Lecture Notes in Computer Science   930

Edited by G. Goos, J. Hartmanis and J. van Leeuwen

Advisory Board: W. Brauer   D. Gries   J. Stoer

W0044080

Springer-Verlag Berlin Heidelberg GmbH

José Mira Francisco Sandoval (Eds.)

From Natural to Artificial Neural Computation

International Workshop on Artificial Neural Networks
Malaga-Torremolinos, Spain, June 7 - 9, 1995
Proceedings

 Springer

Series Editors

Gerhard Goos
Universität Karlsruhe
Vincenz-Priessnitz-Straße 3, D-76128 Karlsruhe, Germany

Juris Hartmanis
Department of Computer Science, Cornell University
4130 Upson Hall, Ithaca, NY 14853, USA

Jan van Leeuwen
Department of Computer Science, Utrecht University
Padualaan 14, 3584 CH Utrecht, The Netherlands

Volume Editors

José Mira
Dept. de Informática y Automática
Universidad Nacional de Educación a Distancia
Senda del Rey s/n., E-28040 Madrid, Spain

Francisco Sandoval
Depto. de Tecnologia Electronica, Universidad de Málaga
Plaza El Ejido s/n., E-29013 Málaga, Spain

CR Subject Classification (1991): I.2, F1.1, C.1.3, C.2.1, G.1.6, I.5.1, B.7.1, J.1, J.2, J.4

1991 Mathematics Subject Classification: 92B20, 68T05, 90C90

ISBN 978-3-540-59497-0   ISBN 978-3-540-49288-7 (eBook)
DOI 10.1007/978-3-540-49288-7

CIP data applied for

Typesetting: Camera-ready by author
SPIN: 10486240   06/3142-543210 - Printed on acid-free paper

Preface

Neural Computation is the clearest complement of Artificial Intelligence (AI) and Knowledge Based Systems (KBS) in a shared attempt to make the most genuine dimensions of human behaviour computational. Born of biology and with a history of periodic ups and downs, it is now at a critical moment. From one side there is a massive interest but, at the same time, standard models of neural computation and learning are nearly exhausted from the computational viewpoint and some fresh air is needed.

This year, the 25th anniversary of the death of Warren S. McCulloch, is a good occasion to remember that neural computation was born around 1943 when W.S. McCulloch and Walter Pitts proposed a formal neural model based on the logic we would presently call minimal sequential circuit, consisting of a combinational function followed by a delay and interconnect with and without feedback.

In the 1960s, both positive and negative events took place for neural computation. On the one hand, the theory of Artificial Neural Networks (ANN) was consolidated with introduction of the Perceptron, the Adalines, the biological models of learning by conditioned reflex, and pioneering studies on biophysical modeling, associative memories, self-organization, pattern recognition, and intentional calculus. The most negative events were Minsky and Papert's criticism of "Linearly Unrecognizible Patterns" as well as the strong growth of AI.

It is interesting to remember that in this initial period neural and symbolic computation moved in unison with computational developments that initially emerged under the umbrella of Neurocybernetics and later gave way to AI, which dominated in the 1970s.

We thus arrive at the powerful revival of connectionism in the 1980s. Unfortunately, since this *fiorello*, neural computation has been usually understood only as layered networks of weighted sums followed by a decision function of the threshold or sigmoid type where external programming is partially substituted by learning. This vision of the field presents at least the following problems:

(1) It distances itself in an almost irreversible fashion from the biological reality from which it originated, using computational modules that are clearly insufficient for describing what is presently know about real neurons.

(2) There is a lack of methodology and an excess of empiricism in the synthesis processes.

(3) It present itself as an alternative which is not integrable with the symbolic perspective of Artificial Intelligence.

The purpose of the third International Workshop on Artificial Neural Networks (IWANN'95) is to partially contribute to posing and resolving these problems. In order to do this, the workshop focuses on biological modeling, the search for theory and design methodologies, and the establishment of a symbolic-connectionist bridge that can make possible the integration of AI and ANNs, taking the best of both paradigms.

The papers presented correspond to the talks delivered at IWANN'95, organized by the University of Málaga and the Spanish Open University at Madrid (UNED) and held in Torremolinos (Malaga), Spain, 7 - 9 of June, 1995. More than 200 papers were submitted and carefully evaluated by the program committee. After averaging the reviewers' score, 143 papers were accepted for oral presentation and are included in these proceedings. Extended papers originating from invited talks related to some of the topics considered are also included as introductions to the corresponding sections.

This workshop has been organized in collaboration with the Spanish RIG IEEE Neural Network Council, the UK&RI Communication Chapter of IEEE, the Spanish Computer Society Chapter of IEEE, and the AEIA (Spanish Association for Computing and Automation).

Sponsorship has been obtained from DG-XII Human Capital and Mobility (EC), Spanish CICYT and DGICYT (MEC), Junta de Andalucía and the organizing Universities (Málaga and UNED).

We would like to thank all the authors as well as all the members of the international program committee for their labor in the production, evaluation, and refinement of the papers. Furthermore, the editors would like to thank Springer-Verlag, in particular Alfred Hofmann, for excellent cooperation.

The papers published in this volume present the current state in neural computation and are organized in nine sections:

- ✧ Neuroscience
- ✧ Computational models of neurons and neural nets
- ✧ Organization principles
- ✧ Learning
- ✧ Cognitive science and AI
- ✧ Neurosimulators
- ✧ Implementation
- ✧ Neural networks for perception
- ✧ Neural networks for communication and control

Turning our eyes once again to neuroscience, we begin with papers related with computational models of cortical neurons, dendro-dendritic processes, synaptic modulation, and self-organization of receptive fields, among others. Neuroscience is an inexhaustible source of inspiration about new styles of computation.

The second topic is related with the formal tools we use to model neurons and neural nets (analogic, logic, or inferential). You will find papers on high-order Boltzmann machines, propositional logic, inferential rules at the synaptic level, stochastic and collective models, and some contributions to learning by reinforcement and pruning.

The study of organizational principles in biology will increase our understanding of living beings and artificial systems. This includes dynamic formulations, autopoiesis, self-organization, cooperative processes and emergent computation, and synergetic, evolutive optimization, and genetic algorithms. Contributions on this topic have not been so numerous as we would have liked. Nevertheless, several papers on cooperation, competition and self-organization, and genetic algorithms are included.

The structural characteristics necessary for the control by learning of the functioning mode of an ANN are decisive for the advance of neural computation as a consolidated engineering field. Inspiration from biological mechanisms of learning and proposals for new "brain-like" algorithms are more than welcome. You will find a good number of papers in this section. Nevertheless, the panorama is not so diverse and innovative as could be desired. Incremental learning, sensitivity analysis, initialization problems, priming, attentional scanning, fuzzy clustering, and schema-based learning, among others, are some of the topics addressed.

Thus far we have examined neuroscience, models, organization, and learning. The next topic is cognitive science and AI. Here we look at hybrid formulations with papers on modeling linguistic problems, extracting rules from ANNs, and dynamic symbol grounding, among others.

Neurosimulators include languages, environments for simulation, development, and prototype evaluation, and the facilities to connect with the implementation stage. Papers on object oriented simulators, hybrid development environments, and biological and applications oriented tools, as well as neural simulation languages, are included in this section.

As is usually recognized, the implementation of neural networks depends directly on which neural model and learning algorithm we seek to implement. The second part is how to do it. For the first part, it is our deepest feeling that the anatomic and physiological bases of natural neurons offer more inspiration than we have been able to formalize until now. For the second part, VLSI technology, FPGAs, parallel architectures, neurodevices, and optical solutions, to name but a few, give us enough alternatives for implementation, once we have clear and complete functional specifications of what we want to implement. Papers in this section make reference to hardware accelerators, Bayesian networks, associative processors, implementation models, analog cellular networks, coprocessor cards, optimal mapping onto FPGAs, and many other related topics.

The last two sections of the proceedings are related to applications in perception, communication, and control. Perception includes low-level processing segmentation, feature extraction, adaptive filtering, textures, motion analysis, and hybrid symbolic-connectionist architectures for artificial vision. Papers are included on image segmentation, pattern recognition, texture classification, and speech processing. Some papers related to active vision and visual feedback in the control of robots as well as parameters estimation and forecasting have also been included in this section.

The applications of ANNs to control and communication systems cover the topics of systems identification, motion planning and control, adaptive, predictive, and model based control, navigation, real-time applications, modems and codecs, network management, and digital communications. In these proceedings we have papers on identification, movement optimization, visuomotor control of mobile robots, forecasting using Kohonen maps, real-time adaptive control, and digital transmission.

Some of the original objectives for IWANN'95 we feel have been fulfilled, with the result that this volume offers a serious, broad, and comprehensive selection of papers ranging from neuroscience to engineering applications.

Other biological and computational problems in the field of ANNs still remain open. Let us mention again the insufficient complexity of our models of biological neurons and the lack of methodology in analysis and design. In the W.S. McCulloch memorial we would like to be provocative in search of the computational "embodiment" of natural computation from which intelligent behaviour emerges.

Madrid, April 1995   J. Mira
                                                                                 F. Sandoval

Contents

1. Neuroscience

Are There Universal Principles of Brain Computation? (Invited Paper) 1
S. Grossberg

Modeling Cortical Networks 7
L. Menéndez de la Prida

Cooperative Organization of Connectivity Patterns and Receptive Fields in the
Visual Pathway: Application to Adaptive Thresholding 15
J. Mira, A. Manjarrés, S. Ros, A.E. Delgado, J.R. Álvarez

Neurobiological Inspiration for the Architecture and Functioning of Cooperating
Neural Networks 24
F. Alexandre, F. Guyot

Synaptic Modulation Based Artificial Neural Networks 31
R.J. Duro, J. Santos, A. Gómez

Self-Organization of Cortical Receptive Fields and Columnar Structures in a Hebb
Trained Neural Network 37
M. Stetter, M. Kussinger, A. Schels, E. Seeger, E.W. Lang

An Analytical Solution of the Compartmental Model for Use in Local Learning in
Artificial Neural Networks 45
J. Hoekstra, M. Maouli

Should ANN be ANGN? 53
J.G. Wallace, K. Bluff

Modeling Retinal High and Low Contrast Sensitivity Filters 61
T. Lourens

Neural Micro-Structures. Three Simple Models 69
J. Miró

Adaptive Hierarchical Structures 76
A. Daffertshofer, H. Haken

Optimal Range of Input Resistance in the Oscillatory Behavior of the Pancreatic β-Cell 85
E. Andreu, B. Soria, S. Bolea, J.V. Sánchez-Andrés

A Computational Model of Periodic-Pattern-Selective Cells 90
P. Kruizinga, N. Petkov

Modeling and Analysis of Some Neural Mechanisms for the Genesis and Control
of Respiratory Pattern 100
I.A. Rybak, J.F.R. Paton, J.S. Schwaber

A Neural Network Model for Plasticity in Adult Striate Cortex   108
F. Morán, M.A. Andrade

Regenerative-Type Neural Interface   114
E. Valderrama, R. Villa, E. Cabruja, P. Garrido, X. Navarro, M. Buti, S. Calvet

New Perspectives in Auditory Coding: Bases for a New Cochlear Behavioural Model   121
R. Villa, J. Aguiló

Nervous System as a Closed Neural Network: Behavioral and Cognitive Consequences   130
J. Mpodozis, J.-C. Letelier, H. Maturana

2. Computational Models of Neurons and Neural Nets

Local Accumulation of Persistent Activity at Synaptic Level: Application to
Motion Analysis   137
M.A. Fernández, J. Mira, M.T. López, J.R. Álvarez, A. Manjarrés, S. Barro

High Order Boltzmann Machines with Continuous Units: Some Experimental Results   144
M. Graña, A. D'Anjou, F.X. Albizuri, A. de la Hera, I. Garcia

An Adaptive Control Model of a Locomotion by the Central Pattern Generator   151
J. Nishii

Self-Consistent Neural Receptive Fields   158
I. Grabec

A Simple Probabilistic Neural Model Producing Multimodal ISHs   166
P.M. Hofman, F.B. Rodriguez, J.A. Sigüenza, V. López, S. Carrillo-Menéndez

An Associative Neural Network to Model the Developing Mammalian Hippocampus   174
M.T. Signes Pont, J.V. Sánchez-Andrés

The Implementation of Propositional Logic in Random Neural Networks   180
M.-D. Weitze, G.L. Hofacker

A Learning Rule for Extracting Temporal Invariances   189
J.V. Stone, A. Bray

The Influence of the Sigmoid Function Parameters on the Speed of Backpropagation
Learning   195
J. Han, C. Moraga

General Transient Length Upper Bound for Recurrent Neural Networks   202
A.M.C.-L. Ho, P. De Wilde

On Some Methods in Neuromathematics   209
R. Moreno-Díaz jr, K.N. Leibovic

Logistic Networks With DNA-Like Encoding and Interactions   216
N.H. Farhat, E. Del Moral Hernandez

A Review on the Stochastic Firing Behaviour of Real Neurons and How it can be Modelled   223
C. Christodoulou, T. Clarkson

The BP-λL1 Algorithm: Non-Chaotic and Accelerated Learning in a MLP Network   231
B. Augereau, T. Simon, J. Bernard, B. Heit

Analysis of Pruning in Backpropagation Networks for Artificial and Real World
Mapping Problems   239
T. Jašić, H.L. Poh

Oscillatory Networks with Hebbian Matrix of Connections   246
M.G. Kuzmina, E.A. Manykin, I.I. Surina

A new Algorithm for Implementing a Recursive Neural Network   252
V. Giménez, P. Gómez-Vilda, E. Torrano, M. Pérez-Castellanos

Visual Information Processing from the Viewpoint of Symbolic Operations   260
J. Barahona da Fonseca, I. Barahona da Fonseca, J. Simões da Fonseca

Dentritic Computation in the Brain   268
I. Barahona da Fonseca, J. Barahona da Fonseca, J. Simões da Fonseca

Stochastic Neuronal Models with Realistic Synaptic Inputs and Oscillatory Inputs   276
P. Hruby

A Neural Paradigm for Controlling Autonomous Systems with Reflex Behaviour
and Learning Capability   283
G. Joya, F. Sandoval

Fast Automatic Architecture Selection in RBF Networks   291
A.M. González, C. Santa Cruz, V. López, J.R. Dorronsoro

Collective Behaviour of a Chain of Hopfield Subnetworks Interconnected Unidirectionally   298
L.Viana

3. Organization Principles

A Comparative Study of Three Neural Networks that use Soft Competition   308
K. Butchart, N. Davey, R. Adams

Neural Networks and Genetic Algorithms for the Attitude Control Problem   315
D.C. Dracopoulos, A.J. Jones

Self-Organising Artificial Neural Networks   322
J.A. Flanagan, M. Hasler

A Distributed Classifier Based on Yprel Networks Cooperation   330
E. Stocker, Y. Lecourtier, A. Ennaji

A Fractal, Selforganizing Map with Partially Chaotic Neurons   338
A. Kosak, K. Goser

Multiple Self-Organizing Maps for Supervised Learning   345
E. Cervera, A.P. del Pobil

An Application of the Saturated Attractor Analysis to Three Typical Models   353
J. Feng, B. Tirozzi

4. Learning

Learning in Evolutive Neural Architectures: An Ill-Posed Problem? (Invited Paper)   361
C. Jutten

Automatic Scaling using Gamma Learning for Feedforward Neural Networks   374
A.P. Engelbrecht, I. Cloete, J. Geldenhuys, J.M. Zurada

Determining the Significance of Input Parameters using Sensitivity Analysis   382
A.P. Engelbrecht, I. Cloete, J.M. Zurada

Multi-Valued Neurons: Learning, Networks, Application to Image Recognition and
Extrapolation of Temporal Series   389
N.N. Aizenberg, I.N. Aizenberg, G.A. Krivosheev

Individual Evolutionary Algorithm and its Application to Learning of Nearest
Neighbor Based MLP   396
Q. Zhao, T. Higuchi

A Practical View of Suboptimal Bayesian Classification with Radial Gaussian Kernels   404
J.-L. Voz, M. Verleysen, P. Thissen, J.-D. Legat

Schema Based Learning & Learning to Detour   412
F.J. Corbacho, M.A. Arbib

Neurons with Continuous Varying Activation in Self-Organizing Maps   419
J. Göppert, W. Rosenstiel

Improving Back-Propagation: Epsilon-Back-Propagation   427
L.A. Trejo, C. Sandoval

Finite State Automata and Connectionist Machines: A Survey   433
M.A. Castaño, E. Vidal, F. Casacuberta

Learning Transformed Prototypes (LTP) - A Statistical Pattern Classification
Technique of Neural Networks   441
Y. Guan, T. Clarkson, J.G. Taylor

Fuzzy Function Estimators as Basis on Learning from Experience   448
R. Ferreiro Garcia, F.J. Perez Castelo

Connectionists and Statisticians, Friends or Foes?   454
A. Flexer

Unsupervised Neural Networks for Speech Perception with Cochlear Implant
Systems for the Profoundly Deaf   462
M. Leisenberg

Obstacle Avoidance by Means of an Operant Conditioning Model   471
E. Zalama, P. Gaudiano, J.L. López Coronado

Qualitative Approach to Gradient Based Learning Algorithms   478
B. Morcego Seix, A. Català Mallofré, N. Piera Carreté

Character Recognition with Neural Assemblies   486
F.J. Vico, F. Ortega, J. Almaraz, F. Sandoval

Learning by Attentional Scanning   492
Z. Schreter

The Synthesis of the Ranked Neural Networks Applying Genetic Algorithm with
the Dynamic Probability of Mutation   498
J. Lis

Global Versus Local Heuristic Terminal Attractor   505
F.J. Marin, F. Garcia, F. Sandoval

Dynamic Learning of Radial Basis Functions for Fuzzy Clustering   513
A. Kanstein, K. Goser

Incremental Learning with a Stopping Criterion - Experimental Results   519
R. Chentouf, C. Jutten

Learning Algorithm with Gaussian Membership Function for Fuzzy RBF
Neural Networks   527
D. Benitez-Diaz, J. Garcia-Quesada

Neural Network Initialization   535
G. Thimm, E. Fiesler

Bidirectional Neural Networks Reduce Generalisation Error   543
A.F. Nejad, T.D. Gedeon

Balancing Bias and Variance: Network Topology and Pattern Set Reduction
Techniques   551
T.D. Gedeon, P.M. Wong , D. Harris

Priming an Artificial Neural Classifier   559
D. Puzenat

5. Cognitive Science and AI

Artificial Neuroconsciousness an Update (Invited Paper)   566
I. Aleksander

Physical and Linguistic problems in the Modelling of Consciousness by Neural
Networks   584
P. De Wilde

A Neural Network Model for the Velocity Vector of an Object and itsConsistency
with Psychological Phenomena   589
K.-i. Miura, T. Nagano

Implementation and Evaluation of a Relevance Feedback Device Based on
Neural Networks   597
F. Crestani

Second-Order Recurrent Neural Networks Can Learn Regular Grammars from
Noisy Strings   605
R.C. Carrasco, M.L. Forcada

Extracting DNF Rules From Artificial Neural Networks   611
H.L. Viktor, I. Cloete

Dynamic Symbol Grounding, State Construction and the Problem of Teleology   619
E. Prem

Analysis of Industrial Economics by means of Neural Nets   627
E. Monte, J.M. Calvet, S. Vilarrubla

Effects of Spatial Frecuency and Stimulus Size on the Orientation Sensitivity of Humans   634
F. Díaz-Otero, A. Caballero, A. Lorenzo, J.A. Sigüenza

6. Neurosimulators

Object Oriented Design of a Simulator for Large BP Neural Networks   642
J.M. Adamo, D. Anguita

Introducing XSim: A Neural Network Simulator that Incorporates Biological Parameters   650
P. Varona, J.A Sigüenza

NETTOOL: A Hybrid Connectionist-Symbolic Development Environment   658
J. Santos, R.P. Otero, J. Mira

EL-SIM: a Development Environment for Neuro-Fuzzy Intelligent Controllers   666
M. Chiaberge, G. Di Bene, S. Di Pascoli, R. Lambert, B. Lazzerini, A. Maggiore,
L.M. Reyneri

Packlib, an Interactive Environment to Develop Modular Software for Data Processing   673
Y. Cheneval

NSL-Neural Simulation Language   683
A. Weitzenfeld

7. Implementation

Low-Cost Accelerator for the Simulation of Cellular Neural Networks 689
A. Torralba

A VLSI System for Neural Bayesian and LVQ Classification 696
P. Thissen, M. Verleysen, J.-D. Legat,, J. Madrenas, J. Dominguez

An Associative Processor Dedicated to Classification by Neural Methods 704
P. Thissen, M. Verleysen, J.-D. Legat

Hardware-Oriented Models for VLSI Implementation of Self-Organizing Maps 712
B. Martin-del-Brio, J. Blasco-Alberto

Hardware Requirements for Spike-Processing Neural Networks 720
U. Roth, A. Jahnke, H. Klar

A VLSI Approach to the Implementation of Additive and Shunting Neural Networks 728
F.J. Pelayo, E. Ros, P. Martin-Smith, F.J. Fernández, A. Prieto

A Low-Power Analog Implementation of Cellular Neural Networks 736
M. Anguita, F.J. Pelayo, F.J. Fernández, A. Prieto

Asynchronously Parallel Boltzmann Machines Mapped onto Distributed-Memory
Multiprocessors 744
J.M. Benitez, J. Ortega, I. Requena

A Coprocessor Card for Fast Neural Network Emulation 752
F. Castillo, J.A. Garcia, J.M. Moreno, J. Cabestany

Digital Hardware Implementation of ROI Incremental Algorithms 761
J.M. Moreno, J. Madrenas, S. San Anselmo, F. Castillo, J. Cabestany

Comparing Implementations of Radial Basis Function Neural Networks on Three
Parallel Machines 771
N. Maria, A. Guérin-Dugué, J.M. Moreno, F. Blayo

Implementing Radial Basis Functions Neural Networks on the Systolic MANTRA Machine 781
F. Blayo, A. Guérin-Dugué, N. Maria

A Mixed Parallel-Sequential SHNN for Large Networks 789
A. Torralba, F. Colodro, L.G. Franquelo

A Modular VLSI Architecture for Neural Networks Implementation 794
O. Vermesan

A Massively Parallel Neurocomputer with a Reconfigurable Arithmetical Unit 800
A. Strey, N. Avellana, R. Holgado, J.A. Fernández, R. Capillas, E. Valderrama

An All-Optical Forward Propagation Multilayer Neural Network 807
I. Saxena, E. Fiesler

A VLSI Current Mode Synapse Chip   815
D.J. Mayes, A. Hamilton

Optimal Mapping of Neural Networks onto FPGA's -A New Constructive Algorithm -   822
V. Beiu, J.G. Taylor

A CPWM Synapsis for Weighted Radial Basis Functions   830
E. Miranda, L.M. Reyneri

Test Pattern Generation for Analog Circuits Using Neural Networks and
Evolutive Algorithms   838
J.L. Bernier J.J. Merelo, J. Ortega, A. Prieto

8. Neural Networks for Perception

About Some Perception Problems in Neural Networks (Invited Paper)   845
J. Hérault

Optimization Neural Networks for Image Segmentation   860
D.L. Vilariño, D. Cabello, A. Mosquera

Segmentation of Range Images: A Neural Network Approach   868
W.P. Cheung, C.K. Lee, K.C. Li

A Neural Architecture for Preattentive Segmentation of Sewage Pipes Video Images   875
J. Ruiz-del-Solar, M. Köppen

A CNN Model For Grey Scale Image Processing   882
M.A. Jaramillo-Morán, F.J. López-Aligué, M. Macías-Macías, M.I. Acevedo-Sotoca

Kohonen's Self Organizing Maps for Contour Segmentation of Gray Level
and Color Images   890
R. Natowicz

A Geometrical Based Procedure for Source Separation Mapped to a Neural Network   898
C.G. Puntonet, M. Rodriguez-Alvarez, A. Prieto

Quasi-Optimum Combination of Multilayer Perceptrons for Adaptive Multiclass
Pattern Recognition   906
A. Ruiz Garcia, F.J. Arcas Túnez

A Text Recognition System Based on a Neural Network and on a Deformed System   913
J. Echanobe, J.R. González De Mendivil, J.R. Garitagoitia

Simultaneous Recognition of Multiple Objects Using the MEM Model   919
S. Shams

On-line Handwritten Character Recognition by a Hybrid Method based on Neural
Networks and Pattern Matching   926
J.-W. Cho, S.-Y. Lee, C.H. Park

Analysis and Application of the STORE Neural Model in Recognizing Handwritten
Symbols   934
J. Pérez Maroto, Y.A. Dimitriadis, J.M. Cano Izquierdo, J.L. López Coronado

An Adaptive Orthogonal Asociative Memory and its Application to Character
Recognition   942
F. Ibarra-Picó, J.M. García-Chamizo, R. Rizo-Aldeguer, D. Corredor Lacha

Texture Classification on Real Time Using Semi-cover Vector and an Orthogonal
Neural Network   948
D. Asensi Muñoz, A. Almagro León, F. Ibarra Picó

An Architecture for Texture Segmentation: from Energy Features to Region Detection   956
P.M. Palagi, A. Guérin-Dugué

A Lattice-Based Time-Delay Neural Network for Speech Processing   963
P. Gómez-Vilda, V. Rodellar, V. Nieto, M.A. Hombrados

Acquisition of Internal Representation by Learning of Identity-mapping Using
Overload Learning   971
I. Noda

A Multiacuity Connectionist Model for Local Speed Estimation   979
C. Bandera, I.M. Conde, J. Jerez, M. González, F.J. Vico, F. Ortega

Using Artificial Neural Networks for Ultrasonic Signals Processing from Simple
Geometric Shapes   987
F. Arroyo, A. Gonzalo, J.R. Hilera

A Hierarchical Neural Network for Mobile Visual Tracking with a Robot Head   993
D. Maravall, L. Baumela

Short Term Load Forecasting Using Neural Nets   1001
R.S. Zebulum, K. Guedes, M. Vellasco, M.A. Pacheco

Image Compression Using Feedforward Neural Networks - Hierarchical Approach   1009
S. Osowski, R. Waszczuk, P. Bojarczak

9. Neural Networks for Communications and Control

Neural Approaches to Robot Control: Four Representative Applications (Invited Paper)   1016
C. Torras, G. Cembrano, J.del R. Millán, G. Wells

On Line Identification of Causal Relationships Between Variables in the Feed Water
System of a Nuclear Power Plant   1036
J.R. Álvarez, J. Mira, R.A. Fernández, L. Sainz, V. Arroyo, A.E. Delgado

Dynamic Neural Units for Nonlinear Dynamic Systems Identification   1045
M. Ayoubi, M. Schäfer, S. Sinsel

Optimal Identification Using Feed-Forward Neural Networks   1052
V. Vergara, S. Sinne, C. Moraga

Recurrent Neural Networks for Identification of Friction 1060
M. Dominguez, J.M. Michelin, J.M. Martinez

Solving an End-Effector Positioning Problem by Hopfield Neural Network 1068
S. Cavalieri, M. Martini

Visuomotor Control using an Artificial Neural Network 1076
P.F. McGuire, G.M.T. D'Eleuterio

Learning the Visuomotor Coordination of a Mobile Robot by Using the Invertible
Kohonen Map 1084
C. Versino, L.M. Gambardella

Neural Networks for Automatic Fuzzy Control System Design 1092
J. Villadangos, J.R. González De Mendivil, C.F. Alastruey, J.R. Garitagoitia

Supervised Classification With Variable Kernel Estimators 1099
P. Comon, Y. Cheneval

Daily Electrical Power Curves: Classification and Forecasting Using a Kohonen Map 1107
M. Cottrell, B. Girard, Y. Girard, C. Muller, P. Rousset

CMAC Real-Time Adaptive Control Implementation on a D.S.P. Based Card 1114
G. Mercier, K. Madani

Neural Networks in Digital Data Transmission 1121
I. Ortuño Ortín, J. Serra Sagristà

A New Neural Network Approach to the Floorplanning of Hierarchical VLSI Designs 1128
M.S. Zamani, G.R. Hellestrand

A Neural Network Approach to Quality Control Charts 1135
T. Stützle

Bankruptcy Prediction with Artificial Neural Networks 1142
E. Fernández, I. Olmeda

Author Index 1147

Are There Universal Principles of Brain Computation?

Stephen Grossberg
Boston University
Department of Cognitive and Neural Systems
and
Center for Adaptive Systems
111 Cummington Street
Boston, Massachusetts 02215 USA

1. Introduction

Are there universal computational principles that the brain uses to self-organize its intelligent properties? This lecture suggests that common principles are used in brain systems for early vision, visual object recognition, auditory source identification, variable-rate speech perception, and adaptive sensory-motor control, among others. These are principles of matching and resonance that form part of Adaptive Resonance Theory, or ART. In particular, bottom-up signals in an ART system can automatically activate target cells to levels capable of generating suprathreshold output signals. Top-down expectation signals can only excite, or prime, target cells to subthreshold levels. When both bottom-up and top-down signals are simultaneously active, only the bottom-up signals that receive top-down support can remain active. All other cells, even those receiving large bottom-up inputs, are inhibited. Top-down matching hereby generates a focus of attention that can resonate across processing levels, including those that generate the top-down signals. Such a resonance acts as a trigger that activates learning processes within the system.

In the examples described herein, these effects are due to a top-down nonspecific inhibitory gain control signal that is released in parallel with specific excitatory signals.

2. Neural Dynamics of Multi-Source Audition

How does the brain's auditory system construct coherent representations of acoustic objects from the jumble of noise and harmonics that relentlessly bombards our ears throughout life? Bregman (1990) has distinguished at least two levels of auditory organization, called primitive streaming and schema-based segregation, at which such representations are formed in order to accomplish auditory scene analysis. The present work models data about both levels of organization, and shows that ART mechanisms of matching and resonance play a key role in achieving the selectivity and coherence that are characteristic of our auditory experience.

In environments with multiple sound sources, the auditory system is capable of teasing apart the impinging jumbled signal into different mental objects, or streams, as in its ability to solve the cocktail party problem. With my colleagues Krishna Govindarajan, Lonce Wyse, and Michael Cohen, a neural network model of this primitive streaming process, called the ARTSTREAM model (Figure 1), has been developed that groups different frequency components based on pitch and spatial location cues, and selectively allocates the components to different streams. The grouping is accomplished through a resonance that develops between a given object's pitch, its harmonic spectral components, and (to a lesser extent) its spatial location. Those spectral components that are not reinforced by being matched with the top-down prototype read-out by the selected object's pitch representation are suppressed, thereby allowing another stream to capture these components, as in the "old-plus-new heuristic" of Bregman (1990). These resonance and matching mechanisms are specialized versions of ART mechanisms.

The model is used to simulate data from psychophysical grouping experiments, such as how a tone sweeping upwards in frequency creates a bounce percept by grouping with a

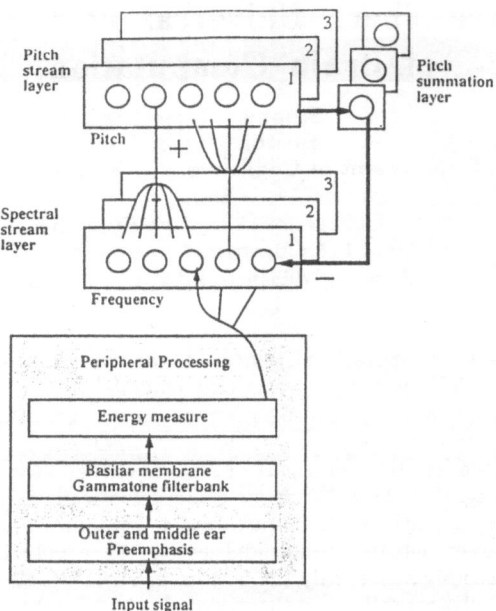

Figure 1. Block diagram of the ARTSTREAM auditory streaming model. Note the non-specific inhibitory feedback from pitch representations to spectral representations.

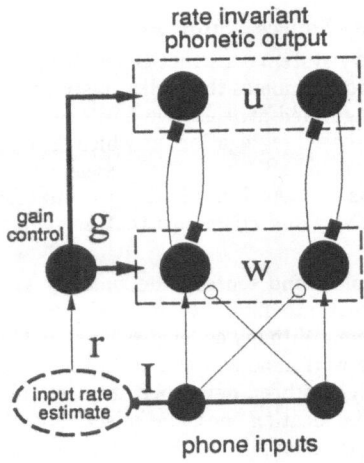

Figure 2. The ARTPHONE model. Working memory nodes (**w**) excite chunks (**u**) through previously learned pathways. List chunks send excitatory feedback down to their item source nodes. Bottom-up and top-down pathways are modulated by habituative transmitter gates (filled squares). Item nodes receive input in an on-center off-surround anatomy. Total input (I) is averaged to control an item rate signal (r) that adjusts the working memory gain (g). Excitatory paths are marked with arrowheads, inhibitory paths with small open circles.

downward sweeping tone due to proximity in frequency, even if noise replaces the tones at their intersection point. The model also simulates illusory auditory percepts such as the auditory continuity illusion of a tone continuing through a noise burst even if the tone is not present during the noise, and the scale illusion of Deutsch whereby downward and upward scales presented alternately to the two ears are regrouped based on frequency proximity, leading to a bounce percept. The stream resonances provide the coherence that allows one voice or instrument to be tracked through a multiple source environment.

3. Neural Dynamics of Variable-Rate Speech Categorization

What is the neural representation of a speech code as it evolves in real time? With my colleagues Ian Boardman and Michael Cohen, a neural model of this schema-based segregation process, called the ARTPHONE model (Figure 2), has been developed to quantitatively simulate data concerning segregation and integration of phonetic percepts, as exemplified by the problem of distinguishing "topic" from "top pick" in natural discourse. Psychoacoustic data concerning categorization of stop consonant pairs indicate that the closure time between syllable final (VC) and syllable initial (CV) transitions determines whether consonants are segregated, i.e., perceived as distinct, or integrated, i.e. fused into a single percept. Hearing two stops in a VC–CV pair that are phonetically the same, as in "top pick," requires about 150 msec more closure time than hearing two stops in a VC_1–C_2V pair that are phonetically different, as in "odd ball." As shown by Repp (1980), when the distribution of closure intervals over trials is experimentally varied, subjects' decision boundaries between one-stop and two-stop percepts always occurred near the mean closure interval.

The ARTPHONE model traces these properties to dynamical interactions between a working memory for short-term storage of phonetic items and a list categorization network that groups, or chunks, sequences of the phonetic items in working memory. These interactions automatically adjust their processing rate to the speech rate via automatic gain control. The speech code in the model is a resonant wave that emerges after bottom-up signals from the working memory select list chunks which, in turn, read out top-down expectations that amplify consistent working memory items. This resonance may be rapidly reset by inputs, such as C_2, that are inconsistent with a top-down expectation, say of C_1; or by a collapse of resonant activation due to a habituative process that can take a much longer time to occur, as illustrated by the categorical boundary between VCV and VC–CV. The categorization data may thus be understood as emergent properties of a resonant process that adjusts its dynamics to track the speech rate.

4. Neural Dynamics of Boundary and Surface Representation

With my colleagues Alan Gove and Ennio Mingolla, a neural network model, called a FACADE theory model (Figure 3), has been developed to explain how visual thalamocortical interactions give rise to boundary percepts such as illusory contours and surface percepts such as filled-in brightnesses. Top-down feedback interactions are needed in addition to bottom-up feedforward interactions to simulate these data. One feedback loop is modeled between lateral geniculate nucleus (LGN) and cortical area V1, and another within cortical areas V1 and V2. The first feedback loop realizes a resonant matching process, as in ART, which enhances LGN cell activities that are consistent with those of active cortical cells, and suppresses LGN activities that are not. This corticogeniculate feedback, being endstopped and oriented, also enhances LGN ON cell activations at the ends of thin dark lines, thereby leading to enhanced cortical brightness percepts when the lines group into closed illusory contours. The second feedback loop generates boundary representations, including illusory contours, that coherently bind distributed cortical features together. Brightness percepts form within the surface representations through a diffusive filling-in process that is contained by resistive gating signals from the boundary representations. The model is used

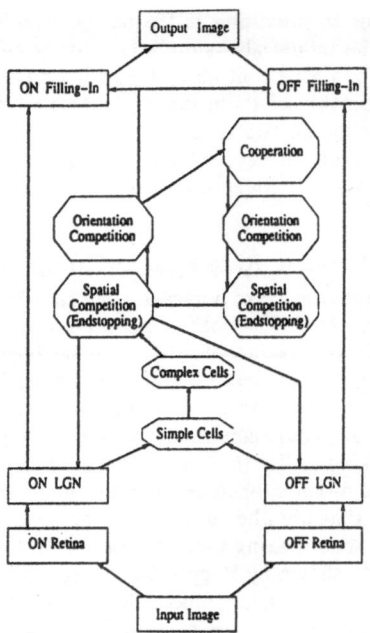

Figure 3. FACADE model macrocircuit. Boundary representation, or BCS, stages are designated by octagonal boxes, surface representation, or FCS, stages by rectangular boxes.

to simulate illusory contours and surface brightnesses induced by Ehrenstein disks, Kanizsa squares, Glass patterns, and café wall patterns in single contrast, reverse contrast, and mixed contrast configurations. These examples illustrate how boundary and surface mechanisms can generate percepts that are highly context-sensitive, including how illusory contours can be amodally recognized without being seen, how model simple cells in V1 respond preferentially to luminance discontinuities using inputs from both LGN ON and OFF cells, how model bipole cells in V2 with two colinear receptive fields can help to complete curved illusory contours, how short-range simple cell groupings and long-range bipole cell groupings can sometimes generate different outcomes, and how model double opponent, filling-in and boundary segmentation mechanisms in V4 interact to generate surface brightness percepts in which filling-in of enhanced brightness and darkness can occur before the net brightness distribution is computed by double opponent interactions. Taken together, these results emphasize the importance of resonant feedback processes in generating conscious percepts in the visual brain.

5. Neural Dynamics for Multimodal Control of Saccadic Eye Movements

Saccades are eye movements by which an animal can scan a rapidly changing environment. While the saccadic system plans where to move the eyes, it also retains reflexive responsiveness to fluctuating light sources. These two types of saccades ultimately result in control of the same set of eye muscles. Visually reactive cells encode gaze error in a *retinotopically* activated motor map. Planned targets are coded in *head–centered* coordinates. When two conflicting commands attempt to share control of the saccadic eye movement system, the system must resolve the conflict and coordinate command of one set of eye muscles.

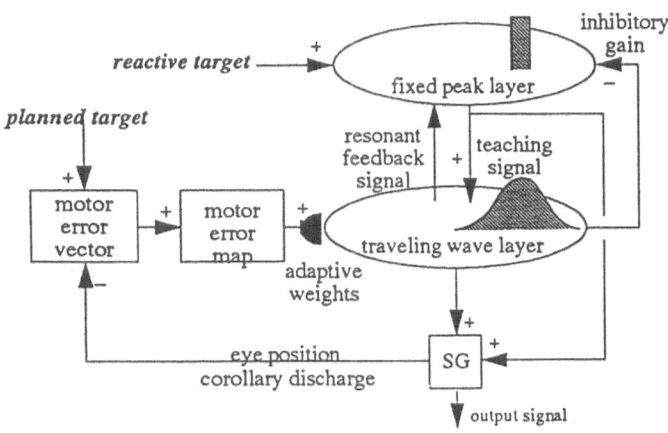

Figure 4. Model for multimodal control of saccadic eye movements by the superior colliculus.

The superior colliculus is a brainstem region that plays a prominent role in both planned and reactive saccades. This region coordinates information to adjust movements of the head and eyes to a stimulus. In order to combine these visual, somatic, and auditory saccade targets in the superior colliculus, the targets in head–centered coordinates are mapped to a gaze motor error in retinotopic coordinates.

How does the saccadic movement system select a target when visual and planned movement commands differ? How do retinal, head–centered, and motor error coordinates interact during the selection process? How are these coordinate systems rendered consistent through learning? Recent data on superior colliculus (SC) reveal a travelling wave of activation whose peak codes the current gaze error (Munoz et al., 1991). In contrast, Waitzman et al. (1991) found that the locus of peak activity in SC remains constant while the activity level at this locus decays as a function of residual gaze error. Why do these distinct cell types exist?

With my colleagues Mario Aguilar, Dan Bullock, and Karen Roberts, a neural network model has been developed that answers these questions while providing a functional rationale for both signal patterns (Figure 4). The model assumes that calibration between visual inputs and eye movement commands is learned early in development within a visually reactive saccade system (Grossberg and Kuperstein, 1989). Visual error signals coded in retinotopic coordinates calibrate adaptive gains to achieve accurate foveation. The accuracy of planned saccades derives from using the gains learned by the reactive system. For this, a transformation between a planned head–centered and a retinotopic target representation needs to be learned. ART matching and resonance control the stability of this learning and the attentive selection of saccadic target locations. Targets in retinotopic and head–centered coordinates are rendered dimensionally consistent so that they can compete for attention to generate a movement command in motor error coordinates. Simulations show how a decaying, stationary activity profile is obtained in the pre–motor layer due to feedback modulation. In addition, a travelling wave activity profile is produced in the motor layer due to modulation from the pre–motor layer and the nature of the local connectivity. The simulations show how this model reproduces physiological data of these two classes of collicular neurons simultaneously during a variety of behavioral tasks (e.g., visual, memory, gap, and overlap conditions), an achievement previously unattained by eye movement models. In addition, the model also clarifies how the SC integrates signals from multiple modalities, and simulates collicular response enhancement and depression produced by multimodal stimuli (Stein and Meredith, 1993).

6. Concluding Remarks

In addition to these systems are the more familiar ART systems for explaining visual object recognition and its breakdown due to hippocampal lesions that lead to medial temporal amnesia (Carpenter and Grossberg, 1993). Taken together, these results provide accumulating evidence that the brain parsimoniously specifies a small set of computational principles to ensure its stability and adaptability in responding to many different types of environmental challenges.

References

Bregman, A.S. (1990). **Auditory scene analysis**. Cambridge, MA: MIT Press.

Carpenter, G.A. and Grossberg, S. (1993). Normal and amnesic learning, recognition, and memory by a neural model of cortico-hippocampal interactions. *Trends in Neurosciences*, **16**, 131–137.

Cohen, M.A., Grossberg, S., and Wyse, L.L. (1995). A spectral network model of pitch perception. *Journal of the Acoustical Society of America*, in press.

Gove, A., Grossberg, S., and Mingolla, E. (1995). Brightness perception, illusory contours, and corticogeniculate feedback. *Visual Neuroscience*, in press.

Govindarajan, K.K., Grossberg, S., Wyse, L.L., and Cohen, M.A. (1994). A neural network model of auditory scene analysis and source segregation. **Technical Report CAS/CNS-TR-94-039**, Boston University.

Grossberg, S., Boardman, I., and Cohen, M.A. (1994). Neural dynamics of variable-rate speech categorization. **Technical Report CAS/CNS-TR-94-038**, Boston University.

Grossberg, S. and Kuperstein, M. (1989). **Neural dynamics of adaptive sensory motor control: Expanded edition**. Elmsford, NY: Pergamon Press.

Munoz, D., Guitton, D., and Pélisson, D. (1991). Control of orienting gaze shifts by the tectoreticulospinal system in the head-free cat, III: Spatiotemporal characteristics of phasic motor discharges. *Journal of Neurophysiology*, **66**(5), 1642–1666.

Repp, B.H. (1980). A range-frequency effect on perception of silence in speech. Haskins Laboratories Status Report on Speech Research, **SR-61**, 151–165.

Stein, B.E. and Meredith, M.A. (1993). **The merging of the senses**. Cambridge, MA: MIT Press.

Waitzman, D., Ma, T., Optican, L., and Wurtz, R. (1991). Superior colliclulus neurons mediate the dynamic characterstocs of saccades. *Journal of Neurophysiology*, **66**(5), 1716–1737.

Modeling Cortical Networks

Liset Menéndez de la Prida

Complex Systems Research Group. Departament de Fisica i Enginyeria Nuclear.
Universitat Politécnica de Catalunya. Sor Eulàlia d'Anzizu s/n. Campus Nord,
Mòdul B4 08034 Barcelona (Spain)

Abstract

A simple model of a microcolumn is developed. Each neuron is taken as a particular oscillator that is strongly connected to several others in a specific network. To model neuronal and network activities non-linear dynamical systems are used. Through stability and bifurcation analysis the dependence of neuronal activities with parameters is studied. A neurophysiological-based heterogeneity among different oscillators could be guarantee selecting different values for parameters of each cell. It is shown that with this simple approach more complex and varied temporal patterns are obtained.

1 Introduction

Real systems are integrated by many interconnected elements that are far to have a similar behaviour. In cerebral cortex for example, electrical responses of cells are subject to change by depending of layer distribution. Pyramidal neurons of layer V are able to generate self-sustained oscillations in a range of 5 to 12 Hz while layers II-III cannot maintain stable periodic behaviour [1]. This neurophysiological-based heterogeneity have important effects in the appearance of rhythms, synchronization, desynchronization and the genesis of several complex patterns that arise from experimental studies.

To develop a model for electrical activity in a nervous circuit these aspects must be considered. In this paper a simple model of a cortical microcolumn is developed (Fig.1). Neurophysiological differences among neurons are included. The variety of temporal patterns obtained is prominent and the complexity that emerges make the study of heterogenous networks a very promising field for the neurosciences.

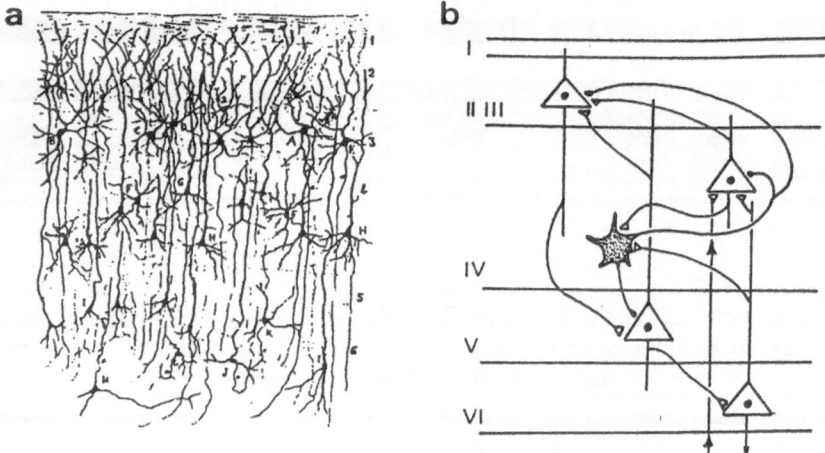

Figure 1: **a.** Cerebral cortex have a laminar structure composed by six layers. Layer I is rich in nervous fibers while the others are mainly populated by neurons of differents sizes and electrical properties. Cells in layer V connect widely with both extremes of the microcolumn, i.e layers II-III and VI. The last one is the principal projecting to thalamus but also interact with layer IV. Thalamic ascending pathways connect mainly with layer IV which is populated by stellate excitatory cells and many inhibitory interneurons that innervates locally (for review see [2]).**b.** Five elements network. Excitatory and inhibitory neurons were connected following previous descriptions. Open triangles represent excitatory synapses, black dots are inhibitory.

2 The model

The electrical behaviour of each neuron is simulated by the same set of ordinary differential equations

$$\frac{d\vec{X}_i}{dt} = \vec{f}(\vec{X}_i, \lambda_i) \tag{1}$$

where $\vec{X}_i$ is the state vector correspondent to i-th cell and $\vec{f}$ is a non-linear function. Equations (1) form a 5-dimensional stiff dynamical system based in the well known Hodgkin-Huxley formalism [3]. X_i^I, i.e the first component of state vector, represents membrane potential of i-th neuron. The remaining components of $\vec{X}_i$ are variables that describe ionic current dynamics. This autonomous system show steady-state solutions or several oscillatory modes as a function of λ_i, a parameter that have a direct biological interpretation [1]. In this sense, each neuron could be an active or a passive oscillator in the network and this behaviour could be "tunned" through a suitable selection of parameter.

---

[1] System (1) can be divided in two subsystems with different time scales (slow and fast). Fast dynamic represents the action of Na^+ and K^+ currents while slow subsytem models Ca^{2+} effects. λ_i is a parameter that controls this last dynamic.

To introduce synaptic effects an aditional term is added to equations (1)

$$\frac{d\vec{X_i}}{dt} = \vec{f}(\vec{X_i}, \lambda_i) + \sum_{j=1}^{N} K_{ij} \sum_{\tau \in Q_j} \Theta(t - \tau)(X_i^1 - V_{syp})(t - \tau)e^{-\frac{t-\tau}{\tau_{syp}}} \qquad (2)$$

where K_{ij} are the elements of a connectivity matrix, Θ the Heavise function and Q_j is the time intances set for which presynaptic cell have fired. Two different time scales (τ_{syp}) were taken into account according to slow and fast character of inhibitory and excitatory connections.

Numerical methods were used to compute the solutions of differential equation systems. A modification to Euler's method is introduced [4] taking the integration step as a variable in the range $(1 \times 10^{-9}, 5 \times 10^{-2})$ ms.

Stability and bifurcation analysis is useful to examine and characterize solutions of dynamical systems when parameters are changed [5]. Autonomous system (1) is studied to analize its intrinsic ability to sustain oscillations. Steady-state points are also investigated and stability is characterized through the analysis of eigenvalues of Jacobian matrix.

3   Characterization of unitary responses

One of the most interesting electrophysiological properties of some neurons is their capacity to fire either burst events or repetitive firing according to the nature of stimulus [6]. In this sense, it was talken about the existence of two firing thresholds (low and high) responsibles of such a dichotomy. If we perturb the system (1) with an inhomogenous term $\vec{I}_{ap}(t)$, two different responses emerge (Fig.2). Negative

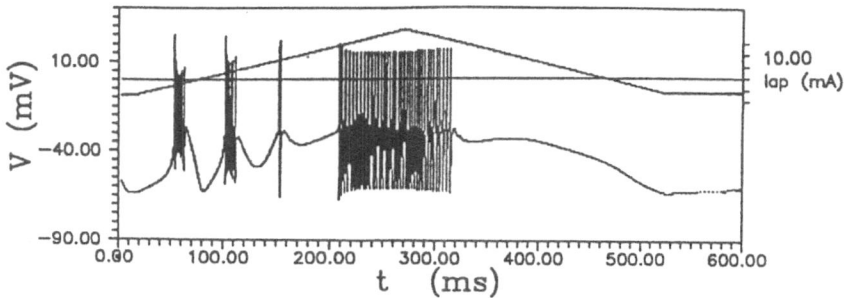

Figure 2: Double-ramp effect. Two differents firing modes are intrinsically present in the neuron model. Controlling parameter values in system (1) it is possible to select burst, fast spiking or both. In right-hand side scale a division correspond to 10 mA.

stimulus provokes what in neurophysiology is known as hyperpolarization, i.e the increase of voltage difference between the inner and outer membrane surface. At this level the slow dynamics is activated driving membrane potential to follow burts pattern. When stimulus depolarize, such a slow dynamics is passive and only fast spikes emerges.

As previously discussed cells throughout cortical layers have different behaviours. Some of them are able to generate bursts while others are spiking cells. According to these elements two groups of different neurons can be formed:

Type A group: Integrated by repetitive firing neurons.

Type B group: Integrated by repetitive bursting neurons.

These groups provide oscillators with two quite different patterns. However, another type of cells (C) must be considered and are those that behave as passive oscillators. These neurons cannot self-sustain any one of the above activities; their responses to stimulus consist only in a short sequence of action potentials.

To characterize the intrinsic dynamics of oscillators a bifurcation diagram of system (1) is presented (Fig.3). As can be seen, for low values of λ_i none oscillatory activity is recorded. The transient for these cases are integrated by two or three bursts followed by a tendency to stable nodus. For $\lambda_i=1$ stationary rhythmic activity emerges associated to the presence of 4 Hz frequency spikes. Following the rising in λ_i

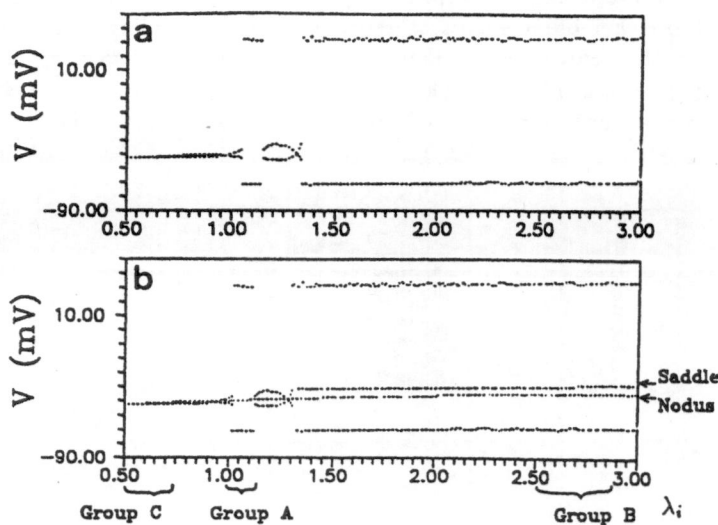

Figure 3: Bifurcation diagram for autonomous system (1).a. Maximum and minimum value of oscillatory behaviour of $\vec{X_i^1}$ (with arbitrary i) are shown. b. Saddle and nodus are superimposed to previous diagram.

spikes sequence is lost and low amplitude passive oscillations appear. This behaviour is presented during a small range at the end of which oscillations are restarted. However, the pattern is very complex for the subsequent parameter interval. The presence of burst discharges increase progressively as λ_i increase and in some cases bursts and spikes are both present. The appearance of a saddle point associated with this transition suggest the homoclinic nature of burst pattern. This complexity in transitions between bursting and continuous spiking have been previously studied as well as the intrinsic mechanism involving burst [7,8]. Increasing λ_i, rhythmic burst pattern is completelly established.

Thus with the use of bifurcation and stability analysis groups could be classified according to λ values. Suitable selection of λ for every neuron guarantee the heterogeneity of oscillators in network model.

4   Cortical network model

To establish the microcolumn model, coupling matrix Kij and parameter λ_i of each cell must be selected. Let us begin with the simplest model of five oscillators (Fig.1). Matrix K was filled based in anatomical data previously summarized (see Fig.1 caption) and neurons of each layer belong to one of the specific groups.

Once dynamical system (2) is established the capacity for syncronization is tested. A brief positive pulse (10 ms and 20 mA of amplitude) stimulate only layer VI excitatory neuron. The synchronized pattern is immediately reached (Fig.4).

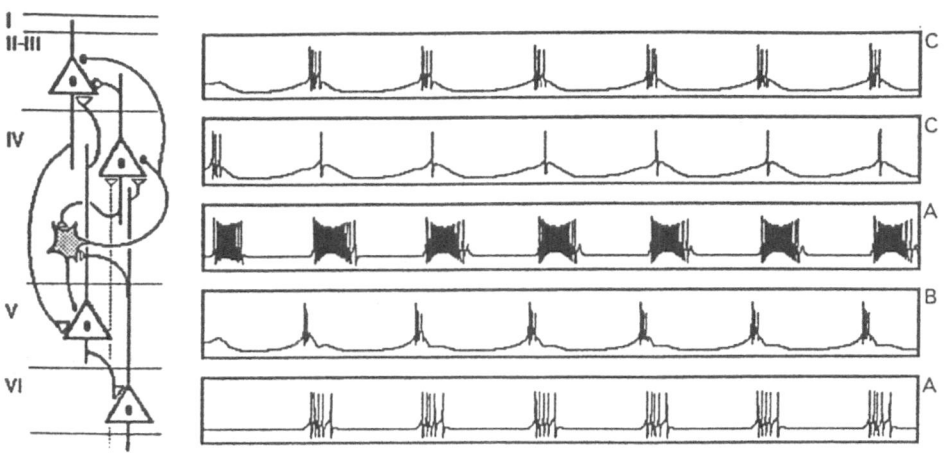

Figure 4: Results obtained from numerical integration of system (2) with N=5. Groups to which each neuron belong are shown in the left-hand side.

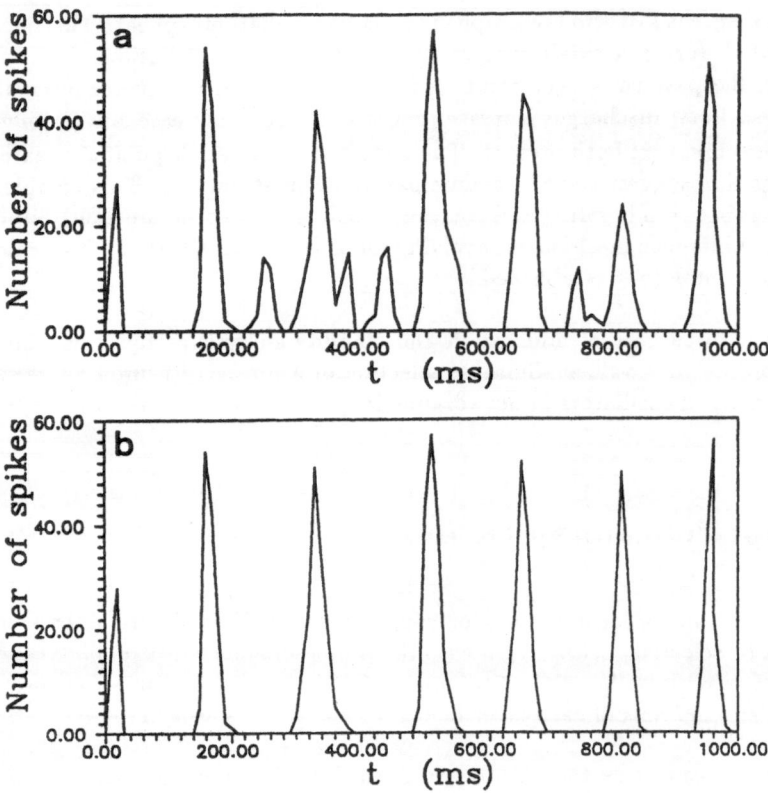

Figure 5: Network with N=10. Two excitatory neurons were added to each layer. A second inhibitory cells was included with wide local connections in layer IV.a. Desynchronization is seen when one of the layer V oscillators is of group A type.b. Synchronization obtained when λ_i is tunned to group B values.

Rhythmic activity is driven by layer V neuron which excite layers II-II and VI. That bursting neurons drive the oscillation within the microcolumn model is consistent with experimental suggestions by Chagnac-Amitai and Connors [6] and a fact that is present for several parameter combinations. The presence of type B neuron is essential for syncronization. Numerical simulations shown that when layer V oscillator was tunned to group C the oscillatory pattern was lost.

Adding more elements to the simple microcolumn model these results become more complex and interesting. First, desyncronization are more frequent but could be avoided selecting other elements in group B. On the other hand, the behaviour of a network with more spiking than bursting cells is quite different. For this last case rhythms are faster and the presence of several spike trains with diferent durations is accentuated.

In Fig.5 results are shown for a 10 elements network. Only one of the oscillators placed in layer V was tunned from group A to B and as can be seen the syncronization of the network is more easily reached. This kind of behaviour is very frequent in this largest network. Again, if we select all active oscillators belonging to type A group, the firing becomes more irregular and the rhythms increase significatively.

5 Conclusions

Several more combinations could be tested. However there is a neurophysiological support to selecting each type of oscillators and connections. It is important to note that with this simple network (less than 10 elements) some interesting experimental findings in cortical tissue could be simulated. The principal fact that allow these results is the consideration of a biological-based heterogeneity among the oscillators. A simple model like this could be theoretically simplified and thus lead to propose a "microcolumn device" that works on the input coming from the thalamus and other cortical regions. The combination of several of such heterogenous-established microcolumns could serve as a more realistic model to study cortical processes like rhythm generation, information proccesing or thalamocortical interactions.

Acknowledgements. I would like to thank Dr. J.V. Sánchez-Andrés for his support and encourage and Dr. R.V. Solé for his useful comments.

6 References

[1] Silva L, Amitai Y and Connors B. 1991. Intrinsic oscillations of neocortex generated by layer 5 pyramidal neurons. Science 251:432-435.

[2] Kandel ER, Schwartz JH and Jessel TM. 1991. Principles of Neural Sciences. Elsevier Science Publishing Co. Inc. 3th Ed.

[3] Hodgkin AL and Huxley AF. 1952. A quantitative description of membrane current and its application to conduction and excitation in nerve. J.Physiol.Lond 117:500-544.

[4] Lambert JD. 1986. A stable sequence of step-lengths for Euler's rule applied to stiff systems of differential equations. Comp. Math with Appls 1213 5/6: 1141-1151.

[5] Guckenheimer J and Holmes PJ. 1983. Nonlinear oscillations, dynamical systems and bifurcations of vector fields. Springer-Verlag: New York, Heidelberg, Berlin.

[6] Chagnac-Amitai Y and Connors BW. 1989. Synchronized excitation and inhibition driven by intrinsically bursting neurons in neocortex. J.Neurophysiol. 62:1149-1162.

[7] Terman D. 1992. The transition from bursting to continuous spiking in excitable membrane models.J.Nonlin.Sci. 2:135-182.

[8] Alexander JC and Cai DY. 1991. On the dynamics of bursting systems. J.Math. Biol. 29:405-423.

Cooperative Organization of Connectivity Patterns and Receptive Fields in the Visual Pathway: Application to Adaptive Tresholding

J. Mira, A. Manjarrés, S. Ros, A.E. Delgado and J.R. Alvarez

Dpto. Informática y Automática. Facultad de Ciencias. UNED
C/Senda del Rey s/n. MADRID, SPAIN
Telf.: (34)-1-3987155 Fax: (34)-1-3986697
e-mail: jose.mira@uned.es

Abstract

A biologically plausible theoretical framework to embody cooperative computation in the visual pathway is proposed. From photoreceptors to ganglion cells, visual processing is properly interpreted by means of linear and nonlinear spatio-temporal filters with center-periphery receptive fields and analogic computation.

At cortical level (mainly recurrent pyramidals) hybrid formulations using a combination of local operators (sum plus sigmoid) and conditionals (inferential rules) are more appropriate. This inferential model is quite general, supports analogic and logic computation as particular cases, and should be applicable to bridge the gap between connectionistic and symbolic artificial intelligence in general and between low level and high level vision, in particular.

To illustrate the possibilities of the model, topographic reorganization of connectivity patterns and receptive fields are considered. Adaptive thresholding as a consequence of cooperative consensus on homogeneity measures in the neighbourhood of each neuron has been simulated. Other properties such as self-organization of columns of contrast, orientation, speed or preferred direction can also be modelled as cooperative processes.

Keywords: Cooperative processes, inferential models, adaptive neighbourhood, functional receptive fields.

1   Biological Problem Statement

In this paper we propose a computational model of cerebral dynamics based mainly on the historical findings of Lashley [1], Luria [2] and Gonzalo [3] concerning the high residual function after traumatic and surgical lesions in animals and men [4, 5] and the more recent neurophysiological results of Gilbert and Wiesel [6] on receptive field dynamics in the adult primary visual cortex. Dynamic changes in the connectivity and receptive field structures may occur continuously during normal vision according to afferent information, selective attention and active search mechanisms. More drastic changes are also found after local lesions (retinal scotoma) with reorganization of the residual tissue increasing the receptive field size for those neurons with receptive fields near the edge of the retinal scotoma [6, 7].

Lashley [1] stated that intact parts of the cortex have the functional capacity to carry out the functions that have been lost by the lesion because all the neurons of a functional area are constantly active and participating in every instance of sensorial information processing. Luria [2] explored the multifactor cooperative theory of cortical tissue where the lesion eliminates "factors" but the perceptual function remains, albeit depressed.

Gonzalo [3], introduces the concepts of cerebral dynamics and cooperative perception based on computational parameters such as excitability, interneuronal permeability and differential sensibility as well as the more important concept of interneuronal reinforcement as an adaptive mechanism of cortical neurons.

The recent results of Gilbert and Wiesel [6] and Merzenich et al. [7] concerning the removal of afferent input to the somato-sensory, auditory, motor or visual cortex and the corresponding changes in cortical connectivity with topographic reorganization, strongly supports the classical proposals of Lashley, Luria and J. Gonzalo.

All these findings on the high reliability, functional stability and continuous reorganization of the cortical tissue suggest that neural processes at cortical level are cooperative since neurons are very complex processors with a cooperative organization and learning whose individual computations are not too relevant. Otherwise, the effect of lesion would be catastrophic, since standard lesions eliminate more than $8 \cdot 10^4$ cortical neurons.

The plexus of long-range horizontal connections and the multiplicity of non linear time dependent dendro-dendritic and dendro-axonic contacts as well as the well-recognized physiological findings on absolute facilitation and inhibition (conditionals), time and activity dependent threshold functions and great diversity of response firing patterns, to name but a few [8], has allowed us to formulate models of neural computation at two levels (figure 1):

(a) Low level (from photorreceptors to ganglion cells and from premotor areas to motoneurons).

(b) High level (from primary cortical areas to motor cortex).

Any specific activity, be it language, movement or perception, is the result of the coordinated and cooperative action of a set of neuronal structures at these levels. Unfortunately, our computational knowledge of the real neurons and neural nets is very limited. The structure-function correlation has only been specified in the low level (from photoreceptors to ganglion cells) where visual information processing, for example, can be experimentally evaluated and properly modeled in terms of spatio-temporal filters (recurrent and not recurrent) with center-periphery receptive fields (ON, OFF, ON-OFF) and other lateral interaction generalized kernels. Within each layer (amacrine, bipolar, horizontal or ganglionar) neural computation is represented by "*context*" and *receptive field* (weighted excitation and inhibition). To understand this analogic computation we also need to know the input and output spaces where these neurons sample information from labelled lines and discharge results. The input space also incorporates the meaning of all previous processes. The output space transports the recoding of this information.

In modelling these low and high level neurons, we are making formal representations of experimental data and theories in terms of formal tools. This means that the properties and limitations of these mathematical tools (integro-differential equations) are imposed on the models. In fact, to be more precise, the conclusions obtained from a model are always implicit from the nature of the mathematics used in its formulation. In the low level case (up to ganglion cells), these tools are more or less appropriate. Per contra, in the cortex, at the first gnostic level, where information has been recodified, integrated with other sensorial modalities and "conceptualized", the usual analytic tools of the low level are not adequate. The correlation structure-computation can not be specified. For this reason, our description of cortical determined behavior can only be at the level of information flow and organizational principles, as well as of possible categories of data and processes (data structures and algorithms) presumably associated with such neurons and anatomical networks. The cooperative model proposed for these high level neurons can only be considered in this sense.

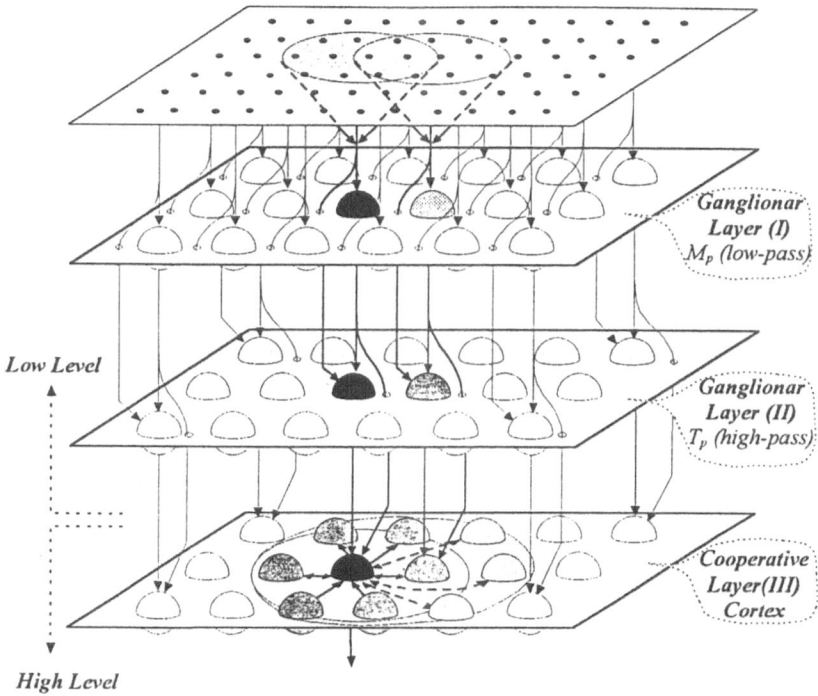

Fig. 1.   Schematic representation of the visual pathway with the distinction between low level analytic
processing (up to ganglion cells) and high level inferential and cooperative processing (from pyramidals
in primary cortex

At cortical level we are making formal representations of pyramidal, stellate or horizontal cells in terms of inferential rules and cooperative processes on neurophysiological symbols. Obviously many components of the model are still purely functional elements.

These symbols would designate things, relations between things, and basic concepts relevant to survival. The accumulation of processes resulting from cultural evolution also add new meanings to these primary-necessity attributes.

In spite of these considerations, it should be made clear that *cortical computation is always connectionistic*. The symbolism is born in the domain of the external observers. *Neurophysiological symbols corresponds to specific patterns of spatio-temporal signals* ("keys"), with an initial referent in the external world, and the synchronous activation of the corresponding anatomophysiological structures ("doors" open by these "keys"). These cortical keys act as transitory dynamical bindings, have been *acquired* by genetics or by learning, *represent* (substitute) the external referent in all the subsequent computations and are *supported* by the long-term structures of the cortex.

2.   Spatially Variable Cooperative Thresholding

To illustrate the usefulness of our proposal on cooperative processing at the cortical level, we use the image-processing problem of adaptive local thresholding based on the dynamic recruitment or rejection of neighbourhood neurons.

The choice of a working scale implies a concrete *description level* at which a specific set of properties are manifest. The concept of *adaptive neighbourhood* has been used in artificial vision to select a working scale at each point of the image and at each level of computing [9][10][11]. Both the size and the shape of this neighbourhood depend on the image data characteristics through parameters defining homogeneity measures in the neighbourhood of each pixel. These homogeneity measures are related to either the grey level of the pixels or to other more complex properties of the image such as texture or local motion parameters (computed by layers I and II). The integration is carried out by a cooperative activity among neurons inside a defined neighbourhood. Thus, if two units are not connected, their local measures do not interfere. Dynamically adapting different neighbourhoods allows properties of different scales to be estimated. Neurons (in layer III) can redefine their neighbourhoods as a function of certain homogeneity measures continuously evaluated over data collected from their receptive fields (apical and basal) at each instant. The problem consists of injecting external knowledge in the design of the net concerning the similarity concept which best identifies the appropriate neighbourhoods.

Self-organizing learning is biologically more plausible than supervised algorithms and implies the use of local criteria on previous data. Supervised learning is a purely functional element without any clear neurological counterpart. Neighbourhood parameters are instantiated by some initial values and the network is executed on a set of training images. After each execution, neighbourhoods are modified using a local error function which compares the portion of image recently processed with the ideally processed image the network uses for feedback.

Adaptive thresholding methods consider grey level distributions around the borders detected in the image as relevant information for threshold estimation. Local thresholding techniques look for a sort of background normalization and in this way compensate the effects of spatial context. The threshold level is varied so as to be adapted to the local grey levels in the neighbourhood of each point, thus preserving local contrasts and resulting in less loss of information. The diverse techniques for local threshold implementation can be grouped into three categories: local thresholding sensible to lines and borders, thresholding based on grey level averages and smooth local thresholding [12,13]. The aim of this simulation is to give a neural solution to the image thresholding problem incorporating the benefits of local and adaptive thesholding in addition to the good characteristics of cooperative computation.

3. Simulation of Layers I, II and III

The network architecture is a particular case of the one presented in figure 1. Each layer consists of 64x64 neurons and the region size is 5x5 pixels on 256x256 pixel images. In this simulation, the property space is bidimensional and the properties measured on pixels are their grey levels and a threshold value estimated in the region to which they belong. The local threshold calculation is distributed over three intercommunicated sublevels: there are two layers (I and II) performing parallel local calculations on the image regions (mean value and distribution type in the zone) and a third cooperative layer (III). The last one integrates the previous results by generating local thresholds which will serve as binary classification criteria for the pixels. Since it is a thresholding, the output intensity space is binary.

Layer I receives as inputs the pixel intensities and calculates the mean grey-level value in the region. *Layer II* receives as inputs the pixel intensities and the values computed by the previous layer. It classifies zones depending on whether they exhibit contrast or not. The functional expression has been formulated by considering the histogram of the zone corresponding to either a unimodal or bimodal pattern. Tonality transitions on real images are never abrupt enough to expect more than two significant modes to appear in an image of the size considered.

The computation in layer III is summarized in figure 2. After receiving two inputs coming from layers I and II, each neuron initiates a data interchange with the neurons inside its neighbourhood.

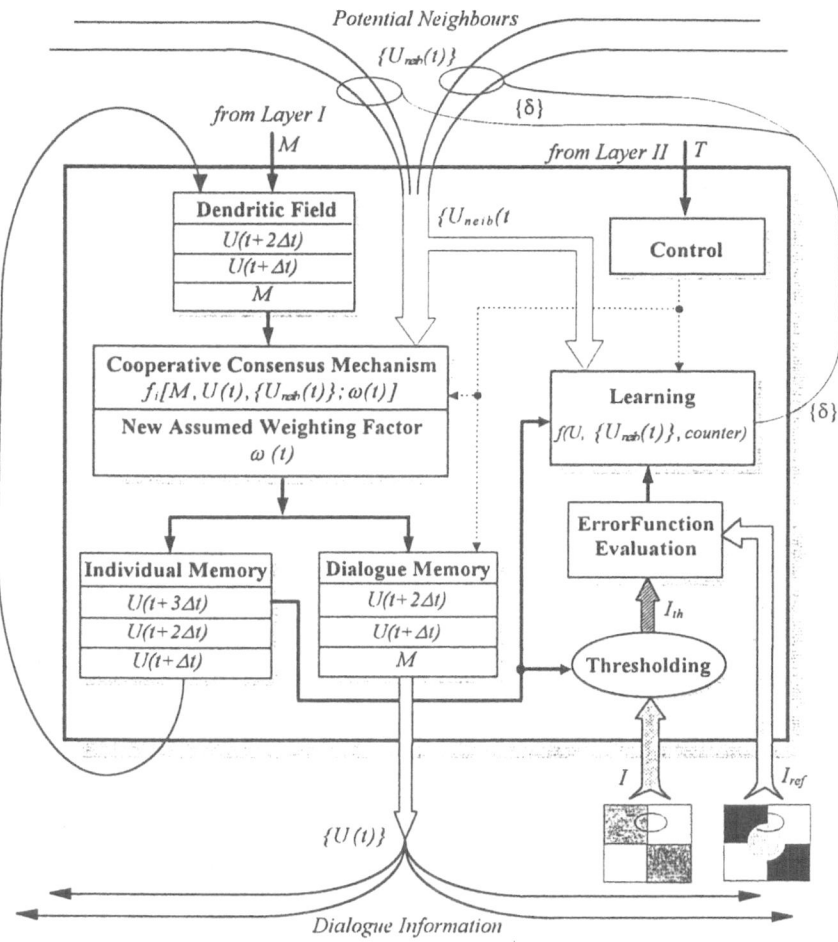

Fig. 2. Cooperative model for cortical neurons. Information from layers I and II is processed, then a dialogue algorithm produces consensus on the values of the property considered (threshold) and, finally, the cooperative learning dynamically reorganizes the connectivity pattern.

The neurons inside a dialogue area that have recieved a "1" value from layer II (visualization of high contrast regions) broadcast the value "M" coming from layer I while the remaining neurons (uniform regions) spread a "NO_OPINION" value, indicating that they cannot contribute any valuable information to the dialogue. In this way, an adaptive thresholding has been performed. Neurons lacking contrast information adopt a threshold resulting from the interpolation of values computed in neighbouring regions. Next, each neuron "reads" the data that comes through its connections and averages the non-zero values giving the same weight to all of them (including its own). This new average value is established as the definite local threshold of the region. A summary of the learning algoritm is shown in figure 3.

LAYER I	Input: $\{I\}$ Output: M

$$M = \sum I / N \qquad N = \#\{I\}$$

LAYER II	Input: $M, \{I\}$ Output: T

$$M_< = \sum_{I < M} I \Big/ \# \{I \mid I < M\} \qquad\qquad M_> = \sum_{I > M} I \Big/ \# \{I \mid I < M\}$$

$\Delta_<$ standard deviation around $M_<$ $\Delta_>$ standard deviation around $M_>$

if $(M_< + \Delta_< + \mu) > M$ OR $(M_> - \Delta_> + \mu) < M$ $T = 1$ else $T = 0$

LAYER III	Step $t = 0$ Input: M, T Output: $U(0)$

     if $T = 1$ $U(0) = M$ else $U(0) = No\_opinion$

     Step $t = 1$ Input: $\{U_{neib}(0)\}$ Output: $U(1)$

     $w(t) = w(t-1) + \Psi, \qquad w(0) = 1$

     if $U(t-1) \neq no\_opinion$:

$$U(t) = \left(\sum_{U_{neib} \neq no\_opinion} U_{neib}(t-1) + w(t) \cdot U(t-1) \right) \Big/ \left(\# \{U_{neib}(t-1) \neq no\_opinion\} + w(t) \right)$$

     else   if $\forall U_{neib} = no\_opinion$ $U(t) = no\_opinion$

     else $U(t) = \sum_{U_{neib} \neq no\_opinion} U_{neib}(t-1) \Big/ \# \{U_{neib}(t-1) \neq no\_opinion\}$

Step $t = 2, \ldots, n$ LEARNING

     Repeat step $t = 1$ with   Input: $\{U_{neib}(t-1)\}$ Output $U(t)$

Step $t = n$ Input: $\delta_{ur}, \delta_{ul}, \delta_{dr}, \delta_{dl}, \{U_{neib}(n-1)\}, I_{th}, I_{ref}$ Output: $\delta_{ur}, \delta_{ul}, \delta_{dr}, \delta_{dl}$

     $count_0(I_{th}, I_{ref}) = $ erroneous 0 pixels, $count_1(I_{th}, I_{ref}) = $ erroneous 1 pixels

     $U_{ur,ul,dr,dl} = \{U_{neib}(n-1)\}$ average in quadrants ur,ul,dr,dl

     $count\_0 > count\_1$ if $U_{ur,ul,dr,dl} > U(n-1)$ $\delta_{ur,ul,dr,dl} = \delta_{ur,ul,dr,dl} + \varepsilon$

                    else $\delta_{ur,ul,dr,dl} = \delta_{ur,ul,dr,dl} - \varepsilon$

     $count\_0 < count\_1$ if $U_{ur,ul,dr,dl} > U(n-1)$ $\delta_{ur,ul,dr,dl} = \delta_{ur,ul,dr,dl} - \varepsilon$

                    else $\delta_{ur,ul,dr,dl} = \delta_{ur,ul,dr,dl} + \varepsilon$

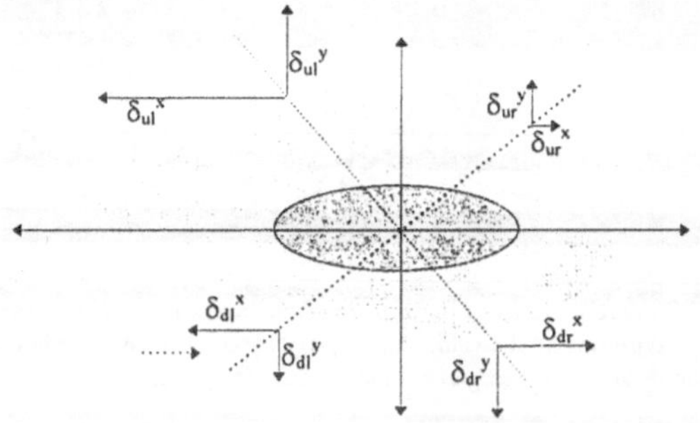

Fig. 3.   A summary of the algorithms in layers I, II and III. In the bottom of the figure the effect of the cooperative learning on the size and form of apical receptive fields is shown.

4.   Dynamic Learning of the Receptive Fields

The objetive of training is to locate zones where global thresholding is apropriate in order to detect contrasts that are interesting to the application. The learning parameters are those described in the dialogue area ("neighbourhood") of the neurons temporarily linked by cooperation. These parameters are 8 integral values that define elliptical sectors as can be seen in figure 3. As the net is being trained, the receptive fields are deformed in such a way that zones where the neurons will dialogue about the more

21

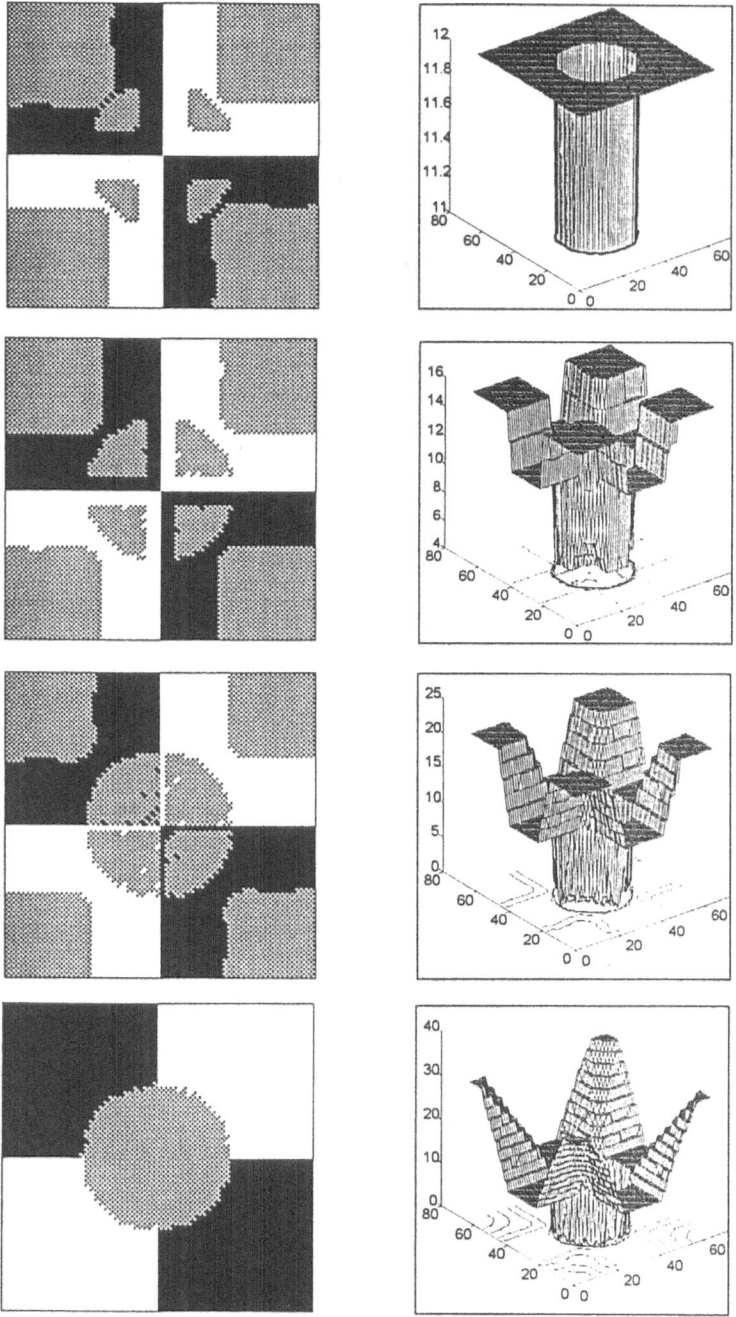

Fig. 4.   Extracting textures with different scales (see description in the text).

apropriated threshold value are defined. The locality or scale of the thresholding is dynamically redefined by the size and shape of the receptive fields.

The training is supervised, based on the ideal thresholding of an image family. The algorithm concludes when the network thresholds the training images with an error under a preestablished threshold value, the error being measured by means of the number of pixels wrongly classified as either object or background.

Possible applications of this cooperative model are the extraction of textures and/or geometrical structures at different levels (e.g. microscopic mineral images and biomedical image processing for diagnosis) as well as recognition of faults in mechanical parts and any other field were it is useful to distinguish among regions which have some property in common which has values in bounded intervals.

In figure 4 we show an example of texture extraction with different scales. The left column shows steps in the evolution of image thresholding during learning. The corresponding pictures in the right hand column represent the parallel evolution of the apical receptive fields (parameter δ_{ni}^x) towards the most suitable values under each texture.

Discussion

We have presented a biologically plausible model of cooperative computation at cortical level. The proposed model supports analogic, logic and inferential computation as well as correlational and supervised learning algorithms. Many components are still purely functional (extrinsic to the biological knowledge).

If the cooperative parameters are the connectivity pattens and the receptive fields, the model proposed is able to explain relevant experimental results on topographic reorganization within the sensorial cortex.

Additionally, this cooperative model is quite general and could be of interest in adaptive thresholding, image segmentation, texture processing and other problems of artificial vision whose formulation requires dynamic binding and integration of low level features.

Acknowledgments

This work has been supported by the Spanish Comisión Interministerial de Ciencia y Tecnología (CCICYT) under projects TIC-92-136 and TIC-94-95.

References

[1]   Lashley, K.S. "In Search of the Engram". *Society of Experimental Biology Symposium* n° 4: *Physiological Mechanisms in Animal Behaviour*, University Press, Cambridge, 1950, pp. 454-480.

[2]   Luria, A.R., *El Cerebro en Acción*. Ed. Fontanella. Barcelona, 1974.

[3]   Gonzalo, J. Las Funciones Cerebrales Humanas según Nuevos Datos y Bases Fisiológicas. *Intituto Cajal de Investigaciones Biológicas*, Vol. XLIV. Madrid, 1952.

[4]   Delgado, A.E., *Modelos Neurocibernéticos de Dinámica Cerebral*. Tesis Doctoral, ETSIT, Madrid, 1978.

[5]   Mira, J., Delgado, A.E. Zapata, E.L. & Cabello, D. "On the Lesion Tolerance Problem for Co-operative Processes". In *Implementing Finctions: Microprocessors and Firmware*. Ed by L. Richter, P. Le Beux, G. Chroust and G. Noguez. pp. 71-80. North-Holland Publishing Company. Amsterdam, 1981.

[6]   Gilbert, C.D. & Wiesel, T.N. "Receptive Field Dynamics in Adult Primary Visual Cortex". *Nature*, Vol.356, 12 March 1992. pp. 150-152.

[7]   Merzenich, M.M., Recanzone, G., Jenkins, W.M., Allard, T.T. & Nudo, R.J. (1988) "Cortical Representational Plasticity". In P. Rakie & W. Singer, eds. *Neurobiology of Neocortex*. Wiley.

[8]   Mira, J., Delgado, A.E., Alvarez, J.R., de Madrid, A.P. & Santos, M. "Towards More Realistic Self Contained Models of Neurons: High-Order, Recurrence and local learning". In J. Mira, J. Cabestany and A. Prieto eds. New trends in neural Computation, LNCS 686. Pp. 55-62. Springer Verlag, 1993.

[9]   Topkar, V., Kjell, B. and Sood, A. "Object Detection Using Scale-space. *In Proceedings of the Applications of Artificial Intelligence* VIII Conference, The Int. Society for Optical Engineering, pp. 2-13, Orlando, Fl, April 1990.

[10]   Witkin, A.P. "Scale-Space Filtering". In *Proc. of the 8th Joint Conf. on Art.I Imtellig.e*, pp. 1019-1022, Karlsruhe, Germany, 1983.

[11] Paranjape, R.B., Rangayyan, R.N., Morrow, W.M. and Nguyen, H.N. "Adaptive Beighborhood Image Processing".
      In *Pro. of Visual Communications and Image Processing*, Boston, Ma, pages 198-207, SPIE, Bellingham, Wa, 1992.
[12] Haralick, R. M. and Shapiro, L. G. *Computer and Robot vision*.Addison-Wesley Pub. Comp.
[13] Sahoo, P.K., Soltani, S., Wong, A.K.C. and Chen, Y.C. "Survey of Thresholding Techniques". *Computer Vision,
      Graphics, and Image Processing*, 41(2):233-260, 1988.

Neurobiological Inspiration for the Architecture and Functioning of Cooperating Neural Networks

Frédéric ALEXANDRE[1] & Frédéric GUYOT[2]

1. CRIN-INRIA
BP 239
54506 Vandoeuvre Cedex
FRANCE
falex@loria.fr

2. Institut des Neurosciences
9, quai Saint-Bernard
75005 Paris
FRANCE

Abstract:

In order to emulate more complex and more realistic human-like functions, it is now well admitted that a single monolithic neural network is not sufficient. Biological data show that the cortex is a set of inter-connected neural networks. Beyond the classical view of one way feedforward neural network guiding an information flow from an input to an output layer, we now have to imagine architectures and functioning rules that permit cooperation and information exchange between such neural networks. Inspired with biological data, we propose here such a scheme bringing into play of complex units like the cortical column, a functional micro-circuit repeated throughout the cortex. This basic unit of treatment gathers the classical weighted sum for feedforward information flow and sigma-pi operations for cooperation between different axis of treatment. We illustrate this model with an application of cooperation between character recognition and localization.

Introduction

Cortical architecture and functioning can be of interest, as one wants to design connectionist models with complex structure allowing typical human tasks. More precisely, the cortex is a very good example of a distributed structure, receiving different information flows and able to make them cooperate in a coherent way and enrich one another. With such a framework of inspiration, our work of modelization consists in progressively integrating the very numerous elements at our disposal from the neurobiological field. Even if we consider neurobiological models such as Burnod's cortical column (Burnod, 1988), the proposed mechanisms, structures and rules are still very complex (even if, it is clear, the neurobiological reality is even much more complex) and cannot be integrated in one go in a computational model. That is why our work has consisted, since 1987, to progressively integrate, adapt and assess these hints from a computer

This work was partly supported by the Centre National d'Etude des Télécommunications, Lannion science point of view. Here we focus on a fundamental issue, namely the cooperation of connectionist models performing complementary tasks. As an illustracion, we will report experiments on the classical application of characters recognition. We first begin with a general presentation of the global model from which we get inspired, with emphasis on the points that we will retain here.

1. Framework of work

1.1. The biological model

The cortical column is a biologically inspired model, proposed by Y. Burnod from the Institut des Neurosciences, Paris (Burnod, 1988). The cortical column is made up of a set of neurons with a specific functionality and it is repeated throughout the cortical sheet.

The main principles of this model can be summarized as follows: we design a high level basic unit (the cortical column), able to perform more sophisticated operations than the classical formal neuron. These operations are performed on specific inputs and outputs which are differentiated and processed with regard to their origin.

These different kinds of input and output are relative to the cortical connectivity. A cortical column can be linked to external world feedforward information flow, to neighbour units, performing different tasks on the same information, to close units, performing the same task on another close piece of information, or to units in other areas. These complex operations consist in tuning a transfer function on the feedforward flow as well as modulating another information flow.

These connectivity principles directly refer to the general architecture of a cortical column based network. (Burnod, 1988) proposes a general plan including all the kinds of cortical area that are classically described. These areas are generally topological maps and are connected with upper and lower areas through feedforward information flows. They are also connected to other areas of other information flows. All these inter-areal connections are not total, but local with limited receptive fields. These information flows are sensorimotor modalities (e.g: visual, auditory pathways) or even functional characteristics inside these modalities (e.g: visual recognition or localization). We will detail below the visual part of this network of areas, in which we are interested here.

A very interesting information representation can be obtained with intra-areal connections. In fact, each of these inter-areal connections listed above does not reach one cortical column, but a set of cortical columns, named maxicolumn. A maxicolumn thus gathers a set of functional units performing different transfer functions on the same information. Hence, connections inside a maxicolumn are generally inhibitory connections.

More interestingly, connections between maxicolumns in a same area can be very fruitful. For low level sensory areas, transfer of learning is thus made possible from a maxicolumn to another.

1.2. The visual flow

The receptive fields, defined as local neighbourhood on the data flow, are well described for visual processing (Hubel & Wiesel, 1962). Some retinal observations show increasing receptive field sizes as one moves away from the visual central axis. This is classically and roughly modelized by a foveal vision, precise but local, and a peripherical one, fuzzy but global. Once in the cortex, the visual flow is split in two complementary pathways. For peripherical vision, the parietal pathway is composed by areas such as V3, whose cells are specialized in 3D movement detection or VIP for ocular movement presentation. On the other hand, the temporal pathway (areas V4, STS, IT) is more precisely responsible for local pattern recognition.

In visual tasks, the parietal cortex can be responsible of eyes movement (to have a good fixation point), but is also responsible of interaction with the temporal cortex where internal shifts are made to have an exact placing of the pattern for recognition. Psychological experiments have shown that, without this mechanism, our recognition system has hardly translation invariance (O'Reagan, 1990).

If the former mechanism has been extensively studied and modeled (e.g: Alexandre et al., 1990), the latter is much harder because it does not refer to an external loop as eye movement does, but to internal influence of one neural network on another. Recently, (Olshausen et al., 1993) have proposed such a model and have related its function to visual attention. Nevertheless, in this paper, only the routing of the information is proposed. We will extend it to functioning and learning principles.

1.3. Previous experiments

From the principles of the model above, we carried out several implementations, including visual (Alexandre et al., 1990), speech (Guyot et al., 1990b) and symbolic (Guyot et al., 1990a) processing to assess that this model was really a general model for human-like functions. From this rather heavy model, we also derived a more simple version and applied it to character recognition (Alexandre et al., 1992).
The goal of this application was to assess the capacities of such a model for classification tasks. We obtained similar results as the ones obtained with a MLP. We can also make other comparisons with this kind of model. Indeed, one of the most efficient MLPs for character recognition (LeCun, 1990) also uses topological maps and weight sharing principle also enables it to transfer learning between units in a same layer.
If we deeply analyse the performances of our model on this character recognition task (Alexandre et al., 1994), we constat that almost all the errors are made because the location of the character was bad. We first tried to learn the corpus with as many as possible translations, but found the same limitations as for MLPs. This technique is interesting to generalize learning on some pixels but gives bad results as we have too much (too many possible places) to learn.
From the general principles of the visual axis that we evoked above, there was obviously another solution. The solution was to consider that the network that we designed for recognition was the temporal axis and that we had to add a parietal axis for localization. We explain below how we build this additional network and also how we connect both networks.

2. Two cooperating networks

From a global point of view, the system that we designed is composed with structures that, from a biological point of view, can be related to a temporal axis (for character recognition) and a parietal axis (for character localization).

2.1 The temporal axis

The structure of this network is very classical. It is a multi-layered network with an input layer (the image), three hidden layers and an output layer with 26 units for the 26 characters. From layer to layer, the connectivity is not complete, but distributed along overlapping receptive fields. The size of the receptive fields and of the overlap have been designed in order to respectively correspond for the three hidden layers to a possible shift of 2, 4 and 8 pixels with regard to the input layer. The forward flow is classically evaluated with a weighted sum of inputs defined by the receptive fields. The principle of weight sharing is used here. A receptive field is received by a set of units specialized by learning on a specific transfer function on this feedforward input. Close receptive fields are received by close set of units. Intra-areal connections between these sets of units enable to share their weights and to propose identical specialized functional units throughout the layer. The original part of functioning takes part in this axis but will be described in subpart 2.3. For the moment, we only perform a classical learning with centered characters (only +/-2 pixels of translation are allowed for generalization), with the classical back-propagation learning algorithm. We report below (fig.1) the results of the convergence of this learning which is quite usual for this kind of task.

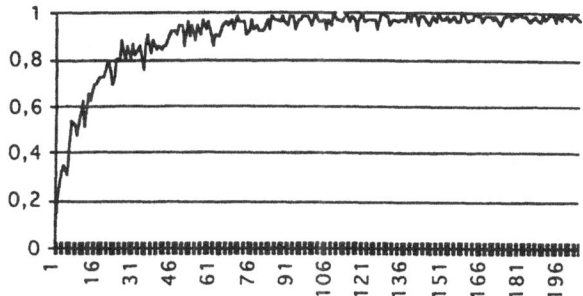

figure 1. Convergence of the temporal axis
(recognition rate on the y axis and number of thousand learned examples on the x axis)

Then, we tested the invariance to translation of this network. We report below (fig. 2) the recognition rate with regard to the maximal shift allowed. It seems thus clear that beyond 2 pixels, the invariance does not exist. So, we need another mechanism to allow this invariance.

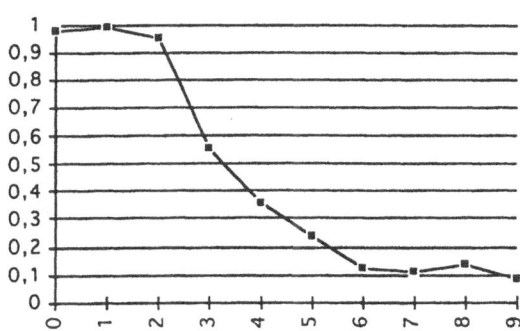

figure 2. Invariance to translation of the temporal axis alone
(recognition rate on the y axis and maximal shift allowed for testing on the x axis)

2.2 The parietal axis

From a scannerized sheet of paper, we first use a classical connected component segmentation to globally detect each character. Then, the goal of this network will be to precisely localize characters (we have seen above that the precision had to be high, close to the pixel, to obtain a good recognition). Its input is the global view of the roughly localized character (30x30 grid). Two hidden layers are then used to learn, on three different output layers, possible shifts of respectively 2, 4 and 8 pixels. (Thus, the maximal shift will be here 14 pixels). The learning technique consists in randomly translating a character along the x and y axis (with a maximal shift of 14 pixels on both axis) and learning the corresponding decomposition in the powers of two (knowing that a shift of one pixel can naturally be learned by the temporal neural network). For this learning task, the back-propagation learning algorithm can also be used.
We report below (fig. 3) the learning of this axis alone

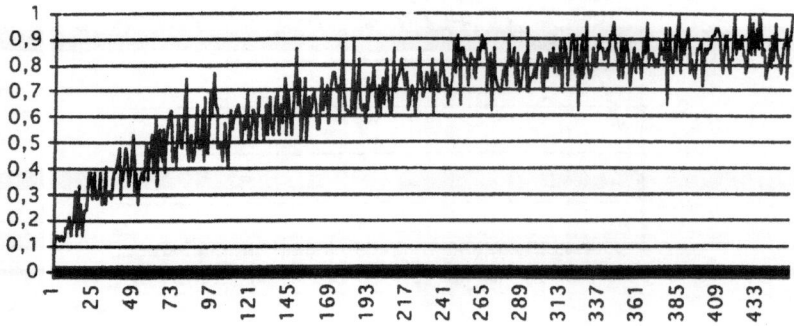

figure 3. Convergence of the parietal axis
(recognition rate on the y axis and number of thousand learned examples on the x axis)

2.3 Design of the cooperation

Here is the central part of our contribution. For the moment, we have two networks at our disposal performing two different tasks. The goal here is not to use the results of the parietal axis in an external way to produce a real shift of information but rather to see how the parietal axis could act upon the temporal axis in an internal way to produce an internal shift. In order to achieve such a goal, we have to introduce the sigma-pi mechanism which is known for a long time (Rumelhart et al., 1986), but is rarely used.

Whereas the classical activation function corresponds to the weighted sum of inputs, the sigma-pi activation function corresponds to weighted sum of product of inputs. We can notice here that the weighted sum is a linear function that enables to tune transfer functions in an additive way. On the other hand, the sigma-pi activation function is multiplicative and can be considered as gating inputs with regard to others. The gating is obtained in that sense that if one input is zero, other inputs will have no effect; if it is 1, other inputs will pass unchanged. Between these two values, this input will modulate (as a weight does) the others. We must also notice here that, if specific learning algorithms are available (Guigon, 1993), it is very simple to derive one from the backpropagation learning algorithm (Rumelhart et al., 1986).

We want to apply here this gating effect between temporal and parital axis and will thus choose conjuncts of size two. We want the parietal axis to modulate the activity of the temporal axis in such a way that it inhibits information when non centered and gates the activity at close centered places. This should be done in an internal way, only on internal activity and not on real information.

Consequently, we connect the three hidden layers of the temporal axis with the corresponding three output layers of the parietal axis. The functioning of the maxi-units in the temporal axis becomes:
-weighted sum of the feedforward inputs
-lateral inhibition inside a maxicolumn (links with units performing different tasks on the same input)
-sigma-pi activation between parietal units and units performing the same task in other maxicolumns of the same area.
As we saw, learning first consists in a classical backpropagation on both axis, then an adapted backpropagation on the sigma-pi links is performed.
We report below (figure 4) the results obtained on the recognition rate with regard to the maximum allowed shift. From bottom to up, the four curves respectively correspond to recognition
-without cooperation (cf. above)
-with cooperation

-with cooperation, with sigma-pi learning allowing +/- 7 pixels of translation
-with cooperation, with sigma-pi learning allowing +/- 1.' pixels of translation

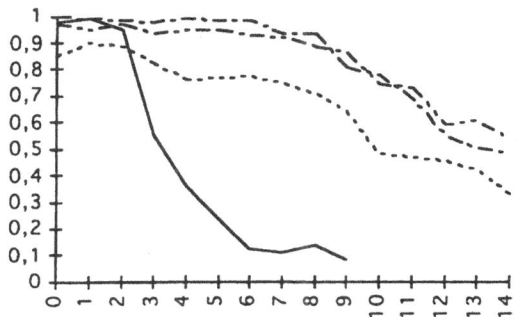

figure 4. Invariance to translation of the temporal axis in cooperation with the parietal axis
(recognition rate on the y axis and maximal shift allowed for testing on the x axis)

Discussion

Results reported in figure 4 indicate that the cooperation designed between the parietal and temporal axis allows for an invariance in translation. Rather than results on this particular application, we prefer to emphasize here the mechanisms that have been introduced to obtain them.

The first one is sigma-pi units. This kind of bilinear model have been known for a long time, but rarely used because of its complexity with regard to classical linear units. Interestingly, biologically oriented works (Burnod et al., 1992; Guigon, 1993) have reported a strong biological plausibility for this model of bilinear combination. This also refers to the distinction between cortical tuning and gating operations. On the other hand, we can also simply use this kind of model for its computational interest to represent multidimensional receptive fields or context-dependent feature detectors (Pican et al., 1993).

This double possibility of biological inspiration or not must also be put forward concerning learning algorithm. We emploied here the backpropagation learning algorithm, which is not very biologically plausible, for the sake of clarity and simplicity and to enlight architectures and functioning rules that were our topic here. Nevertheless, we must mention that it is also possible to integrate other more biologically plausible learning schemes for this kind of treatment (Alexandre et al., 1990; Burnod et al., 1992).

The second point that we want to underline here concerns the architecture of our network. Functionally oriented axis of treatment are frequently described in the cortex and offer a very fruitful basis to build a non-monolithic, understandable artificial neural network. From a more microscopic point of view, the same can be said about the maxicolumn principle which corresponds to the design of a toolbox for information analysis.

On this basis, the internal shift that we propose here is made possible only by the use of weight sharing process. Indeed, this transfer of learning provides the areas with homogeneous transfer functions that can support translation of activity.

We have tried to show here that the adaptation of cortical mechanisms to computational model can lead to good results but above all to very original functions such as cooperation of neural networks. On this basis, we now try to integrate other cortical characteristics in our models to ameliorate their capabilities, but also to provide them with other fundamental properties such as time processing.

References

Alexandre, F., Guyot, F., Haton, J.P. (1990) A connectionist network with two complementary visual processing systems for x-ray image interpretation, INNC'90, Paris.

Alexandre, F., Burnod, Y., Guyot, F., Haton, J. P. (1991) The Cortical Column: a new processing unit for multilayered networks, Neural Networks,Vol 4, n 1, pp. 15-25.

Alexandre, F., Guyot, F. (1992) A connectionist model constrained by an optical implementation, Int. Congress on Artificial Neural Networks, Brighton, Sept. 92.

Alexandre, F., Guyot, F. (1994) Evaluation d'un modèle connexionniste simple pour la reconnaissance automatique de caractères, Colloque National sur l'Ecrit et le Document, Rouen, Juillet 1994.

Burnod, Y. (1988) An adaptive neural network: The cerebral cortex, 2nd Edition, Masson, Paris.

Burnod, Y., Grandguillaume, P., Otto, I., Ferraina, S., Johnson, P., Caminiti, R. (1992) Visuomotor Transformations Underlying Arm Movements toward Visual Targets: A Neural Network Model of Cerebral Cortical Operations, Journ. of Neuroscience, 12, 4, 1435-1453.

Guigon, E. (1993) Modélisation des propriétés du cortex cérébral, PhD Thesis, Ecole centrale de Paris

Guyot, F., Alexandre, F., Haton, J.P. (1990a) Principles and applications of the cortical column symbolic neural model, IJCNN'90, San Diego.

Guyot, F., Alexandre, F., Dingeon, C., Haton, J.P. (1990b) The Cortical Column as a Model for Speech Recognition: Principles and First Experiments, in Speech Recognition and Understanding Recent Advances, Trends and Applications, Springer Verlag.

D. H. Hubel, and T. N. Wiesel, Receptive fields binocular interaction and functional architecture in the cat's visual cortex. J. Physiol. Lond., 160, p. 106-154, 1962.

[LeCun, Y.,] Boser, B., Denker, J., Henderson, D., Howard, R., Hubbard, W., Jackel, L. (1990) Handwritten Digit Recognition with Back-propagation Network, NIPS, vol.2, Morgan Kaufmann.

Olshausen, B., Anderson, C., Van Essen, D. (1993) A Neurobiological model of visual attention and invariant pattern recognition based on dynamic routing of information, The Journal of Neuroscience, 13 (11), pp. 4700-4719.

O'Regan, J. K. (1990) Les "vrais" mystères de la vision, 5èmes journées NSI, Aussois.

Pican, N., Alexandre, F. (1993) Integration of Context in Process Models used for Neuro-Control. Proc. IEEE Systems Man and Cybernetics, Le Touquet.

Rumelhart, D., Mc Clelland, J. (1986) Parallel distributed processing, MIT Press, Cambridge.

Synaptic Modulation Based Artificial Neural Networks

R. J. Duro[1], J. Santos[2] and A. Gómez[2]

[1]Dpto. Ingeniería Industrial
[2]Dpto. Computación.
Universidad de La Coruña
Spain

Abstract

This work introduces complex processing neural network topologies, based on the concept of modulating neuron, which induce higher order terms by means of the modulation of the synaptic weights. These structures present the advantages of being very easy to train, adapting easily to changing contexts and offer very good generalization capabilities along all the dimensions of the problems they are trained to solve. Finally, the function each modulation level or each module performs is very clear, making it simple to extend the model to multilevel hierarchies.

1. Introduction

Different higher order connectionist systems have been proposed in order to endow Artificial Neural Networks (ANNs) of greater capabilities [1]. The objective of these systems is, on one hand, to reduce the number of neurons needed for the implementation of a given function; on the other to introduce higher order correlations among the inputs, leading to richer behaviors and to a speed up in the convergence of the training process, and finally, to rationalize the design process of complex neural networks, allowing their functional modularization and improving the generalization capabilities along all of the dimensions of the problem.

In general, the increase in the order of the ANNs has been achieved through an increase in the complexity of the transfer function of the neurons, leaving the synapses, after the training process, with the fixed weight character they were assigned in the models proposed by McCulloch and Pitts [2].

From the point of view of biological neurons, this is a partial view, as synaptic plasticity is not taken into account except during the training process. If we take a look at the literature, we can find different mechanisms of synaptic plasticity or heterosynaptic regulation, that is, the regulation of the efficiency of a synapsis by means of the activity of a neuron or another synapsis. Some of these mechanisms were proposed by Kandel and Hawkins as a result of their studies on the Aplysia [3], and later by Silva et al. [4] and other authors which classified them into three basic groups:

R. J. Duro, Dpto. Ingeniería Industrial, Universidad de La Coruña, Escuela Politécnica Superior, Esteiro s/n 15403 Ferrol, A Coruña, E-mail: richard@gaes.usc.es, J. Santos and A. Gómez, Dpto. de Computación, Facultade de Informática, Elviña s/n, 15071, A Coruña, E-mail: santos@udc.es and agomez@udc.es.

sensitivation, long term potentiation (LTP) and long term depression (LTD). Dehaene and Changeaux [5][6] proposed a processing model based on neuron triads, ABC, where the efficiency of the synapsis between A and B is influenced by neuron C, which was called modulating neuron (figure 1). This type of synaptic modulation can be thought of as a first order modulation. Peretto and Niez [7] extend the concept of modulating neurons to synapses of any order in their attempt to model multisynaptic contacts reaching a dendrite.

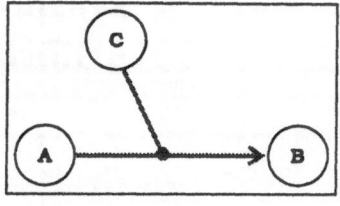

Fig. 1. Dehaene and Changeaux's diagram of synaptic modulation.

In this work we have taken inspiration from the concept of modulating neuron in order to establish complex processing neural network topologies which introduce higher order terms by means of synaptic weight modulation with the basic aim of obtaining neural networks that can adapt to changing contexts and which are capable of modifying their behavior as a function of external stimuli.

2. Synaptic Modulation Based Artificial Neural Networks

We start from an ANN structure divided into two levels, we call it Synaptic Modulation based Artificial Neural Networks (SMANNs). In the bottom level we find a network we call the *modulated network* and whose synaptic weights are partially or totally regulated by a higher level. This network is in charge of performing the base function or functions. In the top level, a second network we call *modulating network* is in charge of regulating all or part of the synaptic weights of the modulated network, and thus, the function it performs.

In figure 2 we display this type of structure. The functional modularity of the network with respect to its inputs can be easily appreciated. This is, inputs $I^m_0,...,I^m_n$, are processed by the modulated network whereas the modulating network determines the type of processing it carries out by means of the appropriate modulation of the synapses as a function of its inputs $I^M_0,...,I^M_N$.

Both the modulating and the modulated network may present any internal connection topology, with the only constraint of being trainable. For the sake of clarity, in the examples we present we have employed a multilayer perceptron type structure for the two levels.

One of the main features of SMANN type topologies is the introduction of higher order terms of the product, sigma-pi or other types in the processing the networks carries out without interfering with their training

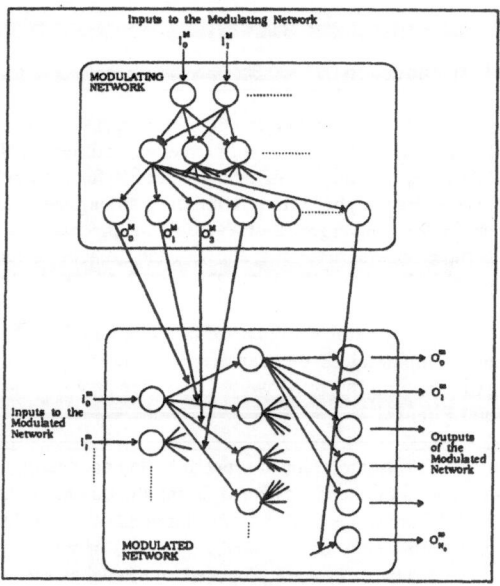

Fig. 2. SMANN Structure.

simplicity. This is due to the fact that the networks preserve their simple topology and these terms arise as a consequence of way they are connected to one another.

If we take into account that the outputs of the modulating network are given by:

$$O_o^M = F[\sum_h w_{ho}^M F(\sum_i w_{ih}^M I_i^M)]$$

where 'h' indicates neurons in the hidden layer, 'i' input neurons, 'o' output neurons, 'm' modulated network, 'M' modulating network, and F the transfer function of the neurons; and that the synaptic weights of the modulated neural network correspond to the values of these outputs,

$$\{w_{ho}^m \cup w_{ih}^m\} = \{O_o^M\}$$

it is easy to see that the outputs include product terms on functions of the inputs $(I_0^M, ..., I_N^M, I_0^m, ..., I_n^m)$.

$$O_o^m = F[\sum_h w_{ho}^m F(\sum_i w_{ih}^m I_i^m)]$$

$$O_o^m = F[\sum_h O_s^M F(\sum_i O_r^M I_i^m)]$$

with $s = N_i N_h + h N_o + o$, $r = i N_h + h$ and $i = 0, ..., N_i - 1$; $h = 0, ..., N_h - 1$ and $o = 0, ..., N_o - 1$, where N_i, N_h, N_o are the number of input, hidden and output neurons respectively for the modulating or modulated network depending on whether its index is 'M' or 'm'.

Some of the advantages derived from this type of modular structure are that both networks can be trained using any well established training algorithm, taking as target outputs for the modulating network, the set of synaptic weights obtained after training the modulated network. That is, each module can be trained by itself, with the consequent reduction in the training complexity for each one of them. On the other hand, the individual networks can be smaller in dimension. Finally, this structure can be extended to as many levels as we desire preserving the advantages we have pointed out. This permits establishing SMANN hierarchies which can be scaled to the dimensionality of the problem.

3. Application Examples

We are now going to present some simple examples where both, the ease of the training process and the generalization capabilities along all the dimensions of the problem of this type of topologies can be appreciated.

The training process was carried out by means of a standard backpropagation algorithm, training the modulating network with the synaptic weights obtained for the modulated network in each case. In order to determine the optimal connection structure between the two networks for each problem we have made use of GENIAL [8], a genetic algorithm based design environment for the generation of trained ANNs which can select optimal topologies as a function of the constraints imposed on the problem.

In the first example we are going to consider a two input, one output SMANN in order to generate variable width gaussian functions. One of the inputs, which determines the width of the gaussian, is assigned to the modulating network, with one input neuron, four hidden neurons and eight output neurons. The second input is assigned to the modulated network, which presents a 1-4-1 structure and which will have to learn a gaussian type function centered in X=0.5 for any input of the modulating network.

In figure 3 we show the results obtained for this problem, where the outputs corresponding to training values are indicated by means of circles and where the lines represent the values obtained for inputs that were not in the training set. We must point out the very good generalization properties of this type of topology, both when interpolating points in a particular gaussian (a task corresponding to the modulated network) and when interpolating intermediate gaussian functions (a task corresponding to the modulating network).

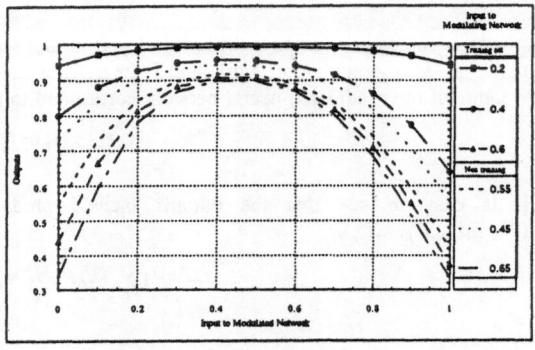

Fig. 3. Results obtained by a SMANN that generates variable width gaussian functions.

The same type of SMANN was trained using GENIAL, but in this case we have allowed the genetic algorithm to determine which synapses of the modulated network were to be modulated by the modulating network and which had to have a fixed weight value. The results obtained show that for this case it is only necessary to modulate 63% of the synapses of the modulated network. Furthermore, the results of executing the network are better than in the first case, probably because in the first case the modulating network was harder to train as it had to control more synapses.

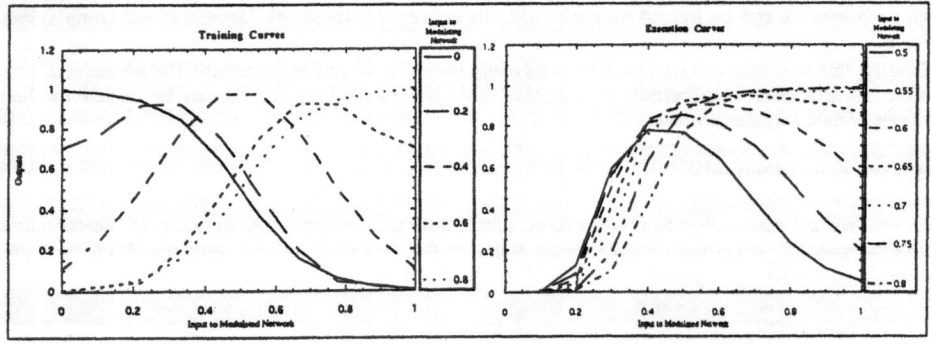

Fig. 4. SMANN processing of evolving curves. a) training curves; b) results obtained for the inputs of the modulating network shown on the right.

With these results in mind, we could question if these types of interpolations can be performed when we evolve between different types of functions where we do not know the functional expression or the parameters that control this variation, having to establish an arbitrary scale. In order to test this, we have trained a two input, one output SMANN with the examples shown in figure 4.a, arbitrarily assigning the values on the right to the second input.

In figure 4.b it can be observed how the network learns the training cases and establishes a "reasonable" interpolation between them (for the sake of clarity we have only presented values for one half of the evolution). In this case, it is clearer that the modulating network determines the function the modulated network will carry out. That is, the modulating network is a network that from a set of examples generalizes to a continuum of functions, allowing for the generation of an infinite number of functions the modulated network can carry out depending on the inputs of the modulator, unlike other types of modular systems, where a discrete set of networks (that is, functions) is established and one of them is selected by a control network.

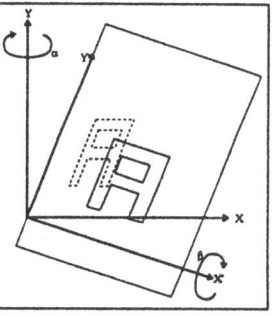

Finally, and in order to consider a more complex problem, we have trained a four input, two output SMANN for reconstructing the original coordinates of any two dimensional object that is located on a plane which is rotated arbitrary angles with respect to the Y and X' axis (see figure 5) from its projection on the XY plane, using as a reference the projection of a control point.

In order to perform this task, we introduce the projected coordinates of the control point as inputs to the modulating network and as the inputs to the modulated network we use the projected coordinates of the object. The outputs of the modulated network will provide the real coordinates of the object on the rotated plane. The training set we employed included projections corresponding to four rotation angles with respect to each axis.

Fig. 5. Rotated 'A' and projection.

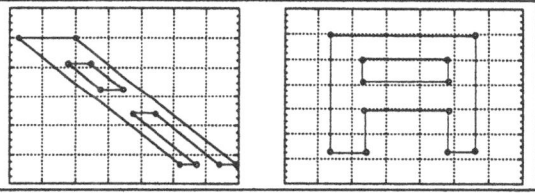

Figure 6 shows an example of this, not included in the training set, corresponding to an A in a plane that was rotated 80 degrees with around the Y axis and 30 around the X' axis. The left side of the figure displays the projection of the 'A', and the right part the reconstruction performed by a 4 input, 2 output SMANN. The results are also satisfactory for any rotation of the plane where the object is with respect to both axes and for any

Fig. 6. Example corresponding to a plane rotation of 80 degrees around the Y axis and 30 around the X' axis. a) projection; b) reconstruction performed by a SMANN.

object. This example shows the ease of training and the versatility of SMANNs with their great generalization capabilities for the reconstruction of objects from their projections.

4. Discussion

In all of the examples we presented it can be appreciated how the tasks to be performed by a SMANNs in order to obtain solutions to a problem are divided into two levels of abstraction. The modulated network learns and generalizes specific functions that act over data, in other words, it maps input data onto output data in the cases we have considered, whereas the modulating network maps its inputs into functions the modulated network must carry out. If we extend this structure to one more level, we would have a modulating network which would determine the way in which the intermediate level modulates the function the modulated network must carry out. This extension process could be extended to as many levels as we wish, each one of them at a higher abstraction level, but with a very clear functionality when applied to specific problems.

As a conclusion we would like to point out the ease of training and a very good generalization capability in each one of the abstraction levels, or from another point of view, along each one of the dimensions of the problem. This provides an indication of the opportunities this type of topologies offer for finding structured ways of approaching complex problems by establishing an abstraction hierarchy which permits finding training sets in a natural way, and above all, to adjust the complexity and topologies of the different levels to the specific functions they must perform within the global scheme. For instance, a modulating network could take temporal elements into account through feedback loops while the modulated network could be a simple multilayer perceptron better suited to the data it receives and the functions it must perform.

References

[1] Giles, C. L., and Maxwell, T., Learning, Invariance, and Generalization in High-Order Neural Networks, *Applied Optics*, Vol. 26, No. 23, December 1987.
[2] McCulloch, W. S., and Pitts, W., A Logical Calculus of the Ideas Immanent in Neural Activity, *Bulletin of Mathematical Biophysics*, Vol. 5, pp. 115-133, 1943.
[3] Hawkins, R. D., Abrams, W.T., Carew, T.J., and Kandel, E.R., A Cellular Mechanism of Classical Conditioning in Aplysia: Activity Depedent Amplification of Presynaptic Facilitation, *Science*, Vol. 219, pp. 400-405, January 1983.
[4] Silva, A. J., Stevents, C.F., Tonehawa, S., and Wang, Y. Y., Defficient Hippocampal Long Term Potentiation in α-Calcium-Calmodulin Kinase-II Mutant Mice. *Science 257*, No. 5067, pp 201-206, 1992.
[5] Dehaene, S., Changeux, J., and Nadal, J., Neural Networks that Learn Temporal Sequences by Selection, Procc. Natl. Acad. Sci. USA, Vol. 84, pp. 2727-2731, May 1987.
[6] Dehaene, S., and Changeux, J., Neuronal Models of Cognitive Functions, *Cognition 33*, pp. 63-109, 1989.
[7] Peretto, P., and Niez, J.J., Long Term Memory Storage Capacity of Multiconnected Neural Networks, *Biol. Cybern. 54*, pp. 53-63, 1986.
[8] Duro, R. J., Santos, J., and, Sarmiento, A., GENIAL, an Evolutionary Recurrent Neural Network Designer and Trainer. In *Proceedings of CAST'94 (Fourth International Workshop on Computer Aided Systems Technology)*, Ottawa, Canada, May 1994.

Self-Organization of Cortical Receptive Fields and Columnar Structures in a Hebb-Trained Neural Network

M. Stetter[1], M. Kussinger[2], A. Schels[2], E. Seeger[2], and E. W. Lang[2]

[1]Dept. of Ophthalmology, University of Regensburg, 93042 Regensburg, FRG
[2]Dept. of Biophysics, University of Regensburg, 93040 Regensburg, FRG

Existing models for the formation of cortical receptive field profiles, orientation maps, and ocular dominance stripes address the emergence of each or some of these features separately. The present work investigates a linear Hebb-trained neural network model for the simultaneous self-organization of receptive field profiles, their arrangement into orientation maps, and the segregation of ocular dominance stripes. Both ON- and OFF-center type input neurons are considered. The requirement of a simultaneous formation of several structures leads to the prediction of additional necessary properties of the input correlation functions. The receptive field- and orientation map formation behaviour predicts, that the range, where ON-ON-correlations exceed ON-OFF-correlations within the LGN, should be about 0.6 times the retinotopic radius of thalamocortical axonal arbors. Additionally, the emergence of ocular dominance stripes requires an asymmetry between ON-ON and ON-OFF correlation functions.

1. Introduction

Simple orientation selective receptive fields in the primary visual cortex of cats and monkeys are sub-divided into two or three elongated subregions, which show either ON- or OFF-type response to small light stimuli (Jones and Palmer 1987). Along the cortical surface, the preferred orientations are arranged into a piecewise continuous orientation preference map with vortices of ±1/2-symmetry (±180° change of the preferred angle around the vortex center) (Blasdel 1992; Bonhoeffer and Grinvald 1991). In monkeys, the vortex centers and linear fractures displayed with differential imaging techniques show a reduced strength of orientation selectivity, which might either result from a dense package of different orientation preferences, or from a reduced orientation selectivity of every single receptive field. This columnar system is superimposed onto a system of ocular dominance stripes (LeVay et al. 1975; Anderson et al. 1988). Most of these features emerge at least partially without visual experience (Sherk and Stryker 1976; LeVay et al. 1978; Hubel and Wiesel 1968), but neural activity seems crucial for their development (Stryker and Harris 1986).

Several models have been made either for the activity dependent development of isolated simple receptive fields (Linsker 1986; MacKay and Miller 1992; Yuille et al. 1989) or for the formation of orientation preference, orientation selectivity, and ocular dominance under a given interaction (Tanaka 1991; Goodhill 1993; Swindale 1992, Obermayer et al. 1992). The latter quantities are treated as primary variables in these models, which thus do not take into account the fact, that in biology orientation map and ocular dominance stripe formation occurs via the coordinated development of cortical receptive fields. The condition of a simultaneous formation of structured receptive fields in models primarily treating columnar systems, however, additionally restricts the parameter space, where biologically observed columnar structures form. Miller investigated a sophisticated neural network model for the formation of receptive fields arranging into orientation maps (Miller 1994), and a similar system for the formation of ocular dominance stripes (Miller et al. 1989). While he showed, that orientation maps can form as a consequence of competing ON- and OFF-type input neurons, his model of ocular dominance stripe formation takes into account only one type of input neurons and thus is valid only for unstructured cortical receptive fields.

The present work investigates a common model for the formation of orientation selective receptive fields and their simultaneous arrangement into orientation maps and ocular dominance stripes, which takes into account both ON- and OFF-center input neurons. The model is implemented as a linear feed-forward neural network guided by a multiplicatively constrained Hebb-type learning rule. From its developmental behaviour, a set of properties of the input correlation structure and the lateral interaction function required for the emergence of biologically observed structures, is formulated. In particular it will be shown, that the presence of two sub-populations of input neurons requires an asymmetry between the shapes of ON-ON (OFF-OFF) and ON-OFF-correlation functions in order to account for the emergence of ocular dominance stripes.

2. Network architecture and learning rule

In the present work, the emergence of cortical functional structures is assumed to result from a correlation based self-organization process, which adjusts thalamocortical synaptic weights. With the investigation of a simple neural network-model for this process, one can formulate required properties which should be present in the immature early visual pathway. In the network model, four sub-populations of input neurons are considered, namely ON-center and OFF-center neurons (denoted by a superscript $\mu = $ ON,OFF) receiving input from the left or right eye respectively (characterized by a superscript $M = L,R$). The input neurons are located within four two-dimensional input layers and send feed-forward projections to a common two-dimensional output layer (Fig. 1). The retinotopic positions of neurons within the input and the output layer are given by vectors r and R respectively. At time t, the synaptic coupling strength between an input neuron located at position r within layer M,μ to an output neuron at position R is given by the synapse function $w^{M,\mu}(R,r,t) \geq 0$. Since thalamocortical axons show finite arborization radii, input and output neurons can only be connected, if their retinotopic distance is not too large. In other words, the synaptic functions are restricted to a finite range of ρ, i.e. $w^{M,\mu}(R,r,t) \neq 0$ only for $|R - r| \leq \rho$ (ρ is the projection radius and may be related to afferent axonal arborization radii). Thus, retinotopy is enforced by the network architecture in the present model. The output neurons are connected by time-independent lateral synaptic couplings, which are assumed to implement a translationally invariant lateral interaction $I(R-R')$ between two neurons at R and R'.

During the training of the network, the four input layers $M=L,R$, $\mu = $ ON, OFF are stimulated to show activity distributions $v^{M,\mu}(r,t) \geq 0$, the statistic behaviour of which is described by the time averages and correlation functions

$$\left\langle v^{M\mu}(r,t)\right\rangle_t \equiv \overline{v}^{M\mu} \quad \forall r,t \qquad \left\langle v^{M\mu}(r,t)v^{N\lambda}(r',t')\right\rangle_t = g(t-t')C^{MN,\mu\lambda}(r-r'), \quad g(0)=1. \tag{1}$$

The symbol $<>_t$ denotes the average over a time interval which is long compared to the time constant τ of the neural dynamics but short compared to the time constant T_l of the learning dynamics.

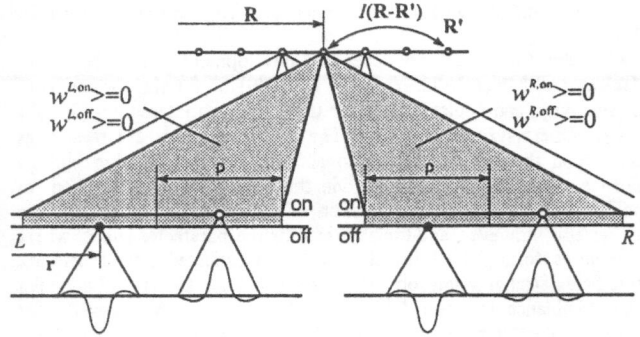

Fig. 1 *Architecture of the neural network used for model simulations. Four input layers (corresponding to ON- and OFF-center sub-populations driven by the left and right eye) send retinotopically ordered, trainable projections to a common output layer (model area 17). The output neurons are connected via time-independent lateral connections.*

The activity $s(\mathbf{R},t)$ of a given output neuron after one time step τ is given by its total (afferent and intracortical) input as

$$s(\mathbf{R},t) = \sum_{\substack{N= \\ L,R}} \sum_{\substack{\lambda= \\ \text{ON,OFF}}} \int_{|\mathbf{R}-\mathbf{r}'|\leq\rho} d\mathbf{r}' \, w^{N\lambda}(\mathbf{R},\mathbf{r}',t-\tau)v^{N\lambda}(\mathbf{r}',t-\tau) + \int d\mathbf{R}' \, I(\mathbf{R}-\mathbf{R}')s(\mathbf{R}',t-\tau) \tag{2}$$

where each integral is taken over the circular area around $\mathbf{R}$ where $w^{N\lambda}(\mathbf{R},\mathbf{r},t)$ is allowed to be nonzero. After each time step, the feed forward weights $w^{M\mu}(\mathbf{R},\mathbf{r},t)$ are changed following a Hebb-type learning rule by the amount $\delta w^{M\mu}(\mathbf{R},\mathbf{r},t)$ with

$$\delta w^{M\mu}(\mathbf{R},\mathbf{r},t) = \frac{\tau^2}{T_l}\Big(s(\mathbf{R},t)v^{M\mu}(\mathbf{r},t-\tau) - F\big(\{w^{N\lambda}(\mathbf{R},\cdot,t)|N=L,R,\lambda=\text{ON,OFF}\}\big)w^{M\mu}(\mathbf{R},\mathbf{r},t)\Big)$$

$$w^{M\mu}(\mathbf{R},\mathbf{r},t) \geq 0, \quad M=L,R; \; \mu=\text{ON,OFF}, \tag{3}$$

where $T_l \gg \tau$ is the time constant of the learning dynamics (i.e. τ/T_l is a small learning rate). $F \equiv F(\mathbf{R},t)$ denotes a general decay functional, which is assumed to increase with increasing values of the synaptic weights to a given output neuron thus limiting their strengths. Iterative replacement of $s(\mathbf{R},t-\tau)$ in Eqn. (2), insertion of the resulting term for $s(\mathbf{R},t)$ into Eqn (3) and time averaging over the learning rule yields

$$\frac{d}{dt}w^{M\mu}(\mathbf{R},\mathbf{r},t) \equiv \int d\mathbf{R}' \, K(\mathbf{R}-\mathbf{R}')\sum_{N,\lambda}\int d\mathbf{r}' \, C^{MN,\mu\lambda}(\mathbf{r}-\mathbf{r}')w^{N\lambda}(\mathbf{R}',\mathbf{r}',t)$$

$$- F(\mathbf{R},t)w^{M\mu}(\mathbf{R},\mathbf{r},t), \quad w^{M\mu}(\mathbf{R},\mathbf{r},t)\geq 0, \quad M=L,R; \; \mu=\text{ON,OFF} \tag{4}$$

where $dw^{M\mu}(\mathbf{R},\mathbf{r},t)/dt \equiv \langle\delta w^{M\mu}(\mathbf{R},\mathbf{r},t)\rangle_t /\tau$ was used and the learning rate τ/T_l is dropped. In addition, the time delays of the synaptic weights have been neglected, $w^{M\mu}(\mathbf{R},\mathbf{r},t-n\tau) \equiv w^{M\mu}(\mathbf{R},\mathbf{r},t)$, $n\tau \ll T_l$, which corresponds to the assumption that the correlation time of the input activities is short compared to the learning time constant, i.e. $g(\tau_c) = 1/e$, $\tau_c \ll T_l$. With Kronecker`s delta function $\delta(\mathbf{R}-\mathbf{R}')$, the coupling function $K(\mathbf{R}-\mathbf{R}')$ is given by

$$K(\mathbf{R}-\mathbf{R}') = \delta(\mathbf{R}-\mathbf{R}') + g(\tau)I(\mathbf{R}-\mathbf{R}') + g(2\tau)\int d\mathbf{R}'' \, I(\mathbf{R}-\mathbf{R}'')I(\mathbf{R}''-\mathbf{R}') + \cdots. \tag{5}$$

In Eqn. (5), the n-th term (starting from $n = 0$) describes the time averaged contribution of input signals, which propagate over n intracortical synaptic connections, to the learning dynamics of the afferent synaptic efficacy $w^{M\mu}(\mathbf{R},\mathbf{r},t)$ considered. This contribution is generated by input patterns, which were applied with a time delay $n\tau$ compared to the current afferent activity pattern $v^{M\mu}(\mathbf{r},t)$.

In the following, only a special case for the the correlation structure within and between the input layers will be treated, namely the case, where ON- and OFF-center neurons tend to show opposite variations of their activities around a common average activity. Under these assumptions, all correlation functions may be written in terms of two covariance functions $G(\mathbf{r})$ and $G^k(\mathbf{r})$, one describing the same-eye covariance structure and one describing the between-eye covariance. It is

$$C^{MM,\text{ON ON}}(\mathbf{r}) = C^{MM,\text{OFF OFF}}(\mathbf{r}) = \bar{v}^2 + G(\mathbf{r}) \tag{6}$$

$$C^{MM,\text{ON OFF}}(\mathbf{r}) = C^{MM,\text{OFF ON}}(\mathbf{r}) = \bar{v}^2 - G(\mathbf{r}) \tag{7}$$

$$C^{MN,\text{ON ON}}(\mathbf{r}) = C^{MN,\text{OFF OFF}}(\mathbf{r}) = \bar{v}^2 + G^k(\mathbf{r}) \tag{8}$$

$$C^{MN,\text{ON OFF}}(\mathbf{r}) = C^{MN,\text{OFF ON}}(\mathbf{r}) = \bar{v}^2 - G^k(\mathbf{r}), \quad M,N=L,R; \; M \neq N. \tag{9}$$

Further, it is convenient to perform a transformation of the weight function into difference weights

$$w^M(\mathbf{R},\mathbf{r},t) := w^{M,\text{ON}}(\mathbf{R},\mathbf{r},t) - w^{M,\text{OFF}}(\mathbf{R},\mathbf{r},t), \quad M=L,R, \tag{10}$$

which are useful for the description of cortical receptive field profiles, and sum weights

$$W^M(\mathbf{R},\mathbf{r},t) := w^{M,\text{ON}}(\mathbf{R},\mathbf{r},t) + w^{M,\text{OFF}}(\mathbf{R},\mathbf{r},t), \quad M = L, R, \tag{11}$$

which average over ON- and OFF- synaptic behaviour and hence are suitable for the description of ocular dominance stripes. For a given output neuron $\mathbf{R}$, the difference weight functions $w^M(\mathbf{R},\mathbf{r},t)$ are henceforth called their (left and right) receptive fields. Note, that the difference weights may change their sign. The time averaged learning rules for difference and sum weights are obtained from (4) and (6) - (9) as

$$\frac{d}{dt} w^M(\mathbf{R},\mathbf{r},t) = \int d\mathbf{R}'\, K(\mathbf{R}-\mathbf{R}')\left[\int d\mathbf{r}'\, G(\mathbf{r}-\mathbf{r}') w^M(\mathbf{R}',\mathbf{r}',t) + \int d\mathbf{r}'\, G^k(\mathbf{r}-\mathbf{r}') w^N(\mathbf{R}',\mathbf{r}',t)\right]$$
$$- f\left(W^L(\mathbf{R},\cdot,t), W^R(\mathbf{R},\cdot,t), w^L(\mathbf{R},\cdot,t), w^R(\mathbf{R},\cdot,t)\right) w^M(\mathbf{R},\mathbf{r},t), \quad M \neq N \tag{12}$$

$$\frac{d}{dt} W^M(\mathbf{R},\mathbf{r},t) = \bar{v}^2 \int d\mathbf{R}'\, K(\mathbf{R}-\mathbf{R}') \int d\mathbf{r}' \left(W^L(\mathbf{R}',\mathbf{r}',t) + W^R(\mathbf{R}',\mathbf{r}',t)\right)$$
$$- f\left(W^L(\mathbf{R},\cdot,t), W^R(\mathbf{R},\cdot,t), w^L(\mathbf{R},\cdot,t), w^R(\mathbf{R},\cdot,t)\right) W^M(\mathbf{R},\mathbf{r},t), \quad W^M(\mathbf{R},\mathbf{r},t) \geq 0 \tag{13}$$

where a factor 1/2 in front of the left hand side was included into the learning rate and dropped to simplify the expressions. The new decay function fulfils $f(\mathbf{R},t) \equiv F(\mathbf{R},t)/2$, but depends on the transformed synapse functions.

3. Results

The behaviour of the model was investigated using three variants of the neural network with increasing complexity:

Monocular model without lateral interactions	=>	undisturbed receptive field formation
Monocular model including lateral interactions	=>	receptive field deformation and arrangement into orientation preference and orientation selectivity maps.
Full binocular model (Fig. 1)	=>	formation of ocular dominance stripes.

3.1 Receptive field formation

Now we consider a network architecture, which results from the system described in the previous section with the assumption of a vanishing lateral interaction function (thus $K(\mathbf{R}\text{-}\mathbf{R}') = \delta(\mathbf{R}\text{-}\mathbf{R}')$), and input layers of one side, for example left eye input layers, only. For a translationally invariant input correlation structure, the neural network then decomposes into independent linear perceptrons with each output neuron $\mathbf{R}$ exclusively driven by its afferent synapse functions $w^L(\mathbf{R},\mathbf{r},t)$ and $W^L(\mathbf{R},\mathbf{r},t)$. It is thus sufficient to consider one of these output neurons, for instance that located at $\mathbf{R} = 0$. The isolated development of its cortical receptive field results from the time development of the difference weight function $w(\mathbf{r},t) \equiv w^L(0,\mathbf{r},t)$, which derives from (12) as

$$\frac{d}{dt} w(\mathbf{r},t) = \int d\mathbf{r}'\, G(\mathbf{r}-\mathbf{r}') w(\mathbf{r}',t) - f(W,w) w(\mathbf{r}',t) \tag{14}$$

For a rotationally invariant covariance function $G(\mathbf{r})$, all eigenfunctions of G factorize into a radial and an angular part and may be characterized by their radial and angular node numbers, (n,l) (Linsker 1990, Stetter et al. 1993). The only stable fixpoints $w(\mathbf{r})$ of equation (14) ly within the subspace spanned by the principal component functions of the covariance function $G(\mathbf{r})$. Therefore, the emerging receptive field profiles can be determined by characterizing the indices (n,l) of the principal components of G as a function of its parameters. Fig. 2 shows parameter domains for the emergence of different receptive field profiles for a spatially oscillating covariance function, as it would result from common signals to input neurons with overlapping mexican hat-shaped (DOG) input receptive fields. The center radius R_0 of these input filters, which relates to the correlation length within the input layer, and their offset z_g (measuring the strength of anticorrelations), are varied.

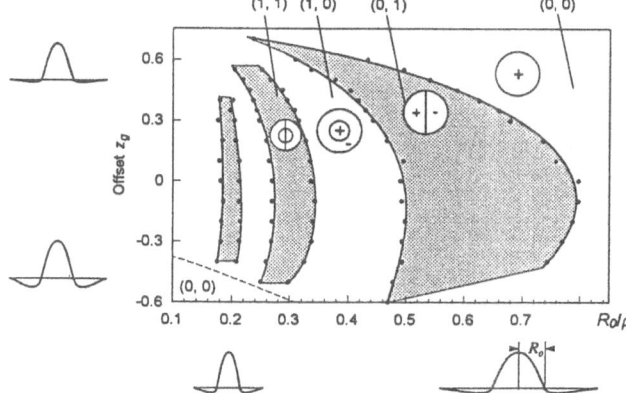

Fig. 2 *Parameter domains for the emergence of cortical receptive field profiles with input correlations originating from overlapping, mexican hat shaped filter functions of the input neurons. The center radius of these filters and the strengths of their antagonistic surrounds (adjusting the strength of anticorrelations) are varied.*

Only rotationally invariant $(n,0)$- and twofold degenerate, orientation selective $(n,1)$-receptive fields emerge. With increasing correlation length, the separation of the principal component eigenvalue and thus the stability of these receptive fields increases, while the number of their lobes decreases. For weak anticorrelations, however, only unstructured, gaussian shaped receptive field functions can emerge. Thus, one necessary property for the input correlation structure to yield structured receptive fields is spatial oscillation of the covariance function $G(\mathbf{r})$. With (6) and (7), this translates into the requirement, that correlations within one sub-population (ON-ON, OFF-OFF) must exceed inter-population- (ON-OFF-) correlations for short distances less than about $1.4\,R_0$, while the reverse must be given for longer distances. In order to obtain the biologically observed simple receptive fields with two or three lobes, the center radius R_0 of the input filters (multiplied by the magnification factor in order to obtain cortical units) should be in the order of magnitude of 0.4-0.7ρ, which lies very close to the correlation lengths generated by overlapping LGN receptive fields (Stetter et al. 1993).

3.2 Orientation maps

Proceeding from these results, a finite lateral interaction was introduced, but the monocularity of the model was maintained in order to investigate the self-organization of orientation maps. Again the difference weights are suitable for the description of the coupled formation of receptive fields. From their final shapes, measures for their preferred orientations and the strengths of their orientation selectivities can be obtained (Stetter et al. 1993) yielding the mature orientation preference and the orientation selectivity map.

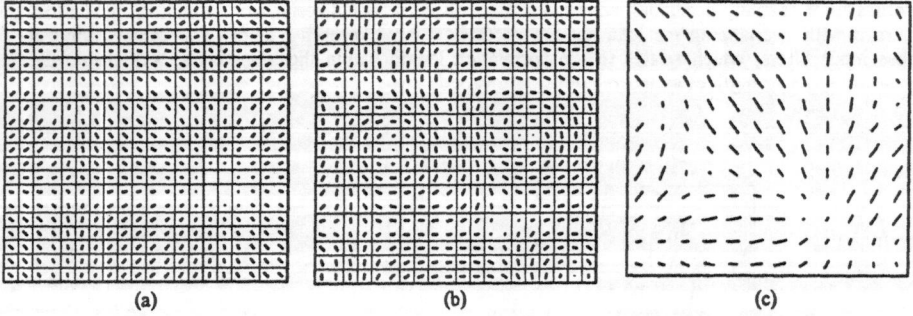

Fig. 3 *Simulated orientation maps (a) for (0,1) correlation parameters and a purely excitatory gaussian lateral interaction (Rc = 2), (b) for (0,1) correlation parameters and a mexican-hat shaped lateral interaction (Rc = 1.2), and (c) for (1,0) correlation parameters with excitatory lateral interaction (Rc = 1). The directions and lengths of the bars denote the preferred orientation and the orientation selectivity strength of the corresponding output neuron.*

Fig. 3 shows simulation results, where for each output neuron both quantities are displayed as the direction and the length of a bar respectively. Note, that orientation preference as well as the bars show 180°-rotational symmetry, while the receptive field profiles themselves often show 360°-symmetry only. For correlation parameters located in the (0,1)-domain (Fig. 2), both radial (Fig. 3a) and tangential vortices (Fig. 3b) with 360°-symmetry (±1-vortices) are observed, while for parameters within the (1,0)-domain, orientation preference maps with the biologically observed ±1/2-vortices emerge, which originate from deformed, phase shifted receptive fields. In order to understand the formation of these structures, it is helpful to consider the lateral interaction as a perturbation of the receptive field learning rule (14) (Stetter et al. 1994). The perturbation term for a given output receptive field depends on the structures of the surrounding receptive fields (and thus on the local structure of the orientation map) as well as on the structure of the lateral interaction. In the (0,1)-domain, the stable bilobed receptive fields are influenced only little by the perturbation term and mostly preserve their 360°-symmetric shape. As a consequence, the orientation maps formed within the (0,1)-domain also show 360°-symmetry. In the (1,0)-domain, the receptive fields are more strongly deformed and phase shifted yielding the opportunity for a ±1/2-vortex system to emerge. Finally, at the vortex centers of the orientation preference maps, the orientation selectivity of the receptive fields is reduced. This can also be understood from the structure of the perturbation, which is nearly rotationally symmetric at vortex centers thus reducing the orientation selectivity of receptive fields near the centers.

3.3 Ocular dominance stripes

In contrast to the previous subsections, the sum weights (11), must be considered in order to investigate the self-organization of ocular dominance stripes. For the correlation structure given by (6)-(9), all stable sum weight distributions must be fixpoints of the corresponding evolution equation (13), which are defined as

$$W_{FP}^M(\mathbf{R},\mathbf{r}) = \frac{\bar{v}^2 \int d\mathbf{R}' \big(K(\mathbf{R}-\mathbf{R}')\big) \int d\mathbf{r}' \big(W_{FP}^L(\mathbf{R}',\mathbf{r}') + W_{FP}^R(\mathbf{R}',\mathbf{r}')\big)}{f_{FP}(\mathbf{R})}, \quad M = L,R \tag{15}$$

The right hand side of (15) neither depends on the input neurons' positions, nor does it contain the lateral index M. Therefore, all fixpoint functions obey the condition

$$W^L(\mathbf{R},\mathbf{r}) = W^R(\mathbf{R},\mathbf{r}) \equiv W_0(\mathbf{R}) \tag{16}$$

Eqn (16) means, that with a balanced correlation structure (6)-(9), all cortical receptive fields remain fully binocular and no ocular dominance stripes can emerge. This is because a balanced correlation structure implies, that ON- and OFF-center input neurons at a given location of one side always tend to show opposite variations of their activities. Hence their feed-forward synaptic weights will on time average always have opposite learning rates, which for this side prevents the simultaneous decoupling of both ON- and OFF-center synaptic weights. In fact, instead of an ocular dominance stripe system, ON-OFF-stripe systems are observed with the current model. Since ocular dominance stripes are observed in a variety of species with a binocular visual field, the present model formulates the requirement of an unbalanced correlation structure, i.e. sign-reversed ON-OFF correlations must be spatially different from ON-ON and OFF-OFF-correlation functions. This result shows, that the consideration of a system with both ON- and OFF-center sub-populations is crucial for a biologically relevant modelling of ocular dominance column formation.

Finally, both ON-OFF-stripe systems and ocular dominance stripes in systems with pure ON-center input are only observed with a spatially oscillating effective lateral interaction function $J(\mathbf{R}\text{-}\mathbf{R}') = K(\mathbf{R}\text{-}\mathbf{R}') - \delta(\mathbf{R}\text{-}\mathbf{R}')$. This states a requirement for the structure of the intracortical couplings.

4. Conclusions

If the present model is valid for the description of activity dependent self-organization processes in the primary visual cortex of higher mammals, certain conditions should be found in the early visual pathway.

- During the period, where simple receptive fields form, the activities of ON- and OFF-center sub-populations must show different statistic structures in order to drive a segregation into ON- and OFF-subregions of the receptive fields.

- During the same time period, the covariance function of the difference weights must show spatial oscillations, i.e. ON-ON-correlations must exceed ON-OFF-correlations for short distances and vice versa for longer distances, in order to yield structured cortical receptive fields (see also Miller (1994) for a model using subtractive constraints within the learning rule). This result follows from a systematic investigation of the correlation parameter space (Stetter et al. 1993), which was done parallel to previous works on systems with subtractive constraints (Linsker 1986, 1990; MacKay and Miller 199) and extending the considerations of Yuille et al. (1989).

- The reversal radius of both correlations (roughly $1.4\,R_0$ in Fig. 2), must lie in the range of 0.6 times the axonal arborization radius ρ, when multiplied with the cortical magnification factor, in order to allow the formation of orientation maps with $\pm 1/2$-vortices. Note, that the correlation length then must vary with the retinal excentricity. Both this variation and the range of the correlation lengths would be even quantitatively given in monkeys, if the correlations of retinal ganglion cells were generated by the overlaps of their receptive fields (Stetter et al. 1993). This restricted range of the predicted correlation length is not given in Miller's (1994) model, which is based on a subtractive constraint of the learning rule and which always leads to the formation of $\pm 1/2$-vortices. The reason for this behaviour might be, that a subtractive constraint does not stabilize the principal component, but traps the synapse functions at hard synaptic boundaries (Miller and MacKay 1994). Thus, bilobed receptive fields are less stable under a subtractive constraint than in the present model, they may be easily deformed and thus allow the development of vortices with 180°-symmetry.

- The spatial oscillation of the covariance function $G(r)$ can not originate from inter-layer and intra-layer correlations as formulated in (6) and (7), the modulations of which are only sign-reversed versions of each other. Instead a different range of both correlations seems to be a suitable situation for the generation of a spatially oscillatiing covariance function in the light of our model. This conclusion can only be obtained using a binocular model with both ON- and OFF-center inputs, which has been considered for the first time in the present model.

- Finally, the model suggests, that the reduced orientation selectivity observed at vortex centers by differential imaging techniques (Blasdel 1992; Bonhoeffer and Grinvald 1991) should originate from a reduced orientation selectivity of every single receptive field rather than by a summation of different orientation preferences inherent to the technique. Therefore, this reduction should also be found with single-cell recordings.

Now, possible mechanisms for the generation of correlations showing the proposed properties, shall be addressed. The requirements for the formation of ocular dominance stripes on the input correlation structure might be fulfilled by the synchronous burst patterns found by Meister et al. (1991) in the retina of fetal cats and ferrets. However, in order to account also for the formation of structured cortical receptive fields, these patterns should generate LGN activity patterns, that are different for ON- and OFF-center neurons and vary in range dependent on the excentric retinotopic location within the LGN. Both properties are not reported so far. On the other hand, simple receptive fields could emerge in a natural way, if the input correlations are generated through overlapping retinal (or LGN-) receptive fields filtering common input. However, within a linear model this common-input-mechanism leads to a balanced correlation structure as given in (6) and (7) and thus prevents the segregation of ocular dominance stripes.

Therefore, in order to obtain a simultaneous formation of structured receptive fields, ocular dominance stripes and orientation maps, either nonlinearities of neural responses, which break the balance of the correlations, are a crucial ingredient for a model candidate, or adequate correlation structures used *ad hoc* in the present model might be generated by some more complicated mechanism in biology. Alternatively one could state, that both columnar structures do not really form simultaneously, but roughly one after the other with each system being driven by its own adequate correlation structure. These questions will be better addressable with a more accurate characterization of required and sufficient properties of correlations and lateral interactions, which allow the simultaneous development of receptive fields, orientation maps and ocular dominance stripes.

References

Anderson PA, Olavarria J and R. C. Van Sluyters, J. Neurosci. **8**, 2183 (1988).
Blasdel GG, J. Neurosci 12, 3115-3138, 3139-3161 (1992)
Bonhoeffer T and Grinvald A, Nature 353, 429-431 (1991)
Goodhill GJ, Biol. Cybern. **69**, 109 (1993).
Hubel DH and Wiesel TN, J. Physiol. (London) 195, 215 (1968).
Jones HP and Palmer LA, J. Neurophysiol. **58**, 1187 (1987).
LeVay S, Hubel DH and Wiesel TN, J. Comp. Neurol. **159**, 559 (1975).
LeVay S, Stryker MP and Shatz CJ, J. Comp. Neurol. 179, 223-244 (1978).
Linsker R, Proc. Natl. Acad. Sci USA 83, 7508-7512, 8390-8394 (1986).
Linsker R, in: Proceedings of the International Joint Conference on Neural Networks (IJCNN), Washington, DC, 1990, edited by M. Caudill (Erlbaum, Hillsdale, NJ, 1990).
MacKay DJC and Miller KD, Network 1, 257-297 (1990).
Meister M, Wong ROL, Baylor DA and Shatz CJ, Science 252, 939-943 (1991).
Miller KD, J. Neurosci. **14**, 409 (1994).
Miller KD, Keller JB and Stryker MP, Science 245, 605 (1989).
Miller KD and MacKay DJC, Neural Comput 6,100-126 (1994).
Obermayer K, Blasdel GG and Schulten K, Phys. Rev. A **45**, 7568 (1992).
Sherk H and Stryker MP, J. Neurophysiol. **39**, 63-70 (1976)
Stetter M, Lang EW and Müller A, Biol. Cybern. **68**, 465 (1993).
Stetter M, Müller A and Lang EW, Phys. Rev. E **50**, 4167-4181 (1994).
Stryker MP and Harris W, J. Neurosci. 6, 2117 (1986).
Swindale NV, Proc. R. Soc. Lond. B **208**, 243 (1980).
Swindale NV, Biol. Cybern. **66**, 217 (1992).
Tanaka S, Biol. Cybern. **64**, 263 (1991).
Yuille AL, Kammen DM and Cohen DS, Biol. Cybern. **61**, 183 (1989).

An Analytical Solution of the Compartmental Model for Use in Local Learning in Artificial Neural Networks

Jaap Hoekstra and Mohamed Maouli
Delft University of Technology, Dept. Electrical Engineering,
P.O. Box 5031, 2600GA Delft, The Netherlands
e-mail: jaap@neuron.et.tudelft.nl

Abstract

Most artificial neural network models do not take into account the extensive structure of the input (dendrite) of a neuron; the neuron is treated as a point at which all inputs meet. For neural network applications in which temporal relations have to be learned, extensive neurons could be interesting because they possess an intrinsic time dependent behavior. Learning in such a network may be done with local-learning models, in which weight updating is based on correlation between inputs and local dendritic parameters, such as the dendritic voltage. Extensive dendrites can be modeled with a compartmental model, a description in terms of electrical circuits. An analytical solution of the time course of the potential in the dendrite is described.

1 Introduction

In most literature on artificial neural network models the node (neuron) is represented as a point, learning rules are based on activity of other nodes and the activity of the node itself. Other models can be developed in which the neuron's extensive dendritic system is taken into account. This seems useful to do because, first, as a consequence of the membrane properties of the dendrite, the spatial structure causes temporal processing of synaptic inputs; the time course of a membrane potential response at any place in the dendrite depends on the location of the synapse, the arrival time of an impulse, and the location and arrival time of impulses on nearby synapses. Second, the mutual distance of synapses are important, for instance, in case excitatory and nearby inhibitory synapses are considered, or in case of (non-Hebbian) local learning mechanisms. Modeling the extensive structure of a dendrite is usually done by compartmental models.

The main contribution of this paper is:

- we describe an analytical solution of the compartmental model in which synaptic input currents and the dendritic parameters can be included in a computational straightforward way.

2 Local Learning

Models incorporating local learning rules allow increase and decrease of weights based on activities of adjacent nodes if the synapses are placed close together on the same dendrite. The spatial structure of dendrites can be taken into account by the use of compartmental modeling techniques. In compartmental modeling the dendrite is subdivided into sufficiently small segments (or compartments), in which the physical properties (e.g. dendrite diameter, specific electrical properties) are spatial uniform and the potential is taken constant. The differences in potential occur *between* compartments rather than within them [8, 7]. The advantage of this modeling is that it places no restrictions on the membrane properties of each compartment. Recently compartmental models received much attention again. A new generation of artificial neural networks extends its models with more (neuro-)biological knowledge [2, 1, 5].

Compartmental models could be used for implementing artificial neural networks if:

- the models can be formulated in such a way that it is possible to consider only a (small) part of the dendrite,

- the models can incorporate synaptic currents and active dendrite phenomena.

The first item is important because compartmental models, in principle, consider the whole dendrite. Consequently for describing local phenomena, parameters of the whole dendrite has to be calculated (which can be very time consuming), unless suitable approximation techniques can be used. The second item is important because for designing networks, different neurons have to be connected. The connection of an axon, through a synapse, onto the dendrite has to be done by a (weighted) current. In addition, general theories on artificial neural networks emphasize the importance of nonlinearities in the models. This points to the necessity to model (nonlinear) active membrane properties, in particular it must be possible to model a maximum voltage in the dendrite. The approximation criterion is described in a previous paper [3]. The second item is the subject of this paper.

3 Compartmental Modeling

In artificial neural network research the dendritic membrane is modeled by the cable equation, see for example [6]. In this model the axial current in the dendritic membrane is assumed to flow along an unbranched cylinder of uniform cross-section [4]. The axial current is then one-dimensional, and the current density changes *only* if current enters or leaves through the cell membrane. The model is based on two assumptions concerning the impedances. These are that the intra-cellular medium acts as an ohmic resistance and that the surface membrane acts as a leaky capacitance.

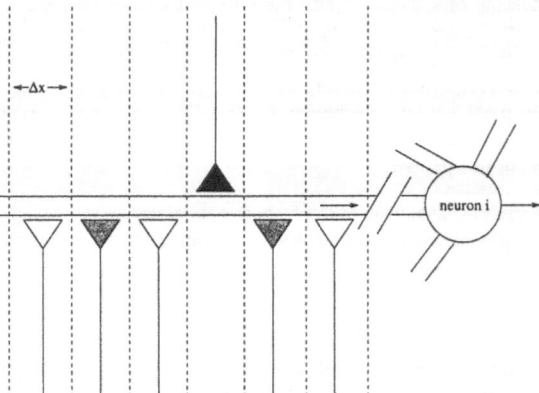

Figure 1: A compartmental model for local learning in artificial neural networks with extensive dendrites

Solutions of an extended model, in which non-uniform cross-sections can be modeled too, can be obtained by using compartmental modeling, in which the membrane is modeled by a cable of equivalent electrical circuits. Synaptic inputs on the dendritic membrane can be translated into (weighted) currents into compartments, therefore the model can be used to describe local learning, and can direct the development of local learning rules.

Figure 1 shows a part of an artificial dendrite divided into six compartments. The compartments with width Δx are chosen such that equally spaced different synapses are placed in each compartment. A local learning rule could now be formulated by stating that the weight on the black synapse will be increased if an impulse arrives at this synapse and, for instance, proportional to the potential in the dendrite underneath the synapse.

At this moment, as a simplification of the models we only consider a flow of information towards the node; such that a potential pulse diffuses only towards the cell body. This simplification, however, still takes into account the influence of signals of neighboring synapses both further away from and closer to the node. Consider the situation in figure 1 in which impulses arrive at the grey synapses just before the impulse on the black synapse that is part of the learning process. The impulse of the grey synapse further away from the neuron will cause a potential pulse in the dendrite that will diffuse towards the black synapse. The actual height of this pulse depends on the properties of the membrane and on the current induced by the impulse on the grey synapse closer to the neuron.

4   Electrical Description

In a single part of this discrete artificial dendrite we can denote the electrical properties of the dendrite. The axial, internal, resistance Ri is divided into two parts both having the value: $1/2 Ri$. In the middle the circuit elements for the leaky capacitance is split of. The radial membrane resistance is denoted as: Rm; and the membrane capacitance as: Cm. An current that represents an active membrane or a synaptic input can be modeled by adding a current Is at the joint between the two axial resistors.

The cable can be analyzed by applying the *Kirchhoff*'s rules, to find all the currents and voltages in the network. To write the necessary equations we apply Kirchhoff's current law (KCL) and Kirchhoff's voltage law (KVL) to the electrical network representing the cable. Figure 2 shows a finite cable of n compartments.

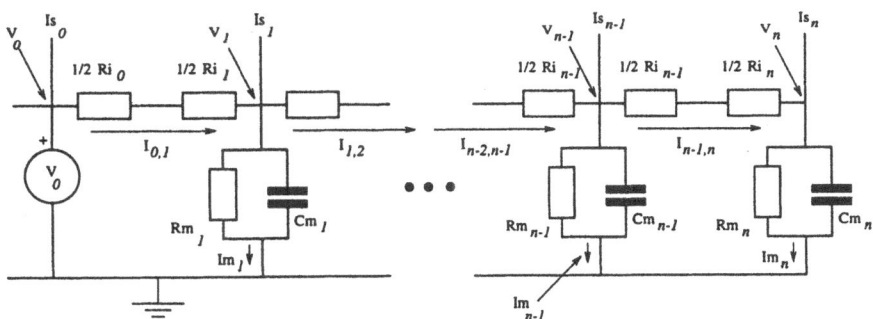

Figure 2: Electrical model of begin and end of a cable

The KCL states that the sum of all the currents towards and from a node in the electrical network is zero. To obtain an description for a finite cable of n compartments, we start from the end of the cable and sum the currents through the different axial resistances:

$$I_{n-1,n} = Im_n - Is_n = Cm_n \frac{dV_n}{dt} + \frac{V_n}{Rm_n} - Is_n \tag{1}$$

$$I_{n-2,n-1} = I_{n-1,n} + Im_{n-1} - Is_{n-1} = Cm_n \frac{dV_n}{dt} + Cm_{n-1} \frac{dV_{n-1}}{dt} + \frac{V_n}{Rm_n} + \frac{V_{n-1}}{Rm_{n-1}} - Is_n - Is_{n-1} \tag{2}$$

$$\bullet$$
$$\bullet$$

Resulting in the current $I_{i-1,i}$ towards compartment i:

$$I_{i-1,i} = \sum_{k=i}^{n} (Cm_k \frac{dV_k}{dt} + \frac{V_k}{Rm_k} - Is_k) \tag{3}$$

At the same time the following voltage relations hold, starting from the beginning of the cable:

$$V_1 = V_0 - I_{0,1}(1/2Ri_0 + 1/2Ri_1) \tag{4}$$

$$V_2 = V_1 - I_{1,2}(1/2Ri_1 + 1/2Ri_2) = V_0 - I_{0,1}(1/2Ri_0 + 1/2Ri_1) - I_{1,2}(1/2Ri_1 + 1/2Ri_2) \tag{5}$$

$$\bullet$$
$$\bullet$$

Resulting in the voltage V_j in the jth compartment:

$$V_j = V_0 - \sum_{i=1}^{j} (1/2Ri_{i-1} + 1/2Ri_i) I_{i-1,i} \tag{6}$$

Combining 3 and 6 we obtain the potential in any compartment.

5 Mathematical Description

We will show a matrix description for a cable of n compartments, in which we assume that $Ri_j = Ri$, $Cm_j = Cm$, and $Rm_j = Rm$ for $j = 0...n$. That is, the axial resistance of all compartments has the same value, and also the membrane resistances and membrane capacitances are taken constant [1]. With the above described simplifications substitution of equation 3 in equation 6 holds:

$$V_j = V_0 - \sum_{i=1}^{j} Ri \sum_{k=i}^{n} (Cm \frac{dV_k}{dt} + \frac{V_k}{Rm} - Is_k) \tag{7}$$

This equation leads to an equation between matrices.

---

[1] This is only done to simplify the description; a similar description can be done without these simplifications. However, in the derivation of the potential as a function of time we assume that the matrices which appear as an argument in the exponential function are diagonalizable.

5.1 Passive membrane with infinite membrane resistance

We first consider an example. Assuming a passive membrane ($Is_j = 0$) with an infinite membrane resistance we obtain the following equation for the potential in compartment j:

$$V_j = V_0 - \sum_{i=1}^{j} Ri \sum_{k=i}^{n} Cm \frac{dV_k}{dt} \tag{8}$$

In order to understand this equation we write down some of the potentials.

$$
\begin{aligned}
V_1 &= V_0 - Ri\,Cm\,(\frac{dV_1}{dt} + \frac{dV_2}{dt} + \frac{dV_3}{dt} + \cdots + \frac{dV_n}{dt}) \\
V_2 &= V_0 - Ri\,Cm\,(\frac{dV_1}{dt} + \frac{dV_2}{dt} + \frac{dV_3}{dt} + \cdots + \frac{dV_n}{dt}) - Ri\,Cm\,(\frac{dV_2}{dt} + \frac{dV_3}{dt} + \cdots + \frac{dV_n}{dt}) \\
&= V_0 - Ri\,Cm\,(\frac{dV_1}{dt} + 2\frac{dV_2}{dt} + 2\frac{dV_3}{dt} + \cdots + 2\frac{dV_n}{dt}) \\
&\quad \bullet \\
&\quad \bullet \\
&\quad \bullet \\
V_j &= V_0 - Ri\,Cm\,(\frac{dV_1}{dt} + 2\frac{dV_2}{dt} + 3\frac{dV_3}{dt} + \cdots + j\frac{dV_n}{dt}) \\
&\quad \bullet \\
&\quad \bullet \\
V_n &= V_0 - Ri\,Cm\,(\frac{dV_1}{dt} + 2\frac{dV_2}{dt} + 3\frac{dV_3}{dt} + \cdots + n\frac{dV_n}{dt})
\end{aligned}
$$

It is clear from this that we are dealing with a set of first order differential equations. We can put this in a matrix equation:

$$
\begin{pmatrix} V_1 \\ V_2 \\ V_3 \\ \vdots \\ V_j \\ \vdots \\ V_n \end{pmatrix}
=
\begin{pmatrix} 1 \\ 1 \\ 1 \\ \vdots \\ 1 \\ \vdots \\ 1 \end{pmatrix} V_0
-
\begin{pmatrix}
Ri\,Cm & Ri\,Cm & Ri\,Cm & \cdots & Ri\,Cm \\
Ri\,Cm & 2Ri\,Cm & 2Ri\,Cm & \cdots & 2Ri\,Cm \\
Ri\,Cm & 2Ri\,Cm & 3Ri\,Cm & \cdots & 3Ri\,Cm \\
\vdots & \vdots & \vdots & \vdots & \vdots \\
Ri\,Cm & 2Ri\,Cm & 3Ri\,Cm & \cdots & jRi\,Cm \\
\vdots & \vdots & \vdots & \vdots & \vdots \\
Ri\,Cm & 2Ri\,Cm & 3Ri\,Cm & \cdots & nRi\,Cm
\end{pmatrix}
\begin{pmatrix} dV_1/dt \\ dV_2/dt \\ dV_3/dt \\ \vdots \\ dV_j/dt \\ \vdots \\ dV_n/dt \end{pmatrix}
\tag{9}
$$

In short form this can be written as:

$$V = BV_0 - A\frac{dV}{dt} \tag{10}$$

To solve equation 10, dV/dt is isolated and we obtain:

$$\frac{dV}{dt} = A^{-1}(BV_0 - V) \tag{11}$$

This differential equation can be solved by integrating both sides:

$$\int_{V(0)}^{V(t)} \frac{dV}{BV_0 - V} = \int_0^t A^{-1} dt \tag{12}$$

Assuming that there are no charges on the capacitances at time $t = 0$, then we can write down:

$$V(t) = (I - e^{-A^{-1}t})BV_0 \tag{13}$$

In which I represents the unit matrix. Because this equation is between vectors and A is a matrix we have to define the matrix exponential term. For simplifying the notation we introduce the matrix Ψ and a function $\eta(t)$:

$$\Psi = -A^{-1}, \qquad \eta(t) = e^{\Psi t}B \tag{14}$$

And obtain the equation:

$$V(t) = (IB - \eta(t))V_0 \tag{15}$$

We now solve this last equation.

For any square matrix Ψ and a scalar t, however, the matrix exponential has to be defined. This can be done through a series analogous to the series expansion for the scalar exponential function:

$$e^{\Psi t} = I + \Psi t + \frac{1}{2!}\Psi^2 t^2 + \frac{1}{3!}\Psi^3 t^3 + \cdots + \frac{1}{k!}\Psi^k t^k + \cdots \tag{16}$$

We can rewrite this expansion by using the *normal form* of the matrix Ψ [2]. The normal form is:

$$\Psi = SAS^{-1} \tag{17}$$

In this equation S is the *eigenvector matrix* and Λ the *eigenvalue matrix*, in which all elements are zero except for the main diagonal, whose elements are the set of eigenvalues of Ψ. Using the normal form we can write any power of Ψ as:

$$\Psi^n = SAS^{-1}SAS^{-1}\cdots SAS^{-1} = S\Lambda^n S^{-1} \tag{18}$$

The series expansion can now be written as:

$$\begin{aligned} e^{\Psi t} &= S(I + \Lambda t + \frac{1}{2!}\Lambda^2 t^2 + \frac{1}{3!}\Lambda^3 t^3 + \cdots)S^{-1} \\ &= Se^{\Lambda t}S^{-1} \end{aligned} \tag{19}$$

To solve eq. 15 a coordinate transformation is defined. If $\{x_1, x_2, \cdots, x_n\}$ is a base of eigenvectors of Ψ, corresponding to the eigenvalues $\{\lambda_1, \lambda_2, \cdots, \lambda_n\}$, then using the transformations:

$$S = [x_1, x_2, \cdots, x_n] \ , \quad y(t) = S^{-1}\eta(t) \ , \quad c = S^{-1}B \tag{20}$$

the equation:

$$\eta(t) = e^{\Psi t}B$$

transforms into

$$y(t) = e^{\Lambda t}c = \begin{pmatrix} c_1 e^{\lambda_1 t} \\ c_2 e^{\lambda_2 t} \\ \vdots \\ c_n e^{\lambda_n t} \end{pmatrix} \tag{21}$$

and we find the *explicit* solution

$$\eta(t) = Sy(t) = c_1 e^{\lambda_1 t}x_1 + c_2 e^{\lambda_2 t}x_2 + \cdots + c_n e^{\lambda_n t}x_n \tag{22}$$

With this equation for $\eta(t)$, eq. 15, and eq. 20, the time evolution of the system is completely determined.

---

[2] The normal form of a matrix is developed by finding the eigenvalues of the matrix; and is treated in most standard textbooks on linear algebra

5.2 Incorporating membrane resistance and synaptic currents

We now consider the basic equation 7 without restrictions on the synaptic current and the membrane resistance.

We obtain the following equation for the potential in compartment j:

$$V_j = V_0 - RiCm \sum_{i=1}^{j} \sum_{k=i}^{n} \frac{dV_k}{dt} - \frac{Ri}{Rm} \sum_{i=1}^{j} \sum_{k=i}^{n} V_k + Ri \sum_{i=1}^{j} \sum_{k=i}^{n} Is_k \tag{23}$$

This set of first order differential equations can transformed in a matrix equation, in the same way as discussed in the previous example:

$$V = BV_0 - A \frac{dV}{dt} - D V + C Is \tag{24}$$

In which the following vector and matrices appear:

$$A = \begin{pmatrix} RiCm & RiCm & RiCm & \cdots & RiCm \\ RiCm & 2RiCm & 2RiCm & \cdots & 2RiCm \\ RiCm & 2RiCm & 3RiCm & \cdots & 3RiCm \\ \vdots & \vdots & \vdots & \vdots & \vdots \\ RiCm & 2RiCm & 3RiCm & \cdots & jRiCm \\ \vdots & \vdots & \vdots & \vdots & \vdots \\ RiCm & 2RiCm & 3RiCm & \cdots & nRiCm \end{pmatrix} \tag{25}$$

$$B = \begin{pmatrix} 1 \\ 1 \\ 1 \\ \vdots \\ 1 \\ \vdots \\ 1 \end{pmatrix} \tag{26}$$

$$C = \begin{pmatrix} Ri/Rm & Ri/Rm & Ri/Rm & \cdots & Ri/Rm \\ Ri/Rm & 2Ri/Rm & 2Ri/Rm & \cdots & 2Ri/Rm \\ Ri/Rm & 2Ri/Rm & 3Ri/Rm & \cdots & 3Ri/Rm \\ \vdots & \vdots & \vdots & \vdots & \vdots \\ Ri/Rm & 2Ri/Rm & 3Ri/Rm & \cdots & jRi/Rm \\ \vdots & \vdots & \vdots & \vdots & \vdots \\ Ri/Rm & 2Ri/Rm & 3Ri/Rm & \cdots & nRi/Rm \end{pmatrix} \tag{27}$$

$$D = \begin{pmatrix} Ri & Ri & Ri & \cdots & Ri \\ Ri & 2Ri & 2Ri & \cdots & 2Ri \\ Ri & 2Ri & 3Ri & \cdots & 3Ri \\ \vdots & \vdots & \vdots & \vdots & \vdots \\ Ri & 2Ri & 3Ri & \cdots & jRi \\ \vdots & \vdots & \vdots & \vdots & \vdots \\ Ri & 2Ri & 3Ri & \cdots & nRi \end{pmatrix} \tag{28}$$

To solve equation 24, dV/dt is isolated and we obtain:

$$\frac{dV}{dt} = A^{-1}(BV_0 - (D+I)V + CIs) \tag{29}$$

This differential equation can be solved by integrating both sides:

$$\int_{V_{(0)}}^{V_{(t)}} \frac{dV}{BV_0 - (D+I)V + CIs} = \int_0^t A^{-1}dt \tag{30}$$

Assuming that there are no charges on the capacitances at time $t = 0$, then we can write down the solution for equation 24:

$$V(t) = (D+I)^{-1}(I - e^{-(D+I)A^{-1}t})(BV_0 + CIs) \tag{31}$$

Or:

$$V(t) = \Delta^{-1}(I - e^{-\Gamma t})\epsilon, \tag{32}$$

$$\begin{aligned} \Delta &= (D+I) \\ \Gamma &= (D+I)A^{-1} \\ \epsilon &= (BV_0 + CIs) \end{aligned}$$

In the equation, Δ^{-1} is the attenuation of the voltage pulse caused by the ratio Ri / Rm; ϵ is the total input vector, consisting of the synaptic input current and the source pulse. The time evaluation of the pulse, or the speed with which the pulse propagates through the dendrite is determined by Γ. The final equation (32) can directly be calculated with the technique described in the previous section.

References

[1] D.L. Alkon, K.T. Blackwell, G.S. Barbour, S.A. Werness, and T.P. Vogl, 'Biological Plausibility of Synaptic Associative Memory Models', *Neural Networks*, Vol. 7, pp. 1005-1017, 1994.

[2] P.C. Bressloff and J.G. Taylor, 'Dynamics of Compartmental Model Neurons', *Neural Networks*, Vol. 7, pp. 1153-1165, 1994.

[3] J. Hoekstra, 'Approximation of the Solution of the Dendritic Cable Equation by a Small Series of Coupled Differential Equations', In: **New Trends in Neural Computation**, LNCS, J. Mira, J. Cabestany, and A. Prieto (Eds.), Springer Verlag, pp. 41-48, 1993.

[4] J.J.B. Jack, D. Noble, R.W. Tsien, **Electric current flow in excitable cells**, Oxford: Clarendon Press, 1975.

[5] A.J. Klaassen, J. Hoekstra, 'Biophysical and Spatial Neuronal Adaptation Modalities: Biological Prerequisite for Local Learning in Networks of Pulse-coded Cable Neurons', In: Proc. Neuro-Nimes 93, Nimes, Oct. 25-29, pp. 75-82, 1993.

[6] **Methods in Neuronal Modeling**, C. Koch and I. Segev (Eds.), Cambridge MA: MIT Press, pp. 63-97, 1989.

[7] W. Rall, 'Cable Theory for Dendritic Neurons', In: **Methods in Neuronal Modeling**, C. Koch and I. Segev (Eds.), Cambridge MA: MIT Press, pp.9-62, 1989.

[8] I.S. Segev, J.W. Fleshman, and R.E. Burke, 'Compartmental Models of Complex Neurons', In: **Methods in Neuronal Modeling**, C. Koch and I. Segev (Eds.), Cambridge MA: MIT Press, pp. 63-97, 1989.

Should ANN be ANGN?

J. G. Wallace* and K. Bluff**

*Information Technology Institute
**Department of Computer Science,
Swinburne University of Technology, P O Box 218, Hawthorn 3122,
Melbourne, Australia

The neural network paradigms currently dominant in ANN derive their biological inspiration entirely from natural neural computation. Inspiration for new and modified paradigms can be derived by extending consideration to glial cells and, in particular, astrocytes. Astrocytes appear to possess a type of intracellular calcium dynamics which provides a basis of excitability for signaling between them. This raises the possibility that astrocytic networks engage in information processing with very different temporal and spatial characteristics from neuronal signaling. Drawing on a wide range of glial experimental evidence a mechanism is described enabling astrocytes to interact with neurons in a fashion which greatly enhances the effectiveness of long term synaptic facilitation. The potential of this mechanism is illustrated in relation to a range of current research issues in natural and artificial neural networks.

Introduction

The neural network paradigms currently dominant in the investigation of ANN derive their biological inspiration entirely from natural neural computation. The time has come in the search for fresh inspiration to consider the potential of other types of brain cell and, in particular, glia as a source of new paradigms and enhancements of neuron based paradigms.

The promise of glia is supported by a rapidly expanding range of experimental evidence. Glia, such as the astrocytes of cortical gray matter, have elaborate dendritic morphologies superficially similar to those of neurons. As Dani et al. (1992) point out, their fine processes mingle intimately with those of neurons throughout the synaptic neuropil and astrocyte membranes juxtapose or ensheathe most synapses in the mammalian CNS. Astrocytes actually outnumber neurons in many brain regions and are interconnected by gap junctions into vast networks. While these structural features hint at signaling or even information processing functions, astrocytes have traditionally been assigned relatively passive, background roles in structural, metabolic and trophic support of neurons. This view arose and persisted due to the absence in astrocytes of evidence of the electrical excitability which provides the basis of signaling between neurons. Recent studies of astrocytes, however, suggest that they possess a type of intracellular Ca^{2+} dynamics which provides an alternative basis of excitability for signaling between them. This raises the possibility that astrocytic networks engage in information processing but with very different temporal and spatial characteristics from neuronal signaling. The remainder of this discussion will explore the evidence in support of this possibility and its implications for natural brain computation and ANN.

Learning and Synaptic Facilitation

Intracellular calcium levels play a prominent part in the neural mechanism underlying the long term potentiation (LTP) produced by Hebbian learning as a result of conjunction of presynaptic and postsynaptic activity. The mechanism for detecting a conjunction of presynaptic and postsynaptic activity appears to have a postsynaptic locus of control and to be based on postsynaptic depolarization. As Figure 1 indicates, the emergence of LTP is accompanied by a postsynaptic increase in intracellular calcium ion concentration. The neurotransmitter glutamate promotes calcium influx via NMDA-receptor linked channels. Calcium also enters via voltage sensitive calcium channels (VSCC).

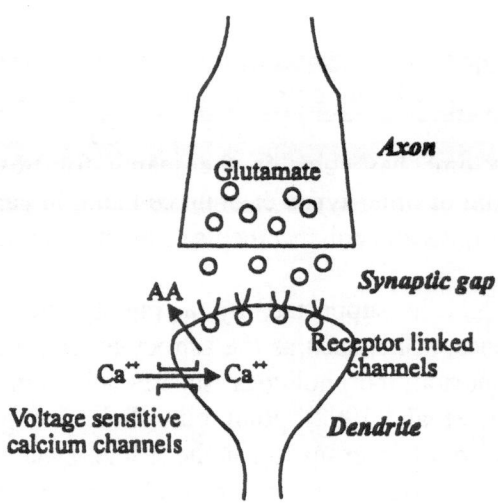

Figure 1: Long Term Potentiation (LTP)

Maintenance of LTP is mediated by presynaptic mechanisms. Williams et al. (1989) postulate the existence of a retrograde messenger to carry information from the postsynaptic side of the synapse to recently active presynaptic terminals. Arachidonic acid (AA) is suggested as a candidate retrograde messenger. Activation of NMDA receptors by neurotransmitter glutamate stimulates release of AA. This maintains LTP by increasing the amount of glutamate available for synaptic facilitation. The mechanism underlying this result introduces glia to the scene. Figure 2 presents a process enabling glial cells to absorb glutamate from a synaptic gap and convert it to glutamine which can be transferred to neurons and reconverted to glutamate. This recycling process enables glia to function as a reservoir for glutamate and has been long known to be a feature of the supportive structure supplied to neurons by glia (McGeer et al., 1978). This function offers a ready explanation of the synaptic facilitation produced by AA. Barbour et al. (1989) have discovered that AA inhibits the uptake of glutamate by glial cells and, thus, produces an increase in the amount of glutamate available for synaptic facilitation. We will revisit this mechanism later.

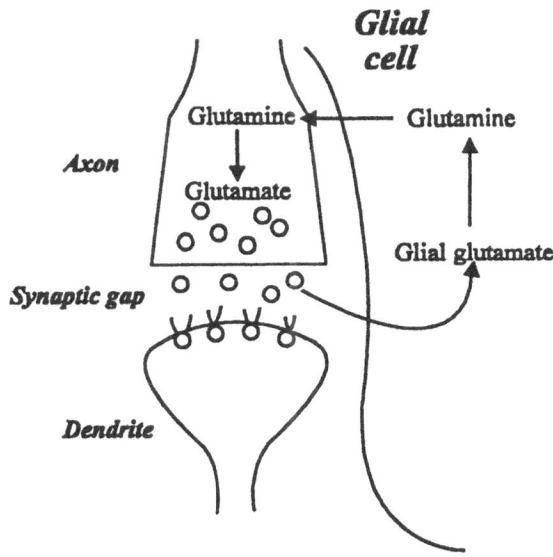

Figure 2: Glutamate Takeup by Glial Cells

Several comments on temporal and operational aspects of LTP will set the scene for our consideration of the role of glia, in general, and astrocytes, in particular, in synaptic facilitation. LTP begins to develop within 10 seconds of the application of stimulation and its effects are fully present within 20 - 30 seconds, (Gustafsson and Wigstrom, 1990). The potentiation can persist without detectable change for weeks, (Staubli and Lynch, 1987). The linking of long-term synaptic facilitation with the results of experience of very short duration must be achieved by means of processes that avoid the obvious dangers of such 'jumping to conclusions' to the survival of a system. Current versions of the neural mechanisms underlying LTP do not take account of this objective.

It is, also, difficult using current LTP mechanisms to account for experimental data such as the results of Martin et al., (1992). Recording from local groups of discriminated hippocampal neurons revealed that while stimulation produced LTP at the level of population evoked field potentials the local, single neuron response was highly variable. The same stimulation produced different LTP outcomes within a local group of simultaneously recorded neurons. Once again, this suggests the need for enhancement of our accounts of synaptic facilitation.

In addition to arguments for the inclusion of additional processes based on the shortcomings of current mechanisms there is direct experimental evidence that supports the involvement of two interacting processes in synaptic facilitation. Williams et al. (1989), in addition to producing LTP in hippocampal tissue, explored the effects of a combination of weak stimulation and added AA. This produced an increase in potentiation comparable with maximal LTP but, in contrast to the rapid onset of LTP, the time course of potentiation revealed a slow onset and a gradual climb to a plateau during a 60 - 120 minute period. As with LTP the enhancement of neural transmission was accompanied by an increase in the release of glutamate.

Astrocytes and Synaptic Facilitation
In this section we will present a necessarily brief description of processes capable of producing temporal and operational characteristics which complement those of LTP and carry us towards a more comprehensive account of synaptic facilitation. The account brings together a wide range of experimental evidence derived from independent lines of investigation of the role and functioning of glial cells.

Figure 3 presents a cycle of sequential and concurrent processes begun by registration on astrocytes of glutamate generated by neural synaptic activity. Evidence from calcium-imaging experiments in hippocampal and cortical astrocytes indicates that glutamate receptor activation in these cells leads to an increase in internal calcium, (Cornell-Bell and Finkbeiner, 1991). Rising intracellular Ca^{2+} results in two highly interesting chains of events. The first provides a mechanism for the appearance of learned associations between astrocytes; the second enables the extended experience of astrocytes to influence synaptic facilitation in the neural system.

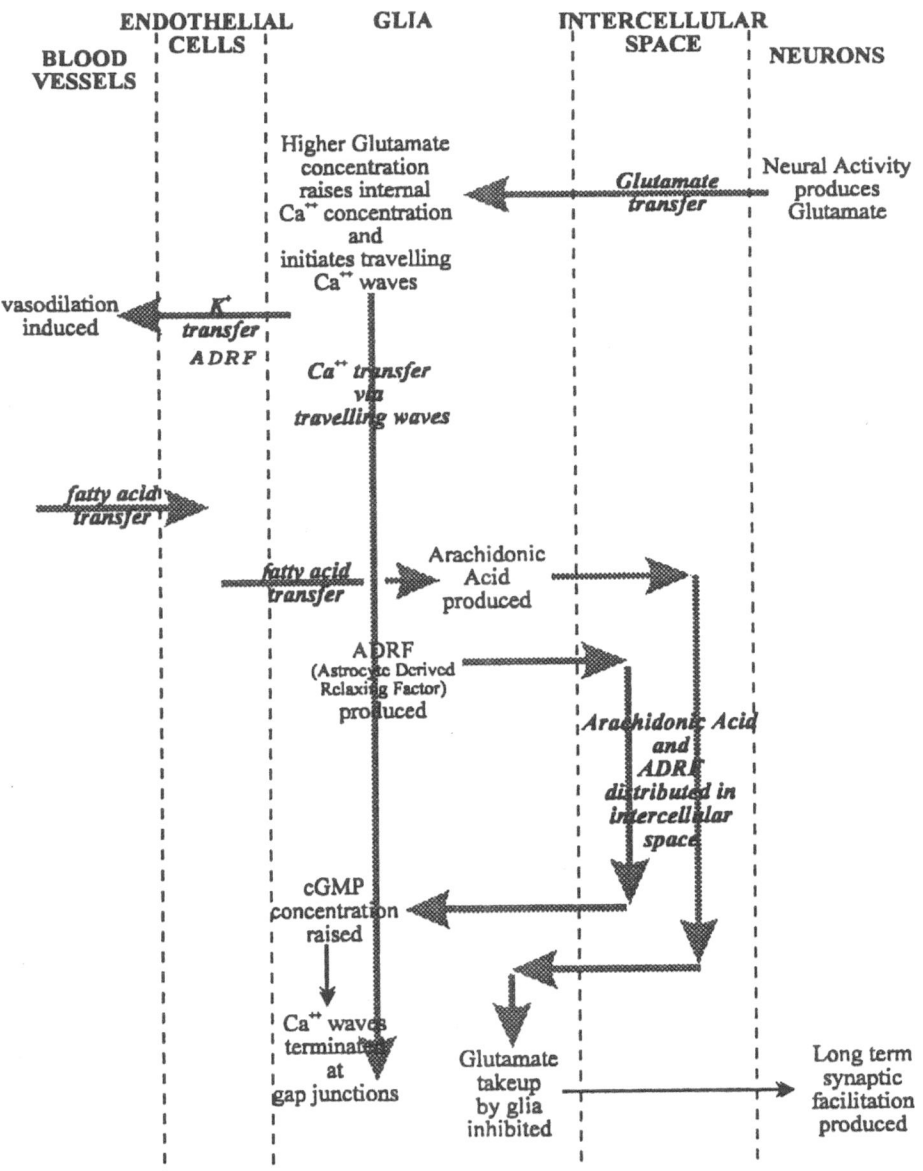

Figure 3: Glial Influence on Synaptic Facilitation

Continued raised levels of intracellular Ca^{2+} frequently propagate as waves, first within the cytoplasm of individual astrocytes and then between adjacent astrocytes, (Dani et al., 1992). Waves have been recorded involving as many as 59 cells before moving beyond the field of view, (Cornell-Bell et al., 1990). The emerging strength of Ca^{2+} based associations between sequences of astrocytes is continuously tested by a chemical reaction chain which reduces conductance across the gap junctions between astrocytes. Decrease in conductance results from a rise in intracellular cyclic GMP (De Vries and Schwartz, 1989). This is produced by the presence of astrocyte derived relaxing factor (ADRF) itself a consequence of raised Ca^{2+} levels in astrocytes.

ADRF provides a link between the processes producing associations between astrocytes and the sequence of events enabling astrocytes to influence synaptic facilitation. Astrocytes are, via specialized end-feet, intimately associated with cerebral blood vessels and have long been regarded as possessing the ability to produce vasodilation for trophic purposes in areas where neural activity is high. Two possible triggers for vasodilation are the vasorelaxant effect of ADRF (Murphy et al., 1990) or of the potassium (K^+) expelled by astrocytes as they absorb Ca^{2+}, (Clark and Mobbs, 1992). The connection between vasodilation and long term synaptic facilitation hinges on blood borne fatty acids which are converted to AA by a two step process involving endothelial cells, lying between blood vessels and astrocytes, and, finally, astrocytes themselves, (Moore et al., 1991). Extracellular AA produced promotes synaptic facilitation by inhibiting glutamate uptake by glia through the recycling mechanism previously described.

Implications and Inspirations

The astrocyte based processes outlined in the previous section are a promising source of a broadened perspective in tackling the shortcomings of current LTP mechanisms. The extended temporal base due to the relative slowness of Ca^{2+} excitability enables state change in astrocytes to reflect the results of neural activity over more extended periods than the brief duration underlying LTP. This offers the possibility of a modified mechanism involving astrocyte - neuron interaction that preserves the advantages of rapid reaction in LTP while avoiding the dangers of long term synaptic facilitation based on short duration experience. It, also, provides a source of variation in the operational context of single synaptic connections that offers an explanation of the different LTP outcomes at the single neuron level produced by the same stimulation.

The extended temporal base and operational characteristics produced by the inclusion of astrocyte mechanisms are potentially highly relevant to a number of current research issues and problems in natural and artificial neural networks. Only brief examples can be provided.

Lynch and Granger (1994) propose that the olfactory-hippocampal circuit raises a sequence of questions involving memories lasting from many weeks to seconds. The search for answers requires "variations of LTP, at least one form of non-LTP plasticity, and a kind of reverberating network." Much of the experimental evidence supporting the existence of the astrocyte mechanism is derived from studies of the hippocampus. It offers an interesting extension at the local level of the sources available to produce the desired effects.

In presenting an account of reinforcement learning based on a natural neural network theory of the striatum, Wickens (1993) highlights the critical significance of maintaining raised postsynaptic calcium levels until the arrival of dopamine produces differential reinforcement. The addition of local astrocyte action to the model would assist in preserving an appropriate temporal and spatial distribution of calcium.

Similar needs for an expansion of the range and temporal basis of associative memory options is evident in research on ANN. Rogers (1994) describes the neural multiprocess memory model (NMMM), "a new artificial synaptic multistage memory system that replaces the single associative long-term memory that normally modulates the efficacy of ANN element inputs. The NMMM enhances learning in an ANN element by supporting multiple time scales in the acquisition, retention and recall of encoded associative memory traces." Further study of astrocyte mechanisms offers a potential source of new ideas on the achievement of temporal flexibility in ANN.

More generally, some attempts to categorize mental functioning into symbolic and subsymbolic forms have drawn heavily on temporal distinctions. Newell (1990), for example, states "There is almost no time available for the neural system to produce fully cognitive behavior." The inclusion of the temporal extension and flexibility provided by astrocyte mechanisms greatly complicates any such attempts to base distinctions on relative duration of processing.

In conclusion, Clarke (1993) predicts that "increased attention to neural plausibility will surely be a hallmark of the next generation of associative engines. The heterogeneity of neural structures (and perhaps learning rules too) will have to be reflected and exploited." This scenario should clearly be extended to include the potential of glial cells and their interaction with the neural system.

References

Barbour, B., Szatkowski, M., Ingledew, N., & Attwell, D., (1989). Arachidonic acid induces a prolonged inhibition of glutamate uptake into glial cells. *Nature* 342, 918-920.

Clark, A. (1993). *Associative Engines: Connectionism, Concepts and Representational Change*. MIT Press, Cambridge, Mass.

Clarke, B. & Mobbs, P. (1992). Transmitter-operated channels in rabbit retinal astrocytes studied in situ by whole-cell patch clamping. *Neuroscience* 12(2), 664-673.

Cornell-Bell, A.H., Finkbeiner, S.M., Cooper, M.S., Smith, S.J. (1990). Glutamate induces calcium waves in cultured astrocytes: long-range glial signaling. *Science*, 247, 470-473.

Cornell-Bell & Finkbeiner (1991). Ca^{2+} waves in astrocytes. *Cell Calcium*, 12, 185-204.

Dani, J.W., Chernjavsky, A., & Smith, S.J. (1992). Neuronal activity triggers calcium waves in hippocampal astrocyte networks. *Neuron*, 8, 429-440.

De Vries, S.H., & Schwartz, E.A., (1989). Modulation of an electrical synapse between solitary pairs of catfish horizontal cells by dopamine and second messengers. *Journal of Physiology*, 414, 351-375.

Gustafsson, B., & Wigstrom, H., (1990). Long-term potentiation in the CA1 region: its induction and early temporal development. *Progress in Brain Research*, 83, 223-232.

Lynch, G. & Granger, R., (1994). Variations in synaptic plasticity and types of memory in corticohippocampal networks. In D.L. Schacter, & E. Tulving, (Eds.) *Memory Systems 1994*, MIT Press, Cambridge, Mass., 65-86.

McGeer, P.L., Eccles, Sir J.C. & McGeer, E.G., (1978). *Molecular Neurobiology of the Mammalian Brain*, Plenum Press, New York.

Martin, P.D., Lake, N., & Shapiro, M.L. (1992). Effects of burst stimulation on neighboring single CA1 neurons in rat hippocampus. *Society for Neuroscience*, 22nd Annual Meeting, Anaheim, CA,.

Moore, S.A., Yoder, E., Murphy, S., Dutton, G.R., & Spector, A.A. (1991). Astrocytes, not neurons, produce docosahexaenoic acid (22:6ω-3) and arachidonic acid (20:4ω-6). *Journal of Neurochemistry*, 56, 518-524.

Murphy, S., Minor, R.L., Welk, Jr., G., & Harrison, D.G., (1990). Evidence for an astrocyte-derived vasorelaxing factor with properties similar to nitric oxide. *Journal of Neurochemistry*, 55, 349-351.

Newell, A. (1990). *Unified Theories of Cognition*, H. U. P., Cambridge, Mass.

Rogers, B.L. (1994). New neural multiprocess memory model for adaptively regulating associative learning. *Neural Networks*, 7, 1351-1378.

Staubli, U., & Lynch, G. (1987). Stable hippocampal long-term potentiation elicited by "theta" pattern stimulation. *Brain Research* 435, 227-234.

Wickens, J. (1993). *A Theory of the Striatum*; Pergamon Press, Oxford.

Williams, J.H., Errington, M.L., Lynch, M.A., & Bliss, T.V.P. (1989). Arachidonic acid induces a long-term activity-dependent enhancement of synaptic transmission in the hippocampus, *Nature*, 341, 739-742.

Modeling Retinal High and Low Contrast Sensitivity Filters

T. Lourens
Department of Computer Science
University of Groningen
P.O. Box 800, 9700 AV Groningen, The Netherlands
E-mail: tino@cs.rug.nl

Abstract

In this paper two types of ganglion cells in the visual system of mammals (monkey) are modeled. A high contrast sensitive type, the so called M-cells, which project to the two magno-cellular layers of the lateral geniculate nucleus (LGN) and a low sensitive type, the P-cells, which project to the four parvo-cellular layers of the LGN. The results will be compared with the ganglion cells as described by Kuffler.

1   Introduction

In the last forty years a lot of work on the mammalian visual system has been done by neuro-physiologists. It was Kuffler [1] who in 1952 recorded the activity from the axons of the retinal ganglion cells that make up the optic nerve. His experiments revealed the type of receptive fields a retinal ganglion cell possess. He found two basic types of center-surround cells. Later Hubel [2] and Wiesel did a lot of pioneering work in the primary visual cortex, its is mainly due to their work that today we understand the functionality of certain parts in the visual system of mammals.

There are two types of ganglion cells that will be modeled in this paper. The first type of cells are the *ganglion magnocellular cells* or short the *M-cells*, this type of cells are retinal ganglion cells which project to the magnocellular layers (ventral layers) in the lateral geniculate nucleus. The second type of cells are *parvocellular cells* or *P-cells* which project to the parvocellular layers (dorsal layers) in the lateral geniculate nucleus.

The two ventral layers in the lateral geniculate nucleus are more sensitive to luminance contrast than the cells in the four dorsal layers of the lateral geniculate nucleus. The differences in these layers is due to the fact that there are differences in the retinal ganglion cells which provide excitatory synaptic input to the LGN-neurons and not because of the organization (pattern of connectivity) in the LGN [3].

The small P-cells in the monkey are color sensitive (wavelength selective) and have small concentric center-surround receptive fields. They are not very contrast sensitive. The large M-cells are not wavelength selective, they also have concentric center-surround receptive fields but these receptive fields are larger than the receptive fields of the P-cells and are highly sensitive to contrast [4, 5, 6].

In section 2 the center-surround receptive field will be modeled by a so called *mexican-hat function*. The function can be modeled by taking the difference of two gaussian functions (DOGs). In this section is also described how the center and surround of the mexican-hat function can be controlled by these two gaussians. In this paper the receptive fields of M-cells and P-cells modeled with the mexican-hat function are only used for comparison. The center and surround of the mexican-hat function in fact are used to calculate respectively the average center and surround luminance in the receptive field of an M-cell or P-cell. This implies that a mexican-hat filter is used which is constant in both center and surround but differs in sign, i.e. the center of the mexican-hat is positive and surround is negative. In section 3 the relation between dendritic tree and visual field is given. In the fourth section a contrast filter for the two types of cells is created and after that the receptive fields of an M-cell and P-cell are modeled. In the last section the experimental results are given.

2   Center-surround receptive fields

In this section the properties of gaussians are described which are used for modeling a center-surround receptive field.

The standard two-dimensional mexican-hat function which is derived from one gaussian function can be defined as follows:

$$G_m(r) = \frac{2\sigma - r^2}{\sigma^2} e^{-\frac{r^2}{2\sigma}} \tag{1}$$

where $r = \sqrt{x^2 + y^2}$. The relation between the center and surround radius is $1 : \sqrt{-\log \varepsilon}$ where ε is a positive constant near zero. If $\varepsilon = e^{-9}$ then we get a ratio of $1 : 3$. This means in fact that one is only able to control either center or surround of a mexican-hat function, for details of (1) and its properties we refer to [7].

2.1 The mexican-hat function with differences of two gaussians

In case one wants to be able to control both center and surround, a mexican-hat function should be modeled with two gaussians. The difference of these two gaussians (DOGs) will give the desired mexican-hat function.

Let us define the *center* of the mexican-hat as that part of the the function for which $G_m(r) \geq 0$ and the *surround* as the parts for which $G_m(r) < 0$. The mexican-hat function can be created by taking the difference between a center and a surround gaussian. Assume that the mexican-hat is modeled by a center gaussian G_c which is subtracted from a surround gaussian G_s, where the center gaussian is defined as:

$$G_c(r) = e^{-cr^2} \qquad (2)$$

and the normalized surround gaussian is defined by:

$$G_s(r) = \frac{1}{m^2} e^{-c\frac{r^2}{m^2}} \qquad (3)$$

where $\frac{1}{m^2}$ is the normalization-factor, $m > 1$. The ratio between center and surround gaussian is $1 : m$. The mexican-hat function is a combination of the two previous equations:

$$G_m(r) = G_c(r) - G_s(r) \qquad (4)$$

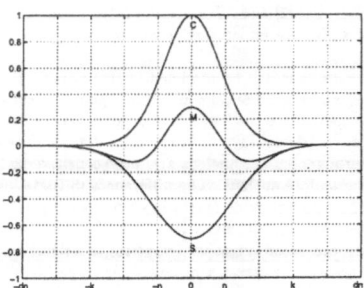

Figure 1: The center gaussian C, surround gaussian S, and the mexican hat function M. For better visualization the one-dimensional functions are used.

For normalization of two-dimensional gaussian filters the following integral is used:

$$\int \int e^{-\gamma(x^2 + y^2)} dx dy = \frac{\pi}{\gamma} \qquad (5)$$

where γ is a constant. This implies that the appropriate normalization factor is γ.

The border of a gaussian is used to control the center or surround of a mexican-hat function. The border of the center gaussian is defined as the radius k for which $G_c(k) = \varepsilon$ (Figure 1). Since a gaussian will never be exactly 0, a small value $\varepsilon > 0$ will be used to estimate the border of a gaussian:

$$G_c(k) = \varepsilon \qquad (6)$$

$$\equiv$$

$$c = -\frac{\log \varepsilon}{k^2} \qquad (7)$$

2.2 Controlling the difference of two gaussians

Let us assume that the center of the mexican-hat function has radius n and that the surround has a radius that is related to the center radius in such a way that the surround radius is d times larger than the center radius (see also

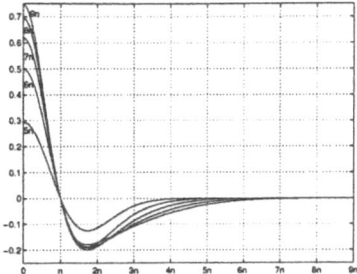

Figure 2: Five different mexican-hat functions all with the same center radius n but different surround radii, the different surround radii are $5n$, $6n$, $7n$, $8n$, and $9n$, respectively. For better visualization again the one-dimensional functions are used.

Figure 1). Given the center and surround gaussians from (2) and (3), we should find out how c and m should be chosen to get a mexican-hat center radius n and a surround radius dn (Figure 2).

For the center radius of the mexican-hat function we obtain:

$$G_m(n) = 0 \tag{8}$$

$$\equiv$$

$$c = \frac{-2\log m}{n^2\left(\frac{1}{m^2} - 1\right)}, \tag{9}$$

from which we conclude that there is a relation between c and m.

For the surround radius of the mexican-hat function we get:

$$G_m(dn) = -\varepsilon \tag{10}$$

$$\equiv$$

$$d = \sqrt{\frac{\log\left(m^2\varepsilon\right)\left(1 - m^2\right)}{2\log m}}. \tag{11}$$

Where $\varepsilon > 0$ is a small constant. Equation (11) gives the relation between m and d.

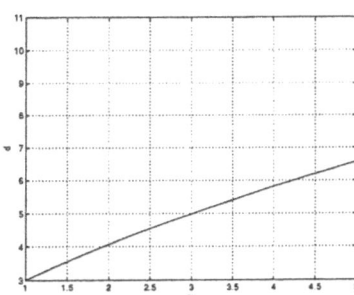

Figure 3: The relation between m and d in the two-dimensional case. Note that the minimum $d = \sqrt{-\log\varepsilon}$. This means that only a relation between center and surround can be realized where the surround radius is at least $\sqrt{-\log\varepsilon}$ times larger than the center radius.

3 The relation between receptive field and dendritic tree of a ganglion cell

The direct connections from a ganglion cell to the bipolar cells is called the dendritic field of a ganglion cell and the maximum size or radius of such a dendritic field is the dendritic field size or radius of a ganglion cell. The dendritic field size of a ganglion cell to the receptors is defined as the maximum dendritic field size that can be reached by the direct pathway, from receptors to bipolars to ganglion cells [2], and is shown in Figure 4.

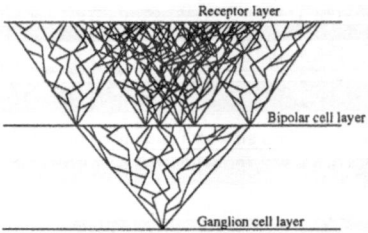

Figure 4: The dendritic field of the ganglion cell itself plus the dendritic fields of the bipolar cells which are directly connected with this ganglion cell is defined as the dendritic field of the ganglion cell to the receptors.

The direct pathway is responsible for the the receptive field centers. The indirect pathway, the path also via the horizontal cells and amacrine cells, is responsible for the the surround response of the receptive field. From this we conclude that the dendritic field radius of a ganglion cell to its receptors is equal to the receptive field center radius of the same ganglion cell. This implies that the variation of the dendritic field size with retinal eccentricity of the cells is equal to the variation in receptive field center size of the cells.

In [4] the dendritic field sizes of the ganglion cells grow in a linear way with retinal eccentricity. We also assume that the dendritic field sizes of the bipolar cells grow in a linear way with increasing retinal eccentricity. It can be easily verified that the dendritic tree between ganglion cell and receptors also grows in a linear way with increasing retinal eccentricity.

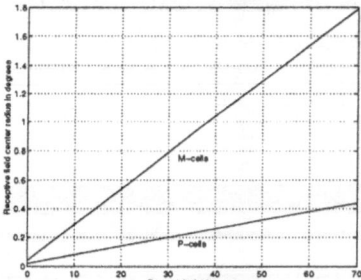

Figure 5: The receptive field center radius grows linear with eccentricity for both M-cells and P-cells. Note that the receptive fields of the M-cells are about a factor four larger than the P-cells.

In the model the assumption is made that given the eccentricity both ganglion cells and bipolar cells have exactly the same dendritic field radius. In [4] the dendritic field size for the ganglion cells to the bipolar cells with retinal eccentricity is given, in the model given the assumption that both ganglion and bipolar cells have the same dendritic field radius then the dendritic field radius from ganglion cells to the receptors is twice the size of the dendritic field radius of the ganglion cells. The receptive field center radius for a given eccentricity α for the M-cells (Figure 5) will be:

$$M_c(\alpha) = 2.5 \cdot 10^{-2}\alpha + 4.0 \cdot 10^{-2} \tag{12}$$

and the smaller receptive field center radius of the P-cells (Figure 5) will be:

$$P_c(\alpha) = 6.0 \cdot 10^{-3}\alpha + 4.0 \cdot 10^{-2} \tag{13}$$

where α is the eccentricity in degrees.

4 Modeling M-cells and P-cells

Instead of using the mexican-hat as a filter for modeling the receptive field of an M-cell or a P-cell, the center and surround of the mexican-hat function are used to calculate the average center and surround luminance in a receptive field respectively. This implies that a mexican-hat filter is used which is constant in both center and surround but differs in sign, i.e. the center of the mexican-hat is positive and surround is negative. Note that the M-cells and P-cells are also modeled with the mexican-hat function but that it is only used for comparison. (For the center-surround (mexican-hat) filter see [7]). In fact the filter uses only the center radius n and the surround radius dn, which is chosen to be about three times larger than the center radius of the mexican-hat function.

If the center of a mexican-hat is placed at position (φ, r) in the visual field then the luminance of the center L_c is defined as the average luminance of that part of the visual field which fits in the center of the mexican-hat:

$$L_c(\varphi, r) = \frac{\int_{\varphi_1=0}^{2\pi} \int_{r_1=0}^{n} I(r\cos\varphi + r_1\cos\varphi_1, r\sin\varphi + r_1\sin\varphi_1) r_1 \mathrm{d}r_1 \mathrm{d}\varphi_1}{\int_{\varphi_1=0}^{2\pi} \int_{r_1=0}^{n} r_1 \mathrm{d}r_1 \mathrm{d}\varphi_1} \tag{14}$$

Note that n is the center radius, for the M-cell $n = M_c(\alpha)$ and for the P-cell $n = P_c(\alpha)$. For the artificial visual field I a two-dimensional image is used.

The luminance of the surround L_s is defined as the average energy in the visual field which fit in the surround of the mexican-hat:

$$L_s(\varphi, r) = \frac{\int_{\varphi_1=0}^{2\pi} \int_{r_1=n}^{dn} I(r\cos\varphi + r_1\cos\varphi_1, r\sin\varphi + r_1\sin\varphi_1) r_1 \mathrm{d}r_1 \mathrm{d}\varphi_1}{\int_{\varphi_1=0}^{2\pi} \int_{r_1=n}^{dn} r_1 \mathrm{d}r_1 \mathrm{d}\varphi_1} \tag{15}$$

where d is about 3, which implies that the center is about three times smaller than the surround.

Because both M-cells and P-cells are contrast sensitive, we introduce the term *contrast*. Contrast is the relative difference between the maximum and minimum luminance and is defined as follows:

$$\text{Contrast} = \frac{L_{\max} - L_{\min}}{L_{\max} + L_{\min}} \tag{16}$$

where $L_{\max}$ and $L_{\min}$ are the maximum and minimum luminances respectively.

From the receptive field sizes (12)-(13) and the definition of contrast (16) the *contrast filter* is rather trivial, it is the contrast between L_c and L_s, given a position (φ, r) in the visual field:

$$C(\varphi, r) = \frac{|L_c(\varphi, r) - L_s(\varphi, r)|}{L_c(\varphi, r) + L_s(\varphi, r)} \tag{17}$$

Note that the size of the center, M_c and P_c of the M-cell and P-cell respectively, gives the difference between a contrast filter for the M-cell or P-cell.

The response of the M-cell or P-cell depends on the contrast. The contrast-response for the M-cell, which is defined as the receptive field of the M-cell or high-contrast sensitive filter is as follows:

$$R_M(C_M) = \frac{aC_M}{0.13 + C_M} \tag{18}$$

where C_M is the contrast difference for an M-cell and a is the impulse amplitude. For the P-cells the contrast-response is defined as the receptive field of the P-cell or low-contrast sensitive filter and is as follows:

$$R_P(C_P) = \frac{aC_P}{1.74 + C_P} \tag{19}$$

where C_P is the contrast differences for a P-cell and a is the impulse amplitude which is identical to the impulse amplitude of the M-cell. Both contrast-response equations are taken from [3].

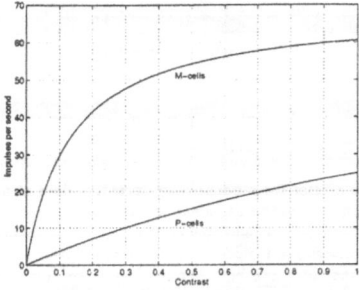

Figure 6: For both type of cells a response-contrast function is given. The magnocellular cells (M-cells) give a high response amplitude on stimulus contrast. This in contrast with the parvocellular cells (P-cells) which give a lower contrast response.

5 Experimental results and future research

The results filtering the face image of Figure 7 with a mexican-hat filter for M-cells and P-cells are shown in Figure 8a and 8b respectively. If we compare the results of the mexican-hat filter (Figure 8a-b) with the contrast filter (Figure 8c-d), it is clear that the results obtained by the contrast filter are better. For example compare the eyebrow of Figure 8a, which is almost invisible, with the eyebrow in Figure 8c. It is remarkable that the differences between Figure 8c and Figure 8e are really small, this means that the contrast filter of a P-cell and the receptive field of the same P-cell have identical properties.

In future research we want to create a model that is able to detect and recognize objects in a visual scene by applying techniques which are also used by mammals. In this model the wavelength selective P-cells give high detailed information which will be used for both color and detailed information description of an object. The non-color selective M-cells which are large will be used together with the simple cells [8, 9] for edge detection and partly for object detection. By using these biologically motivated techniques we hope to get more insights in the functionality of the human brain.

Figure 7: An artificial visual field, which is represented by a face image.

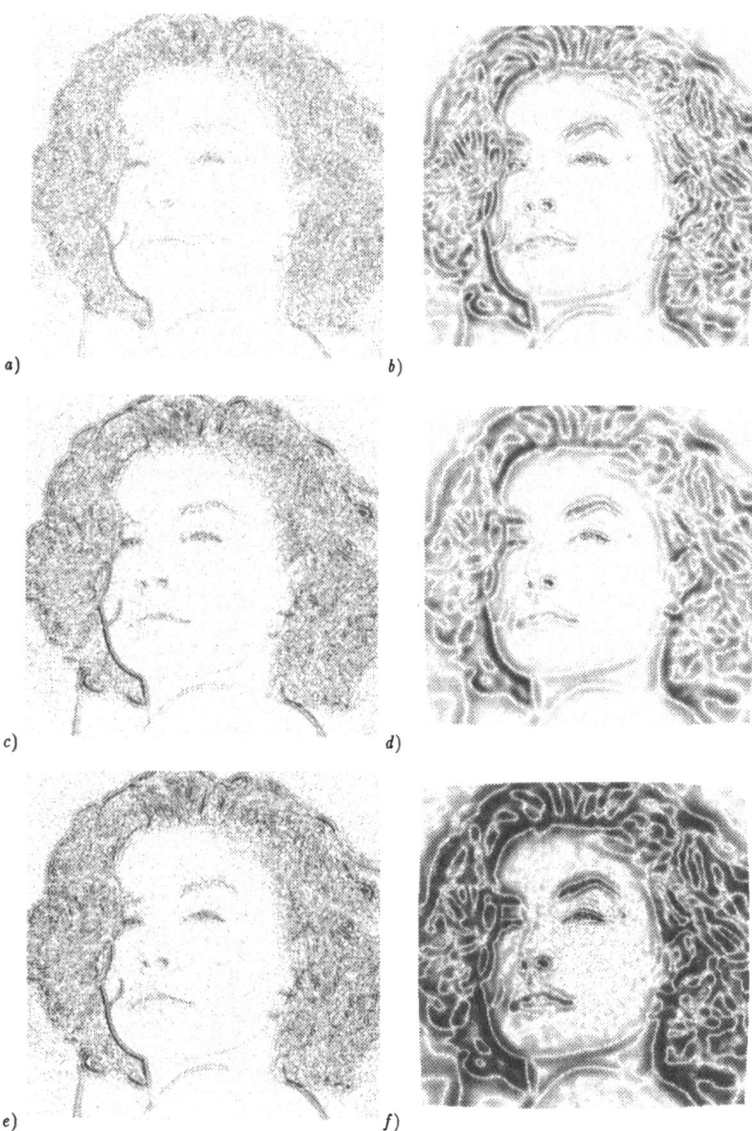

a) b)

c) d)

e) f)

Figure 8: The three left images are related to the parvocellular properties. The three images on the right are related to the magnocellular properties. The original image filtered with a mexican-hat filter *a)* and *b)* . The contrast filters are shown in *c)* and *d)*. The impulse responses on the contrast image are shown in *e)* and *f)*.

References

[1] S. W. Kuffler. Discharge patterns and functional organization of mammalian retina. *Journal of Neurophysiology*, 16:37–68, 1953.

[2] David H. Hubel. *Eye, Brain and Vision*. Scientific American Library, New York, 1988.

[3] E. Kaplan and R. M. Shapley. The primate retina contains two types of ganglion cells, with high and low contrast sensitivity. *Proc. Natl. Acad. Sci. U.S.A.*, 83:2755–2757, April 1986.

[4] Robert Shapley and V. Hugh Perry. Cat and monkey retinal ganglion cells and their visual functional roles. *Trends in Neuroscience (TINS)*, 9:229–235, May 1986.

[5] Margaret Livingstone and David Hubel. Segregation of form, color, movement, and depth: Anatomy, physiology, and perception. *Science*, 240:740–749, 1988.

[6] John G. Nicholls, A. Robert Martin, and Bruce G. Wallace. *From Neuron to Brain -A Cellular Molecular Approach to the Function of the Nervous System*. Sinauer Associates, Inc., third edition, 1992.

[7] T. Lourens. Building a biological filter. submitted to ESANN '95, 1995.

[8] N. Petkov, P. Kruizinga, and T. Lourens. Biologically motivated approach to face recognition. In J. Mira, J. Cabestany, and A. Prieto, editors, *New Trends in Neural Computation, Proceedings of the International Workshop on Artificial Neural Networks, IWANN '93*, volume 686 of *Lecture Notes in Computer Science*, pages 68–77. Springer-Verlag Berlin Heidelberg, Sitges, Spain, June 9-11 1993.

[9] N. Petkov and T. Lourens. Human visual systems simulations - an application to face recognition. In H. Dedieu, editor, *Circuit Theory and Design 93, Proceedings of the 11th Conference on Circuit Theory and Design*, pages 821–826, Davos, Switzerland, Aug. 30 - Sept. 3 1993. Elsevier Science Publishers B.V. Amsterdam.

Neural Micro-Structures
Three Simple Models

Josep Miró Nicolau.
Member of Real Academia de Medicina y Cirugia

Universitat de les Illes Balears,
Ed. A. Turmeda. Campus
07071 Palma de Mallorca SPAIN

Abstract   Discovering neural substructures in the brain is practically impossible. Modules assumed in the brain so far have not been structurally established but *functionally*. To approach the understanding of the performance of the brain in terms of neurons seems over-optimistic. In this paper three simple micro-structures are proposed in terms of neurons, with the purpose of using them in further papers to explain more complex functional modules.

1 INTRODUCTION

This paper is meant to be a a small part of a broader message, in which the author plans to develop his views on a possible brain organization, result of years of introspection, reflection and study. The overall philosophy of this approach will be very briefly resumed in the first Section In the same Section, a condensed outline of the basic structure of knowledge as it is conceived by the author will commented. The main body of this paper can be found from Sections 2 on.

1.1 An hypothesis

The first step in every scientific development is the formulation of an hypothesis. This hypothesis is then described in scientific terms, that is, using a language amenable to scientific methodology of reasoning and testing. If the results of reasoning fit with the results of tests, then this fitness is considered enough reason to say that *observational evidence supports the hypothesis*, and then the scientific community usually agrees in granting it the title of *principle*.

In engineering, in particular, software engineering, if the hypothesis leads to useful programs, then it constitutes the basis of a *method*. If the method has been inspired by an hypothesis concerning the performance of a natural organism, then is usually referred as a *model* of the organism. And so it may happen that a certain model may be useful, as far as computer simulation is concerned, although there might not be any evidential grounds confirming that the natural organism behaves at all as the model

On the other way, an unconfirmed hypothesis together with a whole body of thought deduced from it, if it is accepted by a sub-community, might be referred to as *Theory*. The starting hypothesis may be a very simple sentence or a very sophisticated system. In fact, the overall Greek mythology may be considered as a unique hypothesis, that for centuries was accepted as a correct description of the relations among supernatural powers.

Notice that the formulation of an hypothesis must precede its testing for evidence and its use as a model. Therefore, for every hypothesis it may have existed a period of time when there was no evidence supporting it yet.

This paper contributes to the formulation of an overall hypothesis. Its confirmation from natural science will take a very long time. Its usefulness as a model shouldn't take that long, but it will not be immediate either. For quite a while it will have to be regarded as an unconfirmed hypothesis. If a whole body of thought is eventually developed upon it, it might become a theory, but so far it is only the proposal of a model.

1.2 The brain-mind caesura

The concept of mind is an old one. Three thousand years ago they were already mentioning it. Its performance has been described in terms of theories. Bunge mentions ten different groups of them.

Twenty five hundred years ago, they already knew the brain existence, not its performance. The oldest drawing of the brain Clarke could find (Clarke 1972) is not quite a thousand years old. Practically, all we know about the brain today has been learnt during the last two hundred years. We have theories to deal with the biological performance of the brain. But these theories start with biological data and end with biological result. No available theory gives a good account of the *mental* performance of the *biological* brain. We express this by saying that there is a *caesura* between the biological knowledge of the brain and epystemological knowledge of the mind.

In my opinion, the main reasons behind this caesura are:

• 1) The overwhelming linguistic and functional "*distance*" between the neuron and the brain. Discovering neural substructures in the brain is practically impossible because no clear interfaces can be observed. In the past the existence of modules has been *functionally* assumed, the structure of each module being unknown. On the other hand, to approach the understanding of the behavior of the brain in terms of neurons, seems too optimistic. It would amount to trying to explain the performance of a car in terms of bolts and nuts. It would be more reasonable to try to explain the performance of big systems in terms of smaller subsystems, that could then be understood in terms of bolts and nuts.

• 2) The fact that we are trying to describe the performance of the brain in terms of old concepts of the *mental category* not fit for the task. Once and again, I have proclaimed myself a *reluctant inheritor* of the three thousand year old culture it has been my tough luck to receive in custody.

My contention is:

• 1) I can *think* about the brain. I am doing so now.
• 2) Penrose doubts I can *reason* about the brain. I like to believe I can.
• 3) If the brain is described properly, and the mind is described properly, and reason is an adequate tool for reasoning, then it must be possible to infer mental performance from the biological nature of the brain.
• 4) The brain description is beyond this author's knowledge and research power.
• 5) No fancy reasoning is available to me. Only the usual one.

Therefore:

• 6) The author must either quit, or try to break the impasse by reformulating the description of knowledge.
• 7) If the author wishes to break away from the old culture, he cannot refer to it, therefore he must use his own introspection.

1.3 Structure of knowledge

Introspection has been used in order to announce a general structure for knowledge. Knowledge in itself doesn't exist. It is a concept created by man in order to describe some of those mental activities qualified as *cognitive*. If knowledge doesn't exist, it doesn't have a structure either. The structure is a coordination added to the several elements of the description that helps its overall understanding. This coordination may be original, but usually is taken from previous descriptions. In a sense a structure is a metaphoric relation, that is borrowed in one context to be extended into another. In his description of knowledge the author used a structure borrowed from geometry, that will not be discussed here.

Once the structure was established, it was proper to inquire whether any evidence would suggest that the real brain organization might be operationally described with the help of this structure. Of course there was no direct evidence. Nor grounds for testing it either.

One of the initial requirements that led to the adoption of the geometrical structure was the repetitive nature of the overall system model. Some fundamental components would be repeated again and again. A fundamental question was whether these elementary components could be implemented with natural neurons or with their models. The purpose of this paper is to offer the descriptions of three neuron micro-structures whose expected performance coincides with that of the ideal components that emerged in the conceived structure of knowledge.

2. SEQUENCE MICRO-STRUCTURES

2.1 Fundamental concepts

In the cognitive operation of the brain, a few concepts become relevant to describe it: Recognition, Attention, Consciousness, Simultaneity, Sequentiality, and probably some other. A good model should represent all of them, but at this research stage, it is still premature to try. Pattern recognizers have received most attention and we can conceive them. Today I wish to address sequentiality and simultaneity only.

Sequences are very common in brain operation. A syllable is a sequence of sounds or written characters, a word is a sequence of syllables, a sentence is a sequence of words. A paragraph is a sequence of sentences, and so on. In order to communicate, the brain must be able to detect and generate sequences. An often detected sequence may become a symbol, that is, it is detected as a whole, as if it were a pattern.

For the purpose of this discussion one may imagine that a particular *meaning* is associated to the triggering of one particular m-neuron. Two symbols have the same "meaning" if they are responsible of firing the same m-neuron. Detection of meaning may be a fairly complex process, involving pattern detection, memory, sequence detection and so forth. The process to consider here is simply the detection of a sequence, paying no attention to whether the result of the detection is fed to a m-neuron or to another module.

From the early stages of this research it became apparent that the detection and generation of sequences were crucial steps. It was not so much because of their puzzling nature, because they are not puzzling at all. Computer engineers have been designing sequence detectors since the very first computer was built. We know too well how it can be done. The problem was that the solution we have given in computer engineering, didn't seem plausible. Take syllable detection, for example. Assume further that a syllable can be a combination of up to five characters. If a newborn brain had a syllable detection module, since at birth the brain does not know in what language it will be trained, the module must be in principle able to detect every possible five letter combination. Well, almost all of them, because some combinations like bbbbb are not likely to occur in any language. This means that the detector must be prepared to detect the first element of the sequence then the second and so forth. Assuming that there is one vowel per syllable, the total number of possible syllables is of the order of N

$$N = 5 \times 22 \times 22 \times 22 \times 22 \times 5!$$

a figure far higher than a hundred million. Assuming that one neuron is capable to detect one element of the sequence, it turns that many millions of structures would be devoted to syllable detection, with an average of three or four neurons each. Of course, the number of brain neurons is very high, (they say that 10^{11}) but the brain has many things to do, and devoting several hundred million cells to the task of detecting syllables seemed kind of exaggerated to me.

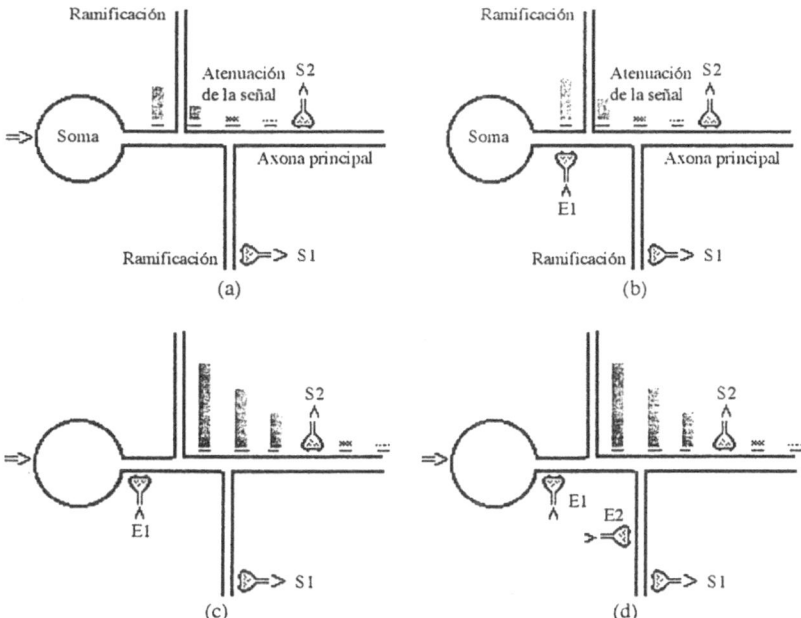

Fig 1. The axo-axonic model

I am indebted to Prof. Simoes da Fonseca of the University of Lisbon for having pointed out to me some research results in neurobiology that turned out to be very helpful. Gray (Gray 1962), had already discovered the axo-axonic synapses, that is, a synapsis delivering a signal from one axon directly to another axon, the group of Rall (Rall et al, 1966) had found evidence of existence of dendro-dendritic synapses, Frank (Frank 1969) had found the anatomically similar pre-synaptic inhibition, and Graubard and Calvin in 1979 suggested that pre-synaptic dendrites were probable candidates for local computation. The physiological computation of these synapses was studied in detail, and the result I want to remark here is that, unlike the classical neuron model, whose firing depended only upon the weighted sum of soma excitations, and such that the neuron realizes only one function, in the studied neurons the output boutons near axonic dendrites do not fire according the typical discontinuous process, but by means of a continuous inner potential change. In fact there are neural systems that do not fire through the discontinuous firing process (Famiglietti 1972). As a result, the output signal depends upon the input synapses near by, and therefore a single neuron might perform the computation described by multiple functions, possibly one different function per output synapsis.

A detail study has been done, mainly for the *Aplysia* and can be found in Graubard (1973) Some details of Graubard 1979 will help to understand the sequel.The reason why the signal propagates toward the narrow end and not backward toward the soma, will not be discussed here.

2.1 The axo-axonic model (Graubard 1979)

Figure 1 (a) represents the soma, the principal axon, and two narrower branches of the axon. It also shows two output synapses O_1 and O_2. The inputs to the soma have produced some excitation that propagates and decays along the principal axon. The decay is shown indicating the places where the signal has been measured. The column size indicates the size of the signal at every measuring place. The farther away from the soma, the smaller the propagated signal. Accordingly no output synapsis is excited.

Fig 1(b) represents the same neuron without somatic excitation, with an additional input synapsis, I_1, that also produces an excitation not reaching the output synapses. Fig 1(c) shows the case in which both somatic and axonic inputs collaborate. In this case the axonic signal is bigger, and it decays along the axon, but but the signal at O_2 is big enough to trigger it, an output signal being produced in the form of neuro-transmitters being released. In the two branchings there is strong enough signal to propagate, but it decays along the branch without any further firing. Fig 1(d) shows the same neuron with another exited axonic input, I_2. In this case both O_1 and O_2 receive enough excitation to deliver an output signal.The understanding of the sequence micro-structures is now immediate.

2.2 Sequence generators

For the purpose of this discussion assume that x represents an action, and *"to have the intention of x"* is the mental interpretation of the particular i-neuron corresponding to x being fired. Once the intention has been in existence then the usual thing is to proceed to generate a sequence of signals. For example, assume that x represents the action of pronouncing the syllable "brown". In this case the brain must generate a sequence of five different signals, address the first signal t_1 to the b-generator, the second t_2, to the r generator and so forth. A simple neuron may do that. Signal propagation in an axon is not very fast, thus five successive synapses along an axon will generate five signals delayed in time that can be addressed one to the b second to the r and so on. Fig 2 (a) shows this simple generator. A more sophisticated micro-structure, shown in Fig 2 (b), would generate the signal addressed to generate the t_2 only after receiving confirmation that the signal addressed to the t_1-generator has been delivered. An even more sophisticated one, shown schematically in Fig 2(c), would confirm that the b has been pronounced, and in the case the b has not been uttered, would insist in pronouncing it. In the case that an unforeseeable trouble would delay the signal confirming the pronunciation of the b, the micro-structure would stutter.The stuttering nature would depend upon the sequence and the cause of the delay. If the sequence were a syllabic one then it should sound like *" I b bb b bb believe"* Should the sequence be a word one, then it should sound *"I be be bebe believe."*

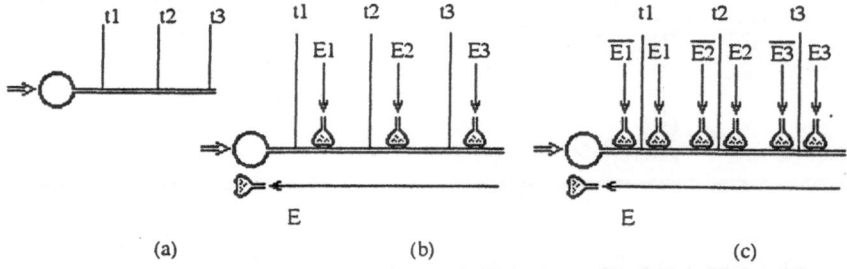

Fig 2. Sequence generator

We certainly do not know how the brain sequence generators are, if there are any, but we can imagine them, and the imagination results are not at all far fetched. For example, assume one neuron per syllabic sequence. I do not know how many syllables we handle, but just for fun I counted how many word initial syllables start with b. I counted less than five hundred. This means that the total number of syllables should be of the order of, say ten thousand. The number of different words is at most thirty thousand, the number of different basic sentences structures is very small. It follows that the total number of sequence generators for language purpose must be bellow fifty thousand, a figure very reasonable in neural orders of magnitude. Of course I should expect that other sequence generators are used for rote learning, muscle control, etc. Typists, actors, musicians, etc possibly singers, generate very long sequences, which are initiated by one single signal. According to this model, many a routine appearing along our life, used for repetitive actions, might be determined by a sequence generated by means of only one neuron each.

2.3 Sequence detectors

A possible design of a basic sequence detector structure might require two complementary operations. The first one may be performed by a single axo-axonic neuron, locally carrying over the computation required for the detection of a sequence.of stimuli. Assuming that the soma responds to some general control signal, then the first stimulus selects on branch, the second stimulus selects a branch of the selected branch, and so on. In this way, not all branches are excited but only those satisfying the input sequence.

An example will illustrate the structure. Fig 3, shows a neuron whose axon is excited by a synapsis driven by a signal generated by an input **b**. This signal might originate in a sound detector organ, or in a vision organ driving a pattern recognizer. After receiving the **b** signal, the axon stem gets exited. Immediately afterwards the signal **a** reaches the corresponding synapsis and the branch **ba** gets excited. Notice that the branch **br** is not exited. Later, **c** reaches the neuron, and the branch **bac** gets exited, but **bad** doesn't. If later the signal **k** reaches its sysnapsis, the branch **back** will be exited, while the branch **bach** is not. Thus, with an adequate branching, an only neuron might be able to detect all the syllables or phonemes starting with b.

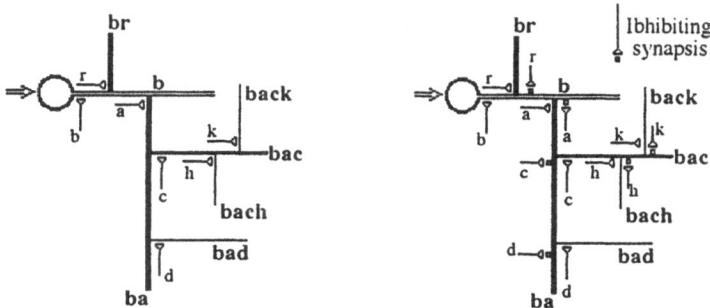

Fig 3 A sequence detector

The need of the second operation is immediate by imagining that the sequence **bach** has reached the neuron. Its effect has been to excite the outputs **b, ba, bac, bach**,while only the branch **bach** should. The solution is very simple, when a branch is excited, the parent branch must be inhibited. This may be accomplished by an inhibiting synapsis placed after the branching. The signal driving might be either the same signal exiting the branch, or the branch signal after being fed back.

Again, number wise, the proposed structure seems reasonable. If a neuron were capable of detecting all the syllables starting with b, as in the presented model, it would have about five hundred outputs, one per syllable, and assuming an average of three inhibitors per syllable, the total number of synapsis would be less than three thousand, quite a reasonable amount as typical neurons go.

In summary: the micro-structures proposed as models for sequence generation and detection doesn't seem unreasonable.

3. SIMULTANEITY

This word is meant to suggest a typical brain property. As soon as one has seen the symbol *grape* he or she knows all the features of the object this word represents. If you see an object you never saw before, simultaneously you become aware of all the attributes you can grasp. There is apparently no delay between the overall view and that of the details.Taking into account that brain signal transmission is not very fast, and that if the name of an object doesn't come to mind simultaneously with its look, it may take quite a while to retrieve it, it seems reasonable to presume that simultaneity is not an accident but the result of the particular nature of the brain. The conclusions born from the knowledge structure research have suggested the possibility of simultaneity being the property of a particular micro-structure, with multiple occurrence. From several possibilities, the following description is chosen.

Consider a classical neuron, with many somatic dendritic inputs and one axonic output. Inputs a_i, b_j, $i, j = 1, 2, \ldots$ proceed directly from other places outside the structure, and therefore are independent of the operation of the cognitive unit. c_k, $k=1, 2,\ldots$ are binary controls, probably indicating overall states of the cognitive module, or general brain states. Let x_r, y_s, $r, s = 1, 2, \ldots$ designate signals generated in the cognitive unit, and therefore are dependent upon the operation that is being carried. Assume further that the output is z and it may be described in term of the inputs by means of the following boolean expression.

$$z = \Sigma a_i + \Sigma x_j + \Sigma c_k \Sigma b_k + \Sigma c_s \Sigma y_s$$

The simplest micro-structure would consist of two such neurons, their equations being

$$z_1 = c_2 z_2 + \Sigma a_i + \Sigma x_j + \Sigma c_k \Sigma b_k + \Sigma c_s \Sigma y_s$$
$$z_2 = c_1 z_1 + \Sigma a_i + \Sigma x_j + \Sigma c_k \Sigma b_k + \Sigma c_s \Sigma y_s$$

In normal operation $c_2 = c_1 = 1$, and as soon the inputs of one unit causes it to fire, the input z to the other unit is enough to make it trigger. Once they are triggered the controls c_1 and c_2 are required to cease the firing.

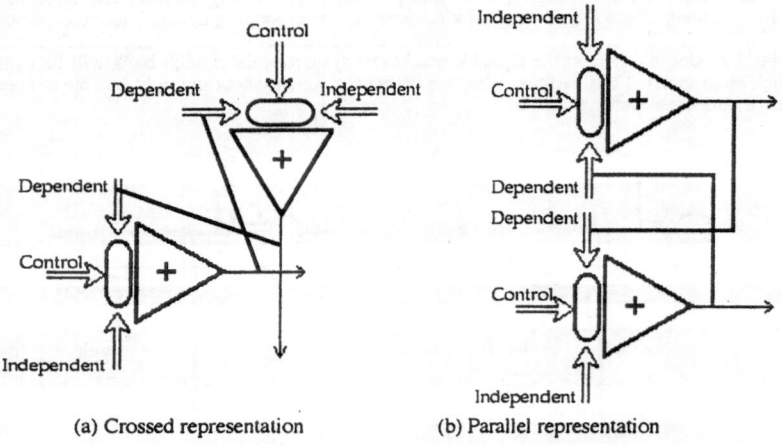

(a) Crossed representation   (b) Parallel representation

Fig 4. A micro-structure for simultaneity

The equations are general enough to fit almost any function, but one can imagine all sorts of variations. The point to remark is that one output triggers the other with practically no delay. Fig 4 illustrates the simplest micro-structure. How this micro-structure may expand or how thousands of them become integrated into a unique module is beyond this introductory paper.

4. FINAL COMMENTS

In this paper three very simple micro-structures are proposed in terms of neurons, with the purpose of using them in further papers to explain more complex functional modules.

It is well known that at birth, brain development is at an early stage. As well as the brain of a species (whatever that is) change with time along successive generations, through a process called *evolution* , also each normal individual brain changes with time through a process of *growth*. Usually the term growth carries the connotation of *increase in size*. In the particular case of the brain it means also a process of adaptive change in the synaptic weights, affecting the modular operation of the brain. In the particular case of the cognitive unit it is specially so. It is therefore reasonable to assume that should any micro-structure exist in the brain, its operation would be determined by the synaptic weights, and they would get their adult values during the process of growth. It is not too difficult to imagine synaptic weight change policies that might result into a mature macro-structure through a process of testing followed by weight change. Still, the author prefers not to make any comment in this regard at this moment, waiting for simulation results

The reader should be aware that the micro-structures considered in this paper were isolated, without being part of anything. Therefore whatever understanding has been reached, must not be considered as a set of blue prints of easily identifiable brain parts, but as useful concepts to gather further understanding of more complex structures. It is the author's contention that important mental functions will be more approachable in terms of micro-structures than in terms of neurons. Those basic ones presented here will serve as a references for future work.

REFERENCES

M. Bunge. *The Mind-Body Problem. A Psychobiological Approach* Pergamon Press Ltd. Oxford 1980

E. Clarke, K. Dewhurst *An Illustrated History of Brain Function* University of California press, Berkeley and Los Angeles 1972

J. Simoes da Fonseca, I.Barahona da Fonseca *De Natura Animae: Ensaio numa Perspectiva Computacional.* (preprint (1993))

E.V. Famiglietti, A. Peters *The synaptic glomerulus and the intrinsic neuron in the dorsal lateral geniculAte nucleus of the cat.* J. Comp. Neurol. vol 144, pp 285-334, 1972E. G. Gray, *A Morphological Basis for Presynaptic Inhibition*, Nature, vol 193, P 82-83, 1962

W. Rall , G.M. Shepherd, T.S.Reese, M.T. Brightman *Dendrodendritic synaptic Pathway for inhibition in the Olfactory Bulb* Exp. Neurol. vol 14, pp 44-56, 1966

K. Graubard W.H. Calvin *Presynaptic Dendrites, Implications of Spikeless Synaptic Transmission and Dedritic Geometry.* en F.D. Schmitt, F.G. Warden Eds. Neurosciences, Fourth Study Progamm, MIT Press Cambridge Mass. 1979

K. Graubard *Morphological and electronic properties of identified nerve cells in the lobster stomatogastric ganglion,*. Ph. D. dissertation, Department of Physics, University of California, San Diego, 1973.

Adaptive Hierarchical Structures

A. Daffertshofer[†] and H. Haken

Institute for Theoretical Physics and Synergetics
University of Stuttgart
Pfaffenwaldring 57/4, D-70550 Stuttgart, Germany

[†] Email: marlow@theo.physik.uni-stuttgart.de

Abstract

The construction of a hierarchical system to distinguish classes of patterns can be improved by the combination of a dynamical elastic matching with a time dependent image resolution. Using the elastic matching as a preprocessing for synergetic computers one achieves an invariant perception by means of arbitrary spatial transformations. On the other hand a time dependent resolution, realized with a Gaussian distribution, can be interpreted as a dynamical change of the available image information. The coupling of these two dynamics results in a hierarchy of locality of pattern transformations. Therefore this system utilizes a dynamical feature extraction and implies the definition several classes of patterns.

1 Introduction

The interpretation of nonlinear dynamical systems as generators of associative memories leads to the nowadays most popular performance of synergetic computers by means of a device for pattern recognition [13]. Pattern recognition may be regarded as a special case of associative mappings, namely, as a process in which classes of patterns are directly mapped on a set of discrete elements [18]. Explicitly, starting with a partly given set of features the synergetic computer is able to reconstruct a pattern as a result of a competition between order parameters. These order parameters belong to the complete set of prototypes that are stored in memory. In the original formulation these concepts do not include further information about the context of patterns, i.e. semantical or syntactical properties which become important in the case of a very large number of prototypes. With regard to human perception it might be necessary that for instance faces should not be compared with characters and vice versa. A first step in overcoming this deficiency can be realized by the introduction of hierarchical structures or classes of patterns during the recognition process [6]. A pattern class is a set of patterns that share some common properties [11]. This leads to the problems of finding general algorithms to combine arbitrary patterns into larger sets or estimating the intrinsic representation of patterns. In other words one has to get some knowledge about relevant features on which a combination into pattern classes can be based. Amongst others the basic question: *"In what form is information stored, or remembered?"* [22] led to numerous approaches to the problem of feature extraction. Usually the given patterns are decomposed into different basic systems. Depending on the loss of information and on the nessecity of invariances a certain number of (projection–) coefficients are used as features, e.g. moments [16],[4], Fourier descriptors [8], principal components [17],[19], or methods based on the topology or geometry of patterns [23]. Whilst on the one hand subject–minded approaches result in a very sufficient performance, on the other hand the algorithms are mostly quite limited to special recognition tasks.

Hence, the aim of this paper is to present a very general way of constructing classes of patterns without using special approaches of extracting the features of the patterns. Further an application on the problem of character recognition shall be presented. For this reason, first off all we will roughly summerize the algorithm which is used as classifier, namely the synergetic computer for pattern recognition (cf. [13] for detailed information).

2 Pattern recognition — synergetic computers

Every procedure of pattern analysis requires some kind of similarity measure, especially the combination of patterns into sets. Such measures have to define a distance between patterns so that a distinction between patterns becomes as unique as possible. Here, patterns are described as elements of an N-dimensional vector space Q which usually refers to $\mathcal{R}^N$. In practice, patterns are digitized images with an image resolution N, and the components of vectors in Q are simply the grey values of each pixel. We denote the time dependent test pattern as $q = q(t)$ whereas $\{v_k\}$ symbolizes the set of M prototypes. Making use of the analogy between pattern recognition and pattern formation which can be observed in a huge variety of (e.g.) open physical or chemical systems, we describe the recognition procedure as a set of nonlinear differential equations [12]. Therefore we assume that the test pattern becomes time–dependent following an evolution given by

$$\dot{q} = \sum_k \lambda_k v_k \left(v_k^\dagger q\right) - B \sum_{k' \neq k} \left(v_{k'}^\dagger q\right)^2 \left(v_k^\dagger q\right) v_k - C \left(q^\dagger q\right) q \qquad \text{with} \quad B, C, \lambda_k \in \mathcal{R}_+ .$$

(1)

The first term on the right hand side of the ordinary differential equation (1) implies an exponential growth of the vector q. It includes parameters for adjustment called attention parameters λ_k and the learning matrix $v_k^\dagger v_k$ known from the Hopfield model [15] in neural networks ($v_{k'}^\dagger$ is implicitly defined by $v_{k'}^\dagger v_k = \delta_{k'k}$). With the second term we obtain a discrimination between different prototypes whereas the last term saturates the growth of $q(t)$. Now, order parameters ξ_k are introduced as scalar products between the test pattern and the adjoint prototype vectors, $\xi_k := v_k^\dagger q$. Thus, the test pattern can be expanded into

$$q = \sum_k \xi_k v_k + w \qquad (w \in \mathcal{R}^N \text{ is a residual vector}) .$$

(2)

By the reformulation of (1) with these order parameters the dimension of the dynamical system decreases enormously from N to M because the number of stored patterns M is usually much smaller than the image resolution N and with an appropriate normalization we obtain

$$\dot{\xi}_k = \left\{ \lambda_k - B \sum_{k' \neq k} \xi_{k'}^2 - C \left[\sum_{k'} \xi_{k'}^2 + \left(w^\dagger w\right) \right] \right\} \xi_k \quad \rightsquigarrow \quad \lim_{t \to \infty} \xi_k(t) = \delta_{kk_0} .$$

(3)

Therefore pattern recognition is the result of competition between order parameters ξ_k which correspond to patterns $\{v_k\}$ in memory [13]. That means every prototype will be compared simultaneously with the input q which finally collapses on a single pattern for the purpose of cognition

[2]. Using the language of common artifical neural networks this synergetic computer works as a "Winner-Takes-All" classifier [3] with the metric given implicitly by the dual base of the feature vectors. One of the advantages of the algorithm is that it contains only soft nonlinearities which results in an elimination of all irrelevant minima which often occur in present days' neural networks. In comparision to other approaches in pattern recognition, especially those based on statistics (e.g. [10], [20] or [1]) or probabilistics (e.g. [3] or [5]) it should be mentioned neither large sets of parameters (synaptic weights) has to be adjusted nor assumptions about possible (conditional) probabilities are necessary.

3 Formal definition of classes

Obviously, (1) or (3) do not include syntactical or semantical information which might be contained within the storage. Every prototype will be dynamically compared with the test pattern. In this section we want to formulate an analogous pattern recognition considering that different sets of patterns can be combined building up a class, subclasses, etc. The resulting system forms hierarchical structures so that firstly the test pattern will be assigned to a certain class. In the next step a recognition takes place only within this extracted class, and so on. Therefore the given vector space $\mathcal{Q}$ (remember that $\dim \mathcal{Q} = N$) is now divided into M disjoint subspaces $\mathcal{U}^\sigma$ each with dimension M^σ. In total analogy to (2) any arbitrary element of $\mathcal{Q}$ can be written as

$$q = \sum_{\sigma=1}^{M} \sum_{i=1}^{M^\sigma} \xi_i^\sigma u_i^q + w := \sum_{\sigma=1}^{M} \xi^\sigma \left\{ \sum_{i=1}^{M^\sigma} \eta_i^\sigma u_i^q + w^\sigma \right\} + \tilde{w} , \tag{4}$$

where $w^\sigma, \tilde{w} \in \mathcal{R}^N$ are residual vectors (cf. (2)) .

Again the dimension of this system should be reduced by introducing order parameters. Therefore we recurrently make use of the orthonormal properties of the dual base: $(u_k^\tau)^\dagger u_i^q = \delta_{\tau\sigma}\delta_{ik}$. Further, we are free to normalize the coefficients η_i^σ with regard to the dimension of the subspace like

$$\forall_\sigma : \sum_{i=1}^{M^\sigma} (\eta_i^\sigma)^2 \equiv M^\sigma , \tag{5}$$

which leads to the formal definition of the *class–coefficients* ξ^σ and representative adjoint prototypes $(r^\sigma)^\dagger$ as

$$\xi^\sigma := \frac{1}{M^\sigma} \sum_{i=1}^{M^\sigma} \eta_i^\sigma (u_i^q)^\dagger q \quad \text{and} \quad (r^\sigma)^\dagger := \frac{1}{M^\sigma} \sum_{i=1}^{M^\sigma} \eta_i^\sigma (u_i^q)^\dagger . \tag{6}$$

In the case of known parameters η_i^σ one can easily see that with the definition (6) we obtain a very simple formula for the class–coefficient, namely $\xi^\sigma = (r^\sigma)^\dagger q$. The values η_i^σ, however, normally depend explicitly on the offered test pattern. Therefore it becomes necessary to approximate these values apriori or at least the difference $|\eta_i^\sigma - \eta_k^\sigma|$ inside a single class. For this reason we assume that the angles between the adjoint prototypes within $\mathcal{U}^\sigma$ are small, i.e.

$$(u_i^\sigma)^\dagger = (u_0^\sigma)^\dagger + (\epsilon_i^\sigma)^\dagger \quad \text{and} \quad \| (\epsilon_i^\sigma)^\dagger \| \leq \epsilon^\sigma \ll 1 . \tag{7}$$

The boundery ϵ^σ in (7) can be interpreted as a measure of the class size and should be chosen sufficiently small. (7) enables us to approximate the individual coefficients ξ_i^σ as

$$|\xi_i^\sigma| = |(u_0^\sigma)^\dagger q + (\epsilon_i^\sigma)^\dagger q| \approx |(u_0^\sigma)^\dagger q| \, . \tag{8}$$

Additionally the coefficients η_i^σ can be fixed, for instance as $\eta_i^\sigma \equiv \eta_0^\sigma \overset{!}{=} const.$ (for all indizes i) and the normalization (5) results in $|\eta_0^\sigma| \equiv 1$. With this estimate the representative prototypes for each class as well as their order parameters can now be written in a very general but simple fashion, namely

$$(\tilde{r}^\sigma)^\dagger := \frac{1}{M^\sigma} \sum_{i=1}^{M^\sigma} (u_i^\sigma)^\dagger \quad \Longrightarrow \quad \tilde{\xi}^\sigma := (\tilde{r}^\sigma)^\dagger q \, . \tag{9}$$

Note, that the definition (9) expresses that $|\tilde{\xi}^\sigma|$ becomes maximal if the sum of overlaps ξ_i^σ within a class σ has the maximal absolute value, at least under the assumption of a small neighbourhood ϵ^σ. The motivation for this notation is given by the following inequality

$$\sum_{i=1}^{M^\sigma} |\xi_i^\sigma| \geq \left| \sum_{i=1}^{M^\sigma} \xi_i^\sigma \right| \geq \left[\sum_{i=1}^{M^\sigma} |\xi_i^\sigma| - \sum_{i=1}^{M^\sigma} |\epsilon^\sigma| \right] \, , \tag{10}$$

where one has to remember that $\tilde{\xi}^\sigma \overset{(9)}{=} \frac{1}{M^\sigma} \sum_{i=1}^{M^\sigma} \xi_i^\sigma$ as well as $\xi_i^\sigma := (u_i^\sigma)^\dagger q$ hold. In other words the sum of the absolute values $|\xi_i^\sigma|$ can be approximated by the absolute value of the sum over all overlaps, and therefore the sum of all $|(u_i^\sigma)^\dagger q|$ is almost equal to $|(\tilde{r}^\sigma)^\dagger q|$. Up to now we have shown how similar prototypes u_i^σ can be combined to a certain class which is represented by a single vector $\tilde{r}^\sigma$. The following step is to make these vectors time dependent. Then, the dynamical evolution will given by (1), or (3) for the overlaps $\tilde{\xi}^\sigma$, respectively. Depending on the number L of *levels* in our hierarchy the complete pattern recognition has to be formulated as an iteration like

$$q \rightsquigarrow \tilde{\xi}_{|1}^{\sigma_0} \to \cdots \to \tilde{\xi}_{|j}^{\sigma_0} \to \cdots \to \tilde{\xi}_{|L}^{\sigma_0} \rightsquigarrow v_{k_0} \tag{11}$$

and each *branch* j of the recognition tree consists of the following dynamical system (compare (3))

$$\dot{\tilde{\xi}}_{|j}^\sigma = \left\{ \lambda_{|j}^\sigma - B_{|j} \sum_{\sigma' \neq \sigma} \tilde{\xi}_{|j}^{\sigma'2} - C_{|j} \left[\sum_{\sigma'} \tilde{\xi}_{|j}^{\sigma'2} + \left((w_{|j}^\sigma)^\dagger w_{|j}^\sigma \right) \right] \right\} \tilde{\xi}_{|j}^\sigma \, . \tag{12}$$

In general it might be a problem to decide which prototypes are that similar to build up a class. The assumption (8), however, can always be fulfilled by an additional preprocessing which reduces the "information" of each pattern. For this reason our preprocessing changes the image resolution dynamically and simultaneously it creates an invariance against spatial transformations (see below), at least if those are not of a special interest.

4 General preprocessing

Pattern recognition tasks usually require algorithms that are invariant against translation, rotation and scaling of the patterns. This can be realized by an adequate preprocessing (see for instance [9] or [21]). Here, however, we want to emphasize the problem of locally deformed patterns. For sake of mathematical simplicity from now on patterns are described as continuous scalar functions depending on the n-dimensional position vector x (here $n = 2$). We denote the test pattern by v and prototype patterns by u_k; k is still the index of prototypes in memory. Further, we assume the following normalization

$$\int q^2(x)\,d^n x = \int u_k^2(x)\,d^n x \equiv 1 \qquad \text{with} \quad q, u_k : \mathcal{R}^n \to \mathcal{R}\,. \tag{13}$$

As already mentioned our preprocessing shall change the image resolution dynamically, and simultaneously it shall create an invariance against spatial transformations. In order to cope with deformations we therefore introduce new coordinates: $x \to \tilde{x}_k := x + S_k(x)$, and define new prototype functions $\tilde{u}_k$ as $\tilde{u}_k(x) := u_k(x + S_k)$. The idea is to find transformation vectors S_k such that $\tilde{u}_k$ become as similar as necessary to q [7]. Hence, we construct potential functionals $V_k = V_k(S)$ which have the property that their minima correspond on the one hand to the best fitting transformations while on the other hand they avoid that patterns which differ too much are transformed into each other. Explicitly we chose the potential as

$$V_k := \underbrace{\frac{1}{2}\int (q - \tilde{u}_k)^2\,d^n x}_{(i)} + \underbrace{\frac{\eta}{4p}\left(1 - \int \tilde{u}_k^2 d^n x\right)^{2p}}_{(ii)} + \underbrace{\frac{1}{2}\int \sum_{\kappa\lambda\mu\pi} \theta_{\kappa\lambda\mu\pi} \frac{\partial S_{k,\lambda}}{\partial x_\kappa} \frac{\partial S_{k,\pi}}{\partial x_\mu}\,d^n x}_{(iii)} \tag{14}$$

where p is an arbitrary constant, $p \in \mathcal{N}$; (e.g. $p = 1$).

The first term (i) of the potential deforms the prototypes into the given test pattern and the second term (ii) is used to keep the prototypes ($\tilde{u}_k$) normalized whereas the third term (iii), lent from elastodynamics, will limit the possible transformations; note that the greek subscripts stand for the n components of x or $S(x)$. With the constant even exponent $2p$ in term (ii) the dynamical normalization can be adjusted (see also eq. (13)); in the numerical simulations below p is always equal to 1. The minima of the potentials V_k can be determined as solutions of a common variational principle which can be solved bye use of its parametrized gradient dynamics $[S_k \to S_k(x, t)]$, i.e.

$$\frac{\delta}{\delta S_k} V_k(S_k) = 0 \longrightarrow \frac{\partial}{\partial t} S_{k,\iota}(x, t) = -\gamma_S \frac{\delta}{\delta S_{k,\iota}} V_k \qquad \text{for } \iota \in \mathcal{N}^n;\ \gamma_S \in \mathcal{R}_+\,. \tag{15}$$

For the ιth component of the transition vector S_k follows in the isotropic case with a constant material tensor θ

$$\frac{\partial}{\partial t} S_{k,\iota} = \left\{ q + \left(\eta \left[1 - \int \tilde{u}_k^2 d^2 x \right]^{2p-1} - 1 \right) \tilde{u}_k \right\} \frac{\partial \tilde{u}_k}{\partial \tilde{x}_\iota} + \mu \sum_\kappa \frac{\partial^2 S_{k,\iota}}{\partial x_\kappa^2} + \nu \sum_\kappa \frac{\partial^2 S_{k,\kappa}}{\partial x_\iota \partial x_\kappa}\,. \tag{16}$$

Note that here the two dimensional case is considered ($n \equiv 2$, the patterns are for instance planar images). Even if we generalize the cost function to an elastic energy depending explicitly on the patterns $\tilde{u}_k$ and their spatial derivatives $\theta \to \theta(\tilde{u}_k, \partial \tilde{u}_k/\partial x_p, \ldots)$ the resulting evolution can be approximated in a very similar fashion [7]. In this case one has the possibility to describe a general cost potential with few constant parameters which are valid for different classes of patterns, such as characters, faces, etc.

In summary the minimization of the absolute difference between prototypes and test patterns under the condition of a minimal deformation leads to the possibility of an invariant pattern recognition. With a simple cost function – a harmonic elastic energy with a constant material tensor – the template matching (prototype → test pattern) can be limited such that only similar patterns are transformed into each other. For the generation of spatial invariance by (16) it is absolutely necessary that the patterns are continuous and differentiable. To guarantee this property we add a smoothing algorithm which enables us to integrate the partial differential equation, at least numerically. In detail, the smoothing is realized by a convolution with a Gaussian function g_{a_k} of width a_k

$$(q, u_k) := \left[g_{a_k} * (q, u_k)_{\text{original}} \right](x) \qquad \text{with} \quad g_{a_k}(x) \propto \exp\left\{ -\frac{x^2}{2a_k^2} \right\} . \qquad (17)$$

This convolution improves a low pass filtering of the images whereby the filter function is again a Gaussian (one may think of convoluting with a Bessel function of the first kind and first order to accelerate the numerical simulation). Adjusting this filter dynamically the Gaussian width is chosen time dependent with the following evolution

$$\dot{a}_k(t) = \begin{cases} -\gamma_a a(t) \exp\left\{ -\frac{\dot{\zeta}_k^2}{2\beta^2} \right\} & \text{for } \dot{\zeta}_k \geq 0 \\ 0 & \text{otherwise} \end{cases} , \qquad (18)$$

where we define the overlap ζ_k as $\xi_k \longrightarrow \zeta_k := \int u_k q \, d^n x$ which is a slightly different measure of similarity compared to ξ_k (see above). The damping factor $\gamma_a \in \mathcal{R}_+$ should be chosen as $\gamma_a \ll \gamma_S$. In other words the smoothing has to be reduced so slowly ($\tau_a \gg \tau_S$) that the deformation is almost finished before the dynamics of $a(t)$ starts. Because of the nonvanishing smoothing during the evolution of $S(x, t)$, measured by ζ_k and adjusted by β ($\ll 1$), its convergence is still guaranteed. This separation of the time scales belonging to two different dynamical systems can be justified by the method of adiabatic elimination [12]. The dynamics (18) implies that the better the fit between test and prototype pattern the faster $a(t)$ will decrease [14]. Now, the combination of the two dynamical systems (16) and (18) results on the one hand in the minimization of the deformation potential or the absolute distance between prototypes and test with respect to a minimal elastic energy while on the other hand this egalization is supported by a dynamical resolution $a(t)$. Therefore this preprocessing enables us to verify the definition of classes by (9).
In the following the applicability of the dynamical preprocessing will be explained using the example of character recognition. Handwritten characters typically show a large variety of local deformations. Certainly, these patterns have to be collected in several classes, e.g. classes a, b, c, and so on. We want to investigate the dynamical behaviour of different prototypes (figure 1, second row) as well as the evolution of the smoothing which is expect to be similar for patterns that should be members of a single class (figure 1, lower part). As visualized in figure 1 the most similar patterns show a drastical reduction of the smoothing $a(t)$ (or a high increase of image resolution) after a little time during the preprocessing. Thus, the step in $a(t)$ can be used to decide which class contains which patterns. The different levels of the hierarchical system are therefore defined by different smoothing values while the representation of the lower branches are still given by the corresponding mean values. On this basis examples of character classes are shown in figure 2 and the corresponding recognition procedures are falsified in figure 3. As one can see in every example the systems firstly distinguishes the correct class before the most similar character is recognized within the set of class members.

test pattern:

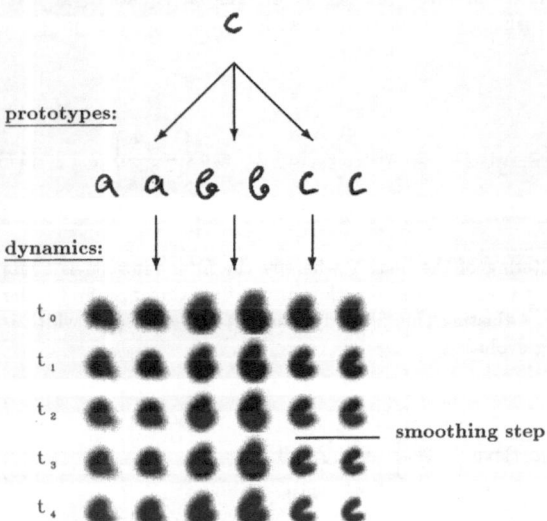

prototypes:

dynamics:

Figure 1: *Dynamical smoothing — the convolution width of the most similar prototype-set is reduced rapidly. Here, of course, similarity is measured as distance between the offered test pattern (first row) and the prototypes (second row); the preprocessing is show below.*

t_0
t_1
t_2
t_3
t_4

smoothing step

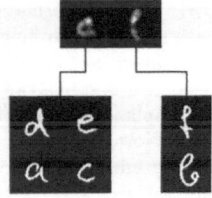

Figure 2: *Recognition tree for handwritten characters: each level corresponds to a certain smoothing.*

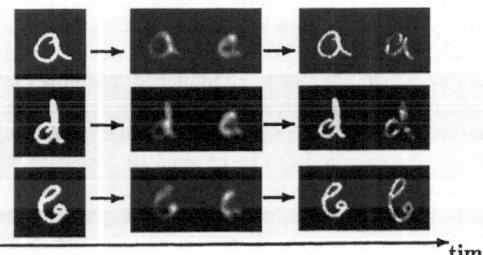

time

Figure 3: *Recognition procedures for 3 different offered test patterns a, d, and b — left: test patterns, middle: class distinction, and right: recognition within the class.*

83

5 Results

Utilizing the farreaching analogies between spontaneous formation and recognition of patterns the synergetic computer can be constructed as a device performing pattern recognition. With an offered set of features (pixels) these features can form orders parameter which will compete with each other. The competition described by a gradient dynamics results in a complementation process which has a complete correspondence to the associative memory during perception. Furthermore we obtain invariance against arbitrary spatial transformations by minimizing the absolute difference between prototypes and the test pattern under the condition of a minimal deformation. Considering several classes of patterns, for instance handwritten characters, a very simple cost function – a harmonic elastic energy – can limit the template matching (prototype → test pattern) such that only similar patterns are transformed into each other. Adding an dynamical smoothing one achieves an automized construction of pattern classes depending on the level of smoothing. The resulting recognition procedure becomes partly iterative by means of a decision tree. Its preformance in the sense of recognition tasks, falsified by several numerical simulations, show a high degree of robustness because firstly the preprocessing is invariant agaisnt spatial transformations and associativity is implied by the choice of our classifier.

References

[1] Abdel-Mottaleb M., Rosenfeld A.: *Inexact Bayesian estimation*, Pat. Recogn. (25)6, 641-646, 1992

[2] Amit D.J.: *Modelling Brain Function*, Cambridge University Press, Cambridge 1992

[3] Beale R., Jackson T.: *Neural Computing: An Introduction*, Hilger, Bristol 1990

[4] Belkasim S.O., Shridhar M., and Ahmadi M.: *Pattern recognition with moment invariants: A comparative study and new results*, Pat. Recogn. (24)12: 1117-1138, 1991

[5] Berger J.O.: *Statistical Decision Theory, Foundations, Concepts and Methods*, Springer, Berlin 1980

[6] Daffertshofer A., Haken H., Lorenz W., and Ossig M.: *Hierarchical Structures in Pattern Recognition*, Proceedings ICASSE '94, Erlangen 1994

[7] Daffertshofer A., Haken H.: *A new approach to recognition of deformed patterns*, published in Pat. Recogn., Dec. 1994

[8] Dougherty, E.R., Loce R.B.: *Robust morphological continuous Fourier descriptors I+II*, Int. J. Pattern Recogn. Artif. Intell. 6 (5) 873–911, 1992

[9] Fuchs A., Haken H.: *Pattern Recognition and Associative Memory as Dynamical Process in a Synergetic System I+II*, Erratum, Biol. Cybern. 60: 17-22, 107-109, 476

[10] Geman S., Geman D.: *Stochastic relaxation, Gibbs distributions, and the Bayesian restoration of images*, IEEE Trans. Pattern Analysis and Machine Intelligence PAMI-6:721-741, 1984

[11] Gonzalez R.C., Thomason M.G.: *Syntactic Pattern Recognition, An Introduction*, Addison-Wesley, Massachusetts 1978

[12] Haken H.: *Advanced Synergetics* (3rd print), Springer, Berlin 1987

[13] Haken H.: *Synergetic Computers and Cognition, A Top–Down Approach to Neural Nets*, Springer, Berlin 1991

[14] Hölle B.: *Erkennung dreidimensionaler Körper aus einer ebenen Projektion mittels einer Potentialdynamik*, PhD-Thesis, University of Stuttgart 1992

[15] Hopfield J.J.: *Neural networks and physical systems with emergent collective computational abilities*, Proc. Nat. Acad. Sci. 79:2554-2558, 1982

[16] Hu M.: *Visual Pattern recognition by moment invariants*, IRE Trans. Inf. Theory 8: 179-187, 1962

[17] Kirby M., Sirovich L.: *Application of the Karhunen–Loéve procedures for the characterization of human faces*, IEEE Trans. Pattern Analysis Mach. Intell. 12(1) 103–108, 1990

[18] Kohonen T.: *Self–Organization and Associative Memory*, (2nd edition), Springer, Berlin 1988

[19] Liu Y., Shouval H.: *Localized principal components of natural images — and analytic solution*, Network 5: 317-324, 1994

[20] Opper M., Haussler D.: *Generalization Performance of Bayes Optimal Classification Algorithm for Learning a Perceptron*, Phys. Rev. Lett. 66(20) pp. 2677-2680, 1991

[21] Reitböck H.J., Altmann J.: *A Model for Size- and Rotation–Invariant Pattern Processing in the Visual System*, Biol. Cybern. 51: 113-121, 1984

[22] Rosenblatt F.: *The perceptron: a probabilistic model for information storage and organization in the brain*, Physiological Review 65:386-408

[23] Yuille A. L., Hallinan P.W., and Cohen D.S.: *Feature Extraction from Faces Using Deformable Templates*, Intern. Journ. Comp. Vision 8:2 99-111, 1992

Optimal Range of Input Resistance in the Oscillatory Behavior of the Pancreatic ß-Cell

E. Andreu, B. Soria, S. Bolea and J.V. Sánchez-Andrés

Dept. de Fisiología. Instituto de Neurociencias. Univ. Alicante. Aptdo. 374, 03080 Alicante (Spain)

Abstract

The pancreatic β-cell produces the hormone insulin in response to the blood glucose levels. This system is damaged in the diabetes. The central region of the β-cell electrical response is oscillatory. In this study, we demonstrate that in this region of the response, the membrane resistance of the cells oscillates in phase with the membrane potential. As the cells input resistance reflects the balance of the ionic conductances, we hypothesize that an optimal range of the input resistance determines the capability of the cells to get into oscillations. Under this scope, we propose an electronic circuit able to reproduce the behavior of the biological system. This circuit would be useful to build an artificial insulin pump able to reproduce the physiological pattern.

1. INTRODUCTION

We have already described the analogies of the pancreatic β-cell with a voltage controlled oscillator (VCO). This system responds to a glucose challenge with a depolarization that drives the cell into a pattern of oscillations[1]. The frequency of the oscillations is modulated by the glucose concentration. There is a good correlation between the membrane potential and the output of the system: the insulin secretion[2,3,4,5]. A detailed analysis permits to observe that the oscillatory pattern takes place only on a rather narrow range of the response[6]. This limitation in the oscillatory capability should be determined by the β-cells electrical properties. If the comparison of the β-cells with a VCO proves to be correct, the described limitation in the oscillatory capability of the β-cells should be explained in terms of their electrical properties. The aim of this study is to analyze the evolution of the cell input resistance in parallel with the voltage events that result of an exposition to different glucose concentrations. The study of the input resistance has been chosen because is representative of the global state of the cell conductances, and, indirectly, can reflect the state of coupling of a given cell with their neighbors.

2. METHODS

Intracellular activity of β-cells from mouse islets of Langerhans was recorded with a bridge amplifier (Axoclamp 2A) as previously described[7]. Recordings were made with thick wall borosilicate microelectrodes pulled with a PE2 Narishige microelectrode puller. Microelectrodes were filled with 3 M potassium citrate, resistance around 100 MΩ. The modified Krebs solution used had the following composition (mM): 120 NaCl, 25 $NaHCO_3$, 5 KCl , and 1 $MgCl_2$ and was equilibrated with a gas mixture containing 95% O_2 and 5% CO_2 at 37°C. Islets of Langerhans were microdissected by hand. Data recorded were directly monitored on an oscilloscope and stored in tape for further analysis.

3. RESULTS

3.1. Input resistance measurements

The major limitation to the input resistance measurements comes for the small size and weakness of the β-cells. This induces to the use of thick-walled, very high resistance (≅200 MΩ) borosilicate microelectrodes, with a very limited linear range of response to current injection. In order to overcome this limitation, we have reduced the resistance of the electrodes to ≅100 MΩ all over the study. In this condition it is possible to obtain complete I-V relationships with a high degree of linearity for the range of ±0.5 nA pulses (Fig. 1).

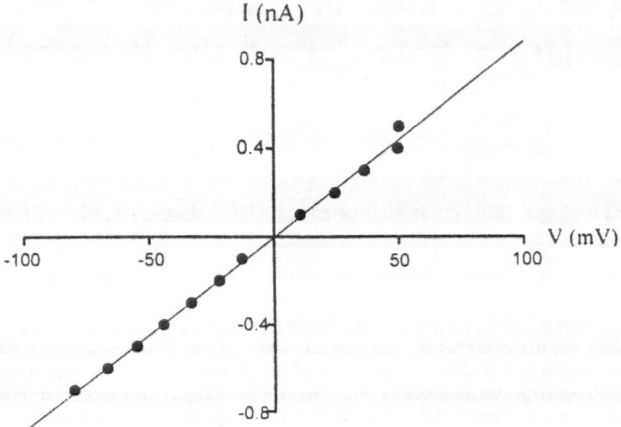

Fig.1 . I-V relationship obtained by current injection through the microelectrode (-0.7-0.5 nA, 25 ms). The input resistance values were measured at the steady-state of the voltage deflection (y=0.003+0.009x, R=0.998).

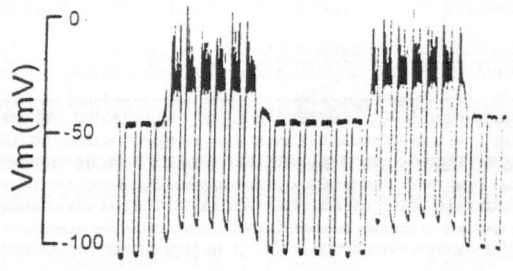

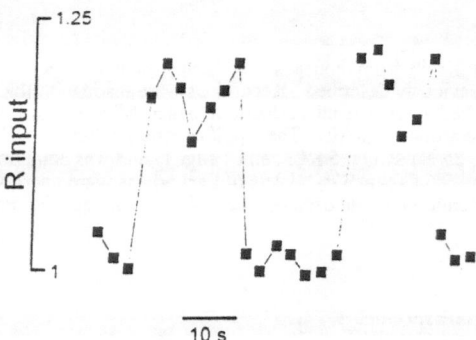

Fig. 2. Top. Intracellular recording of a β-cell exposed to 11.1 mM glucose. The oscillations are representative of the pattern at the steady-state. Downward deflections reflect hyperpolarizing current injection pulses (-0.3 nA, 850 ms). Bottom. Input resistance. Normalized values: 1: 180 MΩ.

3.2. Voltage and input resistance changes in response to glucose

The β-cells maintain their membrane potential hyperpolarized and stable in the absence of glucose. Glucose addition causes a linear depolarization and a concomitant increase in the input resistance (data not shown). When the glucose concentration arrives to a threshold level ($\cong 7$ mM) the membrane potential starts to oscillate (Fig. 2, top). The depolarized levels in the oscillation are called active phases, in opposition to the hyperpolarized levels (silent phases). A quantitative analysis of the input resistance along the oscillation shows an increase as the potential arrives to the top of the active phases. At this level the input resistance stays constant. The repolarization of the membrane towards the silent phase is accompanied by a reduction in the input resistance. Consequently, the input resistance oscillates in phase with the membrane potential (Fig. 2, bottom).

Glucose addition (7-20 mM) increases the duration of the active phases a the expense of the silent phases (Fig. 3). Further addition of glucose (>20 mM) keeps the cell steadily depolarized, canceling both the oscillations in the membrane potential and in the input resistance (data not shown).

[Glucose]

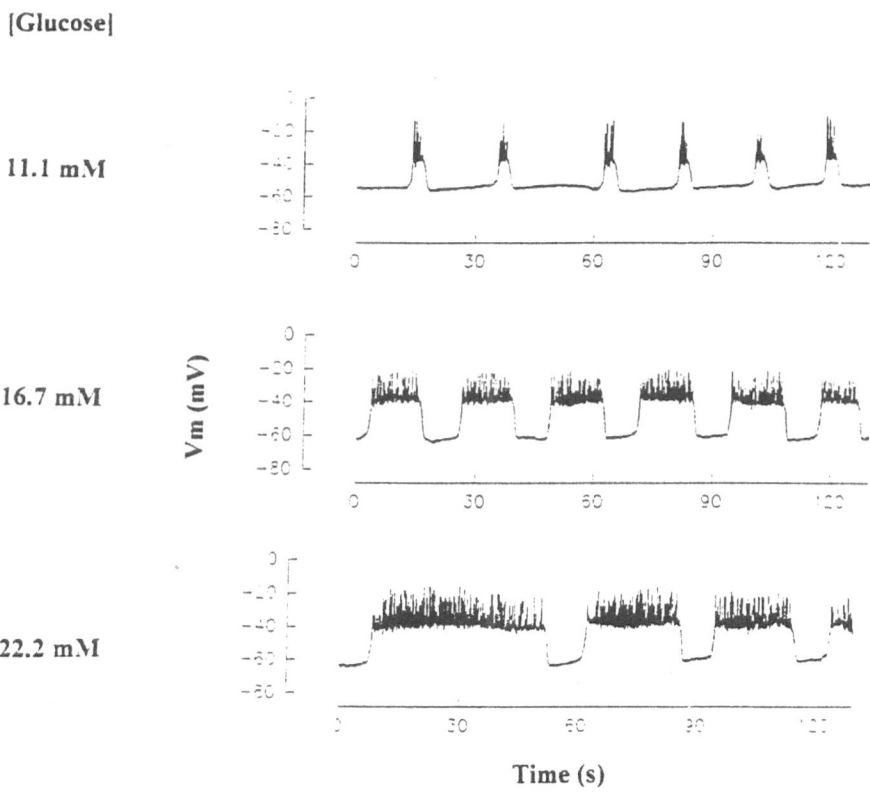

Fig. 3. Effect of the exposition of the β-cells to increasing glucose concentrations (indicated beside every record).

88

4. DISCUSSION

An oscillator is characterized for providing an oscillatory output to a continuous non-oscillatory input, requiring a positive feed-back loop[8]. The fact that the oscillatory behavior appears just into a window of the complete response, between two no oscillatory states, and given the absence of qualitative differences at any level of the input, suggests that the appropriate conditions are only fulfilled in that window. As it was pointed above, in a cell, the input resistance values summarizes the behavior of the membrane and junctional conductances that are responsible of the membrane potential changes. The oscillatory behavior of the membrane potential is only apparent when the input resistance oscillates. The absence of phase lag between the oscillations of the membrane potential and the input resistance points to the possibility of the input resistance determining the appropriate balance of conductances that permits to the cell to work as an oscillator. Out of the appropriate range the cell becomes unable to oscillate.

Under this scope the β-cell system can be compared with an astable (Fig. 4 top) configured as a VCO which oscillatory capability depends on the input voltage (block A, input). If the applied voltage is smaller than that set by R3 the oscillation would be canceled. Block A would play the role of the glucose sensor, driving the cell in or out of the oscillatory range in dependence of the physiologically pre-adjusted glucose reference level. In Fig. 4 bottom is shown a detailed proposed circuitry of an astable able to generate a wave whose active and silent phases are independent. Active phases are set by R1 and R2, while silent phases are regulated by R1 and R3. R1, R2 and R3 would constitute the components of the experimentally measured membrane resistance. Then, only when the appropriate balance between them would be obtained, the cell will oscillate in a physiological manner.

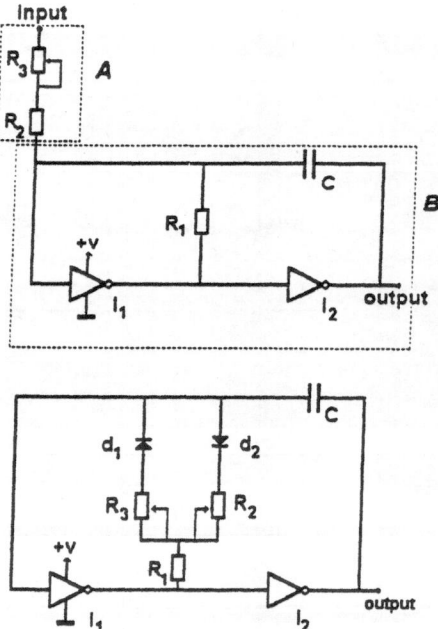

Fig 4. Top. CMOS astable configured as a VCO. Bottom, two gates asymmetrical astable.

The proposed circuit coupled to a glucose sensor can be useful to implement an artificial insulin pump able to reproduce the physiological oscillations in the hormone secretion. This device could have a therapeutical application in patients of diabetes type-I (insulin-dependent)[9].

Acknowledgments: We are indebted to A. Pérez-Vergara and S. Moya for technical assistance and to L. Menéndez de la Prida for helpful comments. This study was supported by a grant from DGICYT (Dirección General de Investigación Científica y Técnica, Spain), PM92-0115.

5. REFERENCES

[1] Dean, P.M. and Matthews, E.K. (1968) *Nature 219*, 389-390.
[2] Santos, R.M.; Rosario, L.M.; Nadal. A.; Garcia Sancho, J.; Soria, B. and Valdeolmillos, M. (1991) *Pflugers Arch. (Eur. J. Physiol) 418*, 417-22.
[3] Wollheim, C.B. and Pralong, W.F. (1990) *Biochem. Soc. Trans. 18*, 111-4.
[4] Rorsman, P., Abrahamsson, H., Gylfe, E. and Hellman, B. (1984) *FEBS Lett. 170*, 196-220.
[5] Soria, B., Martin,.F. and Sanchez-Andres, J.V. (1994). In: *Frontiers in β-cell research* (Flatt, P. and Lenzen, S. eds) Smith-Gordon, London.
[6] Sanchez-Andres, J.V. and Soria, B. (1983) *Lect. Notes Comp. Sci. 686*, 37-42
[7] Sánchez-Andrés, J.V.; Ripoll, C. and Soria, B. (1988) *FEBS Lett. 231*, 143-7.
[8] Reace, E. *Teoría y práctica de los osciladores*, Ed. Técnicas REDE, 1989, pp. 151-166.
[9] Kahn, S.E. and Porte Jr. D. In: *Diabetes Mellitus*, Rifkin and Porte eds., 4°. ed, Elsevier, NY, 1990, pp 436-56.

A Computational Model of Periodic-Pattern-Selective Cells

P. Kruizinga and N. Petkov

University of Groningen, Dept. of Computing Science
P.O. Box 800, 9700 AV Groningen, The Netherlands
Email: peterkr@cs.rug.nl, petkov@cs.rug.nl

Abstract: A computational model of so-called grating cells is proposed. These cells, found in areas V1 and V2 of the visual cortex of monkeys, respond strongly to bar gratings of a given orientation and periodicity but very weakly or not at all to single bars. This non-linear behavior is quite different from the spatial frequency filtering behavior exhibited by the other types of orientation selective cells. It is incorporated in the proposed model by using an AND-like non-linearity to combine the responses of simple cells and compute the activities of so-called grating subunits which are subsequently summed up. The parameters of the model are adjusted to reproduce the results measured by neurophysiologists with different visual stimuli. The proposed computational model of a grating cell is used to compute the collective activation of sets of such cells, referred to as cortical images, induced by natural visual stimuli. On the basis of the results of such simulations we speculate about the possible role of grating cells in the visual system and demonstrate the usefulness of grating cell operators for some computer vision tasks, such as automatic face recognition and document processing.

Keywords: Grating cells, visual cortex, computational model, texture analysis, face recognition, document processing

1 Introduction

Extensive neurophysiological studies in the past fifty years have led to the accumulation of considerable knowledge of the visual system of primates. The functional description of classes of visual neurons makes a major part of this knowledge. The discoveries of the center-surround organization of the receptive fields of retinal ganglion cells [5] and of the orientation selectivity of the majority of neurons in the primary visual cortex [3] are major milestones in the development of this area.

These neurophysiological discoveries were followed by model building aiming at a precise quantitative description of the functional behavior of visual neurons. For instance, the concept of a receptive field function which specifies the response of a visual neuron to a light spot stimulus as a function of position has successfully been applied to describe the behavior of retinal ganglion cells and simple primary cortical cells, enabling one to predict their responses for arbitrary visual stimuli. Such models are the basis of computer simulations in which the collective activation of many cells are computed and visualized giving an opportunity for improved insights into the function of the visual system [10]. They are also important for building artificial vision systems which would perform like natural systems.

Recently Von der Heydt et al. [12] reported on the discovery of a new type of orientation selective neurons in areas V1 and V2 of the visual cortex of monkeys which they named *grating cells*. Similarly to other orientation selective neurons, such as the so-called simple, complex and hyper-complex cells [4], grating cells will respond vigorously to a grating of bars of appropriate orientation, position and periodicity. In contrast to other orientation selective cells, grating cells respond very weakly or not at all to a single bar. This non-linear behavior is quite different from the spatial frequency filtering behavior exhibited by the other types of orientation selective cells. In particular, the behavior of a grating cell cannot be explained by weighted spatial summation (linear filtering), followed by half-wave rectification as in the case of simple cells (see Section 2). Neither can their behavior be explained by three-stage models (linear filtering, rectification, summation) used for complex cells [11].

Most grating cells reported in [12] start to respond when a grating of a few bars (2 to 5) is presented. In most cases the response rises linearly with the number of additional bars up to a given number (4 to 14) after which it quickly saturates and the addition of new bars to the grating causes the response to rise only slightly or not at all and in some cases even to decline. Similarly, the response rises with the length of the bars up to a given length after which saturation and in some cases a decline (end-stopping) is observed. The responses to moving gratings are unmodulated and do not depend on the direction of movement. The dependence of the response on contrast shows a switching characteristic, in that turn-on and saturation contrast values lie pretty close. (The most sensitive grating cells start to respond at a contrast of 1% and level off at 3%.)

The above properties suggest that the primary role of grating cells is to signal periodicity in oriented textures, ignoring other details (such as contrast). On the other hand, their relatively narrow bandwidth for both spatial frequency (median of 1 octave) and orientation (about 20°) causes this type of cells to be activated relatively rarely (as compared to other orientation selective cells) by natural visual stimuli. Therefore, the role of grating cells needs to be clarified. One approach to this problem adopted in this study is to construct a computational model of a grating cell and use it to compute the collective activation of many such cells with different preferred orientations and periodicities, covering the visual field, induced by natural visual stimuli. On the basis of the results of such simulations we speculate about one possible role of grating cells in the visual system and demonstrate the usefulness of grating cell operators for some computer vision tasks.

2 A computational model of grating cells

Grating cells are found in the same cortex area (V1) as simple cells and similarly to simple cells show orientation selectivity. On the other hand they show a more complex non-linear behavior. These facts may suggest that grating cells receive input from simple cells. Although this is a speculation which is not proven experimentally, we propose a model in which the responses of simple cells are used to compute the responses of grating cells. (This is similar to the speculation that complex cells may receive inputs from simple cells [4]). In this model there are three layers where the first layer consists of simple cells and the last one of grating cells (see Fig.1). The intermediate layer plays an auxiliary role; in reality the units of this layer, to be referred to here as *grating subunits*, may correspond to dendrites of grating cells.

We use the following model (for a similar model with a different parametrization, see e.g. [2]) to compute the response r of a simple cell characterized by a receptive field function $g(x, y)$ to a composite visual signal $s(x, y)$, $(x, y) \in \Omega$ (Ω - visual field domain):

(i) An integral

$$\bar{s} = \iint_{\Omega} s(x, y) g(x, y) \, dx dy \tag{1}$$

is evaluated in the same way as if the receptive field function $g(x, y)$ were the impulse response of a linear system and

(ii) the result $\bar{s}$ is submitted to half-wave rectification (more generally to thresholding) and

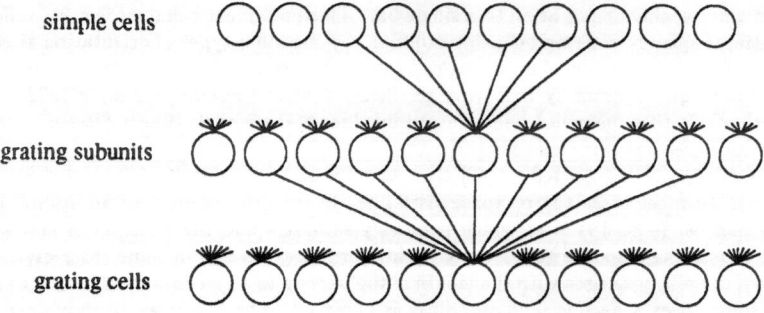

simple cells

grating subunits

grating cells

Figure 1: Schematic representation of a computational model of grating cells: grating cells (bottom layer) carry out spatial summation of inputs which they receive from a layer of intermediate cells, called the grating subunits, which combine inputs from simple cells (top layer) in an AND-like way.

local non-linear normalization as follows:

$$r = 0 \quad \text{if} \quad \tilde{s} \le 0. \tag{2}$$

$$r = \log(1 + \frac{\tilde{s}}{a}) \quad \text{if} \quad \tilde{s} > 0, \tag{3}$$

where a is the average luminance in the receptive field of the concerned neuron.

As to the particular form of the function g, it has been the subject of neurophysiological research for more than thirty-five years. In particular, the receptive fields of simple cells were found to consists of a number of oriented altering parallel excitatory and inhibitory zones. We use the following family of functions to model simple cells:

$$g_{\xi,\eta,\sigma,\gamma,\Theta,\lambda,\varphi}(x,y) = e^{-\frac{(x'^2 + \gamma^2 y'^2)}{\sigma^2}} \cos(2\pi \frac{x'}{\lambda} + \varphi) \tag{4}$$

$$x' = (x - \xi)\cos\Theta - (y - \eta)\sin\Theta$$

$$y' = (x - \xi)\sin\Theta + (y - \eta)\cos\Theta$$

where the arguments x and y specify the position of a light spot in the visual field and ξ, η, σ, γ, Θ, λ and φ are parameters whose effect on the function g is explained in more detail in [10]. In this study, we use functions g with values of the phase parameter $\varphi = 0$ and $\varphi = \pi$ which correspond to symmetric receptive field functions with one central excitatory lobe surrounded by two inhibitory lobes (to be referred to as center-on functions in analogy with retinal ganglion cell responses, Fig.2a) and one central inhibitory lobe and two excitatory side lobes (to be referred to as center-off functions, Fig.2b), respectively.

Substituting a receptive field function $g_{\xi,\eta,\sigma,\gamma,\Theta,\lambda,\varphi}(x,y)$ in eqs.1-2, one can compute the response $r_{\xi,\eta,\sigma,\gamma,\Theta,\lambda,\varphi}$ of a simple visual cortical cell modelled by this function to an input image $s(x,y)$. In the following we skip the parameters γ and σ, since, as neurophysiological research has shown [1], γ varies in a very restricted domain — we take a constant value of $\gamma = 0.5$ — and σ is closely related with λ in that the ratio $\frac{\sigma}{\lambda}$ is relatively constant on the population of all simple cells — in this case we take $\sigma = 0.5\lambda$. Hence, instead of $r_{\xi,\eta,\sigma,\gamma,\Theta,\lambda,\varphi}$ we shall write in the following for short $r_{\xi,\eta,\Theta,\lambda,\varphi}$.

Next, a quantity $q_{\xi,\eta,\Theta,\lambda}$ is assigned to each grating subunit of the intermediate layer. One such subunit is taken for each distinct combination of the parameter values. The concerned quantity is computed as follows:

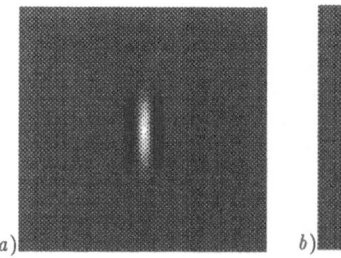 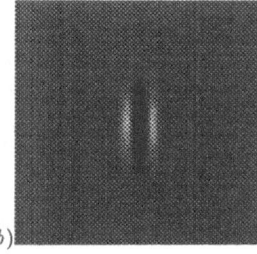

Figure 2: An example of symmetric center-on (a) and a center-off (b) simple cell receptive field functions. The values of the other parameters are $\xi = 0, \eta = 0, \gamma = 0.5, \Theta = 0, \lambda = 2\sigma$ and $\sigma = 0.0625L$ (L - image size). White and black colors indicate excitatory and inhibitory areas, respectively.

Let

$$M_n(\xi, \eta) = Max\{r_{\xi', \eta', \Theta, \lambda, \varphi_n} \mid \xi', \eta' : n \le \frac{2(\xi' - \xi)}{\lambda\cos\Theta} < n + 1, \quad n \le \frac{2(\eta' - \eta)}{\lambda\sin\Theta} < n + 1\} \quad (5)$$

$$\begin{aligned} n &= -3, -2, -1, 1, 2, 3 \\ \text{where} \quad \varphi_n &= 0 \quad \text{for} \quad n = -3, -1, 2 \\ \text{and} \quad \varphi_n &= \pi \quad \text{for} \quad n = -2, 1, 3 \end{aligned}$$

and let

$$M(\xi, \eta) = Max\{M_n(\xi, \eta) \mid n = -3, -2, -1, 1, 2, 3\} \quad (6)$$

The above quantities are related to the activities of simple cells along a line segment of length 3λ passing through point (ξ, η) in orientation Θ. This segment is divided in intervals of length $\frac{\lambda}{2}$ and the maximum activity of one sort of simple cells, center-on or center-off, is determined in each interval. $M_{-3}(\xi, \eta)$, for instance, is the maximum activity of center-on simple cells in the corresponding interval of length $\frac{\lambda}{2}$; $M_{-2}(\xi, \eta)$ is the maximum activity of center-off simple cells in the next interval, etc. Center-on and center-off simple cell activities are alternately used in alternate intervals. $M(\xi, \eta)$ is the maximum among the above interval maxima. The response $q_{\xi, \eta, \Theta, \lambda}$ of a grating subunit in point (ξ, η) is computed as a binary-valued function as follows:

$$\begin{aligned} \text{if} \quad & (M_n(\xi, \eta) \ge \rho M(\xi, \eta) \quad \text{for} \quad n = -3, -2, -1, 1, 2, 3) \\ \text{then} \quad & q_{\xi, \eta, \Theta, \lambda} = 1 \\ \text{else} \quad & q_{\xi, \eta, \Theta, \lambda} = 0 \end{aligned} \quad (7)$$

where ρ is a threshold parameter with a value smaller than but near 1 (e.g. $\rho = 0.9$).

Roughly speaking, the concerned grating cell will be activated if center-on and center-off cells of the same preferred orientation (Θ) and spatial frequency ($\frac{1}{\lambda}$) are alternately activated in intervals of length $\frac{\lambda}{2}$ along a line segment of length 3λ centered on point (ξ, η) and passing in direction Θ. This will be the case if three parallel bars with spacing λ and orientation Θ of the normal to them are encountered. In contrast, the condition is not fulfilled by the simple cell activity pattern caused by a single bar or two bars. At this point, an explanation is due of a question why this condition is applied on responses of simple cells and not on the pixels of an input image. If applied on the pixels of the input image, periodicity of three crests and three troughs along a line with orientation Θ passing through point (ξ, η) will be detected. This periodicity need however not be due to a system of three parallel bars. Experiments with checkerboard patterns (see Fig.12D in [12]) in which the direction of the periodicity of the checks does not coincide with the normal to the diagonals — this is the case when the aspect ratio of the checks is different from 1 — have shown that grating cells detect the periodicity of the diagonals (which evidently resemble bars in the response they elicit) rather than the periodicity of the checks.

Finally, the grating cells sum the responses of the grating subunits in their receptive fields. The response $w_{\xi,\eta,\Theta,\lambda}$ of a grating cell whose receptive field is centered on point (ξ, η) and which has a preferred orientation Θ of the normal to the grating and periodicity λ is computed as follows:

$$w_{\xi,\eta,\Theta,\lambda} = \int G_\lambda(\xi - \xi', \eta - \eta') q_{\xi',\eta',\Theta,\lambda} \, d\xi' d\eta' \qquad (8)$$

where G_λ is a Gaussian function with radius 5λ at half amplitude. This means that the responses of all grating subunits which detect three bars are summed up in the receptive field of a grating cell. This provision is made to model the summing properties of grating cells with respect to the number of bars and their length as well as their unmodulated responses with respect to the exact position (phase) of a grating.

3 Computed cell responses

Von der Heydt et al. [12] describe the responses of grating cells for different visual stimuli. We next turn to the question of how the model presented above performs for the set of visual stimuli used by Von der Heydt et al. The aim is to validate the model and to find the values of its parameters for which it will optimally approximate the behavior of grating cells.

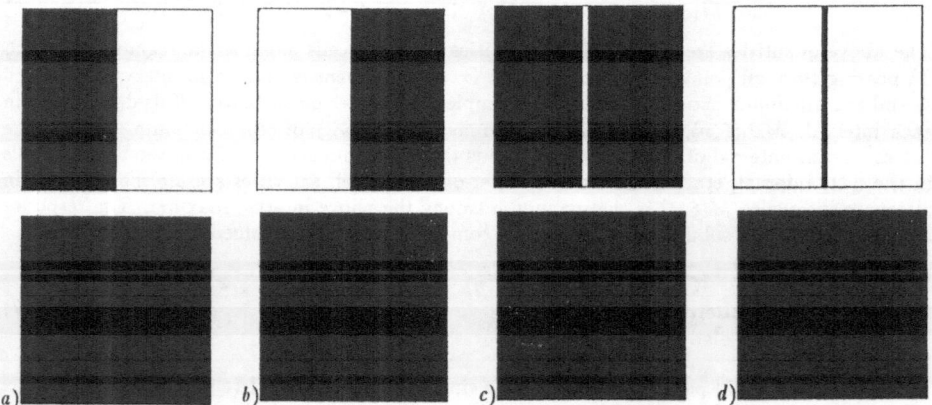

Figure 3: Input visual stimuli (first row) and computed grating cortical images (second row). None of the cells is activated (black and white mean no activity and strong activity, respectively). The simulated grating cells have vertical preferred orientation, $\Theta = 0$, and periodicity of $\lambda = 0.03125L$ (L - image size).

In Fig.3 the upper row of images shows a set of input visual stimuli for which the responses computed according to the above presented model are visualized in the respective images of the lower row. This presentation form of computed grating cell responses needs an explanation, since it differs from the one used in neurophysiological experiments (compare with Fig.1 in [12]). Each pixel (ξ, η) in an image of the lower row of Fig.3 represents the computed activity $w_{\xi,\eta,\Theta,\lambda}$ of a grating cell with preferred orientation Θ of the normal and periodicity λ and a receptive field centered at point (ξ, η). The intensity of the pixel, i.e. its gray value, specifies the computed activity (i.e. firing rate) of the cell for the respective input image. The computed activities of the grating cells which have the same preferred orientation Θ and periodicity λ but differ in the position of their receptive fields are thus represented together in one image. We refer to such images as *cortical images*, or more precisely *grating cell cortical images* to distinguish them from *simple cell cortical images* as used in our previous works [6, 7, 8, 9].

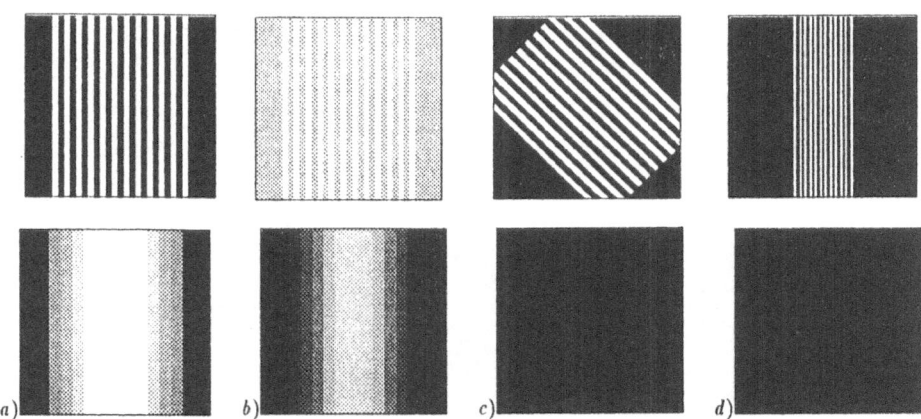

Figure 4: Input visual stimuli (first row) and computed grating cortical images (second row). The simulated grating cells have vertical preferred orientation, (orientation of the normal $\Theta = 0$), and periodicity of $\lambda = 0.03125L$.

In the particular case shown in Fig.3, grating cells with vertical preferred orientation are simulated; although the oriented stimuli in the input images have the same orientation as the preferred orientation of the cells and although they have enough spectral power in the spatial frequency domain for which the cells are selective, none of the cells is activated. In contrast, many cells are activated by a grating of bars with the proper orientation and periodicity (Fig.4a-b). Bar gratings of periodicity (Fig.4c) and orientation (Fig.4d) which differ substantially from the preferred periodicity and orientation of the simulated grating cells fail to activate them.

Fig.5 illustrates the behavior of the grating cell model when a checkerboard pattern (Fig.5a) is presented. In this simulation a model of grating cells with vertical preferred orientation ($\Theta = 0$) and periodicity λ equal to the periodicity of the checkerboard in horizontal orientation is used. The simulated cells would respond to one isolated row of checks, but as can be seen from Fig.5c, the cells do not respond when the checkerboard pattern is presented as a whole. (Real grating cells do not respond in this case either - see Fig.12B in [12].) This is due to the fact that the simple cells whose responses are used in the model integrate the intensity along the columns of the checkerboard and are not activated as shown in Fig.5b. In this way the model is made sensitive for periodicity of bar gratings but not to mere periodicity along a line.

Fig.6 illustrates the behavior of the grating cell model when a rotated checkerboard pattern (Fig.6a) is presented. In this case a model of grating cells with vertical preferred orientation ($\Theta = 0$) and periodicity λ equal to the periodicity of the *diagonals* of the checkerboard in

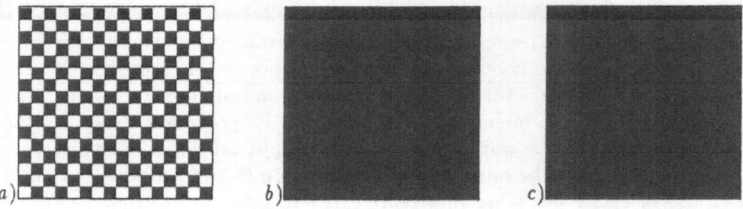

Figure 5: A checkerboard input stimulus (a) and a cortical image (c) comprising the responses of simulated grating cells with vertical preferred orientation and preferred periodicity equal to the periodicity of the checkerboard in horizontal orientation. The middle image (b) shows the corresponding simple cell cortical image used to compute the grating cell cortical image on the right (c).

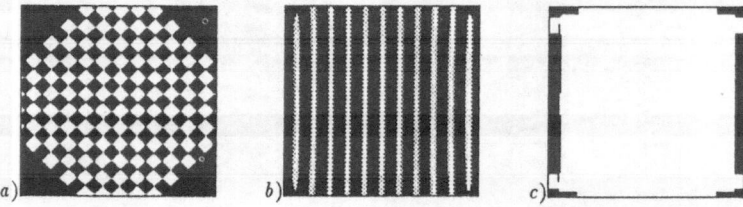

Figure 6: A rotated checkerboard input stimulus (a) and a cortical image (c) comprising the responses of simulated grating cells with vertical preferred orientation and preferred periodicity equal to the periodicity of the checkerboard diagonals in horizontal orientation. The middle image (b) shows the corresponding simple cell cortical image used to compute the grating cell cortical image on the right (c).

horizontal orientation is used. Similar to their biological counterparts (compare with Fig.12D in [12]), the simulated grating cells detect the periodicity of the diagonals, although perceptually one would rather give preference to the periodicity along the rows and columns.

4   Experiments with natural images

We now apply the grating operators proposed above on an image of a human face (Fig.7a), one of the most commonly encountered type of visual patterns in the practice of man.

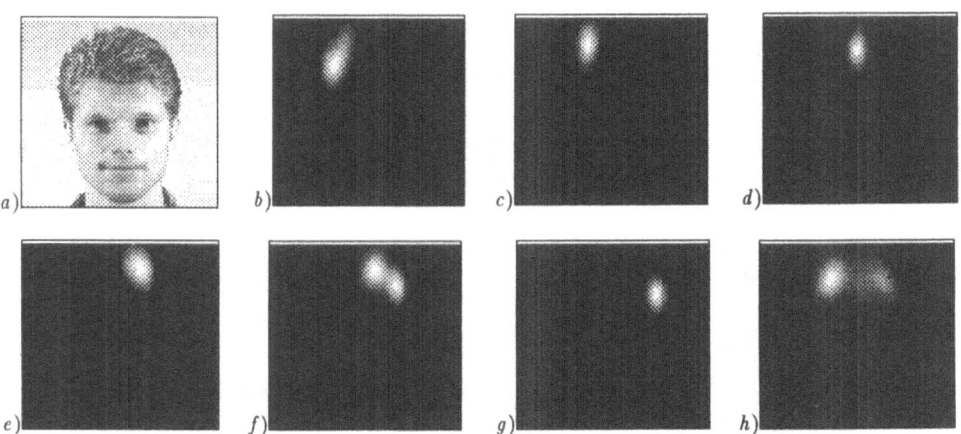

Figure 7: An input image (a) and a number of grating cell cortical images (b-g) obtained from it using different preferred orientations and periodicities. The superposition of 64 such images is shown in bottom-right position (h).

Figures 7b through 7g show grating cell cortical images obtained from the input image shown in Fig.7a by applying grating cell operators with different preferred orientations and periodicities. Different operators produce activity in different areas of the hair according to the orientation and periodicity of the hair bunches. The superposition of 64 grating cortical images as resulting from the use of 16 orientations and 4 periodicities is shown in Fig.7h. As can be seen from this composite cortical image, the area of activity quite well coincides with the texture of the hair. Similar results are obtained with other natural scenes such as forests, fields, etc. — grating cell operators act as detectors of oriented texture areas. Since the orientation and spatial frequency bandwidths of the proposed grating cell operators are not very narrow — 20° and 1 octave at half amplitude, respectively — activity is caused not only by perfectly oriented periodic stimuli: the bandwidth allows for a certain spread in orientation and distance and thickness of the constituting bars and lines.

Fig.8 illustrates the usefulness of grating cell operators for document processing. Two problems often encountered in this area are to separate pictures and figures from text and to determine the orientation of the text. When a bank of grating cell filters is applied on such a document image, the filter which has the proper orientation and spatial frequency will produce the strongest response and allow to identify text areas and the orientation of the text lines. Narrowly tuned grating cell filters allow to determine text orientation with great precision, much more unambiguously than 'classical' methods such as the Hough transform.

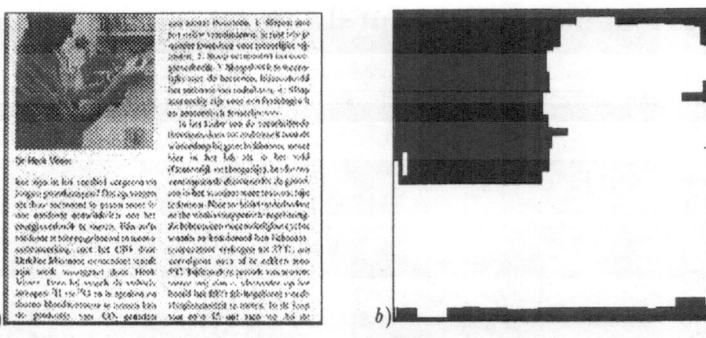

Figure 8: A useful application of grating cell operators in document processing: a filter with appropriate orientation and periodicity allows to identify text areas.

In a future work we will report on the relation of the parameters of the model to the bandwidth properties of grating cells.

5   Discussion

For vision applications it is often necessary to mask out information which is of lesser importance for a given task. Areas of constant illumination, for instance, are less important for many computer vision tasks concerning the form of objects. As a matter of fact such areas are completely characterized by their illumination and contour. Edge detecting operators, such as the operators modelling simple and complex cortical cells as well as retinal ganglion cells, extract intensity transitions, such as edges, from an image, discarding constant illumination information. In this way, edge detection operators separate form information from other visual information.

Similarly to constant illumination, texture is another type of information in natural scenes, which is of lesser interest with respect to form as represented by edges and contours. Perceptually, texture areas are quite similar to areas of constant illumination; in fact fine texture areas appear as areas of constant illumination in low resolution images. One can easily observe this effect by looking at a texture and slowly closing one's eyelids — when the gap between the eyelids becomes small enough the texture area appears as an area of constant color and illumination. In contrast to areas of constant illumination, textures are not suppressed but are rather enhanced by edge detection. As a result, texture information and form information cannot be separated from each other by edge detection. An edge detector applied on the face image shown in Fig.7a, for instance, would enhance the contours of the face and other important form clues, such as the mouth, the eyes, the brows, etc., but at the same time it will enhance a tremendous number of edges in the hair. (This was a major problem in our previous works on biologically motivated face recognition [6, 7, 8, 9].) Additional processing will be needed to separate the form clue edges from the texture edges.

In contrast, a grating cell operator will enhance a system of parallel or semi-parallel lines which are close together, arranged equidistantly or semi-equidistantly. In this way, such an operator can enhance texture areas, in particular when oriented textures are encountered, but will discard isolated lines and edges. This gives the opportunity to effectively separate oriented texture information from form clue information, as present in single lines, edges and contours. Fig.7 is a good illustration of this idea. One may speculate that this is one of the possible roles of grating cells in biological vision. At this point a question may arise of whether it would not be more efficient for form information extraction to take edge and line detection operators which enhance isolated edges and lines but do not respond to gratings of lines. It is worth noting that in their study Von der Heydt et al. have actually found such cells [12], but in the referred study their attention was focussed on grating cells. Another, and possibly the main, role of grating cells is encoding of oriented texture information. With respect to the larger receptive fields of these cells as compared to simple and complex cells, such an encoding would be more efficient than an encoding based on simple cells. This may be an explanation of the fact that grating cells are less frequently found in areas V1 (4%) and V2 (1.6%). Finally, as proposed by Von der Heydt et al. [12] some textures may be of vital importance for an animal (e.g. a bunch of bananas for a monkey) which requires effective and quick visual mechanisms for their proper classification.

References

[1] J.G. Daugman: "Uncertainty relations for resolution in space, spatial frequency, and orientation optimized by two-dimensional visual cortical filters", *Journal of the Optical Society of America A*, Vol.2 (1985) No. 7, pp.1160-1169.

[2] F. Heitger, L. Rosenthaler, R. von der Heydt, E. Peterhans, O. Kübler: "Simulation of neural contour mechanisms: from simple to end-stopped cells", *Vision Research*, Vol 23 (1992) No. 5, pp.963-981.

[3] D. Hubel and T. Wiesel: "Receptive fields, binocular interaction, and functional architecture in the cat's visual cortex", *J. Physiol. (London)*, Vol. 160 (1962), pp.106-154.

[4] D.H. Hubel: "Explorations of the primary visual cortex, 1955- 1978" (1981 Nobel Prize lecture), *Nature*, Vol. 299 (1982) pp.515-524.

[5] S.W. Kuffler: "Discharge patterns and functional organization of mammalian retina", *Journal of Neurophysiology*, Vol.16 (1953) pp.37-68

[6] N. Petkov, P. Kruizinga and T. Lourens: "Biologically Motivated Approach to Face Recognition", *Proc. International Workshop on Artificial Neural Networks*, June 9-11, 1993, Sitges (Barcelona), Spain (Berlin: Springer Verlag, 1993) pp.68-77

[7] N. Petkov, T. Lourens and P. Kruizinga: "Lateral inhibition in cortical filters", Proc. of *Int. Conf. on Digital Signal Processing and Int. Conf. on Computer Applications to Engineering Systems*, July 14-16, 1993, Nicosia, Cyprus, pp.122-129.

[8] N. Petkov, P. Kruizinga and T. Lourens: "Orientation competition in cortical filters - An application to face recognition", *Computing Science in The Netherlands 1993*, Nov. 9-10, 1993, Utrecht (Stichting Mathematisch Centrum: Amsterdam, 1993) pp.285-296.

[9] T. Lourens, N. Petkov, and P. Kruizinga. "Large scale natural vision simulations", *Future Generation Computer Systems, Issue: High Performance Computing and Networking (HPCN)*, 10:351-358, June 1994.

[10] N. Petkov: *Biologically motivated image classification system*, in ed. Ph. Laplante and A. Stoyenko *Real-Time Imaging* (Academic Press, 1995, in print) 31 pages

[11] H. Spitzer and S. Hockstein: "A complex-cell receptive field model", *Journal of Neurophysiology*, Vol.53 (1985), pp.1266-1286.

[12] R. von der Heydt, E. Peterhans and M.R. Dürsteler: "Periodic-pattern-selective cells in monkey visual cortex", *The journal of neuroscience*, April 1992, 12(4), pp.1416-1434

Modeling and Analysis of Some Neural Mechanisms for the Genesis and Control of Respiratory Pattern

Ilya A. Rybak*, Julian F. R. Paton** and James S. Schwaber*

*Central Research & Development, E. I. du Pont de Nemours & Co.
Experimental Station E-0328, Wilmington DE 19880-0328, USA

**Department of Physiology, School of Medical Sciences, University of Bristol
University Walk, Bristol BS8 1TD, UK

Abstract. *Simulations were developed in the framework of the network theory of respiratory rhythmogenesis. The main goals were: (i) to develop computational models of neural mechanisms that provide the genesis and control of both respiratory oscillations and specific patterns of respiratory neurons, and (ii) to test some hypotheses about the neural mechanisms of respiratory rhythmogenesis on the basis of an analysis of these models. Our specific objectives were to understand the mechanisms of integration and specific roles of intrinsic properties of respiratory neurons, network properties of their interconnections, and effects of afferent feedback in the genesis and control of the respiratory pattern. The models of single respiratory neurons were developed in the Hodgkin-Huxley style. The single neuron models produce the specific firing patterns of respiratory neurons recorded experimentally (i.e. adapting and ramping bursts). Different model versions of the respiratory rhythm generator have been considered. They consist of interconnected neurons and vagal feedback from lung stretch receptors. The models demonstrate a stable respiratory rhythm and specific patterns of respiratory neuronal discharges. The performances of the models are compared and analyzed in light of existing hypotheses and physiological data.*

1. INTRODUCTION

The primary respiratory rhythm generator is located in a relatively small area of the ventrolateral medulla. Thus, the genesis of the primary respiratory oscillation is likely to be defined by intrinsic and network properties of neurons within this limited area. In the current work we have considered a network paradigm for respiratory rhythmogenesis. Accordingly, the respiratory neural network is able to generate rhythm without any extrinsic periodical input drive or pacemaker properties of individual neurons. However, the network basis of respiratory rhythmogenesis does not exclude intrinsic properties of neurons as playing an important role for the genesis and control of respiratory rhythm and pattern. To date network models of respiratory rhythmogenesis are based on relatively simple continuous models of a single neuron [1,9,12,13,22,29]. These models principally consider network properties and are too crude to explore the role of individual intrinsic neuron properties in the genesis of the respiratory rhythm. It is also very difficult to use these models for simulating the specific activity patterns of single respiratory neurons or to compare the simulation results with data from intracellular recordings. For example, these models have been unable to reproduce experimentally recorded augmenting firing patterns of inspiratory and expiratory neurons, and to explore the importance of these patterns for network performance. Recent intracellular data from single respiratory neurons [3,5-8,17,19,21,26-28] in concert with the development of new computational approaches to simulation of complex, more realistic neuron models [15,16,20,32] now provide an opportunity to develop more realistic neuronal models of respiratory rhythmogenesis.

Our main objectives were to develop more realistic models of respiratory neurons based on experimental recordings and to understand the roles of specific intrinsic neuronal properties for both the genesis and control of respiratory activity.

Developing our network schemes of the central respiratory generator we have taken as a starting point the theory of a three-phase respiratory cycle developed by Richter and his collaborators [24,25]. In this theory the respiratory cycle consists of three phases: inspiration, post-inspiration and late (stage II) expiration. The respiratory related neurons are usually classified into types or groups, depending on shape of the bursts and their timing in respect to the respiratory circle. In the network models we used the following types of respiratory neurons: early inspiratory (early-I); ramping inspiratory (ramp-I); late inspiratory (late-I); post-inspiratory (post-I); decrementing expiratory (dec-E); decrementing stage II expiratory (dec-E2); constant stage II expiratory (con-E2); augmenting stage II expiratory (E2); preinspiratory (pre-I) neurons. Several versions of the network models based on Richter's theory were published recently [1,13,22]. Some alternative models also were developed, for example the two-phase model of Duffin [9]. These previous models were based on direct and indirect experimental data about connections between different respiratory neurons. The previous models have shown that the key to stable respiratory oscillations is the proposed mechanisms for "switching" between the sequential phases of the respiratory cycle. The main hypothesis about the inspiratory off-switch [4] has some experimental support and has been successfully explored in some models [1,13,22]. However, the mechanism for the expiratory off-switch is still unknown. The problem is to provide the expiratory off-switch and the augmenting pattern of the E2 neurons simultaneously. Another unresolved problem is how respiratory reflexes (i. e. vagal feedback from the lung via the pulmonary stretch receptors (PSR)) participate in phase switching mechanisms.

Our goal was to propose biologically plausible phase switching mechanisms providing: stable respiratory oscillations, specific activity patterns of respiratory neurons, and realistic changes of the respiratory rhythm and pattern under the influence of different reflexes.

2. SINGLE NEURON MODEL AND TYPES OF NEURONS

The basic single neuron model is of the typical Hodgkin-Huxley type. We have employed data from the literature about respiratory and non-respiratory brainstem neurons of mammals (neuronal sizes, membrane capacities, input resistance, resting membrane potentials, types of membrane ionic channels and their kinetics). The neuron model contains the following set of membrane channels: Na_{fast}; K_{dr}; K_A; $K_{AHP}(Ca)$; Ca_L; Ca_T; Leak; Syn_e; Syn_i. The Na_{fast}, K_{dr}, Leak, Syn_e and Syn_i channels form a basic set of ionic channels. The K_A, $K_{AHP}(Ca)$, Ca_L and Ca_T channels have been incorporated into the model because of strong evidence of wide distribution and specific function of these channels in various classes of respiratory neurons [5-8,21,26,27]. Mathematical description of channel kinetics was taken and adapted from the Huguenard and McCormic models [15,16,20]. The description of $[Ca^{++}]$ dynamics was taken from Yamada et al. [32].

Our single respiratory neuron models comprise three types to achieve three types of firing behavior to sustained excitatory drive: constant firing, decrementing/adapting and augmenting/ramping. The maximal conductance of Ca^{++}-dependent potassium channel $\bar{g}(K_{AHP})$ and the maximal conductances of Ca^{++}channels ($\bar{g}(Ca_L)$ and $\bar{g}(Ca_T)$) were tuned to get the required types of behavior of different types of neurons. The differences between the considered three types of neuron are just in $\bar{g}(Ca_L)$ and $\bar{g}(Ca_T)$. In subsequent neural network models each type is used to represent more than one class of respiratory neuron with some tuning to match the individual behaviors of specific neuron classes.

2.1 Type I neuron

This model is considered to be a basic neuron model. The following set of maximal membrane conductunces $\bar{g}$ [μS] was set for this neuron type: $\bar{g}(Na_{fast})$=3.0000; $\bar{g}(K_{dr})$=1.000; $\bar{g}(K_A)$=0.0700; $\bar{g}(K_{AHP})$=0.0700; $\bar{g}(Ca_L)$=0.0003; $\bar{g}(Ca_T)$=0.0003; $\bar{g}$ (Leak)=0.0100. Values of $\bar{g}(Na_{fast})$ and $\bar{g}(K_{dr})$ were recalculated from the classical Hodgkin-Huxley model accordingly to the size of neuron; $\bar{g}(K_A)$ is consistent with that measured experimentally from respiratory neurons [6,7]; $\bar{g}$ (Leak) was derived from experimentally measured input resistance of respiratory neurons. For the type I neuron, $\bar{g}(Ca_L)$ and $\bar{g}(Ca_T)$ are set low and do not influence neuronal firing behavior. The type I model shows a regular spike train of constant frequency in response to constant excitatory synaptic stimulation exceeding some threshold.

2.2 Type II neuron

This neuron differs from the type I neuron by an increased value of the Ca_L conductance: $\bar{g}(Ca_L)$=0.0030. The Ca_L and $K_{AHP}(Ca)$ channels together provide an adaptive frequency response to stepwise excitation [16, 32]. The type II neuron shows an adapting response to a step increase of excitatory synaptic stimulation (Fig. 1A). The mechanism for spike frequency adaptation is the following: synaptic excitation opens the low-threshold Ca_L channels; Ca^{++} concentration inside the cell increases; it causes the slow activation of $K_{AHP}(Ca)$ channels; the potassium current trough these channels causes the slow decrease of spiking frequency. All neurons of type II are the identical in the network models considered except that they may have different time constants of the K_{AHP} channel activation required for different rates of adaptation to match experimental data.

2.3 Type III neuron

Type III neuron differs from the type I neuron by an increased value of Ca_T conductance: $\bar{g}(Ca_T)$=0.0030. We have found that a combination of K_{AHP} (Ca) and Ca_T currents with some participation of K_A current provides a series of types of *spike frequency augmenting/ramping* responses following a period of strong synaptic inhibition under conditions of constant synaptic excitation (Fig. 1B-D). An increase of the preceding inhibition causes the sequential transition in neuron type III response from the "soft ramping" type (Fig. 1B) to the "special ramping" type consisting of an initial single rebound (Ca^{++}+ Na^+) spike, some period of silence, and a train of spikes with ramping frequency (Fig. 1C), and to the "delayed ramping" type (Fig. 1D). The mechanism for augmenting types of responses is the following (Fig. 1, B-D): the abrupt

102

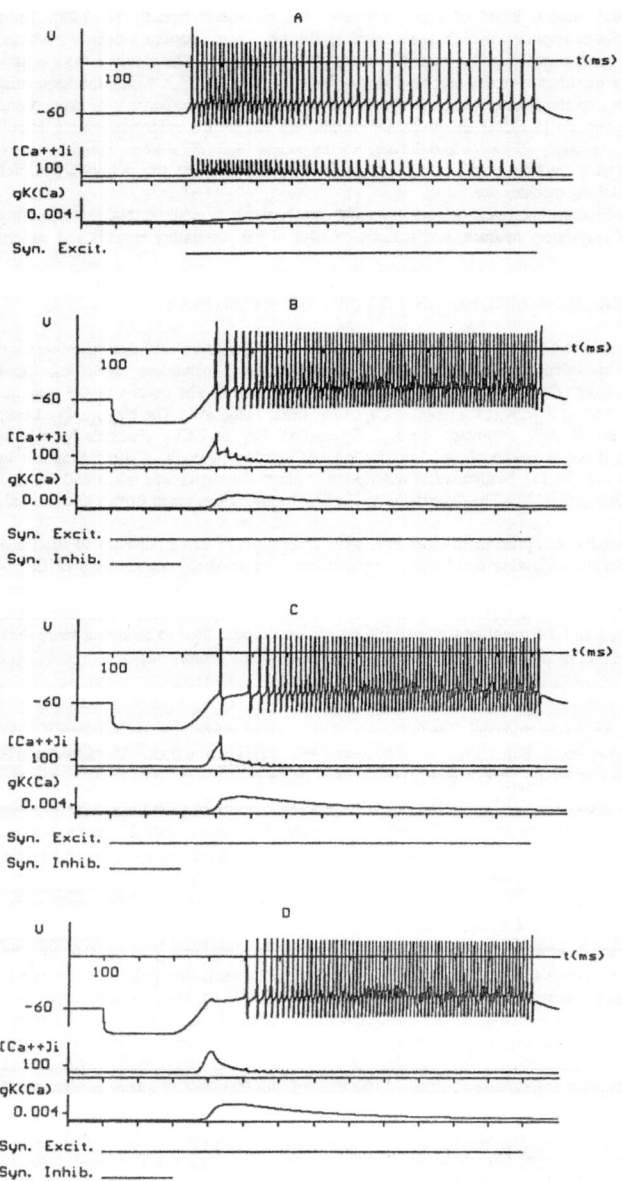

Fig. 1. Behavior of the neuron type II (A) and neuron type III (B-D)

elimination of the preceding hyperpolarization removes the inactivation of the high-threshold Ca_T (and K_A) channels; because of the Ca_T current the concentration of Ca^{++} inside the cell increases; the $K_{AHP}(Ca)$ channels are quickly activated and then slowly decay; this causes a shunting of the membrane which provides a slowly increase of spike frequency. It is interesting that the model can show the "special ramping" type of response (Fig. 1C) even without the K_A channel. The K_A current, slowly inactivating after abrupt disinhibition, together with the Ca_T current, provides "soft ramping" (Fig. 1B) and "delayed ramping" (Fig. 1D) (where the K_A current blocks the rebound Na^+ spike).

It is very important that the three types of ramping responses closely resemble the intracellular recordings from the ramp-I and E2 neurons made in Richter's laboratory [i. e. 22, Fig. 5; 26, Fig. 1]. Taking into account this fact and the strong evidence of wide distribution and specific function of $K_{AHP}(Ca)$, Ca_L, Ca_T and K_A channels in various classes of respiratory neurons [5-8, 21, 26, 27] we have hypothesized that *the ramping bursts of ramp-I and E2 neurons are based on the intrinsic neuron properties described above.* This hypothesis may be experimentally tested *in vivo* as well as *in vitro.*

3. MODELING AND ANALYSIS OF DIFFERENT NETWORK SCHEMES OF THE RESPIRATORY RHYTHM GENERATOR

Three network models of the respiratory rhythm generator have been considered. Their schematics and performance are shown in Fig. 2-4. In developing our models we have used the following criteria: (i) a model should demonstrate the realistic bursting patterns of neurons of different respiratory groups; (ii) a model should be able to generate the stable respiratory rhythm providing switches between the sequential phases of the respiratory cycle; (iii) a model should maintain the respiratory rhythm and show biologically plausible changes of the pattern under different feedback perturbations.

3.1 Neuronal firing behaviors

Since previous respiratory network simulations used simple single neuron models they were forced to use additional mechanisms to provide the augmenting/ramping firing patterns described for some inspiratory and expiratory neurons. This included, for example, recurrent excitation [1,13,22], or positive coefficients of adaptation [13,22]. There is some experimental evidence for the recurrent excitation, but the positive feedback providing the ramping pattern often makes the network model behavior unstable, and the resulting ramping patterns look very artificial. The augmenting firing pattern of ramp-I and E2 neurons in the presented models (Fig. 2B-4B) is based on the intrinsic properties of the neuron type III (described above) together with network properties providing phase dependent inhibition to the neurons and its abrupt elimination (a necessary condition for the ramping bursts in the neuron model type III). In some of our other network schemes the augmenting pattern of the E2 results from disinhibition from dec-E neuron. At first glance these models look less stable.

The adapting firing patterns of neurons in our models are based on the intrinsic properties of the neuron type II described above.

3.2 Phase Switching

All respiratory rhythm generating network models are able to generate the stable respiratory rhythm (Fig. 2A-4A). All models have the same mechanism for the inspiratory off-switch which is based on Cohen-Feldman hypothesis [4]. This mechanism was also included in Richter's three-phase theory and utilized in some previous respiratory network models [1,13,22]. The inspiratory off-switch operates in the following way: the increasing activity of the ramp-I neuron excites the late-I neuron at the end of the inspiratory phase; the latter inhibits the early-I neuron, which in turn disinhibits the post-I (and dec-E in Fig. 3A) neuron.

The switch between post-inspiration and expiration results from reciprocal inhibitory interactions between the post-I and E2 neurons, and due to the adaptation of firing in the post-I neuron.

The switch between expiration and inspiration (so called expiratory off-switch) is most interesting but poorly understood. There is no an accepted point of view for this mechanism. For example, in pervious models based on Richter's hypothesis [13, 22], the expiratory off-switch is based on reciprocal interaction between the E2 and early-I neurons, and on adaptation in the burst pattern of the E2 neuron. But this does not correspond to the convention that the E2 neuron should demonstrate an augmenting but not an adapting burst of activity. To create an opportunity to consider alternative mechanisms we have developed a series of network models with different expiratory off-switch mechanisms.

3.3 Inputs and peripheral feedback

All neurons in all models, excluding the late-I and pre-I neurons, receive constant excitatory drive (not shown in Fig. 2A-4A). The latter two receive a constant inhibitory drive. The inhibitory drive to the late-I neuron is supported by experimental evidence [19]. The motor output from the network is shown by the integrated phrenic (Phr.) activity and derived by integrating the spiking activity of the ramp-I and post-I neurons. All models contain the same scheme of vagal feedback from pulmonary stretch receptors (PSR), which are activated by lung inflation. A simple model of the lung and PSR have been developed and described by 1st degree differential equations. We have considered the following version of vagal feedback: the PSR afferents inhibit the early-I neuron but excite the post-I neuron (Fig. 2A-4A). This version of vagal feedback is based on the following: (i) early-I neurons are inflation negative [9], and (ii) electrical stimulation of the vagus nerve excites post-I neurons [10,23].

3.4 Respiratory network: Model Version 1

The schematic of the Model Version 1 is shown in Fig. 2A. Behavior of this model version is shown in Fig. 2B. This model version is a modification of the model based on Richter's theory [13,22] and proposes the expiratory off-switch mechanism based on the decrementing burst of the dec-E2 neuron and on the reciprocal interaction between the dec-E2 and early-I neurons. However, the model in [13,22] does not contain any neurons showing an expiratory augmenting pattern. The problem was to obtain both an augmenting burst of the E-2 neuron and the expiratory off-switch simultaneously. We have solved this dilemma by the assumption that the expiratory output and expiratory off-switch are provided by different subpopulations of E2 neurons. There are two neurons in the network which show activity bursts in the late inspiratory phase: dec-E2 and E2. They work in parallel. The E2 neuron exhibits an augmenting burst pattern and can provide expiratory output to expiratory motoneurons. In contrast, the E2-dec neuron has an adapting pattern and provides the expiratory off-switch. The other differences of the Model Version 1 vs. model [13,22] are: ramping patterns based on intrinsic properties of the neuron type III; the vagal feedback via PSR; lack of feedback from the reticular activating system.

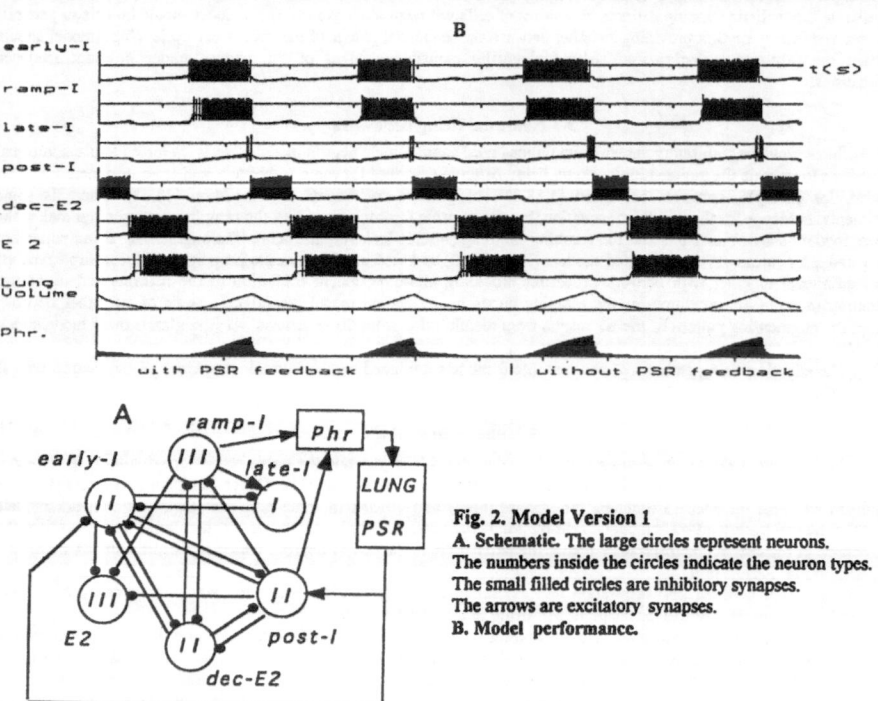

Fig. 2. Model Version 1
A. Schematic. The large circles represent neurons.
The numbers inside the circles indicate the neuron types.
The small filled circles are inhibitory synapses.
The arrows are excitatory synapses.
B. Model performance.

The Model Version 1 demonstrates a stable respiratory rhythm and realistic burst patterns of respiratory neurons under normal conditions, and when the feedback from PSR is disconnected (Fig. 2B). Strong stimulation of the vagal feedback causes postinspiratory apnea (i.e. the cessation of the rhythm at the postinspiratory phase) that is consistent with experimental data [23]. The Model Version 1 also reproduces the Hering-Breuer reflex [4,10,11]. Disconnection of the vagus feedback causes an increase in the duration of the inspiratory phase (T_I) and a decrease in the durations the stage I expiration (T_{E1}) and the whole expiratory phase $(T_E=T_{E1}+T_{E2})$ (Fig. 2B). Short stimuli applied to the PSR feedback in the first part of the inspiratory phase cause a small decrease of T_I. Stimulation during the second part the inspiratory phase causes an increase of T_I. Short stimuli applied to the PSR feedback during the stage I of expiration cause an increase of T_{E1} and T_E. Stimulation during the stage II does not change the duration of expiration. This is consistent with experimental data [2,10,11,17,18,23].

3.5 Respiratory network: Model Version 2

The schematic of the Model Version 2 is shown in Fig. 3A. The model behavior is shown in Fig. 3B. As in Duffin's model [9], the expiratory off-switch emerges as a result of long decrementing of spike frequency of the dec-E neuron and because of the reciprocal mutual inhibition between the early-I and dec-E neurons. The dec-E neuron shows an adapting pattern during both expiratory stages. Neurons of such type are present within the medulla [3]. The Model Version 2 is a two-phase model, since the other expiratory neurons (post-I, con-E2, E2) do not participate in rhythmogenesis, but only provide the postinspiratory (post-I neuron) and augmenting expiratory (E2 neuron) motor output patterns. The con-E2 neuron in the model has the same membrane properties as the E2 neuron (neuron type III), but it is less inhibited by the post-I neuron. When the activity of the post-I neuron decreases, it starts activity earlier than the E2, inhibits the post-I neuron, and provides the abrupt disinhibition of the E2 neuron that is necessary for the ramping pattern.

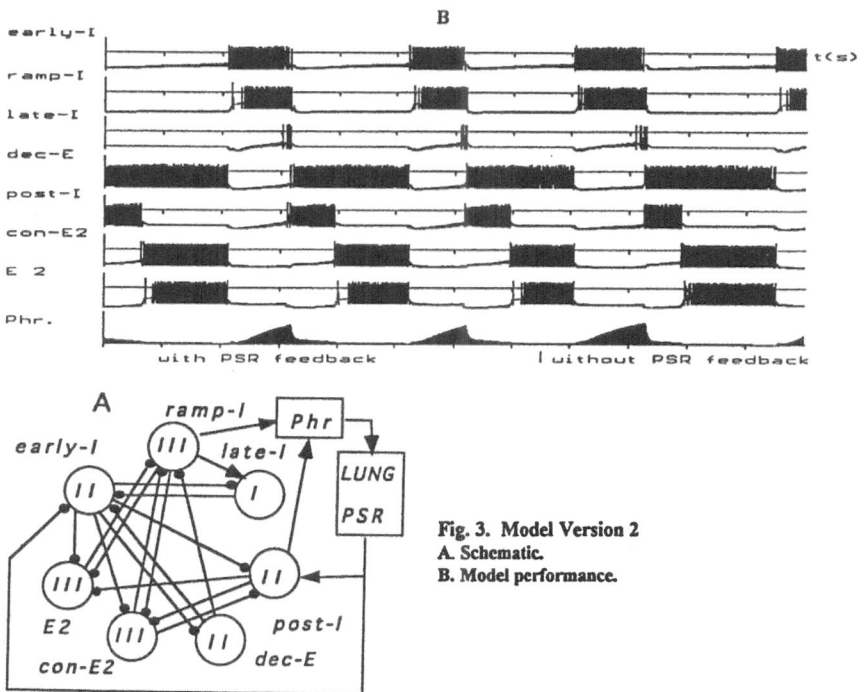

Fig. 3. Model Version 2
A. Schematic.
B. Model performance.

The Model Version 2 also demonstrates a stable respiratory rhythm and realistic activity patterns of respiratory neurons under normal conditions, and when the feedback from PSR is disconnected (Fig. 3B). Strong stimulation of the vagal feedback also causes the postinspiratory apnea. However, the Model Version 2 cannot completely reproduces the Hering-Breuer reflex. Disconnecting the vagus feedback causes an increase of T_I, but does not cause a decrease of T_E (Fig. 3B). The T_{E1} decreases, but the T_{E2} increases correspondingly. We could not find a scheme for the vagal feedback in the frameworks of this model that provided a prolongation of the whole expiratory phase after vagatomy. This is obviously a disadvantage of this model version. Short stimuli applied to the PSR feedback during the first and second parts of the inspiratory phase also cause a small decrease of T_I and its increase correspondingly. Short stimuli applied during the stage I of expiration cause an increase of T_{E1}. However, T_E does not change, which is not consistent with the literature [2,10,11,17,18].

3.6 Respiratory network: Model Version 3

The schematic of the Model Version 3 is shown in Fig. 4A. Behavior of the model is shown in Fig. 4B. The model is based on the use of the pre-inspiratory (pre-I) neuron for the expiratory off-switch for which there is recent experimental evidence [17,22,31]. In the Model Version 3 the expiratory off-switch operates in the same way as the inspiratory off-switch. The pre-I neuron plays an identical role to that of the late-I in the inspiratory off-switch. It is excited by increasing spike frequency of the E2 neuron, and then terminates the firing bursts of the E2 and con -E2 neurons by strong inhibitory feedback. The E2 neuron in turn disinhibits the early-I and ramp-I neurons. The con-E2 neuron plays here the same role as in the Model Version 2. It is interesting that we could develop a modification of this model version in which just one neuron performs the functions of both the late-I and pre-I neurons.

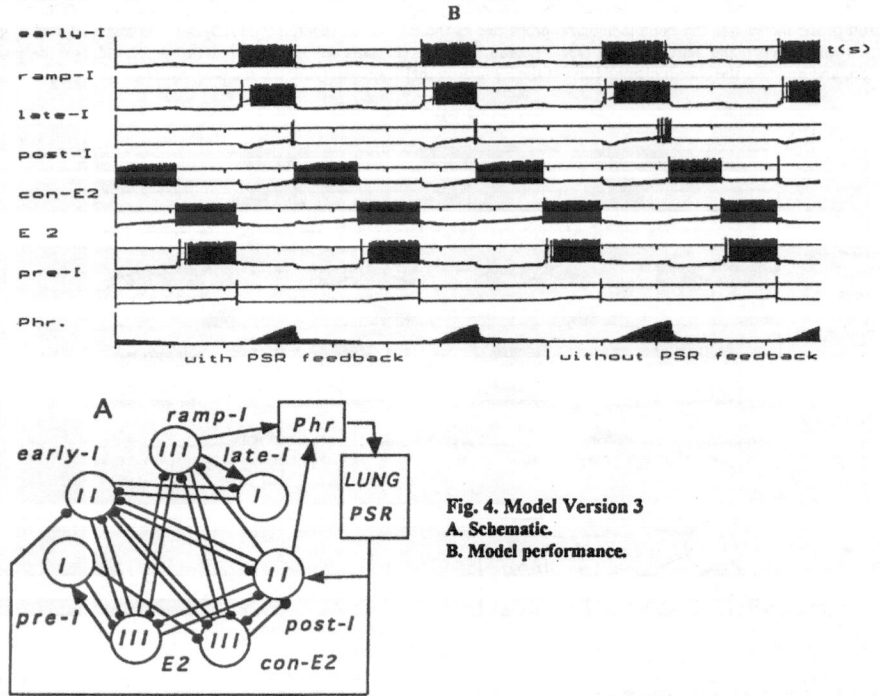

Fig. 4. Model Version 3
A. Schematic.
B. Model performance.

As well as the Model Versions 1 and 2, the Model Version 3 also demonstrates a stable respiratory rhythm and realistic patterns of respiratory neurons under normal conditions as well as when the feedback from PSR is disconnected (Fig. 4B). The strong stimulation of the vagal feedback also causes postinspiratory apnea similar to experimental data. Like the Model Version 1, it reproduces the Hering-Breuer reflex. Disconnecting the vagal feedback causes an increase of T_I and a decrease of T_{E1} and T_E (Fig. 4B). The Model Version 3 shows the consistent with the literature [2,10,11,17,18,23] changes in the duration of the respiratory phases under the conditions of feedback stimulation by short stimuli.

4. CONCLUSIONS

All model versions described above are capable of stable generating respiratory pattern. We tested the models and tried to compare them on the basis of responses to different perturbation with the PSR feedback. All model versions have the same mechanism for the inspiratory off-switch, but different mechanisms for the expiratory off-switch. Thus, the main differences in behavior of the models during perturbation applied to PSR feedback consist in different changes in the duration of the whole respiratory phase (T_E), and especially in the duration of its late stage (T_{E2}). Unfortunately it is difficult to make a final decision in the comparison of the model versions on this basis, because of contradictions in existing experimental data [2,10,11,17,18,30]. Further experimental data are now required in order that these models can be more fully analyzed.

References

1. S. N. Botros & E. M. Bruce. (1990) Neural network implementation of the three-phase model of respiratory rhythm generation. *Biol. Cybern.* 63: 143-153.
2. G. W. Bradley (1986) The effect of CO_2, body temperature and anesthesia on the response to vagal stimulation. In B. Duron (Ed.) *Respiratory Centers and Afferent Systems*, 139-154. Paris: INSERM.
3. T. H. Bryant, S. Yoshida, D. de Castro & J. Lipski (1993) Expiratory neurons of the Botzinger complex in the rat: A morphological study following intracellular labeling with biocytin. *J. Comp. Neurology.* 335: 267-282.
4. M. I. Cohen & J. L. Feldman (1977) Models of respiratory phase-switching *Federation Proc.* 36: 2367-2374
5. J. Champagnat, T. Jacquin & D. W. Richter. (1986) Voltage-dependent currents in neurons of the nuclei of the solitary tract of rat brainstem slices. *Pflugers Arch.* 406: 372-379.
6. J. Champagnat & D. W. Richter. (1994) The Roles of K^+ conductance in expiratory pattern generation in anaesthetized cats. *J. Physiol.* 479: 127–138.
7. J. Champagnat, D. W. Richter, T Jacquin & M. Denavit-Saubie. (1986) Voltage-dependent conductances in neurons of the ventrolateral NTS in rat brainstem slices. In C. von Euler & H. Langercrantz (Eds.) *Neurobiology of the Control of Breathing*, 217-221. New York: Raven.
8. M. S. Dekin & P. A. Getting. (1987) In vitro characterization of neurons in the ventral part of the nucleus tractus solitarius. II. Ionic basis for repetitive firing patterns. *J. Neurophysiol.* 58: 215-229.
9. J. Duffin. (1991) A model of respiratory rhythm generation. *Neuroreport* 2:623-626.
10. C. von Euler (1986) Brainstem mechanism for generation and control of breathing pattern. In *Handbook of physiology. The respiratory system II* (N. S. Chernack & J. G. Widdicombe Eds.), 1-67. Washington: Am. Physiol. Soc.
11. J. L. Feldman (1986) Neurophysiology of breathing in mammals. In *Handbook of physiology*, Section 1. vol. 4.(F. E. Bloom ed.), 463-524. Bethesda, MD: Am. Physiol. Soc.
12. S. Geman & M. Miller (1976) Computer simulation of brainstem respiratory activity. J. Appl. Physiol. 41: 931-938.
13. A. Gottshalk, M. D. Ogilvie, D. W. Richter & A. I. Pack (1994) Computational aspects of the respiratory pattern generator. *Neural Comput.* 6: 56-68.
14. L. Grelot, A. L. Bianchi, S. Iscoe & J. E. Remmers. (1988) Expiratory neurones of the rostral medulla: anatomical and functional correlates. *Neurosci. Lett.* 89: 140-145.
15. J. R. Huguendard & D. A. McCormick (1992) Simulation of the currents involved in rhythm oscillations in thalamic relay neurons. *J. Neurophysiol.* 68: 1373–1383.
16. J. R. Huguendard & D. A. McCormick. Vclamp and Cclamp. A Computational Simulation of Single thalamic relay and cortical pyramidal neurons. Neural Simulation Instruction Manual.
17. S. Klages, M. C. Bellingham, & D. W. Richter (1993) Late expiratory inhibition of stage 2 expiratory neurons in the cat - A correlate of expiratory termination. *J. Physiol.* 70:1307-1315.
18. C. K. Knox (1973) Characteristics of inflation and deflation reflexes during expiration in the cat. *J. Physiol.* 36: 284-295.
19. E. E. Lawson, D. W. Richter, D. Ballantyne & A. Kuhner (1989) Peripheral chemoreceptor inputs to medullary inspiratory and postinspiratory neurons of cats. *Phlugers Arch.* 414:523-533.
20. D. A. McCormick & J. R. Huguenard (1992) A model of the electrophysiological properties of thalamocortical relay neurons. *J. Neurophysiol.* 68: 1384-1400.
21. S. Mifflin, D. Ballantyne, S. Backman & D. W. Richter. (1985) Evidence for a calcium-activated potassium conductance in medullary respiratory neurons. In A. Bianchi & M. Denvait-Saubie (Eds.) *Nerogenesis of Central Respiratory Rhythm*, 179-182. Lancaster, UK: MTP.
22. M. D. Ogilvie, A. Gottschalk, K. Anders, D. W. Richter & A. I. Pack. (1992) A network model of respiratory rhythmogenesis. *Am. J. Physiol.* 263: R962-R975.
23. J. E. Remmers, D. W. Richter , D. Ballantyne, C. R. Bainton & J. P. Klein (1986) Reflex prolongation of stage I of expiration. *Phlugers Arch.* 407: 190-198.
24. D. W. Richter & D. Ballantyne. (1983) A three phase theory about the basic respiratory pattern generator. In M. Schlafke, H. Koepchen & W. See (Eds.) *Central Neurone Environment*, 164-174. Berlin: Springer.
25. D. W. Richter, D. Ballantyne & J. E. Remmers. (1986) How is the respiratory rhythm generated? A model. *News Physiol. Sci.* 1:109-112.
26. D. W. Richter, J. Champagnat , T. Jaquim & R. Benacka (1993) Calcium currents and calcium-dependent potassium currents in mammalian medullary respiratory neurons. *J. Physiol.* 470: 23-33.
27. D. W. Richter, J. Champagnat & S. W. Mifflin. (1986) Membrane properties involved in respiratory rhythm generation. In C. von Euler & H. Langercrantz (Eds.) *Neurobiology of the Control of Breathing*, 141-147. New York: Raven.
28. D. W. Richter, F. Heyde & M. Gabriel. (1975) Intracellular recordings from different types of medullary respiratory neurons of the cat. *J. Neurophysiol.* 38: 1162-1171.
29. J. E. Rubio (1972) A new mathematical model of the respiratory center. Bull. Math. Biophys. 34: 486-481.
30. M. Sammon, J. R. Romaniuk & E. N. Bruce (1993) Bifurcations of the respiratory pattern produced with phasic vagal stimulation in the rat. J. Appl. Physiol, 75: 912-936.
31. S. W. Schwarzacher, J. S. Smith & D. W. Richter. (1991) Respiratory neurons in the pre-Botzinger region of cats (Abstract) *Eur J. Physiol.* 418, Suppl. 1: R17.
32. W. M. Yamada, C. Koch & P. B. Adams (1989) Multiple cannel and calcium dynamics. In C. Koch & I. Segev (Eds.) *Methods in Neuronal Modeling*, 97-133. Cambridge: MIT.

A Neural Network Model for Plasticity in Adult Striate Cortex

Federico Morán[1] and Miguel A. Andrade[2]

[1]Dpto. de Bioquímica y Biología Molecular I, Facultad de Químicas,
Universidad Complutense de Madrid, E-28040 Madrid, Spain
[2]Biocomputing Group, EMBL, D-69012 Heidelberg, Germany

Abstract. Recently, outstanding plasticity in the cat visual cortex after birth following a retinal lesion has been described. Previously, we have formulated a neural network model for the development of receptive fields in the mammal visual cortex prior to coherent visual experience. This model is based on self-organization rules, such as Hebbian and anti-Hebbian learning, spread of neural signal among neighbouring neurons, and limitation of the synaptic growth. Here we present how the same model is able to simulate plasticity after birth just tunning the conditions of the simulation to those of a retinal lesion. Thus, our model accounts for the experimentally described long-term plasticity as well as for the short-term one. The significance of the experimental results are discussed in the context of the neural self-organization theory.

1 Introduction

Plasticity in the nervous system of higher mammals has been considered to be restricted to either the embryonic stages of development or to the period immediately following birth (the *critical period*); in both cases the nervous system is still immature. However, in recent works, Gilbert and colleagues [2,4,12] have reported a topographic remodeling of the visual cortex following retinal lesion (*scotoma*) in adults. This process occurs in two distinct steps, each with a different temporal scale: a fast redistribution of receptive fields in the area of the lesion, and a long-term reorganization that leads to the final receptive field configuration. Although the mechanisms by which this slow rearrangement occurs are becoming clear [2], the first step remains obscure. We think that the use of models of visual cortex self-organization may elucidate the underlying mechanisms of both forms of *plasticity*. Thus, we have developed a model which demonstrates that short-term changes in cortical properties due to retinal lesions are not caused necessarily by structural plasticity.

2 Model

Neurophysiological models using neural networks have shown that the ontogenetic development of the visual nervous system can be understood on the basis of self-organization processes governed by rather elementary rules [5,6,7,8,9,10,13]. With this idea in mind, we have developed a model for the self-organization of variable-sized receptive fields that can account for the ontogenesis of receptive fields in different parts of the visual system [1,11].

The architecture of the network consists of two layers of neurons innervated by weighted excitatory connections representing the thalamo-cortical pathway (see figure 1). Within each layer there is diffusion of neural activity constant with time. Lateral inhibition in the laminae of the visual cortex (that has been shown to influence greatly cortical dynamics [3]) is represented by weighted inhibitory connections between neurons. The evolution of any weight follows Hebbian rules based on activity correlation.

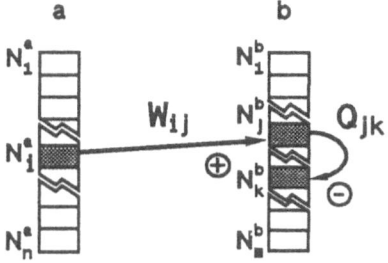

Figure 1. Scheme of the network and their connections.

So, let $\mathcal{W}$ be the activatory connection matrix and $\mathcal{Q}$ the corresponding for lateral inhibition. The time evolution of each individual weight is governed by:

$$\dot{W}_{ij} = <A_i^a(t), A_j^b(t)>_t \tag{1}$$

$$\dot{Q}_{jk} = <A_j^b(t), A_k^b(t)>_t \tag{2}$$

$A_x^a(t)$ for $x = 1, \ldots, n$ being the activity of the source layer neuron x, and $A_y^b(t)$ and for $y = 1, \ldots, m$ being the activity of the target layer neuron y. The activity of every neuron is used to compute the time evolution of the weights. Accordingly to the developmental period simulated, the only source of activity is spontaneous uncorrelated activity that is feeded into the source layer neurons. This activity is propagated along the network to reach the activity values of all the system neurons.

Following the procedure described previously by us [11], the time evolution of activatory and inhibitory connection is described by the following set of differential equations:

$$\dot{W}_{ij} = \beta W_{ij}(t) \left[\sum_{k=1}^{n} D_{ki}^a E_{kj}(t) - \gamma W_{ij}^2(t) \right] + \alpha \tag{3}$$

$$\dot{Q}_{ij} = \beta Q_{ij}(t) \left[\sum_{k=1}^{n} E_{ki}(t) E_{kj}(t) - \gamma Q_{ij}^2(t) \right] + \alpha \tag{4}$$

where

$$E_{ki}(t) = \sum_{p=1}^{n} D_{kp}^a \sum_{l=1}^{m} W_{pl}(t) \left(D_{li}^b - \sum_{o=1}^{m} D_{lo}^b Q_{oi}(t) \right) \tag{5}$$

and similarly for E_{kj}. D_{xy}^a and D_{xy}^b are the signal lateral diffusion terms between neurons x and y, located in the same source and target layer, respectively. In the results presented below, gaussian diffusion of activity has been assumed. α and β are positive constants that regulates the weight growth independent of the activity, and the rate of change of the connections, respectively.

The matrix $\mathcal{E}$, which elements are $E_{xy}(t)$, for $x = 1, \ldots, n$, $y = 1, \ldots, m$, represents the resulting effect that the activation of the different source layer neurons produce in any target layer neurons. Therefore $\{\{E_{1y}(t), E_{2y}(t), \ldots, E_{ny}(t)\}\}$ describes the actual *receptive field* of the y neuron of the tarject layer as a consequence of the activation of neurons of the source layer. This matrix combines activation, inhibition, and diffusion effects.

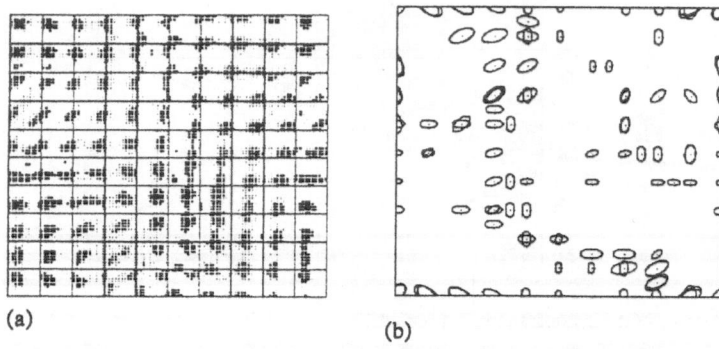

(a)   (b)

Figure 2. Final state at ($t = 400$) of the connections that yield a receptive field structure in a two-layer network composed of 10×10 neurons layers. Parameter values: $\alpha = 10^{-5}$, $\beta = 0.002$, $\gamma = 2 \cdot 10^5$, $h_a = h_b = 4$, and $s_a = s_b = 1$.

(a) Representation of $\mathcal{E}$ weights matrix, corresponding to the receptive field values. Notice that it has four dimensions, and it is shown unfolded in two dimensions. Each 10×10 submatrix indicates the resulting weights arising to one target layer neuron. The maximal square size indicates a value of 4.

(b) Disposition of the receptive fields in the visual space. Each ellipse represents the orientation and position of one receptive field. The size of the ellipse is proportional to the spatial period, and they have been scaled to make the representation clearer.

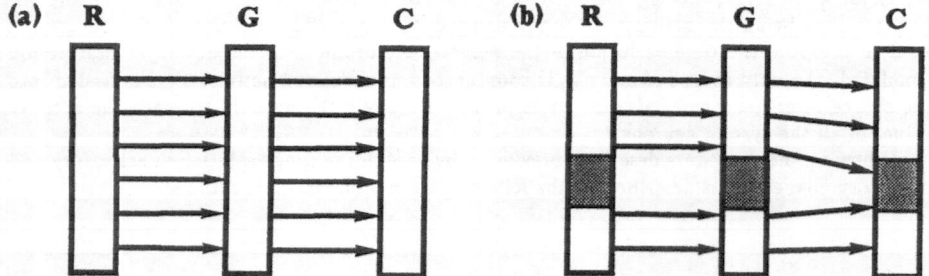

Figure 3. Schematic representation of the proposed effect to simulate the retinal lesion produced in the Gilbert and Wiesel experiment [4]. The left layer represents the retina, the central one the geniculate, and the right layer corresponds to the cortical layer. Connections from retina to geniculate are considered retinotopical and fixed. The lesion in the retina (dark pattern) produces a "shadow region" without activity in the corresponding region of the geniculate layer, that it is correspondent too to a shadow region in the cortical layer. (a) Connection pattern before the scotoma. (b) After the scotoma.

When the set of differential equations (3) and (4) is numerically integrated, the system converges to a distribution of weights, where every target layer neuron receives positive connections from a region of the source layer. Moreover, short range inhibitory weights are developed in the target layer, thus connecting closer neurons. The addition of both effects accounts for on-off receptive fields in the target layer neurons. If two-dimensional layers are used, a great variability of position, shape and orientation of the resulting receptive fields is observed. Figure 2a shows a representation of the matrix $\mathcal{E}$ and figure 2b represents the corresponding receptive field distribution for a tipical experiment.

3 Simulation of the retinal lesion

Assuming that the final state mentioned above reflects the adult state of the visual cortex, the scotoma is simulated by eliminating any input activity for a given region of the source layer. Then, all connections coming from neurons situated in the *shadow* region become inactive, is schematically represented in figure 3. Concretely, the neurons having receptive fields in the shadow region lose them, since they have no stimulus to detect. This causes the instantaneous displacement of the centers of the receptive fields. This effect is not due to neural plasticity since the connections have not been altered yet, but to the changes in the source layer properties. Figure 4a shows the resulting matrix $\mathcal{E}$ and figure 4b the displacements of the centers of the receptive fields.

As an effect of the scotoma, the cortical neurons connected to the shadow region, do not receive activity directly from them. But, at the same time, they do receive lateral activity from neighbour cortical regions and from weak connections from non-affected regions of the source layer. Therefore, as the model allows that every neuron establish connections to every neuron, they can leave their connectivity to the scotoma neurons and strenghen their connectivity to regions around the scotoma, i.e., a slow reorganization of the connectivity occurs.

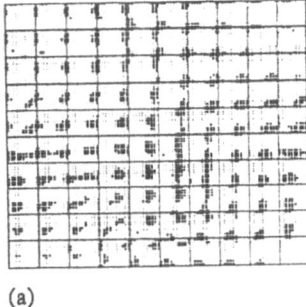 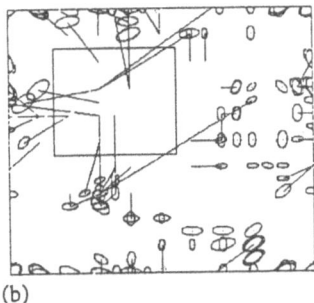

(a)   (b)

Figure 4. (a) Receptive field values immediately after the scotoma. The original receptive field values are those of figure 2a. The diffusion terms are the same as in that simulation except that $D_{ij,kl}^a$ was made null for all $i, j = 2, \ldots, 5$. Thus, the receptive fields with origin in this region are cancelled. The maximal size of the squares indicates a value of 4.

(b) Displacement of the receptive fields immediately after the scotopic lesion. The square represents the shadow region corresponding to the scotoma. The receptive fields are displayed in the visual space as in figure 2b. The grey ellipses show the position before the lesion and the black ones after the lesion. The lines joining the ellipses indicate the displacement of each receptive field. All the receptive fields inside the scotoma exit from it. Those having their center outside the scotoma have also changed since they have part of the receptive field inside the scotoma.

As time progresses, the first effect produced by the absence of spontaneous activity in a region of the source layer is the decreasing of the excitatory weights coming from this region. The vanishing of certain intralayer weights is "detected" by the inhibitory weight matrix. Since they grow by anti-Hebbian mechanisms, those target layer neurons that lose connectivity to the source layer lose activity and lose also inhibitory connections. This is a non-equilibrium situation since the tendency of each neuron is to have the same summatory of strength of connections than the others. Then they displace their connection trees to the active regions of the source layer. As a result, the active part of the source layer is covered by a new distribution of receptive fields. The resulting matrix $\mathcal{E}$ and the corresponding receptive field distribution are shown in figure 5. As observed in [4], there is an average increment in the receptive field size.

4 Discussion

The results reported in the work of Gilbert and Wiesel [4] are explained by means of a theoretical model based on neural network self-organization. These results involve a two step changes. Firstly, the sudden change in the retinal layer properties causes an immediate change in the receptive field positions into an unstable situation. Secondly, as them have pointed out, neural plasticity drives the system to a new steady state. The broad neural plasticity considered in the model has been shown to be crucial for the results obtained. These results suggest that the

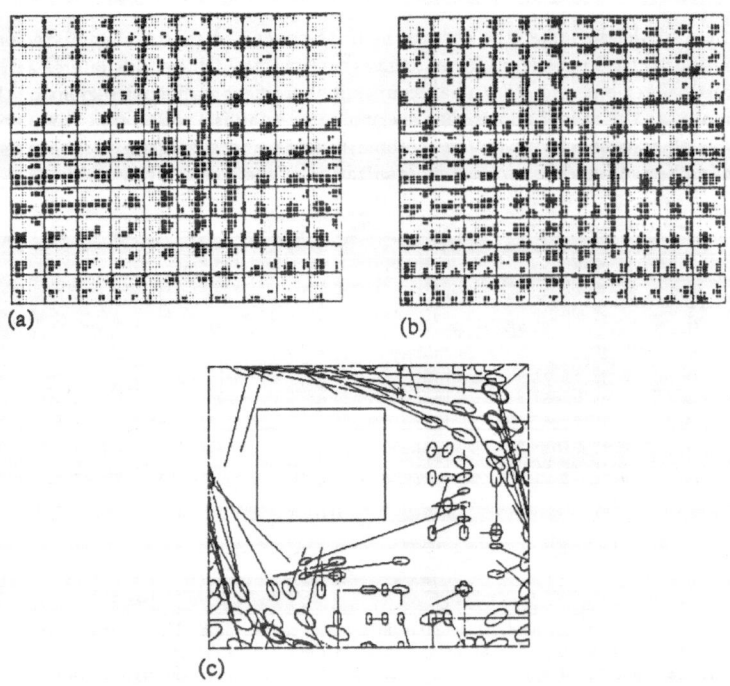

Figure 5. Proof of plasticity in adult stage:
(a) Distribution of resulting receptive field values after the scotoma at $t = 40$. The maximal square size indicates values of 0.02, 0.04 and 4, respectively. (b) Receptive field values at $t = 500$. The numerical integration of the system shown in part (a) was continued. (c) Rearrangement of the receptive fields due to plasticity.

adult plasticity described by Gilbert respond to rules qualitatively similar to that driving the development of the visual system in the critical period.

This model also gives light on the formation, properties, and structure of cortical receptive fileds. A receptive field is like a tip of an iceberg. Behind the actual receptive field there is a wider tree of excitatory connections. Only part of it is expressed as a center of an on-off receptive field due to the existence of inhibitory connections in the second layer. However, the change in the properties of the source layer alter the inhibitory effects and a different part of the activatory tree is expressed. This is reflected as a change in the *center of mass* of those receptive fields partially covering the scotoma. This effect, not due to actual structural plasticity since the connections have not been altered at this point, explains the sudden change in the receptive field position experimentally observed minutes after producing the scotoma.

In this stage, the system is in a non-stable condition, since cortical neurons located in the retinotopic position corresponding to the scotoma do not receive direct stimulation from this region, and thus they have less activation than other cortical neurons. They do receive, however, activity from small direct connections originating from non-affected neurons of the input layer. These small connections are reinforced and the system evolves finally to a new, stable state where again every neuron receives the same amount of activity from the imput layer. Thus, our model accounts for the experimentally described long-term plasticity as well as for the short-term one.

Therefore, artificial neural-network models have been shown to be powerful tools to explain plasticity phenomena, such as those described by Gilbert. This result waits for experimental verification.

Acknowledgements

We would like to thank Prof. Ch. von der Malsburg for his fruitful discussions during the development of the present work. This work has been supported in part by grants PB92-0908 and PB92-0456 from DGICYT (Spain). M.A.A. is a holder of an EC Human Capital and Mobility post-doctoral grant.

References

1. Andrade, M.A. and Morán, F. 1993. A model for the development of neurons selective to visual stimulus size. In *Lecture Notes in Computer Science. New Trends in Neural Computation* J. Mira, J. Cabestany & A. Prieto (Eds.) **686**, pp.24-29. Springer-Verlag, Berlin.

2. Darian-Smith, C. and Gilbert, C.D. 1994. Axonal sprouting accompanies functional reorganization in adult cat striate cortex. *Nature.* **368**, 737-740.

3. Gilbert, C.D. 1992. Horizontal integration and cortical dynamics. *Neuron*, 9, 1-13

4. Gilbert, C.D. and Wiesel, T.N. 1992. Receptive field dynamics in adult primary visual cortex. *Nature*, **356**, 150-152.

5. Häussler, A.F. and von der Malsburg, C. 1983. Development of retinotopic projections: an analytical treatment. *J. Theor. Biol.* **2**, 47-73.

6. Linsker, R. 1986a. From basic network principles to neural architecture: Emergence of spatial-opponent cells. *Proc. Natl. Acad. Sci. USA.* **83**, 7508-7512.

7. Linsker, R. 1986b. From basic network principles to neural architecture: Emergence of orientation-selective cells. *Proc. Natl. Acad. Sci. USA.* **83**, 8390-8394.

8. Linsker, R. 1986c. From basic network principles to neural architecture: Emergence of orientation columns. *Proc. Natl. Acad. Sci. USA.* **83**, 8779-8783.

9. Miller, K.D., Keller J.B. and Stryker, M.P. 1989. Ocular dominance column development: analysis and simulation. *Science* **245**, 606-615.

10. Miller, K.D. 1992. Development of orientation columns via competition between ON- and OFF-center inputs. *NeuroReport*, **3**, 73-76.

11. Morán, F. and Andrade, M.A. 1993. A model for the development of variable sized on-off receptive fields. Proceedings of the World Congress On Neural Networks, 1993 INNS Annual Meeting, Portland, OR (USA); pp. 116-120.

12. Pettet, M.W. and Gilbert, C.D. 1992. Dynamic changes in receptive-field size in cat primary visual cortex. *Proc. Natl. Acad. Sci. USA*, **89**, 8366-8370.

13. von der Malsburg, C. 1973. Self-organization of orientation sensitive cells in the striate cortex. *Kybernetic.* **14**, 85-100.

Regenerative-Type Neural Interface[1]

Elena Valderrama[1,2] , Rosa Villa[1] , Enric Cabruja[1] , Paco Garrido[1,2] , Xavier Navarro[3] ,
Miquel Buti[3] , Santiago Calvet [4]

(1) Centro Nacional de Microelectrónica
(2) Universidad Autónoma de Barcelona, Dpt. Informática
(3) Universidad Autónoma de Barcelona, Dpt. de Biología Celular y Fisiología
(4) Universidad Autonoma de Barcelona, Dpt. de Ciencias Morfológicas

Abstract:

Microdevices capables of in-vivo recording action potentials and, eventually, stimulating axons, can be valuable tools to increase our knowlegde about the organization of the nervous system. We present a regenerative-type, long-term neural interface to interact with the peripheral nervous system, currently under development at our center. In the paper we show different versions of passive interfaces (nterface containing only via-holes) designed for checking the peripheral nerves regeneration capabilities and a first prototype of active interface (interface containing microelectrodes). Nerve regeneration through the dice via-holes has been proved succesfull by implanting passive interfaces in the sciatic nerve of rats. Finally, a discussion about the main challenges, future work and applications is included.

1.- Introduction

Different approaches have been reported to interact with the nervous system; i.e, silicon needles for intracortical stimulation [Campell91]; silicon shafts [Kurpestein81][BeMent86], regenerative-type interfaces [Edell86][Kovacs92][Kovacs94][Akin91][Hetke94] etc. The recent and spectacular growth of micromachining technologies based on silicon sharing processes with microelectronics opens new and powerful approaches. Regenerative-type neural interfaces are microsystems designed to make possible access to a large number of neural signals by using an array of via-holes surrounded by microelectrodes, implanted between the severed ends of a peripheral nerve. When conviniently fixed, a number of axons regenerate through the via-holes, making possible both recording of quasi-individual action potentials and axonal stimulation. Figure 1 shows an artistic drawing of such a regenerative interface.

---

[1] This work is being carried out under the frame of INTER ESPRIT project and FIS grants.

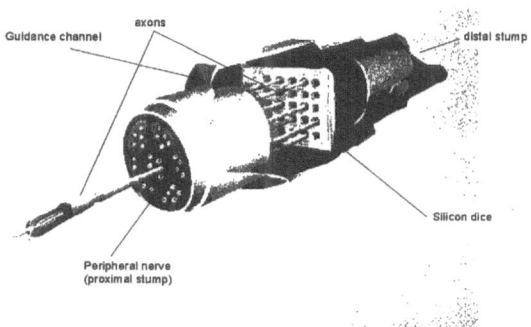

Figure 1: Schematic drawing of the neural interface

The developments made at our institutions are the main concern of this paper. At the first stage, several configurations of passive interface namely interfaces containing only the via-holes array without microelectrodes were designed in order to test, at the very early steps, the possibilities of peripheral nerve regeneration. The structure, design and fabrication of the passive dices are covered in point 2, while the experiments done to check regeneration are explained in point 5. The second step was the development of suitable microelectrodes and the obtention of active interfaces. One active dice has been developed, as shown in point 3. Point 4 deals with package and insulation problems arising when trying to implant in vivo the interface. Finally, in point 6 we try to summarize the main problems that remain unsolved, and give some comments about possible short-term applications of the neural interface.

2.- Passive dice

Figure 2 shows the structure of a passive dice of Si containing 121 (11x11) via-holes on a 10μm thickness membrane. Via-holes are round shaped, with a diameter of 40μm, and separated 70μm center-to-center.

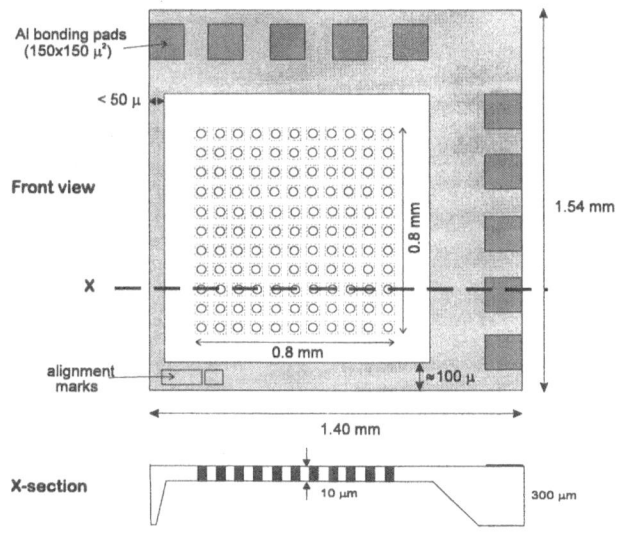

Figure 2: Passive dice containing 121 via-holes

Passive dices are fabricated by using standard <100> oriented Si P-doped wafers. Process begin with an oxidation step, followed by a nitride deposition. The back of the wafers has a photolitography step to remove both layers from the areas where the membrane will be later defined. Aluminium is then sputtered on the front side to create the bonding pads, a layer of undoped Pyrox is deposited, and a photolitography process is carried out to define the areas where the via-holes are immediately after opened using RIE techniques. Finally the back of the wafers is etched in a KOH solution to define the membranes. Etch process is time-stoped, taking about 4 hours. Time is calculated to etch Si up to 5µm before reaching the final membrane thickness. The last 5µm are etched in a RIE system.

Three diferent configurations of passive dices have been fabricated, as summarized in table 1, and shown in figure 3.

Type	Dice size (mm^2)	Dice thickness (µm)	Membrane size (µm^2)	Membrane thickness (µm)	#Via-holes	Via-holes diameter (µm)	Center-to-center (µm)
1	3x3	500	1000x1000	80	5x5	200	400
2	1.4x1.5	300	800x800	10	11x11	40	70
3	1.4x1.5	300	800x800	10	20x20	10	40

Table 1

Figure 3: Picture showing passive dices type 1 (left), 2 (middle) and 3 (right).

Axons succesfully regenerated through some of these dices, as explained in point 5.

3.- Active dice

An active dice containing microelectrodes was also designed and fabricated, and it is shown in figure 4. Traditionally, Ir, Pt, Au or other metals have been used to build the microelectrodes because of their biocompatibility, low impedance and high charge delivery capability. These electrodes are located at the surface of the dice (front side), measuring about 100 µm^2 although different sizes are used. A different approach was used here: Microelectrodes consist of a highly doped (Si n++) silicon area around the via-hole, covering the wall of the microperforation. In this way, a larger area can be assigned to the electrode without causing excessive surface area penalty, allowing enough surface available to fit the connections layers between the electrodes and the bonding pads.

In the active dice shown in figure 3, polysilicon based electrodes were additionally included for modelling porpouses. The left half of the dice contains 66 Si n++ electrodes, 5 of them connected to the upper line of bonding pads. Right half contains 66 polysilicon based electrodes, covering three different configurations: Monopolar (22 lowermost electrodes), bipolar-1 (middle, 22) and bipolar-2 (22, uppermost). Two pads are used to contact the reference electrode, which is distributed along the whole dice.

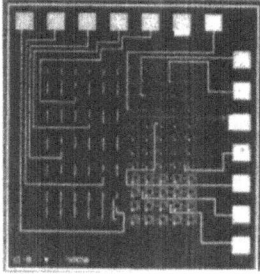

Figura 3: Active dice

Figure 4 shows the internal microstructure of the Si n++. Impedance at 1 KHz. falls within the range considered sufficient to succesfully record extracellular potentials (around 50 to 500 KΩ); however an excesively wide range is found for electrodes in the same dice. This point needs deeper study currently going on. More stable and lower values are found when the electrode is electroplatinized.

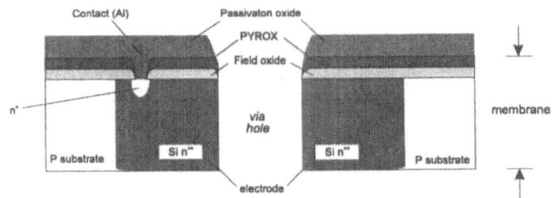

Figure 4: Microstructure of the Si n++ electrode

Regeneration studies are currently going on.

4.- Packaging

The complete interface system will consist on (1) one active dice (an intelligent dice in the future), (2) a support to enable wire bonding and to facilitate surgical manipulation, (3) a guidance channel to provide stability, (4) the connection wires and (5) an external connector. All these parts are shown in figure 5. The whole system must be biocompatible, ocupy a minimun size, be stable, robust and easy to manipulate, be reproductible, provide an stable dice-nerve junction, and make possible the axonal regeneration. These requirements represent a serious challenge for each one of the parts of the system.

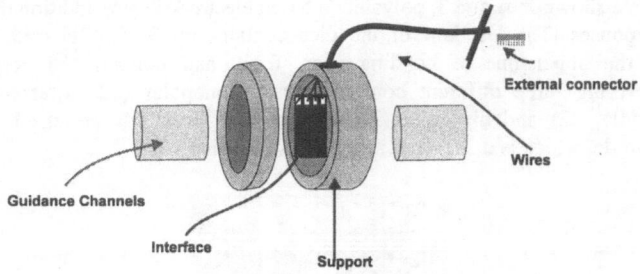

Figure 5: Complete microsystem.

The support is necessary to facilitate wire bonding and to provide mechanical protection to the fragile dice structure. Selection of an adequate material is a crucial issue. Table 2 shows a summary of the different materials studied, together with some of their more relevant properties.

	Biocompat.	Thermal stability	Micromech anization	Type	Electrical insulation	Density (weigh)
Silicone	Yes	Low melting point	Difficult	polimer	Yes	Low
Teflon	Yes	Dilatation coeficient	Difficult	polimer	Yes	Low
Stainless Steel	Yes	Yes	Yes	metal	Worse	High (10)
Titanium	Yes	Yes	Yes	metal	Worse	Lower
Ceramics	Yes	Yes	Yes	ceramics	Yes	Medium
Hidroxipatite	Yes	Yes	Yes	ceramics	Yes	Low

Table 2

Titanium and ceramics were first selected. Hidroxipaptite seems to be a promising material, but we had some availability problems. Figure 6 shows a picture of the titanium support.

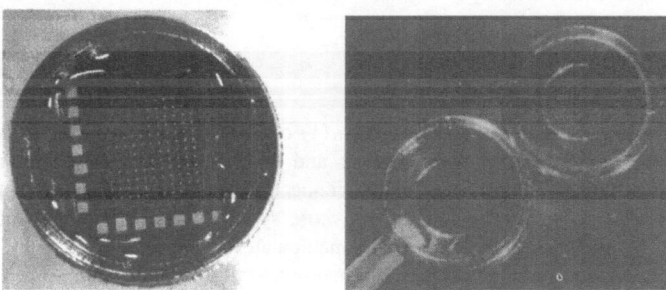

Figure 6: Picture of the titanium support

Gold wires isolated by polyurethane were used to connect the dice to the external connector. The Al bonding pads were covered with epoxy to avoid direct contact between living tissues and aluminium.

Silicone tubes of different diameters, one introduced inside the other, and fixed to the support with a photocurable epoxy, were used as guidance channel. The aim of this guidance channel is to fix the nerve ends and the different parts of the complete interface to avoid undesirable detachment that would prevent axonal regeneration. Wires end in an external connector to provide electrical continuity to the external signal acquisition system.

Other alternatives are currently under study to improve the stability and surgical handling of the complete system.

5.- Axonal regeneration through dices

The passive dices were placed in silicone chambers designed for in vivo implantation. Briefly, one dice is placed into a transverse slit cut in a silicone tube (approx. 8 mm long, 1.5 or 2 mm. i.d.) and the external surface cemented with silicone adhesive to maintain the dice in place and avoid openings to the extraneural space. Operations were performed under anesthesia on several groups of Sprague-dawley rats. The sciatic nerve was exposed at the thigh and transected, the nerve stumps positioned in the chamber facing both sides of the chip with a short gap of 2-3 mm. between the atumps, and secured in place with 10-0 sutures through the tube wall and the epineurium. The chamber was filled with saline solution.

After different postoperative intervals, the animals were again anasthetized and the repaired nerve reexposed. The sural, peroneal and tibial branches distal to the chamber were pinched with fine forceps, and the reflex response recorded to asses the presence of regenerating axons. The nerve was the dissected, fixed and processed for optical and scanning electron (SEM) microscopy studies. Based on gross examination of the chambers and SEM studies, successive regeneration ocurred on 10 of 12 rats implanted with the passive dice type 1 (25 via-holes). The chamber proximal and distal to the dice was filled by a regenerating nerve cable which widened when facing the dice surfaces (figure 7). The size of the intratubular cable and the dice surface covered increased in aninals inspected at longer postoperative times. SEM allowed viewing of regenerating nerves reaching the proximal and distal surfaces of the dice, and small axon fascicles travesing through via-holes. Functional reinnervation, expressed as reappearance of evoked muscle action potentials, reactive swet glands and nociceptive responses in distal target were recorded from 1 month postoperation in some animals and by 3 months in most of those tested.

Figure 7: Nerve regeneration through the implanted dice

6.- Discussion and future work

Regenerative-type neural interfaces constitute an exciting research field where the convergence of different fields of science and technology is completely mandatory. This work is being carried out under the frame of INTER ESPRIT project and FIS grants. Alternatives to almost any point considered here are currently being studied inside the centers belonging to the INTER Consortium[2]: At the IBMT (Germany), active dices containing Pt-based microelectrodes, together with implantable hybrid conditioning circuitry are on-going; at the (Pisa-Italy), alternatives to the silicone guidance channel based on the use of spray/heating techniques are being studied; advanced supports and package techniques are under investigation at IMIT (Germany); the use of artificial neural networks for future signal identification and motor and sensory separation is being addressed, together with the study of in vitro characterization issues at Tuebingen University-IPTC (Germany), and, finally, CHUV (Suisse) is carrying out in vivo studies. Microelectronic circuitry to be included on-chip and the technologies-compatibilization issues associated are being addressed at the CNM. New results are expected in a relatively short time.

Although the control of prosthesis seems to be a clear application for the neural interface in the (far) future, the interface system can represent, in a shorter time, an important tool for neurophysiology research by providing access to a relatively large number of neural signals, in a stable, long-term and bidirectional way. In turn, this research will contribute to a better understanding of the peripheral nervous system behaviour that can make possible the future control of phrostetic devices.

References

[Akin91]: "A micromachined silicon sieve electrode for nerve regeneration applications". T. Akin, K. Najafi.

[Campell91]: "A silicon-based, 3-dimensional neural interface: Manufacturing processes for an intracortical electrode array". P.K. Campbell, K.E. Jones, R.J. Huber, K.W. Horch, R.A. Normann. IEEE Trans. on Biomedical Engineering, vol 38, n 8, 758-767, 1991.

[BeMent86]: "Solid-state electrodes for multichannel multiplexed intracortical neuronal recording". S.L. BeMent, K.D. Wise, D.J. Anderson, K. Najafi, K.L Drake. IEEE Trans. on Biomedical Engineering, vol 33, n 2, 230-241, 1986.

[Edell86]: "A peripheral nerve information transducer for amputees: Long-term multichannel recordings from rabbit peripheral nerves". D.J. Edell. IEEE Trans. on Biomedical Engineering, vol 33, n 2, 203-214, 1986.

[Hetke94]: "Silicon ribbon cables for chronically implantable microelectrode arrays". J.F. Hetke, L. Lund, K. Najafi, K.D. Wise, D.J. Anderson. IEEE Trans. on Biomedical Engineering, vol 41, n 4, 314-321, 1994.

[Kovacs92]: "Regeneration microelectrode array for peripheral nerve recording and stimulation". G.T. Kovacs, C.W. Storment, J.M. Rosen. IEEE Trans. on Biomedical Engineering, vol 39, n 9, 893-902, 1992.

[Kovacs94]: "Silicon-substrate microelectrode arrays for parallel recording of neural activity in peripheral and cranial nerves". G.T. Kovacs, C.W. Storment, M. Halks-Miller, C.R. Belczynski, Ch.C. Della Santina, E.R. Lewis, N.I. Maluf. IEEE Trans. on Biomedical Engineering, vol 41, n 6, 567-575, 1994.

[Kurpestein81]: "A practical 24 channel microelectrode for neural recording in vivo". M. Kurpestein, D.A. Whittington. IEEE Trans. on Biomedical Engineering, vol 28, n 3, 288-293, 1981.

---

[2] INTER Consortium is made up of the following partners: Scuola Superiore Santa Anna-SSSA (leader), Hahn-Schickard Gesellschaft Institute fur Mikro-und-Informationstechnik-IMIT, Institu für Physikalische und Teoretische Chemie-IPTC, Centre Nacional de Microelectrònica-CNM, Centre Hospitalier Universitaire Vaudois-CHUV and Fraunhofer Institute for Biomedical Engineering-IBMT.

New Perspectives in Auditory Coding: Bases for a New Cochlear Behavioural Model

R. Villa, J. Aguiló
Centro Nacional de Microelectrónica
Campus Universitat Autònoma de Barcelona
08193 Bellaterra (Barcelona, SPAIN)

ABSTRACT

In this paper we present a computational model of the auditory system, called "stable pulse correlation", that was developed at the CNM. In the model, the cochlear system has been considered as a black box, incorporating some non very complex devices based on structural elements in the cochlear system itself. First of all, we consider a bandpass filter (basilar membrane) and some transductor units (inner hair cells), whose behaviour could be defined by some specific parameters such as threshold, absolute and relative refractory period, and relaxation function. Our model considers that the output coded response to a particular single tone input signal will be given by the frequencies and the number of the generated stable states of a short number of different output signals and their correlations. A graphic output with 8this information could be extracted from the model (VARG diagram). This graph uses data obtained from the numeric model. The model and associated VARG diagram allows a close prediction of the cochlear behaviour to a specific excitation under certain limits. The results obtained are compared with experimental response of the auditory fibers and the auditory system under normal conditions. From this comparison we have a validate model that could be used as a cochlear system simulation model. In summary, we propose a new theory for auditory system encoding signals that could be applied to cochlear implant systems as well as to speech recognizers.

INTRODUCTION

The purpose of this study was to develop a model to facilitate the study of information encoding and signal pre-processing at the cochlear system level and the input/output behaviour of that system. From the well known input signals our model should simulate the cochlear system behaviour. So, it should generate temporal and spacial discharge patterns in whole auditory nerve fibers (AN) as response to simple and complex stimulus.

Throughout this study, the inner ear has been treated as a "black box" that acts as a transducer of the pressure signals to nervous pulses that fall upon it though the nerve bundles to upper states where they are interpreted as audible signals and the information they contain is "decoded". Our interest is to roughly establish the correlation existing between the two types of signals.

Let us consider the auditory system from the system theory point of view (perhaps more concretely, the electronic systems) without concerning ourselves excessively about particular aspects whose influence on the global system is insignificant from a functional point of view.

The auditory system has been considered a deterministic one, in the sense that its response to excitation is univocal. Consequently, certain parameters have nor been considered: those which undoubtedly influence its behaviour (such as temperature, liquid, viscosity, energy lost by friction, chemical composition, form and size of the cochlea, length of the basilar membrane, binaural audition, etc.) The response to the same stimuli has been considered the same for all cases.

We will treat the system as "analog/digital" (like biological system). The input signals are analog (they vary in a continuous form) and their output signals are electrical pulses.

Our study advances methodologically from lesser to greater system complexity and from lesser to greater input stimuli complexity of the systems. Thus, the response of an individual hair cell to the stimulus of one pure tone (sinusoidal wave of specific frequency and range) is studied. This is followed by a study of the functional behavior of a group of hair cells, and thus successively until completing the study, without a great deal of detail, of the whole system to a complex stimulation signal.

THE MODEL

a. Overview

The function of the outer, middle and inner ears is to transform the pressure fluctuations in the air into the appropriate neural excitation, which is the transmitted to the central nervous system.

The middle ear's primary function is to match the impedance (Pickles 88). Middle ear dynamics are linear over the intensity range of human audition.

The cochlea (Fig. 1) consists of a coiled fluid-filled tube with a still cochlear partition (the basilar membrane and associated structures) separating the tube lengthwise into two chambres (called scalae). At one end of the tube, called the basal end or simply the base, a pair of flexible membranes called windows (oval and round) connect the cochlea acoustically to the middle-ear cavity.

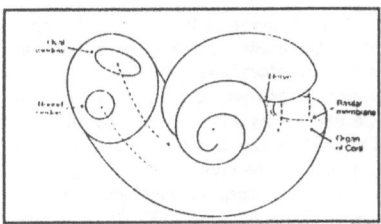

FIGURE 1. Artist's conception of the cochlea. The dashed lines indicate fluid paths from the input at the oval window and back to the round window for pressure relief.

When the oval window is pushed in by a sound wave, the fluid in the cochlea moves and the fluid bulges back out through the round window. The distortion of the cochlear partition by sounds takes the form of a traveling wave, starting at the base and propagating toward the far end the cochlear ducts, known as the apical end, or apex. As the wave propagates, a filtering action occurs in which high frequencies are strongly attenuated. The sensory cells are called "hair cells". This transducers sit alond the edge of the partition in a structure known as the Organ of Corti. There are two groups of hair cells: inner hair cells (IHC)(n=3000) and outer hair cells (OHC) (n=20.000) in the human cochlear.The role of the IHCs is to send messages to the brain about the presence of acoustic vibrations at a specific place in the cochlea. The role of the OHCs is to influence these mechanical vibrations before they reach the IHCs and they are primarily responsible for the sharp tuning observed in cochlear mechanics (amplifies the vibration of the basilar membrane and sharpens the tunning of the mechanical vibrations delivered to the IHCs) .(Neely 1993) From each ear, 30.000 nerve fibers run to the cochlear nucleus. Almost all of these fibers end in a one-to-one fashion on IHCs. Thus, each IHCs has about ten synapses.

b.-Basilar Membrane

Mechanical to electrical transduction in the cochlea is mediated by vibrations of the basilar membrane. Its responses to sound have been measured in only a few laboratories. By far the most celebrated series of investigations were carried out by Georg von Békésy, for which he was awarded the 1961 Nobel Prize for Physiology or Medicina. (Békésy 1961). He showed that the cochlea performs a spatial frequency analysis. Each site of the BM responds to sound stimuli by vibrating.

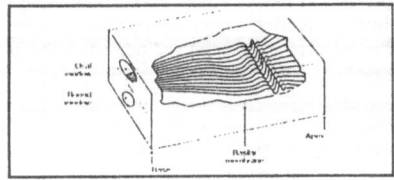

FIGURE 2. A sinusoidal travelling wavw on the basilar membrane, in a simplified rectangular-box model.

A displacement traveling wave propagates on the basilar membrane from the cochlear base toward its apex. The traveling wave grows in amplitude, reaches a maximim and then decays. The location of the peak is a function of stimulus frequency: Vibrations in response to high-frequency sounds peak near the cochlear base, while very low-frequency sounds travel all the way to the cochlear apex.

Recent studies demostrating that all frequency-specific non-linear properties of auditory nerve and IHC responses have mechanical counterparts in the vibration of the basilar membrane.(Ruggero 92).

If we submit the basilar membrane to periodic, prolonged external stimulus, the affected part of the membrane will vibrate with the same periodicity. That is, a wave is generated in the basilar membrane whose equation is:

$$y(x,t) = Ae^{-f(x,\omega)}\cos(Kx - \omega t)$$

where A is the maximum deformation range of the membrane. The second term is the function of the point and frequency, and accounts for the tuning; and the last term represents the periodicity of the vibration. Due to the form, elastic characteristics, and particular boundary conditions of the membrane, evaluation of the second term is not an easy task.

An exact model of the membrane's behavior requires knowing the function, f(x,w). To simplify our study at the initial steps we will assume that each point of the membrane has a particular tuning frequency distribution. To further simplify this study at mean time, we will consider the linear deformation of the membrane from the point of tuning and depending on the excitation range. Finally we consider the sigmoid that will depends from the input frequency and intensity as the function of membrane deformation. We will work with relative deformation amplitudes to ease calculation.

c.-Hair Cell
The physical phenomena occurring when the auditory system is stimulated oblige some hair cells to change their direction. When the deformation is sufficient, a series of processes are triggered, whose cycle ends with the generation of a nervous impulse. The transduction unit is incapable of immediately giving a new impulse. During another period, generation of a new impulse is conditioned by the intensity of the excitation. Possible ranges for these values, and the function of "return to recovery" will be defined, which we call the relaxation function of the hair cells.

The model assumes that interspike intervals consist of a constant-length absolute refractory period (ARP), followed by random-length relative refractory period (RRP) Refractoriness appears as a supression of discharge probability immediately after an action potential

Thus, a behavior model of the hair cell is obtained on four parameters: threshold deformation, UM0, ARP, RRP, and the type of temporal function of return to the situation of equilibrium or relaxation function.

Obviously, ARP determines the maximum firing rate of the hair cell. That is, the maximum firing rate that can be observed, Fdmax, will thus determine the value of ARP, ARP = 1/Fdmax. If the maximum firing rate of a transductor unit is 1kHz, ARP will be 1 ms.

Our study assumed that the responde of a single hair cell will be the same to any stimulus, regardless of its frequency, that is, no specific timing mechanism is assigned to its tuning.

The response of the hair cell will be quantitatively distinct for each group of parameters values. However, we retain the following common property: given a set of parameters which determine a specific behavior, there will be a large number of stimuli, defined by their frequency and amplitude, to which the hair cell will respond periodically and its firing frequency will coincide with some submultiple of the excitation frequency. Thus, for the same excitation frequency, the larger the intensity of the stimulus, the closer this submultiple will be to the maximum possible firing rate, or saturation value.

It can be proven that the value of the minimum deformation threshold does not affect the qualitative results,

124

and its quantitative influence is minimal. We have considered that UM0 = 10% of the maximum possible deformation range.

If we call the threshold relaxation function R(t), where t minds the time since the last fire, there is,

(1) $R(t) \geq UMO, \quad 0 \leq t \leq PRA + PRR$

(2) $R(t) \geq A_{max}, \quad 0 \leq t \leq PRA$

(3) $R(t) = UMO, \quad t \geq PRA + PRR$

(1) Firing period will be a function of the time and the intensity of the stimulus.
(2) So, the system will always be under the threshold and no firing will be produced.
(3) So, after a complete cycle time hair cell will be in their stand-by position.

A new pulse will be produced when,

$$y(x_0, t) = A' \cos(kx_0 - \omega.t) \geq R(t) \quad ; \quad A' = Ae^{-f(x_o, \omega)}$$

where A' is the maximum deformation amplitude at point x_0 due to exitatory signal.

Distinct situations have been simulated with distinct parameter values and relaxation functions (linear, hyperbolic, and logarithmic). The conditions to be met by the functions refer to the limits: they must pass through the point (ARP, Amax) and (ARP+RRP, UM0). In the hyperbolic case it is allowed to pass through the point (RRP, 1,09*UM0) since it tends asymptotically to UM0.

Recent studies have suggested tha teh refractory characteristics of an AN fiber vary with stimulus level or discharge rate. In experiments dates (June Li 1993; Carney 1993) characterize quantitatively the refractory behavior of AN fibers. The ARP is found to be constant , independent of discharge rate, with mean value between 0.56 and 0.86 ms.The RRP decreases in duration as discharge rate increases; RRP mean length is less than 2 ms in most cases.

d.-Correlations between adjacent Hair Cells.
Under a specific excitation and due to the tuning of the membrane there will be a point of maximum membrane deformation. The response of the transducer unit is maximum at this point. As we have assumed that the deformation decreases at both sides of this point, neighbouring transducer units will have a "similar" response. That is, their response will correspond to an excitation of the same frequency with less amplitude.

Information regarding the frequency and excitation range, can be obtained from the correlation between the response of adjacent cells, and more concretely from the distinct submultiples or "stable" frequencies generated. The excitatory frecuency is univocally determined by knowing the firing rate of the two consecutive hair cells.

e.-Cochlear System.
Th global model undoubtedly conserves all the characteristics of his components and relations listed thus far. In other words, the system will be determined according to a series of parameters that define each of its components, plus a series of particular parameters that must be determined to account for specific behavior of the group formed by the membrane and individual transducer units.

An important characteristics of the system is the tonotopic frequency distribution. Different experimental studies of this distribution allow us to obtain the following relation between the position of the transductor unit, x, and its resonant frequency, FCF.

$$x(FCF) = 58.5 - 5.6 * \log(FCF)$$

We will use this equation as the reference for comparison of theoretical tonotopic distributions.

Whichever the system, it has necessarily any maximum of their sensitivity. In particular, for two different signals to be considered as such, the temporal distance between the equivalent points considered for the elaboration of the response must be greater than a threshold which we call Tmin. Let us assume that Tmin is the minimum temporal interval that can be considerer by the system.

If we assume that two units will have not exactly same meaning as part of the system so if F and F' correspond to the tuning frequencies of two adjacent units, the following should be fulfilled,

$$F' \leq F(\frac{PRR}{PRR + T_{min}})$$

In this equation, if we choose the equal sign, the induced tonotopic distribution qualitatively coincides with the reference. If the coincidence must be absolute quantitatively, the following condition must be met:

$$1 + \frac{T_{min}}{PRR} = e^{\frac{x'-x}{5.6}}$$

where x'-x is the distance between hair cells.

Assuming 3000 hair cells uniformly distributed in 35 mm. that is, a distribution of approximately 11 microns, then Tmin/PRR = 0.00164. A reasonable set of values complying with the previous condition would place the value of Tmin around 10 µs. and the PRR between 2 and 10 ms.

The above consideration until now does not account for the discrimination amplitude. The response will deteriorate some what as the excitation frequency decreases, since for low frequencies the number of stable firing rates are much less, and the interval of possible amplitudes between them, much larger. Bearing this in mind, it can be conceived that the tonotopic distribution is proportional to the number of stable firing frequencies for each of them.

The result of this consideration is that the tuning frequency of two adjacent transductor units fulfills,

$$\frac{F'}{F} = 1 + PRA - PRR$$

from which, if the coincidence must be absolute quantitatively,

$$1 + PRA - PRR = e^{\frac{x'-x}{5.6}}$$

where once again, x'- x is the distance between adjacent hair cells. Like the previous case, this implies that 1/(1+PRA-PRR) = 1.000164 ms. If we continue assuming that PRA = 1 ms, then PRR = 2.64 ms. which is perfectly compatible with the range of values obtained previously.

We must still considerer the width of the affected membrane zone on both sides of the unit whose characteristic frequency coincides with that of excitation. This is taken from the tuning curves of the auditory nerve fibers [Kiang, Neely, and Kim]. If FCF is the frequency of excitation, and FMAX and FMIN, those of the transducer units located at the extreme points where the excitation is barely perceptible, it results that the relations FMAX/FCF, FCF/FMIN and thus FMAX/FMIN are constants not depending of FCF, and thus,

$$\log(F_{max}) - \log(F_{min}) = K$$

That is, the width of the affected zone is thus independent of the excitation frequency.

Results based on experimental data (if second filter effect is omitted) show a group of possible values to be: FMAX/FCF=1.5, FCF/FMIN=2.5, for which k=0.6. From these values it can be determined that the width of the area affected by 50 dB incident intensity is around 5.5mm.

If we do not assume a differential behavior for some of the transducer units according to their position, the response of any unit located between FMAX and FMIN will correspond to a harmonic displacement of the same frequency as the center, but with less amplitude.

A diagram of the response os a specific number of hair cells to the frequency excitation FCF results in the following.
We call it the VARG diagram that stands for Variable Auditory Response Graph (FIGURE 6;7).

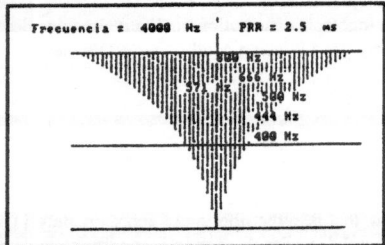

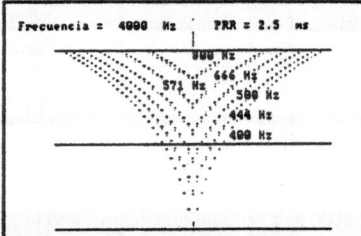

FIGURE 3:   VARG Diagram.   FIGURE 4:VARG Diagram..

In general, an audible complex signal can be seen as the sum of an infinite series signals, each a pure tone of specific range. In reality, we may reduce this sum to a few terms since the range of the rest are practically negligible. Supporting ourselves with this reasoning, we will thus considerer primarily those excitations produced by pure tones, that is by waves of only one frequency and specific range.

RESULTS

We consider that the theoretic obtention of a tonotopic distribution like that obtained experimentally, assuming the existence of a minimum discriminable time interval is a result.

It is also worth citing the obtention of the VARG diagram which gives the details of the auditory code generated by a pure tone as a function of the incident intensity.

This section deals only with results that can be compared with existing experimental data and that have not been used in the development of the model.

Synchronism: It is known that for low frequency excitations, the nervous pulses that are generated are synchronous with respect to the excitation frequency. This synchronism is maintained in some measure until the frequency of 5 kHz, and totally disappears beyond this frequency.

The following figures show the experimental data obtained and the percentage of synchronism obtained with the present simulation model.

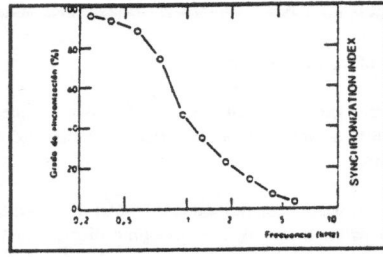

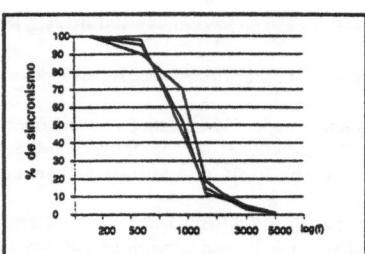

FIGURE 5:   FIGURE 6:

Frequency Discrimination: The following figures show experimental data (Figure 7A) of the frequency discriminations as a function of the central frequency, and (Figure 7 B y C) that obtained from the present simulation model.

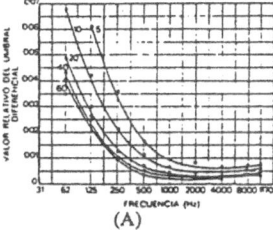
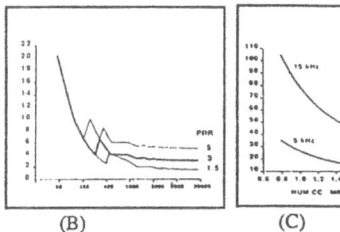

(A) (B) (C)

FIGURE 7: Frequency discrimination

If we observe the graphs of frequency discriminations as a function of the number of hair cells, the simulation results allow predicting a discrimination loss. This loss fundamentally affects high frequencies due to an indiscriminate loss of hair cells.

Incident Phase-Generated Impulse Ratio: The following figures show experimental data and that obtained with the present simulation model.

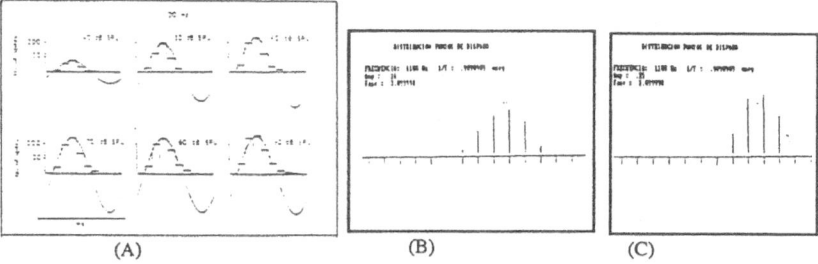

(A) (B) (C)

FIGURE 8: Incident phase-Generated Pulse a) experimental dates, b and c result of model

DISCUSSION

Knowledge of the function of the auditory system indicates that there are only two possible mechanisms for this: the system can use spatial information to detect small displacements in the site of cochlea excitation, know as the place theory, or use temporal information to detect different time intervals between the pulses produced by one signal or another, know as the timing theory. Distinct evidence favours one or another, and neither theory is conclusive, consequently controversy has been maintained for over 100 years.

In the theory presented there, frequency discrimination does not exclusively depend on the place of excitation or the number of pulses per unit of time generated in an axon; it depends instead on the signal correlation between adjacent transducer units. The frequency discrimination depends on a specific correlation of stable firing frecuencies which, as additional information, are produced at a specific location in the cochlea.

The frequency discrimination mechanism proposed coincides exactly with the place theory when the excitation intensity is 0 dB, that is, under quasi-minimum threshold conditions. According to our model, under these conditions we could obtain a sound sensation as response, without discriminating the tone of the signal; this agrees with the acoustic behavior, experience with the human ear.

Following the frequency, the model's behavior is similar to that foreseen in the timing theory until approximately 1 kHz and 140 dB, since the stable frequency will be synchronous with the excitation frequency until that range. At frequencies of less than 1 kHz, a large number of transducer until will transmit pulses synchronous to the excitation frequency. after this value, there will be a transducer until that will give the maximum number of pulses, set by the system, which will coincide with the maximum deformation point;

neighboring transducer units on both sides will transmit pulses corresponding to submultiples of the excitation frequency.

We emphasize that the information to discriminate a given frequency is obtained using the correlation of stable pulses transmitted by various transducer units located in a specific area of membrane. This behavior is independent of the excitation frequency nor use different mechanism (Timing and Place Theory) as eclectics hold true.

-Given the experimentally known model structure, a minimum discriminable time interval between 5 and 10 microseconds is enough to obtain the frequency resolution, without the need to reach values of tenths of microseconds as supported by the timing theory nor count on a precision of less than 10 μm in the tonotopic frequency location.

One aspect validating our model is its possibility to simulate, and consequently predict, the system's behavior under some pathological situations. A typical example of this is the effect produced by the loss of hair cells, as clinically demonstrated in neurosensor deafness. Our model predicts that a random loss of transducer units significantly alters the system's discrimination in the high frequency zone. This is obvious since the model present is much denser stables coding for these frequencies than for low frequencies, moreso at low intensities, making code identification since there is no information available from some transducer units. Of course, the loss of high frequency discrimination is much more acute if the transducer units lost are those located near the oval aperture.

As the excitation intensity increase, our system needs to incorporate or reclute new hair cells to the process. At the same time, stables pulse frequencies detected in the transducer units involved will increase. Considering various units, the transducer unit whose characteristic frequency coincides with that of excitation will always emit the largest pulse frequency, and as the intensity nears the maximum threshold, its pulse frequency will near 1000 c/s.

Consequently, and even though this unit is giving maximum response, or saturation, at a specific intensity, the system continues discriminating intensities for discrete increments of stable impulse frequencies sent by neighbouring transducer units, in addition to adding other units.

This work has analyzed or studied parts of the system in some detail; concretely, those sections whose analysis or comprehension offers concepts or data to discuss. Some of these data simply validate the model regarding its correlation with bibliographic data; others offer ideas to better understand auditory coding; and finally, some serve as nothing more than verification, checks, or the degree of influence on the global model.

Phase: Phase change do not affect the sound sensation, In our model, after a certain transitory period in the order of 1 to 3 ms, the information transmitted to a transducer unit does not depend on this phase. If the phase information is not sent to upper levels, this sensation cannot exist.

Excitation is only produced in the positive phase of the wave. In our model, only the positive rising slope of the excitation signal is susceptible to provoke transducer unit response. Analyzing results of repetitive excitation experiments has shown that these correspond to the sum of responses of all the excitations. Thus, excitation is apparently produced only during the positive phase of the cycle.

Tunning: Basic theoretical studies of the membrane deformation mechanics show that the membrane acts as a tuning filter at distinct frequencies according to its position. On the other hand, the selectivity of the units acts as a new filter, very similar to the first. Knowing the response of these units, the model of the membrane's mechanical response has been adapted accordingly. As a result of the adaptation and simplification, we approach membrane deformation, as an asymmetric triangle with the most abrupt slope directed towards the lower frequencies. Since the deformation contour is compatible with the system mechanics, it may once again be interpreted that the selectivity of the transducer units is conditioned by the changes in the membrane deformation produced by different excitations, rather than a particular selectivity characteristic of the membrane itself.

Absolute Refractory Period (PRA): Based on the model's configuration, 1 ms was assumed as the PRA value, since the fiber cannot transmit pulses of more than 1000 cycles/sec.

Relative Refractory Period (PRR): The need for this parameter and a range of possible values has been studied in this work. Different results obtained have inferred that the PRR has a marked influence on the quantitative response of the model. If this term is eliminated, the possibility of range discrimination would disappear in our model.

So that the results obtained with the model agree with the experimental data for synchronism, integration time discrimination ranges in frequency, tone and/or minimum discriminable time, etc., the possible range of PRR values is between 2 and 6 ms. In view of the results, we could assign a value of 3.3 ms. in a first approximation.
The form the relaxation function has proven to be important in the study performed. The most convincing results were found with hyperbolic functions. Since there is no bibliographic data available that allows direct evaluation of this concept , its validation, like the PAR value, has been on global results of model behaviour.

Minimum Integration Time: In the measure that we have demonstrated the existence of a stabilization time depending on the frequency, we understand that this should coincide with the maximum time interval known as minimum integration time value, according to our model, lies between 3 and 10 ms. These values agrees with diverse experimental studies.

CONCLUSIONS
The conclusions of the developed model may be classified in two groups according to their origin: if they are reached during the development of the behavior model of the cochlear system or they deal with the function of the cochlear system derived from the response of the model itself.

Conclusions regarding the developed model: The coding of excitation frequencies and intensities is determined through the existing relation of stables frequencies. These are submultiples of the input frequency. The VARJ Diagram enables easy obtention of a graph of a known input signal code. This is the final result of the group of algorithms developed in the model. The model can be seen as the link between two auditory theories (Place and Timing Theories).

Conclusions regarding the cochlear system: The tonal distribution is imposed by: 1) the need to discriminate frequencies that are multiples among themselves; 2) the need to discriminate between two very close frequencies; and 3) the existence of a minimum discriminable time interval.
For the response obtained with our model to be correct, the value of the minimum discriminable time interval must range from 5 to 12 μs.
According to our model, the interval of possible PRA values between 0.7 and 1.2ms and for PRR, between 1.5 and 6 ms. The relation function is hyperbolic.
The system self-stabilizes within 2 to 5 ms regardless of the excitation frequency.

BIBLIOGRAPHY
von Békésy G, (1960) Experiments in Hearing. Translated and edited by Wever EG. New York: McGraw-Hill,

Javel E (1986) "Basic response properties of auditory nerve fibers" in Neurobiology of Hearing: The Cochlea, edited by R.A. Altschuler, D.W. Hoffman, and R.P. Bobbin (Raven ; New York) pp.213-245.

Pickles, J.O.(1988). "An introduction to physiology of hearing", 2nd edition, Academic Press,

Ruggero M.A. (1992) "Responses to sund of the basilar membrane of the mammalian acochlea. Current Opinion in Neurobiology" , 2:4449-456.

Li June and Young Eric D. (1993) "Discharge-rate dependence of refractory behavor of cat auditory-nerve fibers", Hearing Research 69 151-162

Neely ST. (1993) A model of cochlear mechanic with outer hair cell motility J.Acoust.Soc. Am 94(1),July 1993

Carney LH, (1993) A model for the responses of low-frequency auditory-nerve fibers in cat J.Acoust. Soc. Am 93 (1), January

Nervous System as a Closed Neuronal Network: Behavioral and Cognitive Consequences

Jorge Mpodozis, Juan-Carlos Letelier and Humberto Maturana

Departamento de Biologia, Facultad de Ciencias, Universidad de Chile
Casilla 653, Santiago, Chile
FAX= (562) 271-2983 tel= (562) 271-2865 ext 260
e-mail= letelier@abello.seci.uchile.cl

Abstract.

We present here a theoretical framework about the nervous system operation that explains the origin of coordinated behavior without violating the structural determinism inherent to the constitutive autonomy of living systems. At the same time we will show that cognition is not the outcome of a computational task, as it is envisaged by the tradicional paradigm that considers the brain an information processing device, but rather is the results of the spontaneous structural coupling that take place ontogenically and phylogenically between a living system and its circumstances of living .

Introduction

Modern neurobiological thinking assumes that the nervous system, moment by moment, senses the external world, recognizes the configuration of stimuli and finally chooses an appropriate response. Under such view, the response of the nervous system is based on the correct determination of the entities present in the outside world, and in the ontogeny and phylogeny of that particular nervous system. This point of view (which we will call the "representationist framework" and wich could be summarized by the dictum "the brain is an information processing machine") underlies most of the research with respect to brain operation that has taken place during the second half of this century. But, in spite of the explosive development of the experimental research in the last two decades, it has become clear that many of the main questions about brain operation arising from such framework (such as object recognition or motor coordination) are still pretty much unsolved. The study of the vertebrate visual system gives a perfect example of such situation. In effect, if we arbitrary define Kuffler's experiments (1953) on the properties of retinal ganglion cells as the starting point of the research effort on this field, we find ourselves in the necessity to explain that 40 years latter, although immersed in a sea of anatomical and physiological details, we cannot imagine the "shape" that the representationist answer to the question "what is it to see?" will finally have.

The representationist framework has also been challenged from a theoretical perspective that considers autonomy and structural determinism as central features of living systems (Maturana, 1969, 1970, Maturana and Varela, 1973, 1980. A review of this growing field could be found in Mingers, 1995). Such theoretical thinking, embodied in the notion of "Autopoietic Systems", considers that the internal dynamics produced by the circular network of processes constituting the living organization cannot be specified by external stimuli. The internal dynamics and the observable states of autopoietic systems are the result of the operations of the living system itself, while external stimuli can only trigger, but never define, the magnitude, kind and direction of the structural changes that the system undergoes. Its seems that this new point of view creates for itself an apparently insurmountable problem, namely, to explain the origin of coherent and adequate behavior. This problem does not exist in the context of the representationist framework, as the observed congruence between behavior and medium is implicitly explained by assuming that such coherence is the outcome of an adequate operation of the nervous system on the representation that it has of the medium.

In this paper we propose a new conceptual framework to understand the operation of the nervous system that: a) uses the autonomy of living systems as a central and starting point and b) avoids the seductive but sterile metaphor of "processing information". Developing this point of view, we shall also show: a) the unexpected relation between the problem of the nature of the living organization and cognition, b) the behavioral, as opposed to the computational, nature of cognition, and, perhaps more fundamentally, c) how the "external world" is not a given problem that the organism has to "solve" but that it is constituted through the interactions between an organism and its circumstances of living. We shall first present our biological fundaments, after that we shall develop our views about nervous system and its operation as component of the living systems, and finally we shall reflect about the behavioral nature of cognition.

I. Living Systems.

1) Living systems, as all systems that we deal with in our daily life and scientific endeavours, are structure determined systems (SDS). That is, they are systems such that all that happens in them and to them arises determined in their structure. Nothing external to an SDS can specify what happens in it. An external agent that incides on an SDS can only trigger in it a structural change determined by it. Moreover, the structure of an SDS determines also what it admits in an encounter as an agent that triggers in it a structural change (Maturana and Mpodozis, 1987). Structural determinism is not an a priori truth, not a principle, and not an ontological assumption, it is an abstraction that observers make of the regularities of theirs experiences as they use them to explain these experiences (Maturana, 1990).

2). Living systems are dynamic molecular structure determined systems, organized as closed networks of molecular interactions that produce the same kinds of molecules that produced them, and specify dynamically at every instant the extension and boundaries of the network. Such a network is closed in terms of its dynamics of states of molecular productions, but is open to the flow of matter and energy through it. Maturana (1970) and Maturana and Varela (1973) have shown that those statements constitute a complete characterization of living systems as molecular systems, specifying their conditions of existence and autonomy. Maturana and Varela (1973) called this organization the autopoietic organization, and claim that living systems are molecular autopoietic systems. According to this notion, cells are first order autopoietic systems and multicellular systems are second order autopoietic systems. A multicellular living system is realized through the autopoiesis of its cellular components, and through its own realization as a multicellular totality, make possible the autopoiesis of these. As autopoietic systems, living systems are in a continuous structural change, both as a result of their intrinsic internal dynamics, and as a result of the changes triggered in them in the course of their recurrent interactions in a medium. A living system lives as long as its structural changes take place in the conservation of its first or second order autopoietic organization.

3). A living system, as a composite cellular and molecular system, exists in two domains: a) in the domain in which its components realize it as a first or second order autopoietic entity, namely in the metabolic or physiological domain, and b) in the domain in which it interacts and relates with the medium that contains it as a totality, namely in the relational or behavioral domain. The phenomena of the metabolic or physiological domain take place in the structural dynamics of the components of the living system, and are totally contained in it. Contrariwise, the phenomena of the behavioral domain arise in the relation living system/medium, and are not determined by the living system or the medium alone. That is, the behavior of a living system is not something that the living system does, not something that the medium specifies of its own, the behavior arises and takes place in the relation living system/medium (Maturana and Mpodozis, 1987, 1992). There are, of course, as many different kinds of physiological and behavioral domains as there are different kinds of living systems with different structures and different manners of living.

4). The two phenomenal domains in which a living system exists cannot be reduced to each other because they take place in non intersecting phenomenal domains, and then, any attempt to explain the phenomena of one domain in terms of the other, is inadequate. There is, however, a recursive dynamic generative relation between them through the structural changes that living system and medium trigger in each other in the course of their interactions: A) as living system and medium interact, they trigger in each other structural changes; B) the structural changes triggered in the living system result in a change in the manner in which the living system encounters the medium in the next interaction, and the same happens with the medium with respect to the living system; C) as a result of what happens in moments A) and B), the relation between living system and medium changes, and the structural changes that living system and medium trigger in each other in their next encounter change too; and D) the process indicated in points A), B) & C), repeats recursively in a manner that appears to an observer both as if the behavior modulated the physiology, and as if the physiology modulated the behavior, even though they take place in phenomenal domains that do not intersect.

5). The ontogeny of a living system from its inception to its death takes place as an epigenetic process that results from a systemic dynamics involving a recursive interplay of physiological and relational phenomena, in the manner indicated above. So, a living system is a systemic entity that: exists as a living being in the physiological domain of its bodyhood and, realizes its manner of living in its domain of relations in recurrent interactions with the medium, through a dynamic interplay of its body dynamics and its behavior. Accordingly, what reproduces when a particular living system reproduces, is a particular systemic entity whose realization takes place in the continuous dynamic interplay of a particular bodyhood and a particular configuration of dynamic circumstances that have arisen in the medium along the phylogenic history of the reproducing living system. At the same time, what is organically passed to the next generation through reproduction is an initial structural configuration that makes possible the epigenetic realization of a particular manner of living that entails the systemic conservation of a particular bodyhood and bodyhood dynamics if it is placed in the proper circumstances of the medium. Inheritance, then, as it consist in the reproductive conservation of an epigenetic manner of living, is a systemic process, and as such is not determined by any particular set of molecular or cellular components, however essential these may be for its occurrence (Maturana and Mpodozis, 1992).

6). When the realization of a manner of living begins to be systemically conserved generation after generation through reproduction, a lineage is constituted and established. Such lineage will last as long as that manner of living remains conserved. Moreover, as a lineage is constituted and conserved in the systemic conservation of the manner of living that defines it, all the features of the physiological domain, as well as all the features of the relational domain of the living systems that realize the lineage, become free to change around that which is conserved, in a way in which both living systems and medium remain in dynamic reciprocal operational congruence (Maturana and Mpodozis 1992). In these circumstances a new lineage arises when some variation in the realization of a particular manner of living becomes part of the manner of living henceforth systemically conserved generation after generation (Maturana and Mpodozis, 1992).

7). As a conclussion, it is possible to say that the recursive ontogenic and phylogenic mutual modulation of behavior and structure that take place trough the interactions between living systems and medium have two fundamental results. The first is that the structure of the living system and the structure of the medium change together, and in congruence, both in ontogeny and philogeny. The second is that all living systems at every moment of their ontogenic an phylogenic histories necessarily have dynamic structures that are adequate for the generation of a behavior adequate for the dynamic medium in which they are alive, or they die.

II. Nervous System and Behavior.

1). Multicellular organisms usually present a nervous system, that is, a closed network of synaptically interacting active cellular components (nerve, muscle and secretor cells), that we shall call neural elements . The nervous system operates as a closed network of changing relations of activity between its neuronal components: any change in the relations of activity holding between some components of the network leads to further changes in the relations of activity holding between other components of it, an so on recursively, in a potentially never ending dynamic (Maturana 1969; Maturana and Varela, 1980; Maturana and Mpodozis 1987). The course that follows these changes of relations of activity is at every moment determined by the state of the activity of the neuronal elements of the network at that moment. At the same time, the state of activity of the cells that compose the neuronal network is at any moment the result of the state of their dynamic structure at that moment, and change as this change through their synaptic operations within the network and through their structural intersection (by means of synaptic, trophic, hormonal and transducer-like effects; see below) with other components of the network and the organism.

2). The nervous system structurally intersects the organism at several body areas that constitute the latter' s internal and external sensory and effector surfaces. The external surfaces constitute the interfaces by which the organism encounters the medium. The internal surfaces constitute the interfaces by which the nervous system, as a component of the organism, encounters the physiological dynamics of the organism. Accordingly, the neural components of the sensory and effector areas have a double identity and a double operation. First, as elements of the nervous system they operate in the closed dynamics of changing relations of activities of the nervous system. Second, as parts of the organism they operate as components of its surfaces of internal and external interactions.

3). As a consequence of this structural intersection, the nervous system through its operations as a closed network of changing relations of activity between its neuronal components, continuously generates in the organism sensory/effector correlations that modulate both the flow of its interactions in the medium, and the flow of its physiological dynamics. The behavior of the organism arises in the dynamic encounter organism-medium through the sensory/effector correlations of the organism and the structural dynamics of the medium. Therefore, the nervous system participates in the generation of the behavior of the organism through the sensory/effector correlations to which it gives rise at any moment, according to its structure at that moment.

4). The nervous system does not interact with the medium, it is the organism that does so through the operation of its effector and sensory surfaces. It is the structure of the organism as a whole that determines wich sensory/effector correlations are possible for it, not the dynamics of the nervous system alone. All that the nervous system can do as it intersects with the external and internal sensory and effector surfaces of the organism, is trigger in these structural changes that result in one or another of the sensory/effector correlations that are possible for the organism according to its present structural dynamics (Maturana and Mpodozis, 1987). Furthermore, the structural changes triggered in the external sensors both, as components of the sensory surfaces of the organism and as neuronal elements, are determined in their structure and not by the circumstances of the interaction that trigger them. In these circumstances, as the organism interacts with its medium, its nervous system undergoes changes in the flow of its synaptic operations that are contingent to the interactions organism-medium, but that are determined by the structure of the nervous system, and not by the characteristics of the medium. As a result, the nervous system does not and cannot operate with representations of the medium, and what it does, it does according to its structure at any moment.

5). The structure of the neural cells is in continuous change, both as a consequence of its own autopoietic dynamics and as a consequence of its participation as components of the nervous system and the organism. Some of these structural changes are specially relevant, because they entail long term changes in the synaptic dynamics of the neural cells. As far as we know, these structural changes happen in four ways: A) Through the so called "transducer effects", that are structural changes triggered at the neural component of the sensors through the encounters of the organism with the medium. These structural changes have been traditionally called "transducer effects", through thinking that what is significant in them is an energy transfer. According to us, what is significant is that these structural changes are those that coupled the activity of the nervous system to the flow of interactions of the organism or to is internal physiological dynamics. B) Through synaptic effects, which are structural changes of different time constants triggered in the neural cells by the actual flow of synaptic interactions. C) Through trophic effects, which are structural changes that arises in the neural cells triggered by substances of neural origin that are produced by processes orthogonal to the synaptic flow (because they involve molecules and cellular interactions which are not proper to the synaptic operation of the nervous system) but contingent to it. D) Through hormonal effects, which are structural changes triggered in the neural cells by substances produced in the organism through physiological process that do not involve directly the operation of the nervous system.

6). If we consider together points II-3, II-4 and II-5, it becomes clear that although the medium does not specifiy what happens in the nervous system, during the ontogeny of an organism the structure of its nervous system (the neural conectivity, the cellular dynamics of production of neurohumors, membrane receptors, molecular channels, etc.) changes in a manner contingent to the flow of interactions of the organism in the medium, to the internal physiological and developmental history of the organism, and to the flow of the operation of the nervous system as a component of the organism.

7). The structure of an organism and the structure of its nervous system are structures that have arisen in a evolutionary history of transgenerational conservation of a manner of living (see I-5). Such history is in fact an epigenetic relational dynamics organism-medium, that consists of the realization of the living of the organism. For these reasons, the structure of the nervous system, as it arises through its development in any particular organism, and the closed dynamics of changing relation of activity that it generates during its development, cannot but be adequate for the generation of the particular interactional behavioral dynamic that the manner of living of the organism entails.

8). The course of the structural changes that the nervous system undergoes through the life history of the organism that it integrates, is de facto constrained by two conditions: A) by the structure that the nervous system has as a component of an organism that belongs to a particular lineage, as its follows from I-7, and B) by the actual contingencies that occur in the living of the organism, through the processes described in I-5, II-5 and II-6. As a consequence of this, every organism has at every moment a nervous system adequate to the generation of the sensory/effector correlations proper to its particular history of realization of its manner of living, precisely because the structure of its nervous system is the present of a history of structural changes contingent to the course of the phylogenetic an ontogenetic history of this organism. In other words, every animal always has a nervous system proper to its biological identity, as this consists and is realized in the relational space of its manner of living.

9). It follows from the previous points that a nervous system operates with different dimensions than those with which the observer sees the organism to operate in the relational and interactional space in which it exists as a totality. The observer sees the organism in its relational and interactional space interacting and relating with entities of different kinds or (in the case of social animals) with relations and symbols as if these were also entities. The nervous system in its internal dynamics, however, operates as a closed network of changing relations of activities between its component elements, and not with the kinds of entities that arise in the domain of relations and interactions of the organism.

10). In summary, the nervous system as a component of a living system, is constituted as a closed network of neuronal elements that operates as a closed recursive network of changing relations of activities (between the neuronal components) in which every change of relation of activity recursively leads to other changes of relations of activities in it. The nervous system as such a neuronal system intersects with the organism at its sensory and effector surfaces, and its closed operation gives rise to sensory/effector correlations in the organism that in the interactions of the organism in its medium, constitute its behavior. The nervous system does not operate making representations of the medium in which the living system that it integrates exists. Nevertheless, it has a plastic structure that changes following the contingencies of the living system while this system maintains its autopoietic organization in a medium. As a result of these structural changes, the closed operation of the nervous system continuously gives rise to configurations of sensory/effector correlations in the organism that realize its living in its changing medium, until this congruence is lost and the organism dies.

III. Final reflections. Cognition or Behavior ?

In general terms, an observer claims that cognition takes place in a system when he or she sees that the system behaves adequately as it operates in a given domain of interactions, in the understanding that a behavior is adequate because the system conserves through it certain features, or gives rise to some results that the observer defines as of importance. When the systems is a living system, and the observer is a biologist, adequate behavior usually means behavior adequate to the survival and to the realization of the manner of living of the living system .

In the representationist framework, cognition cannot but be the result of a process trough wich a system generates an internal representation of its medium, and computes a behavior according to its goals or purposes. In this framework, cognition is a feature adscribed to a system, both to describe the operational congruence of a living system and its medium, and to explain such operational congruence as a result of the operation of some special "cognitive"processes. The existence of these special cognitive processes indeed violates the constituve autonomy of living systems as structure determined systems, since it would requiere that the perturbations that impinge upon a system could specify the characteristic of the internal structural transitions that the system will undergoe. To accept that would be equivalent to state that the finger that dials a telephone number has the property of specifying the subsequent behavior of the thelephone network. Even more, this example makes easy to note that from the point of wiev of the internal telephone mechanism it is absolutely impossible to ascertain wich finger operated it, or even whether the mechanism was triggered by a finger, a pencil, or any other object or circumstance.

What we have said in this article can be used to explain cognition without having to suppose the existence of any special computational or informational process. It is esay to follows from the previous points that cognition is not something that a system has, nor something that a system acquieres trough its interactions. Cognition is the adequate behaviour of a system in a medium, that results from the dynamic structural congruence between such a system and that medium. As such, cognition only entails the dynamic structural congruence which arises spontaneously between systems that do enter in recursive interactions, as a dynamic system and its medium nessesarily does. Cognition, then, will takes place in the relational space or domain of interactions of any dynamic system while it operates in structural coupling with its circumstances, and then cannot be consider as a feature or property of the system itself. Therefore, cognition does not requiere nor does its involve any operation that an observer could treat as generating a representation of the medium to be used to obtain behavioural effectiveness.

Although living systems are autopoietic systems, it is not their autopoiesis which makes them cognitive systems. Rather, it is their manner of operation as dynamic autonomous entities with a plastic structure. In living systems, cognition arises as a result of its phylogenic and ontogenic history of structural coupling with its circumstances of living. All living systems at all moments of the history of the biosphere are born with an initial structure which, if deposited in the proper place, allows them to live an epigenesis in which they become transformed in congruence with the medium while realizing the particular manner of living of their kind. In a strict sense, as a living system lives, it does not encounter a preexisting medium, but this medium arises with its operation, even if for the observer it seems to be already there. Also, the characteristics that the medium acquires as the space of living (niche) of each living system, arise in the realization of the manner of living of the living system (Maturana and Mpodozis, 1992). It is here where the nervous system has a central participation in the generation of a domain of structural plasticity that goes beyond of what the organism does alone, and allows for an enormous expansion of the domain of possible states of the organisms in autopoiesis.

References

KUFLER S W (1953) Discharge patterns and functional organization of the mammalian retina. J Neurophysiol 16: 37 - 68

MATURANA H (1969) Neurophysiology of Cognition. P GARVIN (ed) Cognition, a multiple view. New York, Spartan Books, pp 3 - 24.

MATURANA H (1970) Biology of cognition. BCL Report 9. Biological Computer Laboratory, Departament of Electrical Engineering, University of Illinois.

MATURANA H, VARELA F (1973) De Máquinas y Seres Vivos. Santiago, Editorial Universitaria.

MATURANA H, VARELA F (1980) Autopioieis and Cognition: The Realization of the Living. Boston, D. Riedel Publishing.

MATURANA H, MPODOZIS J (1987) Percepción: configuración conductual del objeto. Arch Biol Med Exp 20: 319 - 324.

MATURANA H (1990) Science and daily life: the ontology of scientific explanations. W. KROHN, G. KUPPERS (eds.). Selforganization: Portrait of a Scientific Revolution. Dodrecht, Kluwer Academic Publisher, pp. 12 - 35.

MATURANA H, MPODOZIS J (1992) El origen de las especies por medio de la deriva natural o la diversificación de los linajes a través de la conservación y el cambio de los fenotipos ontogénicos. Museo Nacional de Historia Natural de Chile, Publicación Ocasional 46.

MINGERS J (1995) Self-producing Systems. Implications and Aplications of Autopoieis. New York, Plenum Press.

Local Accumulation of Persistent Activity at Synaptic Level: Application to Motion Analysis

M.A. Fernández[*], J. Mira[**], M.T. López[*], J.R. Álvarez[**], A. Manjarrés[**]
and S. Barro[***]

[*]Departamento de Informática.Universidad de Castilla la Mancha.
Campus Universitario. Avda. de España s/n. E-02006 Albacete. Spain.
email: miki@info-ab.uclm.es

[**]Dpto. Informática y Automática. Facultad de Ciencias.
UNED. Madrid, Spain.

[***]Dpto. de Electronica y Computación. Facultad de Física.
Universidad de Santiago de Compostela. Spain.

Abstract

Usually we have assumed that the neuron was the "atom" in the architecture of the Neurons System. However, there is such a wealth of drendo-dendritic connections and synaptic mechanisms that it seems essential to distinguish different styles of analog microcomputation.

In this paper we look inside the synaptic structure after a local process of accumulation of persistent activity and their discharge towards the spike trigger zone. To illustrate the usefulness of this information processing behaviour in image motion analysis, and architecture for extraction and selection of length velocity ratio invariants (LVR) is proposed, simulated and partially evaluated.

Introduction

Until we know more about computational neuroscience, a key to the field of modular and self-programming computation by means of nets of artificial neurons lies in the use of fragmental knowledge on the anatomy and physiology of biological neurons as a source of inspiration. This biologically plausible inspiration can also be used to propose local processes looking after the partial reproduction of the logical organization, wealth and emergent complexity of the behaviour of biological neurons.

In this paper we look inside a neuron after two specific aspects of subcellular microcumputation:

1. The existence of two time scales (dendro-dendritic *local time*, $t_n = n \cdot \Delta t$; and axonal *global time*, $T_k = k \cdot \Delta T$, with $\Delta T >> \Delta t$).

2. The existence of a style of analogic computation with *local autonomy* around each synapse, that acts as a non-linear local *permanence memory*.

To illustrate the possible usefulness of these computational processes at synaptic level, a neural network for target identification based on invariant length-velocity ratio (LVR) traces has been simulated. Detection of this LVR invariant is realized by overlapping receptive fields with a micro-structure of synaptic memories working in local time (t_n). Integration at global neuronal level (T_k) of these synaptic memories is made to detect weather a sufficient number of local discharges occur in the receptive field to produce axonal spike activity.

The organization of this paper is as follow. We will first speak of permanence memories at synaptic level. The next section is devoted to introducing the application of these memories for motion analysis.

Then, the network architecture with focus on the layer for the detection and tuning of LVR invariants, is presented. Finally some comment on the experimental results are included.

Physiological Bases of Synaptic Computation

Adult neurons are cells specialized in the integration and distribution of global states of excitation in receptive fields which manifest themselves as action potentials. However, the diversity of contacts (axon-soma, dendrite-dendrite, dendrite-soma,...), along with the specificity and complexity of local computation carried out by those contacts, make one think of authentic subcellular microcomputation [Calvin & Graubard, 1979; Graubard & Calvin, 1979; Koch, 1990; Rall & Segev, 1990; Mira et al., 1993]. That is to say, before arriving at the level of the neuron and considering only the local processes existing within its dendritic field, there is already sufficient experimental evidence to formulate a model of synaptic computation with the following properties.

1. Local convergent processes around each synapses.

2. Semiautonomous functioning, with each synapse capable of spatio-temporal accumulation of local inputs (time scale, t_n) and conditional discharge.

3. Attenuated transmission of these accumulations of persistent coincidences in the different synapses of the dendritic field towards the spike trigger zone of the neuron that integrates at the global time scale (T_k) the results of these charge-discharge local memories.

The generic model of this synaptic structure which carry out the local accumulation of persistent

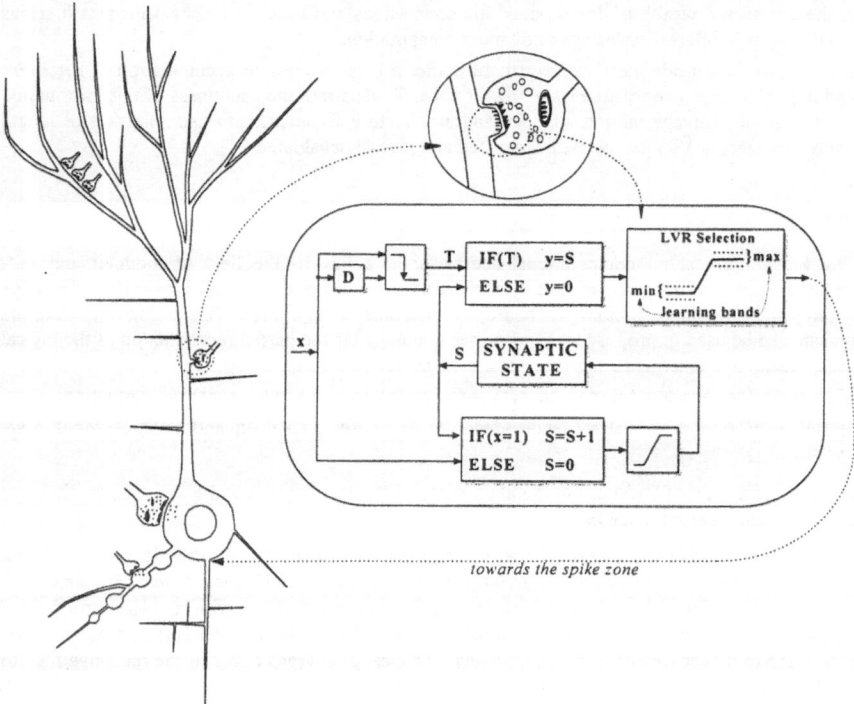

Figure 1: Permanence local memory in synaptic computation. Persistent activity is accumulated, filtered and passively transmitted towards the spike zone.

activity and their subsequent discharge towards the spike trigger zone, and then extinction, is shown in Figure 1. The digital version of a synaptic memory (time dependent conductance in series with a voltage source and in parallel with a membrane capacity and a membrane resistance) is represented by the algorithm included in the box at Figure 1.

Connection with the Motion Analysis

Let us consider now the permanence effect created by an image on vision sensors. This effect can be easily observed when an intense light focus –the torch used by a circus fakir, for example– blinds the vidicon of a video camera for seconds. As the fakir moves the torch, the luminous trail created by the torch motion can be observed on the television set. Thanks to the saturation produced in those pixels of the vidicon which are receiving the blinding light of the torch, we can see the trajectory followed.

The charge-discharge synaptic behaviour previously described is similar to that of the video camera vidicon sensor: it is charged in the presence of afferent activity from the previous layer and discharged in the absence of this element. The synaptic field with such behaviour is composed of as many synapses as sampling pixels are defined in the image. The input value for each synapses is assumed to be one in case the associated image pixel has been occupied by an element of interest, and it is assumed as zero in case this condition is not satisfied. The selection of the processes prior to the synaptic field are not fundamental for the model of accumulative and local synaptic computation proposed in this work. Let us consider, for example, that some segmentation has been accomplished by the previous layer [Fernández & Mira, 1994].

Thus basically each synapses increases its charge value while its input remains active indicating that the associated pixel is occupied by an element of interest, and its charge is decreased when the inactive input indicates that there is no element of interest occupying the associated image pixel. This synaptic behaviour carries out an information when a moving element has passed along as illustrated in Figure 2.

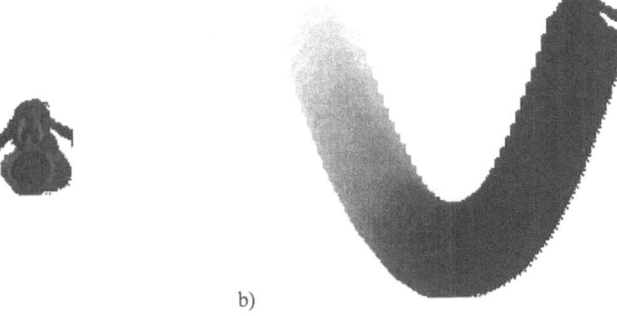

a)   b)

Figure 2: Illustration of the computational process accomplished by a permanence memory. a) Mobil element.
b) Luminous trail created in the receptive field.

We can denominate synaptic local discharges the values that, coming from the synapses are transmitted towards the spike trigger zone of the neuron, computing another function more broadly representative of all the local inputs. [Graubard & Calvin, 1979].

Let us accept in the motion analysis application the assumption that gray level variations in the scene are due to the motion of the composing elements. Then, the analysis of the charge values obtained will provide information both about the period of time an image element has occupied a sampling pixel and about the period of time elapsed after its output. The analysis of these parameters provide enough resources to obtain additional information about the motion of the elements of interest appearing in the scene.

An element that remains constant in a pixel in the sensor will not generate variations in the gray level of such pixel, and therefore the activation or disactivation value it generates will remain constant; this fact implies that the charge value of the associated synapses remains always at a minimum or it is charged to a maximum point and becomes saturated. Thus, to detect motion, the spike trigger zone of a neuron having as input the values of the synaptic field, will only need the capability to detect a sufficient number of value discharges ranging from the saturation value and the synapses minimum charge value. A sufficient number of intermediate discrete discharges will indicate the presence of motion in the elements of interest in the scene. Thus, if we are correct in our interpretation of synaptic computation, motion detection will require very simple neural mechanisms.

As it has already been mentioned, the presence of spike discharges of intermediate frequencies indicates the existence of motion in the elements of interest. Next it is advisable to guess how the discharge value they pass to the spike trigger zone provides information about the characteristics of the motion which generates them. The synaptic state is directly related to the time the input has remained active and an actively constant input is related to the time the same gray level area of interest occupies a pixel in the sensor. A gray level area of interest moving at high velocity will occupy little time the same pixel in the sensor, and in turn, a gray level area of interest with motion will occupy longer a pixel in the sensor the grater its length in the direction of motion is. The value transmitted by a synapses provides information about the length-velocity ratio of a gray level area of interest which has passed over it receptive field.

The neural mechanism necessary for data processing generated by the synaptic field are very simple; it is just necessary to design dendritic fields and spike trigger thresholds capable to detect discharges in a specific interval to be able to distinguish the presence of a gray level area of interest with a specific value of length-velocity ratio (LVR).

The relation between length and velocity is invariant with the distance respect to the projection plane. We have the following relations for the apparent lengths (L_1, L_2) and velocities (v_1, v_2) due to the similar triangles as shown in the Figure 3:

$$\left. \begin{array}{l} L \cdot d_p = L_1 \cdot d_1 = L_2 \cdot d_2 \\ v \cdot d_p = v_1 \cdot d_1 = v_2 \cdot d_2 \end{array} \right\} \frac{L_1}{v_1} = \frac{L_2}{v_2}$$

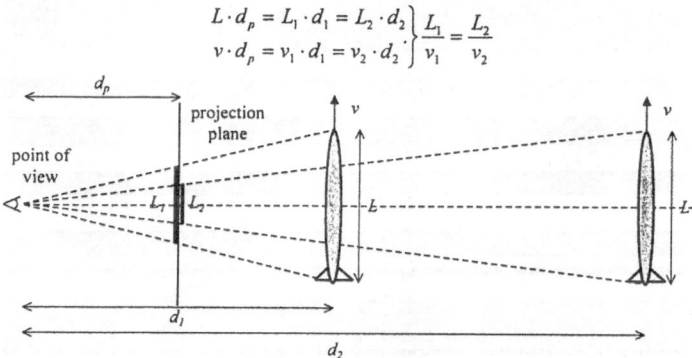

Figure 3: Justification of the usefulness to detect the motion of specific objects by means of the invariant descriptor LVR.

There are also other possibilities, slow or sudden charge and slow discharge, for the detection of motion direction or analysis of trajectory on the trails created by the moving elements [Fernández & Mira, 1992]. The attention of our work is focused on the first type of these mechanisms.

The information obtained by means of the neural mechanism we have mentioned can be introduced in subsequent neural layers in order to carry out higher level interpretations with the motion information extracted from the scene. Subsequent proper processing tasks might be the classification of these moving elements, or the interpretation of specific motion situations (scene understanding).

Neural Network Architecture

A three layers neural network architecture has been used to provide an example of how the use of synaptic computation allows very simple realizations for motion analysis in image sequences. The architecture here proposed has been designed and simulated and works with sequences synthetically generated both with synthetic and real elements. The input space is made up by the data offered by an image sensor of 512x512 pixels, each of which can have values ranging from 0 and 255. The process carried out with the signal has been divided into three stages: segmentation, features extraction and classification (Figure 4). Each stage is accomplished by a specific neural layer.

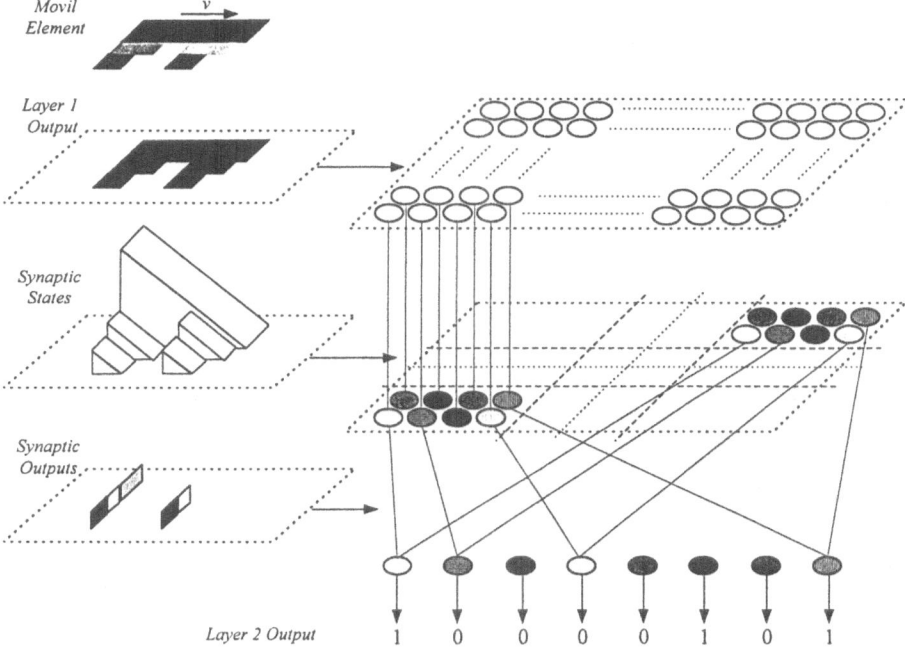

Figure 4: Neural network architecture with an example of layer two processes (accumulative activity in the synapses states and spatio-temporal integration at axonal level).

The first layer carries out segmentation tasks with the input signal converting, in each step of the process, the gray level of the sampling pixels into binary values. This layer then generates the input of the synapses in the second layer [Fernandez & Mira 1994]. In the second layer, denominated LVR detection layer, the detection and motion analysis tasks take place, and here is where the synaptic computation object of this work is located, as shown in Figure 4.

The third layer which forms the architecture proposed corresponds to the classification processes. The characteristics that the previous layers learnt to differentiate are associated in order to classify the elements or events which compose the scene. The network simulated classifies the LVR values appearing in the scene to detect the elements shown during the learning stage, whether they appear with noise or defective.

Let us now describe the layer object of this work which is in charge of the selection and extraction of LVR characteristics. This layer must detect the existence of a number n of LVR values intervals in the image. The learning process in this layer is directed to define which LVR combination will best classify the elements presented during the learning phase.

The connectivity structure of this layer is shown in Figure 4. The receptive fields of each neuron cover the entire output of the previous layer. Neurons are organized in functional groups which are associated to different ranges of LVR values selected in their synaptic field. The operation that must be accomplished locally is simply the summation of synaptic discharges within each specific discharge interval. The parameters to be adjusted (Figure 1) during the learning process are the values which define each specific discharge interval. These parameters are adjusted by using competitive mechanisms supported by the information coming from the third layer, regarding the capability of each LVR combination to differentiate the elements which appear in the learning stage. Each neuron detects a different LVR band. The learning process must decide which neural group classifies best the set of moving elements used to train the system.

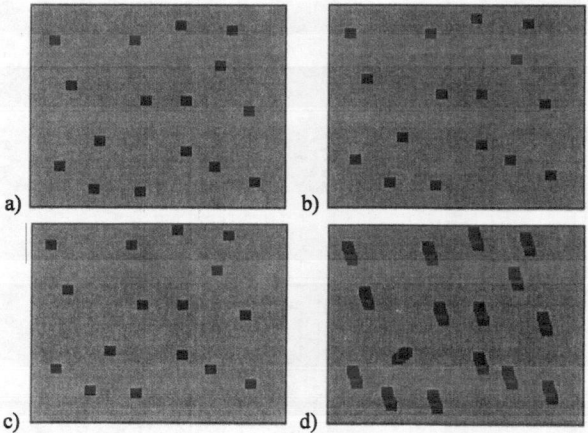

Figure 5: Elements with different directions are detected by identifying gradients in specific-directions synaptic fields. a), b), c) Photograms 1,3,5 are in sequence. d) The associated trail.

In summary each neuron is associated to a discharge band. Each neural body "listens" to the discharge values offered by their synaptic fields and is only activated when , from all the permanence values offered, the one corresponding to its discharge band sounds with enough intensity.

Simulation Results

The operation of the neural layer has been first carried out using synthetic image sequences with different combinations of LVR values. Then several learning sessions using image segments obtained from real images were used. The training process is aimed at the recognition of a set of targets and the detection of such targets even if they appear defective or noisy [Fernandez & Mira, 1994].

In Figure 4 we show the signal pathway from image segmentation to the output of the LVR neurons. To visualize the local computation at synaptic level we include the charge-discharge activity generated by the movement of the targets in the synaptic fields of the different neurons. Finally, the result of the spatio-temporal integration of these discharges in each LVR neuron are also included.

There are other several possibilities for motion analysis using charge-discharge synaptic fields. A specific design for LVR detection has been presented but there are other possible architectures for motion detection in different directions by designing neurons to work as gradient filters in the charge-discharge synaptic field. The Figure 5 a), b), c), d) shows how elements with different directions can be detected by identifying gradients in specific directions in the synaptic field. Another possibility is the detection of the trajectories since the dynamic problem of trajectory calculation is transformed into a static problem if the analysis is realized in a synaptic field of the type suggested in this paper.

Conclusions

From the computational neuroscience viewpoint, the purpose of this paper has been to contribute to the formulation of a model of synaptic computation with semiautonomous local memories and multiple processing paths within a neuron with local time (t_n), different of the usual as axonal level, where the spike trigger mechanisms integrates multiple, specific and fast computations in the line suggested by Calvin, Graubard, Rall, Koch and Poggio, among many others.

From the motion analysis viewpoint, we have proposed the size-velocity ratio traces as an invariant descriptor about the motion in a scene capable of being easily implemented by a field of microstructures of synaptic memories distributed over the extensive dendritic surface of a neuron.

Classical processes for motion analysis in image sequences are image differences, whether accumulated or not, and optical flow. In general, solutions within the field of neural networks consider motion computation based on calculations of optical flow, either by designing networks that work with flow values or by determining optical flow in a neural manner [Fenández & Mira, 1992] [Simpson, 1988] [Zhou & Chellapa, 1990]. These processes imply a rather high computational volume as well as complicated learning mechanisms. Complexity and computational volume of these solutions oblige the possible solutions of hardware implementation to be very specific and, at the same time, very flexible.

The use of synaptic fields with charge-discharge behaviour offers, as it has been shown, the possibility to design neural networks with motion analysis capability in image sequences, with a much lower degree of complexity, both for operational processes and for learning algorithms. This family of neural networks admits simple hardware solutions with capacity to work in real time.

Acknowledgments

We acknowledge the economical support of the Spanish CICYT under projects TIC 92/136 and TIC-94/95. Also the Spanish CAM project 0011/94 is acknowledged.

References

[Calvin & Graubard, 1979] Calvin, W.H. and Graubard, K. "Styles of Neural Computation". In Schmitt and Worden, eds., *The Neurosciences Fourth Study Program*. pp. 513-524. The MIT Press, 1979.

[Fernandez & Mira, 1992] Fernandez, M.A.; Mira, J.; "Permanence Memory: A System for Real Time Motion Analysis in Image Sequences". *IAPR Workshop On Machine Vision Applications*; pp 249 - 252; Mva'92. Tokio. 1992.

[Fernandez & Mira, 1994] Fernandez, M. A.; Mira, J.; "Arquitectura Neuronal Paralela para la Deteccion de Niveles de Gris con Movimiento en Secuencias de Imagen" Fernadez et al. (eds.) *Nuevas Tendencias En La Informatica: Arquitecturas Paralelas Y Programacion Declarativa*; pp 165 - 183;. Servicio de publicaciones de la Universidad de Castilla La Mancha. 1994.

[Graubard & Calvin, 1979] Graubard, K. and Calvin, W.H."Presynaptic Dendrites: Implications of Spikeless Synaptic Transmission and Dendritic Geometry". In Schmitt and Worden, eds., *The Neurosciences Fourth Study Program*. pp. 317-331. The MIT Press, 1979.

[Koch, 1990] Koch, C. "Biophysics of Computation: Towards the Mechanisms Underlying Information Processing in Single Neurons". In E.L. Schwartz, ed. Computational Neurosciences, pp. 97-113. The MIT Press, 1990.

[Mira et al., 1993] Towards More Realistic Self-contained Models of Neurons: High-order, Recurrence and Local Learning. In J. Mira et al (eds): *New Trend in Neural Computation*. LNCS, 686, pp. 55-62. Berlin, Springer-Verlag, 1993.

[Rall & Segev, 1990] Rall, W. and Segev, I. "Dendritic Branches, Spines, Synapses, and Excitable Spine Clusters". In E.L. Schwartz, ed. Computational Neurosciences, pp. 69-81. The MIT Press, 1990.

[Simpson, 1988] Simpson, W.; "Depth Discrimination from Optical Flow". *Perception*, Vol. 17; Pp 497 - 512; . 1988

[Zhou & Chellappa, 1990] Zhou, Y. T.; Chellappa, R.; "A Network for Motion Perception". *Proc. IEEE/Inns Int. Joint Conf. Neural Network*, Vol. II June; pp 875 - 884; IEEE. San Diego Ca.. 1990

High Order Boltzmann Machines with Continuous Units: Some Experimental Results

M. Graña, A. D'Anjou, F.X. Albizuri, A. de la Hera, I. Garcia
Dept. CCIA, UPV/EHU+
Apartado 649, 20080 San Sebastián
e-mail: ccpgrrom@si.ehu.es

Abstract: This work reports the results obtained with the application of High Order Boltzmann Machines without hidden units to classification problems with continuous features. The Boltzmann Machine weight updating algorithm is easily generalised when some of the units can take values inside a continuous interval. The absence of hidden units and the restriction to classification problems allows for the estimation of the connection statistics, without the computational cost involved in the application of simulated annealing. In this setting, the learning process can be speed up several orders of magnitude with no appreciable loss of quality of the results obtained.

0 Introduction

The Boltzmann Machine is a classical neural network architecture [1, 2] that has been relegated from practical application due to its computational cost and the difficulty to tune the several parameters involved, such as the annealing temperatures. We consider [1] the Boltzmann Machine as a maximiser of a consensus function. Our aim in this paper is to show that Boltzmann Machines can be successfully applied to problems with continuous input features and that their training can be much easier than was previously thought whenever two restrictions apply. The first is the avoidance of hidden units, using high order connections to model the high order correlations of the input. High order connections have been referred sometimes as product or sigma-pi units [7, 8, 11, 12, 14] . To our knowledge, our work [6] is the first to report results on training algorithms applied to high order networks. The second is the restriction of the domain of application to classification problems.

For the Boltzmann Machine without hidden units and binary units the Kullback-Leiber distance between the data distribution and that of the machine is convex and has a single global minimum [3]. This convexity is assumed to be a general feature of the learning algorithm regardless of the king of units considered. So, in our experimental works we always set the initial weights to zero. This implies that it is not needed to perform several replications of the learning to estimate the average behaviour of the machine for a given problem. Besides that, because of the absence of hidden units, simulated annealing is not required for the estimation of the equilibrium distribution of their states.

In the case of classification problems, input and output units can be distinguished. Output units are binary and the output vectors are always orthogonal, because only one class assignment is associated with each output unit. Due to the absence of hidden units, the clamped phase can be performed once for all the learning process, and without any stochastic relaxation needed. The free phase of the learning algorithm involves clamping the input units, and a linear-complexity search for the output unit with maximum gain for each input pattern. As the expected response of the network is the class assignment of the input features, the limit behaviour of simulated annealing in the free phase is to obtain a configuration with only the maximum gain output unit set to one. This task can be performed with a simple search. So, we can also avoid the use of simulated annealing in the free phase. Taken together, all these simplifications drastically decrease the computation time for the learning process, and allow to apply Boltzmann Machines to non trivial problems. Despite the simplifications, the quality of the results is comparable to other neural architectures, and the number of learning cycles needed is much less than in the classical approach.

+ Work supported by the Dept. Educación, Univ. e Inv. of the Gobierno Vasco, proyecto PGV9220

The software used for the experiments reported in this paper has been written in ADA, and can be accessed via anonymous FTP at the node ftp.sc.ehu.es in the directory pub/unix/hobm.

In this paper we have considered machines that include continuous units that can take states in arbitrary real intervals. The codification of the problem features into the machine units becomes straightforward. The learning algorithm preserves the main features of the basic binary case. It suffices to consider the mean activation level of the connections, instead of the activation probability. A point of interest is that a necessary condition for the convexity of the Kullback-Leiber measure is that the units have positive (zero included) state spaces. However, the learning algorithm with the simplifications enumerated above seem to work well even in the case of unit state spaces that include negative values, as the results on the vowel recognition problem show. The use of continuous units allows for big reductions on the number of units used to codify the learning problem, and, therefore, of the network complexity. The change in modelling power due to the change in codification (from binary to continuous) is an open problem.

Section 1 introduces the test problems. Section 2 reviews our notation for High Order Boltzmann Machines, and the learning algorithms applied in this paper. Section 3 gathers the definitions of the machines applied to each problem, and the results obtained trying several high order topologies. Section 4 gives some conclusions and directions for further research.

1 The test problems

The learning problems, that we have used in this paper to test the learning power of High Order Boltzmann Machines, have the following common characteristics: 1- The data are in the public domain, and can be accessed by anonymous FTP. 2- Other techniques have been applied to the data, and the results are public. These results play the role of objective references to assert the quality of our own results. 3- The experimental method is clearly defined by the existence of separate train and test data sets. 4- They are classification problems. The output of the network is a binary vector. The class assignment is a vector of zeros with a single component with value 1.

Classification of Sonar Targets

We have used the data used by Gorman and Sejnowski [5] in their study of sonar signal recognition using networks trained with backpropagation. The data has been obtained from the public database of the CMU (node ftp.cs.cmu.edu, directory /afs/cs/project/connect/bench). The goal of the experiment is the discrimination between the sonar signals reflected by rock and metal cylinders. Both the train and test data consist of 104 patterns. The partition between train and test data has been done taking care that the same distribution of incidence angles appears in both sets. Each pattern has 60 input features and a binary output. Input features are real numbers falling in the [0,1] interval.

$$x = \left(x_1,..,x_{60},x_o\right) \in [0,1]^{60} \times \{\text{metal,rock}\}$$

In [5] a set of experiments was performed with a varying number of hidden units, to explore the power of the learning algorithm depending of the topology. Results were averaged over 10 replications of the learning process with varying random initial weights. The best result reported was an average 90.4 per cent of success on the test data, with a standard deviation of 1.8, for a topology with 12 hidden units.

A vowel recognition problem

The data for this problem has also been obtained from the CMU public database. They have been used [4, 9, 10] to test several neural architectures . The best results reported were obtained with a Euclidean nearest neighbour classifier. It attains a 56% success on the test data. Each pattern is composed of 10 input features. Input features are real numbers. The class (vowel) assignment is given by a discrete variable.

$$x = (x_1,..,x_{10},x_0) \in R^{10} \times \text{Vowels}$$

$$\text{Vowels} = \{\text{hid}, \text{hId}, \text{hEd}, \text{dAd}, \text{hYd}, \text{had}, \text{hOd}, \text{hod}, \text{hUd}, \text{hud}, \text{hed}\}$$

The detail of the data gathering and pre-processing can be found in [9, 10]. The train data contains 528 patterns, and the test data contains 462 patterns. The specific characteristics that make this problem worth of study, from our point of view, are three. First, it is a multicategorical classification problem, whereas the Sonar problem involves only two categories. Second, the input features are not normalised in the [0,1] interval. Roughly, they take values in the [-5,5] interval. Finally, the existence of negative estates does not allow to assume the convexity of the Kullbak-Leiber distance. We wish to test the robustness of the approach taken (specially the initialisation of the weights to zero) in this clearly unfavourable case.

2 High Order Boltzmann Machines with continuous units

A High Order Boltzmann Machine is described by a triplet (U,R, L,W), where U is the set of units, $R = \{R_i \subset R \,|\, i = 1..|U|\}$ is the family of the state spaces of the units, L the set of connections between the units (the network topology) and W are weights associated with the connections. A connection $\lambda \in L$ is a subset of U. The order of a connection is its cardinality. The order of the Boltzmann Machine is that of the connection with maximum order. Conventional Boltzmann Machines are of order 2. The weights W can be formulated as a mapping that associates each connection with a real number $W:L \to \mathbb{R}$. The configuration space is the product of the state spaces of the units, so that $k \in R^{|U|}$. The consensus function has the form $C(k) = \sum_\lambda \omega_\lambda \prod_{u_i \in \lambda} k(u_i)$ where $k(u_i)$ is the state of unit u_i when the global configuration is k.

As usual in Boltzmann Machines, learning is defined as the minimisation of the Kullback-Leiber pseudo-distance $D(q'(c)/q(c)) = \int_k q'(c,k) \left(\ln \frac{q'(c,k)}{q'(c,k)} \right) dk$ between the distribution $q'(c)$ of the machine configurations when the visible units (input/output units in the case of classification problems) are set to the values of the patterns that specify the problem to be learnt (clamped phase), and $q(c)$ the distribution of the machine configurations when the units are free to evolve (free phase). In practice, we consider in the free phase that only the output units are free, clamping all the input to the inputs of the training patterns. The parameter c is the temperature. It can be taken as 1 without loss of generality. The minimisation of the Kullback-Leiber pseudo-distance is performed applying gradient descent on the weights. This gradient is of the form:

$$\frac{\partial D\left(q'/q\right)}{\partial \omega_\lambda} = -\frac{1}{c}\left(a'_\lambda - a_\lambda\right)$$

Where $a_\lambda = \int_k q_k \prod_{u \in \lambda} k(u) dk$ is the mean activation level of the connection λ under the free distribution of the machine configurations. The mean activation level of the connection under the clamped distribution is defined analogously. In this paper we have used a weight updating rule with a moment term.

$$\Delta_t\omega_\lambda = \alpha\left(\hat{a}'_\lambda - \hat{a}_\lambda\right) + \mu\Delta_{t-1}\omega_\lambda$$

The values of the learning parameters are $\alpha=1$ and $\mu=0.9$. The estimation of mean activation levels is performed without recourse to the simulated annealing for the reasons discussed in the introduction. In the clamped phase, each of the training patterns was set at the input/output units. The activation state of each connection was recorded, to compute the mean activation level after the presentation of all the training patterns. The free phase is a series of learning cycles. In each cycle, the input components of each pattern were set on the input units. The response of the network was computed searching for the maximum gain output unit, which is set to 1, while the other output units are set to 0. (Remember we deal only with orthogonal binary output vectors). The activation state of each connection is recorded, and the mean activation state after the presentation of all the training patterns is taken as the estimate of the mean activation level in the free phase. This weight updating schedule is often called batch or off-line adaptation. The weights are updated according to the rule employed and a new learning cycle starts. The initial weights were always set to zero.

3 Experimental results

In this section we give a detailed account of the application of High Order Boltzmann Machines to the problems introduced in section 2. In each case we have tested several high order topologies (without hidden units) to explore the behaviour of the learning algorithm.

High Order Boltzmann Machines with continuous [0,1] units for the Sonar problem

The set of units employed to model the patterns is $U=\{u_i \; i=1..60, u_0\}$. The unit state spaces are: $R_1=..=R_{60}=[0..1]$. and $R_0=\{0..1\}$. The mapping of the patterns into the unit states is

$$k(u_i)=x_i. \; i=1..60$$
$$k(u_0)=1 \text{ if } x_0=metal$$

In the experiments two kinds of general topologies were used: The *densely connected topologies of order* r, with all the connections of order r or less that include the output unit.

$$L^r = \left\{\lambda \subset U \middle| (|\lambda| \le r) \wedge (u_0 \in \lambda)\right\}$$

and the *"in line" topologies*, which are densely connected topologies with the additional restriction that the input units in the connection are consecutive:

$$L^r_\ell = \left\{\lambda \subset L^r \middle| (r > 2 \Rightarrow (u_i \in \lambda \Rightarrow (u_{i-1} \in \lambda \vee u_{i+1} \in \lambda)))\right\}$$

Table 1 shows the results obtained for this problem with the moment rule for weight updating. The results are comparable to those obtained by Gorman and Sejnowski. The "in line" topologies appear to be well fitted to this problem, probably due to the sequential nature of the data. The correlation between inputs near in time and the situation of the peaks of the signal seem to be the relevant characteristics for the classification of the signals, and they are well captured by the "in line" topologies. Note that the "in line" topologies are much simpler than the densely connected topologies. The best generalisation for this problem (89%) is obtained with the densely connected topology of order 4. "In line" topologies of order 4 and 5 give also good results. In any case, the overfitting effect increases as the order of the topology grows.

Topology	cycles	%train	%test
L^2	500	94	76
L^3	77	95	89
L^4	48	99	89
L_ℓ^3	205	97	83
L_ℓ^4	97	100	88
L_ℓ^5	97	100	88
L_ℓ^6	180	98	85
L_ℓ^7	103	100	87
L_ℓ^8	110	97	86
L_ℓ^9	101	89	85
L_ℓ^{10}	136	99	88
L_ℓ^{11}	95	91	82

Table 1. Results on the sonar signal recognition using the moment rule.

It can be concluded that the learning algorithm works well with input continuous units. The unit state spaces are normalised to the interval [0,1], so the interest of the next experiment is to verify that our approach works well even in the case of non-normalised input features, and that the algorithm converges even when the units can take negative values.

High Order Boltzmann Machines with continuous units for a vowel recognition problem

The set of units for this problem is $U=\{u_i \ i=1..10, \ u_{oj} \ j=1..11\}$, the input units u_i have state spaces R_i included in the interval [-5,5], the output units u_{oj} have range $R_{oj}=\{0,1\}$. The input units take the value of the input component $k(u_i)=x_i$. The mapping of values to the output units makes $k(u_{oj})=1$ if x_0 takes its j-th value The topologies used for the experiments are the *densely connected topologies of order* r (connections of order r or less and only one output unit in each connection):

$$L^r = \left\{ \lambda \subset U \middle| (|\lambda| \le r) \wedge \left(u_{oj} \in \lambda\right) \wedge \left(\forall k \ne j\left(u_{ok} \notin \lambda\right)\right) \right\}$$

And the "in line" topologies (with the additional restriction of the input units being consecutive):

$$L_\ell^r = \left\{ \lambda \in L^r \middle| \left(r > 2 \Rightarrow \left(u_i \in \lambda \Rightarrow \left(u_{i-1} \in \lambda \vee u_{i+1} \in \lambda\right)\right)\right) \right\}$$

Table 2 shows the results of the experiments using the moment rule. Again, the results are comparable to those of reference, given by Robinson. Surprisingly good results were obtained with the "in line" topologies of order 3 and 4, that improve on the reference results.

Topology	cycles	%train	%test
L^2	100	62	43
L^3	80	98	54
L^4	40	96	46
L^5	60	100	45
L^6	50	100	45
L_ℓ^3	120	87	58
L_ℓ^4	120	87	57
L_ℓ^5	120	84	51
L_ℓ^6	250	97	55

Table 2. Results on the vowel recognition problem. Moment rule

Our conclusion from the results in table 2 is that the learning algorithm proposed for High Order Boltzmann Machines with continuous input units is robust, in the sense that it performs well even in the case that the convexity of the Kullback-Leiber distance is far from clear. We remind the reader that we always start from zero weights, assuming the convexity of this distance. Bad generalisation (poor results on the test set) seems to be inherent to the data, as the reference works give also poor results. Therefore, we do not interpret this experiment as a benchmark of the applicability of High Order Boltzmann to voice recognition. In the case of the densely connected topologies, the overfitting effect can be more clearly appreciated as the order of the topologies grows. Also a reduction of the number of learning cycles needed to fit the data can be appreciated.

4 Conclusions and further work

High Order Boltzmann Machines without hidden units applied to classification problems allow for simplifications of the learning process that speed up it several orders of magnitude, making of practical interest this kind of neural networks. The results obtained were comparable to those found in the reference works, obtained with other techniques or other neural network architectures. We have also found that small number of learning cycles are needed, once a correct guess on the topology has been done. In this paper we have not tried any pruning, weight decay or topological design scheme. Early attempts to apply them have been reported in [6] with poor results. However, we believe that the absence of hidden units in our topologies can lead to interesting and efficient algorithms. Further work is being done in this direction.

High Order Boltzmann Machines without hidden units allow the use of continuous units. The learning algorithm remains essentially the same, regardless of the kind of units used. The main benefits of the use of continuous units are the reduction of the network complexity (and further speedup of the learning and application processes) and the straightforward codification of the problem. The cost of this reduction is the loss of modelling power. Therefore, the formal characterisation of the probability distributions that can be modelled, the study of convexity conditions for the Kullback-Leiber distance and the convergence of the learning process for this generalisation of the binary Boltzmann machines are theoretical works needed to give more light into their practical application.

References

[1] Aarts E.H.L., J.H.M. Korst "Simulated Annealing and Boltzmann Machines: a stochastic approach to combinatorial optimization and neural computing" John Wiley & Sons (1989)
[2] Ackley D.H., G.E. Hinton, T.J. Sejnowski "A learning algorithm for Boltzmann Machines" Cogn. Sci. 9 (1985) pp.147-169
[3] Albizuri F.X. , A. D´Anjou, M. Graña, F.J. Torrealdea, M.C. Hernandez "The High Order Boltzmann Machine: learned distribution and topology" IEEE Trans. Neural Networks en prensa
[8] Deterding D. H. (1989) "Speaker Normalisation for Automatic Speech Recognition", PhD Thesis, University of Cambridge,
[10]. Gorman, R. P., and Sejnowski, T. J. (1988). "Analysis of Hidden Units in a Layered Network Trained to Classify Sonar Targets" in Neural Networks, Vol. 1, pp. 75-89.

[11] Graña M. , V. Lavin, A. D'Anjou, F.X. Albizuri, J.A. Lozano "High-order Boltzmann Machines applied to the Monk's problems" ESSAN'94, DFacto press, Brussels, Belgium, pp117-122

[16] Perantonis S.J. , P.J.G. Lisboa "Translation, rotation and scale invariant pattern recognition by high-order neural networks and moment classifiers" IEEE Trans. Neural Net. 3(2) pp.241-251

[17] Pinkas G. "Energy Minimization and Satisfiability of Propositional Logic" en Touretzky, Elman, Sejnowski, Hinton (eds) 1990. Connectionist Models Summer School

[18] Robinson A. J. (1989) "Dynamic Error Propagation Networks".PhD Thesis Cambridge University Engineering Department,

[19] Robinson A. J. , F. Fallside (1988), "A Dynamic Connectionist Model for Phoneme Recognition" Proceedings of nEuro'88, Paris, June,

[20] Sejnowski TJ. "Higher order Boltzmann Machines" in Denker (ed) Neural Networks for computing AIP conf. Proc. 151, Snowbird UT (1986) pp.398-403

[24] R. Durbin, D.E. Rumelhart "Product Units: A computationally powerful and biologically plausible extension to backpropagation networks" Neural Computation 1pp133-142

[25] M. Graña, A. D'Anjou, F.X. Albizuri, J.A. Lozano, P. Larrañaga, Y. Yurramendi, M. Hernandez, J.L. Jimenez, F.J. Torrealdea, M. Poza "Experimentos de aprendizaje con Máquinas de Boltzmann de alto orden " submitted to Informática y Automática

[27] Lenze B. "How to make sigma-pi neural networks perform perfectly on regular training sets" Neural Networks 7(8) pp.1285-1293

An Adaptive Control Model of a Locomotion by the Central Pattern Generator

Jun Nishii

Department of Mathematical Engineering and Information Physics
Faculty of Engineering
University of Tokyo
7-3-1 Hongo, Bunkyo-ku, Tokyo, 113, JAPAN
E-mail: jun@szklab.t.u-tokyo.ac.jp, Tel: +81-3-3812-2111(ex.6882), Fax: +81-3-5802-2913

Abstract

Most of basic locomotor patterns of living bodies are controlled by central pattern generators (CPGs) which are collective neural oscillators. The CPG sends control signals to muscular systems, and the activity of the CPG is strongly affected by sensory signals from the body. Therefore, it can be said that locomotor patterns are generated by the interaction between the CPGs and the body movements. To control a physical system, such as a leg, by the CPG, it would be necessary to obtain an adequate intrinsic frequency of the CPG and interactions between the CPG and the physical system. In this article, we regard a physical system as a physical oscillator and propose a learning algorithm to acquire these parameters and apply the learning method to the control of a hopping robot. The proposed learning method does not need any information about the dynamics of the controlled physical system and requires only local informations like the Hebbian rule.

1 Introduction

The basic locomotor patterns of most living bodies, such as swimming, walking, and flying, are controlled by central pattern generators (CPGs) which generates rhythmic activities. The CPG is a central nervous system which locates in spinal cords of vertebrates and segmental ganglia of invertebrates. The periodic activities of the CPG which are initiated by a burst from a higher motor center induce muscle activities [Grillner 1985, Grillner et al. 1991]. After the initiation of the locomotion, the activity of the CPG is affected by sensory signals which show the bending of the body and so on [Williams et al. 1990]. In mathematical models, the CPG is considered as collective nonlinear oscillators [Cohen et al. 1982, Rand et al. 1988, Williams et al. 1990], because the CPG consists of collective neural oscillators. The physical systems to produce locomotion, such as leg, can also be considered as physical oscillators. Therefore, it can be said that the locomotor patterns are generated through the mutual interactions between the oscillators, that is, the CPG and the musculomotor system. In order to generate a desired locomotor pattern, it would be necessary that (1) the intrinsic frequency of the CPG and (2) the interactions between the CPG and the musculomotor system must be determined appropriately. These parameters might be achieved genetically but must be tuned according to the body growth and some injury.

In this paper, we propose a learning model for locomotor control. In this model, the appropriate parameters are acquired so as to minimize an evaluation function which shows the performance of the locomotion.

2 A Learning Model of a Locomotion by the CPG

In this section, we consider a system which is composed of an oscillator as the CPG and a physical system with 1-degree freedom which shows a periodic movement. By regarding the physical system as a physical oscillator and defining phases of the oscillator $\theta \in \mathbf{S}^*$ and the physical system $\tilde{\theta} \in \mathbf{S}$, the dynamics of the CPG and the physical system is assumed to take the form,

$$
\begin{cases}
\dot{\theta} &= \omega + \sum_{l=1}^{L} w^l R(\phi - \psi^l) \\
\dot{\tilde{\theta}} &= \Omega + \epsilon_f F(-\phi),
\end{cases}
\tag{1}
$$

where $\phi \equiv \tilde{\theta} - \theta$, ω and Ω are the intrinsic frequencies of the CPG and the physical system, respectively, R shows the effect of feedback signals from the physical system to the CPG, F shows the effect of the control signal from the CPG to the physical system, and $\epsilon_f \ll 1$ is a constant showing the strength of the control signal. Various phase delayed couplings are assumed as the feedback interaction to the CPG and ψ^l ($l = 1, \ldots, L$) represents the phase delay, and w^l ($\sum_l^L w^l \ll 1$) is the coupling strength of each coupling. These phase delayed effects are caused when different cells in a neural oscillator receives a feedback signal and various feedback signals, such as displacement, velocity and acceleration, are obtained. The purpose of the learning is to obtain an adequate phase relation between the CPG and the physical system to realize a desired locomotion by turning parameters ω and w^l. The parameters are changed according to the following learning rule.

$$
\begin{cases}
\dot{\omega} &= \varepsilon \sum_{l=1}^{L} w^l R(\phi - \psi^l) \\
\dot{w}^i &= \varepsilon \gamma E(\phi) \cdot R(\phi - \psi^i),
\end{cases}
\tag{2}
$$

where $E(\phi)$ is the evaluation function which shows the performance of the physical system, and $\varepsilon \ll 1$ and $\gamma \ll 1$ are constants determining the learning rate of the frequency and the weight. These constant values are determined so as that the parameters changes much more slowly than the phase difference ϕ does and the coupling weight w^l changes much more slowly than the frequency ω does.

The above learning rule implies that the intrinsic frequency ω is modulated according to the effect of the input signals to the CPG and the coupling weight w^l is modulated according to the correlation between the evaluation function and the effect of the feedback signal. Concerning this learning rule, the following theorem is obtained.

Theorem:

When θ and $\tilde{\theta}$ have a stable phase locked solution on the dynamics (1) during the learning, $E(\phi) = 0$ is asymptotic stable, if

1. There exist more than two phase delayed sensory feedback signals to the CPG, which give the different zero points of $R(\phi - \psi^l)$. i.e., $\forall \phi \in \mathbf{S}$, $\sum_{l=1}^{L} (R(\phi - \psi^l))^2 \neq 0$.

2. $\forall \phi \neq \phi_E^0$, $E(\phi) \sin 2\pi g(\phi) > 0$ holds, where ϕ_E^0 satisfies $E(\phi_E^0) = 0$ and $g : \mathbf{S} \to \mathbf{S}$, $g \in C^1$, $g' > 0$.

3. For all $w \equiv (w^1, \ldots, w^L) \neq (0, \ldots, 0)$, there exists a function $h_w : \mathbf{S} \to \mathbf{S}$, $h_w \in C^1$, $h'_w > 0$ which satisfies $\forall \phi \neq \phi_w^0$, $\sin 2\pi h_w(\phi) \cdot \sum_{l=1}^{L} w^l R(\phi - \psi^l) > 0$, where ϕ_w^0 satisfies $\sum_{l=1}^{L} w^l R(\phi_w^0 - \psi^l) = 0$.

$\square$

Proof: The dynamics of the phase difference takes the following form from eq. (1),

$$
\dot{\phi} = \Omega - \omega + \epsilon_f F(-\phi) - \sum_{l=1}^{L} w^l R(\phi - \psi^l),
\tag{3}
$$

---

*Here, we define $\mathbf{S} = \mathbf{R}$, (mod 1).

where the assumption that θ and $\tilde{\theta}$ have a phase locked solution implies that the above dynamics has a stable point. Because the changes of parameters are very slow, $\dot{\phi} \simeq 0$ holds. At the stable point ϕ,

$$f + a > 0 \tag{4}$$

holds, where

$$f = \epsilon_f F'(-\phi), \quad a = \sum_{l=1}^{L} w^l R'(\phi - \psi^l). \tag{5}$$

From eq.(3), the following relation is obtained.

$$\begin{cases} 0 = -1 - (f + a)\dfrac{\partial \phi}{\partial \omega}, \\ 0 = -(f + a)\dfrac{\partial \phi}{\partial w^l} - R(\phi - \psi^l). \end{cases}$$

Therefore, we obtain

$$\begin{cases} \dfrac{\partial \phi}{\partial \omega} = -\dfrac{1}{f + a}, \\ \dfrac{\partial \phi}{\partial w^l} = -\dfrac{R(\phi - \psi^l)}{f + a}. \end{cases} \tag{6}$$

Consider the following Liapunov functions

$$V_1 = \frac{1}{\pi} \sin^2 \pi h_w(\phi), \quad V_2 = \frac{1}{\pi} \sin^2 \pi g(\phi).$$

First, we will show $\dot{V}_1 < 0$.

$$\dot{V}_1 = h' \sin 2\pi h_w(\phi) \cdot \frac{\partial \phi}{\partial \omega} \dot{\omega}$$

$$= -h' \epsilon \sin 2\pi h_w(\phi) \cdot \frac{\sum_l w^l R(\phi - \psi^l)}{f + a} < 0$$

holds around $\phi = \phi_w^0$, where we used eq. (4), (6), the condition 3, and the assumption that the change of the frequency ω is much faster than that of the weights w^l. Therefore,

$$h_w(\phi) = 0, \quad \sum_l w^l R(\phi - \psi^l) = 0, \quad \dot{\omega} = 0 \tag{7}$$

is obtained after sufficient time.

Second, we will show $\dot{V}_2 < 0$. In a time scale of w^l, eq. (7) holds.

$$\dot{V}_2 = g' \sin 2\pi g(\phi) \cdot \left\{ \frac{\partial \phi}{\partial \omega} \dot{\omega} + \sum_l \frac{\partial \phi}{\partial w^l} \dot{w}^l \right\}$$

$$= -\epsilon g' \sin 2\pi g(\phi) \cdot E(\phi) \sum_l \frac{(R(\phi - \psi^l))^2}{f + a} \leq 0$$

is obtained around $\phi = \phi_E^0$, where we used eq. (4), (6), (7), the condition 1 and 2. Hence,

$$E(\phi) = 0$$

is asymptotic stable. □

When the functions R, F in eq.(1) are not functions of the phase difference ϕ but depends on θ_1 and θ_2, the dynamics of the CPG and the physical system takes the form.

$$\begin{cases} \dot{\theta} = \omega + \sum_{l=1}^{L} w^l R(\theta, \tilde{\theta} - \psi^l) \\ \dot{\tilde{\theta}} = \Omega + \epsilon_f F(\tilde{\theta}, \theta), \end{cases} \tag{8}$$

When the above dynamics has a phase locked solution, the averaging theory can be applied to obtain the following approximated form [Ermentrout and Kopell 1991]

$$\begin{cases} \dot{\theta} = \omega + \sum_{l=1}^{L} w^l \bar{R}(\phi - \psi^l) \\ \dot{\tilde{\theta}} = \Omega + \epsilon_f \bar{F}(-\phi), \end{cases} \tag{9}$$

$$\bar{R}(\varphi) = \int_0^1 R(\theta, \theta + \varphi) d\theta, \quad \bar{F}(\varphi) = \int_0^1 F(\theta, \theta + \varphi) d\theta. \tag{10}$$

Therefore the learning rule can be given as the form

$$\begin{cases} \dot{\omega} = \epsilon \sum_{l=1}^{L} w^l \bar{R}(\phi - \psi^l) \\ \dot{w}^i = \epsilon \gamma E(\phi) \cdot \bar{R}(\phi - \psi^i). \end{cases} \tag{11}$$

3 Simulation Results

3.1 Control of the Phase Difference Between Oscillators

To ascertain the performance of the proposed learning rule, simulations were done by assuming a physical system as a simple oscillator. The dynamics of the CPG and the physical system follows dynamics (1). The functions take the following form,

$$F(\phi) = \sin 2\pi\phi, \quad R(\phi) = \sin 2\pi\phi, \quad E(\phi) = \sin 2\pi(\phi - \phi_d), \quad \phi_d = 0.8. \tag{12}$$

Therefore, the goal of this simulation is to obtain the phase difference $\phi = 0.8$. The learning algorithm is given by eq.(2). Fig.1 and Fig.2 show the simulation results. The desired phase relation was obtained through the learning in a few cycles although two oscillators do not synchronize in the initial state. The same results were obtained for other desired phase differences and other initial conditions.

3.2 Control of a Hopping Robot

We applied the learning rule to the control of an one-dimensional hopping robot (Fig.3). The robot is composed of a trunk with a mass and a leg which has a spring component, a damping component and a thruster. The thruster generates a force between the trunk and the toe according to the phase of the CPG. The CPG receives the normalized velocity $\hat{v}$ of the trunk of the robot as a sensory feedback signal and its dynamics takes the form

$$\dot{\theta} = \omega + \sum_{l=1}^{L} w^l P(\theta - \psi^l) \hat{v}, \tag{13}$$

where $P(\theta)$ is a function showing the effect of the input signal on the dynamics of the CPG, and $\hat{v}$ is given by

$$\hat{v} = \dot{x}_0/\bar{v}_0, \quad \tau \dot{\bar{v}}_0 = -\bar{v}_0 + |\dot{x}_0|. \tag{14}$$

where x_0 is the position of the trunk of the robot and $\tau = 5.0$ [s] is a time constant. As the function $P(\phi)$, $P(\phi) = \sin \phi$ is given, which is obtained when the dynamics of the CPG can be transformed to the Hopf normal form [Nishii et al. 1995]. The dynamics of the robot is shown in the appendix.

The purpose of the learning is to make the time averaged position of the trunk of the robot $\bar{x}_0$ approach to a desired height x_d, i.e., the evaluation function is given by

$$E = x_d - \bar{x}_0. \tag{15}$$

where

$$\frac{\tau}{\omega} \dot{\bar{x}}_0 = -\bar{x}_0 + x_0. \tag{16}$$

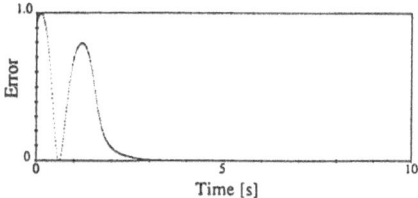

Fig.1. The time profile of the error of the phase difference, i.e., $\sin^2 \pi(\phi - \phi_d)$. Parameters are given as $\Omega = 2.0$ [Hz], $L = 2$, $(\psi^1, \psi^2) = (0, 0.2)$, $\epsilon_f = 0.1$, $\epsilon = 1.0$, and $\gamma = 1.0$. As the initial condition, we put $(\theta(0), \tilde{\theta}(0)) = (0.5, 0.7)$, $(w^1, w^2) = (0.1, -0.1)$ and $\omega = 1.0$ [Hz].

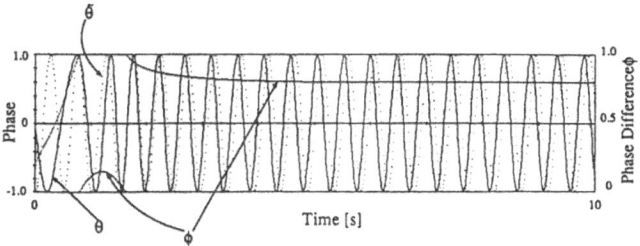

Fig.2. The time profile of the phases, $y = \sin 2\pi\theta$ (solid line) and $y = \sin 2\pi\tilde{\theta}$ (dotted line), and the phase difference, $\phi = \tilde{\theta} - \theta$. The desired phase relation $\phi_d = 0.8$ is acquired by learning, although two oscillators does not synchronize in the initial state because of the large difference of the frequencies.

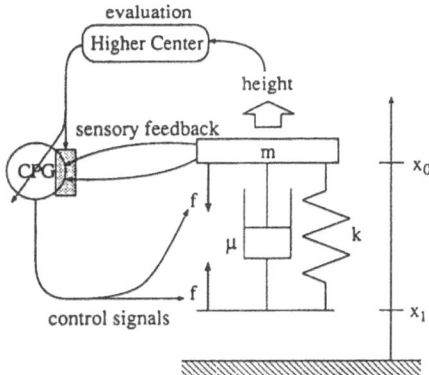

Fig.3. One-dimensional hopping robot. The robot consists of a trunk with mass m and a leg which has a spring component (elastic coefficient : k), a damping component (damping coefficient μ), and a thruster. The CPG sends control signals to the thruster which generates the force f between the trunk and the toe and receives the sensory feedback signals from the robot. According to the evaluation of the performance of the hopping, the parameters, the intrinsic frequency of the CPG and the weights of the sensory feedback, are learned.

The learning rule takes the form,

$$
\left\{
\begin{array}{rcl}
\dot{\omega} &=& \epsilon \sum_{l=1}^{L} w^l \bar{R}^l \\[2mm]
\dot{w}^i &=& \epsilon \gamma E(\phi) \cdot \bar{R}^i,
\end{array}
\right.
\tag{17}
$$

where the time averaged function $\bar{R}^i$ is obtained by

$$
\frac{\tau}{\omega} \dot{\bar{R}}^i = -\bar{R}^i + P(\theta - \psi^i)\hat{v}.
\tag{18}
$$

The simulation results are shown in Fig.4. The one period of the hopping is about one second in a steady state. In the first 20 seconds the learning was not done to achieve a steady relation between the CPG and the robot. The desired height $x_d = 0.6, 0.7, 0.8, 0.9$ [m] were obtained within one hundred periods.

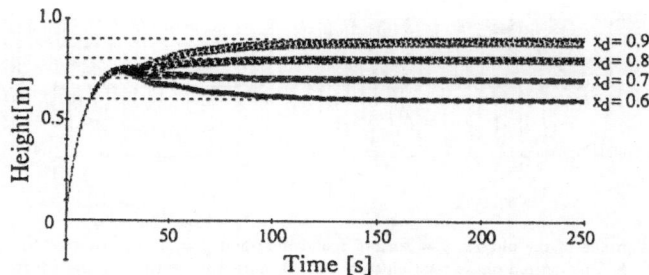

Fig.4. The time profile of the time averaged height $\bar{x}_0$ of the trunk of the robot. Parameters are given as $(\psi_1, \psi_2) = (0, 0.8)$, $\epsilon = 0.3$, $\gamma = 1/3$, and $\tau = 5.0$. As an initial condition, we put $\theta = 0.5$, $(w^1, w^2) = (1.0, -1.0)$, $\omega = 1.0$ [Hz], $x_1(0) = 0.1$ [m], $x_0(0) = x_1 + l = 0.8$ [m].

4 Discussion

We proposed a learning model for the CPG to acquire a desired locomotor pattern and showed that the learning rule can be applied to the control of a hopping robot. The proposed learning rule does not require any information about the controlled object and the learning is done by using local informations, so the back propagation of the error is not necessary. The realized locomotor pattern in our model mainly depends on the oscillatory factor of the physical system and the CPG does not slave the physical system but compensate for the movement of the physical system to obtain a desired performance. The locomotor system of living bodies would behave in such manner as Raibert and Hodgins (1992) mentioned.

In this paper, a physical system with one-degree freedom was treated. We are studying about a learning model for a system composed of multi oscillators and a physical system with multi-degree freedom [Uehara et al. 1994], and about the relation between the proposed learning rule and the learning rule observed in neurophysiology.

Acknowledgement

The author thanks Prof. Ryoji Suzuki, Prof. Kaoru Nakano. and those who gave comments to my research.

References

Cohen AH, Holmes PJ, Rand RH (1982) The nature of the coupling between segmental oscillators of the lamprey spinal generator for locomotion: A mathematical model. J Math Biol 13:345-369

Ermentrout GB, Kopell N (1991) Multiple pulse interaction and averaging in systems of coupled neural oscillators. J Math Biol 29:195-217

Grillner S (1985) Neurobiological bases of rhythmic motor acts in vertebrates. Science 228:143-149

Grillner S, Wallen P, Brodin L (1991) Neuronal network generating locomotor behavior in lamprey: Circuitry, transmitters, membrane, properties and simulation. Annu Rev Neurosci 14:169-199

Nishii J, Uno Y, Suzuki R (1995) Mathematical models for the swimming pattern of a lamprey. I: Analysis of collective oscillators with time delayed interaction and multiple coupling. Biological Cybernetics, in press

Raibert MH, Hodgins JK (1992) Legged robots. In Biological Neural Networks in Invertebrate Neuroethology and Robotics, Academic Press, pp 319-354

Rand RH, Cohen AH, Holmes PJ (1988) Systems of coupled oscillators as models of central pattern generators. In: Cohen AH, Rossignol S, Grillner S (eds) Neural Control of Rhythmic Movements in Vertebrates, John Wiley and Sons, pp 333-367

Uehara M, Nishii J, Suzuki R (1994) Adaptive control of hopping robot by oscillators (in japanese). Tech Rep of IEICE MBE93-143:73-80

Williams TL, Sigvardt KA, Kopell N, Ermentrout GB, Remler MP (1990) Forcing of coupled non-linear oscillators: Studies of intersegmental coordination in the lamprey locomotor central pattern generator. J Neurophysiol 64(3):862-871

Appendix A: Dynamics of a One-Dimensional Hopping Robot

When the robot is jumping, the dynamics is

$$
\begin{cases}
m\ddot{x}_0 &= -mg \\
0 &= k(x_0 - x_1 - l) + \mu(\dot{x}_0 - \dot{x}_1) + f.
\end{cases}
\tag{A.1}
$$

where $m = 0.1$ [kg] is the mass of the trunk of the robot, x_0 and x_1 are the position of the toe and the trunk, respectively, $k = 7.0$ [N/m] and $\mu = 0.05$ [N/(m/s)] are the elastic coefficient and the damping coefficient of the leg, respectively, $l = 0.7$ [m] is the resting length of the spring, $f = f_{max} \sin 2\pi\theta$ is the force generated by the thruster, and $f_{max} = 0.2$ [N] is its maximum force. When the robot is on the ground ($x_1 = 0$ and $N > 0$), the dynamics takes the form

$$
\begin{cases}
m\ddot{x}_0 &= -mg - k(x_0 - x_1 - l) - \mu(\dot{x}_0 - \dot{x}_1) - f \\
0 &= k(x_0 - l) + \mu\dot{x}_0 + f + N \\
\dot{x}_1 &= 0,
\end{cases}
\tag{A.2}
$$

where N is the normal force of the reaction from the ground.

Self-Consistent Neural Receptive Fields

I. Grabec

Faculty of Mechanical Engineering, University of Ljubljana
Pob 394, 61001 Ljubljana, Slovenia

ABSTRACT

The problem of an optimal determination of neural receptive fields is addressed. A principle of self-consistency is formulated which is based on the maximization of receptive field overlapping with the reference data. An iteration formula is derived which yields for the receptive field width approximately the distance to the center of nearest neighbor. Numerical examples show that the probabilistic neural network with self-consistent receptive fields yields better estimataion of continuous probability density than Parzen's estimator on normal, multimodal and exponential distributions. A generalization to multivariate case yields a simple method for the determimnation of ellipsoidal basis function neural network.

INTRODUCTION

A primary task of probabilistic neural networks is a proper estimation of probability density function $f(x)$ (pdf) of input sensory signals that generally represent a continuous random variable X. (Grabec 1990, 1991) For the sake of simplicity we first consider a one-demensional variable X, and later generalize our results to multivariate case. In the following treatment we assume that beside the continuity of random variable X and a set of N independent, identically distributed reference values $\{x_1,...,x_N\}$ no other a priori information about its properties is known. Therefore, a non-parametric presentation of pdf by the kernel estimator is utilized. (Révésh 1991; Nadaraya 1989) It has been introduced as a linear filtering of empirical pdf (Rosenblatt 1956; Parzen 1962):

$$f_e(x)= \frac{1}{N} \sum_1^N \delta(x\text{-}x_n) \tag{1}$$

based upon an appropriately selected kernel or window function $w(x\text{-}q,h)$. In the parlance of neural networks the parameters q and h describe the center and the width of the receptive field of a neuron respectively. The filtered function

$$\hat{f}(x)= \frac{1}{N} \sum_{n=1}^N w(x\text{-}x_n, h_N) \tag{2}$$

is a smooth estimator of probability density provided that the kernel $w(x\text{-}q,h)$ is a smooth approximation of the delta function. The empirical pdf (1) is an unbiased, but non-consistent estimator of probability density. The non-consistency stems from the singularity of delta function. Filtering by a smooth kernel suppresses the singularity but introduces a bias of the estimator. At an increasing number N of reference points a consistent and asymptotically unbiased estimator is obtained by properly decreasing the kernel width h_N towards 0, like for example $h_N = h_1/\sqrt{N}$. However, the problem how to select the receptive field width at fixed number N is generally not solved, because the solution depends on the probability distribution under observation. One of the fundamental problems at the application of probabilistic neural networks is thus how to obtain a method for the determination of receptive fields that could be applicable to various distributions.(Bishop 1994, Masters 1993)

Introduction of linear filtering of pdf leaves freedom in the choice of the kernel function and its width. As a most appropriate kernel function for the presentation of empirical data one often assumes the Gaussian function. A proper selection of the receptive field is more problematic: if it is too narrow the estimated function fluctuates violently, and if it is too wide the estimator is insensitive to variations of pdf at finite N. This problem is outstanding when multimode distribution is estimated from a low number of reference data. In this case a single width h_N generally does not yield a proper smoothing everywhere. Instead of describing the width of receptive field by some ad hoc selected value it is better to adapt it to the data. With this aim nearest-neighbor (Révésh 1991; Fukunaga 1972,1973; Duda & Hart 1937) and the variable kernel width estimators have been introduced. (Loftsgaarden and Quesenbery 1965) In the case of nearest-neighbor estimator the pdf is represented by the expression

$$\hat{f}_K(x) = \frac{1}{N} \sum_{n=1}^{N} w(x-x_n, h_K(x)) \tag{3}$$

in which $h_K(x)$ denotes the width of the receptive field centered at x and containing K values from the sample set $\{x_1,...,x_N\}$. The width is thus not a constant but is determined by the data. It has to be calculated for each point of observation, therefore the method is computationally extensive. Beside this, a disadvantage is that the number of neighbors K and its dependence on N is ad hoc selected and that a couple of functions, $w(x)$ and $h(x)$, is used in the definition of the estimator. We try to improve these deficiencies by using an optimal receptive field around each reference point only. By this modification a simple form of the kernel estimator (2) as well as the favorable properties of the estimator (3) are preserved. For this purpose we introduce a new principle of an optimal selection of kernel width, by which the arbitrariness in the selection of the number of neighbors is avoided. However, if the probability distribution function is not known, one can not strictly avoid indeterminacy in the selection of receptive fields but only reasonably diminish it by assuming that a particular window function is properly spread around a particular reference point when its width is selected in accordance with the position of the other reference points. Spreading is thus needed to establish a continuous transition of probability density between the reference points. As all the reference points represent the same phenomenon, a receptive field associated with a particular point should, in some sense, optimally cover the other reference points. A corresponding selection of receptive field is then considered to be self-consistent.

The overlapping of receptive field with the reference points is mathematically most simply described by a correlation. It is therefore applied in the following section of the article to the formulation of a criterion of self-consistency.

THE PRINCIPLE OF MAXIMAL SELF-CONSISTENCY

Our final goal is to obtain a smooth estimator of pdf which would be more adaptable to local properties of probability distribution than the estimator based on a unique kernel width h_N. For this purpose we assume that the contribution of each reference point can be represented by the same window function but with a consistently selected width :

$$\hat{f}(x) = \frac{1}{N} \sum_{i=1}^{N} w(x-x_i, h_i) \tag{4}$$

As a condition for consistency of i-th window function with the other reference values we apply the maximum of correlation between this window function and a truncated empirical pdf which does not include the i-th reference point :

$$C(h_i) = \int_{-\infty}^{\infty} w(x-x_i, h_i) \frac{1}{N-1} \sum_{\substack{j \neq i}}^{N} \delta(x-x_j)\, dx = \max(h_i) \quad, \quad i = 1,..., N \tag{5}$$

Integration yields the formula

$$C(h_i) = \frac{1}{N-1} \sum_{j \neq i}^{N} w(x_j - x_i, h_i) = \max(h_i) \quad , \quad i = 1,..., N \tag{6}$$

which after derivation leads to the conditions

$$\sum_{j \neq i}^{N} dw(x_j - x_i, h_i)/dh_i = 0 \quad , \quad i = 1,..., N \tag{7}$$

By utilizing the Gaussian window function

$$w(x, h) = \frac{1}{\sqrt{(2\pi)} h} \exp(-\frac{x^2}{2h^2}) \tag{8}$$

we get from Eq. (7) the following system of nonlinear equations

$$\sum_{j \neq i}^{N} \frac{1}{h_i^2} \left[\frac{(x_j - x_i)^2}{h_i^2} - 1 \right] \exp\left[\frac{-(x_j - x_i)^2}{2h_i^2} \right] = 0 \quad , \quad i = 1,..., N \tag{9}$$

Its solution can be obtained by using the iteration formula :

$$h_i^2 = \frac{\displaystyle\sum_{j \neq i}^{N} (x_j - x_i)^2 \exp\left[\frac{-(x_j - x_i)^2}{2h_i^2} \right]}{\displaystyle\sum_{j \neq i}^{N} \exp\left[\frac{-(x_j - x_i)^2}{2h_i^2} \right]} = G(x_1,.., x_N, h_i) , \quad i = 1,..., N \tag{10}$$

Although the iteration depends on transcendental functions, the properties of the solution can be estimated from the analysis of the largest terms in the numerator and denominator in Eq. (10). They are determined by the point x_{jo}, which lies nearest to x_j. The resulting optimal kernel width is therefore approximately given by the distance to the nearest neighbor $d_{jo} = |x_{jo} - x_i|$ while in the case when the sample set includes just two points it is exactly given by this distance. The generating function satisfies the condition $dG(x_1,.., x_N, h_i)/dh_i \approx 0$ for $h_i \approx d_{io}$, therefore the iteration process converges very rapidly. Numerical investigations have shown that the initial rate of convergence of the iteration process does not essentially depend on the starting values of the widths and that the solutions of Eq. (10) satisfy the approximate relation

$$h_i^2 \gtrless \min_j (x_j - x_i)^2 \tag{11}$$

The form of the iteration generating function suggests a simple statistical interpretation of the self-consistent kernel width. If the exponential function taking place in Eq. (10) is treated as a statistical weight assigned to a distance between points of the sample set, then the generating function $G(x_1,.., x_N, h_i)$ can be interpreted as a window average of the square distance from the window center to the neighbor reference points. However, the window width must be determined self-consistently. The number of the nearest neighbor K, is thus not selected ad hoc but determined by an effective number of neighbors inside the window. This number is generally greater than 1, therefore the properties of self-consistent estimators are similar to the properties of estimators defined by a k-nearest-neighbor rule.

The adaptation of the widths $\{h_1,...,h_N\}$ to the reference values $\{x_1,...,x_N\}$ can be interpreted as a mapping of a random variable X to a new random variable H. Due to the fact that this mapping is determined by the iteration process, the properties of the probability distribution of the variable H could hardly be explained analytically and generally. By invoking the approximate relation (11) it can be conjectured that the width of each window diminishes with increasing number N when samples become ever more densely spaced. The kernel functions then tend to the delta function and the estimator bias is evanescent asymptotically. The estimator (4) thus retains the favorable properties of Parzen's as well as of nearest-neighbor estimators automatically without special care about the conditions governing the dependence of the widths on the number of reference points N.

In the adaptation process stemming from the correlation criterion (5) the a particular receptive field is not influenced by the widths of the other fields. The field must therefore effectively cover the complete gap between its center and neighbor reference points. This strong condition can be relaxed if the mutual overlapping of the neighbor fields is considered. In this case the widths of receptive fields are reduced on average for a factor 2, but the corresponding mathematical treatment is more involved. (Grabec 1993)

THE SELF-CONSISTENT METHOD IN A MULTIVARIATE CASE

The self-consistent approach can be simply generalized to n-dimensional example by using multivariate kernels. Let us for this purpose express the window function centered at reference vector x_i by a multivariate Gaussian pdf:

$$w_i(x) = (2\pi)^{-n/2} |B_i|^{1/2} \exp[-\tfrac{1}{2}(x-x_i)^T B_i (x-x_i)] \qquad (12)$$

with $B_i = [b_{i,kl}]$ denoting the inverse of the covariance matrix associated with the i-th point. The conditions of the self-consistency are then

$$C(B_i) = \frac{1}{N-1} \sum_{\substack{j \ne i}}^{N} w(x_j - x_i, B_i) = \max(b_{i,kl}) \quad, \quad i = 1,2,..,N \qquad (13)$$

Derivation with respect to $b_{i,kl}$ and application of the formulas

$$|B| = \sum_{k=1}^{N} b_{kl} B_{kl}^c \quad ; \quad \partial|B|/\partial b_{kl} = B_{kl}^c \quad ; \quad |B_{kl}^c| = B^{-1} = [h_{kl}] \qquad (14)$$

with B_{kl}^c being the cofactor of term b_{kl} in determinant $|B|$, leads to the system of equations for the covariances $h_{i,kl}$

$$\sum_{\substack{j \ne i}}^{N} [h_{i,kl} - (x_j - x_i)_k (x_j - x_i)_l] \, w(x_j - x_i, B_i) = 0 \quad ; \quad i, k, l = 1...N \qquad (15)$$

These equations can be again written in a form applicable for iteration

$$B_i^{-1} = [h_{i,kl}] = \frac{\sum_{\substack{j \ne i}}^{N} (x_j - x_i)^T (x_j - x_i) \, w(x_j - x_i, B_i)}{\sum_{\substack{j \ne i}}^{N} w(x_j - x_i, B_i)} \qquad (16)$$

This expression shows that iteration formula (10) of the one-dimensional case can be directly adapted to multivariate problem by transforming h_i^2 and $(x_j - x_i)^2$ into matrices, and utilizing the multivariate normal distribution. The resulting receptive field is then described by ellipsoidal window function. Unfortunately in this case the result of iteration is the covariance matrix which must be inverted before application in the expression of the window function, which is generally a time consuming task. In the multivariate case a problem can appear at the inversion if the covariance matrix is singular. It can be generally avoided by assuming that no covariance can be lesser than the experimental error of instrumentation : $h_{kl} \ge h_e$ for any pair k,l.

NUMERICAL EXAMPLES

The purpose of this section is to demonstrate numerically some characteristic features of the estimator based on self-consistent determination of receptive fields in one dimension. Like the nearest neighbor estimator, this one is especially appropriate for a low number of reference points, therefore we demonstrate it on examples including only few data. Problems of this kind are often met when pdf is described by a set of reference values obtained by self-organization of neurons.(Grabec 1990) However, an essential difference between representation of a pdf by a few independent random samples or by a set of prototype values obtained by self-organization should be mentioned here. In the latter case the spacing between the prototype points is not determined by chance, as in the case when a few independent samples are measured, but it is rather determined by limiting values proceeding from a quantization process that is driven by a series of random samples. The spacing of reference points thus exhibits a systematic property of pdf. As our procedure has been mainly developed for estimating pdf on the ground of prototype values obtained by the self-organization process,(Grabec 1990) we primarily present the examples with sample spacings exhibiting some well-defined specific property of pdf as is, for example, an equal spacing of samples for the case of uniform distribution or increasing spacing with distance from the center of a normal distribution. The demonstrated cases represent a normal, a uniform, an exponential, and a bimodal pdf by properly spaced reference points. Beside this an example with randomly changing samples is demonstrated for the case of a uniform distribution. For the purpose of comparison Parzen's estimator with a global width selected according to the rule $h_o{}^2 = \text{var}(x)/N$ is also demonstrated. In all cases the self-consistent width was determined by 10 iterations, starting with the global width of Parzen's estimator. During the iteration the value of the width differed from the final value for less than 5 % of the limiting value after 5 iterations already. The results of numerical procedures are shown in Figs. 1-5. In each diagram the vertical strokes on the upper line indicate the positions of the samples or prototypes, while the appertaining table shows their values and corresponding widths. The

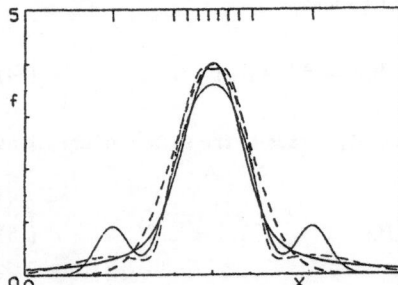
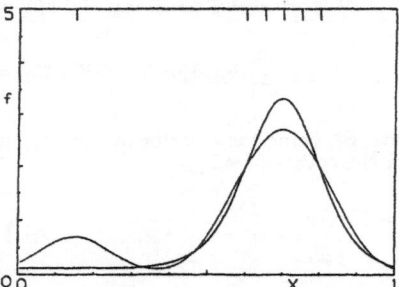

Fig. 1-left. Estimation of normal probability density function. The prototype samples are denoted by strokes on the top line. − bold : self-consistent estimator, --- bold : original distribution, − thin : Parzen's estimator with $h_o{}^2 = \text{var}(x)/N$, --- thin : Parzen's estimator with h_o = distance to nearest-neighbor.

Table 1. - The sample values x_i and the corresponding widths h_i

x_i	0.240	0.400	0.430	0.460	0.490	0.510	0.510	0.570	0.600	0.770
h_i	0.270	0.111	0.056	0.051	0.051	0.051	0.051	0.056	0.111	0.270

Fig. 2-right. Example of estimation of probability density function from a group of closely spaced prototypes and a single separated sample. − bold : self-consistent estimator, − thin : Parzen's estimator with $h_o{}^2 = \text{var}(x)/N$

Table 2. - The sample values x_i and the corresponding widths h_i

x_i	0.150	0.600	0.650	0.700	0.750	0.800
h_i	0.545	0.108	0.069	0.070	0.069	0.108

163

bold line corresponds to the self-consistent estimator while the thin one corresponds to Parzen's estimator.

Fig. 1 shows the case corresponding to a normal distribution. For the purpose of comparison the original normal distribution and the estimators obtained by Parzen's width or distance to the nearest-neighbor are shown too. One can estimate intuitively that the self-consistent estimator adapts better to the original distribution than the other two. Two side lobes of Parzen's estimator indicate that this estimator assigns too large a weight to an individual sample separated from the rest group. Figures 2 and 3 are prepared to point out this property. The example shown in Fig. 2 shows an individual sample that is separated from the group of other samples. The width corresponding to this sample is much larger than the widths of samples in the group, therefore its kernel contributes a lesser amplitude to the estimated pdf than the others in the group. In the case of Parzen's estimator the width corresponding to each sample is the same and consequently the separated sample contributes to the estimator a greater deal as in the self-consistent case.

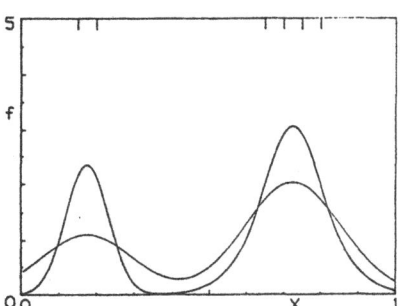 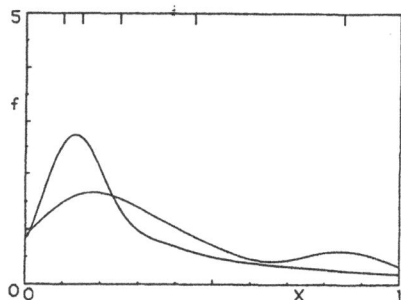

Fig. 3-left. Example of estimation of probability density function from two separated groups of closely spaced prototype samples. – bold : self-consistent estimator, – thin : Parzen's estimator with $h_o{}^2 = var(x)/N$

Table 3. - The sample values x_i and the corresponding widths h_i

| x_i | 0.150 | 0.200 | 0.650 | 0.700 | 0.750 | 0.800 |
| h_i | 0.050 | 0.050 | 0.089 | 0.060 | 0.060 | 0.089 |

Fig. 4-right. Estimation of approximately exponential probability density function. The prototype samples are denoted by strokes with increasing spacing on the top line. – bold : self-consistent estimator, – thin : Parzen's estimator with $h_o{}^2 = var(x)/N$

Table 4. - The sample values x_i and the corresponding widths h_i

| x_i | 0.100 | 0.150 | 0.250 | 0.450 | 0.850 |
| h_i | 0.071 | 0.070 | 0.144 | 0.304 | 0.599 |

The situation is quite opposite in the case when two samples get together, as is for example shown by the distribution in Fig. 3. In this case the distance to the nearest neighbor at the pair of nearby samples is small and the self-consistent estimator gets an expressive peak so that a bimodal distribution emerges. This property corresponds to the better adaptability of the self-consistent estimator to multimodal distributions in comparison with Parzen's one, which is a convenient characteristic property for nearest neighbor estimators too. It becomes especially evident when a distribution with drastically changing spacing between prototypes is represented, as is for example an approximately exponential one shown in Fig. 4. In this case the region with small spacing, as well as the region with large spacing is better described by the self-consistent estimator than by Parzen's one. However, in the case of a uniform distribution, as shown in Fig. 5, Parzen's estimator performs better. This is the consequence of the fact that a larger average distance to the neighbors corresponds to the

164

extreme samples on the edge of the distribution as in the middle of it. As the edge samples influence the width of the samples in their vicinity inside the distribution, the smallest width corresponds to samples close to the edges but inside the distribution. This causes maxima of distribution close to the edges, which resembles a similar effect observed in a representation of a box-car function by a finite number of terms of the Fourier series.

The last example, shown in Fig. 6, is used to represent the applicability of the self-consistent estimator to the estimation of pdf from independent, randomly distributed samples. The samples were generated by a random generator with a corresponding uniform pdf in the interval form 0.1 to 0.9. Due to random acquisition of samples, their spacing is not uniform, but varies randomly. As the self-consistent width is approximately determined by the nearest neighbor, larger fluctuations are generally observable in the corresponding estimator as in Parzen's one, which is based on the global width determined from the variance of samples. This difference has been observed previously in the comparison of the properties of kernel-type and nearest-neighbor estimators and will not be analyzed further. (Duda & Hart 1973) We only mention that in spite of larger fluctuations the nearest-neighbor estimators are, like Parzen's ones, asymptotically unbiased and consistent (Révèsh 1991) and the same is valid for the self-consistent one, which can be treated as a joint version of both types.

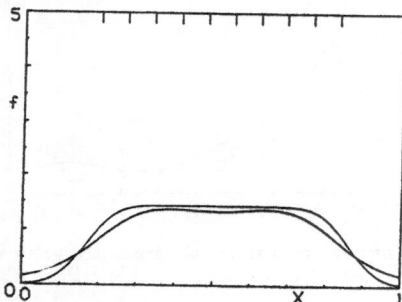 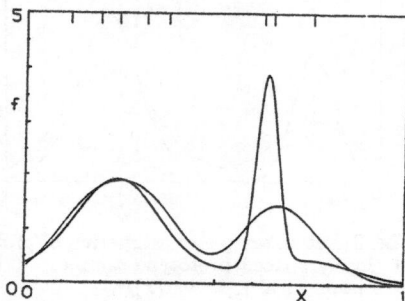

Fig. 5-left. Estimation of a uniform probability density function. The prototype samples are denoted by equally spaced strokes on the top line. − bold : self-consistent estimator, − thin : Parzen's estimator with $h_o^2 = var(x)/N$

Table 5. - The sample values x_i and the corresponding widths h_i

x_i	0.200	0.270	0.340	0.410	0.480	0.550	0.620	0.690	0.760	0.830
h_i	0.231	0.108	0.127	0.149	0.162	0.162	0.149	0.127	0.108	0.231

Fig. 6-right. Estimation of a uniform probability density function in the interval from 0.1 to 0.9 from the random samples. The samples are denoted by randomly spaced strokes on the top line. − bold : self-consistent estimator, − thin : Parzen's estimator with $h_o^2 = var(x)/N$

Table 6. - The sample values x_i and the corresponding widths h_i

x_i	0.110	0.370	0.190	0.240	0.620	0.65	0.75	0.310
h_i	0.148	0.193	0.088	0.086	0.025	0.025	0.121	0.088

CONCLUSIONS

The main purpose of our approach to the estimation of pdf by probability neural networks was to propose a principle by which a receptive field of a neuron that corresponds to a local kernel width could be simply determined. The proposed principle of maximal self-consistency leads to a simple iteration formula, which yields kernel widths approximately equal to the distance to the nearest neighbor. An ad hoc specification of the number of nearest neighbors, or an arbitrary selection of the kernel width in Parzen's approach is thus avoided. The examples demonstrated indicate that our approach is especially appropriate for the estimation of pdf from the reference points obtained by self-organized adaptation or quantization procedures. This is of advantage for the application of a self-consistent approach in the field of neural networks, where the prototypes can be extracted from the parameters of self-organized neurons. (Grabec 1990, 1991) However, for this purpose the generalized multidimensional case has to be considered. Recently, a rule of self-organization in a neural network has been derived on the basis of Parzen-window estimation of pdf. (Grabec 1990) In relation to this object of research there appears a question whether a similar rule could not be derived on the basis of a self-consistent estimator as well. The fundamental problem of how to determine proper receptive fields of neurons, could thus be solved by the principle of self-consistency.

REFERENCES

Bishop, C. M. , (1994). Rev. Sci. Instr., 65, 1830-1832

Duda, R.O., P.E. Hart, P.O.,(1973). Pattern Classification and Scene Analysis, J. Wiley & Sons, NY, Ch. 4.

Fukunaga, K., (1972). Introduction to Statistical Pattern Recognition, Academic Press, NY Grabec, I., Sachse, W., (1991). J. Appl. Phys., v. 69, pp. 6233-6244.

Grabec, I., (1990). Biol. Cyb., v. 63, pp. 403-409.

Grabec, I., (1993). Neural Prallel & Scientific Computations, 1, 83-92

Loftsgaarden, D. O., Quesenbery, C. P., (1965). Ann. Math. Stat., v. 36, pp. 1049-1051.

Masters, T., (1993). Practical NN Rewcipes in C++, AP, Boston

Nadaraya, E. A., (1989). Non-parametric Estimation of Probability Densities and Regression Curves, Kluwer Academic Publ., Dordrecht.

Parzen, E., (1962). Ann. Math. Stat., v. 35, pp. 1065-1076.

Révész, P., (1991). Density Estimation, Handbook of Statistics, Eds.. P. R. Krishnaiah, P. K. Sen, Elsevier Sci. Publ., v. 4, pp. 531-549.

Rosenblatt, M., (1956). Ann. Math. Stat., v. 27, pp. 832-835.

A Simple Probabilistic Neural Model Producing Multimodal ISHs

Paul M. Hofman, Francisco B. Rodríguez
Juan A. Sigüenza*, Vicente López*
Instituto de Ingeniería del Conocimiento, Universidad Autónoma de Madrid, Canto Blanco, Mod. C-XVI, P.4, 28049 Madrid, Spain.
* Departamento de Ingeniería Informática, Universidad Autónoma de Madrid, Canto Blanco, 28049 Madrid, Spain.

Santiago Carrillo-Menendez
Departamento de Matemáticas, Universidad Autónoma de Madrid, Canto Blanco, 28049 Madrid, Spain.

Abstract

Simple probabilistic neural models can be used to study the information processing occurring in the brain. Suitable models have to reproduce complex Interspike Histograms (ISHs) observed experimentally, but have to be simple enough to allow theoretical analysis. The simple probabilistic integrate and fire model we present in this paper can be used to identify the origin of peaks appearing in complex multimodal ISHs.

1 Introduction

In recent years, a large number of neural network models have been proposed . They can be divided into two large categories. Those devised having in mind specific applications, and those concerned with the understanding of information processing in the brain [1]. The model we present in this work falls in the second category. Isolated neurons are represented with a simple model in which only basic biological properties are included, and in which noise plays a relevant role [2]. We seek for a model that being capable of reproducing general patterns of neural firing, is simple enough to allow for the study of information processing mechanisms in the brain. Also, a simple model of neural activity makes easier the search for neural networks architectures in which general mechanisms could be tested in the direction of practical applications.

2 Theoretical Model

We have selected a simple model with few assumptions to study basic unknowns of information processing in the nervous system [3]. In the model, signals from the surrounding sites are received and integrated by the neuron. Membrane potentials are the basic carriers of information and spikes are produced by the neuron after the membrane potential has reached a given threshold.

Due to the nature of biological systems, information processing has to be robust in the presence of (background) noise. The stochastic character is incorporated to the model with the random firing of isolated neurons and with the irregularity of spike trains after stimulus.

2.1 The isolated neuron

Neurons are modeled as stochastic units with a discrete number of states. Time is also considered to be discrete and only two parameters are relevant for the description of the dynamic behavior of every unit: the number of states and the probability of incrementing the state at every time step.

In what follows we use $a_i(t)$ to represent the activity of unit i at time t. Possible states of every unit are in the range from 1 to N_i. In absence of interaction with other neurons, the transition between states is governed by the probabilistic rule:

$$a_i(t+1) = \begin{cases} a_i(t)+1 & \text{with probability } p_i \\ a_i(t) & \text{otherwise} \end{cases} \tag{1}$$

with $a_i(t)\epsilon\{1,..,N_i\}$. The spike is represented by the transition from state N_i to state 1, where the cycle starts again.

According to this model, an isolated neuron behaves like a stochastic oscillator. The elapsed time between consecutive spikes, T_i, has a probability given by the Negative Binomial distribution $P_{N_i}(T_i)$:

$$P_{N_i}(T_i) = \begin{pmatrix} T_i - 1 \\ T_i - N_i \end{pmatrix} p_i^{N_i}(1-p_i)^{T_i - N_i} \tag{2}$$

For such a unit the interspike interval has and expected value τ_i with standard deviation σ_i. The value of both statistic parameters are easily derived from $P_{N_i}(T_i)$, and they are:

$$\tau_i = 1 + \frac{N_i - 1}{p_i}, \quad \sigma_i = \frac{\sqrt{(N_i - 1)(1 - p_i)}}{p_i} \tag{3}$$

respectively [4]. Spike trains produced by the isolated neuron i in the present model are quite simple, and can be adequately described by only two statistical parameters: τ_i and σ_i. We display in Figure 1 the overall behavior of two different isolated neurons.

As the reader will easily infer form the statistical description of ISHs produced with this model, the spike interval spread relative to the mean value will tend to zero for large values of the mean interspike interval. This shortcut of the model implies the use of models in which the activity of two units have to be included to reproduce some observed ISHs. In fact, the simplicity of the selected model is such that it can be used to simulate "active centers" inside a neuron. For instance, a neuron could be modeled as a complex unit in which there are several active sites, each one of the type modeled with the simple stochastic integrate and fire model we have selected. Also, a suitable model for a neuron could be a unit in which two active centers are considered: one for the neural body and the other to simulate the overall source of noise in which it is embedded.

2.2 Units Interaction

Interaction between units is included to model an immediate spike transmission through the neural axon, which produces a postsynaptic change in the receiving unit. More explicitly, neuron j is affected by a spike of neuron i according to:

$$a_j(t) = a_j(t) + \epsilon_{ij} \tag{4}$$

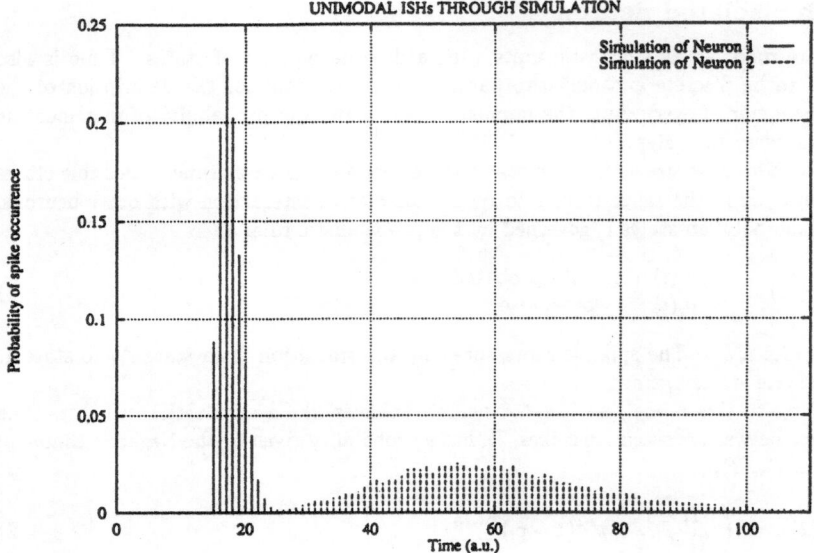

Figure 1: *Simulation of two isolated neurons with parameters:* $N_1 = 15$, $p_1 = 0.85$, $N_2 = 10$, $p_2 = 0.17$. *Neuron 1 fires fast and regularly whereas neuron 2 is slow and noisy.*

where the sign of ϵ_{ij} refers to the type of connection (inhibitory: $\epsilon_{ij} < 0$, excitatory $\epsilon_{ij} > 0$), and the magnitude refers to the strength. If the unit activity reaches the threshold ($a_j(t) > N_j$) due to spike transmission, then it will immediately discharge and reset its state to 1, regardless of the exact value $a_j(t)$ before discharging.

3 Multimodal Interspike Histograms

Interspike Histograms (ISHs) of different kinds have been measured for neurons under several circumstances (see [5] for a general description). Unimodal ISHs can be obviously obtained from the neuron model we have described. However, it is not clear how multimodal interspike histograms can be originated using a simple model as the one described in this work. Interspike histograms of the multimodal type occur in various single-unit recording experiments (such as in shark multimodal sensory cells [6]). Complex models have been considered in order to explain the recordings. For instance, an under-threshold oscillation mechanism has been proposed to be responsible of the observed multimodal ISHs [6].

Simple stochastic neuron models, like the one presented in this work, could be used to study the information processing capabilities of large ensembles of neurons if they can explain, at least, the complex ISHs observed in single unit recording. In order to test the possibilities of the suggested model we have studied in which circumstances it can yield multimodal ISHs.

For this complex interspike pattern to occur it is only needed the interaction between two units. In fact, a multimodal ISH can be measured as the output of a model unit receiving the interaction of another unit that produces a highly random sequence of spikes. Such a simple "network" can be directly thought as the model of a two neurons neural network, or as the model of a single neuron surrounded by a noisy environment. That is, the irregular spike train could as well have the origin in a large number of neurons. In this picture, the sender unit is modeling an irregular environment.

The configuration we have studied essentially consists of a slow noisy neuron acting upon a fast regular neuron by means of an **inhibitors** connection of appropriate strength. In the presence of inhibitors connections model units are allowed to have negative states. Such states may have a realization in the decreasing of membrane resting potentials of biological neurons.

In what follows we use label 1 for the receiving unit and label 2 for the sender. Multimodal ISHs can be obtained in the simple network for appropriate unit parameters. It can be measured in the activity of unit 1 for $\tau_2 > \tau_1$, $\sigma_2 >> \sigma_1$, and ϵ_{21} (the inhibitory connection strength) of the order of N_1. For instance, a multimodal ISH is given by the monitoring of the activity of unit 1 with the following model parameters: $N_1 = 15$, $p_1 = 0.85$, $N_2 = 10$, $p_2 = 0.17$ and $\epsilon_{21} = -35$. Thus, a slow and fuzzy neuron (2) acting upon a fast, regularly firing neuron (1) by means of an inhibitory connection ($\epsilon_{21} < 0$).

3.1 Simulation results

The ISHs "measured" are quite simple when isolated units are recorded. ISHs displayed in Figure 1 are those corresponding to the parameters of the network when the interaction is not present. The recording changes qualitatively when interaction between units is allowed. In Figure 2 we present the ISHs of the faster neuron acted upon by the slower neuron, with the network parameters given above. The ISH is made up with the counting of 10^6 spikes produced in the simulation of the model evolution from a a single initial state.

The simplicity of network and unit models allows for an understanding of the source of the multimodal structure in the ISH. Time $t = 0$ in the ISH graph is defined as the time immediately after the last firing of neuron 1: neuron 1 will be in state $x_1 = 1$, neuron 2 can be in any state x_2 with a probability $\pi(x_2)$. Another 'contribution' to the ISH will be made as soon as neuron 1 fires again.

The first peak in the ISH accounts for all cases in which neuron 1 fires without the occurrence of a spike from neuron 2. Even if neuron 1 would increment its state every time step it cannot fire before $t = N_1$.

The second peak accounts for the cases in which neuron 2 has fired once before neuron 1 fires again. Since the spike transmission caused a descent of the state of neuron 1 by an amount of $|\epsilon_{21}|$, neuron 1 has $|\epsilon_{21}|$ states more to overcome. Obviously, the total number of states to be traversed will be $N_1 + |\epsilon_{21}|$ and in this case, neuron 1 cannot fire before $t = N_1 + |\epsilon_{21}|$. The time neuron 1 is delayed when neuron 2 has fired once is, in average, $< t_{delay} >= 1 + (|\epsilon_{21}| - 1)/p_1$.

An analogous reasonings holds for the $D+1$-th peak. Neuron 2 has fired D times before neuron 1 fires again, therefore, neuron 1 needs to traverse $D|\epsilon_{21}|$ more states. Hence, it will not discharge before $t = N_1 + D|\epsilon_{21}|$. Once again, the average time that neuron 1 is delayed, when neuron 2 has fired D times, is $< t_{delay} >_D = D(1 + (|\epsilon_{21}| - 1)/p_1))$.

This model of two neurons can produce a broad diversity of multimodal ISHs when network parameters are varied. For example, consider the following configuration: $N_1 = 20$,

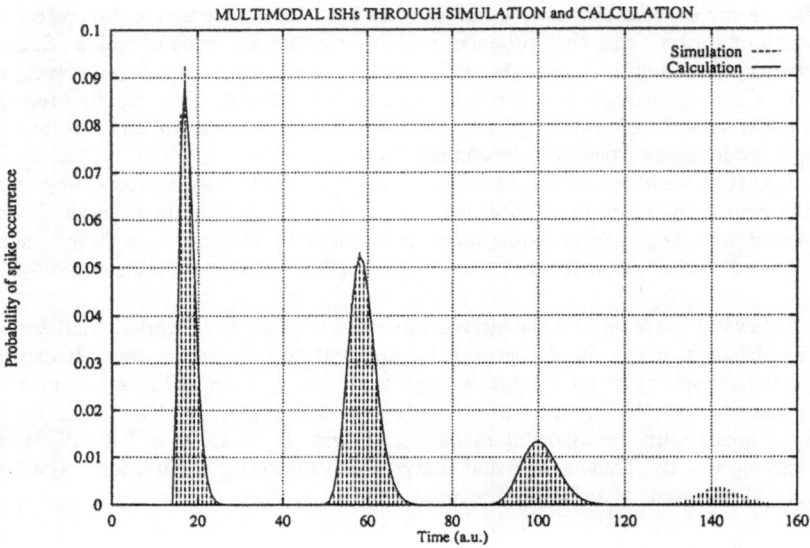

Figure 2: *Simulation and calculation the ISHs of a two neuron network with configuration:* $N_1 = 15$, $p_1 = 0.85$, $N_2 = 10$, $p_2 = 0.17$ and $\epsilon_{21} = -35$. *The figure shows the ISHs of neuron 1. The solid line shows the calculation, whereas the impulses represent the simulation.*

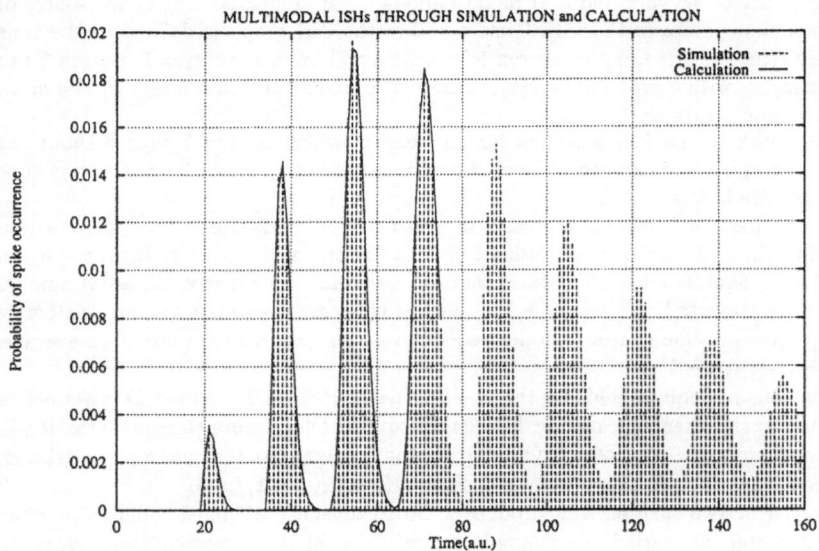

Figure 3: *Simulation and calculation the ISHs of a two neuron network with configuration:* $N_1 = 20$, $p_1 = 0.9$, $N_2 = 6$, $p_2 = 0.3$ and $\epsilon_{21} = -15$. *The figure shows the ISHs of neuron 1. The solid line shows the calculation, whereas the impulses represent the simulation.*

$p_1 = 0.9$, $N_2 = 6$, $p_2 = 0.3$ and $\epsilon_{21} = -15$. The ISH recorded in neuron 1 during simulation with these parameters is shown in Figure 3. This is just an example of the variety observed during simulations performed for different parameters.

3.2 Direct calculation of the ISHs structure

Due to the simplicity of the selected network model it is also possible to approximate the Interspike Histogram of the receiving unit using a direct summation of relevant contributions identified in the simulation analysis.

Time in the ISH graphs is defined as the time elapsed between the last two spikes of neuron 1. The clock is reset after each spike; as soon as the next spike occurs the time is registered and the clock is reset again. Inhibition of neuron 1 consists of pulling down its state by an amount ϵ_{21}. After some time, when it has recovered from the decrement and exceeds its maximum state after all (i.e. it discharges and releases a spike), it will have traversed $N_1 + |\epsilon_{21}|$ states in total. In the case of k times inhibition (k spikes of neuron 2) before the next spike of neuron 1 the total would be $N_1 + k|\epsilon_{21}|$ states.

At time $t = 0$ the state of neuron 1 is $x_1 = 1$, since it has just fired. Neuron 2 can be in any state $x_2 \in \{1, .., N_2\}$, according to a certain distribution $\pi_2(x_2)$. $\pi_2(x_2)$ is non-trivial, since spikes of neuron 2 affect the spiking behavior of neuron 1, since $t = 0$ is defined with respect to the spikes of neuron 1 and $\pi_2(x_2)$ is the distribution of states x_2 at $t = 0$.

For a spike of neuron 1 (#1) and k spikes of neuron 2 (#2), one has to consider or take into account:

- all permutations of states of #1 and k spikes of #2 in time, which lead to state $x_1 = N_1 + 1$ (discharge) at time t.

- distribution $\pi_2(x_2)$ of states x_2 of neuron 2 at $t = 0$.

- restrictions that each and every one of the k inhibitions has to occur before #1 exceeds its maximum potential (#2 has to pull down #1 in time).

- minimum time for #1 to release a spike: with $N_1 + k|\epsilon_{21}|$ states to be traversed in total, it cannot discharge before $t = N_1 + k|\epsilon_{21}|$.

- amount of non-increments: #1 will make exactly $t - (N_1 + k|\epsilon_{21}|)$ non-increments.

Within this approach, the probabilities of unit 1 receiving k inhibitory interactions are given by:

$$P_0(t) = p_1^{Y_0+1} q_1^{X_0} \binom{t}{X_0} Q_2^*(t)$$

$$P_1(t) = p_1^{Y_1+1} q_1^{X_1} \sum_{\xi_1=\xi_0}^{X_1} \sum_{\eta_1=\eta_0}^{Y_0} \binom{t-\tau_1}{X_1-\xi_1} Q_2(t-\tau_1) \binom{\tau_1}{\xi_1} P_2^*(\tau_1)$$

$$P_{k\geq 2}(t) = p_1^{Y_k+1} q_1^{X_k} \sum_{\xi_1=\xi_0}^{X_k} \sum_{\xi_2=\xi_1}^{X_k} \cdots \sum_{\xi_k=\xi_{k-1}}^{X_k} \sum_{\eta_1=\eta_0}^{Y_0} \sum_{\eta_2=\eta_1}^{Y_1} \cdots \sum_{\eta_k=\eta_{k-1}}^{Y_{k-1}}$$

$$\times \quad \binom{t-\tau_k}{X_k-\xi_k} Q_2(t-\tau_k) \prod_{i=1}^{k} \binom{\tau_i - \tau_{i-1}}{\xi_i - \xi_{i-1}} P_2^{(i)}(\tau_i - \tau_{i-1})$$

with the following definitions:

- $\tau_i = \xi_i + \eta_i$, $\xi_0 = \eta_0 = 0$

- $Y_k = N_1 + k|\epsilon_{21}|$, $X_k = t - Y_k$.

- $P_2^{(1)}(\tau) = P_2^*(\tau)$, $P_2^{(i \geq 2)} = P_2(\tau)$.

- $P_2^*(\tau), P_2(\tau)$: the probability that neuron 2 fires at time τ.

- $Q_2^*(\tau), Q_2(\tau)$: the probability that neuron 2 does not fire until time τ.

- $*$: Refers to the precondition of neuron 2. That is, "$*$" indicates that at $t = 0$ neuron 2 will be in state x_2 with a certain distribution $\pi_2(x_2)$. Otherwise (no "$*$") neuron 2 will be in $x_2 = 1$ at $t = 0$. $P_2(t)$ refers to the negative binomial distribution.

- p_1: the probability of incrementing state of neuron 1 ($q_1 = 1 - p_1$).

In verifying the equation for the ISH above, the measured distribution $\tilde{\pi}(x_2)$ (observed in simulation) rather than a theoretically derived one, $\pi(x_2)$, was used. The derivation of $\pi(x_2)$, which is currently in progress, appears to be non-trivial since it relates recursively to $P_{ISH}(t)$.

The final histogram $P_{ISH}(t)$ is obtained by adding contributions for all k. In Figure 1 and 2 histograms obtained with direct calculation and with simulation are compared for major peaks. The agreement observed is excellent.

4 Conclusion

The simple probabilistic integrate and fire model we have presented in this communication is capable of reproducing general characteristics of a large variety of non-trivial ISHs observed in experimental recordings. The model can be used to simulate neurons or individual active sites that can build a more complex model. For instance, the basic model can simulate the noisy environment in which the cell is embedded.

An advantage of this model is that allows for theoretical analysis with a complete identification of the origin of every peak in complex ISHs. Also, due to its simplicity can be used as the building block of Neural Networks in which information processing mechanisms may be analyzed. Work in which learning rules are introduced is under progress.

References

[1] F. Crick and C. Koch. "Some Reflections on Visual Awareness". Cold Spring Habor Symposia on Quantitative Biology. Volume LV.

[2] G.L. Gerstein y B. Mandelbrot (1964). "Random walk models for the spike activity of a single neuron", Biophys. J., 4, pp. 41-68.

[3] V. López, J.A. Sigüenza, J.R. Dorronsoro y S. Carrillo-Menendez. "Stochastic specificity in neural interaction". Proceedings ICANN 93 pp 196–199. Eds. S. Gielen and B. Kappen Springer Verlag. 1993.

[4] W. Feller "An Introduction to Probability Theory and Its Applications". Vol. I, Ed. J. Wiley and Sons.

[5] H. Tuckwell. "Stochastic Processes in the Neurosciencies". SIAM (1989).

[6] H.A. Braun, H. Wissing, K. Schäfer and M.Ch. Hirsch. "Oscillation and noise determine signal transduction in shark multimodal sensory cells". Letters to Nature, Nature – Vol. 367 – 20 January 1994.

An Associative Neural Network
to Model the Developing Mammalian Hippocampus

Maria Teresa Signes Pont and Juan Vicente Sanchez-Andres

Dept. Fisiología, Fac. de Medicina, Inst. Neurociencias
Universidad de Alicante (Spain)

ABSTRACT

Electrophysiological recording of pyramidal hippocampal cells along early postnatal development shows a pattern of maturation consisting of a progressive reduction of the accommodation and increasing excitability. Electrophysiological, pharmacological, behavioural and lesion techniques permit to manipulate cellular,synaptic and connectivity properties in order to explain how cellular and synaptic mechanisms interact with the pattern of connectivity to give rise to a behaviorally important output pattern. These techniques, although powerful, have their limitations in that only some of the potentially important cellular or synaptic properties are amenable to experimentation. We propose a complementary approach using an associative network model based Hebbian laws, able to simulate the biological system, whose sequential output depends on the interference between a slow and a fast components.

INTRODUCTION: NEONATAL RABBITS EXHIBIT CHANGES IN EXPLORATORY BEHAVIOUR AT THE TIME OF EYE OPENING.

Like many mammals, rabbits are born with their eyes closed. In this state, the newborn depends on his mother for all of its needs and is developmentally programmed to stay close to her for a variety of reasons all of which are critical to the animal's survival (e.g. avoiding predation, staying close to the source of food). Neonatal rabbits open their eyes around postnatal day 10 and almost immediately shift their behaviour from the earlier sedentary pattern to an exploratory mode. In this period, the rabbits move around their immediate environment in a more purposeful behavioral pattern that many investigators of the hippocampus have termed 'spatial' or 'cognitive mapping'. This second behaviour probably serves the same survival-enhancing function : here, the visual system provides the raw data for cognitive mapping improving the capability of the young animal to find food and to avoid predation by himself. These behavioral changes have parallel impact in hippocampus, namely in the electrophysiology of CA_1 pyramidal cells and the distribution of membrane-associated protein kinase C.

ACCOMMODATION OF CA_1 NEURONS: FUNCTIONAL DISCONNECTION OF THE HIPPOCAMPUS.

CA_1 pyramidal cells from adult animals typically respond to current injection with a burst of action potentials superimposed on a depolarizing envelop. In contrast, in the neonatal animal (prior to day 10) this response pattern is not present. Instead, the neonatal CA_1 cells show a pattern characterized by non-bursting responses to current injections of the same size as those injected in the CA_1 neurons of adult rabbit (i.e. accommodation in Fig.1, top trace). However, as the developmental program of the rabbit neonate progresses past day 8, this accommodation is quickly replaced by the adult pattern of bursting (Fig. 1, lower traces).

We consider that the net effect of the accommodation is to block the functional transfer of neurally encoded information along the trisynaptic circuit of the hippocampus. This block is removed at precisely the same time that it becomes useful for the animal to exhibit exploratory or spatial behaviors. It would seem that the accommodation is tantamount to a gating function and subserves two important purposes. First, it functionally disconnects the hippocampus, preventing extensive sensory-derived input at a stage when such input would be maladaptive. Second, the electrophysiological block in the hippocampus allows normal neocortical development to be completed in the sense it prevents indexing functional connections between associated cortical modules at some period in development when such modules may not be ready for associations to be formed.

TRYING TO FIND A SUITABLE NETWORK.

Emergent properties of highly connected neural nets are studied in Computational Science. We know successful applications of little neural nets for central pattern generators (CPGs) in invertebrates. These particular nets control rhythmic and stereotyped movements: swimming, walking, breathing, scratching, etc. Our challenge is to match the formal characteristics of the CPGs to vertebrates neural systems. In invertebrates, most of the neurons have their own identity; they can be studied one by one and bilateral interactions can be described. Each artificial neuron in the model can easily correspond to the animal's one. We assume that in vertebrates, in spite of a huge number of neurons, an assumption can be accepted: there are groups of neurons sharing similar physiological properties, so that single neurons appears like representative of their class.

Here we present a theoretical approach based upon a net which evolves like a dynamic system: from an input pattern to another one which has been stored. Such a net is called associative memory. Mathematically, associative memories can be characterized by the transition to a stable solution represented by a local minimum of a 'function of energy'. During recall, an unknown input pattern is submitted, giving an output pattern whose feedback is impinging upon the input. Every neuron receives all feedback's but its own. The process goes on until a stable solution is reached.The net is synchronously updated, a single neuron per iteration so that random delays appear as the propagation proceeds. Along the training, the weight matrix is built based upon Hebbian learning law. In the recall, we can distinguish three terms: a) training pattern which is nearest of unknown pattern; b) crossed correlation between training patterns; and c) correlation between unknown and all training patterns. Convergence is proved by Lyapunov theorem.

MODELING

We choose Hopfield's associative memory. Consider a neural net with N interconnected neurons; each neuron output can take two values, the rest value is 0 and the maximum firing rate is 1.

We define the embedded states:

$V^\mu = \{ V_i^\mu \}$ for $i = 1...N$

$V^{\mu,\nu}$ is the μ-th embedded state in the ν-th pattern.

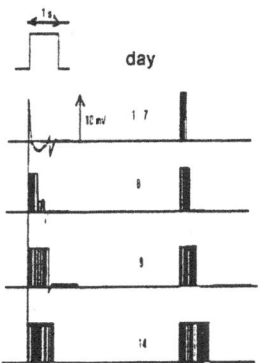

day

Figure 1

Synaptic connections
We consider four states:

$$
\begin{matrix}
& V_1 & V_2 & V_3 & V_4 \\
i=1 & \begin{bmatrix} 1 \\ 1 \\ 1 \\ 1 \end{bmatrix} & \begin{bmatrix} 0 \\ 1 \\ 1 \\ 1 \end{bmatrix} & \begin{bmatrix} 0 \\ 0 \\ 1 \\ 1 \end{bmatrix} & \begin{bmatrix} 0 \\ 0 \\ 0 \\ 1 \end{bmatrix} \\
& j=1 & j=2 & j=3 & j=4
\end{matrix}
$$

We define the synaptic connections:

T_{ij} between post-synaptic i and pre-synaptic j neuron. T_{ij} shares in two functional components T_{ij} S and T_{ij} L. The first has a short time rate τ_S that shows transitions between consecutive embedded states, and the second has a longer one τ_L that means state duration; then $\tau_L >> \tau_S$.

Naturally T_{ij}L synapses are not symmetric T_{ij}L $=/=$ T_{ji} L; but T_{ij}S does because they only depend on activity inside individual states. Weights T_{ij}S y T_{ij}L are successfully computed with these formulas only when there is no recovering between states i.e.:

$$(1/N) \sum_{j=1}^{N} (2 V_j^{\mu,\nu} - 1)(2 V_j^{\mu',\nu'} - 1) = 0$$

for $(\mu, \nu) =/= (\mu', \nu')$
and when in average half of neurons are active in each state.

$$(1/N) \sum_{j=1}^{N} (2 V_j^{\mu,\nu} - 1) = 0$$

Orthogonal embedded states are needed to satisfy these two conditions. If we have a big net with many neurons, we need randomize. In the average, recovering is nearly $(1/N)$ ½.

We define synaptic inputs:

$$h_i^S(t) = \sum_{j=1}^{N} T_{ij}^S V_j(t)$$

$$h_i^L(t) = \sum_{j=1}^{N} T_{ij}^L \overline{V_j(t)}$$

$\overline{V_j(t)}$ neuron average output

If we consider a pattern with a single output :

$V(t) = V^\mu$; averaged output $\overline{V(t)} = V^{\mu-1}$

$$h_i^S(t) = \sum_{j=1}^{N} T_{ij}^S V_j^\mu$$

$$= (J_o/2) \sum_{\eta=1}^{r} (2 V_i^\eta - 1) \{ ((1/N) \sum_{j=1}^{N} (2 V_j^\eta - 1)(2 V_j^\mu - 1) +$$

$$(1/N) \sum_{j=1}^{N} (2 V_j^\mu - 1) \} \quad \text{nearly equal to}$$

$$(J_o/2)(2 V_i^\mu - 1)$$

$$h_i^S(t) = (J_o/2)(2 V_i^\mu - 1)$$

$V_i^\mu = 0$ resting state --> $h_i^S(t) < 0$ inhibition .
$V_i^\mu = 0$ maximum firing rate ---> $h_i^S(t) > 0$ excitation .

$$h_i^L(t) = \sum_{j=1}^{N} T_{ij}^L V^{\mu-1} \text{ nearly equal to}$$

$$(\lambda. J_o/2)(2 V_i^\mu - 1)$$

$$h_i^L(t) = (\lambda J_o / 2)(2 V_i^\mu - 1)$$

Both inputs are going to stabilize the net in its actual state. When t grows, averaged V(t) shifts gradually from state $V^{\mu-1}$ to state V^μ. This shift causes $h_i^L(t)$ grows and $V^{\mu+1}$ grows too. After the network remains in state V^μ during time τ_L input will be:

$$h_i{}^S(t) = \sum_{j=1}^{N} T_{ij}{}^S V_j{}^\mu \sim (J_o/2)(2V_i{}^\mu - 1)$$

$$h_i{}^L(t) = \sum_{j=1}^{N} T_{ij}{}^L V_j{}^{\mu-1} \sim (\lambda J_o/2)(2V_i{}^{\mu+1} - 1)$$

If λ is big enough (> 1), the net has a fast transition to the state $\mu+1$. If the time scale of $T_{ij}{}^S$ and $T_{ij}{}^L$ is the same ($\tau_S \sim \tau_L$) there is no sequential persistent output.

$$\tau_S \frac{du_i}{dt} = h_i{}^S(t) + h_i{}^L(t) + I_{stim\,i.}$$

$$= \sum_{j=1}^{N} [T_{ij}{}^S V_j(t) + T_{ij}{}^L V_j(t)] + I_{stim\,i.}$$

$u_i(t)$ = input to i-th neuron

I_{stim} = extern input to i-th neuron.

$V_i(t)$ = neuron output.

$V_i(t) = g[u_i(t) - \theta_i]$

θ_i = operative level of the neuron.

In order to build the network circuitry each neuron is represented by an operational amplifier working at saturation, with loading time $\tau^N = RC$. R is the input resistance of the neuron. Synaptic connections between neurons are conductances in proportion to $T_{ij}{}^S$ and $T_{ij}{}^L$. The slow synapses $\omega(t)$ (circles) have characteristic time τ_L. For the fast synapses we take $\tau_S = máx(\tau^S, \tau^N)$. To force patterns to emerge τ_L y τ_S must be separated enough (Fig. 2).

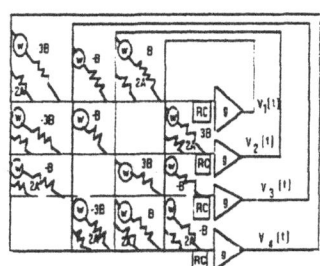

Figure 2.

Short synapse: $T_{ij}{}^S$

$$T_{ij}{}^S = (J_o/N) \sum_{\nu=1}^{q} \sum_{\mu=1}^{r} (2V_i{}^{\mu\nu} - 1)(2V_j{}^{\mu\nu} - 1)$$

$i \neq j$ $J_o > 0$ $T_{ii}{}^S = 0$ for any i

$$T_{12}{}^S = (J_o/4) \sum_{\mu=1}^{4} (2V_1{}^\mu - 1)(2V_2{}^\mu - 1)$$

$$T_{ij}{}^S = \frac{J_o}{4} \begin{bmatrix} 0 & 2 & 0 & -2 \\ 2 & 0 & 0 & 2 \\ 0 & 0 & 0 & 2 \\ -2 & 2 & 2 & 0 \end{bmatrix} \quad \text{(symmetric character)}$$

Long synapse $T_{ij}L$:

$$T_{ij}L = (\lambda J_o/ N) \sum_{\nu=1}^{q} \sum_{\mu=1}^{r-1} (2 V_i{}^{\mu+1\cdot\nu} -1) (2 V_j{}^{\mu,\nu} -1)$$

$i =/= j$ $\lambda > 0$ $T_{ii} = 0$ for any i.

$$T_{12}L = (\lambda J_o/4) \sum_{\mu=1}^{3} (2 V_1{}^{\mu+1} -1) (2 V_2{}^{\mu} - 1)$$

$$T_{ij}^{L} = \lambda \frac{J_0}{4} \begin{bmatrix} 0 & 1 & -1 & -3 \\ 3 & 0 & -1 & -3 \\ 1 & 3 & 0 & -1 \\ -1 & 1 & 3 & 0 \end{bmatrix} \quad \text{(symmetric character)}$$

$$\frac{J_0}{4} = A ; \quad \frac{\lambda J_0}{4} = B$$

Network dynamics:
We must prove the capability of our model to produce stable oscillations. For that we will use two state neurons with synchronous update law ($\delta t = \tau_S$).

$$V_i (t + \delta t) = stp [h_i{}^S (t) + h_i{}^L (t) - \theta_i] =$$

$$= stp [1/2 \sum_{j=1}^{N} (T_{ij}{}^S (2 V_j (t) - 1) + T_{ij}{}^L (2 \overline{V_j (t)} - 1)]$$

where stp (x) = +1 para x > 0
$\qquad\qquad\qquad$ = 0 para x < = 0
and the delay   w(t) = $\delta (t - \tau_L)$.

Immediately after the network has reached state V^1, the i-th neuron output is $V_i{}^1$, but the delayed output will be:

$$\overline{V_i (t)} = V_i (t - \tau_L) = V_i{}^2 .$$

After the next updating, i-th neuron output is:

$$V_i (t + \tau_S) = stp [1/2 \sum_{j=1}^{4} T_{ij}{}^S (2 V_j{}^1 - 1) + (T_{ij}{}^L (2 V_j{}^2 - 1)]$$

$$= stp \left[\frac{J_0}{8} \begin{bmatrix} 0 & 2 & 0 & -2 \\ 2 & 0 & 0 & 2 \\ 0 & 0 & 0 & 2 \\ -2 & 2 & 2 & 0 \end{bmatrix} \begin{bmatrix} 1 \\ 1 \\ 1 \\ 1 \end{bmatrix} + \frac{\lambda J_0}{8} \begin{bmatrix} 0 & 1 & -1 & -3 \\ 3 & 0 & -1 & -3 \\ 1 & 3 & 0 & -1 \\ -1 & 1 & 3 & 0 \end{bmatrix} \begin{bmatrix} -1 \\ 1 \\ 1 \\ 1 \end{bmatrix} \right] = stp \frac{J_0}{8} \begin{bmatrix} -4-3\lambda \\ -5\lambda \\ 2-5\lambda \\ 2-3\lambda \end{bmatrix} = \begin{bmatrix} 0 \\ 0 \\ 0 \\ 1 \end{bmatrix} \quad \frac{2}{5} < \lambda < \frac{2}{3}$$

$$= stp \left[\frac{J_0}{8} \begin{bmatrix} -3\lambda \\ 4-7\lambda \\ 2+\lambda \\ 2+5\lambda \end{bmatrix} \right] = \begin{bmatrix} 0 \\ 0 \\ 1 \\ 1 \end{bmatrix} \quad \lambda > 1$$

$$v_j^{[1+\cdot}s^{\cdot\cdot}t]=\text{s}\phi\left[\frac{J_A}{\theta}\begin{bmatrix}0&2&0&-2\\2&0&0&2\\0&0&0&2\\-2&2&2&0\end{bmatrix}\begin{bmatrix}-1\\-1\\1\\1\end{bmatrix}+\frac{1}{\theta}\begin{bmatrix}0&1&-1&-3\\3&0&-1&-3\\1&3&0&-1\\-1&1&3&0\end{bmatrix}\begin{bmatrix}\cdot\cdot\\-1\\-1\\1\end{bmatrix}\right]$$

Here we have state the V^4. The system finds stability in state 4.

It would be necessary to experiment this model with the theoretical value of λ. To resume the cycle a new pulse is required.

CONCLUSION.

We consider that it is possible to obtain oscillatory patterns for at least some values of λ, as we have indicated. This model does not need any clock: the sequential output depends on the interference between slow and fast synaptic components. The learning Hebbian laws we have used to specify the weights are suitable when the recovering between the embedded states is negligible. But this is the case of many biological systems. It would be necessary to adapt the model with hidden units in order to reduce recovering. Finally, we can explain the evolution of the synaptic weights like a mechanism for dynamic learning: T_{ij}^L means two events separated in time (τ_L and T_{ij}^S).

REFERENCES.

- Koch, K. and Segev, I. (1989) *Methods in neuronal modeling: from synapses to networks*. MIT Press.
- McKenna, T., Davis, J. and Zorn, S.F. (1992) *Single neural computation*. Acad. Press.
- Mira Mira, J. (1993) *Computación neuronal*. UNED. Madrid.
- Traub, R.D. and Miles, R. (1991) *Neuronal networks of the hippocampus*. Cambridge Univ. Press.
- Hush, D.R. and Horne, B.. *An overview of neural networks*. Part I. static networks. Part II: dynamic networks. New Mexico Univ.
- Mira, J. and Delgado, A.E. (1993) *Always trying to write an equation for the brain*. UNED. Madrid.
- Pearlmutter, B.A. (1990) *Dynamic recurrent neural networks*. Carnegie Mellon Univ.
- Rumelhart, D.E., Hinton, G.E. and Williams, R.J. *Learning internal representations by error propagation*.
- Sanchez-Andrés, J.V., Olds, J.L. and Alkon, D.L. (1993) Gated informational transfer within the mammalian hippocampus: a new hypothesis. *Behav. Brain Res.* 54, 111-6.
- Torras, C. (1992) *Relaxation and neural learning: points of convergence and divergence*. CSIC-UPC, Barcelona.

The Implementation of Propositional Logic in Random Neural Networks

Marc-Denis Weitze & G. L. Hofacker

Institut f. Theoret. Chemie, Technische Universität München, 85 747 Garching, Germany
Tel. +49-89-3209-3592, Fax +49-89-3209-3266, Email weitze@theochem.tu-muenchen.de

Describing the evolution of cognition as an informational process, we want to compare the complexity and computational power of connectionist models of some cognitive functions to the maximal information transmitted in brain relevant genes during neocortex evolution.
In this paper we implement propositional logic as Boolean functions in neural networks and investigate what types of functions emerge in nets with specified architecture (feedforward vs. fully backcoupled) and random weights. For $N=2,3$ arguments the relative portions of Boolean functions are given as results of a Monte Carlo simulation.

1. Introduction

If one describes evolution as an informational process, one may get an estimate of human neocortical complexity measured as the maximal information transmitted in brain relevant genes during human neocortex evolution. This information is about $2 \cdot 10^5$ bits (Hofacker et al. 1988:245) and is now to be related to the complexity of neural network (NN) structures. We confine ourselves to the restricted case of propositional logic. "Boolean operations of propositional logics ... allow for the construction of ... logical statements whenever features can be classified by two categories (true/false, friend/foe, feature present/absent)" (Hofacker et al. 1988:249). There is no indication that random NNs self-organize into logical processors, and the world - in particular the primitve world of early humans - is certainly too chaotic to provide training sets for logical processors. So we have to look for other (and we think of genetic) possibilities to implement logic into NNs.

The very small and simple NNs simulated in this investigation are to be read as connectionist models of some special cognitive functions (cf. McCulloch&Pitts 1943). For we regard the processural level, our model neurons and model synapses are by no means to be identified with single biological neurons or synapses, though this would be a possible physical realization .

In chapter 2 we state the notation of propositional logic in terms of BFs and in chapter 3 we describe our simulations of NNs implementing BFs, followed by a brief discussion of our results.

2. Propositional Logic and Boolean Functions

Propositional logic deals with assigning N-tupels of Ts and Fs (T for True and F for False) as T or F. One can regard this assignment as a mapping M: $\{-1,+1\}^N \rightarrow \{-1,+1\}$ in bipolar coordinates or as a Boolean function (BF) with N arguments (there are $2\uparrow 2^N$ different BFs). Besides giving a corresponding propositional calculus formula (PCF) or tabulating the mapping explicitly like a truth-table, there are more descriptions which turn out as convenient at different points of our investigation: i) Each mapping can be described in short by a 2^N-element bipolar string. That is simply the right hand column in the explicit formulation of the mapping given a fixed - say lexicographic[1] - order of the mapped N-element vectors. ii) A quite extensive description is the listing of all vectors mapped to +1. We will call this description 'T-list'. iii) One property of the PCF is that there are many PCFs for one BF. One may prefer using only ANDs, ORs or NOTs; to get some consistency one often agrees upon using normal forms like the disjunctive normal form (DNF)[2].

Because the number of BFs increases very fast with N, it is advisable to look for methods taking some BFs together into types of BFs. For our purpose it turns out to be reasonable to start with a look at the PCF description: for we handle T and F as equal (though opposite) and do not mind the permutation of the literals, we consider those BFs as belonging to the same type, which pass into each other by means of negation and permutation of literals in the PCFs.

The number of BF types for N=1,2,3,4 is as follows:

N	# BF	# BF types
1	4	3
2	16	6
3	256	22
4	65536	402

In the appendix BF-types for N=2,3 literals are given together with the BFs belonging to that type.

3. The Implementation of Boolean Functions in Neural Networks

We will investigate what kind of BF some NN with specified architecture (i.e. number of units, type of connectivity) will implement with weights determined randomly. The NNs investigated consist of units j whose state $s_j(t+1)$ at time step t+1 is determined as a threshold function $\Theta_j(.)$ of the summed activity along its input lines. The input lines are connections from other units i=1,...,N conveying their activity at the previous time-step t, $s_i(t)$ weighted by a factor w_{ij} attached to each connection from unit i to unit j.

In this paper we consider two cases of connectionist architecture: i) the feedforward type NN

---

[1]$x^{(1)}$ is lexicographically before $x^{(2)}$, if $x^{(1)}_i < x^{(2)}_i$ for $i=\min_j\{j|x^{(1)}_j \neq x^{(2)}_j\}$, assuming -1<+1.

[2]The DNF is "the OR of ANDs" (Hampson&Volper 1987:123).

$$s_j(t+1) = \Theta_j(\sum_{i=1}^{N} w_{ij} s_i(t))$$

$$\Theta_j(x) = \begin{cases} +1 \ for \ x \geq \vartheta_j \\ -1 \ for \ x < \vartheta_j \end{cases}$$

consisting of an input layer (N units), one hidden layer (N_{hid} units) and one output unit with connections only between neurons of adjacent layers and ii) the fully backcoupled type with N units, all units connected with each other.

In each case we investigate an ensemble of NNs with a specified architecture but different weights w_{ij}. For each NN we assign all weights randomly from the interval [-1,+1]. The thresholds ϑ_j are generated in the same way. No learning is involved.

i) feedforward NNs (FFNs)
Each FFN with N input neurons and one output neuron can - regardless of hidden layers - implement one of $2\uparrow2^N$ BFs with N arguments (neuron state -1 (+1) corresponding to truth value False (True)). Given a particular FFN and computing the output state for each of the 2^N input vectors one can find out the bipolar string characterizing the BF that is implemented by that FFN.

In our simulation we generated for each of the connectionist architectures with N=2,3,4 input neurons and different numbers of hidden neurons (N_{hid}=0,1,2,3,4,5,9) ensembles of 1,000,000 FFNs. The simulation results yield for each connectionst architecture a list in which for each of the $2\uparrow2^N$ BFs the number of FFNs implementing this BF is given.

Consider as an example the perceptron with two input neurons (N=2, N_{hid}=0). If we only consider *types* of BFs we get tables like this:

BF type	# BF belonging to that type (Y) (Σ=16)	#FFN implementing that type (X) (Σ=1,000,000)	X / Y
1	1	82603	82603
2	4	251169	62792
3	4	333004	83251
4	2	0	0
5	4	249535	62383
6	1	83689	83689

Column 4 is a quotient yielding an average number each BF of the corresponding type is implemented.

As one may see, only BF type 4 is n ot implemented. In our notation this type corresponds to the XOR function (this is just the famous Minsky & Papert (1969) example which represented a severe blow to the perceptron concept). Even from this short example we can learn that 'simple' functions as tautology (type 1) and contradiction (type 6) are implemented easier (i.e. more often) than functions like AND (type 2) and OR (type 5).

In our simulations with FFNs, generally each BF will be implemented with the same portion as its inverse BF^3. Because two BFs that are inverse to each other may not belong to the same type of BFs, type 1 and 6 as well as type 2 and 5 received the same portions in this example.

The simulation results for N=2,3 are visualized as pie diagrams: each pie piece corresponds to the share of a BF type being implemented (type-pieces are ordered counter-clockwise in ascending numerical order). To emphasize the effect of increasing the number of hidden neurons the pie pieces corresponding to BF-types not implemented in NNs with one neuron less are exploded. Alone as a result of network structure, there emerges clearly a preference for implementing special types of BFs (Figs.1,2).

ii) fully backcoupled NNs (FBNs)

Each FBN with N neurons may be considered as implementing one of $2\uparrow2^N$ BFs with N arguments. A bipolar input vector $\{s_j, j=1,...,N\}$ is considered True (False) if the corresponding activity pattern in the FBN is a stable (not stable) state in the parallel dynamics[4].

$\{s_j\}$ is stable, iff

$$s_j = \Theta_j(\sum_{i=1}^{N} w_{ij}s_i) \quad \textit{for } j=1,...,N$$

$$\Theta_j(x) = \begin{cases} +1 & x \geq 0 \\ -1 & x < 0 \end{cases}$$

Note that in this part all thresholds are taken as zero ($\vartheta_j=0$) because the self couplings w_{ii} do the same job as the thresholds and may replace them for formal convenience.

In order to determine the BF implemented by a FBN one has to test the stability of each of the 2^N activity patterns in the FBN. In our simulation we generated FBNs of N=2,3,4 neurons. Ensembles of 100,000 FBNs were generated by determining the connection weights randomly. Because of the chosen dynamics we get within a FBN always *pairs* of attractors: if s is stable, (-s) is stable, too. This of course limits the amount of implementable BFs quite strongly. In the following tables the simulation results for N=2,3,4 are listed:

N=2:

BF type	1	4	6
% FBN	56.2	37.6	6.2

---

[3]This is a result of the symmetry of the interval [-1,+1] we take the synaptic weights from.

[4]We only consider stable attractors and regard oscillatory attractors as 'instable' because this seems to be the most convenient interpretation in the framework of propositional logic.

N=3:

BF type	1	9	14	20	22
% FBN	59.4	32.4	7.2	1.0	0.06

N=4:

BF type	1	46	106	127	191	212
% FBN	61.5	36.3	5.5	2.2	1.5	0.08

218	286	318	355	379	400	402
0.08	0.07	0.01	0.00	0.00	0.00	0.00

4. Discussion

We made out special types of BFs that emerge in NNs in the sense that they are preferably implemented with minimal information about the weights. In feedforward NNs those BFs like the XOR function are hard to implement, whereas in the case of fully backcoupled NNs even some of those BF types emerged (for N=2 arguments BF type 4 and for N=3 arguments types 9, 14, and 20).

As a quantitative measure for the complexity of a single BF i one may regard the Kullback entropy

$$K_i = -\log_2 \frac{number\ of\ NNs\ implementing\ BF\ i}{number\ of\ all\ NNs}$$

(Carnevali&Patarnello 1987:1203). It is noteworthy that, if one considers learning in this model, "it is not always the BFs with largest phase volume that are easy to learn" (Van den Broeck&Kawai 1990:6213, who give a detailled investigation of the entropy landscape for a similiar problem). Rojas (1993) presents an appealing geometric approach to the weight space.

It may be interesting to test the validity of these connectionist models with regard to empirical results, i.e. to compare our results to the ease with which animals or humans perform tasks involving special BFs. Let us state, as a working hypothesis, that logical functions considered as "simple" in natural, bio-logical context are just those functions emerging in NNs with random weights and appropriate architecture. Of course there are many problems in experimental evidence, citing just one of them, "the notoriously poor fit between the meanings of some of the connectives and quantifiers of standard logic and the nearest analogous natural language concepts" (Braine 1978), might exemplify this.

For our work in the context of evolution and cognition the results provide a possibility to estimate the information needed to code for some cognitive functions, which could be compared to the maximal amount of information that was selected during a special period of evolution.

Literature

M.D.S.Braine: On the Relation between the Natural Logic of Reasoning and Standard Logic, *Psychological Review* **85**(1987)1-21.

P.Carnevali & S.Patarnello: Exhaustive Thermodynamical Analysis of Boolean Learning Networks, *Europhysics Letters* **4**(1987)1199-1204.

S.E.Hampson & D.J.Volper: Disjunctive Models of Boolean Category Learning, *Biological Cybernetics* **56**(1987)121-137.

G.L.Hofacker, B.Borstnik & M.Schöniger: Evolutionary Adaption to a Real and an Artificial World, in: *Biological and Artificial Intelligence Systems*, E.Clementi & S.Chin (Eds.), ESCOM, Leiden 1988.

W.S.McCulloch & W.Pitts: A Logical Calculus of the Ideas Immanent in Nervous Activity, *Bulletin of Mathematical Biophysics* **5**(1943)115-133.

R.Rojas: *Theorie der neuronalen Netze*, Springer, Berlin 1993.

C.Van den Broeck & R.Kawai: Learning in Feedforward Boolean Networks, *Physical Review* **42**(1990)6210-6218.

Appendix

For N=2,3 arguments all BF types are listed and described. Note that we do not use a systematic description like normal forms. In each case we introduce the most convenient one to provide the reader with some idea, what BFs belong to what type.

N=2:

BF type	# BF belonging to that type	# elements in T-list	description
1	1	0	contradiction
2	4	1	$x_1 \wedge x_2$
3	4	2	x_1
4	2	2	$(x_1 \wedge \neg x_2) \vee (\neg x_1 \wedge x_2)$
5	4	3	$x_1 \vee x_2$
6	1	4	tautology

N=3:

BF type	# BF belonging to that type	#elements in T-list	description
1	1	0	contradiction
2	8	1	$x_1 \wedge x_2 \wedge x_3$
3	12	2	$x_1 \wedge x_2$
4	12	2	$x_1 \wedge ((x_2 \wedge \neg x_3) \vee (\neg x_2 \wedge x_3))$
5	24	3	$x_1 \wedge \neg(x_2 \wedge x_3)$
6	6	4	x_1
7	8	3	$(x_1 \wedge \neg x_2 \wedge \neg x_3) \vee (\neg x_1 \wedge x_2 \wedge \neg x_3) \vee (x_1 \wedge \neg x_2 \wedge x_3)$
8	8	4	$(x_1 \wedge x_2) \vee (x_2 \wedge x_3) \vee (x_1 \wedge x_3)$
9	4	2	$(x_1 \wedge x_2 \wedge x_3) \vee (\neg x_1 \wedge \neg x_2 \wedge \neg x_3)$
10	24	3	$(x_1 \wedge x_2) \vee (\neg x_1 \wedge \neg x_2 \wedge \neg x_3)$
11	24	4	$(x_1 \wedge x_2) \vee (\neg x_2 \wedge \neg x_3)$
12	24	4	$(x_1 \wedge \neg(\neg x_2 \wedge \neg x_3)) \vee (\neg x_1 \wedge \neg x_2 \wedge \neg x_3)$
13	24	5	$x_1 \vee (\neg x_1 \wedge \neg x_2 \wedge \neg x_3)$
14	6	4	$(x_1 \wedge x_2) \vee (\neg x_1 \wedge \neg x_2)$
15	24	5	$(x_1 \wedge \neg(x_2 \wedge x_3)) \vee (\neg x_1 \wedge x_2)$
16	12	6	$\neg(x_1 \wedge x_2)$
17	2	4	$(x_1 \wedge \neg x_2 \wedge \neg x_3) \vee (\neg x_1 \wedge x_2 \wedge \neg x_3) \vee (\neg x_1 \wedge \neg x_2 \wedge x_3) \vee (x_1 \wedge x_2 \wedge x_3)$
18	8	5	$(x_1 \wedge \neg(x_2 \wedge x_3)) \vee (\neg x_1 \wedge ((x_2 \wedge x_3) \vee (\neg x_2 \wedge \neg x_3)))$
19	12	6	$(x_1 \wedge \neg(x_2 \wedge x_3)) \vee (\neg x_1 \wedge \neg(x_2 \wedge \neg x_3))$
20	4	6	$(x_1 \wedge \neg(x_2 \wedge x_3)) \vee (\neg x_1 \wedge \neg(\neg x_2 \wedge \neg x_3))$
21	8	7	$\neg(x_1 \wedge x_2 \wedge x_3)$
22	1	8	tautology

N=2

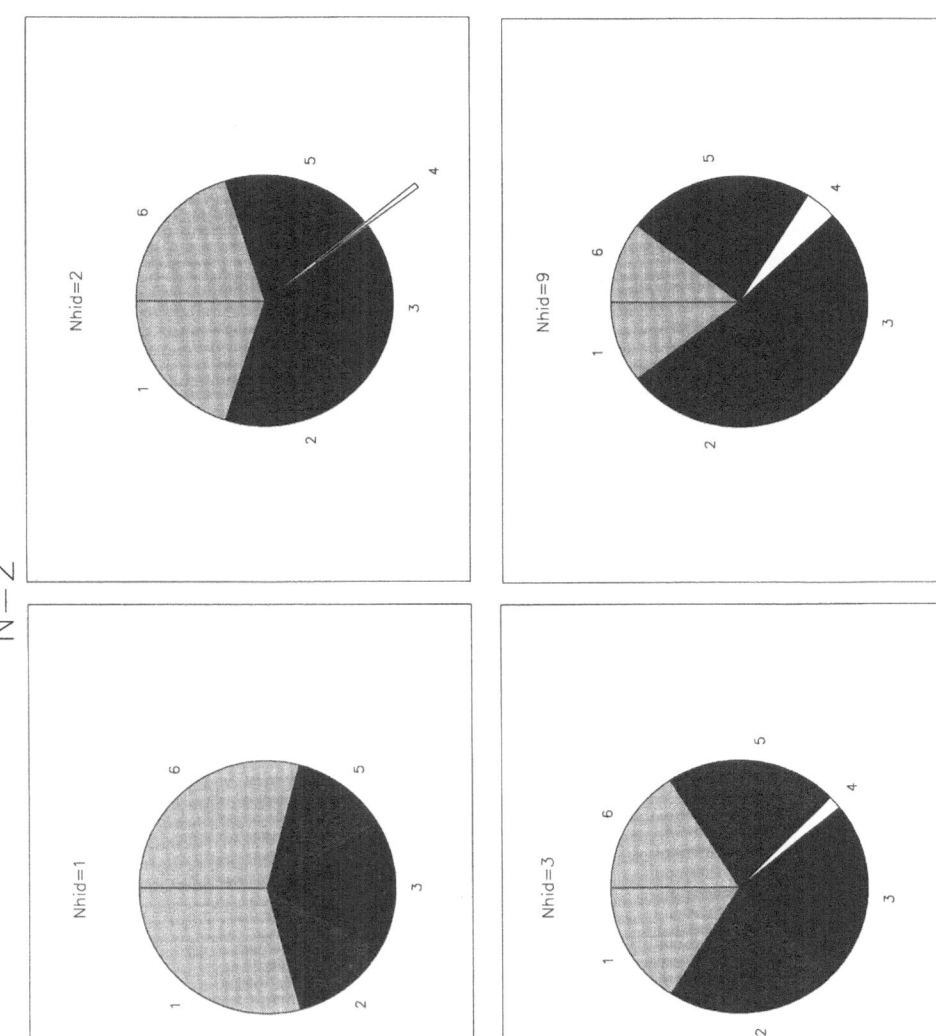

N=3

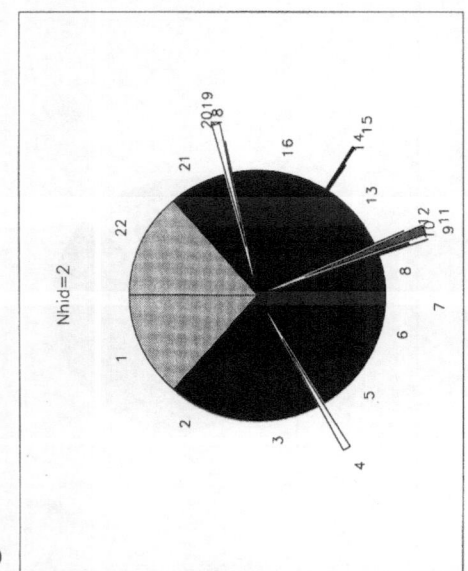

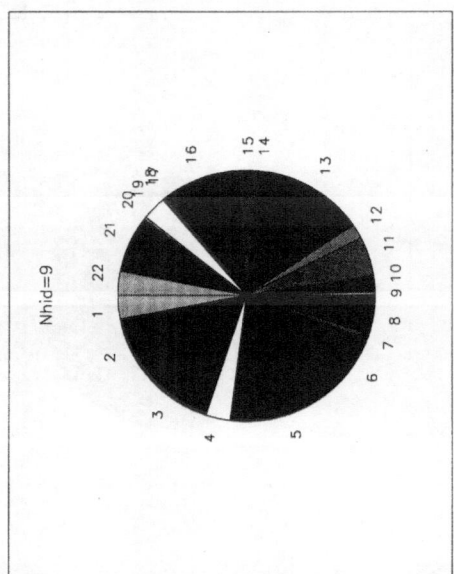

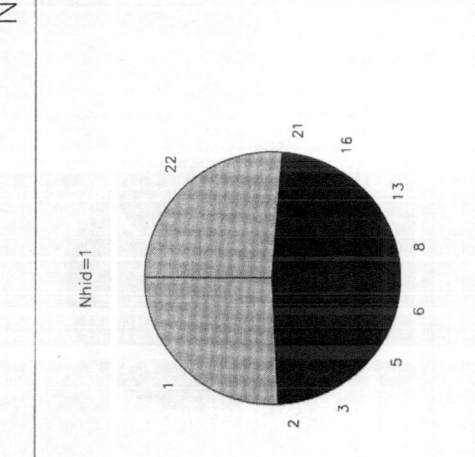

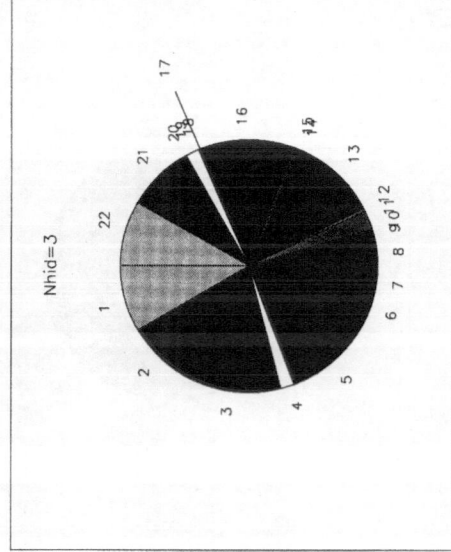

A Learning Rule for Extracting Temporal Invariances

James V Stone & Alistair Bray

School of Cognitive and Computing Sciences,
University of Sussex,
Sussex,BN1 9QH
Email: jims@cogs.susx.ac.uk, alib@cogs.susx.ac.uk
Fax: 44 273 671320
Phone: 44 273 678763

Abstract

Neurons recode information implicit in their inputs. Within a population of neurons, this recoding process is often associated with the removal of redundant information. Clearly, an important objective is to extract a perceptually salient invariance, whilst also specifying its value. A linear combination of Hebbian and anti-Hebbian adaptation (HAH learning), operating simultaneously upon the *same* connection weights but at different time scales, is shown to be sufficient for the unsupervised learning of temporal invariances. A model neuron which implements this rule learns to detect an invariance, and also specifies the value of that invariance.

We present a learning rule that is based upon an assumption which we believe to be quite general: the inputs to sensory receptors (e.g. retinal photoreceptors) tend to change rapidly and discontinuously over time, whereas the physical parameters (e.g. depth of a surface) underlying these changes vary more slowly, and more smoothly. Accordingly, if a unit codes for a physical parameter then its output should also change slowly and smoothly, despite its rapidly fluctuating inputs. We demonstrate that a unit which adapts to make its output vary smoothly over time can learn to code the invariances underlying its input.

1 Introduction

A fundamental property of the visual system is the incremental convergence of information across cortical areas. This convergence ensures that information initially implicit in a large number of neurons with narrow tuning curves can be represented in the outputs of a smaller number of more broadly tuned neurons.

The input to a perceptual system can be characterised as a vector in a high dimensional space. For example, there are approximately 10^6 retinal ganglion cells projecting from primate retina to the cortex; the input over the whole visual field may be viewed as a vector in this space of 10^6 dimensions. Over time, this vector changes rapidly, and the path it follows in the high-dimensional space is discontinuous. However, the vector is a function of a physical world which can usually be described using only a handful of parameters. It is these parameters that are useful to an organism. These parameters tend to vary in a slower, smoother manner than the input vector. A large part of the problem of perception consists of deciphering these underlying physical parameters, given access to the input vector alone or, as Gibson stated [1], obtaining invariant structure from continually changing sensations. This paper addresses the general problem, relevant beyond the domain of perception, of how to extract a small number of smoothly varying parameters from a high-dimensional, rapidly varying data-stream.

Current models exploit temporal correlations in their inputs [2, 3, 4]. These models update synaptic strengths between units according to exponentially weighted time-averages of post-synaptic (or pre-synaptic) activity over the recent past. Using traces with Hebbian synaptic modification allows each unit to make strengthen connections to others with which it is temporally correlated. Essentially, such invariance-seeking units do no more than compute a logical **OR** on a subset of their inputs. For example, a unit that learns to be selective for edge-orientation, whilst invariant to edge-position, forms uniformly strong connections to the subset of all input units having the appropriate orientation, regardless of spatial position, since activity within this subset is temporally correlated. These models are limited because they do not specify the *value* of the invariance.

2 The Learning Rule

The learning rule is based upon an assumption which we believe to be quite general: The inputs to sensory receptors (e.g. retinal photoreceptors) tend to change rapidly and discontinuously over time, whereas the physical parameters (e.g. depth of a surface) underlying these changes vary more slowly, and more smoothly. Accordingly, if a unit codes for a physical parameter then its output should also change slowly and smoothly, despite its rapidly fluctuating inputs. Therefore, *a unit that adapts to make its output vary, but to vary smoothly over time, can learn to code the invariances underlying its input.* The output can be made to reflect both smoothness *and* variability by forcing it to have a small *short-term* variance, and a large *long-term* variance.

This strategy can be implemented for a single linear unit u. The output of u at time t is y_t. The temporally weighted output $\tilde{y}_t$ of u is a short-term average of outputs y. More precisely, $\tilde{y}_t$ is a *short-term* temporal exponentially weighted sum of outputs y_t, and $\overline{y}$ is a similar *long-term* average. We can obtain the desired behaviour in y by altering u's input weights so that y has a large long-term variance V, and a small short-term variance U. Maximising V/U maximises the variability of y over 'long' inter-

vals, whilst simultaneously minimising its variability over 'short' intervals. The output y_t of u is $y_t = \sum_j w_j \, x_j$, where w_j is the value of a weighted connection from input x_j to u. A merit function F can be defined as:

$$F = log \frac{V}{U} = log \frac{\sum_{t=1}^{T}(y_t - \overline{y}_t)^2}{\sum_{t=1}^{T}(y_t - \tilde{y}_t)^2}$$

In order to maximise F it is necessary to determine its derivative with respect to each weight w_j. This can be re-written as:

$$F = log \frac{1}{2} \sum (y - \overline{y})^2 - log \frac{1}{2} \sum (y - \tilde{y})^2$$

Its derivative $\frac{\partial F}{\partial w_j}$ is:

$$\sum \frac{(y - \overline{y})}{V} \left(\frac{\partial y}{\partial w} - \frac{\partial \overline{y}}{\partial w} \right) - \sum \frac{(y - \tilde{y})}{U} \left(\frac{\partial y}{\partial w} - \frac{\partial \tilde{y}}{\partial w} \right)$$

Given that y is a linear function of its inputs, this yields $\frac{\partial F}{\partial w_j}$:

$$\frac{1}{V} \sum (y - \overline{y})(x_j - \overline{x}_j) \; - \; \frac{1}{U} \sum (y - \tilde{y})(x_j - \tilde{x}_j)$$

which can be specified as:

$$\frac{1}{V} \langle (y - \overline{y})(x_j - \overline{x}_j) \rangle - \frac{1}{U} \langle (y - \tilde{y})(x_j - \tilde{x}_j) \rangle$$

This is the direction of steepest ascent in F, which can be used to maximise F.

The rule for synaptic modification combines Hebbian and Anti-Hebbian learning (the HAH-rule), operating simultaneously at two different time-scales. The slow Hebbian adaptation is:

$$\Delta_L w_j = \frac{\eta}{V} (y - \overline{y})(x_j - \overline{x}_j)$$

Where η is the learning rate. The fast anti-Hebbian adaptation is:

$$\Delta_S w_j = -\frac{\eta}{U} (y - \tilde{y})(x_j - \tilde{x}_j)$$

The interpretation of this rule is as follows. If V is small relative to U then learning is principally Hebbian, which has the effect of increasing the variability of outputs over long periods. That is, it prevents the output of the unit being constant. However, if V is large relative to U, then learning is principally anti-Hebbian. This has the effect of decreasing the variability of outputs over shorter periods. The net effect of these changes is to generate an output which has a large range, but which varies smoothly over time.

The BCM rule [5, 6] and the ABS rule [7, 8, 9] provide computational and biological support for combining Hebbian and anti-Hebbian synaptic modification to acquire stimulus selectivity. However, these rules have no significant temporal component, and adaptation is a function of the *instantaneous* pre-synaptic activity. By ignoring temporal structure in the unit's pre-synaptic input, these rules permit units to learn about spatial, but not temporal, correlations.

3 Experiments

We demonstrate this rule in two computer simulations. For the first simulation (see figure 1), consider a vector whose elements are all zero except one, which is unity. Over time, the position of the non-zero element oscillates about the centre, from one end of the vector to the other in simple harmonic motion. Using the above rule, and taking this temporally changing vector as input, the unit learns to code the *position* of the non-zero element along the vector. For the second simulation (see figure 2), we use the two-dimensional analogue of the first. The non-zero element moves independently in two dimensions. Two output units, with a single decorrelation weight between them, learn to code separately for the two independent motion parameters.

In neither of these simulations would previously mentioned synaptic modification rules have been successful. Standard Hebbian/anti-Hebbian rules would ignore the temporal correlations in the input.

4 Conclusion

We have derived a learning rule that allows a linear unit to compute linear invariances that vary smoothly over time. Such a unit is capable of specifying activity over a set of narrowband input units to a value-coding on the output of far fewer broadly-tuned output units. It is known that narrow-band tuning is characteristic of neurons found in the early stages of perceptual systems (e.g. orientation tuning in the primary vsiual cortex). However, in later stages broad spatial tuning is often found (e.g. coding of shape in infero-temporal cortex [10]). Narrow band tuning of a large number of neurons is often the necessary 'expensive' result of extracting high-order, non-linear statistics (e.g. binocular disparity, see [11]) from data. Once this extraction has been achieved, the linear HAH-rule provides a means of extracting linear invariances in these statistics, that are themselves non-linear invariances in the input.

References

[1] **Gibson J J**. *The Ecological Approach to Visual Perception*. Boston: Houghton Mifflin, 1979.

[2] **Földiák P**. Learning invariance from transformation sequences. *Neural Computation*, 1991.

[3] **Wallis G, Rolls E T and Földiák**. Learning invariant responses to the natural transformations of objects. *International Joint Conference on Neural Networks*, pages 1087–1090, 1993.

[4] **Barrow H G and Bray A J**. A model of adaptive development of complex cortical cells. In Aleksander I and Taylor J, editors, *Artificial Neural Networks II: Proceedings of the International Confernece on Artificial Neural Networks*. Elsevier Publishers, 1992.

[5] **Law C C and Cooper L N**. Formation of receptive fields in realistic visual environments according to the bienenstock, cooper, and munro (bcm) theory. *Proceedings National Academy Science USA*, 91:7797–7801, 1994.

[6] **Bienenstock E L, Cooper L N and Munro P W**. Theory for the development of neuron selectivity: Orientation specificity and binocular interaction in visual cortex. *Journal of Neuroscience*, 2:32–48, 1982.

[7] **Artola A and Singer W**. Long-term depression of excitatory synaptic transmission

and its relationship to longterm potentiation. *TINS*, 16(11):480–487, 1993.

[8] **Artola A, Bröcher S and Singer W**. Different voltage-dependent thresholds for inducing long-term depression and long-term potentiation in slices of rat visual cortex. *Nature*, 347:69–72, 1990.

[9] **Stanton P K and Sejnowski T J**. Associative long-term depression in the hippocampus: induction of synaptic placticity of hebbian covariance. *Nature*, 339:69–72, 1990.

[10] **Perret D I**. Viewer-centred and object-centred coding of heads in the macaque temporal cortex. *Experimental Brain Research*, 86:159–173, 1991.

[11] **Stone J V**. Learning spatio-temporal invariances. In *Proceedings of the British Machine Vision Conference*, pages 681–690, September 1994.

Figure 1

Extracting one invariance

The value of an invariance is coded by the position of activation on a set of inputs; the output unit adapts such that its value correlates with the invariance. At each time t a linear unit has input elements x_j ($1 \leq j \leq 101$). All input elements are zero except $x_{j'} = 1$ where $j' = round(51 + 50 * sin(t * 360/\rho))$ and $\rho = 450$. Weighted connections between the unit and its inputs are initially random ($\sum_{j=1}^{101} w_j^2 = 1$). At each time-step the weight vector adapts according to the HAH-rule so that:

$$\Delta w_j = \frac{\eta}{V}(y - \overline{y})(x_j - \overline{x}_j) - \frac{\eta}{U}(y - \tilde{y})(x_j - \tilde{x}_j)$$

Where $\eta = 0.001$, and $w_j^{t+1} = w_j^t + \Delta w_j$.

The trace $\overline{y}$, with a half-life of h_L time-steps, is computed as:

$$\overline{y}^{t+1} = \lambda_L \overline{y}^t + (1 - \lambda_L)y^{t+1}$$

Where $\lambda_L = e^{log(0.5)/\tau_L}$. Similar formulae are used for $\overline{x}$, $\tilde{y}$ and $\tilde{x}$, where the long-term averages have $\tau_L = 2\rho$, and the short-term averages have $\tau_s = (\rho/31)$. Values for V and U are approximated using similar formulae:

$$V^{t+1} = \lambda_L V^t + (1 - \lambda_L)(y^{t+1} - \overline{y}^{t+1})^2$$

$$U^{t+1} = \lambda_l.U^t + (1 - \lambda_l)(y^{t+1} - \overline{y}^{t+1})^2$$

Both V^t and U^t are computed using the same value of $\tau_l = 2\rho$; their values are used, rather than V and U defined earlier, since $V \approx V^t$ and $U \approx U^t$ if the input statistics are approximately constant over a period τ_l. *No weight normalisation is required* because the merit function $F = log(V/U)$ is independent of the magnitude of y.

[a]. The Connectivity. A single output unit has weighted connections to 101 binary-valued inputs; at any time, only a single input is non-zero and its position oscillates over time.

[b]. The Output. After learning, the value of y is plotted against time for 5ρ (2250) time-steps. The correlation r between y and the position j' of the non-zero element is is $|r| > 0.999$.

[c]. The Weights. The value of w_j is plotted against position j. The straight line has a correlation $|r| > 0.999$.

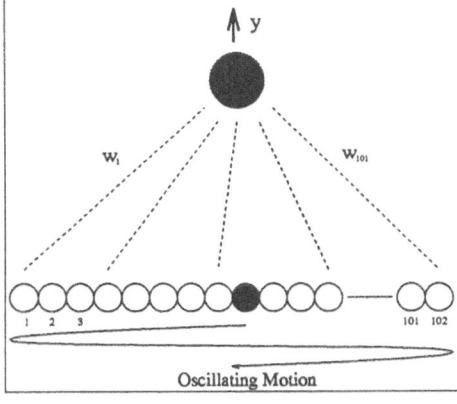

[a]

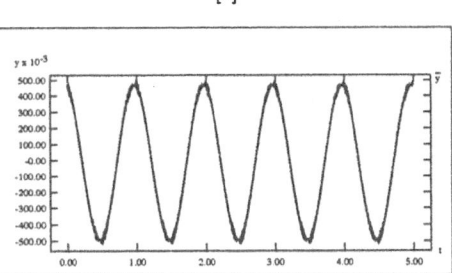

[b]

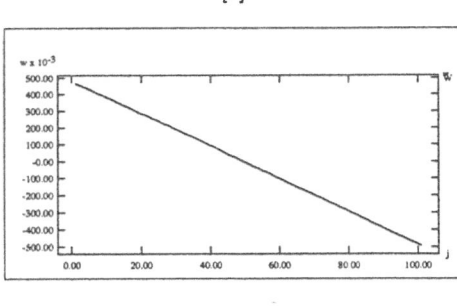

[c]

Figure 1:

Figure 2

Extracting two invariances.

Two invariances are coded by the position of activation on a set of inputs; two output units adapt such that their values correlate with these invariances[1]. Two linear units receive the same inputs x_{ij} $(1 \leq i, j \leq 51)$; at time t all inputs are zero except $x_{i'j'} = 1$ where $i' = round(26 + 25 * sin((t*360/\rho)+\phi))$, $j' = round(26+25*sin((t* 360/\rho) - \phi))$, $\rho = 450$ and $\phi = (t*17)/360$. Weighted connections between each unit and its inputs are initially random $(\sum_{i=1}^{51} \sum_{j=1}^{51} w_{ij}^2 = 1)$. There is a single weight w_{ah} that acts to de-correlate the two outputs y_1 and y_2; it is defined to be the negative of the correlation co-efficient between y_1 and y_2 over the last τ time-steps, where $\tau = min(20\rho, t)$. The outputs are defined as:

$$y_1 = \sum_{i=1}^{51} \sum_{j=1}^{51} w_{1_{ij}} . x_{ij}$$

$$y_2 = \sum_{i=1}^{51} \sum_{j=1}^{51} w_{2_{ij}} . x_{ij} + k.w_{ah}.y_1$$

where $k = 10$. At each time-step the weight vectors are adapted as in the previous simulation, with $\eta = 0.001$. In computing the traces $\tau_l = 2\rho$ and $\tau_s = (\rho/31)$.

[a]. The Connectivity. Two output units each have weighted connections to the same array of 51x51 binary-valued inputs; at any time, only a single input is non-zero and its position oscillates independently in the two spatial dimensions over time. The non-symmetrical anti-Hebbian connection ensures that the output of the second unit is de-correlated from that of the first.

[b]. The Output. After learning, the values for y_1 and y_2 are plotted against time for 5ρ time-steps. The value of y_1 correlates almost perfectly with the position i' of the non-zero element along the i dimension and the value of y_2 with its position j' along the j dimension. In both cases the correlation coefficient $|r| > 0.975$

[c]. The Weights. The two weight vectors w_1 and w_2 are shown as 2D grey-scale arrays. Each

[1]This simulation is analogous to Földiák's if one invariance is considered to be edge-orientation and the other edge-position: one unit becomes spatially invariant, the other orientation invariant.

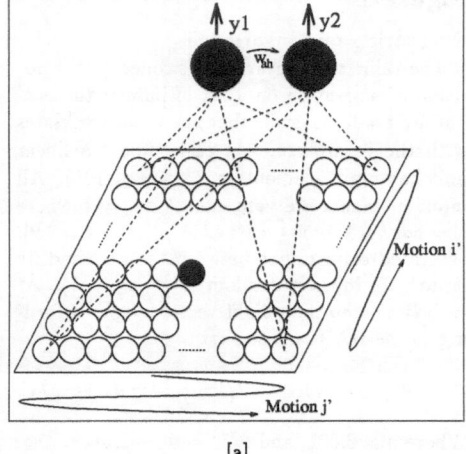

[a]

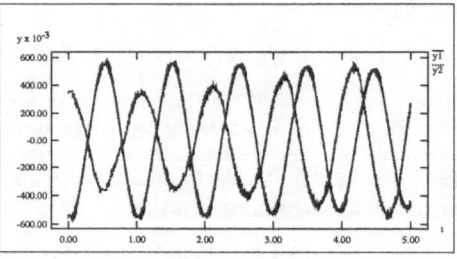

[b]

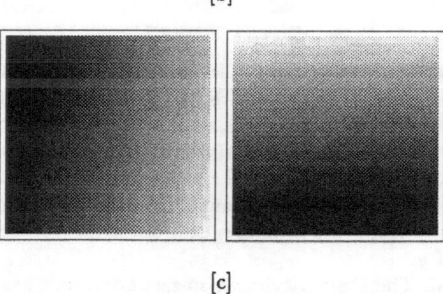

[c]

Figure 2:

is a 2D ramp with weight values rising linearly in one direction but remaining constant in the orthogonal direction. The directions of weight increase for the two units are approximately orthogonal to one another.

The Influence of the Sigmoid Function Parameters on the Speed of Backpropagation Learning

Jun Han   Claudio Moraga

Research Group Computational Intelligence
Dept. of Computer Science, University of Dortmund
D-44221 Dortmund, Germany

Abstract

Sigmoid function is the most commonly known function used in feed forward neural networks because of its nonlinearity and the computational simplicity of its derivative. In this paper we discuss a variant sigmoid function with three parameters that denote the dynamic range, symmetry and slope of the function respectively. We illustrate how these parameters influence the speed of backpropagation learning and introduce a hybrid sigmoidal network with different parameter configuration in different layers. By regulating and modifying the sigmoid function parameter configuration in different layers the error signal problem, oscillation problem and asymmetrical input problem can be reduced. To compare the learning capabilities and the learning rate of the hybrid sigmoidal networks with the conventional networks we have tested the two-spirals benchmark that is known to be a very difficult task for backpropagation and their relatives.

1 Introduction

A sigmoid function is a bounded differentiable real function that is defined for all real input values and that has a positive derivative everywhere. It shows a sufficient degree of smoothness and is also a suitable extension of the soft limiting nonlinearities used previously in neural networks. The backpropagation algorithm also focused attention on it due to the existence of theorems which guarantee the so called "universal approximation" property for neural networks. Hornik and White showed that feedforward sigmoidal architectures can in principle represent any Borel-measurable function to any desired accuracy, if the network contains enough "hidden" neurons between the input and output neuronal fields [1] [2]. However, backpropagation learning is too slow for many applications and it scales up poorly as tasks become larger and more complex. The factors governing learning speed are poorly understood [3]. A number of people have analyzed the reasons why backpropagation learning is so slow, such as step size problem, moving target problem and all the problems that occur with gradient descent (See e.g. [3],[4],[5]). In this paper we introduce a variant sigmoid function with three parameters that denote the dynamic range, symmetry and slope of the function respectively and a hybrid sigmoidal network: *a fully feedforward network with the variant sigmoid function and with different parameters configuration in different layers*. By regulating these three parameters in different layers we can reduce the attenuation and dilution of the error signal problem, the oscillation problem and the asymmetrical input problem that contribute also to the slowness of backpropagation learning. Results on the two-spirals benchmark are presented which are better than any results under backpropagation feed forward nets using monotone activation functions published previously.

2 Three Parameters of the Variant Sigmoid Function

A standard choice of sigmoidal function, depicted in Fig. 2.1, is given by [6],[7]

$$f(h) = \frac{1}{1+\exp(-2\beta h)}$$

which satisfies the condition $f(h) + f(-h) = 1$. Further, a sigmoidal function is assumed to rapidly approach a fixed finite upper limit asymptotically as its argument gets large, and to rapidly approach a fixed finite lower limit asymptotically as its argument gets small. The central portion of the sigmoid (whether it is near 0 or displaced) is assumed to be roughly linear. Among physicists this function is commonly called the Fermi function, because it describes the thermal energy distribution in a system of identical fermions. In this case the parameter β has the meaning of an inverse *temperature*, $T = \beta^{-1}$ [8].

The form of the variant sigmoid function is given as

$$f(x, c_1, c_2, c_3) = \frac{c_1}{1+\exp(-c_2 x)} - c_3$$

where

- c_1: Parameter for the dynamic range of the function; (See Fig. 2.2)

- c_2: Parameter for the slope of the function; (See Fig. 2.3)

- c_3: Parameter for the symmetry (or bias) of the function. (See Fig. 2.4)

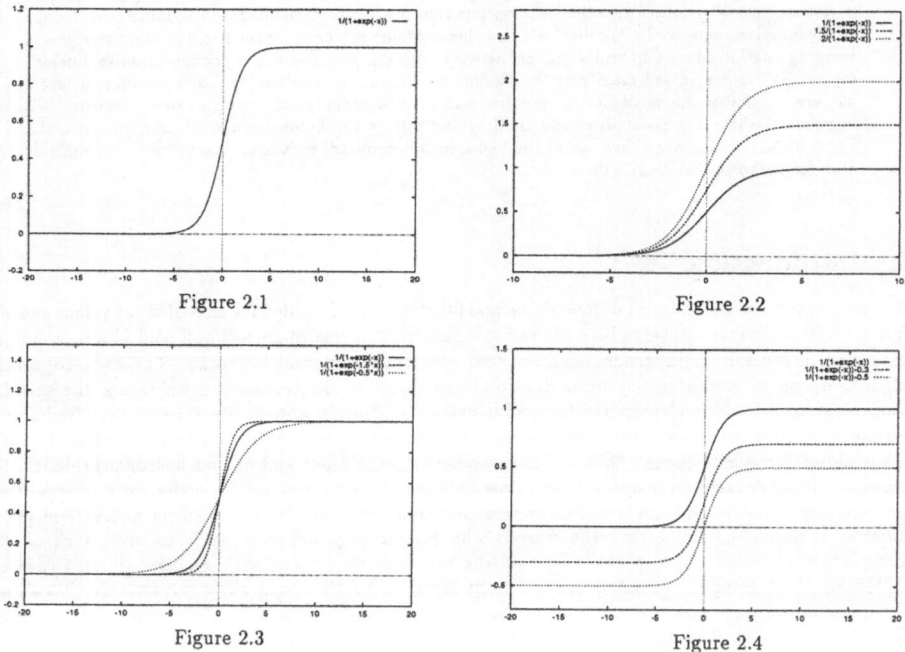

Figure 2.1

Figure 2.2

Figure 2.3

Figure 2.4

Notice that if $c_3 = \frac{c_1}{2}$,

$$f(x, c_1, c_2, c_3) = \frac{c_1}{1+\exp(-c_2 x)} - \frac{c_1}{2}$$

becomes a symmetric (hyperbolic) function. Note that the hyperbolic tangent is just the standard sigmoid function biased and rescaled, as shown by

$$f(x, c_1, c_2, c_3) = \frac{c_1}{1+\exp(-c_2 x)} - \frac{c_1}{2} = \frac{c_1}{2}\left[\frac{1-\exp(-c_2 x)}{1+\exp(-c_2 x)}\right]$$

$$= \frac{c_1}{2}\left[\frac{\exp(+\frac{c_2}{2}x)-\exp(-\frac{c_2}{2}x)}{\exp(+\frac{c_2}{2}x)+\exp(-\frac{c_2}{2}x)}\right] = \frac{c_1}{2}\tanh(\frac{c_2}{2}x)$$

3 The Influence on the Speed of Backpropagation Learning

Many experiments have reported that backpropagation learning slows down dramatically (perhaps exponentially) as we increase the number of hidden layers in a network. This slowdown is partly due to an *attenuation and dilution* of the error signals as they propagate backwards through the layers of the network. The other reason is the so called "moving target effect" and it is well solved through the Cascade-correlation learning architecture [4]. One way to reduce the attenuation and dilution of the error signal problem is to decrease the number of hidden layers and increase the number of hidden units in a given hidden layer of the net, because in a given layer of the net the hidden units cannot communicate with one another directly. The other way is to enlarge the error signals in different hidden layers respectively. That is: *the more a hidden layer is far from the output layer, the more we have to enlarge the error signals.*

First, let us consider the standard backpropagation algorithms [9]. In general, with an arbitrary number of layers, the backpropagation update rule always has the form

$$\Delta w_{pq} = \eta \sum_{patterns} \delta_{output} \times \Gamma_{input}$$

where *output* and *input* refer to the two ends p and q of the connection under consideration and Γ stands for the approriate input-end activation from a hidden unit or a real input. The meaning of δ depends on the corresponding layer. For the last layer of connections δ is given as

$$\delta_i^\mu = f'(h_i^\mu)[\xi_i^\mu - O_i^\mu] \tag{3.1}$$

while for all other layers it is given as shown below

$$\delta_j^\mu = f'(h_j^\mu) \sum_i W_{ij} \delta_i^\mu \tag{3.2}$$

Equation (3.1) allows us to determine δ for a given hidden unit Γ_j in terms of the δ's of the units O_i that it feeds. Meanwhile in equation (3.2) the coefficients are just the usual "forward" W_{ij}'s, but here they are propagating errors (δ's) backwards instead of signals forwards. Both in (3.1) and (3.2) $f'(h_i^\mu)$ denotes the derivative of the activation function of the corresponding unit. We can simply change the parameter c_1 or c_2 of different hidden layers to enlarge the error signals respectively.

C_2 can also be used to reduce the oscillation problem. Because the output layer tends to have larger local gradients than hidden layers, the learning step size should be smaller in the output layer than in the hidden layers. From (3.1) we can see that by changing parameter c_2 in the output layer we can control the learning step size and therefore reduce the oscillation effectively.

C_3 can displace the function up or down along the Y axis and can therefore eliminate the problems arised by the asymmetrical way in which the two states (0's and 1's) of the input are treated.

From (3.2) we can see

$$\delta_i^\mu = f'(h_i^\mu) \sum_i W_{ij} \delta_i^\mu = f(h_i^\mu)[1 - f(h_i^\mu)] \sum_i W_{ij} \delta_i^\mu$$

when $f(h_i^\mu) = 1$ or $f(h_i^\mu) = 0$, we have $f(h_i^\mu)[1 - f(h_i^\mu)] = 0$ and $\delta = 0$.

Thus, for some typical pattern only half of the weights from the input to the hidden layer will be modified, if $c_3 = 0$.

A hybrid sigmoidal hetwork is therefore formed by using different parameter configuration of sigmoid functions in different layers. Let us consider an N-layers hybrid net with $n=1,2,...,N$.

We use Γ_i^n for the output of the ith unit in the nth layer. Γ_i^o will be a synonym for ξ_i, the ith input. Note that superscript n denotes layers, not patterns. Let w_{ij}^n mean the connection from Γ_j^{n-1} to Γ_i^n. Then the backpropagation learning procedure is the following:

1. Initialize the weights with random values to avoid the symmetry breaking.

2. Choose a patten ξ_m^μ and apply it to the input layer ($n=0$) so that

$$\Gamma_m^o = \xi_m^\mu \qquad \text{for all } m \tag{3.3}$$

3. Propagate the signal forwards through the network using

$$\Gamma_i^n = f(h_i^n) = f\left(\sum_j w_{ij}^n \Gamma_j^{n-1}\right) \qquad (3.4)$$

for each i and n until the final outputs Γ_i^N have all been calculated.

4. Compute the deltas for the output layer

$$\delta_i^N = f'(h_I^{n-1})[\xi_i^\mu - \Gamma_i^N] \qquad (3.5)$$

by comparing the actual outputs Γ_i^N with the desired ones ξ_i^μ for the pattern μ being considered.

5. Compute the deltas for the preceding layers by propagation the errors backwards

$$\xi_i^{n-1} = f'(h_i^{n-1})\sum_j w_{ji}^n \delta_j^n \qquad (3.6)$$

for $n = N, N-1, ..., 2$ until a delta has been calculated for every unit.

6. Use

$$\Delta w_{ij}^n = \eta \delta_i^n V_j^{n-1} \qquad (3.7)$$

to update all connections according to $w_{ij}^{new} = w_{ij}^{old} + \Delta w_{ij}$.

7. Go back to step 2 and repeat for the next pattern.

From (3.6) it is straightforward shown that by adopting different parameters in different hidden layers (i.e. tuning the slopes or dynamic ranges of the functions) we can enlarge the deltas for every hidden layer respectively. Similarly, we can also see from (3.5) that in the same way we can adjust the deltas for the output layer and regulate the step size further.

4 Two-Spirals Benchmark Test

The two-spirals problem is a benchmark to test the behavior of a network. The task is to learn to discriminate between two sets of training points (194 points) which lie on two distinct spirals in the x-y plane. These spirals coil three times around the origin and around one another. (See Fig. 4.4a) This benchmark was chosen as the primary benchmark for this study because it is an extremely hard problem to solve for algorithms of the backpropagation family [4].

To show the behavior of the hybrid net clearly, we illustrate a small two-spirals problem that consists of 100 points. (See Fig. 4.1a)

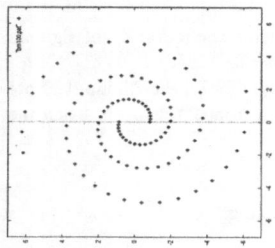

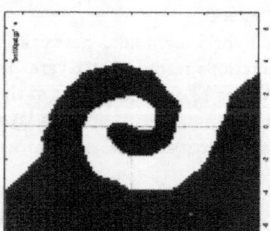

Figure 4.1a

Figure 4.1b

(Figure 4.1: (a) Training points for the small two-spirals problem
(b) Output pattern for one trained hybrid net)

To solve this problem, we build a sigmoidal hybrid net. The structure and the activation functions are given as follows:

$$2 - 65_{N-2} - 65_{N-1} - 1_N$$

where

$$f(h^{N-2}) = \frac{2}{1+exp(-4h)} - 1 \qquad (c_1=2,\ c_2=4,\ c_3=1)$$

$$f(h^{N-1}) = \frac{1}{1+exp(-1.8h)} \qquad (c_1=1,\ c_2=1.8,\ c_3=0)$$

$$f(h^N) = \frac{1}{1+exp(-0.8h)} \qquad (c_1=1,\ c_2=0.8,\ c_3=0)$$

The derivatives of the activation functions for different layers are given as follows:

$$f'(h^{N-2}) = 8 \left(\frac{1}{1+exp(-4h)}\right)\left(1 - \frac{1}{1+exp(-4h)}\right)$$

$$f'(h^{N-1}) = 1.8 \left(\frac{1}{1+exp(-1.8h)}\right)\left(1 - \frac{1}{1+exp(-1.8h)}\right)$$

$$f'(h^N) = 0.8 \left(\frac{1}{1+exp(-0.8h)}\right)\left(1 - \frac{1}{1+exp(-0.8h)}\right)$$

Fig. 4.2 shows that the ranges and slopes of the derivatives by these functions are enlarged correspondingly. This net solves the small two-spiral problem in less than 50 epochs. (See Fig. 4.1b and Fig. 4.3(e)) Now we compare it with another four conventional nets:

- $2 - 65_{N-2} - 65_{N-1} - 1_N$ with $f(h^{N-2}) = \tanh(2h)$ for every layer

- $2 - 65_{N-2} - 65_{N-1} - 1_N$ with $f(h^{N-1}) = \frac{1}{1+exp(-1.8h)}$ for every layer

- $2 - 65_{N-2} - 65_{N-1} - 1_N$ with $f(h^N) = \frac{1}{1+exp(-0.8h)}$ for every layer

- $2 - 65_{N-2} - 65_{N-1} - 1_N$ with standard sigmoid function for every layer

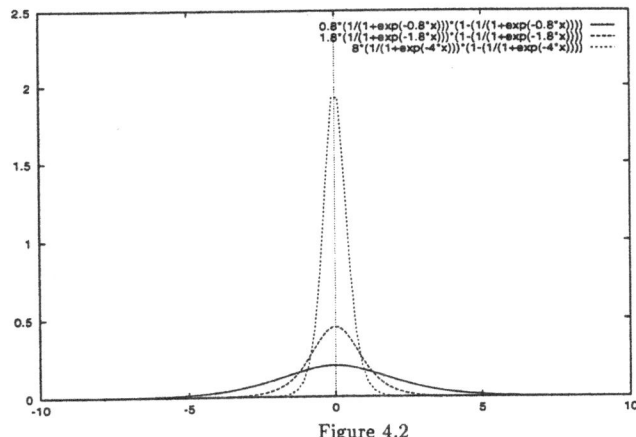

Figure 4.2

(Figure 4.2: Derivative of the activation functions for different layers)

Fig. (4.3) shows the comparison clearly:

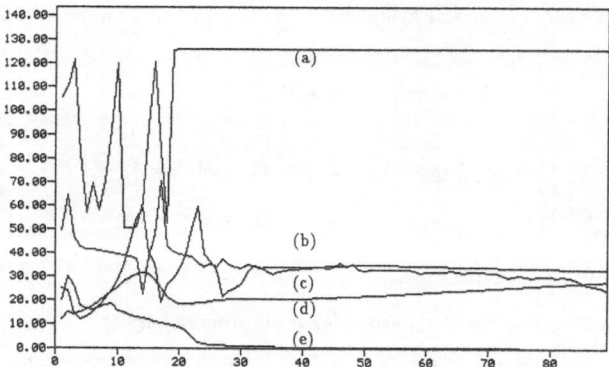

(Figure 4.3: Comparative results for small two-spirals problem from top to bottom: (a) Feed forward net using tanh as activation function (b) Feed forward net using a sigmoid activation function with maximum slope 1.8 (c) Feed forward net using the standard sigmoid function (d) Feed forward net using a sigmoid activation function with maximum slope 0.8 (e) The hybrid net)

Finaly, we build a hybrid net to solve the original two-spirals problem:

$$2 - 130_{N-2} - 130_{N-1} - 1_N$$

where

$$f(h^{N-2}) = \frac{3.2}{1+exp(-2h)} - 1.6 \qquad (c_1=3.2,\ c_2=2,\ c_3=1.6)$$

$$f(h^{N-1}) = \frac{1}{1+exp(-1.2h)} \qquad (c_1=1,\ c_2=1.2,\ c_3=0)$$

$$f(h^N) = \frac{1}{1+exp(-0.7h)} \qquad (c_1=1,\ c_2=0.7,\ c_3=0)$$

This net solves the two-spirals problem in 900 epochs (average over 8 trials by shuffle, all successful) (See Fig. 4.4b and Fig. 4.5). Table (4.1) shows training epochs necessary to solve the two-spiral problem by backpropagation networks and their relatives. It is fair to mention, that the comparison does not include the computational effort of a genetic algorithm to optimize the parameters c_1, c_2 and c_3 for each layer.

network model	number of epochs	reported in
Backpropagation	20000	Lang and Witbrock (1989)
Cross Entropy BP	11000	Lang and Witbrock (1989)
Cascade-Correlation	1700	Fahlman and Lebiere (1990)
Hybrid net under BP	900	This paper

Table 4.1

(Tab.4.1: Training epochs necessary for the original two-spiral problem with backpropagation networks and their relatives)

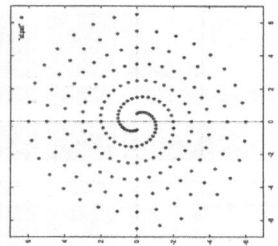

Figure 4.4a

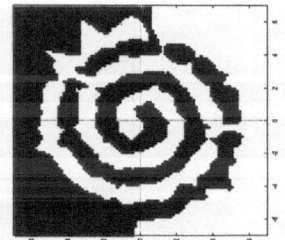

Figure 4.4b

(Figure 4.4: (a) Training points for the original two-spirals problem
(b) Output pattern for one trained hybrid net)

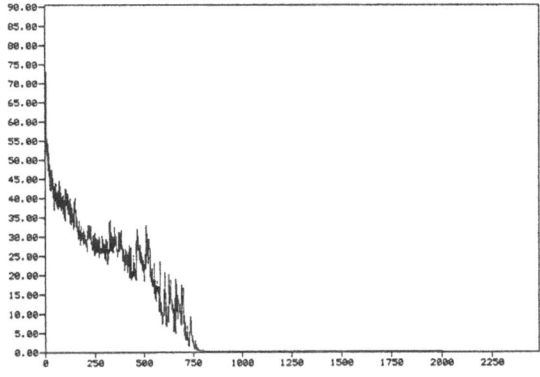

Figure 4.5
(Figure 4.5: Test result of the original two-spirals problem for one hybrid net)

5 Closing Remarks

In this paper we have illustrated how the sigmoid function parameters influence the speed of backpropagation learning and introduced a hybrid sigmoidal network with different parameters configuration in differente layers. In this way it is possible to reduce the attenuation and dilution of the error signal problem, the oscillation problem and the asymmetrical input problem.

Finally it should be mentioned that the goal of this study was the improvement of the learning speed in backpropagation nets with monotone activation functions and not the reduction of the number of nodes. Further improvements are however expected by extending the former strategy from the layer level to the node level. It may be seen that a good parameter configuration may provide a basis for the achievement of greater performance of fast learning under backpropagation.

6 References

[1] Hornik, K., Stinchcombe, M., and White, H.: Multilayer Feedforward Networks are Universal Approximators. Neural Networks, Vol. 2, no. 5, 359-366, (1989)

[2] Kosko, B.: Neural Networks and Fuzzy Systems. Prentice-Hall, INC. (1992)

[3] Fahlman, S.E.: An Empirical Study of Learning Speed in Back-Propagation Networks. Technical Report CMU-CS-88-162, CMU, (1988)

[4] Fahlman, S.E., Lebiere, C.: The Cascade-Correlation Learning Architecture. in Touretzky (ed.) Advances in Neural Information Processing Systems 2, Morgan-Kaufmann, (1990)

[5] Stornetta, W.S., Huberman, B.A.: An improved Three-Layer, Back Propagation Algorithm. IEEE First Int. Conf. on Neural Networks, (1987)

[6] Little, W.A.: The Existence of Persistent States in the Brain. Math. Biosci. 19, 101, (1974)

[7] Little, W.A., Shaw, G.L.: Analytic Study of the Memory Capacity of a Neural Network. Math. Biosci. 39, 281, (1978)

[8] Müller, B., Reinhardt. J.: Neural Networks. Springer-Verlag, (1990)

[9] Hertz, J., Krogh, A., Palmer, R.G.: Introduction to the theory of neural computation. Addison-Wesley publishing Company, (1991)

General Transient Length Upper Bound for Recurrent Neural Networks

A. M. C.-L. Ho
Ph. De Wilde

Imperial College of Science, Technology and Medicine
Department of Electrical and Electronic Engineering
London SW7 2BT, United Kingdom

Abstract

We show how to construct a Lyapunov function for a discrete recurrent neural network using the variable-gradient method. This method can also be used to obtain the Hopfield energy function. Using our Lyapunov function, we compute an upper bound for the transient length for our neural network dynamics. We also show how our Lyapunov function can provide insights into the effect that the introduction of self-feedback weights to our neural network has on the sizes of the basins of attraction of the equilibrium points of the neural network state space.

1   INTRODUCTION

The analysis of the transient length of a discrete recurrent neural network (RNN) operating on discrete states was studied for the synchronous update mode in [9]. For asynchronous update mode, upper bounds were placed on this transient length in [2], [5], [8] and [3] but they excluded the possibilities of negative self-feedback weights for the processing units in an RNN. For a continuous RNN operating on continuous states, a related quantity, the convergence time were studied in [10] and [13]. Their analysis is done essentially by linearisation of the RNN dynamical equation about the origin. Thus they only deal with local convergence in the neighbourhood of an equilibrium point. This approach tends to underestimate the convergence time of the RNN dynamics.

Our present work involves approximating the dynamics of a discrete RNN operating on discrete states with a system of autonomous nonlinear ordinary differential equations with a transfer function with sufficiently high gain. In section 3 we use the variable-gradient method, first to obtain a time derivative of a Lyapunov function for our system of nonlinear differential equations; second, to recover the Lyapunov function from this time derivative. We also show how this method can be used to obtain the Hopfield energy function [7]. In the process, it will be clear why neither our Lyapunov function nor the Hopfield energy function is suitable for studying the dynamics of an asymmetric RNN. Both this Lyapunov function and its time derivative will be used in section 3 to obtain an upper bound on the transient length of a discrete RNN. Finally, in section 5, we use the concepts we developed in previous sections to gain some insights into the effect on the sizes of the basins of attraction of the equilibrium points of our RNN state space. Before all these, we state in the next section the architecture of our RNN and its dynamics.

2 NETWORK ARCHITECTURE

We consider recurrent neural networks (RNNs) which operate in discrete time and discrete space. Each one of the n processing units can be connected to any processing units including itself. We denote the weight of the connection from processing unit j ,which produces a signal to processing unit i, as T_{ij}. The dynamics of the RNN can be described by the following difference equation:

$$x_i(t+1) = sgn[\sum_{j=1}^{n} T_{ij}x_j(t)] \qquad (1)$$

where sgn is the sign function given by:

$$sgn[\sum_{j=1}^{n} T_{ij}x_j(t)] = \begin{cases} +1 & \text{if } \sum_{j=1}^{n} T_{ij}x_j(t) > 0 \\ x_i(t) & \text{if } \sum_{j=1}^{n} T_{ij}x_j(t) = 0 \\ -1 & \text{if } \sum_{j=1}^{n} T_{ij}x_j(t) < 0 \end{cases} \qquad (2)$$

and $x_i(t)$ denotes the state of processing unit i at time t. The state of the network is denoted by $\mathbf{x} = [x_1 \ x_2 \ \cdots \ x_n]^T$. An equilibrium state of our RNN, $\mathbf{x}^{(s)} = [x_1^{(s)} \cdots x_n^{(s)}]^T$ satisfies :

$$x_i^{(s)}(t) = sgn[\sum_{j=1}^{n} T_{ij}x_j^{(s)}(t)] \qquad (3)$$

We only consider random asynchronous update mode in which a processing unit i is randomly chosen and the network dynamics according to equation (1) are iterated once. For the next update episode, another processing unit is chosen randomly and the whole process is repeated.

For reasons that will become apparent in the following sections, we impose the following two extra constraints to our network architecture:

1. The weight matrix $\mathbf{T} = [T_{ij}]$ is to be symmetric : $T_{ij} = T_{ji} \ \forall i, j$;

2. We allow the self-feedback weight, T_{ii} to assume any real value provided that there is no oscillation between different network states.

3 LYAPUNOV FUNCTIONS FOR RNN

In this section, we use the variable-gradient method to construct a Lyapunov function for our RNN. We start by approximating the network dynamical equation (1) by the nonlinear ordinary differential equation:

$$\dot{x}_i = -x_i + g[\sum_{j=1}^{n} T_{ij}x_j(t)] \qquad (4)$$

where $\dot{x}_i$ is the time derivative of x_i. Function $g(h)$ is the function $\tanh(\beta h)$ with $\beta \to \infty$. Writing equation (4) in vector form, we have:

$$\dot{\mathbf{x}} = F(\mathbf{x}) \qquad (5)$$

where $F(\mathbf{x}) = [\cdots - x_i + g[\sum_{j=1}^{n} T_{ij}x_j] \cdots]^T$. Accordingly we can now give a definition of a Lyapunov function ([6]).

Let $\mathbf{x}^{(s)}$ be an asymptotically stable equilibrium point for equation (5). Let V be a single value function define on a neighbourhood U of $\mathbf{x}^{(s)}$, differentiable on $U - \mathbf{x}^{(s)}$, such that:

1. $V(\mathbf{x}^{(s)}) = 0$ and $V(\mathbf{x}) > 0$ if $\mathbf{x} \neq \mathbf{x}^{(s)}$;

2. $\dot{V} < 0$ in $U - \mathbf{x}^{(s)}$ and $\dot{V}(\mathbf{x}^{(s)}) = 0$.

3.1 Variable-Gradient Method

The variable-gradient method ([11], [12] and [14]) for the construction of a Lyapunov function works from $\dot{V}$ back to V. The autonomous nonlinear system given by equation (5) has the time rate of change of V given by

$$\dot{V} = \frac{dV}{dt} = \nabla^T V F(\mathbf{x}) \tag{6}$$

where $\nabla V = [\cdots \frac{\partial V}{\partial x_i}, \cdots]^T$ Instead of assuming a specific form for the Lyapunov function itself, we assume that the gradient function has the following form:

$$\nabla_i V = \sum_{j=1}^{n} a_{ij} x_j \tag{7}$$

where a_{ij} are coefficients to be decided. The Lyapunov function $V(\mathbf{x})$ is recovered from equation (7) by:

$$V(\mathbf{x}) = \sum_{i=1}^{n} \int_{x_i^{(s)}}^{x_i} \nabla_i V dx_i \tag{8}$$

where it is crucial that the curl condition is satisfied:

$$\frac{\partial \nabla_i V}{\partial x_k} = \frac{\partial \nabla_k V}{\partial x_i} \tag{9}$$

$\forall i, k = 1, \ldots, n$ in order that $V(\mathbf{x})$ can be obtained independent of the path of integration of ∇V.

For our RNN from equations (5),(6) and (7) we have :

$$\dot{V} = \sum_{i=1}^{n} [(\sum_{j=1}^{n} a_{ij} x_j)(-x_i + g(\sum_{j=1}^{n} T_{ij} x_j))] \tag{10}$$

Consider the k-th term of equation (10):

$$(\sum_{j=1}^{n} a_{kj} x_j)(-x_k + g(\sum_{j=1}^{n} T_{kj} x_j))$$

For a nontrivial change of network state, neither the expression in the first set of brackets nor the one in the second set of brackets can vanish. This means that x_k and $g(\sum_{j=1}^{n} T_{kj} x_j)$ must have different signs. If $x_k = +1$, then $g(\sum_{j=1}^{n} T_{kj} x_j) = -1$ and the second bracketed term equates to -2. We require $(\sum_{j=1}^{n} a_{kj} x_j)$ to be positive. If $x_k = -1$, then $g(\sum_{j=1}^{n} T_{kj} x_j) = +1$ and the second bracketed term equates to $+2$. We require $(\sum_{j=1}^{n} a_{kj} x_j)$ to be negative. If we choose $a_{ij} = -T_{ij}$, $\forall i, j = 1, \ldots, n$, these two conditions will be fulfilled. It remains to be checked that whether such prescriptions of a_{ij} will satisfy the curl condition:

$$\frac{\partial \nabla_i V}{\partial x_k} = \frac{\partial (-T_{ik} x_k)}{\partial x_k} = -T_{ik}$$

Similarly $\partial \nabla_k V / \partial x_i = -T_{ki}$. Thus we require $T_{ik} = T_{ki}$, $\forall i, k$. Our weight matrix is symmetric in any case and therefore the curl condition is met. To obtain our Lyapunov function we perform the following integration:

$$
\begin{aligned}
V(\mathbf{x}) &= -\sum_{i=1}^{n} \int_{x_i^{(s)}}^{x_i} [\sum_{j, j \neq i} T_{ij} x_j + T_{ii} x_i] dx_i \\
&= -\sum_{i=1}^{n} [\sum_{j, j \neq i} T_{ij} x_j x_i + \frac{1}{2} T_{ii} x_i^2]_{x_i^{(s)}}^{x_i} \\
&= -\sum_{i=1}^{n} \sum_{j, j \neq i} T_{ij} x_j (x_i - x_i^{(s)})
\end{aligned}
\tag{11}
$$

Notice that the self-feedback term, T_{ii}, does not feature in our Lyapunov function because $x_i^2 = 1 \ \forall x_i \in \{-1, +1\}$.

It should be clear from our argument above that our Lyapunov function is only applicable for asynchronous update mode. In synchronous update mode, all the processing units are updated simultaneously and hence the time derivative of a Lyapunov function should take into account the collective behaviour of all the processing units instead of merely the behaviour of one processing unit in isolation as we have done above.

If we take the coefficients a_{ij} to be $-\frac{1}{2} T_{ij}$, perform the integration (11) with the lower limit as the origin $[0 \cdots 0]^T$ instead of $\mathbf{x}^{(s)}$ and if all the self-feedback terms T_{ii} in our RNN vanish, we get the energy function of [7]:

$$
E(\mathbf{x}) = -\frac{1}{2} \sum_{i=1}^{n} \sum_{j, j \neq i} T_{ij} x_j x_i
$$

This energy function behaves in exactly the same manner as our Lyapunov function, except that the first condition in our definition of Lyapunov function is now replaced by: $E(\mathbf{x}) > E(\mathbf{x}^{(s)})$ if $\mathbf{x} \neq \mathbf{x}^{(s)}$ and $\mathbf{x}$ is in the basin of attraction (neighbourhood) of the equilibrium point $\mathbf{x}^{(s)}$;

For an RNN that processes binary $\{0, 1\}^n$ vectors instead of bipolar $\{-1, +1\}^n$ vectors (see [7]), the network dynamics are given by the following difference equation:

$$
u_i(t + 1) = Hv[\sum_{j=1}^{n} T_{ij} u_j(t)]
\tag{12}
$$

where Hv is the Heaviside unit step function given by:

$$
Hv[\sum_{j=1}^{n} T_{ij} u_j(t)] = \begin{cases} 1 & \text{if } \sum_{j=1}^{n} T_{ij} u_j(t) > 0 \\ u_i(t) & \text{if } \sum_{j=1}^{n} T_{ij} u_j(t) = 0 \\ 0 & \text{if } \sum_{j=1}^{n} T_{ij} u_j(t) < 0 \end{cases}
\tag{13}
$$

One can easily verify that using the variable-gradient method, we obtain the following Lyapunov function :

$$
V_{bin}(\mathbf{u}) = -\sum_{i=1}^{n} \sum_{j, j \neq i} T_{ij} u_j (u_i - u_i^{(s)}) - \frac{1}{2} \sum_{i=1}^{n} T_{ii} (u_i^2 - u_i^{(s)2})
\tag{14}
$$

where $\mathbf{u} = [u_1 \cdots u_n]^T$, with $u_i \in \{0, 1\}$ denotes the network state and $\mathbf{u}^{(s)}$ is an asymptotically stable equilibrium point.

4 An Upper Bound for the Transient Length of an RNN

We define a nontrivial transition as a change of RNN state from $x^{(m)}$ to $x^{(m+1)}$ where $x^{(m)} \neq x^{(m+1)}$ as a result of one iteration of equation (1). We can then define the transient length from an initial state x to an equilibrium state $x^{(s)}$ as the number of nontrivial RNN state transitions from x to $x^{(s)}$ with the repeated iteration of the network dynamics.

Intuitively, we can treat the Lyapunov function $V(x)$ we constructed in subsecion 3.1 as a function computing a distance from state x to its equilibrium state $x^{(s)}$. We then can visualise the time derivative of our Lyapunov function at state x, $\dot{V}(x)$, as the velocity (in Lyapunov distance per unit time) of our system approaching the equilibrium state. For our RNN with discrete dynamics, we approximate the change in Lyapunov distance, $\Delta V = V(x(t+1)) - V(x(t))$ by $\dot{V}(x)$, taking $\Delta t = 1$. We compute an estimate for the upper bound for the transient length of our RNN by:

$$\frac{max[V]}{min[-\dot{V}]} \tag{15}$$

This approach was adopted in [3], [5] and [8].

To compute $max[V]$, we work from equation (11) :

$$V(x) = -\sum_{i=1}^{n} \sum_{j,j\neq i} T_{ij}x_j(x_i - x_i^{(s)})$$

We get :

$$max[V] = 2\sum_{i=1}^{n} \sum_{j,j\neq i} |T_{ij}| \tag{16}$$

To compute $min[-\dot{V}]$ we use equation (10) and put $a_{ij} = -T_{ij}, \forall i, j$:

$$-\dot{V} = \sum_{i=1}^{n}[(\sum_{j=1}^{n} T_{ij}x_j)(-x_i + g(\sum_{j=1}^{n} T_{ij}x_j))] \tag{17}$$

Since we are only dealing with asynchronous update mode, in any nontrivial state transition, only one of the n terms of the first summation (from $i = 1$ to n) on the right hand side of equation (17) is nonvanishing. Let this term be term k. Because of our selection of coefficients a_{ij} in subsection 3.1, we know that $\dot{V}$ is always negative for nontrivial transition so $-\dot{V}$ must always be positive accordingly:

$$[(\sum_{j=1}^{n} T_{kj}x_j)(-x_k + g(\sum_{j=1}^{n} T_{kj}x_j))] > 0 \tag{18}$$

In inequality (18), we notice that the second bracketed term can only assume the value of either $+2$ or -2. Therefore, we conclude that if processing unit k is randomly chosen for update and that its state is flipped, the time derivative of the Lyapunov function multiplied by -1 is

$$-\dot{V} = 2 |\sum_{j=1}^{n} T_{kj}x_j| \tag{19}$$

Hence for the upper bound of the transient length of our RNN, we have:

$$\frac{\sum_{i=1}^{n} \sum_{j,j\neq i} |T_{ij}|}{min_{x,k}[|\sum_{j=1}^{n} T_{kj}x_j|]} \tag{20}$$

5 THE ROLE OF SELF-FEEDBACK WEIGHTS

In this section, we will show how the introduction of self-feedback weights to the processing units in an RNN can affect the transient length.

To obtain a time constant at a non-equilibrium state x when the state of processing unit k is to be flipped, we compute :

$$\frac{V}{-\dot{V}} = \frac{-\sum_{i=1}^{n} \sum_{j,j \neq i} T_{ij} x_j (x_i - x_i^{(s)})}{2 \mid \sum_{j=1}^{n} T_{kj} x_j \mid} \tag{21}$$

from equations (11) and (19). This is a constant because once the weights and the dynamics of our RNN are decided, for each state in the network state space, this value will not change ([1]). It is self-evident that the self-feedback terms, T_{ii}, do not contribute to and hence will not affect the numerator of equation (21). If we rewrite the denominator as :

$$2 \mid \sum_{j,j \neq k} T_{kj} x_j + T_{kk} x_k \mid \tag{22}$$

We can see that the time constant is increased if we add a self-feedback term T_{kk} to processing unit k and if both $\sum_{j,j \neq k} T_{kj} x_j$ and $T_{kk} x_k$ have the same sign. This is also true at least for some of the states on the same trajectory and close to our state under scrutiny, state x. This will result in a shorter transient length to an equilibrium point. Conversely, if $\sum_{j,j \neq k} T_{kj} x_j$ and $T_{kk} x_k$ have different signs, our time constant will be decreased. Using a similar argument as above, we can see that the transient length to an equilibrium point will be lengthened. An RNN with both positive and negative self-feedback weights is shown in [4] to have a trajectory that includes every single vertex of a hypercube that depicts the entire state space of the RNN.

An equilibrium state with more of its transient lengths increased than those that are decreased by the self-feedback term(s) will have a larger basin of attraction whereas as an equilibrium state with more of its transient lengths decreased than those that are increased will have a smaller basin of attraction. An equilibrium state with the similar number of transient lengths falling into both categories will have a basin of attraction of similar size.

6 CONCLUSION

The variable-gradient method we use for obtaining the Lyapunov function and its time derivative enables us to give a general form for the upper bound of the transient length in the sense that we only require our RNN to have a symmetric weight matrix and that its states do not oscillate. This upper bound also provides an insight into the way basins of attraction of asymptotically stable equilibrium points are affected with the introduction of non-null self-feedback weights to some of the processing units in the RNN.

References

[1] F. Csáki. *Modern Control Theory*. Akadémiai Kiado Budapest, 1971.

[2] P. Floréen. Worst-case convergence times for hopfield memories. *IEEE Trans. on Neural Networks*, 2(5):533–535, 1991.

[3] F. Fogelman-Soulie *et. al.* Transient length in sequential iteration of threshold functions. *Discrete Applied Mathematics*, 6:95–98, 1983.

[4] E. Goles and S. Martínez. *Neural and Automata Networks*. Kluwer Academic Publisher, 1990.

[5] E. Goles *et. al.* Decreasing energy functions as a tool for studying threshold networks. *Discrete Applied Mathematics*, 12:261–277, 1985.

[6] M. Hirsch and S. Smale. *Differential Equations ,Dynamical Systems and Linear Algebra*. Academic Press, 1974.

[7] J. Hopfield. Neural networks and physical systems with emergent collective computational abilities. *Proc. of Nat. Acad. Sci, U.S.A.*, 79:2554–2558, 1982.

[8] Y. Kamp and M. Hasler. *Recursive Neural Networks for Associative Memory*. Wiley, 1990.

[9] J. Komlós and R. Paturi. Convergence results in associative memory model. *Neural Networks*, 1:239–250, 1988.

[10] A. Michel *et. al.* Qualitative analysis of neural networks. *IEEE Trans. on Circuits and Systems*, 36(2):229–243, 1989.

[11] R. R. Mohler. *Nonlinear Systems :Dynamics and Control*. Prentice Hall, 1991.

[12] P. C. Parks. A. M. Lyapunov's stability theory—100 years on. *IMA Journal of Mathematical Control and Information*, 9:275–303, 1992.

[13] N. Peterfreud and Y. Baram. Second-order bounds on the domain of attraction and the rate of convergence of nonlinear dynamical systems and neural networks. *IEEE Trans. on Neural Networks*, 5(4):551–560, 1994.

[14] J.-J. E. Slotine and Li W. *Applied Nonlinear Control*. Prentice Hall, 1991.

On Some Methods in Neuromathematics (or the Development of Mathematical Methods for the Description of Structure and Function in Neurons)

R. Moreno-Díaz jr and K.N. Leibovic*

Facultad de Informática-Universidad de Las Palmas
35017 Las Palmas, Spain
moreno@edi.ulpgc.es
and
*Biophysics Department
SUNY at Buffalo, Buffalo, NY14214, USA.
bphknl@ubvms.cc.buffalo.edu

Abstract

Success in exploring neural circuitry and its functions depends critically on the availability of data. This determines what kinds of questions can be asked and what analytical tools can most appropriately be used. Biophysical studies have relied heavily on statistics -e.g. as applied to neuronal spike trains- and differential equations and matrix algebra-e.g. as applied to the Hodgkin/Huxley axon and in modeling some networks. Some other approaches have been relatively neglected. These include the search for optimality criteria in relating structure and function and the decomposition of informational processes into simple units.

In this paper we describe how a particular optimaliy criterion has led to new insights and to the classifications of one type of neural cell; and we describe a new family of filters with interesting properties, which serve as simple information processing units and which can be concatenated to provide both high level and low level descriptions. Both methods were developed in connection with visual processing in the retina. But they can be extended with appropriate reformulations to other areas of the nervous system.

0.- Introduction.

It is widely agreed that biological structures have evolved for optimum function. Therefore when we study a biological system we are impelled to ask what optimality principles may be involved.

In this paper we show how the idea of optimality can be applied to understanding the relation between structure and function of a retinal photoreceptor. But as we move from receptors along the nervous system the degree of semantic complexity of the input, and thusof the activity of every cell increases dramatically. It is observed, for example, that there is much convergence and divergence between different levels of the nervous system. This has led us and others to research on the properties of such systems and on the operating features which make them optimal in information processing {3}-{6}. The notion of receptive fields is central to such studies. Receptive fields have been modeled using tools of matrix algebra or integral equations. Here we present an entirely new approach to modeling receptive fields to target cell tranformations by what we have called Newton Filters (NF). In addition to their interesting properties NFs offer great flexibility which makes them attractive for computational implementations.

We motivate our presentation by first discussing photoreceptor optimality and then describing Newton Filters as our candidate for optimality studies in neural networks.

1.- The natural design of optimal visual machinery.

Photoreceptors, rods and cones are transducers that absorb photons and produce electric responses whose amplitude and duration depend on intensity and wavelength of absorbed light. Rods are specially sensitive cells, since they are able to have a response as the consequence of a single photon absorption. The molecules of rhodopsin, the photopigment that captures photons, are placed on discs inside the outer segment (OS) of the rod. A long OS with many layers of rhodopsin increases absorption probability. But there is continuous electrical noise inside the OS due to the maintained transduction biochemistry and this noise increases with OS length. In addition the rhodopsin molecules suffer spontaneous thermal isomerizations which are indistinguishable from photon responses and thus produce "false alarms".

Thus, photon absorption and transduction biochemistry noise present two competing demands to take into account in the design of the length of a rod outer segment: the shorter it is, the less noise is generated and the emitted electrical signal will be clearer, but the probability of photon absorption will decrease. For a longer outer segment this last probability will increase, and the noise as well.

In this situation, let us define the function R(s), Useful Response Signal, as:

$$R(s) = (1-e^{-as}) (1- Ds/TL) A - Ns/L$$

The first bracket in thi expression is the photon absoprtion probability given by the Beer-Lambert law, "a" being the optical density of the medium in which the photon travels and s the length of an outer segment considered as a variable. The second bracket excludes the possibility of a spontaneous isomerization of the rhodopsin (which occurs every T seconds with a duration of D for a length L of O.S.). The product of both brackets by A (amplitude in picoamps of single photon response) gives the expected response excluding false alarms. from this we subtract the continuous thermal noise Ns/L, the noisy background which a reliable signal must overcome. (For the original and extense mathematical treatment the reader is referred to {1} and {2}. Experimental values of the different parameters and all data sources can be found in {2}).

Outer Segment length is optimized for those values of s, Sm, for which:

$$dR/ds = 0$$

When using this criterion to "predict" the length of the O.S. of several species showing well developed duplex retinae one gets results very close to the experimental, actual, length of the cells (it is so for toads, frogs, rabbits and monkeys). Thus, it is possible to state that in those cases, rod O.S. are optimized for the detection of one or two photons and the cells are designed to be reliable detectors of the smallest possible amount of light. This optimality criterion applies only to those species whose cells actually have the basic two characteristics used in the derivation of the equations 1 and 2, that is, sensitive detectors in a noisy environment. It is not as good in some other species like the skate, which has an only-rod retina that has to work over the whole physiological range (both scotopic and photopic levels) and thus the premise of ultimate sensitivity cannot be a strict constraint in its design.

In this sense, the criterion can be used as a clasifier of visual cell, giving a method of distinguishing among families of rod cells. Not all of them (in the animal kingdom) can have the same sensitivity due to the differences in the retinae and the different illumination environments of living forms.

This method of exploring structure-function relationships (showing a result of the old axiom that structure is suited to function) gives a hint of the complexity we should look for at more central levels of the Nervous System. The evolutionary design has produced structures and functions that are inseparably linked and this obliges us to be careful when we use artificial models that we hope include natural characteristics that have been improved generation after generation. At the same time, it shows the kind of deep knowledge it is neccesary to have about nervous functioning to be able to mathematically encapsulate the structure into two or three useful (in the sense of giving accurate real results) equations. But even for the next layer of retinal cells, bipolars and horizontals, it would be very difficult to state such a criterium, not due to the lack of mathematical tools, but due to the lack of data and suitable models.

In the previous example we have started from known function to explain structure. In the following we will simulate an observed structure (e.g. the dendritic tree of ganglion cells) to design an artificial neuron-like machine and explore its properties, showing that the result mimics the observed profile of natural ganglion cell receptive fields.

2.- From computation in dendrites to discrete filters...and back.

The dendritic tree of neurons has always fascinated neuroscientists. Two of their immediately-related properties regarding information processing (e.g. convergence and divergence of lines) have originated lots of research on completeness of computation and reliability of transmission {3,4 for example}. We will concentrate now on a very simplified structure inspired by dendritic connections, the one shown in Figure 1, and for the purpose of calculations will consider a one dimensional receptive field (the extension to two dimensions is quite straightforward).

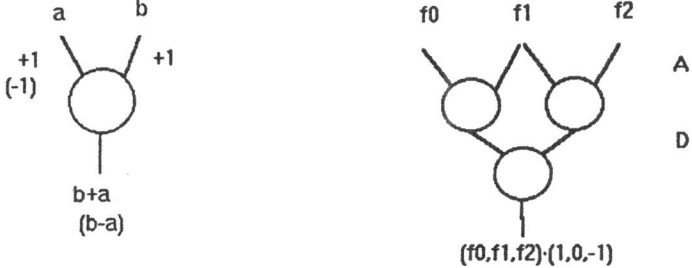

Figure 1: On the left, a simple microcomputational additive unit. On the right, a monodimensional two-layered structure (Newton Filter N(A1,D1)   ) and the computation it performs expressed as vector scalar product. The representation of the global weights vector gives the filter profile.

On Figure 1a we can see a microprocessing unit that can be of one out of two kinds: additive if it performs an addition of the two inputs, or subtractive if it performs a difference. On an additive layer all the are +1, while on a subtractive layer are alternately +1 or -1.On Figure 1b a neuron-like computing unit consisting of layers of such microprocessors is shown. That structure can be seen as a discrete filter acting on a set of input data $f_0,...,f_n$. Let us define the length L of such a filter as the number of total input data contributing to the calculation, L=n-1. It is easy to prove that for a L-length input vector it is necesary to have L-1 layers of microprocessors as shown in Figure 1 to get a neuron-like processing machine that we call a one dimensional Newton Filter (N.F.){5} and we will denote it by N(An,Dm) (where n expresses the number of additive layers and m the number of subtracting layers, and L=n+m+1). Due to the linearity of operations involved in the definition of N.F. the two following relations hold {5}:

N(An,Dm)=N(Dm,An) (Permutation of layers does not change the general performance)
$N(Am_1,...,Am_i,Dn_1,...,Dn_j)=N(Am_1+...+m_i,Dn_1+...+n_j)$

The global weights of the Filters are easy to calculate using a variation of the Pascal triangle (the one used to calculate the factors of the Newton binomial, where the name of the Filters comes from). Thus, a purely additive filter is actually calculated with the Pascal triangle, e.g. N(A3) is

$$
\begin{array}{cccc}
 & 1 & & \\
1 & & 1 & \\
1 & 2 & & 1 \\
1 & 3 & 3 & 1
\end{array}
$$

Adding diference layers means differencing instead of adding. For example, N(A3,D1) is:

$$
\begin{array}{ccccc}
 & & 1 & & \\
 & 1 & & 1 & \\
 & 1 & 2 & 1 & \\
1 & 3 & 3 & 1 & \\
1 & 2 & 0 & -2 & -1
\end{array}
$$

And the Filter N(A1,D1) of Figure 1(right) is calculated:

$$
\begin{array}{ccc}
 & 1 & \\
1 & & 1 \\
1 & 0 & -1
\end{array}
$$

Then, we can write the global weights as vectors, e.g. $N(A3,D1)=(1,2,0,-2,-1)$ and the operation as a scalar product as shown in Figure 1: $F=(f0,f1,...fn)^T \cdot N(Ax,Dy)$.

When the nature of the processes is the same (i.e. there is no mixture of additive and subtractive microprocessors within a layer, all are either additive or subtractive), then, it is an immediate result that if we consider all possible permutations of additive-difference layers for a given L-inputs Newton Filter, there will be L different Newton Filters of length L and we can construct that set of N.Fs. in an orderly fashion:

$$
\begin{array}{c}
N(An) \\
N(An-1,D1) \\
... \\
N(A1,Dn-1) \\
N(Dn)
\end{array}
$$

(where L=n+1)

On Figure 2 the profiles of several such filters of the same length are shown, where it is easy to recognize the discretized gaussian (the pure additive filter) and a discrete center-perifery shaped filter (in general, any N(Ax,D2) N.F. has this profile). The complexity of the profile (that is, the number of zero-crossings of the representation) is directly related to the number of difference layers in the N.F. As Fig. 2 shows, it is possible to generate symmetric and antisymmetric fields, two important calsses found in natural nervous systems and used also in artificial neural networks. Thus, this structure built by connecting very simple microprocessing units can generate truly complicated computational kernel profiles on its input data. Furthermore, the same basic structure can be used in different calculations by only changing the nature of the connections (or, looking at Figure 1, changing the sign of the weight of the connection from + to - or viceversa).

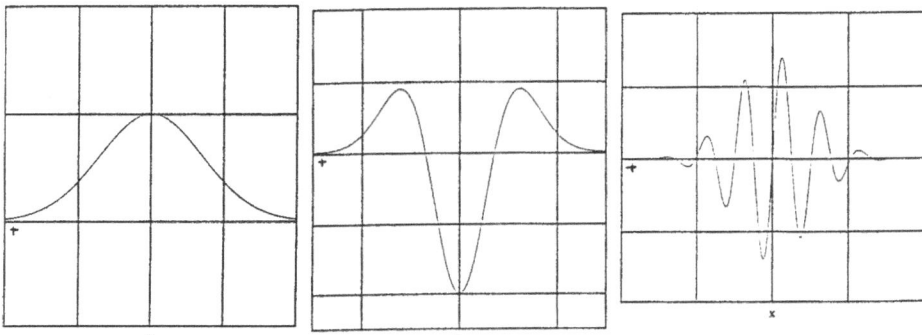

Figure 2: The profiles of N.Fs. N(A20),N(A18,D2),N(A1,D19) plotted as continuous functions. In general, the longer the filter, the smoother the representation of the profile is. As explained in text, any N(Ax,D2) has a center-surround structure. The character (ON-OFF characteristics) of such profiles are easily changed by multiplying by -1. The discrete values of the Filters are calculated as explained in text. The "x" marks the receptive field center, and the "+" the zero line.

Going back to the implications in neuronal modelling, we can recall that hundreds of experiments have justified certain weight functions for cells in the retinal pathway as gaussians or difference of two gaussians (DOGs). Also, more complicated receptive fields, showing several concentric or parallel ON-OFF (excitatory-inhibitory) regions are often used in simulating the performance of the visual pathway as a Fourier analyzer. As we have just seen, building a layered machine such a N.F. one can get the discretization of gaussians, DOGs and parallel inhibitory-excitatory receptive field profiles as the result of the structure of the N.F., not of the fixing of some difficult-to-justify weight functions. In other words, the "dendritic complexity" of the neuron-like machine guarantees the existence of the appropriate kernel functions. In the model, all synaptic contacts have the same weighted absolute value and from the contribution of every microprocessor it is possible to obtain a global weight function that mimics the ones found in natural systems. From the computational point of view, a N.F. is a parallel layered machine.

There is also an implication on the artificial, image proccessing side that is easy to see for everyone involved in that field. Some of the kernels shown in Figure 2 are the most widely used for low-level processing tasks such as edge detection, sharping and contour discrimination. Newton Filters can be used for low-level image processing as fast and easy-to-implement tools {6}.

In the context of converging and diverging information processing {1} we can represent overlapping receptive fields by overlapping input elements to Newton Flters. The target cells are the NF output units. Multivariable, broadly tuned RF can be modeled by the built in properties of a given NF or of a modifiable NF structure. These topics will be considered in a subsequent paper. Suffice is to say that our initial studies have shown that feasability and usefulness of NF in the study of convergence and divergence and optimal performance.

3.- Conclusions.

New points of view and new mathematical tools are needed in the search for answers to many questions in the Nervous Systems. The classical mathematical tools have been useful in modeling certain characteristics but are still unsufficient to fully understand the relation between neural structure and computation. It has been shown how a deep knowledge of the function of a group of cells can lead to a convincing explanation about their structure thorough the formulation of an optimality criterion about their performance. Thus, from function we can derive the structure.

On the other hand, in more complicated neural systems where the function is poorly understood it has been useful to mimic a simplified version of the structure and extract what mathematical characteristics it embraces assuming the local operations are as simple as possible. Again, one can see how the structure determines the function and how a fixed structure can compute (perform) different functions with a slight change in one of the computational properties (not in the structure itself).

These two examples, the design of visual transducers and the computational properties of dendritic-like structures, support the proposition that complex problems on neural elements and networks can be elucidated in the context of optimality relating structure and function.

4.- References.

{1} Leibovic, KN. (1990)"Visual Information: Structure and Function" in KN Leibovic Ed. "Science of Vision", Springer-Verlag, New York.
{2} Leibovic, KN, Moreno-Diaz jr R. (1992)"Rod Outer Segments are designed for optimum Photon Detection", Biological Cybernetics, V66 pp301-306.
{3} Leibovic,KN (1981) "Principles of Brain Function: information processing in convergent and divergent pathways" in Pichler-Trappl Eds. Progress in Cybernetics and Systems, Vol I. Hemisphere Publ. Washington DC.
{4} Moreno-Diaz jr R, Leibovic KN, Bolivar-Toledo O. (1994) "Preservation of Information in Retinal Systems: completeness, structure and function", R. Trappl Ed. Cybernetics and Systems, EMCSR94, World Pub. Co., Singapore.
{5} Moreno-Diaz jr R. (1993) "Computacion paralela y distribuida: relaciones estructura-función en Retinas" PhD Thesis (in Spanish; English version under preparation) , Universidad de Las Palmas, ISBN:84-4090-019-2.
{6} Moreno-Diaz jr R., Correas-Suarez B. (1994) "Newton Filters: discrete computing tools to model Retinal systems", T. Oren Ed. proceedings of CAST94, Univesity of Ottawa, Ottawa, Canada.

Logistic Networks with DNA-Like Encoding and Interactions

Nabil H. Farhat and Emilio Del Moral Hernandez

University of Pennsylvania
The Moore School of Electrical Engineering
200 South 33rd Street
Philadelphia, PA 19104 USA

Abstract

We consider a logistic network consisting of a coupled population of externally driven logistic processing elements (LPEs) or "neurons" with quantized interactions between them. The interactions are modeled after the encoding of genetic information in molecular biology, i.e. as in DNA molecules in terms of four nucleotide bases. A unique and versatile scheme for generating complex spatio-temporal input patterns to drive the the network, that could contain chaotic components, is employed to study the network's behavior. Both coherent (phase-locked) and incoherent input patterns can be generated. We find that DNA-like encoding of interactions causes quantization and clustering to appear in the activity of the network which we represent by limit-set-diagrams (LSDs). Clustering means grouping of processing elements into subpopulations with period-m orbits where m is constant for each cluster, but the values of m are different for different clusters. The clustering is found to characterize the particular input pattern being applied to the network, it changes gradually with gradual change in the input, and appears to persist even when the input pattern contains chaotic components. A striking similarity of bifurcation diagrams of isolated driven LPEs and of the LSDs generated to the bar patterns observed in gel-electropheresis of oligonucleotides* and DNA fragments is observed. This could portend a useful link to molecular biology and serve as basis for introducing a molecular computing paradigm in neural networks.

1. Introduction

The study of logistic nets reported here was inspired by work on molecular computing by Conrad and others [1],[2] and by the recent ground-breaking work of Adelman [3] on solving NP-complete problems with simple protocols carried out with DNA substrates. Evidence is presented here on what could be a useful parallel in the dynamics of neural networks that employ functionally complex processing elements with special forms of interactions matrix, and certain features of encoding of information in molecular biology. A logistic net consists of a coupled population of recursive processing elements whose behavior is described by the logistic map [4]. There are several reasons for choosing logistic processing elements (LPEs). One is their functional complexity exemplified in the bifurcation diagram of the logistic map (see Fig. 1) which exhibits fixed point trajectories ($0 < \mu < 3$), period-doubling route to chaos consisting of a cascade of period-m orbits ($3 \leq \mu < \mu_c = 3.56$), and chaotic trajectories ($3.56 < \mu \leq 4$) depending on the value of the nonlinearity or bifurcation parameter μ. The three regimes of qualitatively distinct behavior are designated R_1, R_2 and R_3 in Fig. 1. Second, is the general expectation that the functional capabilities and the range of computations a network could perform is determined not only by collective behavior but also by the functional complexity of the individual processing elements. Third, is our observation that several researchers [5]-[10] have found the behavior of neuronal groups or netlets, as described by the mean activity (percentage of neurons active (firing) at discrete instants of time), is described by an iterative map of the unit interval bearing close similarity to the logistic map.

---

* An oligonucleotide is a short chain of nucleotide. It is a synthetic single stranded DNA molecule that is made with a chosen sequence of nucleotides selected from the four nucleotide bases G,C,T,A [14]. In the logistic net described here, the four nucleotide bases are replaced by the four numbers 1,2,3,4 and the oligonucleotide-like sequences produced, designated $O_i^s(n)$ and $O_{ij}(n)$, have elements selected from the set 1,2,3,4 equivalent to the four nucleotide bases in molecular biology.

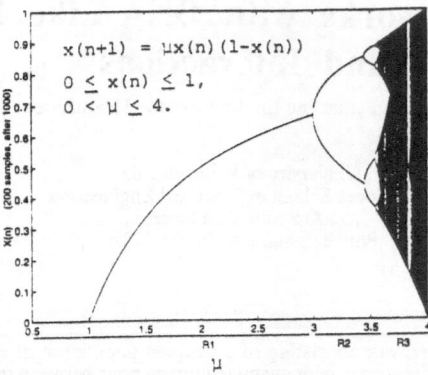

Fig. 1. Bifurcation diagram of the logistic map showing the values assumed by the trajectory X(n) at every value of the nonlinearity or bifurcation parameter μ. The three regions of quantatively distinct trajectories are shown: R_1 (fixed point), R_2 (period-m) and R_3 (mostly chaotic).

Logistic nets have been studied by other workers [11]-[13]. Our network differs from others in two distinct ways. It employs global rather than local interactions and, more importantly, the interactions between neurons are encoded so as to resemble the sequences of nucleotide bases A,T,C,G in oligonucleotides and DNA molecules in molecular biology.

In the following sections we describe first the logistic network and the binning schemes used to quantize the trajectories $x_i(n)$ of its LPEs in order to generate DNA-like or oligonucleotide-like interactions. Then we present the results of numerical simulations illustrating the quantization and clustering observed in the limit-set diagrams used to describe the collective activity of the network especially in response to various spatio-temporal input vectors or patterns and in isolated processing element of the network subjected solely to a driving signal from logistic sensory element. Finally the meaning and ramifications of the observed results are discussed.

2. Logistic Net with Oligonucleotide-Like Interactions

The logistic net we consider, shown in Fig. 2, consists of N LPEs designated L_i i=1,2,...N each of which obeying the logistic map equation,

$$x_i(n+1) = \mu_i(n) \, x_i(n) \, [1 - x_i(n)] \tag{1}$$

where $0 < x_i(n) \leq 1$ is the discrete-time trajectory of the i-th LPE, where n=1,2,3 ..., $\mu_i(n)$ is a discrete-time modulated nonlinearity parameter of the i-th map or LPE confined to the range $0 < \mu_i(n) \leq 4$ and,

$$\mu_i(n) = \frac{1}{2} \, \{O_i{}^s(n) + \frac{1}{N} \, \Sigma \, O_{ij}(n)\} \tag{2}$$

where $O_i{}^s(n)$ and $O_{ij}(n)$ are oligonucleotide or DNA-like sequences derived respectively from $x_i{}^s(n)$ and $x_j(n)$ by binning functions:

$$O_i{}^s(n) = B(x_i{}^s(n)) \tag{3}$$

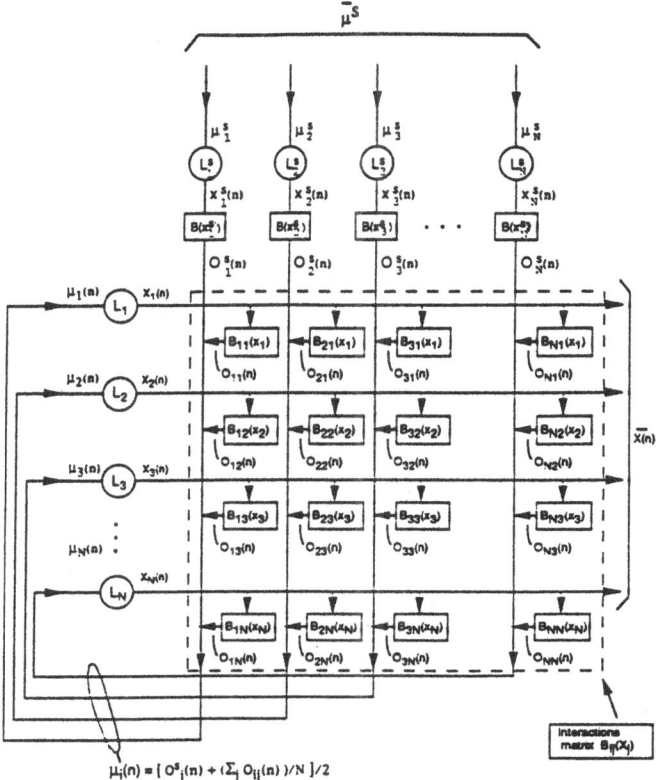

Fig. 2. Logistic net with oligonucleotide-like interactions matrix.

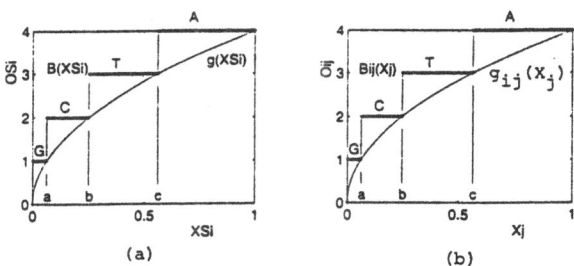

Fig. 3. Binning functions. (a) $B(X_i^S)$ for the logistic sensory elements, and (b) $B_{ij}(X_j)$ for the interactions matrix.

and

$$O_{ij}(n) = B_{ij}(x_j(n)) \qquad (4)$$

where the step-like binning functions $B(.)$ and $B_{ij}(.)$ are as defined in Fig. 3. The thresholds a,b,c of the binning functions are defined by the intersection of the smooth functions

$$g(x_i{}^s) = 4(x_i{}^s)^{C_s} \qquad (5)$$

and

$$g_{ij}(x_j) = 4x_j{}^{C_{ij}} \qquad (6)$$

with the four lines or levels 1,2,3,4 in Fig. 3, corresponding to the four bases G,C,T,A.

Both exponents C_s and C_{ij} range between ε and 1, where ε is small positive number say .1. In the simulations described below we have mostly used $C_s = C_{ij} = .5$. Note that changing C_s or C_{ij} alters the values of the thresholds a,b,c in Fig. 3 and therefore also the step-like binning or mapping of $x_i{}^s(n)$ and $x_j(n)$ into corresponding oligonucleotides $O_i{}^s(n)$ and $O_{ij}(n)$. Note also that C_{ij} = constant represents a uniform interaction matrix. It is also worth noting that the oligonucleotide-like interactions between the LPEs may be adapted for learning purposes through C_{ij}, where $i, j = 1,2,...N$. C_{ij} can therefore be viewed as an interaction parameter matrix of the network.

The upper portion of Fig. 2 constitutes the input generating part of the network. It consists of N logistic maps (or logistic sensory elements (LSEs)) designated $L_i{}^s$ $i = 1,2,...N$ each of which is controlled by a nonlinearity parameter $0 < \mu_i{}^s < 4$ which controls the trajectories $x_i{}^s(n)$ produced by the various LSEs. Thus depending on the values $\mu_i{}^s$, the trajectories $x_i{}^s(n)$, and their binned versions $O_i{}^s(n)$, may be fixed-point, period-m, or chaotic. A wide range of complicated spatio-temporal input or driving patterns, $x_i{}^s(n)$, or equivalent $O_i{}^s(n)$, can therefore be formed by merely selecting the shape of the input generating vector $\bar{\mu}^s$. To produce repeatable input patterns $\bar{x}^s(n)$ or $\bar{O}^s(n)$, we find it is necessary to initiate the iterations of all LSEs from the same initial value $x_i{}^s(o) = x_o$, $0 < x_o < 1$. This also guarantees that trajectories $x_i{}^s(n)$ $(O_i{}^s(n))$ are coherent or phase-locked* and this ensures repeatability of the LSDs from run to run when $\bar{\mu}^s$ is fixed provided the LPEs are also initiated always from the same initial condition. In the simulations presented below we set $x_o = .5$. When, for fixed $\bar{\mu}^s$, random initial conditions are used in iterating the various logistic elements, the resulting driving input vectors $\bar{x}^s(n)$ or $\bar{O}^s(n)$ become incoherent (i.e. their components cease to be phase-locked). We find this introduces fuzziness in the LSDs in the sense that exact repeatability of LSDs from run to run is lost. It is interesting to note that while $O_i{}^s(n)$ in eq. (2) assumes four levels 1,2,3,4, the sum $\frac{1}{N} \Sigma_j O_{ij}(n)$ assumes (3N+1) levels ranging between unity and four. The sum, which is the contribution of all LPEs to the modulation of the nonlinearity parameter $\mu_i(n)$ of the i-th LPE, acts therefore as fine tuning mechanism of the coarser modulation of $\mu_i(n)$ caused by the sensory input $O_i{}^s(n)$.

It can be shown that because of clustering, the possible number of distinct trajectories $x(n)$ of the network can be very large. For example, if $x_i(n)$ is distinguished over L levels, and the order in which the L levels are visited matters, then the number of distinguishable network states or trajectories $\bar{x}(n)$ is

$$N_s = (L! \sum_{m=0}^{L} 1/(L-m)!)^N \qquad (7)$$

---

*The term phase-locking stems from the fact that the variable $0 < x(n) < 1$ of the logistic map can also be translated into a phase-variable, $0 < \phi_n < 2\pi$, by scaling.

This formula takes into account that in our networks the order in which the orbits $x_i(n)$ visit the L levels or bins in the interval [0,1] makes these orbits distinguishable. Equation (7) shows that for a very small network of N=3, and for L=3, the number of distinguishable network states or trajectories is N_s = 4096 and for moderate size networks Ns becomes astronomical. Networks of relatively small size may therefore possess a large number of states because of the inherent temporal nature and phase-locking (clustering) capability of logistic nets. When the order in which $x_i(n)$ visits the L levels is immaterial, equation (7) is replaced by

$$N_s = 2^{NL} \qquad (8)$$

which is the number of possible distinct LSDs of a network of N neurons with trajectories $x_i(n)$ where $x_i(n)$ is measured over L distinguishable levels ranging between 0 and 1. Equation (8) yields N_s = 512 for N=3 and L=3 which is still quite large. It is interesting to compare the predictions of eqs. (7) and (8) with the number of possible states assumed by a network of N sigmoidal neurons where the neuron's state is discernible over L levels. In this case $N_s = L^N$ = 27 for N=3, and L=3, which is significantly lower than N_s in eqs. (7) and (8). The higher numbers of states possible in logistic nets is due to their temporal nature and their phase-locking and clustering capability.

An important point to notice is that the LSDs characterize spatio-temporal input patterns that can also include chaotic components. Thus logistic nets, in contrast to conventional neural networks, can handle complex spatio-temporal input patterns in a natural way.

3. Simulation Results

In this section we present simulation results designed to illustrate: a) how the bifurcation diagram of the driven isolated LPE gets quantized through driving and binning, and b) the way our logistic net classifies complex spatio-temporal input patterns formed by specific input generating vectors $\bar{\mu}$s.

Figure 4(a) shows the bifurcation diagram of a single isolated LPE in Fig. 2. The isolated LPE receives only the sensory input $O_i{}^s(n)$ and no inputs $O_{ij}(n)$ are present. The bifurcation diagram in Fig. 4(a) shows the trajectories, i.e., the values of $0 \leq x_i(n) \leq 1$ visited by the i-th LPE, for every value of the nonlinearly parameter $0 < \mu_i{}^s \leq 4$ of the i-th LSE driving the i-th LPE. Quantization is clearly evident as compared to the bifurcation diagram of the undriven logistic map of Fig. 1. The bar patterns in Fig. 4(a) bear similarity to the patterns observed in gel electropheresis of oligonucleotides and fragmented DNA molecules [15]. Figure 4(b) is a repeat of Fig. 4(a) when the step-like binning function $B(x_i{}^s(n))$ is replaced by the smooth function g(.) expressed in eq. (5) and shown in Fig. 3(a). The quantization observed with binning is seen now to vanish.

In Fig. 5, parts (a) and (b) show two LSDs for a network of N=100 LPEs and for two distinct input generating vectors $\bar{\mu}$s. The LSDs are seen to characterize the distinct spatio-temporal inputs produced by the two $\bar{\mu}$s vectors. The LSDs in (a) and (b) are repeatable because the same initial condition x_o = .5 is used in iterating all logistic elements of the ntwork. Seven clusters, i.e., groups of individual neurons, each group with its own distinct period-m orbit, can be identified in the LSD in (a), and 21 clusters, some of which are single-element clusters, in the LSD in (b). Figure 5(c) shows two LSD runs obtained with the same $\bar{\mu}$s as in (a) but with random initial conditions x(o) used in iterating all logistic elements of the network. Repeatability of fine detail is now lost however gross features of the LSD are preserved. The LSDs shown, and specially that in (a), are seen to have striking similarity to the patterns observed in gel-electropheresis of oligonucleotides and fragmented DNA sequences.

In all simulations previously described, the interaction parameters C_{ij} were selected to be uniform, i.e C_{ij} = .5. Figure 6 shows the effect, on the LSD, of introducing nonuniformity in C_{ij} when the input generating vector $\bar{\mu}$s was as in Fig. 5(b). A C_{ij} = .5+ε where ε is a random variable uniformly distributed between -0.05 and +0.05 is used. The observed dependence of the LSD on the form of the C_{ij} matrix is important for the development of learning algorithm for logistic nets.

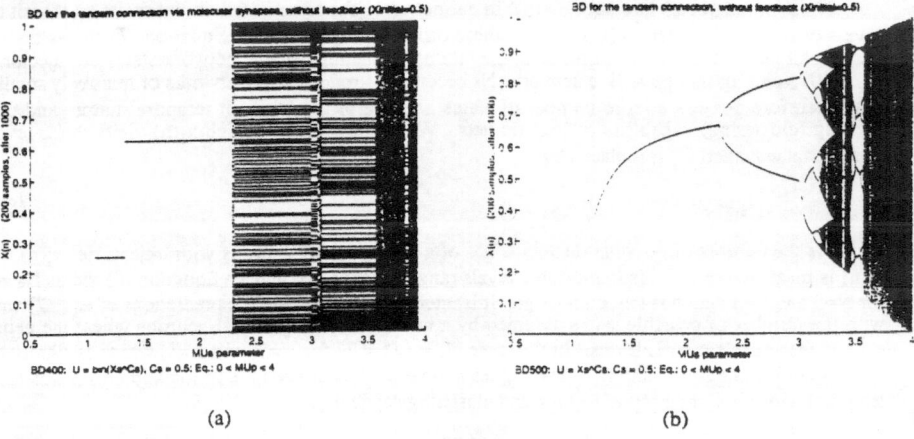

(a)   (b)

Fig. 4. Bifurcation diagram for isolated driven LPE, (a) with binning, (b) without binning.

4. Discussion

The concept of logistic net with DNA-like encoding and interactions introduced here, and the scheme for generating complex spatio-temporal driving (input or stimulus) vectors (patterns) that could contain chaotic components, provide a convenient setting for studying the effect of incorporating concepts from molecular biology and genetic encoding of information in DNA molecules into neural networks. Such endeavor could result in specific ideas, like those presented in this paper, for the development of molecular computing and learning machines. Results of the few key simulations given here furnish useful insight in the behavior of logistic nets and provide intriguing parallels between certain aspects of encoding of information in molecular biology and the behavior of the logistic net as exemplified in LSD and bifurcation diagrams. The stimulus-dependent clustering in the LSDs means that logistic nets, with suitable learning algorithms, may ultimately be useful in the classification of complex spatio-temporal patterns and in the generation of complex functions for motor control. The persistence of clustering when chaotic components are present in the input stimulus pattern, is an intriguing property suggesting that logistic nets may have ability to classify complex spatio-temporal inputs even in the presence of localized interference. This feature seems also to be related to the observation that LSDs satisfy the graduality criterion* [14], and by the persistence of their gross features when random initial conditions for iterating the logistic elements of the network are used.

Returning to the original goal of this work, stated in the introduction, we find now that we could view the inputs or oligonucleotide-like sequences $O_i{}^S(n)$, in the arrangement of Fig. 2, as encoding enzyme exitases, (to borrow a term from Conrad [2]), whose introduction into the system (the network) activates specific "molecular reactions" which result in specific reaction products (outputs) $x_i(n)$ or associated oligonucleotide-like sequences $O_i(n)$ that can be derived from $x_i(n)$ by suitable binning. This rather loose way of looking at the behavior of the network brings to mind enzyme activated reactions in molecular biology and may provide a conceptual frame-work for development of specific ideas for incorporating molecular computing concepts in neural networks.

---

*Graduality refers to gradual change of the LSD with gradual change in the stimulus

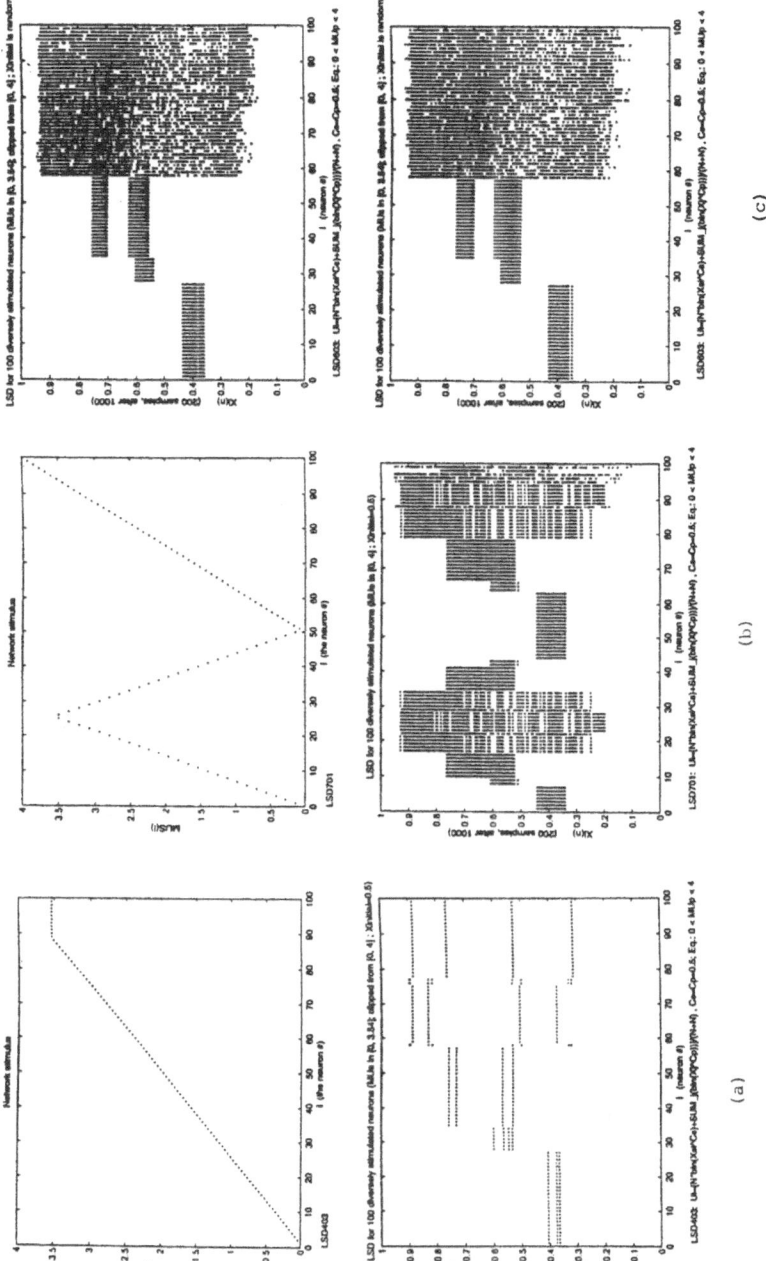

(a)

(b)

(c)

Fig. 5. Limit set diagrams (LSDs), for a network of N=100 LPEs, produced by distinct input-pattern generating vectors μ̄s. (a) and (b) show the μ̄s vectors (top) and the LSDs they produce (bottom) when fixed initial condition $x_0 = .5$ is used to ensure repeatability. Note the stimulus in (b) contains chaotic components because some components of μ̄s exceed $\bar{\mu}_c = 3.56$ (see Fig. 1 and associated narrative in text). In (c) the LSDs produced by the same μ̄s in (a) are shown for two different runs with random initial condition x_0. For random x_0, repeatability of fine detail is lost and only gross features of the LSDs are preserved. Notice the similarity of the LSDs, especially in (a) and gel-electrophoresis patterns produced in the analysis of DNA fragments in molecular biology.

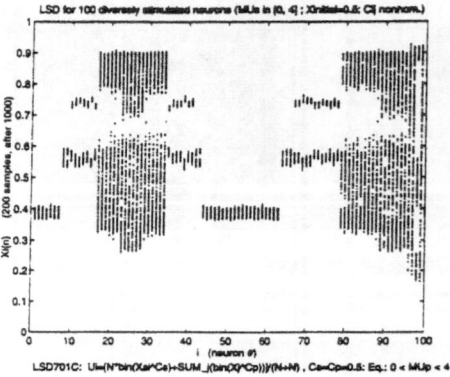

Fig. 6. Limit set diagram of the driven logistic net produced by the input generating vector μs shown in the top with nonuniform interaction matrix $C_{ij} = .5 + \varepsilon$ is a random variable uniformly distributed between -0.05 and +0.05.

5. Acknowledgement

This work was supported in part by a grant from the Office of Naval Research . One of us (E.H.) wishes to acknowledge the support of CNPq of the Ministry of Science and Technology of Brazil.

6. References

1. Computer, Special Issue on Molecular Computing, November, 1992.
2. M. Conrad, In Physics and Mathematics of the Nervous System, M. Conrad et. al. (Eds.), Springer Verlag, Berlin, 1974, pp. 82-127.
3. L.M. Adelman, Science, vol. 266, pp. 1021-1024, November, 1994.
4. R.C. Hilborn, Chaos and Nonlinear Dynamics, Oxford University Press, New York (1994).
5. E.M. Harth, T.J. Csermely, B. Beek and R.D. Lindsay, J. Theor. Biol., vol. 26, pp. 93-120, (1970).
6. P.A. Anninos, B. Beek, T.J. Csermely, E.M. Harth and G. Pertile, J. Theor. Biol., vol. 26, pp. 121-148, (1970).
7. E. Harth and N.S. Lewis, J. Theor. Biol. vol. 55, pp. 201-228, (1975).
8. M. Usher, H.G. Schuster and E. Niebur, Neural Computation, vol. 5, pp. 570-586, July, 1993.
9. G.M. Eldelman, Neural Dawinism: The Theory of Neuronal Group Selection, Basic Books Inc., Publishers, New York, (1987).
10. E. Harth, IEEE Trans. on Systems Man and Cybernetics, vol. SMC-13, pp. 728-789, Sept/Oct., 1983.
11. J.C. Perez and J.M. Bertille, Proc. INNS First Annual Meeting, Bopston, Sept. 1988, p. 121.
12. J.M. Bertille and J.C. Perez, Proc. IJCNN'90, Vol. I, L. Erlbaum Associates Publishers, Hillsdale, N.J., 1990., pp. I-361 to I-364.
13. K. Kaneko, in Theory and Applications of Coupled Map Lattices, K. Kaneko, Ed., J. Wiley, New York, 1993, pp. 1-49.
14. D. Freifelder and G.M. Malacinski, Essentials of Molecular Biology, (Second Edition), Jones and Bartlett Publishers, Boston, (1993).
15. In (gel-electropheresis), fragments of DNA are made to travel through a gel by applying a voltage to it. The speed at which a fragment moves depends on its size or more correctly its charge density - (change to mass ratio) larger fragments more slowly than smaller ones. For further detail see [14].

A Review on the Stochastic Firing Behaviour of Real Neurons and how it Can be Modelled

Chris Christodoulou and Trevor Clarkson

Department of Electronic and Electrical Engineering
King's College London
Strand, London WC2R 2LS
England, UK

ABSTRACT

The types of spike trains recorded in real neurons from different parts of the brain, can either be completely random or bursty. Certainly, at very high firing rates regular spike trains are observed. This paper examines the neurobiological spike trains observed experimentally and analytically and presents how they can be modelled and accounted for by using the biologically inspired Temporal Noisy-Leaky Integrator (TNLI) neuron model, with partial reset. The complete randomness or high firing variability, can be achieved for certain input parameter values at high firing rates, which results from the dendritic temporal summation of postsynaptic responses and the use of random synaptic inputs. It is also demonstrated that bursting behaviour can indeed be achieved, using the TNLI, which is a result of the use of random synapses and distal inputs. The firing variability is demonstrated by calculating the Coefficient of Variation (C_v) of the interspike interval (ISI) distribution and by observing the corresponding ISI histograms.

1. INTRODUCTION

The understanding and explanation of the origin of experimental neurophysiological results has become nowadays a crucial factor in the development of clearer and more solid views of how the brain works. It is therefore essential to construct artificial neurons which incorporate real neuron features, so that they can be used for modelling and understanding real neuron behaviour.

This paper reviews initially the experimental recordings and observations on neuronal firing which show that they can either be bursty or highly irregular. Therefore, an artificial biological model should be able to reproduce both types of these spike trains and account for that behaviour. Such a model is the Temporal Noisy-Leaky Integrator (TNLI) (Christodoulou et al., 1992, 1993). The TNLI is a biologically inspired hardware neuron which models temporal features of real neurons such as the temporal summation of dendritic postsynaptic response currents, which have controlled delay and duration and the decay of the somatic potential due to its membrane leak. In addition, the TNLI models the stochastic neurotransmitter release by real neuron synapses, by using probabilistic RAMs (pRAMs) (Clarkson et al., 1992, 1993) at each input. This close modelling of biological neuronal features makes the TNLI highly suitable for use in reproducing real neuron behaviour and for investigating and understanding real neuron responses. In this area the TNLI has already been successfully used for investigating the computational role of inhibition in real neurons (Christodoulou et al., 1993).

The paper continues by showing that with the TNLI it is possible to achieve near-random firing capability (i.e., $C_v \approx 0.5\text{-}1$), which complies with the high firing variability of real cortical neurons (Smith and Smith, 1965, Burns and Webb, 1976, Softky and Koch, 1993) and cannot be explained by using simple Leaky Integrate-and-Fire (LIF) or compartmental neuron models with total reset. LIF models can only generate irregular spike trains, close to the natural ones produced by real cortical neurons, at low firing rates where they act as coincidence detectors (Bugmann, 1990, 1991). This high firing variability, observed with the TNLI, is further enhanced in the presence of inhibition. It is also demonstrated that large C_v's ($C_v > 1$) which denote bursting behaviour can be achieved for low input frequencies due to the random synapses and the use of distal inputs.

2. FIRING BEHAVIOUR OF REAL NEURONS AND WAYS OF CHARACTERISING IT

2.1 What is noise in spike trains?

Real neuron spike trains are "noisy". If we define "noise" as any undesired departure from a perfectly ordered system, then noise in real neuron firing spike trains is denoted by the departure of the interspike intervals (ISIs) from the refractory period. The refractory period, which is the minimum period between successive firing, sets the limit for maximum firing without noise.

Though a perfect system can survive under constant environmental conditions, in the real world robust noisy systems are needed (like the real neurons) that can adjust themselves to any change. Therefore, artificial neuron models should be able to reproduce this noisy firing behaviour of real neurons.

2.2 Ways of characterising stochastic neuronal firing properties
There are several ways to represent the structure or organisation of spike trains. The most widely used method is the ISI histogram distributions and their related statistics like the Coefficient of Variation (C_v) and the mean variance. This is the method that we are using for spike train analysis in this paper. An ISI histogram distribution can be: (i) exponential which indicates that the spike trains obey a Poisson process and they are therefore completely random, (ii) γ distribution which denotes that spike trains lie somewhere between randomness and regularity, (iii) bimodal distribution and (iv) skewed distribution. According to observations and analysis by Nagai and Ueda, 1981, the bimodal and skewed distributions reflect regular and irregular burst discharges. Distributions can also exhibit a sharp leading hump with a long, but flat and low tail which is an indication of bursting behaviour. In addition, ISI histogram distributions can be multimodal and such distributions have been recorded experimentally by Nakahama et al, 1968, from an analysis of the spontaneous activity of the single neurons of the lateral geniculate body (LG) and the ventrobasal complex (VB) of cats. Multimodal distributions have also been obtained by Pfeiffer and Kiang, 1965 and Rodieck et al., 1962, from the single neurons in the cochlear nucleus of cats.

The Coefficient of Variation (C_v) of the interspike intervals is a measure of spike train irregularity and is defined as the standard deviation divided by the mean, i. e.:

$$C_v \cdot \sigma_{\Delta t} / \Delta t_M \qquad (1)$$

where Δt_M is the mean interspike interval and $\sigma_{\Delta t}^2$ is the variance of the spike train given by:

$$\sigma_{\Delta t}^2 \cdot \frac{1}{N} \sum_{i=1}^{N} (\Delta t_i \cdot \Delta t_M)^2 \qquad (2)$$

where N is the number of spike intervals and Δt_i is the *ith* interspike interval (ISI).

For a random pure Poisson process $C_v = 1$ and the ISI histogram distribution follows an exponential shape. Values of $C_v >$ 1, indicate bursting behaviour and when the C_v values are close to zero it means that the spike train in question approaches regularity. Basically the C_v is a measure of the relative spread of the distribution and its deviation from exponentiality.

A second method of spike train characterisation is the serial correlogram which is a measure of the dependence of the length of any given interval on the length of the preceding interval (first order), interval before that (second order) and so forth. The serial correlogram is the product-moment correlation between adjacent intervals, second-order intervals (alternate), etc. from the first through the higher order (Perkel et al., 1967).

Other methods of spike train analysis are the joint interval histograms, autocorrelograms and scatter plots of adjacent ISIs. The joint interval histogram is a plot of the joint probability of a pair of adjacent intervals as two dimensions: the abscissa representing the duration of the first of the pair of intervals and the ordinate representing the duration of the second of the pair of intervals. The autocorrelogram is an averaged histogram in which each impulse serves as an origin for a histogram of all succeeding impulses, with separate histograms added and divided by the total number of impulses (Perkel et al., 1967).

2.3 Experimentally observed high variability in real neurons
Most of the experimental studies showed that real neuron firing is highly irregular. Smith and Smith, 1965, investigated the spontaneous cortical activity in the biologically isolated forebrain of a cat and pointed out that trains of action potentials in this propagation, sometimes represent a series of events that is almost random with respect to time. They showed that there is a large range of intervals over which the probability of occurrence of any chosen interval can often be predicted from the Poisson distribution. Similarly, Burns and Webb, 1976, have observed that nerve cells in the cerebral cortex of an unanesthetized mammal appear to exhibit spontaneous activity. In other words they discharge in an irregular fashion at times which often bear no obvious relation to the events in the animal's environment. More recently Softky and Koch, 1993, analysed recordings from primary visual cortex (V1; by Knierim and Van Essen, 1992) and extrastriate cortex (MT; by Newsome et al., 1989) of awake behaving macaque monkey and showed that firing in virtually all of these neurons was nearly consistent with a completely random process.

2.4 Bursting behaviour of real neurons
Spike trains classified as bursty are the ones characterised by clusters of short intervals interspersed between irregular long intervals. Such spike trains were recorded in real neurons; Lamarre et al., 1971, observed that neuronal discharges in the ventrolateral nucleus (VL) of the thalamus in cats during sleep exhibit activity characterised by high frequency bursts separated by long silent intervals. Barlow and Fraioli, 1978, observed sustained oscillations (i.e., bursts) in the discharge of optic nerve cells which are produced by strong inhibitory effects at high level of illuminations of the retina *in situ*. Bursting behaviour

has also been recorded by Nagai and Ueda, 1981, in the gustatory impulse discharges in rat chorda tympani fibers. Recent experimental studies by Baranyi et al., 1993, showed that firing activity and synaptic response properties of neurons in the motor cortex of conscious cats is characteristically different and these neurons can be functionally separated into four cell classes that may operate differently in the neocortical circuits. Indeed, two of these classes (inactivating and noninactivating bursting neurons) indicate bursting behaviour while the rest of the neurons are either fast-spiking or regular-spiking ones. Jensen et al, 1994, investigated the firing modes of rat hippocampal pyramidal cells and showed that they form a continuous variation in burstiness that ranges from regular spiking to spontaneous burst firing. This variation is modulated by the level of extracellular potassium. Finally, an interesting experimental discovery indicating the importance of bursts in real neurons is shown by Ramirez and Pearson, 1993, who recorded bursts in the interneurons of a locust and found that they contribute to the generation of flight motor pattern by amplifying the proprioceptive pathways which is crucial for the generation of intact flight motor pattern.

3. MODELLING OF NEURONAL FIRING

There have been several analytical models trying to model neuronal firing. The model of Rospars and Lánský, 1993, with partial reset accounts for bursting behaviour and reports on other previous models of bursting and C_v's greater than one (see for example Tuckwell, 1979, Barbi and Ferdeghini, 1980, Hanson and Tuckwell, 1983, Kohn, 1989, Lánský and Smith 1989, Lánský and Musila, 1991). Softky and Koch, 1993, have used a simple leaky integrate-and-fire neuron and also a compartmental model, in their attempt to reproduce the high firing variability of cortical cells, but they predicted very low variability ($C_v < 1$), and they argued that neuron models with temporal integration of random EPSP's should fire regularly. Below, we present our neuron model and results which indicate that it can be used for reproducing both high variability and bursts.

3.1 The model

For reproducing the neuronal stochastic firing behaviour, we used the Temporal Noisy-Leaky Integrator (TNLI) neuron model (Christodoulou et al., 1993), with the added feature of partial reset. The TNLI is a biologically inspired hardware realisable neuron model. An analogue hardware outline of the TNLI using a pRAM at each input and a Hodgkin and Huxley equivalent circuit (Hodgin and Huxley, 1952) for a leaky cell membrane, is shown in Fig. 1. In the TNLI, the pRAMs model the stochastic and spontaneous neurotransmitter release by the synapses of real neurons (Katz, 1969; Pun et al., 1986).

The postsynaptic response generators (PSR) shown in the diagram of Fig. 1, model the dendritic propagation of the postsynaptic current responses. For every spike generated by the pRAMs, the PSR

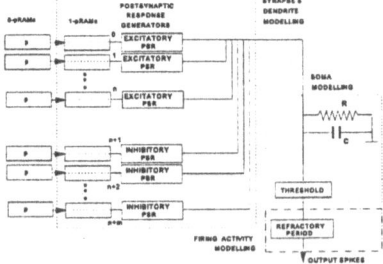

Figure 1: *Block diagram of the TNLI neuron model*

Figure 2: *Postsynaptic Response shapes used in the TNLI*

generators produce postsynaptic current responses ($PSR_{ij}(t)$) of controlled shapes, shown in Fig. 2, which can either be excitatory or inhibitory. Therefore, in the TNLI we have included the temporal function as a property of the spike propagation in the postsynaptic

dendrite as observed in motor-neurons (Redman, 1986). These particular ramp shapes were chosen for the postsynaptic responses (instead of smooth exponential ones) due to the fact that they can easily be implemented and because their defined parameters can be trained. In addition, these shapes result in smoother responses after passing through the leaky integrator circuit, if long rise and fall times (d_r and d_f) are selected, compared to responses produced by rectangular shapes commonly used as inputs to neurons. This enables us to reproduce the smooth postsynaptic potentials produced in distal dendrites of real neurons (Rall, 1964, Shepherd and Koch, 1990). The postsynaptic current responses are summed temporally and the total postsynaptic current response is fed into the RC circuit (Fig. 1) which models the decay of the somatic potential due to its membrane leak. If the potential of the capacitor exceeds a constant threshold (V_{th}), then the TNLI neuron fires. The basic difference of the model used in the current investigation and the one described before (Christodoulou, et al, 1992, 1993) is the resetting of the membrane potential (by resetting the potential of the capacitor) whenever the neuron fires (as in some other leaky integrator models, see for example Lapique, 1907, Stein, 1965, Knight, 1972 and Bugmann, 1991), while the accumulated current due to the temporal summation of the postsynaptic responses (PSRs) on each dendritic input is not reset (as in Kohn, 1989 and Rospars and Lánský, 1993). This is what we call partial reset. The TNLI neuron then waits for a refractory period (t_R) and fires again if the potential is above the threshold. Therefore, the maximum firing rate of the TNLI is given by $1/t_R$.

Further details about the digital hardware structure and the theoretical analysis of the TNLI can be found elsewhere (Christodoulou, et al, 1992, 1993).

3.2 Simulation Details

The parameters used for the postsynaptic responses (Fig. 2) are: t_d (delay time) = 10ms, $d_r = d_f$ = 10ms, t_p (peak period time) = 50ms, h (postsynaptic peak current) = 5pA. The other TNLI parameters are: t_R = 4ms, R (Membrane Leakage Resistance) = 120MΩ, V_{th} = 15mV, C (Membrane Capacitance) = 25pF. The simulation time step used was Δt = 1ms and the system was left to operate for T = 10000ms. It must be noted that the value of the membrane time constant $\tau = RC$ = 3ms might seem to be small compared to realistic values of 13.2 ± 4.0ms (Mason et al., 1991), but due to the fall time (d_f = 10ms) and the peak period time (t_p = 50ms) of the postsynaptic response shape, there is a slower decay which increases the effective value of the membrane time constant. At the TNLI inputs, random spike trains of controlled mean frequency (f_j) were utilised (produced by the 0-pRAMs). These were unaffected by the 1-pRAM action since their memory contents were set to `1' for an input spike and `0' for no spike and thus they fired for each input spike.

3.3 Results and Discussion

Results were taken with 16 excitatory PSR generators and 0 or 10 inhibitory ones respectively. Fig. 3 shows the output frequency of the TNLI neuron and the Coefficient of Variation (C_v) of the output spike trains (eqn. 1) as a function of the Mean Input Current (I_M). Both the cases with and without inhibition are shown in Fig. 3. In addition, Fig. 3 shows the transfer function of the TNLI and the C_v of the output spike trains when regular input spike trains are used, for the case without inhibition only. I_M in the TNLI neuron i is given by:

$$I_M \cdot \sum_{j=0}^{N} f_j \times PSR_{ij}^{*} \tag{3}$$

where f_j is the mean input spike frequency which in our simulations is the same for each input j and PSR_{ij}^{*} is the time integral of the postsynaptic current (PSR_{ij}) produced by a spike arriving on input line j. N is the total number of input lines (or pRAMs).

As we can observe from Fig. 3, the TNLI gives a sigmoidal non-linear transfer function instead of a step function, which this model would produce with continuous input current. This behaviour seems to be similar to that of the formal neuron (McCulloch and Pitts, 1943, Little, 1974, Hopfield, 1982), which has a sigmoid transfer function given by: $y = 1/[1 + exp(-\beta A_i)]$, where β is a constant that determines the slope of the sigmoid and A_i is given by: $\sum_{j} x_j w_{ij}$ where x_j is the jth input to neuron i and w_{ij} is the connection weight value from neuron j to neuron i. A_i is equivalent to I_M in the TNLI (eqn. 3).

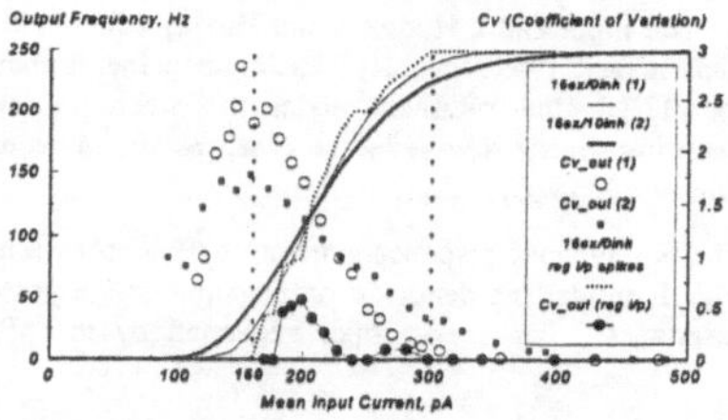

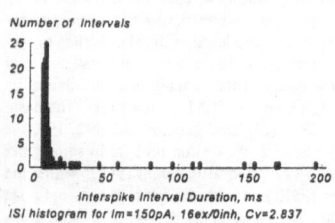

Figure 3: *Transfer function of the TNLI and C_v of the output spike trains with and without inhibition and with regular input spike trains*

The inhibitory inputs increase the fluctuations of the input current around its mean saturation level and therefore reduce the slope of the sigmoidal transfer function (Fig. 3) of the TNLI as explained elsewhere (Christodoulou *et al.*, 1993).

In order to observe the firing variability of the TNLI, the Coefficient of Variation (C_v) of the output spike trains (see eqn. 1) was calculated. As it can be seen from Fig. 3, for low I_M values (up to approx. $I_M \approx$ 230pA), the C_v of the output rises initially above one indicating bursting behaviour. Using t_p = 50ms restricted us to the case of distal inputs which cause a rather continuous input current (i.e., with small fluctuations around I_M). In this case one usually assumes that regular spike trains with a lower C_v are produced (similar to the case of large number of inputs studied by Softky and Koch, 1993, where the input current is also relatively continuous). This is true in the LIF models with total reset. However, in the TNLI, reset does not erase the history of past inputs. Similar input currents are present during few output spikes which will show similar intervals and possibly high frequency bursts followed by longer silences. This causes a non-poissonian distribution of Interspike Intervals (ISI) characterised by very large C_v's (i.e., $C_v \gg 1$). Therefore, the TNLI can reproduce the bursting behaviour that is recorded in real neurons (see section 2.4). An ISI histogram shown in Fig. 4 for I_M = 150pA (p = 0.032) giving C_v = 2.837 (no inhibition case), indicates the bursting

Figure 4: *ISI histogram showing bursting behaviour at low firing rates (f_{out} = 13.58Hz)*

behaviour of the TNLI neuron. For this particular simulation case, an output spike train of 131 spikes within 10000 time steps was obtained giving a very low mean firing rate of 13.58Hz (Δt_M = 73.63ms). These output spikes were distributed over 42 interspike intervals with different lengths, the longest being 1755ms. For graphical clarity the 14 longest intervals are not shown in the histogram of Fig. 4. As it can be seen, the bursting behaviour is characterised by the high C_V and the long tail of the ISI histogram. In this region of I_M values (i.e., up to approx. $I_M \approx$ 230pA) inhibition gives lower C_V's (closer to one; see Fig. 3), because it increases the variability of I_M and therefore reduces the possibility of bursts.

For I_M > 230pA, C_V values around one can be observed (Fig. 3), which are exponentially reduced to zero due to saturation (for I_M > 350pA). These C_V values are increased when the ten inhibitory inputs are introduced, due to the fact that they cause a more irregular input current. Therefore, in the TNLI with reset on the soma only, high firing variability can be achieved (i.e., $C_V \approx$ 0.5-1) for a certain range of I_M values (e.g., $I_M \in$ [230pA, 280pA] for the simulation parameters used above). As it can be seen more clearly from Fig. 5, this high variability is observed at the relatively high mean firing rates of 155Hz to 215Hz ($\Delta t_M \in$ [4.65, 6.45ms]). Near $\Delta t_M = t_R$ = 4ms (i.e., giving maximum possible firing rate of 250Hz), the C_V drops towards zero because we approach saturation, due to the refractory period which induces completely regular patterns. This high variability at high firing rates complies with experimental results from cortical neurons as described in section 2.3.

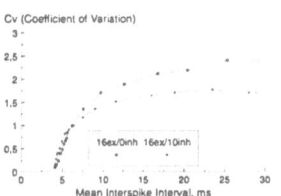

Figure 5: *TNLI firing variability with and without inhibition presented as $\{C_V\}$ vs $\{\Delta t_M\}$*

Fig. 6 shows the ISI histograms of the output spike trains for: (a) I_M = 229pA (p = 0.085), (b) I_M = 256pA (p = 0.095) and (c) I_M = 283pA (p = 0.105) in the case of inhibition. The mean firing rates for these I_M values are 155.3Hz (Δt_M = 6.44ms), 188.7Hz (Δt_M = 5.3ms) and 213.7Hz (Δt_M = 4.68ms) respectively. Part of the histograms with longer intervals have not been included for graphical clarity. The histogram for I_M = 229pA (Fig. 6a) does not include the 17 longest intervals, the longest being 83ms; similarly from the histogram for I_M = 256pA (Fig. 6b), the 11 longest intervals have been omitted, the longest being 65ms and from the histogram for I_M = 283pA (Fig. 6c), the 4 longest intervals have been omitted, the longest being 36ms. As it can be seen, the Δt_i's are exponentially distributed, which justifies near-random behaviour (i.e., $C_V \approx$ 1). The number of occurrences of an ISI = 4 (where 4ms is the value of the refractory period t_R) increases sharply as the output spike train saturates at the rate of $1/t_R$ (i.e., 250Hz) and becomes more regular (i.e., as C_V decreases). The refractory period, t_R = 4ms, is indicated in the histograms of Fig. 6 by the virtual absence of ISI's below that value. Fig. 5 shows also very clearly, the very high C_V values (i.e., $C_V \gg$ 1) obtained at low firing rates. These values are reduced with the introduction of inhibition (i.e., they get closer to one), due to the fact that the possibility of bursts is reduced since inhibition increases the variability of the input current. At high firing rates however (Fig. 5), where $C_V \leq$ 1, higher C_V values are obtained with inhibition (i.e., giving enhanced variability), than with no inhibition. This is a result of the more irregular input current caused by the inhibition.

Figure 6: *ISI histograms of the output spike trains for three I_M values, showing near random behaviour at the high mean firing rates of: (a) f_{out} = 155.3Hz, (b) f_{out} = 188.7Hz and (c) f_{out} = 213.7Hz*

The importance of random synaptic inputs in achieving high firing variability is demonstrated very clearly by the use of regular input spike trains which gives very low C_V values (up to 0.5) as shown in Fig. 3 (case of no inhibition). These low C_V values ($C_V \leq$ 0.5), which are produced when regular input spike trains with the same phase are used, is a result of the long peak period time t_p (t_p = 50ms) used for the postsynaptic responses. This is demonstrated in Fig. 7 which shows that with t_p = 20ms (rest of the parameters as in Fig. 3), the regular input spike trains produce a completely regular output ($C_V \approx$ 0). In addition, the reduction of the peak period time to t_p = 20ms (Fig. 7) reduces slightly the very high C_V values when

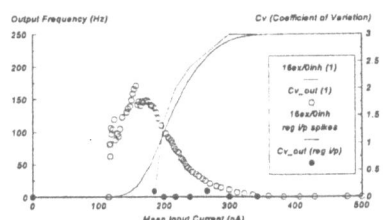

Figure 7: *Transfer function and C_V of the TNLI with random and regular inputs and t_p = 20ms (rest of the parameters as in Fig. 3)*

random inputs are used, because obviously the temporal summation in the dendrites is reduced giving therefore shorter bursts.

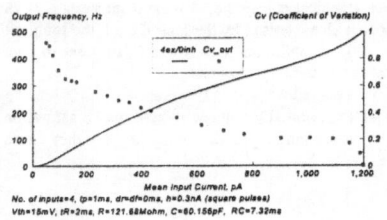

Figure 8: *TNLI transfer function and C_V by using square pulses as postsynaptic responses*

Another indication that the effect of temporal summation in the dendrites is quite strong, is shown in Fig. 8 where by using $t_p = 1$ms (equal to the time step Δt) and $d_r = d_f = 0$ms (i.e., square pulses and no temporal summation in the dendrites), leads to a very reduced C_V. This temporal summation of postsynaptic responses in the dendrites is a temporal summation of noise in terms of irregularity, that leads to a variance $\propto [1-exp(-1/t_p)]^{-1}$, which is nearly equal to t_p (Bressloff and Taylor, 1992). This factor is large when $t_p = 50$ms. Thus the very reduced C_V values for $t_p = 1$ms (Fig. 8), result from the absence of temporal summation of noise in the dendrites. In addition this reduction of the C_V values is facilitated by the increase of the membrane time constant ($RC = 7.32$ms in the simulation of Fig. 8), which increases the effective temporal integration in the soma.

The C_V values around one that can be observed in the simulation results shown in Fig. 8, are due to the fact that the TNLI behaves like a simple LIF model when square input pulses are used and at low mean firing rates it acts as a coincidence detector (Bugmann, 1990, 1991). Fig. 9 shows four ISI histograms for the case of square pulses at the low mean firing rates of (a) 16.4Hz ($\Delta t_M = 61.96$ms), (b) 22.53Hz ($\Delta t_M = 44.38$ms), (c) 34.36Hz ($\Delta t_M = 29.10$ms) and (d) 63.05Hz ($\Delta t_M = 15.86$ms) obtained from the Mean Input Current values of (a) $I_M = 60$pA ($p = 0.025$), (b) $I_M = 73$pA ($p = 0.03$), (c) $I_M = 94$pA ($p = 0.04$) and (d) $I_M = 142$pA ($p = 0.06$) respectively and giving corresponding C_V values of 0.895, 0.825, 0.714 and 0.635. As it can be observed, in none of the histograms of Fig. 9 the Δt_i's follow an exponential shape and therefore high variability is not justified. In fact, it should be noted that a value of $C_V \approx 1$ only is not always in itself substantial for ascertaining a random pure Poisson process. As shown by Keilson, 1979, there is a broad class of ergodic time-reversible models (e.g., Markov chains), for which the first passage times that usually describe interspike intervals, are completely monotonous and thus $C_V \geq 1$. The histograms of Fig. 9(a) and (b), give multimodal distributions while the histogram of Fig. 9(c), gives two rather distinct modes at 11ms and 37ms and the distribution is therefore a bimodal one. The distribution of the histogram of Fig. 9(d) is unimodal, asymmetric with non-exponential decay and with positive skewness. Experimental multimodal distributions have been recorded in real neurons (see section 2.2). Wilbur and Rinzel, 1983, suggested a modified version of Stein's model (Stein, 1965) with reversal synaptic potentials and a time-varying threshold which accounts for the recorded experimental bimodal distributions. From the histograms shown in Fig. 9 we can claim that the TNLI can give multimodal ISI distributions at low firing rates when square input pulses are used and no temporal PSR summation is present in the dendrites.

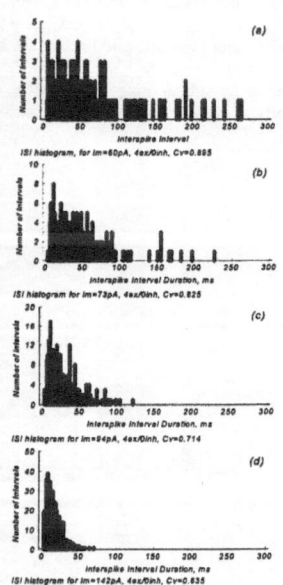

Figure 9: *ISI histograms of the output spike trains for the case of square input pulses at the low mean firing rates of: (a) $f_{out} = 16.4$Hz, (b) $f_{out} = 22.53$Hz, (c) $f_{out} = 34.36$Hz and (d) $f_{out} = 63.05$Hz, showing multimodality and positive skewness*

4. CONCLUSIONS

In this paper we have reviewed the statistical behaviour of the output spike trains of real neurons and demonstrated how this behaviour can be modelled and investigated using the TNLI neuron model with partial reset. Firstly the relatively continuous postsynaptic current caused by distal inputs can cause large C_V's ($C_V > 1$) at low firing rates, when only the soma is reset by an output spike. Such an effect cannot be reproduced by the simpler LIF model with total reset. Therefore, we may note that the TNLI can be used to study combined proximal and distal inputs, which the LIF cannot reproduce, having only one decay constant. These large C_V's also result from the use of random synaptic inputs (0-pRAMs). In addition, high variability can be achieved with the TNLI ($C_V \approx 0.5$-1) at high firing rates, which results from the use of random input spike trains and the dendritic temporal summation imposed as a property of spike propagation (temporal summation of noise in the dendrites). This becomes very significant if sufficiently long postsynaptic responses are used (long t_p). The high variability is enhanced by the use of inhibitory inputs. Finally, we have seen that bimodal and multimodal ISI distributions can also be obtained with the TNLI at low firing rates for the case of using square input pulses and having no temporal PSR summation in the dendrites.

We can therefore claim that in general the TNLI neuron, despite its relative simplicity, can be successfully used for modelling and understanding real neuron behaviour due to the fact that it models biological neuronal features.

ACKNOWLEDGMENT
This work is supported by a grant (Ref. Number F. 40Z) awarded by The Leverhulme Trust to King's College London (University of London).

REFERENCES
Baranyi, A., Szente, M. B. and Woody C. D. (1993). Electrophysiological characterization of different types of neurons recorded *in vivo* in the motor cortex of the cat. *I*. Patterns of firing activity and synaptic responses. *J. of Neurophysiol.*, **69**, 6, 1850-1864.

Barbi, M. and Ferdeghini, E. M. (1980). Relevance of the single ommatidium performance in determining the oscillatory response of the *Limulus* retina. *Biol. Cypern.* **39**, 45-51.

Barlow, R.B. and Fraioli, A. J. (1978). Inhibition in the *Limulus* lateral eye in situ. *J. of Gen. Physiol.*, **71**, 574-592.

Bressloff, P. C. and Taylor, J. G. (1992). Temporal sequence storage capacity of time-summating neural networks. *J. of Physics A: Math. Gen.*, **25**, 833-842.

Bugmann, G. (1990). Irregularity of natural spike trains simulated by an integrate-and-fire neuron. *In: Extended Abstracts,3rd Int. Symposium on Bioelectronic and Molecular Electronic Devices.* R & D Association for Future Electron Devices, Kobe, Japan, 105-106.

Bugmann, G. (1991). Summation and multiplication: two distinct operation domains of leaky-integrate-and-fire neurons. *Network*, **2**, 489-509.

Burns, B. D. and Webb, A. C. (1976). The spontaneous activity of neurons in the cat's visual cortex. *Proc. Royal Soc. London B*, **194**, 211-223.

Christodoulou, C., Bugmann, G., Taylor, J. G. and Clarkson, T. G. (1992). An extension of the Temporal Noisy-Leaky Integrator neuron and its potential applications. *Proc. of the Int. Joint Conf. on Neural Networks*, Beijing, **III**, 165-170.

Christodoulou, C., Bugmann, G., Clarkson T. G. and Taylor J. G. (1993). The Temporal Noisy-Leaky Integrator neuron with additional inhibitory inputs. *New Trends in Neural Computation*, Lecture Notes in Computer Science, ed. by J. Mira, J. Cabestany and A. Prieto, Springer-Verlag, **686**, 465-470.

Clarkson, T. G., Ng, C. K., Gorse, D and Taylor, J. G. (1992). Learning Probabilistic RAM Nets using VLSI structures. *IEEE Trans. on Computers*, **41** (12), 1552-1561.

Clarkson, T. G., Ng, C. K. and Guan, Y. (1993). The pRAM: An adaptive VLSI chip. *IEEE Trans.on Neural Networks*, **4** (3), 408-412.

Hanson, F. B. and Tuckwell, H. C. (1983). Diffusion approximation for neuronal activity including synaptic reversal potentials. *J. of Theor. Neurobiol.* **2**, 127-153.

Hodgkin, A. L. and Huxley, A F. (1952). A quantitative description of membrane current and its application to conduction and excitation in a nerve. *J. of Physiol.* (London) **117**, 500-544.

Hopfield, J. J. (1982). Neural Networks and physical systems with emergent collective computational activities. *Proc. Natl. Acad. Sci.*, USA, **79**, 2554-2558.

Jensen, M.S., Azouz, R. and Yaari, Y. (1994). Variant Firing Patterns in Rat Hippocampal Pyramidal Cells Modulated by Extracellular Potassium. *J. of Neurophysiol.*, **71**, 3, 831-839.

Katz, B. (1969). *The release of Neural Transmitter substance.* Liverpool University Press, Liverpool.

Keilson, J. (1979). *Markov Chain Models - Rarity and Exponentiality.* Applied Mathematical Sciences, **28**. Springer-Verlag.

Knierim, J. and Van Essen, D. (1992). Neuronal Responses to static textual patterns in area V1 of the alert macaque monkey. *J. Neurophysiol.*, **67**, 961-980.

Knight, B. W. (1972). Dynamics of Encoding in a Population of Neurons. *J. of Gen. Physiol.*, **59**, 734-766.

Kohn, A. F. (1989). Dendritic Transformations on Random Synaptic Inputs as Measured From a Neuron's Spike Train - Modelling and Simulation. *IEEE Trans. on Biomedical Engineering*, **36**, 44-54.

Lamarre, Y., Filion, M. and Cordeau, J. P. (1971). Neuronal discharges of the ventrolateral nucleus of the thalamus during sleep and wakefulness in the cat. I. Spontaneous Activity. *Exp. Brain Res.*, **12**, 480-498.

Lánský, P. and Musila, M. (1991). Variable initial depolarization in Stein's neuronal model with synaptic reversal potentials. *Biol. Cybern.*, **64**, 285-291.

Lánský, P. and Smith, C. E. (1989). The effect of a random initial condition in neural first-passage-time models. *Math. Biosci.*, **93**, 191-215.

Lapique, L. (1907). Reserches quantatives sur l' excitation électrique des nerfs traitée commeune polarization. *J. Physiol. Pathol. Gen.*, **9**, 620-635.

Little, W.A. (1974). The existence of persistent states in the brain. *Math. Biosci.*, **19**, 101-120.

Mason, A., Nicoll, A. and Stratford, K. (1991). Synaptic Transmission between Individual Pyramidal Neurons of the Rat Visual Cortex *in vitro*. *J. of Neurosci.*, **11**, 72-84.

McCulloch, W. S. and Pitts, W. (1943). A logical calculus of the ideas immanent in nervous activity. *Bull. Math. Biophys.*, **5**, 115-133.

Nakahama, H., Suzuki, H., Yamamoto, M., Aikawa, S. and Nishioka, S. (1968). A Statistical Analysis of Spontaneous Activity of Central Single Neurons. *Physiol. Behav.*, **3**, 745-752.

Newsome, W., Britten, K., Movshon, J. A. and Shadlen, M. (1989). Single neurons and the perception of motion. In: *Neural Mechanisms of visal perception* (Man-Kit Lam D, Gilbert C, eds), The Woodlands, TX: Portofolio, 171-198.

Nagai, T. and Ueda, K. (1981). Stochastic properties of gustatory impulse discharges in rat chorda tympani fibers. *J. of Neurophysiol*, **45**, 574-592.

Perkel, D. H., Gerstein, G. L. and Moore, G. P. (1967). Neuronal spike trains and stochastic point processes. I. The single spike train. *Biophys. J.* **7**: 391-418.

Pfeiffer, R. R. and Kiang, N. Y-S. (1965). Spike discharge patterns of spontaneous and continuously simulated activity in the cochlear nucleus of anesthetized cats. *Biophys. J.*, **5**, 301-316.

Pun, R. Y. K., Neale, E. A., Cuthrie, P. B. and Nelson, P. G. (1986). Active and Inactive Central Synapses in the Cell Culture. *J. of Neurophysiol.*, **56**, (5), 1242-1256.

Rall, W. (1964). Theoretical significance of dendritic trees for neuronal input-output relations. *In: Neural Theory and Modelling.* Ed. by Reiss,. R. F., Stanford, Stanford University Press, 73-77.

Ramirez, J-M. and Pearson, K. G. (1993). Alteration of bursting properties in interneurons during locust flight. *J. of Neurophysiol.*, **70**, 5, 2148-2160.

Redman, S. (1986). Monosynaptic Transmission in the spinal cord. *News in Physiol. Sci.* **1**, 171-174.

Rodieck, R. W., Kiang, N. Y-S. and Gerstein, G. L. (1962). Some quantitative methods for the study of spontaneous activity of single neurons. *Biophys. J.*, **2**, 351-368.

Rospars, J. P. and Lánský, P. (1993). Stochastic model neuron without resetting of dendritic potential: application to the olfactory system. *Biol. Cybern.*, **69**, 283-294.

Shepherd, G. M. and Koch, C. (1990). Appendix: Dendritic Electrotonus and Synaptic Integration. *The Synaptic Organisation of the Brain*, (3rd edition), ed. by Shepherd, G. M., Oxford University Press.

Smith, D. R. and Smith G. K. (1965). A statistical analysis of the continuous activity of single cortical neurons in the cat unanesthetized isolated forebrain. *Biophys. J.*, **5**, 47-74.

Softky, W. R. and Koch, C. (1993). The Highly Irregular Firing of Cortical Cells Is Inconsistent with Temporal Integration of Random EPSP's. *J. of Neurosci.*, **13** (1), 334-530.

Stein, R. B. (1965). A theoretical analysis of neuronal variability. *Biophys. J.*, **5**, 173-195.

Tuckwell, H. C. (1979). Synaptic transmission in a model for stochastic neural activity. *J. of Theor. Biol.*, **77**, 65-81.

Wilbur, W. J. and Rinzel, J. (1983). A theoretical basis for large Coefficient of Variation and biomodality in neuronal interspike interval distributions. *J. of Theor. Biol.* **105**, 345-368.

The BP-λL1 Algorithm: Non-Chaotic and Accelerated Learning in a MLP Network

Bertrand AUGEREAU[*], Thierry SIMON[*], Jacky BERNARD[*], Bernard HEIT[**]

[*] laboratoire Signaux Images et Communications, Poitiers (France)

[**] Centre de Recherche en Automatique de Nancy, Nancy (France)

Summary

Multilayer perceptrons are learning structures often used in the connectionist approach. A backpropagation algorithm, which enables learning, is a fixed point research algorithm. As such, it induces the various behaviours of chaotic dynamics. One can apply to it the tools of chaos theory. Amongst the useable tools, a measurement of behaviour, or more precisely stability, exists, namely Lyapunov numbers. The obtaining of these numbers comes through awareness of the eigen values of the jacobian matrix associated to the weight modification functions. We give a method of calculation whose efficiency comes from the use of the particularities of the backpropagation. From the calculation of the Lyapunov numbers, the basic backpropagation algorithm is modified. We propose the first part of a new learning algorithm whose originality resides in a strategy of gradient step constraint, arising from the obligation of a stable behaviour. Its values is related to obtaining rapid convergence.

Introduction

Multilayer perceptrons (MLP) are appearing increasingly more frequently in pattern recognition applications [1]. The primary obstacle to the development of these connectionist methods arises from difficulties encountered in the learning phase [2][3]. Among the various algorithms that can be used in this phase, gradient backpropagation (BP) is the most often used [4][5][6]. The present work starts with a BP algorithm to generate a new learning algorithm : BP-λL1. Its originality resides in the strategy used to determine gradient step, that takes into account a measurement of network behavior by the use of Lyapunov numbers, thereby guaranteeing the stability of the learning process. If cover can be assured, BP-λL leads to accelerated convergence. This communication is composed of four parts : the first presents notations used and sets the BP algorithm in the framework of the search for a fixed point; the second involves the elaboration of a measurement of behavior; the third presents the BP-λL1 algorithm. Finally, the last part compares our results with those obtained with two other algorithms, BP and ELEANNE.

Learning by backpropagation of the gradient search for a fixed point

In the course of supervised learning, an MLP network uses examples. The learning sample designates the set of p pairs (X_i, D_i). The X_i vectors are the patterns presented and the D_i vectors are the corresponding desired outputs. Starting with a sample $\{(X_i, D_i)\}_{i=1...p}$ and network architecture A, the learning process seeks weights W such that the overall transition function $F_{W,A}$ covers the sample, $D_i = F_{W,A}(X_i) \quad \forall i = 1...p$. Learning is the result of an iterative optimization of the weights of connections by gradient descent on a function of cost of errors committed $C(W) = \dfrac{1}{p} \sum\limits_{i=1}^{p} |F_{W,A}(X_i) - D_i|^2$ [1][2]. We furnish the results of a demonstration that has become classical and introduce notations given to the nth step, for a neuron j, for a network with c connections, the set of weights forming a vector W_n of $\mathbf{R}^C$, $W_n = \left(w_{j_1 i_1}(n) \quad w_{j_2 i_2}(n) \quad ... \quad w_{j_c i_c}(n) \right)$. We define the total input of the neuron by $a_j(n) = \sum\limits_{i \in A(j)} w_{ji}(n) x_i(n)$, $A(j)$ designating the set of prior neurons and x_i the output of these neurons. During propagation, the output of neuron $x_j(n) = f(a_j(n))$ is the image of the total input by a transfer function $f(a) = [1 + exp(-a)]^{-1}$. During backpropagation, a new weight of a connection is obtained by applying the iterative modification $w_{ji}(n+1) = w_{ji}(n) + \lambda(n) \Delta_{ji}(W_n)$. In this relationship, the parameter λ is the gradient step and the weighting Δ_{ji} is equal to $\dfrac{\partial x_j(n)}{\partial w_{ji}(n)} K_j(n)$. The term K_j depends on the layer to which the neuron belongs. We have $K_j(n) = 2(d_j - x_j(n))$ for the output layer and $K_j(n) = \sum\limits_{k \in P(j)} \dfrac{\partial x_k(n)}{\partial x_j(n)} K_k(n)$ for the other layers. $P(j)$ designates all posterior neurons. Weighting corresponds to a function of $\mathbf{R}^C$ in $\mathbf{R}$ and in order to adopt a more synthetic notation, we consider the modification function $R_{ji}(W) = w_{ji} + \lambda \Delta_{ji}(W)$. The action of the BP algorithm takes iterations into account and is written $w_{ji}(n+1) = R_{ji}(W_n)$. We end up with the relationship

$$(1) \quad W_{n+1} = R(W_n),$$

where the vectorial function R is composed of c $R_{ji}(W)$ functions of $\mathbf{R}^C$ in $\mathbf{R}$. In addition, a scaler, or vectorial value $\bar{x}$ is a fixed point for the function F if we have $\bar{x} = F(\bar{x})$. The search for the fixed point designates the iterative method, based on the recurrence function $x_{n+1} = F(x_n)$ that constructs a series $\{x_n\}$, or trajectory, converging towards $\bar{x}$. This search, extensively described in numerical analysis [7][8], defines an iterative process; in the present case it is a single variable, first order process. Relationship (1) can be used to affirm that a BP algorithm is the search for a fixed point $\bar{W}$ of the vector of weights of an MLP network for the modification function $R(W)$.

Measuring learning behavior

Iterative searches produce trajectories whose potential accumulation enables the fixed point to be obtained, with success depending on adherence of the series generated. The study of behavior enables the convergent or divergent nature of the process used to be determined. Our analysis shows that faults generally attributed to the BP algorithm arise in large part from the fact that behavior during learning is not known. To this end, we propose the use of Lyapunov numbers [9][10]. In the case of MLP networks, however, a preliminary is necessary.

Let us take an example and consider the architecture of *Figure 1*. The chosen conditions for learning associate the input $x_0 = 1.5$ and the required outputs $d_2 = 0.6$ for the neuron 2, and $d_3 = 0.25$ for neuron 3. *Figures 2.a* and *2.b* show the diagrams of bifurcation of outputs x_2 and x_3 for a gradient step varying from 15 to 50. The attractor set, the adhesion of the trajectories is carried there according to λ.

Apparently there are all the characteristics of chaotic dynamics [11][12], the stable, oscillatory and chaotic regimes.

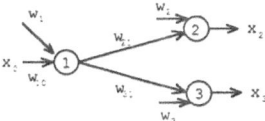

Figure 1 : a network with three neurones with bias.

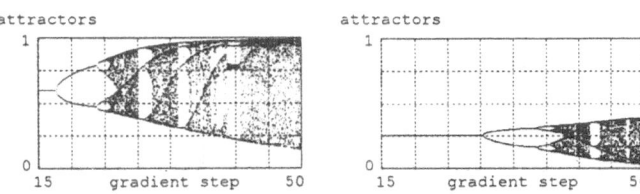

Figure 2.a : diagram of bifurcation of x_2. *Figure 2.b* : diagram of bifurcation of x_3.

This very simple example shows that, in general, nothing can justify a parallelism of output behaviors. In the course of backpropagation, the error committed on an output does not affect the weight of connections associated with the other neurons of the output layer. As a result, we introduce the notion of the partial interconnection graph Γ_s. For a network with architecture A, Γ_s designates the set of connections reached by backpropagation of the error committed at the desired output d_s, s describing the set Σ of subscripts of neurons in the output layer. Let us now suppose that we can determine the Lyapunov numbers $\gamma(s)$, descriptors of learning behavior restricted to Γ_s. The overall Lyapunov number, γ, is obtained by considering that in the case of simultaneity, overall behavior is the reflection of unstable behavior rather than a simple convergence. This overall system results in the relationship $\gamma = sup(\gamma(s))_{s \in \Sigma}$. This reduced convergence enables a restriction hypothesis to be formulated. In order to calculate the $\gamma(s)$ numbers, we will first envision a network whose output layer is reduced to one neuron.

The stability of the process is determined by comparing the change in a trajectory of initial value W_0 with that of a trajectory starting with a similar initial value $\tilde{W}_0 = W_0 + \Delta W_0$. After an initial iteration, we have $\tilde{W}_1 = W_1 + \Delta W_1 = R(W_0 + \Delta W_0)$ and each component of the vector $\tilde{W}_1$ is $\tilde{w}_{ji}(1) = R_{ji}(W_0 + \Delta W_0)$. Via finite increases, the second member is taken to be $R_{ji}(W_0) + \sum_{l,k=1}^{c} \dfrac{\partial R_{ji}(W_0)}{\partial w_{lk}} \Delta w_{ji}(0)$. Since $\tilde{w}_{ji}(1) = w_{ji}(1) + \Delta w_{ji}(1)$, we obtain the various components $\Delta w_{ji}(1) = \sum_{l,k=1}^{c} \dfrac{\partial R_{ji}(W_0)}{\partial w_{lk}} \Delta w_{ji}(0)$ of the deviation ΔW_1. The relationship between deviations in matrix form is written $\Delta W_1 = {}^c J_0 \Delta W_0$, with ${}^c J_0 = \left(\dfrac{\partial R_{ji}(W_0)}{\partial w_{lk}} \right)_{i,j,k,l \in \{1..c\}}$ the jacobian matrix. For the following deviations, we have $\Delta W_2 = {}^c J_1 \Delta W_1$ and $\Delta W_n = {}^c J_{n-1} \Delta W_{n-1}$, *i.e.* $\Delta W_n = \left(\prod_{i=0}^{n-1} {}^c J_i \right) \Delta W_0$. The partial derivatives are calculated with

$$(2) \quad \frac{\partial R_{ji}(W)}{\partial w_{ji}} = 1 + \lambda \frac{\partial \Delta_{ji}(W)}{\partial w_{ji}} \quad \text{or (3)} \quad \frac{\partial R_{ji}(W)}{\partial w_{lk}} = \lambda \frac{\partial \Delta_{ji}(W)}{\partial w_{lk}}$$

(2) for the terms of the principal diagonal, (3) for the other terms. The determination of Lyapunov numbers is expressed in terms of deviations with the hypothesis of a geometric model for a mean convergence or divergence rate of neighboring trajectories. In the multidimensional case, the effective rate is equal to the product of the eigenvalues of the jacobian matrix [13][14]. The elements of the spectrum, designated β, are the roots of the characteristic polynomial $det({}^c J - \beta^c I)$. By the use of the approximation of linearization of derivatives $\dfrac{\partial \Delta_{j_h i_h}(W)}{\partial w_{j_h i_h}} \dfrac{\partial \Delta_{j_k i_k}(W)}{\partial w_{j_h i_k}} \cong \dfrac{\partial \Delta_{j_h i_h}(W)}{\partial w_{j_h i_k}} \dfrac{\partial \Delta_{j_k i_k}(W)}{\partial w_{j_h i_h}}$, where strict equality is verified around a fixed point, we obtain by recurrence on dimension c, the characteristic equation. The spectrum is thus composed of two eigenvalues

$$\text{(4) } \beta_{1,n} = 1 \text{ of order c-1 and } \beta_{2,n} = 1 + \lambda \sum_{k=1}^{c} \frac{\partial \Delta_{j_k i_k}(W_n)}{\partial w_{j_k i_k}} \text{ of order 1.}$$

We obtain a Lyapunov number for the learning process by BP algorithm in an MLP network with c connections given by $\gamma = \lim\limits_{n \to \infty} \dfrac{1}{n} \sum\limits_{p=0}^{n-1} ln\left|1 + \lambda \sum\limits_{k=1}^{c} \dfrac{\partial \Delta_{j_k i_k}(W_p)}{\partial w_{j_k i_k}}\right| = \lim\limits_{n \to \infty} \gamma_n$. The relationship obtained takes into account only type (2) derivatives. Derivation furnishes a first general form

$$\frac{\partial R_{ji}(W)}{\partial w_{ji}} = 1 + \lambda \left[\frac{\partial^2 x_j}{\partial w_{ji}^2} K_j + \left(\frac{\partial x_j}{\partial w_{ji}}\right)^2 \frac{\partial K_j}{\partial x_j} \right] = 1 + \lambda L_{ji}|W. \text{ We have}$$

$$\text{(5) } L_{ji} = 2\left[\frac{\partial^2 x_j}{\partial w_{ji}^2}(d_j - x_j) - \left(\frac{\partial x_j}{\partial w_{ji}}\right)^2 \right] \text{ or (6) } L_{ji} = \frac{\partial^2 x_j}{\partial w_{ji}^2} K_j + \left(\frac{1}{x_j}\frac{\partial x_j}{\partial w_{ji}}\right)^2 \sum_{k \in P(j)} (w_{kj})^2 L_{kj},$$

(5) when neuron j belongs to the output layer, (6) when neuron j belongs to a hidden layer. We have shown above the elements required for elaborating a measurement of learning behavior for a BP algorithm. If we remove the hypothesis of restriction to partial graphs, the Lyapunov number is

$$\text{(7)} \quad \gamma = sup(\gamma(s))_{s \in \Sigma} \text{ with } \gamma(s) = \lim\limits_{n \to \infty} \gamma_n(s) \text{ and } \gamma_n(s) = \frac{1}{n} \sum_{k=0}^{n-1} ln\left|1 + \lambda \sum_{(i,j) \in \Gamma_s} L_{ji}|W_k\right|.$$

Σ is the set of subscripts of output layer neurons. The ordinate pair (i,j) describes the set Γ_s of connections of the partial graph associated with neuron s. L_{ji} values are obtained in the backpropagation mode from (5) or (6). If the iterative process is stable the number is negative. A null number corresponds to the start of a cycle. A positive number is characteristic of unstable behavior. To conclude, let us take up again our example. *Figure 3* has been obtained in similar conditions to those stated above and by using our algorithm of calculation of Lyapunov numbers. The results are satisfying and confirm the observation of various regimes of chaotic dynamics. They illustrate perfectly the validity of Lyapunov numbers as global descriptors of learning behaviour.

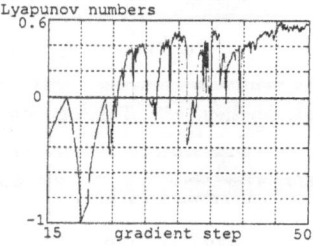

Figure 3 : Lyapunov numbers for process described by *Figures 1, 2.a, 2.b.*

The BP-λL1 algorithm, constrained backpropagation

Lyapunov numbers furnish a measurement of the stability of the series $\{W_i\}$ generated during learning. In order to remain consistent with the hypotheses formulated, an asymptotic visit to the fixed point must be imposed, which induces the geometric model for deviations of similar trajectories. It is a question of imposing stable behavior on the BP algorithm, inducing negative Lyapunov numbers. We know that a necessary condition for convergence towards a simple attractor is the existence of a negative Lyapunov number. We establish a constraint by imposing on all instantaneous Lyapunov numbers γ_n the obligation to be equal to the same negative ν; in this case, the overall Lyapunov number $\gamma = \lim_{n \to \infty} \gamma_n$ is

equal to this ν. Thus, the learning process is subjected to the influence of a stable attractor. In the framework of the restriction hypothesis, we apply the defined constraint. In expression (7), the L_{ji} values are objective data of the process. The constraint can be exerted only on parameter λ and this

dependence is noted by λ_k. We obtain the equality $\nu = \dfrac{1}{n}\sum_{k=0}^{n-1} ln\left|1 + \lambda_k \sum_{(i,j)\in C} L_{ji}|W_k|\right|$, the ordinate pair (i,j)

describing the set C of connections. From the preceding equality, we deduce that for all n,

$\left|1 + \lambda_n \sum_{(i,j)\in C} L_{ji}|W_n|\right| = exp(\nu)$, which enables us to obtain (8) $\sum_{(i,j)\in C} L_{ji}|W_n| < 0$ and

$\lambda_n = -(1 \pm exp(\nu))\left(\sum_{(i,j)\in C} L_{ji}|W_n|\right)^{-1}$. The condition $\sum_{(i,j)\in C} L_{ji}|W_n| < 0$ is caused by the juxtaposition of a

positive step arising from the properties of descent methods of the gradient, and from the extension of conditions of stability. The constant ν can be analyzed in terms of rate of convergence. As $exp(\nu)$ approaches zero, the surface generated by modification functions tends more to locally resemble a tangent hypersurface. This tendency is accelerated with the absolute value of ν and corresponds to the notion of asymptotic visit of fixed points. We may take *a priori* as initial values $\sum_{(i,j)\in C} L_{ji}|W_n| \geq 0$, since

constraint does not have a feedback effect. In order to overcome this difficulty, it was decided to initialize weights at zero. Thus, after the first propagation, the only non-null L_{ji}, those of the output layer, are negative. Finally, λ_n is chosen by the conditions of stability and the properties of descent

methods of the gradient. It is preferable to take $\lambda_n = -(1 - exp(\nu))\left(\sum_{(i,j)\in C} L_{ji}|W_n|\right)^{-1}$. This value belongs to

a zone of monotonic convergence, supporting the hypothesis of the geometric model. If we remove the hypothesis of restriction to partial graphs, we can now propose learning by a BP-λL algorithm :

(9) *Initialize weights to zero.*
At the nth step :
During backpropagation we calculate $\sum_{(i,j)\in\Gamma_s} L_{ji}|W_n|$ *for all s of Σ and with* (5), (6).
We determine $B = sup\left(\sum_{(i,j)\in\Gamma_s} L_{ji}|W_n|\right)_{s\in\Sigma}$.
Before modifying weights, we obtain the gradient step $\lambda_n = -(1 - exp(\nu))B^{-1}$.

When the BP-λL1 algorithm is used, the user no longer has to be concerned with initializing weights and the choice of a gradient step. In addition, when the sample is covered, convergence is notably accelerated. We have brought to the fore a constraint enabling the determination, in an iterative way, of a gradient step which confirms constant and negative Lyapunov numbers, imposing a stable behaviour on the BP algorithm. To obtain a certain simple convergence, K-cycles, oscillatory regimes, must also be avoided. This is the other part of the complete BP-λL algorithm where an arrest condition indicates whether the neuronal structure is capable of processing the information. Either the learning succeeds, the sample is covered and convergence is accelerated, or the learning fails faced with a too large quantity of information. Anyway it is another work, starting from BP-λL1, and unfortunately his correct and complete description is impossible here.

Experimental results

We have compared three learning algorithms on an MLP network : a BP algorithm, a least squares algorithm, ELEANNE [15], and our BP-λL1 algorithm. The three methods seek to minimize a function of error cost. The test involved the changes of a value representative of the error we chose equal to $\frac{1}{2}\sum_{i=1}^{p}|F_{W,A}(X_i)-D_i|^2$. The learning sample included 16 pairs $((x_1,x_2),d)$ (see *Table 1* for nomenclature).

$x_2 \therefore x_1$	-1,5	-0,5	0,5	1,5
-1,5	1	1	1	1
-0,5	1	1	1	1
0,5	1	1	0	0
1,5	0	0	0	0

Table 1 : composition of the learning sample.

The architecture chosen, that of *Figure 5*, is composed of three layers. The input layer is passive and composed of two neurons. The hidden layer includes two neurons and the output layer one neuron. All the neurons, except those of the input layer, have biases.

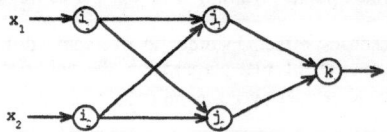

Figure 5 : architecture used.

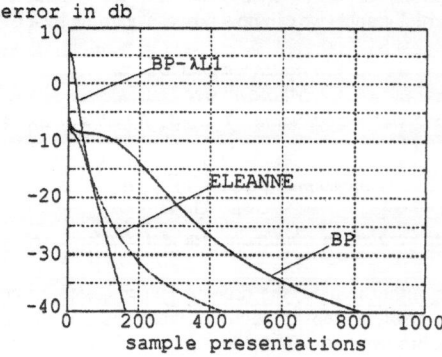

Figure 6 : changes in error committed in learning sample.

The BP algorithm uses a gradient step set at 1 and ELEANNE a learning coefficient of 1. *Figure 6* shows the error committed, expressed in decibels, as a function of the number of sample presentations. Since the BP and ELEANNE algorithms use a random initialization of weights, we conducted 10 runs and calculated a mean in order to obtain significant curves. It is of particular interest to note the different shapes of the graphs. The BP and ELEANNE algorithms have asymptotic behavior, while BP-λ L is quasi-linear. The rapidity of our algorithm results from a constant gain on error, while this gain decreases in the other algorithms. Evidently the rates of convergence of BP and ELEANNE depend on the step chosen, but the asymptotic behaviors remain identical.

Results concerning cover are shown in *Table 2*. The algorithms were run in the conditions described above until the difference between the effective output and desired output for each element in the sample was, in absolute terms, less than 10^{-2}. Despite his calculations increase, BP-λL1 is largely faster than the two others algorithms. The speed comparison is given trough a time ratio independent of the computer type. For example, the coefficient 1 is here about 450 milliseconds for a 486-DX2-66.

algorithm	**BP**	**ELEANNE**	**BP-λL1**
presentations	≈ 45000	≈ 22000	*316*
time ratio	≈ 150	≈ 90	*1*

Table 2 : numbers of sample presentation to obtain covering

Conclusion

The learning algorithm on MLP networks proposed presents a special feature that can be profitably used. By the use of Lyapunov numbers a learning behavior can be imposed. For example, by the use of an iteratively determined gradient step, the algorithm can be forced to remain stable. The BP-λL1 algorithm limits user responsibility to choosing an architecture and a sample, since it initializes weights and determines gradient step. In addition, if learning can succeed, the sample is covered and convergence is considerably accelerated.

References

[1]   **M. Milgram** - *Reconnaissance des formes, méthodes numériques et connexionnistes* - Armand Colin, 1993.
[2]   **Y. Le Cun** - *Modèles connexionnistes de l'apprentissage* - Thèse de doctorat, Université de Paris VI, 1987.
[3]   **B. Widrow, & al.** - *30 years of adaptative neural networks : perceptron, madaline and backpropagation* - Procedings of the IEE, vol. 78, n° 9, p. 1415-1442, IEE, 1990.
[4]   **S. Shah, F. Palmieri, M. Datum** - *Optimal filtering algorithms for fast learning in feedforward neural networks* - Neural Networks, vol. 5, p. 779-787, Pergamon Press, 1992.
[5]   **R.A. Jacobs** - *Increased rates of convergence through learning rate adaptation* - Neural Networks, vol. 1, p. 295-308, Pergamon Press, 1988.
[6]   **Y. Lee, S.H. Oh, M.W. Kim** - *An analysis of premature saturation in backpropagation learning* - Neural Networks, vol. 6, p. 719-728, Pergamon Press, 1993.
[7]   **P. Lascaux, R. Théodor** - *Analyse numérique matricielle appliquée à l'art de l'ingénieur* - Masson, 1986.
[8]   **C. Brezinski** - *Algorithmique numérique* - Ellipses, 1988.
[9]   **P. Frederickson, J.L. Kaplan, E.D. Yorke, J.A. Yorke** - *The Lyapunov dimension of strange attractors* - Journal of Differential Equations, vol. 9, p. 185-207, Academic Press, 1983.

[10]   **A. Grorud, D. Talay** - *Approximation of Lyapunov exponents of non-linear stochastic differential systems* - Rapport de recherche n° 1341, INRIA, 1990.

[11]   **J. Weitkämper** - *A study of bifurcations in a circular real cellular automaton* - Departament de Matematica aplicada i analisi, Universitat de Barcelona, 1991.

[12]   **F. Zou, J.A. Nossek** - *Bifurcation and chaos in cellular neural networks* - IEEE transactions on Circuits and Systems : Fondamental Theory and Applications, volume 40, n° 5, pages 166-173, IEEE, 1993.

[13]   **D. Guegan** - *Notion de chaos, approche dynamique et problèmes d'identification* - Rapport de recherche n° 1623, INRIA, 1992.

[14]   **P. Maneville** - *Structures dissipatives, chaos et turbulence* - Collection Aléa Saclay, 1991.

[15]   **B. Karayiannis, A.N. Venetsanopoulos** - *Efficient Learning Algorithms for Neural Networks* - IEEE Transactions on systems, man and cybernetics, vol. 23, n° 5, p. 1372-1383, 1993.

Analysis of Pruning in Backpropagation Networks for Artificial and Real World Mapping Problems

Teo JAŠIĆ, Hean Lee POH

Department of Information Systems and Computer Science
National University of Singapore
Singapore 0511
E-mail: iscteoj@leonis.nus.sg

Abstract

In this study, the properties of hidden nodes in backpropagation networks with respect to their contribution to the solution of the problem after initial training are examined. Using a pruning method, redundant nodes are removed and weights are redistributed in the pruned network. After reducing the network size, additional retraining may not be needed and the pruned network's generalization performance improves. The results show that the removal of one specific category of redundant nodes leads to improvement in terms of pruned network size, retraining speed, and improved generalization performance. An additional implication is that hidden nodes in a trained network are not evenly fault tolerant and that by changing the order of removal of different categories of redundant nodes we may find different solutions to the problem.

1 Introduction

In this paper, the performance of a pruning method for removing hidden nodes with *redistribution* of weights is examined on examples incorporating artificial and real world data sets. Hidden nodes in backpropagation network can be categorized and removed according to a redundancy criterion, and the network size can be reduced by progressively removing nodes. Removal schemes with updates and without updates of weights are observed and compared in terms of retraining speed and generalization performance of the pruned network. During training, different nodes display varying levels of contribution to the solution and some nodes are more susceptible to pruning. In addition, different solutions in terms of network weights can be achieved by different order of unit removal and subsequent retraining of the network.

In this study, feedforward backpropagation networks [Rumelhart et al, 1986] with one hidden layer and a sigmoid activation function are used. This restriction would not hinder the use of any results derived as both Hornik et al.(1989) and Funahashi (1989) showed that backpropagation networks with one hidden layer which have a sigmoid output function can approximate virtually any function of interest provided sufficiently many hidden nodes are available.

From the pattern space interpretation of neural networks [Pao, 1989; Chung & Lee 1992], it can be postulated that the hidden nodes would converge to one of the two types after training: *solution* nodes and *excessive* or *redundant* nodes. Solution nodes will generate linearly separable hidden patterns and they cannot be removed from the network without significantly impairing its performance. Such nodes are always included in the oversized networks. Redundant nodes can be found to be arbitrarily located in the input pattern space and they are the suitable candidates for pruning because they just enlarge the dimension of hidden patterns which remain linearly separable. After the removal of redundant nodes, the network could be retrained and the original performance restored.

The following categories of redundant nodes have been identified [Sietsma & Dow, 1991; Chung & Lee, 1992]: *noncontributing nodes, duplicate nodes, inversely-duplicate nodes,* and *inadequate nodes.* *Noncontributing nodes* give similar output across all training patterns. The output of a *duplicate node* is correlated with at least one other node, so if two nodes are found to discriminate the input patterns in a similar manner, one of them can be removed. An *inversely-duplicate node* is just the opposite of the duplicate node. The *redundant inadequate nodes* make same class separation rather than different class separation in the hidden pattern space. The removal of a redundant inadequate node should cause the merging of the same class hidden patterns in the reduced hidden pattern space.

In this study, the detection and pruning of three categories of redundant hidden nodes in the context of real to real mapping is conducted. The detection and removal of inadequate redundant nodes is not pursued as their definition is dependent on the pattern class concept. This concept is not so obvious in the real-to-real mapping problems because the interpretations for the dichotomies of decision hyperplanes with an output threshold of 0.5 are not strictly defined. Restricting the categorization of redundant nodes to only three categories in the context of real-to-real mapping does not affect the solution as it has been shown that the application of the node pruning algorithm may yield even better results if there are no specific rules set for the detection of redundant inadequate nodes [Chung and Lee, 1992].

2 Pruning of Redundant Nodes

Following the definitions of three categories of redundant nodes and the methodologies of Sietsma and Dow (1991) and Chung and Lee (1992), the detection and update rules for each category are derived for the case of real to real mapping.

Detection and Removal Scheme for Redundant Noncontributing Nodes. Noncontributing hidden nodes give similar output across all training patterns. These nodes can be removed and the weights given to their outputs redistributed in such a way as to make almost no change to the network's performance over the training set. The rule to detect a redundant noncontributing node i is defined as follows:

$$red_{non}(i) = max_{p_j, p_k}(|o_{p,i} - o_{p_k i}|) < \varepsilon_{non} \quad \forall p_k, p_j \in P \quad k \neq j \tag{1}$$

for respective patterns p_k and p_j in a set of training patterns P, and returns a value in a certain range. The threshold ε_{non} determines the level of trade-off between network size and network performance because adopting smaller thresholds will identify fewer nodes to be pruned and will tend to preserve the performance of the original network after pruning. Taking into account the assumption that the output of a unit is approximately constant, then it is acting like an additional bias to all units to which it is connected. If the average output of unit i is a, and all its outputs (across the training set) fall within some range of $(a \pm d)$, where d is small, then for each unit, j, on the next layer, the bias weight can be substituted by

$$bias'_j = bias_j + w_{ji}a. \tag{2}$$

and unit i can then be removed.

Detection and Removal Scheme for Redundant Duplicate Nodes. One of the hidden nodes will be redundant if the outputs of the two nodes i and j duplicate each other across training patterns, i.e. $o_{p_k j} \approx o_{p_k i} \ \forall p_k \in P$. Thus the redundant duplicate nodes would be detected by

$$red_{dup}(i,j) = max_{p_k}(|o_{p_k i} - o_{p_k j}|) < \varepsilon_{dup} \quad \forall p_k \in P \tag{3}$$

where the function $max_p(|o_{pi} - o_{pj}|)$ returns a value in a certain range and measures the difference between the outputs of the duplicate nodes on each training pattern. The threshold ε_{dup} determines the tolerance in the definition of redundant duplicate nodes. Given two duplicate nodes i and j, only one of them is needed. Each node k in the next layer receives input from both i and j which contains the same information. To remove i without changing the solution, at each node k on the next layer the weights can be updated as follows

$$w'_{kj} = w_{kj} + w_{ki} \tag{4}$$

Detection and Removal Scheme for Redundant Inversely-Duplicate Nodes. Two nodes can convey the same information, not by giving the same output, but by giving opposite outputs. Since an inversely duplicate node is just an inverse of a duplicate node, the detection rule will be identical to that of a duplicate node, after inverting one of the nodes:

$$red_{inv}(i,j) = max_{p_k}(|inv(o_{pi}) - o_{pj}|) < \varepsilon_{inv} \quad \forall p_k \in P \tag{5}$$

where $inv(x) = 1 - x$. Thus the threshold ε_{inv} has similar properties as ε_{dup} for detecting redundant duplicate nodes. The outputs of two inversely-duplicate nodes i and j are related as $a_i \approx 1 - a_j$. (The activation or output function of the units is bounded by zero and one.) For any node k on the next layer which depends on i and j, the weights can be updated as follows:

$$
\begin{aligned}
net_k &= \sum_l w_{kl}a_l + bias_k = ... + w_{ki}a_i + ... + w_{kj}a_j + ... + bias_k \\
&\approx ... + w_{ki}(1 - a_j) + ... + w_{kj}a_j + ... + bias_k \\
&\approx ... + (w_{kj} - w_{ki})a_j + ... + (bias_k + w_{ki})
\end{aligned}
$$

In order to remove node i, w_{kj} is replaced by $w_{kj} - w_{ki}$ and $bias_k$ by $bias_k + w_{ki}$.

2.1 A Node Pruning Algorithm

Based on the detection rules for different categories of redundant nodes described earlier, a node pruning algorithm can be derived. Like existing global pruning algorithms [Le Cun et al, 1990; Mozer and Smolensky, 1989], the proposed algorithm starts pruning with an oversized network. The detection rules are then applied to the converged network and those identified redundant nodes are pruned and weights redistributed depending on the category of nodes removed. After pruning, the network may be retrained to obtain the final network. Three thresholds ε_{non}, ε_{dup}, and ε_{inv} can be set to a certain value in the detection rules and their effects on node pruning examined.

The algorithm is devised in such a way that if one node is found to be correlated with more than one other node, the weights are updated only for the correlated node with lower ordinal index. Similarly, if one node is found to be inversely-correlated with more than one other node, the weights and bias are updated only for the correlated node with lower ordinal index. The reason is that the other hidden nodes can also be correlated themselves (depending on the threshold levels) and the node with a higher ordinal index will be updated accordingly.

There are several pruning strategies which can be pursued. The pruning strategy in this study differs from that of Chung and Lee (1992) who apply detection rules and prune nodes which belong to different categories of redundant nodes simultaneously (depending on the threshold level ε). The method applied in this study, however, removes nodes for only one category of redundant nodes at a time. In addition, the rules for the redistribution of weights are defined separately for each category of redundant nodes so that the effects of pruning can be examined separately. The occurrence of redundant nodes in different categories and the retraining of pruned networks can then be compared for the same level of thresholds ($\varepsilon_{non}, \varepsilon_{dup}, \varepsilon_{inv}$). Because different removal and update schemes can be used for the same problem, the final solution network may actually differ in terms of the hidden nodes remaining after pruning.

Thresholds ε are upper-bounded at 0.9 because of real-to-real mapping properties. By starting with small thresholds ($\varepsilon_{non}, \varepsilon_{dup}, \varepsilon_{inv}$), a correspondingly small number of redundant noncontributing, duplicate and inversely-duplicate nodes are detected and pruned. The pruned network is retrained if the performance of the pruned network is degraded and such a prune-and-retrain process is repeated by increasing the thresholds progressively. The retraining process is terminated when the performance of the current retrained network is degraded or the maximum threshold value is reached.

3 Experiments

In order to investigate the performance of the proposed pruning method, the XOR Problem, the Decoder Problem, and the time series prediction for the Sunspot data were used. One hidden layer networks

ε	Pruned Node Type	Pruned Net Size	after node removal		retraining	
			$MSE_{testing}$ with WU	$MSE_{testing}$ without WU	# of iter. with WU	# of iter. without WU
0.35	4 N	2-3-1	0.0482*	0.1161*	100	80
0.4	3, 4 N	2-2-1	0.0794*	0.1654*	975	500
0.3	3-4, D	2-3-1	0.0121	0.1250*	-	100
0.62	3-4, 1-3 D	2-2-1	0.1817*	0.0626*	**	**
0.7	3-4, ID	2-3-1	0.0457*	0.1250*	50	100
0.8	2-4, 3-4, ID	2-2-1	0.0514*	0.0978*	50	60

Table 1: Pruning of redundant Nodes in Network for the **XOR** Problem: Initial Size 2-4-1 (N - non-contributing, D - duplicate, ID - inversely duplicate, WU - weight updates, * misclassification, ** non-convergent)

were trained using backpropagation and pruned in these simulations. Weights were initialized between -0.5 and 0.5 and the magnitudes of the training patterns were normalized to the range of [0,1]. The objective of pruning is to improve the generalization performance, which is an "effectiveness" criterion for the quality of pruning. Additionally, there are two "efficiency" criteria which refer to the network size and the speed of retraining. For the XOR and Decoder Problems, we are interested in the efficiency of retraining smaller networks for a defined target error. For the Sunspot data, we observe both the efficiency and the effectiveness criteria. In general, the maximum number of retraining iterations was estimated to be smaller than the number of iterations taken to train the initial oversized network.

3.1 XOR Problem

The parity problems can be considered as special cases of real-to-real mapping and therefore they can be used to evaluate the performances of the proposed node pruning algorithm. In this experiment, the error target was set to 0.01 and the learning rate and momentum terms of the backpropagation algorithm were set to 0.5 and 0.7 respectively. The initial 2-4-1 network is trained with 380 iterations. The Mean Squared Error(MSE) is 0.0088 for the training set and 0.010 for the testing set. The testing set consists of 16 patterns - 4 times repeated training patterns with added uniform noise of ± 10%.

A summary of the results of pruning for increasing threshold levels is presented in Table 1. By removing two non-contributing nodes, the number of iterations needed for retraining is 980 which is higher than in the case of removing non-contributing nodes without update of weights (only 500 iterations needed). The worst degradation in performance after pruning to a 2-2-1 network occurs in the case of removing duplicate nodes (0.1817) and it is not possible to retrain such a pruned network. For removal and update scheme of inversely duplicate nodes, the MSE error after node removal is the smallest (0.0514) and the number of iterations for retraining is smaller (50) than the number of iterations when no update scheme for weights is used.

It should be noted that the pruning of different redundant nodes to a 2-2-1 network occurs with different nodes for different categories: nodes 3 and 4 for non-contributing, nodes 2 and 3 for inversely duplicate, and nodes 1 and 3 for duplicate nodes. The highest threshold for removing non-contributing and inversely-duplicate nodes occurs with Node 1. This node can be called a resistant node and its removal may affect the pruned network significantly in terms of the testing errors after node removal and the number of iterations needed for retraining. The results shown in Table 1 confirm this observation.

3.2 Binary One-of-Eight Decoder

This is a simple problem often used in discussions of backpropagation learning. It deals with the decoding of a 3 bit binary code. The initial networks was a 3-9-8 network. There are eight training samples. The testing data set contains 6 repetitions of the training set with added uniform input noise at ±10%. The initial 3-9-8 network is trained at a learning rate 0.5 (0.0 momentum) with ≈ 770 iterations. The training and testing errors for the initial network are 0.0071 (MSE) and 0.00987 (MSE) respectively.

ε	Pruned Node Type	Pruned Net Size	after node removal		retraining	
			$MSE_{testing}$ with WU	$MSE_{testing}$ without WU	# of iter. with WU	# of iter. without WU
0.4	1, D	3-8-8	0.00629	0.0421	-	20
0.6	1,2 D	3-7-8	0.01619	0.1357 *	-	100
0.7	1,2,6 D	3-6-8	0.0176	0.2315 *	-	175
0.82	1,2,6,8 D	3-5-8	0.2687 *	0.00416	375	700
0.87	1,2,4,5,6,8 D	3-3-8	0.4163 *	0.8238	**	12500_{tr}
0.2	1,2,4 ID	3-6-8	0.0095	0.1699 *	-	250
0.5	1,2,3,4 ID	3-5-8	0.0715 *	0.3989	25	**
0.6	1,2,3,4,5 ID	3-4-8	0.1186 *	0.3798	50	1200
0.71	1,2,3,4,5,7 ID	3-3-8	0.4508 *	0.5179	300	2000
0.73	1,2,3,4,5,6,7 ID	3-2-8	0.4642 *	0.5253 *	22100_{tr}	18000_{tr}

Table 2: Pruning of Redundant Nodes for the **Decoder Problem**: Initial Architecture 3-9-8, (N - non-contributing, D - duplicate, ID - inversely duplicate, WU - weight updates, * misclassification, ** non-convergent)

The threshold levels ε_{non} for removing non-contributing units are large ($\varepsilon_{non} > 0.9$ for removing five hidden nodes, and $\varepsilon_{non} \in [0.6, 0.9]$ for removing the remaining 4 or fewer hidden nodes). The threshold levels ε_{dup} for removing redundant duplicate nodes are also relatively high ($\varepsilon_{dup} > 0.8$ for removing six nodes and $\varepsilon_{dup} < 0.6$ for removing the remaining 3 or fewer nodes). For the inversely-duplicate units, thresholds are more evenly distributed ($\varepsilon_{inv} < 0.2$ for removing 3 nodes and $\varepsilon_{inv} > 0.7$ for removing 4 nodes). Since the thresholds for removing non-contributing nodes are too high, pruning is examined for duplicate and inversely-duplicate nodes only.

A summary of the results for pruning redundant nodes for various threshold levels is presented in Table 2. For $\varepsilon_{dup} = 0.82$, there are four pairs of duplicate nodes detected and four nodes are subsequently removed. The weights are updated accordingly. For retraining the pruned 3-5-8 network, 375 iterations are needed for the correct classification of the testing set as compared to 250 for the training set. If the removal of duplicate nodes is performed without weight updates, then the number of retraining iterations needed for the 3-5-8 network is larger at 700 iterations. Likewise, the initial network of 3-8-8 can be pruned to a 3-3-8 network with correct classification, if the inversely duplicate nodes are removed at $\varepsilon_{inv} = 0.71$. Retraining in this case requires 300 iterations with weight updates during node removal and 2000 iterations without weight updates.

Pruning the inversely duplicate nodes with weight updates gives the best results as the correct classification can be achieved after relatively small number of retraining iterations, and the network can be pruned from 3-9-8 to 3-3-8. For each category of redundant nodes removed, retraining is generally much faster if the weights are updated although the threshold level may be high. An additional observation is that there may be significant differences when the performance of the pruned network is measured on the noisy testing set because the retraining may be successful for the correct classification of the training set, but not the testing set.

3.3 Sunspot Data

The Sunspot data is a popular benchmark in the statistics community because of its inherent nonlinearity and periodicity. A wide arrays of statistical methods based on linear and non-linear models have been developed. Connectionist models have also been applied to this problem [Weigend, 1991]. In this experiment, we use the observed yearly Sunspot data from 1712-1911 for training the neural networks (220 patterns) and data from 1931-1980 for the testing set. The twelve input variables correspond to the preceding twelve years of Sunspot data whereas the output is the current year's Sunspot data. A number of authors which use pruning methods for this problem start with an initial 12-8-1 network and end up with 12-3-1 networks [Weigend, 1991]. In this experiment, we also start with an initial 12-8-1 network and then apply the pruning algorithm to remove redundant nodes and observe the generalization capability of the pruned network.

ε_{inv}	Pruned Node Type	Pruned Net Size	after node removal		retraining			
			NMSE w/ WU	NMSE w/o WU	# iter. w/ WU	NMSE w/ WU	# iter. w/o WU	NMSE w/o WU
0.1	4, 6 N	12-6-1	0.1885	0.1866	30/750	0.1800/0.1757	-/800	-/0.176289
0.2	1,3,4,5,6 N	12-3-1	0.4892	0.3315	-/8450	0.2001	-/8500	0.2001
0.2	3-6 D	12-7-1	0.2215	(0.7247)	900/1800	0.1809/0.1798	1050/2000	0.1800/0.1777
0.3	1,3,4 D	12-5-1	0.2985	0.3234	3720	0.1818	3750	0.1817
0.4	1,3,4,5,7 D	12-3-1	1.1518	1.0995	600	0.1961	5000	0.1944
0.3	1,3 ID	12-6-1	0.1290	0.4657	-	-		
0.4	1,3,4 ID	12-5-1	0.4325	0.3234	75	0.1956		
0.5	1,2,3,4,5 ID	12-3-1	2.6141	2.1677	50/125	0.1626/0.1466	4300	0.1926

Table 3: Detecting and Pruning Redundant Nodes in the Network for **Sunspot Data**: Initial Size 12-8-1; (N - non-contributing, D - duplicate, ID - inversely duplicate, WU - weight updates, * misclassification, ** non-convergent)

The initial network is trained at a learning rate of 0.05 (no momentum) and $\approx$ 10000 learning iterations with a normalized testing MSE of 0.18000. The thresholds ε_{non} for removing non-contributing units are small: $\varepsilon_{non} < 0.1$ for two hidden nodes, $\varepsilon_{non} \in [0.1, 0.2]$ for 3 hidden nodes, and $\varepsilon_{non} \in [0.2, 0.4]$ for the remaining 5 hidden nodes). The threshold levels ε_{dup} for removing redundant duplicate nodes are higher than for non-contributing nodes: $\varepsilon_{dup} \in [0.1, 0.2]$ for removing one node and $\varepsilon_{dup} \in [0.2, 0.5]$ for removing the rest of the nodes. For inversely-duplicate nodes, thresholds are the largest: $\varepsilon_{inv} > 0.2$ for removing one node, $\varepsilon_{inv} \in [0.2, 0.5]$ for removing 2-4 nodes, and $\varepsilon_{inv} > 0.5$ for the remaining 4 nodes.

During the pruning process, retraining is monitored with two values: one is the number of iterations needed to obtain the testing error of the initial network (NMSE = 0.1800) and another is the number of iterations taken to improve on the NMSE of 0.1800 until the testing error starts to deteriorate. Table 3 shows the results for pruning redundant nodes in a 12-8-1 network. By detecting and removing non-contributing nodes, it is possible to prune the initial network to the 12-3-1 configuration but the retraining process cannot restore the generalization performance of the initial network and is not efficient in terms of the number of iterations in retraining. By pruning duplicate nodes from the initial 12-8-1 network it is possible to prune the initial network to the 12-3-1 configuration and still preserve the generalization performance after retraining with $\approx$ 900 iterations. By detecting and removing inversely-duplicate nodes from the initial 12-8-1 network, the smaller 12-3-1 and 12-2-1 configurations can also be obtained. For the pruned 12-3-1 network, only 50 iterations of retraining are needed to achieve the $NMSE_{testing}$ of the initial network, and for 125 iterations, the minimal $NMSE_{testing}$ of 0.1464 is obtained.

4   Conclusions

In this study, a pruning algorithm which removes three categories of redundant hidden nodes with weight updates has been examined. The number of hidden nodes removed depends on the threshold level and the category of redundant nodes selected for pruning. This algorithm has been tested on two artificial (the XOR and Decoder Problems) and one real world data sets (Sunspot Data). The criteria for the effectiveness and efficiency of the algorithm are based on the improved generalization error and/or efficient retraining of smaller networks to restore or exceed the initial network performance.

In the case of the XOR and Decoder Problems, the target testing error has been set at 0.01. Therefore, only the efficiency of the algorithm in terms of retraining smaller networks is observed for these two problems. For the Sunspot data, both the effectiveness and the efficiency of the algorithm are observed. Given the XOR problem, the pruning from the initial 2-4-1 to the smaller 2-2-1 network is the most efficient if inversely-duplicate nodes are removed. The number of retraining iterations needed in this case is 50 with weight updates and 60 if there are no weight updates. For the Decoder Problem, the best result for a smaller 3-3-8 network is also obtained by removing and updating inversely-duplicate nodes. The number of retraining iterations is $\approx$ 300 in this case. For the Sunspot data, the best result with respect to the effectiveness (i.e. the best generalization performance) and the efficiency (i.e. the smallest possible network) is also obtained by pruning inversely-duplicate nodes. The best generalization result of $NMSE_{testing} = 0.1290$ is obtained with a 12-6-1 network. This is achieved by removing two inversely duplicate nodes without retraining. With a smaller 12-3-1 network, however, the best possible result obtained by removing inversely duplicate nodes is less favorable at $NMSE_{testing} = 0.1464$. Nevertheless, both the 12-6-1 and the 12-3-1 networks have testing errors which compare favorably with that of the initial network of 12-8-1 at $NMSE_{testing} = 0.18$.

An interesting observation is that across all three problems, the best results are achieved for the pruning of *inversely duplicate nodes* which involves updating both the weights and the biases. Finally, it was shown in the experiments above that the pruning of different nodes would lead to different generalization results in different problems. Hence, the fault tolerance of neural network is not evenly distributed among the nodes and is problem dependent.

5 References

Baum E.B. and D. Haussler (1990), "What size net give valid Generalization?" in: *Neural Computation*, 1:1, 151-160.

Chauvin Y. (1989) "A backpropagation algorithm with optimal use of hidden units" in: *Advances in Neural Information Processing Systems 1*, Ed. D. S. Touretzky, Morgan Kauffman.

Chung F.L. and T. Lee (1992), "A Node Pruning Algorithm for Backpropagation Networks", *International Journal of Neural Systems*, 3, 3, 301-314.

Le Cun Y., J. S. Denker and S. A. Solla (1990), "Optimal Brain Damage," in *Advances in Neural Information Processing Systems 2*, Ed. D. S. Touretzky, Morgan Kauffman.

Funahashi K. (1989), "On the approximate realization of continuous mappings by neural networks", in: *Neural Networks* 2, 183-192.

Hanson S.J. and L. Y. Pratt (1989), "Comparing biases for minimal network construction with back-propagation," in *Advances in Neural Information Processing Systems 1*, Ed. D. S. Touretzky, Morgan Kauffman.

Hornik K., M. Stinchcombe and H. White (1989), "Multilayer feedforward networks are universal approximators," in: Neural Networks 2, 359-366.

Mozer M.C. and P. Smolensky (1989), "Skeletonization: A technique for trimming the fat from a network via relevance assessment," in *Advances in Neural Information Processing Systems 1*, Ed. D. S. Touretzky, Morgan Kauffman.

Pao Y.H.(1989), *Adaptive Pattern Recognition and Neural Networks*, Addison- Wesley, MA.

Rumelhart D.E., G. E. Hinton and R. J. Williams (1986), "Learning internal representations by error propagation," in *Parallel Distributing Processing*, Vol I and II. Eds D. E. Rumelhart and J. L. McClelland, MIT Press, Cambridge, MA.

Sietsma J. and R. J. F. Dow (1991), "Creating Artificial Neural Networks That Generalize", in: *Neural Networks*, 4, 67-79.

Oscillatory Networks with Hebbian Matrix of Connections

Kuzmina M. G.,
Keldysh Institute of Applied Mathematics, Russian Academy of Sciences,
Miusskaya Sq. 4, 125047 Moscow, Russia Phone: 7(095)972-3491, Fax: 7(095)972-0737
e-mail: kuzmina@applmat.msk.su

Manykin E. A., Surina I. I.
Superconductivity and Solid State Physics Institute of Russian Research Center
"Kurchatov Institute", Kurchatov sq. 1, 123182 Moscow, Russia Phone: 7(095)196-91-07,
Fax: 7(095)196 59 73 e-mail: edmany@nlodep.kiae.su

The systems of symmetrically coupled limit cycle oscillators admit the design of recurrent associative memory networks with Hebbian matrix of connections. Unlike the similar neural networks this matrix proved to be the complex-valued Hermitian one with nonzero diagonal. In the case of strong interaction in oscillatory system the memory vectors of the network are slightly perturbed properly normalized eigenvectors of matrix of connections. They can be calculated by perturbation method on the appropriate small parameter. The self-consistent analysis of dynamical system fixed points in the case of homogeneously all-to-all connected oscillators is presented. It is proved that for positive values of connection strength only a single memory vector can be stored. Some questions concerning the "extraneous" memory of the networks are discussed.

1 Introduction

Large systems of coupled oscillators [1-4] in the regime of synchronization (phase locking) have an ability to memorize information [5-8]. So the problem of neural oscillatory system of associative memory design arises. The design includes determination of matrix of connections and proper choice of other modifiable parameters of the corresponding dynamical system to provide effective retrieval characteristics of the network.

One of the most attractive features of oscillatory models is undoubtedly possible numerous physical implementations. In contradiction with well known optoelectronic and nonlinear optical implementations based on the idea of vector-matrix multiplier oscillatory models promise direct - and by this reason much more effective - implementations. When one analyzes neural network implementations based on photon-echo effect [9], it becomes clear that the potentialities of this effect that have been used so far are exceedingly greater than those already used in the known schemes.

As for theoretical study of oscillatory systems from the viewpoint of associative memory modeling there is a number of various ways of the modeling based on systems of coupled oscillators. One of them is the encoding of memory patterns by two subpopulations of oscillatory system in the vicinity of phase transition into synchronized state - subpopulations of synchronized and unsynchronized states - that has been developed in a number of works (see, for instance, [5]).

The modeling of *recurrent* associative memory oscillatory network in the state of *synchronization* is still at the very beginning [6-8]. Up to now only the special kind of oscillatory system with the simplest kind of interaction - limit-cycle interacting oscillators with linear interactions of pairs - have been studied.

It has been found out that such special oscillatory systems are closely related to the systems of magnetic spins on the plane (clock spin glasses or phasor systems). The associative memory networks with Hebbian matrix of connections have been designed for the clock spin systems and the phase transition of memory "overloading", permitting to obtain the retrieval characteristics of the network, has been analyzed [10]. However, the important problem of "extraneous" memory for clock spin networks is not studied at all.

An attempt has been done to study phasor networks with asymmetrical complex-valued matrix of connections and non-zero thresholds (clearly, the Hopfield model is imbedded into it). This model is the natural generalization of the phasor network model studied in [14]. The further study of this model would be quite desirable.

As far as we know, the present work is the first attempt to design and to begin study of recurrent associative memory oscillatory network with Hebbian matrix of connections.

2 The Dynamical Equations of the Model of Phase Oscillators.

We consider the system of N limit-cycle oscillators on the plane with symmetrical nonhomogeneous coupling, the state of each being defined as a complex-value a point $z_j = r_j exp(i\theta_j)$ of complex plane. In appropriate parametric domain the dynamical system governing the dynamics of oscillatory system can be reduced to "phase" dynamical system

$$\dot{\theta}_j = \omega_j + K \sum_{k=1}^{N} \mathcal{W}_{jk} sin(\theta_k - \theta_j + \beta_{jk}), \quad j = 1, ..., N. \quad (1)$$

where $\omega_j, \quad j = 1, ...N$, are the natural frequencies on the cycles and complex-valued Hermitian $N \times N$ matrix $W = [W_{jk}] = [\mathcal{W}_{jk} exp(i\beta_{jk})], \quad W = \bar{W}^{\mathsf{T}} \equiv W^+$ specifies the weights of connections of oscillators in the network, the real value K defines the absolute value and the sign of interaction strengths in the system [10].

The dynamical system (1) defines the model of system of "phase oscillators" which corresponds to the approximation that interaction of network oscillators does not influence on the amplitudes of oscillations, the last being constant. So, the state vector of the network of phase model is

$$z = (z_1, ..., z_N)^{\mathsf{T}}, \quad z_j = exp(i\theta_j),$$

Note that the matrix W in (1) should not have the zero diagonal in difference with the case of neural networks. This is just the consequence of the form of representation of "operator" of interaction of amplitude-phase dynamical system that is reduced to phase system (1).

Any Hermitian matrix W can be represented in a form

$$W = N^{-1} \sum_{m=1}^{M} \lambda^m V^m (V^m)^+, \quad M = rankW, \tag{2}$$

where $\{V^m\}$ is the set of mutually orthogonal eigenvectors of W corresponding to the set of its nonzero eigenvalues [12]:

$$WV^m = \lambda^m V^m, \quad (V^s)^+ V^m = N\delta_{ms}, \quad m = 1, ...N. \tag{3}$$

where δ_{ms} denotes the Kronecker symbol. With the help of expansion (2) the dynamical system (1) can be rewritten in the form

$$\dot{\theta}_j = \omega_j + (K/N) \sum_{m=1}^{M} \sum_{k=1}^{N} \lambda^m sin([\theta_k - \beta_k^m] - [\theta_j - \beta_j^m]). \tag{4}$$

One more form of system (1) can be obtained if one uses the expansion of state vector z in eigenbasis $\{V^m\}$ of matrix W

$$z = \sum_{m=1}^{M} Z^m V^m, \quad Z^m = N^{-1}(V^m)^+ z = N^{-1} \sum_{j=1}^{N} exp(i[\theta_j - \beta_j^m]) = R^m exp(i\psi^m), \tag{5}$$

The variables Z^m, the inner products of current state vector z and the basis vectors V^m, are the macrovariables (in the case of high dimension N of the dynamical system). They are just the "overlaps" which are usually used in asymptotical analysis of retrieval characteristics of associative memory neural networks. For the case of oscillatory networks the "overlaps" have the additional sense: the "order parameters", governing the phase transition of oscillatory system into synchronized state.

Being rewritten in terms of macrovariables $Z_m R^m exp(i\psi^m)$, the system (1) has the form of N independent equations.

$$\dot{\theta}_j = \omega_j + K \sum_{m=1}^{M} \lambda^m R^m sin(\psi^m + \beta_j^m - \theta_j), \tag{6}$$

System (3) provides the "self-consistent field" description of oscillatory network. It proved to be very convenient for the analysis of fixed points of the phase dynamical system.

3 The "Hebbian" Solution to Associative Memory Network Design Problem

As it is very well known from the theory of associative memory neural networks, the matrix of connections that is the sum of outer products by orthogonal set of memory vectors just provides the simplest solution to network design problem. The outer-product matrices of connections themselves are usually regarded as "Hebbian" because of the relation to Hebbian learning algorithm.

As it follows from (2),(3), the "Hebbian" solution to the associative memory design problem exists for the model of system of "phase oscillators". More exactly, we have the following result.

I. The case $\omega_j = 0$ (phasor networks).

Let the set of M, $M \leq N$, of linearly independent vectors $\{\mathcal{V}^m\}$ is given as $\mathcal{V}_j^m = exp(i\beta_j^m)$ is given. Then the "Hebbian" solution to the problem can be realized in the following steps:

• Find the orthogonal system of vectors $\{V^m\}$, corresponding to the set $\{\mathcal{V}^m\}$.

• Define the matrix W by formulas (2),(3), where λ^m are some real values.

Then vectors V^m are just the stable fixed points of phase dynamical system (4). Thus, the memory vectors of oscillatory network coincide exactly with V^m.

The question of proper choice of $\{\lambda^m\}$ should be the subject of special analysis. It's worth to recall from the theory of neural networks with Hebbian matrix of connections that there is serious disadvantage in the choice of equal $\{\lambda^m\}$. Such a choice just leads to drastically great extraneous memory (this is quite natural from the viewpoint of linear algebra). So the choice of close, but different $\{\lambda^m\}$ seems to be preferable.

The existence of "extraneous" memory should be also a subject of special analysis. In any case, it is very plausible that the "extraneous" memory exists in the subspace $kerW$ of complex unitary space of network state vectors [12].

II. The case $\omega_j \neq 0$, $\sum_{j=1}^{N}\omega_j = 0$. It can be shown that at arbitrary ω_j in the case of "strong" interaction in the oscillatory network there exist the set $\tilde{V}^m$, close to V^m, which is the set of memory vectors of oscillatory network with the matrix of connections defined by formulas (2),(3).

To formulate the result more exactly, first of all note that the parameter $\gamma = \Omega/K$, where $\Omega = max_j|\omega_j|$, is the essential parameter of the system. For instance, the simple sufficient condition of synchronization of oscillatory network is $\gamma \leq 1$. The case of $\omega_j = 0$ can be considered as the limit case of infinitely strong interaction of oscillators in the network ($K \to \infty$). When K is great, but finite value, γ is the small parameter, and the perturbation method for the system of equation defining the fixed points of network dynamics can be derived. It is just the system (6) that proved to be the most convenient for this purpose. The perturbation method provides the following result.

At sufficiently small $\epsilon = \gamma = \Omega/K$ the memory vectors $\tilde{V}^m$ of oscillatory network belong to small vicinities of vectors V^m and the following estimations take place:

$$\tilde{V}_j^m = V_j^m + \epsilon(\lambda^m)^{-1}\omega_j + O(\epsilon^2), m = 1, ...M, j = 1, ...N. \tag{7}$$

These facts permit to conclude that the retrieval characteristics of clock spin networks (phasor networks) obtained in [10] (storage capacity $\alpha \sim 0.037$, the limit value of "overlap" equals to ~ 0.9) are simultaneously the retrieval characteristics of oscillatory networks in the case of strong interaction.

The fact that the memory vectors of the network under strong interaction are slightly perturbed eigenvectors of matrix W is confirmed in computer experiments. The last ones also show that in the process of further gradual increase of γ the stable fixed points $\tilde{V}^m$ disappear one after another. This process stops at $\gamma = \gamma^*$ where γ^* is the threshold of synchronization. In small vicinity of γ^* the dynamical system (4) has a single stable fixed point. So, in principle, the parameter γ can be used as the parameter controlling the memory storage abilities of oscillatory networks.

4 The Oscillatory Networks Containing a Single Memory Vector

The oscillatory networks with homogeneous all-to-all connections can be regarded as the networks containing a single memory vector. Indeed, their matrices of connections $[W_{jk}] = N-1$ contain only the single term of the expansion (2):

$$W = N^{-1}VV^{\mathsf{T}}, \quad V = (1,...1)^{\mathsf{T}}, \tag{8}$$

and the dynamical system in the form (6), which one is especially simple, can be written as:

$$\dot{\theta}_j = \gamma\tilde{\omega}_j + Rsin(\psi - \theta_j), \tag{9}$$

where

$$Z = N^{-1} < z, V >= N^{-1}\sum_{j=1}^{N} exp(i\theta_j) = Rexp(i\psi), \tag{10}$$

and $\tilde{\omega}_j \equiv \omega_j/\Omega$. Note that in this case the single macrovariable Z coincides with well known order parameter which was elsewhere used for investigation of phase transition into synchronized state for the systems of uniformly all-to-all coupled phase oscillators [1-4]. The functional self-consistent equation for Z in closed form, which together with the equations for fixed points

$$\gamma\tilde{\omega}_j + Rsin(\psi - \theta_j) = 0 \tag{11}$$

delivers the self-consistent analysis of network equilibria, can be obtained. The self-consistent analysis shows that the phase ψ can be chosen arbitrary (the consequence of rotational symmetry of the system) and the self-consistency equation for R can be written in the forms:

$$R^2 = N^{-1}\sum_{j=1}^{N}(R^2 - \gamma^2\tilde{\omega}_j^2)^{1/2}, \tag{12}$$

or

$$\gamma = N^{-1}u\sum_{j=1}^{N}(1 - \tilde{\omega}^2 u^2)^{1/2}, \quad u \equiv \gamma/R. \tag{13}$$

The analysis of fixed points of phase dynamical system on the base of (11), (12) and (13) gives the following results (for $K > 0$):

1. There exists the single stable fixed point of the network $\bar{z} = \tilde{V}$. At $\gamma = 0$ $\tilde{V} = V$, at $\omega_j \neq 0$ $\tilde{V}$ is slightly perturbed V at small γ in accordance with (7). Gradual increase of γ leads to greater deviations of $\tilde{V}$ from V. The point $\tilde{V}$ exists up to $\gamma = \gamma^*$ being the threshold of synchronization. The later can be calculated exactly from (13) and in the case of three oscillator network is equal to .588.

2. The condition $R = 0$ defines N unstable fixed points of the network that are symmetrical ones and can be complicated equilibrium states of the dynamical system.

5 Concluding Remarks

The following results have been obtained.

- The associative memory oscillatory network with Hebbian Matrix of connections is designed. In the case of strong oscillatory interaction ($\gamma = \Omega/K$ is small) the memory vectors of the oscillatory network are slightly perturbed properly normalized eigenvectors of matrix of connections. The retrieval characteristics (in the same case) coincide with those ones of a clock spin network [10]

- An example of self-consistent analysis of total fixed points of the network is given in the case of the network, containing a single memory vector (of the network with uniform all-to-all connections).

Acknowledgment

The research was supported by Russian fund of fundamental investigations, grant number 94-01-00406.

References

1. Y. Kuramoto, I. Nishakawa, Statistical macrodynamics of large dyamical systems. Case of phase transition in oscillator communities - *J. Stat. Phys.*, Vol. 9, No. 3/4, pp. 569-605, 1987.
2. H. Sakaguchi, S. Shinomoto, Y. Kuramoto, Phase transitions and their bifurcation analysis in a large population of active rotators with mean field coupling - *Progr. Teor. Phys.*, Vol. 79, No. 3, pp. 600-607, 1988.
3. S. H. Strogatz, R. E. Mirollo, Phase-locking and critical phenomena in lattices of coupled nonlinear oscillators with random intrinsic frequencies - *Phys. D*, Vol.31, No.2., pp. 143-168, 1988.
4. L. L. Bonilla, C. J. Perez Vicente, J. M. Rubi, Glassy synchronization in a population of coupled oscillators - *J.Stat. Phys.*, Vol.70, No. 3/4, pp. 921-937, 1993.
5. Sompolinsky H.,Tsodyks M., Processing of sensory information by a network of oscillators with memory - *Intern. J. Neur.Syst.* Vol. **3**, Supp., p.51-56, 1992.
6. A. Y. Plakhov, O. I. Fisun, An oscillatory model of neuron networks - *Matematicheskoe modelirovanie*, Vol. 3, No.3, pp. 48-54, 1991 (in Russian).
7. Kuzmina M.G., Manykin E.A.,Surina I.I., Problem of Associative Memory in Systems of Coupled Oscillators - *Second European Congress on Intelligent Techniques and Soft Computing. Aachen, Germany, September 20-23, 1994, Proceedings*, Vol. 3, EUFIT'94, pp. 1398 - 1402, 1994.
8. Kuzmina M.G., Surina I.I., Macrodynamical approach for oscillatory networks - *Proceedings of the SPIE. Optical Neural Networks*, Vol. 2430 pp. 227-233, 1994.
9. Belov M.N., Manykin E.A. Photon-echo effect in optical implementation of neural network models. - In: *Neurocomputers and Attention, eds. Holden A.V. and Kryukov V.I., Manchester University Press*, Vol. 2, p.459-466, 1991.
10. J. Cook, The mean-field theory of a Q-state neural network model - *J. Phys. Math. Gen.*, Vol.22, pp. 2057-2067, 1989.
11. A. J. Noest, Associative memory in sparse phasor neural networks - *Europhys. Lett.*, Vol.6, No. 6 pp. 469-474, 1988.
12. Kuzmina M.G., Surina I.I., Oscillatory Networks of Associative Memory, submitted to IWANN'95.

A New Algorithm for Implementing a Recursive Neural Network

V. Giménez[1], P.Gómez-Vilda[2], E. Torrano[1] and M.Pérez-Castellanos[2]

[1] *Departamento de Matemática Aplicada*
[2] *Departamento de Arquitectura y Tecnología de Sistemas Informáticos*
Facultad de Informática, Universidad Politécnica de Madrid,
Campus de Montegancedo s/n, Boadilla del Monte, 28660, Madrid, SPAIN
Phone: + 34.1.336.74.29, Fax: + 34.1.336.74.12
E-mail: marga@datsi.fi.upm.es

Abstract - This paper describes a method of designing a procedure based in a new vision of the well known Hopfield algorithm. Our approach is also a Hebb's law based algorithm for describing a Recursive Neural Network. In the training stage we used a *Graph* method for acquiring the data [1], the energy associated to any possible state of the net is represented as a energy point *(a,b)* in the plane R^2. We prove that all the states with similar energy level are on an *hyperbolic surface*, $x.y = k$, when the net changes its state its associated energy point is placed in a utter *hyperbolic surface* $x.y = q$, *(q>k)*; in this way a convergence is proved. When a pattern is called for retrieving, a parameter may be used for controlling the radius of attraction and the number of fixed points in the system; this parameter is related with a *coloring* [2] or partition neighborhood of the *Resulting Graph* obtained after training. As a clear application we have developed an example where we may see the frequency distribution associated with a given state and the incidence of the parameter on the the number of fixed points [3].

1. Introduction.

A Recursive Neural Network is a discrete time, discrete valued dynamic system which at any given instant of time *t* is characterized by a binary state vector. For a recursive network with *n* units, the state vector at time *t* is of the general form

$$x(t) = \left[x_1(t), ..., x_i(t), ..., x_n(t) \right] \in \{1,-1\}^n \tag{1}$$

Let *W* be a real $n \times n$ matrix and θ a real *n-vector*. The behavior of the system along the time axis is described by a *dynamic* equation of the type

$$x_i(t+1) = \text{Sgn}\left[\sum_{j=1}^{n} w_{ij}x_j(t) - \theta_i \right], \quad i = 1,...,n. \tag{2}$$

with

$$\sum_{j=1}^{n} w_{ij} \cdot x_j(t) - \theta_i = 0 \Rightarrow x_i(t+1) = x_i(t) \tag{3}$$

The matrix *W* and the vector θ, are the *parameters* of the network; the threshold vector turns out to be of little importance, so we suppose $\theta = 0$. Let us assume that by some means a copy of *p* binary vectors, $L=\{\xi^1,.., \xi^r,.., \xi^p\}$, has been integrated into the structure of the system. The system will behave as an associative memory if it reproduces at its output one of the vectors, say ξ^r, when triggered at the input by a vector $x(0)$ which is sufficiently close to ξ^r [1].

The building mechanism of the matrix W is based in Hebb's law. At first, a null value is assigned to every w_{ij}, then, when a pattern $\xi^\mu \in L$ is presented to the net, this value is modified by $\Delta w_{ij} = \xi_i^\mu \cdot \xi_j^\mu$. The final value of w_{ij}, when all the vectors in L were presented, represents the relative *gain* in sign-coincidence between the *i-th* and *j-th* components of the vectors in L.

In the new approach we propose the state vector at time t is

$$x(t) = \left[x_1(t), \ldots, x_i(t), \ldots, x_n(t) \right] \in \{0, 1\}^n \tag{4}$$

The network may then be expressed as a *Complete Graph G* [2], with n vertices $\{v_1, \ldots, v_n\}$, and *one* bi-directional *edge* a_{ij} connecting every possible pair of different vertices. At the *Training Process*; initially, a null value w_{ij} is assigned on every edge a_{ij} in the graph; afterwards, when a pattern ξ^μ belonging to $\{0,1\}^n$, is presented to the net, this value is modified as follows:

$$\Delta w_{ij}^\mu = \begin{cases} +1 & \text{if } \xi_i^\mu = \xi_j^\mu = 1, \ i \neq j, \\ -1 & \text{if } \xi_i^\mu = \xi_j^\mu = 0, \ i \neq j, \\ 0 & \text{otherwise.} \end{cases} \tag{5}$$

This procedure is also based in Hebb's law, but with a significant difference. We must interpret that when a learning pattern ξ^μ is acquired by the net, it is superposed over the graph G. This means that its components, $\{\xi_1^\mu, \ldots, \xi_n^\mu\}$ are being mapped, in a bijective correspondence, over the vertices $\{v_1, \ldots, v_n\}$ of the graph G. This mapping may be interpreted as a *coloring* of the edges in G [2]: the edges connecting vertices in correspondence with every component $\xi_i^\mu = 1$ are, for example, *red* colored; reversibly, those edges connecting vertices in correspondence with every component $\xi_i^\mu = 0$, are *blue* colored Then all the edges in the *red* subgraph G_1^μ are positively reinforced and all the edges in the *blue* subgraph G_0^μ are negatively reinforced; those edges connecting both subgraphs remain unchanged. Once acquired the pattern ξ^μ the colors are erased and we repeat the same color assignation with the next pattern to be acquired by the net and so on. When every vector in L has been integrated in the net, the training stage is finished [3], the *Resulting Graph G* has become edge-valued and its adjacency matrix A is the *weight matrix* of the net. Every possible state vector may be interpreted as a *coloring* of the *Resulting Graph G*. We have developed an example taking the 37 eight-bit patterns corresponding with a certain codification of the most relevant phonetic sounds for the *Spanish* language given in [4], as the *training pattern set*. After applying the *training algorithm* [3] we obtained the graph in Fig 1.

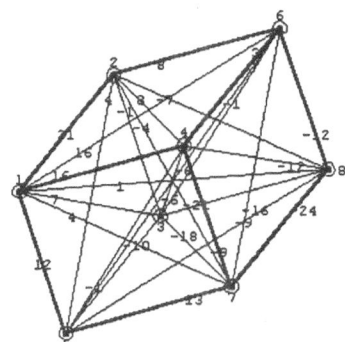

Figure 1. *Resulting Graph G* after applying the *training algorithm*.

When the *training algorithm* is finished, the final value w_{ij} assigned to the edge a_{ij}, represents the same kind of relative *gain* as explained before. The state vector x at time t could also be interpreted as a *coloring* of the edges in G, but now this coloring is going to be used for retrieving the data. The summation of all the edges in the subgraph colored with the same color as the edge a_{ij} may be interpreted as the *gain* of that subgraph. With these considerations in mind, a different *dynamic equation*, taking into account the relation between both of these *gains*, may be defined. In this paper we explain how this system is designed. At last, the advantages of the system are exposed through an already developed application.

2. Dynamics of the system.

If A is the adjacency matrix of the *Resulting Graph G*, we define two numbers, associated with the state vector $x(t)$ of the net,

$$I^x = \frac{1}{2}\sum_{i=1}^{n}x_i(t).x_j(t).w_{ij} = \frac{1}{2}x(t).A.x(t)', \qquad (6)$$

$$O^x = \frac{1}{2}\sum_{i=1}^{n}\overline{x}_i(t).\overline{x}_j(t).w_{ij} = \frac{1}{2}\overline{x}(t).A.\overline{x}(t)', \qquad (7)$$

where, $\overline{x}(t) = \{\overline{x}_1(t),\ldots,\overline{x}_n(t)\}$, is the dual vector of $x(t)$, in the sense that,

$$[x_i(t) = 1 \Leftrightarrow \overline{x}_i(t) = 0] \text{ and } [x_i(t) = 0 \Leftrightarrow \overline{x}_i(t) = 1] \qquad (8)$$

These two numbers can be easily interpreted once the G^x-coloring, associated with $x(t)$, has been applied over the *Resulting Graph G*: I represents the summation of all the values on the edges of the *red (black in figure 2.)* subgraph G_I^x and O represents the summation of all the values on the edges of the *blue (gray in figure 2.)* subgraph G_O^x. We have developed an example taking the 37 eight-bit patterns corresponding with a certain codification [4] of the most relevant phonetic sounds for the *Spanish* language, as the *training pattern set*. The I^x and O^x numbers corresponding with the state vector $(1,1,1,1,1,0,0,0)$ are the ones shown in the figure 2.

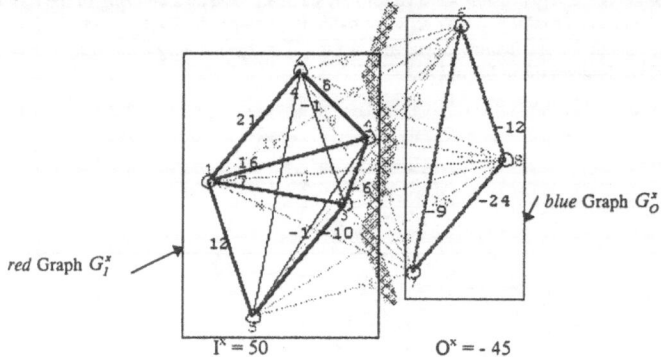

red Graph G_I^x

blue Graph G_O^x

$I^x = 50$ $O^x = -45$

Figure 2. G-coloring associated with $\xi^r = (1,1,1,1,1,0,0,0)$

We define now the *Relative Weight* w_i^x of the vertex v_i in the G-coloring as the contribution of this vertex to I^x. If v_i belongs to a graph with only one vertex is $w_i^x = 0$, otherwise

$$w_i^x = \begin{cases} \dfrac{\displaystyle\sum_{j=1}^{n} x_i(t).x_j(t).w_{ij}}{I^x}, & \text{if } v_i \in \text{red graph } G_I^x \\[2em] \dfrac{\displaystyle\sum_{j=1}^{n} \overline{x_i}(t).\overline{x_j}(t).w_{ij}}{O^x}, & \text{if } v_i \in \text{blue graph } G_O^x \end{cases} \qquad (9)$$

If, for example, $v_i \in red\ graph\ G$, the *Relative Weight* w_i^x, represents the sum of all the values on every *red edge* connecting with v_i divided by the sum of all the values on edges in the *red graph* G_I^x. Then, two numbers w_{iI}^x and w_{iO}^x associated with v_i are defined; the number w_{iI}^x represents w_i^x, changing if necessary, the value of x_i (and consequently its associated coloring), in such a way that $v_i \in red\ graph\ G_I^x$; the number w_{iO}^x represents w_i^x, changing if necessary, the value of x_i in such a way that $v_i \in blue\ graph\ G_O^x$. In our application, as may be seen from the figures, for the vector state $x = (1,1,1,1,1,0,0,0)$, the relative weight w_{3I}^x will be $-10/50$; if x_3 changes to value 0, (vertex v_3 changes to be a vertex in the *blue* graph), then $x = (1,1,0,1,1,0,0,0)$ and w_{3O}^x will be $45/90$.

The *dynamic equation* may be interpreted [5], then, from this new point of view: the output of $x_i(t+1)$ will be equal to "1", not if the summation of the values on edges connecting v_i with every other vertex in the *red* subgraph is bigger than the summation of those connecting v_i with every other vertex in the *red* subgraph, as the usual deterministic approach does; the output of $x_i(t+1)$ will be equal to "1" if w_{iI}^x is bigger than w_{iO}^x. With this consideration, the *dynamic equation* may be defined as follows:

$$x_i(t+1) = f_h\left[w_{iI}^x - w_{iO}^x\right] \quad i = 1,...,n, \qquad (10)$$

where f_h is the Heaviside's step function,

$$f_h(x) = \begin{cases} 1 & \text{if } x \geq 0 \\ 0 & \text{if } x < 0 \end{cases} \qquad (11)$$

and with the convention,

$$w_{iI}^x = w_{iO}^x \Rightarrow x_i(t+1) = x_i(t) \qquad (12)$$

3. Energy.

The trajectory at the initial state $x(0)$, may be defined as the sequence,

$$x(0), x(1), x(2),.......,x(t); \quad \text{where } x(t) = x(t+1). \qquad (13)$$

For determining the conditions under which the network is devoid of cycles so as for ensuring that from any initial state a final state is always obtained, it is sufficient to find a scalar function, somewhat inappropriately called *energy function*, which is strictly decreasing on any trajectory of the system. We shall define the energy associated with some state vector x as

$$E(x) = -\left|I^x . O^x\right| \qquad (14)$$

For proving that $E(x)$ is strictly decreasing on any trajectory, we see that if, for example, in the state vector

$$x(t) = \left[x_1(t),\ldots,x_i(t),\ldots,x_n(t)\right], \text{ with } x_k(t) = 0, \tag{15}$$

the *Relative Weight* w_{kO}^x is bigger than the *Relative Weight* w_{kl}^x , then the vertex v_k must be in the *red* subgraph, or under other interpretation: the updated state vector

$$x(t+1) = \left[x_1(t+1),\ldots,x_k(t+1),\ldots,x_n(t+1)\right], \text{ with } x_k(t) = 1, \tag{16}$$

must be in a lower energy level. So, we must prove that in case that:

$$[w_{kl}^x > w_{kO}^x] \Rightarrow |I^x(t+1) \cdot O^x(t+1)| > |I^x(t) \cdot O^x(t)| \tag{17}$$

Taking into account expression (8) and the relation:

$$x_i(t+1) = x_i(t), \ \forall \, i \neq k; \ i = 1,\ldots,n; \tag{18}$$

expression (17) may be written as:

$$[w_{kl}^x > w_{kO}^x] \Rightarrow \frac{1}{I^x(t+1)} \cdot \sum_{j=1}^{n} x_k(t+1) \cdot x_j(t) \cdot w_{kj} > \frac{1}{O^x(t)} \cdot \sum_{j=1}^{n} \overline{x}_k(t+1) \cdot \overline{x}_j(t) \cdot w_{kj} \tag{19}$$

where

$$\sum_{j=1}^{n} x_k(t+1) \cdot x_j(t) \cdot w_{kj} =$$

$$\left[\sum_{i=1}^{k-1}\sum_{j=1}^{n} x_i(t+1) \cdot x_j(t+1) \cdot w_{ij} + \sum_{j=1}^{n} x_k(t+1) \cdot w_{kj} + \sum_{i=k+1}^{n}\sum_{j=1}^{n} x_i(t+1) \cdot x_j(t+1) \cdot w_{ij}\right] -$$

$$\left[\sum_{i=1}^{k-1}\sum_{j=1}^{n} x_i(t) \cdot x_j(t) \cdot w_{ij} + \sum_{j=1}^{n} x_k(t) \cdot w_{kj} + \sum_{i=k+1}^{n}\sum_{j=1}^{n} x_i(t) \cdot x_j(t) \cdot w_{ij}\right] =$$

$$\left[\sum_{i=1}^{n}\sum_{j=1}^{n} x_i(t+1) \cdot x_j(t+1) \cdot w_{ij} - \sum_{i=1}^{n}\sum_{j=1}^{n} x_i(t) \cdot x_j(t) \cdot w_{ij}\right] = I^x(t+1) - I^x(t).$$

Changing to the dual space, we prove that

$$\sum_{j=1}^{n} \overline{x}_k(t) \cdot \overline{x}_j(t) \cdot w_{kj} = O^x(t) - O^x(t+1) \tag{20}$$

And now the expression (19) can be written as:

$$[w_{kl}^x > w_{kO}^x] \ \frac{I^x(t+1) - I^x(t)}{I^x(t+1)} > \frac{O^x(t+1) - O^x(t)}{O^x(t+1)} \Rightarrow 1 - \frac{I^x(t)}{I^x(t+1)} > 1 - \frac{O^x(t)}{O^x(t+1)}$$

$$\Rightarrow \frac{O^x(t+1)}{O^x(t)} - \frac{I^x(t+1)}{I^x(t)} > 0 \Rightarrow \frac{I^x(t+1) \cdot I^x(t+1) - I^x(t) \cdot O^x(t)}{I^x(t+1) \cdot O^x(t+1)} > 0, \tag{21}$$

if expression (21) is analyzed for every possible signs of the pairs $(I^x(t) \cdot O^x(t))$ and $(I^x(t+1) \cdot O^x(t+1))$, we obtain

$$|I^x(t+1) \cdot O^x(t+1)| > |I^x(t) \cdot O^x(t)| \tag{22}$$

and

$$E(x(t)) > E(x(t+1)). \tag{23}$$

The family of *hyperbolic surfaces*, $|xy| = k$, defined in the plane R^2 may be seen as the energy level lines of the system, we have proven that if energy point (x,y) associated to a given state-vector is in the surface $|xy| = k$, then, when a change of state is produced, the point (x,y) must be placed in other surface $|xy| = q$, $(q>k)$. In our example, if the net is in the state $x(t) = (1,1,1,1,1,0,0,0)$, its energy point, as we saw before, is $(50, -45)$, after applying the *dynamic equation* in expression (12) to the third component we obtain,

$$x_i(t+1) = f_h \left[w_{il}^x - w_{io}^x \right] = f_h \left[(-10/50) - 45/90 \right] = 0. \tag{24}$$

the energy point associated with $x(t+1) = (1,1,1,0,1,0,0,0)$ is the point $(60, -90)$ which is obviously in a higher *hyperbolic surface* with lower energy.

4. Capacity and Recall.

It could be said that the *capacity* of the net is the proportion between the number of fixed and not fixed prototypes in the net. Every possible pattern ξ^r may be represented by its (I', O') coordinates in an (I, O) axis diagram. It can be said that ξ^r is a fixed point if after applying the *training algorithm* to ξ^r the point (I', O') doesn't move. The point (I', O') doesn't move if its corresponding *red* subgraph G_I^r and the *blue* subgraph G_O^r verify a certain condition, which could be interpreted as the *stability condition* of the system. Only those prototypes strongly correlated with the net must verify this condition, but a prototype is strongly correlated when all of its components maintain a high representation in relation to the patterns in training set L. We have, associated with ξ^r the vector of *Relative Weights*, $w^r = (w_1^r, ..., w_i^r, ..., w_n^r)$ and we prove that:

$$\left[2I' = \sum_{i=1}^{n}\sum_{j=1}^{n} \xi_i^r . \xi_j^r . w_{ij} \right] \Rightarrow \left[2 = \frac{\sum_{i=1}^{n}\sum_{j=1}^{n} \xi_i^r . \xi_j^r . w_{ij}}{I'} \right] \Rightarrow \left[2 = \sum_{i=1}^{n} \xi_i^r . \left(\frac{\sum_{j=1}^{n} \xi_j^r . w_{ij}}{I'} \right) \right] \Rightarrow$$

$$\left[2 = \sum_{i=1}^{n} \xi_i^r . w_i^r \right] \Rightarrow \left[2 = \sum_{i=1}^{n} w_i^r, \quad \forall i / \xi_i^r = 1 \right] \Rightarrow \left[2 = \sum_{i=1}^{n} w_i^r, \quad \forall i / v_i \in G_I^r \right]$$

and, in a similar way, we could prove that

$$\left[2 = \sum_{i=1}^{n} \bar{\xi}_i^r . w_i^r \right] \Rightarrow \left[2 = \sum_{i=1}^{n} w_i^r, \quad \forall i / \bar{\xi}_i^r = 1 \right] \Rightarrow \left[2 = \sum_{i=1}^{n} w_i^r, \quad \forall i / v_i \in G_O^r \right], \text{ but}$$

$$v_i \in G \Leftrightarrow v_i \in G_I^r \cup G_O^r, \text{ and consequently } \sum_{i=1}^{n} w_i^r = 4 \tag{25}$$

If we adequately re-scale the components of w^r taking into account the number of vertices in the *red* and *blue* subgraphs, we can force:

$$\sum_{i=1}^{n} w_i^r = 1, \quad \text{for every } G'\text{-coloring.}$$

The w^r-*distribution* for $(1,1,1,1,1,0,0,0)$ and $(1,1,0,1,1,0,0,0)$ may be seen in figure 3. $\qquad$ (26)

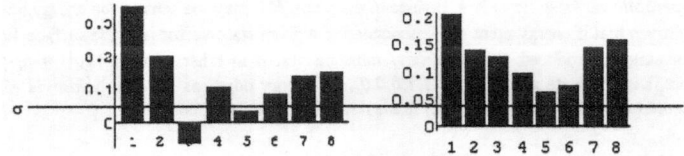

Figure 3. w' *-distribution* for *(1,1,1,1,1,0,0,0)* and *(1,1,0,1,1,0,0,0)*

In this way, we have proved that the more uniform the w' *-distribution* is the more equilibrium or degree of acceptance are sharing the components of ξ' [6]. Once a fixed point is reached, we use an external parameter σ in the sense that the recall algorithm is retrieving all the vectors whose relative weight vector components are lower than σ. For example, in figure 3, we see that if $x(t) = (1,1,1,1,1,0,0,0)$ and $\sigma = 0.4$ then $w_3^x \leq \sigma$, so the neuron number three must change its state. We can say then that the *stability conditions* of a given pattern may be established in relation with a, so to be called, *capacity parameter or threshold* σ [6].

The results of giving different values to the *capacity parameter* σ may be seen in Figures 4 and 5. For a low value of σ, where a high correlation must be required, the five points fixed by the network are represented by the bold points in Fig. 4, the small points representing the non-fixed 251 out of the 256 possible ones. Giving a high value to σ the 18 out of 256 ones represented in Fig 2 will result. This fact shows the influence of σ on the number of patterns fixed by the net

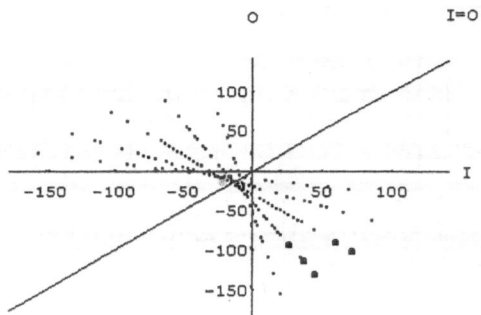

Figure 4. *Points fixed when a low-value parameter (high correlation) is used.*

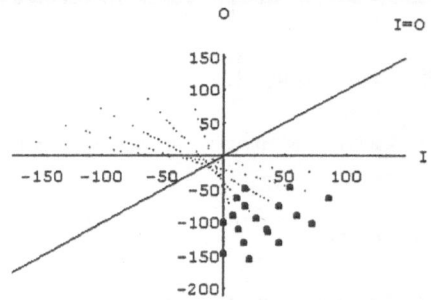

Figure 5. *Points fixed when a high-value parameter (low correlation) is used.*

5. Conclusions.

The conclusions which could be derived from the present research can be seen under the point of view of the re-interpretation of classical *Neural Network Theory* as *Graph Formalisms* [1]. Consequently powerful techniques based in *Graph Theory* could be derived and applied to advantageously extract new knowledge from the classically well-known methods of *Network Training* [6], as for example in the case where an *n-Complete-Graph* is given, and the partition of this graph in two *Subgraphs* is desired, in which the sum of all the edges of both *Subgraphs* obtained as a consequence of the partition are under a pre-established *threshold*. This question may be particularly useful to researchers in *Graph Theory*, where different partitions of a given graph may be obtained and studied according with the *threshold level* chosen, in order to code the degree of similarities or dissimilarities in the information present in the *learning pattern set*.

Acknowledgments: This research is carried out under Grants Ns. TIC 930702C02-01 and AE00347/94

References.

[1] V. Giménez, P. Gómez-Vilda, M. Pérez-Castellanos and V. Rodellar, *A New Approach for Finding the Weights in a Neural Network using Graphs*, Proc. of the 36th Midwest Symposium on Circuits and Systems, Detroit, August 16-18, 1993, pp. 113-116.

[2] F. S. Roberts, *Applied Combinatorics*, Prentice-Hall Inc., New Jersey, 1985.

[3] V. Giménez, P. Gómez-Vilda, M. Pérez-Castellanos and E. Torrano, *A New Approach for improving the capacity limit on a Recursive Neural Network*, Proc. of the AMS'94. IASTED, Lugano, Suiza, June 20-22, 1994, pp. 90-93.

[4] V. Rodellar, P. Gómez, M. Hermida and R. W. Newcomb, *An Auditory Neural System for Speech Processing and Recognition*, Proceedings of the ICARCV'92, Singapore, September 16-18, 1992, pp. INV-6.2.1-5.

[5] Yves Kamp and Martin Hasler, *Recursive Neural Networks for Associative Memory*, Wiley-Interscience Series in Systems and Optimization, England, 1990, pp. 10-34.

[6] R. Shonkwiler, *Separating the Vertice of N-Cubes by Hyperplanes and its Application to Artificial Neural Networks*, IEEE Trans. on Neural Networks, Vol. 4, No. 2, March 1993, pp. 343-347.

Visual Information Processing
from the Viewpoint of Symbolic Operations*

J. Barahona da Fonseca[1], I. Barahona da Fonseca[2] and J. Simões da Fonseca[3]

[1]FCT/UNL, DEPARTMENT OF ELECTRICAL ENGINEERING
Phone: 351-1-294 44 64, EXT 1007, Fax: 351-1-2957786,
E-MAIL:JBF@FCT.UNL.PT,
QUINTA DA TORRE, 2825 MONTE DA CAPARICA, PORTUGAL
[2]FACULTY OF PSYCHOLOGY OFLISBON,
Fax: 351-1- 793 34 08
ALAMEDA DA UNIVERSIDADE, 1600 LISBOA, PORTUGAL.
[3]FACULTY OF MEDICINE OF LISBON,
Fax: 351-1-796 40 59
AV PROF EGAS MONIZ, 1600 LISBOA, PORTUGAL.

Abstract:

Visual information processing at the symbolic level is examined from the point of view of conceptual operations implied in perception, representation and their cognitive components. Visual symbolic representation is considered as being carried by visual particular attributes of the scene independently of linguistic verbal representations. It is hypothetized that visual symbolic representations precede verbal linguistic competence acquisition that generalizes symbolic operations which in the development process occurred in the visual system. A distinction is made between visual mappings and significant-significate relationships carried by high level visual symbols.

Key Words: Visual Processing, Mapping Lines, Symbolic Representation, Icons, Visual Diagrams.

I. Introduction

The analysis of visual information processing in the Nervous System has been deeply influenced by very significant advances in the knowledge of neurophysiological phenomena such as lateral inhibition, feature and object detection, hypercomplex responses to invariant characteristics of stimuli (Hubel, Wiesel, 1965; Jung 1959;. Maturana, Lettvin, McCulloch, Pitts, 1960). An alternative approach that deserves examination is the analysis of significant-significate relationships when the signals which constitute visual mappings are considered as symbolic carriers of information. In one proposal of Charles Sanders Peirce (1966) signs may assume the characteristics of icons, diagrams or symbols. These distinctions may be applied to visual processing and contribute to clarify some issues of high level symbolic visual processing. If we consider a concrete example of a complex visual perception or representation, its firstness can be experienced subjectively in its vividness and sensory freshness projected on the external world. At this level the distribution of some characteristics of coloured surfaces, their texture and some particular shapes for instance proportions present in lines composed by elements of the image produce an immediate aesthesic/aesthetic experience in observers

---

* This work was sponsored by a grant from Bial Foundation, Porto, Portugal.

of the scene perceived or its representation. Some particular attributes of colour and shape are identified upon the background and contribute also for this iconic representation contained in the whole of visual perception. The iconic elements are used in our culture to provoke emotions, feelings or aesthetic or motivational states. Examples may be found in the form of particular identifications: a tower of a church, the colour of the sky, the white, the gray, the green in an ondulation of an extension of land or water, the colour of skin, to mention only a few examples. As we begin to analyze in more detail those signs contained in visual perception or representation, diagrams become apparent and are successively scanned. Diagrams are summary conceptual representations and descriptions written in a visual alphabet and languages contain information about position, dimension, borders, continuities, vicinity, distance, attributed structure, function and possible modes of interaction. In children before they are able to use linguistic competence they can play games, perform acts and intentional plans of action and imitate in a simplified diagramatic form the gestures of some present or past models. When we are driving a car, playing a sport or acting, motivationally or emotionally at a basic level, most of the time it is used this diagramatic system of representation which we share with other species and which allows the use of a rudimentary but efficient system of inference. When we drive a car in routine conditions or else in extreme emergency circumstances the diagramatic system of representation is used in decision making. The same is true concerning sports which require extreme skill and perceptive-motor coordination. The same happens in artistic performances which implies extremely virtuosistic ability. Objects which elicit decision and action, and the matching of action performance and its object and purpose are performed within a non linguistic system of representation. They may be afterwards described using a linguistic code in some of the characteristics they assumed but these linguistic codes have not been used during perception analysis and behaviour performance except for some particular aspects. At this level of analysis visual perception contains information which is represented and analyzed in terms of diagramatic signals. Stimulus detection and evaluation as well as actions or patterns of action in basic motivational acts are performed in other species and probably in many circunstances in humans exclusively within a conceptual system carried by these diagramatic signals.

II. Symbolic Processing

Neurophysiological processing from retina to the brain produces elementary components of the mapping of stimuli in the CNS. Simultaneous contrast, sucessive contrast, position of line segments, angles, colour and colour contrast are representation elements which share, all of them, an almost immediate referential relationship with the stimulus which ellicited them. A similar relationship is present in some object detectors found in neurophysiological experimentation (Lettvin, Maturana, Pitts, McCulloch, 1959). When a mapping of an object is processed with the aim of contour identification, using the neurophysiological paradigm three procedures seem to be likely to occur: (1) definition of a contour by line and angle detectors which define a closed area in a surface; (2) detection of the gravity center of that area and (3) sampling by a delta wave transformation rotating along the line contour with the rotation center coinciding with the gravity center of the area. Autocorrelation may allow the immediate characterization of regular poligonal forms as well as circle, ellipse, oval and other closed curves. A neural net which implements all these sucessive operations may be said to represent in its states the figure. When we examine this problem in a more general situation which occurs when the contour itself has to be found and specified other operations are known to be performed and other cues are used to characterize the stimulus. For instance, binocular angle of convergence contributes significantly to calculate distances. When the CNS scans a scene containing tridimensional objects, a systematic procedure may produce the characterization of the visible surfaces in terms of the position of the points which compose them in a tridimensional system of coordinates. Another processing is then necessary to identify the contour of these surfaces namely which of them may be chosen as characteristic contours of a tridimensional object. A procedure for this purpose might be the scanning, using successive measures of the angle of binocular convergence of distance to chosen points in the object surface. An approach might then be the attribution of the quality of contour point to any position at which distance measure suffers a sudden change, large enough relatively to the mean value of a previous set of measures taken along the same meridian or parallel geodesic lines along the surface limiting the volume exceed a specified threshold α,

$$\left| X_i - X_{i-1} \right| - \frac{1}{k} \sum_{i=n-k+1}^{n} \left| X_{i-k} - X_{i-k-1} \right| > \alpha.$$

This expression defines the decision criterion for surface contour segmentation and may be conceived as associated to an adaptive algorithm which adjusts threshold α as well as the average window size and spatial frequency, k. After the surface contour detection algorithm would be applied, then the gravity center of the volume might be determined and a two dimensional rotational delta transformation might be calculated. A problem of this kind is raised by the experimental finding of visual system neurons for which only the shape of a human face is an adequate stimulus. The algorithm behind this neuronal type of response which occurs only when the stimulus has the shape of a human face raises many interesting questions in symbolic visual processing. First three extreme positions may be considered in which elements of the face are visible- frontal anterior, profile (transversal) and frontal posterior. Between the first and the second position as well as between the second and third position are definable characteristic sets of transformations with preservation of some of the elements of the face, and not others. Considering the frontal anterior position, a binocular scanning algoritm can operate the segmentation of the complete shape defining components such as the external contour, hair, nose, lips, ears, eyes. The contour of the areas corresponding to these components might be calculated and determined their center of gravity. Then after closing open contour the rotational delta transformation would be calculated and its cross-correlation with a template might be obtained. This information for each component and the relative position of the center of gravity of other components would then be represented in a vector which might be compared with a reference vector. The same would apply for the second and third position as well as for characteristic transition positions. These operations would be performed in a neural net and the response of some of its neurons would occur only when the stimulus was the shape of a human face. The visual concept of a human face would correspond then to the visual information used in the decision of the identification net and the decision rules and processing operations used in the identification operation. Summarizing our discussion about visual identification of a complex form (Human Face) we have the following relationships:

$$\forall ML_i, \quad Ref(ML_i) \equiv WL_i$$

in which **Ref** represents the reference relationship between mapping lines, **ML**, which correspond to the result of low level neurophysiological processing and world lines, **WL**, which belong to external objects which act as stimuli. The next relationship defines a second level of processing in which symbols X_i have a significant/significate relationship, **Sig**.

$$X \equiv \{X_1, X_2, ..., X_n\} \ \& \forall X_i, (X_i \ Sig \ ML_i)$$

A transformation **U**

$$\exists ML_i \ \ U(ML_i) \equiv \exists X_i$$

which may be interpreted as the basic symbolisation process. It establishes a correspondence between ML_i and the new symbols X_i which satisfy the relationship X_i **Sig** ML_i- that is the new X_i carries a new significant relationship while ML_i while ML_i carried a referential relationship to WL_i.

X_i **Sig** ML_i means that X_i is an attribute which is related to ML_i as a significant of an attribute which is conveyed by ML_i .Over significant symbols, X_i, there are defined predicates, $P(X_i)$, as relationships over the X_i.

$$P(X_i) \equiv R_i(X_j, X_k, ..., X_l), \ \ \{X_j, X_k, ..., X_l\} \subset X$$

where R_i corresponds to the relationship which specifies $P(X_i)$. $P(CF)$, predicate characteristics of a complex form, **CF**, are defined as higher level relationships over predicates $P(X)$, $P(Y)$, ..., $P(Z)$.

$$P(CF) \equiv R_k(P(X), P(Y), ..., P(Z))$$

such that $R_k \equiv$ boolean matrix **[o]**, $(P(X), P(Y), ..., P(Z)) \equiv$ boolean vector **[u]** such that

$$[o] \ [u]=[v]$$

with Hamming distance

$$dist([v],[r]) \leq \delta$$

in which **[r]** is a vector which expresses the complete set of attributes which **[v]** may possess to satisfy relationship $P(CF)$ represented by **[o]** over the components of **[u]**.

For the representation of the human face we identify it as a complex form **CF(HF)**, defined as a relationship R_{HF}

$$CF(HF) \equiv R_{HF}(P(E), P(N), P(M), P(Er), P(Ey), P(Fh), P(Har), P(Cont.))$$

for each $P(X)$; for instance, for $P(E)$, there are

$$X(E) \equiv \{X_i(E), X_\gamma(E), ..., X_Z(E)\}$$

such that for any $X_k(E)$,

$$X_k(E) \equiv T^n(Rot.^n(Sc^o(Proj^p(X_i(E))))) \subset X(E)$$

in which **T** represents a translation transformation, **Rot**, a rotation transformation, **Sc**, a scaling (dilatation and contraction) transformation such that the next relationships are verified

$$[o_E] \; [u_E] = [v_E]$$

$$dist(\; [v_E],[r_E]) \leq \alpha_E$$

As it was proposed in the introductory text **CF(HF)** does not correspond to the verification of all **P(X)** considered independently one from another. There are "gestalt" relationships such as stereometric distance between the gravity center of all the **ML** which define **X** over which **P(X)** are specified; normalized relatively to the simetry meridian, the individual angular $Rot(ML_i)$ that must be equal for all **ML** characteristic of all X_i from $P(X_i)$. A similar problem occurs relatively to each of the components- namely the eye contain the pupil (circle) enclosed by a ring (the iris) enclosed by an ellipsoid form (the conjuntive) enclosed by the upper and lower eyelids, with the nose between the eyes then the mouth, both at similar distance from the ears and their distance one from another less than the width of the mouth.Other elements of the CF are relevant such as colour, limmits of coloured surfaces, characteristic projective transformations. Another necessary concept is the notion of an incomplete **CF-ICF**. Mirror images, photographic images as well as artistic representations depend for their recognition on the definition of attributes of incomplete forms as well as on some transformations like light-shade modelling transformations, perspective transformations and projected size transformations. Relative size, intersection or interposition as well as iconic identification contribute to the identification of ICF representation under the form of purposefully prepared referents.

A clear distinction appears at this level, in visual processing, if we compare it with the initial visual mappings identified in neurophysiological experimentation. Namely, conceptual visual representation requires that components of the mappings and their transformations belong now to classes of equivalence of representational elements of visual concepts. They are no more simple direct or transformational mappings but rather symbols of particular attributes of the visual concept. Their relationship with the concept corresponding to the stimulus is now specified as one of significant to segnificate similar to the significant-significate relationship which occurs in linguistic systems. An autonomous visual language of information processing at the conceptual level exists and is autonomous and in many aspects independent of the verbal linguistic systems of communication and conceptual representation. During human maturation visual and probably multisensory systems of conceptual representation precede the acquisition of linguistic competence and probably persist as an autonomous system of representation after this acquisition. It is probably due to this fact and also to the refinement of perceptive operations in the auditory system and also to the precision and refinement in sound emission that human language was developed on sound emission and perception. Visual concepts were carried in part to this system of symbolisation and the performance of visual conceptual operations was then incorporated in reading and writing operations.

III. The Sig relationship

The relationship

$$\exists X_i, \; Sig(X_i) \equiv \exists ML_i$$

plays a central role in our theoretical proposal because it implies a second level of symbolization which goes beyond obligatory mapping procedures. Complex cell responses were first described by Hubel and Wiesel in Visual II associative area. These cell responses are elicited by longer line segments than in the case of simple cell responses. These responses are proportional to the length of the line stimulus in Visual I, primary visual projection area cells which detect and identify segments of straight lines with a fixed specific dimension and orientation in space are examples of ML_i representations. Complex responses, CR_i, may be interpreted as X_i which imply the following relationships: (a) CR_i are generated by a set of ML_i detectors; (b) these detectors, ML_i, are colinear with CR_i; (c) for each ML_i there is an origin which coincides with the initial point of CR_i or else with the final point of the preceeding ML_{i-1}, and a final point which coincides with the extreme of CR_i or else with the initial point of ML_{i+1}. Angle detectors, A_{ij}, are related to the CR_i and CR_j through the following relationships: (a) the vertex of the angle, A_{ij}, coincides with one extreme of CR_i and one of the extremes of CR_j; (b) segments i and j which define A_{ij} are respectively coincident with ML_F and ML_{INIT}, the final segment ML_i of CR_i and the initial segment ML_j of CR_j. Cells which detect angles were described in visual association areas and were denoted as hipercomplex cells. While at the ML_i analogical level of representation the mapping ML_i of WL_i has a definite spatial correspondence with its reference, on the contrary at the X_i level this relationship is carried by a representation using rules of a convention and not by simple spatial analogy:

$$X_i \equiv (L(CR_i)\Lambda L(CR_j)\Lambda A_{ij})$$

The following dendritic operative networks embodie these hypotheses.

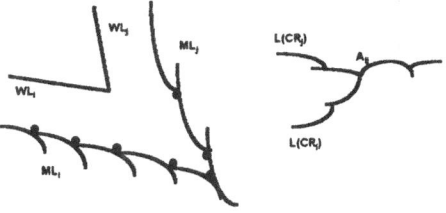

Figure 1a- At the X_i level the relationship may be expressed at a second level of representation by $A_{ij}\Lambda L(CR_i)\Lambda L(CR_j)$. Representation on the left side corresponds to an analogical mapping constructed over ML_i and ML_j. Representation on the right side implies the information about CR_i and CR_j and their respective lengths and the angle A_{ij} between them. Conventions about dendritic computations can be found in "Dendritic Computation in the Brain" in this volume.

While rotation transformation might be represented in analogical form by a dendritic network presented in figure 1b.

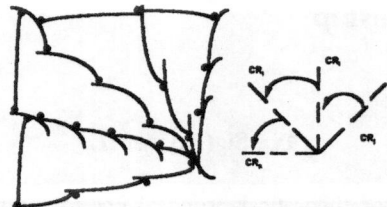

Figure 1b- A rotation of CR_i into CR_k and of CR_j into CR_l. X_i is represented on the right side. At the level of the X_i it is represented as

$$T(CR_i,CR_i) \equiv CR_k,CR_l$$

and angle of rotation

$$V(A_{ij})=V(A_{kl}),$$

$$Rot(A)=A_{ik}=A_{jl}.$$

Concerning translation transformation, Tl, of a CR_i,

$$Tl(CR_i)=CR_j \wedge (CR_i \text{ paralell } CR_j \vee CR_i \text{ colinear } CR_j)$$

is represented at the ML_i level by a dendritic network presented in figure 1c.

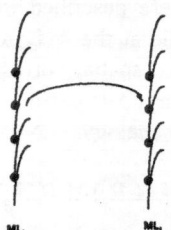

Figure 1c- A dentritic network that performs the translation of CR_i into CR_j.

Scaling, that is, dilatation D_{ij} or contraction C_{ik} may be represented along a dendritic network of a similar type (figure 1d).

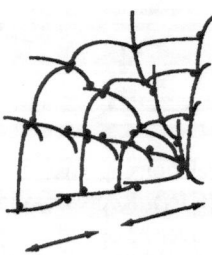

Figure 1d- A network which performs contraction and dilatation of a given CR_i.

As we have shown $\varphi_{ij}(f_i(x),f_j(x))$, cross correlation of $f_i(x)$ and $f_j(x)$ to be used as $\varphi_{ij}(CR_i,CR_j)$ is also representable at the level $P(X_i)$, the next level we hypothetized in figure 1e provides an example for this level of operations which may be interpreted as occurring in Visual III.

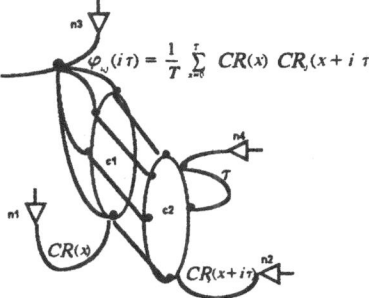

$$\varphi_{\nu}(i\,\tau) = \frac{1}{T} \sum_{x=0}^{\tau} CR(x)\; CR_{\nu}(x + i\,\tau)$$

Figure 1e- The dendritic circuits needed to calculate cross correlation between two functions $CR_i(x)$ and $CR_j(x)$ which specify dendritic messages which are propagated from dendritic trees of neurons n1 and n2 respectively. The lag of $CR_i(x)$ is produced by a delay loop τ. Multiplicative interaction is produced by iterative additions over a set of neurons n3.

The final level of processing implies a dendritic network which may calculate the internal product of a $P(X), P(Y), ..., P(Z) \equiv [u]$:

$$[o]\ [u]=[v]$$

$$dist([v],[r])<\delta$$

As we indicated in the previous section beyond the identification of a **CF** there are identification operations for the attributes of all of its components **P(X)**, **P(Y)** and all the others. These remarks illustrate in a more intuitive form how dendritic computation may be used as an approach to model the way symbolic visual processes occur in the brain in an oversimplified model which nevertheless may provide some guidelines to build models of complex visual processes.

References

HUBEL DH, WIESEL TN (1965): Receptive Fields, Binocular Interation and Functional Architecture in two Non-Striate Visual Areas (18 and 19) of the Cat.*J. Neurophysiol, 28.*

HUBEL DH, WIESEL TN (1965): Binocular Interaction in Striate Cortex of Kittens Reared with Artificial Squint. *J. Neurophysiol., 28.*

JUNG R (1959): Microphysiology of Cortical Neurons and its Significance for Psychofisiology in *Festschrift Prof. C. Estable, An. Fac Med. Montevideo, 44.*

LETTVIN JY, MATURANA HR, PITTS WH, McCULLOCH WS (1959): What the Frog's Eye Tells the Frog's Brain. *Proc Inst. Radio Eng. ,47.*

MATURANA HR, LETTVIN JY, McCULLOCH WS, PITTS WH (1960): Anatomy and Physiology of Vision in the Frog (rana pipers). *J. Gen. Physiol.* 43, II Supplement.

PEIRCE CS (1966): *Collected Papers.* Havard Univ. Press. Cambridge, Mass.

PIAGET, J (1952): *The Origins of Intelligence in Children.* Int. Univ. Press, NY.

Dentritic Computation in the Brain*

Isabel Barahona da Fonseca[1], J. Barahona da Fonseca[2] and J. Simões da Fonseca[3]

[1]FACULTY OF PSYCHOLOGY OFLISBON,
Fax: 351-1- 793 34 08
ALAMEDA DA UNIVERSIDADE, 1600 LISBOA, PORTUGAL.

[2]FCT/UNL, DEPARTMENT OF ELECTRICAL ENGINEERING
Phone: 351-1-294 44 64, EXT 1007, Fax: 351-1-2957786,
E-MAIL:JBF@FCT.UNL.PT,
QUINTA DA TORRE, 2825 MONTE DA CAPARICA, PORTUGAL

[3]FACULTY OF MEDICINE OF LISBON,
Fax: 351-1-796 40 59
AV PROF EGAS MONIZ, 1600 LISBOA, PORTUGAL.

Abstract:

Discrete messages formed by action potentials are processed by cell bodies of neurons which act as time/space integrators of synaptic excitation elicited by this potentials. Data from neurophysiological research suggest that dendritic potential waveforms may be processed in dendro-dendritic circuits largely independent from cell bodies. The spatio-temporal arrangement and spatial frequency of dendro-dendritic synapses allow the implementation of convolution and deconvolution algorithms in such dendro-dendritic circuits. Cross correlation with periodic delta functions implemented under the form of periodic ramification of dendritic trees allows the recovery of periodic waveforms processed in dendro-dentritic networks. Considering the size of dendritic circuits which may attain lengths less than a few hundred microns and a conduction velocity which may attain two to four meters per second or more produces an expectation for the frequency of dendritic circuit much above the 1KHz limit for the frequency of cell body and axon signals. Under these conditions coding of neuronal sequenties of action potentials can only be understood taking into account structural and functional active characteristics of dendritic processes.

Key words: Dendritic Networks, Neural Computation, Graduate Potentials, Brain Models, Periodic Waveforms.

I. Introduction

Data obtained in the last few years have shown that human EEG provided a source of signals which are significantly correlated in a characteristic manner with distinct semantic contents of cognitive states. A method developed by J. Barahona da Fonseca, (1985a,b; J. Barahona da Fonseca et al., 1990), allowed the detection of periodic waveforms during cognitive operations produced independently of any immediate eliciting sensory

---

* This work was sponsored by a grant from Bial Foundation, Porto, Portugal.

stimulus (Isabel Barahona da Fonseca,1991,1993, Isabel Barahona da Fonseca et al., 1988, 1991). Summarizing, cross correlation of EEG waveforms with a periodically repeated delta function allowed the extraction of regular, periodic waveforms which, after an averaging procedure of an ensemble of independent records, using as a synchronizing marker the peak of the wave, was followed by Multivariate Discriminant Analysis of the power spectrum of those waveforms. Four cognitive and cognitive-affective states proved to be adequately classifiable on the basis of the cross-spectral comparisons. As EEG waveforms are most likely to be generated by synaptic and dendritic potentials (Pedley, Traub, 1990) these results raised the need of developing a model for synaptic and dendritic computation which might provide to be a system's approach for the explanation of the experimental data we did mention.

The first problem is to clarify which may be the signals processed in dendritic trees by synaptic transmission in axo-dendritic and dendro-dendritic functions and which is the influence that signals generated in dendritic segments may have on other parts of the dendritic tree of a single neuron, either apical or else proximal to the cell body (J. Simões da Fonseca, 1993). Furthermore which is the reciprocal interaction between dendritic and cell body potentials and also between dendrits belonging to distinct neurons which are linked by a dendro-dendritic synapsis. Finally, which are the computational characteristics of systems formed by dendro-dentritic circuits and those which result from the joint operation of axons ramifications, cell body and dendritic circuits. A neural net approach for all these types of dendritic, axonal and somatic operators will be proposed. The main difference lies in the analogical character of operations in dendritic circuits and the non-relevance of thresholds in these operations.

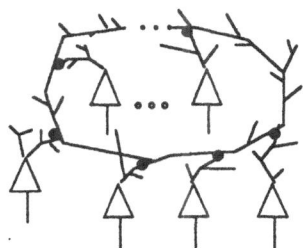

Figure 1- A dendro-dendritic circuit formed by a subset of the dendritic trees of **n** neurons. • signals an excitatory synaptic site and O an inhibitory synaptic site.

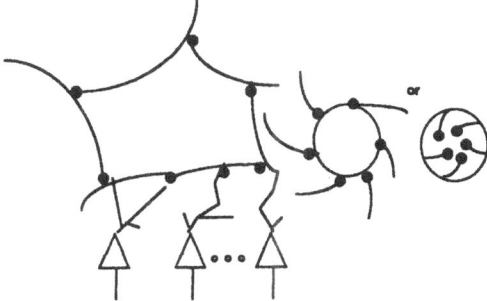

Figure 2- Further simplifications of the representation of dendro-dendritic circuits.

II. Dendritic potentials

Dentritic potentials may result either from local synaptic transmission during activation of axo-dendritic or dendro-dendritic synapses. Dendritic trees may also be activated by propagation of excitation from the cell body along the dendritic tree. The question of dentritic potentials has been extensively examined in the past, most of the conclusions pointing that the modulation of cell body excitation due to the propagation of dendritic depolarization was necessarily of minor signification due to the attenuation of signals conducted along the membrane. Results obtained by Graubard and Calvin, 1979, raised other possibilities. These authors revised extensively both experimental and theoretical data on dendritic signal processing. Examining conditions of propagation, cable transmission equations for steady state conditions yeld

$$V_{PSP} = \frac{g_{PSP}}{g_{PSP} + g_{IN}} E_{PSP},$$

in which V_{psp}, is the size of the postsynaptic potential, which is a function of the synaptic conductance, g_{psp}, driving potential, E_{psp} and the input conductance of the cell computed for that specific location, g_{in} (Graubard, Calvin , 1979). This means that propagation from an excitation site in a dendritic tree is attenuated in its propagation to the cell body so that its contribution may very probably be negligible if we compare it with synaptic sources of excitation directly located in the cell body membrane. On the contrary, propagation of excitation from a dendritic site towards the apical extrem of the tree occurs with minor attenuation of the signal. The amplitude of this signal may be calculated as implying a relevant depolarization amplitude approximately in the range of cell body spike amplitude. Nevertheless dendritic potentials which have been recorded after inhibition of spike production through the use of Tetradoxin show the characteristics of graduated potentials with waveform and duration very different from those of action potentials which have their size and duration constant.

Dendro-dendritic transmission was directly recorded by Graubard and Calvin, 1979. Due to the spikeless characteristics of dendritic potentials and the transmission characteristics of dendro-dendritic synaptic junctions it can be inferred that exclusively dendro-dendritic circuits are possibly relevant for local computations. The signals they receive, convey and process are waveforms which circulate in dendro-dendritic close loops. Complex waveforms which characterize those signals may be modified by synaptic addition or else subtraction in the case of inhibitory synapses. This way sub-cellular local operators can be defined as formed by dendro-dendritic synapses and their adjacent dendritic segment until the next synapse - in the same neuron or from other neuron's dendritic ramifications.

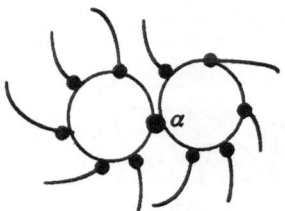

Figure 3-Term by term addition of two ordered sequences of dendritic signals at synapse α.

Axo-dendritic synapses may also activate segments of those dendro-dendritic circuits. Cell body signals may as well propagate into the dendritic branches and put a facilitating bias for synaptic transmission on the dendro-dendritic circuits. As dendro-somatic synapses have been identified they are the putative targets for the attribution of the role of rendering dendritic computations available for cell body processing and axonal transmission. Other consequences are for instance: (1) the possibility of convolution operation between signals defining in its waveforms respectively distinct periodic functions f(x) and f'(x); (2) adequate sampling through periodically spaced dendrits may extract periodic components by cross correlation; (3) the problem of neural coding by pulses which propagate along a single channel or else along parallel channels is here substituted by convolution and deconvolution of waveforms along dendro-dendritic processing networks.

Figure 4- A periodic additive sink. On the dendritic tree of a neuron with periodicity specified by spatial frequency of the dendro-dendritic branching into the neuron.

Another relevant suggestion which results from this approach is that a major part of the problem of neural coding lies probably on anatomic, that is, functional structure of neural channels and operators. Successions of action potentials modulate the activity of the cell body of neuron but also their dendritic trees where specific geometry and distribution of synaptic excitatory and inhibitory weights as well as the characteristic distribution of their thresholds is most relevant to determine the computational functions of the dendritic and neural net. Local time references, delay in synaptic transmission as well as the geometry of the operative net render messages formed by action potentials not interpretable if they are examined out of the context of the generating structure for these messages as well as the receiver's structure and operation and how those two types of operations are related.

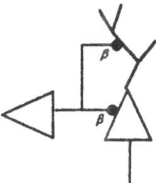

Figure 5- An amplifying factor β which contributes to dendritic signals with a bias β due to somato-dendritic or else to axo-dendritic propagation of excitation. Backward and forward excitation and inhibition may result from axo-dendritic or dendro-dendritic interaction.

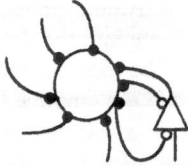

Figure 6- A dendro-somatic interaction of dendrodritic circuits and the cell body of a neuron

III. Some computational hypotheses about dendritic and integrated neural nets

The preceeding considerations can be summarized under a minimal approach by the following structures.

Let $N_1,...,N_n$ be a set of neurons interconnected by axons and dendrits, with connecting activating and inhibiting synaptic junctions with dimension, each one, t_{ij}.

Each neuron cell body state of activation 1 or **rest, 0**, will be determined by

$$V(N_k) = \sum_{j=1}^{s} t_{jk} \ V_j \ > \ L_k$$

with values 1 or 0 for the cases in which $V(N_k)$ is greater than L_k or $V(N_k)$ is smaller than L_k, respectively.

Furthermore let us assume dendritic circuits, DC, with x_i synapses spaced by spatial intervals $I_{i,i+1}$ between two contiguous synapses i and i+1.

The simultaneous excitation at all synaptic nodes with a time limited solicitation defines a waveform with its final limit before the next synaptic junction is attainable through cable propagation, for all of them. After this solicitation the state of the DC circuit will be given by

$$V(x_i) = a_{ij} t_{ij}^{0} + a_{i+1,j+1} t_{i+1,j+1}^{1} + a_{i+2,j+2} t_{i+2,j+2}^{2} + \cdots$$

a polynomial with terms defined for a fixed decay time for each excitation with a delay interval for the conduction of excitation in that dendritic segment. For a sufficiently high spatial frequency of synapases a_{ij}, all periodic function of frequency $F(w)$ will be adequately defined by $F'(a_{ij})$ with $a_{ij} \geq 2w$, w greater than the highest frequency contained in $F(w)$ according with the Nyquist sampling theorem.

Also

$$\varphi_{ij}(\tau) = \int_{0}^{T} f_i(t) \ f_j(t+\tau) \ dt$$

will allow the convolution between two functions which are represented in two dendritic circuits which contact at a single node i. Under inactivation of input synapses the damping factor of conduction will allow the system relax to 0. Furthermore, the independence of **DC** oscillations relatively to cell body excitation will assure the stabilization of cell bodies which will relax into **0**. In case the array $f_{DC}(t)$ assumes a pattern of weights which can activate cell bodies, then a pattern Γ_i of active states of neurons N_i will result until $f_{DC}(t)$ is damped to values in all its components $a_{ij} \, t_{ij}$ less then L_i.

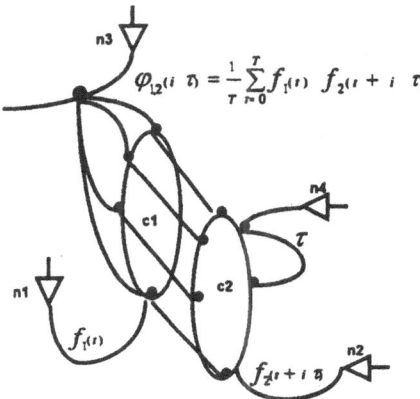

$$\varphi_{12}(i \;\; \tau) = \frac{1}{T}\sum_{t=0}^{T} f_1(t) \; f_2(t + i \;\; \tau)$$

Figure 7- The dendritic circuits needed to calculate cross correlation between two functions f1(t) and f2(t) which specify dendritic messages which are propagated from dendritic trees of neurons n1 and n2 repectively. The lag of f2(t) is produced by a delay loop τ. Multiplicative interaction is produced by iterative additions over a set of neurons n3. Autocorrelation φ_{11} is immediately represented making the input of the two dendritc circuits c1 and c2 equal to f1(t).

A cell body excitation, **B**, which propagates along the dendritic ramifications will assure that the excitation which was circulating along a DC_j

$$V (DC_{\;j}) = B(a_{ij} \, t_{ij}^{0} + a_{i+1,j+1} \, t_{i+1,j+1}^{1} + a_{i+2,j+2} \, t_{i+2,j+2}^{2} + \ldots)$$

for the time which **B** lasts such that **B** $a_{ij} \, t_{ij} > L_i$, where L_i is the minimal excitation required for synaptic transmission.

Memory will depend on the damping factors acting on **B** and $V(DC_i)$ but also in the implementation of structural fixed synaptic connections with strength t_{ij}. The proposal of J. Simões da Fonseca and McCulloch, 1967 and J. Simões da Fonseca, 1968 allow the obtention of a solution for any sequence of values **0, 1** circulating in a closed loop in a way such that the elicitation through a single activating node in a modified resting **DC** will provide a stored pattern of **0s** and **1s** in a structural form.

These examples are intended as an attempt to produce a minimal model which allows other forms of neural processing as well as encoding and decoding which result from present knowledge on dendritic interations in real neurons.

IV. Final Remarks

Another domain of representation results immediately from the use of Pseudo-Boolean equations and inequalities (J. Simões da Fonseca, 1968). In this approach in which variables assume values 1 or 0 and coefficients belong to the field of real numbers it is possible to represent the circuit functioning conditions simultaneously at the analogical and logical levels. Perceptron like circuits result if we consider excitatory and inhibitory synapses distributed along dendritic trees. This subcellular operators have not been included due to the fact that oscillatory behaviour in dendritic closed loops may contribute to represent some psychological operations in a manner close to neurophysiological data and clarify some issues about the biological support for subjective experiences- for instance the distinction between sounds of high pitch which require frequencies not attainable by action potentials in the cell. This problem have been extensively discussed in a communication (Isabel Barahona da Fonseca, J. Simões da Fonseca, In Press) and described in two communications in 1993 (J. Simões da Fonseca, 1993;Isabel Barahona da Fonseca, 1993).

References

BARAHONA DA FONSECA, J (1985a): Technical Note. *Acta Psiq. Port.* 31(3).

BARAHONA DA FONSECA, J (1985b): *Sistema de Aquisição e Processamento de Potenciais Evocados*. Tese de Licenciatura, IST.

BARAHONA DA FONSECA, J; SERRO, J; HORTA, MP; GARCIA FERNANDEZ, I; FERREIRA, MF; SIMÕES DA FONSECA, JL (1990): A Minimal System for the Study of Relationships between Brain Processes and Psychological Events. In F. Pichler, R. Moreno Diaz (Eds) *Computer Aided Systems Theory Eurocast'89*. Springer-Verlag.

BARAHONA DA FONSECA, Isabel (1991): *Os Fenomenos Oscilatorios do EEG como Indicadores de Eventos Psiquicos*. M. Sc. Thesis, Fac. of Psychology and CE, Lisbon.

BARAHONA DA FONSECA, Isabel (1993): High Level Implications of Dendro-Dendritic Computation. Communication presented to the *Fourth Las Palmas Seminar on Computer Sciences (Dir. R. Moreno Diaz)*. Univ. De Las Palmas.

BARAHONA DA FONSECA, Isabel, BARAHONA DA FONSECA, J, FERREIRA, MF, SIMÕES DA FONSECA JL (1988): Indicators of Stimulus Free Cognitive States and Their Proper Time Structure. *Acta Psiq. Port., 34(2,3,4)*.

BARAHONA DA FONSECA, Isabel; BARAHONA DA FONSECA, J; SIMÕES DA FONSECA, JL (1991): As Formas de Onda de Fenómenos Eléctricos Periódicos Cerebrais como Indicadores Distintivos de Características Semânticas. *Actas das I Jornadas de Estudos de Processos Cognitivos,* SPP, Lisbon.

BARAHONA DA FONSECA, Isabel; SIMÕES DA FONSECA, JL (In Press): Modos de Simbolização no Sistema Nervoso Central.

GRAUBARD, R; CALVIN, WH (1979): Presynaptic Dendrites: Implications of Spikeless Synaptic Transmission and Dendritic Geometry. *Neurosciences, Fourth Study Program,* Mit Press, Cambridge Mass:317-331

LEE (1960): *Statistical Theory of Communication*. J. Wiley, NY.

PEDLEY, TA; TRAUB, RD (1990): Physiological Basis of The EEG. In DD Daly, TA Pedley (Eds): *Current Practice of Clinical Electroencephalography* Raven Press, NY.

PERKEL, D. H. ; BULLOCK (1968): Neural Coding - A Report Based on a NRP Session. *Neurosciences Research Program Bulletin (6,3)*.

SIMÕES DA FONSECA, JL (1967): What is the Purpose of Delay in Nets of Real Neurons. *Investigation Progress Report 2B*. Centro de Estudos Egas Moniz, Lisboa.

SIMÕES DA FONSECA, JL (1968): *Invited Discussion in Symposium on Biocybernetics of the CNS*. L. Proctor (Ed) Little Brown, Boston

SIMÕES DA FONSECA, JL (1993): Dendro-Dendritic Computation. Communication presented to the *Fourth Las Palmas Seminar on Computer Sciences (Dir. R. Moreno Diaz)*. Univ. De Las Palmas.

SIMÕES DA FONSECA JL, McCULLOCH WS (1967): Synthesis and Linearization of Non-Linear Feedback Shift Registers - Basis of a Model of Memory. *Quarterly Progress Report, 86* Research Laboratory of Electronics, MIT; June, 15, pp 367.

Stochastic Neuronal Models with Realistic Synaptic Inputs and Oscillatory Inputs

Pavel Hruby

Technical University Brno, Faculty of Electrical Engineering and Computer Science
Bozetechova 2, 612 66 Brno, Czech Republic

Abstract

The Poisson process driven stochastic models of the neural activity and their diffusion approximation are studied. Two main studies are presented here: stochastic models driven by nonhomogeneous Poisson process with oscillatory intensity, and double compartment model with realistic synaptic inputs. The "phase lock" and the "amplitude lock" behaviour was observed in the model with oscillatory inputs and strong dependence on the initial phase after reset the membrane potential. Introducing the realistic synaptic input to the stochastic models opens new class of neuronal models: it has significant influence on all statistic parameters and the model behaviour. The double compartment model with realistic synaptic inputs is able to produce the bursting activity and this mechanism is described.

1. Introduction

The main reason of stochastic models of neural unit activity is to describe or to explain some characteristics of the action potential trains, usually recorded from the live neurone cell. The membrane potential of neurone cell is described by one-dimensional stochastic process X(t), t≥0. The random excitation and inhibition synaptic inputs evoke the increasing and decreasing changes in the membrane potential, what is the trajectory of X(t). When X(t) for the first time exceeds the threshold level S, the neurone produces an action potential. Then the process X(t) is reset and starts again from the value X(0). This is why we can describe neural firing as the first passage time of the stochastic process X(t). Then the interspike interval (ISI) is the random variable T given by relationship:

$$T = \inf\{t \geq 0; X(t) > S\} \tag{1}$$

We can get different stochastic neural models by describing the process X(t).

1.1. Stein's model

Stein's neural model is described by the equation (1), where the stochastic process X(t) is given by the equation:

$$dX = -\frac{1}{\tau} X \, dt + a \, X \, dN^+(t) + i \, X \, dN^-(t) \tag{2}$$

where $X(0)=x<S$ is the initial value, $\tau>0$ is the membrane time constant, a and i, $a>0$, $i<0$, are magnitudes of synaptic activation steps, $N^+(t), N^-(t)$ are two independent homogeneous Poisson processes with initial values $N^+(0) = N^-(0) = 0$ and with intensities λ and ω, that describe level of the excitatory and inhibitory synaptic activation.

The basic Stein's model has constant synaptic activation magnitudes a and i. Synaptic activation appears at random time intervals with exponential distribution and independently for excitatory and inhibitory synapses. The excitatory synaptic input causes the instantaneous increase of the membrane potential by a and the inhibitory input causes the instantaneous decrease by i. During intervals between excitatory or inhibitory synaptic activation the membrane potential decays to zero with relaxation time constant τ.

1.2. Stein's model with reversal potentials

Stein's model with reversal potential is more realistic modification of Stein's model. The magnitudes of synaptic activation steps are dependent on the action potential. The model is given by the equation:

$$dX = -\frac{1}{\tau} X dt + a (V_E - X) dN^+(t) + i (X - V_I) dN^-(t) \tag{3}$$

where $V_E > S, V_I < x_0$ are the excitatory and inhibitory reversal potentials, and other constants have the same interpretation as above. Because of the reversal potentials, the magnitudes of activation steps decrease linearly, as X approaches the boundaries V_E and V_I.

1.3. Diffusion approximation of Stein's model

The Stein's model has discontinuous trajectories, what makes its mathematical treatment difficult. If the intensities λ and ω of input Poisson processes are sufficiently high and magnitudes of synaptic steps a and i are relatively small, the diffusion process is good continuous approximation of Stein's model. The scalar diffusion process X can be described by the equation

$$dX = \mu(X,t)dt + \sigma(X,t)dW \tag{4}$$

where $\mu(X,t)$ and $\sigma(X,t)$ are real-valued functions, $X(t), X(0) = x_0$ represents the membrane potential of neurone cell, W is the standard Wiener process. Stein's model and its diffusion approximation are related by equations:

$$\mu = \lambda a + \omega i$$
$$\sigma = \lambda a^2 + \omega i^2 \tag{5}$$

under conditions $\lambda \to \infty, \quad \omega \to \infty, \quad a \to 0, \quad i \to 0$.

W get different diffusion models of the neural activity by describing the functions $\sigma(X,t)$ and $\mu(X,t)$. Wiener process and Ornstein-Uhlenbeck process are two main diffusion approximations of the Stein's model.

Wiener process is a diffusion approximation of perfect integrator or "integrate and fire" model, or Stein's model without reversal potentials, where $\tau \to \infty$. Wiener process has constant $\mu(X,t) = \mu$ and $\sigma(X,t) = \sigma$. The neurone model is then described by equation (1) and the equation (6):

$$dX = \mu \, dt + \sigma \, dW \tag{6}$$

where symbols have the same meaning as in the equation (4).

Ornstein-Uhlenbeck model is an approximation of "leaky integrator model" or Stein's model without reversal potentials.

Ornstein-Uhlenbeck process has $\mu(X,t) = \frac{1}{\tau} + \mu$ and $\sigma(X,t) = \sigma$. The neurone model is then described by equation (1) and the equation (7):

$$dX = (\frac{1}{\tau} + \mu) \, X dt + \sigma \, dW \tag{7}$$

1.4. Models with oscillatory input

Models with oscillatory input are based on assumption that synaptic inputs are non homogeneous Poisson processes that intensities can vary in time. The intensity of excitation input λ in equation (1) vary according to:

$$\lambda(t) = \lambda_1 + \lambda_2 \cos(2\pi \nu t) \tag{8}$$

where $\lambda_1 > \lambda_2$ and υ are constants and intensity of inhibition input ω is constant.

Numerical simulation reported here was concerned on three models:

1. Model with sinusoidal excitatory input, without the reset the phase of sinusoidal input after each spike.

2. Model with reset the input phase at time instants of action potential.

3. Ornstein-Uhlenbeck process with oscillatory input and with the constant variance σ.

1.5. Models with realistic synaptic potential

The excitatory and inhibitory synaptic inputs in Stein's model are step functions or jumps, that increase or decrease the membrane potential. The realistic synaptic potential is modelled by alpha function $A_i(t - t_i)\exp(-t/\tau)$, where t_i is the time instant of the synaptic activation. The time constant τ is in range from several to several hundred milliseconds, depending on synapse type. This time is longer than the mean interspike interval in many cases. We were interested in the difference between the original Stein's model driven by step functions and the Stein's model driven by realistic synaptic inputs and if the simplification of synaptic input into the step function is not too strong.

In the model submitted here the neurone is divided into two compartments: one is the dendritic tree and other one is the neurone body with axon. The action potential appeared on the body and the axon does not reset the synaptic potentials on the dendritic tree. It means that after each spike some level of synaptic potential can remain on the dendritic tree of the neurone.

2. Methods

All simulation programs were written in Matlab with Simulink and run on PC 486 and HP 9000 machines. Results of each simulation are saved in Matlab file containing spike times, values of amplitude and phase of sinusoidal modulation at spike time points, and the average amplitude and average phase of modulation signal.

Due to increasing of computational speed two special algorithms were used: the "method of thinning" for the non homogeneous Poisson process (Lewis 1978) and the "differential equations method" for the problem of summation of synaptic potentials (Bernard 1994).

3. Results

3.1. Neuronal models with oscillatory input

We used the same basic parameters of neuronal models as in references Musila (1990), Tuckwell (1978, 1979), due to possibility of comparing results with literature. We used sinusoidal modulation of the intensity of excitatory and inhibitory activation inputs with the period in range from 1 to 40 ms and amplitude from 0.08 to 1 mV. Under these settings we observed almost no difference between the Stein's model driven by Poisson processes and it's diffusion approximation by Ornstein-Uhlenbeck process and Ornstein-Uhlenbeck process where σ was constant. The interspike interval histogram influenced by modulation is non monotone and this effect increases with increasing amplitude and period of modulation signal. The period of the histogram peaks approximately corresponds to the signal period and the best correspondence appears when the modulation period is close to the value of mean ISI of the spontaneous activity without modulation, Fig. 1.

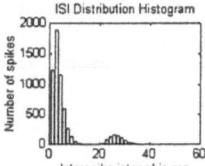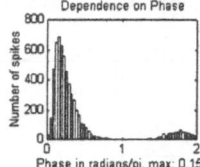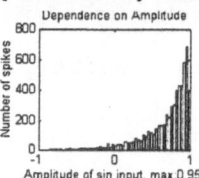

Fig. 1.: Parameters: Ornstein-Uhlenbeck process with cosinus with reset; phase = 0, period = 30 ms; amplitude = 1; file = win61h.mat. Statistics: Mean ISI = 6.565; median ISI = 3.5; std = 8.194; CV = 1.248; peaks at 2.81, 25.9, 55.1±0.7; total number of spikes = 6092.

We can observe two dependencies from histograms, describing the relationship between spike time and phase and amplitude of modulation signal. First, the "continuous" dependency, that expresses the fact that if the amplitude of modulation signal is higher, the probability of spikes is also higher. This effect appears in all frequency ranges and similar result is valid also for the phase histogram. There is no direct relationship between phase and amplitude histograms: the phase histogram is bimodal when the cosinus initial phase is 0 or π, what is not visible on the amplitude histogram, and the maximal histogram values on both histograms do not strictly correspond (Fig. 1.).

Second dependency is the "discrete dependency", appeared when using high modulation frequency (period less then 1/2 of mean spontaneous ISI). Both the phase and amplitude histograms have "discrete character", neurone fires only at certain values of amplitude or phase (Fig. 2.). We can partly explain this fact by existence of relative refractory period in evaluated models.

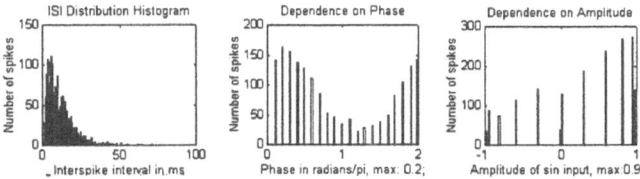

Fig. 2. Parameters: Ornstein-Uhlenbeck process with cosinus without reset; period = 1 ms; amplitude = 1; file = y61614.mat. Statistics: Mean ISI = 11.57; median ISI = 9.3; std = 8.68; CV = 0.75; peak at 4.14±0.35; total number of spikes = 1728, other parameters as above.

The relationship of spike time and phase of modulation signal is reported as the "phase lock" in the literature (Lansky, Tuckwell). According to my opinion it could be better to report on the "amplitude lock" because of following reasons: (1) the amplitude histogram is more sensitive for small amplitudes of modulation signal (Fig. 3.), (2) the amplitude lock is "more invariant": the amplitude histogram has sharp maximum that always corresponds to maximal amplitude of modulation signal, while position of phase histogram maximum depends on modulation frequency and on the initial phase in models with reset. The maximum of phase histogram is not so sharp, too, comparing to amplitude histogram.

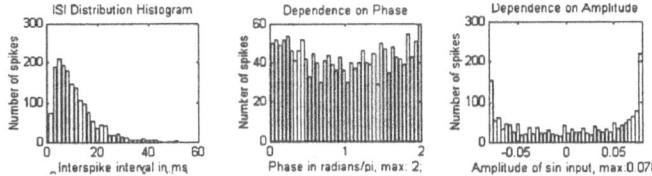

Fig. 3. Parameters: Ornstein-Uhlenbeck process with cosinus without reset; period = 20 ms; amplitude = 0.08 mV; file = y6149.mat. Statistics: Mean ISI = 11.56; median ISI = 9.2; std = 7.77; CV = 0.74; peak at 4.88±0.8; total number of spikes = 1729, other parameters as above.

All models with oscillatory input phase reset after spike are dependent on value of initial phase after reset. This dependence is stronger for modulation signal with long period and this effect decreases with increasing modulation frequency (Fig. 4.). For modulation periods less then one half of mean spontaneous interspike interval there is almost no difference in ISI histogram, mean and CV for different initial phases after reset.

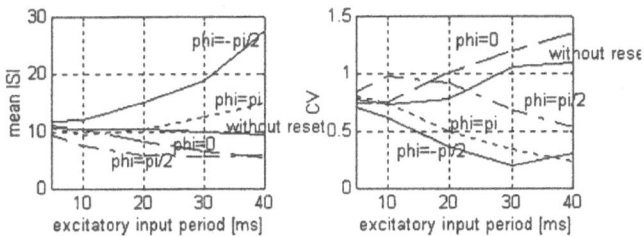

Fig. 4. Dependence of the mean ISI and the coefficient of variance on the modulation signal period and the initial phase after reset.

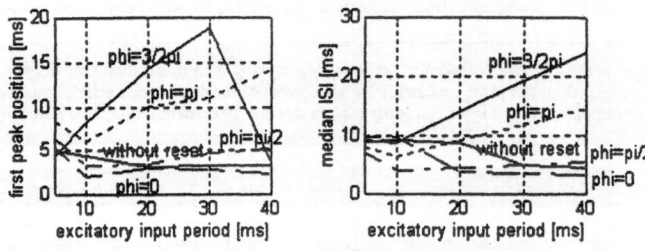

Fig. 5. Dependence of the first peak position and the median ISI on the modulation signal period and the initial phase after reset.

The phase and amplitude histograms refer about the value of phase or amplitude at the spike time instant. The models with the initial phase reset also change the average phase and amplitude over the simulation period that differs from the π, and 0, resp. The Fig. 6. shows the dependence of the average amplitude and phase on the modulation signal period.

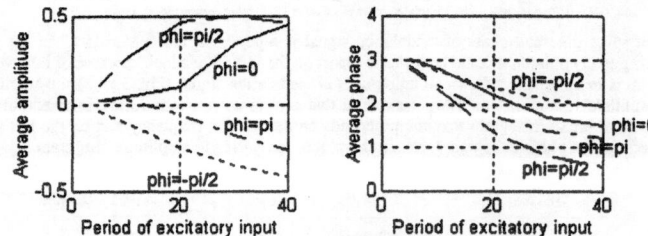

Fig. 6. Dependence of the average amplitude and phase on the modulation signal period.

3.2. Neuronal models with realistic synaptic input

We used the realistic synapse independently for excitatory, inhibitory and both excitatory and inhibitory inputs. In the first study we used the same set of parameters as for the Stein's model above, and the same time constant both for excitatory and inhibitory synapses, as well. The physiological range of chemical synaptic time constants is approximately from 0.2 ms to 150 ms. With model parameters by Tuckwell (1979) we were not able to calculate histograms for synaptic time constants higher than 5 ms (for which the mean ISI was 1260 ms) because of the computational speed, but it is clear that the time constant of realistic synapse is very important parameter of stochastic neurone models (Fig. 7).

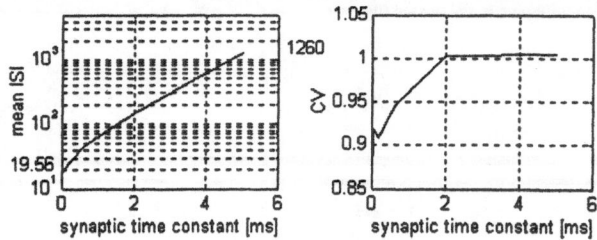

Fig. 7. Dependence of the mean ISI and the coefficient of variance on the synaptic time constant

In the second study we were changing the model parameters and we used the realistic synapse either for excitatory or inhibitory input. The "influence" of excitatory or inhibitory synaptic time constant in the model without reversal potentials depends approximately on the values $a\lambda_e$, $i\lambda_i$, resp.

The most important result following from experimenting with this model is, that under certain conditions we can observe the bursting activity of neurones (Fig. 11.). The bursting activity typically appears if the synaptic time constant is higher than the mean interspike interval and the threshold level is smaller, or very near the number $a\lambda_e - i\lambda_i$ (in the model without reversal potentials). We can explain the mechanism of bursting in the following way: the realistic synapse slows down the changes of membrane potential, so that periods with high excitation on the dendrite membrane are much longer than the mean interspike ISI. In these periods of high excitation the neurone can fire several times. If the number $a\lambda_e - i\lambda_i$ is higher than threshold S, the spikes in bursting periods are almost regular (the same distances). Both excitatory or inhibitory synapses with long time constant can cause the bursting activity. Long synaptic time constants are quite often in real neurones; for example GABA B has time constant about 150 ms. The bursting effect of such a cell depends also on synaptic "weight" $i(X - V_i)$ and on neurone spontaneous firing frequency.

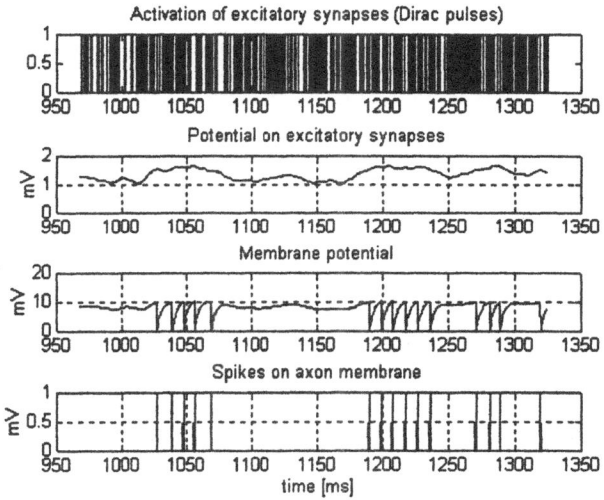

Fig. 8. The mechanism of the bursting activity of neurones caused by realistic synapses

4.   Conclusions

a. The "amplitude lock" and the "phase lock" were observed in models with oscillatory input. The continuous dependency on amplitude, resp. phase changed to the "lock" for the modulation period less than one half of mean spontaneous interspike interval. The "amplitude lock" has higher influence on ISI histogram than "phase lock", mainly when modulation amplitude was very small.

b. There is almost no difference between Stein's model and Ornstein-Uhlenbeck process and Ornstein-Uhlenbeck process with constant σ, when using basic parameters by Tuckwell (1979, 1980) and Musila (1990), for wide range of modulation frequencies and amplitudes.

c. Using realistic synapse instead of step function in the Stein's model has very strong effect on the model behaviour and opens the new class of neuronal models.

d. Stein's model without realistic synapse is sometimes used for estimation of the neurophysiological or neuroanatomical parameters, like level of synaptic activation from the ISI histogram (Tuckwell 1978). Because of the high influence of the realistic synapse time constant this using of Stein's model is questionable.

e. Stein's model with realistic synaptic potentials can be set to produce the bursting activity. In literature the mechanism of bursting was modelled either by using the conductance models with dependent ionic currents, or by certain neuronal connections in simple neural networks. This model shows that the realistic synapse itself can also cause the bursting activity, if we separate the dendrite compartment and do not reset it after each spike as the axon membrane.

f. Bursting behaviour can also be observed in the model without exponential decay of membrane potential ("perfect integrator"). This result presages that the most important change is the separation of the neurone to the dendrite compartment and the body and axon compartment. This fact opens possibilities for further simplification of the model, or, on the other hand, for the more exact membrane description with using conductance neuronal models, derived from Hodgkin and Huxley equations.

References

Bernard C., Ge Y. C., Stockley E., Willis J. B., Wheal H. V.: Synaptic integration of NMDA and non-NMDA receptors in large neuronal network models solved by means of differential equations, Biol. Cyb. 70, 267-273 (1994)

Holden A. V.: Models of the Stochastic Activity of Neurons, Springer Verlag 1976

Kloeden E. P., Platen E.: Numerical Solution of Stochastic Differential Equations, Springer Verlag 1992

Lansky P., Rospars J. P., Valliant J.: Some Neuronal Models with Oscillatory Input, 1994

Lansky P., Smith C. E., Ricciardi L. M.: One-dimensional stochastic diffusion models of neuronal activity and related first passage time problems, Trends in Biological Cybernetics 1

Lewis P. A. W., Shedler G. S. :Simulation methods for Poisson processes in nonstationary systems, IBM research report 1978

Longtin A., Bulsara A., Pierson D., Moss F.: Bistability and the dynamics of periodically forced sensory neurons, Biol. Cyb. 70, 569-578, 1994

Musila M., Lansky P.: Simulation of a diffusion process with randomly distributed jumps in neuronal context. Int. J. Biomed Comput 31, 233-245, 1992

Musila M., Stochastic models of neuronal activity, Czech Technical University, PhD thesis 1990 (in Czech language)

Tuckwell H. C., Richter W.: Neuronal interspike time distributions and the estimation of neurophysiological and neuroanatomical parameters, J. Theor. Biol 71, 167-183, 1978

Tuckwell H. C.: Synaptic transmission in a model for stochastic neural activity J. Theor. Biol 77, 65-81, 1979.

Tuckwell, H. C.: Stochastic Processes in Neurosciences, Monash University 1989

Wilbur J. W., Rinzel J.: A theoretical basis for large coefficient of variation and bimodality in neuronal interspike interval distributions J. Theor. Biol 105, 345-368, 1983.

A Neural Paradigm for Controlling Autonomous Systems with Reflex Behaviour and Learning Capability

G. Joya(*) and F. Sandoval(**)

(*) Dept. Arquitectura y Tecnología de Computadores y Electrónica.
Universidad de Málaga, Plaza El Ejido s/n, 29013 Málaga, SPAIN
E-Mail: joya@tecmal.ctima.uma.es.

(**) Dept. Tecnología Electrónica.
Universidad de Málaga, Plaza El Ejido s/n, 29013 Málaga, SPAIN
E-Mail:sandoval@tecmal.ctima.uma.es

Abstract

In this paper we present a neural paradigm for controlling the reflex behaviour of autonomous systems which are able to modify their behaviour by interaction with the environment. This paradigm incorporates the ideas expressed by Russell [1] about how to model the living being's reflex behaviour. In this paradigm a new type of connection is introduced: the so called high order Or connection. Learning is local and unsupervised,i.e., the change in the weight of a connection takes place as a consequence of its activation. We present two functions to update the weights which incorporate the forgetting capability. Some topologies have been simulated to provide the basic capabilities such as inhibition, stimuli association an reinforcement.

1.- INTRODUCTION.

In 1913, S.Bent Russell [1] presented the design of a hydraulic device which simulated the reactions of a nervous system. The scheme of this system consists of sensitive and motor terminals interconnected by nervous channels. The signal, a discharge sent by a sensitive terminal, is propagated through a channel until it excites a motor terminal.

This model incorporated some basic characteristics of the living being's reflex behaviour, among which we can point out the following:

1.- Reflex behaviour is modified as a result of individual experience, in such a way that the same sensation can produce two different answers in different moments and vice-versa.

2.- Nervous system's susceptibility to a certain stimulus is related to the stimulus presentation frequency. The longer a stimulus takes to reappear, the smaller the nervous system's reaction capability will be.

Besides, this model allows the simulation of basic behaviour change mechanisms as inhibition, association of ideas or reinforcement.

In this paper we adapt the Russell's ideas to the field of neural networks, elaborating a neural paradigm specifically oriented to generate the reflex behaviour of an autonomous system with capability to modify its conduct as the result of the interaction with its environment.

This paradigm presents the following characteristics, which we consider crucial when distinguishing it from other neural paradigms already known [2].

1.- A distinction is not established between connections regarding its function as exciting or inhibiting. All the weights are positive and all the connections have the same nature.

2.- The capability for behaviour change is determined by the weight change algorithm as well as the network topology itself; so that a certain distribution of connections will allow the change of certain conducts and will prevent the change of other ones.

3.- The weight change algorithm is unsupervised, independent from the external behaviour of the system, local for every connection, and the same for all connections. Let us remark that the learning process of a neural network is often dealt with in terms corresponding to a student-teacher learning process. This treatment is not altogether correct, since when the teacher corrects the student, he is not acting directly over the student's neurons, and the connection between the teacher's stimulus (external) and the neuron stimulus (internal) is not known. In our neural system, the relation between an external stimulus and the weight change of a connection is perfectly defined: the weight of a connection changes whenever that connection is activated.

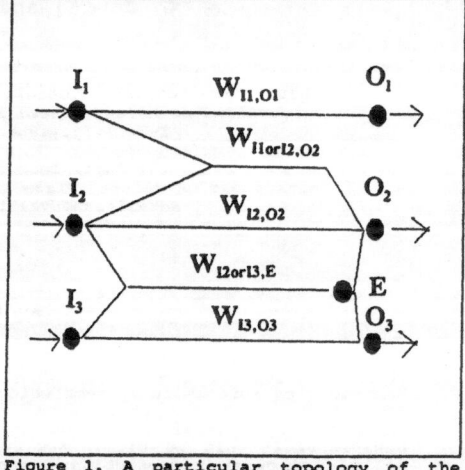

Figure 1. A particular topology of the neural paradigm

4.- The activation signal of a neuron is not transmitted as a whole to each one of its output connections, but it is distributed among these ones proportionally to their weight value.

In the remaining of this paper we describe each of the elements of our paradigm, as well as its learning algorithm (section II). We study, in particular, some topologies, which provide some basic learning capabilities such as inhibition, stimuli association and reinforcement (section III). Finally, we carry out a summary of results and conclusions.

2.- DESCRIPTION OF THE NEURAL PARADIGM.

2.1.- Description of the topology.

The neural network in figure 1 contains the most relevant elements of our paradigm, and will be useful for its description.

Three types of neuron are used:

a) Input neurons (I_1, I_2, I_3): they receive the signal coming from a sensorial terminal. Depending on the type of information collected, the activation function of these neurons will be threshold or sigmoid.

b) Output neurons (O_1, O_2, O_3): They produce the control module response. Its output is connected to the system motor organs. Its activation function will be sigmoid.

c) Intermediate neurons (E): they are used as nodes in a connection bifurcation. Its activation function will be linear.

The connections are of two kinds:

a) First order connections ($W_{I1,O1}$, $W_{I2,O2}$, $W_{I3,O3}$, $W_{E,O3}$, $W_{E,O2}$): they directly connect an output neuron to another input neuron.

b) High order OR connections ($W_{I1orI2,O2}$, $W_{I2orI3,E}$): they constitute the confluence of the outputs of two neurons in just one connection. The signal sent through this connection will be the addition of the signals sent by both neurons. These connections are essentially different from the high order connection Sigma-Pi which have appeared in the literature [3],[4], since in these ones the propagated signal is the product of the outputs of every neuron, in such a way that if one of them is not active the signal sent will be zero. With our nomenclature, these connections Sigma-Pi could be called High Order AND connections.

Each connection has an assigned weight W. The signal sent by a neuron is distributed between its output connections, proportionally to the weight of each one. For example, if $S(I_1)$ is the activation value of the neuron I_1, the signal sent by its connection with O_1 will be

$$\frac{W_{I1,O1}}{W_{I1,O1} + W_{I1orI2,O2}} \, S(I_1) \tag{1}$$

and the signal sent by its connection with O_2 will be

$$\frac{W_{I1orI2,O2}}{W_{I1,O1} + W_{I1orI2,O2}} \, S(I_1) \tag{2}$$

When two output signals are connected to motor organs, which provide opposite movements, the resulting movement will be the corresponding to the neuron with the greatest activation level.

2.2.-Description of the learning process

We consider learning as a change in the reflex conduct of the autonomous system, and it is a consequence of its individual experience.

In our systems, the reflex conduct before a certain set of stimuli is established by two factors: On one hand, the connections between input and output neurons (particular network topology). On the other hand, the weight value of each of these connections. Consequently, a change in the reflex conduct of the system will be possible if an adequate experience and network topology co-occur. Thus, as we will further see, in the network in figure 1 the connections between neurons I_1, I_2 and O_1, O_2, allow a change in the system conduct by inhibition, and the connections between neurons I_2, I_3 and O_2, O_3, allow a change in conduct by stimuli association.

The variation in the connection weights takes place in an unsupervised way. In general, the connection weight sharply increases if there is signal conduction, and decreases along the time of inactivity of the connection (forgetting capability).

We propose two functions describing the evolution of the weight of a connection from the moment this has been excited by a signal.

$$W_i(t) = W_i(t_0) + C_1 e^{-a(t-t_0)} \tag{3}$$

and

$$W_i = W_i(0)(1 - e^{-b(t-t_0)}) + w_i(t_0) e^{-b(t-t_0)} + C_1 e^{-a(t-t_0)} \tag{4}$$

$$\text{with } b << a$$

where $W_i(t)$ is the weight of the connection t seconds after activation, $W_i(t_0)$ is the weight of the connection at the time of its activation, C_1 is the weight increase produced by the activation, and a and b are the weight decay coefficients for connection inactivity.

The algorithm modelled in equation (3) produces an updating mechanism without return, i.e, the forgetting effect is partial, since each stimulus consolidates the weight value in that instant. Thus, even if the characteristics of the environment are modified, the system will not go back to its initial behaviour (see figure 2.a).

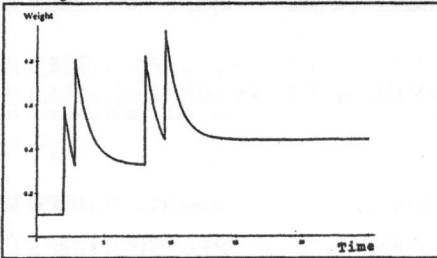

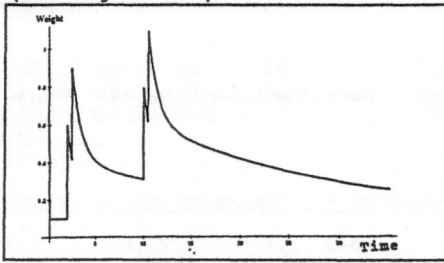

Figure 2.a: Weight evolution produced by the learning function (3).

Figure 2.b: Weight evolution produced by learning function (4)

The algorithm modelled in equation 4 allows a reversible updating mechanism. The forgetting mechanism is of such a kind that if the circumstances that motivated the change stop, the weight of the connection, and consequently the behaviour of the system, will be again the initial one (see figure 2.b).

3.- MODELLING SOME BASIC REFLEX CAPABILITIES.

In this section we study three elementary neural topologies. Each allows a basic behaviour change: inhibition, conditioned and unconditioned stimuli association, and reinforcement.

3.1.-Inhibition.-

The neural network in figure 3 represents a basic control module, capable of changing the system reflex behaviour by inhibition. We have the following initial conditions: $W_{I1,O1} >> W_{I1orI2,O2}$, $W_{I2,O2} >> W_{I1orI2,O2}$; in addition, we assume that O_1 and O_2 produce two opposite responses R_1 and R_2 (forward-backward movement, holding-loosing).

The environment will be of such a kind that the reaction R_1 produces the activation of I_2. Then, an activation of I_1 originates the following sequences:

The activation of I_1 --> signal conduction through $W_{I1.O1}$ and $W_{I1orI2.O2}$ --> reaction R_1 --> activation of I_2 --> signal conduction through $W_{I2.O2}$ and $W_{I1orI2.O2}$.

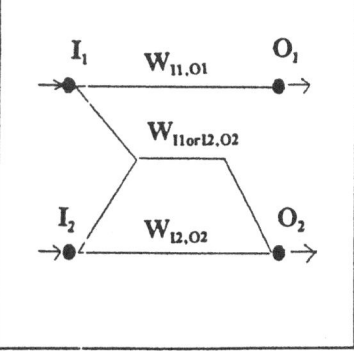

As we see, the connection $W_{I1orI2.O2}$ has been activated two consecutive times. Therefore, its weight will increase more than $W_{I1.O1}$ and $W_{I2.O2}$. If the process is repeated several times, $W_{I1orI2.O2}$ will exceed $W_{I1.O1}$ and $W_{I2.O2}$ and the activation of I_1 will produce the activation of O_2 (R_2) instead of O_1. In other words, the response R_1 has been inhibited. Figures 4.a and 4.b show the evolution of the outputs O_1 and O_2 for a system using the learning rule in equation (3). It could happen that a change in the environment make R_1 not to drive the activation of I_2. In this case, since the weight $W_{I1orI2.O2}$ remains greater than $W_{I1.O1}$, the system will not go back to its initial behaviour.

Figure 3. Neural topology for a system with inhibition capability.

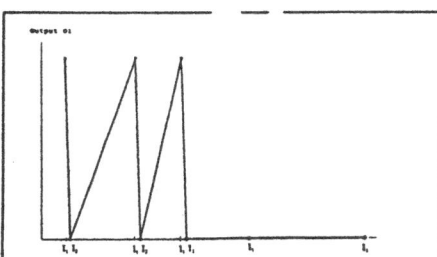

Figure 4.a: Evolution of output O1 for a particular sequence of stimuli.

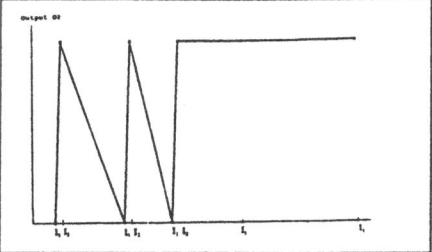

Figure 4.b: Evolution of output O2 for a particular sequence of stimuli

3.2. Conditioned and unconditioned stimuli association.

The neural network in figure 5 represents the control module of an autonomous system which is able to change its behaviour when the association of a conditioned stimulus (CS) with an unconditioned stimulus (US) happens. For the initial conditions $W_{I1.O1} \gg W_{I1orI2.E}$ and $W_{I2.O2} \gg W_{I1orI2.E}$, the activation of I_1 (CS) will produce the activation of O_1, and the activation of I_2 (US) will produce the activation of O_2 (response R_2). But if the occurrence of CS is always followed by the occurrence of US, the connection $W_{I1orI2.E}$ will be excited twice for each excitation of $W_{I1.O1}$. After a sufficient number of excitations, $W_{I1orI2.E}$ will be the greatest weight, so that the activation of I_1 will produce the response R_2 even if I_2 is not activated. In other words, both stimuli have been associated (see figure 6).

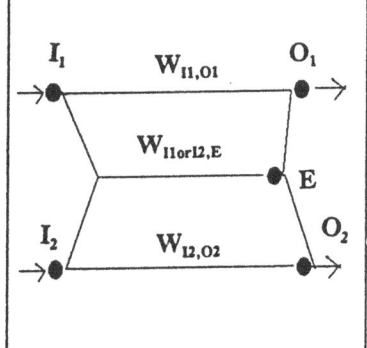

Figure 5. Neural topology for a system with stimuli association capability.

3.3.-Reinforcement.

Figure 7 shows the neural topology for a system with reinforcement capability. Let us assume that I_1 and I_2 are neurons connected to sensory terminals which are excited by the same kind of stimuli, but for reasons of physical location in the system, the stimulus will some times be received by the sensory terminal of I_1 and some others by the sensory terminal of I_2. In that case, for the initial conditions $W_{I1,O1}, W_{I2,O2} >> W_{I1orI2orRe1}, W_{I1orI2orRe2}$ the system response will be R_1 or R_2, according to the neuron has been activated, I_1 or I_2. However, if the activation of O_1

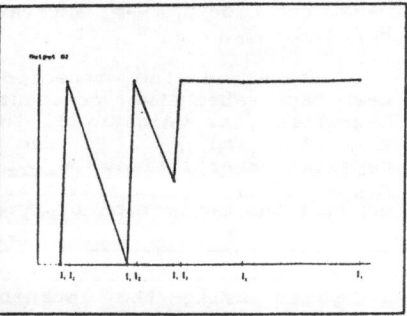

Figure 6: Evolution of output O2 for a particular sequence of stimuli.

(response R_1) is followed by the activation of reinforcement neuron Re1, the connection $W_{I1orI2orRe1}$ will be excited two consecutive times for every activation of I_1, and, after some repetitions of the process, it will reach the greatest weight. At this moment, the presentation of the stimulus will activate the neuron O_1 (response R_1), regardless of the sensory terminal activated. That is, the behaviour R_1 has been reinforced. Analogously, if the activation of O_2 leads to the activation of reinforcement neuron Re2, the reinforced behaviour will be R_2 (see figure 8).

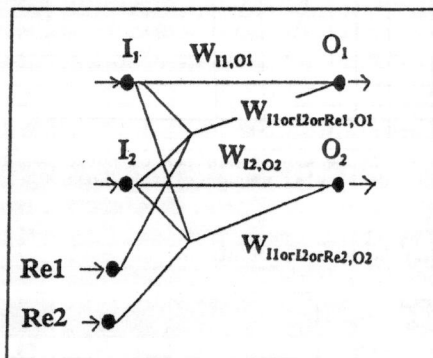

Figure 7. Neural topology for a system with reinforcement capability.

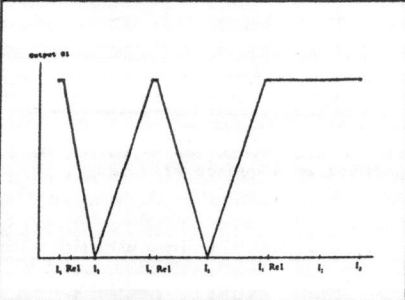

Figure 8: Evolution of output O2 for a particular sequence of stimuli.

4.- CONCLUSIONS

From Russell's ideas (1913) about the characteristics that a system simulating a natural nervous system must verify, we have generated a neural paradigm specifically oriented to the control of an autonomous system with reflex behaviour, which is able to change its conduct as a result of its individual interaction with the external environment.

Among the elements configurating the topology of this paradigm we point out a new type of connection that we call High Order OR connection, which means a generalization of the concept of high order connection that appears in the literature. Thus, the connection usually considered of high order, where the signal conducted is the product of two or more neuron outputs would become a particular case of high order connection; this connection could be called high order AND connection.

Other remarkable characteristics of this paradigm are:

All the connections are of the same nature, since it is not necessary to differentiate between inhibiting and exciting connections for achieving the inhibiting or reinforcing of a behaviour.

Learning is unsupervised, and we could call it "unconscious", since the modification of the weights has a physical cause located at the microscopic level of the neural system: the excitation of a connection. This modification of the weight neither can be controlled by an external agent nor by the system itself.

The paradigm we show fulfils at least five of the six organizational requirements of an artificial neural system attempting to be similar to a biological neural system [5], namely:

1.-Learning is local, since the modification of the weight of each connection is caused only by the excitation of that connection and there is not an external module in charge of producing that variation.

2.-The connections more frequently used increase their weights, whereas these decrease with the time of inactivity.

3.- The network does not start its learning from scratch, but in its topology has innate knowledge. The reflex behaviour of the system, once it is born, will be determined by its initial topology.

4.-The information the neural system receives arrives as signals from the external world, collected by sensory terminals, and the outputs of the neural network are connected to the motor organs of the autonomous system to put out the responses.

5.- Learning in each neuron takes place in an independent and heterogeneous way, since each connection increases its weight in a different instant time, as a consequence of a certain stimulus.

The reconfigurability and fault tolerance requirement will not be fulfilled if we design a network with a minimum connectivity, however, redundant connections could achieve a certain degree of fault tolerance.

We have described two functions for the learning process. Both incorporate the forgetting capability, but whereas in the first function (equation 3) it is partial, in such a way that the system will never go back to the initial behaviour, even if the causes that motivated its change of conduct would definitely disappear, in the second function (equation 4), the forgetting can become complete, i.e., if there is not a stimulus long enough in time, the system will recover its initial behaviour and will be able to be adapted to a new change in its environment.

Simulations of the behaviour of different topologies have been carried out.

Acknowledgement.

This work has been partially supported by the Spanish Comisión Interministerial de Ciencia y Tecnología (CICYT), Project No. TIC92-1325-PB

References.

[1].-   S. B. Russell, "A practical device to simulate the working of
          nervous discharges", Journal of Animal Behaviour, 3, (15)
          (1913), pp. 15-35.
[2].-   R.P. Lippmann, "An Introduction to computing with neural
          nets", IEEE ASSP Magazine, April 1987, pp.4-22.
[3].-   C.L. Giles and T. Maxwell, "Learning, invariance, and
          generalization in high-order neural networks", Applied Optics,
          26, (23), (1987), pp. 4972-4978.
[4].-   M.L. Minsky and S.Papert, *Perceptrons*, MIT Press,
          Cambrige,MA,1969.
[5].-   J. Mira, A.E. Delgado, J.R. Alvarez, A.P. Madrid, and M.
          Santos, "Towards more realistic self contained models of
          neurons: high-order, recurrence and local learning", in
          J. Mira, J. Cabestany, A. Prieto eds, *New Trends in Neural
          Computation*. Lecture Notes in Computer Science 686, Springer-
          Verlag, 1993, pp. 55-62.

Fast automatic architecture selection in RBF networks

A. M. González, C. Santa Cruz, V. López and J. R. Dorronsoro
Instituto de Ingeniería del Conocimiento (IIC) and
Departamento de Ingeniería Informática
Universidad Autónoma de Madrid
28049 Madrid, Spain

Abstract

Fast automatic methods of architecture selection are of great interest for use in local, dynamical modeling. A general procedure to select an optimal network architecture is proposed in the case of RBF nets. Taking as a starting point the universal approximation properties of RBF networks and the natural interpretation of their weights, a search for the simplest, best generalizing network is done. This requires some measure of goodness of fit, which allow to discard initial overweighted nets and to stop once too lean ones do not provide good generalization; in our case we will analyze both the error evolution and the statistical nature of the network residues. Since when going down in the number of network units fast network retraining is needed, we will also discuss this issue.

1 Introduction

Feedforward neural networks have known in the past years a widespread use in a variety of fields, specially in areas such as modeling, classification or prediction. The key of the applicability of neural networks in such problems is their excellent approximation properties. In fact, in mathematical terms, each one of the above problems can be ultimately stated as that of finding an unknown function f that maps a D dimensional vector of known data X into a target value y. The problem is then to build a good approximation $F(X, W)$ to f in such a way that we have $y = f(X) \approx F(X, W)$. In this formula, F is chosen usually within a family of functionally related maps, and W denotes a set of parameters (weights in neural network parlance) that establish the choice of a concrete function from that family.

The success of neural networks in this setting is due to the fact that they provide universal approximators, by which we mean that, given a reasonable f and a properly chosen family $F(X, W)$ of neural functions, no matter what a tolerance ϵ is given, we can find a concrete weight set W_0 such that, over a concrete modeling region $\mathcal{R}$, $\int_{\mathcal{R}} (f(X) - F(X, W_0))^2 dX < \epsilon$.

Most of these neural families are of the form $F(X, W) = \sum_0^N w_i \Phi(X, \Omega_i)$, where the weight set has the form $W = (w_0, \Omega_0, \ldots, w_N, \Omega_N)$, N any positive integer. The best known example of such a family are the perceptrons [10], for which we have $\Omega = (\omega, \beta)$, with $\omega \in R^D$, $\beta \in R$, and $\Phi(X, \Omega) = \sigma(\omega^t X + \beta)$, and where by σ we denote the sigmoidal activation function $\sigma(x) = 1/(1 + e^{-x})$.

Another well known and widely used family is that of the Radial Basis Functions (RBF) networks [7, 8], where now Φ is a D dimensional gaussian, and $\Omega = (C, s)$, with C the gaussian center and s is standard deviation. More precisely,

$$\Phi(X, \Omega) = e^{-\|x - C\|^2 / 2s^2}.$$

Both kind of networks have been widely used and both give universal approximation families $F(X, W)$ (see [3] for the perceptrons and [6] for the RBF networks). However, most of these applications deal with what we can call static networks, in the sense that the concrete network to be used is built once and for all; in particular, it usually takes a rather long time to settle upon a concrete network architecture, a decision that requires a fair amount of trial and error, and a rather subtle analysis to make the finally prevailing architectural choice.

Nevertheless, there is also a high number potential neural network applications in which these static networks are bound to fail. To choose a particularly interesting field, consider the modeling and forecasting of industrial plants and processes. A well known situation in that setting is [11, 12] the drifting of working conditions due often to noncontrollable factors. This drifting means, in particular, that a concrete data vector X can correspond to different y values in different time intervals. Thus, we no longer have to model a single unknown function f but rather a time varying such family.

The most natural way of coping with this situation is to shift to **dynamical**, or **local** neural models [13]: on each concrete setting of the plant a particular neural network transfer function F is used. But this means that two choices have to be made: first we have to decide in the concrete architecture to be used, which for the above examples essentially means fixing the number N of network units, and then we have to train the resulting networks to get the optimal weight set W_0. Moreover, for the networks to be really useful, these choices have to be made in a fast way. This clearly precludes the traditional approach to network training: choose different sets of random starting weights, train them and compare extensively the resulting networks. This approach has two drawbacks: the time it takes and the skill it requires on the persons supervising the training. The time requirements are usually unacceptable in industrial settings, and the needed supervising skills would hamper the extensive use of such neural systems and limit them to a very few installations.

The only acceptable alternative is to use automatic methods of network architecture determination coupled with fast training procedures. Then perceptron networks turn out to be poor candidates for use in local neural networks. The reason is that, although very impressive in modeling and generalization capabilities, it is very hard to interpret the role of individual units, and therefore, very difficult to decide upon the choice of initial weight values before training, and after it is complete, upon the advantages or the influence of a particular unit.

On the other hand, the natural probabilistic interpretation of the gaussian units in a RBF network makes both the choice of initial weights and the analysis of the influence of individual units easier to adapt to automatic procedures of network selection. This is the approach that we will follow here, the details of which we are going to sketch. First, the most natural way of arriving to a good neural architecture is by means of **network pruning**, that is, to start with nets of rather large size and to diminish their number of units until a good network is obtained. The reason why this is so is the nonconstructive nature of the universal approximation theorems for neural networks: they simply state the if a large number of units is provided, good approximation will result.

We will thus start with a large network and take advantage of the gaussian nature of a RBF network units to decide which units could be removed and how the remaining ones can be grouped, if possible, to obtain, after the appropriate retraining, a smaller network with better generalizing properties. This sequence of steps will be repeated until a no longer acceptable network is obtained.

Clearly, if this is to be done in a fast way, efficient training methods have to be used. On the other hand, the probabilistic interpretation of these gaussian units also suggests how to make good estimates for the initial weight values, which in particular, makes possible to perform just one training procedure to obtain a good network within a fixed architecture. In what follows, we will deal with both issues for RBF networks: parameter initialization and fast training will be discussed in section 2, and network pruning in section 3. Finally, in section 4 we will apply these considerations to a particular significative example.

2 RBF network training

Our objective in RBF training is to obtain good behavior of the very first network we train. For that to be so, it is clear that weight initialization is a crucial issue: we cannot simply make a random choice and hope for the best. Instead, we follow the ideas already proposed by Moody and Darken [4]; we assume that a concrete network size N has been decided. In the first place, notice that the gaussians making up a RBF network can be seen as providing a particular covering of the data region; in fact, the combination (C, s) of their parameters define a family of balls which concentrate the influence of their gaussian, and that, in turn, reflects a particular clustering of the data points.

Thus, a reasonable first choice for the C values are the centers provided by a particular clustering procedure. Many such procedures appear in the literature; one good candidate in our case is the so called K–means algorithm [2], whose aim is to divide the data set into K clusters for which the within–cluster sum of squares is minimized.

In any case, K–means is only used in the construction of the first network. For the other smaller ones, it would not make sense to throw away the knowledge about the data that prior larger networks have. Thus, for the selection of its initial centers, we take advantage of the work already done in a fashion that will be described below.

Once we have the gaussian's centers C_i, there also are a variety of methods to compute their deviations s_i. A good procedure is the P–nearest heuristic, in which s_i is chosen as the average of the distances from C_i to the P nearest centers, that is, $s_i = \left(\sum_j \| C_i - C_j \| \right) / P$. Finally, to complete the initial parameter choice, we have to select the initial coefficients ω_i, for which we just perform a linear regression of the target values y_i against the basis of gaussians $\Phi(X, C_i, s_i) = \exp(- \| x - C_i \|^2 / 2s_i^2)$.

Although a reasonable first choice, these parameters C_i, s_i and ω_i may not be optimal (even if in Moody and Darken's paper they are taken as such). Therefore, we have to train the network of N gaussians taking these values as initial weights. Again, a course of action could be to apply the traditional method of gradient descent. But, since we already start from a kind of minimizing weight set, we can expect to be close to an optimizing set. We can thus use a faster minimizing procedure, and a particularly well suited to our situation is the well known method of Levenberg–Marquardt.

A good description of this method can be found in [9]; we simply note that its good suitability for neural training stems from the fact that it alternates between what essentially is pure gradient descent when the procedure is still far from a minimum, and a faster second order algorithm when closer to it. In fact, in our experiments, either the method already started in its second order version, or just needed a few iterations to shift to it.

Summing things up, the combination of the initial weight selection with the fast training of the Levenberg–Marquardt algorithm, allows, once the network size is decided, the easy construction of a good transfer function. Hence, while pruning is being done, once a network size reduction has been decided, we can easily obtain a network of that size well suited to the data to be modeled. Now, it remains to establish criteria to size up a network's performance; this, and the procedures to go from a given network size to a smaller one will be discussed next.

3 RBF network pruning

We can now give what could be the pseudocode of our pruning algorithm:

```
while network performance is good:
```

1. throw away redundant units;

2. regroup remaining units if necessary;

3. reset initial values;

4. retrain the resulting network;

5. measure network performance;

Up to this point, we know how to perform steps 3 (for the initial net) and 4; we will deal with all others now.

First by redundant units we mean units that do not contribute significantly to the transfer function. Observe that during retraining some units could evolve to this kind of situation. Three such cases can be considered:

- units whose influence regions, as defined by their centers and deviations, are far from the training region $\mathcal{R}$;
- units with very small deviations, which correspond to very sharp gaussians and to locally distorted features of the training data (such as large pointwise noise components);
- units with very small a_i coefficients, which clearly correspond to noise contributions.

In the first case we discard those units Φ_i such that the distance δ_i between C_i and $\mathcal{R}$ is smaller than 2 times s_i. To get rid of sharp gaussians, we compare their deviation s against the quantity $\rho d(\mathcal{R})/K$, where $d(\mathcal{R})$ is the diameter of the training region, K is the number of actual units and ρ is a regularization parameter; we choose $\rho = 0.1$ in our experiments. Finally, we take the current mean square error ϵ as the threshold against which we compare the unit's coefficients; if they are smaller than 1.5 times ϵ, they are removed. Clearly, any unit verifying one of these conditions can be discarded without any hampering of the network performance.

In the initial stages of the pruning procedure, several units may be discarded in the above way, but this becomes rarer as smaller network sizes are considered. It is then necessary to regroup units, for which we select a distance threshold of the form $\eta d(\mathcal{R})/K$ (with η again a regularization parameter) and successively apply it to each center. If for a given one a number of other centers are at a smaller distance, all the units are coalesced into a single gaussian, whose center, deviation and coefficient is chosen in such a way as to minimize its square distance to the units it will replace. All the centers involved are then marked out and we proceed to deal with the remaining ones.

Here the η parameter does not have a fixed value but instead varies in fixed increments from a starting smallest value η_m. Notice that this unit regrouping should be done in a smooth fashion; this is precisely the role of the η parameter. We first try to regroup units within a starting minimal distance η_m and only if this fails to produce the desired effect is η actually incremented. The regrouping stops as soon in a pass over all the centers some units coalesce, and it yields for each gaussian starting centers, deviations and coefficients (either the old parameters or those computed during the regrouping). We have thus a new starting set of weights with which the new, smaller network can be retrained as described above.

In order to complete all the phases of the above pseudocode, we have to measure the performance of the resulting network $\mathcal{N}$. Perhaps the most obvious quantity is its mean square error

$$ e(\mathcal{N}) = \frac{1}{M} \sum_{1}^{M} |\hat{y}_i - F_{\mathcal{N}}(X_i)|^2, $$

where M denotes the number of training points and $F_{\mathcal{N}}$ the network transfer function.

Initially, with a large number of units, $\mathcal{N}$ not only approximates the correct values $y_i = f(X_i)$ but its noisy versions $\hat{y}_i$ in the data set. Thus, $e(\mathcal{N})$ will be quite small, falling below the noise variance σ_n. While the network reduces its size, the error will slowly grow up, even surpassing the noise variance but still being of the same order of magnitude. The corresponding networks will adequately approximate f even if their size can still be reduced without hampering their performance. However, after a critical size is passed, the network's performance sharply deteriorates.

A stopping criterion can thus be obtained by observing the square error evolution and detecting this change. Of course it is convenient to normalize in some way the successive error values so as to homogenize them. A possibility is to consider a kind of derivative in the following way. Suppose that during pruning we obtain from a size N network another one with N' units, and the associated errors are e and e'. We define then the quantity $\Delta = |e - e'|/|N - N'|$. Then Δ should remain quite small during most of the pruning procedure and then shot up sharply as soon as the network size is too small to offer good generalization. We can thus fix a threshold value Δ_0 and stop the pruning as soon as $\Delta > \Delta_0$.

Although it won't be developed here, a more subtle alternative stopping criterion can be derived by taking into account the nature of the noise present in the $\hat{y}$ values. Most likely it will be due to random factors affecting either the process itself or the measuring instruments. It is thus reasonable to suppose it gaussian in nature, that is $\hat{y}_i = y_i + \epsilon_i$, where the ϵ_i are samples of independent gaussians with 0 mean and a standard deviation σ_n.

In these circumstances, if a network transfer function F_N approximates correctly the function f to be modeled, the square error $e(N)$ associated to the data should be a good estimate of the variance σ_n^2 of the gaussian noise, and the random variable $Z_N = \frac{\hat{Y} - F_N}{\sqrt{e(N)}}$ should behave as a $(0,1)$ normal distribution.

Of course, when the network N has too many weights, the above approximation won't work, and the distribution Z_N will be far from being normal. Moreover, if the network has instead too few units, the same phenomenon will happen. Thus, if in each network size reduction we register the degree of fitting between Z_N and a $(0,1)$ normal, it should evolve from a starting minimum to another ending minimum and thus present a maximum somewhere in the middle of its evolution, associated to an optimal network architecture. Thus, we can complement the above presented analysis of the Δ values with the results of an appropriate statistical test (either a χ^2 or a Kolmogorov–Smirnov) of the fitting between Z_N and a $(0,1)$ normal, stopping the pruning either when the Δ values shot up or we pass through a maximum fitting and it starts to degrade.

We are already working on this approach, and more concrete conceptual and numerical discussion of this approach will appear in a forthcoming paper. However, as it stands, this approach would certainly result rather expensive computationally.

In any case, an analysis along these lines can be used to confirm the sound workings of the Δ approach. First, notice that each network's output is linear in the gaussian functions, and also optimal. Thus it may be regarded as the output of a linear regression procedure with respect to the data $\hat{y}_i$ and a basis of, say, N gaussians. In this setting it is well known [1] that, writing $\hat{D}_N = \sum_1^M \|\hat{Y}_i - F_N(X_i)\|^2$, $\hat{D}_N/(M - N)$ is an unbiased estimator for the noise variance σ_n, and that $\hat{d}_N = \hat{D}_N/\sigma_n$ also has a χ^2 distribution with $M - N$ degrees of freedom (recall that σ_n denotes the true noise variance).

Since in our examples (and in any realistic situation if overfitting is to be avoided), we will have that the number of points M is much larger than that of units N, the normalized version of the above $\hat{d}_N$ distribution, $Z_N = (\hat{d}_N - M + N)/\sqrt{2(M - N)}$, will approximately be a $(0,1)$ normal. If so, we can test the likelihood of getting concrete $\hat{z}_N$ values of the Z_N distribution by evaluating the probabilities

$$P_{\hat{z}_N} = P(|Z_N| > |\hat{z}_N|) = \frac{2}{\sqrt{2\pi}} \int_{|\hat{z}_N|}^{\infty} e^{-x^2/2} dx$$

Thus we can confirm the correctness of the behavior of the Δ values by comparing them against those of the above defined $P_{\hat{z}_N}$. For these, we can expect initial low values of $P_{\hat{z}_N}$ for large N networks, bigger values for networks with the appropriate size, and a progressive deterioration when that size is much too small. Our numerical examples of the coming section will confirm just that.

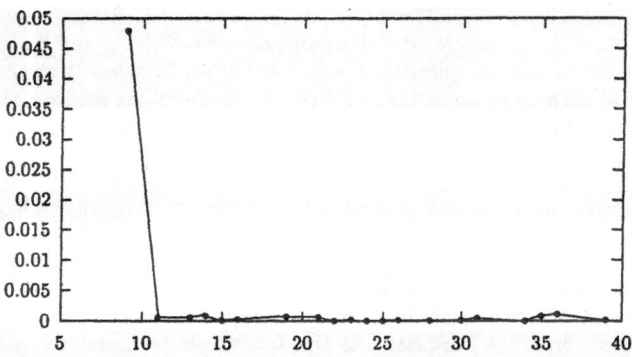

Figure 1: Evolution of the Δ values: they are almost zero for most of the evolution of the pruning procedure, and shoot up for too small networks (less than 11 units).

4 A numerical example

In order to test these ideas, we will work with a f function of the form $f(X) = \sin(2\pi x/25) + \sin(2\pi x/10)$, to which we have added a $(0,0.2)$ gaussian noise ν. This test function has been used as a benchmark in some other problems [5]. We obtained a 200 point test set by choosing uniformly randomly distributed points X_i in the interval $[0,100]$, computing the associated $y_i = f(X_i)$ and finally obtaining the $\hat{y}_i$ adding to them random samples of the distribution ν.

The starting network had 50 units, and therefore 150 free parameters. As mentioned above, after an initial clustering was performed, all the remaining networks were obtained by pruning of previous ones. As it can be seen of the evolution of the Δ values, they remain constant and very small for most of their evolution (figure 1).

Thus, these Δ values tell us that networks of size larger than 11 units will approximate well the unknown function. Of course, the smallest of these networks should be preferred.

However, when this is compared with the evolution of the $P_{\hat{z}_N}$ values, it is seen that the degradation starts somewhat earlier, when the network goes below 13 units. Clearly this is a reasonable situation: the too sharp shoot up of the Δ tells us degradation must have started earlier, even if we are not yet able to detect it. Seen under this light, the $P_{\hat{z}_N}$ behavior is quite reasonable and it suggests, as a rule of thumb, not to choose just the networks after which the rise of the Δ takes place but may be the network obtained just one or two steps before that jump.

Before finishing, we briefly point out that the relative low values (around 0.7) of the $P_{\hat{z}_N}$ quantities; usual values for χ^2 tests are around 0.90 or 0.95. This is explained by the fact that in their computation enters not a true gaussian error, that in this case would be $|\hat{Y} - f|$, but a somewhat different one, $|\hat{Y} - F_N|$. Of course, we don't thus get a true χ^2 distribution, but rather an approximation. This fact shows itself in the $P_{\hat{z}_N}$ values depicted above, which in any case are reasonably large when the best number of centers are used.

As a conclusion, we have proposed a simple method for automatic architecture selection, in which starting from a relatively large network, smaller ones are successively constructed and tested until a sharp degradation of network performance is detected. The underlying algorithms can be made to run with adequate speed and, although very easy to implement and to interpret, related statistical analysis of the performance of the networks constructed during the algorithm's evolution show that, under standard assumptions on the noise involved, the method shows a good ability to select adequate architectures.

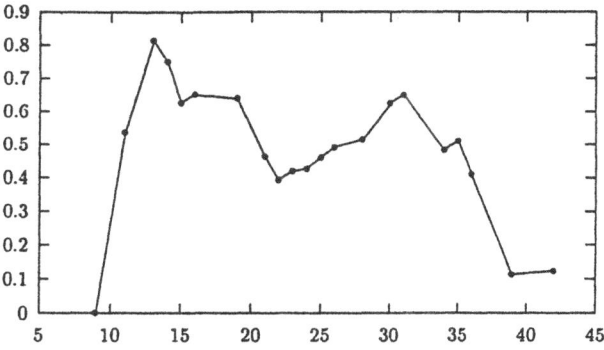

Figure 2: Evolution of the P_{i_N} values: they reach a maximum when 13 units are used and deteriorate sharply afterwards.

References

[1] H. Cramer, "Mathematical methods of statistics", Princeton University Press, 1946.

[2] R. O. Duda, P. E. Hart, "Pattern classification and scene analysis", Wiley, 1973.

[3] K. Hornik, M. Stinchcombe, H. White, "Universal Approximation of an Unknown Mapping and Its Derivatives Using Multilayered Feedforward Networks", Neural Networks, 3, 551-560, 1990

[4] J. Moody and C. J. Darken, "Fast Learning in Networks of Locally–Tuned Processing Units", Neural Computation, vol. 1, pp. 281–294, 1989.

[5] K. S. Narendra, K. Parthasarathy, "Identification and Control of Dynamical Systems Using Neural Networks", IEEE Trans. on Neural Networks, 1, 4–27, 1990.

[6] J. Park, I. W. Sandberg, "Universal Approximation Using Radial–Basis–Function Networks", Neural Computation 3, 246–257, 1991.

[7] T. Poggio and F. Girosi, "Regularization Algorithms for Learning That Are Equivalent to MultiLayer Networks", Sciences, vol. 247, pp. 978–982, 1990.

[8] M.J.D. Powell, "The Theory of Radial Basis Function Approximation in 1990", Numerical Analysis Reports DAMTP 1990/NA11, December 1990, Cambridge University.

[9] William H. Press, Brian P. Flannery, Saul A. Tekolsky, William T. Vetterling, "Numerical Recipes in C. The Art of Scientific Computing", Cambridge University Press, 1990.

[10] D.E. Rumelhart, G.E. Hilton, R.J. Williams, "Learnings Internal Representation by error propagation" In: Parallel Distributed Processing, 1,2, Cambridge, MA:MIT Press, 1986.

[11] C. Santacruz, V. López , R. Huerta and J. R. Dorronsoro, "Neural forecasting in real time industrial control", Proc. ICANN 1994, 1193-1198.

[12] C. Santacruz, R. Huerta, F. Rodríguez, J. R. Dorronsoro and V. López, "Local neural modelling in chemical reactor control systems", Neurocomputing (1995, in press).

[13] L. Bottou, V. Vapnik , "Local Learning Algorithms", Neural Computation 4, 888-900, 1992.

Collective Behaviour of a Chain of Hopfield Subnetworks Interconnected Unidirectionally

L. VIANA

Lab. de Ensenada, Instituto de Física, UNAM
A. Postal 2681, 22800 Ensenada, B.C., MEXICO
e-mail: laura@ifuname.ifisicaen.unam.mx

Abstract. A review of recent results on the performance of a string of n Hopfield subnetworks interconnected by means of strong unidirectional low density connections is presented. Such networks store p random unbiased patterns $\{\xi_i^\mu\}$, composed by n subpatterns, each of them to be recalled sequentially, by one of the subnetworks.

1. INTRODUCTION.

Neural Networks (NN) have been the subject of extensive research in recent years. Efforts in this field have been done from many points of view: bilogical, computational, etcetera. Physicist have also contributed to this subject, as they have made use of an analogy between the behaviour of an array of simple, formal neurons, interacting by means of excitatory and inhibitory synapsis, and some disordered magnetic systems called Spin Glasses. In this way, statistical mechanics methods have been used to study these systems assuming they are composed by a high number of elements [1].

One of the best studied models of NN is that of Hopfield [2], which consists of an array of N two-state elements completely interconnected according to the Hebb rule [3]. The use of this "learning" prescription allows to store p random unbiased patterns $\{\xi_i^\mu = \pm 1\}$, with $i = 1, \ldots, N$ and $\mu = 1, \ldots, p$, in the sense that these states become attractors for the dynamics of the system. From the biological point of view, this is a very crude model where neurons are two-state elements denoted by $S_i = \pm 1$ and interactions between them are symmetric; nevertheless it has proved to work well as a robust associative memory. It has been found that in the thermodynamical limit $(N \to \infty)$ such a network can store exactly a finite number p of patterns $(\alpha = (p/N) \to 0)$ [4], so the attractor states are identical to the nominated patterns. On the other hand, by means of computer simulations it has been found that its dynamical evolution depends on the initial state of the system, characterized by the overlap $m_o = (1/N) \sum_i S_i \xi_i^\nu$ with each of the ν-th patterns, as follows: If $m_o < m_{low}^c$, the probability of the system to recall this pattern is zero, while it is one for $m_o > m_u p^c$; finally, for m_o within those two values, the probability increases with m_o from zero to one. The width of this "band of criticallity" diminishes as N grows, tending to zero in the thermodynamical limit [5].

The dynamical behaviour of a string of $n \geq 2$ Hopfield networks interconnected unidirectionally has been studied [6,7], and many interesting characteristics have been found. The idea is to store a group of p patterns in the whole network, each composed by n subpatterns, in a way that each of the subpatterns is stored in one of the n subnetworks. It is an interesting question to find out how many of these subnetworks can be connected such that a given pattern can be correctly recovered by putting the first subnetwork within the basis of attraction of its corresponding subpattern (independently of the initial state of the other subnetworks).

Studying this kind of systems is of interest as we know that complex behaviours are constituted by a series of simpler operations. On the other hand, eventhough this could be a simple, unrealistic model, from the biological point of view, it can give us a clue about the reason for which neurons in the periferic nervous system have evolved to have very long axons. In this way, we find that in most real neural circuits there are just a few relays between ganglia. A similar model has also been used to model semantic memory [8].

2. TWO INTERACTING SUBNETWORKS.

We consider two N-element Hopfield subnetworks interconnected by means of unidirectional connections where each neuron in the first subnetwork sends information exclusively to another one in the second. The connections inside each of the subnetworks are given by the Hebb rule

$$J_{ij} = \frac{1}{N} \sum_{\mu=1}^{p} \xi_i^\mu \xi_j^\mu (1 - \delta_{ij}), \tag{1.a}$$

with i and j belonging to the same subnetwork, and $\xi_i^\mu = \pm 1$ with the same probability. The connections between nodes build up the relationship between subpatterns belonging to the same pattern. They are given by

$$W_{ij} = \sum_{\mu=1}^{p} \xi_i^\mu \xi_j^\mu \delta_{i,j+N}. \tag{1.b}$$

The dynamical evolution of the system is assumed to occur synchronically, at zero noise ($T = 0$), according to the updating rule

$$S_i(t + 1) = sign(h_i^n(t)), \tag{2}$$

where $S_i = \pm 1$ denotes the dynamical state of i–th neuron, and $n = 1, 2$ refer to the first and second nodes, respectively. The "fields" $h_i^n(t)$ are given by

$$h_i^1(t) = \sum_{j=1}^{N} J_{ij} S_j, \tag{3.a}$$

in the first subnetwork, and

$$h_i^2(t) = \sum_{j=N+1}^{2N} J_{ij} S_j + \sum_{j=1}^{N} W_{ij} S_j \delta_{i,j+N}, \tag{3.b}$$

in the second. Notice that the first node is a typical isolated Hopfield NN, while the second can be seen as a Hopfield NN in the presence of an external field.

It is convenient to give a macroscopical description of the system, allowing us to quantify how similar the states of the subnetworks are to a given subpattern (to be called subpattern1). To that end, it is introduced m^n, the overlap between the dynamical state of subnetwork$n$ and subpattern1. This is given by

$$m^n = \frac{1}{N} \sum_{i=(n-1)N+1}^{nN} \xi_i^1 S_i,$$ (4)

where, if subscripts f or o are used, this accounts for final or initial values, respectively. The maximum value for this overlap has been normalized to 1, so the fraction of correct bits in subnetwork n is given by $\frac{1}{2}(1 + m^n)$. Therefore, the overlaps $\{m^n\}$ are related to the total overlap q, between the complete pattern1 and the state of the whole system, as follows

$$q = \frac{1}{2N} \sum_{i=1}^{2} N\xi_i^1 S_i = \frac{1}{2}[m^1 + m^2].$$

Computer simulations of this system show that if $m_o^1 > m_{low}^c$, then the network is capable to recall the complete pattern, or it is not, with an α-dependent probability [6]. But if it does so, the basin of attraction of this pattern covers the whole space of states, since any initial state in subnetwork2 (any m_o^2), will take the system to the same attractor [9]. However, the position of this attractor, in the space of states, is displaced with respect to the nominated subpattern (subpattern$_2^1$).

Figure (1) shows a statistical study of the displacement of the attractors as a function of p for $N = 100$, over a sample of 100 different networks[10]. The x-axis shows the value of the overlap m_f^2 between the nominated subpattern and the final state, and the y-axis shows the percentage of times this value was obtained; the figure in the right upper corner indicates the percentage of cases in which the pattern was recovered. As we can see, the parity of p has an important effect in the results. This can be clearly seen in figure (2), where the first graph shows the percentage of successes in the recall, and the second one gives the average value for the final overlap $< m_f^2 >$. In both graphs, the results are shown as a function of p; with solid (dotted) line joining odd (even) values for p.

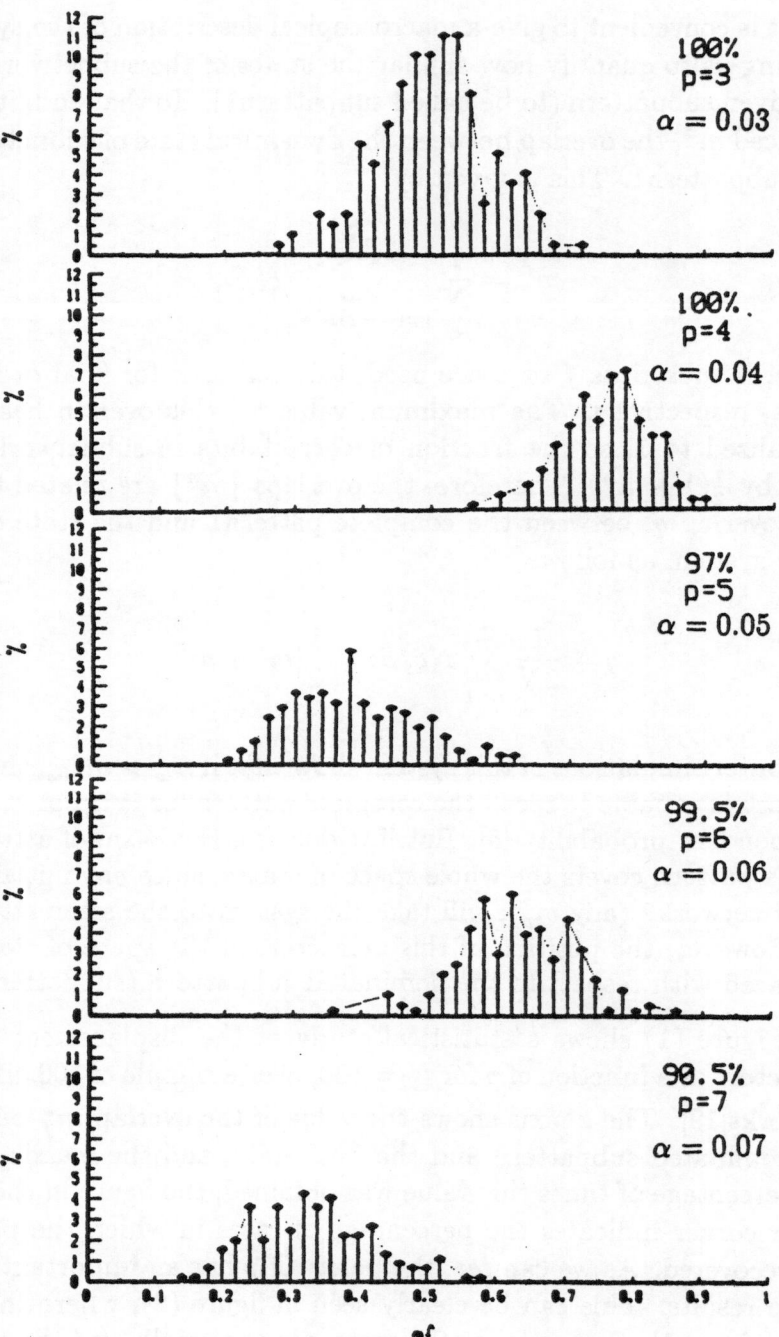

Figure (1)

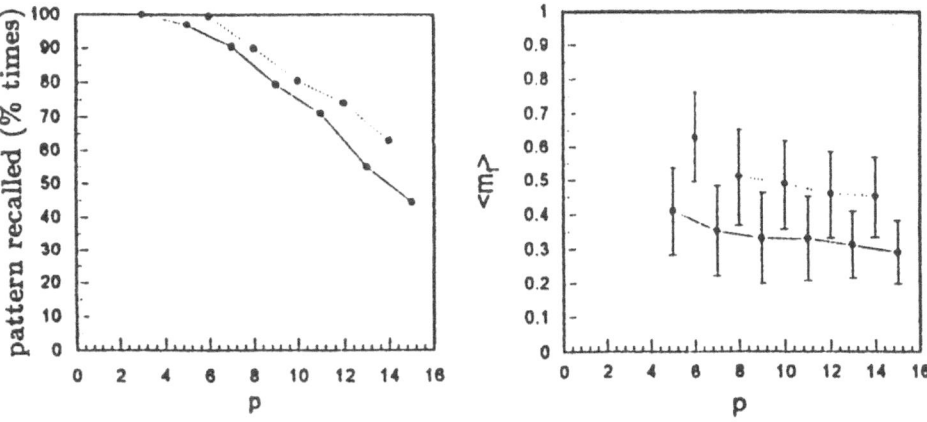

Figure (2)

Summarizing the previous results: if the first subnetwork is set initially in a state within the basis of attraction of a given subpattern, there is a very high probability for the second subnetwork to recover its corresponding subpattern, independently of its initial state. That is, the first subnetwork sends to the next one the information of which of the learnt states in the second one has to show up. However, the position of this attractor will be displaced from the nominated pattern. A similar behaviour is observed for higher values of N: there is a gradual deterioration, in both, the number of times the pattern was recalled, and its quality [6].

2. CHAIN OF HOPFIELD SUBNETWORKS

The results in the preceeding section, suggest a simple procedure to increase the quality of the information recovered by the second subnetwork. This consists in cutting the connection between subnetworks, once both of them have reached their equilibrum state. This would modify the space of states of the second subnetwork, and it would allow it to search for a new attractor taking as a initial state the one obtained when still connected to its feeding partner. If this second subnetwork is able to correctly recall the pattern, it could be the first subnetwork in a new pair. Therefore, another interesting question is, how many of such subnetworks can be connected, so a given pattern can be recovered by putting the first subnetwork within the basin of attraction of its corresponding subpattern. In this way it is considered a dynamical process with as many stages as subnetworks are in the chain, so during l-th dynamical stage, subnet l sends information to subnetwork $l+1$ until both reach an stationary state, then the connections between them are cut and the second subnetwork becomes the first one in a new pair. This is repeated until one of the subnetworks is unable to recall its corresponding subpattern (bad quality in the recovery by the first subnetwork or no recovery by the second).

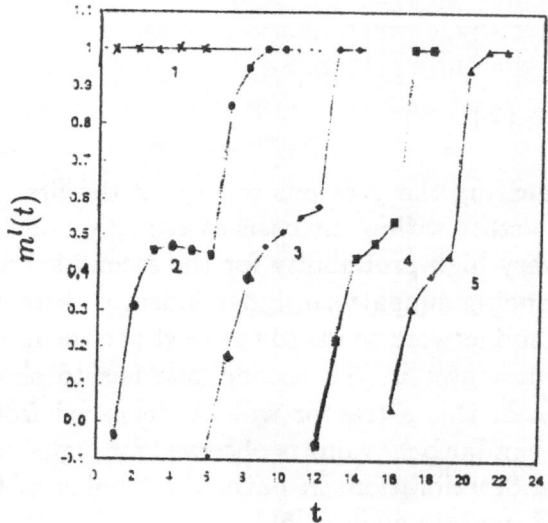

Figure (3)

Figure (3) ilustrates this process by showing the value for each of the 5 partial overlaps m^l as a function of time t: The first subnetwork is set in its corresponding subpattern ($m_o^1 = 1$), and the rest of the subnetworks are set in a random state ($m_o^l \sim 1/\sqrt{N}$, for all l). Then, the network is updated synchronically by pairs according to equation (2). The discontinuity in the value of m_l reflects the point at which the connection to its feeding pattern is cut. Notice that all patterns were exactly recalled ($m_f^l = 1$, for all l).

Figure (4) shows (a) the average number $< k >$ of subnetworks in a chain that were able to recall correctly up to k subpatterns (k-th subpattern was not recovered), as a function of the load $\alpha = p/N$, for

$N = 500$. The simulation included 100 different cases. These results have a very high dispersion $\sigma(k) = \{< k^2 > - < k >^2\}^{1/2}$, of the same order of magnitude than $< k >$, this is shown in (b) ; however, the distribution is not symmetrical and small values of k are safe. Notice that for low values of p, its parity has an important effect on $< k >$.

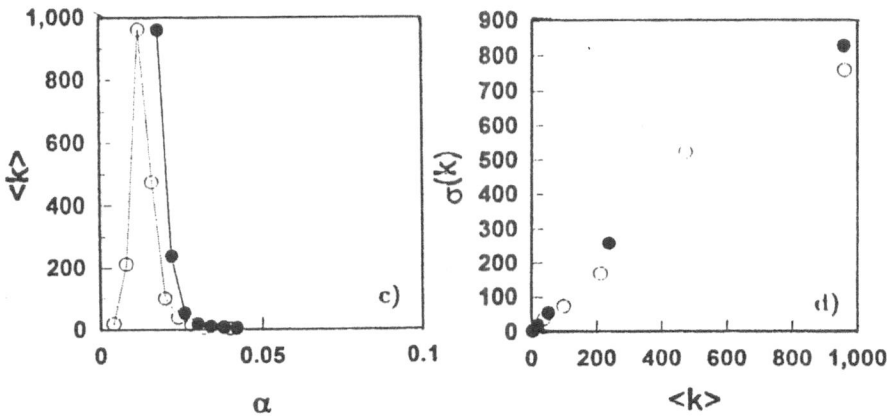

Figure (4)

Notice in figure (4), that for low values of p there is an important dependecy on the parity of $< p >$. For even values (o), the length of the average chain $< k >$ is very low; then it increases to a maximum, for $\alpha = \alpha^*$, and then it decreases. For odd values for p ($\bullet$), this lenght diverges for $p = 1$ and decreases as p grows. For $\alpha > \alpha^*$ there is not an important difference in the performance of networks with an even or odd number of stored patterns. Figure (5) shows the value of α^c for networks of different sizes. As we can see from these results, the load capacity α^* of a chain of k Hopfield subnetworks is much below than load capacity $\alpha_c \approx 0.138$ of the Hopfield network. So the maximum number of collective behaviours is much below the maximum number of individual ones. This could be the reason for which real neural circuits have only a few relays, as nature cannot afford the possibility of having, only sometimes, the necesary response.

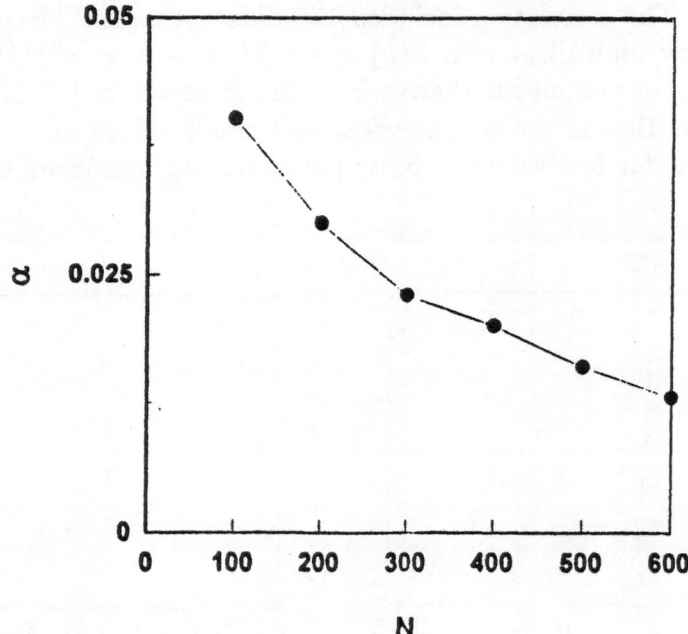

Figure (5)

ACKNOWLEDGEMENTS.

The author wishes to thank C. Martínez for her collaboration in preparing the diagrams. This work was partially suported by DGAPA Project No. IN100294 of the National University of Mexico (UNAM).

REFERENCES

[1] AMIT DJ, GUTFREUND H and SOMPOLINSKY H, *Phys. Rev.* **A32** 1007 (1985).

[2] HOPFIELD JJ, *Proc. Natl. Acad. Sci. USA* **79** 2554 (1949).

[3] HEBB DO, The Organization of Behaviour, Willey, New York, (1949).

[4] AMIT D.J., GUTFREUND H. and SOMPOLINSKY H., *Phys. Rev. Lett.* **55**, 1530 (1985).

[5] FORREST BM, *J. Phys.* **A21** (1988) 245.

[6] VIANA L, and MARTINEZ C., to be published.

[7] VIANA L, to be published.

[8] VIRASORO MA, REICH S. and LAUROGROTTO R., to be published.

[9] This behaviour has a few exceptions for some values of N and p, as discussed in [6].

[10] MARTINEZ C., M.Sc. Thesis, CICESE, México (1994).

A Comparative Study of Three Neural Networks that Use Soft Competition

Kate Butchart *, Neil Davey and Rod Adams
comrkb@herts.ac.uk ; comqrnd@herts.ac.uk;comqrga@herts.ac.uk
School of Information Science
University of Hertfordshire
Hatfield, Herts., UK. AL10 9AB

Abstract - This paper provides a comparative study of three proposed self organising neural network models that use forms of soft competition. The use of soft competition helps the neural networks to avoid poor local minima and so provide a better interpretation of the data they are representing. The networks are also thought to be generally insensitive to initialisation conditions. The networks studied are the Deterministic Soft Competition Network (DSCN) of Yair et al. , the Neural Gas Network of Martinetz et al and the Generalised Learning Vector Quantisation (GLVQ) of Pal et al. The performance of the networks is compared to that of standard competitive networks and a Self Organising Map when run over a variety of data sets. The three proposed neural network models appear to produce enhanced results, particularly the Neural Gas network, but in the case of the Neural Gas network and the DSCN this is at the cost of greater computational complexity.

1.0 Introduction

A standard Competitive Learning Neural Network (CLNN) [1] uses gradient descent on a cost function to settle into an equilibrium state. By using gradient descent such networks are prone to finding local minima, the quality of which is highly dependent upon the initial random state of the network. Soft competition, where there is more than one winner, has been used to try to prevent some of the problems of finding local minima. Networks that also use a temperature coefficient to reduce the "softness" of the competition over time, employing a simulated annealing type approach, may be the most successful in avoiding local minima. This report analyses and evaluates three CLNNs [2,3,4] which use forms of soft competition. The three CLNNs all claim to produce solutions independent of the initial state of the network, and two claim to find near global solutions.

Section 2 of the report looks at clustering, standard CLNNs and the theory of simulated annealing. In section 3 the three new network models are described. The three network models were run over different types of data and the results of these tests are reported and discussed in section 4.

---

*correspondence to: K. Butchart , comrkb@herts.ac.uk;

2.0 Clustering and Competitive Learning Neural Networks

Most clustering algorithms reduce a cost function by iteratively altering the prototype weight vectors in response to the inputs. The cost function that is being reduced determines the form of the update rule used. The most common cost function is based on the Mean Squared Error (MSE) measure that is given as

$$E = \sum_{x} \left\| x - y_{i*} \right\|^2 \qquad (1)$$

where x is input vector
y_{i*} is the winning node

and

$$\left\| x - y_{i*} \right\|^2 \leq \left\| x - y_i \right\|^2 \qquad \forall i \qquad (2)$$

A standard CLNN performs clustering on the input data by reducing the MSE given in (1). The MSE is reduced by updating the winning node for each input using the rule

$$y_i = y_i + a(x - y_i) \qquad (3)$$

where a is the learning rate and is global to the whole network.

The update rule in (3) uses gradient descent to minimise the MSE and as such may well find local minima. The quality of the minima found is very dependent upon the initial random starting weight values of the competitive units.

2.1 Simulated Annealing

Annealing is a process whereby a collections of atoms is reduced to its lowest energy state. Simulated annealing was first introduced by Metropolis et al.[5] who proposed a simple algorithm that can be used to simulate a collection of atoms in equilibrium at a given temperature. In each step of this algorithm an atom is given a small random displacement and the resulting change ΔE, in the energy of the system is computed. If $\Delta E <= 0$ the displacement is accepted and the state is changed. If $\Delta E > 0$ then the state is accepted with probability

$$P(\Delta E) = e^{-\frac{\Delta E}{K_b T}} \qquad (4)$$

where K_b is the Boltzman constant, and T is the temperature

This choice of $P(\Delta E)$ means that the system evolves to a Boltzman Gibbs distribution, where the probability of the system being in a state s is

$$P(s) = \frac{1}{Z} e^{-\beta E(s)} \qquad (5)$$

where $\beta = 1/T$ and Z is a normalisation factor defined so that the sum of P(s) over all possible states is unity, and so conforms to a probability density function.

By gradually reducing T over time simulated annealing is performed [6]. For high values of T there is a an element of random movement in the system, as states that increase E have a fairly high probability of being accepted. This random movement allows the system to escape local minima. At low temperatures a state that increases E is less likely to be accepted. As T is reduced so is the probability of accepting states that increase E and the system gradually evolves towards gradient descent. If the temperature schedule is correct for the problem the global minima should be reached[6].

3.0 The Three Networks

The neural networks that are analysed and evaluated in this report are :
* Yair et al's Deterministic Soft Competition Network (DSCN) [4]
* The Neural Gas network of Martinetz et al[2]
* Generalised Learning Vector Quantisation (GLVQ) of Pal et al[3].

3.1 The Deterministic Soft Competition Network

This network provides a deterministic form of Soft Competition derived from the principles of simulated annealing. It does not produce winners and losers, but updates all of the nodes for each input. The nodes are updated at differing rates, the rate of learning for each node being dependent upon: its distance from the input, the current temperature of the network and the current size of its individual counter.

The update rule for each vector is

$$y_i = y_i + a_i P_n(i)[x - y_i] \qquad (6)$$

$P_n(i)$ is the probability of output node y_i winning according to a Gibbs distribution

$$P_n(i) = \frac{e^{-\beta|x-y_i|^2}}{Z} \qquad (7)$$

where $\beta = 1/T$ $\qquad (8)$

and $Z = \sum_i e^{-\beta|x-y_i|^2}$ $\qquad (9)$

a_i is inversely proportional to the node's individual counter.

$$a_i(n) = \frac{1}{n_i(n)} \qquad (10)$$

n_i is the counter for each unit that is updated for each input by the rule

$$n_i(n) = n_i(n-1) + P_n(i) \qquad (11)$$

Initially when the temperature is high, $P_n(i)$ values are approximately uniform so the output nodes all migrate towards each input vector. As the temperature lowers those near to the input take a larger step size and those further away a smaller one. It should be noted that nodes which are close together will have a similar $P_n(i)$ value and are likely to have a similar a_i value. Hence nodes that become close have very similar learning rates and so can become "stuck" together. Since at high temperatures the nodes all tend to migrate to the same point it is very important the network is not given too high an initial temperature or all the nodes may become merged.

The value of T is reduced at the start of each pass through the input data set. The secondary temperature a_i is set to unity each time the number of passes through the data set equals a perfect square. The periodic increase in the secondary temperature makes the network's movements more random and so can help to jog the network out of local minima.

3.2 Neural Gas Network

Martinetz et al [2] propose the Neural Gas network that uses soft competition and a temperature coefficient to effect the softness of the competition over time. The network minimises the cost function

$$E = \sum_{x} \sum_{i=1}^{N} e^{-\beta(ki(x,y))} . \|x - y_i\|^2 \qquad (12)$$

where

$\beta = 1/T$

$k_i(x,y)$ is a function which represents the ranking of each weight vector y_i i with each input vector x . If y_i is closest to input x then k = 0, for the second closest k = 1 and so on.

At high temperatures many of the nodes in the network add to the cost (error) of the network. As the temperature is reduced only the nodes close to the input contribute significantly to the error, and at very low temperatures effectively just the winning node produces an error. Thus the cost function is reduced to the MSE cost function in (1) for very low temperature values.

The update rule, that is applied to all nodes and which reduces the cost function in (12) is

$$y_i = y_i + \alpha e^{-\beta(ki(x,y))}(x - y_i) \qquad (13)$$

where α is in the range 0 to 1 and is global to the network.

3.3 GLVQ network

A Generalised LVQ (GLVQ) method is proposed by Pal et al [3]. It uses soft competition as all of the nodes are updated unless the input is classified perfectly. The cost function being minimised is

$$E = \sum_{x} \sum_{i} g_{ji} . \|x - y_i\|^2 \qquad (14)$$

where

j is the winning unit

and $g_{ji} = 1$ if i=j   otherwise

$$g_{ji} = \frac{1}{\sum_{k=1}^{N} \|x - y_k\|^2} \qquad (15)$$

where N is the number of nodes

There is just one learning rate for all of the losers, that is partially dependent upon the classification error produced by the winning node.

The update rules for the network are

for the winner

$$y_i = y_i + \alpha(x - y_i)\frac{D^2 - D + \|x - y_i\|}{D^2} \qquad (16)$$

for the losers

$$y_i = y_i + \alpha(x - y_i)\frac{\|x - y_i\|}{D^2} \qquad (17)$$

where $D = \sum_{i} \|x - y_i\|^2 \qquad (18)$

α is global to the network and reduces over time

When the match between winner and input is perfect then losing nodes are not updated. As the mismatch between the winning weight vector and input vector increases the impact on the losing nodes also increases.

The GLVQ performs gradient descent on a cost function other than the MSE . In so doing it may be able to find a solution that provides a lower MSE than can be found by performing gradient descent on the MSE function itself, particularly if the initial starting conditions are poor.

4.0 The Comparative tests.

The DSCN, Neural Gas network and the GLVQ network were implemented and run over a variety of data sets. A standard CLNN , a SOM[7], and CLNNs with conscience[8] and leaky learning [1] mechanisms were also run over some of the data sets to provide performance comparisons. The CLNN with a conscience mechanism is equivalent to the Frequency Sensitive Competitive Learning method (FSCL) [9,10].

The networks were run over a model problem where the global minima could be calculated, data consisting of more than one Gaussian cluster and data from a single Gaussian cluster.

For the sake of comparison when an error measure was required the same MSE function was used for each network type.

4.1 Squares data - a model problem

To test the performance of the networks in minimising the MSE, a data distribution was chosen for which the MSE global minimum can be calculated for a large number of vectors. This data distribution was used in [2].

The clusters of data points are of a square shape and are separated from each other. The number of nodes was chosen to be four times the number of clusters, hence the final optimal configuration of

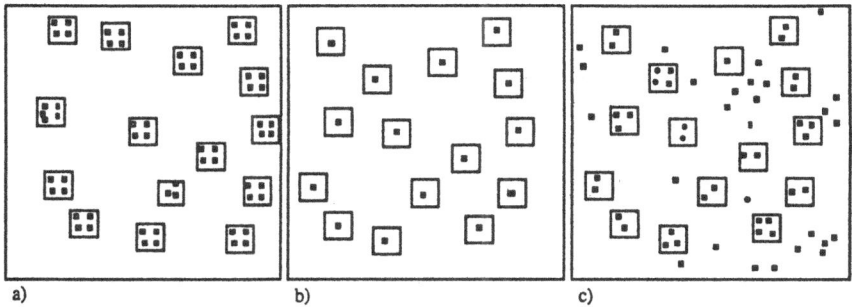

Figure 1 Network representations of the Squares data
The large squares represent the 15 regions where the input vectors are generated with uniform probability. The small squares represent the weight vectors of the nodes.
a) shows the most frequently produced Neural Gas network representation of the data distribution after 60000 adaptation steps. The network has come close to the optimum state.
b) shows the best solution the DSCN network found over this data .
c) shows a typical GLVQ network performance for this data.

the weight vectors is four vectors for each cluster arranged in the known optimal configuration for a single square.

4.1.1 Networks Performance

The Neural Gas network performed well over this data. For some configurations of the clusters it was able to find a global minimum, and for all configurations found a solution that was close to the optimum. On average the Neural Gas performance was 12 % worse than the optimum as shown in Figure 1 a).

The DSCN network was originally designed as a vector quantiser for single source data and did not prove to be very suited to clustered data such as this. The problem encountered with the network is finding the correct temperature schedule. If the temperature is too high the weight vectors of the nodes tend to merge together, if it is too low the network freezes before it has found a good solution.

The authors of the network provide a heuristic for the start temperature of the network over single source data but not for clustered data. Suitable temperature schedules for the network were investigated empirically. Once the weight vectors become very close the network is not able to separate them due to its method for calculating the learning rate for the nodes. (Nodes that are close together will have the same learning rate).

Figure 1 b) shows the best solution the network found over this data, groups of the nodes have become very close together, with each square

having a group of nodes at its centre. Inevitably with nodes so close together the network is not able to reduce the distortion error as would be hoped. This solution represents an ideal solution given just 15 of the 60 nodes. The local clustering of the other 45 nodes added nothing to the functionality of the network. The average distortion values for the network were 370% greater than the optimum.

The states that the network converged to were similar and independent of the starting positions.

The GLVQ network performed reasonably well over this data, but it was not able to bring all the nodes into the competition as shown in Figure 1c). On average one third of the nodes did not move into a position to be able to classify any of the data points. This prevented the network from getting close to the optimum value. The value of E was for this network on average 180% greater than the optimum.

4.1.2 Comparisons with Other Networks

Figure 2 shows the average performance of all seven types of network used. The Neural Gas network was certainly the best performer over this data and proved to be resilient to poor initial conditions. The GLVQ was clearly better than any of the others. Its distortion measure doubled from 147% greater than the optimum to 299% greater under poor starting positions which was a little surprising given its claim to be position independent. However even taking the worst figure it still outperformed all the networks except the Neural Gas.

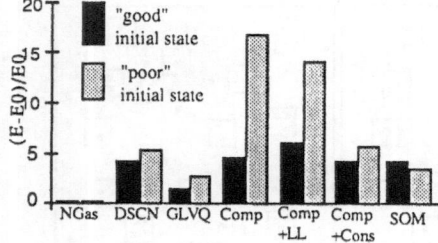

Figure 2 : Average performance of the networks
The average performance of each network is shown from both good and poor initial states. E0 is the known global minimum of E. A good initial state exists where the initial positions of the output nodes are evenly spread across the input space. Poor initial states are defined as ones where the output nodes are initialised in a small area of the input space away from any of the inputs.

The DSCN network performed comparably with the Competitive network with a conscience mechanism, and a little worse than the SOM. All three were not badly effected by poor starting positions. The standard competitive networks, and the competitive network with a leaky learning mechanism produced acceptable performances over the data with good starting positions. However both networks did very badly with poor initial conditions thus emphasizing the random nature of the solution found with such networks, unless good initial positions can be guaranteed.

4.1.3 Convergence Rates

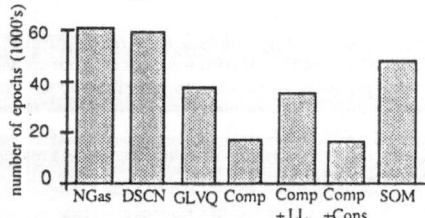

Figure 3: the number of presentations of inputs required for each network to converge to its optimum solution.

Figure 3 shows the number of input presentations required by the networks to converge to their optimum solutions. It should be noted that the networks that took longer to find their optimum solutions , such as the Neural gas network and the SOM would still converge to a reasonable solution with a smaller number of presentations.

4.2 Gaussian Clusters

The three networks were also run over more standard data consisting of Gaussian clusters.

4.2.1 Test 1

The first test consisted of four clusters, three of which contained 100 data points and the fourth 50 data points. All clusters had a normal Gaussian distribution. The networks had seven competitive nodes. The optimal solution is for each of the clusters consisting of 100 data points to be represented by two of the output nodes , and the smaller cluster by just one.

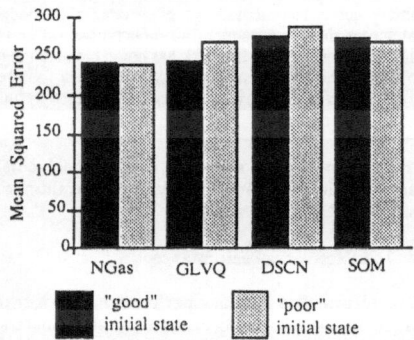

Figure 4: Networks Mean Squared Error Performance for Test1.

For good initial state of the weight vectors both the Neural Gas and GLVQ networks were able to find a solution close to the optimal configuration. The DSCN did not find a solution as close to the optimal configuration and consequently produced a poorer MSE.

For poor initial weight vector state the Neural Gas performed the best with an MSE almost the same as for good initial state data. The SOM network performed well in comparison with the other networks and outperformed the DSCN network.

4.2.2 Tests 2 and 3

The networks were also run over two dimensional Gaussian data with more noisy inputs . For the second test there were three dense clusters of 200 points each , and one sparse cluster covering the whole input area giving the effect of a lot of noisy inputs. The networks had 64 competitive units.

The data was extended to four dimensions for test 3.

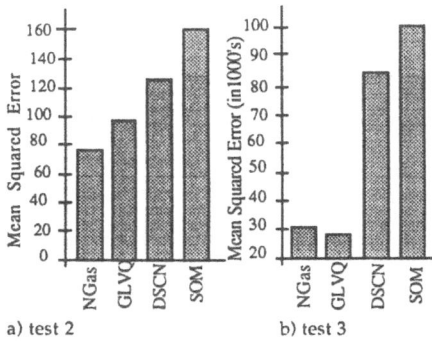

a) test 2   b) test 3

Figure 5: Mean Squared Error performance for tests 2 and 3

As in all the previous tests the Neural Gas network performed the best in test 2. In test 3, where the input data was in four dimensions and effected by quite a lot of noisy inputs the GLVQ network outperformed the Neural Gas network, although both were much better than the DSCN and SOM.

The problem the networks had to solve was determining how many of nodes should cover the clusters and how many the sparse noisy inputs. GLVQ tended to spread the nodes across the whole area as did the SOM. The Neural Gas network provided significantly more nodes in the dense clusters but still provided cover for the noisy inputs.

Over this type of data the NGas and GLVQ again consistently outperformed the other network types.

4.3 Gauss Markov Source

All of the previous tests used data with the inputs clustered into groups. The tests in this section were run over data with just one cluster of inputs produced from first order Gauss-Markov processes of the form $x_n = \alpha x_{n-1} + w_n$, where w_n is an independent identically distributed Gaussian process with zero mean and unit variance. The coefficient α defines the level of correlation between inputs. The tests were run with data of 2, 4 and 8 dimensions, with differing degrees of correlation between inputs.

Tables 1 , 2 and 3 show the MSE for each network over 2, 4 and 8 dimensional data respectively. The lowest MSE for each correlation is underlined in each table.

Corr	NGas	GLV	DSC	SOM	FSCL
0.0	162	159	198	259	227
0.2	175	173	218	272	241
0.6	229	214	287	412	321
0.9	792	790	986	1356	1134

Table 1 MSE for networks over two dimensional data from a Gauss-Markov Source, with varying correlation between inputs.

Corr	NGas	GLV	DSC	SOM	FSCL
0.0	356	341	376	420	385
0.5	459	446	472	579	478
0.9	1357	1380	1442	1773	1422

Table 2 MSE for networks over four dimensional data from a Gauss-Markov Source, with varying correlation between inputs.

Corr	NGas	GLV	DSC	SOM	FSCL
0.0	1713	1717	1728	1908	1720
0.5	2282	2340	2398	2608	2263
0.9	6166	6215	6295	6958	6288

Table 3 MSE for networks over eight dimensional data from a Gauss-Markov Source, with varying correlation between inputs.

Over this single source data the GLVQ and Neural Gas networks produced very similar results. The DSCN did not perform as well as had been hoped over this data, and, in fact, the Competitive network with a conscience mechanism produced results that were similar and sometimes better than the DSCN over this data.

5.0 Network Evaluations and Conclusions

5.1 Does Soft Competition provide enhanced performance

The Neural Gas and GLVQ networks consistently performed better than the standard networks in terms of the MSE produced over all the tests performed.

The relative performance is improved most noticeably when the data consists of more than one cluster and there are a large number of nodes. Over single source data the enhancements are not as great, especially when the dimensionality is increased.

5.2 How the networks compare with each other

The GLVQ and Neural Gas networks outperformed all the other networks over all the tests run. Over data with more than one cluster Neural Gas is certainly the best performer, in terms of the MSE reached. For single source data Neural Gas and GLVQ produced fairly equal performances. The DSCN network generally

314

outperformed the standard networks but did not do as well as the Neural Gas and GLVQ networks. The main problem encountered by the DSCN was that the nodes converged to the same point if the temperature was too high. The network therefore had to be started with a low temperature that caused it to be too static and unable to find the optimum solution.

5 . 3 Sensitivity to parameters and initial conditions

The DSCN is very sensitive to the initial temperature value. If this value is too high the network provides very poor performance, if it is too small the network again performs badly. The GLVQ and Neural Gas appear to be relatively resilient to alterations in parameter values.

A major strength of both the Neural Gas and DSCN networks is the fact that empirically they appear to be resilient to the quality of the starting positions. The other networks tended to be quite badly effected by poor starting conditions.

5 . 4 Computational load and Speed of Convergence

The DSCN and Neural Gas networks both take up to twice as long to converge than the other networks. This is caused by the need to allow time to operate a suitable temperature schedule.

Computationally the DSCN and Neural Gas are again the most expensive, the Neural Gas network needs to rank all of the nodes for every input and calculate their individual update rates. The DSCN has to calculate the probability of each node winning according to a Gibbs distribution, then calculate the update value for every node before updating them all. The GLVQ network is not significantly more computationally expensive than a FSCN or SOM.

5.5 Conclusions

The Neural Gas network provides a reliably good MSE over many different data types, and is resilient to initial starting positions and variations in parameter settings. This is however gained at the expense of an increased computational load and slower convergence rate. The GLVQ network does not perform quite as well as the Neural Gas network over certain types of data and is not as resilient to poor starting conditions. It does, though, perform better than the standard networks, converging at a similar rate and with a computational load equivalent to all but the standard (winner update only) CLNN.

The DSCN network is capable of better performance than the standard networks but is overly sensitive to the temperature schedule implemented.

References

[1] Hertz, J.,Krogh A., Palmer, R., "Introduction to the theory of Neural Computation", Addison Wesley, 1991 .
[2] Martinetz , T. , Berkovich , S. and Schulten , K. "Neural-Gas Network for Vector Quantisation and its Application to Time-Series Prediction." IEEE transactions on Neural Networks. July 1993, vol. 44 (4) .
[3] Pal , N . , Bezdec , J. and Tsao, E., "Generalised Clustering Networks and Kohonen's Self-Organising Scheme." IEEE transactions on Neural Networks , July 1993, vol. 44 (4).
[4] Yair , E. , Zeger , K. and Gersho, A. , "Competitive Learning and Soft Competition for Vector Quantiser Design." IEEE transactions on Signal Processing, 1992 . vol. 40 (2)
[5] Metropolis , N . , Rosenbluth , A. W. , Rosenbluth , M.N. ,Teller , A.A . and Teller , E. "Equations of state calculations by fast computing machines," Journal Chemical Physics ., 1953, vol. 21, pp. 1087-1091.
[6] Kirkpatrick , S. , Gelatt C . D . and Vecchi , M. P. , "Optimisation by simulated annealing," Science, 1983, vol. 220 , pp 671-680.
[7] Kohonen,T. , "Self Organisation and Associative memory". Third Edition Springer Verlag, 1989.
[8] DeSeino, D . "Adding a conscience to competitive learning," in Proceedings IEEE International Conference Neural Networks, 1988,pp .1117-1124.
[9] Ahalt, S.C. , Krishnamurthy, A. K., Chen, P., and Melton , D.E., "Competitive learning algorithms for vector quantization," Neural Networks, vol. 3,no. 3, pp. 277-290, 1990.
[10] Krishnamurthy, A. K., Ahalt, S.C. , Melton , D.E. and Chen, P., "Neural Networks for Vector Quantisation of Speech and Images,"IEEE Journal on Selected Areas in Communications, vol. 8, no. 8, pp. 1449-1457,1990.

Neural Networks and Genetic Algorithms for the Attitude Control Problem

Dimitris C. Dracopoulos
Brunel University
Department of Computer Science
London, UK

Antonia J. Jones
Imperial College
Department of Computing
London, UK

Abstract

A general adaptive control method using genetic algorithms and neural networks is proposed and applied to a highly nonlinear problem, the attitude control problem. Examples are given where the method successfully control a rigid body satellite with unknown dynamics, including an example where the satellite is subject to external forces trying to lead it into a chaotic motion.

1 Introduction

The orientation control of a rigid body has important applications from pointing and slewing of aircraft, helicopter, spacecraft and satellites, to the orientation control of a rigid object held by a single or multiple robot arms. The problem considered in this paper is the attitude control problem.

Attitude control is the process of achieving and maintaining an orientation in space by performing attitude maneuvers (an attitude maneuver is the process of reorienting a satellite or spacecraft from one attitude to another). The attitude of the body is relative to some external reference frame. This reference frame may be either inertially fixed, or slowly rotating, as in the case of Earth-oriented satellites. This paper describes a general control architecture, using genetic algorithms and neural networks, which can be used for

- reaching a particular state of the dynamic system, even when the system dynamics are chaotic.

- trajectory tracking

- finding an inverse model (even when there is ill-posedness)

The proposed control architecture is applied to the attitude control problem. The goal of the described control application is to detumble a satellite (reduce its kinetic energy) while achieving a desired orientation. It is assumed that the satellite is a rigid body, so that the motion is described by the Euler equations [4]. The system is equipped with reaction thrusters which provide control torques about the three principal axes. The above control problem has no complete general solution [7], it is highly nonlinear, exhibits chaotic behaviour under certain circumstances [6] and therefore is a challenging problem for testing the proposed control method.

In addition, the plant may change its characteristics. This could happen due to to damage, malfunction, or a change in the principal moments of inertia. In this event an adaptive controller is required. The proposed control architecture is adaptive to the changes of the plant dynamics. As shown in following examples, the proposed controller is able to operate successfully even in the presence of external forces which try to lead the system into a chaotic motion.

2 The Euler equations

The Euler equations describe the rotational motion of a rigid body about its mass center. They are given by the following equations [4]:

$$
\begin{aligned}
L &= I_x \dot{\omega}_1 - (I_y - I_z)\omega_2\omega_3 \\
M &= I_y \dot{\omega}_2 - (I_z - I_x)\omega_3\omega_1 \\
N &= I_z \dot{\omega}_3 - (I_x - I_y)\omega_1\omega_2
\end{aligned}
\tag{1}
$$

where I_x, I_y, I_z are principal moments of inertia, $\omega_1, \omega_2, \omega_3$ are the angular velocities of the rigid body about the x, y, z body axes and L, M, N are the torques about the x, y, z axes respectively. In addition to equations (1) the following equations describe the orientation of the body relative to an inertial frame:

$$
\begin{aligned}
\dot{\Phi} &= \omega_1 + \omega_2 \sin \Phi \tan \Theta + \omega_3 \cos \Phi \tan \Theta \\
\dot{\Theta} &= \omega_2 \cos \Phi - \omega_3 \sin \Phi \\
\dot{\Psi} &= (\omega_2 \sin \Phi + \omega_3 \cos \Phi) \sec \Theta
\end{aligned} \tag{2}
$$

where Φ, Θ, Ψ are the angles given by three consecutive rotations about the x, y, z body axes respectively. The two systems (1) and (2) of differential equations are the dynamic equations required for the attitude control problem.

There are several cases where freely rotating rigid bodies can exhibit chaotic behaviour [6, 8]. The first case is when one of the moments (L, M or N), varies periodically in time. The second case is where one has parametric excitation through time-periodic changes in the principal inertias, for example, $I_y = I_0 + B \cos \Omega t$.

The third case comes from the schemes based on linear or quadratic feedback, which have been proposed to stabilize rigid body spacecraft attitude, described by the Euler equations (1). A simple such scheme employs jets to impart torque according to suitable linear combinations of the sensed angular velocities ω about the body-fixed principal axes.

If the torque feedback matrix for the nonlinear system is denoted by $\mathbf{A}$, so that $\mathbf{G} = (L, M, N) = \mathbf{A}\omega$, the equations (1) become

$$
\begin{aligned}
(\mathbf{A}\omega)_1 &= I_x \dot{\omega}_1 - (I_y - I_z)\omega_2 \omega_3 \\
(\mathbf{A}\omega)_2 &= I_y \dot{\omega}_2 - (I_z - I_x)\omega_3 \omega_1 \\
(\mathbf{A}\omega)_3 &= I_z \dot{\omega}_3 - (I_x - I_y)\omega_1 \omega_2
\end{aligned} \tag{3}
$$

It has been noticed [6, 10] that for certain choices of I_x, I_y, I_z and $\mathbf{A}$ equations (3) exhibit both strange attractors and limit cycles. Since these linear feedback rigid body motion equations are slightly more complicated than Lorenz's equations [6], this conclusion is not surprising. The interpretation of strange attractor motion, in this case, is that the body executes a wobbly spin first about one, then about the other of a conjugate pair of directions fixed relative to the body axes (and symmetric about an axis). The limiting spin magnitudes and directions define rest points of the system, called eye-attractors because of their appearance. Figure 2 shows the Poicaré map of the double strange attractor of equations (3) in the $x - z$ plane for $I_x = 3$, $I_y = 2$, $I_z = 1$ and

$$
\mathbf{A} = \begin{pmatrix} -1.2 & 0 & \sqrt{6}/2 \\ 0 & 0.35 & 0 \\ -\sqrt{6} & 0 & -0.4 \end{pmatrix} \tag{4}
$$

The attractor of an orbit is determined by the location of the initial point of that orbit.

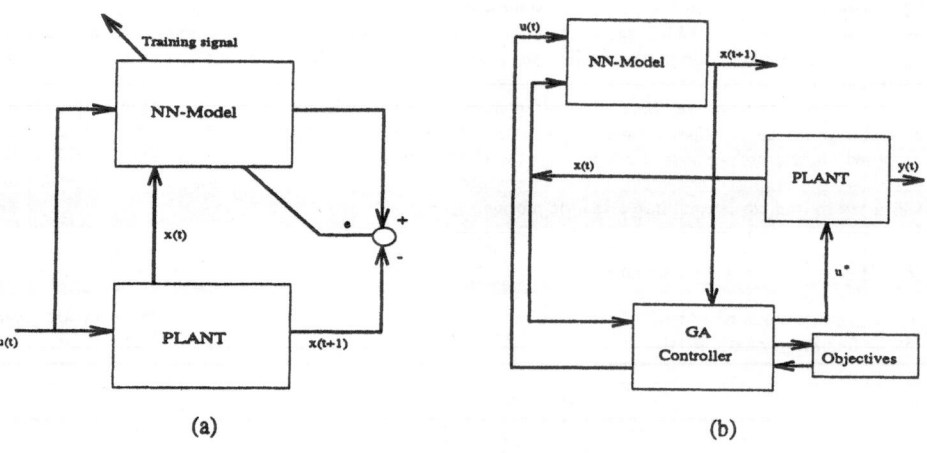

(a)   (b)

Figure 1: (a) Building the neuromodel for the plant. (b) The Genetic controller.

3   The Adaptive Control Architecture

The rigid body satellite is described by the Euler equations (1) and (2). The problem faced here is, its detumbling and reorientation in a prespecified way. This is desired to be done, even in cases where its dynamics are unknown, due to damage, malfunction or a change in the moments of inertia.

Assume that the dynamics of the satellite are unknown or have changed for an unknown reason (which could be one of the above). The proposed adaptive control architecture is shown in Figures 1a and 1b. As soon as the satellite dynamics change, a model of its dynamics is constructed. This is done by adapting (training) a neural network model to identify the plant dynamics (Figure 1a). At each time instant, the difference between the correct plant output, and the network's estimated plant output, is used as a training error for the neuromodel (as described in the block diagram of Figure 1a).

Recent work [1, 2] has shown that feedforward neural networks are able to model dynamic systems of different complexity (including chaotic systems) over a large part of the phase space by being trained only on a few hundred data of a single trajectory. For example in Figures 2a and 2b are shown the Poincaré map of the chaotic dynamic system described by equations (3), (4) and the Poincaré map of a 12-10-5-3 neural network trained to predict for time $\Delta t = 0.5$. The neural network was trained using the backpropagation algorithm [11] on 500 data taken from a *single trajectory* starting at $(\omega_1, \omega_2, \omega_3) = (3, 2, 1)$. For the requirements of the adaptive control architecture described in this paper and for the following control examples it will be assumed that such a neuromodel describing the plant dynamics locally, has already been trained.

The main module of the controller proposed here is based on genetic algorithms. Assuming that the goal of the control application is to detumble the satellite, and spin it about one of the body axes, while bringing it in a desired orientation, the following objective function can be defined:

$$U(t) = |\dot{\Phi} + \Phi| + |\Phi| + |\dot{\Theta} + \Theta| + |\Theta| \tag{5}$$

The Genetic controller uses the constructed plant model as a predictor of the future values of the plant (Block diagram of Figure 1b). According to the above objective function, which has to be minimized, the goal is to spin the satellite about the z body axis, whilst detumbling it about the x, y axes (make ω_1, ω_2 equal to zero), and reorient it to the position $\Phi = 0, \Theta = 0$. It must be noted, that if the goal was a different one, then it suffices to change the objective function. The Genetic controller minimizes (5) by maximizing the following adjusted fitness function:

$$E(t) = \frac{1}{1 + U(t)} \tag{6}$$

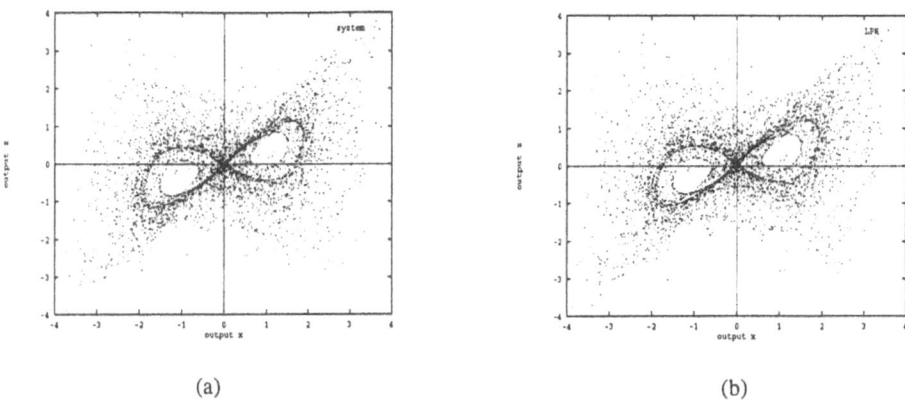

(a)   (b)

Figure 2: Euler system. Poincaré sections through the $x - z$ plane. (a) system (b) neural network model.

Thus at each time instant, the Genetic controller seeks for individuals (chromosomes) of the form in Figure 5, which lead the system to a "good" state, according to the objectives (5). The genetic algorithm used for *the Genetic controller is shown in Figures 3, 4.*

Although, different fitness functions based on an objective function, have been used in the genetic algorithm literature, all the simulations for this work showed that the choice of (6) leads always to good solutions (as long as the other parameters of population size, individual size, etc. are chosen carefully), while others, like the raw fitness function do not. As it was noticed in the simulations for this work, the adjusted fitness exaggerates the importance of small differences in the value of the standard raw fitness usually used, as the standard fitness approaches 0 (something which occurs in later generations of a run). The individuals (control inputs) which have the binary form of Figure 5 take integer values in the range $[-2^{15}, 2^{15}]$. The length of each chromosome was 48 bits, with 16 bits dedicated to each of the control variables L, M, N respectively (Figure 5). Mutation (alternation of a gene) with probability $P_m = 0.0333$ was used. A 1-point crossover operator with probability $P_c = 0.6$ was used to exchange genes between individuals (more details on these two operators can be found in [3, 5].

A population size of 50 individuals was used and the two best individuals of each generation were copied in the next generation. The termination criterion for the genetic algorithm was to reach generation 100.

4 Spin Stabilization of a Satellite from a random initial position

Assume, that for some unknown reason (damage), a satellite with specified dynamics, changes its characteristics. Its moments of inertia, become $I_x = 1160, I_y = 23300$ and $I_z = 24000$.

During the period, where the system changes dynamics, unknown forces lead it to the $(\omega_1, \omega_2, \omega_3) = (3, 2, 1)$ and $(\Phi, \Theta, \Psi) = (2, 1, 3)$ state. The goal is to detumble the satellite about the x, y body axes, spin it about the z body axis, and reorient it, so that $\Theta = 0, \Phi = 0$.

The application of the Genetic adaptive controller architecture, described by Figures 1a, 1b leads to the situation described by Figures 6a, 6b. Figure 6a shows the evolution in time, of the angular velocities ω_1, ω_2 about the x, y body axes respectively. It is easily shown that the Genetic controller, soon leads both of these angular velocities to the prespecified value of zero. The third angular velocity is arbitrary, since it was not in the control objectives. Consequently the control torque N is not subject to evolutionary pressure.

Figure 6b show the reorientation of the satellite for the angles Φ, Θ after the application of the controller. While these are becoming zero, the satellite is rotating about its z axis. It should be noted, that the controller not only leads the system to a desired state, but it maintains this state afterwards.

```
population_size := 50;
generation := 1;
Initialize population with random binary strings;
while generation ≤ 100 do
        Find the two best individuals of current population;
        Copy the two best individuals to new population;
        for i = 1 to population_size − 2 step 2 do
            Select two individuals based on fitness;
            Probabilistically mutate the two individuals;
            Probabilistically perform crossover;
            if crossover_performed then
                Copy the two offspring into new population;
            else
                Copy the two (mutated) individuals into new population;
            endif
        endfor
        generation := generation + 1;
endwhile
```

Figure 3: Pseudocode of the Genetic Algorithm for the Attitude Control Problem.

```
Select()
        j := 1; sector := 0;
        number := random();
        while sector ≤ number and j ≤ population_size do
            sector := sector + fitness_of_individual_j;
            j := j + 1;
        endwhile
        return individual_{j-1};
end
```

Figure 4: Select procedure of the Genetic Algorithm for the Attitude Control Problem.

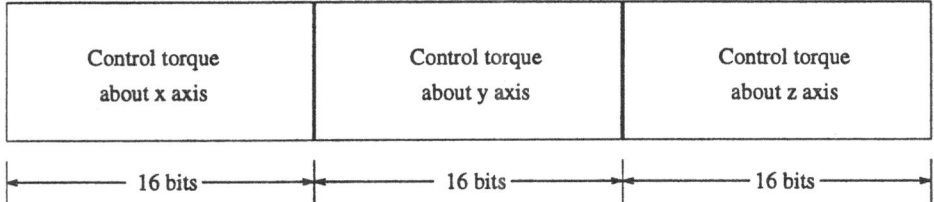

Control torque about x axis	Control torque about y axis	Control torque about z axis
←——— 16 bits ———→	←——— 16 bits ———→	←——— 16 bits ———→

Figure 5: The individual chromosomes for the Genetic controller

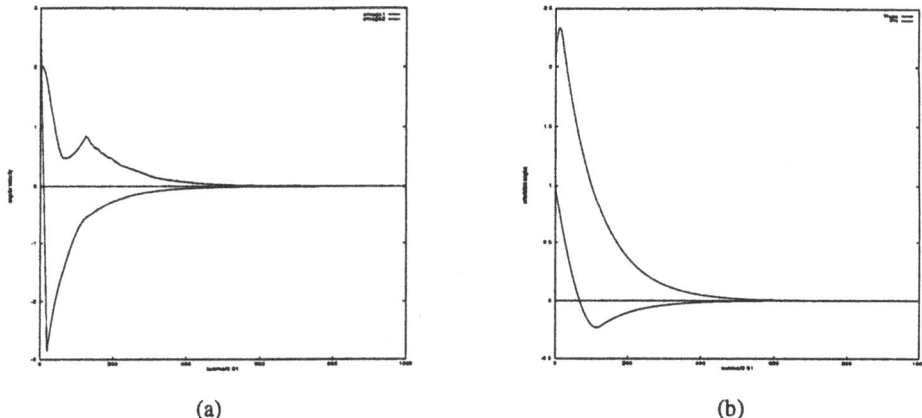

(a)   (b)

Figure 6: (a) Angular velocities ω_1, ω_2 for the satellite after the application of the Genetic Controller. (b) Orientation angles Θ, Φ for the satellite after the application of the Genetic Controller. Initial conditions $(\omega_1, \omega_2, \omega_3) = (3, 2, 1)$ and $(\Phi, \Theta, \Psi) = (2, 1, 3)$.

5 Satellite Attitude Control subject to external forces leading to chaos

Assume that a satellite changes its characteristic moments of inertia to $I_x = 1160, I_y = 23300$ and $I_z = 24000$. External forces are asked upon it for $t > 0$. The system is described by dynamic equations (2) and

$$I_x\dot{\omega}_1 - (I_y - I_z)\omega_2\omega_3 = N_1 + L$$
$$I_y\dot{\omega}_2 - (I_z - I_x)\omega_3\omega_1 = N_2 + M \tag{7}$$
$$I_z\dot{\omega}_3 - (I_x - I_y)\omega_1\omega_2 = N_3 + N$$

$$(8)$$

where the torques N_1, N_2, N_3 produced by the external forces are given by the equations

$$
\begin{aligned}
N_1 &= -1200\omega_1 + 1000 \cdot \frac{\sqrt{6}}{2}\omega_3 \\
N_2 &= 350\omega_2 \\
N_3 &= -1000 \cdot \sqrt{6}\omega_1 - 400\omega_3
\end{aligned}
\tag{9}
$$

The above is a system in which the externally imposed torques N_1, N_2, N_3 would left to themselves result in a chaotic motion, while the thrust vector $G = (L, M, N)$ is trying to control this chaotic motion and lead the system into a prespecified state (control of chaos). In this case, it is assumed that the initial state is $(\omega_1, \omega_2, \omega_3) = (1.3, 3.0, 2.8)$ and $(\Phi, \Theta, \Psi) = (2, 1, 3)$ while the desired target state is $(\omega_1, \omega_2, \omega_3) = (0, 0, 0)$ and $(\Phi, \Theta, \Psi) = (0, 0, 0)$. Since the goal from the previous simulation is different, a new objective function is defined as follows:

$$
U(t) = |\dot{\Phi} + \Phi| + |\Phi| + |\dot{\Theta} + \Theta| + |\Theta| + |\dot{\Psi} + \Psi| + |\Psi|
\tag{10}
$$

Figure 7a show the evolution in time, of the angular velocities $\omega_1, \omega_2, \omega_3$ about the x, y, z body axes respectively. The Genetic controller, soon leads these angular velocities to the prespecified value of zero despite the fact that during the control process the external perturbing torques were as large as 44% of the maximum available control torques. As soon as this is achieved, the controller maintains these angular velocities.

In Figure 7b the reorientation of the satellite for the angles Φ, Θ, Ψ after the application of the controller is shown. Again it can be noticed, that the controller not only reorients the satellite, but it maintains this reorientation.

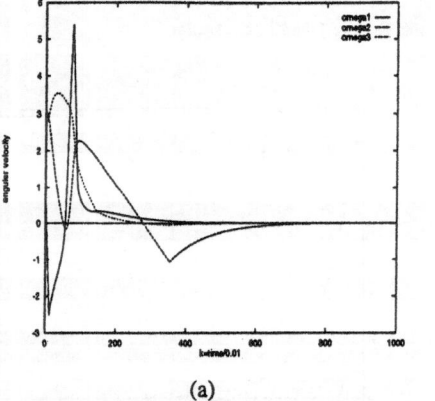
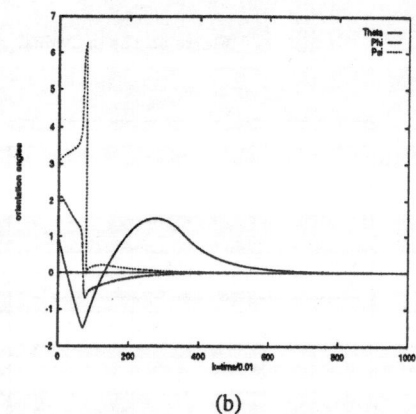

(a)   (b)

Figure 7: (a) Angular velocities $\omega_1, \omega_2, \omega_3$ for the satellite, in the presence of forces trying to lead it in a chaotic motion, after the application of the Genetic Controller. (b) Orientation angles Θ, Φ, Ψ for the satellite, in the presence of forces trying to lead it in a chaotic motion, after the application of the Genetic Controller. Initial conditions $(\omega_1, \omega_2, \omega_3) = (1.3, 3.0, 2.8)$ and $(\Phi, \Theta, \Psi) = (2, 1, 3)$.

6 Conclusions

An adaptive control architecture was proposed using the attitude control problem as a testbed. This highly nonlinear problem is complicated and as shown becomes more complicated, when the dynamics of the system are chaotic under certain conditions. The adaptive control architecture proposed in this paper is able to attitude control a rigid body. The method deals directly with the underlying nonlinearities describing the motion of the rigid body, rather than with linearized equations (an approach usually adapted by other attitude control methods), which can lead to inaccurate or incorrect description of the system dynamics.

In contrast with most other attitude control approaches, it deals with the adaptive problem. In cases where the system dynamics change due to damage or malfunction, the proposed control architecture adapts itself to the new plant dynamics, and where possible it tries to control the new system by constructing a neural network model of the plant dynamics. An exceptional case of difficulty in the control of the highly nonlinear system was introduced by adding external forces acting upon the system and leading it into a chaotic motion. Even in this case, the control architecture behaved extremely well. The chaotic motion of the body was controlled. The satellite was completely detumbled, it was reoriented in the desired orientation and then it was maintained in that target position.

Compared with the OGY method [9] (which currently is the only other available method for controlling chaos), the control architecture proposed in this paper can effectively control a chaotic system starting from any system state. In contrast the OGY method in order to start control has to wait until the system trajectory passes through an unstable periodic orbit [9]. In addition the OGY method requires some knowledge of the plant dynamics (i.e. unstable periodic orbits) while the method proposed in this paper requires no minimum knowledge of the plant dynamics.

References

[1] Dimitris C. Dracopoulos and Antonia J. Jones. Modeling dynamic systems. In *1st World Congress on Neural Networks Proceedings*. INNS/Erlbaum Press, 1993.

[2] Dimitris C. Dracopoulos and Antonia J. Jones. Neuromodels of analytic dynamic systems. *Neural Computing & Applications*, 1(4):268–279, 1993.

[3] David E. Goldberg. *Genetic Algorithms in Search, Optimization and Machine Learning*. Addison Wesley, 1989.

[4] Herbert Goldstein. *Classical Mechanics*. Addison Wesley, second edition, 1980.

[5] John J. Greffenstette. Optimization of control parameters for genetic algorithms. *IEEE Transactions on Systems, Man and Cybernetics*, SMC-16:122–128, 1986.

[6] R. B. Leipnik and T. A. Newton. Double strange attractors in rigid body motion with linear feedback control. *Physics Letters*, 86A:63–67, 1981.

[7] George Meyer. On the use of Euler's theorem on rotations for the synthesis of attitude control systems. Technical Report TN D-3643, NASA, 1966.

[8] Francis C. Moon. *Chaotic and Fractal Dynamics*. John Wiley and Sons, 1992.

[9] E. Ott, C. Grebogi, and James Yorke. Controlling chaos. *Physical Review Letters*, 64(11), 1990.

[10] George E. Piper and Harry G. Kwatny. Complicated dynamics in spacecraft attitude control systems. *Journal of Guidance, Control and Dynamics*, 15(4):825–831, July-August 1992.

[11] David Rumelhart, James McClelland, and the PDP research group. *Parallel Distributed Processing - Explorations in the Microstructure Cognition*, volume 1. MIT Press, 1986.

Self-Organising Artifical Neural Networks

John A. Flanagan and Martin Hasler
Department of Electrical Engineering, Swiss Federal Institute of Technology,
Lausanne, EL-Ecublens, CH-1015 Lausanne, Switzerland,
fax: +41 21 693 67 00, e-mail: flanagan@circnxt.epfl.ch

Abstract

Self-organisation is displayed during the unsupervised learning of the Kohonen Neural Network (KNN) algorithm. Classical Markov techniques, the ordinary differential equation (ODE) method being included among these, have so far not yielded a complete general analysis of self-organisation in the KNN. In order to obtain a more general understanding of self-organising behaviour which could then be applied to the analysis of self-organisation in the KNN two simpler self-organising algorithms are described. The first algorithm is based on a simple intuitive understanding of self-organisation. The second is based on a simplification of the KNN algorithm. Using the ODE method general results on the self-organising abilities of the two algorithms are given.

1 Introduction

Self-organisation is a type of behaviour exhibited by non-linear dynamical systems, and examples of this behaviour are abundant in many areas of science [1], [2]. One particular biological example is seen in the self-organisation of the neural connections between cells in the retina of the eye and neurons in the visual cortex. Self-organisation in this respect means that adjacent neurons in the cortex respond to adjacent stimuli in the retina. T. Kohonen has developed an Artifical Neural Network (ANN) model of this self-organisation process, the Kohonen Neural Network (KNN), which is generally recognised as a gross simplification of the real biological process. This simplified model does however retain the self-organising ability of the real process, and indeed has found many practical applications given that after an unsupervised training phase the neuron weights form a topological map of the input signal space [3], [4].

The general KNN algorithm can be described as follows. There are a total of N neurons labelled with the integers $1, \ldots, N$. Each neuron k has an associated position vector $\mathbf{I}_k = (i_{k1}, i_{k2}, \ldots, i_{kM}) \in \mathbb{N}^M$, which describes the position of the neuron in an M dimensional grid. This position vector does not change. The number of neurons B in any column of the neuron grid is the same [1].

The neuron weight vector for neuron i at iteration t is denoted by $\mathbf{X}_i(t) \in \mathbb{R}^D$, and where it only makes sense if $M \leq D$. Neuron weight j of the weight vector of neuron i is denoted X_{ij}. The neuron weights are initialised randomly and then the algorithm proceeds as follows. At iteration t an input $\omega(t) \in [0,1]^D$ with probability distribution function $\mu(\omega)$ is presented to the network. The *'winner'* neuron v is chosen such that

$$\|\omega(t) - \mathbf{X}_v(t)\| \leq \|\omega(t) - \mathbf{X}_i(t)\| \qquad \forall \, i \tag{1}$$

In the case where there is more than one possible choice of winner (i.e $\|\omega(t) - \mathbf{X}_r(t)\| = \|\omega(t) - \mathbf{X}_s(t)\| < \|\omega(t) - \mathbf{X}_i(t)\| \, \forall \, i \neq r, s$) then a predefined rule is used to choose the winner, for example in the one dimensional case the neuron with the smallest index could be chosen. The neuron weights are then updated as,

$$\mathbf{X}_i(t+1) = \mathbf{X}_i(t) + \alpha(t) \, h(d(\mathbf{I}_i, \mathbf{I}_v)) \, \{\omega(t) - \mathbf{X}_i(t)\} \tag{2}$$

The function h is a *'neighbourhood'* function $h : \mathbb{R}^+ \to \mathbb{R}^+$. In practise it is a decreasing function about zero, such that for $|a| \leq |b| \in \mathbb{R}^+$ then $h(a) \geq h(b)$.

The function $\alpha(t)$ is a gain $\alpha(t) \in (0,1)$ such that $\alpha(t)$ is constant or $\alpha(t) \to 0$, $t \to \infty$. The function d is a metric function which gives a measure of the distance between $\mathbf{I}_i$ and $\mathbf{I}_j$, which for $M = 1$, is generally chosen such that $d(\mathbf{I}_i, \mathbf{I}_j) = |i - j|$. The *width* of the neighbourhood function is taken to be the largest value of $d(\mathbf{I}_i, \mathbf{I}_j)$ for which $h(d(\mathbf{I}_i, \mathbf{I}_j)) \neq 0$. In an attempt to simplify equations $h(d(\mathbf{I}_i, \mathbf{I}_v))$ will be written as $h(i,v)$.

As seen the training phase of the algorithm itself is quite easy to implement however its self-organising ability has so far defied a general rigorous analysis. The main problem is to know when the neuron weights will converge to an organised configuration. Part of the difficulty also lies in defining what an organised configuration is. In a KNN

---

[1] note that $N = B^M$.

where $M = D = 1$ this organised configuration corresponds to the neuron weights being in one of the following configurations,

$$\mathcal{D}' = \{\mathbf{X} : X_1 < X_2 < \ldots < X_N\}$$

$$\mathcal{D}'' = \{\mathbf{X} : X_1 > X_2 > \ldots > X_N\}$$
(3)

Where the vector $\mathbf{X} = (X_1, \ldots, X_N)^T$. The reason for defining these configurations as organised is that they are *absorbing*, that is to say if $\mathbf{X}(t) \in \mathcal{D}'$ then $\mathbf{X}(t + T) \in \mathcal{D}'. T > 0$. This fact facilitates the analysis of the organising phase and has been analysed by several authors for the $M = D = 1$ KNN. By treating $\mathbf{X}$ as a stochastic process [5], or more precisely a Markov process, Cottrell and Fort [6] first demonstrated rigorously that for a width one neighbourhood function, and a uniform distribution μ that $\mathbf{X}$ converges to one of the configurations $\mathcal{D}', \mathcal{D}''$ with probability one in a finite time, and will remain there. This result has been further extended by Bouton and Pagès [7] for a more general μ. Erwin *et al* [8] have also given the outline of a proof for a uniform distribution μ and a neighbourhood function of width greater than or equal to $N/2$. The case of discrete inputs ω and discrete neuron weights has been treated by Thiran and Hasler [9], also for the $M = D = 1$ KNN and a general neighbourhood function. A more complete summary of all these results can be found in Cottrell *et al* [10].

Due to the absence of known absorbing configurations in KNNs where $M = D > 1$, the analysis of self-organisation is more difficult. Flanagan [11] has however defined an organised configuration in higher dimensions and shown that the first entry time of the neuron weights into this organised configuration is finite with probability one, given certain conditions.

All of these above mentioned proofs are based on the same technique of showing the existence of sets of inputs which will take the neuron weights from any initial state to an organised configuration in a finite time. This technique gives no great insight into the mechanism of self-organisation in the KNN. However there exists another technique for analysing stochastic processes called the Ordinary Differential Equation (ODE) method, described by Ljung [12], Kushner and Clark [13]. This technique has been applied to the analysis of the final convergence phase of the neuron weights in the KNN [14] (i.e. after self-organisation) but not in a rigorous manner to the analysis of self-organisation in the general KNN. The reason for this simply being that the technique is based on associating a set of determininstic differential equations to the update equations of the neuron weights, the neuron weights can then be shown [12] to converge to stable stationary points of these differential equations. The problem is however that these associated differential equations are formed by appropriately averaging the update equations for the neuron weights and the resultant ODEs are non-linear and complicated to analyse even for the simplest of cases.

The aim of this work is to analyse self-organisation in ANN inspired algorithms using the ODE method. As just explained the KNN is a self-organising ANN but difficult to analyse with the ODE method so now two self-organising algorithms are presented and the results obtained on their self-organising ability using the ODE method are summarised.

The first algorithm is the A algorithm which can be shown to have only stationary points to its ODEs which are in an organised configuration. The second algorithm is the AA algorithm which is a simplification of the KNN algorithm similar but more general than that described by Cottrell and Fort [15]. When analysing it with the ODE method however it turns out that its associated ODEs are linear, which makes its analysis easier compared to that of the KNN. Before describing these two algorithms a brief introduction to the ODE method is given.

2 The ODE Method

Consider a stochastic process generated by a recursive algorithm of the form

$$\mathbf{X}(t + 1) = \mathbf{X}(t) + \alpha(t)\Delta\mathbf{X}(\mathbf{X}(t), \omega(t))$$
(4)

where $\Delta\mathbf{X}(\mathbf{X}(t), \omega(t))$ is the update at time t and ω is a random variable. The associated ODEs are given by averaging the update $\Delta\mathbf{X}$ to give

$$\frac{d\mathbf{X}}{d\tau} = E(\Delta\mathbf{X}(\mathbf{X}(\tau), \omega(\tau)))$$
(5)

Where E is an ensemble average over the input space of ω and τ a pseudo-time variable related to the gain function α. A stationary point $\mathbf{x}_\infty$ of this set of ODEs satisfies,

$$\frac{d\mathbf{X}}{d\tau} = 0$$
(6)

Ljung has shown that the stochastic process $\mathbf{X}$ can only converge to stable stationary points $\mathbf{x}_\infty$ and the Kushner Clark theorem states that if the stochastic process visits the *basin of attraction* of a $\mathbf{x}_\infty$ infinitely often then $\mathbf{X}(t) \to \mathbf{x}_\infty, t \to \infty$ with probaility one. Using these ideas which have been very briefly described self-organisation is analysed in the A and AA algorithms which are now presented.

3 The A Algorithm and Self-Organisation

In this section the A algorithm is presented and then self-organisation is analysed using the ODE method. The A algorithm is quite similar to that of the KNN (i.e. structure of the neuron array, the notation etc). There is a difference however in that there is no explicit neighbourhood function and the update equation is different. New neuron indices $i'_{kj}, 1 \leq j \leq D$, are defined so that if $M = D$ we let $i'_{kj} = i_{kj}$. If $D > M$ then for $j \leq M$ we let $i'_{kj} = i_{kj}$ and for $D \geq j > M$ we arbitrarily let $i'_{kj} = i_{kM}$. The algorithm proceeds as follows. At iteration t an input $\omega(t)$ is presented to the network and the winner neuron v is chosen with probability p_v in a random fashion which may or may not depend on $\omega(t)$ and/or the $\mathbf{X}_j$'s. Each neuron weight is then updated as,

$$X_{kj}(t+1) = X_{kj}(t) + a(t)(\omega_j(t) - X_{kj}(t))$$
$$+ a(t)(i'_{kj} - i'_{vj})/N^{1/M} \tag{7}$$

where $k = 1, \ldots, N$, $j = 1, \ldots, D$.

Note there are two parts to the update. The first part $\omega_j(t) - X_{kj}(t)$ is essentially an attractive term which draws the neuron weights towards the input. The second part $(i'_{kj} - i'_{vj})/N^{1/M}$ is a term which forces the neuron weights apart. To prove that this algorithm organises, it is first necessary to define the organised configuration for this algorithm,

Definition 1 *Neuron weight vectors $\mathbf{X}_t$ and $\mathbf{X}_s$ are organised if for every j such that $i'_{tj} - i'_{sj} > 0$ then,*

$$X_{tj} - X_{sj} > 0 \tag{8}$$

The network is in an organised configuration when every pair of neuron weight vectors is organised.

Note that there is only one organised configuration, which in the $M = D$ case we normally accept as being similar to one of the organised configurations of the KNN. In fact we now show that if the input ω has a finite average value then in fact the only stationary states of this algorithm are in an organised configuration. This is shown using the ODE method.

Applying the ODE method to equation (8) as described in section 2 gives the set of associated ODEs,

$$\frac{dX_{kj}}{d\tau} = \int_\Omega (\omega_j - X_{kj}) d\mu(\omega) + \sum_{v=1}^N p_v \frac{(i'_{kj} - i'_{vj})}{N^{1/M}} \tag{9}$$

$\forall k, j$. If p_v is independent of the $\mathbf{X}_j$'s then the system of ODEs is linear with a single stationary point which is asymptotically stable, in the case where p_v depends on the $\mathbf{X}_j$'s this is not necessarily so. However for either case any stationary point must satisfy,

$$x_{kj} = \bar{\omega}_j + \sum_{v=1}^N p_v \frac{(i'_{kj} - i'_{vj})}{N^{1/M}} \quad \forall k, j \tag{10}$$

Thus we see that each equilibrium value of the neuron weight is the average value of the input plus a term which for constant p_v depends only on the indices. The difference $x_{tj} - x_{sj}$ is then given by,

$$x_{tj} - x_{sj} = \frac{i'_{tj} - i'_{sj}}{N^{1/M}} \tag{11}$$

Thus if $i'_{tj} - i'_{sj} > 0$ then $x_{tj} - x_{sj} > 0$. Note that this is true independent of whether p_v depends on ω or the $\mathbf{X}_j$'s and thus any stationary state is in an organised configuration. Figure 1 shows a plot of the initial state of the network for $M = D = 2$ and $N = 400$ and figure 2 shows the final stage after training with $\Omega = [0, 1]^2$ and ω and p_v uniformly distributed. We see that the result is what might be expected from the above analysis. With 400 neurons, the above algorithm with a gain factor of 0.4, reached an organised state in 15 iterations. This algorithm is strictly a self-organising algorithm and does not reveal anything about the distribution of the input signal. What it does however is centre the neuron weights in an organised fashion about the average value of the input. The algorithm in itself is trivial in its application, and its analysis is made trivial by the use of the ODE method. On its own this algorithm could only serve as a simple completely analysable self-organising algorithm. However it has been shown [11] how this algorithm can be combined with the KNN to accelerate the self-organising phase of the KNN algorithm.

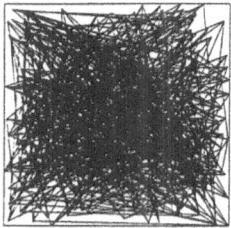

Figure 1: Initial state of the neuron weights for the A algorithm.

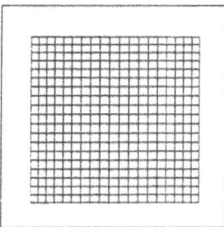

Figure 2: Neuron weights after training for the A algorithm.

4 Self-Organisation in the AA Algorithm

As an introduction to the AA algorithm consider the KNN algorithm described in the introduction. Consider the case where the Voronoi tessellation cell for each neuron [2] is *small* and thus the neuron weight is close to the centroid of its cell. For a given input then the update equation for weight j of neuron i can be approximated by,

$$X_{ij}(t+1) \approx X_{ij}(t) + a(t) h(i,v) (X_{vj}(t) - X_{ij}(t)) \tag{12}$$

Consider the case where this was the update equation for the KNN. What would happen? On closer examination it is seen that the whole map would collapse or rather all the neuron weights would converge to the same value. Thus this algorithm would not be very useful. To avoid this collapse consider a fixed set of boundary neurons which can be chosen as the winner neuron but whose weight values are constant. It is possible to then imagine that the neuron weights will not converge to the same values, but what would they converge to? Finally consider the case where there is no input as such to the algorithm but that the winner neurons are chosen randomly according to some probability law. This is essentially the new self-organising algorithm which is now proposed in a more formal manner.

In terms of the neurons the chief difference between the standard KNN and the AA algorithm is the use of fixed boundary neurons. In the KNN there are B^M neurons which are updated, the same convention will be used here, where these neurons are indexed by $1 \leq i \leq B^M$. These neurons are on an M dimensional grid with B neurons along any axis of the grid. The position vector of neuron k is given by $\mathbf{I}_k$ and is such that if i_{kj} is the jth component of this vector then $1 \leq i_{kj} \leq B$. The neuron weight vector for neuron k is as before given by $\mathbf{X}_k = (X_{k1}, X_{k2}, \ldots, X_{kD})^T$. The other set of neurons, the boundary neurons, of which there is a total of $(B+2)^M - B^M$ are indexed by $B^M + 1 \leq k \leq (B+2)^M$ and their position vector $\mathbf{I}_k$ is such that for at least one $1 \leq j \leq D$ that i_{kj} is either equal to 0 or $B+1$. The weight vector for a boundary neuron k is given by $\boldsymbol{\theta}_k = (\theta_{k1}, \theta_{k2}, \ldots, \theta_{kD})^T$. For the complete network there is a total of $N' = (B+2)^M$ neurons. For the other parameters everything else is the same as the KNN algorithm (i.e. neighbourhood function h, gain function a, and the distance function d).

The AA algorithm proceeds as follows. The neuron weight vectors $\boldsymbol{\theta}_k$ are initialised to values on the boundary of the unit hypercube in $\mathbb{R}^D$. For example,

$$\theta_{kj} = \frac{i_{kj}}{B+1} \qquad B^M + 1 \leq k \leq (B+2)^M, \ 1 \leq j \leq D \tag{13}$$

This initialisation is of course very important in determining the direction of organisation of the final state of the network, and an initialisation of this form will be assumed in the following. The neuron weights $\mathbf{X}_k$ are initialised

---

[2] the set of inputs ω for which the neuron is the winner neuron

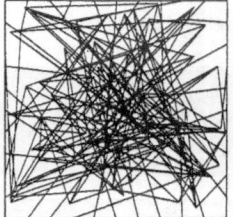
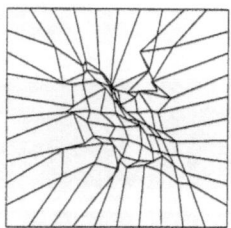

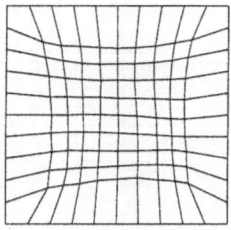

Figure 3: These three pictures show plots of the neuron weights for different parts of the training phase of the AA algorithm. The top left picture shows the initial state. The top right is after 250 iterations. The bottom picture shows the state after 500 iterations.

randomly. At iteration t a neuron $1 \leq v \leq N$ is chosen as the winner neuron, there is a probability $0 \leq p_v \leq 1$ of this happening for each neuron. The neuron weights are then updated, for v not a boundary neuron,

$$X_{kj}(t+1) = X_{kj}(t) + \alpha(t)\, h(k,v)\, (\, X_{vj}(t) - X_{kj}(t)\,) \tag{14}$$

for v a boundary neuron,

$$X_{kj}(t+1) = X_{kj}(t) + \alpha(t)\, h(k,v)\, (\, \theta_{vj}(t) - X_{kj}(t)\,) \tag{15}$$

The algorithm continues as in the KNN. The results of a simulation of this algorithm are shown in figure 3 for a $M = D = 2$ net with $B = 9$. The gain function was of the form $\alpha(t) = 150/(500 + t)$, and $p_i = p_j$, $\forall\, i, j$. In running this algorithm it is noticeable how the organised state is very quickly reached. One other point about this algorithm is that the width of the neighbourhood function can never be decreased to zero. Finally the algorithm is much quicker to simulate than the KNN because it is not necessary to calculate the Euclidean distance between the input and the neuron weights to determine the winner neuron, an operation which becomes time consuming in higher dimensions.

4.1 The ODE Method and the AA Algorithm

The ODE method is now applied to the analysis of the AA algorithm and its self-organising ability. Only the main results are given without their proofs which are lengthy. A fuller description is given in [11].

The update for neuron k weight j is given by,

$$\Delta X_{kj} = \alpha(t)\, h(k,v)\, (\, X_{vj}(t) - X_{kj}(t)\,) \tag{16}$$

The ODE for this equation is then given by the average of this update for every neuron winner, thus,

$$\frac{dX_{kj}}{d\tau} = \sum_{v=1}^{N} p_v\, h(k,v)\, (\, X_{vj} - X_{kj}\,) + \sum_{v=N+1}^{N'} p_v\, h(k,v)\, (\, \theta_{vj} - X_{kj}\,) \tag{17}$$

or collecting all the variable terms together gives,

$$\frac{dX_{kj}}{d\tau} = -X_{kj} \sum_{v=1}^{N'} p_v\, h(k,v) + \sum_{v=1}^{N} p_v\, h(k,v)\, X_{vj} + \sum_{v=N+1}^{N'} p_v\, h(k,v)\, \theta_{vj} \tag{18}$$

A differential equation like this results for all $1 \leq k \leq N$ and all $1 \leq j \leq D$. So to reduce the complexity of the notation the index j will be dropped and what follows applies to any $1 \leq j \leq D$. In effect there is no coupling between the different components of the neuron weight vectors, something which occurs in the ODEs for the standard KNN. From now on equation (18) will be written as,

$$\frac{dx_k}{d\tau} = -x_k \sum_{v=1}^{N'} p_v\, h(k,v) + \sum_{v=1}^{N} p_v\, h(k,v)\, x_v + \sum_{v=N+1}^{N'} p_v\, h(k,v)\, \theta_v \qquad 1 \leq k \leq N \qquad (19)$$

So the vector $\mathbf{x} = (x_1, x_2, \ldots, x_N)^T$ is the vector of the jth components of the weight vectors $\mathbf{X}_i$, $1 \leq i \leq N$, the vector of random variables $\mathbf{X}_j$ is then given by $\mathbf{X}_j = (\mathbf{X}_{1j}, \mathbf{X}_{2j}, \ldots, \mathbf{X}_{Nj})^T$. The stationary point of the system of equations is given as the solution of $\frac{dx_k}{d\tau} = 0$, $1 \leq k \leq N$. Which results in the set of equations,

$$-x_k \sum_{v=1}^{N'} p_v\, h(k,v) + \sum_{v=1}^{N} p_v\, h(k,v)\, x_v = -\sum_{v=N+1}^{N'} p_v\, h(k,v)\, \theta_v \qquad 1 \leq k \leq N \qquad (20)$$

Written in matrix notation as,

$$\mathbf{A}\mathbf{x} = -\mathbf{b} \qquad (21)$$

where the constant $N \times N$ matrix,

$$\mathbf{A} = \begin{pmatrix} -a_{11} & a_{12} & a_{13} & \cdots & \cdots & a_{1N} \\ a_{21} & -a_{22} & a_{23} & \cdots & \cdots & a_{2N} \\ \vdots & \ddots & \ddots & \ddots & \ddots & \vdots \\ \vdots & \ddots & \ddots & \ddots & \ddots & \vdots \\ a_{N-11} & \cdots & \cdots & \cdots & -a_{N-1N-1} & a_{N-1N} \\ a_{N1} & \cdots & \cdots & \cdots & \cdots & -a_{NN} \end{pmatrix} \qquad (22)$$

with $a_{kk} = \sum_{v=1}^{N'} p_v\, h(k,v)$, and $a_{kj} = p_j\, h(k,j)$, $k \neq j$. The constant $N \times 1$ vector $\mathbf{b}$ is given by,

$$\mathbf{b} = \begin{pmatrix} b_1 \\ b_2 \\ \vdots \\ \vdots \\ b_{N-1} \\ b_N \end{pmatrix}, \qquad b_k = \sum_{v=N+1}^{N'} p_v\, h(k,v)\, \theta_v \qquad (23)$$

Where once again $x_k = X_{kj}$ for any $1 \leq j \leq D$, and $\mathbf{X}_j = (\mathbf{X}_{1j}, \mathbf{X}_{2j}, \ldots, \mathbf{X}_{Nj})^T$. It is this system of equations which will be used to analyse self-organisation for the newly proposed algorithm. The next theorem forms the basis for the analysis of the algorithm using these equations,

Theorem 1 *If* $\mathbf{A}$ *is nonsingular the system of equations (21) has the unique solution* $\mathbf{x}_\infty = -\mathbf{A}^{-1}\mathbf{b}$. *Thus for any* $\mathbf{X}_j(0) \in [0,1]^D$ *this solution is the single asymptotically stable stationary point of the stochastic process described by equations (14) and (15), assuming the Robbins-Monro conditions on the gain function* α.

Proof : The proof of this theorem is a direct result of the Kushner-Clark Theorem [13]. Given that the ODEs for the algorithm are linear, and that the set $[0,1]^D$ is an invariant set for the process, it is quite easy to check that the conditions for the Kushner-Clark Theorem are satisfied. If $\mathbf{x}_\infty$ is the stable point then as it is the unique solution of the linear system then it can be shown that the matrix A is dominantly diagonal (i.e. $\sum_{i=1, j\neq i}^{N} a_{ij} < a_{ii}$, $\forall i, j$) with negative diagonal terms. Thus from Gersgorhin's theorem [16] all eigenvalues of $\mathbf{A}$ must have a negative real part. Thus the system is asymptotically stable. $\square$

Given this linear matrix equation $\mathbf{A}\mathbf{x} = -\mathbf{b}$ we would like to known when it has a unique solution. The existence of a unique solution is given in terms of the *connectedness* of the neurons which is defined as follows,

Definition 2 *A neuron* $0 \leq i_0 \leq N$ *is said to be 'connected' to a neuron* i_k *if there exists a set of neurons,* $\{i_1, i_2, \ldots, i_k\}$ *such that,*

$$\prod_{r=1}^{k} p_{i_r} h(i_{r-1}, i_r) > 0 \qquad (24)$$

We note that as a result of this definition that a neuron i cannot be connected to a neuron j if $p_j = 0$. The existence of a unique solution is given in terms of the connectedness of all the neurons in the network to a boundary neuron.

Theorem 2 *There exists a unique solution* $\mathbf{x}_\infty$ *to the set of equations* $\mathbf{Ax} = -\mathbf{b}$, *'iff' every neuron* $i, 1 \le i \le N$ *is connected to at least one boundary neuron* $k, N < k \le N'$.

Having obtained a result on the existence of a unique solution it is interesting to analyse the form of this solution, or rather the ordering of the neuron weights in this stationary state. In the $M = D = 1$ case for a Gaussian type neighbourhood function h such that,

$$h(a) < h(b), \mid a \mid > \mid b \mid$$

$$h(d(1, N+2)) > 0 \tag{25}$$

it is quite trivial to show the following,

Theorem 3 *If* $p_{N+1} > 0$ *and* $p_{N+2} > 0$ *then the equilibrium point solution* $\mathbf{x}_\infty = (x_{1\infty}, \ldots, x_{N\infty})$ *is such that,*

$$0 < x_{1\infty} < x_{2\infty} < \ldots < x_{N\infty} < 1 \tag{26}$$

What is interesting about the result of this theorem is that the organised state has not been pre-specified but rather it is a natural result of the ODEs associated with the system. Also unlike other proofs of self-organisation this proof is not specific to a particular type of neighbourhood function or probability distribution function. The restrictions on the neighbourhood function and the probability distribution function are included in the connectedness property of the neurons to the boundary neurons.

Unfortunately this result is not readily generalisable to the case of $M = D > 1$, however it is possible to analyse the equations and it becomes obvious that if there is a *regular* [11] distribution of the probabilites p_v then the stationary state will be in what we would consider to be an organised configuration. From this analysis an appreciation of the self-orgainising ability is obtained.

The analysis of this AA algorithm has also been applied to the analysis of self-organisation in the KNN [11] by assuming that a KNN with small gain factor and a large number of neurons is approximated by the AA algorithm.

5   Conclusion

In this paper the question of self-organising artifical neural networks has been presented. The first ANN algorithm to demonstrate self-organisation, the KNN, is presented. This is followed by a discussion which gives a brief overview of the state of the art in the analysis of the self-organising ability of the KNN. It is pointed out that the Markovian technique used so far to analyse self-organisation in the KNN is not suitable for very general cases. Another promising analysis tecnhique, the ODE method, is presented which allows the convergence of stochastic processes to be analysed using a set of associated deterministic ordinary differential equations. However when it is applied to the analysis of the KNN a set of complicated non-linear ODEs are obtained.

It is however still desirable to use the ODE method to analyse self-organisation, and to this end two other self-organising algorithms are presented, the A and AA algorithms. The A algorithm is a trivial self-organising algorithm but it is possible to make a general analysis of its self-organising behaviour and from it we obtain a more intuitive idea of the requirements an algorithm must satisfy for self-organisation to occur (i.e assymetrical attraction and repulsion of the neuron weights). The AA algorithm turns out to be a simplification of the KNN algorithm and its main feature is the linearity of its ODEs (despite the nonlinearity of the algorithm itself). Given these linear equations it is possible to determine conditions which must be satisfied for the algorithm to converge to a stationary state, and in the one dimensional case this is easily shown to be an organised state.

In presenting the A and AA algorithms we see that self-organisation is not just a characteristic of the KNN algorithm and that it is possible to generate other simpler self-organising algorithms. These algorithms are less practically useful than the KNN algorithm, but they still retain the essential features for self-organisation to occur. It also turns out that their self-organising ability is more readily analysable in a general case using the ODE method. Given these results a greater appreciation of the general conditions necessary for self-organisation to occur can be obtained, which allows for a better understanding of the behaviour of the KNN algorithm.

Acknowledgements

This project is financially supported by the Swiss National Science foundation SPP-IF grant no. 5003-34353.

References

[1] G. Nicolis and I. Prigogine. *Exploring Complexity*. W. H. Freeman, New York, 1989.

[2] E. Jantsch. *The Self-Organising Universe*. Pergamon Press, Oxford, New York, 1980.

[3] T. Kohonen. *Speech Recognition based on topology preserving neural maps*. Kogan Page, London, 1989.

[4] F. Blayo and P. Desmartines. Kohonen algorithms : Application to the analysis of economic data. *Bulletin des Schweizerischen Eletrotechnischen Vereins und des Verbandes Schweizerischer Elektrizitaetswerke*, 85(5):23–26, 1992. (in French).

[5] E. Parzen. *Stochastic Processes*. Holden-Day, Inc., San Francisco, London, Amsterdam, 1962.

[6] M. Cottrell and J.C. Fort. Étude d'un processus d'auto-organisation. *Ann. Inst. Henri Poincaré*, 23(1):1–20, Jan. 1987.

[7] C. Bouton and G. Pagès. Self-organisation of the one-dimensional Kohonen algorithm with non-uniformly distributed stimuli. *Stochastic Processes and their Applications*, 47:249–274, 1993.

[8] E. Erwin, K. Obermayer, and K. Schulten. Self-organising maps: Ordering, convergence properties and energy functions. *Biol. Cybern.*, 67:47–55, 1992.

[9] P. Thiran and M. Hasler. Self-organisation of a one dimensional Kohonen network with quantized weights and inputs. To appear in *Neural Networks*.

[10] M. Cottrell, J.C. Fort, and G. Pagès. Two or three things we know about the Kohonen algorithm. In *Proceedings of ESANN*, 1994.

[11] John A. Flanagan. Self-organising neural networks, 1994. Thèse no 1306(1994) Electricity Department, EPFL, Lausanne, Switzerland.

[12] L. Ljung. Analysis of recursive stochastic algorithms. *IEEE Trans. on Automatic Control*, AC-22(4):551–575, Aug. 1977.

[13] H. J. Kushner and D. S. Clark. *Stochastic Approximation Methods for Constrained and Unconstrained Systems*, volume 26 of *Applied Mathematical Sciences*. Springer-Verlag, New York, Heidelberg, Berlin, 1978.

[14] C. Bouton and G. Pagès. Convergence in distribution of the one-dimensional Kohonen algorithm when the stimuli are not uniform. *Advances in Applied Probability*, 26(1), March 1994.

[15] M. Cottrell and J.C. Fort. A stochastic model of retinotopy : A self organising process. *Biol. Cybern.*, 53:405–411, 1986.

[16] W. Kaplan. *Advanced Mathematics for Engineers*. Addison-Wesley, New York, 1981.

A Distributed Classifier Based on Yprel Networks Cooperation

Emmanuel STOCKER, Yves LECOURTIER, Abdel ENNAJI

Université de Rouen, La3i
UFR des Sciences et Techniques
F-76821 Mont Saint Aignan Cedex, France
e-mail : stocker@la3i.univ-rouen.fr

Abstract : In this paper we present a scheme of classification based on a particular processing element ("neuron") called Yprel. The main characteristics of the approach are: (i) an Yprel classifier is a set of Yprels networks, each network being associated with a particular class; (ii) the learning is supervised and conducted class by class; (iii) the structure of the network is not a priori chosen, but is determined step by step during the learning process; (iv) the learning process is incremental : each network improves its own learning base with the errors of the previous test; (v) networks cooperate : each network can use the outputs of the previously builded networks. Preliminary results are given on a well-known classification task (recognition of typographic characters).

I. INTRODUCTION

Classification is a key task in any problem of pattern recognition. Several classical textbooks developing the main concepts and algorithms (Bayes approach, K-Nearest-Neighbours, Parzen-Windows etc ...) are available (see e.g. [BAL82, BEL92, BRE84,...]). A renewal of interest has arisen during the last decade due to the emergence of neural network methodology [FOG87, HEC90, KAM90, KOH88, RUM86,...]. However, even if some papers give comparisons between classical and neural approach [IDA92], some fundamental problems remain not satisfactorilty solved. Three of them are dealt with this paper: (i) the building of a network structure adapted to a given goal; (ii) the ability to easily retrain the classifier with an improved learning set, and (iii) the way to make cooperate a set of networks. These points have been recognized as key problems for neural network methodologies by several authors during the last years [FAH90, FRE90, GEN93, HUN93, PER93].

In the first part of the paper, we briefly recall the main points of the Yprel methodology. Details are developed in [LEC93a-b] and [LEC94]. Then we present (i) the implemented strategy to produce an incremental learning, and (ii) a scheme of networks cooperation in learning which improve the performances of the classifier. The last part of the paper gives some results obtained from the proposed methodology.

II. YPREL NETWORKS AND CLASSIFICATION

Yprel methodology has been developed to perform a classification task after a supervised learning phase. An Yprel classifier is a set of Yprels networks, where each network is linked to a particular class. Each network must recognize the elements of the learning set which belong to its class and must reject all the elements from the other classes. For a few elements, a network can come to no conclusion, it means that these learning data belong to a region of the input space, for which the learning process cannot take a decision. Thus, a network acts as a specialized classifier for a particular two-classes problem: the class it has to identify and the set of all the other classes. Each network of the classifier can be considered as an independent agent or an expert which tries to solve its own goal. In fact, this goal is a sub-problem of the whole classification task. Such an approach presents several advantages. Since the goal to be solved by a network is simpler than the general problem, we have a real reduction of the task for each network, so its structure and its learning are simplified. Moreover, we can improve the classifier by retraining only a few bad networks without decreasing the performances of the others. Finally, the networks being independent from each other, an implementation on a parallel machine is possible with a distribution of the networks on several processors.

Each network needs a specific knowledge to solve its particular goal. So, the learning data used must be different according to the network to build. An Yprel network uses a particular subset of the whole training set as its own learning set. In this part of the paper, we will just assume that it is possible to build such a learning set for each network.

The elaboration of the network decision is based on the following assertions: (i) each Yprel of a network tries to come to a conclusion, and (ii) if a previous Yprel in the network take a decision, every following Yprels just transmit it. Thus, the decision mechanism is an entire part of the functionning of the network. This differs from other neural networks like the Multi Layer Perceptron which does not take any decision, but calculate likelihoods: thus, the classification mechanism is an extra mechanism, the simpler of them is: the "winner take all". An example of Yprel network associated with a class is given on Figure 1 for a problem with only two input features.

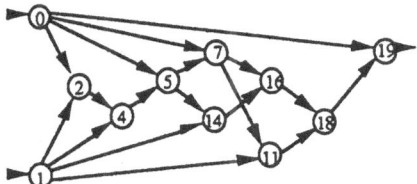

Figure 1: An example of Yprel network

A first layer of Yprels is built, each Yprel is associated with a feature (Yprel 0 and 1). Such Yprels work mainly as a normalizing factor on the inputs data. All the other Yprels in the network have two inputs which are the outputs of two previous Yprels. Each Yprel can have multiple outputs. In this kind of networks, there is no real layers structure, since each Yprel output can become an input for any other Yprel. Usually, in neural methodologies the network structure is chosen from an empirical knowledge, and the neural network user has just to adapt the weights on each connexion. With Yprel methodology, the structure of the network is not a priori chosen: the structure determination is a part of the learning process. This learning process is based on cooperative and competitive schemes. Such an approach could be compared to the ones used in Genetic Algorithms [GOLD92], but without requiring an explicit genes encoding.

During the learning phase of each network, we generate a set of new Yprels. Each Yprel becomes the terminal Yprel of a sub-net which tries to come to a conclusion about all the elements of the network learning set. All the created sub-nets have the same goal, but each is candidate to be the best answer to the two-classes problem of the network. Only the sub-net with the best performances will become the final network. This principle of competition is illustrated on Figure 2.

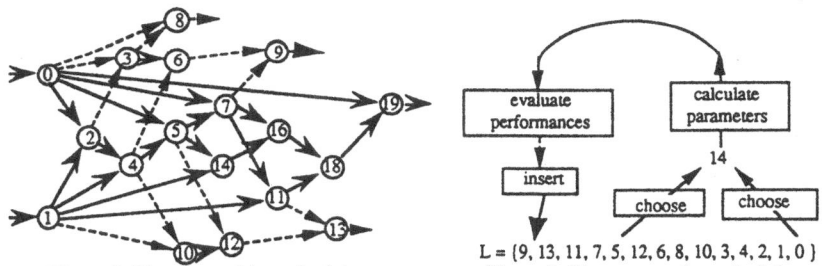

Figure 2: The competition principle. Figure 3: The cooperation principle.

The Figure 3 illustrates the way to generate sub-nets. The list L represented on figure 3 corresponds to the list of created Yprels classified by performances. The output of each Yprel of the list is the output of a particular sub-net which can conclude about some elements of the network learning set. One step of the learning algorithm consists in making cooperate the classification capacities of two of these sub-nets by choosing the outputs of their terminal Yprels as the two inputs of a new Yprel (principle of cooperation). The created Yprel is inserted in the list L, only if its number of non-classified elements is lower than the number of non-classified elements of both inputs. The number of non-classified elements acts as the Yprels performance criterion. The generation of new Yprels stops when all the elements of the network learning set are classified by a new Yprel, or when the maximum number of trials is reached. Then we select the best Yprel of the list L (the Yprel with the minimal number of non-classified elements) and all the Yprels used to build its inputs to become the new network for the class under study.

With this learning algorithm based on competitive and cooperative schemes, the parameters of each Yprel are calculated only once, when this Yprel is used as a terminal element of a new sub-net. They are never modified later, so the calculation to generate a new sub-net is reduced to calculate the parameters of only one Yprel. It is a very fast procedure. Yprels parameters calculation, action of an Yprel network in the features space and algorithm to choose the inputs of a new Yprel are detailed in [LEC93b] and [LEC94].

III. INCREMENTAL LEARNING

The performances of a network depend highly on its learning set (LS). However, it is impossible to know which samples are the most useful to collect from LS, in order to obtain a network with the best possible performances. The way used to solve this problem is to let the network itself select its own learning set. The algorithm used to build the network associated with the class k is the following:

1 - Select randomly few samples of each class in the learning set LS, and call this subset LB_k: the Learning Base of the network associated with the class k;
2 - Use the learning base LB_k to determine the best network associated with the class k by applying the methodology of section 2;
3 - Test the whole Learning Set LS and select all the samples for which the network gives an erroneous answer; call the set of these samples the Error Base EB_k associated with the network;
4 - while performances are not considered good enough,
    4.1 - Select among the samples of the error base EB_k a subset of elements (belonging either to class k or to other classes) and add them to the learning base LB_k;
    4.2 - Use the learning base LB_k to determine a new network associated with the class k by trying to use again the informations of the previous network at first, and by applying the methodology of section 2;
    4.3 - Test the whole learning set LS and build an error base EB_k as in step 3.

The incremental learning algorithm is illustrated by the following Figure:

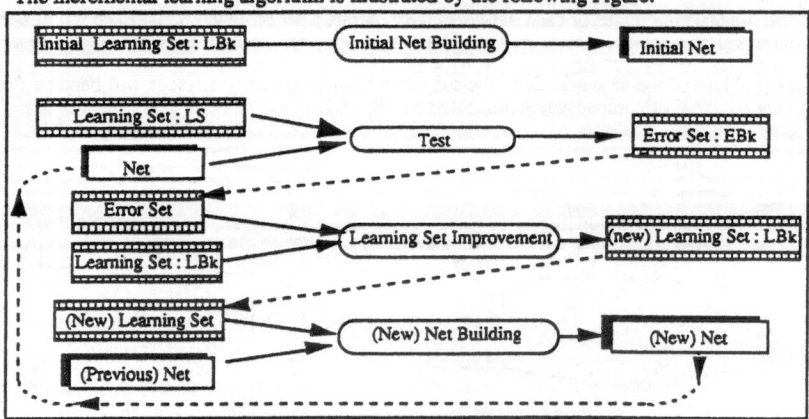

Figure 4 : Incremental Learning Principles

Some comments can be made about this algorithm:

- First, if the initial learning base LB_k can be the same for all the classifier networks, the resulting LB_k at the end of learning of each class will be different. Indeed, the goal of the network being different for each class, the errors made by the network will be different. So the samples added to LB_k depend on class k and on the performances of the previous network. Moreover, it is possible to change the whole learning set LS. Indeed, all the useful elements of the previous learning set are kept in the set LBk, then we can always improve the network learning with the samples selected inside a new learning set LS.

- Two of the key ideas of the methodology are to obtain for each class a learning base LB_k as reduced as possible, and to have in this network learning base a good representation of the class k and all the nearest other classes. For that purpose, we add only a particular subset of the error base EB_k which is erased and re-calculated at each iteration.

- Having a reduced learning base LB_k leads to learning phases faster than learning phases on the whole learning set LS. Reduction factor is important: usually after the first learning, networks are able to correctly reject more than 98% of the elements to be rejected.

- As the used methodology concentrates on the definition of borders of convex domains, the useful elements to add to the network learning set are those close to such borders. These elements are the most difficult to classify whatever the scheme of classification used. So, an efficient learning must enable to select them. In this way, we associate to each element of the error base EBk a Hamming distance which is calculated in the features space. If the EBk elements belong to the class k, we calculate the distances with the furthest LBk elements of the class k. On the contrary, if the EBk elements belong to another class, the distances are calculated with the nearest LBk elements of the class k. The EBk elements subset we add to the previous learning set contains the class k elements which have the longest distances, and for each other class we add a particular number of EBk elements which have the shortest distances. These numbers are linked to the numbers of errors made by the networks in each class.

- On step 4.2 of the algorithm, it seems interesting to re-use the previous learning informations and performances. Indeed, the network obtained during the last iteration can correctly conclude about all the elements of the learning set, except about the new selected ones. So, this network can become an efficient sub-net of the new network. The way used to take into account such information is to rebuild an initial list L with all the Yprels of the network obtained at the previous step by re-calculating the Yprels parameters with the new learning set.

- The step 4 of the algorithm requires the definition of a performance criterion. The current criterion is that the network under construction can conclude about all the elements of the whole learning set. Such a criterion often leads to an over-learning network which increases the Yprels number without improving the network performances. Among the information that such a criterion must integrate, the improvement of number of well classified elements of the learning set LS is one of the most important

IV.COOPERATION OF YPRELS NETWORKS

An Yprel classifier is a distributed classifier, where each network works as an independent agent which attempts to reach its own goal: to identify the elements of its class and reject all the others. The networks are built class by class. One of the network sub-goals associated with the class k, is to reject all the elements which belong to the classes 0 to k-1. No input feature enables to easily conclude if an element belongs to these classes. Nevertheless, the networks linked to the classes 0 to k-1 are built and have solved their own goals. They can conclude if an element belongs to these classes. So, the necessary informations to solve this particular sub-goal of the class k network can be found in the re-use of knowledges of the networks associated with the classes 0 to k-1. A direct re-use of the networks knowledges associated to the previous classes must simplify the learning phase of the network under construction. The way used to take into account such knowledges during a learning phase leads to make cooperate a set of previously built networks with the network in learning. For this purpose, we consider the set of built networks outputs as a set of supplementary possible inputs features for the network in learning. These feature values are obtained by simulating all the elements of the current learning set in each network. Then, the network in learning associated to the class k must choose its inputs among the set of the initial features and the set of the networks outputs linked to the classes 0 to k-1.

In this part of the paper, we call sub-net a network which is used as an input of another network. This cooperation process in learning is made class by class for all the classifier networks. Then, the sub-nets chosen by a particular network used their own sub-net sets. The different cooperation links between all the networks can be represented by a tree structure where each node is a particular network. The following figure gives an example of cooperation links which are obtained for the network associated with the class 36. This network directly uses as sub-nets the networks linked to the classes: 11, 15, 17, 18, 20 and 22.

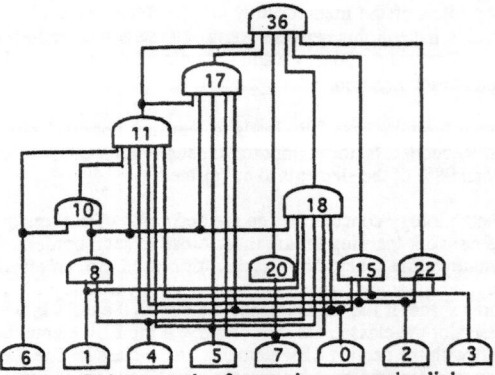

Figure 5: An example of networks cooperation links.

The simulation of a particular network requires to simulate all the sub-nets present in this graph. To implement this cooperation scheme, we have created a global supervisor. Its role is to manage the different simulations by having a representation of all the networks cooperation links. During a learning or a testing phase, each network communicates with the agent supervisor by sending a message to ask for a simulation of a particular sub-net.

As we can see on the figure 5, a particular sub-net can be used several times in the same network. On one hand, it can be directly used as an input of the network. On the other hand, it can also be used as a sub-net inside other sub-nets. Then, to avoid simulating many times the same sub-net, the results of the first simulation are recorded on a particular module working as a BlackBoard (according to the A.I meaning). If other simulations of the same sub-net are useful, the cooperation process between the different networks will be made through the BlackBoard. A network communicates with this module by sending messages to read the sub-net simulation values.

One of the most interesting problem which is raised by this cooperation methodology is that we have introduced in the construction of the Yprel classifier the notion of Temporality in learning. Indeed, a network in learning can only use the knowledges of previously built networks. So, all the possible cooperation links between several networks depend on the order in which the networks are built. Then, it becomes necessary to plan the learning process of each class with a view to build the most useful cooperation links to improve the classifier performances.

Now, other ways to use this Temporality notion and this cooperation scheme are under study: (i) to build for the same classification task a second set of Yprels networks associated with the same classes by making cooperate these new networks with the previous set of networks. (ii) To associate with one class a set of cooperating networks, whose goals are more and more complicated according to the Temporality.

V. PRELIMINARY RESULTS

Some preliminary results have been obtained with the previous methodology. The problem under consideration was the classical problem of multifont optical character recognition. The samples of the learning set LS are of eleven different fonts as illustrated on Figure 7, of various size, some of them being slightly rotated.

A A A A A G G G G 6 6 6 t t t t
Figure 6: examples of characters of the learning set

A total of 8700 samples has been collected (upper-case and lower-case letters, and digits). The size has not been used as one discriminate feature. So upper an lowercase letters such as V and v, or U and u have been merged in the same class. A total of 45 different classes is considered. Moreover, a set of additional samples (like little square or star...) not belonging to a useful class has been added :such elements must be rejected by all the networks. From the image of each sample, a set of 50 features has been extracted. The features used are classical features in the field of OCR, namely the dimensions of the surrounding rectangle, the number of intersections with lines, the distances between the surrounding rectangle and the first pixel of the character, the densities of pixels in some regions etc...

The following results must be considered as preliminary results obtained without any optimisation. Their goal is to validate the key points of the learning methodology.

The following histograms point out the values of some parameters obtained for the set of 45 networks when the learning process is stopped. The first one gives the number of steps in the learning process before reaching a learning without error on the whole training set. The second histogram gives the number of Yprels of the obtained networks. It can be seen that all of them have less than 200 Yprels and 2/3 of them have less than 75 Yprels. Then the third histogram shows that, among the 50 external features proposed, each network selects the usefull ones for its particular task. None of them uses the whole set of features, most of them using only half of these features. As we have seen with the cooperation scheme, besides the external features, the outputs of previous builded networks can also be used as internal features. The second part of the histogram gives us the total number of features (both internal and external) used by the networks. It must be noticed that the networks have to select their inputs among an increasing set of features, the last one having to choose among 100.

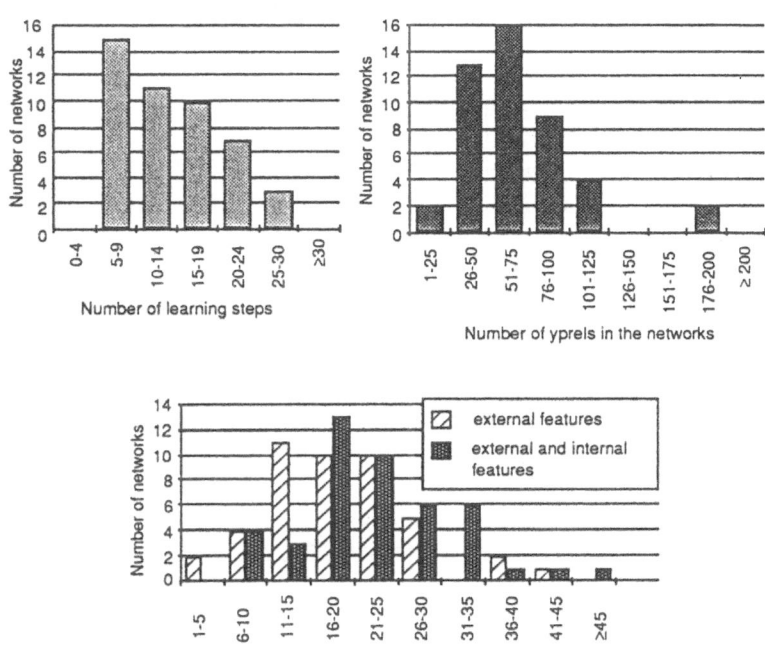

The next histogram displays the number of elements selected among the 8700 ones of the whole training set as the usefull elements in order to precise the borders of the class to be identified by the network. Each network selects between 3% and 10% of the training set. Moreover, the selected samples are well-suited for the task: as an example, among the 491 samples selected by the network "1" there are 115 "1" but also 82 "I" and 23 "t", and 38 classes have no sample in the selected set. Similarly, among the 373 samples selected by the network "5", there are 115 "5" but also 32 "S" and 37 "6", and 37 classes have no sample in the selected set.

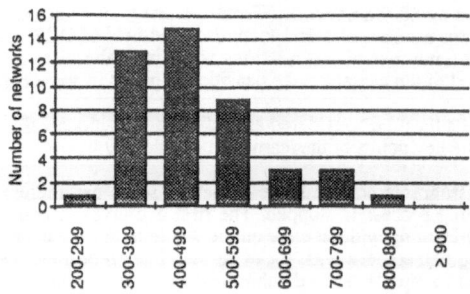

Number of elements selected in the whole learning set

The properties of the classifier have been tested using a test set of 3888 samples (among them 162 do not belong to a class and must be rejected by all the networks) although, with incremental learning, there is no difference between a training set and a test set because any error made can be used to improve learning set of the faulty network. The following histogram shows that, if the rejection process works well for all networks, the recognition process must be improved. However, it must be noticed that the whole training set have only between 100 and 200 samples of each class. So, it can be hypothetized that this number is too small to reach a good recognition rate according to the number of classes.

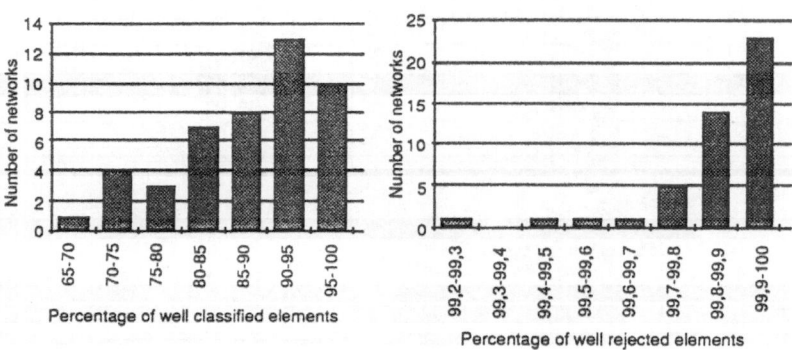

Percentage of well classified elements

Percentage of well rejected elements

The cooperation scheme is illustrated by the next histogram.

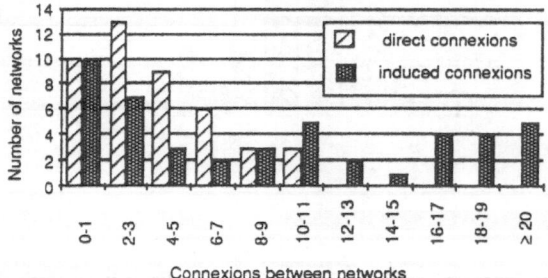

Connexions between networks

This histogram shows that 70% of the networks use less than 5 direct links to other networks. As it is illustrated by the cooperation tree of the Figure 5, a particular network used to take a decision a set of direct sub-nets (direct connexions) which have their own sub-nets (induced connexions of the network). As we can see, even with quite direct connexions the networks access to informations of a great number of networks.

The global performance of the classifier (set of networks) can be summarized by the following table :

Recognition	3221	86,5%
Correct rejection	144	88,9%
Ambiguity 2 networks	101	2.6 %
Other ambiguities	2	0.05 %
Total of correct decisions	3468	89,2%
Rejection	299	7.7 %
Confusion	121	3.1 %
Total base	3888	100 %

"Ambiguity k networks" means that the network associated to the class of the analysed sample gives a right answer but that k-1 other networks do not reject the sample. A confusion arises when the network associated to the class rejects the sample while one or several other networks accept it. The table shows that the main drawback of the low recognition rate of some networks is to generate rejection : all the networks (among which the one associated to the class of the sample) reject the sample. So an improvement of the learning set with more samples for each class would be improve the results.

VI CONCLUSION

This paper shows that, for supervised classification, good performances can be obtained with a particular network associated to each *a priori* classe. The structure and the training of each network is based on an incremental learning algorithm. This incremental algorithm enables to increase the classifier performances by improving the networks learning with several differents training sets. Moreover, the networks can use the knowledges of the previously built networks. This cooperation process between the networks introduce the notion of the temporality during a learning phase. The consequences of the temporality on the classifier performances and the scheme of cooperation between several networks associated with the same class in order to learn a strategy of recognition are currently under study.

REFERENCES

[BAL82] BALLARD A. H., BROWN C.M., *Computer vision*, Prentice-Hall, 1982.

[BEL92] BELAID A, BELAID Y, ., *Reconnaissance des formes*, InterEditions, 1992.

[BRE84] BREIMAN L., FRIEDMAN J.H., OLSHEN R.A., STONE C.J., *Classification and Regression Trees*, Wadsworth&Brooks, Pacific Grove, CA, 1984.

[FAH90] FAHLMAN, S.E., LEBIERE, C., The Cascade-correlation learning architecture, *Advances in Neural Information Processing Systems*, D.S. Touretsky Ed., 2, 524-532, Morgan Kauffmann, 1990.

[FOG87] FOGELMAN-SOULIE, F., ROBERT, Y., TCHUENTEE, M. *Automata Networks in Computer Science*. Manchester Univ. Press, 1987.

[FRE90] FREAN, M. The Upstart algorithm: a method for constructing and training feed-forward neural networks, *Neural Computation*, 2, 198-209, 1990.

[GEN93] GENTRIC P., WITHAGEN H., Constructive Methods for a new Classifier Based on a Radial-basis-function Neural Network accelerated by a Tree, *New Trends in Neural Computation*, J. Mira, J. Cabestani, A. Prieto Eds.,125-130, Springer-Verlag, Berlin, 1993.

[HEC90] R. HECHT-NIELSEN, *Neurocomputing*, Addison-Wesley, Reading, MA, 1990.

[HUN93] H. HÜNING, A node Splitting Algorithm that reduces the Number of Connections in a Hamming Distance Classifying Network, *New Trends in Neural Computation*, J. Mira, J. Cabestani, A. Prieto Eds.,102-107, Springer-Verlag, Berlin, 1993.

[IDA92] Y. Idan, J.M. Auger, N. Darbel, M. Sales, R. Chevallier, B. Dorizzi, G. Cazuguel, Comparative study of neural networks and non parametric statistical methods for off-line handwritten character recognition, *Proceedings ICANN 92*, Brighton, September 1992.

[KAM90] KAMP, Y., HASLER, M. *Réseaux de neurones récursifs pour mémoires associatives*. Presses Polytechniques Romandes, Lausanne, 1990.

[KOH88] KOHONEN, T. *Self organisation and Associative Memory*. Springer series in Information Sciences, Springer Verlag, Berlin,1988.

[LEC93a] LECOURTIER Y., DORIZZI B., SEBIRE P., ENNAJI A., MLP Modular versus Yprel Classifiers, *New Trends in Neural Computation*, J. Mira, J. Cabestani, A. Prieto Eds., 569-574, Springer-Verlag, Berlin, 1993.

[LEC93b] Y. LECOURTIER, A. ENNAJI, F. GILLES, P. CHAVY, Yprel networks and classification. IEEE SMC Conf, 3, 463-468, Le TOUQUET 1993.

[LEC94] Y. LECOURTIER, A. ENNAJI, E. STOCKER, F. GILLES, Réseaux d'Yprels, classification et apprentissage incrémental. Actes du 3ème colloque CNED, 109-118, Rouen 1994.

[MAR91] G. L. Martin, J. A. Pittman, Recognizing hand printed letters and digits using backpropagation learning, *Neural Computation* 3, 258-267, 1991.

[PER93] PEREZ J.C., VIDAL E., Constructive Design of LVQ and DSM Classifiers, *New Trends in Neural Computation*, J. Mira, J. Cabestani, A. Prieto Eds., 334-339, Springer-Verlag, Berlin, 1993.

[RUM86] RUMELHART, J , Mc CLELLAND, J. *Parallel Distributed Processing*, MIT Press, Cambridge MA, 1986.

A Fractal, Selforganizing Map with Partially Chaotic Neurons

A. Kosak and K. Goser
University of Dortmund, Faculty of Electrical Engineering
D 44221 Dortmund, Germany
email: kosak@luzi.e-technik.uni-dortmund.de
goser@luzi.e-technik.uni-dortmund.de

Abstract

A new type of a chaotic neuron is proposed which allows a new, simplified architecture of selforganizing maps. The chaotic neurons are examined with respect to their application in self-organizing maps. A new concept of self-organizing maps with fractal architecture is proposed and its suitability for a VLSI-implementation is examined. It is shown that the efficiency of conventional selforganizing maps can be overcome while the expenditure, with regard to a hardware-realisation, is notably reduced.

1. Introduction

Due to observations that different areas of the human brain are specialized in particular tasks, Kohonen proposed self-organizing maps in 1981 [1] [2]. The simulation of self-organizing maps on conventional computers is restricted by the immense demand of computing power, so even with simplified algorithms only the implementation of small maps is possible so far. On the other hand, even simple applications require large numbers of neurons, so a hardware realization is expected to solve this dilemma. For that reason several attempts for a hardware realization were made, some of which will be mentioned in the following.

Among others, one publication develops the concept of a neural coprocessor [4], where a conventional computer controls the procedure and a neural coprocessor handles the operations which are sensitive to computing time. Another publication reports the successful hardware realisation of a 2x2 map with binary input vectors BISOM [5]. Although the hardware realization could be the appropriate way to solve the problem of computing power, it reveals many new problems, many of them insoluble for the current technology.

In summary the present status quo prevents practical applications because on the one hand the implementation of self-organizing maps on computers is by far too slow and on the other hand a hardware realisation in today's technology is impracticable.

Our proposal of a new architecture in combination with partially chaotic neurons makes not only fast simulations of complex maps on common computers possible but it also solves a lot of problems referring to a hardware realisation.

2. Algorithms for a self-organizing Map with chaotic Neurons

The introduction of chaos into artificial neurons provides a remarkable increase of performance, and it especially enables the later proposed fractal architecture. For that reason we apply a modified version of the logistic equation [7] [8] as transfer function of the chaotic neurons. Usually the logistic equation is used for predictions of population dynamics, and despite of its simplicity it has a surprisingly complex solution. With modifications the final transfer function of the chaotic neuron follows as:

$$y_{n+1} = 1 - 4(1 - In) \cdot y_n \cdot (1 - y_n) \tag{2.1}$$

The iteration has to be performed for a sufficient number of times (N ~ 50) until the solution has lost all memory of its initial condition. Dependent on the input value In the sequence shows periodic or chaotic behaviour. The output values of such a neuron as a response on particular In-values are shown in Figure 1 a. Chaotic behaviour is observed above input values of about 0.15. Only In-values in the range of]0:1] are permitted. The performance can be improved additionally if a part of the periodic section of the function is suppressed in favour of the constant range. If N is the total number of iterations, then the average between the two last iteration results represents the final output value of the chaotic neuron:

$$y^{out} = \tfrac{1}{2}(y_{N-1} + y_N) \tag{2.2}$$

For this reason the final chaotic stage of the neuron consists of two parts. The first part performs the iteration and transfers each value of the sequence to the second part where each value is added to the previous one to form the definitive output value. The resultant plot from the modified iteration function is shown in Figure 1 b.

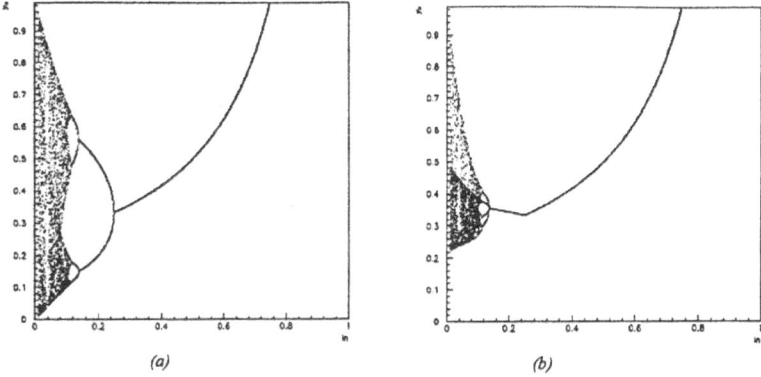

(a)   (b)

*Figure 1 : Initial (a) and modified (b) chaotic iteration function. The modified iteration function
was chosen as the final transfer function of the chaotic neuron*

The proposed chaotic iteration function is quite difficult to implement on integrated circuits. A considerable simplification could be achieved by some non-linear electrical networks with similar characteristic, such as a simple electrical circuit [9], consisting of a series RCL circuit driven by a controlled oscillator, as shown in figure 2 which is described by the following differential equation:

$$L \cdot \ddot{q} + R \cdot \dot{q} + V_c = V_d(t) = V_0 \cdot \sin(2 \pi f t) \qquad (2.3)$$

Here V_c is the voltage across a Si varactor diode, that represents the non-linear element. In this system the input value In corresponds to the driving voltage V_0, while the voltage measured across the varactor diode is the chaotic output value y_n.

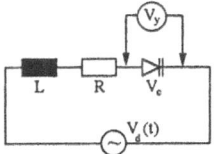

Figure 2: simple electronic circuit with chaotic characteristic

The plot in figure 3 shows the experimentally determined behaviour of this system. With the previously discussed modifications, such circuits could replace the iteration function in the chaotic stage of the neuron.

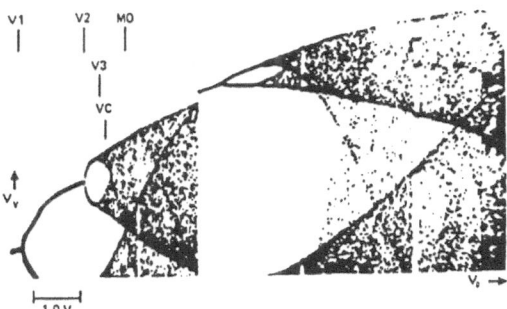

*Figure 3: Experimentally determined correspondence to the logistic equation, from the study of a simple
non-linear electric network. From Testa, Pérez and Jeffries, Phys. Rev. Lett. 48, 714 (1982)*

Independent of the final implementation of the transfer function, the chaotic behaviour of the proposed neurons makes the usually applied best-match approximation no longer practicable, because in this case the winning neuron does not inevitably need to be the one with the most similar weight vector. For that reason the more complicated original model must be applied. With respect to the limited range of valid input values, the calculation of the *In*-value requires new procedures as well. After all the scalar product between weight vector $\vec{w}$ and input vector $\vec{x}$ turned out to be an adequate quantity which holds the required specifications.

$$\vec{x} \cdot \vec{w} = |\vec{x}| \cdot |\vec{w}| \cdot \cos(< \hat{x}, \hat{w}) \qquad (2.4)$$

For this purpose the vectors have to be standardized to unit length:

$$\hat{x} = \frac{\vec{x}}{|\vec{x}|} = \frac{1}{\sqrt{\sum_i x_i^2}} \cdot \vec{x} \qquad (2.5)$$

In this way the *In*-value reduces to the angle between weight and input vector:

$$\hat{x} \cdot \hat{w} = |\hat{x}| \cdot |\hat{w}| \cdot \cos(< \hat{x}, \hat{w}) = \cos(< \hat{x}, \hat{w}) \qquad (2.6)$$

The scalar product between two vectors represents the projection of the one vector onto the other one. Its value reaches its maximum if both vectors are equal and decreases the more they differ. At large *In*-values the neuron works in its constant range and produces large output values, corresponding to the transfer function shown in Figure 1 b. If the weight vector of a neuron is similar enough to the input vector, which causes the scalar product to be above about 0.15, the neuron will most likely produce larger output values than almost every other neuron that works in its chaotic range. In this case the neuron with the most similar weight vector will win the competition with high probability. However, if in the beginning of the training no neuron with a sufficient similar weight vector exists, a neuron working in its chaotic range will have the largest output value and therefore win the competition.

By this way, a complete participation of all neurons at the learning process is achieved, because by presenting an unknown vector, not always the neuron with the most similar weight vector wins the competition. If there is no neuron on the map to know the input vector sufficiently well, the winner will probably be located in an area of the map which has not yet participated at the learning process and therefore has not yet stored any input vector.

Already at the beginning of the training the input vectors are distributed more evenly onto the whole map, a circumstance which effects that fewer training cycles are needed and also a smaller number of neurons is sufficient to store the whole set of input vectors.

The additional coupling decreases rapidly with the distance between the neurons, a reason for that in the following only the coupling of adjacent neurons is considered:

$$f_C = \underbrace{\alpha}_{\substack{\text{coupling} \\ \text{strength}}} \cdot \underbrace{\left(\frac{1}{\sum\limits_{i\pm1, j\pm1} 1} \right)}_{\substack{\text{number of} \\ \text{direct neighbours}}} \cdot \underbrace{\sum\limits_{i\pm1, j\pm1} y_{i,j}^{Out}}_{\substack{\text{sum over outputs} \\ \text{of direct neighbours}}} \qquad (2.7)$$

The total number of neighbours varies with the location of the neuron (i,j) within the map. Whether it is located in the centre, at the margin or in a corner of the map, the number of direct neighbours can be 8, 5 or 3, which is considered by the factor $N_{neighbours}$. An additional factor α fixes the strength of the interaction, which generally should be weaker than the scalar product. Best results so far have been achieved with values about $\alpha \sim 5\% - 10\% = 0,05 - 0,1$. An additional factor $\frac{1}{2}$ keeps the *In*-value within the admissible range, so that the final equation follows as:

$$In = \frac{1}{2} \cdot \left\{ \hat{x} \cdot \hat{w} + f_c \right\} \qquad (2.8)$$

3. Basic Architecture of the Neuron

The sketch in Figure 4 describes the operation of the chaotic neuron. It consists of three independent units. In this case, the neuron operates with five-dimensional vectors $(x_1,...,x_5)$. The final concept is based on three different stages. The first stage is responsible for the calculation of the input value for the chaotic stage. The second stage carries out the given number of iterations of the chaotic transfer function and passes the sequence of chaotic numbers out to the third stage, where the final output value of the neuron is calculated and given out.

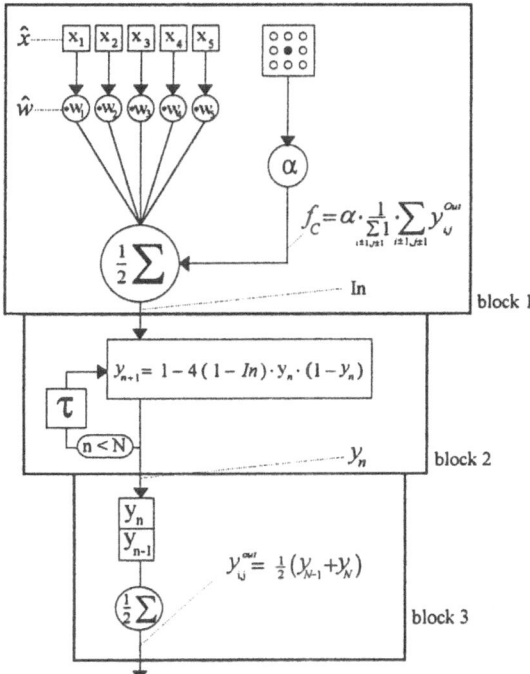

Figure 4: The functional blocks of the chaotic neuron

- The input stage consists of two independent units. Their task is to produce an adequate real number which is transferred to the chaotic stage. Every component of the input vector is multiplied with its correspondent weight component. Simultaneously the output values of the direct neighbours are added up and standardized by the factor α. The resultant six numbers are summed up and standardized with the factor ½ to keep the required range between 0 and 1.

- The following chaotic stage carries out the N iterations, as described before. The result y_n of every iteration step is fed back with the time constant τ until the given number of iterations is reached. The result is a sequence of output values changing with the period τ.

- In the output stage always the latest output value y_n of the sequence is stored. Being added to the next value y_{n-1} and standardized with ½ it forms the next output value of the neuron, leading to a sequence of output values, which are, dependent on the operation state, either remaining constant or changes chaotically.

Experimental observations confirm the assumption that biological neurons behave analogously to our artificial neurons. Familiar input patterns are responded by ordered firing, whereas strange patterns lead to chaotic activity. If the output sequence of the artificial neuron is identified with the firing of its biological equivalent, the different reaction to known or respectively strange data input indicates that like the proposed neurons, biological neurons have different operation modes which depend on the familiarity of the input.

4. Conceptual Architecture

4.1. The Chaotic self-organizing Map

The chaotic self-organizing map is assembled analogously to conventional self-organizing maps, but with the previously introduced chaotic neurons. During one training step a randomly chosen input vector is presented to all neurons of the map. The input stage combines the two independent input terms to the adequate input value for the chaotic stage. The chaotic stage carries out the fixed number of iterations and passes each result to the output stage. Each two latest output values of the chaotic stage are combined to the final output value of the neuron. After the given number of iteration steps is completed, the neuron with the largest output value assimilates its weight vector a bit towards the input vector, while its direct neighbours adapt with a reduced rate. After that another vector is chosen, and the procedure is repeated successively for the given number of training steps.

4.2. Fractal Design

Fractals are considered to be self similar. Each time, a self-similar structure is magnified, another layer of finer detail is revealed, which looks almost similar to the previous structure. Geometrical fractals are easily constructed [5] by repeatedly applying a generator to an initiator. Each time it is applied, the generator is scaled appropriately to provide structure on another scale length. The left side of figure 6 gives an example of such a construction in which the initiator is a equilateral triangle and the generator is an excising operation which removes a smaller inverted triangle. At each iteration stage of growth, the generator is reduced to accommodate the remaining smaller triangles.

A fractal self-organizing map with chaotic neurons can be realized in the same way as illustrated on the right side of figure 5. A square as an initiator is repeatedly divided by a generator of the shape of a cross. The number of neurons within a level and therefore the complexity of the map increases with the number of iterations. The particular generator illustrated in Figure 5 divides the initiator into four parts, but divisions into 9, 16, 25, 36, etc. are also possible. Each square represents a neuron within the map. The different plains are arranged hierarchically, so that every neuron has its own number of associated neurons in its subordinate level, depending on the generator applied and respectively on the number of parts it creates in one step.

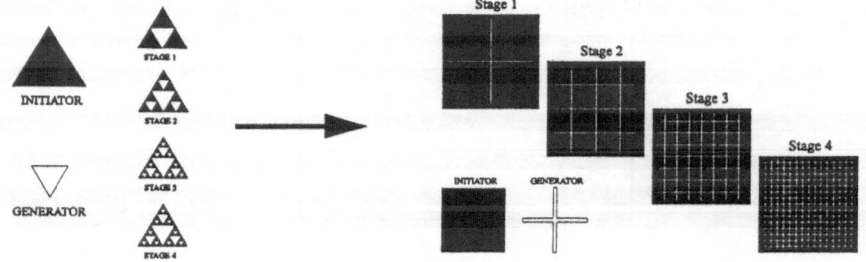

Figure 5: Example for the generation of two-dimensional fractals and transition to generation of fractal self-organizing maps

The different modules are all identical, each of them being a complete, small self-organizing map. The method, how the different modules cooperate, is explained by the example of 2x2 modules in figure 6.

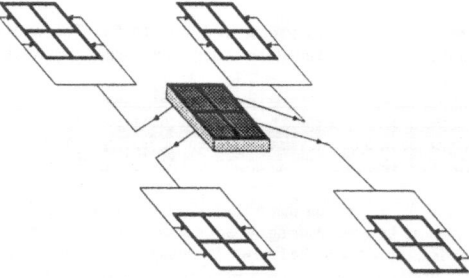

Figure 6: First level neurons and associated subordinate modules

The grey squares represent the neurons of the first-level module. Each neuron of this level has an associated module in the next subordinate level, represented by white squares. Every input vector is presented to the first-level module. Supposing that the neuron in the left upright corner wins this cycle, the input vector is transferred only to its associated second level module in the upper left corner. The other three modules of the second level remain inactive until the end of this cycle. The transfer of the input vector to a third or higher level module goes on analogously to the steps before. By successive addition of further levels, self-organizing maps of nearly any complexity can be realised.

Imagine a fractal map with five levels, build up by 3x3 map modules. Starting with 9 neurons in the first level, the fifth level contains already 59.049 neurons. Since it has the fractal architecture, even such a complex map remains easy to handle. One reason is that there is no communication between modules of the same level, so if a single module requires the time τ to carry out one step, the time required by the complete map is $T = N \cdot \tau$, where N is the total number of levels. Because τ is only dependent on the number of neurons within a module, T grows linearly with the number of levels, whereas the number of neurons increases by square. Referring to our simulations, a single 3x3-module requires around $\tau \approx 13ms$ (a PC 486/50 MHz was used). As illustrated in figure 7, each additional level increases the total processing time by 13ms while the number of neurons by factor 9.

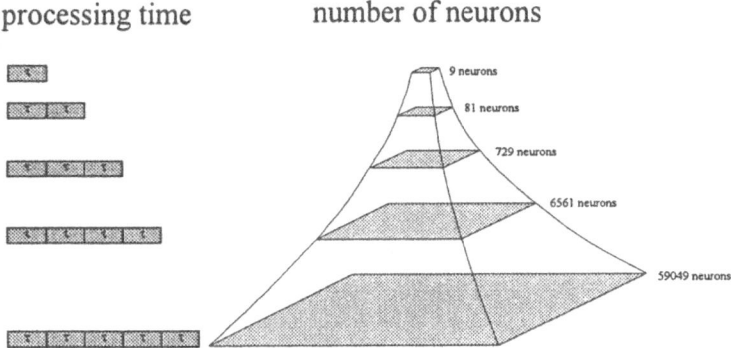

Figure 7: development of processing time in comparation with increasing complexity, for a fractal map, assembled of 3x3 modules

Each level is based on the previous one, so that for comparison to conventional maps only the number of neurons in the last level should be considered. Therefore, the correspondent conventional map would have the size 243x243 or 59049 neurons. But while the fractal map needs about 65 seconds for 1000 steps, the simulation of the corresponding conventional map is more than 1160 times slower and already needs 20.9 hours.

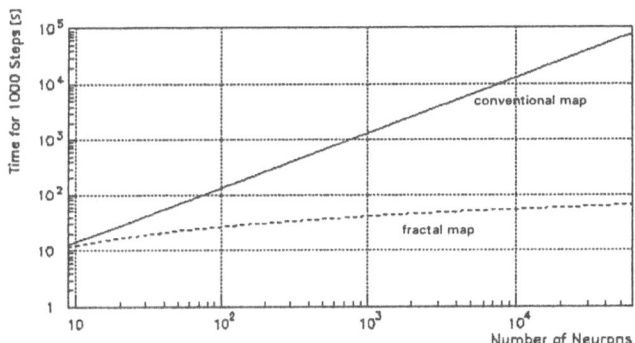

Figure 8: Computing time [seconds] required by self-organizing maps in fractal architecture in comparison with correspondent conventional maps

The development of computing time versus increasing complexity of the map is plotted in figure 8. With an increasing number of neurons, the difference between conventional and fractal maps rises up to several orders. Even maps of high complexity are practicable, if the fractal architecture is applied.

Besides, the resolution of the fractal map can additionally be improved by the training of single modules. Self-organizing maps generally tend to store very similar vectors in the same neuron, that means the map cannot distinguish them. With conventional maps only an extended training can reduce this incompetence. In the case of fractal maps further training can be restricted respectively to the affected last level module. In this second training stage, an isolated training is performed at which only those input vectors participate which have already been stored within the affected module.

5.   Advantages for the VLSI-Implementation

The proposed concepts effect several advantages with regard to a hardware realization in VLSI-technology:

- The chaotic behaviour of the neurons avoids the initialisation of the weight vectors with random numbers. So there is no need of data exchange between a random generator and each neuron. On the one hand the number of wires required can be reduced, but on the other hand there is no need of an internal random generator, which is difficult to realize in VLSI-techniques.
- In the initial part of the training, the winner is preferentially located in areas of the map which operate in the chaotic state and therefore have not yet participated at the training. This causes the preferential selection of untrained neurons, so that the distribution of the whole set of input vectors is reached faster and covers the map more evenly.
- Separate training of single modules allows on the one hand the improvement of the resolution and on the other hand it is possible to include some new input vectors without having to repeat the whole training.
- Within a self-organizing map every neuron has to be connected with each of its direct neighbours. The complexity of the interconnection increases with the number of neurons. Up to now one reason which has prevented the realization of complex maps has been the difficult and often insoluble wiring problem which is solved by this architecture automatically. A single module of the proposed map consists typically of 10 to 100 neurons, so that wiring remains feasible. Supposing that a single chip contains a complete module, then the different chips can be connected sequentially. The input vector can be passed through each module of the map in turn. Only an additional activation index has to be transmitted, which is responsible for the activation of the map, corresponding to the winner of the previous level. Therefore the total number of wires, which are required to connect the modules remains orders below those of a comparable conventional map.
- The realization of different VLSI-chips is expensive. Because of the fractal design, the final size of the map need not to be considered already at the layout of the chip, a reason why different sized maps can be realised with a single serial manufactured and therefore low cost chip.

6.   Conclusion

We have demonstrated that the new concept is capable of simplifying the realization of a self-organizing map in VLSI-technology. The introduction of chaos into neurons improves their power and shortens the time for training. Furthermore the distribution of the complete set of input vectors becomes more evenly. Together with the fractal structure, the efficiency of the map is increased by several orders. By the way, some important problems occurring with the VLSI-realization are solved automatically. On the one hand the modular principle solves the difficult wiring problem and on the other hand single modules are possible to be realised with common VLSI-techniques. The final assembly of the map can be handled flexibly and does not already need to be considered at the design of the module, therefore many different types of fractal maps can be assembled with a single type of module.

References

[1] Kohonen, T.: Automatic Formation of Topological Maps of Patterns in a Selforganizing System, Proceedings of the 2nd Scandinavian Conference on Image Analysis, Helsinki (June 15-17, 1981), pp. 1-7.

[2] Kohonen, T.: Self-Organised Formation of Topology Correct Feature Maps, Biological Cybernetics 43 (1982), pp. 59-69.

[3] Dingle, Allison, A., Andreae, John, H., Jones, Richard, D.: A Chaotic Neural Unit, IEEE 5/93, 1993, pp 335-340.

[4] Tryba, Viktor: Selbstorganisierende Karten: Theorie, Anwendung und VLSI-Implementierung, Fortschrittsberichte VDI, Reihe 9: Elektronik, Nr. 151.

[5] Jaggard, Dwight, L.: Special Section on Fractals in Electrical Engineering, Proceedings of the IEEE, Vol. 81, No. 10, October 1993, pp 1423-1427.

[6] S. Rüping, U. Rückert, K. Goser: Hardware design for a Self-Organizing Feature Map with binary Input Vectors, Proceedings for IWANN (1993), Sitges (Spain), pp 488-493.

[7] R.W. Leven, B.-P. Koch, B. Pompe: Chaos in dissipativen Systemen, Vieweg1989, pp 13, 53, 102

[8] Eryk Infeld, George Rowlands: Nonlinear waves, solitons and chaos, Cambridge University Press, 1990

[9] James Testa, José Pérez, Carson Jeffries: Evidence for Universal Chaotic Behaviour of a Driven Nonlinear Oszillator, Physical Review Letters, 1982, Volume 48, pp 714-717

Multiple Self-Organizing Maps
for Supervised Learning

Enrique Cervera and Angel P. del Pobil

Computer Science Dept.
Jaume I University
Campus Penyeta Roja
E-12071 Castelló. SPAIN
{ecervera, pobil}@inf.uji.es

Abstract. A scheme for supervised learning based on multiple self-organizing maps is presented and its performance is compared with other methods in several pattern classification benchmarks using both synthetic and real data. The advantage of this approach is that the learning method is simplified because the problem is divided into several SOMs, which are trained in the standard unsupervised way. The resulting network preserves the SOM properties like dimensionality reduction and cluster formation, while classifying with an accuaracy comparable to other supervised methods on a wide range of problems.

1. Introduction

Self-organizing networks (like Kohonen's maps [1]) perform unsupervised learning. Frequently, they generate ordered mappings of the input data onto some low-dimensional topological structure. In other cases, they are used to partition the input data into subsets (or clusters) such that data items inside one subset are similar but items from different subsets are dissimilar.

In many situations, however, input data together with corresponding output data are provided. The problem is then to learn the underlying relationship from a limited number of examples. Supervised learning methods are in these cases used to train networks to generate the desired output, when they are presented with the input part of a specific data pair. Although this is not very useful in itself it is hoped that, after finishing the training, the network will be able to generate "reasonable" output values also for unknown input data. This is often denoted as *generalization*. It is a commonplace today that to achieve good generalization the number of free parameters of the network must be kept small. Otherwise there is the danger of "over-fitting" which denotes a situation where the network still improves on the training data, but already has a decreasing performance on the test data. Typical application areas for supervised learning include pattern classification or function approximation.

In the rest of this paper, a scheme for supervised learning based on multiple self-organizing maps is presented and its performance is compared with other methods in several pattern classification benchmarks using both synthetic and real data. The advantage of this approach is that the learning method is simplified because the problem is divided into several SOMs, which are trained in the standard unsupervised way. The resulting network preserves the SOM properties like dimensionality reduction and cluster formation, while classifying with an accuaracy comparable to other supervised methods on a wide range of problems.

2. Supervised Learning with Multiple Self-Organizing Maps

The process in which a SOM is formed is an unsupervised learning process. There are N continuous-valued inputs, ξ_1 to ξ_N, defining a point ξ in an N-dimensional real space. The output units O_i are arranged in an array (generally one- or two-dimensional), and are fully connected via w_{ij} to the inputs. A competitive learning rule is used, choosing the winner c as the output unit with weight vector closest to the current input ξ:

$$\|w_c - \xi\| \leq \|w_i - \xi\| \quad \text{(for all } i\text{).}$$

And the learning rule is:

$$\Delta w_{ij} = \Lambda(c,i) \, (\xi_j - w_{ij})$$

for *all* i and j. A typical choice for $\Lambda(c,i)$ is

$$\Lambda(c,i) = \eta \, \exp(-\|r_c - r_i\|^2 / 2\sigma^2),$$

where $r_c \in \mathbf{R}^2$ and $r_i \in \mathbf{R}^2$ are the radius vectors of nodes c and i, respectively, in the array, and both η and σ are some monotonically decreasing values.

The rule drags the weight vector w_c belonging to the winner towards ξ. But it also drags the w_i's of the closest units along with it. Therefore we can think of a sort of elastic net in input space that wants to come as close as possible to the inputs; the net has the topology of the output array (i.e., a line or a plane) and the points of the net have the weights as coordinates.

This training algorithm is unsupervised, since only input values are used. The resulting map, however, can be used in classification tasks if its units are *labeled*. This can be done by introducing a number of known examples into the map, and looking where the winners lie. The map units are then labeled according to the majority of examples of a given class 'hitting' a particular map unit. Then, an unknown example can be classified by computing the winner unit, and its label will give us the class of that example.

Though classification accuracies obtained by this method are not very high, it is very useful for monitoring and visualization of high-dimensional data. We have used it successfully in sensory mapping in robotic applications [2].

2.1 Probability Density Approximation and Bayesian Classification

Our approach is an extension of this scheme. In order to increase the accuracy, we might use the output information of the training data during the learning process, thus performing *supervised* learning.

The key is that a SOM approximates the probability density of the training set. If the probability densities of each class (denoted by $p(\xi|C_i)$ for class i) were known, we could use Bayesian theory to minimize the average rate of misclassifications, assuming that the a priori probabilities of each class (denoted by $P(C_i)$) are also known [3]. Then a sample ξ would be assigned to class C_i iff:

$$p(\xi|C_i) \, P(C_i) > p(\xi|C_j) \, P(C_j) \qquad \text{for all } j \neq i$$

Although the distribution of the SOM units is not exactly the same as the distribution of the underlying training density (for the one-dimensional case the density of output units is proportional to $P(\xi)^{2/3}$ around point ξ [4]), this does not matter in Bayesian classification where only ratios of class densities are important. Furthermore, dimensionality of the SOM is less than that of the training set, thus a SOM is a "nonlinear projection" of the probability density function of the high-dimensional input data onto the one- or two-dimensional array of units. For the sake of simplicity, we will use the distance of the input vector ξ to the best matching unit as a measure of the probability density at that point.

2.2 Supervised Learning Procedure

Our simple yet powerful idea is based on *training a different SOM for each class*. Thus, output data from the training set is used to divide all data in several subsets, each one containing only data from one class. This is the supervised part of the procedure. After this, the SOMs will be trained as usual, in an unsupervised way. A similar idea was used by Kurimo in [5], where the system is initialized with little SOMs, one for each phoneme class, which are combined and applied to other methods. Another similar system was explained in [6] where one SOM is trained for each speaker. In our approach, however, classification is performed with only one vector of input data. In the experiment reported in [6], a whole sequence of inputs were tested on each SOM and the SOM which gave the smallest quantization error was the 'winner'. Our approach is a generalisation of this scheme which should be able to classify properly not only sequences but single data samples.

The complete learning process has three phases:

1) Division of the training set into several subsets, as many as different classes.

2) Training of different SOMs, each with only a subset of data of the same class, following the unsupervised learning procedure defined in previous section.

3) Labelling of units of all the SOMs. This can be done in two ways. Each SOM is separately labeled, thus all its units will be labeled as belonging to the class it was trained with. Or all SOMs are labeled at once with all training data. We have found the latter method more accurate when there are overlappings between classes.

Now, classification of unknown data is done with all the SOMs together. When an example is presented, the best matching unit of **all** maps will give the correspondent class.

3. A Classification Example

This is a problem introduced by Kohonen [7] that consists of two normally distributed classes in a d-dimensional Euclidean space with equal class a priori probabilities. In the "easy" test the first class has an $N(0,I_d)$ distribution and the second class an $N(\mu,4I_d)$ distribution with $\mu=(2.32,0,...,0)\in R^d$. the "hard" test has the two classes distributed as $N(0,I_d)$ and $N(0,4I_d)$, respectively. Figure 1 depicts the normal distributions in the one-dimensional case, and the optimal decision boundaries for classification.

The dimensions considered are d=2,4,6. The SOM sizes are 6x6, 8x8, and 10x10, thus the number of units increases approximately linearly with the number of dimensions. Two independent sets for training and testing were used. Each set consisted of 2000 randomly chosen samples from each class.

Data and the trained SOMs for each class in the two-dimensional "easy" case are shown in figure 2. A summary of the results is given in table I. The second column gives the theoretically optimal classification errors. The third and fourth columns give the results for our method with separate labelling -MSOM(1)- and common labelling -MSOM(2)- respectively. The next two columns give the results for backpropagation and LVQ as reported in [7].

Classification results are not better than those obtained by purely supervised methods like backpropagation and LVQ, but keep rather close, specially in the two-dimensional case. This is due to the fact that we are using a two-dimensional SOM, with which is hard to approximate a normal distribution of more dimensions, since it is quite homogeneous.

Furthermore, we are using more units than in the other methods, but, on the other hand, algorithm complexity is somewhat smaller, specially than backpropagation. Moreover, training a SOM in this experiment only takes 25 epochs (presentations of the complete training set). However, we recognise that training a SOM usually requires fixing the values of some parameters in order to get good convergence, and maybe a large number of units would be necessary in order to get a good approximation of a tricky class, but these are not major drawbacks since the algorithm is computationally cheap. Moreover, our scheme can be easily parallelised.

Theoretically, one should get near-optimal results with a great number of units in each SOM, but SOMs are best suited to data which is distributed in a subspace of the high dimensional original space. Ideally, the SOM number of dimensions should be equal to the dimension of this subspace.

The main advantage of our method is that it simplifies the original problem. Training smaller networks with subsets of data is easier and faster than training a bigger network with all the data. Furthermore each independent SOM might be reused in other problems of classification of its class with other different classes. Thus we could get a collection of SOMs and pick the relevant ones depending on the classes involved in a given problem.

Another advantage is that this approach gives a reply to a criticism frequently made of neural networks, regarding the lack of significance of the weights of the units, and *how* the network solves the problem. In our method, each SOM is an approximation of the probability density of a class, and classification is done by Bayesian theory.

4. Simulation Results

We have tested our approach in other real problems, with electronically available data from the Carnegie-Mellon University connectionist benchmark collection.

4.1 Classification of Sonar Signals

This is the data used by Gorman and Sejnowski in their study of the classification of sonar signals using a neural network [8]. The task is to train a network to discriminate between sonar signals bounced off a metal cylinder and those bounced off a roughly cylindrical rock. The data set contains 208 samples. Half of them were used for training, and the other half for testing. Each sample has 60 components.

We used two SOMs of 6x6 size each. Units had 60 inputs. And the SOMs were trained for 500 epochs each. After training and labelling, we got 76.2% and 78'6% right classifications on the test set with separate and common labelling respectively. In

[8], results for different backpropagation networks with variable number of hidden units range from 77.1% to 84.7%.

One should note that the input data has many more dimensions than the SOM. However, due to the dimensionality reduction of the algorithm, classification results are comparable to those of other methods, and they would be considerably improved if there were more training data available.

4.2 Speaker Independent Vowel Recognition

These data are recorded examples of the eleven steady state vowels of English spoken by fifteen speakers for a speaker normalization study. After processing, input space to the network was 10-dimensional. Robinson used these data in his thesis [9] to investigate several types of neural network algorithms. He used 528 frames from four male and four female speakers to train the network and used the remaining 462 frames from four male and three female speakers for testing the performance.

In our simulations we use 11 SOMs of size 3x3 each, thus using a total of 99 units. Each SOM was trained during 2200 epochs, and the test results of the commonly labeled SOMs, together with other methods reported by Robinson, are shown in table II. The percentage of correct classifications obtained by our method stands among the best ones in a real hard problem like this.

5. Conclusions

A method for supervised learning using multiple Self-Organizing Maps has been presented. The learning procedure is straightforward, since it only requires the division of the training set in separate subsets, one for each data class. Then a different SOM is trained in a unsupervised way with the subset of a data class. Classification is based on the SOM aproximation of the probability density of each class and Bayesian decision. We know not only *what* computes the network but also *how* it does.

Training is simplified since input data is simpler than the complete set, thus a complex problem is split in several easier problems, and the significance of the units is shown as aproximators of the densities of data. Furthermore, SOM properties like dimensionality reduction and cluster formation are preserved and allow the integration of this method with other AI techniques. We are currently working in that direction [10], combining self-organizing maps with qualitative reasoning.

Test results have been presented on both synthetic and real classification problems, which demonstrate that this method is comparable to other pure supervised methods widely used -like backpropagation networks- on hard problems, and should be tested on other tasks. Further studies are also required in order to get better aproximations of probability densities with SOMs.

Acknowledgements

This work has been funded by the CYCIT under project TAP92-0391, and by a grant of the FPI Program of the Spanish Department of Science and Education. We wish to thank Prof. Kohonen and Prof. Oja for making possible a stay of one of the authors at the Laboratory of Computer and Information Science at Helsinki University of Technology, and specially Dr. Kangas, from the mentioned laboratory, and Prof. Holmström, from Rolf Nevanlinna Institute at University of Helsinki, for helpful discussions.

References

[1] Kohonen, T., Self-Organized Formation of Topologically Correct Feature Maps, *Biological Cybernetics*, 43, pp. 59-69, 1982.

[2] Cervera, E., del Pobil, A.P., Marta, E., Serna, M.A., Unsupervised Learning for Error Detection in Task Planning, *EEE Transactions on Robotics and Automation, Special issue on§ Assembly and Task Planning*, (submitted for publication).

[3] Devijver, P.A., Kittler, J., *Pattern recognition: A statistical approach*. Prentice Hall, London, 1982.

[4] Hertz, J., Krogh, A., Palmer, R.G., *Introduction to the Theory of Neural Computation*, Addison-Wesley, 1991.

[5] Kurimo, K., Hybrid training method for tied mixture density hidden Markov models using LVQ and Viterbi estimation, *Proc. IEEE Workshop on Neural Networks for Signal Processing*, pp. 362-371, 1994.

[6] Naylor, J., Higgins, A, Li, K.P., Schmoldt, D., Speaker recognition using Kohonen's self-organizing feature map algorithm, *Neural Networks*, v. 1, pp. 311, 1988.

[7] Kohonen, T., Barna, G., Chrisley, R., Statistical Pattern Recognition with Neural Networks: Benchmarking studies, *Proceedings of the IEEE International COnference on Neural Networks, San Diego*, v. 1, pp. 61-68, 1988.

[8] Gorman, R.P., Sejnowski, T.J., Analysis of Hidden Units in a Layered Network Trained to Classify Sonar Targets, *Neural Networks*, v. 1, pp. 75-89, 1988.

[9] Robinson, A.J., *Dynamic Error Propagation Networks,* Cambridge University, PhD Thesis, Cambridge, 1989.

[10] Cervera, E., del Pobil, A.P., Perception-based Qualitative Reasoning in Manipulation with Uncertainty, *International Joint Conference on Artificial Intelligence*, Montreal 1995 (submitted for publication).

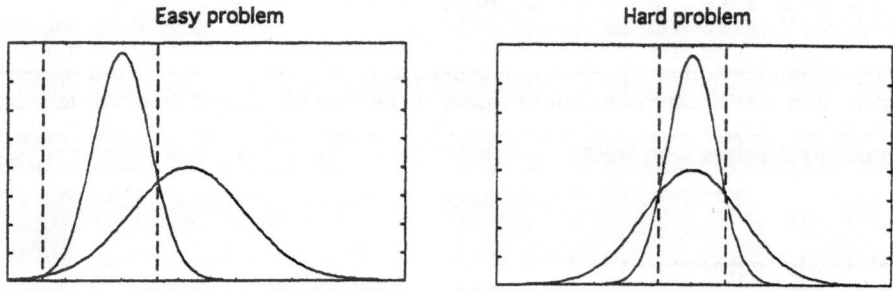

Figure 1. Two normally distributed classes, and optimal class boundaries.

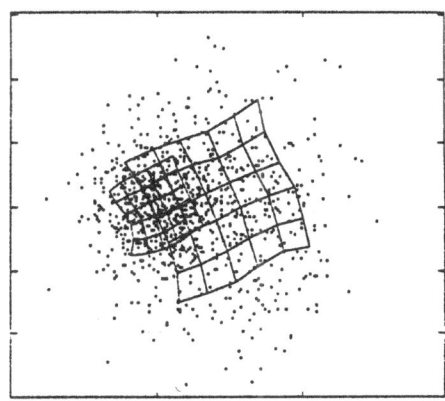

Figure 2. Data and SOMs trained in the two-dimensional "easy" problem.

Table I. Test results of the two class problem.

Easy problem

D	theoretical	MSOM (1)	MSOM (2)	BP	LVQ
2	16.4	21.7	18.4	16.4	17.0
4	11.6	19.3	16.3	12.5	13.1
6	8.4	16.4	15.9	10.8	10.7

Hard problem

D	theoretical	MSOM (1)	MSOM (2)	BP	LVQ
2	26.4	30.3	26.6	26.3	26.5
4	17.6	24.3	23.7	19.4	18.8
6	12.4	20.9	20.8	20.7	15.3

Table II. Test results on the vowel recognition problem.

Classifier	No. of hidden units	% correct
Single-layer perceptron	-	33
Multi-layer perceptron	88	51
Multi-layer perceptron	22	45
Multi-layer perceptron	11	44
Nearest neighbor	-	56
Modified Kanerva Model	528	50
Modified Kanerva Model	88	43
Radial Basis Functions	528	53
Radial Basis Functions	88	48
Multiple SOM	**99**	**52**

An Application of the Saturated Attractor Analysis to Three Typical Models

J. FENG   B. TIROZZI

Mathematical Department
University of Rome "La Sapienza", P.le A. Moro, 00185 Rome

Abstract

The saturated attractor analysis, an approach proposed first in [FP] for a comprehensive study of the dynamics of the Linsker model and then successfully applied to the dynamic link model[FT1], is further developed here. By a unified approach to the Hopfield model, the Linsker model and the dynamic link model, three typical models in the field of the neural networks, we show a way to choose the parameters of these dynamics in order to obtain any chosen saturated attractor which is general enough in most applications. We generalize our previous results for the Linsker model and the dynamic link model with the clipping function to the case of the sigmoid like function. Our results allow us for the first time to understand the underlying mechanism among these models and thus to furnish a useful guidance in the further possible applications.

§1 INTRODUCTION

The past decade has seen an explosive growth in the studies of neural networks, the theory underlying learning and computing in networks has developed into a mature subfield existing somewhere between mathematics, physics, computer science and neurobiology. In part this was the result of many deep and interesting theoretical exposition in physics and mathematics, for example, the application of the spin glass theory to the Hopfield model allows us to understand clearly the phase transition from the retrieval to non retrieval state. Another major impulse was provided by the successful explanation of some biological phenomena, at least in a primitive level, for example, the Linsker model mimics the ontogenesis development of the primary visual system[Lin]. Of course, the most important impulse comes from the learning techniques successfully applied to some practical problems which were traditionally thought of as some of the hardest problems in the AI. One of the recent examples of such an application is the face recognition using the dynamic link model, a model proposed by von der Malsburg first in 1981[KMM].

However, at this moment, the theoretical treatment of these models is obviously far away from being satisfactory, mainly due to the lack of theoretical tools to deal with the nonlinearity exploited in most of the models reported today. In the present paper, in terms of our previous work on the Linsker model and the dynamic link model we develop a unified theoretical framework for tackling the Hopfield model, the Linsker model and the dynamic link model.

Our approach allows us to reformulate many problems studied in the Hopfield model. A concrete criterion to check whether a stored pattern is an attractor of the network is given. The capacity, a quantity which plays a central role in the spin glass approach to the Hopfield model, is naturally introduced here. One advantage of the present approach is that we do not impose the restriction of the symmetry of the connection matrix. Our results also reveal the role of different parameters in the Hopfield model. We consider the Linsker model with the sigmoid like function in the updating dynamics of its synaptic connections(a definition of the sigmoid like function is in section 2). All conclusions in [FP][FPR] are reobtained, where the clipping function, a special case of the sigmoid like function and so a special case of the present paper, is used for the development of the synaptic connections. The present paper tells that the appearance of the structured receptive fields is independent of the choice of the clipping function, which is thought of as one of the drawback of the Linsker model. Furthermore, we also take into account on the reason for the appearance of the oriented receptive field in the further layers of the Linsker network. For the dynamic link model, a principle to choose all five parameter employed in the model is furnished, which confirms our previous claim that all results contained in [FT1] for the clipping function are true for a more general class of function, i.e. for the sigmoid like function.

Although here we confine ourselves to the models on which we worked before [ATYD] [FP] [FPR] [FQ] [FT1], the essential part of our approach is to analyze the dynamics with the sigmoid like function, and it is possible to adopt our method here to analyze other models in the field of the neural networks such as the B.P. and the recurrent network.

The general idea behind the saturated attractor analysis is quite straightforward. Consider a dynamics defined on the space $[-1,1]^N$, where N is either the number of neurons (the Hopfield model and the dynamic link model) or the number of connections (the Linsker model). It is reasonable to confine ourselves to a subset of all the fixed points of the dynamics, i.e. to all saturated fixed points in $\{-1,1\}^N$ since in the Hopfield model all the stored patterns take values on $\{-1,1\}^N$, while in the Linsker model and the dynamic link model this confinement has been confirmed by the numerical simulations[Lin][KMM]. In particular, the fast dynamic link model is proposed in terms of this observation[KMM]. As we all know, it is relatively easy to determine the *whole* region of the dynamic parameters, say $\Gamma(w)$, in which a given pattern w of particular interest (in the Hopfield it is one of the stored patterns, in the Linsker model it is the structured receptive field, in the dynamic link model it is the on-center configuration) is a fixed point. If we are further able to prove the stability of the fixed point w, we assert that if and only if as the dynamic parameters are in the region $\Gamma(w)$, w is an attractor of the dynamics. Fortunately, due to the special form of the sigmoid like function and we restrict ourselves to all saturated fixed points, the idea above can be carried out as in [FP][FPR][FT1], but for a more general and more significant class of functions, the sigmoid like functions. We call such an approach *the saturated attractor analysis*.

Due to the limitation of the space, we are only able to briefly report our results. For a full exposition and detailed proofs, we refer the reader to our whole paper[FT2].

§2 GENERAL MODEL AND NOTATION

For a given positive integer N, an $N \times N$ matrix $Q = (q_{ij}, i, j = 1, \cdots, N)$ and an N dimensional vector $r = (r_i, i = 1, \cdots, N)$, consider the following dynamics

$$w_i(\tau + 1) = f(w_i(\tau) + k_1 + \sum_{j=1}^{N} [(q_{ij} + k_2)r_j w_j(\tau)]) \tag{1}$$

where $\tau = 1, 2, \cdots$ is the discrete time, $w(\tau) = (w_i(\tau), i = 1, \cdots, N) \in \mathbb{R}^N$, (k_1, k_2) are two parameters of the dynamics, and f is a continuous function defined on $\mathbb{R}^1$ satisfying

(f1). $f(x) = 1$, if $x \geq 1$, $f(x) = -1$, if $x \leq -1$,

(f2). $f(x)$ is a strictly increasing and continuous function for $x \in [-1, 1]$, $f(x) \geq x$
if $x \in (0, 1]$ and $f(x) \leq x$ if $x \in [-1, 0)$.

We call a function with the properties (f1) and (f2) a *sigmoid like function*.

Note that for the sigmoid function σ_β with range between -1 and 1, $\sigma_\beta(x) = \frac{2}{1+\exp(-\beta x)} - 1$, both conditions (f1) and (f2) are approximately satisfied when β is large. It is reasonable to expect that in the numerical simulation, both (f1) and (f2) are true for the sigmoid function σ_β with large β. Due to this reason, we believe that our results on the dynamics (1) with the sigmoid like function below reflect the exact properties of dynamics (1) with $f = \sigma_\beta$ (β large) which are mostly observed by numerical simulation. The function termed as *the clipping function* and used in the dynamics of the development of the synaptic connection of the Linsker network is defined by $f_c(x) = x$ if $|x| < 1$, and $f_c(x) = 1$ if $x > 1$, $f_c(x) = -1$ if $x < -1$, which of course fulfills both (f1) and (f2) [Lin][FP][FPR]. In the dynamic link model, fast dynamic link model, or the discrete version of it, the function f adopted for the dynamics is either the clipping function or the sigmoid function [KMM][FT1].

It is easily seen that the conditions on the range of the function (f1) is not essential and can be relaxed.

Let us now introduce three functions which will play a crucial role in our later development. Let $w \in \{-1,1\}^N$ be a given configuration then $J^+(w) = \{i, w_i = 1\}$, $J^-(w) = \{i, w_i = -1\}$ are (respectively) the set of all sites with $w_i = 1$ and all sites with $w_i = -1$.

First we introduce the *slope function* $c(w)$ on $\{-1,1\}^N$ defined by

$$c(w) = \sum_{i \in J^+(w)} r_j - \sum_{i \in J^-(w)} r_j. \tag{2}$$

Then we consider the *intercept functions* $d_1(w)$ and $d_2(w)$:

$$d_1(w) = \max_{i \in J^+(w)} [\sum_{j \in J^-(w)} q_{ij} r_j - \sum_{j \in J^+(w)} q_{ij} r_j] \tag{3}$$

and

$$d_2(w) = \min_{i \in J^-(w)} [\sum_{j \in J^-(w)} q_{ij} r_j - \sum_{j \in J^+(w)} q_{ij} r_j]. \tag{4}$$

The reason why we call them slope function and intercept functions will be clear after the Theorem 1 below.

§3 THE SET OF ALL SATURATED ATTRACTORS

The set of all fixed points of the dynamics (1) is $FP = \{w; w_i = f(w_i + \sum_{j=1}^{N}(q_{ij} + k_2)r_j w_j + k_1),\ i = 1, \cdots, N\}$. From the compactness of the range of the function f and the continuity of f, we get that the set FP is nonempty by the Brouwer's fixed point theorem. A fixed point is called an attractor if it is a stable fixed point. We will confine ourselves to a subset of all attractors in $\{-1,1\}^N$ which is general enough in most of applications.

Definition 1. *A configuration in the set* $\Omega := \{w \in \{-1,1\}^N;$ *there exists a nonempty neighborhood* $B(w)$ *of* w *in* $\{-1,1\}^N$ *such that* $\lim_{\tau \to \infty} w(\tau) = w$ *if* $w(0) \in B(w)$ *and* $k_1 + \sum_{j=1}^{N}(q_{ij} + k_2)w_j \neq 0, \forall i = 1, \cdots, N\}$ *is called a saturated attractor of the dynamics (1).*

The following theorem establishes that, for the case of the dynamics (1), the condition (5) below is strong enough to ensure that w is an attractor of the dynamics.

Theorem 1. *If w is a saturated attractor of the dynamics (1), $\lim_{\tau \to \infty} w(\tau) = w$, then there exists a $T > 0$ such that $w = w(T + \tau), \forall \tau \geq 0$. Furthermore w is a saturated attractor of the dynamics (1) if and only if*

$$d_1(w) < k_1 + c(w)k_2 < d_2(w). \tag{5}$$

For a given configuration w, Theorem 1 tells that w is a saturated attractor of the dynamics (1) if and only if (k_1, k_2) lies in between the two parallel lines(see Fig. 1) $k_1 + k_2 c(w) = d_1(w)$ and $k_1 + k_2 c(w) = d_2(w)$. Hence $c(w)$ is the slope function of the lines above, and d_1, d_2 are the two intercept functions. If $d_2(w) > d_1(w)$, the parameter region $\Gamma(w) := \{(k_1, k_2)$ such that w is a saturated attractor of the dynamics (1)$\}$ is a nonempty set. If $d_2(w) < d_1(w)$ $\Gamma(w)$ is an empty set. So in this sense the larger is the difference between $d_2(w)$ and $d_1(w)$, the more stable is the attractor w. From Theorem 1 we can derive some interesting consequences which are shown in the following corollaries:

Corollary 1. *(Fig. 1)*
1) The parameter region of (k_1, k_2) *in which* $(1, \cdots, 1)$ *is a saturated attractor of the dynamics (1) is*

$$k_1 + \sum_j r_j k_2 > d(+) := -\min_{i=1,\cdots,N} \sum_{j=1}^{N} q_{ij} r_j \tag{6}$$

2) The parameter region of (k_1, k_2) *in which* $(-1, \cdots, -1)$ *is a saturated attractor of the dynamics (1) is*

$$k_1 - \sum_j r_j k_2 < d(-) := \min_{i=1,\cdots,N} \sum_{j=1}^{N} q_{ij} r_j \tag{7}$$

Corollary 2. *(Fig. 1)*
1) If q_{ij} depends only on j, then only the configuration $(1, \cdots, 1)$ and $(-1, \cdots, -1)$ are saturated attractors of the dynamics (1).
2) If $q_{ij} = \delta_{ij}$, and $\min\{r_j, j = 1, \cdots, N\} > 0$, then any configuration $w \in \{-1,1\}^N$ is a saturated attractor of the dynamics (1).

§4 APPLICATION TO THE HOPFIELD MODEL, THE LINSKER MODEL AND THE DYNAMIC LINK MODEL

The Hopfield model, to which most of the theoretical investigations in the field of the neural network have been devoted so far, is defined by $q_{ij} = T_{ij} = \frac{1}{N} \sum_{\mu=1}^{p} \xi_i^\mu \xi_j^\mu$, $i, j = 1, \cdots, N$ and by the equalities $k_1 = \theta$, the threshold, $k_2 = h$, the external field and $r_i = 1, i = 1, \cdots, N$. $w_i(\tau)$ is the neural activity at time τ of the i-th neuron, and $\xi^\mu = (\xi_i^\mu, i = 1, \cdots, N)$ is the μ-th pattern to be stored in the network. The dynamics (1) now reads

$$w_i(\tau+1) = f(w_i(\tau) + \sum_{j=1}^{N}(T_{ij} + h)w_j(\tau) + \theta), i = 1, \cdots, N. \tag{8}$$

In most of the theoretical investigations, in particular in the statistical physics approach, ξ_i^μ is assumed to be i.i.d. and $p(\xi_i^\mu = 1) = p(\xi_i^\mu = -1) = \frac{1}{2}$, $\forall i, \mu$.

The dynamics (8) is a discrete time version of the continuous Hopfield model. Next we apply our results of section 3 to the Hopfield model. Since the stored patterns take values $+1$ and -1, it is enough general for us to restrict ourselves to the space $\{-1, 1\}^N$. $d_2(w)$ and $d_1(w)$ may be expressed using the overlap parameters $m(w, \xi^\mu) := \frac{1}{N} \sum_{i=1}^{N} w_i \xi_i^\mu$

$$d_1(w) = -\min_{i \in J^+(w)} \sum_{\mu=1}^{p} \xi_i^\mu m(w, \xi^\mu) \tag{9}$$

and similarly,

$$d_2(w) = -\max_{i \in J^-(w)} \sum_{\mu=1}^{p} \xi_i^\mu m(w, \xi^\mu). \tag{10}$$

Combining (9), (10) and Theorem 1, we see that the criterion for the existence of a saturated attractor of the Hopfield model is that

Theorem 2. *For the dynamics (8), a configuration $w \in \{-1, 1\}^N$ is a saturated attractor of the Hopfield model if and only if*

$$-\min_{i \in J^+(w)} \sum_{\mu=1}^{p} \xi_i^\mu m(w, \xi^\mu) < \theta + c(w)h < -\max_{i \in J^-(w)} \sum_{\mu=1}^{p} \xi_i^\mu m(w, \xi^\mu). \tag{11}$$

In the practical applications, we are mainly interested to establish if $w = \xi^\mu, \mu = 1, \cdots, p$ is a saturated attractor of the dynamics (8), a fact which can be easily checked by using Theorem 2. Note that this criterion is not based on the independence of the patterns ξ^μ. Many interesting examples can be constructed using this approach[FT2].

In spite of the extensive investigation of the Hopfield model, a little attention was paid to the parameter (θ, h) until now. Our theorem allows us for the first time to have a clear understanding of the role played by the two parameters in the dynamics (8) as explained below. The Hopfield model is described by a picture of the type of Fig. 1 , which is redrawn in Fig. 2. It is easily seen from the Fig. 2 that the number of stored patterns, i.e, of saturated attractors, of the Hopfield model depends on the parameters (θ, h). There is one region in which many saturated attractors coexist(see Fig. 2). In this region, the network will have the highest capacity, a quantity studied extensively in the literature. Outside this region, the capacity will become lower and lower. When h, the external field, is negative, there will be only one saturated attractor corresponding to one of the stored patterns if $c(\xi^\mu) \neq c(\xi^\nu)$ for $\mu \neq \nu$. Thus the capacity for the network is only $1/N$ in this case. However this region is good for retrieving a specific memory w if it is a saturated attractor. This remark suggests a way to recall an information avoiding the spurious states[FQ]. The above example suggests us the following definition of the critical capacity of the Hopfield model which can be applied to more general models also with dependent patterns:

Definition 2. *The critical capacity α_c of the dynamics (8) is*

$$\alpha_c := \inf\{\alpha = p/N, \langle d_2(\xi^\mu)\rangle - \langle d_1(\xi^\mu)\rangle = 0 \text{ for any } \mu = 1, \cdots, p\}, \tag{12}$$

where $\langle \cdot \rangle$ represents the expectation with respect to the distribution P of ξ^μ.

Further discussion on the relation of α_c defined above and the critical capacity founded in the spin glass approach is contained in [FT2] and the numerical simulation of α_c is shown in [FT2] also.

Now we consider the Linsker model.

The Linsker's model [Lin][FP][FPR] resembles the visual system, with an input feeding onto a number of layers corresponding to the layers of the visual cortex. The units of the network are linear and are organized into two-dimensional layers indexed L_0(input), $L_1, \cdots$ and so on. We suppose that each layer has N neurons and periodic boundary condition(wrapped up). There are feed-forward connections between adjacent layers, with each unit receiving inputs decreasing monotonically with the distance from the neurons belonging to the underlying layer. Let $w_{ki}^{(n)}(\tau)$ be the synaptic connection between the neuron i of the $(n-1)$-th layer and the neuron k of n-th layer, $r_{ki}^{(n)}$ is the synaptic density function between the $(n-1)$-th layer and the n-th layer, $(\sum_{i=1}^{N} r_{ki}^{(n)} = 1, \ \forall \, n, k)$. Making the averages with respect to the neuron activities we get

$$w_{ki}^{(n)}(\tau+1) = f(w_{ki}^{(n)}(\tau) + k_1 + \sum_{j=1}^{N} (q_{ij}^{(n)} + k_2) r_{kj}^{(n)} w_{kj}^{(n)}(\tau)). \tag{13}$$

(see the above bibliography for more details). The dynamics (13) is the updating process of the synaptic connections and characterizes the Linsker network. The index k can be dropped from the equation (13) since the appearance of a structured receptive field does not depend on it. We change our notation a little bit in order to apply theorem 1 of the section 3 to the dynamics (13). Let

$$d_1(w, n) = \max_{i \in J^+(w)} [\sum_{j \in J^-(w)} q_{ij}^{(n)} r_j^{(n)} - \sum_{j \in J^+(w)} q_{ij}^{(n)} r_j^{(n)}],$$

$$d_2(w, n) = \min_{j \in J^-(w)} [\sum_{i \in J^-(w)} q_{ij}^{(n)} r_j^{(n)} - \sum_{j \in J^+(w)} q_{ij}^{(n)} r_j^{(n)}].$$

Theorem 3. *If w is a saturated attractor of the Linsker model, then there exists $T > 0$ such that $w = w(T + \tau), \forall \tau \geq 0$. Furthermore w is a saturated attractor of the dynamics (13) if and only if*

$$d_1(w, n) < k_1 + c(w)k_2 < d_2(w, n), \quad n = 1, \cdots. \tag{14}$$

In the Linsker model a structured (an on-center or an oriented) receptive field is of particular interest. We know from the simulations that these receptive fields appear between the L_1 and L_2 layers if all the synaptic connections between the L_0 layer and the L_1 layer are kept positive. Recently this kind of structured receptive field has been founded important in the application of similar networks to the image recognition (see next section). Theorem 3 gives results which agree with what has been discovered by Linsker in its numerical simulations for the third layer (L_2) [FPR]. These results are shown in Fig. 3. The application of this theorem becomes more important as we encounter the necessity, in the practical application, to control the size of the on-center receptive field configuration by selecting the parameter of the dynamics (13). All results in [FP] [FPR] are true for the dynamics (13), we will not repeat them here and refer the reader to them for more details.

An interesting problem for us to do in the future is to check which model is the most optimal one in the sense to have the biggest $d_2(w) - d_1(w)$ among the models proposed to describe the ontogenesis of the visual system[Mal] here w is a structured receptive field. Our approach here makes this comparison possible.

Next we are going to consider the dynamic link model.

The power of the dynamic link network, a model proposed by von der Malsburg first in 1981, is demonstrated and developed in recent years in different applications, see for example [KMM]. In [FT1], a discrete version of the dynamic link model is proposed and a principle for choosing parameters used in the dynamic link model is given for the clipping function defined as in section 2. Here, by our results of section 3, we are able to reobtain all results in [FT1]. The dynamic link network is essentially a two layers network, say layer X and layer Y with both inter-layer connections and intra-layer connections. There are N neurons in the two layers and they are arranged in a two dimensional torus (i.e with periodic boundary conditions). For the details of the model the reader can refer to the paper [KMM]. In this model there are two time scales, one (τ) varied rapidly corresponding to the neural dynamics and the other one (ν) associated to the characteristic time scale of the synaptic dynamics. Making analogous transformations as those in [FT1] the neuron dynamics can be written in a form which is similar for the two layers

$$\begin{cases} \eta_i(\tau+1,\nu) &= f(\sum_{j=1}^{N} k_{ij}\eta_j(\tau,\nu) + I_i(\tau,\nu)) \\ \eta_i(0,\nu) &= 0 \end{cases} \tag{15}$$

and which is the dynamics we will focus on. The input $I_i(\tau,\nu)$ is different for the two layers and contains all the information connected with the image recognition problem. We suppose here that $I_i(\tau,\nu)$ is independent of i and τ denoting it as $I(\nu)$. The case of the dependence of I on i and τ is considered in [FT1]. The matrix k_{ij} is defined by $k_{ij} = \gamma e^{-\|i-j\|^2/s} - \mu = \gamma p_{ij} - \mu$ where p_{ij} is the weight interaction function inside the layer X or Y, $\gamma > 0$ and $\mu > 0$ are the intensities of excitatory or inhibitory connections respectively. Let $c(w)$ be defined as in Section 3 and $e_1(w)$ and $e_2(w)$ be two functions of the configuration w defined similar to $d_1(w)$ and $d_2(w)$

$$e_1(w) = \max_{i\in J^+(w)}[\sum_{j\in J^-(w)} p_{ij} - \sum_{j\in J^+(w)} p_{ij}], \quad e_2(w) = \min_{i\in J^-(w)}[\sum_{j\in J^-(w)} p_{ij} - \sum_{j\in J^+(w)} p_{ij}]. \tag{16}$$

Then we can apply Theorem 1 of section 3 to the dynamic link model. The proofs are similar to that in [FT1].

Theorem 4.
(1). For $\forall w \in \{-1,1\}^N$, $w \neq (1\cdots,1),(-1,\cdots,-1)$, a necessary and sufficient condition ensuring that there exists a nonempty set of (μ,γ,s,I) in which w is a saturated attractor of the dynamic (15) is $e_2(w) > e_1(w)$ and $\gamma > \gamma_0 := 2/[e_2(w) - e_1(w)]$. Furthermore the larger the γ, the bigger the parameter region ensuring that w is a saturated attractor of the dynamics (15).
(2). In the circumstances of (1), there exists a positive number μ_0 such that when μ is in the set $\{\mu,\mu \geq \mu_0\} \cap \{\mu, \gamma e_1(w) + 1 < I(\nu) - c(w)\mu < \gamma e_2(w) - 1\}$ then w is a saturated attractor of the dynamics (15) and $(1,\cdots,1)$, $(-1,\cdots,-1)$ will no longer be attractors of the dynamics (15).
(3). If s is large enough so that p_{ij}, $i,j = 1,\cdots,N$ are constants independent of i,j, then only $(1,\cdots,1)$ and $(-1,\cdots,-1)$ are the only possible saturated attractors of the dynamics (15).
(4). If s is small enough so that $p_{ij} = \delta_{ij}$ with $\gamma > 1$, then any state $w \in \{-1,1\}^N$ is an attractor of the dynamics (15).
(5). w is a saturated attractor of the dynamics (15) if and only if $I(\nu) \in [\gamma e_1(w) + c(w)\mu + 1, \gamma e_2(w) + c(w)\mu - 1]$.

For an explanation of Theorem 4, we refer the reader to Fig. 4. Theorem 4 shows that the effect of the lateral inhibition is to have some non trivial pattern as an attractor of the dynamics (15) and to avoid the region in which the trivial configuration $(1,\ldots,1)$ or $(-1,\ldots,-1)$ is the global minima. Theorem 4, (5) establishes the good fluctuation region of the input signal. If $I(\nu)$ is in the region of $[\gamma e_1(w) + c(w)\mu + 1, \gamma e_2(w) + c(w)\mu - 1]$, w will remain as an attractor of the dynamics (15) (Fig. 4). The interval in which $I(\nu)$ changes can be taken as an estimate for an effective interval in the case when the input signal is not translation invariant and depends on the time τ of the neural dynamics. From all the above arguments one derives a useful way for choosing the four parameters μ,γ,s,I^X in order to have an on-center configuration which is an attractor of the dynamics (15).

From Theorem 4 one can show that if the input $I(\tau)$ increases, the size of the on-center of the configuration which is an attractor of the dynamics (15) will increase also(see Fig. 4) if $\mu > 0$, the opposite will happen if μ is negative. One important open question here is how to ensure the convergence of the algorithm. A simple way to achieve it, as one may suggest similar to that in the Kohonen network , is to shrink the size of the on-center field. This can be done by decreasing the input $I(\tau)$ as well(Fig. 4). However from our Theorem 4 it follows after some estimates that in general it is impossible to shrink arbitrarily the size of the on-center pattern if s is fixed. This is a main difference between the Kohonen network and the dynamic link model[KMM].

§5 Conclusions

This paper unifies the approach to the Hopfield model, the Linsker model and the dynamic link model, three typical networks arising from three typical areas in the study of the neural network. Since most of models proposed so far in the field of the neural networks use the sigmoid function in their dynamics of learning or retrieving procedure, as discussed in section 2, we are able to characterize the attractors of these models in terms of the present approach. So the power of the present method is not restricted to these three models reported here.

In the Hopfield model, we give a sufficient and necessary condition in order to check if a given pattern is an attractor of the network. The capacity of the network is reconsidered from a point of view different from the usual statistical physics approach. It is also obvious that we could apply our method here to analyze generalizations of the Hopfield model.

The present approach becomes more efficient if we are mainly interested in one or a few kind of patterns. This is the case in the Linsker network and in the dynamic link network. For the former network, the appearance of the on-center and oriented receptive field is the core of its dynamics. For the latter, the on-center structured pattern is an important one in its dynamics. The present paper asserts that the appearance of the structured receptive field in the Linsker model is universal in the sense that the appearance of the structured receptive field is independent of the specific choice of the clipping function used in the numerical simulation in the Linsker's network. For the dynamic link network, we propose a principle for the selection of these parameters employed in the model.

The significance of this unification approach is obvious. It helps us to understand the mechanism underlying each model more deeply. It furnishs a useful guidance in the practical application of these models, in particular in choosing the parameters of the dynamics. For example, the results in the Linsker network suggest a way to shrink the size of the on-center receptive field, which may help the convergence of the algorithm used in the pattern recognition based upon the dynamic link network; essentially the self-organization in the Linsker model and the dynamic link model is a procedure of retrieving 'memory', the structured receptive field ('memory') is stored in the model already; and so on. Besides these findings we report here and in [FT2], there is still a lot of work to be done further.

Acknowledgement. This paper is partially supported by the CNR of Italy.

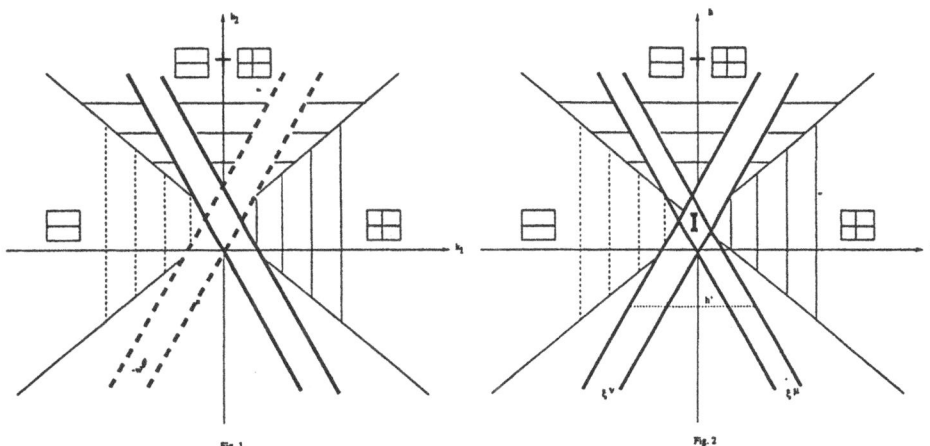

Fig. 1   Fig. 2

Fig. 1. The parameter region of different saturated attractors of the dynamics (1). ⊞ is the parameter region of the all positive attractors $w_i = 1, i = 1, \cdots, N$ (Corollary 1). ⊟ is the parameter region of the all negative attractors $w_i = -1, i = 1, \cdots, N$ (Corollary 1) and ⊞ + ⊟ is the parameter region of the all positive and the all negative attractor. The region of (k_1, k_2) between two dark lines is the parameter region in which $w \neq (1, \cdots, 1), (-1, \cdots, -1)$ is an attractor of the dynamics (1). The region of (k_1, k_2) between two dash dark lines is the parameter region in which $-w$ is an attractor of the dynamics (1)(see also [FT2]).

Fig. 2. The parameter region of (θ, h) in which w is a saturated attractor of the Hopfield model (see Fig. 1 also). In the region I enclosed by dark lines, the Hopfield model has the highest capacity. In this region, for example, ξ^μ, ξ^ν are both attractors of the Hopfield model. When $h = h'$(horizontal dash line), the capacity of the model becomes lower.

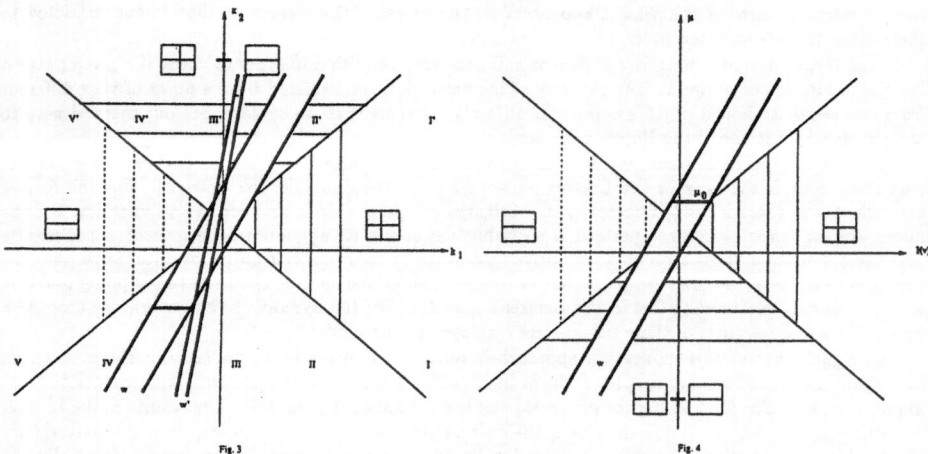

Fig. 3

Fig. 4

Fig. 3. The smaller is the size of the on-center, the narrower is the band in which that the on-center receptive field is an attractor([FPR]). The same conclusion is true for the off-center receptive field. When k_1 decreases($k_2 < 0$), the system passes through region I(all-excitatory), II(on-center), III(several attractors coexist), IV(off-center), V(all-inhibitory). If k_1 decreases($k_2 > 0$), we go from region I'(all-excitatory), II'(off-center), III'(several attractors coexist), IV'(on-center), V'(all-inhibitory). If we fix $k_1 > 0$, decrease $k_2 < 0$ the size of the on-center of the configuration which is an attractor of the Linsker model becomes smaller and smaller.

Fig. 4. The parameter region of $(I(\nu), \mu)$ in the dynamic link model. The slope of the dark lines is $c(w)$. So for fixed $\mu = \mu_0$, the smaller is the $I(\nu)$, the smaller is the radius of the on-center of the configuration which is an attractor of the dynamics (15). As r is small enough, the configuration with on-center radius r will no long be an attractor of the dynamics. For fixed $I(\nu) > 0$, when $\mu > 0$ becomes small, the size of the on-center of the configuration which is a saturated attractor of the dynamics (15) will become large.

References

[ATYD] Antonucci, M. & Tirozzi, B. & Yarunin, N.D. & Dotsenko, U.S.(1994), *Numerical Simulation of Neural Networks with Translation and Rotation Invariant Pattern Recognition*, Inter. Jour. of Modern Physics B, vol. 8, nos. 11 & 12, pp. 1529-1541.

[FP] Feng, J. & Pan, H.(1993),*Analysis of Linsker-type Hebbian Learning: Rigorous Results*, Proc. 1993 IEEE Int. Conf. on Neural Networks - San Francisco Vol. III, (Piscataway, NJ: IEEE) pp. 1516-1521.

[FPR] Feng, J. & Pan, H. & Roychowdhury, V. P.(1995), *A Rigorous analysis of Linsker's Hebbian learning network*, to appear in Advances of Neural Information Processing Systems, Vol .6 (NIP*S 94).

[FQ] Feng, J. F. & Qian, M. P.(1993), *Two-Stage Annealing in Retrieving Memories I*, In: Badrikian, A.; Meyer P-A, and Yan, J-A (ed.), Probability and Statistics, Singapore: World Scientific, pp. 149-176.

[FT1] Feng, J. & Tirozzi, B.(1995), *A Discrete Version of the dynamic link model*, to appear in Neural Computation.

[FT2] Feng, J. & Tirozzi, B.(1994), *An analysis of the dynamics with the sigmoid like function*, preprint.

[KMM] Konen, W. & Maurer, T. & von der Malsburg, C.(1994), *A fast dynamic link matching algorithm for invariant pattern recognition*, Neural Network, Vol. 7, Nos, 6/7, pp.1019-1030.

[Lin] Linsker, R.(1986), *From Basic Network Principle to Neural Architecture (series)*, Proc. Natl. Acad. Sci. USA, 83, 7508-7512, 8390-8394, 8779-8783.

[Mal] von der Malsberg, C.(1995),*Network self-organization in the ontogenesis of the mammalian visual system*, An Introduction to Neural and Electronic Networks, Second Edition, S.F. Zornetzer, J. Davis and C. Lau, Eds. (Academic Press).

Learning in Evolutive Neural Architectures: An Ill-Posed Problem?

Christian JUTTEN

INPG-TIRF 46, avenue Félix Viallet 38031 Grenoble Cedex
Tel. + 33 76 57 45 48 Fax + 33 76 57 47 90 Email chris@tirf.inpg.fr

Abstract. Basically, evolutive architectures are networks able to be modified by adding or pruning neurons or connections. In the paper, by a synthesis a various works, we point out that evolutive architectures involve a lot of tricks because of indeterminacy of solutions and suboptimality, which are characteristic of ill-posed problems. We also emphasize on interest of stopping criteria, essential to control adding as well as pruning procedure and to avoid overfitting. Finally, we suggest another formulation of learning in evolutive architectures based on more realistic "hardware" constraints.

1. INTRODUCTION

1.1. EVOLUTIVE ARCHITECTURES

Since beginning of 80's, interest of researchers increases toward evolutive architectures, that is architectures which may grow. Most attempts have been done for multi-layer perceptrons (MLP) but the first growing algorithm seems to be the one of Reilly *et al.* [1], in 1982 which can be viewed as a radial basis function (RBF) neural network.

Another way to modify architectures is to simplify them: such a pruning concept is very well known, and for long time, to simplify binary trees. Direct methods for neural networks pruning have been proposed in 90's: Le Cun *et al.* [2] proposed in 1989 a method based on weights pruning, Siestmas and Dow [3] showed how it is possible to cancel redundant neurons. Indirect methods, based on a modification of the cost function minimized by the algorithm, have been also studied from 1986 [4].

Finally, other questions can be related to evolutive architectures. How to learn new data ? How to do permanent learning ? Up to now, these questions are not addressed in the literature, except with recent works on active data selection [5].

Of course, evolutive architectures will be efficient if architecture modifications do not call into question remaining parameters, previously optimized.

1.2. WHY EVOLUTIVE ARCHITECTURES ?

Theoretical results on function approximation are existence theorems which gave justifications in using neural networks. But they provide any information neither on the architecture of the networks, especially on the number of neurons, nor on the optimization process.

In fact, these theorems claim more or less that it exists a 2-layer neural network (with sigmoidal neurons) able to approximate any continuous and bounded function. But perhaps, the network needs an infinite number of cells. For instance, we can define functions which can be approximate by a 2-layer network with an infinite number of neurons in the hidden layer, or by a 3- (or more) layer network with a small number of neurons.

We also know that adding more and more units is not good because it leads to overfitting. Moreover, network parameters are optimized from empirical data (typically noisy data) and the optimal number of free parameters (number of neurons and connections) is related to the number of samples and to the data dimension.

Therefore, evolutive architectures could be an efficient and attractive answer in chosing the network size.

1.3. MAIN PROBLEMS

Three main problems must be solved to build efficient evolutive architectures:

• after adding or pruning, the global structure of the network must not change. This constraint guarantees the adding or pruning procedure can be easily repeated.

• after adding or pruning, starting from the previous mapping, the network optimization can be computed with a complexity significantly smaller than starting again the learning. Otherwise, evolutive architectures have no interest.

• To have good enough mapping, and avoid overfitting, pruning as well as adding procedures must be controlled by a stopping criterion.

1.4. CONTENT OF THE PAPER

In section 2, we analyse architectures proposed in the literature for evolutive networks. Section 3 is devoted to stopping criteria, both for growing and pruning. Remarks concerning permanent learning are added in section 4, before a conclusion.

2. ARCHITECTURE CONSTRAINTS

In this section, we focus on architecture constraints. We study constraints related both to MLP architectures and to RBF architectures, first when adding neurons to improve the mapping, and then when adding a new class in classification. For each method, we also discuss briefly on retraining aspects after adding neuron or pruning. Finally, architecture problems due to pruning are discussed.

2.1. MAPPING IMPROVEMENT

2.1.1. MLP Architectures

Consider a classical MLP. Input-layer and output-layer sizes are defined according to data dimension and coding. For sake of simplicity, we assume n input units (data dimension is n) and 1 output unit (scalar mapping).

To improve the network ability, it seems natural to add new units in the hidden layers. How to determine in which layer ? How to compute the weights of the new unit and to merge it with the current network ? Is the network optimal after the merger ?

To evade the first question, consider a single hidden layer (with N neurons) MLP. We want to add a unit to this layer. Mappings before and after merger are:

$$
\begin{aligned}
y_N(\mathbf{w}, \mathbf{x}) &= \phi\left[\sum_{j=1}^{N} w_{ij}^{(1)} y_j^{(1)} \right] = \phi\left[\sum_{j=1}^{N} w_{ij}^{(1)} \phi\left(\sum_{k=1}^{n} w_{jk} x_k \right) \right] \\
y_{N+1}(\mathbf{w}, \mathbf{x}) &= \phi\left[w_{i,N+1}^{(1)} \phi\left(\sum_{k=1}^{n} w_{N+1,k} x_k \right) + \sum_{j=1}^{N} w_{ij}^{(1)} \phi\left(\sum_{k=1}^{n} w_{jk} x_k \right) \right]
\end{aligned}
\tag{1}
$$

MLP with linear outputs

If the output of the new unit, say $y_{N+1}^{(1)}$, is trained to fit a target function, after merging, its contribution on network output is distorted by the non linear function ϕ. Then, the target function of the new unit is not easy to define, except if the output unit is linear. In that case, network output reduces to:

$$y_{N+1}(\mathbf{w},\mathbf{x}) = w_{i,N+1}^{(1)}\, \phi\!\left(\sum_{k=1}^{n} w_{N+1,k}\, x_k\right) + \sum_{j=1}^{N} w_{ij}^{(1)}\, \phi\!\left(\sum_{k=1}^{n} w_{jk}\, x_k\right). \qquad (2)$$

This observation has been done by Jutten *et al.* [6, 7], who proposed to teach the new unit on the current network error. Assuming the target outputs are noisy data $y_d(\mathbf{x}) = y(\mathbf{x}) + n$, then current mapping $y_N(\mathbf{w},\mathbf{x})$ leads to the error $\varepsilon_N(\mathbf{w},\mathbf{x}) = y_d(\mathbf{x}) - y_N(\mathbf{w},\mathbf{x})$. The new unit is teached to approximate this error. Then, after merging with the previous network, the approximation is improved by $\hat{\varepsilon}_N(\mathbf{w},\mathbf{x})$. However, after merging, the mapping $y_N(\mathbf{w},\mathbf{x}) + \hat{\varepsilon}_N(\mathbf{w},\mathbf{x})$ is not optimal: the non linear mapping is a differential variety in $\mathbb{R}^n$, and residuals have no particular properties. On the contrary, in linear mapping, the error is orthogonal to the solution space spanned by the data (projection theorem). However, the approximation is generally close to the optimal and some learning iterations are sufficient to achieve the optimum mapping $y_{N+1}(\mathbf{w},\mathbf{x})$.

This scheme is very simple, and can only be applied on the last hidden layer provided that output unit is linear. Experimental results on function approximation [7] point out 2 essential advantages with respect to classical backpropagation:

- the incremental procedure is 20 to 30 % faster than classical backpropagation,

- it avoids local minima: for simple function approximation, classical backpropagation failed 6 times over 10, while this never occured with the incremental learning.

This algorithm has been designed for approximation, and at first glance, generalization for classification seems evident, but surprisingly it comes up against problem of error definition, of output coding and of homogeneousness of the network and of the new unit.

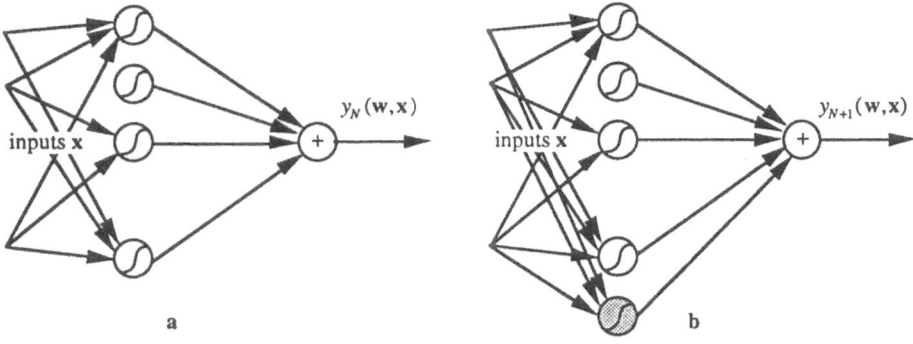

Figure 1. The new hidden unit (grey in b) is teached to map what the current mapping (a) cannot do, that is the error. Simple fusion is possible because the output unit is linear.

Cascade networks

A well-known scheme of constructive MLP architecture is the so-called Cascade-Correlation learning architecture proposed in 1989 by Fahlman and Lebiere [8]. The idea consists in teaching new hidden units which will not modified after merging to the current network. At each cycle, the new unit is computed as a non linear mapping of inputs and of outputs of still existing hidden neurons:

$$y_1(\mathbf{w}, \mathbf{x}) = \phi\left[\sum_{k=1}^{n} w_k x_k\right],$$

$$y_2(\mathbf{w}, \mathbf{x}) = \phi\left[w_{n+1} y_1(\mathbf{w}, \mathbf{x}) + \sum_{k=1}^{n} w_k x_k\right],$$

$$y_3(\mathbf{w}, \mathbf{x}) = \phi\left[w_{n+2} y_2(\mathbf{w}, \mathbf{x}) + w_{n+1} y_1(\mathbf{w}, \mathbf{x}) + \sum_{k=1}^{n} w_k x_k\right], \tag{3}$$

$$y_{N+1}(\mathbf{w}, \mathbf{x}) = \phi\left[w_{n+N} y_N(\mathbf{w}, \mathbf{x}) + ... + w_{n+1} y_1(\mathbf{w}, \mathbf{x}) + \sum_{k=1}^{n} w_k x_k\right].$$

The equations are more complex than (2), and each new neuron corresponds to a new hidden layer in the network (Fig. 2). Fahlman *et al.* suggested to teach the new unit, at each step, in order to maximize a covariance between the residual error and the activity of the new unit. Then, the weights of this unit are frozen, and the unit is merged to the networks. The weights of the network are then optimized to minimize the mean square error. And so on. Results on simple classification tasks (XOR and 2-spiral problems) especially prove improvement of convergence speed.

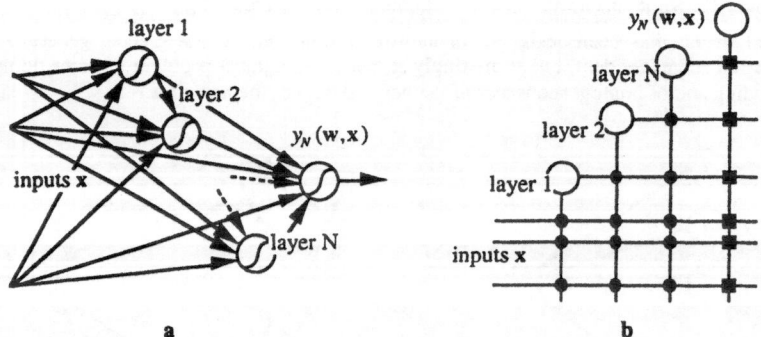

Figure 2. Cascade Correlation network. Each new unit is added in a new layer: classical representation (a), and equivalent scheme (b). Only weights corresponding to squares are updated in (b).

An alternative of the Cascade-Correlation scheme, studied by Littmann and Ritter [9], consists in replacing each hidden unit $y_N(\mathbf{w}, \mathbf{x})$ by a pool of hidden units which are powers of hidden unit. The idea is not new, and is even extensively used in non linear mapping, because it has the advantage to be linear with respect to the parameters, and then it allows a very easy and efficient optimization.

Neural trees

The concept of neural tree, introduced by Sirat and Nadal [10] in 1990, also allows efficient growing MLP algorithms and architectures [11], but devoted to classification. In that case, the network is a neural implementation of a binary classification tree, which implies very specific tasks for each layer. Consider the 2-D distributions of Fig. 3.a. It is easy to build a binary classification tree by splitting first the space, with respect to classes, in two parts with a hyperplan (simple straight line, in 2-D space), say H1. In each part defined by H1, we again split classes by hyperplans H2 and H3. And so on, up all the classes are completely (or optimaly, according to a given criterion) separated. We may associate to this procedure a classification binary tree (Fig. 3.b). This tree can also be implemented by a 3-layer neural network. Neurons of the first layer compute the hyperplans Hi: output of the neuron Hi is positive for any point located on one side and negative for any point of the other side. Then, there are as many neurons in this first layer as hyperplans in the tree. Neurons of the second layer compute the different areas corresponding to the different class clusters. There are then as many neurons in this layer as leaves on the classification tree (here 5). The computation involves simple logical AND of first layer outputs. Last layer neurons provide the decision, by simple logical OR of second layer outputs: in fact, they do merge points (regions) having the same class label. Their number is egal to the class number (3 on Fig. 3.c).

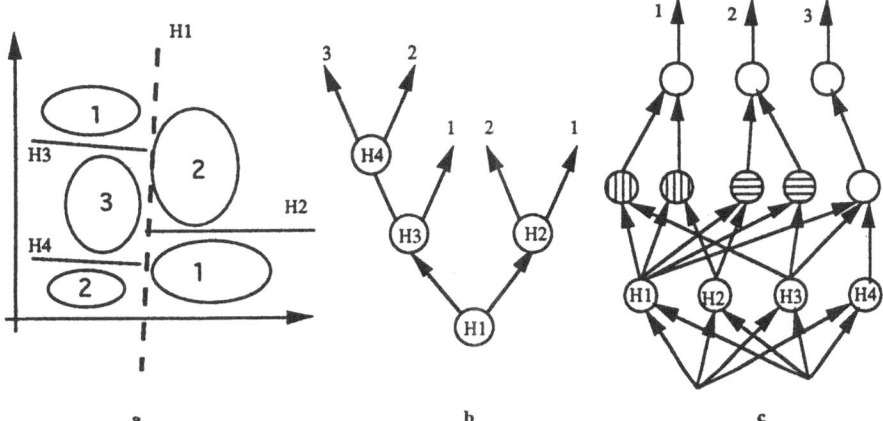

Figure 3. Classification tree and neural network. The separation of classes by hyperplans (a) can be represented by a binary classification tree (b), and implemented in a 3-layer neural network (c).

Assume now that a new hyperplan H5 must be add to provide a better separation. Then, a neuron corresponding to H5 is added in the first layer. Because there is one leave more, one neuron is also added in the second layer. Connections of one neuron (if we except the new one) of this layer will be modified by adding a connection coming from output H5. In the last layer, the neuron number remains unchanged, but a connection coming from the new neuron of the second layer to the output neuron corresponding to its class (Fig. 4). Clearly, this simple example proves the growing procedure does not question the previous connections. It only adds a few neurons and a few connections, according to simple rules which can be automatically implemented. Neural tree is an interesting way to set up initial architecture and weights, close to a good solution, which can be easily optimized afterwards.

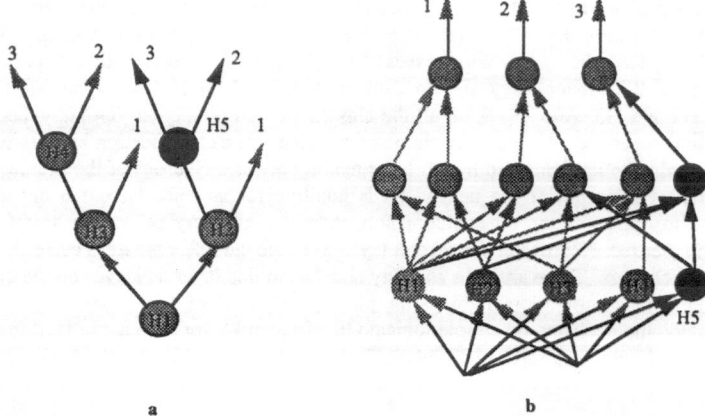

Figure 4. Modification of the classification tree (a) are easily implemented on the neural network (b). Previous connections and neurons are in grey, modifications are in black.

2.1.2. RBF Neural Architectures

We now consider RBF neural networks, which can also be viewed as approximators. Assuming for sake of simplicity a scalar mapping $y(\mathbf{x})$, where $\mathbf{x}$ is n-dimension vector. Using N noisy samples $(\mathbf{x}, y(\mathbf{x}) + n)$, the RBF approximation is:

$$\hat{y}(\mathbf{x}) = \lambda_0 + \sum_{j=1}^{P} \lambda_j K\left(\left\|\mathbf{x} - \mathbf{c}_j\right\|\right), \tag{4}$$

where $K(u)$ is a monovariate even function, which decreases as $|u| \to \infty$. In (4), parameters (λ_j and c_j) are generally computed by supervised learning on the N data base samples.

Although Gaussian kernel are frequently used, other kernels with various shapes are proposed in the literature. See for instance Girosi *et al.* [12] for a discussion on RBF networks as stabilizers in regularization schemes.

For a very long time, normalized RBF networks are also used as density estimators, whose Parzen window estimator is a well known example [13, 14]. Consider N samples of class k, $\mathbf{x}_j, 1 \le j \le N$, the density of class k is estimated by:

$$\hat{p}_N(\mathbf{x}) = \frac{1}{N} \sum_{j=1}^{N} K_h\left(\left\|\mathbf{x} - \mathbf{x}_j\right\|\right) = \frac{1}{Nh} \sum_{j=1}^{N} K\left(\frac{\left\|\mathbf{x} - \mathbf{x}_j\right\|}{h}\right), \tag{5}$$

where h is a width factor, and K_h is a positive kernel, whose L_1 norm is equal to 1. In (5), each kernel has the same weight ($1/N$), but refined density estimator involves kernels with variable widths and then weights are different [15, 16].

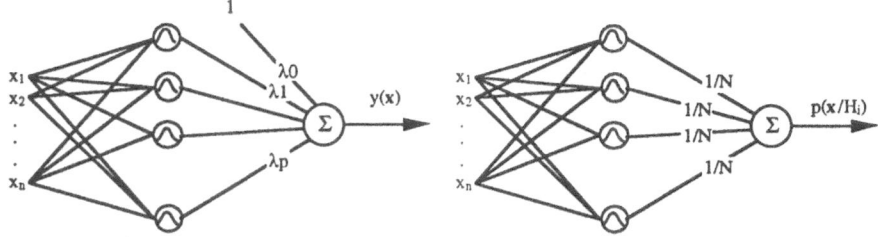

Figure 5. RBF and normalized RBF networks point out very close architectures, but very different learning procedures.

In both case, the estimation corresponds to a very simple 2-layer RBF network (Fig. 5), whose output units are linear. The density estimator can be used as a basic bloc for classification neural networks which implements Bayesian Neural Classifiers, as suggested first by Specht [17], and Comon [18]. Figure 6 represents such a Bayesian neural classifier. Outputs of the first layer are kernel conditional density estimators. It is very simple to prove [19] that the Bayesian decision can be obtained by comparing weighted (by prior probabilities and costs) sums of the densities.

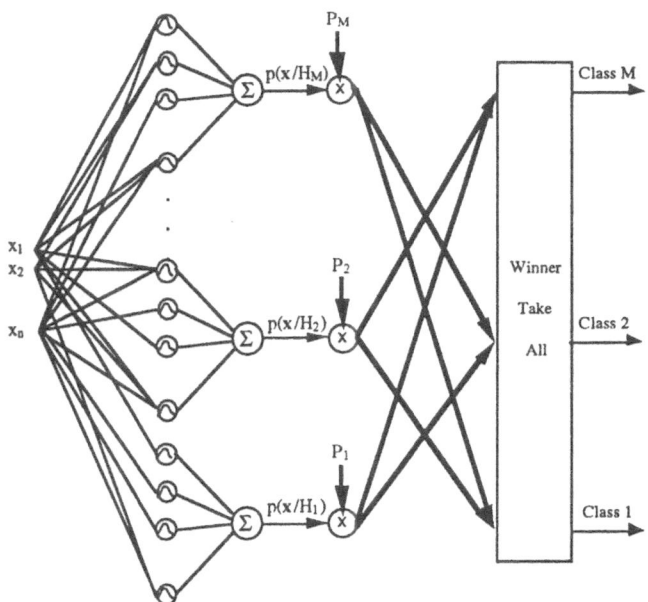

Figure 6. Bayesian Neural Classifier

Basically, RBF networks emphasize local approximation properties, because kernels have local influence. Then, local corrections should be very easy to do. In fact, in (4), adding a kernel centered in c_i, only influences the approximation for points x close to c_i, and consequently only requests to adjust weights of neighbouring kernels. Exploiting this properties, a quasi-orthogonal incremental algorithm has been derived by $ [20]. According to (5), the density estimator can be improved for each new sample, say x_{N+1}, of class k, by adding a new kernel centered on x_{N+1}, and by reducing the weight to $\frac{1}{(N+1)h}$. Concerning learning, estimators (4) and (5) point out very different behaviours: an infinite number of data can be used to computes (by least square algorithm for instance) parameters of (4). On the contrary, an infinite number of data provides an infinite number of kernels in (5), which becomes untractable. Practically, in (5), vector quantization techniques are essential to reduce data number to a given and reasonable number, say N. Then, the weight remains equal to $1/N$, but the location of each kernel center is modified at every step by the vector quantization algorithm.

2.2. PRUNING THE NETWORK

In this section, we only take into account brute-force pruning methods which modify the networks by removing neurons or synapses.

Sietsma and Dow's idea [3] consists in merging redundant neurons (neurons strongly correlated with positive or negative coefficients, close to ± 1) or cancelling useless neurons (for instance, neurons with constant outputs). Optimal Brain Damage [2] (OBD) is based on the computation of an average sensitivity, the so-called saliency, on the quadratic cost function (square error) with respect to each parameters (weights). Then, the parameter (synapse) with the smallest saliency is remoted. In these two cases, after pruning, the network mapping is no more optimal, although very close, and must be retrained during a few iterations. To avoid retraining after pruning in OBD, Hassibi and Stork [21] introduced a new algorithm, called Optimal Brain Surgeon (OBS), which, by constrained minimization, computes directly the parameter to cancel and the one-shot updating of remaining parameters. However, saliency in OBD and OBS is based on a second order expansion of the cost function. Therefore, the one-shot updating of OBS, or the retraining during few iterations with OBD are efficient provided that parameters cancelled are small enough.

In RBF networks, one only can cancel kernels, and it is very easy to compute the error due to kernel removal. With RBF, balancing the removal should be achieved after a few iterations. With normalized RBF (density estimators), it is more complex because we must define a set of samples representative of the density, smaller by one unit. This can be done by vector quantization on the whole learning data base, but it generally takes a long time. In the case of monovariate kernel estimators, Fambon et al. [22] apply the Hassibi & Stork's idea, and balance removal of a few kernels by updating, in one-shot, locations of remaining kernels. However, up to now, pruning of RBF (normalized or not) networks has not been intensively studied.

3. STOPPING CRITERIA

As we said in Introduction, a crucial role of evolutive architectures is to avoid overfitting. In fact, this can be achieved if the adding procedure is controled by a stopping criterion.

3.1. STOP GROWING

Most of evolutive neural architectures do not address this main point. Stopping criteria are completely fuzzy and consist of stopping network growing "when the approximation is good enough". Nevertheless, the problem is well known in regression and in identification, and for a long time. A practical method consists in computing Akaike's B information criterion (BIC) [23]

$$BIC = \log\left(\frac{\sigma_r^2}{N}\right) + n_p \frac{\log N}{N}, \tag{6}$$

where σ_r^2 is the residual error variance, N the number of samples, and n_p the number of independent parameters. The first term of (6) measures data closeness, the second one is a complexity term which points out the relation between the number of parameters and the number of samples in the learning data base. In the case of evolutive architectures, at each step, one must compute BIC and stop when BIC is minimum. Notice that n_p is the independent number of parameters and is not equal to the total number of network parameters (weights): parameters corresponding to redundant or strongly correlated neurons and synapses must be cancelled.

This procedure is very general and may also be used to control pruning as done by Cottrell *et al.* [24].

In the case of approximation from noisy data, a new criterion has been recently proposed by Jutten *et al.* [6, 7]. The idea consists in remarking that if the approximation is perfect, the residual error becomes equal to the noise. Assuming noise samples are zero mean and i.i.d., the adding procedure is stopped if the approximation by one neuron of the residual error leads to a function very close to 0. To decide is the function is close enough to 0, we may use a simple hypothesis test based on Gaussian asymptotical behaviour of least square estimators. We currently study other tests, more sensitive based on methods used in non linear regression [25].

3.2. STOP PRUNING

The problem of stopping criteria is also crucial in pruning procedures. Otherwise, the network maybe will not have enough parameters to insure a consistent approximation.

In OBD [2] and OBS [21], authors proposed to prune the weight which has the smallest sensitivity. In fact, if the smallest sensitivity is too large, the weigh must not be pruned, but how to define a threshold ? It can be shown [26] that a statistical nullity test, as suggested by Cottrell *et al.* [24], is equivalent to OBD and OBS. It seems surprising because one often argues that small parameters can have important effect in the network, and must not be pruned. However, remember that OBS and OBD are based on a second order expansion of the cost function with respect to the parameters, which is true if parameter variations are small, that is if pruning only concerns small enough weights. Practically, based on Gaussian asymptotical properties of least square estimators, Cottrell *et al.* suggest a nullity test with a given confidence interval, which defines a threshold for the saliency, at the same time.

Notice that all these criteria are statistical variables typically computed on the whole learning data base. Of course, stopping criteria (for both growing and pruning architectures) are cost consuming, but essential.

4. PERMANENT LEARNING

4.1. LEARNING NEW DATA

Permanent learning looks like a difficult problem which includes learning of new data. Assuming a current mapping, what does one do when a new data arrives ? Basically, the answer is twofold depending of the learning scheme.

In the case of supervised learning, learning and recall phases are distinct, and do not use the same informations. Especially, during the recall phase, target outputs are not known, and the network cannot be trained. The problem of permanent supervised learning is then an unusual scheme.

On the contrary, permanent learning is easy with unsupervised algorithms. Typically, there is no distinction between learning and recall phases. Learning may be permanent because criteria used to update the parameters can always be computed.

4.2. ADDING A NEW CLASS

Consider another aspect of permanent learning: addition of a $(N+1)$th classe in a neural classifieur.

Clearly, the Bayesian neural classifier (Fig. 5) will be easily expanded by adding a new kernel estimator, and a new row and a new column to the cost matrix. Neural tree also will be easily modified to take into account the new class.

With a classical MLP, assuming "canonical" coding on outputs, it will be necessary to add a new neuron on output layer and to connect it to each unit of the last hidden layer. A backpropagation learning process will adjust the new weights, but also modify all the weights before the last hidden layer.

4.3. PRUNING DURING LEARNING

Contrary to brute-force pruning methods of section 2.2., which alternate learning and pruning, another (and soft) way to control network complexity consists in adding complexity term E_w to the classical mean square error term E_d. The cost function which then becomes:

$$E = E_d + \lambda E_w \tag{7}$$

where λ controls the trade-off between the two terms. In the literature, a few complexity terms have been proposed, but without explanation. Recently, Williams [27] proposed a statistical interpretation of the complexity term, based on estimation theory. In absence of prior information on the weight distribution, estimation on network parameters will be based on simple Maximum Likelihood estimator. If noise on data are zero-mean i.i.d. samples, it is well known that maximize the likelihood is equivalent to minimize mean square error. On the contrary, if we assume a weight distribution $p(\mathbf{w})$, the optimal solution is given by the Maximum A Posteriori equation, which consists in chosing parameters $\mathbf{w}$ which maximize *a posteriori* conditional density

$$p(\mathbf{w} / y_d) = \frac{p(y_d / \mathbf{w}) p(\mathbf{w})}{p(y_d)}. \tag{8}$$

It is easy to see that maximize log of (8) is then equivalent to minimize a cost function like (7). If parameter distribution $p(\mathbf{w})$ is assumed Gaussian, or Laplacian respectively, it leads to complexity term equal to $\sum_i w_i^2$ or $\sum_i |w_i|$, respectively (see [28] and [29] for reviews).

Unfortunately, parameters being computed from unknown data, it would be surprising that weight distributions are Gaussian or Laplacian. Then, the optimal choice of a complexity term in (7) is still an ill-posed problem, and the statistical interpretation can explain practical problems encountered when minimizing (7).

5. CONCLUSION

In previous sections, I tried to list and discuss problems encountered when designing evolutive architectures. First, I prove that evolutive algorithms demand *ad hoc* architectures. Moreover, it seems that the architectures cannot be well-suited both for approximation and classification tasks. Secondly, interest of evolutive methods leads in the speed enhancement in learning with respect to trial/error procedures, and in the ability to avoid local minima during the learning. Finally, I recomment to turn a special care on stopping criteria in order to avoid overfitting.

Now, I wonder if the problem of evolutive architectures, as it is currently (and above) addressed, is not an ill-posed problem, and if it would not be better to address other questions, strongly related to hardware constraints: optimal ressources allocation and reliability:

Given a classification or a mapping problem, and finite ressources (neurons, synapses), what are ressources allocations which lead to best performances ?

Once an architecture is chosen, how to train the network to get a robust solution, that is unsensitive as possible to a random pruning ?

The backpropagation algorithm with mortality, designed by Kerlirzin *et al.* [30], gives partial answer to the second question. The first question is completely open.

In this paper, I never spoke about what brains do. It is time yet. In Neurobiology, it is well known that neuron number is decreasing during the life. More precisely, we know that, during the first months after birth, synaptic plasticity in neural architectures is driven by environment stimuli and especially corresponds more or less to simple synaptic selection, regression and reinforcement schemes. In other situations, for instance after lesions of certain parts of the brain, it has been observed that neighbouring parts can balance and functionally replace destroyed structures. Basically, every day, our neural networks are randomly pruned.

Robust learning and dynamic ressources allocation would it not be essential tasks, but of everyday purpose, leading to efficient evolutive architectures ?

Acknowledgements

This work is partially funded by Esprit project ELENA (6891).

REFERENCES

[1] **Reilly D., Cooper L., Erlbaum C.**, "A Neural Model for Category Learning", *Biological Cybernetics*, Vol. 45, pp. 35-41, 1982.

[2] **Le Cun Y., Denker J., Solla S.**, "Optimal Brain Damage", NIPS Conf., Denver (USA) 1989. In *Advances in Neural Information Processing*, 2, pp. 598-605, 1990.

[3] **Sietsma J., Dow R.**, "Creating artificial networks that generalize", *Neural Networks*, Vol. 4, n° 1, pp. 67-79, 1991.

[4] **Plaut D., Nowlan S., Hinton G.**, "Experiments on learning by back propagation", Tech. rep. CMU-CS-86-126, Carnegie-Mellon Univ., 1986.

[5] **Plutoski M., White H.**, "Selecting Concise Training Sets from Clean Data", *IEEE Trans. on Neural Networks*, Vol. 4, n° 3, pp. 305-318, 1993.

[6] **Jutten C., Chentouf R.**, "A New Scheme for Incremental Learning", *Neural Processing Letters*, Vol 2, n° 1, pp. 1-4, 1995.

[7] Chentouf R., Jutten C., "Incremental learning with a stopping criterion. Experimental results", IWANN 95, Malaga (Spain), June 1995.

[8] Fahlman S., Lebiere C., "The Cascade-Correlation Learning Architecture", in *Advances in Neural Information Processing Systems II*, Touresky *et al.* (eds), pp. 524-532, 1989.

[9] Littmann E., Ritter H., "Cascade Network Architectures", *Proc. Intern. Joint Conf. on Neural Networks*, Baltimore, Vol. II, pp. 398-404, 1992.

[10] Sirat J., Nadal J.-P., "Neural Trees: a New Tools for Classification", *Networks*, 1, pp. 423-438, 1990.

[11] Moreno J. M., Castillo F., Cabestany J., "Hardware implementation of piecewise linear separation incremental algorithms", Proceeding of NeuroNîmes 93, Nîmes (France), October 1993, pp; 199-208, 1993.

[12] Girosi F., Jones M., Poggio T., "Regularization Theory and Neural Neetworks Architectures", *Neural Computation*, Vol. 7, n° 2, pp. 219-269, 1995.

[13] Parzen E., "On Estimation of a probability density function and mode", *Ann. Math. Statist.*, 33, pp. 1065-1076, 1962.

[14] HärdleW., *Smoothing techniques with implementations in S*, Springer-Verlag, 1990.

[15] Silverman B., *Density estimation for statistics and data analysis*. Chapman and Hall, 1986.

[16] Comon P., "Supervised classification: a probabilistic approach", Invited paper, ESANN 95, Brussels (Belgium) April 1995. D facto publisher.

[17] Specht D., "Probabilistic Neural Networks", *Neural Networks*, Vol. 3, 1, pp. 109-118, 1990.

[18] Comon P., Bienvenu G., Lefebvre T., " Supervised design of optimal receivers", *NATO Advanced Study Institute on Acoustic Signal Processing and Ocean Exploration*, Madeira (Portugal), July 1992.

[19] Jutten C., Comon P., "Neural Bayesian Classifier", IWANN 93, Barcelona (Spain), June 1993. In "News Trends in Neural Computation", Mira J., Cabestany J., Prieto A. (Eds), Lecture Notes in Computer Science n° 686, Springer-Verlag, pp. 119-124, 1993.

[20] Chen S., Cowan C., Grant P., "Orthogonal least squares learning algorithm for radial basis function networks", *IEEE Trans. on Neural Networks*, Vol. 2, n° 2, pp. 302-309, 1991.

[21] Hassibi B., Stork D., "Second Order Derivatives for Network Pruning: Optimal Brain Surgeon". *Neural Information Processing Systems*, 1992.

[22] Fambon O., Jutten C., "Pruning kernel density estimators". ESANN 95, Brussels (Belgium) April 1995. D facto publisher.

[23] Akaike H., "Statistical predictor identification", *Ann. Inst. Statist. Math.*, 22, pp. 203-217, 1970.

[24] Cottrell M., Girard B., Girard Y., Mangeas M., Muller C., "Neural modeling for time series: a statistical stepwise method for weight elimination". To appear in *IEEE Trans. on Neural Networks*.

[25] Antoniadis A., Berruyer J., Carmona R., *Régression non linéaire et applications*, Economica, Paris, 1992.

[26] Fambon O., Jutten C., "A comparison of two weight pruning methods", ESANN 94, Brussels (Belgium), April 1994, D facto publisher, pp. 37-42, 1993.

[27] **Williams P.**, "Bayesian regularization and pruning using a Laplace prior", *Neural Computation*, Vol. 7, n° 1, pp. 117-143, 1995.

[28] **Reed R.**, "Pruning algorithm - a survey", *IEEE Trans. on Neural Networks*, Vol. 4, n° 5, pp. 740-747, 1993.

[29] **Jutten C., Fambon O.**, "Pruning methods: a review". Invited paper, ESANN 95, Brussels (Belgium) April 1995. D facto publisher.

[30] **Kerlirzin Ph., Vallet F.**, "Robustness in multilayer perceptrons", *Neural Computation*, 5, pp. 473-482, 1993.

Automatic Scaling Using Gamma Learning for Feedforward Neural Networks

AP Engelbrecht   I Cloete   J Geldenhuys   JM Zurada*
ian@cs.sun.ac.za

Computer Science Department, University of Stellenbosch, Stellenbosch 7600, South Africa

Abstract

Standard error back-propagation requires output data that is scaled to lie within the active area of the activation function. We show that normalizing data to conform to this requirement is not only a time-consuming process, but can also introduce inaccuracies in modelling of the data. In this paper we propose the gamma learning rule for feedforward neural networks which eliminates the need to scale output data before training. We show that the utilization of "self-scaling" units results in faster convergence and more accurate results compared to the rescaled results of standard back-propagation.

1 Introduction

Many artificial neural networks trained with the popular error back propagating training algorithm, also called the *delta rule*, contain units having the well-known sigmoid activation function where λ is a positive constant [Zurada, 1992a]:

$$f(\lambda, y) = \frac{1}{1 + e^{-\lambda y}} \qquad (1)$$

A problem with this squashing function is that its output is always in the range $[0, 1]$, thus requiring scaling of the desired output before training to fit into this range. In addition to scaling of the output data, the input data is normally scaled to lie within the active area of the sigmoid activation function (e.g. to the range $[-\sqrt{3}, \sqrt{3}]$). In this paper we investigate the effects that scaling of the output data has on the learning process.

In practice, the data set presented to a neural network often contains values which lie outside the active range of the sigmoid activation function. If the delta learning rule with sigmoid activation functions is used to learn the data set, the data must be pre-processed before training. During pre-processing, the data is compressed to fit into the active range of the sigmoid function. This scaled data set is then used for training purposes. To interpret the results obtained from the neural network, the outputs must be rescaled to the original range. From the user's viewpoint the accuracy obtained by the neural network refers to this rescaled data set. We show that the scaling of outputs into a smaller range than the original unscaled range leads to longer training times to reach a specified accuracy on the rescaled data.

In this paper we extend the delta rule to the so-called *gamma learning rule* which adjusts the output range of the sigmoid activation function during learning. Thus, the gamma rule is effectively performing automatic scaling – a property applicable to almost all applications. Recently, Zurada proposed the *lambda learning*

---

*University of Louisville, Louisville, Kentucky 40292, USA

rule where the constant λ in (1) is treated as a variable and also adapted during training [Zurada, 1992b]. We base the derivation of the gamma rule on the lambda rule and denote the combination as the *lambda-gamma learning rule*, which is more general than the delta rule.

The gamma rule is reminiscent of biological neurons which are able to adjust to signals of various natures through transmitter depletion and contrast enhancement. For instance, cells in the auditory system exhibit "stimulus selectivity" [Morgan, 1991], becoming attuned to a characteristic frequency.

In the next section we investigate the effects of scaling of the output data, and show the advantages of self-scaling output units. The lambda-gamma rule is derived for a single neuron in section 3, and extended to single layer learning in section 4. Section 5 generalizes the rule for hidden layer learning. A complete general learning algorithm is presented in section 6, and experimental results are reported in section 7.

2   Effects of scaling

For the purpose of this exposition assume an output layer which consists of one neuron[1]. Without loss of generality, assume that the desired output data is scaled into the range $[0, 1]$ using linear scaling:

$$d_s = c_1 d_u + c_2 \qquad (2)$$

where d_u is the original unscaled desired output data (i.e. raw data), and d_s is the corresponding scaled desired output data to be used for training. To scale data to the range $[0, 1]$ the scaling factors c_1 and c_2 are the following:

$$c_1 = \frac{1}{\max_{p=1,\ldots,P}\{d_u^{(p)}\} - \min_{p=1,\ldots,P}\{d_u^{(p)}\}} \qquad c_2 = \frac{-\min_{p=1,\ldots,P}\{d_u^{(p)}\}}{\max_{p=1,\ldots,P}\{d_u^{(p)}\} - \min_{p=1,\ldots,P}\{d_u^{(p)}\}} \qquad (3)$$

with P the total number of patterns. Then, from (2) the rescaled desired output d_r is

$$d_r = \frac{1}{c_1}d_s - \frac{c_2}{c_1} \qquad (4)$$

Let o_s denote the actual output of output neuron o. Then, similarly to (4), o_r is the actual output rescaled to the original output range. Assume it is possible to learn original unscaled data, and let o_u denote the actual unscaled output of neuron o. Since the values of desired output data are not changed during training, it is clear that $d_u = d_r$, and under ideal conditions we will also have that $o_u = o_r$. However, this will require perfect learning with zero error which is in practice not realizable. Let MSE_s and MSE_r respectively denote the mean square error for the scaled and rescaled data over the entire training set. Then,

$$MSE_s = \sum_{p=1}^{P}(d_s^{(p)} - o_s^{(p)})^2/P \qquad (5)$$

$$MSE_r = \sum_{p=1}^{P}(d_r^{(p)} - o_r^{(p)})^2/P \qquad (6)$$

By substitution of equation (4) in (6) we obtain

$$MSE_r = \sum_{p=1}^{P}[(\frac{1}{c_1}d_s^{(p)} - \frac{c_2}{c_1}) - (\frac{1}{c_1}o_s^{(p)} - \frac{c_2}{c_1})]^2/P = (\frac{1}{c_1})^2 MSE_s \qquad (7)$$

Equation (7) illustrates a clear relation between the scaled and rescaled error. If

$$|c_1| < 1 \qquad (c_1 \neq 0) \qquad (8)$$

---

[1]The derivations in this section can easily be extrapolated to an output layer with more than one neuron.

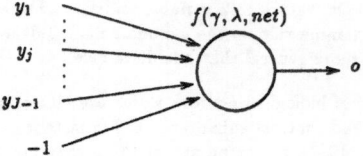

Figure 1: Lambda-gamma learning for a single neuron

then (7) indicates that the rescaled error is a factor of $(\frac{1}{c_1})^2$ larger than the scaled error, where condition (8) corresponds to the compression of data into a smaller range than the original range.

For the following, assume it is possible to learn the original unscaled data. Let MSE_u denote the mean square error for the unscaled data. The relationship illustrated above indicates that in order to obtain a rescaled accuracy MSE_r which is equal to MSE_u, the network must be trained longer until

$$MSE_s = (c_1)^2 MSE_r \tag{9}$$

On the other hand, if

$$|c_1| > 1 \tag{10}$$

we have from (7) that the rescaled error is a factor of $(\frac{1}{c_1})^2$ smaller than the scaled error. This corresponds to our claim that training on data which is expanded over a wider range will lead to faster convergence, since a scaling factor c_1 which conforms to condition (10) represents the scaling of data to a larger range than the original unscaled range.

From this investigation into the effects of scaling we conclude that it is preferable to use "self-scaling" output units and to learn original unscaled data when the range is greater than $[0, 1]$. This will significantly decrease the number of training cycles compared to learning scaled data, especially when $|c_1|$ is very small. Currently only linear self-scaling output units are available. In the next sections we propose the use of sigmoid self-scaling output units where the output range of the sigmoid activation function is dynamically adapted to span the original output range. The online adjustment of the output range enables the learning of unscaled data.

3 Lambda-gamma single neuron learning

The customary sigmoid function (1) for a single neuron is modified to include a *range coefficient* γ and a *steepness coefficient* λ, to give:

$$f(\gamma, \lambda, net) = \frac{\gamma}{1 + e^{-\lambda net}} \tag{11}$$

where the activation value is $net(\vec{w}, \vec{y}) = \vec{w}^t \vec{y}$, the augmented input vector is $\vec{y} = [y_1 \ y_2 \ \ldots \ y_{n-1} \ -1]^t$ and the weight vector is $\vec{w} = [w_1 \ w_2 \ \ldots \ w_n]^t$. In the classical delta rule the neuron therefore learns in $(n-1)$-dimensional non-augmented weight space in which n weights are adjustable. The lambda and gamma learning rules expand the learning space to $(n+1)$ dimensions, while the lambda-gamma learning rule expands it to $(n+2)$ dimensions. In addition to weight learning, both the steepness λ and the range γ undergo adjustments in the negative gradient direction.

Referring to Figure 1 and using the customary expression for error between the desired value d and the actual output of the neuron o,

$$E(\gamma, \lambda, \vec{w}) = \frac{1}{2}[d - o(\gamma, \lambda, \vec{w})]^2$$

where

$$o(\gamma, \lambda, \vec{w}) = f(\gamma, \lambda, net(\vec{w}, \vec{y}))$$

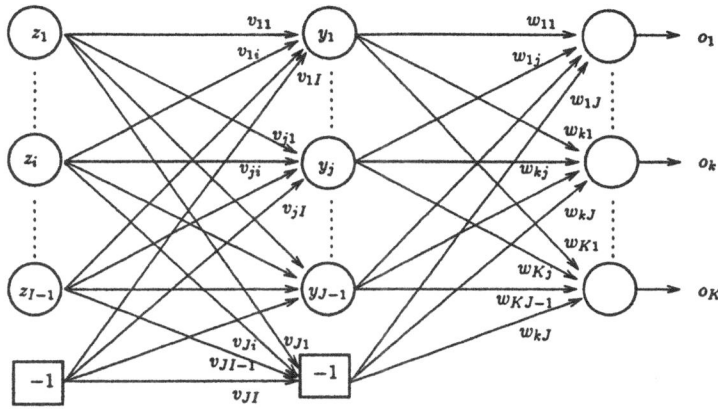

Figure 2: Layered feedforward network

we obtain the following weight adjustments for single neuron learning:

$$\Delta w_j = -\eta_1 \frac{\partial E}{\partial w_j} = -\eta_1 \frac{\partial E}{\partial o}\frac{\partial o}{\partial net}\frac{\partial net}{\partial w_j} = \eta_1 (d-o)\frac{\lambda}{\gamma}o(\gamma - o)y_j \tag{12}$$

$$\Delta \lambda = -\eta_2 \frac{\partial E}{\partial \lambda} = -\eta_2 \frac{\partial E}{\partial o}\frac{\partial o}{\partial \lambda} = \eta_2 (d-o)\frac{1}{\gamma}o(\gamma - o)net \tag{13}$$

$$\Delta \gamma = -\eta_3 \frac{\partial E}{\partial \gamma} = -\eta_3 \frac{\partial E}{\partial o}\frac{\partial o}{\partial \gamma} = \eta_3 (d-o)\frac{1}{\gamma}o \tag{14}$$

where η_1, η_2 and η_3 are positive learning constants usually selected as arbitrarily small values. Inspection of expression (12) coincides with the delta learning rule [Zurada, 1992a] when $\lambda = \gamma = 1$, and the lambda learning rule [Zurada, 1992b] when $\gamma = 1$. Expression (13) similarly reduces to the lambda rule when $\gamma = 1$. The extension to the lambda-gamma learning rule where a neuron's activation value can be "self-scaling" to the desired range is repesented by expression (14).

The next section illustrates single layer learning for the lambda-gamma learning rule.

4 Single layer learning

Assume that the neurons in the output layer $\vec{o}$ undergo training. In addition to a net input and activation value, each neuron o_k $(k = 1, \ldots, K)$ has a range coefficient γ_{o_k} and a steepness coefficient λ_{o_k}. The range and steepness coefficients are trained along with the weights w_{kj} for all hidden neurons y_j $(j = 1, \ldots, J)$ shown as the rightmost two layers in Figure 2. The usual definitions for error and error signal terms are used:

$$E = \frac{1}{2}\sum_{k=1}^{K}[d_k - o_k(\gamma_{o_k}, \lambda_{o_k}, net_{o_k})]^2 \tag{15}$$

$$\delta_{o_k} = \frac{\partial E}{\partial net_{o_k}} = -\frac{\lambda_{o_k}}{\gamma_{o_k}}(d_k - o_k)o_k(\gamma_{o_k} - o_k) \tag{16}$$

where $\vec{d}$ and $\vec{o}$ are respectively the desired output and the actual output vectors, and $\vec{\delta}_o$ and $\vec{\delta}_y$ are respectively the error signal term vectors for the output and hidden layers. The adjustments to the learning variables are given below:

$$\Delta w_{kj} = -\eta_1 \frac{\partial E}{\partial w_{kj}} = -\eta_1 \frac{\partial E}{\partial o_k}\frac{\partial o_k}{\partial net_{o_k}}\frac{\partial net_{o_k}}{\partial w_{kj}} = -\eta_1 \delta_{o_k} y_j \tag{17}$$

$$\Delta\lambda_{o_k} = -\eta_2 \frac{\partial E}{\partial \lambda_{o_k}} = -\eta_2 \frac{\partial E}{\partial o_k} \frac{\partial o_k}{\partial \lambda_{o_k}} = -\eta_2 \delta_{o_k} \frac{net_{o_k}}{\lambda_{o_k}} \tag{18}$$

$$\Delta\gamma_{o_k} = -\eta_3 \frac{\partial E}{\partial \gamma_{o_k}} = -\eta_3 \frac{\partial E}{\partial o_k} \frac{\partial o_k}{\partial \gamma_{o_k}} = \eta_3 (d_k - o_k) \frac{1}{\gamma_{o_k}} o_k \tag{19}$$

where

$$o_k = f(\gamma_{o_k}, \lambda_{o_k}, net_{o_k}(\vec{w}_k, \vec{y})) \quad \text{and} \quad net_{o_k}(\vec{w}_k, \vec{y}) = \sum_{j=1}^{J} w_{kj} y_j$$

Hidden layer learning for the lambda-gamma learning rule is described in the next section.

5 Hidden layer learning

To train the neurons y_j $(j = 1, \ldots, J)$ in the hidden layer, the weights v_{ji} $(i = 1, \ldots, I)$, the steepness coefficient λ_{y_j} and the range coefficient γ_{y_j} must be adjusted during each iteration of the learning algorithm, using respectively

$$\Delta v_{ji} = -\eta_1 \frac{\partial E}{\partial v_{ji}} \tag{20}$$

$$\Delta\lambda_{y_j} = -\eta_2 \frac{\partial E}{\partial \lambda_{y_j}} \tag{21}$$

$$\Delta\gamma_{y_j} = -\eta_3 \frac{\partial E}{\partial \gamma_{y_j}} \tag{22}$$

Using the error (15) and error signal term

$$\delta_{y_j} = \frac{\partial E}{\partial net_{y_j}} = \frac{\lambda_{y_j}}{\gamma_{y_j}} (\gamma_{y_j} - y_j) y_j \sum_{k=1}^{K} \delta_{o_k} w_{kj} \tag{23}$$

we obtain

$$\frac{\partial E}{\partial v_{ji}} = \frac{\partial E}{\partial net_{y_j}} \frac{\partial net_{y_j}}{\partial v_{ji}} = \delta_{y_j} z_i \tag{24}$$

$$\frac{\partial E}{\partial \lambda_{y_j}} = \frac{\partial E}{\partial y_j} \frac{\partial y_j}{\partial \lambda_{y_j}} = \delta_{y_j} \frac{net_{y_j}}{\lambda_{y_j}} \tag{25}$$

$$\frac{\partial E}{\partial \gamma_{y_j}} = \frac{\partial E}{\partial y_j} \frac{\partial y_j}{\partial \gamma_{y_j}} = \frac{1}{\gamma_{y_j}} f(\gamma_{y_j}, \lambda_{y_j}, net_{y_j}) \sum_{k=1}^{K} \delta_{o_k} w_{kj} \tag{26}$$

where we have from (15)

$$\frac{\partial E}{\partial y_j} = \sum_{k=1}^{K} \delta_{o_k} w_{kj} \tag{27}$$

Substitution of equations (24), (25) and (26) into equations (20), (21) and (22) respectively yields the following adjustments for the hidden layer neurons:

$$\Delta v_{ji} = -\eta_1 \delta_{y_j} z_i$$

$$\Delta\lambda_{y_j} = -\eta_2 \delta_{y_j} \frac{net_{y_j}}{\lambda_{y_j}}$$

$$\Delta\gamma_{y_j} = -\eta_3 \frac{1}{\gamma_{y_j}} f(\gamma_{y_j}, \lambda_{y_j}, net_{y_j}) \sum_{k=1}^{K} \delta_{o_k} w_{kj}$$

where

$$y_j = f(\gamma_{y_j}, \lambda_{y_j}, net_{y_j}(\vec{v}_j, \vec{z})) \quad \text{and} \quad net_{y_j}(\vec{v}_j, \vec{z}) = \sum_{i=1}^{I} v_{ji} z_i$$

with δ_{o_k} and δ_{y_j} respectively the error signal term of the k-th output neuron and the j-th hidden neuron by equations (16) and (23). As mentioned previously, the adjustments reduce to that of the delta rule when λ and γ are constants equal to 1, the lambda rule when γ is equal to 1 and the gamma rule when λ is equal to 1. For training, the complete back-propagation algorithm in [Zurada, 1992b] is updated to reflect the changes given above. The updated algorithm is presented in the next section.

6   Complete lambda-gamma learning algorithm

The algorithm presented in [Zurada, 1992b] is modified below to reflect the lambda-gamma learning rule which is more general than the delta and lambda learning rules. Changes correspond to the adjustments to weights, steepness and range coefficients.

Begin: Given P training pairs of vectors of inputs and desired outputs $\{(\vec{z}_1, \vec{d}_1), (\vec{z}_2, \vec{d}_2), \ldots, (\vec{z}_p, \vec{d}_p)\}$
where $\vec{z}_i$ is $(I \times 1)$, $\vec{d}_i$ is $(K \times 1)$ and $i = 1, \ldots, P$; $\vec{y}$ is $(J \times 1)$ and $\vec{o}$ is $(K \times 1)$.

Step 1:   Choose the values of the learning rates η_1, η_2 and η_3 according to the learning rule:

Delta learning rule	$\eta_1 > 0, \ \eta_2 = 0, \ \eta_3 = 0$
Lambda learning rule	$\eta_1 > 0, \ \eta_2 > 0, \ \eta_3 = 0$
Gamma learning rule	$\eta_1 > 0, \ \eta_2 = 0, \ \eta_3 > 0$
Lambda-gamma learning rule	$\eta_1 > 0, \ \eta_2 > 0, \ \eta_3 > 0$

Choose an acceptable training error E_{max}. Weights W $(K \times J)$ and V $(J \times I)$ are initialized to small random values. Initialize the number of cycles q and the training pairs counter p to $q = 1$, $p = 1$. Let $E = 0$ and initialize the steepness and range coefficients

$$\lambda_{y_j} = \gamma_{y_j} = 1 \quad \forall\, j = 1, \ldots, J \quad \text{and} \quad \lambda_{o_k} = \gamma_{o_k} = 1 \quad \forall\, k = 1, \ldots, K$$

Step 2:   Start training. Input is presented and the layers' outputs are computed using $f(\gamma, \lambda, net)$ as in equation (11):

$$\vec{z} = \vec{z}_p, \quad \vec{d} = \vec{d}_p \quad \text{and} \quad y_j = f(\gamma_{y_j}, \lambda_{y_j}, \vec{v}_j^t \vec{z}) \quad \forall\, j = 1, \ldots, J$$

where $\vec{v}_j$, a column vector, is the j-th row of V and

$$o_k = f(\gamma_{o_k}, \lambda_{o_k}, \vec{w}_k^t \vec{y}) \quad \forall\, k = 1, \ldots, K$$

where $\vec{w}_k$, a column vector, is the k-th row of W.

Step 3:   The error value is computed:

$$E = E + \frac{1}{2}(d_k - o_k)^2 \quad \forall\, k = 1, \ldots, K$$

Step 4:   The error signal vectors $\vec{\delta}_o$ $(K \times 1)$ and $\vec{\delta}_y$ $(J \times 1)$ of both the output and hidden layers are computed

$$\delta_{o_k} = -\frac{\lambda_{o_k}}{\gamma_{o_k}}(d_k - o_k)o_k(\gamma_{o_k} - o_k) \quad \forall\, k = 1, \ldots, K$$

$$\delta_{y_j} = \frac{\lambda_{y_j}}{\gamma_{y_j}}y_j(\gamma_{y_j} - y_j)\sum_{k=1}^{K}\delta_{o_k}w_{kj} \quad \forall\, j = 1, \ldots, J$$

Step 5:   Output layer weights and gains are adjusted:

$$w_{kj} = w_{kj} + \eta_1\delta_{o_k}y_j \quad \lambda_{o_k} = \lambda_{o_k} + \eta_2\delta_{o_k}\frac{net_{o_k}}{\lambda_{o_k}} \quad \gamma_{o_k} = \gamma_{o_k} + \eta_3(d_k - o_k)\frac{1}{\gamma_{o_k}}o_k$$

for all $k = 1, \ldots, K$ and $j = 1, \ldots, J$.

Step 6: Hidden layer weights and gains are adjusted:

$$v_{ji} = v_{ji} + \eta_1 \delta_{y_j} z_i \quad \lambda_{y_j} = \lambda_{y_j} + \eta_2 \frac{1}{\lambda_{y_j}} \delta_{y_j} net_{y_j} \quad \gamma_{y_j} = \gamma_{y_j} + \eta_3 \frac{1}{\gamma_{y_j}} f(\gamma_{y_j}, \lambda_{y_j}, net_{y_j}) \sum_{k=1}^{K} \delta_{o_k} w_{kj}$$

for all $j = 1, \ldots, J$ and $i = 1, \ldots, I$.

Step 7: If $p < P$ then let $p = p + 1$ and go to Step 2; otherwise go to Step 8.

Step 8: One training cycle is completed. If $E < E_{max}$ then terminate the training session. Output the cycle counter q and error E; otherwise let $E = 0$, $p = 1$, $q = q + 1$ and initiate a new training cycle by going to Step 2.

7 Experimental results

We have used a simple function approximation experiment to substantiate our claims to the effects of scaling. A 1-10-1 network architecture was used to approximate the function $f(z) = |z|$. For the experiments described below, we have used the lambda-gamma learning algorithm to train on original unscaled data, and the delta learning algorithm to train on the scaled data. For illustration purposes, Figure 3 also shows the learning profile on the rescaled output data. The mean square error MSE_r on the rescaled output data is calculated form equations (4) and (6) after each epoch. Both experiments use the same initial weights, which are initialized as random values in the range $[\frac{-1}{\sqrt{fanin}}, \frac{1}{\sqrt{fanin}}]$.

- **Experiment 1:** For this experiment we have $z \in [0.4, 0.6]$, and $d \in [0.4, 0.6]$. The desired outputs d are linearly scaled to $[0, 1]$ using (2). From (3) we have $|c_1| = 5$, which illustrates the effect when output data is scaled to a larger range than the original. Figure 3(a) shows that the mean square error for the rescaled data is smaller than the mean square error of the scaled data for each epoch when $|c_1| > 1$. For example, from Figure 3(a) we see that a required error of 0.0005 on the scaled data has already been reached at epoch 55 on the rescaled data compared to epoch 150 on the scaled data.

- **Experiment 2:** For this experiment we have $z \in [-5, 5]$ and $d \in [0, 5]$. The desired outputs d are linearly scaled to $[0, 1]$ using (2). Then, from (3) we have $|c_1| = \frac{1}{5}$ which corresponds to the compression of data. From Figure 3(b) we observe that an error of 0.02, which is reached at epoch 20 using lambda-gamma learning on unscaled data, is reached at epoch 140 on the rescaled data. Longer training is therefore required on scaled data to obtain a specified error equivalent on the original data.

Figure 3 also shows the learning profile for the lambda-gamma rule on unscaled data. Table 1 shows that condition (7) holds for arbitrarily selected epochs: for any given epoch, the error MSE_r on the rescaled data is a factor of $(\frac{1}{c_1})^2$ larger than the error MSE_s on the scaled data when $|c_1| < 1$, and a factor $(\frac{1}{c_1})^2$ smaller when $|c_1| > 1$.

The results presented in this section confirm our conclusion that training on output data that is scaled into a smaller range causes longer training times to reach a required accuracy on the rescaled output data (training accuracy is normally specified in terms of the rescaled data). The problem escalates as $|c_1|$ becomes very small. With the lambda-gamma learning rule, the same accuracy is obtained in less training cycles.

8 Conclusions

We have derived a relationship between the mean square errors for scaled and rescaled data when output data is linearly scaled. This relationship has indicated that the compression of data causes longer training

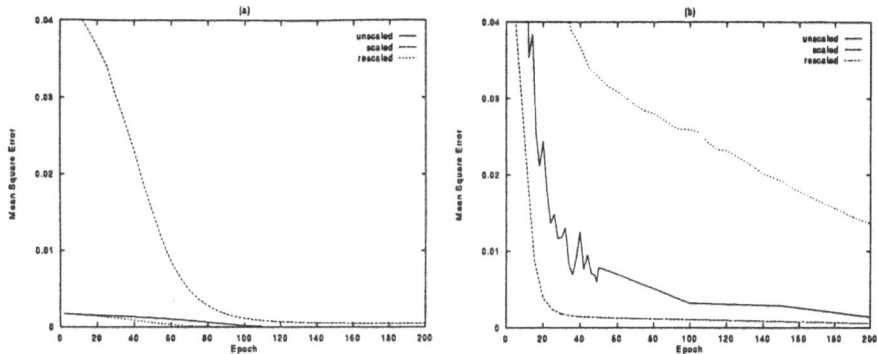

Figure 3: Learning profiles for unscaled, scaled and rescaled data.

Epoch	Experiment 1			Experiment 2		
	MSE_s	MSE_r	$(\frac{1}{c_i})^2 MSE_s$	MSE_s	MSE_r	$(\frac{1}{c_i})^2 MSE_s$
5	0.043914	0.001757	0.0017566	0.037712	0.94279	0.9428
50	0.013873	0.000555	0.0005549	0.001315	0.03288	0.032875
100	0.001086	0.00043	0.00004344	0.001039	0.025983	0.025975
150	0.000489	0.00002	0.00001956	0.000766	0.01915	0.01915
200	0.000455	0.000018	0.0000182	0.000542	0.013551	0.01355

Table 1: Comparison of MSE_s with MSE_r.

times compared to training on the unscaled data. We have presented the lambda-gamma learning algorithm which utilizes self-scaling sigmoid output units.

In order to perform scaling, the maximum and minimum ranges of the input and output must be known. For incremental learning systems this is difficult to obtain, since all training pairs are not available before training. Upper and lower bounds need to be determined beforehand. Gamma learning eliminates this problem since the output range of the sigmoid activation function is dynamically adjusted during training. The lambda-gamma learning rule further seems to eliminate the need for internal rescaling within the units as reported by Rigler, Irvine and Vogl [Rigler, 1991].

References

[Morgan, 1991] DP Morgan and CL Scofield, *Neural Networks and Speech Processing*, Kluwer Academic Publishers, 1991.

[Rigler, 1991] AK Rigler, JM Irvine and TP Vogl, *Rescaling of Variables in Back Propagation Learning*, Neural Networks, 4, pp 225–229, 1991.

[Zurada, 1992a] JM Zurada, *Introduction to Artificial Neural Systems*, West Publishing Company, 1992.

[Zurada, 1992b] JM Zurada, *Lambda Learning Rule for Feedforward Neural Networks*, Proceedings of the IEEE International Conference on Neural Networks, March 28–31, 1992, San Fransisco, California.

Determining the Significance of Input Parameters Using Sensitivity Analysis

AP Engelbrecht   I Cloete   JM Zurada*

ian@cs.sun.ac.za

Computer Science Department, University of Stellenbosch, Stellenbosch 7600, South Africa

Abstract

Accompanying the application of rule extraction algorithms to real-world problems is the crucial difficulty to compile a representative data set. Domain experts often find it difficult to identify all input parameters that have an influence on the outcome of the problem. In this paper we discuss the problem of identifying relevant input parameters from a set of potential input parameters. We show that sensitivity analysis applied to a trained feedforward neural network is an efficient tool for the identification of input parameters that have a significant influence on any one of the possible outcomes. We compare the results of a neural network sensitivity analysis tool with the results obtained from a machine learning algorithm, and discuss the benefits of sensitivity analysis to a neural network rule extraction algorithm.

1 Introduction

Machine learning and neural network algorithms are increasingly being used to solve problems in a variety of domains. Some of the best known real-world problems are in the medical (cancer, diabetes, thyroid), agriculture (soybean) and financial (credit approval, stock market prediction) domains. These problems are frequently being used to perform benchmark tests on new algorithms [Prechelt 1994]. Machine learning algorithms, for example CN2 and C4.5, have been used successfully in these domains to extract production rules from data sets [Theron 1994]. The extracted production rules can be used by the respective domain experts to aid in their decision-making. Recently, neural networks have been used for rule extraction [Craven 1993]. In addition, neural networks are also used to learn the functional relationship between inputs and outputs, which is then used to predict the outcome given a set of inputs.

Machine learning and neural network algorithms learn from data. With many real-world problems domain experts find it difficult – if not impossible – to identify all those input parameters that have an influence on any one of the outcomes of the problem. For large problems, it becomes difficult to see all the relationships inherent to the data set. Consequently, the expert may include redundant input parameters, or neglect to include one or more influential parameters.

Automatic tools that have the ability to identify all relevant input parameters from a set of potentially relevant parameters will therefore be beneficial. The expert merely includes all potential input parameters, and the automated tool will determine which parameters are relevant to each one of the outcomes using the statistical relationships among these parameters.

By definition, machine learning algorithms possess the ability to ignore input parameters with little, or no, significance to any of the outcomes. By proper adjustments of the pruning parameters, the machine

---

*University of Louisville, Louisville, Kentucky 40292, USA

learning algorithms will generate those rules which contain only input parameters that have a significant influence on any outcome of the problem. Despite this ability of machine learning algorithms, we have found that it is still necessary to pre-process data sets in order to remove obvious irrelevant input parameters (we show in section 3 that the inclusion of such parameters influences the quality of the generated rules).

Recent research has shown that a trained neural network can be used to extract symbolic rules from the learned weights. Craven and Shavlik show that the rules extracted from a neural network using a clustering and soft weight-sharing algorithm generalize better than rules obtained from C4.5 [Craven 1993]. However, a neural network does not have the ability to distinguish between relevant and irrelevant input parameters. In this paper we describe a sensitivity analysis tool which can be applied to a trained neural network in order to automatically identify all input parameters which have a significant influence on any one of the possible outcomes. The identified irrelevant input parameters can be pruned, thus effectively reducing the dimension of the input space. A reduction of the number of input parameters simplifies the neural network architecture, which will lead to the extraction of a reduced number of general rules. We show that a neural network rule extraction algorithm in conjunction with sensitivity analysis eliminates any need to pre-process the original data set in contrast to C4.5 which requires pre-processing of the data set to produce sensible rules.

Section 2 gives an overview of sensitivity analysis as a means to establish the significance of each input parameter. We also present a pruning algorithm to eliminate irrelevant input variables. Section 3 shows that sensitivity analysis correctly identifies relevant input parameters. We also compare the results from sensitivity analysis to that of C4.5. We conclude in section 4 that sensitivity analysis is an efficient tool to determine the significance of each input parameter to the outcome of the problem.

2 Sensitivity analysis

In most real-world applications it is difficult to identify all input variables that have an influence on the outcomes of the problem. Sensitivity analysis provides a neural network tool to automatically identify all relevant parameters from a set of potential parameters. The neural network is trained on the entire data set after which sensitivity analysis is applied. Additionally a pruning algorithm can be used to eliminate irrelevant parameters using the significance measures obtained from the sensitivity analysis tool.

In this section we summarize the sensitivity analysis technique developed by Zurada, Malinowski and Cloete [Zurada 1993]. We first define a metric for the individual input-output sensitivities, and then show how this metric can be used to determine the sensitivities of each input to each output over the entire training set. Lastly, we summarize the sensitivity analysis and pruning algorithm.

2.1 Input-output sensitivities

In the following exposition we consider only a three layer network, although the extension of sensitivity analysis to neural networks with more layers is straightforward.

Consider a three layer feedforward neural network, where $\vec{z} = (z_1, \cdots, z_i, \cdots, z_I)$, $\vec{y} = (y_1, \cdots, y_j, \cdots, y_J)$ and $\vec{o} = (o_1, \cdots, o_k, \cdots, o_K)$ respectively denote the input, hidden and output layers. Define a training pair p as the tuple $p = (\vec{z}^{(p)}, \vec{t}^{(p)})$, where $\vec{t} = (t_1, \cdots, t_k, \cdots, t_K)$ denotes the target vector. Given any training pair p, define the sensitivity $S_{ki}^{(p)}$ of a trained output o_k with respect to an input z_i as

$$S_{ki}^{(p)} = \frac{\partial o_k}{\partial z_i} \tag{1}$$

$$= o_k' \sum_{j=1}^{J} w_{kj} \frac{\partial y_j}{\partial z_i} \tag{2}$$

$$= o'_k \sum_{j=1}^{J} w_{kj} y'_j v_{ji} \qquad (3)$$

where y_j denotes the output of the j-th hidden neuron of the hidden layer $\bar{y}$, o'_k is the value of the derivative of the output layer activation function

$$o_k = f(\sum_{j=1}^{J} w_{kj} y_j) \qquad (4)$$

and y'_j is the value of the derivative of the hidden layer activation function

$$y_j = f(\sum_{i=1}^{I} v_{ji} z_i) \qquad (5)$$

where w_{kj} denotes the weight value between hidden neuron y_j and output o_k, and v_{ji} is the weight value between input z_i and hidden neuron y_j.

Equation (3) only defines the sensitivity of one output o_k with respect to one input z_i for pattern p. For training pattern p, define the pattern sensitivity matrix $S^{(p)}$, which consists of entries $S_{ki}^{(p)}$, as

$$S^{(p)} = O'WY'V \qquad (6)$$

where W $(K \times J)$ and V $(J \times I)$ are respectively the output and hidden layer weight matrices, and O' $(K \times K)$ and Y' $(J \times J)$ are defined as

$$O' \doteq \mathrm{diag}(o'_1, \cdots, o'_K) \qquad (7)$$

$$Y' \doteq \mathrm{diag}(y'_1, \cdots, y'_J) \qquad (8)$$

2.2 Sensitivity measures over entire training set

Equation (6) defines the sensitivity matrix for a specific training pattern p. However, each training pair p produces a different sensitivity matrix $S^{(p)}$. In order to apply sensitivity analysis, the sensitivity matrix $S^{(p)}$ must be evaluated over the entire training set. Zurada, Malinowski and Cloete define three different metrics over the entire training set [Zurada 1993]:

- The *mean square average sensitivity* matrix S_{avg} defined as

$$S_{ki,avg} \doteq \sqrt{\frac{\sum_{p=1}^{P} [S_{ki}^{(p)}]^2}{P}} \qquad (9)$$

- The *absolute value average sensitivity* matrix S_{abs} defined as

$$S_{ki,abs} \doteq \frac{\sum_{p=1}^{P} |S_{ki}^{(p)}|}{P} \qquad (10)$$

- The *maximum sensitivity* matrix S_{max} defined as

$$S_{ki,max} \doteq \max_{p=1,\cdots,P} \{S_{ki}^{(p)}\} \qquad (11)$$

Any one of the sensitivity measure matrices defined in (9) – (11) is sufficient to assess the relative significance of each input to each output.

To allow accurate comparison among inputs, it is necessary to scale inputs and outputs to the same range. If the original inputs and outputs were not scaled to the same range before training, additional scaling is necessary using (12)

$$S_{ki,avg} = S_{ki,avg} \frac{(\max_{p=1,\cdots,P}\{z_i^{(p)}\} - \min_{p=1,\cdots,P}\{z_i^{(p)}\})}{(\max_{p=1,\cdots,P}\{o_k^{(p)}\} - \min_{p=1,\cdots,P}\{o_k^{(p)}\})} \qquad (12)$$

2.3 Pruning

In the previous section, we have defined three metrics to determine the significance of each input to each output. In this section we show how these significance measures can be used to identify and prune irrelevant input parameters.

Using one of the measures defined in (9) – (11), the dimension of the input vector $\vec{z}$ can be reduced by pruning those inputs z_i which bear a low significance to each output. Define the significance Φ_i of input z_i over all outputs as

$$\Phi_i = \max_{k=1,\cdots,K}\{S_{ki}\} \tag{13}$$

where S represents any one of the matrices S_{avg}, S_{abs} and S_{max}. The vector $\vec{\Phi}$ is then sorted in descending order such that

$$\Phi_i \geq \Phi_{i+1} \tag{14}$$

for $i = 1, \cdots, I - 1$.

The significance vector $\vec{\Phi}$ is then parsed to find two consecutive inputs z_i and z_{i+1} such that the gap between their respective sensitivities Φ_i and Φ_{i+1} is large enough. Therefore, if the gap $g_{i,i+1}$ is such that

$$g_{i,i+1} > \text{constant} \tag{15}$$

all inputs $z_{i+1}, \cdots, z_I$ have little significance to any of the outputs and can therefore be pruned.

The reader must note that sensitivity analysis is a neural network post-processing technique. To prevent the erroneous pruning of relevant input variables, sensitivity analysis can only be applied to well trained networks. The next section presents a complete sensitivity analysis algorithm including the pruning of irrelevant inputs.

2.4 Complete algorithm

This section summarizes the sensitivity analysis and pruning algorithm:

Step 1: Train the neural network on the original training set.

Step 2: Calculate all input-output sensitivities for each training pair using (6).

Step 3: Calculate the sensitivity matrix over the entire training set using one of (9), (10) or (11).

Step 4: Calculate the significance of each input over all outputs using (13).

Step 5: Apply the pruning algorithm as described in section 2.3.

Step 6: Retrain the neural network on the pruned training set.

Step 7: Repeat steps 1 to 6 until no irrelevant inputs remain.

3 Experimental results

In this section we apply the sensitivity analysis technique described in section 2 to the breast cancer problem from the PROBEN1 data set for neural networks [Prechelt 1994]. We show that sensitivity analysis correctly identifies irrelevant input parameters. We also show that none of the input parameters indentified through sensitivity analysis for pruning occur in any C4.5 generated rule. We further show that the robustness of C4.5 degrades when trained on the original training set which includes an obvious irrelevant input parameter.

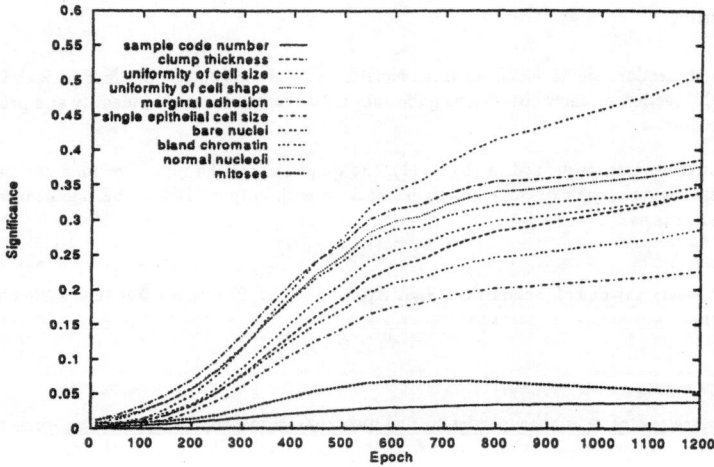

Figure 1: Significance profile for original training set

The breast cancer database was compiled at the University of Wisconsin Hospitals by Dr WH Wolberg [Mangasarian 1990]. The database consists of 699 instances, where each instance consists of 10 input parameters and a class parameter.

For the purposes of this exposition the data set is divided into a training set of 489 instances and a test set of 210 instances. The class parameter is modelled as two separate parameters, respectively for the *benign* and *malignant* outcomes. Without further ado, we can immediately identify the *sample code number* input parameter as being irrelevant to the class parameters.

We have conducted the following experiments with a feedforward neural network and C4.5, using the same training and test sets:

- **Neural network experiments:** For the first experiment we have trained a 10-25-2 neural network with the original training set. The learning rate and momentum term are respectively set to 0.1 and 0.5. The weights are initialized with random values in the range $[-\frac{1}{\sqrt{fanin}}, \frac{1}{\sqrt{fanin}}]$. Both the inputs and outputs are linearly scaled to the range $[0, 1]$. Hidden and output units use the sigmoid activation function. The network is trained for an average output unit error of 0.00028, defined as

$$E = \frac{\sum_{p=1}^{P} \sum_{k=1}^{K} [t_k^{(p)} - o_k^{(p)}]^2}{2PK}$$

where P and K are respectively the number of patterns and the number of output units.

The significance profile in Figure 1 illustrates the evolution of the significance Φ_i of each input z_i to the outputs during training. The learning profile is depicted by Figure 2. Figure 1 shows that input parameters *sample code number* and *mitoses* have a very low relevance to the outputs. Consequently, these two input parameters are removed from the training and test sets.

In a second experiment, we have trained an 8-25-2 neural network on the pruned training set (i.e. the original set without the *sample code number* and *mitoses* input parameters). The learning profile depicted by Figure 2 shows that the neural network converges to the same error on the pruned training set as on the original training set. We can therefore conclude that pruning has been done correctly.

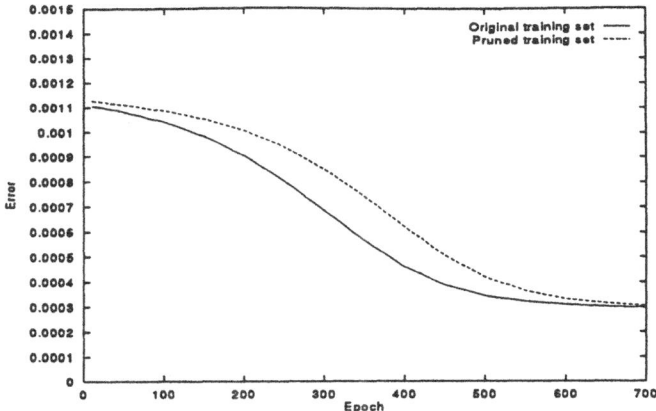

Figure 2: Learning profile for original and pruned training set

Original data set		Sample code number excluded	
Training set error	Test set error	Training set error	Test set error
2.45%	6.19%	1.84%	3.81%

Table 1: Performance results for C4.5

- **C4.5 experiments:** For the first C4.5 experiment we have used the original training set for rule induction. The results have shown that the inclusion of the irrelevant input parameter *sample code number* degrades the robustness of C4.5. The following unsensible rule is generated:

```
sample code number > 1.02612e+06
sample code number <= 1.21409e+06
uniformity of cell shape > 1
normal nucleoli > 3
-> class malignant
```

We have further observed that input parameter *mitoses* does not occur in any rule.

For the second C4.5 experiment we have removed the *sample code number* parameter from the training set. Table 1 illustrates a higher degree of accuracy on both the training and test sets for the processed data set compared to the lower accuracy on the original data set. Again, we observe the exclusion of the *mitoses* input parameter from the generated rules, which corresponds to the elimination of this input parameter through sensitivity analysis.

The experiments above show that sensitivity analysis correctly identifies those input parameters which have little relevance to the outputs, and can therefore be pruned. For the breast cancer problem, we have further observed that the robustness of C4.5 degrades when the user neglects to identify and remove the irrelevant input parameter *sample code number*. For this problem, a neural network rule extraction algorithm which utilizes sensitivity analysis will prove to be a more robust rule induction tool, since sensitivity analysis correctly identifies *sample code number* to be irrelevant. We also observe from Figure 1 that pruning can start at an earlier stage, since we have a clear grouping among the less significant and more significant parameters already at epoch 500.

4   Conclusion

We have shown in this paper that sensitivity analysis is an efficient technique to determine the relevance of each input to the outputs. Experiments on the breast cancer problem have shown that sensitivity analysis leads to the correct pruning of irrelevant input parameters. We have indicated that a neural network rule extraction algorithm in conjuction with a sensitivity analysis and pruning tool will prove to be more robust than the C4.5 algorithm. The experimental results presented in this paper also suggest that the pruning of input parameters using sensitivity analysis can be done at an earlier stage. However, further research is needed in order to define when a neural network has learned enough to effectively apply sensitivity analysis.

References

[Craven 1993] MW Craven and JW Shavlik, *Learning Symbolic Rules using Artificial Neural Networks*, Proceedings of the Tenth International Conference on Machine Learning, 1993, pp 73–80.

[Mangasarian 1990] OL Mangasarian and WH Wolberg, *Cancer Diagnosis via Linear Programming*, SIAM News, 23(5), September 1990, pp 1 & 18.

[Prechelt 1994] L Prechelt, *PROBEN1 – A Set of Neural Network Benchmark Problems and Benchmarking Rules*, Technical Report 21/94, Fakultät für Informatik, Universität Karlsruhe, Germany, September 1994.

[Quinlan 1993] JR Quinlan, *C4.5: Programs for Machine Learning*, Morgan Kaufmann Publishers, San Mateo, California, 1993.

[Theron 1994] H Theron, *Specialization by Exclusion: An Approach to Concept Learning*, PhD dissertation, Department of Computer Science, University of Stellenbosch, 1994.

[Zurada 1993] JM Zurada, A Malinowski and I Cloete, *Sensitivity Analysis for Minimization of Input Data Dimension for Feedforward Neural Network*, IEEE International Symposium on Circuits and Systems, London, May 30 – June 3, 1994.

Multi-Valued Neurons: Learning, Networks, Application to Image Recognition and Extrapolation of Temporal Series

Naum N. Aizenberg [*], Igor N. Aizenberg [**],

Georgy A. Krivosheev [***]

[*]University of Uzhgorod, Professor of the Department of Cybernetics, Minaiskaya (Geroev Stalingrada) 28, kv. 49, UZHGOROD,294015, UKRAINE. Tel (+7 03122) 23908; Fax (+7 03122) 36120; E-mail: igor@pgd.uzhgorod.ua
[**] Company "INFORM RTG Ltd", Chief of the Research Department, dr. of sc. math., Grushevskogo (Engelsa) 27, kv.32, UZHGOROD,294015, UKRAINE. Tel (+7 03122) 33471; Fax (+7 03122) 36120; Fax (+7 095) 261-84-27; E-mail: igor@pgd.uzhgorod.ua
[***] Company "INFORM RTG Ltd", Scientific Manager, dr. of sc. techn., Paustovskogo 3, kv. 430, MOSCOW, RUSSIA. Tel (+7 095)261-56-20; Fax (+7 095)261-84-27; E-mail: georgy@dkl.msk.su

This work is supported by Russian Company "INFORM RTG Ltd", Tel. (7 095)261-49-77; Fax (7 095)261-84-27;E-mail: prog@dkl.msk.su

Abstract

In this paper we consider in the developing conception of multi-valued neurons. First of all significant reinforcement of the learning algorithm which led to the 20-30 - times acceleration of the convergence of learning is proposed. Then neural network based on multi-valued neurons where each neuron is connected with restricted number of other ones (function of connections is defined as random function) is considered. Application of such an network to image recognition is proposed. Then approach to extrapolation of the temporal series based on the representation of the series as multiple-valued function, learning of the single neural element and furtheron forecasting of the function's values is also considered.

1.INTRODUCTION

Conception of multi-valued neurons based on the notion of multiple-valued threshold function of k-valued logic [1] has been introduced in [2] and then developed in [3] and [4]. The main feature of this conception is representation of the signals processed in neurons and of the weights by complex numbers. From the first point of view such an representation is so complicate, but as has been shown [1-4] it gives possibility to define output function of the neuron by way which lead to the very efficient learning algorithm and therefore to solution of the different problems on the neural networks based on the considered neurons. I.g. in [2] associative memory for storing of gray-scale images based on cellular network from multi-valued neurons has been proposed. In [3] and [4] application of the same cellular network (also of the network based on universal binary neurons which are similar to multi-valued but implement arbitrary Boolean functions) to solution of some image processing problems (edge detection, impulsive noise filtering) has been considered. The main goal of this paper is further developing of the conception of multi-valued neurons. We will consider learning algorithm presented in [2] and [3], but with significant reinforcement which makes it possible 30-times acceleration of the convergention of the learning algorithm and also makes it possible to increase number of the levels of signals which are processed on considered neural networks (from hundreds to thousands). This modernization of the learning algorithm also gives possibility to consider very interesting problem of the extrapolation of temporal series which may be represented by multiple-valued functions. Wonderful feature of approach presented here is solution of such an problem on the single neuron. Also we would like to propose new type of the neural network based on multi-valued neurons as alternative to both of the Hopfield-like and

Cellular-like networks. In the network proposed here each neuron is connected with restricted number of other ones, but function of connections is defined as some random function. Wonderful application of such an network to image recognition and restoring also will be considered below.

2.MATHEMATICAL MODEL

First of all we have to remind some basic notions of the theory of multi-valued neurons. These neurons perform multiple-valued (k-valued) functions of n variables $f(x_1,\ldots,x_n)$ by their representation through $n+1$ complex-valued weights $w_0, w_1, \ldots, w_n$:

$$f(x_1,\ldots,x_n) = P(w_0 + w_1x_1 + w_nx_n) \qquad (1)$$

where $x_1,\ldots,x_n$ -variables of which performed function depends (values of function and of variables are also coded by complex numbers which are k-th power roots of a unit: $\varepsilon^j = \exp(i2\pi j / k)$, $j \in [0, k-1]$, i- is an imaginary unit, in another words values of the k-valued logic are represented as k-th power roots of a unit: $j \rightarrow \varepsilon^j$) and P - is output function:

$$P(z) = \exp(i * 2\pi * j / k),$$
$$\text{if } 2\pi * (j + 1) / k > \arg(z) \geq 2\pi * j / k \qquad (2)$$

where j=0,1,...,k-1 - values of the k-valued logic, i- is an imaginary unit, $Z = W_0 + W_1X_1 + W_nX_n$ - weighted sum , $\arg(z)$ is the argument of the complex number z. (2) is illustrated by Fig.1.

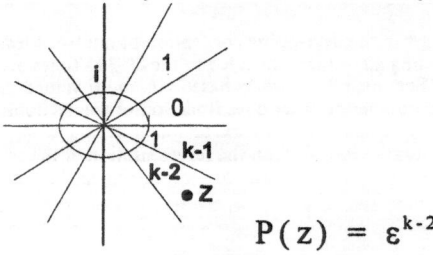

$$P(z) = \varepsilon^{k-2}$$

Fig.1

For the first time idea of such an threshold-like representation of the multiple-valued functions has been proposed in [1] and then developed in [2-4].

3.LEARNING ALGORITHM AND IT'S REINFORCEMENT

In our case learning consist in direction of the weighted sum of the neuron's input signals into the need sector (from k ones) on the Complex plane (Fig.1). Approach to solution of this problem has been proposed in [1] and some kinds of learning algorithm for multi-valued neuron (also for universal binary neuron) have been proposed in [2-4]. It has been shown by many experiments that time which is need for learning in the case considered here is much less in comparison with time which is need for solution of learning problem for Hopfield's, Chua's and some other popular types of neurons including multi-layer perceptron (insted of last it is always possible to use universal binary neuron [3-4] which based on the mathematical model simillar to presented here and can to perform arbitrary Boolean function of arbitrary number of variables).

Procedure of the direction of the weighted sum into the need sector on the Complex plane is implemented by the next formula:

$$W_{m+1} = W_m + \omega\varepsilon^q \overline{X}, \qquad (3)$$

where W_m and W_{m+1} - current and next weighting wectors, ω - correction coefficient, $\overline{X}$ - vector of the neuron's input signals with comlex-conjugated components, ε^q - is the value of neuron's output signal and therefore q - is number of the need sector (in another words q - is an integer value of the k-valued function implemented by neuron). ω have to be choosen from the consideration that weighted sum $z = W_0 + W_1 x_1 + W_n x_n$ have to be closer as soon as possible to the need sector (or have immediately get to the need sector) - sector number q - after step of the learning defined by (3). As a rule it is sufficient to consider two cases for choosing of the ω - $\left| \arg(\varepsilon^q) - \arg(z) \right| \leq \pi / 2$ or $\left| \arg(\varepsilon^q) - \arg(z) \right| > \pi / 2$. If first of this conditions is true as a rule it is possible to get to the need sector by 1-2 steps. For the second condition way is so longer, but not very. Iteration of the learning is sequential checking of the (1) for all elements from the learning (training) set and in the case (1) is false for some element from this set we have to apply learning procedure (3). This process have to continue till moment when we will obtain weighted vector $W = (w_0, w_1, \ldots w_n)$ satisfied (1) for all elements from the learning set. As has been shown by many experiments learning procedure based on (3) is convergenced for lot of partial-defined functions of k-valued logic and for full-defined multi-valued threshold functions (precise proof for the last case is given in [1]). Even in the case of non-convergenced process for some function (especial for partial-defined functions) practicaly always it is possible to change value of k and reduce given function of k-valued logic to the function of $\tilde{k}$-valued logic (where $\tilde{k} \leq k$ or $\tilde{k} \geq k$) for which learning process will be convergenced.

Learning algorithm based on (3) has been implemented by direct application of complex arithmetic with floating-point operations. On the base of such an implementation some interesting problems (such as restoring of the gray-scale images in associative memory, edge detection on gray-scale images, etc.) have been solved [2-4]. But with increasing of k and n (number of variables) evaluations became more and more slower and i.g. for learning of the network from 4096 neurons (each of them has 30 inputs, k=256, learning set for each neuron consist of 30 elements) 70 hours on the 486-66 Mhz IBM-PC computer are requested. The most difficult moment is evaluation of the argument of weighted sum when k is more than 300.This dufficulty involves restriction on value of k (in previous experience case k>300 did not consider). Here we would like to propose modification of the algorithm which gives possibility to overcome these disadvantages also as significant increasing of values of k and n for which learning algorithm is working without any computational problems.

First of all we have to reject of evaluations with floating-point operations and therefore floating point data. Now all weights will be presented by integer 32-bit numbers per real and imaginary part respectively and weighted sum will be presented by integer 64-bit numbers per real and imaginary part respectively. This approach is very efficient also because makes it possible to use internal formats of the standart 386/486 processors. For the fast evaluation of the output of neural element, in another words, for the fast perfomance of (2) it is necessary to put one restriction on the possible values of k. It have to devide on 8. It is not hard restriction, because if this condition is not satisfied it is always possible to increase k to get need value. But thanks to this restriction we can evaluate j (neuron's output in integer form corresponds to the complex ε^j)by very simple way. Let $z = (Re(z), Im(z))$ - complex number which represent weighted sum and

$$\phi = \begin{cases} abs(Re(z)) / abs(Im(z)) & \text{if } abs(Re(z)) > abs(Im(z)) \\ abs(Im(z)) / abs(Re(z)) & \text{if } abs(Im(z)) < abs(Re(z)) \end{cases} \tag{4}$$

Instead of evaluation of the $arg(z)$ and comparisons of ungles for evaluation of the neuron's output j we will directly evaluate value of j using the table which always may be created in advance:

$$T(l) = tg(2\pi l / k), \text{ where } l = 0, 1, \ldots, (k / 8) - 1. \tag{5}$$

Now, from (4), (5) and Fig. 2 it is evident that it is easy may be found such value of l that $T(l) \leq \phi \leq T(l + 1)$. Taking to account sign and value of the ratio $Im(z)/Re(z)$ it is easy to find number s of the subsector to which z is got. Now it is evident that

$$j = s(k / 8) + l, s = 0, 1, \ldots, (k / 8) - 1, \tag{6}$$

where j - is neuron's output.

Modification of the learning algorithm presented here makes it possible 30-times acceleration of the evaluations in (1) - (3), i.g., of all evaluations concerning multi-valued neural element and breaks restrictions on values of k. If earlier we considered in applications k<300, now there are no problems

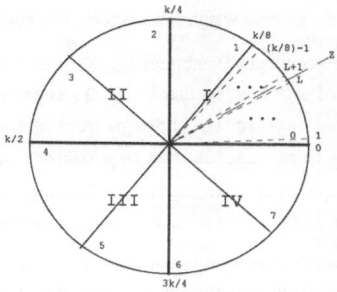

Fig.2

with, i.g. k=1024 (even in the case k=2048 it is all right, but need more time for all evaluations, first of all for learning). To compare the speed of learning we can return to example with learning of the network which contains 4096 30-inputs neurons on the learning set consisted of 30 elements (we will also consider the same network below). Evaluations based on (4)-(6) need only 2 hours 25 minutes on the 486-66 Mhz IBM-PC computer (one may compare with 70 hours for complex floating-point evaluations on the same computer).

3. IMAGES RECOGNITION

We considered in [2-3] cullular network based on multi-valued neurons and it's applications. Especially in [2] using of such an network as associative memory for gray-scale images has been proposed. Despite the fact that such memory can to correct lot of impulsive distortions, it has significant disadvantage - cellular structure of the network makes impossible correction of errors in the case of full distortion of some regular region of image. To break this restriction we would like to propose another structure of the network. Network proposed here has local connections feature also as cellular one, i.g. each nuron is connected with limited number of other ones . But, if in cellular network each neuron is connected only with neurons from it's nearest neighborhood (Fig.3a), in network considerd here function of connections for each neuron is generated as some random function (one may compare with

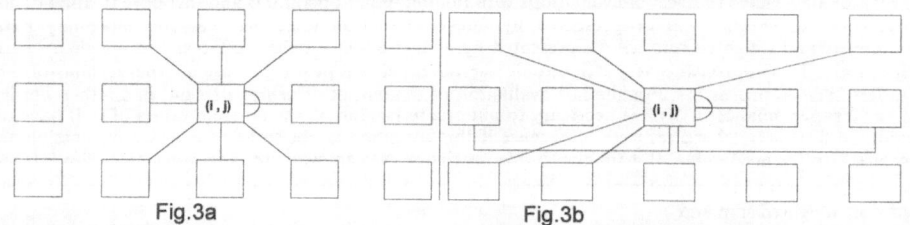

Fig.3a   Fig.3b

Hopfield's-Kohonen's network [5], but they are fully-connected) On the Fig.3b one can see example of such network where (i,j)-th neuron is connected with 8 neurons and with itself simillar to previous model but numbers of neurons which are connected with given one are defined by random function.

For simulation of the image recognition system on the neural network based on multi-valued neurons and with random connections the next experiment has been provided.

Fig.4

Our goal was to train network to recognize 40 images of the faces of different people (4 of these images are presented on Fig.4) with sizes 64*64 and 256 gray-scale levels.

For solution of this problem we used software simulator of the neural network which contains 4096=64*64 neurons. Each neuron has been connected with 29 other ones (numbers of them has been generated as values of random function) and with itself. Then learning algorithm based on (3)-(6) sequentially has been applied to all neurons of the network. Learning set for each neuron contained from the values of brightness in the pixels from all images which had the same coordinates that given ((i,j)-th) neuron. Therefore, learning of the network in our case is learning of each neuron to implement of the partial-defined multiple-valued threshold function which express value of brightness in (i,j)-th pixel of the picture through values of brightness in another 29 pixels of the same picture.

Only 2 hours 45 minutes has been requested for convergention of this learning procedure for all neurons of the network on the PC-486-66Mhz computer. Results of images recognition after learning are presented on Fig. 5 and 6. On the Fig. 5a one can see original image (one of the 40 images from learning set). On the Fig. 5b the same image but with 75% of information replaced with uniform noise is

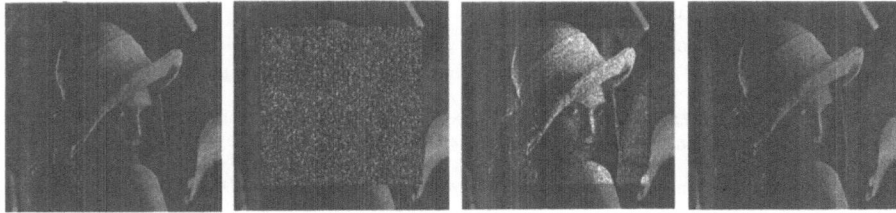

| Fig. 5a | Fig. 5b | Fig. 5c | Fig. 5d |

presented. On the Fig.5c result of the recognition (restoring) after 20 iterations of the neural network action (1 iteration is simultaneous action of all neurons of the network) is presented. On the Fig. 5d result of the recognition (restoring) after 40 iterations is presented. This image differs from original one only in some separate pixels. On the Fig. 6a original image from the learning set is presented. On the Fig. 6b one can see the same image distored by impulsive noise (probability of distortion 0.35, range of the noise signal-[0.255]). On the Fig. 6c - result of the recognition (restoring) after 35 iterations and on the Fig. 6c -

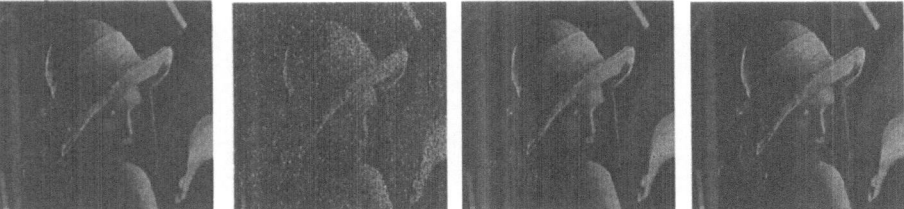

| Fig. 6a | Fig. 6b | Fig. 6c | Fig. 6d |

result of the restoring after 70 iterations. Last image again differs from original only in some separate pixels. One may compare results presented here with experiments presented in [6]. Learning set in that work has been consisted only of 8 patterns (in our case 40 - and this number is restricted only by speed of conventional computer - greater number of images in learning set involves increasing of time which need for learning) and only small destorted region (window around eyes) in comparison with image sizes (ratio of sizes 1:8) was possible to restore.

4.EXTRAPOLATION OF THE TEMPORAL SERIES

Solution of the extrapolation problem on the neural networks is very popular now. We would like to propose here original approach of extrapolation of the temporal series on the single multi-valued neuron.

Let we have temporal series

$$X_0, X_1, X_2, \ldots, X_{n-1}, X_n, X_{n+1}, \ldots, X_i, \ldots \qquad (7)$$

where range of X is $[a, b]$, $a, b \in R$. Let we also have k-valued multi-valued neural element with n inputs. To create multi-valued function for learning and extrapolation we have first of all transform data from the range $[a, b]$ to the range $[0, k-1]$. This may be reached by different means (normalization, linear transformation and so on, it is not on principle, as shown by our experience - in each partial case suitable approach may be always found). For simplification let assume that (7) represent series with transformed range of values. Now we have to create function f for learning. Let assume that each X_i is depended of previous n values of X. Therefore we have:

$$\left.\begin{array}{l} x_n = f(x_0, x_1, \ldots, x_{n-1}) \\ x_{n+1} = f(x_1, x_2, \ldots, x_n) \\ \cdots \quad \cdots \quad \cdots \quad \cdots \quad \cdots \\ x_i = f(x_{i-n+1}, x_{i-n+2}, \ldots, x_{i-1}) \end{array}\right\} \qquad (8)$$

Let learning set contains s values: X_n, X_{n+1}, . . . , X_i, i=n+s-1. To represent function f according to (1) we have to apply learning procedure (3)-(6) for obtaining of the weights W_0, W_1, . . . , W_n. It have be noted that function defined by (8) is partial-defined multi-valued function and always we can find such value of k for which learning procedure (3)-(6) for this function will be convergenced, therefore this function will be partial-defined multiple-valued threshold function for given value of k. After finishing of the learning we can extrapolate values of X_{i+1}, X_{i+2}, . . . using (1).

This approach has been successfully tested on different examples which are presented below.

1.FORECASTING OF THE CURRENCY COURSE

Temporal series which has been extrapolated - values of the USD/DM course. Successful forecasting of the course is very pressing problem and as has been shown this problem may be successfully solved by approach presented here.

For the forecasting with high precision data have been presented with 4 digits after decimal point. Course in each current day has been presented as function of 1036 variables (number of the working days in a 4 years) in 1024-valued logic (k=1024 - such precise is necessary for correct presentation of the range of data). Input data have been normalized to change range of their values to [0, 1923] (after learning and forecasting inverse transformation to change range of data to natural form has been done). Values of course have been presented as (7) and function (8) has been defined. In our case X_0 corresponds to the value of course on Aprl, 23, 1984. Then learning process has been began.

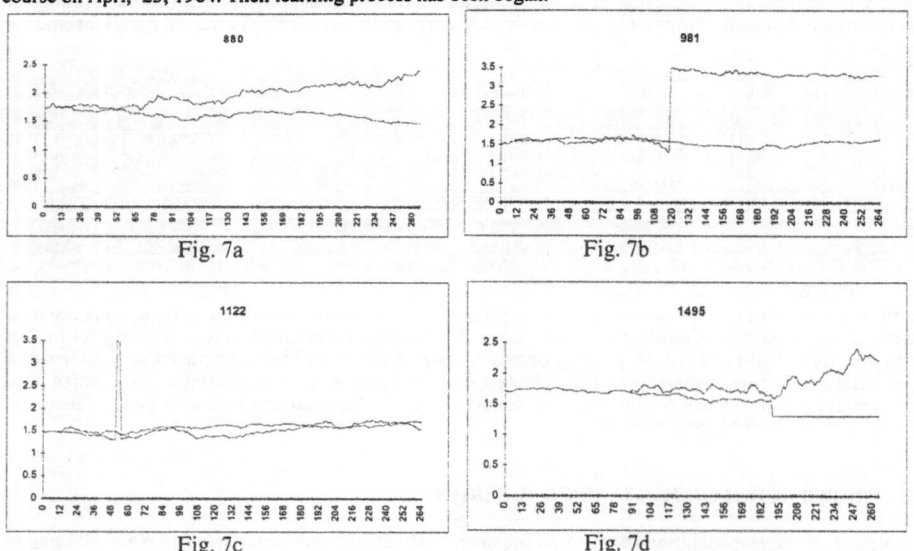

Fig. 7a   Fig. 7b

Fig. 7c   Fig. 7d

On the Fig. 7 results of the forecasting (extrapolation) after learning on the different number of the input data are presented (on horisontal axis - numbers of working days (0 corresponds to the next day after finishing of learning), on vertical axis - price of the 1 USD in DM, more smooth curve corresponds to real course, more breaked - to extrapolated one): Fig. 7a - after learning on the 330 values, Fig. 7b - after learning on the 981 values, Fig. 7c - after learning on the 1122 values and Fig. 7d - after learning on the 1495 values. It is evident that each next forecast is better than previous one. That means that increasing of number of data in the learning set involves increasing of the precise of extrapolation. It has be noted that time which requested for learning is incresed lineary (not exponentially) with increasing of the number of data in the learning set. Last result presented on the Fig.7d is from our point of view wonderful - 96 values of the series have been predicted with absolute precise (that means that precise forecast of the course is known on futheron 96 days) - some differences between real and predicted values were only in 4-th digit after decimal point.

2.EXTRAPOLATION OF THE TRIGONOMETRIC FUNCTIONS.

The second example of the extrapolation - prediction of the values of trigonometric functions. On the Fig. 8 one can see comparison of the 259 extrapolated values of the function

$$sin(2\pi * j / 1024), \quad j=573, ...,832 \qquad (9)$$

(extrapolated function of 1024-logic (k=1024) is depended of 259 variables and learning set consist of 313 values - 313+259=572, therefore for the first extrapolated value j=573) with real values evaluated by (9):

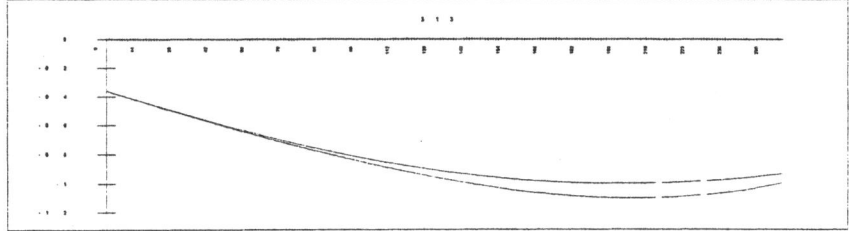

Fig.8

It is evident from the Fig. 8 that predicted values almost equals to real values after learning on half of period. Despite of some very small differences behavior of the function SIN is represented by extrapolation with very high precise.

CONCLUSIONS

The main conclusion which we would like to do - high efficiency of multi-valued neurons for solution of the different problems - from designing of the networks for image recognition to extrapolation of the temporal series on the single neuron. The main feature of the multi-valued neuron - complex-valued arithmetic (complex-valued weights and complex-valued signals). This arithmetic togather with original output function of the neuron made possible to elaborate high effective fast convergenced learning algorithm for multi-valued neuron and therefore to solve by learning different kinds of problems .

REFERENCES

[1] N.N.Aizenberg Multiple-Valued Threshold Logic. Kiev: Naukova Dumka Publisher House (1977) (in Russian).
[2] N.N.Aizenberg , I.N.Aizenberg : CNN Based on Multi-Valued Neuron as a Model of Associative Memory for Gray-Scale Images, Proc. of the 2-d International IEEE Workshop on Cellular Neural Networks and their Applications, Munich, Germany,   October 14-16 (1992) IEEE 92TH0498-6, ISBN 0-7803-875-1, pp.36-41 .
[3] N.N.Aizenberg , I.N.Aizenberg "Fast Convergenced Learning Algorithms for Multi-Level and Universal Binary Neurons and Solving of the some Image Processing Problems", Lecture Notes in Computer Science, Ed.-J.Mira, J.Cabestany, A.Prieto, v.686, Shpringer-Verlag, Berlin-Heidelberg (1993) pp. 230-236.
[4] N.N.Aizenberg , I.N.Aizenberg "Neural Networks based on Universal and Multi-Valued Neurons and their application to solving of the some problems of Image Processing and Pattern recognition", Proc. of the "COST 229" Int. Workshop on Adaptive Systems, Intelligent Approaches, Massively Parallel Computing and Emergent Techniques in Signal Processing and Communications", Bayona (Vigo), (1994) Publication of the University of Vigo and Politec. de Madrid, pp. 223-228.
[5] T.Kohonen "Content-Addressable Memories" Springer -Verlag, Berlin - Heidelberg - New-York, 980.
[6] U.Ramaher, W.Raab, J.Anlauf, U.Hachmann, J.Beichter, N.Bruls, M.Weseling, E.Sichender, R.Manner, J.Glass, A.Wurz "Multiprocessor and Memory Architecture of the Neurocomputer Synapse-1" Proceedings of the 3-d International Conference on Microelectronics for Neural Networks. April 6-8, 1993, Edinburgh, UK, pp. 227-231.

Individual Evolutionary Algorithm and Its Application to Learning of Nearest Neighbor Based MLP

Qiangfu Zhao and Tatsuo Higuchi

Graduate School of Information Sciences
Tohoku University, Sendai, Japan 980
Tel: +81-22-222-1800 ext. 4288
Fax: +81-22-263-9406
Email: zhao@higuchi.ecei.tohoku.ac.jp

Abstract

A society $S(I, T)$ is defined as a system consisting of an individual set I and a task set T. This paper studies the problem to find an efficient S such that all tasks in T can be fulfilled using the smallest I. The individual evolutionary algorithm (IEA) is proposed to solve this problem. By IEA, each individual finds and adapts itself to a class of tasks through evolution, and an efficient S can be obtained automatically. The IEA consists of four operations: *competition*, *gain*, *loss* and *retraining*. *Competition* tests the performance of the recent I and the fitness of each individual; *gain* increases the performance of I by adding new individuals; *loss* makes I more compact by removing individuals with very low fitness; and individuals are adjusted by *retraining* to make them better. An evolution cycle is: *competition* $\land$ *(gain* $\lor$ *loss)* $\land$ *retraining*, and the evolution is performed cycle after cycle until some criterion is satisfied. The performance of IEA is verified by applying it to the learning of nearest neighbor based multilayer perceptrons.

Keywords : Evolutionary algorithm, genetic algorithm, individual evolutionary algorithm, multi-individual-multi-task problem, nearest neighbor based multilayer perceptron

1 Introduction

In the literature, the term genetic algorithm (GA) is almost a synonym of evolutionary algorithm (see [1],[2] and all references therein). GA is a direct computer simulation of the neo-Darwinian evolutionary theory, and consists of four operations: *competition, reproduction, selection* and *mutation*. All individuals in a population *compete* for surviving; those with high fitness will get more chance to *reproduce*; individuals with very low fitness will be *selected* and removed; and *mutation* gives more chances to get the fittest individual. If we define an evolution cycle as: *competition* $\land$ *(reproduction* $\lor$ *selection)* $\land$ *mutation*, the evolution can be performed cycle after cycle until a good individual is obtained. Here, $\land$ and $\lor$ represent logic "and" and "or", respectively. There are four features in GA:

1) All individuals compete for one task or in one environment;
2) The life span of an individual is only one evolution cycle;
3) The individuals have no learning ability, and the fitness is determined only by chance;
4) The final evolution result is one of the fittest individuals.

In practice however, we often meet the problem that many individuals compete for many tasks or environments. For example, many kinds of creatures compete for different living environments on the earth; many factories compete for producing different kinds of products needed by a community; many workers compete for different job positions in a factory, and so on. In such problems, each individual finds a class of tasks (environments) to fulfill (to live) by evolution, and many kinds of individuals can live in the same "society" peacefully. For example, many kinds of creatures can live on the earth together, with each kind of creatures living in their own environment; many factories can exist in the same community, with each factory making its own products; many workers can work in the same factory, with each worker working in his own position; and so on. Thus, the multi-individual-multi-task (MIMT) problem is an important evolution problem in the real world.

The MIMT evolution is an individual based evolution. That is, an individual must live for a relatively long period of time, and must have some learning ability so as to find a class of tasks to fulfill, and adapts itself to these tasks. After each individual finds a class of suitable tasks, the society in which they live will become a relatively stable and efficient system. For example, when each worker in a factory gets a suitable job position for him, and works hard to keep that position, the factory will function well; when each factory finds some suitable products, and produces them to meet the needs of the community, the community will become prosperous; and so on. Note however, it does not mean that "all individuals" can survive. In fact, if an individual can not find a suitable task to fulfill or can not adapt itself to the task it already found, this individual will be selected by nature, and removed

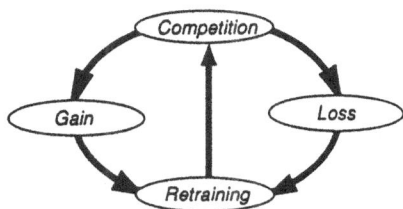

Figure 1: The individual evolutionary algorithm

from the society. Thus, the result of the MIMT evolution is an efficient system in which all tasks can be performed using as less individuals as possible.

2 The Individual Evolutionary Algorithm (IEA)

Let's define a society $S(I, T)$ as a system consisting of two sets: an individual set I and a task set T. Then, the MIMT problem is to find an efficient society S such that all given tasks in T can be fulfilled using the smallest I. To get some ideas for solving this problem, let us first look at some practical problems in the real world.

First, let us consider the evolution of a factory. In a factory (society), a lot of job positions (tasks) are opened for many workers (individuals). At the beginning, the individual set I is empty, and no task can be fulfilled by it. After employing some workers, some positions are filled and some are still opening. Before employing some new workers, the employer would try to train the employed ones to make them more skillful and do as many jobs as they can. After training, the employer would test the ability of all workers. If they can not do all the jobs, some new workers would be employed. On the other hand, if all jobs can be done with easy, the employer would try to find some workers who are not so skillful, and dismiss them. When some new workers are employed or some old workers are dismissed, the employer would train the workers once more, so that more jobs can be done by them. The four operations *test*, *employ*, *dismissal* and *training* would be performed successively until all jobs can be done with the least number of workers. Clearly, the result of the evolution will be the desired factory.

Another example is the evolution of a community. In a community (society), many kinds of goods must be provided (tasks) to the citizens, and many kinds of factories (individuals) must be constructed to produce these goods. The first thing the governor (the actual governor here is the market) must do is to make a plan to construct some factories. After these factories are constructed, they can provide some kinds of goods, but may not be able to provide all goods. However, before planing to construct some new factories, the ability of existing ones should be improved so that more goods can be provided by them. After improving, the productivity of existing factories would be tested by the governor. If there still exist some kinds of goods that can not be provided, the governor would make another plan to construct some new factories. On the other hand, if all goods can be provided with easy, some useless factories would be withdrawn or changed to produce more useful goods. After some new factories are constructed or some old factories are withdrawn, the factories are improved again, so that more and better goods can be provided by them. There are also four operations in this case: *test*, *construction*, *withdrawal* and *improvement*, and the evolution is performed cycle after cycle until the community become prosperous enough.

Inspired by the above evolution processes, we find that the MIMT problem in general can be solved in an evolutionary manner using four basic operations: *competition*, *gain*, *loss* and *retraining*. *Competition* tests the performance of the recent individual set I and the fitness of each individual; *gain* increases the performance of I by assigning tasks to new individuals (some individuals get their jobs); *loss* makes I more compact by removing some individuals with very low fitness (some individuals loss their jobs); and individuals are adjusted by *retraining* to make them perform better. An evolution cycle can be defined as: *competition* $\wedge$ (*gain* $\vee$ *loss*) $\wedge$ *retraining* (Fig. 1), and the evolution is performed cycle after cycle until some criterion is satisfied. This is an evolutionary algorithm based on the evolution of the individuals, and is thus called the individual evolutionary algorithm (IEA) in this paper. Note however, although the evolution is emphasized on the individuals, the final result is the whole society $S(I, T)$ — an efficient system.

3 Learning of the Nearest Neighbor Based MLP

To make the idea of IEA more concrete and to show its efficiency, the evolutionary learning of the nearest neighbor based multilayer perceptron (NN-MLP) is considered in this paper. NN-MLP is a neural network realization of the nearest neighbor classifier (NNC). It is a one-hidden-layer network as shown in Fig. 2(a), where each hidden neuron

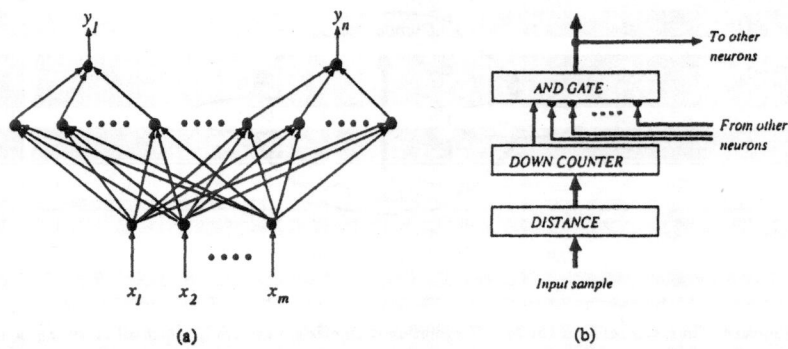

Figure 2: (a) The nearest neighbor based multilayer perceptron and (b) a conceptual realization of the nearest neighbor neuron (small circles represent invert operations)

corresponds to a training sample and each output neuron corresponds to a pattern class. The hidden neurons are realized using the model shown in Fig. 2(b). In this neuron model, the distance (any distance measure can be used) between the input sample and the weighting vector of the neuron is first computed, and the result is then put into a count-down counter. The neuron fires (outputs a 1) when its counter becomes 0, and its output is then used to inhibit firing of all neurons with different class labels. In the NN-MLP, all hidden neurons compute the distances and count-down simultaneously. An output neuron fires if at least one of its inputs is 1.

From pattern recognition theory, NN-MLP can be used for any decision making if the number of hidden neurons is large enough [3]. Further, since patterns are represented locally, NN-MLP is very suitable for self-organization and real-time learning. This is the main reason that NN-MLP has been studied by many authors in different forms [4]–[19]. However, if the hidden neurons correspond directly to all samples in the training set Ω, the network will be extremely large. The basic idea to reduce the network size is to find a prototype set P which consists of much less elements than Ω, and use the prototypes in P as the connection weights of the hidden neurons. The prototype set P must satisfy the following conditions:

$$
\begin{aligned}
\cup_{x_{ij} \in P} D(x_{ij}) &= \Omega \\
D(x_{ij}) \cap D(x_{kl}) &= \Phi, \qquad i \neq k \\
M(P) &= min
\end{aligned}
\tag{1}
$$

where $\cup$ and $\cap$ represent the union and intersection of sets, respectively, Φ is the empty set, $x_{ij} \in P$ is the j-th prototype of the i-th pattern class, $D(x)$ is the set of samples with the same class label as x, and $M(\bullet)$ is the cardinality of a set.

Two approaches can be used to find the prototype set. In the first approach, the prototypes are selected directly from the training set Ω. This approach is relatively simple, and has been studied by many authors. For example, the condensed nearest neighbor (CNN) rule [21], the reduced nearest neighbor (RNN) rule [22], the restricted coulomb energy (RCE) algorithm [18] and its modified version [19], all belong to this approach. The main problem in these methods is that the prototypes may not be in the optimal positions in the feature space, and thus $M(P)$ is usually much larger than necessary.

In the second approach, prototypes optimal in some sense can be found iteratively using training samples. Algorithms of this approach include: the vector quantization (VQ) algorithm [20], the competitive learning algorithms [10]–[16], and so on [23],[24]. A common problem in these algorithms is that there is no efficient way to determine $M(P)$. For example, in the self-organization feature map (SOFM) and the learning vector quantization (LVQ) algorithms of Kohonen, a large enough $M(P)$ must be assumed, and all prototypes are actually used after learning. In the adaptive resonance theory (ART) of Grossberg, each prototype has a certain "effective region", and can be used to represent a certain subspace of the feature space. A new prototype is added to the prototype set when a sample is "far" from any existing prototypes. Thus, $M(P)$ can be determined automatically. However, in practice, it is hard to know the effective region of each prototype. A small predetermined effective region may result in too many prototypes, while a large effective region may produce a poor recognition rate.

Therefore, it is difficult to get a prototype set that satisfies (1) using existing algorithms. However, if we consider the training set Ω as the task set T, and the prototype set P as the individual set I, the problem for finding the smallest P is in fact a typical MIMT problem, and can be solved using the IEA algorithm. In this case, the operation *competition* tests the ability of the recent prototype set, and the importance (fitness) of each prototype; the operation *gain* creates some new prototypes for unknown samples; the operation *loss* removes some unimportant prototypes when necessary; and the prototypes are adjusted by the operation *retraining* to achieve better performance after some prototypes are created or removed. After many evolution cycles, the prototype set is expected to be able to recognize all training samples, with the least number of prototypes.

Note that if we consider the prototype set as an NN-MLP, and the prototypes as the hidden neurons, all algorithms discussed above can be adopted directly to the learning of NN-MLP. Thus, in the following discussions, hidden neurons and prototypes will be used interchangeably, and a prototype set will mean the same thing as an NN-MLP.

4   Evolutionary Learning of NN-MLP Based on IEA

In the IEA, each operation is realized by a process or a subroutine. The operation *competition* is a process to test the ability of the recent network (individual set) and the importance of each hidden neuron (individual) using training samples (tasks). For these purposes, two special parameters λ and ρ are assigned to each hidden neuron. λ is the class label of a hidden neuron, and ρ is a measure of importance of that neuron. Both parameters are assigned to the neuron automatically in the evolution process.

To test the ability or the recognition rate of the recent network is relatively easy, and can be performed using the nearest neighbor rule as usual. To test the importance of the hidden neurons is one of the key points. This is performed also by competition, but in a different way from conventional methods. Conventionally, a hidden neuron is a winner whenever its weighting vector is nearest to the input sample x, or in other words, if it fires first. In our algorithm however, a hidden neuron y is a winner only if it satisfies

$$\lambda(y) = \lambda(x) \tag{2}$$
$$d(y, x) < d(y', x), \ for \ \lambda(y') \neq \lambda(x) \tag{3}$$

and has the largest ρ value among all fired neurons. A hidden neuron is a loser if it satisfies the above two conditions, and has the smallest ρ value of all fired neurons. That is, if two or more hidden neurons fire for a given sample, the most important one will be the winner, and the most unimportant one will be the loser. Suppose y_w is the winner and y_l is the loser, their ρ values are changed as follows:

$$\rho(y_w) = \rho(y_w) + \delta$$
$$\rho(y_l) = \rho(y_l) - \delta \tag{4}$$

where δ is a sufficiently small positive number. In practical applications, it is necessary to normalize the value of ρ to prevent it going too large or too small. In the experiments given in this paper, the above simple rule is used directly because only a few learning cycles are always sufficient.

After changing ρ, the winner will become more important, and the loser will become more unimportant. Thus, if a hidden neuron frequently wins in the competition, its importance ρ will become larger and larger. On the other hand, if a hidden neuron often loses, its ρ will be smaller and smaller.

After *competition*, we can know how the present network performs and how important each hidden neuron is. If the recognition rate r is too low, some hidden neurons should be added to the network. On the other hand, if r is very high, some of the unimportant neurons should be removed to make the network more efficient.

We may add a hidden neuron whenever a sample is mis-classified (as is done in ART or RCE), but this is not an efficient method because the network may soon become too large. In our algorithm, only one hidden neuron is added for each mis-classified class in the operation *gain*, and this is done only when the recognition rate is smaller than a certain threshold r_0. A neuron is added by recovering an unused neuron, or by creating a new neuron. Thus, it is not necessary to specify the number of hidden neurons before learning.

We can remove a neuron whenever its ρ becomes negative. However, to make the learning process stable, we do not remove any neurons in the operation *competition*, but do it in the operation *loss*. In *loss*, only one neuron with negative ρ is removed, and this neuron is selected randomly. If reduction of a neuron results in too many recognition errors, this neuron will be returned to the network. Removing of a neuron is performed by making it *unused*, or deleting it physically from the network.

Retraining is necessary when some hidden neurons are removed or added. It is the process to rearrange the knowledge learned up to now, and to make them more abstract and simpler. In *retraining*, network parameters are readjusted so that the network can achieve the highest ability using recent neurons. Any supervised competitive learning method can be used for this purpose. In this paper, the algorithm given in [17] is adopted due to its desired convergent property, and it is given as follows:

<u>Step 1</u>: $\forall x \in \Omega$, test if x can be recognized, If yes, no change; otherwise, update the connection weights as follows:

$$y_0 = y_0 - \alpha(x - y_0)$$
$$y_1 = y_1 + \alpha(x - y_1) \tag{5}$$

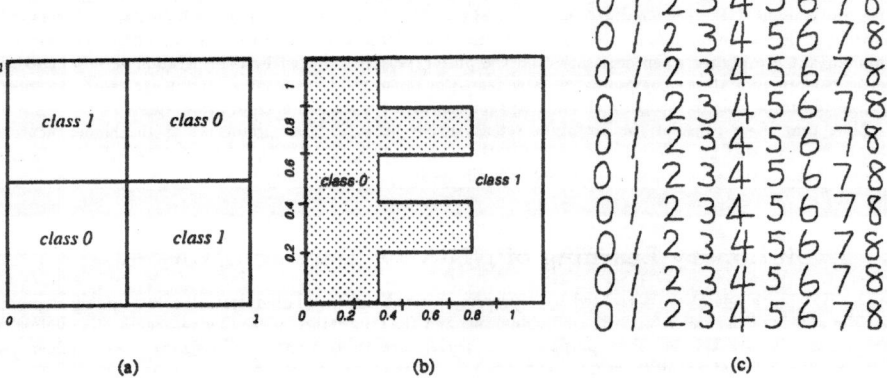

(a)   (b)   (c)

Figure 3: (a) The generalized XOR problem, (b) the straight line class boundaries problem, and (c) part of the samples for digit recognition

where y_0 is the nearest neuron of different pattern classes, y_1 is the nearest neuron of the same pattern class, and α is the convergent ratio;

$\underline{Step\ 2:}$ Terminate if the number of iterations reaches to the given value, or all samples have been correctly recognized; otherwise, return to Step 1.

5 Experimental Results

To demonstrate the performance of IEA, we have conducted experiments with three pattern recognition problems. The first two problems are artificial problems adapted from [17](see fig. 3(a)(b)), and the third one is a practical handwritten digit recognition problem. As in [17], both training set and test set for the artificial problems consist of 6400 samples taken at random. For the digit recognition problem, 80 samples were written by one of the authors in an ordinary manner for each pattern class. The training set consists of the first 40 samples of each class, and the test set consists of the remained samples. In the experiment, every digit was written using Mouse in a 256 × 256 frame. No normalization of any kind was performed. Part of the samples are shown in Fig. 3(c). Note that 9 is not used because features employed in our experiment are the crossing numbers of the image on some concentric circles. These features are rotational invariant, and 6 and 9 belong to the same class in this case.

A. The Generalized XOR Problem

The generalized XOR problem is a two-class problem as depicted in Fig. 3(a). There are four perfect prototypes corresponding to the centers of the four subregions. Tab. 1(a) gives the experimental results, where $M(P)$ is the number of prototypes, and the error rate is obtained using the test set. As shown in the table, the NNC uses all 6400 training samples as prototypes, the CNN, RNN and RCE use about 200, and the NN-MLP obtained by IEA uses only 4 (hidden neurons). Further, the error rate of the 4-hidden-neuron NN-MLP is much smaller even than that of the NNC. Thus, the generalization ability has also been increased for this problem.

In the experiment, the parameter δ in Eq. (4) is 0.001 and the initial value for α in Eq. (5) is 0.1. α is decreased linearly in the retraining process. Further, each training sample is presented once in *competition*, and 20 times in *retraining*. The number of evolution cycles is 25. The four prototypes obtained after the 25-th cycle are as follows:

$$(0.267457,\ 0.739306,\ 1)\qquad(0.734149,\ 0.262727,\ 1)$$
$$(0.733432,\ 0.738035,\ 0)\qquad(0.266190,\ 0.262267,\ 0)$$

where the 1-st and 2-nd numbers are the feature values, and the 3-rd one is the class label. Clearly, they are very close to the perfect prototypes.

B. The Straight Line Class Boundaries Problem

The straight line class boundaries problem is also a two-class problem, and is depicted in Fig. 3(b). Similar to the generalized XOR problem, we can easily find out that there are ten perfect prototypes for this problem. Using

Table 1: Results for (a) the generalized XOR problem, and (b) the straight line class boundaries problem

(a)	M(P)	Error rate (%)	(b)	M(P)	Error rate (%)
NNC	6400	0.91	NNC	6400	0.97
CNN	231	1.11	CNN	307	1.17
RNN	162	1.27	RNN	216	1.14
RCE	183	0.69	RCE	243	0.56
New	4	0.13	New	10	0.48

the same parameters as in the generalized XOR problem, the prototypes obtained by IEA after the 25-th evolution cycle are as follows

$$
\begin{array}{ll}
(0.847078,\ 0.356892,\ 1) & (0.551071,\ 0.427418,\ 1) \\
(0.048296,\ 0.827414,\ 0) & (0.551475,\ 0.827253,\ 1) \\
(0.844268,\ 0.778659,\ 1) & (0.228646,\ 0.248717,\ 0) \\
(0.430515,\ 0.183015,\ 1) & (0.550853,\ 0.771827,\ 0) \\
(0.429797,\ 0.220361,\ 0) & (0.554753,\ 0.364483,\ 0)
\end{array}
$$

The experimental results are summarized in Tab. 1(b). Again, the smallest prototype set has been obtained by IEA, and the performance of this prototype set is the best.

C. The Handwritten Digit Recognition Problem

As stated previously, the handwritten digit recognition problem considered here is a 9-class problem. Features used in the recognition are crossing numbers on some concentric circles, with the center of the concentric circles being the gravity center of the image. For comparison, the number of concentric circles (features) has been changed from 12 to 32, with increment of 4. All training samples are presented once in *competition*, and 10 times in *retraining*. The initial value of α in Eq. (5) is 0.1, and the number of learning cycles is 25. The parameter δ in Eq. (4) is 0.01 here, so that unimportant hidden neurons can be detected more quickly.

Tab. 2 gives the experimental results. Clearly, for each case, the new algorithm always produces the smallest or nearly smallest prototype set, and the error rate is comparable with that of the NNC.

6 Comparison with The Genetic Algorithm

From the above discussions we can see that there is an interesting similarity between IEA and GA. Both of them consist of four basic operations. In fact, if we speak in a GA language, the operation *competition* in IEA is also a competition process in which all individuals compete for surviving; the operation *gain* (re)produces new members for the population (individual set); the operation *loss* selects individuals with low fitness and removes them from the population; the operation *retraining* mutates existing individuals to make them perform better. However, there are also important differences between IEA and GA, and the main differences are given as follows:

(1) The fitness of individuals in IEA is determined by a process, while the fitness of individuals in GA is often given in one-step by calculating a fitness function;

(2) Only one generation of individuals is considered in IEA, and every individual evolves from its initial state to an expert for fulfilling a certain class of tasks. In GA however, better individuals are reproduced by cross-over of existing individuals, and the learning ability of individuals is totally ignored;

(3) The *retraining* operation in IEA is "purpose-controlled", so that all individuals become more useful, and the individual set becomes more efficient. However, in GA, mutation is performed randomly, and there is no assurance that the individual can become better after mutation;

(4) The final result of IEA is a functional harmonic society although it is based on the evolution of the individuals, while the result of GA is one of the fittest individuals although the evolution is based on the families of individuals;

Table 2: Results for the handwritten digit recognition problem

Number of features		12	16	20	24	28	32
NNC	Error rate (%)	6.9	3.9	3.6	3.6	5.0	4.4
	M(P)	360	360	360	360	360	360
CNN	Error rate (%)	6.7	8.3	6.4	3.1	10.3	7.2
	M(P)	50	45	49	45	53	45
RNN	Error rate (%)	5.3	8.3	6.9	3.9	10.8	8.1
	M(P)	41	37	41	35	45	39
RCE	Error rate (%)	11.1	5.8	6.7	5.3	6.7	1.7
	M(P)	80	63	62	61	54	55
New	Error rate (%)	8.1	3.3	5.3	5.6	6.7	4.7
	M(P)	15	13	11	11	12	10

(5) In IEA, many individuals compete for many tasks, and each individual will finally find a class of tasks suitable for him (it). In GA however, many individuals compete for one task, and only one of the fittest individuals can get it.

At present, it is still hard to say which approach is better. However, rather than talking about which one is better, it is more profitable to combine these two different approaches to obtain a more powerful evolutionary algorithm. This seems to be a very interesting topic for future research.

7 Concluding Remarks

In this paper, a new evolutionary algorithm has been proposed. This algorithm, called the individual evolutionary algorithm (IEA) is suitable for solving multi-individual-multi-task (MIMT) problems such as learning of NN-MLP. Using IEA, the smallest or nearly smallest NN-MLPs can be obtained from given initial random networks by successively performing four operations: *competition*, *gain*, *loss* and *retraining*. The algorithm is very simple and suitable for parallel realization. Its efficiency has been verified using experimental results.

Many topics are remained for future studies. For example, how can we do if the task set changes with time; what shall we do if each task can not be fulfilled by a single individual; what kind of evolutionary algorithm can we obtain if we combine IEA with GA, and so on. In fact, finding a good evolutionary algorithm itself is an evolution process, and requires different ideas from different areas.

Acknowledgment

This research is supported in part by the telecommunications advancement foundation (TAF) of Japan.

References

[1] X. Yao, "Evolutionary artificial neural networks," International Journal of Neural Systems, Vol. 4, No. 3, 203-222, Sept. 1993.

[2] D. B. Fogel, "An introduction to simulated evolutionary optimization," IEEE Trans. on Neural Networks, Vol. 5, No. 1, pp. 3-14, Jan. 1994.

[3] T. M. Cover and P. E. Hart, "Nearest neighbor pattern classification," IEEE Trans. on Information Theory, Vol. IT-13, No. 1, pp. 21-27, Jan. 1967.

[4] Q. F. Zhao and T. Higuchi, "A study on the determination of MLP structures," Proc. IEICE Karuizawa Workshop, pp. 121-126, Karuizawa, Japan, April 1994.

[5] Q. F. Zhao, "Neural network realization of the nearest neighbor methods," Technical Report of IEICE, NC93-65, pp. 83-87, Dec. 1993.

[6] Q. F. Zhao and T. Higuchi, "Efficient learning of nearest neighbor multilayer perceptrons." Proc. International Conference on Fuzzy Logic, Neural Nets and Soft Computing, pp. 77-78, Iizuka, Japan, Aug. 1994.

[7] Q. F. Zhao and T. Higuchi, "Supervised organization of nearest neighbor MLP," Proc. International Conference on Neural Information Processing, pp. 1398-1403, Seoul, Korea, Oct. 1994.

[8] O. J. Murphy, "Nearest neighbor pattern classification perceptrons," Proc. IEEE, Vol. 78, No. 10, pp. 1595-1598, Oct. 1990.

[9] N. K. Bose and A. K. Garga, "Neural network design using Voronoi diagrams," IEEE Trans. on Neural Networks, Vol. 4, No. 5, pp. 778-787, Sept. 1993.

[10] T. Kohonen, "Self-organized formation of topologically correct feature maps," Biolog. Cybern.,Vol. 43, pp. 59-69, 1982.

[11] T. Kohonen, "The self-organizing map," Proc. IEEE, Vol. 78, No. 9, pp. 1464-1480, Sept. 1990.

[12] J. A. Kangas, T. Kohonen and J. T. Laaksonen, "Variants of self-organizing maps," IEEE Trans. on Neural Networks, Vol. 1, No. 1, pp. 93-99, Mar. 1990.

[13] B. Kosko, "Unsupervised learning in noise," IEEE Trans. on Neural Networks, Vol. 1, No. 1, pp. 44-57, Mar. 1990.

[14] B. Kosko, "Stochastic competitive learning," IEEE Trans. on Neural Networks, Vol. 2, No. 5, pp. 522-529, Sept. 1991.

[15] G. A. Carpenter and S. Grossberg, "ART 2: self-organization of stable category recognition codes for analog input patterns," Applied Optics, Vol. 26, No. 23, pp. 4919-4930, Dec. 1987.

[16] G. A. Carpenter and S. Grossberg, "The ART of adaptive pattern recognition by a self-organizing neural network," IEEE Computer, Vol. 21, No. 3, pp. 77-88, Mar. 1988.

[17] S. Geva and J. Sitte, "Adaptive nearest neighbor pattern classification," IEEE Trans. on Neural Networks, Vol. 2, No.2, pp. 318-322, Mar. 1991.

[18] D. L. Reilly, L. N. Cooper and C. Elbaum, "A neural model for category learning," Biol. Cybern. 45, pp. 35-41, 1982.

[19] Y. Okamoto, "Neural network model for real time adaptation to rapidly changing environment," IEICE Trans., Vol. J73-D-II, No. 8, pp. 1186-1191, Aug. 1990.

[20] Y. Linde, A. Buzo and R. M. Gray, "An algorithm for vector quantizer design," IEEE Trans. on Communication, Vol. COM-28, No. 1, pp. 84-95, Jan. 1980.

[21] P. E. Hart, "The condensed nearest neighbor rule," IEEE Trans. on Information Theory, Vol. 14, No. 5, pp. 515-516, May 1968.

[22] G. W. Gates, "The reduced nearest neighbor rule," IEEE Trans. on Information Theory, Vol. 18, No. 5, pp. 431-433, May 1972.

[23] Q. B. Xie, C. A. Laszlo and R. K. Ward, "Vector quantization technique for nonparametric classifier design," IEEE Trans. on Pattern Analysis and Machine Intelligence, Vol. 15. No. 12. pp. 1326-1330, Dec. 1993.

[24] C. L. Chang, "Finding prototypes for nearest neighbor classifiers," IEEE Trans. on Computers, Vol. 23, No. 11, pp. 1179-1184, Nov. 1974.

A Practical View of Suboptimal Bayesian Classification with Radial Gaussian Kernels

Jean-Luc Voz, Michel Verleysen, Philippe Thissen, Jean-Didier Legat

Université catholique de Louvain,
Laboratoire de Microélectronique - DICE,
3 Place du Levant, B-1348 Louvain-La-Neuve, Belgium

For pattern classification in a multi-dimensional space, the minimum misclassification rate is obtained by using the Bayes criterion. Kernel estimators or probabilistic neural networks provide a good way to evaluate the probability densities of each class of data and are an interesting parallel implementation of the Bayesian classifier [1]. However, their training procedure leads to a very high number of neurons when large datasets are available; the classifier then becomes too complex and time consuming for on-line operation. Suboptimal Bayesian classifiers based on radial Gaussian kernels [2] uses an iterative unsupervised learning method based on vector quantization to obtain a significant simplification of the network structure, while keeping sufficiently accurate estimations of probability densities. In this paper, we study the vector quantization problem and the effects of codebook size and data space dimension on the optimal width factors of the radial Gaussian kernels used in the estimation.

1 Introduction

For multi-dimensional classification tasks, the use of the Bayesian classification theory permits to minimize the misclassification probability given the a priori probabilities of the classes and their probability density functions. For real-case problems, these functions are never known and the only data available are a finite set of observations with known classes (the training set). The challenge is thus to "learn" the spatial distribution of each class on this training set and then to take the best classification decision for any new vector to classify.

The principle of Parzen windows [3] or kernel estimators is to estimate the probability density functions through the training vectors, and then to use these estimates in the Bayes law to classify a given vector u. The use of these estimators to build classifiers is very interesting because they also provide a useful way to estimate the probabilities of each class for any point to classify. However, from a practical point of view, kernel classifiers imply a computational load in the recognition mode that is unrealistic in practical situations : the calculation of the estimates of the probability density functions of each class at one given point u requires to evaluate at location u as many Gaussian kernel functions as there are vectors in the training set.

To avoid these problems while keeping the advantages of the kernel Bayesian approach, different reduction methods where proposed [4, 5, 6, 7, 2, 8], most of them being based on clustering techniques to replace the initial design set by another one having a strongly reduced number of samples.

In [2, 8], we proposed a reduction method to choose the reduced set but also the widths of the kernels in an optimal way. The theory leading to this reduction method is based on two main hypotheses: we first suppose that the vector quantization process (to decrease the number of samples) converges to centroids having the same distribution as the initial points, and secondly, we derivate the optimal values of the kernel function width factors from the hypothesis that all clusters will be small compared to local variations of the true densities.

In this paper, we first provide a brief introduction to the Bayesian classification theory and its approximation by the use of kernel classifiers. We then present a reminder of our method to build an efficient suboptimal Bayesian classifier and the hypotheses that are used in this purpose. Through extensive experiments, we then study the effect of these hypotheses in real case problems, how they are respected in function of the data space dimension, and the codebook size and how it is possible to enhance the performances of our classifier. The simulation results give a qualitative view of how the hypotheses of [6] and [2] must be applied in different situations.

2 Statistical classification: theory and practice

The problem consists in classifying an observed vector u of $\mathcal{R}^d$ among c known classes denoted ω_i, $1 \leq i \leq c$. In the Bayesian context, it is assumed that any vector u belonging to a given class ω_k is drawn from a single conditional density $p(u|\omega_k)$ and that the occurrence of any class ω_i has a constant probability denoted $P(\omega_i)$. With these assumptions, if all wrong decisions are given the same penalty, the Bayes classification decision will be to select the most probable class, i. e. the class for which the product $p(u|\omega_i)P(\omega_i)$ is maximum.

2.1 Bayes-like classification with kernel density estimate

According to the Bayes law, the knowledge of the conditional densities $p(u|\omega_i)$ and of the a priori probabilities $P(\omega_i)$ of each class is needed to take the decision which minimizes the probability of misclassification for an observed vector u. But these values are never known in real case problems: we only have at our disposal a finite set A_N of observations $x(n)$, $1 \leq n \leq N$ having known classes $\omega_{x(n)} : A_N = \cup\{A_{N_i}\}$ with $A_{N_i} = \{x(n), \omega_{x(n)} = \omega_i, 1 \leq n \leq N_i\}$ and $N = \sum_{i=1}^{c} N_i$. The a priori probabilities $P(\omega_i)$ may be simply estimated by the relative frequency of the class occurrences in the learning set $\hat{P}(\omega_i) = N_i/N$.

A consistent estimate of a multivariate probability density function can be obtained by a kernel density estimator [3, 9]. Using such estimator, the probability density in each class ω_i can be estimated by

$$\hat{p}(N_i, u|\omega_i) = \frac{1}{N_i} \sum_{n=1}^{N_i} K\left(\frac{u - x(n)}{h(n)}\right) \tag{1}$$

where $\{x(n), 1 \leq n \leq N_i\}$ denote the available patterns in class ω_i and $K(\cdot)$ a radial kernel function depending only on the norm of its argument. Parameter $h(n)$ is called the *width factor* of the kernel. The estimator is said to be "variable" if $h(n)$ depends of $x(n)$ and "fixed" otherwise. Variable estimators always provide better estimates, but it is very difficult to locally compute the optimal value of $h(n)$.

Due to their nice analytical properties, radial Gaussian kernels in dimension d are often used:

$$K\left(\frac{u - x(n)}{h(n)}\right) = \frac{1}{\left(h(n)\sqrt{2\pi}\right)^d} exp(-\frac{1}{2}\left(\frac{\|u - x(n)\|}{h(n)}\right)^2), \tag{2}$$

So, a classifier based on kernel density estimation require an extremely light computational cost during the learning (a simple storage of the training patterns) and have very good Bayes-like classification performances. Unfortunately, for large training sets the required memory size and the computational cost of the classification become incompatible with hardware constraints and real time classification tasks. The purpose of the suboptimal Bayesian classifier presented here below is to drastically reduce the number of kernels N_i in each class, in order to use (1) in realistic situations, avoiding to reduce the quality of the density estimation.

2.2 The suboptimal Bayesian classifier

The principle of the proposed method is to use a vector quantization technique to split into clusters the portion of the space where vectors can be found. The aim is thus to approximate the sets of patterns A_{N_i} by sets of so-called centroids $B_{M_i} = \{c(m), \omega_{c(m)} = \omega_i, 1 \leq m \leq M_i\}$, where $M_i << N_i$, roughly keeping the same probability density of vectors for sets A_{N_i} and B_{M_i}.

For the estimation of probability densities in each class, we then use the reduced sets B_{M_i} to build variable kernels estimators of each class instead of the original sets A_{N_i}; this strongly decreases the number of operations involved in (1).

The vector quantization used is an iterative version of the "Generalized Lloyd Algorithm" [10], the "Competitive Learning" (CL); the iterative character of this rule is used to set the position of the centroids and to evaluate the inertia of each cluster in order to obtain an approximation of the optimal variable width factors associated to each cluster. The principle of this method is the following in each class ω_i.

First, the M_i centroids $c(m)$ are randomly initialized to any of the N_i patterns, keeping the same a priori probabilities of classes for both sets A_{N_i} and B_{M_i}. The inertia coefficients $i(m)$ associated to each cluster are initialized to zero. Then, each of the N_i patterns $x(n)$ is presented to the set B_{M_i}, and the centroid $c(a)$ closest from $x(n)$ is selected and moved in the direction of the presented pattern while its inertia coefficient is updated:

$$c(a) = c(a) + \alpha(x(n) - c(a)) \tag{3}$$

$$i(a) = i(a) + \alpha(\|x(n) - c(a)\|^2 - i(a)) \tag{4}$$

where a is the index of the closest centroid to a learning vector $x(n)$ and α is an adaptation factor ($0 \leq \alpha \leq 1$) which must decrease with time during the learning to ensure the convergence of the algorithm.

After several presentations of the whole set of patterns A_{N_i}, the distribution of centroids $c(m)$ in B_{M_i} is supposed to reflect this of the training set A_{N_i}, and the inertia coefficients $i(m)$, $1 \leq m \leq M_i$, converge to the average inertia of points in the clusters associated to $c(m)$ ((4) being a kind of convex combination at each iteration between the previously estimated value of $i(a)$ and a new contribution $\|x(n) - c(a)\|^2$ due to the input vector $x(n)$):

$$i(m) \simeq \frac{1}{n(m)} \sum_{v \in C(m)} \|v - c(m)\|^2 \tag{5}$$

where the sum goes on every point v of the original training set belonging to $C(m)$, the cluster associated to the centroid $c(m)$ in the Voronoi tessellation obtained after the vector quantization, and $n(m)$ is the number of these points.

At the end of the learning, and under the hypothesis of a sufficiently large number of centroids for a good coverage of the partition of the space where the classes are present, the clusters will be sufficiently small so that the true probability density inside each cluster can be approximated by a constant. We use this hypothesis to set the width factors of the Gaussian kernel function in order to keep the estimate (1) of the density as constant as possible over two consecutive clusters (clusters sharing the same border). Under this hypothesis, if we consider that the local arrangement of the centroids of consecutive clusters will be as the vertices of an hypercube, it may be shown [8] that the relation between $h(m)$, the optimal width factor of the Gaussian kernel function to set on $c(m)$ and the estimated inertia $i(m)$ is:

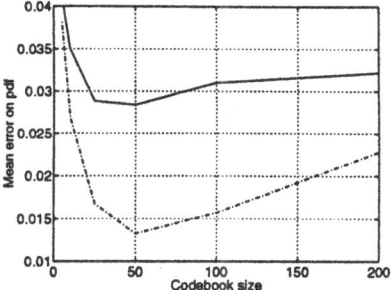

Figure 1: Mean probability density estimation error of the suboptimal kernel estimator for the estimation of a Gaussian mixture with three modes using the number of points per cluster (dashed line) or not (solid line).

$$h(m)^2 = \frac{3}{2\ln 2} \frac{i(m)}{d} \tag{6}$$

where d is the data space dimension.

Finally, the estimation of probability density in each class will be calculated through (1), applied on a set of centroids fixed by (3), the width of the kernels being fixed by (6). Bayesian classification is then realized by using the Bayes law where the probability densities are replaced by their estimates $\hat{P}(\omega_i)$ and $\hat{p}(M_i, u|\omega_i)$ (7).

3 Empirical results and discussion

3.1 Vector quantization effect on the codebook distribution

The first main hypothesis of the method we use to build the suboptimal Bayesian classifier is that the vector quantization process will lead to a distribution of centroids $c(m)$ in B_{M_i} similar to this of the training set A_{N_i} for each class. This hypothesis would be well verified if $n(m)$, the number of points belonging to $C(m)$ (the cluster associated to centroid $c(m)$ in the Voronoi tessellation obtained after the vector quantization) would approximately be constant for each cluster.

Several experiments on artificial and real distributions showed us that this hypothesis is verified for large codebook sizes, but if we desire to drastically reduce the complexity of the estimator, the codebook size must be sufficiently small. In this case $n(m)$ can be locally approximated by a constant (over a few consecutive clusters), but will globally depend on the clusters position in the initial distribution. So, in order to keep the best approximation of the probability density function in each class, the estimator proposed in [6] will provide better results, and the equation of the kernel estimator based on the reduced design set B_{M_i}

$$\hat{p}(M_i, u|\omega_i) = \frac{1}{M_i} \sum_{m=1}^{M_i} K\left(\frac{u - c(m)}{h(m)}\right) \tag{7}$$

has to be replaced by:

$$\hat{p}(M_i, u|\omega_i) = \frac{1}{N_i} \sum_{m=1}^{M_i} n(m) K\left(\frac{u - c(m)}{h(m)}\right) \tag{8}$$

To illustrate this, we used the reduced estimator on a two-dimensional Gaussian mixture distribution with three modes containing 2500 training patterns $p(x) = p_1(x)/2 + p_2(x)/2 + p_3(x)$, where $p_1(x)$ and $p_2(x)$ are radial Gaussian functions of standard deviation $\sigma_x = \sigma_y = 0.2$ and of

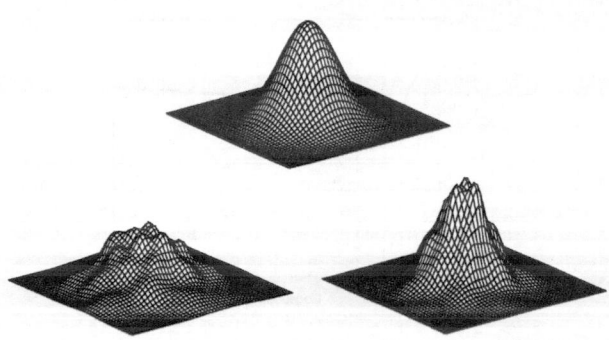

Figure 2: Estimation of a simple Gaussian distribution (top) with a reduced kernel estimator of 25 centroids using the number of points per cluster (bottom right) or not (bottom left).

respective mean $(0,0)$ and $(0,2)$ while $p_3(x)$ is centered on $(2,1)$ and has a diagonal covariance matrix with $\sigma_x = 0.2$ and $\sigma_y = 1$.

The estimator was built with a codebook size varying from 5 to 200; the CL learning consisted of 10 presentations of the 2500 training patterns with a α adaptation factor linearly decreasing from 0.3 to 0.001. Figure 1 shows the evolution of the mean error on the probability density function (pdf) estimation (the square root of the mean square error computed over a 50x50 grid covering more than 99.9% of the distribution) for estimators built with (7) and (8) using width factors provided by (6).

This is also illustrated in figure 2 where a simple Gaussian distribution is approximated with a codebook of 25 centroids using the number of points per cluster $n(m)$ or not.

The vector quantization process leading to values of $n(m)$ which are "locally constant", the hypothesis used to obtain the "optimal" value of the $h(m)$ width factor (equation 6) is still verified, even if the values of $n(m)$ are not "globally constant". But, as we will see in the following, the actual optimal value of $h(m)$ will also depend on the data space dimension and on the codebook size.

3.2   The optimal width factor

As said in section 2.2 the hypothesis leading to the "optimal" value of the $h(m)$ width factor (6) is that the number of centroids is sufficiently large so that the CL learning leads to clusters small enough in order to allow to approximate the true probability density inside each cluster by a constant.

On the other hand, as the codebook size decreases, the vector quantization will lead to larger clusters which do no more have the above mentioned property of being "small"; we can thus guess that (6) will be no more valid and that the optimal width factor $h(m)$ will decrease. In fact, if the codebook size exactly corresponds to the number of modes in the learning distribution the optimal value of $h(m)$ will corresponds to the maximum likelihood estimate of the standard deviation of an isotropic Gaussian centered on centroid $c(m)$ and modeling the mode of the distribution centered on $c(m)$ [6, 9]. This minimum value of the optimal $h(m)$ is linked to the averaged inertia coefficient $i(m)$ by:

$$h(m)_{min}^2 = \hat{\sigma}^2 = \frac{i(m)}{d} \tag{9}$$

So, depending on the codebook size, the width factor providing the best approximation will be

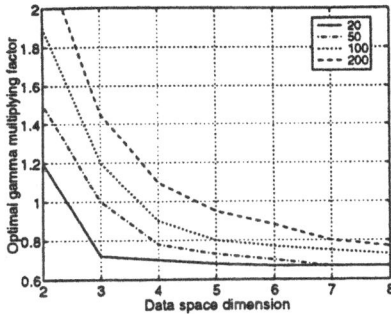

Figure 3: Optimal γ multiplying factor in function of the database dimension with 20, 50, 100 and 200 centroids.

$$h(m)_{opt} = \gamma \sqrt{\frac{3}{2\ln 2} \frac{i(m)}{d}} \qquad (10)$$

where γ is a multiplying factor depending on the codebook size, on the number of modes in the initial distribution and on the data space dimension (γ being egal to 0.6798 when $h(m)_{opt}$ egal $h(m)_{min}$ and to 1.0 when the codebook size become sufficient).

To illustrate this, we tested the suboptimal Gaussian classifier on a set of seven databases corresponding to the same problem, but with dimensionality ranging from 2 to 8. For these databases, class 0 is represented by a multivariate normal distribution with zero mean and standard deviation 1 in all dimensions, and class 1 by a normal distribution with zero mean and standard deviation 2 in all dimensions. There are 5000 patterns, 2500 in each class. In order to test only the influence of an increase of dimension, the databases were generated in the same way for of them. The vectors presentation order is thus the same and for a given vector, all the shared attributes in the 7 databases are the same [1].

For the different dimensions, the γ multiplying factor corresponding to the minimum misclassification error was computed for a codebook size of 20, 50, 100 and 200 centroids (Averaged Holdout test over five partitions of the database in a learnset and a testset of the same size). The results are reported in figure 3. The importance of the data space dimension on the value of the γ parameter corresponding to the optimal width factor is well illustrated on this figure. This phenomenon may be explained as follows: for a given codebook size, when the dimension increases, the coverage of the initial distribution by the codebook will be worst and the optimal width factor will tend to reach the value corresponding to the maximum likelihood estimate of the standard deviation (equation 9). This is due to the "empty space phonomenon" problem appearing for kernel estimators built on finite datasets in large dimension [12].

3.3 A real-world problem

Tests have been carried out on a real-world classification database used in the European ROARS ESPRIT project [13]: "phoneme". Its aim is to distinguish between the classes of nasal and oral vowels. The database contains 5404 vowels coming from isolated syllables (for example: *pa, ta, pan,...*). Five different attributes characterize each vowel: the amplitudes of the five first harmonics, normalised by the total energy (integrated on all frequencies).

Simulations consisted in measuring the performances of the suboptimal Bayesian classifier (8) built with a total number of 20, 50, 100 or 200 clusters (for all classes together). The reported error

---

[1]Similar databases were already used by Kohonen in [11]

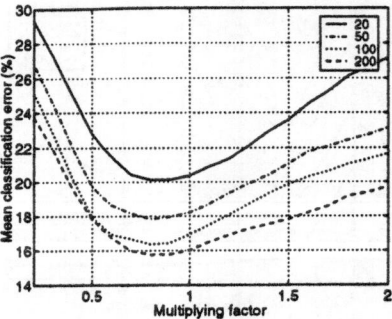

Figure 4: Mean classification error on the "phoneme" database with 20, 50, 100 and 200 centroids.

percentages were obtained by a Averaged Holdout test method over five different partitions in a learnset and a testset of equivalent size (2702 patterns) and the Competitive Learning consisted of 10 presentations of the 2702 training patterns with the α learning factor linearly decreasing from 0.3 to 0.001. The errors were computed for a γ multiplying factor varying from 0.2 to 2; value 0.67 corresponds to the maximum likelihood estimate (9) and 1.0 to (6).

Figure 4 clearly shows a minimum in the value of the error for a multiplying factor $\gamma \simeq 0.8$. This value of γ may be compared to the optimal values reported for gamma in figure 3 (0.68 to 0.95). The small differences may be due to the differences of the distributions in the two cases (number of modes,...)

It is important to mention that a large number of simulations carried out on other databases showed similar qualitative results.

4 Conclusion

The use of kernel estimators with reduced design sets provided by vector quantization techniques enables to approach the Bayesian classification solution with a minimum amount of computations.

While the vector quantization process is deemed to have converged to centroids having the same distribution as the initial points, experiments showed that this process leads to clusters including different number of points. The solution we use to increase the quality of approximation is to take into account the number of points associated to each cluster.

Another problem is the evaluation of the appropriate optimal widths factors for the kernels used in the estimation of probability densities; in this paper, we proposed the use of a γ multiplying factor which could take into account the effects of the data space dimension, of the codebook size and of the particularities of the distributions to approximate. With the hypothesis of small clusters, verified with large codebooks γ tends to 1, which can be seen as an experimental check of the hypothesis used in [2, 8]. When the codebook size decreases, γ decreases too, verifying the results of [6] valid when the number of classes decreases to reach the number of modes of the distribution. The experiments presented in this paper may thus be seen as an unified way to present the optimal width kernel factors of radial Gaussian kernel estimators, depending on the hypotheses on the size of the clusters and the dimension of the space.

5 Acknowledgments

Part of this work has been funded by the ESPRIT-BRA project 6891, ELENA-Nerves II, supported by the Commission of the European Communities (DG XIII). Michel Verleysen is a Senior

Research Assistant of Belgian National Fund for Scientific Research (FNRS). Philippe Thissen is working towards the Ph.D. degree under an IRSIA (Institut pour l'Encouragement de la Recherche Scientifique dans l'Industrie et l'Agriculture) fellowship. All simulations where run on the Packlib simulator developed at EPFL (Lausanne, Switzerland) in the framework of the ELENA project.

References

[1] M. Verleysen, P. Thissen, J.L. Voz, and J. Madrenas, "An analog processor architecture for neural network classifier", *IEEE Micro*, vol. 14, no. 3, pp. 16–28, June 1994.

[2] J.L. Voz, M. Verleysen, P. Thissen, and J.D. Legat, "Handwritten digit recognition by suboptimal bayesian classifier", in *Neural Networks and their applications 94*, Marseille, December,15-16 1994, IUSPIM.

[3] T. Cacoullos, "Estimation of a multivariate density", *Annals of Inst. Stat. Math.*, vol. 18, pp. 178–189, 1966.

[4] K. Fukunaga and R.R. Hayes, "The reduced Parzen classifier", *IEEE Transactions on Pattern Analysis and Machine Intelligence*, vol. 11, no. 4, pp. 423–425, Apr. 1989.

[5] Q. Xie, C. A. Laszlo, and R. K. Ward, "Vector quantization technique for nonparametric classifier design", *IEEE Transactions on Pattern Analysis and Machine Intelligence*, vol. 15, no. 12, pp. 1326–1330, december 1993.

[6] P. Comon, G. Bienvenu, and T. Lefebvre, "Supervised design of optimal receivers", in *NATO Advanced Study Institute on Acoustic Signal Processing and Ocean Exploration*, Madeira, Portugal, July 26-Aug. 7 1992.

[7] P. Burrascano, "Learning vector quantization for the probabilistic neural network", *IEEE Transactions on Neural Networks*, vol. 2, no. 4, pp. 458–461, July 1991.

[8] J.L. Voz, M. Verleysen, P. Thissen, and J.D. Legat, "Suboptimal bayesian classification by vector quantization with small clusters", in *ESANN95-European Symposium on Artificial Neural Networks*, M. Verleysen, Ed., Brussels, Belgium, April 1995, D facto publications, Submitted.

[9] P. Comon, "Supervised classification: a probabilistic approach", in *ESANN95-European Symposium on Artificial Neural Networks*, M. Verleysen, Ed., Brussels, Belgium, April 1995, D facto publications.

[10] Y. Linde, A. Buzo, and R.M. Gray, "An algorithm for vector quantizer design", *IEEE Transactions on Communications*, vol. 28, pp. 84–95, January 1980.

[11] T. Kohonen, G. Barna, and R. Chrisley, "Statistical pattern recognition with neural networks: Benchmarking studies", in *IEEE Int. Conf. on Neural Networks*, San Diego, CA, 1988, vol. 1, SOS Printing.

[12] P. Comon, J.L. Voz, and M. Verleysen, "Estimation of performance bounds in supervised classification", in *ESANN94-European Symposium on Artificial Neural Networks*, M. Verleysen, Ed., Brussels, Belgium, April 1994, pp. 37–42, D facto publications.

[13] P. Alinat, "Periodic Progress Report 4", Tech. Rep., ROARS Project ESPRIT II- Number 5516, February 1993, Thomson report TS. ASM 93/S/EGS/NC/079.

Schema Based Learning
&
Learning to Detour

Fernando J. Corbacho & Michael A. Arbib

Center for Neural Engineering, U. S. C.

Abstract

Corbacho & Lee (1993) used a schema-based model of Anuran Detour Behavior to show how interacting schemas may analyze a complex environment to generate an appropriate course of behavior. However, having explicitly constructed the model on the basis of neuroethological data, we now turn to the question: how are such schema assemblages acquired. In particular, we address the issue of learning to detour to provide a case study in the evolution of adaptive behaviors. We consider the main sensory, motor and adaptive components of a prototype model for anuran detour behavior. To place the detour analysis in the more general context we introduce the more general framework of Schema Based Learning (SBL).

1. Introduction

SBL is a new learning paradigm whose strength is based on the following three intertwined aspects:
i) Cope with new experiences based on old experiences. The agent comes with a variety of schemas to perform certain tasks. These schemas can be reconfigured to solve novel tasks by incrementally growing more refined structures by interacting with the environment.
ii) Use of gross primitive *a priori* knowledge/principles to perform a broad-brush analysis of the *scene* -e.g. prey detection, rude grasping in babies. They point out what is relevant thus reducing the estimation space and allowing for the bootstrapping of the system -learning in the right space.
iii) Construction/prediction based system. The system is able to construct aspects of the future based on its past experiences. Organisms become more and more predictive to afford for better survivability.

In a typical environment the agent is not given explicit specifications of the task to be accomplished -e.g. detour around a fence, but drives and the capabilities to interact with the environment. In our particular case study the goal is the construction of the Detour schemas based on more primitives schemas through self-organization. Schema-based learning will combine currently available schemas to construct new schemas which then may be tuned by experience into a well-adapted integrated unity. Without this, given a new task, the system would start *ab initio* to train a new unstructured network dedicated to the task.
The SBL architecture includes mechanisms for modulating, adapting, and constructing -aggregate, encapsulate-schemas for encoding knowledge about experiences. Schemas start as gross approximations of what are relevant patterns of action-perception and thus reduce the combinatorial space a lot (initial islands of reliability) so that the system learns with minimum *scene* statistics.

2. Learning to Detour

To test SBL we have constructed a simulated environment and a simulated *agent*. The agent is an evolving model of anuran visuomotor coordination (c.f. *Rana computatrix*; Arbib, 1987; Corbacho & Arbib, 1993). The simulated environment contains several entities placed on a grid. These different entities as well as the agent are allowed to move and interact in the environment. The simulated agent consists of many integrated[1] structures arranged from sensorial maps to perceive its environment to motor outputs which can affect the environment.

Ingle (1983) and Collett (1982) have observed that a frog or toad's approach to prey or avoidance of a threat are also determined by the stationary objects in the animal's surround. A frog or toad, viewing a vertical paling fence barrier (see top down view in Fig. 3) through which it can see a worm, may either approach directly and snap at the worm, or detour around the barrier. However, if no worm is present, the animal does not move. Thus, it is the worm that triggers the animal's response but when the barrier is present, the animal's trajectory to the worm changes in a way that reflects the relative spatial configuration of the worm and the barrier.
Arbib and House (1987) modeled this using a "potential field" approach. We have used a schema-based model of Anuran Detour Behavior to show how interacting schemas may analyze a complex environment to generate an appropriate course of behavior (Corbacho & Lee, 1993; Lee, 1994). However, having explicitly constructed the

---

[1] Proper integration of perception, adaptation, and action - *embedded learning* reduces estimation.

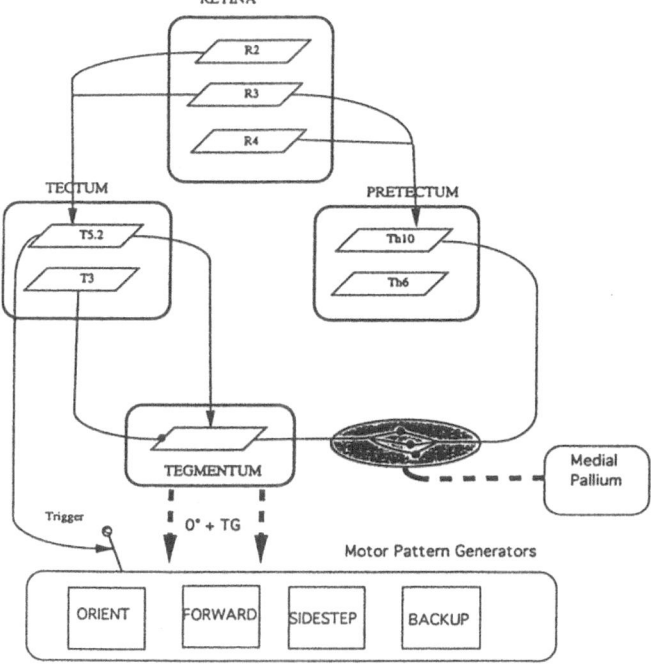

RETINA

Figure 1. Prototype neural architecture for simple detour behavior.

model on the basis of neuroethological data, we now turn to the question: how are such schema assemblages acquired. In particular, we address the issue of learning to detour to provide a case study in the evolution of adaptive behaviors.

The main components of this prototype model are (Corbacho & Arbib, 1993): Retina as the front-end visual perception system. Tectum as the subsystem for prey recognition. Pretectum for stationary object recognition (barrier recognition). Tegmentum as the place of integration of the sensory maps. It receives the signals corresponding to the prey and barrier internal pattern representations, as well as projection from the tactile map. The tegmentum contains the representation of the "heading" map, that is, body angle w.r.t. target position (see Fig. 2 for patterns of activity in tegmentum). Modulation of the signals from the perceptual maps (tectal, pretectal, tactile) by a plastic structure (e.g. telencephalic: medial pallium). Lastly, *winner take all* dynamics (more in section 3.2) over the tegmental array activity will decide which of the available motor synergies will be activated.

Pure reflexive navigation may include bumping avoidance, gap seeking, and so on. In these systems the agent produces a detour behavior by simply avoiding visible obstacles and going for the passable visible gaps. More complex detour is a step beyond reflexive detour, since there might not be visible passable gaps within the visual field of the agent, that is, no visible "passage" through which navigate towards the goal. Nevertheless, a passable gap does exist in the environment -namely the edge of the barrier-. Thus, there is a solution to the problem based on another strategy, but the animal will have to construct it. Bumping avoidance might be far from efficient in this case -e.g. following the perimeter of a wall versus making a diagonal trajectory towards the end of the wall. In some animals, amphibia in particular, more efficient detour behavior seems to be learned not innate. One of the purposes of this paper is to show how this behavior can be learned, namely how are the detour schema -e.g. sidestep detour schema- constructed. We will also show how systems have evolved to be more and more *predictive*, since this affords for better survival of the species, as they become more efficient.

3. Schema Based Learning Underpinnings

3.1. Schemas

We will extend Piaget's definition of schema as a unit of behavior and knowledge which interacts and evolves with its physical environment, and with other schemas. A *perceptual schema* (Arbib, 1992) embodies the process that allows the organism to recognize a given domain of interaction. Various schema parameters represent properties such as size, location, and motion. A *schema assemblage* (an assemblage of perceptual schema instances) provides an estimate of environmental state with a representation of goals and needs. New sensory input as well as internal processes update the schema assemblage. The internal state is also updated by knowledge of the state of execution of current plans made up of *motor schemas*, which are akin to control systems but distinguished in that they can be combined to form *coordinated control programs* which control the phasing in and out of patterns of coactivation, with mechanisms for the passing of control parameters from perceptual to motor schemas.

Let us now define another kind of schema. A *relational schema* is a unit of perception, action and adaptation since it relates percepts, and actions in an adaptive manner. Their structure is similar to that described by Becker (1973). Relational schemas have a context, an action and a (expected/predicted) result. For instance the sidestep detour schema consist of a context which is the perception of gaps -but none of them passable, an action which corresponds to the sidestep motor schema, and an expected result which is: passable gap in the animals visual field.[2]

Our relational schemas are *prediction-value* versus *situation-action:* the system does not directly learn what action to take in a given situation, but rather learns what would happen next for each possible action. It may then select what action to take based in part on the value of an achievable result. The animal may face an almost infinite combination of situations due to the complexity of the environment. Nevertheless there is a much lesser number of potential goals that the animal should achieve. This greatly reduces the *search* space since the action selection will be based on the expected results, and the context/situation will simply help trimming the space a bit more. Predictive: not just where you are but where you want to be.

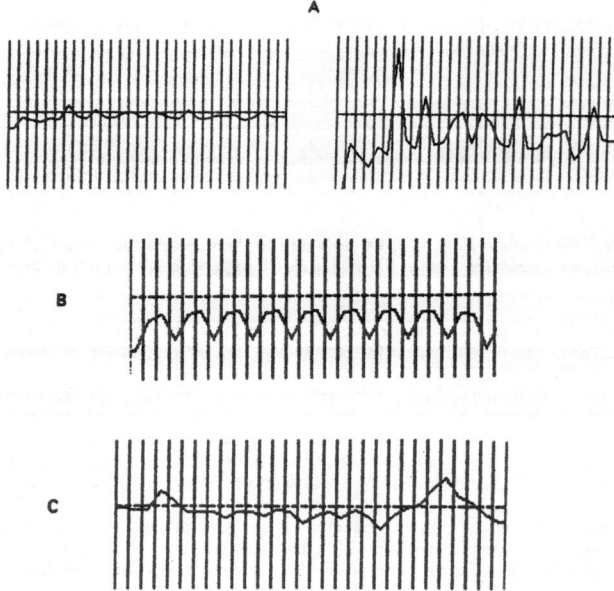

Figure 2: Patterns of activity in the tegmental map. The figures represent a horizontal cross section over the 2D tegmental array. A) Winner take all dynamics over tegmental activity: reset time on the left; after 10 timesteps on the right. B) Impasse on tegmentum. C) Peak of activity in *tegmentum* constructed by the sidestep detour schema.

---

[2] The sidestep motor schema corresponds to a synergy of muscle activations. The detour sidestep schema is a relational schema which has as its action the sidestep motor schema.

3.2. Schema Dynamics

First let us emphasize that the dynamics of schemas are not controlled by any centralized executive control but it is a distributed controlled system where stable solutions emerge from interactions of many local variations in a selforganizing way (Malsburg, 1993). Compare explicit executive control structures such as the ones used in traditional AI systems with the *competition and cooperation* dynamics used here and other brain models (Amari & Arbib, 1977), which provide the basis for selforganization. Cobas and Arbib (1992) postulated the construction of motor activity through the interaction of different motor schemas via a process of *competition and cooperation* wherein there is no need for a unique schema to win the competition (although that might well be the case) since several schemas may simultaneously be active and interact to yield a more complicated motor pattern. They focused mainly on the interaction in the heading map, SBL will extend and formalize the theory to many more schemas integrated in many more maps. This brings much more complex behavior/dynamics since the interactions are far richer.

The interactions between two schemas are governed by a set of interconnection variables.

$$H = \{ W_{ij} \mid i,j \in N \}$$

where i refers to the incoming schema element and j to the outcomming one. The connection set may be defined explicitly -enumeration of the elements- or implicitly -function to generate the elements-. The implicit definition allows for cheaper system adaptation by tuning the parameters of the function. The elements of the set may also have independent or dependent update.[3] In this paper we use an example of the implicit, dependent update connectivity. Difference of Gaussians (DOG) interconnection pattern where system adaptation is produced by tuning the amplitude parameter of the positive gaussian thus causing the update all of all single interconnection variables in a dependent manner.

Our system is composed of a finite number of schemas:

$$S = \left\{ S^1, S^2, ..., S^n \right\}$$

The temporal evolution of schemas is governed by a system of differential equations of the following form:

$$\tau^i \frac{d s^i(t)}{d t} = -\alpha^i s^i(t) + \sum_j S^i(t) \times W^{i,j} + \rho^i(t) \qquad \text{lumped model}$$

$$\tau^i_{x,y} \frac{d s^i_{x,y}(t)}{d t} = -\alpha^i s^i_{x,y}(t) + \sum_j \sum_{x,y} S^i_{x,y}(t) \times W^{i,j}_{x,y} + \rho^i(t) \qquad \text{distributed model}$$

where $s^i_{x,y}(t)$ is the activity variable for schema i, and $S^i_{x,y}(t) = \sigma(s^i_{x,y}(t))$ is the effect of this schema on other schemas, σ is one of the popular nonlinear threshold functions, and $\rho^i(t)$ is white noise.

3.3. Seed Schemas

Evolution must provide a basic repertoire of survival mechanisms, and initial guidelines for finding what is salient. That is, general - coarse but flexible - innate mechanisms: primitive broad schemas which are rough approximations/vague descriptions of aspects of the world as well as "organizational" mechanisms. All these will give raise to more refined behaviors/structures through the agent's "own" experience.

For the Detour problem the following constitute some of the possible (non-exclusive) schemas. Some of them can be defined as primitives and some others may actually be built by the individual. It is a matter of design analogous to the evolution/development design dichotomy.

---

[3] An example of an algorithm with independent update is classical backpropagation.

Perceptual Schemas to sense and interpret the environment:
Recognition of moving objects: Prey and Predator recognition schemas. Recognition of stationary objects. Tactile/pain Perception implemented by a primitive "tactile" system.

Motor Schemas allow that agent to interact in/with the environment.
Approach as well as Avoidance related motor patterns. These are decomposed into more elementary schemas -synergies of muscles: forward, orient, snap, back up, sidestep, freeze, tilt, ... which in turn are decomposed in muscle patterns. These schemas activate corresponding pools of motor neurons in the spinal cord that actually carry out the action. It is not an exhaustive list and also they may internally intersect in the sense that they use overlapping sets of resources.

Relational schemas e.g. Bumping avoidance schema to avoid the damage of the organism. Prey Acquisition: *Approach food* "implements" the evolutionary principle that food "affords" survival. This schema is triggered by the presence in the sensorial field of stimuli of certain characteristics. Predator Avoidance: implements the evolutionary principle that avoiding predators affords survivability versus destruction.

4. Learning Schemas

During the first trials the naive animal will bump into the barrier (see Fig. 2A describing the evolution of the heading map as it produces a target position within a non passable gap). This causes the bumping avoidance schema to trigger the adaptation of the pretectal projection over tegmentum. After the adaptation non-passable gaps will not be able to *fire* activity in tegmentum -animal has learned that some gaps are not passable. Thus, an *impasse* (Fig. 2B) is produced on the tegmental array (array activity can not reach the tegmental output threshold). Since there is no output activity in tegmentum no definitive motor action can be taken. The sidestep motor action will then eventually be tried -due to the *competition and cooperation* among different schemas. This schema projects activity to tegmentum (Fig 2C) as if there was a passable gap at that retinotopic position. At this point the influence of the gaps on the heading map is very low, thus, the prey location as well as the bump and sidestep locations become the driving forces in the tegmental array.

Sidestep detour schema constructs pattern in Heading map (tegmentum) for "ghost" gap. The system is predicting (thus constructing) the existence of a passable gap upon completion of the sidestep action. Thus the schema triggers the action as well as the construction of the "ghost" gap activity in the heading map.

    context: gaps, fences
    action: motor sidestep schema
    prediction/result: passable gap on visual field

Model prediction: Tegmental cells firing even if no "passable" gap is on view (Fig. 2C). That is, the brain is constructing a "ghost" gap. This is a reflection of the reality in the environment, namely, there was open gaps in the boundaries outside of the initial visual field of the animal.

How is the schema constructed in the first place?
The system initially does not know the results of some of its actions before it takes them. Nevertheless, as the system takes an action it already constructs the (expected/predicted) results of it; so that upon termination of the action the system can compare the actual result with its predicted result and calculate the *error*. The system must become increasingly better in predicting the results of its actions. Learned schemas may not be perfect but they are the most reliable given the past experiences. Given the same context the same action should produce a similar result.

When the schema is activated for the first time the activity in the result (slot) array may be null -no prediction has been built. Then the system takes the action and *observes* the result. Upon completion of the sidestep motor action the heading map has got out from its previous *impasse*, since now there is a clear passable gap on sight. This is a very desirable partial goal. So the motor synergy is linked to the context (impasse on heading map) and to the result (projection of activity from sidestep schema to heading map). When the animal is again in the same situation it will try again this schema. On the long run, the reliability of the prediction for this action upon this context increases (is reinforced).

5. Results

In this short paper we will only include a small sample of some of the observed behaviors which we think elucidate some of the main capabilities of anurans for the detour behavior, as well as the learning capabilities of the model.

Experiment I: Barrier 10 cm wide, 20 cm away from the frog.
Case 1) For small barriers such that the boundaries of the barrier are clearly perceived, the frog innate gap detection mechanism suffices. This is a reflex towards the "most-passable" gap. A winner take all mechanism over the gaps -with a bias for worm location- produces perfect detour behavior towards the boundaries of the barrier.
Thus, in the model the behaviors of the animal have been replicated by using winner take all dynamics over the
       displays model results. A) Naive animal: produces several small sidesteps. B) Partially trained animal: produces
       fewer wider sidesteps. C) Trained animal: produces a *perfect* sidestep.
tegmental array activity. Also in the model the visual receptive field has a 50 deg area within which stimuli are perceived with great clarity and beyond which the stimuli are perceived with increasingly lower clarity.

Experiment II: Barrier 20 cm wide, 20 cm away from the frog.
Case 1) The barrier boundaries are not clearly perceived as they fall beyond the 50 visual degrees border line. Thus, this is a harder problem since there is no clear feature in the environment to hint as a possible target for the generation of a detour behavior. The animal eventually tries out the sidestep motor pattern and gets closer to the goal - two consecutive sidesteps may let the animal see the edge of the barrier. From that on, the animal constructs a detour sidestep schema with wider and wider angles (Fig. 3) so as to achieve efficient detour by sidestepping all the way close to the edge of the barrier.
Case 2) If we introduce an open wide gap in the middle of the barrier the frog goes for that gap. Again this is due to the winner take all dynamics over the tegmental array representation.

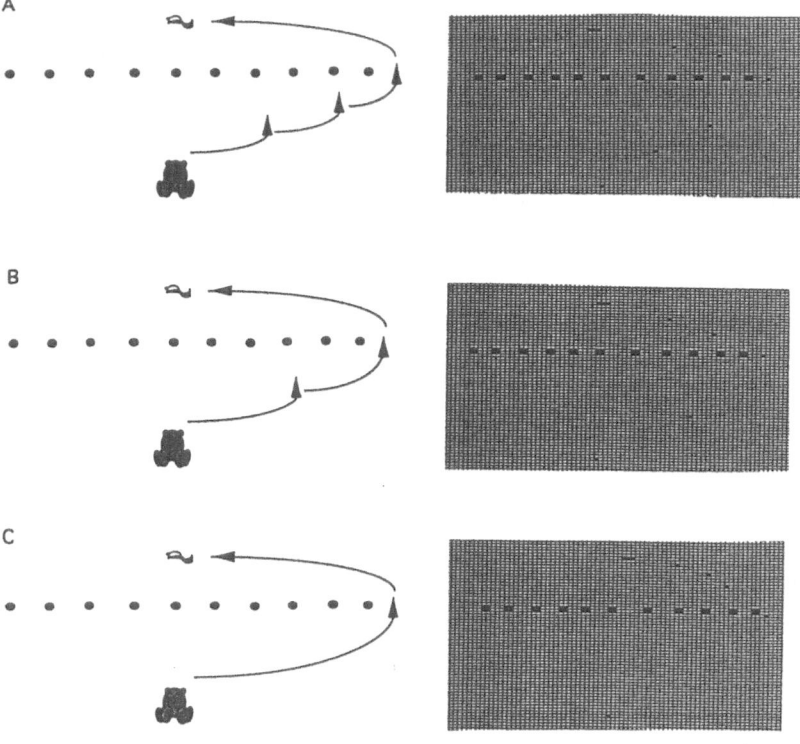

Figure 3: Evolution of the Sidestep Detour Schema. This figure shows the improvement in performance as the agent gets increasingly exposed to the task. Left column displays experimental results. Right column

Selected References

Amari, S-I. & Arbib, M. A. (1977). *Competition and Cooperation in Neural Nets*. In: Systems Neuroscience (ed.) pp. 119-165. New York: Academic Press.

Arbib, M. A. (1987). Levels of modeling of visually guided behavior. *Behav. Brain Sci.* 10, 407-465.

Arbib, M. A. (1992). Schema Theory. In: *The Encyclopedia of Artificial Intelligence*. (Shapiro, S., ed.) 2nd Ed pp. 1427-1443. New York, NY: Wiley Interscience.

Arbib, M. A. & House, D. H. (1987). Depth and detours: an essay on visually guided behavior. In:*Vision, Brain and Cooperative Computation* (Arbib, M. A. & Hanson, A. R. eds) pp. 129-163. A. Bradford Book/The MIT Press.

Becker, J. (1973). A model for the Encoding of Experiential Information. *Computers Models of Thought and Language*, eds. Shanck, R. and Colby, K. pp 396-434. San Francisco: Freeman.

Cobas, A. & Arbib, M. A. (1992). Prey-catching and Predator-Avoidance in Frog and Toad: Defining the Schemas. *J. Theor. Biol.*

Collett, T. (1982). Do toads plan routes? A study of detour behavior of B.viridis. *J. Comp. Physiol.* A, 146. 261-271.

Corbacho, F. & Arbib, M. A. (1993). Integrated Learning in Rana Computatrix. *Proceedings of IWANN 93*. Lecture Notes in Computer Science 686. Springer Verlag: Berlin.

Corbacho, F. & Lee, H. B. (1993). Schema Based Learning and Anuran Detour Behavior. *Proceedings of the Workshop on: Neural Architectures and Distributed AI: From Schema Assemblages to Neural Networks*. Los Angeles.

Grobstein, P. (1988) Between the retinotectal projection and directed movement: Topography of sensorimotor interface. *Brain Behav. Evol.* 31, 34-48.

Ingle, D. (1983). Brain mechanisms of visual localization by frogs and toads. (*Advances in Vertebrate Neuroethology*, J. -P. Ewert, R.R.Caprinica and D.J.Ingle, Eds), 177-226.

Lee, H. B. (1994). *A Neural Network and Schematic modeling of anuran visuomotor coordination in Detour Behavior*. Ph. D. Thesis. University of Southern California.

Malsburg, C.v.d. (1981). *The correlation theory of brain function*. Internal report, 81-2, Max-Planck-Institut für Biophysikalische Chemie, Postfach 2841, 3400 Göttingen, FRG.

Neurons with Continuous Varying Activation in Self-Organizing Maps

Josef Göppert, Wolfgang Rosenstiel

Lehrstuhl für Technische Informatik, Universität Tübingen
Sand 13, 72076 Tübingen, Germany

Abstract

A new training and recall method for self-organizing maps (SOM) is developed by comparison of SOM to the human information processing system. As neurons and cortical columns do not change their activity instantly, it is increased or decreased in a smooth way. This fact is introduced in SOM-neurons. In a same way, recognition of objects is supposed to be a task of analysing complete sets of feature vectors and finding the region in the SOM which represents the current inputs best. This method especially allows the evaluation of ambiguous feature vectors and of objects which are decomposed in sets of basic feature vectors or which are aquired in a continuous temporal flow.

1 Introduction

The SOM takes into account cooperative aspects of neighbouring neurons by using individual activation for all neurons. The activation reflects the distance of one unit to the winner in the grid of neurons and is used for the calculation of the adaptation strength. So, similar adaptation of neighbouring units leads to similar features and in a global view of the map, to a topology preserving mapping of the training vectors. From a biological point of view the activation of a neuron raises its attention to the next stimulus. But how to introduce this notion into the SOM-model? In artificial neural networks this type of cognitive supervision is not possible, but further information of training data may be known, which can be used in the same way. Such information may be the class of input pattern, or the circumstances of the acquisition of the pattern.

2 An extention of the distance

In SOM the similarity between the sensory input $\mathbf{X}$ and stored item $\mathbf{W}_i$ of neuron i is defined by the euclidean distance (Sensory distance D_i^S):

$$D_i^S = \sqrt{\sum_j (x_j - w_{ij})^2} \tag{1}$$

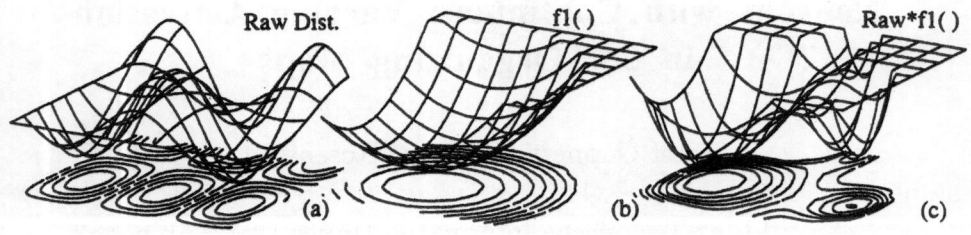

Figure 1: Influence of the attention onto the Distance.
(a) Sensory distance D^S, (b) attentional distance D^E, (c) overall distance D.

A smaller distance represents a bigger similarity and defines the probability of winning. Why shouldn't we use this distance to model a change of attention of a neuron. This can be done by modifying the distance with an expectation value.

The winner position from one step to the next is not to change abruptly which leads to a more continuous variation of the excitation or attention center on the map. This may assign to each neuron i (placed at the position $\mathbf{C}^i$ in the two-dimensional map) an expectation ("attentional distance") D_i^E that influences the sensitivity of these neurons to the input pattern. This expectation value represents the distance of neuron i to the center of attention $\mathbf{C}_A$. The parameter σ allows to ballance the sensory distance and the attentional distance and represents the size of the attention region. This leads a modified distance:

$$D_i^E = 1 + \frac{\sum_{j=0,1}(\mathbf{C}_{Aj} - \mathbf{C}_j^i)^2}{\sigma^2} \tag{2}$$

$$D_i = D_i^S D_i^E = \left(1 + \frac{\sum_{j=0,1}(\mathbf{C}_{Aj} - \mathbf{C}_j^i)^2}{\sigma^2}\right) \sqrt{\sum_j (x_j - w_{ij})^2} \tag{3}$$

Figure 1 visualizes this new distance graphically. Notice that figure 1 a shows a ambiguous configuration with 3 neurons having the same minimal distance. In combination with the attentional distance a unique winner can be found, the one which is closest to the attention center (See figure 1 c).

Two methods can be applied to define the attention center $\mathbf{C}_A$. Both methods are used and described in the following sections:

- Training and recognition can be described as a dynamic process. So, continuous varying processes with a continuous flow of input vectors result in a continuous variation of the activation center over the neuron-grid. Only one activation center is present.

- Training and recognition can be seen as the task of recognition sets of input vectors. Attention acts in two ways. First by focussing sensory attention onto the target, and next by restricting mental activation onto the region that best represents the set of sensory input. Here several class specific activation centers are present.

These ideas are inspired by psychological research, and lead to modifications for artificial neural net training. Comparision of the basic structure of the human cortex has a lot in common with the self-organizing map. For further information see [Cow88, Koh82, Mer84, Rit90].

3 Time continuous flow of data vectors

It is supposed that from a biological point of view, a neuron does not change its activation abruptly and already activated neurons will have a higher sensitivity to new input patterns. Thus active neurons have a higher probability to win and the activation center varies from one step to the next in a more continuous way.

Continuous variation of the attention center also necessitates an ordered presentation of the input vectors. In fact: A random choice of input stimuli is biologically not plausible. Most types of training data have any type of one- or multidimensional ordering (temporal order, spatial position, continuous varying parameters, etc..). Normal SOM do not take advantage of this fact, even if it would help in exploring the data space. Adapted methods of presentation order would be to go back and forth (right and left) in the ordered list (matrix) of training vectors, to follow predefined paths or to perform some kind of random walk.

The excitation center C_A is initialized by the grid position C_{w_0} of the winning neuron w_0 of the first pattern. The center for the following patterns is set to the position of the previous winner:

$$C_{A_1} = C_{w_0} \tag{4}$$

$$C_{A_{l+1}} = C_{w_l} \tag{5}$$

The training principle of the self organizing map stays unchanged [Koh84]. Adaptation of neurons around the winner is defined by the adaptation function. One supplementary parameter is the width σ of the attentional function. Good results have been obtained with big starting values in order to allow global organization of the map and a smooth reduction during the training.

The training patterns have to be presented in roughly ordered sequence. Thus they get trained together with neighbouring vectors in a given context. Due to the modified training, these vectors are also represented by neighbouring neurons, and the contextual order will lead to a *context preserving organization* of the map. As in SOMs similar vectors will be represented by neighbouring units, but only, if they have been trained in a common context. Dynamic training can be seen as an extension from topological feature space into spatio-temporal space of continuous varying stimuli.

Context leads also to new principles for the recognition of ambiguous data. If the same stimulus occurs in several ways and in different contexts, the use of some neighbouring inputs may allow to build up a context sensitive recognition procedure, by finding an activation center which best represents a set of input vectors.

4 Sets and classes of input vectors

Input analysis is the task of finding an activation center on the map that represents best possible the actual "context" by a recursive process of pattern recognition that may help to find the best representation area. In the recognition of a complex object, attention acts in two ways. First by focussing sensory attention onto the target, and next by restricting mental activation onto the region that best represents the set of sensory input. This can be seen as a task of recognizing a whole set of pattern. One single input may be sufficient, but the more pattern are taken into account, the better this object will be recognized. This recursive process of finding the activation center C_A of a set of pattern is initialized by the grid position C_{w_0} of the winning neuron w_0 of the first pattern. The center is adapted iteratively, by applying randomly other patterns of the same set, finding the winning neurons w_l for this step $l \in [1 \ldots L]$ and moving the excitation center towards its grid position C_{w_l}:

$$C_{A_1} = C_{w_0} \tag{6}$$
$$C_{A_{l+1}} = C_{A_l} + \gamma(C_{w_l} - C_{A_l}) = (1 - \gamma)C_{A_l} + \gamma C_{w_l} \tag{7}$$

Herein the mobility of the activation center is defined by the adaptation factor γ. In order to reduce the risk of ending in a local minimum, the mobility of the cluster center and the cluster size are changed during the retrieval. In the first phase of recognition, the best region has to be found; the aim in the second phase is to find the exact position of the activation center. Both stages are attained by continuous change of mobility (Equ 7, λ) and the size of the cluster (Equ: 3, σ). In the beginning the mobility and the cluster size will be chosen big ($\lambda \approx 0.4$; σ >biggest size of the map). During the training phase the activation center should stabilize by reducing the mobility towards 0 and the cluster size towards its final value.

The new recognition method for SOM allows the coherent evaluation of more than one training pattern and the retrieval of the most adapted region on the map for this set. Applying this method to topological maps, ordered by the standard SOM-algorithm, may not lead to the desired result. The reason is that normal training arranges the stored items on the map according to euclidean distances. Any relationships between features and classes are neglected. The next step is to find an adapted SOM-Training method.

It is supposed, that every class has its own center. Class specific centers are adapted according to equation 7, but only if the input vector appertains to the corresponding class (Equ. 2, k). During the self-organizing process, the cluster centers are moved according to the winner position. Starting from random initial position they are forming more and more an organization which reflects the similarities of classes.

5 Results for time continuous data flow

To show the performance of this new training method, some topological difficult Lissajous' figures have been chosen. A circle can be represented by two sine curves with a phase shift of $\frac{\pi}{2}$. Presented to a one dimensional SOM, it will be approximated in a

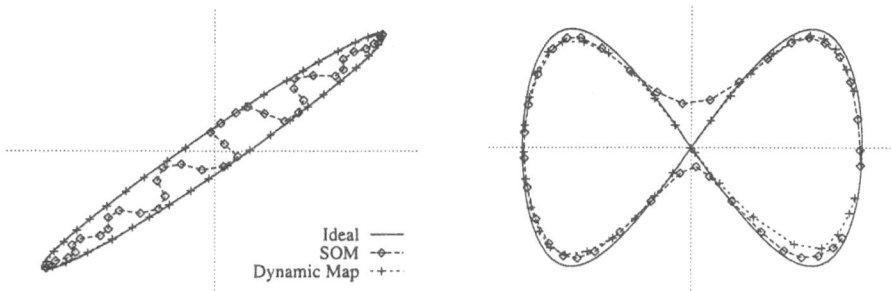

Figure 2: Approximation properties of the dynamic training methode. (a) Clustering properties in an Ellice. (b) Topological vs. contextual order.

Figure 3: Decomposition of the digit "1" into a set of 15 local 5 × 5 sub-images. Center pixels are emphased in grey colour.

good way. If the phase shift is reduced to, say $\frac{\pi}{8}$, the circle is distorted into an ellipse. Trying to approximate this figure by a SOM results either in a stretched map (big neighbourhood) or in a meandered map inside the ellipse (figure 2). Dynamically trained maps lead to a context preserving approximation.

In the second example, two sine waves with different frequencies produces an "∞ like" shape. Here standard SOM produces a quite good approximation of the shape, which changes the direction near the crossing point. This might be correct from a topological point of view, but is a contextual error. A dynamically trained map is able to follow the shape in a contextual correct manner. The crossing point is an ambiguous point, which might also be correctly recognised if presented in its context.

In both examples the training-neighbourhood was not reduced to zero, in order to come to a more discriminative graphical representation. A reduction to zero would push the prototypes closer to their optimal position, but would not change the topological aspects at all.

6   Results for classes of input vectors

A very easy way of feature extraction is to decompose an image in small sub-images. These windows contain local properties like direction and curvature of the line. Other information like the exact position of the sub-image gets lost. In a character recognition task the exact position of the character in the pixel image does not contain important information and loss of position increases the aptitude of the method for position and scale invariant character recognition.

he pixel image of the 10 digits of 12 different screen fonts were combined to a

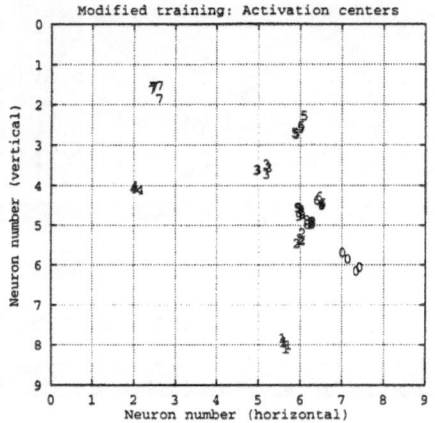

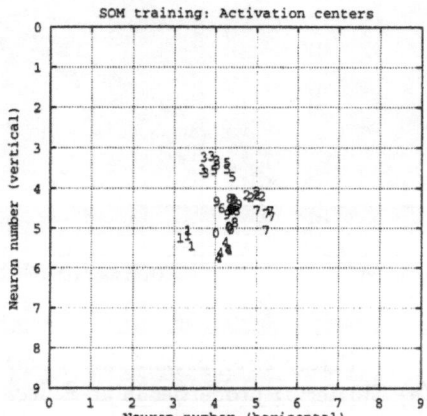

Figure 4: Modified training. Position and stability of the activation centers of different retrieval cycles.

Figure 5: SOM training. Position and stability of the activation centers of different retrieval cycles.

training data set. Height of these digits varies from 9 to 13 pixel and width from 5 to 7 pixel. The style was either normal or italic. These digits are decomposed into all possible 5×5 sub-images with a black coloured pixel in the center. An example of such a decomposition is shown in figure 3.

The sub-images are trained to a self-organizing map according to the algorithm presented. Good results have been obtained with a map with 10×10 neurons, 10000 iterations of training ($\sigma^k = 2$ Grid positions, $\gamma = 0.1$).

One example of a resulting map and a comparision to normal SOM is shown in figure 4. If training was successful, the activation centers of specific digits should converge always to the same position in the grid of the map, allowing the recognition of the digit in this way. The cluster centers of five independent retrieval cycles are indicated in figure 4. For each of these retrieval processes, 800 feature vectors out of the 2277 training vectors are presented at random, while the activation center mobility (γ) decreases ($0.4 \ldots 0$) and the excitation width σ^k decreases from 20 to 1 grid positions.

After 800 steps, the mobility was reduced to zero and the activation center was stabilized and compared. The position of the classes seem to be particular stable. Most classes can be distinguished by the grid position, but some classes are situated in a small neighbourhood. Especially the digits "8" and "9" are placed quite near to the digits "2" and "6". This is not very surprising, because these four digits share a large set of common features. Observing the mutual distances of class centers on the map, it can be constated that the distance is reflecting the similarity of the features and the size of the subset of common feature vectors.

A comparision to the standard SOM (Figure 5) shows big differences. The features of one class are distributed over the whole map. The clusters have big sizes, overlap in wide ranges and stabilize near the middle of the map.

Statistical comparison of the two training methods confirms the hypothesis of different organization. The standard SOM training and modified SOM training were performed 10 times. For each training 15 retrieval cycles were calculated. Five of these evaluation cycles are middled in order to find the reference class centers. The activation centers of the other 10 cycles are used for the evaluation. The set of windows from one input pattern was assigned to the class of the nearest reference center. This procedure was performed for trained digits, as well as for the digits of a new font, which was not contained in the training set:

	Correct recognition	
	Std. SOM training	Mod. SOM training
Training set	75.9%	95.2%
New font	48%	81.5%

The mean distance of the class centers on the map can be seen as an indicator for the distinctivity of classes. In the modified training, the mean distance between centers is 2.6 grid positions, while standard SOM place the activation center mainly in the center of the map in a small proximity of 1.04 grid positions. This is the main reason, why the evaluation result of SOM-training is worse than the result of modified training.

Notice that the result in terms of correct recognition is still not good enough. It must be constated that digit recognition, using the presented type of pre-processing is an especially hard evaluation problem. More evaluated pre-processing like replacement of the sub-windows by more evaluated feature vectors may increase the recognition result considerably. The new aspect is the coherent evaluation of sets of input-vectors which are extracted from one digit. A very important property is that the number of vectors may vary and that the ordering of these vectors does not matter. This is not the case if all features are composed to one single feature vector.

7 Conclusion

A new training algorithm for self-organizing maps was presented. Inspired by natural neurons and cortical columns an aspect of attention was introduced which produced "Dynamic behaviour". Properties were introduced which modify the topology preservation towards a context preserving training of the map. Some simple examples show the basic difference to standard SOM training. We expect that these properties allow a better representation of context by neural maps and show higher performance in the evaluation of ambiguous or noisy data. The algorithm is adapted to problems of ordered data (e.g. time series: an ellipse is a sampled sine curve with a window size of 2) and may provide a new contextual storage and recognition principle.

Another interesting aspect in neural networks and especially in self-organizing maps is the possibility to evaluate complete sets and varying numbers of training vectors. The idea is that specific regions are representing the various facets of a complex object. This behaviour was achieved by introducing an aspect of selective attention through class-specific activation centers. It is supposed that such extention may raise the aptitude of SOM for the evalutaion of difficult data sets like described in [Bog94, Göp92]. They are moving dynamically on the map in order to find the best adapted region.

The presented work is in a very early stage of research. Nevertheless it seems to be promising for several kinds of data evaluation, especially for the evaluation of problems with a varying number of features. Further research will cope with theoretical aspects of this new kind of training and empirical investigations of the influence of training and map parameters onto the self-organization and recognition results.

References

[Bog94] M. Bogdan and W. Rosenstiel. Artificial neural nets for peripherical nervous system - remoted limb prostesis. In *Proceedings of Neural Networks and their applications*, pages 193 – 202, Marseilles, France, 12 1994.

[Cow88] N. Cowan. Evolving conceptions of memory stage, selective attention, and their mutual constraints within the human information-processing system. *Psychological Bulletin*, 104(2):163–191, 1988.

[Göp92] J. Göppert, H. Speckmann, W. Rosenstiel, W. Kessler, G. Kraus, and G. Gauglitz. Evaluation of spectra in chemistry and physics with Kohonen's self-organizing feature map. In *Proceedings of Fith International Conference Neuro-Nimes 92*, pages 405–416, Nanterre France, 10 1992.

[Koh82] T. Kohonen. Self-organized formation of topology correct feature maps. *Biological Cybernetics*, 43:59–69, 1982.

[Koh84] T. Kohonen. *Self-Organization and Associative Memory*. Springer Verlag Heidelberg New York Tokyo, 1984.

[Mer84] M. M. Merzenich, R. J. Nelson, M. P. Stryker, M. S. Cynader, A. Schoppmann, and J. M. Zook. Somatosensory cortical map changes following digit amputation in adult monkeys. *J. Comp. Neurol.*, 224:591 – 605, 1984.

[Rit90] H. Ritter, T. Martinez, and K. Schulten. *Neuronale Netze: Eine Einführung in die Neuroinformatik Selbstorganisierender Netzwerke* . Addison Wesley, 1990.

Improving Back-Propagation: Epsilon-Back-Propagation

Luis A. Trejo, Carlos Sandoval

Computer Science Department,
ITESM-CEM, Carretera a Lago de Guadalupe KM 3.5, Atizapán de Zaragoza,
C.P. 52926 Edo. de México, México

Abstract. *A modified version of back-propagation learning algorithm is introduced. This new algorithm called epsilon-back-propagation allows a neural network to learn faster or at least as good as back-propagation. Experimental data is given in order to compare both methods.*

1 Introduction

Back-propagation is one of the most successful learning algorithms for neural networks . However it has some drawbacks. The main one is that training takes too much time. Our idea is to present a new algorithm that allows to reduce the number of training steps when a network is learning a training set. Since this new algorithm is based on back-propagation we have named it epsilon-back-propagation. The organization of the article is as follows: in section 2 and 3 we review the back-propagation algorithm as well as the momentum term. In the next section we present the proposed algorithm. In section 5 we give some experimental results and finally we give our conclusions.

2 Back-Propagation algorithm

We are presenting here the back-propagation algorithm for a neural network with L layers, n inputs and 1 output, in order to compare with the algorithm we are proposing. You can refer to [3] for the derivation of the back-propagation learning rule. This algorithm is used when training a back-propagation network with the pattern by pattern case. This procedure has two parts: the feedforward part (steps 3 - 4 in the algorithm below) and the back-propagation of the errors and updating of the weights (steps 5 - 7). It is performed until a given approximation value is reached for all patterns, that is whenever the difference between the target output y' and the actual output y is less than or equal to a given ϵ for all patterns.

Back-propagation algorithm

1. Initialize all the weights w_{ij}^l to small random values (typically between -0.1 and +0.1), where w_{ij}^l is the weight in layer l which connects unit i in layer $l-1$ with unit j in layer l.

2. Initialize the activations for the thresholding units. The values of these units will never change.
$$x_0 = 1, h_0^l = 1 \ \text{ for } \ 1 \le l < L.$$

3. Choose a pattern P_k, where $P_k = (p_1, p_2, \ldots, p_n)$ and apply it to the input layer so that:
$$x_i = p_i \ \text{ for } \ 1 \le i \le n.$$

4. Propagate the signal forwards through the network, using for the first layer:
$$h_j^1 = \frac{1}{1+e^{-\text{NET}_j^1}} \ \text{ for } \ 1 \le j \le H_1 \text{ and where } \text{NET}_j^1 = \sum_{i=0}^{n} w_{ij}^1 x_i$$

For layers $1 < l < L$:
$$h_j^l = \frac{1}{1+e^{-\text{NET}_j^l}} \ \text{ for } \ 1 \le j \le H_l \text{ and where } \text{NET}_j^l = \sum_{i=0}^{H_{l-1}} w_{ij}^l h_i^{l-1}$$

For the output layer:
$$y = \frac{1}{1+e^{-\text{NET}_1^L}} \ \text{ and where } \text{NET}_1^L = \sum_{i=0}^{H_{L-1}} w_{i1}^L h_i^{L-1}$$

Where H_l is the number of units in the hidden layer l.

5. Compute the error for the output layer with:

$$\delta_1^L = (y' - y)y(1 - y)$$

by comparing the actual output y with the desired one y' for pattern P_k that is being considered.

6. Compute the errors for the preceding layers by propagating the errors backwards. For hidden layer $L - 1$:

$$\delta_i^{(L-1)} = h_i^{(L-1)}(1 - h_i^{(L-1)})\delta_1^L w_{i1}^L \quad \text{for} \ 1 \leq i \leq H_{L-1}$$

For hidden layers $L - 1 > l \geq 1$:

$$\delta_i^l = h_i^l(1 - h_i^l) \sum_{j=1}^{H_{l+1}} \delta_j^{l+1} w_{ij}^{l+1} \quad \text{for} \ 1 \leq i \leq H_l$$

7. Update weights with the following formula:

$$w_{ij}^l = w_{ij}^l + \Delta w_{ij}^l$$

Where Δw_{ij}^l for the first hidden layer:

$$\Delta w_{ij}^1 = \eta \delta_j^1 x_i \quad \text{for} \ 0 \leq i \leq n \ \text{and} \ 1 \leq j \leq H_1$$

For weights between two layers of hidden units:

$$\Delta w_{ij}^l = \eta \delta_j^l h_i^{l-1} \quad \text{for} \ 0 \leq i \leq H_{l-1}, 1 \leq j \leq H_l, \ \text{and} \ 1 < l < L$$

For weights between the last hidden layer and the output layer:

$$\Delta w_{i1}^L = \eta \delta_1^L h_i^{L-1} \quad \text{for} \ 0 \leq i \leq H_{L-1}.$$

8. Go back to step 3 and repeat for the next pattern.

3 Adding a momentum term

Selecting the proper learning rate η used in step (7) is difficult. A small learning rate may result in very slow learning. A large learning rate may lead to oscillations. One way to avoid oscillation at large η (see [2, p. 123]) is to add a *momentum term*, which will make the change $\Delta w_{ij}^l(t)$ in the weight dependent of the previous change $\Delta w_{ij}^l(t - 1)$:

$$\Delta w_{ij}^l(t) = \eta \delta_j^l x_i + \alpha \Delta w_{ij}^l(t - 1) \tag{1}$$

The *momentum term* α must be between 0 and 1; many authors recommend a value of 0.9 (see [2, p. 123], [4, p. 506], [6, p. 54]). A reasonable value for η is 0.35 (see [4, p. 505]). So, in the algorithm presented in the previous section we must use formula (1) in step 7 to add the momentum term.

4 Proposed algorithm

There are two mayor considerations in the learning process. The first one is that when a net has already learned some of the patterns (but not all of them), it is important that these patterns are not forgotten when learning new ones; it makes no sense to learn pattern B if in the process pattern A is forgotten. However, if the net has already learned pattern A, it will be desirable to put more attention to the other patterns that have not been learned yet. It might seem that this criteria is contradictory, but later we will show a method that has both features.

A second factor is how much time do we spend in the different steps of the learning process. Back-propagation consist of two mayor phases: the feedforward step and the back-propagation of the error (with the corresponding adjusting of the weights in the different layers). The feedforward step is straightforward: there are no difficult operations involved and only a fraction of the time is spent in this part. In contrast, the second step comes to be more complex and time consuming, because of the need to compute for each layer the error involved, and then adjust the weights. If we were able to skip this when possible, i.e. for patterns that have already been learned, and concentrate in other patterns, we will be going in the right direction for reducing the time consumed in the learning process.

The goal is therefore to obtain an algorithm that allows to train a neural network. This algorithm should reduce the number of training steps performed, thus reducing the number of operations involved in the training. If we are using a computer to train the network, we will reduce in consequence the many times expensive CPU usage.

When analyzing step (7) which gives us the modification in the weights for the output units, we can see that this modification is proportional to $y' - y$, which is the error that the output unit has for a given pattern P_k. Now, suppose that the network has already learned pattern P_k, or in this iteration is very close to y'. In that case, the error for pattern P_k will be very close to zero, along with the contribution for the overall error of the network. The weight modification in this step will also be very close to zero. If the change is sufficiently small, performing or not this back-propagation step it really does not matter, because the network remains almost the same. Taking this into consideration, we proposed the following modification to back-propagation, when the network only consist of one unit (a generalization for more output units can be made):

Epsilon-Back-propagation algorithm

1. Initialize all the weights w_{ij}^l to small random values (typically between -0.1 and $+0.1$), where w_{ij}^l is the weight in layer l which connects unit i in layer $l - 1$ with unit j in layer l.

2. Initialize the activations for the thresholding units. The values of these units will never change.

$$x_0 = 1, h_0^l = 1 \text{ for } 1 \leq l < L.$$

3. Choose a pattern P_k, where $P_k = (p_1, p_2, \ldots, p_n)$ and apply it to the input layer so that:

$$x_i = p_i \text{ for } 1 \leq i \leq n.$$

4. Propagate the signal forwards through the network, using for the first layer:

$$h_j^1 = \frac{1}{1 + e^{-\text{NET}_j^1}} \text{ for } 1 \leq j \leq H_1 \text{ and where NET}_j^1 = \sum_{i=0}^{n} w_{ij}^1 x_i$$

For layers $1 < l < L$:

$$h_j^l = \frac{1}{1 + e^{-\text{NET}_j^l}} \text{ for } 1 \leq j \leq H_l \text{ and where NET}_j^l = \sum_{i=0}^{H_{l-1}} w_{ij}^l h_i^{l-1}$$

For the output layer:

$$y = \frac{1}{1 + e^{-\text{NET}_1^L}} \text{ and where NET}_1^L = \sum_{i=0}^{H_{L-1}} w_{i1}^L h_i^{L-1}$$

Where H_l is the number of units in the hidden layer l.

5. If $|y' - y| > \epsilon$ then

   (a) Compute the error for the output layer with:

   $$\delta_1^L = (y' - y)y(1 - y)$$

   by comparing the actual output y with the desired one y' for pattern P_k that is under consideration.

   (b) Compute the errors for the preceding layers by propagating the errors backwards. For hidden layer $L - 1$:

   $$\delta_i^{(L-1)} = h_i^{(L-1)}(1 - h_i^{(L-1)})\delta_1^L w_{i1}^L \text{ for } 1 \leq i \leq H_{L-1}$$

   For hidden layers $L - 1 > l \geq 1$:

   $$\delta_i^l = h_i^l(1 - h_i^l) \sum_{j=1}^{H_{l+1}} \delta_j^{l+1} w_{ij}^{l+1} \text{ for } 1 \leq i \leq H_l$$

   (c) Update weights with the following formula:

   $$w_{ij}^l = w_{ij}^l + \Delta w_{ij}^l$$

   Where Δw_{ij}^l for the first hidden layer:

   $$\Delta w_{ij}^1 = \eta \delta_j^1 x_i \text{ for } 0 \leq i \leq n \text{ and } 1 \leq j \leq H_1.$$

   For weights between two layers of hidden units:

   $$\Delta w_{ij}^l = \eta \delta_j^l h_i^{l-1} \text{ for } 0 \leq i \leq H_{l-1}, 1 \leq j \leq H_l, \text{ and } 1 < l < L.$$

   For weights between the last hidden layer and the output layer:

   $$\Delta w_{i1}^L = \eta \delta_1^L h_i^{L-1} \text{ for } 0 \leq i \leq H_{L-1}.$$

6. Go back to step 3 and repeat for the next pattern, until no weight update is made in one epoch, that is $|y' - y| \leq \epsilon$ for all patterns.

As explained in section 3 the momentum term can be readily included in step 5c.

This slight modification of back-propagation might seem not to do any good or even to damage the overall performance of the net in the learning process, but if it is carefully analyzed we can see the following:

- At the beginning, when the network has not yet reached the desire approximation to any of the patterns, back-propagation and the previous algorithm behave exactly the same.

- As the net begins to learn some patterns with more and more precision, the two methods begin to diverge: plain back-propagation will spent time with all patterns, no matter how close to the desired output the outcome of the net has become. In contrast, epsilon-back-propagation will begin to skip some backward steps for the patterns that the network already recognize and gives a satisfactory response for them. This will result in the net spending more time in the patterns that have more interest at that time, because they have not been learned yet.

- What happens if the net forgets one of the patterns that has been previously learned? This happens when the weights are adjusted for other patterns and the net no longer gives a "good enough" answer for pattern A. In this case, the error is greater than the ε we have selected. When computing the error of the net for pattern A (remember that the feedforward step is never skipped), the method behaves as plain back-propagation, which assures the net will train again for that value as many times as necessary (and the rest of the values that have not been learned yet): we achieve part one of our considerations, spending more time with patterns that have not been learned.

- It is reasonable to expect that the number of forward steps needed to train the network may be greater if we use this epsilon modification than if we are not using it at all, but hopefully, the number of back-propagation skips will compensate for it, since these last steps are more complex, we may indeed use less mathematical operations in the overall process, and consequently less CPU time: part two of our considerations, and giving as result the success in our goal.

In the next section we give some experimental results comparing back-propagation with epsilon-back-propagation.

5 Experimental results

Here we present the data obtained from the tests performed. A C program was used to simulate and test both back-propagation and epsilon-back-propagation in a RS/6000. Both algorithms were tested with the same set of training samples, the same topology of the network, the same values for $\eta = 0.35$, $\alpha = 0.9$ as well as the same set of initial random weights for each problem. The criterion for stopping back-propagation and epsilon-back-propagation was that for all patterns, the approximation were less or equal to the selected ϵ value. The tests were performed for several ϵ values for each problem. Four problems were selected:

1. AND. Two binary inputs and one binary output. The output must be the logic AND of the inputs. All possible patterns were in the training sample. The topology selected was a network with two layers, two inputs, one hidden layer with eight hidden units and one output unit. The data is shown in table 1.

	Epsilon-Back-propagation			Plain Back-propagation		
ϵ	Epochs	Feedforward	Backward	Epochs	Feedforward	Backward
0.200	66	264	201	68	272	272
0.100	103	412	312	118	472	472
0.050	225	900	620	271	1,084	1,084
0.025	633	2,532	1,620	793	3,172	3.170
0.010	3,148	12,592	7,613	4,013	16,052	16,052
0.005	11,481	45,924	27,156	14,698	58,792	58,792
0.001	258,549	1,034,196	593,581	339,832	1,359,328	1,359,328

Table 1: Statistics for the AND problem

2. XOR. Two binary inputs and one binary output. The output must be the logic XOR of the inputs. All possible patterns were in the training sample. The topology selected was a network with two layers, two inputs, one hidden layer with eight hidden units and one output unit. The data is shown in table 2.

	Epsilon-Back-propagation			Plain Back-propagation		
ϵ	Epochs	Feedforward	Backward	Epochs	Feedforward	Backward
0.200	363	1,452	1,380	367	1,468	1,468
0.100	412	1,648	1,592	424	1,696	1,696
0.050	559	2,236	2,132	603	2,412	2,412
0.025	1,080	4,320	4,043	1,233	4,932	4,932
0.010	4,440	17,760	16,373	5,271	21,084	21,084
0.005	15,993	63,972	58,640	19,108	76,432	76,432
0.001	381,292	1,525,168	1,399,553	454,770	1,819,080	1,819,080

Table 2: Statistics for the XOR problem

3. SQUARE ROOT. One real input and one real output. The output must be the *square root* of the input. The training sample was conformed with 17 values, from 0.10 up to 0.90 with increments of 0.05. The topology selected was a network with three layers, two inputs, two hidden layers with four hidden units each layer and one output unit. The data obtained is given in table 3.

	Epsilon-Back-propagation			Plain Back-propagation		
ϵ	Epochs	Feedforward	Backward	Epochs	Feedforward	Backward
0.040	141	2,397	1,700	2,718	46,206	46,206
0.035	143	2,431	1,731	4,596	78,132	78,132
0.030	595	10,115	3,298	5,251	89,267	89,267
0.025	2,634	44,778	11,800	5,680	96,560	96,560
0.020	9,003	153,051	45,171	6,135	104,295	104,295
0.015	9,517	161,789	67,433	6,631	112,727	112,727
0.010	9,787	166,379	96,691	7,505	127,585	127,585

Table 3: Statistics for the SQUARE ROOT problem

4. SQUARE. One real input and one real output. The output must be the *square* of the input. The training sample was conformed with 17 values, from 0.10 up to 0.90 with increments of 0.05. The topology selected was a network with two layers, two inputs, one hidden layer with eight hidden units and one output unit. Results are presented in table 4.

	Epsilon-Back-propagation			Plain Back-propagation		
ϵ	Epochs	Feedforward	Backward	Epochs	Feedforward	Backward
0.15	36	612	262	34	578	578
0.10	51	867	353	56	952	952
0.08	167	2,839	601	202	3,434	3,434
0.07	310	5,270	895	5.629	95,693	95,693
0.06	866	14,722	2,152	13,190	224,230	224,230
0.05	2,323	39,491	5,547	60,647	1.030,999	1,030,999
0.04	10,840	184,280	26,263	373,411	6,347,987	6,347,987

Table 4: Statistics for the SQUARE problem

6 Conclusions

The following conclusions can be drawn from the data in the tables:

1. Epsilon-back-propagation was better in all cases for the AND problem. The number of feedforward steps was smaller and for the backward phase, the number of steps performed was about half of the backward steps in plain back-propagation, as the epsilon value decreases.

2. For the XOR problem, we can see that the epsilon approach was always better, even not as much as the previous problem. The number of feedforward steps was also always better than plain back-propagation, but the number of backward steps which were skipped, were not so high. This can be explained as the XOR problem is harder for the network to learn than the AND problem.

3. The data obtained for the SQUARE ROOT problem was not so encouraging. With a large value of ϵ (i.e. 0.04), the epsilon approach is a lot better. The number of feedforward steps in the epsilon approach is about 5% of those required in plain back-propagation, and the backward steps about 4%. The advantage is maintained with values of $0.035, 0.03, 0.025$ for the ϵ parameter. With values smaller than 0.020 the epsilon algorithm performs even more feedforward steps. In this case, we have to take into consideration that a backward step is more costly than a feedforward step. If we consider that both steps would take the same time, both algorithms come out to be more or less the same: consider the sum of feedforward and backward steps and an epsilon value of 0.01, we have for the epsilon approach $263,070$ total steps against $255,170$ total steps in plain back-propagation. For this only case, we tested both algorithms with the absolute time, and it came out that both were almost the same: epsilon approach took 36.98 seconds against 37.35 seconds taken by plain back-propagation (the time given includes system overhead). We can see that the difference is not significative.

4. The best performance obtained was for the SQUARE problem. With a large epsilon, both algorithms are more or less the same, but as the epsilon decreases, the speed up obtained is amazing. For example, with an epsilon value of 0.04, the number of feedforward steps performed in the epsilon algorithm is about 3% the number of feedforward steps in plain back-propagation, and the number of backward steps is about 0.4% !

5. We can say that there are problems very well suited for this approach, as the SQUARE problem, others not as good, but still the performance would be as acceptable as plain back-propagation.

7 Future research

Future research can be done. Specifically, we have in mind to make a generalization of epsilon-back-propagation for multiple outputs. We also are currently working on a parallel version of the algorithm, in order to take advantage of the parallelism involved in a neural network.

One variation of the epsilon approach not taken into account in the present work, is to consider a dynamic epsilon value, i.e. begin the algorithm with a large value of ϵ and, once the net has converged, reduce the ϵ value and train the network again with the previously computed weights, repeating this until the desired ϵ value has been reached.

References

[1] Hecht-Nielsen, R., *Neurocomputing*, Addison-Wesley Publishing Co., Reading, MA, 1990.

[2] Hertz, J., A. Krogh and R. G. Palmer, *Introduction to the Theory of Neural Computation*, Addison-Wesley Publishing Co., Redwood City, CA, 1991.

[3] Kröse, B.J. and P. van der Smagt, *An introduction to Neural Networks*, Dept. of Computer Systems, University of Amsterdam, 1993.

[4] Rich, E., and K. Knight, *Artificial Intelligent Systems*, McGraw-Hill, New York, 1991.

[5] Sandoval, Carlos A., *A Modified Learning Algorithm for Backpropagation Networks*, Master Thesis, University of Texas at el Paso, 1993.

[6] Wasserman, P.D., *Neural Computing Theory and Practice*, Van Nostrand Reinhold, New York, 1989.

Finite State Automata and Connectionist Machines: A Survey

M.A. Castaño[†], E. Vidal[††] , F. Casacuberta[††]

[†]Dpto. de Informática. Universitat Jaume I de Castellón.
Campus de Penyeta Roja.12071 Castellón (Spain)
[††]Dpto. Sistemas Informáticos y Computación. Universidad Politécnica de Valencia
Camino de Vera s/n. 46071 Valencia (Spain)

Abstract

Work in the literature related to Finite State Automata (FSAs) and Neural Networks (NNs) is review. These studies have dealt with Grammatical Inference tasks as well as how to represent FSAs through a neural model. The inference of Regular Grammars through NNs has been focused either on the acceptance or rejection of strings generated by the grammar or on the prediction of the possible successor(s) for each character in the string. Different neural architectures using first and second order connections were adopted. In order to extract the FSA inferred by a trained net, several techniques have been described in the literature, which are also reported here. Finally, theoretical work about the relationship between NNs and FSAs is outlined and discussed.

1. INTRODUCTION

Work presented in the literature about Finite State Automata (FSAs) and Neural Machines or Neural Networks (NNs) has focused on two topics: How to *infer* the Regular Language (RL) associated to some positive and/or negative samples through a neural model; and how to *represent* the FSA corresponding to a Regular Grammar (RG) through a NN.

The history of Grammatical Inference (GI) and Neural Models began when [Minsky,67] proved that "Every State Machine is equivalent to and can be simulated by some neural machines [McCulloch,43]". However, later work on GI and NNs goes back to the end of the eighties. Most of these studies have dealt with FSAs, though neural machines have been recently designed to learn Context-Free Grammars [Sun,90] [Das,93] [Lucas,93]. In order to infer RGs, different neural architectures were adopted, and are briefly described and compared in section 2.1. To be able to determine the internal structure that these connectionist models had exactly learned (that is, to "extract" the inferred automaton from the net) different techniques were proposed in the literature, which are described in section 2.2.

With reference to represent FSAs through connectionist models, section 3 presents two techniques for building a NN that behaves like a given automaton. The possible equivalence of both tools is also discussed.

Taking into account that the behavior of a trained net is deterministic (two identical input situations provide the same output activations), NNs can only learn deterministic FSAs. So, in what follows, we will only refer to deterministic FSAs.

2. INFERENCE OF FINITE STATE AUTOMATA THROUGH NEURAL NETS

The inference of FSAs was considered for either *recognition* or *prediction* purposes. The recognition task consists of training connectionist architectures through positive and negative strings belonging or not belonging to a RL, respectively, so that after all characters of a test string have been sequentially processed, the network should accept or reject it [Pollack,91] [Giles,92a-c] [Watrous,92] [Omlin,92-93] [Giles,93a-b] [Manolios,93] [Miller,93] [Chen,95]. On the other hand, the objective of the predicting task was to forecast the possible successor(s) that may follow each input symbol of the string generated by the target grammar [Servan,88] [Cleeremans,89] [Smith,89] [Servan,91] [Fahlman,91] [Sopena,93] [Castaño,93a-b]. In this case, only positive samples are employed in the training process.

2.1. ARCHITECTURES AND METHODS EMPLOYED

The difficulty of inferring RGs through NNs arises from the fact that a substring can be followed by different characters and that the length of the strings is variable. Thus, inference problems have been approached through Recurrent NNs, in which the concept of time is implicit and past events are somehow remembered. Two generic recurrent connectionist models have been employed in the literature to infer RGs: First Order Recurrent Networks (FORNs) and Second Order Recurrent Networks (SORNs). Mixed models with first and second order connections have also been used.

2.1.1. INFERENCE THROUGH FIRST ORDER RECURRENT NETS

Different Simple Recurrent Networks (SRNs) such as the Elman or Jordan models were adopted to induce RGs; other complex FORNs such as Williams' net or Fahlman's Recurrent Cascade Correlation architecture were also employed.

2.1.1.1. ELMAN'S SIMPLE RECURRENT NET

A basic *Simple First Order Recurrent Net* adopted in the literature to induce RGs was introduced in [Elman,88]. This SRN was trained through an adequate modification of the standard *Backward-Error-Propagation algorithm* [Rumelhart,86] which truncates the gradient computation of the backward recurrence and does not strictly minimize the error achieved by the net.

Although this connectionist approach can be employed for both prediction and recognition processes, it has only been used for predicting successive elements in a sequence [Cleeremans,89] [Servan,88] [Servan,91]. To do so, at each time cycle, the network was presented with an element of the current (positive) string and was required to predict the possible successor(s) on the output layer. Thus, the learned SRN behaves like a Moore machine [Maskara,92].

The performance of the net was tested on the RG shown in Figure 1-a and called *Reber grammar* [Servan,88] [Cleeremans,89] [Servan,91] which many GI researchers have used in their work. Another grammar with the Reber RG embedded in it twice and with identical transition probabilities for both embedded Reber grammars (see Figure 1-b) was also tried. Table 1 reports the features and performances for both (simple and double Reber RG) learning processes. The net which was trained to learn the simple grammar correctly predicted a test data of 20.000 strings; however, only 75% of the 20.000 test strings considered for the double Reber RG were predicted. The recurrent layer of an Elman's SRN was also trained through the Real Time Recurrent Learning (RTRL) algorithm [Williams,89] which, in contrast to the one previously employed, minimizes the error function. The prediction of the preceding RGs was again approached [Sopena,93] and both of them were correctly learned; i.e., the trained nets were tested on one thousand transitions each without error, outperforming those results obtained with Cleeremans et al.'s technique (see Table 1).

On the other hand, SRNs were not only able to predict the next symbol(s) of strings randomly generated by a SRN, but also seemed to be capable of accurately estimating the probability distribution corresponding to the string generation, as is shown in the experiments that were reported in [Castaño,93a] and [Castaño,93b].

2.1.1.2. WILLIAMS' FIRST ORDER RECURRENT NETWORK

A different first order connectionist approach to infer FSA consisted in a fully-connected recurrent architecture trained through the RTRL algorithm [Williams,89]. The only inference works carried out with this model [Smith,89], were focused on predicting the simple and double symmetric Reber RG. A 100% recognition rate (on one million test strings) was achieved for both grammars. The features of the net and training (summarized in Table 1) were not identical for Williams', Cleeremans' and Sopena's experiments and so, the obtained numerical performance results can not be exactly compared; however, the convergence results provided with this approach do not seem to improve those obtained by Sopena.

2.1.1.3. FAHLMAN'S RECURRENT CASCADE CORRELATION

Fahlman designed another second order neural model called *Recurrent Cascade Correlation* (RCC) [Fahlman,91], with which the induction of RGs was also considered. As in the two previous first order methods, both the simple and symmetric double Reber grammar were tried through this architecture. In both experiments, several trials using different training sets were run. All the learned nets for the simple Reber RG correctly predicted the test data (256 strings). However, prediction performances (on a set of

256 test strings) were always above 92% for the double Reber grammar. Moreover, [Chen,95] showed that Fahlman's model was not capable of representing all RLs (such as those with sequences of period two or less).

EXPERI MENT	TRAINING ALGORITHM	SYMBOL PREDICTION CRITERION	HIDDEN UNITS	TRAINING STRINGS
Reber Grammar				
Elman's net [Cleeremans,89] [Servan,91]	Continuous training until 20.000 validation strings are predicted	The activation of the target symbol is above .3	3 5	20.000 60.000
Elman's net + RTRL [Sopena,93]	Continuously training 2.000 grammar transitions until 1.000 new test transitions are predicted	The activation of the possible successors is above .25 and the remaining outputs less than .25	6	1.000
Williams' net [Smith,89]	Continuous learning until 10.000 successive strings are predicted	The same as that just employed above with a threshold of .3	2	19.000
Fahlman's net [Fahlman,91]	Continuous training until the error function is above a certain value	The highest activations are for the possible successors	2-3	25.000
Symmetric Double Reber Grammar				
Elman's net* [Cleeremans,89] [Servan,91]	Continuous training (the criterion to stop learning was not specified in the papers)	The highest activation is for the target and its *Luce ratio* (ratio of this activation to the sum of all outputs) is greater than .6	Not specified	900.000
Elman's net + RTRL [Sopena,93]	Continuously training 2.000 grammar transitions until 1.000 new test transitions are predicted	The activation of the possible successors is above .25 and the remaining output units less than .25	12 12+12	26.000 16.000
Williams' net [Smith,89]	Continuous learning until 1.000 successive strings are predicted	The highest activations is for the target and its Luce ratio is above .6	12	25.000
Fahlman's net* [Fahlman,91]	Continuous training until the error function is above a certain value	The highest activations are for the possible successors	5-7	200.000

Table 1. Comparative behavior of different FORNs on the simple and the symmetric double Reber grammar. The training algorithm adopted, the criterion to assess the correct prediction, the number of hidden units of the trained net and the number of training strings employed, are specified for every experiment. All of them assumed a local representation of the grammar alphabet and considered a string as correctly predicted when the corresponding prediction criterion (specified in the Table) was verified for every symbol in the string. 100% recognition rates were obtained for all experiments, except for those labeled with the * symbol.

2.1.1.4. JORDAN'S SIMPLE RECURRENT NETWORK

In contrast to the previous approaches, the following mechanism was employed to tackle recognition problems, that is, to accept or reject strings of RGs. The simple recurrent FORN considered was introduced in [Jordan,88], and Manolios and Fanelly spread it out through time in a similar way as that adopted in [Rumelhart,86] to deal with this inference task [Manolios,93].

In order to make comparisons on the performance of different networks and induction methods, another benchmark set of seven problems introduced in [Tomita,82] (and shown in Figure 2) has also been adopted by many GI researchers. Manolios and Fanelli considered a subset of seven Tomita grammars to train their model. However, more than obtaining numerical recognition performances, they analyzed the behavior of their nets through graphic techniques. Therefore, the most interesting of these graphical studies revealed that during the learning phase the net first *decides* the structure of the FSA and *reinforces* the acquired information later.

2.1.2. INFERENCE THROUGH SECOND ORDER RECURRENT NETWORKS

GI has also been approached through recurrent nets with high order connections among their units. Three second order models (Giles', Pollack's and Chen's SORNs) and the corresponding works with which induction was carried out are briefly described in the next sections.

2.1.2.1. GILES' SECOND ORDER RECURRENT NETWORK

Due to the promising advantages of high order nets [Maxwell,89] [Miller,93], a second order version of the Williams' architecture was used to learn RLs by Giles et al. for recognition purposes. The net was initially trained on a small fraction of the training set (called *working set)* until the error on every sample of the working set was less than an *error tolerance*. The obtained net was then tested on the remaining training set, so that if the net correctly classified it, the training ended; otherwise, 10 misclassified strings were added to the working set and training resumed.

The first studies about Giles' SORN were focused on the 7 Tomita grammars [Giles,92a] and, for comparison purposes, Table 2 shows the performances obtained for the fourth grammar. A larger randomly generated RG corresponding to the FSA displayed in Figure 3 was also satisfactorily learned [Giles,92b] and the results are reported in Table 2.

Giles' learning method was modified by himself and Omlin in order to inject *a priori knowledge* about the FSA (defined as *hints*) into the net. They demonstrated [Omlin,92] [Giles,92c] that by inserting hints into the initial net, the training time needed to converge was improved (although not necessarily the generalization performance), even when weak hints were provided. On the other hand, Chen et al. showed that networks took advantage of the previous knowledge they acquired [Chen,95].

Omlin and Giles also tried a simple method for pruning Giles' SORN [Omlin,93] which consisted in repetitively reducing the size of the architecture by eliminating the smallest weight and retraining the net. Comparing the convergence results obtained for the random FSA referred above (see Table 2) with those reported in [Giles,92b] for the same grammar, this pruning network seems to improve convergence times.

[Miller,93] presented an experimental study on the effect of the connectivity order of a net, comparing Giles' SORN and Williams' FORN as applied to the task of inferring RLs. First, Miller and Giles showed that for simple RGs such as Tomita's benchmarks, these two architectures had comparable learning power and a comparable generalization performance. However, for the larger randomly-generated grammar of

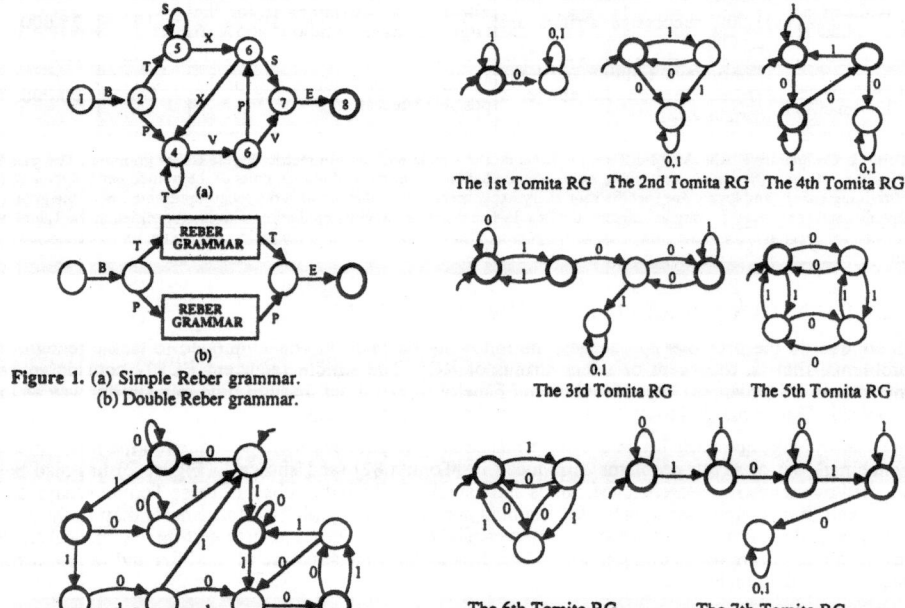

Figure 1. (a) Simple Reber grammar.
(b) Double Reber grammar.

Figure 3. Randomly generated FSA.

The 1st Tomita RG   The 2nd Tomita RG   The 4th Tomita RG

The 3rd Tomita RG   The 5th Tomita RG

The 6th Tomita RG   The 7th Tomita RG

Figure 2. The seven Tomita grammars.

Figure 3, the SORN significantly outperformed the FORN, both in convergence time and generalization capability. Finally, they observed that while high order networks converged more quickly to a solution and could find it more reliably than FORNs, high order solutions seemed to be of poorer quality than those of first order.

2.1.2.2. POLLACK'S SECOND ORDER RECURRENT NETWORK

Pollack proposed another second order approach which consisted of a standard feed-forward *slave network* and a *master network* [Pollack,91] which were unrolled following the Back-Propagation through time technique [Rumelhart,86]. Pollack tackled the problem of recognizing strings generated by a RG, by considering these networks as *dynamical recognizers*. Studying his second order network from this perspective, he discovered a new mechanical inference method called *Induction by Phase Transition*, which consisted in periodically analyzing the output of the network through graphical processes. Based on the fractal dynamics he studied, he empirically established that depending on whether the attractors of the state-space in the limit of a dynamical recognizer were limit points, self-similar regions or chaotic regions, the recognized language was regular, context-free or context-sensitive.

2.1.2.3. CHEN'S MODEL

A simple scheme to *dynamically construct* a recurrent net was proposed in [Chen,95]. The model consists in an Elman SRN (with first or higher ordered connections) in which inputs can be connected to output units. The mechanism repeatedly adds a fully-connected HU after a specific fixed number of training strings are presented to the net, until the net learns to correctly classify all the training samples· This constructive approach was tested on the seven Tomita grammars [Chen,95] but the obtained numerical results were not reported in the original paper.

2.1.3. INFERENCE THROUGH MIXED MODELS

Architectures with mixed first and second order connections between their units have also been employed in the literature to infer RGs; such as the one suggested in [Watrous,92], with which the 7 benchmark Tomita grammars were induced. Several trials were made to learn each grammar and for most of the seven languages, at least one network converged to a very low mean squared error value although there was no correlation between this error and the number of iterations until termination or the generalization of the language. Nevertheless, the performances provided were not above those achieved with Giles' SORN (Table 2 shows the performances obtained for the fourth Tomita grammar).

EXPERIMENT	HIDDEN UNITS	TRAINING ITERATIONS	ERROR TOLERANCE	RECOGNITION RATE
The fourth Tomita Grammar				
Giles' net [Giles,92a]	3	94	.2	100% on all strings of length up to 15
Watrous' net [Watrous,92]	Not specified	60-10.700	.1	21-61% on all strings of length up to 10
Random Grammar				
Giles' net [Giles,92b]	6	1.000	.2	99% on all strings of length up to 15
Giles' net + Pruning method [Omlin,93]	7	104	.2	99.8% on all strings of length up to 20

Table 2. Comparative behavior of different SORNs on the fourth Tomita RG and a randomly generated RG. The number of HUs of the trained net, the number of iterations to termination, the error tolerance and the network generalization performance are specified for all experiments.

2.2. EXTRACTION OF FINITE STATE AUTOMATA THROUGH NEURAL NETS

Although the previous connectionist learning models have shown considerable empirical success in order to infer RGs, it is hard to exactly characterize *what they learned*. To be able to determine how the grammar structure is accurately approached, the inferred automaton should be extracted from the internal representation of the net. Four different techniques of extracting FSAs have been employed in the literature, which are described immediately below. They are based on *hierarchical clustering, dynamic state partitioning, neighborhood criteria* and *networks with direct extraction*, respectively.

The first studies on the nature of the internal representations that the network developed on its HUs, was introduced in [Elman,88] in the following way: Assuming a trained NN, its connection weights are frozen and strings are passed through the net submitting the HU activations for each input character to a **hierarchical clustering** analysis on the basis of Euclidean distance. The resulting clusters are assumed to be the *states* of the automaton inferred by the net. In order to obtain the *state transition diagram*, strings are again presented to the net, so that every symbol of a string and its corresponding current state is connected to another state with the minimum Euclidean distance to the new computed hidden activations. Applied to the task of predicting RGs, this cluster analysis revealed that the activations corresponding to a cluster were first grouped according to the successor(s) predicted for a given node and later to the previous path of the current string [Cleeremans,89] [Smith,89] [Servan,88] [Servan,91]. Nevertheless, a clear disadvantage of this method lies in the fact that the size of the automaton is required for the hierarchical clustering procedure, when it is assumed to be unknown in a GI process.

The structure of the automaton represented by the trained net can also be extracted through another procedure called **dynamic state partitioning**. The algorithm initially divides each hidden neuron range into q partitions of equal size and consequently, the output space for N hidden neurons is split into q^N possible partition *states*. In order to construct the *state transition diagram*, strings are presented to the net, so that, the input symbol is associated with the current partition state and the next partition state they activate. Later, the extracted FSA is reduced to its equivalent minimal state representation using a standard minimization algorithm. The studies carried out on this issue [Giles,92a] [Giles,92b] [Miller,93] revealed that when the network had successfully learned the grammar, the minimum FSA obtained did not depend of q, the number of HUs or the initialization of the net. On the other hand, regarding the classification of unseen strings, the extracted FSAs often outperformed the NN from which they were built [Giles,92a] [Giles,92b]. However, despite these advantages, this method is computationally intensive for large numbers of neurons, because of the employed partitioning process.

Manolios and Fanelli proposed another FSA extraction technique based on **neighborhood criteria** [Manolios,93]. This method consists in representing the successive activations of the HUs for the strings presented to the net, into an n-dimensional unit hypercube where n matches the size of FSA in terms of states. The centroids of the activations in the hypercube should represent the *states* of the FSA. In order to obtain them, a number of markers are randomly distributed within the hypercube and every marker is then moved towards that point a distance equal to the distance between them divided by the times the marker was moved plus one. The evolution of the successive activations for each string in the hypercube, indicates the *state transitions*. As in the first extraction method, a priori information about the automaton is required to obtain the structure learned by the net, which can be an obstacle for real-world inference tasks.

A new technique on the internal representation of the net was explored in [Zeng,94], where the learned FSA was **directly extracted**. Zeng adopted a first order architecture for binary sequences, which consisted in two subnetworks with shared HUs, so that only one of them is enabled at a time, depending on the binary value of the input. Assuming that this technique forces the learning of discrete states in the HU activation space, each point in the discretized hidden activation space was automatically defined as a *state*. Experimental results showed that the extracted FSAs had a small number of states. However, the time required to learn was higher with discretized nets than that employed for analog machines.

3. REPRESENTATION OF FINITE STATE AUTOMATA THROUGH NEURAL NETS

In spite of the large body of experimental work related to GI and NNs, only a few theoretical works about the relationship between NNs and FSAs have been reported. This section tries to answer how to build a NN recognizer for a given RL, as well as whether or not there such a recognizer always exists. Two different models to represent FSAs were proposed in the literature, (which are briefly described in the following sections). On the other hand, looking at the discrete nature of FSAs and the analog feature of NNs, it is not clear that any NN can be represented through a FSA. In fact, no work in the literature is related to how to build NNs through FSAs. Consequently, the equivalence between both machines will probably not be established.

3.1. REPRESENTATION THROUGH SINGLE LAYER RECURRENT NEURAL NETS

The first proposed approach to represent FSAs through connectionist models is focused on a simple architecture called *Single Layer Recurrent Neural Network* (SLRNN), which consists of a first or second order Elman net without output layer. Starting from this net, Goudreau and Giles "demonstrated" that a second order SLRNN could implement any Finite State Recognizers (by using a simple constructive method reported in [Goudreau,94]) while a first order SLRNN could not; however, they "suggested" that if the first order SLRNN is augmented with output layers of feed-forward neurons, it could probably implement any "non-minimal" Finite State Recognizer.

3.2. THE FINITE STATE RECURRENT NETWORK MODEL

A different approach for representing FSAs through connectionist machines is presented in [Sanfeliu,92] and [Alquezar,93]. This method starts from both first and second order Williams architectures, in which the current state of the FSA was coded. The model, called *Finite State Recurrent Network*, formulates the problem as a system of linear equations, where the unknown variables correspond to the network weights, which have to satisfy a set of constraints posed by the RL to be represented.

4. CONCLUSIONS AND FUTURE WORK

The inference of regular grammars (RGs) through Neural Networks (NNs) has traditionally focused on either learning to *accept or reject* strings generated by the grammar or learning to *predict* the possible successor(s) for each character in the string. *Prediction tasks* have been approached though different first order architectures such as Elman's SRN, Jordan's SRN, Williams' net or the constructive Recurrent Cascade Correlation model. Comparative studies among them considered both a simple and a complex benchmark problem centered on the Reber grammar. Although the experimental conditions did not exactly coincide for the four neural models, an Elman net, whose recurrent layer was trained with the Real Time Recurrent Learning algorithm, seemed to provide the best convergence performances.

Regarding the *recognition tasks*, learning machines with first order connections (such as Jordan's SRN) and second order models (such as Giles' or Pollack's nets) were employed. On this occasion, the experiments were focused on the seven Tomita grammars and a longer randomly generated grammar. The inference of these grammars through Jordan's and Pollack's models attempted to understand the behavior of NNs during both training and testing through graphic techniques, more than obtaining numerical recognition results. On the other hand, Giles et al. developed extensive studies centered on their model. In fact, a simple pruning mechanism of this net provided the best performances to simulate both the Tomita RGs and the random grammar. They also showed that, by inserting into the net a priori knowledge about the Finite State Automaton (FSA) to be learned, convergence speed increased.

Four different techniques employed in the literature to extract the deterministic FSA embedded in a training net were also described in the paper. And finally, in order to study the relationship between RGs and NNs, two different methods to represent FSAs through two specific NNs were proposed. The opposite task, how to build a NN through FSAs, is still an pending problem which will probably not be solved, due to the continuous and discrete nature of the corresponding learning machines.

Nevertheless, most of these works are based on heuristics, and conclusions on a specific model are perhaps too rapidly generalized to generic NNs after a slight experimentation was carried out. Deeper and more systematic experimental research as well as theoretical work about the capabilities and power of a *generic NN* and the corresponding learning strategies to infer RGs are desirable. Theoretical relationships between generic NNs and FSAs also constitute another interesting problem to be solved.

5. BIBLIOGRAPHY

[Alquezar,93] *Representation and Recognition of Regular Grammars by means of Second-Order Recurrent Neural Networks.* R. Alquézar, A. Sanfeliu. In New Trends in Neural Computation. Eds. J.Mira, J.Cabestany, A.Prieto. Springer Verlag. Lecture Notes in Computer Science, Vol. 686, pp. 143-148. 1993.
[Castaño,93a] *Simulation of Stochastic Regular Grammars through Simple Recurrent Networks.* M.A. Castaño, F. Casacuberta , E. Vidal. In New Trends in Neural Computation. Eds. J. Mira, J. Cabestany, A. Prieto. Springer Verlag. Lecture Notes in Computer Science, Vol. 686, pp. 210-215. 1993.
[Castaño,93b] *Inference of Stochastic Regular Languages through Simple Recurrent Networks.* M.A. Castaño, E. Vidal, F. Casacuberta. In Procs. of the First International Conference on Grammatical Inference. 1993.
[Chen,95] *Constructive Learning of Recurrent Neural Networks: Limitations of Recurrent Cascade Correlation and a Simple Solution.* D. Chen, C.L. Giles, G.Z. Sun, H.H. Chen, Y.C. Lee, M.W. Goudreau. IEEE Transactions on Neural Networks. 1995. In press
[Cleeremans,89] *Finite State Automata and Simple Recurrent Networks.* A. Cleeremans, D. Servan-Schreiber, J.L. McClelland. Neural Computation, no. 1, pp. 372-381. 1989.
[Das,93] *Using Hints to Successfully Learn Context-Free Grammars with a Neural Network Pushdown Automaton.* S. Das, C.L. Giles, G.Z. Sun. In Advances in Neural Information Processing Systems 5. Eds. C.L. Giles, R.P. Lipmann. 1993.

[Elman,88] *Finding Structure in Time*. J.L. Elman. Technical Report No. 8801. Center for Research in Language. University of California. La Jolla. 1988.

[Fahlman,91] *The Recurrent Cascade-Correlation Architecture*. S.E. Fahlman. Technical Report CMU-CS-91-100, School of Computer Science, Carnegie Mellon University, Pittsburgh. 1991.

[Giles,92a] *Learning and Extracting Finite State Automata with Second-Order Recurrent Neural Networks*. C.L. Giles, C.B. Miller, D. Chen, H.H. Chen, G.Z. Sun, Y.C. Lee. Neural Computation, no. 4, pp. 393-405. 1992.

[Giles,92b] *Extracting and Learning an Unknown Grammar with Recurrent Neural Networks*. C.L. Giles, C.B. Miller, D. Chen, G.Z. Sun, H.H. Chen, Y.C. Lee. In Advances in Neural Information Processing Systems 4. Eds. J.E. Moody, S.J. Hanson, R.P. Lipmann. 1992.

[Giles,92c] *Inserting Rules into Recurrent Neural Networks*. C.L. Giles, C.W. Omlin. In Procs. of the 1992 IEE Signal Processing, pp. 13-22. 1992.

[Giles,93a] *Rule Refinement with Recurrent Neural Networks*. C.L. Giles, C.W. Omlin. In Procs of the 1993 IEE International Conference on Neural Networks. 1993.

[Giles,93b] *Extraction, Insertion and Refinement of Symbolic Rules in Dynamically-Driven Recurrent Neural Networks*. C.L. Giles, C.W. Omlin. Connection Science, vol. 5, no. 3, pp. 307-337. 1993.

[Goudreau,94] *First-Order vs. Second-Order Single Layer Recurrent Neural Networks*. M.W. Goudreau, C.L. Giles, S.T. Chkradhar, D. Chen. IEEE Transactions on Neural Networks, vol. 5, no. 3, pp. 511-513. 1994.

[Jordan,88] *Serial order: A parallel distributed processing approach*. M.I. Jordan. Technical Report No. 8604. Institute of Cognitive Science. University of California. San Diego. 1988.

[Lucas,93] *Algebraic Grammatical Inference*. S.M. Lucas. In Procs. of the First International Conference on Grammatical Inference. 1993.

[Manolios,93] *First Order Recurrent Neural Networks and Deterministic Finite State Automata*. P. Manolios, R. Fanelli. Technical Report NNRG-930625A, Department of computer Science and Physics, Brooklyn College of the City University of New York. Brooklyn. 1993.

[Maskara,92] *Forcing Simple Recurrent Neural Networks to Encode Context*. Procs. of the 1992 Long Island Conference on Artificial Intelligence and Computer Graphics. 1992.

[McCulloch,43] *A logical Calculus of the Ideas Imminent in Nervous Activity*. W.S. McCulloch, W. Pits. Bulletin of Mathematical Biophysics, vol. 5, pp. 115-133. 1943.

[Maxwell,89] *Generalization in Neural Networks: The Contiguity Problem*. T. Maxwell, C.L. Giles, Y.C. Lee. In Procs. of the International Joint Conference on Neural Networks, vol. 2, pp. 41-46. 1989.

[Miller,93] *Experimental Comparison of the Effect of Order in Recurrent Neural Networks*. C.B. Miller, C.L. Giles. International Journal of Pattern Recognition and Artificial Intelligence. 1993.

[Minsky,67] *Computation: Finite and Infinite Machines*. M.L. Minsky. Chap. 3.5. Ed. Prentice-Hall, Englewood Cliffs, New York. 1967.

[Omlin,92] *Training Second-Order Recurrent Neural Networks using Hints*. C.W. Omlin, C.L. Giles. In Procs. of the Ninth International Conference on Machine Learning. 1992.

[Omlin,93] *Pruning Recurrent Neural Networks for Improved Generalization Performance*. C.W. Omlin, C.L. Giles. Technical Report No. 93-6. Computer Science Department, Rensselaer Polytechnic Institute, Troy, N.Y. 1993.

[Pollack,91] *The Induction of Dynamical Recognizers*. J.B. Pollack. Machine Learning, no. 7, pp. 227-252. 1991.

[Rumelhart,86] *Learning sequential structure in simple recurrent networks*. D.E. Rumelhart, G. Hinton, R. Williams. Parallel distributed processing: Experiments in the microstructure of cognition, vol. 1. Ed. Rumelhart,D.E. McClelland,J.L. and the PDP Research Group. MIT Press. Cambridge. 1986.

[Sanfeliu,92] *Understanding Neural Networks for Grammatical Inference and Recognition*. A. Sanfeliu, R. Alquézar. In Advances in Structural and Syntactic Pattern Recognition, pp. 75-948. Ed. H.Bunke. 1992

[Servan,88] *Encoding sequential structure in simple recurrent networks*. Servan-Schreiber, D.A. Cleeremans, J.L. McClelland. Technical Report CMU-CS-183. School of Computer Science. Carnegie Mellon University. Pittsburgh, PA. 1988.

[Servan,91] *Graded State Machines: The Representation of Temporal Contingencies in Simple Recurrent Networks*. D. Servan-Schreiber, A. Cleeremans, J.L. McClelland. Machine Learning, no. 7, pp. 161-193. 1991.

[Smith,89] *Learning Sequential Structure with the Real-Time Recurrent Learning Algorithm*. A.W. Smith, D. Zipser. International Journal of Neural Systems, vol. 1, no. 2, pp. 125-131. 1989.

[Sun,90] *Connectionist Pushdown Automata that Learn Context-Free Grammars*. G.Z. Sun, H.H. Chen, C.L. Giles, Y.C. Lee, D. Chen. In Procs. of the International Joint Conference on Neural Networks, vol. 1, pp. 577-580. 1990.

[Tomita,82] *Dynamic Construction of Finite-State Automata from Examples using Hill-Climbing*. M. Tomita. In Procs. of the Fourth Annual Cognitive Science Conference, pp. 105-108. 1982.

[Watrous,92] *Induction of Finite-State Languages Using Second-Order Recurrent Networks*. R.L Watrous, G.M. Kuhn. Neural Computation, no. 4, pp. 406-414. 1992.

[Williams,89] *Experimental Analysis of the Real-time Recurrent Learning Algorithm*. R.J. Williams, and D. Zipser. Connection Science, vol. 1, no.1, pp. 87-111. 1989.

[Zeng,94] *Discrete Recurrent Neural Networks for Grammatical Inference*. Z. Zeng, M. Goodman, P. Smith. IEEE Transactions on Neural Networks. Vol. 5, no. 2, pp. 320-330 1994.

Learning Transformed Prototypes (LTP) – A Statistical Pattern Classification Technique of Neural Networks

Y Guan, T G Clarkson and J G Taylor*

Department of Electronic and Electrical Engineering
*Department of Mathematics
King's College London
Strand, London WC2R 2LS, UK

ABSTRACT

A statistical pattern recognition algorithm called learning transformed prototypes (LTP) is developed for probabilistic RAM (pRAM) neural networks. With LTP the pRAM net learns to map statistically the input sets to the output prototypes, or codebook vectors, in the binary domain. The method allows the pRAM net to self-organise the codebook vectors in the output space of arbitrary dimension. The similarities and differences of LTP with those algorithms such as LVQ (learning vector quantisation), SOFM (self-organised feature maps) and pRAM reinforcement learning are discussed. The training data processed in the method is the input-output spike series of the neural net, therefore the technique can be built into a hardware system with the currently available pRAM learning chips.

INTRODUCTION

This paper presents a statistical pattern recognition method, learning transformed prototypes (LTP), for pRAM neural networks. Developed from the pRAM reinforcement learning rule (e.g., Gorse and Taylor 1989, Guan, Clarkson, Taylor & Gorse 1992, 1993, 1994), the LTP algorithm trains a neural net to perform a stochastic mapping of input sets to the self-organised output *codebook vectors*, or *prototypes*, in the binary domain. Unlike the algorithms in conventional neural networks such as LVQ (learning vector quantisation) and SOFM (self-organising feature maps) (e.g. Kohonen 1982, 1989 and 1992), where the network structure is limited to a single layer, LTP allows multi-layer networks. The favoured responses of the networks (or "winners") in LTP are not selected as the winning neuron nodes, as they are in SOFM, but chosen as the winning output vectors (or prototypes) of the network which fall into the output hyperspace of 2^n vector elements but created from only n output neurons. The way of selection of the winners in LTP is also quite different from those methods of direct comparison of input vectors with winner-associated weights. The dimension of output prototypes in LTP may be different from that of input sets. With considerations on its implementation, however, the LTP algorithm can be regarded as a statistical learning algorithm comprising features from both LVQ and SOFM. Some pattern classification tasks have been performed with LTP and initial results have shown that the training process speeded up dramatically compared with the previous pRAM training algorithms while maintaining the "sub-optimal" classification which maximises rewards. This will be described later in the paper, though the method of optimising the classification is still under investigation.

REVIEW OF RELATED PATTERN RECOGNITION METHODS

Most pattern recognition tasks involve statistical classification activities. Based on classical statistical decision-making theory, pattern recognition methods with neural networks have been developed and successful application results have been achieved. This section described several classification methods that are related to the proposed LTP algorithm in one or more aspects.

Bayes classifier

As a classical statistical pattern classification method, Bayes classifier was widely discussed and regarded as the theoretical foundation for many classification methods. We take the symbol representations used in, Kohonen 1992 for instance, and assume $x \in R^n$ is the vector of an input set, and $(C_i, i=1,2,....,L)$ is the set of classes to which x may belong. Let $p(x|C_i)$ be the probability density function of x in class C_i, and $P(C_i)$ the a priori probability of occurrence of samples from class C_i. Then $d_i(x) = p(x|C_i)P(C_i)$ corresponds to the class distribution of those samples of x which belong to class C_i. The classification rule is defined as:

$$x \text{ is classified as belonging to } C_i \text{ if } d_i(x) > d_j(x) \text{ for all } j \neq i$$

and the decision surface (decision border) of neighbouring classes C_i and C_j for input x is defined by

$$p(x|C_i)P(C_i) = p(x|C_j)P(C_j).$$

The main problem existing in Bayes classifier is that the analytical expressions for $d_i(x)$ are usually not available, while in neural network algorithms the decision surface can be learned adaptively.

Learning vector quantisation (LVQ)

Learning vector quantisation (LVQ) was originated from vector quantisation (VQ) (Gray 1984) which approximates continuous functions of vectorial variables by a finite number of codebook vectors. The approximation criterion is that the distance of the input sample from the closest codebook vectors in some metric is minimised on the average.

If $x \in R^n$ is the input vector, and the codebook vectors are denoted by $m_i \in R^n$, i = 1,2,...,c,..., then the following iterative process leads to the minimisation of the distance:

$$m_c(t+1) = m_c(t) + \alpha(t)[x(t)-m_c(t)]$$
$$m_i(t+1) = m_i(t) \text{ for } i \neq c. \tag{1}$$

where t = 0,1,2,... and $0 < \alpha(t) < 1$.

The unsupervised learning process of VQ was extended by Kohonen to LVQ and three types of LVQ were introduced: LVQ1, LVQ2 and LVQ3. Each of the input samples is belong to one of the classes and each codebook vector selected represents one of these classes. During learning, the codebook vectors are updated so as to optimally represent the assigned classes, whereas during testing, an unknown input x is classified as belong to a certain class if the chosen codebook vector of that class is nearest to x. This is a supervised learning method since the class-affiliation of input samples is known before training.

The novelty of LVQ lies in the optimisation for the decision surfaces of class distributions of input samples. LVQ is subdivided into three algorithms LVQ1, LVQ2 and LVQ3, in which the codebook vectors are updated with different strategies.

LVQ1: In LVQ1 algorithm, the selected codebook vectors that correctly classify the input samples are moved so as to reduce the classification error, while the codebook vectors that lead to misclassifications are moved in the opposite direction. The learning process of LVQ1 is formulated as the following:

$$m_c(t+1) = m_c(t) + \alpha(t)[x(t)-m_c(t)]$$
$$\text{if } x \text{ and } m_c \text{ belong to the same class}$$
$$m_c(t+1) = m_c(t) - \alpha(t)[x(t)-m_c(t)]$$
$$\text{if } x \text{ and } m_c \text{ belong to different classes}$$
$$m_k(t+1) = m_k(t) \text{ for } k \neq c. \tag{2}$$

where $0 < \alpha(t) < 1$.

LVQ2: The updating rule of LVQ2 is defined as the following:

$$m_i(t+1) = m_i(t) - \alpha(t)[x(t)-m_i(t)]$$
$$m_j(t+1) = m_j(t) + \alpha(t)[x(t)-m_j(t)] \qquad (3)$$

where m_i is supposed to be the nearest codebook vector to x and m_j the next-to-nearest one, respectively, and x belongs to C_j but not C_i. Two codebook vectors are updated at a time. One is called the "winner", and the other, the "runner-up", which represent C_i and C_j respectively. The midplane of there two vectors is shifted towards the place where the Bayes decision surface is supposed to exist so that the misclassification rate is reduced.

LVQ3: This is an improved algorithm of LVQ2 and its updating rule is defined as the following:

$$m_i(t+1) = m_i(t) - \alpha(t)[x(t)-m_i(t)]$$
$$m_j(t+1) = m_j(t) + \alpha(t)[x(t)-m_j(t)]$$
where m_i and m_j are the two closest codebook vectors to x, and if x and m_j belong to the same class, while x and m_i belong to different classes
$$m_k(t+1) = m_k(t) + \varepsilon\alpha(t)[x(t)-m_k(t)]$$
for $k \in [i,j]$, if x, m_i and m_j belong to the same class. $\qquad (4)$

Compared with LVQ2, it is noted that LVQ3 is obtained by adding correction factors multiplied by a small parameter ε to LVQ2. With this correction, the codebook vectors can be prevented from staying at non-optimal stationary places.

Self-organising feature maps (SOFM)
Self-organising neural nets create abstract representations for input features in the form of ordered mappings. For a given input x, one of the neuron nodes is defined as the "winner" if its codebook vector m_c matches best with the input. Then updating is performed of that node, as well as those that lie in the topological neighbourhood N of node c. That is, if node c is the "winner" satisfying

$$\| x(t) - m_c(t) \| = \min \{ \| x(t) - m_i \| \} \text{ for all } i \qquad (5)$$

in Euclidean distances, then the updating rule is

$$m_i(t+1) = m_i(t) + \alpha(t)[x(t)-m_i(t)] \qquad \forall i \in N_c(t)$$
$$m_i(t+1) = m_i(t) \qquad \forall i \notin N_c(t) \qquad (6)$$

The pRAM LEARNING TRANSFORMED PROTOTYPES (LPT) ALGORITHM
The pRAM neuron is a RAM-based neuron model. The weights of a pRAM neuron are the 2^N memory contents α_u (where u is an n-bit binary vector) which take on a continuous range of values in $[0, 1]$ (Gorse and Taylor 1989).

The output $a \in \{0, 1\}$ of the pRAM is 1 with probability dependent on the binary input vector i and given by

$$Prob(a = 1|i) = \sum_u \alpha_u \prod_{j=1}^{N} \left(i_j u_j + \bar{i}_j \bar{u}_j \right) \qquad (7)$$

where for any real number z, $\bar{z}$ is defined to be equal to $1 - z$, and where $i = (i_1, i_2, ..., i_N)^T$, $u = (u_1, u_2, ..., u_N)^T$ (N-dimensional binary vectors), and $\alpha = (\alpha_{00...00}, \alpha_{00...01}, ..., \alpha_{11...11})^T$ (2^N-dimensional real vector). It is clear that this exhibits a response which is of maximal non-linearity in the components of i. In the deterministic case (when $\alpha \in \{0,1\}^{2^N}$) pRAM can compute any of the 2^{2^N} possible Boolean functions of its inputs. In the language of parallel distributed processing (PDP), a pRAM is the ultimate Σ–Π unit in the binary domain, since products up to order N enter on the right hand side of Eq.(7). A pRAM is also maximally stochastic, in that the probabilistic aspects of its behaviour are governed by the 2^N random variables α_u rather than a single stochastic threshold variable.

Reinforcement training of pRAM networks

The pRAM reinforcement learning [Gorse and Taylor 1991] is an extension of Barto's A_{R-P} algorithm (Barto and Anandan 1985) and can be written as:

$$\Delta\alpha_u(t) = \rho[(a - \alpha_u)r + \lambda(\bar{a} - \alpha_u)p](t) \times \delta_{u,i} \qquad (8)$$

where $r, p \in \{0, 1\}$ are global success and failure signals emitted with a probability dependent both on the binary input i the pRAM and the pRAM's output a:

$$Prob(r = 1|a, i) = \rho_{a,i}$$
$$Prob(p = 1|a, i) = \pi_{a,i}$$

Note that $\pi_{a,i} \neq 1 - \rho_{a,i}$. Independent reward and penalty signals allow the possibility of "neutral" actions which are neither punished nor rewarded but which may correspond to a useful exploration of the environment. The aim of the reinforcement rule is to update the α_u so as to increase the probability of actions which lead to a reward $r = 1$, while discouraging actions which lead to a punishment $p = 1$. This can be achieved by the update rule Eq.(8).

The constant λ represents the ratio of punishment to reward; a nonzero value of λ is necessary in order to prevent the system converging on false minima. When $r = 1$ (success) the probability α_u changes so as to increase the chance of emitting the same value, a, from that location in the future, while if $p = 1$ (failure) the probability of emitting the other value, $1 - a$, when addressed increases. The probabilities of reward and penalty are independent in this model; this allows the possibility of "neutral" actions which are neither punished nor rewarded but may correspond to a useful exploration of the environment. The presence of the Kronecker delta in Eq.(8) ensures that the update only occurs at the location which was accessed at time t; in the continuous-input extension to be presented below the role of the $\delta_{u,i}$ will be played by a more general distribution function.

The noisy reinforcement learning rule (e.g. Guan, Clarkson, Taylor & Gorse 1992, 1993, 1994) is an extension of the above algorithm. With dynamically injected noise during training, the generalisation property of the pRAM nets can be improved and the optimal representation of input classes can be achieved in the weight space of the networks. The noisy processing theory and weight space analysis of simple system prototypes are given in Guan, Clarkson, Gorse and Taylor 1994.

The learning transformed prototypes (LTP) algorithm

As it is seen in the reinforcement learning algorithm, a pRAM net is usually assigned a set of binary target output codes, or vectors, for the input sets, where binary output coding is normally preferred, and the network is then trained to map each input class to its output vector that represents the class to which it belongs. This is a kind of supervised learning and has been proved to be very effective for classification tasks that requires strong generalisation. It has been realised, however, that the pre-assignment of the output target vectors for input classes may not be optimal in the output vector space and this will affect the classification rate. For the optimal assignment of target vectors, the vector distribution in the output space should be able to best represent the input classes in a certain manner. For example, the target vectors could have maximum distance (e.g. in Hamming distance) between each other; the vectors could be in such positions that over-lapping parts of input classes can be mapped into some non-target vectors that are nearest to their target vectors, rather than any other vectors. It is also realised that the previous pRAM learning algorithms takes much longer time to converge in the cases where the number of classes and the number of layers of the net are large. The increase of training time in respect to large numbers of pre-assigned target vectors and network layers is the inherent problem which exists in reinforcement learning rules. Two, three or more layers seems inevitable in RAM-based neural networks since otherwise increasing the input number to a RAM neuron will exponentially increase the memory space and this is obviously impractical. Then it is natural to think of improving the way of assigning the target vectors for input classes.

The output coding methods discussed before (e.g. Guan, Clarkson, Taylor & Gorse 1994) are basically unary coding and binary coding. Like many other neural networks, unary coding in a pRAM net allow only one of the output nodes to respond to one class of inputs. If n is the number of output nodes, or the dimension of output vector space, and l is the number of input classes, then it must satisfy

$$n \geq l.$$

In binary coding, however, the target vectors can take any of the whole $\{0, 1\}^n$ space and the number of the output nodes can be greatly reduced by just satisfying

$$n \geq \log_2(l).$$

The basic idea of learning transformed prototypes (LTP) is to allow a pRAM net to create and "self-organise" its target output vectors, or prototypes in this sense, in $\{0,1\}^n$ output space, while the network is being trained to perform this mapping. If the transform is denoted as F_p, and y is the output vector produced after the transformation, the mapping relation can be written as

$$x \in \{0, 1\}^m \rightarrow F_p(x) \rightarrow y \in \{0, 1\}^n \qquad (9)$$

where the subscript p specifies that the transform is a stochastic process with elements of the transform matrix being probabilities. In fact, Eq.(9) is another form of Eq.(7). $x \in \{0,1\}^m$ is the input set of dimension m, and $y = [y_1, y_2, ..., y_n]^T \in \{0,1\}^n$ is the output vector. Every prototype obtained from LTP should be able to represent one of the input classes, and every input should be classified by its assigned prototype with a reasonable probability. Due to its stochastic property, a pRAM network may generate different output vectors for an input x at different running times, depending on the weight distribution of the net. Suppose y^i is the prototype, or codebook vector, of input $x^i \in C_i$, where $\{C_i, i=1, 2, ...,l\}$ is the set of classes, the prototype y^i selected should represent the input class. Since the input space and the output space may have different dimensions in LTP, it is impossible to measure directly the input and output with Euclidean distance as it is done in LVQ and SOFM. Therefore, the selection of prototype y^i ought to be preferably by some sort of self-organisation mechanism rather than by direct comparison between the input and output vectors; the prototype chosen should be capable of best representing the input class C_i.

Following the above ideas it is then suggested that y^i should be selected from the set of output vectors that are generated from the network driven by input x^i. Furthermore, it is reasonable to chose the output vector that appears most often, in a fixed number of runs driven by the input, as the prototype. The prototype is then the largest response of the net for the input in each iteration. That is

$$y^i = \{y : \max prob(output = y | x^i \in C_i) \quad for\ all\ y \in \{0,1\}^n\} \qquad (10)$$

and what follows is the update phase in which the reinforcement learning rule (8) is used as in the previous pRAM training algorithms. Thus reward or penalty is assigned for other pattern of C_i if input of that new pattern leads to the prototype y^i or not.

LTP is regarded as an unsupervised learning algorithm. Unlike LVQ, where the codebook vectors are directly compared with input and chosen for the purpose to approximate the input classes (see Eqs. 1-6) , the prototypes of LTP are chosen according to the output responses or firing rates, i.e., how frequently a vector appears as the response to the input. In general, LTP is a process of learning of mapping of the input sets to the output prototypes.

It is noted that the nomination of the prototype y^i in LTP is an adaptive process and the prototype for x^i is expected to change during training. Suppose the initial prototype is y^i for input x^i , then the prototype sequence is $y^{i(0)}$, $y^{i(1)}$, ..., $y^{i(k)}$, ..., which is due to the stochastic property of the pRAM net and competitive effects between input samples. This property provides the possibility of optimising the LTP learning in pattern recognition. If we assume the error function in LTP is:

$$E = \sum_{C_i, x^i \in C_i} -\ln p(y^i | x^i \in C_i) \qquad (11)$$

where $p(y^i | x^i \in C_i)$ is the probability that the network generates prototype y^i for input x^i, then the process of learning to get stabilised prototypes is the process of minimising the error E.

Pattern classification

LTP learning is a unsupervised learning process of self-organising the prototypes of input classes in output vector space. On the other hand, LTP is also an application of the pRAM reinforcement learning rule as discussed early in this paper. Within each training epoch, which consists of a comparison phase and a updating phase, LTP works like the reinforcement learning as shown in Eq.(8). That is, once a prototype is selected in the comparison phase, it is kept constant as the target vector in the updating phase, though it may be different in the next comparison phase, and Eq.(8) is applied for memory modification. As LTP learning approaches convergence, i.e., the prototype of every input class remains constant for all training epochs at this time and the time after and is of reasonably high appearance frequency, LTP settles to be the normal reinforcement learning.

It has been proved in Guan, Clarkson, Gorse and Taylor that training noise is crucial in optimisation of the pRAM neural networks. Optimal representation of class structures has been achieved in pRAM nets with pre-designated target output vectors by using the noisy reinforcement learning algorithm. Then it is expected that, beside speeding up the learning process, the LTP algorithm maintains at least the optimisation features of pRAM reinforcement learning, i.e., optimal representation of input classes in network's weights space. This might be called "sub-optimisation" in the context of LTP, which can be easily realised by noise injection to the network at the time when the LTP learning process starts to converge. The overall optimisation of the LTP algorithm, which includes the optimisation of prototype distributions in the output vector hyperspace, may still lie in the appropriate use of injected noise in the training process.

LTP has been used for a character recognition task and a speaker voice identification task. In the first task, the training sets are the ten digits displayed as binary images. In the second task the training sets consist of 148 speech feature vector samples belonging to four speakers and the testing set are 12 samples, and these samples are digitised into binary images before use. In both applications, the image sizes are 16 x 16 pixels and the pRAM network consists of three layers: 32, 8, 4 8-input nodes in the first, second and output layer, respectively. With LTP, the network converges at about 20 epochs for the first task and about 30 epochs for the second. With normal pRAM reinforcement learning, however, the network does not converge even after 300 epochs for both tasks with the same training parameters such as ρ and λ. This is because the pre-assigned target output vectors act as a sort of hard limitation on the network and the prototypes of the LTP algorithm are assigned naturally in accordance with the network responses to the input sets. The trained network classified the training and testing sets with 100 percent success rate.

The initialisation of LTP is quite simple. The weights of the neural network are set to be neutral, i.e., at the value of 0.5, and the initial choice of prototypes are therefore completely random. The final formation of prototypes at the stable stage are obtained from the self-organising process of output vectors.

DISCUSSION

This paper has described the learning transformed prototypes algorithm and its applications in pattern classification. At each operation step, the input and output signals of neurons are 0, 1 signals which enable direct hardware realisation with pRAM learning chips (e.g. Clarkson, T. G., Gorse, D., Taylor, J. G. & Ng., C. K. 1992), though the weights (or firing probabilities) take real values.

The LTP algorithm is an efficient pattern recognition technique. The initial results have demonstrated that the pRAM net trained with LTP possesses good generalisation abilities due to the inherent features from pRAM reinforcement learning introduced before. The speed of convergence is greatly increased compared with the previous pRAM reinforcement learning algorithms. LTP has the features of LVQ and SOFM in that its prototypes are self-organised in the output vector hyperspace and the prototypes are updated according to the statistical properties of the input sets. In order to realise optimal pattern classification, LTP has to be used in conjunction with training noise and this research is in progress.

Based on the previous discussions of LTP, LVQ and SOFM, the comparisons are listed in the following table:

	LVQ	SOFM	LTP
Codebooks:	In input space; real domain; finite number (= number of input classes).	In input space; class boundary (no prototypes); infinite number (as many as the number of nodes in the net).	In transformed space (reduced dimension); binary domain; finite number (= number of input classes).
Learning:	$LVQn$ movement by direct comparison (Eqs 2, 3 and 4).	By direct comparison (Eqs 5, 6).	Learning better transformation by reinforcement so leading to better prototypes (Eqs 8 - 11).

Table 1. The comparisons of LVQ, SOFM and LTP algorithms.

REFERENCES

Barto, A. G. & Anandan, P. (1985). Pattern recognising stochastic learning automata. *IEEE Transactions on Syst. Man Cyb.*, **15**, 360-374.

Clarkson, T. G., Gorse, D., Taylor, J. G. & Ng., C. K. (1992). Learning probabilistic RAM nets using VLSI structures. *IEEE Transactions on Computers.* **41**, 1552-1561.

Clarkson, T. G., Guan, Y., Gorse, D. & Taylor, J. G. (1993). Generalisation in probabilistic RAM nets, *IEEE Transactions on Neural Networks*, **4**, 360-364.

Gorse, D. & Taylor, J. G. (1989). An analysis of noisy RAM and neural nets. *Physica D.* **34**, 90-114.

Gorse, D. & Taylor, J. G. (1991). Universal associative stochastic learning automata. *Neural Network World*, **1**, 193-202.

Gray, R. M. (1984). Vector quantisation. *IEEE ASSP Mag*, **1**, 4-29.

Guan, Y. Clarkson, T. G., Taylor, J. G. & Gorse, D. (1992). The use of encoded outputs and reinforcement training in pRAM nets. In I. Aleksander & J. G. Taylor (Eds.). *Artificial neural networks, 2* (pp.653-656), London: North Holland.

Guan, Y. Clarkson, T. G., Taylor, J. G. & Gorse, D. (1994). Noisy reinforcement training for pRAM nets". *Neural Networks*, Vol. 7, No. 3, pp.523-538, 1994.

Kohonen, T. (1982). Self-organized formation of topologically corrected feature maps. *Biological Cybernetics*, **43**, 59-69.

Kohonen, T. (1989). *Self-organization and associative memory*. Third ed. Berlin, Heidelberg: Spring-Verlag.

Kohonen, T. (1992). Learning vector quantisation and the self organising map. In J. G. Taylor & C. L. Mannion (Eds.). *Theory and applications of neural networks*, pp.235-242.

Fuzzy Function Estimators as Basis on Learning from Experience

R. Ferreiro Garcia and F.J. Perez Castelo
Dep. Electrónica y Sistemas. Universidad de La Coruña
E.S. Marina Civil, C/ Paseo de Ronda 51, 15011. La Coruña. SPAIN
Tel.: Int+81 25 67 00. Fax.: Int+81 25 15 68

Abstract: This paper describes an alternative method to neural networks for description of high nonlinear systems identification as well as systems in which being nonlinear there are variables with very high time varying rate with respect to the other system variables.It consists in a learning algorithm to be applied in process control, covering several topics of control applications such as system identification, observer design and adaptive control in a simple and useful way which make the method reliable to be applied on industrial process control. Knowledge acquired by means of a proposed learning algorithm is stored into a DAM or FAM (deterministic or fuzzy associative memory) for finally be applied on controller mapping, state observer mapping or model parameter mapping. With such mappings, control design techniques may be applied included the adaptive/learning or hybrid control algorithms.

Keywords Fuzzy associative memory, Learning algorithm, Membership-function set, Universe of discourse.

Introduction

The fundamental process involved in controlling a dynamical system include the mathematical modeling, identification of the system based based on experimental data, processing of the outputs, and using them in turn to synthesize control inputs to achieve the desired behavior. In spite of the advances carried out last years, our understanding of the fundamental theoretical issues involved in the control of nonlinear multivariable systems is not well developed, and the best developed aspect of control theory continues to be that related to systems defined by linear operators.

Function estimators on the basis of learning algorithms have found their way into standard practice, opened to a wide spectrum of complex applications. Such systems are characterized by poor models, high dimensionality of the decision space, hierarchies, multiple performance criteria, distributed sensors and actuators, high noise levels and complex information patterns. All that system's characteristics may be summarised in three categories (Narendra, Astrom 1992): computational complexity, presence of nonlinearities and uncertainty. In recent years, motivated by the superior abilities of biological systems to cope with complex environments, of the type described avobe, researchers have been exploring the use of fuzzy function estimators for information processing including some subcategories like the one defined as fuzzy associative memories (Bart Kosko, 1992).

There are basically two alternatives to machine learning: Neural networks and fuzzy function estimators. While the neural approach requires the especification of a nonlinear dynamical system, usually feedforward, the acquisition of a suficiently representative set of numerical training samples, and the encoding of such training samples in the dynamical system by repeating learning cycles, fuzzy systems requires only to fill in a linguistic rule matrix, which is much simpler than design and train a neural network.

The massively parallel nature of the fuzzy associative memories, FAM, permits computations to be performed at high rates; the FAM's which contains nonlinear components can be used to approximate nonlinear maps to any desired degree of accuracy. The fact of adjusting the clustered information contained into a FAM using input-output data implies that they can be used as adaptive or learning systems under different conditions of uncertainty. Using these attractive feactures, FAM's have been shown to be extremely efficient in pattern recognition problems, system identification and system control in a greater number of satisfactory applications than rule-based expert systems.

The problem of mapping I/O information from a dynamic or static system differs from a system to another because it depends strongly not only on the nonlinear characteristics of the system to be mapped but also in the ammount of data to be clustered into a limited space. As much information as it is being achieved from a learning process much more laborious with the inherent time cost, is the task of clustering and retrieving into and from the FAM such information. Mapping information from a system to be clustered by a learning process is the aim of this work and it is based in the fuzzy function estimation theory (Kosko, 1992), (Ferreiro, 1993).

Basic problem formulation on function estimators

Learning process may be defined as the capacity to store data related to expirience. From this point of view, function estimators are learning systems

Learning from samples of syncronous input/output data means mapping a reaction model by storing data into a FAM or product clustering space. Funtion estimators include dynamical systems defined by transfer functions which implies plant models and controller models.

An interesting feature of learning-fam algorithms is that they provide a generic class of nonlinear functions. This is used in the following ways when modelling nonlinear dynamical systems (Astrom and Thomas J. McAvoy., 1992). Consider, for example, the linear model defined by the phase state space variable method

$$dx/dt = Ax + Bu \qquad (1)$$

$$Y = Cx \qquad (2)$$

Expression (1) and (2) can be replaced with

$$D^n(x) = f[K_{n-1}D^{n-1}(x), K_{n-2}D^{n-2}(x),..$$

$$...K_0x,u] \qquad (3)$$

$$Y = g[C_{m-1}D^{m-1}(x), C_{m-2}D^{m-2}(x),.C_0x] \text{ with } m < n$$

where the function f is a set of input variables to the hypercube which defines the fuzzy associative memory (FAM) and will be used to get the system model by the application of the proposed real time training-learning algorithm, which stores the knowledge on a FAM system, and the $D^n x$ is the n order derivative of variable x. Figure 1 illustrates the mapping task in which input and output data from the system is processed under a training learning phase, and is being accumulated into a fuzzy associative memory.

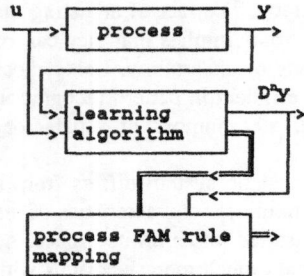

Fig. 1.Learning task from system samples.

Learning from system behaviour

The problem of learning from samples or expirience requires that all available data be stored in order to map the most relevant chraracteristics of that particular model. In doing so, the neccessity of accuracy in problem representation demand a variable space clustering.

The training/learning task (Ferreiro, 1994) is a previous part of the identification procedure and is carried out in fixed clustering space by following the steps described below as,

-Find the universe of discourse of any input and output variables.
-Initialise the DAM by filling it in the training learning process with the nominal values at the meadle of its range.
-Begin an inner loop until time for arithmetic mean expires
-Read I/O values
-Compute the values of membership-function for each input variable
-Store the result of adding actual to past data as,

$$U_T(j,k) = U_T((j-1,k-1)$$
$$N(j,k) = N(j-1,k-1)+1$$

where j and k are the membership-function values which belong to input variables.
-Come back on inner loop until the time for evaluations is reached.
-Compute arithmetic mean under an hypercube rule-base and update DAM

$$U_{To}(j,k) = .5*[U_{To}(j,k)+U_T(j,k)/N(j,k)]$$

-wait for update the FAM or DAM information.
-Come back on outer loop, repeating training /learning process loop.

Some practical function estimation algorithms

The proposed method is centred in one of the popular concepts in approximation theory: least-square curve fitting, or regression (the statistical determination of parametric dependencies, e.g., the polynomial coefficients of a least-square curve fit to a set of data),[4]. The simplest and most common fitting function is the straight line.

If we are given a set of data values (x_i, y_i) where x is defined as the independent variable and y the dependent variable, we have

$$y_i = a + b \cdot x_i + E_i \qquad (1)$$

where E_i represents the measurement error. Therefore, a and b cannot be obtained from equation 1. The least squares method can be stated as follows: The "best" values of a and b are those that minimize S(a,b):

$$S(a,b) = \sum_{i=1}^{i=n} (y_i - a - b \cdot x_i)^2 \qquad (2)$$

The above least-square demonstration example can be generalized to an Mth-degree polynomial in which M <= N-1.

As the parameters a and b are to be chosen so that S(a,b) is minimized, at this minimum, it is the case that

$$\frac{\partial S}{\partial a} = 0, \frac{\partial S}{\partial b} = 0 \qquad (3)$$

Thus, we have the two equations:

$$\sum_{i=1}^{i=n} (y_i - a - b \cdot x_i) = 0 \quad , \quad \sum_{i=1}^{i=n} x_i \cdot (y_i - a - b \cdot x_i) = 0 \qquad (4)$$

The solutions for the coefficients are then

$$a = \bar{y} - b\bar{x} \qquad (5)$$

$$b = \frac{\sum (x_i - \bar{x})(y_i - \bar{y})}{\sum (x_i - \bar{x})^2} \qquad (6)$$

where

$$\bar{x} = (\frac{1}{N}) \sum_{i=1}^{i=n} x_i \quad , \quad \bar{y} = (\frac{1}{N}) \sum_{i=1}^{i=n} y_i \qquad (7)$$

The linear least-squares fitting algorithm is a simple and special case of the general least-square polynomial-approximation procedure.

Function estimation based on multidimensional least-squares method

The generalization of least-square fitting by means of the multidimensional approach, will be used in the process ofpolynomial regression defuzzification (PRD).

Extending an application to the two independent variable function, the lest-square solution will be the following polynomial regression given as,

$$Y = A_1 + A_2 x_1 + A_3 x_1^2 + A_4 x_2 + A_5 x_2^2 + A_6 x_1 x_2$$
$$+ A_7 x_1^2 x_2 + A_8 x_1 x_2^2 + A_9 x_1^2 x_2^2 \qquad (8)$$

Function estimation task by on line Lagrange Interpolation

Lagrange interpolation is based on the simple premise that for every set of sequential table values (every one-dimensional rulbase), there exists a unique polynomial curve that passes through each and all those table values. The Lagrange polynomial for the cubic case is, [4],

$f(x) = L_3(x) =$
$[f1(x-x2)(x-x3)(x-x4)/(x1-x2)(x1-x3)(x1-x4)]+$
$[f2(x-x1)(x-x3)(x-x4)/(x2-x1)(x2-x3)(x2-x4)]+$
$[f3(x-x1)(x-x2)(x-x4)/(x3-x1)(x3-x2)(x3-x4)]+$
$[f4(x-x1)(x-x2)(x-x3)/(x4-x1)(x4-x2)(x4-x3)]+$

Then, given a single input crisp value x, the interpolation function supply the output value f(x) for the control variable. At figure 2 it is shown the cubic polynomial fitted to four table values.

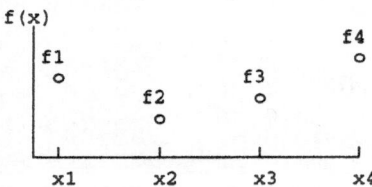

Fig. 2 Three membership labels defined by Lagrange interpolation.

It is to be remarked that the x coordinates of the table values need not be equally spaced. Such property used to define the membership function labels of an universe of discourse defines the x coordinate. The cubic polynomial fitted through these four points is unique, that is, there exists no other cubic polynomial having different coefficients that also passes through the same points. In the figure 2 it is shown how three membership labels of an universe of discourse of any variable can be processed by means of Lagrange interpolation, where the distance between each x are the labels of the fuzzy input variable and the f(x) is the output variable.

Conclusions

Some alternatives to neural networks as a mean to learn the behaviour of a dynamic process has been shown. Extensive test on such methods has been carried out and it is to be remarked that the learning task is condensed under a dynamic process mapping (traslation into a FAM or DAM), which can be updated evrey time we want. Such copy is then used as the acquired knowledge which is applied in illustrating the system dynamics by means of one of the described methods. Continuous nonlinearities are well mapped. High gradient discontinuities are not satisfactory mapped if variable space clustering is not used. So that when a priori knowledge is not available, the training/learning procedure must be carried out by using variable clustering space in order to store more information as a compression method. Finally, the learning speed depends only on the necessary time to scan all input/output process variables because it independent of the perceptron dynamics.

References

Bart Kosko, (1992).
  Neural Networks and Fuzzy Systems. Chap. 8, pp 299-339.
  Prentice Hall International Editions. U.S.A.
Ferreiro García, R. (1994).
  Associative memories as lerning basis applied to the roll control of an
  aircraft. Proceedings of the 5th Int. Symp. AMST'94. Application of
  Multivariable System Techniques, pp 27-32. University of Bradford. West
  Yorkshire, U.K.
Ferreiro Garcia R., (1994).
  FAM rule as basis for poles shifting applied to the roll control of an aircraft,
  Proceedings of the IMACS International Symposium on Signal Processing,
  Robotics and Neural Networks SPRANN'94, pp 375-378, Editors: P. Borne
  T. Fukuda S.G. Tzafestas, E.C.Lille, France
K.J. Astrom and Thomas J. McAvoy, 1992.
  Intelligent control:An overview and evaluation.
  Handbook of Intelligent Control, Chap. I, pp. 3-21, Ed. David A. White and
  Donakd A. Sofge, Ed. Van Nostrand Reinhold. U.S.A
K.S. Narendra, 1992.
  Adaptive control of dynamic systems using neural networks
  Handbook of Intelligent Control, Cap. V, pp. 141-181, Ed. David A. White
  and Donakd A. Sofge, Ed. Van Nostrand Reinhold. U.S.A
Sinha N. K. and Kuszta B. (1983)
  Modeling and Identification of Dynamic Systems. Chap. 3, pp 27-50
  Ed. Van Nostrand Reinhold Company. N.Y. U.S.A.

Connectionists and Statisticians, Friends or Foes?

Arthur Flexer

The Austrian Research Institute for Artificial Intelligence
Schottengasse 3, A–1010 Vienna, Austria
arthur@ai.univie.ac.at

Abstract:

This investigation on relationships between the field of artificial neural networks (connectionism) and statistics starts with a look on relevant work based on a classification of possible points of contact. Then follows a distinction between connectionism seen as a tool for data analysis (engineering connectionism) and seen as a model for human thinking or, as one might say, a tool for cognitive or biological modeling (explanatory connectionism). It will be argued that statistics will have a major impact on the former but a rather minor on the latter. As a consequence, the gap between applied neural network research and research concerning cognitive modeling with artificial neural networks will become even bigger than it already is. Statistics will be adopted as the theory of engineering connectionism and therefore entail its development fom a purely empirical to a fullgrown theoretical science. Explanatory connectionism has its own problems and will have to undergo its own changes. Consequently, it will finally be seen as a science of its own independent from mere data analysis purposes.

1.) Introduction

Already in the early work that started the second rise of connectionism in the mid–1980's, analogies between neural networks and statistical methods were drawn (e.g. Stone 1986). However, it took quite a few years to bring such commonalities to the center of (connectionist) awareness. Statisticians realized, that "neural networks [...] are now beginning to be used in a wide range of subject areas traditionally thought by statisticians to be their domain" (Ripley 1993). Out of this situation of competition emerged an ever growing discussion and argument between statisticians and connectionists about the relationship of the two fields of science. The claim that connectionism is in fact nothing but statistics, is brought forward more and more often both in scientific publications (e.g. "In one sense neural networks are little more than non–linear regression and allied optimization methods" Ripley 1993) and in more informal media (e.g. the comp.ai.neural–nets newsgrpoup on the internet). Since at the same time the 'media hype' about neural networks vanishes and in consequence lots of interest and funding is going back to statistics, the question on the connectionist side seems to arise as to what a statistician is to a connectionist: a friend or a foe?

Since such criticism on connectionism is often expressed generally without any further discrimination, it is highly overdue to try shed some light on the proper treatment of the relationship between statistics and connectionism. Such a systematic account is even more urgent, because the already very sharp argument seems to lead to a condemnation of the idea of connectionism in its entirety.

2.) Points of contact between connectionism and statistics

The following overview is divided into paragraphs based on a classification of possible points of contact. The first section is concerned with the theoretical comparison of artificial neural networks and statistics based on a formalization in a joint famework. The second section deals with the empirical comparison of connectionist and statistical methods in the routine of data analysis and is therefore about competition. The third section is concerned with the role that statistics can play in neural network experimentation, especially in the evaluation of such studies.

2.1) Theoretical comparison

To allow for a comparison of neural network models and statistical models on a theoretical basis, both kinds of models must be described in one joint theoretical framework. Up to now, most of the work in this line of research has been done by statisticians (see e.g. Ripley 1992, Sarle 1994). Naturally, the theory used in such endeavours is that of statistical mathematics.

Ripley (1993) concentrates on "the engineer's viewpoint" in connectionism that tries to find "new computing and pattern recognition paradigms" in contrast to the biologist's view. He gives an overview of the statistical aspects of a variety of neural network algorithms and compares their performance on one data set to statistical procedures (see also 2.2).
Concerning the theoretical comparison, his emphasis is on feed forward neural networks. He comments on problems like the rather unexplored convergence behaviour of the backpropagation algorithm. He elaborates on the problem of getting stuck in local minima instead of reaching the global minimum whilst minimizing the total squared error (i.e. the sum of the squared differences between network outputs and target outputs).
According to Ripley, this minimization of the squared error could be achieved more effectively by a nonlinear least squares algorithm or by general minimization algorithms such as quasi−Newton and conjugate gradient methods and simulated annealing than by relying on the backpropagation algorithm.

Sarle (1994) provides a translation from connectionist to statistic terminology and points out the common underlying principles. Because connectionist researchers "routinely reinvent methods that have been known in the statistical or mathematical literature for decades or centuries", Sarle is able to provide equivalents in the form of statistical models for most of the more common neural network models.
The following overview is a compilation of Sarle's comparisons:

Neural Network Model	Equivalent Statistical Model
Simple Linear Perceptron	Multivariate Multiple Linear Regression
Simple Nonlinear Perceptron	Logistic Regression
Adaline	Linear Discriminant Function
Multilayer Perceptron	Simple Nonlinear Regression, Multivariate Multiple Nonlinear Regression
Unsupervised Hebbian Learning	Principal Components
Adaptive Vector Quantization	Least−squares Cluster Analysis
Learning Vector Quantization	Variation of Nearest−neighbour Discriminant Ananlysis
Radial Basis Functions	Kernel Regression Methods

He also discusses the statistical view of a few other network models, for which he does not find precise equivalents such as counterpropagation, self−organizing maps or ART networks.
However, Sarle continues stating that the training algorithms for the network models are very often inefficient. One reason being that "they are designed to be implemented on massiveley parallel computers but are, in fact, usually implemented on common serial computers". Another more severe argument is that connectionist researchers "often fail to understand how these methods work" because of their lack of understanding of the underlying statistical principles. For instance, multilayer perceptrons can be trained more efficiently with general purpose nonlinear modeling or optimization programs. Sarle even names the statistical procedures of a common statistical software package to be used for training of the various connectionist models.

The reformulation of neural network models in terms of a statistical framework is often followed by modifications of those models themselves:

Röscheisen et al. (1991) are able to achieve better performance of their neural controller by recasting the problem in Bayes' decision theory leading to a specific network model and training regime. Tresp et al. (1994) deduce an improved algorithm for the training of feedforward neural networks with incomplete data based on Maximum Likelihood considerations.

So the conclusion is that neural network models can be described properly within a statistical framework because of the inherent similarities between statistics and neural networks. What often follows is that statistical procedures are better suited for network training than the originally proposed connectionist learning algorithms. Moreover, the neural network models themselves often have to be altered to achieve optimal performance.

2.2) Empirical comparison

The most common use of neural networks for engineering purposes is that of data analysis. In both classification via supervised learning and clustering via unsupervised learning the main goal is optimal performance of the chosen method for a given problem. Therefore, working on a data analysis problem, the applied neural networks have to compete with appropriate other, e.g. statistical, methods. Only by not restricting oneself to the methods of one single technology, one can hope for optimal solutions.

Which statistical methods are to be chosen for such a comparison should be a result of previous theoretical studies (see 2.1). Because statistical methods are far better understood and there is a general lack of theory within the field of connectionism (see e.g. Dorffner et al. 1991), one should stick to statistical methods as long as they perform at least as good as their connectionist competitors.

Generally, one must say that the quality of research in the area of connectionism definitely needs improvement. In his study, Prechelt (1994a) examines 113 articles from two connectionist "top journals" (*Neural Computation* and *Neural Networks*) for the amount of experimental evaluation they contain. 34% of all articles feature no comparison with other algorithms at all and only 19% compare to more than two known algorithms.

An extensive empirical comparison between 23 statistical, machine learning and neural network methods on 21 different data sets is given in Michie et al. (1994). The different methods employed include algorithms like linear and quadratic discriminants, nearest neighbour and Bayes approaches on the statistical side, various decision tree— and rule based methods on the machine learning, and multi—layer perceptrons, Kohonen and radial basis networks among others on the neural networks side.

Unsurprisingly, no general best method can be identified for the diverse data sets in this study. With the help of statistical and information—based measures of the datasets, it is tried to find some connection between the features of the data sets and the performance of the different methods. Regarding neural networks, the authors conclude that they perform quite well on most of the problems. A radial basis network is the winning algorithm on one data set and for 15 of the 21 data sets at least one neural network algorithm ranks in the top five. However, the authors also point out certain difficulties like computational inefficiency, laborious adjusting of parameters and hard to understand behaviour of the completed networks because of their high degree of nonlinearity.

Ripley (1993) compares various statistical and neural network methods and a decision tree classifier on one data set. Looking at the classification accuracy, some of the networks performed as well as the statistical methods. However, the author also reports about the "degree of frustration" involved with parameter tuning and getting the back—propagation algorithm to converge.

Most of the other empirical studies found in neural network literature do not include statistical methods. Some of them are restricted to comparison of connectionist to machine learning methods (e.g. Schaffer (1993), Shavlik et al. (1991)), most of them remain within the field of connectionism (e.g. Prechelt 1994b who provides performance figures for a number of neural networks as a starting point for further exploration using his benchmark collection of datasets). A comprehensive review of such previous empirical comparisons is provided in Michie et al. (1994).

So the bottom line of this section is that in general comparison of connectionist and statistical methods does not happen as often as it should. If such a performance comparison is undertaken, connectionist methods often perform quite well, although there are a lot of difficulties in applying the algorithms.

2.3) Statistics as a tool for connectionist experimentation

Because most of the neural network learning algorithms remain too complex for rigorous formal treatment, connectionism is mostly performed as an experimental science (see Kibler & Langley (1988) for a discussion within the framework of the related field of machine learning). Such experiments should involve systematic variation of independent variables (e.g. learning rate, number of hidden units) and examination of their effects on dependent variables (e.g. accuracy, learning behaviour). Statistics can be of great help in the proper design and evaluation of those experiments.

When measuring the accuracy of any classification rule, it is widely known that the error rate estimated from the same set of data as that used for construction of the classificator is usually over−optimistic (see e.g. Michie et al. (1994)). In statistical terms it is said that such error rates tend to be *biased*. Therefore it is at least necessary to use different sets of data for training and testing.
An overview of other statistical procedures like the bootstrap algorithm or cross−validation to minimize such a bias is given in Michie et al. (1994).
Paass (1993) discusses the use of bootstrap especially in the context of feedforward neural networks. By applying bootstrap, confidence intervals for the prediction distribution of inputs unknown to the network during training can be derived.

With neural network algorithms, it is necessary to tune some parameters (e.g. learning rate, number of layers, numbers of units, etc.) to get the best performance. As Michie et al. (1994) point out, a division of the available data into three different sets is recommended. One should hold back approximately 20% of the data and divide the remaining data in a set for training and a set for testing, and then tune the parameter using those two sets. The final network should use both training and test data for learning with the now optimized parameters and should then be finally tested with the remaining, never before used, 20% of the data. If the use of such a third independent data set is omiited, the obtained error rates will again be biased and over−optimistic.

Connectionist experimentation involves the comparison of many different test runs. Only by means of such multiple test runs, the effects of the varying of parameters, of different algorithms employed, different kinds of data used and so forth can be observed.
Often such comparisons are made by looking only at absolute numbers like the accuracy achieved in a certain classification task. But only by statistical testing it can be ensured that the observed performance differences are caused by the varied independent variables (e.g. network parameters, kind of network method, etc.) and not by mere chance (i.e. whether the observed phenomena are *significant* in a statistical sense or not).

The following papers use some sort of statistics to compare performance results.
Finnoff et al. (1992) and Hergert et al. (1992b) use "a (robust) modification of a t−test statistic" for comparison of different algorithms. Hergert et al. (1992a) compare some network pruning methods with a four field table $\chi2-$test. Shavlik et al. (1991) investigate different algorithms on a variety of data sets and employ a two−way analysis of variance (ANOVA) to isolate the source of variation in the results. Additionally, a t−test for paired differences is used.
Egmont−Petersen et al. (1994) give an excellent overview on statistical evaluation of neural network classifiers especially for the medical domain, including e.g. standard errors and confidence intervals for sensitivity and specificity.

All the problems connected with empirical research and experiment design are well known to statisticians. A closer look at connectionist literature reveals that there is only little awareness of such issues within the neural network community.

3) What kind of impacts on what kind of connectionist research?

In the previous section, the relationship between statistics and connectionism has been elaborated based on a survey of possible points of contact. From that, an overview of possible impacts of statistics on connectionism that can be expected or that can already be observed is straightforward.
The possible "benefit" from statistical methods is threefold:
i) They can make some neural networks obsolete because these are less well suited to solve certain problems.
Statistics can show that some networks are not the optimal choice for certain problems by means of theoretical analysis (see 2.1) or through direct empirical competition (see 2.2).

ii) They can lead to better training algorithms and better models for neural networks.

This can be achieved by reformulation of neural networks in a statistical framework (see 2.1).

iii) They can help to improve the quality of connectionist research.

The quality of neural network research can profit from statistical knowledge about design and evaluation of empirical research (see 2.3).

Before further elaborating on these three points, it is necessary to consider what kind of connectionist research will "benefit" in these ways in order to avoid mistakes due to unsystematic and undifferentiated treatment of connectionism. The field of connectionism is usually divided into two subfields (see e.g. Winston 1992 for a discussion in the more general framework of artificial intelligence):

The first one persecutes an engineering goal trying to solve real–world problems using connectionist methods. The emphasis here is on the construction of hopefully good solutions and not so much on how these are achieved. From now on, we will call this subfield *engineering* connectionism.

The second subfield is concerned with the understanding and explanation of human intelligence, either with an emphasis on biological or cognitive explanations. This is attempted by building computational models and examination of their behaviour, where the emphasis is clearly on how certain performances are achieved. In what follows, we will call this subfield *explanatory* connectionism.

Concerning engineering connectionism, there can be impacts from all three "benefits".

ad i): If there exists a (statistical) method that solves a given problem better than neural networks, then there is no reason to stick to connectionist methods.

ad ii): As the main objective of engineering connectionism are optimal engineering solutions, every help from statistics that leads to the improvement of algorithms and models is welcome.

ad iii): Again, any help from statistics towards proper empirical connectionist research should be highly appreciated.

Concerning explanatory connectionism, the possible impacts of statistics are rather minor.

Explanatory connectionism is trying to build computational models of cognitive or biological phenomena. Therefore, each part of such a model, like the architecture (units, layers, weights, etc.) or the algorithms (updates of weights and activations), represents something of what is being modelled. For instance units may stand for biological neurons, weights for synapses, activations for membrane potentials, update rules for certain biochemical processes. Even if the intention is to model something on a higer cognitive level of abstraction, there still is the connection of each part of the model to the phenomenon which it represents. This implicates strong constraints on what impacts from statistics can be awaited on that type of research.

ad i): Connectionist models cannot be replaced by statistical models just because of inferior performance. To warrant a replacement, a statistical model must be shown to be a model of the phenomena one is interested in.

ad ii): As has been outlined above (see 2.1), the reformulation of connectionist models in statistical terms can give great insight into mechanics and performance of neural networks. This, of course, is of great value also to explanatory connectionism. But again, changes within the connectionist models cannot be based solely on such performance considerations. Because there exists a mapping from all the parts of the connectionist model to the phenomena that are being modelled, changes in the connectionist model must always be plausible from the point of view of those modelled phenomena.

ad iii): This is the only point in which explanatory connectionism can profit from statistics in the same way as engineering connectionism or any other empirical investigation.

Even if such a division into the two subfields is employed in comparisons of statistics and connectionism, statements are nevertheless often applied generally to the whole field of connectionism (e.g. "This paper is concerned with artificial neural networks for data analysis", "If artificial neural networks are intelligent, then many statistical methods must also be considered intelligent", both Sarle 1994).

4) On the road to ... where?

In the previous sections a survey of likely impacts of statistics on connectionism has been given. In what follows, possible future directions and changes of connectionism will be sketched, again based on the division into engineering and explanatory connectionism proposed above.

The fact that already an increasing amount of research in the engineering subfield includes statistics in its endeavours, is an indication of an impact that is already taking place. Therefore it can be expected that in the near future such references and links to statistics will become obligatory and, hopefully, a must for acknowledgement of scientific soundness.

This will become a difficulty for all those connectionist researchers who want to pursue engineering goals with neural networks and do not have the rigorous statistical background that could become a must—have very soon. Those people involved in connectionism that stem from very diverse scientific fields "and are engineers, physicists, neurophysiologists, psychologists, or computer scientists who know little about statistics" (Sarle 1994) will then become what some think that they already are: laymen or amateur statisticians.

Another impact that can already be observed is that many neural network models will simply not be used for data analysis any more, or at least not in their original form. Models that are simply not suited for engineering tasks will either vanish completely or be severly altered with the help of statistical knowledge (see 2.1).

Hopefully, because of the improvement of the quality of connectionist experimentation (see 2.3) and the increasing competition (see 2.2), the steady growth of the number of different models and algorithms in connectionism will be stopped. This will happen because the majority of these methods will be recognized as what they really are: Highly specialized solutions to mostly single (and often artificial) problems.

Therefore, only those connectionist models and algorithms will remain in the engineering subfield, that survive the increased competition or that contain new and interesting ideas even from the point of view of a statistician.

For instance, Multi Layer Perceptrons have been recognized as being "especially valueable because [...] the complexity of the model" can be varied "from a simple parametric model to a highly flexible, nonparametric model" (Sarle 1994). Recurrent networks are already being examined for their commonalities with NARIMA (nonlinear autoregressive—moving average) models. The underlying statistical principles of such networks are still unexplored, especially in the case of more complicated models (e.g. multi—recurrent networks, see Ulbricht 1994).

An often especially by statisticians noticed drawback of neural networks is the conceptual opaqueness of the representations in the hidden units. This has been termed the "explanation problem" (Partridge 1987) and addresses the fact that it is often very difficult for the designer of a network to understand the network's behaviour. For instance it is usually very hard to describe what kinds of inputs have what kinds of influence on the network's weights and outputs. This, of course, is a direct consequence of the distributed nature of the representation in the hidden units, which is a notion that lies at the very heart of connectionism. Prem (1995b) proposes the use of "symbol grounding" to enable the networks to explain their behaviour by themselves. Therefore, such networks can be said to employ some kind of infering statistics without giving up their distributedness.

To sum things up for engineering connectionism, there is an increasing probability that statistics will become the theoretical basis of this subfield of connectionism. This will entail the development of engineering connectionism from a purely empirical to a fullgrown theoretical science.

But what about what we have termed explanatory connectionism? This subfield of connectionism has its own problems and is consequently already undergoing major changes of its paradigms (see Dorffner 1995 for an overview).

Prem (1995a) argues that in the study of what has been termed "new artificial intelligence" (including fields like artificial life, behaviour based robotics and also neural networks) a move towards the original aims of artificial intelligence is being made. Such a "new artificial intelligence" stresses the physical embodiment of the designed systems (i.e. the necessity to build robots), the aspect of cognitive modeling and a greater proximity to biology. As a consequence, connectionist paradigms within this "new wave" of artificial intelligence employ new architectures and techniques that have less and less in common with their former engineering twins and therefore with statistical methods as well.

As has been outlined above, statistics can give insight into mechanics and performance of neural networks and help to improve the quality of experimentation even for explanatory connectionism. But this will not stop the widening of the gap between engineering and explanatory connectionism.

5) Conclusion

The starting point of the debate about the relation between statistics and connectionism was a lot of justified criticism on connectionism brought forward by statisticians. This criticism has often been stated very bluntly: "The marketing hype claims that neural networks can be used with no experience and automatically learn whatever is required; this, of course, is nonsense." (Sarle 1994). It is the belief of the author, that the whole field of connectionism can only profit from such sharp criticism.

In accordance with the three possible "benefits" from statistics outlined in the previous section, every connectionist working within the engineering subfield should keep the following three questions in mind during all of his research work and consequently relate to them in all his publications:
Is there a statistical method that can solve my problem more optimal?
How is my neural network model related to statistical methods?
What is the proper design and evaluation of my neural network experiment?
The engineering part of connectionism will either mature to a fullgrown independent theoretical science or simply be seen as a part of statistics.

As a consequence, the modeling of cognitive or biological phenomena with neural networks will finally be realized as being an endeavour on its own only remotely linked to engineering purposes. A lot of research in the short history of connectionism has been handicapped by the following double strategy: People originally interested in biological or cognitive modeling had to show some additional engineering payoff of their neural networks (e.g. usability for data analysis) in order to keep the funding agencies and institutions interested. As a consequence they ended up with neural network models that are neither plausible in terms of biology or cognition nor are they really suited for engineering purposes. The current criticism from statisticians will help to clarify those confusions and stop further misuse of the original ideas of connectionism.

Therefore, statisticians should rather be seen as friends that give connectionists some good advice and help them to untangle what has been muddled years ago, and not as competing foes.

Acknowledgements

The author wishes to thank Erich Prem for ongoing discussions about the idiosyncrasies of connectionist research. The following persons made valuable comments on previous versions of this work: Johannes Fürnkranz, Paolo Petta, Lutz Prechelt and Erich Prem. The Austrian Research Institute for Artificial Intelligence is supported by the Austrian Federal Ministry of Science and Research.

Literature:

Dorffner, G., Prem, E., Ulbricht, C., Wiklicky, H. (1991), Theory and Practice of Neural Networks, in Brauer, W. & Hernandez, D. (eds.), Verteilte künstliche Intelligenz und kooperatives Arbeiten, Springer Berlin, Informatik—Fachberichte 291.

Dorffner, G. (ed.) (1995), Neural Networks and a New AI, Chapman & Hall, London.

Egmont—Petersen, M., Talmon, J.L., Brender, J., McNair, P. (1994), On the quality of neural net classifiers, Artificial Intelligence in Medicine 6, 359—381.

Finnoff, W., Hergert, F., Zimmermann, H.G. (1992), Improving Generalization Performance by Nonconvergent Model Selection Methods, in Aleksander, I. & Taylor, J. (eds.), Artificial Neural Networks, 2, North—Holland, Amsterdam, pp.233—236.

Hergert, F., Finnoff, W., Zimmermann, H.G. (1992a), A Comparison of Weight Elimination Methods for Reducing Complexity in Neural Networks, in IJCNN International Joint Conference on Neural Networks, Baltimore, IEEE, pp.980—987.

Hergert, F., Zimmermann, H.G., Kramer, U., Finnoff, W. (1992b), Domain Independent Testing and Performance Comparisons for Neural Networks, in Aleksander, I. & Taylor, J. (eds.), Artificial Neural Networks, 2, Nort—Holland, Amsterdam, pp.1071—1076.

Kibler, D. & Langley, P. (1988), Machine Learning as an Experimental Science, Machine Learning, 3(1), 5—8.

Michie, D., Spiegelhalter, D.J., Taylor, C.C. (eds.) (1994), Machine Learning, Neural and Statistical Classification, Ellis Horwood, England.

Paass, G. (1993), Assessing and Improving Neural Network Predictions by the Bootsrap Algorithm, in Hanson, S.J., et al. (eds.), Neural Information Processing Systems 5, Morgan Kaufmann, San Mateo, CA.

Partridge, D. (1987), What's Wrong With Neural Architectures, Computer Science Dept., Univ. of Exeter, Research Report 142, 1987.

Prechelt, L. (1994a), A Study of Experimental Evaluations of Neural Network Learning Algorithms: Current Research Practice, Fakultät für Informatik, Universität Karlsruhe, Technical Report 19/94.

Prechelt, L. (1994b), PROBEN1 — A Set of Neural Network Benchmark Problems and Benchmarking Rules, Fakultät für Informatik, Universität Karlsruhe, Technical Report 21/94.

Prem, E., (1995a), New AI: Naturalness Revealed in the Study of Artificial Intelligence, in Dorffner (1995).

Prem, E., (1995b), Understanding Complex Systems: What can the speaking lion tell us?, in Steels, L. (ed.), The Biology and Technology of Autonomos Agents, Springer, NATO ASI Series.

Ripley, B.D. (1993), Statistical Aspects of Neural Networks, in Barndorff—Nielsen, O.E, Jensen, J.L., Kendall, W.S., (eds.), Networks and Chaos: Statistical and Probabilistic Aspects, London: Chapman & Hall.

Röscheisen, M., Hofmann, R., Tresp. V. (1992), Neural Control for Rolling Mills: Incorporating Domain Theories to Overcome Data Deficiency, in Moody, J.E., et al. (eds.), Neural Information Processing Systems 4, Morgan Kaufmann, San Mateo, CA.

Schaffer, C. (1993), Technical Note: Selecting a Classification Method by Cross—Validation, Machine Learning, 13(1), 135—143.

Shavlik, J.W., Mooney, R.J., Towell, G.G. (1991), Symbolic and Neural Learning Algorithms: An Experimental Comparison, Machine Learning, 6(2).

Sarle, W.S. (1994), Neural Networks and Statistical Models, Proceedings of the Nineteenth Annual SAS Users Group International Conference, Cary, NC: SAS Institute, pp 1538—1550.

Stone, G.O. (1986), An Analysis of the Delta Rule and the Learning of Statistical Associations, in Rumelhart D.E. & McClelland J.L., Parallel Distributed Processing, Explorations in the Microstructure of Cognition, Vol 1: Foundations, MIT Press, Cambridge, MA.

Tresp, V., Ahmad, S., Neuneier, R. (1994), Traininig Neural Networks with Deficient Data, in Cowan, J.D., Tesauro, G., Alspector, J. (eds.), Neural Information Processing Systems 6, Morgan Kaufmann.

Ulbricht, C. (1994), Multi—recurrent Networks for Traffic Forecasting, Proceedings of the AAAI'94 Conference, Seattle, Washington, Volume II, pp. 883—888.

Winston, P.H. (1992), Artificial Intelligence, Addison—Wesley, Reading, MA.

Unsupervised Neural Networks for Speech Perception with Cochlear Implant Systems for the Profoundly Deaf

Manfred Leisenberg

University of Southampton, Inst. for Sound and Vibration Res., Southampton, UK-SO9 5NH
Email: *leisen@mail.soton.ac.uk*

ABSTRACT: **Recently we have proposed a new speech processing concept for Cochlear Implant (CI) - systems. The concept is based on speaker independent signal representation and a neural net classifier which can be combined with the well known CI- speech- coding- strategies. This paper describes some new simulation results: For every speech input frame a 4- dimensional feature vector has been extracted by employing a relative spectral perceptual linear predictive (RASTA-PLP) technique. To classify the feature vectors into so called "auditory related units (ARU)" we applied the self-organizing Kohonen neural net. The best matching ARU's will directly control the synthesis of a "alphabet" of patient adapted stimulus patterns. Simulation results show that the Kohonen algorithm finds representative clusters in the feature vector space for different net dimensions. A discussion of the results and a overview of present experiments with deaf patients will be given.**

1. Introduction

Consider the problem of the huge amount of information a cochlear implant (CI) patient has to process shortly after being implanted with one of the common CI- systems. Wouldn't it be a good idea to supply the patient first with a limited "alphabet" of stimulus patterns and increase the number of "characters" step by step during the rehabilitation process until he will be able to recognize the full continuous information stream sufficiently?

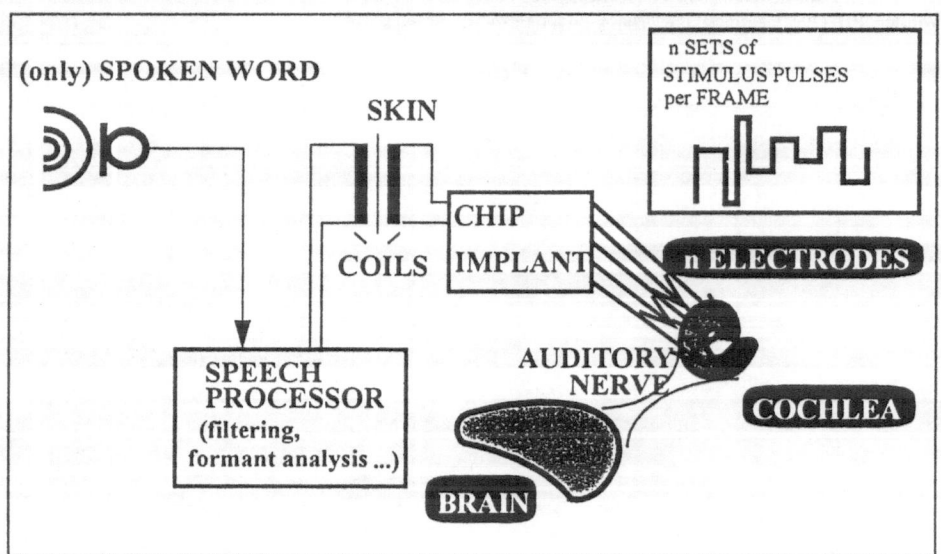

Fig.0: Fundamental scheme of a CI- System

CI- systems are based on the principle of coding acoustical information into electrical stimulus patterns. Fig.0 illustrates the principle of a common CI- system: First, the acoustical input signal will be preprocessed (filtering, FFT) in the speech processor and some feature extraction will be conducted. Then, the extracted features are applied to speech processing algorithms like formant analysis etc. and finally the data are coded into stimulus parameters or patterns in order to produce the stimulus pulses appropriate to the related speech input frames. The creation of the stimulus pulses is done by the special circuits of the chip implant while the parameter data will be transferred to the implant transcutanieously by RF- signals. Finally, the mentioned pulses will stimulate the auditory nerve of the deaf patient in order to provide him some sort of sound impression.

Existing systems are providing patterns nearly unlimited in their variety. The same word spoken twice by one person will never produce exactly the same pattern sequences. The influence of different speakers or inadequate acoustical conditions produces often an unrecognizable stream of information at the auditory nerve. Especially during the patients first rehabilitation phases it seems to be promising to limit the provided information to a relatively small amount of different stimulation patterns. A limited number of such patterns, the "alphabet", and their sense might be learnt by the hearing impaired quickly and easily. For this purpose we have designed a new CI- speech processing concept in order to improve existing systems. The concept is based on speaker independent signal representation and neural net classification [1] and could be combined with the known methods. Reminiscently, this paper first briefly describes the basic idea behind the proposed concept. In the second part we will explain the selected classification method and it's advantages. Then, recent results of the ongoing computer simulations will be given and discussed. A prospect on further investigations will conclude the present paper.

2. Technical idea behind the concept

The technical idea (see Fig.1) of the concept can be described as follows:

*One **utterance** on the CI-system's input must always produce by processing the input signals the **same sequence of stimuli patterns** on the systems output which will be characteristic to that particular utterance and adapted to the patients impairment, **independent** of characteristics of different speakers, influences of background noise etc.*

Employing the simplified bloc scheme of the proposed CI- system (Fig. 2) the processing of acoustical information in our example (Fig.1) will be explained: An utterance, maybe part of a word, has been sampled into 4 frames, F1 - F4. The features describing these frames have been extracted by employing the relative spectral perceptual linear prediction (RASTA-PLP) method [2]. This feature extraction algorithm has proved properties like speaker independence and robustness against inadequate acoustical conditions [3] and has been evaluated to be suitable for the purpose of CI- speech- processing [1]. The result of the feature extraction are 4 coefficients describing each frame sufficiently. Now, these coefficients are assigned to so called "auditory related units (ARU)" by classification. The mapping between acoustical information and stimulus pattern will be done in this bloc. It is the most important part of the processor. The classifier is based on an unsupervised neural net algorithm which will be able to learn properties of the relation between acoustical input features and the ARU's. Obviously, the definition of the ARU's as classification targets is a very important issue. The classification method will be discussed in next chapter. The subsequent synthesizer generates stimulus patterns by processing the ARU's. Since the classifier produces a limited number of different ARU's the synthesized pattern could be seen as "characters" of a limited "pattern-alphabet".

464

Please note that this technique can be applied to electrical stimulus patterns as well as to patterns for tactile actors or visual representation.

3. Classification of speech features employing unsupervised artificial neural nets

Neural networks offer specific processing advantages, which make it the technology of choice in multiple applications areas. These advantages include adaptive learning, self-organization, fault tolerance via redundant information coding and real- time operation. Neural networks have been applied successfully to solve several speech processing tasks [4,5], especially for speech recognition purposes. Adaptive learning is one of the most attractive features of neural networks; that is, they learn how to perform certain tasks by undergoing training with illustrative examples. Because neural networks are able to learn discriminating patterns based on training, we do not have to have elaborate *apriori* models. To solve the classification task within the proposed CI- system neural net is the selected method because of its capability of learning and adaptability.

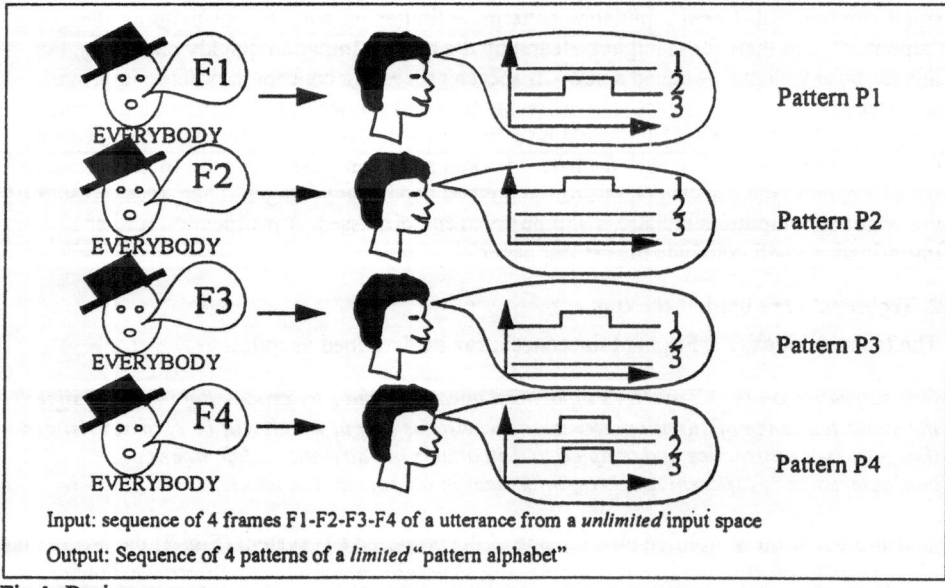

Input: sequence of 4 frames F1-F2-F3-F4 of a utterance from a *unlimited* input space

Output: Sequence of 4 patterns of a *limited* "pattern alphabet"

Fig.1: Basic concept

Neural nets consist of different layers of processing nodes: input, output and hidden layers, processing training or test mode. During the learning phase training patterns are presented to the input layer. Artificial neural nets are working supervised or unsupervised. In case of supervised neural networks during the learning process the connectionist system has to know which nodes of the output layer have to be activated related to a specific input pattern. In terms of our classification problem we need to know the targets in order to classify every frame of acoustical input information. For that purpose we would have to incorporate a supporting phonetic speech model consisting of phonemes, syllables or words. Such models have been applied to solve speech recognition tasks. Our investigations show that the supervised method is not very effective to process speech in CI- systems. Since we don't want to recognize speech automatically we do not need the overhead of a phonetic model.

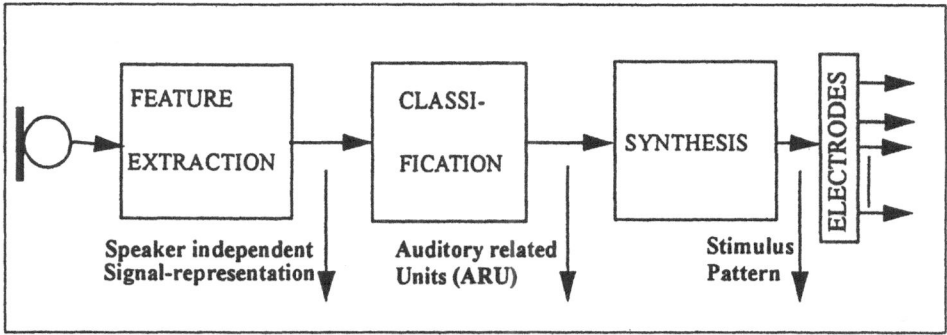

Fig.2: Block scheme of the proposed CI-system

Unsupervised neural nets do not need any external adjustment to determine the desired input- output transformation. Because of it's special property of effectively creating spatially organized "internal representations" of various speech features we choose the unsupervised Kohonen feature map [5] for our experiments. During the training process this net creates clusters closely related to the distribution of the input feature vectors. The resulting clusters of classified features very closely resemble the topographically organized maps found in the cortices of more developed animal brains[7].

The basic concept underlying the Kohonen map is that we can distribute a set of vectors across a space so that the way they span the space mimics the probability distribution of a set of training data. This is an efficient data compression scheme which can be used for codebook creation and accessing like the above mentioned classification problem.

Fig.3 shows the topological structure of the Kohonen map. During the training process sequences of feature vectors c_i are presented to the map via input nodes. In our case, the feature vectors c_i are 4- dimensional;

$$c_i = \{F1_i, F2_i, B1_i, B2_i\} \quad i = 1, 2, ...n. \qquad (1)$$

The components are representing the extracted speech features with $F1_i, F2_i$ peak frequencies

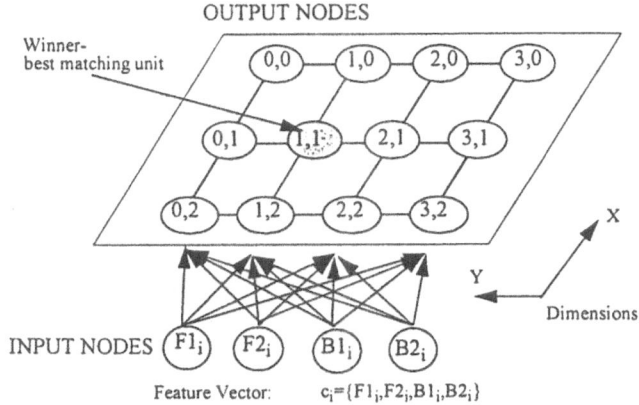

Fig.3: Structure of the Kohonen map

of the first and second formant, $B1_i$, $B2_i$ related bandwidths and i denotes the number of the sampled speech frame. The output layer consists of a 2- dimensional array. Every input node is connected to every output node (full connectivity) via weighted links. During the training process the weights w_{xyk} are adapted to the input vectors. At the end of the learning process the density function of the weights closely represents the probability density function of the input vectors c_i [7]. In other words: the weight vectors w_{xy} of the output nodes are representing clusters of the feature vector input space. After finishing the training process we label the relevant output nodes according to the classification. These labels denote the "characters" of the mentioned "alphabet". The resulting "alphabet" is part of a codebook which will be used to code the acoustical input information during the following test steps. Each "character" of the "alphabet" relates to one cluster extracted from the input vector space.

During testing the map the minimal Euclidean distance between the presented input feature vector c_i and all nodes represented by their weight vectors $w_{x,y}$

$$w_{xy}=\{w_{xyF1}, w_{x,y,F2}, w_{x,y,B1}, w_{x,y,B2}\} \quad x=1,2,..X, \; y=1,2,...Y \quad (2)$$

denotes the best matching unit - the winner node. The mentioned Euclidean distance is measure of quantization error Q_{err}.

4. Experiments

4.1. Simulation Results

The experimental environment is shown in Fig.4. The input acoustical information was provided by the TIDIGIT speaker-independent connected-digit database (Linguistic Data Consortium). The binary coded speech waveform files were divided into training and test data and have been processed by the RASTA-PLP-algorithm [2]. Then, the resulting sequences of speaker independent coefficient, the feature vectors c_i, have been fed to the Kohonen map simulator [8]. The actual simulation including a graphic control desktop, all the speech preprocessing, the neural net simulator and data visualization has been implemented in "C" on a UNIX workstation. In the following section we will describe and discuss the results of two experiments:

First, a map consisting of 15*20 nodes has been initialized randomly and, then trained with sequences of connected- digit utterances and tested with the utterance "nine". Fig.5 shows the simulation results according to different training states. The left diagrams of fig.5 show the reciprocal of the maximum quantization error for each matching node of the map. The right diagrams show the distribution of the related weight vectors w_{xyi} (indicated with [2]) and the input feature vectors c_i (indicated with +) in a 3- dimensional coordinate system. The fourth dimension is the colour and, unfortunately, can not be seen here. The Euclidean

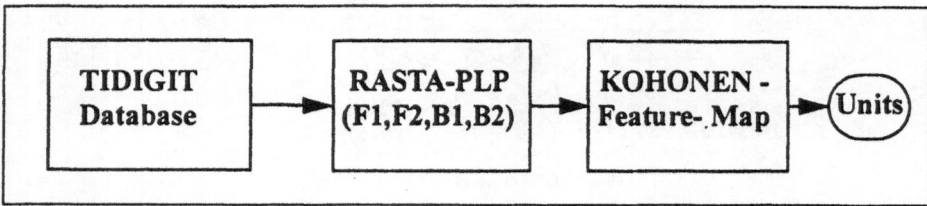

Fig.4: Experiment

distances between the w_{xyi} and c_i are shown as connecting lines. The upper two diagrams illustrate a test with "nine" after 77 training steps. The distances between weight and feature vectors are still long, the average quantization error is high and there are no significant clusters to be discriminated. The test results after 462 training steps are shown in the lower part of fig.5: The quantization errors are slightly smaller now and in the lower left diagram the creation of clusters can be seen

The second simulation experiment has been conducted with a map consisting of 40*60 nodes. The upper 2 diagrams of fig.6 are maps tested with the utterance "nine-five-one". It is very impressive, how the reciprocal of the maximum quantization error for each matching node rises with the number of training steps. The lower 2 diagrams of fig.6 are the 40*60 maps tested with the utterance "nine". Obviously, the number of matching nodes is smaller than above because of the length of the test utterance. Also in this case, the quantization error decreases with the number of training steps. If we compare these two diagrams with the related diagrams in fig.5 we can conclude that the quantization error strongly depends on the dimensions of the map.

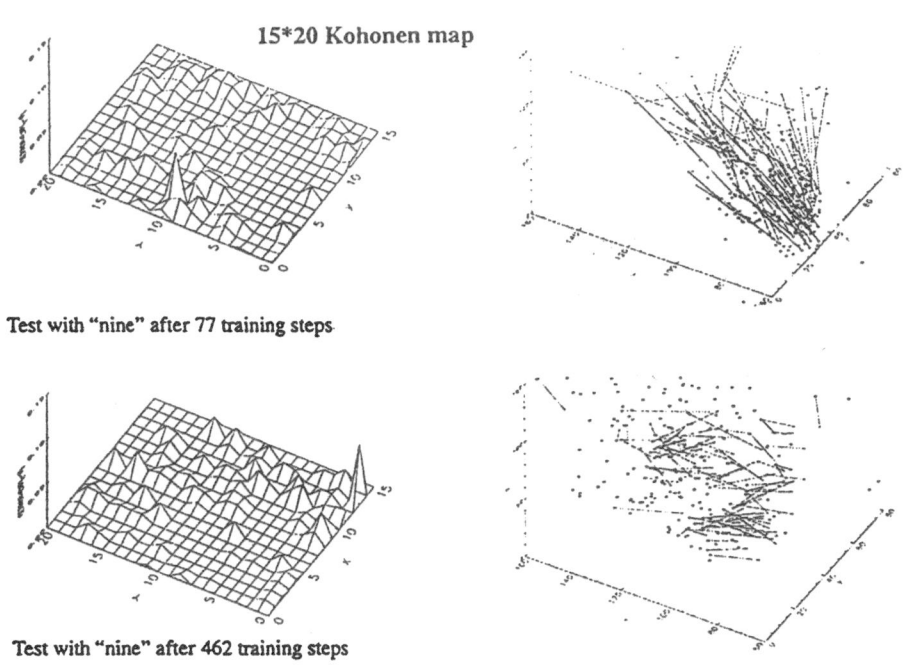

15*20 Kohonen map

Test with "nine" after 77 training steps

Test with "nine" after 462 training steps

Fig.5: Results of simulation 1

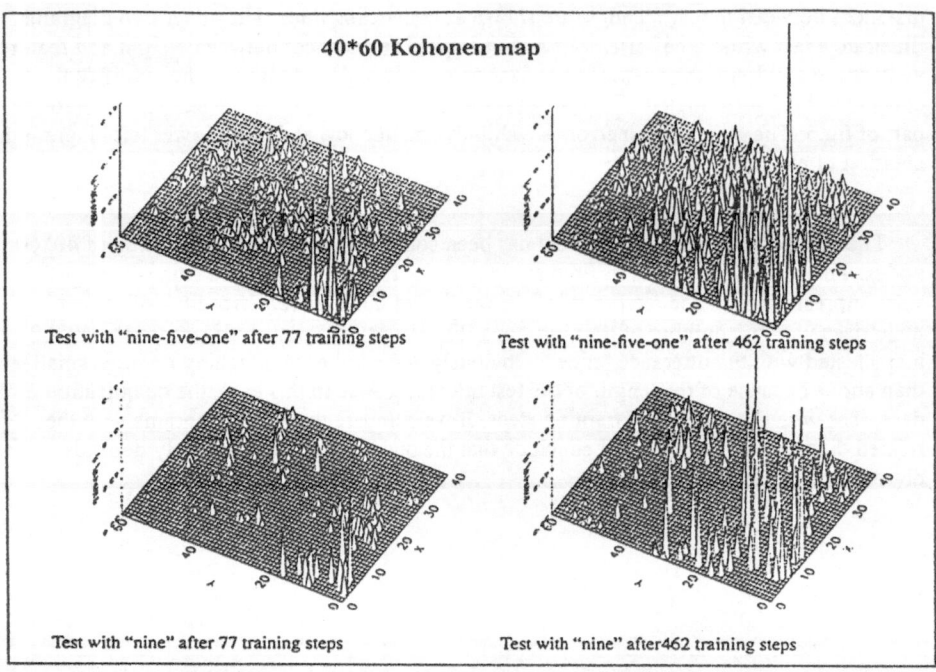

Fig.6: Results of simulation 2

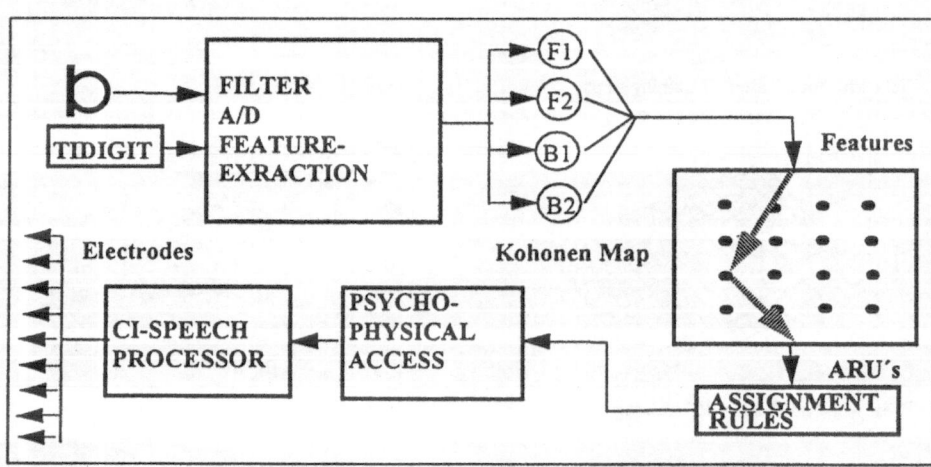

Fig.7:Experimental Environment of present psychophysical experiments

4.2. Experiments with profoundly deaf patients

The experimental environment of the psychophysical experiments we are presently conducting is shown in fig. 7. To input speech we currently employ the TIDIGIT database or a microphone. The speech processing front end and the Kohonen neural network algorithm processing work like described above. For this experiment, the neural net has already been trained on a fast UNIX workstation by using utterances of 20 speakers of the TIDIGIT database. Each ARU at the neural net output denoting a particular cluster extracted from the feature vector space $\{c_i\}_M$ is assigned to particular sets of stimulation parameters by using assignment rules. The mentioned assignment rules represent our resent knowledge concerning highly distinguishable stimulus patterns. They are experimentally obtained and will be extended and validated within our ongoing patient tests. Actually, these rules are the most important part of the experiment in terms of the expected psychophysical results. For example, the eligibility of speech transformed into "letters" of the above mentioned "pattern alphabet" strongly depends on the assignment rules.

The stimulus parameters are input to the psychophysical access facilities. These facilities consist of a special interface board which links a PC to a commercial CI- processor and a software library of functions to access to the internal coding circuits of the mentioned CI-processor. Finally, by using the stimulus parameters the coding circuit produces the stimulus pattern which are then sent to the chip implant in order to create the appropriate stimulus pulses.

5. Conclusions

An alternative concept for speech processing in CI- systems has been proposed. Computer simulation results validate the feature extraction and neural net classifier principle. Classification employing an unsupervised Kohonen map seems to be a promising clustering method in order to derive ARU's automatically. A limited "stimulus pattern alphabet" will be created under control of the mentioned ARU's. The "letters" of the "alphabet", and their sense might be learnt by the hearing impaired quickly and easily.

The quantization error of the classification strongly depends on the number of nodes of the Kohonen map. The mentioned number represents the number of ARU's and finally the number of "characters" of the "stimulus pattern alphabet". Therefore it is necessary to find a optimum between these parameters which have to be adapted to the patients needs. Probably it will be possible to derive "alphabets" with different numbers of "characters" from one basic map. This would support a strategy to increase the size of the "alphabet" during the rehabilitation phase of the hearing impaired patient.

The experimental facilities to conduct the necessary psychophysical investigations have been described. They have already delivered the first patient test results. The concept has been implemented with CINSTIM V2.0 - The Southampton Cochlear Implant/ Neural network STIMulation framework [9]. First experimental results confirm the new CI speech processing strategy.

References

[1] Leisenberg, M., A concept of an adaptive, neural net based cochlear- implant- system using speaker independent signal representation, Proc. International Symposium on cochlear implants, Toulouse, Fr.1992

[2] Hermansky, H.,Perceptual linear predictive analysis of speech, J.Acoust.Soc.Am 1990; 87

[3] Morgan, N., Compensation for the effect of the communication channel in auditory-like analysis of speech, Proc.Eurospeech'91, Genova,1991

[4] Windheuser,C., Competitive Sequence Learning, Diplomarbeit Rheinische Friedrich-Wilhelms- Universität Bonn, Bonn 1991

[5] Parten CR (ed), Handbook of neural computing applications. San Diego: Academic Press, 1990

[6] Kohonen, T.: The self-organizing map, Proc. IEEE, 78(1990)9, p. 1464 ff.

[7] Tavan, P., Grubmüller, H., Kühnel, H., Self- organization of associative memory and pattern classification: Recurrent signal processing on topological feature maps, Biological Cybernetics, 64, p. 95-105, 1990

[8] Kohonen,T.,Kangas,J.,Laaksonen,J.,SOM-PAK- The Self-Organizing Map Program Package, TR, Helsinki University of Technology, Helsinki 1992

[9] Leisenberg,M. , Downes,M.: CINSTIM: The Southampton Cochlear Implant/ Neural network STIMulation framework - implementation advances of a new, neural net based speech processing concept,International Cochlear Implant,Speech and Hearing Symp.,October 1994, Melbourne, Australia

Obstacle Avoidance by Means of an Operant Conditioning Model

Eduardo Zalama†, Paolo Gaudiano‡, Juan López Coronado†

†Departamento de Ingenieria de Sistemas y Automatica, Universidad de Valladolid
Paseo del Cauce s/n, Valladolid 47011, Spain
‡Department of Cognitive and Neural Systems, Boston University
111 Cummington Street, Boston, MA 02215, USA

Abstract

This paper describes the application of a model of operant conditioning to the problem of obstacle avoidance with a wheeled mobile robot. The main characteristic of the applied model is that the robot learns to avoid obstacles through a learning-by-doing cycle without external supervision. A series of ultrasonic sensors act as Conditioned Stimuli (CS), while collisions act as an Unconditioned Stimulus (UCS). By experiencing a series of movements in a cluttered environment, the robot learns to avoid sensor activation patterns that predict collisions, thereby learning to avoid obstacles. Learning generalizes to arbitrary cluttered environments. In this work we describe our initial implementation using a computer simulation.

1 Introduction

One of the aspects to consider when an animal or intelligent machine has to operate in an unknown environment is that it must learn to predict the consequences of its own actions. By learning the causality of environmental events, it becomes possible to predict future and new events.

Models of classical and operant conditioning have emerged from the field of psychology in order to try to explain how an organism can achieve autonomous behavior in a constantly changing environment (Rescorla & Wagner, 1972; Sutton & Barto, 1981; Grossberg, 1982).

In the classical conditioning paradigm, learning occurs by repeated association of a Conditioned Stimulus (CS), which normally has no particular significance for an animal, with and Unconditioned Stimulus (UCS), which has significance for an animal and always gives rise to an Unconditioned Response (UCR). For example, a dog that repeatedly hears a bell before being fed will eventually begin to salivate when the bell is heard. The response that comes to be elicited by the CS after classical conditioning is known as the Unconditioned response (CR).

Another related form of learning is known as *operant conditioning*. In this case an animal learns the consequences of its actions. More specifically, the animal learns to exhibit more frequently a behavior that has led to a reward, and to exhibit less frequently a behavior that has led to punishment. For example, a hungry cat placed in a cage from which it can see some food will learn to press a lever that allows it to escape the cage to reach the food. In this situation, the animal cannot simply wait for things to happen, but it must generate different behaviors and to learn which are effective. This kind of learning has also been referred to as *reinforcement learning* (Sutton & Barto, 1981). The main problem in modeling this form of learning is how to learn which of a large array of behaviors has produced the reward.

We have used a model of classical and operant conditioning proposed by Grossberg (1971, 1986), Grossberg and Levine (1987) to train a wheeled robot to avoid obstacles by learning the patterns of sensor activation that predicts an imminent collision.

The remainder of this paper is structured as follows. In section 2 we describe briefly the conditioning model (Grossberg & Levine, 1987), describing its functionality and its applicability to obstacle avoidance. In section 3 we describe our simplified implementation of the model. In section 4 we show some results and performance of the model, and finally section 5 is dedicated to conclusions and future work.

2 Grossberg's conditioning model

Grossberg & Levine's (Grossberg & Levine, 1987) implementation of Grossberg's conditioning circuit is in figure 1. This model was used by Grossberg and Levine to explain a number of phenomena from classical conditioning.

In this model the *sensory cues* (both CSs and UCS) are stored in Short Term Memory (STM) within the population labeled S, which includes competitive interactions to ensure that the most salient cues are contrast enhanced and stored in STM while less salient cues are suppressed. At the bottom of the figure the *drive node D* is represented, and conditioning can only occur when the drive node is active. This node and the nodes in the population labeled P,

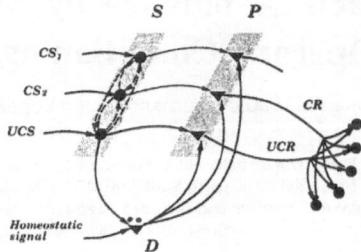

Figure 1: Conditioning circuit proposed by Grossberg & Levine. See text for details.

are represented as triangular nodes and are polyvalent cells: Polyvalent cells require the convergence of two inputs in order to become active. Finally, the neurons at the far right of the figure are represent the response (conditioned or unconditioned), and are thus connected to the motor system.

The design of figure 1 satisfies a number of fundamental constraints. Most important are the requirements that: (1) initially, only the UCS must be able to cause a response (the UCR); (2) after conditioning, the CS must be able to elicit a response similar to the UCR; (3) in order for learning to take place, the CS and UCS must be presented nearby in time.

The system satisfies these requirements as follows: initially it is presumed that only the UCS has a strong connection to the drive node D. When the UCS is turned on (e.g., a shock), it activates the drive node, which in turn sends activation up toward the polyvalent cells P. The P that receives joint input from the UCS and D nodes becomes active, and "reads out" the UCR on the motor cells. When an CS (e.g., a light) is presented by itself prior to conditioning, it cannot activate the D node, and thus it cannot activate its P cell, and no action is generated. However, if the UCS is turned on shortly after the CS has become active, then the D node becomes active, and the CS has a brief opportunity to sample the D node's activity through an associative learning rule. At the same time, the D node will also briefly activate *all* P cells that are receiving sensory input, and thus the P cell corresponding to the CS will be active, and it will learn the pattern of motor activity being generated by the UCS. If the pairing is repeated several times, the CS will develop strong connections to the D node, while it polyvalent cell will learn to imitate the UCR. Eventually activation of the CS generates a large enough signal to activate the drive node D, which in turn helps to activate the polyvalent cell P, and reads out a response similar to the UCR, that is, the CR.

Notice that in figure 1 the drive node must also receive a *homeostatic signal* to become active. This reflects the observation that a motivated behavior, such as eating, will not be released unless the animal sees food *and* is hungry. When the CS and UCS represent a fearful cue, it is assumed that the homeostatic signal corresponds to some form of survival drive, which is presumably always active in normal animals. In this case, as we do below, on e can simply assume that the drive node only needs a strong sensory cue to become active.

One of the main characteristics of the model is its dynamical nature and temporal causality. Temporal causality refers to the fact that the association between stimulus and response can be learned even though they are presented at different times in different trials. For example when an animal presses a lever to get food, the animal will learn the effect of its action regardless of exactly how quickly the food is presented (within a window of several seconds). The present model is able to reproduce other paradigms like blocking second order conditioning (Grossberg & Levine, 1987). Second order conditioning is useful as it allows a CS previously paired with a UCS, to act as a UCS for other conditional stimuli. For instance, a bell repeatedly paired with shock eventually comes to elicit fear. If a light is now repeatedly presented shortly before onset of the bell, even though the shock is never turned on, the light will also come to elicit fear. This form of "higher order" conditioning is extremely useful as it allows animals to learn early predictors of important events.

3 Conditioning and obstacle avoidance

In this section we describe how we have applied Grossberg's conditioning model to the problem of obstacle avoidance with a mobile robot. The mobile robot we have used in the simulations is a differential-drive robot, as is shown in figure 2.

We have previously introduced a neural network controller for this type of mobile robot, which learns the robot's forward and inverse odometry through a learning-by-doing cycle (Zalama, Gaudiano, & López-Coronado, 1995). That model, which we called NETMORC, includes two neural populations that code the distance and angle to a target as registered by the robot's sensory system. The distance and angle information is used to generate the

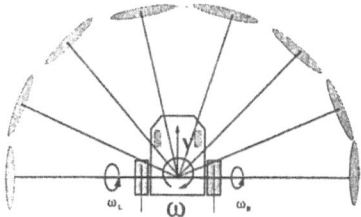

Figure 2: Mobile robot structure. The mobile robot has a set of 8 ultrasonic sensors uniform and frontal distributed through the robot.

wheel velocities required to move the robot toward the target. However, the NETMORC is not capable of avoiding obstacles.

In the present work we combine the conditioning model of figure 1 with the NETMORC in such a way that the robot can learn to avoid obstacles by modifying its angular velocity when it encounters obstacles in its path. Figure 3 illustrates the scheme we use to represent angular velocities. In this figure the leftmost node represents an angular velocity of $-w_m \, rad/s$, the right node represents an angular velocity of $w_m \, rad/s$ (where w_m is the maximum angular velocity developed by the robot), and the central node corresponds to a straight line movement. The activation pattern over the population is used to determine the wheel velocities that will move the robot. The map includes a sigmoidal transformation, whereby angular velocities close to zero are represented by a greater number of nodes. The sigmoidal function which selects the most active node in the map as a function of the angular velocity is given by:

$$
n_d = \begin{cases} \frac{N}{2} + \frac{N(a_w + 0.5 w_m)w}{w_m(a_w + w)} & \text{if } w > 0 \\ \frac{N}{2} + \frac{N(a_w + 0.5 w_m)w}{w_m(a_w - w)} & \text{otherwise} \end{cases} \tag{1}
$$

where n_d represents the most active node, w is the angular velocity of the robot. N is the number of nodes in the map, w_m is maximum angular velocity and, a_w controls the steepness of the sigmoid.

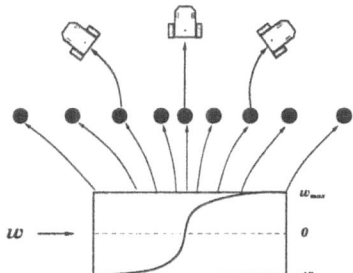

Figure 3: Angular velocity map. Each spatial position represents an angular velocity developed by the robot. The transformation has been performed by means of a sigmoidal function which permits more density of nodes for velocity values close to zero.

In our simulations, instead of only activating a single node in the population, we activate a neighborhood of nodes, with the position of maximal activation corresponding to the preferred angular velocity or direction of movement, and the activation of nearby nodes falling of as a Gaussian. We will show later that this form of distributed activity lends itself well to the problem of obstacle avoidance.

The mobile robot used in the simulations has eight ultrasonic sensors angularly distributed every $25.5°$, covering the frontal plane of the robot as figure 2 shows. In the simulations we have assumed a maximum range of $5m$ for each sensors, and that a collision occurs when any of the sensors measures a distance under $1m$. An alternative solution in a practical situation could utilize information from bump sensors positioned on along the robot's perimeter.

However, it is important to point out that the conditioning model has no knowledge of the robot's geometry or of the location of the sensors, as we will show below.

In figure 4 the proposed model for obstacle avoidance is shown. In this case each sensory cue (or CS) corresponds to the signal from an ultrasonic sensors, subtracted from the maximum value of each ultrasonic sensor, so that closer obstacles are represented by larger activity. The unconditioned stimulus (US) in this case corresponds to a collision detected by the robot. For simplicity, rather than treating the collision detector as a UCS with a strong, fixed connection to the drive node, we let the collision signal activate directly the drive node. This corresponds to the situation discussed above in which an internal "survival" signal is always impinging upon the drive node. This simplification is reasonable as long as a single kind of UCS and drive are necessary.

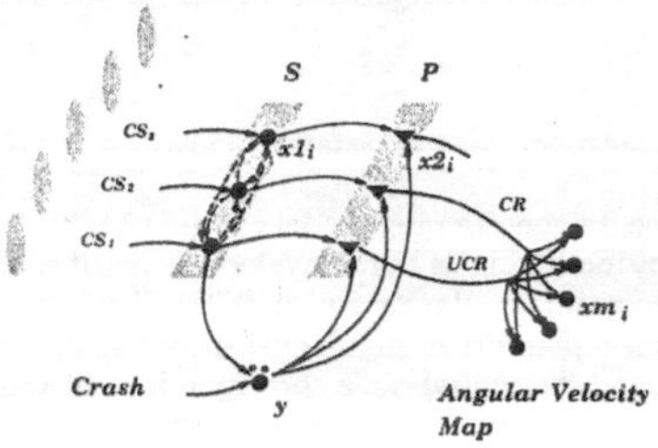

Figure 4: Conditioning model for obstacle avoidance. The ultrasonic range data represents the conditioned stimuli; the crash is the unconditioned stimulus. After conditioning, the pattern of activity across the ultrasonic sensors can predict a collision and change the angular velocity to avoid the obstacle.

The overall idea behind the model in figure 4 is that whenever the robot collides with an obstacle, learning in the circuit will create a connection between the current pattern of ultrasonic sensor activity and the angular velocity that the robot had at the time of collision. Later activation of a similar sensor pattern will cause inhibitory activation of those angular velocities that would have caused a crash, causing the robot to change direction and steer away from an obstacle.

One of the main properties of the model is its real-time function, in the sense that it is not necessary to separate explicitly learning from normal operation. However, in order to achieve a faster learning we have performed an initial learning phase where the robot performs random exploratory movements in a cluttered, activating sequentially each node from the angular velocity map, and thus sampling various patterns of activity that lead to collision.

As the robot travels, it takes measures from the ultrasonic sensors. The complementary values are contrast-enhanced and stored within the STM field S. The population S, which was originally modeled by Grossberg as a *recurrent competitive field*, removes the inherent noise from the ultrasonic sensors and enhances the activity of those sensors receiving maximal activation, that is, those registering the closest obstacles. A more complete description of the properties of this kind of network can be found elsewhere (Grossberg, 1973, 1982). We have used a simplified discrete time version of the recurrent competitive field, which quickly and efficiently normalizes and contrast-enhances the sensor activations. Specifically, the activation x_{1i} of each neuron in population S is given by

$$x_{1i}(t) = M_x \frac{[R - I_i(t)]^+}{\sum_i x_{1i}(t)} - (1 - M_x)x_{1i}(t-1) \tag{2}$$

where I_i is the "raw" ultrasonic measures, R is the maximum range for each ultrasound, M_x is a constant that determines how much the previous activation is weighted relative to the current input, and the notation $[x]^+$ represents half-wave rectification (returns only those values of $x > 0$). The summation is taken over all ultrasonic activations, thus ensuring normalization over the entire population S.

Prior to conditioning the drive node D is activated when the robot collides against obstacles. After conditioning, sufficient activation of a pattern of sensors can also activate the drive node. This permits second order conditioning, so that after the initial learning stages the ultrasonic measures can themselves predict the collision, and lead to conditioning of other sensor patterns. The activation y of the drive node is given by:

$$y(t) = \sum_i x_{1i}(t)z_{1i}(t) - T_y + U_{CS}(t) \tag{3}$$

where U_{CS} represents the collision of the robot ($U_{CS} = 1$ if collision just occurred, and $U_{CS} = 0$ otherwise), z_{1i} the adaptive weight connecting the sensory node x_{1i} to the drive node, and T_y is a threshold that controls how easily the drive node is activated, and thus indirectly controls how easily second order conditioning will occur.

The activation x_{2i} of polyvalent cells is given by:

$$x_{2i}(t) = x_{1i}(t)f(y(t)) \tag{4}$$

where $f(y(t))$ is given by:

$$f(y(t)) = \begin{cases} 1 & \text{if } y(t) > 0 \\ 0 & \text{otherwise} \end{cases} \tag{5}$$

Two different kinds of learning take place: the learning that couples sensory nodes (ultrasounds) with the drive node (the collision), and the learning of the angular velocity pattern that existed just before the collision. The first type of learning follows an associative learning law with decay:

$$z_{1i}(t) = Lz_{1i}(t-1) + Px_{1i}(t)f(y(t)) \tag{6}$$

where P is the learning rate, L is the weight decay.

The equation which learns the velocity map is also of an associative form:

$$zm_{i,J}(t) = zm_{i,J}(t-1) - Mx_{2i}(t)[xm_J + zm_{i,J}(t)] \tag{7}$$

where $zm_{i,j}$ represents the adaptive weights from the polyvalent cells to the nodes within the angular velocity map, M is the learning rate, and J is the winner node in the angular velocity map. However, in this case the learned weights are negative, thus when the robot collides against obstacles, the above rule learns to inhibit the actual direction of movement. Note that this learning rule is equivalent to learning a pattern of activity over a map of neurons that are coupled through mutual inhibitory connections with the angular velocity map. The use of negative connection weights is computationally more parsimonious.

Once learning has occurred, the activation of the angular velocity map is given by two components. A first excitatory component reflects the angular velocity required to reach the target without the influence of obstacles. The second, inhibitory component (because the weights are negative), moves the robot away from the obstacles as a result of the activation of ultrasound signals in the conditioning circuit. The equation that reflects this behavior is given by;

$$xm_j(t) = \exp\left[-(j - n_d(t))^2/\sigma\right] + \sum_i x_{2i}(t)zm_{i,j}(t) \tag{8}$$

where $n_d(t)$ is the index of the node that represents the desired angular velocity to reach the target without obstacles, and σ is the variance of the Gaussian.

The reason for using a Gaussian distribution of activity is depicted in figure 5. When an excitatory Gaussian is combined with an inhibitory Gaussian at a slightly shifted position, the resulting net pattern of activity exhibits a maximum peak that is shifted in a direction *away* from the peak of the inhibitory Gaussian. Hence the presence of an obstacle to the left causes the robot to shift to the right, and *vice versa*.

In an earlier paper we have described an adaptive neural controller that utilizes distance and angle information to move the robot. In this example for simplicity we use the kinematic equations directly to move the simulated robot. Specifically, for a given pattern of activity of the angular velocity nodes, we select an angular velocity according to:

$$w = \begin{cases} \frac{w_m a_v[\max_j(xm_j(t)) - N/2]}{N(a_w + 0.5w_m a x) - w_m[\max_j(xm_j(t)) - N/2]} & \text{if } \max_j(xm_j(t)) > N/2 \\ \frac{w_m a_v[\max_j(xm_j(t)) - N/2]}{N(a_w + 0.5w_m a x) + w_m[\max_j(xm_j(t)) - N/2]} & \text{otherwise} \end{cases} \tag{9}$$

This function is essentially the inverse of the sigmoid described by equation 1 above, where $\max_j(xm_j(t))$ represents the node number with the largest activity in the angular velocity map.

4 Experimental results

In this section we show our preliminary results on the model's performance. In a first stage, we let the model develop a set of weights by letting the robot perform movements at different angular velocities in an environment cluttered with obstacles, by activating sequentially the nodes in the angular velocity map. After this initial learning phase the robot is able to avoid obstacles in arbitrary positions.

Figure 6 shows the projections of the adaptive connections between the sensory nodes x_2 and the angular velocity map xm. In the figure you can see for example how angular velocities that make the robot turn to the right

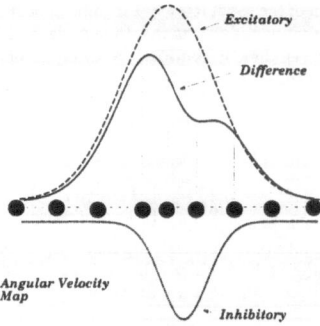

Figure 5: Positive Gaussian distribution represents the angular velocity without obstacles, and negative distribution represents activation from the conditioning circuit. The difference represents the angular velocity that will be used to drive the robot. Notice how the maximum of the excitatory Gaussian is shifted by the inhibitory Gaussian.

(nodes close to 20 in the figure) are inhibited when the robot receives ultrasonic sensor signals to the right (values close to 7).

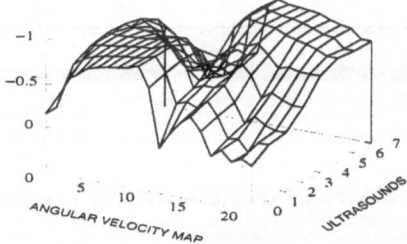

Figure 6: Representation of the adaptive connections $x_2 \rightarrow xm$. This map intrinsically codes the location of the ultrasonic sensors, in such way that when echoes are received in a given direction, that direction is inhibited in the angular velocity map.

Figure 7 illustrates the the robot's performance in the presence of several obstacles. The robot starts from the initial position labeled 1 and reaches a sequence of points 2-7 along the path shown by the dashed line. During the movements, whenever the robot is approaching an obstacle, the inhibitory profile from the conditioning circuit changes the selected angular velocity and makes the robot turn away from the obstacle. The presence of multiple obstacles at different positions in the robot's sensory field cause a complex pattern of activation that steers the robot between obstacles.

5 Conclusions

In this article we have described preliminary results with a model that learns obstacle avoidance for a wheeled mobile robot by means of ultrasonic information learned in a conditioning paradigm. The robot progressively learns to avoid the obstacles without the necessity of external supervision, but by negative reinforcement signals produced by the collision of the robot. One of the main properties of the model is that it is not necessary to know the robot's geometry nor the configuration of ultrasonic sensors on the robot's surface, because the robot learns from past experiences to avoid directions of movement that make the robot collide against the obstacles.

We are extending this models of conditioning to develop more complex behaviors. In particular, we are investigating conditioning circuits that permit the robot to choose among different behaviors (avoid, escape, wall following, etc.) depending on the moment-by-moment combination of sensorial information and its internal necessities. For example, a more complex system of sensory and drive nodes could be used to modulate how much the robot will try to avoid obstacles depending on its necessity to recharge its batteries.

477

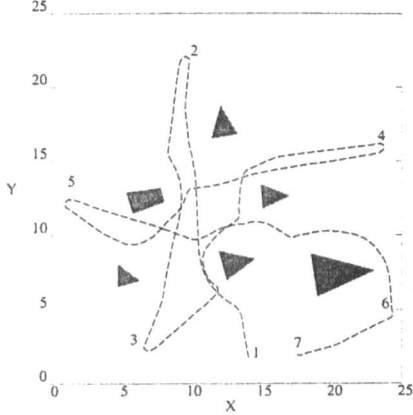

Figure 7: Trajectory followed by the robot in presence of six obstacles (shaded polygons). The robot travels from point 1 to point 7. Distances are expressed in meters. Parameters employed in simulations: $N = 21, w_m = 2.0, a_w = 0.3, M_x = 0.6, R = 5.0, T = 0.02, T_y = 0.2L = 0.9, P = 0.1, M = 0.05\sigma = 80$

References

Grossberg, S. (1971). On the dynamics of operant conditioning. *Journal of Theoretical Biology, 33,* 225–255.

Grossberg, S. (1982). A psychophysiological theory of reinforcement, drive, motivation and attention. *Journal of Theoretical Neurobiology, 1,* 286–369.

Grossberg, S., & Levine, D. (1987). Neural dynamics of attentionally modulated Pavlovian conditioning: blocking, interstimulus interval, and secondary reinforcement. *Applied Optics, 26,* 5015–5030.

Grossberg, S. (1973). Contour enhancement, short-term memory, and constancies in reverberating neural networks. *Studies in Applied Mathematics, 52,* 217–257.

Grossberg, S. (Ed.). (1982). *Studies of Mind and Brain: neural principles of learning, perception, development, cognition and motor control.* Reidel, Boston.

Grossberg, S. (Ed.). (1986). *The Adaptive Brain I: Cognition, Learning, Reinforcement, and Rhythm.* Elsevier/North-Holland, Amsterdam.

Rescorla, R. A., & Wagner, A. R. (1972). A theory of pavlovian conditioning: variations in the effectiveness of reinforcement and nonreinforcement. In Black, A. H., & Prokasy, W. F. (Eds.), *Classical Conditioning II,* chap. 3, pp. 64–99. Appleton, New York.

Sutton, R. S., & Barto, A. G. (1981). Toward a modern theory of adaptive networks: Expectation and prediction. *Psychological Review, 88,* 135–170.

Zalama, E., Gaudiano, P., & López-Coronado, J. (1995). A real-time, unsupervised neural network for the low-level control of a mobile robot in a nonstationary environment. *Neural Networks, 8,* 103–123.

Qualitative Approach to Gradient Based Learning Algorithms

Bernardo Morcego Seix[1], Andreu Català Mallofré[1], Núria Piera Carreté[2]

(1) Systems Engineering Department, Univesitat Politècnica de Catalunya, c/ Pau Gargallo 5, 08028 Barcelona, Spain.

(2) Applied Mathematics II, Univesitat Politècnica de Catalunya, c/ Pau Gargallo 5, 08028 Barcelona, Spain.

Abstract

This work is concerned with the establishment of a relation between the fields of qualitative reasoning (QR) and neural networks. We explore how well-known backpropagation learning algorithm can be studied from the point of view of QR. Qualitative models are based on the discretization of their parameters and the use of closed operators on the sets induced by the discretization. Henceforth, a qualitative version of backpropagation is an algorithm in which the variables involved in it belong to one of the finite classes defined. We analyse the algorithms resulting from this transformation and test their performance with a set of four problems. The results are encouraging and provide an empirical basis for a deeper, theoretical study, which can be very useful to realize physical implementations of the algorithm or as a starting point for the development of reinforcement learning algorithms.

Introduction

The task of establishing a relation between the fields of qualitative reasoning (QR) and neural networks is not an obvious one because the treatment data receives in most neural networks, from beginning to end, is purely numerical: weights interconnecting neurons are real numbers, activation functions are (usually) bounded real numbers, and in general all the variables involved in the process of learning and test are real numbers.

In spite of this, there have been three remarkable approaches of QR techniques applied to neural networks. The first one is concerning the use of qualitative data as input or output to neural networks: a well-known application of this kind is the recognition of speech sounds using as input articulatory features of letters, which is done in [1]. The second is reinforcement learning [2], a supervised method in which there is no teacher but a critic because the error signal is a reward or a penalty rather than a real valued quantity. For example, Barto, Sutton, and Anderson developed a neural network that learned associations from the states of a dynamic system to control its actions using reinforcement learning [3]. The third approach to QR is the use of discretized weights [4]. Different attempts to discretize weights for digital implementations have been proposed, allowing binary {0, 1} and ternary {+1, -1, 0} values, [5]. At the same time many efforts have been put in developing learning algorithms for discrete weights [6].

On the other hand, it is evident that learning is much more explainable at a qualitative level. It seems feasible and justifiable to study the learning algorithms used to configure neural network systems from a qualitative point of view. Take, for example, gradient algorithms. They intend to follow the steepest gradient descent to reach a global optimum. The main objective of this kind of algorithms is to compute the direction (positive or negative) a weight must follow to minimize the difference between the output and the desired output. Some of these algorithms (backpropagation, for example) do not even try to find the size of the increment but they just add the weight a quantity proportional to the gradient. One can then say that, to some extent, backpropagation follows a qualitative scheme and we will try to prove it throughout this article.

Qualitative Reasoning

One of the main objectives of AI is the machine-reproduction of man-made models when reasoning. Considering that many mental processes are thought to follow a qualitative nature it is easy to understand the increasing interest the scientific community is putting on the theory of qualitative reasoning.

QR is concerned with all kinds of scientific disciplines that use mathematical models to represent their phenomenae. Qualitative modelization tries to build models up allowing system behavioural description in terms of significant magnitudes, tendencies, change rates, etc., where the representation of causality (the expert knowledge is usually expressed with cause/effect relations) and imprecision are both essential points.

The meaning of the term qualitative is related to reasoning about continous properties via discrete abstractions. One method of dicretization is the representation of any quantity by its sign and it is reasonable because different things tend to happen when signs change. Moreover, as stated by Forbus [7], the sign quantization verifies the *relevance principle* which states that this quantization is relevant to the kind of reasoning performed.

The formalism involved in the use of sign quantities is a calculus based on the *algebra of signs*. It is not the aim of this work to describe with full details the algebra of signs but a small overview is needed in order to understand the following derivation. The set SD = {[-], [0], [+]} contains the equivalence classes induced by signs and S = {[-], [0], [+], [?]} expands SD with class [?], the meaning of which is that nothing can be said about the sign of that magnitude. Product and addition are commutative and associative operators. Product is a distributive operation with respect to addition and their tables are:

⊕	[-]	[0]	[+]	[?]
[-]	[-]	[-]	[?]	[?]
[0]	[-]	[0]	[+]	[?]
[+]	[?]	[+]	[+]	[?]
[?]	[?]	[?]	[?]	[?]

Addition Table

⊗	[-]	[0]	[+]	[?]
[-]	[+]	[0]	[-]	[?]
[0]	[0]	[0]	[0]	[0]
[+]	[-]	[0]	[+]	[?]
[?]	[?]	[0]	[?]	[?]

Product Table

The previous tables depict symbol [?] grayed to remark its appearance when applying the basic operators. We will soon see that obtaining [?] as a result is a situation one should better avoid. For the rest of this work, when we talk about qualitative values we mean magnitudes classified into the classes in SD or results of additions and products of them.

The Backpropagation Algorithm

Backpropagation [8] is a learning algorithm for multilayer feedforward neural networks and it is based on gradient descent on the hypersurface of errors. Fig. 1 shows layers and indexes in a multilayer feedforward neural network and following there is a synthetic derivation of the algorithm.

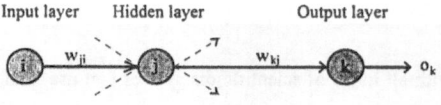

Fig. 1

The input to node j in the hidden layer is $s_j = \Sigma\, o_i\, w_{ji}$ and its output is $o_j = f(s_j)$ where f is the activation function (usually a sigmoidal function, but sometimes also linear or gaussian).

A gradient search of the minimum system error is based on the minimization of the sum of the squared errors of each pattern presented to the network and weight updates are done according to:

$$\Delta w = -\eta\, \frac{\partial E}{\partial w} \tag{1.1}$$

where η is a user supplied constant: the learning rate. The derivatives of weights have different expressions for the weights of the hidden layer and for those of the output layer. The final update formulae are the following:

$$\frac{\partial E}{\partial w_{kj}} = e_k \cdot f_k{}'(s_k) \cdot o_j = \delta_k \cdot o_j \tag{1.2}$$

$$\frac{\partial E}{\partial w_{ji}} = f_j{}'(s_j) \cdot \sum_k (\delta_k w_{kj}) \cdot o_i = \delta_j \cdot o_i \tag{1.3}$$

In the context of this article, the most remarkable characteristic of backpropagation is that it can *only* obtain the direction in which a weight must be altered to decrease the errors made by the network. The amount of weight increment results from the heuristic that the step size must be proportional to the error.

This heuristic is the main drawback of the algorithm and many authors tried to reduce its effect adding second order derivative terms (a review can be found in [9]). These approaches usually lead to much faster and accurate learning at a high computational cost and the question of when will second order learning algorithms globally outperform first order ones still remains open.

The Qualitative Version of Backpropagation

Before beginning with the derivation of the qualitative version of backpropagation two questions need to be answered: is it feasible to derive a qualitative version of backpropagation? and, what do we understand by a qualitative version of backpropagation?

The answer to the first question was partially explained in the previous sections and is clearly yes. The principal reason to say so is because the only *true* information available after a training cycle is the sign of Δw. Therefore, if using the algebra of signs we are able to obtain the sign of the error derivative (without obtaining a [?] result) and we modify the constant η to counterbalance the effect of working with products of ± 1 we will have obtained what we looked for.

A remark must be done before answering the second question. We do not want to alter the structure and functioning of the network. This means we can only treat the variables involved in the learning process, which are e, f' and w. Therefore, a qualitative version of backpropagation can be different things depending on which variables are classified into SD, i.e. which ones are turned into qualitative.

Now we can proceed with the derivation of the qualitative version of backpropagation. The main objective of such a derivation will be to obtain δ_{ok} and δ_{oj} in SD using the algebra of signs.

In expression (1.2) one can easily notice that if the values e and $f' \in$ SD the resulting δ_{ok} will always belong to SD. This is because the product is a closed operation in SD.

Equation (1.3) needs to be analysed more carefully. On one side, addition is not a closed operation in SD (for example, [+] + [-] = [?]) and on the other side, some attention must be paid if we want to convert w to a qualitative value because this parameter is intrinsically real valued and making such a conversion may be meaningless.

In order to solve the first problem and not alter the behaviour of backpropagation we used the 'majority' operator, which is a version of the qualitative addition. This operator carries out a sort of voting between the addenda and the result is the winning sign, or [0] if there is no winner.

There are two versions of the majority operator and each can take different forms depending on the addenda. Version 1 quantifies the voting and its result is a real number (positive or negative) indicating the difference between [+] and [-]. Version 2 normalizes this result giving only a sign. It is also noticeable that when the addenda are signs (i.e. both f' and w are qualitative) the voting is equitable, while when either f' or w is real valued the voting is not really fair because voters have different number of votes.

As one can easily see, there is not only *one* qualitative version of backpropagation but quite a few. There are two operators and three variables to convert to qualitative, which results in 14 versions of the algorithm. A brief description of the meaning of turning each variable into qualitative follows:

- e: if the error is a sign (belongs to SD) the network only knows if it gets closer or further to the minimum rather than how closer or further. Approaches similar to this one are the basis for most reinforcement learning algorithms.

- f': this variable is used to obtain the direction pointing to a minimum. It is, therefore, very reasonable to choose only the sign of the derivative of the activation function. On the other hand, it could allow the use of different activation functions with hard to calculate derivatives.

- w: conversion of real valued weights to signs (in order to have a fair voting) is controversial. On one side, having $w \in$ SD would normalize the contribution of all weights and δ would be the most important learning factor. On the other side, the error backpropagating should somehow be proportional to the degree with which a certain neuron contributed to generate it: the weight.

Experimental Results

In order to validate and compare the different versions of the learning algorithm obtained, we set a test bed consisting of 4 problems, each one representing a general class of problem. They are:

XOR: $\{0,1\}^2 \longrightarrow \{-1,1\}$. The well known non-linear separable combinatorial function.

DECODE: $\{0,1\}^n \longrightarrow \{-1,1\}$. A combinatorial function with n binary inputs and n binary outputs. Each input vector contains all zeros and a single 1, and the network must be able to codify in the hidden layer (of $\log_2 n$ neurons) the n different inputs.

DONUTS: $\Re^2 \longrightarrow \{0,1\}$. (x,y) real pairs are classified into two groups: the ones inside a donut

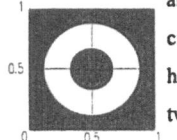

and the ones outside it. When this task is simplified reducing the internal circle radius to zero, it can be solved using a small network of 2 inputs, 2 hidden neurons and 1 output. To solve the DONUTS task one can compose two of these networks, obtaining a 4 layer network.

SQRT: $\Re^3 \longrightarrow \Re$. Analytical function: it is the square root of the division of two different linear combinations of the inputs. This is the saturation factor of an image's pixel calculated with the red, green and blue levels of the pixel. Generally speaking, analytical functions mapping real

numbers to real numbers can be solved, but the accuracy depends on the number of hidden neurons. Sometimes, though, it is possible to find non-standard configurations that work accurate enough, like the one shown.

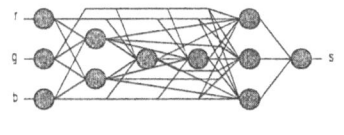

A remark must be done before proceeding with the results of the tests: the implementation of the qualitative version of backpropagation induces new parameters to the algorithm. These parameters determine the intervals of discretization of each variable (e, f, and w). Parameter setting allowed all possible tests for XOR and SQRT but some tests could not be realized for DECODE and DONUTS. The complete statistical analysis was, therefore, only performed for those tests which allowed it.

Each test consisted of 45 runs of a network learning the task at hand. Tests were realized with backpropagation and its 14 different qualitative versions. In each run the network had to attain a previously fixed degree of precision for the function to be learned.

After concluding the tests, only versions 7, 8, and 14 of the qualitative algorithm were comparable with backpropagation on each of the four problems. Algorithms 8 and 14 use the majority operator version 2 while algorithm 7 uses version 1 of the operator. The qualitative variable of algorithm 8 is *e* and they are *e*, *f'* and *w* for algorithms 7 and 14. The following table shows 99% confidence intervals for the mean of the learning time (computed in learning cycles) and those shaded indicate the mean can be considered statistically smaller than that of backpropagation with more than 1% of probability:

	XOR	DECODE	DONUTS	SQRT
Backpropagation	89.2 ± 51.6	106.68 ± 7.88	783.5 ± 260.8	984.2 ± 235.5
v7: Maj. v1 + *e*, *f'*, *w*	9.22 ± 1.26	90.4 ± 25.6	747 ± 326	2534 ± 682
v8: Maj. v2 + *e*	28.2 ± 2.26	149.89 ± 21.79	1688 ± 528	316 ± 86.3
v14: Maj. v2 + *e*, *f'*, *w*	7.76 ± 1.06	49.1 ± 41.2	605.9 ± 220.7	300 ± 120

Two facts can be observed when examining the previous table: first is that version 14 of the qualitative algorithm performed better than backpropagation in 3 of the four problems and second, problem DONUTS was very hard for all qualitative versions.

It should be noticed that version 14 of the qualitative algorithm is the "most" qualitative of all and its performance shows that it cannot be stated that backpropagation makes better use of the information provided by the error than it does.

On the other hand, DONUTS is a difficult task because it needs accuracy, in the sense that the boundaries of the donut can be assigned either class if the network is not accurate enough. We observed that qualitative versions of backpropagation got very close to a solution but failed to attain it, i.e. the learning rate needed to be adaptive.

Conclusions

This work is a first attempt to deal with the intrinsic qualitative aspects of the learning processes employed by neural network models. The results obtained using qualitative operators and qualitative parameters modifying backpropagation show that such gradient based learning algorithms follow an intrinsic qualitative scheme, i.e. they find the sign of weight increments. This is no discovering, but what is noticeable is that some qualitative versions of backpropagation performed significantly better than it, which means that backpropagation might be managing more information than it needs, considering its aim of calculating a sign. This results are encouraging and might be very interesting for the development of learning algorithms, as well as for their use in VLSI implementation.

It is also noticeable that a more sophisticated qualitative description of parameters would be possible and that might lead to more efficient algorithms, although such finer partitions of the real line rise new problems to take into account.

Bibliography

[1] Jeffrey L. Elman. Finding structure in time. CRL *Technical Report 8801*, 1988.

[2] Barto A.G., Sutton R.S., and Brower. Associative Search Network; A Reinforcement Learning Associative Memory. *Biological Cybernetics*, 40: 201-211, 1981.

[3] Barto A.G., Sutton R.S., Anderson C.W. Neuronlike elements that can solve difficult learning control problems. *IEEE Trans. on Systems, Man, and Cybernetics*, 13: 835-846, 1983.

[4] Carrabina, J. Xarxes Neuronals VLSI d'alta velocitat. *Doctoral dissertation*. Universitat Autònoma de Barcelona, 1991.

[5] Sivilotti M.A., Emerling M.R., Mead C.A. VLSI Architectures for Implementation of Neural Networks. *American Institute of Physics conference Proceedings 151, Neural Networks for Computing*, pp 408 - 413, 1986.

[6] Pérez Vicente C., Learning algorithm for binary synapsis. *Statistical Mechanics of Neural Networks*, Springer Verlag 1990.

[7] Forbus K.D., Qualitative Process Theory, *Artificial Intelligence 24*, 1984.

[8] Rumelhart D.E., Hinton G.E., Williams R.J. Learning internal representations by error propagation. *Parallel Distributed Processing: Exploration in the Microstructures of Cognition*, vol.1, pp 318 - 362, Mit Press, 1986.

[9] Buntine, W.L. and Weigend, A.S., Computing Second Order Derivatives in Feed-Forward Networks: a Review. *IEEE Transactions on Neural Networks*, (revised February 1993).

Character Recognition with Neural Assemblies

Francisco J. Vico , F. Ortega , J. Almaraz & F. Sandoval

E.T.S.I. de Telecomunicación, Dpto. Tecnología Electrónica.

Facultad de Psicología, Dpto. Psicobiología

Universidad de Málaga. Plaza El Ejido s/n, 29013 Málaga, Spain.

voice: +34-5-213-13-52; fax: +34-5-213-14-47.

E-mail: fjvico@ctima.uma.es

Abstract

In living beings, any pattern recognition task involves complex processes of concepts formation. We propose a model based on the principle of neural assemblies to develop internal representations of characters. Neural assemblies self-organize themselves by extracting the relevant information of the characters set depending on the previously stored characters. The principal consequences of this are that less storage capacity is needed, and it makes easer the clasiffication with a delta rule. Examples of this task are exposed, where it can be seen the way in which the assemblies store the characters, and how the classification of the internal representations is performed.

I. Introduction

A great variety of paradigms have found a ground of application in recognizing handwritten characters. *Backpropagation* (Burr, 1986; Khotanzad, 1988; Pawlicki, 1988; Shimoara, 1988) and *Learning Matrix* (Steinbuch, 1963) have been widely used. Biologically inspired models, like *Neocognitron* (Fukushima, 1983) and *ART* (Carpenter, 1987), were also successfully applied in this field.

The main part of the model we proposed is inspired in psychobiological considerations. Living beings perform any pattern recognition task in two stages: features extraction, and concepts formation. In the development of the model we have represented characters in a two-valued, 5×5 grid. We can consider these 25 binary values as features of handwritten characters.

Our goal has been to find a method to form internal representations of characters. The 25 particular features of a character, when seen by a human being, are grouped according to their spatial relations (simultaneous occurrence), and their affinity with other characters. This affinity is given in terms of similarity, and belonging to the same or different classes.

The approach of neural assemblies has been proposed as a good method for modelling cortical dynamics (Gerstein, 1989). Consequently, it has found application in image processing tasks (Gerstein, 1990). We have studied the dynamics of neural networks with a wide family of associative learning rules (Vico, 1994). Interesting psychological processes (categorization and classical conditioning phenomena), that are commonly verified in animal learning experiments, where observed in this dynamics.

II. Description of the model

The architecture of the neural network is shown in *Figure 1*.

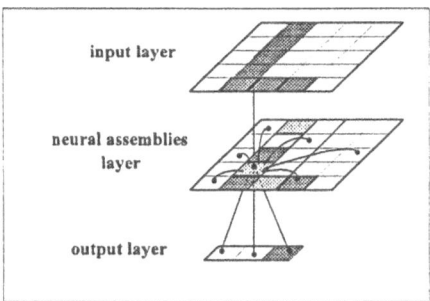

Figure 1. Neural network architecture, and basic pattern of connections.

The character is presented in the input layer, a grid of 5×5 neurons. Each neuron in this layer is connected to the corresponding in the neural assemblies layer, that has the same dimensions. From now on, neurons in this layer will be termed NA-neurons, as they are to form the stable neural assemblies that constitute the internal representations of the characters. This is made according to the similarities with other characters, and to external supervision. NA-neurons are fully connected among them.

The output layer represents the different categories in which the characters will be classified. There is also a full connectivity from each NA-neuron to the output layer. The neuron model that we have used is the ramp function. The activity of a neuron, a_j, is given by equation (1),

$$a_j = \sum_i w_{ij} x_i \tag{1}$$

where w_{ij} is the synaptic weigth between neurons i and j. x_i is the output of neuron i, whose value is given by equation (2).

$$x_j = \begin{cases} 0 \ , & \text{if } (a_j < 0) \\ a_j \ , & \text{if } (0 \leq a_j \leq 1) \\ 1 \ , & \text{if } (a_j > 1) \end{cases} \qquad (2)$$

The character recognition process is performed in three stages:

- Presentation of the character in the input layer.
- Self-organization of the neural assemblies layer for this input.
- Correction of the network output.

The 5×5 points character is presented at the input layer by setting the outputs of these neurons to the corresponding value in the character: 1 if the character fills this point, and 0 if it is blank.

This value is maintained until the internal representation is formed. This is made by self-organization of the neural assemblies layer. Equation (3) is a Hebbian-like learning rule that is applied to the synapses that interconnect the NA-neurons:

$$\Delta w_{ij} = \alpha x_i x_j (x_i - x_j)(1 - x_j) D \qquad (3)$$

where α is the learning rate, and D is the discrepancy between the output layer and the desired outputs. Its value is computed by equation (4).

$$D = \sum_j |e_j| \qquad (4)$$

$$e_j = x_j - p_j^n \qquad (5)$$

In this expression, x_j represents the activity of the jth output neuron, and p_j^1 is the value of the jth element of pattern n. e_j is the error computed at neuron j.

This learning rule has the following meaning: a neural assembly is reorganized when there is an error in the classification of a character. The Hebb rule is expressed in the first two terms. The third term makes the weight change in the sense of reducing the difference between pre- and post-synaptic activity. The function of the last term is to stop learning when the post-synaptic neuron has reached the maximum activity (1 in our neuron model).

When a character is presented, the input neurons propagate the activity to the neural assemblies layer. The self-organizing process joins them together in a stable neural assembly. The self-organization evolves for a number of steps. When this stage is finished, the neural assembly propagates its activity to the output layer. Finally, the output layer is compared with the desired output, and the connections are corrected according to equation (6).

$$\Delta v_{ij} = \beta x_i e_j \qquad\qquad (6)$$

This is the delta rule. It adjusts the weight between the ith NA-neuron and output neuron j, v_{ij}, depending on the error. β is a learning rate.

III. Simulations and Results

Simulations have showed stable formation of internal representations in the neural assemblies layer, and their correct matching with the desired output.

The following experiment shows how the assemblies store the most relevant part of the character, depending on the similarities with other characters. The experiment consists in learning the correct classification of characters I, i, and T.

First of all, character I was presented until the corresponding assembly was built, and the correct output was selected. *Figure 2* shows the activity in the network when I is presented.

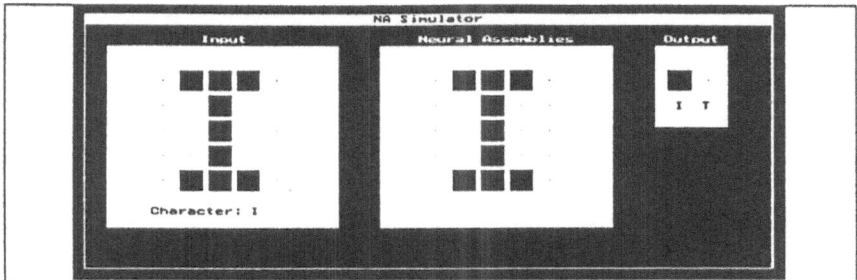

Figure 2. Self-organization of the neural assembly that represents character I.

The assembly is formed with all the active neurons, and the correct output neuron is selected. When T is presented, if it has coincident inputs with previously stored characters, then there is a competition and the assembly for this character is formed with the relevant inputs. In this case with the neurons that make T different from I. The final assembly for T is presented in *Figure 3*.

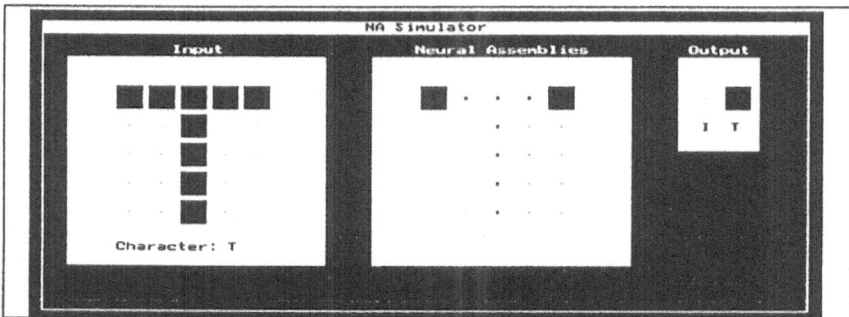

Figure 3. The internal represenation of T consists in the relevant NA-neurons.

Finally, when we present a character belonging to a previously learned class, then the assembly of this class is reorganized to store the minimum number of neurons that represents the class. This effect can be seen in *Figure 4* when character *i* is presented.

The processes that are present in the reorganization of the assemblies have a correlation with psychological phenomena. Acquisition, blocking, overshadowing, and conditioned inhibition are classical conditioning phenomena that can be observed in this kind of self-organization.

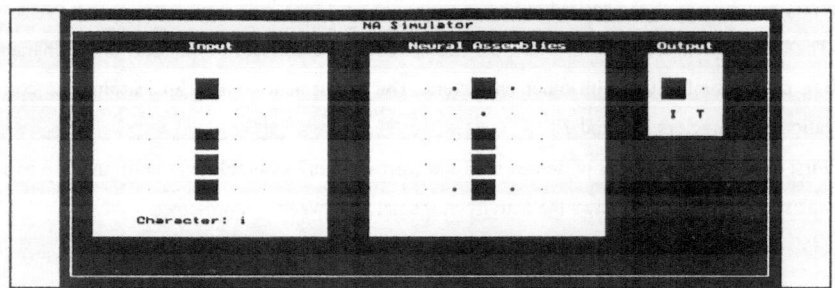

Figure 4. Presentation of character *i* makes five neurons irrelevant.

IV. Future Work

Presently we are expanding the model for a bigger neural assemblies layer, to overcome the limited storage capacity of the model.

References

Burr, D. (1986). A neural network digit recognizer. In *Proceedings of the 1986 IEEE International Conference on Systems, Man, and Cybernetics*, pp. 1621-1625. Atlanta: IEEE.

Carpenter, G. & Grossberg, S. (1987). A massively parallel architecture for a self-organizing neural pattern recognition machine. *Computer Vision, Graphics and Image Processing*, **37**, pp. 54-115.

Fukushima, K., Miyake, S. & Ito, T. (1983). Neocognitron: A neural network for a mechanism of visual pattern recognition. *IEEE Transactions on Systems, Man and Cybernetics*, **SMC-13**, (5), pp. 826-834.

Gerstein, L. G., Bedenbaugh, P., & Aertsen, M. H. J. (1989). Neuronal assemblies. *IEEE. Trans. on Biomedical Engineering*, **36**, (1), pp. 4-14.

Gerstein, G. L., & Turner, M. R. (1990). Neural assemblies as building blocks of cortical computation. In E.L. Schwartz (Ed.), *Computational Neuroscience*, pp. 179-191. MIT Press.

Khotanzad, A., & Lu, J. (1988). Distortion invariant character recognition by a multi-layer perceptron and backpropagation learning. In *Proceedings of the IEEE International Conference on Neural Networks*. Vol. I, pp. 625-632. San Diego: IEEE.

Pawlicki, T., Lee, D. S., Hull, J., & Srihari, S. (1988). Neural networks and their application to handwritten digit recognition. In *Proceedings of the IEEE International Conference on Neural Networks*. Vol. II, pp. 63-70. San Diego: IEEE.

Shimoara, K., Uchiyama, T., & Tokunuga, Y. (1988). Backpropagation networks for event-driven temporal sequence processing. In *Proceedings of the IEEE International Conference on Neural Networks* Vol. I, pp. 665-672. San Diego: IEEE.

Steinbuch, K., & Piske, U. (1963). Learning matrices and their applications. *IEEE Transactions on Electronic Computers*, **EC-12**, pp. 846-862.

Vico, F.J., Sandoval, F. & Almaraz, J. (1994). A Hebb-like learning rule for cell assemblies formation. In *Proceedings of the ICANN'94*, Vol. 1, pp. 781-784.

Learning by Attentional Scanning

Zoltan Schreter
Department of Psychology
University of Tasmania
GPO Box 252 C
Hobart, TAS 7001, Australia
Phone: +61-02-202887, Fax: +61-02-202883
Email: zoltan@psychnet.psychol.utas.edu.au

Abstract: In contrast to many Artificial Neural Network models, the human visual system operates in a mixed parallel-sequential mode: Signals are transmitted in parallel from many areas of the retina towards the brain. In addition, the retina - by moving the eyes or the whole head or the whole body - is directed towards different parts of the environment in sequence. The aim of this project was to create a neural network learning model that has both parallel and sequential aspects. The network learns to focus an attentional window at important, discriminative areas of patterns it is learning to recognise. One advantage of the model is that it is able to learn to discriminate or to categorise patterns that differ from each other only in small areas. In such tasks it often performs better than straightforward backpropagation networks.

Introduction

In humans and animals, attending to selected features can easily combine a *parallel* and a *sequential* mode of operation. Signals are transmitted in parallel from many areas of the retina towards the brain. In addition, however, the retina can be directed — by moving the eyes (or the head or the whole body) — towards different parts of the environment in sequence. This sequential aspect of visual processing allows for quickly gathering information about the environment, without having a very large retina that would allow seeing everything at once.

One important problem for any such system is how to integrate the results of individual glimpses into a whole, recognised image. In particular, do we achieve this integration in *iconic memory* (Sperling, 1960) or using some more abstract, higher level representational system, like *schematic maps* (Hochberg, 1968)? There is much evidence that the latter is true (Bruce & Green, 1990), and the model to be described later is based on the idea of a higher level integration as well.

Another question posed by a sequential scanning system is: how should a decision be made about *where* to scan *when*? The answer to this question obviously depends on the task given the system. One very usual task is to recognise a scene or picture and/or to react to it differentially. This involves discrimination of the scene from other scenes. Humans and animals do this by preferentially scanning the discriminative parts or features of the scene. How do they know which features are discriminative? Ideally, they should be able to learn it. There is a long history of ideas about attentional discrimination learning. In psychology, Lovejoy (1965), and Mackintosh (1965), and more recently Schreter (1990), Bowles (1992), and Kruschke (1992) developed mathematical and neural network models of attentional learning and feature selection processes.

A mixed parallel-sequential operation would be a desirable feature of artificial visual recognition systems as well. It is easy to realise why this is so if we think of a neural network that we would like to train to recognise pictures. A straightforward method sometimes used is to let the units in the input layer correspond to individual pixels of the pictures, and to let their activation value correspond to the brightness of the pixels. The output layer would correspond to different categories of the pictures. A hidden layer would be located between the input- and output layers, and the network would be trained by some version of backpropagation. If we want this network to recognise even moderately sized pictures, we need a large number of input layer units — eg. 250.000 input layer units for pictures of the size 500x500 pixels. However, the real problem is the number of connections necessary. Even if we had only 10 hidden layer units in a network of 250.000 input layer units, with total interconnectivity between the input and hidden layers we would end up with 2.500.000 connections between those layers, making the network's operations excessively slow except if it is implemented on special purpose hardware or on very fast computers.

One solution that neural network researchers had for this problem was to decrease the resolution of the picture, thereby decreasing the number of pixels used. Although low resolution versions of some types of images contain enough information so that people are quite good at recognising them (Bruce, 1988), this solution always carries the risk of losing important information.

Another approach to dealing with the problem of too many inputs and connections could be incorporating into the model the sequential aspect of human/animal visual recognition. The network could have a relatively small input layer, say 50x50 pixels. It could use this layer as a *window* through which it could 'see' a part of the picture. For the system to see the whole picture it would have to move the window to different parts of the picture, necessitating a sequential scanning process[1]. Also, for it to be a model of attentional discrimination learning, the network would have to be able to learn to direct the window preferentially to discriminative areas of the picture.

Such a network will be described in this paper. It is an improved version of the model described in Schreter and Latimer (1992) and of the one described in Schreter (1994). It has an attentional learning mechanism that has both a sequential and a parallel aspect: it contains an attentional window that covers a limited area of the input pattern at any one time and that is allocated in a sequential manner. The limited amount of information covered at any one time is categorised, stored, and can be used to identify or categorise the whole pattern.

The Attentional Scanning Network (ASN) Model

The structure of the model is shown in Figure 1. Each unit in the *Attentional Window* (*AW* from here on) represents an element of a pattern (eg. a pixel of a picture). Each unit in the *Attentional Window Position Control Layer* (AWPC from here on) represents a position of the input window.

Each unit in the *Category Layer* (CL from here on) represents a category of the patterns the network "sees" in the AW. Each unit in the *Output Layer* represents a pattern or pattern category.

A unit in the *CL* receives connections from each of the units in the *AW*. The weights of these connections change during learning. The connections to the *CL* from the *AWPC* are only implicitly there (more about this later).

Attentional Learning of Patterns in the Model

This process occurs in *scanning cycles*, which consist of the steps outlined below. In the first step of a scanning cycle, one of the AWPC units gets activated, by the following equation:

$$N_k = w_k + I_k$$

$$o_k = \begin{cases} 1 & \text{if } N_k = \max_n \{N_n\} \\ 0 & \text{otherwise} \end{cases}$$

where w_k: weight of the location k

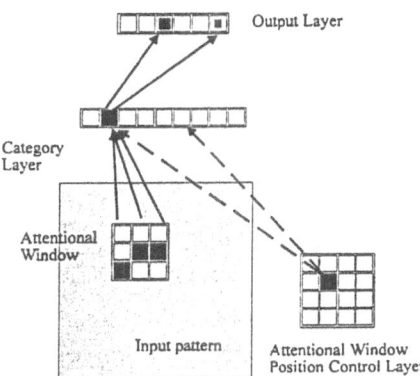

Figure 1. The structure of the ASN model. The arrows with non-broken lines mean changeable connection weights, the arrows with broken lines mean non-changeable connection weights set to 1. See the text for further explanation.

---

[1]   A *further method for* dealing with this problem can be connecting the input layer units only in a relatively small area to each hidden layer unit, mimicking the limited receptive field of ganglion cells of the retina (LeCun, 1989). These two methods are not necessarily contradictory: they could be complementary, just as they are in the human visual system.

o_k: output of AWPC unit k
N_k: net input to AWPC unit k
I_k: inhibitory input to AWPC unit k

That is, the AWPC unit that gets activated is the one that receives the maximal net input. This net input consists of a 'weight' value — corresponding to the current estimate of the system about the importance of the location for discriminating the patterns form each other — and of an inhibitory value. This inhibitory value is set to zero for each AWPC unit, before a pattern is scanned. It is set to a large negative value after the unit is activated. This prevents the same unit from becoming activated - that is, the same location to be scanned - twice during each scanning cycle.

In the second step, the *AW* is allocated into a position above the pattern indicated by the currently active *AWPC* unit. Then, that part of the pattern is copied into the *AW*.

In the third step, a unit in the *CL* becomes activated, according to the following equations:

For all *CL* units i associated with current *AWPC* position: $D_i = \sum_j abs(w_{ij} - o_j)$

$If D_i w_k > \omega$

$$o_i = \begin{cases} 1 & if D_i = \min_n\{D_n\} \\ 0 & otherwise \end{cases}$$

else create new *CL* unit i, set $o_i = 1$, and connect it to both the *AW* and the *Output Layer*, and associate it with current *AWPC* position.

For the *CL* unit with $o_i = 1$, set the $w_{ij} = o_j$.

where w_{ij}:   weight of the connection between *AW* unit *j* and *CL* unit *i*
        o_j: output of *AW* unit *j*
        o_i: output of *CL* unit *i*
        D_i: measure of the distance between the vector of weights of the
                connections between *AW* unit *j* and *CL* unit *i* and the vector of *AW*
                unit outputs
        ω threshold
        w_k: weight of the current location k

'Associating' *CL* units with *AWPCV* positions is represented in Figure 1 by the broken lines. No new *CL* unit is created to represent the current *AW* pattern, except if either the weight-pattern distance or the importance of the current location are relatively large. The latter condition ensures that relatively important locations are associated with a larger number of categories than relatively unimportant ones. This results in a *finer discrimination* of *AW* patterns in important locations.

The value of ω is slowly decreased during learning. This gives a chance for creating new *CL* units if necessary (for example because a particular pattern area contains a lot of variation). The starting value of ω is determined after the first epoch to be the largest $D_i w_k$ value found.

In the fourth step of the scanning cycle, the *Output Layer* units become activated, according to the *logistic output function*. Their activity is determined by the *CL* units active and by the weights coming from those units.

This completes the *activation* phase of the scanning cycle. It is followed by the *learning* phase. There are three parts of the network where learning occurs: in the weights between the *AW* and the *CL*, in the weights of the *AWPC* units, and in the weights between the *CL* units and the *Output Layer*.

Changing the weights coming to the currently active CL unit is achieved by the following equation:

$$\Delta w_{ij} = c(o_j - w_{ij})$$

where w_{ij}:   weight of the connection between AW unit *j* and
Category Layer unit *i*
        o_j: output of AW unit *j*
        c: constant

The above equation is the same as used in many self-organising type weight change algorithms, as in Kohonen (1988). It ensures a move of the weight vector of the currently active *CL* unit towards the activation vector in the *AW*. It results in creating, by the network, a number of *categories* to represent *AW* patterns. Together with the ability of the network to create new categories if the available ones are not adequate, this part of the network is similar to Adaptive Resonance Theory (ART; Carpenter and Grossberg, 1988). The threshold parameter ω corresponds to the 'vigilance' parameter in ART.

Changing the weights between the *CL* units and the *Output Layer* units is achieved by the delta rule:

$$E_i = (t_i - o_i)o_i(1 - o_i)$$
$$\Delta w_{ij} = cE_i o_j$$

where E_i: current error in Output Layer unit i
  t_i: target output of Output Layer unit i
  o_i: output of Output Layer unit i
  o_j: output of Binder Layer unit j
  c: constant
  Δw_{ij}:   change of the weight of the connection between Binder Layer unit j
        and Output Layer unit I

The error of the whole *Output Layer* is propagated backwards to the currently active *CL* unit, using the error backpropagation procedure of Rumelhart, Hinton, and Williams (1986). This error is used to change the weight of the currently active *AWPC* unit:

$$E_j = o_j \sum_i E_i w_{ij}$$

$$\Delta w_k = E_j$$

where E_j: current error in the just activated *CL* unit j
  E_i: current error in *Output Layer* unit i
  w_{ij}:   weight of the connection between the just activated *CL* unit j
        and *Output Layer* unit i
  o_j: output of the just activated *CL* unit j
  Δw_k: change of the weight of the *AWPC* unit k

Changing the weights in the *AWPC* units results in changing the sequence of *AW* allocations in the next scanning cycle in a way that the network tends to allocate the *AW* to *more discriminative* locations first. A larger *AWPC* weight for a particular location means an earlier movement of the *AW* to that location.

After the learning phase is finished, the next allocation of the *AW* follows. This is repeated until either of the following two events (*break-off criteria*) occurs:

1.   A maximal number of *AW* allocations is reached (this is normally the
      number of *AWPC* units);

2.   The sum of error squares in the *Output Layer* sinks below a threshold.

In either of these two events, the scanning cycle finishes, the next input pattern is presented, and the next scanning cycle starts. This is repeated until the network learns to identify or categorise all patterns according to the targets.

The output of the *Output Layer* units can change with each new *AW* allocation. That is, it might be that the network can not identify the pattern based on the first *AW* allocation, but it can do it after the second or third one. Typically, after the network learned the task, only a few *AW* allocations — to some of the most discriminative locations — are necessary to make a decision about the input pattern.

Testing the network's performance

Testing how well the network can identify the learned patterns or how well it can generalise is done similarly as described above, except for two changes:

1.   The network's weight structure does not change;

2.   There is a different break-off criterion: scanning is broken off as soon as one of the *Output Layer* units becomes substantially more active than all other *Output Layer* units. This is interpreted to be the network's identification/categorisation decision.

Experiments with ASN

Experiment 1:

In this experiment the task was to learn to identify eight different patterns (associating each of the patterns with one of eight *Output Layer* units). The patterns consisted of 10x12 matrices (10 columns, 12 rows). In most locations, the pattern matrices were exactly the same for all eight patterns. The patterns differed only in 3 out of the 120 locations , all of which were in the upper left part of the matrices. Figure 2 shows the eight input patterns.

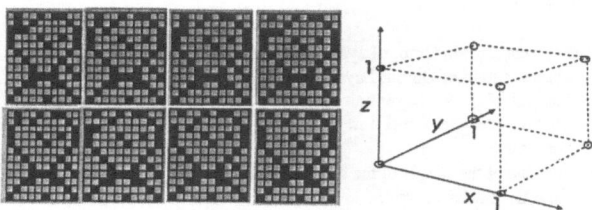

Figure 2. The input patterns used in Experiment 1. The figure on the right shows the 3 important locations and their values, corresponding to 8 value patterns.

The *AW* had a size of 3x3. The size of the *AWPC* was 5x6. That is, the *AW* could be allocated above every second pixel of the input, resulting in some overlap between the *AW* contents.

The learning performance of the ASN network was compared to that of a simple 3-layer backpropagation (BP) network with different numbers of hidden units. The learning rate of the BP network was set to 0.5. Learning of this task is much faster in the ASN than in the BP: perfect learning in about 50 epochs versus about 300 epochs. In fact, this is overly optimistic for the BP: it often happened that the BP network did not learn the task at all during up to 500 epochs.

The ASN used 37 *CL* units, resulting in a network size of 9x37 + 8x37 + 8 + 37=674. The size of the BP network was: 120x20 + 8x20 + 8=2568. That is, for this problem ASN proved to be more efficient, both with regard to speed of learning and to space requirements.

Experiment 2:

In this experiment the task was similar to the one in the previous experiment (identification of all the patterns in the above 3 dimesional feature space with a background that is the same for all patterns), with the following differences: 1) The important locations were not neighbouring each other; 2) The patterns could appear in *two* different locations. Each of the *Output Layer* units were associated not with one pattern but with the same pattern in two different locations. The 16 patterns consisted of 20x20 matrices (Figure 3 shows two of them). That is, this task involved identifiying 6 important pixels in an input containing 400 pixels.

The *AW* size was the same as in Experiment 1, but the *AWPC* size was increased to a 10x10 matrix.

ASN learned the task perfectly in 56 epochs, while BP did not learn it in up to 190 epochs (the weights seemed to be oscillating, without improvement).

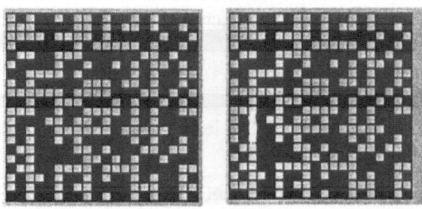

Figure 3: Two of the sixteen input patterns used in Experiment 2.

Discussion

A neural network model has been described that learns to recognise scenes by sequentially scanning them. It 'sees', at any one time, only a relatively small proportion of the whole pattern: 7.5 % of it in Example 1 and 2.25% in Example 2. It can learn to categorise the results of single glimpses, and it stores these categories together with information on the location it saw them in a higher level representation format — thus it has knowledge about *what* it saw *where*.

In the Examples, it demonstrated its ability to learn to discriminate patterns from each other that differed only in a small area. It was able to do this by learning to focus its 'attention' to parts of the scene that discriminate it from other scenes.

This task was very difficult for straightforward backpropagation networks. This is because in this task such networks initially have to deal with a lot of irrelevant information, at the same time as with the relevant data. They try to build up distributed representations combining these two types of information. If there is a relatively large amount of irrelevant information, this can lead to decrements in learning performance. In contrast, ASN can build separate representations for important and non-important parts of the input.

This task is probably relatively easy for humans as well, and for the same reason that it is for ASN: they can easily recognise the low discriminative importance of large areas of the images, and focus their attention into the important area.

After learning, the ASN is very fast at recognising the patterns it was trained on because it allocates its 'attentional window' only to a few areas of discriminative importance.

Bibliography

Bowles, A. (1992). Machine learns which features to select. In A. Adams & L. Sterling (Eds.), *Proceedings of the 5th Australian Joint Conference on Artificial Intelligence* (pp. 127-132). Singapore: World Scientific.

Bruce, V. (1988). *Recognising faces*. Hillsdale:Lawrence Erlbaum Associates.

Bruce, V., and Green, P.R. (1990). *Visual perception. Physiology, psychology and ecology*. Hillsdale: Lawrence Erlbaum Associates.

Carpenter, G. & Grossberg, S. (1988). The ART of adaptive pattern recognition by a self-organizing neural network. *Computer, 21 (3)*, 77-88.

Hochberg, J. (1968). In the mind's eye. In R.N.Haber (Ed.), *Contemporary theory and research in visual perception*. London: Holt, Rinehart, and Winston.

Kohonen, T. (1988). The "neural" phonetic typewriter. *Computer, 21*, 11-24.

Kruschke, J.K. (1992). ALCOVE: An exemplar-based connectionist model of category learning. *Psychological Review, 99*, 22-44.

LeCun, Y. (1989). Generalization and network design strategies. In Pfeifer, R., Schreter, Z.,Fogelman-Soulie, F.,& Steels, L. (Eds.). *Connectionism in perspective*. Amsterdam: Elsevier, 277-282.

Lovejoy, E.(1965). An attention theory of discrimination learning. *Journal of Mathematical Psychology, 2*, 342-362.

Mackintosh, N.J. (1965). Selective attention in animal discrimination learning. *Psychological Review*, 64,124-150.

Rumelhart, D., Hinton, G., & Williams, R.J. (1986). Learning internal representations by error propagation. In Rumelhart, D.E., McClelland, J.L. and the PDP Research Group (1986). *Parallel Distributed Processing. vol.1, Foundations*. Cambridge: MIT-Press, 318-362.

Schreter, Z. (1990). *Modelling with connectionist networks: Interactions between cognition and arousal*. Unpublished PhD Thesis, University of Zurich.

Schreter, Z. (1994). ASN, an attentional scanning network. *Proceedings of ISANN 94, Tainan, Taiwan*, 259-264.

Schreter, Z., and Latimer, C.R. (1992). A connectionist model of attentional learning using a sequentially allocatable "spotlight of attention". *Proceedings of the Third Australian Conference on Neural Networks (ACNN'92), Canberra, 1992*, 143-146.

Sperling, G. (1960). The information available in brief visual presentations. *Psychological Monographs, 74*, whole no. 498.

The Synthesis of the Ranked Neural Networks Applying Genetic Algorithm with the Dynamic Probability of Mutation

Joanna Lis

Institute of Biocybernetics and Biomedical Engineering, Polish Academy of Sciences, Trojdena 4, 02-109 Warsaw, Poland.

ABSTRACT. The real problem connected with the construction of neural networks is to appoint the number of elements (neurons) and the connection structure. The algorithms, which construct the classification networks using ranked layers allow to establish the number of neurons and weights of their connections on the basis of training data set. In this paper, the method of ranked layer construction, using genetic algorithm has been presented. The applying the genetic algorithm leads to obtaining the network with less number of elements, hence the investigated network is of more ability to generalize information. Such a smaller network preserves its given classification property referring to the data of training set. In addition, applying of the genetic algorithm allows to use in the construction process any neuron activation function, e. g. rectangular activation function. This feature results in considerable reduction of the network size.

1. INTRODUCTION. The main problem in researches on the neural networks is fitting of the topology and the parameters of the network to the environment or to the set of data. To obtain a good generalization ability it is necessary to built into the network as much as possible of the knowledge about the problem and to limit the number of the connections in respective way. Therefore, it is required to find such algorithms, which not only perform the optimization of the weights for the given architecture, but they also optimize the architecture itself. In particular it means the optimization of the number of layers and the number of units in the layer. Of course, there exist the different optimization criteria e.g. generalization ability, learning time, number of units etc. But in practice, where different technical limits exist, the cost function - for the architecture only - can be some complicated. Mainly, one focuses on applying as small as possible number of units. Ones of more promising are the approaches, where for a given job the construction or modification of the architecture is performed in the incremental way beginning from the small network and developing it up to appropriate size. To the networks constructed such a way belongs the ranked network of formal neurons [Bobr92].

2. FORMAL NEURON. Formal neuron - the central element of the Perceptron - is the network with several inputs and one output which can be found in tuned or in neutral state. The decision rule of the formal neuron depends on weights being modified in the process of learning the weights. That neuron can be treated and used as a linear classifier: the input signals are classified by the neuron to the tuned group or to the neutral group.

$$r_1(x) = r_{w,\theta} = \begin{matrix} 1 & \Leftrightarrow & \langle w,x \rangle \geq \theta \\ 0 & \Leftrightarrow & \langle w,x \rangle < \theta \end{matrix} \qquad (1)$$

$$\mathbf{x} = [x_1, ..., x_N] \in \mathbf{R}^N \quad \text{- input vector}$$

$$\mathbf{w} = [w_1, ..., w_N] \in \mathbf{R}^N \quad \text{- weigth vector}$$

where:

$$\theta \in \mathbf{R} \quad \text{- threshold}$$

$$<\mathbf{w}, \mathbf{x}> = \sum_{i=1}^{N} w_i x_i \quad \text{- inner product}$$

The bounded applicability of the Perceptron arised from the linearity of its modifiable decision rule. The networks built from the formal neurons modified in several layers don't have that bound. Such network can be treated as the universal tool for generation linear as well as non-linear decision rules. It is essential when the multiclass classifiers are being designed.

3.THE RANKED NEURAL NETWORKS. The ranked network of formal neurons is the two-layers classifying network. Its job is to classify the input vectors (data) to one of K classes. On the stage of construction of the network the classes are represented by the finite separable training sets of vectors: D_k. The first layer of neurons transforms the input vectors into the binary vectors with such property, that the set of transformed vectors from one of classes D_j can be separated with the hiperplane from the set of transformed vectors from whole remaining classes. The weights of the neurons of the second layer obtaining such transformed vectors are selected such a way, that at the input of the neuron of that layer appears 1 for the input vector belonging only to the set D_j, and 0 for the input vector belonging to any other set. The way of construction of ranked network of formal neurons have been described in the work [March90] in the case of 2 classes (2 training sets) and in [Bobr92] in general case of N classes (N training sets). In such network the first layer of the neurons is called the ranked layer. The algorithm of construction of that layer appoints - on the basis of the training sets - both the number of the neurons and the weight coefficients of the layer. The algorithm consists of several stages and on each stage the successive neuron is being added. to the layer. The weights of the added neuron are appointed in such a way, that it can be tuned by the several or by all vectors of one of any training set, and it can't be tuned by any vector of any other training set. Then, from the training set are being removed the vectors tuning the added vector. The training sets modified in such way are used to add the next neuron in the next stage. The procedure repeats until the training sets are exhausted. The number of neurons in the second layer equals the number of classes separating the network. The weights of the neurons of the second layer can be appointed after finishing the construction of the first layer, applying one of standard algorithms of learning of formal neuron. The number of obtained neurons in the ranked layer is not greater than the number of vectors in all training sets, because during the process of construction of each of neuron of that layer at least one vector is being removed from the training sets. In general, it is desirable to use the network which consists of as small as possible number of neurons and which performs the appropriate classification of the vectors from given training sets, because the network of smaller number of neurons has usually greater abilities of generalization of the information obtained on the synthesis stage. To obtain such a network it is necessary, in successive stages of the construction of the ranked layer, to select the weights of added neuron in such a way so that it would be tuned by as many as possible vectors from one of training sets, i.e. as much as possible number of vectors would be removed from the training sets in successive stages of the construction of the layer. The algorithms applied thus far [Bobr94] appointed the weights of the neuron fulfilling only approximately above condition; it sometime led to the network developed too much. For that reason in this paper the applying of Genetic Algorithms (GA's) to appointment the weights of the neurons in the ranked layer have been proposed. GA's, because of their large resistance on the staying in the local extremes, are good in performing above job [Mich92]. Furthermore, the proposed method of appointment the weights of neurons in the ranked layer can be applied to any binary activation function of those neurons e.g. to the rectangular function (sinusoidal rule):

$$r_2(w,x,\theta) = \begin{array}{ll} 1 & \text{for } \sin(<w,x>-\theta) \geq 0 \\ 0 & \text{for } \sin(<w,x>-\theta) < 0 \end{array} \qquad (2)$$

The construction of the ranked layer from such elements can let to decrease the size of the network considerably. In extreme case ranked layer consisting of formal neurons, presented in Fig. 1, has to consist of n-1 neurons (where n - the number of vectors in the training sets), while for the construction of such layer from the elements with rectangular activation function the one element is enough.

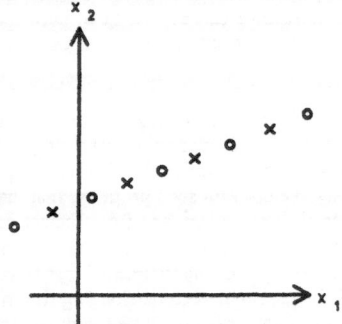

Figure 1. Example of the data from two groups situated alternately.

4.GENETIC ALGORITHMS. The idea of genetic algorithms arises from inspiration by processes observed in the nature i.e. selection of individuals and evolution of species, rules of multiplication (reproduction) and connected with them the inheritance of features [Holl75], [DeJo75]. As a result of activity of natural mechanisms the new species appear; they better fit the environment they are used to live in, and at the same time they push out the worse fitted species. The task of genetic algorithms is to fit the demands, i.e. to find the global extreme of the function. They are the algorithms which, starting from the initial set of probable solutions of given problem, create the sets of more and more better solutions. New solutions are generated applying genetic operators, which simulate the natural processes of reproduction. Each of elements of space of solutions (populations) with constant size is named the chromosome, the components of chromosome code are named genes, and then the genetic operators refer to following names: selection, crossover (cross-matching), mutation. During the process of searching of the domain with the GA the proposals of solutions (i.e. the individuals) are being generated and evaluated by the environment. In the case of genetic algorithms a part of the environment plays so called "fitness function " (accommodation, adaptation) defined in each point of searched domain and measuring the degree of adaptation of a given chromosome to needs of the problem. The task of genetic algorithm is a continuous improvement of value of fitness function of the whole population until the global extreme is reached by that function. To construct the GA it is necessary to settle its five components: the genetic representation of solutions of a given problem, the method of a generation of primary population of solutions, the shape of fitness function (in the sense of evaluation of potential solutions), defining and selection the genetic operators which change the genes in the chromosomes, the values of control parameters of genetic algorithm (e.g. the size of population, the probability of application of a given operator).In practical applications of genetic algorithm the respective stages of its activity are going cyclically. The probabilities of arising of cross-matching, mutation, and a size of population are being defined at the beginning. During the execution of the algorithm, after selection of the population the coupling of the chromosomes takes place, then these couples are being crossovered (due to the probability of the crossover), and on created such a way new couples of chromosomes the mutation is being performed (due to a respective probability). Only after execution above operations the new values of fitness function for each individual of the population are being defined [Gold89]. After certain number of generations, if any other improvements are not observed, the best chromosome represents the optimal solution. In practice, the algorithm stops after given number of iterations depending on time criteria and existing (possessed) resources.

5. THE GENETIC ALGORITHM APPLIED TO THE CONSTRUCTION OF RANKED NEURAL NETWORKS..

5.1. Representation of solutions: The floating point representation of solutions in form of the sequence of real numbers defining vector $(-\theta, w)$ from the neuron weights space has been applied. Not each of solutions represented in the form (θ, w) is admissible one, i.e. not each solution fulfils the condition that the neuron - element is tuned only by the training vectors from the same class. Therefore, the representation of solutions has been taken, where it is possible to present not only admissible solutions, whereas the fitness function has been constructed to act as a filter not giving the inadmissible solutions any chance for survival

5.2. The proposed form of fitness function: The fitness function put it down to the admissible solution the certain percentage of vectors from selected training set; these vectors tune the neuron. On the contrary, to the inadmissible solution that function put it down the percentage of points from behind the selected training set which do not tune the neuron; that percentage is being divided by the number of vectors of the selected training set

$$fittness = \begin{cases} \dfrac{\#(J)}{\#(D_1)} & for \quad (J - D_1) = \varnothing \\[2ex] \dfrac{\#(D_2 - J)}{\#(D_1) \cdot \#(D_2)} & for \quad (J - D_1) \neq \varnothing \end{cases} \qquad (3)$$

where:
- $\#(X)$ — number of elements in set X
- D_1 — selected group of trening vectors
- D_2 — all others learning vectors
- J — vectors activating the constructed neuron

That form of the function lets to eliminate the inadmissible solutions and to distinguish, among the admissible solutions, the better from the worse ones. At the same time the information is being carried, which of inadmissible solutions is better i.e. closer to the admissible one. The initial population selected by lot usually consists of inadmissible solutions only, thus the evolution towards admissible solution is possible.

5.3. Crossover operator: The GA, based on the floating point representation of solutions, most frequently use the crossover operator performing some kind of averaging. Most frequently is applied the crossover operator, which works in this way, that such chromosome is calculated, which individual elements are the arithmetic average from analogous regions of parental chromosomes; this chromosome becomes one of two descendent chromosomes the crossover operator has to produce. On the contrary, the second descendent chromosome is the better one, in the sense of the value of function of fitting of the chromosome from the couple of parental chromosomes. In fact, such operator contains a certain mechanism of selection, and this selection is being performed in deterministic way where worse of chromosomes is always being abandoned. In the elaborated method the mechanism of selection is applied, where worse chromosome is not automatically abandoned, but only obtains less chance of survival or transition to the following generation. The applied crossover operator produces on the basis of two parental chromosomes two new descendent chromosomes and due to assumed strategy of creating of the next generation the parental chromosomes become forgotten, and on their place come the descendent chromosomes.

Descendent chromosomes x' and y' are being calculated on the basis of the pair of parental chromosomes x and y due to the following rule:

$$x'=px+(1-p)y$$
$$y'=py+(1-p)x, \qquad\qquad (4)$$

where p -the random number drawn out by lot from monotonous distribution along the sector [0,1]

5.4. Mutation operator: The applied mutation consists in addition to the mutated vector the vector randomly selected (so called mutating vector) with length not exceeding the half of the length of mutated vector. The mutating vector is produced in following way: each of its component is being drawn by lot from monotonous distribution in the sector [-1,1] and then is being also drawn by lot the length of the vector from the monotonous distribution in the interval [0,1/2 of the length of the mutated vector]. To obtain that length the random vector is being multiplied by the respective real number. So, the operations of mutations of vector W one can write as follows:

$$ w' = w + p\frac{r\cdot\|w\|}{2\cdot\|r\|} \tag{5} $$

where: w   - vector before mutation

      w '   - vector after mutation

      r -   - random vector, where each of its components comes from the monotonous distribution in the interval [-1,1]

      p   - real number drawn from the monotonous distribution in the interval [0,1]

5.5. Initial population. Initial population consists of the random vectors. The random vector is selected in such way, that each of its components is being drawn by lot from the monotonous distribution in the interval [-1,1]. The vectors obtained as a result of such selection are of similar length depending on dimension of vectors. That phenomenon is not profitable, because the nearly equal length of vectors W doesn't let to exist in the initial population the vectors representing the partition of the solutions space on the large number as well as on the small number of stripes. Vector w of the small length yields the partition of the training vectors space on wide stripes with positive or negative values of the function sin(<w, x>-θ). On the contrary, the large value of the vector w leads to the partition of the space on large number of small stripes. Because of that it isn't known in advance what kind of partition of the space is appropriate for a given problem, so it is useful to set in the initial population both the long and short vectors w. Therefore, after drawing out by lot the components of each of vectors of initial population there is saved the direction of each vector, but its length, as a real number is additionally being drawn out by lot from the monotonous distribution in the interval (0,100). The initial population constructed in this way lets the fast beginning of effective searches by means of GA.

5.6. The control parameters of genetic algorithm:
In this work the method of the dynamic establishing of the value of **mutation probability** has been applied. It consists in the evaluation of degree of concentration of the values of fitness function in the population near the maximal value and due to that evaluation increasing or decreasing the mutation probability in following steps to obtain the population concentration degree recognized as an optimal. To this effect it is necessary to define the measure of concentration degree of fitness function values of the population in the way which:
-displays us the described features of the histogram of the fitness function,
-additionally makes possible to select the optimal value of that criterion as much as possible independently on concrete sets of separated points, among others on the spatial dimension.
It has been experimentally established [Lis94], that such requirements are fulfiled by the following measure of degree of population concentration: it is the standard deviation of fitness function value among the half of the number of chromosomes with the largest value of fitness function. That measure will be called the scattering degree of population and will be signed in n-th iteration as ρ(n):

$$ \rho(n) = \frac{\sqrt{\frac{1}{\lfloor N/2 \rfloor}\sum_{i=1}^{\lfloor N/2 \rfloor}\text{fittness}(ch_{n,i}) - \frac{1}{\lfloor N/2 \rfloor}\left(\sum_{i=1}^{\lfloor N/2 \rfloor}\text{fittness}(ch_{n,i})\right)^2}}{\max_{i=1...N}\text{fittness}(ch_{n,i})} \tag{6} $$

where: $ch_{n,i}$ - chromosom with number i from n-th iteration (generation)

         $\lfloor \ \rfloor$ - integer part of the number

and the chromosomes are ordered due to decreasing value of the fitness function i.e.:/

$$\forall n, i, j \quad i < j \Rightarrow \text{fittness}(ch_{n,j}) \geq \text{fittness}(ch_{n,j}) \tag{7}$$

The standard deviation measured within the whole population was formed mainly by the chromosomes with the fitness function considerably different from the maximal one and it reacted much weaker to change of mutation probability than the given criterion.

It has been observed, that formerly appointed optimal fixed probabilities of mutation for individual examples corresponds the average value of given criterion within 20 successive iterations amounted about 0.1-0.15 of the maximal value of fitness function in the population.

The **crossover probability** was constant and equal 0.25.

6. EXPERIMENTS, RESULTS: Below have been presented the results describing how decreased the number of elements, necessary to construct the ranked layer, as the effect of the application the elements with sinusoidal decision rule, in relation to the layer of neurons consisted only of neurons with threshold decision rule.

Set of the data	Number of elements in the ranked layer	
	threshold rule	sinusoidal rule
The example containing 44 two-dimensional (2D) vectors - the components from two groups.	5	4
The set of 176 3D vectors describing the results of laboratory tests of two groups of chemical substances.	8	6

Table 1. Results of experiments.

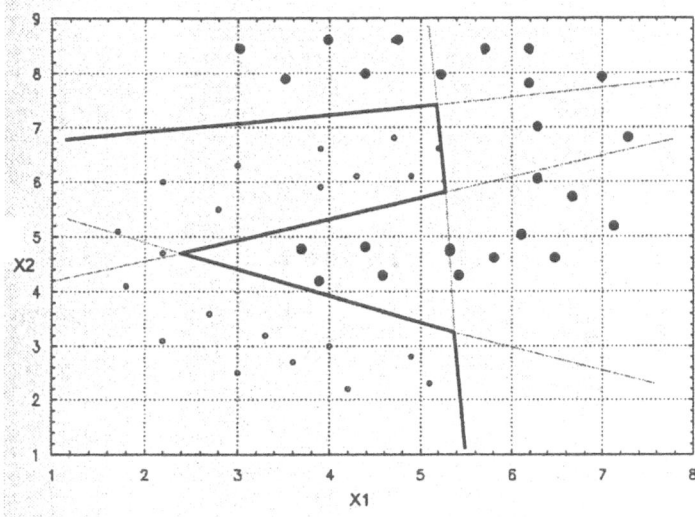

Fig.2. The partition of set of data (44 two-dimensional vectors) obtained as a result of applying the formal neurons with threshold decision rule. That partition is being performed by the ranked layer consisted of 5 elements.

504

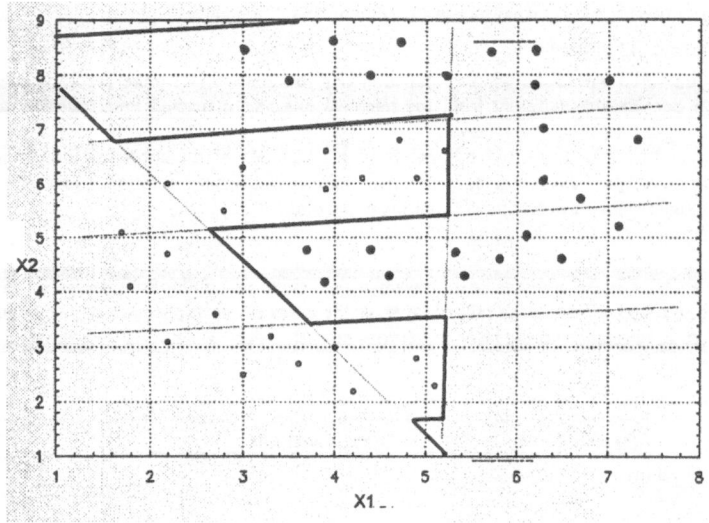

Fig.3. The partition of the same set of the data obtained as a result of applying the elements with the sinusoidal decision rule. That partition is being performed by the ranked layer consisted of 4 elements.

7. REFERENCES.

[Bobr92] Bobrowski L.,"The ranked networks of formal neurons", Biocybernetics and Biomedical
    Engineering, vol.12, no.1-4, 1992.
[Bobr94] Bobrowski L., "Ranked layer of formal neurons and their applications in features extraction"
    (in polish), Proc. First National Conference on Neural Networks, Częstochowa, 1994.
[DeJo75] De Jong K. A.,"An analysis of the behavior of a class of genetic adaptive systems", Ph.D.
    dissertation, Univ. Michigan, 1975.
[Gold89] Goldberg D. E., "Genetic Algorithms in Search, Optimization and Machine Learning",
    Reading, MA: Addison-Wesley, 1989.
[Holl75] Holland J. H. "Adaptation in Natural and Artificial Systems", Ann Arbor, MI: University of
    Michigan Press, 1975.
[Lis94] Lis J., "The Algorithms of neural networks for Clssification Purposes", Ph.D. dissertation,
    Institute of Biocybernetics and Biomedical Engineering of Polish Academy of Sciences, Warsaw,
    1994.
[March90] Marchand M., Golea M., Rujan P., "A convergence theorem for sequential learning in two-
    layer Perceptrons", Europhysics Latters, vol.11, no.6, 1990.
[Mich92] Michalewicz Z., "Genetic Algorithms + Data Structures = Evolution Programs". Springer-
    Verlag.1992.

Global Versus Local Heuristic Terminal Attractor

F.J. Marín[1], F. García[1] and F. Sandoval[2]

(1) Dpto. Arquitectura y Tecnología de Computadores y Electrónica.
(2) Dpto. Tecnología Electrónica.
Universidad de Málaga, Plaza El Ejido s/n, 29013 Málaga, (SPAIN)
Email: marin@tecmal.ctima.uma.es

Abstract

The Heuristic Terminal Attractor (H.T.A) [2,3] is one of the most widely used algorithms for training feedforward neural networks. This algorithm ensures the completion of the learning process in finite time, as well as reaching the global minimum of the error function. The H.T.A. implementation introduces a local adaptive gain factor, in the way that it only affects the weights which belong to a given neuron. The weights actualization rule works with partial information from the weights vector, making the local algorithm slower in learning time. In this paper, we introduce the global gain strategy for H.T.A., which updates the matrix weights using all the weights of the network, instead of partial information of the matrix weights. This global strategy provides shorter learning times than the local one. A theoretical computational study is given to compare the viability of global versus local algorithm, for an only processor and in the case in which a processor is available for every neuron. The results are shown for the encode/decode problem and for the pattern recognition of the alphabetical capital letters.

1.- INTRODUCTION

Multilayer neural networks are being successfully applied to the resolution of problems such as function optimization, signal processing and pattern recognition. Backpropagation (B.P.) [1] is the most frequently used learning algorithm. It presents some disadvantages such as it is slow and falls into local minima. Learning algorithms based on the terminal attractor concept [4] reach the global minimum of the error function in finite time and are faster than backpropagation.

The Heuristic Terminal Attractor is one of the most known from these types of algorithms. It introduces an adaptive gain factor in the weight update rule so that the algorithm is easily implemented in parallel, because the information that a neuron needs to update its weights is local. The most important drawback of all these algorithms is its sequential nature, i. e., the learning process requires 4 steps that must be sequentially carried out: first, computing errors for each neuron, starting from the input layer and backpropagating the error until the first hidden layer; then, computing the weight increment once the errors for each neuron are known; third, computing the output layer errors and, finally, the weights actualization.

In this paper, we introduce the global gain strategy for H.T.A., which updates the matrix weights using all the weights of the network, instead of partial information of the matrix weights. Also, we measure the improvement introduced by the parallel implementation in case that local information is available, and the inconvenience of updating the weights only with partial information from the weights vector.

2.- BASIC CONCEPTS

In a multilayer feedforward neural network the output $U_{l,j}$ of a neuron j-th of the layer l-th is expressed as:

$$U_{l,j} = f\left(\sum_{i=0}^{N_{l-1}} W_{l,j,i} U_{l-1,i}\right) \tag{1}$$

where $W_{l,j,i}$ is the weight that connects units $U_{l-1,i}$ and $U_{l,j}$. For notational convenience, layer 0 will contain the input vector, i. e., $U_{0,p} = X_p$ with X_p the p-th learning pattern, and the last layer L will represent the output vector. We will also suppose, in order to treat the weight bias θ as another weight, that the 0-th component of the input vector to a layer is 1.0, i. e., $U_{l,0} = 1.0$, so $W_{l,j,0}$ is the weight bias of unit $U_{l,j}$. N_l is the number of neurons in layer l. f() is the activation function of the neuron.

The learning process, i. e., finding the optimal set of weights that exactly produces the desired output pattern for every input pattern, is done with the backpropagation learning algorithm. This algorithm defines a cost or error measure function for each pattern p as:

$$E(W) = \frac{1}{2}\sum_{q=1}^{N_l} (U_{L,q}(X_p) - d_q(X_p))^2 \tag{2}$$

where $d_q(X_p)$ is the desired output for the q-th output neuron for the pattern X_p. The cost function is a cuadratic function in the network weights, being 0 when the solution is satisfied. This function depends on the patterns and the weights $W_{l,i,j}$ and is independent from the neuron activation function.

The global error is expressed as:

$$E = \sum E_p \tag{3}$$

The net weights are determined by iteration according to the descending gradient method that changes each weight by a quantity proportional to the gradient of the cost function E :

$$W_{l,j,i}(k+1) = W_{l,j,i}(k) - \mu\left[\frac{\partial E(w)}{\partial W_{l,j,i}}\right]_{W(k)} \tag{4}$$

where μ is a small and positive constant called learning rate. Convergence gets very slow when μ is too small, and oscillates when μ is too big. A momentum term, $\alpha(W_{l,j,i}(K) - W_{l,j,i}(K-1))$, is added so that every connection weight tends to make the current search direction an exponentially ponderated average of previous directions, helping to maintain the weights movement through plain regions of the error surface, once a deep descent has been produced.

In practice, the B.P. algorithm is used incrementally, i. e., a training pattern p is assigned to the input and all the weights are updated before presenting a new pattern. This makes to decrease the cost function that adapts itself to a local gradient, in the sense that it is only necessary to evaluate values in the two terminals of a connection to compute the actualization of a weight associated to that connection (output errors and input neuron activation). This makes it suitable for biological implementation because of its high paralellization grade. If weights updating is done when all the patterns have been presented (batch mode), it is necessary an additional local storage for each connection. Effectiveness of these two approaches depends on the problem, but for very regular or redundant training sets it seems that the incremental one is better.

Computing the cost function for a network with n connections has n operations, evaluating n derivatives. Therefore, the computational complexity is n^2 order operations. The B.P. algorithm allows to compute all the derivatives in order n operations.

The most important disadvantages are the long convergence time in a multilayer network, its poor generalization capacity and getting into local minima when the gradient approximates 0 and the error function has not reached a global minimum, a typical problem with gradient based algorithms.

3.- HEURISTIC TERMINAL ATTRACTORS

To correct these inconveniences, one the most widely applied algorithms is the heuristic terminal attractor. The main idea behind terminal attractor algorithms is based upon a violation of the Lipschitz condition at an equilibrium point [4], which guarantees that the solution can be reached in finite time.

The Heuristic Terminal Attractor is implemented as the B.P. with the difference that it adds an adaptive gain factor γ in the equation (4) of the weight update rule. It ensures the completion of the learning process in finite time, and reaching the global minimum of the function error, with a convergence faster than the B.P. algorithm. The gain factor is defined as the quotient between the error function and its derivative with respect to the square of the weights matrix (global information):

$$\gamma = \frac{E^k}{|\nabla_w E|^2} \qquad (5)$$

where k is constant and 0<k<1. If the system gets into a local minimum, the denominator approximates 0, but the numerator does not, which makes γ very large and the system will get out of the local minimum, and reduce the learning process time. When it falls into a global minimum, $E \geq 0$, its derivative is ≤ 0, furthermore, it is 0 only when E=0, which satisfies the Liapunov condition that guarantees that the global minimum of the error function is found in finite time.

When implementing the H.T.A. algorithm, the momentum term, which is present in B.P., is not considered and the gain factor is modified so that it is not necessary to know all the net weights. According to this, the gain concept is extended, being applied only to the vector of weights belonging to a given neuron c (local information). Thus equation (5) can be expressed as:

$$\gamma_c = \frac{E^k}{|\nabla_{w_c} E|^2} \qquad (6)$$

The system still benefits from the terminal attractors properties with the advantage that when computing γ_c it is only needed local information, which allows an efficient parallel implementation. Besides, weight updates are only permitted when adaptive gain is equal or greater than 1, ensuring faster convergence than B.P. If gain is less than 1, weight updates just follows B.P. algorithm with momentum term.

4.- COMPUTATIONAL STUDY

Our aim is to study the feasibility of the global algorithm versus local one in terms of time required to complete a weight update (equation (4)) and number of iterations needed to obtain a given error.

Each iteration consists of the following steps: computing the unit error, calculating weights updates, evaluating the output layer global error, calculating the gain and, finally, updating the weights. In our notation, the time required by an operator to perform an operation is: s for addition and substraction, m for multiplication, d for division, p for exponentiation, and e to evaluate derivatives.

- COMPUTING EACH NEURON ERROR

For an only processor, the following must be computed: equation $E_o = U_L - D_L$ for the output layers, which requires $s*N_L$ time units; equation $E_i = \Sigma_j (U_j)' E_j W_{ji}$ for each hidden unit i with j from 0 to N_{I+1} (Multilayer perceptron with total connectivity), which requires $(s+2m+e)[\Sigma_{k=2,L} N_K N_{K-1}]$ time units. In total, evaluating the error of all the neurons requires the following time units, for both local and global algorithms:

$$T_{err} = (s+2m+e) \left[\sum_{K=2}^{L} N_K N_{K-1} \right] + sN_L \tag{7}$$

In case that a processor is available for each neuron, the calculation of layer l errors must wait for the calculation of the errors of layer l+1. First, must be evaluated the error of the output layer L ($E_o = U_L - D_L$) which requires s time units. Then the error of layer L-1 is evaluated ($E_i = \Sigma_j (U_j)' E_j W_{ji}$) which takes $(s+2m+e)*N_L$ time units, and so on until layer 1. In total, evaluating the error of all the neurons requires the following time units, for both local and global algorithms:

$$T_{err} = s + (s+2m+e) * \sum_{K=2}^{L} N_K \tag{8}$$

- COMPUTING WEIGHT UPDATES

For an only processor, and for each weight, we need to compute the equation $\Delta W_{ij} = - \mu * E_i * (U_i)' * U_j$; $E_i * (U_i)'$ has been previously evaluated for each neuron. The first step will be to multiply this value times μ for each hidden and output neuron ($m*\Sigma_{K=1,L} N_K$ time units). Then, we need to multiply each weight by U_j. In total, evaluating the weight updates for all the neurons in the net takes the following time units, for both global and local algorithms:

$$T_{\Delta W} = m * \sum_{K=1}^{L} N_K + m * \sum_{K=1}^{L} N_K N_{K-1} \tag{9}$$

In case that a processor is available for each neuron, for each weight we need to evaluate the equation $\Delta W_{ij} = - \mu * E_i * (U_i)' * U_j$; $E_i * (U_i)'$ has been calculated in the previous step. The data is directly available to each unit, i. e., it does not have to wait for data processed in other layers. For neuron i the value $\mu * E_i * (U_i)'$ is a constant, we need to evaluate it once (m time units) and then multiply this value by the appropriate U_j. Thus, it will take $(m+m*N_{l-1})$ time units. The amount of time needed to perform this operation will be the longest time taken by any processor. In total, the calculation of the weight updates for all the neurons takes the following time units, for both local and global algorithms:

$$T_{\Delta N}=m + m*MAXIMO(N_0,N_1,N_2,\ldots,N_{L-1}) \tag{10}$$

- COMPUTING THE OUTPUT LAYER TOTAL ERROR

For an only processor it will be enough to add the errors stored in each output layer neuron. Thus, we require the following time units for both local and global algorithms:

$$T_{ERR-SAL}=s*N_L \tag{11}$$

In case that a processor is available for each neuron, it will be enough to add the errors stored in every output layer neuron. In order to accomplish this, a neuron must fetch all the error data and add then (or each processor should increment a variable by E_i but it is equivalent for timing purposes). So, computing the total output layer, error is defined by equation (11), for both global and local algorithms.

- COMPUTING THE GAIN FACTOR

For an only processor, in the global algorithm the gain is computed with equation (5). The following operations must be carried out: a multiplication, a division, an exponentiation and, for every weight an addition because $\nabla_w E$ it is equivalent to the values of ΔW_{ij}. The following time units are needed:

$$T_\gamma=d+m+p+ s*\sum_{K=1}^{L} N_K N_{K-1} \tag{12}$$

For an only processor, in the local algorithm the gain is computed with equation (6). The term E^K is evaluated only once and for each neuron a division, a multiplication and s additions must be performed. The following units are needed:

$$T_\gamma=p+\sum_{K=1}^{L} N_K(d+m+s*N_{K-1}) \tag{13}$$

In case that a processor is available for each neuron in the global algorithm, every processor is in charge of performing the partial sums of its own weight increments ($s*$MAXIMUM N_K time units). A processor out of each two performs the addition of their two partial sums. A processor out of each four performs the addition of their four partial sums, and so on until all the weight sums have been carried out ($s*\lceil \log_2 N\rceil$ time units). Finally, a multiplication and a division must be performed, because E^K is evaluated with the first processor in idle state. The following time units are required:

$$T_\gamma=s*MAXIMO(N_0,N_1,\ldots,N_{L-1}) + s*\sum_{K=1}^{L} \lceil \log_2 N_K \rceil +m+d \tag{14}$$

In case that a processor is available for each neuron, in the local algorithm the gain is computed with equation (6): an addition must be performed for every weight (the number of weights of a neuron equals the number of neurons in the previous layer), and a multiplication, a

division and a exponentiation. Again, the time required for all the
gains to be evaluated will be the time taken by the neuron with more
weights. In total, computing the gain in the local algorithm takes the
following time units:

$$T_\gamma = m+d+p+ s*MAXIMO(N_0,N_1,N_2,\ldots,N_{L-1}) \tag{15}$$

- UPDATING THE WEIGHTS

For an only processor, and since the weight updates have been already
computed, we will only need to evaluate $W_{ij} = W_{ij} + \gamma*\Delta W_{ij}$, where γ is the
gain computed in the previous step. That is, an addition and a
multiplication for each weight in the network. Updating the weights
requires the following time units, for both local and global algorithms:

$$T_{ACT} = (s+m) * \sum_{K=1}^{L} N_K N_{K-1} \tag{16}$$

In case that a processor is available for each neuron, and since the
weight updates and gain have been already calculated, each processor
will perform an addition and multiplication for each weight of a neuron.
Updating the weights, for both local and global algorithms, requires the
following time units:

$$T_{ACT} = (s+m) * MAXIMO(N_0,N_1,N_2,\ldots,N_{L-1}) \tag{17}$$

As can be seen in this study, the only difference between global and
local implementation (in the best case of a processor for each neuron)
is that the former takes $s*\Sigma_{K=1,L}\lceil \log_2 N_K \rceil$ time units longer than the latter
and one less derivative. This, as the experiments will show, does not
implies a better performance if we take into account the information
lost in the weight update rule as a consequence of using a partial
weight vector.

5.- EXPERIMENTS AND RESULTS

The hyperbolic tangent has been employed as activation function in all
the experiments, showing the average of 25 runs. In figures 1 and 2, the
ordinates reflect the normalized error ($\Sigma E/$(number of patterns * number
of output units) and the abscissa shows the number of iterations, which
correspond to the weight update performed when a training pattern is
presented. Figures 1 and 2 have been obtained using the optimal values
for μ, α and K.

The first experiment is the encode/decode 10-5-10. The neural network
has 10 inputs with only 10 possible input patterns, 5 hidden neurons
which must code those patterns and 10 output neurons which must decode
the hidden units to give the same input vector. This problem is one of
the most widely used as a benchmark for neural network learning
algorithms.

In figure 1 the normalized error is represented with values of $\mu=0.1$,
$\alpha=0.1$ and K=0.33. It may be observed that the global algorithm requires
less iterations than the local and back-propagation algorithms to reach
a given error. If we compare the number of iterations and learning time
of the global versus local algorithm, it is concluded that it is
preferable to use global information to shorten the learning time. For
example, to reach a normalized error of less than 0.02 the global
algorithm requires 767 iterations, while the local takes 833 iterations.
In the global one each iteration only needs to perform 7 additions more

and a exponentiation less than the local algorithm, compared to the 51 additions, 52 multiplications and 1 division per iteration needed in both implementations (note that evaluating the derivative of the activation function is equivalent to perform an addition and a multiplication).

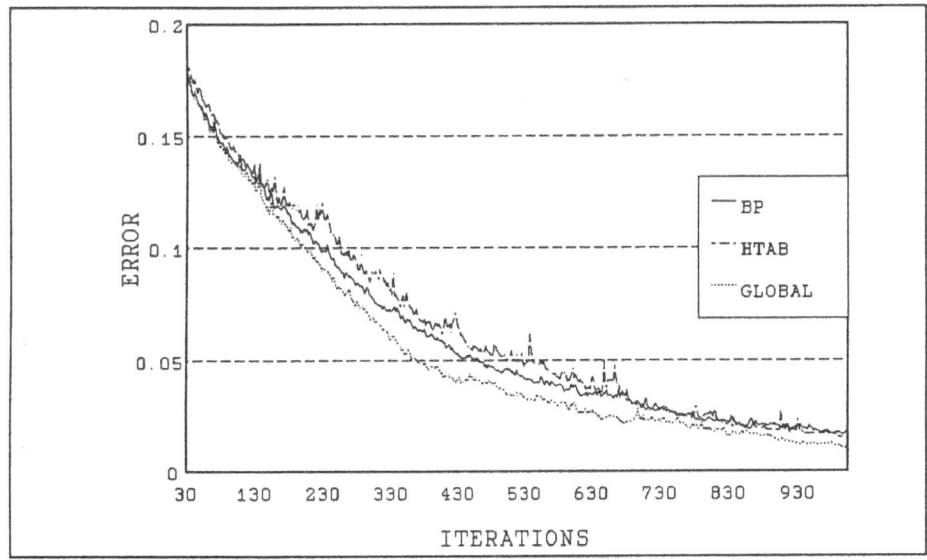

Figure 1. Encode/Decode problem

The second experiment is the recognition of the upper case letters from the alphabet. There are 27 different letters and each one is represented in a 6x7 matrix, and the output is binary coded, so 5 bits are needed to distinguish the 27 patterns. The neural network has 42 input layers, 10 hidden units and 5 output neurons. The normalized error can be seen in Figure 2, with the values $\mu=0.01$, $\alpha=0.1$ y $K=0.1$. It is observed that the global algorithm takes less iterations than the local and back-propagation algorithms to reach a given error. If we comparete number of iterations and learning time of the global versus local algorithm, again, it is clear that is preferable to use the global information to speed the learning time. In the global algorithm each iteration takes only seven additions more and an exponentiation less than the local one, compared to the total of 100 additions, 101 multiplications and one division per iteration needed in both implementations.

6.- CONCLUSIONS AND FUTURE DIRECTIONS

In this paper we show, in a theoretical way and for various problems, that the global implementation is better than the local algorithm in terms of learning time, with better performance with larger net sizes.

In learning algorithms based upon gradient methods, because of their sequential nature, the parallel approach using local information to update the weights will only have positive effects when the net size is large enough; in this case $\Sigma_{K=1,L} \lceil \log_2 N_K \rceil$ should compensate the additional slowness when information on weights that do not belong to that neuron is lost.

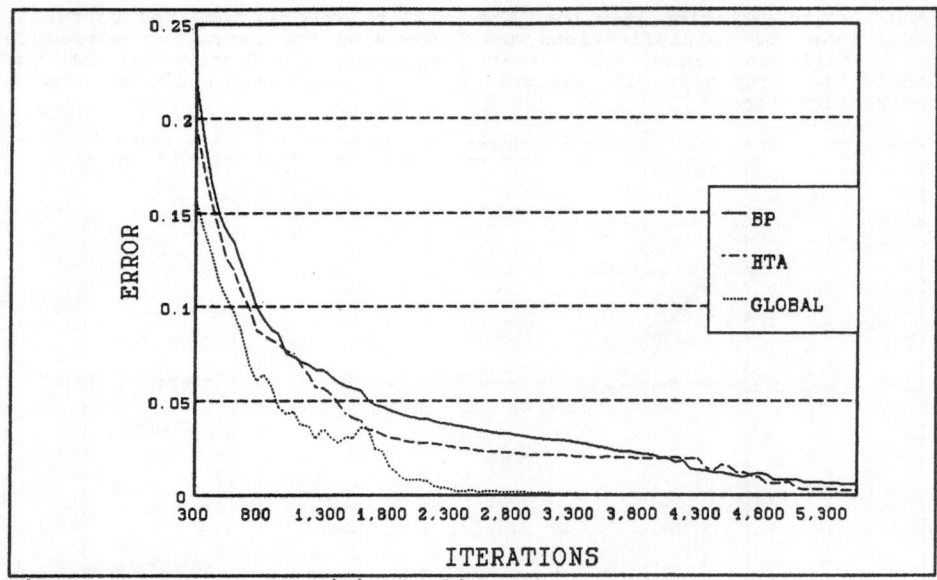

Figure 2. Pattern recognition problem.

The parallel implementation of the global algorithm that we have carried out needs more study as for optimizing the process time of each processor. We are currently comparing the global algorithm with other learning algorithms known in the literature. We are also working on using the global algorithm as a neural network topology constructor.

Acknowledgements This work has been partially supported by the Spanish Comisión Interministerial de Ciencia y Tecnología (CICYT), Project Nº. TIC92-1325-PB.

REFERENCES

[1] Rumelhart, D.E., Hinton, G.E. & Williams, R.J. (1986), "Learning internal representations by error propagation", in D. E. Rumelhart, J.L. McClelland and the PDP Research Group (Eds), Parallel Distributed Processing: Explorations in the Microstructure of Cognition, Vol. 1, pp. 318-362, Cambridge, MA: MIT Press/Bradford Books.

[2] Wang, S. and Hsu, C. (1991), "Terminal Attractor Learning Algorithms for Back Propagation Neural Networks", International Conference on Neural Networks, pp. 183-189, Singapore.

[3] Wang, S. and Hsu, C. (1991), "A Self Growing Learning Algorithm for Determining the Appropriate Number of Hidden Units", International Conference on Neural Networks, pp. 1098-1104, Singapore.

[4] Zak, M. (1989), "Terminal Attractors in Neural Networks", Neural Networks, vol. 2, pp. 259-274.

Dynamic Learning of Radial Bases Functions for Fuzzy Clustering*

Andreas Kanstein and Karl Goser

University of Dortmund, Faculty of Electrical Engineering,
D-44221 Dortmund, Germany.
E-mail: kanst@luzi.e-technik.uni-dortmund.de

Abstract

We present a new architecture of a radial basis function network for fuzzy classification. This architecture supports supervised classification learning with a dynamic adaptation of the number of hidden neurons. The radial basis functions are adapted using a statistically motivated learning function to learn fuzzy clusters of the input patterns. The network provides a classification method that comes from fuzzy control systems.

1 Introduction

Radial Basis Function Networks (RBFN) are neural networks for classification and function approximation. The activation of the hidden layer neurons is computed with radial basis functions. These functions describe fuzzy points in the input space, i.e., membership regions around center vectors. The functionality of RBFN's is very similar to that of fuzzy inference systems [JS93] but is different in regard to their output representation [BG94]. This is discussed in section 2.

For classification tasks, the network structure can be simplified. We group the RBF units together which do the clustering of the inputs of one specific class. These units are connected to a special output pattern unit that does the evaluation for the concerning class. This greatly reduces the number of connections between RBF and output units, a feature which is very important for hardware implementations of neural networks. The motivation is to design an architecture with efficient learning of the RBF's to provide a concept for a VLSI implementation of this classification network. Our architecture is presented in section 2.

Under these constraints the learning of the network reduces to the adaptation of the RBF's. A widely accepted method is the learning vector quantisation [Spe92]. This method is also used in other recent neural network models which are very similar to RBFN's, but are derived from Self-Organizing Maps, e.g., the Vector Quantisation and Projection neural network [DH93], the Fuzzy SOM [Vuo94], and the differential equation SOM [KG94]. Here we will follow this idea. Our update functions are derived from simple statistics and presented in section 3. In section 4 we demonstrate the classification features by means of a well known example and section 5 gives conclusions and an outlook to future work.

2 Network Architecture

Our network architecture (Fig. 1) is motivated from RBF networks and from fuzzy controllers. Both models do a very similar kind of input mapping, but have a different representation of output values. Fuzzy controllers map the fuzzified input onto fuzzy sets of the output variables. In our architecture these fuzzy sets are stored in the output pattern units that are not part of standard RBF networks. To get a crisp output value the output units combine the fuzzy sets of a

---

*This work is supported by the "Deutsche Forschungsgemeinschaft (DFG)", grant number Go-379/11

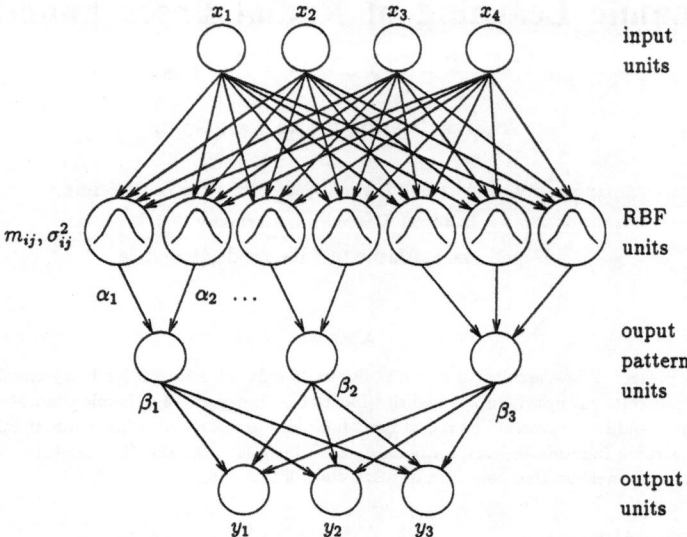

Figure 1: Architecture of the proposed fuzzy classification network. The output pattern units reduce the number of connections between RBF and output units by means of combining the RBF units in groups.

variable and then defuzzify to compute the outputs y_i. Standard RBF networks do not have fuzzy output sets. In addition, their output units compute the weighted sum of the RBF's activations and calculate the output with a sigmoid function, i.e., the decision boundaries are hyperplanes. This gives a different type of reasoning [BG94].

The original Radial Basis Function Network resembles to a three layer feed-forward neural network with special neurons in the hidden layer. These neurons calculate a distance of input and weight vector and compute a locally limited activation, in contrast to the normal feed-forward network. The activation $\alpha_i(\vec{x})$ of the hidden neurons is defined as

$$\alpha_i(\vec{x}) = \exp\left(-\sum_{j=1}^{n_{in}} \frac{\|x_j - m_{ij}\|}{2\sigma_{ij}^2}\right) \quad i = 1 \ldots n_{RBF} \tag{1}$$

with n_{in}: number of input units, n_{RBF}: number of RBF units, m_{ij}, σ_{ij}^2: center and variance of RBF i in input dimension j. Usually the norm $\|\cdot\|$ is the Euclidean norm, but other metrics are also possible [Spe92]. We use individual variances σ_{ij}^2 for each input dimension to increase the flexibility in the shape of the activation regions.

The similarity of the input mapping of RBF networks and fuzzy controllers results from the local activation computation. Each RBF is a composition of one-dimensional fuzzy sets A_{ij} with gaussian shapes, combined using the arithmetic product (as t-norm). We can rewrite (1) as

$$\alpha_i(\vec{x}) = \prod_{j=1}^{n_{in}} \exp\left(-\frac{\|x_j - m_{ij}\|}{2\sigma_{ij}^2}\right) = \prod_{j=1}^{n_{in}} \mu_{ij}(x_j)$$

with $\mu_{ij}(x_j)$: membership value of input x_j to A_{ij}. So the RBF's are premises of rules combining individual fuzzy sets in each input variable. The activation α_i of a RBF then is the activation of the rule, or the truth value of its implication. This is shown on the left side of Fig. 2.

In our architecture each RBF unit is connected to only one output pattern unit. This greatly reduces the number of connections between RBF units and outputs, which is very important for

IF $X_1 = A_{11}$ AND $X_2 = A_{21}$ THEN $Y_1 = yes$ AND $Y_2 = no$

IF $X_1 = A_{12}$ AND $X_2 = A_{22}$ THEN $Y_1 = no$ AND $Y_2 = yes$

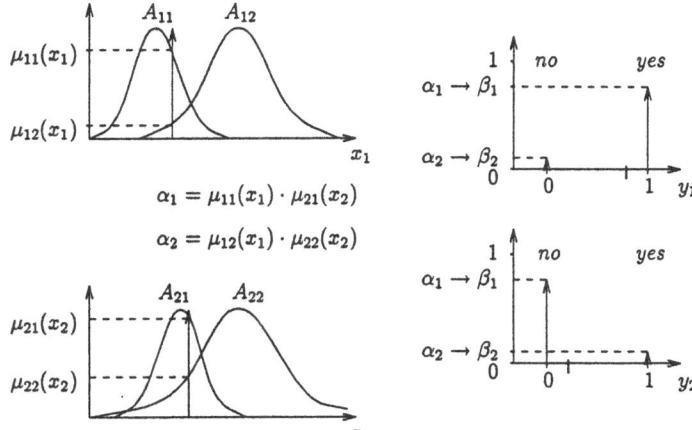

$$\alpha_1 = \mu_{11}(x_1) \cdot \mu_{21}(x_2)$$

$$\alpha_2 = \mu_{12}(x_1) \cdot \mu_{22}(x_2)$$

Figure 2: Example of the fuzzy control mapping of the network. Here $\beta_i = \alpha_i$ because of the small number of rules.

hardware implementations. We call an output pattern unit and the RBF units connected to it a *group*. Every group represents a class of input patterns. The output pattern units combine the activation of the group's RBF units using the arithmetic sum (as t-conorm). This gives the truth value β_i of the output fuzzy sets *yes* and *no* of the rule's conclusions (see Fig. 2). Here output fuzzy sets are singletons, i. e., crisp values, representing the class memberships of the input.

The output units evaluate the output fuzzy sets. Every unit i calculates the possibility of the input belonging to class i. The activation of the output fuzzy set *yes* is the activation of the fuzzy sets *no* of the other groups. Then the defuzzified output value is

$$y_i = \frac{\beta_i}{\sum_{j=1}^{n_{out}} \beta_j} \quad , \quad i = 1 \dots n_{out} . \tag{2}$$

The calculation of a weighted average is the common method for defuzzification in fuzzy control and, as already mentioned, is the main difference to standard RBF networks.

3 Adaptation Strategy

The architecture of the network also simplifies the adaptation process. We use a 1-out-of-k coding of class membership of the input patterns and do only adapt the group neurons which represent the specific class. The update rules do a competitive learning of the RBF centers and are similar to Kohonen's learning vector quantisation (LVQ) algorithm [Koh90]. Of the group's RBF units only that unit is adapted which center is closest to the input vector, i. e., the center $\vec{m}_c$ with

$$\|\vec{x} - \vec{m}_c\| = \min_i \{ \|\vec{x} - \vec{m}_i\| \} . \tag{3}$$

Then, this center is adapted in the direction of the input vector. In our network the variances $\vec{\sigma^2}_c$ of that RBF are adapted, too.

The adaptation formulas for $\vec{m}_c$ and $\sigma^2{}_c$ are derived from simple statistics. From the formulas of mean and variance of a standard probability distribution

$$m(t) = \frac{1}{t} \sum_{i=1}^{t} x_i \qquad \text{and} \qquad \sigma^2(t) = \frac{1}{t} \sum_{i=1}^{t} (x_i - m)^2 \tag{4}$$

follow the iterative expressions

$$m(t+1) = m(t) + \frac{1}{t+1}(x(t+1) - m(t))$$
$$\sigma^2(t+1) = \sigma^2(t) + \frac{1}{t+1}\left((x(t+1) - m(t))^2 - \sigma^2(t)\right) - \frac{1}{(t+1)^2}(x(t+1) - m(t))^2 .$$

The last term in the iterative calculation of σ^2 can be ignored. The total error is

$$\epsilon(t) = \frac{1}{t^2}(m(t+1) - x(t+1))^2$$

and gets very small with large t. In n_{in} dimensions, the adaptation formulas then become

$$\vec{m}_c(t+1) = \vec{m}_c(t) + \frac{1}{t+1}(\vec{x}(t+1) - \vec{m}_c(t)) \tag{5}$$
$$\sigma^2{}_{cj}(t+1) = \sigma^2{}_{cj}(t) + \frac{1}{t+1}\left((x_j(t+1) - m_{cj}(t))^2 - \sigma^2{}_{cj}(t)\right) \quad j = 1 \ldots n_{in}. \tag{6}$$

It should be noted that the adaptation of the center vectors is identical to Kohonen's adaptation formula of self-organization and vector quantisation [Koh90]. Here, the learning rate is $1/(t+1)$ and decreases with every adapted vector. The counters t_i of the number of adapted vectors are stored individually in each RBF unit and incremented only if $\vec{m}_i$ and $\sigma^2{}_i$ are adapted. This increases the stability of the learned clustering, because "old" RBF units that did a large number of adaptations have a very low learning rate and change very slowly. This is important because these units represent a large number of training patterns.

The number of RBF units of a group is determined dynamically, i. e., as desired to cover the clusters of input patterns. We start with a single RBF unit in each group. During learning, we add a RBF unit when the activation β_i of the group unit in response to an input $\vec{x}(t)$ is lower than a threshold θ_α. The new unit is initialised with the input as center vector. Then the adding of units is suppressed for some time to let the group's RBF units settle for some adaptation steps. Therefore we count the number of adaptations since the last change of the group and add a new unit only when this number increased enough. The number of RBF units in a group influences σ^2. Therefore we multiply σ^2 with the number of RBF units in the group to held the activation region in a reasonable range.

Like in Kohonen's LVQ algorithms we also do supervised learning in the sense of reducing the influence of RBF's that do belong to the wrong group. If the activation of RBF units of other groups is bigger than the threshold θ_α, these units are adapted in the opposite direction of the input vector. This reduces the overlap of activation regions and produces a fuzzy system with lower entropy, that is, with less overlap of the fuzzy sets [SKG93], and yields more plausible fuzzy clusters.

4 Example

We want to demonstrate the learning of fuzzy clusters with Anderson's Iris data [And35]. These are real measurements of flowers of three species of Iris, Iris Setosa (SE), Iris Virginica (VI) and Iris Versicolor (VE). The measured dimensions are Sepal length (SL), Sepal width (SW), Petal length (PL) and Petal width (PW). The clustering of the 150 four-dimensional patterns is a very

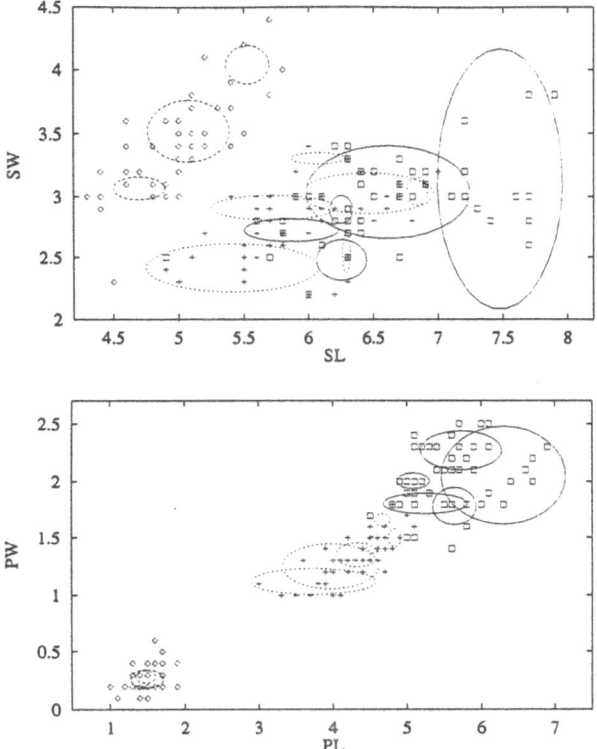

Figure 3: Activation regions of the RBF's and input patterns of Iris Setosa (◇), Iris Virginica (□), and Iris Versicolor (+). The plot shows the different dimensions of the variances σ_{ij}^2. The intersection of activation regions results from the projection of the four-dimensional sets onto two dimensions.

well known and often performed task [Bez74], because this data has nice properties. Two classes intersect in all dimensions, while the third class can easily be separated (SE). In one dimension (SW) all three classes intersect totally, and in a second dimension (SL) at least heavily.

Here classification is performed in order to demonstrate the clustering properties of the proposed method. With an adaptation threshold $\theta_\alpha = 0.3$ the network settles after 26 learning epochs. Four RBF units describe the fuzzy classification region of SE, 6 units that of VE and 7 that of VI. With these RBF's the net classifies all but one training pattern correctly.

More interesting than the performance is the visualisation of the RBF's activation regions in the input dimensions (Fig. 3). It can be seen that the individual activation regions fit the data nicely, mainly because of the individual variances σ_{ij}^2. The intersection of activation regions results from the projection of the four-dimensional sets onto two dimensions.

A final pruning removed the RBF's which had a very small activation region. These units obviously represented an overtrained identification. Pruning removes a RBF unit i when

$$\sum_j \sigma_{ij}^2 < \theta_\sigma \,. \tag{7}$$

For this example we used a threshold of $\theta_\sigma = 0.005$. It is important not to change the scaling of the variances when units are removed. The scaling factor remains the former number of RBF units, even when this number decreased because of pruning. In our example pruning removes two RBF units from VC and one RBF unit from SE. Even now the net classifies all but two training patterns correctly.

This easy pruning strategy again stresses the transparency of the network functionality. The locally restricted activation of the RBF units let overtrained units be found easily. These units can be removed without a change of the classification ability of the other units.

5   Conclusions and Outlook

We presented a special RBF network architecture that implements a fuzzy decision system. The RBF's do a fuzzy clustering of the input data. The proposed network features

- reduction in connectivity

- dynamic adaptation for a low number of RBF units

- competitive learning derived from simple statistics for centers m_{ij} and variances σ_{ij}^2

- very few learning parameters

- reduction of overlap for a low entropy of the fuzzy rule base.

This leads to a network with high transparency and a learning strategy that facilitate a continuous adaptation of the radial basis functions. The network architecture with output pattern units and groups of RBF units yields a reduction of the number of connections. This is a promising approach and inherits a high potential for further investigations. Simplifications of the architecture and the adaptation rules are very important for future VLSI implementations of this system.

References

[And35]   E. Anderson. The Irises of the Gaspe Peninsula. *Bulletin of the American Iris Society*, 59:2–5, 1935.

[Bez74]   J. C. Bezdek. Numerical taxonomy with fuzzy sets. *Journal of Mathematical Biology*, 1:57–71, 1974.

[BG94]   H. Bersini and V. Gorrini. MLP, RBF, FLC: what's the difference. In *Proceedings of EUFIT'94*, pages 19–26, Aachen, 1994.

[DH93]   P. Demartines and J. Hérault. Vector Quantization and Projection neural network. In J. Mira, J. Cabestany, and A. Prieto, editors, *New Trends in Neural Computation*, pages 328–333. Springer-Verlag, 1993.

[JS93]   J.-S. Roger Jang and C.-T. Sun. Functional equivalence between radial basis function networks and fuzzy inference systems. *IEEE Transactions on Neural Networks*, 4:156–159, 1993.

[KG94]   Andreas Kanstein and Karl Goser. Self-organizing maps based on differential equations. In *Proceedings of ESANN'94*, pages 263–269, Brussels, 1994.

[Koh90]   Teuvo Kohonen. The self-organizing map. *Proceedings of the IEEE*, 78:1464–1480, 1990.

[SKG93]   Hartmut Surmann, Andreas Kanstein, and Karl Goser. Self-organizing and genetic algorithms for an automatic design of fuzzy control and decision systems. In *Proceedings of EUFIT '93*, pages 1097–1104, Aachen, 1993.

[Spe92]   Donald F. Specht. Enhancements to Probabilistic Neural Networks. In *Proceedings of the IEEE International Conference on Neural Networks*, volume 1, pages 761–768, Baltimore, 1992.

[Vuo94]   Petri Vuorimaa. Fuzzy Self-Organizing Map. *Fuzzy Sets and Systems*, 66:223–231, 1994.

Incremental Learning with a Stopping Criterion Experimental Results

Rachida Chentouf, Christian Jutten

Laboratoire de Traitement d'Images et Reconnaissance de Formes
Institut National Polytechnique de Grenoble
46 Av. Félix Viallet, F-38031 Grenoble Cedex, France
Tel. + 33 76 57 45 50 Fax. + 33 76 57 47 90 Email : chentouf@tirf.inpg.fr

Abstract. We recently proposed a new incremental procedure for supervised learning with noisy data. Each step consists in adding to the current network a new unit (or small 2- or 3-neuron networks) which is trained to learn the error of the network. The incremental step is repeated until the error of the current network can be considered as a noise. The stopping criterion is very simple and can be directly deduced from a statistical test on the estimated parameters of the new unit. In this paper, we develop experimental comparison between few alternatives of the incremental algorithm and classic backpropagation algorithm, according to convergence, speed of convergence and optimal number of neurons. Experimental results point out the efficacy of this new incremental scheme especially to avoid spurious minima and to design a network with a well-suited size. The number of basic operations is also decreased and gives an average gain on convergence speed of about 20%.

1. Introduction

Using incremental neural networks procedures to perform learning tasks is certainly a very attractive idea. These methods allow automatic tuning of the network size what one generally does empirically with a risk of overestimation or underestimation, which implies computer consuming trial/error procedure. The idea of incrementality is not new and many growing algorithms have been proposed [1], [2], [3], [4]. Unfortunately, these procedures often involve overfitting because their stopping criterion is not well defined. The main questions concerning incrementality that must be adressed are : how to estimate added unit parameters to improve the current network ? how to connect added units to the current net so that it is possible to carry on learning without restarting ? when must one stop the adding process ?

Another aspect of evolutive architectures is the pruning approach which can be applied to oversized networks. In this case, the network can be considered as a target model and one removes units which keep the model as unchanged as possible. An error criterion is then available and few theoretical approaches based on second order expansion of the error [5, 6] and [7, 8] or on intercorrelation of hidden units outputs [9, 10] have been proposed, and can sometimes be explained by statistics [11].

We recently proposed a new scheme for incremental learning in multilayer perceptrons [12]. The procedure includes weights precalculation for the added unit, the fusion with the existing net and a stopping criterion.

In this paper, a practical refinement of the idea is done, leading to different versions of the method. In order to test their efficiency, these versions are compared to classic backpropagation and

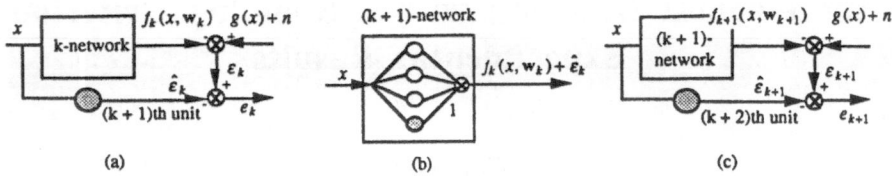

Figure 1. Description of the adding procedure

(a) Before network fusion ; (b) Network fusion (c) After network fusion

to each other. This comparison is done in terms of convergence, speed of convergence and network complexity.

This paper is organized as follow : In section 2, we briefly present the method and explain the interest of adding several new units rather only one. In section 3, the experimental method is described. In section 4, we experimentaly compare classic backpropagation algorithm and 3 alternatives of our incremental algorithm. Finally, we conclude in section 5.

2. The Incremental Scheme

Consider the approximation of a function $y = g(x)$, where $y \in \Re'$ and $x \in \Re'$, by a 1-hidden layer network with sigmoidal hidden units, and linear output units. Let us denote $f_k(x, w_k)$ the k-network approximation obtained with the weight vector w_k. We assume the data are corrupted by a zero mean i.i.d. noise, that is samples are pairs :

$$(x^i, y^i) = (x^i, g(x^i) + n^i). \tag{1}$$

In the following, for the sake of clarity, we consider the approximation of $y = g(x)$ with $x \in \Re$ and $y \in \Re$, but generalization to any dimension is straightforward.

2.1 Adding procedure

Suppose we already have a k-network providing the estimation $f_k(x, w_k)$ of $g(x)$. For each x^i, the error is :

$$\varepsilon_k(x^i) = y^i - f_k(x^i, w_k). \tag{2}$$

If we want to improve the current approximation of $g(x)$, we may add a new unit which must estimate what $f_k(x, w_k)$ did not, that is the error $\varepsilon_k(x)$. The output of the new unit is then :

$$\phi(w_{k+1} x - \theta_{k+1}) = \hat{\varepsilon}_k(x), \tag{3}$$

where $\phi(.)$ is an odd sigmoidal function. In (3), w_{k+1} and θ_{k+1} are adjusted in order to minimize :

$$E = \frac{1}{2} \sum [\phi(w_{k+1} x^i - \theta_{k+1}) - \varepsilon_k(x^i)]^2. \tag{4}$$

If the stopping criterion (see next subsection) is not satisfied, the approximation will be improved simply by adding the new unit to the hidden layer of the network and linking its output to the linear output unit with a unity weight (see Fig.1).

The residual error, just before fusion of the networks, reduces to :

$$e_k(x) = \varepsilon_k(x) - \hat{\varepsilon}_k(x). \tag{5}$$

Because optimizations of the k-network and of the new unit have been done separately, this does not give an optimal solution, although good enough, and it is necessary to train again the (k + 1)-network. After a few learning steps, the approximation is $f_{k+1}(x, w_{k+1})$ with a residual error $\varepsilon_{k+1}(x)$.

And we may repeat the procedure by adding a new (k + 2)th unit which must learn the error $\varepsilon_{k+1}(x)$, and so on.

2.2. Stopping criterion

This adding procedure must be controled to avoid overfitting. The k-network will give the optimal approximation if:

$$\forall i, \quad f_k(x^i, w_k) = g(x^i). \tag{6}$$

In that case, the residual error reduces to :

$$\forall i, \quad y^i - f_k(x^i, w_k) = n^i. \tag{7}$$

If we try to learn, using one neuron, the function $\varepsilon_k(x^i) = n^i$, where n^i are zero-mean i.i.d samples, we then adjust w_{k+1} and θ_{k+1} to minimize:

$$E = \frac{1}{2}\sum_i \left[\phi(w_{k+1}x^i - \theta_{k+1}) - n^i\right]^2. \tag{8}$$

Because n^i are zero-mean i.i.d samples, it is simple to show that minimizing (8) leads to $w_{k+1} = \theta_{k+1} = 0$. Thus, the adding procedure will be stopped if the estimated parameters of the new unit are statistically equal to zero. The estimated parameter vector $\hat{P} = (\hat{w}_{k+1}, \hat{\theta}_{k+1})$ of the new unit, obtained after a Least Square (LS) optimization, is the realization of a random vector P. As LS estimator, it is well known that P is asymptotically Gaussian in the linear case. The asymptotic behaviour remains true also in more general cases, especially in non linear case, under weak conditions [13]. Then, we can test the nullity of $\hat{P}$ using a standard hypothesis test on each component. For a significance level of 5%, the parameters $\hat{w}_{k+1}$ and $\hat{\theta}_{k+1}$ will be considered equal to zero if:

$$t_p = \left| \frac{\hat{p}_{k+1}}{\hat{\sigma}_\varepsilon / \sqrt{N}} \right| < 1.96, \tag{9}$$

where N is the number of samples in the learning set, and $\hat{\sigma}_\varepsilon$ is the estimated standard deviation of residual errors $\varepsilon_k(x^i)$.

In fact, the condition above is necessary but not sufficient : it can be verified if the error is a zero mean white noise, or unfortunately if the approximation $\hat{\varepsilon}_k(x) \approx 0$ is a local minimum of the error $\varepsilon_k(x)$ (see Fig.2). It is then important to detect the 2 situations. We currently test various methods based on local residual analysis and on jump detection suggested in [14].

Moreover, adding only one unit may lead to very poor approximation. Especially, such an approximation is monotonous. We propose in this paper alternative solutions which consist in adding small 2- or 3-neuron networks. The stopping criterion is then a little bit more different, because there are more weights in the added network, for which some non zero combinations can lead to $\hat{\varepsilon}_k(x) \approx 0$. As above, such a situation may occur in 2 cases, and we must be able to detect local minima as suggested previously.

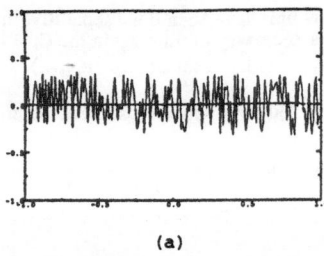 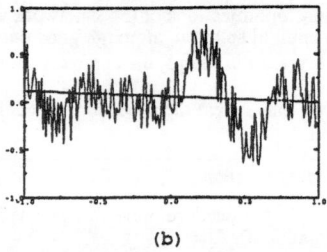

(a)   (b)

Figure 2. The stopping criterion can be satisfied if $\varepsilon_k(x)$ is a zero mean noise (a) or if the estimation $\hat{\varepsilon}_k(x) = 0$ is a local minimum of $\varepsilon_k(x)$ (b).

2.3. The algorithm

The initial algorithm is very simple :

```
Initialization, k := 1, Stop := 0
REPEAT
    Optimization of the k-network
    Optimization of the new unit
    Stopping test of adding new unit
    IF the test is false
        THEN
            Fusion of the 2 networks, k := k + 1
        ELSE
            Stop :=1
    ENDIF
UNTIL Stop = 1
END
```

It may be refined by optimizing simultaneously 1-, 2- or 3-neuron networks. If the test is true for the 1-neuron network, the algorithm stops, otherwise one adds to the main network, the network which computes the best approximation (leading to the smallest residual error). And so on.

3. Experimental Method

Experiments concern approximation of a real function $g(x)$ of the real variable x. The learning data base consists in 200 noisy samples $(x, \ g(x)+n)$. The zero-mean noise has a uniform distribution with a variance equal to 0.3. In the following experiments, the network was trained using simple backpropagation algorithm (BP) but the method can be applied to any supervised algorithm minimizing a square error. The BP algorithm is stopped if the average variation (computed on 20 epochs) of the mean square error becomes less than 10^{-3}.

A good approximation of $g(x)$ can be achieved with a 2-layer perceptron having 6 neurons in its hidden layer. The same function has been estimated using 3 alternatives of the incremental algorithm. The first alternative denoted 1-IA consists in adding neurons one by one. The second alternative consists in adding neurons 2 by 2 and is denoted 2-IA. The last alternative consists in performing optimizations of $g(x)$ and $\varepsilon_k(x)$ "simultaneously" and is denoted 1-AIA for Alternative Incremental Algorithm with adding neurons one by one.

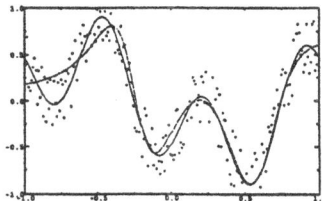

Figure 3. Exemple of estimation provided by the classic BP algorithm.

Are shown the set of training samples corrupted by a zero mean noise of variance 0.3 (dots), the function $g(x)$ and the estimation $f_k(x, w_k)$ provided by a 6-network.

Ten trials have been done for each algorithm, and always using the same BP parameters, and with different initial conditions. Weights are generated randomly in the interval [-M, M] where M is a function of the variance of learning samples, which forces network sigmoides to work in their linear parts at the beginning of the learning.

For each run, we have computed the mean square error (MSE), the number of epochs (one epoch consists in one presentation of the entire data base, here 200 patterns) and the number of operations. This last number is proportional to $3k + 1$ where k is the number of hidden units.

4. Experimental Results

4.1. Classic BP

The results obtained with ten trials on a 6-neuron networks, with different initial conditions, and with same BP parameters are summarized in Table 1. Only 6 runs over 10 lead to convergence. The 4 remaining cases lead to spurious minima. The averaged MSE obtained in this last case was 0.153 while the averaged MSE for convergence was 0.054. The average number of iterations before reaching the local minima is about 155, which implies that the network was rapidly trapped in these minima. We also notice that, in the case of convergence, the BP algorithm needs an important number of operations (about 2,7 millions). An example of estimation provided by a 6-network is given in Figure 3.

run	MSE	# epochs	# Op 10^6
1	0.046	540	2.1
2	0.05	1120	4.3
3	0.057	1260	4.8
4	0.064	300	1.1
5	0.051	480	1.8
6	0.055	520	2
Average	0.054	703	2.7

Table 1. Results obtained with classic BP on a 6-network. Only 6 runs over 10 lead to convergence and are shown in the table.

4.2. Adding new neurons one by one (1-IA) and 2 by 2 (2-IA)

With 1-IA, we start with one hidden unit and add units one by one during incremental process. We notice that no local minimum occured which proves the efficacy of the method in avoiding bad minima. When compared to classic BP, we notice that the number of iterations increases by a factor 112% but the number of operations decreases by a factor 24%. So 1-IA procedure has the double benefit of avoiding local minima and reducing the number of operations and consequently time of convergence. The optimal network complexity is practically the same as classic BP (respectively 5 and 6).

With 2-IA, we start with 2 hidden units and add units 2 by 2 during incremental process. When compared to 1-IA procedure, we notice that the number of operations increases by a factor 15%. This number represents a reduction by a factor 12.5% when compared to classic BP.

run	# neurons	MSE	# epochs	# op 10^6
1	6	0.047	1420	2
2	4	0.049	1320	1.6
3	5	0.051	1580	2.2
4	4	0.048	1340	1.7
5	6	0.048	1800	3.1
6	7	0.044	2220	2.4
7	5	0.046	1180	1.7
8	4	0.050	1560	2
9	5	0.048	1220	1.6
10	5	0,047	1280	2.1
Average	5	0.048	1492	2

Table 2. Results of incremental procedure when adding neurons 1 by 1.

run	# neurons	MSE	# epochs	# op 10^6
1	4	0.047	640	9.5
2	6	0.050	1600	3.2
3	9	0.047	1480	4
4	6	0.047	960	2
5	8	0.045	1200	3.4
6	8	0.048	1600	3.4
7	8	0.047	940	2
8	4	0.046	680	1.3
9	6	0.048	980	1.7
10	5	0,045	740	1.5
Average	6	0.047	1082	2.3

Table 3. Results of incremental procedure when adding neurons 2 by 2.

4.3. Alternative Incremental Algorithm (1-AIA)

This version consists in performing alternatively the two optimizations that are $g(x)$ and $\varepsilon_k(x)$. We train the two networks until the error corresponding to one of them can not be reduced more. In fact, this algorithm consists in learning during one epoch $g(x)$ on the main network, and then during one epoch $\varepsilon_k(x)$ on the new unit, and so on. This procedure can be used with adding neurons either 1 by 1 or 2 by 2. Following results have been obtained when using "alternative learning" with adding neurons 1 by 1 and the procedure is denoted 1-AIA.

When compared to classic BP, we notice that the number of operations decreases by a factor 29%, which is better than both 1-IA and 2-IA procedures (24% and 12.5% respectively).

run	# neurons	MSE	# epochs	# op 10^6
1	5	0.042	1840	1.3
2	5	0.049	1760	1.8
3	5	0.042	1740	2
4	5	0.052	1800	1.8
5	6	0.044	1800	2.3
6	5	0.049	1680	1.7
7	5	0.047	1460	1.8
8	4	0.051	1600	2.2
9	5	0.042	1820	2
10	5	0,041	1800	2.2
Average	5	0.046	1730	1.9

Table 4. Results of alternative incremental learning algorithm. Neurons are added 1 by 1

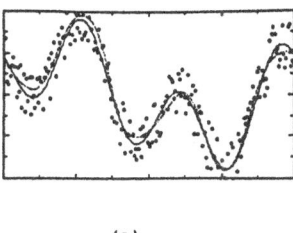 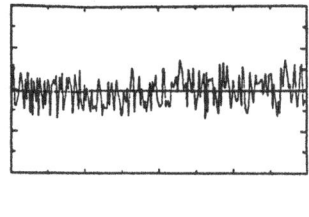

(a) (b)

Figure 4. Experimental results for a function estimation problem using 1-IA procedure

(a) shows the set of training samples corrupted by a zero mean noise of variance 0.3 (dots), the optimal function $g(x)$ and the estimation f_k (x, w_k) provided by a k-network (k = 6) ; (b) is a plot of the corresponding residual error and its 1-neuron network estimation.

5. Discussion

It is well known that the classic BP algorithm can converge to spurious local minima. In our experiment, this situation occurs 4 times over 10 runs. Using the incremental procedure we proposed in this paper, gives us the possibility to avoid bad minima and offers an automatic tuning of the network complexity. The incremental method we proposed is based on 2 new ideas:

• the added unit (or units) must approximate what the current approximation cannot do, that is, the residual error.

• the stopping criterion is a simple nullity test on the new unit parameters.

All alternative of our incremental algorithm are very efficient to avoid local minima. No spurious local minimum has been observed over all experiments we did. The 1-IA procedure reduces by a factor 24% the number of operations. For 2-IA procedure, number of iterations is reduced by only 12.5%, while for 1-AIA, it is reduced by a factor 29%. Experiments prove the interest and the efficacy of this new incremental method (see Fig.4), the alternative learning (1-AIA) seems to be the most efficient procedure.

The method is very simple and can be applied for various optimization algorithms. Especially, it is not restricted to the backpropagation algorithm. The procedure can be easily extended to function approximation from $\mathfrak{R}^r$ to $\mathfrak{R}^s$ and also adapted for classification.

Acknowledgements

This work has been partly funded by the ESPRIT project ELENA (6891).

References

[1] D.R. Reilly, L.N. Cooper and C. Erlbaum. A Neural Model for Category Learning, *Biological Cybernetics* vol. 45, pp. 35-41, 1982.

[2] E. Alpaydin. Grow and Learn : An Incremental Method for Category Learning, *Proc. Int. Neural Network Conf. (Paris)* vol. II, pp. 761-764, 1990.

[3] M. Mézard and J.P Nadal. Learning in Feedforward Layered Networks : the Tiling Algorithm, *J. Phys. A : Maths. Gen.* vol. 22, pp. 2191-2203, 1989.

[4] M. Frean. The Upstart Algorithm : A Method for Constructing and Training Feedforward Neural Networks, *Neural computation* vol. 2, pp. 198-209, 1990.

[5]   Y. Le Cun, J. Denker and S. Solla. Optimal Brain Damage, *Advances in Neural Information Processing* vol. 2, pp. 598-605, 1990.

[6]   M. Cottrell, B. Girard, Y. Girard and M. Mangeas. Time Series And Neural Network : A Statistical Method For Weight Elimination, *1st European Symposium on Artificial Neural Networks (Brussels)*, pp. 157-164, 1993.

[7]   M. Cottrell, B. Girard, Y. Girard, M. Mangeas, C. Muller. Neural Modeling For Time Series: A Statistical Stepwise Method For Weight Elimination. To appear in *IEEE Trans. on Neural Networks*.

[8]   R. Reed. Pruning Algorithm - A Survey, *IEEE Trans. on Neural Networks* vol. 4, n°5, pp. 740-747.

[9]   J. Sietsma, R.J.F. Dow. Creating artificial networks that generalize. *Neural Networks* , vol. 4, n°1, pp.67-79, 1991.

[10] C. Louis, B. Gittler and F. Moutarde. Un algorithme de dimensionnement pour atteindre un réseau d'architecture minimale, *Neural Networks and their applications (Marseille)*, December 15-16, 1994.

[11] O. Fambon and C. Jutten. A Comparison of Two Weight Pruning Methods, *European Symposium on Artificial Neural Networks, (Brussels)* , pp. 147-152, 1994.

[12] C. Jutten. and R. Chentouf. A New Scheme for Incremental Learning, Accepted for *Neural Processing Letters (Brussels)* , 1995.

[13] T. Ameniya. *Advanced Econometrics*. Basil Blackwell, 1986.

[14] A. Benveniste, M. Métivier, P. Priouret. *Adaptive Algorithms and Stochastic Approximations*. Springer-Verlag, Berlin, 1990.

Learning Algorithm with Gaussian Membership Function for Fuzzy RBF Neural Networks

D. Benitez-Diaz, J.Garcia-Quesada

Universidad de Las Palmas G.C., Edificio Departamental de Informática y Matemáticas,
Campus Universitario de Tafira, 35017 Las Palmas (Spain). E-mail: domingo@fobos.ulpgc.es

Abstract

In this paper a new learning algorithm for Fuzzy Radial Basis Function Neural Networks is presented, which is characterized by its fully-unsupervising, self-organizing and fuzzy properties, with an associated computational cost that is fewer than other algorithms. It is intended for pattern classification tasks, and is capable of automatically configuring the Fuzzy RBF network. The methodology shown here is based on the self-determination of network architecture and the self-recruitment of nodes with a gaussian type of activation function, i.e. the center and covariance matrices of the activation functions together with the number of tuned and output nodes. This approach consists in a mix of the "Thresholding in Features Spaces" techniques and the updating strategies of the "Fuzzy Kohonen Clustering Networks" introducing a Gaussian Membership function. Its properties are the same as those of the traditional membership function used in Fuzzy c-Means clustering algorithms, but with the membership function proposed here it lets a nearer relationship exist between learning algorithm and network architecture. Data from a real image and the results given by the algorithm are used to illustrate this method.

1. INTRODUCTION.

Radial Basis Function (RBF) Neural Networks can be interpreted within the framework of the *Mapping Neural Networks* (MNN) that allow to learn a probability density function *p(x)* which is estimated from the observed inputs patterns[9] x_i, i=1,...,r,

$$p_r(\mathbf{x}) = \frac{1}{r} \sum_{i=1}^{r} \frac{1}{h_r^n} \psi\left(\frac{\mathbf{x} - \mathbf{x}_i}{h_r}\right) \quad,, \quad \mathbf{x}_1 \in \Re^n \tag{1}$$

where ψ represents a bounded no-negative potential function of the n-dimensional input vector **x**, and h_r is a sequence of positive numbers such that $lim_{r \to \infty} h_r = 0$, $lim_{r \to \infty} r\, h_r^n = \infty$, and $p_r(x)$ converges to $p(x)$ as r tends towards ∞.

Equation (1) has an associated problem related with the number of potential functions (ψ) required for implementing an unknown function ($p_r(x)$). In some methodologies it is proportional to the number of input patterns[4,10,11]. This is due to the fact that this equation is based on the shifted summation of potential functions with prespecified shapes assigned to individual input samples[9].

A generalized form of (1), incorporating the adjustment of shape parameters and the self-recruitment of potential functions, can be expressed as,

$$f(\mathbf{x}) = \sum_{i=1}^{M} \alpha_i \, \psi(\mathbf{x}, \mathbf{p}_i) \tag{2}$$

where M is the number of potential functions to be recruited, α_i represents the summation weight and $\mathbf{p}_i$ is a parameter vector including both, the position shift and shape parameters of the i-th potential function. M, α_i and $\mathbf{p}_i$ are subject to the adjustment through learning.

Fuzzy Kohonen Clustering Networks[3,16] give a formal framework for determining *f(x)*. Let *c* be an integer, $1 < c < r$, and let $X = \{\mathbf{x}_1, \mathbf{x}_2, ..., \mathbf{x}_r\}$, $\mathbf{x}_k \in \Re^n$, $\forall k$. We say that c-fuzzy subsets $\{u_i : X \to [0,1]\}$, $1 \le i \le c$ are a fuzzy c-partition of X in case $u_{ik} = u_i(x_k)$, $1 \le k \le r$, satisfy[2]:

$$(i)\; 0 \le u_{ik} \le 1, \qquad \forall\, i,k;\; (ii) \sum_i u_{ik} = 1, \qquad k;\; (iii) \sum_k u_{ik} > 0,\; \forall i \tag{3}$$

u_{ik} is interpreted as the membership of x_k in the i-th cluster (partitioning subset) of X. These type of clustering networks try to find out the cluster centers (v_i, $1 \leq i \leq c$) in an unsupervised or self-organizing manner. But this approach is not fully unsupervised so the c parameter is given beforehand.

We propose to get M and p_i in (2) from a fuzzy c-partition of the input data set with a learning algorithm which combines the Thresholding in Feature Spaces techniques with the updating strategy of the Fuzzy Kohonen Clustering Networks. It accommodates input patterns to a Fuzzy RBF Neural Network into classes depicted in a feature representation space. We suppose that the number of classes is unknown and every $\alpha_i=1$. A similar approach has been applied to color image segmentation in Statistical Pattern Recognition[17], but this is not intended for Neural Networks implementations.

This work is divided as follows. Section 2 gives an overview of the Fuzzy RBF network architecture and the Gaussian Membership Function is defined. In section 3 the learning algorithm steps are developed. An experiment made with data from a real problem in pattern classification appears in section 4. And finally section 5 concludes the paper.

2. FUZZY RBF NEURAL NETWORK ARCHITECTURE.

The *Fuzzy Radial Basis Function* Neural Network (FRBF) architecture to be specified by the training algorithm is illustrated in figure 1. It is intended for pattern classification tasks and is composed of three layers: the *Input Layer*, the *Tuning Layer*, and the *Output Layer*. The Input Layer has as many nodes as components have the vector representing to the input pattern (x_k). This is full connected to the Tuning Layer, which has the same number of nodes (c) as different classes have been learned in the net training process that will be explained in the next section.

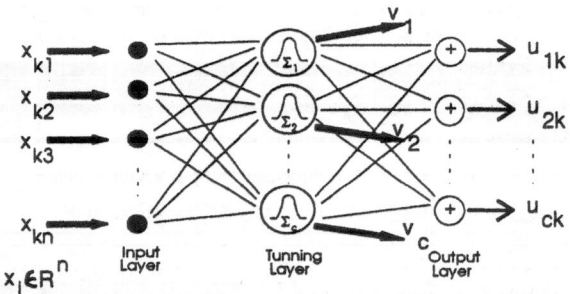

FIGURE 1. Fuzzy Radial Basis Function Neural Network architecture.

Tuned Node is the name of every node in the Tuning Layer. The activation function of this node shall be taken as a gaussian function and represents a pattern class,

$$\psi(x, p_i) = \exp\left(-\gamma \|x - v_i\|_M^2\right) \tag{4}$$

$$\|x - v_i\|_M^2 = (x - v_i)^T [\Sigma_i]^{-1} (x - v_i) \quad ,, \quad i = 1,...,c \tag{5}$$

$p_i=(v_i,\Sigma_i)$ is a vector whose components determine the position and shape parameters of the i-th activation function. Their values are determined at the end of the learning algorithm. $\|x-v_i\|_M$ is the Mahalanobis distance between x and v_i in $\Re^n$, and γ is a scale parameter depending on the value range of x_j.

Each output processing unit computes the membership function of an input pattern to a determined class. In the Output Layer there exist also c nodes, as many as patterns classes have been characterized. We propose to take as the operation performed in this stage, the provided by the following Gaussian Membership Function,

$$u_{ik} = \left(\sum_{j=1}^{c} \left(\frac{\exp\left(-\gamma \left\| x_k - v_i \right\|_M^2 \right)}{\exp\left(-\gamma \left\| x_k - v_j \right\|_M^2 \right)} \right)^{\frac{-2}{m_t - 1}} \right)^{-1} \qquad \forall i,k \qquad (6)$$

where $m_t \in [1,\infty)$ is the membership weighting exponent. Equation (6) satisfies the three conditions associated with a fuzzy c-partition mentioned in (3). Subsequently as it can be noted in (6), our network provides the output with a relative pertinence degree of the input vector to a pattern class through the fuzzy Gaussian Membership Function.

At the Output Layer it can be applied the probabilistic Winner-Take-All rule[8,13] for getting a conventional crisp or hard c-partition of X. On the other hand, the fuzzy information in every output node relative to some input pattern set can be used by a higher level processor, as for instance a knowledge-based recognition system, in order to decide the class of one pattern. This situation may appear in Image Processing when we are segmenting an image. Sometimes every window of pixels is analyzed or transformed to obtain a result that is used to classify the central pixel[1].

3. LEARNING ALGORITHM WITH GAUSSIAN MEMBERSHIP FUNCTION.

In what follows, we are going to give the proposed strategy of training the Tuning and Output Layers that will automatically configure the FRBF classifier. Suppose that we are given a training set $X = \{x_1, x_2, \dots, x_r\}$, $x_k \in \Re^n$ where x_{kj} is the j-th component, $1 \le j \le n$. x_{kj} can take discrete values $\{z_m\}$, $m = 0, 1, \dots, L$.

FRBF1. Calculate a histogram in each coordinate: $\mathbf{h}_j = \left(h_{j0}, h_{j1}, \dots, h_{jL} \right)$, $j = 1, \dots, n$,

$$h_{jm} = \sum_{q=1}^{r} F_{jm}\left(x_{qj} \right) \qquad m = 0,1,\dots,L \qquad F_{jm}\left(x_{qj} \right) = \begin{cases} 1 & x_{qj} = z_m \\ 0 & x_{qj} \ne z_m \end{cases} \qquad (7)$$

Finally we have n histograms $h = \left\{ \mathbf{h}_j \right\}_{1 \le j \le n}$.

FRBF2. Now some 1-D convolutions are done on every histogram (h_j) with an optimized gaussian operator[5] (g), $H_j = \mathbf{h}_j \otimes g$, $j = 1,\dots,n$

FRBF3. Next find the peaks occurring in each filtered histogram H_j, $\{max_{js}\}$, s=1,...,MT$_j$.

$$\left. \frac{\partial H_j}{\partial x} \right|_{x=max_{js}} = 0 \qquad \left. \frac{\partial^2 H_j}{\partial x^2} \right|_{x=max_{js}} < 0 \qquad \forall j,s \qquad (8)$$

and then take a pair of boundary valleys around every peak:

$$\left(minA_{js}, minB_{js} \right), minA_{js} < max_{js} < minB_{js}, \forall s$$

FRBF4. Select a pair of thresholds $\left(ThA_{js}, ThB_{js} \right)$ for each peak location. They are defined as follows, $\forall j,s$

$$ThA_{js} = minA_{js} + (max_{js} - minA_{js}) * fuzzy\_index \qquad (9.1)$$
$$ThB_{js} = minB_{js} - (minB_{js} - max_{js}) * fuzzy\_index \qquad (9.2)$$
$$fuzzy\_index \in [0,1]$$

FRBF5. In the j-th histogram there exist MT_j regions bounded by a pair of thresholds. Data contained outside the bounded regions are further processed in next steps. Thus the input space, $\mathfrak{R}^n$, is partitioned into several regions, identifying each of them to a patterns class,

$$class_i = \left\{ \mathbf{x}_k \in \mathbf{X} \mid \forall j \, \exists s \quad ThA_{js} \leq x_{kj} \leq ThB_{js} \right\}$$

We name Nc_i the cardinal number of $class_i$. If $Nc_i < 0.05\,r$, then $class_i$ is not a valid class, and their patterns will be processed in next steps with patterns outside the bounded regions. "c" coincides with the final number of classes. The recruitment of tuning and output nodes is reached at this step. Their number coincides with c.

From the beginning to this point it has been applied the Thresholding in Features Spaces technique[15]. The learning algorithm we are describing is fully-unsupervised in the sense that the number of pattern classes is also determined, along with the activation function parameters. This is one of the differences with other fuzzy c-Means clustering methods used in Artificial Neural Networks.

FRBF6. Initially the center of masses and covariance matrices are calculated in each class,

$$\mathbf{v}_i = \left(v_{i1}, v_{i2}, \cdots, v_{in}\right) \quad ,, \quad \Sigma_i = \begin{pmatrix} \sigma_i^{11} & \cdots & \sigma_i^{1n} \\ \vdots & \vdots & \vdots \\ \sigma_i^{n1} & \cdots & \sigma_i^{nn} \end{pmatrix} \quad i = 1, \cdots, c$$

$$v_{ij} = \frac{1}{Nc_i} \sum_k x_{kj} \quad ,, \quad \sigma_i^{js} = \frac{1}{Nc_i - 1} \sum_k \left(x_{kj} - v_{ij}\right)\left(x_{ks} - v_{is}\right) \quad ,, \quad \mathbf{x}_k \in class_i \tag{10}$$

FRBF7. Fix the convergence threshold $\varepsilon > 0$, the iteration number limit $t_{max} > 0$, and the initial exponent factor $m_o > 1$.

FRBF8. Now apply the strategy of a Fuzzy Kohonen Clustering Network[16] to the unclassified set of patterns $\mathbf{X}' = \{\mathbf{x}'_1, \mathbf{x}'_2, \dots, \mathbf{x}'_r\}$, $\mathbf{X}' \subset \mathbf{X}$, on the basis of the initial cluster centers and the respective covariance matrices worked out in step FRBF6,

$$\mathbf{V}_o = \left\{\mathbf{v}_{1,o}, \mathbf{v}_{2,o}, \cdots, \mathbf{v}_{c,o}\right\} \quad ,, \quad \Sigma_o = \left\{\Sigma_{1,o}, \Sigma_{2,o}, \cdots, \Sigma_{c,o}\right\}$$

For $t = 0,1, \dots, t_{max} - 1$
{

FRBF8.1 Update all cr' memberships $\{u_{ik,t}\}$ with equation (6) and $m_t = m_o - t\,\Delta m$, where being

$$\Delta m = \frac{m_o - 1}{t_{max} - 1}.$$

FRBF8.2 Update each center of masses with the next formula,

$$\mathbf{v}_{i,t+1} = \mathbf{v}_{i,t} + \frac{r'}{r} \frac{\sum_{i=1}^{r'} a_{ik,t}\left(\mathbf{x}'_k - \mathbf{v}_{i,t}\right)}{\sum_{i=1}^{r'} a_{ik,t}} \qquad a_{ik,t} = \left(u_{ik,t}\right)^{m_t} \qquad i = 1, \dots, c \tag{11}$$

FRBF8.3 Update the covariance matrices $\Sigma_{i,t+1}$. First, every unclassified pattern $(\mathbf{x}'_k)$ is temporary assigned to $class_i$ if $u_{ik,t} \geq u_{i*k,t}$; $\forall i \neq i*$; $i, i* = 1, \dots, c$. $Nc_{i,t+1}$ is the new number of patterns belonging to $class_i$. With the patterns assigned to every class in FRBF5 and the new assignments, calculate

$$\sigma_{i,t+1}^{js} = \frac{1}{Nc_{i,t+1} - 1} \sum_k \left(x_{kj} - v_{ij,t+1}\right)\left(x_{ks} - v_{is,t+1}\right) \tag{12}$$

FRBF8.4 Compute E_t, $E_t = \sum_{i=1}^{c} \left\| \mathbf{v}_{i,t+1} - \mathbf{v}_{i,t} \right\|_M^2 = \sum_{i=1}^{c} \left(\mathbf{v}_{i,t+1} - \mathbf{v}_{i,t}\right)^T \left[\Sigma_{i,t+1}\right]^{-1} \left(\mathbf{v}_{i,t+1} - \mathbf{v}_{i,t}\right)$

FRBF8.5 If $E_t < \varepsilon$ the learning algorithm stops, else t=t+1.

}

3.1 Gaussian Updating Rule Properties.

It can be proved that the Gaussian Learning Rates $\{a_{ik,t}\} = \{(u_{ik,t})^{m_t}\}$ satisfy the same properties as those proposed in other Fuzzy c-Means clustering algorithms[2,3,16].

Property a. $\ell im_{m_t \to +\infty}\{a_{ik,t}\} = 0 \quad \forall i,k$

For large values of m_t (near m_O) all c nodes are updated with lower individual learning rates.

Property b. $\ell im_{m_t \to +1}\{a_{ik,t}\} = 0 \ or \ 1 \quad \forall i,k$

The lateral distribution of learning rates is a function of t as $m_t \to +1$. In the limit $m_t=1$, the gaussian update rule transforms the learning algorithm to the Hard c-Means model[2], where the winner is only updated.

Property c. The gaussian learning rates calculated with equation (11) are, for fixed c, $\{v_{ik,t}\}$, and m_t, of the following form for each x'_k,

$$(u_{ik,t})^{m_t} = \frac{\left(\exp(-\gamma \, d_{ik,t})\right)^{\frac{2m_t}{m_t-1}}}{\beta} \tag{13}$$

where β is a positive constant and $d_{ik,t} = \|x'_k - v_{i,t}\|_M$. The same as with the learning rates that have been used in other algorithms, the effect of (13) is to distribute the contribution of x'_k to the next update of all tuning node vectors directly proportional to their respective gaussian activation function responses. The $v_{i,t}$ closest to x'_k will be moved farther along the line connecting $v_{i,t}$ to x'_k than any other vectors. It can be proved that $\sum_i a_{ik,t} \leq 1$, so these amounts will distribute partial updates across all c nodes for each $x'_k \in X'$. In this way we say that the learning algorithm is also self-organizing.

4. TEST RESULTS.

The learning algorithm for Fuzzy RBF Neural Networks is applied to a chromatic patterns classification problem in images taken from real scenes. Every pixel in a digitized color image has three components in the tridimensional R-G-B space. As it has been proved[12], this is not a good representation of chromatic features, so it is used to make a linear or no-linear transformation from R-G-B to another color space. We choose the constant luminance plane proposed by Yuichi Otha named I2-I3 (I2=R-B, I3=G-0.5(R+B)). Figure 2 illustrates the two-dimensional distribution of points representing pixel colors belonging to an 435 x 462 image that has been taken from a scene, where there are five objects, each of them with a unique color. The test consists in giving to every point several pertinence degrees to chromatic classes, then making a clustering process.

The patterns are depicted by a vector set

$$X = \{x_1, x_2, ..., x_r\} \quad ,, \quad x_k \in \Re^2 \quad ,, \quad x_k = (x_{k1}, x_{k2}) \quad ,, \quad x_{k1} = I2_k, x_{k2} = I3_k$$

where r=200970. In this way the learning algorithm developed in previous section is started. First of all (step FRBF1), the I2 and I3 histograms are worked out (h_1 and h_2 respectively) and the resulting graphic representations are shown in figure 3. At the next algorithm step (FRBF2) some one-dimensional convolutions are done on both histograms with the following optimized gaussian operator, $\frac{1}{70}[6,17,24,17,6]$, to get a smooth version of them (see figure 4).

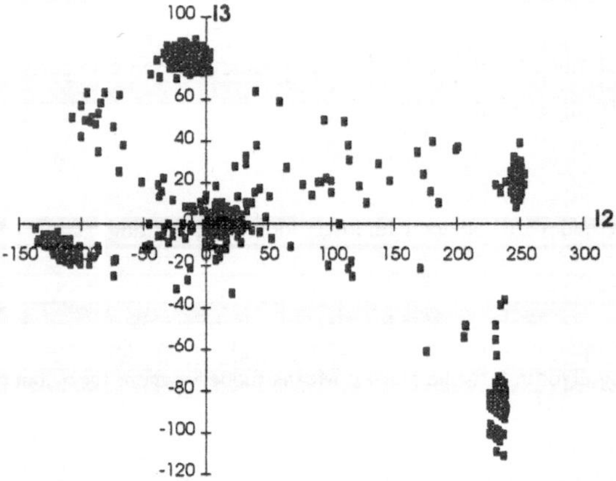

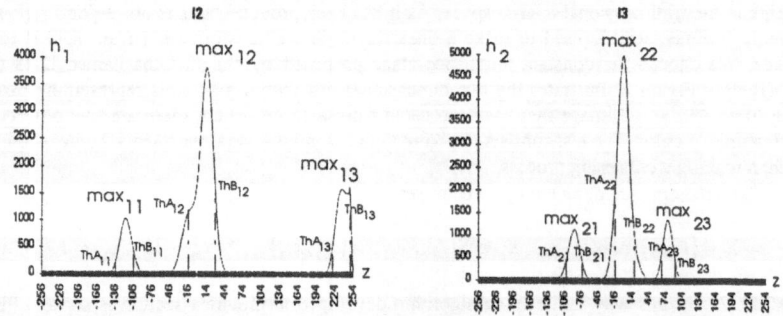

FIGURE 2. Distribution of 200970 points that represent in the I2-I3 space to pixel colors of an image taken from a real scene where there are five objects.

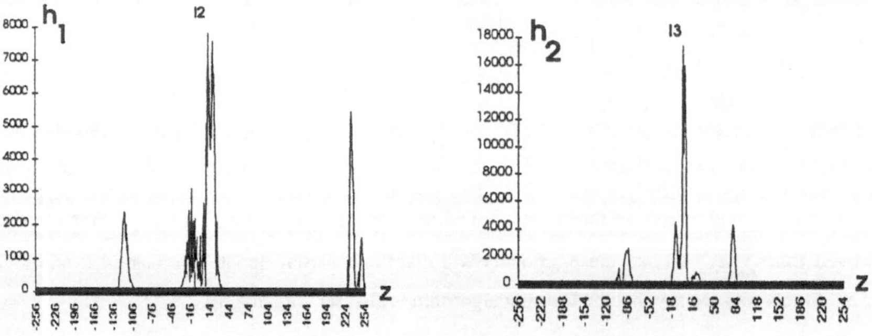

FIGURE 3. Original I2 and I3 histograms.

FIGURE 4. Smoothed versions of the I2 and I3 histograms.

Now the peaks and valleys are found out in each smoothed histogram (step FRBF3). As it can be seen in figure 4, there are 4 peaks in I2 histogram and 3 peaks in I3 histogram. Therefore several thresholds bounding every peak are taken (step FRBF4), which are also depicted in figure 4. We have taken *fuzzy_index*=0.5. In this way the I2-I3 plane is initially divided into regions (gray square regions in figure 5) that can belong to different pattern classes. After the points number in each of these areas is calculated, it has been found out that in five of them the amount of patterns is higher than 5 % of the total number of input patterns that will coincide with the final number of classes (step FRBF5). Whence it is concluded that there will exist five units in the Tuning and Output layers. Thus, the number of unclassified patterns has been reduced to 12094 that will be further processed in the following algorithm steps.

In the next step (FRBF6) five centers of masses and covariance matrices are calculated. These are initial parameters of an iterative loop where it will be obtained the gaussian activation functions of the Fuzzy RBF network Tuning Layer. In step FRBF7 we assess ε=0.1, $t_{max} = 10$, and $m_o = 10$. Finally the algorithm goes to step FRBF8, and in our experiment after four iterative processes the training stopped and the final gaussian activation function positions (v_i) and covariance matrices (Σ_i) are given in Table I. The computing time spent in the training process was two minutes in a HP APOLLO 425 workstation.

The ellipses that enclose the 95 % of points with the same and highest activation function are obtained from standard statistical tables, taking into account that the mentioned points are distributed as χ^2 (chi-square) for two degree of freedom[6]. They are depicted in figure 5. Each ellipse clusters the majority of chromatic patterns that are assigned to a class.

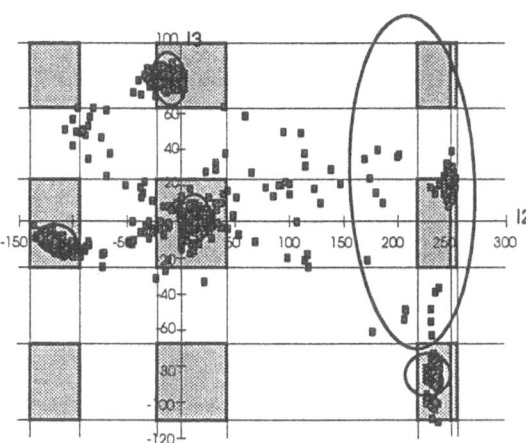

FIGURE 5. Initial square regions (in gray) pointing out probable pattern classes and the final ellipses, each of them representing to a final class, that cluster the 95% of points with the highest membership function.

TABLE I. Final center of masses and covariance matrices that result of applying the Learning Algorithm of the Fuzzy RBF Neural Network to the two-dimensional distribution of points shown in figure 2.

Class number(i)	v_i=(I2$_i$,I3$_i$)	σ_i^{11}	σ_i^{12}	σ_i^{22}
1	(-109.1,-9.9)	66.1	-13.4	9.7
2	(21.9,2.9)	70.0	2.0	12.9
3	(-6.4,77.5)	129.6	12.9	14.6
4	(230.9,-82.6)	22.9	-6.8	69.0
5	(241.4,-18.3)	4325.9	178.6	219.4

534

In our opinion the algorithm developed here gives better performance for classification tasks than others like Back-Propagation[7,14] at a much lower computational cost because it utilizes local representation in a feature space. Furthermore in comparison with some learning algorithms specially thought for classifier Neural Networks[4,9,10,11], so the number of hidden and output units are reduced in basis on the fact that each gaussian radial basis function represents only one class. And moreover in this case, the computational cost is also lower because not all input patterns participate individually in the updating rule (11).

5. CONCLUSION.

We have developed an efficient training algorithm for Fuzzy RBF Neural Network classifiers. It has been proved that it is fully-unsupervised, self-organizing, fuzzy and the associate computational cost is fewer than the spent by other algorithms used for getting similar objectives. Up to now the membership function in a fuzzy c-Means clustering algorithm has been built on a ratio of distance functions. In this paper, the Gaussian Membership Function is introduced in order to take into account the cluster fuzzy sizes and positions in a feature space. For us the value of the sum of gaussian function ratios in equation (6) measures the activation of a hidden unit relative to the others. Whence the Gaussian Membership Function and the Gaussian Updating Rule allow to connect the network learning algorithm with the architecture performance. One test has been conducted to verify the validity of the techniques used. Results have verified the effectiveness of the proposed approach.

Acknowledgments.

The authors gratefully acknowledge helpful comments from and discussions with Professor Roberto Moreno-Díaz, Dr. M.M. González, Dr. Jordi Carrabina and Professor Pedro Gómez, furthermore of Mrs. Carolina Rodríguez Juárez for kindly advising the English version of the paper.

References.

[1] Benítez-Díaz D., Carrabina J., González M.M.; "Neural-like Network Model for Color Images Analysis Systems"; Proceedings of the IEEE International Conference on Neural Networks, Vol.3, pp.1415-1420, 1994.
[2] Bezdek J.C.; Pattern Recognition with Fuzzy Objective Function Algorithms; Plenum Press, New York, 1981.
[3] Bezdek J.C.; "Computing with Uncertainty"; IEEE Communications Magazine, pp.24-36, September 1992.
[4] Chen S., Cowan C.F.N., Grant P.M.; "Orthogonal Least Squares Learning Algorithm for Radial Basis Function Networks"; IEEE Transaction on Neural Networks, Vol.2, No.2, pp.302-309, 1991.
[5] Davies E.R.; "Design of optimal gaussian operators in small neighborhoods"; Image and Vision Computing, Vol.5, No.3, pp.199-205, 1987.
[6] Hald A.; Statistical Tables and Formulas; Wiley, New York, 1952.
[7] Hush D.R., Horne B.G.; "Progress in Supervised Neural Networks"; IEEE Signal Processing Magazine, pp. 8-39, January 1993.
[8] Kohonen, T.; Self-Organization and Associative Memory, third edition, Springer Verlag, 1989.
[9] Lee S., Kil R.M.; "A Gaussian Potential Function Network with Hierarchically Self-Organizing Learning"; Neural Networks, Vol.4, pp.207-224, 1991.
[10] Moody J, Darken C.J.; "Fast Learning in networks of locally-tuned processing units"; Neural Computation, Vol.1, pp.281-294, 1989.
[11] Musavi M.T., Ahmed W., Chan K.H., Faris K.B., Hummels D.M.; "On the Training of Radial Basis Function Classifiers"; Neural Networks, Vol.5, pp.595-603, 1992.
[12] Ohta Y.; Knowledge-Based Interpretation of Outdoor Natural Color Scenes"; Pitman Advanced Publishing Program, 1985.
[13] Osman H., Fahmy M.M.; "Probabilistic Winner-Take-All Learning Algorithm for Radial-Basis-Function Neural Classifiers"; Neural Computation 6, 927-943, 1994.
[14] Pham D.T., Bayro-Corrochano E.J.; "Self-Organizing Neural-Network-Based Pattern Clustering Method with Fuzzy Outputs"; Pattern Recognition, Vol.27, No.8, pp.1103-1110, 1994.
[15] Pratt W.K.; Digital Image Processing, second edition; John Wiley and Sons Interscience, 1991.
[16] Tsao E.C., Bezdek J.C., Pal N.R.; "Fuzzy Kohonen Clustering Networks", Pattern Recognition, Vol.27, No.5, pp.757-764, 1994.
[17] Won Lim Y., Uk Lee S.; "On the Color Image Segmentation Algorithm Based on the Thresholding and the Fuzzy c-Means Techniques"; Pattern Recognition, Vol.23, No.9, pp.935-952, 1990.

Neural Network Initialization

G. Thimm and E. Fiesler

IDIAP, Case Postale 592, CH-1920 Martigny, Switzerland

Abstract

Proper initialization is one of the most important prerequisites for fast convergence of feed-forward neural networks like high order and multilayer perceptrons. This publication aims at determining the optimal value of the initial weight variance (or range), which is the principal parameter of random weight initialization methods for both types of neural networks.

An overview of random weight initialization methods for multilayer perceptrons is presented. These methods are extensively tested using eight real-world benchmark data sets and a broad range of initial weight variances by means of more than 30,000 simulations, in the aim to find the best weight initialization method for multilayer perceptrons.

For high order networks, a large number of experiments (more than 200,000 simulations) was performed, using three weight distributions, three activation functions, several network orders, and the same eight data sets. The results of these experiments are compared to weight initialization techniques for multilayer perceptrons, which leads to the proposal of a suitable weight initialization method for high order perceptrons.

The conclusions on the weight initialization methods for both types of networks are justified by sufficiently small confidence intervals of the mean convergence times.

1 Introduction

The learning speed of multilayer and high order perceptrons[1] depends mainly on the initial values of its weights and biases, its learning rate, its network topology, and on learning rule improvements like the momentum term. The optimal values for these parameters are usually unknown *a priori* because they depend mainly on the training data set used. In practice it is not feasible to perform a global search for obtaining the optimal values of these parameters, as the convergence behavior of the network might change significantly for small changes in the initial weights, as was demonstrated by J. F. Kolen and J. B. Pollack [Kolen-90]. An extensive search for the optimum values requires therefore much more overhead than performing a relatively small number of simulations using non-optimal values. Furthermore, current mathematical techniques are insufficient for a complete theoretical study of the learning behavior of these neural networks. Nevertheless, it is important to have a good approximation of the optimal initial value of these parameters; or with the words of J. F. Kolen and J. B. Pollack: to start the learning process in the "eye of the storm," to reduce the required training time.

Several weight initialization methods for multilayer perceptrons have been suggested. The simplest method among them is random weight initialization, which is often preferred for its simplicity and its ability to produce multiple solutions, as the weights may, due to their initial randomness, converge to various attractors [Kolen-90]. Other methods involve extensive statistical and/or geometrical analysis of the data and are therefore very time consuming. The most rigorous among those is the pseudo-inverse method for perceptrons, which, besides being limited to linear separable data, has several other drawbacks (see [Hertz-91]). Some other weight initialization methods are based on special properties of a network that can not be applied to high order or multilayer perceptrons, as for example the weight initialization technique for radial basis function networks by J. C. Platt [Platt-91].

D. E. Rumelhart et al. observed that if all weights in a neural network are initialized with zero, they have the tendency to assume identical values during training. They therefore proposed random weight initialization to avoid this undesired situation by *breaking the symmetry* [Rumelhart-86]. However, the efficiency of this method depends much on the initial weight distribution. Several researchers therefore proposed random weight initialization methods. An overview of these methods is presented in section 2, and their performance is evaluated in section 4.2.1.

In order to obtain a thorough insight in the initialization characteristics of high order networks, which have not been studied before, numerous experiments were performed, varying the following parameters:

- the shape of the initial weight distribution: uniform, normal or Gaussian[2], and a novel distribution which is uniform over the intervals $[-2a, -a]$ and $[a, 2a]$, and zero everywhere else, (for a motivation of the later see section 4.3),

- the variance of the initial weight distribution,

- the order and topology of the network, and

- the activation function.

---

[1] For a definition of high order neural networks and associated references see [Fiesler-94] and [Thimm-94.1].

[2] Neural network weights are often assumed to be normally distributed [Bellido-93].

The results of these experiments is a simple weight initialization method using an application independent variance.

2 Weight Initialization Techniques for Multilayer Perceptrons

S. E. Fahlman performed an early experimental study on the random weight initialization scheme for multilayer perceptrons. Based on this study, he proposed to use a uniform distribution with a range of $[-1.0, 1.0]$, but found that the best weight range for the data sets in his study varied from $[-4.0, 4.0]$ to $[-0.5, 0.5]$ [Fahlman-88].

Other researchers tried to determine the optimal weight range using network parameters:

L. Bottou uses an interval $[-a/\sqrt{d_{in}}, a/\sqrt{d_{in}}]$, where a is chosen is such way that the weight variance corresponds to the points of the maximal curvature of the activation function (which is approximately 2.38 for a standard sigmoid), and d_{in} is the fan-in (or in-degree) of a neuron, without justifying this interval further in a theoretical manner. L. Bottou trains the neural network only on speech data and does not compare this method with others [Bottou-88].

J. W. Boers and H. Kuiper initialize weights using a uniform distribution over the interval $[-3/\sqrt{d_{in}}, 3/\sqrt{d_{in}}]$, without any mathematical justification. They state that this interval performed the best on their speech data [Boers-92].

F. J. Śmieja uses uniformly distributed weights which are normalized to the magnitude $2/\sqrt{d_{in}}$ for each node. The thresholds of the hidden units are then initialized to a random value in the interval $[-\sqrt{d_{in}}/2, \sqrt{d_{in}}/2]$ and the thresholds of the output nodes are set to zero. He obtained these values from reasoning about hyperplane spin dynamics, and did not validate his method by experiments [Smieja-91].

L. F. A. Wessels and E. Barnard describe two weight initialization methods. The first method sets the initial weight range to a value which assumes that the output of the network and the output patterns have the same variance. The second method puts equally distributed decision boundaries in the input space (without considering input or output patterns), which produces initial weights for the first interlayer weight matrix. The weights of the second interlayer weight matrix are set to 1.0. They compared both methods on generalization for three data sets. They found that the second method performed better in terms of generalization. However, they did not compare convergence speeds [Wessels-92].

An approach similar to the first method of L. F. A. Wessels and E. Barnard was introduced by G. P. Drago and S. Ridella [Drago-92]. They aim at avoiding flat regions in the error surface by restricting the number of neurons with absolute activations greater than 0.9. They developed a simple formula to estimate the best weight initialization scheme for multilayer perceptrons and showed for three data sets that this scheme uses satisfactory good initial weight ranges. The weights are uniformly distributed over the interval $[-a, a]$, with $a = 1.3/\sqrt{1 + E[x^2]}$ for the input layer and $a = 1.3/\sqrt{1 + 0.3d_{in}}$ for the output layer (assuming that all input values x have the same expected value E).

Y. Lee, S.-H. Oh, and M. W. Kim showed theoretically that the probability of prematurely saturated neurons (small weight changes cause only negligible changes of the neuron output) in multilayer perceptrons increases with the maximal value of weights. They conclude that a smaller initial weight range increases the learning speed of multilayer perceptrons. Simulations performed using two data sets confirm their reasoning, but they disregard that learning speed also decreases for weight ranges that are too small. Y. Lee et al. do not suggest an optimal weight range [Lee-93].

P. Haffner, A. Waibel, H. Sawai, and K. Shikano use a normal initial weight distribution. Unfortunately they do not compare their approach to others, give details, or justify it mathematically [Haffner-88].

R. L. Watrous and G. M. Kuhn compared a Gaussian distribution to a uniform distribution and found differences on the conditioning of the Jacobian matrix of a neural network, but found no relation to the convergence speed [Watrous-93].

D. Nguyen and B. Widrow use a multilayer perceptron with piecewise linear activation functions as an approximation of a network with logistic activation functions. Based on this simplification, they calculated an optimal length of $^{d_1}\sqrt{N_2}$ for the randomly initialized weight vectors and an optimal bias range of $[-^{d_1}\sqrt{N_2}, ^{d_1}\sqrt{N_2}]$ for neurons in the hidden layer, where N_2 is the number of hidden nodes. The weights of the neurons in the output layer are randomly initialized in the interval $[-0.5, 0.5]$, without any justification given [Nguyen-90].

Y. K. Kim and J. B. Ra calculated a lower bound for the initial length of the weight vector of a neuron to be $\sqrt{\eta/d_{in}}$, where η is the learning rate [Kim-91].

Besides these random weight initialization methods, some non-random methods are described here for completeness.

A mixture between a random weight initialization scheme and the pseudo inverse method was developed by C.-L. Chen and R. S. Nutter for perceptrons with one hidden layer. First, the weights in the first interlayer weight matrix of the network are initialized with random values. Then, the weights in the second interlayer weight matrix are calculated using the pseudo inverse method applied to the activation values of the hidden layer. C.-L. Chen et al. refined this technique further by alternating the adjustment of the first interlayer weight matrix in a backpropagation-like process with the mentioned method of calculating the second interlayer weight matrix. These adjustments are repeated until a convergence criterion is reached, after which the backpropagation training begins. The authors report faster training in number of backpropagation cycles [Chen-91], but they disregard the computational complexity of the matrix inversions.

T. Denoeux and R. Lengellé initialize a one hidden layer perceptron with prototypes. This method requires a transformation of the input patterns to vectors of unit length and increased size. Additionally, prototypes have to be found by a cluster analysis. The authors reported improvements in training time, robustness versus local minima, and better generalization [Denoeux-93].

3 High Order Perceptrons

High order perceptrons are high order neural networks [Lee-86], having only unidirectional interlayer connections. They are also a generalization of Sigma-Pi networks, which are multilayer perceptrons having high order connections [Rumelhart-86], and functional link networks [Pao-89].

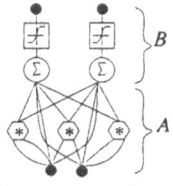

Figure 1: A two-layer high order perceptron.

A high order connection connects a set of neurons in one layer to neurons in the next layer (marked A in figure 1). Each connection applies its specific *splicing function* to the activation values of the lower layer. The number of activation values combined by the splicing functions determines the order of the connection and the connection with the highest order determines the order ω of the network. A network is called a *full (n-th order) network*, if all possible interlayer connections up to this order are present. The network shown in figure 1 is, for example, a full two layer second order network.

The results of the splicing functions are fed, together with the activation values of the lower layer, into the next layer of neurons (marked B). Each neuron consists of a summation unit and an activation function (depicted by a Σ and a symbolized function, respectively).

High order perceptrons can be trained using the backpropagation algorithm, with possible extensions such as a momentum term [Rumelhart-86] or flat spot elimination [Fahlman-88].

From now on in this publication, only two layer high order perceptrons are considered. The splicing function used in this study is multiplication, but other functions are also conceivable.

4 The Simulations

A simulation consists of initializing a neural network and applying the online backpropagation algorithm, alternated by convergence tests, until (non-)convergence is observed. A number of simulations starting with the same initial conditions is called an experiment.

The experiments are performed with two major aims: firstly, to see whether the performance of a network changes for different types of initial weight distributions, and secondly to find the optimal initial weight variance, depending on the activation function of the output neurons and the network order.

Each experiment consists of at least 50 simulations. The number of simulations per experiment was increased until the size of the 95% confidence interval for the mean convergence time permitted a sound conclusion. The confidence intervals were calculated under the assumption that the mean convergence time is student-t distributed.

For the simulations performed, a suboptimal learning rate is used, as it is too laborious and computing time consuming to find the optimal learning rate for each combination of data set and network, and as the learning rate and initial weight variance seem to have an independent influence on the learning speed. Because of the suboptimal learning speed, the results do not necessarily allow a comparison between different activation functions and experiments reported elsewhere, as the maximal possible learning rate may differ largely from the one actually used. For example, a third order network has, for the solar data with a shifted/scaled logistic output function, an optimal learning rate of about 0.05. In contrast, the same network and data set, except for using now a standard logistic output function, has an optimal learning rate of about 4.0, and converges in about the same number of iterations.

Some of the convergence criteria chosen in these simulations are rather crude and not related to the task to be solved. This is done in the aim to reduce the high computational expense, which still was several months of Sparc 10 CPU time.

4.1 The Data Sets

Most öf the data sets used, and shortly described below, are obtained (if not stated otherwise) from an anonymous-ftp server at the University of California [Murphy-94], which also contains further references and documentation. In the following list of data sets, the two numbers in brackets behind the name of the data set are the number of input and output values, respectively.

Solar (12,1) contains the sun spot activity for the years 1700 to 1990. The task is to predict the sun spot activity for one of those years, given the activity of the preceding twelve years ($\hat{=}12$ real valued inputs). The data are scaled to the interval $[0, 1]$.

Wine (13,3) is the result of a chemical analysis of wines grown in a region in Italy derived from three different cultivars. The analysis determined the quantities of 13 constituents found in each of the three types of wines. A wine has to be classified using these values, which are scaled to the interval $[0,1]$. The output patterns use boolean values, encoded as $+1$ and -1.

CES (2,1) is the output of the constant elasticity of a substitution production function for thirty pairs of labor and capital input (see [Judge-85], pages 195 and 210). The patterns have two real valued inputs and one real valued output, none of them scaled.

Servo (12,1) was created by Karl Ulrich (MIT) in 1986 and contains a very non-linear phenomenon: predicting the rise time of a servomechanism in terms of two (continuous) gain settings and two (discrete) choices of mechanical linkages. The input is coded into two groups of five boolean values each, and two discrete inputs, one assuming four, the other five values. The output is real valued, and like all real valued inputs, scaled to the interval $[0,1]$.

Vowels (20,5) is a subset of 300 patterns of the vowels data set, obtainable via ftp from cochlea.hut.fu (130.233.168.48) with the LVQ-package (lvq_pak). An input pattern consists of 20 unscaled cepstral coefficients obtained from continuous Finnish speech. The task is to determine whether the pronounced phoneme is a vowel, and, in the case it is, which of the five possible ones. The boolean output values are encoded as $+1$ and -1.

Auto-mpg (7,1) concerns city-cycle fuel consumption of cars in miles per gallon, to be predicted in terms of 3 multi-valued discrete and 4 continuous attributes. All values are scaled to the interval $[0,1]$ (incomplete patterns have been removed).

Glass (9,1) consists of 8 scaled weight percentages of certain oxides and a 7 valued code for the type of glass (window glass, head lamps etc.). The output is the refractive index of the glass, scaled to $[0,1]$.

Digits (256,10) consists of 500 handwritten digits (50 patterns for each of the ten digits) of the *NIST Special Database 3* [Garris-92]. Each digit was scaled to fit into an image of 16x16 points, and each pixel is represented by an eight bit value. The input values are scaled to the interval $[-1,1]$ and the boolean output values are encoded as $+1$ and -1.

4.2 The Experiments for Multilayer Perceptrons

In the aim to validate and compare the performance of the random weight initialization techniques for multilayer perceptrons mentioned in section 2, a large number of experiments has been performed using the data sets listed in the previous section. The network topology used has one hidden layer which is fully interlayer connected to both input and output layer. The network has no intralayer or supralayer connections, and all activation functions in the hidden and output layer are hyperbolic tangents. No optimization technique was used for training.

For each data set, a sequence of experiments with uniform initial weight distributions of a varying variance were performed (100 simulations per experiment). The outcome of these experiments was used to determine the overall best weight variance as a reference for comparing the random weight initialization schemes of L. Bottou, F. J. Śmieja, G. P. Drago et al., and D. Nguyen et al. It should be noted that D. Nguyen et al. do not seem to use a bias in the output layer. However, neither [Nguyen-90] nor the references mentioned in it state why. To make the simulations fair (leaving out the bias makes learning more difficult), a bias is used in the all simulations reported in this publication.

4.2.1 The Results for Multilayer Perceptrons

The outcome of the experiments for the multilayer perceptrons are shown in tables 1 and 2. These tables list in the first column the name of the data set, the number of neurons in the hidden layer N_2, the convergence criterion ϵ (the notation '$< a$' means that the mean square distance between network output and target pattern has to be smaller than a, and '$b\%$' means that at least b percent of the patterns must be classified correctly), and the learning rate α. The subsequent columns labeled with the initial weight variances in table 1 and names in table 2, respectively, contain the outcome of the experiments. The names in table 2 refer to the random weight initialization schemes described in section 2. A single number in these columns corresponds to the mean number of required online learning cycles until convergence (an online learning cycle is a presentation of all patterns with a weight update after each presentation). A number printed in bold face marks the best result in a row and an entry p/c signifies that the network did not converge in p percent of the online learning cycles, where a trial is judged as non-convergent if the number of cycles exceeds c. The rightmost column in table 2 shows the maximal radius t_{max} of the confidence intervals for the mean number of required learning cycles[3] for the methods listed in this table and a for random weight initialization with a variance of 0.2 (see table 1).

4.2.2 Analysis of the simulations for Multilayer Perceptrons

The average convergence behavior of a multilayer perceptron is depicted in figure 2, where region A indicates the optimum initial weight variances that have been encountered and region B non-convergence. As the curve is flatter on the left side of the optimal initial weight variance than on the right, the loss in performance is much more tolerable for

---

[3]The radius of a confidence interval is the difference between the mean and the upper limit of the interval.

	N_2	ϵ	α	10^{-6}	0.0001	0.001	0.005	0.01	0.05	0.1	0.2	0.5	0.7	1.0	2.0	3.0	4.0
											The initial weight variance σ_w^2						
Solar	5	<0.06	0.3	202	202	195	176	174	165	167	157	148	142	1/500	157	2/500	2/500
Wine	6	90%	0.2	109	105	104	103	104	99.2	100	98.5	97.0	94.3	94.6	92.1	4/500	5/500
CES	4	<0.14	0.1	248	171	131	108	97.9	75.1	63.1	49.7	36.9	35.4	46.9	1/500	1/500	1/500
Servo	3	<0.08	0.1	129	99.0	84.9	74.9	71.0	63.8	64.0	62.8	68.5	73.5	81.3	119	173	4/500
Vowels	20	90%	0.05	158	143	133	125	122	117	113	109	101	102	113	16/800	49/800	53/800
Auto-mpg	3	<0.06	0.3	35.7	33.6	33.9	34.0	33.6	31.9	31.5	31.5	34.7	39.2	44.4	1/500	1/500	1/500
Glass	6	<0.04	0.6	20.9	17.2	15.6	14.1	13.7	12.2	12.0	12.5	16.7	20.0	27.2	48.1	82.8	120
Digits	30	95%	0.3	12.7	12.1	11.6	11.4	11.2	11.6	12.8	15.1	32/100	46/100	25/100	98/100	100/100	100/100

Table 1: Performance of multilayer perceptrons for a uniform weight distribution over the interval $[-a, a]$, with $a = \sqrt{3\sigma^2}$.

data	N_2	ϵ	α	Bottou	Wessels	Šmieja	Drago	Nguyen	t_{max}
Solar	5	<0.06	0.3	152	146	151	153	162	3.1
Wine	6	90%	0.2	98.0	96.9	96.0	98.4	99.3	1.3
CES	4	<0.14	0.1	40.4	32.5	40.8	44.1	42.7	3.2
Servo	3	<0.08	0.1	56.6	55.6	62.2	59.9	63.2	1.4
Vowels	20	90%	0.05	110	105	111	111	115	1.0
Auto-mpg	3	<0.06	0.3	31.8	30.9	31.4	33.1	31.4	1.2
Glass	6	<0.04	0.6	11.0	11.9	11.7	12.1	11.9	0.4
Digits	30	95%	0.1	11.3	11.7	12.8	12.9	11.4	0.2

Table 2: Random weight initialization with the methods of other authors

initial weight variances smaller than the optimal value as compared to bigger variances. Moreover, non-convergence was only encountered for simulations using initial weight variances bigger than the optimal value.

The rather small differences obtained for an initial weight variance of 0.2 as compared to the optimal result, suggests to use this value for a simple weight initialization method. A comparison between the results for this simple method and table 2 shows that the weight initialization method of L. F. A. Wessels et al., which uses the same weight variances as the method of J. W. Boers, performs the best.

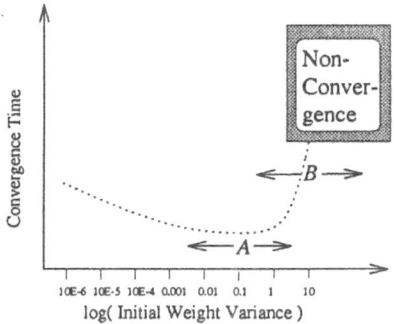

Figure 2: The average behavior of a multilayer perceptron in convergence speed for changing the initial weight variance.

Remark: Some of the weight initialization methods presented in section 2 scale the upper and lower bound of the initial random weight interval by the reciprocal square root of the fan-in. This corresponds to scaling the initial weight variance by the reciprocal of the fan-in. Hence, these methods assume a negative correlation between the fan-in of a neuron and the best initial weight variance. This correlation can not be confirmed or rejected by the results of the experiments; more experiments with other data sets are necessary for this.

4.3 The Experiments with High Order Perceptrons

The networks used in the simulations are usually full and the biases are initialized with a random value of the same distribution as the weights. The only exception is the network trained on the digits data set. This network includes all first order connections and only second order connections with both inputs corresponding to different pixels in the same row or the same column in the image. This configuration should allow the extraction of sufficient features to learn the digits. Training sessions on the in section 4.1 described digits data set gave an acceptable recognition of untrained digits, despite the small training set used.

The three different initial random weight distributions used are: uniform on the interval $[-a, a]$ (with $a = \sqrt{3\sigma^2}$), normal (restricted to an absolute value of $3\sigma^2$), and uniform over the intervals $[-2a, -a]$ and $[a, 2a]$ (with $a = 3\sigma^2/7$) while zero everywhere else. The three types of activation functions used are: a linear f_l, a hyperbolic tangent f_t, and a scaled/shifted hyperbolic tangent f_{st}, shown in figure 3. The use of the function f_{st} was motivated by several ideas: the scaling in the direction of the y-axis prevents the weights from becoming very big and thus cause the same effect as for example scaling the output data to $[-0.9, 0.9]$ and the change of the steepness and the shifting of the sigmoid in the direction of the x-axis were used to force the outcome of the summation step in the neurons to be in the interval $[0, 1]$. Also, experiments with this activation function where performed to see, whether a relation between a deformation of the activation function and the optimal initial weight range exists. The only optimization technique applied to speed up learning is flat-spot elimination.

The detailed results of the high order neural network simulations are not included in this publication due to space constraints.

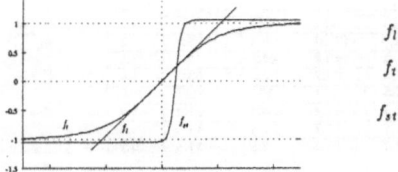

$$f_l(x) = x$$

$$f_t(x) = \frac{2}{1 + e^{-x}} - 1 = tanh(\frac{x}{2})$$

$$f_{st}(x) = \frac{2.1}{1 + e^{-10.1x + 5.07}} - 1.05$$

$$= 1.05\, tanh(5.05x - 2.54)$$

Figure 3: The activation functions

4.3.1 Analysis of the Simulations for High Order Perceptrons

The minimal convergence times for all three initial weight distributions show no difference of statistical significance. The average behavior of the learning time as a function of the initial weight variance, which is depicted in figure 4, is explained as follows. The main difference for the three distributions is the value of the weight variance where the convergence time starts increasing drastically. This "edge" (point A) is roughly at the same location for both the uniform distribution and the uniform distribution over two intervals, but slightly shifted to higher variances for the normal distribution. As the optimal weight variance (point B) is similarly displaced, the performance of two networks, initialized with two different weight distributions of the same variance, is difficult to compare. This might explain the better performance for a Gaussian initial weight distribution in the report of P. Haffner. For the various combinations of data set, network order, etc., the optimal weight variance was encountered in region D, whereas non-convergence was, if at all, only observed in region C.

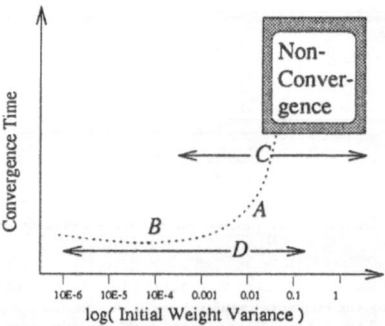

Figure 4: Average behavior of a higher order perceptron in convergence time for changing initial weight variance

As the three different initial weight distributions yield no significant difference in network performance, only the commonly used uniform distribution is considered from now on. For the shifted/scaled logistic and the linear activation functions, the best fixed weight variance is about 10^{-4} (which corresponds to an interval $[-0.017, 0.017]$). For the logistic activation function, the best value for the weight variance depends a lot on data set and network order. In general, the performance with optimal initial weight variance differs not much more than about 10% from the results obtained with a variance of 10^{-4} or even smaller. Therefore a variance of 10^{-4} may be used as a simple application independent random weight initialization scheme. This initialization scheme is also justified by a smaller risk: a network performs nearly as good for an initial weight variance smaller than the optimum. The loss in performance for choosing the initial weight variance too small is much less significant than it is for multilayer perceptrons.

The experiments confirm also that the data set itself has a large influence on the optimal initial weight variance: for the solar, wine, and servo data sets, the networks have about the same size for the same order, but the optimal value for the weight variance differs a lot for the network with the logistic activation function. Further, the optimal value for the initial weights remained for some data sets nearly unchanged for different net orders or even different activation functions, while it changes greatly for other sets. It remains the question, which attribute of the data sets causes this behavior.

5 Conclusion

The experiments show that a suitable and convenient weight initialization method for high order perceptrons[4] with identity activation function[5] is a random initialization with a rather small variance of about 10^{-4} (which corresponds to a weight range of $[-0.017, 0.017]$). If a hyperbolic tangent is used as activation function, the best performance is obtained for the interval $[-a/\sqrt{d_{in}}, a/\sqrt{d_{in}}]$, where a is chosen such that the weight variance corresponds to one third of the distance between the points of the maximal curvature of the activation function[6] (this is approximately 0.8 for an unscaled hyperbolic tangent with steepness one). The "rules of thumb" which perform well for multilayer perceptrons are not suitable for high order perceptrons (which can be explained by their different topologies).

The steepness (and/or the horizontal shift) of the activation function has a big influence on the convergence time of high order perceptrons. A mathematical study, partly inspired by this research, showed that this is indeed the case if the steepness of the activation function is changed (assuming the initial weight range is adapted) [Thimm-94.2].

On the other hand, the shape of the initial weight distribution of three rather different distributions showed no or only very little effect on the optimal convergence time of high order perceptrons. The main effect observed is a dislocation of the optimal value for the initial weight variance. There is consequently no preference for one of the three distributions as the optimal learning speeds are similar.

For multilayer perceptrons with one hidden layer, the weight initialization method of L. F. A. Wessels et al. performed on average the best, but the performance of the methods proposed by J. W. Boers et al., L. Bottou, F. J. Śmieja, G. P. Drago et al., and D. Nguyen et al. are nearly as good. An fixed weight variance of 0.2, which corresponds to a weight range of [-0.77, 0.77], gave the best mean performance for all the applications tested in this study. This performance is similar or better as compared to those of the other weight initialization methods. Despite the fact that the method of L. F. A. Wessels et al. and L. Bottou apply the same initial weight variances for the experiments performed for this publication, the first should be preferred, as it scales the initial weight variances depending on the activation function (due to the calculation of the network output variance).

The experiments show that the best initial weight variance for both types of neural networks is determined by the data set. Consequently, some reasoning on the data set has to be included in the determination of this value, if better values than those proposed in this publication are desired. On the other hand, an initial weight variance close to the optimal value is often acceptable, as the impact on the number of required learning cycles is not too big for small deviations in variance. In general, the loss in convergence speed for both types of neural networks is bigger when too high a variance is chosen than when too small a variance is chosen, as compared to the optimal value.

The evaluation of the experiments performed in order to find the best weight initialization scheme for high order and multilayer perceptrons includes the calculation of confidence intervals for the mean convergence time. This is a much more reliable measure than simply counting the number of simulations performed, as used in most other publications. Moreover, the simulations showed that some data sets require a bigger amount of simulations for a sufficiently small size of the confidence interval than others. In the experiments performed in this research, these numbers varied between 50 and 2,000. The authors encourage other researchers to report their results in a similar way.

Acknowledgments

The authors want to thank the Institute for Logic, Complexity, and Deduction Systems at the University of Karlsruhe for the supply of computing power, which allowed the performance of the simulations.

References

[Bellido-93]   I. Bellido and E. Fiesler. **Do Dackpropagation Trained Neural Networks Have Normal Weight Distributions?** In Stan Gielen and Bert Kappen (eds.), *ICANN '93; Proceedings of the International Conference on Artificial Neural Networks*, pp. 772–775, London, U.K., 1993. Springer-Verlag.

[Boers-92]   E. J. W. Boers and H. Kuiper. **Biological Metaphors and the Design of Modular Artificial Neural Networks.** Master's thesis, Leiden University, Leiden, The Netherlands, Aug. 1992.

---

[4] The results of this study apply in part to standard (first order) perceptrons, since high order perceptrons are a generalization of standard perceptrons.

[5] A linear activation function of steepness one.

[6] This results in about one tenth of the weight variance.

542

[Bottou-88] L.-Y. Bottou. **Reconnaissance de la Parole par Reseaux Multi-Couches.** In *Neuro-Nimes'88; Proceedings of the International Workshop on Neural Networks and Their Applications,* pp. 197–217, 1988. ISBN: 2-906899-14-3

[Chen-91] C. L. Chen and R. S. Nutter. **Improving the Training Speed of Three-Layer Feedforward Neural Nets by Optimal Estimation of the Initial Weights.** In *International Joint Conference on Neural Networks,* vol. 3, pp. 2063–2068. IEEE, 1991.

[Denoeux-93] T. Denoeux and R. Lengellé. **Initializing Back Propagation Networks with Prototypes.** *Neural Networks,* vol. 6, pp. 351–363, Pergamon Press Ltd., 1993.

[Drago-92] G. P. Drago and S. Ridella. **Statistically Controlled Activation Weight Initialization (SCAWI).** *IEEE Transactions on Neural Networks,* vol. 3, num. 4, pp. 627–631, Jul. 1992.

[Fahlman-88] S. E. Fahlman. **An Empirical Study of Learning Speed in Backpropagation Networks.** Technical Report CMU-CS-88-162, School of Computer Science, Carnegie Mellon University, Pittsburgh, PA, Sep. 1988.

[Fiesler-94] E. Fiesler. **Neural Network Classification and Formalization.** In J. Fulcher (ed.), *Computer Standards & Interfaces,* vol. 16, num. 3, special issue on Neural Network Standardization, pp. 231–239. North-Holland/Elsevier, 1994. ISSN: 0920–5489

[Garris-92] M. D. Garris and R. A. Wilkinson. **NIST Special Database 3.** National Institute of Standarts and Technology, Advanced System Division, Image Recognition Group, Feb. 1992.

[Hertz-91] J. Hertz, A. Krogh, and R. G. Palmer. **Introduction to the Theory of Neural Computation,** vol. I. Addison Wesley, 1991. ISBN: 0-201-51560-1

[Haffner-88] P. Haffner, A. Waibel, H. Sawai, and K. Shikano. **Fast Back-Propagation Learning Methods for Neural Networks in Speech.** Technical Report TR-I-0058, ATR Interpreting Telephony Research Laboratories, 1988.

[Judge-85] G. G. Judge, W. E. Griffiths, R. Carter Hill, and T.-C. Lee. **The Theory and Practice of Econometrics.** Wiley Series in Probability and mathematical statistics. John Wiley and Sons, 2nd edition, 1985.

[Kolen-90] J. F. Kolen and J. B. Pollack. **Back Propagation is Sensitive to Initial Conditions.** Technical Report TR 90-JK-BPSIC. Laboratory for Artificial Intelligence Research, Computer and Information Science Department, 1990.

[Kim-91] Y. K. Kim and J. B. Ra. **Weight Value Initialization for Improving Training Speed in the Backpropagation Network.** In *International Joint Conference on Neural Networks,* vol. 3, pp. 2396–2401. IEEE, 1991.

[Lee-86] Y. C. Lee, G. Doolen, H. Chen, G. Sun, T. Maxwell, H. Lee, and C. L. Giles. **Machine Learning Using a Higher Order Correlation Network.** *Physica D: Nonlinear Phenomena,* vol. 22, pp. 276–306, 1986. ISSN: 0167-2789

[Lee-93] Y. Lee, S.-H. Oh, and M. W. Kim. **An Analysis of Premature Saturation in Back Propagation Learning.** *Neural Networks,* vol. 6, pp. 719–728, 1993.

[Murphy-94] P. M. Murphy and D. W. Aha (Librarians). **UCI Repository of machine learning databases** [Machine-readable data repository], anonymous-ftp access ics.uci.edu: pub/machine-learning-databases, 1994.

[Nguyen-90] D. Nguyen and B. Widrow. **Improving the Learning Speed of 2-Layer Neural Networks by Choosing Initial Values of the Adaptive Weights.** In *Proceedings of the International Joint Conference on Neural Networks (IJCNN) San Diego,* vol. III, pp. 21–26, Edward Brothers, 1990.

[Pao-89] Y.-H. Pao. **Adaptive Pattern Recognition and Neural Networks.** Addison-Wesley Publishing Company, Inc., Reading, Mass., 1989. ISBN: 0-201-12584-6

[Platt-91] J.C. Platt. **Learning by Combining Memorization and Gradient Descent.** In R. P. Lippman et al. (eds.), *Advances in Neural Information Processing Systems,* vol. III, pp. 714–720. Morgan Kaufmann, San Mateo, 1991.

[Rumelhart-86] D. E. Rumelhart, J. L. McClelland, and the PDP Research Group. **Parallel Distributed Processing: Explorations in the Microstructure of Cognition,** vol. 1: Foundations. The MIT Press, Cambridge, Mass., 1986. ISBN: 0-262-18120-7

[Smieja-91] F. J. Śmieja. **Hyperplane "Spin" Dynamics, Network Plasticity and Back-Propagation Learning.** GMD report, GMD, St. Augustin, Germany, Nov. 28, 1991.

[Thimm-94.1] G. Thimm, R. Grau, and E. Fiesler. **Modular Object-Oriented Neural Network Simulators and Topology Generalizations.** In M. Marinaro and P. G. Morasso (eds.), *Proceedings of the International Conference on Artificial Neural Networks (ICANN 94),* vol. 1, pp. 747–750, London, U.K., 1994. Springer-Verlag. ISBN: 3-540-19887-3

[Thimm-94.2] G.. Thimm, P. Moerland, and E. Fiesler. **The Learning Rate and the Gain of the Activation Function in Backpropagation Neural Networks are Exchangeable.** Submitted to Neural Computation. See also P. Moerland, G. Thimm, and E. Fiesler. **Results on the Steepness in Backpropagation Neural Networks.** In Marc Aguilar (ed.), *Proceedings of the '94 SIPAR-Workshop on Parallel and Distributed Computing,* Inst. of Informatics, University Pérolles, Chemin du Musée 3, Fribourg, Switzerland, pp. 91–94, Oct. 1994. SI Group for Parallel Systems.

[Wessels-92] L. F. A. Wessels and E. Barnard. **Avoiding False Local Minima by Proper Initialization of Connections.** *IEEE Transactions on Neural Networks,* vol. 3, num. 6, pp. 899–905, Nov. 1992.

[Watrous-93] R. L. Watrous and G. M. Kuhn. **Some Considerations on the Training of Recurrent Neural Networks for Time-Varying Signals.** In M. Gori (ed.), *Second Workshop on Neural Networks for Speech Processing,* pp. 5–17, Trieste, Italy, 1993. Università di Firenze, Edizioni LINT Trieste S.r.l.

Bidirectional Neural Networks Reduce Generalisation Error

A.F. Nejad[1] and T.D. Gedeon[1,2]

[1] School of Computer Science and Engineering
The University of New South Wales
Sydney, 2052, AUSTRALIA

[2] Department of Telecommunications and Telematics
The Technical University of Budapest
Sztocek u. 2, H-1111, HUNGARY

Abstract:

BiDirectional Neural Networks (BDNN) are based on Multi Layer Perceptrons trained by the error back-propagation algorithm. They can be used as both associative memories and to find the centres of clusters.

One of the major challenges in neural network research is data representation. We have used cluster centroids obtained by BDNNs and some heuristic techniques to achieve good representations. This is the key factor in reducing generalisation error. Evaluation of these methods is done by statistical learning theory supported by experimental results. A variety of data sets from real-world problems have been used to support the results of our methods.

The results are consistent with the Vapnik-Chervonenkis bounds. Our methods can be considered as efficient means of designing the required bias in solving dynamic and complex learning systems and to increase their expected performance.

1. Introduction

Neural networks, in particular multi layer perceptrons trained by the error back-propagation learning algorithm (Rumelhart et al, 1986), have been studied extensively by researchers. This learning method can be considered as a type of non parametric model-free estimator.

Two of the major recent challenges for researchers in neural networks have been improving learning models, and data representation. For the former, we designed BiDirectional neural networks (Nejad and Gedeon, 1994), which are powerful models based on feed-forward neural networks. We have shown that BDNNs are effective tools for classification, prediction, associative representation, and finding the centroid of a cluster.

The data representation problem is concerned with appropriate form of input or output transformation. A good representation is often more important than the learning algorithm and does most of work. This is emphasised by many researchers (eg. Anderson, 1988; Stuart, 1992). However this has been studied much less than learning models.

This paper is concerned with data representation techniques to reduce generalisation error and balance bias and variance of the learning algorithms. The experiments have been done both the standard error back-propagation algorithm and BiDirectional Neural Networks on four real-world data sets and some artificial data sets. Vapnik-Chervonenkis (VC) dimension is used to study the consistency between average generalisation performance and the worst case bounds obtained from formal learning theory (Blumer, 1989; Haussler, 1990). The representational capacity of BP and BDNN are measured by the VC dimension and support the results of our experiments by the suggested methods. We show that BDNNs are appropriate tools for enhancing data representation techniques and partitioning effectively the feature space and reducing generalisation error.

The organisation of this paper is as follows. Section 2 introduces bidirectional neural networks. A brief description of the data sets used in our experiments will appear in section 3. Section 4 very briefly reviews the data representation problem and popular methods of input and output scaling and transformations. In section 5 the concept of the distance based learning method is introduced, and together with the methods of centroid shaking, biased classification, and principal components learning are described and the experimental results are reported. In section 6 the latest work on analysing learning system VC-dimension and network generalisation are discussed. Section 7 is the conclusion and discussion.

2. BiDirectional Neural Networks

A new method of training neural networks is suggested that enables them to remember input patterns as well as output vectors, given either of them. They may be considered to be associative memories, and are capable of classification, prediction, and finding cluster centroids.

Bidirectional associative memories (BAM) (Kosko, 1988, 1992) and the bidirectional version of Counter-propagation networks (Hecht-Nielsen, 1986) have been developed to take advantage of the possible advantages of bidirectional mappings. Remaining faithful to the simplicity and capability of the error Back-Propagation algorithm (BP), however, BDNNs (Nejad and Gedeon, 1994) avoid many of the serious problems these earlier models of hetero-associative bidirectional attractors or hybrid networks demonstrate such as capacity, efficiency, or multiple learning. Moreover, BDNNs have the power to determine expected cluster centroids, at a desired level of optimisation for each class, while minimising the effects of outliers.

To use BDNNs as content addressable memories, it has been assumed that a complete set of training data with a one-to-one relation between input and output vectors is available, otherwise some data preparation techniques may be used to create a one-to-one relationship between existing input-output data, unless the problem is inherently not appropriate for such transformations of the vector spaces. A one-to-one relationship is not necessary for finding cluster centres.

We apply the error back-propagation technique in both reverse and forward directions to adjust the weight matrix of the network. In our experiments we did not need to use more hidden units or more hidden layer weights in training a bidirectional network in comparison to the case of training a network in the traditional way. Therefore, preventing any increment in network size, we avoid any increment in VC-dimension of the network which would lead to a decrease in generalisation ability of the learning system.

As mentioned, the networks have been trained bidirectionally with the same weights in each direction. For each input and output node a threshold is assigned, but assigning extra threshold units to hidden layer units is not necessary. That is, the same threshold weights are used in both directions. Due to a flatter search space, usually a higher number of epochs will be necessary for the network to converge in comparison to those of traditional single direction back-propagation networks.

3. Data Sets

The following data sets have been used in our experiments. Due to space constraints, only the results on the first data set are quoted where these results are consistent with the other data sets.

3.1. Students Final Mark Prediction (SFMP) data set consists of 150 samples. Each pattern has fourteen input attributes. Four output have been used to classify the marks. Each record comprises student information, semester assessments without the final exam component and the subsequent final mark for a sample of students from a first year Computer Science subject at the University of New South Wales (Gedeon and Turner, 1993).

3.2. Geographical Information Series (GIS) data set consists of some satellite information from 190 samples from a rectangular grid of 24494 points. Each pattern has seventeen input attributes and one output. 143 records have been used in the training set and 47 records have been used as unseen data. The satellite information has been collected, augmented, and preprocessed in the School of Geography in University of NSW to classify a large geographical area in some forest supra-type categories (e.g. dry sclerophyll and wet sclerophyll) (Bustos and Gedeon, 1995).

3.3. Flight Simulation (FS) data set consists of 1800 records. Each records has 20 attributes. Outputs should predict the appropriate command in each time step to control automatic flight.

3.4. Gross Domestic Product (GDP) data set consists of 160 patterns. Each record has fourteen socio-economic indicator input attributes. They are used to predict the amount of GDP for a specified developing country (Gedeon and Good, 1993).

3.5. Artificial data sets with three inputs and one output to estimate the function $Y = \alpha X_1 + \beta X_2 + \gamma X_3$. Training samples of size 1331 have been deliberately generated to cover all the possible range of independent variables for 343 multivariate equations (Nejad and Gedeon, 1995).

4. Introduction to Data Representation

Each problem requires its own representation. Neural networks with more than one output should estimate an appropriate value for each output. Multi-Task Learning (MTL) will be done if there is more than one output. Input and output data are either real or binary depending on the problem and learning model. For unique concepts binary attributes are usually more appropriate. Representations are either local or distributed. In a local representation, only one attribute is on at a time. In a distributed representation a set of attributes could be on at the same time. A distributed representation has the advantage of using less attributes, thus smaller size network. Normalisation and scaling are common approaches to encode the raw data. The most common scaling technique used is (x - LowerBound) / Range. Mathematical transformations such as Fast Fourier Transform (FFT) and Wavelets are strong tools to capture the complexity of large size inputs. This is done frequently in tasks such as image processing.

5. Suggested Representation Methods

In this section we discuss a number of techniques for data representation. The first two methods use the cluster centres found using our BiDirectional method.

5.1 Distance Based Learning

5.1.1. Methodology

BDNNs were used to find the expected cluster centroids of the desired data set. These centroids are more reliable than the mean vector of input patterns for a specific class (Nejad and Gedeon, 1994). Outliers can significantly diverge the centroids. BDNNs are more reliable because they will adapt to the desired level of optimisation; thus, minimising the effects of the outliers.

Suppose c_k is the cluster centroid of class k, and p_i is the ith input pattern. $p_{ik} = (p_i - c_k + 1)/2$ was used to measure and scale the distance of each record from cluster centroids of SFMP data set. Two extra units were used to determine the class of p_i and the related centroid of k. This increased four times the number of input examples. We trained the network to learn these distances. In the recall phase, new patterns were produced for each test pattern using the same formula to measure distances from cluster centroids. According to the level of generalisation appropriate thresholds were used to decide the class of unseen patterns under uncertainty.

5.1.2. Experimental Results

We applied this method on SFMP data set. We randomly selected 70 records as training data. Fifty unseen records were selected to measure the performance. Applying this distance bases learning method reduced generalisation error of the test data set from 34% to 18%. That is, the performance increased from 66% to 82%.

5.2. Centroid Shaking Method

5.2.1. Methodology

Abu-Mostafa shows how we can take advantage of prior knowledge about the relationship between the input features and the function we want to learn. BDNNs can be used to extract some implied knowledge (e.g. invariance, monotonicity, and approximation hints) about the function we want to estimate. Once these hints are extracted using the positive and negative instances, it is possible to use these kinds of information to enhance learning system ability.

BDNN's cluster centroids were used to determine the most expected value of the attributes of input patterns for each class. Then some small amount of noise was added to each attribute. When pre-knowledge of the problem is not available other methods should be used to confirm that the changed attributes do not exceed the allowed ranges. Causal index, hidden index, or sensitivity analysis methods can be used as appropriate methods to assure this. These methods assign a significance factor to each input. The actual amount of noise should be inversely related to the significant of each attribute. Then these new records will be added to our training set to cover a wider variety of the problem space.

Some statistical methods may be used to produce the new exemplars. For example, we could calculate the standard deviation and mean value of the attributes. Then, *shaking* the records may be done by perturbing each variable with respect to its standard deviation.

5.2.2. Experimental Results

An experimental study on SFMP data set, by producing 20 extra *shaken* centroid exemplars, showed 10% increment in generalisation ability of the trained network, from 66% to 76%, which is not as good as the previous method. Figure 1 shows the input units significant indexes of SFMP data set obtained by three methods. These indexes have been used to determine the shaking bounds of each independent variable.

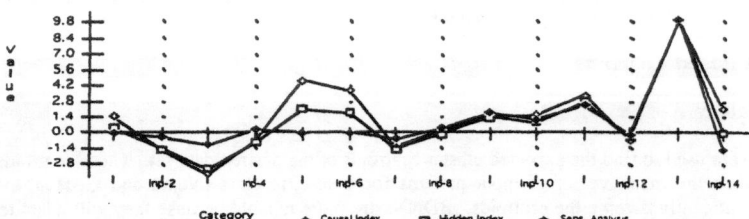

Figure 1. Comparing Hidden Indexes vs CI and SA approaches applied to a trained neural network with five hidden units to predict student final marks.

These methods are in general agreement, the simple perturbation based sensitivity analysis provides the least discrimination. The hidden index method is descibed elsewhere (Nejad and Gedeon, 1995).

5.3. Biased Classification Method

5.3.1. Methodology

Although most researchers have reported the dependency of learning system generalisation on sample size, relatively little attention has been paid to the appropriate distribution of positive and negative examples among output classes. Even if we had a large number of input patterns, but not distributed appropriately in the feature space to cover all the possible states of input patterns of separate classes, the training network will be biased toward the classes with higher density. This section provides a solution to optimise the classification in this situation.

Suppose $t_j^k \in \{0.1, 0.9\}$ is the desired output of the jth dependent variable of the kth class, and x_i is the ith input vector. We argue that in order to have an unbiased learning system t_j^k should be substituted by $t_j^k * \dfrac{p(c_k \mid x_i)}{p(c_j \mid x_i)}$, for all k and i. $p(c_k \mid x_i)$ is the posterior probability of class c_k given a sample x_i. This Baysian projection of output vectors will reduce the classification bias toward the classes with higher density.

When the negative examples of a class belong to more than one class, computing baysian probability needs to determine prior and conditional probability of each class by using expert domain knowledge. To be faithful to the notion of simplicity of programming in neural computation we suggest computing $t_j^k * \dfrac{p(c_k)}{p(c_j)}$ instead of $t_j^k * \dfrac{p(c_k \mid x_i)}{p(c_j \mid x_i)}$ only for *off* output units. To avoid meaningless values in the target and the slow convergence of fermi function in upper and lower bounds, it is preferable to assign lower bound .1 and upper bound .5 to target values (except where $k = j$). Therefore, if the output vector is [.9, .1, .1, .1] and the prior probability of the classes are .50, .20, .20, and .10 respectively, the transformed output vector will be [.9, .25, .25, .5].

5.3.2. Experimental Results

We transformed the output vectors of SFMP by the proposed method and repeated the training of the network. Evaluating the generalisation ability of the new method showed 82% against the 66% of traditional representation. This result is as good as the distance base learning method. It worth noting that in the cases of classification of continuous categories the contingency restriction may enhance this method. For example in the GDP data set we can restrict farther units from having a higher value than closer units. for example, if the transformed output vector is [.5, .2, .2, .9], the restricted transformation will be [.2, .2, .2, .9].

5.4. Principal Components Learning

5.4.1. Methodology

Principal Component Analysis (PCA) is a statistical method to solve the problem of multi-colinearity among a set of variables by extracting a small number of uncorrelated variables (principal components). The new variables (factors) are a linear combination of the original variables. That is, for j^{th} factor of the kth sample is estimated as $\hat{F}_{jk} = \sum_{i=1}^{p} W_{ji} X_{ik}$, where X_{ik} is the original value of the jth variable for sample k and W_{ji} is the factor score coefficient for the jth factor and i^{th} variable.

This is the same task which is down by the hidden layer of a multilayer perceptron. Unfortunately, for the statistical method of PCA as well as feed-forward neural networks, goodness-of-fit tests for the adequacy of these factors are directly related to the number of examples. Thus, deliberately selecting the weighted combinations using pre-knowledge of the problem domain is required. Inputting expert knowledge into a system, by defining appropriate partitions on the input vector space, may reduce the input dimension. Other kinds of transformations are also known to reduce the input dimension. FFT and wavelets are among these,. which are frequently used in image processing tasks. These techniques are able to reduce the complexity of the problem by capturing some part of the complexity in the preprocessing phase. This will let the network capture better and more efficiently the remained complexity.

5.4.2. Experimental Results

To show the multi-colinearity effect on the trained neural networks, we did some experiments with our artificial data sets. We studied a few networks which were trained to estimate the output of some multivariate equations by correlated inputs. The experiments showed that not only was the convergence slowed down, but the causal index and sensitivity analysis were unable to determine the correct significant effect of each variable in all cases.

Linear combinations were used to reduce the input dimension of the SFMP data set to five. We dropped the first and the second variables. A weighted linear combination of the lab marks, which are highly correlated and with large variance, and two other simple linear combinations reduced our input dimension from 14 to 5. Then a neural network with 3 hidden units and the same outputs as before was used to classify the classes. The result was an increase in performance to 82% from 66%. This was a great reduction in generalisation error, size of network, and convergence speed. One of the major advantages of this method is the smaller size of the network leads to easier analysis and meaningful and understandable rules can be more readily extracted.

Note that this method again increased the performance from the base of 66% to 82%, like two of the previous methods.

6. Generalisation and Vapnik-Chervonenkis Dimension

To understand better why our representation methods are able to have better performance on the patterns which have not been seen during training, we will use the concept of Vapnik-Chervonenkis (VC) dimension. VC dimension and Valiant's model for Probably Approximately Correct (PAC) learning have been utilised to measure the amount of capacity or complexity of representational methods.

These models agree that when sample complexity and computational complexity, and therefore learning system VC dimension increases, more examples will be needed to be learned accurately as training data to enable the system to represent a wide variety of approximation functions (Valiant, 1984; Blumer, 1989).

It has been shown that for the N-input single-layer network, the VC dimension of the network is $N+1$ (Baum and Haussler, 1989). If $m \geq O\left(\frac{W}{\varepsilon}\log\frac{N}{\varepsilon}\right)$, $0 < \varepsilon \leq \frac{1}{8}$, random examples can be loaded on a feed-forward network using linear threshold function with N nodes and W weights, so that at least a fraction $1 - \frac{\varepsilon}{2}$ of the examples are correctly classified, then one has confidence approaching certainty that the network will correctly classify a fraction $1 - \varepsilon$ of the future test examples drawn from the same distribution.

The Vapnik upper bound for the worst-case generalisation error is less than or equal to ε, for $\varepsilon \leq O\left(\frac{d}{m}\ln\frac{m}{d}\right)$ (Vapnik, 1982). Ehrenfeucht (1988) showed that ε for some classes is $\varepsilon \geq O\left(\frac{d}{m}\right)$. Haussler (1990) showed that for binary classification $\varepsilon \leq \frac{d}{m}$. Recently, it was shown that in some cases, the average generalisation is significantly better than the VC bound (Cohn and Tesauro, 1992). There exist some tasks for which increasing network size improves generalisation (Amirikian and Nishimura, 1994). In fact, the VC-dimension of a feed-forward neural network with linear threshold gates is $\Omega(w.\log w)$ instead of $O(w.\log w)$ (Maass, 1994).

7. Discussion and Conclusion

The question of how complex learning a problem of interest by a given network remains an interesting problem still to be closely investigated. However, there is no doubt that an increase in W will usually result in an increase in VC-dimension, and generalisation performance significantly depends on the data representation method and problem specification.Therefore, in order to get better performance from a trained network we should either reduce W or increase the sample size.

Due to the nature of the problem and other restrictions, increasing the sample size is not always possible. Thus, it is suggested to reduce the network size and focus on data representation. There exist different methods of reducing the network size. A method of weight elimination is shown in Weigend (et al, 1991). Some researchers have suggested to reduce the number of hidden nodes and therefore the W (eg. Kruschke, 1988; Gedeon and Harris, 1991). Balancing bias and variance by determining the optimal (not necessarily the minimal) number of hidden units will result in the best performance of a specified network (Geman et al, 1992, Slade and Gedeon 1993).

We have used our BiDirectional neural networks model which we have shown elsewhere to be effective tools for classification, prediction, associative representation, and finding the centroid of a cluster. Here we have used the BDNNs to find the centres of clusters to address some parts of the data representation problem. We have used the centroids with our distance based learning approach, and with our centroid shaking approach to achieve good representations, which is a key factor in reducing the generalisation error. The biased classification and PCA methods require significantly more effort to use, and only achieve the same level of results. For example, for best results, the PCA approach may require expert domain knowledge. Nevertheless, the approach we have presented provided a different benefit – the smaller network can be more easily analysed in terms of rules.

In summary, we have used a variety of data sets from real-world problems to support the benefits of our methods, and evaluated the relationship of these methods to statistical learning theory. Our results are consistent with the Vapnik-Chervonenkis bounds, thus our methods can be considered as efficient means of designing the required bias in solving dynamic and complex learning systems and to increase their expected performance. This is by enhancing data representation and partitioning effectively the feature space and reducing generalisation error.

Acknowledgments

We would like to thank Claude Sammut from the A.I. Group in the School of Computer Science and Engineering at the University of NSW for giving access to the flight simulation data, and A.K. Skidmore from The Centre for Remote Sensing in the School of Geography at the University of NSW for giving access to the GIS data used.

References

Amirikian, B., and Nishimura, H. 1994. What Size Network Is Good for Generalisation of a Specific Task of Interest? Neural Networks, Vol. 7, No. 2, 321-329.

Anderson, J.A. 1988. Cognitive and Psychological Computation with Neural Models. *IEEE Transactions on Systems, Man, and Cybernetics 13*, 799-815.

Baum, E.B., and Haussler, D. 1989. What Size Net Gives Valid Generalisation? *Neural Computation 1*, 151-160.

Blumer, A., Ehrenfeucht, A., Haussler D., and Warmuth M. 1989. Learnability and the Vapnik-Chervonenkis dimension. *Journal of the ACM, 36(4)*, pp.929-965.

Bustos, RA and Gedeon, TD Decrypting Neural Network Data: A GIS Case Study, *Proceedings International Conference on Artificial Neural Networks and Genetic Algorithms (ICANNGA)*, Alès, 1995.

Cohn, D. and Tesauro, G. 1992. How Tight are the Vapnik-Chervonenkis Bounds? *Neural Computation 4*, 249-269.

Ehrenfeucht, A., Haussler, D., Kearns, M., and Valiant, L. 1988. A General Lower Bound on the Number of Examples Needed for Learning. In *Proceedings of the 1988 Workshop on Computational Learning Theory*. San Mateo, CA, Morgan Kaufmann.

Gedeon, T.D. and Bowden, T.G. 1992. Heuristic pattern reduction. *International Joint Conference on Neural Networks*, Beijing, pp. 449-453.

Gedeon, TD and Good, RP Interactive modelling of a neural network model of GDP, *Proceedings International Conference on Modelling and Simulation*, pp. 355-360, Perth, 1993.

Gedeon, TD and Harris, D Network Reduction Techniques, *Proceedings International Conference on Neural Networks Methodologies and Applications*, AMSE, vol. 1, pp. 119-126, San Diego, 1991.

Gedeon, T.D. and Turner, H. 1993. Explaining student grades predicted by a neural network. *Proceedings International Joint Conference on Neural Networks*, pp. 609-612, Nagoya.

Geman, S., Bienenstock, E., and Doursat R. 1992. Neural Networks and the Bias/Variance Dilemma. *Neural Computation 4*, 1-58.

Haussler, D., Littlestone, N., and Warmuth, M. 1990. Predicting {0,1}-Functions on Randomly Drawn Points. *Technical Report UCSC-CRL-90-54*, University if California at Santa Cruz.

Hecht-Nielsen, R. 1987. Counterpropagation Networks. *Applied Optics*, vol. 26, no. 3, 4979-4984.

Kosko, B. 1988. Bidirectional Associative Memories. *IEEE Transactions on Systems, Man, and Cybernetics*, vol. SMC-18, 49-60.

Kosko, B. 1992. Neural Networks and Fuzzy Systems.

Kruschke, J.K. 1988. Creating Local and Distributed Bottlenecks in Hidden Layers of Back-Propagation Networks. *Processing of the 1988 Connectionist Summer School*. Carnegie-Mellon University, Pittsburgh, PA: Morgan Kaufmann.

Maass, W. 1994. Neural Nets with Superlinear VC-Dimension. *Neural Computation 6*, 877-884.

Nejad, A.F., Gedeon T.D. 1994. BiDirectional MLP Neural Networks. *International Symposium on Artificial Neural Networks*. Tainan, Taiwan, pp. 308-313.

Nejad, A.F., Gedeon T.D. 1995. Analyser Neural Networks. *International Workshop on Applications of Artificial Neural Networks to Telecommunications*. Stockholm, Sweden.

Rumelhart, D.E., Hinton, G.E., Williams, R.J. 1986. Learning internal representations by error propagation. in Rumelhart, D.E., McClelland, *Parallel Distributed Processing*, Vol. 1, MIT Press.

Slade, P. and Gedeon, T.D. 1993. Bimodal Distribution Removal, *Proceedings IWANN International Conference on Neural Networks*, Barcelona. also in Mira, J., Cabestany, J. and Prieto, A. 1993. *New Trends in Neural Computation*, pp. 249-254, Springer Verlag, Lecture Notes in Computer Science, vol. 686.

Valiant, L.G. 1984. A Theory of the Learnable. *Communications of the ACM*, 27, 1134-1142

Vapnik, V.N. 1982. Estimation of Dependencies Based on Empirical Data. Spring-Verlag, New York.

Weigend, A., Rumelhart, D., and Huberman, B. 1991. Generalisation by Weight Elimination with Application to Forecasting. In *Advances in Neural Information Processing Systems 3*, R. Lippmann, J. Moody, and D. Touretzky, eds. Morgan Kaufmann, Denver, Co.

Balancing Bias and Variance: Network Topology and Pattern Set Reduction Techniques

T.D. Gedeon [1,3], **P.M. Wong** [2] and **D. Harris** [4]

[1] School of Computer Science and Engineering
[2] Centre for Petroleum Engineering
The University of New South Wales
Sydney, 2052, AUSTRALIA

[3] Department of Telecommunications and Telematics
The Technical University of Budapest
Sztocek u. 2, H-1111, HUNGARY

[4] Centre for Neural Networks
Kings College, London, UK

Abstract

It has been estimated that some 70% of applications of neural networks use some variant of the multi-layer feed-forward network trained using back-propagation. These neural networks are non-parametric estimators, and their limitations can be explained by a well understood problem in non-parametric statistics, being the "bias and variance" dilemma. The dilemma is that to obtain a good approximation of an input-output relationship using some form of *estimator*, constraints must be placed on the structure of the estimator and hence introduce bias, or a very large number of examples of the relationship must be used to construct the estimator. Thus, we have a trade off between generalisation ability and training time.

We overview this area and introduce our own methods for reducing the size of trained networks without compromising their trained generalisation abilities, and to reduce the size of the training pattern set to improve the training time again without reducing generalisation.

Model Assumptions

In this paper, we will assume a multi-layer feed-forward network trained using back-propagation [1], and will use the general expression "neural network" to mean such a network. All connections are from units in one level to units in the next level, with no lateral, backward or multi-layer connections. Each unit is connected to each unit in the preceding layer by a simple weighted link. The network is trained using a training set of input patterns with desired outputs, using the back-propagation of error measures. The network is tested using a validation set of patterns which are never seen by the network during training and thus can provide a good measure of the generalisation capabilities of the network. The separation of the total set of patterns into training and test sets is generally at random to avoid introducing experimenter bias.

By back-propagation we mean the general concept of developing the error gradient with respect to the weights, and not restricted to the original gradient descent method. In the examples we use here, we have used the basic sigmoid logistic activation function, $y = \left(1 + e^{-x}\right)^{-1}$, though this is not essential to the substance of our results.

Statistical Background

Non-parametric statistics is concerned with model free estimation. When employed for classification both parametric and non-parametric statistics seek to construct decision boundaries between the various classes using a collection of training samples. Non-parametric methods differ from parametric methods in that there is no particular structure assigned to the decision boundaries, *a priori*.

The obvious advantage of parametric techniques is efficiency. By setting the structure of the decision boundary before estimation begins then fewer data points (or training examples) are required. This is because there are (hopefully) a small number of parameters in the parametric model that require estimation. Non-parametric or model free estimation potentially requires the estimation of an infinite number of parameters and hence needs a much larger number of training examples. However, the efficiency of parametric methods comes at a cost. If the actual form of the decision boundary departs substantially from the assumed form, then parametric methods can result only, in the 'best' approximation for the decision boundary from within the adopted class of decision boundaries. Non-parametric methods place no restriction on the class in which the decision boundary used in estimation must reside.

Informally, *consistency* is the asymptotic convergence of the estimator to the object of estimation. In this context asymptotic refers to the sample size or the number of patterns in the training set approaching infinity. Most non-parametric algorithms are consistent for any regression function $E[y|x]$[2]. Indeed it has been shown [2, 3] that feed forward networks are consistent, under appropriate conditions relating to the architecture of the network. Consistent in the sense that the weights in the network will, in the limit of training set size approaching infinity, converge to the optimal weights w^*. Although this is an encouraging property, non-parametric methods can be very slow to converge, and this has indeed been observed in the training times for neural networks.

Non-parametric estimators are guaranteed to perform optimally in the limit. In the context of neural networks, they are only guaranteed to outperform other parametric estimators when the size of the training set approaches infinity. For a finite sample, non-parametric estimators can be very sensitive to the actual realisations of (x,y) contained in the sample. This sensitivity results in an estimator that is high in what is known as *variance*. The only way to control this variance is to introduce some *a priori* structure into the estimator, that is to use parametric methods. This approach also has its pitfalls. In complex classification problems, it is difficult to know the structure to impose on the estimator. As mentioned above, this can result in estimators that converge to an incorrect solution. This creates models that are high in what is known as *bias*. The performance function used in back-propagation can be readily decomposed into a bias and a variance term [4, 5].

It is this dilemma between bias and variance that can explain the limitations of non-parametric learning. Low bias and low variance requires large numbers of training examples. In situations where it is not possible to obtain sufficiently large numbers of training examples it is necessary to allow some bias into the training procedure. In the next section we provide a brief overview of the decomposition of neural network training into bias and variance terms.

Bias-variance decomposition

The neural network is solving a regression problem, so we construct a construct function based on the training set:

$$f(x;D) \qquad f \text{ depends on training data } D$$

We use the mean squared error of f as an estimator of regression:

$$E_D\left[(f(x;D) - E[y|x])^2\right]$$

The bias-variance decomposition of the estimator is:

$$[E_D[f(x;D)] - E[y|x]]^2 \qquad \text{"bias"}$$
$$+ E_D\left[(f(x;D) - E_D[f(x;D)])^2\right] \qquad \text{"variance"}$$

The interpretation of the above is as follows:

◊ If the expected value of the function is different on average from the regression, then it is biased.

◊ If the function has large differences from the expected value it has high variance.

We do not in general have access to the estimator function learned by the neural network, the derivation of the function encoded by neural network weights is an entire area of research. Nevertheless, the estimator function can be approximated, over a number of test patterns over a number of trained neural networks. We chose the test patterns to be 1/3 at random of the available patterns.

The bias variance decomposition estimated bias (over the set of test patterns) is:

$$\lozenge \qquad Bias\,|x| \approx \frac{1}{\text{test}} \sum_{i=1}^{\text{test}} \overline{f}\,|x_i,\,w|-y_i\ ^2$$

The estimated variance is calculated over the test patterns over 50 different trained neural networks, with each network being trained on a randomly selected 3/4 of the patterns from the remaining 2/3 of the original pattern set available. Each network is thus trained using 1/2 of the original patterns. The estimated variance is:

$$\lozenge \qquad Variance\,|x| \approx \frac{1}{\text{test}} \sum_{i=1}^{\text{test}} \frac{1}{50} \sum_{k=1}^{50} f\,|x_i,w;\,D_k|-y_i\ ^2$$

Network topology reduction – pruning

The introduction of bias into a trained neural network by reducing the number of processing units is generally referred to as pruning. Although the output units perform computations, the elimination of one or more of these units modifies the nature of the output function being learnt by the network, hence we will restrict ourselves to considering the case of hidden units only.

The seminal work on pruning trained networks [6] uses the outputs of units in a two stage pruning process, which operates by inspection. Such pruning by inspection is difficult even on small examples, some automatable process would be ideal. Properties such as *relevance* [7, 8], *contribution* [9], *sensitivity* [10], *badness* [11], and *distinctiveness* [12] have been described in detail elsewhere. We will briefly describe *distinctiveness* here.

The *distinctiveness* of hidden units is determined from the unit output activation vector over the pattern presentation set [12]. That is, for each hidden unit we construct a vector of the same dimensionality as the number of patterns in the training set, each component of the vector corresponding to the output activation of the unit. This vector represents the functionality of the hidden unit in (input) pattern space. In this model, vectors for clone units would be identical irrespective of the relative magnitudes of outputs and recognised. Units with short activation vectors in pattern space are recognised as insignificant and removed.

Pattern	1	2	3	4	5	6
p.000	1.000	1.000	1.000	1.000	1.000	0.000
p.001	0.000	1.000	1.000	0.000	0.000	0.000
p.002	0.000	0.000	0.000	1.000	0.000	1.000
p.003	0.000	0.000	0.000	1.000	1.000	0.000
p.004	0.000	0.000	1.000	1.000	0.000	0.000
p.005	1.000	0.439	1.000	1.000	0.999	0.706
p.006	0.000	1.000	0.000	1.000	1.000	0.000
p.007	1.000	0.000	0.000	1.000	0.000	0.000
p.008	1.000	0.000	0.000	1.000	0.000	0.000
p.009	0.000	0.000	0.000	0.000	0.000	0.000
p.010	0.000	0.000	1.000	0.000	0.167	0.000
p.011	0.000	0.000	1.000	0.000	0.000	0.000
p.012	0.000	0.000	0.000	0.000	0.000	0.000
p.013	1.000	0.000	0.000	0.989	1.000	1.000
p.014	0.000	0.000	0.000	0.000	0.000	0.000
p.015	0.000	0.000	1.000	1.000	1.000	0.000

Table 1. Six hidden unit activations by pattern.

The recognition of similar pairs of vectors is done by calculation of the angle between them in pattern space. Since all activations are constrained to the range 0 to 1, the angle calculations are normalised to 0.5, 0.5. Angular separations of up to about 15° are considered too similar and one of them is removed. The weight vector of the unit which is removed is added to the weight vector of the unit which remains. With low angular separations, the averaging effect is minor and the mapping from weights to pattern space remains adequate in that the error measure is no worse subsequently. This produces a network with one fewer unit which requires no further training. Similarly, units which have an angular separation over about 165° are complementary, and both can be removed. It would be possible to discover pairs of similar units by inspection of the above table, but would be difficult for much larger numbers of patterns or units. The vector angles for the six hidden units range from 61.6° to 90°. Clearly, none of the units are similar to any other.

A further category of undesirable units is also discovered and included in the distinctiveness analysis, which is not found by the other methods. Groups of three or more units which together have no effect, or two or more units with a constant effect can be recognised. That is, in the pattern space, the sum of their vectors is zero or constant. The discovery of such groups is done by a sorted Gaussian vector pivot on the cumulative rectangular matrix of pattern space vectors. This produces the reduced row echelon matrix. For the case of groups of units with jointly no effect, the entire group can be removed. If the joint effect is constant, a bias can replace the entire group. Because of the inaccuracies incurred by banding and subsequent use of this information, it may be necessary to retrain the network. Fortunately, such groups are not common in our experience, therefore retraining is not often required.

Reducing the size of training sets

During training, frequency distributions of the errors for all patterns in the training set show that very early in training, the distribution of the pattern errors is approximately normal with a large variance. The network then dramatically reduces the errors for the majority of the training set. There remain patterns with relatively high error, creating an almost bimodal error distribution, with the low error peak containing patterns the network has learnt well, and the high error peak containing the outliers. From the two peaks in the error distribution it is clear that the network can identify outliers itself. It is difficult and time consuming to identify a bimodal error distribution during training. Fortunately, a measure of the variance achieves the same effect.

Patterns should not be removed too quickly, as those patterns with midrange errors could eventually be learnt by the network. Bimodal Distribution Removal (BDR) is intentionally conservative in its removal of patterns to give the network opportunity to learn these *slow coaches* [5].

Should BDR be continued indefinitely, eventually all the patterns would be removed from the training set. This is undesirable because as the training set becomes smaller the network is devoting a large number epochs of training to a reduced set of examples, thus potentially dramatically increasing the overfitting effect. Removal of the outliers from the training set causes the high error peak to shrink – this could be used as a halting condition for training.

BDR attempts to address the usual weaknesses of outlier detection and removal methods in that:

* pattern removal does not start until the network itself has identified the outliers,
* the number of patterns removed is not hard wired, but instead is data driven, and
* patterns are removed slowly, to give the network time to extract information from them.

This method was tested by calculating the estimated bias and variance terms over the test set, over a number of networks being trained. Two other methods were tested as controls. The first of these is the Least Trimmed Squares (LTS), which minimises only the lowest mean square errors, hence the outliers never used which leads to better generalisation, however we must know how many outliers exist [13]. The second method is the Heuristic Pattern Reduction (HPR) method, which is not an outlier reduction method *per se*, but attempt to reduce the overall size of the training set by mapping the individual training patterns onto a one-dimensional adjacency measure using their contribution to the total sum of squares. This adjacency is used as the base to make arbitrary assumptions as to sizes of clusters of training patterns. Thus, assuming an homogenous distribution of clusters of pairs of patterns, the training set can be reduced to half its size [14].

One third of the patterns were set aside as a test set and were never used in training. The remaining patterns were used to create 50 sets of pattern training sets at random. Each of these used about 3/4 of the remaining 2/3 of the original patterns. Fifty networks for each of methods BDR, LTS and HPR as described above were trained, as well as normal back-propagation. The estimated bias and variance were then calculated.

Least Trimmed Squares provided some control of variance at the expense of higher bias. This control is significant if training is continued for a long time (since the overfitting effect increases). The LTS method to control variance requires *a priori* knowledge of the amount of noise in the training set. Heuristic Pattern Removal produces an surprising result. We have discussed that neural performance becomes optimal as the size of the training set approaches infinity. Yet, measurements of bias and variance for training on a half size training set showed that the Heuristic method performs almost as well as the Bimodal Distribution Removal method. Bias and variance are very sensitive to the complexity of the data and by how much the training set is reduced. Eventually, the Heuristic method lead to the most uncontrolled increase in variance of all the methods. Bimodal Distribution Removal provides a similar control of variance to LTS. It is an improvement over LTS since both bias and variance are lower during training, and we do not need the *a priori* knowledge of the number of outliers. The problem remains in finding the 'correct' time to halt removal of further points.

The values for bias in the LTS and BDR methods for this data set in comparison to normal back-propagation indicates that even though these methods perform well, the noisy data points are being useful in this case. This means our implicit assumption that the probability law γ is approximately degenerate is barely valid. This points to the requirement for appropriate choice of a data set.

Both the HPR and BDR methods require some means of determining when to terminate the training set reduction. The benefits of these methods derive from the use of the neural network behaviour to help determine which points are outliers. Both of these methods use static measures, hence we developed a method to utilise the dynamics of the network behaviour during training on training patterns, called error sign testing.

Error Sign Testing

During the training phase each input vector of the training set is presented to the network and an output vector is obtained. An error vector is then calculated by taking the difference between the desired and the output vectors. The magnitude of the error vector can be used as a measure of how easily the input pattern is learnt in a given epoch, the lower the value the easier to learn. For successful learning of a good pattern, the error magnitude for a particular pattern in the nth epoch of batch training, say E_n, should be smaller than the previous one, E_{n-1}. Therefore, counting the number of negative signs of the expression E_n-E_{n-1} for K epochs can be used to define a *good* pattern. In order words, a pattern with a large percentage of negative signs will be a good pattern. Similarly, a small percentage of negative error signs (or large percentage of positive error signs) will be a noisy pattern or outlier. This method of detecting outliers is called Error Sign Testing (EST).

The value of K can be determined by the root-mean-square-error (rmse) of all the training patterns presented to the network. In most applications, the rmse starts from a high value and drops very quickly in a small number of epochs. In this example study, K was determined at an epoch when oscillation of rmse began, say 200. This is due to the presence of the outliers, and hence the learning of good patterns slows down and the network starts to overfit the outliers.

After K epochs, the percentage of negative signs of each pattern is calculated. In this study, a pattern which has less than 10% of the negative error signs is defined as an outlier. The outliers are then removed and further training is performed using a reduced pattern size. A validation data set is also used to test the performance of the trained network, but is not used for training, nor are any outliers removed from this test set. An example study using data from an oil well in a real reservoir will be used to demonstrate our method.

Example Study

Our technique for pattern reduction was used on a data set from the wireline logs and core data in an oil well located at the North West Shelf, Australia. The data set consists of points along the well with 4 wireline log measurements. The log variables in this well included deep resistivity (RLLD), bulk density (RHOB), sonic travel time (DT) and gamma ray (GR). The rock type of each of the points was identified through the careful examination by a geologist of the core sample taken at that location. In this study, five dominant rock types were found and were then named as rock types 1 to 5 for simplicity purposes. A crossplot of RHOB versus DT is shown in Figure 1. The core sampling process is not usually done on every well and the predictions of rock types must then rely on the available log data. Neural network methods have found some recent applications in well log analysis [15, 16].

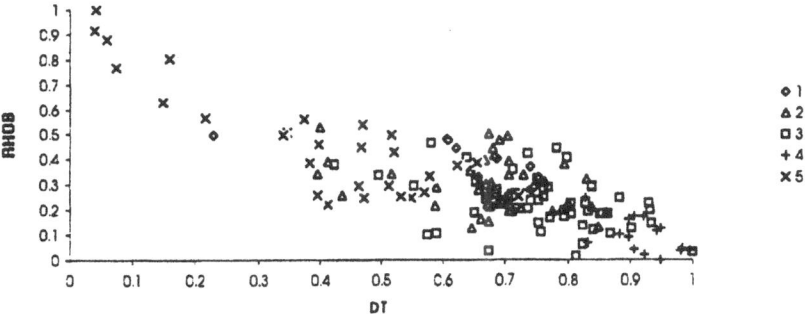

Figure 1. Relationships of RHOB and DT for different rock types

In this study, the four log measurements and the corresponding rock types were used as the input patterns and the desired output patterns respectively. The network architecture used has four input units corresponding to each of the log variables, and four units in the hidden layer. Each rock type was represented by one output unit, and hence five output units were used. The classification was considered to be correct if the outputs of the network were within 0.1 of the desired value of 0 or 1. The network configuration used produced acceptable predictions of the rock type. By acceptable, we meant the results from neural network was a consistent improvement compared to those obtained from the standard statistical techniques.

Experiments and Results

The log measurements usually contain noisy data, especially in a lithologically complex reservoir. However, the outliers are not easily recognised in most of the cases (refer to Figure 1). Note that this does not include potentially unbounded outliers, which are clearly markedly different to the normal data and are excluded using standard methods. This exclusion of markedly different outliers is necessary for our method, as it is specialised to the location of any remaining outliers, and the sum of squares error term is not sufficiently sensitive to compensate for very extreme values which could pull the network towards them and still reduce the overall squared error. In this study, the proposed error sign testing method was used to detect and remove any remaining outliers. Four experiments with different training sets were trained and the performance of each set (i.e. classification accuracy) was evaluated using the validation test data. The structure of each training set is described as follow:

(1) whole data set - this data set consisted of the original data which was used as a control experiment.
(2) half of the whole data set - this data set was selected based on the percentage of negative error signs during training using each of the original patterns. The patterns were then sorted in ascending order and the new training set was formed using every second sorted pattern. The aim of this training set was to remove half of the good and noisy data, in analogy with the HPR method described earlier.
(3) clean data set - the outliers, defined by the EST method, were removed from the original data set. After doing this, the data set was then considered to be 'clean'.
(4) half of clean data set using error sign testing method - this data set was formed using the clean data set in (3) followed by the half reduction method using the EST method as in (2). This experiment was designed to further simplify the set of patterns in (3).

Each of the above experiments was performed with 10 different sets of initial weights, and was trained for 10,000 epochs. In each run, the same initial weights are used for training the original data set (1), and the corresponding sets of reduced patterns (2), (3) and (4). This was done in order to minimise the effects of the initial random functionality of the network unit weights. The validation set was also used to record the highest classification accuracy (in percentage) during the training phase.

The results of the comparison study are tabulated in Table 1. Note that the classification accuracy shown are the maximum values of the number of runs done and can be from different runs. The average training times (measured in number of epochs) required to achieve the maximum accuracy in each experiment are also shown. The success of the method is shown by the high classification accuracy on the validation set. Statistical evaluation of the same data using discriminant analysis was also done on the whole training set. The classification accuracy was only 57% on the validation data. Therefore, the results using neural network methods showed better performance in generalisation.

Training Set	Average Training Time (epochs)	Classification Accuracy (%)
(1) Whole Data Set	3500	66
(2) Half of Whole Data Set	3400	64
(3) Clean Data Set	2900	66
(4) Half of Clean Data Set	1400	68

Table 1: Performance of Different Training Sets on the Validation Data.

The results on this example also showed that a smaller set of training patterns did not necessarily degrade the generalisation ability of the trained network. Pattern reduction techniques aim to simplify the error surface of the pattern space, however simple elimination of half of original training patterns did not show significant improvement. This is most likely due to the presence of half of the original outliers in the remaining training data set, and these outliers may affect the process of error surface simplification. When the outliers were defined and removed from the training set, the same classification accuracy was obtained compared to the whole data set.

The clean data set was also reduced to half its size and the results showed better performance in generalisation. This improvement was probably due to the simplification of the error surface in the pattern space. Figure 2 shows the outliers identified by the EST method in the run with maximum classification accuracy. In this case, 12 outliers were found in rock types '2' and '3', and they were coded as '2x' and '3x'. As displayed in Figure 2, the outliers defined tend to lie within the clusters of other rock types, and therefore including these data in the training set will significantly increase the training time, and overfitting of these outliers will also occur.

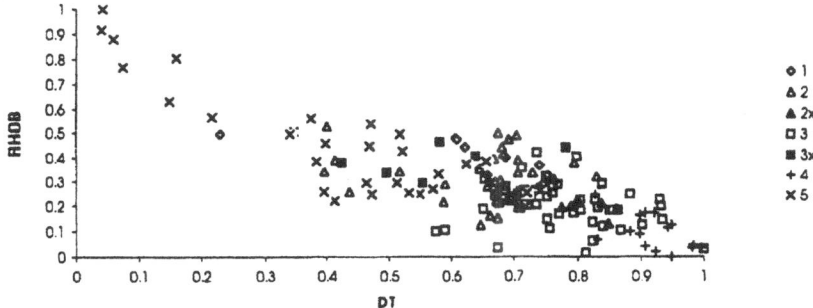

Figure 2. Outliers (coded as '2x' and '3x') identified by the Error Sign Testing Method

The removal of outliers did not seem to improve generalisation, however the clean data set took less training time to achieve the same results as obtained from the original data set. Therefore removing outliers in the training set does reduce the amount of training time. Note that the above time are shown in epochs, yet the set (4) is less than half the size of set (1), and hence the training time of set (4) versus set (1) is reduced by a factor of over 5. This result can be explained in terms of the bias-variance balance. The removal of half of the patterns increases the bias, as the network will now be more sensitive to the particular patterns that happen to be represented in the smaller training set, rather that to the underlying function we are attempting to approximate. This is balanced by the reduction in variance we obtain from correctly removing the outliers in the training set. Since the results are statistically the same between sets (1) and (4), we can therefore be convinced that the EST method has indeed removed the correct outliers.

Finally, it is worth mentioning that the neural network method in this study performed better than the standard statistical approach. However, the classification accuracy obtained from discriminant analysis can be used to provide a minimum baseline level of accuracy for the neural network method to achieve.

Conclusions

We have discussed neural networks in term of non-parametric estimators, and how their limitations can be explained by a well understood problem in non-parametric statistics, being the "bias and variance" dilemma, which for neural networks is essentially a trade off between generalisation ability and training time.

We have described some approaches to control the bias and variance balance in a neural network during training, methods for reducing the size of trained networks without compromising their trained generalisation abilities, and to reduce the size of the training pattern set to improve the training time again without reducing generalisation. We have also discussed how to measure the bias and variance in a number of neural networks during training. We have introduced a method of outlier detection, and indicated how the balance of bias and variance can be used to come to some qualitative understanding of the usefulness and correctness of the outlier detection algorithm.

References

1. Rumelhart, DE, Hinton, GE, Williams, RJ, "Learning internal representations by error propagation," in Rumelhart, DE, McClelland, *Parallel distributed processing*, Vol. 1, MIT Press, 1986.
2. White, H, "Learning in artificial neural networks: A statistical perspective," *Neural Computation*, vol 1., pp. 425-464, 1989.
3. Gallant, AR and White, H, *A Unified Theory of Estimation and Inference for Nonlinear Dynamic Models*, Basil Blackwell, Oxford, 1988.
4. Geman, S, Bienenstock, E and Doursat, R, "Neural networks and the bias/variance dilemma," *Neural Computation*, vol. 4, pp. 1-58, 1992.
5. Slade, P and Gedeon, TD "Bimodal Distribution Removal," *Proceedings IWANN International Conference on Neural Networks*, Barcelona, 1993. also is Mira, J, Cabestany, J and Prieto, A, *New Trends in Neural Computation*, pp. 249-254, Springer Verlag, *Lecture Notes in Computer Science*, vol. 686, 1993.
6. Sietsma, J, & Dow, RF, "Neural net pruning - why and how," *Proc. Int. Joint Conf. on Neural Networks*, vol. 1, pp. 325-333, 1988.
7. Mozer, MC, Smolenski, P, "Using relevance to reduce network size automatically,", Connection Science, vol. 1, pp. 3-16, 1989.
8. Segee, BE, & Carter, MJ, "Fault Tolerance of Pruned Multilayer Networks," *Proc. Int. Joint Conf. on Neural Networks*, vol. 2, pp. 447-452, Seattle, 1991.
9. Sanger, D, "Contribution analysis: a technique for assigning responsibilities to hidden units in connectionist networks," *Connection Science*, vol. 1, pp. 115-138, 1989.
10. Karnin, ED, "A simple procedure for pruning back-propagation trained neural networks," *IEEE Transactions on Neural Networks*, vol. 1, pp. 239-242, 1990.
11. Hagiwara, M, "Novel back propagation algorithm for reduction of hidden units and acceleration of convergence using artificial selection," *IJCNN*, vol. 1, pp. 625-630, 1990.
12. Gedeon, TD, Harris, D, "Network Reduction Techniques," *Proc. Int. Conf. on Neural Networks Methodologies and Applications*, AMSE, San Diego, vol. 2, pp. 25-34, 1991.
13. Joines, M and White, M, "Improving generalisation by using robust cost functions," *IJCNN*, vol. 3, pp. 911-918, Baltimore, 1992.
14. Gedeon, TD and Bowden, TG, "Heuristic Pattern Reduction," *Proc. Int. Joint Conf. on Neural Networks*, vol. 2, pp. 449-453, Beijing, 1992.
15. Rogers, SJ, Fang, JH, Karr, CL and Stanley, DA "Determination of Lithology from Well Logs using a Neural Network," *AAPG Bulletin*, 76, p. 731-739, 1992.
16. Wong, PM, Gedeon, TD and Taggart, IJ "An Improved Technique in Porosity Prediction: A Neural Network Approach," *IEEE Trans. Geoscience and Remote Sensing*, (in press), 1994.

Priming an Artificial Neural Classifier

Didier Puzenat

LIP - URA 1398 du CNRS

Ecole Normale Supérieure de Lyon

69364 Lyon Cedex 07, France

dpuzenat@lip.ens-lyon.fr

January 12, 1995

Abstract

Repetition priming capacity enables biological systems to manage easily with recently met situations. Priming an artificial neural network is of great interest in some modeling tasks. The network is an incremental neural classifier. This system creates units when it is not able to recognize a pattern correctly. Repetition priming is introduce through a priming function, by reinforcing the recognition of recently seen categories. Characteristics of this function are discussed in order to find the more suitable shape. Experiments are performed on handwritten recognition application. Methods described enable to detect easily priming with low computation (computation of a simple linear regression). More computation enables to measure the phenomenon (difference between the slope of the regression lines, with and without priming).

1 Introduction

Repetition priming capacity enables a quicker, or easier, answer if a *stimulus* has already been met ([KK92] page 374). This phenomenon is often used by neuro-psychologists to understand neural systems architecture. In a modeling task of a cognitive science project, we have met the need of a neural classifier with such capacities, to model perceptive memory. This article presents a way to prime an incremental neural classifier network.

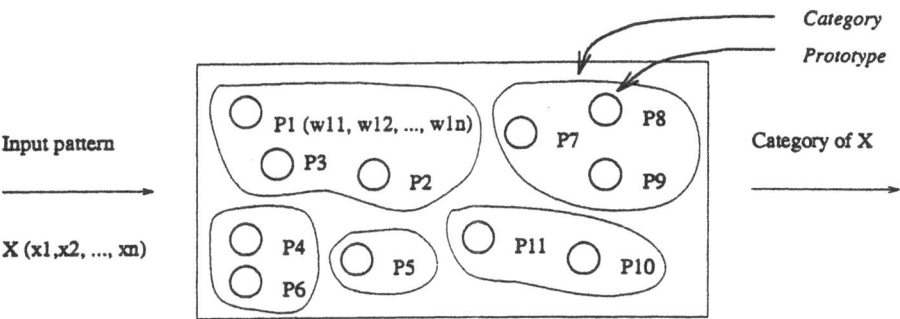

Figure 1: Architecture of an incremental classifier

2 Incremental classification

2.1 Architecture

The network is an incremental classifier (figure 1) [AG92]. The knowledge is stored in weighted connections with the input. It is modeled by a vector named prototype. A prototype is created, or modified, when the classifier is not able to recognize a pattern correctly. The modification of a prototype is performed according to a Grossberg learning rule [CG88].

2.2 Learning and generalization

A classifier associates patterns with categories. During the *recognition phase*, the system gets a pattern and replies by the associated category. During the *learning phase*, patterns are presented to the classifier, the system clusters similar patterns into categories (unsupervised learning). Then the user labels categories which have emerged (after the learning process ends). A classifier can also learn patterns presented with a label giving the category (supervised learning). An incremental classifier always learns, increasing the number of prototypes if necessary. However, a *generalization* phase enables to test the recognition performances on new patterns.

2.3 Algorithm of the classifier

When an input pattern X is presented to the classifier, the algorithm computes the similarity s_{P_j} between each prototype P_j and the input pattern. Formula 1 computes the similarity for all the prototypes, w_{jk} is the component k of the prototype P_j, x_k is the component k of the input pattern, and n is the number of prototypes at this time.

$$s_{P_j} = \frac{2 \sum_k w_{jk} x_k}{(\sum_k w_{jk}^2) + (\sum_k x_k^2)} \qquad \forall j \in [1, n] \tag{1}$$

Step 1 The algorithm finds the best prototype P_{best}, that is the prototype with the best similarity s_{best}. The prototype P_{second} (similarity s_{second}) is the second best prototype from another category. Thus $C(P_{best}) \neq C(P_{second})$ where $C(P_j)$ is the category of P_j. In other words, P_{second} is the best prototype among prototypes that do not belong to $C(P_{best})$. P_{best} must be close enough to X and far enough from P_{second} (equations 2). If both conditions are not verified, P_{best} does not exist.

$$\begin{cases} s_{best} > \Theta_{influence} \\ s_{best} - s_{second} > \Theta_{confusion} \end{cases} \tag{2}$$

Step 2 The algorithm modifies or creates prototypes.

Presentation of a pattern X	Presentation of a pattern X with a label $C(X)$
- If P_{best} exists : - X is recognized to belong to $C(P_{best})$, - upgrade P_{best} to take X into account. - Else : - X is not recognized, - Create P_{new} from X, in a new category.	- If P_{best} exists and $C(P_{best}) = C(X)$: - X is recognized, - upgrade P_{best} to take X into account. - Else : - X is not recognized, - Create P_{new} from X, with $C(P_{new}) = C(X)$.

3 Priming method

3.1 Principle

The classifier has to answer easily if a pattern belong to the category of the input pattern has already been presented. A method is to promote prototypes from categories that have been seen recently. The *help* must be as high as the last presentation is recent. The key is to add a *bonus* to categories which have been recently presented.

3.2 Priming function

Let $t_{C(P_j)}$ denote the elapsed time (*i.e.* the number of input patterns presented) since an input pattern of the category $C(P_j)$ has been presented. The score S_{P_j} of the prototype is computed by adding a priming function f (equation 3). Step 2 of the primed algorithm deals with scores instead of similarity. Step 1 is unchanged.

$$S_{P_j} = s_{P_j} + f(t_{C(P_j)}) \qquad \forall j \in [1, n] \tag{3}$$

The *priming function* f must be positive and strictly decreasing. Figure 2 gives a good candidate for f. This function promotes categories seen since the last 10 presentations (figure 3).

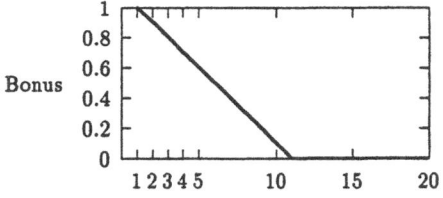

Elapsed time (number of presentations)

Figure 2: Priming function f

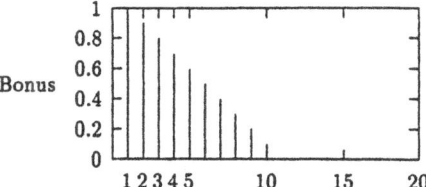

Elapsed time (number of presentations)

Figure 3: Priming function f (discrete)

4 Detecting and Measuring priming

4.1 Principle

There is a duality between a quick answer, and an easy answer. The higher the system trust in an answer, the quicker its answer. Therefore, the priming is measured by monitoring the trust that the classifier has in its answers.

4.2 Trust, a definition

Let T denote the trust that the classifier puts in its answer. T is a function of the best scores. It comes from the new algorithm that :

$$T = S_{best} - S_{second} = \underbrace{s_{best} - s_{second}}_{\Delta_s} + f(t_{C(P_{best})}) - f(t_{C(P_{second})})$$

4.3 Trust variations

The following table studies the influence of the priming function on trust. Since the algorithm is *winner takes all*, $i \neq j \Rightarrow t_{C(P_i)} \neq t_{C(P_j)}$.

$$
\begin{cases}
t_{C(P_{best})} < t_{C(P_{second})} & \Rightarrow & f(t_{C(P_{best})}) - f(t_{C(P_{second})}) > 0 \\
t_{C(P_{best})} > t_{C(P_{second})} & \Rightarrow & f(t_{C(P_{best})}) - f(t_{C(P_{second})}) < 0
\end{cases}
$$

Trust T	$t_{C(P_{best})} < t_{C(P_{second})}$	$t_{C(P_{best})} > t_{C(P_{second})}$
$\Delta_s > 0$	Priming confirms the classifier in its choice. Trust is **great**.	Priming makes another category to be chosen. Trust in this new choice is **small**.
$\Delta_s < 0$	Priming helps the second best choice, trust in the first choice is **small**.	**Impossible** since $T > \Theta_{confusion}$ (formula 2), hence $T > 0$.

An analysis of trust level is difficult. However, the important point is that in all the cases, a smaller $t_{C(P_{best})}$ increases the trust. Hence, it is possible to detect and measure the priming effect by an analysis on the slope of trust as a function of the occurred delays. An occurred delay is the number of input patterns presented between two presentations of patterns labeled by the same category. A negative slope would be a sign for priming.

5 Experiments

Experiments are performed on handwritten recognition application [AA92] [APMP94]. The classifier learns 500 characters, and tries to generalize with 500 others patterns. Each pattern is presented (only one time) according to a random order. The input of the classifier comes from a module that extracts oriented segments. Generalization gives the good category for about 95 % of the patterns. There is no interest in priming learning. Hence only the generalization algorithm has been changed.

5.1 Monitoring priming

5.1.1 Measure

Priming must be detectable and measurable. Monitoring trust gives the distribution of figure 4. The abscissa (named *delays*) gives the experimental occurred delays (sorted). Figure 5 presents the average trust for each delay (the thick curve). The thin curve gives the average trust without priming. Both curves are almost illegible, but they clearly differ.

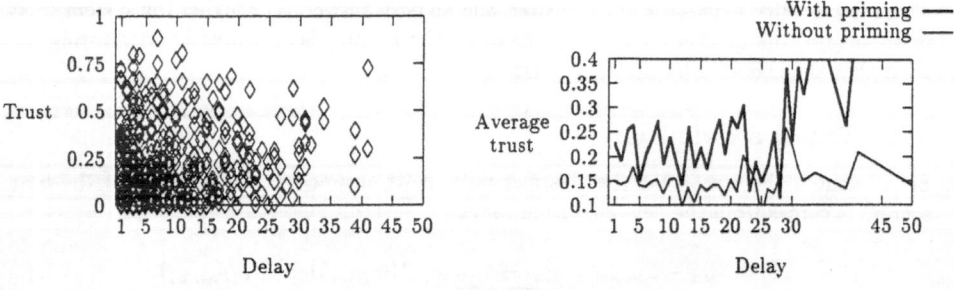

Figure 4: Trust

Figure 5: Average trust

5.1.2 Meaningful delays

Figure 6 gives the number of occurrences for each delay. Most of the delays are too rare to be meaningful. It seems to be reasonable not to take into account delays that appear less than 5 % of the time (figure 7). Computing linear regression on delays which percentage are higher than 5 % give :

With priming   : y = -0.000789 x + 0.222
Without priming : y = -0.000377 x + 0.152

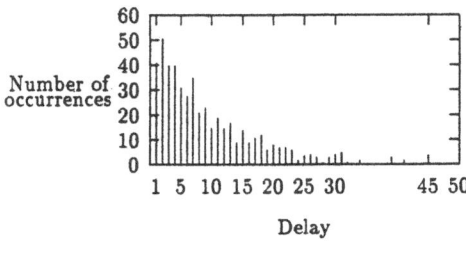
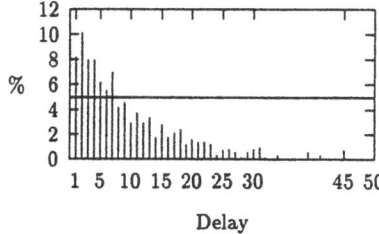

Figure 6: Occurrence of delays   Figure 7: Percentage of occurrence of delays

The slope of the linear regression on trust for the primed classifier is smaller. However, results are not very conclusive. Moreover, other similar experiments have shown that various slopes are obtained from varying the presentation order of the data sets.

5.1.3 Meaningful regression lines

An answer for reducing the influence of learning is to compute the average linear regression over several experiments. An equivalent method is to increase the trust distribution (figure 4) over several learning and generalization cycles and to compute the regression line on the large distribution. This last method has been chosen. Figure 8 presents the average trust after 150 learning-generalization cycles, with priming (the thick curve) and without priming (the thin curve). Figure 9 gives the percentage of occurrences for every delays. Seven delays are meaningful, they represent 54.7 % of all the delays. Figure 10 shows the evolution of the slope as a function of the number of cycles. Priming is easily detectable after 5 cycles. Results are meaningful since about 50 cycles. The slope reaches its final value since about 100 cycles. A very meaningful linear regression is computed after 150 cycles :

With priming   : y = -0.00341 x + 0.253
Without priming : y = -0.000147 x + 0.147

The slope of the regression line gives evidence of the priming phenomenon. The difference between the slopes of the regression lines, with and without priming, measures priming level. This measure is stable and accurate.

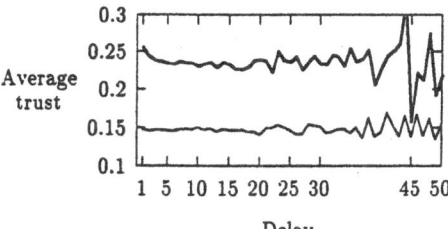
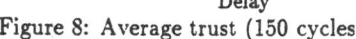
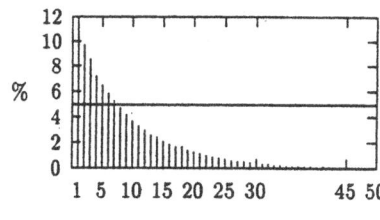

Figure 8: Average trust (150 cycles)   Figure 9: Percentage of occurrence of delays

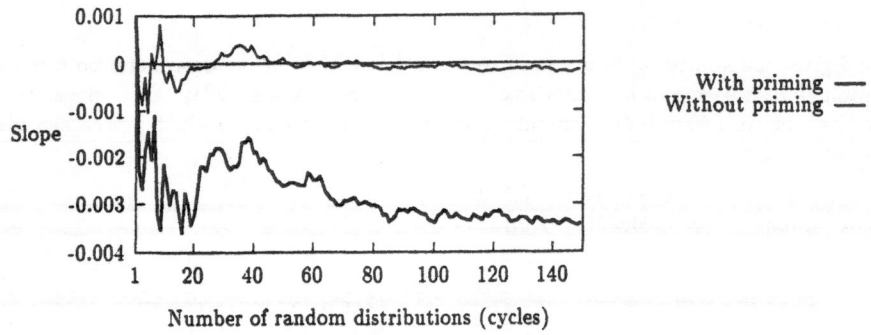

Figure 10: Evolution of regression slope

6 Priming function

6.1 Moving base

The choice of the priming function is arbitrary. Figure 11 presents various shapes, by modifying the *base* of the function (2, 5, 10, or 15 presentations). Figure 12 gives the corresponding evolution of the regression line slope. The best slope comes with 5 for the base (i.e. the 5 last presentations are promoted). This result can be linked with the number of meaningful delays (figure 9) : the function must promote the more frequent delays.

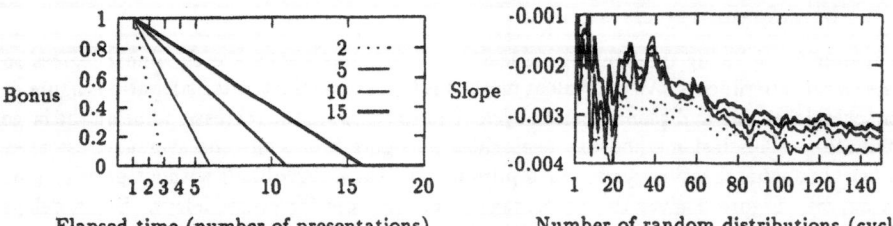

Figure 11: Different bases for f Figure 12: Slopes for different bases

6.2 Moving height

Another way to increase priming is to increase the *height* of f. Experiments has been performed with the value 5 for the base, heights equal to 0.1, 0.5, 1, and 2. Figures 13 and 14 show that a higher f increase priming.

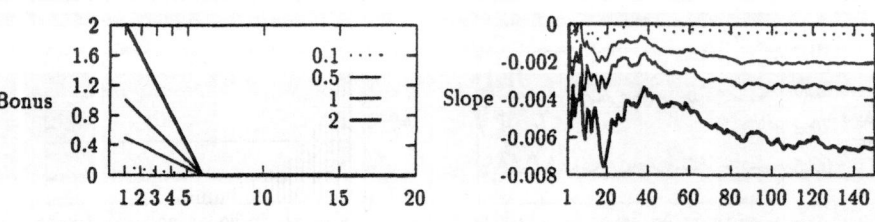

Figure 13: Different heights for f Figure 14: Slopes for different heights

7 Classification

Increasing the height of f appears to be the best way to produce priming. But this method has *a cost*. The classifier frequently recognizes an input pattern as belonging to the category of the last presented patterns. Therefore many answers are wrong. Such a drawback does not matter for many tasks (e.g. modeling). However, if classification performances must be kept, the height of f must stay small. Table 15 and figure 16 show that a height of 0.1 reconciles priming detection and performances.

Height	Success (%)	Slope (150 cycles)
0.1	93.3	-0.000526
0.5	80.5	-0.002045
1.0	58.6	-0.003674
2.0	41.7	-0.006570

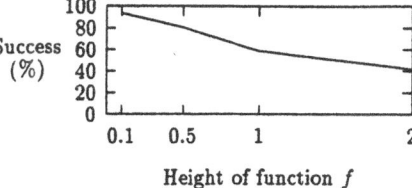

Height of function f

Figure 15: Percentage of recognition   Figure 16: Percentage of recognition

8 Conclusion

It is possible to prime an incremental neural classifier by reinforcing the recognition of prototypes from recently seen categories. The classifier gives easier responses to recently seen patterns. Numerous cycles are needed to study the phenomenon accurately; mainly for analysising the influence of the shape of the priming function. Since this work have been done, it is easy to find a good priming function for quick results. Methods described enable to detect easily priming in a few cycles (computation of a simple linear regression). More cycles enable to measure the phenomenon (difference between the slope of the regression lines, with and without priming). These measurements are of great interest since they enable to do quantitative analysis of systems.

References

[AA92]   A. Azcarraga and B. Amy. An incremental neural classifier of configurations of active orientation-specific line detectors. In *Artificial Neural Networks 2*, 1992.

[AG92]   A. Azcarraga and A Giacometti. A prototype-based incremental network model for classification tasks. In *Proc. Neuro-Nîmes*, pages 121–134, November 1992.

[APMP94] A. Azcarraga, H. Paugam-Moisy, and D. Puzenat. An incremental neural classifier on a mimd parallel computer. In C. Girault, editor, *Applications in Parallel and Distributed Computing*, volume A-44 of *IFIP Transactions*, pages 13–22, Caracas, April 1994.

[CG88]   G. Carpenter and S. Grossberg. The art of adaptive pattern recognition by a self-organizing neural network. *IEEE Computer*, 21(3):77–88, March 1988.

[KK92]   S. M. Kosslyn and O. Koenig. *Wet Mind : The new cognitive Neuroscience*. The Free Press (ISBN 0-02-917595-X), 1992.

Artificial Neuroconsciousness
an Update

Igor Aleksander
Department of Electrical and Electronic Engineering, Imperial College
London, UK

Abstract

The concept of a theory of artificial neural consciousness based on neural machines was introduced at ICANN94 (Aleksander, 1994)[15]. Here the theory is developed by defining that which would have to be synthesized were consciousness to be found in an engineered artefact. This is given the name "artificial consciousness" to indicate that the theory is objective and while it applies to manufactured devices it also stimulates a discussion of the relevance of such a theory to the consciousness of living organisms. The theory consists of a fundamental postulate and a series of corollaries. In this paper the series of corollaries is extended and illustrated by means of characteristic state structures. Studies of artificial neuroconsciousness aim at two results: first to provide a single perspective on many mechanisms which perform cognitive tasks; and second, it provides an explanation of consciousness which stands alongside the many discussions found in the literature of the day[1-4].

1 Theory

The theoretical framework used in this work has one fundamental postulate from which follow 12 corollaries. This framework has been inspired by Kelly's[5] theory of "personal constructs"which explains the causes of personality differences in human beings.

The Fundamental Postulate: Consciousness and Neural Activity.

> *The personal sensations that lead to the consciousness of an organism are due to the firing patterns of some neurons, such neurons being part of a larger number which form the state variables of a neural state machine, the firing patterns having been learned through a transfer of activity between sensory input neurons and the state neurons.*

The words of this postulate are intended to have specific meanings which need to be stressed so that the corollaries which follow should make sense.

Personal sensation: Much of the controversy surrounding consciousness comes from the problem of infinite regress. Here it is implied that neural activity leads directly to personal sensation so dismissing the problem of infinite regress.

Firing patterns: Neurological terminology has been adopted to refer to the output activity of a group of neural elements. In an artificial system 'firing patterns' could refer to any measurement of the output quantity of the elements which constitute that system.

Neurons: This adoption of this neurological term is used to indicate that the theory is that of a cellular system where "neuron" is the name given to a basic cell.

Neural state machine: A state machine is the most general model of a finite computing process - it calls on the concept of an inner state which is a function of input sequences. Neural versions assume that neurons generate the variable values which, when taken together, form a state. (Corollary 1 formalises this notion and the generality of neural state machines has been argued elsewhere[6]).

Learned: Neurons are assumed to be plastic and it is this plasticity which allows them to learn meaningful, representational, firing patterns.

Iconic Transfer: This key property relates to the source of information which controls the learning of the neurons. It will be seen that distal, sensory information is postulated to impose output patterns on neurons so that these may be learned and recalled in the absence of input. It is this transfer that creates inner perception in the conscious organism.

Sensory Neurons: These are transducer neurons that transform energy from environmental input into the distal, sensory signals which control iconic transfer.

Corollary 1: The brain is a state machine.

> *The brain of a conscious organism is a state machine whose state variables are the outputs of neurons. This implies that a definition of consciousness be developed in terms of the elements of state machine theory.*

Corollary 1 is a consequence of the intent in the fundamental postulate that the theory of artificial consciousness be based on state machine theory. State machines can model any system with inputs outputs, internal states and input-dependent links between such states. The states and their links form a state structure. Such machines can be probabilistic where links between states are defined as probabilities, they can have a finite or an infinite number of states. The fact that any conscious organism must have something called a brain with an attendant state structure is evidently true and not controversial. The key question is whether enough can be said about the nature of the state structure of organisms that are said to be conscious which distinguishes consciousness itself. This becomes the task for the corollaries which follow - to define the characteristics of state structure that are necessary for and specific to organisms that are said to be conscious.

Formalization of Corollary 1.
In any state machine, five items need to be defined:

i) The total *input* to the neural state machine is a vector i of input variables i_1, i_2 ...
$$i=[i_1,i_2...]$$
The i_1, i_2 ..variables are the outputs of sensory neurons.
In living brains the number of such variables, being the number of neurons involved in the early layers of all sensory activity, is very large but finite. There is also some debate about whether it is important for these variables to be considered as binary (firing or not) or real (firing intensity per unit time). While it will be seen that this decision does not alter the course of the theory, it is assumed here that these variables are binary. This is done without loss of generality but with the gain that, using the methods of automata theory, it becomes possible to develop non-linear models.

Also, I is defined to be the set of all possible input vectors.

ii) The total *output* of the neural state machine is a vector z of output variables z_1, z_2 ...
$$z = [z_1, z_2 ...]$$
The z_1, z_2 .. variables are the outputs of 'actuator' neurons.

Again the variables z_1, z_2.... are considered to be binary, and, in living brains, would be seen as the output parts of the brain which are responsible for muscular action .

Also, Z is said to be the set of all possible output vectors.

iii) The *inner state* of the neural state machine is defined as a vector q of variables q_1, q_2 ..
$$q = [q_1, q_2 ...]$$
The q_1, q_2 .. variables are the outputs of 'inner' neurons.
Again, variables q_1, q_2 ... are binary, and, in brains, would be the states of neurons neither involved in input sensing nor output generation.

Also, Q is said to be the set of all possible input vectors.

iv) The *state dynamics* of the neural state machine are determined by the equation
$$q' = \beta(q,i)$$
where q' is the "next" state, q is the current state and β a function, which in the case of a finite number of binary variables may be expressed as the mapping,
$$\beta : \quad Q \times I \rightarrow Q.$$
where x is the Cartesian product. (In the general case this mapping is considered to be probabilistic in the sense that every pair (q,i) of $Q \times I$ maps into every element of Q with some probability.)

v) The *output function* of the neural state machine is determined by the equation
$$z = \omega(q)$$
where ω is a many-to-many mapping which in the general case is probabilistic.
$$\omega : \quad Q \rightarrow Z.$$

In addition to the above group of five properties which are required in the definition of any state machine, the definition of a *neural* state machine contains the following key property which relates to the generalization of the neurons.

vi) The state dynamics and output functions of the neural state machine *generalize* in the sense that:

given a stable state $\quad q_a = \beta(q_a, i_a)$,

and $\quad z_a = \omega(q_a)$

there exists a set $\quad (Q \times I)_a$, a subset of $(Q \times I)$

such that for any $\quad (q_j, i_k)$ from $(Q \times I)_a$

$\quad q_a = \beta(q_j, i_k)$

and $\quad z_a = \omega(q_j)$

That is, there is an equivalence class of pairs (q_j, i_k) which includes (q_a, i_a) for which the next state q_j is the same and the output z_a is the same. Note that the case for the stable state has been quoted here, although any state transition has similar equivalent state-input pairs. More is said of generalization in corollary 7 and appendix 1. The basic notion expressed in this corollary is expressed in fig.1.

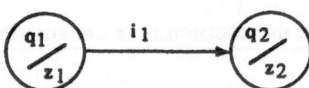

Fig.1 A classical element of state structure. It is stressed that the transition is learned and that the state, input and output vectors are composed of neural state variables.

The following three corollaries are stated together because their justifications are interleaved.

Corollary 2: Inner Neuron Partitioning

> *The inner neurons of a conscious organism are partitioned into at least three sets:*
> *Perceptual Inner Neurons : responsible for perception and perceptual memory;*
> *Auxiliary Inner Neurons : responsible for inner 'labelling' perceptual events.*
> *Functional Inner Neurons : responsible for 'life-support' functions - not involved in consciousness.*

Corollary 3: Conscious and Unconscious States

Consciousness in a conscious organism resides directly in the perceptual inner neurons in two fundamental modes:
Perceptual : which is active during perception - when sensory neurons are active;
Mental : which is active even when sensory neurons are inactive.
The activity of the inner perceptual neurons ranges over the same states in both these modes.

The same perceptual neurons can enter semi-conscious or unconscious states that are not related to perception.

Corollary 4: Perceptual Learning and Memory

Perception is a process of the input sensory neurons causing selected perceptual inner neurons to fire and others not. This firing pattern on inner neurons is the inner representation of the percept - that which is felt by the conscious organism. Learning is a process of adapting not only to the firing of the input neurons, but also to the firing patterns of the other perceptual inner neurons. Generalisation in the neurons (i.e. responding to patterns similar to the learnt ones) leads to representations of world states being self-sustained in the inner neurons and capable of being triggered by inputs similar to those learned originally .

Comment on corollaries 2, 3, 4.

i. All three corollaries stem from the statement in the fundamental postulate that a conscious organism is conscious through *owning* the sensation-causing firing patterns of its inner neurons.

ii. All three corollaries meet the requirement that an organism could not be said to be conscious unless sensations due to sensory input may be sustained in the absence of such sensory input, albeit in reduced detail. (The organism is conscious even with its eyes closed and other senses shut off).

iii. Allowing for unconscious function in a brain-like organism, corollary 2 indicates that perceptual states occur in a subset of inner neurons. That is, not all inner neurons store perceptual memories - some may be encode concepts such as duration, ordinality or even 'mood', while others just keep the organism "alive".

iv. Corollary 3 indicates that the fundamental postulate leaves open the possibility that not all the states of the perceptual inner neurons have direct sensory correlates. This allows the model to account for effects such as sleep or anaesthesia.

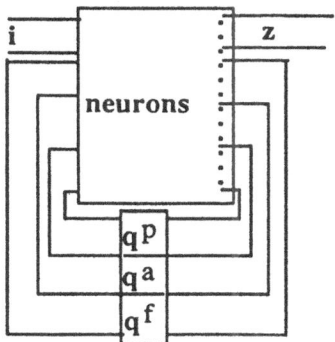

Fig.2 The classical state machine has its state vector partitioned into perceptual, auxiliary and functional variables.

v. Corollary 4 suggests that the formalization of the fundamental postulate should account for the creation of perception-related states by reference to the learning properties of the neuron. This includes a formalization of the process of retrieval of inner perceptual states. *Formalization of corollary 2*

Let {q} denote the set of variables that make up vector q.
{q} is partitioned into three subsets:

    {q^P} the **perceptual** set with a corresponding vector q^P,
    {q^a} the **auxiliary** set with a corresponding vector q^a
and   {q^f} the **functional** set with a corresponding vector q^f.

Related to these partitions are the sets of all states on the variables:

    Q^P is the set of all states on the **perceptual** variables,
    Q^a is the set of all states on the **auxiliary** variables,
    Q^f is the set of all states on the **functional** variables.
Also $Q = Q^P \times Q^a \times Q^f$
Fig. 2 shows what is intended physically by this partition.

Formalization of corollary 3
A particular perceptual input i_w has a state correlate q_w on the set of variables {q^P} so that:

  $q_w = ß(i_w , q_w)$, the perceptual mode.

Also if i_w is replaced by some "neutral" input $i_\emptyset$, due to generalization (corollary 1 [vi])

  $q_w = ß(i_\emptyset , q_w)$, the mental mode.

(more will be said about the nature of "neutral" inputs in conjunction with corollary 10 relating to "will").

Let the set of all input-related states such as q_w be Q^i, a subset of Q^P. The remainder of Q^P contains states not related to perception: unconscious and semi-conscious states.

An assumption has been made in the above, and that is that
  $ß:$ $Q^P \times I \rightarrow Q^P$ rather than $ß:$ $Q \times I \rightarrow Q$ as in corollary 1.
This effectively assumes that vector q^P is independent of the rest of the state variables. This has been done to simplify the formal notation in this and subsequent corollaries and to retain a clear line of explanation. It is stressed that this assumption should not be made in general, as dependence between state variable sets provides a way of explaining links between perceptual states on one hand and auxiliary or functional states on the other. The general idea is illustrated in fig. 3.

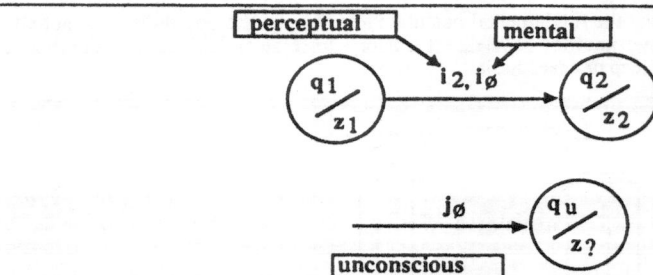

Fig. 3. The upper example suggests that a meaningful transition may be sustained even in the absence of perceptual input. The lower example suggests that the same state variables as those involved in conscious perception are at work in a less controlled way when an unconscious sensory environment $J_\emptyset$ is present.

Formalization of corollary 4

Learning is the process of first associating an input i_w and an arbitrary state $q_\emptyset$ to form the following element of the forward network function ß,

$$q_w = ß(i_w, q_\emptyset).$$

The key factor is that there is μ, a fixed sampling mapping which transfers some of i_w into q_w, causing q_w to be defined by i_w:

$$q_w = \mu(i_w).$$

To create the stable representation (attractor) for i_w, ß is augmented by:

$$q_w = ß(i_w, q_w).$$

(This is the "iconic" training methodology fully described elsewhere[6]).
Further, the generalization of the neurons as formalised in corollary 1, ensures that the requirement of corollary 3:

$$q_w = ß(i_\emptyset, q_w), \text{ the mental mode,}$$

is satisfied. Figure 4 illustrates this as a state structure.

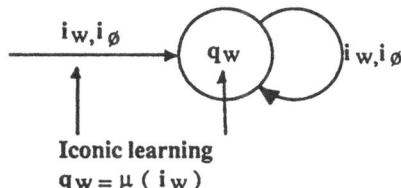

Iconic learning
$$q_w = \mu(i_w)$$

Fig. 4. State q_w is formed to retain the spatial characteristics of i_w which may be retrieved even in the mental mode.

Corollary 5: Prediction

> *Relationships between world states are mirrored in the state structure of the conscious organism enabling the organism to predict events.*

Prediction is one of the key functions of consciousness. An organism that cannot predict would have a seriously hampered consciousness. It can be shown formally that prediction follows from a deeper look at the learning mechanism of corollary 4.

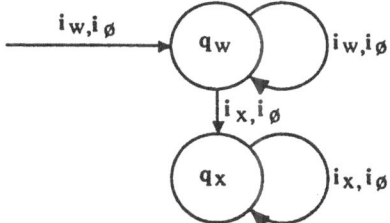

Fig. 5. Planning is seen to be effective for mental input $i_\emptyset$

Formalization of corollary 5

Say that i_x follows i_w as a result of world state changes. Say that the organism is in state q_w in response to i_w. If the input changes to i_x and iconic learning is taking place, the following element

of ß will be added:
$$q_x = ß(i_x, q_w), \quad \text{followed by}$$
$$q_x = ß(i_x, q_x), \quad \text{where} \quad q_x = \mu(i_x).$$
In the mental mode, the following two transitions become equally probable:

$q_w = ß(i_\emptyset, q_w)$, $q_x = ß(i_\emptyset, q_w)$ .This means that, in time, state q_w will lead to q_x in the mental mode, completing the prediction as shown in Figure 5.

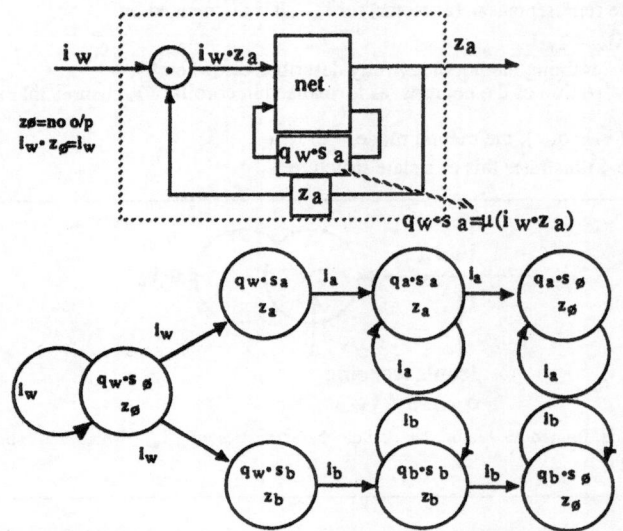

Fig.6. Internal and external loops which show how the effect of two actions is learned.

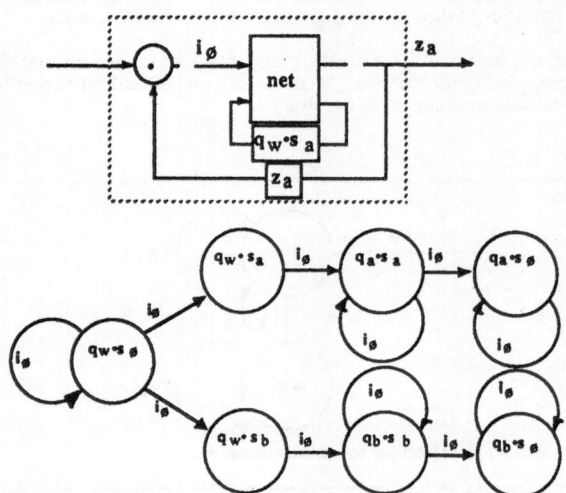

Fig.7. Internal and external loops which show how the effect of two actions is learned.

Corollary 6: The Awareness of Self

As a result of iconic learning and feedback between physical output and the senses, the internal state structure of a conscious organism carries a representation of its own output and the effect that such an output can have on world states. This includes a representation of what can and cannot be achieved by the organism itself.

Awareness of self is the ability to distinguish between changes in world states that are caused by the organism's own actions and those that occur in a way that is not controlled by the organism. Here it is argued that this ability follows from the prediction corollary and implies that the organism stores the knowledge.

Formalization of Corollary 6

The objective of this formalization is to involve the output of the organism as part of the input.

Let $z_\emptyset$ be a special "no output" condition and let all other output actions perceivable by the input be:
$$z_1, z_2 \ .... \ z_j$$
An input to the neural state machine therefore contains at least two components:
$$\{ i_w \bullet z_j \}$$
$[p = \{r \bullet s\}$ should be interpreted as "p is made up of two *fields* r and s]

(note that if the output is $z_\emptyset$ then $i_w \bullet z_\emptyset = i_w$, and that is the situation assumed in earlier corollaries)
As iconic learning defines $q_w = \mu(i_w)$, this can be extended to
$$\{ q_w \bullet s_j \} = \mu \{ i_w \bullet z_j \} = \mu \{ j_w \} \bullet \mu \{ z_j \} \ (say) \ .$$
Hence iconic learning leads to parts of states such as s_j which are internal representations of the organism's own actions.

Now suppose that the world is in some state i_a and that this has been learned with $z_\emptyset$, producing therefore a direct iconic representation $q_a = \mu(i_a)$, suppose that action z_1 changes the world state to i_1 where it remains even if the action ceases (i.e. the output reverts to $z_\emptyset$). As in the prediction corollary, 5, the system will learn the following linked internal representations.
$$q_a \rightarrow \{ q_a \bullet s_1 \} \rightarrow \{ q_1 \bullet s_1 \} \rightarrow q_1$$
Hence iconic learning leads to representations of the way in which the organism's own actions achieve changes in the world state which is an existence proof for a representation of the awareness of self.

Figure 6 shows how the internal and external loops create representations of the way in which action z_a changes the world state to i_a and that this is represented as q_a. A similar representation is shown for the consequence of action z_b . Figure 7 shows how, in the mental mode, this "awareness of self" may be retrieved as a probabilistic event.

Corollary 7: Representation of Meaning.

When sensory events occur simultaneously or in close time proximity in different sensory modalities, iconic learning and generalization of the neural state machine ensures that one can be recalled from the other.

When it is said that a conscious organism knows the meaning of input (e.g. words or events) this refers to the creation of an internal representation that is related to that input. It could be the word related to an event or an entire internal scenario related to a word. In this theory, the word "knows" translates to the retrieval of complete iconic internal representations from partial inputs. Association of this kind is a basic property of an iconically trained neural state machine and is formally stated below. However, this is also the area in which much of the controversy of whether a machine can or cannot "know" the meaning of words and phrases arises[7]. While linguistic output and the understanding of language are discussed in the corollaries that follow, it may be stated here that internal representations have precisely the property of subjectivity that Searle[3] has advocated. This results from such representations being iconic and dynamic "images" of world states which include the representation of world behaviour due to the actions of the organism itself. This is not the case with the preprogrammed symbolic representations of conventional Artificial Intelligence where the symbols are only arbitrarily related to sensory input.

Formalization of Corollary 7

The corollary has been worded so as to embrace two major forms of association of meaning: i) spatial and ii) temporal. These are treated separately.

i) Spatial association.

Here two input events, say i_v and i_u, (e.g. vision and utterance) occur together so as to be the components of an overall input i_w . Say that, using the notation of corollary 6:

the input is represented by $\{ i_v \cdot i_u \}$

then $\qquad q_v \cdot q_u = \mu \{ i_v \cdot i_u \} = \mu \{ i_v \} \cdot \mu \{ i_u \}$

and for iconic learning
$$q_v \cdot q_u = \beta(\{i_v \cdot i_u\}, \{q_v \cdot q_u\})$$

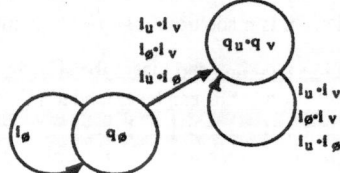

Fig. 8. Spatial association.

It is shown in appendix 1 that there exists a generalization such that
$$q_v \cdot q_u = \beta(\{i_v \cdot i_\emptyset\}, q_\emptyset) \text{ or } q_v \cdot q_u = \beta(\{i_v \cdot i_\emptyset\}, q_\emptyset) \text{, where } i_\emptyset \text{ and } q_\emptyset \text{ are arbitrary}$$
values.
This leads to the stable states
$$q_v \cdot q_u = \beta(\{i_v \cdot i_\emptyset\}, \{q_v \cdot q_u\}) \text{ or } q_v \cdot q_u = \beta(\{i_\emptyset \cdot i_u\}, \{q_v \cdot q_u\})$$

This means that inner representation of the pair of associating events is recalled from the presence of one input event from the pair as illustrated in Figure 8.

ii) Temporal association.
In this case, input events such as i_v and i_u occur in sequence, and the retrieval of the second could be regarded to be the prediction made by the system as a result of the occurrence of the first. The formalization is therefore the same as that in the Prediction Corollary, 5.

Corollary 8: Learning Utterances.

> *The feedback loop responsible for the creation of "self" representations is also responsible for the creation of state representations of the basic utterances of the organism which are retrieved in response to the utterances of other, "adult" , organisms and may be used by the adult to teach the organism more complex utterances such as the words of a language.*

A salient measure of the level of sophistication of an artificially conscious organism is the complexity of the means of communication that the organism has with others and the way that others may be responsible for teaching the learning organism to harness its abilities to communicate. In the spectrum of living organisms there appears to be a large gap in this sophistication between humans and animals which makes this and the next corollary appear to be particularly directed at human-like behaviour. In artificial systems it is possible to imagine a greater degree of continuity along this scale of sophistication than that which exists in nature. This is reflected in the formalization of this corollary.

The corollary involves the assumption that the organism is capable of a basic set of utterances, which, before any learning takes place, are generated arbitrarily. However, due to the output - input long-routed feedback, these, in the manner of corollary 6, form representations in the state space of the organism. If an "adult" wishes to key in to these representations it will have to imitate the utterances of the organism. This, among humans, is called baby talk.

The corollary extends to the way in which an adult, having keyed in to these representations, can use them to label events with words. In corollary 10 it will be shown that the organism can use such representations when it "wishes" to call for objects described by these words.

Formalization of Corollary 8

Concentrating on a specific set of the actions found in corollary 6, assume that the organism is capable of a finite set of elementary "utterances".

$$Z_u = \{ z_{u\emptyset} , z_{u1}, z_{u2} \dots z_{um} \}$$

There is again an elementary utterance $z_{u\emptyset}$ which corresponds to no utterance or silence.

Say that the organism emits these arbitrarily and in the absence of other input, they are perceived at the input and learned as stable states as

$$s_{uj} = \beta (z_{uj}, s_{uj})$$

where $s_{uj} = \mu (z_{uj})$.

However, as learning affects the output neurons too, a link is established in the output mapping ω (see corollary 1):

$$z_{uj} = \omega (s_{uj})$$

In summary, this means that an utterance z_{uj} has an internal representation s_{uj} which is both retrieved by z_{uj} through mapping β and sustained by it through mapping ω. Hence a stable state has been established in both the internal feedback in the neural state machine *and* the long-routed feedback via output and input.

The implication of the above is that once state s_{uj} is entered, the organism cannot exit the state and will repeatedly and uncontrollably output z_{uj}. This problem vanishes if it is assumed that all utterances except $z_{u\emptyset}$ are of short duration and that the normal sustained state is that of silence.

Now consider that another organism (which we call an 'adult') is capable of mimicking z_{uj} as, say, z_{uj}'

If, $\sigma(z_{uj}' , z_{uj}) > \sigma(z_{uj}' , z_{uk}), k \neq j$ (σ is 'similarity' as defined in appendix I),

then z_{uj}' can 'induce' s_{uj} and the utterance z_{uj} in the organism through the generalization property stated in appendix I.

Finally, a 'word' is defined as consisting of a sequence of utterances taken from the set of elementary utterances Z_u. Say that the adult utters the word represented by some sequence

$$z_{ua}' , z_{ub}' \dots$$

with the object in question in view at the time, causing (say) i_v . As a combination of corollaries 5, 6 and 7 the transition between states

$$\{ q_v \cdot s_{ua} \} \rightarrow \{ q_v \cdot s_{ub} \} \dots$$

may be learned, where q_v is the state representation of i_v linked spatially to s_{ua} and s_{ub} which are the state representations related to the utterance sequence z_{ua}' , z_{ub}' ... (i.e. the "meaning" label which is word-like rather than a single symbol as in corollary 7).

It may be seen that being given the stimulus i_v, the entry into states $\{ q_v, s_{ua} \} \rightarrow \{ q_v, s_{ub} \} \dots$ can give rise to the organism 'naming' the stimulus as $z_{ua} , z_{ub} \dots$ through a generalization of the output link $z_{uj} = \omega (s_{uj})$. This effect will be elaborated in corollary 10 and is illustrated in Figure 9.

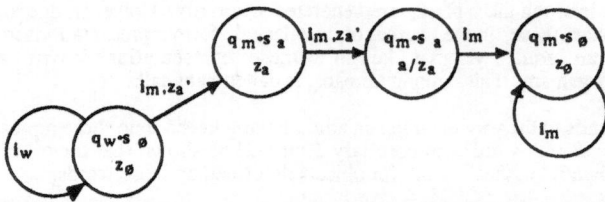

Figure. 9   Utterance learning with i_m being   mother's face and $z_a{'}$ baby talk for utterance z_a.
        Mental interpretation: organism expects mother if $z_a \, z_a$ is produced.

Note that if the same utterance is repeated (as in *mah mah* for example) this would bring into play auxiliary states (see corollary 2) and the above transition would be of the form

$$\{ q_v{\cdot}s_{ua}{\cdot}a_1 \} \; \text{->} \; \{ \; q_v{\cdot}s_{ua}{\cdot}a_2 \} \ldots$$

where a_1 and a_2 are the auxiliary state labels that distinguish between repeated utterances.

Corollary 9: Learning Language.

> *Language a result of the growth process of a societal repository from which it can be learned by a conscious organism, given the availability of knowledgeable 'instructors'.*

It is not the intention here to give a full account of language learning by a neural state machine. It is, however, necessary to stress the three points that are central to this corollary. First, it accepts the concept (as accepted by many commentators on language and consciousness[1,3,11]) that language is the result of a process of growth of a repository to which new words and structures are being added by the users of the language and which resides collectively in the brains of such users. Second, it calls for the presence of an instructor. It has been argued[6] that an organism that merely explores or observes an environment cannot learn the rules that govern phrase-structured languages. This is overcome through the intermediacy of a teacher who selects progressively more complex examples of phrases from which the rules can be learned. The role of such a teacher is demonstrated in the formalization of this corollary.

The third point, while not explicitly expressed in the corollary, is central to the nurture-nature debate in language. It is the "poverty of stimulus" argument[8]. This suggests that deep language structure needs to be innate as, otherwise, more stimuli than appear to be available to developing children, would be required. What needs to be demonstrated formally is that adequate rules may be developed by a learning state machine from a reasonable number of examples.

The role of the instructor is more complex than that implied in corollary 8. This corollary relies on the instructor not only to provide word labels for objects and actions, but also to create the opportunity for the organism to build up the experience that underlies phrases and sentences. For example the phrase "a bag of beans" has to be learned as being not only grammatically correct with respect to "bag a beans of", but also semantically correct with respect to "a bean of bags". The training should also distinguish the use of the word "of" in "the sound of thunder" and "a bag of beans". The formalization of this corollary concentrates these phrases as a particular example, while the learning of other linguistic objects (such as anaphora[9]) are the subject of current studies.

Formalization of corollary 9.

It follows from corollaries 7 and 8 that state representations can, spatially, within a state, represent both the appeearence and the meaning of objects. In order to learn the rule for accepting "a bag of beans" and rejecting "a bean of bags", the neural state machine has to learn the difference between a "container" and the "contained". This follows from an extension of corollaries 7 and 8, in the sense that the representation of an object as a state can carry a large list of attribute labels that represent the experience of the organism, either first hand or through manipulative experience. In fact, the state of a neural state machine can partition its state variables into "fields" which (as in classical databases) carry attributes such as "non-porous", "hard", "concave", "open at the top", "container". In contrast with classical databases, however, given the first four, the fifth could be retrieved without it being given through the generalization of the system.

It follows that in the time domain, state q_{a1} ($= \mu(i_{a1})$, the input word for, say, "bag") is followed by input word (hence iconic state) <of> which is followed by q_{b1} ($= \mu(i_{b1})$, the input word for ,say, "bag"), and this is repeated several times for

$$q_{a2} <of> q_{b2} \quad , q_{a3} <of> q_{b3} \quad ....$$

and if $\quad q_{a1}, q_{a2}, q_{a3} ....$ learn "container", q_{cr}, in common in some specific field (corollary 7),

i.e. $\quad \{ q_{a1} \cdot q_{cr} \}, \{ q_{a2} \cdot q_{cr} \}, \{ q_{a2} \cdot q_{cr} \}, ...$

and if $\quad q_{b1}, q_{b2}, q_{b3} ....$ learn "contained", q_{cd}, in common in some specific field (corollary 7),

i.e. $\quad \{ q_{b1} \cdot q_{cd} \}, \{ q_{b2} \cdot q_{cd} \}, \{ q_{b2} \cdot q_{cd} \}, ...$

then the sequence <a> < { $q_{aj} \cdot q_{cr}$ } > <of> < { $q_{bj} \cdot q_{cd}$ } >

will be learned while any phrase of the form <a> < { $q_x \cdot q_{cd}$ } > <of> < { $q_y \cdot q_{cr}$ } > will be rejected.

Note that in the above, only one attribute has been appended, but this attribute could be inferred from the presence of others through the generalization of the net. It is this that circumvents "poverty of stimulus" problem through the process of the dependence and independence of state fields, which can be established with a relatively small number of examples.

Corollary 10: Will

The organism , in its mental mode, can enter state trajectories according to need, desire or in an arbitrary manner not related to need. This gives it its powers of acting in a seemingly free and purposeful manner.

The effect which is modelled by this corollary is that which would lead the organism to describe a sensation as "I want X". This corollary follows from the representation of "self" in corollary 6. Suppose that a state, in the mental mode, can lead to several subsequent states. There are only two ways in which one of these transitions can be favoured. One is a purely arbitrary transition, and the other is some "desire" which sets the internal representation into the appropriate part of the state. The "desire" can be induced from some external stimulus such as someone's statement "do you want a chocolate or a vanilla ice cream?" or from some internal stimulus such as hunger. Whatever the case, the result is that the automaton in its mental state (as in corollary 5) predicts the result associated with each of the states constrained by the stimulus. This may evoke emotional links (see corollary 12: emotion) such as greater pleasure associated with chocolate as opposed to vanilla or instinctive links (see corollary 11: instinct) that lead to crying to alleviate hunger (interpreted by an observer as "baby *wants* food").

In an organism that has learned to utter words (as in the discussion of corollary 8), an example of two actions as discussed above, could be uttering *mah-mah* or not with the result of bringing "mother" into view or not. Therefore the sense in which an organism could learn to express will by saying "I want X" is precisely a "field" generalization learned from an adult as in corollary 9. The interpretation of such a statement is that the organism, for an unspecified reason, has X in a specific mental field, and takes the learned action to control the environment in a way that in past experience brought X to the fore.

Formalization of Corollary 10

Assume that the neural state machine is in some state q_w from which it has learned that its own actions z_a, z_b, ... lead to world states i_a, i_b, ... (corollary 6) leading to mental representations

$$(q_a \cdot s_a \cdot a_{a1} \cdot a_{a2} ...), (q_b \cdot s_b \cdot a_{b1} \cdot a_{b2} ...) \quad ...$$

where q_a is a representation of world state i_a, s_a a representation of action z_a and $a_{a1} \cdot a_{a2} \dots$ are additional attributes of the state learned as in corollary 7.
The learned transitions from q_w are of the form

$$(q_a \cdot s_a \cdot a_{a1} \cdot a_{a2} \dots) = \text{ß}(z_a , q_w)$$

and $\quad (q_a \cdot s_a \cdot a_{a1} \cdot a_{a2} \dots) = \text{ß}(i_a, \ q_a \cdot s_a \cdot a_{a1} \cdot a_{a2} \dots)$

Similar learning takes place for other states such as $(q_b \cdot s_b \cdot a_{b1} \cdot a_{b2} \dots)$.

Corollary 5 has explained that in the mental mode these states will be visited either in some arbitrary fashion, or (as in appendix 1) can be retrieved in response to the presence of a subset of attributes.

That is, given state q_w and the absence of all other attributes the system, in the mental mode, can transit to any of the states $(q_a \cdot s_a \cdot a_{a1} \cdot a_{a2} \dots)$ or $(q_b \cdot s_b \cdot a_{b1} \cdot a_{b2} \dots) \dots$ etc.
This would be described as the system "knowing" that it can do
$\qquad z_a$ with world consequences i_a, and mental consequences $(q_a \cdot s_a \cdot a_{a1} \cdot a_{a2} \dots)$
or $\qquad z_b$ with world consequences i_b, and mental consequences $(q_b \cdot s_b \cdot a_{b1} \cdot a_{b2} \dots)$ etc..

The existence of a preferred attribute will lead to one transition becoming more probable than others through learning. That is, a state label such as a_{a1} can be the distinguishing label which causes a learned transition bias to $(q_a \cdot s_a \cdot a_{a1} \cdot a_{a2} \dots)$ in preference to others . For example, if pleasure is preferred to pain (as will be seen in corollary 11) or the presence of the mother's face is preferred to the absence due to a bound attribute such as the alleviation of hunger, this will determine the more

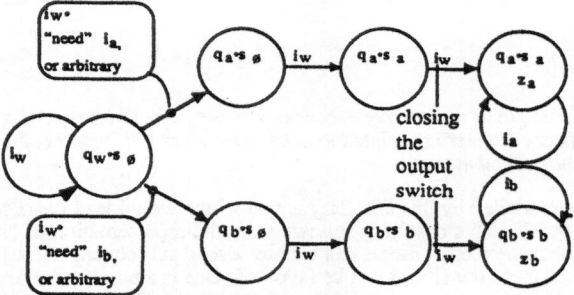

Fig. 10.   Knowing that the world can be changed to i_a or i_b.

likely mental transition and consequent action. Figure 10 illustrates the state structure which represnts knowledge which allows either a "free" choice or a choice according to need. The right-hand side of the figure shows how the mental model leads to predicted action.

Corollary 11: Instinct

To enhance survival, an organism needs a substrate of output actions that are instinctively l inked to inputs. These form anchors for the growth of state structure which develops as a result of learning.

There is no doubt that living organisms are born with instinctive reactions to some inputs. This is particularly true of animals who need these instincts for early survival. Typical of such links is the suckling reaction in response to a feeling of hunger and the presence of the mother's breast. In an artificially conscious organism, such reactions are to be determined by the designer. The thrust of this corollary is that such instinctive reactions influence the development of learned state structure in the neural state machine.

Formalization of Corollary 11.
An instinctive reaction is of the form

$z_a = \lambda(i_a)$.

The way in which this affects state structure, soon after "birth" is (as in corollary 6) through the iconic creation of a state

$(q_a \cdot s_a)$ where $q_a = \mu(i_a)$ and $s_a = \mu(z_a)$

which becomes linked to the action z_a through learning as

$z_a = \omega(qa \cdot sa)$

This means that a representation of such instinctive links makes an early entry into the system's state structure. This influences the development of semantic learning, utterance and language acquisition as described in corollaries 7, 8 and 9, above. Figure 11 shows the $z_a = \lambda(i_a)$ effect.

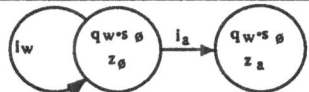

Fig. 11.   A "instinctive" output reaction to a world input happens without a change of 'mental' state.

Corollary 12:   Emotion.

Auxiliary state variables encode emotions which develop from built-in instincts.

The presence of auxiliary neurons has been outlined in corollary 2. In living organisms instinct can account for basic sensations such as fear and pleasure. As such sensations are not representative of world events they must be encoded in auxiliary neurons, primarily in an instinctive way as defined in corollary 11. However, as learning progresses, these encodings become associated with mental representations of world events. This process is responsible for both the spreading of such codes to states that are not only reactions to world events but are related to such events through the process of prediction. Subtle, mood-like emotions can be modelled in this way. Also it may be possible to model pathological cases where the emotion labels have spread to inappropriate mental states. This is a vast topic which requires a great deal more attention than is provided here. The formalization below merely indicates a way in which theory could be developed in this area.

In artificial systems, in much the same way as instincts, useful, basic emotions would have to be determined by the designer.

Formalization of Corollary 12.

A primary emotion is instinctively linked to some input as

$a_f = \varepsilon(i_f)$

where a_f is a state of auxiliary neurons.

For example, a_f could be "fear" and i_f the image of a looming object.

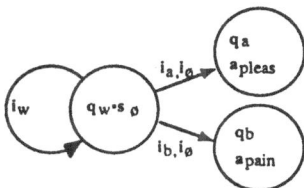

Fig. 12.   Auxiliary emotional states representing pleasure and pain.

As in the case of instincts, iconic learning will lead to internal representations of the world events to be labelled by the emotional states, for example, if

$$q_f = \mu(i_f)$$

iconic training will lead to internal representation of the kind

$$(q_f \cdot a_f)$$

which can be accessed in both perceptual and mental modes. Figure 12 shows the linking of two different auxiliary messages with two world states. From this point onwards the prediction and self awareness processes formalized in earlier corollaries come into play, associating the auxiliary states with state trajectories (thoughts) related to world events. The mixtures of auxiliary states and mental representations of world states leads to subtler forms of emotions. A simple example is the transfer of the pleasure of the alleviation of hunger to the representation of the mother's face. It is this property that, through the will corollary (10) endows the organism with "pleasure seeking" or "pain avoidance" behaviour.

Consider now the example of a pathological case (say, depression) which may be linked to the instinctive emotion of "unhappiness", say, a_u. It is possible for the a_u sensation to become linked to a large number mental states on the perceptual neurons. Neural theory can show that under some conditions, the a_u code can become self-sustaining on the auxiliary variables and be present even for unrelated states on the perceptual neurons. An observer would ascribe "chronic depression" to this organism. The theraputic act of a counsellor could be described as trying to recreate the links of a_u to appropriate mental states and liberate the inappropriate ones from it.

2   Technical critique

In this section and the one that follows the theory presented above will be given the shorthand title of ACT (Artificial Consciousness Theory). ACT is by no means rigorously complete. Its aim is to establish, through what could be described as "ramshackle" formality, that there exists a theoretical framework that is appropriate to a description of consciousness. Stress has been laid on the word "artificial" to keep in mind the constructional nature of the exercise: given an engineered artefact, what properties must it have to capture consciousness? Particular care has been taken not to fall into the trap of saying that only a biological brain can have this property as this endows the brain with mystical properties that set it apart from the physical world and which require analytical methods that have as yet not been discovered and, indeed, may never be discovered. The diametrically opposite point of view has been taken - the brain is merely one of a large class of conscious mechanisms, some of which may be manufactured. These operate according to a set of principles that may be described using existing analytical methods. However, it is the existence of the biological brain that inspires and informs the content of ACT and ACT itself can be applied to a detailed analysis of what is known about biological brains. ACT also asks a more general question: say that the universe of mechanisms could be divided between those that can be said to possess consciousness and those that do not, what is the difference between the two and what is a canonical model for the class of mechanisms that have consciousness?

It is the fundamental postulate that contains the belief the consequences of which lead to the required distinction. That is, consciousness is an inward sensation which can be attributed to the activity of a subset of constituent components called neurons. In living organisms the processing machine made up of the totality of neurons is called the brain. While this assertion is shared by many contemporary views of consciousness (see 3, below) the consequences of such a belief have not been previously analysed, and this is the objective of the 12 corollaries developed in this paper. The language of probabilistic automata and a general ,physically realisable model, the neural state machine, have been used to show in the corollaries that attributes of consciousness can be captured with a constructional methodology. Clearly, the list of corollaries need not be limited to 12. In fact it is important to keep the list open to encourage the further development and refinement of the theory. It is evident that the corollaries are ordered, the earlier ones establishing properties on which the later ones can build.

A question that remains to be answered here is whether the approach in this paper explains the richness of conscious experience which we as humans enjoy in the course of introspection.   It is difficult to be quantitative about this, but it can be shown that the artificial model discussed here would certainly provide both states and a state structure of remarkable richness due to the fact that the number of states of an automaton increases exponentially with the number of state variables (neurons). Assuming that states can be visited at the rate at which live neurons fire (about a 100 times a second) over a period of 100 years, the automaton would be able to change state approximately 4×10^{11} times. Now, as an automaton with n state variables has 2^n states, and 4×10^{11} is approximately 2^{36}, it would take over a hundred years for a modest automaton with 36 neurons to visit all its states. Of course the nature of neural state machines in ACT is such that only a small subset of all states is used to represent world events. Despite this, the key point is that ACT can represent mental states in a very rich way which should be contrasted with models in traditional artificial intelligence that are highly parsimonious. Also, the richness allows for complex functions to be built up from absorbing a large number of events, which provides an explanation of how it is that a mass of brain cells can learn to behave in ways that are modelled by the parsimonious, symbolic representations created by a programmer's efforts in traditional artificial intelligence.

3 Conclusion: philosophical critique

ACT appears to clash specifically with Penrose's contention[2] that consciousness is such an important phenomenon that it is hard for anyone to believe that it is "something just accidentally conjured up by computation". While ACT is undoubtedly couched in terms of a computational framework, there is no appeal to an accidental emergence of consciousness. Indeed, the main aim of the theory is to show that the complex mixture of properties normally attributed to a conscious organism are the properties associated with some computing structures and may be described through appropriate formalisms. Part of the problem lies with the word "computer" as this is taken by Penrose[4] to mean "machine that operates according to rules designed by a programmer". But that is only one interpretation of computation. Here the alternative of "neural computation" forms the basis of a theory in which the programmer merely plays the role of creating machinery that is capable of absorbing and representing its environment. The programmer does not feature in the theory itself which is about the state structures that the computational organism creates for itself or through the aid of other conscious organisms. Therefore while it is possible to agree that consciousness cannot be captured by a programmer's recipe (algorithm), the door should at least be kept open for computational models of consciousness based on systems that are capable of building up their own processing structures.

At the other end of the spectrum of belief in computational models, Dennett's explanation of consciousness[1] relies very heavily on elements of computer engineering. The theory is expressed in terms of consciousness being the product of a multi-agent program running on a virtual machine where the autonomous "bossless" agents are continually reassessing sensory information ("revising drafts"). In comparison with ACT, this outlook appears strongly metaphorical. All the concepts are metaphors drawn from computer engineering, but as metaphors they do not necessarily constitute a theory. While the Cartesian ghost in the machine has been expunged, the ghost of the programmer is still there, and this does little to explain how the machine components come into being and do what they do. This is left to a briefly stated appeal to evolution. There is agreement between Dennett's description and ACT in what regards language being held in an evolving societal repository - an idea borrowed from Dawkins[10].

Evolutionary models of consciousness such as those of Edelman[11] abound in the literature. The mechanics of the brain are seen as being the result of a very finely tuned evolutionary process which requires an accelerated evolution of neural structures during the course of life. This contrasts with ACT in the sense that learning in a neural state automaton does not demand the physical changes associated with lifetime evolution. The plasticity and generalization of the neurons is sufficient. An evolutionary standpoint is also part of the work of Humphrey[12], who sees conscious representations as evolving from the internalization of instinctive reactions to the environment. This seems an unlikely process as such internalizations have subsequently to be externalized in order to develop a sense of self and to use language as indicated in corollaries 6-12 of ACT.

The philosophical standpoint which does not clash with ACT is that of Searle[3] as, at least, it leaves open the possibility of finding the causal nature of the brain (i.e. causing consciousness through neural activity) and the subjective nature of consciousness (intentional relationship with the environment) in materials other than biological brains as we know them. While he goes on to believe that it is most unlikely that such alternative chemistries could be found, his philosophy does not exclude a theoretical analysis of alternatives as being of relevance to an understanding of biological consciousness. ACT is precisely such an analysis. It is based on an abstraction of what is understood to be the function of the brain as it attempts to capture both the causal (fundamental postulate) and subjective (corollaries 6-12) character of consciousness. Even if some artificial system built on ACT principles were to leave skeptics unsatisfied about it "actually being conscious", it is likely that this judgement would be based on the belief that only biological systems could be conscious. This would not reflect on the appropriateness or otherwise of ACT which addresses what it is to be conscious whether this is said of organisms that are biological or not.

Contemporary debate on consciousness has a much wider participation than that of the authors quoted so far. For example, Nagel's suggestion that it is necessary to say what *it is like to be* a particular conscious organism[13], can, in ACT be expressed in terms of a taxonomy of state structures (i.e. how does the state structure of a bat differ from that of a human?). Also the fundamental postulate in ACT contradicts McGinn's contention[14] that the link between introspective sensation and the function of the brain cannot be understood as a scientific object . It has been shown that the alternative view leads to a comprehensive theory which should be evaluated against the pessimistic result McGinn's proposal that consciousness cannot be understood through logical enquiry. ACT creates an opprotunity for a much longer exposition that includes the views and nuances of the many who have expressed opinions on the nature of consciousness. All that has been done here is to draw attention to the fact that this can be done in a principled way.

Appendix I: Generalization

Corollary 7 has asked for an existence proof for a generalization which causes an input pattern where one part is the same as a training input and the others are arbitrary to retrieve the appropriate response.

The existence proof of such a generalization makes use of a *similarity measure* $\sigma(a,b)$ that can be used to compare any two patterns a and b. This is a value which is at a maximum when the two patterns are identical and a minimum when the patterns are the inverse of one another. The measure could be expected to be half way between maximum and minimum for two arbitrarily chosen patterns. Also when pattern a is made of patterns k and l, i.e. $a=(k \cdot l)$ and $b=(m \cdot n)$
then $\sigma(a \cdot b) = \sigma(k \cdot m) + \sigma(l \cdot n)$ In neural technology, the negative of Hamming distance - the difference between two patterns measured in number of bits - is used as such a measure, and has the required properties, which indicates an existence of the assumed measure. Hamming distance also seems a good measure to describe similarity in living neural systems.

Assertion 1: Ideal generalization is expected to yield q_w in response to neural state machine input/state pair (i_x, q_x) if (i_w, q_w) is the most similar pattern (among a training set of such pairs) to (i_x, q_x) and $q_w = \beta (i_w, q_w)$.

Referring to the situation in corollary 7 where the input is a mixed pattern of the form $(i_v \cdot i_u)$, contradiction of assertion 1 would require that there be some $(i_x \cdot i_y, q_z)$ that is more similar to $(i_v \cdot i_u, q_w)$ than is $(i_v \cdot i_\emptyset, q_\emptyset)$. Remembering that $q_w = (q_v, q_u)$this would lead to the erroneous retrieval of q_z in place of the correct retrieval q_w. But this would require that

$$\sigma(i_x, i_v) + \sigma(i_y, i_\emptyset) + \sigma(q_z, q_\emptyset) > \sigma(i_v, i_v) + \sigma(i_u, i_\emptyset) + \sigma(q_w, q_\emptyset)$$

As it can be expected that $\sigma(i_y, i_\emptyset) \approx \sigma(q_z, q_\emptyset) \approx \sigma(i_u, i_\emptyset) \approx \sigma(q_w, q_\emptyset)$ ($\approx$ is used to indicate "roughly the same as") .The above inequality implies that $\sigma(i_x, i_v) > \sigma(i_v, i_v)$ which cannot be as $\sigma(i_v, i_v)$ is maximum. It is noted that this is an expectation and not a guarantee, based on the expectation of the similarity of two arbitrarily chosen patterns.

References

1. Dennett, D. C. *Consciousness Explained* (Allan Lane/Penguin, London, 1991).
2. Penrose, R. *The Emperor's New Mind* (Oxford U. Press, 1989)
3. Searle, J.R. *The rediscovery of the mind.* (MIT Press, Boston,1992).
4. Penrose, R. *The Shadows of the Mind* (Oxford U. Press, 1994)
5. Kelly, G. A. *The Psychology of Personal Constructs* (Norton, New York, 1955)
6. Aleksander, I. and Morton, H.B.*Neurons and Symbols: the Stuff that Mind is Made of.* (Chapman and Hall, London, 1993).
7. Searle, J.R. *Behavioural and Brain Sciences* 3, 417-424, 1980.
8. Chomsky, N. *Language and Mind* (Harcourt Brace Jovanovich, 1972)
9. Parfitt, S.H. *Proc. ICANN94*, 188-194 (1994)
10. Dawkins, R. *The Selfish Gene* (Oxford University Press, Oxford, 1976)
11. Edelman, G. M. *Bright Air, Brilliant Fire.* (Allen Lane, London, 1992)
12. Humphrey, N. *A History of Mind* (Vintage, London, 1993)
13. Nagel, T *The View from Nowhere* (Oxford University Press, Oxford, 1982)
14. McGinn, C. *The Problem of Consciousness* (Basil Blackwell, Oxford, 1991)
15. Aleksander I., Towards a Neural Model of Consciousness, *Proc ICANN 94*, Springer (1994)

Physical and Linguistic Problems in the Modelling of Consciousness by Neural Networks

Philippe De Wilde

Imperial College of Science, Technology and Medicine
Department of Electrical and Electronic Engineering
London SW7 2BT, United Kingdom

Abstract

We consider consciousness as a large scale phenomenon that can be reduced to small scale interactions between neurons, and hence to quantum mechanics. We show that this has consequences for a reductionist approach in modelling consciousness. An alternative to this is a metaphorical approach. We draw a parallel with the alchemists language in describing his experiments. Finally, we investigate the formalist viewpoint that consciousness is modelled and described by its use in language.

1 Is a model of attention an element of reality?

Ever since Niels Bohr advanced the principle of complementarity [3], it has been debated whether quantum mechanics provides a model of reality, or merely correlates experimental results. Most scientists adhere to the latter, formalist, viewpoint. This so-called Copenhagen interpretation states that quantum mechanics provides a *language* to describe and predict observations. This language has quantum numbers and wave functions as variables, and laws as primitive entities. It forms a subset of English. Not everybody knows this subset, but it is nevertheless a recognized part of the English language.

Quantum mechanics is a definitive theory. There are no hidden variables [9]. The behaviour of systems of any complexity can, in theory, be accurately described and predicted by quantum mechanics. This is the case for ions, simple and complex organic molecules, and biological cells. Hence, quantum mechanics *does* play a role in the description of macroscopic systems.

Everybody who accepts the implications of the Copenhagen interpretation for macroscopic systems experiences sooner or later an uneasiness, as in the paradox of Schrödingers cat. It is difficult to accept that an assertion about reality cannot be claimed true until an observation has taken place. This problem is related to epistemology, see [5].

Einstein, Podolsky, and Rosen [8] put forward as an essential requirement for any physical theory that *every element of the physical reality must have a counterpart in the physical theory*. Intuitively, this sounds right. However, it is well known that this gives rise to paradoxes in quantum mechanics, e.g. non-locality. For this reason, it makes no sense to *believe* that an electron is an element of physical reality. It is characterised by its quantum numbers, and the theory predicts outcomes of measurements of these numbers. This is as far as modelling can go. A model cannot decide whether something is an element of physical reality or not. These consequences of quantum mechanics are not confined to sub-atomic phenomena, but are also playing a role in more complex systems, e.g. in the brain.

Any model of consciousness, if it has to coexist with modern science, has to stand attempts at reduction. It either reduces correctly to physics, or it contains irreducible parts that do not contradict physics.

We will discuss the reduction of higher brain functions in Section 3. Here we consider *attention* as an intermediate level brain function. It can be reduced to the level of biological neurons [11, 14]. Via the dendrites this can be further reduced to microscopic quantum mechanical systems [6, 7].

This leads to the conclusion of this Section. Because attention can be reduced to quantum mechanics, it cannot be considered an element of physical reality when it is modelled. This is contrary to most "beliefs", but has to be accepted in order to escape quantum mechanical paradoxes.

2 Modelling as alchemy

Many thinkers reject a reductionist approach for modeling consciousness. An example of this is the maxim that mind is more that its constituent matter. Consciousness is seen as a metaphor for emergent properties of some system. A computer program, for example, may be made more and more complex. At a certain stage, some of its properties can be called consciousness [10]. This involves two levels, the computer programme on the one hand, and a human observer on the other.

If the computer programme is complex, the human observer may *understand* individual small parts of it, but not the whole. This is often the case with artificial neural networks and cellular automata. They are easy to scale by using more artificial neurons or more cells, so it is easy to make complex systems. These systems can then show emergent properties.

This problem of understanding emerging properties in complex systems has appeared before, namely in alchemy [13]. The alchemist was working with chemical substances. He could not reduce the reactions he observed to chemical reactions. He had no molecular model to serve as ground for the language he was using to describe his observations.

The alchemist developed an anthropomorphic language for chemical reactions. The hermetic alchemist literature [13] uses the word healing for the change of lead into gold, and a metal was said to be killed, then revived. Hot and cold were male and female. Dilution was baptism, gas was spirit (anima). King and queen had a chemical marriage, a retort was a uterus where the philosophers son was born. Fire was gaining insight (exaltatio intellectus). The black work (nigredo) was the meditation.

The work of the alchemist had become a metaphor for gaining insight and for the awakening of consciousness. The search for gold was secondary in the hermetic tradition.

All this is happening again. On a simple level, it can be an automaton or a programme that does not accept a particular input. The user says and thinks "he doesn't want it". For a complex system, e.g. Conway's "game of life" cellular automaton [2], the user can observe and describe the "aggression" of certain moving patterns. The user sees a more and more complex virtual reality and describes it using ever more abstract terms. He does not reduce the phenomena in terms of commands in a programming language any more, but uses complex words. Consciousness, as well as doggedness can be ascribed to a computer.

An observer who ascribes consciousness to a computer programme uses language in the same way as the alchemists did.

3 Consciousness irreducible

How is it possible to speak about consciousness, as it is not part of reality (Section 1), and without sounding as an alchemist (Section 2)?

Philosophers can speak about consciousness assuming that it is a primitive concept, as point and line are in Euclides system of geometry. This is not possible for the scientist who *models* consciousness.

The scientific model is built up gradually. Observations of the neuronal pathways from the retina in the eye to parts of the brain lead to models of attention and visual processing. These models are quite well established by now. Researchers try to model different aspects of perception, memory, control, group behaviour, and recognition. These are concepts on a higher level than attention. Also, they cannot be reduced to properties of individual neurons, because of the large numbers of neurons involved. Any calculation that is more than just schematic would be practically impossible.

The interactions between the neurons are on a *microscopic level*. Phenomena as attention and memory are manifestations on a *macroscopic* level of the microscopic interactions between the neurons. The macroscopic phenomena are emergent abilities of the neural networks in the sense of [12]. Some of the brain's emergent abilities are, in increasing order of complexity, muscle coordination, attention, memory, planning, intelligence, language, understanding, consciousness.

In simulating some of these macroscopic phenomena, prototypes of networks with a small number of neurons and a typical connectivity, modelled on part of the brain, are used. Some models of the thalamus or the cerebellum use only a few thousand neurons. It is here that an important phenomenon occurs in the description of the simulations. Emergent properties of the model are described in terms from cognitive psychology. There is no reduction to biological neurons possible. The language used in describing the observations has broken away from biological reality. A shift of language [16] has occurred, from biology to cognitive psychology.

Another view [1] holds that, if consciousness can be talked of in formal terms, it could be modelled as a macroscopic phenomenon of a mathematical automaton. Such an automaton exhibits transitions between states. This is a path in the middle. It does not reduce consciousness to neurons and thence to physics. It explains conscious behaviour in terms of some other "lower" concepts. There is a language shift here, too. It is not from biology to psychology, but from mathematics to psychology. Similar shifts are those from mathematical catastrophe theory to human evolution [17], and from thermodynamics to behaviourism [16].

Still another view on consciousness is advanced by Searle [15]. Consciousness is defined as what happens when awakening after a sleep, when becoming aware of yourself and the world. Via a subtle and well-founded argument, he decides that consciousness can't be reduced, because it is all *appearance*. So, consciousness is like pain, while in the same neuronal state, one person will have pain, another will not. As Wittgenstein has shown for pain, it is important how one talks *about* it. This will be important for consciousness, too.

All these views point to consciousness as an important class in language.

4 Consciousness as a part of language

Consciousness seems to be elusive. If it can be reduced, it is not an element of physical reality, and if it cannot be reduced, it is only an appearance. Why then do most people believe consciousness is real, as folk psychology points out? Why do most people adhere to a "folk materialism"?

As the language used in describing consciousness has become uncoupled from biology, how can we prevent it from becoming an alchemists language? By being open for the dynamics of language itself. Language is not only driven by observations, but has a set of internal rules that allow it to develop on its own [18]. Language has its own dynamics, independent of physical reality. Aristotelian physics, for example, often contradicts the experiments, but still offers a consistent philosophical system.

These internal rules are simple if one describes them without the use of linguistic categories. Texts consist of fragments. Fragments can be words, part of a sentence, a sentence, several sentences, several sentences and part of sentences, parts of an encyclopedia, a number of books in several libraries, etc. Fragments can also be part of a word.

Written language is only the shadow of spoken language. Some written language is the shadow of the language of thoughts. This will be specified later. Spoken language also has its fragments, they are phonemes, words, sentences, speeches.

Spoken language is only one of the languages in which we express ourselves. There is also body language, sign language, breathing, gym, driving a car, and typography, hypertext, etc. Our *acts* too form a language, e.g. belonging to a party, a religion, a bird-watchers group, a weight-watchers group. All these languages have their fragments. The fragments can overlap and can include each other. They can form hierarchies and long serial lists.

Writing is combining fragments. They can be concatenated or placed in a hierarchy, as in hypertext. They can genetically influence each other, for example when inferring the past tense of a new verb from a list of past tenses. Expressing oneself in one of the other languages works in the same way.

We can even consider a language of thought. Its fragments are the chemical reactions in the brain, the states of the neurons, the states of large groups of neurons just before some physical act is performed. After this, we have fragments of body language or of writing, if the arm is moving while holding a pen.

Some combinations are causal, e.g. chemical reactions, action and reaction of the body, or words combined according to a grammatical rule. Other combinations are random. Some remain at a low frequency for no obvious reason to their users, for example poetry.

Are there any rules in these languages composed of fragments? On the lowest level, the usage of words is governed by their frequency in the language [4]. More structure in language can be explained and predicted from Markov models of orders 1, 2, 3, etc. Additional structure can be brought in via laws of logic, as syllogisms, and grammar. Sometimes the rules can't be specified, because our thinking, the ultimate cause of any linguistic expression, is not linked to language. In that case one can talk about intuition or the subconscious.

For this reason, if there are any emergent abilities in language, it is *structure* that is emergent. Rules are not necessarily structure, they can be just statistics, frequent occurrences of a pattern. This is consistent with learning by example. If rules are statistics, they can not be used to ground language in physical reality.

The words for higher brain functions, as memory, anxiety etc., that can't practically be reduced to a lower, biological level because of complexity, are generated inside a language. One should not believe, as in alchemy, that they are legitimated or grounded by the underlying physical matter. They are only fragments of a language.

The conclusion is that it is not possible to understand consciousness, or make predictions about it, but that we can only show how people *talk* about consciousness.

A model of consciousness has to be built up from simple words, well defined by their common usage. The complexity of the model then arises from the complexity of the language game played with the basic terms. This is essentially the programme that the later Wittgenstein followed when he discussed psychology, perception of colours, mathematical concepts, sense data, private language, etc. The scientist should clean his language from such terms as "qualia", "intentionality", "mind", "dream", "personality", and build up a model of consciousness from simple terms and their combinations. Their meaning derives not from a grounding in physical reality nor from a ritual use, but from practice. This programme may take long, but such is the nature of an organic growth of meaning.

The author would like to thank Dr. V. I. Kryukov for discussions about modelling attention and higher brain functions, and Prof. I. Aleksander for interesting remarks.

References

[1] Igor Aleksander and Helen Morton. *Neurons and Symbols*. Chapman and Hall, London, 1993.

[2] Elwyn R. Berlekamp and John H. Conway. *Winning Ways for your Mathematical Plays*, volume I&II. Academic Press, London, 1982.

[3] Niels Bohr. Can quantum-mechanical description of physical reality be considered complete? *Physical Review*, 48:696–702, 1935.

[4] Colin Cherry. *On Human Communication*. MIT Press, Cambridge, Massashusetts, 1982.

[5] Bernard d'Espagnat. *Une Incertaine Réalité*. Gauthier-Villars, Paris, 1985.

[6] John C. Eccles. A unitary hypothesis of mind brain interaction in the cerebral cortex. *Proceedings of the Royal Society of London Series B - Biological Sciences*, 240(1299):433–451, 1990.

[7] John C. Eccles. Evolution of consciousness. *Proceedings of the National Academy of Sciences of the USA*, 89(16):7320–7324, 1992.

[8] Albert Einstein, Boris Podolsky, and Nathan Rosen. Can quantum-mechanical description of physical reality be considered complete? *Physical Review*, 47:777–780, 1935.

[9] Stuart J. Freedman and John F. Clauser. Experimental tests of local hidden-variable theories. *Physical Review Letters*, 28:938–941, 1972.

[10] John Haugeland, editor. *Mind Design*, Cambridge, Massachusetts, 1987. MIT Press.

[11] Arun V. Holden and Vitaly I. Kryukov, editors. *Neurocomputers and Attention*, Manchester, 1991. Manchester University Press.

[12] John J. Hopfield. Neural networks and physical systems with emergent collective computational abilities. *Proceedings of the National Academy of Sciences of the USA*, 79:2554–2558, April 1982.

[13] Carl Gustav Jung. *Erlösungsvorstellungen in der Alchemie*, volume 6 of *Grundwei C. G. Jung*. Walter-Verlag, Freiburg im Breisgau, 1985.

[14] V. I. Kryukov, G. N. Borisyuk, R. M. Borisyuk, A. B. Kirillov, and E. I. Kovalenko. *The Metastable and Unstable States in the Brain (in Russian)*. Academy of Sciences of the USSR, Pushchino, 1986.

[15] John R. Searle. *The Rediscovery of the Mind*. MIT Press, Cambridge, Massachusetts, 1992.

[16] Isabelle Stengers. *D'une science à l'autre, des concepts nomades*. Seuil, Paris, 1987.

[17] René Thom. *Stabilité Structurelle et Morphogénèse*. InterEditions, Paris, 1977.

[18] Ludwig Wittgenstein. *Philosophische Untersuchungen*, volume 1 of *Ludwig Wittgenstein Werkausgabe*. Suhrkamp, Frankfurt am Main, 1984.

A Neural Network Model for the Velocity Vector of an Object and Its Consistency with Psychological Phenomena

Ken-ichiro Miura and Takashi Nagano

College of Engineering, Hosei University
3-7-2, Kajino-cho, Koganei, Tokyo, 184, JAPAN

Abstract

A biologically plausible neural network model is proposed that can detect the actual velocity vector of an object by using local motion signals detected by many local motion detectors. First, the computational theory to find the actual velocity vector of a rigid object correctly is described. Then a neural network is shown that implements the theory. The neural network model is constructed by two layers: a local velocity vector extraction layer and an integration layer. Many velocity vectors of points on a moving object are detected by local detectors in the local velocity vector extraction layer. Then these local velocity vectors are integrated in integration layer in order to obtain the actual velocity vector of the object. The computational processing pathway of the proposed model is well fit to the motion processing path of visual stimulus in the actual nervous system (from LGN to MT via V1). We also show the model can explain motion perceptual phenomena in the case where a moving stimulus is observed through an aperture.

1. Introduction

Neurobiological findings tells us that early stages in motion detection process are performed by cells with local receptive fields in, for example, V1 and MT areas in the visual cortex[1][2][3]. It is well known that velocity vector detected through a local window (or a receptive field) is usually different from the actual velocity vector of an object. However we can perceive the actual velocity vector of an object as a whole accurately despite that early motion signals from the object are detected by detectors with local receptive fields. Therefore, we must have the mechanism in our brains that integrates various signals detected by many local motion detectors to obtain the velocity vector of an object.

In this paper, we first propose a computational theory that gives a way of computing the velocity vector of an object by using many locally detected velocity vectors of the object under the assumption that the object is rigid and that all the possible velocity vectors can be detected. Then, a biologically plausible neural network model that implements the theory is proposed. The proposed model is constructed with two layers: the local velocity vector extraction layer and the integration layer. One of the authors previously proposed a model that can detect the motion of a point stimulus[4]. The model has the selectivity for the velocity vector of a moving point stimulus. In addition, the response properties of the model are well fit to some neurobiological and psychophysiological findings. The models are used as local motion detectors in our model. They are allocated at all the position in two dimensional visual field to detect all the possible velocity vectors of every point on an object. Finally, the local velocity vectors detected are integrated in order to obtain the velocity vector of an object. In addition, we explain some perceptual phenomena[5] in the case where a moving object is observed through an aperture.

2. Velocity vector of an object

In the section, we propose a computational theory for the velocity vector of a rigid object and a neural network model that implements the computation. If we can know which point on a rigid object at time $t_1+\delta t$ corresponds to each point on the object at time t_1, the velocity vector of the object can be easily obtained. However, a difficult problem called "the correspondence problem" have to be solved with several restrictions and iterative computation in order to know the corresponding point[6]. Detecting of the velocity vector of an object is done very quickly and therefore is thought to be done with simple computation at an early stage of visual processing. Therefore, it is hard to imagine that such complex computation is performed in the actual visual system. Therefore, it is necessary to develop the method to obtain the actual velocity vector of a rigid object without solving the correspondence problem. A computational theory satisfying this requirement is described in the following.

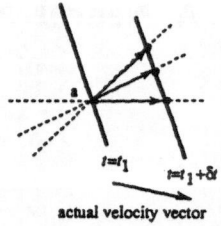

actual velocity vector

figure1. Possible velocity vectors of a point in the object

2.1 Computational theory for the velocity vector of a rigid object

Here, we develop a theory for the velocity vector of an object on the assumption that the object is rigid in the two dimensional space. Any other assumptions, for example, the shape of an object are not made. For simplicity, a two dimensional image is represented by an array of binary signals: 1 and 0. Consider the case where a rigid object shown in figure1 translates in the two dimensional space. A point on the object at time t_1 may correspond to all of the points on the object at $t_1+\delta t$ if motion is detected locally. Therefore, the velocity vector of every point on the object can not be decided uniquely. We consider the case where basic local motion processing units to detect the direction and velocity of each point are allocated at every position in the two dimensional space. Each unit responds when a point moves with a specific velocity vector. When the object shown in figure1 moves from time t_1 to $t_1+\delta t$, velocity vectors from each point of the object at time t_1 to each point at time $t_1+\delta t$ are detected. If the object is constructed with n points, each point on the object has n different velocity vectors. Since the moving object is rigid, the set of n different velocity vectors detected at each point on the object at t_1 necessarily contains the correct velocity vector given by the correct corresponding point which shows the actual velocity vector of the object. All the other (n-1) velocity vectors are spurious. When the number of detected velocity vectors is counted for each velocity vector, the number of the actual velocity vector is equal to n, the number of points constructing the object. It is easily understood that the numbers of the spurious velocity vectors are less than that of the actual velocity vector. Therefore, the velocity vector with the maximum number gives the actual velocity vector of a rigid object. This fact does not depend on the shape of a rigid object. The number of a detected velocity vector, called "the amount of a velocity vector" hereafter, is equal to the area of the overlapped part of the image of the object at t_1 and that given by translating the image at time $t_1+\delta t$ by $-V_i \cdot \delta t$, where V_i denotes a velocity vector. Let $f(p)$ and $g(p)$ denotes the images at time t_1 and $t_1+\delta t$ respectively and that $f(p)$ and $g(p)$ have 1 when p gives the position where an object exists and 0 otherwise. Because of the assumption of the rigidity of an object, the relation between $f(p)$ and $g(p)$ is described by the following equation.

$$f(\mathbf{p}) = g(\mathbf{p} + \mathbf{V}_r \cdot \delta t) \qquad (1)$$

Here, V_r denotes the actual velocity vector of a rigid object. The area of the overlapped part, which shows the amount of a velocity vector V_i, $N(V_i)$ is given by the following equation.

$$N(\mathbf{V}_i) = \int_\mathbf{p} f(\mathbf{p}) \cdot g(\mathbf{p} + \mathbf{V}_i \cdot \delta t)) d\mathbf{p} \qquad (2)$$

$N(V_i)$ has the range from 0 to the area of a rigid object because $f(p)$ and $g(p)$ are represented by 0 or 1. $N(V_i)$ has the maximum value when $f(p)$ equals to $g(p+V_i \delta t)$. Therefore, the following relation can be obtained.

$$N(\mathbf{V}_r) > N(\mathbf{V}_i) \qquad (\mathbf{V}_i \neq \mathbf{V}_r) \qquad (3)$$

Therefore, the actual velocity vector of a rigid object is given by the following equation irrespective of the shape of a rigid object.

$$\mathbf{V}_r = \arg\left(\max_i \left(N(\mathbf{V}_i) \right) \right) \qquad (4)$$

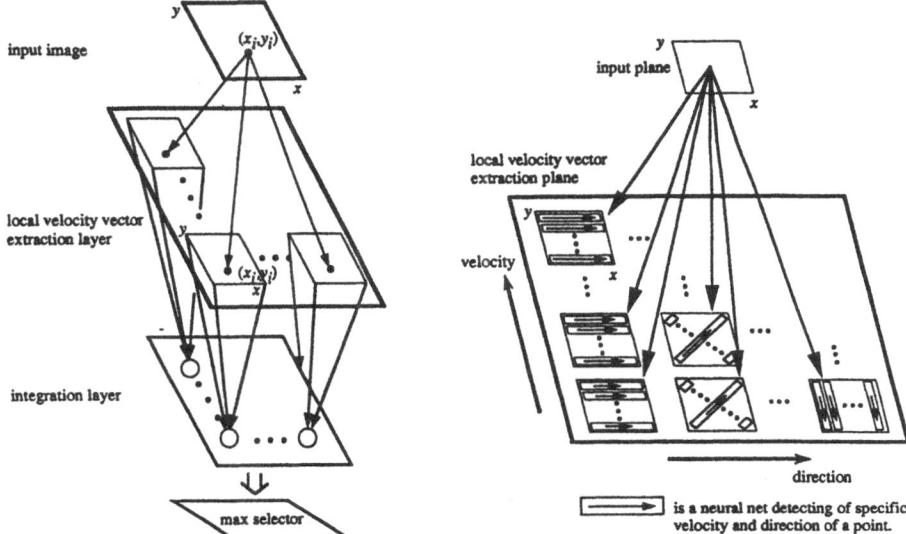

figure2. Schematic diagram of the model   figure3. Local velocity vector extraction layer

2.2 Neural network model

In the section, a neural network model is proposed that can compute the actual velocity vector of an object by using the proposed theory. The schematic diagram of the proposed model is shown in figure2. The model is constructed with two layers: "local velocity vector extraction layer" and "integration layer". The former layer extracts all the possible local velocity vectors of all the points on an object. The latter layer integrates those local velocity vectors in order to obtain the velocity vector of a rigid object.

2.2.1 Local velocity vector extraction layer

The layer is composed of many basic module, each of which responds selectively to a specific velocity vector when a point stimulus moves across its receptive field. From among various velocity vector selective models proposed up to now[4][7][8], we chose the model proposed by Hirahara and Nagano[4] as a basic module to construct the layer because its response properties are well fit to neurobiological and psychophysical findings. The layer is constructed by allocating a set of basic modules each of which responds selectively to each velocity vector at every position in the two dimensional space. The schematic diagram of the layer is shown in figure3. Each square in the large square corresponds to each spatiotemporal velocity vector extraction plane, in which many basic modules with selectivity the same velocity vector are allocated for all the positions corresponding to those of the input plane. In the layer, all the possible local velocity vectors of each point on an object are extracted by basic modules.

2.2.2 Integration layer

In this layer, local velocity vectors are integrated to obtain the actual velocity vector an object. Cells in this layer perform the computation described in section2.1. Each of the cells in the integration layer corresponds to each velocity vector and receives the outputs of all the basic modules existing in each extraction plane of the local velocity vector extraction layer. The outputs $I(V_i)$ of cells in the layer are defined by the following equation.

$$I_{V_i} = \sum_{x,y} O_{V_i, xy} \qquad (5)$$

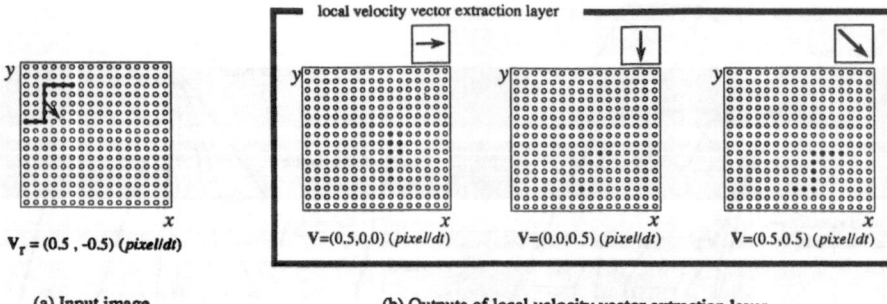

(a) Input image   (b) Outputs of local velocity vector extraction layer

$V_r = (0.5, -0.5)$ (*pixel/dt*)

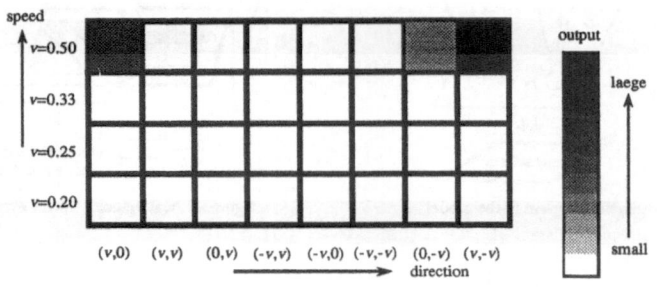

(c) Outputs of integration layer

figure4. result of numerical simulation

where, $O_{V_{ixy}}$ denotes the output of basic module selective to velocity vector V_i at the position (x,y). The cell which gives the maximum output in this layer shows the actual velocity vector of a rigid object. The cell can be detected by the maximum selector.

2.3 Numerical simulations

Some numerical simulations are conducted on a computer to confirm that the behavior of the proposed model is the expected one. In all the simulations, we assume that input images are represented by arrays of binary signals: 0 or 1 and that movements of objects in input images are only translation. 32 (8 directions and 4 speeds) kinds of basic modules are allocated for every position on the two dimensional space in the local velocity vector extraction layer. One of the results is shown here. The input stimulus is shown in figure4(a). In the simulation the object moves with a velocity vector $V_r=(0.5,-0.5)$ *pixel/&t*. *&t* denotes are clock time in simulated movement. Circles and an allow in the figure show pixels of the image and the actual velocity vector of the object respectively. The outputs of local velocity vector extraction layer after some clocks in figure4(b). Although the model has 32 velocity vector extraction planes, the output patterns of planes which do not have any activated calls are omitted in the figure. From the result we can see the output pattern of the plane extracting V_r is the same as the input pattern. The outputs of cells in the integration layer are shown in figure 4(c) . The result shows that the cell corresponding to V_r gives the largest output. This means the model can detect the actual velocity vector of the object. We conducted some other simulations using the other types of objects and confirmed that the model could detect the actual velocity vectors of objects in all the simulations.

3. Improvement of the computational theory
3.1 Psychophysiological phenomena

It is well known that Wallach H investigated the perceived direction of movement when a moving object was observed

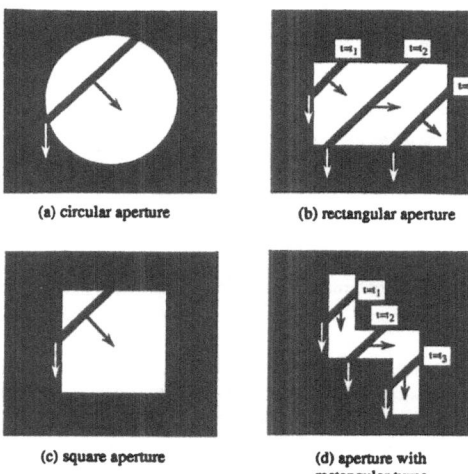

(a) circular aperture

(b) rectangular aperture

(c) square aperture

(d) aperture with
rectangular turns

⟶ and ⟹ denote a perceived
direction and an actual direcion.

figure5. Perceived direction in observations through various apertures

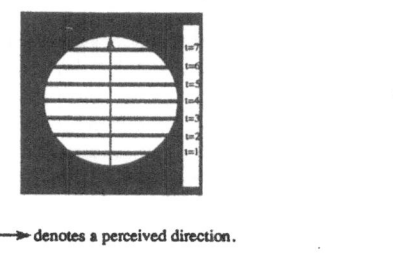

⟶ denotes a perceived direction.

figure6. Moving bar observed through
a circular aperture

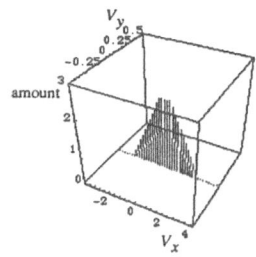

figure7. Amounts of velocity vectors in observations
through a circular aperture

through various apertures[5]. According to the results of his experiments, the perceived direction of object movement depends on the shape of the aperture through which the object is observed. Some of his experimental results are shown in figure 5. These examples show the perceived direction when an oblique bar moving downward is observed through various apertures. In the case(a), for example, in which an aperture is a circle, the perceived direction is orthogonal to the orientation of the bar.

In 2.1, we have described the computational theory that gives the actual velocity vector of a rigid object on the assumption that the whole object is visible. In this section, we will show that some psychological phenomena obtained by observing an object through an aperture can be explained by the theory described in 2.1. We will consider the two cases as example in which a moving bar is observed through a circular (a) and a rectangular (b) apertures.

A moving bar observed through a circular aperture at several clocks is shown in figure6. Here, the orientation of the bar is horizontal. First the bar is perceived to moves upward magnifying its length quickly. When the bar arrives around the center of the circular aperture, it moves upward almost without changing its length. Then it moves to the same direction reducing its length quickly until it disappears. Consider the case where the bar magnifies its length quickly. More than two velocity vectors have the maximum amount given by equation(2). We consider the direction range of velocity vectors whose amounts are the maximum. At the time when the bar touches the circular aperture, the range is from 0 deg to 180 deg. As the stimulus moves on, the range gradually becomes narrow. When it reaches the center of the circular aperture, the velocity vector that has the

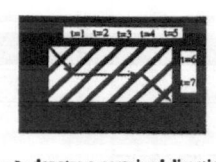

—→ denotes a perceived direction.

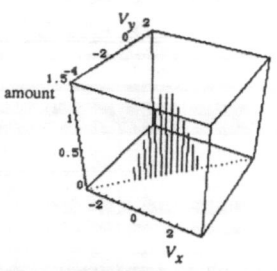

figure8. Moving bar observed through
a rectangular aperture

figure9. Amounts of velocity vectors in observations
through a rectangular aperture

maximum amount is only one since the bar can be regarded as rigid. After the stimulus passes through the center of the aperture, the range gradually becomes broad again. The speed range of velocity vectors changes the same as the direction range. The amounts of each velocity vector at $t=2$ are shown in figure7. The bottom plane of the box and each vertical bar in the figure denotes the velocity vector space and the amount of each velocity vector respectively. The theory described in 2.1 can explain movement perception only when the bar moves on near the center of the aperture because the bar can be assumed rigid at the time, but can not do in the other cases.

As an another example, we consider the case in which a bar is observed through a rectangle aperture. figure8 shows the case in which the orientation of the bar is 45 deg. In this case, three kinds of velocity vectors can be observed depending on the position of the bar. First we perceive the bar moves diagonally right-downward magnifying its length until the lower endpoint of the bar reaches the left bottom corner. Next, rightward movement is perceived until the upper endpoint reaches the right top corner. Finally the bar is perceived to move toward the same direction as the first period reducing its length. A perceived velocity vector can be explained by the theory in 2.1 in the case of the second period since the bar is rigid in the case. However, perceptual phenomena can not be explained in the first and the final periods because the visible length of the bar changes. The amounts of velocity vectors in the first period is shown in figure9. We can see that the range of velocity vectors with the maximum amount is always from 0 to 90 deg. Consequently the theory can not explain perceptual phenomena in the case where the rigidity assumption does not hold. A improved theory is described in the following that can explain those perceptual phenomena.

3.2 Improved theory

When a rigid object is observed through an aperture, the object moves changing generally the shape or size of its visible. Therefore, it can be said that movement observed through an aperture is more ambiguous than that observed without aperture. However we perceive a specific velocity vector even when a moving object is observed through an aperture. In order to explain this fact, we introduce the spatiotemporal interaction of the amounts of velocity vectors into the theory. The amount of a velocity vector at time t_1 is given by the weighted sum of the amount of itself at time t_1-δt, those of neighboring velocity vectors at time t_1-δt and the total of outputs of basic modules selective for the velocity vector at time t_1. By this interaction of the amounts of velocity vectors, the range of velocity vectors with the maximum amount gets narrower. As the interaction among the amounts of velocity vectors is propagated to the computation at the next clock by temporal interaction, the range become narrower gradually. Consequently only one velocity vector has the maximum amount after sufficient interactions.

This computation can be implemented by adding excitatory self-recurrent and mutual connections to the integration layer of our model. The more similar velocity vectors corresponding to cells are, the larger the connection weights between them become. The self-recurrent connection has, therefore, the largest weight. The schematic diagram of the improved model is shown in figure10. The output function of a cell in the integration layer is given by the following equation.

$$I_{V_i}(t) = \sum_j M_{ij} \cdot I_{V_j}(t - \delta t) + W \cdot \sum_{x,y} O_{V_i xy}(t) \qquad (6)$$

where, M_{ij} and denotes the connection weight between cells corresponding to V_i and V_j and W is denotes the connection

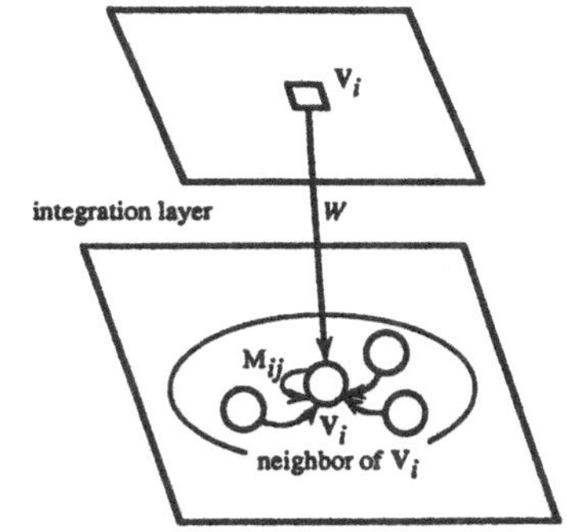

local velocity vector extractionlayer

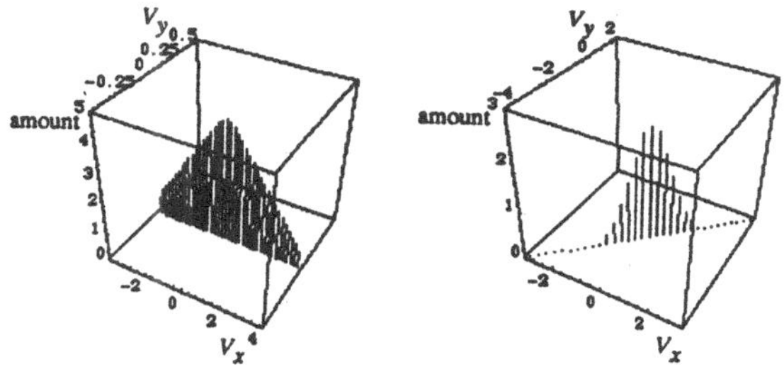

(a) circular aperture

(b) rectangular aperture

figure10. Schematic diagram of the improved model

figure11. Amounts of velocity vectors
given by improved theory

weight from the output cell in the local extraction layer to the cell in the integration layer. The amounts of velocity vectors given by equation(6) are shown in figure11. (a) and (b) in the figure show the amounts in the cases of observations through a circular aperture and rectangle one. In both the cases, it is understood that the model can well explain the perceptual phenomena.

4. Discussion

4.1 Comparison with other model

Marshall JA had proposed a model that can explain perceptual phenomena in the case where a moving bar is observed through a rectangular aperture[9]. He uses in his model many feature detectors selective to each length and orientation. Although it is reported that complex cells (end-stop type) have the selectivity to the length of a bar or an edge, it is hardly thought that cells with the selectivity to quite long length exist in V1 because the receptive fields of cells in the area are limited . Therefore his model does not seem to be plausible in the actual visual system. Our model, on the other hand, does not use cells with length and orientation selectivities. Therefore our model can explain the perceptual phenomena irrespective of the shape and size of a visual stimulus. In addition, it is suggested that simple motion perception might not require the information on the shape of an object because our model does not require any other selectivities than that to velocity vectors.

4.2 Correspondence to neurobiological findings

It is well known that motion informations of visual stimuli are processed mainly on the pathway from LGN to MT via V1[10]. We might consider that basic models in the local velocity vector extraction layer of the proposed model correspond to the pathway to V1 and the integration layer corresponds to MT. It is reported that response properties of most cells in MT do not depend on the shape of a visual stimulus but depend on its velocity vector[11]. Cells in the integration layer are similar to MT cells in that they respond strongly when an input stimulus moves with velocity vectors preferred by them.

4.3 Psychophysical findings

In 3. we described some of the experimental results given by Wallach H and compared the behaviors of the proposed model with these experimental results. It was confirmed that the output velocity vectors of the improved model were the same as the perceived ones in all the cases shown in figure5. When an object moves changing its own shape and/or size, the improved model needs some time to decide the velocity vector of the object. We also need some time to perceive the velocity vector of an object in such case. In this point the behavior of the improved model is said to be similar to human perception. It is necessary to investigate the interval between the time when a stimulus starts to move and the time when the velocity vector of a stimulus is perceived in the case where a stimulus moves changing its size and/or length.

Recently perceptual phenomena in the case where two moving sine gratings are observed through a circular aperture have been studied by a few researchers[12][13][14]. The phenomena are that coherent motion is perceived in the case where two

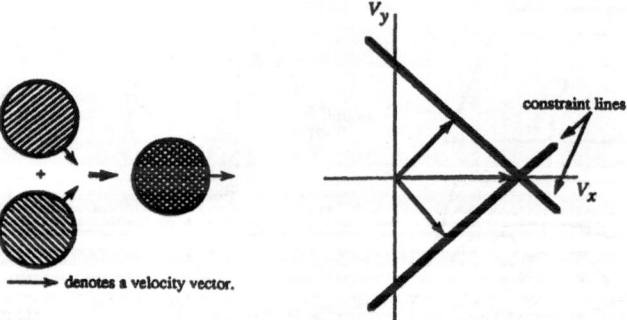

figure12. Perceptual phenomena in the case of observations of two sine gratings through a circular aperture

sine grating with similar contrast and spatial frequency move on independently in a circular aperture (figure12). It was shown that when velocity vectors of the two gratings are satisfy a certain relative condition, perceived direction of the whole visual stimulus is the same as that given by the theory called IOC (intersection of constraint) construction[12]. This theory tells that the velocity vector of the stimulus is decided by the intersection of the "constraint lines" perpendicular to the two velocity vectors of sine gratings in the velocity vector space. The proposed model has the possibility to explain these perceptual phenomena because this case may be considered as the extension of the case where a moving bar is observed through a circular aperture.

5. Conclusion

In this paper, we propose a neural network model that integrate local motion signals to obtain the actual velocity vector of an object. It is shown that the proposed model can explain some psychological phenomena on movement perception. The neural network model is simple and its process is well fit to the pathway of visual motion processing in the cortex. Therefore, it may be thought that the computational process is similar to the model exists in the actual brain.

References
[1] Allman J, Miezin F, McGuinness E (1985) Direction- and velocity-specific responses from beyond the classical receptive field in the middle temporal visual area (MT). Perception, 14, 105-126
[2] Maunsell JHR, VanEssen DC (1983) Functional properties of neurons in middle temporal visual area of the macaque monkey.I.Selectivity for stimulus direction, speed and orientation. J. neurophysiol.,49[5], 1127-1147
[3] Hubel DH, Wiesel TN (1959) Receptive fields of single neurons in the cat's striate cortex. J. Physiol. 148, 574-591
[4] Hirahara M, Nagano T (1993) A neural network model for visual motion detection that can explain psychophysical and neurophysiological phenomena. Biol.cybern., 68, 247-252
[5] Wallach H (1976) On Perception. New York: Quadrangle.
[6] Ullman S (1979) The Interpretation of Visual Motion. Cambridge, Mass.: MIT Press.
[7] Barlow HB, Levick WR (1965) The mechanism of directionally selective units in rabbit's retina. J.Physiol., 178, 549
[8] Poggio T, Reichardt W (1973) Considerations on models of movement detection. Kybernetik 13, 223-227
[9] Marshall JA, (1990) Self-Organizing Neural Network for Perception of Visual Motion., Neural Networks, 3, 45-74
[10] Maunsell JHR, Nealey TA, DePrist DD (1990) Magnocellular and parvocellular contributions to responses in the middle temporal visual area(MT) of the macaque monkey. J. Neuroscience 10, 3323-3334
[11] Zeki S (1974) Functional organization of a visual areas in the superior temporal sulcus of the rhesus monkey. J.Physiol. 236, 460-468
[12] Adelson EH, Movshon JA (1982) Phenomenal coherence of moving visual patterns. Nature, 300, 523-525
[13] Welch L (1989) The perception of moving plaids reveals two motion-processing stages. Nature, 337, 734-736
[14] Ferrera VP, Wilson HR (1987) Direction specific masking and the analysis of motion in two dimensions. Vision Res, 27, 1783-1796

Implementation and Evaluation of a Relevance Feedback Device Based on Neural Networks

Fabio Crestani
Dipartimento di Elettronica e Informatica
Universita' di Padova
I-35131 Padova - Italy

Abstract

This paper presents the results of an experimental investigation into the use of Neural Networks for implementing Relevance Feedback in an interactive Information Retrieval System. The most advance Relevance Feedback technique used in operative Interactive Information Retrieval systems, Probabilistic Relevance Feedback, is compared with a Neural Networks based technique. The latest uses the learning and generalisation capabilities of a 3-layer feedforward Neural Network with the Backpropagation learning procedure to learn distinguishing between relevant and non-relevant documents. A comparative evaluation between the two techniques is reported using an advance Information Retrieval System, a Neural Network simulator, and an IR test document collection. The results are reported and explained from an Information Retrieval point of view.

1 Information Retrieval

Information Retrieval (IR) is a science that aims at storing and allowing fast access to a large amount of information. This information can be of any kind: textual, visual, or auditory. An *Information Retrieval System* (IRS) is a computing tool which stores this information to be retrieved for future use. Most actual IR systems store and enable the retrieval of only textual information or documents. However, this is not an easy task, just to give a clue to its size, it must be noticed that often the collections of documents an IRS has to deal with contain several thousands or even millions of documents.

A user accesses the IRS by submitting a query, the IRS then tries to retrieve all documents that "satisfy" the query. As opposed to database systems, and IRS does not provide an exact answer but produce a ranking of documents that appear to contain information relevant to the query. Queries and documents are usually expressed in natural language and to be processed by the IRS they are passed through a query and a document processors which breaks them into their constituents words. Non-content-bearing words ("the", "but", "and", etc.) are discarded, and suffixes are removed, so that what remains to represent query and documents are lists of terms that can be compared using some similarity evaluation algorithms. Good

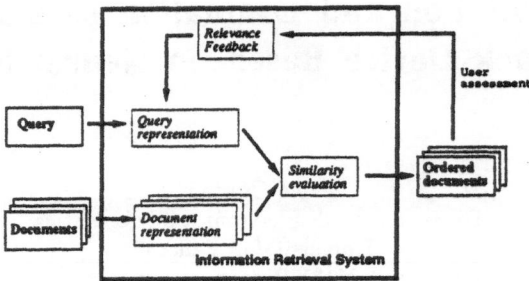

Figure 1: Schematic view of and Information Retrieval System

IR systems typically rank the matched documents so that those most likely to be relevant (those with the higher similarity with the query) are presented to the user first. An example of an IR system is depicted in Fig. 1.

Some retrieved documents will be relevant (with varying degree of relevance) and some will, unfortunately be irrelevant. In some advanced IRS, the user can appraise the documents that he considers relevant and feeds them back through a process called *Relevance Feedback* (RF). RF modifies the original query to produce a new improved query by adding additional query terms (see Fig. 2). As a consequence a new ranking of documents is produced. If the IR system is interactive this feedback process will go on until the user is happy of the resulting list of relevant documents.

In recent years big efforts have been devoted to the attempt to improve the performance of IR systems and research has explored many different directions trying to use with profits results achieved in other areas. In this paper we will investigate the possibility of using Neural Networks (NN) in IR and in particular we will concentrate on the RF process. RF is not always present in operative IR systems, but it has been widely recognised to improve considerably IR systems' performance. The purpose of this work is therefore to investigate the possibility of using NN to implement a Neural RF. Previous research (see [1] for a survey) shows that, though giving encouraging results, NN cannot be effectively used in IR at the current state of the NN technology. The scale of real IR applications, were hundreds of thousands of documents are at stake, makes it impossible to use NN in an effective way, unless we make use of very poor document and query representations. However, recent results [2] proves that it may be still possible to use NN in IR for very specific tasks were the numbers of patterns involved (and therefore the training) are reduced to a manageable size. In this paper we show that it is indeed possible to use NN to implement a RF device. However, the results here reported also show that a Neural RF device is not effective in real world's IR applications, where it performs worse than traditional RF techniques.

2 Relevance feedback and query adaptation

Relevance Feedback is a technique that allows a user to express in a better way his information requirement by adapting his original query formulation with further information provided

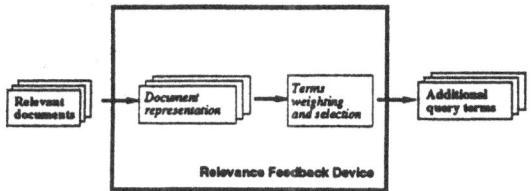

Figure 2: Schematic view of a Relevance Feedback Device

by indicating some relevant documents. When a document is marked as relevant the RF device analyses the document text, picking out terms that are statistically significant to the document, and adds these terms to the query. RF is a very good technique of specifying an information requirement, because it releases the user from the burden of having to think up lots of terms for the query. Instead the user deals with the ideas and concepts contained in the documents. It also fits in well with the known human trait of "I don't know what I want, but I'll know it when I see it". Obviously the user cannot mark documents as relevant until some are retrieved, so the first search has to be initiated by a query. The IRS will return a list of ordered documents covering a range of topics, but probably at least one document in the list will cover, or come close to covering, the user's interest. The user will mark the document(s) as relevant and starts the RF process performing another search. If RF performs well the next list should be closer to the user's requirement. A schematic view of a RF device is depicted in Fig. 1.

We may also think at a RF device as a filter that receives as input a query and a set of relevant documents (or considered so by the user) and that gives as output a modified or adapted query. This process of query adaptation is supposed to alter the original user formulated query to take into consideration the information provided by features of relevant documents. More formally the input of the RF device has the following form:

$$(q_i, d_1^i, d_2^i, d_3^i, \ldots d_l^i)$$

where q_i is the representation of the query i, and d_j^i is the representation of a document relevant to the query q_i. The set is composed of only a subset (l documents) of all the documents (k documents, with $k > l$) that are relevant to that particular query. The aim of the RF is to enable the retrieval of the other $(l - k)$ relevant documents.

In IR we can use different RF techniques, depending on the IR model being preferred. In the following Section we will illustrate a technique called Probabilistic RF, which is based on the Probabilistic IR model.

3   Probabilistic relevance feedback

Probabilistic Relevance Feedback (PRF) is one of the most advance technique for performing RF in operative IR systems. For a better and more complete explanation of the mathematics

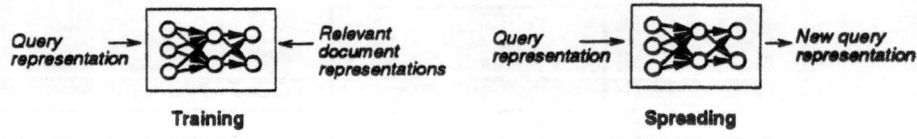

Figure 3: Training and spreading phases of Neural Relevance Feedback

involved see [3] and [4]. Briefly, the technique consists in adding a few other terms to those already present in the original query. The terms added are chosen by taking the first m terms in a list where all the terms present in relevant documents are ranked according to the following weighting function:

$$w_i = \log \frac{r_i\,(N - n_i - R + r_i)}{(R - r_i)\,(n_i - r_i)}$$

where: N is the number of documents in the collection, n_i is the number of documents with an occurrence of term i, R is the number of relevant documents pointed out by the user, and r_i is the number of relevant documents pointed out by the user with an occurrence of term i. Essentially what this rather complex function does is compare the frequency of occurrence of a term in the documents the user marked as relevant with the term frequency of occurrence in the whole document collection. So if a term occurs much more frequently in the documents marked as relevant than in the whole document collection it will be assigned a high weight. In the experiments reported in this paper the number of terms added to the original query has been experimentally set to 10.

4 Neural relevance feedback

In this work we intend to investigate the possible use of NN in designing and implementing RF. In [2, 5] an Adaptive IR System using NN was presented. We refer back to those papers for the architecture, the algorithms, and the learning performance of the system. Here we will describe only how that system can be used to perform *Neural Relevance Feedback* (NRF).

We can perform NRF using a RF device based upon a NN. This device learns from training examples to associate new terms to the original query formulation. The NRF device acts in a way similar to the classical RF. The main difference is that the weights used to order and select the terms are obtained from the output of a 3–layer feedforward NN trained using the Back Propagation (BP) learning algorithm. We use such a simple NN structure because we intended to consider the NN as a black box to investigate the feasibility of the approach. After this first investigation we will go on into tuning the structure to make it more effective to the problem at hand. We use as many input nodes as the terms that can be used to formulate queries, and as many output nodes as terms that can be used to represent all the documents in the collection. Each node in the NN input layer represents a query term and each node in the output layer represent a document term. We use the BP learning algorithm because its definition itself, as back propagation of errors, suggests its use in the RF context (see [6, 7]). In fact, what BP does is to back propagate the difference between the desired output (the representation of a relevant document) and the actual output (the representation of a non-relevant document). The BP learning algorithm is directed to make this difference the smallest.

Fig. 3 shows that the NRF acts in two phases. During the *training phase* the input and the output layer of the NN are set to represent a training example made of a query representation q_i and a relevant document representation d_j^i. The BP algorithm is used for learning, and is repeated for every relevant document d_j^i, with $j = 1, \ldots, l$. The learning is monitored by the NN control structure and when some predetermined conditions are met (such as, for example, a mean error below a certain threshold) the training phase is halted. During a *spreading phase* the activation produced on the input nodes by the query spreads from the input layer to the output layer using the weight matrices produced during the training phase. A new query is produced by adding to the terms already present in the query the first m (here $m = 10$) higher activated terms (nodes) on the output layer.

5 Experimental Settings and Evaluation

In order to perform an experimental analysis on the performance of PFR and NRF these tools are necessary: a document collection with relevance assessment, a probabilistic IRS with PRF, and a NN or a NN simulator on which implement a NRF.

The document collection chosen for the investigation is the *ASLIB Cranfield test collection*. This collection was built up with considerable effort in the 60s as the testbed for the ASLIB-Cranfield research project, aimed at studying the "factors determining the performance of indexing systems". They produced two test collections of documents about aeronautics, comprehensive of documents, requests and relative relevance judgements. For a full description of the collections see [8]. In the investigation reported in this paper the 200 document collection, with 42 requests and relevance judgements, has been used. It is of course understood that this limits the generality of the results obtained. However, the main purpose of this investigation is to demonstrate the feasibility of the proposed approach. There are still many open issues in the application of NN to IR. and the problem of scaling the results is one of the major ones. Nevertheless, it must be noticed that the dimension of the data set used in the these experiments is one of the largest used in any application of NN techniques to IR.

The choice of IR systems with PRF was quite limited. In fact, there are not many operative IR systems with such an advance feature. One of the best systems is *News Retrieval Tool* (NRT) [9], developed at the University of Glasgow for the Financial Times. NRT implements PRF as specified in Section 3.

For the investigations reported in this paper a NN simulator, *PlaNet 5.6* [10], running on a fast conventional computer has been used. The architecture and the algorithm's details of the implementation are reported in [2].

For the evaluation, the main retrieval effectiveness measures used in IR are Recall and Precision. These two measures have been used in evaluating and comparing the performance of PRF and NRF. *Recall* is the proportion of all documents in the collections that are relevant

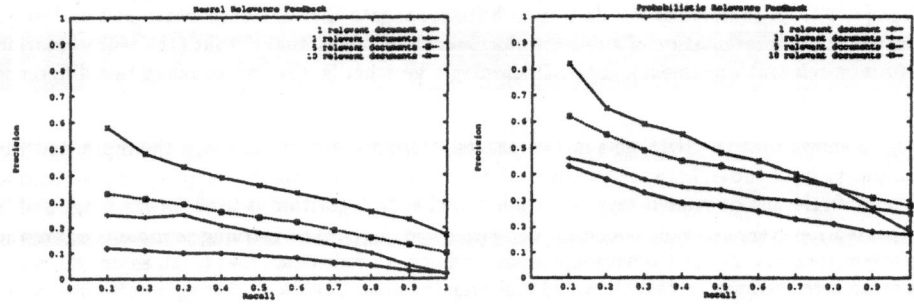

Figure 4: Performance of NRF and PRF w.r.t. the training sets

to a query and that are actually retrieved. *Precision* is the proportion of the retrieved set of documents that is also relevant to the query. In order to give a measure of the generalisation performance of NRF and PRF, Recall and Precision have been evaluated with different dimension of the set of training examples. If some learning of the domain knowledge has taken place, and if the NN can generalise it, then an improvement in the retrieval performance has to be expected.

6 Experimental results

The results here reported refer to the PRF implemented in NRT and to a NRF implemented on PlaNet using a 3–layer feedforward NN with BP learning. The numbers of input nodes was set to 195, the number of hidden nodes to 100, and the number of output nodes to 1142. The number of input and output nodes correspond respectively to the number of terms used in all the queries and to the number of terms used by the documents of the Cranfield collection. The mean number of documents relevant to a query is about 6.8.

Fig. 4 shows graphically how the performance of the NRF and PRF increases when the RF device is fed with increasing amount of relevant documents. This shows that both PRF and NRF act like a pattern recognition device and the more information they receives the more they can discriminate between patterns of relevant and not relevant documents. The performance has been evaluated averaging over all the set of queries in the relevance assessment at different values of the number of relevant documents given as feedback (the training examples). The graph shows that the NRF increases more rapidly in performance than the PRF. This is due to the better characteristics of non linear discrimination of NRF, that enables it to separate better the two patterns. However, the performance of PRF are better at any level of training and especially at lower levels. This makes PRF more useful in real world applications where the percentage of relevant documents used in the training over the total number of relevant document is usually very low.

Tab. 1 reports performance figures for the case of 10% training, i.e. when 10% of the number of relevant documents totally present in the document collection are fed to the RF device for

Recall	Precision NRF	Precision PRF
0.10	0.245	0.440
0.20	0.234	0.363
0.30	0.225	0.332
0.40	0.201	0.301
0.50	0.179	0.275
0.60	0.160	0.259
0.70	0.132	0.239
0.80	0.111	0.221
0.90	0.051	0.210
1.00	0.022	0.130

Table 1: Comparison of the performance of Neural and Probabilistic RF for 10% training

training. It shows numerically how PRF outperforms NRF in a realistic situation where the number of training examples, i.e. the number of relevant documents fed to the RF device is low compared to the total number of relevant documents present in the collection. This confirms what Fig. 4 already shows, and gives a stronger argument for preferring PRF to NRF in IR.

7 Conclusions

The results of this investigation, briefly summarised in this paper, demonstrate that a RF device based on NN (NRF) acts in a similar way to classical RF techniques. However, a comparison with one of the most advance RF techniques presently used in IR (PRF) shows that for low levels of training, which is the most common case in IR, NRF does not perform as well as PRF. Though it is necessary to proceed with further research using different NN architectures, and different learning algorithms, we think that NN are currently not suitable for use in IR applications where the number training examples is usually very little and the number of nodes and connections is extremely high.

References

[1] F. Crestani. A survey on the application of Neural Networks's supervised learning procedures in Information Retrieval. Rapporto Tecnico CNR 5/85, Progetto Finalizzato Sistemi Informatici e Calcolo Parallelo - P5: Linea di Ricerca Coordinata Multidata, December 1991.

[2] F. Crestani. Learning strategies for an Adaptive Information Retrieval System using Neural Networks. In *Proceedings of the IEEE International Conference on Neural Networks*, pages 244–249, S.Francisco. California, USA, March 1993.

[3] C.J. van Rijsbergen. *Information Retrieval*. Butterworths, London, second edition, 1979.

[4] D. Harman. Relevance feedback and other query modification techniques. In W.B. Frakes and R. Baeza-Yates, editors, *Information Retrieval: data structures and algorithms.*, chapter 11. Prentice Hall, Englewood Cliffs, New Jersey, USA, 1992.

[5] F. Crestani. An Adaptive Information Retrieval System based on Neural Networks. In J. Mira, J. Cabestany, and A. Prieto, editors, *New trends in Neural Computation*, volume 686 of *Lecture Notes in Computer Science*, pages 732–737. Springer-Verlag, June 1993.

[6] D.E. Rumelhart, J.L. McClelland, and PDP Research Group. *Parallel Distributed Processing: exploration in the microstructure of cognition*. MIT Press, Cambridge, 1986.

[7] J. Hertz, A. Krogh, and R. Palmer. *Introduction to the theory of Neural Computation*. Addison-Wesley, New York, 1991.

[8] C. Cleverdon, J. Mills, and M. Keen. *ASLIB Cranfield Research Project: factors determining the performance of indexing systems*. ASLIB, 1966.

[9] M. Sanderson and C.J. van Rijsbergen. NRT: news retrieval tool. *Electronic Publishing*, 4(4):205–217, 91.

[10] Y. Miyata. *A user's guide to PlaNet version 5.6: a tool for constructing, running and looking into a PDP network*. Computer Science Department, University of Colorado, Boulder, USA, December 1990.

Second-Order Recurrent Neural Networks Can Learn Regular Grammars from Noisy Strings

Rafael C. Carrasco and Mikel L. Forcada
{carrasco,mlf}@dtic.ua.es
Departament de Tecnologia Informàtica i Computació
Universitat d'Alacant, E-03071 Alacant (Spain)

Abstract

Recent work has shown that second-order recurrent neural networks (2ORNNs) may be used to infer deterministic finite automata (DFA) when trained with positive and negative string examples. This paper shows that 2ORNN can also learn DFA from samples consisting of pairs (W, μ_W) where W is a noisy string of input vectors describing the degree of resemblance of every input to the symbols in the alphabet, and μ_W is the degree of acceptance of the noisy string, computed with a DFA whose behavior has been extended to deal with noisy strings.

1   Introduction

A number of recent papers have explored the ability of second-order recurrent neural networks (2ORNNs) to learn simple regular languages[1, 2, 3, 4, 5]. As noted in [1, 2], it is possible to obtain a symbolic representation of the language —i.e., the states and transitions of a deterministic finite automaton (DFA) accepting the language— from the trained network. For this purpose, a complete sample consisting of a collection of strings classified as belonging or not to the language is used to train the network. After training, the network behaves as a DFA, that is, a syntactic pattern recognizer.

However, some empirical situations are better described in terms of properties which are difficult to discretize, but are still tractable under the syntactic pattern recognition paradigm. Ideal patterns consist of primitives in the prototype set (for example, strings are made of symbols from an alphabet), but often times real patterns contain primitives which cannot be, in general, clearly identified with a single prototype. Instead, one has (1) a degree of resemblance between each primitive and each of the prototypes (for instance, a probability that the primitive is a noisy version of that prototype), and (2) a degree of acceptance of the whole pattern in a given pattern class (or language). For example, the first part of a speech recognizer could output a *noisy string* of phonemes (or the so-called microphonemes) representing a spoken word, in which each position holds a vector of probabilities with one component for each phoneme. These noisy strings could be then fed into a neural recognition system that classifies the string as being, for instance, a verb or not a verb with a given degree ranging from 0 to 1.

In this paper, we explore the capability of 2ORNNs to reconstruct deterministic finite automata (DFA) from noisy samples. These samples consist of strings not made of ideal, noiseless symbols from an alphabet, but instead, of vectors describing the degree of resemblance of every input to each symbol in the alphabet. The degree (or probability) of acceptance (as computed using a DFA in an extended way) of each string is given instead of a Boolean value indicating whether the string is accepted or not.

2 Preliminaries

Let $\mathcal{A}$ be an alphabet of size $|\mathcal{A}|$. A noiseless string w of length $|w|$ is a sequence of $|w|$ symbols in the alphabet. A noisy string W of length $|W|$, however, consists of $|W|$ input vectors $\mathbf{I}^{[t]}$ $(1 \leq t \leq |W|)$ of dimension $|\mathcal{A}|$. Each component is labeled with a symbol from $\mathcal{A}$: component a of $\mathbf{I}^{[t]}$ has a real value $I_a^{[t]} \in [0, 1]$ representing the *degree of resemblance* of the t-th input to class or symbol a in $\mathcal{A}$. The concepts of noisy string concatenation and empty noisy string are analogous to their noiseless counterparts.

When the symbols in the alphabet $\mathcal{A}$ represent disjoint classes (*i.e.* a partition), the restriction that all components add to one,

$$\sum_{a \in \mathcal{A}} I_a^{[t]} = 1 \quad \text{for all } t, \tag{1}$$

may be imposed. Therefore, when the alphabet is $\mathcal{A} = \{0, 1\}$, it suffices to give the 1-component of the vector in each position, as $I_0^{[t]} = 1 - I_1^{[t]}$. For instance, the noisy string $W = (0.8)(0.1)$ indicates that the input at the second position of W is a 1 with degree of resemblance 0.1 (or equivalently, a 0 with degree 0.9).

In all of our experiments we assume that the language to be learned is regular, that is, a DFA contains all the information needed in order to evaluate the degree of acceptance of the noisy strings. Indeed, all the experiments reported in this paper are simulations on strings produced by known DFA.

The following straightforward generalization allows a DFA to work with noisy strings. Let $M = (Q, \mathcal{A}, \delta, q_1, F)$ be a DFA, where Q is the set of states, $\mathcal{A}$ the input alphabet, $q_1 \in Q$ the initial state, $F \subseteq Q$ the subset of accepting states, and $\delta : Q \times \mathcal{A} \to Q$ the transition function. The occupancy of state q after reading the noisy string W will be called $n(q, W)$. When W is the empty noisy string Λ,

$$n(q, \Lambda) = \begin{cases} 1 & \text{if } q = q_1 \\ 0 & \text{otherwise} \end{cases} \tag{2}$$

Once the occupancy of each state after reading a noisy string W is known, one may compute the occupancy of state q after reading one more input I, *i.e.*, after reading string WI:

$$n(q, WI) = \sum_{q' \in Q} \sum_{a \in \mathcal{A}} I_a \, n(q', W) \, \kappa(q', a, q), \tag{3}$$

where

$$\kappa(q', a, q) = \begin{cases} 1 & \text{if } \delta(q', a) = q \\ 0 & \text{otherwise} \end{cases} \tag{4}$$

The *degree of acceptance* (or simply *acceptance*) μ_W of a noisy string W in the noiseless language $L = L(M)$ accepted by the automaton M is computed then as follows:

$$\mu_W = \sum_{q \in F} n(q, W). \tag{5}$$

It has to be noted that this formulation is very similar to the kind of *fuzzy automata* defined in [6], but with the product as a norm, the sum as a co-norm, and a crisp initial state.

The noisy-string representation of a noiseless string $w = a^{[1]}a^{[2]}...a^{[l]}$ is $W = \mathbf{I}^{[1]}\mathbf{I}^{[2]}...\mathbf{I}^{[l]}$ where $I_a^{[t]} = 1$ if $a = a^{[t]}$ and 0 otherwise. The definition of degree of acceptance has been chosen so that, for these strings, $\mu_W = 1$ if $w \in L(M)$ and 0 otherwise.

For instance, the set of binary strings with an odd number of ones is accepted by the automaton $M = (\{q_1, q_2\}, \{0, 1\}, \delta, q_1, \{q_2\})$ with $\delta(q_1, 0) = q_1$, $\delta(q_1, 1) = q_2$, $\delta(q_2, 0) = q_2$, $\delta(q_2, 1) = q_1$. As the automaton reads the length-two noisy string $W = (0.8)(0.1)$ (remember that only component I_1 is needed in each position), the occupancies of states evolve as follows: $n(q_1, \Lambda) = 1$, $n(q_2, \Lambda) = 0$; $n(q_1, (0.8)) = 0.2$, $n(q_2, (0.8)) = 0.8$; $n(q_1, (0.8)(0.1)) = 0.26$, $n(q_2, (0.8)(0.1)) = 0.74$. The degree of acceptance of the string is then $\mu_{(0.8)(0.1)} = 0.74$. Clearly, the noisy-string representation of a noiseless 011 would be $(0)(1)(1)$ and its acceptance, $\mu_{(0)(1)(1)} = 0$.

A noisy sample over the binary alphabet will consist of pairs (W, μ_W) where μ_W is the degree of acceptance of W in the language, and W is given as a sequence of resemblance vectors.

3 Architecture and training

For our experiments we have used a second-order recurrent neural network as in [1, 2, 3, 7]. The basic features of the architecture are shown in fig. 1. The network has N hidden recurrent neurons plus $|\mathcal{A}|$ input neurons; hidden neuron number 1 is arbitrarily chosen to represent the degree of acceptance. In a way that parallels the function of a DFA, the states $S_i^{[t]}$ of the hidden neurons $(1 \leq i \leq N)$ after reading t inputs depend on the states $S_j^{[t-1]}$ $(1 \leq j \leq N)$ of the hidden neurons after reading $t - 1$ inputs and the values of the $|\mathcal{A}|$ components of the t-th input vector $\mathbf{I}^{[t]}$, as presented through each one of the input neurons. The products (hence the name *second-order*) of the states of each input and each hidden neuron are weighted, summed, and mapped through the sigmoid function $g(x) = 1/(1 + \exp(-x))$ to determine the new states of the hidden neurons:

$$S_i^{[t]} = g(\sigma_i^{[t]}), \tag{6}$$

with

$$\sigma_i^{[t]} = \sum_{j=1}^{N} \sum_{a \in \mathcal{A}} C_{ija} \cdot S_j^{[t-1]} I_a^{[t]}, \tag{7}$$

where C_{ija} $(i, j = 1, ..., N, a \in \mathcal{A})$ are the real-valued weights. After $|W|$ iterations, a noisy string W has been completely processed and the degree of acceptance is given by $S_1^{[|W|]}$ (recall that, due to the sigmoid function, the values of all states are always in the range $[0, 1]$).

The network is completely defined by the set of weights $\{C_{ija}\}$ and the set of initial activations of the hidden neurons, i.e., the set $\{\sigma_i^{[0]}\}$ before any input[1]. Optimization of the $N^2|\mathcal{A}| + N$ parameters characterizing the network is the goal of the learning process, and proceeds basically as described in [1, 2, 7]. At the very beginning, the initial value of every parameter is selected randomly, but this is only done once during the whole learning process. From that point on, the values are changed, trying to minimize the error function for the learning sample S

$$E = \frac{1}{2} \sum_{W \in S} \left(S_1^{[|W|]} - \tilde{\mu}_W \right)^2 \tag{8}$$

where $\tilde{\mu}_W$ is the customary projection of $\mu_W \in [0, 1]$ onto the range $[\epsilon, 1 - \epsilon]$ (we take $\epsilon = 0.1$). The search of the minimum is simply accomplished by means of gradient descent[1, 2, 7] using an

---

[1] The values of $\sigma_i^{[0]} = g^{-1}(S_i^{[0]})$ are preferred to the actual state values $S_i^{[0]}$ because, unlike the latter, they are not restricted to any range of values. This simplifies the optimization of the whole set of parameters[7].

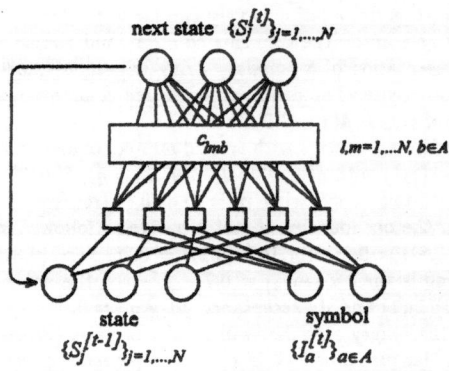

next state $\{S_j^{[t]}\}_{j=1,\dots,N}$

C_{lmb} $l,m=1,\dots N,\ b\in A$

state
$\{S_j^{[t-1]}\}_{j=1,\dots,N}$

symbol
$\{I_a^{[t]}\}_{a\in A}$

Figure 1. The architecture of the recurrent network

adapted version of the RTRL method[8]. The sample S is repeatedly fed and, after each noisy string is completely processed, weights and initial states are updated:

$$\Delta P = -\alpha \frac{\partial E}{\partial P} + \eta \Delta P \qquad (9)$$

where P stands for any of the parameters C_{ija} or $\sigma_i^{[0]}$. On the right-hand side of equation (9), ΔP represents the change computed in the previous iteration; α is the *learning rate* and η is called the *learning momentum.* The derivatives

$$\frac{\partial E}{\partial P} = \left(S_1^{[\|W\|]} - \mu_w\right)\, g'(\sigma_1^{[\|W\|]})\, \frac{\partial \sigma_1^{[\|W\|]}}{\partial P} \qquad (10)$$

are recursively calculated:

$$\frac{\partial \sigma_i^{[t]}}{\partial C_{lmb}} = \delta_{il} S_m^{[t-1]} I_b^{[t]} + \sum_{j=1}^{N} \sum_{a\in A} C_{ija}\, g'(\sigma_j^{[t-1]})\, \frac{\partial \sigma_j^{[t-1]}}{\partial C_{lmb}} I_a^{[t]} \qquad (11)$$

(where δ_{il} is the Kronecker delta) for $P = C_{lmb}$, and

$$\frac{\partial \sigma_i^{[t]}}{\partial \sigma_j^{[0]}} = \sum_{j=1}^{N} \sum_{a\in A} C_{ija} I_a^{[t-1]} g'(\sigma_j^{[t-1]}) \frac{\partial \sigma_j^{[t-1]}}{\partial \sigma_i^{[0]}} \qquad (12)$$

for $P = \sigma_j^{[0]}$, with the initial values:

$$\frac{\partial \sigma_j^{[0]}}{\partial C_{lmb}} = 0 \qquad (13)$$

and

$$\frac{\partial \sigma_i^{[0]}}{\partial \sigma_j^{[0]}} = \begin{cases} 1 & \text{if } i = j \\ 0 & \text{otherwise} \end{cases} \qquad (14)$$

for $i, j, l, m = 1, \dots, N$ and $a, b \in A$.

4 Experiments

Samples of noisy strings over $A = \{0, 1\}$ have been generated for three regular languages:

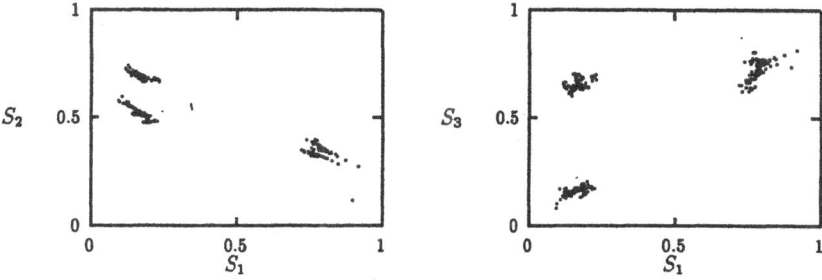

Figure 2. Clusters in the state space of a three-hidden-neuron 2ORNN after training with L_2 (see text).

- L_1, the strings having an odd number of ones: $0^*1(0 + 10^*1)^*$;

- L_2, the binary numbers that are multiples of three: $(0 + 1(01^*0)^*1)^*$;

- and L_3, Tomita's[9] fourth language (strings not containing 000 as a substring): $(1 + 01 + 001)^*(\epsilon + 0 + 00)$.

The strings in the samples are generated as follows: first, a length in the range 1 to 10 is randomly selected; then a noisy string of that length with totally random input vectors (a very noisy string) is built, and finally, its degree of acceptance is computed (as described in section 2) for the target automaton.

The network is first trained with a small subset (40 strings) of the sample. Every 100 iterations, a new group (40 more strings) is added to the training set. The process stops when either convergence is reached or the sample is exhausted. The learning rate and momentum are $\alpha = 4.0$ and $\eta = 0.5$. The training process is said to converge when the DFA extracted from the network computes exactly the degree of acceptance of every string in the sample (in practice, we checked that the extracted DFA was equivalent to the one used to generate the sample). DFA extraction is carried out by running noiseless strings through the network to form the set of "pure" states, and then extracting the DFA by a method similar to the one described in [1, 2].

A network having $N = 2$ hidden units was sufficient to learn L_1. Ten runs were performed; a fresh sample of 400 strings and a random set of values for all learnable parameters were used for each run. Convergence was reached after presentation of 40 strings (the initial set) in 3 cases and after 240 strings in 7 cases. That gives an average of 180 strings.

For L_2, a larger network was needed, $N = 3$. Nineteen out of twenty runs converged, using 400-string samples. The number of strings needed were: 80 (in 5 runs), 120 (6 runs), 160 (3 runs), 200 (2 runs), 240 (1 run), 320 (1 run), and 400 (1 run). This shows that a larger number of strings may sometimes be needed to learn this, more complex language. Figure 2 shows two views of the hidden-neuron state space for this network. The points visited by the network as it processes the 255 noiseless strings of length up to (and including) 7 are clearly shown to form three clusters corresponding to the states of the minimal automaton that accepts the language. Two of the states are non-accepting (low values of S_1 and the other is an accepting state (high value of S_1).

Learning L_3 is harder. A network with $N = 3$ was unable to learn the DFA from 400-string samples in a preliminary suite of ten runs. The training of a $N = 4$ network on 400-string samples converged only in four runs out of twenty; the number of strings needed were 240 in three cases and 280 in the fourth case. The results clearly indicated the need for a larger sample. In a test suite using 800-string sets, the training converged in four out of ten runs, having used 760, 800, 200 and 800 strings.

5 Concluding remarks

We have shown that second-order recurrent neural networks (2ORNN) can learn simple deterministic finite automata from rather small samples consisting of pairs of very noisy strings and their degrees of acceptance, computed according to an extension of the behavior of a DFA to noisy strings. This learning task may be related to many interesting, practical syntactic pattern recognition problems, and happens to be more difficult than the customary task of inferring DFA from noiseless strings. These experiments with recurrent neural networks presented in this paper give a clear indication about the learnability of finite-state recognizers from noisy samples and a motivation to search for other language-theoretical methods (such as [10]) to accomplish the same learning task.

Acknowledgements: The authors wish to thank the Dirección General de Investigación Científica y Técnica of the Government of Spain for support through project CICYT/TIC93-0633-C02.

References

[1] Giles, C.L., Miller, C.B., Chen, D., Chen, H.H., Sun, G.Z., and Lee, Y.C. (1992a) "Learning and extracting finite state automata with second-order recurrent neural networks" *Neural Computation* 4, 393–405.

[2] Giles, C.L., Miller, C.B., Chen, D., Sun, G.Z., Chen, H.H., and Lee, Y.C. (1992b) "Extracting and learning an unknown grammar with recurrent neural networks", *Advances in Neural Information Processing Systems*, vol. 4 (J. Moody *et al.*, eds; Morgan-Kaufmann, San Mateo, Calif., U.S.A.), 317–324.

[3] Siegelmann, H.T., Sontag, E.D., and Giles, C.L. (1992) "The complexity of language recognition by neural networks" *Information Processing 92*, vol. 1 (Elsevier/North-Holland), p. 329–335.

[4] Watrous, R.L. and Kuhn, G.M. (1992a) "Induction of Finite-State Automata Using Second-Order Recurrent Networks", *Advances in Neural Information Processing Systems*, vol. 4 (J. Moody *et al.*, eds; Morgan-Kaufmann, San Mateo, Calif., U.S.A.), 306–316.

[5] Watrous, R.L. and Kuhn, G.M. (1992b) "Induction of Finite-State Languages Using Second-Order Recurrent Networks", *Neural Computation* 4, 406–414.

[6] Steimann, F. and Adlassnig, K.-P. (1994) "Clinical monitoring with fuzzy automata", *Fuzzy Sets and Systems* 61, 37–42.

[7] M.L. Forcada and R.C. Carrasco (1994) "Learning the initial state of a second-order recurrent neural network during regular-language inference", *Neural Computation*, in press.

[8] Williams, R.J. and Zipser, D. (1989) "A learning algorithm for continually running fully recurrent neural networks" *Neural Comp.* 1, 270.

[9] Tomita, M. (1982) "Dynamic construction of finite-state automata from examples, using hill-climbing" *Proceedings of the Fourth. Annual Cognitive Science Conference*, (Ann Arbor, Mich., U.S.A.) p. 105–108

[10] Carrasco, R.C. and Oncina, J. (1994) "Learning stochastic regular grammars by means of a state merging method", in *Grammatical Inference and Applications, Proc. of the 2nd. Intl. Colloq. on Grammatical Inference ICGI-94 (Alicante, Spain, September 1994)* (Carrasco, R. and Oncina, J., eds.) Lecture Notes in Artificial Intelligence 862 (Springer-Verlag) p. 139–152.

Extracting DNF Rules
from Artificial Neural Networks

HL Viktor   I Cloete
{hlv, ian}@cs.sun.ac.za

Computer Science Department, University of Stellenbosch, Stellenbosch 7600, South Africa

Abstract

Artificial neural networks are powerful classification mechanisms. Neural networks encode knowledge in a set of numerical weights and biases. This data driven aspect of neural networks allows easy adjustments when change of environments or events occur. Numeric weights, however, are difficult to interpret in terms of rules, making it difficult for a human to understand what the neural network has learned.

One approach to understanding the representations formed by neural networks is to extract symbolic rules from networks, since concepts represented by symbolic learning algorithms are more easily understood by humans. It has been shown that most concepts described by humans usually can be expressed as production rules in disjunctive normal form (DNF) notation. Rules expressed in this notation are therefore highly comprehensible and intuitive.

A method that extracts production rules in DNF is presented. The extracted rules are accurate and results compare favourably with traditional symbolic rule extraction methods. Since the rules are in a logically manipulatable form, significant simplifications in the structure thereof can be obtained, yielding a highly comprehensible set of rules.

Introduction

Artificial neural networks (ANNs) have been proven to be successful general machine learning techniques for, amongst others, classification. However, one disadvantage of the neural network approach is that it is very difficult to determine why a particular conclusion was reached. This is due to the 'black box' nature of the neural network, as well as the fact that neural networks encode knowledge in a set of numerical weights and biases. Although this data driven aspect of neural networks allows easy adjustments when change of environments or events occur, it is difficult to interpret numeric weights, making it difficult for humans to understand. One approach to understanding the representations formed by neural networks is to extract symbolic rules from networks, since concepts represented by symbolic learning algorithms are intuitive and are therefore more easily understood by humans.

Over the last few years, a number of methods to extract symbolic rules from ANNs have been published. Since it is known that the task of rule-extraction from an arbitrary-configured ANN is extremely difficult [Towell 1992], every method utilizes its own dedicated architecture. There are, however, some general assumptions that these algorithms adhere to. In the next section, these general assumptions underlying rule extraction algorithms are presented. A short overview of algorithms discussed in the literature follows.

A previously reported algorithm that extracts NofM-style rules [Towell 1992, Craven 1993] is discussed in some detail. A main drawback of this algorithm is the style of its rules. Individual rules are not highly comprehensibly [Towell 1992], which may lead to an incomprehensible rule set. Moreover, it is known that humans usually express concepts in simple DNF-notation, rather than the complex NofM style [Wnek 1994]. An extension to the NofM algorithm that generates production rules in DNF notation is presented. A comparison with other methods show that the extracted rules are intuitive, accurate and highly comprehensible.

1 Preliminaries

The first assumption that most rule extraction algorithms makes, is that a non-input unit is either maximally active (activation near 1) or inactive (activation near 0). This Boolean valued activation is approximated by using the standard logistic activation function $f(x) = 1/(1 + e^{-sx})$ and setting $s \geq 5.0$. The use of the above function also guarantees that non-input units always have non-negative activations in the range [0,1] [Towell 1992].

The second underlying premise of rule extractions is that each hidden and output unit implements a symbolic rule. The concept associated with each unit is the consequent of the rule, and certain subsets of the input units represent the antecedents of the rule. The output a_i of a unit is a function f of the input weights w_j and a bias θ, where $a_i = f(Input - \theta) = f(\sum_{j/a_j \approx 1} w_j - \theta)$. Here, θ acts as a threshold. Rule extraction algorithms search for those input weights that guarantee that $(\sum_{j/a_j \approx 1} w_j > \theta)$.

2 Rule Extraction Algorithms

Fu [1993, 1994], as well as Saito and Nakamo [1991], developed similar methods to extract crisp rules from general ANNs. Saito and Nakamo's algorithm forms rules that maps directly from inputs to outputs. The algorithm is limited to units with only four inputs, whereas Fu extracts rules with k connectives. Towell and Shavlik [1992, 1994] refer to the method as the Subset method. The method works on a simple premise: combinations of input weights to a particular unit is exhaustively searched for combinations that exceed the bias of the unit. Yoon [1994] implemented a similar approach that employs weight and unit pruning prior to rule extraction. These algorithms tend to extract a large number of rules, even when considering networks of moderate complexity. In addition, the process of searching for rules is problematic because of the combinatorics involved, making it unsuitable for real-world problems.

Other research includes RuleNet, developed by McMillan, Mozer and Smolensky [1991], which generates both symbolic (rule) and subsymbolic (category) learning. The algorithm assumes that there are only two weight clusters: one with large positive weights, and one with weights near zero. This limits the usefulness of their algorithm to a small number of problems.

Research into fuzzy rule extraction includes work by Hayashi [1991], who extracts IF-THEN form rules, Berenji [1991] and Tazaki and Inque [1994], who extract rules from ANNs with planar lattice architectures. Most fuzzy rule extraction methods do not assume near-Boolean activations, but rather activations in the range [-1, 1].

2.1 The NofM Algorithm

The NofM algorithm represents the state of the art in published literature. The algorithm was originally developed for Knowledge-Based ANNs (KBANNs) by Towell and Shavlik [Towell 1992]. They presented an algorithm NofM, named after the sentence N of M, to extract crisp rules from a restricted case of trained KBANNs. (In a KBANN, the topology and initial weights of the network are specified by a domain theory consisting of symbolic inference rules.)

Craven and Shavlik [Craven 1993] extended the NofM algorithm by using soft weight-sharing training [Nowlan 1992a, Nowlan 1992b] instead of backpropagation. This training method was used to encourage clustering during training. They proceeded to extract NofM-style rules from general ANNs. In addition, rules of the form *N more than (Pos-set, Neg-set)* (where the predicate returns true if the number of satisfied antecedents in Pos-set minus the number of satisfied antecedents in Neg-set is larger than N) were also extracted, since it was found that general ANNs generate a large number of these type of rules. The basic idea of the NofM algorithm is the use of equivalence classes. Similarly weighted links are grouped into clusters, the assumption being that the individual weights do not have unique importance.

The algorithm consists of 6 steps. See [Towell 1992, Towell 1994] for a detailed description.

Step 1: Clustering. With each hidden and output unit, group similarly weighted links into clusters. Stop when no pair of clusters is closer than a set distance (default 0.25).

Step 2: Averaging. Set link weights of all the members in a cluster to the average of the group.

Step 3: Eliminating. Eliminate any groups that do not significantly affect whether the unit will be active or inactive.

Step 4: Retrain the Biases.

Step 5: Extraction. Form a single rule for each hidden and output unit. The rule consists of a threshold given by the bias and weighted antecedents specified by the remaining links.

Step 6: Simplification. Where possible, simplify rules to eliminate weights and thresholds. The final rules are of the form *if (N of the M antecedent are true) then the consequent is true.* If the elimination of weights and biases requires rewriting a single rule with more than five rules, then the rule is left in its original state.

3 Extension using Disjunctive Normal Form (DNF)

The clustering of link weights considerably reduces the combinatorics of the NofM algorithm as opposed to, for example, the Subset algorithm. In addition, the NofM algorithm generates accurate rules that approximate the network.

A large number of learning problems, however, does not learn NofM-style concepts [Wnek 1994]. This considerably limits the applicability of the NofM algorithm. In addition, it was found that concepts generated by human subjects when asked to create classes of entities and to express them linguistically, usually fall into the DNF category [Wnek 1994].

Another significant drawback of the NofM algorithm is that individual rules generated are often not easily comprehensible [Towell 1992, Towell 1994], especially if the elimination of the weights and biases does not take place due to the limitation placed on the number of rules generated during simplification. Redundant rules that should have been subsumed, may be generated. Even though the NofM algorithm performs some subsumption, the structure of the rules is not ideal for the detection and resolution of all such redundancies. It would be preferable to write the rules in some form that is mathematically sound and manipulatable.

One of the earliest discoveries regarding propositional logic was its similarity with algebra [Van Dalen 1989]. Because of the Boolean nature of the NofM rules, it follows that these rules may be expressed in propositional logic. Propositions may be manipulated using standard algebraic laws, providing us with convenient simplification mechanisms.

In particular, propositions may be written in normal forms, a powerful mechanism to manipulate logical predicates. Two normal forms are of importance: The *disjunctive normal form* is a disjunction of conjuncts as follows: $(p_{11} \wedge p_{12} \wedge ... \wedge p_{1n}) \vee ... \vee (p_{m1} \wedge p_{m2} \wedge ... \wedge p_{mn})$. The *conjunctive normal form*, on the other hand, is a conjunction of disjuncts: $(p_{11} \vee p_{12} \vee ... \vee p_{1n}) \wedge ... \wedge (p_{m1} \vee p_{m2} \vee ... \vee p_{mn})$.

The DNF approach extends the NofM algorithm by rewriting the NofM-style rules into normal form rules. The method consists of the following steps:

Step 1: NofM Steps 1-4. Execute Steps 1-4 of the NofM algorithm for general ANNs. The clustering of weights into equivalence classes significantly reduces the number of rules generated and the number of computations.

Step 2: Normalization. Form DNF production rules. The weighted antecedents within each cluster that is guaranteed to exceed the bias form conjuncts, and a disjunction exists between the various clusters.

Step 3: Simplification. Simplify the rules generated in step 2. In some cases the conjunctive normal form is more suitable, since many rules include more AND than OR predicates. The disjunctive normal form may be transformed to conjunctive normal form. If such a transformation, however, leads to replication for rules involving many disjunctions and few conjunctions, it should be avoided.

4 Evaluation

Various measures should be taken into account when evaluating the quality of the rules extracted.

The **comprehensibility** is an important measurement, since this was the initial motivation for rule extraction research. That is, *does the algorithm encode the information it learns in such a way that it may be inspected and understood by humans?* Firstly, the number of rules serves as a good indication of the overall understandability of the algorithm. Second, the complexity of the individual rules should be considered. The complexity of a rule may be measured by the number of tests within a rule [Wnek 1994].

The **accuracy** of the rules should approximate the accuracy of the trained ANN from which it was extracted. The accuracy of the rule extraction should be comparable to other traditional symbolic methods to motivate the extraction process.

The **similarity** of the rules generated by the various algorithms is related to both the accuracy and comprehensibility. Two rule sets are similar if they contain approximately the same tests within the rules.

For example, the following two rules are similar:

if (a = 1) and (b ≥ 0) then c.   (1)

if (b > 1) and (a = 1) then c.   (2)

Finally, the rules should be meaningful and reflect the properties of the data set it symbolizes.

5 Experimental Results

This section presents a set of experiments to determine the strengths and weaknesses of the DNF method.

The digit recognition example was used in the experiments. Digits are ordinarily displayed on electronic watches and calculators using seven horisontal and vertical lights in *on-off* combinations. The goal of the exercise is to determine which digit is currently being displayed. Let i denote the ith digit, that is we have

($i = 0, 1, 2, ...,9$) and take $(x_{i1}, ..., x_{i7})$ to be a seven-dimensional vector of zeros and ones, corresponding to positions *(top, left-top, right-top, middle, left-bottom, right-bottom, bottom)*. For example, to display the digit 1, the lights of *right-top* and *right-bottom* will be set to *on*, and the lights of the rest of the positions will be set to *off*, e.g. $x_{im} = 1$ if the light in the mth position is *on* for the ith digit, and $x_{im} = 0$ otherwise.

The data for the example were generated from a faulty calculator with respectively 0%, 5%, 10% and 20% noise. Two output classes, recognising digits 1 or 2, were identified. The database consisted of 63 instances. For the purpose of the evaluation the data set was divided into 48 training instances and 15 testing instances. Prior to rule extraction, the ANN was trained using patterns reflecting the various levels of noise, until, in each instance, an accuracy of 98.2% was reached. A 7-5-2 configuration was found to be optimal, since more hiddens leads to redundant rules and less hiddens to overfitting.

Two sets of experiments were done. Firstly, the rules obtained by employing the DNF extension was compared with that obtained from the NofM algorithm. Secondly, the quality of the DNF rules was compared to that of symbolic rule extraction methods.

DNF versus NofM

In the first set of experiments, the **comprehensibility** of the DNF extension was compared to that of the NofM algorithm. Since the two algorithms use the same extraction steps, the accuracy and the similarity of rules are implicit.

Table 1 presents a summary of the results:

Algorithm	Noise	# of Rules	Max # Antecedents	Average # Antecedents
NofM	0%	6	7	3.2
	5%	9	5	3
	10%	12	5	2.3
	20%	6	7	6
DNF	0%	3	1	1
	5%	3	2	1.6
	10%	7	4	2.4
	20%	3	5	2.6

Table 1: A comparison of the comprehensibilities of the NofM and DNF approaches.

The number of rules extracted by the NofM algorithm is significantly higher than those extracted by the DNF method. This is due to the fact the NofM algorithm may generate redundant tests that are eliminated by the methods's simplification step. The following example illustrates the reduction that may be achieved.

Assume that the NofM algorithm extracts the following rules:

$Hidden_1 \leftarrow 1$ *of* $\{a,b\}$ *and 1 of* $\{d\}$. (1)

$Hidden_2 \leftarrow 2$ *of* $\{a,d\}$. (2)

$Hidden_3 \leftarrow 2$ *of* $\{b,d\}$. (3)

$Output \leftarrow 1$ *of* $\{Hidden_1\}$. (4)

$Output \leftarrow 1$ *of* $\{Hidden_2, Hidden_3\}$. (5)

In DNF notation, we have the following rules:

$Hidden_1 \leftarrow (a \lor b) \land d.$ *(6)*

$Hidden_2 \leftarrow (a \land d).$ *(7)*

$Hidden_1 \leftarrow (b \land d).$ *(8)*

$Output \leftarrow Hidden_1 \lor (Hidden_2 \lor Hidden_3).$ *(9)*

Rule 9 may be rewritten, by substitution, to the following form:

$Output \leftarrow (a \lor b) \land d) \lor ((a \land d) \lor (b \land d)).$ *(9a)*

$Output \leftarrow ((a \land d) \lor (b \land d)) \lor ((a \land d) \lor (b \land d)).$ *[Distributivity] (9b)*

$Output \leftarrow ((a \land d) \lor (b \land d)).$ *[Idempotency] (9c)*

The result is a single DNF rule with only two conjuncts: this is a significant reduction of the initial rules.

Employing normal forms are only worthwhile if the normalization process does not significantly increase the number and size of the rules [Towell 1992]. The results of the experiments as illustrated by the example show that the normalization does indeed lead to a reduction of both the number and size of the rules.

DNF versus CN2 and C4.5

In the second set of experiments, the quality of the DNF approach was compared to two symbolic machine learning methods CN2 [Clark 1989] and C4.5 [Quinlan 1993]. CN2 used set-covering whereas the C4.5 algorithm uses decision trees. The DNF rules were written into an unordered CN2-format rule file. The rule file was evaluated using the CN2 evaluation mode.

Table 2 shows the results of the comparisons:

Algorithm	Noise	Accuracy (train)	Accuracy (test)	# of Rules	Max # Antecedents	Average # Antecedents
CN2	0%	100%	100%	2	1	1
	5%	100%	100%	5	2	1.4
	10%	100%	85.7%	5	2	1.8
	20%	93.8%	86.7%	10	3	2.2
C4.5	0%	100%	100%	3	1	1
	5%	97.9%	93.3%	3	1	1
	10%	95.8%	92.9%	3	1	1
	20%	91.7%	66.7 %	7	2	1.8
DNF	0%	100%	100%	3	1	1
	5%	95.8%	86.7%	3	2	1.6
	10%	96.5%	92.9%	7	4	2.4
	20%	93.8%	86.7%	3	5	2.6

Table 2: A comparison of rules extracted by CN2, C4.5 and DNF.

Comprehensibility. C4.5 generated rules with few antecedents, even with 20% noise added. These rules does not accurately reflect the properties of the inputs, but rather presented a generalized view of the data and relationships. The CN2 rules were more descriptive, but also failed to identify the relationships between the inputs and the outputs as clearly as the DNF algorithm.

The number of rules and antecedents of the DNF rules are comparable to that of the symbolic methods. It extracts the least rules when the noise level is 20%. Moreover, it is the only method that extracts the rule (with 20% noise): *IF (right-bottom = 1) and (right-left = 1) THEN digit 1.* This rule accurately describes the characteristics of the digit 1.

Accuracy. The CN2 algorithm generates the most accurate rules for the training sets, while the C4.5 results are the lowest. The DNF method yields the best results for the test sets with 10% and 20% noise. The results show that the DNF method compares favourably with the traditional machine learning methods. It also shows that the accuracy approximate, and in some cases extend, that of the original trained ANN (98.2%). This is a promising result, since it indicates that the use of trained ANNs to obtain rules proves to be a worthwhile exercise.

Similarity. The rules that are generated by the two symbolic methods differ considerably. As mentioned above, the C4.5 rules are more general since a considerable number of rules contains only one test. These rules generally subsume the more fine-tuned rules generated by the DNF and CN2 methods. Although the rules extracted by CN2 and DNF are similar, the DNF rules tend to include more tests. These descriptive DNF rules reflect the properties of the data sets more accurately than the other methods. This verifies an observation by Towell (1994), namely that rules extracted from ANNs are usually more fine-tuned than traditional symbolic rules.

Conclusion and Future Extensions

Rule extraction from ANNs is an important topic of research, since the extraction of symbolic rules will increase our understanding and confidence in neural network decisions.

Humans usually express concepts in a way that is expressible in a simple DNF-notation. A method that extracts symbolic DNF production rules was presented. It extracted considerably less rules than the previously reported NofM-style algorithm. The DNF rules also tend to be simpler and therefore more easily comprehensible. The comprehensibility and accuracy of the rules were comparable to, and in some cases extended, that of symbolic machine learning methods. The DNF rules reflected the properties of the experimental data sets accurately and were more fine-tuned than any other approach.

Future research should include the extraction of fuzzy rules from arbitrarily configured ANNs with non-Boolean activations. The employment of pruning algorithms and/or dimensionality reduction prior to extraction might also prove worthwhile.

References

[Berenji 1991] HR Berenji, Refinement of approximate reasoning-based controllers by reinforcement leaning, Proceedings of the Eight International Machine Learning Workshop, Evanston, IL: Morgan Kaufman, p475-479.

[Clark 1989] The CN2 Induction Algorithm, Machine Learning, 3, 1989.

[Craven 1993] MW Craven and JW Shavlik, Learning Symbolic Rules using Artificial Neural Networks, Proceedings of the Tenth International Conference on Machine Learning, Amherst, Morgan Kaufmann, 1993.

[Fu 1993] LM Fu, Knowledge-Based Connectionism for Revising Domain Theories, IEEE Transactions on Systems, Man and Cybernetics, Vol 23, No 1, p173-182.

[Fu 1994a] LM Fu, Rule Generation from Neural Networks, IEEE Transactions on Systems, Man and Cybernetics, Vol 24, No 8, p1114-1124.

[Hayaski 1991] Y Hayaski, A Neural Expert System with Automated Extraction of Fuzzy If-Then Rules and Its Applications to Medical Diagnosis, Advances in Neural Information Processing Systems, III, (editors RP Lippmann, JE Moody and DS Touretzky), 1991, p578-584.

[Mcmillan 91] C McMillan, MC Mozer and P Smolensky, The Connectionist Scientist Game: Rule Extraction and Refinement in a Neural Network, Proceedings of the 13th Annual Conference of the Cognitive Science Society, Chicago, IL, Erlbaum, Hilsdale, NJ, 1991, p420-430.

[Mcmillan 92] C McMillan, MC Mozer and P Smolensky, Rule Induction through Integrated Symbolic and Subsymbolic Processing, Advances in Neural Information Processing Systems, IV, (editors JE Moody, SJ Hanson and RP Lippmann), 1992.

[Nowlan 1992a] SJ Nowlan and GE Hinton, Adaptive Soft Weight Tying using Gaussian Mixtures, Advances in Neural Information Processing Systems, IV, (editors JE Moody, SJ Hanson and RP Lippmann), 1992.

[Nowlan 1992b] SJ Nowlan and GE Hinton, Simplifying Neural Networks by Soft Weight-Sharing, Neural Computation, Vol 4, No 4, 1992, p473-493.

[O'Neal 1994] MB O'Neal and WE Edwards, Complexity Measures for Rule-Based Programs, IEEE Transactions on Knowledge and Data Engineering, Vol 6, No 5, October 1994.

[Quinlan 1993] JR Quinlan, C4.5: Programs for Machine Learning, Morgan Kaufman Publishers, San Mateo, California, 1993.

[Saito 1988] K Saito and R Nakamo, Medical diagnostic expert system based on PDP model, Proceedings of the IEEE International Conference on Neural Networks, Vol 1, p255- 262, IEEE Computer Society, Washington. DC, 1988.

[Shavlik 1994] JW Shavlik, Combining Symbolic and Neural Learning, Machine Learning 12, 1994, p321-331.

[Towell 1992] GG Towell and JW Shavlik, Interpretation of Artificial Neural Networks: Mapping Knowledge-Based Neural Networks into Rules, Advances in Neural Information Processing Systems, IV, (editors JE Moody, SJ Hanson and RP Lippmann), 1992.

[Towell 1994] GG Towell and JW Shavlik, Refining Symbolic Knowledge using Neural Networks, Machine Learning 12, 1994, p321-331.

[Tazaki 1994] E Tazaki and N Inque, A Generation Method for Fuzzy Rules using Neural Networks with Planar Lattice Architectures, The 1994 IEEE International Conference on Neural Networks, Vol III, Orlando, Florida, 1994, p1743-1748.

[Towell 1992] GG Towell and JW Shavlik, Extracting Refined Rules from Knowledge-Based Neural Networks, Machine Learning, 13, 1993, p71-101.

[Towell 1994] GG Towell and JW Shavlik, Refining Symbolic Knowledge using Neural Networks, Machine Learning, Vol 4, (editors RS Michakski and G Tecuci), Morgan Kaufmann, 1994, p489-519.

[Tazaki 1994] E Tazaki and N Inoue, A Generation Method for Fuzzy Rules using Network Networks with Planar Lattice Architectures, The 1994 IEEE International Conference on Neural Networks, Orlando, Florida, 1994.

[Yoon 1994] B Yoon and RC Lacher, Extracting Rules by Destructive Learning, The 1994 IEEE International Conference on Neural Networks, Orlando, Florida, 1994.

[Van Dalen 1989] D van Dalen, Logic and Structure (Second Edition), Springer-Verlag, 1989.

[Wnek 1994] J Wnek and RS Michalski, Comparing Symbolic and Subsymbolic Learning: Three Studies, Machine Learning, Vol 4, (editors RS Michakski and G Tecuci), Morgan Kaufmann, 1994, p489-519.

Dynamic Symbol Grounding, State Construction and the Problem of Teleology

Erich Prem
erich@ai.univie.ac.at
The Austrian Research Institut for Artificial Intelligence*
Schottengasse 3, A-1010 Wien, Austria
Tel. +43 1 5336112, FAX +43 1 5320652

January 1995

Abstract

Symbol grounding has originated within the connectionist-symbolic debate so as to gap the bridge between the two approaches. This paper provides an overview about recent results concerning symbol grounding, which is critically reviewed here. A thorough analysis reveals that symbol grounding parallels transcendental logic and is best viewed as automated model construction. If this diagnosis is true, the necessary next question must be which sort of models are generated in symbol grounding systems. The answer to this question very much depends on the kind of network architecture employed for grounding. An illustrative neural network architecture is used to explain how a dynamic symbol grounding system generates a formal model. It is argued that, depending on the architecture, very different kinds of signs ranging from input to goal descriptions can be grounded.

1 Symbol Grounding

1.1 The Symbol Grounding Problem

Based on a fundamental criticism of symbolic models and Searles "Chinese room" argument [Searle 80], Stevan Harnad introduced "The Symbol Grounding Problem" in his 1990 paper. Symbol grounding (SG) tries to answer the question as to how it is possible for a computer program to use symbols which are not arbitrarily interpretable. Whereas the meaning of signs in conventional programs is just "parasitic on the meaning in our heads", grounded symbols should possess at least some "intrinsic meaning" [Harnad 90]. The problem can also be formulated as how it is that formal symbol systems can acquire a semantics which is not based on other symbol structures but on the system's own sensory experience.

It should be noted here that Searle—as opposed to Harnad—is more concerned with *intentionality* than with *meaning*. However, as of today, it is quite obvious that symbol grounding cannot solve the problem of original intentionality. The reason for this lies in the observation that intentionality is connected to consciousness, cf. [Frixione & Spinelli 92]. But Harnads proposal is obviously directed towards the generation of correlational semantics. (For the more philosophical discussion see [Christiansen & Chater 92, Christiansen & Chater 93]. Harnads answers can be found in [Harnad 94].)

Consequently, symbol grounding has mainly been performed in the context of linguistic research so as to clarify the question of semantics or the origin of reference in words ("word meaning"). The general idea is to equip a symbol system with some kind of measurement device. The goal of the system is to find invariances in the sensory data whenever a specific symbol is shown to the system. A perfect symbol grounding system would then be able to use a specific symbol in order to describe the sensory input.

Such a system would, of course, also contribute to gap the bridge between symbolic and subsymbolic models of cognition, since it would use symbols for communication but, on the other hand, possess rich semantic meaning for these symbols which would not consist in other symbols, cf. [Dorffner & Prem 93].

*The Austrian Research Institute for Artificial Intelligence is supported by the Austrian Federal Ministry of Science an Research.

1.2 Symbol Grounding Systems

A group of connectionist proposals for network architectures addresses the problem of grounding object categories in perception. Typically, the models possess two types of input: one for (simulated) sensory data, another one for symbolic descriptions of the data. The networks are usually chosen so that the sensory input is categorized trhough un- or self-supervised categorization algorithms like Kohonen networks. The resulting category representations are then associated with some sort of symbolic description of the static input. Such architectures have been suggested by [Chauvin 89, Cottrell et al. 90, Schyns 91, Dorffner 89]. A few models have been suggested which try to ground symbols in dynamic input data, e.g. [Cottrell et al. 90, Nenov & Dyer 94].

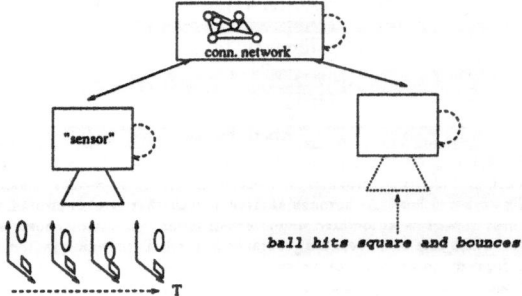

Figure 1: (Simulated) sensory data from visual scences is mapped on a symbol space by means of (recurrent) connectionist methods.

Figure 1 shows the "generic" SG-architecture (generalized to sequences of input). In most of these models categorization of input and naming of these categories is clearly separated in two parts, i.e. in two neural network modules. The motivation for this modularization either consists in assumptions concerning cognitive aspects of categorization (e.g. that children form categories without always having names for the categories [Schyns 91]) or in aspects concerning features of typical AI symbols (e.g. in the observation that "symbols are arbitrary" [Dorffner 89]).

Techniques for generating the categories include Kohonen nets, auto-associative backpropagation networks, and "winner-take-all" strategies. In systems with dynamic input data recurrent auto-associative backpropagation networks are used to find compressed category representations in the hidden layer [Cottrell et al. 90]. The second module connects categories to labels (symbols) and must therefore be trained with a supervised algorithm, e.g. backpropagation.

In the next section I shall explain why and how SG models do a bit more than just labeling inputs.

2 Symbol Grounding and Model Construction

There are three main recent results concerning symbol grounding, which fundamentally change our view of what SG is and can be used for. These results are: (i) SG is, essentially, automated model construction; (ii) SG implements aspects of transcendental logic; (iii) dynamic SG can generate state transition models.

2.1 SG is automated model construction

In SG models the only purpose of symbols is their reference to external objects, respectively to the sensory data generated by them. A theory of SG which largely adopts such a conception (basically a form of correlational semantics) can only result in a specifically scientific model (in the sense of Fig. 2) which is designed to reflect nature for epistemic purposes. This process is equivalent to the automated development of a formal model of a natural system. ("Natural" because we are still talking about *grounded* systems, which are connected to the world through measurement devices.)

This interpretation of SG is not changed by the observation that SG models construct some kind of *subjective* model of their environment—based on the inherently statistical nature of neural network algorithms, because the system itself still generates a symbolic model of this environment. SG systems try to bring a manifoldness

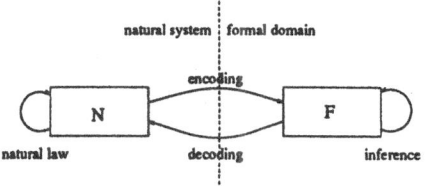

Figure 2: Model of a natural system.

of sensory data into the unity of concepts and *express* these concepts by means of arbitrarily chosen forms. It is essentially this feature which can be found in any mathematical model of a natural system: Some kind of encoding procedure maps "objects" of the environment (usually through employing meters) onto symbols in a formalism. Implications in the formalism are then used to predict what natural law does to the natural system. This situtation is depicted in Figure 2. SG systems try to map sensory data from N on the symbol in F such that this symbol is understood (i.e. used, arbitrarily chosen) by a human teacher.

2.2 SG implements transcendental locic

I have elsewhere argued in detail [Prem 94b, Prem 95a] that the grounding of signs (symbol, index, icon) in SG systems exhibits strong similarites with the three classical forms of reasoning (de-, in-, abduction). SG searches for the conditions which enable signs (specifically symbols) to refer to objects. Therefore it is very close to Kants program of transcendental logic. It can be argued that SG systems are not only grounding *symbols,* but—depending on the concrete architecture—sometimes ground *icons* in the Peircean sense (due to the continuous character of connectionist mappings) and also exhibit features of *indexes* because they are working according to physical laws. (I.e. in the same way as a thermometer signifies temperature could grounded symbols signify what they stand for.)

Let us now concentrate on how the sign becomes connected to its object. In the case of an icon some similarity of representation and represented negotiates between sign and significatum. This reference to a common property takes exactly the form of a logical inference, namely abduction. (M is P. Q is P. → M is Q.) The position of P, the middle term, corresponds to Aristotle's second figure. Reference of an index to what it stands for is made possible through natural law. Therefore, it is the objects themselves which inform about the set of things referred to. Viewed logically, this process is similar to induction, which informs about the total set of objects with a specific property.

Finally, the symbol is a general and arbitrarily chosen representation of an object. Therefore, it is *the sign alone* which negotiates its meaning. In other terms, the symbol represents in- and extension, it alone ensures that one is connected to the other. This perfect double reference to the singular (extension) and the general (intension) is what makes symbolic negotiation similar to deduction. Fig. 3 tries to support these similarites by showing hierarchical concept trees and how different types of signs relate to the objects they stand for.

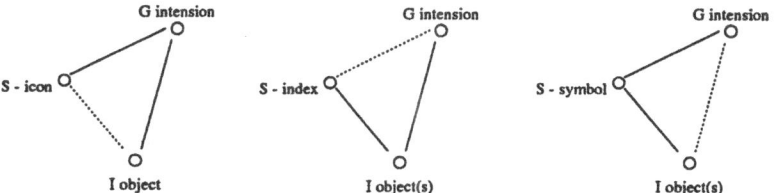

Figure 3: Different views of hierarchical concept trees (arbor porphyriana) and corresponding types of signs. Straight lines show relations given, dottet lines represent conclusions drawn.

In Kantian terms, it are the three logical principles of *sameness* (A is B), *separation* (if B then NON-B), and *opponentship* (A is either B or NON-B, law of the exluded middle) which are realized in SG systems. This observation will allow us to extend this analysis of static SG to dynamic SG. In static ones, states (objects) of the environment are only formalized into symbols. Synamic SG systems should also generate predictive rules. The extension of this discussion to space and time is quite natural and has also been performed by Kant and his successors.

Figure 4: Schematic representation of three transcendental principles.

Fig. 4 tries to appeal to intuition and to suggest as to how the three logical principles can be extended to space and time. *Sameness* creates the permanency of objects in time, *separation* serves as the (inductive) source of creating causal relations (Kant: "The real whereupon something follows."). *Opponentship* puts the "objects" into relation with each other.

2.3 Dynamic SG can generate state transition models

In order to show that SG really is automated model construction it must now be demonstrated how SG systems can generate state transition rules (in addition to state formation). We do this on the basis of the principles of transcendental logic outlined above. Our idea has it been to identify sameness with the notion of a state in a formal model, separation with the timely ("causal") sequence of states.

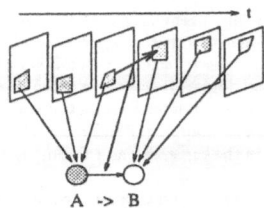

Figure 5: Implication in SG-models. Consecutive measurements of objects are interpreted as belonging to states, generate state labels and formulate state-transitions.

In Fig. 5 the function of such a system is depicted. Firstly, a flow of input data which is supposed to contain measurements of objects is mapped on formal symbols (A and B). Secondly, the timely sequence of these objects is mapped on a formal connection (implication arrow) between the two signs. The names of the states (e.g. A) are, of course, arbitrarily selected by a supervisor.

Thirdly, opponentship is identified with the fact that only one state can be active at a time. This is what we would expect from a typical Newtonian state model, where the states are arbitrarily labeled, see Fig. 6.

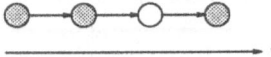

Figure 6: State transition sequence in Newtonian models. At one instant in time only one state is active. Arrows represent state transitions. (States could be labeled $s_1, \ldots$)

3 Dynamic Symbol Grounding: An example

3.1 Architecture

The following neural network example has been implemented to illustrate the theoretical argumentation, it is *not* supposed to serve as yet another SG architecture. (Details of the architecture and the algorithms can be found in [Prem 94b].)

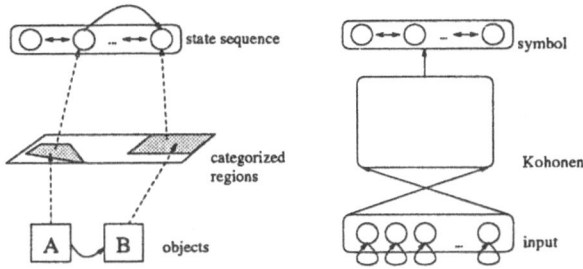

Figure 7: Left: Two consecutive objects are categorized in a Kohonen network and generate a state sequence. Right: The network consists of recurrent input units, a Kohonen layer and a IAC-output layer.

Fig. 7 shows the basic idea and architecture employed in our experiments. Two consecutive objects are categorized by means of a Kohonen network, i.e. their measurements (which can be distorted) map onto different regions of the map. These regions can then be labeled through another layer, the symbol layer. Whereas the formation of categories happens without a teacher, the name for a category (i.e. which unit of the symbol layer should represent the category) is defined by a human supervisor.

The input layer which receives signals from the (simulated) measurements consists of self-recurrent units. The weights of the self-recurrent connections range from 0.0 to 0.7 so as to only slowly change their activation through time and to capture differnt (historical) aspects of the time sequence, cf. [Ulbricht 94].

The Kohonen layer was trained using a variant of the common Kohonen algorithm [Zell 94].

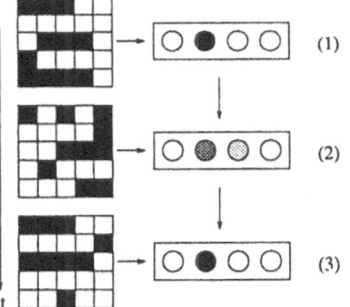

Figure 8: Clear recognition of states in the symbol layer despite of distorted inputs. *Interactive activation* serves to suppress short-term variation of the symbol units, see text.

The symbol layer implements the *principle of the excluded middle* to ensure that only one symbol unit is active per time. This is done by means of *interactive activation* (lateral inhibition [McClelland & Rumelhart 81]) of symbol units. One advantage of this technique is to suppress short-time variations of the symbol due to noise in the input data, see Fig. 8. In addition to the inhibitory connections in the input layer there are also trainable connections which try to capture which active symbol unit follows which.

The symbol layer connections to the Kohonen layer were trained according to a simple Hebbian learning scheme. The intra-layer connections should capture the sequential aspect, i.e. which symbol (state) follows which. In order to capture only really "causal" sequences in the input Hebbian learning between two consecutively active symbol units only occured if (i) the activation of the preceding unit is above a threshold and (ii) if the follwing symbol unit is above a threshold in the very next time step. Condition (i) serves to make sure that only safely recognized states can imply others, condition (ii) ensures that only really implicatons are learned. In order to forget wrongly learned implications all these weights in the symbol layer are reduced by a small

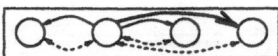

Figure 9: The symbol layer which captures the state transitions. Besides of the IAC connections (dottet lines) other intral-layer connections exist that learn the state graph. The figure shows the connections of the second unit which has obviously often been active immediately before the fourth and so generated a strong connection.

amount in each time step.

It should be mentioned that this architecture is basically a regular SG system (except for the recurrent connections in the input and the intra-layer connections in the symbol layer).

3.2 Experiments

In the experiments with the architecture up to 5 prototypical 5×5 pixel input vectors were used. These pictures were artificially distorted (up to 25%) and presented to the network in a predefined merry-go-round fashion, see Fig. 10. A simple input sequence consisted of showing all 5 noisy prototypes, one after the other, to the network. Each prototype was presented for 3 consecutive time steps ("permanency of objects in time"). It was easy for a network presented with such an input and the sequence of $1, 2, 3, 4, 5$ as the desired output labels to (i) map distorted inputs on symbol units and (ii) to generate the desired state transition graph.

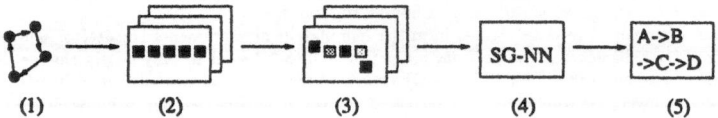

Figure 10: Experimental design. (1) a hypothetical sequence of objects is modeled as a sequence of prototypes (2), which are measured by noisy meters (3). A SG architecture (4) tries to formulate the sequence into a state-transition graph (5).

A more difficult input consists of the following sequence of input patterns: $\alpha \to \beta \to \gamma \to \delta \to \epsilon \to \gamma$. The network was trained successfully to label these inputs thruogh the symbol units number $1, 2, 3, 4, 5, 6$. One interesting feature of this architecture is that the recurrent input units allowed the Kohonen net to distinguish between the two same input patterns γ which can only be distinguished through their different context in time. In this case, the state-transition sequence generated in the output layer was as shown in sequence 1.

$$1 \xrightarrow{0.95} 2 \xrightarrow{0.95} 3 \xrightarrow{0.7} 4 \xrightarrow{0.95} 5 \xrightarrow{0.7} 6 \xrightarrow{0.85} 1 \tag{1}$$

It can be seen that the SG system very well models the input sequence, however, has some difficulties in distinguishing between the two same inputs. (Numbers on top of the arrows represent weights between corresponding symbol layer connections.) If the network is trained so as to generate the labels $1, 2, 3, 4, 5, 3$ then object γ is easier to recognize for the symbol layer, but then the state transition sequence becomes something like $1 \to 2 \to 3 \to 1$ and additionally $4 \to 5 \to 3$ and another transition $3 \to 4$. Which obviously reflects that the teacher of the system used too few states for clear and unambigous transitions.

Many variations of this architecture could be studied (but need a more practical context). The aim of the above experiments was only to show how a slightly modified general SG architecure can be used as a automated generator of formal models of informal domains where the state labels are chosen by a supervisor.

4 Teleology and "groundable" Signs

If it is correct that SG is model construction, the necessary next question must be, which sort of models are generated in SG systems. Today's SG architectures can be partitioned into two groups depending on whether they label unsupervised categorizaton or autosupervised backpropagation networks (Fig. 11).

SG has sometimes been suggested to have a system label its internal states (not necessarily found through unsupervised categorization) and to thereby support finding explanations for the system's actions (e.g. [Prem 95b]).

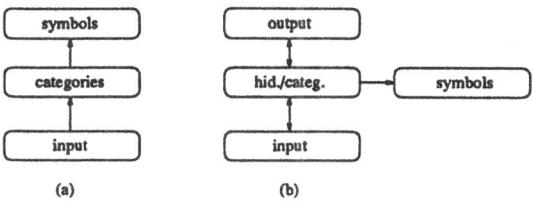

Figure 11: Variants of SG (a) input- and (b) hidden layer categorization.

This would mean to create the system's desired output by means of a supervised learning technique, use representations which generate the desired output for categorization and to then label these categories with arbitrary signs.

It is easy to see that the categories which are generated are now not only dependant on the input, but also on the output. Depending on which task is to be learned by a network with architecture (b) in Fig. 11, other kinds of symbols can be successfully grounded. In another experiment with the modified architecture we trained a backpropagation network to distinguish between the two instances of the same input pattern γ using two output units (01 first occurence of γ, 10 second occurrence, 00 other inputs). The patterns of the hidden layer of this backpropagation network have now been used to generate categories in the Kohonen network and to be labeled by symbol layer units. This time, the architecture could not successfully ground descriptions of the input.

In other terms, in the case of the architecture which categorizes the hidden states of the backpropagation network grounding of the very same set of symbols fails (if the categorization component is forced to find a small set of categories, otherwise some symbols can successfully be grounded). In this case, only labels for the *output* can be grounded, i.e. only labels for the output of the backpropagation network can be correctly produced.

Employing networks which solve another task makes it necessary to switch from input to output descriptions. The symbols which are now grounded do not only describe objects in the measurements, but notions which have to do with the task, which express some kind of "in order to". The reason lies in the fact that the representation which the network has constructed is more a model of the task to solve than of the input. This will alsways happen, when some sort of error minimization training procedure is used. The backpropagation network finds a representation in the hidden layer which is useful for itself, i.e. for its task. Consequently, it now becomes difficult to use these very same encodings of the input as the generator for symbolic descriptions which would make sense to the human observer.

The practical consequence of this result is that if one tries to use the representations generated in the hidden layer of a backpropagation net which is not auto-associative, arbitrary input descriptions (symbols) cannot be grounded. Only those symbols are "groundable" which are in accordance with the goal of the backpropagation network. Without knowing this goal, grounding is hardly possible.

5   Conclusion

The SG architecture presented in this paper can be extended in many ways: One could study differenct ways of processing the input, other categorization methods can be used, and the generation of the state-transition sequence could well be achieved by other means, too. However, such a study does not seem to make sense without a concrete problem, i.e. without a concrete practical domain which is to be modeled automatically. The results presented here do not make claims about the practicality of the presented architecture. They serve to support my proposal that symbol grounding as it is pursued today and as it is argued by Harnad and others is nothing else but the (semi-)automatic construction of a formal model of a non-formal domain.

This need not be a merely negative result. Of course, it suggests that SG cannot solve problems of intentionality (Chinese room) or original meaning. It is also doubtful that it can contribute very much to the semantics of natural language, since it is only based on a very primitive view of correlational word meaning. It can, however, turn out to be a useful and interesting field of research on its own. There are many cases where an automated construction of a world model that remains, at least in parts, understandable for a human, would be desirable. Consider, for example, a robot who is moving in an unknown territory. Having this robot produce a correct set of labels for the environment it experiences or report about an expedition to new territory would certainly be desirable. These are reasons why symbol grounding still deserves further attention.

References

[Chauvin 89] Chauvin Y.: Toward a Connectionist Model of Symbolic Emergence, in Proceedings of the Eleventh Anual Conference of the Cognitive Science Society, Lawrence Erlbaum, Hillsdale, NJ, 580–587, 1989.

[Christiansen & Chater 92] Christiansen M.H., Chater N.: Connectionism Learning and Meaning, Connection Science, 3(4), 227–252, 1992.

[Christiansen & Chater 93] Christiansen M.H., Chater N.: Symbol Grounding– the Emperor's New Theory of Meaning?, Proc. of the 15th Annual Conference of the Cognitive Science Society, Boulder, CO, 1993.

[Cottrell et al. 90] Cottrell G.W., Bartell B., Haupt C.: Grounding Meaning in Perception, in Marburger H. (Hrsg.), GWAI-90, Springer, Berlin, 1990.

[Dorffner 89] Dorffner G.: A Sub-Symbolic Connectionist Model of Basic Language Functions, Indiana University, Computer Science Dept., Dissertation, 1989.

[Dorffner & Prem 93] Dorffner G., Prem E.: Connectionism, Symbol Grounding, and Autonomous Agents, Proceedings of the 15th Annual Meeting of the Cognitive Science Society, Boulder, CO, pp. 144–148, 1993.

[Frixione & Spinelli 92] Frixione M., Spinelli G.: Connectionism and Functionalism: The Importance of Being a Subsymbolist, JETAI Journal of Experimental and Theoretical Artificial Intelligence, 4(1), 1992.

[Harnad 90] Harnad S.: The Symbol Grounding Problem, Physica D, 42, pp. 335–346, 1990.

[Harnad 94] Harnad S.: The Origin of Words, in Durham, W. & Velichkovsky B. (Hrsg.) Naturally Human: Origins and Destiny of Language, Nodus Pub., Muenster, 1994.

[McClelland & Rumelhart 81] McClelland J.L., Rumelhart D.E.: An Interactive Activation Model of Context Effects in Letter Perception: Part 1. An Account of Basic Findings, Psychological Review, Vol.88, 375–407, 1981.

[Nenov & Dyer 94] Nenov V.I., Dyer M.G.: Perceptually Grounded Language Learning: Part 2 – DETE: a Neural/Procedural Model, Connection Science, 6(1), 3–42, 1994.

[Prem 94a] Prem E.: Symbol Grounding Revisited, Oesterreichisches Forschungsinstitut fuer Artificial Intelligence, Wien, TR-94-19, 1994. Presented at the Workshop for Combining Symbolic and Subsymbolic Approaches, 11th European Conference on AI, Amsterdam, 1994.

[Prem 94b] Prem E.: Symbol Grounding: Die Bedeutung der Verankerung von Symbolen in reichhaltiger sensorischer Erfahrung mittel neuronaler Netzwerke, PhD Thesis, Faculty of Computer Science, University of Technology, Wien, 1994.

[Prem 95a] Prem E.: Symbol Grounding and Transcendental Logic, Proc. of the Second Swedish Conference on Connectionism, Skoevde, Sweden, 1995.

[Prem 95b] Prem E.: Understanding Complex Systems–What Can the Speaking Lion Tell Us?, to appear in Steels L. (ed.), The Biology and Technology of Autonomous Agents, Springer, Berlin Heidelberg New York, NATO ASI Series F, 1995.

[Schyns 91] Schyns P.G.: A Modular Neural Network Model of Concept Acquisition, Cognitive Science, 15(4), 1991.

[Searle 80] Searle J.R.: Minds, Brains and Programs, Behavioral and Brain Sciences, 3, 417–457, 1980.

[Ulbricht 94] Ulbricht C.: Multi-recurrent Networks for Traffic Forecasting, Proceedings of the AAAI'94 Conference, Seattle, Washington, 883–888, 1994.

[Zell 94] Zell A.: Simulation Neuronaler Netze, Addison-Wesley, Bonn-Paris-Reading, 1994.

Analysis of Industrial Economics by Means of Neural Nets

E.Monte, J.M.Calvet,S.Vilarrubla

E.T.S.I.Telecomunicació
P.O.Box. 30002
08080 Barcelona.Spain
E-mail:enric@tsc.upc.es

Abstract

This work was has been done in order to see the feasibility of prediction of the performance of firms in industrial economics, using two different methods. One was the a multilayered perceptron and the other the a linear prediction filter. A data base of industrial corporations of Spain was used for the experiments. The predicted variables that were the benefits of the corporations and sales. The information used for the prediction was related to the internal variables of each corporation, the relative position in the sector, and additional macroeconomic data. Also some statistical tests were done in order to ascertain the reliability of the results. It was found that the results with the neural net based predictor were statistically more reliable than linear prediction in the sense that results were more accurate with a better confidence margin.

1-Introduction

The objective of this work was designing a system based on neural nets, which aims at predicting the evolution of several economical variables of a set of corporations [1]. The tool for prediction was a multilayer perceptron, which is known to be able to learn arbitrary functions [2] (with some restrictions) between $R^m \rightarrow R^n$. We make use of this property and we assume that there is an unknown function between the actual value of some internal factors of each corporation, the relative rank of the corporation in it´s sector and the results obtained the following year. And that this unknown function can be computed if the multilayer perceptron is well trained. We think that the hypothesis that we assume are correct, because of the experience in the so called "performance analysis" [3] in the field of the industrial economics, that aims to find or test empirically if there is a relation (in the basis of observed data) between the observed values of the variables that represent the structure and strategic behaviour concepts of the previous model and the observed variables of performance. In formalised terms if it exists a empirical function between the performance which is a function of the kind $f(s_1,..,s_1)$, with s_i; the strategical variable i.. This is based in the well stablished model of the "industrial economics" by authors like Mason, Baim,Scherer, etc. is based on the sequence's structure of the sectors, the strategic behaviour of the enterprises/firms that operate in those sectors and the performance attained by the firms.

In order to see if a multilayer perceptron can give improvements a comparison is made between the neural net and a linear predictor filter which is the classical method based on regression.

2-Economical Variables

The variables that were used for characterising each corporation were taken from a database that has information about largest and medium sized corporations of Spain with annual sales greater than 10 million \$ [4] and information about economical sectors. The data consisted on the market power, the concentration variables, dimension variables, barriers of entry variables, other barriers, variables related with the conditions of the demand, strategic behaviour variables, variables related to the efficiency of the corporation, and variables related to the results/performance of the corporation. This information was used in order to describe each corporation and it's context. It was found that each group of variables was highly correlated, and that the observation vector for each corporation was of length that was too long for being tackled correctly by a neural net. Because of this, some prepossessing was done on the data in order discrimate the useful information and to be able to estimate correctly the weights of the net.

3-Processing of the data

The data base consisted of six years from 1980 to 1986. The training set was of the years 1980 to the 1985, enchaining the prediction form one year to the following one. Due to the correlations of some of the variables of the database prepossessing was done on the input data before the prediction was done by the net. We observed that the correlation was high between the variables that belonged to certain groups. Also it was noted that the dimension of the global observation vector over which the prediction had to be done was high and it would be a cause of problems in the training of the neural net. The prepossessing of the data consisted on a principal component analysis (PCA) [5] over the data of some of the groups of variables. This is a statistical method that consists on the projection of the data on the eigenvectors associated with the highest eigenvalues of the autocorrelation matrix of the data. The dimension of the projection was selected in all cases to be such that the 95% of the total variance explained was preserved. The PCA made possible to reduce the number of dimensions of the concentration variables from 30 to 2 preserving 97% of the variance, on the dimension variables the reduction was done from 9 to 4 preserving 95%, on the group of entrance variance the reduction was done from 9 to 5 preserving also the 95%, the other variables were not needed to be compressed. These prepossessing gave an observation vector upon which the prediction was done of dimension 38. The compression matrix was only computed using the training database, and the same projection matrix was used for the testing.

It was also found that a beneficial information for the prediction would be the use of the derivative of the input vector. These derivative was computed as shown in equation (1), where x^n_{in} is the input vector at the moment n . The use of the derivative gave information about the tendency of the time series.

$$Dx^n_{in} = \frac{x^{n+1}_{in} - x^{n-1}_{in}}{2} \tag{1}$$

The test database with the data to be predicted was also processed in order to adapt these data to the characteristics of the output of a multilayer perceptron with a sigmoidal nonlinearity. First of all the data was normalized to the range [0,1]. We also did an equalisation of the histograms in order to give more resolution to the output of the net. This equalisation gives more resolution to the more probable margins of the variables to be predicted. An example of the equalisation method can be seen in figure 1.

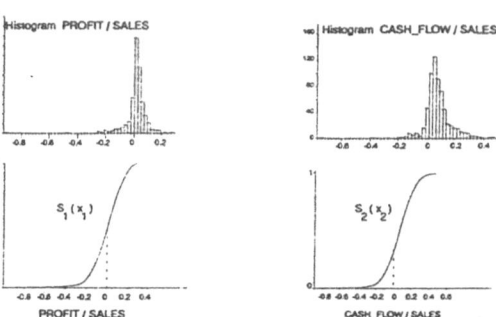

Figure 1. Equalisation of the data.

4-Structure of the prediction net

The structure of the net was taken so that at the input information could be of three different kinds: a) the observation of a given year in order to predict the next one, b)could have a context, which would be the observation of a given year and the previous one, c) the observation of a given year and the temporal derivative of the observation vector. The prediction variables are the sales and the benefits of the corporation for the following year. Thus the function to be implemented would have as an argument a vector of dimension either 38 or 76 depending on the input and as output a vector of dimension 2. This was done using one multilayer perceptron with a hidden layer in one case and in the other using two linear prediction filters. In the case of the neural nets a sigmoidal nonlinearity was used and different number of hidden nodes where taken (from five to seventeen).

5-Statistical tests

Several tests were done in order to ascertain the statistical reliability of the results [6]. Due to the limited amount of data of the database . The training database for the structure a) [see section 5] had 690 corporations. The training database for the structures b) or c) had 505 corporations. The test database used only 147 corporations. The reason that justifies the different number of corporations per database was that there was missing information for some corporations at certain years. The reliability of the prediction was done by means of the confidence interval of the regression line between the predicted data and the real data. We also computed the confidence margin of the derivative of the regression line and the residual analysis.

6-Comparison of the prediction results of both methods.

We will present in this section the results obtained when the prediction is done with a neural net and when it is done with a model based on linear prediction. Four cases will be presented, where the comparison is made taking into account several statistical parameters, such as:

RMS Error in the test and validation database: This information will give us an idea of the training point where a neural net has the best generalisation.
Confidence Margin of the regression line between the predicted data and the real data.
Dispersion of the points with respect to the regression line
Confidence margins of the regression line and the relative position of the corporations.

631

Experiment 1: Prediction using an a) type structure [see section 5]. Data to be predicted: Cash_flow/Sales (o) and Profits/Sales (x). The neural net that was used to do the prediction had an input vector of dimension 38, nine hidden units and two output units. The results are shown in figure 2.

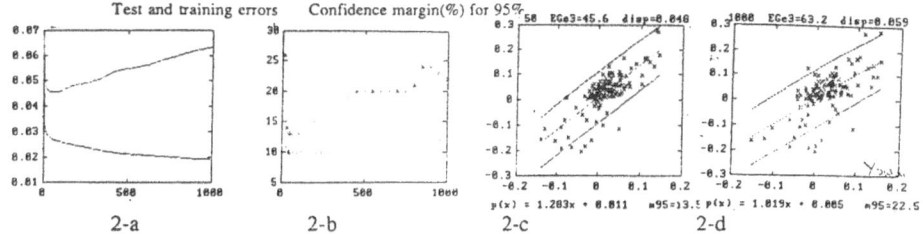

2-a   2-b   2-c   2-d

Figure 2. Prediction results with a neural net.
2-a Training and Test Errors as a function of the training epoches.
2-b Confidence margin (95%) of the residual rule
2-c and 2-d Confidence margin (95%) of the residual line with the output of the net as a function of the desired value. Cash_flow/Sales (2-c)and Profits/Sales (2-d).

Experiment 2: Prediction using an a) type structure. Data to be predicted: Cash_flow/Sales and Profits/Sales. A linear predictor was used which had as input a vector of dimension 38. The results are shown in figure 3.

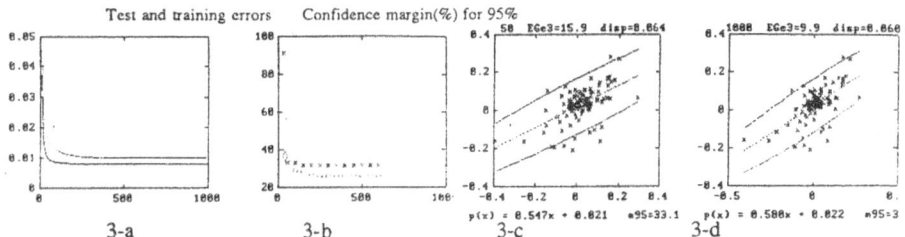

3-a   3-b   3-c   3-d

Figure 3. Prediction results with a linear prediction filter.
2-a Training and Test Errors as a function of the training epoches.
2-b Confidence margin (95%) of the residual rule
2-c and 2-d Confidence margin (95%) of the residual line with the output of the net as a function of the desired value. Cash_flow/Sales (2-c)and Profits/Sales (2-d).

Experiment 3: Prediction using an b) type structure. Data to be predicted: Cash_flow/Sales (o) and Profits/Sales (x). The neural net that was used to do the prediction had two input vectors of dimension 38, fifteen hidden units and two output units. The results are shown in figure 4.

632

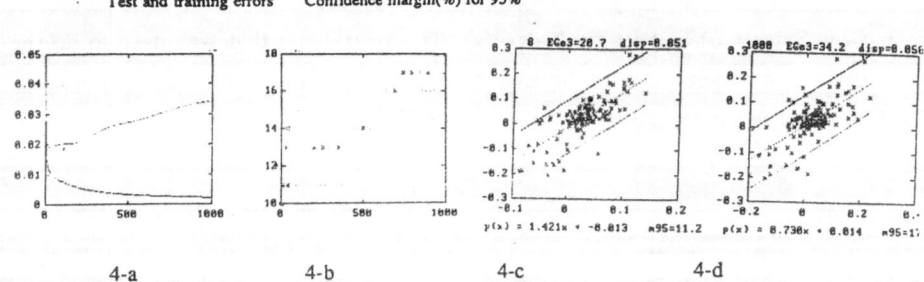

Test and training errors   Confidence margin(%) for 95%

4-a   4-b   4-c   4-d

Figure 4. Prediction results with a <u>neural net.</u>
2-a Training and Test Errors as a function of the training epoches.
2-b Confidence margin (95%) of the residual rule
2-c and 2-d Confidence margin (95%) of the residual rule with the output of the net as a function of the desired value. Cash_flow/Sales (2-c)and Profits/Sales (2-d).

Experiment 4: Prediction using an b) type structure. Data to be predicted: Cash_flow/Sales (o) and Profits/Sales (x). A <u>linear predictor</u> was used which had as input two vectors of dimension 38. The results are shown in figure 5

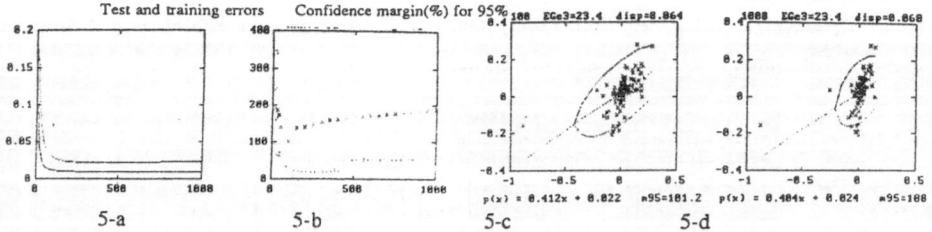

Test and training errors   Confidence margin(%) for 95%

5-a   5-b   5-c   5-d

Figure 5. Prediction results with a <u>linear predictor</u>.
2-a Training and Test Errors as a function of the training epoches.
2-b Confidence margin (95%) of the residual rule
2-c and 2-d Confidence margin (95%) of the residual rule with the output of the net as a function of the desired value. Cash_flow/Sales (2-c)and Profits/Sales (2-d).

The results obtained when the prediction was done by means of the temporal derivative (structure c) are not presented because the results were inferior and less reliable. Probably this result was due to the fact that the temporal derivative of the observation vector was computed using only two observations, which gave a noisy estimate.

7-Conclusion

In this paper we have shown that the use of a neural net for the prediction of corporation results can give better and more reliable results than the use of a linear predictor filter using a database in applied economics. These results were shown to be more reliable in all the statistical measures. We have also shown a way for prepossessing the data in order to a compression so that the data can be processed and the problems related to the estimation of parameters are reduced. The experiments could not be done without this compression that took away the redundancy from the data. The training time for the neural net was higher than the training time for the linear predictor, nevertheless the test phase was of the same order of magnitude in relation to the CPU time. An open question is the application of the neural nets methodology to the determination of the merits of the input variables to the net.

8-References

[1] S.Vilarrubla,*Anàlisi d'economia aplicada industrial amb xarxes neuronals*. PFC. ETSIT Barcelona.1994.

[2] Hetch Nielsen, *Neurocomputing*. Addison Wesley. 1989

[3] D.A. Hay y D.Morris. *Industrial economics and organization* . Theory and evidence). Oxford University press. 1993.

[4] J. Costa. Las bases y el soporte software para una gestión integral cualitativa de empresas. PFC. ETSIT Barcelona 1994.

[5] R.Duda and Hart. Pattern, *Classification and Scene Analysis*. John Wiley and Sons. 1973.

[6] Daniel Peña Sánchez, *Estadística, Modelos y Métodos*. Alianza Editorial.Textos 1989.

Effects of Spatial Frequency and Stimulus Size on the Orientation Sensitivity of Humans

F. Díaz-Otero, A. Caballero, A. Lorenzo and J.A. Sigüenza[1]

Instituto de Ingeniería del Conocimiento, Mód. C–XVI, P. 4, Universidad Autónoma de Madrid. 28049 Madrid, Spain.
Facultad de Medicina, Universidad Autónoma de Madrid. 28029 Madrid, Spain.

Abstract

Threshold for grating detection have been measured in six human observers for different orientations (vertical, horizontal, oblique 45° and oblique 135°). Gratings of fifteen different spatial frequencies were presented monocularly to the observers through a circular window. The area of the window was different under two experimental conditions ($A = 3.14°^2$ and $B = 0.785°^2$). In all cases, the sensitivity was higher for the vertical orientation than for the other ones. Moreover, the sensitivity was lower for all the orientations when the B window was used, in this case, the sensitivity for the oblique orientations was higher than the horizontal one.

1 Introduction

The existence of orientation selectivity in the neurons of the visual cortex as well as their columnar organization, would be expect to reflect the orientation sensitivity present in the human visual system. Previous studies in human subjects using psychophysical methods, have shown that the sensitivity is greater for the main orientations (vertical and horizontal), than for the obliques (5, 8, 15) These studies are supported by electrophysiological results obtained from human subjects (3, 11)

Several authors (5), have shown that the so called "oblique effect" (2), seems to have a neural substrate. In relation with that, other authors (18) have pointed out that this effect is not an artifact due to the optics of the eye. At least three hypothesis have been formulated attempting to explain the oblique effect:

- There is a larger amount of neurons selective for vertical and horizontal orientations than for oblique ones (12, 13). This hypothesis is supported by the findings of Blakemore and Cooper (1970) who tested the orientation properties of visual cortex neurons from reared kittens in vertical an horizontal environments. However, other authors, after recording single cells in the cat's visual cortex, report no vertical or horizontal preferences (10)

---

[1]This study has been supported by Grant DGICYT PB91-0045

- There is not a larger amount of neurons selective to any orientation, but neurons selective to the vertical and horizontal orientations would have stronger response properties than the oblique ones. However, Rose and Blakemore (1974) have provided strong evidences against this hypothesis.

- Neurons with preferred vertical or horizontal orientation are very selective to a narrow band of spatial frequencies (1, 14). In the case of simple cells of the cat visual cortex there is a close relationship between the increase of the spatial frequency and the narrowness of the band (17).

These theories reflect the disparity in the interpretation of results. Moreover, other authors have found a specific pattern of 60 degrees of alternate changes in the orientation selectivity for a spatial frequency ranging between 4 and 4.5 cycles/degree (9). This study disagrees with the proposal of the vertical and the horizontal as the two principal orientations. Tootle and Berkley (1983) studying the contrast sensitivity for vertically and obliquely oriented gratings as a function of the grating area, showed an increase for both vertically and obliquely oriented grating with increases in stimulus area, the sensitivity growing more slowly for oblique gratings. More recently work by Essock (1990), reported an increase in contrast sensitivity for a high spatial frequency (20 cycles/degree) grating at all orientations with increasing stimulus length, although that increase was greater for horizontal than for oblique stimuli.

The aim of the present study has been to analyze the combined effect of the stimulus size and grating orientation in the visual detection of spatial frequencies, in human subjects. Therefore, we have measured the contrast sensitivity function of a balanced set of spatial frequencies, combined with two different stimulation areas, when the grating stimuli were presented at four main orientations: Vertical, Horizontal, Oblique 45 degrees and Oblique 135 degrees.

2 Methods

Subjects

Experiments were performed on six adult subjects (2 male and 4 female). Two of them had normal vision, and the other four wore correction lenses. No subject had an astigmatism bigger than 0.5 D.

Apparatus

The stimuli consisted of sinewave gratings generated in an oscilloscope Tektronix 608 (green phosphor). The screen subtended 3.05° wide and 2.45° height at a viewing distance of 228 cm. The oscilloscope was connected to a Picasso CRT Image Synthesizer (Innisfree Inc.) which was controlled by an IBM personal computer. The mean luminance of the screen was $20 cd/m^2$. Gratings were viewed monocularly through a circular window generated at the oscilloscope screen. Two window sizes were used: $3.14°^2$ and $0.785°^2$ for two different experimental series. Fifteen different spatial frequencies ranging 1.44 to 12.1 cycles/degree were tested for four different orientations: vertical, horizontal, oblique 45 and oblique 135.

Procedure

The subjects were seated comfortably in a chair and viewed the screen monocularly, which had a fixation point of 0.1° diameter. The measurement of the contrast sensitivity function was performed following the method of Campbell et al. (1966) with two main differences:

- While the subject looked at the screen one experimenter controlled the computer.

- The stimulus changed in orientation instead of changing the point of view of the subject.

In each case the detection sensitivity was measured instead of the discrimination. The subjects knew the grating orientation, but not the spatial frequency of each grating.

Each experimental series consisted in the presentation of fifteen different spatial frequencies at the four orientations for two different window areas. This gives 15*4*2=120 single stimulus presentations for each subject. Each subject was tested in at least four experimental series given a total of 480 single stimulus presentations per subject.

Each single stimulus presentation was developed as follows: A grating of a random spatial frequency and of a given orientation was presented through a circular window at zero contrast to the subject. From here the contrast was increased by discrete steps, until the subject detected the grating and stopped the contrast increase. The inverse of the contrast was plotted against the corresponding spatial frequency, obtaining the contrast sensitivity curves for each orientation and window. All the contrast sensitivity curves presented here are the mean values of four individual experiments.

The statistical analysis of the data was performed by means of Statview for Apple Macintosh. After an ANOVA test we use the t-test for comparison between the means of the correspondent sensitivities.

3 Results

Despite each of the six subjects had a specific pattern in their contrast sensitivity curves, some general aspects, are commonly present. For instance, the maximum sensitivity was found to be around 4-5 cycles/degree, with a decrease for the sensitivity towards the lower and higher spatial frequencies. The curves were similar in shape to previously described ones (5).

Figure 1 shows the contrast sensitivity curves obtained from the left eye of F.D. for the four tested orientations using a window of 3.14°². In this case the sensitivity was higher for the vertical orientation, and no significant differences between the horizontal and oblique orientations were found. Subjects A.C. and I.G. showed a similar behavior. The other two subjects (P.Z. and R. S.) showed a completely different pattern with a clear preference for the horizontal orientation. In any case the best orientations corresponded with the oblique ones.

The contrast sensitivity curves using a smaller window 0.785°² were also determined in four of the six subjects. Figure 2 shows the comparison between the curves obtained for subject F.D. for the two windows and the two orientations. In the four orientations the sensitivity was lower (p<0.05) for the small window and all the spatial frequencies.

Although the relationship between the area of both windows is 4:1, the change in sensitivity is in the range 2:1. Those results are in agreement with Campbell and Maffei (1970) in spite of these authors only showed this fact for a spatial frequency of 2.5 cycles/degree.

A way of measuring of the oblique effect intensity is though the ratio between the sum of sensitivities for the vertical and horizontal orientations and the sum of sensitivities for the oblique orientations. Mansfield and Ronner (1978) have shown that the mean values of that relationship for spatial frequencies between 1.89 and 4.70 cycles/degree and window areas from $1.82°^2$ to $2.44°^2$ was of 1.47 ± 0.07. Table I shows the ratios found for two subjects (F.D. and A.L.) for three spatial frequencies (2.08, 5.28 and 9.52 cycles/degree) in both windows. In the same table the ratios between the horizontal and the oblique (45°) orientations are shown (the different behavior of the vertical and horizontal orientations has been shown in Figure 1). The values corresponding to the subjects A.C. and R.S. have also been calculated and commented in the text.

The analysis of the data in Table I reveal the following facts: the oblique effect increases in parallel with the spatial frequencies in both windows. The other fact is that the ratio for a spatial frequency of 2.08 cycles/degree and a window of $0.785°^2$, stays below 1 in all the subjects.

The comparison between V+H/O45+O135 and H/O45 suggests that for high spatial frequencies and small windows, the oblique effect is due to the sensitivity for the vertical orientation because the relationship H/O45 is lower than 1 (with the exception of subject F.D.). The results showed in Table I, suggest the existence of a change in the oblique effect depending on the spatial frequency and the window size. Representing the sensitivity against the orientation for each spatial frequency (Figures 3 and 4), is possible to describe two different behaviors, depending on whether we consider, low (1.44-4.2 cycles/degree) or high (4.56-12.12 cycles/degree) spatial frequencies, also with a clear dependence of the window size.

In Figure 3A we can observe the different behavior for high spatial frequencies in subject A.L. at the largest window size. For low spatial frequencies there are not significant differences between horizontal and oblique orientations, while for high spatial frequencies the sensitivity for the horizontal orientation is larger than that for the oblique ones (H/O45 $p<0.085$ and H/O135 $p<0.05$). Figure 3B shows the data for the same subject but using the smallest window size. In this case, there is not a predominant orientation for low spatial frequencies (including the vertical), but for spatial frequencies larger than 5 cycles/degree, the oblique orientations had higher sensitivity than the horizontal ones. This behavior was also observed for subject A.C. and for subject F.D. (in this last case only for the large window size). Subject R.S. showed no difference between low and high spatial frequencies.

4 Discussion

Is it possible that the sensitivity in the detection of a grating with an oblique orientation is higher than that for the so-called main orientations (vertical or horizontal)?

From our results, the answer to that question seems to be 'yes' in some circumstances. This is possible if we consider that the oblique effect, as given by the relationship V+H/O45+O135 showed in Table I, could be due to the fact that the increase in the oblique effect corresponding with the increase in spatial frequency, arises from the vertical orientation sensitivity but not for the horizontal one. This finding is pointed out in Table I when by comparing column V+H/O45+O135 and column H/O45.

If we compare our Figure 1 with Figures 2 and 5 from Campbell et al. (1966), we see that both cases show that anysotropy rises with the increase in spatial frequency. But it is necessary for spatial frequencies higher than 4-5 cycles/degree for the horizontal orientation sensitivity to be better than the oblique ones. Similar findings were reported by Campbell et al. (1966) for the vertical and oblique 45 orientations.

From the above mentioned facts, it is not possible to assume that a greater sensitivity to the main orientations is only due to a large amount of receptors for these orientations (12, 13).

Electrophysiological experiments cannot support the fact of a higher sensitivity for the main orientations than for the oblique ones (6). These authors have not found significant differences in the visual evoked potentials obtained in the cat's visual cortex using low spatial frequency gratings ranging 0.26 to 1.7 cycles/degree for any orientation. Other authors (14) searching for cells implied as orientation detectors, have found no differences between the main orientations and the oblique ones. By the other hand, the results of Blakemore and Cooper (1970) dealing with plasticity in the visual system seem to support the fact that the environment determines the orientation specificity of the neurons.

If we increase the stimulus area, we can assume a proportional increase in the total amount of receptors implied in the detection of that stimulus. If the total amount of receptors is the same for each orientation, then, it would be the same increase in sensitivity for all the orientations. We do not find this in our results but we do find an increase in the window area evoke the increase in sensitivity for the vertical and horizontal orientations in comparison with the oblique ones. A similar effect has also been observed by Tootle and Berkley (1983). However, these authors reported that this behavior is independent of the spatial frequency of the grating (although they only used high spatial frequencies: 10, 14 and 20 cycles/degree). In our case the increase in sensitivity is proportionally inverse to the spatial frequency increase. At the same time Tootle and Berkley have considered that the area increase affects the oblique effect, as has been also pointed out by Essock (1990).

We have found a considerable influence of the window area on the oblique effect. For instance, with the small window we use $(0.785°^2)$, the oblique effect could disappear and even a higher sensitivity for the oblique orientations than the horizontal ones could occur. Other authors have also shown, that the increase in the spatial frequency, increases the anysotropy in the sinewave gratings detection (8)

Our visual environment has a considerable amount of "vertical" stimulus, this fact is present in the orientation test in a majority of authors, where that orientation shows the best sensitivity. From this point of view any other orientation seems to be "tilted", and even the horizontal orientation could be considered as an oblique 90. When the stimulus size is not very small, then the representation of the horizontal orientation is bigger than the other oblique orientations, because of their relative abundance in the environment. However, when the stimulus area is very small and the spatial frequencies high, it seems to be important to have a good representation of all the orientations in order to have a detailed visual spatial representation.

5 References

1. Andrews, D.P. (1967). Perception of contour orientation in the central fovea. Part I: short lines. *Vision Research*, 7, 975-997.

2. Appelle, S.(1972). Perception and discrimination as a function of stimulus orientation: the "oblique effect" in man and animals. *Psychological Bulletin*, 78, 266-278.

3. Blakemore, C. & Campbell, F.W. (1969). On the existence of neurones in the human visual system selectively sensitive to the orientation and size of retinal images. *Journal of Physiology, 203*, 237-260.

4. Blakemore, C. & Cooper, G.F.(1970). Development of the brain depends on the visual environment. *Nature, 228*, 477-478.

5. Campbell, F.W., Kulikowski, J.J. & Levinson, J. (1966). The effect of orientation on the visual resolution of gratings. *Journal of Physiology, 187*, 427-436.

6. Campbell, F.W., Maffei, L. & Piccolino, M. (1973). The contrast sensitivity of the cat. *Journal of Physiology, 229*, 719-731.

7. Essock, E.A. (1990). The influence of stimulus length on the oblique effect of contrast sensitivity. *Vision Research, 30*, 1243-1246.

8. Heeley, D.W. & Timney, B. (1989). Spatial frequency discriminations at different orientations. *Vision Research, 29*, 1221-1228.

9. Hirsch, J. & Hylton, R. (1984). Orientation dependence of hiperacuity contains a component with hexagonal symmetry. *Journal of the Optical Society of America, A, 1*, 300-308.

10. Hubel, D.H. & Wiesel, T.N. (1962) Receptive fields, binocular interaction and functional architecture in the cat's visual cortex. *Journal of Physiology, 160*, 106-154.

11. Maffei, L. & Campbell, F.W. (1970). Neurophysiological localization of the vertical and horizontal coordinates in man. *Science, 167*, 386-387.

12. Mansfield, R.J.W. (1974). Neural basis of orientation perception in primate vision. *Science, 186*, 1133-1135.

13. Mansfield, R.J.W. & Ronner, S.F. (1978). Orientation anisotropy in monkey visual cortex. *Brain Research, 149*, 229-234.

14. Rose, D. & Blakemore, C. (1974). An analysis of orientation selectivity in the cat's visual cortex. *Experimental Brain Research, 20*, 1-17.

15. Timney, B. & Muir, D.W. (1976). Orientation anisotropy: Incidence and magnitude in Caucasian and Chinese subjects. *Science, 193*, 699-700.

16. Tootle, J.S. & Berkley, M.A. (1983). Contrast sensitivity for vertically and obliquely oriented gratings as a function of grating area. *Vision Research, 23*, 907-910.

17. Vidyasagar, T.R. & Sigüenza, J.A. (1985). Relationship between orientation tuning and spatial frequency in neurones of cat area 17. *Experimental Brain Research, 57*, 628-631.

18. Weymouth, F. W. (1960) Stimulus orientation and threshold: an optical analysis. *Journal of the Optical Society of America, 63*, 763-765.

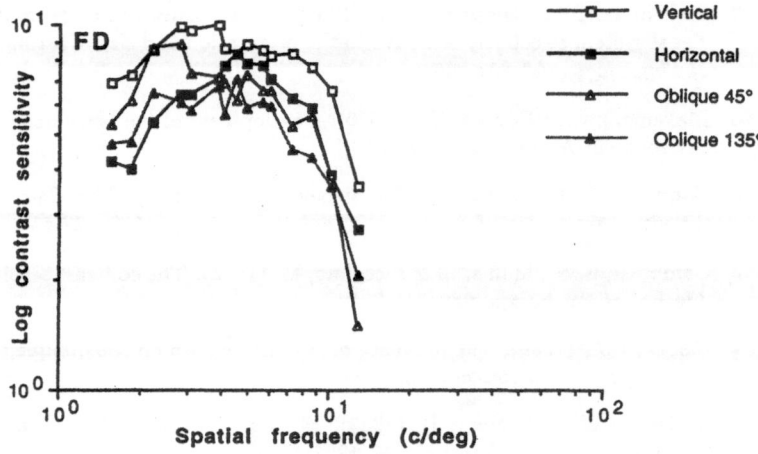

Figure 1: Contrast sensitivity curve obtained from subject FD, for the four orientations we analyzed: vertical, horizontal, oblique 45 and oblique 135. Each curve represented the mean value from four single experiment.

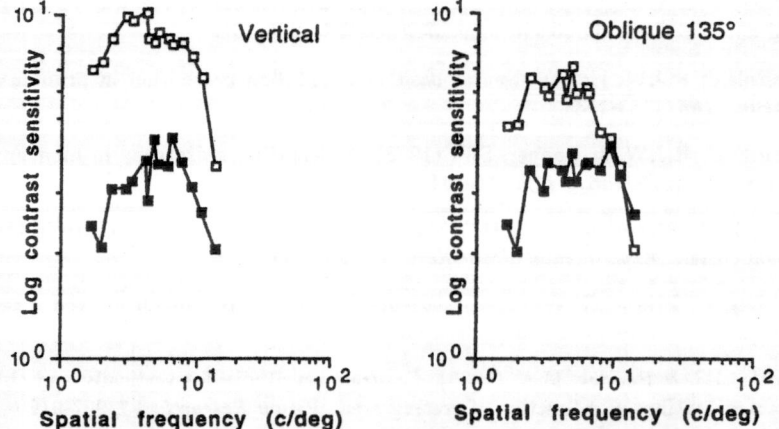

Figure 2: This figure shows differences between contrast sensitivity curves obtained from subject FD, for twop different stimulus size: A=3.14°² (open squares) and B= 0.785°² (filled squares) and for two orientations: Vertical (left) and Oblique 135 (right).

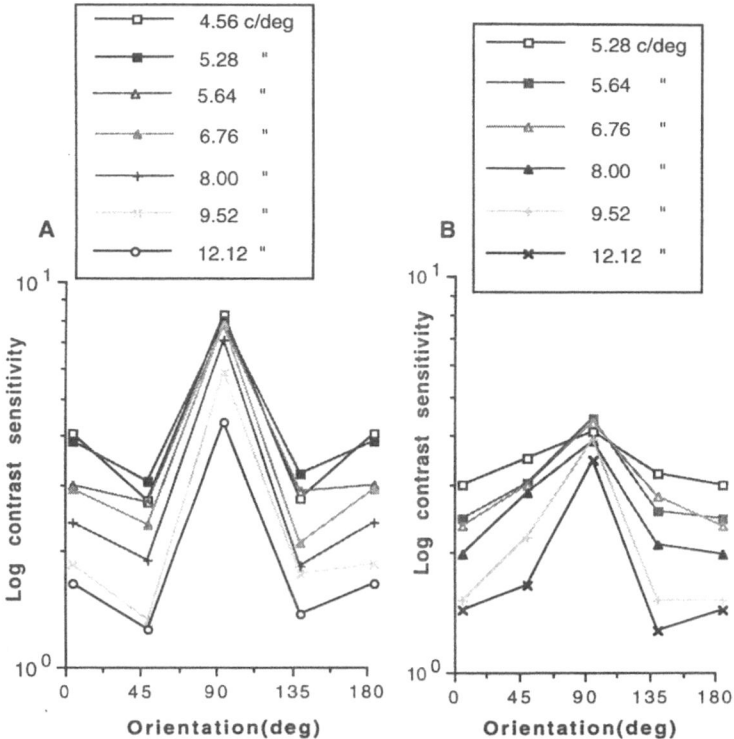

Figure 3: Values of contrast sensitivity obtained from subject AL for a stimulus size of $3.14°^2$ (A) and $0.785°^2$ (B) and represented as a function of the orientation: 0°= Horizontal, 45°=Oblique 45, 90°=Vertical, 135°=Oblique 135, 180°=Horizontal.

Subject	SF	Area	V+H	O45+O135	V+H/O45+O135	H/O45
FD	2.08	3.140	13.225	14.425	0.92	0.62
FD	5.28	3.140	15.475	12.075	1.28	1.17
FD	9.52	3.140	9.925	7.075	1.40	0.99
FD	2.08	0.785	5.175	6.475	0.80	0.71
FD	5.28	0.785	6.825	6.775	1.01	1.01
FD	9.52	0.785	5.175	5.975	0.87	0.95
AL	2.08	3.140	6.900	5.140	1.34	0.90
AL	5.28	3.140	11.340	5.970	1.90	1.27
AL	9.52	3.140	7.340	2.940	2.50	1.39
AL	2.08	0.785	4.400	4.500	0.89	1.07
AL	5.28	0.785	6.750	6.420	1.05	0.85
AL	9.52	0.785	5.175	3.550	1.45	0.69

Table 1: This table shows the measurementof the oblique effect for subjects FD and AL, as the ratio between the sum of sensitivities for the vertical and horizontal orientations (V+H) and the sum of sensitivities for the oblique orientations (O45+O135). SF: Spatial frequency (c/deg). Area: Window area (square degrees).

Object Oriented Design of a Simulator for Large BP Neural Networks

J.M.Adamo [†],   D.Anguita [‡]

[†]Université Claude Bernard & LISA CPE-Lyon, 25 rue du Plat, 69288 Lyon cedex 02, France
e–mail: adamo@cpe.ipl.fr
[‡]University of Genova, D.I.B.E., Via Opera Pia 11a, 16145 Genova, Italy
e–mail: anguita@dibe.unige.it

Abstract

In this paper we describe the implementation of the backpropagation algorithm by means of an object oriented library (ARCH). The use of this library relieves the user from the details of a specific parallel programming machines and at the same time allows a greater portability of the generated code.

To provide a comparison with existing solutions, we survey the most relevant implementations of the algorithm proposed so far in the literature, both on dedicated and general purpose computers.

Extensive experimental results show that the use of the library does not hurt the performance of our simulator, on the contrary our implementation on a Connection Machine (CM–5) is comparable with the fastest in its category.

1 Introduction.

Since its introduction, the backpropagation algorithm and its variants have been implemented on an innumerable amount of general purpose machines. As the demand for computational power increased, the implementation shifted towards dedicated architectures. Both the purpose and the result of these implementations are varied but the basic idea is to provide the user with an effective tool for fast NN learning.

To compare our implementation to existing solutions, we present here a survey of what is available in literature so far, hoping to point out their advantages and disadvantages. We will deliberately mix general purpose and dedicated machines (or, in other words, software and hardware implementations) without addressing the issue of which solution is the most convenient for a particular application. This choice depends upon too many factors and is beyond the scope of this paper. The model addressed will be the Multi Layer Perceptron and its classical learning algorithm (BP).

The effectiveness of an implementation can be measured in several ways: our choice in this paper has been to use some widely accepted parameters. The most commonly used (and abused) is the efficiency of the implementation or, more precisely, the speed at which the weights of the network can be updated. This speed is usually measured in MCUPS (or Millions of Connection Updates per Second).

Some implementations differ in the arithmetic used for the computations. Many dedicated architectures take advantage of the greater speed of fixed–point operations (compared to floating–point) to achieve high performances to the detriment of precision. Obviously, all the general purpose machines use floating–point arithmetic instead.

One of the important characteristic of an implementation is the type of backpropagation algorithm used. In general the use of the batch version allows a better utilization of the hardware because of greater parallelism. On the other hand, the on–line version shows a faster convergence in some cases expecially with large databases.

In the following section we will compare briefly some of the implementations proposed in the literature. In section 3 the details of our implementation are presented. Experimental results regarding its performance are reported in section 4.

2 Implementations of bp: the state of the art.

In Table 1 some of the best–known solutions proposed in literature are presented.

Table 1: General purpose and dedicated implementations of the backpropagation.

Computer	MCUPS	Problem size	Alg.	FP	Ref.
CNS–1 (128/1024)	$22000^1/166000^1$	–	P	N	[5, 22]
Adapt. Sol. CNAPS (512)	2379	$1900 \times 500 \times 12$	P	N	[13]
Sony GCN–860 (128)	1000^1	$256 \times 80 \times 32,5120$	–	Y	[12]
Sandy/8 (256)	$118/567^1$	NETtalk/peak	P	Y	[15]
TMC CM–2 (64k)	350	$128 \times 128 \times 128, 65536$	E	Y	[27]
HNC SNAP (16/64)	80.4/302	$512 \times 512 \times 512$	–	Y	[1]
MUSIC (60)	247	–	P	Y	[20, 21]
MANTRA I (1600)	133^1	–	P	N	[28]
RAP (40)	102	$640 \times 640 \times 640$	P	Y	[19]
SPERT	100^1	$512 \times 512 \times 512$	P	N	[29, 6]
TMC CM–5 (512)	76	$256 \times 256 \times 131072, 111$	P	Y	[16]
FUJITSU VP–2400/10	60	NETtalk	P	Y	[25]
A.C.A. (4225)	51.4^1	NETtalk	P	N	[10]
Cray Y–MP (2)	40	$256 \times 256 \times 131072, 111$	P	Y	[16]
TMC CM–2 (4k/64k)	$2.5/40^1$	$256 \times 128 \times 256, 64$	B	Y	[30]
Cray X–MP (4)	18	$256 \times 256 \times 131072, 111$	P	Y	[16]
IBM 6000/550	17.6	$500 \times 500 \times 1, 1000$	E	Y	[4]
Intel iPSC/860 (32)	11	NETtalk	B	Y	[14]
Cray 2 (4)	10	$256 \times 256 \times 131072, 111$	P	Y	[16]
DEC Alpha	3.2	–	P	Y	[20]
Sun SparcStation 10	1.1	–	P	Y	[20]
Inmos T800 (16)	0.7	192 units (3 layers), 128	B	Y	[23]
PC486	0.47	–	P	Y	[20]
MasPar MP–1	0.3	–	–	Y	[11]

In the first column, the name of the system on which a backpropagation implementation has been realized is reported. The number of processors (if greater than one) is reported in parenthesis. Note that most of the dedicated systems use massive parallelism in order to exploit the native parallelism of the algorithm, while general purpose computers are based on more conventional architectures (with the most notable exception of the Connection Machine).

All general purpose systems are commercially available, while only three of the dedicated machines come from a non–academic environment (Adaptive Solution CNAPS, HNC SNAP and Siemens SYNAPSE). The last system is not included in the table because the MCUPS figure is not published (we could find only references to MCPS). The building block of SYNAPSE is a systolic fixed–point matrix–matrix multiplier (MA–16) with a peak performance of 800 MCPS. A system with eight MA–16 has been reported to perform at 5.3 GCPS [24].

In the second column the performances of the systems are reported (in MCUPS). Note that some of the values are not actual runs of the implementation but estimates[1]. The top lines of the table are occupied by dedicated systems while conventional workstations and super–computers lie in the bottom part. The fastest implementation on a general purpose (super)computer reaches 350 MCUPS using a Connection Machine 2 with 64k processors.

Scanning the table one can make surprising comparisons. For example, the fastest implementation on a large–grain supercomputer (FUJITSU VP–2400/10) outperforms the a conventional workstation (IBM 6000/550) only by a factor of three. A single dedicated microprocessor (SPERT) is an order of magnitude faster than a Cray–2 supercomputer. Note that the performance ratio between the fastest and the slowest implementation is $\sim 500,000$: in other words, a run that takes one hour on the fastest neurocomputer (CNS-1) would require more than fifty years on a conventional personal computer (PC).

The difference in terms of raw computing power between the systems showed in Table 1 is not sufficient to justify such a huge difference in terms of neurocomputing power. Part of the explanation lies in columns 3 and 4 of the table.

Column 3 shows the size of the problem tested on a particular implementation. The first numbers refer to the size of the network and the last one (if present) to the number of patterns in the training set. In some cases the problem is the well–known NETtalk [26]: this is a common benchmark to measure the learning speed of an implementation and allows a fair comparison between different systems. As the size of the problem influences heavily the performances, it is not easy to compare other systems.

Column 4 shows which version of the algorithm has been used. P stands for *by pattern* (or on–line backpropagation), E for *by epoch* (or batch) and B is an intermediate version *by block* [23]. In the first case the weights of the network are updated after each pattern presentation, while in the second the gradient is accumulated through the entire database before doing a learning step. The consequences of the different versions are both on the computational requirement and on the speed of convergence. The last one is outside of the scope of this paper and won't be addressed here (see for example [18]). The effect on the computation affects directly the maximum speed achievable by an implementation. As mentioned before, the on-line version can be implemented through matrix–vector multiplication, while the batch version requires matrix–matrix multiplications [7, 8]. As showed in [9] the latter can be implemented more efficiently on the majority of systems.

Finally, the column labeled FP indicates the arithmetic used by an implementation. As can be seen, the fastest system can obtain such astonishing performances using fixed–point math instead of floating–point. For the same reason, a single dedicated microprocessor (SPERT) can be the fastest single processor system and outperform all the conventional systems (except the CM-2 with 65536 processors). Obviously, all the general purpose systems make use of the floating–point format.

3 Implementation of the simulator with ARCH.

3.1 The ARCH library.

Our implementation is based on an object–oriented library for parallel computing (ARCH) developed by one of the authors. Greater detail about ARCH can be found in [2, 3].

The library provides a layered programming environment that can be accessed at different levels according to the user needs and allows various styles of parallel programming available in a common syntax (that of C++). No programming-primitive insertion was performed in the C++ compiler: one of our purposes was to check how far it would be possible to go, by simply using the extension mechanisms provided by C++.

The library was built from scratch. Threads were introduced first, from which the process class was derived. A thread corresponds to an autonomous line of control. Processes are structured threads with more disciplined activation and termination protocols. Threads and processes are well suited to asynchronous-systems programming. Threads and processes can communicate and synchronize via message passing. A tool- set for message passing has been developed, based on the "Concurrent Sequential Processes" model. Advanced features of C++ were used (templates, etc.) so the C++ compiler can check the correct use of message-passing functions (type checking, send-recv consistent matching).

Global read/write functions are also proposed as an alternative to message-passing functions. Together with barrier-synchronization such functions provide the right tools for loosely-synchronous system programming.

One feature, among the most interesting in the library, lies in its facilitation of efficient parallel programming abstractions. we started with developing the spread pointer and array classes [2]. These are very useful abstractions that allow the user to specify how arrays are spread over the distributed memory, and then use these familiar data structures regardless of data location: the data are accessed the same as in conventional sequential languages.

Building the neural-net simulator, it was quite easy to derive the required spread matrix and vector classes (parallel BLAS) from spread arrays [3]. One of the purposes was to take advantage of the specific architecture of the CM5 nodes (vector units) in order to achieve full efficiency. As matrix and vector operators are loosely-synchronous processes, implementing the spread matrix and vector classes made extensive use of global read/write functions together with barrier-synchronization.

3.2 The algorithm.

Before detailing the implementation of the backpropagation, we will summarize here the algorithm in terms of matrix and vector operations. In particular we will examine both the on–line and batch versions. In the following text we'll use bold letters to indicate vectors and matrices (lower case and capital respectively) and normal letters to indicate scalars.

Let's consider a Multi Layer Perceptron (MLP) with L layers of neurons (in our notation, an MLP with one hidden layer has $L = 2$). The l-th layer is composed by N_l neurons (N_L being the number of outputs and N_0 the number of inputs). The weights of each layer can be stored in a matrix $\mathbf{W}_l$ of size $N_l \times N_{l-1}$. For convenience we will store the biases in a separate vector for each layer $\mathbf{b}_l$ of size N_l.

The learning database is composed by two sets of vectors, the input patterns $I_P = \{s^1, s^2, \ldots, s^{N_P}\}$ and the target patterns $T_P = \{t^1, t^2, \ldots, t^{N_P}\}$. They can be organized in two matrices: $\mathbf{S}_0$ of size $N_P \times N_0$ and $\mathbf{T}$ of size $N_P \times N_L$.

There are only few differences between the two versions. The on–line backprop deals with one vector at a time, choosing it randomly from the database, while in the batch case the whole matrix is involved. The intermediate results for each layer are stored in vectors $\mathbf{s}_1, \ldots, \mathbf{s}_{L-1}$ (or matrices $\mathbf{S}_1, \ldots, \mathbf{S}_{L-1}$ in the batch case). The output of the network is stored in $\mathbf{s}_L$ (or $\mathbf{S}_L$). Vectors ($\mathbf{d}_i$) or matrices ($\mathbf{D}_i$) store the error. The operator $\times$ denotes the element–wise product and the function $sgm'()$ is the first derivative of $sgm()$. Function $rnd()$ returns a value between 1 and N_P and function $sgm()$ computes the sigmoidal activation function for each element of its argument. Vector $\mathbf{e}$ is a column of 1.

On–line version	Batch version
$n := rnd(); \mathbf{s}_0 := \mathbf{S}_0^n$	
for $i := 1$ to N_L	for $i := 1$ to N_L
$\quad \mathbf{s}_i^t := sgm(\mathbf{s}_{i-1}^t \cdot \mathbf{W}_{i-1} + b_i)$	$\quad \mathbf{S}_i := sgm(\mathbf{S}_{i-1} \cdot \mathbf{W}_{i-1} + b_i \cdot e^t)$
$\mathbf{d}_L := (t - \mathbf{s}_L) \times sgm'(\mathbf{s}_L)$	$\mathbf{D}_L := (\mathbf{T} - \mathbf{S}_L) \times sgm'(\mathbf{S}_L)$
for $i := N_{L-1}$ to 1	for $i := N_{L-1}$ to 1
$\quad \mathbf{d}_i^t := \mathbf{d}_{i+1}^t \cdot \mathbf{W}_{i+1}$	$\quad \mathbf{D}_i := \mathbf{D}_{i+1} \cdot \mathbf{W}_{i+1}$
for $i := 1$ to N_L	for $i := 1$ to N_L
$\quad \Delta \mathbf{W}_i = \mathbf{d}_i \cdot \mathbf{s}_{i-1}^t$	$\quad \Delta \mathbf{W}_i = \mathbf{S}_{i-1}^t \cdot \mathbf{D}_i$
$\quad \Delta b_i = \mathbf{d}_i$	$\quad \Delta b_i = \mathbf{D}_i^t \cdot e$
for $i := 1$ to N_L	for $i := 1$ to N_L
$\quad \mathbf{W}_i \mathrel{+}= \eta_k \Delta \mathbf{W}_i^{new} + \alpha_k \Delta \mathbf{W}_i^{old}$	$\quad \mathbf{W}_i \mathrel{+}= \eta_k \Delta \mathbf{W}_i^{new} + \alpha_k \Delta \mathbf{W}_i^{old}$
$\quad b_i \mathrel{+}= \eta_k \Delta b_i^{new} + \alpha_k \Delta b_i^{old}$	$\quad b_i \mathrel{+}= \eta_k \Delta b_i^{new} + \alpha_k \Delta b_i^{old}$

3.3   The implementation with ARCH.

In this section we will detail the implementation of the algorithm using the ARCH library.

The following code is the declaration of the class *Neural_net* for the batch algorithm. Whenever possible, the same notation of the algorithm described in the preceding section has been used. The template of the class defines the type of the variables $\{float, double\}$ and the number of layers of the network (L).

```
template<class T, int L>
class Neural_net{

//forward
SpreadMatrices<T, 4, 8> *S[L];          // Matrices S_i
SpreadMatrices<T, 4, 8> *W[L-1];        // Matrices W_i
SpreadVectors<T, 4, 8>  *B[L-1];        // Vectors b_i

//backward
SpreadMatrices<T, 4, 8> *TG_ptr;        // Matrix T
SpreadMatrices<T, 4, 8> *D[L-1];        // Matrices D_i
SpreadMatrices<T, 4, 8> *DW[2][L-1];    // Matrices \Delta W_i

SpreadMatrices<T, 4, 8> *TG_ptr;

DSpreadMatrix<T, 4, 8> *aux_ptr;

int SWITCH;
T eta, alpha, e_new, e_old;

//const
T c_alpha, c1_eta, c2_eta;

   public:

       //constructor
       Neural_net(int *N, char* in_file, char *t_file);

       //learning
       void back_prop(int *N);

       //keep_on_learning condition (user defined function)
       int keep_on_learning(int iter);

       //matrix and vector initialization (user defined functions)
       void data_file(SpreadMatrices<T,4,8> *Matrix, char *data_file);

       void matrix_initialization(SpreadMatrices<T,4,8> *Matrix, T Range);

       void vector_initialization(SpreadVectors<T,4,8> *Vector, T Range);

};
```

Two copies of the matrices ΔW_i and vectors Δb_i are kept in memory because the Vogl's algorithm requires a backtracking if the current learning step is not correct.

The constructor *Neural_net()* requires the number of neurons in each layer and the number of patterns (N[]), the learning data file (in_file) and the test set (t_file) data file. Method *back_prop()* implements the learning algorithm. It is executed until the exit condition is satisfied (supplied by the user with the method *keep_on_learning()*).The last three methods are for data initialization.

Here is the code for the constructor *Neural_net()*. It performs all the basic initializations including the input from data files and the random initialization of weights.

```
template<class T, int L>
Neural_net<T, L>::Neural_net(int *N, char* in_file, char *t_file){

    for(int i=0; i<L; i++){
        S[i] = new SpreadMatrices<T, 4, 8>(N[L], N[i]);
        if(!i) data_file(S[i], in_file);
    }

    for(i=0; i<L-1; i++){

        W[i] =   new SpreadMatrices<T, 4, 8>(N[i], N[i+1]);
        matrix_initialization(W[i], (T) N[i]+1);

        B[i] = new SpreadVectors<T, 4, 8>(N[i+1]);
        vector_initialization(B[i], (T) N[i]+1);

        D[i] = new SpreadMatrices<T, 4, 8>(N[L], N[i+1]);

        DW[0][i] = new SpreadMatrices<T, 4, 8>(N[i], N[i+1]);
        DW[0][i] = new SpreadMatrices<T, 4, 8>(N[i], N[i+1]);

        DW[1][i] = new SpreadMatrices<T, 4, 8>(N[i], N[i+1]);

        DB[0][i] = new SpreadVectors<T, 4, 8>(N[i+1]);

        DB[1][i] = new SpreadVectors<T, 4, 8>(N[i+1]);
    }

        TG_ptr =    new SpreadMatrices<T, 4, 8>(N[L], N[L-1]);
        data_file(TG_ptr, t_file);

        aux_ptr = new DSpreadMatrix<T, 4, 8>(N[L], N[L-1]);

        SWITCH = 0;
        c_alpha = ALPHA;
        c1_eta  = ACCELERATION;
        c2_eta  = DECELERATION;
        eta = ETA;
}
```

The main part of the simulator lies in the *back_prop()* method:

```
template<class T, int L>
void Neural_net<T, L>::back_prop(int *N){

    int iter = 0;
    e_old = (T) 1.0e6;

    barrier_synchronization();

    while (keep_on_learning(iter)){
        iter++;
        int i;
        for(i=1; i<L; i++){
            S[i]->Vector_to_Matrix_Vexpansion(B[i-1]);
            S[i]->C_equal_u_D_plus_v_A_mult_B(1, 1, S[i], S[i-1], W[i-1]);
            S[i]->TANH();
}
```

```
D[L-2]->C_equal_A_minus_B(TG_ptr, S[L-1]);
e_new = (T)1.0/(N[L]*N[L-1])*(D[L-2]->glob_norm());
if(e_new < e_old){
    eta *= c1_eta;
    alpha = c_alpha;
    S[L-1]->C_equal_A_evmult_B(S[L-1], S[L-1]);
    S[L-1]->C_equal_u_minus_A(1, S[L-1]);
    D[L-2]->C_equal_A_evmult_B(D[L-2], S[L-1]);
    D[L-2]->C_equal_u_A((T)2.0/(N[L]*N[L-1]), D[L-2]);

    //back
    for(i=L-2; i>0; i--){
        D[i-1]->C_equal_A_mult_tB(D[i], W[i]);
        aux_ptr->on(N[L], N[i]);
        aux_ptr->C_equal_A_evmult_B(S[i], S[i]);
        aux_ptr->C_equal_u_minus_A(1, aux_ptr);
        D[i-1]->C_equal_A_evmult_B(D[i-1], aux_ptr);
        aux_ptr->off();
    }

    int SWITCH_new = (SWITCH+1)%2;
    for(i=0; i<L-1; i++){
        DW[SWITCH_new][i]->C_equal_u_D_plus_v_tA_mult_B
                          (alpha, eta, DW[SWITCH][i],S[i],D[i]);
        D[i]->Matrix_to_Vector_Vreduction(DB[SWITCH_new][i]);
        DB[SWITCH_new][i]->C_equal_u_A_plus_v_B
                          (alpha, eta, DB[SWITCH][i], DB[SWITCH_new][i]);
        W[i]->C_equal_A_plus_B(W[i], DW[SWITCH_new][i]);
        B[i]->C_equal_A_plus_B(B[i], DB[SWITCH_new][i]);
    }
    SWITCH = SWITCH_new;

}else{
        eta *= c2_eta;
        alpha = 0;
        for(i=0; i<L-1; i++){
            W[i]->C_equal_A_minus_B(W[i], DW[SWITCH][i]);
            B[i]->C_equal_A_minus_B(B[i], DB[SWITCH][i]);
        }
    e_old = e_new;
    }
  }
}
```

The code follows exactly the structure of the algorithm detailed in the previous section.

4 Experimental results.

In Tables 2 and 3 the speed in MCUPS is reported for different problem sizes. In particular, a two–layer network with the same number of neurons in each layer was used. For the batch/block version of the algorithm, the number of patterns of the learning set was varied (obviously, the speed of the on-line version is not particularly affected by this parameter). The configuration of the CM-5 is 32 processors with four vector units each.

Table 2: Speed (in MCUPS) for the batch/block version.

Np	Neurons per layer						
	32	64	128	256	512	1024	2048
512	3.98	9.90	21.02	43.26	78.38	123.4	177.4
2048	6.44	13.13	28.17	56.39	94.16	161.5	245.7

The block/batch version, as expected, is far more efficient than the on-line version of the algorithm. Note that the performance is less affected by the size of the training set than by the size of the network.

Table 3: Speed (in MCUPS) for the on-line version.

Neurons per layer					
32	64	128	256	512	1024
0.1	0.36	1.15	2.57	3.95	4.75

In Table 4 some results with networks derived from real–world applications are showed.

Table 4: Speed (in MCUPS) for some real–world applications.

Application	Size	MCUPS
NETtalk	$203 \times 80 \times 26$	18.33
Data compression	$512 \times 64 \times 512$	36.87
Speech recognition	$234 \times 1024 \times 61$	72.42

5   Conclusions.

An implementation of the back-propagation algorithm based on the ARCH parallel object-oriented library has been presented. Any neural network of any size can be instanciated and executed.

Experimental results show that the implementation is well–suited for problems with large networks and when the batch/block version of the algorithm is used. In this case, our solution outperforms most of the currently available implementations on non–dedicated architectures.

6   Acknowledgments.

We would like to thank Prof. J.Feldman for letting us experiment with the Connection Machine 5 at Univ. of California, Berkeley, USA; Gerd Aschemann (Institut für Systemarchitektur, Darmstadt, Germany) for his contribution in debugging the latest version of ARCH and Eric Fraser (University of Berkeley) for his effective administration of the CM-5. Most of the work was developed while the authors were visiting researchers at the Int. Computer Science Institute, Berkeley, USA.

References

[1] *SNAP – SIMD Numerical Array Processor*. HNC, 5930 Cornerstone Court West, S.Diego, CA, 1994.

[2] J.M.Adamo. *Object-Oriented Parallel Programming: Design and Development of an Object-Oriented Library for SPMD Programming*. ICSI Technical Report TR-94-011, February 1994.

[3] J.M.Adamo. *Development of Parallel BLAS with ARCH Object–Oriented Parallel Library, Implementation on CM-5*. ICSI Technical Report TR-94-045, August 1994.

[4] D.Anguita, G.Parodi, and R.Zunino. *An Efficient Implementation of BP on RISC-based Workstations*. Neurocomputing 6, 1994, pp. 57–65.

[5] K.Asanović, J.Beck, T.Callahan, J.Feldman, B.Irissou, B.Kingsbury, P.Kohn, J.Lazzaro, N.Morgan, D.Stoutamire and J.Wawrzynek. *CNS-1 Architecture Specification*. ICSI Technical Report TR-93-021, April 1993.

[6] K.Asanović, J.Beck, B.E.D.Kingsbury, P.Kohn, N.Morgan, J.Wawrzynek. *SPERT: A VLIW/SIMD Microprocessor for Artificial Neural Network Computations*. ICSI Tech. Rep. TR-91-072, January 1992.

[7] A.Corana, C.Rolando, S.Ridella. *A Highly Efficient Implementation of Back-propagation Algorithm on SIMD Computers*. In High Performance Computing, J.-L.Delhaye and E.Gelenbe (Eds.), Elsevier, 1989, pp.181–190.

[8] A.Corana, C.Rolando, S.Ridella. *Use of Level 3 BLAS Kernels in Neural Networks: The Back–propagation algorithm*. Parallel Computing 89, 1990, pp.269-274.

[9] J.Dongarra. *Linear Algebra Library for High-Performance Computers*. Frontiers of Supercomputing II. K.R.Ames and A.Brenner (Eds.), University of California Press, 1994.

[10] B.Faure, G.Mazare. *Implementation of back-propagation on a VLSI asynchronous cellular architecture.* Proc. of Int. NN Conf., July 9–13, 1990, Paris, France, pp. 631–634.

[11] K.A.Grajski, G.Chinn, C.Chen, C.Kuszmaul, S.Tomboulian. *Neural Network Simulation on the MasPar MP-1 Massively Parallel Processor.* Proc. of Int. NN Conf., July 9–13, 1990, Paris, France, p.673.

[12] A.Hiraiwa, S.Kurosu, S.Arisawa, M.Inoue. *A two level pipeline RISC processor array for ANN.* Proc. of the Int. Joint Conf. on NN, January 15–19, 1990, Washinghton, DC, pp.II137–II140.

[13] P.Ienne. *Architectures for Neuro-Computers: Review and Performance Evaluation.* Technical Report 93/21, Swiss Federal Institute of Technology, Lausanne, January 1993.

[14] D.Jackson, D.Hammerstrom. *Distributing Back Propagation Networks Over the Intel iPSC/860 Hypercube.* Proc. of the Int. Joint Conf. on NN, July 8–12, 1991, Seattle, WA, pp. I569–I574.

[15] H.Kato, H.Yoshizawa, H.Iciki, K.Asakawa. *A Parallel Neurocomputer Architecture towards Billion Connection Update Per Second.* Proc. of the Int. Joint Conf. on NN, January 15–19, 1990, Washinghton, DC, pp.I147–I150.

[16] X.Liu and G.L.Wilcox. *Benchmarking of the CM-5 and the Cray Machines with a Very Large Backpropagation Neural Network.* Proc. of IEEE Int. Conf. on NN, June 28 – July 2, 1994, Orlando, FL, pp.22–27.

[17] R.Means, L.Lisenbee. *Extensible Linear Floating Point SIMD Neurocomputer Array Processor.* Proc. of the Int. Joint Conf. on NN, July 8–12, 1991, Seattle, WA, pp. I587–I592.

[18] M.Moller. *Supervised Learning on Large Redundant Training Sets.* Int. J. of Neural Systems, Vol.4, No.1, 1993.

[19] N.Morgan, J.Beck, P.Kohn, J.Bilmes, E.Allman, J.Beer. *The Ring Array Processor: A Multiprocessing Peripheral for Connectionist Applications.* Journal of Parallel and Distributed Computing, Vol.14, N.3, March 1992, pp.248–259.

[20] U.A.Müller. *A High Performance Neural Net Simulation Environment.* Proc. of IEEE Int. Conf. on NN, June 28 – July 2, 1994, Orlando, FL, pp.1–4.

[21] U.A.Müller, M.Kocheisen, A.Gunzinger. *High-Performance Neural Net Simulation on a Multiprocessor System with "Intelligent" Communication.* Advances in Neural Information Processing Systems 6, J.D.Cowan, G.Tesauro, J.Alspector (Eds.), Morgan Kaufmann Publ., 1994, pp.888–895.

[22] S.M.Müller. *A Performance Analysis of the CNS-1 on Large, Dense Backpropagation Networks.* ICSI Technical Report TR-93-046, September 1993.

[23] A.Petrowsky, G.Dreyfus, *Performance Analysis of a Pipelined Backpropagation Parallel Algorithm.* IEEE Trans. on Neural Networks, Vol.4, No.6, Nov. 1993, pp.970–981.

[24] U.Ramacher. *SYNAPSE – A Neurocomputer That Synthesizes Neural Algorithms on a Parallel Systolic Engine.* Journal of Parallel and Distributed Computing, Vol.14, N.3, March 1992, pp.306–318.

[25] E.Sànchez, S.Barro, C.V.Regueiro. *Artificial Neural Networks Implementation on Vectorial Supercomputers.* Proc. of IEEE Int. Conf. on NN, June 28 – July 2, 1994, Orlando, FL, pp. 3938–3943.

[26] T.J.Sejnowsky and C.R.Rosenberg. *Parallel Networks that Learn to Pronounce English Text.* Complex Systems 1, 1987.

[27] A.Singer. *Exploiting the Inherent Parallelism of Artificial Neural Networks to Achieve 1300 Million Interconnets per Second.* Proc. of Int. NN Conf., July 9–13, 1990, Paris, France, pp.656–660.

[28] M.A.Viredaz. *MANTRA I: An SIMD Processor Array for Neural Computation.* Proc. of the Euro-ARCH '93 Conf., München, October 1993.

[29] J.Wawrzynek, K.Asanović, and N.Morgan. *The Design of a Neuro-Microprocessor.* IEEE Trans. on NN, Vol.4, No.3, May 1993.

[30] X.Zhang, M.Mckenna, J.P.Mesirov, D.L.Waltz. *An Efficient Implementation of the Back-Propagation Algorithm on the Connection Machine CM-2.* Advances in Neural Information Processing Systems 2, D.S.Touretzky (Ed.), Morgan Kaufmann Publ., 1990, pp.801–809.

Introducing XSim: A Neural Network Simulator that Incorporates Biological Parameters

Pablo Varona, J. A. Sigüenza.
varona@iic.uam.es, siguenza@iic.uam.es

Instituto de Ingeniería del Conocimiento, Mód. C–XVI, P. 4, Universidad Autónoma de Madrid. 28049 Madrid, Spain.
Departamento de Ingeniería Informática, Mód. C–XVI. Universidad Autónoma de Madrid. 28049 Madrid, Spain.

Abstract

XSim is a neural network simulator that allows to construct a wide variety of neural network architectures. It is specially designed to implement neural networks that incorporate any kind of biological parameters. XSim supports different levels of biological modeling, from basic single-cell models to networks with thousands of units. XSim's principal features include: a command level interface and a graphical interface to facilitate user interaction with the system and the network; interactive and batch processing of commands and data; temporal and spatial displays of the network's performance; library of cell types with different kinds of compartments; easy and flexible construction of networks in three dimensions with complex connectivity patterns and weight distributions; and a methodology for extending the command and cell type library.

1 Introduction

Theoretical modelers of biological neural networks often spend a great amount of time implementing their models in a computational language. The need for tools that facilitate the implementation of the models has been increasing during the last years as new models appear. There are several neural network simulators available for the neuroscience community [1, 2, 3]. Some of them are of very specific use and often require a great amount of training. We introduce XSim Neural Network Simulator as an easy-to-use tool to implement neural networks incorporating biological parameters. XSim has an extensive library of neuronal paradigms and includes an easy methodology to incorporate new models.

XSim has been developed at the Instituto de Ingeniería del Conocimiento, Universidad Autónoma de Madrid. Initially, XSim was intended to be an expansion of the Rochester Connectionist Simulator (RCS)[4]. Finally, and although XSim inherits some of the ideas and structures of RCS, it ended as a new neural network simulator with its own internal and external features. XSim has been implemented using the C programming language and has been successfully tested in several platforms[1] supporting UNIX and X Window System (Motif required to compile source code).

---

[1]IBM RS-6000, Silicon Graphics, i486 under Linux.

2   XSim's compartmental design

In order to carry out its task, XSim makes use of the standard computational elements of artificial connectionist networks which are units and connections. In addition, XSim provides a rather detailed cell design that allows to perform complex information processing inside each single cell, and to establish complex connectivity patterns among the cells. This architecture is useful to incorporate biological parameters that are thought to have an active roll in the task of information processing performed by real neurons.

Many neurosimulators –specially those developed for single-neuron modeling– divide the neuron into a number of compartments. The compartmental approach [5] represents a spatial discretization of the neuron so that each compartment has its own activity and contributes to the overall operation of the neuron. This approach facilitates the mathematical and computational treatment of the model.

XSim has not been thought to be used for complex single-neuron modeling, there are other tools more suitable for this task –see [1]–. Yet, the simulator makes use of a compartmental approach to take advantage of a detailed cell design while constructing networks with several units. XSim can act as a single-neuron simulator when testing new paradigms and when adjusting the cell's parameters. During this task, the simulator allows to monitor the behavior of each compartment. Later on, the compartments can also be used to create a complex pattern of links among units, allowing the cell to receive connections in different compartments.

Compartmental structures in XSim

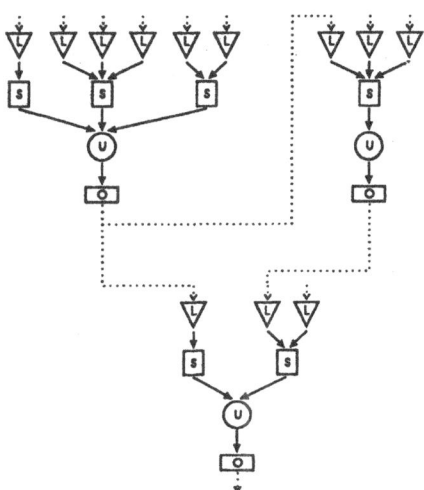

Figure 1: XSim's compartments and mechanism of action. **L**: Link (synapsis), **S**: Site (dendrite), **U**: Unit (cell body), **O**: Output (axon).

The cell design in XSim is built out of three kinds of compartments: *unit*, *sites* and *links* –see figure 1–. Each neuron has a single unit compartment, but it can have several sites and links. A cell made out in this way behaves like a little network. Each compartment can receive weighted information from the preceding compartments, and after making some computation on it, will send an output to the next compartment. There is a clear biological resemblance in these structures that helps to understand the reason and advantages of this compartmental design. Unit would act as the neuron body, sites would be the neuron's dendrites and links would act as the axon's synapses. The inputs coming to the cell through the links are gathered and weighted by the different sites and then sent to the unit. The unit compartment then evaluates the input from all its sites and produces the cell's output. This output is sent through a simulated axon and received by other cells through their respective links.

XSim's compartments follow an object oriented philosophy for their internal design. Each compartment has associated two structures: a structure which is common to all compartments of the same type, and a private structure that contains the parameters specific to that particular compartment. In XSim, one begins creating the type structure out of a neuronal template, and then one builds the compartment attaching this type to the compartment's private structure. Thus, a typical neuron consists of the following structures: unit, unit type, site, site type, link and link type. The type structures contain the functions that the compartment performs, and the set of parameters whose values are common for all neurons of that particular type. The private structures contain the set of parameters whose values are specific for each neuron and are updated during each simulation step by the compartment's function. This design is useful to build different neuronal types just by setting different values for the parameters in the type structures. Although all these neuronal types share the same compartment function, the behavior is different due to the particular values of their type parameters.

3   Building neurons and networks

In XSim, units and networks are created using a reduced number of commands that allow the specification of plenty of options. The building of networks in XSim comprises the following steps:

- Define cellular types (unit, site, link): First step to build a compartment is to define a cellular type. This is done by specifying a neuronal template (contains the initialization, behavior and learning functions for the cellular type) and a set of values for the type parameters.

- Create layers: This is the way of creating unit compartments out of unit type structures. Layers can consist of a single or several unit compartments. The number of units in a layer can be specified or calculated by XSim if space restrictions are applied. XSim's commands to create layers allow the specification of a three-dimensional space where the layers will be enclosed. Several distributions are available to set the positions of the units. All units in a layer are labeled with a name and indexes that allow the operation of XSim's commands on a specified group of cells inside a layer.

- Create sites: A site compartment is created specifying a unit compartment to which it will be attached, the site type that will be used to build the compartment and its name. A neuron can have several sites, and one can assign different site type and private structures for each site of the cell.

- Create links: Links are created out of a link type structure and established from the unit compartments through a simulated axon. Target units receive these links at the specified sites. A delay time can be set for the arrival of the synapses.

Building cells out of these three compartments provides flexibility and the possibility to perform complex neural computation to the models. However, it is not necessary to endow the three kinds of compartments with behavior. Some models may not need this detailed cell design and one can set a null behavior for site or link compartments.

Once the network is constructed, the compartments remain as flexible entities. At any time, it is possible to modify their type structures or the values of the parameters so as to achieve a desired behavior.

4   Connections

Some of XSim's most outstanding features are the facilities provided by the link commands to establish connections among units. With XSim one does not have to specify each connection on a one by one basis. The simulator is able to create realistic connectivity patterns dealing with plenty of units at the same time.

XSim's link commands can act on particular units or on layers of units. These commands allow to link units from layer to layer using a specified weight distribution. One can also specify a probability distribution to establish the links. As mentioned above, XSim handles positions for cell bodies, and dimensions for dendrites and axons (including patch displacements) –see figure 2– which are used to condition the links.

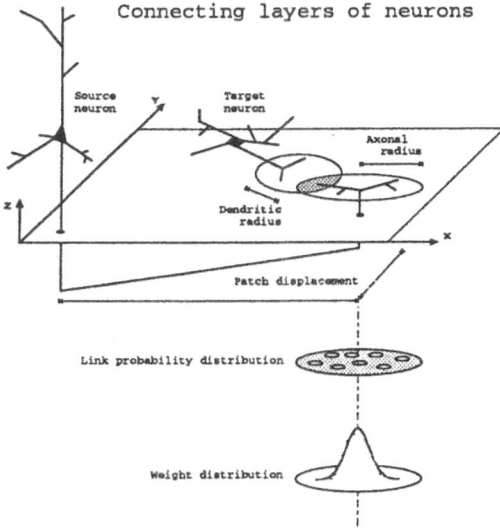

Figure 2: XSim's link commands provide facilities to connect units using different probability distributions, weight distributions and patch displacements.

Facilities are provided to change the weight values and to prune links once the initial connections are established.

5 Neuron models

XSim version 2.0 incorporates several neuron models which range from a perceptron like neural paradigm to a modified Hodgkin-Huxley neuron model with four ionic channels. XSim allows to incorporate new models into the simulator's library (through an assisted methodology to define new templates with the compartment functions and parameters). New templates are defined separately from the core of the program (knowledge of the C programming language is required). The possibility of compiling the program with a selected group of templates is provided so as to reduce system requirements.

There are currently three classes of neural models in XSim:

- Artificial neural models:

  - fpertron: a perceptron type neural model. The implementation of this model shows that it is also possible to build artificial neural paradigms using XSim.

  - fstoch: a probabilistic neural model [6]. This model does not incorporate biological parameters either. Neurons have a finite number of discrete states and a probability to jump to the next state in each simulation step. The firing of the neuron is the jump from the last state to the first and always occurs with probability 1. Synapses contribute to change (either increase or decrease) the current state of the neuron. Neurons built out of this model show a behavior that favors the synchronization of the spikes.

- Integrate and fire models:

  - fp11: an integrate and fire model based on behavior functions derived from the work of R. J. MacGregor [7, 8]. The model includes several membrane-related variables such as: membrane potential, ion conductances, equilibrium potentials, time constants, etc. The family of behavior functions built out of this model can reproduce realistic repetitive firing patterns by tuning these parameters.

  - fanton: an integrate and fire model with shunting and driving force effects based on the theoretical work by P. Antón [9]. This model approximates the membrane property and driving-force effects via the interactions of simple PSP functions in order to simulate physiological responses. The simplicity of the equations of the model maintains the computational efficiency in networks with large numbers of neurons.

- H-H neural models:

  - fliset1: A simplified Hodgkin-Huxley model with Na and K channels [10]. The behavior of the model is highly realistic although requires more computational time than the integrate and fire models since the solution of two highly nonlinear differential equations is involved. This model accounts for particularly optimum responses to the arrival of consecutive inhibitory synapses.

- fliset2: A modified Hodgkin-Huxley model with Na, K, Ca and Ca dependent K channels [10]. This model adds two more ion channels to fliset1 and, as a consequence, neurons can exhibit bursting action potentials. The model has a wide range of tuning through the change of the parameter values (conductances for all channels, membrane capacitance, resting potentials, time constants...).

There is also available a statistical neural model that makes statistics (number of spikes arrived simultaneously, spiking frequency, total number of spikes, etc) on individual units or on groups of cells. This is a powerful tool that facilitates the analysis of the network's behavior. Different ways to implement stimulus on the neurons, such as current injections, are provided. Learning functions to modify weights or synaptic conductances as the network evolves are also available for most models. These functions include hebbian and anti-hebbian learning rules.

6 Graphical interface

XSim's graphical interface allows for interactive graphical control over the compartments and neurons. With XSim it is possible to view the spatial setup of the network in bidimensional projections and monitor the network's behavior as a whole or in individual units. XSim allows for the plot of the parameters in real time and the display of the connections inside the network. XSim has a set of parameter windows available for each neuron from which one can edit, record or monitor the values of the parameters. These windows are also used to navigate through the neuronal compartments and to give access to their internal structures. All these graphic tools ease the analysis of the different neural network architectures.

See figures 3 and 4 for an outlook of XSim's graphical interface.

7 Miscellaneous features

Commands are introduced in XSim through a command interface. Most of them support default values and are also available through buttons and other gadgets from the graphical interface. There is an on-line help for the use of commands and their syntax. Script files to construct neurons and to build networks and graphic setups are supported.

XSim allows to save the current state of the simulation at any time (including the graphical setup). XSim generates record files that are compatible with most tools for data analysis available in the public domain. It is possible to keep a log file of the simulation sessions with an echo of commands used and their output.

The simulator has a reference manual [11] where the facilities provided are explained along with examples that show how to use them. Details on the implementation, the neuronal models available and on how to expand the simulator are also included.

Full source code is provided and there is a methodology to change or expand the command library.

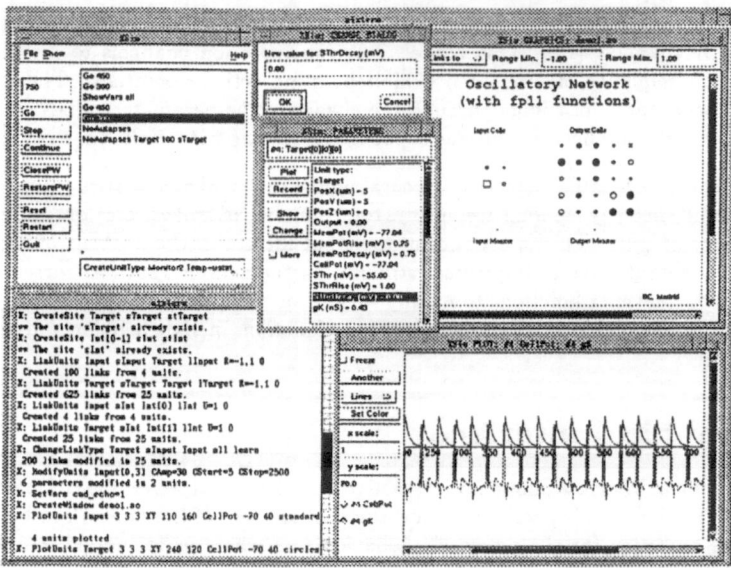

Figure 3: XSim's graphical interface. First row, from left to right: main window, dialog and parameter windows, graphic window. Second row: xterm window, plot window.

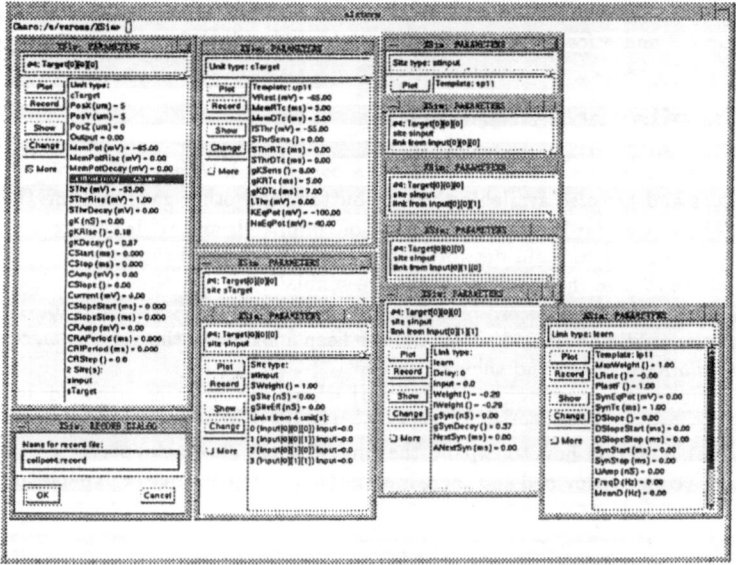

Figure 4: The six classes of parameter windows: unit (1st column), unit type (2nd column, 1st window), site (2nd column, 2nd and 3rd windows), site type (3rd column, 1st window), link (3rd column, 2nd-5th windows) and link type (4th column).

8 Discussion

XSim is a useful tool for building neural networks suitable for teaching and research. The simulator has a flexible architecture that allows the implementation of numerous neural paradigms, though its design has been thought to take advantage of a compartmental approach in order to incorporate biological parameters. XSim has powerful tools for constructing complex networks with large numbers of units and connections. The graphical interface allows to monitor the network behavior and find the best tune ups for the different models.

XSim is public domain software. It can be freely distributed and is available via ftp through internet service.

Acknowledgements:

We are grateful for the support of DIGICYT/PB91/0045. Pablo Varona also acknowledges M. E. C. for financial support and Alberto Cobas for his encouragement to develop XSim.

References

[1] De Schutter, E., (1992). *A consumer guide to neuronal modeling software.* Trends in Neuroscience, 15: 462-464.

[2] Bergdoll, S., (1990-1993). *Biosim. Ein biologisch orientierter Simulator für neuronale Netze.* BASF Inc., Department of Software Engineering (ZXA/US). D–67056 Ludwigshafen, Germany.

[3] Ekeberg Ö., Hammarlund P., Levin B., Lansner A., (1993). *SWIM – a simulation environment for realistic neural network modeling.* In J. Skrzypek, editor, *Neural Network Simulation Environments.* Kluwer.

[4] Goddard N. H., Lynne K. J., Mintz T., Bukys L., (1989). *Rochester Connectionist Simulator.* University of Rochester, Computer Science, Technical Report 233 (revised).

[5] Koch C. and Sergev I., eds., (1989). *Methods in Neuronal Modeling: From Synapses to Networks.* MIT Press.

[6] Hofman P. M., Rodríguez F. B., Sigüenza J. A., López V., Carrillo-Menendez S., *Probabilistic Neural Networks,* submitted to IWANN'95.

[7] MacGregor R. J., (1987). *Neural and Brain Modeling.* Academic Press.

[8] MacGregor R. J., (1993). *Theoretical Mechanics of Biological Neural Networks.* Academic Press.

[9] Antón P. S., (1991). *Simulations of Information Processing. Control, and Plasticity Effects in the Olfactory Bulb.* Technical Report No. 91-63. University of California. Irvine, California.

[10] Menéndez de la Prida, L., *Modeling Cortical Networks,* submitted to IWANN'95.

[11] Varona P., Cobas A., Sigüenza J. A., (1994). *XSim Neural Network Simulator.* IIC Technical Report B03/94. Instituto de Ingeniería del Conocimiento, Universidad Autónoma de Madrid. 28049 Madrid.

NETTOOL: A Hybrid Connectionist-Symbolic Development Environment

J. Santos[1], R.P. Otero[1], and J. Mira[2]

[1]Dpto. de Computación. Facultade de Informática. Univ. da Coruña. Spain.
[2]Dpto. de Informática y Automática. Facultad de Ciencias. UNED. Spain.

ABSTRACT.

In this work we present an environment, NETTOOL, for the development of hybrid connectionist-symbolic systems, in which the connectionist representation is based on the same knowledge representation model as that of symbolic systems. The hybridation between the knowledge elements is local, the connectionist training algorithm is also localized in each element of the network so that the knowledge required for the learning process is a part of it. There is also the possibility of including capabilities for inferential level processing in the elements of the network.

1. INTRODUCTION.

An effort has been made in the last few years towards benefiting from the advantages of the connectionist and symbolic paradigms. The aim has been to make the most outstanding characteristics of each one of them help the other where it performs worse. Thus there are several systems in which the learning capabilities of the networks are used when imprecise knowledge is found in the symbolic art. This makes an equivalent network refine the knowledge of a problem provided by a symbolic part. Whenever this happens we talk about *"expert networks"* or *"connectionist expert systems"* [1][2][3]. There are very few articles [4][5] that we can consider as using true hybrid connectionist-symbolic systems, where the connectionist networks and the symbolic parts are intermingled, act as pre or post processing stages of each other or one modifies the processing of the other to some extent.

In our work we try to make both the connectionist and symbolic parts fit the same representation, more specifically that of micro-frames, which we have called Generalized Magnitudes. This representation of knowledge has the structural and functional connectivity that is appropriate for the representation of connectionist networks, as well as the possibility of including inferential capabilities and localization of the training in each element of the network. In this line, we have developed an environment, NETTOOL, for handling hybrid systems under this representation, as an extension of MEDTOOL, an expert system development environment [6].

The distinction between the connectionist and symbolic paradigms becomes diffuse when both work under the same representation. The differentiation based on the way information is handled in both systems, "value passing" between nodes of a connectionist system or "symbol or message passing" in a symbolic system, is not good enough if we contemplate it from the point of view of the processors or the elements which ultimately execute the required operations, or we contemplate it from an external reference system of an observer that only perceives inputs and outputs of a computational system [7].

2. THE KNOWLEDGE REPRESENTATION MODEL OF GENERALIZED MAGNITUDEs

The model for the representation of knowledge of the MEDTOOL expert system development tool is structured around a basic entity which we have called Generalized Magnitude (GM). Consequently the Knowledge Base (KB) is a set of GMs.

A GM is defined from the facts and events which an expert system (ES) must know. Each and every one of the possible facts the ES is going to handle must be associated forming GMs. This association of several possible facts in this unit must verify that in every application of the knowledge only one of the facts which are grouped in each GM is to be verified, while the remaining facts are false.

The *Knowledge Module* slot of a GM express the relationship between the true fact of a GM and the true facts of others. The syntax is a logical-relational expression:

        <assignment 1> if <condition 1>;

        ....

        <assignment n> if <condition n>;

If the <condition i> part is verified, the fact calculated or assigned in the <assignment i> part is the true fact assigned to the GM.

The operators that can be used with each <assignment> are the relational, mathematical, logical and conditional operators, between numeric and symbolic facts, temporal operators that return the associated time of a true fact, and an operator of access to the past (ant(GM)) that returns the previous true fact of a GM.

3. IMPLEMENTATION METHODOLOGY OF CONNECTIONIST MODELS.

In the implementation of the connectionist mechanisms based on the concept of GM, the knowledge modules must include all the necessary processing and training mechanisms for each particular model of neural network. We consider two types of nodes: "transfer nodes" and "learning nodes". The transfer nodes are implemented with GMs that incorporate the transfer function and input combination function in their KMs. The learning nodes include in the KM of their corresponding GM the particularized or localized training algorithm. It is the case of the connections between nodes. In order to distinguish the processing mode of a network we will employ a "control GM" we will call MODE and which can take the alphanumeric values of *initiation, recall or training*.

In the calculation stages, the connections act as numerical constants. In the initial stage we usually take a random collection of values, and in the learning stage, the connection GMs must locally account for the corresponding part of the value change, depending on the particular training algorithm. In order to take these considerations into account, when using the GM concept, we will make use of the power of the conditional operator in the specification of the knowledge module in order to distinguish in which mode all the GMs implementing connections are in a particular instant of time. A particular connection GM will have a KM such as:

actual value of C if MODE = recall
modification of C according to learning algorithm if MODE = training
initial value if MODE = initiation

Thus, the control GM MODE, according to its current true fact, "controls" the operation of all the GMs of the network through the conditional dependence on the KMs of the remaining GMs.

Once all the necessary GMs for the specification of a network are available, the general operation is the following: The GMs of the network make up a KB which will use the *Mechanism for the Application of the Knowledge (MAK)* -Inference Engine in Production Rule terminology-. The MAK will apply the expressions of knowledge (the KMs) which reside in the KB to a series of starting facts (of primitive GMs), obtaining all the conclusional facts (of derived GMs). The starting facts take the form of a true fact for a particular GM. It will therefore appear in the Base of Facts (BF) before applying the knowledge. Then the MAK interprets or applies, to each of the derived GMs, its associated KM, obtaining as a result the true fact for this GM, in this application of the knowledge, depending (this result) on the true facts of the other GMs included in its KM.

The results inferred by the MAK in the derived GMs will correspond, during the calculation stage, among others, to the values of the output nodes in the corresponding GMs. During the training stage, the values inferred by the MAK will be the new values of the connection GMs (Fig 1). With an external control of the values assigned to the MODE GM we have a control over all the elements or GMs that make up a network.

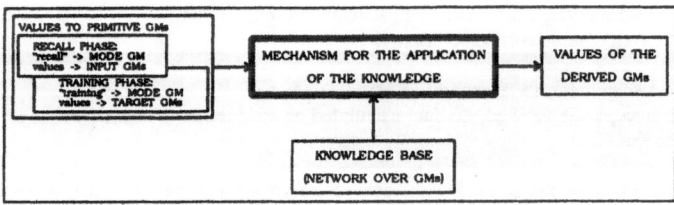

Fig. 1. Processing of a Neural Network by the MAK.

For example, for a multilayer perceptron structure with supervised training through backpropagation [8] in the transfer nodes, a very usual function is a sigmoid acting on the sum of the product of inputs times weights, although the model permits the incorporation of any other more complex function. Thus, for a hidden node H0 its KM would be:

GM   H0
KM $1/(1+\exp(-1*(I0*CI0H0+I1*CI1H0+I2*CI2H0+...)))$

where the primitive GMs I0, I1, I2, ... correspond to the input nodes to the network and the GMs CI0H0, CI1H0 and CI2H0 represent the connections between input and hidden nodes. We could obtain the output GMs in the same way.

In the connection nodes, their values in the calculation stage are the connection weights between two nodes. In the training stage, we must implement the backpropagation algorithm, which consists in the formalization of the least squares learning procedure over networks with hidden layers, making use of the MODE GM for conditionally specifying the knowledge module these GMs need to infer depending on the operation mode. For instance, for the CH0S0 node:

ant(CH0S0) if MODE = recall
ant(CH0S0) + COEF*DS0*ant(H0)
if MODE = training
rand() if MODE = initiation

where COEF is a GM which represents a numerical value in the weight modification process, and the DS0 GM contains the difference between the target and the real output (T0) corresponding to node S0 (Figure 2).

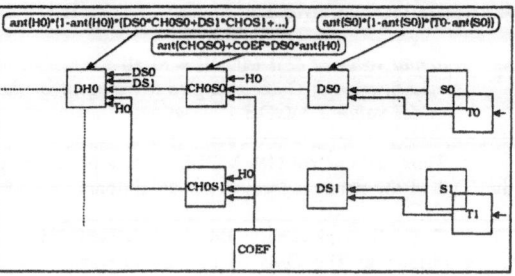

Fig. 2. Inferences in the GMs during the training phase.

The use of operator ant() is necessary in order to force the MAK to obtain the value of those derived GMs appearing in the KM of another GM, whenever the knowledge is executed again, as in the processing changes occurring between the calculation and learning modes.

In unsupervised training there is no need for target outputs for each input of the network and the modification of the connections is carried out so that it produces similar outputs for very similar inputs. A simple example are Kohonen´s selforganizing maps, where for a given input vector, only one neuron (winning neuron) or group of neurons in a region present a logical 1. Each weight associated with the winning node is modified by an amount that is proportional to the difference between its value and the input value to which it is connected. In this case we need a GM that detects which is the winning node. We call it "largest output" (LO) and for a connection CI0S0 between input node I0 and an output node S0 its KM would be:

ant(CI0S0)+ALFA*(I0-ant(CI0S0) if ant(LO)=ant(S0) and MODE = training

so that not all the connections are modified during the training stage, only those that are connected to the node or nodes with the highest activation level in the output, as specified in the additional condition of the KM. This is an example of how the inferential processing capabilities are included in the elements of the network.

Even though the processing in our networks is essentially asynchronous as the MAK infers the value of a GM when all the GMs referenced in its KM have inferred a fact as true, it is possible to simulate synchronous processing, also valid for recurrent networks, in which all the nodes or GMs infer at the rhythm established by a master clock. This is only possible if all the nodes infer from the previous value of the nodes connected to it, making use of the ant(GM) operator. Each clock cycle corresponds to each execution of all the knowledge.

The following table summarizes how the processing and training mechanisms are expressed in the KMs of the GMs for this and other models:

CONNECTIONIST MODEL	KM EXPRESSION
Multilayer Perceptron	A set of GMs incorporate the combination and transfer functions of the nodes, and another group of GMs the propagation of the error (figure 2) for its minimization.
Competitive Learning	A GM that detects the output node with higher activation level (LO) is necessary. In a connection between an input node I0 and an output node S0, its KM will be: ant(CI0S0)+ALFA*(I0-ant(CI0S0) if ant(LO)=ant(S0) and MODE = training
Dynamic Networks	The feedback loops are achieved through the ant (GM) operator.
Stochastic Networks	A GM that indicates which connection GMs must be modified by a random amount is necessary CIxOy=CIxOy+rand()
Fuzzy Networks	The nodes incorporate the membership functions of the fuzzy sets and other GMs the usual operators of the fuzzy rules max, min, not.

The model permits the connection of different types of networks with different topologies and training methods. There is the possibility of defining individual processing of each node or particular training in each connection, as well as an immediate local hybridation to symbolic modules implemented through other GMs.

4. NETTOOL ENVIRONMENT.

The environment we have called NETTOOL can be defined as a hybrid connectionist-symbolic development environment built around a GM representation model and the knowledge application mechanism of the MEDTOOL system.

Now the KBs are a mixture of networks and symbolic networks. The representation is unique. There are only GMs which differ in their KM, some processing functions that are typical of neural networks, and others processing symbolic knowledge which acts as a preprocessing stage, post processing stage, or modifies the behavior of the neural network or networks.

Thus, unlike other hybrid environments [10] that only provide a programmable connection between connectionist and symbolic systems built in different environments, NETTOOL incorporates the local hybridation idea between basic knowledge elements which process numbers or symbols and consequently handles hybrid KBs, but with a uniform representation.

In its orientation towards handling connectionist models [11], NETTOOL incorporates a series of modules which are usual in the different connectionist development environments, in addition to its own particularities as a hybrid system (figure 3):

1. An automatic generator of knowledge bases for architectures and training methods that are usual in networks (adalines, multilayer perceptrons, competitive networks, associative memories, etc...), following the methodology seen in the previous section. Symbolic parts are added to the users desire and the KMs of all the GMs are compiled in a notation that can be interpreted in an easier way by the MAK for its execution. This compilation stage also detects errors in the edition of the KMs of the GMs.

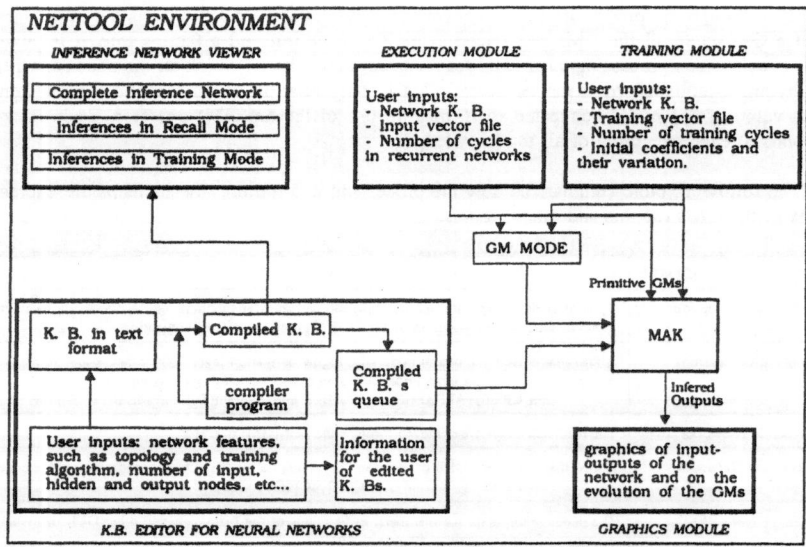

Fig. 3. Modules of the NETTOOL Environment.

2. A network execution module. A network defined in a KB and previously trained can be executed after loading its weights. The KB must be taken as consisting of, in addition to the definition of the GMs, the trained weight configuration. The inputs to the network can be the training data or testing data specified by the user. In recurrent architectures, the user also specifies the number of processing cycles desired or, in other words, the number of times the MAK must execute all the knowledge from the input data true facts saved in the Base of Facts in each execution cycle.

A graphical representation of the value of the input and output GMs of the network is generated each execution (figure 4), and the value of all the numerical and alphanumerical GMs the user desires is shown when they are updated in the execution stage.

3. Training module. The information with the training data (network inputs or desired inputs-outputs in supervised training) is kept in a file read by the environment for training a previously edited

network. The user specifies the number of training stages needed as well as the initial value and the increment of the GMs participating in the training process, such as the different weight change coefficients.

The environment is in charge of performing two knowledge executions each cycle, one execution stage with the current weight configuration (MODE GM with the value of "recall") and a training stage, updating the MODE GM to a value of "training". As in the execution stage, the values of all the GMs specified by the user are shown when they are updated in either stage. The last values of the connection GMs are stored.

4. Graphic module for the evolution of training. A graphic representation of the evolution of a given GM is generated during the training process, GM or group of numerical GMs specified by the user.

5. Knowledge visualization module. The inference network corresponding to a KB is graphed in order to clearly visualize the set of interrelations among all the GMs. The network is presented as a tree structure, labelling the nodes with the symbol of the GM and the arcs with the relationship between GMs operator. This way, the knowledge engineer, will graphically have a focalizing image of the knowledge he introduces.

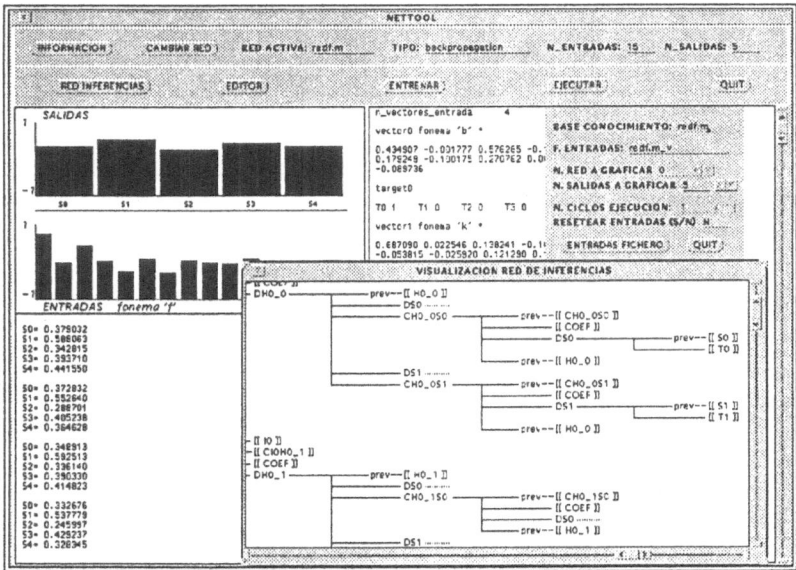

Fig. 4. NETTOOL's XWindow Interaction Screen.

In handling KBs that constitute connectionist systems and in order to make the knowledge dependence relationships more explicit, we include the possibility of generating the inference network separately in the calculation and training modes. In the calculation mode the network is generated without considering the dependencies from the training algorithm so that the final visualization expresses the dependencies between node GMs in the usual way, that is, the topology of the network. The feedback loops are specified by the dependencies of a GM with its previous value or that of other GMs. The representation in the training mode denotes how the connections of a network are modified as a function of the value of the GMs from which it depends, appreciating this way how the training algorithm is localized. As the inference network can be of considerable size, the user can choose to represent the dependencies of a single GM or of the whole network in either mode (lower window in figure 4).

Figure 4 shows the main interaction screen with NETTOOL under XWindow. We present the different subwindows of the environment. The top one displays basic information of the currently active network, such as name and topology. Under this we find the different options of the tool. A submenu is open and presents the different options. The central left subwindow graphically represents the inputs and outputs or the evolution of training. The lower left subwindow shows the different GMs and the values inferred after executing the knowledge. The right window shows the training vectors. In order to represent the inference network we generate a new window that can be enlarged and has scrolling capabilities (central window).

5. APPLICATION EXAMPLE.

We have built a hybrid system with the NETTOOL environment for the recognition of Spanish phonemes [12]. In this domain additional knowledge to the phonetic recognition a neural network is able to produce is necessary. In this line, we have integrated different backpropagation modules trained for the recognition of groups of consonantal phonemes, those that sobremodulate the vowel core, differentiating between plosive and non plosive phonemes, and those whose influence falls mainly outside the vowel nucleus (Fig 5). Therefore a first level of GMs corresponding to properties like existence of segment, energy level of the segment, length, existence of silence, increment of energy, etc., implements decision rules of what backpropagation submodule must be activated, propagating ignorance over the GMs of the other submodules. The user edits these GMs and compiles them with the other submodules created automatically by the tool in an unique KB.

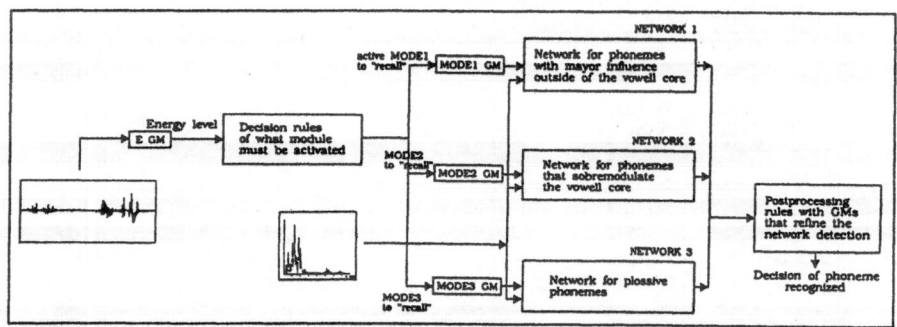

Fig. 5. The hybrid system for the recognition of the Spanish phonemes.

With the same idea of local hybridization a final module performs a lexical and grammatical analysis that postprocesses what the previous modules have recognized.

For example, a network decision GM has rules of the type:
if(EXIST EXPLOSION OF ENERGY) then active GMs of the plosives network in the first interval of the segment

and a postprocessing rule would be:
if the network has detected 'p' and (THERE ISN'T ENERGY PREVIOUS TO THE EXPLOSION (over 20 msec.) then is more probable that the networks detection is correct

because it is a distinctive characteristic of that phoneme over the other plosives, and thus itpermits the refinement of the knowledge between the symbolic and connectionist parts.

6. CONCLUSIONS AND FUTURE WORK.

We have presented an environment for the development of hybrid connectionist-symbolic systems making the artificial neural networks fit the same model for the representation of knowledge as the symbolic systems. The training algorithm is localized in each element of the network so that the training is a part of the network, and is not imposed as an algorithm of the external host. The hybridation with symbolic parts makes both processing modes undistinguishable. The GMs pass either values or symbols in both paradigms.

We are following this methodology for implementing different connectionist architectures and training methods. As an application we are constructing a speech recognition system at the phonetic level in which symbolic parts cooperate with connectionist modules.

ACKNOWLEDGEMENTS

This investigation was supported by CICYT projects TIC-92-0136 and TIC-94-0095 and the grants XUGA10502A92 and XUGA10503B94.

REFERENCES

[1] Gallant, S., Connectionist Expert Systems, *Communications of the ACM*, Vol. 31, No. 2, February 1988.
[2] Latcher, R.C., Hruska, S.I., and Kuncicky, D.C., Backpropagation Learning in Expert Networks, *IEEE Transactions on Neural Networks*, Vol. 3, No. 1, pp. 62- 72, January 1992.
[3] Shavlik, J.W., and Towell, G.G., An Approach to Combining Explanation-Based and Neural Learning Algorithms, *Connection Science*, Vol. 1, No. 3, pp. 231- 253, 1989.
[4] Ciesielski, V., and Spicer, J., Embedding Neural Nets and Expert Systems in Diagnostic Microbiology Laboratories, *IEEE Expert*, pp. 42-48, Jun 1994.
[5] Handelman, D.A., Lane, S.H., and Gelfand, J., Integrating Neural Networks and Knowledge-Based Systems for Intelligent Robotic Control, *IEEE Control Systems Magazine*, Vol. 10, No. 3, 1990.
[6] Otero, R.P., *MEDTOOL: Una Herramienta para el Desarrollo de Sistemas Expertos*, Tesis Doctoral, Univ. de Santiago, Dpto. Electrónica y Computación, Santiago de Compostela, 1991.
[7] Mira, J., and Delgado, A. E., Always Trying to Write an Equation for the Brain. In *Lecture Notes in Computer Science, Vol. 540, Artificial Neural Networks*. A. Prieto ed., pp. 93-100, Springer-Verlag, Berlin, 1991.
[8] Rumelhart, D.E., Hinton, G. E., and Willians, R. J., Learning Internal Representations by Error Propagation. In *Parallel Distributed Processing, Explorations in the Microestructure of Cognition*, MIT Press, Cambridge, Vol. 1, pp. 318-362, 1986.
[9] Kohonen, T., Self-organization and Associative Memory. *Series in Information Sciences*, Vol. 8, Springer-Verlag, Berlin, 1984.
[10] Khebbal, S., and Treleaven, P., An Object-Oriented Hybrid Environment for Integrating Neural Networks and Expert Systems, *Proceedings of The First New Zealand International Two-Stream Conference on Artificial Neural Networks and Expert Systems*, pp. 210-213, Dunedin, New Zealand, November 1993.
[11] Murre, J.M., Neurosimulators, *Handbook of Brain Research and Neural Networks*, M.A. Arbib (Ed.), MIT Press, in preparation for 1995.
[12] Santos, J., and Otero, R.P., Connectionist Models for Syllabic Recognition in the Time Domain, in *New Trends in Neural Computation*, J. Mira, Cabestany and Prieto (Eds.), *Lecture Notes in Computer Science*, Vol. 686, 149-154, Springer-Verlag, Berlin, 1993.

EL-SIM: a Development Environment for Neuro-Fuzzy Intelligent Controllers

M. Chiaberge*, G. Di Bene**, S. Di Pascoli***,
R. Lambert*, B. Lazzerini***, A. Maggiore*** and L.M. Reyneri***

* Dip. Elettronica, Politecnico di Torino
** I.S.E. Ingegneria dei Sistemi Elettronici s.r.l.
*** Dip. Ingegneria della Informazione: Elettronica, Informatica, Telecomunicazioni
E-Mail: {beatrice,lmr}@iet.unipi.it

Abstract

[1] This paper presents a new technique for the design of real-time controllers based on a hybrid approach which integrates several control strategies, such as intelligent controllers (e.g., artificial neural networks, fuzzy systems), traditional linear controllers, finite state automata. An integrated programming environment, called EL-SIM, is also presented, suited for developing high-performance intelligent controllers for industrial applications. EL-SIM provides general tools to support the development and optimization of control systems based on the aforementioned approach, by means of several cognitive or hybrid algorithms, which allow also improvement of environmental performance indexes, like power consumption or toxic waste emission. EL-SIM permits both the study of new experimental techniques in research application and the design, tuning and testing of widely used control architectures for industrial applications.

1 Introduction

The use of traditional linear control theory can often lead to the necessity of expensive high performance hardware to obtain the required system performances. For example, in a typical industrial robot application, a high positioning precision is usually obtained by means of a highly rigid and heavy structure, which requires complex high power actuators. As a consequence, typical robot systems are heavy, energy consuming and expensive. In this situation and, in general, in the case of complex nonlinear systems, or "ill-behaved" systems, including systems with low rigidity links (distributed flexibility), loose joints, friction, nonlinearity, time or temperature-dependent effectors (e.g., muscle-like actuators), pneumatic actuators, etc, an alternative more sophisticated control strategy would be desirable, possibly based on "intelligent" control, aimed at improving the capability of controlling the target system over time through accumulation of experience.

Two well-known intelligent control techniques are neural networks and fuzzy systems. Neural networks are characterized by their capability of learning from examples, their intrinsic parallelism and fault-tolerance; whereas fuzzy systems implement a linguistic control strategy suited to poorly modeled target systems. Neural networks and fuzzy systems are commonly integrated into the neuro-fuzzy approach, which has proven well adapted to nonlinear control applications, and particularly efficient for low-power embedded applications, where a small size, associated with a reduced power consumption and a high reliability are often a must.

[1] This work has been partially supported by ASI contract 249/3 neural architecture for spaceborne manipulators and EEC CAPRI initiative Elect.

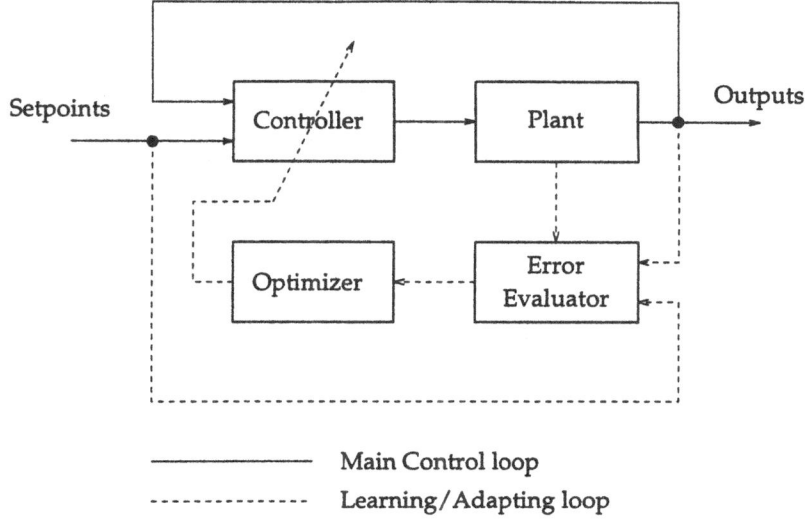

Main Control loop

Learning/Adapting loop

Figure 1: General structure of an adaptive control architecture

The fundamental reason for mixing the neural and fuzzy approaches is to integrate the ease of controller description, typical of fuzzy systems, and the efficiency of controller implementation offered by neural networks. A well-known method to merge fuzzy systems with neural networks consists in directly mapping the fuzzy rules into the network structure [1]. More precisely, we propose a control strategy which integrates several intelligent architectures by appropriately combining and optimizing them.

However, although the neuro-fuzzy approach alone provides good performance for nonlinear controllers, it cannot always satisfy all the requirements of a real plant. Indeed, in some applications the plant has a number of well-defined states in which the static and dynamic characteristics of the plant may differ from one state to the other. Consider, for instance, the forward and backward steps of autonomous walking machines, or the optimal trajectory of a manipulator with two largely different load conditions. In such cases, the chosen control strategy must be integrated with finite state automata to keep track of the different plant states and vary accordingly some controller characteristics (e.g., synaptic weights, fuzzy inference rules).

Unfortunately, there are some major limitations to an extensive use of the neuro-fuzzy approach in the field of real-time control, mainly due to the high technological risks that can arise from the choice of a non-traditional control methodology. Given this fact, and also taking into account the very short time-to-market required by a lot of industrial applications and the high cost of researches in the field, we believe that a development and simulation environment is an effective means to evaluate possible benefits of embedded control solutions based on a neuro-fuzzy approach, in order to assess the potential for the development of new products and solutions. For this reason, we have developed an integrated programming environment, called EL-SIM, to investigate, optimize and test several control architectures.

EL-SIM is basically composed by i) a set of predefined operational blocks, which provide control and optimization functions, ii) an inter-block interface which, among others, allows an easy integration of user-supplied blocks, such as the plant simulator and pre-defined hardware drivers or blocks which implement new control functions, and iii) a user interface which allows the user to build the final control system.

2   Analysis of Control Strategies

Generally, an "intelligent" control system can be modeled as shown in Fig. 1. The *Main Control loop* consists of the plant and the controller (e.g., a neural network, a fuzzy inference system, a traditional linear controller, or even a combination of them); the controller generates the control variables (namely the plant inputs) from the setpoints (the target of the control action) and the plant outputs. The *Learning/Adaptive loop* includes an Error Evaluator block, which evaluates an appropriate performance index, and an optimizer, which tunes the parameters of the controller to improve overall performance. The most interesting approaches to intelligent control are the following:

- Fuzzy control: the know-how of a human expert is mapped, using linguistic variables, into a set of control rules, which are processed by a fuzzy inference mechanism [2].

- Simple neural control: a neural network controller can be trained with the backpropagation algorithm by a set of input-output training pairs supplied externally, maybe reproducing an existing reference controller. As an alternative to backpropagation, other neural paradigms such as Radial Basis Functions can also be used [1].

- Passive Learning Neural control: a neural network learns directly from the plant by on-line minimization of an appropriate cost function (often the square sum of errors). The main drawback of this control strategy is the problem of overfitting around the actual trajectories in the state space [3].

- Active Learning Neural control: as for the previous technique, a neural network learns directly from the plant, but now it learns also from untrained regions in the state space, thus improving performance.

- Neuro-fuzzy control: a neural controller is either trained to imitate a given fuzzy logic controller, or built to map directly a fuzzy rule base [2, 4]. In particular, Weight Radial Basis Functions [1] have been developed to simplify this process.

- Inverse Neural control: two neural networks can be trained to emulate both the direct and the inverse plant characteristics [4, 5]. The approximate inverse model of the plant can then be used either as a stand-alone controller (possibly in an open loop architecture), or to improve stability, linearity and performance of another controller (Fig. 2).

- Traditional Linear control: a traditional linear controller can be used to generate the training set data for the neural network or as a part of a fuzzy hybrid controller.

EL-SIM provides a set of basic blocks which allow to implement all previously listed control techniques, namely: blocks representing traditional controllers, fuzzy inference systems, and several kinds of neural networks. Furthermore, to simulate control systems of a wider class of plants, which may need analog control and yet have well defined discrete states requiring different control parameters, EL-SIM provides basic blocks implementing Finite State Automata. For example, a finite state automaton block can be used to change the weights of a neural controller as the plant changes state.

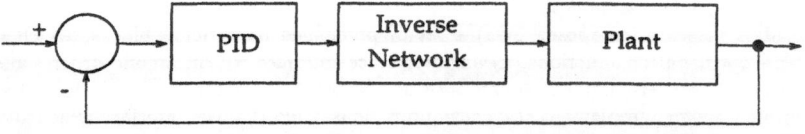

Figure 2: Canceling nonlinearities with a trained neural network

As already mentioned, it is also possible to add data acquisition boards and driver blocks to interface hardware devices (the real plant, a hardware neural network, etc.) with the rest of the simulator, therefore introducing the real plant in the optimization loop. Indeed, an implementation of the neuro-fuzzy paradigm with a hardware neural network based on the Weight Basis Radial Functions [6] can be notably efficient in real-time applications. In such a way the final optimization of the controller can be done acting directly on the real plant. This avoids the introduction of errors due to the use of a simulation of the plant. Moreover, the software controller can be used as the actual controller of the plant.

3 Optimization Strategies

To increase the overall performance of a control strategy, an optimization algorithm can be used to automatically tune the parameters of a basic control block. EL-SIM provides basic optimizers which can minimize (or maximize) some performance index, generally an error function, possibly calculated inside the optimizer. This performance index can be a function of the plant outputs and the setpoints, or a function of the outputs only (e.g., toxic chemical emissions), or it can be provided by the plant itself (e.g., power consumption). A brief description of some optimization methods used by EL-SIM follows.

One of the algorithms for general minimization is gradient descent, which uses derivatives of the error with respect to parameters to move towards the minimum. The need for this derivatives is the main obstacle to the use of this algorithm and adds some constraints to the design of the blocks to be optimized. In many cases, indeed, the derivatives of the output error with respect to plant inputs are not known. The gradient descent block can use heuristics by adopting, for instance, the momentum or an adaptive learning rate in order to reduce the probability of getting stuck in a local minimum.

A different solution to the optimization problem is implemented by simulated annealing. Simulated annealing is a probabilistic gradient descent algorithm which typically converges to the absolute minimum. This algorithm does not require derivatives of the error and can be used when the plant outputs derivatives are not known.

A different approach to the problem is to use an optimizer based on genetic algorithms [7]. Such block does not need any derivative of the error and may be faster. Alternatively, a hybrid approach which combines simulated annealing and genetic algorithms can be used [8]. This approach keeps the advantages of both its components, that is, low memory requirements of simulated annealing and good performance of genetic algorithms.

These optimization algorithms allows to improve performance with respect to a wide range of indexes, ranging from traditional ones, like dynamic specifications (settling time, bandwidth, delay), errors (e.g., position, temperature, etc.), or more "advanced" ones: component wear, mechanical stresses, and environmental issues, such as concentration of toxic chemicals in waste output, power consumption or every other quantity for which it is possible to define a numerical performance index to feed to the optimization algorithm.

4 The Programming Environment

The integrated development environment is a set of tools for the design, simulation and field testing of control algorithms. It is a windows-like environment with menus, pop-up panels, and selection lists. The user develops its own project, by using pre-defined objects from a built-in library. The elements selected can be configured for the specific purpose (e.g., number of layers for neural networks, number of rules for fuzzy systems, etc.) and connected to each others, then the complete system can either be simulated or tested on the real plant.

A specific configuration process is built using an editor which asks the user for the parameters required by each block and the connections between the selected blocks. Apart from controllers, optimizers and the plant simulator, EL-SIM also provides several other utility blocks, such as signal generators, signal analyzers with graphic capabilities, mathematical primitives, etc., and driver blocks for hardware devices, like data acquisition boards, RS-232 lines, etc.

Connections and coherency of the desired configuration are tested by a connection inspector, which analyses the interconnections between blocks and warns the user whenever a problem is found. The user can plot signals either during the simulation/testing process or off-line. Statistical analyses can also be done automatically by using histograms, pie charts and other graphic representations of data. EL-SIM has been written in the C/C++ language and has been designed to be easy to upgrade and maintain.

C++ is naturally adapt to implement a structure organized in functional blocks, each one implementing a basic control operation. The blocks communicate with each other by standard signals, which can be scalars (e.g., the performance index), vectors (e.g., setpoints, control variables and plant outputs), or matrices (e.g., the weight matrix of a neural network) of integers, booleans or floats. It is worth noting that every information exchange among blocks is done through signals, including both communications, for example between the controller and the plant, and parameter tuning that occurs between the optimizer and the controller. Graphics and the interface with data acquisition boards and other hardware devices have been made using the LabWindows libraries.

The simulation process is time-discrete (and not event-driven) as all blocks update their output signals "simultaneously" at regular intervals.

5   A Practical Application

The first practical application of EL-SIM has been developed to implement a software for controlling the thickness of aluminium deposited in a vacuum roll coating machine (Fig. 3). The process of evaporation of aluminium is highly nonlinear with respect to the temperature and it is fairly complex to be described in an analytic manner, due to the dependency of the parameter values on the constructive details of the coating machine.

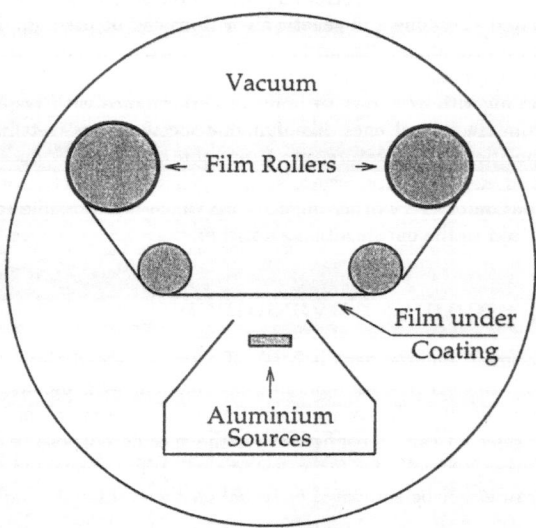

Figure 3: Simplified representation of a coating machine

Notwithstanding, the process can be described in terms of linguistic variables, because the thickness is clearly dependent from the temperature and the area of the aluminium surface. Another important condition to be achieved is that of avoiding disturbance to the liquid surface, in order to prevent "spitting". Based on these considerations we decided for the implementation of a fuzzy controller to be tested on the machine. The controller has to be developed under the form of software to be run by the computer already present to manage the process. The development of the controller was undertaken basically in three steps by using EL-SIM: 1) development of the simplified plant simulation; 2) development and optimization of the fuzzy controller; 3) testing of the controller on the real plant.

The plant simulation was kept as simple as possible, though keeping in mind the necessity of representing the basic phenomena as the dependency of the aluminium evaporation on the temperature and on the evaporating area, which in turn depends on the feeding speed of the aluminium wire. After the plant simulator showed to perform according to the real plant, the fuzzy controller was developed, based on two inputs (relevant to the state of the aluminium bath and the evaporation condition) and two outputs (aluminium feeding rate and heating power) for each of the 18 aluminium sources. The controller implemented was a 25 rules one (we decided it was not worth to optimize the number of rules). The parameters of the controller were first chosen according to the operator's experience and to the knowledge of the involved phenomena, and then optimized by means of a genetic optimizer, were the fitness indicator was the settling time after a disturbance. The main goal of the testing phase was to verify the installation of the software controller on the computer of the plant. This was accomplished without major problems, but we had to optimize the execution times of the controller, in order to permit the existing computer to withstand the new load.

The results we obtained were very good so that we are now investigating about the possibility of automating completely the process of roll coating in order to achieve a machine which is virtually operator independent.

The results of this very first real application were very exciting, considering that EL-SIM was developed having primarily in mind the world of robotics and servo-controls, where we intend to investigate how to tackle the control of moving parts with backlash and low stiffness. This is a very wide class of problems, which show up whenever there is the need for fast and precise movements, as in the case of robots and machine tools for example.

6   Conclusion

A new approach to intelligent control has been presented. In this new paradigm, several nonlinear control techniques and some cognitive optimization algorithms have been used as basic blocks to build an integrated and hybrid architecture, which can exploit the positive aspect of each technique, while it is effective in avoiding their drawbacks. The integration of general purpose optimization algorithms into this hybrid approach gives the possibility of improving system performance with respect both to traditional indexes, such as dynamic characteristics (settling time, steady state error, etc...) and to other performance indexes, like power consumption, component wear, toxic chemical emission, or to any quantity for which it is possible to develop a suitable performance index.

EL-SIM, a development environment for intelligent controllers, has been outlined. EL-SIM provides a set of tools for the design, simulation and testing of control algorithms. It provides the basic blocks to build several intelligent and traditional control architectures.

An application of the proposed approach has been described: EL-SIM was used to control the thickness of aluminium deposited in a vacuum roll coating oven. The results obtained confirmed the effectiveness of the proposed control strategy.

References

[1] L.M. Reyneri: "Weighted Radial Basis Functions for Improved Recognition and Signal Processing", submitted to *Neural Processing Letters*.

[2] J.R. Jang: "Self-Learning Fuzzy Controllers Based on Temporal Back Propagation", in *IEEE Transactions on neural networks*, Vol. 3, No. 5, September 1992.

[3] P. D. Wassermann: "Advanced Methods in Neural Computing", Van Nostrand Reinhold, 1993.

[4] D.A. White and D.A. Sofge: "Handbook of Intelligent Control", Van Nostrand Reinhold, 1992.

[5] L.M. Reyneri, M. Chiaberge, L. Zocca: "CINTIA: A Neuro-Fuzzy Real Time Controller for Low Power Embedded Systems", in *Proc. of MICRONEURO 94, Fourth Int'l Conf. on Microelectronics for Neural Networks and Fuzzy Systems*, Torino (I), September 1994, IEEE Computer Society Press, pp. 392-404.

[6] E. Miranda, L.M. Reyneri: "A CPWM Synapsis for Weight Radial Basis Functions", submitted to *IWANN 95*, Malaga.

[7] L. Davis: "Handbook of Genetic Algorithms", Van Nostrand Reinhold, New York, 1991.

[8] M. Chiaberge, J.J. Merelo, L.M. Reyneri, A. Prieto, L. Zocca: "A Comparison of Neural Networks, Linear Controllers, Genetic Algorithms and Simulated Annealing for Real Time Control", in *Proc. of ESANN 94, European Symposium on Artificial Neural Networks*, Brussels (B), April, 1994.

Packlib, an Interactive Environment to Develop Modular Software for Data Processing

Yves Cheneval[1]

Ecole Polytechnique Fédérale de Lausanne
Laboratoire de Microinformatique
Bâtiment INF - Ecublens
CH-1015 Lausanne, Switzerland
E-Mail: cheneval@di.epfl.ch
URL: http://lamiwww.epfl.ch/ymosaic/YC.html

Abstract: In this paper, we present a software concept and implementation allowing to develop modular software, i.e. a single software composed of modules written by different people and then joined together. We begin by establishing a set of rules that should be verified to obtain a powerful environment: modular code, powerful graphic capabilities, compiled code to reach maximum speed, mechanism to easily modify the parameters of the modules. We apply these rules, introduce the concepts of the environment called Packlib and describe a graphical tool, grapher, which implements the Packlib concepts. As a demonstration of the capabilities of Packlib, we show an example/tutorial of a Kohonen simulator and describe briefly the modules needed: data generation module, then data processing (the kohonen module) and finally graphic output (cgraph). To control all the parameters of the simulation, a module (panel) allows to design a graphical user interface with control panels, sliders, buttons, menus, etc.

Introduction

Ten years ago, powerful and affordable Unix workstations began to appear. On a high-resolution screen (typically 1000x1000 pixels), they were able to display windows containing text and/or graphics. In the beginning, most of these machines had a proprietary windowing system. As the years passed, we assisted in the emergence of a standard for windowing systems: this standard is X-Windows, developed originally by the MIT and handled now by the X Consortium. X-Windows is in the public domain, which means that all the source code is available and freely distributable. Because of this and because X-Windows is an open system with a revolutionary way of handling the screens (application running on machine A can display its window on machine B), it ensured a large success among users. It is now accepted as a standard internationally by the leading manufacturers of Unix workstations.
X-Windows offers a well-documented and powerful platform to design applications with graphical windows, control panels containing sliders, buttons, etc. The plain X-Windows system contains some basic buttons, sliders (called widgets), but usually with an awkward look. To obtain a better look-and-feel, some companies are providing toolkits featuring more powerful and user-friendly widgets. These toolkits must be used with X-Windows, but they are not part of it. The most widespread toolkits are XView (Sun Microsystems, public domain), and Motif (OSF, under license).
However, it is a mistake to believe that X-Windows plus these toolkits offer an ideal environment for using "off-the-shelf" graphics. A lot of programming has still to be done to obtain the desired result. Although these toolkits are more or less easy to program for experienced people, they are difficult to use for the scientist who uses his computer as a tool to help him solve actual problems, and whose main job is not to program a computer. What this person needs is a computer environment which allows, with a minimal waste of time, to design, simulate and analyse the models he has imagined. The environment should let him focus on the resolution of the problem instead of wasting time on the details such as the coding of the problem or the visualisation of results. Ideally, the scientist should load a configuration file in his environment and begin immediately to modify parameters to monitor changes in the results. This ideal environment does not exist, because no environment can contain in advance all the models

---

[1] This work has been funded by OFES project #93.0048 for the ELENA ESPRIT-BRA project #6891

imagined by scientists. It is therefore unavoidable that some code has to be written.
But the code corresponding to the problem itself is usually simple. For example, here are the tasks
needed in a typical learning loop for a neural network:
  i)   Acquire data (from sensors, from disk, from memory, etc...)
  ii) Train network
  iii) Output results, either while learning or while testing (to disk, to memory or to graphic frame)
The task ii) is usually the easiest to program. To illustrate this article we will take the example of the
coding of a simple Kohonen Self-Organising Map algorithm. Such a code possesses only approximately
one hundred lines of C code. So, the problem is not to code the central part of the algorithm but, rather,
to provide good input-output (tasks i) and iii)) capabilities. Neural networks generally output a lot of
numbers, usually grouped as matrices. The problem is to interpret these matrices correctly, and very often we need to display the matrices as lines, coloured squares, curves, etc...

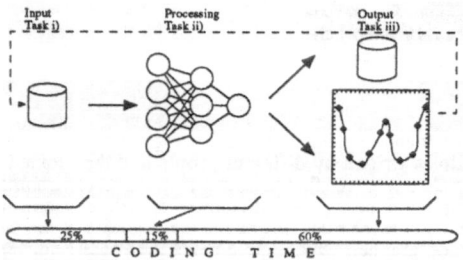
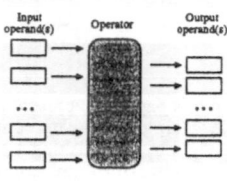

Fig. 1 Coding time for a neural network iteration   Fig. 2: The operator-operands model

If we examine the time needed to program the three tasks (see fig. 1), we discover that 85% of the time
is needed to program the tasks i) and iii). Moreover, this percentage is valid only for simple graphic outputs. It must be increased for complicated outputs.
The percentages of fig. 1 are still valid for other topics such as mobile robotics, image processing, where
the problems can be decomposed into these tasks. As a result, if we can provide an environment that a)
contains code for tasks i) and iii), b) minimises as much as possible the coding time and difficulty of
task ii) and c) maximises the global performances of the environment, a large amount of time will be
saved for the scientist. More precisely, we established a set of rules to ensure that the resulting environment will be interactive and user-friendly:

1) Code written by one user should be fully usable by others. Code written by different users should be easily "merge able" into a single software.
2) User should not write a single line of graphic code. However, he should be able, if need be, to supply additional code for input-output capabilities
3) Powerful graphic output should be available
4) Code should be compiled to reach maximum speed
5) User should be able to easily change parameters of the model and monitor the changes immedi-ately on the outputs, even while the simulation is running.

Table 1: Rules for the Packlib environment

As it can be noticed, these rules are not linked with any implementation (except the second part of rule
5). In the next section of this paper, we will present the Packlib environment in more details and describe the technical choices we have made to stick to these rules.

Description of Packlib

In this section, we will describe the concepts of Packlib which results from the rules 1 to 5 of Table 1.
The first part of rule 1 forces us to provide common input and output formats. The first, second and third
rule imply that we must provide a mechanism to display graphically the outputs of the task ii). Second
part of rule 2 tells us that the user should be able to write code for any task.
Therefore, we propose to introduce the **modular design** mechanism: each task will be implemented as a
self-contained module, with common inputs and outputs types, so that the output of a module can be
linked to the input of the next. For example, the processing module can output a matrix containing the

weights of the Kohonen module. The graphic module will then take this matrix and display it in its own window. Of course, we must also have a sequencer, a mechanism that organises the temporal sequence of the tasks: first, task i), then task ii) and finally task iii).

Rule 4 implies that all the module will be written in a compilable language. Because we work on Unix machines, the natural choice is C or C++. The C language was chosen because future users of the environment were more likely to know C than C++ (actually, modules can be programmed in C++ because object files of C++ and C are compatible).

The problem we want to address deals clearly with data processing (the modules themselves) and data flow (how to organise the sequencer) paradigms. In summary, we can say that we want to design a software environment allowing data processing **inside** the modules and data flow **between** the modules. Data processing must be defined once and for all, but data flow is implementation-dependant.

Data processing

Data processing can be defined with the help of **operators** and **operands** (fig. 2)

Operators are like "filters" that accept input operands (left), process them and give back output operands (right). To **implement** data processing, we have to define these basic concepts:
  i) The operands, i.e. input or output data.
  ii) The operators.

i) Implementation of operands

To simplify the processing, operands have been **typed**; each operator must accept one of these types as inputs and outputs. In Packlib, there are six basic types available, plus a special one (the ANY type). Each basic type has three representations: a C typedef (formal C type definition), a TAG (integer value used to identify them) and a NAME (a textual form) (See Table 2)

ii) Implementation of operators

Operators are implemented as C routines. Operators with N input operands and M output operands are called ND-MD operators. With Packlib, N>=1 and M >=1 (hence the void type when an operator has no inputs or no outputs)

<u>Remark</u>: Because of this implementation, "operator" and "routine" (respectively "operands" and "arguments") are considered synonyms in the rest of this paper.

C typedef	TAG	NAME	Description	Example
typedef char *string	STRING	"string"	Data consists of a group of character, terminated by '\0'	"Hello"
typedef long int32	INT32	"int32"	Data is a 32bit integer number.	34
typedef double real64	REAL64	"real64"	Data is a 64bit real number.	-0.456
typedef struct mxh mxh	MXH	"mxh"	Data is a matrix defined in our matrix library	{{0.1, 0.2}, {0.3, 0.4}}
typedef char bool	BOOLEAN	"bool"	Boolean. Data can be True or False	TRUE or FALSE
C intrinsic	VOID	"void"	Implements a case where data is non-existent. Usefulness of this type is the ability to have a function with no argument	----------
----------	ANY_TYPE	"any"	Any one of the type above	{1} or 3.1 or "hello"

Table 2: Packlib data types

Modules & Classes

It is often useful to join related operators together. We define a **module** as a *set* of ND-MD operators *coupled with a C data structure* (see fig. 3). Modules are implemented as a collection of C routines forming a single source file. For example, in the Kohonen example, operators exist to **learn** a new vector, **recall** the activity of neurons, **initialise** the networks, etc... It is natural to join these operators together in the same module.

These operators acts on the same data: a Kohonen module must store the weight and activation matrices, the net structure, the parameters for neighbourhood and learning rate functions, etc... In the Packlib model, all these fields **must be** joined in a single C structure. The operators set, the C structure plus the module class name forms what we call a **class**. Classes are a programming abstraction. Before using a module, **instances** must be created. From an implementation point of view, we must provide a mechanism able to instantiate a class and destroy an instance. This kind of approach is very near (and was inspired) from object-oriented programming. Each instance is identified by its **instance name**. The instance name is generally derived from the class name as user instantiates his module.

The basic idea behind this is to allow to have multiple instances of the same module. Because the code is the same between the two modules, this mechanism that allows to have re-entrant code: when a new instance is created, each module dynamically allocates memory to hold a new instance of the structure. So, although the actual code is the same, each module instance is unique in the sense that it possesses its own memory. This memory is also shared by all the operators of a given module. Each operator can thus read (or write) any field of the structure.

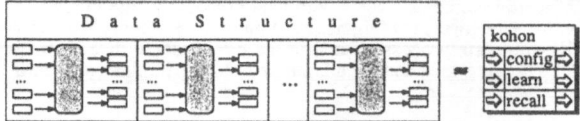

Fig 3: A module as a set of operators with the C structure

Inside a module, we will distinguish between the private routines set (i.e. the routines which are callable only from inside the module and which are implemented to simplify the coding of the module) and the public routines set. Packlib is interested only by the latter set as those routines are the only ones callable from an external tool[2]. To distinguish between private and public sets, the routines belonging to the latter set must be **registered**. Only registered routines may be "seen" and executed by a tool. To register operators, the user needs to fill a table (the registration table) inside the source code of the module, stating clearly the name of exported operators, the arguments each one takes as input and output. This table is static, because it is compiled with the module and cannot be modified at run-time.

Exported parameters (resources)

We often need to configure a module using parameters. For example, in the Kohonen module, we would like to change the dimension of the network, the shape of the neighbourhood, etc... A mechanism called **resources** have been designed for that purpose. The idea is to export variables so that they can be seen and manipulated outside the module. In a way, the method is the same than the registration of the operators (public routine set), except that here we want to export data itself instead of exporting methods to manipulate data.

For the registration of operators, the user needs to fill a static table (the registration table, see above). But the situation with the exportation of data is different, because we want to be able to dynamically export data (export or unexport parameters at run-time).

To take into account this new feature, we will use library calls (instead of a static registration table) to register at run-time the exported parameters. This method is very flexible: each time we want to export a parameter, we call a routine whose parameters will describe totally the exported parameter. Parallelly, we can unexport a parameter when needed.

Before continuing further, we must define what are the resources:

i) A *single resource* (or only *resource*) is a string composed of a *key* (or name) and a existent *value*. Parentheses separate keys from values. *value* must be the string representation of one of the type described in table 2.

ii) An *empty resource* is a resource where the *value* is non-existent. (See e.g. the BETA resource in the table 3).

iii) A *nested resource* is a resource where *value* is itself a resource (See e.g. the CALL resource in the table 3).

iv) A *multiple resource* is a string composed of one or more single, empty or nested resources separated by spaces. (See the "ALPHA(0.65) NEURONS(80)" resource in the table 3).

Resources provide a useful way of <u>associating</u> keys with values.

Resource	Key/Name	Value	TAG type of value
"ALPHA(0.65)"	"ALPHA"	"0.65"	REAL64
"BETA"	"BETA"	NULL	VOID
"ALPHA(0.65) NEURONS(80)"	-----	-----	-----
"WEIGHTS({{0.3, 0.4}})"	"WEIGHTS"	"{{0.3, 0.4}}"	MXH
"CALL(MODE(cross))"	"CALL"	"MODE(cross)"	STRING

Table 3: Examples of resources

---

[2] See the Implementation details section to know what is a "tool"

We also want to give access to exported parameters (i.e. resources) by an external tool. The basic operations we need to do are to set a resource with a new value or get the existing value of the resource. We have provided, inside Packlib, automatic methods to give access to the resources. There are two main method of access: graphical and textual. They both need the registration (inside each module) of a special operator called the **config routine**, whose function is to process the resources given as input. It is beyond the scope of this article to explain in more details the exact implementation of this mechanism. It is sufficient to know that each module is notified when a parameter has been modified, can perform tests on the new value of the parameter (valid range checks), and can take appropriate decisions, for example to reset a parameter to a correct value.

For example, if the programmer want to give the possibility to modify the network width (total number of neurons = width × height) inside the Kohonen module, he will register inside the kohonen module a resource with key = "NET_WIDTH", and specify a **pointer in memory** where the value will be stored. When the user changes the number of neurons (using textual or graphical access), the module will be notified and will be able to take the necessary actions (such as, in this case, probably reset the weights of the new network).

Textual access

Textual access is achieved by building a **resource string** as defined in Table 3 and giving this string as argument to the config routine of the module. Therefore, each module should have a config routine. (see "Configuring using the textual method" section below)

Graphical access

Graphical access can be used only if the tool provide graphical access. In this version of Packlib, only grapher gives such an access. (You can see an example of the graphical access in the "Grapher as a Packlib tool" section).

Implementation details

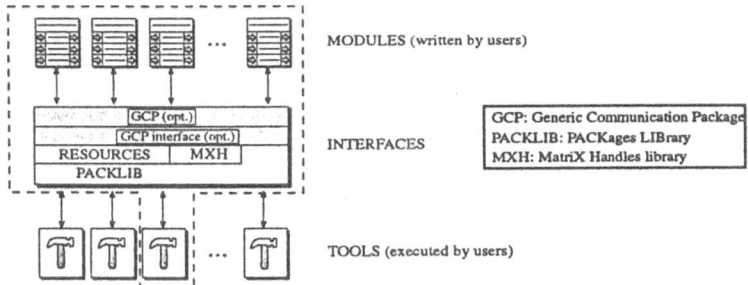

Fig. 4 Packlib block diagram

Packlib has been written on a Sun workstation using the XView toolkit provided by Sun Microsystems. This decision was taken in 1992, and at this time the Motif toolkit didn't exist for Suns.

You can see in Fig. 4 a block diagram of Packlib.

At the top are the MODULES, written by all users, described above.

At the bottom are the TOOLS that use the modules (e.g. grapher[3] is one of these tools), by calling the different operators of instantiated modules. The user can only access the modules by using a tool.

In the middle lies the INTERFACE between the modules and the tools composed of the GCP (optional), MXH, RESOURCES and PACKLIB libraries:

• The GCP library is used to run a specific module on a different machine. It is completely transparent from the programmer of the module point of view; user can decide at run time which module runs on which machine.

• The MXH library handles the matrices (memory allocation and current operations such as multiplication, inversion, etc.).

• The PACKLIB library contains functions to register a module, to create or destroy an instance of a module, and to call the operators of each module.

• The RESOURCES library handles the exported parameters of each module as described in the

---

[3] grapher is the main Packlib tool. It is a graphical front-end allowing to visually see you modules and to link them together. See the "Grapher as a Packlib tool" section.

"Exported parameters" section.

The area enclosed in dashed line of Fig. 4 is one tool. A tool is an executable file which must be compiled and linked with all the modules it needs.

The tools are the applications actually run by the users, although they are not written by the users. They are provided by the designer of the Packlib environment. A few tools exist for various purposes (a tool able to use the configurations file created by grapher with no possibility of modification (called runtime) is used for demonstrations).

A number of other user-callable C libraries are provided to the programmers of the modules. These libraries can be freely called from within the module; the programmer can also design his own libraries callable from his modules. These libraries are not shown in Fig. 4.

We will now present briefly a tool, called grapher, which has been used for two years by the partners of the ELENA (Enhanced Learning for Evolutive Neural Architectures), Nerves II, ESPRIT-BRA project #6891.

Grapher as a Packlib tool

grapher is the most intuitive and user friendly tool of the packlib environment (fig 5). It is used to quickly design applications involving neural networks or any other kind of data processing, by creating a data flow architecture. grapher is using a custom made multitasking kernel, which means that in most cases, the user will be able to perform more than one action at the same time (for example, when grapher is performing lengthy computations, you can still modify the environment, start a second data-flow loop, etc...).

Obviously, data processing requires data storage and processing abilities. These two concepts are the key of packlib, and thus the key of grapher. Grapher uses the surface of the computer screen as a flat desk. The grapher frame is separated in three parts: the top panel containing the "Grapher" and "Actions" menu and the running mode, the *dock* (left) and the *desktop* (right). The top panel elements are used to control the grapher. The right side of the window is called the *desktop*, where data and processing are created, moved and destroyed using the mouse. The *dock* is located to the left of the window. It contains base objects that can be instantiated by dragging them on the *desktop*. The dock is used to hold root operands, controls and packages object, along with the trashcan. You may consider the dock as a place from were you will fetch all objects that you need to build your application.

Only the six types of data elements of packlib (void, boolean, integer number, real number, matrix, string) can be stored on the desktop, where they are called *operands*. (A special type, the "Any" type, is a meta type: it can hold any one of the basic six types). Processing abilities are also stored on the desktop, where they are simply called *modules* or *packages*. Both standard modules (modules provided by the grapher designer) and your own packages reside on the desktop.

grapher uses a standard user interface look-and-feel: either operands or modules (that is: *objects*) may be selected with a simple click on the mouse button, opened with a double click, moved with a drag. Actions are initiated with a "drag and drop" movement: an object can be selected, dragged over another object, and dropped inside.

You can see on fig. 5 a setup allowing to run the Kohonen maps algorithm we have used as illustration for this article. The numbers below refer to figure 5. We will first describe the basic elements of grapher in the dock, where you see five groups of objects:

1 A *trashcan* that is used to destroy objects.
2 Seven operands, each one is an *instance* (an example) of a void value, a boolean flag, an integer and a real value, a matrix, a string and the special Any type.
3 *Flow control cells*, used to control how information will flow between modules. Loops , type conversions, "If-then-else" structures can be created using flow control cells. These cells are unique to the grapher tool.
4 A *subgrapher*, i.e. a grapher (child) frame within another (parent) frame. A subgrapher is like the main grapher frame: it contains a dock, a desktop and so on. However, it differs from the main grapher frame because it contains input and output pads, corresponding to respectively the input and output of the subgrapher frame.
5 Some *root packages* (or *modules*) ready to be used. Here you have all the classes that have been linked with the tool. The packages are *closed* on the dock (as opposed to *open* on the desktop, see below), which means that only the first line with the class name is visible. You can choose which packages you want and which you don't before compiling the tool by using a package selection utility.

On the desktop are more objects that have been instantiated by taking them from the dock.

9 Four packages have been created onto the desktop. They are represented as the rectangular boxes, whose first line of the name of the instance of the module; next lines are reserved for each operator. Each operator interacts with the grapher through input and output *pad(s)*, represented as arrows. By convention, the left pad is for input, the right is for output. A ND-MD operator will have N input pads and M output pads. To activate an operator, the basic operation is to drop an operand inside its input pad[4].
The first module is a distrib module, (1_distrib is the instance name); the second is the kohonen module (4_kohon); the third is a cgraph module, labelled 2_cgraph. In this example, they implement the task i), ii) and iii) respectively. The goal of the distrib module is to generate learning vectors that will be used by the kohon module. The cgraph module will then display the weights (and activities) of the network.

6 These are operands. To modify operands, simply click on it to get a line editor displayed above the desktop (see "Location of the line editor" in Fig. 5). You can double-click on string and matrices to obtain a graphical text or matrix editor respectively.

7 When we need to see the result of an operator, a *drop* object has the effect of "dropping" the result on the desktop, to its right. Each time it is activated, the drop object creates a new operand on the desktop. For example, the output of the config routine of the 4_kohon package has a drop connected to it: each time a result is produced, it will appear on the desktop. If some previous result already lies there, it will be updated. If no drop object was here, the user would not see the result of the operator .

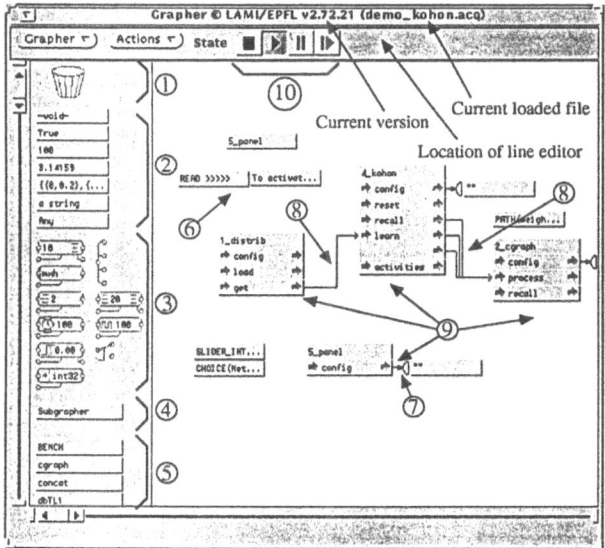

Fig. 5: Snapshot of the grapher screen

8 When we need to pass the result of an operator as argument to another operator, we use a *link* object. The link connects an output pad to an input pad, realising the data flow mechanism.
In this example, there are links between the 1_distrib and 4_kohon modules, and links between the 4_kohon and 2_cgraph modules. Each time the get operator is activated, by dropping a void into get, a matrix containing the vector to be learned is forwarded to the kohon module. The kohonen module adapts his weights according to the Kohonen algorithm. The outputs of the learn operator are double: the top operand is the weights matrix, the bottom is the activation ma-

---

[4] If N>1, all operands must be presented (in any order) for the operator to be activated. Only the last presentation will be taken into account

trix, i.e. the matrix of the activities (function of the distance of each neuron to the input vector). These outputs both converge to the 2_cgraph module, where they are displayed.

10 This is the grapher running mode display. There are 2 main modes available, the PLAY (right-pointing arrow) mode and the PAUSE (two vertical bars) mode. We won't describe these modes in this article.

To start one loop of the simulator, we must start the data-flow machine. This is achieved by dropping a void operand into the get operator of the 1_distrib module. The key concept of grapher is that every operation can be achieved by a drag-and-drop operation. It is a matter of seconds to build the setup shown on fig. 5. It is important to note that there is not a single line of graphic code inside the source of the kohonen module (see rule 2).

Suppose now that you want to modify the network width of the kohon module (a resource called "NET_WIDTH" has been registered in the kohon module). You want to change it to 17 neurons. You have many solutions to do this:

i) Configuring using the textual method

This is the most basic method. Instantiate a new string in the grapher frame, edit it and put "NET_WIDTH(17)". Then, drop this string in the config routine of the 4_kohon module.

ii) Configuring using the configuration panel of the module

Double-click on the kohon module name with the left mouse button. A window, looking like the one of Fig. 6, should appear. This window contains all the parameters (resources) of the module. To change the network width, simply click on the arrows of the NET_WIDTH resource to raise or lower it (or else, type directly the correct number).

iii) Configuring using a panel module

There is another, very important, way of setting resources of the modules. This is the panel module. that is, like any other module, located in the module list (5). When you create a new panel, a frame, like the one in Fig. 7 appears (with no objects inside).

The panel module is used to control the resources of other modules by providing a consistent user interface. We could for example want to provide a graphical interface to control the network width of the 4_kohon module, but without using the configuration panel of the module, because we want to:

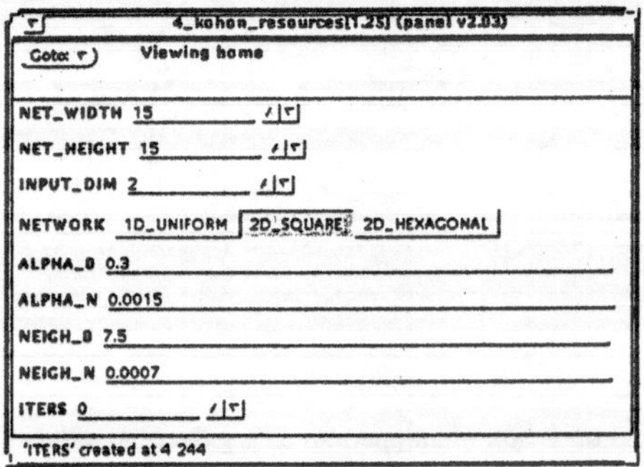

Fig. 6: Configuration panel of the 4_kohon module

- Provide simplified control over the module, i.e. hide unwanted parameters
- Concentrate all the simulation parameters on one (or a small number of) frame only, so that the user doesn't need to have all the configuration panels open to modify desired parameters (for example to build a simulator that will run with the runtime tool)

The panel frame is divided into 2 parts. The upper part contains icons of available objects. The bottom part is where the actual objects will be located. You can place any panel object on the panel frame and assign actions that will be executed each time the user interacts with the object. Everything that can be done with the mouse on the grapher frame can be executed by panel objects. You can configure any resource of any module, modify any operand on the desktop and simulate the drop of any operand in any operator.[5]

You can configure all the panel objects. You can see on fig. 8 the parameters of the "Width" slider of the fig. 7. You can change the slider name, its position on the frame etc... The most important resource is the CALL resource. It allows to associate an action with the object. In this example, the slider is configured to modify the network width of the 4_kohon module:

4_kohon,config(NET_WIDTH(%))[6]

This line means that each time the slider is moved, the slider will replace the '%' by the exact value of the slider and call the config operator of the 4_kohon module with the resulting string (exactly like the textual method of configuring). If the configuration panel of the kohon module is open, it will reflect the change as the slider is moved.

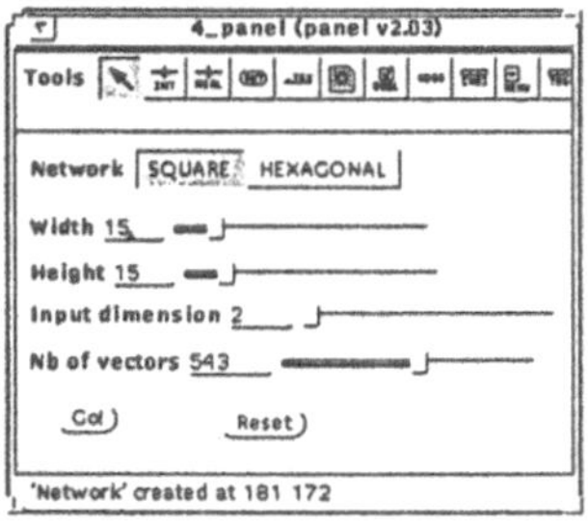

Fig. 7: Panel to control the kohonen simulator 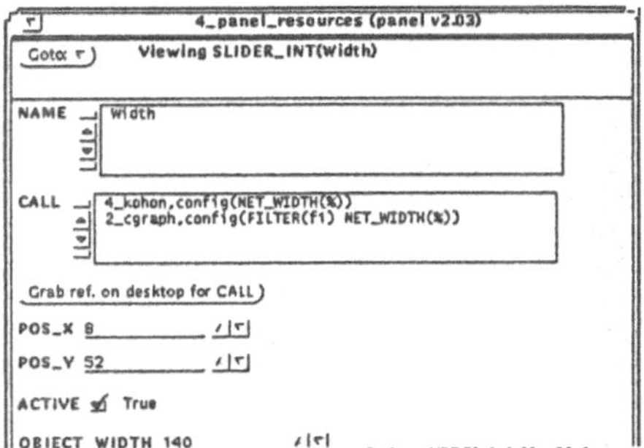   Fig. 8: Some parameters of the "Width" slider

The cgraph module

We will use the cgraph module to display the results of the kohonen network on a graphic frame.
The cgraph module allows to display on the same frame:
- Independent data, i.e. data coming from different matrices
- Linked data, i.e. data contained in a single matrix, using different visuals, each visual being a different representation of the same matrix.

Each visual is displayed in a sub window inside the cgraph frame. User can dynamically add filters between input matrix and each visual. These filters allows to select portions of matrices and/or apply transformations to the input matrix content. They may be easily written by the user if a predefined filter does not exist for the desired task. Filters are programmed like the modules.
The cgraph frame shown on fig. 9 has two views (actually, there are three internal views, but two of them are superposed, giving the illusion of two views only).
On the left, we see the representation of the neurons in the weight space. Each cross corresponds to the neuron position. The lines indicate the neighbourhood relations between neurons.
On the right is the representation of the neuron activities, ranging from black (low activity) to white (high activity).
Before going further, here is the internal cgraph configuration that led to this graphic frame (fig. 10)
As you can see, this figure contains boxes linked together. Each box represents a visual or a filter and contains two lines (except the OUT box). The top line shows the instance name of the filter/visual; the bottom one displays the class name of this object. The visuals are represented as the rightmost boxes on fig. 10.
With such a configuration, each time a matrix is dropped in the process routine of the cgraph module, the correct path is chosen according to the **name** of the matrix:
- If the matrix is called weights, the top path is chosen.
- If the matrix is called activities, the bottom path is chosen.

---

[5] You cannot, however, create new objects or link objects. This can be done with the mouse only.

[6] There are automatic methods to create this string. It is possible to create it by a simple mouse click in the configuration panel of the kohon module.

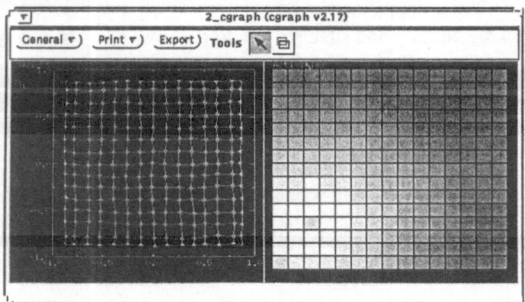
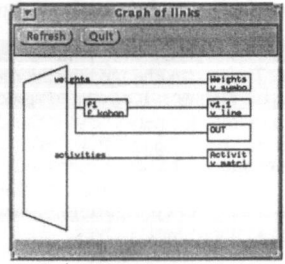

Fig. 9 The cgraph frame displaying kohonen network   Fig. 10: Internal configuration of cgraph

Suppose the matrix is called weights. In this case, the path is split. First, the matrix goes directly to visual Weights of class v_symbol and is displayed. Second, the matrix goes through f1 filter (this filter builds the neighbourhood lines associated with the Kohonen network) and then is displayed in visual v1.1 of class v_line (lines). Third, it goes to the OUT box, which means the output of the process routine.

This configuration means that each time a matrix whose name is weights is dropped in the process routine, it
* is shown through 2 visuals, named Weights and v1.1
* is the output of the process routine (so that it can be processed by other modules).

Here, the two visuals (v1.1 and Weights) are superposed to show the symbols (the neurons) and the lines (connections between nearest neighbours) as if they were inside the same view (see fig. 9).

Each visual is specialised for one kind of graphical representation. There are visuals to display symbols, lines, circles, rectangles, images, matrices as small coloured squares, etc. Each visual can be configured for user needs.

The cgraph module offers also the possibility to modify the displayed data directly on the cgraph frame, by selecting elements with the mouse and move them. If you move the neurons, the neighbourhood lines follow the neuron. This is possible because of the backtracking mechanism built inside cgraph. As you can see in the fig 10 above, the visuals Weights and v1.1 have the same source matrix (weights). When you modify the weights matrix by dragging elements, cgraph automatically backtrack from right to left in the graph figure. It finds the link going to f1, where new lines are computed and finally displayed in the v1.1 visual.

Conclusion & Acknowledgements

We have described the concept and presented a software environment allowing to design modular software. This environment has been used for two years by the partners of the ELENA project. There are approximately 80 modules (all with online help) for Packlib, written by all the partners and implementing mainly neural networks models used for data classification. Packlib is used also in applications involving mobile robotics in our lab. A demonstration version of Packlib will be distributed in the public domain. Contact the author for a copy.

I would like to thank L. Tettoni (LAMI-EPFL), co-designer and programmer of the first version of grapher in Nov. 92, for numerous discussions about Packlib; J.-L. Voz (DICE-UCL) for intensive beta-testing and suggestions; and finally all the people who developed modules for the Packlib environment. Packlib wouldn't exist without its modules!

References

Y. Cheneval, *"Deliverable R2-B2-P: Unified Graphic Environment"*, ELENA ESPRIT-BRA project #6891, July 1994

NSL – Neural Simulation Language

Alfredo Weitzenfeld

Departamento Académico de Computación
Instituto Tecnológico Autónomo de México
Rio Hondo #1, San Angel 01000
México D.F., México

NSL, Neural Simulation Language, is a general purpose simulation system providing a high-level language with many constructs and libraries developed to ease the specification of large neural networks. NSL integrates *object-oriented* programming methodologies in its design and implementation, providing a simulation environment for users with little programming background, as well as those with more extensive programming expertise, who can use C++ as an extension to NSL's modeling language. NSL is widely used in research and teaching, having lead to many different neural network models, both in the artificial and biological domains. NSL enables the simulation of models with different levels of neural details, with special support for the *leaky integrator*.

1. INTRODUCTION

In the area of neural network simulation many tools have been built to facilitate the scientist in the task of modeling and simulating neurons at different levels of detail. The more detailed neuronal models, such as the *Hodgkin-Huxley* model (Hodgkin and Huxley, 1952), and the *compartmental* model (Rall, 1959), permit only the modeling of a few neurons at a time, and are supported by simulation systems such as GENESIS or NEURON (De Schutter, 1992). The coarser neural models, such as the *leaky integrator* model (Arbib, 1989), permit the modeling of thousands of neurons, and are supported by simulation systems such as NSL (Weitzenfeld, 1991; Weitzenfeld and Arbib, 1994).

2. NSL SYSTEM

NSL, Neural Simulation Language, is a general purpose simulation system providing a high-level language with many constructs and libraries developed to ease the specification of large neural networks. NSL integrates *object-oriented* programming methodologies (Wegner, 1990) in its design and implementation, providing a simulation environment for users with little programming background, as well as those with more extensive programming expertise, who can use C++ (Stroustrup, 1991) as an extension to NSL's modeling language. NSL is offered as public domain software (*anonymous ftp* from usc.edu).

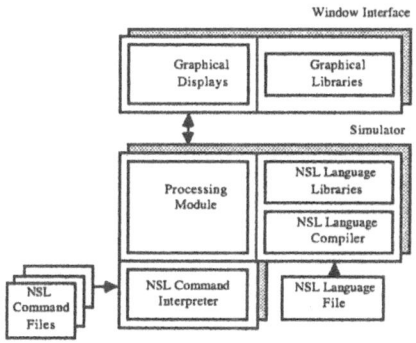

Figure 1. The NSL simulation system is composed of two units: (1) the Simulator, containing the Processing Module, NSL Language Compiler, NSL Language Libraries, and NSL Command Interpreter; (2) the Window Interface, containing the Graphical Displays, and the Graphics Libraries.

The system, whose architecture is shown in Figure 1, includes a command interpreter for interactive and batch processing, an X windows graphical interface, and temporal and spatial displays, including 2D and 3D graphics.

3. NSL LANGUAGE

In order to model neural networks with NSL (NSL Language Compiler shown in Figure 1) it is necessary to describe (1) the neurons making up the network, (2) the neurons' interconnections, and (3) the dynamics of the neurons and their interconnections.

3.1. Neurons

The basic neural model in NSL is the single-compartment neuron, having one output and many inputs, as shown in Figure 2. The internal state of the neuron is described by a single scalar quantity, its membrane potential m, which depends on the neuron's inputs and its past history. The output is described by another single scalar quantity, its firing rate M, and may serve as input to many other neurons, including itself. As the input to a neuron varies, the membrane potential and firing rate also vary.

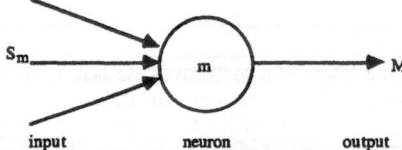

Figure 2. The single-compartment neuron model is represented by one value m corresponding to its membrane potential, and one value M corresponding to its firing rate. S_m represents the set of inputs to the neuron. There is a single output.

The *membrane potential* for m is described by the differential equation

$$\tau_m \frac{dm(t)}{dt} = f(S_m, m, t)$$

which depends on the neuron's input S_m, previous values of m, and time parameter t. τ_m is the time constant. The choice of f defines the particular neural model utilized. In particular the *leaky integrator* model is described by $f(S_m, m, t) = -m(t) + S_m(t)$, or

$$\tau_m \frac{dm(t)}{dt} = -m(t) + S_m(t)$$

The *firing rate M*, the output of the neuron, is obtained by applying a *threshold function* to the neuron's membrane potential,

$$M(t) = \sigma(m(t))$$

where σ is usually a non-linear function. Some of the most common threshold functions, such as *ramp*, *step*, *saturation* and *sigmoidal*, are shown in Figure 3.

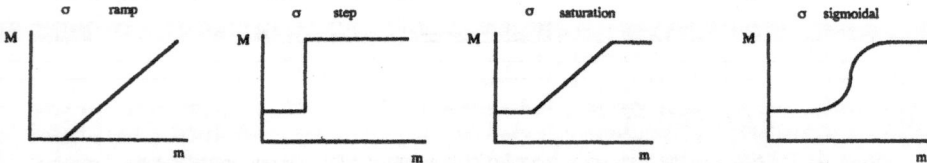

Figure 3. The figures shows some typical threshold functions.

In NSL two **DATA** structures are required to represent a single neuron, one structure corresponds to the membrane potential and the other one to the firing rate. The notation is as follows (with a semicolon at the end of each statement):

```
DATA m;
DATA M;
```

The membrane potential m is represented by a differential equation

```
DIFF (m, ᴛₘ) = f (Sₘ, m);
```

where **DIFF** defines a first order differential equation for m with time decay t_m (time parameter t is implicit in the equation). The leaky integrator model corresponds to

```
DIFF (m, ᴛₘ) = -m + Sₘ;
```

The firing rate M is represented simply by

```
M = σ(m);
```

where σ represents the choice of threshold function.

3.2. Interconnections

When building neural networks, the output of a neuron serves as input to other neurons. Links among neurons carry a connection weight which describes how neurons affect each other. Links are excitatory or inhibitory depending on whether the weight is positive or negative. The most common formula for the input to a neuron v is

$$S_v = \sum_{i=1}^{n} w_i M_i(t)$$

where $M_i(t)$ is the firing rate of neuron m whose output is connected to the ith input of neuron v, and w_i is the weight on that link.

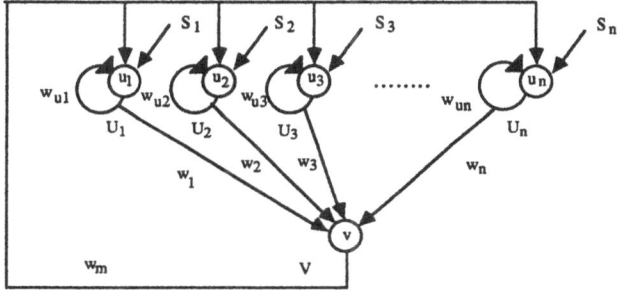

Figure 4. The neural network shown corresponds to the architecture of the Maximum Selector model (see Arbib, 1989), where ui and v represent membrane potentials, Ui and V represent firing rates, Si represent inputs to the network, and wi represent connection weights.

For example, a neural network architecture corresponding to the *Maximum Selector* model (see Arbib 1989) is shown in Figure 4. u and v represent membrane potential (analogous to m), and U and V represent firing rate (analogous to M).

The input to neuron v is given by

$$S_v = w_1 U_1 + w_2 U_2 + w_2 U_2 + \ldots + w_n U_n$$

while the input to the u_i neuron is (there is n such equations)

$$S_{ui} = w_m V + w_{ui} U_i + S_i$$

In NSL, these expressions describing interconnections among neurons in the neural network are described in a similar way. For example, the input to neuron v, represented by S_v, would be the summation of the outputs of all the neurons u multiplied by the corresponding connection weights w:

$$S_v = w_1 * U_1 + w_2 * U_2 + w_3 * U_3 + \ldots + w_n * U_n$$

The input to neuron u_i, is represented by S_{ui} (there is n such equations)

$$S_{ui} = w_m * V + w_{ui} * U_i + S_i$$

3.3. Layers and Masks

When modeling thousands of neurons and their interconnections it becomes extremely difficult to name every single one of them. Since in the brain we often find neural networks structured into two-dimensional homogeneous neural layers, with regular connection patterns between various layers, we extend the basic neuron abstraction into neural layers and connection masks.

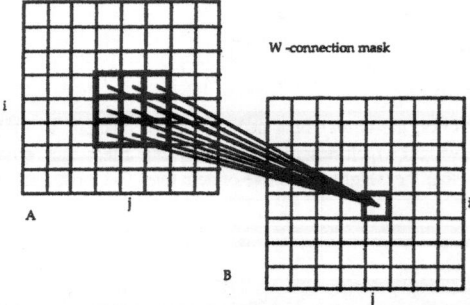

Figure 5. W represents the connection or convolution mask between layers A and B corresponding to the equation $B=W*A$. In this example W is a 3x3 mask which is overlapped over a window of A to obtain a single value in B.

The computational advantage of introducing such concepts when describing a neural network is that neural layers and interconnection masks can then be concisely described as higher level data structures. Instead of describing neurons on a one by one basis, a layer can be described as an array and, similarly, the connections between layers can be described by a mask storing synaptic weights. An interconnection among neurons would then be processed by computing a spatial convolution of a mask and a layer. For example, as shown in Figure 5, if A represents an array of outputs from one layer of neurons, and B represents the array of inputs to another layer, and if the mask $W(k,l)$ (for $-d \leq k, l \leq d$) represents the synaptic weight from the $A(i+k,j+l)$ (for $-d \leq k, l \leq d$) elements to $B(i,j)$ element for each i and j, we then have

$$B = \sum_{k=-d}^{d} \sum_{l=-d}^{d} W(k,l) A(i+k, j+l)$$

which can be described by a simple expression

$$B = W*A$$

In order to support layers and masks, the basic **DATA** structure in NSL is extended with two layers types, **VECTOR** and **MATRIX**, differing according to the number of dimensions they have. To simplify matters, masks, which may have any rectangular shape, are also defined as layers whose values are interpreted in a different way.

For example, the layers of neurons shown in Figure 4 would be described by

```
VECTOR (S, n) ;
VECTOR (u, n) ;
VECTOR (U, n) ;
DATA (v) ;
DATA (V) ;
```

3.4. Sample Model

The complete set of equations describing the *Maximum Selector* model, shown in Figure 4, are:

$$\tau_u \frac{du_i(t)}{dt} = -u_i + w_u f(u_i) - w_m g(v) - h_1 + s_i, \ 1 \le i \le n$$

$$\tau_v \frac{dv}{dt} = -v + w_n \sum_{i=1}^{n} f(u_i) - h_2$$

where w_{ui} is the connection weight for the self connection of u_i, and $w_{u1} = w_{u2} = ... = w_{un} = w_u$ and $w_1 = w_2 = ... = w_n$ (in this particular model these connection weights are the same), h_1 and h_2 are constants, and the threshold functions are

$$f(u_i) = \begin{cases} 1 & u_i > 0 \\ 0 & u_i \le 0 \end{cases}, \ g(s_i) = \begin{cases} s_i & s_i > 0 \\ 0 & s_i \le 0 \end{cases}$$

In NSL the description of the complete model is arranged into modules, the **INIT_MODULE** containing re-initialization statements, and the **RUN_MODULE** containing statements which are continuously executed as part of the simulation.

The above equations correspond to the following code arranged in two **RUN_MODULE**s:

```
RUN_MODULE (U)
{
    DIFF(u,tu) = - u + wu*U - wm*V - h1 + S;
    U = step(u);
}
```

```
RUN_MODULE (V)
{
    DIFF(v,tv) = - v + SUM(wn*U) - h2;
    V = ramp(v);
}
```

Note that S, u, and U are vector layers and all operations are applied to the layer as a whole. $SUM(w_n*U)$ first multiplies the connection weight w_n by the firing rate U, and the returns a single value corresponding to the vector summation of the expression. This is necessary since v is a single element layer.

4. DISCUSSION

NSL has been successfully utilized as a simulation tool for both biological and artificial neural networks, where various types of applications have been developed, such as the *visuomotor coordination* model (Arbib and Lee, 1993), and the *generation of saccades* model (Dominey and Arbib, 1992). The main challenge in the development of NSL, as well as with other simulation tools, is on one hand to provide a general purpose user-friendly simulation environment, while at the same time being as efficient as possible in the time consuming process of neural network simulation.

As NSL keeps on evolving, it will offer a distributed and parallel framework for the simulation of neural networks (Weitzenfeld and Arbib, 1991) integrating with *schema* models, as described in ASL, Abstract Schema Language (Weitzenfeld, 1993), to enable the development of hierarchical and distributed neural networks, such as needed in robotics applications (Fagg et al., 1992).

5. REFERENCES

Arbib, M.A., 1989, The Metaphorical Brain 2: Neural Networks and Beyond, Wiley.

Arbib, M.A., and Lee, H.B., 1993, Anuran Visuomotor Coordination for Detour Behavior: From Retina to Motor Schemas, in From Animals to Animats 2: Proc. of 2nd International Conference on Simulation of Adaptive Behavior (J.-A. Meyer, H.L. Roitblat, and S. Wilson, Eds), A Bradford Book/MIT Press :42-51.

De Schutter, E., 1992, A Consumer Guide to Neuronal Modeling Software, in Trends in Neuroscience, 15(11):462-464.

Dominey, P.F., and Arbib, M.A., 1992, A Cortico-Subcortical Model for Generation of Spatially Accurate Sequential Saccades, Cerebral Cortex, 2:153-175.

Fagg, A.H., King, I.K., Lewis, M.A., Liaw, J.S., Weitzenfeld, A., 1992, A Neural Network Based Testbed for Modeling Sensorimotor Integration in Robotics Applications, Proc. of IJCNN '92, Baltimore, MD.

Hodgkin, A.L., and Huxley, A.F., 1952, A quantitative description of membrane current and its application to conduction and excitation in nerve, J. Physiology, London, 117:500-544.

Rall, W., 1959, Branching dendritic trees and motoneuron membrane resistivity, Exp. Neurol., 2: 503-532.

Stroustrup, B., 1991, The C++ Programming Language, 2nd. Ed., Addison-Wesley.

Wegner, P., 1990, Concepts and Paradigms of Object-Oriented Programming, SIGPLAN, OOPS Messenger, 1(1):7-87, Aug.

Weitzenfeld, A., 1991, NSL: Neural Simulation Language, Version 2.1, CNE-TR 91-05, University of Southern California, Center for Neural Engineering, Los Angeles, CA.

Weitzenfeld, A., 1993, A Hierarchical Computational Model for Distributed Heterogeneous Systems, TR 93-02, Center for Neural Engineering, University of Southern California, Los Angeles, California, May.

Weitzenfeld, A., and Arbib, M., 1991, A Concurrent Object-Oriented Framework for the Simulation of Neural Networks, Proceedings of ECOOP/OOPSLA '90 Workshop on Object-Based Concurrent Programming, SIGPLAN, OOPS Messenger, 2(2):120-124, April.

Weitzenfeld, A., and Arbib, M.A., 1994, NSL Neural Simulation Language, in Neural Network Simulation Environments, Ed. J. Skrzypek, Kluwer.

Low–Cost Accelerator for the Simulation of Cellular Neural Networks

A. Torralba

Dpto. de Ingeniería Electrónica

Escuela Superior de Ingenieros, Universidad de Sevilla,

Avda. Reina Mercedes s/n, SEVILLA–41012, SPAIN

e–mail: torralba@gtex02.us.es

December 1994

Abstract

As proposed by L.O.Chua and L.Yang in [1], if we discretize the time in the equations that describe a Cellular Neural Network (CNN), we obtain a difference equation that recalls the dynamics of cellular automata. In this paper a hardware accelerator for CNN's with the structure of a cellular automaton is proposed. Its simple architecture allows the implementation of large arrays with high efficiency.

1 Introduction.

Neural computing has certain properties analogous to those of the human brain, i.e., association, generalization, parallelism, adaptation and learning. Hence, Artificial Neural Networks (ANN's) have been successfully used in areas such as pattern recognition, language processing and optimization problems.

Cellular Neural Networks (CNN's) are a class of ANN's whose local interconnection feature are attractive for VLSI implementation [1]. Numerous applications of the CNN's have been reported.

An XY CNN is defined by the circuit equations:

$$C\frac{dx_{(i,j)}(t)}{dt} = -\frac{1}{R}x_{(i,j)}(t) + \mathbf{A}*y_{(i,j)}(t) + \mathbf{B}*u_{(i,j)} + I \qquad (1)$$

$$y_{(i,j)}(t) = \frac{1}{2}(|x_{(i,j)}(t)+1| - |x_{(i,j)}(t)-1|) \qquad (2)$$

$1 \leq i \leq X; 1 \leq j \leq Y$

where $*$ denotes the two–dimensional *convolution operator* that includes the effects of the adjacents cells within a neighborhood of radius r, as defined in [1]. A and B are $(2r+1)$ x $(2r+1)$ matrices that are assumed to be independent of i and j (that is, *space invariants*). If the central element of the matrix A satisfies $\mathbf{A}(r,r) > \frac{1}{R}$, a stable equilibrium with binary–valued outputs is guaranteed.

In [2] an inmediate op amp implementation of a simple CNN is proposed. In [3]–[5] other actual implementations are presented. In general, the continous–time feature of analog implementations allows high cell density and promises impressive processing power. However they still present some problems that are currently being solved [6].

As suggested in [1], if we discretize the time t in the cell equation (1), we would obtain

$$x_{(i,j)}(k+1) = x_{(i,j)}(k) + \frac{T}{C}[-\frac{1}{R}x_{(i,j)}(k) + \mathbf{A}*y_{(i,j)}(k) + \mathbf{B}*u_{(i,j)} + I] \qquad (3)$$

that recalls the dynamic behaviour of cellular automata, except for the real nature of the state variables. In this equation, T is the integration step. Small values of T guarantee a good precision of the solution at the expense of a slow convergence. $T < 2RC$ is required to guarantee the convergence in most cases [7].

In this paper a hardware accelerator for cellular neural networks with structure of a cellular automaton is proposed. The objective of the proposed hardware is to serve as a co–processor in personal computers and workstations. Then, a simple, flexible linear systolic array is selected. Section 2 deals with mapping details. In section 3 analytical expressions for the proposed hardware are derived. Finally, section 4 presents experimental results measured on a propotype board.

2 Network Mapping.

Cellular automata are highly regular structures with local connectivity. First introduced by J.V.Neuman in the early 1950's, different prototypes and comercial cellular machines have been proposed since then ([8]–[9]). Most of them are high–priced, high–performance, dedicated systems with complex hardware and software. In this paper we face the design of a simple systolic coprocessor capable of an efficient mapping of large arrays of neurons onto a reduced number of Processing Elements (PEs). In order a coprocessor to be effective, some requirements have to be fulfilled:

1. *Simplicity*. It should provide an efficient implementation of the equation (3), avoiding complex array initialization, programming and data loading.

2. *Flexibility*. It should be capable of efficient reconfiguration to deal with large networks of variable size.

3. *Expansion capability*. The array size should be able to grow to cope with growing user requirements. Therefore, it should build efficient systems ranging from only a few to hundreds of PEs.

4. *Low cost*. Cost requirements force the selection of simple data and control paths.

In this paper a linear systolic array rather than a two–dimensional mesh of processing elements is selected due to the following reasons:

1. Simpler array initialization, programming and data loading.

2. Flexibility. A simple reconfiguration of external storage buffers allows large networks of arbitrary size to be mapped on the array.

3. Expansion capability. The proposed hardware can be expanded both by increasing the number of PEs in the array or by pipelining.

4. Low cost. Control and data paths are simple in a linear array.

An extense list of works related to parallel simulation of ANN's can be found in the literature ([10]–[16]). Regular structure and local connectivity are inherent features of cellular automata. Therefore, the methods which have been devised for mapping algorithms on systolic arrays can be applied to map cellular automata on systolic arrays. ([17]–[18]).

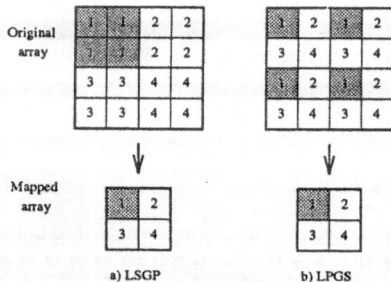

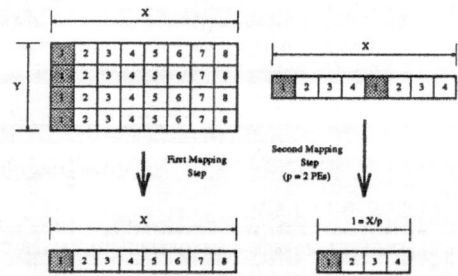

Figure 1. Basic Mapping strategies. Numbers inside cells indicates processor assignment. Shaded cells are assigned to processor number 1.

Figure 2. Proposed mapping. Shaded cells are assigned to processor number 1.

For a systematic mapping, the cellular array is regularly partitioned into blocks. In this case, only rectangular clusters will be considered. Following [18], there are two methods for mapping clusters to the array:

1. Locally Sequential Globally Parallel (LSGP). See figure 1a.

2. Locally Parallel Globally Sequential (LPGS). See figure 1b.

The second method (LPGS) shows some advantages with respect to flexibility. In the LPGS method, intermediate data can be stored outside the PEs in separate buffers. Hence, local memory size on the PE can be kept constant, independent of the neural network size. Besides, *framming* (a technique to reduce computation time by considering only the active part of the whole array) is only effective in the LPGS method.

In this paper, the LPGS method is hierarchically applied in two steps. In the first step, the whole two–dimensional array of neurons is clustered by columns and each cluster is assigned to a PE (figure 2a), leading to a linear structure. In the second step, the array size is further reduced by folding (figure 2b).

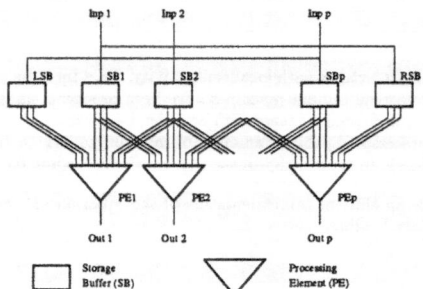

Figure 3. The systolic accelerator.

Figure 3 shows the final architecture, including external Storage Buffers, in the case of a neighborhood of radius $r = 1$. In the figure, p is the number of PEs and XY is the neural network size. In a favorable case, X/p is an integer. In figure 3, SB stands for Storage Buffer.

The architecture in figure 3 reduces to the systolic array proposed in [19] in the case $p = 1$.

3 The Systolic Accelerator.

In figure 3, an SB receives a sequential stream of input cells. Each cell (i, j) contains the state variable $x_{(i,j)}(k)$ of the neuron (i, j) in time step k. As input matrix u is constant, it can be preloaded on internal memory in each PE. In the case of storage limitation it can flow through the systolic array along with the matrix x.

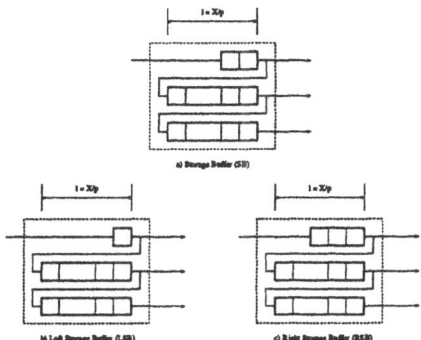

Figure 4. Internal structure of the Storage Buffers.

An SB has the internal structure of figure 4a. It consists of three linked shift registers of cells whose lenghts are 2, X/p and X/p, respectively. Two additional SBs are placed on the extremes of the array to account for its borders. Figures 4b and 4c present the internal structure of the Left and Right Storage Buffers (LSB and RSB, respectively). The upper shift register has three cells in the RSB and only one in the LSB. In periods of $l = |X/p|$ cycle times, the LSB and the RSB ignore their normal input and store a cell with a value $x = 0$ which represents the left or right border of the network.

In cycle time k, the nine inputs to a PE are one central cell and its eight neighbors (figure 5). The PE evaluates the expression (3) on the central cell (i, j) producing $x_{(i,j)}(k + 1)$.

Let's now consider how the array in figure 3 does work. A XY input CNN is represented in figure 5, where $X = 12$ and $p = 4$.

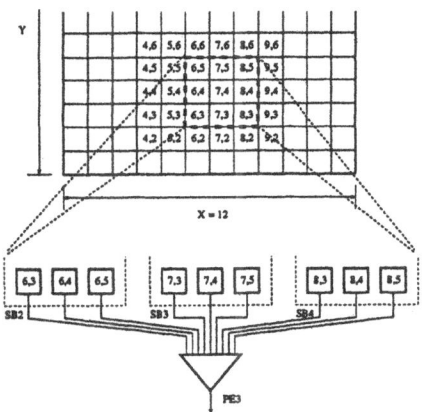

Figure 5. Inputs to the Processing Element.

We define the array cycle time t_c, as the time between the output of two consecutive cells from the array. Each cycle time, p adjacent cells of a row are entered to the array. The Storage Buffers of the extremes of the array (the LSB and the RSB) make the link between two succesive groups of p cells entering in two succesive cycle times. In the example, the nine inputs to the PE number 2 are: a central cell $((7, 4))$ and its eight neighbors $((6, 3), (6, 4), (6, 5), (7, 3), (7, 5), (8, 3), (8, 4)$ and $(8, 5))$. This situation arises in the 15th cycle time (figure 6b). A computation step takes place on the PE and the new cell $(7, 4)$ appears in the output of the array. In a similar way, PEs number 1, 3 and 4, output the new cells $(8, 4), (6, 4)$ and $(5, 4)$, respectively. All this process takes one cycle time, t_c.

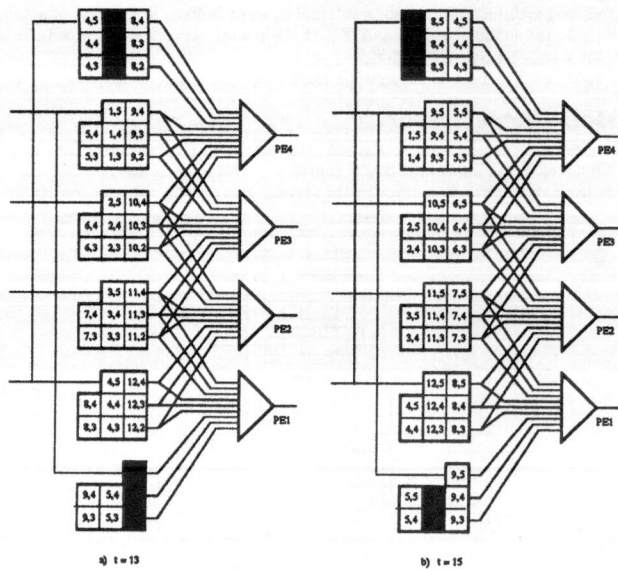

a) t = 13   b) t = 15

Figure 6. Storage Buffers contents in cycle times: a) 13 and b) 15.

Figure 6 depicts the contents of the Storage Buffers in cycle times 13 and 15 with the example network of figure 5. In this figure, a black cell denotes the left or right border of the CNN.

The architecture of figure 3 can be extended to the case $r > 1$ at expense of higher complexity. Figure 7 shows the internal structure of an SB and a PE in the case $r = 2$.

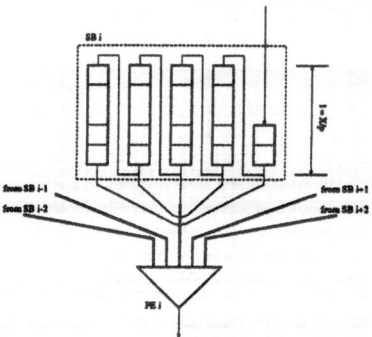

Figure 7. The systolic accelerator for a neighborhood of radius $r = 2$.

The time required to pass a XY CNN through the array is:

$$T = [(\frac{X}{p} + 2) + \frac{XY}{p}]t_c \tag{4}$$

p is the number of PEs in the array.

The first term is the latency time, that is, the time required to fill the storage buffers to start obtaining cells in the output of the array. The second term is the time required to process the full XY network.

Several stages, each composed of the array of figure 3 can be cascaded. That is, the outputs of stage i can be used to fed the inputs of the stage $i + 1$. Therefore, further concurrency is attained via pipelining at expense of higher latency time (figure 8).

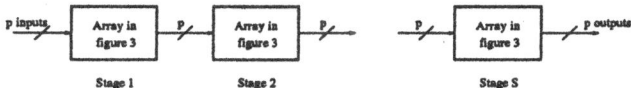

Figure 8. Pipelining the array of figure 3.

Consider an XX input neural network requiring L iterations of the equation (3) to converge. We use S stages of the hardware of figure 3. Assuming $L >> S$, L/S passes through the array are needed in all. The total time is:

$$T = \frac{L}{S}[S(\frac{X}{p} + 2) + \frac{X^2}{p}]t_c \tag{5}$$

4 Experimental Results.

The array of figure 3 has been mapped onto a prototype board that uses 16 transputers $T800$ as Processing Elements. Transputers are 32-bit microprocessors specially designed for concurrent processing [20]. They have been used in ANN's implementations (see [21] and references therein). Figure 9 shows network mapping in the case $r = 1$.

A cycle time for the transputer m in stage n consists of three steps:

1. Step Horizontal communications.

   shift the registers in the SBs
   send $x_{(i-p,j)}(k+1)$ to transp. m in stage $n+1$
   and receive $x_{(i+p,j+1)}(k)$ from transp. m in stage $n-1$

2. Step Vertical communications.

   send $[x_{(i,j-1)}(k), x_{(i,j)}(k), x_{(i,j+1)}(k)]$ to
   and receive $[x_{(i-1,j-1)}(k), x_{(i-1,j)}(k), x_{(i-1,j+1)}(k)]$ from

   transp. $m + 1$ in stage n

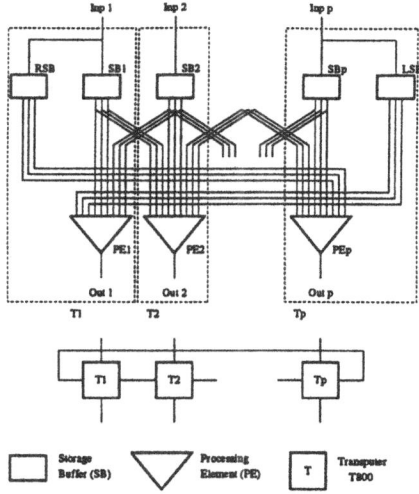

Figure 9. Mapping the accelerator on a transputer array.

send $[x_{(i,j-1)}(k), x_{(i,j)}(k), x_{(i,j+1)}(k)]$ to

and receive $[x_{(i+1,j-1)}(k), x_{(i+1,j)}(k), x_{(i+1,j+1)}(k)]$ from

transp. $m - 1$ in stage n

3. Step Evaluate expressión (3).

Operations in steps 1 and 2 above are done in parallel.

Note that one iteration of the equation (3) involves two matrix–vector multiplications and three additions. A dependence–graph for these operations can be generated and a dedicated array can be built to map it following well-known techniques describes in [18]. Relations (4) and (5) still mantain with a reduced cycle time t_c. For simplicity, in our prototype board, all the operations involved in one iteration of the equation (3) are sequentially done in one transputer.

In the case $r > 1$, the mapping in figure 9 can be also used, although operations in step 2 above would require r steps. As the transputer $T800$ has only four serial links to communicate with its neighbors, a processing element with more communication links would allow a more efficient mapping in this case.

To test the architecture we have implemented the edge detector proposed in [1], with an squared XX initial image. Matriz u was preloaded on transputers memory. Table I resumes some of the experimental results with different values of X (network size), S (number of stages) and p (number of PEs per stage). To obtain the results of Table I, the array was forced to do 100 iterations of the equation (3) (i.e., $|100/S|$ passes through an S stage array), although actual convergence was achieved in only 21 iterations. Note that the number of iterations that are necessary to converge depends on the value of the constants T/C and R used in the equation (3).

A graphical representation of Table I is given in Figure 10, where the efficiency of parallelism η has been used as a figure of merit

$$\eta = \frac{t_1}{n * t_n} \tag{6}$$

t_i is the time that i processors take to complete.

Figure 10 depicts the efficiency as a function of the number of stages S for different values of X and p. As expected, the highest efficiency is achieved in the case of $S = 1$ (near 100 %) due to the near negligible latency of the first stage of the array. The excellent behavior of the case $p = 1$ is due to the absence of vertical communications. This result indicates that the speed of the transputers communication links becomes a bottleneck.

In a natural VLSI implementation of the proposed hardware, a Processing Element will be assigned to a chip. In the case of a neighborhood of radius r, each PE requires $2r + 1$ bidirectional connections with the two neighboring PEs of the same stage. Besides, one unidirectional connection is required with its companin SB and another with an SB of the following stage. Several Processing Elements can reside in the same chip. This design will not run into pin limitations, since in addition for power, ground and control, the set of m adjacents PEs of the same stage will only require $2 * m + 2 * (2r + 1)$ pins for offchip communication.

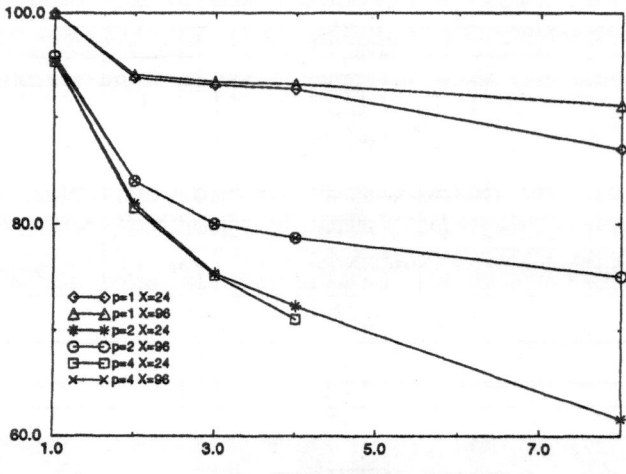

Figure 10. Measured efficiency.

S	$p=1$			$p=2$			$p=3$		
	$X=8$	24	96	$X=8$	24	96	$X=8$	24	96
1	3.02	28.11	440.44	1.59	14.73	229.64	1.06	9.83	153.57
2	1.62	15.00	234.28	1.03	8.58	131.08		5.73	87.39
3	1.11	10.08	157.36	0.82	6.22	91.76			
4	0.87	7.59	118.27	0.69	4.87	70.00		3.27	46.64
8	0.46	4.04	60.43	0.49	2.85	36.70			

S	$p=4$			$p=6$			$p=8$		
	$X=8$	24	96	$X=8$	24	96	$X=8$	24	96
1	0.79	7.37	115.30	0.53	4.92	76.87	0.40	3.69	57.65
2	0.52	4.31	65.54	0.36	2.88	43.75	0.27	2.18	32.81
3	0.45	3.12	45.88						
4	0.39	2.47	34.98						

Table 1: Time required to complete 100 iterations (seg.).

5 Conclusions.

Neural computing is synonimous of massively parallel processing. Special–purpose hardware is a solution to optimize cost and response times. In this paper a hardware accelerator for CNN's has been presented. Its efficiency has been calculated by means of analytical expressions. The proposed hardware has been simulated using a transputer network and experimental results are reported.

References

[1] Chua, L.O. and Yang. L. Cellular Neural Networs: Theory. *Trans. on Circuits and Systems*, 35 (1988), pp. 1257–1272.

[2] Chua, L.O. and Yang, L. Cellular Neural Networs: Applications. *Trans. on Circuits and Systems*, 35 (1988), pp. 1273–1290.

[3] Yang, L., Chua, L.O. and Krieg, K.R. VLSI Implementation of Cellular Neural Networks. *Proc. Int. Symp. on Circuits and Systems*, IEEE, 1990, pp. 2425–2427.

[4] Roska, T. A Hardware accelerator board for cellular neural networks: CNN–HAC. *Proc. First Cellular Neural Networks & Applications International Workshop*. IEEE, 1990, pp. 160–168.

[5] Harrer, H., Nossek, J.A. and Stelzl, R. An Analog Implementation of Discrete–Time Cellular Neural Networks. *Trans. on Neural Networks*, 3, (1992), pp. 466–477.

[6] Vittoz, E. Analog VLSI Implementation of Neural Networks. *Proc. Int. Symp. on Circuits and Systems*, IEEE, 1990, pp. 2524–2527.

[7] Rodríguez Vázquez, A., Domínguez–Castro, R. and Huertas, J.L. Accurate Design of Analog CNN in CMOS Digital Technologies. *Proc. First Cellular Neural Networks & Applications International Workshop*, IEEE, 1990, pp. 273–280.

[8] Preston,K. and Duff, M.J.B. *Modern Cellular Automata. Theory and Applications*. Plenum Press, N.Y., 1984.

[9] Toffoli, T. and Margolus, N. *Cellular Automata Machines*. MIT Press, Cambridge, Massachussets, 1987.

[10] Forrest, B.M., Roweth D., Stroud N., Wallace, D.J., and Wilson, G.V. Implementing neural network models on parallel computers. *Comput. J.* 30 (1987), pp. 413–419.

[11] Pomerleau, D.A., Gsciora, G.S., Touretzky, D.S., and Kung, H.T. Neural network simulation at warp speed: How we get 17 million connections per second. *Proc. Int. Conf. on Neural Networks*, IEEE, July 1988, Vol. II, pp. 143–150.

[12] Kung, S.Y., and Hwang, J.N. A unified systolic architecture for artifical neural nets. *Proc. Int. Conf. on Systolic Arrays*, IEEE, 1988, pp. 163–174.

[13] Ghosh, J., and Hwang, K. Mapping neural networks onto message–passing multicomputers. *J. Parallel Distrib. Comput.* 6, (1989), 221–230.

[14] Lin, W.M., Prasanna Kumar, V.K., and Wojtek Przytula, K. Algorithmic mapping of neural network models onto parallel SIMD machines. *Trans. Comput.*, 40, (Dec. 1991), pp. 1390–1401.

[15] James, Mark, and Hoang, D. Design of low–cost, real–time simulation systems for large neural networks. *J. Parallel Distrib. Comput.*, 14, 221–235.

[16] Chu, L.-C., and Wah, B.W. Optimal mapping of neural–network learning on message–passing multicomputers. *J. Parallel Distrib. Comput.*, 14, 319–339.

[17] Moldovan, D.I.M. and Fortes, J.A.B. Partitioning and mapping of algorithms into fixed size systolic arrays. *Trans. on Comput.*, 35 (1986), pp. 1–12.

[18] Kung. S.Y. *VLSI Processor Arrays*. Prentice-Hall, Englewood–Cliffs, N.J., 1987.

[19] Lougheed, R.M. and McCubbrey, D.L. The Cytocomputer: A practical pipelined image processor, in: *Proc. 7th. Ann. Int. Symp. on Computer Architecture*, 1980, pp. 271–277.

[20] INMOS. The transputer family 1987. 1987.

[21] Nordström, T. and Svensson, B.. Using and Designing Massively Parallel Computers for Artificial Neural Networks". *J. Parallel Distrib. Comput.*. 14, pp. 260–285.

A VLSI System for Neural Bayesian and LVQ Classification

Philippe Thissen*, Michel Verleysen*, Jean-Didier Legat*, Jordi Madrenas[†], Jordi Domínguez[†]

* Université catholique de Louvain,
Laboratoire de Microélectronique - DICE,
3 Place du Levant, B-1348 Louvain-La-Neuve, Belgium
[†] Universitat Politècnica de Catalunya,
Departament d'Enginyeria Electrònica
Gran Capita, s/n, E-08071 Barcelona, Spain

Various types of neural networks may be used in multi-dimensional classification tasks; among them, Bayesian and LVQ algorithms are interesting respectively for their performances and their simplicity of operations. The large number of operations involved in such algorithms may however be incompatible with on-line applications or with the necessity of portable small-size systems. This paper describes a neural network classifier system based on a fully analog operative chip coupled with a digital control system. The chip implements sub-optimal Bayesian classifier and LVQ algorithms.

1 Introduction

Many neural network methods have been developed recently to solve the problem of multi-dimensional data classification; among these methods, we can select sub-optimal Bayesian classification (see for example [1] and [2]) for its performances and its convergence to the (optimal) Bayesian boundaries between classes, and LVQ (Learning Vector Quantization) [3] or 1-Nearest-Neighbor for its simplicity of operations.

Without going into the details of the equations, we can mention the principle of these two classes of algorithms. In the first one, the principle is to approximate the true probability density of learning vectors, by superposition of Gaussian (for example) kernels [4][5]; the idea is to place on each vector of the learning set a radial-basis kernel, which is usually chosen as Gaussian, and to sum all such kernels class by class; it may be proven that if the number of learning points is large, this approximation will converge to the true probability density of vectors in each class. Once the probability densities have been estimated, the Bayes law can be used to select the most probable class to attribute to a vector to classify, by choosing the class having the largest product between the estimate of probability density at the location of this vector, and the a priori probability of the class, this last one being estimated by the ratio between the number of learning points in each class and the total number of points. Such a Bayesian classifier however requires a large number of Gaussian functions to obtain acceptable approximations of probability densities. In order to reduce the number of computations, sub-optimal methods have been developed [1][2][6];

they mainly consist in selecting a reduced number of kernels, their locations being fixed by a vector quantization method, and their width by some function of the variance of each cluster.

The LVQ principle is even more simple: once the prototype vectors (equivalent to the centers of the radial functions in the Bayesian classifier) have been located by some adaptive method of vector quantization, the principle of the classification by LVQ [3] is to attribute to a vector to classify the class of the nearest prototype, using either an Euclidean either a Manhattan distance.

2 Architecture of the ANKC processor

Because of the high number of computations involved in sub-optimal Bayesian and in LVQ classifiers, it may be interesting in some applications to have a VLSI specialized chip implementing these two kinds of algorithms; the advantage of a specialized chip in this context is not only to offer a high speed of computation (see evaluations in section 6), but also to realize portable classification systems, without the need for cumbersome general-purpose computers.

The general architecture of the ANKC (Analog Neural Kernel Classifier) processor, designed to implement sub-optimal Bayesian and LVQ classifiers, is detailed in figure 1.

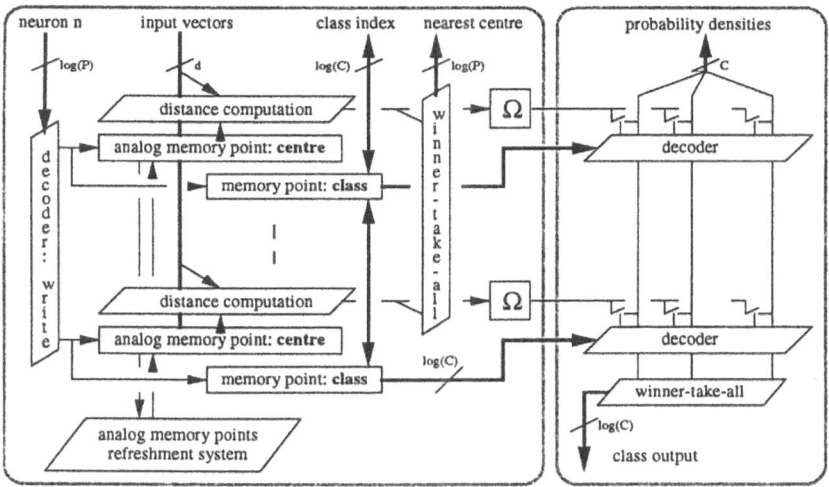

Figure 1: Functional description of the analog processor

The core of the system is built around P identical cells, each of them being composed of memory points to store the coordinates of the centroid (prototype vector) and its class, together with a distance calculator to compute the distance between this centroid and an input vector. Shortly, the system will work as follows. A set of P centroids will be stored in the processor; in the case of the Bayesian algorithm, the coordinates of the centroid correspond to the center of the kernel function Ω. Then, when an input vector is presented to the circuit for classification, all distances between this input vector and each of the centroids are computed in a parallel way; this is the purpose of the P mentioned distance computation cells.

The P computed distances are then used in two ways. On one hand, they are compared to find the smallest one, in order to select the closest centroid from the input vector; this is used in LVQ-like algorithms, in the purpose of selecting the winning centroid. On the other hand, the distances serve as inputs to P Gaussian-like kernel functions, used in Bayesian algorithms.

In the case of the LVQ algorithm, the selection of the winning centroid completes the recognition phase of the algorithm. In the case of the Bayesian algorithm, the P kernel outputs are summed class by class, in order to estimate the probability density of each class. The parameters of the kernels, namely their widths and shapes, may be adjusted by external commands. According to the Bayes law, classification of the input pattern is then realized by selecting the largest probability density among the different classes. In the Bayes law however, the probability density estimates must be multiplied by the a priori probabilities of the classes, before selecting the largest value; this is verified in our circuit if the number of centroids in each class is proportional to these a priori probabilities, which is the usual condition required on the centroids in all vector quantizations algorithms used for classification.

3 Test processor

A test version of the ANKC processor has been designed according to the above architecture. The chip has 16 centroids in dimension 8, each of them belonging to a maximum of 8 different classes.

The locations of the centroids are memorized in analog memory cells; all cells are detailed in reference [7]. The principle of the analog memory cells is to memorize a voltage on a capacitor by a cascode current copier cell. Leakage currents and charge injection may disturb the memorized value; the circuit has been designed in order to make the charge injection negligible in comparison with one LSB of the dynamics, and to make the leakage currents small enough so that a value is stable on the capacitors during approximately 1/10 sec., at the precision of one LSB. The accuracy of each memory point is 7 bits.

Each regulated cascode current copier used as analog memory point is connected to an input circuitry, also implemented as a regulated cascode cell. The principle of the two algorithms described above is to compute the distances between an input vector and each of the prototypes; to simplify the operations, the Manhattan distance is used in the system (the use of one type of distance or another only influences the performances of the classification in a negligible and non-systematic way). The input circuitry will thus produce a current in the same range as the current memorized in the cell, and their subtraction component by component and summation on the 8 components per vector by Kirchoff's law will produce the Manhattan distance between the input vector and one of the memorized ones. Before summation, positive and negative currents are separated in order to sum their absolute values.

The result of the Manhattan distances computation is fed either in a looser-take-all circuit for LVQ purposes (to select the smallest of these distances), either in a Gaussian kernel for the Bayesian classifier. The analog cells implementing the kernel functions have been designed to realize Gaussian-like functions with both adjustable slope and width, allowing thus two parameters that can be used to improve the performances of the classifier. Slopes and widths are identical for all kernel functions on a chip.

We will describe into more details the analog memory points refreshing system. This system converts the current memorized in a memory point by a successive approximation scheme (SAC) based on the propagation of a token in a finite-state automata [8]. In figure 2, at the first step of the SAC the memorized current is first compared to a current equal to 2^6 LSBs; the two (opposite) currents are serially connected at point C, and the voltage at this point rapidly goes high or low respectively if the memorized current is smaller or higher than the source. Voltage C is then compared with the medium point of the dynamics, and the result of the comparison determines if the source of 2^6 LSBs will remain connected during the next phases of the SAC or not (if C is higher than M, it means that the source current of 2^6 LSBs is smaller than the memorized current, and thus the source will remain connected during the next phases; if C is less than M, it means the source current is too high, and it will thus be disconnected for the next phases).

Latches in figure 2 will not only transfer the token at each step, but will also memorize the status of each successive comparison. Once the decision about the MSB has been taken, the

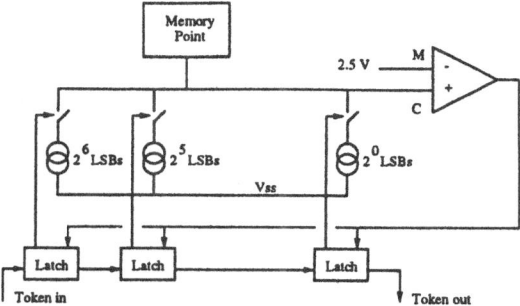

Figure 2: Principle of the refreshment system for analog memory points

token will be transmitted to the second cell, and the decision about the switching on or off the 2^5 LSBs source will be taken, and so on... After 7 iterations, the sum of all the sources of 2^i LSBs ($0 \leq i \leq 6$) which has been switched on during the approximation process will constitute a lower estimation of the memorized current.

The memory cell has been designed to ensure that the leakage currents will always have the same sign; in other words, it ensures that the current memorized in the cell will always decrease. Refreshing the memorized current by a value equal to the above approximation plus one LSB will thus ensure to keep the memorized current fixed at the precision of one LSB, if the leakage currents between two successive refreshments are kept under this limit. Simulations showed that a delay of 0.1 sec. between two successive refreshments may be chosen, for a capacitor value in the cascode memory point equal to 1 pF, and a precision of 7 bits in each cell, even with large overestimations of leakage currents and parasitic effects. This means that the whole circuit has to be switched in refreshing mode only each 0.1 sec.; since the refreshing time for the whole circuit (128 cells) will be around 1 msec. (with a clock frequency of the refreshment system of 1 MHz), this means that the percentage of time that the circuit has to be switched in refreshment mode is about 1%, which is more than acceptable; during the other 99% of the time, the circuit can be used in classification mode. The two modes cannot be mixed, since the refreshment modifies the voltage at node C, and the current flowing from a memory point will thus be corrupted; one mode or another must be chosen through a set of switches connected to an external control line.

Connection of the refreshment system to one memory point or another is assumed by two multiplexors connected to switches not represented in figure 2. The state of these multiplexors (line and column of the memory points array) is determined through two shift registers; the first one (row) changes its state each time the token in the refreshment system goes out, the second one (column) each time the row shift register achieves a complete cycle. By this way, all memory points will be successively connected to the successive approximation analog-to-digital convertor, and refreshed. The end of the cycle of the column shift register will signify the end of a complete refreshing cycle.

Let us finally mention that the use of the refreshment system is not the only possible way of refreshing the values stored on the capacitors; another way is to duplicate all memorized values in a digital external memory, and to periodically refresh the internal memory points with the values stored in the external ones. Because of the strongly reduced number of steps involved in a refreshment of the complete memory array, the global refreshment time is now reduced to approximately 0.02 sec., which is an obvious advantage of this solution, the drawback being the more complex peripheral components and control system to be used.

4   External control and setup

A setup for the prototype system is shown in figure 3. To keep the system portable, an embedded specific controller is used, but connection to a host could be done as well. The ANKC test chip is digitally controlled by means of an FPGA and an programmable read-only memory (E)PROM. The former provides the required control signals to the ANKC processor, and the latter contains the programming information of the analog processor. Two switches (that could be physically hardwired to reduce components) determine if the system performs a previous digital test, and the internal or external refreshing mode.

The functions performed by the external digital control of the ANKC processor are the following: clock generation, digital test, processor initialisation and analog memory point refreshing in external mode.

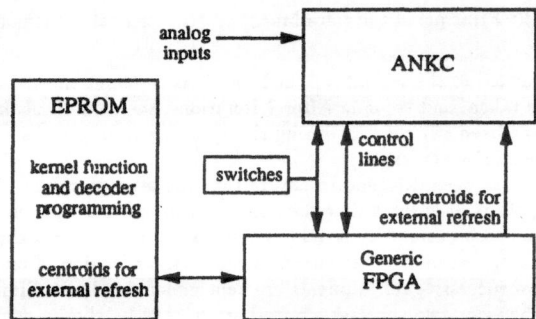

Figure 3: Control architecture of the ANKC system

- Digital test: The ANKC internal latches form in a test mode scan-path chains that allow to feed test vectors through a serial line and thus bring the digital part of the processor to a known state. The output stream of the scan-path chain is compared by the controller with the previously entered data in order to verify if any bit has been altered. In the prototype system only the test of the processor digital part is considered, but it can be easily extended to test the analog function, provided analog inputs and output measurement are available. The test is performed if the associated switch is active.

- Processor initialisation: After testing the digital part, the decoder switches and the width and slope of the kernel functions (figure 1) are programmed. Then, the row and column registers are reset and the analog memory points are programmed through the D/A converter input lines. All the information needed in this step is stored in the (E)PROM.

- Refreshing: During normal processing, as indicated in the previous section, classification and refreshing are interleaved. During refreshing, if internal mode, the controller just counts the refreshing interval and activates the refresh signal. In external refreshing mode, it also enters the digital value of the memory points.

5   Accuracy of computations

Determining the necessary accuracy for all computations in an analog chip is not obvious. Limitations on accuracy mainly come from charge injection, leakage currents, mismatching between devices and noise (like power supply noise). In the ANKC processor, charge injection and leakage currents affect the values memorized on the capacitors in the analog memory points; we showed however in a previous section how this problem has been addressed, by periodic refreshment of the stored values. Mismatching between devices is also much more critical than possible noise; the following shows how mismatching has been modeled and how it affects classification results in a suboptimal Bayesian classifier (the IRVQ [2] algorithm was used for the simulations).

Mismatching between transistors may be modeled by a circuit simulator as HSPICE. However, in the case of a complex analog circuit as ANKC, it is out of question to try to simulate the whole chip by a circuit simulator; electrical simulations are only carried out at the cell level, and to verify the connections between cells. To test the influence of mismatchings on the classification task itself, a more reasonable way on a computational point-of-view is to model the parasitic effects in a simulation of the classification algorithm itself.

Three mains steps are achieved in the ANKC processor to classify a vector: Manhattan distance between the vector and the centroids, non-linear (Gaussian) functions of the distances, and winner-take-all to select the winning class. In the Manhattan distance computation, the crucial point resides in the current mirrors which have to be used to sum the currents and to take their absolute values; mismatching at this level has been modeled by adding a white noise at the output of the each distance computation block the amplitude of the noise being determined by a constant fraction of the total dynamics, expressed in bits (a precision of b bits corresponds to noise dynamics equal to $1/2^i$ of the total dynamics).

Concerning the Gaussian functions, simulations showed that the exact shape of the function has no influence on the classifications results. What is more important is to select the width and the slope factors of the curve; these parameters have been chosen in these simulations as those giving the best classification results by the IRVQ algorithm implemented on the chip without noise and mismatchings (ideal implementation of the algorithm).

Finally, the mismatching in the current mirrors used in the winner-take-all has also been modeled by adding a white noise value at the input of this device, as after the distance computation. We may consider that the influence of these two noise sources will be approximately equal to the influence of mismatchings in the distance computation and winner-take-all devices, with an identical number of bits defining the dynamics of the range and the percentage of mismatching in current mirrors.

Simulations have been made with the IRVQ [2] algorithm. They are illustrated here on the IRIS database (150 points in dimension 4, and 3 classes), which is a standard in classification studies; simulations were however carried out on a large number of databases (reals ones coming from the industrial world and artificial ones generated for special tests), and all qualitative results are similar.

Figure 4 illustrates the classification error on this database in function of the dynamics of the two sources of noise described above; we can see that the performances of an ideal circuit are very closely approximated if the precision of the mismatched devices is around 7 bits, while acceptable performances are already obtained for 3-4 bits accuracy; these results were used in the implementation of the ANKC processor, by taking the necessary precautions to design the critical current mirrors with an accuracy of about 7 bits.

6   Performances

Evaluating the performances of such analog processor greatly depends on the type of measure used for the evaluation; comparing serial digital and parallel analog processors in terms of FLOPs is not fair, because of the very different nature of operations between processors. In the following, the performance of a processor is thus measured in terms of number of operations that should be used on a serial digital processor to realize an identical set of computations (GOPS $= 10^9$ operations per second).

Giving this definition, we can evaluate the number of these operations that are computed in

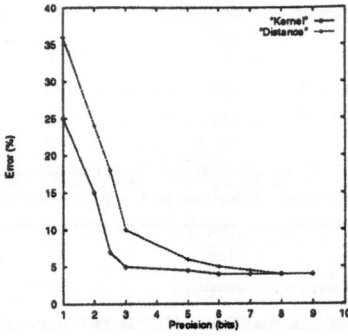

Figure 4: Influence of the two main sources of mismatching in the ANKC circuit on the classification error by the IRVQ algorithm on the IRIS database

one *clock cycle* (period between the presentation of two successive input vectors) in the ANKC processor. If N is the dimension of the vectors, P the number of centroids in the ANKC processor, and C the number of different classes, we have:

Task	Number of operations
P distance computations	$3.P.N$ additions
P kernel functions	$P.\log2(N)$ comparisons
P additions	P additions
1 winner-take-all over C values	C comparisons
1 winner-take-all over P values	P comparisons
Total	$3.P.N + P.\log2(N) + 2.P + C$

In our implementation of the ANKC test chip, we have $P = 16, N = 8$ and $C = 8$; with a clock of $10MHz$ (this frequency is used as the main clock of the circuit, so that one vector is classified at each cycle), this leads to a number of operations (as defined above) per second equal to 4.72 GOPS. The chip size in MIETEC 2.4 μm technology is 5x4 mm.

To compare the performances of the ANKC processor with another up-to-date digital processor implementing similar algorithms, the *Ni 1000 Recognition Accelerator* from INTEL, we have to take the technology and size of the circuits into account. The *Ni 1000* chip is fabricated in a $0.8\mu m$ technology, and occupies 12x12 mm. It has been evaluated that an ANKC processor with $P = 256, N = 32$ and $C = 16$ can be implemented in the same technology on the same silicon area. Evaluations of the performances according to the above definition then give 264 GOPS for the ANKC processor, and 16 GOPS for the *NI 1000*, which shows the strong advantages obtained with a fully analog implementation of such kind of classification algorithms.

7 Conclusion

The analog implementation of a neural network classifier, based on suboptimal Bayesian or LVQ algorithms, has been described in this paper. The analog operative chip is coupled to a digital control part, to form a complete, portable (small size), powerful classification system. It implements the classification phase of the algorithms, while learning has to be performed on an external computer and the results of the learning downloaded in the system. A test chip has been designed in MIETEC 2.4 μm technology, with a reduced number of cells. Evaluations show that the performances of this classification system surpass those of an up-to-date digital chip implementing similar algorithms by a factor around 16.

8 Acknowledgments

Part of this work has been funded by the ESPRIT-BRA project 6891, ELENA-Nerves II, supported by the Commission of the European Communities (DG XIII). Philippe Thissen is working towards the Ph.D. degree in microelectronics under an IRSIA (Institut pour l'Encouragement de la Recherche Scientifique dans l'Industrie et l'Agriculture) fellowship. Michel Verleysen is a Senior Research Assistant of Belgian National Fund for Scientific Research (FNRS). Jordi Domínguez holds a Research fellowship (FPI) from the Spanish Ministry of Education and Science. We thank Olivier Herbeuval and Yves Van Daele for their contribution in the design of the ANKC test chip.

References

[1] P. Comon, G. Bienvenu, and T. Lefebvre, "Supervised design of optimal receivers", in *NATO Advanced Study Institute on Acoustic Signal Processing and Ocean Exploration*, Madeira, Portugal, July 26-Aug. 7 1992.

[2] J.L. Voz, M. Verleysen, P. Thissen, and J.D. Legat, "Handwritten digit recognition by suboptimal bayesian classifier", in *NeuroNîmes94 (Neural Networks and their applications)*. October 1994, EC2, submitted.

[3] T. Kohonen, *Self-Organization and Associative Memory*, Springer-Verlag, Berlin, 1989, 3rd Edition.

[4] T. Cacoullos, "Estimation of a multivariate density", *Annals of Inst. Stat. Math.*, vol. 18, pp. 178–189, 1966.

[5] E. Parzen, "On the estimation of a probability density function and the mode", *Ann. Math. Stat.*, vol. 27, pp. 1065–1076, 1962.

[6] Q. Xie, C. A. Laszlo, and R. K. Ward, "Vector quantization technique for nonparametric classifier design", *IEEE Transactions on Pattern Analysis and Machine Intelligence*, vol. 15, no. 12, pp. 1326–1330, december 1993.

[7] M. Verleysen, P. Thissen, J.L. Voz, and J. Madrenas, "An analog processor architecture for neural network classifier", *IEEE Micro*, vol. 14, no. 3, pp. 16–28, June 1994.

[8] D. Macq, J.D. Legat, and P. Jespers, "Analog storage of adjustable synaptic weights", in *Proceedings of the SPIE conference on Applications of Artificial Neural Networks, Orlando (USA)*, April 1992, pp. 712–718.

An Associative Processor Dedicated to Classification by Neural Methods

Philippe Thissen, Michel Verleysen, Jean-Didier Legat

Université catholique de Louvain,
Laboratoire de Microélectronique - DICE,
3 Place du Levant, B-1348 Louvain-La-Neuve, Belgium

Multi-dimensional classification tasks by neural methods are interesting for their performances and their simplicity of operations. However, the number of computations needed by these algorithms is very impressive and drastically limits practical applications of such methods. This paper describes a digital bit-serial, massively parallel associative processor to speed up neural-like classification tasks. To achieve this goal, each block of this associative processor is optimized to the main operations involved in classification algorithms.

1 Introduction

This paper describes the architecture of a massively parallel processor dedicated to classification tasks. This processor belongs to the bit-serial, word-parallel SIMD associative machines family. The bit-serial, word-parallel configuration offers a powerful way of realizing parallel computations with a relatively low complexity hardware design.

The processor implements the operative part of the architecture which is under full custom design. Coupled with a control part made of programmable logic devices, the whole forms a complete high performance and high speed classification system. The massively parallel operative part is in fact well adapted to classification algorithms by neural methods. The advantage of a specialized chip is not only to offer a high speed of computation but also to realize portable classification systems, without the need for host computers.

The paper is divided into three parts. The first one concerns the functional description of the massively parallel architecture. The different constitutive basic blocks are briefly explained and performances for arithmetic operations are given. In the second part, the hardware realization of this processor is presented. The third part gives information about performances of such a parallel processor for classification algorithms by neural methods like the LVQ, RCE or other ones.

2 Functional description of the architecture

The architecture described in this section implements the data path (or operative part) of a global classification system. This system is composed of a lot of simple processing units working in parallel; each unit is formed by a memory, a elementary processor and a status register. An external control part manages all the processing units. To have the parallel structure working into one common goal, some interactions exist between the elementary processors.

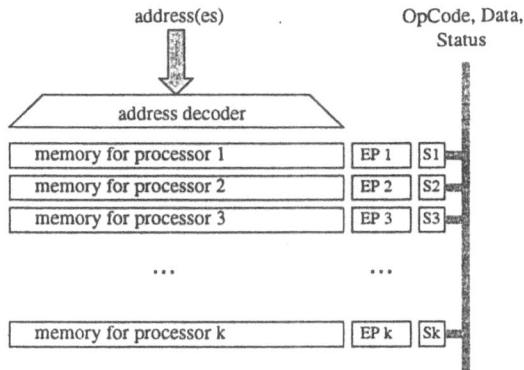

Figure 1: Architecture of a bit-serial, word-parallel associative processor.

Figure 1 represents the global architecture of a bit-serial and word-parallel associative processor [1, 2]. Each elementary processor (EP) exchanges information with its associated memory (on the same horizontal line). A special 1 bit register (S) fixes the status of the elementary processor (see below). All data transfers are made through a 1 bit data bus.

The original part of this approach resides in the structure of this associative processor well suited to neural-like classification. The associative memory and the elementary processing unit are designed according to the most frequent and complex operations involved in this task. Moreover, specialized hardware has been incorporated to achieve collective calculations. In the following, a description of each part of the processor is given in more details.

- The memory used is a random access memory with two separate bus : one for the reading and another one for the writing into the memory cells. At each clock period, two addresses are thus needed. Both addresses, coming from outside of the circuit, are propagated to all rows of memory; this two bus architecture permits to execute two different operations on two different data at the same clock period : the memory reading and writing.

- The arithmetic and logic unit (EP on figure 1) is built to process simple data in a bit serial way. This part of the elementary processing unit is able to execute addition, subtraction and comparison between a data coming from the memory and an other one coming from the outside of the circuit, in one clock period; it also has the possibility to carry on arithmetic and logic operations on two data coming both from memory in two clock periods. The inputs are taken in the input buffer of the ALU and the result is written in its output buffer. The operations are identical for all elementary processing units and are executed depending on the value of the status register.

- The status register (S) allows the current instruction to be processed or not. This status information can be loaded from and to the memory. The result of a computation (arithmetic or logic) can be directly written into the status register. Another possible working mode is the activation of all the processing units independently of the value stored in the status register.

Figure 2 gives a schematic representation of the elementary processing unit. Three different operations can be realized at the same time on different data in one clock period :

- the reading of 1 bit from the memory and its loading into the input buffer of the ALU;

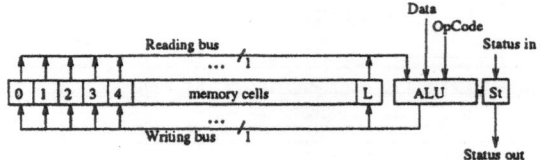

Figure 2: Schematic representation of the elementary processing unit.

- the computation of an arithmetic or logic operation on the input of the ALU and the storage of the result in its output;

- the writing of the contents of the output buffer in the memory.

This pipelining multiplies by 3 the number of operations realized by unit of time in comparison with the clock frequency.

Collective functions, taking the outputs of all elementary processing units as inputs, have been added to the associative structure. These functions, dedicated to speed up classification tasks, come from the study of classification algorithms. Two functions are particularly useful in these tasks : *the selection of the first active EP* and *the summing of all their outputs*.

- The first function is very powerful and frequently used in classification algorithms. It is a combinatory block (without synchronization) : a token goes through the structure (all processors) and changes its state only when it finds the first active status register. The associated elementary processor becomes then active and all others are turned off. To avoid large delays due to a very long chain of elementary processors, a method, similar to the carry look-ahead for adders, is used to bypass the normal way of the token.

- The second collective function sums all outputs of the elementary processors. This function is useful for computing the number of active processors but also for summing the contributions of all processors as explained in the case of suboptimal Bayesian classification (section 4.3). This function is built with simple 1 bit full-adders. The adders are laid in a tree structure and signals go through the tree in a pipelined way. At each clock period, all data are processed to the next level in the tree; the sum is obtained in a bit-serial way at the output of the last adder of the tree.

This architecture is multi-purpose and expandable. In fact, the processing unit executes basic but fundamental functions (addition, subtraction and comparison). It is expandable because the number of processing units is not fixed to a predefined value. If, for specific applications, we need a larger number of processors, one can connect together more than one VLSI circuit (of fixed size) and build, by this way, a large system composed of many elementary processors working in parallel.

Each elementary processor is able to compute bit serial additions without any interaction with the other ones. By extension, the three others basic integer operations (subtraction, multiplication and division) are possible on this device [3]. The execution of such operations on a floating point representation is also possible by separating the number into two parts (mantissa and exponent) and processing the adequate function to each part. Table 1 illustrates the number of operations that can be executed on this architecture composed of 1024 elementary processors clocked at a frequency of 100 MHz; the results are given in million of operations (addition, multiplication ...) per second (MOPS). For floating point operations, we suppose that numbers are given in a 32 bits IEEE format with a 24 bits mantissa and an 8 bits exponent.

The number of comparisons that can be computed by time unit is also very impressive; these operations act only on integers and have the same complexity as a integer subtraction. Let us insist on the fact that each chip inputs/outputs are realized through a 1-bit bus.

| | integer | | | real |
	8 bits	16 bits	32 bits	32 bits
addition				
$X + A = C$	10240 MOPS	5687 MOPS	3012 MOPS	80 MOPS
$B + A = C$	5687 MOPS	3012 MOPS	1552 MOPS	80 MOPS
subtraction				
$X - A = C$	9310 MOPS	5390 MOPS	2925 MOPS	80 MOPS
$B - A = C$	5390 MOPS	2925 MOPS	1527 MOPS	80 MOPS
multiplication				
$X \times A = C$	397 MOPS	122 MOPS	35 MOPS	57 MOPS
$B \times A = C$	317 MOPS	92 MOPS	25 MOPS	42 MOPS
division				
$X/A = C$	132 MOPS	40 MOPS	12 MOPS	20 MOPS
$B/A = C$	112 MOPS	35 MOPS	10 MOPS	17 MOPS

Table 1: Computational power of the massively parallel processor formed by 1024 elementary processing units clocked at 100 MHz; one MOPS is equivalent to one million of operations per second.

3 Hardware realization of the processor

3.1 The data path

The main part of the data path consists in the elementary processors. The arithmetic part of these ALU must be able to execute the following operations

$$X + A, \quad X - A, \quad B + A, \quad B - A,$$

where A and B represent numbers stored into the memory and X an external data.

The main part of the arithmetic unit is a 1 bit full adder. We must add to this part the possibility to invert one data for subtraction, and to store a data in a latch for operations concerned with two operands coming from the memory. A schematic representation of this arithmetic part of the ALU is illustrated at figure 3. The *FF_sign* register and the *Xor* gate are used to invert one input of the full adder; we have chosen this solution to easily implement operations like $X + |A|$.

The logic part of the ALU is built in such a way that comparisons can be executed as fast as subtractions. A specialized hardware, design to speed up drastically the comparison operations and the research of extremum value with a low silicon area, has been chosen for this part of the circuit; it works like a finite state automata, comparing the two numbers from the more significant bit (MSB) to the less one (LSB). By this way, the result of a comparison can eventually be known directly after the comparison of the two MSBs.

The operative part of this associative processor is currently under design in a 1 μm ES2 technology. The elementary processor and the static memory have been successfully simulated at frequency up to 100 MHz. The density of integration of this processor is about 1 memory point per 560 μm^2, including the contributions of the elementary processors and others peripheral devices. The parallel processor is of course expandable and large networks can be realized by connecting several circuits together.

3.2 The control unit

The control part of the associative processor is realized through a programmable logic device. The main features of the mircosequencer are : an incremental conditional structure, a stack used with

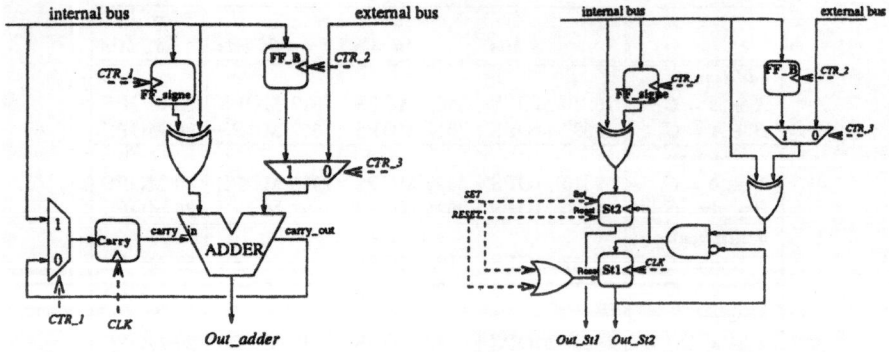

Figure 3: A standard cells representation of the arithmetic part (left) and logic part (right) of an elementary processor.

subroutines and a counter dedicated to the implementation of loops. The architecture is adapted to classification algorithms and optimized for such instructions. To avoid skew and others clock problems, the control part is built according to the basic logic blocks available in the chosen programmable device and special attention is paid to the synchronization of the different parts.

4 Implementation of classification algorithms

Some neural-like classification algorithms have been slightly adapted to fit into this processor. In such algorithms, a lot of computations can be realized in parallel without any interaction between the different processing units. During the learning phase, the network is progressively built according to some rules to represent the input distribution of patterns; then the classification phase compares an input pattern to all stored patterns and makes some computations on these distances.

Three types of neural-like algorithms have already been partially or totally implemented into this parallel processor : RCE, LVQ, and suboptimal Bayesian classifiers.

4.1 The RCE algorithm

The RCE algorithm proposed by Reilly, Cooper and Elbaum [4] is an incremental model of neural classifiers. During the learning, an input vector is presented to the network and neurons fire if this input vector belongs to their region of influence. Neurons leading to misclassification are modified (the size of their region of influence is decreased) and a new neuron is added if the input element is not well classified.

Each elementary processing unit in the architecture represents a neuron and processes all the computation of this neuron; the center and the radius of the concerned neuron are stored in the memory associated to the processing unit. The radius of elementary units leading to misclassification are modified and new elementary units are added if necessary. To simplify the computation, the Manhattan distance metric is used. The operations of the RCE algorithm, implemented in our architecture, are thus :

1. To compute the Manhattan distance between the input pattern and all patterns stored in the memory (one pattern by elementary processor).

Learning

	Simulation		Theory	
	Distance	Total	Distance	Total
5 bits	51%	1 000 000	41%	800 000
16 bits	58%	450 000	43%	332 000

Classification

	Simulation		Theory	
	Distance	Total	Distance	Total
5 bits	65%	1 278 000	61%	1 200 000
16 bits	69%	536 000	63%	490 000

Table 2: Number (theoretical and simulated) of clock periods needed to achieve the learning or classification of one input vector with the RCE algorithm.

2. To select the neurons or elementary processors whose influence region includes the input pattern. This is done by switching active only the processors having a region of influence radius higher than the calculated distance.

3. To modify the radius of the influence region for neurons which misclassify the input pattern.

4. To added, if necessary, one neuron (one elementary processor) centered on the input and with an a priori fixed radius.

Simulations of our architecture on the RCE algorithm have been carried out on a bidimensional database containing 1000 elements divided into two separated classes (a circle and a external ring). The data accuracy has been truncated respectively to 5 and 16 bits. Table 2 gives the number vectors that can be learned per second, and the percentage of time used for the distance computation on it. Values illustrated in the *"simulation"* column represent the mean value for 1000 presentations of the whole learning set; those indicated in the *"theory"* column represent a maximum value of the number of clock cycles (some operations depending on the result of the others).

During the generalization phase of the algorithm, three operations occur. First, the distances between the input pattern and all the stored patterns are computed and only the neurons which include the input pattern are set active. Then, the class of the vector to classify is randomly set to the class of one of the active patterns. Finally, it must be verified that all active processors proposed the same class; if not, no solution exists for the given input. Table 2 gives the theoretically computed and simulated numbers of vectors that can be classified per second.

4.2 The LVQ algorithm

The Learning Vector Quantization (LVQ) method, proposed by Kohonen [5, 6] is an adaptative vector quantization algorithm which is used for classification purposes by adding an information of class to each quantized centroid.

During the learning phase, centroids are associated to elementary processors. At each presentation of an input pattern, the nearest centroid - in terms of Manhattan distance - is moved toward or away from the input stimulus depending on the concordance or not of their classes. Because of the move of only one centroid at each learning step, only one elementary processor will be active while all others are turned off; the parallel computation power of this associative processor is thus spoiled and the performances drastically decrease. Table 3 illustrates the number

Learning

	Simulation		Theory	
	Distance	Total	Distance	Total
5 bits	18%	358 000	16%	308 000
16 bits	10%	80 000	6%	43 400

Classification

	Simulation		Theory	
	Distance	Total	Distance	Total
5 bits	58%	1 142 000	46%	909 000
16 bits	51%	398 000	30%	238 000

Table 3: Number (theoretical and simulated) of vectors that can be learned or classified per second with the LVQ algorithm, and percentage of time spent in distance computation.

of learning cycles (presentation of an input pattern and adaptation of the network) that can be achieved per second, and the percentage of time spent in the distance computation.

For the classification phase (once the representation of the input space has been built), the power of this parallel architecture is much more exploited and the results obtained are very impressive as shown in table 3. The classification algorithm consists in computing all distances between the input pattern and the stored centroids, and choosing the smallest of these distances. The class of the input vector is then selected as the one of the nearest centroid.

4.3 Suboptimal Bayesian classification and other algorithms

Bayesian and suboptimal Bayesian classification [7] algorithms evaluate the probability densities of the classes and classify an input vector by selecting the most probable class according to the Bayes criterion. The algorithm first selects appropriate locations in the space by a vector quantization technique on the data from each class, and then sums class by class Gaussian-like (bell shaped) functions centered on these locations. The most probable class is chosen as the one giving the maximum of these sums among the different classes. In our architecture, the Gaussian-like functions are evaluated in the processing unit; the sum and the detection of maximum are implemented by the collective functions of the chip.

Of course, it is possible to implement a lot of other algorithms (for classification or not) on this massively parallel processor since it is basically built on a general SIMD architecture with simple and powerful functions; performances will however be optimally exploited in neural-like classification tasks.

5 Conclusion

We described in this paper a parallel SIMD architecture dedicated to classification and pattern recognition applications. The SIMD architecture is based on a bit-serial word-parallel associative processor. One bit-column of each word is processed simultaneously and only one ALU is implemented by word. This particular SIMD architecture and the ALU have been designed in order to speed up the main operations involved in standard classification algorithms (distance, extremum computations, ...); some specialized resources have been added for this purpose too. This architecture is programmable and many algorithms can be implemented only by writing their associated microprogram. The computational power of the architecture is fully exploited by neural-like algorithms for classification.

Such a processor, operating at a frequency of 100 MHz and composed of 1024 EPs, is able to process up to 10 billions of 8 bits additions per second and to class 400 000 vectors with neural network algorithms like LVQ or RCE. This associative processor is implemented in 1 micron VLSI technology with full custom design.

6 Acknowledgments

Philippe Thissen is working towards the Ph.D. degree in microelectronics under an IRSIA (Institut pour l'Encouragement de la Recherche Scientifique dans l'Industrie et l'Agriculture) fellowship. Michel Verleysen is a Senior Research Assistant of Belgian National Fund for Scientific Research (FNRS).

References

[1] B. Parhami, "Associative memories and processors: an overview and selected bibliography", *Proceedings of the IEEE*, vol. 61, no. 6, pp. 722–730, 1973.

[2] S.S. Yau and H.S. Fung, "Associative processor architecture—a survey", *Computing Surveys*, vol. 9, no. 1, pp. 3–27, 1977.

[3] I. Scherson, D. Kramer, and B. Alleyne, "Bit-parallel arithmetic in massively-parallel associative processor", *IEEE Transactions on Computers*, vol. 41, no. 10, pp. 1201–1210, 1992.

[4] D.L. Reilly, L.N. Cooper, and C. Elbaum, "A neural model for category learning", *Biological Cybernetics*, vol. 45, no. 1, pp. 35–41, 1982.

[5] T. Kohonen, "Improved versions of the learning vector quantization", in *Proceedings of the International Joint Conference on Neural Networks, San Diego*, June 1990, vol. I, pp. 545–550.

[6] T. Kohonen, "Statistical pattern recognition revisited", in *Advanced Neural Computer*, 1990, pp. 137–144.

[7] C. Jutten and P. Comon, "Neural Bayesian classifier", in *New Trends in Neural Computation*, Mira Cabestany Prieto, Ed., Berlin, June 1993, number 686 in Lecture Notes in Computer Sciences, pp. 119–124, Springer-Verlag.

Hardware-Oriented Models for VLSI Implementation of Self-Organizing Maps

Bonifacio Martín-del-Brío[1], Javier Blasco-Alberto[2]

[1] Tecnología Electrónica. EUITIZ, Universidad de Zaragoza. 50009 Zaragoza, Spain
[2] Dpto. de Ingeniería Eléctrica e Informática. Universidad de Zaragoza. 50009 Zaragoza, Spain

Tel. +34 76 351609. Fax +34 76 555638. E-mail: nenet@cc.unizar.es

Abstract

In this work, two hardware-oriented models for the Self-Organizing Feature Map (SOFM) are introduced. The first model, based on a dot-product neuron, is suitable for analog VLSI implementations. For the second one, we choose minimal expressions, from the viewpoint of the digital VLSI, for the different aspects of the conventional SOFM, which lead to a digital-oriented SOFM. Both hardware-oriented models are verified by means of computer simulations. Finally, a SIMD coprocessor based on the digital SOFM introduced is designed and simulated by means of a well known hardware description language, VHDL.

1. Introduction.

The first step towards the hardware implementation of a neural architecture should be the modification of the formal model in order to fit the features of the specific implementation technology. Thus, in the present work we introduce two different hardware-oriented models for the Self-Organizing Feature Map (SOFM) [Kohon90]. The first model, based on a new dot-product Kohonen neuron, is more suitable for analog VLSI implementations. This new Kohonen neuron is formally similar to the conventional one used in many other artificial neural architectures, as backpropagation (BP) or Hopfield networks, and does not require vector normalization.

Next, we propose a second model oriented to the digital hardware. For this purpose, we choose minimal expressions, from the viewpoint of the digital VLSI implementation, for the different aspects of the conventional SOFM: distance measure, learning rule, parameter updating, etc. While some implemented techniques are original contributions, others are taken from previous works, but all these aspects conform a novelty digital SOFM.

Both hardware-oriented models are verified by means of computer simulations in two different cases, an academic and a real data analysis problem. Finally, for demonstrating the capabilities of the digital-oriented SOFM, a SIMD digital architecture is designed and simulated by means of a hardware description language, VHDL. This architecture acts as a coprocessor attached to a host computer, of conventional von Neumann type.

The paper is organised as follows. In Sect. 2 we give a brief introduction to the SOFM model. In Sect. 3 we introduce the analog-oriented SOFM. In Sect. 4 the digital SOFM is discussed. In Sect. 5 the neural coprocessor quoted above, based on the digital model, is described. The paper ends with the conclusions of our work. The updating rule for the bias of the first neuron model is derived in one Appendix.

2. Self-Organizing Map models.

The SOFM (or Kohonen map) is a competitive unsupervised neural system, used as a classifier, vector quantifier [Kohon90] or as a tool for exploratory data analysis [Martf93]. The SOFM projects a high dimensional input space onto a usually one or two dimensional output space, represented by a discrete lattice of neurons or processing elements (PE), usually arranged in a rectangular layout. Thus, this model transforms relations between input patterns in neighbourhood relations on the map, in such a way that similar input patterns are mapped onto near neurons, preserving the topology of the input space.

Let us suppose a rectangular Kohonen network composed by n_x x n_y neurons, every one labeled with a pair of indices $i \equiv (i,j)$ which give their location (x,y) on the map. In the recall phase, every neuron (i,j) computes the similarity between the input vector x, $\{x_k / 1 \leq k \leq n\}$, and its synaptic weight vector w_{ij}; the neuron g whose w_{ij} is more similar to x is called 'the winner'. In the learning phase, the weights of the neurons belonging to a neighbourhood of the winner are adjusted following a certain learning rule.

Some similarity measures, and their associated learning rules, are showed in Table 1 [Martf94]. Each learning rule is derived from its corresponding distance measure by gradient descent, in such a way that the similarity measure and the training algorithm are metrically compatibles [Martf94, Kohon93].

The most common Kohonen neuron is based on the Euclidean distance [Kohon90] (model 3 of Table 1); notice that its associated learning rule is very simple[1]

$$\Delta w_{ijk}(t) = \alpha(t).\left(x_k(t) - w_{ijk}(t)\right)$$
(1)

Nevertheless, the dot product has been often used as a similarity measure because of several reasons:

a) Its natural neural network implementation [Sutto94], because it is the most common neuron model used in many neural systems (BP, Hopfield, adaline, etc.) [Zurad92].

b) From the VLSI point of view, the cost in operators is against the Euclidean distance [Demar92, Vitto90].

c) It is more biologically plausible.

Notice (see Table 1) that when the dot product is used as a similarity measure, if we want to use the simple learning rule (1) derived from the Euclidean distance, we must normalized the input vectors (model 2.a). This is so because if the vectors are normalized, then Euclidean distance and dot-product are equivalent and, therefore, metrically compatibles, as can be seen from the following expansion

$$d^2(\mathbf{w}_{ij},\mathbf{x}) = \sum_{k=1}^{n}\left(w_{ijk} - x_k\right)^2 = \|\mathbf{w}_{ij} - \mathbf{x}\|^2 = \|\mathbf{w}_{ij}\|^2 + \|\mathbf{x}\|^2 - 2\mathbf{w}_{ij}^T.\mathbf{x} = 2(1 - \mathbf{w}_{ij}^T.\mathbf{x})$$
(2)

NEURON MODEL	DISTANCE MEASURE	DERIVED LEARNING RULE
1) Manhattan	$d(\mathbf{w}_{ij},\mathbf{x}) = \sum_{k=1}^{n}\left|w_{ijk} - x_k\right|$	$\Delta w_{ijk}(t) = \alpha(t).sign\left(x_k(t) - w_{ijk}(t)\right)$
2) Dot product		
2a) Normalized vectors ‖x‖=constant=1	$d(\mathbf{w}_{ij},\mathbf{x}) = \sum_{k=1}^{n} w_{ijk}x_k$	$\Delta w_{ijk}(t) = \alpha(t).\left(x_k(t) - w_{ijk}(t)\right)$
2b) Non-normalized vectors ‖x‖≠constant	$d(\mathbf{w}_{ij},\mathbf{x}) = \sum_{k=1}^{n} w_{ijk}x_k$	$w_{ijk}(t+1) = \dfrac{w_{ijk}(t) + \alpha(t).x_k(t)}{\left\|\mathbf{w}_{ij}(t) + \alpha(t).\mathbf{x}(t)\right\|}$
3) Euclidean	$d^2(\mathbf{w}_{ij},\mathbf{x}) = \sum_{k=1}^{n}\left(w_{ijk} - x_k\right)^2$	$\Delta w_{ijk}(t) = \alpha(t).\left(x_k(t) - w_{ijk}(t)\right)$

Table 1. Several models of the Kohonen neuron [Kohon90, Kohon93].

---

[1] In this work, we do not take into account the neighbourhood function h(.), for notation simplicity.

hence, a maximum dot-product gives a minimum Euclidean distance, and both lead to equivalent results.

Although the dot-product neuron (2.a from Table 1) is frequently preferred for the reasons exposed above [Demar92, Sutto94], it has one additional (heavy) computational cost due to the vector normalization requirement, that in VLSI implementations is time and area consuming. Thus, it would be very interesting eliminating that constraint. In [Demar92] it is shown that the dot-product could be applicable without normalization only when the dimension of the inputs is high. In [Sutto94] is studied a simpler normalization: Instead of using a equal-length constraint, it is suggested an alternative equal weight-sum normalization. Nevertheless, some distortion of results occurs with this procedure [Sutto94].

3. A biased Kohonen neuron for analog VLSI implementation.

In this section, we propose a new Kohonen neuron model which uses the dot-product as similarity measure and is equivalent to the Euclidean distance neuron, but without any normalization requirement. For this purpose, let us start with the Euclidean distance as a similarity measure, we look for the winning neuron $g \equiv (g1, g2)$, whose weight vector w_g is most similar to the input pattern x

$$d(w_{g1g2}, x) = \min_{ij}(d(w_{ij}, x)) = \min_{ij}\left(\left\|x - w_{ij}\right\|^2\right) = \min_{ij}\left(x^T.x + w_{ij}^T.w_{ij} - 2w_{ij}^T.x\right) = \tag{3}$$

As $x^T.x$ has the same value for all neurons (i,j), it is not necessary for computing the similarity, hence

$$\min_{ij}\left(\left\|x - w_{ij}\right\|^2\right) = \min_{ij}\left(w_{ij}^T.w_{ij} - 2w_{ij}^T.x\right) = \tag{4}$$

that can be rewritten as

$$= \max_{ij}\left(w_{ij}^T.x - \frac{1}{2}w_{ij}^T.w_{ij}\right) \tag{5}$$

Let us define the bias θ_{ij} of the (i, j) Kohonen neuron as

$$\theta_{ij} \equiv \frac{1}{2}w_{ij}^T.w_{ij} = \frac{1}{2}\sum_{k=1}^{n} w_{ijk}^2 \tag{6}$$

then, (3) can be expressed as

$$\min_{ij}(d(w_{ij}, x)) = \max_{ij}(w_{ij}^T.x - \theta_{ij}) \tag{7}$$

Therefore, we have obtained a dot-product unit, whose weight vector is that of the Euclidean neuron, and whose bias θ_{ij} is given by (6). Thus, finding the winning neuron g according to the Euclidean distance, is equivalent to find the unit whose output y_{ij} given by

$$y_{ij} = \sum_{k=1}^{n} w_{ijk} x_k - \theta_{ij} \tag{8}$$

is maximum. We can also apply to y_{ij} a common monotone increasing activation function, as is the case with the sigmoid of BP networks, because the neuron whose state y_{ij} is maximum and that whose output $f(y_{ij})$ is maximum are the same.

This new Kohonen neuron is very simple, because is based on the dot-product and does not require normalization. Its dot-product form makes easy its analog implementation, and it can be realized by means of the electronic circuits already used for the neurons of BP of Hopfield nets [Zurad92], without extra circuits for vector normalization (Fig. 1).

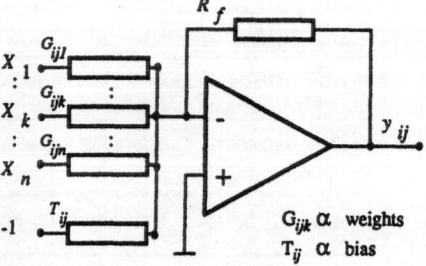

Figure 1. Example of a simple analog implementation of a Kohonen neuron with bias, based on op-amp.

For the training of a SOFM based on that new Kohonen unit, it can be used the conventional rule (1) derived from the Euclidean one, because both neurons are equivalent. The biases could be recalculated in every iteration by means of (6), but this is a costly procedure. In the Appendix we show that the following incremental updating rule for the biases can be derived from (1)

$$\Delta\theta_{ij}(t) = \alpha(t).\left(y_{ij}(t) - \theta_{ij}(t)\right) \tag{9}$$

that is formally similar to the Euclidean rule for the weights (1), replacing w_{ijk} by θ_{ij}, and x_k by y_{ij}. Thus, the learning phase of that new Kohonen neuron is only slightly complex than that of the Euclidean, while its recall phase is more simple.

We have verified the proposed SOFM neuron by means of computer simulations for two different cases. The first one is the classical, squared, uniform, two-dimensional probability density function (p.d.f.) [Kohon90]. We have carried out simulations for different network sizes, from 5x5 to 30x30, attaining similar results to those of the conventional SOFM. We have found in the simulations that the biases can be initially set at random (as occurs with the weights): Due to the updating rule (9), the biases gradually tend to fulfill condition (6), and the neuron finally becomes Euclidean. See [Martí94] for more details.

The model has been also verified in a real problem of data processing: The analysis of the Spanish banking crisis of 1977-85 [Martí93]. The results achieved by using the proposed Kohonen neuron are similar to those obtained in [Martí93] by using the conventional SOFM model (Fig. 2).

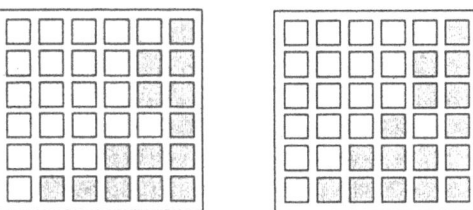

Figure 2
Analysis of the data from the Spanish banking crisis. a) Conventional SOFM model; b) SOFM with biased neurons. Neurons with bankrupt banks are represented in grey.
Notice that the results for both models are similar.

4. A SOFM model for digital VLSI implementation.

In the present Section we re-write the conventional Kohonen model in order to make its digital implementation easier. Our aim is to find simple expressions (from the viewpoint of the digital hardware) for the different issues of the model: data coding, distance measure, learning rule, neighbourhood function, parameter updating and winner-takes-all. This new digital SOFM combines original contributions, techniques previously reported [Melto92, Rüpin93], and the simple learning rule recently introduced in [Kohon93] (model 1 from Table 1).

Number representation. Data are coded by using the binary two's complement representation. If we denote by b the precision (register size in bits), the permitted numbers range from -2^{b-1} to $+2^{b-1}-1$. The inputs, real numbers, must be adjusted to that range by using a linear transformation, and later, discretized to the nearest allowed integer. The required precision depends on several factors, such as the p.d.f. of the input space and the number of neurons on the network. The theoretical study of this question is difficult [Thira94]; for instance, in [Thira94] there is pointed a way to find a lower bound. By means of computer simulations (see the end of this section), we find that $b=8$ is a reasonable number of bits.

Distance function and learning rule. In digital, the simplest similarity measure for computing the distance between x and w_{ij} is the Manhattan distance (Table 1), because it only carries out additions, subtractions and absolute values. Note that the Euclidean distance, the most commonly used, requires computing products, of greater hardware complexity in digital systems.

If we rewrite the learning rule derived from the Manhattan distance (recently introduced in [Kohon93]), it is easy to see that it is the simplest one:

$$\Delta w_{ijk}(t) = \begin{cases} +\alpha(t) & \text{if } x_k(t) > w_{ijk}(t) \\ 0 & \text{if } x_k(t) = w_{ijk}(t) \\ -\alpha(t) & \text{if } x_k(t) < w_{ijk}(t) \end{cases} \tag{10}$$

This rule does not require products, but only additions, subtractions and comparisons: Thus, its hardware realization is very simple. As an additional issue, in our digital Kohonen model, if overflow appears on additions (subtractions), the values will be saturated to the maximum (minimum) allowed integer.

Neighbourhood function and parameter updating. In the case of continuos (real) weights, it has been shown that the convergence of a Kohonen map becomes faster and of greater quality by using smooth neighbourhood functions, such as the Gaussian [Lo91]; a similar conclusion is reached in [Thira94] for the case of discretized weights. The problem with smooth functions is their hardware complexity in a digital implementation. For this reason, we use in our model the simplest one, the step function: Only neurons whose distance to the winner is smaller than the present $R(t)$ (neighbourhood radius) are updated, and in a same amount, moreover. Nevertheless, in this case, we should be careful in properly choosing the parameters and their updating to reach a final map correctly developed [Thira94]. Later in this Section we will show the values we have used in our simulations.

The neighbourhood function depends on the present iteration and on the distance between the coordinates of every neuron $\mathbf{i} \equiv (i, j)$ and those of the winner $\mathbf{g} \equiv (i_g, j_g)$. As in the previous Section, we use the Manhattan distance, because of its simplicity and because it can be carried out by the same hardware device [Melto92]:

$$d(\mathbf{i}, \mathbf{g}) = |i - i_s| + |j - j_s| \tag{11}$$

The learning parameters are the learning rate α and the neighbourhood radius R. Both parameters are time decaying, from their initial values (α_0 and R_0) towards their final ones ($\alpha_f = 0.01$ and $R_f = 1$, with real numbers). We select a linear updating

$$R(t) = R_0 + (R_f - R_0)\frac{t}{t_{Rf}} \tag{12} \qquad \alpha(t) = \alpha_0 + (\alpha_f - \alpha_0)\frac{t}{t_{\alpha f}} \tag{13}$$

being t_f the iterations until reaching the final values. In our case, we deal with integer values, thus $R(t)$ and $\alpha(t)$ are integers belonging to the range seen in Sect. 2.1. If we compute the iterations needed for $\alpha(t)$ and $R(t)$ to decrease in one unit, T_α and T_R:

$$T_R = \frac{t_{Rf}}{R_0 - R_f} \tag{14} \qquad T_\alpha = \frac{t_{\alpha f}}{\alpha_0 - \alpha_f} \tag{15}$$

then, updating $R(t)$ and $\alpha(t)$ consist of noticing when T_R and T_α iterations have passed, respectively, and then decrement their values in one unit. The digital implementation of this scheme is very simple, because it only deals with comparisons and decrements. The following are the values we have used in our simulations: $\alpha_0 \cong (2^{b-1}-1)/2$, $R_0 \cong max(n_X, n_y)$, $\alpha_f = 1$, $R_f = 1$

Searching for the winning neuron. One should not search for the minimum distance by pairwise comparing the outputs of the neurons, because the process will be slow and dependent upon network size. Instead, we use an iterative process of successive approximations [Melto92], similar to the one used in some analog to digital converters, completed with a system of lateral inhibitions between neighbours, implemented with OR gates (see Sect. 5). By means of this procedure, the time needed for finding the winner is the number of resolution bits (8 in our case), and does not depend upon the network size. If there are several winning neurons, the neuron with lower indices will be declared as the only winner.

We have checked the proposed model by means of computer simulations for the same two cases of Sect. 3. We have carried out numerous simulations for the two-dimensional p.d.f., for different network sizes (from 5x5 to 30x30) and by using different resolutions b. For 5x5 networks, we find that 5 bits are enough for obtaining a perfectly unfolded and ordered final map. For larger networks (up to

30x30), we find that more bits are required, always attaining good results with 8 bits. Fig. 3 depicts the final results for a 10x10 net, using 5, 6, 7 and 8 resolution bits. With 5 bits, the map is not correctly ordered; with 6, it is, but a more satisfactory aspect is reached by using 7 and 8 bits, similar to that obtained by using 32 bits in floating point format.

In the second problem (the Spanish banking crisis), the results attained by using the proposed digital SOFM with 8 bits are similar to those obtained in [Martí93], by using the conventional Kohonen model and 32 bits floating point arithmetic. Thus eight resolution bits seem to be a good choice.

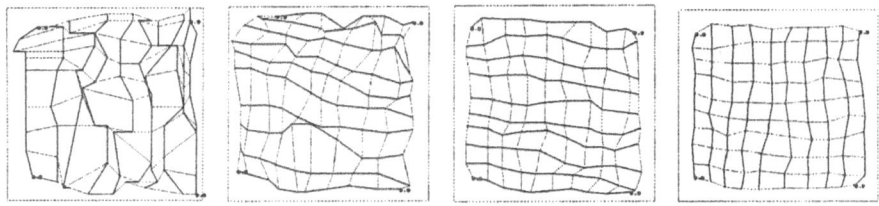

Figure 3. Final appearance of a 10x10 map representing an uniform p.d.f., by using 5, 6, 7 and 8 bits of precision.

5. VHDL simulation of a SIMD coprocessor based on the digital SOFM.

In this Section, we explain the main features of a neural digital coprocessor designed and simulated making use of VHDL, which implements the digital Kohonen model described in the preceding Section. This processor consists of a SIMD machine (single instruction, multiple data), which acts as a coprocessor of a conventional computer system (host).

Coprocessor description. The coprocessor uses integer arithmetic in two's complement format, and it is managed from the host by means of a computer program, written in assembler or C. The architecture of the coprocessor consists of the following main blocks (Fig. 4a): A set of processing elements (PE), each one implements one neuron, and stores its own weight vector w in a 16 byte local RAM; the control unit (CU) or controller, which carries out the communications with the host, and controls the operation of the different blocks of the coprocessor; the internal 10-bit bus, to whom are attached the CU and the PEs. The OR-gates of Fig. 4a implement the lateral inhibitions between the PEs in the winner-takes-all phase.

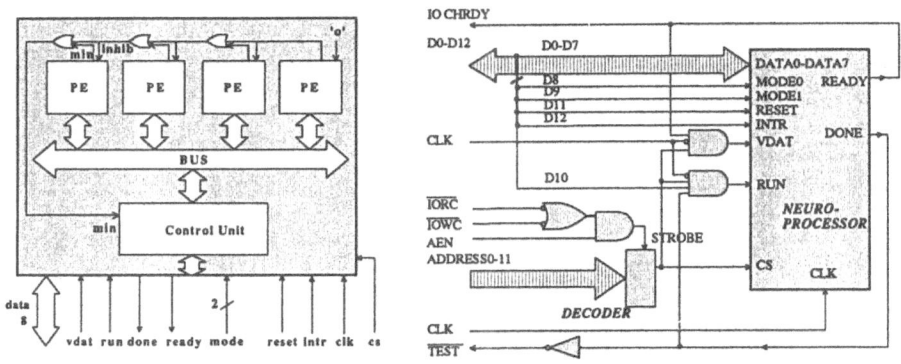

Figure 4. a) Block diagram of the neural coprocessor. b) Example of its connection to a PC computer.

System operation. The host sends sequentially the components of the input vector x_k (8 bits) to the coprocessor; the maximum number of components allowed is 16. The controller sequences the actions of every component of the system, depending on the present operating mode (MODE) selected (setup, read, recall or learning). One of the actions carried out is setting the present operation mode of the PEs (store, show_weights, compute, distance, update, hold, compare, winner_is). The parallel operation of the system lies in that all the PEs carry out the same operation with the input x_k and its own local weights; thus, the operating speed does not depend on the map size. Moreover, in the critical phases (learning and recall), a one stage pipeline is implemented: While a new data is fetched, the previous one is computed in parallel. The maps that the coprocessor can emulate are limited to a size of 16x16 neurons (because there are 8 address lines), and the inputs have a maximum of 16 components. It is not very usual working with larger Kohonen networks; nevertheless, the VHDL description can be easily modified, if necessary, and the architecture is easily expandable.

The system has four operating modes (set by MODE): setup, read, recall and learning. In the setup mode, the host stores in the coprocessor the net parameters (net dimensions, starting synaptic weights and learning parameters). The CU receives and stores the parameters in their right place (its own registers or in the PEs). In the read mode, the coprocessor's CU sends to the host the synaptic weight stored in the PEs; thus, the host can monitor a learning process. In the recall mode, the host initialize the coprocessor (store the starting parameters), next, sequentially sends the input vector components, which the controller drives to the PEs. Every PE, in parallel, subtracts and accumulates (computing the Manhattan distance), then, the winner is searched (under control of the CU), and its label is sent to the host. In the learning mode, first, a recall phase is carried out, the label of the winning neuron is sent to all the PEs, and every one in parallel computes its distance to the winning neuron, and accordingly update its weights. Notice that in one recall or learning phase only one input pattern is processed.

VHDL simulation of the system. By means of VHDL, a hardware description language of widespread use, we have described the technical specifications of the coprocessor, and simulations of the whole digital system have been carried out. VHDL allows us to describe, simulate and optimize the architecture at different description levels, from behavioral, to structural or data-flow levels. By means of VHDL, every block (entity) can be described (architecture) at the definition level more interesting at every moment. Thus, VHDL is a powerful tool for top-down design.

We have carried out the VHDL description of a first version of the neural processor. The entities *ProcElem* (Fig. 5a) and *Controller* (Fig. 5b), which implement the PE and the CU, respectively, have been defined. The description of the architectures associated to both entities has been developed at a behavioral level. The entity *Neuroprocessor* (Fig. 4a), which implements the neural coprocessor, has been declared. Its architecture has been defined at a structural level, by connecting the PEs and the CU by means of the internal bus. Finally, the entity *System* has been described (Fig. 4b), its architecture implements the connection neuroprocessor-host, and the operations of the host. In Figure 4b, an example of a possible connection scheme of the coprocessor to a conventional PC is shown. The simulations of both, the neuroprocessor architecture and its connection to the host, have been fully satisfactory. We estimate that about 300 MCUPS could be reached in its VLSI implementation and near to 10 PEs could be integrated in one chip, making use of the available present submicron technologies.

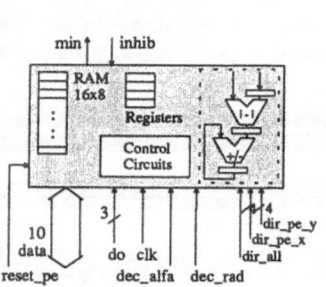
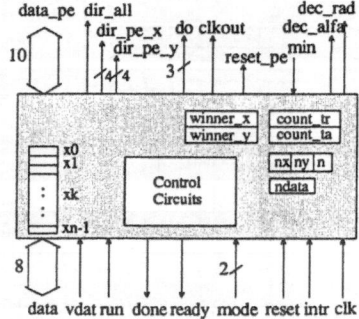

Figure 5. a) Processing Element. b) Neuroprocessor Control Unit.

Conclusions.

Two hardware-oriented Kohonen models have been proposed, one suitable for analog VLSI implementation, and the other one for digital. From the digital SOFM, a neural coprocessor has been described and simulated by using VHDL. In a future work, we will address the electronic implementation of the architecture.

References.

[Demar92] Demartines, P., Blayo, F. *Kohonen Self-organizing Maps: Is the normalization necessary?*. Complex Systems, 6, 105-123, 1992.

[Kohon90] Kohonen, T. *The Self-Organizing Map*. Proc. of the IEEE, 78, 9, pp. 1464-1480, 1990.

[Kohon93] Kohonen, T. *Things you haven't heard about the Self-Organizing Map*. IEEE 1993 Int. Conf. on Neural Networks, San Francisco, pp. 1147-1156, 1993.

[Lo91] Lo, Z.P., Bavarian, B. *On the rate of convergence in topology preserving neural networks*. Biological Cybernetics, 65, 55-63, 1991.

[Martf93] Martín-del-Brío, B., Serrano, C. *Self-organizing neural networks for analysis and represen-tation of data: Some financial cases*. Neural Computing and Applications, 1, 193-206, 1993.

[Martf94] Martín-del-Brío, B. *Procesamiento Neuronal con Mapas Autoorganizados: Arquitecturas Digitales*. PhD Thesis (in Spanish), University of Zaragoza, Spain, December, 1994.

[Melto92] Melton, M.S., Phan, T., Reeves, D.S., Ven den Bout, D.E. *The TInMANN VLSI Chip*. IEEE Trans. on Neural Networks, 3, 3, 375-384, 1992.

[Rüpin93] Rüping, S., Rückert, U., Gosser, K. *Hardware desing for self-organizing feature maps with binary input vectors*. Proc IWANN'93, Sitges (Spain), 488-493, 1993.

[Sutto94] Sutton III, G.G., Reggia, J.A. *Effects of normalization constraints on competitive learning*, IEEE Trans. on Neural Networks, 5, 3, pp. 502-504, 1994.

[Thira94] Thiran, P., Peiris, V., Heim, P., Hochet, B. *Quantization effects in digitally behaving circuit implementations of Kohonen networks*. IEEE Trans. on Neural Networks, 5, 3, 450-8, 1994.

[Vitto90] Vittoz, E. *VLSI implementations of Kohonen maps*. Int. Rep., ESPRIT prj. NERVES, 1990.

[Zurad92] Zurada, J. M. *Artificial Neural Systems*. West Publishing Company, 1.992.

Appendix. Incremental updating rule for the bias.

We are looking for an incremental rule for updating the bias of the Kohonen neuron of Sect. 3

$$\theta_{ij}(t+1) = \theta_{ij}(t) + \Delta\theta_{ij}(t) \tag{A.1}$$

Let us start with an Euclidean neuron, whose bias fulfills condition (6). Writing (A.1) in $t+1$, we have

$$\theta_{ij}(t+1) = \frac{1}{2}\sum_{k=1}^{n} w_{ijk}^2(t+1) = \frac{1}{2}\sum_{k=1}^{n}(w_{ijk}(t) + \Delta w_{ijk}(t))^2 = \frac{1}{2}\sum_{k=1}^{n} w_{ijk}^2(t) + \frac{1}{2}\sum_{k=1}^{n}(\Delta w_{ijk}(t))^2 + \sum_{k=1}^{n} w_{ijk}(t)\Delta w_{ijk}(t) \tag{A.2}$$

Noticing that the first term is the bias, and rejecting terms of second order, we have

$$\theta_{ij}(t+1) \cong \theta_{ij}(t) + \sum_{k=1}^{n} w_{ijk}(t)\Delta w_{ijk}(t) \qquad \text{or} \qquad \Delta\theta_{ij}(t) = \sum_{k=1}^{n} w_{ijk}(t)\Delta w_{ijk}(t) \tag{A.3}$$

and replacing the weight increment by (1)

$$\Delta\theta_{ij}(t) = \sum_{k=1}^{n}\left[w_{ijk}(t).\alpha\left(x_k(t) - w_{ijk}(t)\right)\right] = \alpha\left[\sum_{k=1}^{n} w_{ijk}x_k(t) - \sum_{k=1}^{n} w_{ijk}^2(t)\right] \tag{A.4}$$

Taking into account expressions (8) and (6), we finally have

$$\Delta\theta_{ij}(t) = \alpha.\left(y_{ij}(t) - \theta_{ij}(t)\right) \tag{A.5}$$

That is the incremental rule for the bias (9).

Hardware Requirements
for Spike-Processing Neural Networks

Ulrich Roth, Axel Jahnke, and Heinrich Klar

TU-Berlin, Institut für Mikroelektronik, Jebensstr. 1, D-10623 Berlin

Experimental results suggest that the time structure of neuronal spike trains can be relevant in neuronal signal processing. In view of these results, a shift of interest from analog neural networks to spike-processing neural networks has been observed. For tasks like image processing the simulation of these networks has to be performed with the speed of biological neural networks. We investigated the performance of available hardware and showed, that the required performance for large networks could not be achieved. According to these results we formulated concepts for the design of dedicated hardware for spike-processing neurons. For an efficient hardware implementation it is necessary to know the requisite precision for computations. Through simulations with fixed-point numbers we examined the effects of word length limitation on the behaviour of a spike-processing network. The network was able to perform its basic task as long as the word length does not fall below a certain limit. On this basis we derived conditions for the lower bound of the requisite word length.

1 Introduction

In the eighties interest in artificial neural networks was revived by the incorporation of statistical methods and analogies in physical systems, e.g. the back-propagation algorithm and the Hopfield model. This led to the well-known growth of this field. For a few years there has been a strong tendency towards a return to biology and towards including more details of neuronal signal processing. The background of this shift of interest is the experimental proof of stimulus-induced synchronized brain activity [Eckh88] [Gray89]. Together with the Correlation Theory by von der Malsburg [Mals86] this results in the assumption, that temporal correlation of activity might be used by the brain as a code to bind features to one object and to segregate one object from others.

The synchronised firing of neuronal assemblies could serve as a versatile and general mechanism for feature binding, pattern segmentation and figure/ground separation. How the brain accomplishes these fundamental tasks for invariant object recognition is an old questions of brain science. Invariant object recognition is a natural and easy task for human beings and animals. But it is a difficult and intricate problem in pattern processing for machine vision and until now largely unsolved in a real world environment.

Various model neurons and network architectures have been presented which allowed to reproduce the essential phenomena of synchronized activity in simulation studies. To mention only a few: [Eckh89] [Hart92] [Gers93] [Somp90] [Horn91] [Koen91] [Spor89], see also the references therein. The majority of these model neurons share the following properties: spike trains as incoming signals which induce time-varying potentials at a synapse and outgoing spike trains which originate from a threshold function processing the combination of the post-synaptic potentials. Some models incorporate a time-varying threshold modelling the relative refractory period and/or include axonal delay times. Recently Maas presented a stimulating theoretical analysis of the computational power of spiking neurons [Maas94].

In section 2 we describe the particular model neuron and the model network we chose for our investigations. Unfortunately simulations of these more complex model neurons demand more computer resources. Our first question was whether there the need for specific hardware if someone wants to simulate large networks with thousands of neurons in order to tackle technical problems or to model brain areas. To answer this question we investigated the simulation times on workstations and state of-the-art neurocomputer, like the CNAPS and the SYNAPSE. We present the results in section 3. On this basis we investigated the requirements for an efficient design of a digital accelerator for spike-processing neurons.

Through simulations with fixed-point numbers we examined the effects of word length limitation. In section 4 we present the results and derive conditions for the lower bound of the requisite limitation value and the upper bound of the quantization step.

Finally, in section 5 we summarize the requirements for an efficient architecture of an accelerator for spike-processing neurons.

2 Spike-Processing Neurons

2.1 Model Neuron

The particular model neuron we chose for our investigation is shown in Fig 1.

This neuron is basically the same as the one introduced by Reitboeck, Eckhorn et al. [Eckh89]. Incoming spikes $x(t) \in \{0, 1\}$ are weighted and induce time-varying potentials $u(t)$ at a synapse which change on a time scale much longer than a single spike. There are two types of input potentials: a feeding potential $u_f(t)$, used for direct inputs, and a linking potential $u_l(t)$, used for lateral and feedback connections. The linking potential acts modulatory onto the feeding potential. In this way it shifts the temporal occurrence of the spikes without affecting the mean firing rate of the neuron and, thus, supports synchronization in interconnected model neurons. Furthermore there is a time-varying threshold potential $u_t(t)$ which enables the model neuron to act as a local nonlinear oscillator.

Figure 1: Block diagram of the model neuron

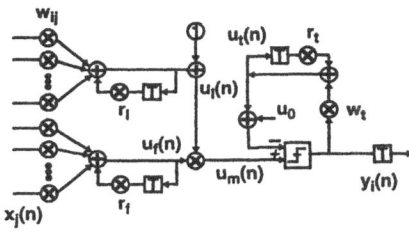

The response functions model a postsynaptic potential and are realized as first-order recursive digital filter in this discrete version with a relaxation factor

$$r = \exp((-T)/\tau) \qquad (2.1)$$

where T denotes the basic time unit for one simulation step, normally $T = 1$ ms.

The feeding, the linking and the threshold potentials are described by the following equations:

$$u_f(n) = \left(\sum_j w_j x_j(n) + r_f u_f(n-1) \right)$$

$$u_l(n) = \left(1 + \left(\sum_j w_j x_j(n) + r_l u_l(n-1) \right) \right)$$

$$u_t(n) = (w_t y(n) + r_t u_t(n-1)) \qquad (2.2\text{-}2.4).$$

The combination of the input potentials yields the membrane potential

$$u_m(n) = u_f(n) \cdot u_l(n) \qquad (2.5)$$

The neuron emits a spike if the membrane potential becomes greater than the sum of threshold potential and a static offset u_o, i.e.

$$y(n) = H(u_m(n) - (u_t(n) + u_0)) \qquad (2.6).$$

2.2 Model network

As model network for the study of simulation times we chose a enlarged version of the neural network presented by Reitboeck, Stoecker et al. [Reit93]. Our network consists of a two-dimensional layer of 128x128 neurons. Each neuron receives an input signal to its feeding input from its corresponding pixel in the input image. The input image has the same size as the layer of neurons. Furthermore each neuron is connected to its eighty nearest neighbours via linking inputs in a square of 9x9 neurons. The linking weights decrease exponentially with the distance between the neurons. This type of interconnection is identical for all neurons in the network, except for those close to a border of the layer.

Moreover, there is a global inhibitory neuron to which all neurons of the layer are connected. This neuron sums up all outgoing spikes and feeds back the resulting inhibitory potential $u_i(n)$ to all neurons. The membrane potential $u_m(n)$ of each neuron of the layer is, thus,

$$u_m(n) = u_f(n) \cdot u_l(n) - u_i(n) \qquad (2.7)$$

with

$$u_i(n) = \left(\sum_i w_i y_i(n) + r_i u_i(n-1) \right) \qquad (2.8).$$

If the network is presented an input image with two or more objects it is able to bind together pixels which belong to one object and to separate one object from others. All neurons which receive input from one object are forming one assembly that is defined by the synchronized firing of its member. Neurons which receive input from an other object also form an assembly. But the synchronized assemblies of different objects fire with different relative phases to each other, thus, separating one object from others. For a throughout explanation how the network accomplishes this task see [Reit93].

3 Simulation of the model network

3.1 Available Hardware

Despite simulating spike-processing neural networks on a general-purpose computer such as a workstation, we can use some of the existing neurocomputer. On the one hand there exist the so-called general-purpose neurocomputers allowing the simulation of a broad class of neural networks. Examples are the CNAPS from Adaptive Solutions, the SNAP from Hecht-Nielsen and the SYNAPSE-1 from Siemens AG. Otherwise, the so-called neuroemulators are designed for specific neural network with the intention to gain more computational power. Recently only one neuroemulator for networks with spiking neurons is known, which has been developed at the TU Berlin.

3.2 Reference network

In order to compare different hardware approaches, we choose the network of section 3 as reference. Before continuing, some special properties of this network should be discussed concerning the simulation.

When simulating spike-processing neurons on digital computers, the time t proceeds in discrete basic time units. A *time slice* denotes the simulation of one basic time unit. In order to compare the simulated networks with biological neural networks, we have to compute the artificial neurons with the speed of biological neurons. Therefore the simulation of one time slice have to be finished in less than one millisecond. This could roughly be defined as *minimal realtime requirements*.

A conventional method for storing the network topology is the use one matrix with dimension N x N (N denotes the numbers of neurons). This, however, leads to a very large but sparsely occupied matrix for the reference network due to both the local connectivity and the high number of neurons. To reduce the memory requirements, the topology of sparsely connected networks is commonly stored in two related matrices C and W, each with dimension N x N_C (N_C denotes the maximum number of connections per neuron). Each row i of C contains several numbers c_m cor-

responding to the neurons connected to neuron n_i. Each cell of W contains the weight w_m corresponding to a connection c_m in C. Matrix C could be ordered in two opposed ways:

1. The number c_m denotes the neuron n_j, to which neuron i sends a spike.
2. The number c_m denotes the neuron n_j, from which the neuron receives to which neuron n_i sends its spikes.

The first method should be called *sender oriented* or *spike distributing*, the second *receiver oriented* or *spike collecting* [Hart93]. Let the network activity be a measurement of the average number of neurons spiking (the so-called *active neurons*) per one time slice. Usually a low network activity has been observed in networks with the model neuron described in 2.1 [Reit93]. Thus, the spike-distributing method has to be faster than the spike-collecting method by order of log N.

In the reference network, we can also exploit the fact, that the connection between two neurons depends solely on the distance vector d. Therefore we only need to store a small weight vector containing N_C values. Calculating the distance vector on-line can also be done with little effort when referencing the neuron by its Cartesian coordinates.

Examining the simulation of the network, each time slice can be divided in two steps:

1. In *step 1*, the numbers of the active neurons spiking the previous time slice have to be distributed or to be collected. According to the incoming spikes, each neuron will increment its potentials u_f, u_l and u_s by the corresponding weight.
2. During *step 2*, each neuron has to relax its potentials and to calculate $u_m(n)$.

Obviously the time needed for step 1 depends on the network activity, while the time for step 2 is rather independent of network activity.

3.3 Experimental results

First of all we simulated the network on a conventional computer (Sparc-10, f_t = 40 Mhz) Times reported in table 1 are measured by a profiler while simulating and averaging over a large number of time units. Activity of the network has been controlled by changing the value of the stimulus. The spikes are distributed via on-line calculating the connectivity in a spike distributing manner. As one would expect, most of the time is spent executing step 2, even for high activity. Furthermore, the time needed for one time slice exceeds the value of 1 millisecond by order of 40.

Table 1: Simulation Sparc-10
- activity : network activity (percent)
- step1: increment of potentials (ms)
- step2: relaxation & calculate u_m (n) (ms)
- time slice : Σ (step1, step2) (ms)

activity	0.01	0.21	0.8	1.5	2.5
step1	1	4	7.4	13.5	17.7
step 2	39	39.2	39.2	39.3	39.3
time slice	40	43.2	46.6	52.8	57

Table 2: Simulation CNAPS: CP1

activity	0.01	0.21	0.8	1.5	2.5
step1	0.4	41.5	98.2	160	184
step 2	1.6	1.7	1.8	2	2
time slice	2	43.2	100	162	186

Table 3: Simulation CNAPS: CP2

Activity	0.01	0.21	0.8	1.5	2.5
step1	0.1	2.4	4.2	4.2	4.2
step 2	1.4	1.6	1.8	1.8	1.8
time unit	1.5	4	6	6	6

To reduce the simulation time, we implemented the network on a CNAPS-256 from Adaptive Solutions [Hamm90]. The SIMD parallel computer has 256 processing elements (PE's) and local memory of 4 KB each. There are two 8b-buses accessible, for input and output, respectively.

We simulated two different implementations of the network. It should be noted in particular, that we have to calculate the connectivity on-line. Otherwise a network of this complexity could not be simulated on the CNAPS-256 due to the roughly limited size of available memory. A more irregular network of the same size would require about 5 MB for C and W.

The two implementations - CP1 and CP2 - are mainly different with respect to the method to calculate the connectivity:

1. CP1 is a straightforward implementation of the network. The 16384 neurons are distributed to all PE's in such a way, that every second neuron of row k of neuron mesh is allocated to PE_{2*k}, while the resting 64 neurons of row k are allocated to PE_{2*k+1}. The simplified algorithm for step 1 is shown here:

1. Distributes number of active neurons
2. -- Parallel code:
   for all neurons
      if connected to active neuron
         increment potential
3. If more active neurons, go to 1.

Note, that the spikes are distributed over the bus, but each PE have to collect the spikes. The results are shown in table 2. As one would expect, the time needed for step 1 is dominant even for moderate activity. Furthermore, the overall performance is worse than these of the serial implementation. This behaviour is caused by both the unfavourable distribution of the neurons and the spike collecting on each PE.

2. Therefore, we chose a different approach for CP2 and divided the network in 64 rectangular sectors with 16 * 16 neurons each. Neurons with an identical position p relative to a sector are assigned to the same PE. Position p is stored in its local memory. The algorithm for step 1 is shown here:

1. PE_j with active neurons distributes p_j
2. -- Parallel code on PE_i:
   calculate distance vector $d_i = p_i - p_j$
3. for all active neurons
      distribute index of active neuron
      -- Parallel code on PE_i:
      incr. potential of neuron with same index
      -- end of parallel code
4. Go to 1. for next PE_j

Note, that now each PE distribute the spike to its local neurons. Table 3 shows the computation times. The time needed for step 1 has been reduced by the order of 10 and is also nearly independent of the activity A. The results indicates, that the realtime requirements are almost met CP2. Further improvement should be possible. Nevertheless, the good results were only achieved due to the strong regularity of the used network. Otherwise, the needed memory for the matrixes C and W would lie beyond the available local memory, even when switching to larger CNAPS systems.

3.4 Some additional remarks

Another neurocomputer, the SYNAPSE from

Siemens AG, exhibits a high-performance for a broad class of conventional neural networks [Rama93]. One of the key points of SYNAPSE is the reusing of weights from a small subset of neurons for several consecutive pattern before computing the next subset. When simulating e.g. networks with spike processing neurons, one time slice has to be computed after the other. Thus, the SYNAPSE is unsuitable for our reference net or, generally spoken, for the simulation of neural networks processing dynamic pattern.

As mentioned above, a neuroemulator has been developed at the institute of microelectronic at the TU Berlin [Pran93]. The underlying neuron model is the model originally presented by Eckhorn et al. [Eckh89], which includes a response function for each synapse. The chip incorporates 16 identical filter blocks and one output block, each containing all parameters required for 16 neurons, allowing therefore the simulation of 16 neurons with 16 synapses each. Due to the full cascadability of this chip, the reference network could be calculated with a network of these chips. The time needed for one time slice is independent of network size and activity. One time slice could be simulated in less than 0.001 ms. However, the unacceptable number of *5184 chips* is needed.

3.5 Summary of the results

Our investigation indicates, that the considered hardware do not meet the above formulated real-time requirements for large spike-processing neural networks. The reasons are:

- missing parallelism (Sparc-10),
- inefficient spike-collecting (CP1),
- limitation to small networks (CP2),
- limitation to static neural networks (SYN-APSE)
- unreasonable amount of hardware due to inflexibility concerning the neuron model (neuroemulator)

These points have to be taken into consideration for the development of dedicated hardware for spike-processing neurons.

4 Resolution analysis

For our study of the requisite precision we used a simplified one-dimensional version of the network described in section 2. In that way we could reduce the simulation times but the essential effects of word length limitation remains the same. In a chain of 128 neurons each neuron is connected to sixteen neighbours to the left and

sixteen neighbours to the right with linking weights of equal magnitude. The input image consists now of two static lines of 32 pixels. Table 4 lists the settings of the weights and the relaxation factors .

Table 4: Parameter Settings of the Chain Network

	weight	relax. fac.
feeding pot.	1.3	0.904837
linking pot.	0.1	0.367879
inhibitory pot.	0.3	0.904837
threshold pot.	40	0.920044

Before each simulation run we initialized the potentials of each neuron with randomly chosen values of a prior simulation run. The network needs a mechanism which breaks the initial symmetry in order to separate the two objects. The network forms now two assemblies which fire about every 28 simulation steps. The firing of the two assemblies occurs with a phase difference of π, i.e. about every 14 simulation steps fires one assembly. If the basic time unit for each simulation step is defined as 1 ms, the firing interval of the assemblies corresponds to an oscillation of nearly 36 Hz.

An input signal arriving with a fixed period k leads to a peak value of the potential which can be calculated as the limit value of the sum over a geometric series according to the following equation

$$u_{peak} = \frac{w}{1 - r^k} \qquad (4.1).$$

At the perfectly synchronized network the linking, the inhibition and the threshold potential reach their minimum value at the firing time and their peak value one time step later. The minimum value can be computed as

$$u_{min} = r^k u_{peak} \qquad (4.2).$$

Table 5 displays the resulting values for these potentials together with the feeding potential which reach a constant value and the membrane potential which is calculated according to eqn. *(2.7)*. The value of the static offset was 5. These values have been verified with the simulation results.

For our investigation of the effects of word length limitation we proceeded as follows: we changed for one type of potential at a time the

Table 5: Peak and minimal values at the perfectly synchronized chain network

	u_{peak}	u_{min}
feeding pot.	13.66	13.66
linking pot.	4.2	1
inhibitory pot.	12.74	3.14
membrane pot.	44.63	10.52
threshold pot.	44.30	4.35

Figure 2: Performance measure Q vs. limited word length for the potentials of the chain network in fixed-point representation

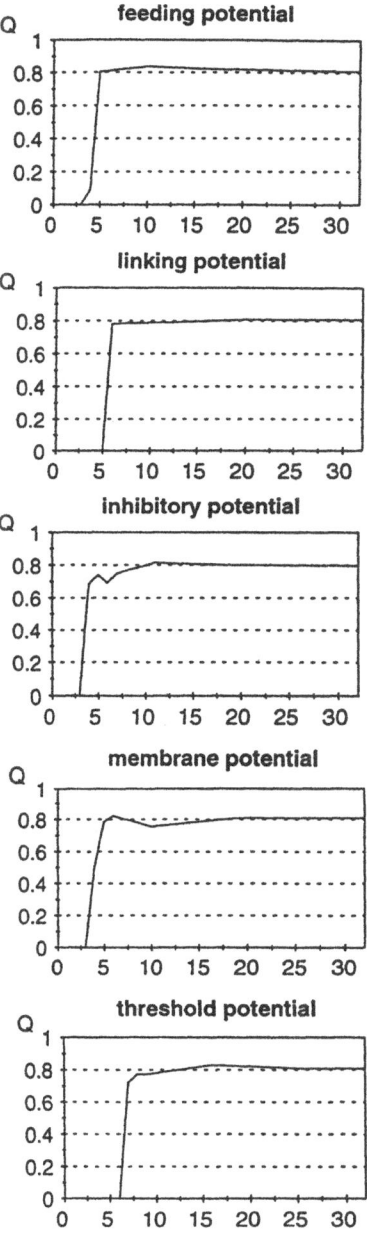

representation from floating-point to unsigned fixed-point number. These fixed-point numbers - i-bit integer part and f-bit fraction part - were used to store the results of the computations in finite length registers. A value exceeding the maximum representable number, the limitation value $v_l = 2^i \text{-} 2^{-f}$, was set to this limitation value. Quantities smaller than the minimum representable number, the quantization step $q = 2^{-f}$ were truncated. For every determined word length i+f and scaling (i,f) we performed 20 simulation runs with different initialization.

We judged the performance of the network by two valuation criteria: the maximum quantity of neurons firing at one time step N_f an the standard deviation of the firing times of one assembly σ. The scaled values are combined in the following performance measure

$$Q = \sigma \cdot N_f \qquad (4.3).$$

We reduced the word width and - if necessary - changed the scaling as long as the performance measure exceeded a certain value and the network was able to perform the basic task of binding and separation. We applied this procedure to each type of potential.

The results (Fig. 3) indicate that finite resolution has no significant influence on the performance of the network as long as the word length does not falls below a certain limit. This proves again, that the network performance does not depend critically on accurate parameter settings, or - assuming the influence of finite resolution as additive noise - that the network is astonishing robust against noise. The limit of the word length is between 5 and 7 bits for the various types of potentials. Going beyond the limit results in a break down of the network performance.

The network was able to perform its basic task with a minimal word length of 5 bits and a scaling of (3,2) for the feeding potential, 6 bits and

(1,5) for the linking potential, 4 bits and (2,2) for the inhibitory potential, 5 bits and (5,0) for the membrane potential and finally 7 bits and (6,1) for the threshold potential. As we see, there is no serious problem for all potentials except the threshold potential, if the limitation value is smaller than their peak value. The limitation of the various potentials has, however, different effects on the network performance. Now we will examine the effects of finite resolution for the various potentials in detail.

A limited feeding potential lowers the oscillating frequency, but the network is able to form two assemblies as long as the feeding potential has the chance to become greater than the static offset.

The peak value of the linking potential is only reached when nearly all connected neighbours fire at the same time step. This happens, however, only if one assembly is perfectly synchronized, and the linking influence is no longer necessary. A limited linking potential reduces the quantity of neighbours which can exhibit a synchronizing influence at one time step. The simulation results indicate that this quantity should be greater than a distinct value in order to allow the neurons of one assembly to synchronize themselves. This value lies between 9 and 19 neighbours for the model network, corresponding to a scaling of (0,5) resp. (1,5).

A limited inhibitory potential shortens the firing periods of the assemblies and the firing period between two assemblies. The two assemblies are no longer firing with a phase difference of π. These effects impair the network performance, but the network is able to separate two objects as long as the limitation value is greater than the minimum value of the inhibitory potential. A limited membrane potential effects the network performance during the initial synchronizing phase. At this stage large values of the membrane potential are necessary to force a neuron which fired a few time steps earlier to fire again, now synchronized with his assembly. Therefore, the limitation value of the membrane potential should be in the same range as the sum of maximum threshold potential and static offset.

The only potential which must never be limited is the threshold potential, because a limited threshold potential results in a lasting firing of the neuron.

Summarizing, we can deduce the lower bound for the limitation value of the feeding, the linking, the inhibitory and the membrane potential. The limitation value needs only to be in the range of 2^n with

$$n = \lfloor ld(u_{max}) \rfloor. \qquad (4.4)$$

The limitation value of the threshold potential has to be greater than 2^{n+1}. The maximum value of the various potentials can be computed in advance if the sum of the weights and the minimal period k_{min} of incoming spikes are fixed:

$$u_{max} = \frac{\Sigma w}{1 - r^{k_{min}}}. \qquad (4.5)$$

Now we will analyse the influence of the quantization step q on the network performance. Computation with truncation has two effects: weights with smaller magnitude are added and the values of the potentials are lowered after each multiplication with the relaxation factor r. Therefore, smaller values are reached for the various potentials compared to the floating-point case. Furthermore, the potentials are decaying faster and arrive at zero, when their value falls below q. Computing with fixed-point numbers leads to an impulse response of finite length, if no rounding is applied.

Smaller values for the feeding, linking and inhibitory potential give rise to almost the same effects as the limitation of the maximum values. The firing period is changed and the capacity to bind and separate is reduced in nearly the same way as we described before. At the threshold potential a greater quantization step results in a shorter firing period of the neuron due to the steeper decay of the potential.

The network performance is only slightly impaired as long as the quantization step does not become greater than a certain limit value. In the following we will derive a upper bound for the requisite quantization step.

At every potential the minimal input quantity, i.e. the minimal weight w_{min}, should at least have an influence over a distinct time. We will call this time influence length k_{inf}, measured in discrete time steps. Now we demand as upper bound for q that

$$q < \left[w_{min} r^{k_{inf}} \right]_t \qquad (4.6)$$

where $[\ ]_t$ denotes the computation with truncation. Calculating the right expression yields

$$q < w_{min}\frac{r^{k_{inf}} - r^{k_{inf}+1}}{2 - r - r^{k_{inf}+1}}, \qquad (4.7)$$

if we assume that the maximum error q occurs at each computation step. The simulation results validate this upper bound for the quantization step.

5 Conclusion

The investigation of the existing hardware shows, that there is still a need for an specific neurocomputer dedicated for spike processing neural networks. An architecture for such a neurocomputer should include following features:

- parallelism
- distribution of spikes
- efficient algorithm to compute the connectivity for regular networks
- flexibility concerning the neuron model
- simulation of one time slice in less than 1 ms (independent of the network activity)

For an efficient hardware design it is necessary to know the effects of word length limitation. The results of our examination indicates, that there is no need for floating-point precision. The network was able to perform its basic task of binding and separating objects as long as the word length of the potentials does not falls below a certain limit. On this basis we derived conditions for the lower bound of the request word length. We checked our results with different parameter settings and verified the derived conditions for the requisite resolution: The minimal word length depends only on the minimal weight, the sum of the weights and the firing characteristics of each potential

Acknowledgment

We would like to thank R. Eckhorn and G. Hartmann for stimulating discussions, M. Stoecker for providing us the source code of his network and M. Pfister for supporting our work with the CNAPS. Special thanks to L. Bala, who did a lot of the programming work. This work has been supported in part by the Deutsche Forschungsgemeinschaft (DFG) under Grant No. Kl 918/1-1.

References

[Eckh88] Eckhorn R, Bauer R, Jordan W, Brosch M, Kruse W, Munk M. Reitboeck HJ (1988) Coherent oscillations: A mechanism of feature linking in the visual cortex ? Biol Cybern 60:121-130

[Eckh89] Eckhorn R, Reitboeck HJ, Arndt M, Dicke P (1989) Feature linking via stimulus-evoked oscillations: Experimental results from cat visual cortex and functional implication from a network model. Proc ICNN I:723-730

[Gers93] Gerstner W, Ritz R, Hemmen JL van (1993) A biologically motivated and analytically soluble model of collective oscillations in the cortex. Biol Cybern 68:363-374

[Gray89] Gray CM, Singer W (1989) Stimulus-specific neuronal oscillations in orientation columns of cat visual cortex. Proc Natl Acad Sci USA 86:1698-1702

[Hamm89] Hammerstrom D (1990) A VLSI Architecture for High-Performance, Low-Cost, On-Chip Learning. Proc. IJCNN II: 537-543.

[Hart92] Hartmann G (1992) Hierarchical neural representations by synchronized activity: a concept for visual pattern recognition. In: Neural Network Dynamics, edited by j. G. Taylor et al. (Springer, Berlin et al.) pp 356-370

[Hart93] Hartmann G (1993) 1. Workshop zum Förderungsschwerpunkt „Elektronisches Auge". Summary: 10-19.

[Horn91] Horn D, Usher M (1991) Segmentation and Binding in an oscillatory neural network. Proc IJCNN II:243-248

[Koen91] Koenig P, Schillen W (1991) Stimulus-dependent assembly formation of oscillatory responses: I. Synchronization. Neural Comput 3(2):155-166

[Maas94] Maas W (1994) On the computational complexity of networks of spiking neurons. NeuroCOLT Technical Report NC-TR-94-021

[Mals86] Malsburg C von der, Schneider W (1986) A neural cocktail-party processor. Biol Cybern 54:29-40

[Pran93] Prange SJ, Klar H (1993) Cascadable Digital Emulator IC for 16 Biological Neurons. Proc ISSCC: 243-244.

[Rama92] Ramacher U, Beichter J, Brüls N (1991) Architecture of a General Purpose Neural Signal Processor. Proc IJCNN I: 443-446.

[Reit93] Reitboeck HJ, Stoecker M, Hahn C (1993) Object separation in dynamic neural networks. Proc IEEE Int Conf Neural Networks II:638-641

[Somp90] Sompolinsky H, Golomb D, Kleinfeld D (1990) Global processing of visual stimuli in a neural network of coupled oscillators. Proc Natl Acad Sci USA 87:7200-7204

[Spor89] Sporns O, Gally JA, Reeke GN, Edelman GM (1989) Reentrant signalling among neuronal groups leads to coherency in their oscillatory activity. Proc Natl Acad Sci USA 86:7265-7269

A VLSI Approach to the Implementation of Additive and Shunting Neural Networks

F.J.Pelayo[1]; E.Ros; P.Martin-Smith; F.J.Fernández; A.Prieto

Departamento de Electronica y Tecnologia de Computadores
Universidad de Granada, 18071-Granada Spain. Email fpelayo@ugr.es

Abstract: Biologically inspired VLSI circuits are proposed which can be particularized to approximate the real-time dynamics of either additive or shunting neural models. Analog inputs to these circuits are represented by short spikes and, both, their transient and steady-state behaviours depend only on process-independent local ratios. The paper includes simulation results and experimental measures of a CMOS prototype, which illustrate the utility and the feasibility of the proposed VLSI approach.

I. INTRODUCTION

The mathematical descriptions used to compute the neural activity in many artificial neural network (ANN) models can be considered as particular cases of equations in the form:

$$\frac{dx}{dt} = - A(x-x_R) + \sum (Weighted\ EXC.inputs) - \sum (Weighted\ INH.inputs) \qquad (1)$$

which describe the dynamics of the post-synaptic activity **x**, or short-term memory (STM) trace, for each neuron in the so called **additive** neural networks [GRO88].

Alternatively to the additive ANNs, several researchers have concentrated upon network models, known as **shunting** or multiplicative [GRO88, ÖGM93, CAR88, CAR90, BAL91, GAU91], in which the neural post-synaptic activity is described by an equation similar to the membrane potential in biological neurons according to the model of Hodgkin y Huxley [HOD52].In a shunting neural network, the equations describing the STM traces are in the form [GRO88]:

$$\frac{dx}{dt} = -A(x-x_R) + (B-x)[\sum (Weighted\ EXC.inputs)] - (x-D)[\sum (Weighted\ INH.inputs)] \qquad (2)$$

$$with: \quad B > x_R \geq D \geq 0$$

Both, additive and shunting equations include a passive decay term $-A.(x-x_R)$ that makes the post-synaptic activity tend to a resting value x_R in the absence of external inputs, with a rate given by the constant **A**. The excitatory term can include positive feedback and external excitatory inputs, and the last term includes the negative feedback and external inhibitory inputs. The feedback terms will be usually non-linear functions of the neural activity of others neurons, weighted by fixed connection strengths and/or by variable long-term memory (LTM) traces representing the synaptic weights.

---

[1] Partially supported by the EC Project: Fellowships for Research on Artificial Neural Networks (ERBCHBGCT920027)

In neurons whose dynamics is described by eqn.(2), the STM trace is restricted to the interval **[D,B]** due to the multiplicative terms **(B-x)** and **(x-D)**, which act as an automatic gain control for the neural states and, as can be proved mathematically, regulate the total activity in networks of neurons interconnected forming an on-centre off-surround topology [GRO88]. This is an interesting feature that makes the neurons respond to the ratio between its input and the total input activity (built-in normalization). The shunting terms also produce interesting effects on the transient activity of networks whose nodes are described by eqn.(2), which have different temporal responses to the on-set and the off-set of the input pattern. In fact, due to the shunting multiplicative terms, the response rates of different neurons in a network will depend on their respective input strengths.

The above features, in some cases combined with mechanisms for continuous adaptability, are being explored and used by several groups in problems such as adaptive pattern recognition, visual perception, and visuo-motor control. Some examples can be found in [ÖGM93, CAR88, CAR90, BAL91, GAU91].

Solving the continuous model given by the differential equations (1) and (2) for the set of neurons in an additive or a shunting network is a difficult task when performed in a digital computer, particularly when real-time operation is desired. In the more general case, for a set of **n** neurons, a system of **n** non-linear coupled differential equations should be solved. Many currently popular ANN models can be derived from equation (1), but using discretized forms to be simulated on digital computers and, often considering only steady-state solutions. However, temporal interaction mechanisms in neural networks can be better implemented if equations describing the continuous or "real time" dynamics of neurons are used, rather than considering particular steady-state solutions of such equations. The circuits presented in the next section allow a real-time emulation of networks described by either the additive or the shunting equations. As it occurs in biological systems, the time is used in these circuits to encode analog inputs and inter-neuron signals, which will be represented by voltage spikes. Instantaneous values of signals will depend only on the spike rate, irrespective of the duration and amplitude of such spikes.

Codification of analog signals by spike streams has been used by several authors, in the so called pulse-stream circuits, to implement ANNs [COR93,CHU93,MUR88,MAS91]. The main limitations of these implementations are that they usually maintain dependencies on the width of pulses, and/or on process-dependent parameters; and sometimes synchronism is explicitly introduced. Moreover, most proposals deal with circuit implementations for "computer versions" or algorithms describing additive feed-forward ANNs. In this case, the time codification of signals is seen as a problem limiting the computing speed, rather than as a potential advantage.

Spikes produced by biological neurons are transferred by their axons, which extend their connections to the dendritic trees of other neurons (wired interconnections). An efficient way of implementing interconnections in VLSI neural systems where signals are represented by asynchronously produced short pulses is by means of digital buses in which the address of each output neuron is written (during a very short period) each time it produces a spike. Without loss of parallelism, the whole neural system can be split in several chips, each one implementing a group of neurons belonging to one or several layers, and the connectivity between groups is defined by the personalization of the address encoders and decoders that communicate the neural groups with the bus (virtual interconnections). This communication scheme drastically reduces the number of physical interconnections and is being used in neural networks like VLSI perception systems [MOR93, LAZ93, MAH92, ARR93].

For the circuits described in this paper, a non-arbitrated event-driven communication scheme [MOR93] is assumed, in such a way that fast digital buses carrying short duration addresses (tens of nanoseconds) can serve to communicate, with a small collision probability, relatively large networks in which inter-spike times can vary in the range of milliseconds.

With typical inter-spike times from tens to hundreds of milliseconds, living organisms can produce actions or take decisions involving several neural areas in a few seconds and even less than a second; that is, a time in which a neuron can produce only a few spikes. In a neural population, if neurons that receive input spikes with highest strength (highest synaptic weight), not only become the most active but also respond the fastest, they could prevent, by inhibition, from other neurons to produce active responses (temporal competition [NAK93]). As a result, the responses of a whole neural system could be based on the specialization of different neurons or groups of neurons to respond before to certain input stimuli. In order to implement similar temporal interaction mechanisms in VLSI we need circuits able to reproduce both the steady-state and transient responses of neural models. This implies to reproduce the continuous dynamics of the model and to have a good control on their response times.

II. CIRCUIT IMPLEMENTATION OF STMs:

Fig.1 shows a schematic representation of the basic circuits to implement the neuron models described by eqns.(1) and (2). For simplicity, only one excitatory and one inhibitory synapse are shown in the figure. Further synapses can be added by direct connections to the summing capacitor C_x, which represents the membrane capacitance of the neuron. For each input spike (represented by a very short voltage pulse), local synaptic capacitances (C_i^+ or C_j^-) are suddenly discharged during the spike, followed by a constant-current charge. Due to the action of the voltage comparators, locally scaled replicas of such currents are added to, or subtracted from, the membrane capacitor C_x during a time which depends on the synaptic weight; i.e. on the values of the pre- and post-synaptic capacitances and the current scaling factors, K_i^+ or K_j^-. For a single excitatory connection, assuming an initial membrane potential $V_x = V_{x0}$, the charging time in response to an input spike is given by:

$$t_{ch} = \frac{C_i^* . C_X}{I_{bias} . C_X + K_i^* . I_{bias} . C_i^*} \cdot (V_B - V_{x0}) \tag{3}$$

and the resulting change in the membrane potential by:

$$\Delta V_X = \frac{K_i^* . C_i^*}{C_X + K_i^* . C_i^*} \cdot (V_B - V_{x0}) \tag{4}$$

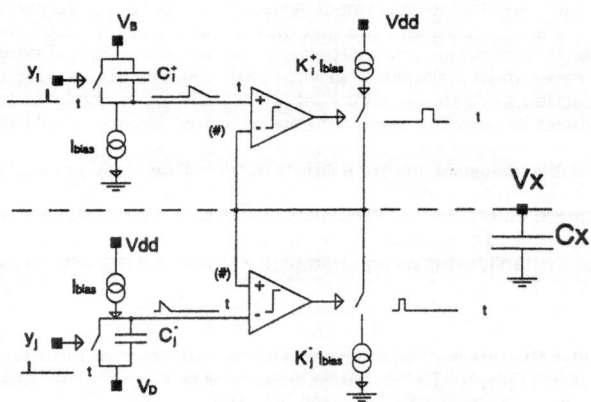

FIGURE 1: *Schematic representation of the circuits to implement excitatory (upper) and inhibitory (lower) shunting synapses. Voltages V_B and V_D correspond respectively to the parameters B and D in eqn.(2). V_x represents the STM trace x in eqns.(1) and (2). To implement additive synapses, the comparator inputs marked (#) must be set to a voltage reference ($V_B > V_{ref} > V_D$) instead of to the summing node.*

According to eqn.(4), if we assume that $K_i^* . C_i^* \ll C_X$, i.e. the changes in V_x produced by individual spikes are sufficiently small to consider a near-continuous variation, the following expression is a good approximation when the excitatory synapse receives an instantaneous spike frequency F_i^*:

Shunting excitatory synapse: $\dfrac{dV_X}{dt} \approx w_i^{\cdot} . F_i^{\cdot} . (V_B - V_X)$; $w_i^{\cdot} = \dfrac{K_i^{\cdot}.C_i^{\cdot}}{C_X}$ (5)

In a similar way, assuming that $K_j^{-} . C_j^{-} \ll C_X$, an instantaneous spike frequency F_j^{-} at the inhibitory input y_j produces a variation of V_X which can be described by:

Shunting inhibitory synapse: $\dfrac{dV_X}{dt} \approx - w_j^{-} . F_j^{-} . (V_X - V_D)$; $w_j^{-} = \dfrac{K_j^{-}.C_j^{-}}{C_X}$ (6)

To implement the decay term in eqns.(1) and (2), constant conductances could be added to the summing node of the circuit in Fig.1; but if we want to have a good control on the decay rate towards the resting potential, which can be accurately defined for several neurons (even when they are implemented in different chips), then a pair of synapses, one excitatory and one inhibitory, receiving a constant spike frequency, can perform the same function. If an excitatory connection (defined by K_R^{+} and C_R^{+}) and an inhibitory one (K_R^{-} and C_R^{-}) both receive the same input spike frequency F_R , the change in V_X they produce can be approximated by:

Passive decay term: $\dfrac{dV_X}{dt} \approx - A . (V_X - V_R)$

(7)

with: $V_R = \dfrac{K_R^{\cdot}.C_R^{\cdot}.V_B + K_R^{-}.C_R^{-}.V_D}{K_R^{\cdot}.C_R^{\cdot} + K_R^{-}.C_R^{-}}$; $A = \dfrac{K_R^{\cdot}.C_R^{\cdot}.F_R + K_R^{-}.C_R^{-}.F_R}{C_X}$

That is, a pair of synapses receiving a spike frequency F_R makes the post-synaptic potential tend to the resting value V_R with a rate governed by such a frequency.

As can be seen, expressions (5), (6), and (7) approximate excitatory, inhibitory, and passive decay terms that drive the neural potential towards V_B, V_D, and V_R, respectively, with rates modulated by the corresponding input spike frequencies. The aggregation of several excitatory and inhibitory terms and the pair of synapses implementing the decay term would make V_X vary as the STM trace of a shunting neuron (eqn.(2)). Moreover, both the steady-state and the transient values of V_X are defined by local current and capacitance ratios, which can be controlled much better than absolute values in VLSI. In other words, process-independent values for the neural activity and response times are obtained.

If the comparator inputs marked (#) in Fig.1 are connected to a common reference V_{ref} instead of to the summing node, then individual excitatory and inhibitory synapses produce respectively the following contributions to the membrane potential V_X:

Additive excitatory synapse: $\dfrac{dV_X}{dt} = w_i^{\cdot}.F_i^{\cdot}$; $w_i^{\cdot} = \dfrac{K_i^{\cdot}.C_i^{\cdot}}{C_X}.(V_B - V_{ref})$

(8)

Additive inhibitory synapse: $\dfrac{dV_X}{dt} = - w_j^{-}.F_j^{-}$; $w_j^{-} = \dfrac{K_j^{-}.C_j^{-}}{C_X}.(V_{ref} - V_D)$

Thus, synapses of this kind, together with a resting element, can be used to approximate the dynamics of additive neurons (see eqn.(1)). In this case the synaptic weights also depend on the voltage reference.

The circuits in Fig.1 can be efficiently implemented in CMOS. A simplified version of excitatory and inhibitory synapses is shown in Fig.2. The comparators are implemented by the differential pairs **mn1-mn2** and **mp1-mp2**, and the local current ratios (K_i^{+} and K_j^{-}) are defined by the relative size ratio of transistors **mp3-mp4** and **mn3-mn4** respectively.

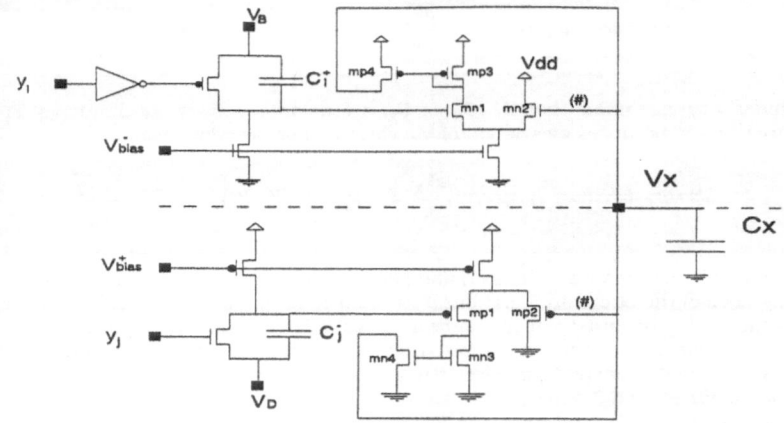

FIGURE 2: A simplified CMOS implementation of the circuit in Fig.1.

III. EXPERIMENTAL RESULTS:

The simulation results included in this section have been obtained with a functional simulator of the circuits described above. This programme implements either the shunting or the additive neural models using the equations (5) to (8), and assuming that analog inputs and outputs are codified by the rates of asynchronously produced spikes.

On the other hand, the experimental measures correspond to CMOS chips that include different STM prototypes, with or without added activity to spike frequency converters. In the present prototype, each synapse includes a section of the total membrane capacitance C_x, and occupies about $60 \times 70 \mu m^2$. Cascode current mirrors have been used instead of the simple mirrors shown in Fig.2.

Figure 3 shows experimental measures of elementary shunting and additive STMs. Fig.3.a corresponds to a shunting firing neuron with a single excitatory synapse. Output spikes are generated with a circuit similar to the Axon-Hillock circuit proposed in [MEA89], with self-reset after each output spike. The input spikes (not shown in the figure) have the frequencies: $F_i^+=5.28$KHz for the bottom trace, $2.F_i^+$ (middle), and $4.F_i^+$ (top). The resulting exponential shaped waveforms have the expected relative time constants. The measures in Fig.3.b correspond to an additive STM with one excitatory and one inhibitory synapses which receive alternate streams of 50 spikes. The linear behaviour is maintained up to values close to GND.

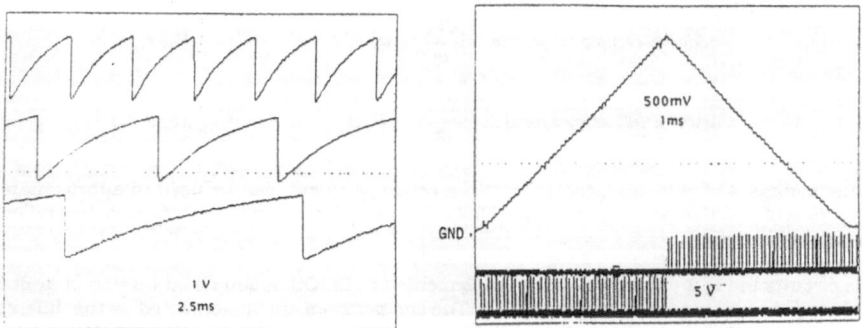

FIGURE 3: Experimental measures of CMOS shunting (a) and additive (b) STMs.

For the measures in Fig.4, a shunting neuron is alternatively excited and inhibited with a spike frequency of 12KHz. In the absence of inputs stimuli, it recovers its resting state with a rate three times slower (F_R is 4KHz and all the synapses, including those implementing the passive decay term have the same weight).

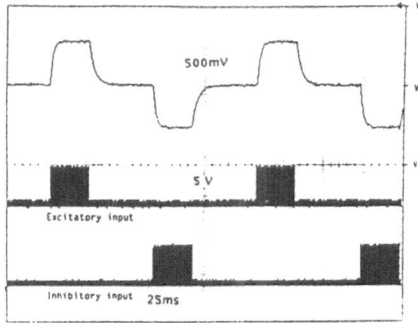

FIGURE 4: Experimental measure of a CMOS shunting STM with excitatory, inhibitory, and decay terms.

Figure 5 shows the simulation results and the corresponding experimental measures for a neural circuit useful to perform some early vision tasks like transient contrast and light adaptation in real-time neuro-vision systems [FAY91]. A neuron (N1) and an inhibitory inter-neuron (N0) receive the same input (E0) but with different weights. As a result of the different response rates and the delayed inhibition of N1 produced by the inter-neuron N0, N1 shows an initial transient response to the input on-set followed by a decaying activity plateau. The simulated outputs are also represented by spikes (S0 and S1 in the figure) whose frequency, over a given threshold, is a linear function of the corresponding neural potentials. When the input stimulus E0 is suppressed, the neural potential of both neurons tends towards the resting value, set equal to V_D in this experiment ($V_R=V_D=1.5$Volt). A stronger inhibition from N0 to N1 would reduce the plateau, and the neural circuit will behaves as an input transient detector.

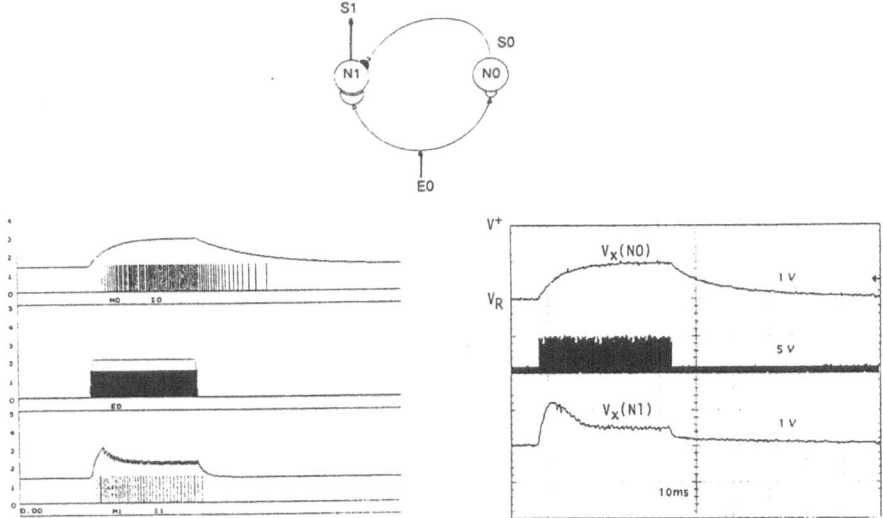

FIGURE 5: Simulation (a) and experimental results (b) of a two-neuron circuit (see text).

Figure 6 illustrates the regulation of the total activity in an on-centre off-surround topology. This behaviour, which has been also verified experimentally, can not be obtained with additive neurons. The inputs to the network (E0, E1, and E3) are made to vary but maintaining their relative ratios with respect to the total input activity. Due to the shunting terms, the total activity of the three neurons is self-regulated, producing the same relative responses while the total input changes. Due to the small decay rate used in this simulation, the STMs tend very slowly towards their resting state when the input are suppressed. That is, the network elements behave as analog temporal buffers.

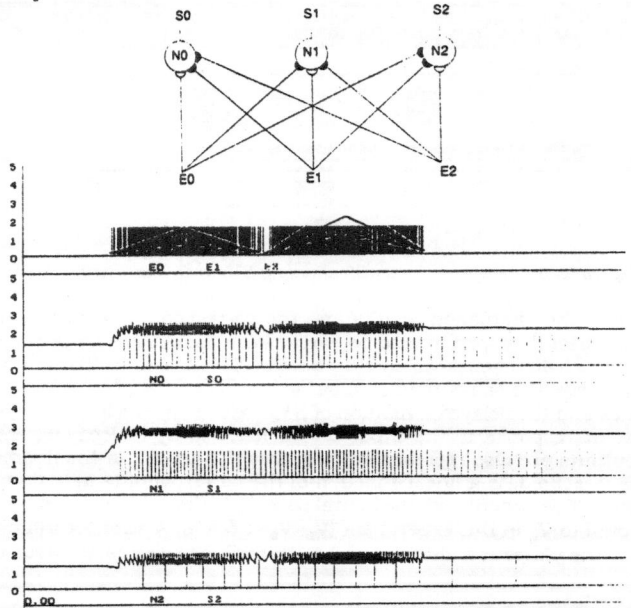

FIGURE 6: *Simulation of a three-neuron on-centre off-surround shunting network. The output activity values depend on the relative input values. If the input spikes are sufficiently separated in time, the set of inhibitory synapses of each neuron can be replaced by a single synapse receiving all the inhibitory inputs, thus simplifying its circuit implementation.*

IV. CONCLUDING REMARKS: Analog VLSI circuits implementing the continuous dynamics of neural models with well controlled time constants at different time scales provide the opportunity to explore temporal interaction mechanisms in a variety of neural systems. In particular, for visual perception VLSI systems [ARR94], these circuits can be efficiently combined with artificial retinas to perform temporal processing tasks like novelty detection, selective attention, and motion processing; tasks requiring many computing resources when they are implemented in real-time with digital computers [BAL91].

Moreover, with the proposed circuits, the synaptic weights are modeled by process-independent local ratios, thus facilitating the analog VLSI implementation of off-line trained neural networks. In-chip learning is not mandatory (as it occurs in process-dependent analog VLSI approaches) unless continuous adaptivity would be required. To implement networks with continuous learning, short and medium-term memories can be easily obtained with the proposed circuits. Long-term memories (LTM) based on non-volatile analog memory devices are being investigated. For some network models [GRO88], the dynamics of the LTMs can be also approximated by circuits with the structure described in this paper but working at a very different time scale. Another possibility is to use digital circuits (and digital weight memories) acting on the communication buses and on the address encoders/decoders, in order to implement the weight adjusting by changes of the pre-synaptic spike rates.

ACKNOWLEDGEMENTS: Authors express their gratitude to the Commission of the EC, to the Centre Suisse d'Electronique et de Microtechnique, to EUROCHIP, and to the Fakultat für Elektrotechnik of Dortmund. Many thanks for their or indirect collaboration and/or assistance to Xavier Arreguit, Eric Vittoz, Pierre Marchal, Michel Verleysen, Jochen Mueller, and to Francisco J. Corbera.

REFERENCES:

[ARR93]   Arreguit,X.; Vittoz,E.A.; van Schaik,F.A.; Mortara,A.: Analog implementation of low-level vision systems. Proc. of the ECCTD'93, Davos. Aug.30-Sept.3, 1993.

[ARR94]   Arreguit,X.; Vittoz,E.A.: Perception Systems Implemented in Analog VLSI for Real-Time Applications. Proc. of PerAc'94. Lausanne, Switzerland, Sep 5-9, 1994.

[BAL91]   Baloch,A.A.; Maxman,A.M.;: Visual Learning, Adaptive Expectations, and Behavioral Conditioning of the Mobile Robot MAVIN. Neural Networks. Vol.4. pp.271-302, 1991.

[CHU93]   Churcher,S.; Baxter,D.J.; Hamilton,A.; Murray,A.F.; Reekie,H.M.: Generic Analog Neural Computation - The EPSILON Chip. Advances in Neural Information Processing 5. Morgan Kaufmann Publishers, Inc. 1993. pp.773-780.

[CAR88]   Carpenter,G.A.; Grossberg,S.: The ART of Adaptive Pattern Recognition by a Self-Organizing Neural Network. Computer, March.88, pp.77-88.

[CAR90]   Carpenter,G.A.; Grossberg,S.: ART 3: Hierarchical Search Using Chemical Transmitters in Self-Organizing Pattern Recognition Architectures. Neural Networks. Vol.3, pp.129-152, 1990.

[COR93]   Del Corso,D.; Gregoretti,F.; Reyneri,L.: An Artificial Neural System using Coherent Pulse Width and Edge Modulations. Proc. of Third International Conference on Microelectronics for Neural Networks. Edinburgh, April 1993. pp.105-114.

[FAY91]   Fay,D.A.; Maxmam,A.M.: Real-Time Early Vision Neurocomputing. Proc. of the IEEE IJCNN'91, pp 1621-1626, 1991.

[GAU91]   Gaudiano,P.; Grossberg,S.: Vector Associative Maps: Unsupervised Real-Time Error-Based Learning and Control of Movement Trajectories. Neural Networks, Vol.4, pp.147-183, 1991.

[GRO88]   Grossberg,S.: Nonlinear Neural Networks: Principles, Mechanisms, and Architectures. Neural Networks, Vol.1, pp.17-61, 1988.

[HOD52]   Hodgkin,A.L.; Huxley,A.F.: A quantitative description of Membrane current and its application to conduction and excitation in Nerve. Journal of Physiology. 117:500-544.

[LAZ93]   Lazzaro,J.; Wawrzynck,J.; Mahowald,M.; Sivilotti,M.; Gillespie,D.: Silicon Auditory Processors as Computer Peripherals. Advances in Neural Information Processing 5. Morgan Kaufmann Publishers, Inc. 1993. pp.820-827.

[MAH92]   Mahowald,M.: Computation and Neural Systems. Ph.D.dissertation. California Institute of Technology.1992.

[MAS91]   Massengill,Ll.W.: A Dynamic CMOS Multiplier for Analog VLSI based Exponential Pulse-Decay Modulation. IEEE Journal of Solid State Circuits, vol.26, no.3. March.1991. pp.268-276.

[MEA89]   Mead,C.: Analog VLSI and Neural Systems. Addison-Wesley. 1989.

[MOR93]   Mortara,A.; Vittoz,E.A.: A Communication Architecture Tailored for analog VLSI Artificial Neural Networks: Intrinsic Performance and Limitations. IEEE Trans. on Neural Networks.

[MUR88]   Murray,A.F.; Smith,A.V.W.: Asynchronous VLSI Neural Networks Using Pulse-Stream Arithmetic. IEEE Journal of Solid State Circuits, vol.23, no.3, pp.688-697.

[NAK93]   Nakamura, K.: Temporal Competition as an Optimal Parallel Processing of the Cerebrohypothalamic System. Proc. of IEEE International Conference on Neural Networks. San Francisco, March-1993.

[ÖGM93]   Ögmen,H.: A Neural Theory of Retino-Cortical Dynamics. Neural Networks. Vol.6, pp.245-273, 1993

A Low-Power Analog Implementation of Cellular Neural Networks

Mancia Anguita, Francisco J. Pelayo, Francisco J. Fernandez, and Alberto Prieto

Departamento de Electrónica y Tecnología de Computadores
Facultad de Ciencias. Universidad de Granada. 18071-Granada, Spain

Abstract: Low-power CMOS analog circuits to implement programmable Cellular Neural Networks (CNN) are proposed, which are based on sub-threshold and bipolar operated MOS transistors. The circuits implementing CNN cells process images directly captured by embedded image sensors. The feasibility of the VLSI approach here described is proved by electrical simulations taking into account electrical parameter variations, and by experimental measures of a fabricated CNN prototype based on the proposed circuits.

I. INTRODUCTION

The Cellular Neural Network model proposed by L.Chua [1] has a similar structure to that found in cellular automata; the evolution of this network is based on the dynamics of locally connected cells working in parallel as analog processing elements with a functionality defined by a set of configurable parameters.

As defined in [1], in a CNN with M rows and N columns (see Fig.1), the state equation for the cell with the index $c \in \{1,2,..,M.N\}$ is:

$$\tau \frac{d\, x_c(t)}{dt} = -x_c(t) + \sum_n A_{n-c}\, y_n(t) + \sum_n B_{n-c}\, u_n + I \tag{1}$$

$$1 \leq n,c \leq N.M \; ; \quad n \in N_R(c); \quad 1 < A_0; \quad |x_n(0)| \leq 1; \quad |u_n| \leq 1; \quad \tau = C_x R$$

where:
- **-n** denotes a generic cell belonging to the neighbourhood of cell c, $N_R(c)$, with radius equal to R. Thus, $N_1(c)$ is the set of 3x3 cells centred in c, $N_2(c)$ the set of 5x5 cells centred in c, and so on.
- $-x_c$ is the state of cell c,
- $-y_n$ is the output of cell n, defined in terms of a nonlinear function f (see Fig.2):

$$y_n = f(x_n) = \frac{1}{2}(|x_n + 1| - |x_n - 1|) \tag{2}$$

- u_n is the input to the cell n,
- I is an offset term to shift the cell nonlinear function, and
- A and B are the **feedback** and **control cloning templates**, respectively.

Depending on the values of the cloning templates, the offset term, and the initial states, the same CNN can be configured to perform different processing tasks. CNNs have been successfully tested by simulation in solving mainly image processing tasks. Examples of the application of CNNs to several image processing tasks, such as noise removal, edge and corner detection, hole filling, connected component detection, etc., have been reported in several publications [2,3]. Since the publication of the paper of L.Chua and other related theory works, several authors have proposed alternative analog VLSI implementations of CNN cells [4-12], which can serve to build CNNs under different constraints concerning the size of the network, the kind of inputs (analog/digital), and the programmability features of the cloning templates.

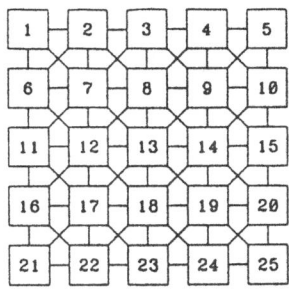

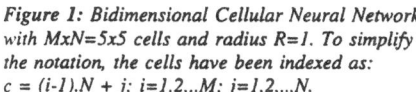

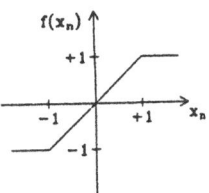

Figure 1: Bidimensional Cellular Neural Network with MxN=5x5 cells and radius R=1. To simplify the notation, the cells have been indexed as: c = (i-1).N + j; i=1,2,...M; j=1,2,...,N.

Figure 2: Pseudo-linear function defining the output nonlinearity.

Circuits to perform specific processing tasks can produce denser VLSI implementations. Programmable CNN circuits can either be used to perform concrete tasks on the input images, or they can be sequentially configured to carry out compatible successive tasks.

In order to obtain analog VLSI CNNs that can be efficient for image processing, the design effort must not only concentrate on optimizing the operation and circuit implementation of isolated CNN cells, but also on solving some key problems related to the implementation and functioning of a whole VLSI CNN based on a great amount of such cells interconnected and operating in parallel. These problems may be summarized in the following points:

1.- *Input of data to be processed:* to exploit the parallel operation of the CNN components, the initial inputs and states must be supplied to the network at a speed sufficiently high to prevent the initialization process from spending more time than that required for processing the data in parallel by the CNN cells. The best choice for image processing is to integrate the image sensors in the CNN cells.

2.- *Process dependent non-idealities:* due to the distributed processing that must be carried out by a set of cells integrated over a large chip area, some self-compensation techniques must be applied in order to guarantee that the processing is carried out in such a way that the effects of manufacturing process variations and other non-idealities are reduced to a minimum.

3.- *Programmability of cloning templates:* the feedback and control cloning templates and the offset terms in a programmable CNN must also be either distributed or locally generated in such a way that their values (when they are interpreted by the different CNN cells) remain independent from mismatches inherent to the manufacturing process.

With the CMOS approach here described we have attempted to cope with the above problems in order to obtain a continuous time CNN which makes compatible a high integration density with features of programmability, low power consumption, and integrated image sensors. This implementation is based on sub-threshold and bipolar operated MOS transistors and is described in Section II. Simulation results illustrating the functionality and robustness against parameter variations of CNN cells and networks based on the proposed circuits are shown in Section III, which also includes experimental results of a fabricated CNN prototype. Section IV provides some conclusions.

II. CMOS IMPLEMENTATION OF THE CNN CELLS

Figure 3 shows a block diagram of a CNN cell, where m, and in consequence the number of multipliers included in the summing blocks, depends on the number of cells considered in the neighbourhood of the cell c. The input block includes the image sensor and provides the initial state $x_c(0)$ and the input u_c, which also has to be transmitted to the neighbour cells.

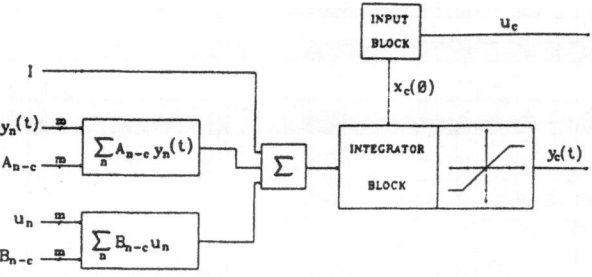

Figure 3: Schematic block diagram of a CNN cell.

Figure 4 shows the basic CMOS circuits to implement the building blocks of Fig.3. We use MOS transistor operating in weak inversion for the differential pairs. Parasitic lateral bipolar transistors, available in standard CMOS processes, are used to implement the current sources because they exhibit better matching properties than MOS transistors [16]. This matching is required between neighbour cells. MOS transistor in strong inversion implement the current mirrors that collect the outputs of all the multipliers in a cell.

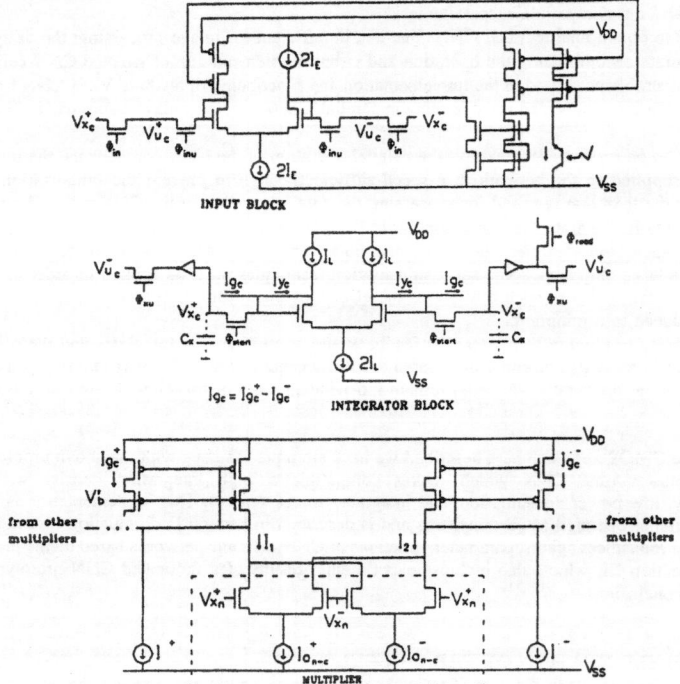

Figure 4: Circuits to implement the building blocks in Fig.3.

For n-channel MOS devices in saturation (that is when $V_{DS} = V_S + 3$ to $6\ U_T$), neglecting the Early effect, the sub-threshold current is given by [14,15]:

$$I_{DS} = I_0\ e^{\frac{kV_G - V_S}{U_T}} \qquad (4)$$

where I_0 is:

$$I_0 = I_S \, e^{\frac{(1-k)V_B - kV_{TO}}{U_T}} \quad ; \qquad I_S = \frac{2\mu C_{OX} U_T^2}{k} \frac{W}{L} \tag{5}$$

V_{TO} is the zero-bias threshold voltage; V_G, V_S and V_B are the gate, source and substrate potentials respectively; $U_T = KT/q$ is the thermal voltage; and k measures the effectiveness of the gate potential in controlling the channel current. For a p-channel device the signs of all voltages are reversed.

Symbols representing voltages and currents in Fig.4 are associated with the respective variables and parameters in the cell; thus:
- $Vx_c^+ - Vx_c^-$ and $Vu_c^+ - Vu_c^-$ represent the state $x_c(t)$ and the input u_c of the cell c respectively.
- I_L is the current that limits the cell output $y_c(t)$.
- Ig_c is the total current representing the two weighted sums of the neighbourhood and the offset term I.
- $I^+ - I^-$ corresponds to the offset term and is obtained from an external differential voltage.
- $Ia_{n-c}^+ - Ia_{n-c}^-$ represents a component of the feedback template, it is also obtained from an external differential voltage by lateral bipolar transistor. The components of the control template (not appearing in the figure) are applied and multiplied in a similar way.

The input block in Fig.4.a includes the vertical bipolar transistor that acts as sensor and a shifter/limiter of the current I_{in} that the sensor produces. The switches controlled by Φ_{in} are used to load the initial cell state ($x_c(0)$). When the state has been initialized Φ_{in} falls and, with $\Phi_{start}=1$ in the integrator block, the cell starts its computation. The final output (+1 or -1) will be given by a voltage close to V_{DD} or V_{SS} at the capacitor nodes of the integrator block (see Fig.4.b). The inverter and the switch controlled by Φ_{read} are used to load the final outputs as binary signals on shared (row or column) lines.

One of the advantages of using externally programmable values for the parameters that define the "programming" of the CNN is that several processing tasks can be sequentially performed on the same initial input image; the most usual is a noise removal as a previous step to other tasks. To include this feature in the circuits of Fig.4, two additional switches are required to load as the new input the output of the cell generated in the previous processing task instead of the signal coming from the sensor.

The following expressions can be derived for the circuit in Fig.4:
Integrator block (Fig.4.b):

$$(6) \quad \tau \, \frac{d \, I_{x_c}(t)}{dt} = -I_{y_c}(t) + Ig_c(t)$$

where: $\qquad \tau = \frac{C_x U_T}{k} \frac{I_L}{(I_L)^2 - (I_{y_c}(t))^2} \quad ; \quad for \quad -I_L < I_{x_c}(t) < I_L \tag{7}$

The output of each multiplier (see Figs.4.b and 4.c) is given by:

$$I_1 - I_2 = \frac{I_{y_n}}{I_L} \, (\, I_{a_{n-c}}^+ - I_{a_{n-c}}^- \,) \tag{8}$$

From Fig. 4 and from the last expression it can be derived that Ig_c is :

$$Ig_c = \sum_n \frac{I_{y_n}(t)}{I_L} \, (\, I_{a_{n-c}}^+ - I_{a_{n-c}}^- \,) + \sum_n \frac{I_{u_n}}{I_L} \, (\, I_{b_{n-c}}^+ - I_{b_{n-c}}^- \,) + (\, I^+ - I^- \,) \tag{9}$$

Thus, the state equation for the entire CMOS cell is:

$$\tau \frac{d \, I_{x_c}(t)}{dt} = -I_{y_c}(t) + \sum_n I_{y_n}(t) \, A_{n-c} + \sum_n I_{u_n} \, B_{n-c} + I =$$

$$-I_{y_c}(t) + \sum_n \frac{I_{y_n}(t)}{I_L} \, (\, I_{a_{n-c}}^+ - I_{a_{n-c}}^- \,) + \sum_n \frac{I_{u_n}}{I_L} \, (\, I_{b_{n-c}}^+ - I_{b_{n-c}}^- \,) + (\, I^+ - I^- \,)$$

(10)

which coincides with expression (1) except in the first term on the right hand side, which refers now to the output instead of the cell state. Moreover, the time constant (τ in expression (1)) depends, as occurs for all CT-CNN cell implementations proposed to date, on process-dependent parameters. The time constant also varies with the instantaneous cell output; the state evolution being slower when the output approaches its final values. Similar dependencies have been obtained by other authors for different CNN implementations based on strong inversion CMOS circuits. In the simulations we have carried out for corner and border extraction, connected component detection, noise removing, and shadow detection, the time- and cell-dependence of the time constant given by expression (7), only produces a variable delay for the different cells in the CNN in reaching their final state, but it does not affect its functionality (see next Section for simulation examples).

III. SIMULATION AND EXPERIMENTAL RESULTS

Using the electrical parameters of the LEVEL=2 model of AMS (Austria Micro Systems) for its 1.2μm analog CMOS process, which have been tested experimentally for MOS transistors operating in weak inversion, several HSPICE simulations of transistor-level descriptions of CNNs have been carried out for different image processing tasks. The influence of process parameter variations has been tested by Monte Carlo analysis, assuming worst-case parameter tolerances for transistors in different cells. For simplicity, the simulation results in the figures below only show three and two of the Monte Carlo trials respectively.

Figure 5 shows the simulated transient response of a simple 7-cell CNN configured for connected component detection. The first 7 plots correspond to the differential voltages between the capacitors C_x (see Fig.4), which represent the cell states $x_c(t)$. The two bottom plots show the signals Φ_{start} and Φ_{in} that temporize the network operation, and the total power consumption. The mean worst-case consumption is less than 60 μwatts per cell.

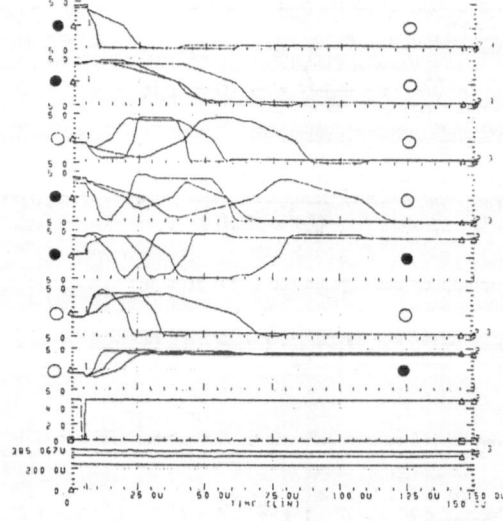

Figure 5: Electrical simulation of a connected-component detection CNN. Black and white dots represent the output values +1 and −1 respectively.

The simulation results in Fig.6 correspond to a CNN configured for vertical shadow detection. The tables included in the figure represent the input image, the programming parameters and initial conditions, and the final output obtained by the network.

A test chip with 8x8 CNN cells has been designed and fabricated, which is based on the circuits in Figure 4 but using their complementary versions. An integration density of 10.7 cells per square millimetre has been obtained, using the $1.2\mu m$ CMOS process of AMS. Figure 7 shows the layout of a cell. The chip is being successfully tested for different image processing tasks. As an example, Figure 8 shows the experimental measure of the 8 outputs of a column of cells when the network perform a connected component detection similar to that shown in Figure 5.

IV. CONCLUSION

An analog CMOS implementation of continuous time cells for Cellular Neural Networks (CNN) has been presented, which has the following features:

-    Low-current operation; which has important consequences on power consumption, one of the key issues for building large single-chip CNNs.
-    Embedded image sensors implemented by vertical bipolar transistors for direct capturing of input images, thus exploiting the potential advantage of CNNs for parallel image processing.
-    Robustness against process parameter variations between cells, due to the use of equal dynamic ranges for the common and inter-cell signals, which are converted to different inner units for each separate cell.
-    The parameters defining the processing task the CNN is to carry out are modifiable by means of external signals, which enable different processing tasks to be performed on the same or different input images.

Simulation results have been provided illustrating the functioning of the proposed circuits in the presence of worst-case parameter variations, as well as some experimental measures of an integrated CNN prototype based on the proposed circuits.

ACKNOWLEDGEMENTS: This work has been partially supported by project MIC-89-04-15, financed by the Spanish National Plan for Microelectronics (CICYT-SPAIN).

REFERENCES

[1]   Chua, L.O.; Yang,L.: "Cellular Neural Networks: Theory", *IEEE Trans. on Circuits and Systems*, Vol.35,No.10, pp.1257-1272, Oct.1988.

[2]   Proceedings of the IEEE First and Second *International Workshops on Cellular Neural Networks and their Applications*. IEEE Cat. No. 90TH0312-9 and 92TH0498-6.

[3]   *IEEE Transaction on Circuits and Systems*, Special Issue on Cellular Neural Networks. Vol.40, No.3, March 1993.

[4]   Varrientos,J.E.; Sánchez-Sinencio,E.; Ramírez-Angulo,J. : "A Current-Mode Cellular Neural Networks Implementation", *IEEE Trans. on Circuit and Systems. Special Issue on Cellular Neural Networks*, Vol.40, No.3, pp. 147-156 March 1993.

[5]   Harrer,H.; Nossek, J.A.; Stelzl, R.: "An Analog Implementation of Discrete-Time Cellular Neural Networks", *IEEE Transactions on Neural Networks*, Vol.3, No.3, pp. 466-476, May 1992.

[6]   Halonen,K.; Porra,V.; Roska,T.; Chua,L.O.: "VLSI Implementation of a Reconfigurable cellular Neural Network", *Proceedings of the IEEE Cellular Neural Networks & Applications International Workshop*, IEEE Cat. No. 90TH0312-9, pp. 206-215, Budapest, 16-19 Dec. 1990.

[7]   Baktir, I.A.; Tan, M.A.: " Analog CMOS Implementation of Cellular Neural Networks", *IEEE Transaction on Circuits and Systems*, Special Issue on Cellular Neural Networks. Vol.40, No.3, pp. 200-206 March 1993.

[8]   Cruz,J.M.; Chua,L.O.: "A CNN Chip for Connected Component Detection", *IEEE transactions on Circuits and Systems*, Vol.38, No.7, pp. 813-817, July 1991

[9]   Anguita,M.; Pelayo,F.J.; Prieto,A.; Ortega,J.: "*Analog CMOS Implementation of a discrete-time CNN with Programmable Cloning Templates*". IEEE Trans. on Circuit and Systems. Special Issue on Cellular Neural Networks, Vol.40, No.3, pp. 215-219, March 1993.

[10]   Dalla Betta,G.F.; Graffi,S.; Masetti,G.; Kovacs,Zs.M.:"CMOS Implementation of an Analogically Programmable Cellular Neural Network", *IEEE Trans. on Circuit and Systems. Special Issue on Cellular Neural Networks*, Vol.40, No.3, pp. 206-215, March 1993.

[11]   Espejo,S.; Rodriguez-Vazquez, A; Huertas,J.L.: "*Design and Testing Issues in Current-Mode Cellular Neural Networks*",*Proceedings of the IEEE Second International Workshop on Cellular Neural Networks & Applications*, IEEE Cat. No. 92TH0498-6. pp.169-174. Munich, October 1992.

[12]   Espejo,S: *Redes neuronales celulares: modelado y diseño molotítico*. PhD.Dissertation. Universidad de Sevilla, Spain. 1994.

[13]   Andreou,A.G.; Boahen,K.A.; Pouliquen,Ph.O.; Pavasovic,A; Jenkins,R.E.; Strohbehn,K.: "*Current-Mode Subthreshold MOS Circuits for Analog VLSI Neural Systems*". IEEE Trans. on Neural Networks, Vol.2, no.2, pp.205-213. March.1991.

[14]   Vittoz,E.A.; Wegmann,G.: "*Dynamic Current Mirrors*". In "Analogue IC Design: The Current Mode Approach". C.Ioumazov, F.J.Lidgey, and D.G.Haigh (Eds.). *IEE Circuits and Systems Series 2*. pp.297-326, 1990.

[15]   Mead,C.: *"Analog VLSI and Neural Systems"*. Addison-Wesley Publ.Comp. Inc, 1989.
[16]   Arreguit, X.: "Compatible Lateral Bipolar Transistors in CMOS technology: model and applications. *These No. 817. Ecole Polytechnique Federale de Lausanne*, 1989.

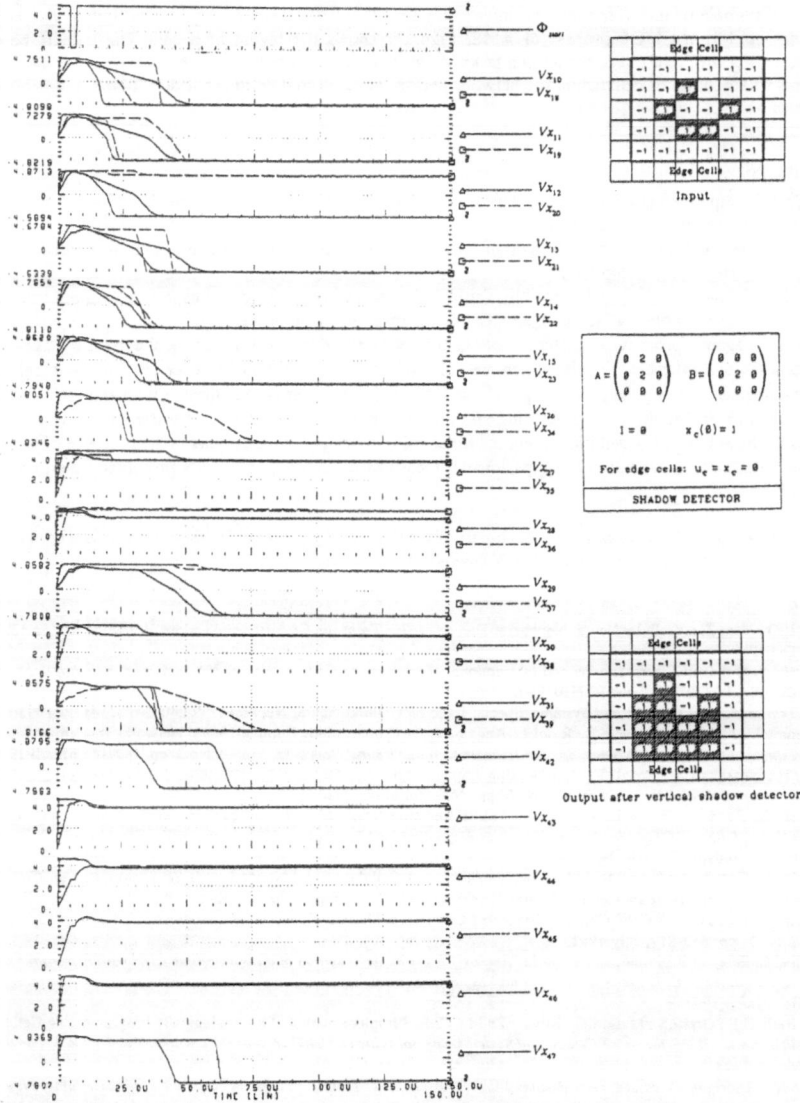

Figure 6: Simulation results of a 6x5-cell CNN (8x7 cells including the boundary) configured to perform a vertical shadow detection.

Figure 7: *Layout of a programmable continuous-time cell based on the circuits of Fig.4. The photosensor and the bipolar transistors implementing the current sources are placed in the central row; whose size can be greatly reduced using a BICMOS process. The whole cell, except the sensor, is covered with a second-level metal not shown in the figure.*

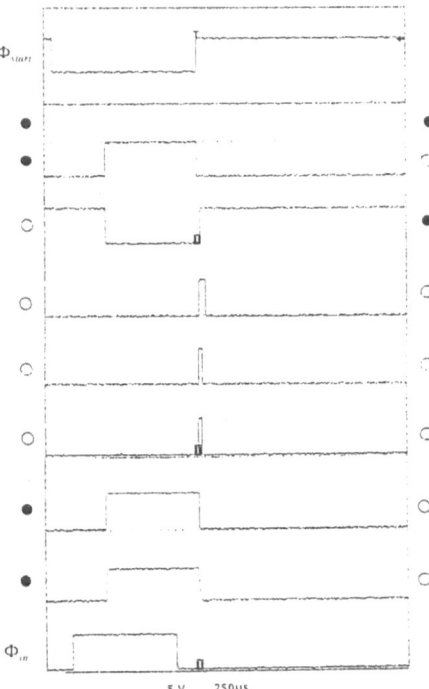

Figure 8: *Experimental measures of the outputs $y_i(t)$ of a column of the 8x8 integrated CNN, the top plot and bottom plots correspond to the signals Φ_{start} and Φ_{in} respectively.*

Asynchonously Parallel Boltzmann Machines Mapped onto Distributed-Memory Multiprocessors[1]

J.M. BENÍTEZ[a]; J. ORTEGA[b]; I. REQUENA[a]
[a]Departamento de Ciencias de la Computación e Inteligencia Artificial
[b]Departamento de Electrónica y Tecnología de Computadores
Universidad de Granada (Spain)

Abstract
This paper provides an experimental analysis of the convergence properties of the asynchronously parallel Boltzmann Machine (APBM) implemented on a 2-D Mesh Multicomputer. A trade-off between the speed-up achieved with multiple processors and the quality of the optimal solution determined is also observed. The results have been obtained by using a simulator, based in the discrete-event model, which allows an accurate modelling of the hardware elements that have a relevant influence in the behaviour of the APBM.

1. INTRODUCTION.

The Boltzmann Machine (BM) [1] is a stochastic neural network in which each neuron determines its next state in a probabilistic way, according to the state of the neurons connected to it, and the weights of their connections. A global state of the BM is called a **configuration**, k, and can be represented by a vector with N components $(k(1), ..., k(N))$, where $k(i) \in \{0,1\}$ is the state of the ith neuron. Let k be a configuration, the **consensus**, $c(k)$, associated to the configuration is given by:

$$c(k) = \sum_{i=1}^{N} \sum_{i \leq j}^{N} s_{i,j} \, k(i) \, k(j) \qquad (1)$$

where $s_{i,j}$ is the weight of the connection between the neurons i and j, verifying that $s_{i,j}=s_{j,i}$. In the BM, a **state transition mechanism** establishes the way the neurons change their states in order to reach a final configuration that maximizes the consensus function. If the state of neuron i is changed, the increment in the consensus function is

$$\Delta c_k(i) = (1 - 2k(u))[\sum_{j=1}^{N} s_{i,j} \, k(j) + s_{i,j}] \qquad (2)$$

If $\Delta c_k(i)>0$, the transition in neuron i is accepted because a better consensus is obtained, and otherwise, the transition is accepted with a probability

$$P_k(i,T) = 1 \, / \, (1 + \exp(-\Delta c_k(i) \, / \, T)) \qquad (3)$$

where T is a parameter that controls the acceptance probability. Initially, T has a high value, which is decreased according to a **cooling schedule**. This results in a number of nonimproving transitions

---

[1]This work has been supported by projects TIC94-0506 (CICYT, Spain) and PB-0945 (DGICYT, Spain)

accepted, which is high at the beginning and is gradually lowered. This way, local maxima traps can be avoided.

Due to its behaviour, the BM is asymptotically able to solve combinatorial optimization problems as it is described in [1]. Nevertheless, the main problem in the application of BM is that it requires a large amount of computational time if the number of neurons is high, or when the BM is used in a real-world problem in which a slow decrease in T is required to obtain a good enough solution. To speed up the computations required, two main approaches to the implementation of the BM have been used. The first one assigns a physical processing element to each neuron in the BM [2], thus allowing to get very high computing speeds but supporting only a specific BM. The other way to accelerate the computations is to use a parallel computer, which allows to program different structures and cooling schedules.

As the neurons in a BM evaluate their state transitions locally, only influenced by the state of its neighbours, it could be easy to get an efficient parallel implementation of the BM previously described. Nevertheless, as the new state of a neuron has to be communicated to all the neurons connected to it before these neurons could evaluate their new state, and as the neurons in a BM usually have a lot of connections to other neurons, the synchronization overhead is usually quite high. So to get an acceptable speed up from a parallel implementation of a BM, either the architecture of the machine is able to support a fine-grain parallelism, or the synchronization overhead is reduced.

In order to reduce the synchronization overhead, there are two main approaches to implement parallel state transitions in a BM [1]. The first one is called **synchronous parallelism**, and is based in the scheduling of state transitions in successive trials and in the communication of the state transitions after each trial, so that all neurons could have updated information about the states of their neighbours. This kind of parallelism requires a synchronization scheme that controls the end of each trial.

The other approach is the **asynchronous parallelism**, in which neurons evaluate state transitions simultaneously and independently, using information about the states of their neighbours that is not necessarily up-to date. This type of parallelism allows to reduce the synchronization overhead because a neuron has not to wait for the new states of its neighbours, and so a global synchronization scheme is not necessary. But it has the drawback of the absence of a model to describe the parallel state transitions which allows to prove the asymptotic convergence of the BM.

What respect to fine-grain parallel architectures, several distributed-memory multiprocessors have been proposed recently [3-5] providing low-overhead primitive mechanisms for communication and synchronization by using presence tags, fast task switching, and wormhole routing of messages. There are several proposals [6, 7] of implementation of a BM in a distributed-memory multiprocessor. Nevertheless, they use the synchronously parallel BM, mainly because many evidences of its convergence have been found by simulation [8] and that it is theoretically proved in part [9]. In this paper, we study the convergency and the speed-up of the **asynchronously parallel BM** implemented in a distributed-memory multiprocessor with a bidimensional mesh interconnection network.

The rest of the paper is organized in five sections. Section 2 provides the details of the asynchronously parallel BM (APBM) we propose, and Section 3 gives the description of the simulator we have developed to analyze the convergence of the APMB. The results obtained and the conclusions of the paper are presented in Sections 4 and 5, respectively.

2. BRIEF DESCRIPTION OF THE IMPLEMENTATION.

In the Asynchronous Parallel Boltzmann Machine (APBM), to evaluate a possible state transition, each neuron considers the present values of the states of its neighbours. These values are not necessarily up-to date with the actual values of the neighbour states. Thus, the implementation of an APBM would decrease the synchronization overhead. Moreover, it is easy to map the neurons in the nodes of the multiprocessor because each neuron need not to know where have been mapped its neighbours to send them its new state.

The APBM with N neurons is implemented, in a 2-D mesh multicomputer with n processors, by spawning a process that emulates the behaviour of each neuron. The number of those concurrent processes running in each processor is equal to the number of neurons allocated to that processor and the load (the processes implementing the neuron behaviour) is balanced among all the processors. In the local memory of each processor there is a word corresponding to each neuron connected to any of the neurons which have been mapped in that processor. This word stores the weight of the connection from that neighbour neuron and its state, represented by using three bits, b1, which is the activation state, and b2 and b3 that will be explained bellow. The behaviour of the neuron j is described in Procedure 1.

```
{Procedure 1: Neuron Behaviour Algorithm}
NEURON_EVOLUTION (j,N)
begin
  T:=T_0;
  t:=1;
  k(j):=U({0,1})
  /* Uniform Distribution in {0,1} */
  while (not STOP_CONDITION) do
  begin
    while (not EQUILIBRIUM_CONDITION) do
    begin
      if (TRANSITION_PROPOSED) then
      begin
        Compute Δc_k(j);  /* Using  (2) */
        if ((Δc_k(j)>0) or (P_k(j,T)>rand())) then
        begin
          k(j):=(1-k(j));
          Broadcast(TRANS,j,k(j))
        end;
      end;
      Compute(EQUILIBRIUM_CONDITION)
    end;
    Broadcast(EQUIL,j,k(j));
    Wait(THERMAL_EQUILIBRIUM);
    T:=g(T_0,t);
    /* g is the cooling schedule */
    t:=t+1
    Compute(STOP_CONDITION)
  end;
  Broadcast(STOP,j,k(j))
end.
```

```
{Procedure 2: A specific implementation of Procedure 1}
NEURON_EVOLUTION (j,N)
begin
  T:=T_0; t:=1;
  /* Stop, Change and m are variables */
  /* to compute the STOP_CONDITION */
  Stop:=0; Change:=0; m:=0;
  STOP_CONDITION:=false;
  /* Uniform Distribution in {0,1} */
  k(j):=U({0,1})
  while (not STOP_CONDITION) do
  begin
    for i:=1 to L do
    begin
      if (TRANSITION_PROPOSED) then
      begin
        Compute Δc_k(j);
        if ((Δc_k(j)>0) or (P_k(j,T)>rand())) then
        begin
          k(j):=(1-k(j));
          Change:=1;
          Broadcast(TRANS,j,k(j))
        end;
      end;
    end;
    /* The two following sentences can be */
    /* eliminated for a complete  asynchronous */
    /* evolution */
    Broadcast(EQUIL,j,k(j));
    Wait(THERMAL_EQUILIBRIUM)

    /* g is the cooling schedule */
    T:=g(T_0,t);
    t:=t+1;
    /* STOP_CONDITION is computed */
    if (Change=0) then m:=m+1 else m:=0;
    Change:=0;
    if (m=M) then STOP_CONDITION:=true
  end;
  Broadcast(STOP,j,k(j))
end.
```

In Procedure 1, the Broadcast() function means that a message is sent from the processor where the neuron is mapped to all the processors of the multicomputer. It can be implemented without deadlocks as it is indicated in [10]. There are three possible types of messages that a neuron can broadcast. The message of type TRANS is sent when the neuron changes its activation state, the message EQUIL indicates that the neuron has reached the thermal equilibrium at a given value of T, and the message STOP is generated when the neuron has ended its evolution.

Besides the bits codifying the type (TRANS, EQUIL, or STOP), the message also includes bits to codify the neuron index and its activation state. The neuron index, i, has three fields, i=(i1,i2,i3), i1 and i2 codify, respectively, the row and column of the processor where the neuron is mapped, while i3 is an index that identifies the neuron among the other neurons in the same processor. The routing of the messages in the network is done by a router module associated to each processor [4]. This module receives the messages from its own processor or from other nodes in the network and determines, as in

[10], where to send the message in order to broadcast it to all the processors.

The control flow in Procedure 1 is determined by the values of the conditions: STOP_CONDITION, EQUILIBRIUM_CONDITION, THERMAL_EQUILIBRIUM, and TRANSITION_PROPOSED. They are set by events and procedures that can be specified in some different ways, thus allowing implementations of APBM with different characteristics.

The **EQUILIBRIUM_CONDITION** is reached after a number of state transitions at a given temperature T are rejected. In our implementation, this condition is set after a fixed number of trials, L. When a neuron reaches its EQUILIBRIUM_CONDITION it communicates this event to all the neurons by broadcasting an EQUIL message.

After reaching its EQUILIBRIUM_CONDITION, a neuron must wait the other neurons to reach that condition too. This is the meaning of **Wait(THERMAL_EQUILIBRIUM)** in Procedure 1, where the **THERMAL_EQUILIBRIUM** condition is locally computed in each processor thanks to the EQUIL messages, broadcasted from remote processors. They allow to update the neurons that have set its EQUILIBRIUM_CONDITION as it is explained bellow.

The process implementing the neuron behaviour is suspended until the THERMAL_EQUILIBRIUM has not been reached, and the control of the processor is transferred to other processes. When an **EQUIL** message arrives to the processor, it interrupts the current process and the process that manages the THERMAL_EQUILIBRIUM condition takes the control of the processor. This process sets to 1 the value of the bit b2 in the word of memory where is stored the state and weight of the neuron which is the source of the EQUIL message, and determines if all the neurons have its bit b2 equal to one. If so, the THERMAL_EQUILIBRIUM condition is set, and the processes waiting for it can follow their asynchronous evolution using a new value for the temperature, T, obtained from $g(T_0, t)$.

The THERMAL_EQUILIBRIUM condition establishes a synchronization point for all the connected neurons and produces an overhead associated to it. So the primitive **Wait()** acts as a barrier where each process stops until all the processes have arrived to their respective Wait(). It is possible to eliminate that overhead if the APBM is implemented allowing the neurons to continue their asynchronous evolution when they arrive to their local EQUILIBRIUM_CONDITION. In this case, no EQUIL messages are necessary and they are not used. In this paper, the convergence properties for both APBM implementations (with and without synchronization at thermal equilibrium) have been analyzed.

The **TRANSITION_PROPOSED** condition determines when a neuron evaluates a state transition. So if TRANSITION_PROPOSED is set, the process which corresponds to Procedure 1 can take the control of the processor, otherwise, this process is suspended. Two possibilities for the generation of this condition have been considered in this paper:

TP1: when a processor receives a message of type TRANS. This is a reasonable policy that makes a neuron to evolve when there are changes in other neurons connected to it. In extreme cases, it could generate a high number of transitions and messages or such a small number of them that the machine cannot evolve properly.

TP2: by using a random function either implemented in hardware or evaluated by a process that can interrupt the processor. The values taken by that function determine the time when the TRANSITION_PROPOSED condition is set. In this second possibility, the reception of a TRANS message only produces a control transference to a process that actualizes the local memory with the state of the sender neuron.

The **STOP_CONDITION** indicates that a neuron has finished its processing. We have considered that a neuron stops when it has not changed its state since a number, M, of consecutive values of the temperature. Then, the STOP_CONDITION is set and a STOP message is broadcasted to all the processors. When a STOP message arrives to a processor, it is interrupted and the control is transferred to another process, which sets to 1 the bit b3 in the memory word where it is stored the state of the neuron that broadcasted the message.

The characteristics of the specific APBM implementation here proposed are shown in Procedure 2, given above.

3. THE SIMULATION MODEL.

To analyze the behaviour of an APBM implemented on a 2-D mesh multicomputer, we have developed a simulator. This simulator uses a description of the functional level of the multicomputer, and models the computation and communication processes described in the previous section, which are the elements that determine the convergence and speed-up properties of the APBM implementation.

In our experiments, we have considered a multicomputer with n processors, in which an APBM with N neurons has been mapped. Each processor executes a copy of the same program, which simulates the behaviour of a neuron, so $N=n$, and it has a router module which is able to send, receive, and route messages in the network. This way, the network management and the neuron computation are done concurrently. The simulated tasks for each processor are the following ones:

1. To receive messages from its router, coming from the neurons connected to it in the BM. These messages update the states stored in the local memory of the processor.
2. To evaluate state transitions and decide whether these transitions are accepted or rejected.
3. To send a message of type TRANS to its router with information of its new state. The router will broadcast this message to all the active processors of the multiprocessor.
4. To keep updated counters that hold information about when the equilibrium has been reached (EQUILIBRIUM_CONDITION) and the APMB evolution has concluded.
5. To send messages of type EQUIL to its router when the neuron is in its equilibrium state for a given temperature.
6. To send messages of type STOP indicating when the neuron has definitively stopped.

The simulator we have implemented is based in a well-known, general model called discrete-event model [11]. It has been designed to reflect the main characteristics of both the APBM model itself and the hardware system on which the neural network is implemented, i.e. a 2-D mesh multicomputer.

The behaviour of the simulator is very flexible and it can be controlled by giving proper values to its input parameters. These are categorized in two classes, either related to the APBM or related to the hardware. Parameters in the first class include the number of neurons, the connection topology and weights, the initial temperature, the cooling schedule, the criterium for thermal equilibrium, stop condition and whether there is synchronization at thermal equilibrium or the evolution is purely asynchronous. The second class of parameters corresponds to the number of processors and their distribution, the propagation time (time that a message takes to propagate from a processor to a neighbour one, t_p), the transmission time (the time that needs a router to read a message and send it, t_t) and the computation time (time a processor requires to execute an iteration of its process, t_c).

The implemented simulation model is accurate enough to provide an approximation for the actual time that evolution of the neural network would take on a real multiprocessor. Moreover, by giving different values to t_p, t_t, and t_c it is possible to characterize the influence of the hardware elements which are responsible of each parameter, on the performance of the APBM implementation.

4. EXPERIMENTAL RESULTS.

While the asymptotical convergence in probability for the sequential BM is formally proved [1], no similar result has been obtained yet for the APBM. In this section, it is presented the experimental analysis of the implementations of an APBM on a bidimensional mesh multicomputer.

The definition of the APBM model is quite general and allows several particularizations. According to the possibilities indicated in Section 2, different specific implementations of the APBM has been analyzed in this paper:
- APBM1, with synchronization at thermal equilibrium and transitions proposed after the arrival of a TRANS message (TP1).
- APBM2, with synchronization and transitions scheduled by a random function uniformly distributed in the interval $[0, 2t_c]$ (TP2).

- **APBM3**, without synchronization and transitions proposed after a TRANS message has been received.
- **APBM4**, without synchronization and transitions randomly scheduled (by using the same function as in APBM2).

For each APBM implementation, the experiments have been done by using the same test sets of Boltzmann machines: a set BM1 of BMs randomly generated (BM1), and a set BM2 corresponding to different instances of the Max-Cut problem [1]. Each of these sets consist of BMs with sizes of 20, 50, and 100 neurons, which also correspond to the number of processors in the multicomputer where they have been implemented. The experimental results are summarized in Tables I and II, where they can be compared with the behaviour of a Sequential Boltzmann Machine (SBM).

All the executions of the SBM and the different APBMs have been done with an initial temperature T_0=5, the acceptance criterium expressed in (3), an exponential cooling schedule [1] with parameter α=0.9, and values L=3 and M=2. Regarding to the parameters representing the characteristic of the multicomputer, we have selected the values t_c=(Number of operations) · 0.05 μs, t_p=5 μs, and t_t=0.5 μs obtained from [3].

Table I shows the values corresponding to the **consensus index**, I, which is a measure of the coincidence between the consensus reached by the SBM and the APBM implementations. The index is defined as I=C(APBM)/C(SBM), where C(APBM) and C(SBM) are the consensus yielded by APBM and SBM, respectively, in the optimization of the same neural network. In the table, we provide the values of $I(C_{max})$=C_{max}(APBM)/C_{max}(SBM) and $I(C_{avg})$=C_{avg}(APBM)/C_{avg}(SBM), where C_{max} stands for the maximum value of the consensus in all the executions (25 in our experiments) of the same problem, and C_{avg} corresponds to the average value of the consensus. In Table I, the values of the variation coefficient, Var_Coeff (computed from the standard deviation σ as Var_Coeff=σ/C_{avg}) are also provided. The smaller the values of Var_Coeff, the more homogeneous the behaviour of the APBM. The highest values for $I(C_{avg})$ and $I(C_{avg})$ are printed in boldface. Table II, gives the values of the speed-up, S, and the efficiency, E, computed for the average values of the execution times in the different APBMs, t_{avg}(APBM), and in the SBM, t_{avg}(SBM). This way, S=t_{avg}(APBM)/t_{avg}(SBM) and E=S/n.

Two facts are shown in Table I. First, in most cases the consensus obtained by the APBMs are quite similar to those obtained by the SBM. Second, the Var_Coeffs are quite small so they exhibit an homogeneous behaviour. Hence, we can conclude that the APBM converges.

The best figures of convergence, i.e. the smallest values of Var_Coeff correspond to APBM2 which also reaches the highest values for the consensus: it reaches values of $I(C_{max})$ equal to 1.0 and higher than 0.9 in most cases. The worst values for the indices of convergence in APBM2 correspond to the BM2 problem with N=100. Nevertheless, it is possible to improve these figures, as we have checked with additional experiments, by increasing the average time intervals among transitions. This causes an increase in the time to reach the final configuration and so a reduction of the speed-up and efficiency. We have found that, when the convergence results are the same for the SBM and the APBM, the efficiency decreases from 0.44 to 0.35.

As the best results in convergence correspond to the APBM2 and APBM4, when compared to APBM1 and APBM3, respectively, it is clear that transitions scheduled by a random function offer better performances than transitions fired after the arrival of a TRANS message, specially as n increases. Moreover, by comparing APBM2 against APBM4, and APBM1 against APBM3, APBM2 and APBM1 are better, so an advantageous influence in the convergence of the synchronization at thermal equilibrium is observed. Nevertheless, this synchronization reduces the values of the speed-up and efficiency of APBM1 and APBM2 with respect to APBM3 and APBM4, which sometimes reach superlinear speed-ups, as it is shown in Table II.

Although the convergence of APBM2 is better than the convergence of APBM4, the differences between their respective values of the consensus indices (I) are not very important. Besides this, the speed-up obtained for APBM4 is higher than that for APBM2 because, as in APBM4 there is not synchronization at thermal equilibrium, it allows to obtain efficiencies close to one or even higher than one (a superlinear speed-up). This way, sometimes it could be advantageous to use the APBM4 instead of APBM2, mainly when the goal is to get a fast solution with good enough quality although not being the best one.

Table I. Consensus Indices (I)

Size	BM	Indices	APBM1	APBM2	APBM3	APBM4	SBM (Consensus)
20	BM1	$I(C_{max})$	1.0000	1.0000	0.9861	0.9484	629
		$I(C_{avg})$	0.9675	0.9210	0.7517	0.7718	
		Var_Coeff.	0.0891	0.0851	0.1862	0.1210	
	BM2	$I(C_{max})$	0.9856	0.9936	0.9692	0.8049	380
		$I(C_{avg})$	0.6943	0.6867	0.5924	0.5377	
		Var_Coeff.	0.5344	0.3331	0.3456	0.3612	
50	BM1	$I(C_{max})$	0.9875	0.9976	0.6360	0.9608	2065
		$I(C_{avg})$	0.9180	0.8996	0.2975	0.7052	
		Var_Coeff.	0.1405	0.1007	0.4108	0.1829	
	BM2	$I(C_{max})$	0.8777	0.9255	0.7359	0.9337	883
		$I(C_{avg})$	0.5607	0.7124	0.3881	0.6363	
		Var_Coeff.	0.3541	0.1651	0.3141	0.1923	
100	BM1	$I(C_{max})$	0.9249	0.9599	0.3714	0.9083	4509
		$I(C_{avg})$	0.7703	0.8809	0.2284	0.7965	
		Var_Coeff.	0.2515	0.0500	0.6511	0.0757	
	BM2	$I(C_{max})$	0.7094	0.8485	0.4615	0.7404	2949
		$I(C_{avg})$	0.3277	0.6760	0.2127	0.6310	
		Var_Coeff.	1.4016	0.1480	0.5403	0.1288	

Table II. Speed-ups (S) and Efficiencies (E).

Size	BM		APBM1	APBM2	APBM3	APBM4	SBM (time, μs)
20	BM1	Speed-up	4.58	5.60	16.17	17.93	640
		Efficiency	0.23	0.28	0.81	0.90	
	BM2	Speed-up	6.50	7.92	21.25	22.54	896
		Efficiency	0.32	0.40	1.06	1.13	
50	BM1	Speed-up	30.61	16.03	79.82	37.09	3795
		Efficiency	0.61	0.32	1.60	0.74	
	BM2	Speed-up	28.36	12.16	67.37	23.39	3450
		Efficiency	0.57	0.24	1.35	0.47	
100	BM1	Speed-up	65.56	49.44	307.42	100.31	20800
		Efficiency	0.66	0.49	3.07	1.00	
	BM2	Speed-up	53.48	43.99	250.40	85.15	15600
		Efficiency	0.53	0.44	2.50	0.85	

5. CONCLUDING REMARKS

The experimental results presented in the previous section show that the asynchronously parallel implementation of a BM on a 2-D mesh multicomputer is able to converge. Nevertheless, the performances obtained depend on some specific characteristics of the implementation. The best values for the consensus correspond to an APBM that synchronizes its neurons at thermal equilibrium and proposes state transitions at time intervals determined by a random distribution.

In relation to the efficient use of multiple processors, the absence of any synchronization in some APBMs allows to get high values for the speed-up and the efficiency. So it seems to be a trade off between the quality of the solution obtained and the parallelism achieved.

Nevertheless, it has to be into account that the same annealing scheduling, initial temperature, and criterium of acceptance of a transition have been used for both the SBM and APBM. So it is possible to improve the performances of the different APBM implementations either by increasing the initial temperature, or by increasing the size of the time interval in the random distribution, or both. Having a higher time interval in the random distribution, there is more time among transitions, thus using a more updated information of the states of the neighbours. These possibilities to get better results will produce an increase in the time to reach the final equilibrium, which reduces the speed-up. We have made some experiments that show a 30% reduction in the efficiency when similar average consensus to the SBM has been reached.

The test sets of BMs we have used to evaluate the convergence and speed-up of the APBM correspond to fully connected BMs. We think that the results for the APBM would be even better in those more realistic applications that could be mapped onto BMs with less connections among their neurons. This way, in a future work, we plan to analyze the use of APBMs in order to speed-up the solution of problems related with the design and test of circuits.

Finally, it has to be considered that although we have used specific values for t_c, t_t, and t_p, our simulator allows to change them to whatever values corresponding to other system characteristics. This way, it would be possible to analyze the influence of different hardware modules on the performance of the APBM.

6. REFERENCES.

[1]   Aarts, E.H.L.; Korst, J.M.H.: *Simulated Annealing and Boltzmann Machines*. John Wiley & Sons, 1990.

[2]   Tomberg, J.; Rattinen, H.; Kaski, K.: *VLSI architecture of the Boltzmann Machine algorithm*. Proc. Int. Conf. on Neural Network, pp.274-277, 1990.

[3]   Athas, W.C.; Seitz, C.L.: *Multicomputers: Message-Passing Concurrent Computers*. IEEE Computer, Vol.21, No.8, pp.9-24. August, 1988.

[4]   Noakes, M.D.; Dally, W.J.: *System Design of the J-Machine*. Proc. Sixth MIT Conf. Advances Research in VLSI, MIT Press., pp.179-194, 1990.

[5]   Dally, W.J. *et al.*: *The Message-Driven Processor: A Multicomputer Processing Node with Efficient Mechanisms*. IEEE Micro, pp.23-39. April, 1992.

[6]   Oh, D.H., Nang, J.H; Yoo, H.; Maeng, S.R.: *An efficient mapping of Boltzmann Machine computations onto distributed-memory multiprocessors*. Microprocessing and Microprogramming, 33, North Holland, pp.223-236, 1991/92.

[7]   Nang, J.: *A Parallel Implementation of Boltzmann Machine for Solving Combinatorial Optimization Problems on a Network of Transputers*. Research report, IIAS-RR-92-17E, Fujitsu Ltd.. December, 1992.

[8]   Korst, J.M.H.: *Combinatorial Optimization on a Boltzmann Machine*. J. Parallel Distribut. Comp., 6, pp. 331-357, 1989.

[9]   Zwietering, P.J.; Aarts, E.H.L.: *The convergence of Parallel Boltzmann Machines*. In 'Parallel Processing in Neural Systems and Computers' by R. Eckmiller *et al.* (Ed.), Noth Holland, pp.277-280, 1990.

[10]   Lin, X.; McKinley, P.K.; Ni, L.M.: *Deadlock-Free Multicast Wormhole Routing in 2-D Mesh Multicomputers*. IEEE Trans. on Paral. and Dist. Syst., Vol.5, No.8, pp.793-804. August, 1994.

[11]   Bank, J.; Carson, J.S.: *Discrete-event System Simulation*. Prentice-Hall, 1984.

A Coprocessor Card
for Fast Neural Network Emulation[1]

F.Castillo[*], J.A.García[**], J.M.Moreno[**], J.Cabestany[**]

Universidad Politécnica de Catalunya
Dept.Eng. Electrónica
[*]c/ Víctor Balaguer s/n
08800 Vilanova i la Geltrú
SPAIN

[**]c/ Gran Capità s/n
Modul C4, 08034 Barcelona
SPAIN

Abstract:

In this article we present a coprocessor card for PC, designed for the fast emulation of neural network-based systems. The card is composed of 6 custom neural processors arranged using a special parallel architecture. The processor is briefly presented and the structure of the card discussed, along with the architecture's basics. Also presented are the software tools developed around the card, the graphical interface used for drawing the net and the compiler used to translate the drawing into a series of instructions understood by the processor. Last, we present some results regarding a sample application in which the card was tried: in the control of an inverted pendulum.

I. Introduction

In this article we present a neural coprocessor card based on a custom-made processor designed in our laboratory, the UTAK1 [Cast91]. This card is a first prototype which is aimed to be used in high-speed control and pattern recognition applications. At present, several products exist which simulate neural networks, useful if the intended application does not require a real-time response. On the other hand, some products have begun to appear in the market in the recent years which are specifically-designed hardware for neural networks [Inte91] [Nest94] [Adap91]. We present our hardware solution as another alternative, a simple yet versatile way to emulate neural networks.

II. Card Structure

As mentioned in the introduction, the card is based on the UTAK1 processor. This processor was specifically designed for the architecture of its intended use, the Dynamic Ring Architecture (DRA) [Cast92]. Though space does not allow us to explain this architecture exhaustively, we shall try to describe here its main features. It is proved in the reference, that a Multi-layered Perceptron (MLP) type of network can be emulated using a ring topology for the architecture, both in the forward phase and during backpropagation learning. The MLP is emulated layer by layer, the number of neurons in the layer determining the size of the ring communication path, thus the name Dynamic Ring for the architecture. Figure 1 shows a block diagram of the architecture in simplified form. Not shown in this figure is the bus common to all processors, which is used to send the instructions common to all

---

[1] This work has been partially supported by SIDSA Corp.

processors as the architecture is that of the SIMD type (see [Haye88] for example), in other words, all processors execute the same instructions but on different data.

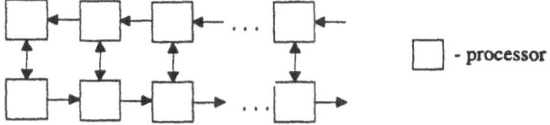

Figure 1. DRA Architecture

During the forward phase, the number of processors used at each time step depends only on the number of neurons in the current and previous layer. Concretely, the expression which determines the number of processors needed in emulating layer l of the network is given by:

$$N_l = max\left\{ g\left(\left\lceil \frac{n_l}{v_l} \right\rceil\right), N_{l-1} \right\} \tag{1}$$

$$\text{where } g(x)=\begin{cases} x+1 & \text{if } x \text{ is odd} \\ x & \text{if } x \text{ is even} \end{cases}$$

and n_l is the number of neurons in layer l and v_l the number of virtual neurons per processor. Virtual neurons refers to the number of neurons each processor is able to emulate, which in the case of the UTAK1 processor is v=3 for each layer.

Thus, if we use a 4-8-5 neuron, 3-layered network, and assume that v=1, the resulting architecture mapping would be that shown in figure 2. Take note of the inactive processors, in charge only of taking the outputs from the previous layer (which had 8 neurons) and closing the ring correctly.

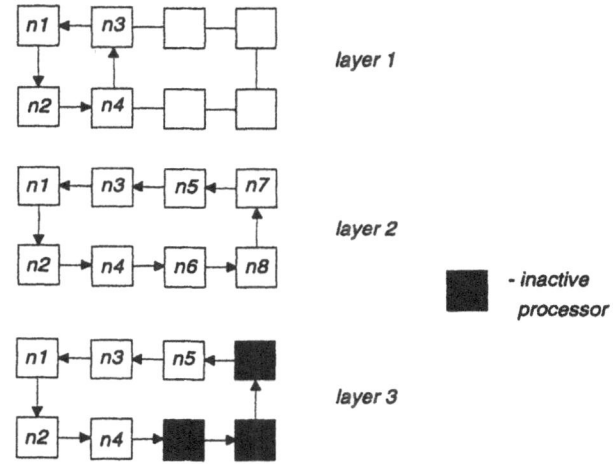

Figure 2. 4-8-5 MLP being emulated using v=1.

One of the advantages of the DRA architecture is that the speed of the emulation depends linearly on the number of processors used, without any degradation in speed occurring even for very large arrays. Additionally, it is studied in [More94] that if incremental algorithms are to be mapped to hardware, the DRA is one of the more efficient solutions which may be used. As the architectural structure is a modular one, more processors may be added to the architecture so as to make it faster. Inherent hardware redundancy, as well as the ease in which faulty processors may be isolated makes us think that this architecture could also be used in integrating various processors in a single chip.

instr #	Name
0	no operation
1	load threshold
2	sequential write to RAM
3	operate
4	calculate activation
5	transfer activation
6	write status register
7	read status register
8	load activation
9	read RAM
10	explicit write to RAM
11	synchronous reset
12	select layer
13	save accumulator
14	increment RAM pointer
15	read activation register

Figure 3. Photograph of the UTAK1 and instruction set

In our realization, we used the custom-designed chip which we have called the UTAK1, also shown on figure 3, with its 16 instructions on the right. The processor contains an internal RAM to store the weights and thresholds, and a ROM which serves as a Look-Up Table (LUT) that implements the non-linear activation function. Precision is limited to 8 bits for its internal registers. Also, aside from a bus which is used for receiving the instructions and data from the interface, it has 3 communication ports as required by the DRA. 4 programmable status registers, one for each possible layer to be used, are in charge of configuring the I/O settings, flagging the processor as in/operative and telling the processor how many neurons are to be emulated at that particular layer.

In the case of the PC card developed, a maximum of 6 processors were used, the actual number being flexible, as the user can configure the software in accordance with the number of processors actually placed on the board. Figure 4 shows the layout of the card. The card has been designed for interface with a PC, and is intended to function in parallel with the PC. Figure 5 also shows a photograph of the actual board.

The card has an on-board clock functioning at 10MHz, while the processors are functioning at half this speed. This has been done even if the processors can work up to 10MHz, however, the limitation are the RAMs currently used, which are much slower than the processors. This could be solved by using faster RAMs, or interleaving them.

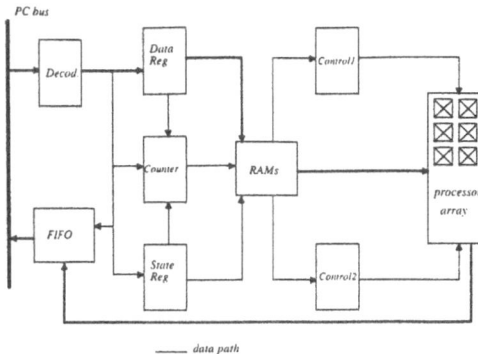

Figure 4. Block diagram of the board.

Figure 5. Photograph of the PC neural card.

Two banks of RAMs are used: instruction (opcode) RAMs, and control (ct1, ct2) RAMs. The first is in charge of storing all the instructions which are to be sent to the processor array, while the second stores control signals used by the board which must be synchronized with the instructions. The latter includes the signals which control the different board buses, the chips' enable signals, as well as the enable signals to the FIFO, counter, etc. The processor array is completely synchronous, which means that all access to the RAMs and buses is done in accordance with the instruction cycle. This means that at each clock pulse both the processor and the card controller should be aware of both the instruction being executed, and the cycle which it is performing. Thus, for example, when the processor writes to the bus, the card controller should anticipate the action by disabling access of all other devices to the bus. In order to do this, we opted for encoding many of the control bits in the control RAM, and generating these bits along with the instructions, using the compiler. In this way, the card's controller need not decode the instruction, and determine all the control signals as a function of the array configuration and clock cycle, but only handle the simpler control functions which are realized by 3 high-density EPLDs present on the card.

The two RAM types may in turn be differentiated into initialization (_ini) and execution (_exe) RAMs. The first refers to the instruction and control RAMs which are in charge of initializing the whole processor array. This means loading the weight values to the processors and configuring their status register for proper operation. This is done sequentially for every processor and is performed only once for a particular neural network. Execution RAMs store all the instructions needed during the actual emulation of the neural network. In this second phase of operation, all the processors work in parallel, save for the writing of inputs to the net, and the reading of outputs.

In order to facilitate the use of the board, a control register is contained which is programmed via address 300h of the PC. One may think of this register as a primitive instruction register, which determines the actions the card is to take as shown on table 1.

Description	Bit	Description	Bit
Select RAM for writing	0	Select opcode_ini	3,4,5
Select RAM for reading	0	Select ct1_ini	3,4,5
Load counter with address	1,2	Select ct2_ini	3,4,5
Reset counter	1,2	Select opcode_exe	3,4,5
Set counter for counting	1,2	Select ct1_exe	3,4,5
Enable RAM	7	Select ct2_exe	3,4,5
Disable RAM	7	Select initizalization	3,4,5
Enable neuron decoder	6	Select execution	3,4,5
Disable neuron decoder	6		

Table 1. Board actions taken by programming the card's control register.

Instructions and control values are initially loaded to the RAMs. Afterwards, initialization is performed by selecting the corresponding RAMs and thereafter, execution by choosing the other bank. Once the array has calculated the outputs, these are loaded sequentially to the FIFO. On the last storage, the card interrupts the PC and waits for it to read the results. It then waits until the next set of inputs are supplied and the process is repeated.

III. Software

One of the things we devoted considerable time to was the development of a good S/W for interfacing the board. This was needed for two reasons: (1)user-friendliness; (2) the complexity in programming the processors. Both of these are in fact related as a user thinks of a neural network structure he wishes to implement, the processors however, function at a much lower level of abstraction: they understand simple, 8-bit instructions just as a common microprocessor. While it is true that these processors have been specifically designed for Neural Network emulation, the translation of the neural network structure to the processors is neither immediate, nor trivial. We therefore realized different levels of abstraction and specification, along with a compiler for each level, as shown in figure 6.

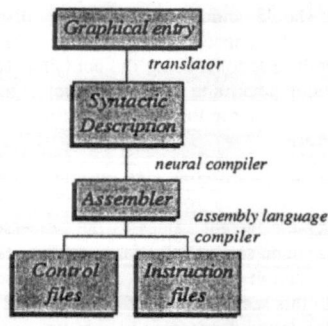

Figure 6. Different hierarchical levels of the software

Using the graphical interface (figure 7), the user is actually able to draw the neural network. The program aids the user in defining the number of layers and neurons, and edit the values of the weights and thresholds. Before compiling this neural network, he may also visualize the resultant mapping of the neural network to the different processors. The program shall map the NN so as to maximize speed in accordance with the number of available processors as shown in the example of figure 8.

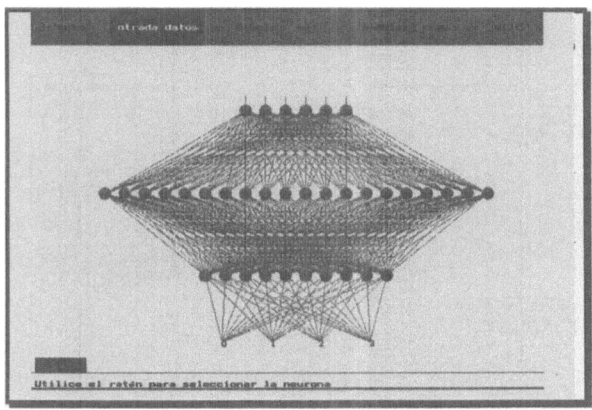

Figure 7. The board's graphical interface.

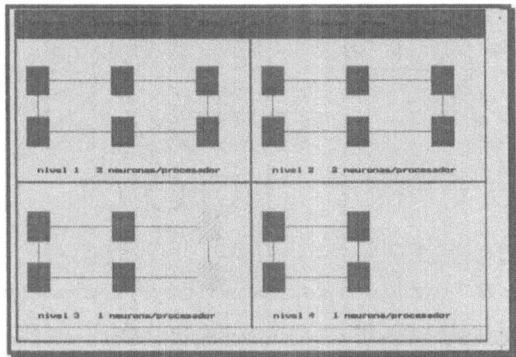

Figure 8. Automatic neural network mapping as done by the interface.

Figure 9 shows in a simplified flowchart form, the steps taken so as to generate the instructions. The first three steps are part of the initialization process: loading weights to the processors, initializing the configuration of each processor, and resetting the registers to be used for the calculations. Afterwards, the instruction sequence is generated layer by layer. Starting with the input layer, activations are loaded to the processors. After, the threshold is loaded into the accumulator and the first multiply and accumulate operation is effectuated. The activation is transferred, simultaneously receiving one. This process is repeated for each activation until all are processed. When all activations are processed, the next layer's activations are calculated by applying the accumulator to a look-up table (LUT). The sequence is then repeated for each of the layers.

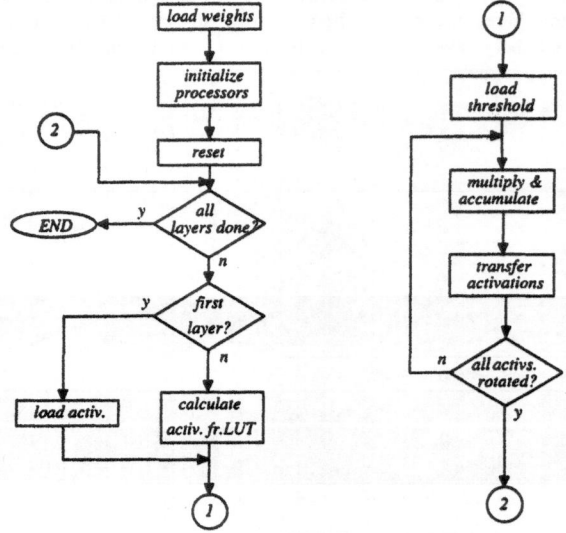

Figure 9. Flowchart showing compiler procedure.

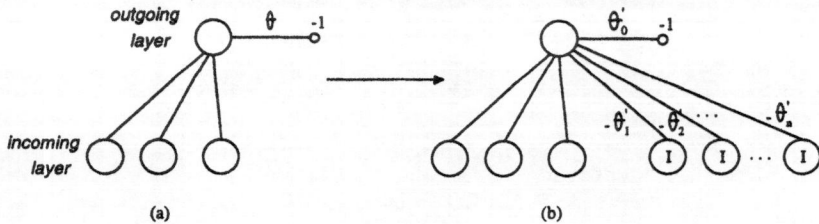

Figure 10. Normalization procedure: new dummy neurons are added.

It is possible that during learning, weights and/or thresholds of a value larger than the maximum codeable value (± 7.9375) are generated. In order to emulate such nets, a normalization procedure is included, which is in charge of reproducing the same function using an equivalent, encodeable net. Take as an example figure 10(a) where θ is a threshold value greater than the maximum codeable value, θ_{max}. So as to create the same effect as θ, dummy neurons are created as shown in figure 10(b). The sum of the new thresholds and weights shall then have to satisfy:

$$\theta = \theta_0' + I \sum \theta_n' \tag{2}$$

where I is the maximum possible output coded by the processor (0.9928185). Thus, it is easy to calculate for the number of new neurons, n, which need to be created in order realize the weights θ_n':

$$n = \left\lceil \frac{\theta - \theta_0'}{I\theta_{max}} \right\rceil \tag{3}$$

where $\theta_0{}'$ may be fixed to θ_{max}. In order to realize a constant I output for the newly-created neurons, the easiest way is to fix their threshold to $-\theta_{max}$ and all of their weights to zero. As for the values of the new weights created, any values may be used for as long as expression (2) is satisfied.

The process involved in normalizing nets with overflowed weights is similar, the concept being much simpler even. First, let us define the terms incoming and outgoing layer as shown in figure 10, to identify the two layers involved. An overflowed weight may be substituted by replicating neurons in the same way as for the thresholds. The sum of the new weights created to the implicated outgoing neuron should be equal in value to the old weight.

In order to determine how many new neurons need to be reproduced in the incoming layer, the normalizer first makes a first pass in order to determine the largest weight in each incoming layer neuron and the largest threshold from among all the outgoing layer neurons. These shall impose the requirements of how many times a neuron needs to be replicated.

IV. An Application Example

In this section, we shall discuss an example which we tried with the board: the control of an inverted pendulum (see [Hech90] for example), in which the inverted pendulum was modeled and simulated. Figure 11 shows the set-up used.

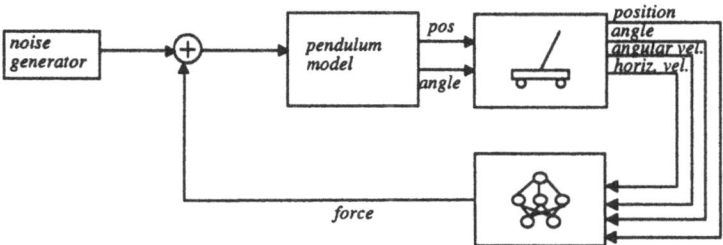

Figure 11. Sample problem used to test the board.

The pendulum model calculates, given a force, the resulting cart position and the angle the bar makes with the vertical axis. These two parameters are then passed to the drawing routine, which updates the screen with the position of both cart and bar. The neural network, implemented using the PC card, takes the position, angle and the first derivative of these two as its inputs and calculates the force needed in order to maintain the bar in vertical position, as well as maintain the cart near the middle of the platform.

With a 486-PC the simulation appears on the screen in real time. The network used consisted of a very simple 5 neuron network, 4 neurons at the input layer, one at the output. In this example, the compiler generated 58 instructions for the initialization, 38 instructions for the execution per se. In terms of speed, the initialization takes $42.8\,\mu s$, while the time elapsed from the presentation of inputs until an output was obtained took $19.8\,\mu s$. Indeed, the time it takes the PC to model the inverted pendulum takes more time than that for the board to compute for the force.

In terms of performance, there was practically no difference between the simulated neural network and the performance of the board. We tabulated the results at different time steps and saw that there were differences in their values due to the limited fixed-point precision of the processors, yet both managed to stabilize the cart equally well.

V. Conclusions

In this article a PC-based board was presented as a practical solution for emulating neural nets in high-speed. This experience has served to see that, indeed, the idea of developing specialized hardware for neural networks which could be applied to diverse applications (we envision applications in control and pattern recognition) is feasible. It has also served to see the limitations of the current system developed. On-chip learning would certainly make the product more attractive, as well as the possibility of using diverse types of activation functions.

From the use of the hardware, we believe that perhaps one of the most important features is the ease-of-use and user-friendliness of the software. When developing applications, one merely wants to plug-in and try, and not have to worry about the particularities of the hardware.

VI. Bibliography

[Adap91]   Adaptive Solutions, Inc., "CNAPs Neurocomputing", 1991.

[Cast91]   F.Castillo, J.Cabestany, J.M.Moreno, "An Integrated Circuit for Artificial Neural Networks", in Lecture Notes in Computer Science: Artificial Neural Networks, Ed. A.Prieto, Springer-Verlag, 1991.

[Cast92]   F.Castillo, J.Cabestany, J.M.Moreno, "The Dynamic Ring Architecture", in Artificial Neural Networks,2 , Eds.I.Aleksander, J.Taylor, North-Holland, 1992.

[Haye88]   J.P.Hayes, "Computer Architecture and Organization", 2nd Ed., McGraw-Hill, 1988.

[Hech90]   R.Hecht-Nielsen, "Neurocomputing", Addison-Wesley, 1990, p. 342..

[Inte91]   Intel Corporation, "80170NX Electrically Trainable Analog Neural Network", Experimental Data Sheet, June 1991.

[More94]   J.M.Moreno, "VLSI Architectures for Evolutive Neural Network Models", PhD.Thesis, Univ. Politècnica de Catalunya, December 1994.

[Nest94]   Nestor Inc., "Ni1000 Development System", May 1994.

Digital Hardware Implementation of ROI Incremental Algorithms

J.M. Moreno, J. Madrenas, S. San Anselmo, F. Castillo, J. Cabestany

Departament d'Enginyeria Electrònica
Universitat Politècnica de Catalunya

Abstract: In this paper we address the problem of constructing efficient hardware solutions for Region of influence (ROI) incremental algorithms. First we shall review the main features associated with these neural models, paying special attention to the basic operations required in order to fulfil the data flow imposed by their training and recall phases. Taking into account the resource organization demanded by this data flow, we shall propose an efficient digital realization which is capable to convert into a physical implementation the organization principles stated previously. The proposed realization is composed of a bidimensional array of processing units, which have been developed as RISC processors. After explaining the emulation sequence to be used for ROI incremental models on the proposed realization, we evaluate the performance (measured in terms of processing speed) attainable by the system when real world classification tasks have to be handled. Our results shown that the proposed realization considerably outperforms recent commercial developments.

1. Introduction

Evolutive neural models have concentrated an increasing research interest during the last years, motivated mainly by the efficient solutions they provide for the problems associated with classical artificial neural network models. Among others, these problems are related to the lack of deterministic methods for establishing the proper network structure to be used in order to solve a particular task, and to the non-incremental learning scheme able to adapt the network parameters so as to meet the requirements imposed by the problem to be handled.

In order to overcome the above stated problems, evolutive neural models offer the possibility to construct the proper network structure for solving a particular task. In this way, the training phase associated with these models is in charge of adapting the internal network parameters (i.e., the strength of the connection between units), but it also evolves the network structure (i.e., number of units and/or layers) appropriate for the stated problem. Furthermore, some evolutive neural models admit also incremental learning capabilities, thus avoiding the necessity of performing an exhaustive training process when new knowledge about the considered problem is available.

If a hardware realization suitable for facing real world tasks is considered, incremental evolutive neural models are preferable to the decremental evolutive ones, since the pruning methods characteristic of the latter ones impose very high computational and memory loads. Among the different types of incremental evolutive neural models [1], [2], the ROI (Region of Influence) incremental models provide an efficient alternative for a physical realization, due to the very simple data flow imposed by both the learning and the recall phases which define their behaviour. These neural models are used mainly for solving classification tasks, so that they construct the corresponding discriminant function by estimating the regions of the input space where the different categories are dominant. The RCE (Restricted Coulomb Energy) [3] or the GAL (Grow and Learn) [4] algorithms may be considered, among others, as belonging to the category of ROI incremental evolutive neural models.

In this paper we shall present an efficient digital realization for the ROI incremental models which significantly improves the performance offered by recent hardware developments [5], [6]. First we shall briefly review the main features of the ROI evolutive neural models, especially those concerning the network organization and the basic operations to be performed by the units generated by the algorithm. Then we shall present a digital realization, inspired in digital systolic architectures, able to map efficiently the data flow imposed by some ROI evolutive models. After explaining the emulation process for these models, we shall provide estimates of the attainable performances expected for this realization when handling real-world tasks. Finally, the conclusions and our future work will be presented.

2. Architecture proposal

The network structure generated by ROI algorithms is organized in general in the two layers depicted in figure 1. The first one, the *coding* layer, obtains a similarity measure between the input vector and the sample stored by each of the units constituting the layer. The similarity measures are then used by the *decision* layer to assign a category to the input vector. The decision layer can be a Winner-Take-All (WTA) system combined with a lookup table, but more complex schemes are also possible.

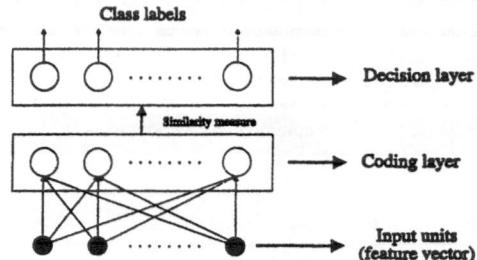

Figure 1. Organization of ROI-based networks

2.1 Coding and Decision layers

Coding layer. The similarity measure is generally obtained by calculating the distance from the stored sample to the input vector. No significant performance differences have been observed for different distance measures, so the Euclidean distance is used. Since the result obtained by the decision layers is the same for the squared Euclidean distance, the output of the coding layer units is:

$$o_i = \left\| x^p - \omega_i \right\|^2 = \sum_{j=1}^{N} \left(x_j^p - \omega_{ij} \right)^2 \tag{1}$$

where o_i is the output given by the i-th coding unit, x^p is the p-th input vector, ω_i the sample vector stored in the i-th coding unit, and N the number of dimensions (components) of the input vectors. A parallel architecture that performs efficiently this data flow is the BBA (Broadcast Bus Architecture) [7], which consists of a linear array of processing units (PUs) accessing a common input and a common output bus (figure 2).

Decision layer. This layer differs for the GAL and RCE algorithms considered in this work. In the GAL algorithm, the decision layer just selects the most similar sample, that is, that which provides the smallest distance. This WTA behavior is digitally performed by comparing the distance measures. The category associated with the most similar (smallest distance) sample is thus obtained. An efficient architecture to perform this calculation is the SRAGB (Systolic Ring Architecture with Global Bus) [7], whose organization is depicted in figure 3.

In the case of the RCE algorithm, each distance measure is compared to a radius associated with each sample of the coding layer. If the distance measure is smaller than the radius, the input vector is inside the hypersphere centered on its sample vector. If the vector is inside only one hypersphere, the category to which the sample belongs is selected, otherwise rejection or further learning is done. For the RCE decision layer, the proposed architecture is again the BBA.

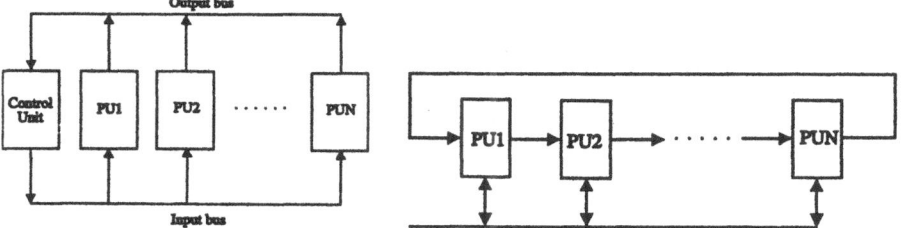

Figure 2. Broadcast Bus Architecture   **Figure 3.** Systolic Ring Array with Global Bus

2.2. The Dynamic Ring Array (DRA)

This architecture was originally developed to emulate Multi-Layer Perceptrons [8] in the recall phase, but the interesting properties it shows makes it very suitable for ROI emulation as well. The DRA architecture organization is shown in figure 4. In this systolic architecture, the processing units which form the ring emulate layer by layer the neural network. Because of the connection topology, the network size is easily reconfigurable to fit different layer sizes and thus to avoid dead cycles. The DRA architecture performs operations on a SIMD (Single Instruction Multiple Data) basis, being the instructions dispatched to the processing units by an array control unit.

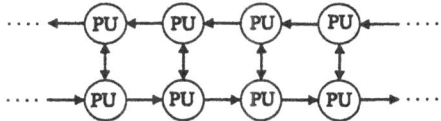

Figure 4. DRA architecture organization

To enhance efficiency during the training phase, a modified DRA architecture that provides also the synapse parallelism processing scheme has been proposed [2]. It includes the addition of a local memory to each PU and a multiplexer that allows also the control unit access (figure 5). The array works as a SRAGB with synapse parallelism during the training phase, and as a BBA with neuron parallelism during the recall phase to perform PLS algorithms emulation.

The internal organization of the processing units composing the modified DRA structure is depicted in figure 6. The four basic blocks are:

- *Communications block.* It consists of the communications registers and bus, and is used to configure the physical array.

- *Register file block.* It is in charge of storing the synaptic vector during the recall phase and the synaptic components during the learning phase.

- *Arithmetic block.* It performs the basic arithmetic operations. Includes a multiplier, adder, accumulator, some auxiliary registers and a status register.

- *Control block.* It coordinates the other system blocks. Includes the control unit, instruction register and control bus.

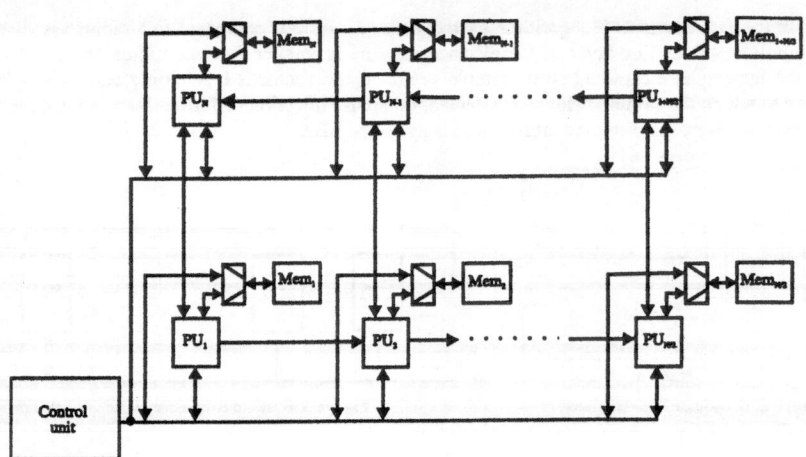

Figure 5. Modified DRA architecture

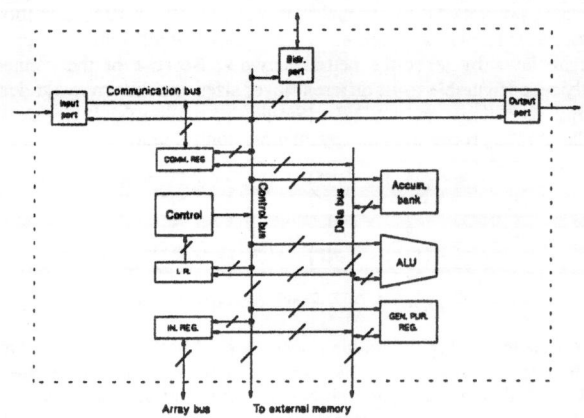

Figure 6. Internal organization of the processing units

A realization of the processing unit has been described and simulated in hardware description languages (Verilog and VHDL). Automatic synthesis is currently being undertaken and a VLSI implementation will be fabricated in the near future.

The ASIC is a 16-bit RISC (Reduced Instruction Set Computer) processor with three pipeline stages (instruction decode, execution and write result) and a execution rate of 1 instruction per clock cycle. It includes a 128-register file and addressing capability for a 64 K x16-bit external local memory. From the 128 registers contained in the register file, only three (count0, count1 and a register connected to an internal random number generator) are devoted to specific task, being the rest available as general purpose registers. The Wallace-tree multiplier included in the arithmetic block yields an execution delay of 17 ns, and thus the product is performed without extra clock cycles.

The working clock frequency is 40 MHz for a 1-micron digital CMOS technology. Estimated area overhead of the prototype is about 20 mm^2, but memory cell optimization can reduce it significantly. This would enable multiple processor integration in one chip.

In addition to MLP and PLS, the modified DRA architecture can be efficiently used to emulate the GAL and RCE algorithms. We shall consider emulation separately for coding and decision layers, the former being common to both algorithms.

3. Emulation of ROI incremental neural models

3.1. Emulation of the recall phase

Coding layer. Taking into account the operations of the ROI layers described in the previous section, the basic steps required to emulate the coding layer are shown in figure 7, namely loading, subtraction, multiplication, and accumulation.

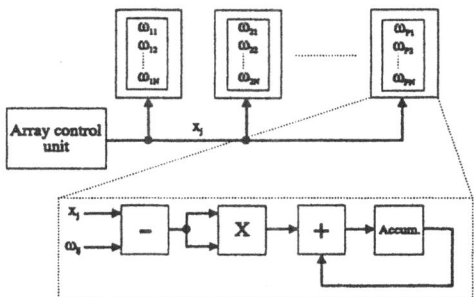

Figure 7. DRA architecture emulating a ROI coding layer

Emulation is performed in a BBA configuration with neuron parallelism. Each PU stores a sample vector and the control unit broadcasts sequentially the input vector components. For N components, the coding layer is emulated in 4N instructions (or clock cycles).

Decision layer for the GAL algorithm. Taking the previously calculated distances as inputs for this layer, the minimum distance value has to be calculated, in order to output the category associated with it. This is done by configuring the system as a SRAGB (figure 8). The PUs transfer the distance values through the local communication links as many times as processing units are in the array, so that each distance value reaches all the PUs. At each cycle, each PU compares its own stored distance with the latest received distance value. If the own value is greater than the received distance, a ">" status flag is set, and a subsequent instruction resets conditionally a register. Finally, the control unit determines the winner by accessing sequentially each PU and detecting the only count0 or count1 registers containing a value different than zero. The category is determined by the control unit using a lookup table. A rejection mechanism may also be implemented, as will be explained later.

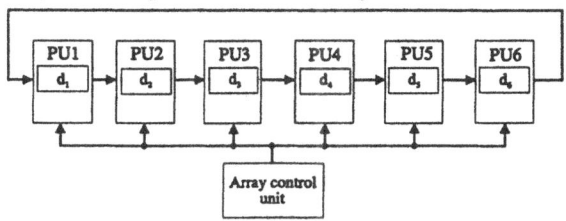

Figure 8. DRA emulating the GAL decision layer

Decision layer for the RCE algorithm. In this case, the decision layer is emulated in a much simpler way (figure 9). The distances are compared in just three cycles. In the first cycle a specific register of all the PUs is set, in the second one, the distance values are compared by each PU with the corresponding radius values, and in the third step the specific registers are conditionally reset if the distance is greater than the radius. Thus, the BBA organization is suitable for this layer emulation. After the distance comparisons, the control unit accesses sequentially the specific registers contained in each PU in order to determine the register containing a nonzero value. A lookup table indicates the category of the input vector. Of course, if more than one (or none) specific registers are nonzero, the result should be rejected or a learning process re-started.

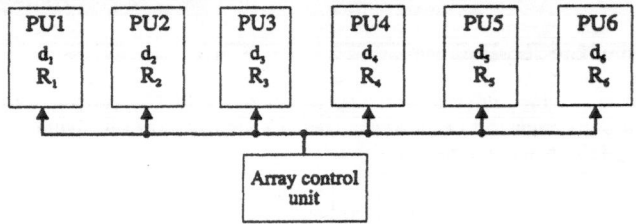

Figure 9. DRA emulating the RCE decision layer

3.2. Emulation of the training phase

Regarding the emulation of the training phase associated with the GAL algorithm, the array control unit should be in care of handling a counter and a lookup table. The counter indicates the number of units generated currently, and the lookup table consists of a memory which stores at each position the category to which coding unit whose number points to this position belongs. Therefore, upon presentation of an input vector x, three possibilities can arise:

1) The category associated with x is not represented in the network. In this case, the control unit writes in its lookup table the code representing the category, and then writes on the corresponding processing unit in the array the components of x, which shall act thereafter as its weight vector.

2) The category associated with vector x is represented in the network. In this case, the emulation of the *coding* and *decision* layers is performed in the DRA structure. If x is classified correctly, no action is taken.

3) The category associated with vector x is represented in the network, but x is classified as belonging to a different class. In this case, a new coding unit is created and initialized as in case 1).

Bearing in mind that a 128x16 register file is contained in each processing unit, and recalling that only five registers must be reserved to keep the internal data flow (one register for the final distance, two registers reserved for count0 and count1, one register for intermediate results, and the register 20, which is connected to the internal random number generator), the modified DRA architecture presented in previous sections is capable of emulating directly GAL networks for up to 123-dimension input vectors (higher dimensions are also allowed if the components over the 123rd dimension are stored in the local memory banks of each processing unit, but at the expense of longer execution cycles, since the access to the external memory bank is performed by means of two consecutive instructions). Alternatively, if the number of samples to be stored exceeds the number of processing units in the DRA structure, each processing unit should perform the function of several coding units, being the maximum number of coding units per processor limited by $124/(N+2)$, where N is the dimension of the input vectors (the two additional storage elements are required to hold the distance value and a copy of the count0 or count1 registers).

Similar data flow and control structure are required in order to implement the training phase associated with the RCE algorithm. In this case, however, when a misclassification occurs (meaning that the input vector lies in the hyperspherical region of a coding unit associated with a different category), the radius of the coding unit activated has to be reduced by a constant factor in order to avoid overlapping. As two additional registers are needed (one register for the hypersphere's radius and one register for the radius reduction factor), it can be deduced that each processing unit is capable of handling input vectors belonging to an input space of dimension up to 121 (higher dimensional input vectors can be processed if the components corresponding to dimensions over the 121st one are stored in the local memory bank). In the case the number of coding units to be emulated exceeds the number of available processing units in the DRA structure, each processing unit is able to perform the function of up to $123/(N+3)$ coding units (the three additional registers are used to store the distance value, the radius of the associated hypersphere and a copy of the count0 or count1 registers).

As can be deduced, learning capabilities for ROI incremental models can be handled by the proposed architecture at the expense of a little amount of additional complexity for the array control unit (basically, just a counter)

4. Performance evaluation

In this section we shall estimate the performances (basically, the maximum allowable processing speed) provided by the proposed architecture. For this purpose, and bearing in mind that the proposed architecture is composed of RISC processing units able to execute one instruction per clock cycle, we shall first evaluate the number of cycles required to emulate the *coding* and the *decision* layers generated by ROI incremental algorithms. Then we shall obtain the real performances of the system for two real world tasks. Since the basic operations to be carried out in order to complete the training phase depend on the particular problem to be handled, we shall provide only the indicators corresponding to the emulation of the recall phase.

The emulation of the coding layer is performed in the same way for all the ROI models, as was indicated in section 2. It consists basically in determining the Euclidean distance from the input vector to all the weight vectors stored in the coding units. Therefore, bearing in mind the basic steps indicated previously (load input vector component, subtract from weight component, multiply and accumulate), the number of cycles required to emulate the coding layer is given by:

$$N_{op} = K \cdot [(4 + 4 \cdot E) \cdot N]$$ (2)

where:

- N_{op} : Number of cycles.
- K : Number of coding units emulated by each processing unit in the DRA structure.
- E : Fraction of the K coding units whose weight vectors are stored in the local memory bank of the processing units. The 4 additional cycles required for the emulation of these coding units are due to the read process of the components (2) and to the write cycle of the resulting distance (2).

We shall consider separately the emulation of the *decision* layer for the RCE and GAL algorithms, due to the different data flows imposed by each model. Taking into account the emulation sequence explained in section 2 for the emulation of the *decision* layer of the GAL algorithm, and assuming the DRA structure is composed of P processing units, we can deduce the following basic steps:

- P cycles in order to configure the communication links between the processing units. P cycles in order to initialize the count0 or count1 registers.
- $(K-1) \cdot (2 + 4 \cdot E) + 4 \cdot (P-1)$ cycles in order to compare and transfer the distance values.
- $K \cdot (1 + 3 \cdot E) \cdot P$ cycles in order to access the count0 or count1 registers in each processing unit (stored either internally or in the local memory bank).

If the rejection mechanism (i.e., the possibility to mark an input vector as "unable to classify" if the two coding units which yield the lowest distance values are associated with different categories and the difference of these distances falls below a certain threshold) of the GAL algorithm is implemented, then $(K-1) \cdot (2 + 4 \cdot E) + 5 \cdot P + 1 + K \cdot (1 + 3 \cdot E) \cdot P$ additional cycles are required (which correspond to the repetition of the whole process indicated previously plus two compare instructions).

Regarding the emulation of the *decision* layer corresponding to the RCE algorithm, and bearing in mind the data flow explained in section 2, the following steps have to be performed:

- $K \cdot (3 + 8 \cdot E)$ cycles in order to compare the distance values obtained after the emulation of the *coding* layer with the radii associated with the coding units.

- $K \cdot (1 + 3 \cdot E) \cdot P$ cycles to access the count0 and count1 registers of each processing unit, so as to determine the coding unit whose hyperspherical region contains the input vector.

As can be deduced from the previous expressions, two main problems may arise if the proposed architecture must handle real world classification tasks by means of ROI incremental algorithms:

- If the number of coding units generated by the ROI algorithm is low, but the input dimension is larger than 123 (GAL) or 121 (RCE), some components of the weight vectors have to be stored in the local memory banks of each processing unit. It means that the emulation of the recall phase is slightly slowed down (terms affected by the E parameter in previous expressions), due to the fact that each memory access requires to consecutive cycles (instead of just one cycle for the accesses to the internal register file).

- If the number of coding units generated by the ROI algorithm is larger than the number of processing units constituting the DRA architecture, then each processing unit has to emulate several coding units during the recall phase. It means also a penalty factor (terms affected by the K parameter in the previous expressions) for the maximum processing speed attainable by the proposed realization, due to the loss of the parallelization efficiency.

However, as we shall demonstrate through real world classification examples, the high throughput processing units developed for the proposed architecture permit very high a processing speed even if there exist I/O problems due to the size of the database. The databases we have used in our performance analysis are:

- *Wine database*, from the Institute of Pharmaceutical and Food Analysis and Technologies in Salerno, Italy [9]. It consists of 178 input patterns belonging to three different wine categories. Each input pattern is represented by a 13-dimensional vector, whose components indicate the quantitative value of the respective constituents found in the chemical analysis of the wine samples.

- *Phoneme database*, provided by THOMSON-SINTRA, and used in the ROARS Esprit project (no. 5516). It is composed of 5427 5-dimensional vectors, each of them corresponding to a nasal or to an oral vowel. The components of each vector are the first five harmonics of each sample divided by the total energy.

Our simulations with the RCE and GAL algorithms have shown that a mean number of 27 coding units are generated after the training phase for the *Wine* database (150 learning vectors) is completed, while 921 coding units are generated for the *Phoneme* database (4000 training vectors). As can be easily deduced, if we assume a DRA structure composed of 128 processing units (assuming 4 processing units can be integrated on the same chip, a conservative assumption with state-of-the-art technology), in the first case (*Wine* database) each processing unit is in charge of emulating only a coding unit, while for the *Phoneme* database each processing unit has to emulate the function of 7 coding units, whose weights are stored in the internal register file. A system frequency of 40 MHz has been assumed in the calculations.

By substituting the proper values in the parameters constituting the formulas presented in previous paragraphs, it will be then possible to evaluate the performance provided by the proposed hardware realization when it is used for emulating ROI incremental models. Table 1 provides these performance indicators (presented as number of cycles, time required to process an input vector and processing speed). for the *Wine* database.

Algorithm	Number of cycles	Processing time per vector	Patterns per second
GAL (without reject)	174	4.35 μsec.	229885
GAL (with reject)	301	7.525 μsec.	132890
RCE	76	1.9 μsec.	526316

Table 1. Performance indicators for the *Wine* database

Table 2 shows the same performance indicators for the *Phoneme* database.

Algorithm	Number of cycles	Processing time per vector	Patterns per second
GAL (without reject)	1684	42.1 μsec.	23752
GAL (with reject)	3233	80.825 μsec.	12372
RCE	1057	26.425 μsec.	37843

Table 2. Performance indicators for the *Phoneme* database

As can be deduced from the previous tables, the proposed system outperforms significantly the performances provided by recent commercial developments (1500 patterns/sec. are attainable by the realization [6], assuming 5-bit input vectors). Furthermore, the system behaves well even when handling real world tasks. Finally, due to the scalable properties inherent to the proposed hardware realization, it is possible to adapt it (in terms of number of processors constituting the DRA structure) to meet the requirements imposed by the particular application to be handled.

5. Conclusions

In this paper we have presented an efficient digital realization for implementing ROI incremental neural models. After reviewing the main features associated with these neural models, we have studied in detail the data flow required for the emulation of the training and recall phases which define their behaviour.

Bearing in mind the resource organization demanded by the emulation of these models, an efficient digital hardware realization has been developed, with the aim to provide an alternative for handling real world tasks. The main features of the proposed realization are:

- It is organized as a bidimensional array of processing units, which can be structured dynamically so as to match the organization requirements imposed by the problem to be solved.

- Since the processing units are connected by means of local links and only a global bus is required, the proposed realization is scalable, so that the number of processing units will be imposed only by the throughput factor to be attained by the target system developed using the organization principles associated to this realization.

- The processing units have been developed as RISC processors, for which a quite flexible instruction set has been designed.

- Due to the flexible instruction set developed for the processing units, on-chip learning capabilities are also possible for the proposed realization, at the expense of a little more complex array control unit.

- The performance analysis carried out for real world classification problems has shown that the proposed realization outperforms the throughput offered by recent commercial products.

Our current work is concentrated in the last design stages for the first prototype system built around the principles proposed in this paper. Furthermore, we are also envisaging the possibility of using the same realization for the physical implementation of another incremental neural models.

Acknowledgements

This work has been partially funded by the Esprit III Project Elena-Nerves 2 (no. 6891) and by the Spanish CICYT action TIC92-629.

6. References

[1] F. Castillo, "Incremental Neural Networks: A Survey", Technical Report, Institut National Polytechnique de Grenoble, France, 1991.

[2] J.M. Moreno, "VLSI Architectures for Evolutive Neural Models", Ph.D. Thesis, Universitat Politècnica de Catalunya, Spain, December 1994.

[3] D.L. Reilly, L.N. Cooper, C. Elbaum, "A Neural Model for Category Learning", Biological Cybernetics 45, pps. 213-225, 1991.

[4] A.I.E. Alpaydin, "Neural Models for Incremental Supervised and Unsupervised Learning", Ph.D. Thesis, Ecole Polytechnique Federale de Lausanne, Switzerland, 1990.

[5] F. Castillo, J. Cabestany, J.M. Moreno, "Region of Influence (ROI) Networks. Model and Implementation", New Trends in Neural Computation, J. Mira, J. Cabestany, A. Prieto (eds.), pps. 96-101, Springer-Verlag, 1993.

[6] Nestor Inc., "Ni1000 Recognition Accelerator User's Guide", 1994.

[7] P. Ienne, "Architectures for Neuro-Computers: Review and Performance Evaluation", Technical Report no. 93/21, Ecole Polytechnique Federale de Lausanne, Switzerland, 1991.

[8] F. Castillo, "VLSI Architectures for Artificial Neural Networks", Ph.D. Thesis, Universitat Politècnica de Catalunya, Spain, September 1992.

[9] S. Aeberhard, D. Coomans, O. de Vel, "Comparison of Classifiers in High Dimensional Settings", Technical Report no. 92-02, Dept. of Computer Science and Dept. of Mathematics and Statistics, James Cook University of North Queensland, 1992.

Comparing Implementations of Radial Basis Function Neural Networks on three Parallel Machines [*]

N. Maria[†]   A. Guérin-Dugué[*]   J.M. Moreno[‡]   F. Blayo[§]

Abstract

In this paper we compare the implementations of Radial Basis Function (RBF) Neural Network on three parallel Neuro-Computers : the DRA machine (1D), the SMART machine (1D) and the MANTRA machine (2D). RBF networks can be used as probability density function estimators in a classification framework. The amount of calculation required for the simulation of such networks grows rapidly with the size of the learning database. Due to the highly parallel nature of RBF networks, parallel architectures are ideal candidates for such simulations. In this work we have tried to make a comparison of the three architectures based on the efficiency measure. We conclude this paper by outlining the different algorithmic constraints imposed by the particularities of each of the three architectures. We also discuss the I/O limitations for real time classification. Finally, we consider two real data-bases examples on which we compare the different machines.

1 Introduction

Function approximation is a fundamental problem in numerous applications as, for example, classification or adaptive non linear control. Therefor, kernel estimators are commonly used [4] (probability density function estimation or non linear process function approximation). It is the same purpose for Artificial Neural Networks (ANN) based on Radial Basic Functions (RBF) ([19], [17], [9],[3]).

In this article, we discuss parallel implementations of such ANN, for 1D and 2D target architectures in the classification framework. First we describe the algorithm principles and its computational complexity (section 2). We describe then the effective 1D and 2D target architectures (section 3) . Afterwards we give the different implementation alternatives (section 4 for 2D and section 5 for 1D). Finally we compare the efficiency of these different implementations on real databases (section 6).

2 Radial Basis Functions Network

2.1 Principle

Suppose we have C different classes ($[\omega_c, c = 1..C]$) where class ω_c contains N_c samples. Implementing a classifier in the Bayesian framework (equation 2), requires the knowledge of the probability density function (*pdf*) $p(\vec{x}|\omega_c)$.

$$p(\omega_c|\vec{x}) = \frac{P_c p(\vec{x}|\omega_c)}{p(\vec{x})} \qquad (1)$$

$$class(\vec{x}) = arg(max[P_c p(\vec{x}|\omega_c), c = 1..C]). \qquad (2)$$

2.1.1 Radial basis function (RBF) estimator

An estimation of the *pdf* $p(\vec{x}|\omega_c)$ can be obtained using the RBF estimator given by [4]:

$$\hat{p}(\vec{x}) = \frac{1}{N} \sum_{n=1}^{N} \frac{1}{h(n)^d} . K(\frac{d(\vec{x}, \vec{x}(n))}{h(n)}) \qquad (3)$$

where P_c is the a priori probability that class ω_c occurs.

This equation shows that the value $\hat{p}(\vec{x})$ is the sum of N contributions. Each contribution depends on $\vec{x}$ and $\vec{x}(n)$ via a radial kernel function $K(.)$ of width $h(n)$.

[*]Part of this work has been funded by the ESPRIT-BRA project number 6891, ELENA-Nerves2, supported by the Commission of the European Communities (DG XIII)

[†]Laboratoire de Traitement D'Images. et Reconnaissance de Formes. Institut National Polytechnique de Grenoble. 46, Avenue Félix-Viallet, 38031,Grenoble, FRANCE.

[‡]Departament d'Enginyeria Electrónica. Universitat Politècnica de Catalunya, 08034, Barcelona, SPAIN.

[§]LGI2P-EMA-EERIE. Parc Scientifique G. Besse, 30000, Nimes, FRANCE. Address the correspondence to : SAMOS - Université Paris 1 Panthéon-Sorbonne, 90 Rue de Tolbiac, F-75634 Paris Cedex 13.

2.1.2 A Neural Network Architecture

The classifier described by the above given equations can be seen as a neural network (figure 1.a) formed of three different layers : an RBF estimation layer, a risk evaluation layer (in case decisions are affected with different penalties) and a winner take all (WTA) layer ([9]). In fact, equation 3 can be rewritten as :

$$\hat{p}(\vec{x}) = \frac{1}{N} \sum_{n=1}^{N} \phi_{ij}(\vec{x}) \tag{4}$$

where $\phi_{ij}(\vec{x}) = \frac{1}{h(n)^d}.K(\frac{d(\vec{x},\vec{x}(n))}{h(n)})$.

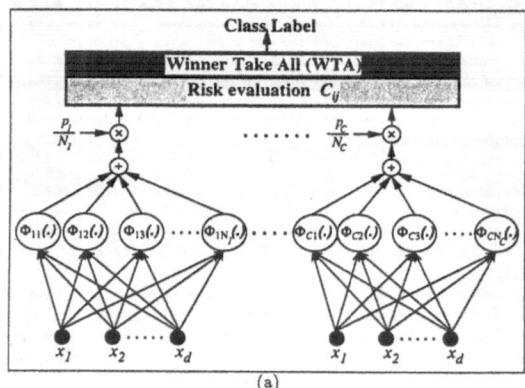
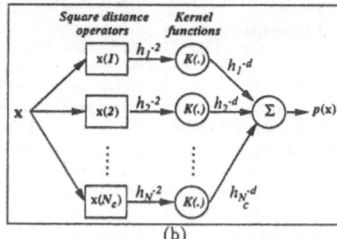

Figure 1: (a) Complete architecture of the Radial Basic Network for Bayesian Classification including prior probabilities and cost matrix multiplications . (b) Parallel form of the per class *pdf* estimator. The parameters h_n^{-2} and h_n^{-d} act as connexion weights.

The learning phase essentially produces the width factors $h(n)$ for the learning set. The initial width factors $h(n)_{t=0}$ can be obtained by various procedures. The k-NN (k-Nearest Neighbour) rule can be used to get an initial guess of the width factors (distance to the k-nearest sample inside the same class). This involves, for each class, calculating the intra-class mutual distances matrix, then sorting the rows of this matrix to obtain the smallest distance of rank k. The value k depends on the number of samples N_c of class ω_c, through an empirical relation like $k = round(N_c^{0.4})$ ([18], [3]). These factors can then be refined [3] in order to get more optimal values.

2.2 Vectorizing the test phase

The bottleneck to this algorithm is the probability estimation phase. In fact, while practical applications may contain a huge number of samples they usually have a small number of classes ($C \ll N_c$) and the decision phase needs relatively very little calculation. Fortunately, the test phase is highly parallel. As we can see in figure 1.b, for a class ω_c with N_c samples, the estimation of the probability density function, when a new input sample $\vec{x}$ is presented, involves the following vector operations (all of dimension N_c) :

- process the distance vector : $\vec{v_d}$
- update vector : $\vec{v_1} = \vec{v_d} * v_{h^{-2}}$
- apply the kernel function $\vec{v_2} = K(\vec{v_1})$
- evaluate the dot product : $s = < \vec{v_2}, v_{h^{-d}} > = \vec{v_2}^T.v_{h^{-d}}$

where $\vec{v_d} = \{d(\vec{x}, \vec{x}(i)), i = 1..N_c\}$ is the distance vector of $\vec{x}$ with the N_c samples, $v_{h^{-2}} = \{h(i)^{-2}, i = 1..N_c\}$ and $v_{h^{-d}} = \{h(i)^{-d}, i = 1..N_c\}$ are constant vectors, $\vec{v_1}$ and $\vec{v_2}$ are two intermediate vectors, and s is the probability density estimate. We can then evaluate the product $P_c.s$ before proceeding to the decision phase.

2.2.1 Implementing the kernel function

Different types of kernel functions can be used. The gaussian function is commonly used mainly when the number of samples is small. For high dimensions and in order to avoid the "empty space phenomenon" [18] a generalised form of the Gaussian kernel can be used :

$$K(r) = B.e^{-[A.r^2]^g} \tag{5}$$

where g is a parameter controlling the decrease rate of the Gaussian. The parameter g is chosen as a growing function of the dimension d. For $d = 2$, we obtain the Gaussian kernel by setting $g = 1$.

An accurate implementation of such kernels requires conditional operations which are unavailable within a strict SIMD model. One solution is to use an approximation (for example a piecewise linear approximation or a series development) of the kernel function. The accuracy of the approximation might compromise the good behaviour of the algorithm. Another solution is to use a lookup table. If such a table is to be replicated over several processors its size might become prohibitive. In order to keep the size of the table reasonable, we can take advantage of the fact that the kernel is practically non zero only in a limited and rather small interval around the origin.

A simple approximation can be obtained by using the polynomial developpement of the exponential function : $e^{-x} = \lim_{n \to \infty} (1 - \frac{x}{n})^n$. By taking $n = 2$, for example, we can write the simple approximation [14]:

$$e^{-x} = \begin{cases} (1 - \frac{x}{2})^2 & \text{if } x < 2 \\ 0 & \text{otherwise} \end{cases}$$

which yields the final result $K(r^2) \approx (1 - \frac{r^2}{2})^2$.

3 Models for 1D and 2D architectures

Systolic architectures ([10]) are cost-effective, high-performance, special-purpose solutions for exploiting massive parallelism in computationally intensive applications. Technological advances allow the adaptation of the elementary systolic elements to a variety of applications ([8], [11]) yielding General-Purpose Systolic Arrays. The three architectures presented in this paper are such examples.

3.1 2D systolic model

3.1.1 General principles

The architecture considered here is a 2D regular array of processing elements (PE). Each PE is simple, composed of arithmetic resources and registers. The communication is only local between neighbour cells. To reach optimal performances, the data flow fed in and out the array must be a regular and pipeline one.

3.1.2 MANTRA architecture

The MANTRA machine is an SIMD array processor dedicated to the implementation of neural networks. The kernel of this machine is a 2D-systolic array of simple processing elements (PE) named GENES IV. A complete description will be found in [13].

A PE is composed of a simple arithmetic and logic unit, local multiplexing data paths and registers. Two registers are connected in a "ping pong" way providing parallelism between processing and transfer. These two registers are used to store the synaptic weights, providing two memory resources. Four communication registers are available to communicate with the neighbour cells.

3.2 1D SIMD model

3.2.1 General principles

The basic 1D SIMD (Single Instruction Multiple Data) architecture is composed of a chain of PEs. This architecture is the simplest one and allows pipeline neural implementations ([7]). The instruction flow passes sequentially through the PEs. At each cycle, a new instruction can enter the processing pipeline. The associated data flow is synchronized with the instruction flow.

Several architectures can be derived according the type of the communication networks linking the PEs. A complete description is available in [6]. We consider here two kinds of communication networks. The first one uses a simple global broadcast bus linking all the PEs. This architecture is called "BBA" (Broadcast Bus Architecture). In the second one, a systolic ring links the PEs in a chain ("SRA" Systolic Ring Array). By this way, each PE is linked to its two direct neighbors. A third architecture is available combining the two previous communication links, the global bus and the systolic ring ("SRAGB" Systolic Ring Array with Global Bus). This architecture is more flexible. We will show that parallel and pipeline neural algorithm implementation are possible. The DRA architecture ([16]) and the SMART one ([12]) are two realisation examples of this architectural class. The main difference between the two realisations is that SMART uses a rather small number (≈ 16) of complex PEs while the DRA architecture uses a great number (≈ 128) of simple PEs.

3.2.2 DRA architecture

The DRA (Dynamic Ring Array) is composed of an array of processing units, each of them provided with a local memory bank [16]. These memory banks can be accessed either by the associated processing units or the control unit. Figure 2.a shows two "lines" of PEs. Each line is scalable with more integrated PEs. A global bus links these two lines and the connected PEs. Local communications are provided by systolic links between neighbour PEs on a line, and neighbour PEs between the two lines (in an orthogonal way). The PEs have been designed as RISC processors, with an arithmetic and logic unit, a register file, an instruction controller and an input/output controller.

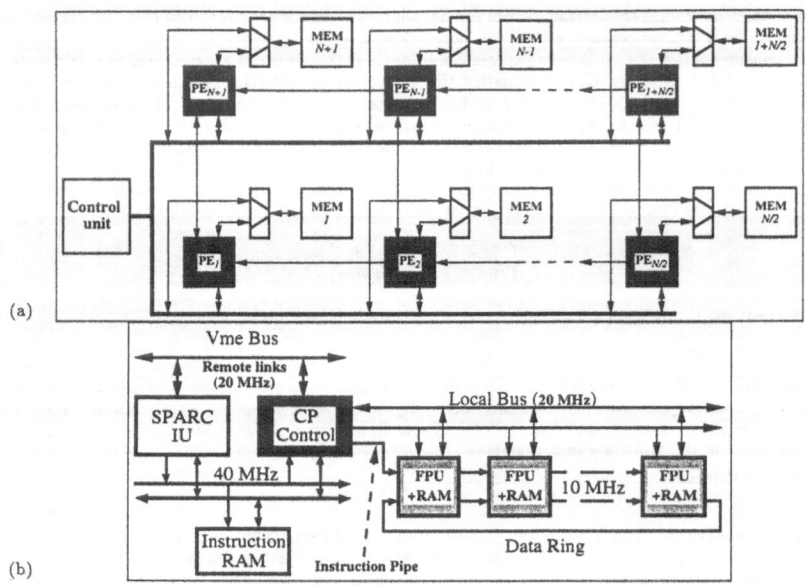

Figure 2: General overview of the DRA (a) and SMART (b) architectures.

The register file is large enough to reduce transfers with the memory bank. The presence of several communication paths makes the architecture very flexible and facilitates the implementation of several algorithms. The architecture may be viewed either as BBA, or as SRAGB or as a multiple SRAGB.

3.2.3 SMART architecture

The SMART (Sparse Matrix Adaptive Recursive Transforms) architecture has been studied for a "General Purpose Neuro Computer" ([12], [1]). Main characteristics are the generality of the approach, the flexibility of the architecture and the computational efficiency by supporting sparse matrix processing. This last characteristic will not be developed here [12]. Figure 2.b shows the chain of PEs (SMART) as a vectorial coprocessor linked to control a unit and the whole is connected to a host computer via a VME bus. The programming model is the *RISC model*. The control unit provides decoded instructions which can enter the instruction pipeline at each operating cycle. There are two data paths (local ring and global bus) providing both parallel and pipe line implementations.

4 2D systolic implementation

For this theoretical performance estimation, we have only simulated the efficiency of the most time consuming phase (distance computation to obtain individual contributions to the *pdf*). A detailed description of the implementation of the whole classification process (including the initialisation phase) can be found in [2] and [15].

Let us here recall the main characteristics of this implementation in order to compare the strong and weak points of the different 1D and 2D implementations.

1. For the initial phase, in order to use the k-NN rule, the intra-class mutual distances matrix must be first
   . calculated. An optimal implementation is achieved if the number of processors on a row p (or column since the grid is square) is taken equal to the input dimension d. This implementation takes into account the symmetry of the mutual distances matrix. As a consequence of the systolic data flow, the efficiency grows as the number of learning samples grows. The input distances are evaluated on blocks of d samples.

2. The initial widths are fixed by sorting each line of the distances matrix, by a bubble sort algorithm.

3. For the test phase, only distances between the input test sample $\vec{x}$ and each of the learning samples, are to be calculated. The learning database is split into blocks of p samples. The input vector keeps cycling in the array. Once the first block is loaded into the array, the calculation of the required distances starts. While a block is being treated in the array the next block is charged into the ping-pong registers. The general optimal case is where d is a multiple of p, like the number of samples per class.

4. As the distances come out of the array they pass through dedicated kernel function operators in order to obtain the contributions to the *pdf* estimate.

5. The remaining phases from the contributions accumulation, up to the decision by WTA, will be poorly implemented in a 2D structure. So, these tasks are devoted to a particular 1D architecture (inspired from the SMART one).

5 1D SIMD implementation

We shall consider two schemes used for mapping 2D algorithms on a 1D chain of PEs. For multilayer perceptron implementations, multiple possibilities have been studied in ([11]). We can use here these results because the data paths for matrix by vector multiplication (main operations in MLP) and distance computation are equivalent.

5.1 Partitioning by the individuals

This classical approach consists in distributing the different samples on the SIMD array : one or more samples per PE. Zeros, if needed, are added to balance the calculation charge and thus keep all the processors synchronous. Each processor calculates the distances between the samples it holds and the input vector which is diffused to all the PEs via the global bus. Once calculated, the resulting distance vector is thus distributed over all the processors. Partitioning the vector $\bar{h}$ and all the other vectors of section 2.2 in the same way permits all of the remaining steps of the algorithm to be carried out in parallel. The kernel function must also be replicated on each PE. In this way, each PE calculates a partial sum of the *pdf* estimate. In general, a gather operation is provided in order to sum up all the partial sums. Both the DRA and the SMART architectures realize this operation through the ring pipeline. One should however note that, if mutual distances are to be calculated (initial phase) this partitioning scheme will lead to poor performance since each PE must diffuse the patterns it holds to all the other PEs.

5.2 Partitioning by the dimension

The second approach consists in partitioning each pattern among the p PEs. In this scheme, each PE holds one or several components of of a given pattern. The input test vector is distributed in a similar manner. Each PE then calculates a partial sum of the distance and the final result is collected through the pipeline. Once calculated this distance vector can be distributed over all the processors in order to perform the remaining steps in the same way as before. The kernel operator can be duplicated over all the processors.

This partitioning scheme avoids the copy of the input vector over all the processors. A major advantage of this scheme is that it permits the calculation of the mutual distances without any communication cost. Furthermore, it is possible to take advantage of the fact that the mutual distance matrix is symmetric and consequently divide the amount of the required computation by two.

6 Available machines and environment constraints

6.1 Available machines : Brief technical descriptions

6.1.1 MANTRA

The MANTRA I system includes four different boards : a *processor board*, a *Control board* based on a TMS 320C40, an *input/output board* and a *GENES IV array board*. The chip GENES IV integrates a 2D systolic grid composed of 4×4 PEs. The last board includes 100 GENES IV chips (CMOS $1\mu m$, is $6.3 \times 6.1 mm^2$ die size), and can reach a size configuration up to 1600 PEs (32×32).

The complexity of the processing performed by the GENES operators, requires the power of a DSP to generate the control sequence of input data.

6.1.2 DRA

As has been indicated in previous sections, the DRA architecture is organized as a unidimensional array of processing units, being able of emulating the data flow associated with both SRAGB and BBA architectures. The processing units constituting the DRA architecture have been developed [16] as specific RISC processors, whose internal organization is depicted in figure 3.a. The ALU building block consists of a parallel 16-bit multiplier, a 16-bit adder and a comparator. The communication ports (2 unidirectional and 1 bidirectional port) required in order to permit the resource organization demanded by the DRA architecture are included in the External IO building block. The register file depicted in figure 3.a contains 128 16-bit registers, from which only three are dedicated to specific tasks. The external memory bus integrated in each processing unit is capable of handling a local memory with a capacity of 64K 16-bit words. The RISC processor constituting each PE has been designed so as to permit a maximum clock frequency of 40 MHz, being capable to execute one instruction per cycle, which finally yields a total performance of 40 MIPS (Millions of Instructions per Second). Being the DRA architecture modular and scalable, the current technology limitations allow for a physical realization composed of 128 processing units, and this is the architecture size we shall consider hereafter in our performance estimations.

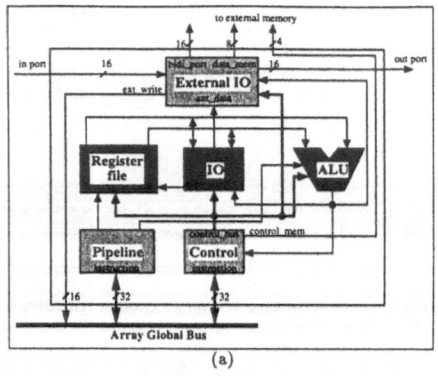 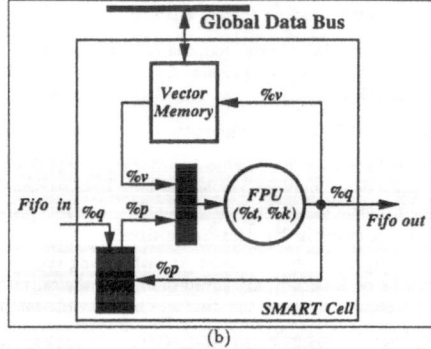

Figure 3: (a) Architecture of the DRA PEs, (b) Architecture of SMART's cells.

6.1.3 SMART

All the arithmetic cells composing the SMART coprocessor are identical. Each cell is based on a fast local memory which serves a general purpose Floating-Point-Unit 3.b. Each cell has also a FIFO memory, and a local indirection table for handling sparse matrices.

The instruction set of the FPU has more than 60 arithmetic and logical operations, and uses 32-bit data (integer or floating point format). Each FPU can deliver a peak of 20 MFLOPS.

FIFO memories are viewed as vectorial registers (like all the resources). They are shared between neighbouring cells. For each cell, the %p registers (4 registers) refer to the local FIFO buffers and the %q registers refer to the corresponding FIFO buffers of the following cell. Hence, the cells work *serially* if %q is used as a destination argument, *independently* otherwise. The FIFO model both masks pipeline delays and supports sequences of basic instructions on vector components. The vectorial registers (8 registers) %v are in fact local pointers to the memory cases on each cell.

By a suitable architecture, SMART can handle efficient processing on sparse matrices ([12]). In [15], we have shown the link of this capability and the dynamical learning in neural networks, allowing neurons growing or pruning. This point will not be discuss here.

6.2 Algorithmic constraints

6.2.1 MANTRA implementation

From an algorithmic point of view, the pipelined nature of the MANTRA machine imposes to study the epoch updating of the learning algorithm. Here, the whole learning database is split into blocks of p samples. Typically, to sustain an utilization rate around 100%, the number of examples must be greater than the physical array size. This makes the MANTRA architecture well tailored for sufficiently large problems, as it is shown in the figure 6.b., for a huge database ("Satimage").

6.2.2 DRA implementation

The DRA architecture has been designed in order to provide very integrated PEs. So this machine will be available with a great number of PEs. This justifies the choice of the partition by individuals for the learning database. So, the implementation of the distances computation is very classical. For the implementation of the accumulation phase of the contributions to the *pdf*, two alternatives are possible. We have called these possibilities, "clusters of classes" and "sequential classes" method.

The first one takes advantage of the array flexibility as a multiple systolic ring array. Figure 4 shows the principle of this strategy. Here the number of samples allocated per PE, is the same for all the classes, that is to say, the number of PEs per class follow the prior probabilities. So for a given number of PEs and a given total number of samples, this strategy will be more or less efficient according to the equilibrium of the prior probabilities. The optimal efficiency is obtained with equal prior probabilities.

With the "sequential classes" method, the accumulation of the *pdf* estimate in each class is processed serially in the pipeline ring. If one sample is allocated per PE, the efficiency is minimum and equals $= 1/p$ ($Efficiency = \frac{T_{seq}}{p.T_{par}}$). The efficiency increases with the parallelism degree, that is, as the number of samples per PE increases. So with few PEs, the efficiency of this strategy is comparable with the efficiency of the first one implemented with more PEs.

In the first strategy, the degree of parallelism is equal to the number of classes times the number of samples per class. In the second strategy, the degree of parallelism is the number of samples per class. So, even if the prior

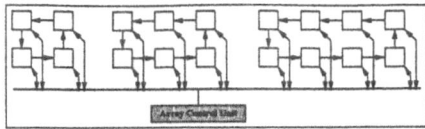

Figure 4: DRA organisation as a multiple systolic ring array, one ring per class for the accumulation of the contributions to the probability density function.

probabilities are not quite equal, for the same number of PEs higher efficiency will be always achieved by the "clusters of classes" strategy.

Let us consider a simple example, with 128 samples distributed in three classes (80, 20, 28) and two hardware configurations (128 PEs, 32 PEs). The efficiency of the "clusters of classes" method on 128 PEs is almost four times smaller than the efficiency of the second method implemented on 32 PEs, and greater than the efficiency of this second method implemented on 128 PEs. The calculated efficiency only concerns this pipe line phases, that is to say here, the pipe line latency. Table 1 shows this comparison. The loss of efficiency due to this phase (serial phase) might have a dramatic influence on the overwhole performance (Amdahl's law).

		PEs / Class			Samples / PE			Nb cycles	Efficacity
		ω_1	ω_2	ω_3	ω_1	ω_2	ω_3		
128 PEs	Clusters of classes	80	20	28	1	1	1	80	$\frac{1}{80}$
	Sequential classes	80	20	28	1	1	1	128	$\frac{1}{128}$
32 PEs	Clusters of classes	20	5	7	4	4	4	24	$\frac{4}{24} = \frac{1}{6}$
	Sequential classes	32	20	28	3	1	1	83	$\frac{4}{83} \approx \frac{1}{20}$

Table 1: . Comparing the efficiency of the contributions accumulation phase for the two strategies with 32 and 128 PEs.

6.2.3 SMART

The partitioning by the individuals on SMART has the same proprieties as those described above in the case called "sequential classes". However three main differences must be noted : (i) data access time is always constant on SMART, (ii) Memory can attain 1M 32-bit words per cell, and (iii) the number of PEs in SMART is at maximum 32.

In the case where $d \gg N_c$, the partitioning by the dimensions scheme has two main advantages over the previous one :

1. it simplifies the problem of the kernel function implementation. In fact, a dedicated operator can be inserted at the end of the array. In this way, as each component of the distance vector comes out of the last PE it passes through the kernel operator before being fed back into the FIFO register of the first cell. Moreover, the operator can also hold the width factor vectors and thus perform the accumulation of partial contributions. Since all the SMART instructions take one cycle (RISC model), the operator must finish its computation in at most $p - 1$ cycles. Otherwise, the instruction pipeline must be frozen until the computation is done.

2. if d is a multiple of p the charge will be always balanced no matter the repartition of the samples among the different classes.

In order to minimize the pipeline latency on SMART, the components of the distance vector are serially stacked (as they come out of the last cell) onto the FIFO register of the first cell. In this way, a single cycle is lost per distance component calculation, that is a total loss of N_c cycles for class ω_c.

6.3 Hardware constraints

6.3.1 MANTRA

For the MANTRA machine, the memory resources inside the systolic array are not a limitation in the case of large databases. Two registers per PE are needed, in order to have both processing and transfer. So the constraint is on the external memory size connected to the DSP controller.

From the input/output point of view, MANTRA machine is well adapted if the input test samples are presented to the network by burst. By this way, the lost time in pipeline latency becomes negligible comparing to the computation time. Figures 6 illustrate this fact, the efficiency grows with the length of the burst.

6.3.2   DRA and SMART

The main hardware constraints for the DRA or SMART architectures when emulating RBF networks are imposed by the size of the problem to be handled (i.e., number of samples to be stored by the network and number of attributes characterizing each sample). In both machines, it is due to the limited capacity of the memory linked at each PE. For example, the DRA machine will be available with 128 PEs and so 128×64 Kwords (16 bits). This memory limitation is only linked to the current realization, it can be removed by extending the memory capacity or by increasing the number of PEs.

For the DRA machine, the fixed number of the internal registers is not viewed as a limitation for the implementations. If all the samples cannot be stored in registers, the extra samples will be stored in local memories. The database memorisation remains distributed not only on registers but also on the local memories. Consequently, the processing speed is only penalized due to the slower local memory bank access.

7   Synthesis - Relation with databases

7.1   Data bases examples

We consider here two examples of real databases, called "Iris' ([4]) and "Satimage" ([5]). The first one is a very popular database in pattern recognition. The data set contains 3 classes of 50 instances each, where each class refers to a type of the iris plant. The attributes associated to each sample, represent the size of sepals and petals, in 4 dimensions. So for architectural comparison, this database is a little one with equal priors.

The second one is a huge database (6435 samples) in 36 dimensions. This database was generated from Landsat Multi-Spectral Scanner. Each sample represent a sub-area of the scene viewed from four spectral bands. A decision have been done in 6 classes for 6 types of areas. The repartition of the samples among the 6 classes is as follows : 1533, 703, 1358, 626, 707, 1508.

7.2   Comparison between DRA and SMART

First let us compare a DRA machine of 128 PEs with a SMART machine of 16 PEs. For each of the above mentioned databases, we study the behaviour of the efficiency by varying d. This will give an idea on how much robust the implementation is, in function of the input space dimension.

On DRA, and for the Iris database, we suppose that the samples have been loaded in the registers of the different PEs according to the "clusters of classes method". For the Satimage database we suppose, since the registers are not able to hold all the samples, that all the samples are stored in external memory (there is an overcost of two cycles per memory access).

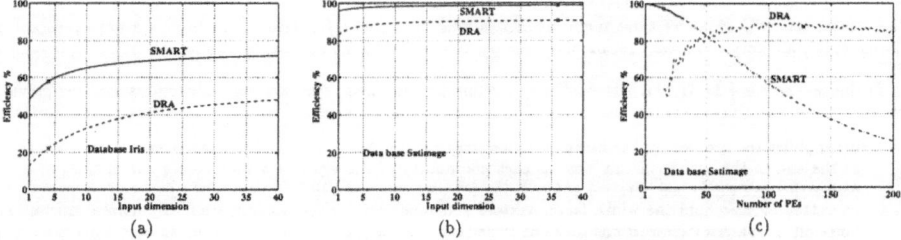

(a)   (b)   (c)

Figure 5: Comparing the efficiency of the DRA and SMART architectures on two real databases. (a) Iris base with d as parameter, (b) Satimage base with d as parameter, (c) Satimage base with p as parameter.

Figure 5 (a) compares the behaviour of the DRA-128 and SMART-16 on the Iris as a function of d. The efficiency for $d = 4$ corresponding to the Iris database is marked with a dot for both machines. When the total number of samples is small the DRA performs poorly since only a part of the PEs participate in the effective calculation. For example, for the iris data base, the efficiency of the DRA is about 20% because only 75 (2 samples per PE, 25 PEs per class) out of the 128 PEs participate in the effective calculation. As for SMART the efficiency is much better since all the PEs participate in the calculation. The loss of efficiency is only due to the serial pipeline phase needed to sum up the contributions from the PEs.

The SMART architecture behaves better for small d, the asymptotic value is reached faster than on the DRA. This is due to the fact that the degree of parallelism is higher on SMART since PEs on SMART contain more samples than those of DRA. When the number of samples is huge the two architectures behave in a similar manner. In such cases the load is much better balanced over the PEs and the efficiency is very good. For the Satimage data base SMART slightly outperforms DRA because the pipeline latency of SMART is smaller than that of DRA (which is equal to the latency of the biggest cluster).

However as the number of PEs increases, the DRA architecture is much more scalable than the SMART one as shows figure 5.c. This is explained by the "clusters of classes" method used on the DRA which keeps the latency of the pipeline phase rather small where as it increases in function of p on SMART. We must however keep in mind that on the shelf technology allows a maximum number of PEs (very complex PEs) of about 32 for SMART value for which the efficiency is still about 90%. Any way the DRA architecture by allowing a bigger number of simpler PEs permits to attain processing speed that SMART can't reach. A VLSI implementation of the DRA architecture will allow even for more processors. The scalability of the DRA architecture is then an essential characteristic which allows to reach very high processing speed.

7.3 Performance of the MANTRA machine

For this theoretical performance estimation, we have only simulated the efficiency on the key phase, the most time consuming one (distance computation to process individual contributions to the *pdf*). Detailed explanations on the formula can be found in [2]. The MANTRA machine uses here 32×32 PEs. So for the two databases, we consider that the input dimension d is equal to 32. For the Iris database, 28 zeros are filled per learning samples. With the Satimage database, we have taken the 32 first dimensions (issued from a Principal Component Analysis for example).

We can see on figures 6.a and .b :

1. For the Iris database, the efficiency is really poor, this is mainly due to the weak utilisation rate of the PEs ($\frac{4}{32}$).

2. For the Satimage database, good efficiency is achieved if the test bloc size is greater than 40. The asymptotic efficiency is not reached, because a correct partition of the learning database into blocks of 32 samples requires the addition of extra null learning vectors.

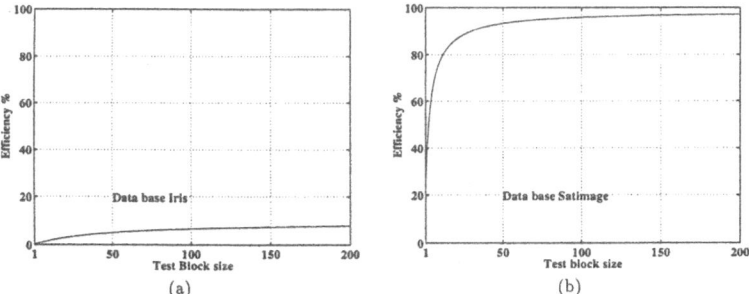

(a)   (b)

Figure 6: Efficiency of the test phase on MANTRA 32x32 versus the number of test samples. (a) Iris with $d = 4$, (b) Satimage with $d = 32$,

8 Conclusion

In this paper we have compared the three presented architectures for the implementation of RBF like networks. Both the DRA and SMART machines present quite similar behaviours. SMART and DRA machines follow two different strategies : few complex PEs for SMART and much more simpler PEs for DRA. As we observe the theoretical efficiency estimation by varying the number of PEs, a good efficiency is held on DRA, contrary to SMART. The two machines are well designed, each in a different objective : the maximum number of PEs in SMART must be small compared to the number of PEs in DRA.

Implementation on MANTRA is particular due to the systolic processing into the 2D square array, and an extra constraint is imposed by the fact that the grid must be square. The matching between the hardware size and the input dimension is very critical, as it is shown by the simulation on the Iris database. The efficiency is maintained as the number of samples gets larger the hardware size.

For RBF networks, we have noticed two main phases, a parallel one (distance computations) and a pipeline one (contributions accumulation). For the three machines, the efficiency of the parallel phase implementation is good though slightly modulated by the proportionality between the number of PEs and the number of samples. For MANTRA, this efficiency increases with the size of the test database. The efficiency of the pipeline phase is dependent on the number of PEs. As the number of PEs increases, the time devoted to this phase increases and so the efficiency decreases. This tendency is well limited in the DRA machine by the choice of the "clusters of classes" strategy.

In conclusion, we see that a cost-effective choice of a hardware solution consists in finding a compromise between the power speed and the scalability needed. This work will serve as a basis in establishing guidelines on how to choose the best hardware solution to a given classification problem.

780

References

[1] P. Bessiere, A. Chams, and T. Muntean. A Virtual Machine Model for Artificial Neural Network Programming. In *Int. Neural Network Conf.*, Paris, July 1990.

[2] F. Blayo, A. Guérin-Dugué, and N. Maria. Implementing Radial Basis Function Neural Networks on the Systolic MANTRA machine. *Submitted to IWANN 95*, June 1995.

[3] P. Comon, J.L. Voz, and M. Verleysen. Estimation of Performance Bounds in Supervised Classification. In Michel Verleysen, editor, *ESANN : European Symposium on Artificial Neural Networks*, pages 37–42, Brussels, Belgium, April 1994.

[4] R Duda and P. Hart. *Pattern Classification and Scene Analysis*. John Willey & Sons, 1973.

[5] C. Feng, A. Sutherland, and S. King. Comparison of machine learning classifiers to statistics and neural networks. In *AI and Statistics Conference*, 1993.

[6] P. Ienne. Architectures for Neuro-Computers : Review and Performance. Technical Report 93/21, LAMI-EPFL, Lauzanne, Switzerland, 1993.

[7] A. Johannet, L. Personnaz, G. Dreyfus, J.D. Gascuel, and M. Weinfeld. Specification and implementation of a digital Hopfield-type associative memory with on-chip training. *IEEE Transactions on Neural Networks*, 3(4):529–539, July 1992.

[8] K. T. Johnson and A.R. Hurson. General-Purpose Systolic Arrays. *IEEE Comp.*, pages 20–31, November 1993.

[9] C. Jutten and P. Comon. Neural Bayesian Classifier. In *New Trends in Neural Computation*, number 686 in Lecture Notes in Computer Science, pages 119–124. Springer-Verlag, 1993.

[10] H. T. Kung. Why systolic architectures. *IEEE Comp.*, 15(1):37–46, January 1982.

[11] S. Y. Kung and J. N. Hwang. Parallel Architectures for Artificial Neural Nets. In *IEEE Int. Conf. On Neural Networks*, volume 2, pages 165–172, San Diago, July 1988.

[12] J.C. Lawson, N. Maria, and Hérault J. SMART : A Neuro-computer Using Sparse Matrices. In *Euro Micro PDP Workshop*, Canari (Spain), January 1993.

[13] C. Lehmann, M. Viredaz, and F. Blayo. A Generic Systolic Array Building Block for Neural Networks with On-Chip Learning. *IEEE Transactions on Neural Networks*, 4(3):400–407, May 1993.

[14] P. Maffezzoni and P. Gubian. VLSI Design of Radial Functions Hardware Generator for Neural Computations. In *Fourth International Conference on Microelectronics for Neural Networks and Fuzzy Systems*, pages 252–259. IEEE Computer Society Press, September 1994.

[15] N. Maria, A. Guérin-Dugué, and Blayo N. 1D and 2D systolic implementations for Radial Basis Functions Networks. In *Fourth International Conference on Microelectronics for Neural Networks and Fuzzy Systems*, pages 34–45, Turin, Italy, September 1994. IEEE Computer Society Press.

[16] Moreno. *VLSI Architectures for Evolutive Neural Models*. PhD thesis, Universitat Politècnica de Catalunya, Spain, December 1994.

[17] R.M. Sanner and J.J.E. Stoline. Gaussian Networks for Direct Adaptive Control. *IEEE Transactions on Neural Networks*, 3(6):837–863, November 1992.

[18] B.W. Silverman. *Density Estimation for Statistics and Data Analysis*. Chapman and Hall, 1986.

[19] D. Specht. Probabilistic Neural Networks. *Neural Networks*, 3(1):109–118, 1989.

Implementing Radial Basis Functions Neural Networks on the Systolic MANTRA Machine*

F. Blayo[†]   A. Guérin-Dugué[‡]   N. Maria[‡]

Abstract

The development of neural network models requires the study of dedicated hardware architectures. In this paper, we propose an implementation of Radial Basis Function networks, derive an architecture based on an already existing 2D-systolic machine (MANTRA). A systolic algorithm is described to implement the required functions and the suitable sequence of operations. Theoretical efficiencies are estimated on the key tasks and some guidelines are given for a best usage of the Mantra machine in the studied framework.

1 Introduction

Numerous 2D implementations of neural networks, has been already proposed in [5], [4], as the multi layered perceptron, the self organizing features map, ... Radial Basis Functions based networks present an alternative tool for classification by the estimation of the density probability functions ([3],[2]).

We describe an implementation of such a network, on a 2D systolic grid. The architectural model used here, is the MANTRA architecture ([8]). After a brief description of the RBF algorithm (section 2), and the MANTRA architecture (section 3), we detail (section 4) the implementation for each algorithmic phase, and suggest dedicated architecture if needed. Finally, theoretical performance evaluation are done, by the mean of the efficiency (section 5), in order to evaluate the matching between the algorithmic problem size and the hardware size.

2 Radial Basis Functions Network

The general organization of the RBF based network is depicted in figure 1. In this case, we consider a problem with C classes. Class ω_c has N_c samples. An input vector z is presented to the network. For each class ω_c, the conditional probability density function $\hat{p}(z|\omega_c)$ is estimated by accumulation of the contributions of the N_c samples :

$$\hat{p}(z|\omega_c) = \frac{1}{N_c} \sum_{n=1}^{N_c} \frac{1}{h(n)^d} . K(\frac{d(z, z(n))}{h(n)}) \tag{1}$$

The function $K(.)$ is the kernel function whose the argument is the distance between the learning sample $z(n)$ and the input vector divided by the width factor $h(n)$. The discriminant function is obtained by the multiplication by the a priori probability P_c. To implement a classifier in the Bayesian framework, such a processing is duplicated for each class. A Winner Take All function follows these structures in order to make class decision (equation 3) :

$$p(\omega_c|\vec{x}) = \frac{P_c p(\vec{x}|\omega_c)}{p(\vec{x})} \tag{2}$$

$$class(\vec{x}) = arg(max[P_c p(\vec{x}|\omega_c), c = 1..C]) \tag{3}$$

For this implementation, we will discuss on :

1. The initialisation phase, which consist of the determination of the initial width factors. An empirical procedure has been proposed to determinate this factor by the k nearest distance ([2]). The parameter k depends on the number of samples. For this phase, all the mutual distances between the learning samples of each class will be computed (section 4.1).

2. The initial width factors $h(n)$ are initialized by a bubble sort algorithm (section 4.2).

3. The test phases which consists of the distances processing between the learning samples and the input vector (section 4.3.1).

4. The last phase (class decision) is not time consuming, the parallelism degree is low (number of classes). This one will be briefly describe in section 4.3.2.

---

*Part of this work has been funded by the ESPRIT-BRA project number 6891, ELENA-Nerves2, supported by the Commission of the European Communities (DG XIII)

[†]Laboratoire pour les Etudes et la Recherche en Informatique. Parc Scientifique G. Besses, 30000, Nîmes, FRANCE. Address the correspondence to : SAMOS - Université Paris 1 Panthéon-Sorbonne, 90 Rue de Tolbiac, F-75634 Paris Cedex 13.

[‡]Laboratoire de Traitement D'Images et Reconnaissance de Formes. Institut National Polytechnique de Grenoble. 46, Avenue Félix-Viallet, 38031,Grenoble, FRANCE.

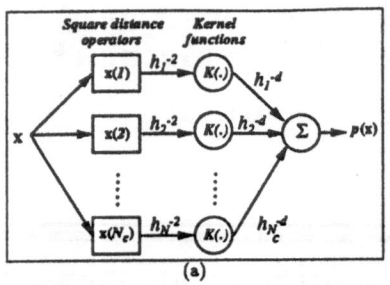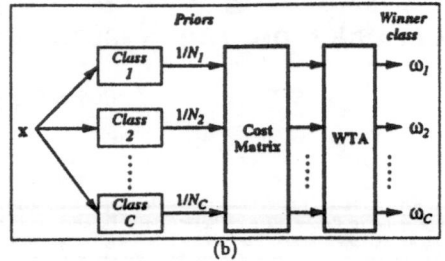

Figure 1: (a) The basic per class density estimator. The parameters h_n^{-2} and h_n^{-d} act as connexion weights. (b) Parallel implementation of a complete Bayesian classifier. The parameters P_c are connexion weights (prior probabilities). Dark blocks are the per class probability estimators as described in a.

3 MANTRA as a 2D systolic model

The MANTRA machine is a fine-grain systolic SIMD processor array dedicated to the implementation of neural networks. The kernel of this machine is a 2D-systolic array of simple processing elements (PEs) named GENES IV. A first study has shown [1] that four basic scalar and matrix operations are sufficient to implement the most widely-used neural networks on the 2D-systolic array. Figure 6.a shows the data phaths across the PEs. We can notice the suitable short-circuits on all the borders for data cycling, and the I/O via diagonal cells. This analysis has been extended by [5], resulting in nine operations. An hardware machine has been completely developed by [8] who discussed the performance measures and applied them to this machine.

At the present time, the MANTRA machine is dedicated to a finite set of models, including Perceptron, Adaline, Delta Rule, Back Propagation rule, Kohonen self-organizing maps. The MANTRA I system includes four different boards : a *Processor board*, a *Control board* based on a TMS 320C40, an *input/output board* and a *GENES IV array board*. This last board includes 100 GENES IV chips, and can reach a configuration size up to 1600 PEs (32 × 32).

From an algorithmic point of view, the pipeline nature of the MANTRA machine imposes to study the epoch updating of the learning algorithm, so as the partitioning of the learning databases when if it greater than the dimension of the array. Typically, to sustain an utilization rate around 100%, it requires to have a number of examples equal to the double of the physical cells. This makes a GENES-based computer well tailored for sufficiently large problems. Then, we propose to use the MANTRA machine to implement a neural network based on the Bayesian theory for classification tasks.

4 2D systolic implementation of Radial Basis Functions network

In this section, we propose a systolic implementation of the RBF algorithm on a square lattice of processing elements (PE). The implementation constraints are analyzed, and a discussion on the pipelining possibility is developed.

4.1 Mutual distances computation

This part is dedicated to the presentation of a simple procedure to compute the mutual distances between points belonging to the same class. This procedure will be executed on a 2D-systolic architecture. We consider a 2-D square array of processing elements (PE). The dimension of this array is supposed to be exactly equal to $(d \times d)$. The value d is the dimension of the points in the database. For sake of simplicity, we consider that the number of vectors for which we have to compute the mutual distances is equal to the space dimension d. This restrictive condition will be discussed in the future explanation of the algorithm.

4.1.1 Principle of computation

For each vector $x(i) = (x_1(i), x_2(i), ...x_d(i))$, belonging to the class ω_c, we need to compute all the distances :

$$d(x(i), x(j)), j = 1..N_c, i \neq j \tag{4}$$

where N_c is the number of points belonging to the class ω_c.

The idea of the computation is to load the d first $x(i)$ points in the array and to compute immediately in a systolic way the distances with all the other vectors of the same class. In a first part, we will describe the data flow between the elements, and in the second part we will bring up the architecture of each processing element.

4.1.2 Data flow

At the beginning of the computation, the 2D-square array is empty. We load the first component of the $x(1)$ vector in the first PE of the array, as indicated in the figure 2.a. The input vectors are transmitted through the processing

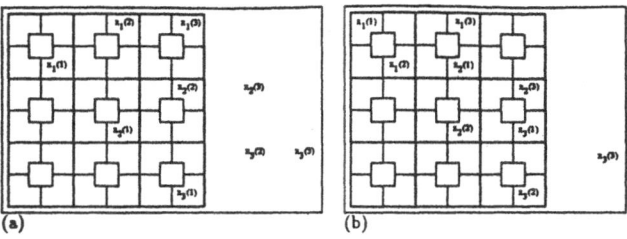

Figure 2: (a) Data loading, (b) Data transmission.

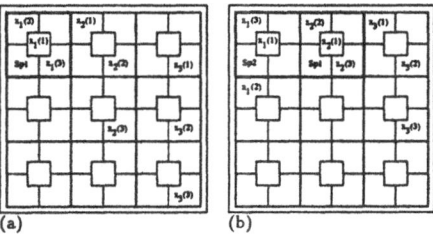

Figure 3: (a) Memory point loading, (b) Local computation and accumulation.

elements, without performing any computation. When a diagonal cell is reached, the vector component in transmitted to the upper cell, where it moves until a north cell is reached. This data flow is shown in the figure 2.b.

When a value is entered in a cell, coming from the north, and if the local memory point is empty (contains a null value), then this value is stored in the local memory point, as shown in figure 3.a. This value is never transmitted to the lower cell. Such a procedure has the advantage of avoiding the computation of the distance between the coordinates of a point with itself. Furthermore, it allows an easy loading of the input vector coordinates, without the requirement of a dedicated data path.

All the following steps are based on this behavior. During the next step, shown in the figure 3.b, two identical computations are performed in two different PEs (PEs performing a significant computation are surrounded in boldface). As usual, the result of the local computation is transmitted to the nearest East neighbor, to take part in the next computation.

After this step, as indicated in the figure 4.a, three PEs are active, for three different computations. Now, for the first value reaching the East border of the array, the accumulation of the values is completely achieved.

If we consider that each PE is able to compute locally a quadratic difference between the local memory point value and the value coming from the North border of the cell, then the first $Sp1$ value will be :

$$Sp1 = \sum_{i=1}^{3}(x_i(1) - x_i(2))^2 \qquad (5)$$

The effective computation of the Euclidean distance between $x(1)$ and $x(2)$ requires the application of the square root function to this value. However, this function can be applied at the East border of the cell using a dedicated operator. But for the RBF networks, we do not need this.

For the second value $Sp2$, the process is identical, and the computed value is equal to the Euclidean distance between $x(1)$ and $x(3)$. This step is shown of the figure 5.a.

On figure 5.b, the last value corresponding to the Euclidean distance between $x(2)$ and $x(3)$ is computed.

To end up the loading of the values in local memory, we have to transfer the last value in the last memory point. This step should not be useful in such a simple case, but we will see, in the following, that it is necessary in order to chain the computations with other input vectors.

4.1.3 Architecture of the processing elements

As mentioned before, each processing element must be able to compute a local quadratic difference between the values of two components of vectors. A local accumulation of the result must be done, and the correct transfer between the cells, depending on their location in the array, must be provided. We propose a simple architecture, well adapted for the computations and the data flow required. It is depicted on figure 6.b.

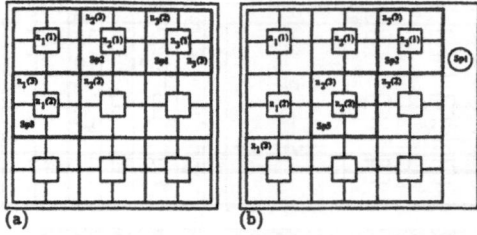

Figure 4: (a) First complete accumulation, (b) Transformation of the accumulation through the dedicated operator.

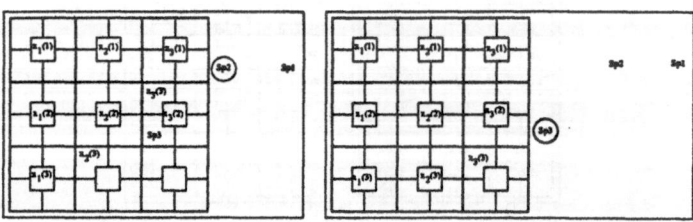

Figure 5: (a) Distance between $x(1)$ and $x(3)$, (b) Distance between $x(2)$ and $x(3)$.

This systolic PE is quite similar to the GENES IV cell presented in previous works [5]. The main difference comes from the additional data path provided between the D and W registers. It is used during the initialization phase to move the components of the $x(k)$ vector into the local W register. This is simply implemented by a demultiplexer, which redirects the output of the D register between the input of W or the SouthOut data path. The operation performed in the cell is the difference between the local D register and the local W register, accumulated and squared, lastly transmitter to the neighboring cell. At the present time, this operation is implemented on the GENES chip. The only modification comes from the data path required. Nevertheless, it must be mentioned that the direct loading of the W values is possible on the GENES board. It can be used to compute the same systolic algorithm on GENES, but some useless computations will be done, decreasing the maximum performance that can be reached by the hardware. Anyway, it can be a good solution for a first prototype because it does not require a complex and costly development of a new chip.

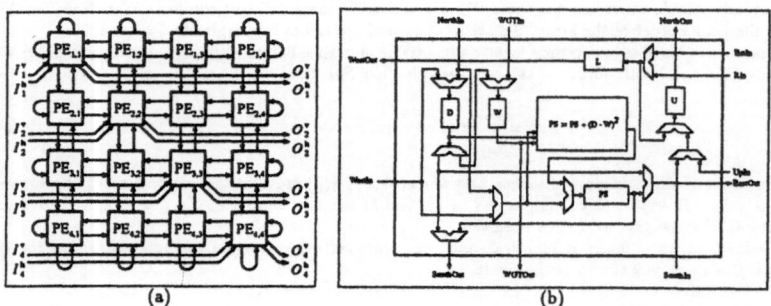

Figure 6: (a) Systolic data path across the MANTRA 2D systolic grid (from Lehmann, 93). (b) Basic architecture of one diagonal PE. The non-diagonal PE have neither the input/output signals Uin, Rin, Uout and Rout, nor the corresponding multiplexers. (From Viredaz).

The efficiency of this implementation is poor ($\approx 16\%$) because the cell activity is low. We show, in the next section, that pipelining the computation is possible, and we analyze the possible database partitioning.

4.1.4 Pipelining the computation

The previous explanations are valid for a problem where the number of vectors in the class ω_c is exactly equal to the dimension of the vector space d. This is a too much restrictive condition, and we show in the following that it can be overcome by a suitable arrangement of the vectors transformed by the 2D-square systolic array.

Now, we consider that the number of vectors associated to class ω_c is much larger than d. We have to compute all the mutual distances between vectors.

$$d(x(i), x(j)), j = 1..N_c, i \neq j \tag{6}$$

This problem can be split in smaller subproblems. First of all, it is not necessary to compute the distances $d(x(i), x(j))$ and $d(x(j), x(i))$ twice. Then, for each $x(i)$, it is sufficient to evaluate :

$$d(x(i), x(j)), j = i + 1..N_c, i \neq j \tag{7}$$

With a systolic array containing exactly d vectors, the first evaluation will concern all the N_c vectors of the class ω_c. Thus, after a first pass of all the vectors, the following distances will be computed :

$$d(x(i), x(j)), i = 2..N_c, j = 1..d \tag{8}$$

The second pass requires the loading of the vectors $x(d+1)$ to $x(2d)$ into the memory points of the array. After, a second evaluation pass can be done. It concerns all the vectors such as :

$$d(x(i), x(j)), i = (d+1)..2d, j = (d+2)..N_c \tag{9}$$

The entire class ω_c will be processed after a splitting in d subsets, and the remaining vectors processed individually. The worst case will be for :

$$Rest[\frac{N_c}{d}] = d - 1 \tag{10}$$

In this case, the last subset is filled with $d - 1$ null learning vectors. A first set of N_c vectors is presented to the systolic array, followed by a set of $N_c - (d + 1)$ vectors, and so on. During the computation, a flow of distances is delivered by the systolic array on its East border. The second operation required by the learning the algorithm is the computation of the initial width factors. In the following, we show that this operation can be also implemented on a 2D systolic array.

4.2 Determination of the initial width factor $h(n)_{t=0}$

The initial width factor $h(n)_{t=0}$ can be fixed by various procedures, but the simplest one consists in using the previously computed distances, in order to find, for each $x(u)$, the k-nearest point.

The same systolic array can be used to evaluate all the k-nearest distances from one point $x(u)$ to all other points of the same class. Practically, the network will provide the k-distances, with $k = 1..d$, where d represents the space dimension. This is essentially due to the architecture of the array, whose dimension is $(d \times d)$.

First of all, we consider that the mutual distances previously computed are represented in the memory as a $(d \times N_c)$ matrix, containing all the distances. This requires a dedicated mechanism to address the memory during the extraction of the mutual distances, in order to duplicate the distances which have not been explicitly computed. For example, $d(x(1), x(2))$ is computed for the point $x(1)$, but the distance $d(x(2), x(1))$ is necessary for the $x(2)$ point, but not computed. Then, it must be duplicated at the adequate position into the distances matrix. This operation produces a square matrix, from an upper triangular one, composed of d lines and k columns. It is thus possible to compute the first, the second, up to the dth nearest neighbor associated to each point $x(i)$ by a bubble sort algorithm, applied to each line of the distance matrix. We obtain the values of the k-nearest neighbor, in the decreasing order, but we loose the index of the point associated to each value. The systolic array used for this operation is depicted on figure 7.a. After completion of the bubble sort, the learning phase is achieved, because we have at disposal all the values h, required for the initialization of the RBF algorithm. Note that a restriction can appear if we consider that the value k depends on the number of samples N of the database. These variables are dependent through the empirical relation $k = round(N^{0.4})$. Then, if the number of samples is very large, the dimension of the systolic array can to be smaller than k, and thus, a virtual bubble sort computation must be considered. For practical reasons, this operation will not be described in this paper. After learning, the evaluation phase can be applied to classify unknown vectors. In the next section, we propose a systolic implementation of this phase, and estimate performances on real databases.

4.3 Evaluation phase

The evaluation phase is applied when all the width factors are computed either by the dedicated hardware, or by an external computer. In our case, we consider that the initial width factor $h(n)_{t=0}$ is estimated by the dedicated systolic array as described before, and the refinement procedure is executed on an host computer. The first reason is the complexity of this procedure. Furthermore, such a procedure is not well stabilized from a theoretical point of view ([2]), and a hardware implementation is probably too much specialized for a long lifetime.

Then, we consider that an external hardware is able to compute a table containing all the width factors, from the initial ones delivered by the systolic array, as seen before. Each point $x(n)$ is associated to a width factor $h(n)$, used for the estimation of conditional pdf by a suitable Radial Basis Function. After the learning phase, which has essentially produced the width factor, the evaluation phase consists on the choice of the most probable class ω_c to be associated to the point x. The first step is the estimation of $p(x|\omega_c)$, the conditional probability of x, when the class ω_c is known. Then, this value must be multiplied by P_c, which is the a priori probability for the class ω_c to appear.

The value $p(x|\omega_c)$ is estimated by the sum of distances between x and all the points belonging to the same class according to equation 1.

4.3.1 Evaluation of individual contributions

The first part of this computation is the processing of whole the N_c distances with the input sample x. This is performed in the systolic array in a similar way as the previously described procedure. The values of the reference points are loaded in the systolic array, they are not computed in the array. Partition is needed if N_c is greater than p.

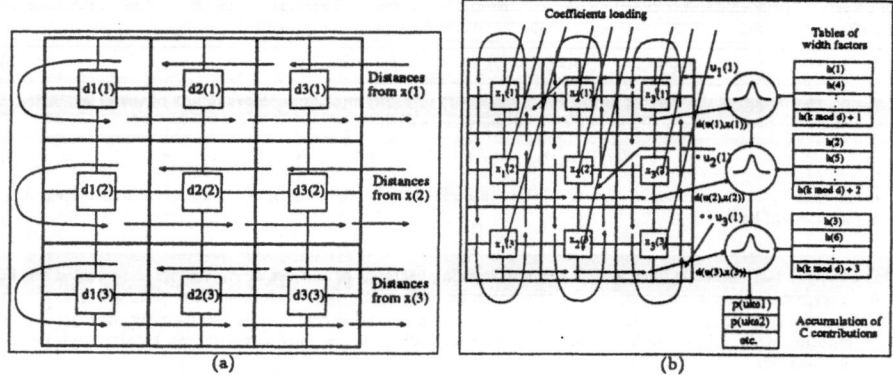

Figure 7: (a) Systolic array for bubble sorting. (b) Architecture of the 2D systolic array for pdf estimation.

So, in this case, after the computation of the d distances between x and $x(1), x(2), .., x(d)$, the other distances corresponding to the other reference points of the same class must be computed. Thus, while the first computations are performed, the x value is cycling in the array, being fed back from the South of the array, to be available in the diagonal after exactly $2.d$ time steps. An exchange of the memory points, correctly preloaded with the values $x(d+1), ..., x(2d)$ allows the next computation. This is repeated until the whole dataset for a same class is examined. Note that such a process can be performed with a pipelining, allowing a maximum occupation rate of the systolic array.

In a second part, a kernel function (parameter $h(n)$) is applied to each distance to obtain the contributions, and then all these contributions are accumulated to obtain $p(x|\omega_c)$. The computation of the kernel function is realized by a tabulation or a specialized operator. This second part is implemented in a dedicated hardware, connected to the East border of the array. The whole architecture is depicted in figure 7.b.

4.3.2 Determination of the class label

The last part concerns the determination of the class label associated to the test samples. This part is itself divided into two operations : the multiplication of the contributions by the priors P_c associated to each class, and the multiplication of the result by a cost matrix. This matrix represents the cost associated to the decision. The priors are computed by the number of points belonging to each class, divided by the number of examples in the dataset. We consider that this value is available for each example. The cost matrix W is also available, given by the user.

The choice of the class is done, after the two multiplications, through a winner take all operator which determines the highest value computed amongst the C available values. The *matrix $\times$ vector* product used to compute the C values can be realized by a systolic operator. In fact, it is exactly the operations performed for the Kohonen algorithm, and has been described in previous papers [8]. This operator has a $C \times C$ dimensions, and each register contains the C^2 values corresponding to the costs associated to the classes.

5 Theoretical efficiency of the distances computation

We notice here only the theoretical estimations achieved in the mutual distances computation (initialisation phases) and in the distance computations for the test phases. These estimations are useful to give ideas on the matching between the algorithms size of the problem and the target hardware size.

5.1 Efficiency on the mutual distances computation

We must consider three main cases :

1. The input dimension is less than p for a $p \times p$ array. So the whole learning samples are completed with zero values to reach the dimension d. A great loss of efficiency occurs.

2. The input dimension is equal to the p. This is the optimal case.

3. The input dimension is greater than p. If d is not a multiple of p, extra dimensions are filled with zero. With this new database, the computations are made sequentially by subsets of d components. The accumulation of the partial distances are implemented on the DSP. So, here also, the efficiency is lower.

For the number of samples par class, we have also the three same cases, comparing the number of samples N_c with p. The same approach is used for the cases 1 and 2. For the case 3 applied to the number N_c of samples $(N_c > p)$, the database is split into Q_c subsets with $Q_c = ceil(N_c/p)$ of p samples. Let us notice N'_c, as $Q_c \times p$. For the parallel implementation, the class ω_c has N'_c samples, and the input dimension is to p by filling with zero values. The process is sequentially carrying on (the first block with all the blocks, the second block with all the blocks except the block 1, and so on). For one class ω_c, the sequential time can be estimated by $T_{seq_c} = dN_c(N_c - 1)/2$ computing cycles. The efficiency $(T_{seq_c}/p/p/T_{par_c})$ follows equation 11.

$$T_{par_c} = latency + bias + \sum_{q=1}^{q=Q_c}(p-1) + \sum_{q=1}^{q=Q_c-1} p(Q_c - q) = 3p - 1 + \frac{N'^2_c + N'_c d - 2N'_c}{2p}$$

$$Eff_c = \frac{d}{p} \cdot \frac{N_c(N_c - 1)}{N'^2_c + N'_c p - 2N'_c + 6p^2 - 2p} \approx \frac{d}{p} \cdot \frac{N_c(N_c - 1)}{N'^2_c + N'_c p - 2N'_c} \tag{11}$$

Theoretical simulations are shown in figure 8. The efficiency rapidly increases as the number of samples increases, the pipeline latency becomes negligible, and this increase is faster as the hardware size is small (the reason is the same).

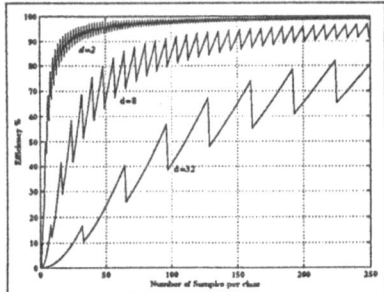

Figure 8: Efficiency of the mutual distances calculation on MANTRA for one class versus the number N_c of samples.

5.2 Efficiency of the distances computation for the test phases

For this estimation, let us consider two databases examples. The first database is called "Iris". It is a famous database in pattern recognition. The dimension is 4, the number of classes is three with 50 samples per class. The second database is called "Satimage". This database was generated from Landsat Multi-Spectral Scanner. Each sample is in 36 dimensions. The number of classes is 6. The mean number of samples per class is about 1000, the minimum a priori probability is 0.9, the maximum is 0.23.

For this estimation, we consider that N_T test samples are presented to the network, for classification. With the same notations as previously, in the parallel implementation, the computing time T_{par} and the efficiency can be estimated according the equation 12. The sequential time is $N \times N_T \times d$, where N is the total number of samples. The T_{par} is defined according to the equation 12. where $N' = \sum_{c=1}^{c=C} N'_c$.

$$T_{par} = \sum_{c=1}^{c=C}(latency + bias + \frac{N'_c N_T}{p}) = C(3p - 1) + \frac{N_T}{p} \cdot \sum_{c=1}^{c=C} N'_c$$

$$Eff = \frac{N.N_T.d}{(3.C.p^2 - C.p + N'.N_T).p} \approx \frac{N}{N'} \cdot \frac{d}{p} \tag{12}$$

Efficiency for the Iris database is very bad (since each sample is padded with 28 zeros). With only one test sample, the efficiency in this worst case, is 0.20%. For the database Satimage, if the size of the test block is greater than 40, the efficiency is good. With only one sample, the minimum efficiency is 1.6%.

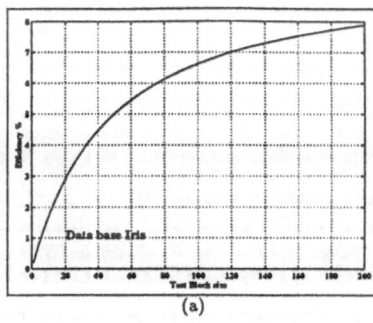 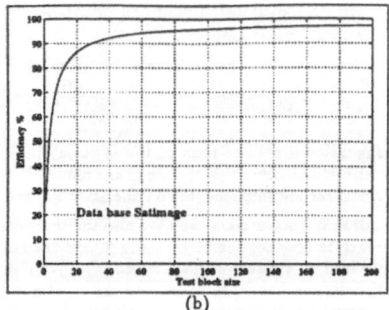

(a)   (b)

Figure 9: Efficiency of the test phase on MANTRA 32x32 versus the size N_T of the test block. (a) Iris with $d = 4$, (b) Satimage with $d = 32$,

6   Conclusion

In this paper, we have presented a possible systolic implementation of a kernel based algorithm for classification tasks. Implementations have been proposed for all the time consuming tasks. These tasks involve distances computations. Other tasks are less time expensive and can be efficiently implemented in a 1D structure.

By theoretically studying the efficiency in two cases (small and huge database), we have clearly outlined three conditions for a good usage of the Mantra machine : (i) The input dimension d must be quite equal or a multiple of the hardware size p, (ii) The size of the database must be almost proportional to the hardware size. The exact proportionality is not critical as the number N_c of samples increases. (iii) Better efficiencies are achieved if N_T test samples are presented in burst, and if the number N_T grows with the hardware size.

This work is currently compared ([7] [6]) to other existing implementations on 1D machines (SMART and DRA). In the same framework, it will provide user guides to help a designer in the choice of a machine for the best possible implementation depending on various criteria, such as the portability, the integration and the computation performances.

References

[1] F. Blayo. *Une implantation systolique des algorithmes connexionnistes*. PhD thesis, Ecole Polytechnique Fédérale de Lausanne, Suisse, 1990. *Thèse N. 904*.

[2] P. Comon, J.L. Voz, and M. Verleysen. Estimation of Performance Bounds in Supervised Classification. In Michel Verleysen, editor, *ESANN : European Symposium on Artificial Neural Networks*, pages 37–42, Brussels, Belgium, April 1994.

[3] C. Jutten and P. Comon. Neural Bayesian Classifier. In *New Trends in Neural Computation*, number 686 in Lecture Notes in Computer Science, pages 119–124. Springer-Verlag, 1993.

[4] S. Y. Kung and J. N. Hwang. Parallel Architectures for Artificial Neural Nets. In *IEEE Int. Conf. On Neural Networks*, volume 2, pages 165–172, San Diago, July 1988.

[5] C. Lehmann, M. Viredaz, and F. Blayo. A Generic Systolic Array Building Block for Neural Networks with On-Chip Learning. *IEEE Transactions on Neural Networks*, 4(3):400–407, May 1993.

[6] N. Maria, A. Guérin-Dugué, J.M. Moreno, and Blayo F. Comparing Implementations of Radial Basis Function Neural Networks on three parallel machines. *Submitted to IWANN 95*, June 1995.

[7] N. Maria, A. Guérin-Dugué, and Blayo N. 1D and 2D systolic implementations for Radial Basis Functions Networks. In *Fourth International Conference on Microelectronics for Neural Networks and Fuzzy Systems*, pages 34–45, Turin, Italy, September 1994. IEEE Computer Society Press.

[8] M. A. Viredaz. MANTRA I : An SIMD Processor Array for Neural Computation. In *Proceedings of Euro-ARCH'93*, Munich, Germany, October 1993.

A Mixed Parallel-Sequential SHNN for Large Networks

A.Torralba, F.Colodro and L.G.Franquelo

Dpto. de Ingeniería Electrónica, de Sistemas y Automática
Escuela Superior de Ingenieros
Avda. Reina Mercedes, s/n, SEVILLA–41012
Tlf: (95) 4556851, Fax: (95) 4556849
e–mail: carrasco@obelix.cica.es

Abstract

This paper presents an architecture for the implementation of a large Stochastic Hopfield Neural Networks (SHNNs). The sequential SHNN, originally proposed in [1], takes a long time to convergence. On the other hand, the connection between chips limits to one hundred the number of neurons of the fully parallel SHNN proposed in [2]–[3]. A multichip approach proposed in this paper overcome both problems. The architecture, using a mixed parallel–sequential strategy, reduces the number of interconection lines to k while accelerates the convergence time of the network in [1] by a factor k. A partitioning problem is simulated to evaluate the behavior of the network.

I. INTRODUCTION

Hopfield Neural Networks (HNNs), composed of one–layer neurons and fully connected feedback weights, are adequate to solve optimization problems. Different strategies based on analog and digital technologies have been used for a hardware implementation of HNNs. Recently, architectures employing stochasticism have been used. Stochasticism has a number of advantages over preceeding analog and digital implementations, such as signal multiplication using a simple AND gate.

In [1], a Stochastic Hopfield Neural Network (SHNN) was presented, where the architecture is depicted in figure 1. This circuit implements the time discretized charging equation of a HNN

$$u_i(t + \delta t) = u_i(t) + (\sum_{j=0}^{n-1} G_{ij} v_j(t) + I_i) \times \delta t \qquad (1)$$

$$u_i(t) = f(u_i(t)) \quad i = 0, 1, ..., n-1 \qquad (2)$$

A stochastic signal s_{ij} that pulses with probability proportional to $G_{ij} v_j$ is obtained by ANDing two stochastic signals that pulse with probability proportional to G_{ij} and v_j, respectively. In this way, it is possible to calculate simultaneously a lot of products $G_{ij} v_j$ by means of simple AND gates. Nevertheless, in [1] a neuron accumulates only one synapse input $G_{ij} v_j$ per clock cycle, due to the difficulty of a parallel summation of stochastic signals; therefore, n cycles are spent in the evaluation of $\sum_{j=0}^{n-1} G_{ij} v_j(t)$.

To reduce the convergence time of the network, a digital circuit (called circuit F) was proposed in [2]–[3] to carry out the parallel summation of the synaptic input pulses s_{ij} to the neuron i. This approach reduces the convergence time of a parallel SHNN by a factor of n. Note that only the stochastic signals that pulse with a probability proportional to the neuron state are neccesary to interconect the neurons. Then, a multichip approach of an n neuron, fully parallel SHNN, requires n input/output pins for neuron interconnection. With present technology limitations in the number of I/O pins, a multichip network with no more than one hundred neurons can be implemented. Even so, such networks are small compared to the types of optimization problems found in real applications.

In this paper, a mixed sequential–parallel SHNN for the implementation of a network with a large number of neurons is presented. This approach represents a compromise between convergence time, area cost and feasibility of the implementation.

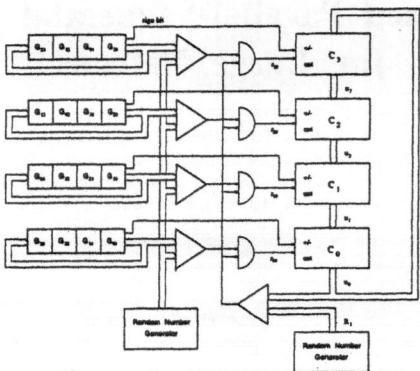

Figure 1: Stochastic Hopfield Neural Network (SHNN).

II. THE CIRCUIT F

The main problem to calculate $\sum_{j=0}^{n-1} G_{ij} v_j$ using stochastic logic is the summation. Several approaches have been described in the literature:

- In [1] the authors implement the summation by multiplexing stochastic pulses in time

- In [6], an analog summation circuit is proposed

- In [7] the stochastic signal is transformed by means of a exponential function. The summation is then replaced by a multiplication, which is carried out using a AND gates and some auxiliary logic.

In [2]–[3] the authors proposed a digital circuit to carry out the parallel summation: the circuit F. Circuit F is a combinatorial network which receives n bits of equal weight as input and produces a $(d = log_2 n)$ bit–word

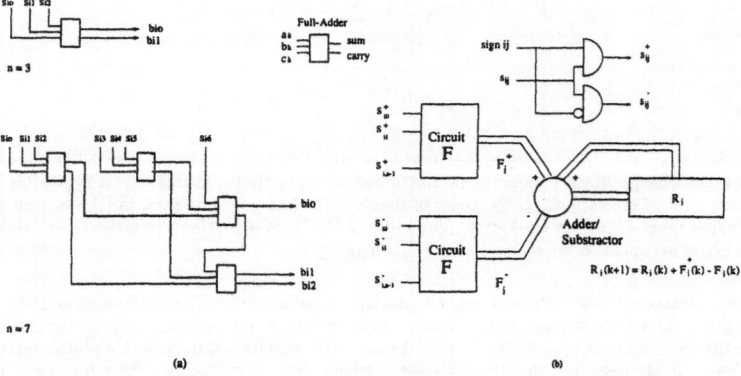

Figure 2: a) Circuit F using full–adders. b) Parallel SHNN.

corresponding to their sum, as output. This kind of circuit was termed a (n, d) counter by Dadda [8]. A direct implementation of the circuit F using two–level logic, such as a PLA, would lead to a circuit whose complexity would increase combinatorially with the number of synaptic input pulses n, making this approach impractical even for moderate values of n. In [3] two implementations of the circuit F with complexity $O(n)$ are presented. The so called parallel implementation of the circuit F using full–adders (figures 2.a and 2.b) takes only one clock cycle time to calculate

$$F^+ = \sum_{G_{ij}v_j > 0} G_{ij}v_j \tag{3}$$

$$F^- = \sum_{G_{ij}v_j < 0} G_{ij}v_j \tag{4}$$

The maximum propagation delay and the number of full–adders the circuit F requires are presented in table 1.

n	No.adders	Max.prop.delay (two–gate levels)
3	1	1
7	4	3
15	11	5
31	26	7
63	57	9
127	120	10

Table 1

III. MIXED SEQUENTIAL–PARALLEL SHNN

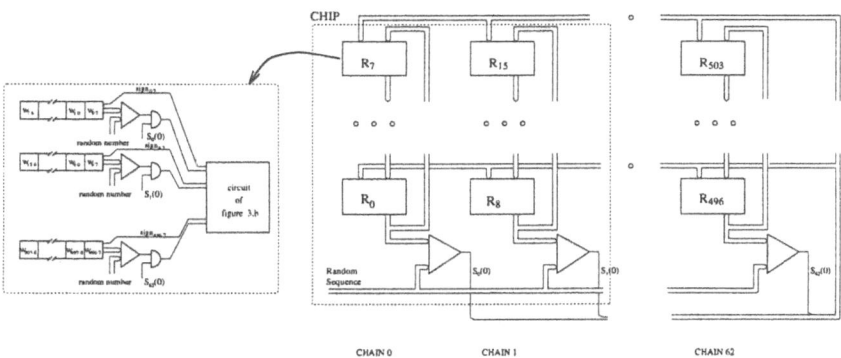

Figure 3: A mixed (sequential–parallel) implementation for large networks.

Let's now consided a chip for a large circuit (say 504 node) partitioning. A sequential implementation of the chip would require 504 cycles to make the summation $\sum_j G_{ij}v_j$.

Using the parallel circuit of figures 2.a and 2.b, only 1 cycle would be required, (two if positive and negative pulses are accumulated in different cycles). An acceleration factor of 504 is achieved. Unfortunately a multi–chip implementation of such a fully parallel hardware would require more than 504 I/O pins.

A new sequential–parallel mixed network is proposed here. For the example above, the structure is composed by 63 parallel chains of 8 neurons each (figure 3). Registers $(R_0,R_1,...,R_{503})$ contain the neural state vector $(u_0,u_1,...,u_{503})$. For the sake of simplicity, synaptic weight storage is not shown in figure 3. Inputs I_i can be consided to be the weights $G_{n+1,i}$ of a $(n+1)$–th neuron (the bias neuron), whose output is forced to 1. A possible multichip implementation of the proposed architecture would place 2 chains on a chip, as depicted in figure 3. Note that only 64 I/O external pins are required in each chip for neuron interconnection.

To evaluate the behavior of the proposed circuits in a real case, one prototype of mixed sequential–parallel SHNN with 504 neurons is being designed using 1.0 μm standard CMOS process. The network will be implemented as a multichip architecture. The chip has 16 neurons and the structure of one neuron is depicted in figure 3.

Once initialized, the register's contents of the m–th chain are continously rotated, so that the bottom register contains the state of the neuron $k_{mod(N)} + m \times N$ at the cycle k, where N is the number of states per chain (8 in this example). This value is compared with a random number to generate the stochastic signal $S_m(k)$. $S_m(k)$ is fedback to the neurons and multiplied by the weight w_{ij} to accumulate the content of R_r, where i and j are

$$i \;=\; k_{mod(N)} + m \times N \tag{5}$$

$$j \;=\; (k+r)_{mod(N)} + \lceil \frac{r}{N} \rceil \times N \tag{6}$$

Synaptic weights in the chip are also arranged in chain which are continously rotating. For the example above there are 63 synaptic weight chains per register, with 8 weights each.

Due to the stochastic nature of the signals, the circuit (figure 3) behaves similarly to the system described in equations 1, provided that it runs for a significant number of cycles. The neural transfer function f can be modified by adjusting the probability distribution function of the random sequency.

For the 504 neurons case, the proposed network is able to update each neuron in $O(8)$ cycles. Mixed sequential–parallel architectures are a promising way to build large networks with a reduced updating time at resonable cost.

IV. RESULTS

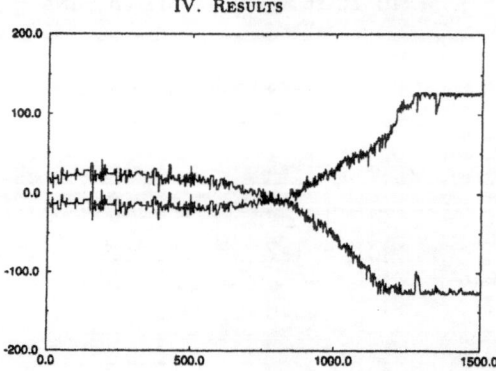

Figure 4: Simulation results. Two neuron dinamic of the proposed example

The dynamics of the slowest neurons are depicted in figure 4 when the network is used to cluster with 504 nodes. The results were obtained by simulation. All neurons were initialized with a random numbers in the range of [-20..20]. Note only 1200 cycles are neccessary to convergence. The sequential approach of [1] would require 64×1200 cycles in the same case.

The area estimated for one neuron is $1.2mm^2$ using $1.0\mu m$ CMOS process. Nevertheless, $12.5mm^2$ are required for synaptic storage. Therefore, a chip with 16 neurons (figure 3) uses approximately $82mm^2$.

V. CONCLUSIONS

A strategy to design large Hopfield Neural Network has been presented. Stochasticism was used for area efficiency. Precedent architectures were limited by long convergence times and/or limitations in the number of I/O pins. A mixed parallel–sequential SHNN, using the circuit F proposed in [2], is presented to overcome these problems. The main limitation of this new architecture is due to the area required to store the synaptic weights,which is a bottleneck of all Neural Network architectures. A n–neuron network with the proposed architecture can be designed in a multichip approach. If the number of interconnection lines between chips is N, $\frac{n}{N}$ clock cycles are spent in the summation of equation (1).

REFERENCES

[1] D.E. van den Bout and T.K.Miller III, "A digital architecture employing stochasticism for the simulation of Hopfield neural nets". *IEEE Trans. Circuits and Systems*, vol. 36, pp. 732–738, May 1989.

[2] A.Torralba, F.Colodro. "Towards a fully parallel Stochastic Hopfield Neural Network". *Proc. of the ISCAS'93*, pp. 2741–2743, May 1993.

[3] A.Torralba, F.Colodro. "Two digital circuits for a Fully Parallel Stochastic Neural Network". IEEE. *Trans. of Neural Network (to appear)*,

[4] D.E. van den Bout and T.K.Miller III, "TInMANN: The Integer Markovian Artificial Neural Network".

[5] M.S.Melton, T.Phan, D.S.Reeves, D.E. van den Bout, "The TInMANN VLSI chip". *IEEE Trans. Neural Networks*, vol.3, no. 3, May 1992.

[6] Y.Kondo and Y.Sawada, "Functional abilities of a stochastic logic neural network". *IEEE Trans. Neural Networks*, vol. 3, no. 3, May 1992.

[7] C.Janer and J.M.Quero, "Fully parallel summation in a new Stochastic Neural Network architecture". *IEEE Trans. Int. Conf. in Neural Network*, San Francisco, 1993.

[8] L.Dadda. "Some schemes for parallel multipliers". *Alta Freq.*, vol. 19, pp. 349–356, May 1965.

A Modular VLSI Architecture
for Neural Networks Implementation

O. Vermesan

Microelectronics Group, Department of Physics
University of Bergen, Allégaten 55, 5007 Bergen, Norway

ABSTRACT

This paper describes a modular analog VLSI architecture for the implementation of artificial neural networks. Analog neural network implementations are faster and smaller than their digital counterparts, but the problem of smaller dynamic range of the analog weight memory and the linearity of the synapses based on analog multipliers increases the need for design effort at the circuit level. We suggest that a complex neural network system can be implemented in a single chip if a modular architecture design using simple analog circuits is followed. To demonstrate the VLSI implementability of the neural network system, a description of each analog circuit block is provided.

1. Introduction

Neurocomputing is the term used to characterise the analog computation based on artificial neural networks that are designed to simulate the functioning of the brain and tries to implement intelligent machines based on artificial neurons. Neurocomputing could be seen conceptually as analog computation systems that simulate dynamic systems defined through synaptic connections.

Artificial neural networks are massively parallel computer systems that have the ability to learn from experience, adapt to new situations and process data very fast, once they have been properly trained. The basic building blocks for artificial neural networks are neurons and synapses. Each synapse multiplies an input by a stored weight, and each neuron takes the outputs of some number of synapses, sums them and passes them through a transfer function [12].

Over the last few years different techniques ranging from software and hardware/silicon/VLSI to optics and optoelectronics have been employed to implement artificial neural networks [11].

We present in this paper a modular analog neural network implementation which includes the analog weight storage, the synapses and the neurons.

Analog VLSI circuit techniques offers area-efficient implementation of the functions required in a neural network such as multiplication, summation and Sigmoid transfer function, but are much more sensitive to the problems of cascability, process variation and device matching, than digital logic. Special attention must be given to the limitations of the MOS transistor and to the design techniques. As result, in analog neural hardware the high accuracy found in digital implementations is traded off for the simplicity and interconnectivity found in analog circuits. This means that the analog neural networks, for less silicon area, allow for sufficient redundancy at the system level to compensate lower accuracy at the process level [9].

2. Circuit Description

The block diagram of a modular neuron and the corresponding modular synapses connected to the neuron is presented in Figure 1.

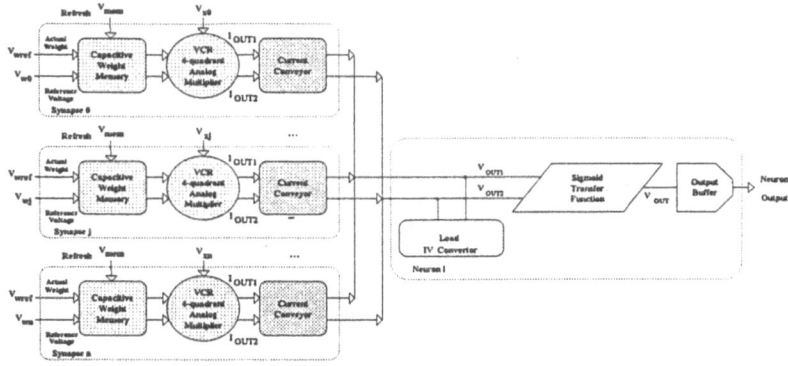

Figure 1 Block diagram of modular architecture for neuron and synapses

Each synapse is realised using a simple multiplier based on voltage controlled resistors (VCR). The differential output current from the analog multiplier is passed to a current conveyor that has the role of maintaining the virtual short between V_1 and V_2. The differential output current from the current conveyor of each synapse is summed into a common bus and transformed into a differential voltage output using an active load circuit.

This differential output voltage is applied at the input of an operational transconductance amplifier which implements the actual Sigmoid transfer function. The output of the neuron is buffered and applied to the synapses of other neurons.

Differential operation is maintain from the input to the output in order to improve device matching, power supply noise rejection and multiplication accuracy. The circuits described above were implemented in a 1.2 μm n-well CMOS process. The process is a mixed analog digital process with two layers of metalization and two layers of polysilicon.

3. Synapse Characteristics

An analog multiplier approach using MOS transistor voltage controlled resistors multiplier circuit has been chosen to implement the core of the synapse circuit.

An analog multiplier based on proportionality of the channel conductance to the gate voltage was described in [6][5]. The scheme presented, uses two junction field-effect transistors in a balanced (bridge) configuration. One variable (a voltage) is applied with opposite sign to appropriately biased gates, increasing the conductance of one channel and decreasing the conductance of another one. Another variable is applied as a voltage to the channels. The difference in channel currents is proportional to the product of the two variable V_A, V_B and a constant value C and is given by:

$$\Delta I = C \cdot V_A \cdot V_B \qquad (1)$$

An improvement to the circuit can be realised by using a parallel compensation technique [2], by simply attaching in parallel to the voltage controlled resistor a diode-connected MOS transistor (M_{D1}, M_{D2}) as presented in Figure 2.

The four transistors M_1, M_2, M_{D1} and M_{D2} are operated in strong inversion in the non saturated region and the current $I_1 = I_{OUT1}$ and $I_2 = I_{OUT2}$ for the equivalent transistors M_{1e}, M_{2e} can be expressed after a fair amount of algebra as:

$$I_1 = I_{OUT1} = \frac{k_1' W_1}{2 \cdot L_1} \cdot \left[2 \left(V_{GS1} - 2 V_{Th1} \right) V_{DS1} \right] \qquad (2)$$

and

$$I_2 = I_{OUT2} = \frac{k'_2 W_2}{2 \cdot L_2} \cdot \left[2 \left(V_{GS2} - 2 V_{Th2} \right) V_{DS2} \right] \qquad (3)$$

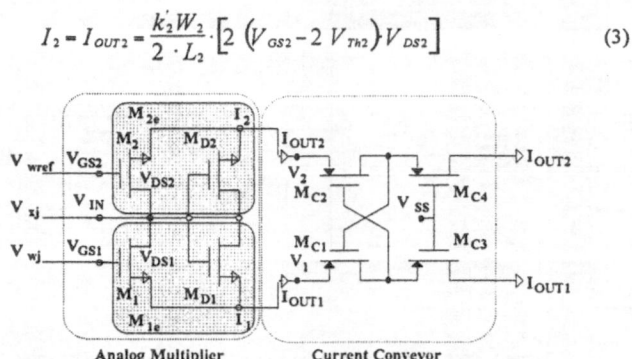

Figure 2 Four transistor four quadrant multiplier

Where we have considered that M_1 is matched with M_{D1}, and M_2 is matched with M_{D2}. Further more if M_{1e} and M_{2e} are identical and $k'_1 = k'_2 = k'$, $W_1 = W_2 = W$, $L_1 = L_2 = L$, $V_{Th1} = V_{Th2} = V_{Th}$, and if we replace the expressions for $V_{DS1} = V_{IN} - V_1$, $V_{DS2} = V_{IN} - V_2$, and consider the voltages $V_1 = V_2 = V = $ constant, the difference between I_{OUT1} and I_{OUT2} can be expressed as:

$$\Delta I = k \cdot \left[\left(V_{IN} - V \right) \left(V_{GS1} - V_{GS2} \right) \right] \qquad (4)$$

In our design we applied the synapse input value x_j to the neuron i, as $V_{IN} - V$ and the weight value w_{ij} as $V_{wj} - V_{wref} = V_{GS1} - V_{GS2}$. The linearity of the circuit is a function of the input amplitudes. A suitable set of V_{xj} (1.5-2.5V), V_{wj} (0-1V), and V (2V) reduces the non linearity to less than 1.5%.

The condition $V_1 = V_2 = V = $ constant is achieved in practice by using a current conveyor. A simple four transistor current conveyor is presented in Figure 2. The circuit presents interesting features such as a virtual short for voltages V_1, V_2. The area occupied by multiplier and current conveyor is 3,400μm².

However mismatch of the transistors is the most serious problem in the design. The threshold voltage variations are the major cause of transistors mismatch while the current factor variations (which includes variations in W/L area) are a secondary factor. In the recent years problems associated with mismatches, offsets and inaccuracy in determining exact device characteristics have been tackled by using several circuit techniques. The transistors from the multiplier have to be matched with almost identical transconductance parameters and threshold voltages. For matching, the transistors must be in close proximity in the layout. A common-centroid geometry layout has help us to minimise the offset. At the same time it is necessary to keep the number of bends and corners in the layout to a minimum for the devices that must match.

There are a number of second-order effects that can have a negative effect on the circuit such as:

♦ Threshold voltage modulation determined by the body effect coefficient for the case when the source and the body are not at the same potential. This is seen for example when using n-channel MOS transistors and when we have a n-well process where the substrate of all these transistors is connected to V_{SS}.

♦ Channel length modulation which will determine the modification of the current expression in order to include the effect of the channel length modulation parameter λ.

These two effects, and especially the channel length modulation, are significant for devices with large areas operating in strong inversion in the non saturated region. At the same time the linearity of the circuit will depend on the device matching and the length of the transistors must be large in order to obtain acceptable input voltage and dynamic range. As result a trade off was made between silicon area, linearity and dynamic range.

4. Capacitive Analog Weight Storage

In artificial neural networks implementation the values of the synaptic weights are adjusted during the learning process until an optimal weight set is determined. During this process the weights are memorised in the corresponding synapses and are updated as the learning process progresses until the optimal set of weights is obtained. This optimal weight set should be memorised by the network and preserved for every use of the network in solving the specific problem for which the network was trained [10].

An active capacitor structure is used to implement the weight storage or weight memory element in our circuit. The active capacitor structure is formed by using a MOS transistor as a capacitor when biased in non saturation region, the gate forming one plate and the source, drain and channel forming the other . The weight memory element and the switch transistor are presented in Figure 3 [7][8][10].

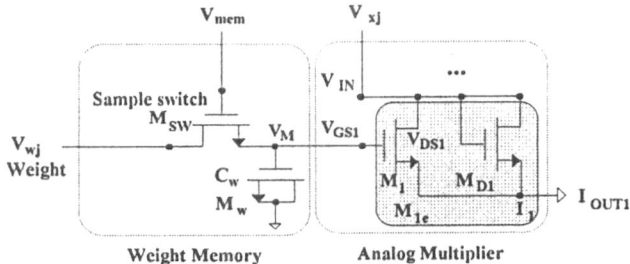

Figure 3 Analog memory element using an active capacitor structure

The value of the active capacitor formed C_w can be expressed as [4]:

$$C_w = C_{ox} \cdot W \cdot L = \frac{\varepsilon_{ox}}{t_{ox}} \cdot W \cdot L \qquad (5)$$

where the C_{ox} is the oxide capacitance per unit area between the gate and the channel, W the width and L the length of the MOS transistor M_w. From this formula is easy to determine the transistor M_w dimensions (W, L) for a certain capacitance C_W needed as a memory element. The capacitance of the weight storage capacitor was chosen $C_W = 1pF$ ($840\mu m^2$).

The switch transistor M_{SW} is an n-channel MOS transistor which was designed with a minimum size (W/L= $2.4\mu m/1.2\mu m$) in order to keep the charge injection at a minimum, and increase the retention time of the storage element. The switch introduces a noise voltage equal to square root from kT/C_w which adds to the actual weight voltage. Since we use a double capacitor storage approach the noise is cancelled. The weight resolution is approximately 6 bits and it is limited by the clock feed through on the sampling switches.

5. Neuron Characteristics

The differential output current from each synapse of the neuron is summed into a common bus and converted into a differential voltage by a simple current to voltage converter (IV Converter) as presented in Figure 4. The converter is realised using the same voltage controlled resistors configuration as for the analog multiplier. All four transistors M_1, M_2, M_{D1} and M_{D2} are operated in strong inversion in the non saturated region.

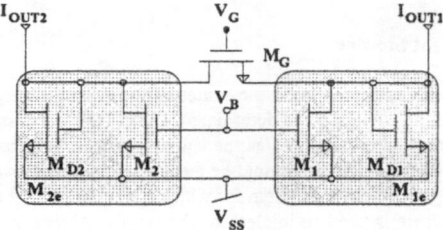

Figure 4 Current to voltage converter

The converter is linear over a 1 V_{P-P} swing in output voltage. The voltage V_B is used to bias the voltage controlled resistors. The transistor M_G is used to provide a variable gain to optimise the dynamic range for desired weights w_{ij} and input values x_j, or to initialise the network. The neuron gain can be controlled by the gate voltage of M_G. When the gate voltage V_G is decreased and the M_G turns off the gain of the neuron increases to its maximum value.

The Sigmoid transfer function that characterise the behaviour of the neuron's body is implemented using a simple n-channel input unbuffered two stage CMOS operational transconductance amplifier as presented in Figure 5 [1][3].

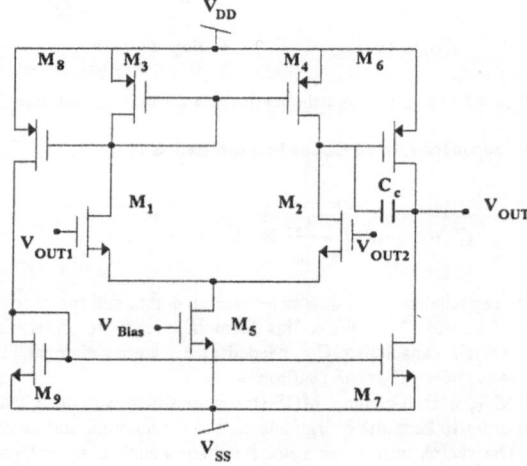

Figure 5 OTA Sigmoid Transfer Function Implementation

In the design, the transistors M_1 and M_2 are identical, M_3 and M_4 have identical small aspect ratio in order to reduce the noise, while M_6 has large aspect ratio in order to increase the frequency response. The bias voltage of the operational transconductance amplifier was chosen to 1.6 V.

To overcome the loading problems for the next synapses to which is connected, the output of the neuron is buffered using a simple voltage follower. The buffer circuit is implemented with a low output impedance in order to supply sufficient current to low input impedance node of the drain of the MOS transistors from the analog multiplier of other synapse.

6. Conclusions

In this paper we have discussed a modular analog VLSI architecture for implementing artificial neural networks. We have chosen this modular design approach in order to increase the size and the complexity of a neural network implemented on a single chip.

A description of each module of the actual implementation is presented and trade-offs are made among silicon area, linearity, computation precision, device matching and layout design.

The architecture is intended to work with descent based learning algorithms such as Back Propagation and Weight Perturbation.

7. Bibliography

[1] P.E. Allen and D. R. Holberg, *CMOS Analog Circuit Design*, New-York, NY, Holt, Rinehart and Winston, Inc., 1987.

[2] L.N.M. Edward, "Comment on "Voltage-Controlled Linear Resistor by Two MOS Transistors and its Application to Active RC Filter MOS Integration"", in *Proc. IEEE*, Vol. 74, No. 5, pp. 753-755, 1986.

[3] R.L. Geiger, P. E. Allen, and N. R. Strader, *VLSI Design Techniques for Analog and Digital Circuits*, New-York, NY, McGraw-Hill, 1990.

[4] P.R. Gray and R. G. Meyer, *Analysis and Design of Analog Integrated Circuits*, 3rd ed., New-York, NY, John Wiley & Sons, 1993.

[5] K.K. Moon, F.J. Kub and I.A. Mack, "Random Address 32×32 Programmable Analog Vector-Matrix Multiplier for Artificial Neural Networks", in *Proc. IEEE CICC '90*, pp. 26.7.1-26.7.4, 1990.

[6] V. Radeka, " Fast Analogue Multipliers With Field-Effect Transistors", *IEEE Trans. Nuclear Science*, Vol. 11, No. 1, pp. 302-307, 1964.

[7] F.M.A. Salam and M. R. Choi, "Analog MOS Vector Multipliers for the Implementation of Synapses in Artificial Neural Networks", *Journal of Circuits, Systems, and Computers*, Vol. 1, No. 2, pp. 205-228, 1991.

[8] S. Satyanarayana, Y. P. Tsividis, and H. P. Graf, " A Reconfigurable VLSI Neural Network", *IEEE J. Solid-State Circuits*, Vol. SC-27, No. 1, pp. 67-81, 1992.

[9] O. Vermesan, *The MOS Transistor as the Basic Building Block for Analog VLSI Implementation of Neural Networks*. Scientific/Technical Report No. 1994-11, ISSN 0803-2696, University of Bergen, Norway, 1994.

[10]O. Vermesan, *Memory Units for Analog VLSI Implementation of Neural Networks*. Scientific/Technical Report No. 1994-12, ISSN 0803-2696, University of Bergen, Norway, 1994.

[11]O. Vermesan, *Neural Networks Implementation - Issues and Techniques*. Scientific/Technical Report No. 1994-20, ISSN 0803-2696, University of Bergen, Norway, 1994.

[12]O. Vermesan, and A.I. Vermesan, "The Use of Hybrid Intelligent Systems in Telecommunications" In J. Liebowitz and D.S. Prerau, Eds., *Worldwide Intelligent Systems-Approaches to Telecommunications and Network Management*, Amsterdam, IOS Press, Chapter 10, pp. 186-226, 1995.

A Massively Parallel Neurocomputer with a Reconfigurable Arithmetical Unit

Alfred Strey

Abteilung Neuroinformatik
Universität Ulm
D-89069 Ulm, Germany

Narcis Avellana

Abteilung Allgemeine Elektrotechnik und Mikroelektronik
Universität Ulm
D-89069 Ulm, Germany.

Raul Holgado, J. Alberto Fernández, Ramon Capillas,
Elena Valderrama

Departemento de Diseño de CIs
Universidad Autonoma de Barcelona – C.N.M.
08193 Bellaterra, Spain

Abstract

This paper presents a massively parallel neurocomputer system which is mainly based on a new reconfigurable arithmetical unit optimized for the simulation of neural networks. The system offers a very high performance for all typical neural network operations combined with a high flexibility to adapt the available hardware resources to the requirements of a user-selected neural network model. The main system features are the support of many different bitlengths, a high memory bandwidth, a good scalability and a dynamic reconfigurability.

1 Introduction

In recent years several neurocomputers have been built and can be used for the high speed simulation of neural networks. Some neurocomputers are available as a commercial product ([3], [6]), some have been realized as a research prototype ([5], [9]). They all have in common that the basic operations are performed with a limited and fixed precision of mostly 16 bit. Also the size of the weights is limited to 16 bits. For the case of multi-layer perceptrons with error backpropagation learning it was shown by Holt [4] that a precision of 16 bit is mostly sufficient. However for many neural network models neither a theoretical analysis nor an experimental study about the required precision are available. Furthermore, the required precision strongly depends on the data set used in the learning phase and it is not clear at all if a general statement about the minimal precision can be done.

So we dicided to implement a neurocomputer which supports many different bitlengths in all basic operations required for the simulation of neural networks. The maximum precision of all data elements is 24 bit. If less than 24 bits are used, the degree of parallelism in the arithmetical units is increased, so that the same hardware resources are optimally utilized.

2 Basic Operations

The basic operations required for the simulation of neural networks can be divided into two classes. One class contains all the *serial operations* which are performed only once for each neuron i (e.g. the calculation of the neuron output y_i from the local field x_i by using a sigmoidal or Gaussian function f: $y_i = f(x_i)$). The other class is composed of some *massively parallel operations*. Let w_{ij} denote the neural weight connecting input/neuron i with neuron j and δ_j some error measure related to the output y_j of neuron j. The parameter ϵ represents the learning rate. Some of the massively parallel operations to be realized are the following ones:

- product of input/state vector and weight matrix W: $x_j = \sum_i y_i \cdot w_{ij}$
- product of error vector and transposed weight matrix W^T: $u_i = \sum_j \delta_j \cdot w_{ij}$
- outer vector product for weight updating: $\Delta w_{ij} = y_i \cdot \delta_j$
- squared Euclidean distance between input vector $\mathbf{u}$ and weight vector $\mathbf{w}$:
  $d(\mathbf{u}, \mathbf{w}) = \sum_i (u_i - w_i)^2$
- parallel adaptation of a parameter matrix $\Psi = \psi_{ij}$ (which can be the learning rate matrix (ϵ_{ij}) [8] or the weight update matrix (Δ_{ij}) [7]):

$$\psi_{ij}(t) = \begin{cases} \eta^+ \cdot \psi_{ij}(t-1) & \text{if } y_i(t-1)\delta_j(t-1) \cdot y_i(t)\delta_j(t) > 0 \\ \eta^- \cdot \psi_{ij}(t-1) & \text{if } y_i(t-1)\delta_j(t-1) \cdot y_i(t)\delta_j(t) < 0 \\ 0 & \text{else} \end{cases}$$

3 System Architecture

The main idea of the system architecture consists in using special reconfigurable arithmetical units (see next section) for the massively parallel operations and a standard DSP for the serial operations and for system control.

One board of the neurocomputer system (see figure 1) consists of 4 arithemtical units
AU 1 to AU 4 which perform synchronously the same operation on different data elements read from the local weight memories WM 1 to WM 4 (SIMD operation principle).
All incoming weights of a neuron are mapped onto the same WM and are processed
by the same arithmetical unit. Figure 2 presents the mapping of a typical feedforward
neural network layer onto a system with 4 AUs. The weights stored in WM 1 and
processed by AU 1 are shown by soild lines. Weights of 4,6,8,12,16 and 24 bits are supported, they are stored as 48-bit words in the weight memories WM. Thus, one memory
word contains either two 24-bit weights, three 16-bit weights, four 12-bit words, ..., or
twelve 4-bit weights.

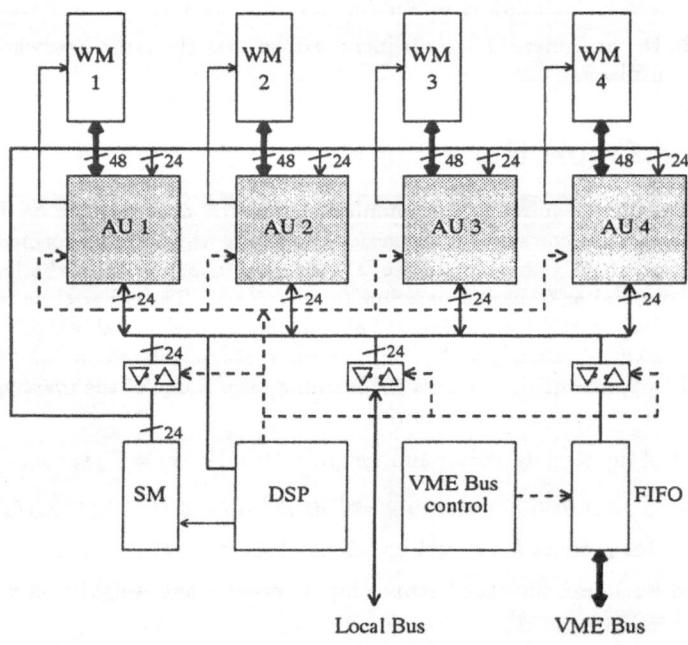

Figure 1: Architecture of one neurocomputer board

In each clock cycle the four arithmetical units read in parallel one memory word from the
corresponding local weight memories and perform the massively parallel operations on
all weights coded in one memory word. The required input patterns and neuron outputs
are centrally stored (by using 2,3,4,6,8,12,16 or 24 bit data) in the state memory SM
and are broadcasted to the four AUs. The results of the arithmetical units are collected
by the DSP which performs some serial operations and writes the new neuron outputs y
back to the state memory SM. All memories are realized by fast static memory chips.

Several boards can operate in parallel to emulate large neural networks. Here all neurons and their incoming weights weights are distributed uniformly over all boards; all
necessary communication between different boards (especially the broadcast of neuron
outputs) is managed by the DSPs via a special local bus. Each board is a slave VMEbus

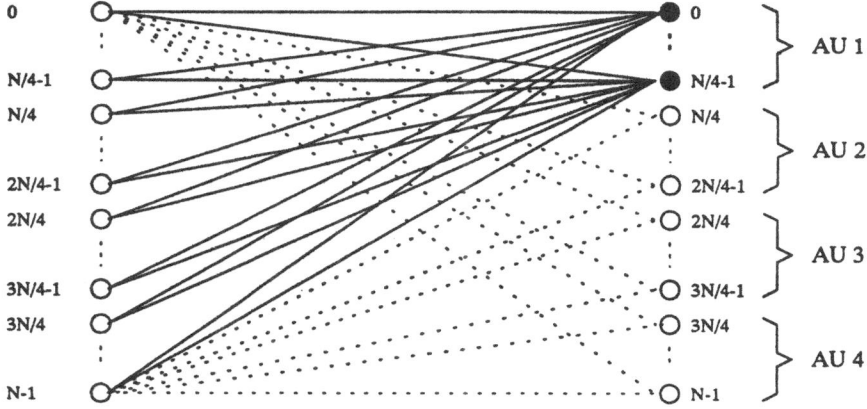

Figure 2: Mapping of one neural network layer with N neurons onto 4 AUs

board; the VMEbus master (host) can write a new input vector to the boards while the last input vector is still being processed. This is realized by buffering all I/O vectors in a FIFO memory.

4 The Arithmetical Unit (AU)

Based on our first experimental design of a neuroemulator system [1] and [2] we developped a new reconfigurable arithmetical unit especially suited for the massively parallel operations described above.

The central part of the AU (see figure 3) is a configurable multiplier with a 48-bit input A and a 48-bit input B. Both inputs A and B can be splitted into $k \in \{2, 3, 4, 6, 8, 12\}$ different data values $a_1, \ldots, a_k$ or $b_1, \ldots, b_k$ with a bitlength of $48/k$ bits. The configurable multiplier can operate in one of the following two modes:

I $(a_1, a_2, \ldots, a_k) \cdot (b_1, b_2, \ldots, b_k) \rightarrow a_1 \cdot b_1 + a_2 \cdot b_2 + \ldots + a_k \cdot b_k$

II $(a_1, a_2, \ldots, a_k) \cdot B_{23-0} \rightarrow (a_1 \cdot B_{23-0}, a_2 \cdot B_{23-0}, \ldots, a_k \cdot B_{23-0})$

In operation mode I the multiplier can be configured to compute either two 24x24 bit, three 16x16 bit, four 12x12 bit, six 8x8 bit, eight 6x6 bit or twelve 4x4 bit multiplications in parallel. This mode can be used for multiplying a vector $\mathbf{y}$ with the weight matrix W. Here the $k = 2$ to 12 data elements $b_1, \ldots, b_k$ are read from the weight memory port W and represent k different weights coded compactly in one 48-bit memory word. The k operands $a_1, \ldots, a_k$ represent neural states or inputs y_i and are read from the state memory port S, which has a width of 24 bit. The up to $k = 12$ parallel products are internally added so that only the sum of all products is presented at the output of the multiplier. Thus, in each clock cycle the the multiplier computes the partial inner product $y_i \cdot w_{ij} + y_{i+1} \cdot w_{i+1,j} + \ldots + y_{i+k-1} \cdot w_{i+k-1,j}$, the j-th element $x_j = \sum_i y_i w_{ij}$

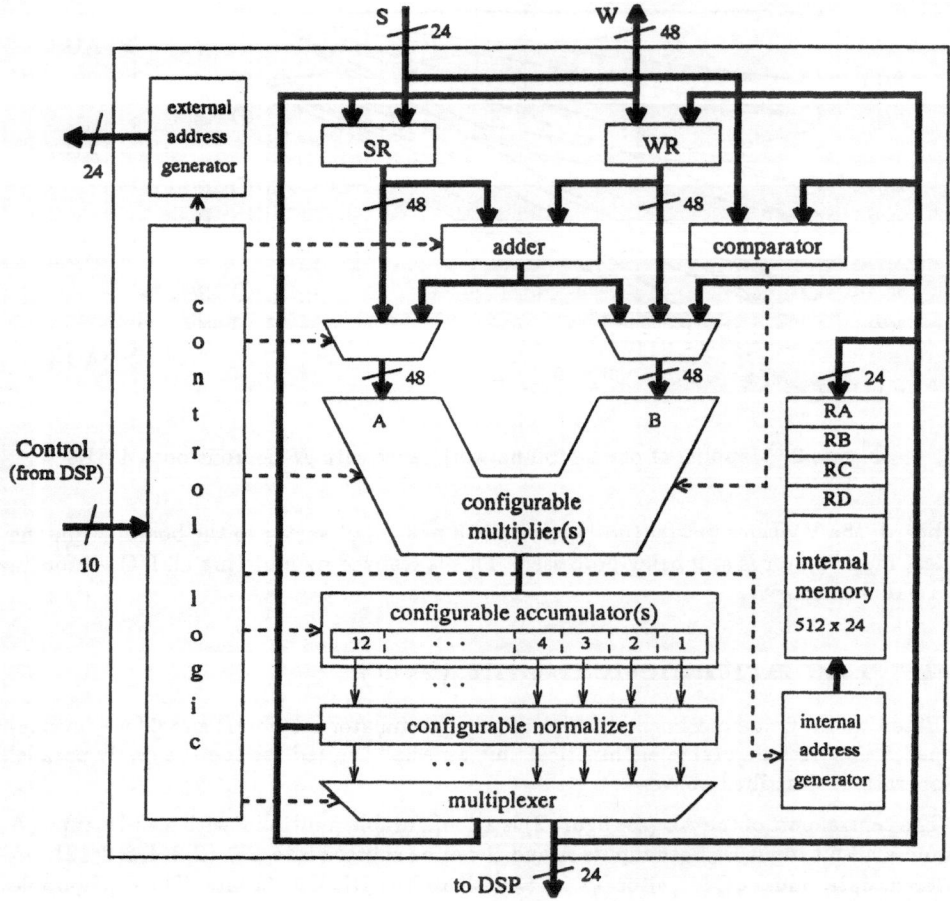

Figure 3: The Arithemetical Unit (AU)

of the result vector x can be calculated by accumulating the partial inner products. The operation mode I can also be used for computing in parallel the squares of up to k different values (e.g. for calculating Euclidean distances). Here the same input values must be presented at both inputs A and B of the multiplier.

For operation mode II the multiplier can be configured to compute in parallel two 24x24 bit, three 16x24 bit, four 12x24 bit, six 8x24 bit, eight 6x24 or twelve 4x24 bit multiplications. Here $k = 2$ to 12 different data elements a_i constitute the 48-bit value at input A. The second operand B however is a fixed 24-bit data value which is *identical* for all parallel multiplications and is expected to be presented at bits 23 to 0 of the multiplier input B. This mode is especially suited for computing the product of the error vector δ and the transposed weight matrix W^T in the backward phase of error backpropagation algorithms. Here the data elements $a_1, \ldots, a_k$ represent the weights w_{ij}

read from the weight memory port W. The operand B represents the error measure δ_j of a certain neuron j and is read from the internal memory (the state memory port cannot be used here because each AU needs a different δ value). The multiplier calculates in parallel the k products $w_{ij} \cdot \delta_j, w_{i+1,j} \cdot \delta_j, \ldots, w_{i+k-1,j} \cdot \delta_j$ and delivers the k results to the configurable accumulator (see below). Then the next locally stored value δ_{j+1} is presented as input B and the weigths $w_{i,j+1}, \ldots, w_{i+k-1,j+1}$ are read from the weight memory port W. The new k partial products are added to the old ones in k separate accumulators. Finally, the i-th, $(i+1)$-th, $\ldots$, $(i+k-1)$-th element of the result vector $\mathbf{u} = \delta \cdot W^T$ can be computed externally by the DSP which adds in sequence all corresponding partial products computed by the 4 AUs. The operation mode II is also useful for computing the outer vector product $\Delta w_{ij} = y_i \cdot \delta_j$. Here k states $y_i, \ldots, y_{i+k-1}$ are presented at input A and multiplied in parallel by δ_j read from the internal memory and presented at input B.

The accumulator can be configured as two 60-bit accumulators, as up to six 36-bit accumulators or as up to twelve 24-bit accumulators. So the maximum of 12 products computed in parallel by the configurable multiplier can be accumulated in up to 12 separate accumulators.

The other components of the AU chip are

- a configurable adder (e.g. for computing the differences $w_{ij} - u_i$ of Euclidean distances) which can add/subtract in parallel k different data elements coded in two 48-bit words

- a comparator (used e.g. for learning algorithms with adaptive learning rates)

- a configurable normalizer for selecting the relevant result bits from the accumulators

- an internal memory of size 512x24 which is used for storing the error measures δ_j in the backward phase of some learning algorithms and also for storing some parameters (like learning rate ϵ or momentum term μ)

- an external memory address generator for addressing the local weight memory.

The AU is realized by a standard cell ASIC design and was developed and simulated by using VHDL and the Synopsys CAD tool.

5 System Software and Applications

The 4 AUs are controlled by the DSP which broadcasts the required microinstructions. The corresponding microprograms and also the DSP procedures for the serial operations of the neural network simulation are downloaded into the internal RAMs of the DSP by the host. All popular neural network models (error backpropagation network with several variations proposed in the literature, radial basis function networks, self-organizing feature maps, learning vector quantization, ...) can efficiently be mapped onto the system architecture. A high-level neural network description language supporting the various bitlengths of all data elements has been designed; a compiler for automatically generating the DSP code and the microprograms is under development.

6 Conclusion

We presented a new general neurocomputer which is not dedicated to a specific neural network model. It can be configured for many neural network models with different precision requirements to simplify the selection of the optimal neural network architecture for a specific application.

Parallelism on three different levels is supported: on system level (by using several boards), on board level (the 4 AUs emulate in parallel 4 different neurons) and on chip level (each AU operates in parallel on 2 to 12 different weights of one neuron).

The theoretical performance of a system with 4 AUs (one board) is 600 MCPS and 150 MCUPS in the case of the standard error backpropagation algorithm, if a precison of 16 bits for the weights and 8 bits for the states is assumed.

At the moment of writing this paper (January 1995) the AU chips are specified and simulated by using VHDL. The schematics have been synthesized by the Synopsys CAD tool and the layout was generated by using Cadence Design Framework II. The complete system is planned to be operational in summer 1995.

References

[1] Avellana, N., Carrabina, J., Lisa, F., Reyes, M., and Valderrama, E. "Unidad aritmetica con paralelismo configurable para emular redes neuronales. In *IX Congreso de diseño de Circuitos Integrados* (Gran Canarias, Spain, 1994).

[2] Avellana, N., Carrabina, J., Rabaneda, L., and Valderrama, E. Experience on the development of a neuroemulator system. In *Fifth Eurochip Workshop on VLSI Training* (Dresden, Germany, 1994).

[3] Hammerstrom, D. A VLSI Architecture for High-Performance, Low-Cost, On-chip Learning. In *Proc. IJCNN* (San Diego, 1990), pp. 537–543.

[4] Holt, J., and Hwang, J. Finite precision error analysis of neural networks hardware implementations. *IEEE Transactions on Computers* 42 (1993), 281–290.

[5] Morgan, N., Beck, J., Kohn, P., Bilmes, J., Allman, E., and Beer, J. The Ring Array Processor: A Multiprocessing Peripheral for Connectionist Applications. *Journal of Parallel and Distributed Computing* 14 (1992), 248–259.

[6] Ramacher, U. Synapse – A Neurocomputer that Synthesizes Neural Algorithms on a Parallel Systolic Engine. *Journal of Parallel and Distributed Computing* 14 (1992), 306–318.

[7] Riedmiller, M., and Braun, H. A direct adaptive method for faster backpropagation learning: The RPROP algorithm. In *IEEE International Conference on Neural Networks (ICNN)* (1993), H. Ruspini, Ed., pp. 586–591.

[8] Tollenaere, T. SuperSAB: Fast adaptive backpropagation with good scaling properties. *Neural Networks* 3, 5 (1990), 561–573.

[9] Viredaz, M. MANTRA I: An SIMD Processor Array for Neural Computation. In *Euro-ARCH '93* (1993), P. Spies, Ed., Springer-Verlag, pp. 99–110.

An All-Optical Forward Propagation Multilayer Neural Network

I. Saxena and E. Fiesler

IDIAP, Case Postale 592, CH-1920 Martigny, Switzerland
Electronic mail address (Internet): ISaxena@IDIAP.CH

Abstract

An adaptive multilayer optical neural network with optical thresholding by a liquid crystal light valve (LCLV) and all-optical forward propagation is described. It has a large number of modifiable optical interconnections which are implemented by liquid crystal television screens, and it has a modular structure allowing the cascading of layers, each layer with its own light source.

Sigmoid fits to response curves of four LCLVs are evaluated and their suitability as optical thresholding functions is examined on the basis of neural network simulations.

1 Introduction

The promise of massive parallelism in optics has led to the investigation of several optical neural network implementations. Amongst these, Hopfield neural networks have been most frequently implemented optically due to their effective single layeredness. However, they are mainly limited to associative memory applications, whereas multilayer neural networks are capable of addressing a broad class of problems. Multilayer networks, in contrast, require a differentiable thresholding function, usually chosen to be the sigmoidal function $\frac{1}{1+e^{-x}}$ [Rumelhart-86].

Previous multilayer opto-electronic neural networks (MONNs) performed thresholding electronically (in software [Jang-93, Yu-90, Kranzdorf-89] or hardware [Ohta-90, Yuk-93]), and it is primarily the vector-matrix multiplication which was done optically. To progress towards all-optical recall and to reduce the overhead involved in conversion of optical signals into electronic, and re-conversion (after electronic thresholding) into optical signals for the next MONN layer, *optical* thresholding is essential. A liquid crystal light valve (LCLV) is a potential device for optical thresholding as its non-linear response is sigmoid-like. To evaluate the usefulness of LCLVs for optical thresholding in MONNs, the closeness of their response curves to sigmoidal functions is studied here.

In the following section, a MONN is described in which optical thresholding is implemented by an LCLV. Next, the properties of an ideal optical thresholding function for all-positive incoherent optical systems are examined. In the light of these properties, sigmoid curve-fits of four LCLV response curves are compared with ideal sigmoids, and their thresholding characteristics discussed. Finally, neural network simulations with the sigmoid-fits are analyzed to make a qualitative assessment of the most suitable LCLV for optical thresholding.

2 Multilayer Optical Neural Network

In optical neural networks where information is coded in light intensity, all variables including the interconnection weight matrix (IWM) elements have to be positive. (In general, IWMs contain both negative and positive weights, which are often referred to as bipolar weights.) An application independent, adaptive MONN architecture without negative weights is described here, in which optical thresholding is implemented by an LCLV. Three possible solutions allow working with unipolar weights in MONNs. In two-layer[1] systems (without hidden layer), the second and final layer can be electronic, and bipolar weights may be separated into two sets of unipolar weights, which, after being photodetected separately, either (i) *spatially* [Kasama-90] or (ii) *temporally* [Yu-90], are subtracted electronically. Bipolar weights may alternatively (iii) be transformed to unipolar weights by adding a bias term and compensating accordingly. In the MONN described here, the third possibility is adopted, and the argument of the activation function is biased at the hidden layer (with the aid of a personal computer), before optical thresholding.

Our three layer adaptive MONN has 256 input neurons, 256 hidden layer neurons, and 16 output neurons. This size enables addressing a wide range of applications. The information flow through the MONN can be described

---

[1]In this report, a *layer* of a neural network is defined as a layer of neurons (cf. [Fiesler-94.1]). In the three layer neural network described in the text, the input layer is considered number one, the hidden layer number two, and the output layer as layer number three.

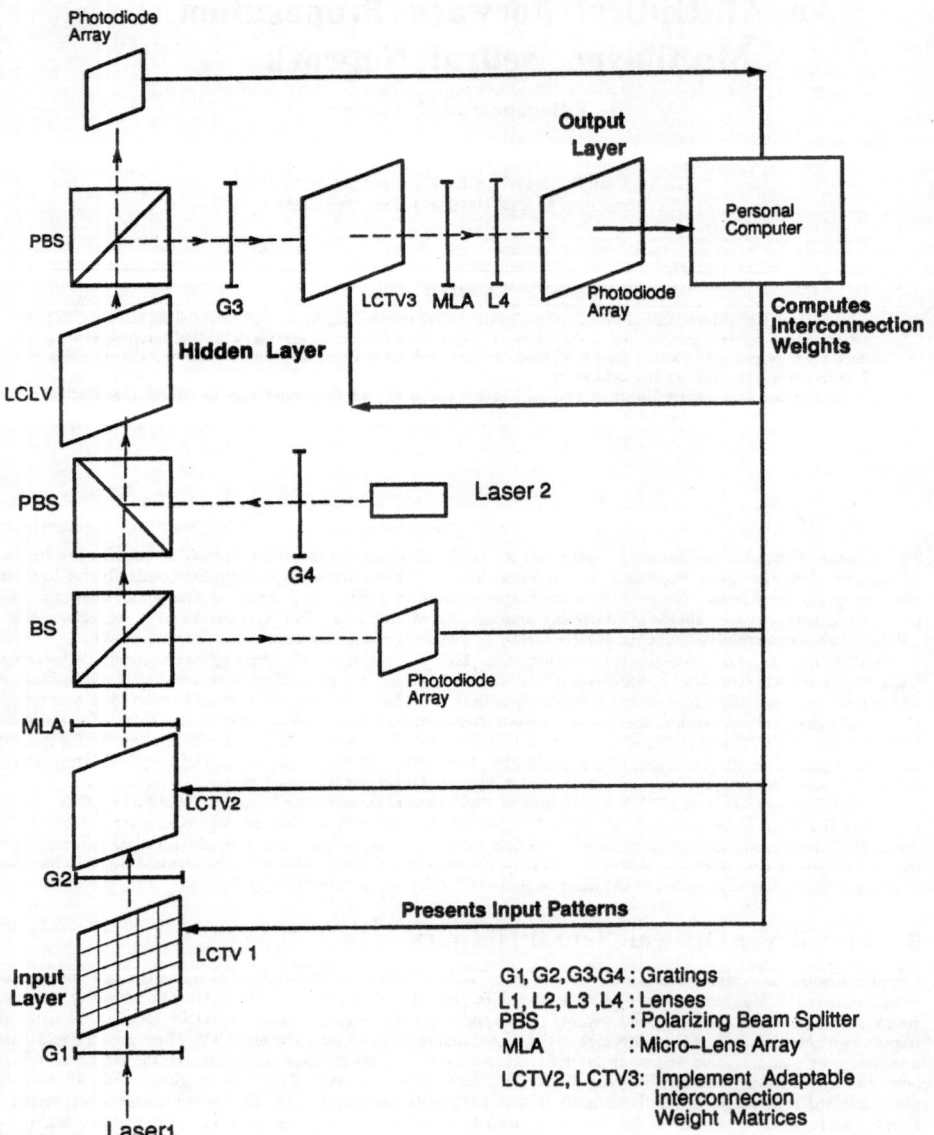

Figure 1: Schematic of the Multilayer Optical Neural Network.

Table 1: Sigmoid parameter definitions

Properties	Standard Sigmoid	Generic Sigmoid
Y-axis intercept	$\frac{1}{2}$	$\alpha + \frac{\gamma}{1+e^{-\delta}}$
Inflection Point (x,y)	$(0, \frac{1}{2})$	$(\frac{\delta}{\beta}, \alpha + \frac{\gamma}{2})$
Slope at inflection point	$\frac{1}{4}$	$\frac{\gamma\beta}{4}$
Steepness	1	$\gamma\beta$
$S_{max}(x)$ = asymptote for $x \rightarrow +\infty$	1	$\alpha + \gamma$
$S_{min}(x)$ = asymptote for $x \rightarrow -\infty$	0	α

as follows. Figure 1 shows the optical system whereby the inputs are presented by a Liquid Crystal Television (LCTV1) [Collings-94] to the fanned-out laser beam, and the 16×16 input array is replicated 256 times onto the transmissive LCTV2, following the fan-out design of Weible et al. [Weible-90, Weible-92]. The IWMs between layers are represented by the transmission values of the pixels of the LCTVs. The light transmitted by groups of pixels of LCTV2 is integrated by a micro-lens array to represent the vector-matrix product of inputs and weights[2]. This product, after biasing, is thresholded by the LCLV to produce the activation values of the hidden layer which are read by a new light source. The hidden layer outputs are similarly imaged onto LCTV3, which represents the 256×16 IWM between the hidden and output layer. Next, the micro-lens array following LCTV3 produces the second vector-matrix product. The final neural network outputs are detected by a 4×4 photodetector array. The hidden layer non-linearity, concerning 16×16 neurons and implemented optically by a transmissive LCLV, shall be further discussed in section 3.

In the MONN described above, light emitted by an Argon-ion laser at 480nm transmits through LCTV1, the fan-out optics, LCTV2, and the fan-in optics, and then writes to the LCLV. A helium-neon (633nm) light source reads the optically thresholded hidden layer output from the LCLV. Provided that it has sufficient gain, this method can amplify the light intensity to compensate for light losses in the first two layers of the neural network. Subsequently, each additional layer, with its own light source, can be modularly stacked onto the previous module to build an optical neural network with multiple layers.

3   Ideal Optical Thresholding

A sigmoid of the form $S(x) = \frac{1}{1+e^{-x}}$, to be referred to as the *standard sigmoid* in this paper, is the most often used thresholding or activation function in multilayer neural network simulations based on the backpropagation learning rule [Rumelhart-86]. The most interesting part of this function is centered about the y-axis for which the argument (of neural inputs) can assume both positive and negative values and can generate only positive outputs, $S(x)$, in a range from 0 to 1. However, in an incoherent optical processor based neural network, all variables (inputs and outputs) can only be represented by all-positive light intensities. This constraint requires the interesting part of an optical thresholding function to be in the all-positive quadrant, which can be obtained by translating a standard sigmoid along the x-axis by an amount δ. The corresponding sigmoid function is $S(x) = \frac{1}{1+e^{-x+\delta}}$. The shift, δ, could be determined by the range of neural network inputs for the problem at hand, akin to dynamic or input-dependent thresholding [Jang-88] in all-positive neural networks.

To allow a comparison with what real devices have to offer, a generic sigmoid, translated along the x- and y-axes with respect to the standard sigmoid by amounts δ and α respectively, and where both the range, γ (= $S_{max}(x) - S_{min}(x)$), and slope at the inflection point can have any value, is defined:

$$S(x) = \alpha + \frac{\gamma}{1 + e^{-\beta x + \delta}}. \qquad (1)$$

Table 1 provides a comparison of the characteristics of this generic sigmoid with those of the standard sigmoid. Note that the *shift* of a generic sigmoid is defined as the x-coordinate of its inflection point, and depends on the parameters δ and β.

To satisfy the all-positive nature of thresholding in an incoherent optical system, the response of an optical device should approximate a generic sigmoid satisfying the following conditions:
(i) $S_{min}(x)$ should be non-negative, that is, for all $x \geq 0$, the corresponding $S(x) \geq 0$, and,
(ii) the magnitude of the difference between the y-intercept and α is small, typically within a few percent of γ.

For sigmoids having unity range γ, that are most commonly used in neural network simulations, the slope at the inflection point differs only by a numerical constant from β, which is usually taken as the *steepness* and plays an important role in the learning rule (see section 5). However, the steepness of a generic sigmoid (equation 1) should be explicitly defined to encompass the vertical scaling factor γ, thereby rendering it invariant to vertical (and horizontal) scaling, as follows:

$$\text{Sigmoid Steepness} = \gamma\beta. \qquad (2)$$

---

[2] When light of unit intensity transmits through a material, a fraction of it is transmitted corresponding to the product of the incident intensity with the transmittance of the material.

Table 2: Sigmoid curve fit parameters for LCLV transfer curves

LCLV #	α	$\alpha' = \frac{\alpha}{\alpha+\gamma}$	β	γ	$\gamma' = \frac{\gamma}{\alpha+\gamma}$	δ
1	-28.2	-0.41	8.7×10^{-2}	97.7	1.41	0.93
2	-82.7	-2.34	6.2×10^{-3}	118.0	3.34	-0.82
3	0.46	0.015	4.3×10^{-2}	31.0	1.0	3.20
4	-0.43	-0.019	1.40	23.1	1.0	4.57

Table 3: Properties of scaled sigmoid-fits of LCLV response curves.

Properties	LCLV1	LCLV2	LCLV3	LCLV4
Y-axis intercept	-0.01	-0.024	0.05	-0.01
Inflection Point (x,y)	$(10.79, 0.3)$	$(-132.71, -0.67)$	$(75.26, 0.51)$	$(3.26, 0.49)$
Slope at inflection point	0.03	0.005	0.01	0.36
Steepness	0.123	0.021	0.042	1.4
$S_{max}(x)$	1.0	1.0	1.0	1.0
$S_{min}(x)$	-0.41	-2.34	0.015	-0.019

This definition also holds for the hyperbolic function $tanh(\beta x)$, (which is sometimes used instead of the standard sigmoid as the activation function), whose slope at the origin is β, and whose range is 2. If its range is multiplied by a factor γ, its slope becomes $\gamma\beta$.

4 LCLV thresholding

In addition to offering benefits of optical over electronic thresholding, LCLVs enable MONNs to be scalable in size, offering very large numbers of effective pixels [Clark-92], as compared to devices for electronic thresholding which are hardware limited by the number of pixels, as in 'smart' spatial light modulators or photodetector arrays.

Response curves or transfer characteristics for three LCLVs were obtained by Xue [Xue-94]. In order to find their useful functional approximations for neural network implementation, they were fitted to a generic sigmoid (equation 1). The three transfer characteristics and their generic sigmoid fits [Xue-94] are shown in figures 2 to 4, where the y-coordinate represents the actual optical read-out efficiency. In general, these transfer characteristics are truncated and asymmetric in comparison to generic sigmoids. The corresponding curve-fit parameters, both with optical coordinates, and with scaled y-coordinates, are given in table 2. The scaling, obtained by dividing α and γ by $(\alpha + \gamma)$, normalizes $S_{max}(x)$ to unity, and provides a comparison with the standard sigmoid in addition to facilitating their integration into neural network simulations. Characteristic properties of the sigmoid fits are given in table 3, together with those of another LCLV, LCLV4 [Micro-Optics]. The response curve of LCLV4 and its sigmoid fit is shown in fig 5. The useful sections of these curve-fits are the truncated parts for which $x \geq 0$ and $y \geq 0$. These truncated parts closely approximate the actual LCLV transfer characteristics, as can be seen in figures 2 to 5. It should be noted, however, that the y-coordinate of the lower asymptote, $S_{min}(x)$, of the LCLV2 curve-fit is decidedly negative due to the asymmetry of its transfer characteristic (see table 3).

The response curve of a given LCLV varies significantly with electrical and optical operating parameters like the frequency and magnitude of applied voltages and the wavelength of the write light [Xue-94, Davis-91]. Desired response curves can theoretically be generated by combining various available response curves [Caulfield-91], at the cost of increased hardware and speed. Alternatively, the operating parameters may be tuned to obtain a more desirable transfer characteristic to optimize system performance. Transfer curves at a write light wavelength of 480nm were obtained for the higher write sensitivity LCLV4 by varying the applied voltage. It was found that while the transfer characteristic of LCLV4 at a lower voltage could not be fit properly to a generic sigmoid, a good sigmoid fit could be obtained at a higher voltage of 15 volts.

An important aspect concerning the suitability of an LCLV as an optical thresholding device for an MONN is its write sensitivity which is defined as the light intensity per unit area required to switch it on. The transmissive LCLV4, operating at a wavelength of 480nm, has the lowest intensity requirement, and hence, the highest write sensitivity of $0.1\mu W/cm^2$. The minimum gain requirement [Psaltis-88] at each neural layer is $\sqrt{N_l}/2$, for incoherent optical systems, (N_l being the number of neurons in that layer). A layer with 256 neurons would then require a gain of approximately 8, which corresponds to a read-out beam having $0.8\mu W/cm^2$ for LCLV4, which is easy to achieve. For LCLV1, with its write sensitivity of $45\mu W/cm^2$ at 555nm, the read-out light must have $360\mu W/cm^2$ over the entire area of the LCLV. For even less sensitive LCLVs the read-out beam power requirement increases to more than one mW/cm^2 for LCLV3, and more than $4mW/cm^2$ for LCLV2; both at 515 nm [Xue-94].

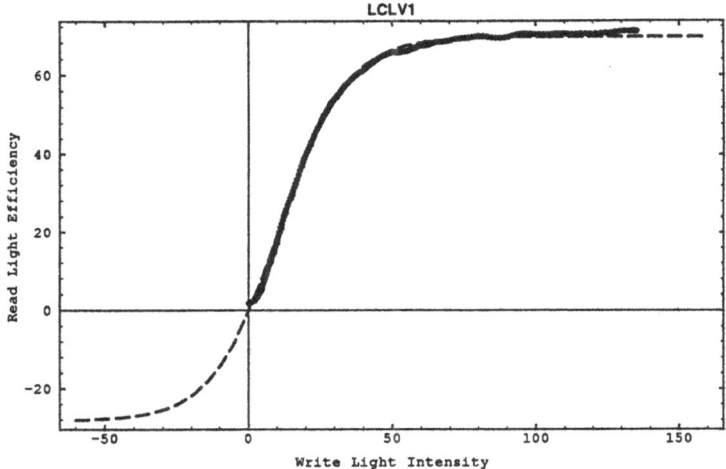

Figure 2: Transfer Curve of LCLV1 and curve fit.

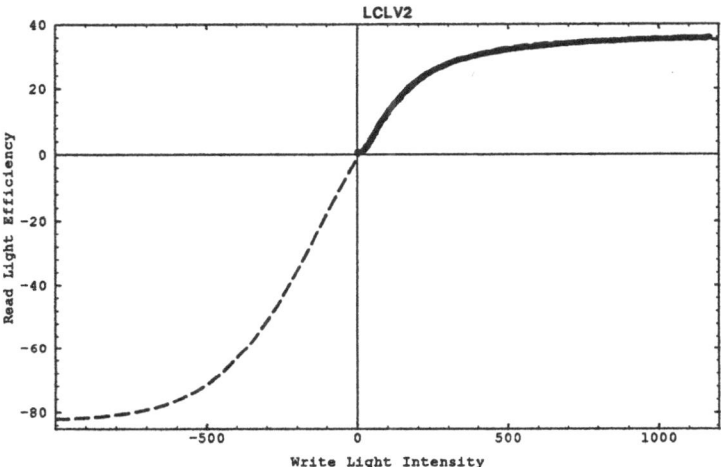

Figure 3: Transfer Curve of LCLV2 and curve fit.

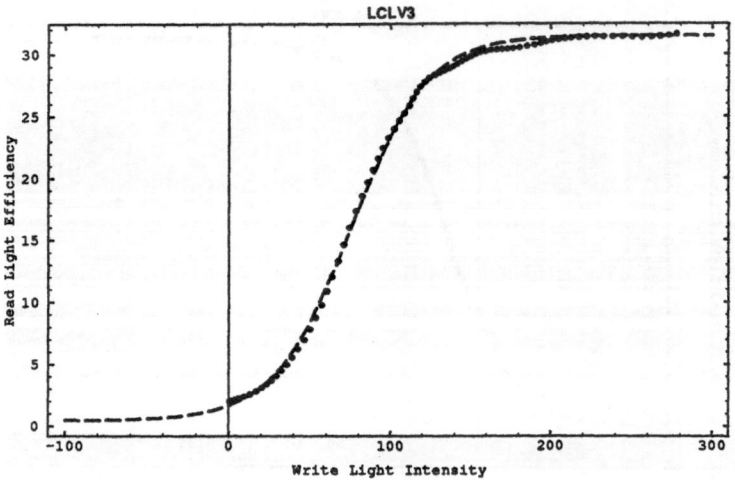

Figure 4: Transfer Curve of LCLV3 with curve fit.

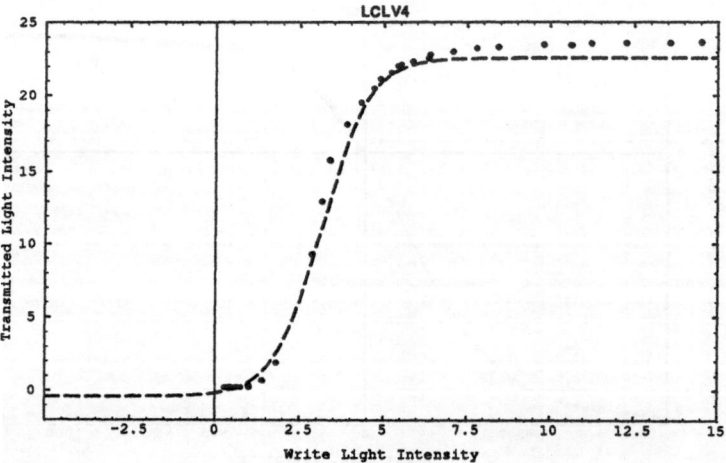

Figure 5: Transfer Curve of LCLV4 with curve fit.

5 Simulation Results

In order to assist in the selection of the most appropriate LCLV for optical thresholding, an extensive set of computer simulations was performed in which the LCLV transfer curves are represented by the sigmoid functions obtained from their curve fits.

The training of the optical neural network is performed by the weight discretization and update algorithm described in [Fiesler-94.2]. Besides a functional representation of the transfer curve itself, in this case the LCLV curve fit, the information needed for incorporating it into this training procedure are its derivative, its steepness, and its shift. The simulated neural networks are fully interlayer connected and initialized with small random weights.

Four well-known benchmark data sets, exclusive OR, parity, addition, and the two output sonar problem, were used for training; see [Fiesler-94.2] for details concerning these benchmarks. The first stage of each simulation consists in pre-training the neural network with continuous weights using the backpropagation learning rule. The convergence of this pre-training is a useful criterion for determining whether a given transfer function is suitable, since this is an important indicator for successful training of the network with discrete weights.

The results of a complete simulation were labeled *good* if the network was able to learn all the benchmarks with a maximum error of ten percent, using a maximum of twenty neurons in the hidden layer, and less than ten weight discretization levels. As a point of reference, a set of simulations has been performed using the standard sigmoid under conditions identical to those to be used for LCLV curve-fits, whose results were good for all the four benchmarks [Fiesler-94.2].

The first step towards adapting the training algorithm to enable the use of LCLV threshold curve fits was translating the standard sigmoid four units along the positive x-axis, such that the relevant part of the sigmoid was shifted into the all-positive quadrant. It was subsequently truncated at the y-axis to render it an all-positive sigmoid with a small y-intercept of about 0.018. It was possible to adapt the learning rule in such a way that it compensated for this translated and truncated sigmoid with only a negligible loss of performance.

Next, this adapted learning rule was used to train the neural network using each of the four LCLV sigmoid-fits. For each set of simulations with a given benchmark data set, the same learning rate, weight discretization step size, and initial weight range were used for all four sigmoids.

The results of the pre-training using the sigmoid fits for LCLV1, LCLV2, and LCLV3 did not converge and those of the complete simulation were also not satisfactory. Only the results of the simulations based on LCLV4, whose transfer characteristic is almost symmetric and closely approximated by its curve fit, were good.

It was then noted that the steepness of LCLV4 was closest to the standard value of one (see table 3) and that the steepness could therefore play an important role in the simulations. Analysis of a series of subsequent experiments yielded the following heuristic:

> *Divide the neural network learning rate, the weight discretization step size, and the initial weight range by the parameter β of the activation function.*

Using this heuristic, which renders each LCLV sigmoid fit with its own set of parameters, the simulation results for all LCLVs improved considerably, especially those for LCLV1 and LCLV3. The results based on LCLV1 and LCLV3 are now good for three out of the four benchmarks, failing only on the sonar problem.

6 Summary and Conclusions

A large and application independent adaptive MONN, based on the LCLV as a thresholding device and using available opto-electronic hardware, is presented. Sigmoidal thresholding in MONNs has been discussed and practical LCLVs for performing the thresholding have been evaluated based on sigmoid-fit approximations to their response curves. Translated and truncated sigmoids have been successfully incorporated as activation functions in our weight discretization learning rule for training multilayer neural networks. With this adapted learning rule, in which sigmoid fits have been used as thresholding functions, a large number of simulations have been performed to enable the selection of a useful optical thresholding LCLV. Simulations show that this adapted learning rule performs well with truncated generic sigmoids like the curve fit of LCLV4.

The adapted learning rule has been further extended by a heuristic to improve its capability of handling activation functions with a non-standard steepness, like the LCLV1 and LCLV3 transfer curve fits. Based on these simulations, the LCLVs can be arranged in order of suitability for optical thresholding, as follows: The LCLV4 response at 15 volts is the most desirable because of its similarity to a shifted standard sigmoid, and its high write sensitivity for providing sufficient gain. Good simulation results are obtained using its curve fit for all four benchmark problems. LCLV3 and LCLV1 are both a close second, with good results for three of the four benchmarks, with LCLV1 having the additional asset of a high write sensitivity at 555nm. The highly asymmetric LCLV2 response curve did not yield a desirable sigmoid fit for optical thresholding in neural network simulations, which does not perform well on the benchmarks.

7 Acknowledgements

We would like to thank Neil Collings for his contribution towards the optical design, Perry Moerland and Jeroen van Valburg for performing neural network simulations, and Wei Xue and Christoph Berger for their assistance concerning the data-acquisition and curve-fitting software.

This work is partially supported by the 'Fonds national suisse de la recherche scientifique', project number 21-36497.92, awarded to IDIAP in collaboration with the 'Institut de Microtechnique' (IMT) of the University of Neuchâtel.

References

[Rumelhart-86] David E. Rumelhart, James L. McClelland, and the PDP Research Group, 'Parallel Distributed Processing: Explorations in the Microstructure of Cognition', Volume 1: Foundations, The MIT Press, Cambridge, Massachusetts, 1986, ISBN: 0-262-18120-7.

[Jang-93] Jang, Shin, Yuk, Shin and Lee, 'Dynamic Optical Interconnections Using Holographic Lenslet Arrays for Adaptive Neural Networks', *Optical Engineering*, Vol. 32, No. 1, p. 80–87, January 1993.

[Yu-90] F. T. S. Yu, T. Lu, and X. Yang, and D. A. Gregory, 'Optical Neural Network with Pocket-Sized Liquid-Crystal Televisions', *Optics Letters*, Vol. 15, No. 15, pp. 863–865, August 1990.

[Kranzdorf-89] M. Kranzdorf, B. J. Bigner, L. Zhang, and K. M. Johnson, 'Optical Connectionist Machine with Polarization-Based Bipolar Weight Values', *Optical Engineering*, Vol. 28, No. 8, pp. 844–848, August 1989.

[Ohta-90] J. Ohta et al., 'Optical Neurochip Based on a Three-layered Feed-Forward Model', *Optics Letters*, Vol. 15, 1362–1364, 1990.

[Yuk-93] Seong-Won Yuk, Hyuek-Jae Lee, Soo-Young Lee, and Sang-Yung Shin, 'Optical Neural Networks Based on Error Back Propagation Learning for Hetero-Association of Two-Dimensional Patterns', *International Journal of Optical Computing*, Vol. 2, pp. 397–407, 1991.

[Kasama-90] N. Kasama, Y. Hayasaki, T. Yatagai, M. Mori, and S. Ishihara, 'Experimental Demonstration of Optical three Layer Neural Network', *Japanese Journal of Applied Physics*, Vol. 29, No. 8, p. L1565–L1568, August 1990.

[Collings-94] Both LCTV1 and LCTV2, having 440 × 480 pixels, are from a VPJ-200 Seiko Epson video projector, see: N. Collings, 'Design Considerations for a Useful Two-Layer Neural Network', presented at *Euro-American Workshop on Optical Pattern Recognition* at La Rochelle, France, June 1994.

[Weible-90] K. J. Weible, G. Pedrini, W. Xue, and R. Thalmann, 'Optical Implementation of a Neural Network Associative Memory Using Diffraction Gratings', *Japanese Journal of Applied Physics*, Vol. 29, No. 7, pp. L1301–L1303, July 1990.

[Weible-92] K. J. Weible, N. Collings, and A. Pourzand, 'Initial Results of a Fully Interconnected Neural Network with Modifiable Interconnects', *8th. Workshop on Optics in Computing*, Paris, 8th.–9th. September 1992, European Optical Society: Topical Meetings Digest series.

[Jang-88] Ju-Seog Jang, Su-Won Jung, Soo-Young Lee, and Sang-Yung Shin. 'Optical Implementation of the Hopfield Model for Two-Dimensional Associative Memory.' *Optics Letters*, Vol. 13, No. 3, pages 248–250, March 1988.

[Clark-92] N. A. Clark and K. M. Johnson, 'Applications of Liquid Crystals in Optical Computing', in Birendra Bahadur (editor), 'Liquid Crystals Applications and Uses', Vol. 3, World Scientific Publishing Co., 1992.

[Xue-94] Wei Xue, 'Characterization of Liquid Crystal Light Valves for Neural Network Applications', Ph.D. dissertation at the University of Neuchâtel, Switzerland, 1994.
Also in W. Xue, N. Collings, and K. J. Weible, 'The Characteristics of Three Kinds of LCLVs: Measurement and Comparison', IMT Report, January 1992.
And in, N. Collings and W. Xue, 'Liquid-Crystal Light Valves as Thresholding Elements in Neural Networks: Basic Device Requirements', *Applied Optics*, Vol. 33, No. 14, p. 2829–2833, May 1994.

[Micro-Optics] Micro-Optics Technologies, 8608 University Green # 5, Middleton, WI 53562, U.S.A.

[Davis-91] Jeffrey A. Davis, 'Binary Operation of a Liquid Crystal Light Valve', *Applied Optics*, Vol. 30, No. 3, p. 267, January 1991.

[Caulfield-91] H. John Caulfield, Janine Reardon, and Bahram Javidi, 'Simulating Arbitrary Response Curves with Available Response Curves for SLMs', *Proceedings of the SPIE*, Vol. 1562, p. 103–105, 1991.

[Psaltis-88] Demetri Psaltis, David Brady, and Kelvin Wagner, 'Adaptive Optical Networks Using Photorefractive Crystals', *Applied Optics*, Vol. 27, No. 9, May 1988.

[Fiesler-94.1] E. Fiesler, 'Neural Network Classification and Formalization', *Computer Standards & Interfaces*, Vol. 16, No. 3, special issue on Neural Network Standards, John Fulcher (editor), pp. 231–239, North-Holland/Elsevier Science Publishers B. V., Amsterdam, The Netherlands, 1994, ISSN: 0920-5489.

[Fiesler-94.2] E. Fiesler, A. Choudry, and H. J. Caulfield, 'A Universal Weight Discretization Method for Multi-Layer Neural Networks', *IEEE Transactions on Systems, Man, and Cybernetics (IEEE-SMC)*, (Accepted for publication). See also: E. Fiesler, A. Choudry, and H. J. Caulfield,' A Weight Discretization paradigm for Optical Neural Networks', in *Proceedings of the International Congress on Optical Science and Engineering*, volume SPIE 1281, pp. 164–173, SPIE, Bellingham, Washington, 1990, ISBN: 0-8194-0328-8.

A VLSI Current Mode Synapse Chip

David J Mayes and Alister Hamilton
Dept. of Electrical Engineering, University of Edinburgh,
King's Buildings, Mayfield Rd.,
Edinburgh, EH9 3JL.

Jean E Louvet
Electrical, Electronic and Computer Engineering Dept.,
Napier University,
219 Colinton Rd.,
Edinburgh, EH14 1DJ.

Abstract

Two–quadrant multipliers are required for several neural network architectures. The efficient implementation of these architectures in silicon requires the development of small, compact, reliable and accurate hardware multipliers. This paper details simulation and hardware results for the DYnamic Mirror PuLsed Experimental Synapse (DYMPLES) Chip. DYMPLES is an analogue current mode chip which utilises dynamic current mirrors and current matching to implement two quadrant multiplication based on the pulse stream approach[1]. HSPICE simulations indicate that the DYMPLES circuits produce excellent current matching and this is reinforced by the hardware results.

Introduction

Implicit in many hardware neural architectures (e.g. MLPs, RBFs [2]) is the need for robust, reliable and accurate multipliers. These are used to produce the weight X neural state ($T_{ij}S_j$) product terms required to calculate the output neural states.

In order to realise massively interconnected hardware neural networks, large numbers of $T_{ij}S_j$ synaptic multipliers are required. Analogue multipliers require less silicon area and dissipate less power than their digital equivalents allowing more multipliers, and hence more synapses, to be implemented per mm^2 of silicon real estate. Furthermore, analogue neural chips allow a direct interface to the real world (useful for sensor fusion applications) and perform truly parallel neural computations in real time.

The DYnamic Mirror PuLsed Experimental Synapse (DYMPLES) is based on dynamic current mirrors and has been designed to be robust, reliable and cascadable. The DYMPLES chip consisting of an 8x8 array of these synapses along with the necessary support circuitry has been fabricated and tested and some of the significant initial results are presented here for the first time.

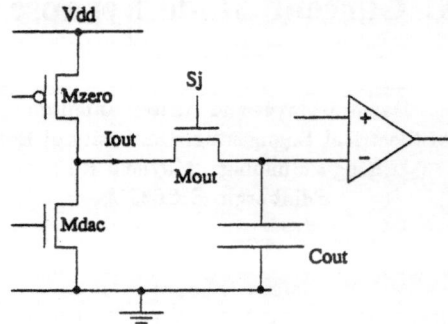

Figure 1: Pulsed synapse circuit.

Pulsed Synapse Operation

The principle of operation of the pulsed synaptic multiplier is well established[1] and is illustrated in Figure 1. Two transistors, M_{dac} and M_{zero}, are configured to act as an electronically programmable current sink and a fixed current source respectively. The currents through these transistors combine to produce an output current $I_{out} = I_{zero} - I_{dac}$ which linearly charges or discharges the output integration capacitance, C_{out}, under the control of the pass transistor, M_{out}. By applying a signal pulse, whose width represents the input neural state (S_j), to the gate of M_{out}, the resultant change in charge on C_{out} represents the required product term $T_{ij}S_j$. By applying the final voltage on C_{out} to one input of a comparator, and comparing it to a symmetrical ramp voltage on the other, a variable-width output pulse is produced.

Ideally, for large fabricated arrays of such circuits, the current sources and sinks and the output capacitors should be matched across the whole synapse array. However, transistor performance varies across a silicon die due to process variations and therefore, in reality, synaptic performance also varies.

Previous examples of pulsed synapses[1] stored a voltage on the gate capacitance of a transistor and used the transistor transconductance to generate the required currents. Weight voltages were refreshed from an off-chip RAM via a digital to analogue *voltage* convertor. This method makes the explicit assumption that either all the transistors can be matched across an array of multipliers or that techniques have been used to compensate for transistor mismatch.

With the DYMPLES design presented here, the transistors are operated in current mode. Values stored in off-chip RAM are converted to currents using a simple on-chip digital to analogue *current* convertor. The use of dynamic current mirrors allows the derivation of transistor gate voltages locally at each synapse, the resultant gate voltages being stored dynamically at the synapse site. This method implicitly accounts for across-chip transistor variation provided the currents are accurately controlled.

Whilst others have used dynamic current mirror techniques in the context of analogue neural networks[3], they have not yet been used for pulsed synapses.

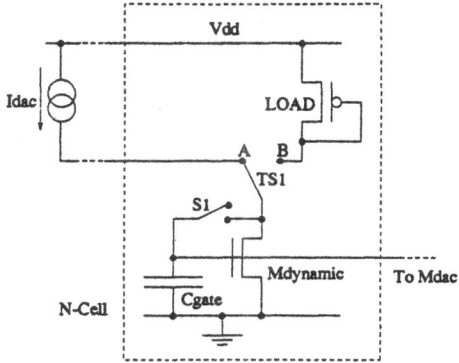

Figure 2: NMOS dynamic current mirror circuit.

Dynamic Current Mirrors

The principle of operation of a simple dynamic current mirror [4] is illustrated in Figure 2. With the toggle switch **TS1** connected to node **A**, current I_{dac} flows through transistor $M_{dynamic}$ provided it has a sufficient gate voltage. When switch **S1** is closed, $M_{dynamic}$ is forced into saturation and the voltage required by $M_{dynamic}$ to conduct I_{dac} is established on C_{gate}. When the appropriate gate-source voltage has been set for $M_{dynamic}$, the transistor has effectively been programmed to sink current I_{dac} and **S1** can be opened. When **TS1** is subsequently connected to node **B**, a current of value I_{dac} is pulled through the load due to the gate-source voltage of $M_{dynamic}$.

The complete dynamic mirror pulsed synapse is shown in Figure 3. As alluded to previously for this implementation, the current I_{dac} is generated by reading a digital word from RAM and using it to switch binary weighted currents onto the I_{dac} line. This technique allows a direct interface from a digital RAM to the analogue neural chip.

The PMOS constant current source from Figure 1 (the P-Cell in Figure 3) is implemented in a similar manner to the programmable N-Cell of Figure 2, with the NMOS transistors replaced by PMOS transistors and vice versa, and the input current line, I_{zero} (corresponding to I_{dac}), connected to a constant current sink of value $I_{dac(max)/2}$. Under a common timing control scheme, both dynamic mirrors can be programmed simultaneously.

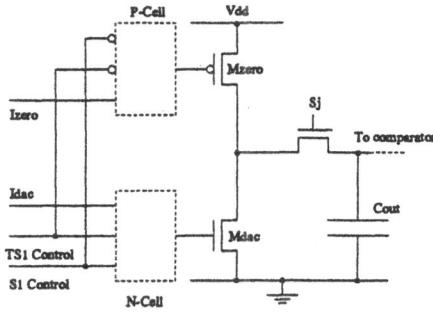

Figure 3: Dynamic current mirror pulsed neural synapse.

The output stage of the synapse is formed by matching the output transistors M_{zero} and M_{dac} to the $M_{dynamic}$ transistors of the PMOS and NMOS dynamic current mirrors respectively (this is achieved by closely juxtaposing the transistors in silicon) and their gates are connected to the appropriate C_{gate} capacitances. An advantage of this circuit configuration is that it allows refresh of the stored voltage on C_{gate} to be achieved, for little loss in overall performance, whilst network calculations are being performed. Thus the data throughput of the chip is independent of the refresh time.

Simulations

Our simulations were performed using HSPICE and the transistor models for ES2's 1.5um CMOS fabrication process. Three transistor models were available corresponding to the typical transistor threshold voltage (V_T) produced by the process (TYPICAL transistor model) plus the two extremes of high process V_T (SLOW transistor model) and low process V_T (FAST transistor model). These three model types were used to construct typical, slow and fast synapses for the simulation trials.

Since our circuit relies on the use of the equation

$$I_{out}.\Delta t = C_{out}.\Delta V_{out} = T_{ij}.S_j \tag{1}$$

to produce the correct multiplicative operation, its ultimate performance depends on the capacitance, C_{out}, which in turn depends on the characteristics of the fabrication process. For this reason, and because the total output capacitance is distributed, and hence averaged, across the array, the output capacitor for the simulations was configured as a typical 1.5um NMOS transistor. This ensured output load consistency and thus allowed the operational performance of the different synaptic multipliers to be compared. The first simulation involved establishing 16 equally distributed currents, corresponding to binary currents of 0000_2 to 1111_2, in the dynamic mirror for each of the three synapse types and measuring the voltage established on C_{gate} in the NMOS current sink. These results are shown in Figure 4. This graph confirms that different gate-source voltages are indeed required, due to threshold voltage variation, to conduct the same current through fast, slow and typical synapses.

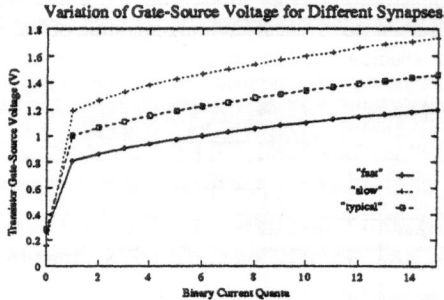

Figure 4: Variation of V_{gs} of an NMOS dynamic mirror transistor with binary current (in quanta) for fast(bottom curve), slow(top curve) and typical synapses.

The second set of simulations investigated the two quadrant multiplication characteristic of the synapse. Each of the sixteen currents were individually re-established in all three synapses and various pulse widths were applied to the gate of M_{out}. The final output voltage on C_{out} was then noted and some of the obtained results (for binary weights of 0000_2 (top graph), 0101_2, 1010_2 and 1111_2(bottom graph)) are shown in Figure 5. As can be seen, the output voltage varies linearly with the applied pulse width for the different binary weightings. This result confirms that the multiplier works as intended[1]. Furthermore, since the multiplication characteristic is virtually identical for each type of synapse, this result also indicates the potential current matching ability of the synapse circuits.

Results

The fabricated DYMPLES chip is being tested at the time of writing and some significant initial results are presented here. The chip will be fully tested in the coming months and a full set of results will be presented at the conference.

Having designed suitable output test transistors for some of the synapses and the current DAC, it was possible to test these circuits, verify their correct operation and show that

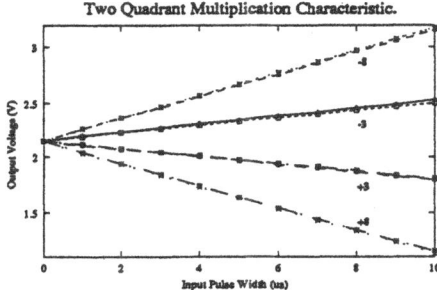

Figure 5: Simulated multiplication characteristic of fast, slow and typical synapses corresponding to currents of -8(top), -3, +3 and +8(bottom) current quanta.

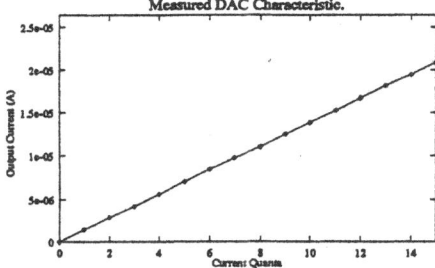

Figure 6: DAC Performance.

---

[1] A further inversion is performed by the output comparator to re-invert the final product term.

DYMPLES was able to perform Pulse Width Modulation calculations. Circuit operation was tested by connecting the output of each test transistor, in turn, to the inverting input of a JFET OP-AMP configured as an inverting amplifier and monitoring the output voltage of the OP-AMP for various DAC inputs. Clearly for this chip the necessary indicators of overall performance will be the linearity of the current DAC and the current matching ability of the synapses. Initial results for both are presented in Figures 6 and 7.

Figure 6 shows the variation in the output current from the OP-AMP connected to the DAC output as the 4-bit input word size is varied. This graph gives a clear indication that the DAC performance is linear and it is worth highlighting that the slope of the graph can be varied by varying the DAC's bias current.

Figure 7 shows the excellent current matching ability of the NMOS dynamic current mirror (Figure 2) within the synapse circuit. The results shown were obtained from the end synapses of the second, fourth, fifth and seventh rows in the 8x8 array by scaling the OP-AMP output voltage data.

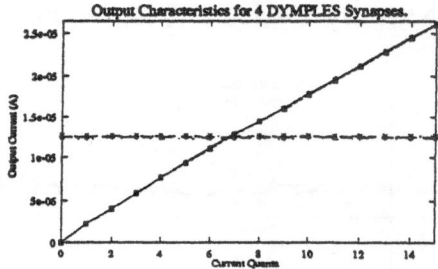

Figure 7: Performance of 4 DYMPLES Chip Synapses showing the currents flowing in the NMOS and PMOS dynamic current mirrors.

Some initial quantitative information has been obtained regarding DYMPLES' ability to perform Pulse Width Modulation two–quadrant multiplication calculations[1] as per Figure 5. These highly encouraging results are presented in Figure 8. However, the results are neither full nor complete and merely prove the chip's operational and functional performance. A complete set of results will be presented at conference.

Possible Improvements

Ideally the synapse cell should be as small as possible as this will allow many such circuits to be fabricated on-chip. At the moment, however, the size of the synapse cell is dictated by three large capacitors.

Each of the dynamic mirror capacitors was made large to compensate for the gate voltage variations due to charge injection from the switches of the dynamic mirrors. By employing suitable compensation circuits to account for this phenomenon[4], it is hoped to significantly reduce the size of two of these capacitors and hence also decrease the circuit's set-up time.

The dimensions of C_{out} are determined by the size of the maximum output current from the synapse circuit and the maximum desired voltage change required in a given time. However, because the final operation of the circuit will ultimately depend upon how accurately this capacitor can be fabricated, some alternative form of integrator may be better. One possibility would be to use an integrator whose operation is dependent on the ratio of two capacitors since this will be fabricated to a greater accuracy.

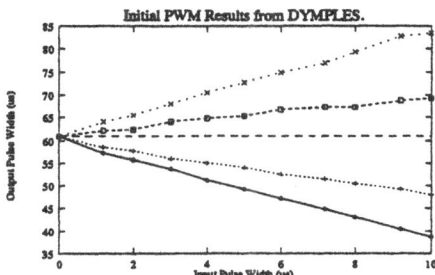

Figure 8: Initial Pulse Width Modulation (PWM) multiplication performance for DAC inputs 1111 (top), 1010, 0011 and 0000 (bottom).

Conclusions

We have presented a pulsed circuit whose simulated performance shows it is capable of reproducing the $T_{ij}S_j$ multiplication as required by hardware ANNs. The DYMPLES circuit was designed to be reliable, accurate, robust and cascadable and the HSPICE simulations highlighted its potential for process tolerance. Its current matching potential and ability to perform two–quadrant multiplications have been demonstrated by the initial hardware trials. The circuit is applicable to many pulse-stream neural hardware implementations and it is intended that the circuit will feature as the basic synapse circuit in our analog pulse stream RBF chips since it has several advantages over the EPSILON synapse in terms of biasing, control circuitry and simplicity of operation[5].

References

[1] HAMILTON, A., MURRAY, A. F., BAXTER D. J., CHURCHER, S., REEKIE, H. M., and TARASSENKO, L.: 'Integrated pulse-stream neural networks - results, issues and pointers', *IEEE Transactions on Neural Networks*, Vol. 3, No. 3, 1992, pp. 385-93.
[2] LIPPMANN, R. P.: 'An introduction to computing with neural nets', *IEEE ASSP Magazine*, 1987, Vol. 4, pp. 4-22.
[3] VERLEYSEN, M., THISSEN, P., VOZ, J-L., and MADRENAS, J.: 'An analog processor architecture for a neural network classifier', *IEEE Micro*, Vol. 14, No. 3, June 1994, pp. 16-28.
[4] VITTOZ, E. A., and WEGMANN, G.: 'Dynamic current mirrors' in TOUMAZOU, C., LIDGEY, F. J., and HAIGH, D. G. (Eds.), 'Analogue IC design: the current-mode approach' (Peter Peregrinus Ltd., London, 1990), pp.297-326.
[5] HAMILTON, A.: 'Synaptic weight representation in pulsed neural network VLSI: Voltage or Current ?', *Proc. of Int. Conf. on Neural Information Processing.*, Seoul 1994, pp.353-8.

Optimal Mapping of Neural Networks Onto FPGAs[1]
– A New Constructive Algorithm –

Valeriu Beiu[†,‡] and John G. Taylor[†]

[†]*King's College London, Centre for Neural Networks, Department of Mathematics Strand, London WC2R 2LS, United Kingdom*

[‡]*on leave of absence from "Politehnica" University of Bucharest, Computer Science Department Spl. Independentei 313, RO-77206 Bucharest, România*

Abstract — The paper shows how a feedforward neural network defined by a set of m binary examples of n bits each, can be determined by a novel constructive algorithm (which determines the number of layers, the number of neurons in each layer and the synaptic *weights* of a particular neural network). For doing that, the optimisation criteria of the new algorithm can be chosen from the following: (i) the *area* of the circuit A; (ii) the AT^2 complexity measure of VLSI; (iii) the *delay* T; or (iv) the maximum *fan-in* for a gate Δ. As a result the neural network which is build can be optimised for mapping onto a FPGA. By considering the maximum *fan-in* of one neuron as a parameter, we proceed to show its influence on the *area*, and suggest how to obtain a full class of solutions. We also compare our results with other constructive algorithms and benchmark it on the classical "two spirals problem." Conclusions and some open problems are closing the paper.

1. INTRODUCTION

In this paper we shall consider only feedforward neural networks (NNs) made of threshold gates (TGs):

$$Z_k = (z_0, \dots, z_{n-1}) \in \{0,1\}^n, \; k = 1, \dots, m, \text{ and } f(Z_k) = sgn(\textstyle\sum_{i=0}^{n-1} w_i z_i + \theta) ,$$

with w_i called the synaptic *weights*, θ known as the *threshold*, and sgn the signum function. A feedforward NN will be a feedforward TG circuit having as cost functions:(i) *depth* (i.e. number of layers), and (ii) *size* (i.e. number of TGs). These are linked to $T = delay$ and $A = area$ of a VLSI chip. Still, TGs do not closely follow these proportionalities as: (i) the *area* of the connections counts; (i) the *area* of one TG is related to its associated *weights*. That is why the *size* and *depth* complexity measures are not the best criteria in ranking different solutions when going for silicon. Several authors have taken into account the *fan-in* [12], the total number of connections, the total number of bits needed to represent the *weights* or even more precise approximations like the *sum of all the weights and the thresholds* [4]:

---

1   This research work has been started while Dr. Beiu has been with the Katholieke Universiteit Leuven, Belgium, and has been supported by a grant from the *Concerted Research Action of the Flemish Community* entitled: *"Applicable Neural Networks."*

The research has been continued under the *Human Capital and Mobility* programme of the European Community as an *Individual Research Training Fellowship*: ERB4001GT941815 hosted by King's College London.

The training project is entitled *Programmable Neural Arrays* and is financed by the Commission of the European Communities under contract ERBCHBICT941741.

$$area \; \propto \; \sum_{all\,TGs} \left(\sum_{i=0}^{n-1} |w_i^{TG}| + |\theta^{TG}| \right). \tag{1}$$

Equation (1) has also been used to define the minimum-integer TG realisation of a function. The *depth* will be assumed to be a good approximation of the *delay* (we neglect the *delay* introduced by long wires [16]). There are also several sharp limitations for VLSI like: (i) the maximal value of the *fan-in* cannot grow over a certain limit; (ii) the maximal ratio between the largest and the smallest weight.

Two natural questions arise when using NNs: (i) *"How to determine the synaptic weights?"*; and (ii) *"How many neurons should one use in each hidden layer?"*. These questions can either be solved by learning techniques or by constructive algorithms. *Learning algorithms* [22] suffer from the inherent error correction function which does not guarantee that the global minimum will ever be reached. That deficiency often stands beside the very long time needed for solving the problem. Other aspects, like the precision of the *weights* [13, 29] or the high *fan-in* of neurons, are completely neglected as such networks are simulated by software. These preclude their VLSI implementation. *Constructive algorithms* [3, 18] are error prone, so they will find the correct set of *weights*. They can be divided into *algebraic* [14], *network-based* [25], or *geometric* [6, 20]. Some of them are for particular functions, or use queries and/or hints [1]. An excellent overview of seven constructive algorithms (sometimes called "direct design" or "growth" algorithms) can be found in [25]. There exist *alternate solutions* developed for overcoming some of the above mentioned deficiencies: *learning algorithms with quantified weights* [2, 8, 9, 10, 11, 16], and *polynomial time direct design* [26].

In section 2 we shall shortly present some results concerning the *depth*, the *size* and the *area* of $F_{n,m}$. This is the class of Boolean Functions (BFs) of n bits having m groups of ones in their truth table. Based on already established *depth*, *size* and *area* values for implementing the COMPARISON of two n-bit numbers [4], we review a constructive theorem [5] for implementing any function belonging to $F_{n,m}$.

In section 3 we detail the mathematical basis of the new constructive algorithm. The m given examples are represented on n bits each (either all the examples are Boolean, or they are reals which have been quantified on at most n bits). The output will be represented on k bits. Thus the learning problem is equivalent to simultaneously implementing k BFs defined on m points. By extending the results from the previous section to the case of implementing k functions belonging to $F_{n,m}$, we determine the *depth*, the *size* and the *area* of such an implementation for a maximum given *fan-in*. We present the same measures for a classical *distributed normal form* (DNF) implementation (suggested in [3]). For taking the best part of both alternatives COMPARISONs should be used only when they give a smaller *area* than the AND-equivalent gates needed to implement all the examples bounded by the COMPARISONs. This theoretical basis helps us to describe the algorithm.

In section 4 we compare and benchmark the algorithm on the well known "two spirals" problem. The results are supporting our claims of VLSI-optimality and simplicity when mapping the NN onto FPGAs. Conclusions as well as open problems for research are pointed in the last section.

2. PREVIOUS RESULTS

A BF is symmetric if and only if it remains unchanged by any permutation of its input variables [7]. More than thirty years ago Muroga [19] has shown that any symmetric function of n inputs can be realised by a *depth*-2 TG circuit of *size* = $n+1$. This result has been improved to *size* = $2\sqrt{n} + O(1)$ in *depth*-3 [24]. Both constructions have polynomial *fan-in*. The *fan-in* can be reduced by tree decompositions [4]. Symmetric functions can be build by using only threshold elementary symmetric function (MAJORITY functions, having only ± 1 *weights*). Any symmetric function can be decomposed in Δ-ary trees [4] having $size_{MAJ}(n,\Delta) = \lceil n-1/\Delta-1 \rceil$, $depth_{MAJ}(n,\Delta) = \lceil \lg n / \lg \Delta \rceil$ and area $A_{MAJ}(n,\Delta) = \Delta \lceil n-1/\Delta-1 \rceil$. In this paper Δ is the maximum *fan-in* [23], $\lceil x \rceil$ is the smallest integer greater or equal than x (or "ceiling"), and $\lg x$ is the base two logarithm of x ($\log_2 x$).

Apart from the useful class of symmetric functions another known class is $F_{n,m}$ which has been defined by Red'kin [21] as *"the class of Boolean functions $f(x_1, x_2, \ldots, x_n)$ that have exactly m groups of ones."* For constructing a TG circuit for $f \in F_{n,m}$, Red'kin uses $\varphi_i(\tilde{\sigma})$ (see [21]) which are COMPARISONs. It achieves an excellent *size* $2\sqrt{m} + 3$ in *depth*-3 (the existence lower bound being $\Omega(\sqrt{m})$ [3, 24]), but with: (i) exponential *weights* and *thresholds*, and (ii) polynomial *fan-in*. From the VLSI point of view these rule out his solution for $n > 8$.

We propose a solution which should use $2m$ COMPARISONs in a hidden layer to determine the beginning and the ending of each of the m groups, and one MAJORITY gate as output which, by taking alternate signs for the *weights*: $+1, -1, +1, -1, \ldots$ (see Fig.1), will "ADD" the results from this hidden layer (the COMPARISON gates). We have proven [4] that there are NNs for computing a 2 n-bit COMPARISON having *size* $O(^n/_\Delta)$ and *depth* $O(^{\lg n}/_{\lg \Delta})$ for all the Δ in the range $O(1)$ to $O(\lg n)$. These results have improved on the previous known *size* $O(n)$ [24]), while still having polynomial bounded *weights* (for $\Delta \le \lg n$, *weights* $\le \sqrt{n}$). We have also shown [4] that by alternating the signs of the *weights* one can achieved the lowest possible *threshold* (-1) for all TGs.

Because for VLSI the chip *area* represents a much more important cost measure than the conventional *size*, we have firstly use the rough estimate *area* $\propto$ *size*. In this case our solution achieves:

$$AT_{COMP}^2(n,\Delta) = O\left(\frac{n\lg^2 n}{\Delta \lg^2 \Delta}\right).$$

This ranks **BPVL_4** (having constant *fan-in* TGs $\Delta = 4$), **BPVL_lg** (having logarithmic *fan-in* $\Delta = \lg n$ TGs [4]), and **SRK** (the best solution, from [24]) as:

$$AT_{SRK}^2 = O(n) < AT_{BPVL\_lg}^2 = O\left(\frac{n\lg n}{\lg^2(\lg n)}\right) < AT_{BPVL\_4}^2 = O(n\lg^2 n).$$

In spite of the fact that AT_{SRK}^2 has the lowest complexity, the larger constant ($27 > 4$) makes **BPVL_lg** the best solution for any practical values ($n \le 2^{564} \cong 6 \cdot 10^{169}$).

For an even better approximation of the *area* we have used eq. (1) which leads to:

$$AT_{COMP}^2(n,\Delta) = \frac{2^{\Delta/2}}{\Delta} \cdot \frac{8n\Delta - 6n - 5\Delta}{\Delta - 2} \cdot \frac{\lg^2 n}{\lg^2 \Delta} = O\left(2^{\Delta/2} \cdot \frac{n\lg^2 n}{\Delta \lg^2 \Delta}\right),$$

and clearly points to **BPVL_4** as the best solution, reversing the ranking:

$$AT_{SRK}^2 = O(n^2) > AT_{BPVL\_lg}^2 = O\left(\frac{n\sqrt{n}\lg n}{\lg^2 \lg n}\right) > AT_{BPVL\_4}^2 = O(n\lg^2 n).$$

The ranking is (again) more subtle: the **BPVL_lg** solution is the best for $n < 6 \cdot 10^4$. Only for $n > 8 \cdot 10^4$ the lower AT^2 values are those of **BPVL_4**.

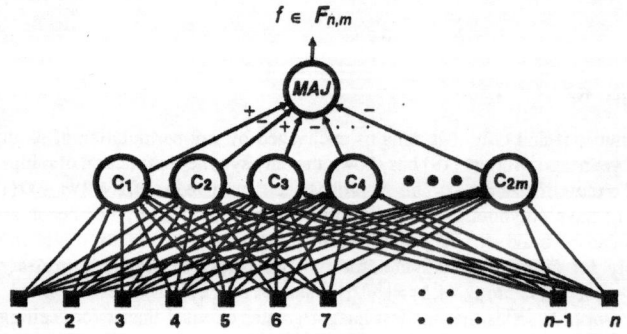

Figure 1. A two layer TG circuit used for implementing any $f \in F_{n,m}$.

By linking these results for COMPARISON with the ones for implementing MAJORITY functions, a systematic solution for decomposing $F_{n,m}$ functions has been developed [4]:

Theorem 1: *Any function $f \in F_{n,m}$ can be computed by a NN with polynomially bounded integer weights in $depth_f(n,m,\Delta) = O(\lg(mn)/\lg\Delta)$ and*

$$size_f(n,m,\Delta) = \begin{cases} O\left(mn/\Delta\right) & if \quad 2m \le 2^\Delta \\ O\left(mn/\Delta^2\right) & if \quad 2m > 2^\Delta \end{cases},$$

and occupying an area of $A_f(n,m,\Delta) = O(mn \cdot 2^\Delta/\Delta)$ if $2m \le 2^\Delta$ for all the fan-in values in the range 3 to $O(\lg n)$.

Varying Δ from 3 to $c \lg n$, the full set of solutions with *area* ranging from $O(mn)$ to $O(mn^{1+c}/\lg n)$, and *delays* from $O(\lg(mn))$ to $O(\lg(mn)/\lg n)$ can be obtained. The AT^2 can also be computed by taking $T \propto depth$, leading to $O\left(mn \cdot \lg^2(mn) \cdot \dfrac{2^\Delta}{\Delta \lg^2\Delta}\right)$.

Remark 1: *These results also suggest that the minimum area A should be reached for $\Delta_{opt\_A} = 3$, while for minimising AT^2 one should use $\Delta_{opt\_AT^2} = 6$. The values are not exact, but the optimum should be in very close vicinity (± 1 or ± 2). This is explained by the fact that the optimum Δ_{opt} have been obtained by estimating the derivatives (we have neglected the ceilings) and numerically solving the resulting transcendental equations (see [4, 5]).*

For a better understanding a small example is presented in Fig. 2. The function has 16 inputs and 3 groups of ones, while the *fan-in* has been limited to $\Delta = 4$. We have chosen the starting of the first group of ones at 0, while the ending of the third group of ones is at $2^{16} - 1$. The overall structure has 4 levels: (i) the first 3 levels represent the decomposition of the "hidden layer" of COMPARISONs; (ii) the 4th level is the "output layer" (in this case it is just one MAJORITY gate). The *size* is 105, but with a drastically limited *fan-in*: some gates having 3 inputs while others having 4 inputs. This leads to very small *weights* (|weights| $\in \{1, 2\}$) and extremely simple Boolean functions are now implemented by the nodes. This is an important feature when thinking to implement such structures in VLSI by means of FPGAs.

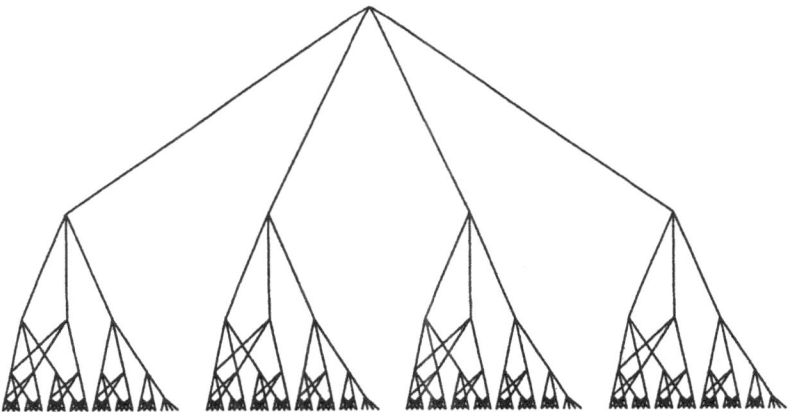

Figure 2. Area-efficient tree decomposition for $F_{16,3}$ ($n = 16$, $m = 3$) when $\Delta = 4$.

3. LEARNING FROM EXAMPLES

The following theorem shows that the decomposition method used for $F_{n,m}$ functions can be directly used to "learn" k functions from m examples (k $F_{n,i}$ functions, $i \leq m$).

Theorem 2: *Any set of k functions $f \in Fn,i$, $i = 1,2,...,m$, $i \leq m \leq 2^{\Delta-1}$ can be computed by a NN with polynomially bounded integer weights having $size_{kf} = O\,(\,^{m\,(2n+k)}\!/_\Delta\,)$ and $depth_{kf} = O\,(\,^{\lg(mn)}\!/_{\lg\Delta}\,)$ and occupying an area of $O\,(\,^{m\,n\cdot 2^\Delta}\!/_\Delta + mk\,)$ for all the values of the fan-in in the range 3 to $O\,(\lg n)$.*

Remark 2: For $2m > 2^\Delta$, many TGs needed in the first level of the $2m$ trees for decomposing COM-PARISONs are redundant. Not wanting to complicate the theorem we have omitted this case.

This result could still be improved as has been suggested by Baum [3]: directly implement the *disjunctive normal form* (**DNF**) of the functions. Instead of 2 COMPARISONs we might use just one TG, thus reducing the *size* by a factor of 2. The TG is an AND-equivalent gate having as inputs the direct value of the variables corresponding to ones and the complemented values for the variables corresponding to zeros. As the minterms are common to all the k functions, the worst case will require m AND-equivalent TGs in the hidden layer instead of $2m$ COMPARISONs. Also of interest is that an AND-equivalent gate is simpler than a COMPARISON gate both with respect to *area*, and to the decomposition into a limited *fan-in* tree.

Theorem 3: *Any set of k functions $f \in Fn,i$, $i = 1,2,...,m$, of n variables specified by m examples can be computed by a NN with ± 1 weights having $size_{kf}^* = O\,(\,^{m\,(n+k)}\!/_\Delta\,)$ and $depth_{kf}^* = O\,(\,^{\lg(mn)}\!/_{\lg\Delta}\,)$ and occupying an area $O\,(\,m\,(n+k))$ for all the values of the fan-in (Δ) in the range 2 to n.*

Remark 3: The area estimate does not depend on the fan-in.

The improvements given by *Theorem 3* over *Theorem 2* are: (i) a possible reduction with respect to *depth*; (ii) a reduction by a factor of almost 4 in *size*; (iii) a large reduction of the *area* by $8\cdot 2^\Delta/\Delta$. We do not know how to constructively reach the $\sqrt{m}$ existence *size* complexity for implementing any BFs with TGs [3, 24]. Still, we can combine together the idea of using COMPARISONs with the one of using AND gates such as to take the best part of both solutions and reduce as much as possible the *area*.

Theorem 4: *It becomes more area efficient to use 2 COMPARISONs instead of α MAJORITY (AND) gates only if the group of ones has a length of:*

$$\alpha \geq \frac{6\cdot 2^\Delta}{\Delta} + \frac{10\cdot 2^{\Delta/2}}{\Delta(\Delta-2)}.$$

Remark 4: The length of the group of ones —starting from which it is more area-efficient to use COMPARISONs— grows exponentially with the fan-in. Still, for minimising the AT^2 complexity measure, small constant values of the fan-in ($\Delta_{opt} = 6 \ldots 8$) have to be used (cf. Remark 1).

Theorem 4 is properly valid only for the one dimensional case. There is a simple way to extend it to the d-dimensional case: work on each of the d axes independently. This is equivalent to *finding hyperplanes which are parallel to the axes*. These hyperplanes are COMPARISONs with certain values and represent the first hidden layer of the network. All the results from this first hidden layer are binaries and the function can be synthesised by a classical AND-OR structure. The network has thus two more layers: (i) a second hidden layer of $2d$–inputs AND gates (each gate corresponds to a d–dimensional hypercube, and at most 2 COMPARISONs are needed for each of the d dimensions); and (ii) a third output layer of k OR gates. For optimising the *area* some COMPARISONs from the first hidden layer should be replaced by AND gates (as proven in *Theorem 4*). This reduces the *area* of the chip but the generalisation capabilities of the network are also degraded. Using AND gates the network will "recognise" only the training examples, while using COMPARISONs the network will *"generalise"* to hypercubes around the given examples.

The algorithm can now be described:

1. Find the maximum and the minimum on each axis.
2. Find the steps required on each axis (the initial steps are maximum-minimum and are subsequently reduced until the classification of the *m* examples becomes possible).
3. Determine the constants for COMPARISONS (this is done by scanning the space with the steps previously determined).
4. Reduce the number of COMPARISONS (by grouping together as many hypercubes as possible).
5. Use a "generalisation" parameter to decide if either COMPARISONS or AND gates should be used in the final circuit.

4. COMPARISON OF RESULTS

It is not very easy to compare our algorithm with others as our main aim was to minimise the *area*, and no other algorithms for doing that are known to the author. Even if we consider *size* as criterion there are very few algorithms to compare with. Most of the direct design algorithms are based on a "learning" phase; as a result of which the *weights* are updated or: (i) a new neuron is added; (ii) a new layer is added. That is why it is not possible to bound the *size* of the network they "learn." Still, there is one algorithm we can compare with: that of Tan and Vandewalle [26]. While the *size* of the network they build is of the same order of complexity O (*mn*) as ours for constant *fan-in* O ($^{mn}/_{\Delta}$), the *depth* complexity is reduced from O (*mn*) – in their case – to O ($^{\lg mn}/_{\lg \Delta}$) – in our case. Moreover, the maximum *fan-in* is reduced from n – in their case – to Δ.

For benchmarking the algorithm we have tested it on the classical "two spirals" problems. The task to be learnt is to classify two sets of training points belonging to two distinct spirals in the plain. It is known that this appears to be a very difficult task for backpropagation networks. In Fig. 3.a., 3.b., 3.c. and 3.d. we can see the step being reduced till the problem becomes solvable by using only COMPARISONS. There are in fact two steps: one for the X axis and one for the Y axis, but in this particular example they are equal. In general, each axis might have its own step.

The results obtained by our algorithm are presented in figure 4.a. (before reduction) and 4.b., 4.c. (after reduction – step 4 of the algorithm). The maximum value for *generalisation* has been used. Figure 4.d. shows in white the space which cannot be classified by the NN as belonging to one spiral or the other.

We compare these results (see Table 1.) with those obtained using backpropagation as well as with a fresh geometric constructive algorithm [6]. For a fair comparison we have included the number of *weights* (i.e. the sum of the *fan-in*) in the last column. Both the *size* and the number of *weights* are comparable with [43] and by far smaller than those obtained by [6]. Anyhow, our solution is better fitted for VLSI as having a small range of integer *weights*, thus simpler functions have to implemented.

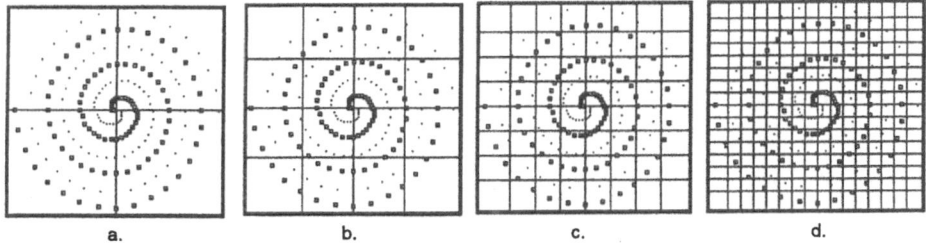

      a.   b.   c.   d.

Figure 3. The two spirals problem and the division of the space by smaller steps (a., b., c., d.) till the two spirals can be separated only by vertical and horizontal lines (parallel to the two axes).

TABLE 1.

Comparison of several solutions for the "two spirals" problems.

Authors	Number of layers	Number of neurons	Type of neurons	Type of operations	Number of weights
D. Waker [29] (using backpropagation)	3	20 10 1	sigmoid sigmoid TG	multiplication, addition multiplication, addition multiplication, addition	281 (reals)
Lang and Witbrock [15] (using backpropagation)	4	5 5 5 1	sigmoid sigmoid sigmoid TG	multiplication, addition multiplication, addition multiplication, addition multiplication, addition	138 (reals)
Bose and Garga [6]	3	193 97 1	TG AND OR	multiplication, addition logical logical	867 (386 reals and 481 integers)
This article (with "generalisation")	3	27 30 1	COMPARISON AND OR	logical logical logical	254 (small integers on 4 bits; ≤ 15)
(without "generalisation")	2	96 1	AND OR	logical logical	1056 (on 1 bit; ± 1)

The algorithm can be directly used to map NNs onto FPGAs. The description of the circuit which is obtained as output of this constructive algorithm can be translated (by software) into an input file for one of the classical software packages for FPGAs. Very few macros would be needed as the circuit is using only COMPARISONs, ANDs and ORs. Finally, the FPGAs programming software package would then select the proper size FPGA and program the chip accordingly.

5. CONCLUSIONS AND OPEN PROBLEMS

In this paper we have shown how to extend the applicability of a decomposition algorithm for functions belonging to $F_{n,m}$ to the problem of "learning from examples." While doing that we have proven that both the *size* and the *area* of a VLSI implementation of the NN grow linearly with the "problem" (i.e. n, m, k) for the two solutions discussed (with COMPARISONs or implementing the DNF). We have also discussed the influence of the *fan-in* on *size*, *depth* and *area*, and presented a better alternative by suggesting how to combine the two solutions proposed such as to minimise *area* even more.

Further research should be pursued to try to lower the *area* of COMPARISON and $F_{n,m}$ functions. Closer estimates could be achieved if the *area* of interconnections would be included in the cost function. For even better estimates the additional *delay* introduced by long wires should not be neglected.

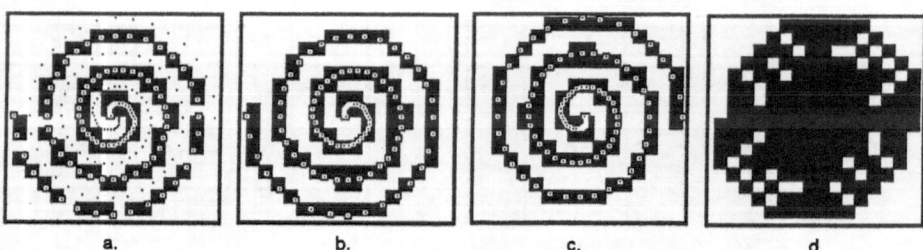

a.   b.   c.   d.

Figure 4. A solution given by our algorithm to the two spirals problem: (a.) before and (b.) after reducing the number of COMPARISONs; (c.) the other spiral; (d) white space cannot be classified.

As has been seen from the two-spiral example presented, an interesting aspect which clearly deserves more attention is the balance between the *generalisation* capabilities and minimum *area* of the chip. The software program is still under development and the *generalisation–area* tradeoff is one of the topics currently under investigation.

REFERENCES

1. Y.S. Abu-Mostafa, "Learning from Hints in Neural Networks," *J. Compl.*, **6**, 192-198 (1990).
2. W.W. Armstrong and J. Gecsei, "Adaption Algorithms for Binary Tree Networks," *IEEE Trans. on Systems, Man and Cybernetics*, **SMC-9**, 276-285 (1979).
3. E.B. Baum, "On the Capabilities of Multilayer Perceptrons," *J. Compl.*, **4**, 193-215 (1988).
4. V. Beiu, J.A. Peperstraete, J. Vandewalle and R. Lauwereins, "Area-Time Performances of Some Neural Computations," in P. Borne, T. Fukuda and S.G. Tzafestas (eds.): *Proceedings of the IMACS International Symposium on Signal Processing, Robotics and Neural Networks SPRANN'94* (Lille, April), GERF EC, Lille, 664-668 (1994).
5. V. Beiu and J.G. Taylor, "VLSI Optimal Neural Network Learning Algorithm" accepted for published in *Proc. ICAN-NGA'94* (Alès, April), Alès, France (1995).
6. N.K. Bose and A.K. Garga, "Neural Network Design Using Voronoi Diagrams," *IEEE Trans. on Neural Networks*, **NN-4**(5), 778-787, 1993.
7. J. Bruck and R. Smolensky, "Polynomial Threshold Functions, AC^0 Functions and Spectral Norms," *SIAM J. Comput.*, **21**(1), 33-42 (1992).
8. S. Diederich and M. Opper, "Learning of Correlated Patterns in Spin-Glass Networks by Local Learning Rules," *Phys. Rev. Lett.*, **58**(9), 949-952 (1987).
9. E. Fiesler, A. Choudry and H.J. Caulfield, "A Universal Weight Discretization Method for Backpropagation Neural Networks," *IEEE Trans. on System, Man, and Cybernetics* (to appear).
10. J.F. Fontanari and R. Meier, "Evolving a Learning Algorithm for the Binary Perceptron," *Network*, **2**, 353-359 (1991).
11. S.I. Gallant, "Perceptron-Based Learning Algorithms," *IEEE Trans. on Neural Networks*, **NN-1**(2), 179-191 (1990).
12. D. Hammerstrom, "The Connectivity Analysis of Simple Association –or– How Many Connections Do You Need," *Proc. NIPS'87* (Denver, November), Amer. Inst. Phys., 338-347 (1987).
13. J.L. Holt and J.-N. Hwang, "Finite Precision Error Analysis of Neural Network Hardware Implementations," *IEEE Trans. on Comp.*, **C-42**(3), 281-290 (1993).
14. J.E. Hopcroft and R.L. Mattson, "Synthesis of Minimal Threshold Logic Networks," *IEEE Trans. on Electr. Comp.*, **EC-6**, 552-560 (1965).
15. K.J. Lang and M.J. Witbrock, "Learning to Tell Two Spirals Apart," *Proc. of the 1988 Connectionist Models Summer School*, Morgan Kaufmann (1988).
16. C.A. Mead and L. Conway, *Introduction to VLSI Systems*, Addison-Wesley, Reading (1980).
17. M. Mézard and J.-P. Nadal, "Learning in Feedforward Layered Networks: the Tiling Algorithm," *J. Phys. A: Math. Gen.*, **22**, 2191-2203 (1989).
18. M.L. Minsky and S.A. Papert, *Perceptron: An Introduction to Computational Geometry*, The MIT Press, Cambridge (1969).
19. S. Muroga, "The Principle of Majority Decision Logic Elements and the Complexity of Their Circuits," *Proc. Intl. Conf. Inform. Processing*, Paris (1959).
20. U. Ramacher and M. Wesseling, "A Geometrical Approach to Neural Network Design," *Proc. IJCNN'89* (Washington, January), IEEE Press, vol. **2**, 147-153 (1989).
21. N.P. Red'kin, "Synthesis of Threshold Circuits for Certain Classes of Boolean Functions," *Kibernetika*, **6**(5), 6-9 (1970).
22. D.E. Rumelhart, J.L. McClelland and the PDP Research Group (eds.), *Parallel Distributed Processing: Explorations in the Microstructure of Cognition*, A Bradford Book, The MIT Press, Cambridge (1986).
23. J.S. Shawe-Taylor, M. H.G. Anthony and W. Kern, "Classes of Feedforward Neural Nets and Their Circuit Complexity," *Neural Networks*, **5**(6), 971-977 (1992).
24. K.-Y. Siu, V. Roychowdhury and T. Kailath, "Depth-Size Tradeoffs for Neural Computations," *IEEE Trans. on Comp.*, **C-40**(12), 1402-1412 (1991).
25. F.J. Śmieja, "Neural Network Constructive Algorithm: Trading Generalisation for Learning Efficiency?" *Circuits, System, Signal Processing*, **12**(2), 331-374, 1993.
26. S. Tan and J. Vandewalle, "Efficient Algorithm for the Design of Multilayer Feed-forward Neural Networks," *Proc. IJCNN'92* (Baltimore, January), IEEE Press, vol. **2**, 190-195 (1992).
27. D. Walker, unpublished result reported by S.A. Frostrom of SAIC (posted to "connectionists" mailing list by Alexis Wieland of MITRE Corporation; data-base maintained by neural-bench@cs.cmu.edu).
28. J. Wray and G.G.R. Green, "Neural Networks, Approximation Theory, and Finite Precision Computation," *Neural Networks*, **8**(1), 31-37 (1995).
29. *CORTEX-PRO*, **Unistat Ltd.** (Unistat House, 4 Shirland Mews, London W9 3DY), England.

A CPWM Synapsis
for Weighted Radial Basis Functions

E. Miranda*
Dipartimento di Elettronica
Politecnico di Torino
C.so Duca Abruzzi, 24 - 10128 Torino - ITALY
e.mail miranda@polimage.polito.it

L.M.Reyneri**.
Dip. di Ingegneria dell'Informazione
Univerità di Pisa
Via Diotisalvi, 2 - 56126 Pisa - ITALY
e.mail lmr@iet.unipi.it

Abstract

This paper describes a silicon synapsis designed to implement Weighted Radial Basis Functions. The synapsis is based on Pulse Stream computation principles, which offer interesting performance, especially for what power dissipation and computation speed concerns. Weighted Radial Basis Functions integrate the advantages of Multi-Layer Perceptrons and Radial Basis Functions alone, therefore the silicon neural networks which results may find applications in several pattern recognition and classification tasks, especially in low power environments. Furthermore it can also be used as a method to map Fuzzy Inference Systems on silicon Artificial Neural Networks.

1 Introduction

Multi-Layer Perceptrons (MLPs) [1] and Radial Basis Functions (RBFs) [2] are two Neural computing paradigms widely used in several applications, ranging from pattern recognition, to industrial control, Neuro-Fuzzy emulation, function approximations, etc. Fuzzy Systems (FSs) [1] represent another computational paradigm widely used in several intelligent controller applications. Each of the three paradigms has its own advantages and drawbacks [3], which are not discussed here

Although apparently the three paradigms mentioned above are very different from each other, they can be unified into the so-called Weighted Radial Basis Functions (WRBF) [5]. This new paradigm described below finds applications in a number of different fields, and in particular in robotic and control [4]. The major drawback of this new paradigm is that at present it can only be emulated on general purpose processors, as dedicated silicon devices are not yet available.

Several chips are available from literature [7], which are limited to implement only some forms of MLPs, Kohonen Maps, or Hopfield nets, which are though not well suited to real-time control applications and often do not provide optimal performance in pattern recognition applications.

This paper describes a silicon implementation of a WRBF synapsis based on the Coherent Pulse Width Modulation (CPWM) technique [6, 8], which may find several applications in low-power equipment, flexible robotic control [4], etc. The same technique has already been used by the authors to develop other neural chips for MLPs [6], as a method to implement very low-power neural computation for real-time control applications [4].

The paper describes only the synaptic circuit which has been designed to implement two major cases of the generic WRBF paradigm, as the rest of the neuron, the I/O interfaces, and the Analog to/from CPWM converters have already been described elsewhere [6]. The VLSI chip proposed below is still under development. It has successfully been simulated and the layout is currently under design. For the neuron and interface circuits the same circuits already developed [6] have been used.

The paper first presents the WRBF algorithm, then it describes the circuits of the synapsis which has been designed to implement such paradigm. Simulation results are also presented, as the final chip is not yet available. For further details on the operation principle and the performance of Coherent Pulse Width Modulation, see [4, 6].

2 Weighted Radial Basis Functions

For the proposed algorithm, each neuron $j \in [1 \ldots M]$ is associated to a pair of vectors, namely a *location center* $\vec{C}^j = \{c_1^j, c_2^j, \ldots, c_N^j\}$ (as for RBFs) and a *weight vector* $\vec{W}^j \{\Theta^j, w_1^j, w_2^j, \ldots, w_N^j\}$ (as MLPs). The output y^j of the neuron is a function of the input vector $\vec{X} = \{x_1, x_2, \ldots, x_N\}$:

$$y^j = \mathcal{H}_n^{F(z)}\left(\vec{X}; \vec{C}^j, \vec{W}^j\right) \tag{1}$$

N	Membership function normalized width	RMS error (%)
	0.5	14.9
2	0.2	6.0
	0.1	3.0

N	Membership function normalized width	RMS error (%)
	0.5	18.0
3	0.2	4.6
	0.1	1.6

Table 1: RMS errors approximating a Fuzzy System with a WRBF. The width of the bell-shaped membership function is expressed as a fraction of the size of the input universe.

where $\mathcal{H}_n^{F(z)}(\cdot)$ is the characteristic of a *Weighted Radial Basis Function of order n*, which is defined as:

$$\mathcal{H}_n^{F(z)}\left(\vec{X}; \bar{C}^j, \vec{W}^j\right) \triangleq F\left(\Theta^j + \sum_{i=1}^{N} \mathcal{D}_n(x_i - c_i^j) \cdot w_i^j\right), \tag{2}$$

where the factor $\mathcal{D}_n(x_i - c_i^j)$ is a function of the distance between the i-th component of the input vector $\vec{X}$ and of the location center $\bar{C}^j$, respectively:

$$\mathcal{D}_n(x_i - c_i^j) \triangleq \begin{cases} (x_i - c_i^j) & \text{for } n = 0 \\ |x_i - c_i^j|^n & \text{for } n > 0 \end{cases} \tag{3}$$

$F(\cdot)$ is one of several possible monotonic *activation functions* such as, for instance:

$$F(z) \triangleq \begin{cases} \dfrac{1 - e^{-z}}{1 + e^{-z}} & (sigmoidal) \\ e^{-z} & (exponential) \\ z & (linear) \end{cases} \tag{4}$$

One or more WRBF layers can be cascaded to build a Multi-Layer WRBF, by connecting all the outputs of a layer to the inputs of the next one, as for MLPs [2]. For instance:

$$\vec{Y} = \mathcal{H}_{n_2}^{F_2}\left(\mathcal{H}_{n_1}^{F_1}\left(\vec{X}; \bar{C}_1, \vec{W}_1\right); \bar{C}_2, \vec{W}_2\right) \tag{5}$$

where $\vec{Y} = \{y_1^j, y_2^j, \ldots, y_N^j\}$ is the vector of outputs, while $\bar{C}_k$, $\vec{W}_k$, n_k and F_k are the location centers matrices, the weight matrices, the order and the activation function of the k-th network layer.

Note that the standard RBF and MLP paradigms can be seen as two cases of WRBFs, while FSs can be approximated by an appropriate WRBF. Further details on WRBF and its learning rule can be found in [5]:

- MLPs [2] are equivalent to $\mathcal{H}_0^{sigm}\left(\vec{X}; \vec{0}, \vec{W}^j\right)$ with sigmoidal activation function, where $\vec{0}$ is a vector of 0s.

- RBFs [1] are equivalent to $\mathcal{H}_2^{exp}\left(\vec{X}; \bar{C}^j, \vec{1}\right)$ with exponential activation function, where $\vec{1}$ is a vector of 1s. With weight vector components equal to one, the input/output characteristic of the neuron is equivalent to that of a RBF [1], since the exponential function (4) behaves as a gaussian, due to the exponent $n = 2$ in (2), (3).

- FSs [1] can be approximated by a two-layer RBF. This statement can be explained by giving an example of a set of Fuzzy inference rules:

```
IF (x1 IN R1A) AND (x2 IN R2A) AND ... (xN IN RNA) THEN SA
IF (x1 IN R1B) AND (x2 IN R2B) AND ... (xN IN RNB) THEN SB
```

where R1A, R2A, ... are appropriate *membership functions* [1]. In case of *bell-shaped* (i.e. gaussian-like) functions, the above rules can be computed analytically as:

$$y_{RBF} \approx T_A \cdot \min\left\{\mathcal{R}_1^A(x_1), \mathcal{R}_2^A(x_2), \ldots \mathcal{R}_N^A(x_N)\right\} + T_B \cdot \min\left\{\mathcal{R}_1^B(x_1), \mathcal{R}_2^B(x_2), \ldots \mathcal{R}_N^B(x_N)\right\} + \ldots \tag{6}$$

where T_A and T_B are the *centers of gravity* of membership functions SA and SB, respectively, while $\mathcal{R}_i^A(x_i)$ and $\mathcal{R}_i^B(x_i)$ are the membership value of input x_i to the rules RiA and RiB, respectively. Under the assumption that (valid for several bell-shaped functions):

$$\mathcal{R}_i^j(x_i) \approx e^{(x_i - c_i^j)^2 \cdot w_i^j}, \tag{7}$$

equation (6) can be approximated by:

$$y_{RBF} \approx T_A \left(e^{(x_1-c_1^1)^2 \cdot w_1^1} \cdot e^{(x_2-c_2^1)^2 \cdot w_2^1} \ldots \right) + T_B \left(e^{(x_1-c_1^2)^2 \cdot w_1^2} \cdot e^{(x_2-c_2^2)^2 \cdot w_2^2} \ldots \right) + \ldots \qquad (8)$$

where the operator "min" in formula (6) has been approximated by a multiplication. Such an approximation introduces a small error, as indicated in Tab. 1, for $N = 2$ and $N = 3$ inputs. This error is also function of the *normalized width* of the membership function. Finally, formula (8) can be modified as:

$$y_{RBF} \approx \sum_{k=1}^{M} \left(T_k \cdot e^{\sum_{i=1}^{N} (x_i - c_i^k)^2 \cdot w_i^k} \right), \qquad (9)$$

which is the characteristic of a two-layer WRBF (from (2), (3), and (4)):

$$\mathcal{H}_0^{\text{lin}} \left(\mathcal{H}_2^{\text{exp}} \left(\vec{X}; \vec{C}, \vec{W} \right); \vec{0}, \vec{T}^1 \right), \qquad (10)$$

or, in other words, a WRBF on the first layer plus a MLP with linear activation on the second layer.

2.1 Applications and Advantages

The proposed algorithm and the corresponding synapsis described in Sect. 3 find several applications, some of which have already been mentioned previously. The algorithm has been deeply tested in several practical cases [4]. Due to space limitations, this section only discusses the relevant advantages of the proposed paradigm in two relevant application areas.

Pattern Classification: RBFs are known to provide interesting advantages with respect to MLPs, especially for what training concerns. Adding a different weight to each input improves learning performance, especially when dealing with noisy pattern, blurry patterns and not-well defined patterns, as can be found in a lot of practical problems (e.g. handwriting and speech recognition).

Intelligent Control: Fuzzy and Neural controllers [3, 4] have different advantages and drawbacks. WRBF can take the advantages of both of them, in the sense that it allows the design of *learning controllers* by either human-friendly Fuzzy Inference methods (therefore reusing the know-how of human experts) or more computer-friendly Neural Learning methods (therefore learning from samples measured directly from the controlled plants). The similarity between WRBFs and FSs allows a straightforward learning of Neural Networks from Fuzzy Controllers and, viceversa, the easy translation into human-readable Fuzzy rules of trained WRBF Neural Networks.

3 Synaptic Circuit Description

The circuit presented here is the synaptic part of a neural cell; this synapsis can implement both MLP and RBF algorithms, as two special cases of the WRBF general algorithm. For the MLP algorithm the parameters considered are $n = 0$ and $\vec{C}^j = \vec{0}$, while for the RBF algorithm, they are $n = 1$, $w_i^j > 0$ and $c_i^j > 0$. Only the synaptic circuit is described here and not the neuron, because the neuron is identical to the one used in a previous work [4, 6], where it is described in detail.

As shown in Fig. 1, the synaptic circuit consists of two *Weight Memories* MEM1 and MEM2, two *Half Current Mirrors* CM1 and CM2 and some other circuits needed for the evaluation of synaptic contributions. All synapses are connected to a *Common Refresh System* CRS, but part of the refresh circuits are implemented directly within the memories MEM1 and MEM2, as shown in Sect. 3.1. The circuit uses the CPWM pulse stream methods developed in previous works [6, 8].

The complete synaptic circuit is shown in Fig. 7, and the following sections describe the working principles of the various blocks. The same circuit can be used either as an MLP synapsis or as a RBF or a WRBF synapsis, and therefore it can also be used to approximate Fuzzy Systems, as indicated in Sect. 2.

3.1 Weight Memories

We used the *Current Copier Cell* described in [7] to implement a memory that is written and read using currents instead of voltages. This *Current Memory* has several advantages more than the previously used voltage memories [6], because it is independent of several circuit parameters. Fig. 2 shows the circuit diagram of a Current Memory and its associated refresh circuits. The transistor M_4 is not part of the current memory, but it is drawn because is necessary for the explanation of operation (in practice transistor M_4 is part of the Common Refresh System and is shared by all Current Memories).

The Current Memory has two operation cycles: The Write Cycle and the Refresh Cycle, but in practice the write cycle is a part of the refresh cycle and is not an independent cycle, but we present it separately because its operation can be described more easily than if analyzed in the refresh process.

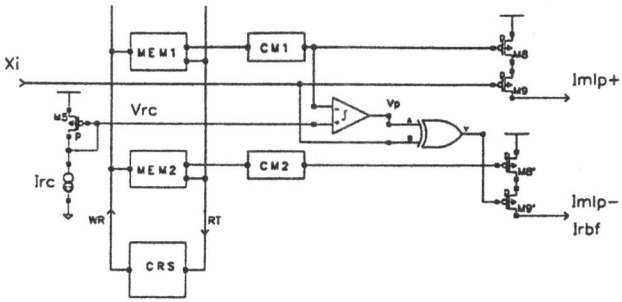

Figure 1: Block diagram of a synapsis

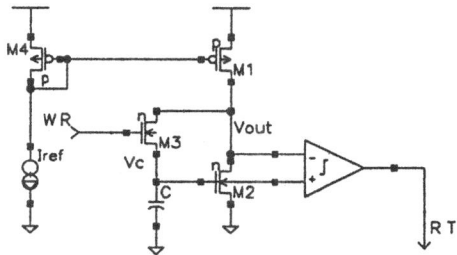

Figure 2: Circuit diagram of a weight memory;

3.1.1 Write Cycle

To store a value into the memory, the current I_{REF} must be set to a specific value (named I_M) and the signal WR must be switched on; in this case, transistor M_2 is connected as a diode (through M_3), therefore it stores its gate voltage V_C on capacitor C, when WR=0. The voltage V_C is a function of I_M, but it is not the output value, being the actual memory output another current generated from V_C by another transistor (M_7 in Figs. 4 and 7) as will be discussed in Sect. 3.2 and 3.3. Considering that M_1 and M_4 are identical, the voltage V_C can be obtained from:

$$V_C = \sqrt{\frac{2 \cdot I_M}{\beta_2}} + V_{TN} \tag{11}$$

where

$$\beta_x = \mu_n \cdot C_{OX} \cdot \frac{W_x}{L_x} \tag{12}$$

and V_{TN} is the threshold voltage of a NMOS transistor. The time constant τ for the write process can then be obtained from:

$$\tau \approx \frac{C}{g_{m2}} = \frac{C}{\sqrt{2 \cdot \beta_2 \cdot I_M}} \tag{13}$$

And for the lowest I_M (≈ 500nA) we obtain $\tau < 230$ns; this means that for a 6 bits memory, the lowest write time must be

$$t_{WR} > -\tau \cdot \ln\left(\frac{1}{64}\right) = 798\text{ns} \tag{14}$$

In order to guarantee a correct storage, the lowest store time has been chosen $t_{WR} = 1\mu$s.

3.1.2 Refresh Cycle

The task of the Refresh System is to maintain the stored voltage V_C unchanged for an unlimited period of time. During a Refresh Cycle, the Refresh System periodically refreshes the stored value V_C.

A Refresh Cycle consists in the application of a current $I_{REF}(t)$ that is a staircase signal, where the number of levels defines the number of memory bits ($N^{\circ}levels = 2^{(N^{\circ}bits)}$). When $I_{REF}(t)$ is close to the current I_M stored during the write cycle, the synaptic memory must fire a signal called *Refresh Trigger* RT. This signal is processed by the Common Refresh System that refreshes the memory, by turning on the WR signal for a short period, and thus storing the instantaneous staircase signal value. In other words, as this value is close to the stored value I_M, the Refresh System restores the original value again.

The optimal operation point of the of the current memory is when the V_{DS} voltage of M_2 is the same as during the Write Cycle, therefore when $V_{DS} = V_C$. In other words, the Refresh Trigger RT signal must be generated when V_{OUT} is equal to V_C, then, the Refresh Trigger RT must be generated by a comparator that compares V_{OUT} with V_C and its output RT is used by the Common Refresh System as shown in Fig. 1. Figure 3 shows the simulation results of two consecutive Refresh Cycles, with different weight values.

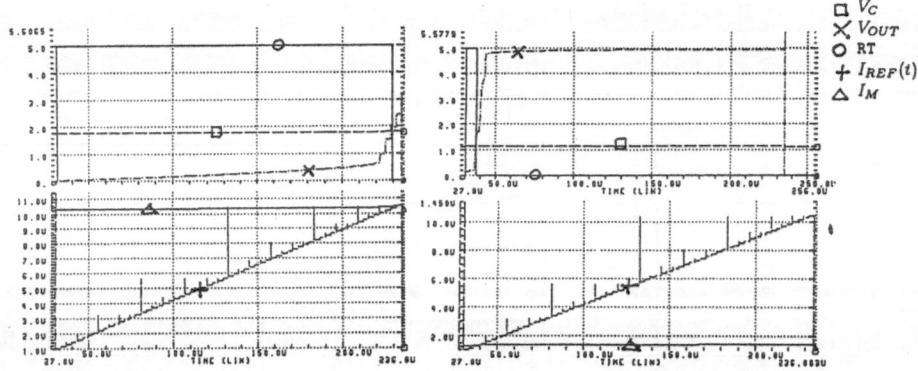

Figure 3: Simulation results of two Refresh Cycles, with two different weight values.

An important aspect of the refresh process is derived from the analysis of the form of capacitor discharge; in our case, the leakage current attempts to discharge the capacitor (negative slope) through the Drain-Bulk diode of transistor M_3 (see Fig. 2) and to compensate this, the slope of staircase signal $I_{REF}(t)$ must be positive.

3.2 Operation of a MLP Synapsis

A MLP synapsis can be obtained using the two Current Memories where one memory stands for the *Excitatory Weight* and the other for the *Inhibitory Weight*. The two parts are identical and are shown in Fig. 4.

Each Weight Memory has a Current Mirror (M_6, M_8) that reads the memory current and divides it by a constant factor in order to obtain an output current I_{MLP} in a range compatible with the neuron circuit $[0 \ldots 2\mu A]$. The transistor M_9 evaluates the product of output current I_{OUT} and the input CPWM pulse x_i to guarantee that the synaptic contribution supplies to the neuron a correct charge quantity [6]. The expression for I_{MLP} is obtained

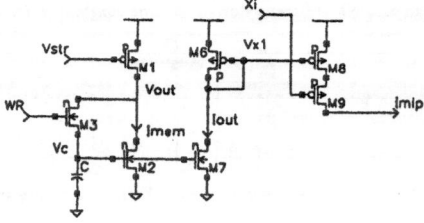

Figure 4: One half of a MLP synapsis

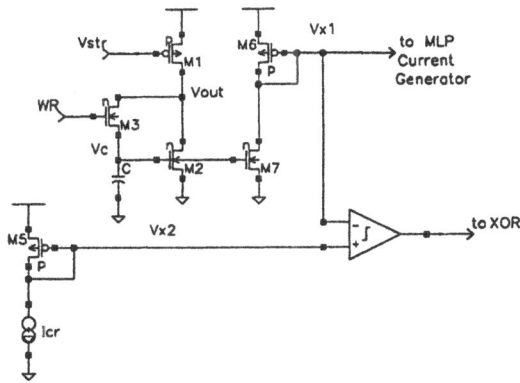

Figure 5: Current to CPWM converter for operation as a WRBF

considering equation 11:

$$I_{MLP} = \left(\frac{\beta_7 \cdot \beta_8}{\beta_2 \cdot \beta_6}\right) \cdot I_M \tag{15}$$

where I_M is the value stored in the Current Memory cell.

The last expression is identical for the two synaptic currents (Excitatory and Inhibitory currents), the only difference is that I_M may be different for the two synaptic weights. The inhibitory and excitatory currents are subtracted in the neuron body (not shown [6]), and then integrated by the neuron circuits. The obtained voltage (neuron output) is proportional to the charge transferred from the synapsis to the neuron, thus:

$$y_i^j = V_N = \frac{1}{C}(I_{MLP}^+ - I_{MLP}^-) \cdot x_i \tag{16}$$

where x_i is the width of the input CPWM pulse and I_{MLP}^+ and I_{MLP}^- are the excitatory and inhibitory currents, respectively. Synaptic weight is therefore represented by the following equation:

$$w_i^j = \left(\frac{\beta_7 \cdot \beta_8}{\beta_2 \cdot \beta_6}\right) \cdot (I_{MLP}^+ - I_{MLP}^-) \tag{17}$$

It is obvious that since the two memories have 6 bits of resolution, the total MLP synaptic weight has a 7 bits resolution.

3.3 Operation as a WRBF Synapsis

A WRBF synapsis must evaluate the expression $w_i^j \cdot |x_i - c_i^j|$ where the first synaptic weight MEM1 of the CPWM synapsis is used as a center value c_i^j while the second (MEM2) is used as a positive weight w_i^j. In order to evaluate the WRBF synaptic contribution we implement an algorithm that generates the absolute value of the difference of two CPWM signals; this algorithm consists of the XOR of two CPWM signals, provided that they all begin at the same time. To use this algorithm, the weight MEM1 that we use as a center value must be first converted into a CPWM pulse and then used.

The conversion is made by a comparator that compares the voltage stored in MEM1 with a *Continuous Ramp Signal* I_{CR} as shown in Figs. 5 and 6, where:

$$V_{X1} = V_{dd} - \sqrt{\left(\frac{2 \cdot \beta_7}{\beta_2 \cdot \beta_6}\right) \cdot I_M} + V_{TP} \tag{18}$$

and:

$$V_{X2} = V_{dd} - \sqrt{\frac{2 \cdot I_{CR}(t)}{\beta_5}} + V_{TP} \tag{19}$$

and since I_{CR} has the same amplitude of I_{REF}, the transistor M_5 must have:

$$\beta_5 = \frac{\beta_2 \cdot \beta_6}{\beta_7} \tag{20}$$

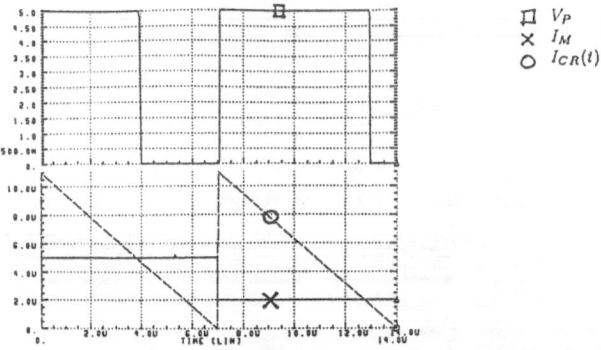

Figure 6: Simulation results of the current to CPWM conversion.

The comparator output will be active when $V_{X1} < V_{X2}$, therefore when $I_{CR}(t) > I_M$. Therefore, being $I_{CR}(t)$ a linear ramp, the output pulse width is directly proportional to the current I_M stored in the memory. Such a signal is a CPWM signal c_i^j that represents a center value and is equivalent to the weight MEM1 stored in the synapsis. Fig. 6 shows the simulation results of this current to CPWM converter.

After the conversion process, the converter output pulse c_i^j is XORed with the input pulse x_i obtaining a CPWM pulse equivalent to $|x_i - c_i^j|$ and then this signal is used to activate the multiplier of MEM2 where a charge quantity current proportional to $w_i^j \cdot |x_i - c_i^j|$ is generated (by transistors M_8 and M_9 in Fig. 7).

3.4 Overall Operation

The complete synaptic circuit is shown in Fig. 7, from which it is clear that the proposed circuit operates either as a MLP or as a WRBF synapsis, according to the status of signal V_{X2}. In fact, if V_{X2} is connected to ground, the comparator output V_P will always be zero, and the XOR output equal to input x_i; in such a case, the two Current Memories will work as a MLP synapsis generating one excitatory current and one inhibitory current.a On the other hand, if V_{X2} is the a linear ramp signal, the two memories operate as described in Sect. 3.3, therefore as a WRBF synapsis.

In this section was presented a synaptic circuit that is compatible with a previously implemented neuron circuit [6]; this synapsis can operate with two different sub-cases of the WRBF algorithm, also, the other important characteristics of this synapsis are the unlimited weight retention time (due to self-refresh) and a low current consumption.

4 Conclusion

This paper has described a novel Neuro-Fuzzy algorithm and its VLSI implementation. The system, which is still under development, finds applications in several low-power real-time control applications, as an improved version of another existing device.

References

[1] P.D. Wasserman, "Advanced Methods in Neural Computing", New York, *Van Nostrand Reinhold*, 1993.

[2] P.D. Wasserman, "Neural Computing: Theory and Practice", New York, *Van Nostrand Reinhold*, 1989.

[3] D.A. White and D.A. Sofge, "Handbook of Intelligent Control", Van Nostrand Reinhold, 1992.

[4] L.M. Reyneri, M. Chiaberge, L. Zocca, "CINTIA: A Neuro-Fuzzy Real Time Controller for Low Power Embedded Systems", in *Proc. of MICRONEURO 94*, Torino (I), IEEE Computer Society Press, September 1994, pp. 392-404.

[5] L.M. Reyneri, "Weighted Radial Basis Functions for Improved Pattern Recognition and Signal Processing", submitted to *Neural Letters*.

[6] L.M. Reyneri, M. Chiaberge, D. Del Corso, F. Gregoretti, "Using Coherent Pulse Width and Edge Modulations in Artificial Neural Systems", *Int'l Jour. Neural Systems*, Vol. 4, no. 4, December 1993, pp. 407-418.

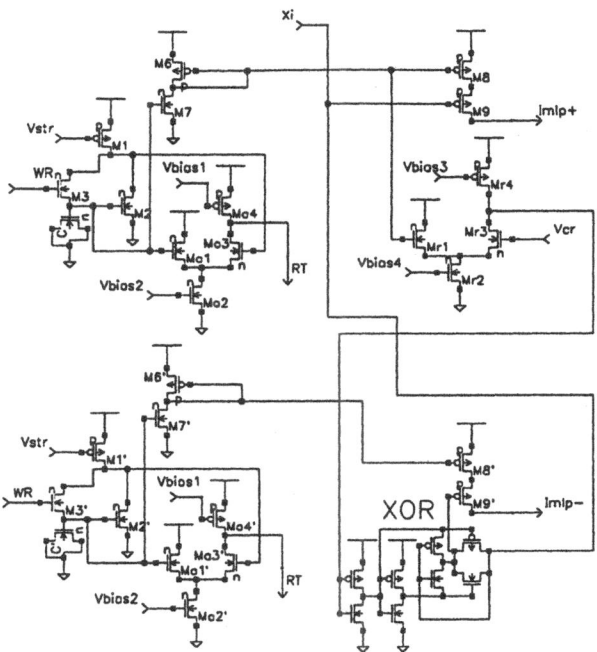

Figure 7: Complete synaptic circuit

[7] -, "Analog VLSI and Neural Networks", Special Issue of *IEEE MICRO*, June 1994.

[8] L.M. Reyneri, "A Performance Analysis of Pulse Stream Neural Networks", accepted for publication on *IEEE Trans. on Circuits and Systems*, 1994.

Test Pattern Generation for Analog Circuits Using Neural Networks and Evolutive Algorithms

J.L. Bernier, J.J. Merelo, J. Ortega and A. Prieto
Dpto. Electrónica y Tecnología de computadores
Campus Fuentenueva s/n. Facultad de Ciencias
18071 Granada. Spain

Abstract

This paper presents a comparative analysis of neural networks, simulated annealing, and genetic algorithms in the determination of input patterns for testing analog circuits. The problem has been modelled as an optimization problem in which the objective is to determine a test signal that maximizes the quadratic difference between the nominal response and the faulty one due to a defect in the circuit. This approach makes possible the search of the test pattern space by using techniques based on neural and evolutive algorithms.

I. INTRODUCTION.

The ever increasing capabilities of VLSI technology have allowed to include complex analog and digital circuits in a single chip. As the size of that circuits grows, the need for tools to automate designing and testing of circuits is higher.

One of the most important step in the process of manufacturing a circuit is its test. It allows to assure that the circuit will work according to the specifications. As more and more circuits are included in a single chip, the cost of testing it is increased not only because its size has grown, but also because its controllability and observability have decreased. Moreover, it is difficult to find accurate solutions to some important problems which appear in the test of circuits because they are NP-complete problems [1].

Focusing on digital circuit testing, there are several efficient procedures to generate test patterns and even some standards [2] about procedures to increase the testability of the circuits. The situation is worst in the analog testing field due mainly to the following two reasons:

1. The size of the majority of analog circuits was small two decades ago, so it was possible an efficient manual generation of the test patterns.

2. The specific difficulties of analog testing compared with digital testing. More precisely, the lack of simple and accurate models for the possible defects of the analog circuits, and the tolerance of the analog elements that in some cases allows a circuit to function according to the specifications, even when its parameters have some deviation with respect to its nominal values.

A possible procedure to detect faults in an analog circuit can be implemented by using the following steps:

Test procedure

1) Generation:
   1.1) Define a set of potential faults.
   1.2) Obtain a set of test patterns for these faults.

2) Application:
    2.1) Apply the test patterns to the circuit under test.
    2.2) If the difference in the output is greater than a predetermined threshold the circuit is diagnosed as faulty.

As it is shown from step 1.2) in the previous procedure, one of the problems which appears in the test of analog circuits is the determination of a set of patterns to detect if the circuit is faulty. It belongs to the class of NP-complete problems which means that a high amount of computing time would be required to find patterns for testing analog circuits with high, or even medium size.

The problem of automatic test pattern generation for analog circuits has been discussed in some previous papers [3,4]. In [4] the problem is formulated as an optimization problem where the goal is to obtain the set of input patterns which maximize the "difference" of responses from normal and faulty circuit. Thus, for a given fault the input stimulus $x(t)$ selected as test pattern should maximize the difference between the outputs $y(t)$ and $y^{*}(t)$, corresponding to the correct and the faulty circuit, respectively.

In what follows, it is provided a function which allows to quantify the difference between responses of circuits with and without faults [4]. It can be obtained by representing the behaviour of the circuits by the discrete version of their impulse response, h_n for the correct circuit, and h^{*}_{n} for the faulty one, and using the convolution to describe the outputs in term of the input stimulus, x_i:

$$y_n = \sum_{k=0}^{n} x_k \, h_{n-k} \tag{1}$$

$$y_n^{*} = \sum_{k=0}^{n} x_k \, h_{n-k}^{*} \tag{2}$$

and, this way, $y_n - y^{*}_{n}$ is

$$\Delta y_n = y_n - y_n^{*} = \sum_{k=0}^{n} x_k \, \Delta h_{n-k} \tag{3}$$

where

$$\Delta h_{n-k} = h_{n-k} - h_{n-k}^{*} \tag{4}$$

As the input x_i corresponds to a set of instants, $i=0,...,n$, it is possible to define a function which represents the overall difference between y and y^{*}:

$$D = \sum_{i=0}^{n} \Delta y_i^2 = \sum_{i=0}^{n} \left(\sum_{k=0}^{i} x_k \, \Delta h_{i-k} \right)^2 \tag{5}$$

This way, the pattern generation for testing a fault in an analog circuit is expressed as a search for the values of x_i, $i=0,...,n$ that maximize D, given h^{*}_{n} to model the effect of the fault in the circuit represented by h_n.

It is possible to show [4] that the function, D, to maximize is semipositive quadratic in terms of x_i, thus the problem to solve is a quadratic programming problem. This way, the maximum of D must occur for $x_i = \pm V_{max}$, and since V_{max} would be a common factor in (5), the problem is to find an optimal assignment of $+1$ or -1 to each x_i.

As it has been said, the optimization problem to solve is a quadratic programming problem, which is NP-complete. There are some numerical methods that allow to determine a local maximum in an iterative way [5] or by using a heuristic algorithm [4] but they do not assure to find a global maximum. In this paper, it is studied the use of the Hopfield neural network, Simulated Annealing, and Genetic Algorithms to maximize (5), thus providing the test pattern for a given analog circuit. These methods, besides introducing new ways to guide the search towards a global maximum in the solution space, provide a methodology that has the potential to exploit fine-grain parallel computing [6]. Thus, they would give way and make easy to use the massively parallel computers in compute-intensive CAD applications.

The rest of the paper is structured in six sections as follows. The description of the set of analog circuits used to make the experimental comparison among the different methods is described in Sect. II. Sections III to V deal with the characteristics and performances evaluated for each method used to optimize (5). Finally, the comparative analysis of the different procedures and the conclusions are provided in Sect. VI., while the references are given in Sect. VIII.

II. BENCHMARK CIRCUITS.

As benchmark circuits we have used a set of lowpass analog Chebyshev filters [7] with order n=1, ...,10, a cutoff frequency at w_o=300 and a 0.5db ripple. The transference function for this kind of circuits is

$$H(s=jw) = \frac{H_o}{a_1 s^n + a_2 s^{n-1} + ... + a_{n+1}} \qquad (6)$$

where n is called the order of the filter. Fig. 1 shows the typical gain response for such circuits in the frequency domain. Higher values for the filter order implies a better approximation to an ideal filtering function (step).

By using a bilinear transform the frequency response (Fig. 1) can be mapped into the digital plane in order to obtain the discrete impulse response $\mathbf{h}$ (Fig. 2).

The set of defects that can modify the right behaviour of the filter has been modelled as a +50% deviation in the value of each a_i, with i=1...n+1. Different patterns have been generated for different circuits with order n=1,2,..,10 using the MATLAB package [7] to calculate the coefficients a_i for each filter and the matrix Δh defined in (4).

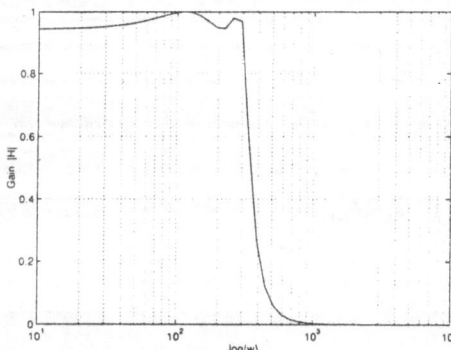

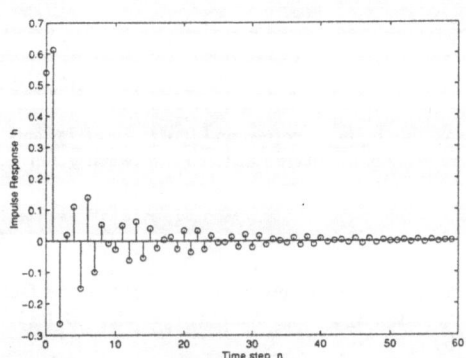

Figure 1. Gain response of a fourth-order filter.

Figure 2. Discrete impulse response of the filter.

III. TEST GENERATION WITH THE HOPFIELD NETWORK.

The Tsai's algorithm [4] is quite similar to a Hopfield network. In this algorithm, for convenience, (5) is expressed as:

$$D = \sum_{i=0}^{n} x_i \left(\sum_{j=0}^{n} \Delta h_{i-j}^2 x_j \right) = \sum_{i=0}^{n} x_i P_i \qquad (7)$$

This way, the algorithm in [4] works according to the following steps:

1) Do $x_i=1$ for i=0,...,n
2) Calculate the minimum $P_i = P_{min}$.
3) If $P_{min}<0$ then $x_{min} = -x_{min}$ else end.
4) Goto 2).

So, it is possible to use a Hopfield network to optimize (7) if the previous algorithm is slightly modified. This way a neuron is associated to each x_i and a candidate neuron is randomly selected to change in each iteration instead of using the heuristic selection. The condition for a change in the state of a neuron is

$$x_i = -x_i \quad if \quad P_i < 0 \tag{8}$$

We have considered that the network is stable when there have been no changes in a selected number of iterations, N. In our case we have used N=50 and have run the algorithm three times for each simulated fault. Tables I and II show the average quadratic difference D, and computing time, respectively.

IV. TEST GENERATION WITH SIMULATED ANNEALING.

We have used the expression (5) as the cost function to maximize. So, the algorithm randomly chooses a x_i that turns into $-x_i$ if the following condition is satisfied:

$$\Delta D>0 \quad OR \quad \exp(\Delta D/T) \geq random[0,1) \tag{9}$$

where T is the temperature and ΔD is the difference between the cost due to the new value of x_i and the cost due to the previous value. The temperature is decreased by a factor α every L iterations [8].

The initial temperature has been set to 1000 and L was always 10. We have made three experiments for each fault, using decrement factors α of 0.95, 0.92 and 0.9, respectively. Tables I and II show the average values for cost and time in the columns SA1.

We have also defined an alternative cost function defined as

$$D' = \sum_{k=0}^{n} |x_k \Delta h_{n-k}| \tag{10}$$

that is computationally cheaper and produces similar results as we can see in Table I. Using that cost function we have ran the same algorithm. The results are shown in column SA2 of Tables I and II, where in order to compare, the cost for each solution has been calculated by using expression (5).

V. TEST GENERATION WITH GENETIC ALGORITHMS.

The Genetic Algorithm (GA) used to generate test patterns is a classical GA [9], and indeed it has been written by using the standard library GAGS [10]. Test patterns are represented as bit strings, with -1 represented as 0 and 1 left as such. Thus, the length of the genotype is (length of the pattern)/8.

The GA parameters we have used are: population size, 400; number of generations, 300. Mutation rate is 1%, crossover used is 2-point crossover. It is a steady state algorithm, so that only 20% of the population is substituted each iteration. Fitness function is eq. (5). One experiment has been run for each fault. GA columns in Tables I and II contain the obtained results.

VI. COMPARATIVE ANALYSIS AND CONCLUDING REMARKS.

We have injected single faults by changing only one selected a_i for i = 1,..,n+1 in all transference functions with n=1,...,10, and for each faulty circuit, 200 temporal samples of its impulse response have been used.

Tables I and II contains some representative results that allow to make a comparative analysis among the different procedures. While Table I shows the quadratic difference of the solution, table II gives the computing time required to reach it. In both tables, the column titled as Tsai provides the results corresponding to the algorithm proposed in [4], and the first column identifies the simulated circuit with an index, where the first number corresponds to the order, n, of the filter and the second, to the subindex, i, of the faulty parameter a_i.

From these tables, we can conclude that the Tsai's algorithm is the fastest, although it is not always able to obtain the best solution. The other procedures provide better solutions, i.e. higher quadratic difference, at cost of longer computing time. It can be seen that by using the function proposed in (9) for the SA algorithm, while the obtained solutions provide test patterns with similar performances, the time to get these solutions is lesser. In most cases, GA provides the best solution, but the computing time is too long compared with the other procedures although the number of generations could be decreased. We conclude that algorithm SA2 is a good election because provides good enough results in a moderate amount of time.

Circuit	Tsai	Hopfield	SA1	SA2	GA
1.1	4.195	4.357	4.072	4.401	5.112
2.3	17.256	17.256	16.527	16.933	18.597
3.4	17.054	17.054	24.713	24.374	28.474
4.3	37.094	31.993	33.409	35.368	37.059
4.5	15.741	15.741	31.250	27.641	37.986
5.1	170.456	151.154	172.513	188.634	170.264
5.4	56.738	51.016	63.884	63.537	67.075
5.6	13.737	13.737	25.002	24.518	25.032
6.7	14.185	14.185	21.656	24.776	26.111
9.3	105.795	132.823	129.6	137.085	145.063

Table I. Quadratic difference obtained by the different algorithms (eq. (5)).

Circuit	Tsai	Hopfield	SA1	SA2	GA
1.1	16.5	33.37	78.86	36.48	~ 4 hours
2.3	11.02	13.22	78.17	36.47	"
3.4	10.98	28.30	78.29	36.56	"
4.3	14.74	45.85	78.77	37.17	"
4.5	11.15	16.60	79.08	36.21	"
5.1	13.74	32.10	78.24	36.25	"
5.4	13.94	27.27	78.43	36.78	"
5.6	11.07	12.66	78.61	36.37	"
6.7	11.65	16.17	78.32	36.61	"
9.3	13.96	30.33	78.59	36.65	"

Table II. Computing time (seconds) used for each algorithm.

Figs. 3-6 show the deviation, Δy, between the nominal and faulty responses versus the time samples for the circuits 6.7 and 9.3, they represent the values of Δy when the test patterns obtained by the different procedures here compared are applied. As the values of Δy increase, the quality of the test is better. While in Figs. 3 and 5 the continuous curve corresponds to the Tsai's solution, the dashed curve to SA1 solution, and the plotted curve to the SA2 solution, in Figures 4 and 6 the continuous and the dashed lines correspond to the test patterns obtained with the Hopfield network and the genetic algorithm, respectively.

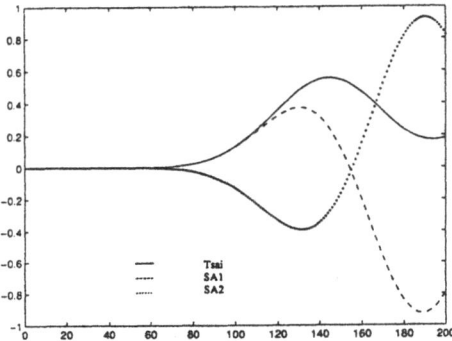

Figure 3. Difference in the output using Tsai, SA1 and SA2 solutions in the circuit 6.7

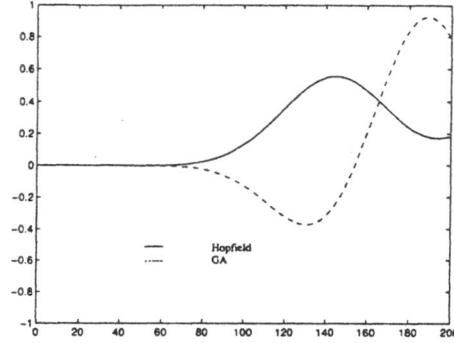

Figure 4. Difference in the output using Hopfield and GA solutions in the circuit 6.7

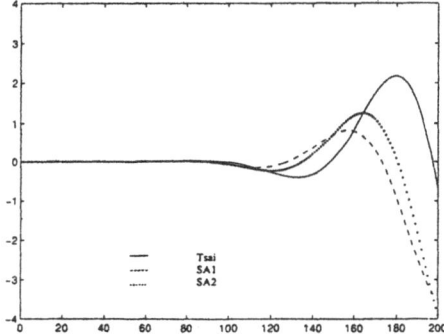

Figure 5. Difference in the output using Tsai, SA1 and SA2 solutions in the circuit 9.3

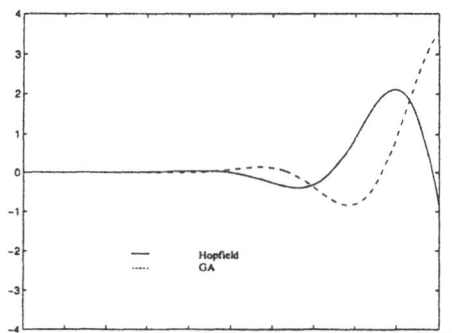

Figure 6. Difference in the output using Hopfield and GA solutions in the circuit 9.3

As a future line of research, the generation of a minimum set of patterns that detects all the defined faults will be investigated; using that results as a basis, multiple faults can be simultaneously simulated, and a minimal set of canonical vectors could be determined. The long-range objective is to develop a environment for test pattern generation from a description of the circuit and a set of faults that are more likely to occur. In that tool, one of the main goals will be to investigate the use of neural networks and other evolutive/soft computing algorithms as pattern generators, in order to take advantage of their inherent parallelism to speed up the testing process.

VII. ACKNOWLEDGEMENTS

This work has been partially supported by project TIC92-603 financed by the "Comisión Interministerial de Ciencia y Tecnología" (CICYT, Spain).

VIII. REFERENCES.

[1]   Garey, M.R.; Johnson, D.S.: "Computers and Intractability: A Guide to the Theory of NP-Completeness". W.H. Freemen and Co, 1979.

[2]   "IEEE Standard Test Access Port and Boundary-Scan Architecture" (IEEE Std.1149). IEEE Computer Society (Test Technical Committee), 1990.

[3]   Schreiber, H.H.: "Fault Dictionary based upon Stimulus Design". IEEE Trans. on Circ. and Syst., Vol. CAS-26, No.7, pp. 529-537. July, 1979.

[4]   Tsai, S.J.: "Test Vector Generation for Linear Analog Devices". Proc. Intern. Test Conf., pp.592-597, 1991.

[5]   Cooper, L.: "Applied Nonlinear Programming for Engineers and Scientist". Aloray Publisher, 1974.

[6]   Chakradhar, S.T.; Bushnell, M.L.; Agrawal, V.D.: "Toward Massively Parallel Automatic Test Generation". IEEE Trans. Comp.-Aid. Des., Vol.9, No.9, pp. 981-994. September, 1990.

[7]   Krauss, T.P.; Shure, L.; Little, J.N.: "MATLAB: Signal Processing Toolbox User's Guide". The MathWorks Inc., 1994

[8]   Aarts, E.; Korst, J.: "Simulated Annealing and Boltzmann Machines". John Wiley & Sons, 1990.

[9]   Goldberg, D.E.: "Genetic Algorithms in Search, Optimization and Machine Learning". Addison Wesley, 1989.

[10]   Merelo, J.J.: "GAGS 0.94: Users's Manual". Available at ftp://kal-el.ugr.es/GAGS. 1995.

About some Perception Problems in Neural Networks

Jeanny HÉRAULT

INPG - TIRF, 46 Avenue Félix Viallet, F-38031 GRENOBLE Cedex, FRANCE

Abstract: Perception is a major problem which is studied as well in life science as in engineering science: This paper concerns a reflection about some of the early mechanisms which underlie perception. Examples are taken in the field of vision in biology and in computer vision, showing the necessity of some adequate pre-processing of the signals. Then, perception appears as a process of representation of signals, that is, as a process of data analysis aimed at finding the structure of the data. Two examples of artificial neural networks are presented to illustrate the problem of data representation. The first one called "Independent Component Analysis" is close to the signal level, the second one, called "Curvilinear Component Analysis" can be seen as a smooth transition between the aspects of Signal Processing and those of Data Analysis.

1. Introduction

The word "perception" has been so widely used, primarily in psychological science, then in cognitive and engineering sciences, that a number of definitions can be found, depending on the subject of the discourse and on the background of the speaker. For example, a dictionary would tell: "to perceive, it is to construct a representation of an object (wide sense) after the impression of our sensory receptors", whereas David Marr [15] said in 1982 "to perceive, it is to know, by looking, what is where". Even if engineers speak of machine perception, they refer to the model of animal perception [15], [19]. The title of this paper has been made voluntarily ambiguous: do we speak of Natural, or of Artificial neural networks? Both of them should be considered. For the natural ones, the problem lies in the discovering of which processes [24] and functional structures [23] in the brain could lead to perception, for artificial ones, it is related to the construction of some functional building blocks able to lead to perceptual properties [25]. Hence, for the second case, it is quite natural to try to get some inspiration from the biological model, as far as it can be understood.

Starting from the preceding definitions, it is clear that between the sensory input signals and the operation of perception itself, there should be some stages of data pre-processing. Before recognising an object, the pixels of the image (bi-dimensional data) must be processed in order to detect the presence of some features, or the amount of some characteristics among all possible ones (orientations of edges, angular points, colours, shades, energy of spectral bands...): the initial 2-D information is converted into a *multi-dimensional data vector* by means of low-level signal processing. Then the object can be represented by a region in a multi-dimensional space, i. e. perceived. The quality of this representation (invariance in position, rotation,...) is due to the nature of the applied pre-processing. Further, for the perception of a scene, the various recognised objects should be characterised with respect to their location, orientation, respective position ..., which constitutes then a new vector of data, used to build an other kind of representation, for example the one of the environment.

The operation of representation from a multi-dimensional space consists in finding the structure of the data, i. e. the relations between the components, or the sub-spaces or sub-manifolds they span, which are characteristic of the analysed object or of the environment. At this stage, two kinds of problems may arise according to the fact that the nature of these relations is known or not. When the nature of the relations is known, we are facing a problem of parameter identification. When it is not known, the problem is to construct a representation space, often with much lower dimensionality than the input data space. In this paper, we will present two examples of self-learning artificial neural networks the use of which reveals interesting properties in perception tasks. The first one is of the parameter identification type: the Source Separation Network or "Independent Component Analysis", which discovers by self-learning the (unknown) independent variables underlying an observed process. The second one is of the space representation type: it proceeds first a Vector Quantization of the input space, then finds, again by self-learning, a non-linear projection into a low-dimensional space, it is named "Curvilinear Component Analysis". Before this presentation, we will give some examples of signal and data representations in biology and in engineering science. They will be taken from the field of vision because this domain has been widely studied in biological science [2], [3], [7], [14], [17], [23], in traditional computer science[1], [4], [15], [21], and more recently in artificial neural networks [3], [11], [22], [25].

2. Some Examples

2. 1 Example of Data reconstruction in the Vestibulo-Ocular Loop

The mechanism of stabilisation of the gaze during the rotation of the head implies a special loop: the Vestibulo-Ocular reflex recently studied by Darlot and Droulez in [7]. In this system, two kinds of sensory inputs are used (figure 1): the *angular velocity H* of the head, which is measured by the semi-circular channels of Vestibule through a frequency high-pass system, and the velocity of the *retinal sliding S*, which is measured by the visual cortical areas through a frequency low-pass system. These systems are formalised in the Laplace Transform space by the following transfer function:

For the head velocity $H \Rightarrow \frac{s\tau}{1+s\tau}$, and for the retinal sliding $S \Rightarrow \frac{1}{1+s\tau}$

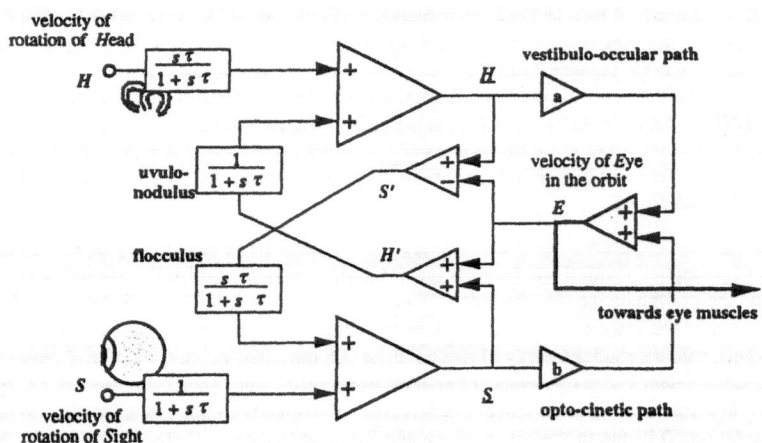

Figure 1. *Vestibulo-Ocular loop and the reconstruction of missing sensory information by the central nervous system. See text. (After [26]).*

As it can be seen, both of the resulting signals are incomplete with respect to spectral considerations. The angular velocity of the Head presents a lack of low-frequency spectrum, and the retinal Sliding, a lack of high-frequency spectrum. We will show that the nervous system manages to approximate and *reconstruct* the missing information in both channels in order to elaborate a signal able to stabilise the gaze during head movements. The retinal sliding S of the gaze is the opposite of the head velocity plus the velocity of eye movement in the orbit V_O: $S = H + V_O$.

Let us name $\underline{H}$ the central representation of the head velocity, it results from the addition of the sensory input signal H high-pass filtered by the sensor, and an efferent copy H' low-pass filtered by a structure of the central nervous system: the "Uvulo-nodulus".

$$\underline{H} = H \frac{s\tau}{1 + s\tau} + H' \frac{1}{1 + s\tau} \tag{1}$$

Symmetrically, the central representation of the retinal sliding velocity $\underline{S}$ results from the addition of the visual-originated signal S low-pass filtered by the sensory pathway, and an efferent copy S' high-pass filtered by an other central structure: the "Flocculus".

$$\underline{S} = S \frac{1}{1 + s\tau} + S' \frac{s\tau}{1 + s\tau} \tag{2}$$

The control signal of the eye velocity in the orbit E is obtained by adding $\underline{H}$ and $\underline{S}$, the two central representations: $E = a \underline{H} + b \underline{S}$, and is used, by closing the loop, to generate the efferent copies $H' = E + \underline{H}$ and $S' = E + \underline{H}$ as shown figure 1. By solving the system for $\underline{H}$, we obtain:

$$\underline{H} = \frac{H s\tau[1 + (1+b) s\tau] + S (1+b)}{(1-a) + s\tau[1 + (1+b) s\tau]},$$

which reduces to: $$\underline{H} = H + \frac{(1+b)}{s\tau[1 + (1+b) s\tau]} S$$

when a tends towards 1, that is, mainly the reconstruction of the head velocity H plus a very low-pass version of S. Similarly, the $\underline{S}$ signal reduces to: $\underline{S} = \frac{S}{[1 + (1+b) s\tau]}$, and when b tends towards -1, we have the exact copies of H and S. Hence, E becomes equal to $H - S$. Then, combined with $S = H + V_O$, the E signal can be used directly to control the direction of gaze and maintain it constant when the head is rotating.

This analyse shows that the signal E is a filtered version of the full $(H - S)$ signals: the central nervous system has *reconstructed the missing part of the sensory information* by making use of redundancy and complementarity of the input channels. Hence, it is directly suitable for the control of eye muscles in order to stabilise the direction of the gaze. The stability of this system has been studied, in an extended and more detailed version, by Zupan in [26].

2. 2. Example of Vision and motion

When a camera is moving, all the objects in the visual field are submitted to different time-varying elementary translations, homotheties and rotations which are all linked to the camera's motion. It is a classical problem of computer vision to determine the ego motion or the shape of objects from the flow field of velocity vectors in the image [1], [4], [21], [22]. Figure 2 shows the motion of an observer (camera) which consists of six parameters: a translation vector, the three components of which are parallel to the X, Y and Z axes and a rotation vector, the three components of which are with respect to the same axes. The projective geometry with unit focal distance gives the co-ordinates (x, y) and velocity vectors (u, v) in the image plane of the image of any point P of the rigid objects of the scene at co-ordinates (X, Y, Z) with respect to the translation velocities (V_X, V_Y, V_Z) and the angular velocities $(\Omega_X, \Omega_Y, \Omega_Z)$:

$$u(x, y) = \left\{ x \, \frac{V_Z}{Z} - \frac{V_X}{Z} \right\} + \left[xy \, \Omega_X - \left(1 + x^2\right) \Omega_Y + y \, \Omega_Z \right] \tag{3}$$

$$v(x, y) = \left\{ y \, \frac{V_Z}{Z} - \frac{V_Y}{Z} \right\} + \left[\left(1 + y^2\right) \Omega_X - xy \, \Omega_Y + x \, \Omega_Z \right] \tag{3'}$$

There is a first step to be considered before exploiting these equations. The components $u(x, y)$ and $v(x, y)$ of the velocity vector at each pixel must be estimated. It is not a simple task of signal processing: local calculations based on time and space derivatives lead to dramatic errors due to the noise and even can only give velocity components normal to the edges of the objects (the famous aperture problem) and not the velocity itself. But let us suppose we have succeeded in getting the real u and v components from suitable image processing techniques.

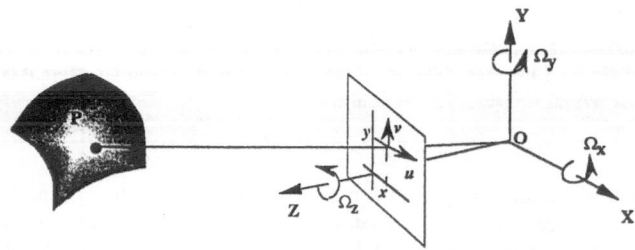

Figure 2. *Projection of an object on the image plane of a moving camera and the resulting components of velocity $u(x, y)$ and $v(x, y)$.*

These equations give the *structure of the observed data*: the components of the velocity vectors in the image depend on the co-ordinates and on the parameters of motion in a known manner. There are now two ways of considering these equations. First, we can be interested in the local information in the image. In this case we consider that neighbour pixels are belonging to the same object. Then, in the motion equations, neighbour pixels are linked by a relation between their depth, this augments the number of equations and parameters but diminishes the ratio between the number of parameters and the number of data. The solution in this case is geometrical or algebraic.

A second way of considering these equations can be more global. Let us suppose that we pave the image by means of a number N of "velocity sensors" whose outputs are u_i and v_i, just as we can find in the retina of rabbits or in our cortical area V_1, or in the brain of insects [8] as well. A given scene is then represented by a N-dimensional space, in some sub-manifold spanned by the data. Finding this sub-manifold could be the task of a self-learning neural network, as for example the CCA network we present hereafter at section 4.

2.3. Discussion

As it can be guessed, starting from the two examples above, the problem of perception may take many aspects, from low-level signal processing to higher level of data representation. Even in this case, it is always preceded by an important stage of signal pre-processing. In the example of the vestibulo-ocular reflex, the nervous system represents the missing part of the information in one signal by modelling it from the other available signals. In the problem of ego-motion, all the data concerning the flow field of the image are correlated by the six components of the camera's motion. But the noise and the lack of accuracy, also the possibility of missing signals, lead to bad results if algebraic methods are used. This is typically a case where the signals should be considered on a statistical point of view, in the frame of data analysis in high dimensions. We will present now two artificial neural networks which can be used in signal and data analysis, with the objective to discover the structure of the data. Both of them are relevant of a class of problems named "Blind identification". This term means that we suppose that very little is known about the signals we are facing. In fact, it means that the minimum possible assumptions are made.

3. Separation of sources

3.1. The problem

In many natural situations, the signals which are received are mixtures of more basic other signals. One example has been given in the formulation of flow field in terms of ego-motion variables. In biology, a number of signals also appear as mixtures of basic variables. For example, in the coding of movements by the proprioceptive fibres of a muscle, there are two kinds of neuro-muscular receptors: the so-called "primary fibres", which are known to be dynamic, coding for the velocity of lengthening of the muscle. But, unfortunately, in its signal a non-negligible part of position is found. In the second kind of receptors, the "secondary fibres", the information is more static, their signal codes for the position, but also contains a certain amount of velocity. So, the static fibres convey more position than velocity and the dynamic ones convey more velocity than position. How the pure information of velocity and position can be retrieved, without knowing the relative amounts of each signal in the observed mixtures? We have firstly solved this problem in 1984, and now it has given rise to an important activity in the field of signal processing (see a review in [10]). In the next section we will show how it can be solved by a self-learning neural network.

3.2. Principle of the network

Suppose we are observing the signals of N sensors e_i, which are in fact N different linear mixtures of M sources x_k. Neither the mixtures nor the sources are known. The problem of finding the sources can be formulated as the solution of the following system:

$$e = A \ x.$$

The vector e of observables is the product of the vector of sources x by the matrix A of the mixtures. In order to go further, we need now to make some hypotheses:

1/ The number N of observables is at least equal the number M of sources

2/ The matrix A is of rank M

3/ The sources are statistically independent signals

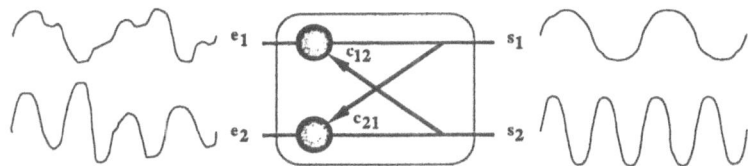

Figure 3. Separation of sources. Two signals e_1 and e_2 are observed. They are in fact two different mixtures of two other unknown independent signals. The neural network learns from the statistics of its output signals how to drive the synaptic coefficients in order to extract the unknown signals.

To solve this problem, we take a fully connected neural network as described in dimension 2 in figure 3. The goal is to make the network learn the inverse model of the mixture by adjusting the synaptic coefficients of its connection matrix **C**. The neurones are not self-connected, so the connection matrix **C** is square and has zeros on its principal diagonal. Then the output vector *s* of the network writes:

$$s = (I + C)^{-1} e \qquad (4)$$

I being the identity matrix. Combining with the formula of mixture we have:

$$s = (I + C)^{-1} e = (I + C)^{-1} A x \qquad (5)$$

Let us now assume for simplicity that the matrix **A** is square, but it is not mandatory in the general case, see [10] and [11]. The most general solution of the problem is that the matrix **C** verifies:

$$(I + C)^{-1} A = D P \qquad (6)$$

D being a diagonal matrix, and **P** a permutation matrix. So, it will ensure that any source x_k will appear, only one time, on any output s_i, that is the minimum we can ask for the separation of the sources. In this case, it has been shown that $s_i = a_{ik} x_k$, the coefficient a_{ik} being an element of **A**.

Now, how can we drive the synaptic coefficients of the network? Starting from the idea that the sources are independent, a criterion would be to drive the coefficients c_{ij} of **C** in order to minimise the mean output power of each neurone: $E[s_i^2]$. This leads to a learning law of the hebbian type:

$$\frac{d}{dt} c_{ij} = - \mu \, s_i \, s_j \qquad (7)$$

In fact, such a law would lead to a symmetric variation of matrix **C** and, at convergence, the outputs would be simply decorrelated and not independent as required for separation. A better criterion would be to minimise the mean fourth power of each output $E[s_i^4]$, which leads to the following learning law:

$$\frac{d}{dt} c_{ij} = - \mu \, s_i^3 \, s_j \qquad (8)$$

This law aims at zeroing the fourth order cross-moment of each pair of the outputs of the network, thus leading to a more accurate search of independence. In many cases this is sufficient, a better law was obtained by replacing s_j by Arctan(s_j) as in [11] or by fourth order cross-cumulants [10].

Then, with such laws, the network converges towards a value of **C** depending on the coefficients of **A**, which leads to the separation of the sources:

$$s = D P x \qquad (9)$$

This network has been applied in a number of problems of additive mixtures in Signal Processing, and has been extended more recently to time-convolutive mixtures of signals for speech enhancement purposes [16].

3.3. Example of application

In the following example, the inputs of the network are the four (x, y) co-ordinates of image pixels of a pair of stereo-cameras, as shown in figure 4, the unknown signals to be discovered are the three dimensions of the object seen by the cameras.

As it can be seen, the dimension of the observable vector is four and the intrinsic dimension of the data is three. There are four neurones in the network and they work as presented above. After convergence of the learning, we remark that one of the outputs if the network gives an almost zero signal whereas the other outputs provide signals which are linked to the dimensions of the analysed object in front of the cameras: the network has performed *by itself* a dimension reduction on the input data and has "discovered" the structuring signals: the co-ordinates linked to the object. It is worth mentioning that despite the mixing process is non-linear, the network has been able to find the best linear approximation of this process, giving thus an exploitable information.

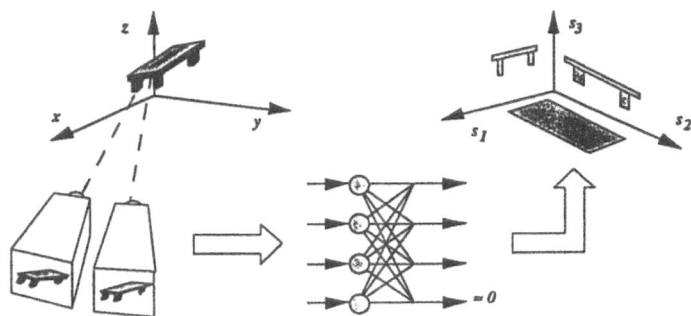

Figure 4. *Separation of sources applied to a stereo signal (four coordinates). The networks finds the structure of the analysed object and the necessary dimension reduction from four to three.*

The Source Separation network has also been successfully applied to more difficult kinds of signals as the R-G-B signals of a colour camera [13]. Here the observables are the three colour channels which can be seen as mixtures of various colours of the objects of the image. In this case, the result is a transformation of the signals so that, by simply thresholding the outputs, one can identify several objects in the image whereas they were not separable in the original R-G-B signals.

3.4. Discussion

In this problem, the signals of observables have been transformed into a *unique* representation, the one of the independent sources that were at the origin of the observed phenomenon. For this reason, this neural algorithm has also been called "Independent Component Analysis" or ICA, making reference to the classical "Principal Component Analysis", or PCA, well known in statistical data analysis. Though also a linear transformation, it differs fundamentally from PCA in the fact that it does not provide a simple decorrelation by means of a rotation of the space of observables, but really leads to independent variables. Thus, it can be given as a mean to identify the structure of the observed data, according to a linear model of mixture. For highly non-linear data structures, the models are rarely known and such a procedure of identification would not be adequate. The next section will treat this case.

4. Curvilinear Component Analysis

4.1. Problem statement

When there is no available model of the structure of the data, the problem is more complicated. Let us suppose that we are observing a process by a series of measurements of n variables ξ_k, the space of observables is R^n, n being possibly very high, say 10 to 1000. In fact this process is driven by a relatively low number of independent variables, say p, which can be combined in various unknown non-linear manners to produce the observables ξ_k such that:

$$\xi_k = f_k(t_1, t_2, ..., t_p) \tag{10}$$

The problem is, firstly to find the intrinsic dimension p of the sub-manifold V^p of R^n where the data are lying, and secondly to find a suitable mapping of this sub-manifold in R^p. In such a problem, there exist some classical techniques related to "Non-Linear Mapping" (NLM), [18] and [20], and a more recent one, very popular in the field of neural networks: SOM, the Kohonen Self-Organising Map [12]. Though having been first stated over more than 30 years and widely used, the Non-Linear Mapping suffers from a dramatic low speed of convergence and gets often trapped in local minima. The well known Self-Organised Map has proved to be useful in numerous examples. However, there remains some cases where the difficulty of convergence should not be ignored, especially in cases of highly folded data structures, or when the output mapping should be of greater dimension than 2.

I would like to present here a recent neural network [5], [6], that was partly inspired by the two previous techniques, but overcomes their drawbacks. It has been called initially "VQP" for "Vector Quantization and Projection", and as it will be seen, a more suitable denomination would be "Curvilinear Component Analysis" (CCA), following the spirit of the preceding network.

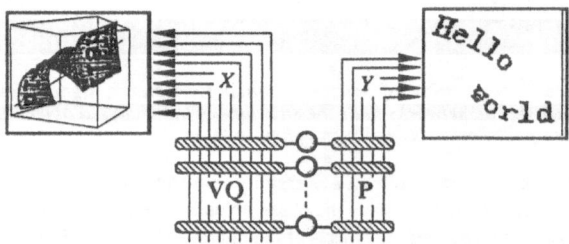

Figure 5. *The VQP or CCA network: a set of N neurones are provided, each with a n-dimensional input vector x_i and a p-dimensional output vector y_i. The input vectors quantify the input data space, and the output vectors y_i map the local topology of x_i's in the output space.*

4.2. The CCA network.

Just as in the SOM, let us suppose that we have a network of N neurones N_i, each of them having a n-dimensional input vector x_i, pointing in the input data distribution (figure 5). But these neurones have no predefined topology. However, they have each a p-dimensional output vector y_i pointing in the output space. Before running the network, two preliminary operations are required: 1) a Vector Quantization (VQ) of the n-dimensional input data sub-manifold, and 2) definition of the output space dimension p according to the dimension of the data sub-manifold.

The first operation can be done by any well known method, classical [9] or neural network-based (review in [5]). The second one can be done on the data space by a Karuehnen-Loeve transform, or better by a determination of fractal dimension, or by successive trials of CCA itself (we will see how later).

Now the CCA network can operate. In fact, the N prototype vectors of the input VQ are the input vectors x_i of the network. At the beginning, the output vectors y_i of the neurones are chosen randomly in the output space. The goal of the network is to learn the local topology of the data and reproduce it in the output space. To do so, without any prior knowledge, the network can only compute the distance $X_{ij} = \| x_i - x_j \|$ between any two given input prototype vectors x_i and x_j, and copy it to the distance between the two corresponding output vectors y_i and y_j : $Y_{ij} = \| y_i - y_j \|$. This copy is made under a learning process of energy minimisation, realised by means of a stochastic gradient approximation.

Let be E the energy function to be minimised according to the matching of prototypes distances in the input- and output spaces. If the output space has the same dimension as the input one, E takes the very simple form:

$$E = \frac{1}{2} \sum_{i,j} E_{ij} = \frac{1}{2} \sum_{i,j} (X_{ij} - Y_{ij})^2$$

(11)

This means that all the possible distances between pairs of input vectors are to be matched to the distances between the corresponding output vectors. A perfect match leads to $E = 0$. Let us notice that the projection is free in translation, rotation or inversion of axes, which allows sufficient freedom for incorporating some other constraints in the mapping process.

When the output space is of lower dimension than the input one, and when the structure of data is highly non-linear, this mapping is no longer possible because of the folding of the input space. We can only hope to perform a *local mapping*, then the energy function E takes the form:

$$E = \frac{1}{2} \sum_{i,j} E_{ij} = \frac{1}{2} \sum_{i,j} (X_{ij} - Y_{ij})^2 \, F(Y_{ij})$$

(12)

$F(Y_{ij})$ being a positive, monotonously decreasing function of the distance Y_{ij}. This means that, in this case, the matching should be perfect only for small distances *in the output space*. Such a function will allow the unfolding of the data sub-manifold and its projection in the output space, even if the input distribution lies in a closed n-dimensional hyper-sphere: during the projection process, the sphere will blow-up and "flatten" in a $(n-1)$-dimensional output space.

The minimisation of E is processed in the following manner: choose, at random or not, a neurone i, and move its output vector y_i according to opposite of the partial gradient of E with respect to y_i, with a step μ:

$$\Delta y_i = -\mu \, \nabla_i E = -\mu \sum_{j \neq i} \nabla_i E_{ij}$$

(13)

The vector y_i will evolve according to the current position of all other vectors y_j and, after having cycled among all the neurones, the energy ceases to decrease and the output is a p-dimensional map of the data sub-manifold.

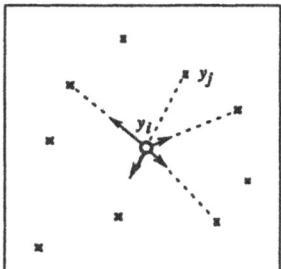 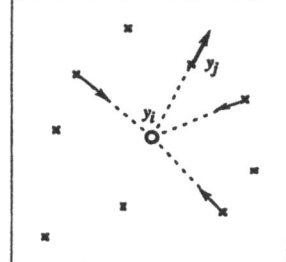

Figure 6. *Learning the prototypes in the output space. Left: y_i moves according to the opposite of the components of the gradient of E with respect of y_i. Right: y_i is fixed, it is the y_j's that move, in the opposite direction.*

In fact, this kind of stochastic gradient descent can easily be trapped in some local minima, as many neural network learning procedure do. In [5] we find an interesting alternative which reveals faster, more accurate and able to escape from local minima. The rule is very simple: after having chosen the neurone i, instead of moving its output vector y_i according to the opposite the gradient of E (see figure 6, left), keep it fixed and move all the other vectors y_j according to the partial gradient of E_{ij} (see figure 6, right):

$$\forall j{\neq}i \quad \Rightarrow \quad \Delta y_j | i = - \mu \, \nabla_j E_{ij} = \mu \, \nabla_i E_{ij} \tag{14}$$

It has been stated in [5] that the mathematical expectation of the energy variation $E[\Delta E]$ over all the iterations is the same as in the previous learning procedure, thus leading to the same global minimisation of energy. But what happens, is that *locally*, this variation can be positive, that is the energy can temporarily grow, and thus *escape from local minima*. This is an important result which can be applied to any problem of mapping. More, in this case the learning is an order of N faster, and the minimum reached by the energy is lower. We will see hereafter some examples of the performances of this new self-organising neural network.

We have said here above that the CCA network could be used to determine empirically the structuring dimension of the data. By observing the statistical joint distribution of distances of prototypes in the input space and in the output space, we have a means to determine if the mapping is correct or not. A strict proportionality indicates a perfect mapping. When the dimension of the output space is too small, the joint distribution scatters around the first diagonal. The optimal output dimension is then the lowest which leads to joint distributions near the first diagonal, at least for small output distances.

4.3. Example of a complex data structure

As an example of what this network can do, we have chosen a fictive problem of clustering and classification. Let us imagine a data structure like the one of the figure 7 (left). It is clear that there are two clusters in this data distribution. But these clusters are intermixed in such a manner that it is impossible to find a compact separating surface for a purpose of classification. Only neural networks with radial basis functions could be used for such a purpose, but we will see that our VQP network is more suitable: it can learn by itself, and more, it can provide a true interpolation in the input and output data spaces.

In this problem, we apply the formula (12) for the energy and the formula (14) for the learning. The network converges in O(N) iterations and provides a 2-dimensional projection in the output space (figure 7, right). In this projection where each data point corresponds to an input data, it is easy to find a separating line for a purpose of classification. This network which can follow and project trajectories in the sub-manifold of the input data onto a space of reduced dimension can be called "Curvilinear Component Analysis". This closes the loop opened with "Principal Component Analysis" and continued above by the "Independent Component Analysis" neural network.

 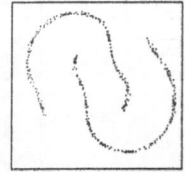

Figure 7. *Example of a difficult 3-dimensional input data distribution. There are two imbricated classes, each on a torus (left). The network makes a 2-dimensional projection where we can find a separating curve.*

More, once the network has converged, it is possible to *interpolate* the projection. For this, we have just to create a new prototype neurone, with an x input vector, pointing at a desired place in the input distribution, and an output vector *y pointing* at random in the output space. This new neurone has just to learn, by the same algorithm, the corresponding place of its output vector. Note that by this procedure, the interpolation is *continuous*. This is called the "forward interpolation problem".

A *backward interpolation* as well is possible. In this case, the procedure is also very simple: we have just to exchange the names of input and output spaces. Then the network learns to map the p-dimensional projection space into the n-dimensional original one. The energy function to be minimised sould be slightly different now:

$$E' = \frac{1}{2} \sum_{i,j} E'_{ij} = \frac{1}{2} \sum_{i,j} (X_{ij} - Y_{ij})^2 F(X_{ij}) \qquad (15)$$

As we have exchanged the input and output spaces, the weighting function F in the energy is now taken with respect to input distances: $F(X_{ij})$. This ensures the full symetry of the algorithm and provides a good estimation of the reverse projection.

4.4. Application in a perceptual task

Let us now consider a discontinuous and non-linear structure of data, as it often happens in the representation of the environment for visually-guided robots. The problem for a navigating robot is to construct some representation of its environment, for example for finding its route, based on the (changing) visual information. In its visual field, different objects appear and disappear, with respect to its motion. Let us suppose that the visual system is capable of evaluating the distance from each viewed object. These distances constitute the data space used for the representation of the environment.

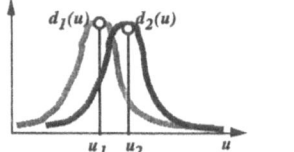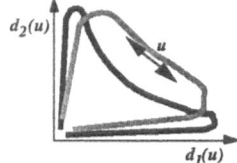

Figure 8. *The signals $d_i(u)$ of sensors sensitive to the distance between the robot location u and objects situated at location u_i (left). Corresponding signal in the space of two sensors when the robot is moving, for two different shapes of the $d_i(u)$ function.*

For more simplicity, let us consider a uni-dimensional motion of a robot along an axis, u. Various objects are located at different places $u_1, u_2,... u_i$. The distance sensors of the robot give a signal for the distance between position u and location u_i of the form (figure 8, left):

$$d_i = \exp[-(u - u_i)^2/\lambda^2] \qquad (16)$$

The data space to be processed by the robot is the one of the observables: the sensor signals. The data are then in a multi-dimensional space, the elements of which $\{d_i\}$ are distances from the robot to the objects. The distribution of the data in this space when the robot is moving is given figure 8 (right) for two objects: it is a particular trajectory, depending on the position of the robot, of location of the objects and. of the shape of the distance-measuring function.

In fact, the robot has a limited field of vision: a distance sensor produces a non-zero signal only when an object is in its field of vision. So, the d_i signal respects the formula (16) only for $(u - u_i)$ greater than a minimum distance λ_0. This results, in the data space, in some discontinuities of the trajectories, as it can be seen figure 9 (bottom left).

In this figure, we see the robot with its limited visual field and three objects unequally spaced at locations u_1, u_2, u_3. When the robot is moving, according to its current position, one, two or three sensors are activated, but not progressively. When a sensor becomes visible, its signal jumps from zero to the current value depending on the distance. If we represent the trajectory (distribution) of the input data for a robot moving along the axis u, we see some discontinuities (figure 9, bottom left).

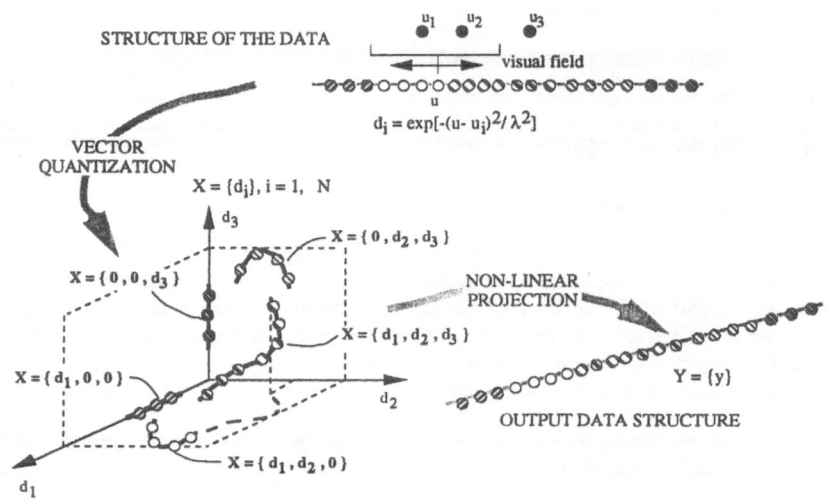

Figure 9. *The problem of input space discontinuities for a moving robot (see text). Top: the structure of the data. Bottom left: the resulting input space for three objects. Bottom right: the representation space obtained by the CCA network.*

By applying such data to a CCA network, after the vector quantization of the input space, the projection in a uni-dimensional space is not strait forward. The distances between prototypes in the input space are not continuously varying: between two prototypes x_i and x_j corresponding to the appearance or the disappearance of an object, the distance jumps, according to the presence or not of a new co-ordinate. For example, when the robot position is on the left-hand side one vector has components $x_i = \{d_1, 0, 0\}$, the next one has $x_j = \{d_1, d_2, 0\}$. The direct mapping of these data to the output space would provide undue discontinuities. For this reason, we need to compare distances in the input space with the same visible objects. The distance between x_i and x_j will be:

$$X_{ij} = \text{Sqrt } [(d_{1i} - d_{1j})^2 + 0 + 0],$$

instead of:
$$X_{ij} = \text{Sqrt } [(d_{1i} - d_{1j})^2 + (d_{2i} - d_{2j})^2 + 0].$$

This means that, for a suitable representation of the data, there is a need to determine *what components are significant* for our problem. This is a general procedure we use in current life.

After this new kind of pre-processing of the data, the CCA network is able to learn the continuous non-linear projection of the input data space (figure 9, bottom right), disregarding the occasional jumps of the trajectories of the data in the input space.

5. Conclusion

With the few examples of signal and data processing we have seen here in the various domains of biological systems, computer vision, and with some solutions provided by artificial neural networks, we have been faced with different aspects of the problem of perception. Let us review them briefly.

In the example of the Vestibulo-Occular reflex, the signals provided by the sensory receptors are incomplete and not directly usable for the control of the direction of the gaze. The central nervous system takes benefit from the redundancy among the various sensors and combine their signals in order to reconstruct the missing part of their frequency spectra. This can be seen as an elementary (?) problem of perception: the representation of hidden information in a group of signals.

For the determination of ego motion from the flow field of an image, the problem seems very simple: the parameters of ego motion are linearly combined in the expressions of the velocity flow field estimation. However, these estimations are highly corrupted by noise, many of them are not significant, due to appearing or disappearing objects. The algebraic methods, even with the use of regularisation techniques, generally fail to find consistent results. It appears that for this problem, there would be a need of a higher level of processing, in the scope of statistical data processing, by selecting the well correlated data and disregarding the remainder.

Now with the artificial neural networks, the Source Separation network appears as a possible transition between Signal Processing (separation of sources) and Data Analysis. It puts forward the concept of Independent Component Analysis as a means to reach the structure of the observed data. It works on the basis of a linear combination of signals, which is often the case for low level processing.

The VQP network deals really with the structure of data, it can find highly non-linear relationships between data and can represent them in a reduced-dimension space. It outperforms a number of other techniques in this task. Working on a continuous projection basis, it allows a smooth representation of continuous phenomena by true interpolation. However, it also suffers from noisy and corrupted data and needs some adequate pre-processing.

This pre-processing seems necessary at any level of signal or data processing. It can be made by some elementary operations in a very limited number of cases. Otherwise, the problem remains open, it is mostly solved by means of ad-hoc solutions and there is a lack of general approaches. I guess that it could be a goal for the next generation of artificial neural networks: selecting the most relevant information and disregarding the least significant data for the problem under consideration. In fact, it is an operation we are commonly doing in everyday's life, why not translating it into some artificial neural network?

6. References

[1] **Adiv G.**, (1985) Determining 3-D Motion and Structure from Optical flow generated by Several Moving Objects. *IEEE Trans. PAMI*, vol. 7, N° 4.

[2] **Atick J.**, (1992) What does the Retina Know about Natural Scenes. *Neural Computation*, 4, 196-210.

[3] **Beaudot W, Palagi P, Hérault J.**, (1993) Realistic Simulation Tool for Early Visual Processing including Space, Time and Colour Data. *International Workshop on Artificial Neural Networks* IWANN'93, Barcelona, Spain

[4] **Cappellini V., Mecocci A.**, (1993) Motion Analysis and Representation in Computer Vision. *Journal of Circuits, Systems and Computers*, Vol. 3, N° 4.

[5] **Demartines P.** (1994) *Analyse de données par réseaux de neurones auto-organisés*, Thèse de Doctorat de l'Institut National Polytechnique de Grenoble, 26 Nov. 1994.

[6] **Demartines P., Hérault J.**, (1993) Representation of non-linear data structures through fast VQP neural networks. *Proceedings of Neuro-Nîmes'93*, Nîmes (France).

[7] **Droulez J., Darlot C.**, (1989) The geometric and dynamic implications of the coherent constraints in three-dimensional sensorimotor interactions. *Attention and Performance*, 14, 4105-526.

[8] **Franceschini N., Pichon J. M., Blanes C.** (1992) From insect vision to robot vision. *Phil. Trans. Roy. Soc.* London. B 337, 283-294.

[9] **Gersho A., Gray R. M.** (1992) *Vector quantization and signal compression*. Kluwer Academic Publishers, London.

[10] **Hérault J., Jutten C.** (1994) *Réseaux neuronaux et traitement du signal*. Traité des nouvelles technologies, Hermes, Paris.

[11] **Jutten C., Hérault J.**, (1991) Blind separation of sources. Part I: an adaptive algorithm based on a neuromimetic architecture. *Signal Processing*, Vol. 24, 1-10.

[12] **Kohonen T.**, (1984) Self-organisation of topologically correct feature maps. *Biological cybernetics*, 43 , 59-69.

[13] **Liu X., Hérault J.**, (1991) Colour image processing by a neural network model. *International Neural Network Conference* INNC'91, July 9-13, Paris (France).

[14] **Livingstone M., Hubel D. H.**, (1988) Segregation of form, colour, movement and depth: anatomy, physiology and perception. *Science*, 240, 740-749.

[15] **Marr D.**, (1982) *Vision: a computational investigation into the human representation and processing of visual information*. WH Freeman & Co, San Francisco

[16] **Nguyen Thi H. L., Jutten C.**, (1995) Blind source separation for convolutive mixtures. *Signal Processing*, (to appear).

[17] **Pettet M. W., Gilbert C. D.**, (1992) Dynamic changes in receptive-field size in cat primary visual cortex. *Proceedings of National Academy of Science*, USA, Vol. 89; 8366-8370.

[18] **Sammon W. J.**, (1969) A non-linear mapping algorithm for data structure analysis. *IEEE Trans. on Computers*, Vol. C-18, N° 5, 401-409.

[19] **Schölkopf B, Mallot H P.**, (1994) *View-based cognitive mapping and path planning*. Technical report N° 7, Max Plank Institut für biologische Kybernetik, Tübingen, Germany.

[20] **Shepard R. N, Carrol J. D.**, (1965) Parametric representation of non-linear data structures. In *International Symposium on Multivariate Analysis*, Krishnaiah P. R, editor, Academic Press.

[21] **Sull S. , Ahuja N.**, (1994) Integrated 3-D Analysis-Guided Synthesis of Flight Image Sequences. *IEEE Trans. PAMI*, Vol. 16, N° 4.

[22] **Sune J. L., Puget P., Samy R.**, (1993) Computation of the depth-from-motion problem from neural networks. *Proceedings of Neuro-Nîmes'93*, NÎMES (France).

[23] **Tononi G., Sporns O., Edelman G. M.**, (1992) Reentry and the problem of integrating multiple cortical areas: Simulation of dynamic integration in the visual system. *Cerebral Cortex*, Vol. 2, N° 4, 316-335.

[24] **Treisman A.**, (1988) Features and objects: the fourteenth Barlett Memorial Lecture. *Quarterly Journal of Experimental Psychology*, Vol. 40 A, 201-237.

[25] **Walter J. A, Schulten K. J.**, (1993) Implementation of Self-Organising Neural Networks for Visuo-Motor Control of an Industrial Robot. *IEEE Trans. on Neural Networks*, Vol. 4, N° 1.

[26] **Zupan L.**, (1995) *Modélisation du réflexe vestibulo-oculaire et prédiction des cinétoses*. Thèse de Doctorat, École Nationale Supérieure des Télécommunications, Paris (France).

Optimization Neural Networks
for Image Segmentation

D.L. Vilariño, D. Cabello and A. Mosquera.

Departamento de Electrónica y Computación. Facultad de Física.
Universidad de Santiago de Compostela.
15706 Santiago de Compostela. SPAIN.
E-mail: eldiego@usc.es

Abstract. In this work we describe the implementation of an artificial neural network, an extension of Hopfield's model, for the segmentation of images. The problem is approached in terms of the minimization of an objective function which integrates statistical and spatial information and which is projected onto the network. It provides a locally optimal solution to the problem of the classification of $N_1 * N_2$ pixels into M classes. The experimental results obtained show the validity of the architecture we propose.

1. INTRODUCTION.

Segmentation is one of the most important stages of an image analysis system. It consists on the division of the image into a set of disjoint elementary regions which are characterized by the fact that some feature (grey level, color, texture, etc...) is constant. In an image interpretation system there will always be a previous segmentation stage. Obtaining a correct interpretation will greatly depend on the quality of the results from the segmentation process. In the image processing literature we find a wide variety of strategies and algorithms for approaching the segmentation problem. The article by Pal and Pal (1993) is a good review of these methods. However, there is no method that provides results that can be considered optimal in every type of image. In fact, the methods that can correctly segment one type of image may not be applicable to other types. Thus, the most adequate method must be employed in each particular application.

Among the classical segmentation procedures we can distinguish between 'edge oriented procedures' and 'region oriented procedures'. The former seek local discontinuities through differential operators, whereas the latter pursue the detection of local areas with homogeneous properties, either in the image space itself, with region growth techniques, or in a feature space by means of pixel classification techniques (histogram thresholding or clustering algorithms). In classification techniques, each pixel of the image is characterized by a feature vector. In these cases, the segmentation problem becomes a classification or partition problem in the feature space. The elementary regions arise from the transportation of the information on the classes in the feature space to the image space: neighboring pixels belonging to the same class will constitute an elementary region. In the methods we mention, the segmentation is carried out following two strategies: based on information on the spatial distribution of the feature, which generally leads to incomplete segmentations of the image, requiring a subsequent stage for analyzing and tracing the edges in order to obtain a complete segmentation, or based on statistical information, which, even though it leads to complete segmentations, does not take into account the spatial distribution of the pixels. A relevant aspect consists in the development of strategies that permit carrying out the segmentation taking both types of information into account.

With the implementation of these ideas in mind, the parallelism found in image processing problems as well as in artificial neural networks in addition to the processing characteristics of the latter (learning capabilities, which in many cases reduces the problems to a correct training of the network, their generalization capabilities, power and robustness with respect to noise, easy computational implementation, possibility of real time outputs through their hardware implementation, etc...) have led to the use by many authors of artificial neural networks for the computational processing of images. The literature in this line is very extensive.

Constraining our discussion to segmentation problems, some solutions based on selforganizing topological maps which implement classical clustering algorithms (Bezdeck et al., 1992; Hall et al., 1992) have been proposed. On the other hand, as the most widely used clustering algorithms (c-means algorithms or any of their variations) are based on he classification of the M feature vectors of the input space into C classes on the grounds of the minimization of an objective function, some authors suggest formulating them in terms of a Hopfield neural network model (Kamgar-Parsi et al., 1990). Thus, Amartur et al. (1992) implement a K-means algorithm by means of an M*C neuron modified Hopfield network and apply it to the segmentation of nuclear magnetic resonance images. Other authors have followed this line of approaching segmentation as an optimization problem, designing both deterministic and stochastic modified Hopfield networks in order to minimize different objective functions. These functions solve the problem of segmenting an image into a prespecified number of classes and in almost all of the works, their characteristic parameters are known (Manjunath et al., 1990 ; Huang, 1992; Mosquera et al., 1993).

In this work we approach the problem of the segmentation of an $N_1 * N_2$ pixel image into M classes as an optimization problem. The segmentation must respond to statistical information on the classes, which must be extracted from the image, as well as to the spatial distribution of the feature. To this end we design an objective function that integrates both terms. For its minimization we employ a multilayer Hopfield network with N_1*N_2*M neurons, which we model by means of non linear amplifiers with a sigmoid type monotonic input/output relationship. The projection of the objective function we have designed onto the general expression of the energy minimized by the network will determine the connection scheme and weights as well as the inputs. Finally, the solution of the set of differential equations that control the dynamics of the different neurons will lead to a state of the network that will represent the desired segmentation.

The article is structured into the following sections. After this introduction we present a brief description of Hopfield networks. We then describe the problem and the design of the network that solves it, as well as its dynamics in detail. In a final section we present different segmentation examples based on grey levels and texture features which demonstrate the validity of the structure we propose.

2. HOPFIELD NETWORKS.

Hopfield networks are one of the most widely used neural structures and there are a large number of implementations of them, both for binary inputs and for continuous inputs. Their weights are usually fixed and their applications are basically oriented towards optimization problems or as associative memories. These networks belong to the category of dynamic networks; the equations of the nodes are described by differential or finite difference equations (Hush and Horme; 1993).

A Hopfield network consists of a totally interconnected neuron layer. If we consider an N neuron network with continuous inputs, the node equations will be:

$$\frac{du_i}{dt} = -\frac{u_i(t)}{\tau_i} + \sum_{j=1}^{N} T_{ij} V_j(t) + I_i \qquad (1a)$$

$$V_i(t) = f(u_i(t)) \qquad (1b)$$

where $u_i(t)$ is the internal state or potential of the i-th neuron, τ_i a time constant, I_i its external input and V_i ($0 \leq V_i \leq 1$) its activation level or output. $f(\cdot)$ is the activation function, usually a sigmoid. T_{ij} represents the weight associated with the connection between the j-th neuron and the i-th neuron. The Hopfield network can thus be taken as a non linear dynamic system with an input vector I, a state vector U(t) and an output vector V(t). Due to the sigmoid type non linearity, the output vector will be inside an N-dimensional unitary hypercube.

The behavior of the network depends on the value of its parameters, and it can result in anything from a stable system to an oscillator, or even a chaotic system. Hopfield proves that a network with a symmetric connection matrix converges to a stable point that is a local minimum of function

$$E = -\frac{1}{2} \sum_{ij} T_{ij} V_i V_j - \sum_i I_i V_i + \sum_i \frac{1}{\tau_i} \int_0^{V_i} u_i \, dV_i \qquad (2)$$

The non linear nature of the Hopfield network generates multiple equilibrium points. This result tells us that for any set of initial conditions a Hopfield network with symmetric connections converges to a fixed equilibrium point, located inside the $(0,1)^N$ hypercube. The exact number of equilibrium points and their locations are determined by the parameters of the network in terms of the connection weight matrix, inputs and neuron gains. As we will later see, in our problem, the inputs and the connection matrix are fixed for the solution we propose and consequently, we will only be free to modify the gain of the neurons. In a low gain situation there are very few equilibrium points located inside the hypercube. However, as the gain is increased, the number of equilibrium points grows and their positions are shifted towards the vertices of the hypercubes, reaching them for a high gain limit situation. In this case, the expression of the network energy is simplified, eliminating the integral in equation (2) (Hopfield; 1984). The characteristics of the Hopfield network in a high gain regime are very interesting due to the fact that many problems that must be solved with this network require binary solutions. In our case we are going to consider high gain neurons.

3. STRUCTURE OF THE PROPOSED NETWORK

The network we want to design has the objective of segmenting an $N_1 * N_2$ pixel image into M classes. For this we propose the use of a multilayer neural structure with $N_1 * N_2 * M$ neurons, associating one neuron to each pixel and class to which it can belong. The first problem to solve will be to determine the appropriate connection structure for these neurons in order to be able to perform a correct segmentation of the image. We can imagine the set of neurons as distributed in M $N_1 * N_2$ processing element planes, with each plane associated to each one of the classes. This way, the value of the output potential of the (ijl)-th neuron, V_{ijl}, which, because of the definition of a sigmoid type activation function, will take values from interval [0,1], will indicate the degree of membership of pixel (i,j) to group or class l.

For the design of the network we impose three constraints that summarize what we could define as 'optimal segmentation': 1) A pixel must belong to the group whose characteristic vector is closer to the feature vector of the pixel, 2) A pixel will tend to belong to the same group as its neighboring pixels, 3) The sum of the memberships of a pixel to all the groups must be equal to one. The first constraint could be called class information, and refers to the classification of the pixels in statistical terms. The second integrates information on the spatial distribution of the pixels and the third one permits interpreting the output of a neuron as a degree of reliability of the membership of the pixel to the class indicated by the neuron in the sense of a classical fuzzy classifier. By integrating these three terms we want to make the system evolve to a point that implies an equilibrium between the three constraints.

As we have already indicated, we work with a system that will evolve towards the minimum of an energy function E. Consequently, the strategy to follow will be to find a function that, for a given input, evolves towards a minimum that implies the desired output for the input. After constructing this target function, we will have to restructure it so that we can equate its terms to those of the energy function that minimizes the Hopfield network, which, for a network such as the one proposed, with $N_1 * N_2 * M$ high gain initially fully connected neurons, corresponds to the expression:

$$E = -\frac{1}{2}\sum_{i=1}^{N_1}\sum_{j=1}^{N_2}\sum_{l=1}^{M}\sum_{i'=1}^{N_1}\sum_{j'=1}^{N_2}\sum_{l'=1}^{M} T_{ijl;i'j'l'}V_{ijl}V_{i'j'l'} - \sum_{i=1}^{N_1}\sum_{j=1}^{N_2}\sum_{l=1}^{M} I_{ijl}V_{ijl} \qquad (3)$$

where $T_{ijl;i'j'l'}$ represents the weight associated to the connection from the (i'j'l')-th neuron to the (ijl)-th. The terms I_{ijl} and V_{ijl} represent the external input and the output potential of the (ijl)-th neuron.

Observing equation (3) we can infer that those terms of the objective function we have designed that are multiplied by an output potential will be identified with inputs that are external to the neurons. On the other hand, the terms that are multiplied by the square of output potentials or products of potentials will be identified with synapses. Finally, constant terms and those that do not depend on the output potentials of the neurons can be ignored, as their presence or absence will not modify the position of the minima of the E function we have constructed. The objective function E will have three terms, each one representing one of the three constraints we have imposed.

3.1 Constraint on class information.

This constraint is the one that introduces the characteristics based upon which we want to segment the image in the segmentation process. We start by describing each pixel by a feature vector. For the construction of the first term of E we use the idea of a minimum distance classifier, that is, assume that a pixel must belong to the group whose characteristic vector is closer to its feature vector. One possible term that is minimized when this criterium is satisfied is the following:

$$E_1 = -\sum_{i=1}^{N1}\sum_{j=1}^{N2}\sum_{l=1}^{M} [V_{ijl} - k^2 |\theta_{ij} - \psi_l|^2]^2 \tag{4}$$

where θ_{ij} represents the estimated feature vector for the (i,j)-th pixel and Ψ_l the characteristic vector of the l-th class (l-th neuron layer). In order to calculate the distance between vectors we have chosen the euclidean distance, but it could well be the one induced by any other norm. This term, E_1, will be minimized when its absolute value is maximized. For this, V_{ijl} will tend to 1 when $\theta_{ij} \approx \psi_l$ and will tend to zero when both are very different. Constant k calibrates, in a certain sense, the sensitivity of the system.

Once a term that responds to the first constraint is constructed, we must rearrange it so that it is easy to identify it with the general expression for the energy given by equation (3). Expanding the squared terms and ignoring the terms that are not related to any potential, the expression becomes:

$$E_1 = -\sum_{i=1}^{N1}\sum_{j=1}^{N2}\sum_{l=1}^{M} (V_{ijl})^2 + 2.k^2.\sum_{i=1}^{N1}\sum_{j=1}^{N2}\sum_{l=1}^{M} |\theta_{ij} - \psi_l|^2.V_{ijl} \tag{5}$$

The first term is multiplied by V_{ijl}^2, which will be translated into a feedback loop. Finally, the remaining term is the one involving the properties of the pixel and is multiplied by V_{ijl}, which means that it will, in fact, be introduced into the system as an external input. By equating the terms in (3) and (5) we obtain the external inputs and the weights of the synapses in the network due to term E_1 as:

$$I_{ijl} = -2.k^2 |\theta_{ij} - \psi_l|^2 \tag{6}$$

$$T_{ijl;i'j'l'} = 2\delta(i-i',j-j',l-l') \tag{7}$$

where $\delta(a,b,c) = 1$ if a=b=c=0 and null otherwise. Consequently, this first term, apart from the external inputs, only generates feedbacks.

3.2 Constraint on the spatial distribution of labels.

The second class of constraint imposes that a pixel should belong to the same class as it neighbors. This constraint arises from the assumption that the objects of the image are represented by homogeneous regions. However, the presence of 'noise' both during the image formation process and the sensing process, might break that homogeneity in the representation of the image, producing discontinuities in the segmentation. The idea of this constraint is that small variations in the spatial distribution of the feature we are using for the segmentation do not generate multiple elementary regions. With this term we implicitly assume a model for the formation of the regions and we make a pixel's label depend on those of its neighbors. Thus, what the second term of E does is that if pixels (i'j'), neighbors of (i,j) in a neighborhood N_s, belong to a given class, then pixel (i,j) also belongs to that class. An adequate term for representing this second constraint could be something like

$$E_2 = \sum_{l=1}^{M}\sum_{i=1}^{N1}\sum_{j=1}^{N2}\sum_{i',j'\in N_s} (V_{ijl} - V_{i'j'l})^2 \tag{8}$$

where the first sum is over all the classes, the second over all the pixels of the image and the third one over the set of neighbors of each pixel. As definition of neighbors of a pixel we have chosen the neighborhood structure that is commonly used in stochastic image models (Derrin and Elliot, 1987). Thus, N_1 includes the four nearest neighbors of each pixel, N_2 includes the eight nearest, etc.... We must point out that near the edge, the pixels will have a smaller number of neighbors than inside, adopting a 'free border' model instead of a toroidal model for the network.

Following a similar process to that carried out for the term E_1, we now rearrange E_2 in order to compare it with equation (3). E_2 will finally be :

$$E_2 = 2.N.\sum_{i=1}^{N1}\sum_{j=1}^{N2}\sum_{l=1}^{M} (V_{ijl})^2 - 2.\sum_{i=1}^{N1}\sum_{j=1}^{N2}\sum_{l=1}^{M} \sum_{i',j'\in N_s} V_{ijl}V_{i'j'l} \qquad (9)$$

where N indicates the number of neighbors of each pixel. Equation (9) is already similar to equation (3). If we compare both, we obtain:

$$T_{ijl;i'j'l'} = -4 (N+1) \delta(i-i',j-j',l-l') + 4\delta(i-i'+\Delta, j-j'+\Delta, l-l') \qquad (10)$$

where Δ generates the set of neighbors. Thus Δ takes the values +1, 0 and -1 for s=2. The connections generated by this term are limited to neurons within the same layer and to the neighborhood considered. On the other hand this term does not induce any external input to the network.

3.3 Constraint on the membership functions.

The last term to be taken into account contains the third constraint, which specifies that the total sum of the memberships of a pixel to the different classes must be one. A term that is minimized when this case occurs would be:

$$E_3 = \sum_{i=1}^{N1}\sum_{j=1}^{N2} (\sum_{l=1}^{M} V_{ijl} - 1)^2 \qquad (11)$$

which, if we expand and identify with equation (3) leads to inputs and synaptic weights given by

$$I_{ijl} = 2 \qquad (12)$$

$$T_{ijl;i'j'l'} = -2\delta(i-i',j-j',0) \qquad (13)$$

This term originates an inhibitory feedback among the neurons of the same column, that is, among all the neurons associated with the same pixel as well as a constant external input.

3.4 Global expression for the energy:

The global energy function that must be minimized by the network in order to solve the segmentation problem we have presented is constructed as the sum of the three terms described above. Its final expression is:

$$E = A.E_1 + B.E_2 + C.E_3 \qquad (14)$$

where A, B, C are constants that will permit the modification of the influence of each term, in other words, they will permit the variation of the weight associated to each one of the constraints. The connection topology, weights and inputs of the network are given by

$$I_{ijl} = 2.C - 2.A.k^2 \| \theta_{ij} - \psi_l \|^2 \qquad (15)$$

$$T_{ijl;i'j'l'} = (-4(N+1)B+2A)\delta(i-i',j-j',l-l') + 4B\delta(i-i'+\Delta,j-j'+\Delta,l-l') - 2C\delta(i-i',j-j',0) \qquad (16)$$

It can be observed that this network satisfies the symmetry condition for the connection matrix, that is, $T_{ijl;i'j'l'} = T_{i'j'l';ijl}$ and consequently, the system it generates will be stable.

4. SIMULATION.

For the simulation of the network we must solve the set of differential equations that describe the dynamics of its neurons. In order to obtain them it is enough to differentiate the global energy equation with respect to the output of each neuron and invert the sign of the result. Thus, if we take into account the sigmoidal relation of the activation function

$$V_{ijl} = \frac{1}{1 + \exp\left(\frac{-2.u_{ijl}}{u_o}\right)} \qquad (17)$$

where u_o is the gain term, the resulting equation system is:

$$\frac{du_{ijl}}{dt} = -u_{ijl} + 2C - 2A.k^2 \parallel \theta_{ij} - \psi_l \parallel^2 -$$

$$-(4B - 2A)\frac{1}{1 + \exp\left(\frac{-2.u_{ijl}}{u_o}\right)} + 4B \sum_{i',j' \in N_s} \frac{1}{1 + \exp\left(\frac{-2.u_{i'j'l}}{u_o}\right)} - 2C\sum_{l'=1}^{M} \frac{1}{1 + \exp\left(\frac{-2.u_{ijl'}}{u_o}\right)} \qquad (18)$$

for $i=1,..,N_1$, $j=1,..,N_2$ and $l=1,..,M$. It is thus a system of first order non linear equation that we will solve by means of a routine for solving differential equations that uses the classical Runge-Kutta method on a vector computer. We would finally like to point out that equation (18) is not rigorously correct as ψ_l is a function of the activities of the neurons and consequently its derivatives should also be included. This dependence is in the form of

$$\psi_l^m(t+1) = \psi_l^m(t) + \frac{\sum_{i,j} (\theta_{ij}^m - \psi_l^m(t)) * V_{ijl}(t)}{2 . \sum_{i,j} V_{ijl}(t)} \qquad (19)$$

where $\psi_l^m(t)$ is the m-th component of the characteristic vector (centroid) of the l-th neuron layer in time t. θ_{ij}^m represents the m-th component of (i,j)-th pixel feature vector. However, when solving the equations numerically using small time increments, the effects of this dependence will be negligible, and the approximation will be valid. Consequently, in the numerical simulation we start from random characteristic vectors. After each iteration, the new centroids that will act on the inputs of the next iteration will be recalculated by means of (19).

5. EXPERIMENTAL RESULTS

In order to show the validity of the structure we propose, we have carried out different trials consisting on the segmentation of different images considering different feature spaces. Thus, in the first place we have attempted the segmentation of images based on their grey levels. In this case both the feature vectors and the centroids of the classes become scalar variables. After this we have approached the segmentation of texture images, which we have characterized through second order Markov fields, making the measurement space four dimensional. Figures 1.a, 1.b and 1.c show the first set of images we processed. All of them are 64*64 pixel binary images that have been corrupted with increasing noise levels. Thus, in figure 1.a, we present the binary image without any type of noise; its grey levels are 96 and 160. The remaining images have been obtained from it by adding a gaussian noise component with an increasing standard deviation. If we define a signal to noise ratio (SNR) for these binary images as

$$SNR = \frac{|q_2 - q_1|}{\sigma} \qquad (20)$$

where q_1 and q_2 are the grey levels of the image and σ the standard deviation of the noise signal, the noisy images present decreasing values for the signal to noise ratio: 2 and 0.5 for figures 1.b and 1.c respectively.

In order to segment these images we have implemented a 64 x 64 x 2 neuron neural network. For the gain of the amplifiers we have chosen $U_0=1/100$, an adequate value for working in a high gain situation. In addition, we have chosen the following values for the constants in the energy expression: A=0.1, B=0.05, C=0.5, K=5 and as N_s a fifth order neighbor system (s=5; N=24). For solving the differential equations the inputs to the network were normalized to the [0,1] interval, and initial conditions for U_{ijl} were established ($U_{ijl} = 0$). After the convergence of the network, the activation levels of the neurons will indicate the membership degree of the pixels to the classes, producing a fuzzy segmentation. As we work with high gain neurons, their potentials will be close to 0 or 1 and consequently, in order to defuzzify this segmentation we apply the rule of the maximum membership function: we assign the pixel to the class to which its membership degree is highest. The regions, on the image plane, are made up of all those adjacent pixels that are assigned to the same class. Figures 1.d , 1.e and 1.f show by means of masks the regions extracted from the images presented in figures 1.a, 1.b and 1.c respectively. The segmentation is totally correct.

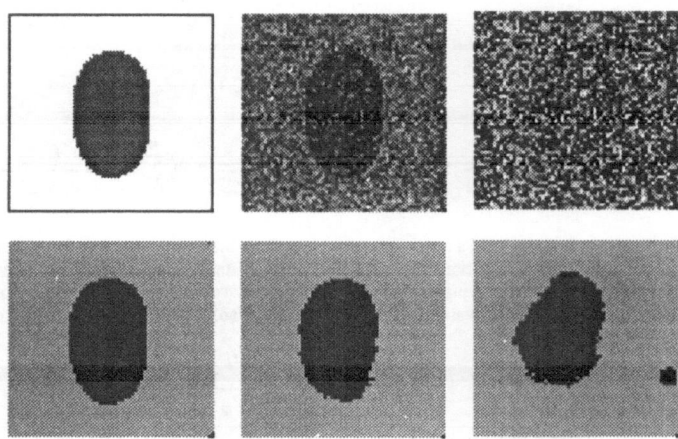

Figure 1. (form left to right and top to bottom). (a): binary image, (b) and (c): images with SNR = 2 and SNR = 0.5 respectively. (d), (e) and (f): results form the segmentation of the previous images.

We have also approached the segmentation of images that are compositions of different textures. Texture is an important characteristic in many types of images, despite the fact that no precise definition for it exists. Its interpretation usually implies the construction of models. Several techniques try to describe a texture through the number and type of primitives it uses and their spatial organization, considering as primitives the regions of the image with specific tonal properties. A real texture may present a large randomness in its primitives and/or their spatial organization, which leads to stochastic models that consider textures as a sample of a probability distribution in the image space (Cross and Jain, 1983). The grey levels of the pixels are random variables, and the grey level value for a pixel of an image is highly dependent on the levels of the remaining pixels, unless the image is simple random noise. Markov's random field represents a particular model for this dependence. In it, the intensity of the pixels of an image is a linear combination of the intensity of the neighboring pixels plus a noise term. Let us consider an image of dimension N*N, let us call its set of points Ω, $\Omega = \{ (i,j); 0 \le i,j \le N-1 \}$, and let s be a 2 dimensional vector such that $s \in \Omega$. If we assume that observations y(s) of a texture are gaussians with a null mean, the Markov random field can be described by means of the following equation (Woods, 1972):

$$ y(s) = \sum_{r \in N_s} \theta_r \, y(s+r) + e(s) \tag{21} $$

where e(s) represents a gaussian type sequence of null mean and variance of υ. N_s is the set of neighbors, given by the order of the markov field and θ_r the parameter vector corresponding to the model. In order to model textures we have considered a gaussian and symmetric second order markov field. The set of neighbors to be considered is $N_2 = \{(0,1), (1,0), (0,-1), (-1,0), (-1,1), (1,1), (1,-1), (-1,-1)\}$. If we assume symmetry, we have that $\theta_{0,1} = \theta_{0,-1} = \theta_h$, $\theta_{1,0} = \theta_{-1,0} = \theta_v$, $\theta_{1,1} = \theta_{-1,-1} = \theta_{ds}$ and $\theta_{-1,1} = \theta_{1,-1} = \theta_{dp}$; the parameter vector of the field will be $\Theta = (\theta_h, \theta_v, \theta_{dp}, \theta_{ds})^t$, where $\theta_h, \theta_v, \theta_{dp}, \theta_{ds}$ are the parameters for the horizontal, vertical, main diagonal and secondary diagonal respectively.

In order to demonstrate the capacity of the network for analyzing this type of images, we have constructed an image as a composition of several textures. These textures were generated artificially as a visualization of

Markov fields with different parameter vectors. The image including four textures we present in figure 2.a has 64*64 pixels and 256 grey levels. In these images, before approaching the segmentation, it is necessary to obtain an estimation of the parameter vector for each pixel. The process is carried out using the maximum probability estimation method proposed by Kashyap and Chellappa (1983). These would be the features used for the segmentation of the image. For solving the problem we have implemented a four layer network and chosen as values for the constants A=0.1, B=0.01, C=0.5, s=2 and k=1. Figure 2.b and c show the results obtained: edges extracted after defuzzifying the segmentation and edges over the original image. The good concordance between these edges with those perceived on the original image can be appreciated and from it we can infer the validity of the tool we have designed for the segmentation of this type of images.

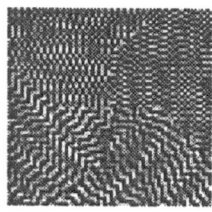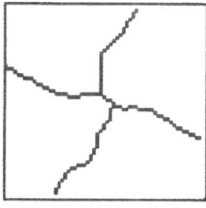

Figure 2. (a): image that is a composition of 4 synthetic textures. (b) and (c): edges resulting from the segmentation process.

Finally, we would like to point out that the constants were chosen experimentally in every case, both for binary and texture images. There is no analytical relationship that indicates the appropriate values of the constants for the segmentation of a generic image.

REFERENCES

1. Amartur, S.C.; D. Piraino and Y. Takefuji (1992). **Optimization Neural Networks for the Segmentation of Magnetic Resonance Images.** IEEE Transactions on Medical Imaging, Vol. 11, N° 2, pp. 215-220.
2. Bezdeck, J.C.; E. Chen-kuo Tsao and N.R. Pal (1992). Fuzzy Kohonen clustering network. Proc. first IEEE Conf. on Fuzzy Systems, pp. 1035-1043. San Diego, 1992.
3. Cross, G.R. and A.K. Jain (1983). **Markov Random Field Texture Models.** IEEE Trans. Pattern Analis. Machine Intell., vol. PAMI-5, no.1, pp. 25-39.
4. Derin, H. and H. Elliot (1987). **Modeling and Segmentation of Noisy and Textured Images Using Gibbs Random Fields.** IEEE Trans. Pattern Analis. Machine Intell., vol. PAMI-9, no.1, pp. 39-55.
5. Hall, L.O.; A.M. Bensaid; L.P. Clarke; R.P. Velthuizen; M.S. Silbiger and J.C. Bezdeck (1992). **A comparison of Neural Networks and Fuzzy clustering tecniques in Segmenting Magnetic Resonance Images of the Brain.** IEEE Trans. Neural Network, vol. 3, n. 5, pp. 672-682.
6. Hopfield J. J. (1984). **Neurons with a grade response have collective computational properties like those of two state neurons.** Proceedings of the National Academy of Science USA, 81, pp. 3088-3092.
7. Huang, Ch. L (1992). **Parallel image segmentation using modified Hopfield model.** Pattern Reconigtion Letters Vol. 13 Number 5, pp. 345-353.
8. Hush, D.R. and B.G. Horne (1993): **Progress in Supervised Neural Network, What's news since Lipmann ?.** IEEE Signal Processing Magazine, pp. 8-39.
9. Kamgar-Parsi, B.; J.A. Gualtieri; J.E. Devaney and B. Kamgar-Parsi (1990) **Clustering with Neural Networks.** Biological Cybernetics, pp. 201-208.
10. Kashyap, R. L. and R. Chellappa (1983). **Estimation and Choice of Neighbors in Spatial-interaction Models of Images.** IEEE Trans. Inform. Theory, vol. IT-29, no.1, pp. 60-72.
11. Manhunath, B.S.; T. Simchon and R. Chellapa (1990) **Stochastic and Deterministic Networks for texture segmentation.** IEEE Trans. Acoustic, Speech and Signal Processing, vol. 38, N° 6, pp. 1039-1049.
12. Mosquera, A.; D. Cabello; M. J. Carreira and M.G. Pencdo (1993). **Texture Image Segmentation Using a Modified Hopfield Network.** In New trends in Neural in Neural Computation (J. Mira; J. Cabestany y A. PRieto, eds.). Lecture notes in Computer Science, vol. 686, pp. 657-663. Springer-Verlag, Berlin.
13. Pal, N.R and S.K. Pal (1993). **A review on image segmentation techniques.** Pattern Recognition. 26,9, pp.1277-1294.
14. Woods, J.W. (1972). **Two-dimensional discrete Markovian Fields.** IEEE Trans. Information Theory, vol.18, pp.232-240.

Segmentation of Range Images:
A Neural Network Approach

W.P.Cheung, C.K.Lee, and K.C.Li

Department of Electronic Engineering, Hong Kong Polytechnic University,
Hung Hom, Kowloon, Hong Kong.

Abstract

In this paper we present a neural computation model for histogram based range image segmentation. An optimal thresholding vector for the range histogram is determined. The number of elements in the vector is characterized by the histogram. Since our model is the parallel implementation of maximum interclass variance thresholding, the time for convergence will be much faster. Together with a real-time histogram builder, real time adaptive range image segmentation can be achieved.

The multithresholding criterion is derived from maximizing the interclass variance and hence the average of the c.g. (center of gravity) of two neighboring class pixel values should be equal to the interclass threshold value. The learning (weight matrix evolution) procedure of the neural model is developed based on the above condition. We use a three-layer neural network with binary weight synapses. The number of neurons in the first layer equals to that of the level of the range image and complex number inputs are used because the arguments of second layer outputs represent the c.g. of the class. The third layer neurons receive the argument output of the second layer and give an indication of the reach of the optimum condition.

1 Introduction

Segmentation is one of the important and critical procedures in pattern recognition and classification, which applies to 2D images, 2½ (range) images and 3D images. This paper mainly focuses on the range images. The quality of the segmentation usually affects the success rate of the recognition. There are many segmentation techniques [1,8,9] and histogram-based thresholding is the most common tool because of direct and one dimensional operations. Hence, many existing real time pattern recognition systems use histogram-based thresholding as the preprocessing element.

Many techniques have been derived to obtain an optimal threshold value to separate the object and background (binary thresholding) and they assume that the image histogram is bimodal. These techniques included fuzzy divergence [2], minimum error [3], maximum likelihood [4] and simulated annealing [5]. However, in many industrial applications such as pick and place, the range image patterns are complex and multimodal. In order to reduce the information loss for recognition process, the reduced range level instead of binary level is used. As a result the multithresholding [10] is required which divides the range histogram into several regions and each be extended their idea to multithresholding, major drawbacks occur in (a) demand of heavy computation and (b) difficulties to perform parallel processing of the algorithms. For these reasons we developed a multithresholding scheme which can be implemented using a three-layer neural network (NN). Actually the multithresholding scheme is a kind of unsupervised competitive learning which adapts the weight value of the threshold neurons to an optimal condition according to the histogram information.

The following section describes the single value threshold selection and the optimal multithresholding condition. The NN implementation is presented in section 3. Section 4 provides the result of multithresholded range images. Also the rate of convergent, computation time comparison and the effect of noise are briefly discussed.

2. Binary and multithresholding

Consider a typical bimodal histogram, $h(g)$ shown in Fig.1b. It can be seen that there are two bell shape regions which indicates the background (lower range level region) and the object (the higher range level region). Here we confine our range level from zero to 255. The objective of the binary thresholding is to select an optimum threshold value such that the classification error is minimum. It can be shown [6] that this value is the valley point between the bell shape regions. However this valley point is usually not well-defined and small ripples may superimpose on it, so simple concavity analysis may not be suitable although the implementation is straight forward.

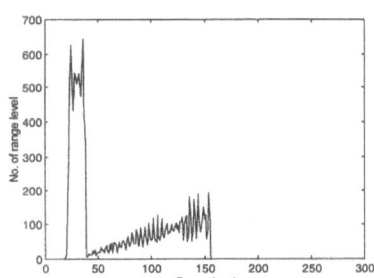

Fig.1a An bimodal range image

Fig. 1b The range histogram of Fig. 1a

Another more robust approach is the maximization of interclass variance [7]. This approach selects the threshold value T so that the interclass variance between the object and background is maximized. The interclass variance between them can be defined as

$$\sigma^2(T) = p_b(\mu_b - T)^2 + p_o(\mu_o - T)^2 \tag{1}$$

where σ = interclass variance at threshold value T,

p_b = the sum of background pixel whose value is less than T

$$= \int_0^T h(g)dg \tag{2}$$

p_o = the sum of object pixel whose value is greater than T

$$= \int_T^\infty h(g)dg \tag{3}$$

μ_b = mean of background pixel values

$$= \frac{\int_0^T gh(g)dg}{p_b} \tag{4}$$

μ_o = mean of object pixel values

$$= \frac{\int_T^\infty gh(g)dg}{p_o} \tag{5}$$

In order to find the value T, Equ. (1) is differentiated with respect to T and then set to zero. The optimum value is determined as

$$T = \frac{1}{2}(\mu_b + \mu_o)$$ (6)

In discrete form, the threshold value is determined as

$$T = \frac{1}{2}\left(\frac{\sum_{0}^{T} gh(g)}{\sum_{0}^{T} h(g)} + \frac{\sum_{T+1}^{\infty} gh(g)}{\sum_{T+1}^{\infty} h(g)}\right)$$ (7)

In Fig. 1b the value obtained by iterating Equ.(7) is 50 which is approximately the valley point of the histogram. Fig.2 shows the segmented image of Fig.1a.

Fig. 2 Segmented image of Fig. 1a

In order to extend the above derivation to the multithresholding case, the histogram is divided into $n+1$ of bell shape regions by n threshold values $(T_1,.., T_n)$ as shown in Fig. 3.

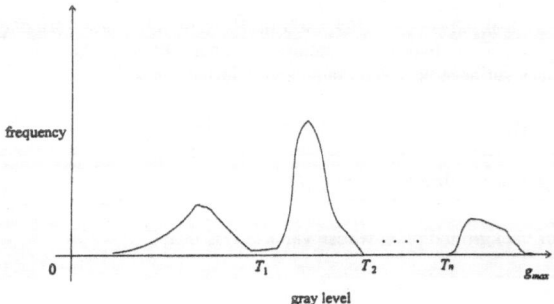

Fig.3 A multimodal histogram

Here we also use the maximum interclass variance criteria. The total interclass variance is defined as

$$\sigma^2(T) = \sum_{i=1}^{n+1} p_i(\mu_i - T_i)$$ (8)

where $T = (T_0, T_1,.., T_n, T_{n+1})^T$ is the threshold vector,

p_i = the sum of pixel values which belong to region i

$$= \int_{T_{i-1}}^{T_i} h(g)dg \qquad \text{for } i = 1,..,n$$ (9)

μ_i = the mean of pixel values in region i

$$= \frac{\int_{T_{i-1}}^{T_i} gh(g)dg}{P_i} \qquad \text{for } i = 1,..,n \tag{10}$$

In the threshold vector, the initial and final element are fixed at

$$T_0 = 0 \text{ and } T_{n+1} = g_{max} \tag{11}$$

and g_{max} is the maximum range level. Similar to the binary thresholding, we differentiate Equ. (8) and set it to zero. The optimum threshold condition is determined by the following equation

$$T_i = \frac{1}{2}(\mu_i + \mu_{i+1}) \tag{12}$$

and in discrete form Equ. (12) becomes

$$T_i = \frac{1}{2}\left(\frac{\sum_{T_{i-1}}^{T_i} gh(g)}{\sum_{T_{i-1}}^{T_i} h(g)} - \frac{\sum_{T_i}^{T_{i+1}} gh(g)}{\sum_{T_i}^{T_{i+1}} h(g)} \right) \qquad \text{for } i = 1,..,n \tag{13}.$$

In fact Equ. (12) and (13) imply that the interclass or region threshold is the average of the c.g (center of gravity) or the centroid of two neighboring class pixel values.

3. NN implementation of multithresholding

In this section we describe our three-layer NN model to perform the iterations of Equ. (13) to find the optimum threshold vector T. The NN consists of input, hidden (second) and output layers as depicted in Fig.4. In the first layer the number of neuron is equal to $g_{max} + 1$ which accept the histogram information. Since the calculation of Equ. (13) involves both range level and frequency data, the input of each neuron is a two element vector and is represented by a complex number as follow

input to the 1st layer ith neuron, $x(i) = (h(i), ih(i))^T$

$$= h(i)e^{j \tan^{-1}(i)} \qquad \text{for } i = 0,..,g_{max} \tag{14}$$

where $j = \sqrt{-1}$.

The second layer computes the moment of pixel values of each region or class. Hence the number of neurons in this layer is $n+1$ with the neuron output

$$y(k) = \sum_{i=0}^{g_{max}} w(k,i)x(i) \tag{15}$$

for $k=1,..,n+1$,

where the weight $w(k,i) \in [0,1]$. The third layer is the output layer which indicates whether the optimum condition is reached. The output of neurons in the layer depends on the 2 adjacent outputs of the second layer and is defined as

$$z(k) = f(y(k)y(k+1)) \qquad \text{for } k = 1,..,n \tag{16},$$

where f is a nonlinear function defined by

$$f(x) = \frac{1}{1 + e^{-\alpha(\delta - |\tan(Arg(x)) - 2T(x)|)}} \tag{17},$$

in which α is a positive constant and δ is the error of the network. Also $Arg(x)$ is the argument of x and $T(x)$ is the threshold value which is calculated by

$$T_k = T(x(k))$$
$$= T(x(k-1)) + \sum_{i=0}^{g_{max}} w(k,i) \quad \text{for } k=1,..,n \text{ with } T(x(0)) = T_0 \tag{18}$$

As a result when the optimum condition is reached, all the neuron outputs will be close to one.

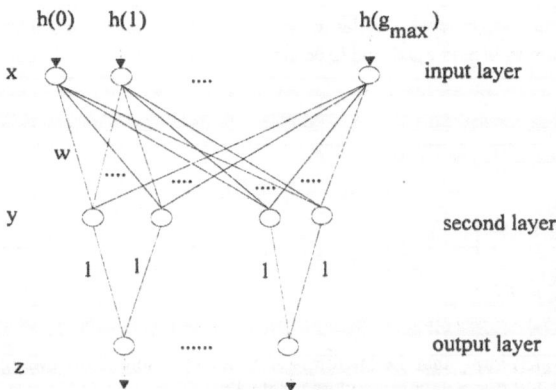

Fig. 4 A NN model for multithresholding

Initially the $n+1$ weight vectors $(w^0(k) = (w^0(k,0),..,w^0(k,g_{max}))^T$ for $k=1$ to $n+1$ are evenly distributed according to

$$w^0(k,i) = \begin{cases} 1 & \text{for } (k)[T_{n+1}/n+1] \le i < (k+1)[T_{n+1}/n+1] \\ 0 & \text{otherwise} \end{cases} \tag{19}$$

Then at each iteration, adjacent neurons compete with each other. The number of 1 appended or removed at the ends of the 1's string of the weight vectors depends on the among of threshold error $\frac{1}{2}\left(2T_k^t - \tan(Arg(y^t(k)y^t(k+1)))\right)$ where the superscript t is the iteration index.

4. Results of Simulation

We simulate the NN model in a Sparc 2 machine and use the histogram of a complex range image (shown in Fig.5) as our input data. Here the size of the image 256 x 256 x 8 bit range image. Hence there are 256 range levels and $T_{n+1} = g_{max} = 255$. Also from Fig. 5 the histogram is trimodal (the background corresponding to zero range level is removed) so n is equal to 2. According to Section 3, we have 256 neurons in the first layer, 3 neurons in the second layer and 2 neurons in the output layer. From Equ. (19) the initial weight vectors between 1st and 2nd layer are

$$w(1) = (1,..,1,0..0)^T$$
$$w(2) = (0,..,0,1,..,1,0,..,0)^T$$
$$w(3) = (0,..,0,1,..,1)^T$$

with each w has 85 number of 1. Therefore the initial threshold vector, $T = (0,85,170,255)^T$. The network error δ is set to 0.1 pixel and α is set to 5. After 30 iterations the threshold vector converges to $(0,135,200,255)^T$ and Fig.6 illustrates different segments of the multithresholded image. If we allow larger network error ($\delta = 1$ pixel) the network converges more rapidly (within 10 iterations) but the threshold vector will have 1 to 2 pixels shift. Furthermore, we have added additive white Gaussian noise (1 to 20dB) to the image and only 2 to 3 pixels shifts occurs which may be accounted by the averaging effect of Equ. (13). However under high noisy environment, the thresholded image segmented is highly corrupted by the noise even though the threshold values come close to the no-noise situation. Bearing in mind that in this NN model, only weights updating and neurons output calculation require computation time and hence they are the same for any number of range levels because they can be done concurrently.

Fig 5a A complex range image (a telephone hand set)

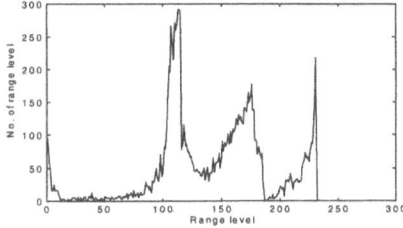

Fig. 5b Histogram of Fig. 5a

Fig. 6a The handle portion

Fig. 6b The component corresponds to the middle region of the histogram

Fig. 6c The region corresponds to the highest bell shape region in the histogram (the mount)

5 Conclusion and discussion

In this paper a NN computation model for histogram multithresholding for range image segmentation is proposed. It is a 3-layer higher order NN with competitive learning capability. The optimum threshold condition is based on the maximum interclass variance and the optimum interclass threshold value is the average of the centroid (or c.g) of the pixel values of the classes. Hence, the model has high white Gaussian noise immunity. Also it is computational efficient and together with a real-time histogram builder as the input module of the NN, real time adaptive range image segmentation can be achieved. Since we basically focus on smooth and normal (objects with normal parallel to the view direction) range images, sharp edges will affect the classification of the surface object. In this case range edge detection should be performed first and segmentation should then be proceeded on the regions bounded by the edges.

6. References

1.   P.K.Sahoo, S.Soltani, A.K.C. Wong and Y.C.Chen, "A survey of thresholding techniques," CVGIP vol. 41, pp233-260, 1988.
2.   D. Bhandari, N.R.Pal and D.D.Majumder, "Fuzzy divergence, probability measure of fuzzy events and image thresholding," Pattern Recognition Letter, vol.13, pp857-867, 1992.

3.   J.Kitter and J.Illingworth, "Minimum error thresholding," Pattern Recognition, vol.19, No.1, pp41-47, 1986.

4.   T.Kurita, N.Otsu and N.Abdelmalek, "Maximum likelihood thresholding based on population mixture models," Pattern Recognition, vol. 25, No.10, pp1231-1240, 1992.

5.   R.Brunelli, "Optimal histogram partitioning using a simulated annealing technique," Pattern Recognition Letter, vol.13, pp581-586, 1992.

6.   A. Rosenfeld and P.De La Torre, "Histogram concavity analysis as an aid in threshold selection," IEEE Trans SMC., vol.13, pp231-235, 1983.

7.   N. Otsu, "A threshold selection method from gray level histograms," IEEE Trans SMC, vol-9, No.1, pp62-66, 1979.

8.   S. Ghosal, R. Mehrotra, "Application of neural networks in segmentation of range images," IJCNN, vol.3, p297-302, 1992.

9.   R. Hoffman and A.K.Jain, "Segmentation and Classification of range images," IEEE Trans. PAMI, vol.9, no. 5, pp608-620, 1987.

10. W.P.Cheung, C.K.Lee and K.C.Li., "Multithresholding for arithmetic logic pattern matching," Electronics Letters, Vol.29 No.5 pp. 481-483, 1993.

A Neural Architecture for Preattentive Segmentation of Sewage Pipes Video Images

Javier Ruiz-del-Solar[1] and Mario Köppen

Fraunhofer-Institut IPK Berlin
Department of Pattern Recognition
Pascalstr. 8-9, 10587 Berlin, Germany
e-mail: javier@ipk.fhg.de

Abstract

This article describes a neural architecture for real time preattentive segmentation of sewage pipes video images, whose mechanisms are based on the mammalian early visual system. The architecture corresponds to a modified and simplified version of the *Boundary Contour System* tuned to take advantage of the circular symmetric characteristics of the pipes images. Remarkable aspects of the proposed architecture are the application of a spatial complex logarithmic mapping stage, and the use of cooperative receptive fields with noncollinear branches.

1. Introduction

The periodical inspection of sewage pipes is necessary to avoid the ecological damage produced when the transported substances are leaked into the environment. That leakage is produced by corrosion, fissures, and split of pipe sections. The small pipes diameter (60 centimeter) does not allow a direct inspection of them. The visual inspection through the processing of inner images is a good possibility to perform that labour. The present work is part of a research project to automate the pipes visual inspection process [1]. Automating the visual inspection process saves human time and effort and can provide accurate, objective, and reproducible results. Additionally, automation can eliminate human errors resulting from fatigue or lack of concentration.

The inspection is performed through the processing of a video signal. That signal is taken by a CCD camera mounted on a remote controlled camera-car, which moves through the inner parts of the pipes. The processing system consists on two independent stages that work alternately. The stage A or *on-line stage* localizes pipes unions (where almost all the faults are produced). The stage B or *off-line stage* performs an analysis of pipes unions, and searches, detects and classifies the faults. A description of stage B can be found in [1].

The proposed neural architecture implements the processing stage A, whose block diagram is showed in fig. 1. The architecture is based on mechanisms motivated and justified by evidence from psychophysics and neurophysiology, which were adapted and simplified to tune the process features. The main process features are: real time analysis, variable illumination conditions of the pipes inner surfaces, and a *priori* knowledge of the physical system properties (geometry of the pipes, CCD

---

[1] The author is supported by a grant from the DAAD (Deutscher Akademischer Austauschdienst).

camera, and camera-car). Preattentive segmentation [2] is one of the basic tasks to perform in stage A. The remainder of this article describes exclusively the Preattentive Segmentation Subsystem, and presents some preliminary results and conclusions.

2. Description of the Preattentive Segmentation Subsystem (PSS)

The Preattentive Segmentation Subsystem (PSS) is formed by three modules (Fig. 2), called: SCLM (Spatial Complex Logarithmic Mapping), DOI (Discount of Illuminant), and SBCS (Simplified Boundary Contour System).

The PSS has two inputs, the Video Input Signal (VIS) and the Parameter Feedback Signal (PFS). The VIS corresponds to the succession of images to be processed. The PFS is a feedback signal coming from the Object Recognition Subsystem (ORS) that allows to adjust the local parameters taking into account the global system state. The global system states are S_0 and S_1. In the initial state S_0 or *predetection state* a fast processing of image circular segments is performed to localize a possible pipe union area. In the state S_1 or *detection state* a processing of the area predetected in S_0 is carried out.

The PSS outputs are the Segmentation Output Signal (SOS) and the Spatial Mapping Signal (SMS), being the inputs to the ORS and Foveation Subsystem (FS) respectively. The following paragraphs describe each module of the PSS.

2.1. SCLM (Spatial Complex Logarithmic Mapping)

The SCLM module performs a complex logarithmic mapping of the VIS (Fig. 3). Studies about optical nerves and visual images projection in the cerebellar cortex show that the global retinotopic structure of the cortex may be characterized in terms of the geometric properties of that mapping [3]. The SCLM module takes advantage of the circular system symmetry focalizing the analysis into circular image segments, which allows a big diminution of the data to be processed and provides an invariant representation of the objects. When the origin of the mapping coordinate system is not adequately selected, the mapped image has a large distortion. If the camera focus is centered in relation to the pipes axis, then the image central position can be used as origin. The Foveation Subsystem performs the adaptive camera centering.

In the state S_0 the processing is performed in the following four circular segments:

$$\{\{r_{00}, r_{01}\}, \{\varphi_{00}+n\frac{\Pi}{2}, \varphi_{01}+n\frac{\Pi}{2}\}\} \qquad n=0,1,2,3 \qquad (1)$$

In the state S_1 the processing is performed in the following circular segment:

$$\{\{r_{10}, r_{11}\}, \{\varphi_{10}, \varphi_{11}\}\} \qquad (2)$$

where r_{xy} and φ_{xy} correspond to radius and angles respectively (see fig. 3). Complete circular segments (360°) are not used because normally there are water or solid sediments in the lower pipe region, which disturb the segmentation.

2.2. DOF - Discount of Illuminant

In this stage variable illumination conditions are discounted by a shunting on-center off-surround network (network defined in [4]), which models the receptive fields response of the ganglions cells of the retina. Image regions of high relative contrast are amplified and regions of low relative contrast are attenuated as a consequence of the discounting process.

2.3. SBCS - Simplified Boundary Contour System

The SBCS module corresponds to a simplified and modified version of the Boundary Contour System (BCS) developed at the Boston University [5][6]. The BCS model is based primarily on psychophysical data related to perceptual illusions. Its processing stages are linked to stages in the visual pathway: LGN Parvo->Interblob->Interstripe->V4 [7]. The BCS model generates emergent boundary segmentations that combine edge, texture, and shading information. The BCS operations occur automatically and without learning or explicit knowledge of the environment. In general the standard BCS algorithm requires a significant amount of execution time that does not allow its utilization in a real time application.

The SBCS implementation uses: monocular processing [8], only the "ON" processing channel [8], a single spatial scale [8], three orientations, and takes as input the SMS (Spatial Mapping Signal). Each processing stage of the model is explained as follows:

2.3.1. Oriented Filtering Stage

Two-Dimensional Gabor Filters, introduced by Daugman in [9], are used as oriented filters. That filters models the receptive fields of simple and complex cells in the visual cortex. We use only odd-symmetric filters, which respond optimally to differences of average contrast across its axis of symmetry. By taking the image circular symmetry into account we use only three oriented filters (see Fig. 4) and we take the absolute response of each filter as output.

2.3.2. First Competitive Stage

Cells in this stage compete across spatial position within their own orientation plane. This is done in the form of a standard shunting equation with two additional terms (see [8]), a tonic input (T) and a feedback signal (V) that comes from a later stage (Feedback Stage). Our modified dynamic shunting equation is given by:

$$\frac{d}{dt} W_{ijk} = -A W_{ijk} + (B - W_{ijk})(J_{ijk} + C V_{ijk} + T) - W_{ijk} \sum_{(p,q) \neq (i,j)} G_{pqij} J_{pqk} \qquad (3)$$

At equilibrium, this stage is defined by:

$$W_{ijk} = \frac{B(J_{ijk} + C V_{ijk} + T)}{A + C V_{ijk} + T + \sum_{(p,q)} G_{pqij} J_{pqk}} \qquad (4)$$

where J is the output of the *Oriented Filtering Stage*; G is a gaussian mask; k is the orientation index; p, q, i, j are position index; and A, B, C are constants.

2.3.3. Second Competitive Stage

At this stage competition takes place only across the orientation dimension, i.e. cells compete with other cells that have the same position but different orientation. Like in [10], the equilibrium condition for this dynamic competition is simulated by finding the maximal response across the orientation planes for each image localization and by multiplying all non-maximal values by a suppression factor (0.5 in our implementation).

2.3.4. Oriented Cooperation Stage

The oriented cooperation is performed in each orientation channel by bipole cells that act like long-range statistical AND-gates. Unlike the standard BCS model we use bipole cells whose receptive fields have collinear and noncollinear branches (Fig. 5). These receptive fields have properties consistent with the *spatial relatability* property [11], which indicates that two boundaries can support a interpolation between themselves when their extensions intersect in an obtuse or right angle. The cooperation in the left-half receptive field (L_{ijk}) is performed among neighbouring cells with the same orientation, and the cooperation in the right-half receptive fields (R_{ijk}) among neighbouring cells across all the orientations.

Because that stage works with only three orientations we do not use a dipole field (see [8]) as input to this stage.

At equilibrium, the output from this stage is defined by:

$$Z_{ijk} = \frac{SL_{ijk}SR_{ijk}}{D+SL_{ijk}SR_{ijk}} \qquad k=0,1,2. \tag{5}$$

with:

$$SL_{ijk} = \sum_{(p,q) \in L_{ijk}} Y_{pqk}F_{pqijk}$$

$$SR_{ijk} = \sum_{r=r_{0_k}}^{r_{max_k}} \sum_{(p,q) \in R_{ijk}} Y_{pqr}F_{pqijr} \tag{6}$$

and:

$$F_{pqijk} = \exp(-E(|B_{pqijk}|-P)^2 + C_{pqijk}^2)) \ [|\cos(\Omega_{pqij}-\theta_k)|]^R$$
$$B_{pqijk} = (p-i)\cos(\theta_k)-(q-j)\sin(\theta_k)$$
$$C_{pqijk} = (p-i)\sin(\theta_k)+(q-j)\cos(\theta_k) \tag{7}$$
$$\Omega_{pqij} = \arctan\left(\frac{q-j}{p-i}\right)$$
$$\theta_k = (k-1)\Delta\theta$$

with F_{pqijk} receptive field kernel; B_{pqijk} and C_{pqijk} the rotated coordinates; Ω_{pqij} the direction of position (p,q) with respect to the position (i,j); Y the output of the Second Competitive Stage; r_{0_k} and r_{max_k} the ranges of relatable orientations; D, E, P, and R constants; and $\Delta\theta$ the angle between orientations.

2.3.5. Feedback Stage

Before cooperative signals are sent to the first competitive stage, a competition across the orientation dimension and a competition across spatial position take place in order to pool and sharpen the signals that are feed back. Both competitions are homologous to Competition 1 and Competition 2, and are implemented in one processing step.

3. Results and conclusions

We present only preliminary results because the full architecture has not been implemented yet. Actually, we are working at the module connection, the algorithm optimization, and the system parameter selection.

As a preliminary example of the system processing, figure 6 shows a sewage pipes input image, figure 7 the spatial mapping signal, and figure 8 the segmentation output signal. In those images we can see that the spatial mapping allows a great data reduction (more than 10 times), that produces an equivalent processing time reduction.

We think that our processing system can also be used to process other images with circular symmetry (images from tubes, tunnels, wheels, etc). Additionally, we believe that more research must be performed in the *Oriented Cooperation Stage* to find a better cooperation algorithm.

References

[1] Lohmann, L. (1993). Untersuchung der Einsatzmöglichkeiten der Bildverarbeitung zur Automatisierung der Rohr- und Kanalanalyse. *Diplomarbeit Technische Universität Ilmenau*, Germany, 1993.

[2] Preattentive segmentation refers to the ability of mammalians to perceive textures without any sustained attention.

[3] Schwartz, E.L. (1980). Computational anatomy and functional architecture of striate cortex: A spatial mapping approach to perceptual coding. *Vision Research*, **20**, 645-669.

[4] Grossberg, S. (1983). The quantized geometry of visual space: The coherent computation of depth, form, and lightness, *The behavioral and brain sciences*, **6**, 625-657, 1983.

[5] Grossberg, S., and Mingolla, E. (1985a). Neural dynamics of form perception: Boundary completion, illusory figures, and neon color spreading, *Psychological Review*, **92**, 173-211.

[6] Grossberg, S., and Mingolla, E. (1985b). Neural dynamics of perceptual grouping: Textures, boundaries, and emergent segmentations, *Perception and Psychophysics*, **38**, 141-171.

[7] Nicholls, J., Martin, A., and Wallace, B. (1992). The Visual Cortex. In: *From Neuron to Brain: A cellular and molecular approach to the function of the nervous system*. 3rd Edition, 1992, Sinauer Associates, Inc., Sunderland, Massachusetts, USA.

[8] Grossberg, S., and Mingolla, E. (1987). Neural dynamics of surface perception: Boundary webs, illuminants, and shape-from-shading. *Computer Vision, Graphics, and Image Processing*, **37**, 116-165.

[9] Daugman, J.G. (1980). Two-Dimensional spectral analysis of cortical receptive field profiles. *Vision Research*, **20**, 847-856.

[10] Lehar, S., Worth A., and Kennedy D. (1990). Application of the Boundary Contour/Feature System to magnetic resonance brain scan imagery. *Proceedings of the International Joint Conference on Artificial Neural Networks*, **I**, 435-440, San Diego, June 17-21, 1990.

[11] Kellman, P.J., and Shipley, T.F. (1991). A theory of visual interpolation in object perception. *Cognitive Psychology*, **23**, 141-221.

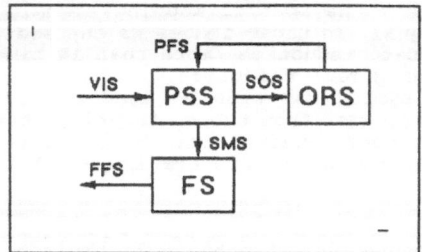

Figure 1: System Stage A.

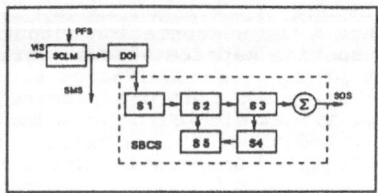

Figure 2: Preattentive Segmentation Subsystem (PSS).

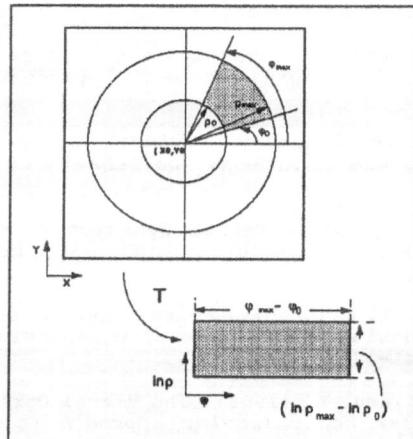

Figure 3: Complex logarithmic Mapping.

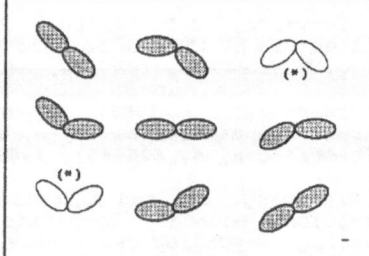

Figure 5: Bipole cells with collinear and noncollinear branches. The bipole cells (*) are not used.

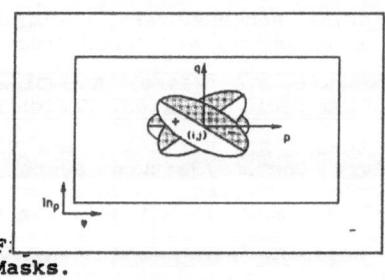

F:
Masks.

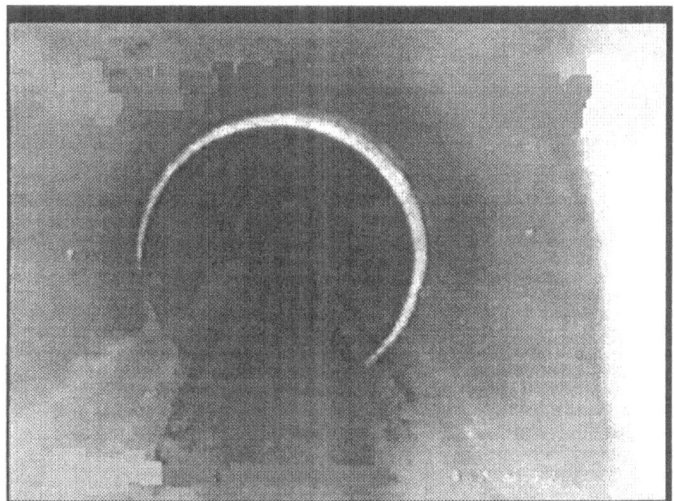

Fig. 6: Input Image (376x288 pixels).

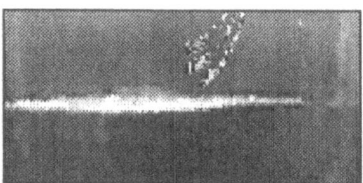

Fig. 7: Spatial Mapping
Signal (128x64 pixels).

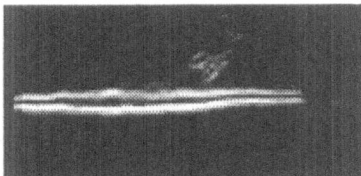

Fig. 8: Segmentation Output
Signal (128x64 pixels)

A CNN Model for Grey Scale Image Processing

Miguel A. Jaramillo-Morán, Francisco J. López-Aligué, Miguel Macías-Macías, María I. Acevedo-Sotoca

Departamento de Electrónica e Ingeniería Electromecánica. Facultad de Ciencias. Universidad de
Extremadura. Badajoz. Spain

Abstract

A modification of the CNN model is proposed in this work. An iterated-map is defined instead of
the original differential equation while a sigmoid function is taken as the cell output. Modifications in the
structure and values of the synaptic scheme allow the use of the model in different tasks. The network´s
behaviour is mainly determined by its feedforward term, while feedback simply adds optimization to the
network´s output. The network´s stability is discussed. Edge detection and contrast enhancement are
performed, stressing the fact that they are only different aspects of the same network property: the
enhancement of brightness gradients. Some examples are presented in which grey scale images are
processed, revealing the capabilities of the model both in detecting edges and in enhancing contrasts.

Introduction

One of the main applications of Neural Networks (NN) is the recognition of previously learned
patterns. A great amount of work has been done on studying their fundamental characteristics. One can
find very efficient learning strategies that ensure optimal pattern storage, neuron models with well-defined
recognition capabilities, theoretical studies that ensure the stability of the network, and analyses of the net
information capacity. Much effort has been put into reproducing two of the brain´s main abilities, learning
and recognition, and into understanding the principles governing them. The difficulty involved has driven
many researchers´ efforts in a slightly different way: the application of neural networks to solving
technical problems. One is image processing with no pattern recognition. This work is based mainly on
certain structures known as Cellular Neural Networks (CNN) [1,2]. They are very simple NN in which,
with the principles of cellular logic, images are processed. The basic network structure is described by the
equations:

$$C\dot{z}(t) = -Rz(t) + Ay(t) + Bu + I \qquad (1.1)$$

$$y_i = \frac{1}{2}(|z_i(t) + 1| - |z_i(t) - 1|) , \qquad (1.2)$$

where C is an input capacitor that is the same for every cell,
R is a diagonal matrix whose elements are the reciprocals of the input resistors, i. e., $1/R_i$, which may be
different for each cell,
I is a vector whose elements are input bias currents that are equal for every cell,
$z(t)$ is the vector that describes the cellular activity,
$y(t)$ is the network output vector,
u is a constant external input to the network,
B and A are connection matrices that describe respectively the input and feedback connectivity.

Equation (1.1) describes the cell activity, and (1.2) its output. The matrices A and B are usually,
although not always, symmetric. In many cases $B_{ij} = 0$, with no input connection assumed. In this case,
the initial conditions of differential equation (1.1) are taken to be the inputs to the network.

Every cell has the same synaptic scheme with fixed values, the so-called "cloning template". The connections can either be excitatory or inhibitory, and give the model its main characteristic: a very precisely defined synaptic structure developed to perform a certain image processing task. The difference from other models is that the synaptic weights do not store any previously learned pattern, but are merely a convenient arrangement of connections suited to carrying out a particular task. An appropriate modification of the values and structure of these "cloning templates" can endow the network with different capabilities such as hole filling [3], image thinning [4] or motion sensitive filtering [5]. It is even possible to simulate some properties of the subcortical visual pathway [6].

Proposed model

Considering the equations (1) one can say that CNN have the structure of Hopfield Networks with two particular features. The first is the CNN synaptic scheme, defined as a fixed distribution of connections for each cell. This is suited for the performance of a specific image processing task and has no previously computed value. The other difference appears in the output function, which is a sigmoid in Hopfield's model and an approximation to sigmoid in CNN, being piecewise-linear with constant values -1 for $z_i(t) \leq -1$ and 1 for $z_i(t) \geq 1$ and a ramp with slope 1 for $-1 < z_i(t) < 1$.

We propose a modification of the CNN model in which the differential equation (1.1) is replaced by a iterated-map, better adapted to computer simulation, and the output (1.2) is a sigmoid function. The "cloning template" structure is maintained. So (1) becomes

$$Rz(t+1) = Ay(t) + Bu + I \qquad (2.1)$$

$$y_i = \sigma(z_i) = \frac{1}{1+\exp{-Sz_i}}, \qquad (2.2)$$

where S is the sigmoid function gain. Rewriting (2.1) in component form with (2.2) included for convenience, we have

$$\frac{1}{R_i} z_i(t+1) = \sum_{j=1}^{N} a_{ij}\sigma(z_j(t)) + \sum_{j=1}^{N} b_{ij}u_j + I. \qquad (3)$$

In order to obtain a more satisfactory set of variables, we consider x_i, $i = 1, ..., N$, defined by:

$$\sum_{j=1}^{N} a_{ij}x_j(t) = \frac{1}{R_i} z_i(t) - \sum_{j=1}^{N} b_{ij}u_j - I. \qquad (4)$$

Setting (4) into (3), we have a more useful equation describing the cell behaviour which includes both the neural activity and the output:

$$x_i(t+1) = \sigma(z_i(t)) = \frac{1}{1+\exp{-SR_i(\xi_i + T)}} , \qquad \xi_i(t) = \sum_{j=1}^{N} a_{ij}x_j(t) + \sum_{j=1}^{N} b_{ij}u_j. \qquad (5)$$

In this equation, we can take R_i as equal for each cell (R) and regard it as an implicit factor in both a_{ij} and b_{ij} without regard to its value. This may be considered as an adjustment factor to give a_{ij} and b_{ij} appropriate values to ensure a good response of function (5). Now, taking $T = I/R$, the last equation can be written in the form:

$$x_i(t+1) = \sigma(\xi_i) = \frac{1}{1+\exp{-S(\xi_i + T)}} , \qquad \xi_i(t) = \sum_{j=1}^{N} a_{ij}x_j(t) + \sum_{j=1}^{N} b_{ij}u_j. \qquad (6)$$

Up to now we have dealt with cell function with no mention of synaptic structure, which is the main characteristic allowing the definition of CNN as a differentiated model because it is responsible for their image processing capabilities. Thus, while we have made some modifications in the CNN cell functions to obtain a new model, the basic structure of synapses defined by "cloning templates" has been kept. There will only be modifications in the distribution and values of synapses in our model that, while of minor theoretical significance, will endow the network with different behaviour. Therefore, as different schemes of "cloning templates" give different image processing capabilities, we must define those tasks before studying the network response.

The applications we have selected for the network are contrast enhancement and edge detection. They have the same "cloning template" scheme and only differ in the values of certain cell parameters. As we will see below they are but two aspects of the same behaviour: the enhancement of brightness gradients.

As has been seen in (6), inputs and feedback are separated to differentiate their nature, and the two terms will be studied separately. In our model inputs define the network´s properties, while the feedback leads to only minor modifications.

Input weights

As previously mentioned, these parameters are usually omitted in CNN so that there is no input to the network considered as a part of the cell equation. The initial conditions of the differential equation (1.1) are taken as input. Nevertheless this term has been kept in the model to define a feedforward structure between the net and its input. Such a structure provides a simplified dynamical behaviour that allows an easier understanding of the network´s properties. When feedforward is included and no feedback is allowed the network is stable and converges to a fixed point in one time step, so that its behaviour may be described easily and with great precision. We can then add the feedback to obtain a slight modification of the network´s properties. Another important reason for including a feedforward term is the possibility of defining a multilayer network in which the output of each layer is taken as the input of the following one.

Before defining the structure of the feedforward term it is important to stress that the input image and the cell layer have the same form, i. e., they have the same number of elements with a one-to-one correspondence between input pixels and cells. The i-th cell feedforward term then has the form:

$$\sum_{j=1}^{m} b_{ij}u_j = b_{ii}u_i - \sum_{j\neq i}^{m} b_{ij}u_j \, , \tag{7}$$

where b_{ij} is the connection between the i-th neuron and the j-th input and "m" the number of input connections of each cell. All b_{ij} (i≠j) are equal and form a square centred on the i-th input, so that we shall call the "radius" of the neighbourhood the distance from the i-th input to the centre of any side of the square. Its value depends on the particular task we want to dedicate the network to. In any case, its size will also depend on the size of the network.

One may take the values of b_{ii} and b_{ij} may be related by the relation $b_{ij} \approx b_{ii}/M$ (i≠j), where M is the number of connections for each cell. Clearly, this is only an approximate relationship that allows the user to make a fine adjustment of the two parameters based on the results provided by numerical simulation. Nevertheless, this approximation is good enough to define a precise behaviour of the network´s response. We may therefore assume that (7) computes the difference between an input and its neighbours, providing a measure of the change of brightness in the area surrounding a pixel. To understand this better, let us consider two cases:

(i) u_i has a small value: the i-th pixel is dark.- Then $b_{ii}u_i$ has a low value. If surrounding pixels are brighter, then $\sum_{j\neq i}^{m} b_{ij}u_j$ is greater than $b_{ii}u_i$. (7) is negative in value, and the cell output (6) is close to zero. If the surrounding pixels are also dark, $\sum_{j\neq i}^{m} b_{ij}u_j$ is close to zero and so is (7). The cell output (6) then has a greater value than in the previous case: the cell retain a small activity that depends on the values of the sigmoid function parameters. Therefore, the i-th cell has an output very close to zero if its surrounding inputs are brighter than the i-th input. A somewhat higher output appears if all the inputs are approximately equal. Thus, the i-th cell output undergoes little variation if there is no gradient in its surrounding inputs, but a sharply lowered when there is a gradient.

(ii) u_i has a high value: the i-th pixel is bright.- Then $b_{ii}u_i$ has a high value. If the surrounding are

darker, then $\sum_{\substack{j \neq i}}^{m} b_{ij}u_j$ is small compared with $b_{ii}u_i$, and (6) gives an high output with a value that depends

on the relationship between b_{ii} and b_{ij} $(i \neq j)$. If the surrounding pixels are also bright, $\sum_{\substack{j \neq i}}^{m} b_{ij}u_j$ is close to

$b_{ii}u_i$ and the output cell (6) has a low value that depends on the values of the sigmoid function parameters.

As has been seen, contrast enhancement is performed because a pixel with a value different from those surrounding it produces a large variation in its corresponding output cell. I. e., bright pixels cause their corresponding cell output to rise, while dark ones cause a fall. This effect is more significant at large gradients, so that the sharper the change of brightness, the more extreme are the cell outputs. Therefore, the edges of images appear in the output as the junction of one bright and one dark line. Smaller changes in brightness produce smoother transitions in the output, so that cells in those areas with uniform brightness have changes of little significance in their outputs that depend on their input values and those of b_{ii} and b_{ij} and the sigmoid function parameters.

Sigmoid function parameters

The network´s response, as was pointed out above, may be slightly modified by changes in the values of the sigmoid function parameters. Therefore a detailed analysis of the effect of those changes must be carried out. Instead of precise numbers, we shall only distinguished approximately in terms of high or low values. Nevertheless they will be precise enough to give an exact understanding of how changing that modifies the network´s response.

S.- This parameter controls the slope of the sigmoid function: the higher it is, the closer is the approach to the step function. A moderate value will provide more detailed output because of the slower change in the function´s output. There thus appear two types of behaviour: the first, with small values of the slope, provides general contrast enhancement with no sharp transition, while the second, with higher values of S, brings out those zones with a sharp change in brightness such as lines or profiles. These transitions appear as a very dark area joined to a very bright one as results from the rise of contrast promoted by the high value of A. Those zones with no change in their brightness give a bright or dark output depending on the value of B, as will be seen in the following paragraphs. So we can state that a small value of S will provide contrast enhancement and a large value, edge detection capability.

T.- This parameter is the threshold of the sigmoid function. Its meaning is clear in the case of high S, when the function approaches the step function. With a low value of A, however, T is only the input for which the output takes the value 1/2. Nevertheless, we can state that it allows the sigmoid function to adapt itself to the input values. This is its application when we want contrast enhancement -low S. But if edge detection is required its variation may produce some changes in the output.
The values of T may be decreased to obtain an output in which only those cells with a very low (negative) value in (7) will remain dark while all the other cells give a bright output. The dark ones represent those pixels on the edge of an input with a low value of brightness. Thus, the output of the network is a white image in which the edges of the input image appear as black lines.
On the other hand, an increase in the value of T will generate a high threshold that will make the sigmoid function give a very bright output for those cells with a high value in (7) and a dark one for the others. Only those cells corresponding to pixels on the bright edge of an input image have a high output while the others are forced to stay in a state close to zero. Thus, a black image with white lines marking the edges of the input appears as the network response.
We have obtained an edge detector in both cases, although we must stress that even though we talk of lines in the output, this is only a rough expression because the images do not usually have sharp edges.

Feedback

As previously mentioned, this term was added to enhance some aspects of the network´s response without further modifications to its properties. Therefore, both its feedforward structure and its synaptic values will be defined to maintain the aforementioned properties of the network.

Feedback can destabilize the dynamical evolution of the network. So stability criteria must be established to ensure a fixed response of the network's output. These conditions can be found in [7], where a system such as that proposed here is considered. Two properties are proved: i) The only attractor of systems such as (6) are fixed points or period-two cycles provided that A is a real symmetric matrix (in [7] a general sigmoid function is assumed instead of the particular one proposed in (6)). ii) Systems such as (6) have only fixed-point attractors when the matrix $(A + \Gamma^{-1})$ is positive definite, where Γ^{-1} is a diagonal matrix in which each element is 1/SR, with S and R as defined above (different R_i and S_i are allowed for each cell in [7]). A sufficient condition for $(A + \Gamma^{-1})$ to be positive definite and thus for (6) to have only fixed-point attractors is $SR_i < |1/\lambda_{min}|$ for all i, where $\lambda_{min} < 0$ is the most negative eigenvalue of the real symmetric matrix A.

Thus if we define A as a symmetric matrix, we have that system (6) is stable and approaches a fixed point or a period-two cycle. But, if we want to ensure that (6) has only fixed-point attractors, it must satisfy the aforementioned relationship for λ_{min}. This eigenvalue can not be obtained analytically because of the high dimension of system (6), and numerical calculation is necessary. Once the feedback is defined, λ_{min} may be calculated and appropriate values given to S to ensure the desired fixed-point stability. But since this stability criterion only provides a particular condition for each system, and its calculation is difficult, one may be satisfied with the symmetry of matrix A. If a period-two appears in a certain system, we can reject that feedback structure and try another until a fixed-point stability is obtained. All the structures we have used to perform the above tasks had a fixed-point stability.

We noted in the previous sections the need for feedback to obtain sharper lines when the network is devoted to edge detection tasks. To achieve this line sharpening effect, we must generate a rise in the activity of those cells defining the edges when they are bright against a dark background, and a decline of black lines against a white background. This may be obtained with a combination of excitatory and inhibitory connections, where the former is the closer and the latter the more distant from the considered cell:

$$\sum_{j=1}^{n} a_{ij}x_j(t) = \sum_{j=1}^{n'} a_{ik}^{+}x_k(t) - \sum_{l=n'}^{n} a_{il}^{-}x_l(t) , \qquad (8)$$

Where "n" is the number of feedback connections of each cell and n' the number of excitatory ones. We must make the first contribution greater than the second to obtain a rise in the output of those cells within the lines, and a decline in those on or close to their borders. Cells inside a line have a high output and are surrounded by others with the same high value. Therefore as the excitation exceeds the inhibition, a significant rise in their values appears. Those cells on or close to the border of the lines have a less bright output and are surrounded by a combination of cells with both higher and lower outputs, but only those cells inside the line make a significant contribution to (8). Now the inhibition exceeds the excitation and those cells have a diminished activity. So the lines appear brighter or darker depending on the selected background.

If we are dealing with contrast enhancement rather than with edge detection the feedback provides a general rise in the brightness of those zones of the image with uniform brightness that is less effective in darker areas - the lower activity generates a smaller feedback contribution. Those zones with a small change in brightness have a small rise in the brighter areas and a small decline in the darker ones. Thus, with contrast enhancement, we can say that only a slight improvement in performance results from the addition of feedback, and we can either leave it out or balance the excitatory/inhibitory ratio. This equilibrium would provide a null contribution everywhere except those zones where there appears a small change in brightness.

Finally, it is important to note that, as the network's properties are determined by the feedforward connectivity with the feedback added to optimize them, the later's contribution must be less than that from the feedforward. Therefore, values must be given for a_{ij}^{+} and a_{ij}^{-} in order to fulfil the conditions outlined above for a certain task. A good approximation is $a_{ij}^{+} = 1/r^{+}$ and $a_{ij}^{-} = 1/r^{-}$ where "r⁺" and "r⁻" are respectively greater than "n'" and "(n-n')" in (8) and are adjusted to fit the above conditions.

Simulation

The two proposed tasks, contrast enhancement and edge detection, were simulated to outline the model's capabilities. The simulation was performed on a VME bus rack with four CPU boards working in parallel.

To carry out the parallel processing, we took into account that input image and layer have the same structure. Then it is easy to divide the whole neural network system into four subsystems, each processed by one of the four boards. Therefore the input image was taken with a square form, and then divided into four equal squares. These squares were augmented with overlap strips taken from the others to allow cells lying on the borders to take account of those input pixels within the input "radius" belonging to the other subsystems. This structure is shown in Fig. 1 (a). Each subsystem was sent to the corresponding board where the simulation was performed.

Once each iteration is completed, each CPU board sends a copy of its part of the output image to a shared memory to form the net output. When another iteration begins, each CPU board takes from this common memory only those outputs from a different board that are considered as feedback by its own border cells. This output structure may be seen in Fig. 1 (b).

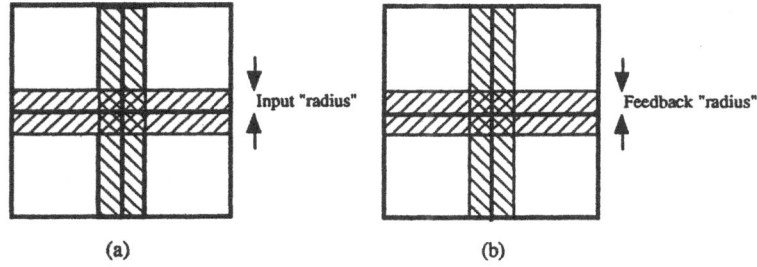

(a)   (b)

Fig. 1

As we set two tasks to be performed by the network, a simulation was carried out for each. The contrast enhancement was performed with the parameters listed in Table 1 and input and output are shown in Fig. 2. As can be seen, those zones where a change in brightness is present have augmented their contrast, leading to easier detection. Edges have also been detected. Those areas with no change in illumination have a uniform background.

Fig. 3 shows edge detection performed with the parameters listed in Table 1. One can see the input, the output with no feedback, and the output with feedback. The last is presented in two ways: white lines against black background and black lines against white background. One can see that this last option provides a clearer response with no noise corrupting the output. The introduction of feedback provides better defined lines, but some details have disappeared. Only the sharper profiles appear in the output. Of course, an appropriate selection of parameters will allow adaptation to different kinds of inputs in which there may be no sharp profiles.

	N	m	n	n'	b_{ii}	b_{ij}	a_{ij}^{+}	a_{ij}^{-}	A	T
Contrast enhancement	10^4	25	25	9	4	0.17	0.125	0.06	3.3	-0.25
Edge detection										
Black background	$4 \cdot 10^4$	168	25	9	3	0.0179	0.135	0.06	12.5	-0.3
White background	$4 \cdot 10^4$	168	25	9	3	0.0179	0.135	0.06	12.5	0.3

Table 1

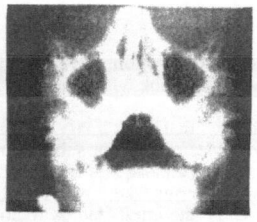 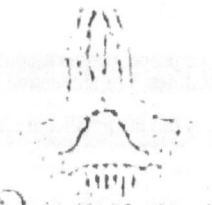

Fig. 2

Fig. 3

Conclusions

We have proposed a modification of the CNN model in which an iterated-map algorithm is used while the "cloning template" structure is retained. It is able to perform image processing of analogue grey scale images, whereas CNN process mainly digital images. Thus a simplification of the CNN model has been obtained and an extension of its capabilities to grey scale image processing provided. Moreover, new tasks may be studied as possible applications of the network.

Acknowledgement

The authors wish to express their gratitude to the Spanish CYCIT for their support of this work through grant TIC 92-054.

References

[1] L. O. Chua, L. Yang. "Cellular Neural Networks: Theory". IEEE Trans. on Circuits and Systems. Vol. 35, No. 10. October 1988. pp. 1257-1272.

[2] L. O. Chua, L. Yang. "Cellular Neural Networks: Applications". IEEE Trans. on Circuits and Systems. Vol. 35, No. 10. October 1988. pp. 1273-1290.

[3] T. Matsumoto, L. O. Chua, R. Furukawa. "CNN Cloning Template: Hole-Filler". IEEE Trans. on Circuits and Systems. Vol. 37, No. 5. May 1990. pp. 635-638.

[4] T. Matsumoto, L. O. Chua, T. Yokohama. "Image Thinning with a Cellular Neural Network". IEEE Trans. on Circuits and Systems. Vol. 37, No. 5. May 1990. pp. 638-640.

[5] B. E. Shi, T. Roska, L. O. Chua. "Design of Linear Cellular Neural Network for Motion Sensitive Filtering". IEEE Trans. on Circuits and Systems. II: Analog and Digital Signal Processing. Vol. 40, No. 5. May 1993. pp. 320-331.

[6] T. Roska, J. Hámori, E. Lábos, K Lotz, L Orzó, J. Takács, P. R. Venetianer, Z. Vidnyánszky, A. Zarándy. " The use of CNN Models in the Subcortical Visual Pathway". IEEE Trans. on Circuits and Systems. I: Fundamental Theory and Applications . Vol. 40, No. 3. March 1993. pp. 182-195.

[7] C. M. Marcus, R. M. Westervelt. "Dynamics of Iterated-Map Neural Networks". Physical Review A. Vol. 40, No. 1. July 1989. pp. 501-504.

Kohonen's Self-Organizing Maps for Contour Segmentation of Gray Level and Color Images

René Natowicz

École Supérieure d'Ingénieurs en Électrotechnique et Électronique
Cité Descartes –BP 99– 93162 Noisy le Grand cedex France
phone/fax: 33-1-45926714/45926699 email: natowicz@esiee.fr

Abstract

This paper is concerned with contour segmentation of gray level and color images. For segmenting gray level images, the set of gray levels is, in a first step, quantized by a one dimensional self-organizing map. Contour segmentation is the second step: one computes the set of spatially close pixels mapped onto distant cells. Noise reduction in the segmentation can be achived when spatial and gray level pixel components are quantized as a whole: the image is considered a set of points in a three-dimensional space and quantized with a three dimensional map. This way, any two pixels close in gray levels and positions are mapped onto two close map cells. These two methods have straightforward extention to color image segmentation.

1 Introduction

Self-organizing feature maps [1,2] have been used in the field of image processing for coding through vector quantization [3,4], texture based segmentation [5,6], early visual processing modeling [7], emission computed tomography [8], contour segmentation of images in gray levels [9,10]. The current paper presents the extension of this last method towards contour segmentation of color images and a new method of contour segmentation from "region-like" segmentation of gray level images.

The method of gray level contour segmentation presented in [9] has two stages:

1. vector quantization of the image gray levels by a one dimensional Kohonens map

2. computation of spatial discontinuities of the gray level quantization function.

Because the dimension of the set of gray levels and the dimension of the map are equal, "tonotopy" property holds: after step 1, any two close gray level values are mapped onto two close cells. In step 2. any couple of pixels spatially close on the image and whose gray levels are coded by two distant cells are labeled as segmentation pixels. Hence, this method is the detection of spatial discontinuities of the gray level quantization function defined by the map.

In this segmentation method, two distant pixels are mapped onto two close cells when their gray level values are close. This situation is irrelevant for the situations detected, but one could be interested in having a mapping through which furthermore, any two distant pixels are mapped onto two distant cells. This way, the inverse mapping of any cell would be a set of spatially close pixels with close gray level values, i.e. a "region-like" set of pixels. Such a mapping can be computed if one considers the image as a set of points in the three dimensional space of spatial and gray level components; tonotopy property holds when quantizing the so considered image with a three-dimensional Kohonen's map: any two close pixels –spatially close with close gray level values– are now mapped onto two close cells. One will see that this second method can be used to compute far less noisy contour segmentations.

Segmenting color images is a straightforward extention: as colors are three component values, in the first method of contour segmentation one needs to make use of a three-dimensional Kohonens map whilst in the second method of contour segmentation through region-like mapping, one needs to make use of a five-dimensional map for quantizing the image considered as a set of five –spatial and color– coordinate points.

2   Self-organizing maps for contour segmentation of gray level images

For processing contour segmentation of gray level images, one considers a one dimensional map (composed of chain connected cells) and the set of gray level values to be quantized. In this case map cells hold scalar values.

Let P be the pixel set of an image in gray levels and $M = \{m_1, ..., m_n\}$ a one-dimensional self-organizing map composed of n linear chain connected cells. Any pixel of the image is characterized by its spatial position and its gray level, i.e. $p = (x, y, g)$. When quantizing the set of gray level values G, the self-organizing map M defines a coding function K over this set, $K : G \longrightarrow M, g \in G \mapsto K(g)$, where $K(g)$ is the cell holding the scalar value closest to gray level g. The coding function K defined by map M can be extended to a coding function over the set of pixels P by assigning any pixel the cell coding for its gray level:

$$K : P \longrightarrow M, \forall p = (x, y, g) \in P, K(p) = K(g).$$

Having defined the coding function K over the set of pixels P, any pixel $p \in P$ is assigned a spatial neighborhood $N_I(p)$ on the image (9-neighborhood or 5-neighborhood including central pixel p) and any cell $m_i \in M$, is assigned a spatial neighborhood $N_M(m_i, r) \subseteq M$, defined as the set of cells whose path length to cell m_i is less than integer r: $N_M(m_i) = \{m_k \in M, |i - k| \leq r\}$.

Segmenting the image for map's neighborhood radius r is computing the set $S(r)$ of pixels p in the neighborhood of which lies a pixel q whose coding cell $K(q)$ is outside the r-neighborhood of coding cell $K(p)$:

$$S(r) = \{p \in P, \exists q \in N_I(p), K(q) \notin N_M(K(p), r)\}$$

The neighborhood defined on the map is parametrized by radius r and, for the same map, one obtains included segmentation sets depending upon radius values: $S(r)$ and $S(r')$ being the sets of segmentation pixels computed for respective radii r and r', one has the following property:

$$r < r' \Longrightarrow S(r) \supseteq S(r').$$

Because of this property, the segmentation is multiresolution; radius r stands for its gray level resolution. Letting $d(M)$ be the diameter of the map[1], one has the two following limit cases.

- $r = 0$: any two neighbor pixels with gray levels mapped onto different cells are segmentation pixels (on common images, almost every pixel is a segmentation one)

- $r > d(M)$: as no couple of cells are at a distance greater than $d(M)$, $S(r) = \emptyset$.

Figure 1. is an example of multiresolution segmentation where the image to be segmented is the classical 256 gray level image "Lena". On this segmentation series, one can observe that lowering the resolution (i.e. increasing radius r) brings noise segmentation pixels to vanish progressively but, at the same time, relevant contour points too.

---

[1]longest path length between two map's cells

3 Contour segmentation of color images using self-organizing maps

In a celebrated application [1] p.142, T. Kohonen illustrates how self-organizing maps approximate in an orderly fashion various vector set probability distributions. In a first situation under consideration, the map has a dimensionality smaller than the distribution's one: a one dimensional map was used to approximate a uniform distribution over a triangular area (map's cells held two component vectors). On this beautiful application, one can notice that tonotopy property does not hold: two close points on the area can be coded by two distant cells on the map.

In the second situation, map and distribution dimensionalities are equal and tonotopy property holds. This matter of fact is general and when one wants tonotopy to be a property of the quantization computed by a self-organizing map, dimension of the map must be that of the distribution.

Keeping this result in mind, the former method for segmenting gray level images on a contour basis can be extended to contour segmentation of color images in a straightforward way : as the only notions involved in the segmentation process are tonotopy of the quantization and neighborhoods on image and map, one must quantize the set of colors present on the image by building a mapping that owns this tonotopy property. This last point imposes the dimension of the map which must be set to three, map's cells holding three component color vectors. In what follows, one decided the map to be a cubic grid with map's cells located at its vertices.

Let P be the pixel set of a color image. Any pixel $p \in P$ is characterized by its spatial position (x, y) and its color vector (r, g, b), i.e. $p = (x, y, r, g, b)$. When quantizing the set C of color values, self-organizing map M defines mapping K over this set. One extends this mapping over the set of pixels P by assigning any pixel $p \in P$ the cell coding for its color vector:

$$K : P \longrightarrow M, p = (x, y, r, g, b) \mapsto K(r, g, b).$$

Here again, any pixel p is assigned a neighborhood $N_I(p)$ on the image and any cell m is assigned a spatial neighborhood $N_M(m, r)$ defined as the set of cells whose path length on the cubic grid to cell m is less than r, interger r standing for the color resolution parameter. Contour segmentation $S(r)$ of the color image for map's neighborhood radius r is, with no change from the gray level case defined by

$$S(r) = \{p \in P, \exists q \in N_I(p), K(q) \notin N_M(K(p), r)\}.$$

One obtains a multiresolution segmentation according to the color component, parametrized by map's neighborhood radius r.

Figure 2. is an example of multiresolution segmentation where the image to be segmented is the 16,000 color image "Lena". On this segmentation series, one can observe (just as in figure 1.) that lowering the resolution brings noise to vanish progressively and, at the same time, relevant contour points too; that some contours are detected in this series which were not in gray levels; that relevant contours are disconnected at lower resolutions; that the segmented images are more noisy.

4 Region-like mapping and contour segmentation of gray level images

It appears from segmentation results depicted in figures 1. and 2. that the segmentations computed at high resolution include a lot of points which are not part of a contour: these segmentation are noisy. These noise points vanish at low resolution but to the detriment of real contour points. One would like to compute far less noisy segmentations preserving real contour points.

The reason of such a level of noise in the method of contour segmentation presented in sections 2. and 3. is that the quantization was computed disregarding the spatial component of the pixels: the difference of gray levels between two pixels is not balanced by their spatial closeness during the map organizing process. One would like the map to organize so that any two pixels close according to their spatial and gray level components are to be mapped onto the same cell. This way, the inverse mapping of any cell would be a set of spatially close pixels with close gray levels (almost a region) but, as this set would not necessary be connex, one could only speak of a region-like segmentation.

To achieve this mapping the image will be considered a set of points in a three-dimensional space and quantized with a three dimensional map. This way, any two pixels close in gray levels and positions will be mapped onto two close map's cells. At this point we are lying between two extreme situations:

- quantizing set of gray level values without taking into account pixel spatial components. In this situation the map is one dimensional and the inverse mapping of any cell is a "connex" set of gray levels, i.e. an interval of gray level values.

- quantizing spatial coordinates without taking into account pixel gray level components. In this situation the map is two dimensional and the inverse mapping of any cell is a connex set of positions. The image is divided into regions according to a regular tesselation because the density of probability of spatial positions is uniform (cf. [1] p. 146).

4.1 Region-like mapping

To achieve this region-like mapping, let P be the set of pixels of an image considered as a three-dimensional set of points. One quantizes this set using a three-dimensional map M organized in a cubic grid, so defining a mapping K, $K : P \longrightarrow M$, $p = (x, y, g) \in P \mapsto K(p)$. According to this mapping and thanks to tonotopy property, any two close pixels are mapped onto two close cells and for any cell m, $K^{-1}(m)$ is the region-like set of pixels mapped onto cell m.

Let m_i and m_j be two connex map's cells[2] and let $B(m_i, m_j)$ be the set of border pixels of these region-like sets $K^{-1}(m_i)$ and $K^{-1}(m_j)$. The union of all such sets of border pixels $B(m_i, m_j)$ defines a region-like segmentation $R(P)$ of the image.

On figure 3. are four instances of region-like segmentations obtained from four different $6 * 6 * 6$-cell maps quantizing the same image in 256 gray levels. One can observe that these segmentations are far less noisy than the contour segmentations of figures 1. and 2, that real contours are present and that gray level homogeneous regions are divided in a tesselation-like fashion.

4.2 Contour computation from region-like mapping

The problem to address now is the one of discarding tesselation-like artefacts from region-like segmentations, while preserving real contour points. This problem is currently under study and we would like to present here a perhaps unexpected possible solution.

It appears from [4] and from our own experiments that the quantization computed by Kohonen's self-organizing maps are less sensitive to initial codebook than the other methods of quantization approximation. Nevertheless, they are sensitive, even though only "a little". One experimented to compute region-like segmentations of the same image differing only by the initial values given to map's cells (i.e the initial codebook). From these experiments it appeared

---

[2]i.e. cells m_i and m_j are at distance 1 on the map

that tesselation-like artefacts are far more sensitive to initialisation than real contour points. We decided to take advantage of this fact for computing contour segmentations $S(P)$ out of these different region-like segmentation in a direct way:

- compute n region-like segmentations $R_1(P), R_2(P), ... R_n(P)$

- compute contour segmentation $S(P)$ as $S(P) = \bigcap R_i(P)$

In figure 3. is the contour segmentation computed as the intersection of the four region-like segmentations. One can observe that tesselation-like artefacts vanished while real contour points are preserved and that little noise segmentation points are present.

5 Discussion

We have presented two methods of contour segmentation of gray level and color images. The first method is the detection of spatial discontinuities of the mapping computed by a quantizing self-organizing map. The multiresolution segmentation property allows to reduce the amount of noise segmentation points, but to the detriment of real contour points. Our experiments tends to show that color images brings more contour details but, at the same time, more noise points in the segmentation.

A "region-like" segmentation has been defined which permits to reduce the amount of noise points in the segmentation while preserving real contour points. We are at present time devoting some effort to compute "classical" region segmentation from self-organizing maps, specially for color images as color is an information that one "naturally" attaches to regions [11].

References
1. T. Kohonen, "Self-organization and associative memory" , Springer-Verlag Berlin, 1984.
2. T. Kohonen, "The self-organizing feature map", proceedings of the I.E.E.E., vol. 78, n. 9, September 1990.
3. N.M. Nasrabadi, Y. Feng, "Vector quantization of images based upon the Kohonen self-organizing feature map", I.E.E.E. Int. Conf. on Neural Networks, pp. 101-108, San Diego California, 1988.
4. E. le Bail, A. Mitchie, "Quantification vectorielle par le réseau neuronal de Kohonen", Traitement du Signal, vol. 6, n. 6, 1989.
5. A. Visa, "Identification of stochastic textures with multiresolution features and self-organising maps", Int. Conf. on Pattern Recognition, pp. 518-522, Atlantic City, 1990.
6. O. Simula, A. Visa, "Self-organising feature maps in texture classification and segmentation", Int. Conf. on Artificial Neural Networks, pp. 1621-1628, Brighton, 1992.
7. W. Beaudot, P. Palagi, J. Hérault, "Realistic simulation tool for early visual processing including space. time and colour data", I.W.A.N.N.'93, lecture notes in computer science, vol. 686, Springer-Verlag, 1993.
8. C. Manhaeghe, I. Lemahieu, D. Vogelaers, F. Colardin, "Automatic initial estimation of the left ventricular myocardial midwall in emission tomograms using Kohonen maps", I.E.E.E. transactions on P.A.M.I., vol. 16 n. 3, march 1993.
9. R. Natowicz, R. Sokol, "Self-organizing feature maps for image segmentation", I.W.A.N.N.'93, lecture notes in computer science, vol. 686, Springer-Verlag, 1993.
10. R. Natowicz, "Segmentation d'images par cartes de Kohonen", colloque Gretsi, Juan les Pins, 1993.
11. Quang-Tuan Luong, "La couleur en vision par ordinateur: une revue", rapport de recherche I.N.R.I.A. (Institut National de Recherche en Informatique et Automatique) n. 1251, June 1990.

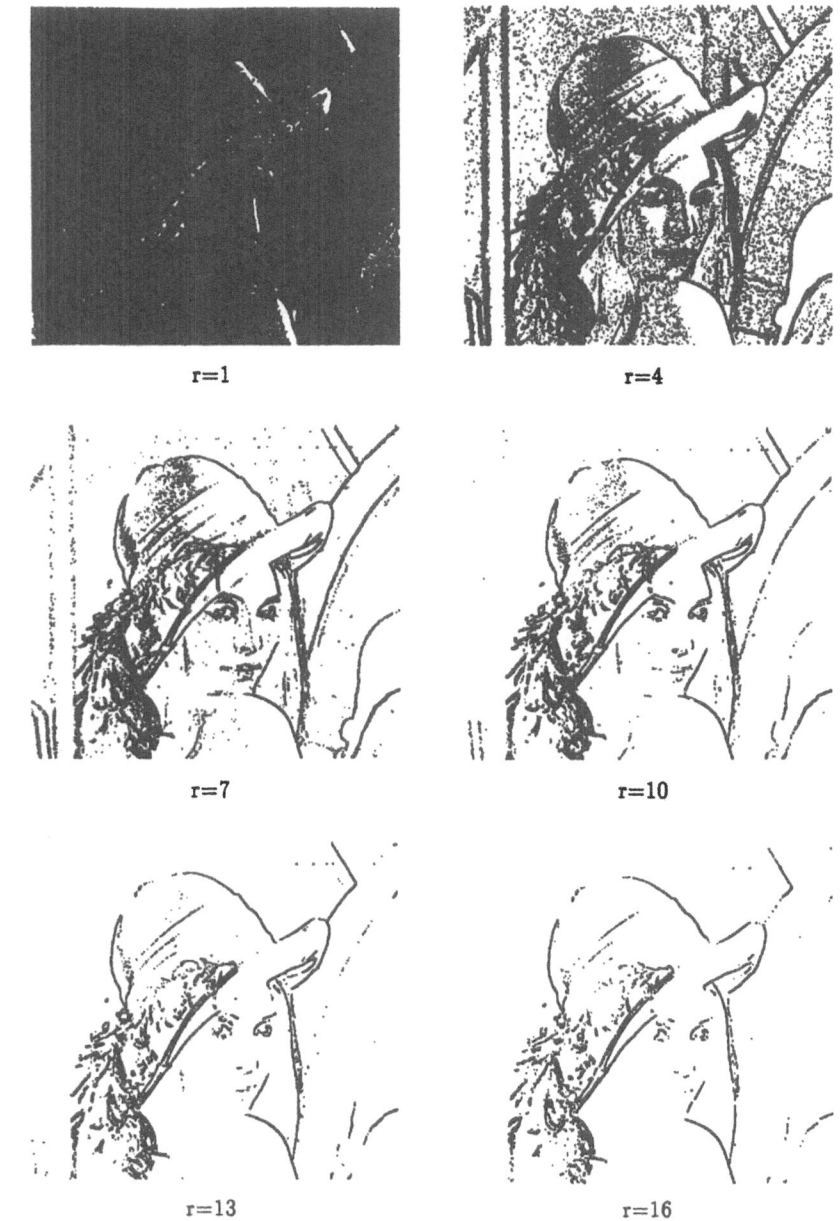

r=1

r=4

r=7

r=10

r=13

r=16

Figure 1: **Contour segmentation of a gray level image.** Original image in 256 gray levels and five contour segmentations at decreasing resolution (highest resolution is for map's neighborhood radius r=1). Quantizing map composed of 50 linearly chained cells.

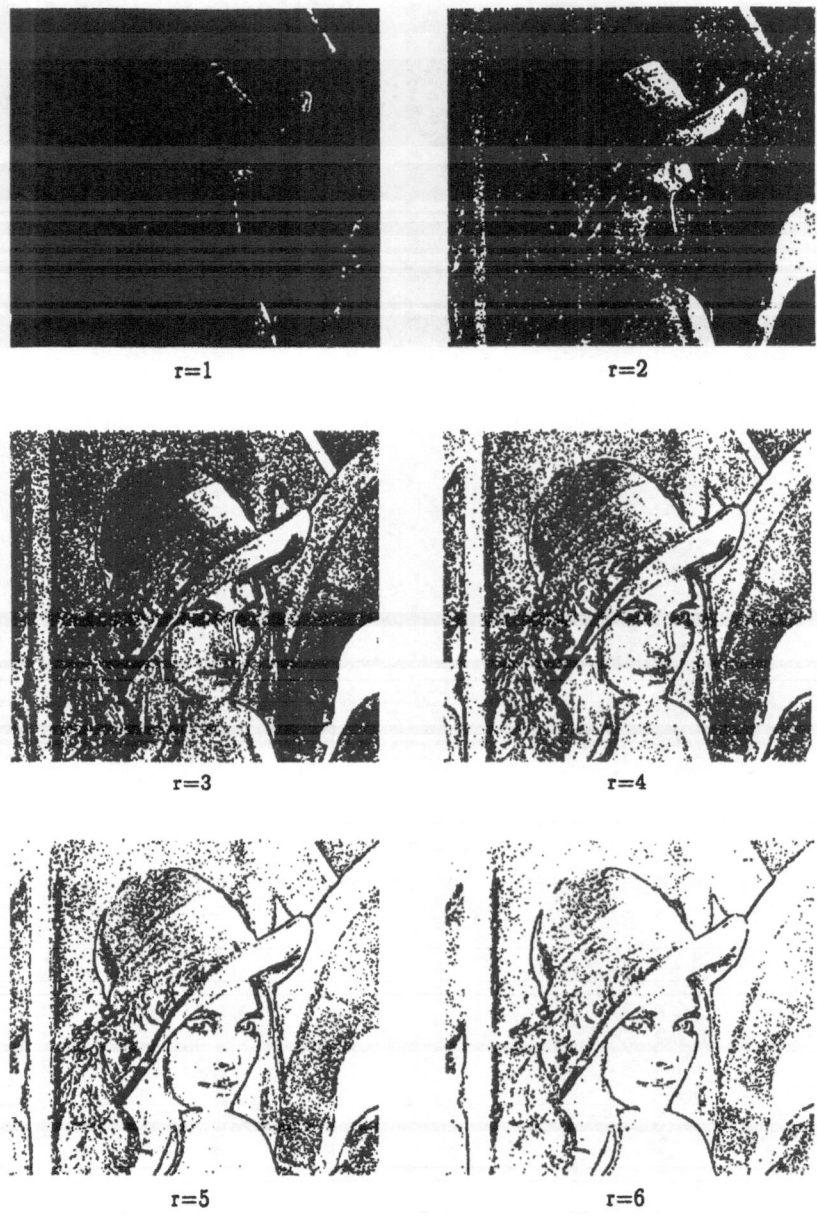

Figure 2: **Contour segmentation of a color image.** Original image in 16,000 colors. Five contour segmentations at decreasing resolution (highest resolution is for map's neighborhood radius r=1). Quantizing map composed of 10 * 10 * 10 cells organized in a cubic grid.

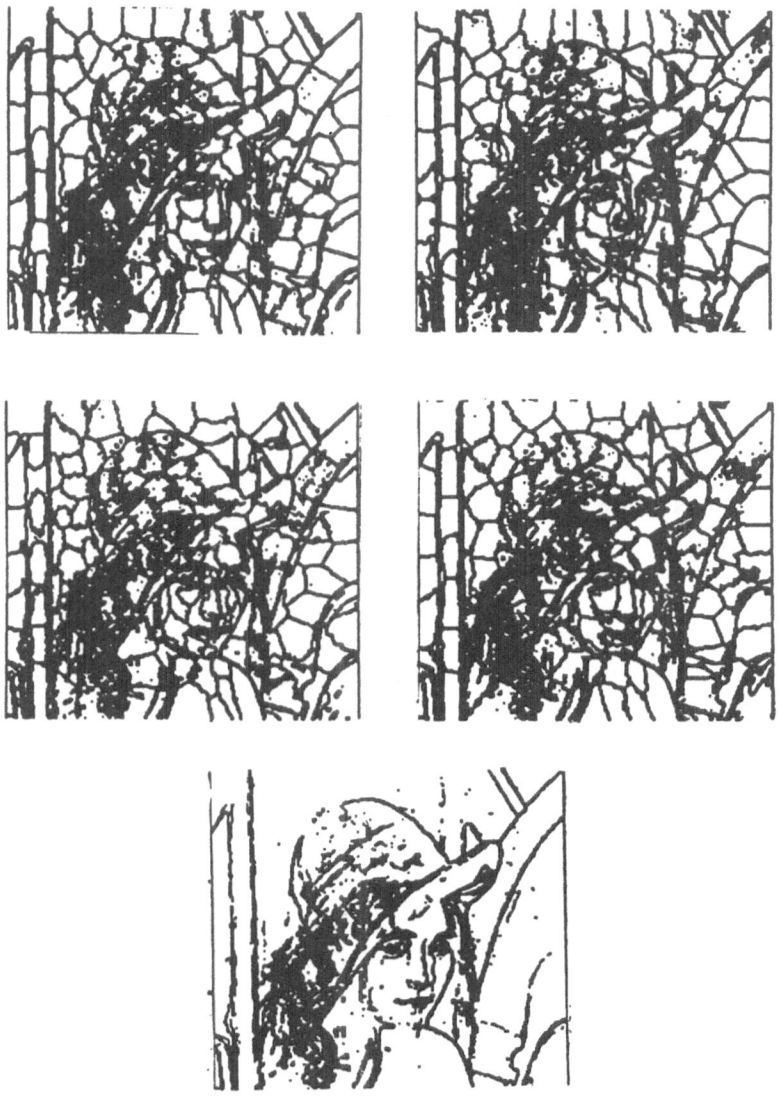

Figure 3: **Region-like and contour segmentations.** Original image in 256 gray levels. Four region-like segmentations $R(p)$ each obtained from a separate $6 * 6 * 6$-cell map organized in cubic grid. Contour segmentation $S(P)$ is the intersection of the four region-like segmentations.

A Geometrical Based Procedure for Source Separation Mapped to a Neural Network

C.G. Puntonet; M. Rodríguez-Alvarez; A. Prieto

Departamento de Electrónica y Tecnología de Computadores.
Universidad de Granada. 18071 Granada (Spain)

Abstract : In many Signal Processing applications, data sampled by sensors comprise a mixture of signals from different sources. The problem of separation lies in the reconstruction of sources from the mixtures. In this paper a new method is proposed for the separation of sources, based on geometrical considerations. After a brief introduction, we present the principles of the new method and provide a description of the algorithm and map this on an artificial neural network. Finally we give examples with synthetic and real signals to illustrate the efficiency and utility of the network.

1. Introduction

In many Signal Processing applications, data sampled by sensors comprise a mixture of signals from different sources. The problem of separation lies in the reconstruction of sources from the mixtures. Let us suppose a medium in which there are q sensors measuring the signals $e_{ok}(t)$ produced by p sources, $s_{oi}(t)$. In fact, the signal $e_{ok}(t)$ observed at the output of the sensor k, is a superimposition of all the source signals, $s_{oi}(t)$, i=1,...,p. Such situations occur in Radar, Sonar, Speech and Array processing [NGU92], and even in Biology, for example, in the coding of odours or movement [JUT91]. The separation of sources consists of reconstructing the sources from the mixtures [HER85], thus eliminating the effect introduced by the environment. In this paper, it is assumed that:

- The signals $s_{oi}(t)$, i=1,...,p, are unknown.
- The number of sensors is equal to the number of sources, p=q.
- The mixing is linear, i.e.:

$$e_{ok}(t) = \sum_{i=1}^{p} w_{oki} \cdot s_{oi}(t) \qquad k=1,2,...,p \tag{1}$$

or, in vector notation:

$$e_o(t) = W_o \cdot s(t)$$
$$e_o, s_o \subset \Re \quad , \quad W_o \subset \Re^{p \times p} \tag{2}$$

where $e_o(t)$ and $s_o(t)$ denote vectors of components $e_{ok}(t)$ and $s_{oi}(t)$ respectively, which are included in Eq.2 as column matrices; W_o is an unknown pxp matrix *(mixing matrix)*, which we assume to be regular. Eq. (1) is known to be an instantaneous linear mixture.

The problem of separation consists of *retrieving the unknown sources $s_o(t)$, only from the observations $e_o(t)$*. With a known matrix W_o, and from the observed signals, $e_o(t)$, we may obtain the original sources, $s_o(t)$, as from Eq.2 it is deduced that:

$$s_o(t) = W_o^{-1} \cdot e_o(t) \qquad (3)$$

In the problem of separation of sources it is standard to assume certain indeterminacies, and thus it is possible to resolve the problem without having to exactly determine the original mixture matrix W_o, but instead another, W, transformed from A_o, from which we obtain the values $s_i(t)$, i=1,..,p, given by

$$s(t) = W^{-1} \cdot e_o(t) \qquad (4)$$

and that correspond to the original sources $s_{oi}(t)$, i=1,...,p, except with an amplification or attenuation (scale factor), or with an offset or phase shift, or with an index permutation (because the order of the sources in the vector $s_o(t)$ is arbitrary), irrelevant for practical purposes. We use the term the *class of matrices similar to W_o* to denote the set of matrices transformed from W_o valid to obtain the original sources. W denotes a matrix belonging to the class W_o, i.e.:

$$W \in class(W_o)$$

In particular, given that:

$$W_o = \begin{pmatrix} w_{11} & w_{12} \\ w_{21} & w_{22} \end{pmatrix} = \begin{pmatrix} 1 & w_{12}/w_{22} \\ w_{21}/w_{11} & 1 \end{pmatrix} \cdot \begin{pmatrix} w_{11} & 0 \\ 0 & w_{22} \end{pmatrix} = W.D \qquad (5)$$

D being a diagonal matrix, then $W \in$ class (W_o).

In this paper we present a neural network which is applied to obtain the original signals only from the instantaneous mixtures, $e_{ok}(t)$. It has the following interesting properties:

- It is applied to digital or analogic signals.
- It does not require that the signals to be separated be statistically independent.
- It is purely based on geometrical considerations concerning the mixtures, and so is easy to understand and to apply.
- The algorithm proposed can easily be mapped on a recurrent ANN.
- The procedure works on-line and is continously adaptive.

In Section 2, we present the principles of the new method. The description of the new adaptive algorithm is shown in Section 3. Section 4 deals with the neural networks that resolve the problem of separation of sources. The behaviour of the algorithm, with synthetic and true signals, is shown in Section 5, and Section 6 contains some conclusions.

2. Principles of the new method: Analytical and Geometrical considerations.

From (2), the w_{ij} coefficients of the W matrix may be obtained with the following expression [PUN94]:

$$w_{ij} = \lim_{(s_{o1},...,s_{op}) \to 0, \, s_q \neq 0} e_i \cdot (e_j)^{-1} = w_{oij} / w_{ojj} \qquad \forall i,j \in \{1,..,p\} \qquad (6)$$

In other words, when the vector $s=(0, 0,..,s_{oj},.,0,.,0)$ is present e_i/e_j provides exactly the coefficient w_{ij} of an W matrix belonging to the class of matrices similar to the unknown matrix W_o, i.e.:

$$w_{jj} = 1 \qquad w_{ij} = w_{oij} / w_{ojj} \qquad \forall i,j \in \{1,...,p\} \qquad (7)$$

Then, the reconstructed signals will be:

$$s_i = w_{ii} \cdot s_{oi} \qquad\qquad \forall i \in \{1,...,p\} \tag{8}$$

From a geometrical point of view [PUN94], the detected signals $e(t)$ are contained within a hyperparallelepiped because the original sources s_{oi} are bounded in a range $[s_{mi}, s_{Mi}]$. Fig. 1 shows the projection of the hyperparallelepiped in the (e_i, e_j) plane when $s_i \in [-|s_{mi}|, +|s_{Mi}|]$, and Fig. 2 shows the projection when $s_i \in [+|s_{mi}|, +|s_{Mi}|]$. To apply the proposed method it is necessary to translate the hyperparallelepiped, such that one of its vertices is located at the origin, with a vector $T=\min(t)$ where $t=e_{o1}+e_{o2}+...+e_{op}$, (Figs. 1-3). In the case of two sources $(p=2)$, once the hyperparallelepiped has been translated, all the values (e_1, e_2) verify the condition:

$$if \quad e_1 > 0 \quad , \quad \frac{e_2}{e_1} \geq \left\lfloor \frac{e_2}{e_1} \right\rfloor_{min} \equiv p_{21m}$$

$$if \quad e_2 > 0 \quad , \quad \frac{e_1}{e_2} \geq \left\lfloor \frac{e_1}{e_2} \right\rfloor_{min} \equiv p_{12m} \tag{9}$$

where p_{21m} and p_{12m} are the slopes of the edges of the polyhedral cone containing the parallelepiped. In general, the points $(e_1,...,e_p)$ of the edges of the polyhedral cone verify the following condition:

$$p_{krm} = \min\left(\frac{e_k}{e_r}\right) \quad , \quad if \quad e_r > 0 \tag{10}$$
$$\forall k,r \in \{1,...,p\} \quad with \quad k \neq r$$

It is easy to see that the coefficients of W matrix can be obtained as follows:

$$w_{ji} = w_{oji} / w_{oii} = \min\{e_j / e_i\} \equiv p_{jim} \qquad if \ e_i > 0$$

$$w_{ij} = w_{oij} / w_{ojj} = \min\{e_i / e_j\} \equiv p_{ijm} \qquad if \ e_j > 0 \tag{11}$$

Then, we obtain a W matrix belonging to class W_o, as was shown in (4).
If the system has more than two mixed signals, corresponding to more than two unknown sources, this procedure obtains all the minors of the W matrix as a projection of the hyperparallelepiped into the plane (e_i, e_j).

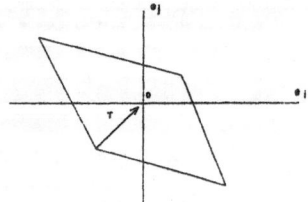

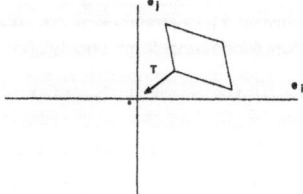

Fig.1. Projection of the hyperparallelepiped in the (e_i,e_j) plane when $s_i \in [-|s_{mi}|, +|s_{Mi}|]$.

Fig.2. Projection of the hyperparallelepiped in the (e_i,e_j) plane when $s_i \in [+|s_{mi}|, +|s_{Mi}|]$.

This conclusion will subsequently be used as the base to obtain the algorithm.

3. Adaptive algorithm

For digital processing, at each instant of time, nT, a point E(n) will be considered.

The main idea of the procedure is to compute the slope of the polyhedral cone edges, once the hyperparallelepiped has been translated to the origin.

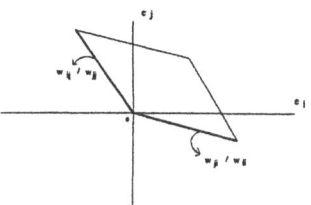

For this purpose the algorithm computes iteratively the points E(n) obtained by sampling the sensed signals. The points E(n) with a minimum (t_o) sum of coordinates (t) are taken as translation vectors, T_o. Then the succesive samples become E(n)-T_o. The algorithm obtains, for each plane (i,j), the

Fig.3. Projection of the hyperparallelepiped in the (e_i,e_j) plane after translation T.

minimum slope w_{ij} of E(n)-T_o. We denote the vector that obtains w_{ij} as ε=E-T_o. When a new translation vector, T, is detected, the coefficients w_{ij} must be adapted as follows:

$$w_{ij} = \frac{e_i - \Delta_i}{e_j - \Delta_j} \equiv \frac{e_i - T_i}{e_j - T_j} \qquad (12)$$

where Δ=T-T_o

Step 1 initializes w_{ij}, p_{ij} and ε_i.
Step 2 obtains the first sample E(n) of the measured signals and considers this as the initial translation vector, T_o.
Step 3 reads iteratively the measured samples, E(n). Then, the algorithm checks whether the new point should be considered a new translation vector, T. If this is not so the translation vector, T_o, remains unchanged, and if $p_{ij}<w_{ij}$ the coefficient w_{ij} is adaptively computed according to the decreasing slope p_{ij}. Note that, in this case, the translated vector ε is stored. Otherwise, with the arrival of a new T, the w_{ij} coefficient is adapted to this vector, taking into account the components of ε and the distance $\Delta\equiv$(T-T_o). **Step 4** separates recursively the signals s_i according to the w_{ij} coefficients computed above, and continues by reading a new vector E (Step 3).

Table 1 includes the algorithm that runs on-line and makes it possible to obtain adaptively the coefficients w_{ij} of an W matrix belonging to the class of matrices similar to W_o. Then, the procedure may be used with time-dependent media.

4. Mapping the algorithm on an Artificial Neural Network.

The algorithm described in Sect. 3 could be mapped in the recursive neural network of Fig.4. The inputs (measured signals) and outputs (reconstructed signals) are analogic signals. The output of each neuron is:

$$s_i = e_i - \sum_{\substack{j=1 \\ j \neq i}}^{p} w_{ij} \cdot s_j \qquad \forall i \in 1,...,p \qquad (13)$$

This represents the desired values of s_i because it satisfies Eq.(1) with w_{ii}=1.

Table 1: Algorithm for on-line separation of sources

STEP	Description	
1	$i,j=1,...,p$	;*initialize* p_{ij}, a_{ij}, ε_i

2	Read $T_o=E\equiv(e_1,...,e_p)$
	$i,j=1,...,p$ $i\neq j$
	{
	  if $(e_i.e_j)\leq 0$ {Go to STEP 2}
	}
	$t_o=e_1+...+e_p$
		
		;*initial translation vector*

3	Read $E\equiv(e_1,...,e_p)$
	  $t=e_1+...+e_p$
	  $i,j=1,...,p$
	  { if $t<t_o$ then
	    { $T=E$, $T-T_o=\Delta\equiv(\Delta_1,...,\Delta_p)$, $T_o=T$
	    $a_{ij}=(\varepsilon_i-\Delta_i)/(\varepsilon_j-\Delta_j)$
	    }
	    else
	    { $p_{ij}=(e_i-T_{oi})/(e_j-T_{oj})$
	      if $p_{ij}<a_{ij}$ and $(e_j-T_{oj})>0$ then
	      { $a_{ij}=p_{ij}$, $\varepsilon_i=e_i-T_{oi}$, $\varepsilon_j=e_j-T_{oj}$
	      }
	    }
	  }
		;*observed vectors*
		
		;*new transl. vector*
		;*distance* Δ
		;*adapting* a_{ij} *to new T*
		
		;*no new translation vector*
		
		;*adapting* a_{ij}
		;*storing E*

4	$i,j=1,...,p$ { $s_i = e_i - \sum_j w_{ij} \cdot s_j$ } Go to STEP 3	 ;*separating sources* ;*reading new observation*

According to the algorithm described in Sect.3 the weights must change as follows:

$$w_{ij}(t+1) = \min\left[w_{ij}(t),(e_i-T_{oi})/(e_j-T_{oj})\right] \qquad if\ \Delta\geq 0$$
$$w_{ij}(t+1) = (\varepsilon_i(t)-\Delta_i)\ /\ (\varepsilon_j(t)-\Delta_j) \qquad if\ \Delta<0 \tag{14}$$

i.e. this equation describes the changes needed for the learning of the network.

Fig.5 shows another neural network, of the feedforward (non-recursive) type, for the separation of two mixed signals.

In this network learning must be implemented as described in the following expressions:

$$w_{ij}(t+1) = \min\left[w_{ij}(t),(e_i-T_{oi})/(e_j-T_{oj})\right] \qquad if\ \Delta\geq 0$$
$$w_{ij}(t+1) = (\varepsilon_i(t)-\Delta_i)\ /\ (\varepsilon_j(t)-\Delta_j) \qquad if\ \Delta<0 \tag{15}$$

The outputs (in the recall phase) are:

$$s_1 = e_1 - w_{12}\cdot e_2$$
$$s_2 = e_2 - w_{21}\cdot e_1 \tag{16}$$

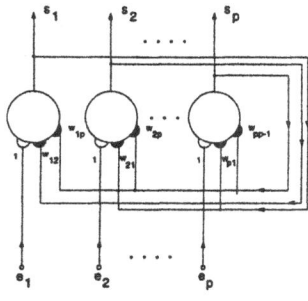

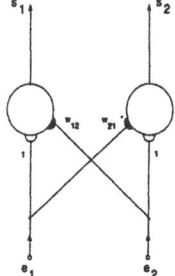

Fig. 4. Recursive network for separation of p sources.　　　Fig.5. Feedforward network for modular separation.

The neuron of Fig.5 can be used as a block for a modular implementation. As an example, with two layers of 3 and 2 blocks (running 4 and 3 neurons) it is possible to separate 3 signals (Fig.6), and with 3 layers of 8, 4 and 4 neurons, respectively, we may separate 4 signals.

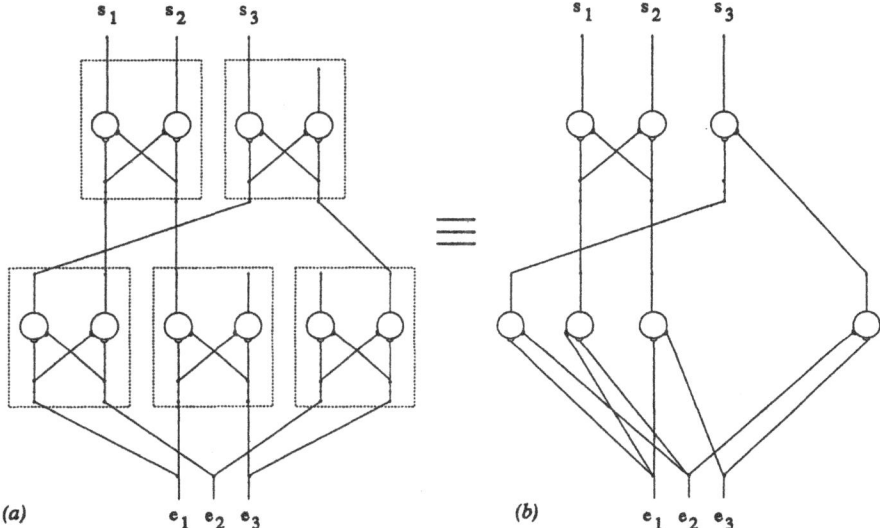

Fig.6 : (a) Modular implementation of a feedforward neural network separating 2 signals;
(b) Simplified version of the network in which unused neurons have been eliminated.

5. Results with simulated and true signals

We have implemented the network in an AT-486 and the coefficients of the mixing matrices were chosen at random. Some results are shown in Fig.7. In this figure, and for each simulation, the upper part shows the original sources (s_{oi}); the middle part corresponds to the signals (e_{oi}) mixed according to the W_o

matrix (neural network inputs); and the lower part corresponds to the neural network outputs (s_i), i.e. the separated signals. The weighting matrix obtained during learning is **W**. The effectivity of the network may be tested by comparing the upper (s_o) and lower (s) parts. It may be seen that during an initial period the network provides an erroneous response; this is because the learning process is then being carried out.

The simulations correspond to the following signals:

Simulation A. Three binary mixed sources.
Simulation B. Two modulated random sources.
Simulation C. Two sinusoidal sources.
Simulation D. Two real signals: the spanish word "tres" ("three") and a randomly generated noise.

6. Conclusions

This paper introduces a procedure to resolve the problem of blind separation of p original signals or sources, reconstructing the mixture matrix. The new method of separation of sources we propose is based on geometric considerations. According to the methods previously reported, this is a new approach which does not need estimation of moments or cumulants [CAR91,COM89,LAC88], and the procedure is purely algebraic.

The algorithm is mapped on two neural networks (one recursive, and the other feedforward and modular), and runs on-line and can be applied to time-variable media, as it is continuously self-adaptive. Some simulation results with synthetic and real signals have been included to illustrate the efficiency of the method.

Acknowledgements

This work has been supported in part by the 1993-1994 French-Spanish Integrated Action HF92-273B and HF93-222B. The authors are grateful to C. Jutten and A. Mansour for their constructive comments and helpful discussions.

References

[CAR91]   J.-F. Cardoso, "Super-symmetric decomposition of the fourth-order cumulant tensor, blind identification of more sources than sensors", Proc. International Conf. Acoust. Speech Signal Process '91, 14-17 May 1991.

[COM89]   P. Comon, "Separation of sources using high-order cumulants", SPIE Conference on Advanced Algorithms and Architectures for Signal Processing, Vol. Real-time Signal Processing XII, San Diego, California, Aug. 8-10, 1989, pp.170-181.

[HER85]   J. Hérault, C. Jutten and B. Ans, "Detection de grandeurs primitives dans un message composite par une architecture de calcul neuro-mimetique en apprentissage non supervisé." 10th GRETSI, Nice, France, pp. 1017-1022, 1985.

[JUT91]   C. Jutten, J. Hérault, P. Comon and E. Sorouchiary, "Blind separation of sources", Parts I, II and III, Signal Processing, Vol. 24, No. 1, July 1991, pp.1-29.

[LAC88]   J.-L. Lacoume and P. Ruiz, "Sources identification: A solution based on the cumulants", Proceedings of the 4th ASSP Workshop on Spectral Estimation and Modelling. Minneapolis, USA, Aug. 1988, pp. 199-203.

[NGU92]   L.H. Nguyen Thi, C. Jutten and J. Caelen, "Speech enhancement: Analysis and comparison of methods on various real situations", Proceedings of EUSIPCO-92, Brussels (Belgium), Aug. 24-27, 1992, pp. 303-306.

[PUN94]   C.G. Puntonet, "Nuevos algoritmos de separación de fuentes en medios lineales", Ph. D. Thesis, University of Granada, Spain, September, 1994.

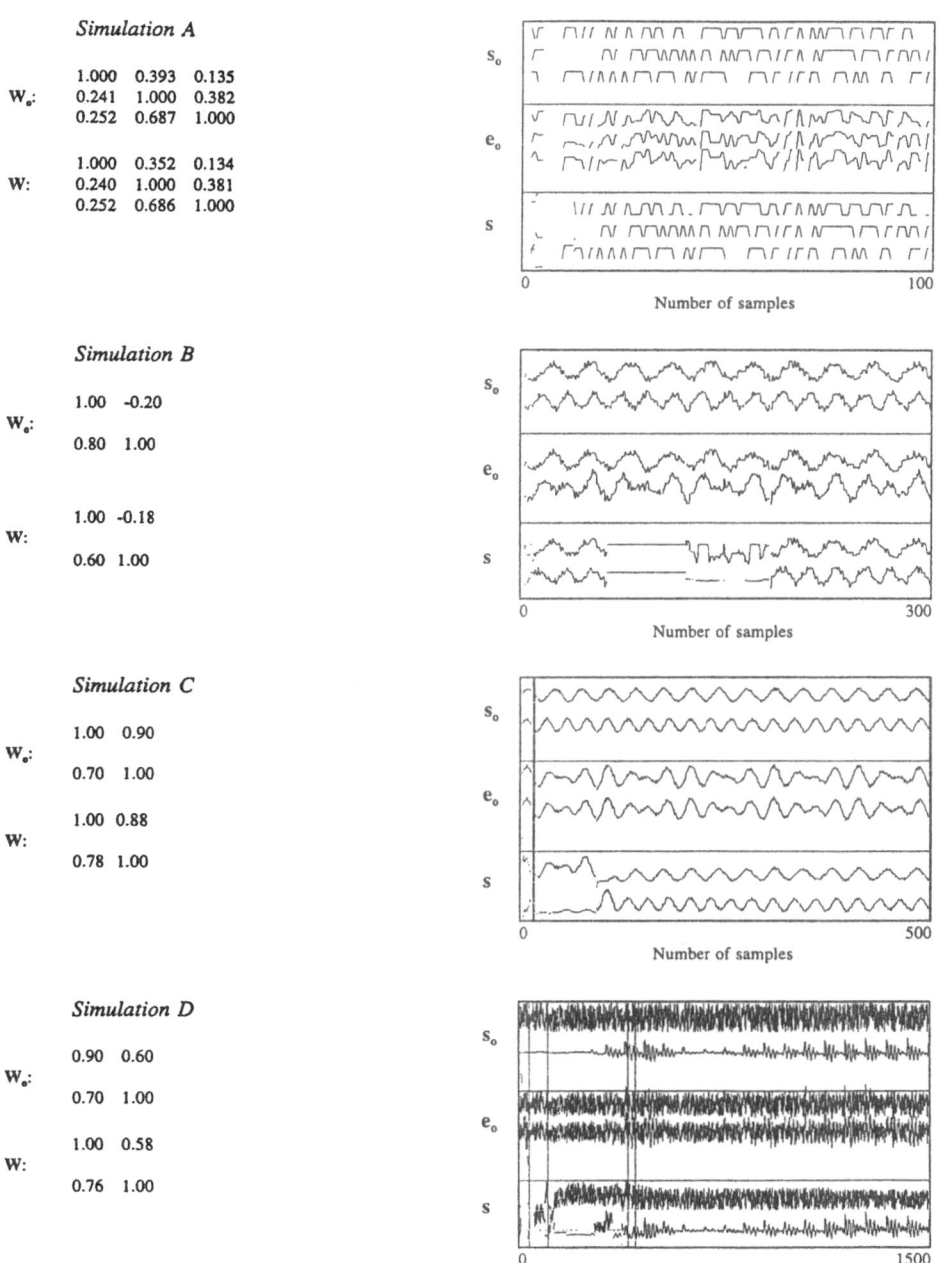

Simulation A

W_0:
$$\begin{array}{ccc} 1.000 & 0.393 & 0.135 \\ 0.241 & 1.000 & 0.382 \\ 0.252 & 0.687 & 1.000 \end{array}$$

W:
$$\begin{array}{ccc} 1.000 & 0.352 & 0.134 \\ 0.240 & 1.000 & 0.381 \\ 0.252 & 0.686 & 1.000 \end{array}$$

Simulation B

W_0:
$$\begin{array}{cc} 1.00 & -0.20 \\ 0.80 & 1.00 \end{array}$$

W:
$$\begin{array}{cc} 1.00 & -0.18 \\ 0.60 & 1.00 \end{array}$$

Simulation C

W_0:
$$\begin{array}{cc} 1.00 & 0.90 \\ 0.70 & 1.00 \end{array}$$

W:
$$\begin{array}{cc} 1.00 & 0.88 \\ 0.78 & 1.00 \end{array}$$

Simulation D

W_0:
$$\begin{array}{cc} 0.90 & 0.60 \\ 0.70 & 1.00 \end{array}$$

W:
$$\begin{array}{cc} 1.00 & 0.58 \\ 0.76 & 1.00 \end{array}$$

Figure 7. Some results with synthetic and real signals.

Quasi-Optimum Combination of Multilayer Perceptrons for Adaptive Multiclass Pattern Recognition

Alberto Ruiz García and Francisco J. Arcas Túnez

Departamento de Informática y Sistemas
Facultad de Informática, Universidad de Murcia, Spain
aruiz@dif.um.es
May 12, 1994

Standard multiclass pattern recognition requires frequent re-learning stages when the set of categories of interest evolves in time. In order to minimize the computation costs of class incorporation and removal, we divide the global multiclass recognizer into a collection of class pairwise neural dichotomizers. When a new class appears, an adequate set of dichotomizers is created and trained to discriminate the new class from the rest. If a class disappears, its associated dichotomizers are eliminated. In both cases previously learned knowledge is not disturbed. The properties of neural recognizers and pairwise modularization allow an analytic quasi-optimum method for combining network outputs to obtain the global multiclass response. An incremental and distributed pattern recognition architecture is presented and its performance experimentally evaluated, obtaining better error rates and learning times than conventional multiclass recognizers using similar resources. The design is highly parallel and asyncronous, adequate for dynamic real time applications.

Keywords: pattern classification, neural networks, density estimation, data fusion, classifiers combination.

I. INTRODUCTION

Multiclass pattern recognizers [5,6] usually implement a common decision strategy for the whole set of classes. In particular this is true for the Back-Propagation Multilayer Neural Networks [7], currently one of the most suitable and flexible solutions to complex pattern classification tasks.

When an automatic system operates in a dynamic environment the categories of interest may change in time. A great computational effort in re-learning is needed every time a new class appears. It is also convenient to forget the knowledge associated to classes which are not very likely to appear in the future. We are interested in a pattern recognition architecture in which most of the acquired knowledge can be retained both at incorporating and at eliminating classes. This paper shows that this goal can be efficiently achieved through the modularization of the global classifier into a set of feedforward multilayer neural networks trained to discriminate pairs of classes. When a new class appears, an adequate set of dichotomizers is created and trained to discriminate the new class from the rest. If a class disappears, its associated dichotomizers are eliminated. In both cases previously learned knowledge is not disturbed. This approach also simplifies training because all the subtasks are easier than the global one. These advantages require a satisfactory combination strategy for computing the global response from the module outputs.

There are two essential problems in multiple classifier combination. On one hand, we have a problem of competence: when the input to a recognition module belongs to a class not used in learning the output should not be, at least in principle, taken into account. But of course we do not know in advance the class of the example in order to enable only the qualified modules. On the other hand, some kind of credit assignment problem appears when several modules trained over the same set of classes do not agree in their responses. Actually, the above questions are among the most inportant ones to be solved in any distributed information processing system involved in data fusion.

Knerr et al. propose [12] a constructive algorithm for neural architectures. First they try to separate every class from the rest by a linear discriminant function. If there are classes for which this is not possible, additional linear functions are trained for pairwise class discrimination. A specific piecewise linear classifier is finally implemented for non linearly separable classes by a decision tree or a multilayer neural network.

Xu et al. [16] combine multiple classifiers by a back-propagation neural network trained to select the most promising module for classifying each example. Instead of using an aggregation algorithm, the network learns to switch alternative solutions depending on certain features of the examples.

Jacobs [10] proposed modular back propagation neural networks. The global classifier is divided into several sub-networks and a master nework. The master network learns to send each example to the proper subnet, which automatically learns to solve a subtask of the whole problem. Both the master and the subnets are trained back propagating the overall error. The number of subtasks and module architecture are set *a priori* by the designer.

Xu *et al.* [17] exhaustively review combination methods for multiple classifiers independently of their specific decision algorithms. According to the amount of information they use the combination methods may operate at three levels: a) using only the class labels provided by each module (abstract classification); b) using the class preference order (rank classification) and c) using cuantitative measures of belief (e.g. *a posteriori* probabilities in bayesian classifiers). In category c) they study output averaging for nearly bayesian classifiers. In category a) they study several variants of voting, then a bayesian approach based on the confusion matrices of the modules and finally evidence agregation by Dempster-Shafer Theory using the success rate of each module to compute the basic probability assignments.

Ho *et al.* [8] combine the class rankings obtained by multiple classifiers for class reduction, using intersection and union methods, and for class reranking, using the highest rank, the Borda count and logistic regression.

Kimura *et al.* [11] combine a statistical and a structural method for handwritten character recognition outperforming the results of each individual approach.

Baxt [3] combine two neural networks for simultaneous reduction of the two kind of errors (related to false alarm probability and sensibility) in disease detection. Each network is trained with majority of examples in one class. The global decision is made according to the certainty degree (approximation of the actual outputs to the desired values) of the networks.

Other approaches to classifier combination can be found on [13] and [15]. Abidi and Gonzalez [1] review data fusion and integration in a more general context.

II. CLASS PAIRWISE MODULARIZATION

Consider an arbitrary L-class pattern recognition problem, with $L > 2$. Usually, each pattern x to be classified is a feature vector extracted from the raw data acquired by the sensors. Select a set of D discriminant functions $f_i(x)$ devoted to learn specific subsets $T_i \subset \{C_1, ..., C_L\}$ of the classes. These functions may instantiate any recognition technique and implicit or explicit rules exist for translating the numeric output values into symbolic class indices. In principle, $f_i(x)$ is only meaningful for $x \in T_i$. Let G denote a global recognizer, trained with all the classes.

In general, the complexity of the optimum assignation regions increases with the number of classes to recognize. Therefore, under reasonable assumptions on the recognition method, f_i should obtain higher accuray than G over the subset T_i; the extra classes learned by G may disturb the optimal regions for T_i.

Let $f(x) = \{f_1(x), f_2(x), ..., f_D(x)\}$ denote the outputs of all the discriminant functions. Conceptually, the components of f can be considered as "second level" features extracted from the patterns. They are expected to be very discriminant

because are trained to separate classes. From a probabilistic point of view the optimum combination strategy is the Bayes rule: decide the class $c \in \{1, 2, ..., L\}$ with the highest value of

$$P\{c|f\} = \frac{p_f(f|c)P\{c\}}{p(f)} \quad (1)$$

Unfortunately, the multivariate statistical distribution of f is not known in advance and a reliable estimation of $p_f(f|c)$ is not feasible in general.

There is a special case in which expression (1) can be simplified. If the components of f are statistically independent, the joint distribution can be factorized into the marginal distributions over each component [14]. Therefore the multivariate rule can be replaced by the sequential application of scalar Bayes rules for every discriminant output:

$$P\{c|f\} = \frac{p(f_1|c)}{p(f_1)} \frac{p(f_2|c)}{p(f_2)} ... \frac{p(f_n|c)}{p(f_n)} P\{c\} \quad (2)$$

The optimum combination method can be easily applied if the discriminant functions have independent outputs following known class conditional probability densities.

Xu *et al.* [17] suggest that discriminant functions trained with different examples will show statistically independent outputs. Hence the classifying modules must be trained with different sets of examples, $T_i \neq T_j$ for $i \neq j$. There are many ways of dividing L-classes into different subsets; for our purposes the proper method is using a module for every pair of classes. This structure leads to minimum re-learning overhead upon creation and destruction of classes and high recognition accuracy.

The final step towards a complete specification of expression (2) is selecting some powerful pattern recognition technique with known and easy to estimate output distributions. Denker and leCun [4] suggest that multilayer neural network output unit values (before squashing) will be nearly normal because they are the sum of many different apportations (Central Limit Theorem). On the other hand, it is well known that, under certain conditions, this kind of neural model is able to approximate any assignation function [9] and estimates [18] the *a posteriori* probabilities required for optimal classification.

In conclusion, the outputs of a set of neural dichotomizers can be combined in a optimum way if the normality and independence assumptions actually hold. These two facts will be experimentally verified in Section IV.

III. A MODULAR RECOGNITION SYSTEM

We have built a general purpose recognition system to evaluate the actual advantages of the proposed modularization of global multiclass recognition into class pairwise neural dichotomizers. In this section we describe the functional blocks of the system (Fiig. 1) and their interaction.

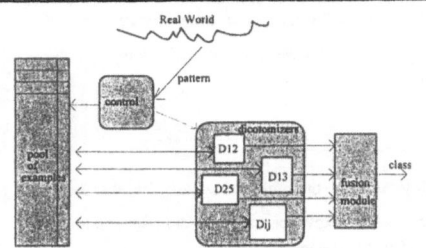

Figure 1. Overview of the Modular Recognition System. It contains a pool of examples, a controler, the set of class pairwise neural dichotomizers Dij and a data fusion module.

The real world provides examples (patterns) from an unrestricted and dynamicaly changing set of classes. Any required raw data preprocessing is assumed to be completed at this point. The most recent examples are stored in a FIFO pool to be available to the dichotomizers. The controler detects new classes as well as obsolete ones, creating and destroying the required dichotomizers.

Independent and asyncronously, every dichotomizer module is always learning from labelled examples in the pool belonging to its assigned pair of classes, incrementally improving its recognition accuracy.

At the same time, the modules compute the sufficient statistics of its outputs in *all* the classes (even for classes not assigned for learning) for data fusion purposes.

When an unlabelled example arrives, the data fusion module asks all the dichotomizers about the likelihoods of the different classes to compute the global output of the system.

The data fusion algorithm requires minimun modifications after creation or destruction of discriminant modules when a class enters or leaves the system. The "forgetting" schema is direct and without side effects and new classes are easily incorporated without any conflict with the available knowledge learned from the older classes.

The system dynamically adapts to changes in the number and nature of categories because of the incremental nature of learning and the schema of creation and destruction of modules according appearance and elimination of classes. A detailed description of the blocks follows.

The Pool of Examples. We store a statistically significant experimental data base for inductive learning, using a first-in-first-out list containing the most recently presented labelled examples. The pool can be read at any time by every module in the system.

The Controler. This block supervises the addition and deletion of examples into the pool and keeps a list of the current class labels. When a pattern from a new class appears, a new set of discriminants is created associated with the new class and all of the older ones. When a class disappears from the pool, all of the discriminants associated to that class may be destroyed. The list of labels is known to the fusion module, allowing it to ask the dichotomizers for the likelihood of every class present in the system.

The Elemental Neural Dichotomizer. The structure of the basic recognition module is shown in Fig. 2. It contains a neural dichotomizer trained to recognize a specific pair of classes and a *likelihood estimator*, which maintains models (mean value and variance) of the statistical distribution of its output over the whole set of current classes.

A reasonable architecture for the neural dichotomizer is a two layer feedforward network with N units in the hidden layer and one output linear unit. The parameter N controls the flexibility of the network for building complex decision regions.

Let $f_{ij}(x)$ denote the output value of the network in the module associated to classes Ci and Cj. The module is always reading labelled examples {x,c} from the pool. If the example belongs to one of the learning classes (Ci or Cj) the network executes one elemental learning step (a back propagation iteration) with example x, trying to obtain different values of $f_{ij}(x)$ in Ci and Cj.

For examples belonging to every class c, the likelihood estimator updates the mean and variance of $f_{ij}(x)$ in class c.

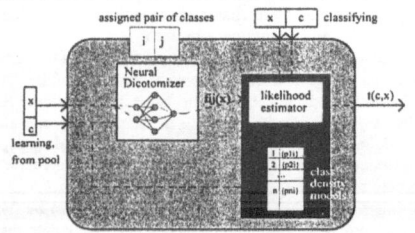

Figure 2. The elemental building block of the modular classifier is essentially a dichotomizer connected to a likelihood estimator.

When an unlabelled example comes from the *classifying* line, the likelihood estimator computes

$$t_{ij}(c,x) = 2\ln\sigma_{ijc} + \left(\frac{f_{ij}(x) - \mu_{ijc}}{\sigma_{ijc}}\right)^2 \tag{3}$$

monotonously related to $p(f_{i,j}(x)|c)$, the relevant magnitude required for the optimum data combination expression (2):

$$p\big(f_{i,j}(x)\big|x\in c\big) = \frac{1}{\sqrt{2\pi}\sigma_{ijc}} e^{-\frac{1}{2}\left(\frac{f_{ij}(x)-\mu_{ijc}}{\sigma_{ijc}}\right)^2} = \frac{1}{\sqrt{2\pi}} e^{-\frac{1}{2}t_{ij}(c,x)} \tag{4}$$

$t_{ij}(c,x)$ is a probabilistically meaningful distance between $f_{ij}(x)$ and the population of $f_{ij}(x)$ in class c. Given a fresh unlabelled example, the data fusion module will ask every module Dij for the values of $t_{ij}(c,x)$ for every class c.

The output distributions change after each learning step. Several strategies can be used in order to maintain acceptable approximations:

- Recompute means and variances after stabilizaton of learning and always before the likelihoods are requested.

- For each example, after the learning step, incrementally update the means and variances using a discounted sum in such a way that, in the long run, the discriminants will be trained and the class models will converge to the actual values.

The Fusion Module. When an unlabelled example x arrives, the fusion module asks all the pairwise recognizers for the likelihood of every class given x. The output values are added up and the class with minimum "probabilistic" distance $d(c)$ is selected as global output:

$$d(c) = \sum_{i>j} t_{ij}(c, x) - \ln P\{c\} \qquad (5)$$

This decision rule is equivalent to (2) as can be easily verified taking logarithms and removing non discriminant magnitudes. The fusion module only requires a list of the current set of classes and access to the likelihoods provided by the discriminant modules. The algorithm is completely analitic and requires no extra learning.

The implemented system [2] has actually some other features: it allows different learning algorithms in the modules; each module automatically switches the learning method (or increments the number of hidden units) if the success rate remains low for a long time; it is possible to compare the combination method with other standard approaches such as votation; etc.

IV. EXPERIMENTAL RESULTS

The proposed modular recognition architecture has been evaluated over a special benchmark and over a real application.

Normality and Independence Assumptions

First, we have built a four synthetic classes recognition problem with complex assignation boundaries (Fig. 3). It has been designed to verify the normality and independence asumptions required for optimum data fusion.

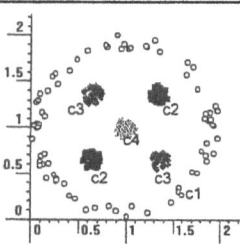

Figure 3. Synthetic recognition problem to verify normality and independence of module outputs. There are four classes with complex boundaries.

This problem has a bayes (optimum) error probability of 0%, because the class conditional probability densities do not overlap. The topologic complexity of the assignation regions (the classes are not linearly separable and some of them are

not connected) makes learning hard for global multiclass algorithms.

However, all the six pairwise discriminant networks can be easily trained in a few hundred iterations with five hidden units. The tasks are perfectly learned as shown in the assignation regions obtained by the dichotomizers (Fig. 4).

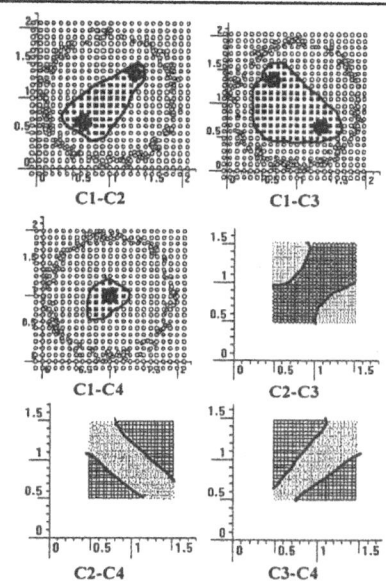

Figure 4. Assignation regions built by the dichotomizers for solving their particular two-class recognition tasks. The regions are less constrained, so class pairwise is easier than global discrimination.

The global assignation regions are satisfactorily inferred (Fig. 5) by the proposed combination algorithm, obtaining a 100% success rate over an indepentent collection of examples (hold out estimation). This result clearly improves the accuracy and learning time of global multiclass recognition algorithm using similar resources.

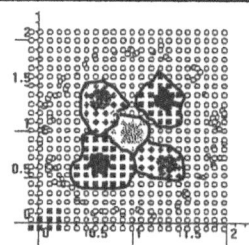

Figure 5. Global assignation regions obtained by the proposed combination method.

We sampled the outputs of all the dichotomizers in each class and observed the following facts: First, the class conditional outputs of each module are not strictly gaussian. They would not pass standard normality tests. However, most of the distributions are unimodal and, therefore, for discrimination purposes a gaussian is an adequate model for the densities (Fig. 6). Note that the neural nature of the discriminant functions radically simplify the output distribution of classes which are far from normal in the original two-feature space. Less flexible (e.g. linear or cuadratic) discriminant functions cannot usually separate complex classes. But, even when they can, data fusion is not trivial because the outputs are less constrained and, therefore, density estimation is harder.

specially when most of the modules achieve high discrimination over their assigned classes. Therefore, although the fusion algorithm reveals to be sub-optimum, the overall recognition schema clearly outperforms standard global approaches to multiclass recognition.

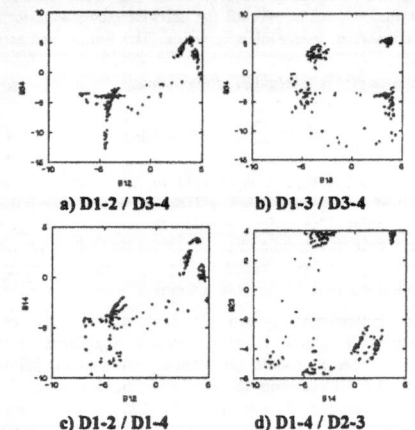

a) D1-2 / D3-4   b) D1-3 / D3-4

c) D1-2 / D1-4   d) D1-4 / D2-3

Figure 7. Some typical scatter plots of module outputs. Cases a) and c) show high correlation. For b) and d) the independence assumption may be acceptable for combination purposes.

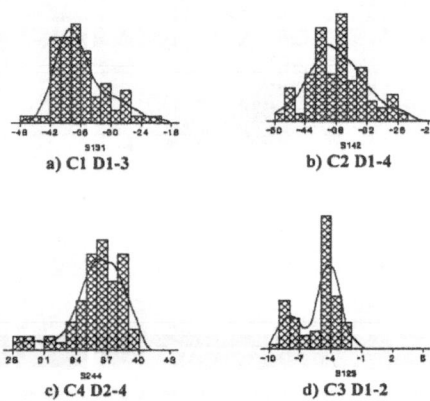

a) C1 D1-3   b) C2 D1-4

c) C4 D2-4   d) C3 D1-2

Figure 6. Some typical histograms of the outputs of the neural dichotomizers on each class. Ck Di-j denotes output of examples from class k on the dichotomizer trained to separate Ci from Cj. Only 5 out of 24 outputs showed clearly bimodal distributions as in example d)

While deviations from normality are essentially irrelevant, the statistical independence violation can be attenuated, opening an interesting line for system operation. In fact, if two modules are highly dependent, the fusion algorithm integrates redundant information. Therefore, either of the modules in a correlated pair (preferably the worst one in recognition accuracy) could be eliminated with small performance degradation.

We studied the effects of dependent module elimination. A set of modules with absolute Pearson correlation below 0.8 (D23, D24, D34) was selected for output combination. The success rate remained remarkably high: 96.3%.

Second, statistical independence of module outputs cannot be assumed in general. The nonparametric Pearson correlation coefficient is high between some modules even if they do not share any learning class. See Table 1 and Figure 7. Conversely, there are uncorrelated modules even when they learn with a common class.

	D12	D13	D14	D23	D24
D13	-.074				
D14	.869	.292			
D23	-.066	.995	.294		
D24	.325	.648	.641	.657	
D34	.891	.045	.909	.064	.488

Table 1. Matrix of Pearson correlations among the six module outputs.

The experiment shows that the requeriments for optimum data fusion do not hold in a strict statistical sense. However, they are sensible approximations for practical purposes,

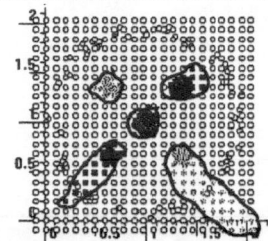

Figure 8. The assignation regions removing the modules trained to discriminate class C1 are essentially correct.

Note that it is possible to correctly discriminate classes (in this case C1, see Fig. 8) without explicit use of the modules trained to recognize those classes!

Restricted data fusion, discarding dependent modules, can be easily implemented maintaining a dynamic correlation matrix among all the current network outputs. This matrix can be incrementally updated or recomputed after learning stabilization. The fusion module should only request the likelihood values to a discriminant and uncorrelated module subset.

Handwritten Character Recognition

Our modular approach to multiclass pattern recognition has been also evaluated on a hand-written character recognition application. We prepared two sets of examples from the small vowels. The master set had 40 patterns / class with high variability and the test set had 20 patterns / class with moderated variability. Typical exemplars from the five classes are shown in Figure 9.

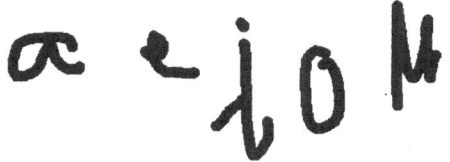

Figure 9. Typical hand-written characters.

We extracted very simple features from the raw bitmap data. Each character was represented by 12 values, the average gray level along five rows, five columns and the two diagonals (Fig. 10). These features are invariant to position and size.

Figure 10. Features extracted from a character (diagonal values are not shown).

Figure 11 shows the success rate of the dichotomizers after an early learning iteration. Finally, after a reasonable amount of time all the ten dichotomizers achieved a high recognition rate using with five hidden units.

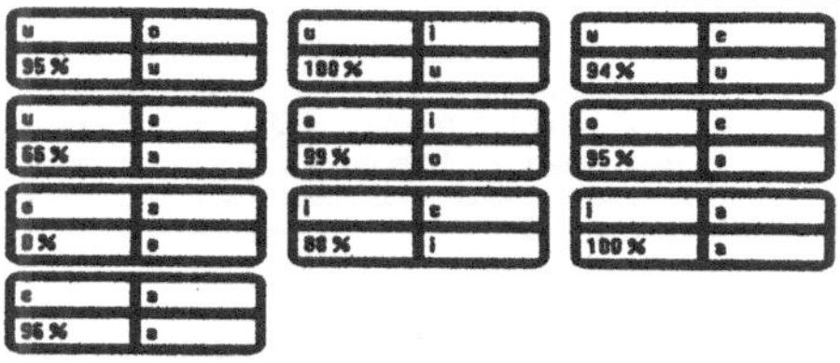

Figure 11. On line information about learning evolution in all the discriminant modules.

The global recognition results of the proposed combination method were quite good: success rate of 94.5% over the independent test set. A simpler voting strategy obtained only 83.5% on the same test set.

A normality and independence study was performed also for this problem, confirming the results found with the synthetic classes. Normality was acceptable but statistical independence among all module outputs cannot be assumed in general. In this case the best set of dichotomizers with absolute correlation below 0.8 included "u-i", "u-a", "o-e", "i-a" and "e-a". Fifty percent of modules were eliminated and the success rate over the test examples only decreased to 90.0%, less than a 5% degradation.

A multiclass network was trained to solve this problem for comparison with the modular approach. In the best case we obtained a success rate of 82.0% over the test set with 12 hidden units. Learning time was one order of magnitude higher.

CONCLUSIONS

This paper presents a modular multiclass pattern recognition architecture for efficient adaptation to changes in the set of categories of interest. Multiclass recognition is achieved combining the outputs of a group of neural class pairwise dichotomizers which are dynamically created and destroyed as new classes appear or obsolete classes are no longer required. The knowledge is distributed, minimizing re-learning costs.

Pairwise neural modularization allows, in principle, an analytic and optimum data combination strategy which requires no extra learning. Our experiments show that the requeriments for optimum combination are not strictly verified. However, normality deviations are small and statistical dependences can be easily managed providing a quasi-optimum fusion algorithm which has proved satisfactory for every practical purpose.

While this approach is adequate for recognition tasks with a moderated number of classes (there are L(L-1)/2 pairs of L classes), dependent module elimination usually keeps the number of active modules near to the number of classes.

The proposed modular architecture is an efficient solution for supervised recognition in applications with a dynamically changing universe of classes. Its distributed nature also improves the performance of comparable standard multiclass pattern recognizers in static recognition tasks.

BIBLIOGRAPHY

[1] A. Abidi, R. C. Gonzalez (eds.), *Data Fusion: In Robotics and Machine Intelligence*, Academic Press, 1992.

[2] F. J. Arcas Túnez. *Descomposición de Reconocedores de Patrones Multiclase*, Term Project, Facultad de Informática, Universidad de Murcia, Spain, December 1993 (in Spanish).

[3] W. G. Baxt, "Improving the Accuracy of an Artificial Neural Network Using Multiple Differently Trained Networks", *Neural Computation*, 4, 772-780 (1992)

[4] J. S. Denker, Y. leCun, "Transforming Neural-Net Output Levels to Probability Distributions" *Adv. Neur. Inf.* 3, Morgan Kaufmann 1991

[5] R. O. Duda, P. E. Hart, *Pattern Classification and Scene Analysis*, John Wiley & Sons, 1973

[6] K. Fukunaga, *Introduction to Statistical Pattern Recognition*, Academic Press, 1990

[7] J. Hertz, A. Krogh, R.G. Palmer, *Introduction to the Theory of Neural Computation*, Addison Wesley 1991

[8] T. K. Ho, J. J. Hull, S. N. Srihari, "Decision Combination in Multiple Classifier Systems", *IEEE T. on Pattern Analysis and Machine Intelligence*, V16 N1 Jan 1994, pp. 66-75

[9] K. Hornik, M. Stinchcombe, H. White, "Multilayer Feed Forward Networks are Universal Approximators", *Neural Networks* N2, 1989

[10] R. A. Jacobs "Task Decomposition Through Competition in a Modular Connectionist Architecture: The What and Where Vision Tasks", *COINS Tech. Rep.* 90-27, March 1990

[11] F. Kimura, M. Shridhar, "Handwritten numerical recognition based on multiple algorithms", *Pattern Recognition*, V24 N10 pp969-983, 1991

[12] S. Knerr, L. Personnaz, G. Dreyfus, "A New Approach To The Design Of Neural Networks Classifiers and its Applic. to the Aut. Recog. of Handwriten Digits", *Adv.Neur.Inf.* 3, Morgan Kauffmann, 1991

[13] N. Morgan, H. Bourlard, "Factoring Networks by a Statistical Method", *Neural Computation*, 4, 835-838 (1992)

[14] A. Papoulis, *Probability, Random Variables and Stochastic Processes* 3rd ed. McGraw-Hill, 1991

[15] L. I. Perlovsky, M. M. McManus, "Maximum Likelihood Neural Networks for Sensor Fusion and Adaptive Classification", *Neural Networks*, V4 pp89-102, 1991

[16] L. Xu, A. Krzyzak, C. Y. Suen, "Associative Switch For Combining Multiple Classifiers", *IEEE*, 1991

[17] L. Xu, A. Kryzak, C. Y. Suen, "Methods of Combining Multiple Classifiers and Their Applications to Handwritten Recognition", *IEEE T Systems, Man and Cybernetics*, V22 N3, 1992

[18] E. A. Wan, "Neural Networks Classification: A Bayesian Interpretation". *IEEE T. Neural Networks*, V1 N4 Dec. 1990

A Text Recognition System Based on a Neural Network and on a Deformed System

Javier Echanobe, José R. González de Mendívil, José R. Garitagoitia.

Department of Electricity and Electronics, University of the Basque Country,
P.O. Box 644, 48080 BILBAO, SPAIN; e.mail:webecarj@lg.ehu.es

Abstract: This paper shows a text recognition system based on a Neural Network which is used as Isolated Character Classifier (ICC), and on a Deformed System that incorporates the contextual knowledge defined by a dictionary. The Neural Network provides for every input character a fuzzy character built up with the ouput unit values. The fuzzy characters are the inputs for the Deformed System which is defined as an automaton representing the dictionary and whose behaviour is fuzzily constrained by fuzzy inputs. Therefore the classification and contextual processes are computed together. Experimental results show good performance for the system.

1. Introduction

The necessity of Contextual Postprocessing (CP) in a Text Recognition System, is justified by the fact that an Isolated Character Classifier (ICC) is able to recognize only up to a certain limit because of several error sources in texts [1]. Different methods for handling the CP can be classified as [2]: (i) Statistical methods, based on the *a priori* knowledge of transition probabilities between characters in words and the use of the *Bayes' Rule* [3]; (ii) Dictionary methods, based on the knowledge of the lexicographical constraints (in a dictionary form) of words in texts [4]; and (iii) Hybrid methods, based on the use of both Statistical and Dictionary methods [1][5]. All of these methods take as inputs the observed words which are provided by the ICC, as the result of the classification of a sequence of isolated letters; that is, $I(x_1)$... $I(x_m)$ being x_j a letter and $I(.)$ a function which, for each x_i, provides a unique character in an alphabet Σ. The classification of a letter depends on the proximity values (in terms of similarity) between a group of prototype letters and the input letter. Consequently, it is possible to take into account those proximity values (instead of a unique classification), and to build a fuzzy character for each input letter x_j: $\{(\text{prox}(x_j,\alpha),\alpha) \mid \alpha \in \Sigma \wedge 0 \leq \text{prox}(x_j,\alpha) \leq 1\}$, where $\text{prox}(.,.)$ represent such proximity values.

In order to use a sequence of fuzzy characters (from an input word) together with a dictionary which represents the lexicographical context, we propose the use of Deformed Systems [6] as it has been introduced by the authors in [7][8]. A first approach for a Text Recognition System with such a method was exposed in [7], and a formal study of Deformed Systems for the Text Recognition Problem has been recently presented in [8]. Furthermore, due to the flexibility of Deformed Systems, it is possible to propose several state equations, which constrain the behaviour of the system using different fuzzy composition strategies: (i)*max-min*; (ii)*max-prod*; (iii)*global average*; (iv)*square average*; (v)*time varying average*.

In this paper, we show a Text Recognition System formed by a Neural Network as Isolated Character Classifier and a Deformed System for Contextual Postprocessing. We also develop a text recognition experiment in order to test the performances of the system. The experiment allows to compare our method with one of the best Hybrid methods [1]. The rest of the paper is organized as follows: Section 2 introduces a Text Recognition System with two main components: An ICC and a CP. The CP method based on a Deformed System is formally presented in Section 3. Section 4 deals with experimental results. Finally, Section 5 is devoted to concluding remarks and future perspectives.

2. A Text Recognition System

This Section is devoted to present a simple formulation for a Text Recognition System which comprises two main coupled components: an ICC and a CP. In the following paragraphs, we introduce the preliminary notation for the problem.

Let Σ be a finite alphabet of characters; $\Sigma = \{\alpha_1, \alpha_2, ...\}$. A subset $D \subseteq \cup_{i:1..N} \Sigma^i$ is called a dictionary, being N the length of the largest dictionary word, where a dictionary word ω is an element of D. Dictionary words are not associated with descriptive information such as meanings or derivations. Define an abstract text, T, as a finite sequence of dictionary-words: $T = \langle .., \omega, .. \rangle$. A representation of an abstract text T given by a physical medium, is called a text and denoted X_T. Therefore, each text is a representation of an abstract text. Assume that each text is a sequence of words (representations of dictionary words): $X_T = \langle .., X_\omega, .. \rangle$, X_ω the representation of ω. By the previous assumption, the lengths of both a text and its corresponding abstract text are the same, $|X_T| = |T|$.

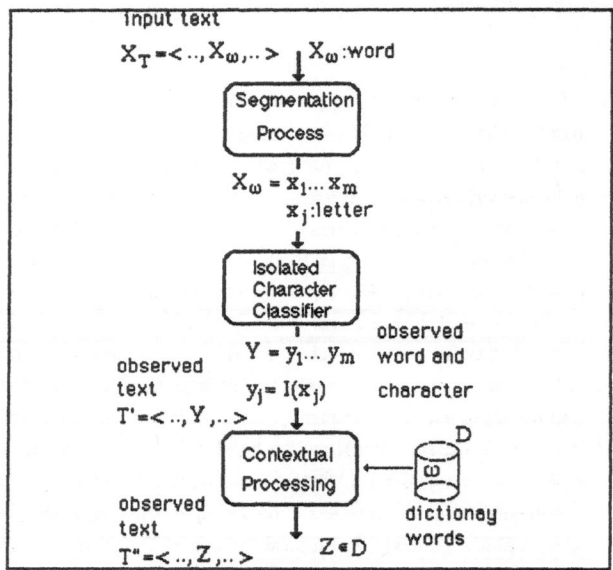

Figure 1. A Text Recognition System.

The goal of a Text Recognition System (figure 1) is to find an observed text for an input text; it can be represented by a function $S(X_T) = T'$. If $T = T'$ for each text X_T, the system is ideal. Let $T' = \langle .., Y, .. \rangle$ be the observed text provided by the system, where $Y \in \Sigma^*$. Assume that $S(.)$ is distributive, $S(X_\omega) = Y$ being Y the observed word provided by the system for the word X_ω. If $\omega = Y$ then the word is *well recognized* and if $\omega \neq Y$ the word is *erroneously recognized*. The performance of the system is defined by means of the Recognition_Rate $\equiv \dfrac{|T'/T|}{|T'|} \times 100$, where T'/T is the subsequence of T' of well recognized words. The error of the system is: Error_Rate $\equiv 100$ - (Recognition_Rate).

In the following, a process for making $S(.)$ is provided. Consider $S(.)$ to be composed by two subprocesses (figure 1): a Segmentation process and an ICC.

Let $\omega = \alpha_1...\alpha_n$ be a dictionary word and X_ω its word in the text. The goal of a Segmentation process is to divide the word X_ω in a sequence of letters (representations of characters), that is, $X_\omega = x_1...x_m$. In this formulation, assume that the Segmentation process preserves the length, so that m = n (no

Segmentation errors are considered). Each one of those letters will be an input for the ICC. The ICC is the process that provides for each letter a unique character in the alphabet. It can be represented as a function $I(x_j) = y_j \in \Sigma$. Therefore, for each word $X_\omega = x_1...x_m$ processed by the ICC, an observed word $Y = y_1 ...y_m$ is obtained. Note that $\omega \neq Y$ would be due to the fact that one or more letters had been badly classified by the ICC; in such a case, change errors have been produced (the present method only deals with change errors due to the ICC).

In order to reduce the number of erroneous recognized words, many Text Recognition Systems use the fact that only certain letter combinations are allowed in the words. This restriction is called the lexicographical context. The CP based on a dictionary is a process that receives as input the observed word provided by the ICC, $Y = y_1...y_m$, and generates the dictionary word $Z \in D$ which has the minimum distance (in some sense) to Y (figure 1). The improvement of the recognition capability due to the incorporation of the lexicographical context, is expressed by comparing the system with and without an CP. Thus, the Error_Reduction_Rate is defined as: $100 - \left(\dfrac{\text{Error\_ Rate(ICC + CP)}}{\text{Error\_ Rate(ICC)}} \times 100 \right)$.

3. Deformed Systems for CP

In order to introduce our method, our attention will be focussed on how the ICC takes a decision when an input letter is processed by it. Many of the existing ICCs (Nearest Neighbour Classifiers [9], Bayesian Classifiers [10], Neural Network based Classifiers [11], etc.), compute the *proximity* between the letter x with each one of the prototype letters representing the characters in Σ. Thus, a collection of pairs $\{ (\text{prox}(x, \alpha), \alpha), \alpha \in \Sigma \}$ is obtained. Then, the ICC executes a decision function and provides as observed character, the character with the maximum proximity to the letter: $I(x)=y$, being $y \in \Sigma / \forall \alpha \in \Sigma$: $\text{prox}(x, y) \geq \text{prox}(x, \alpha)$.

The CP in the Dictionary-based methods is clearly separated from the ICC and starts from the observed word given by it. However, we propose a method that uses all information available from the ICC (the collection of pairs for each input letter) and processes it together with the dictionary. Thus, the decision function is omitted and the choice of the observed character is postponed until the context is taken into account. Therefore, a special class of systems is needed in order to handle, not with sequences of single characters, but with sequences of collection of pairs together with the dictionary. To achieve this goal, we propose the use of a Deformed System [6] which is defined as an automaton whose behaviour is fuzzily constrained by fuzzy inputs. In our case, the automaton represents the dictionary and the fuzzy inputs are the collection of pairs. In the following paragraphs, such concepts are formalized.

Let x be an input letter and $\{ (\text{prox}(x, \alpha), \alpha), \alpha \in \Sigma \}$ its associated collection of pairs provided by the ICC. By making $\text{prox}(x, \alpha) \in [0, 1]$, a fuzzy set can be built in the universe Σ as $\tilde{y} = \{ (\mu_{\tilde{y}}(\alpha), \alpha), \alpha \in \Sigma \}$ where $\mu_{\tilde{y}}(\alpha) = \text{prox}(x, \alpha)$. The fuzzy sets, $\tilde{y}$, are called fuzzy characters. Therefore, for each word $X_\omega = x_1...x_m$, a sequence of fuzzy characters $\tilde{y}_1...\tilde{y}_m$ is obtained

A way for implementing a dictionary is by means of a State Automaton. As the dictionary D is finite, then the proposed automaton is finite and deterministic [12]. Let $A \equiv (\Sigma, Q, q_0, \delta, \lambda, \Delta)$ be a Moore finite deterministic automaton which accepts the dictionary D; the elements of the automaton are defined as: (i) Σ the alphabet; (ii) Q the set of states; (iii) $q_0 \in Q$ the initial state; (iv) δ the transition function defined as δ: Q $\times \Sigma \times Q \to \{0, 1\}$, being $\delta(q, \alpha, q') = 1$ if there exits the transition from q to q' by the character α, and $\delta(q, \alpha, q') = 0$ if there does not exits; (iv) Δ the output alphabet, where $\Delta = D \cup \{\varepsilon\}$ (ε denotes the empty string); (v) λ the output function defined as λ: Q $\to \Delta$. Given a particular dictionary D, the elements above presented can be calculated as follows: Let Q be a finite set of states; initially $\forall q, q' \in Q$: $\delta(q, \alpha, q') = 0$ and $\lambda(q) = \varepsilon$. Given a dictionary word $\omega = \alpha_1..\alpha_t..\alpha_m$, we define via construction a sequence of states

$q_0..q_t..q_m$ being q_0 the initial state, $q_t \in Q$ with $1 \le t \le m$, and update $\delta(q_{t-1}, \alpha_t, q_t) = 1$ with $1 \le t \le m$. Finally, $\lambda(q_m) = \omega$. The method is constrained for obtaining a deterministic automaton.

When the inputs are sequences of fuzzy characters, it is possible to handle the finite deterministic automaton above define by using the fuzzy interpretation. The Deformed System [6] for the automaton A is defined as the tuple $Ad \approx ((\Sigma, \tilde{\mathfrak{F}}), (Q,S), q_0, \delta, \lambda, (\Delta,W))$ where Σ, Q and Δ are fuzzily constrained by the fuzzy sets $\tilde{\mathfrak{F}}$, S and W respectively; δ is the transition function whose domain is fuzzily constrained by $S \times \tilde{\mathfrak{F}} \times S$; λ is the output function whose domain is fuzzily constrained by S and its range is restricted by W. Consider the generic fuzzy sets $\tilde{\mathfrak{F}}$, S and W defined as $\tilde{\mathfrak{F}} = \{(\mu_{\tilde{\mathfrak{F}}}(\alpha), \alpha), \alpha \in \Sigma\}$; $S = \{(\mu_S(q), q), q \in Q\}$; $W = \{(\mu_W(\omega), \omega), \omega \in \Delta\}$ with universes Σ, Q and Δ respectively. The state equations of the Deformed System are the following:

Let $(S_t, \tilde{\mathfrak{F}}_t, S_{t+1})$ be the t-th step of the system (once in the equations $\mu_{S_t}(q_t)$ is simplified by $\mu_S(q_t)$);

(1) Given $\delta(q_t, \alpha_t, q_{t+1})$ then $\mu_S(q_{t+1}) \ge \min(\mu_S(q_t), \mu_{\tilde{\mathfrak{F}}}(\alpha_t), \delta(q_t, \alpha_t, q_{t+1}))$.

(2) Given $\omega = \lambda(q_t)$ then $\mu_W(\omega) \ge \mu_S(q_t)$.

The computation of the fuzzy set W which is the output of the Deformed System, is achieved simultaneously from the equations (1) and (2). Therefore, once the end of the sequence of fuzzy characters for an input word is reached, the observed word provided by the Deformed System will be the dictionary word with the maximum membership function value in the fuzzy set W: $Z \in D / \forall \omega \in D$: $\mu_W(Z) \ge \mu_W(\omega)$. However, the state equation (1) can be very restrictive due to the fact that characters whose values of $\mu_{\tilde{\mathfrak{F}}}(\alpha_t)$ are small, dominate in the computation of the set W. In order to overcome that restriction it is possible to modify the equation (1) in such a way that it accepts other composition functions [13]. This is possible due to the flexibility of the system. In the following, the composition functions which will be used to evaluate the Deformed System, as CP method in a problem of change error recuperation of printed texts, are formulated.

(I) Max-Min. $\mu_S(q_{t+1}) \ge \min(\mu_S(q_t), \mu_{\tilde{\mathfrak{F}}}(\alpha_t), \delta(q_t, \alpha_t, q_{t+1}))$

(II) Max-Prod. $\mu_S(q_{t+1}) \ge \mu_S(q_t) \cdot \mu_{\tilde{\mathfrak{F}}}(\alpha_t) \cdot \delta(q_t, \alpha_t, q_{t+1})$

(III) Global average. $\mu_S(q_{t+1}) \ge \dfrac{\mu_S(q_t) \cdot t + \mu_{\tilde{y}}(\alpha_t)}{t+1} \cdot \delta(q_t, \alpha_t, q_{t+1})$

(IV) Square average. $\mu_S(q_{t+1}) \ge \dfrac{\mu_S(q_t) \cdot t + (\mu_{\tilde{y}}(\alpha_t))^2}{t+1} \cdot \delta(q_t, \alpha_t, q_{t+1})$

(V) Time varying local average. $\mu_S(q_{t+1}) \ge \tau \cdot (\mu_S(q_t) + \gamma \cdot \mu_{\tilde{y}}(\alpha_t)) \cdot \delta(q_t, \sigma_t, q_{t+1})$,

where $0 \le \tau \le 1$ and $\gamma = 1/\sum\limits_{n=0}^{\infty} \tau^n$

4. Experimental Results

In this Section, we present an experiment where the goal is to reduce the number of change errors introduced by an ICC in the recognition of printed texts. This reduction is performed by the incorporation of the context through a Deformed System. Furthermore, the different composition functions for modelling the state transitions which have been proposed in the previous Section, are evaluated and compared.

In the experiment, the ICC is a Neural Network: a Perceptron with two layers. The output layer has 26 units and each one is in correspondence with a character in the alphabet. The Neural Network was previously trained with 26 prototype letters representing the characters in the alphabet. When an input letter is processed by the Neural Network, the value of an output unit (in the real interval [0, 1]) represents the proximity between the input letter and the character associated with that unit. According to this fact, the Neural Network provides as observed character, the character associated with the largest value output unit. However, instead of taking such an output, we build a fuzzy character with all the output unit values as it has been explained in the previous Section; that is, for an input letter x, the Neural Network produces a fuzzy set $\tilde{y} = \{(\mu_{\tilde{y}}(\alpha), \alpha), \alpha \in \Sigma\}$ where Σ is the alphabet and $\mu_{\tilde{y}}(\alpha)$ is the value of the output unit associated with the character α.

Note that a change error is produced by the ICC when for an input letter x which is a representation of the character α, the value $\mu_{\tilde{y}}(\alpha)$ is not the largest membership value in the fuzzy character $\tilde{y}$.

For this experiment, we have selected computational domain texts that contain 6396 words, and a dictionary with 1700 words. The results obtained for the texts are shown in Tables I to V where each Table is obtained with a different composition function for the Deformed System. In the Tables:

(A) denotes the Error_Rate produced by the Neural-Network (only 3 values have been selected: 32.88%, 42.00% and 50.39%).

(B) denotes the Error_Reduction_Rate achieved by the Deformed System.

(C) denotes the Recognition_Rate obtained with the complete system (ICC + Deformed System). Note how the best results are achieved with the max-prod composition (Table II) and the worst results with the max-min composition (Table I).

(A)	32.88%	42.00%	50.39%
(B)	80.18%	81.33%	82.09%
(C)	93.48%	92.15%	90.97%

Table I. Results with the Max-min.

(A)	32.88%	42.00%	50.39%
(B)	95.55%	95.36%	95.68%
(C)	98.54%	98.05%	97.76%

Table II. Results with the max-prod.

(A)	32.88%	42.00%	50.39%
(B)	97.26%	97.48%	95.40%
(C)	99.10%	98.94%	97.93%

Table III. Results with the Global average.

(A)	32.88%	42.00%	50.39%
(B)	95.53%	95.67%	95.55%
(C)	98.53%	98.18%	97.75%

Table IV. Results with the square average.

(A)	32.88%	42.00%	50.39%
(B)	91.98%	92.60%	92.68%
(C)	97.36%	96.89%	96.31%

Table V. Results with the Time varying average.

In the Hull's experiment [1], the texts were of 6372 words and the dictionary of 1724 words. The best result by using his Hybrid method was: Error_Reduction_Rate 87% when the Error_Rate of the ICC was of 31% (81% with single change error, 16% with double and 3% with triple).

In our experiment with similar sizes of texts and dictionary, and also with a similar change error distribution, a 97% of Error_Reduction_Rate was obtained in the case of 32% of Error_Rate (Table II).

5. Conclusions

A Text Recognition System with a Neural Network as Isolated Character Classifier and with a Deformed System for Contextual Postprocessing, has been introduced. Different composition functions for modelling the state transitions in the system have been formulated. The obtained results in the experiment for printed text recognition, show good performances of the method. A comparison with an Hybrid method has been presented, and the results have shown an improvement of a 10% in the Error_Reduction_Rate for a problem of change-error recuperation.

Due to the flexibility of the Deformed System, it is possible to develop another composition functions for particular problems. Furthermore, it is possible in a simple way to join the method with syntactic context based on grammars. A future perspective is the use of the proposed method to deal with the problem of insert and delete errors in printed texts.

Acknowledgements

The authors wish to thank the Basque Government for providing the financial support of this work.

References

[1] J. J. Hull, S.N. Srihari and R. Choudhari, "An Integrated Algorithm for Text Recognition: Comparison with a Cascaded Algorithm", *IEEE Trans. Pattern Analysis Mach. Intell.*, vol. PAMI-5, no. 4, pp. 384-395, 1983.

[2] D. G. Elliman and I.T. Lancaster, "A Review of Segmentation and Contextual Analysis Techniques for Text Recognition", *Pattern Recognition*, vol. 23, no. 3/4, pp. 337-346, 1990.

[3] R. Shinghal and G.T. Toussaint, "Experiments in Text Recognition with the Modified Viterbi Algorithm", *IEEE Trans. Pattern Analysis Mach. Intell.*, vol. PAMI-1, no. 2, pp. 184-193, 1979.

[4] G.M. Landau, "Fast string matching with k differences", *J. Comput. Syst. Sci.*, vol. 37, pp. 63-78, 1988.

[5] R. Shinghal, "A Hybrid Algorithm for Contextual Text Recognition", *Pattern Recognition*, vol. 16, no. 2, pp. 184-193, 1983.

[6] C.V. Negoita, D.A. Ralescu, *Application of Fuzzy Sets to System Analysis*, Birkaeuser, Basilea 1975.

[7] R. Reina, José R. González de Mendívil, José R. Garitagoitia, "Improved Character Recognition System based on a Neural Network incorporating the context via Fuzzy Automata" *2nd International Conference on Fuzzy Logic & Neural Networks*, Izuka (Japan), July 17-22, 1992.

[8] J. Echanove, R. Reina, J.R. Garitagoitia and J.R. González de Mendívil, "Deformed Systems in Text Recognition", *International Conference On Artificial Neural Networks.*, Sorrento (Italy), May 26-29, 1994.

[9] T.M. Cover, P.E.Hart, "Nearest Neighbor Pattern Classification", *IEEE Transactions on Information Theory*, IT-13, January 1967, 21-27.

[10] R. O. Duda, P.E. Hart, "Pattern Classification and Scene Analysis", Addison-Wesley, New York, 1973.

[11] K.Fukushima, S. Miyake, T. Ito, "Neocognitron: A Neural network for a Mechanism of Visual Pattern Recognition" *IEEE Trans. Sys. Man and Cyber.*, SMC-13, no. 5, pp. 826-834, 1983.

[12] J. Hopcroft, J. Ullman, *Introduction to Automata Theory, Languages and Computation*, Addison-Wesley Publishing Company, Reading Massachusetts, 1979.

[13] E. Vidal, F. Casacuberta, E. Sanchís, J. M. Benedí, "A General Fuzzy-Parsing Scheme for Speech Recognition", *NATO ASI series*, vol. F16, pp. 427-446, 1985.

Simultaneous Recognition of Multiple Objects Using the MEM Model

Soheil Shams
Hughes Research Laboratories
3011 Malibu Canyon Rd.
Malibu, CA 90265
shams@maxwell.hrl.hac.com

Abstract

In this paper, a new self-organizing neural network model, called the Multiple Elastic Modules (MEM), is described. The model is applied to the task of invariant object recognition through labeled graph matching, where topologically ordered feature labels of an object (e.g., oriented edges, textures, colors, etc.) are matched to the features present in the input scene. The matching process is accomplished through the use of elastic graph modules, with nodes encoding feature labels and edges encoding expected spatial arrangement of the features. The elastic modules stochastically searches the entire input scene for a correspondingly similar arrangement of features, while being tolerant to small deformations in the match. The key attribute of the MEM model is the use of independent and adaptive receptive fields for each neuron. The receptive field dynamics of this model implement a new type of adaptive annealing mechanism which avoids spurious local minima. This feature also allows for efficient simultaneous search for multiple objects using multiple elastic modules. The performance of this model is demonstrated through simulations on real data.

I.   Introduction

In this paper I will describe a new neural network model, called the Multiple Elastic Modules (MEM), which successfully addresses many of the issues faced by current neural approaches to object recognition. Namely, the MEM model utilizes a dynamically varying focus-of-attention mechanism to search the entire input scene. This approach couples the segmentation and the recognition operations into one coherent process. It also couples the top-down expectation-driven processing with the bottom-up stimulus-driven operations. Both of these mechanisms are conjectured to be utilized by biological visual systems [1]. The MEM model should be contrasted with purely feed-forward models, such as the multi-layer perceptron [17] and the neocognitron [6], in that the synaptic weight values are not learned through a process of repetitive learning on examples. Rather, the MEM model uses an abstract representation, an elastic graph, to encode various features of the object along with the specific topological arrangement of these features. The MEM model is a rapid optimization method which associates previously stored elastic modules to the pattern of activity in the visual field. This optimization process is done collectively by all the neurons in the network, similar to processing performed by recurrent neural network architectures such as the Hopfield net [9], elastic net [5], Kohonen's SOM [12], dynamic link architecture [22], and the like.

The principle attributes of the MEM model is the use of locally controlled focus-of-attention and related variables which enable the model to rapidly search a large field of view for recognition of any of the stored objects in memory. The model also offers a specific mechanism for the system dynamics to "realize" when it has recognized a specific object. This information can be combined with other data, such as recognition of other objects, to form a more complete picture of the scene. If this processing is performed in a hierarchical fashion, it allows for the direct implementation of "schemas" as described by Arbib [1], where a number of sub-schemas are combined to instantiate a more complex schema. In Section 2, I will describe the basic graph matching approach to object recognition and define some of the nomenclature associated with the memory modules. In Section 3, I will present the MEM model and discuss its biological analogies as well as its relation to similar neural models. Simulation results are given in Section 4, followed by concluding remarks in Section 5.

II.   Object Recognition through labeled graph matching

Labeled graphs can be an efficient and robust mechanism for object representation [23]. An object can be defined by a set of features (node labels) and the specific topological arrangement of these features (links of the graph), see Figure 1. In the MEM model, described in detail in the next section, one neuron is associated with each node of the module graph. A node of the module graph is denoted by a neuron i which is sensitive to feature type f_i. The links of the graph are represented by g_{ij}, where $g_{ij} = 1$ if neurons i and j are connected, otherwise $g_{ij} = 0$. The set L_i is defined to represent all the neurons connected to neuron i. In addition to the simple connectedness relation given by g_{ij}, the MEM model also utilizes information of the expected relative arrangement of features. This information can be a valuable hint to limit the search space of possible matches [14]. Although the expected arrangement of features can be defined to cover any space (e.g., spatial, temporal, color, etc.), in this paper we limit our treatment only to spatial relationships. The same principles can be used to extend the application to other dimensions. The expected spatial arrangement between neuron i and j is represented by the vector $\bar{\delta}_{ij}$, with $j \in L_i$. It is assumed that these values are stored in the links between the neurons.

Object   Labeled Graph Representation

Figure 1 - A triangular object with a specific texture can be represented by three oriented edge cells and a texture sensitive cell. The expected spatial organization of the features are encoded in the interconnection links between the cells.

In addition to restricting the search space of the graph matching problem, the use of a separate connectedness g_{ij} and expected spatial arrangement $\bar{\delta}_{ij}$ information allows for simple feature (or sensor) data integration. For example, a "green vertical line" can be represented by two neurons i and j, where f_i="vertical edge" and f_j="green", with $g_{ij}=1$ and $\bar{\delta}_{ij}=0$. Of course, other type of sensory input (such as motion, depth, etc.) can be fused in a similar manner.

In a visual pattern recognition task, the object is to associate each of the nodes i in the module graph, having feature labels f_i, to a particular point in the input image exhibiting a similar feature. Additionally, this matching should be performed such that the expected arrangement of the features, as specified by $\bar{\delta}_{ij}$, is satisfied as well as possible. In other words, if neuron i is bound to a similarly labeled feature in the input at spatial location $\bar{x}$, we should expect to find a feature f_j, at location $\bar{x}+\bar{\delta}_{ij}$, if neuron i is connected to neuron j ($g_{ij}=1$). Finding a match is therefore equivalent to solving a constrained optimization problem. The MEM algorithm, described in the following section, is an efficient method for solving this problem.

III. Multiple Elastic Modules Model

The MEM model consists of several layers of neurons, see Figure 2. Sensory information flows from the input towards higher levels of abstractions (starting from simple features such as oriented edges, to simple shapes, objects, and so on). The first layer L0 is analogous to the retina which is used to represent the input scene. The next layer, L1 is analogous to the Lateral Geniculate Nucleus (LGN) and the simple cells of area V1 of the visual cortex [10, 11]. The neurons in layer L1 have inputs from layer L0 and are used to detect specifically oriented edges at specific spatial location in the input scene. In other words, neurons in L1 have fixed receptive fields projected onto layer L0. Neurons in layer L1 are topologically ordered so neighboring neurons in L1 project to neighboring points in the input image. The final layer in the information flow direction is L2. The neurons in this layer correspond to the nodes of the module graph as described in Section 2. Each neuron in layer L2 is sensitive to a particular feature at any spatial location in the input scene, analogous to the complex cells of area V1 in the visual cortex. Therefore, the neurons in L2 are invariant to spatial translations of features in the input image and can thus be used for translation invariant object recognition.

The task of translation invariant object recognition in this architecture will be to bind each neuron i of layer L2, with feature f_i, to one of the many neurons of layer L1 with similar labels, while preserving the spatial arrangement of features as specified by $\bar{\delta}_{ij}$s. A similar approach for invariant object recognition has been previously proposed by von der Malsburg [22, 24] in which fast synaptic modulation, implemented through temporal correlation of firing patterns, is used to implement the binding mechanism. The characteristic difference between the Dynamic Link Architecture (DLA) [22] and the MEM model is the use of explicit and adaptive receptive fields for neurons in the L2 layer, along with the use of explicit enforcement of a priori spatial arrangement information encoded in the $\bar{\delta}_{ij}$ links. The special dynamics of the receptive fields, along with their auxiliary variables, strongly enforce a number of constraints on the matching criteria [19] and enable the system to escape shallow local minima. In addition, specific constructs of the MEM model allow for assemblage of higher level concepts (such as schemas [1]) from low level concepts. The details of the receptive field dynamics, will be discussed later in this section.

A different approach for implementing fast synaptic modulation for invariant object recognition has recently been proposed using special "control" neurons to modulate the synaptic connection between a pair of neurons [16]. In the model of Olshausen et. al. [16], the dynamic links are used purely for the purpose of visual attention and scale invariance. This model requires a separate associative memory module, such as the Hopfield net [8], for recognition of the object in the attention window. In effect, this system separates the segmentation process from the recognition process. The Olshausen et. al. model does not address the

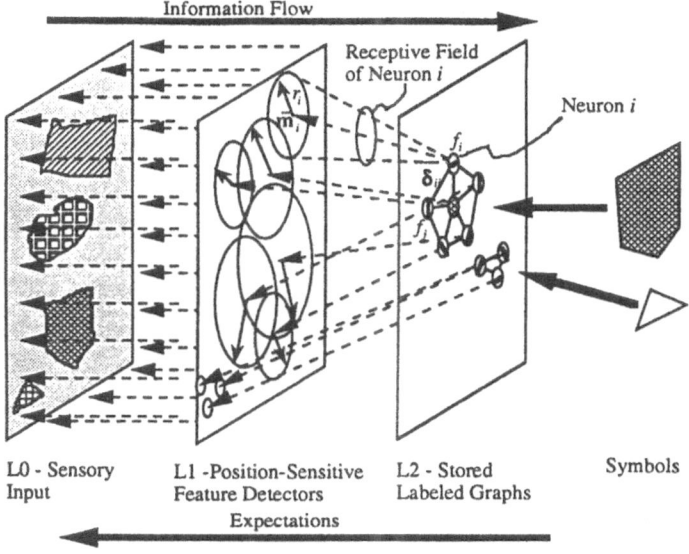

Figure 2 - The architectural organization of the MEM model. Sensory information is propagated bottom-up (from L0 to L2) while expectations, as defined by the receptive field location and size, are propagated top-down. An object in the "symbols" memory is recognized when its corresponding L2 neurons' receptive fields are aligned with similarly labeled features in the L1 layer (such as the small triangle at the lower-left corner of the input image).

figure ground separation problem and will have difficulty recognizing overlapping or deformed objects. In addition, that model is inherently sequential in recognition of objects. That is, the model successively moves the focus of attention, from one object to the next. In contrast, the MEM model performs the segmentation and recognition processing concurrently. Therefore, the MEM model can readily deal with overlapping and occluded objects. In addition, the L2 neurons of the MEM model collectively determine the focus-of-attention of a module graph. Thus, multiple objects can be concurrently recognized. This feature is particularly important as it aids in the recognition of overlapping objects. Due to the interaction between various competing elastic modules (or hypothesis of objects), the features associated with two overlapping objects can be "claimed" by their corresponding module graphs, and thus reduce the search space for each module. This fact leads to an experimentally verifiable predication that it is easier to recognize a familiar object if it is occluded by another familiar object than when it is occluded by an unfamiliar object (e.g., noise).

The fundamental principle of the MEM model is self-organization. It has been previously shown that simple self-organizing models, such as the elastic net [5] and the Kohonen SOM [12] models, can be used to describe certain neural organizations, such as retinotopic mapping, ocular dominance stripe patterns, and cortical orientation maps [7, 20, 21]. The common element in all of these models involves the preservation of order in transformation from one space to the next. For example, neighboring points in the visual field-of-view are mapped to neighboring neurons of the visual cortex. Similarly, neighboring points in orientation space are mapped to neighboring neurons in the columnar structure of the visual cortex. In the MEM model, a similar self-organizing mechanism is used, not for organization, but for detection of topology preserving transformations from the visual world space to the abstract object or schema space.

As stated earlier, the MEM Model associates a neuron from layer L2 to one of the similarly labeled neurons in L1. This association is accomplished by directing the focus-of-attention of the L2 neuron to the area most likely for a good match and continuing to refine the focus-of-attention. The focus-of-attention of an L2 neuron i is represented by a receptive field projected onto layer L1 (see Figure 2) with a center at spatial coordinate $\bar{m}_i$ and radius r_i. The optimization process for finding a good match between a module graph in symbol memory and the input image is done through an iterative constrained stochastic search outlined as:

1. Randomly select an L1 neuron at spatial location $\bar{x}$ with feature label $\mathcal{F}(\bar{x})$.

2. Select the "closest" L2 neuron i^* to the randomly selected neuron of layer L1, whose receptive field covers location $\bar{x}$. The closeness measure is designed to combine both spatial proximity (distance between the receptive field center of all neurons $\bar{m}_j$s to location $\bar{x}$) and feature label similarity (defined by a function $R(\mathcal{F}(\bar{x}), f_j)$ with small R representing high degree of similarity). In the simulations I describe in the next section, the following formula was used:

$$i^* = \min_{j \in A}\{|\bar{m}_j - \bar{x}|R(\mathcal{F}(\bar{x}), f_j)\}, \qquad \text{where} \qquad (1)$$

set A includes all neuron in L2 which have receptive fields covering location $\bar{x}$.

3. Adjust the receptive field center of the winning L2 neuron, $\bar{m}_i^*$, to be closer to the selected location $\bar{x}$ according to

$$\bar{m}_i^* \leftarrow \bar{m}_i^* + \alpha[\bar{x} - \bar{m}_i^*], \qquad \text{where} \qquad (2)$$

α is the update rate of the winning neuron. In the MEM model, the update rate is scaled proportional to the similarity of the feature labels between the winning neuron and the selected L1 neuron. The closer the similarity, the higher the update rate. This is accomplished formally by having $\alpha = \alpha_*[1 - R(\mathcal{F}(\bar{x}), f_i^*)]$, where $\alpha_* < 1$ is the maximum update rate. In the simulations I describe in the next section, a relatively large $\alpha_* = 0.8$ value was used to achieve rapid recognition.

4. Based on the self-organizing principle described earlier, all the L2 neurons connected to the winning neuron ($j \in L_{i^*}$) are adjusted such that the expected spatial arrangement of features, as defined by $\bar{\delta}_{ij}$s, is preserved. Formally,

$$\bar{m}_j \leftarrow \bar{m}_j + \alpha_j'[\bar{m}_i^* + \bar{\delta}_{ij} - \bar{m}_j]. \qquad (3)$$

This updating mechanism, in effect, "pulls" all the neurons connected to the winning neuron i^* closer to their expected position with respect to i^*. The extent of the pulling force is uniquely determined for each neuron j according to the updating parameter α_j'. In other elastic graph matching models [13, 25], a constant updating parameter is used to define the link elasticity. In the MEM model, however, the link elasticity is a direct non-linear function of the local deformation of the module. The local deformation around neuron i is calculated as

$$p_i = \frac{\sum_{j \in L_i}|(\bar{m}_i - \bar{m}_j) - \bar{\delta}_{ij}|}{|L_i|}. \qquad (4)$$

The effect of this updating scheme is that, the more a neuron receptive field diverges from its expected location, as defined by its connected neighbors, the larger the retracting force to pull the neuron back into its expected position. The updating step of equation (4) is repeated consecutively, in a breadth-first manner, to propagate the updating to all neurons of the module graph.

5. The receptive field size of all neurons in L2, whose center locations $\bar{m}_i$s were updated, are adjusted according to

$$r_i = \lambda p_i h_i + e_i + \varepsilon, \qquad \text{where} \qquad (5)$$

λ is a scaling factor, p_i is the measure of local deformation given by eq. (4), h_i is the level of binding or locking, e_i is the expectation value, and ε is the minimum size of the receptive field. The locking value has a range $0 \le h_i \le 1$. The dynamics of the locking value are designed to shrink the receptive field size when a similarly labeled feature is found in the receptive field of neuron i and the receptive field center $\bar{m}_i$ is updated only slightly at each iteration ($d\bar{m}_i/dt$ is small). The expectation value e_i is used to enlarge the receptive field size if a similarly labeled feature is not found in the receptive field of neuron i. The dynamics of the locking and expectation variables are coupled to the dynamics of the receptive field centers and are therefore data-dependent. The details of these dynamics are given elsewhere [18].

6. Go to step 1.

The dynamics of the receptive field centers, as defined by eqs. (1) and (2), are designed to spatially align the L2 neurons' receptive fields, associated with a single object, to satisfy the expected spatial arrangements imposed by $\bar{\delta}_{ij}$s. The dynamics of the receptive field size r_i, as defined by eq. (5), are designed to focus attention more tightly (reduce r_i) when receptive field center $\bar{m}_i$ is located in the desired spatial position with respect to its neighbors (small p_i) and a similarly labeled feature has remained in the receptive field region for some length of time (small h_i). An object is recognized at a specific spatial location when the majority of its constituent L2 neurons have small locking values (very fine focus on the object). This unique ability of the MEM model to "know" when it has recognized the object and also to what degree (value of locking parameter) it is certain of the presence of the specific parts of the object allows for construction of abstract objects from their constituent parts. The details of this mechanism are deferred to a later publication.

IV.   Simulation Results

A cluttered scene with multiple partially occluded objects was captured using a video camera in order to apply the MEM model to a real-world object recognition task. The input scene contained an image of a toy tank and a partially-occluded image of a toy jeep, among a number of other objects, see Figure 3c. A rough line drawing of the two toy objects where drawn using a Computer Aided Design (CAD) package. A simple heuristic algorithm was used to construct the associated module graphs using line orientations as node feature labels and spatial proximity of points for generating the node interconnection patterns. Two types of nodes were used in these simulations. The first corresponds to coarse-grain features (longer lines), indicated in Figure 3a and 3b by larger asterisks, and the second correspond to fine-grain features (shorter lines and subcomponents of larger lines), which are indicated by smaller asterisks. Spatially-variant feature detecting cells of layer L1 were implemented through the use of oriented gabor filters having small constant receptive fields [2]. Gabor filters of two different scales where used to detect the presence of coarse- and fine-grain features. The receptive field center locations $\bar{m}_i$s, for all neurons in L2, (associated with both jeep

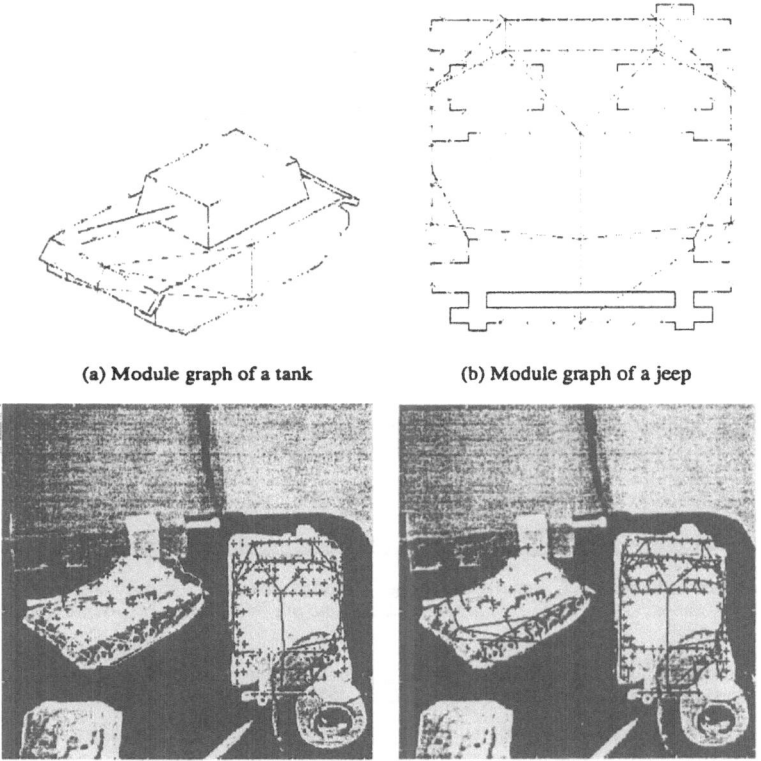

(a) Module graph of a tank   (b) Module graph of a jeep

(c) State of the network after 1000 iterations (d) State of the network after 4000 iterations

Figure 3 - Simultaneous recognition of multiple objects in a cluttered scene. a) Module graph of a tank generated from a CAD line drawing. b) Module graph of a jeep generated from a CAD line drawing. c)

and tank modules) were initialized to the center of the image. Shortly after the start of the self-organization process (after approximately 100 iterations) the receptive fields of both module graphs self-organize into a "jeep-like" and "tank-like" shape by the elastic force imposed through the δ_{ij}s constraints. After approximately 1000 iterations, the receptive field centers, and therefore the focus-of-attention of each module, is localized to the correct region of the image. The network will then proceed to fine-tune. The locking value associated with each neuron indicates the degree of certainty in correct binding between the L2 neuron and the L1 neuron at the center of its receptive field. The network can thus "know" to what degree it has recognized an object. The "recognition confidence level" is depicted in Figure 3 by dark interconnection lines for each neuron having $h_i < 0.1$. In these simulations, the tank module takes longer to be recognized since it considerably differs in proportion to the stored module graph. When the same module graphs are used in scenes without any jeep-like or tank-like object, the network does not converge.

After 1000 iterations, the receptive field centers (depicted by small "+" symbols) have self-organized to corresponding points in the image. The dark interconnection lines indicate a locking value of less than 0.1.
d) State of the network after 4000 iterations

V. Conclusion

In this paper I described a new self-organizing neural network model for concurrent recognition of multiple objects. Although a multitude of other neural network models (e.g., neocognitron [6] and Hopfield net [15]) have been used for visual object recognition, the MEM model is distinct in that it combines various aspects of visual processing, namely segmentation, focus-of-attention, and recognition, to arrive at a system capable of concurrently and cooperatively recognizing multiple objects. Furthermore, the MEM model offers a precise mechanism, coupled in the recognition and focus-of-attention generation, for deciding when and what part of an object is recognized. Examining the MEM model from a statistical optimization framework, the use of independent receptive field sizes for the L2 neurons implements a non-uniform temperature field dependent on the optimality of the local match [4, 19]. This unique feature of the MEM model enables the network to automatically increase the local temperature value in the areas of poor match and escape from undesired local minima.

The mechanism for achieving self-organization in the MEM model, as described in this paper, is similar in spirit to the Kohonen SOM model [12]. However, the basic principles of the MEM model can be applied to other, more biologically plausible self-organizing methods. In fact, research currently in progress shows great promise in this direction. In addition, the MEM model can be used to interpret the experimental data obtained from the lateral intraparietal area (LIP) of alert monkeys in which receptive field of neurons are adjusted to keep retinal space aligned with visual space [3]. The model described in this paper, lays down the fundamental principles of the MEM approach to invariant object recognition. I have demonstrated its use in successfully recognizing multiple objects in a cluttered scene while being invariant to translations and minor distortions in the image plane. The extensions of the MEM model to include scale and orientation invariance can be simply added by enforcement of self-organizing constraints in scale and orientation spaces, respectively. Preliminary results have successfully demonstrated the feasibility of simultaneous search in spatial and scale space. Future work will include the addition of orientation invariance and formulation of hierarchical structure in the L2 layer to implement abstract schemas [1].

Acknowledgments: I would like to thank Patricia Keaton for developing the gabor filter code used in the feature extraction process and Ladan Shams and Bernard H. Soffer for many stimulating discussions and review of this paper.

References:

[1] M. A. Arbib, The Metaphorical Brain 2. Neural Networks and Beyond. John Wiley & Sons, 1989.

[2] J. Buhmann, J. Lange and C. von der Malsburg, "Distortion Invariant Object Recognition by Matching Hierarchically Labeled Graphs," *Proc. of the Inter. Joint Conf. on Neural Networks*, Washington D.C., Vol. 1, pp. 155-159, 1989.

[3] J.-R. Duhamel, C. L. Colby and M. E. Goldberg, "The Updating of the Representation of Visual Space in Parietal Cortex by Intended Eye Movements." *Science.* 255: 90-92, 1992.

[4] R. Durbin, R. Szeliski and A. Yuille, "An Analysis of the Elastic Net Approach to the Traveling Salesman Problem." *Neural Computation.* 1: 348-358, 1989.

[5] R. Durbin and D. Willshaw, "An Analogue Approach to the Traveling Salesman Problem Using an Elastic Net Method." *Nature.* 326: 689-691, 1987.

[6] K. Fukushima, "Neocognitron: A Hierarchical Neural Network Capable of Visual Pattern Recognition." *Neural Networks.* 1: 119-130, 1988.

[7]   G. J. Goodhill and D. J. Willshaw, "Elastic Net Model of Ocular Dominance: Overall Stripe Pattern and Monocular Deprivation." *Neural Computation.* **6**(4): pp. 615-621, 1994.

[8]   J. J. Hopfield, "Neural Networks and Physical Systems with Emergent Collective Computational Abilities." *Proceedings of the National Academy of Science USA.* **79**: 2554-2558, 1982.

[9]   J. J. Hopfield and D. W. Tank, ""Neural" Computation of Decisions in Optimization Problems." *Biological Cybernetics.* **52**: 141-152, 1985.

[10]   D. H. Hubel and T. N. Wiesel, "Receptive Fields, Binocular and Functional Architecture in the Cat's Visual Cortex." *Journal of Physiology.* **160**: 106-154, 1962.

[11]   E. R. Kandel, J. H. Schwartz and T. M. Jessell, Principles of Neural Science. 3rd Edition ed., Norwalk, CT, Appleton & Lange, 1991.

[12]   T. Kohonen, Self-Organization and Associative Memory. second ed., Springer Series in Information Sciences. Springer-Verlag, 1987.

[13]   M. Lades, J. C. Vorbruggen, J. Buhmann, J. Lange, C. von der Malsburg, R. R. Wurtz and W. Konen, "Distortion Invariant Object Recognition in the Dynamic Link Architecture." *IEEE Transactions on Computers.* **42**: 300-311, 1993.

[14]   T. K. Leen, "From Data Distributions to Regularization in Invariant Learning," *Proceedings of the Neural Information Processing Systems,* Denver, CO, 1994.

[15]   N. M. Nasrabadi and W. Li, "Object Recognition by a Hopfield Neural Network." *IEEE Transactions on Systems, Man and Cybernetics.* **21**(6): 1523-1535, 1991.

[16]   B. A. Olshausen, C. H. Anderson and D. C. Van Essen, "A Neurobiological Model of Visual Attention and Invariant Pattern Recognition Based on Dynamic Routing of Information." *The Journal of Neuroscience.* **13**(11): 4700-4719, 1993.

[17]   D. E. Rumelhart, G. E. Hinton and R. J. Williams. "Learning Internal Representations by Error Propagation." In Parallel Distributed Processing: Explorations in the Microstructure of Cognition. Vol. 1, Ed. D. E. Rumelhart and J. McClelland, Cambridge, MIT Press, pp. 318-364, 1986.

[18]   S. Shams, "Multiple Elastic Modules for Object Recognition." Submitted for publicaiton, 1994.

[19]   P. D. Simic, "Statistical Mechanics as the Underlying Theory of 'Elastic' and 'Neural' Optimisations." *Networks.* **1**: 89-103, 1990.

[20]   J. Sirosh and R. Miikkulainen, "Cooperative Self-Organization of Afferent and Lateral Connections in Cortical Maps." *Biological Cybernetics.* **71**: 65-78, 1994.

[21]   C. von der Malsburg, "How to label nerve Cells so that they can interconnect in an ordered fashion." *Proceedings of the National Academy of Science.* **74**(11): 5176-5178, 1977.

[22]   C. von der Malsburg, "Nervous Structures with Dynamical Links." *Physical Chemistry.* **89**: 703-710, 1985.

[23]   C. von der Malsburg, "Pattern Recognition by Labeled Graph Matching." *Neural Networks.* **1**: 141-148, 1988.

[24]   C. von der Malsburg and E. Bienenstock, "A Neural Network for the Retreival of Superimposed Connection Patterns." *Europhysics Letters.* **3**(11): 1243-1249, 1987.

[25]   A. L. Yuille, P. W. Hallinan and D. S. Cohen, "Feature Extraction from Faces Using Deformable Templates." *International Journal of Computer Vision.* **8**(2): 99-111, 1992.

On-line Handwritten Character Recognition by a Hybrid Method based on Neural Networks and Pattern Matching

Jung-Wook Cho, Soo-Young Lee, and Cheol Hoon Park

Computation and Neural Systems Laboratory
Department of Electrical Engineering
Korea Advanced Institute of Science and Technology
373-1 Kusung-Dong Yusung-Gu
Taejon 305-701, KOREA

Abstract

A hybrid system is developed for on-line recognition of hand-written Korean characters. Each syllable consists of several Korean alphabets and is written in a rectangular box to result in a 2-dimensional compositions of alphabets. Although only 24 alphabets exist, the number of syllables easily exceeds 3000. Therefore, instead of the syllables, the 24 alphabets are used as basic recognition unit, which causes difficult segmentation and recognition problems. Each Korean alphabet has at most 4 strokes, and 4 neural networks are trained separately for alphabets with different number of strokes. To improve the recognition preformance, after neural network classifications, classical pattern matching is also applied to check validity of the decisions made by the neural network classifiers. The hybrid systems is robust on distortion and rotation of the characters.

I. Introduction

As computers become more popular and portable recently, more natural Man-Machine interface by speech and character recognition is required. Character recognition is divided into printed character recognition[1] and handwritten character recognition[2]. In the case of Korean characters(Hangul) the printed character recognition is usually performed by the unit of a character[1]. In the handwritten Hangul character recognition[3], the distortion of characters is more severe and complex, since each character is composed of combination of consonants and vowels. So, handwritten Hangul recognition is usually performed by unit of a consonant and a vowel[4,5].

Hangul basically consists of 14 consonants and 10 vowels as phonemes. Some compound vowels are made of these 10 vowels, and a character is composed of consonant(s) and vowel(s), or consonant(s), vowel(s), and consonant(s) in order. For example, '학'(pronounced as *hak*) is composed of a consonant 'ㅎ'(*h*), a vowel 'ㅏ'(*a*), and a consonants 'ㄱ'(*k*). And the number of possible characters is over 10,000 by composition of consonants and vowels, which makes it difficult to recognize characters, especially handwritten characters[6], at once.

927

Recognition can be either on-line[3] or off-line[7]. Generally, on-line recognition is easier because the information on strokes like the number of strokes and the order of strokes can be used. But a fast process is indispensable in on-line recognition[1].

In character recognition, the traditional pattern-matching technique is much used[3]. However, it has many problems, because the features of all characters must be collected manually and recognition time can be long. Moreover it cannot be certified that manual feature extraction is always exact and proper for recognition. The more complex the feature is for accuracy, the longer time recognition takes. HMM(Hidden Markov Model) is used effectively in both off-line and on-line handwritten character recognition[4,5], but has slow speed. Also, fuzzy recognition is relatively fast and easy to design, but the developer has to make all rules in some cases[6].

In order to overcome these problems, we propose a simple feature extraction technique for real-time processing, a neural network for recognition, and a simple pattern-matching method as a postprocessing for improving recognition rate[9]. Since neural networks have capabilities of approximation, classification, and generalization[10], they work well with distorted data and can learn the patterns without developer's dedicated directions. In consideration that people write the letters differently, the neural network shows its excellent ability in on-line recognition.

In this paper, we will propose that Hangul recognition system with the pre-defined order and number of strokes which is unaffected by scaling and small rotation. One of its merits is that only consonants and vowels are used in the learning process and all the Hangul characters are recognized by the construction algorithm.

In the following section, the structure of the recognition system is explained. The algorithm of recognizing consonants and vowels and constructing characters is illustrated in section 3 and the experimental results in section 4. Finally, conslusions are made in section 5.

II. Structure of the recognition system

The smallest units for recognition are shown in Table 1. Each of 'ㄹ', 'ㅂ', 'ㅈ' and 'ㅊ' in this table has two kinds of forms as shown in Fig.1. 'ㄹ' can be written by one stroke or three strokes as shown in Fig.1(a) according to that person who writes, so two different patterns are defined. 'ㅂ' is one of the most versatile consonants in Hangul, and only two- and four-stroke 'ㅂ' is defined for simplicity as shown in Fig.1(b). 'ㅈ' and 'ㅊ' are also defined as two kinds of patterns as shown in Fig.1(c) and 1(d). Other patterns written differently can be added easily by learning.

# of strokes	corresponding consonants	corresponding vowels
1	ㄱ, ㄴ, ㅡ, ㅇ	ㅡ, ㅣ
2	ㄷ, ㅂ, ㅅ, ㅈ, ㅋ, ㄲ	ㅜ, ㅗ, ㅓ, ㅏ
3	ㄹ, ㅁ, ㅈ, ㅊ, ㅌ, ㅎ, ㅃ	ㅠ, ㅛ, ㅕ, ㅑ, ㅖ, ㅒ
4	ㅂ, ㅊ, ㅍ, ㄸ, ㅆ, ㅉ	ㅔ, ㅐ

Table 1. Basic recognition units of consonants and vowels

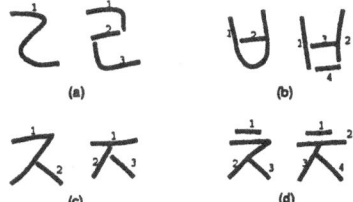

Figure 1. Characters which can be written in two ways

Compound-vowels are not defined, because they are recognized by composition of single-vowels in our system. Therefore, the complexity and the number of patterns to be recognized are decreased even though separation between consonants and vowels becomes a little more difficult. In the case of letter '관'(pronounced as kwan), if compound vowels can be recognized as a vowel, '관' is separated by 2 consonants 'ㄱ' and 'ㄴ' and a vowel '와'. But since only single-vowel can be recognized, 'ㅗ' and 'ㅏ' are recognized separately.

Our Hangul recognition system uses the following as the inputs of neural networks:

1) starting point of each stroke;
2) end point of each stroke;
3) three vectors obtained by cutting out the route of the stroke with equal length.

The features of a character are composed of these five features of each stroke as shown in Fig.2 where 'ㄹ' is written in one stroke. Since these features have x and y values, each stroke has 10 features in total. Recognition is not affected by size of characters, since these features are extracted from stroke after normalization.

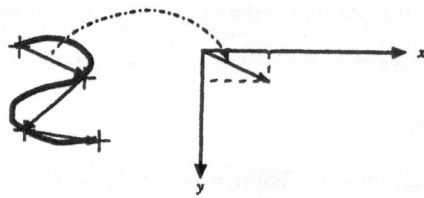

Figure 2. Extraction of vectors from a consonant 'ㄹ'

Recognition system is composed of four multilayer neural networks. One-stroke consonants and vowels are learned only at "one-stroke MLP"(=MLP1). And two-stroke consonants and vowels are learned only at "two-stroke MLP"(= MLP2), and so on. The inputs of neural networks are the features mentioned above and the outputs are the consonants and the vowels which corresponds to each MLP. As an example, MLP1 has 10 inputs for 10 feature values for one stroke, and has 6 output neurons corresponding to 'ㄱ', 'ㄴ', 'ㄹ', 'ㅇ', 'ㅡ', 'ㅣ'.

III. Separation of consonants and vowels and construction of characters

We adopt simple and efficient algorithms for separation and recognition of consonants and vowels. First, if the total number of strokes of a character is less than five, all strokes except the last one is the input of the corresponding MLP because there must be a vowel. If the total number of strokes of a character is more than five, the first four are the inputs of MLP4, because no vowels and consonants have more than four strokes. As an example, since '밝' has 10 strokes as shown in Fig.3, 'ㅂ'(the first four strokes)is an input of MLP4 and will be classified as 'ㅂ' by MLP4. Then, since the number of the rest of strokes after 'ㅂ' is also more than four, the first four strokes except 'ㅂ'('ㅏ' and the first two strokes of 'ㄹ') are the inputs of MLP4. But these patterns of strokes are not taught, so MLP4 will not produce any output. Then, the first three('ㅏ' and the first one stroke of 'ㄹ') are the input of MLP3, but there will be no output, either. Then the first two('ㅏ') are the inputs of MLP2. Eventually, 'ㅏ' can be separated and recognized. This process is shown in Fig.3.

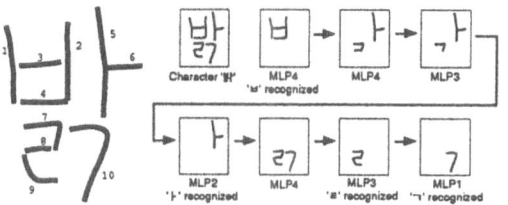

Figure 3. Recognition process in the case of a character '밝'

Basic algorithm is similar to the process explained above, but the separation of consonants and vowels did not work correctly in real situations, sometimes. The reason is that neural networks produce relatively high output value which is undesirable, even if the direction and position of one or two strokes are different. That is, even if those strokes that represent important characteristics of a character is different in direction or starting position, they can be selected in some cases. This is one of the advantages of neural networks, but this works as a disadvantage in this case. High-valued threshold level or more intensive learning on various patterns will be helpful, but it will become more difficult in recognizing the distorted characters. In order to solve this problem, we use a pattern matching method. Neural networks choose possible characters, and pattern matching method picks out totally different characters. Eventually, we use both merits of a conventional method and a neural network together.

As an example, if a character '믈' comes in MLP4 as inputs, the output of '밤' or '왜' can be high since '믈' is not taught in the learning phase of MLP4. In this case, we can correct the disqualified output of neural networks by pattern matching method. For each stroke, two kinds of pattern matching are used with rule tables.

1) Starting and end directions of a stroke

Every stroke of characters has the shape which can be changed in some limitation, and there are the shapes which cannot construct a correct character. As shown in Fig.4, the starting vector and the end vector of '기' have to exist in some angle even if distortion is considered. First, the allowable angles of the starting and end vectors of each stroke of each consonant and vowel are memorized numerically. If some consonant or vowel is selected by neural network, its starting and end vectors are compared with standard data. This comparison consists of OR operation, so it has no effect on recognition time.

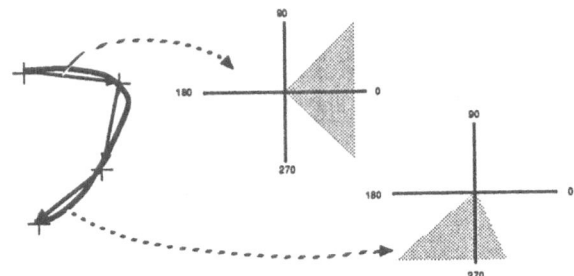

Figure 4. Angle limitation of the starting vector and the end vector

2) Starting and end points of strokes

There is another criterion on position. For example, the starting point of the first stroke of '⊏ '(point a) must be at the left-hand and upper side of the end point of the second stroke(point b) as shown in Fig.5. This comparison also consists of OR operation.

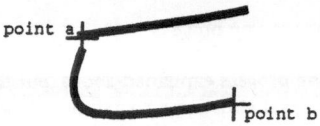

Figure 5. Position-restriction in a consonant '⊏'

All strokes of every consonant and vowel have standard data for these two kinds of rules. These data are binary and thus easily compared by OR operation with efficient coding. We made standard binary codes about position and direction limit. The code of the consonant or vowel selected by neural network is compared with standard binary code by OR operation. If the result of OR operation is different from standar binary code, the selection of neural networks might be wrong.

Another problem comes with final consonants. For example, in the case of '릴', the first stroke of ' ㅁ' with '一' is mistaken as 'ㅜ' as in Fig.6. To solve such a problem, we rearranged the order to separate consonants and vowels. At first, the first consonant is selected out. Then the final consonant is picked out, and at last the vowel is selected out. This process is shown in Fig.7. This method makes recognition time a little longer as shown in Fig.7(b).

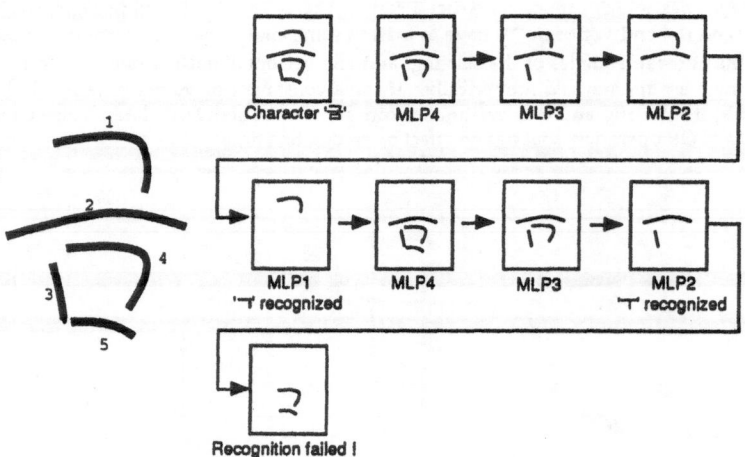

Figure 6. '一' is mistaken as 'ㅜ'

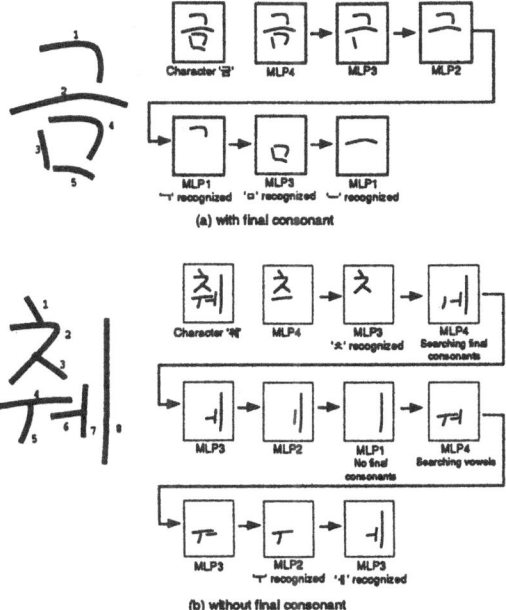

Figure 7. Process of Recognition

IV. Experimental results

A sample set of Hangul written by one person were learned by recognizer. Nine samples for each consonant and vowel were learned 1000 times for each by neural networks. We used error back propagation learning algorithm to train neural networks. It took about 3 hours for neural networks to complete the learning on Sparc 2 machine. We used Macintosh IIcx(CPU:68030, 16Mhz) for recognition.

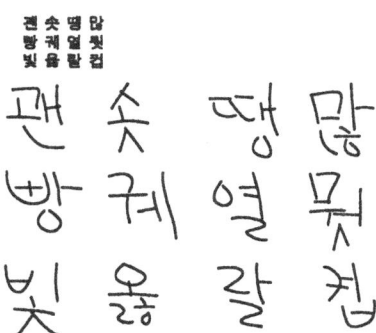

Figure 8. Examples of characters used in recognition experiments

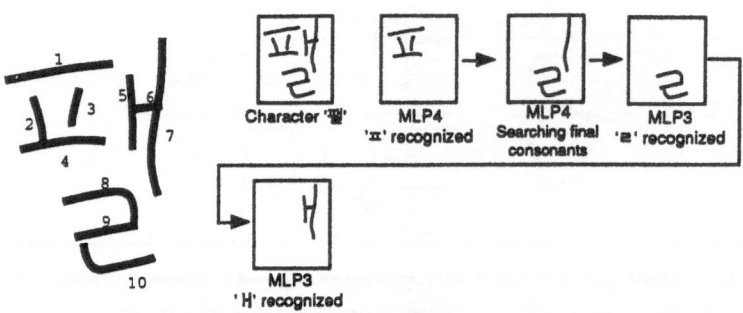

Figure 9. Recognition process in the case of '퍫'

Fig.8 shows the examples of characters which are recognized successfully. Average recognition time is below 1 second. The recognition time has almost nothing to do with the number of strokes but something to do with the type of combination of consonants and vowels. In the case of '퍫', the recognition is completed through MLP4, MLP4, MLP3, and MLP3, as in Fig.9. In spite that '체' has less strokes than '퍫' has, it takes more time to recognize, since the recognition of '체' is completed through MLP4, MLP3, MLP4, MLP3, MLP2, MLP1, MLP4, MLP3, MLP2, and MLP3 as in Fig.7(b).

The recognition rate was measured by three other persons. We selected 300 characters by the order of frequent usage and they wrote 100 letters each for testing. Failure of recognition was only 12 characters out of 300 characters. So, we got 96% recognition rate.

Most of recognition failures are caused by 'ㄹ' written by one stroke. 'ㄹ' written by one stroke can be misconceived easily as 'ㄱ' or 'ㄴ'. Five 'ㄹ' were misconceived. Another problem is that some character like '진' was misconceived as '건'. In this case, human can also mistake it, sometimes.

A possible way of improving performance of this system is to subdive a stroke into more than three vectors. But it needs more input neurons, so it takes longer time to recognize without neural network hardwares.

V. Conclusions

In this paper, we showed the recognition system of handwritten Korean consonants and vowels and the algorithm to compose a character.

The merit of this recognition system is that all Korean characters can be recognized by training of only consonants and vowels. And somewhat distorted and rotated characters can be recognized well by using neural networks. The problem we have to solve is that this recognizer can not recognize successively or cursively written characters. The research on increasing the features of strokes and developing optimal neural network structure are required. We consider the application such as stenography recognition.

Reference

1. S.-B. Cho and J.H. Kim, Recognition of large-set printed Hangul(Korean script) by two-stage backpropagation neural classifier, *Pattern Recognition* 25(11), 1353-1360 (1992)
2. J.S. Jang, M.W. Kim, C.-D. Lim and Y.-S. Song, Recognition of handwritten Korean characters based on constraint relaxation, *Proc. Pacific Rim Int. Conf. on Artificial Intelligence*, pp.1209-1215, Seoul, Korea, Sep. (1992)
3. Y.H. Huh and H.L. Beust, On-line recognition of hand-printed Korean characters, *Pattern Recognition* 15(6), 445-453 (1982)

4. Jin-Young Ha, Se-Chang Oh, Jin-Hyung Kim and Young-Bin Kwon, Unconstrained Handwritten Word Recognition with Interconnected Hidden Markov Models, *Proce.of 3rdInternational Workshop on Frontiers in Handwriting Recognition,*, pp. 455-460, Buffalo, USA, May (1993).

5. Hee-Soen park and Seong-Whan Lee, Off-Line Recognition of Large-Set Handwritten Hangul with Hidden Markov Models, *Proc.of 3rd International Workshop on Frontiers in Handwriting Recognition*, pp.51-61, Buffalo, USA, May (1993).

6. U.J. Seong, Online Hand-written Hangul Recognition Using Hierarchical Curve Representation, *M.S.Dissertion*, Dept. of Computer Science, Korea Advanced Institute of Science and Technology (1991)

7. Yoon-Seon Song, Myung Won Kim, Jong Moon Kim, A Neural Network Approach to Continuous Handwritten Character Recognition without Segmentation, *Internation Conference on Neural Information Processing*, pp.122-127, Seoul, Korea, Oct. (1994)

8. Hendrawan, A. C Downton, C. G. Leedham, A Fuzzy approach to handwritten address verification, *Proc.of 3rd International Workshop on Frontiers in Handwriting Recognition,*, pp.300-311, Buffalo, USA, May (1993).

9. Eduard Sackinger, Bernhard E. Boser, Jane Bromley, Yann LeCun and Lawrence D. Jackel, Application of the ANNA Neural Network Chip to High-Speed Character Recognition, *IEEE Trans. on Neural Networks* 3(3), pp.498-505 (1992).

10. Richard P. Lippmann, An Introduction to computing with neural nets, *IEEE ASSP Magazine*, , pp.4-22, April (1987).

Analysis and Application of the Store Neural Model in Recognizing Handwritted Symbols

J. Pérez Maroto [†], Y. A. Dimitriadis [‡], J. Manuel Cano Izquierdo[†] and J. López Coronado [†] [1]

[†] Department of Systems Engineering and Control,
School of Industrial Engineering, University of Valladolid,
Paseo del Cauce, s/n, Valladolid, 47011, Spain
[‡] Department of Signal Theory, Telecommunications and Telematics Engineering,
School of Telecommunications Engineering, University of Valladolid
Calle Real de Burgos s/n, Valladolid, 47011, Spain

Abstract

On-line handwriting recognition is considered here as a problem of classification of temporal sequences of symbol components. A neural-fuzzy architecture that extends and enhances Fuzzy ARTMAP for the categorization of sequences is proposed in this paper. A new local distance measure permits an efficient processing of the spatial patterns produced by the STORE memory model. This neural network is extensively analyzed in both mathematical and experimental terms, leading to important suggestions for the engineering design of this biologically inspired model. Finally, experimental results of the global handwriting recognition architecture are presented.

1 Introduction

The problem of handwriting recognition has always been one of the favourite problems of the pattern recognition community because of the intrinsic difficulty of the handwriting variability as well as because of its huge importance for the man-machine communication. Although lots of research approaches have been reported, its solution for a reasonably unconstrained environment is still to be reached [14]. Recently several solutions based on artificial neural network models were proposed [12], although in general without biological or psychological foundations.

In on-line handwriting recognition, dynamic information is used as opposed to off-line recognition. The components and strokes of each handwritten symbol provide important features, such as direction, duration and type. In our system, on-line discrete handwritten symbols are considered as sequences of component allographs.

The models derived from the Adaptive Resonance Theory [5, 6] seem to have solid foundations on the competitive nature of the biological neural networks, while at the same time offer an alternative for the traditional clustering algorithms. In previous papers, an ART-2 based neural architecture was proposed [7, 8] with satisfactory results. The need to enhance its performance in terms of speed and recognition rates motivated us to design and study a new architecture that incorporates fuzzy ART networks and integrates them in a temporal sequence processing problem. This new neural fuzzy architecture is described in section 2.

A fundamental aspect of this architecture is the conversion of the temporal list of symbol components to spatial pattens. The problem of storage and recall of temporal lists is of a wider interest and has been studied and applied in various fields such as speech recognition and grammatical inference [13]. The STORE (Sustained Temporal Order REcurrent) model [2, 3] employed in our architecture lies also within the framework of S. Grossberg's theory on human memory [9, 10]. Although the general structure and characteristics of this working memory model were reported in the original papers, its importance led us to study, mathematically and experimentally, its properties and propose practical guidelines for the engineering design of such a model, when embedded in real-life applications. The mathematical analysis and simulation results are shown in section 3.

Finally, experimental results for the global neural-fuzzy architecture are shown in section 4.

2 Description of the on-line handwriting recognition architecture

As stated in the introduction, the problem of on-line run-on discrete handwritten symbol recognition can be decomposed to two subproblems, namely:

- Classification of the individual components (sequence of points between two consecutive pen lifts).

- Classification and labeling of the symbols, where each symbol is considered as a sequence of the component categories, that were obtained in the previous phase.

---

[1]We would like to thank the members of the Neural Networks Group (J.F. Díez-Higuera, F.J. Díaz-Pernas, C. García, F. Merino, A. Muñoz, E. Sanchez, G.Mendez, E. Zalama) for their contributions to the preparation of this paper.

The first subproblem is equivalent to the unsupervised classification of the analog patterns, that correspond to the components. Since a fixed length is required for the input patterns of the categorization module, the normalization of the number of the elements of the input pattern is performed, using the linear elasting matching algorithm that corresponds to a linear interpolation.

After the natural classification of each component to a node of the component classification module, a spatial pattern is required for each symbol that could represent the ordered list of nodes that were activated. This conversion is obtained by a STORE network, that is extensively analyzed in the next section. Also, a signal is necessary that would indicate the beginning and the end of the list for each symbol. This signal is obtained during the segmentation phase of the preprocessing module.

The proposed integrated fuzzy-neural architecture, presented in Figure 1, uses the same philosophy of a previous architecture [7, 8]. Its first innovation lies in the application of Fuzzy ART modules instead of ART-2 based networks. Nevertheless, architectures such as Fuzzy ARTMAP [5] are not prepared for temporal sequence processing, as in our problem.

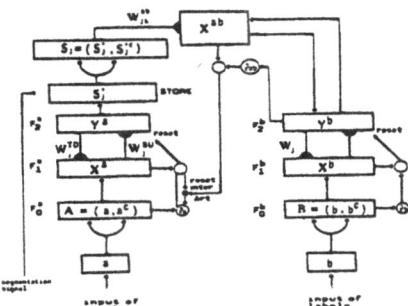

Figure 1: A diagram of the proposed integrated fuzzy-neural architecture

In the **learning phase** of the proposed architecture, the vectors a (sequence item patterns) and b (sequence labels) are presented to ART_a and ART_b respectively. The corresponding patterns A and B are calculated through complement coding. The algorithm employed in Fuzzy ART is used here in order to code the input patterns to the nodes j and k respectively. Following the convention of Fuzzy ART, only the bottom-up weights W_j^{BU} are used. When the input signal that refers to the end of the sequence is activated, then the corresponding spatial pattern S_j^* will be produced by the STORE module, that in turn is transformed to S_j through complement coding. At the end of the cycle, the spatial pattern S_j and the coresponding node k of F_2^b should be associated.

A new *local distance* LD_j, inspired in the concept of fuzzy subset, measures the similarity of the STM pattern S_j amd and the inter-ART LTM vector W_{jk}^{ab}

$$LD_j = |S_j \wedge W_{jk}^{ab}| = \sum_{j=1, S_j \neq 0}^{N} |\min(S_j^*, W_{jk}^{ab})| + \sum_{j=1, S_j \neq 0}^{N} |\min((1 - S_j^*), (1 - W_{jk}^{ab}))| \qquad (1)$$

Note that only the nodes of the STM pattern S_j^* that are not zero are considered in this local distance measure, since they are the ones that carry information, thus reducing greatly the computational cost. This local distance reduces in turn the complexity of the created clusters.

Then the ratio $\frac{LD_j}{|S_j|}$ is calculated and compared to the inter-ART vigilance parameter ρ_{ab}. There exist two cases depending on the result of the above comparison:

First Case: If $\frac{LD_j}{|S_j|} < \rho_{ab}$, then a reset inter-ART is produced indicating that the sequence items were not correctly classified, and therefore a new cycle of categorization has to be initiated. In that case $W_j^{BU} == W_j^{TD}$, i.e. the changes that were produced during the election of the winning node j are suppressed. This process corresponds to the fact that the hypothesis made before was not correct and therefore, we should go back to the situation before the choice of the node j at F_2^a. Then a temporary increase takes effect on ρ_a making it: $\rho_{ai} = \frac{|A_i \wedge W_j|}{|A_i|}$, where A_i is the i-th sequence item, W_j is the LTM for the node that had won for the item i in the previous cycle, and ρ_{ai} is the new vigilance parameter for the item i.

This way, we can influence the unsupervised categorization of the symbol components and enhance the overall recognition rates.

Second Case: If $\frac{LD_i}{|S_j|} \geq \rho_{ab}$, then the classification is considered correct and therefore:

$$W_{jk}^{ab} = \begin{cases} \beta(W_{jk}^{ab} \wedge S_j) + (1 - \beta)W_{jk}^{ab} & \text{if } S_j \neq 0 \\ W_{jk}^{ab} & \text{otherwise} \end{cases} \tag{2}$$

Note: The weights W_{jk} are initialized to zero.

At the **prediction phase**, the general process is similar, but we have to estimate the winning node k at F_2^b. We choose it following the winner-take-all rule: $k : T_k = \max_i(T_i)$ where: $T_i = \frac{LD_i}{\alpha + |W_{ij}^{ab}|}$. Finally the following test is made: $\frac{LD_i}{|S_j|} \geq \rho_{ab}$ and in the case that it is true, the bottom-up weights of the node k of F_2^b are used as the prediction code for the symbol. If the test is not true, then no decision is made.

3 Analysis of the STORE neural network model

3.1 Introduction

As already indicated in section 2 each incoming component vector is classified by a Fuzzy-ART module, thus obtaining a list of nodes at the F_2 level of the ART module. Any further processing and classification of this list would require a model for storage and recall of arbitrary sequences. This problem could also occur in the following levels of handwriting processing, such as the sequence of symbols, words, phrases etc. Many similar problems that treat the processing of sequences could also be mentioned [13].

The STORE neural network model [2], presented in figure 2, is based on the theory of human memory proposed by S. Grossberg [9, 10]. According to it, the individual list items are stored in the working memory as transient activity, thus coding the information of the list items as well as their order of appearance. This type of memory storage contrasts with serial memory models and is supported by psychological data [1]. The following principles are basic for a performance that comply with the human system:

- **The invariance principle**, where the temporal list should be continuously coded as new list items appear, maintaining at the same time the stability of the sublists learned previously e.g. *Transition* and *Transition*-ing.

- **The partial normalization principle**, that expresses the limited capacity of the transient memories.

These two principles correspond to the dilemas of stability-plasticity and noise-saturation for the permanent memory.

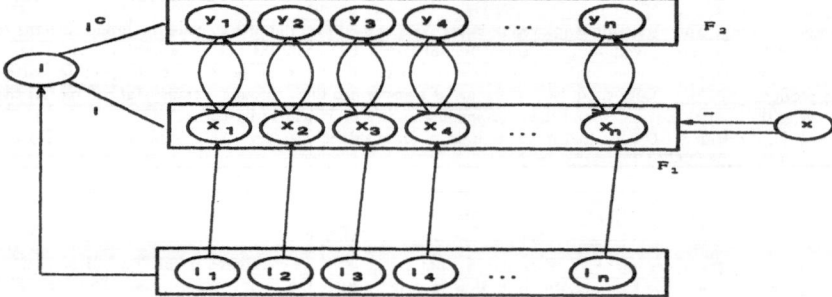

Figure 2: STORE model with a transient memory level F_1 and a permanent memory level F_2

Reviewing its strucure, for an input sequence $I_i(t)$ of

$$I_i(t) = \begin{cases} 1 & \text{if } t_i - a_i < t < t_i \\ 0 & \text{otherwise} \end{cases} \tag{3}$$

its differential equations are:

$$\frac{dx_i}{dt} = [AI_i + y_i - x_i x]I \tag{4}$$

$$\frac{dy_i}{dt} = [x_i - y_i]I^c \tag{5}$$

where we have $x = \sum_k x_k$, $I^c = 1 - I$, and

$$x_i(0) = y_i(0) = 0 \qquad (6)$$

3.2 Properties and limits of the model

In this subsection we present the analytical and experimental results with respect to STORE, that we obtained.

Property 1 Compliance to the invariance principle: *When a new item x_i is presented, the relative activities of the sublist $(x_1, ..., x_{i-1})$ are preserved.*

This is easily demostrated taking into account the equations 4, 6 y 3, and $I_i = 1$, $y_i = 0$. Then the present item will be:

$$x_i \to \frac{A}{x} \qquad (7)$$

while for the previous items x_k: $k < i$, $I_k = 0$ and

$$x_k \to \frac{y_k}{x} = \frac{x_k(t_{i-1})}{x} \qquad (8)$$

□

Property 2 1. Compliance to the partial normalization principle: *The sum of activities at the first level of STORE increments as a function of the number of list items in an asymptotic way towards a maximum value S.*

From 4 y 5, we deduce that the sum of activities x and y at levels F_1 and F_2 respectively obey the following equations:

$$\frac{dx}{dt} = [A + y - x^2]I \qquad (9)$$

$$\frac{dy}{dt} = [x - y]I^c \qquad (10)$$

Since $y(0) = 0$, at $t = t_1$ the equation 9 implies that $x(t_1) = \sqrt{A}$, i.e. its critical point. At stable state and at the instant $t = t_i$, $i > 1$, the equation 9 implies

$$x(t_i) = \sqrt{A + y(t_i)}$$

, while at the same time the equation 10 implies that $y(ti) = x(t_{i-1})$. Then since $S_1 = x(t_1) = \sqrt{A}$, the total activity S_2 at the instant t_2 will be:

$$S_2 = x(t_2) = \sqrt{A + S_1} \qquad (11)$$

At the next step $y(t_2) = x_2 = S_2$ and therefore

$$S_3 = x(t_3) = \sqrt{A + S_2} = \sqrt{A + \sqrt{A + S_1}} \qquad (12)$$

Using inductive reasoning we can prove the equivalent expression for i. Taking the singular point of this expression we resolve the equation $S = \sqrt{A + S}$ whose positive solution is:

$$S = 0.5[1 + \sqrt{1 + 4A}] \qquad (13)$$

□

In figure 3, examples are shown for STORE performance with respect to the above two principles.

Property 3 *The gradient of the spatial pattern produced by STORE can be assigned through an additional parameter B.*

In [3] the equation describing F_2 was modified to:

$$\frac{dx_i}{dt} = [AI_i + y_i - x_i \sum_{k \leq i} x_k - Bx_i]I \qquad (14)$$

The parameter B permits us to control the gradient of the STORE output, as it can be seen by its stable state solution:

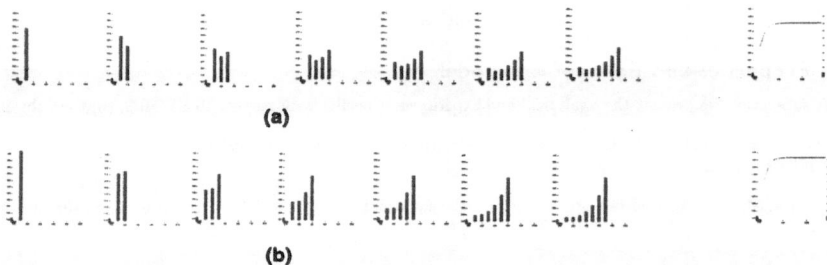

(a)

(b)

Figure 3: Examples of STORE output. In **(a)** $A = 0.6$ and the number of list items varies from 1 to 7. In **(b)** $A = 1$. In last column the total activity S is shown.

$$x_i(t_i) = \frac{A}{\sum_{k \leq i} x_k(t_i) + B} \qquad (15)$$

while the nodes that correspond to previously presented list items will have an activity of:

$$x_{i-1}(t_i) = \frac{x_{i-1}(t_{i-1})}{\sum_{k \leq i-1} x_k(t_i) + B} \qquad (16)$$

i.e. it will be reduced as an effect of the parameter B.
The new asymptote will be:

$$S = 0.5[(1 - B) + \sqrt{(1 - B)^2 + 4A}] \qquad (17)$$

□

An important observation indicates that the choice of the design parameter B is difficult, since the activities of the output level could decrease to a point that it could be confused with noise levels.

Property 4 *Empirical rules are provided for the selection of the design parameters A and B.*

Depending on the form of the gradient of the STORE output, three types of storage and recall of the temporal lists can be acieved, namely primacy, recency and bowing. These forms comply to the ones observed in psychological experiments [1].

Analytical as well as simulation studies have been performed by us in order to find the relation between the form of the output and the design parameters A and B. The results of the simulation of the equation 14, for different number of list items and 4000 responses of the model, are shown at the figure 4 .

The results can be summarized in the following engineering resomendations:

- If $A > 1$ and/or $B > 1$ we will have *recency* independently of the number of list items.

- As we can observe in figure 4, the frontiers between *bowing* and *primacy* depend on the number of list items. It can be proven analytically that if $x_1(t_1) > A$ we will have primacy untill $x_i(t_i) \leq A$ when *bowing* starts to be produced. This means that for short lists the model produces *primacy*, while for longer lists we have *bowing*. This result also confirms the data reported in psychological experiments.

Property 5 *An analytical expression is established in order to obtain independence of the model with respect to the stimulus and interstimulus intervals*

In [2, 3] it was argued that the correct model performance is independent of the stimulus and interstimulus intervals. Since the model is a dynamic system governed by differential equations, the necessary time for the stabilization of the equations has to be considered, as shown in figure 5.

If we perform a linear regression on the data obtained for the minimum required time for the stabilization of the differential equations we obtain:

$$x_{min} = 14.44 - 8.7A \qquad (18)$$

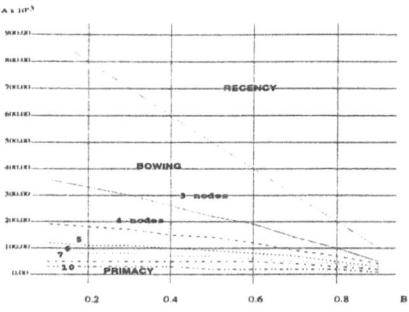

Figure 4: Frontiers among the different types of the model output as a function of A, B and the number of the list items.

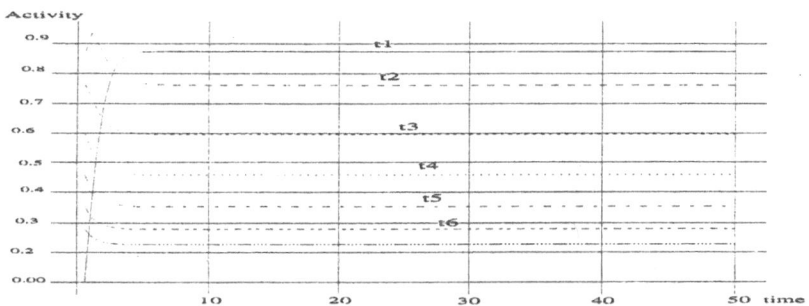

Figure 5: Evolution of the activity of the node 1 for a sequence of 5 items.

It can be observed that x_{min} depends only on the design parameter A.
□

Taking into account the above expression, we can choose the optimal initial values in order to reach the stability point.

Property 6 *The maximum permitted list length is limited by the magic number 7*

Due to the limited capacity of the transient memory, it was observed that human subjects were able to recall only a finite number of list items, when they had been presented to them once. Biological as well as psychological data [11] indicate that the magic number 7 ± 2 can describe the limit of the immediate memory for storage and recall of lists.

The results shown in figure 6 imply that STORE confirm the above data.
□

4   Results and discussion of the global recognition process

The proposed fuzzy neural architecture was tested in the context of a mathematical editor. In this system the symbols are introduced through a digitizing tablet and the information of the pen lifts permits us to extract the components of a symbol. Then the symbols are represented as sequences of their components, that in turn use a measure of the angular velocity as feature vector. The input vector length is normalized to 25 using a linear elastic matching algorithm.

An extensive study of the performance of the system was made, although in the present paper a worst-case example is presented. The users were asked to enter a full page of each symbol and the first 100 samples were used for the learning set, while the rest of them were used for the prediction phase. The parameters used in this experiments are : $\beta_a = 1$, $\alpha_a = 0.001$, $\rho_b = 0.98$, $\beta_b = 1$, $\alpha_b = 0.001$ and $\rho_{ab} = 1$. The vigilance parameter ρ_a of

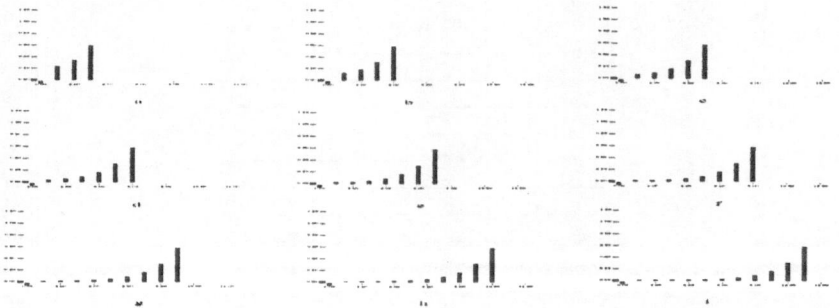

Figure 6: The effect of normalization is translated to an abrupt decrease of the activity as the number of list items is increased. In this figure, we can observe the existence of the magic number 7 ± 2, since after ir the activity of the previously learned items is lost.

ART_a was set to 0.8 and 0.7 for the learning and prediction phases respectively. The number of created nodes at F_a^2 is 215.

Symbol	Correct	Erroneous	Uncertain
+	53	0	15
$\sum$	29	0	10
0	44	0	0
8	43	0	0
H	13	0	15
P	17	8	4
X	35	3	1
h	33	0	0
p	14	12	2
x	11	18	0

Table 1: Recognition rates of the symbol recognition for the proposed fuzzy neural architecture.

The results in Table 1 indicate a robust performance of the proposed architecture and compare favorably to the ones obtained by a previous neural architecture [7, 8]. We should note that the errors or uncertainties are due to factors like bad quality of the input data, inadequate representation of the symbol components by the angular velocity feature vector, incompleteness of the learning set and finally confusion of symbols that are similar in shape but not in size (x and X).

Besides the efficient performance of the new architecture, there is a significant enhancement in terms of computational cost as compared to architectures that use ART-2 modules, since no differential eauations are solved, only sums and comparisons are made avoiding the multiplications of ART2, and the complement coding of Fuzzy ART is less complex than the normalization of ART2. Also, due to the fast-learning slow-recoding mode, the system was stabilized in few learning cycles. In any case, learning is much faster for similar performance than in equivalent modules that employ the feed-forward networks with the backpropagation learning algorithm, as it was reported in a benchmark for off-line character recognition [6].

An interesting study was made with respect to the vigilance parameter selected at the Fuzzy-ART module. It was shown [4] that the increment of the number of nodes created at the F_2 level was linear, till $\rho = 0.7$ while after that value of ρ the increment was exponential. This indirect study of the statistical properties of the learning set can help us considerably to enhance the overall performance of the architecture, since a great number of uncertainties is due to the "misclassification" at the component level.

5 Conclusions

An integrated fuzzy-neural architecture for the recognition of handwritten symbols was presented at this paper. Its modules, inspired in the Adaptive Resonance Theory, form an architecture that extends the Fuzzy ARTMAP architecture for the processing of temporal sequences. The experimental results were shown to be satisfactory in both terms of recognition rates and computation requirements.

Since the decomposition of a handwritten symbol as a sequence of its components is fundamental in this architecture, the employed STORE neural network model was studied analytically and experimentally. Several properties proposed in the original STORE paper, such as the compliance to the principles of invariance and partial normalization were mathematically demonstrated. On the other hand the influence of the design parameters A and B on the STORE output was studied, and decision criteria were proposed depending on the recency, bowing or primacy output characteristics. The limits of the independence of the stimulus intervals were also shown, as well as the existence of the magic number 7 ± 2 for the maximum length of the sequences.

References

[1] R.C. Atkinson and R.M. Shiffrin, "The control of short term memory", *Scientific American*, pp. 82-90, August 1971.

[2] G. Bradski, G.A. Carpenter, and S. Grossberg, "Working memory networks for learning temporal order, with application to three-dimensional visual object recognition", *Neural Computation*, vol. 4, no. 2, pp. 270-286, March 1992.

[3] G. Bradski, G.A. Carpenter, and S. Grossberg, "Working memories for storage and recall of arbitrary temporal sequences" Proc. of the *IJCNN92*, Baltimore, vol. II pp. 57-62, 1992.

[4] J.M. Cano Izquierdo, "Development of a neural architecture for the recognition of handwritten symbols" *Engineering degree thesis*, University of Valladolid, November 1993 (in spanish).

[5] G.A. Carpenter, S. Grossberg, N. Markuzon, J.H. Reynolds and D.B. Rosen, "Fuzzy ARTMAP: A neural network architecture for incremental supervised learning of analog multidimensional maps", *IEEE Trans. on Neural Networks*, vol. 3, pp. 698-713, 1992.

[6] G.A. Carpenter, S. Grossberg, and K. Iizuka, "Comparative performance measures of Fuzzy ARTMAP, Learned Vector Quantization, and Backpropagation for handwritten character recognition", Proc. of the *IJCNN92*, Baltimore, vol. I, pp. 794-799, 1992.

[7] Y.A. Dimitriadis, J. López Coronado, C. García Moreno and J.M. Cano Izquierdo, "On-line handwritten symbol recognition, using an ART based neural network hierarchy", *Proceedings of the IEEE Conference on Neural Networks*, vol. 2, pp. 944-949, March 28- April 1, 1993, San Francisco.

[8] Y.A. Dimitriadis, and J. López Coronado, "Towards an ART-based mathematical editor, that uses on-line symbol recognition", *Pattern Recognition*, in press, 1995.

[9] S. Grossberg, "A theory of human memory: Self-organization and performance of sensory-motor codes, maps, and plans", in *Studies of Mind and Brain*, D.Reidel, Boston, MA, pp. 498-639, 1982.

[10] S. Grossberg, "Behavioral contrast in short-term memory: Serial binary memory models or parallel continuous memory models?", in *Studies of Mind and Brain*, D. Reidel, Boston, pp. 425-447, 1982.

[11] G.A. Miller, "The magic number seven, plus or minus two", *Psychological Review*, vol. 63, pp. 81-97, 1956.

[12] P. Morasso, A. Pareto and S. Pagliano, "Neural models for handwriting recognition", in *From Pixels to Features III: Frontiers in Handwriting Recognition*, S. Impedovo and J.C. Simon (eds), Elsevier, pp. 423-440, 1992, .

[13] R.F. Port, "Representation and recognition of temporal patterns", *Connection Science*, vol. 2, pp. 151-176, 1990.

[14] C.C. Tappert, C.Y. Suen, and T. Wakahara , "The state of the art in on-line handwriting recognition", *IEEE Trans. on Pat. Anal. and Mac. Intel.*, vol. 12, pp. 787-808, August 1990.

An Adaptative Orthogonal Asociative Memory and its Application to Character Recognition

Ibarra-Picó, F.; García-Chamizo J.M.; Rizo-Aldeguer, R.; Corredor Lacha, D.

Departamento de Tecnología Informática y Computación.
Universidad de Alicante.
Apdo. 99. Alicante. Spain.
email: ibarra@dtic.ua.es

Abstract. It is presented an adaptative orthogonal memory model which is based in bidirectional associative orthogonalized memory models. This adaptaive neural system has a higher capacity that others memories (as BAM) and uses and special backpropagation algorithm to get a correct answer in a forward way without bidirectional iteration. In this model, the computational cost is put in the learning phase and is lower than other pervious nets (perceptrons and self-organizative maps); and the response in pattern recognition (without bidirectional iteration) is very quick. This model of associative memory is used as a clasification system, in a simple example of application, to analysis and clasifications of alphabetic characters.

I Introduction

The problem of pattern recognition has been received special attention by several neural approaches : multilevel perceptrons [Rumelhart,86], self-organizative maps [Kohonen, 88] associative memories [Hopfield,82] [Hopfield,84] [Kosko,88a] [Kosko,88b] [Hao, 92] [Hoffmann, 90] and so on. Multilevel perceptrons and self-organizative maps have an high learning capacity, but it offten needs accurate weights adjustments and long times to converge. In other way, associative memories can be used to pattern recognition with a quick learning phase (in general, the weight matrices are built by hebbian rules or algebraic methods); but it has a low storage capacity ($\sqrt{min(n,m)}$) for the BAM model [Wang,90a] being n and m the size of the patterns; and only the recalling of a limited number of patterns could be guaranteed [Wang,90a] [Wang,90b] [Wang,91]). We propose a new adaptative orthogonal memory which is based in Bidirectional Associative OrthoGonalized Memory model (BAOGM) [Ibarra,93] [Ibarra,94] [Ibarra,95] that has a higher capacity that others memories and uses

and special backpropagation algorithm to get a correct answer in a forward way without bidirectional iteration. In this model, the computational cost is put in the learning phase and is lower than other pervious nets (perceptrons and self-organizative maps); and the response in pattern recognition (without bidirectional iteration) is very quick. As a simple example of application, we use the Adaptative BAOGM to analysis and clasifications of alphabetic characters.

II Bidirectional Associative Orthogonal Memories

Let a set of q patterns pairs (a_i, b_i) for i=1,..,q belonging to the vector spaces R^n and R^m, where the patterns a_i are normalized in some way so $|a_i| = k$ $where$ $k \in R$ and i=1..q. The memory is built as a neural network (see figure 1) with two synaptic matrices (containing hebbian correlations) W and V, which are computed $W = AQ^t$ and $V = QB^t$, where A and B are built as :

$$A = \begin{bmatrix} a_{11} & a_{21} & \cdots & a_{q1} \\ a_{12} & a_{22} & \cdots & a_{q2} \\ \cdots\cdots\cdots \\ a_{1n} & a_{2n} & \cdots & a_{qn} \end{bmatrix} \quad B = \begin{bmatrix} b_{11} & b_{21} & \cdots & b_{q1} \\ b_{12} & b_{22} & \cdots & b_{q2} \\ \cdots\cdots\cdots \\ b_{1m} & b_{2m} & \cdots & b_{qm} \end{bmatrix}$$

Q is an intermediate orthogonal matrix (Walsh, Householder, and so on) of dimensions qxq. The q_i vectors of Q are an orthogonal base of the vector space R^q. This characteristic of the qi vectors is very important to make accurate associations including noise patterns [Pao, 89].

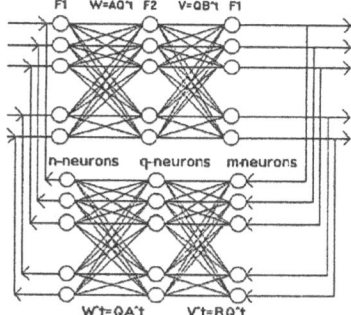

Figure 1. BAOGM topology

The input and output layer uses the clasical biporar filter f_1 (if patterns are coded in pibolar mode). In the hidden layer the BAOGM computes the filter f_2 (where $q^1{}_1$ and q^2 are the two possibles values of the Q components) :

$$f_2(x) = \begin{cases} q^1 & x \geq 0 \\ q^2 & x < 0 \end{cases}$$

The associations between patterns could be done in only one-step or in bidirectional mode. In One-step Recall and forward and backward mode we have :

• Let a_i the input pattern, the output b_i is

$$b_i = f_1\left[f_2\left(a_i^t \cdot W\right) \cdot V\right] = F\left(a_i\right)$$

• Let b_i the input pattern, the output a_i is

$$a_i = f_1\left[f_2\left(b_i^t \cdot V^t\right) \cdot W^t\right] = F^{-1}\left(b_i\right)$$

In the bidirectional mode, the patterns are fed forward and backward (feedback) into the BAOGM in a symilar BAM style while the energy is falling in a minimum of its energy surface. The process continue until a máximum number of iterations or a convergence desired grade is reached :

$$q_i^t \rightarrow F\left(q_i^t\right) \rightarrow b_i^{tl}$$

$$b_i^{tl} \rightarrow F^{-1}\left(b_i^{tl}\right) \rightarrow q_i^{tl} \qquad \text{Convergence condition } k < K$$

$$\cdots\cdots\cdots \qquad o \left| 1 - \frac{q_i^{k-1} \cdot q_i^{k}}{\left|q_i^{k-1}\right| \cdot \left|q_i^{k}\right|} \right| < e$$

$$q_i^{k-1} \rightarrow F\left(q_i^{k-1}\right) \rightarrow b_i^{k} \approx b_i^t$$

As we demostrate in [Ibarra,93] [Ibarra,94] the BAOGM converges to a minimum of its energy surface

$$E(a, q, b) = -\left(q_i^t \cdot W \cdot q_i + q_i^t \cdot V \cdot b\right)$$
$$= -(n + m)$$

and it's a general class of the BAM and MAHON [García, 92] [García, 94] systems.

III Adaptative Model

Forward operation, logically, is quicker than bidirectional mode in BAOGM. However, the linear dependences between the learning patterns make that the intermediate vector is

$$\overline{q} = a_i^t W = q_i + \sum_{j=1}^{q} \cos(a_i^t, a_j) q_j$$
$$j \neq i$$

where $\overline{q}$ has a noise component over q_i, so we have to apply extended [Ibarra,93] [Ibarra,94] or iterative [Ibarra,95] filters in the hidden layer. In this adaptative model, we propose to use backpropagation methods to reduce the linear dependence effect in $\overline{q}$. In the adaptative model a simple f_2 filter is applied in parallel to get an accurate response. The adaptative learning is made in two phases. First, we build the synaptic matrices W and V by hebbian rules. Then, we apply a backpropagation algorithm into the hidden layer to reduce linear dependences between patterns and so we can use the f_2 filter.

III.I Synaptic Ajustment

We built the synaptic weight matrices W and V as $W=AQ^t$ and $V=QB^t$, where A and B have the patterns to learn. A typical intermediate matrix Q is a Householder matrix built as $Q=I-2uu^t$ where is $u^t=[2/q \ 2/q \ ... \ 2/q]$, then the intermediate filter (f_2) in the hidden layer is

$$f_2(x_i) = \begin{cases} 1 - 2/q & x_i \geq 0 \\ -2/q & x_i < 0 \end{cases}$$

Where only the winner neuron x_i which is associated to the a_i pattern rises to $1-2/q$.

We apply an algorithm similar to the multilayer perceptron reduced to the hidden layer. In this case, we can use the activation function

$$f(x) = \frac{1}{1 + e^{-x}} - \left(1 - \frac{2}{q}\right)$$

where q is the number of diferent patterns to learn. The local error associates with an input a_i which an intermediate response $\overline{q} = a_i^t \cdot W$, can be calculated as

$$E_i = \begin{cases} 0 & \overline{q_{ii}} \geq 0 \\ \frac{1}{2} \sum_{j=1}^{q} |q_{ij} - \overline{q_{ij}}| & \overline{q_{ii}} < 0 \end{cases}$$

If we have a local error of zero then synaptic weights are not change; in other case we apply a delta rule which is defined as

$$\delta = \frac{(q_i - \overline{q_i})\left(1 - \overline{q_i} + \frac{2}{q}\right)\left(\overline{q_i} + \frac{2}{q}\right)}{2q}$$

The algorithm is shown in the figure 2. In this algorithm η is the learning constant and E_{max}, the maximum error which is allowed.

Input
  Matrix A, B, Q, E_{max}, η
Output
  Matrix W, V
Method

$W=AQ^t$ and $V=QB^t$
$E=0$
While $(E<E_{max})$ **do**
  $E=0$
  For i=1..q **do**
    $\overline{q} = a_i^t W_j$
    $$E_i = \frac{1}{2} \|q_i - \overline{q_i}\|$$
    if $(E_i>0)$ **then**
    $$\delta = \frac{(q_i - \overline{q_i})\left(1 - \overline{q_i} + \frac{2}{q}\right)\left(\overline{q_i} + \frac{2}{q}\right)}{2q}$$
    $W = W + hda_i^t$
    endif
  endfor
  $E=E+E_i$
Endwhile

Figure 2. Algorithm I

III.II Cluster Discrimination

Synaptic ajustment (algorithm I) has a higher convergence rate than backpropagation methods in perceptrons, because W and V matrices are built in a direct and controled way (without schocastic inicialization). However, if input patterns have some strong linear dependences then convergence time can be reduced by adding cluster discrimination to the process. Here, we build the synaptic matrices from a subset of the training patterns which has low dependences (A_{sub} and B_{sub}) in form diferent clusters. In this case, the matrices are made by outer product (hebbian rule) of the vectors in A_{sub} and B_{sub}; with q_i^t vectors, that have $1 - \frac{2}{q}$, value in the i dimension

$$q_i^t = \left[-\frac{2}{q} - \frac{2}{q} ... 1 - \frac{2}{q} ... - \frac{2}{q} - \frac{2}{q}\right]_{1xq}$$

So, we can built the matrices as

$$W = \sum_{i=1}^{sub} A_{sub,i} q_i^t \qquad V = \sum_{i=1}^{sub} q_i B_{sub,i}^t$$

When we have the V and W matrices, now all patterns of the training (a_i with i=1,..,q) are shown to the net. If output in the hidden layer is bad (another cluster is activated and the associate neuron has a negative value, $\overline{q_{ii}} < 0$) then we update the weights by adding a new cluster

$$W = W + a_i q_i^t \qquad V = V + q_i b_{ii}^t$$

Later, we use algorithm I (section III.I); so the general adaptative method (figure 3) is

Input
  Matrix A, B, Q, E$_{max}$, η, sub
Output
  Matrix W, V
Method

$$W = \sum_{i=1}^{sub} A_{sub,i} q_i^t \qquad V = \sum_{i=1}^{sub} q_i B_{sub,i}^t$$

For i=1..q do
  $\overline{q} = a_i^t W_j$

  if ($\overline{q_{ii}} < 0$) then
  $W = W + a_i q_i^t \qquad V = V + q_i b_{ii}^t$
  endif
endfor
E=0
While (E<E$_{max}$) do
  E=0
  For i=1..q do
  $\overline{q} = a_i^t W_j$

  $E_i = \frac{1}{2} \left\| q_i - \overline{q_i} \right\|$

  if (E$_i$>0) then

  $$\delta = \frac{\left(q_i - \overline{q_i}\right)\left(1 - \overline{q_i} + \frac{2}{q}\right)\left(\overline{q_i} + \frac{2}{q}\right)}{2q}$$

  $W = W + \eta\delta a_i^t$
  endif
  endfor
  E=E+E$_i$
endwhile

Figure 3. Algorithm II

IV Experiments

The adaptative orthogonal memory model can be used as a general classification system too. However, it has a lower convergence time than other classical models. As an example, it has been used to classify characters with a previous processing phase to get an invariant isometric representations [Yüceer, 93]. In RS/6000 workstations and programming in C++ and X_windows, we built a general character classification system. The net is trained to learn a set of characters (figure 4) and its associations with proper cardinal numbers (the class). Each character is represented in a 80x80 binary grid. For recognition test we make several transformations (translation, rotation, scale) over the characters (figure 5 and figure 6) to test and teach the net.

A	B	C	D	E	F
G	H	I	J	K	L
M	N	O	P	Q	R
S	T	U	V	X	Y
Z					

Figure 4. Set of characters for learning

Figure 6 shows convergence time (number of iterations) versus local error to learn a simple set of distorting characters by differents methods . There is a quick convergence rate in algorithm I (2 iterations) and algorithm II (1 iteration); versus multiperceptron with backpropagation (83 iterations). In others experiments with a lot of patterns the results are similar.

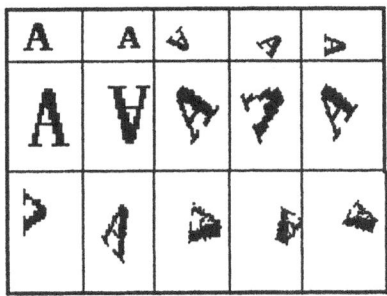

Figure 5. Character A under several transformations used to test

Other advantage of adaptative orthogonal models is that it don't need parameter adjust. In other way, the correct classification was 96% in translation patterns, 91% in rotation with a medium rate of 93.5%.

V Conclusion

Adaptative Orthogonal memories can be a good link between diferents models of neural networks (Associative memories and classifications systems), from a functional wiew. In the learning methods, the model combines hebbian rule in a first phase of training and later it uses backpropagation methods without accurate parammeter adjustment. So, it gets a high capacity of discrimination in a low learning time as we demostrate in experiments.

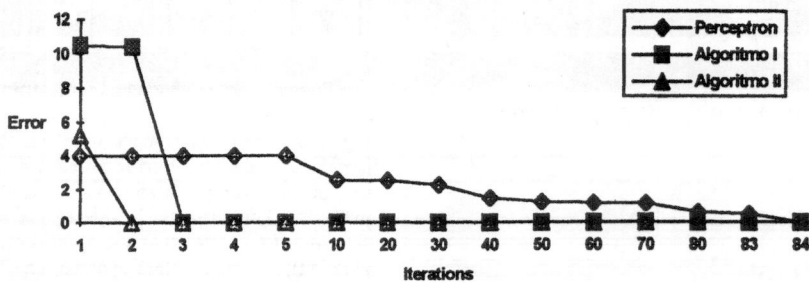

(a) Patterns for training

(b) Error versus iterations

Figure 6. Training set and learning time

References

[García, 92] Garcia-Chamizo J.M., Crespo-Llorente A.: "Orthonormalized Associative Memories". Proceeding of the IJCNN, Baltimore, vol 1, pg. 476-481 (1992).

[Garcia,94] Garcia-Chamizo, J.M. Thesis "Semicoberturas Heterogeneas de Regiones Bidimensionales morfológicamente no restringidas. Modelado Conexionista Aplicado". Servicio de publicaciones, U. de Alicante. 1994.

[Hao, 92] Hao J., Wanderwalle J.: "A new model of neural associative memoriy" Proceedings of the JJCNN92, vol 2, pg. 166-171 (1992)

[Hoffmann, 90] Hoffman G. W. , Davenport M.R. : "A network that uses the outer product rule , hidden neurons, and peaks in the energy landscape". IEEE Tans. on Neural Networks, pg 196-199 (1990).

[Hopfield, 82] Hopfield J.J.:"Neural Networks and physical systems with emergent collective computational abilities". Proceedings of the National Academy of Science, vol 79, pg. 2554-2558 (1984)

[Hopfield, 84] Hopfield J.J.:"Neural networks with graded response have collective Computational properties like those of two-state Neurons". Proceedings of the National Academy of Science, vol 81, pg. 3088-3092 (1984)

[Ibarra,93] Ibarra,F.; Garcia-Chamizo,J.:"Memorias Asociativas Bidireccionales Ortonormalizadas ". Actas AEPIA, vol 1, pg 20-30.

[Ibarra,94] Ibarra F. y Garcia-Chamizo J.M. "A generalized bidirectional associative memory with a hidden orthogonal layer" Proceedings of ICANN.1994.

[Ibarra,95] Ibarra, F. ;Garcia, J.M; Satorre-Cuerda, R. "An Orthogonal

947

Associative Memory with Parallel Iterative Filters". Proceedings of the International Conference in Neural Networks and Genetic Algoritms". France. 1995. Aceptado y pendiente de publicación.

[Kohonen, 88] Kohonen, T. " Self-Organization and Associative Memory". Ed. Springer-Verlag. 1988.

[Kosko, 88a] Kosko, B.: "Bidirectional Associative Memories". IEEE Tans. on Systems, Man & Cybernetics, vol 18, n 1 (1988)

[Kosko, 88b] Kosko, B. : "Competitive adaptative bidirectional associative memories". Procedings of the IEEE first International Conference on Neural Networks, eds M. Cardill and C. Butter vol 2. pp 759-66 (1988).

[Pao, 89] Pao You-Han : "Adaptative Pattern Recognition and Neural Networks". Addison-Wesley Publishing Company, Inc. pg 144-148 (1989).

[Rumelhart,86] Rumelhart, D.E; Hinton G.E.; Williams R.J. "Learning Internal Representations by Error Propagation". Ed. Rumelhart, D.E and Mc Clelland J.L. Parallel Distributing Procesing: Explorations in the Microestructures of Cognition.MIT Press, Cambridge. 1986.

[Wang, 90a] Wang , Cruz F.J., Mulligan: "On Multiple Training for Bidirectional Associative Memory ". IEEE Tans. on Neural Networks, 1(5) pg 275-276 (1990).

[Wang, 90b] Wang , Cruz F.J., Mulligan: "Two Coding Strategies for Bidirectional Associative Memory ", IEEE Tans. on Neural Networks, pg 81-92 (1990).

[Wang, 91] Wang , Cruz F.J., Mulligan: "Guaranted Recall of All training Pairs for Bidirectional Associative Memory". IEEE Tans. on Neural Networks, 2(6) pg 559-567 (1991).

[Yüceer,93] Yüceer, C; Oflazer, K. " A Rotation, Scaling and Translation Invariant Pattern Classification System". Pattern Recognition. Vol. 26. No. 5. pg 687-710. 1993.

Texture Classification on Real Time Using Semi-Cover Vector and an Orthogonal Neural Network

Asensi Muñoz, D.;Almagro León, A; Ibarra Picó, F

Departamento de Tecnología Informática y Computación
Universidad de Alicante
Apdo 99. Alicante. Spain.

Abstract.- A texture image classification system based on the semi-cover vector (SCV) and the bi-directional associative orthonormalized memory (BAOM) is described. The SCV is an statistic extraction method of texture features derived from the fractal geometry. It is invariant under geometric transformations. The BAOM is a neural network with a hidden layer of neurons that increases the learning capacity and reduces the noisy effect. This classifier works in real time and produces about 95.13% of correct classification rate.

1.-INTRODUCTION

Texture classification is one of the most important task in the analysis of texture images. It is at this stage that different texture regions within an image are isolated for subsequent processing, such as texture recognition.

Texture analysis is a particular case on image processing. The presence of texture causes difficulty to segment using the general approaches[1] (classical approach and fuzzy mathematical approach) to image analysis since texture do not has the main properties which the methods that use these approaches (histogram thresholding, edge detection, relaxation,...) are based, such as homogeneous surfaces and correspondence between intensity changes and object boundaries.

The major problem of texture analysis is the extraction of texture features. Texture features should have the followings properties: be invariant under the transformations of translation, rotation, and scaling; a good discriminating power; and take the non-stationary nature of texture account. There are two basic approaches for the extraction of texture features: structural and statistical[2]. The structural approach assumes the texture is characterized by some primitives following a placement rule. In this view, to describe a texture one needs to describe both the primitives and the placement rule. This approach is restricted by complications encountered in determining the primitives and the placement rules that operate on these primitives. Therefore, textures suitable for structural analysis have been confined to quite regular textures rather than more natural texture in practice. In the statistical approach, texture is regarded as a sample from a probability distribution on the image space and defined by stochastic model or characterized by a set of statistical features. The most common features used in practice are based on the pattern properties. They are measured from first and second order statistics and have been used as discriminators between textures.

Nowadays, most extraction methods of texture features provide high correct classification rate and they are easy to compute but are not enough fast for real time environments such the inspection of defects in textile fabric[3]. In this paper, a new method and a neural classifier that provide a high correct classification rate and real time response are present. Neural networks have proved to be powerful tools capable of mapping a finite set of input data onto a set of output data in a parallel way, allowing a significant reduction of computational complexity[4] that would otherwise have to be encountered using a conventional computation mechanism such nearest neighbor classification[5].

The semi-cover vectors (SCV) are a new statistic method for the extraction of texture features that has to do with Hausdorff dimension[6] which is derived from the fractal geometry. The major properties of this approach are: (i) it is fast and easy to compute, (ii) it is invariant under the transformation of translation, rotation, and scaling. Using the SCV joined to a neural classifier based on a simplified bi-directional associative orthonormalized memory (BAOM)[7-9] we obtain a texture segmentation on real time.

2.-SEMI-COVER VECTOR

In this section we propose a new method to represent texture. Texture is described by the distributions properties of the pixel with gray level greater than or equal to some threshold. The pixels may be taken as a set of points on a two-dimensional plane. The set of points is treated as a fractal set such that it can be characterized by the dimension of the set. Theoretically, the Hausdorff dimension is one of the most useful tools for characterizing the set; but it is difficult to estimate for most cases. The entropy dimension[10] and the intersections of a semi-cover are similar definitions to the Hausdorff dimension that are easy to compute. In this paper we present the later definition.

Several sets of points will be obtained if the thresholding of the image is done for several thresholds, and each set has its own number of possible intersections of the semi-cover. Taking this number as texture feature we obtain a vector, called semi-cover vector (SCV), as many components as thresholds.

Mathematic Concepts

Let R^2 be a MxM window. Points in R^2 will be denoted by lower case letters **a**, **b**, etc. If **a** and **b** are points in R^2, the distance between them is the Euclidean distance, i.e.

$$|a-b| = (|x_a - x_b|^2 + |y_a - y_b|^2)^{1/2}$$

where $a=(x_a,y_a)$ and $b=(x_b,y_b)$.

Let S be a set of finite number of binary pixels in R^2.

Definition 1. The diameter of a set $S \subset R^2$ is defined as the greatest distance apart of pairs of points in S, i.e.

$$|S| = \sup\{|a-b| \,/\, a,b \in S\} \qquad \text{sup = supremum}$$

Definition 2. If $\{A_i\}$ is a finite collection of sets of diameter at most δ that cover S, i.e. $S \subset \cup A_i$ with $0 < |A_i| < \delta$ $\forall i$, then $\{A_i\}$ is called a δ-cover of S.

Definition 3. Let h be a non-negative number. The h-dimensional Hausdorff measure of a set $S \subset R^2$ is defined as

$$H^h(S) = \lim_{\delta \to 0} H^h_\delta(S) \tag{1}$$

with

$$H^h_\delta(S) = \inf\left\{ \sum_i |A_i|^h \,/\, \{A_i\} \text{ is a } \delta\text{-cover of S} \right\} \tag{2}$$

$$\text{inf = infinimum}$$

From (2) it is clear that for any set S and $\delta < 1$, $H^h_\delta(S)$ is non-increasing with h, and so is the $H^h(S)$. In fact, letting δ be small enough, we have $H^r(S)=0$ for $r>h$, if $H^h(S)<\infty$.

Definition 4. The critical value of h at which $H^h(S)$ "jumps" from ∞ to 0 is called the Hausdorff dimension of S. Formally

$$D_H(S) = \inf\left\{ h \,/\, H^h(S) = 0 \right\} = \sup\left\{ h \,/\, H^h(S) = \infty \right\} \tag{3}$$

If $h=D_H(S)$, then $H^h(S)$ may be zero, infinite, or may be some finite constant. In general, most sets satisfy the last condition, i.e. $H^h(S)=c$ with c a constant such that $0<c<\infty$.

Proposition 1. The Hausdorff dimension of a set S is invariant under a isometric transformation (translation, rotation, and reflection).

Proposition 2. The Hausdorff dimension of a set S is invariant under a similar transformation (scaling).

Proof. See[10].

In general, it is difficult to measure the dimension of the set directly from the definition. Alternatively, we will use a fast and easy method to compute the dimension.

Intersections of a semi-cover

Although a textured image is irregular and has an heterogeneous nature it also has a certain frame. We can describe texture measuring this frame. To get this information we take a geometric piece with a given shape and size, then we calculate the number of intersections between the piece and texture frame such that the piece is within the frame. The piece is called *semi-cover*.

It is trivial that if we take a small semi-cover then the number of intersections will be large. On the other hand, with a great semi-cover the number of intersections will be small. When the size tends to infinite then the intersections tend to zero, and when the size tends to zero the intersections tend to infinite.

This conclusion it is not a coincidence between the Hausdorff dimension and the intersections of a semi-cover. There is a relation that it is demonstrable. Let S be the set of points that form texture frame. Let $N_\delta(S)$ be the number of intersections of a square semi-cover in S; the semi-cover has the following form

$$[m\delta, (m+1)\delta] \times [n\delta, (n+1)\delta], \quad m, n \in Z$$

The set of semi-covers in S provides a collection $\{A_i\}$ of $N_\delta(S)$ sets of diameter $\delta\sqrt{2}$ that approximately cover S. The δ-cover is an approximation since we consider only the semi-covers into texture frame.

From $H_\delta^h(S)$

$$H_\delta^h(S) = \inf\left\{ \sum_i |A_i|^h / \{A_i\} \text{ is a } \delta\text{- cover of S} \right\} \approx$$

$$\approx \sum_i^{N_\delta(S)} |A_i|^h = \sum_i^{N_\delta(S)} (\delta\sqrt{2})^h = N_\delta(S) \cdot (\delta\sqrt{2})^h$$

we can say that $N_\delta(S)$ is a factor of Hausdorff dimension.

Semi-cover Vector (SCV)

The intensities of an image $I(x,y)$, can be viewed as forming a surface in a three-dimensional Euclidean space (x,y,I).

Definition 5. A thresholded image of an image I at threshold interval $[th1, th2]$ is defined as

$$I_{[th1,th2]}(x,y) = \begin{cases} 1 & \text{if } th1 \leq I(x,y) \leq th2 \\ 0 & \text{otherwise} \end{cases}$$

We can think of the thresholded image $I_{[th1, th2]}(x,y)$ as being a two-dimensional plane with a set of points $S_{[th1,th2]}$ on it, where

$$S_{[th1,th2]} = \{(x,y) / I_{[th1, th2]}(x,y) = 1\}$$

Thus, the number of intersections of the semi-cover into the thresholded image can be computed.

Definition 6. The semi-cover vector (SCV) of an image I is defined as

$$SCV = (N_1, N_2, \ldots, N_i, \ldots, N_m)$$

where N_i is the number of intersections of the thresholded image $I_{[th1, th2]}$ and m is the number of intervals.

Unitarian SCV

If we use SCV computed from the number of intersections of an unitarian square semi-covers, that is, a 1x1 square then we can proof the following properties:

(1) The set of semi-covers in S provides a collection $\{A_i\}$ of $N_{\sqrt{2}}(S)$ sets of diameter $\sqrt{2}$ that cover S. In this case we have a δ-cover of S that it is not an approach.

(2) Unitarian SCV are invariant under isometric transformations. The transformation $f:R^2 \to R^2$ is isometric if it preserves distances, i.e. if $|f(a)-f(b)| = |a-b|$ for all a,b in R^2. Rotations and translations are isometric transformations since they preserve the texture frame. It is trivial that the number of intersections of a unitarian semi-cover is the same when the image is rotated or translated. In general it is the same for any isometric transformation. The unitarian SCV from two isometric transformations have the same modulus and direction.

(3) Unitarian SCV are invariant under similar transformations. The transformation $f:R^2 \to R^2$ is similar if it there is some constant λ such that $|f(a)-f(b)| = \lambda|a-b|$ for all a,b in R^2. Scaling and linear gray-level transformations do not preserve the texture frame but they conserve the linear ratio. In scaling, for example, the number of intersections of a unitarian semi-cover is proportional to the scaling factor. In general it is proportional to λ. The unitarian SCV from two similar transformations have the same direction and proportional modulus. If the classification algorithm is based on distance measurement then we have to normalize the SCV to obtain the any vectors. However, if the algorithm is based on angle measurement then it is not necessary some additional processing.

(4) It is easy and quick to compute unitarian SCV of a textured image since we only have to move the semi-cover through the image and testing the pixels value to increase the intersection counter.

In figure 1 we show some of these properties.

Texture 1

(1:1),0°   (1:1),20°   (2:1),0°   (2:1),20°

Texture 2

(1:1),0°   (1:1),70°   (2:1),0°   (2:1),120°

Texture 3

(1:1),0°   (1:1),90°   (3:1),90°   (2:1),60°

(a)

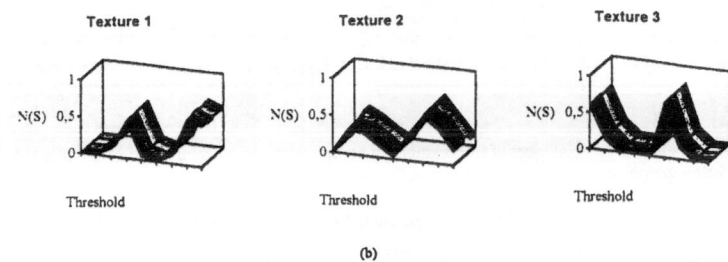

(b)

Fig. 1: (a) Samples of the rotated/scaled variants of three texture classes. (b) The SCV curves of (a)

3.-BI-DIRECTIONAL ASSOCIATIVE ORTHONORMALIZED MEMORY

The Bi-directional Associative Orthonormalized Memory (BAOM) is a special kind of an heteroassociative bi-directional memory with the capacity to store q pairs of patterns (a_i, b_i) i=1,...,q that can be defined into different vectorial spaces R^n and R^m. After this learning step this network is capable to restore the associated pattern b_i from the original a_i or from a noise version of a_i.

The main characteristic of this net is the hidden level of neurons. This hidden level increases the learning capacity and leaves us to introduce orthogonality that reduces the noisy effect that arises from the interdependencies between the learning patterns.

Like most network this have two steps: training and classification. In the training step two matrix V and W are constructed from the matrix A and B which contains the input and output patterns respectively. We can see this process like this:

$$W = AQ^t$$
$$V = QB^t$$

where Q is a qxq matrix

$$Q = I - 2uu^t$$

and u is a vector of q components $u_i = 1 / q^{1/2}$. Note that the q^i vectors of Q are an orthonormal base of R^q.

Then this net have n neurons at the input layer, q neurons at the hidden layer and m neurons at the output layer. All the patters of input an output are filtered into a bipolar domain. The filter used at the hidden layer of neurons is:

$$f_2(x) = 1 - 2u_i^2 \quad si \quad x \geq 0$$
$$f_2(x) = -2u_i^2 \quad si \quad x < 0$$

When the network is trained, we ask to the net which is the response b_i associate with the input a_i, and this is how the net obtain the response:

$$b_i = f_1\left[f_2(a_i^t W)V\right]$$

If this pattern is the lowest value of a energy function E, the finally answer is b_i. If b_i is not the minimum then we have to feedback the net in the opposite direction to obtain a'_i associate with b_i. The feedback answer is given by:

$$a_i = f_1\left[f_2(b_i^t V^t)W^t\right]$$
$$E(a,q,b) = -(a^t W_q + q^t Vb)$$

where E(a,q,b)=-(n+m) is the minimum value of convergence. All this process is repeat until obtain the minimum or a close value.

In figure 2 we show the structure of the network:

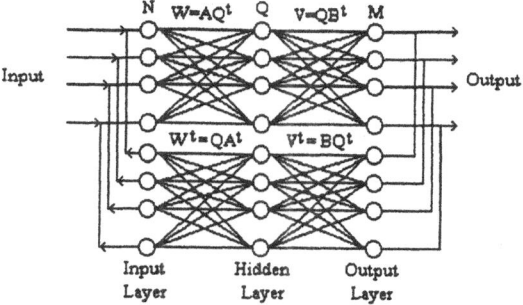

Fig 2: BAOM Network

In our system we try to obtain the fastest response since we want a real time response of a given pattern. Therefore we make a simplification on the BAOM algorithm, that consist of take the first response an do not feedback the net. Note that now we do not obtain the minimun value of the energy function. Instead we obtain one closed. All the process is the same at training and classification steps. We can see the net in figure 3:

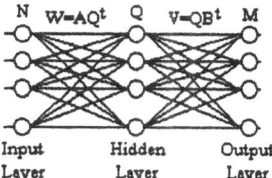

Fig 3: Simplified BAOM.

4.-TEXTURED IMAGE CLASSIFICATION

This section presents the structure of a textured image classification system based on the simplified BAOM that use the unitarian SCV as input vectors. Our main purpose is evaluate the ratio "correct classification rate / response time".

Our original texture images were $512*512*256$ bit , of which a central 40x40 portion was used as the $(1:1)^2$ training sample. In the learning step the BAQM was trained with six pairs of the form (SCV(t),LABEL(t)) where t is a texture, SCV(t) is the vector of unitarian semi-covers of t, and LABEL(t) is the label associated to t. The input space is R10, that is, the SCV has ten elements since we used the following threshold intervals: [40-80], [80-120], [120-160],[160-200], [200-240], [40-120], [80-160], [120-200], and [160-200]. Note that the gray-level range 40..240 was sampled twice.
In the recognition step the BAOM follows the diagram shows in Figure 4.

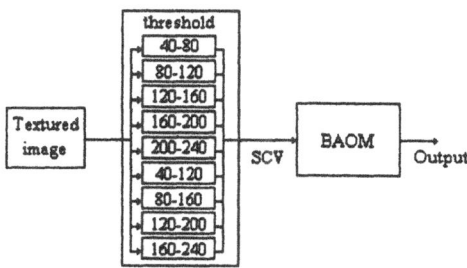

Fig 4: The recognition system.

The input images are windows of 40x40 pixels. In the classification test we use six different textured images of 256x256 pixels, that is 144 windows. The classification results are showed in Table 1:

TEXTURE	CORRECT	INCORRECT	CORRECT (%)
TEXTILE 1	135	9	93.75
BAKELITE	137	7	95.13
FOAM RUBBER	139	5	96.52
TEXTILE 2	134	10	93.05
SKIN	138	6	95.83
LEATHER	139	5	96.52
TOTAL:			95.13

Table 1

In the segment test the input images are windows of 40x40 pixels too, that form the mosaics that are showed in Figure 5(a). Texture mosaic 1 contains portions of the original 512*512 images that are different from the training portion. Texture mosaic 2 has the same textures with some rotation and scaling. Due to the purpose of this system is the classification of textured images and it is not the image segmentation the algorithm sometimes performs badly near the borders between the textures.

Figure 5 (b) is the gray level coded ground truth of Fig 5(a) In Figure 5 (c) we can see the output images provided by the BAOM. The number of windows processed, the classification rate and the response time are presented in Table 1.

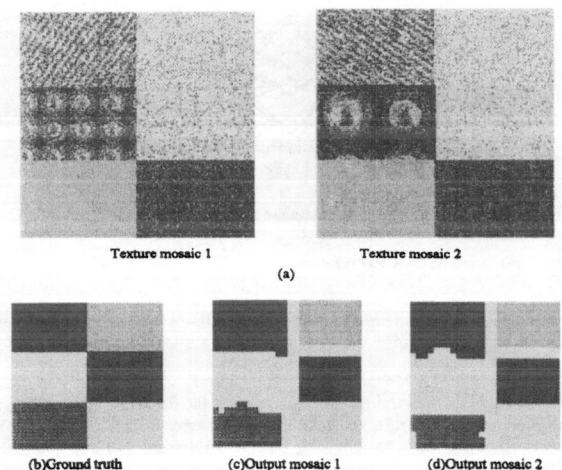

Texture mosaic 1   Texture mosaic 2

(a)

(b)Ground truth   (c)Output mosaic 1   (d)Output mosaic 2

Fig 5:(a) Texture mosaics. (b) Gray Level coded ground truth of (a).
(c)-(d) Classification results; Misclassified windows in white.

In our application the system needs 0.027 seg. in every window of 40x40 pixels, this time was taken on a 80486DX-33MHz.

5.-CONCLUSION

A real time system for texture analysis is successfully applied to texture classification. The system presents a statistic method for feature extraction and a neural classifier.

The method for the extraction of texture features is based on the semi-cover vectors (SCV). The SCV is related to the Hausdorff dimension and its most important properties are: it is easy to compute and it is invariant under geometrical mapping such as rotation, translation and scaling.

A Bi-directional Associative Orthonormalized Memory (BAOM) is used as a classifier; the BAOM has a hidden layer of neurons that increases the learning capacity and reduces the noisy effect that arises from the interdependence between the learning patterns, that is, from SCV.

This system works in real time since the SCV is easy to compute and the BAOM responds quickly. We have obtained a response time of 0.027 sec coding the SCV and the BAOM by a software simulator in C language. The importance of this neural system is that it could be coded in hardware by a VLSI architecture.

On the order hand, the system produces about 95.13% of correct rate. This suggest that this system based on the SCV and the BAOM is a good tool for texture analysis.

REFERENCES

1. N.R. Pal and S.K. Pal, A review on image segmentation techniques, Pattern Recognition, Vol. 26, No.9, pp. 1277-1294, 1993.

2. R.M. Haralick, Statistical and strucctural approaches to texture, Procc. IEEE, Vol. 67, pp. 786-804, 1979

3. C. Neubauer, Segmentation of defects in textile fabric, Procc. IEEE, pp. 688-691, 1992.

4. W.Huang and R.Lippmann, Comparison between neural net and conventional classifier, IEEE First Int. Conf. Neural Networks, Vol. IV, pp. 485-492, 1987.

5. R.O. Duda and P.E. Hart, Pattern recognition and scene analysis, New York, Wiley, 1973.

6. B.B. Mandelbrot, Fractal geometry, Freeman, New York 1993.

7. J.M.G. Chamizo and A.C.Llorente, Orthonormalized associative memories, Procc. IJCNN, Baltimore, vol. 1, pp. 476-481, 1992

8. J.M.G. Chamizo and F.I. Picó, Bidirectional associative orthonormalized memory, Procc. AEPIA, Madrid, 1993.

9. F.I. Picó and J.M.G.Chamizo, Genralized Bidimensional Associative memory with a hidden orthonormalized layer, Procc. ICANN, 1994.

10. C-M. Wu and Y-C. Chen, Multi-threshold dimension vector for texture anlysis and its application to liver tissue classification, Pattern Recognition, 1993.

An Architecture for Texture Segmentation: from Energy Features to Region Detection

P. M. Palagi & A. Guérin-Dugué

Laboratoire de Traitement d'Images et Reconnaissance de Formes,
Institut National Polytechnique de Grenoble, 46 Avenue Félix Viallet,
38031 Grenoble Cedex, France

Abstract. This paper presents a texture segmentation model realised with image treatment processing and artificial neural network techniques. Gabor oriented filters are used to extract texture features and Self-Organising Feature Maps to group these features. In order to decrease the number of filters needed to better extract features, we use a multiresolution procedure and a learning rule property to features interpolation. Two main axes of texture recognition are exploited with this model; one by the interpolating capabilities of the feature space, leading to a segmentation based on the number of chosen filters; and the other by the unsupervised segmentation of textured regions, leading to the number of textured objects on the image. Also presented, is a proposition to improve the model by an active learning for features fusion, and where the detection of contours in low resolution levels are used to control the focus on the textured regions of the immediately posterior resolution level.

1. Introduction

Filtering images successively with Gabor functions that best cover the frequential spectrum, makes it possible to extract the characteristics belonging to different textures, and by consequence, to segment those textures. Some texture segmentation models have already shown the integration of Gabor functions and artificial neural networks [17, 21, 24], and especially the self-organising feature maps (SOFM)[14]. Usually, the computational cost is very high due the number of filters and their kernel size. In this paper, we present two main optimisations while using a multiresolution method of image processing and an adapted learning rule of SOFM.

Another proposition to enhance our texture segmentation model is also introduced based on some psychological experimental data [23]. Sequential vision treatment is realised in the visual cortex: visual information from low to high frequencies is transmitted sequentially through the cortical areas. With an image decomposition on multiple resolution levels, we can simulate this sequential process. The proposed architecture takes into account this spatio-temporal processing even for analysing static images. Segmented images are obtained at each resolution level, but the final results will use all levels. The intermediate results allow to increase the learning speed by controlling the focus in the immediately superior resolution level.

In sections 2 and 3, we present the use of Gabor functions to treat and extract the texture characteristics under the form of energy features, and the multiresolution method implemented. Section 4 shows a general description of the model architecture including the integration of the spatio-temporal process. Section 5 shows the results obtained in textured image segmentation when using this model through two different aspects: interpolating capabilities on the features space and the unsupervised segmentation of the textured regions. In the last section, some conclusions and perspectives are drawn.

2. Texture features: energy of oriented filters

Human texture segmentation is possible due to differences in sensitivities to local graphic configurations such as line segments, crossings or terminations of line segments [13]. In neurophysiological terms, this differentiation could be due to responses of complex and hypercomplex cortex cells, based on earlier received responses of simple cortical cells [12]. Measures from the receptive fields of the visual cortex simple cells of mammalians yielded the evidence of their analogy with band-pass filter functions sensitive to frequency bandwidths and to different orientations [5]. Differences of offset Gaussians (DOOG) or Gabor functions expanded in two dimensions were proved to aptly model orientation sensitive cells [19, 26]. We have chosen to model the receptive fields with Gabor functions, in order to realise texture segmentation. Gabor functions were chosen because of their good local sensitivity both in the spatial and frequential domains and due to their straightforward mathematical formalism and computational simplicity.

In the spatial domain, one Gabor function is defined as a sinusoidal wave modulated by a Gaussian function. In the frequency domain, its response shape is given by the Fourier transform of the Gaussian function and its location (the maximal value of the function) is given by the frequency of the sinusoidal wave, as:

$$\phi(u,v) = \exp\left[-2\pi^2\left(\alpha^2(u-u_0)^2 + \beta^2(v-v_0)^2\right)\right]$$

where α and β are the space constants of the spatial Gaussian envelope in x and y respectively. So, such filters enable the detection of local configurations according to:

  a. their position given by a frequency f_0 and an orientation θ and;

  b. their spatial variances given by α^2 and β^2. In the frequency domain, these parameters are linked to the frequency bandwidths.

Figure 1 represents symbolically the location and the spread of such oriented pass-band filter.

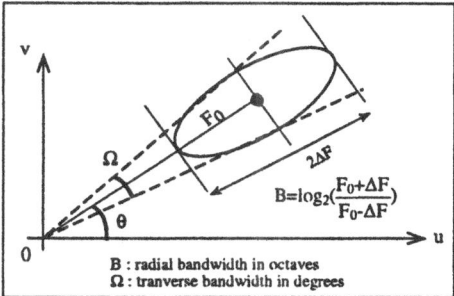

$$B = \log_2\left(\frac{F_0 + \Delta F}{F_0 - \Delta F}\right)$$

B : radial bandwidth in octaves
Ω : tranverse bandwidth in degrees

Figure 1: Symbolic representation of the frequency response of the Gabor filter in the frequency domain.

One texture is characterised by a certain spectral content within the 2D frequential space. In order to best characterise textures, it is necessary to cover the whole frequency space by choosing a certain family of Gabor filters by means of different orientations and frequencies. This family can be generated by changing the spatial frequency and orientation location of a Gabor filter through homothety or rotation.

Let us define the result of the convolution of an image with an even or odd Gabor kernel at the frequency f_0 and the orientation θ, by $G_{f_0,\theta}^e(x,y)$ and $G_{f_0,\theta}^o(x, y)$ respectively . According to Polen and Ronner [22], a pair of simple cells in the same cortical column have a phase difference of 90° and the same sensitivity to frequency and orientation which confirms their modelisation by even and odd Gabor filters. As a consequence, we calculate texture features as local energies, estimated from the odd and even Gabor transforms by:

$$E_{f_0,\theta}(x,y) = \sqrt{G_{f_0,\theta}^e(x,y)^2 + G_{f_0,\theta}^o(x,y)^2}$$

which also allows a better localisation of intensity discontinuities in different textures [11].

Nevertheless, the idea of convoluting one image with a family of Gabor filters leads to the treatment of a great amount of information and to a high computational cost. It is desirable to reduce the kernel sizes and their number. The first reduction is achieved through the multiresolution framework. The second can be realised by interpolation in the energy feature space : more orientations can be obtained with less filters.

3. Multiresolution Principles

Usually, texture segmentation models which integrate Gabor functions need a large number of filters to reach high frequency and orientation selectivities. For instance, they can use 16 filters (4 frequencies and 4 orientations) or, more frequently, 24 filters (4 frequencies and 6 orientations) or even more [7, 18, 20, 25].

With an implementation that directly uses the Gabor transforms, the spatial size of a convolution kernel may double its dimension in the x and y directions as the spatial frequency decreases by a factor of two. For instance, the 11x11 pixel convolution kernel of a Gabor filter localised at $f_0 = 0.25$ (filter example with a certain selectivity [10]) will be transformed into an 81x81 kernel at the frequency $f_0/8$. This means that the computation cost increases 8^2 times between these two filters.

In order to optimise our model, we used a pyramidal multiresolution method of image processing [3, 10]. The pyramidal decomposition is obtained by low-pass filtering and sub-sampling. This method allows the reduction of the image at each level. In this way, instead of using the filters localised at the frequencies $f_0/2$ (21x21 pixels) or $f_0/4$ (41x41 pixels) or $f_0/8$, we use, for all the levels of the pyramid, the same filter at f_0 (11x11 pixels) for the orientations being considered.

With the pyramidal multiresolution method, there is not only an economy of the computational cost, but also a possibility of improvement of the model's performance with biological plausibility. We will discuss these different properties in the next section.

4. General architecture

4.1. General description

The proposed architecture (figure 2) may be described in two parts. The first part (I) is the pre-processing for features extraction. These are estimated from a combination of pyramidal decomposition and Gabor filtering. The features correspond to local energies characterising the images. We obtain with different frequency bands and orientations, vectors of energy features (EF). By using a self-organising features map inside the feature generation process, interpolation capabilities increase the orientation selectivity handled by the original energy features estimated from the filters bank. The new features are called "interpolated energy features" (IEF). Section 5.1 will describe the obtained results. So, for the following architecture, these two features spaces ("IEF", "EF") can be indistinguishably used.

For the detection of textured regions, two approaches may be considered. The first (part II) is a fusion of the spatial locations with the energy features, before the decision. In the second (part III), a simple vector quantifier can provide unsupervised region detection. This part will be described in section 5.2. By this way, a global detection may be done with the features provided by all the frequency bands, or a partial detection in each frequency band. In this case, combining the different decisions is a quite difficult task, original data are lost in the decision space. We think that this approach is limitted due to the poor performance of vector quantisation algorithms for this task. The performance can be enhanced by fusion of other features, as it is proposed on several applications [4, 16].

Here, we consider that the decision task is the ultimate one, it means that fusion is realised in the features space in order to take benefit of features correlations. The part II of the figure 2, proposes at this stage, fusion with spatial locations. This principle is general and it can be applied with multiple feature categories. We use a self organising features map (not necessarily a Kobonen's map) [6] trained with the concatenation of the enegy features (real or interpolated ones) and the associated spatial location. Correlations between them are learned in order to permit boundaries detection. This part of the study is now under on test.

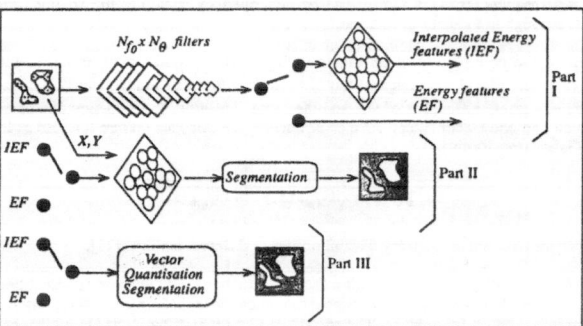

Figure 2: Model architecture description

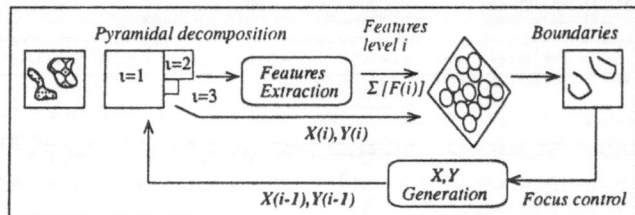

Figure 3: Integration of the boundary detection in each different frequency channel and control of the focus in the immediately superior resolution level.

4.2. Fusion of energy features and spatial locations

We consider here part II of figure 2. The energy features are obtained at each different resolution level. The self-organising features map is trained by associating the energy features and the spatial locations. At the beginning of the learning phase, the weights of the spatial information are important in order to impose the topological structure of the 2D image. The first boundaries can be obtained on the lower resolution level by sequentially scanning the map. At this level, this boundary information can be used to control the new focus locations in the immediately superior resolution level, in order to increase the learning rate. The learning is realised with all the available energy features. This process is repeated until the higher resolution level is reached.

5. Feature detection

Two parts of the model architecture were already implemented (parts I and III in figure 2), and some results are presented here. After the image pre-processing realised with the multiresolution and Gabor filters, the energy features may be used in two ways: either by training a neural network to organise itself according to the number of chosen filters (f_0, θ), or directly as a data base to a vector quantisation in order to segment textured objects without supervision. In the first case of features extraction, one neural unit will be sensitive to just one frequency, in the second case one neuron will be sensitive to one frequential composition (i.e. a particular texture).

5.1. Frequency and orientation detection

In order to have each neuron of a neural network sensitive to a certain frequency and a certain orientation, we used a learning data base composed of an image constructed from concentric rings at frequency $f_0 = 0.25$ (figure 4a). This image was filtered with 12 oriented Gabor filters localised between 0° and 165° (at $f_0 = 0.25$), with a transverse band $\Omega = 15°$. This learning image has been choosen in order to represent all the possible orientations. Learning with this image will represent permanent learning on all the orientations in this frequency band. When a SOFM with 12 neurons is organised with this learning base, we obtain each neuron sensitive to one orientation sector, and the neurons are ordered (figure 4b). We can observe this regularity due to the colour gradation presented in the image segmentation. The same image filtered with 6 filters localised between 0° and 150° (at $f_0 = 0.25$, $\Omega = 30°$) also has a good segmentation when using a SOFM with only 6 neurons (figure 4c).

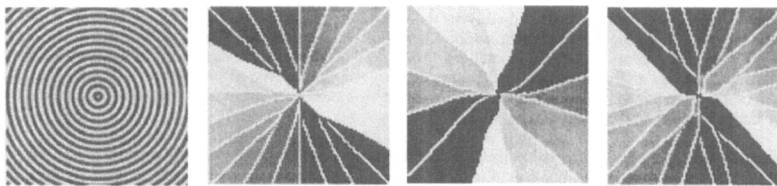

Figure 4: From left to right (a) artificial image, (b) segmentation with 12 filters ($f_0 = 0.25$, θ from 0° to 165°, $\Omega = 15°$) and 12 neurons, (c) segmentation with 6 filters ($f_0 = 0.25$, θ from 0° to 150°, $\Omega = 30°$) and 6 neuron; (c) segmentation with 6 filters ($f_0 = 0.25$, θ from 0° to 150°, $\Omega = 30°$) and 12 interpolated neurons. (b), (c) and (d) result from superposition of texture segmentation and contour detection.

Interpolation capability in weight space has been enhanced by using a gaussian neighbourhood in the classical Kohonen learning rule [15]. By this way, if we use more neurons than filters (i.e. more neurons than input clusters), the extra neurons will be located inside the input distribution, and not considered as "dead units". In our case, we can use less orientations on the chosen filters and still have a good selection in terms of the number of orientations. For example, with six oriented filters, we can obtain at least twice the number of different orientation selectivities as when using 12 neurons (twice the number of filters). This is obtained by redundant features built from overlapped filter configurations explained in [9]. In this case, we can use the same 6 filters localised between 0° and 150° (at $f_0 = 0.25$, $\Omega = 30°$) and still have a good spectral representation of 12 orientations. This result is shown in figure 4d where 12 neurons were obtained from 6 filters, and we still have one winner neuron for each filter and one neuron sensitive to intermediate orientations not presented in the set of filters. The final segmentations are similar in figure 4b and 4c.

The same 12 neurons used to segment figure 4d were used to segment a real image. Figure 5a was obtained by scanning a publicity poster from ICANN-94. In its segmentation (figure 5b), we can find the same oriented regions coloured by the same neurons. The background of the image composed of horizontal lines is coloured by the same neuron. At right, the woman's dress, that is mainly composed by vertical lines, is also coloured by only one neuron. The man's hat and back are formed by curved lines. In their segmentations we can see a degradation of colours due to different neurons sensitivities to justaposed orientations. This means that there is one neuron sensitive to each frequency-orientation composition.

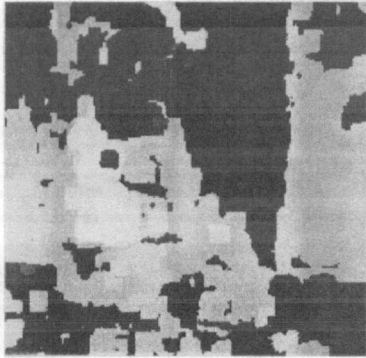

Figure 5: (a) scanned image, (b) segmentation with 6 filters ($f_0 = 0.25$, θ from 0° to 150°, $\Omega = 30°$) and 12 interpolated neurons.

5.2. Region detection

When we select the number of neurons of a SOFM based on the number of regions, we want in fact to compress the image in the number of textured objects to make pattern recognition. In this case, the artificial neural network used realises a vector quantisation, one neuron will represent one different region/class. This operation will succeed if each object (textured region) is well identified in the features space, represented by a well separated cluster. Optimally, if the samples distribution of each cluster is supported by a hyper-sphere region, one prototype will be enough to represent it. If not (i.e. if the shape of the cluster is more complex) more prototypes will be needed. The sensitivity to the probability density must be the main characteristic of the vector quantisation algorithm. With few prototypes, they must first reach the different modes of the distribution, where each mode corresponds to an object. With the growth of the number of prototypes, they will be placed according to the number of samples in each cluster (i.e. the size of the textured object).

We have chosen to evaluate the quantisation quality by the global distortion (mean squared error between input vectors and final prototypes). When training a vector quantisation algorithm with a growing number of prototypes, we can observe a decrease of the distortion criterion (figures 6c and 7c). The shape of the obtained curve can give some indications on the intrinsic data structure. For example, the curve in figure 6c shows a plateau from 3 prototypes. In figure 7c, we only see a continuous decrease of distortion, fast decrease in the beginning and slow decrease with the growth of number of prototypes (the frontier between these two regions is not so evident).

We have tried two classical algorithms to do this. The first one is a modified version of the Competitive Learning algorithm [8] where the sensitivity to the distribution is explicitly implemented: the Frequency Sensitive Competitive Learning (FSCL) [1]. The second one is a SOFM, where the property of vector quantisation of the input space is used, with a learning rule enhanced with a Gaussian neighbourhood, which we call here KG. We noticed that both algorithms have the same performance to this task as can also be observed on the graphs of average distortion.

To realise the pattern recognition, a bank of 6 filters ($f_0 = 0.25$, θ from 0° to 150° with $\Omega = 30°$) is applied to each of the 3 levels of a pyramid transform of one target image (the image to be segmented). The learning data base is created from the original image (figures 6a or 7a). The samples are randomly chosen in the features space.

The segmentation task in figure 6 is quite easy. The evolution of the distortion gives the optimal number of prototypes which exactly corresponds here to the number of textured regions.

A much more difficult test can be observed with the image composition of Brodatz [2] textures of figure 7a. In this composition, the 7 textured regions are not easy to visually discern. Its segmentation is still not very easy when using a SOFM with 12 neurons (fig.7b). The lecture of the distortion curve in figure 7c is not so evident to impose an optimal number of prototypes. The clusters in the feature space are not so separated and have a more complex shape. With 12 prototypes, the different textured regions are detected but also the frontiers between them. The fusion of energy features of Gabor filters and their spatial locations will be realised in future implementations, as discussed in section 4, in order to improve this kind of result, to process precise boundaries.

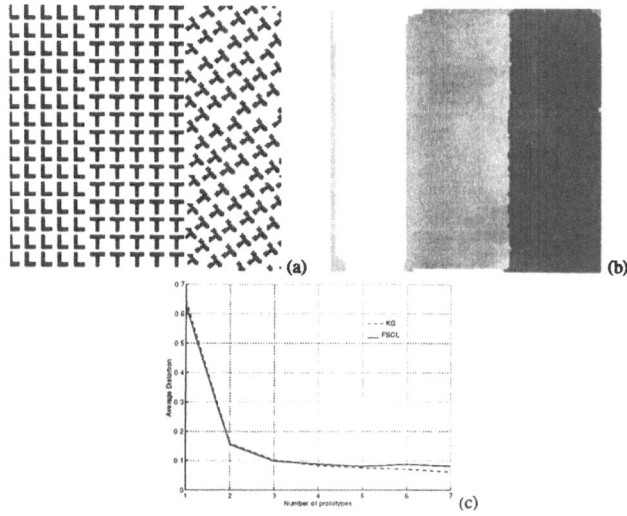

Figure 6: (a) artificial image, (b) segmentation with 3 neurons (one for each region), image filtered with .6 filters ($f_0 = 0.25$, θ from 0° to 150°, Ω = 30°) and 3 pyramid levels; (c) average distortion measured with different numbers of neurons (prototypes).

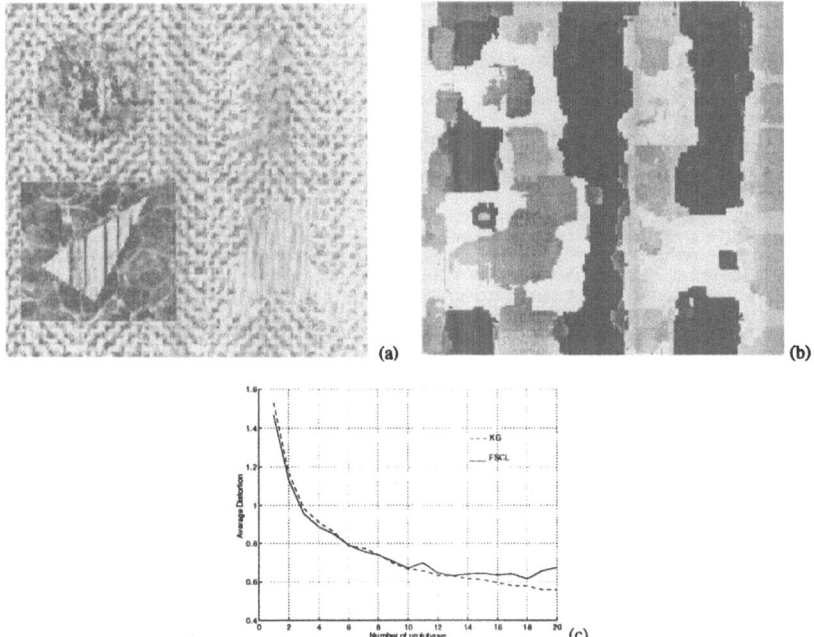

Figure 7: (a) Image composed of Brodatz textures, (b) segmentation with 12 neurons (more neurons than the number of regions), image filtered with 6 filters ($f_0 = 0.25$, θ from 0° to 150°, Ω = 30°) and 3 pyramid levels. (c) average distortion measured with different numbers of neurons (prototypes).

6. Conclusion and Perspectives

This model of segmentation is based on the analysis of frequential characteristics of textured images. The multiresolution framework and the SOFM performed interpolation allow an economical implementation of the filters. The concatenation of a pre-processing of Gabor filters energy features and self-organising features map lead to a good performance of segmentation on images composed of not very complex textures. To achieve better results of segmentation on complex, natural textures, and to have at the same time biological plausibility, we need other types of cortical treatments like temporal transmission of information, control of focus attention, and fusion in features spaces of all the resolution levels. We propose to enhance the model with a spatio-temporal treatment of the frequential information where the contours of the low resolution levels are re-injected in upper levels in order to control the "visual attention" to segmented regions.

Acknowledgements

The authors would like to thank Phillippe Gaussier from the ENSEA ETIS who sent us the Brodatz images.

References

1. Ahalt, S. C. et al. Competitive Learning Algorithms for Vector Quantization, Neural Networks, Vol. 3, 277-290 (1990).
2. Brodatz, P. Textures: A photographic Album for artists and designers. Dover Publications Inc., New York (1966).
3. Chehikian, A. Algorithmes optimaux pour la génération de pyramides d'images passe-bas et laplaciennes. Traitement du signal, Vol. 9, No. 4, 297-307 (1992).
4. Chu, C. C. & Aggrawal, J. K. Image interpretation using multiple sensing modalities. IEEE Trans. Patt. Anal. Mach. Inteligence, Vol. 14, No. 8, August (1992).
5. Daugman, J. G. Uncertainty relation for resolution in space, spatial frequency, and orientation by two-dimensional visual cortical filters. J. Opt. Soc. Ame. A, Vol.2, No.7, 1160-1169 (1985).
6. Demartine, P. Analyse de données par réseaux de neurones auto-organisés. Ph.D. Dissertation, Institut Nationale Polytechnique de Grenoble, December (1994).
7. Ghosh, J. & Bovik, A. C. Neural Networks for Textured Image Processing, in Artificial Neural Networks and Statistical Pattern Recognition: Old and News Connections. Elsevier Science Publishers, 55-73 (1991).
8. Grossberg, S. Competitive learning: From interactive activation to adaptive resonance. Cognitive Science, Vol. 23, 23-63 (1987).
9. Guérin-Dugué, A. & Palagi, P.M. Texture segmentation using pyramidal Gabor functions. Neural Processing Letters, Vol. 1, No. 1, 25-29 (1994).
10. Guérin-Dugué, A. & Palagi, P. M. Implantations de Filtres de Gabor par Pyramide d'Images Passe-Bas. Submitted to Traitement du Signal (1995)
11 Heitger, F. & al. Simulation of Neural Contour Mechanisms: from Simple to End-stopped Cells. Vision Research 32(5) 963-981 (1992).
12. Hubel, D. H. & Wiesel, T. N. Functional architecture of macaque monkey visual cortex. Proc. R. Soc. Lond. B. Vol. 198, 1-59 (1977).
13. Julesz, B. Texton Gradients: The Texton Theory Revisited. Biol. Cybern. Vol. 54, 245-251 (1986).
14. Kohonen, T. Self organisation and Associative Memories. Springer Verlag, Berlin (1984).
15. Kohonen, T. Self-Organisation maps: Optimization approaches. In T. Kohonen, K. Mäkisara, O. Simula, and J. Kangas, editors, Artificial Neural Networks: Proc. ICANN-91, II 981-990. North-Holland (1991).
16. Lovell, R. et all. A model of visual texture discrimination using multiple weak operators and spatial averaging. Pattern Recognition, Vol. 25, No. 10, 1157-1170 (1992).
17. Lu, S. & al. Texture Segmentation by Clustering of Gabor Feature Vectors. IJCNN I-683 (1991).
18. Malik, J. & Perona, P. Preattentive texture discrimination with early vision mechanisms. J.Opt.Soc.Ame.A, Vol.7, No.5, May (1990).
19. Marcelja, S. Mathematical description of the responses of simple cortical cells. J.Opt.Soc.Ame.A, Vol.70, No.11, 1297-1300 (1980).
20. Navarro, A. & Tabernero, A. Gaussian Wavelet Transform: Two Alternative Fast Implementations for Images. Multidimensional Systems and Signal Processing, Vol. 2,ï 421-436 (1990).
21. Oja, E. Self-Organising Maps and Computer Vision. Neural Networks for Perception. Ed. Harry Wechsler. Volume 1 - Human and Machine Perception. Academic Press Inc. San Diego (1992).
22. Polen, D. & Ronner, S. Phase Relationships Between Adjacent Simple Cells in the Visual Cortex. Science, Vol.212, 1409-1411 (1981).
23. Schyns, P. G. & Oliva, A. From Blobs to Boundary Edges: Evidence for Time- and Spatial-Scale-Dependent Scene Recognition. Psychological Science, Vol. 5, No. 4, July (1994).
24. Tomasini, L. Apprentissage d'une Représentation Statistique et Topologique d'un Environnement. Ph.D. Dissertation, Ecole Nationale de l'Aeronautique et de l'Espace, February (1993).
25. Turner, M. R. Texture Discrimination by Gabor Functions. Biol. Cybern, Vol. 55, 71-82 (1986).
26. Young, R. The Gaussian derivative theory of spatial vision: analysis of cortical cell receptive field line-weighting profiles. Tec. Rep. GMR - 4920 General Motors Research, Warren, Mich, USA (1985).

A Lattice-Based Time-Delay Neural Network for Speech Processing

P. Gómez, V. Rodellar, V. Nieto, M. A. Hombrados

Departamento de Arquitectura y Tecnología de Sistemas Informáticos
Universidad Politécnica de Madrid
Campus de Montegancedo s/n, Boadilla del Monte, 28660 Madrid, SPAIN
Tel: +34.1.336.73.84   Fax: +34.1.336.74.12   e-mail: pedro@pino.datsi.fi.upm.es

Abstract

The use of Time-Delay Neural Networks (TDNN's) in *Continuous Speech Recognition* has not been as relevant as it was expected due to the computational costs implied by Time-Delay orders, as it was taken for granted that the bigger the orders, the better the representation of the dynamic essence of Speech. This paper focuses on the true differential nature of this representation, and proposes to see TDNN's as devices working on differential relations among delayed versions of Speech Spectra, using Lattice Predictors as processing delay lines, which de-correlate the information which is presented to the computing nodes. This results in optimally compact structures (minimum number of delays), and better convergence rates. Convergence experiments show that reductions in the global computational costs as low as 1:5 may be achieved using structures based on this method as compared with traditional TDNN's.

1. Introduction.

The precise *Identification and Recognition of Continuous Speech* demands a great ability from *Recognition Algorithms* to detect *Non-stationary Processes*, due to the intrinsic nature of *Speech Generation*. Typically, Hidden Markov Models and Time Delay Neural Networks (TDNN's) have been used for such purposes. Each of these methods represent time variability in *Speech Spectra* using different paradigms. HMM's use probability relations to determine the transitions on a *State Diagram*, while TDNN's make use of weighted associative relations among time-aligned spectral templates, to establish the detection of a given pattern. In some sense, both methods differ in that HMM's use estimations of the statistics of the templates in the *Ensemble Domain*, and TDNN's rely on estimations in the *Time Domain*. TDNN's behave in some way as *Adaptive Filters*, tracing the relations among input and output variables depending on the *Second Order Statistics* of the processes involved. When TDNN's are fed with spectral vectors of a given stochastic process, they can detect the changes in the second order statistics of the process, and as such, are very accurate detectors of the dynamic behavior of Speech. TDNN's have produced good results in applications for phonemic detection, either alone or in collaboration with HMM's [Dug.94]. However, there have been complaints on their relatively high computational complexity. The increase in computational complexity in TDNN's when compared to other Neural Networks comes from the fact that they must use several time-delayed samples of the spectral vectors representing the non-stationary process. This implies a proportional increment in the number of weights, and consequently, in the need for storage and computational costs. The relation between the number of delayed copies of the spectral vectors and the detecting capabilities of the TDNN has not been explored in deep, as it has been generally assumed that the bigger the number of delay stages, the better the tracing of the non-stationary nature of Speech [Mor.91]. In a preliminary research, the basis for such study using the general theory of Adaptive Filtering, has been established [Rod.94a]. The aim of the present paper is to propose a new structure of TDNN to exploit the behavior of dynamic detection of non-stationary processes with the optimal number of delay stages, determining the order of non-stationarity of the processes involved. To adequately dimension TDNN's the number of delayed versions of the activation vectors, the number of hidden layers, and the dimensionality of each layer must be taken into account. The number of delays included is one of the most influential parameters in relation with the computational complexity of the TDNN. In the structure represented in Fig. 1, used for *Phonemic Detection* [Rod.94b] with p input nodes, q nodes in the hidden layer, and r nodes in the output layer, and k and m the number of delay levels in the input and hidden layers respectively, the resulting computational complexity measured as the number of floating point products C, is:

$$C = 2p(k+1)q + 3q(m+1)r + 3q + 4r \qquad (1)$$

and the amount of memory required to store the weights may equally be established as:

$$M = p(k+1)q + q(m+1)r + q + r \qquad (2)$$

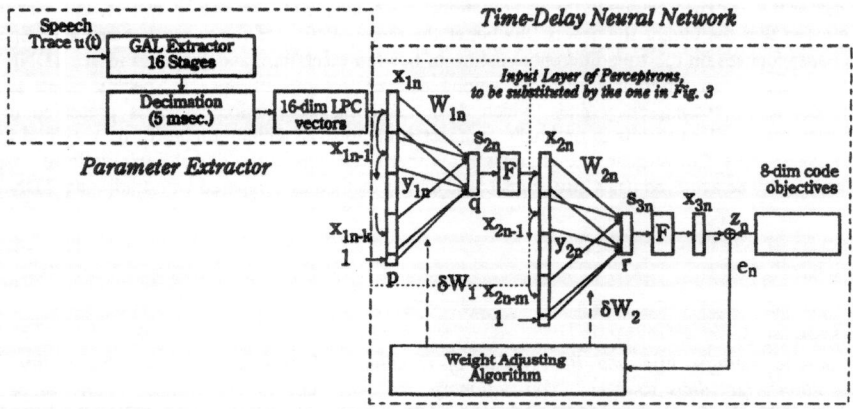

Figure 1. Global Framework for Phonetic Decoding and Architecture of a TDNN typically used for such purposes. The dimension of the input LPC vectors is given by $p=16$. The depth of the delay lines is $k=3$ for the input layer and $m=3$ for the hidden layer. The number of outputs is $q=8$ for the input layer and $r=8$ for the hidden layer. F is a sigmoid-like limiting function. The *Weight Adjusting Algorithm* is classical *Back-Propagation*.

In *Eqs. 1* and *2* the relative importance of k and m may be appreciated, the global computational complexity and memory requirements depending linearly on these parameters. The orders of p and r in a practical case are given by the dimensionality of input data and output objectives, as in the case mentioned above, where $p=16$ and $r=8$. The case for q is slightly different, as this parameter is related to the convergence of the Neural Network [Mor.91], and must be fixed experimentally most of the times. For many Phoneme Detection applications it may be fixed to $q=8$ [Rod.94b]. To establish optimal values for such parameters, reducing the computational expenses to a minimum, the role played by time-delays must be carefully understood. The aim of the present paper is to explore this role, and to give practical measurements of its adequate value for different speech traces. Finally, a generalized structure for a TDNN will be presented to better exploit the hybrid nature of TDNN's amidst Digital Filters and Neural Networks.

2. TDNN's as detectors of the order of Non-Stationarity.

TDNN's were first proposed by Waibel et al. [Wai.89] as a means for detecting the time-varying features of *Mel-Scale Speech Spectra in Phonemic Decoding*. The main reason for their success was that these networks introduced a way for representing the slow time-varying characteristics of Speech Spectra, which are due to the intrinsic non-stationary nature of the Speech Traces. The innovative aspect of the approach was to present ordered sequences of Speech Spectral Vectors to the input and hidden layers of a *Back-Propagation Architecture*. The reported success stimulated other researchers to investigate the applications of such network to Speech Recognition, either as stand-alone detectors, or as cooperating algorithms with Hidden Markov Models [Dug.94]. That Speech is a Non-Stationary process is a well-known fact, due to the dynamic nature of the process of Speech Production, based in the continuous movement of *Articulatory Organs* among certain pre-established positions. These dynamic changes in the Vocal Tract continuously modify the spectrum of the resulting signal, rendering its characterization a rather complicated process.

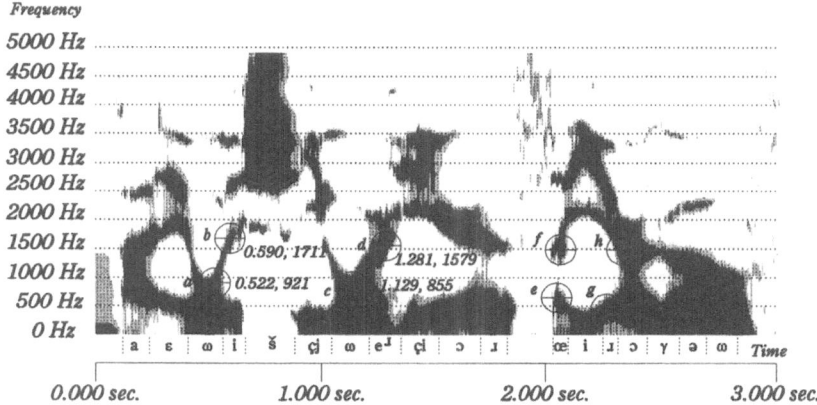

Figure 2. Spectrogram of the utterance *I wish you were here a year ago*, spoken by a male speaker. The horizontal axis is graded in sec., and the vertical axis is given in Hz.

To illustrate this dynamic nature in a practical case, Fig. 2 represents the spectrogram of the sentence *I wish you were here a year ago*, where the positions of the formants (maxima) in the spectrum with time is continuously changing, this being the case in *Continuous Speech Recognition (CSR)*. Formant positions depend on the resonances of the *Vocal Tract*, and they may be seen as dark bands raising and falling with time. Where formant positions are stable, phonemic detection could be accomplished using standard Neural Networks, such as *Backpropagation*. But certain phonetic structures present rapid movements in formant positions, with continuous increase/decrease, showing slopes of many Hz/msec. To adequately characterize such slopes, TDNN's associate sets of weights to time-contiguous samples of spectral vectors x_{1n-i}, where i is the delay index, which when adequately adjusted, are in charge of capturing this temporal evolution. One of the most frequent cases in Speech for non-stationarities takes place in diphthongs. In Fig. 2, the regions (*a-b*) between 522 msec. and 590 msec., corresponding to the diphthong /ɑi/ in /wish/, and (*c-d*), between 1129 msec. and 1281 msec., corresponding to /æe/ in /were/ show slopes of *11.61* Hz/msec. and *4.76* Hz/msec. respectively, illustrating the nonstationary nature of the process. The duration of both processes is of 68 msec. (*a-b*) and 152 msec. (*c-d*). According to classical TDNN theory, to detect a given dynamic process, a delay order covering the duration of the whole process would be required [Mor.91]. This implies that if 5 msec. time delays were used, delay orders of 14 (*a-b*) and 31 (*c-d*) would be required to adequately represent such phonemes. But this assumption is inconsistent, as the rule it implies is *the slower the slope the higher the delay order*, which does not make any sense, because in the limit, stationary processes (negligible slope) would require the highest delay orders. Our approach assumes that delay orders must be related to the orders of curvature observed in formant trajectories. In regions such as *a-b* or *c-d*, where formant trajectories are well represented by straight lines, a first order delay stage (for samples x_{1n} and x_{1n-1} separated 5 msec.) should be enough. Otherwise, in regions *e-g* and *f-h*, a higher delay order could be expected, as the curvature of the spectral formants seems to be of higher order (parabolic). It should be reasonable, then, to dimension the order of delays in a TDNN according to the order of curvature in the formants of the Speech Spectrum being characterized.

3. Measuring the order of Non-Stationarity of Speech using Lattice Networks.

The behavior of a generic node of a TDNN as detector of the non-stationary characteristics in Speech Spectra, can be represented by the Transversal Joint-Process Estimator (TJPE) given in Fig. 3.a, as shown in a preliminary work [Rod.94a]. This result is quite important, as it is well known that this kind of structures may be substituted by Lattice-Based Joint Process Estimators (LJPE), as the one given in Fig. 3.b [Hay.91, pg. 231].

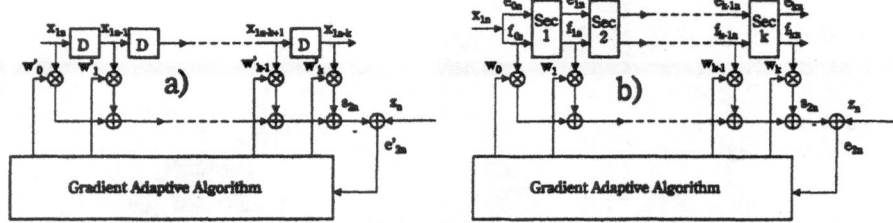

Figure. 3.a). Transversal Joint Process Estimator. b). Equivalent Lattice-Based Joint Process Estimator.

LJPE's present the advantage over TJPE's of optimally estimating the order of Autoregressiveness of the input process from the filter order k, in the mean-square sense. In [Rod.94a] it was concluded that the order of non-stationarity of the Speech Trace may be represented by the order of Autoregressiveness of the components of the spectral vectors fed as inputs to the TDNN's.

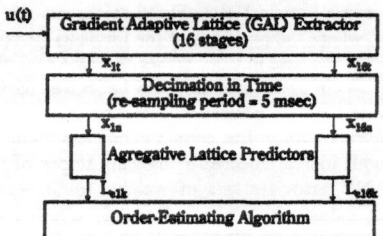

Figure 4. Algorithmic structure for estimating the Autorregressive Order of Spectral vectors x_n

As a consequence, it may be inferred that LJPE's can then be used to measure the orders of non-stationarities of Speech Traces, for the correct dimensioning of TDNN's. In the practical cases which will be discussed later, the structure shown in Fig. 4 was used for such measurements. It should be expected that most of the spectral traces would exhibit a first, or at most, a second order of non-stationarity, according to the conclusions derived from the analysis of Fig. 2. The Speech Trace $u(t)$ is sampled at a given frequency (in our case 10 KHz.), and its spectrum is LPC-estimated using a Gradient Adaptive

Lattice Extractor of order $p=16$, thus producing a set of spectral traces $x_t = \{x_{tl}\}$; $1 \leq l \leq p$. These traces are redundant in excess, as they may be considered stable over time intervals under 10 msec. For this reason, they are re-sampled at a frequency of 200 Hz., which grants that their non-stationary behavior is preserved, thus resulting in the new spectral vectors $x_n = \{x_{ln}\}$. These vectors may then be used as inputs to the TDNN's. To determine their order of Autoregressiveness, which will be related to their order of non-stationarity, a set of Aggregative Lattice Predictors will independently be tuned for each trace x_{ln}. These Lattice Predictors may be specified algorithmically as:

$$e_{0n} = f_{0n} = x_n; \tag{3}$$
$$\text{for } (j=0;j<k;j++)$$
$$\{ c_j = - E\{e_{j-1n} \cdot f_{j-1n-1}\}/[\{E\{e^2_{j-1n}\}E\{f^2_{j-1n-1}\}]^{1/2}; \tag{4}$$
$$e_{jn} = e_{j-1n} + c_j f_{j-1n-1}; f_{jn} = c_j e_{j-1n} + f_{j-1n-1} \} \tag{5}$$

where c_j is the *Partial Correlation Coefficient* of the *j-th Lattice Section*, and e_{jn} and f_{jn} are the *Forward and Reverse Prediction Errors*. The order of Autoregressiveness sought (k) may be inferred from the *Variance* of e_{jn}, $1 \leq j \leq k$, given as:

$$L_{ej} = (1 - c^2_j) L_{ej-1} \tag{6}$$

by tracing the degree of descent in this variance according with a certain criterion. To give a precise idea on how this order influences the decrease of the *Error Variance* with the increasing order of j, the *Variance* of trace x_{16n} of /ca/ (one of the irregularly behaving traces studied) is presented in Fig. 5.

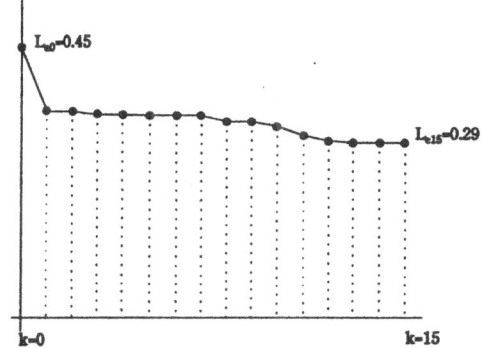

Figure 5. Evolution of the *Error Variance* of x_{16n} from /ca/.

The criterion to determine the optimal order k from the *Error Variance* may be formulated in quite different ways. The one used in the cases discussed in the present paper is given by the following expression:

$$k = \arg\min \{\frac{L_{ej}-L_{e\infty}}{L_{e0}-L_{e\infty}} < 1-\alpha\}; \quad 1 \leq j < \infty \tag{7}$$

where $0 < \alpha < 1$ is the grading parameter. This criterion establishes the optimal order as the lowest integer reducing the *Residual Differential Error Variance* $L_{ej}-L_{e\infty}$ to a fraction of the *Initial Differential Error Variance* $L_{e0}-L_{e\infty}$, given by $1-\alpha$. $L_{e\infty}$ is the asymptotic value of L_e, taken as L_{eN} for $N \gg k$. In Table 1 the

968

different orders of non-stationarity for speech traces corresponding to different sounds of Spanish of the kind C-V or V-C-V are given as a reference for discussion ($\alpha=0.9$).

4. Proposed modifications to the classical TDNN Architecture.

From the results exposed in Table 1 two important consequences may be derived. On one hand, most of the traces may be represented by first order systems (straight lines).

N° Coeff.	/pa/	/ta/	/ca/	/ka/	/ba/	/da/	/ɣa/	/ga/	/aβa/	/aδa/	/aζa/	/aɣa/
1	1	1	1	1	1	1	1	1	1	1	1	1
2	1	1	1	1	1	1	1	1	1	1	1	1
3	1	1	1	1	1	1	1	1	1	1	1	1
4	1	1	1	1	1	1	1	1	1	1	1	1
5	1	1	1	1	1	1	9	1	1	1	1	15
6	2	1	1	1	1	1	1	1	1	1	1	2
7	1	1	1	3	1	1	1	1	1	1	1	3
8	2	1	1	1	1	1	1	1	1	1	1	1
9	1	1	16	2	1	1	1	1	1	1	1	1
10	1	1	1	1	1	3	1	2	1	1	1	1
11	1	1	4	1	1	3	1	1	1	1	1	2
12	1	1	5	4	1	1	1	1	1	1	1	1
13	1	1	8	1	1	1	1	8	1	1	2	1
14	6	1	20	1	1	1	1	1	1	1	1	1
15	1	1	1	1	1	30	1	4	1	1	1	1
16	1	1	13	1	1	1	1	1	1	1	1	1

Table 1. Orders of non-stationarity for different sets of sounds C-V and V-C-V from Spanish ($\alpha=0.9$).

As a consequence, it is evident that dimensioning TDNN's with an arbitrary number of delays will not introduce any advantage under the representation point of view, while increasing substantially the computational complexity of the network. On the other hand, it re-enforces the argument in favor of using Lattice Joint Estimators instead of Tapered Delay Lines in the input stages of TDNN's.

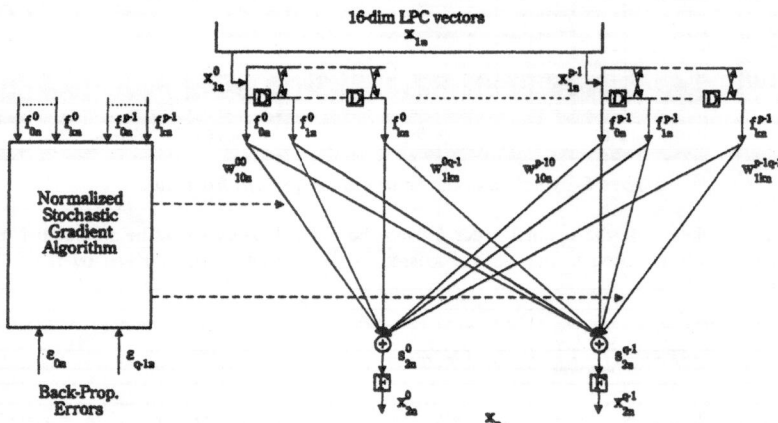

Figure 6. Alternative Input Layer to the structure in Fig. 1, based in Lattice Joint Process Estimators.

This last suggestion seems to be a very promising one, as Lattice Predictors present interesting properties in de-correlating reverse-propagating signals f_{jn}, which when fed to the TDNN may produce a faster and smoother convergence. The approach, would then be to substitute the input layer of perceptrons fed from tapered delay lines in Fig. 1, by its equivalent optimal LJPE's given in Fig. 6. This means that the TDNN will be fed with the prediction errors f_{jn} resulting of de-correlating the *information shared in common by the delayed versions of spectral vectors*, adjusted to an optimal order of estimation, in the sense defined in part 3. The weights of the new Input Layer, are adjusted using the *Normalized Stochastic Gradient Algorithm* [Hay.91, pp. 352-9].

5. Results and Discussion.

The performance of the structure exposed in the previous section has been compared with that of a standard TDNN. For such, a *Code-Book* of 24 *Speech Traces* consisting in 6 utterances of /aβa/, /aδa/, /aζa/, and /aγa/, spoken by 6 different speakers (3 male and 3 female) was prepared. The traces were pre-emphasized, LPC extracted, and the resulting 16-dim PARCOR coefficients were used as inputs in the following experiments. Each trace was given a label for *Phonemtic Decoding* [Rod.94b]. Two TDNN's were implemented in C++ on a 20 MIP's Workstation. The first one, labeled as WTDNN followed Waibel's structure, with $p=16$, $k=3$, $q=8$, $m=3$ and $r=8$. The second one, labeled LTDNN was based on the proposed alternative structure, using Predictor Lattices, with $p=16$, $k=1$, $q=8$, $m=0$, and $r=8$. Their respective computational costs are compared in Table 2.

Task	WTDNN		LTDNN	
Pre-processing Input Data	-	*0*	*11.p.k*	*176*
Data Propagation (IL)	*p.(k+1).q*	*512*	*p.(k+1).q*	*256*
Data Propagation (HL)	*q.(m+1).r*	*256*	*q.(m+1).r*	*64*
Error Backpropagation	*r.(q.(m+1)+1)*	*264*	*r.(q.(m+1)+1)*	*72*
Symmetric Balancing (IL)	*2.q*	*16*	*2.q*	*16*
Symmetric Balancing (HL)	*2.r*	*16*	*2.r*	*16*
Weight Adjustment (IL)	*q.(p.(k+1)+1)*	*520*	*q.(p.(k+1)+1)*	*264*
Weight Adjustment (HL)	*r.(q.(m+1)+1)*	*264*	*r.(q.(m+1)+1)*	*72*
Total # of Products	-	*1848*	-	*936*
Total (%)	-	*100*	-	*50,65*

Table 2. Comparison between the Computational Complexities involved

In two sets of similar experiments, both networks were required to associate the set of 6 spectral vectors corresponding to each utterance, with the labels assigned to each phoneme. The results of forcing both networks to converge to a normalized RMS error under 0.001 in each case are presented in Table 3. The number of required training steps to arrive to such results are given in the second and third columns.

Speech Trace	No. Train. Steps (WTDNN) #1	No. Train. Steps (LTDNN) #2	NTS(#2)/ NTS(#1) (%)	Comparing Effective Cost (%)
/aβa/	6144	1984	32,29	16,3549
/aδa/	6336	2560	40,4	20,4626
/aζa/	5952	2240	37,63	19,0596
/aγa/	5312	2112	39,76	20,1384

Table 3. Relative computational costs to achieve the same RMS convergence errors.

The fourth column gives the ratio between columns 3 and 2, and the fifth column takes into account the effective computational expenses, when the respective computational complexities of each training step, as given in Table 2, are taken into account. According with these results, it seems that the LTDNN achieves a faster convergence in all the cases studied, at half the computational cost of a WTDNN. Two reasons account for this result. On one hand, it seems evident from Table 1 that first order delays would suffice to trace the non-stationarities involved the cases studied (except for some minor irregularities); on the other hand, Predictor Lattices de-correlate the input traces (in a time averaged sense), thus improving convergence rates [Hay.91, pp. 620-3]. It may be concluded then, that controlling the depth of delay lines adaptively for each problem, according to the time-varying dynamics of second-order statistics of input data seem to be a good method for improving convergence. Finally, it should be mentioned that an interesting line to continue this research area is the pre-processing of input signals by means of other Optimal Filtering Methods, as Kalman Filtering [Ruc.92, Con.94] to improve robustness. The applications of this research are being sought in the *Phonemic Detection* of the *Speech Trace* for *Real-Time Correction of Speech* in *Computer-Aided Language Learning*.

Acknowledgments.
This research is being funded by Comisión Interministerial de Ciencia y Tecnología, Grant No. TIC 93-0702-C02-01, and Grant No. C0AE00347/94 from Plan Regional de Investigación de la Comunidad Autónoma de Madrid.

References.
[Con.94] J. T. Connor, R. D. Martin, and L. E. Atlas, "Recurrent Neural Networks and Robust Time Series Prediction", *IEEE Trans. on Neural Networks*, Vol. 5, No. 2, March 1994, pp. 240-253.

[Dug.94] Dugast, C. Devillers, L. and Aubert, X., "Combining TDNN and HMM in a Hybrid System for Improved Continuous-Speech Recognition", *IEEE Trans. on Signal Processing*, Vol. 2, No. 1, January 1994, pp. 217-223.

[Hay.91] S. Haykin, *Adaptive Filters*, Prentice-Hall, Englewood Cliffs, NJ, 1991.

[Mor.91] D. P. Morgan and C. L. Scofield, *Neural Networks and Speech Processing*, Kluwer Academic Publishers, Boston, MA, 1991.

[Rod.94a] V. Rodellar, P. Gómez, M. Pérez, V. Nieto, R. Romera and J. Muruzábal, "Dimensionality and Dynamic Temporal Structure of a Time-Delay Neural Network for Speech Processing", *Proc. of EUSIPCO '94*, Edinburgh, September 13-16, 1994, pp. 816-819.

[Rod.94b] V. Rodellar, V. Nieto, P. Gómez, D. Martínez and M. Pérez, "A Neural Network for Phonetically Decoding the Speech Trace", *Proc. of the Intern. Conf. on Spoken Language Proc. '94*, Yokohama, Japan, 1994, pp. 1575-1578.

[Ruc.92] D. W. Ruck, S. K. Rogers, M. Kabrisky, P. S. Maybeck and M. E. Oxley, "Comparative Analysis of Backpropagation and the Extended Kalman Filter for Training Multilayer Perceptrons", *IEEE Transactions on Pattern Analysis and Machine Intelligence*, Vol. 14, No. 6, June 1992, pp. 686-691.

[Wai.89] A. Waibel, T. Hanazawa, G. Hinton, K. Shikano and K. J. Lang, "Phoneme Recognition Using Time-Delay Neural Networks", *IEEE Trans. on ASSP*, Vol. 37, No. 3, March 1989, pp. 328-339.

Acquisition of Internal Representation by Learning of Identity-Mapping Using Overload Learning

NODA Itsuki

Electrotechnical Laboratory
1-1-4 Umezono, Tsukuba, Ibaraki
JAPAN

Abstract

Acquisition of internal representation by neural networks that learn identity-mappings is one of important methods for feature-abstraction. It is, however, difficult to decide the number of hidden units of the networks. In this article, I show a way to apply the *overload learning* (OLL) technique to overcome this problem. Because OLL causes to reduce the number of effective dimensions of hidden patterns, networks can get reduced internal representation by learning. Moreover, I show that this technique is useful to get spatial representation from symbolic representation, and to integrate various types of information.

1   Introduction

Acquiring good internal representation of information is an important problem in AI research, because the performance of an AI system depends on internal representation used in the system. On the other hand, neural networks have a function to acquire internal representation through their learning: When a multi-layered neural network with hidden layers learns a certain task, activation patterns on the hidden layers represent information that is required for the task. Using this property, many researchers have been trying to analyze characteristics of tasks to learn [2; 7]. They also have been trying to use such internal representation acquired by networks for input data of the higher level of processing [1, 5]. *Sandglass-type networks* are one of remarkable models in such works. A sandglass-type network is a feed-forward layered network with some hidden layers, and is trained to learn an identity mapping of a set of patterns. After learning, patterns of one of the hidden layers become to represent information about input patterns. Therefore we can apply the network for data compression, feature abstraction, representation change, etc. [3, 8, 6]. The most important feature of the sandglass-type networks is that we need not heuristic teachers for its learning, because the network is trained only to output the same patterns as input patterns.

However, there is a problem, how to decide the number of hidden units of the network. In order to guarantee performance of representation, we must use the enough number of units. On the other hand, if the number of units is too enough, pattern representation on the hidden layer becomes redundant. Such redundancy makes analysis and reuse of the patterns difficult.

In order to overcome such a problem, I have proposed a method, *overload learning*[4], by which active dimensions of patterns on a hidden layer are reduced through learning of a network. One of merits of this method is that it is simple enough so that it is easy to apply the method to various models of networks and learning procedures.

In this article, I show how to apply the overload learning to sandglass-type networks, and report how effective it is through some experiments.

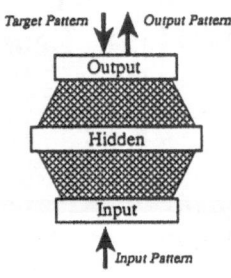

Figure 1: Multi Layered Neural Network

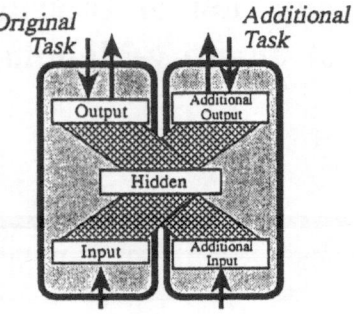

Figure 2: Network for Overload Learning

2 Overload Learning

When we train a network with hidden layers to achieve a given task, we usually give the network more hidden units than that the task will require. In this case, a part of hidden units become superfluous, so that pattern representation on the hidden layers becomes redundant. *Overload learning*[4] is a technique to overcome this problem.

In the overload learning, the number of *active* dimensions [1] are reduced by learning an additional task in the same time of learning an original task. The actual procedure is as follows:

1. Suppose the case when a given network shown in Fig. 1 is trained to achieve a given task (called 'original task'). We add two new layers, the additional-input and the additional-output layers, to the network, and connect the layers with the hidden layer (Fig. 2). [2]

2. We prepare a new task (called 'additional task') that is independent of the original task, and train the network to achieve the additional task using the additional-input and additional-output layers. This training is done in the same time of training of the original task. In other words, the network is trained to minimize the following error by the back-propagation method.

$$E = E_{\text{org}} + \alpha E_{\text{add}} \qquad (1)$$

where E_{org} and E_{add} are the average error of the original task and the additional task respectively, and α is a small positive constant.

3. After training, we remove the additional-output layer, and set an average pattern on the additional-input layer. [3]

In the step 2 of this procedure, patterns on the hidden layer must represent information about both of the original task and the additional task. Therefore the redundant part of the pattern representation for the original task becomes to be used for the additional task. As the result, the pattern representation on the hidden layer is reduced by controlling the parameter α.

3 Learning Identity-mapping by sandglass-type networks

A *sandglass-type network* consists of an input layer, an output layer and some hidden layers, one of which is called the internal-representation layer (Fig. 3). The size of the output layer [4] is same as the input layer, and

---

[1] We say a dimension is *active* when patterns on a hidden layer change along the dimension in the pattern space for various inputs.

[2] Here we use a network with one hidden layer, but we can apply the technique to a network with multi hidden layers. In this case, we connect additional layers to a hidden layer whose pattern-representation we want to reduce.

[3] As the result, inputs from the additional-input layer to the hidden layer become constant. We also can remove the additional-input layer by add the constant inputs to threshold values of units of the hidden layer.

[4] The *size* of a layer means the number of units in the layer.

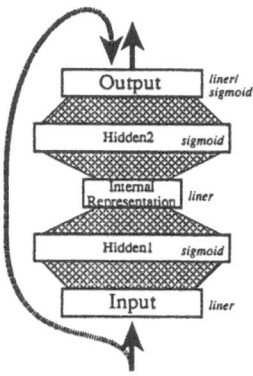

Figure 3: Sandglass-type Network

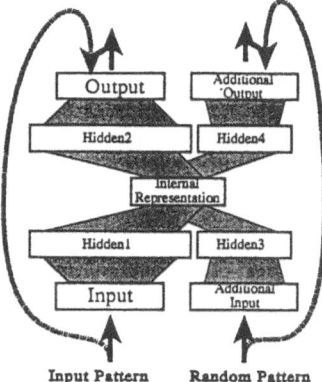

Figure 4: Sandglass-type Network with Overload Learning

the size of the internal-representation layer is usually smaller than the input layer. This network is trained to output the same pattern on the output layer as patterns on the input layer. After training, patterns on the internal-representation layer become to represent information about inputted patterns. Therefore we can use the network for data-compression, feature abstraction and so on: For example, we can use the lower half of the network, between the input layer and the internal-representation layer, as an encoder to compress input patterns, and the upper half, between the internal-representation layer and the output layer, as a decoder to uncompress the encoded patterns.

As mentioned in the introduction, one of the problem about this type of networks is how to decide the size of the internal-representation layer. When the size is too small, patterns on the internal-representation layer can not represent information about inputted patterns sufficiently. On the other hand, when the size is too large, the pattern representation becomes redundant. Redundant representation is obstacle to feature abstraction.

In order to overcome the problem, we apply the overload learning to sandglass-type networks. In order to do this, we add new layers to a sandglass-type network, and construct a network shown in Fig. 4. In this network, the internal-representation layer and additional-output layer consist of liner units [5]. The other layers consist of standard sigmoid units. As mentioned in section 2, the original task (the identity-mapping of given patterns) is learned between the input layer and the output layer. In the same time, the additional task is learned between the additional-input layer and the additional-output layer. As the additional task, we use an identity-mapping of random patterns. In this task, the network receives a randomly generated pattern (called random pattern) to the additional-input layer as input in each time, and is trained to output the same pattern on the additional-output layer as the random pattern. Each value of a random pattern is between -1 and 1. Ultimately, the network is trained to minimize the following error.

$$E = \langle |o_{\text{org-out}} - o_{\text{org-in}}|^2 \rangle + \alpha \langle |o_{\text{add-out}} - o_{\text{add-in}}|^2 \rangle \qquad (2)$$

where $o_{\text{org-out}}$, $o_{\text{org-in}}$, $o_{\text{add-out}}$ and $o_{\text{add-in}}$ are respectively activation patterns of the output, input, additional-output and additional-input layers. $|x|$ means the norm of the vector x, and $\langle x \rangle$ means the average of x. α is a positive value between $0.1 \sim 1.0$.

As mentioned in section 2, the overload learning reduces the number of active dimensions for the original task (the identity mapping of the given patterns), so that we can get simple pattern representation on the internal-representation layer.

4 Experiments and Discussion

Here, we will demonstrate how the proposed network works well through some experiments.

---
[5] each of which outputs the value of the weight-sum of its inputs.

Table 1: Eigenvalues of Principle Components of Patterns on the Internal-Representation Layer

	PC1	PC2	PC3	PC4	PC5	PC6	PC7
Ex.1.1 (NO)	1.000	.651	.000	.000	.000		
Ex.1.1 (OLL)	1.000	.002	.000	.000	.000		
Ex.1.2 (OLL)	1.000	.853	.000	.000	.000		
Ex.2 (NO)	1.000	.738	.581	.195	.146	.143	.045
Ex.2 (OLL)	1.000	.756	.611	.002	.000	.000	.000
Ex.3.1 (NO)	1.000	.572	.304	.225	.141	.020	.010
Ex.3.1 (OLL)	1.000	.869	.454	.000	.000	.000	.000
Ex.3.2 (OLL)	1.000	.683	.619	.000	.000	.000	.000

Each value is normalized by the eigenvalue of the PC1.
(NO) and (OLL) mean the cases without and with overload learning, respectively.

4.1 Ex.1: Learning of Values of Sine and Cosine

In the first experiment (Ex.1), a network was trained to achieve an identity mapping of a pair of values of sine and cosine.

In this experiment, each of the input and output layers of the network consists of 2 units. Units in the input layer receive values of $\sin(x)$ and $\cos(x)$ respectively, and units in the output layer are taught to output the same values.

After training, the activation pattern of the additional-input layer is fixed. Then I recorded patterns of the internal-representation layer for the various sin-cos patterns. I analyzed the recorded patterns by principle component analysis (PCA), and determine an eigenvalue of each principle component. The eigenvalue shows how *active* the principle component is. In other words, the number of non-zero eigenvalues means the number of active dimensions of patterns of the internal-representation layer.

In the first setup of Ex.1, the domain of x for sin and cos is $0° \sim 240°$ (Ex.1.1). The internal-representation, additional input and additional output layers consist of 6, 12 and 12 units respectively, and each of other hidden layers consists of 40 units. Tab. 1 shows the result of PCA [6].From the table, we can find that two eigenvalues remain positive in the case without the overload learning (Ex.1.1-NO), and on the other hand only one eigenvalue remains positive in the case with the overload learning (Ex.1.1-OLL). This means that the network acquires 2-dimensional representation in Ex.1.1-NO, and 1-dimensional representation in Ex.1.1-OLL. In other words, the network with the overload learning represents values of $\sin(x)$ and $\cos(x)$ by a 1-dimensional manifold that is homeomorphic with a part of a circle.

In the second setup, the domain of x is set to $0° \sim 360°$ (Ex.1.2). Other settings are the same as Ex.1.1. In this case, the second eigenvalue becomes positive even in the case with the overload learning. This means that the network uses 2 dimensions for representation the full circle, because the full circle can not be embedded in a 1-dimensional Euclid-space.

From these results, we can say that the network with overload learning can adapt the suitable number of dimensions of patterns of the internal-representation layer flexibly to represent structures of given patterns.

4.2 Ex.2: Acquiring Spatial Representation

In the second experiment (Ex.2), I will show that the proposed network can acquire suitable spatial representation, which reflects topographic character of given patterns. Initially I made patterns for the identity mapping in the following way:

1. I considered a cube whose edges, apexes and sides are labeled like Fig. 5, and also considered an array of 26 units each of which corresponds to each of the labels.

2. I generated activation patterns of the array (Fig. 6). In each pattern, units of an edge and neighboring two apexes and two sides are 'on' (activation values are 1.0), and other units are 'off' (0.0). For example, in the first pattern in Fig. 6, units '00x', '0xx', 'x0x', '000' and '001' are 'on' because edge '00x' connects to sides '0xx', 'x0x' and edges '000', '001'.

3. In the same way, I generated patterns of apexes-and-neighbors and sides-and-neighbors.

---

[6] Values in this table do not show the averages, but typical results of the experiments.

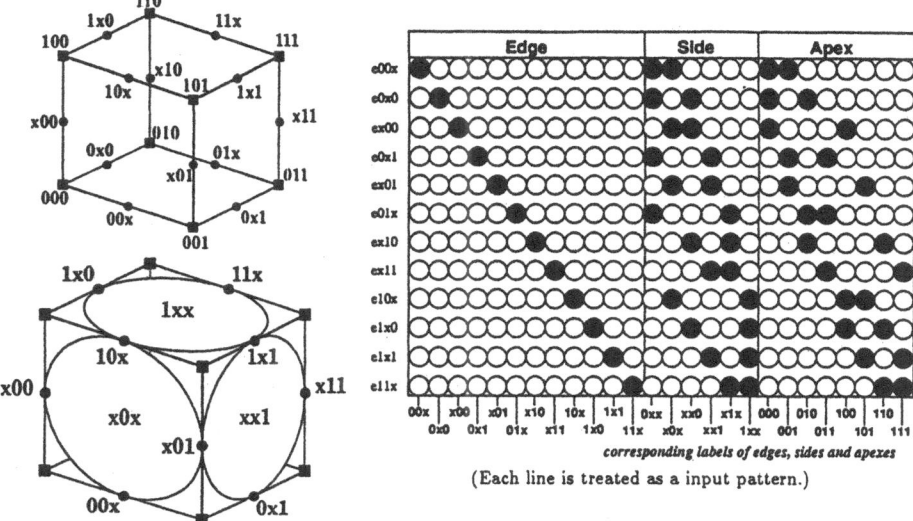

Figure 5: Labeled Cube

Figure 6: Patterns used in Ex.2 (part)

corresponding labels of edges, sides and apexes

(Each line is treated as a input pattern.)

Then I trained a network to achieve the identity mapping of these patterns. In the network, the internal-representation layer consists 10 units, each of additional-input and -output layer consists of 20 units, and each hidden layer consists of 20 units.

After training, I analyzed patterns of the internal-representation layer in the same way as Ex.1. From results of the analysis (Tab. 1), we can find that the network uses only 3 dimensions that is the same as the number of dimensions of the cube used to generate input patterns. Moreover, I investigated an arrangement of the internal representations. For example, Fig. 7 shows positions of the internal-representation patterns for input patterns of apexes in the pattern space. The figure means that the network acquired an internal representation whose arrangement is the same as the original cube.

The result of this experiment shows that the proposed networks have the ability to find primitive dimensions represented in input patterns implicitly, and acquire suitable spatial representation.

We can discuss the result from another point of view, symbols and patterns. I generated input patterns in the symbolic manner: Each unit represents a label (localist representation), activations of the units are binary ('on'/'off'), and each pattern represents a relation of neighborhood-ness of labels discretely. Then the network acquired continuous and spatial representation. This means that the proposed networks have possibility to offer a way to overcome the problem of combining symbol- and pattern-processings.

4.3 Ex.3: Stereo Vision

The third experiment is the stereo vision. Initially I made patterns that have stereo parallax (Fig. 8). Each pattern consists of two arrays of 10 units that corresponds to left and right retinas. In each array, only units in an area of length l and position p (or $p + d$ in the right retina) are 'on'. Domains of parameters are $p = 0 \sim 9$, $l = 1 \sim 4$ and $d = -2 \sim 2$. Then I trained a network like Fig. 9 to achieve the identity mapping of the patterns. In the network, each of the input and output layers consists of 20 units, each of the additional-input and the additional-output layers consists of 40 units, the internal-representation layer consists of 20 units and each hidden layer consists of 40 units.

In the first setup of Ex.3, I trained the network to achieve the identity mapping of the stereo patterns only [7]. From the result of the PCA (Tab. 1), we can find that the network uses 3 dimensions, which is the same as the number of the free parameters, p, l and d. This means that the network determined the suitable

---

[7]The network was also trained to achieve the additional task during learning.

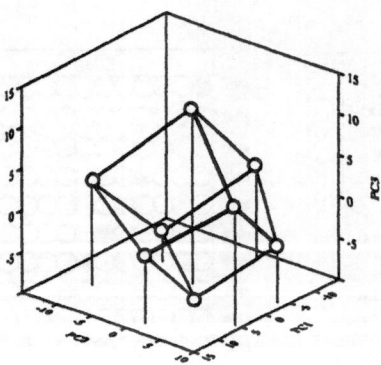

Figure 7: Acquired Internal Representation in Ex.2. (Representation for Each Apex)

number of dimensions for representing the stereo patterns. Moreover, I analyzed the arrangement of the internal representation (Fig. 10, Fig. 11). In Fig. 10, internal representations for stereo patterns of a certain l are plotted on each curve. From this figure, we can find that representation of parameter l and parameter p forms a polar coordinate system. In the coordinate system, l corresponds to the radius component and p corresponds to the angle component. On the other hand, Fig. 11 shows internal representations for stereo patterns of a certain d and $l = 3$. In this case, it is difficult to find a simple structure of representation of d.

In the second setup, I trained the network using the stereo patterns with additional information about the shift parameter d (Ex.3.2). The additional information is 'near-or-not' information: It is represented on one unit that becomes 'on' only when $d < 0$. The motivation of the introduction of the information is that when the object is near (this is the case of $d < 0$), human or robots can get such information through non-visual sensors like touch-sensors.

After training, I analyzed internal representations by PCA, and found that the network uses 3 dimensions, that is the same as Ex.3.1 (Tab. 1). This means that the representation of the additional information is embodied in the representation about stereo-patterns. Fig. 12 shows the result of the same analysis as Fig. 11. As compared with Ex.3.1, curves are arranged in parallel. This means that parameter d is represented as an independent dimension in the internal representation. Note that though the additional information indicates parameter d by only 2 levels ('on' or 'off'), the acquired representation indicates it by 5 levels ($d = -2 \sim 2$) clearly.

From these results, we can say that the proposed networks can reduce the number of dimensions of internal representations, but can not find all primitive dimensions as independent dimensions that are easy to analyze. Moreover, the networks can fuse various types of patterns that represent information redundantly into reduced representation. Especially, as shown in Ex.3.2, there are cases where such fusion of various types of patterns makes internal representations easy to analyze. This effect may offer a way to solve the sensor fusion problem of robotics.

5 Conclusion

In this article, I described a way how to apply the overload learning to sandglass-type networks that learn identity mapping. Because the overload learning cause to reduce the number of active dimensions of patterns on a hidden layer, we can get simple representation of input patterns for the sandglass-type networks. It is easy to analyze such representation, and easy to find primitive dimensions that reflect important features of information. Moreover, I found that the proposed model has the following effects through discussions about result of the experiments:

- The network can acquire spatial representations. When input patterns are concerned with spatial relations, the network can acquire representation that reflects the spatial relations. This effect is important when input patterns represent information in symbolic manners, for example, 'localist representation', because it will offer a way to combine pattern information and symbolic information.

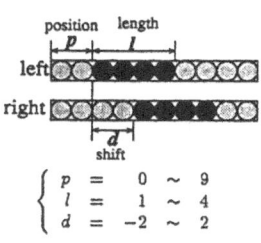

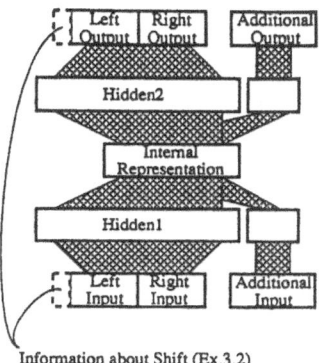

Figure 8: Stereo Parallax Pattern

Information about Shift (Ex.3.2)

Figure 9: Network for Stereo Parallax Pattern.

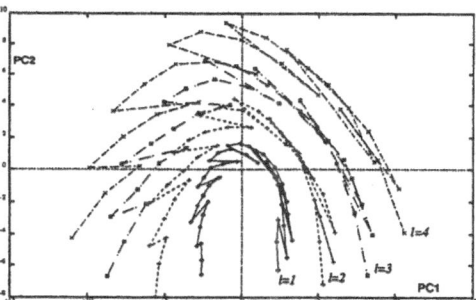

Marks on each curved line are representations for input patterns of the same length l.

Figure 10: Internal Representation of Length and Position(Ex.3.1).

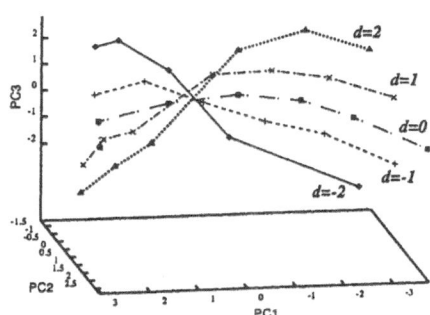

Marks on each curved line are representations for input patterns of the same shift d.

Figure 11: Internal Representation of Shift($l = 3$)(Ex.3.1).

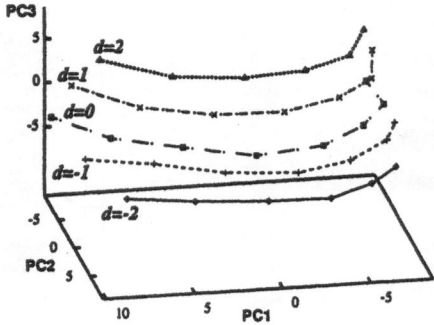

Figure 12: Internal Representation of Shift in the Case with Information abount Shift($l = 3$)(Ex.3.2)

- The network can fuse various types of information into reduced representation. When two parts of a pattern represent the same information in different ways, the network can unify them in the internal representation. Such situation will occur on robots who has many sensors. This effect may be useful for the sensor fusion problems in such multi-sensor robots.

References

[1] Brian T. Bartell and Garrison W. Cottrell. A Model of Symbol Grounding in a Temporal Environment. In *IJCNN91*, pages I–805–810, June 1991.

[2] Jeffrey L. Elman. Finding Structure in Time. Technical Report CRL-TR-8801, Center for Research in Language, University of California, San Diego, April 1988.

[3] Bunpei Irie and Mitsuo Kawato. Acquisition of Internal Representation by Multi-Layered Perceptrons (in Japanese). Technical Report NC89-15, The Institute of Electronics, Information and Communication Engineers, 1989.

[4] Itsuki Noda. Overload Learning (in Japanese). Technical Report NC93-28, The Institute of Electronics, Information and Communication Engineers, 7 1993.

[5] Itsuki Noda. A Model of Recurrent Neural Networks that Learn State-Transitions of Finite State Transducers. In *WCNN'94-San Diego*, pages IV–447–452, Jun. 1994.

[6] Jordan B. Pollack. Recursive distributed representations. *Artificial Intelligence*, 46(1–2):77–105, 1990.

[7] David Servan-Schreiber, Axel Cleeremans, and James L. McClelland. Learning Sequential Structure in Simple Recurrent Networks. In *NIPS1*, pages 643–652. Morgan Kaufmann, 1989.

[8] Shiro USUI, Shigeki NAKAUCHI, and Masae NAKANO. Analysis of Munsell Color Space by Multi-Layered Neural Network (in Japanese). Technical Report NC89-40, The Institute of Electronics, Information and Communication Engineers,, Dec. 1989.

A Multiacuity Connectionist Model
for Local Speed Estimation

C. Bandera*, I. M. Conde**, J. Jerez**, M. González**, F. J. Vico**, F. Ortega**

* Amherst Systems Inc., Buffalo (NY). U.S.A.
E-mail: cba@amherst.com
** Dpt. Tecnología Electrónica. E.T.S.I. Telecomunicación.
Universidad de Málaga. P. El Ejido, s/n. 29013. Málaga. SPAIN
Voice: +34 5 213 13 52
Fax: +34 5 213 14 47
E-mail: martingg@ctima.uma.es

ABSTRACT

Multiresolution foveal imaging offers a field-of-view×acuity×frame rate product much greater than that of uniform acuity imaging, and can reduce the computational complexity of vision in dynamic scenarios with localized relevance. The latter requires space variant retinotopic processing with results that are consistent between different resolutions. A physiologically inspired connectionist model is presented which yields local speed estimates consistent between coarse resolution peripheral vision and fine resolution central vision. A non-uniform acuity profile across the field-of-view is compensated by an inverse profile of weights in the model. This model demonstrates the efficiency of multiresolution vision when processing and retinotopology are properly matched.

INTRODUCTION

Foveal vision, prevalent in vertebrates, features imaging with graded acuity coupled with context sensitive sensor gaze control. It can operate more efficiently than uniform acuity vision in non-deterministic scenarios with localized relevance because resolution is treated as a dynamically allocatable resource. Wide field-of-view (FOV) and localized high acuity are simultaneously supported while minimizing sensor data to that which is relevant. A foveal multiresolution retinotopology also reduces the computational complexity of vision, when compared against a system with high acuity throughout the FOV.

Research into foveal machine vision focuses on two retinotopologies: log-polar sampling and quad-tree tessellation (Kreider, 1990; Bandera 1989). The former is more biomimetic and varies acuity "gracefully", while the latter varies acuity in octave steps and lends itself more to digital implementations. Both can be exploited to reduce the computational complexity of vision, but only if processing uses the same topology (*i.e.*, mapping from the polar domain back to the traditional uniform Cartesian domain is self-defeating). Connectionist systems for foveal machine vision must thus reconcile homogeneous structure with heterogeneous spatial resolution. They must also provide homogeneous function, limited only by acuity, because visual tasks in general cannot be designated to strictly delineated retinotopic regions. For example, local speed estimation, essential in dynamic scenarios, supports the tasks of unresolved cue detection, target tracking, and navigation. It is particularly sensitive to retinotopology, unlike other features such as color, because perceived speed is scaled by spatial resolution, and discontinuities in resolution can lead to motion singularities if processing and retinotopology are not well matched.

A connectionist model is presented which estimates local motion from foveal imagery with a quad-tree tessellation topology. The model, an extension of a 1-D model developed for uniformly sampled imagery (Vico, 1994), assigns neuron-like processing elements commensurably with acuity so as to reduce overall hardware and computational complexity. This new foveal model is inspired by known functions of horizontal and amacrine cells in the retina, which perform spatial averaging and decimation (*i.e.*, acuity reduction) and temporal differencing.

DESCRIPTION OF THE FOVEAL RETINOTOPOLOGY

The retinotopology of the model is based on the exponential (quad-tree) lattice, which tessellates square receptive fields called resolution cells, or *rexels* (*figure 1*). The root lattice contains a 4×4 fovea array of uniformly sized rexels (the size of each fovea rexel is normalized to 1×1). The fovea is surrounded by a ring of rexels of size 2×2, another with rexels of size 4×4, 8×8, and so forth. The size of a rexel is proportional to its L_{∞} distance to the lattice center, so acuity tapers off linearly. Each ring consists of twelve rexels, and adding an additional ring increases the FOV area by a factor of four.

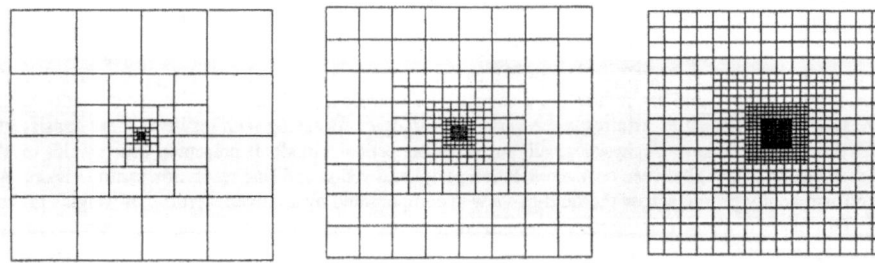

a. Root lattice with five rings about the fovea (d=1, r=5)   b. Subdivided lattice with half the acuity gradient (d=2, r=4)   c. Subdivided lattice with 1/4 the acuity gradient (d=4, r=3)

Figure 1.Examples of foveal retinotopologies

Most applications require a flatter acuity profile because that of the root lattice is very steep and produces a peripheral acuity which is too coarse to be of any value in wide FOV systems. Derivative lattices are obtained by uniformly subdividing each rexel in the root lattice by a subdivision factor *d*, and scaling the geometry upward by the same factor *d* to preserve the maximum acuity. The subdivided lattice has a fovea of size 4*d*×4*d*, and *d* rings of a given rexel size (the collection of *d* rings with the same sized rexels is called a major ring). The acuity of a subdivided exponential lattice is approximated outside the fovea by

$$acuity = \frac{1}{rexel\ size} = \frac{3}{2} \times \frac{d}{L_{\infty}\ distance\ from\ lattice\ center} \tag{1}$$

Acuity profiles have been shaped through natural evolution so as to optimize vision performance during life critical tasks. Rexel subdivision permits a design engineer to likewise optimize foveal machine vision to a particular application by selecting the distribution of resolution without altering the basic topology of the lattice or signal processing algorithms. Higher subdivision improves peripheral acuity, but at the expense of a greater number of rexels. The bandwidth compression factor f_c, defined as the ratio between the number of pixels in a uniform acuity image to the number of rexels in a foveal image with the same FOV and maximum resolution, is an indication of the savings in data and computational bandwidth. It is expressed as

$$f_c = \frac{4^{r+1}}{4+3r} \cdot \frac{w^2}{d^2 12 \log_2(w/d) - 8}$$ (2)

where w is the FOV width in terms of fovea rexels (*i.e.*, pixels), and r is the number of major rings in the foveal lattice (*table 1*).

r	0	1	2	3	4	5	6	7	8	9
acuity	1:1	1:2	1:4	1:8	1:16	1:32	1:64	1:128	1:256	1:512
f_c	1	2.3	6.4	19.7	64	216	745	2621	9,36	33,82

Table 1. Exponential lattice bandwidth compression and central-to-peripheral acuity ratio for different number of rings about the fovea

There are two general types of connectionist models for foveal imaging. The first uses variably sized sensor elements that generates rexel values directly. The second uses a conventional uniform array whose pixel values are averaged into a much smaller number of rexels, except for the fovea which is comprised directly of pixels. This averaging process is called *rexelization*, and is analogous to the convergence of horizontal cells in the retina.

DESCRIPTION OF THE MODEL

The connectionist model estimates the local speed of an edge moving along a preferred direction. The model is one-dimensional, along this direction, although multiple models can be combined in a straightforward fashion to estimate motion along any arbitrary direction in a focal plane. The neuron model uses the classical threshold element defined as

$$a_j = \sum_{i=1}^{n} w_{i,j} x_i$$ (3)

$$x_j = \begin{cases} 0, & if \ (a_j < \theta) \\ a_j, & if \ (a_j \geq \theta) \end{cases}$$ (4)

where a_j represents the activation level of the jth element, computed as the sum of its n inputs weighted by

w_{ij}, x_j represents the output of the element to other neurons in the network, and θ is the neuronal threshold, set to zero for simulations so that the neuronal output is a continuous function in contrast to the McCullough-Pitts two-valued neuron.

The model architecture uses a one-dimensional version of the exponential lattice and consists of four layers (*figure 2*). Layer R represents a uniform linear array of receptors (*i.e.*, pixels) onto which a moving edge is projected. Layer F represents an vector of horizontal cells (*i.e.*, rexels), whose values are computed by a simple averaging rexelization network connecting R to F. Assuming a lattice subdivision of one, the rexel vector consists of a fovea with four rexels at the center. On either side of the fovea is a rexel with twice the receptive field coverage of a fovea rexel, next to which there is another rexel with four time the coverage, and so forth. Layer S stores information about the previous edge position. Finally, layer L computes the local speed of an edge moving along a preferred direction (left to right) integrating the information of R and S. R-neuron, F-neuron, S-neuron, and L-neuron will denote a neuron in the R, F, S, or L layer, respectively.

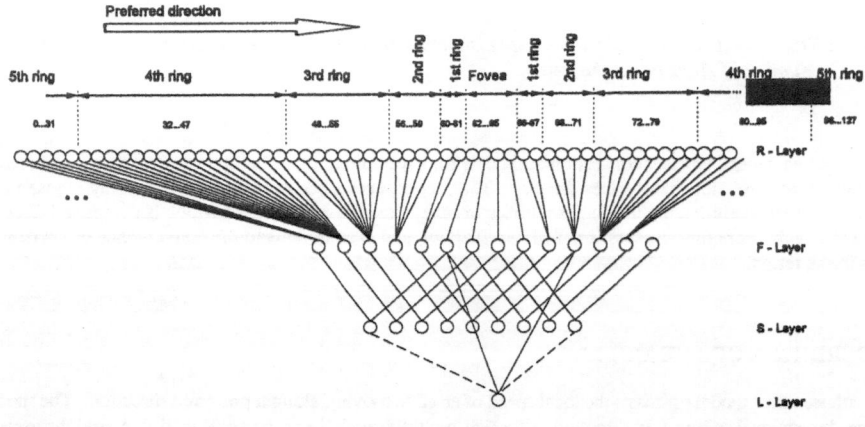

Figure 2. *Foveal network architecture with rexelization*

S- and L-neurons receive edge information from the F neurons (solid lines in *figure 2*) through the weighting function

$$ f = I \cdot S^r \tag{5} $$

where S is the speed about which the linear response of the network is centered (*i.e.*, the speed to which the network is "tuned"), r is the integer ring index of the rexel (r=0 at the fovea), and I is a response scale (*figure 3*). This weighting performs two functions simultaneously. First, it implements a directional edge detector through spatial differentiation such that an S-neuron reaches maximum activity when an edge is centered in its receptive field. Second, it implements an inverse acuity normalization that emphasizes edges detected with lower resolution. Layers R and F together form a space variant bandpass filter, with lowpass rexelization followed by highpass filtering.

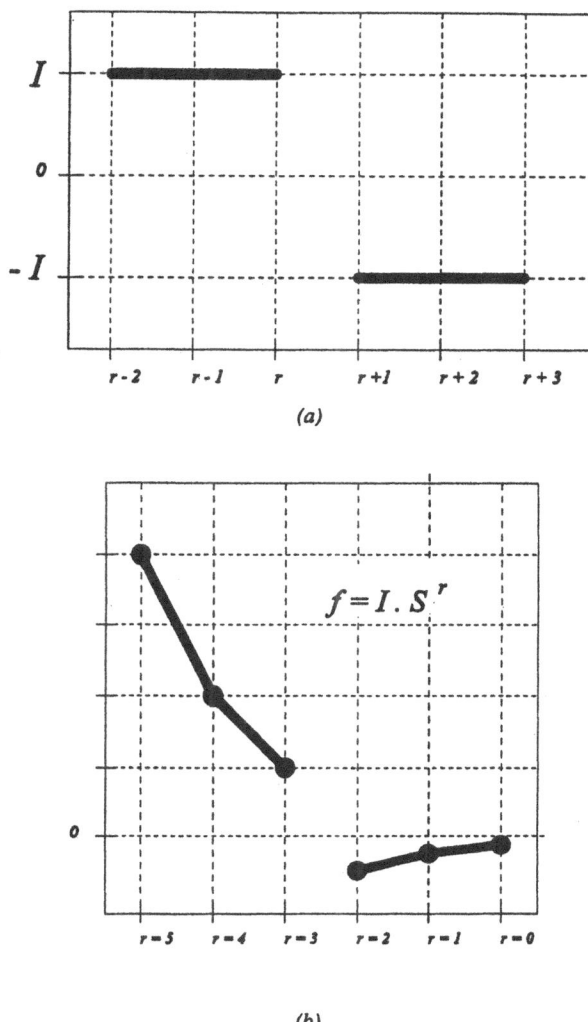

Figure 3. Rexel weighting. (a) Basic edge detector at fovea (I=0.45). (b) Modified edge detector.

The L-neuron receives stimulus from the F-neurons. The L-neuron also receives inhibition from the S-neurons (dashed lines in *figure 1*) according to the weighting shown in figure 4, where inhibition strength decreases with distance. These inhibitory connections reach the L-neuron only from its left side, establishing left to right as the preferred direction of motion. Because the S layer outputs delayed edge position information, the L-neuron computes a temporal derivative of edge location which serves as an estimate of local speed.

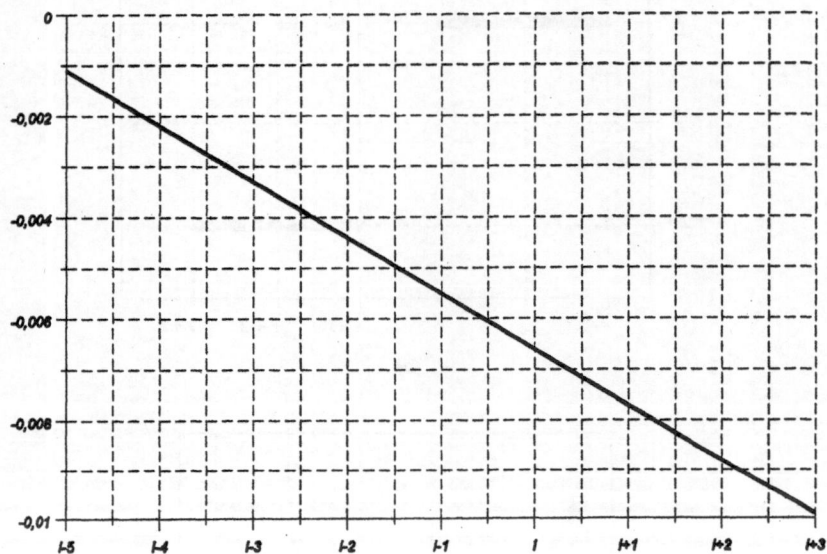

Figure 4.Receptive field weighting from layer S to layer L

SIMULATION RESULTS

A connectionist model for local speed estimation which rexelizes 128 pixels into 14 rexels was simulated. This topology corresponds to a one dimensional root exponential lattice (d=1) with five rings about the fovea, giving a bandwidth compression factor of f_c=216.

Figure 5 shows the speed estimation obtained by the model when a moving edge at different speeds was presented. Comparison of both figures highlights the role of parameter S in equation (5). A high value tunes the network to high speeds, moving the linear part of the curve to this range. Low values of S move the linear part of the estimation curve to the left. In this simulations the speed of the edge is measured in terms of a constant displacement along the retinal cells, so the longer the displacement, the higher the speed.

The low computational peak of the model, and the linear estimation of the object's speed it performs, are desirable features for tracking applications.

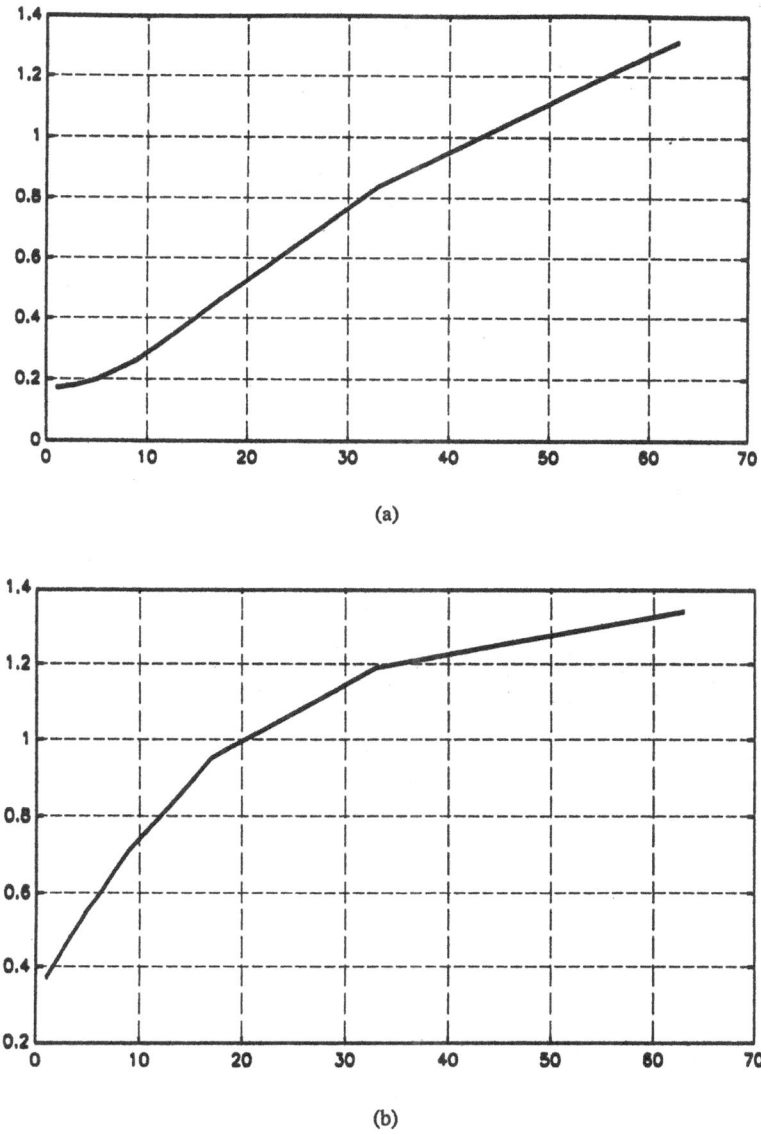

Figure 5. Speed estimation. (a) Tuned to high speeds (S=4). (b) Tuned to low speeds (S=2).

Note that the number of F- and S-neurons is much smaller in this foveal network than if acuity were uniformly maximum throughout the FOV. The limited convergence and lack of divergence in the network between the

R and F layers maintains the overhead of rexelization to a minimum. The size of the S layer in a uniform acuity network is related to that of the foveal network by the bandwidth compression factor f_c, which is determined by the number of rings in the topology. As a reference, consider that human vision has a central to peripheral acuity ratio of over 200:1 and a bandwidth compression factor of 16,000 (Yeshurun, 1989).

CONCLUSIONS

The presented connectionist model exploits a foveal retinotopology to reduce the computational complexity of local motion estimation. It is physiologically inspired in that it incorporates space variant spatiotemporal properties of amacrine and horizontal cells. However, by adopting a quad-tree tessellation of receptive fields, the model also lends itself to implementability in the machine setting.

While the main objective of the model is the estimation of motion, its inverse acuity weighting may also lead to increased peripheral sensitivity to temporal events, such as flashes, which is characteristic of primate vision and supports important preattentive behavioral responses in both biological and machine systems.

REFERENCES

Bandera, C. & Scott, P., "Foveal Machine Vision Systems," *Proceedings of the IEEE International Conference on Systems, Man, and Cybernetics*, Cambridge, MA, November 1989, pp. 596-599.

Kreider, Van der Spiegel, J., Born, I., Claeys, C., Debusschere, I., Sandini, G. & Dario, P., "The Design and Characterization of a Space Variant CCD Sensor," *SPIE vol. 1381 Intelligent Robots and Computer Vision IX: Algorithms and Techniques*, pp. 242-249, 1990.

Vico, F.J., Garrido, F.J., Sandoval, F. & Leibovic, K.N., "A Connectionist Model for Local Speed Estimation," *Proceedings of the ICIP94*, 1994.

Yeshurun, & Schwartz, E.L., "Shape Description With a Space-Variant Sensor: Algorithms For Scan-path, and Convergence Over Multiple Scans," *IEEE Trans. PAMI*, vol. 11, no. 11, pp. 1217-1222, November 1989.

Using Artificial Neural Networks for Ultrasonic Signals Processing from Simple Geometric Shapes

F. Arroyo, A. Gonzalo, J.R. Hilera·⁝·

Dpto. de Lenguajes, Proyectos y Sistemas Informáticos
E.U.I. de la Universidad Politécnica de Madrid
Ctra. de Valencia Km 7,400
D.P. 28031 MADRID
E-mail:farroyo@eui.upm.es

⁝ Dpto. de Matemáticas
Escuela Universitaria Politécnica
Universidad de Alcalá de Henares
Ctra. Madrid-Barcelona Km. 33,600
D.P. 28871 Alcalá de Henares
E-mail:mthilera@alcala.es

Abstract

The aim of this paper is the recognition of simple plane geometric shapes using Artificial Neural Networks and ultrasound echoes obtained from appropriate objects situated in front of one ultrasonic sensor. For us, the two much more simple plane geometric shapes are: segments and corners. Once the echoes were obtained by the sensor and stored in the computer, we preprocessed them by Fast Fourier Transform Algorithm and then we fed to an Artificial Neural Network with the first one hundred coefficients of the Fourier Serie of each echo obtained from the sensor. The Network is trained by Backpropagation algorithm.

Introduction

Our first working point with ultrasonic sensor was application feasibility of Artificial Neural Network for autonomous mobile robot control [1]. In that case we worked with two ultrasonic sensor situated on a mobile robot in order to obtain environmental information -distances from the contour to be followed by the robot. Some questions were formulated, could we obtain more environmental information with ultrasonic sensor?, and if we could, what kind of information would be?. A contour can be broken down into elementary geometric shapes, segments and corners. So the task was served.

At the beginning of the geometric shapes recognition task by means of echoes obtained from an ultrasonic sensor, we were thinking that Artificial Neural Networks were being appropriate devices for the processing of complex signals obtained from the sensor. In order to facilitate the recognition task to the Network, we thought it would be interesting to make the signal preprocessing using Fast Fourier Transform Algorithm, which are widely used in complex and noisy signal processes [2]. So, once the echoes were obtained, signal preprocessing was done as a previous step to Network's training. In this way we transformed each signal into a hundred of its Fourier serie coefficients, offering us an important reduction in the Network learning time.

We got 160 signals from different objects and transformed them into their corresponding coefficients of the Fourier serie, and then we formed two files with 80 or 120 records and 40 records respectively (half segments and half corners Fig. 1). The first file was used for the Network training, second file was used as a performance test in order to prove the Network performance.

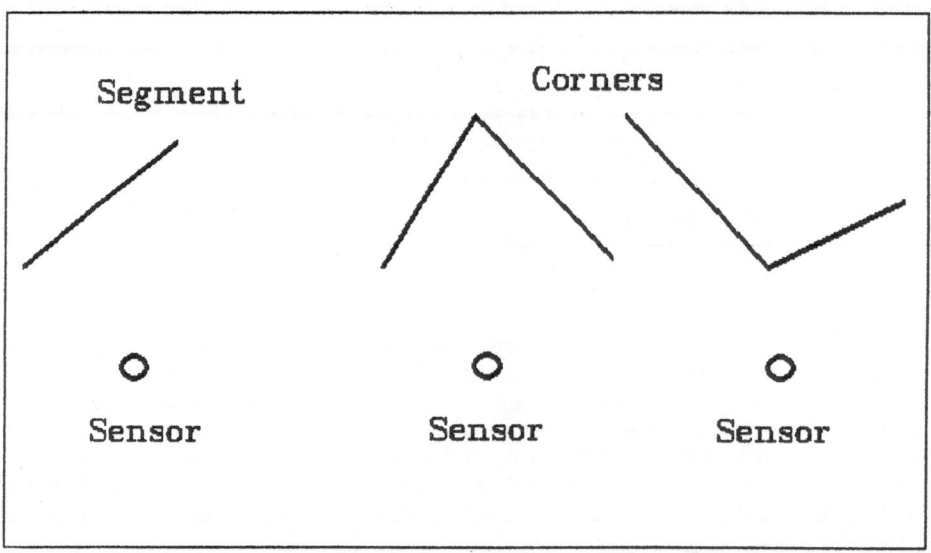

Fig. 1 Shapes presented to the ultrasonic sensor

Data Acquisition

Data acquisition system main component has been an ultrasonic transducer [9], 38 mm. of diameter capable to obtain an echo within a distance range from 0.3 m. to 10.6 m.; the transducer transmits a pulse toward a target and detects resulting echo converted into an electrical signal. Ultrasonic emission is controlled from a computer (Fig.2) by an Input/Output port, where a digital pulse is sent in order to activate a circuit oscillator that generates a pulse composed by 56 μpulses, which applied to the transducer generates the ultrasound. After ultrasound generation the sensor operating mode is shifted in order to behave like a microphone ready to receive the returned echo and transform it into an electrical signal. Signal digitalization was the next step in preprocessing, so we needed an analogical/digital converter board. The DT2811-PLG of Data Acquisition allowed us to read the signal values arrived at the sensor and then load them into computer memory on blocks of 1024 integer values of 12 bits.

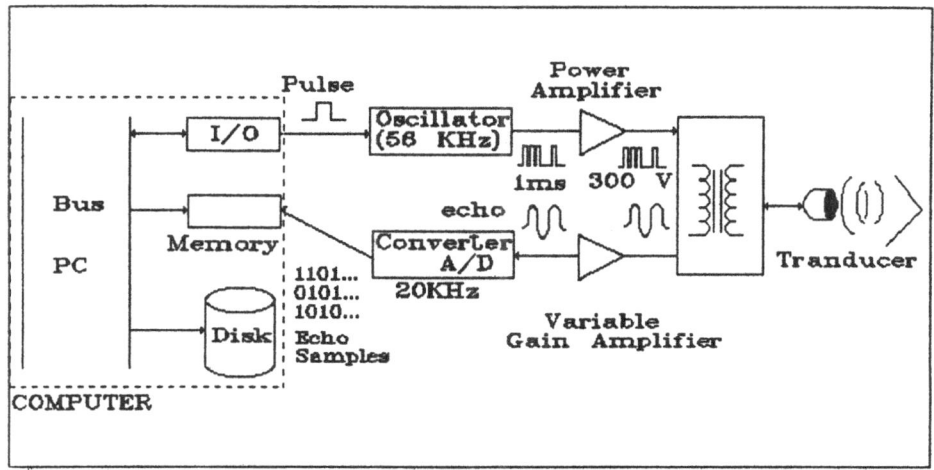

Fig. 2 Ultrasonic Transmit/Receive System

Learning and trial pattern sets were obtained from shapes situated at different distances and orientations, obtaining two files (one for learning phase and one for trial phase). To find desired data sets, the echoes obtained from the sensor were preprocessed using Sandey-Tukey algorithm to compute the Fast Fourier Transform [3][4] for each signal received into the sensor. A record is composed for the first one hundred coefficients of the Fourier serie of each signal and the desired output we expect at the network output units.

Network Architecture

We were working with feedforward multilayer networks like Multilayer Perceptrons ones, trained with classical backpropagation algorithm [6][7][8]. We were varying hidden units number between 10 - 24 and output units number between 1 - 2. Input units number were 100 corresponding to the Fourier coefficients number obtained to each signal echo received in the sensor. Finally we tested networks with two hidden layers.

Networks behaviour having one output unit works like this: When a pattern belonging to segment class was presented to the network, output unit network must be activated and remained non active in other case.

In networks having two output units, each output unit must recognize a specific class of pattern, e.g. if a pattern belonging to segment class is presented to the network only the output unit which recognizes it must be active and the other output unit not.

Results

Even though, system error is greater than one over learning pattern set, we got results between 87% and 97.5% of accuracy on the learning pattern set and 70% and 82.5% of accuracy on the trial pattern set. That can be produced by a lot of environmental conditions in the data acquisition process and by the similarity among pattern which networks were trained. We show different tables below to summarize obtained results.

TABLE I: Results obtained with one output unit networks

Pattern Number	Output Number	Hidden Units Number	Iterations Number	System Error	% Accuracy on learning pattern set	% Accuracy on trial pattern set
80	1	4	5500	1.972	95	75
80	1	6	2450	2.84	90	77.5
80	1	8	7500	1.002	97.5	75
80	1	13	300	3.323	91.25	77.5
80	1	14	600	2.378	95	72.5
80	1	15	1000	2.688	93.75	80
80	1	18	1000	2.542	93.75	80
80	1	19	600	2.793	92.5	77.5
80	1	20	600	3.103	92.5	75
80	1	21	1000	2.675	93.75	77.5
80	1	22	3100	1.213	97.5	72.5
80	1	24	4900	1.001	97.5	77.5
80	1	26	350	2.257	95	70

TABLE II: Results obtained with two output unit networks

Patterns Number	Outputs Number	Hidden Units Number	Iterations Number	System Error	% Accuracy on learning patterns set	% Accuracy on trial patterns set
80	2	5	6650	3.383	96.25	60
80	2	6	12500	3.352	96.25	62
80	2	7	1250	7.457	91.25	75
80	2	11	4000	1.0003	98.75	70
80	2	12	1600	2.628	96.25	62.5
80	2	13	1100	5.296	93.75	82.5
80	2	14	2350	2.029	97.5	70
80	2	22	6300	1.748	97.5	65
120	2	5	10500	1.01	99.16	60
120	2	22	9925	1.027	99.16	60

TABLE III: Results obtained with two layer networks

Patterns Number	Outputs Number	Hidden Units Number Layer 1	Hidden Units Number Layer 2	Iterations Number	System Error	% Accuracy on learning pattern set	% Accuracy on trial pattern set
80	1	2	3	12000	7.012	73.75	72.5
80	1	5	3	14000	1.496	96.25	60
80	1	5	5	2225	8.346	72.5	72.5
80	2	5	3	13050	5.29	93.75	75
80	2	6	6	2850	4.758	93.75	75
80	2	20	10	1150	11.376	81.25	75
120	2	3	3	5000	12.423	87.5	77.5

Conclusions

As we can see in tables I, II and III global system error is not important in system performance, so in two layer networks, with a system error of 5.29 we get a 93.75% of accuracy over learning patterns set and a 75% over trial pattern set. And if we force the system to decrease the global error over learning pattern set, it loses its ability to generalize. In this way we are trying to decrease the global system error without losing its capability of generalizing, even increasing it; exploring other architectures and dynamics. On other hand we are working in order to provide the networks with bigger capability in plain geometric shapes task increasing the pattern class number they can recognize.

Finally, we think Artificial Neural Networks can be used in recognition of ultrasonic echoes from simple plane geometric shapes, but if we want to apply them to autonomous mobile robot control we must try to get better performances with networks.

References

[1]   F. Arroyo, A. Gonzalo, E. Moreno. **Neural Network Desing for Mobile Robot Control Following a Contour**. Lectures Notes in Computer Science N° 540. International Workshop IWANN'91 pp 430-436 1991.

[2]   R.Paul Gorman & T.J.Sejnowski. **Learned Classification of Sonar Targets Using a Massively Parallel Network**. IEEE Trans. on Acoustic Speed and Signal Processing Vol 7 pp 1437-1440 July 1988.

[3]   J.C.Montaño, Mª.C.Florido, M.Castilla, A.Lopez y J.Gutierrez. **Realización de la FFT en PC**. Mundo Electrónico. pp 73-81 Diciembre 1989.

[4]   Kenji Nakayama. **An Improved Fast Fourier Transform Algorithm Using Mixed Frecuency and Time Decimation**. IEEE Trans. on Acoustic Speed and Signal Processing Vol 36 N° 2 February 1988.

[5]   Polaroid Ultrasonic Ranging System Handbook Application Notes/Technical Papers.

[6]   J.J.McClelland & D.E.Rumelhart. **Parallel Distributed Processing**. Vol I MIT Brodford Press, 1986.

[7]   J.Hertz, A.Krogh, R.G.Palmer. **Introduction to the Theory of Neural Computation**. Lecture Notes Vol I Santa Fe Institute. Addison-Wesley 1991.

[8]   Data Translation INC User Manual for DT2811.

[9]   Polaroid Corporation. (1984)
       "Polaroid Ultrasonic Ranging System Handbook".

A Hierarchical Neural Network for Mobile Visual Tracking with a Robot Head

D. Maravall, L. Baumela
Universidad Politécnica de Madrid
Dept. Inteligencia Artificial
Boadilla del Monte
28660 Madrid (SPAIN)

Abstract

Active vision and visual feedback intend to solve some of the problems arising in the development of robots with increasing autonomy. The so–called robot heads are excellent devices to implement and experiment the visual percepcion needed by these advanced robots. This paper presents some results obtained by using a robot head with two degrees of freedom for the automatic visual detection and tracking of mobiles. First a general description and setup of the problem is made, afterwards, a hierarchical neural network arquitecture for the solution of the visual tracking problem is proposed and described. Finally, the experimental results obtained with this novel solution are discussed and compared with the conventional recursive least–squares algorithm.

1 Introduction and general discussion

The emergence during the eighties of a new generation of mobile robots has helped the development of the so–called active vision paradigm [1, 2]: the concept of processing and analysis of digital images as an active task with one or several agents –usually autonomous robots –; i.e. robots which are able to interact with their environment in order to understand it, and modify certain parameters of their sensors as a result of that understanding.

Another concept used in our Laboratory, as early as in the mid eighties, is visual feedback [3, 4], interpreted as the symbiosis between computer vision and control engineering. Obviously, the concept of visual feedback does not have the globality of the active vision one –to which it is common to add the qualifiers intentional and qualitative– because visual feedback means the capacity of the visual system to change its position in the space and its internal parameters as a consequence of the visual information and, of course, the feedback this information to the system for its interpretation and understanding. Active vision is, or intends to be, visual feedback plus other things like environment understanding, planning and learning, which are topics belonging to the artificial intelligence field. In summary, if visual feedback is the result of the union between machine vision and control engineering, active vision should be considered the union of these disciplines with artificial intelligence.

In this paper we are going to focus primarily on one specific problem in visual feedback: automatic mobile tracking. Without entering into too much detail –for which other references should be consulted [5, 6]– we shall introduce first our previous work in the so–called robot heads field.

Robotic heads have appeared very recently [7] and are the result of the inmense effort currently devoted to improve the visual perception in the new generation of mobile robots. They are excellent platforms for the development and test of theories and algorithms in machine vision. We have been working in our Laboratory with a robot head, particulary for the detection and tracking of mobiles, which is a typical problem in dynamic scenes analysis.

One vital aspect of the correct solution to the above-mentioned problem –the visual tracking of mobiles– is the existence of a good model of the mobile's dynamics, essential for the continous presence of the target within the camera's field of view. The block diagram shown in figure 1 will help to understand this.

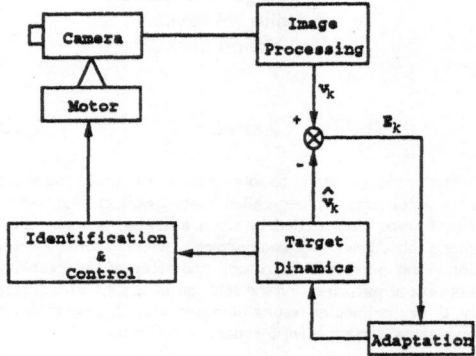

Figure 1: Symbolic scheme usign blocks of the mobile visual tracking robot head

From this figure it is obvious the importance of a good model of the mobile's dynamics for the correct operation of the robot head and accurate positioning of the video–camera.

In our work the basic information needed for target tracking are the cartesian coordinates of the mobile's centroid.

$$\vec{v} = x\hat{a}_x + y\hat{a}_y + z\hat{a}_z \tag{1}$$

where (x, y, z) are cartesian coordinates.

Usual methods for signal and systems estimation and prediction are based on extremely precise and accurate mathematical models. For instance, this is the case of the well known Kalman filter, which handles a model expressed as a system of equations in the so–called state variable representation. Another common method is the Luenberger observer. It is an equivalent approach for the mathematical description of the system under study, although with a different design, as observers are based on the eigenvalues of the system's dynamical matrix in order to estimate and predict the variables of interest. For a comparative study of both methods see [8].

The main disadvantage of these methods lies in the poor flexibility with regards to changing dynamics of the targets, as both the Kalman filter and the Luenberger observer are constrained to use a specific mathematical model for each possible dynamics.

To avoid this problem we introduced in 1988 a new method [9], to our knowledge for the first time in the technical literature, for the estimation of a vision–guided mobile robot's dynamics. This new method is based on a very different approach to the one usually taken in the Kalman filter and Luenberger observer designs, where a mathematical model is obtained after posing physical specifications. On the contrary, with the novel method, the dynamics of the variables are obtained using their Taylor series expansion. Then, we can express the centroid's cartesian coordinates as

$$\vec{v}[(k+1)T] = \vec{v}(kT) + T\dot{\vec{v}}(kT) + \ldots + \frac{T^n}{n!}\vec{v}^{(n}(kT) + \ldots \tag{2}$$

It is elementary to show that the three coordinates follow the same law. Thus, for coordinate x we can write

$$x[(k+1)T] = x(kT) + T\dot{x}(kT) + \frac{T^2}{2}\ddot{x}(kT) + \ldots + \frac{T^n}{n!}x^{(n}(kT) + \ldots \tag{3}$$

Then, taking into account only one of the three coordinates, the following simplifications can be introduced

$$\dot{x}_k \approx \frac{(x_k - x_{k-1})}{T}$$
$$\ddot{x} \approx \frac{(\dot{x}_k - \dot{x}_{k-1})}{T} \approx \frac{(x_k - 2x_{k-1} + x_{k-2})}{T^2} \tag{4}$$

where it has been adopted a more handy notation for the time–sampled variables. Therefore, it is possible to predict coordinate x at instant $t = (k+1)T$ with the following difference equation

$$x_{k+1} \approx w_o x_k + w_1 x_{k-1} + w_2 x_{k-2} + \cdots + w_N x_{k-N} + \cdots \tag{5}$$

The coefficients $w_0, w_1, \ldots, w_N$ fix the target's dynamics. The greater is N, the more accurate is the corresponding model in regards to the real dynamics. For typical target dynamics it is good enough to use a small number of coefficients.

Another factor to be considered, apart from the model order given by N, is the fact that the coefficients $w_0, w_1, \ldots, w_N$ depend on the sampling period T under which the mobile's dynamics is observed or measured. Obviously, as T is fixed by the design of the whole system, the coefficients do not depend sensu stricto on T, although the sampling period determines the stability of the model convergence to the real target dynamics. A correct sampling period choice is one of the crucial aspects in the design and implementation of any control system algorithm.

In summary, the models of a particular mobile's dynamics rely on two elements: a) the order of the model given by N and b)the sampling period T. Naturally, for a tracking system claiming to be valid for any kind of mobile it is necessary to leave open these two elements, N and T, so that the system learns in real time the optimum values for a given mobile. It is in this particular aspect where the neural networks approach plays an important role.

2 Kinematic model of the robot head

Target visual tracking, as posed above, is three dimensional (3D) by its very nature. This would suggest working with stereo vision in order to recover the three dimensions of the real world. Stereo vision based visual tracking poses serious difficulties with regard to computation times, something which makes it almost unpratical for tracking real targets. In previous publications [5, 10], we have proposed a simplified solution based on hypothesizing the center of gravity of the mobile's projection onto the image plane as the correspondence point in the stereo pair. Another approach is the one recently developed in our Laboratory [11, 12], in which the essential aim is to continuosly maintain the target within the field of view of a robot head controlled camera, rather than computing the three coordinates of the target's centroid.

This aim is more efficiently achieved working directly with the camera rotation angles, θ and ϕ, which correspond to the azimuth and elevation degrees of freedom of the robot head. In the above mentioned reference [12] the kinematic model used to transform the (x, y) target coordinates on the image plane into the (θ, ϕ) coordinates of the head axis movement can be consulted. Notice that only two coordinates are used; i.e. the tracking is 2D.

Obviously, the transformation of cartesian coordinates into the θ and ϕ head axis coordinates does not make invalid the linear approximation introduced earlier for the cartesian coordinates. Then it is

perfectly correct to postulate that the dynamics of any target expressed in the head axis coordinates are
also linear

$$\begin{aligned}
\theta_{k+1} &= w_0\theta_k + w_1\theta_{k-1} + w_2\theta_{k-2} + \cdots + w_N\theta_{k-N} + \cdots \\
\phi_{k+1} &= w_0'\phi_k + w_1'\phi_{k-1} + w_2'\phi_{k-2} + \cdots + w_N'\phi_{k-N} + \cdots
\end{aligned} \tag{6}$$

3   A hierarchical neural network architecture for mobile visual tracking

The visual tracking system that we have implemented is based on a hierarchical arquitecture of linear
neural networks organized in two levels. In the first level there are M prediction filters, each one tuned
to a specific dynamic model –bear in mind the previous discussion concerning the parameters N and T-.
In a higher level there is a decision–making and learning system that chooses, at each sampling instant
of the global tracking system, the optimum lower level prediction filter. As the final and basic aim of the
robot head is to permanently maintain the target within the camera field of view, it is vital to guess the
mobile's position, in order to track it accurately. Figure 2 shows the schematic diagram of this hierarchical
architecture.

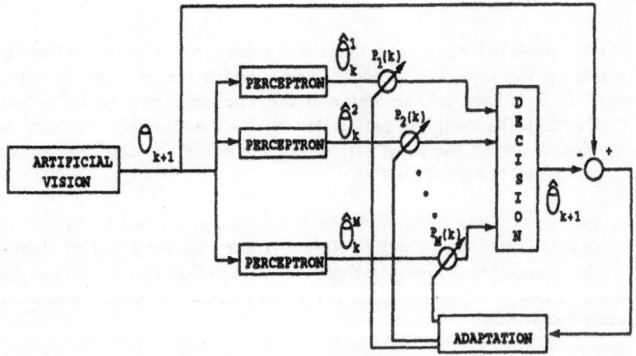

Figure 2: Schematic diagram of the hierarchical neural network architecture

Each prediction filter is implemented by means of a linear neural network with a perceptron–like learning algorithm. Concerning the high–level structure that coordinates the low–level structure, it consists
as well of a learning algorithm, in this case different from the perceptron, as we shall show in the sequel.
Notice that only one of the coordinates will be considered in the next paragraphs, as the other one has
exactly the same behavior.

3.1   Individual perceptron–like prediction filters

We have previously postulated and justified that the dynamics of any mobile can be modeled as a linear
combination of mobile position past values –with a finite memory depth–. Considering exclusively the
spherical coordinates of the head axis, θ and ϕ, we can rewrite the expresion (6) in a compact form

$$\theta_{k+1} = \vec{W}_\theta^T(k)\vec{\theta}_k; \quad \phi_{k+1} = \vec{W}_\phi^T(k)\vec{\phi}_k \tag{7}$$

where

$$\vec{W}_\theta^T(k) = [w_0(k)\ w_1(k)\ \ldots\ w_N(k)]_\theta \ ; \quad \vec{W}_\phi^T(k) = [w_0(k)\ w_1(k)\ \ldots\ w_N(k)]_\phi$$

and

$$\vec{\theta}_k = \begin{bmatrix} \theta_k \\ \theta_{k-1} \\ \vdots \\ \theta_{k-N} \end{bmatrix} ; \quad \vec{\phi}_k = \begin{bmatrix} \phi_k \\ \phi_{k-1} \\ \vdots \\ \phi_{k-N} \end{bmatrix}$$

With this compact notation and restricting ourselves to one of the two dynamical variables, we can express the prediction of a generic individual filter as follows

$$\theta_i(k+1) = \vec{W}_i^T(k)\vec{\theta}_k \tag{8}$$

in which, to shorten the notation, we have removed the reference to the w coefficients of coordinate θ and instead we have introduced in these coefficients the subscript i which make reference to a generic prediction filter.

It can be defined a performance index for this generic filter, J_i, based on the square prediction error

$$J_i(k+1) = \frac{1}{2}\epsilon_i^2(k+1) = \frac{1}{2}[\hat{\theta}_i(k+1) - \theta(k+1)]^2 \tag{9}$$

where $\theta(k+1)$ is the coordinate obtained from the visual tracking system at instant $t = (k+1)T$.

As it is well known, a procedure to minimize this performance index is the so-called perceptron algorithm, based on the error gradient descent. That is

$$\vec{W}_i(k+1) = \vec{W}_i(k) - \mu(k)\left\{\nabla_{\vec{W}_i(k)}J_i(k)\right\}_{\vec{W}_i(k)} \tag{10}$$

Later on we shall present the results obtained with the hierarchical neural network architecture based on this prediction algorithm, and their comparison with the results from the recursive least–squares algorithm, one of the most frequently used prediction filters.

3.2 Competitive and parallel learning

In figure 2 it can be clearly observed the hierarchical structure of the global system implemented for our visual tracking system.

Thereby, the output of all the prediction filters based on the perceptron algorithm are evaluated by a higher level. This evaluation or decision–making can be expressed in a generic way for each dynamical coordinate as follows

$$\begin{aligned} \hat{\theta}_{k+1} &= p_1(k)\hat{\theta}_k^1 + p_2(k)\hat{\theta}_k^2 + \cdots + p_M(k)\hat{\theta}_k^M \\ \hat{\phi}_{k+1} &= p_1'(k)\hat{\phi}_k^1 + p_2'(k)\hat{\phi}_k^2 + \cdots + p_M'(k)\hat{\phi}_k^M \end{aligned} \tag{11}$$

It must be noticed that these expressions coincide with the output of a linear neural network with synaptic weights $p_1(k)$, $p_2(k),\ldots, p_M(k)$ and $p_1'(k)$, $p_2'(k),\ldots, p_M'(k)$. Therefore, the higher level of the global tracking system is composed of two linear neural models.

The physical interpretation of this higher level is straightforward: the synaptic coefficients weigh up the importance of each individual prediction filter made by the perceptron–like neural networks of the lower level. Thus, if the following condition holds

$$\sum_{i=1}^{M} p_i(k) = 1 ; \quad 0 \le p_i(k) \le 1 \quad \forall i = 1,\ldots, M \tag{12}$$

i.e. if the $p_i(k)$ are positive and normalized to unity –which is equivalent to be considered as probabilities–
the one dominant prediction would be that corresponding to the maximum synaptic weight. If some of the
synaptic coefficients were equal to unity, as the rest had to be null, it would mean that the corresponding
prediction would be the only one to be considered.

Posed in this way the higher level of the global tracking system, several alternatives can be followed as
far as the updating or learning of the synaptics weights is concerned. An all–or–nothing solution can be
adopted, in which only one prediction filter is considered at each sampling instant $t = kT$ –the filter with
the maximum synaptic wheight–, or, on the contrary, a less drastic solution can be chosen by taking into
account all the predictions and weighing them up with their respective synaptic weight. In the sequel we
present and discuss some comparative experimental results.

4 Experimental results

One negative feature of this visual tracking system rests on the high computation times requiered by
the machine vision algorithms: digitization, image processing, scene segmentation, mobile identification,
computation of the cartesian coordinates (x, y) and, finally, computation of the spherical coordinates
(θ, ϕ) needed to control the robot head.

These high computation times produce relevant delays in the information flow coming from the video
camera to the robot head actuators. Unless these times were compensated, the visual tracking would be
unpractical. Therefore, the aim of the prediction subsystem is to compensate that time lag by predicting
at each instant $t = kT$ the mobile position at the following instant $k = (k + 1)T$. As previously stated,
it is supposed that the target position is determined by the cartesian coordinates (x, y) of the mobile
projected contour's center of gravity on the image plane. Of course, these cartesian coordinates have to
be afterwards transformed into the spherical coordinates (θ, ϕ).

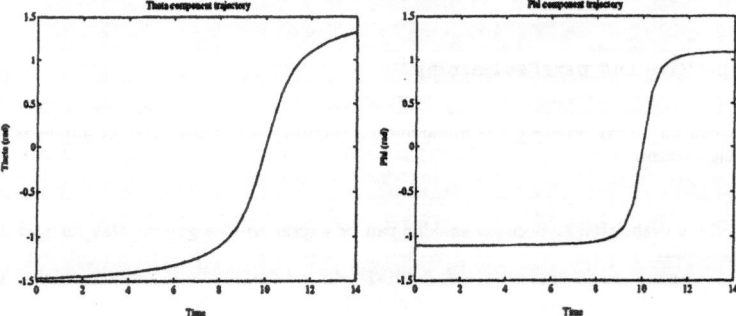

Figure 3: Azimuth and elevation components of the spherical coordinates –i.e. θ and ϕ– corresponding
to a linear trajectory in cartesian coordinates

One problem arising in the target's prediction is the high nonlinearity of the trajectories $\theta(t)$ and $\phi(t)$.
In figure 3 are shown the dynamics of $\theta(t)$ and $\phi(t)$ for a linear $x(t)$ and $y(t)$ trajectory in the cartesian
space. It can be clearly noticed the strong nonlinearities of the spherical coordinates, even though the
cartesian coordinates representation is linear.

Concerning the performance of the hierarchical neural network proposed in this paper, figure 4 shows
the comparative results of this architecture and the recursive least–squares algorithm. For the results

999

displayed in figure 4 the neural network learning procedure is very simple: it is an all–or–nothing type of learning with the synaptic weights $p_i(k)$ and $p_i'(k)$ of expression (11) taking only the values zero or unity. More specifically, at each sampling instant $t = kT$ the perceptron–like prediction filter of the lower level with the minimum square error is the dominant one. So its synaptic coefficient is equal to unity, and the remaining synaptic weights are obviously equal to zero, due to the normalization restriction. In figure 4, it can be noticed the improvement obtained by the neural network as compared with the prediction of the recursive least–squares algorithm.

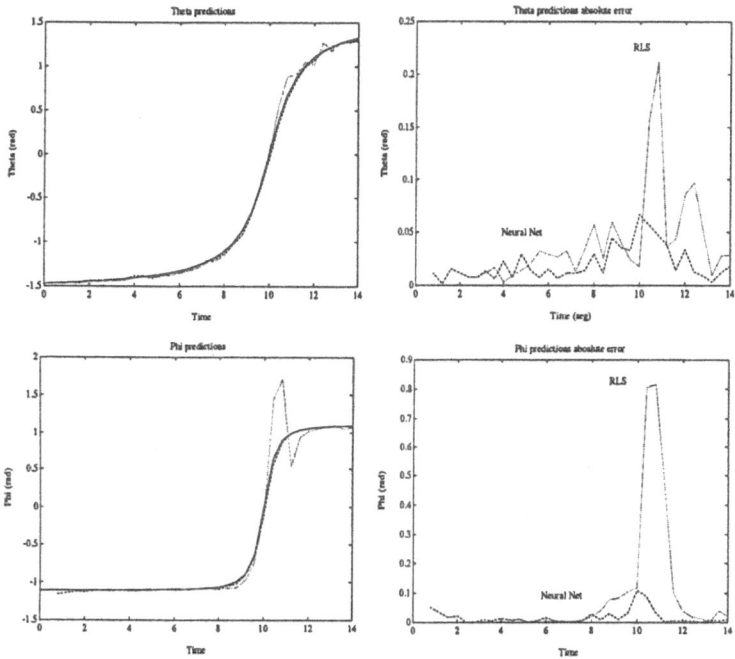

Figure 4: Comparative results of the hierarchical neural network architecture trained with an all–or–nothing algorithm and the recursive least–squuares one.

Finally, in figure 5 are displayed the results of the least–squares algorithm and the hierarchical neural network architecture for a nonlinear trajectory in the cartesian space, which is clearly a more difficult prediction test than the linear one previously considered. In figure 5a can be observed the switching in the hierarchical neural network architecture between two perceptron–like neural networks of the lower level. This switching means that the global system always chooses the prediction filter with minimum square error. In figure 5b the prediction errors of the recursive least–squares algorithm and the hierarchical neural network architecture are shown together, with a remarkable better performance for the last procedure.

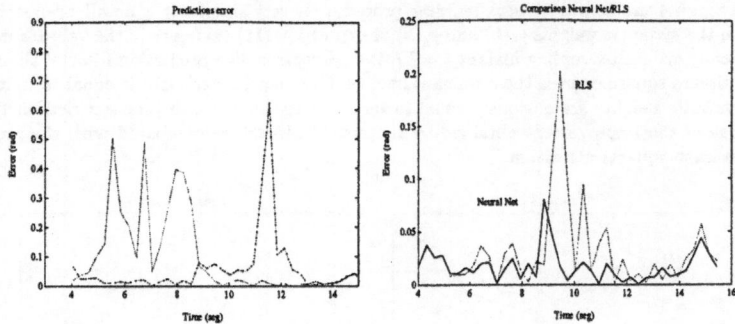

Figure 5: Comparison of predictors for a nonlinear trajectory in the cartesian co ordinates.

References

[1] J. Aloimonos, I. Weiss, A. Bandyopadhyay (1988). "Active vision". *International Journal of Computer Vision*, vol. 1, pp. 333–356.

[2] R. Bajcsy (1988). "Active perception". *Proceedings of the IEEE*, vol. 76, n. 8, pp. 996–1005.

[3] D. Maravall, M. Mazo et al. (1990). "Guidance of an autonomous vehicle by visual feedback". *Cybernetics & Systems*, vol. 21, pp. 257–266.

[4] M. Mazo (1988). "Contribution to the control and guidance of an autonomous vehicle with visual feedback". (In Spanish). PhD. dissertation. Technical University of Madrid.

[5] D. Maravall (1992). "Automatic tracking of mobile objects with self–controlled camera using stereo vision", *Proceedings of the IFAC Symposium on Intelligent Components and Intruments for Control Applications*, Málaga, pp. 633–636.

[6] D. Maravall (1994). "Adaptive systems and learning automata". (In Spanish). In D. Maravall Casesnoves (Ed.). *Frontiers of Computer Science*. (In Spanish). Real Academia de Ciencias, pp. 41–59.

[7] H. I. Christensen, K. W. Bowyer, H. Bunke (Eds). *Active Robot Vision*. World Scientific. Singapore, 1993.

[8] D. Maravall, F. Monasterio-Huelín (1985). "Estimation versus observation in a soaking pit". (In Spanish). *Proceedings of the Sixth Congress on Informatics and Automatica*. Madrid, pp. 447–451.

[9] D. Maravall (1993). "Visual feedback of an autonomous vehicle". (In Spanish). In D. Maravall Casesnoves (Ed.) *Artificial Intelligence. Theory and Practice*. (In Spanish). Real Academia de Ciencias, pp. 59–77.

[10] D. Maravall (1993). "Adaptive control for the automatic tracking of mobiles using a vision–based robot". *Modelling, Measurement & Control – B*, vol. 51, n. 4, pp. 55–63.

[11] J. F. Peters. L. Baumela, D. Maravall, S. Ramana (1994). "Logical design of neural controllers". *Proc. IFAC Symposium on Artificial Intelligence in Real Time Control*. Valencia, pp.201–206.

[12] L. Baumela, D. Maravall (1994). "Geometric model for camera–based object tracking". *Proceedings of the SPIE Conference on Photonics for Industrial Applications: Vision Geometry*. Boston (in press).

Short-Term Load Forecasting Using Neural Nets

Ricardo S. Zebulum † ‡
Marley Vellasco †

Karla Guedes †
Marco Aurélio Pacheco †

† ICA : Núcleo de Inteligência Computacional Aplicada
Departamento de Engenharia Elétrica
Pontifícia Universidade Católica
Rio de Janeiro - Brasil
e-mail: ICA@ele.puc-rio.br

‡ CEPEL - Centro de Pesquisas de Energia Elétrica
Caixa Postal: 2754   CEP : 20.001 - 970
Rio de Janeiro - Brasil
Tel : 598-2134 Fax:260-1340/260-6211

ABSTRACT

Load forecasting is decisive in the operation of power systems, for economic and security reasons. Many techniques have been proposed in the last two decades [1]. This work presents a short-term load forecasting system (whose main objective is to maintain the generation-load balance) using Neural Networks. Neural Networks have demonstrated to be a very efficient technique to time series forecasting, particularly in load series [2]. In the application shown in this paper, a Neural Network is used to learn the daily load behaviour of a real electrical system (CEMIG, Brazil, 1993). The network inputs are : past load data, the forecasting hour and the type of day (weekday or weekend). The windowing technique [3] is used to identify the series characteristics. Many neural nets with different architectures were tested and the results evaluated in terms of forecasting errors. We achieved an average forecasting error close to 1.5%. The forecasting system was developed in C programming language and includes the pre-processing of the input data, the network training and the forecasting. This system offers to the user options such as : tuning of some network parameters (learning rate, momentum term, number of processors in any layer), usage or not of the forecasted values as network inputs, adjustment of the size of the training window etc.

1 - INTRODUCTION

Load forecasting is of great importance to the electric energy companies. There are basically two types of forecasts, depending on the leading time of the prediction: *Short Term* and *Long Term*. In the first case, the objective is to forecast the load from 1 to 24 hours ahead. Short Term Load Forecasting is important to assure an economic and secure operation of the power system. In the second case, the aim is to forecast the load months or years ahead and it is essential to the system's expansion planning.

Time series and regression techniques have for long been used in Short Term Load Forecasting (STLF) as standard techniques [1].

In the last few years, Computational Intelligence (CI) techniques have also been applied to the load forecasting problem. Expert Systems for short term load forecasting have been suggested [4], but Neural Networks (NNs) have offered better results in this field. The non-linear structure of NNs allows the identification of complex properties of a particular series, which is impossible with standard techniques. Many works in this area can be found in the literature. Few of them are summarised bellow.

K.Y. Lee and J. H. Park [5] from the Pennsylvania University developed a Neural Network for hourly load prediction. They used two different networks for week-days and weekends. The network inputs are the past values of the load series for the day of the forecast and for the previous days. Using a one hidden layer architecture and the Backpropagation learning algorithm, Lee and Park obtained a 2% average error.

Park, Sharkawi, Marks, Atlas and Damborg [6] from the Washington University applied NNs in total daily load, peak daily load and hourly daily load forecasting. All networks have one hidden layer and implement the Backpropagation learning algorithm. In the first two cases, Park and his group used only the temperature series to forecast the electric load, achieving 2.04% and 1.68% average error for total daily load and peak daily load forecast, respectively. In the hourly daily load forecasting, the network inputs are the previous load values, the previous temperature values and the predicted temperature for the hour of the forecasting. They obtained a 1.40% average error for an one hour ahead forecast. In this work, they have also investigated the increase in the forecasting error as a result of an increase in the leading time of the forecast.

Peng, Hubele and Karady [7], researchers from the Arizona University, applied NNs to total daily load forecasting. Initially, the network is trained with 52 weeks load data. Then, after each day, the network is trained again. The training set is selected from the total set of available data, using the minimum distance criterion to reduce the network learning time. This work shows the influence of the input data normalisation in the network's performance. Besides, the negative effect of the holiday inclusion in the training set is investigated.

Kun-Long-Ho and his fellows [8] from the Electrical Engineering Department of Taiwan University developed an hybrid intelligent system for hourly load forecasting. An expert system identifies the day type (week-day, weekend or holiday) and a Neural Network forecasts the peak and valley loads of the day. The Expert System gets these values and computes the hourly load values for the day. This system uses a NN with adaptive momentum in order to decrease training time.

The work presented in this article describes an hourly load forecasting system, based on Backpropagation NNs, developed at ICA (Research Centre for Applied Computational Intelligence) of PUC-Rio, Brazil. The system consists of three main parts: the load series pre-processing, the network training and the load prediction. In contrast with the majority of the works published in this area, this work does not make use of temperature as a network input. Forecasting is made using only past load values, hour of the day and type of the day. Our objective is to show that if an adequate training set is chosen, it is possible to obtain good results even without considering temperature or any other weather data. In this work, we also investigate the forecasting performance for different implementations: using single or separate networks for different types of day, the effect of using forecasted load values as NN inputs, the effect of the momentum term and the effect of varying the window size.

The next section briefly discusses the role of load forecasting in electrical power systems. Section 3 presents a selection of the main simulations performed in this work, describing the networks architecture, its parameters and the training sets used in each implementation. The results are shown in section 4 . Section 5 presents some final conclusions.

2 - SHORT TERM LOAD FORECASTING IN POWER SYSTEMS

Short Term Load Forecasting predicts a system load demand for a period ranging from 1 to 24 hours. It is fundamental to the power system operation, as it grants an efficient calibration between generation and demand.

In a period of few seconds, when load variations in a power system are small and random, the *Automatic Generation Control (AGC)* operates making the generation couple with this small variation of the system load. In periods of minutes, when greater variations may occur, the economic load dispatch is done, assuring that the generation changes due to the load variations are economically distributed among the power systems' generators. In the period of hours, Short Term Load Forecasting (STLF) plays the major role, since it allows the analysis of future power exchanges between electrical areas, leading to an economic operation. In addition, investigations of the system behaviour in the presence of a large disturbance, greatly depends on the previous knowledge of the future load demand of the power system [2].

Overall, the main variables that have to be considered for short term load prediction are: economical factors, seasonal effects, day of the week, hour of the day, temperature and other weather parameters, and random effects.

3 - NEURAL NETWORK APPROACH FOR STLF

3.1 - Network Architecture and Analysis Method

The short term load forecasting considered in this work makes use of a load series with a leading time that varies from 1 to 24 hours. The network inputs are typically the past load values, the type of the day and the forecasting hour. The general NN topology is presented in figure 1.

The method for data acquisition and series analysis used in this work is based on the windowing technique [9]. The idea is to use two windows, Wi and Wo , of fixed sizes WI and WO respectively, to look into the dataset (Figure 2). For a given window size, the assumption is that the sequence $Wi_0, ..., Wi_n$ is somehow related to the following sequence $Wo_0, ..., Wo_n$, and this relationship, although unknown, is defined entirely within the dataset. In the case of Neural Networks, Wi → Wo can be used as a training vector. Both windows are shifted along the time series using a fixed step size. In the example of the NN shown in figure 1, we have Wi = n e Wo = 1.

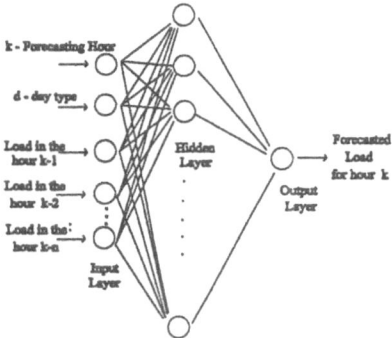

Fig. 1 - A Neural Network Topology Used For Load Forecasting.

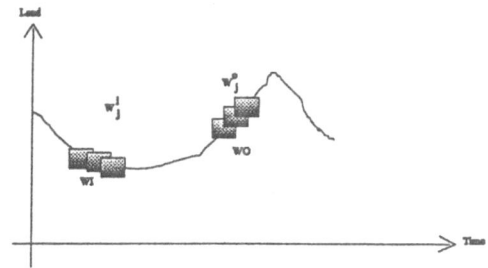

Fig. 2 - The "Windowing" Technique

3.2 - Case Studies

In this work we have used real data from the Brazilian Electrical System. The data contains hourly generation values of the Electric Company of the state of Minas Gerais in 1993. For the sake of simplicity, we will refer to this values as load values, as they represent real load demand. The shape of the load series is shown in Figure 3, for the months January, February and March.

The first step of the work included an extensive analysis of the load series. Through this analysis, we concluded that a week could be divided in four groups of days, each with a particular load shape. The first group includes Tuesdays, Wednesdays, Thursdays and Fridays, while the second, the third and the fourth groups include Saturdays, Sundays and Mondays respectively. Group 1 , made up of week-days, presents the highest load values and group 3, which includes only Sundays, contains the lowest load values. The other two groups (Saturdays and Mondays) are in an intermediate stage. Figure 4 shows the load curves for each day of a week in January 1993. As one could expect, there is a strong relationship between economic activities and the load shape for each day. For instance, Monday morning loads are relatively low, due to the start up of the industrial activities. On the other hand, there is a great decrease in the load shape in the second half of the Saturday, due to the end of the week activities.

Considering these load nuances, we adopted the following input sets to the network:

L $(k\text{-}m)$ → Set of the m past load values before the prediction. These inputs represent the load in the hour $(k\text{-}m)$, where m is a positive integer.

k → Forecasting hour.

d → Type of the day in the week.

The set "L" gives the m load values immediately before the hour of prediction, known as k. The variable m varies from 1 to the window size and constitutes a network parameter, to be chosen before training. The forecasting hour k is also a network input, while d gives to the network information about the week-day. Using such inputs, the network is able to recognise the correct load shape of a given day. However, when different networks are used for different types of day, the input d is not necessary.

The network topology consists of one hidden layer, using Backpropagation algorithm[10] for training (Figure 1). Given the load series regularity, we considered that one hidden layer would perform better. Two hidden layer network would be suitable if the series had a higher degree of randomness. The hidden layer is composed of processors with sigmoid activation function. The output layer is made up of only one processing element, also with sigmoid activation function. This output corresponds to the forecasted value.

The number of processing elements in the network input layer is directly related to the window size chosen, as shown in figure 1.

The number of processors in the hidden layer is not directly bound to the inputs or outputs of the network, but it is closely related with the network's performance. If it is too large, the training time will increase and the network may loose its generalisation capability. Otherwise, if it is too small, the network will not be able to converge. We made many tests, varying the number of hidden processors in the range between N to 3 x N, where N is the number of processors in the input layer. We observed that the best forecasting results occurred when the number of hidden processors were approximately between 1.5 x N and 2 x N.

3.3 - Network Training

The construction of the training set is an important aspect in a NN forecasting system. The load series must be pre-processed, that is, it must be turned into a training set to feed the NN. Pre-processing does not mean simply applying the windowing technique to the series, but also normalising the values that will be provided to the network. Data normalisation to the range between 0 and 1 is important to avoid network saturation due to the sigmoid shape. The network's performance is strongly dependent on the kind of normalisation used. In this work, linear normalisation was used. The following equation illustrates this normalisation:

$$\text{Normalised Load} = \frac{L - MIN}{MAX - MIN}$$

where:

$L \rightarrow$ Load value not normalised.

$MIN \rightarrow$ Minimum value of the series.

$MAX \rightarrow$ Maximum value of the series.

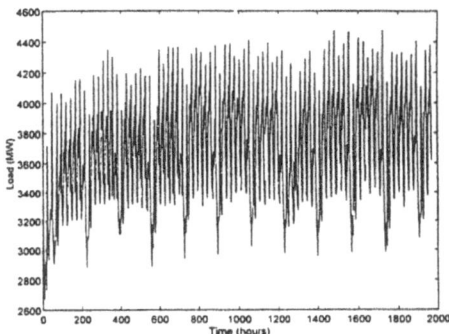

Fig. 3 - Load Series Values   Fig. 4 - Load shape of each day of the

(Jan. , Feb. and Mar.,1993)   week (Jan. 10 to Jan.16, 1993)

3.4 - Implementations

A large number of tests were realised with the available data. We selected four cases to present in this article. Their characteristics are summarised in Table 1. The size of the training set varied from 3 weeks to 2 months, in order to consider the seasonal effects.

The network's performance was investigated for two types of Forecasting: *single-step* and *multi-step*. In the *single-step* forecasting, the network inputs are always the load series values. In the multi-step forecasting, the network inputs are fed with forecasted values, which generally tends to increase the error. In the *multi-step* tests, we used forecasted values as inputs for a period of 24 hours. At the beginning of each day, the network is presented with the correct load series values again. One key result of this work is that we have obtained similar results in both kinds of forecasting.

4 - RESULTS

The results, in terms of daily average errors, for each of the 4 cases illustrated in table 1, are shown in table 2. Each day error is computed taking the average of the 24 hourly errors.

Firstly, it can be verified that case 1 yielded the worst results (3.33% average error - Table 1). In this case, the first three weeks of January were used as the training set and the last ten days of the month were used for test. We concluded that the small size of the training set led to a poor network performance.

In case 2, a single network was trained (as in case 1), but the training set was increased to two months (January and February). The entire month of March was used to test the network. The average error for the forecasting was reduced to 2.49% (Table 1). We could verify that the errors were bigger in the Saturdays and Sundays series, meaning that the use of different networks could yield better results. Additionally, the behaviour of the hourly forecasting error was very irregular, meaning that the type of codification for representing the hour input was not suitable for the network training.

Case	Training Set	Test Set	Nets	Window Size	Type of day	Forecasting Hour	Input Layer Size	Hidden Layer Size	Output Layer	μ	α	f	<e>
1	1/1/1993 to 21/1/1993	22/1/1993 to 31/1/1993	1	10	d = 0 (Sundays and holidays) e d = 1 (Other days)	hour/24	12	20	1	0.6	0.9	Sigmoid	3.33 (Single -Step)
2	1/1/1993 to 28/2/1993	1/3/1993 to 31/3/1993	1	10	d = 0 (Sundays and holidays) e d = 1 (Other days)	hour/24	12	20	1	0.6	0.9	Sigmoid	2.49 (Single -Step)
3	1/1/1993 to 28/2/1993	1/3/1993 to 31/3/1993	4	5	N/A.	Hour binary value (5 bits)	10	18	1	0.6	0.9	Sigmoid	1.68 (Single -Step) e 1.71 (Multi-Step)
4	1/5/1993 to 30/6/1993	1/7/1993 to 31/7/1993	4	5	N/A.	Hour binary value (5 bits)	10	18	1	0.6	0.9	Sigmoid	1.51 (Single -Step)

Table 1 - Four Studied Cases (μ is the learning rate , α is the momentum, f is the activation function and <e> is the average (%) error).

The best results were obtained when we implemented two changes in case 2: first, one network was trained for each of the four day groups (week-days, Saturdays, Sundays and Mondays) and, second, the forecasting hour was codified as a binary number. In case 3, we executed forecasts for March, obtaining a 1.68% average error (Table 1). For week-days (group 1), the average error was only 1.47%.

In case 4, we made forecasts for July of the same year, using May and June to train the network and obtaining 1.51 % (Table 1) average error.

All the tests mentioned above were 1 hour ahead forecasts. To predict the load shape of the entire day (1 to 24 hours ahead), we used the *multi-step* forecasting (Case 3 (*) in Table 2). The forecasted values were used as inputs for the period of one day. It can be seen that the network's performance was almost the same as the observed in case 3 for 1 hour ahead forecasting, which is a very stimulating result.

Figure 5 shows a comparative graphic between hourly forecasted and actual loads for the last week of March, 1993, in the *single-step* case. Figure 6 shows the relative histogram. Figures 7 and 8, respectively, shows the same results for the *multi-step* case.

It is important to mention that the momentum term introduced a great speed-up in the training time of the NNs, particularly when the training set was increased (cases 2, 3 and 4). Less then 1000 iterations were sufficient for the networks to converge. The learning time was about 1 hour, running the NNs in SPARC workstations.

Comparing our results with other works, it can be verified that they are quite reasonable. Lee e Park [5] obtained an average error of 2.2% for the Summer period and 1.8% for the winter period. The main difference between Lee and Park's work and this work is that in the first the forecasting hour is not used as a network input. Park, Sharkawi and others [6] obtained 1.4% of average error for a one hour ahead forecasting in week-days. This value can be compared with the average error of 1.47 % we obtained in case 3 for group 1. It is worthwhile mentioning that Park and Sharkawi's work also used temperature data in their forecasting system.

Day	Case 1	Case 2	Case 3	Case 3 (*)	Case 4	Day	Case 1	Case 2	Case 3	Case 3 (*)	Case 4
1	-------	2.0294	1.5109	1.4603	1.2592	16	---------	2.4407	1.5908	1.6502	1.4674
2	-------	2.1626	1.4287	1.3473	1.1425	17	---------	2.6463	1.5650	1.6045	1.0306
3	-------	2.2388	1.3117	1.2010	0.8129	18	---------	2.4166	1.5192	1.6856	1.6367
4	-------	2.3580	1.2248	1.3137	0.9093	19	---------	2.3872	1.4456	1.3272	1.3637
5	-------	2.5827	1.4725	1.4161	1.1652	20	---------	2.4291	2.5397	2.6194	1.3591
6	-------	3.0090	2.3016	2.0931	1.4611	21	---------	2.7822	2.2890	2.4622	1.6624
7	-------	3.4598	1.2628	1.3456	1.6192	22	2.1614	2.3106	1.9012	1.6771	1.5784
8	-------	2.3897	2.5289	2.4159	1.5092	23	4.1586	2.6900	1.7670	1.6681	1.9818
9	-------	2.5949	1.2671	1.3650	2.1740	24	4.3118	2.5093	1.4578	1.5882	1.5840
10	-------	2.1362	1.3559	1.6335	2.1532	25	2.8389	1.9397	1.1308	1.0984	1.6605
11	-------	2.3616	1.7177	2.1316	1.8128	26	2.7388	2.2129	1.6358	1.6241	1.2766
12	-------	2.6058	1.2853	1.2660	1.4938	27	3.6410	2.3194	2.2438	2.2454	1.9580
13	-------	2.3578	1.7959	1.8676	1.3195	28	2.5401	2.9803	2.2565	2.3783	1.5391
14	-------	3.0602	1.5743	1.5683	1.3248	29	2.5518	2.1611	1.5363	1.5454	1.8151
15	-------	2.4047	2.0850	1.9450	1.4515	30	4.0664	2.8342	1.9855	2.2247	2.0014
						31	4.3607	2.4403	1.3424	1.4338	1.1676

Table 2 - Average daily errors (%) to the 4 cases (* → multi-step)

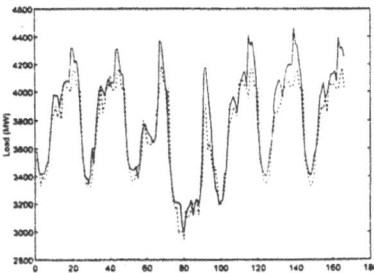

_____ →Actual -------------- → Forecasted

Fig. 5 - Comparison between hourly forecasted and actual load values. (single-step)

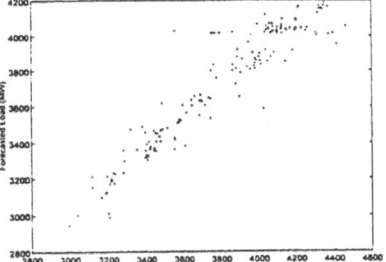

Fig. 6 - Histogram relative to Figure 5.

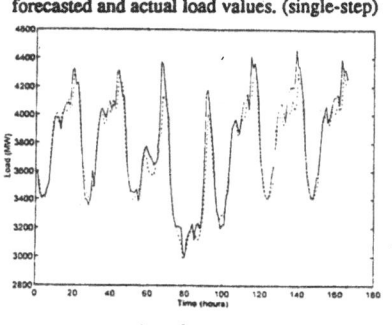

_____ →Actual -------------- → Forecasted

Fig. 7 - Comparison between hourly forecasted and actual load values. (multi-step)

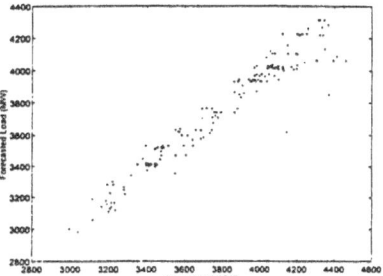

Fig. 8 - Histogram relative to Figure 7.

5 - CONCLUSIONS

This work has investigated the application of a Neural Network to load forecasting, using real data from the Brazilian Electrical System (state of Minas Gears). The main aspects of this work are the absence of temperature data and the relatively small training sets used. However, the results were as satisfactory as the results of other published works. A complete programming environment that includes from the load series pre-processing as well as the forecasting system was built using C language.

The integration of this system with other tools used in Energy Management Systems is the next step of the work.

6 - ACKNOWLEDGEMENTS

The authors wish to thank the Electrical Energy Research Centre of Brazil (CEPEL), for helping in the acquisition of the load data used in this work.

7 - REFERENCES

[1] - I. Moghram, S. Rahman, "Analysis And Evaluation Of Five Short-Term Load Forecasting Techniques", IEEE WM 171-0 PWRS, 1989.

[2] - G. George. R., F. D. Galiana, "Short-Term Load Forecasting", IEEE Proceedings, Vol. 75, No. 12, December 1987.

[3] - Antônio J. G. Abelém, " Redes Neurais Artificiais na Previsão de Séries Temporais", Dissertação de Mestrado, Departamento de Engenharia Elétrica, PUC/RJ, Set. 1994.

[4] - S. Rahman, R. Bhatnagar, "An Expert System Based Algorithm for Short Time Load Forecast", IEEE Trans. on Power System, Vol. 3, No. 2, pp. 392-399, May 1988.

[5] - K. Y. Lee, Y. T. Cha, J. H. Park, "Short-Term Load Forecasting Using an Artificial Neural Network", IEEE Trans. on Power System, Vol. 7, No. 1, pp. 124-132, Feb. 1992.

[6] - D.C.Park, M. A. El Sharkawi, R. J. Marks II, L. E. Atlas, M. J. Damborg, "Electric Load Forecasting Using an Artificial Neural Network". IEEE Trans. on Power Systems, Vol. 6, No. 2, pp. 442-449, May 1991.

[7] - T. M. Peng, N. F. Hubele, G. G. Karady, "Advancement in the Application of Neural Network for Short-Term Load Forecasting" . IEEE Trans. on Power Systems, Vol. 7, No. 1, pp. 250-257, Feb. 1992.

[8] - Kun-Long Ho, Yuan-Yih Hsu, Chien-Chuen Yang, "Short-Term Load Forecasting Using a Multilayer Neural Network with an Adaptative Learning Algorithm". IEEE Trans. on Power Systems, Vol. 7, No. 1, pp. 141-149, Feb. 1992.

[9] -Refenes , A. N. "Currency Exchange Rate Prediction and Neural Network Design Strategies". Neural computing & Applications Journal. London, v.1, n.l, p. 46-58,1992.

[10] - Wasserman, P.D. "Neural Computing: Theory and Practice", New York: Van Nostrand Reinhold, 1993.

Image Compression Using Feedforward Neural Networks – Hierarchical Approach

Stanisław OSOWSKI, Robert WASZCZUK, Piotr BOJARCZAK
Institute of the Theory of Electrical Engineering and Electrical Measurements
Technical University 00-661 Warsaw, pl. Politechniki 1, POLAND

Abstract

The paper presents the hierarchical approach to the problem of image compression using feedforward neural networks. In this approach smaller frames are used in the regions containing more details and larger, when the grey level is uniform. Thanks to this the number of data is reduced, learning speed accelerated and the quality of compression improved. The numerical results, confirming the efficiency of the proposed approach and good generalization properties are presented and discussed.

1   Introduction

Application of the feedforward multilayer neural networks to image compression has been subject of several recent contributions appearing in technical literature [2, 3, 4, 5, 6, 7, 8]. It has been shown, that the autoassociative neural network performs in an indirect way the Karhunen - Loeve transformation of the data, resulting into the reduction of the information to be stored or transmitted. Although the usefulness of the neural networks to obtain the compression has been proved, the reported compression ratios, practically obtained, do not exceed the value of 10 [5], and it is difficult to increase it without additional techniques, applied at the stage of preparation of data.

This paper will show one of the methods of the improvement of the neural approach to the compression by using the hierarchical technique of splitting image into frames. The results of numerical experiments are presented, and they confirm the effectiveness of the proposed solution.

2   Neural architecture and learning algorithm

The image compression/decompression architecture based on the application of the feedforward neural network is shown in Fig. 1. We use 3-layer neural network of linear activity function of the neurons. The input and hidden layers form the compression of data, while the hidden plus output layers perform the decompression. The hidden layer plays double role: the output for compression and input for decompression. The number of input and output neurons is the same (autoassociative mode). Learning of the optimal weights for the compression (weights A_{ij}) and decompression (weights B_{ij}) is done at the same time in the process of training the whole network, i.e. minimizing the least square error between the input data vector x and the output data vector $\hat{x}$, where $\hat{x} = BAx$. The matrices A and B represent weights of the input-to-hidden and hidden-to-output layers, respectively. If we split the given image into N subpictures, arranged in the form of n-dimensional vectors, the LSE error is defined in the form

$$E = \frac{1}{2} \sum_{i=1}^{N} \sum_{j=1}^{n} (x_{ij} - \hat{x}_{ij})^2 \qquad (1)$$

This is the quadratic function with respect to the adjusted weights and its minimization, performed at the learning stage, results into optimal weights of the network.

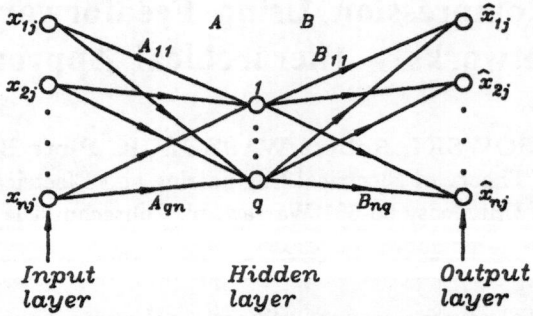

Figure 1: *The 3-layer neural network used for the compression of data*

The learning process of the multilayer feedforward neural network has been performed in the work using conjugate gradient method, associated with the directional minimization. The optimization cycle is composed of 2 steps: the determination of the direction $p(W)$ and the directional minimization, used to find the optimal learning rate η. The update of the weight vector W, represented by the weights A_{ij} and B_{ij}, is done according to

$$\Delta W = \eta p(W) \tag{2}$$

in which W - the weight vector and $p(W)$ is the direction of search. The search direction in this algorithm is given as follows

$$p(k+1) = -g(k+1) + \beta p(k) \tag{3}$$

in which $p(k)$ is the last search direction and β is the conjugacy coefficient adjusted in each step according to the Polak - Ribiere rule

$$\beta = \frac{g(k+1)^t[g(k+1) - g(k)]}{[g(k)]^t[g(k)]} \tag{4}$$

After getting the direction p, the calculation of optimal learning rate η is performed in the second step. In developing it we have applied the efficient mixed third and second order polynomial approximation of the energetic function on the direction p, using the information of the values of the energetic function and its directional derivatives in composition with the nongradient search techniques called bisection. The line search is done only to some limited accuracy and and in practice the directional minimization is performed very limited number of times; normally it is done two or three times in each cycle. This method of adjusting the learning rate has been proved to be extremely efficient in comparison to the constant or adaptive way of choosing it.

The exact value of the gradient vector is generated using the concept of so called adjoint graph [10]. These techniques implemented in the program $NETTEACH$, written in C++, have resulted into quick convergence learning process [11].

However, due to extremely large number of learning data to be presented at every cycle of learning, the training process is relatively long. Any reduction of the length of data accelerates the learning phase and can also improve the quality of compression, as well as increase the value of compression ratio.

3 Hierarchical splitting of the image into frames

The technique of dividing the image into uniform-sized subpictures, usually applied in neural network approach to the compression, does not allow to reduce the number of subpictures. Since different regions of the image may vary in details, it is expected better coding results by allowing the blocks size to vary. Smaller blocks should be used in the regions of the image containing more details, and larger - when the grey level is uniform.

The hierarchical approach begins by initially dividing the image into certain number of equal blocks (in our solution 32 × 32 pixels) and by measuring the grey level scale contrast, i.e., the difference between pixels with the highest and lowest grey scales within the block. A block with excessive contrast is divided into subblocks, which are assigned to lower layers. The typical division of the 32 × 32 block into 16 × 16 and eventually 8 × 8 subblocks is shown in Fig. 2a, and the corresponding structure tree with three levels of division in Fig. 2b. The highest layers contain

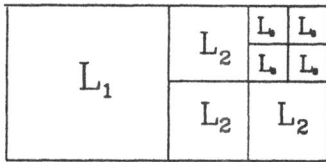

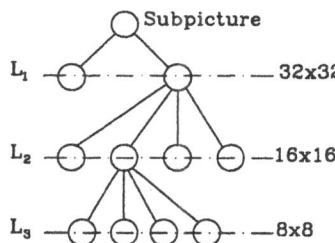

Figure 2: *The hierarchical division of the subpicture into smaller blocks: a) typical division of blocks, b)structure tree corresponding to the division*

blocks in the smooth areas, and the lower layers contain blocks with more details. The algorithm of hierarchical division developed in this work may be presented in the following form

- Initial division of the image into 32 × 32 pixels blocks.

- Calculation of the mean value of the grey level within each block

$$x_m = \frac{1}{h \times v} \sum_{i=1}^{h} \sum_{j=1}^{v} x_{ij} \tag{5}$$

where h and v denote the horizontal and vertical dimension of the blocks (in pixels), and x_{ij} is the grey level associated with each pixel.

- For each pixel of the block the following inequality is checked

$$\mid x_{ij} - x_m \mid \leq R \tag{6}$$

where R is the predefined constant radius (threshold) usually assumed as a certain percentage of $\mid x_{ij,max} - x_{ij,min} \mid$ or as some percentage of the value represented by bits used for coding the grey level. At the same time the number L of pixels, not satisfying relation (6) is counted.

- If the relation (6) is satisfied for all pixels of the block, the process of splitting this block is stopped. In opposite case the number L is compared to the so called tolerance TOL, i.e., the maximum ratio of data for which block splitting is stopped even in the case when some pixels do not satisfy the relation (6). If $\frac{L}{\lambda \times \nu} > TOL$ the block is subject to further splitting.

- The splitting process is continued until the standard size of blocks is achieved (in our case it has been assumed 8×8). Then the larger blocks are reduced to standard size, to get the uniform size of all blocks. The grey level of the pixels of the reduced size are calculated as the mean value of the neighbouring pixels subject to reduction.

As a result of such process the series of standard size 8×8 frames are created. Some of these frames represent original set of pixels of the image and some represent reduced 16×16 or 32×32 subpictures. All frames are coded using binary values, where code 01 means that the frame is obtained by reduction of 32×32 subpicture, code 10 - when the frame has been obtained by reduction of the 16×16 subpicture and the code 11 is the result of spliting the subpicture into the standard 8×8 size.

The opposite to coding is decoding process, i.e., reproduction of the whole image on the basis of subpicture information. Using the information written earlier in the code file, the appropriate subpictures are either left unchanged (the original 8×8 frames) or enlarged to the original sizes 32×32 or 16×16. The information of the grey level of the lacked pixels is obtained through the process of the bilinear interpolation in two dimensions.

To recover higher dimension subpictures on the basis of the known values f_i of pixels of the reduced 8×8 frames, we have applied the interpolation scheme described as follows

$$f(x,y) = (1-t)(1-u)f_1 + t(1-u)f_2 + tuf_3 + (1-t)uf_4 \tag{7}$$

$$t = \frac{x - x_1}{x_2 - x_1} \tag{8}$$

$$u = \frac{y - y_1}{y_3 - y_1} \tag{9}$$

where $x_1, x_2, x_3, x_4,$ y_1, y_2, y_3, y_4 and f_1, f_2, f_3, f_4 are respectively, the x-coordinates, y-coordinates and the grey levels, associated with 4 known pixels (placed on the vertices of a square) subject to interpolation. The variables x and y denote the coordinates of the pixels beeing recovered.

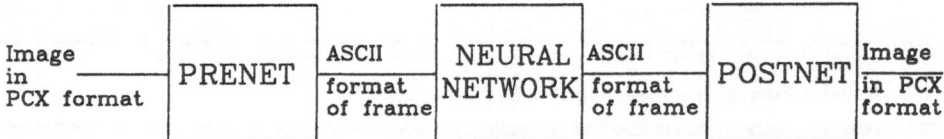

Figure 3: *The image compression/decompression scheme using hierarchical approach*

This procedure of hierarchical preparation of data for compression has been implemented in C++ as a program $PRENET$, and the decoding procedure as the program $POSTNET$.

The image compression/decompression process applied here is shown in Fig. 3. The original image in PCX format is read by the program $PRENET$ and the data are splitted hierarchically into frames, and converted into ASCII format recognized by the neural network learning program $NETTEACH$. After transformation of the image data by the neural network, the output frames in ASCII format are read by the $POSTNET$ program and converted back into PCX format of the resulting transformed image. The original image and the image resulting from the compression/decompression process are compared and the errors of the compression are assessed and calculated.

4   Results of numerical experiments

The hierarchical approach described above has been tested on many examples of images of different sizes, from 128×128, through 256×256 up to 512×512 pixels. Here we will present only the results obtained for 512×512 images. At this size of image, the number of data points is equal 262144. At standard, equal size subpictures the number of frames is very large and the learning stage is lengthy. Application of hierarchical approach has reduced this number in typical cases up to 30%.

Figure 4: *The original image used for training*

Figure 5: *The reproduced images at $K_r = 12.8$ (a) and $K_r = 21.3$ (b)*

Fig. 4 presents the 512×512 image used for training the compressing neural network. At 8 bits description of pixels we have assumed the values of R=10 and TOL=5%. At these values the number of frames generated by program $PRENET$ has been reduced by 28%. This saving has resulted in approximately the same saving of the learning time. Taking into account that the compression ratio is equal approximately $K_r \approx \frac{n}{q}$, where n - dimension of input vector, q - the number od hidden neurons, different compression rates have been tested by using different number of hidden neurons. Fig. 5a presents the obtained results at $K_r = 64/5$ (5 hidden neurons) and Fig. 5b at $K_r = 64/3$ (3 hidden neurons). As it is seen the results deteriorate with the increse of R_r, but even at $K_r = 21.3$ the reproduced image is fully recognizable. The maximum discrepancy between the grey level of original and reproduced images has been found equal 0.601 and this is only very small fraction of the dynamic range of 2^8, represented by 8 bits grey level coding of pixels.

The network trained on the example of one image has been tested on different types of images, and the generalization abilities have been found good. Fig. 6a presents the results of testing the network on the example of the forest and Fig. 6b on the example of mountain landscape. In both

Figure 6: *The reproduced images used for testing the generalization properties of the network*

cases the compression ratio was the same and equal $K_r = 12.8$ (5 neurons in hidden layer). Although the types of pictures used for generalization testing were different from that used for training, the results are acceptable and there is no visible difference between the original and reproduced images (the maximum discrepancy between the grey levels of pixels of original and reproduced images were equal 0.45, less than at the trained image).

It should be pointed here that application od hierarchical approach to image compression has resulted in not only reduction of the learning time, but also increase of compression ratio. Application of hierarchical strategy reduces the differences of contrast within the frame and this allows to decrease the overall error at the same compression rate, or increase the compression rate at the same overall error.

5 Conclusions

The hierarchical neural network image compression algorithm, which is superior to the conventional uniform sized frames, has been presented in the paper. The block sizes used in the proposed approach vary with the features of the different parts of the image, so that a lot of redundancy information is removed from further processing. Small blocks have been employed in high detail regions and large blocks in low detail regions. Digital decimation and interpolation techniques have been used to convert between different vector dimensions. Thanks to the proposed solution the time required for learning the neural compression/decompression networks has been significantly reduced and the quality of compression increased. The program was tested on many images including these presented in the paper, allowing to increase the compression ratios (up to above 20) and to obtain good generalization abilities of the trained neural network.

References

[1] J. Hertz, A. Krogh, R. Palmer, Introduction to the Theory of Neural Computation, Addison Wesley, 1991, Amsterdam

[2] Cottrell G., Munro P., and Zipser D., Image compression by back propagation: an example of extensional programming, Technical Report ICS report 8702, ICS-UCSD, San Diego, California, USA, February 1987.

[3] Kunt M., Image compression using neural networks, Technical Report EPFL Lausanne, (CAR-NAC), 1993

[4] Mougeot M., Azencott R., Angeniol B., Image compression with backpropagation: improvement of the visual restoration using different cost functions, Neural Networks, 1991, vol. 4, pp. 467 - 476

[5] Oja E., Wang L., Image compression by MLP and PCA neural networks, SCIA-93, Tromso, 1993, pp. 1317 - 1324

[6] Sonehara N., Kawato M., Miyake S., Nakane K., Image data compression using neural networks, IJCNN, Washington,1989, pp. II35- II40

[7] G. Vines, M. Hayes III, Map search strategies for IFS image compression algorithms, 14 Colloque Gretsi, Juan-Les-Pins, 1993, pp. 843 - 846

[8] G. Martinelli, L. Prina Ricotti, G. Marcone, Neural ckustering for optimal KLT image compression, IEE Trans. Signal Processing, 1993, vol. 41, pp. 1737 - 1739

[9] P. Yu, A. N. Vanetsopoulos, Hierarchical multirate vector quantization for image coding, Signal Processing: Image communication, 1992, vol. 4, pp. 497 - 505

[10] S. Osowski, J. Herault, Signal flow graphs as an efficient tool for exact gradient and hessian determination, Complex Systems (accepted for publication)

[11] S. Osowski, Fast learning algorithms for feedforward multilayer neural networks, IEEE 1994 IMACS Int. Symp. Signal Processing and Neural Networks, Lille, 1994, pp. 27 - 30

Neural Approaches to Robot Control:
Four Representative Applications*

Carme Torras[1], Gabriela Cembrano[1], José del R. Millán[2], Gordon Wells[1]

(1) Institut de Cibernètica (CSIC-UPC)
Diagonal, 647
08028-Barcelona. SPAIN
e-mail: torras@ic.upc.es

(2) Institute for System Engineering and Informatics
European Commission, Joint Research Centre
TP 361. 21020 Ispra (VA). Italy
e-mail: jose.millan@jrc.it

Abstract

This paper reviews neural network techniques for achieving adaptivity in both manipulator and mobile robots. It is structured in two parts. First, the different learning approaches are classified according to the amount of training information they require: quantitative (supervised approaches), qualitative (reinforcement-based approaches) and none (unsupervised approaches). Afterwards, the adequacy of each approach for solving specific problems in robot control is illustrated through four working industrial prototypes developed by the authors in the frame of two Esprit projects. The problems tackled are the inverse kinematics and inverse dynamics of robot manipulators, visual robot positioning and mobile robot navigation.

1 Introduction

One of the main goals in present-day robotics research is to provide robots with a greater degree of autonomy, so that they can perform tasks efficiently in unstructured or changing environments. To this end, it becomes essential to endow robots with adaptive capabilities, allowing them to modify their mappings from sensory patterns to motor commands as they move in those environments.

Similarly, a leading trend in nonlinear control systems design is concerned with the use of behavioural approaches to control, where the relationship between sensory information and control actions is learned, as opposed to classical model-based techniques, which require explicit mathematical representations of the processes to be controlled and their relationship with the environment.

Neural networks are especially appropriate to tackle this type of problem because of their ability to learn complex nonlinear mappings. Furthermore, the adaptive properties of neural networks and their robustness to noisy or corrupted input data are of great importance in control applications, where measurement errors may appear in sensors and actuators.

*The support from the ESPRIT III Program of the European Union under contracts No. 6715 (project CONNY) and No. 7274 (project B-LEARN II) is gratefully acknowledged. The authors wish to thank all the partners involved in these projects for their cooperation and especially, Dr. Christophe Venaille, for his contribution in the field of visual positioning, Mr. Jesús Sardá for his help in developing dynamic control schemes and Mr. Conor Doherty for his work in the implementation of the inverse kinematics update.

In order to assess the adequacy of neural control in robotics, the Esprit project CONNY[1] is being developed, which involves three different real applications in industrial domains. The goal of the project is to demonstrate the neural control concepts on three real robots, belonging to the industrial partners of the project team. To this goal, a general neurocontrol architecture has been devised, which has been subsequently tailored to each of the three applications. The architecture consists of several modules, among them those for *visual positioning, inverse kinematics update*, and *inverse dynamics*, which have been developed at the Institut de Cibernètica in close relation with the corresponding industrial partners.

A related Esprit project is B-Learn II[2], which has been set up to study the application of Machine Learning techniques in three robotic domains: assembly, machining and monitoring, and navigation. Kaiser et al. [25] describe the integration of different machine learning methods into the control architecture of autonomous mobile robots and report experimental results. In this paper, only the use of neural learning for *mobile robot navigation* will be described. The working prototype has been developed in close collaboration with the Joint Research Centre in Ispra.

The remainder of this paper is structured as follows. Section 2 provides an overview of the neural learning approaches most widely used in the field of Robot Control, classified as supervised, unsupervised and reinforcement-based. Section 3 describes the application of a supervised off-line learning approach to the problem of visual robot positioning, while in Section 4 a supervised on-line learning scheme is applied to the dynamic control of a telescopic robot arm. The use of an unsupervised approach for updating the inverse kinematics mapping of a robot arm after some degradation due to wear has occurred, is explained in Section 5. The next section presents a reinforcement-based approach to the generation of safe trajectories for a mobile robot. Finally, Section 7 outlines the main conclusions obtained so far in these studies.

2 Neural Learning

Neural learning approaches differ in the type of training information they use. In one extreme of the spectrum one finds *supervised approaches*, which require complete target information —in the form of input/output pairs— and whose goal is to build a mapping from inputs to outputs that generalizes adequately. Sections 3 and 4 describe applications developed under approaches of this type.

Unsupervised approaches are placed in the opposite extreme of the spectrum, since they use no problem information at all and their goal is to carry out feature discovery or clustering, i.e. to build a mapping from inputs to statistically salient features of the input population that permit establishing clusters of input patterns with similar features. In a robot control setting, these approaches are often used to represent a given state space in a compact and topology-preserving manner. The application presented in Section 5 relies on a network representation of this type for the robot workspace.

Somewhere in between both extremes lie *reinforcement-based approaches*, which make use only of a reward/penalty signal —i.e. an assessment of how good is an output to a given input — and whose goal is to build a mapping that maximizes reward. As trial-and-error procedures, they have an stochastic component that permits exploring the suitability of different outputs in response

---

[1]Project ESPRIT-III 6715 "Robot Control based on Neural Network Systems" (1992-95). Partners are: Daimler-Benz Aerospace (D), THOMSON CSF (F), Mimetics S.A.(F), Framentec-Cognitech (F), CRAM-AID (I), University College London (UK) and Institut de Cibernètica (E).
[2]Project ESPRIT-III 7274 "Behavioural Learning: Combining Sensing and Action" (1992-95). Partners are: Universität Karlsruhe (D), Universität Dortmund (D), Università di Torino (I), Università di Genoa (I), Universidade Nova de Lisboa (P), Katholieke Universiteit Leuven (B) and Institut de Cibernètica (E).

to the same input. Reinforcement-based approaches have been applied mainly to the learning of sensorimotor mappings, as is the case in the application described in Section 6.

Learning in all these approaches relies on the uniform application of an unineuronal learning rule throughout a network of formal neurons, each being in a state x_j governed by the following generic equation:

$$x_j(t + \delta t) = f[\sum_{i \in I_j} w_{ij}(t) x_i(t)], \tag{1}$$

where I_j is the set of inputs to neuron j —which can come from the environment or be the outputs of other neurons, w_{ij} is the synaptic weight of the connection from neuron i to neuron j, t is the time instant, and f takes usually the form of either a deterministic or a stochastic threshold function.

Several surveys of the various neural learning models proposed can be found in the literature [5, 14, 20, 21, 30, 57, 58]. In what follows we will concentrate on those that have been applied in the field of Robot Control.

2.1 Supervised Approaches

The learning rules used within these approaches work by comparing the response to a given input pattern with the desired response, and then modifying the weights in the direction of decreasing error. Two such rules have been applied to robot control, namely the LMS rule and back-propagation.

The **LMS learning rule** [63] is expressed by the equation:

$$w_{ij}(t + 1) = w_{ij}(t) + c(x_j^*(t) - x_j(t)) x_i(t), \tag{2}$$

where both the desired response $x_j^*(t)$ and the actual response $x_j(t)$ of neuron j take real values, and the f in (1) is the identity function. To prevent the unbounded growth of weights, different normalization procedures have been used, the most common one being that based on the euclidean metrics.

The repeated presentation of pairs (X_l, z_l), $l = 1, \ldots k$, with $X_1, \ldots X_k$ linearly independent, to a two-layer network equipped with this rule, called *Madaline*, causes the convergence of the synaptic weights toward the proper configuration for the response to each stimulus X_l to be the desired real number z_l. The rule thus provides an iterative procedure to find the solution of a system of linear equations. If the patterns $X_1, \ldots X_k$ are not linearly independent, the rule can be slightly modified (converting the parameter c into a variable that goes to zero with time) so that the convergence of the weights minimizes the mean squared error between the actual outputs and the desired ones (this is why this rule is called the LMS rule, for "least mean squares"). With this slight modification, the LMS rule thus computes a linear regression iteratively.

An extension of this rule to the case where the desired response is specified only for a subset of neurons (those whose outputs constitute the output of the network) is **back-propagation** [31, 53]. As its name indicates, it proceeds by propagating error signals from the output neurons back to the sensory neurons, through all intermediate layers, so that appropriate corrections can be applied to all connection weights:

$$w_{ij}(t + 1) = w_{ij}(t) - c \frac{\partial E}{\partial w_{ij}}, \tag{3}$$

where E is the mean squared error between the actual and the desired responses to all input patterns.

Observe that, in fact, the LMS rule is a particular instance of this generic rule, since for the former the f in equation (1) is the identity and thus $\partial x_j / \partial w_{ij} = x_i$, leading to:

$$\frac{\partial E}{\partial w_{ij}} = \frac{\partial E}{\partial x_j} \cdot \frac{\partial x_j}{\partial w_{ij}} = (x_j^* - x_j) x_i. \tag{4}$$

By incorporating this result into (3), equation (2) is obtained.

Back-propagation is the most well-known and widely used neural learning algorithm and, therefore, we are not going to describe it further. Let us just mention that it has the drawbacks of all gradient descent techniques, namely the possibility of getting stuck in local minima and a slow convergence rate. Because of this, numerous acceleration procedures have been proposed. Moreover, back-propagation suffers from *catastrophic forgetting* of the previously learnt patterns when trained with a new pattern. Thus, techniques to prevent forgetting by introducing noise [50] and by explicitly minimizing degradation while encoding a new pattern [51] have been devised.

2.2 Unsupervised Approaches

These approaches usually rely on some variation of the classical **Hebbian learning rule** [19], whose expression in terms of the generic neuron model (1) is:

$$w_{ij}(t+1) = w_{ij}(t) + cx_i(t)x_j(t), \tag{5}$$

where c is a positive constant that determines the speed of learning. The same consideration about normalization made for the LMS rule applies also here.

The Hebbian rule has been incorporated into *competitive learning models* [52], which consist of a set of hierarchically layered neurons, each neuron receiving excitatory input from the layer immediately above. Futhermore, the neurons in each layer are grouped into disjoint clusters, each neuron in a cluster inhibiting all other neurons within the cluster. The name "competitive learning" comes from the fact that the neurons within a cluster "compete" with one another to respond to the pattern appearing on the layer above; the more strongly any particular neuron responds, the more it shuts down the other members of its cluster, which therefore becomes a winner-take-all network. A cluster containing n neurons can be considered an n-ary feature, every stimulus pattern being classified as having exactly one of the n possible values of this feature. It has been proved that, if the stimulus patterns naturally fall into classes, the system will find exactly these classes and the attained classification will be very stable. However, when presented with arbitrary input environments, competitive learning models can become very unstable and the need appears of stabilizing their response through the use of specialized mechanisms, leading to *adaptive resonance models* [16]. These are recurrent modular networks able to form a new cluster whenever they are presented with an input pattern that is very different from the patterns previously seen.

Following the same line of competitive learning, Kohonen [27] has proposed to use *self-organizing feature maps* to learn mappings that preserve topography (i.e. neurons that are spatially close in the network learn to be maximally activated by input vectors close according to the euclidean metrics). Essentially, this is realized through two-layer networks with intralayer lateral inhibition and interlayer plastic excitatory connections. A self-organized map is thus a winner-take-all network, where neuron k wins if it satisfies:

$$\sum_i w_{ik}x_i \geq \sum_i w_{ij}x_i, \ \forall j. \tag{6}$$

The learning rule for this type of network is:

$$w_{ij}(t+1) = w_{ij}(t) + ch_k(j)(x_i(t) - w_{ik}(t)), \tag{7}$$

where $h_k(.)$ is a Gaussian function centered at k used to modulate the adaptation steps as a function of the distance to the winning neuron.

2.3 Reinforcement-based Approaches

These approaches do not require being supplied with the desired responses, either at the single neuron or at the overall network levels, but instead a measure of the adequacy of the emitted responses suffices. This measure is reinforcement, which is used to guide a random search process to maximize reward. Here we will describe only the most widely used reinforcement rule, which is that used in the application presented in Section 6. This is the **associative search learning rule** [6], which incorporates the required source of randomness in the input-output function of the neuron model used, i.e. a noise with gaussian distribution is added to the weighted sum of inputs in equation (1).

The simplest expression of this rule is:

$$w_{ij}(t+1) = w_{ij}(t) + cx_i(t)x_j(t)r(t) \qquad (8)$$

where $r(t)$ is the reinforcement signal.

A neuron model equipped with this rule learns to maximize $r(t)$ for each stimulus situation. If $r(t)$ is a random variable, its mathematical expectation is instead maximized. The neural networks that incorporate the above rule are called Associative Search Networks (ASN) and, if certain conditions are satisfied, they learn to respond to each stimulus situation $X_l = (x_{l1}, \ldots x_{ln})$ of a set $\{X_1, \ldots X_k\}$ repeatedly presented, with the vector $Y = (y_1, \ldots y_m)$ that maximizes the reinforcement function r. The conditions that have to be satisfied are: (a) the function r has to be unimodal, and (b) for each neuron, the subset of stimulus situations in which the optimum response is 0 has to be linearly separable from the corresponding subset in which the optimum response is 1.

Depending on whether the reinforcement signal is provided at the overall network level or is particularized for each single neuron, the structural credit-assignment problem does or does not arise. This is the problem of correctly assigning credit or blame to the action of each neuron that contributed to the overall evaluation received [5]. When dynamical situations need to be considered, because what is of interest is a temporal sequence of events, then a temporal credit-assignment problem also arises. Instead of assigning credit or blame to the action of each neuron in the network, credit or blame must here be assigned to each action in a sequence. In [56] it has been proposed to use *temporal-difference methods* for this purpose. Methods of this type have been embodied, for example, in "critic modules" used in conjunction with reinforcement learning approaches. The goal of these modules in this setting is to produce an heuristic reinforcement signal which, by predicting future outcomes, is more informed than that directly supplied by the environment.

3 Learning Visual Positioning

The inspection application proposed by THOMSON Broadband Systems (F) is primarily concerned with the fine positioning of a robot arm, based on images captured by a camera mounted on the wrist. The robot is a commercial 6-dof articulated manipulator. The inspection task must be performed in a specified relative pose between the viewed object and the robot end-effector, even though the objects may appear in arbitrary positions and orientations. Therefore, the goal of the positioning system is to bring the end-effector to the desired pose relative to the object, based on the information extracted from the camera images.

The learning of fine positioning is not restricted in this application to one specific object for inspection, but rather it is expected to be applicable to a broad class of objects, with very few limitations on their image characteristics. Furthermore, the learning method is obviously not restricted to the inspection task, but is relevant to many other industrial object-handling tasks.

3.1 Previous Work

Many authors have tackled the problem of visual positioning of robot manipulators, for applications ranging from inspection, grasping and assembly of parts to docking and navigation of autonomous guided vehicles. Most of the existing works have tended to rely on simple geometrical features extracted from images, such as points, lines or circles, together with projection transformations to analytically derive the mapping from 2D image space to the robot cartesian coordinates [1, 11, 13, 24, 34].

Although existing approaches based on analytical methods have produced useful results, they all depend on simple geometric features which are assumed to be always visible and extractable in the camera image. In addition, the physical relationship between all object features must be known in order to derive the projection transformations. Furthermore, since more than one cue is typically used, matching must be performed to find the correspondence between each feature in the observed image and a feature in the reference image or physical scene. Finally, in order to derive the complete transformation from image features to robot coordinates, the precise relationship between the camera coordinates and those of the end effector must be known, requiring calibration of the camera as well as knowledge of its intrinsic parameters. Obviously, most of these assumptions place strong restrictions on the observed scene, and the operating conditions and robustness to noise of the resulting system.

The mapping between 2D image feature deviations and robot position or joint angles is a highly nonlinear relationship which depends on the type and relative positions of the object features, the camera-robot relationship, the intrinsic camera parameters, and the robot kinematics. A neural network may be used to learn the entire transformation implicitly based on training examples, thus avoiding explicit computation of all intermediate transformations. A first work in this area was that of Hashimoto [18] who used *back-propagation* networks to learn the mapping between the image deviations of 4 projected points of a viewed object with respect to a "reference" or desired image, and the corresponding joint angles of a robot with a camera mounted on its end effector. Hashimoto's work showed the capability of a neural network to perform visual positioning with a degree of accuracy similar to that obtained using analytical techniques.

However, many possibilities by which neural networks can be exploited to produce more general-purpose visual positioning systems remain open for research. For example, training neural networks to map from more global image descriptors to robot commands could make visual servoing methods much less sensitive to the presence and extractability of a few simple geometric features in the observed scene. Similarly, the need for explicit feature matching may be avoided and robustness to obstructed features, noisy images, and changing reference positions may be increased.

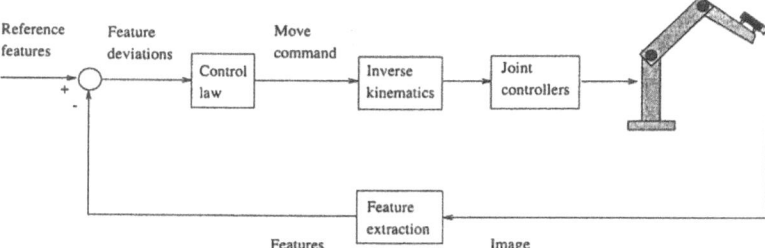

Figure 1: General scheme for visual positioning.

3.2 Supervised Off-line Learning Approach Taken

The goal of this work was to develop more flexible and robust visual positioning methods based on neural networks. In all the experiments performed, the overall objective was to train a neural network to represent the mapping between the variations, with respect to a desired "reference" image, of several prespecified image features or descriptors as seen from a camera mounted on the robot end effector, and the 3D cartesian position and orientation of the end effector relative to the reference pose. Unlike in earlier neural approaches [18], inverse kinematics of the robot were not included in the mapping. In this way aspects related to the visual mapping problem could be studied independently from the robot kinematics and the additional benefit was achieved that the learned transformation is completely independent from the chosen reference position, and may therefore be used to correctly position the robot relative to the object regardless of their initial locations. The general scheme for image-based robot control is shown in Fig. 1.

Two sets of experiments were performed, in which a neural network was used to learn the mapping between a chosen set of extracted image features and the 6D movements required to approximate the camera to a prespecified reference position. In initial experiments, four point features were used, allowing comparison of results with similar existing works based on analytic and neural methods. The differences between the x, y coordinates of the four points in the reference and observed images were used as inputs to the network. In the second set of experiments, the features used consisted of 32 Fourier descriptors used to encode the shape of the extracted silhouette of an observed object on a uniform background. As before, the differences between the descriptors in the reference and observed images comprised the network inputs. In both cases, training sets were constructed by moving the robot-mounted camera to 1000 random positions in the vicinity of the reference pose, and the extracted feature deviations for each image, along with the applied 6D movement, were used as training examples. *Back-propagation* networks were trained using these data sets and tested in a closed-loop visual positioning system to test their ability to guide the camera back to the reference position from any given initial position and orientation within a limited range. Fig. 2 shows the results of visual positioning in an example of a water valve. See [62] for a more detailed exposition of these experiments.

In both sets of experiments, the neural visual positioning systems were capable of converging on the reference position to an accuracy of 2 or 3 millimeters in an average of 2 to 10 movements, depending on the range of initial displacements used. Even when the camera was displaced to three times the range of movements used to generate the training set, the networks generalized quite well and were able to achieve the same final positioning accuracy in just a few more approach movements. Thus, this work demonstrated the capability of using neural networks to accurately position a robot manipulator based on image information even when complex image features are used for which it is not possible to derive explicit transformation relationships or perform feature matching.

4 Learning Inverse Dynamics

This application involves an orange-harvesting robot, a custom-made prototype of which has been developed by the Italian research institute Consorzio per la Ricerca dell'Agricoltura del Mezzoggiorno (CRAM). The robot is composed of a large 3-dof arm structure which holds two smaller 3-dof telescopic harvesting arms with a cutting end-effector for the harvest operation. A camera is mounted on each harvesting arm. The larger arm is heavy and slow and is conceived mainly for approximating the harvesting arms to the correct height on the tree. Conversely, the harvesting arm is a light-weight structure which must perform very fast movements, under strict requirements on positioning accuracy and precision, even with high accelerations. This is an extremely complex problem for classical control techniques. In this case, neural techniques are used for learning the

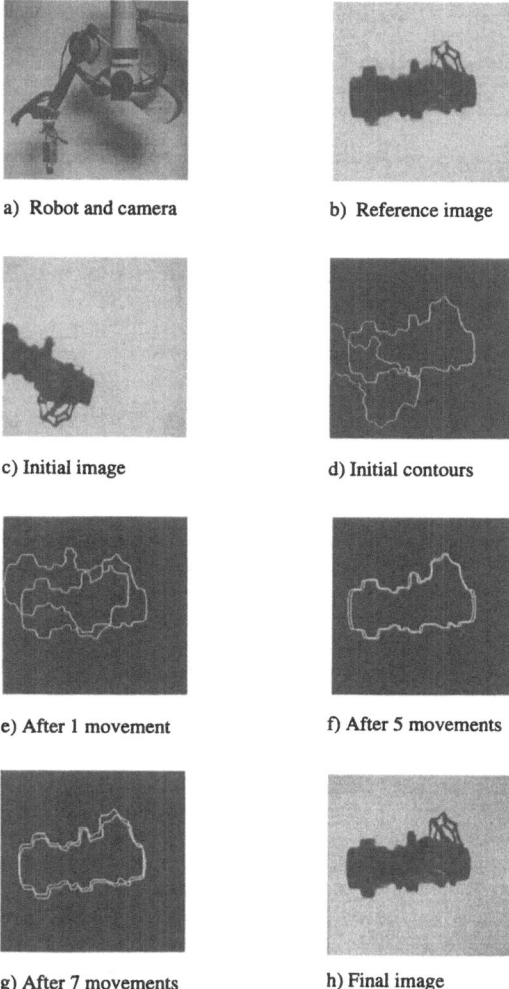

a) Robot and camera   b) Reference image

c) Initial image   d) Initial contours

e) After 1 movement   f) After 5 movements

g) After 7 movements   h) Final image

Figure 2: Results of visual positioning for a water valve.

inverse dynamics of the robot and learning to adaptively control the motor commands of the harvesting arm, so as to minimize the positioning errors.

4.1 Previous Work

The problem of controlling the dynamics in a robot arm is concerned with generating appropriate motor commands at its joints, so that the end-effector follows a desired trajectory, with a high degree of accuracy and precision. In state-space-representation terms, the state variables of this problem are the position, velocity and acceleration of the end-effector and its control variables are the motor commands (currents or voltages) at each joint.

The efficient solution of the dynamic control problem through conventional control schemes would require a deep knowledge of the system behaviour, translated into a very accurate nonlinear mathematical model (see for example [64]), which are usually very hard to obtain. The effects of friction and the inertia moments, for example, depend on the system state and the addition of payloads to the system may greatly affect the overall dynamic behaviour.

For these reasons, neural identification and control are very relevant to the problem of the robot dynamics. On the one hand, the ability to learn a nonlinear behaviour through appropriate examples of inputs and outputs may overcome the modelling difficulties. On the other, the on-line learning capabilities of neural networks provide a good framework for the use of adaptive control schemes.

A general approach to adaptive control, proposed in [45], consists of an indirect model-reference adaptive scheme with a series-parallel structure. Similarly, several concepts related to neural adaptive control are treated in [23] and [9]. A number of more refined versions of *back-propagation* have been proposed in the literature for adaptive control and proved in a variety of applications e.g. [61, 22, 54].

Similarly, the CMAC neural network model (Cerebellar Model Articulation Controller), originally developed by Albus [3], has also been used in neural adaptive control schemes. It has the advantage over back-propagation networks of much faster convergence, with excellent model tracking capabilities [4].

The CMAC network, based on the cerebellar model for neuromuscular control, is basically a nonlinear table look-up technique which maps each n-dimensional input state-space vector to a corresponding output vector of the same or different dimension. Each input vector activates exactly c input neurons with overlapping receptive fields, where c is a variable parameter representing the extent of generalization within the state space. The potentially very large virtual state space is mapped onto a smaller physical weight table using a fixed random hashing function. A supervised training method, resembling the *LMS rule*, is used to adjust the CMAC memory values based on these observations. The learned information is used to predict the command signals required to produce desired changes in the sensor outputs.

Miller et al. [40, 42] have studied a neural-based learning control system for the dynamic control of robot manipulators with multiple feedback sensors and multiple command variables. In their scheme, a neural network is used, in place of an explicit system model, to adaptively learn an approximate dynamic model of the controlled robot in appropriate regions of the system state space. The CMAC neural control module adaptively learns the unknown nonlinear mapping between the sensor outputs and the system command variables from on-line observations of each during system operation. The CMAC neural controller is implemented in a closed-loop control system, where it is used to predict the actuator torques required to make the robot follow a desired trajectory. These torques are a function of the current joint positions, velocities and accelerations, and are used as feedforward terms in parallel with a fixed-gain linear feedback controller. The control signal input to the robot is the sum of the terms from the CMAC module and the feedback controller

4.2   Supervised On-line Learning Approach

The concepts of CMAC have been applied to the control of the orange-harvesting robot in simulation. The simulator of the robot dynamics is a variable structure state-space model of the telescopic harvesting arm of the robot, which takes into account variations of inertia moments and friction coefficients with the elongation. In order to generate acceptable training data, a PID controller is added to the robot model, since it would otherwise produce unstable responses.

The implementation in this project differs from those of the aforementioned works, especially in the complexity of the dynamic behaviour of the robot (variable structure), which poses serious difficulties to the learning process. Firstly, a discrete-time state-space representation of the dynamic process was preferred to the original input-output scheme and the output error signals were used as feedback signals. Therefore, the inputs to the CMAC network include information on the current state and next desired state, as well as tapped delays of both.

In the control scheme adopted for this application (Fig. 3), the linear controller is initially used for a first phase of training the CMAC controller; during this phase, the CMAC module is not connected to the robot. When both controllers provide control signals which are equal, to a certain degree of accuracy, the CMAC controller is switched on, and a period of on-line training starts, where the neural controller can improve its performance with respect to the PID controller results. The linear controller is a backup system which takes over in case of malfunction of CMAC.

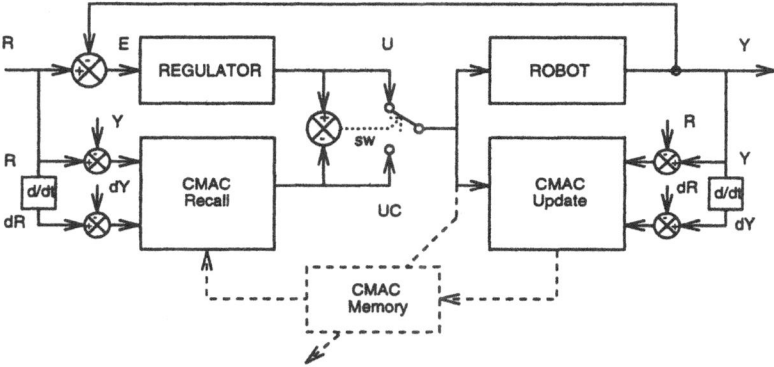

Figure 3: Scheme of CMAC control of robot dynamics.

The results of this implementation show that CMAC can efficiently learn the inverse dynamics of this complex robot model and that it can improve the results of applying a classical PID controller, as shown in Fig. 4. Moreover, it is important to take into account that the PID controller was specially designed for the mathematical model of the simulator, so that the PID control results are the best-case situation, while poorer results are to be expected in real conditions, where deviations from the model will undoubtedly appear. Conversely, the CMAC learning is not model dependent, so that the real robot conditions should not pose additional difficulties for control.

5   Learning to Update Inverse Kinematics

This application, proposed by the German aerospace company Daimler-Benz Aerospace, refers to a 6-degree-of-freedom robot arm to be used in a space station. The robot is intended for two types of tasks, namely, those internal to the station, such as maintenance operations and external ones, such as exploratory missions for reconaissance and sample-collection. The environment in both instances is unstructured; this is readily apparent in the case of external missions, where the environment may be unknown. In the internal maintenance tasks, although geometrical models of the space station may be used, it is reasonable to expect changes in geometry and instrument calibration in long-term operation. Since no manned missions are planned to recalibrate instruments and models, the robot control systems must be endowed with the capability of performing correctly even if changes in operating conditions arise.

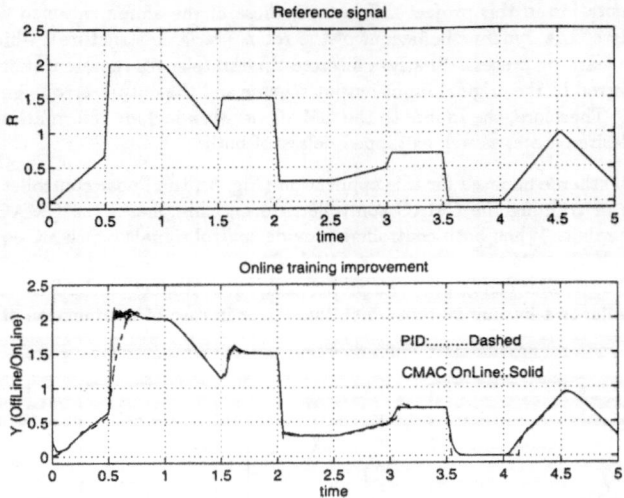

Figure 4: Results of PID and of CMAC control after on-line training with a reference signal composed of steps and ramps.

The need for some degree of autonomous operation stems from the increasing effort to close the feedback control loops in the space station, as opposed to doing so through earth-based operators, which involves considerable time delays for communications. The main autonomous behaviours expected from this robot are related to its ability to react to sensory information in-situ. They involve learning sensorimotor control for positioning after changes or degradations occur and learning how to grasp and handle objects, based on image and force-torque information.

One of the most important functions required for the autonomous behaviour of this robot is the self-calibration of inverse kinematics, which is the module developed at the Institut de Cibernètica, as described below. Other functions for which neural learning is being investigated elsewhere are: modelling of the environment, selecting poses for grasping objects and some examples of object handling, related to the peg-in-hole insertion problem.

5.1 Previous Work

The knowledge of inverse kinematics, which relates the end-effector coordinates with the required coordinates in the robot joints, is essential in any robot control application. Most commercial robots provide explicit mathematical models of their inverse kinematics, which are implemented in their execution controllers. However, robots frequently undergo mechanical changes during operation causing miscalibrations with respect to the original kinematic model. It is reasonable to expect that a robot arm with a certain degree of autonomy be capable of relearning or recalibrating its inverse kinematic relationship.

Most of the neural approaches to inverse kinematics have involved supervised learning schemes. The two learning rules most widely used have been *LMS* and *back-propagation*. Among these works, we find [15, 17, 26, 41]. The most significant result in these papers is the demonstration of the ability of neural networks to learn this complex mapping. Other works, like [2] have studied the combination of *back-propagation* networks and a conventional method, or the use of specialized neural structures [28] in order to improve the overall accuracy and convergence.

A different approach to inverse kinematics is to deal with the problem from a more behavioural point of view: relating images of the end-effector position and orientation to the required coordinates in the joints. Consequently, in this approach, it is not required to know the actual cartesian coordinates of the end-effector, but the goal is to relate sensory information directly to joint coordinates. This approach is, in fact, very appropriate for visuomotor control and is sometimes referred to as the hand-eye coordination problem.

5.2 Unsupervised Learning Approach Taken

The approach adopted for this application follows that by Ritter et al. [35, 48], with its special form of Kohonen self-organizing maps, thus based on unsupervised learning. The target position of the end-effector is defined as a spot registered by two cameras looking at the workspace from two different vantage points.

Neurons are arranged in a 3D lattice to match the dimensionality of physical space. The learning process makes this lattice converge to a discrete representation of the workspace. Each neuron i has an associated four-dimensional vector $\mathbf{w}_i$ representing the retinal coordinates of a point of the workspace. The response of the network to a given input $\mathbf{u}$ is the vector of joint angles θ_k and the 3×4 Jacobian matrix $\mathbf{A}_k$ associated with the winning neuron k, as shown in Fig. 5.

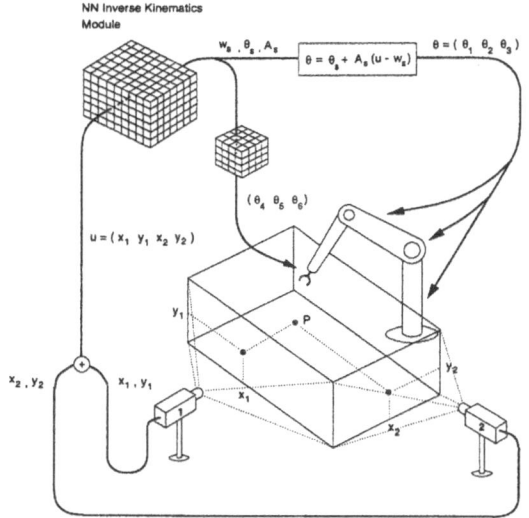

Figure 5: Scheme of hand-eye coordination using self-organizing maps.

The joint angles produced for this particular input are then obtained with the expression:

$$\theta(\mathbf{u}) = \theta_k + \mathbf{A}_k(\mathbf{u} - \mathbf{w}_k). \tag{9}$$

A learning cycle consists of the following four steps:

1. First, the classical Kohonen rule is applied to the weights:

$$\mathbf{w}_i^{new} = \mathbf{w}_i^{old} + c\, h_k(i)\, (\mathbf{u}(t) - \mathbf{w}_k(t)), \tag{10}$$

where c is the learning rate and $h_k(.)$ is a Gaussian function centered at k used to modulate the adaptation steps as a function of the distance to the winning neuron.

2. By applying $\theta(\mathbf{u})$ to the real robot, the end-effector moves to position $\mathbf{u}'$ in camera coordinates. The difference between the desired position $\mathbf{u}$ and the attained one $\mathbf{u}'$ constitutes an error signal that permits applying an error-correction rule, in this case the LMS rule:

$$\theta^* = \theta_k + \Delta\theta = \theta_k + \mathbf{A}_k(\mathbf{u} - \mathbf{u}'). \tag{11}$$

3. By applying the correction increment $\mathbf{A}_k(\mathbf{u} - \mathbf{u}')$ to the joints of the real robot, a refined position $\mathbf{u}''$ in camera coordinates is obtained. Now, the LMS rule can be applied to the Jacobian matrix by using $\Delta\mathbf{u} = (\mathbf{u} - \mathbf{u}'')$ as the error signal:

$$\mathbf{A}^* = \mathbf{A}_k + (\Delta\theta - \mathbf{A}_k\Delta\mathbf{u})\frac{\Delta\mathbf{u}^T}{\|\Delta\mathbf{u}\|^2}. \tag{12}$$

4. Finally, the Kohonen rule is applied to the joint angles:

$$\theta_i^{new} = \theta_i^{old} + c' \ h'_k(i) \ (\theta_i^* - \theta_k(t)), \tag{13}$$

and the Jacobian matrix:

$$\mathbf{A}_i^{new} = \mathbf{A}_i^{old} + c' \ h'_k(i) \ (\mathbf{A}_i^* - \mathbf{A}_k(t)), \tag{14}$$

where again c' is the learning rate and $h'_k(.)$ is a Gaussian function centered at k used to modulate the adaptation steps as a function of the distance to the winning neuron.

The previously cited works showed this method to converge and to self-organize into a reasonable representation of the workspace in about 30.000 learning iterations. This should be taken as an experimental demonstration of the powerful learning capabilities of this scheme, because the conditions in which it is made to operate are the worst possible ones: no a priori knowledge of the robot model, random weight initialization, and random sampling of the workspace during training.

In the application described here, this approach has been modified to suit a more practical setting. In particular, the problem of interest is not to learn the complete inverse kinematics mapping as in the Ritter et al. application, where no a-priori model of the robot is assumed. Rather, the goal is to correct a mapping which has degraded due to joint wear or malfunction. In this case, the topological map of the space relates camera images of the end-effector position and orientation to appropriate corrections of the joint angles and their Jacobians, with respect to the existing inverse kinematics module. Moreover, it has been extended to 6 degrees-of-freedom with success. The implementation in the real robot is now underway and is scheduled to be concluded in mid 1995.

A number of decalibration tests have been performed, ranging from slight variations in parameters of the kinematic model to structural changes in the links or the joints. In all cases, the system has been able to reconstruct its kinematic mapping in a limited number of learning iterations. As an illustration of how the error in position and orientation decreases as learning proceeds, Fig. 6 shows the results obtained for the case in which 3 robot links were extended by $1\,cm$ and the gripper orientation axes were twisted by between 3 and 6 degrees. For the worst-case problems, involving a complete re-learning of the inverse kinematics transformation, this has been achieved in 3000 learning cycles, to an accuracy of less than $1\,mm$ in position. The accuracy of the inverse kinematics correction is dependent on the discretization of the space chosen for the 3-D Kohonen map, but, in principle, the method should converge to any desired accuracy, if the appropriate number of Kohonen cells is provided.

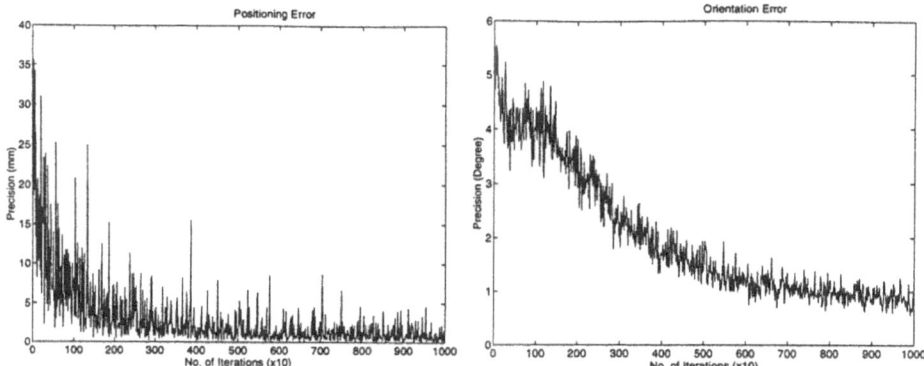

Figure 6: Evolution of the error in position and orientation as learning proceeds.

6   Learning to Navigate

Efficient navigation is critical for robots operating in hazardous environments, which are usually unknown the first time robots face them. This application deals with the problem of controlling an autonomous mobile robot so that it reaches efficiently a goal location in an unknown indoor environment. Some critical aspects of this problem are the following. First, the robot must always make decisions in real time. A robot operating in a hazardous environment cannot afford to stop long times to select the best course of action and/or to update its current knowledge about the task. Second, the robot's knowledge has to be grounded. In fact, since the robot does not have any a priori knowledge of the environment, it can only work upon information that is extracted from its sensors. Third, the robot's controller (and all related modules) has to deal with noisy sensory data.

6.1   Previous Work

In the case of unknown environments, a common approach to the control of autonomous mobile robots is that of reactive systems [8, 10, 49, 55]. They have shown robust real-time navigation capabilities. However, purely reactive systems suffer from two shortcomings. First and most important, reactive controllers may generate inefficient trajectories since they select the next action as a function of the current sensor readings and the robot's perception is limited. Second, they are difficult to program.

To address the first shortcoming of reactive systems some approaches combine planning and reaction, relying on a coarse global map of the environment. However, building and maintaining consistent global maps of the environment is far from trivial since noisy sensory data will introduce errors into the maps. Although we agree that global maps may be very helpful for navigation, we believe that the addition of learning capabilities to reactive systems is sufficient to allow a robot to generate efficient trajectories after a limited experience. In addition to surmounting the first shortcoming of reactive systems, a learning approach like ours even overcomes the second one. As some researchers have recently shown, the robot programming cost is considerably reduced by letting the robot learn automatically the appropriate navigation strategies [7, 12, 29, 32, 33, 36, 38, 44, 47].

In all these works, robots improve their performance through *reinforcement learning*. Briefly, a reinforcement-learning robot learns by doing and does not require a teacher who proposes correct actions for all possible situations the robot may find itself in. Instead, the robot simply tries different actions for every situation it encounters and selects the most useful ones as measured by a reinforcement or performance feedback signal. In the reinforcement learning literature, the situation-action mapping is customarily called "policy".

Thus, reinforcement-learning robots can self-improve their performance continuously and can self-adapt to initially unknown environments without needing extensive previous knowledge about the task. This is not only an advantage over non-learning robots, but also over other kinds of learning robots. Examples of alternative learning approaches to the acquisition of the appropriate reactive navigation strategies are supervised learning [46] and explanation-based learning [43]. In the former approach, robots require a teacher as discussed above. In the latter approach, robots require a perfect (or good enough) model of the task; robots use the model to plan a sequence of actions that achieve the goal, then they generate an explanation that is compiled into reactive rules.

6.2   Reinforcement-based Approach Taken

TESEO [37, 39] is an autonomous mobile robot controlled by a neural network. The neural controller maps the current perceived situation into the next action. A situation is made up of sensory information coming from physical as well as virtual sensors.

The physical robot is a commercial *Nomad 200* wheeled cylindrical platform that has three independent motors. The first motor moves the three wheels of the robot together. The second one steers the wheels together. The third motor rotates the turret of the robot. The robot has 16 infrared sensors and 16 sonar sensors that provide distances to the nearest obstacles the robot can perceive, and 20 tactile sensors that detect collisions. The infrared and sonar sensors are evenly placed around the perimeter of the turret and the tactile sensors cover all the perimeter of the robot below the turret. Finally, the robot has a *dead-reckoning* system that keeps track of the robot's position and orientation.

The input to the neural network consists of a vector of 40 components, all of them real numbers in the interval [0, 1]. The first 32 components correspond to the infrared and sonar sensor readings. In this case, a value close to zero means that the corresponding sensor is detecting a very close obstacle. The remaining 8 components are derived from a virtual sensor that provides the distance between the current and goal robot locations. This sensor is based on the dead-reckoning system. These 8 components correspond to a coarse codification of an inverse exponential function of the virtual sensor reading. The main reason for using this codification scheme is that, since it achieves a sort of interpolation, it offers three theoretical advantages, namely a greater robustness, a greater generalization ability and faster learning.

The output of the neural network consists of a single component that controls directly the steering motor and indirectly the translation and rotation motors. This component is a real number in the interval [−180, 180] and determines the direction of travel with respect to the vector connecting the current and goal robot locations. Once the robot has steered the commanded degrees, it translates a fixed distance ($25cm$) and, at the same time, it rotates its turret in order to maintain the front infrared and sonar sensors oriented toward the goal.

It is worth noting that a relative codification of both the physical sensor readings and the motor command enhances TESEO's generalization capabilities.

The reinforcement signal is a real number in the interval [−3, 0] which measures the cost of doing a particular action in a given situation. The cost of an action is directly derived from the task definition, which is to reach the goal along trajectories that are sufficiently short and, at the same time, have a wide clearance to the obstacles. Thus actions incur a cost which depends on both the step clearance and the step length. Concerning the step clearance, the robot is constantly updating its sensor readings while moving. Thus the step clearance is the shortest distance provided by any of the sensors while performing the action.

TESEO's aim is to learn to perform those actions that optimize the total reinforcement in the long-term (i.e., from the moment an action is taken until the goal is reached). TESEO improves its performance through *reinforcement learning* and acquires efficient navigation strategies in a rapid, safe and incremental way. To do so, it overcomes three critical limitations of basic reinforcement neural learning that prevent its application to autonomous robots operating in the real world. The first and most important limitation is that reinforcement learning might require an extremely long time. The main reason is that it is hard to determine rapidly promising parts of the action space where to search for suitable reactions. The second limitation has to do with the robot's behavior during learning. Practical learning robots should be operational at any moment and, most critically, they should avoid catastrophic failures such as collisions. Finally, the third limitation concerns the inability of "monolithic" neural networks —i.e., networks where knowledge is distributed over all the weights— to support incremental learning. In this kind of standard networks, learning a new rule (or tuning an existing one) could degrade the knowledge already acquired for other situations.

TESEO learns on top of basic reflexes. These reflexes correspond to sensory-motor knowledge of general applicability to a given task (e.g., obstacle avoidance, wall following, target tracking, etc.). These initial sensory-motor rules are transferred to the neural controller, which modifies them in order to adapt this general knowledge to the specific environment faced by the robot. The adaptation seeks to discover a policy that optimizes the robot's performance.

Our learning architecture allows to have an operational robot from the very start that improves its performance rapidly and incrementally as it safely explores the environment. The proposed neural controller is a modular network which is built automatically. Each module codifies a consistent set of reaction rules. Modularity guarantees that improvements on a module will not negatively alter other unrelated modules. The basic reflexes are preprogrammed as simple reactive behaviors. They are used every time the neural network cannot generalize its previous experience to the current sensory situation. The neural controller associates the selected reflex with the sensory situation in one step. The sensory situation is represented by a new neuron of the network and the selected reflex is codified into the weights of the controller. This new reaction rule is tuned subsequently through *reinforcement learning*. In this way, the neural network gets control (and thus suppresses the activation of the basic reflexes) more often as the robot explores the environment.

To demonstrate the appropriateness of our approach, we have chosen the task of reaching a fixed goal specified in cartesian coordinates. The environment is unknown and consists of a corridor with offices at both sides. The task is to generate a safe and efficient trajectory from inside one of the offices to a point at the end of the corridor. TESEO reaches the target location every time and it never gets lost or trapped into malicious local maxima. TESEO learns suitable navigation strategies after travelling 10 times from the starting location to the desired goal (see Fig. 7).

7   Conclusions

The four robotic applications of neural learning described in this paper involve challenging characteristics in terms of required autonomous behaviour in unstructured environments. In choosing the learning solutions, the authors have sought to combine the criteria of efficiency in learning and amenability for real-time implementation with an orientation towards the most behavioural models, i.e., avoiding, as much as possible, the need for explicit coding of workspace, objects or task models.

From the point of view of demonstrating the adequacy of neural learning in robotics, the solutions to the issues of inverse kinematics, inverse dynamics, visual positioning and navigation provide

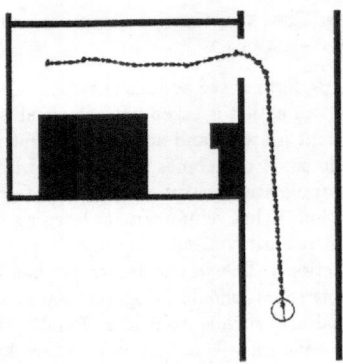

Figure 7: Trajectory found after 10 trials.

an excellent testbed for different concepts in learning. All four problems have been solved by adapting and extending known neural learning approaches so as to meet the specific requirements of the particular applications tackled in the projects CONNY and B-Learn II. Moreover, the neural techniques used constitute a good sample of the current technological "offer", ranging from entirely unsupervised to supervised approaches, and including off-line and on-line training. For a wider coverage of applications of neural learning to the robot control field, the reader is referred to [60].

The implementation of the first three modules in the real robots is currently underway at the industrial sites and is scheduled to be concluded in mid 1995. The navigation module is already installed in the mobile robot NOMAD 200, as described in the preceding section. A number of extensions of this research work are foreseen. On the subject of inverse kinematics, the issues of redundancy and singularities will be dealt with. Furthermore, a thorough study on the stability conditions and the robustness of the dynamic control with CMAC must be carried out. Similarly, the research work on the selection of global descriptors and neural network structures to optimize the convergence of the visual positioning module will be continued. Finally, the extension of the reinforcement approach to handle collision avoidance for manipulator robots will be investigated.

References

[1] M. Abidi and R. Gonzalez, 'The use of multisensor data for robotic applications , *IEEE Trans. on Robotics and Automation*, **6(2)**, (1990).

[2] Z. Ahmad and A. Guez, 'On the solution to the inverse kinematic problem , *Proc. IEEE Conf. on Robotics and Automation*, (1990).

[3] J.S. Albus, 'A new approach to manipulator control: The cerebellar model articulation controller (CMAC)', *Transactions of the ASME, Journal of Dynamic Systems, Measurement and Control*, **97**, 220–227, (1975).

[4] S. Ananthraman and D.P. Garg, 'Training backpropagation and CMAC neural networks for control of a SCARA robot', *Engineering Applications of Artificial Intelligence*, April, (1993).

[5] A.G. Barto, 'Learning by statistical cooperation of self-interested neuron-like computing elements', *Human Neurobiology*, **4**, 229–256, (1985).

[6] A.G. Barto, R.S. Sutton, and P.S. Brouwer, 'Associative Search Network: A reinforcement learning associative memory', *Biological Cybernetics*, **40**, 201–211, (1981).

[7] K. Berns, R. Dillmann and U. Zachmann, 'Reinforcement-learning for the control of an autonomous mobile robot', *Proc. of the IEEE/RSJ Intl. Conf. on Intelligent Robots and Systems*, 1808–1815, (1992).

[8] R.A. Brooks, 'A robust layered control system for a mobile robot', *IEEE Journal of Robotics and Automation*, **2(1)**, 14–23, (1986).

[9] G. Cembrano and G. Wells, 'Neural networks for control', in *Artificial Intelligence in Process Control*, Pergamon Press, 1992.

[10] J.H. Connell, *Minimalist Mobile Robotics: A Colony-Style Architecture for an Artificial Creature*, San Diego, CA: Academic Press, 1990.

[11] F. Chaumette, P. Rives and B. Espiau, 'Positioning of a robot with respect to an object, tracking it and estimating its velocity by visual servoing , *Proc. IEEE Int. Conf. on Robotics and Automation*, Sacramento, (1991).

[12] M. Dorigo and M. Colombetti, 'Robot shaping: Developing autonomous agents through learning', *Artificial Intelligence*, **71(2)**, 321–370, (1994).

[13] B. Espiau, F. Chaumette and P. Rives, 'A new approach to visual servoing in robotics , *IEEE Trans. on Robotics and Automation*, (1992).

[14] F. Fogelman-Soulié, 'Le connexionnisme', *Support de cours MARI 87 - COGNITIVA 87*, Paris, May, (1987).

[15] K. Goldberg and B. Pearlmutter, 'Using a neural network to learn the dynamics of the CMU direct-drive arm II, *Technical Report* CMU-CS-88-160, Computer Science Department, Carnegie-Mellon University, (1988).

[16] S. Grossberg, 'Competitive learning: from interactive activation to adaptive resonance', *Cognitive Science*, **11**, 23–63, (1987).

[17] A. Guez and J. Selinsky, 'A trainable neuromorphic controller , *Journal of Robotic Systems*, **5(4)**, 363–388, (1988).

[18] H. Hashimoto, K. Takashi, M. Kudou, and F. Harashima, 'Self-organizing visual servo system based on neural networks', *IEEE Control Systems*, 31–36, (April 1992).

[19] D.O. Hebb, *The Organization of Behavior*, Wiley, New York, 1949.

[20] R. Hecht-Nielsen, *Neurocomputing*, Addison-Wesley, 1990.

[21] G.E. Hinton, 'Connectionist learning procedures', *Artificial Intelligence*, **40**, 185–234, (1989).

[22] D.A. Hoskins, J.N. Hwang and J. Vagners, 'Iterative inversion of neural networks and its applications to adaptive control', *IEEE Trans. on Neural Networks*, **3(2)**, (1992).

[23] K.J. Hunt, D. Sbarbaro, R. Zbikowski and P.J. Gawthrop, 'Neural networks for control systems - a survey', *Automatica*, **28(6)**, (1993).

[24] M. Kabuka and A. Arenas, 'Position verification of a movile robot using standard pattern , *IEEE Journal of Robotics and Automation*, **13(6)**, (1987).

[25] M. Kaiser, V. Klingspor, J. del R. Millán, M. Accame, F. Wallner and R. Dillman, 'Achiving intelligence in mobility: incorporating learning capabilities in real-world mobile robots', *IEEE Expert*, to appear.

[26] M. Kawato, Y. Uno, M. Isobe and R. Suzuki, 'A hierarchical model of voluntary movement and its application to robotics , *Proc. IEEE 1st Intl. Conf. on Neural Networks*, San Diego, (1987).

[27] T. Kohonen, *Self-Organization and Associative Memory* (second edition), Springer-Verlag, Berlin Heidelberg New-York Tokyo, 1988.

[28] C. Kozakiewicz, T. Ogiso and N. Miyake, 'Partitioned neural network for inverse kinematic calculation of a 6 dof robot manipulator', *Proceedings IEEE INNS*, (1991).

[29] B.J.A. Kröse and J.W.M. van Dam, 'Adaptive state space quantisation for reinforcement learning of collision-free navigation', *Proc. of the IEEE/RSJ Intl. Conf. on Intelligent Robots and Systems*, 1327–1332, (1992).

[30] B.J.A. Kröse and P.P. van der Smagt, *An Introduction to Neural Networks*, 5th edition, University of Amsterdam, 1993.

[31] Y. LeCun, 'Une procedure d'aprentissage pour reseau au seuil assymetrique', *Proceedings of COGNITIVA*, 599–604, (1985).

[32] L-J. Lin, 'Programming robots using reinforcement learning and teaching', *Proc. of the 9th Natl. Conf. on Artificial Intelligence*, 781–786, (1991).

[33] S. Mahadevan and J. Connell, 'Automatic programming of behavior-based robots using reinforcement learning', *Artificial Intelligence*, 55(2), 311–365, (1992).

[34] K. Mandel and N. Duffie, 'On-line compensation of mobile robot docking errors , *IEEE Journal of Robotics and Automation*, 3(6), (1987).

[35] T.M. Martinetz, H.J. Ritter and K.J. Schulten, 'Three-dimensional neural net for learning visuomotor coordination of a robot arm , *IEEE Trans. on Neural Networks*, 1(1), (1990).

[36] J. del R. Millán, 'Reinforcement learning of goal-directed obstacle-avoidance reaction strategies in an autonomous mobile robot', *Robotics and Autonomous Systems*, to appear.

[37] J. del R. Millán, 'Rapid, safe, and incremental learning of navigation strategies', *IEEE Trans. on Systems, Man and Cybernetics*, to appear.

[38] J. del R. Millán and C. Torras, 'A reinforcement connectionist approach to robot path finding in non-maze-like environments', *Machine Learning*, 8(3/4), 363–395, (1992).

[39] J. del R. Millán and C. Torras, 'Efficient reinforcement learning of navigation strategies in an autonomous robot', *Intl. Conf. on Intelligent Robots and Systems (IROS'94)*, (1994). (To appear also in a special issue of the journal *Robotics and Autonomous Systems*).

[40] W.T. Miller, 'Real-time neural network control of a biped walking robot', *IEEE Control Systems*, February, (1994).

[41] W.T. Miller, F.H. Glanz and L.G. Kraft, 'Application of a general learning algorithm to the control of robotic manipulators , *Intl. Journal of Robotics Research*, 6(2), 84–98, (1987).

[42] W.T. Miller, R.P. Hewes, F.H. Glanz, and L.G. Kraft, 'Real-time dynamic control of an industrial manipulator using a neural-network-based learning controller', *IEEE Trans. on Robotics and Automation*, 6(1), 1–9, (1990).

[43] T.M. Mitchell, 'Becoming increasingly reactive', *Proc. of the 8th Natl. Conf. on Artificial Intelligence*, 1051–1058, (1990).

[44] T.M. Mitchell and S.B. Thrun, 'Explanation-based neural networks learning for robot control', in C.L. Giles, S.J. Hanson and J.D. Cowan (eds.), *Advances in Neural Information Processing Systems*, 5, 287–294, San Mateo, CA: Morgan Kaufmann, 1993.

[45] K.S. Narendra and K. Parthasarathy, 'Identification and control of dynamical systems using neural networks', *IEEE Trans. on Neural Networks*, 1(1), (1990).

[46] D.A. Pomerleau, 'Efficient training of artificial neural networks for autonomous navigation', *Neural Computation*, 3, 88–97, (1991).

[47] T.J. Prescott and J.E.W. Mayhew, 'Obstacle avoidance through reinforcement learning', in J.E. Moody, S.J. Hanson and R.P. Lippmann (eds.), *Advances in Neural Information Processing Systems*, **4**, 523–530, San Mateo, CA: Morgan Kaufmann, (1992).

[48] H. Ritter, T. Martinetz, and K. Schulten, *Neural Computation and Self-Organizing Maps*, New York: Addison Wesley, 1992.

[49] S.J. Rosenschein and L.P. Kaelbling, 'The synthesis of machines with provable epistemic properties', *Proc. of the 1986 Conf. on Theoretical Aspects of Reasoning about Knowledge*, 83–98, (1986).

[50] V. Ruiz de Angulo and C. Torras, 'Random weights and regularization', *Proceedings of the International Conference on Artificial Neural Networks (ICANN'94)*, Sorrento, May, (1994).

[51] V. Ruiz de Angulo and C. Torras, 'On-line learning with minimum degradation in feedforward networks', *IEEE Transactions on Neural Networks*, **6(3)**, May, (1995).

[52] D.E. Rumelhart and D. Zipser, 'Feature discovery by competitive learning', *Cognitive Science*, **9**, 75–112, (1985).

[53] D.E. Rumelhart, G.E. Hinton, and R.J. Williams, 'Learning representations by back-propagating errors', *Letters to Nature*, **323**, 533–535, (1986).

[54] D. Sbarbaro-Hofer, D. Neumerkel and K. Hunt, 'Neural control of a steel rolling mill', *IEEE Control Systems*, **13(3)**, (1993).

[55] M.J. Schoppers, 'Universal plans for reactive robots in unpredictable environments', *Proc. of the 10th Intl. Joint Conf. on Artificial Intelligence*, 1039–1046, (1987).

[56] R.S. Sutton, 'Learning to predict by the methods of temporal differences', *Machine Learning*, **3**, 9–44, (1988).

[57] C. Torras, *Temporal-Pattern Learning in Neural Models*, Lecture Notes in Biomathematics No. 63, Springer-Verlag, 1985.

[58] C. Torras, 'Relaxation and neural learning: points of convergence and divergence', *Journal of Parallel and Distributed Computing*, **6**, 217–244, (1989).

[59] C. Torras, 'From geometric motion planning to neural motor control in robotics', *AI Communications*, **6(1)**, 3–17, (1993).

[60] C. Torras, 'Neural learning for robot control'. *Proc. 11th European Conf. on Artificial Intelligence (ECAI'94)*, edited by A. Cohn, 814–819, Amsterdam, August, (1994).

[61] E. Tzirkel-Hancock and F. Fallside, 'Stable control of nonlinear systems using neural networks', *Intl. Journal of Robust and Nonlinear Control*, **2**, (1992).

[62] C. Venaille, G. Wells, and C. Torras, 'A neural network approach to image-based robot positioning', *Proceedings of SICICA*, Budapest, June, (1994).

[63] B. Widrow and M.E. Hoff, 'Adaptative switching capatibility and its relation to the mechanisms of association', *Kybernetik*, **12**, 204–215, (1960).

[64] A.Y. Zomaya and T.M. Nabhan, 'Centralized and decentralized neuro-adaptive robot controllers', *Neural Networks*, **6**, 223–244, (1993).

On Line Identification of Causal Relationships Between Variables in the Feed Water System of a Nuclear Power Plant

J.R. Álvarez[*], J. Mira[*], R.A. Fernández[**], L. Sainz[***], V. Arroyo[***], A.E. Delgado[*]

[*]Departamento de Informática y Automática. Facultad de Ciencias. UNED.
Senda del Rey, s/n. E-28040 Madrid (Spain). E-mail: jras@uned.es
[**]Nuclenor, S.A. Hernan Cortés, 26. E-39003 Santander (Spain)
[***]Departamento de Ingenieria de Sistema. E.T.S.I. Caminos. Universidad de Santander.
Av. Los Castros, s/n. 39005-Santander (Spain)

Abstract

On line identification of nonlinear causal relationships between variables in the feed water control loop of a nuclear power plant is reported. The knowledge about the observable variables of the application has been used in the design of the architecture for the network, in the local function of each elemental processor (quadratic expansion of inputs and recurrence) and, finally, in the selection of the supervised learning algorithm. This learning algorithm is based on the local evaluation and propagation of individual output errors for each sample in the training set. This nonlinear model with delays and quadratic expansion of inputs is compared with the more usual linear dynamic network and a clear improvement is observed. Some preliminary conclusions on the influence of signal noise relationships and the criteria for the selection of the appropriated sampling period are also included.

Introduction

Artificial neural networks are not always the best solution to all computational problems. Consequently when facing a concrete problem it is necessary to first evaluate its computational demands as well as to carry out a viability study of the neural solution. The biological systems have concrete neural networks solutions to specific problems [Mira, Delgado, 1991].

As a general norm, an analytic solution is preferable to an algorithmic one, an algorithmic one is preferable to a heuristic one, and this latter to a neural one. It is also a general norm that the variable and unpredictable character of a process, along with the need for adaptation in "real" time, points to a neural solution. There are a few questions which could provide guidance in the design of a neural solution to a specific problem.

A first step is to state which are the computational demand of the problem. That is to say, what type of task should be executed (classification, decision, control, spatio-temporal association, etc.) and how can we make it computational, independently of the "grain" size and of the programming source (external or autonomous).

We will next look at the distinction between a "blind net" and a "net based on knowledge". The initial design should be based on all the knowledge about the problem. The neural net is a learning system which can be described in two domains. The observer's own domain (OD) and the neural processor's own domain (PD). In the OD the description of the problem is found using natural language and the metalanguage of the specific task. Here, we must define the functionality expected from the neural net. The representational space (input signals and their meanings) and the injected knowledge, which is not inherent in the neurons but is essential to understand the meaning of its computation, must also be specified.

The following pass is to check if it is possible and advisable to use neural nets for the problem resolution. In other way, if the problem is segmentable and therefore it admits a distributed solution. This means whether the nature of the problem possesses: modularity, real time properties, changing environment, parameter adjustment and processor autonomy requirements.

All of the previous requirements make the neural solution advisable. But at this point we need to define the initial architecture of the net. The architecture includes the type of the elemental processors (artificial neurons) and the connectivity between them. At a first stage there is no limitations on the type and number of neurons to be used. But applying again the principle of a "net based on knowledge" it is needed to restrict the types of neurons to those which best can cope with the specific problem we are trying to solve. There are of course more restrictions based on the subsequent implementation.

In the process of designing the elemental processors we must specify the type of specialized local computation to use. The key lies in deciding which computation we are to call "artificial neuron". From the most common suggestion (adder and sigmoid) to the other extreme (self-programmable microprocessor), there is a whole repertory of possible solutions.

As a final point, but the most important one, there is no neural computation without learning. Consequently, the calculus structure proposed in the previous steps must be able to self program. Since the modules are autonomous, self-programming must be local.

The control systems of nuclear power plants require the identification of complex nonlinear dynamic systems. The standard models of dynamic systems are of two kinds: linear dynamic systems or specific nonlinear dynamic systems. The linear models are not adequated for very complex systems. The nonlinear models are specific for each task and requires different techniques for each case [Cellier, 1991]. This field is appropriated for the use of artificial neural networks because it provides a way of solution to problems with little knowledge.

The use of information and knowledge about the problem allow us to design a more efficient and precise solution to a given problem. This work tries to apply the available

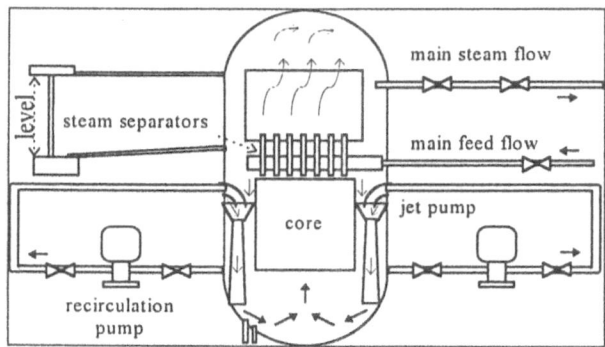

Figure 1: Schematic representation of the physical system being identified. The control circuit is not represented explicitly but is associated with the valves and the sensors.

knowledge about the problem and high order recurrent neurons in the continuous identification of causal relationships between the observable variables of a complex control system.

Problem Statement:

The main goal in this paper is embedded into a project to renovate de Control Systems in the Nuclear Power Plant of Santa María de Garoña in Spain. The identification problem is restricted to the case of the feed-water control system. The real system is not very accessible and it is a delicate part of the plant. This properties of the problem force us to identify the system in closed loop of control and during the normal operation of the plant. In the Figure 1 we can see a schematic diagram of the physical system under investigation. This part represents only the plant but the control system is attached and modifies the operation of the plant.

The system under analysis is nonlinear with multiple dependent variables. In the Figure 2 we can see the diagram of causality dependencies for some of the variables involved in the feed-water control system.

The target is to design a digital control to substitute the old analog control. The digital restriction points to a *digital* neural network solution. The way to deal with nonlinear dynamic systems with difficult analysis is to use a *dynamic* digital neural network with *nonlinear* elemental processors.

Functional Specifications for the Neural Net

A requisite for the correct design of the artificial neural network for a specific problem is to apply all the knowledge about the problem. The structure of the elemental processors and the connectivity is selected according to the task assigned to the net. We want to apply the network to identify a dynamic system, so some degree of recurrence is needed. This requirement is satisfied by the use of digital *fifo* memories in the inputs of each elemental processors [Fonseca, McCulloch, 1967].

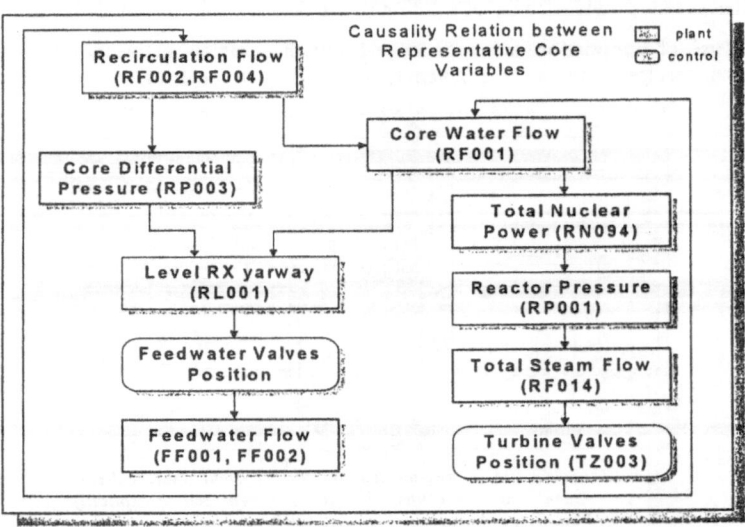

Figure 2: Schematic diagram for causality relation between variables involved in the plant.

To cope with the problem of nonlinear systems some kind of nonlinearity is needed in the elemental processors. The usual nonlinearity used in perceptron like artificial neurons is the sigmoid function at the output. This solution provides only a limited nonlinearity. In this work we use the quadratic expansion of inputs, including the delayed values in the fifo memories. We try with second order polynomials because the nature of dependencies in the simplified models of the plant are of second order [General Electric].

The sigmoid saturation of output is used only as a limiter function. This limit is useful when using quadratic expansion of inputs because of the instability in the dynamic difference equations. This same instability make necessary the use of scaled inputs when using the quadratic expansion. This is not a problem because the knowledge about the system provides the bounds for the signals.

The learning algorithm selected for this problem is based in the local error propagation. This kind of algorithms are well known [Rumelhart, Hinton, Williams, 1986], [Werbos, 1988], [Mira et all, 1994]. They are applicable for this case because the training is supervised with exact values. We want to minimize each output error not the global error, because different output can have different importance and cannot be mixed in a collective cost function. So the applied algorithm makes use of the locality in the propagation of error. All the information needed for the parameter adjustment is taken from the same elemental processor and from the neighbors connected to it.

The training mode selected is by sequential examples because a complete set of training is not available to be passed again and again until the network learns to identify the specific sequence. On the other hand, each sample is fed to the network and influences in the performance of the response. The sequence of examples must be guaranteed in order to maintain the time dependencies coherence. This form of training is appropriated to the changing environment of a control system in a nuclear power plant.

Knowledge and trial interaction

The selection of variables to feed the network as inputs and the desired outputs is done according to the scheme presented in Figure 2. We select some of the more representative variables involved in the system. The kind of relation tested is many to many, because there are no one to one clear relations.

This system includes the control of the feed water which is not analytic for all the range of operation. On the contrary it has different modes of operation that use different variables to calculate the control action. The usual simplified models of this system involve second order terms in the frequency domain. This is translated to the architecture design, for a first approach, in that there is one delay for each of the two layers previous to the quadratic expansion of inputs.

During the phase of design of the neural network there is an interaction between the model, from the knowledge of the system, and the results in trials of the network defined by the topology and structure of elemental processors.

The original data measured for the system under consideration are very noisy because de sensors are embedded in a very aggressive environment due to high pressures, temperatures, radiation, etc. This problem is related with the sampling period selection. The first test done over the data with the initial proposed structure for the neural network gives as a result that the sampling period was too short for this system. We take the time constants from the knowledge of the system which are around 2 seconds. The original data was reduced from 0.5 seconds sampling period to 2 seconds by the mean value of every 4 points measured. This action has the secondary result of reducing the noise of the signals.

Further test in the system leads to more changes in the structure of the network. The instability of the quadratic terms when the coefficients were fed to the difference equation model pointed to a reduction of the original structure to the one presented in the following section of results. This process of trial and adjust produces an incremental design of the neural network solution to the given problem.

Identification of dynamic systems

In the case of identification of systems we need an explainable model of the system in order to allow the effective control of the plant [Barto, 1990]. The use of blind neural networks, that classify and correlate the inputs with the outputs, is useless about the process model to be identified as some parameter values [Mira, Delgado, 1991].

It is clear that any model of a dynamic system must include delays to replicate the dynamics and the relation with the time. There are some ways to construct an elemental processor with delays. In this work the delays are inherent to the input space to each element. Each input and its delays are weighted and summed to give the output. This is given by the following difference equation

$$y(t) = \sum_{k=0}^{m} a_k \cdot u(t - k\Delta t) + \sum_{k=1}^{n} b_k \cdot y(t - k\Delta t) \tag{1}$$

The equation (1) can be expressed in terms of the z^{-1} operator [Fukunaga, 1964], [Narendra, 1990], resulting in a linear digital filter of the form

$$\frac{Y(z^{-1})}{U(z^{-1})} = \frac{\displaystyle\sum_{k=0}^{m} a_k \cdot z^{-k}}{1 - \displaystyle\sum_{l=1}^{n} b_k \cdot z^{-l}} \tag{2}$$

The coefficients $\{a_k, b_k\}$ form the parameters to be learned by the neuron, i.e. the weights. The function that represent the learning in the elemental processor is given to modify

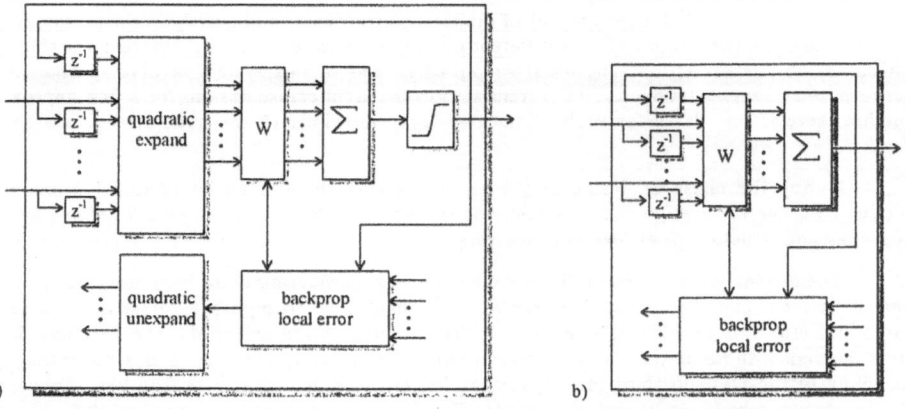

Figure 3: Structures of elemental processors for dynamic systems using delays of the inputs and of the self output. a) Element with quadratic expansion of inputs. b) Element with linear inputs.

the weights used in the output function. The algorithm used is the backpropagation of the local error. This basic model of linear system is represented in Figure 3,b).

Quadratic expansion of inputs

One way to obtain a better approach to the structure of the problem of identification of complex nonlinear systems is the use of one type of nonlinearity in the elemental processors. The traditional way is to add a sigmoid function as a local nonlinearity to the output. The use of a quadratic expansion of the inputs to each elemental processor has been probed to be better solution to fit complex nonlinearities. The sigmoid transfer function is used here as a saturation of the output level when using quadratic expansion of inputs to avoid the excessive grow of output with the square of inputs greater than 1.

This expansion consists in the transformation of the input space in other space including all the cross product as well as the original inputs [Nilsson, 1965], [Röckmann, Moraga, 1991], [Mira et all, 1993a]. This is expressed in the following equation

$$\tilde{x} = \left\{ x_i \cdot x_j; i,j = 0..n; j \leq i; \text{where } x_0 = 1 \text{ and } (x_1,\cdots,x_n) = (u_1,\cdots,u_n) \right\} \qquad (3)$$

The expanded space $\tilde{x}$ is used in the elemental processor as the true input space. The number of parameters (weights) is incremented as needed. An unexpander is used in order to add the appropriated partial derivative of the error to each true input of the neuron for the backpropagation algorithm

$$\frac{\partial}{\partial u_i}\left(\sum \tilde{x} \cdot \tilde{w}\right) = w_{0,i} + 2w_{i,i} \cdot x_i + \sum_{k>i} w_{k,i} \cdot x_k + \sum_{k<i} w_{i,k} \cdot x_k \qquad (4)$$

The schematic of this model with quadratically expanded inputs and the unexpander block is represented in Figure 3, a).

Application

Those techniques proposed in the previous sections has been applied to part of the feed water control systems of the nuclear power plant of Santa María de Garoña (Spain). We have selected some variables from those involved in the feed water system, proposed by the engineers of the plant [Mira, Fernández, Álvarez, 1993b].The design of an appropriated neural network for this problem need additional knowledge about the signals and the relation between them.

A structure based in dynamic perceptrons with one delay for each input has been selected. The connectivity is full between the two groups of neurons. The first group receive the inputs to the network and its self delayed output. The elements of the first group have quadratic expansion of inputs and one delay for each input. The schematic representation of the structure can be seen on Figure 4.a)

The number of elements in the first group is selected as the minimum equal to the number of inputs to the network. This means that the first group performs a combination and a nonlinear transformation of the inputs into a new space of the same dimension.

The second group is formed with linear elements with inputs from the previous group, plus the same delayed and from its self output delayed. The elements of the second group form the output of the network. These elements have linear output without any transfer function.

The training algorithm is a type of backpropagation using local error propagation. The training is done by examples in sequential order. This is very important in dynamic neural networks because the information is order dependent (past data values). The use of training by

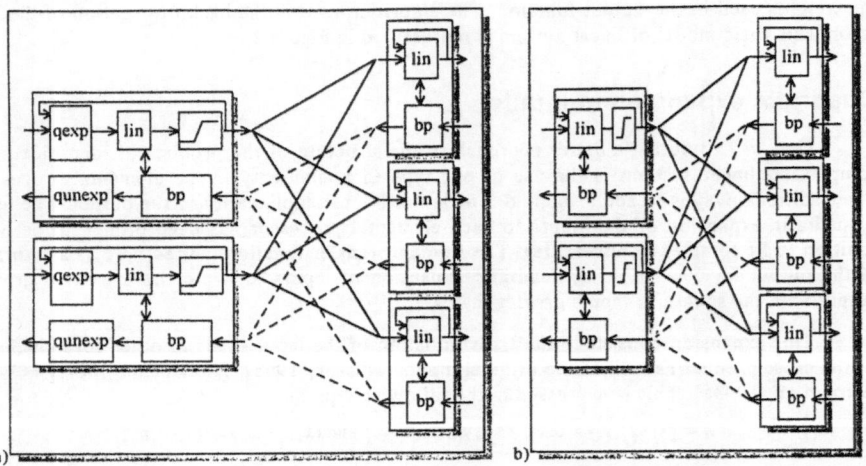

Figure 4: *Connections structure of the neural networks used in the application. a)Hidden layer of elements with quadratic expansion of inputs and output layer of linear inputs elements. b) Hidden and output layers of linear inputs elements.*

epochs is valid only for the case of a restricted set of examples to be learned. In this latter case a repeated continuous adaptation would be needed to identify the system.

Experimental results

The data from a transient in the plant has been used to train the network. We intented to identify the dynamics involved in the feed water control system of the plant. So, representative variables for this part have been selected. The time period of sampling is 2 seconds which is the corresponding to the time constants involved in the system.

The selection of input and output variables to be identified is done based on the causality of the observable variables. The selected variables for input to network are: core water flow and differential core pressure. The measured values for these variables are represented in Figure 5,a). The selected variables for output are: total power, steam flow and feed water flow. The measured values for these variables are represented in Figure 5,b). All the measured values have some fluctuations due to spurious error in the sensors because extreme conditions of the environment. The fluctuations in the measures improves the learning process adding some noise to the signals.

The results of training of the network with quadratic expansion of inputs are presented in Figure 5,c) for the first pass of training. We can observe that the response is very near to the real values. The network output improves as the training progresses and after the 30 first seconds the response is of the same kind that of the real values. Subsequent passes of the training set get a little better response only in the first 30 seconds as can be seen on the Figure 5,d) representing the output of network in the 3rd pass of training.

A comparison is done with another network using the same structure but replacing the elements with quadratic expanded inputs by another elements with linear inputs only. The schematic of this network is represented in the Figure 4,b). The equivalent experimental results are given in Figure 5,e) for the first pass of training and Figure 5,f) for the 3rd. pass of training.

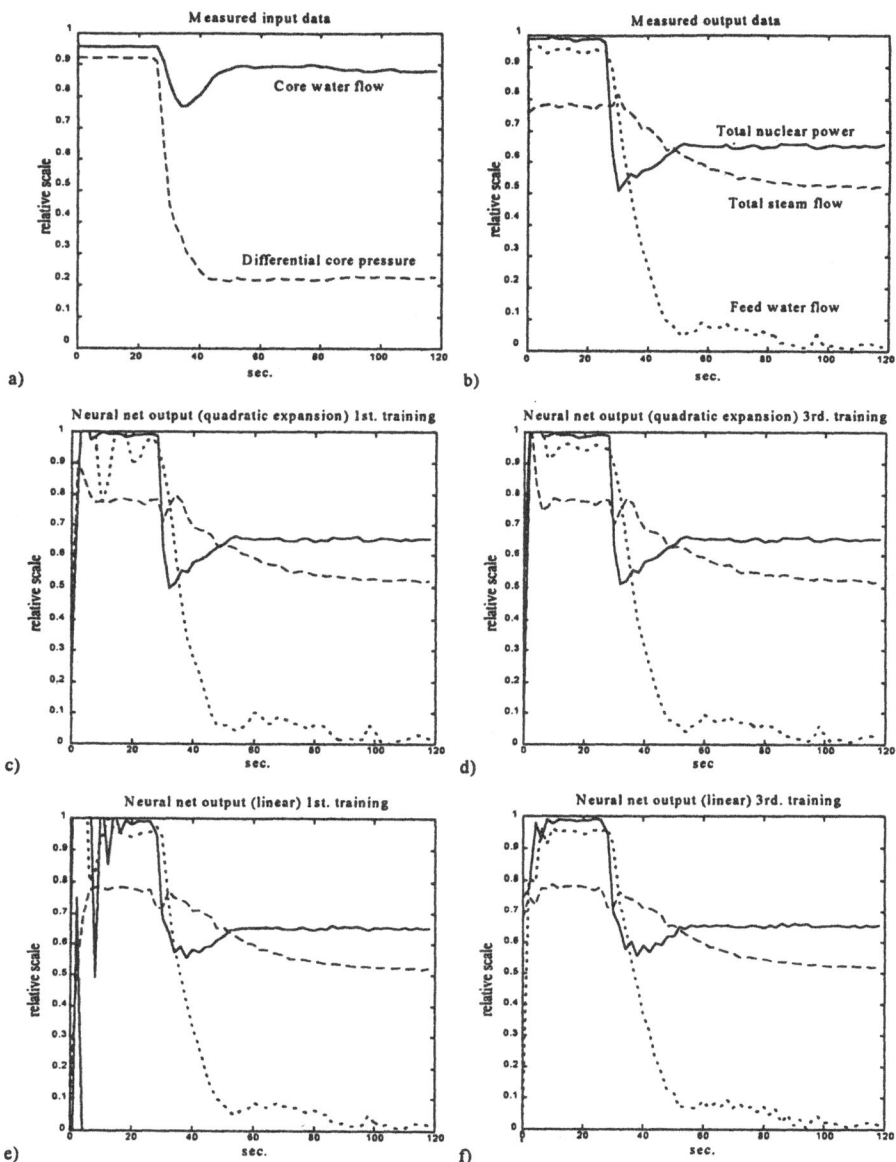

Figure 5: Results of application to real data extracted from the Nuclear Power Plant of Santa María de Garoña (Spain). a) Measured input data: core water flow and differential core pressure. b) Measured output data: total nuclear power, total steam flow and feed water flow. c) Output of neural network using elements with quadratic expansion of inputs at the first training. d) Same as c) but at the third training. e) Output of neural network using all elements with linear inputs at the first training. f) Same as e) but at the third training.

The response in this case is worst than in the previous case. For example, the abrupt change in the total nuclear power cannot be followed by linear inputs elements.

Acknowledgments

The staff of Santa María de Garoña, the Spanish CICYT project TIC-94-095 and the Spanish Comunidad de Madrid project I+D 0011/94, which supported this work.

References

Barto, G. A. (1990): "Conectionist Learning for Control". *Neural Networks for Control*. Miller III, Thomas W.; Sutton, Richard S.; Werbos, Paul J. (Eds.) pp. 5-58. The MIT Press.

Cellier, F. E. (1991): "Continuous System Modeling". pp. 659-671. Springer-Verlag.

Fonseca, J.S. da; McCulloch, W.S. (1967): "Synthesis and Linearization of Non-Linear Feedback Shift Registers: Basis of a Model of Memory". Q.P.R. n° 86. pp. 355. MIT.

Fukunaga, K. (1964): "A Theory of Non-Linear Autonomous Sequential Nets Using Z-Transformes". IEEE Trans., 1964, vol. EC-13, pp. 310-313.

Gallant, S.I. (1993): "Neural Network Learning and Expert Systems". The MIT Press.

General Electric: "Design Specification" (Nuclenor). Plant documentation of Santa María de Garoña Nuclear Power Plant.

Mira, J. (1992): "Computación Neuronal". UNED. Madrid.

Mira, J.; Delgado, A.E.(1991): "Always Trying to Write an Equation for the Brain", Lecture Notes in Computer Science, 540, *Artificial Neural Networks*. A. Prieto (de.) pp 93-100. Springer-Verlag.

Mira, J.; Delgado, A.E.(1991): "Linear and Algorithmic Formulation of Co-operative Computation in Neural Nets." Lecture Notes in Computer Science, 585, *Computer Aided Systems Theory*, *EuroCast '91*. F. Pichler and R. Moreno Díaz (Eds.) pp 2-20. Springer-Verlag.

Mira, J.; Delgado, A.E.; Álvarez, J.R.; P. de Madrid, A.; Santos, M. (1993a): "Towards More Realistic Self Contained Models of Neurons: High-Order, Recurrence and Local Learning" *New Trends in Neural Computation*. pp. 55-62. J.Mira, J.Cabestany, A.Prieto (Eds.) Springer-Verlag.

Mira, J.; Fernández, R.A.; Álvarez, J.R. (1993b): "Evaluación, Análisis y Modelación de los Sistemas de Control de la Central Nuclear de Santa María de Garoña". (Project Report) Madrid y Santander.

Mira, J.; Delgado, A.E.; Santos, M.; P. de Madrid, A.; Álvarez, J.R. (1994): "Local Learning in Networks of Universal Analogic Neurons". *Cybernetics and Systems* 25: pp. 259-273.

Narendra, K.S. (1990): "Adaptive control usign neural networks". *Neural Networks for Control*. Miller III, Thomas W.; Sutton, Richard S.; Werbos, Paul J. (Eds.) pp. 5-58. The MIT Press.

Nilsson, N. (1965): "Learning Machines".pp. 27-30. McGraw-Hill.

Lettvin, J.Y. (1962): "Form-function Relations in Neurons", Q.P.Q. n° 66, pp. 333. Research Lab. of Elect. MIT.

Röckmann, D.; Moraga, C. (1991): "Using Quadratic Perceptrons to Reduce Interconnection Density in Multilayer Neural Networks". Lecture Notes in Computer Science, 540, *Artificial Neural Networks*, A. Prieto (de.) pp. 86-91, Springer-Verlag.

Rumelhart, D.E.; Hinton, G.E.; Williams, R.J. (1986): "Learning internal representations by error propagation", *Parallel Distributed Processing: Explorations in the Microstructure of Cognition*, Vol 1 Foundations. D.E. Rumelhart and J.L. McClelland (Eds.). The MIT Press.

Werbos, P.J. (1988): "Generalization of back propagation with applications to a recurrent gas market model" *Neural Networks*, 1. pp 339-356.

Dynamic Neural Units
for Nonlinear Dynamic Systems Identification

M. AYOUBI, M. SCHÄFER and S. SINSEL

Technical University of Darmstadt, Inst. of Automatic Control, Laboratory of Control Engineering and Process Automation, Landgraf-Georg 4, 64283 Darmstadt, Germany. e-mail: ayou@irt2.rt.e-technik.th-darmstadt.de

Abstract An attempt has been made to establish a time-discrete neuron model which is applied to build Radial Basis Function and Multilayer Perceptron networks with distributed dynamics. The well-known delta-rule is extended to the dynamic delta-rule in order to optimize network parameters. Both network types were used to identify empirical, parametrical models of a turbocharger of a Diesel engine which comply with the demanded accuracy properties to a high degree. The performance of both network types is compared according to required number of parameters, approximation accuracy and computational effort.

1. INTRODUCTION

System analysis describes the task of investigating models of dynamical processes. These models can be applied for analytic redundancy in signal validation applications, fault diagnosis and adaptive control. One time-consuming but important approach when dealing with system analysis is the physical modeling of the system which involves the derivation of mathematical equations describing the various processes within the system based on the conservation laws of physics. Even though very accurate models can be obtained by physical modeling of well-understood processes, major drawbacks are inherent, e.g. the need for considerable human resources for the development of physical models, the difficulty in accurately modeling poorly understood physical phenomena and furthermore the requirement of empirically obtained process characteristics. Another important way for system analysis is the identification task (also called experimental analysis). System identification involves the investigation of the unknown parameters of an appropriate model structure which is chosen a-priori or partially motivated by physical analysis. The main advantages are the fast development and verification of models and the adaptability for on-line changes. However, the model parameters do not have any direct relation to the physical variables. In addition, the model structure selection is an important and crucial step in the overall identification procedure because it places some inherent limitations on the accuracy of the identified model.

In the last decade, parameter estimation methods have been elaborated for the identification of linear dynamical systems (Isermann, 1988; Ljung, 1987) although most complex processes are nonlinear. Unlike linear systems, there is no uniform approach to nonlinear system identification applicable to a broad class of systems. This is mainly due to the inherent complexity of nonlinear systems. Functional series methods such as identification of Volterra kernels can be applied to the class of processes with mild nonlinearities, but are usually limited to second order expansions due to excessive computations and the poor convergence. If a great deal of a-priori information concerning the system and nonlinearities structure is available, then parameter estimation methods based on the underlying differential equations or block-oriented methods in which a linear system is coupled with nonlinear blocks can be considered (Haber and Unbehauen, 1990; Billings and Voon, 1986). These methodes, however, can be very restrictive in general and require many assumptions using physical insight as a guide (Donne and Özgüner, 1992).

On the other hand, Artificial Neural Networks (ANN) offer a technology for processing signals without requiring a specified model structure. The most commonly applied ANN are the feed forward Multilayer Perceptron (MLP) and the Radial Basis Function networks (RBF). Both networks are proved to be universal approximators of static nonlinearities (Cyleenko, 1989; Funahashi 1989) and can fit the function:

$$ f: \quad X \in \mathfrak{R}^n \quad \rightarrow \quad Y \in \mathfrak{R}^m \qquad (1) $$

MLP and RBF are capable of identifying any nonlinear unique static function to arbitrary desired accuracy. Dynamic nonlinear processes, however, map their input $\underline{U}(t)$ time dependent to the output $\underline{Y}(t)$. Such systems therefore perform a spatiotemporal mapping task of the form:

$$ f: \quad \underline{U}(t) \in (\mathfrak{R}^n, t) \rightarrow \underline{Y}(t) \in (\mathfrak{R}^m, t) \qquad (2) $$

Leontaritis and Billings (1985) proved that under some mild assumptions, a wide class of discrete-time, nonlinear, and time-invariant systems can always be represented by the following simplified version of a NARMAX model (Nonlinear Auto Regressive Moving Average with eXogenous inputs):

$$ y(k) = f[y(k-1), ..., y(k-n), \\ u(k), ..., u(k-m)] \qquad (3) $$

where $y(k-i)$ and $u(k-i)$ are output and input past values at instant $[k-i]$ for $i = 1, ..., m, ..., n$. Because the derivation of the NARMAX model was independent of the form of the nonlinear functional, other choices of expansion like ANN were investigated. Recently, several approaches were proposed to introduce dynamics to ANN. Werbos (1990) used a form of the so-called "Back Propagation through time" for training networks that model dynamic systems. Chen *et al* (1990) and Narendra *et al* (1990) fed the network with current and delayed values of the process inputs and outputs which leads to quasi dynamical models, in that the used neural net remains a static approximator and the dynamics is lumped into an external system. The input space dimension of the network therefore increases depending on the number of available process inputs and outputs and the number of used past values. In contrast, other researchers supplied the neuron model with discrete or continuous time dynamics (Ayoubi, 1994a; Gupta *et al*, 1993; Chassiakos, 1991). These approaches differ in networks structure and dynamics implementations. However, dynamic neuron models do not require explicit past values of the process measurements as network inputs.

Within this approach, an attempt has been made to extend the nonlinear, dynamic, time-discrete perceptron model proposed by Ayoubi (1994a) to the Radial Basis Function neuron. Both dynamic neuron models are used to build Dynamic Multilayer Perceptron networks (DMLP) and Dynamic Radial Basis Function networks (DRBF) which were applied to identify empirical, parametrical, multi-input single-output (MISO) models for a turbocharger.

2. DYNAMIC NEURAL NETWORKS

The underlying idea of the dynamic neuron model, the so-called Dynamic Elementary Processor (DEP), was to make the neuron's activity depend on internal neuron states by integrating an ARMA-filter within the model. The neuron acts like a nonlinear, infinite impulse response filter (IIR) and thus processes past values of its own activity and input signals. Within this approach, the idea of DEP is introduced to the peceptron and to the RBF-neuron to build DMLP- and DRBF-networks, respectively.

2.1 Dynamic Perceptron Model based on DEP

Figure 1 represents the structure of the dynamic perceptron model.

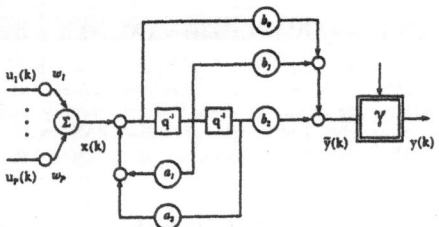

Fig. 1   Dynamic Elementary Processor (DEP) in state space representation with P inputs and 1 output.

The neuron transfer function is described by (4). $\underline{u}(k)$ is the

neuron input vector of the dimension [Px1], y(k) is the neuron output at time instant k. γ is a nonlinear activity function of the neuron with a threshold c.

$$
\begin{aligned}
y(k) &= \gamma(\tilde{y}(k), c) , \\
\tilde{y}(k) &= \underline{\xi}(k)^T \underline{\theta} , \\
x(k) &= \underline{w}^T \underline{u}(k) = \sum_{i=1}^{P} w_i u_i(k)
\end{aligned}
\tag{4}
$$

$\underline{\xi}(k)$ is the data vector of the dimension [5x1]:

$$
\underline{\xi}(k) = \begin{pmatrix} x(k), & x(k-1), & x(k-2), \\ & -\tilde{y}(k-1), & -\tilde{y}(k-2) \end{pmatrix}^T
\tag{5}
$$

$\underline{\theta}$ is the filter coefficients vector of the dimension [5x1]:

$$
\underline{\theta} = \begin{pmatrix} b_0 & b_1 & b_2 & a_1 & a_2 \end{pmatrix}^T
\tag{6}
$$

x(k) is the filter input at time instant k, and w_i is the weight of the i-th neuron input.

2.2 Dynamic Delta-Rule for Optimal Parameters

Widrow and Hoff (1960) proposed the delta-rule which is a Least Mean Squared Error (LMS) method to optimize the weights of their ADALINE (ADAptive LINear Element). We extended the delta-rule to the more general case of dynamic systems, the so-called dynamic delta-rule. The objective of the algorithm is to adjust the neuron parameters (both the weights and filter coefficients), based on a given set of input-output pairs and to determine the optimal parameter set which minimizes the performance index J :

$$
J = \frac{1}{2} \left\{ \sum_{k=0}^{N} (y_d(k) - y(k))^2 \right\}
\tag{7}
$$

where N is the size of the training set. The error signal defined as e(k) is the difference between the desired response $y_d(k)$ and the actual neuron response y(k).
To simplify the derivation of the adaptation algorithm, a linear time shifting operator can be defined by equation (8).

$$
\begin{aligned}
[\tilde{y}(k)] &= \frac{B(q)}{A(q)} [x(k)] \\
q^{-i}[x(k)] &= x(k-i) \\
A(q)[\tilde{y}(k)] &= \tilde{y}(k) + a_1 \tilde{y}(k-1) + \\
& \quad + a_2 \tilde{y}(k-2) \\
B(q)[x(k)] &= b_0 x(k) + b_1 x(k-1) + \\
& \quad + b_2 x(k-2)
\end{aligned}
\tag{8}
$$

Note that the time shifting operator may looks similar to the z-domain operator, but it is more simple in that it just shifts its argument without requiring its existence in the z-domain.

The optimal parameters which minimize J are iteratively approximated by moving in the direction of steepest descent on the cost function surface:

$$\vartheta_{new} = \vartheta_{old} + \eta \left\{ \sum_{k=1}^{N} e(k) \frac{\partial y(k)}{\partial \vartheta} \right\} \qquad (9)$$

where ϑ denotes the network parameter to be adapted and η is the learning rate. It is obvious that

$$\frac{\partial y(k)}{\partial \vartheta} = \frac{\partial y(k)}{\partial \bar{y}(k)} \frac{\partial \bar{y}(k)}{\partial \vartheta}$$
$$= \gamma' \frac{\partial \bar{y}(k)}{\partial \vartheta} \qquad (10)$$

Therefore, the used activity function has to be differentiable. Now, using the time shifting operator, four cases can be distinguished:
ϑ is a filter coefficient of the numerator $B(q)$

$$\frac{\partial [\bar{y}(k)]}{\partial \vartheta} \bigg|_{\vartheta = b_i} = [S_b(k)]$$
$$= \frac{q^{-i}}{A(q)} [x(k)] \qquad (11)$$

ϑ is a filter coefficient of the denominator $A(q)$

$$\frac{\partial [\bar{y}(k)]}{\partial \vartheta} \bigg|_{\vartheta = a_i} = [S_b(k)]$$
$$= \frac{-q^{-i}}{A(q)} [\bar{y}(k)] \qquad (12)$$

ϑ is a neuron input weight

$$\frac{\partial [\bar{y}(k)]}{\partial \vartheta} \bigg|_{\vartheta = w_i} = [S_w(k)]$$
$$= \frac{B(q)}{A(q)} [u_i(k)] \qquad (13)$$

ϑ is a neuron threshold

$$\frac{\partial y(k)}{\partial \vartheta} \bigg|_{\vartheta = c} = \frac{\partial \gamma}{\partial c} \qquad (14)$$

$S_b(k)$ and $S_w(k)$ are parameter states within the dynamic filters described on the right side of (11), (12) and (13). Thus, the change of the neuron activity according to any parameter is determined by the filtered gradient which explains the name of the procedure: the dynamic delta-rule.

2.3 Dynamic Multilayer Perceptron (DMLP)

Now, to make use of the connective power of ANN, those nonlinear, MISO, dynamical, time-discrete neurons can be distributed to build a dynamical Multilayer Perceptron DMLP (see Fig. 2).

Equations (15), (16) and (17) describe the inference in the network beginning by the input layer <K> through the hidden layer <M> to the output layer <L> respectively.

$$X(k)^{<J>} = W^{<J>} U(k) ,$$
$$\bar{Z}(k)^{<J>} = diag\big(\Psi(k)^{<J>} \Theta^{<J>}\big) ,$$
$$Z(k)^{<J>} = \gamma \big(\bar{Z}(k)^{<J>}, C\big) \qquad (15)$$

$$X(k)^{<J>} = W^{<J>} Y(k)^{<J>} ,$$
$$\bar{Z}(k)^{<J>} = diag\big(\Psi(k)^{<J>} \Theta^{<J>}\big) ,$$
$$Z(k)^{<J>} = \gamma \big(\bar{Z}(k)^{<J>}, C\big) \qquad (16)$$

$$X(k)^{<J>} = W^{<J>} Y(k)^{<J>} ,$$
$$\bar{Z}(k)^{<J>} = diag\big(\Psi(k)^{<J>} \Theta^{<J>}\big) ,$$
$$Z(k)^{<J>} = \gamma \big(\bar{Z}(k)^{<J>}, C\big) \qquad (17)$$

where
$X(k)^{<J>}$ is the state vector of the layer J
$W^{<J>}$ is the weight matrix of the layer J consisting of the weight vectors w for each neuron.
$\Psi(k)^{<J>}$ is the data matrix consisting of the data vectors $\xi(k)$ for each neuron in layer <J>
$\Theta^{<J>}$ is the parameter matrix consisting of the parameter vectors $\theta(k)$ for each neuron in layer <J>

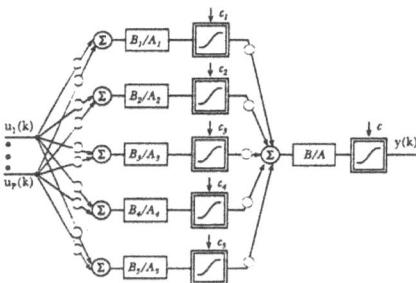

Fig. 2 3-layer DMLP with P inputs and 1 output consisting of 5 dynamic neurons in the <L>-layer and one in the <M>-layer

However, the dynamic delta-rule equations of one neuron can be extended by means of the time shifting operator to the case of a three layer network. This yields the dynamic, generalized delta-rule. The parameter state value of any parameter is calculated within the layer and propagated to the output layer through the dynamic filters of the layers inbetween. Figure 3 presents the scheme for the adaptation of a filter coefficient within the input layer of a 3-layer net.

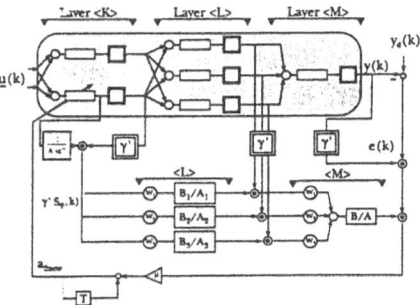

Fig. 3 Basic scheme of the dynamic generalized delta-rule for the adaptation of a 3-layer DMLP

It is important that the proposed DMLP does not require past values of the process measurements. Instead, it processes the system measurements at current instant [k]. This reduces the dimension of the network input space.

2.4 Dynamic Radial Basis Function Neuron based on DEP

Although RBF networks have an excellent approximation ability, their application was strongly limited by its drawback in that the number of used basis functions (number of neurons) increases exponentially with the dimension of the input space such that the approach becomes practically infeasible if the input space dimension is high (Narendra, 1992). This feature limits the application of RBF-networks for the identification of dynamic systems using past measurements values. However, the idea of DEP could be extended to supply the neuron with local memory and thus resign the lagged measurements at neuron input. Figure 4 shows the modified stucture of the dynamic Radial Basis Function neuron.

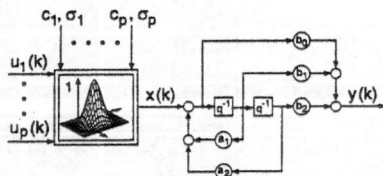

Fig. 4   Dynamic Elementary Processor (DEP) with Radial Basis Function (RBF) in state space representation with P inputs and one output

Similar to the DEP used for Multilayer Perceptrons, a second order ARMA-filter is integrated within the neuron as a dynamic element. The filter input x(k) is calculated as a function of the P neuron inputs $u_p(k)$ by a multidimensional Radial Basis Function with different centres c_p and standard deviations σ_p for each input. The Radial Basis Function is described by (18):

$$x(k) = e^{-\sum_{p=1}^{P} \frac{(u_p(k)-c_p)^2}{2\sigma_p^2}} \qquad (18)$$

The filter transfer function is as for the DMLP-network:

$$[y(k)] = \frac{B(q)}{A(q)}[x(k)]$$
$$= \frac{b_0 + b_1 q^{-1} + b_2 q^{-2}}{1 + a_1 q^{-1} + a_2 q^{-2}}[x(k)] \qquad (19)$$

2.5 Adaptation Algorithm for Optimal Parameters

In order to minimize the performance index J in eq. (7), the neuron parameters a_i, b_i, c_p and σ_p are optimized iteratively using the described dynamic delta-rule. The following equations state the partial derivatives of neuron output y(k) according to neuron parameters (the parameter states).

For a coefficient of the filter numerator B(q)

$$\left.\frac{\partial[y(k)]}{\partial\vartheta}\right|_{\vartheta=b_i} = [S_{b_i}(k)]$$
$$= \frac{q^{-i}}{A(q)}[x(k)] \qquad (20)$$

For a coefficient of the filter denominator A(q)

$$\left.\frac{\partial[y(k)]}{\partial\vartheta}\right|_{\vartheta=a_i} = [S_{a_i}(k)]$$
$$= \frac{-q^{-i}}{A(q)}[y(k)] \qquad (21)$$

For a Radial Basis Function centre c_p

$$\left.\frac{\partial[y(k)]}{\partial\vartheta}\right|_{\vartheta=c_p} = [S_{c_p}(k)]$$
$$= \frac{B(q)}{A(q)}[\frac{u_p(k)-c_p}{\sigma_p^2}x(k)] \qquad (22)$$

For a Radial Basis Function standard deviation σ_p

$$\left.\frac{\partial[y(k)]}{\partial\vartheta}\right|_{\vartheta=\sigma_p} = [S_{\sigma_p}(k)]$$
$$= \frac{B(q)}{A(q)}[\frac{(u_p(k)-c_p)^2}{\sigma_p^3}x(k)] \qquad (23)$$

Equations (22) to (25) require past parameter states to calculate the actual one. Thus, it is necessary to save the last two values of each parameter state during DRBF training procedure.

2.4 Dynamic Radial Basis Function Network (DRBF)

The structure of a dynamic neural network consisting of M Dynamic Elementary Processors with Radial Basis Functions is shown in figure (5).

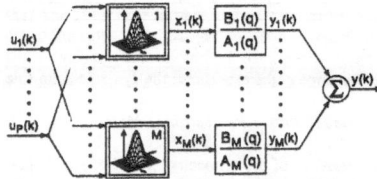

Fig. 5.   DRBF-network with P inputs and 1 output consisting of M Dynamic Elementary Processors (DEP)

The number of network inputs is identical to the number of inputs of each neuron. Thus, the DRBF-network is a multi-input single-output (MISO) system with P inputs and one output. The network output y(k) is the sum of the M neuron outputs $y_n(k)$:

$$[y(k)] = \sum_{m=1}^{M} [y_m(k)]$$

$$= \sum_{m=1}^{M} \frac{B_m(q)}{A_m(q)} [x_m(k)] \quad (24)$$

where $x_m(k)$ is the filter input of the mth neuron

$$x_m(k) = e^{-\sum_{p=1}^{P} \frac{(v_p(k)-c_{mp})^2}{2\sigma_{mp}^2}} \quad (25)$$

Because of the linear superposition of the neuron outputs at the network output, the parameter adaptation algorithm and equations for all neurons are similar.

3. APPLICATION OF DMLP- AND DRBF-NETWORKS TO THE IDENTIFICATION OF A TURBOCHARGER

3.1 Physical Modeling of the Turbocharger

Figure 6 schematically represents the charging process of a Diesel engine by an exhaust turbocharger. The exhaust enthalpy is used by the turbine to drive the compressor which aspirates and precompresses the fresh air in the cylinder. The turbocharger allows thus a higher compresion ratio increasing the power of the engine while its stroke volume remains the same. This is important in the middle speed range. The charging process has a nonlinear input/-output behaviour as well as a strong dependency of the dynamic parameters on the operating point.

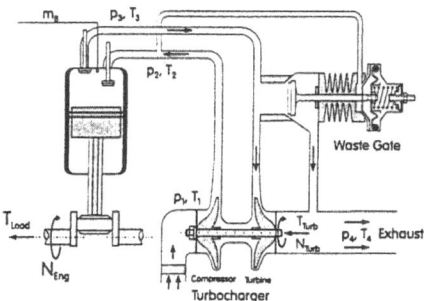

Fig. 6   Basic structure of a turbocharged Diesel engine

In general, the static behaviour of the turbocharger may be sufficiently described by characteristic maps (Look-Up tables) of compressor and turbine. Now, if the dynamics of the turbocharger need to be considered basic mechanical and thermodynamical modeling is required (Zinner, 1985; Boy, 1980; Pucher 1975). Practical applications have shown that these methods are capable of reproducing the characteristic dynamic behaviour of the turbocharger. The model quality, however, essentially depends on the accurate knowledge of several process parameters which have to be hardly derived or estimated in most cases by analogy considerations. An disadvantage also is the considerable computation effort due to the complexity of those methods.

For these reasons, such methods are considered inconsistent with the requirements of typical control engineering applications such as controller design, fault diagnosis and hardware-in-the-loop-simulations. Here, easy identifiable input/output-models suitable for real-time simulations are required. The dynamic neural network are therefore employed to develop a dynamic model of the turbocharger complying with the demanded properties to a high degree. Only the recorded input/output measurements of the real process are required and no theoretical knowledge of the process is necessary.

3.2 Empirical Modeling of the Turbocharger

Since the models identified by the dynamic neural networks should be integrated within a hardware-in-the-loop Diesel engine simulator, only measurements which are commonly availabe within engines like engine speed N_{Eng}, injection rate m_B and charging pressure (loading pressure) p_2 were selected as network input and output variables. The required tests have been performed on a dynamic engine test stand consisting of a 1.6 l VW Diesel engine (4 cylinders, 55 kNm engine torque), a turbocharger, an Electronic Diesel Control Unit EDC and a DC-motor to simulate vehicle load.

System Excitation

An important prerequisite for satisfactory system identification results is the use of suitable excitation signals which sufficiently excite the system in all relevant states of operation. The input cycle has to excite relevant frequencies of system dynamics as well as to cover the nonlinearity range for which the model is required. For a multi-input system, in particular, it is necessary to take care of a sufficient excitation of every individual system input.

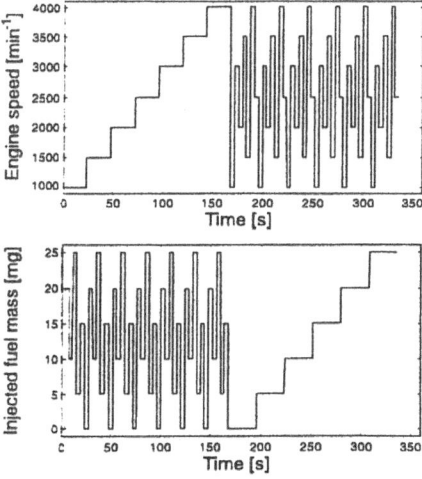

Fig. 7   Engine speed and injected fuel mass stimulation cycle for identification of an ANN model of a turbosupercharger

Figure (7) presents the training cycle used for identification of the turbocharger. As regards the fact that excitation by the engine speed and the vehicle load (fuel mass) should be decoupled, the process is dynamically excited by only one input at a time while the other input is set to different static states. The covered operation ranges are 1000 to 4000 rpm for engine speed and 0 to 25 mg for injected fuel mass. The step width of each dynamic excitation is chosen long enough to allow the loading pressure to reach its new steady state. This guarantees good identification results for both dynamics and static states of operation of the turbocharger. The presented training cycle lasts 340 s at a sampling time of 10 ms. This results in 34000 measurements of which only each 20th was used for neural network training to reduce the amount of data and to increase the sampling time while holding the Shannon sampling condition.

Training Results with DMLP- and DRBF-Networks

The excitation cycle of Fig. 7 was driven. The measured engine speed and the injected fuel mass were selected as input variables of the required model which should predict the loading pressure as desired system output. Several DMLP- and DRBF-networks with different configuration were trained with the measurements. Best results have been achieved for DMLPs with two layers consisting of five dynamic neurons in the input and one neuron in the output layer. Input weights and activity function biases are randomly initialized according to a uniform distribution density function symmetric to zero, while the gradient of the activity functions is set to one. For the DRBF-network, a 36 neuron structure proved to be good. The Radial Basis Functions of the network were positioned on a regular 6x6 grid in the two-dimensional input space. The initial value for the standard deviation of the RBFs is a function of the input ranges covered by the used training cycle. At the beginning of the training procedure, all filter coefficients except b_0 are initialized to zero. b_0 is set to one, so that the filter transfer function is static and supports stable learning. Figure 8 presents the measured turbine pressure versus the response of the trained 5.1-DMLP. The relative error which is also shown, is defined as:

$$e_{rel} = \frac{p_{estimated} - p_{measured}}{p_{measured}} \cdot 100\% \qquad (26)$$

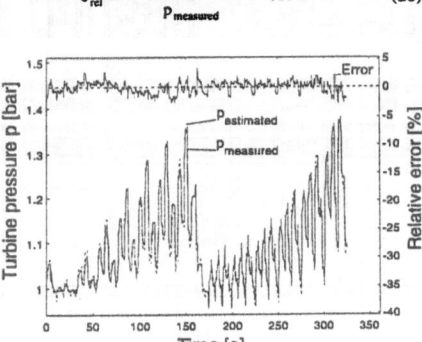

Fig. 8   Identification results of the loading pressure p_2 of the turbocharger using a 5.1-DMLP

Figure 9 presents the corresponding results achieved by the DRBF network.

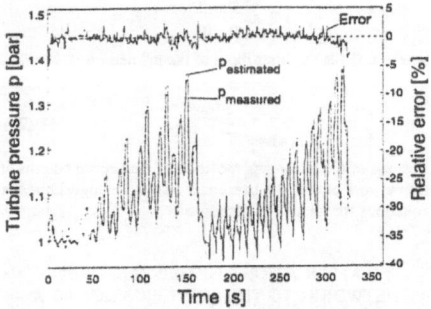

Fig. 9   Identification results of the loading pressure p_2 of the turbocharger using a 36 neuron DRBF-network

Table 1 shows some statistic information concerning training time and quality of approximation for both applicated network types. The mean squared errors relate to the difference between the measured turbine pressure and the network response at the end of the training procedure.

	DMLP (5.1-structure)	DRBF-Network (36 neurons)
Training Duration	10 h	10 h
Training Cycles (PC-AT 486-66)	10450	8330
Mean Square Error	$2.50 \cdot 10^{-4}$	$2.03 \cdot 10^{-4}$
Parameters	51	252

Table 1.   Comparison of training statistics for DMLP- and DRBF-network

Generalization Results with DMLP- and DRBF-Networks

To validate the models identified by the neural networks, the trained DMLP- and DRBF-networks can be tested with new measurements which have not been introduced to the network before. This is usually called generalization. For this purpose. the Diesel engine has been driven in a special vehicle simulation mode in which the dynamic engine test stand simulates vehicle load with the possibility to perform acceleration, braking and gear changing. For a test cycle which contains all these elements, Fig. 10 and 11 present the generalization results of the trained DMLP and DRBF networks. respectively. Again. the response of the trained neural networks is presented versus the measured loading pressure as desired system output.

1051

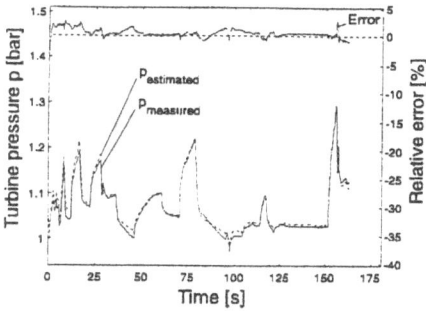

Fig. 10   Generalization results of the 5,1-DMLP network predicting the loading pressure p₂ of the turbocharger

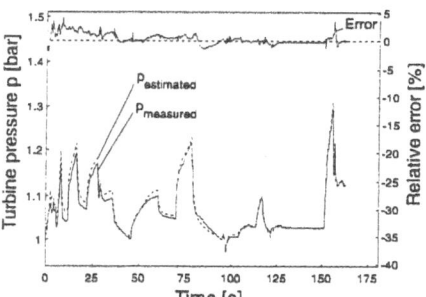

Fig. 11   Generalization results of the 36 neuron DRBF-network predicting the loading pressure p₂ of the turbocharger

The identified models have shown very good generalization results for both the dynamics as well as the steady states. Table 2 presents the generalization mean squared errors for both network types. The fact that these errors hardly differ from the training mean square errors given in table 1, proves good system identification quality.

	DMLP (5,1-structure)	DRBF-Network (36 neurons)
Mean Square Error	$1.54 \cdot 10^{-4}$	$2.49 \cdot 10^{-4}$

Table 2.   Generalization quality for DMLP- and DRBF-network

4. DISCUSSION AND CONCLUSIONS

Within this approach, the dynamic, time-discrete neuron model DEP was introduced to build dynamic Multilayer Perceptron (DMLP) and dynamic Radial Basis Function (DRBF) networks for nonlinear dynamic systems identification. As regards the topology of both networks, two main differences can be obtained. First, a DRBF-network always consists of only one single layer. Second, the static nonlinearity (Radial Basis Function) is placed at the network input in case of DRBF-networks, while the nonlinear activity functions of DMLPs are distributed to the

individual neurons in the different layers of the network. However, other studies have shown that the DMLP is more flexible and thus able to identify processes with unknown structures while the DRBF requires the assumption that the nonlinearity of the process is a function of the input signals. Furthermore, the DRBF still needs more parameters within the network than the DMLP (for the results of this paper: 252 versus 51, respectively).

On the other hand, the authors developed an initialization procedure utilizing powerful linear parameter estimation methods to initialize the filter coefficients of the DRBF-network. This was possible because of the DRBF-structure. The training time was herewith strongly shortend (only 5 min) which would be the subject of a further paper.

5. REFERENCES

Ayoubi, M., 1994a). Fault Diagnosis with Dynamic Neural Structure and Application to a Turbocharger. IFAC Symp. on Fault Detection, Superv. and Safety for Tech. Processes. SAFEPROCESS'94. Finland.
Ayoubi, M. 1994b). Nonlinear Dynamic Systems Identification with Dynamic Neural Networks for Fault Diagnosis in Technical Processes. IEEE Int. Conference on Systems, Man and Cybernetics SMC'94. USA.
Ayoubi, M. and Isermann, R. (1994). Model-Based Fault Detection and Diagnosis with Neural Nets and Application to a Turbocharger. Symp. on Artificial Intel. in Real Time Control AIRTC'94. València, Spain.
Billings S. and Voon W. (1986). Correlation based Model Validity Tests for Nonlinear Models. Int. Journ. of Control, 44, pp. 235-244.
Boy P. (198). Beitrag zur Berechnung des instationären Betriebsverhaltens von mitte schnellaufenden Schiffsdieselmotoren. Diss., TU Hannover.
Chassiakos A., Kosmatopoulos E. and Christodoulou M. 1991. Identification of Robot Dynamics by Neural Networks with Dynamic Neurons. Workshop on NN in Robotics. CA.
Chen S., Billings S. and Grant P. (1990). Nonlinear System Identification using Neural Networks. Int. J. Control. Vol. 51, No. 6. pp 1161-1214.
Cybenko. G 1989). Approximations by Superpositions of a sigmoidal Function. Mathematics of Control, Signals and Systems, 2. 303-314.
Donne J. Ozguner Ü. (1992). A Comparative Study of Neural vs. Conventional Methods for Modeling and Prediction. IEEE Int. Symp. on Intel. Control. Glasgow.
Funahashi, K. (1989). On the Approximat Realization of Continuous Mappings by Neural Networks. Neural Networks. 2. 183-192.
Gupta, M. Rao D. and Nikiforuk P. (1993). Neuro-Controller with Dynamic Learning and Adaptation. Jour. of Intelligent and Robotic Systems7 pp 151-173. Kluwer Publishers. Netherlands.
Haber R. and Unbehauen H. (1990). Structure Identification of Nonlinear Dynamic Systems- A Survey on Input/Output Approaches. Automatica. Vol. 26, pp. 651-677.
Hecht-Nielsen R. (1990). Neurocomputing, Addison-Wesley Publishing Company. Reading.
Leontaritis, I J., Billings, S.A.(1985). Input-Output Parametric Models for Nonlinear Systems, Part 1: Deterministic Nonlinear Systems, Int'l J. of Control, 41, 303-344.
Ljung, L. (1987). System Identification, Theory for the User. PTR Prentic Hall Information and System Sciences Series. Ed. Thomas Kailath, Englewood Cliffs, New Jersey 07632.
Masri, S., Chassiakos A. and Caughey T. 1990. Structure-Unknown Nonlinear Dynamic Systems: Identification through Neural Networks. Smart Mater. Struct. 1, pp 45-56, UK.
Narendra, K.S., Kumpati S. and Parthasarathy K. (1990). Identification and Control of Dynamical Systems Using Neural Networks, IEEE Trans. Neural Networks, vol. 1, no. 1, pp. 4-27.
Pucher H. 1985). Aufladung von Verbrennungsmotoren. Expert Verlag, Sindelfingen
Ludwig, C. Leonhardt, S. and Ayoubi, M. (1994). Real Time Supervision for Diesel Engine Injection. IFAC Symp. on Fault Detection, Supervision and Safety for Technical Processes, SAFEPROCESS'94. Helsink.
Werbos, P. 1990). Backpropagation through Time: What it does and how to do it. Proc. IEEE, Special Issue on Neural Networks 78. October.
Widrow, B. and Hoff, M. (1960) Adaptive Switching Circuits. IRE WESCON Convention Record. 96-104, New York.
Zinner K. A. (1985). Aufladung von Verbrennungsmotoren. Springer-Verlag, Berlin.

Optimal Identification Using
Feed-Forward Neural Networks

V. Vergara, S. Sinne, C. Moraga
Dept. of Computer Science I, University of Dortmund
D-44221 Dortmund, Germany

Abstract.- In this work we present new approaches for the optimal identification of nonlinear systems. We optimize different parameters of feedforward neural networks and of the learning schedule backpropagation by the use of global search methods like genetic algorithms and simulated annealing. We achieve a global increment of their learning capability thereby enlarging the generalization capability and reducing the amount of learning speed.
The result is a more reliable and robust model for nonlinear systems.

I. INTRODUCTION

Dynamic systems in control theory can be seen as black boxes transfering time-dependent input signals to time-dependent output signals. Under computational aspects it is necessary to restrict the time variable to discrete values. SISO systems can be mathematically described by a difference equation (eq. 1). This is a simplified form of the well-known NARX-model, leaving the error term beside. The time variable k is normally chosen to be a series of natural numbers $(0,1,2,...)$.

$$y(k+1) = f\big(y(k), ..., y(k-p), x(k), ..., x(k-q)\big) \qquad (1)$$

It is the task of identification [5] to reveal the proper form of the unknown function f, where the accuracy mainly depends on the application the model is used for. For control and prediction purposes it is often strongly important to find a model simulating the system's behaviour in a very precise manner. Parametric models in combination with parameter-estimation techniques have been found to be good means for this job, especially in the case of highly nonlinear systems.

For parametric models the identification process can be divided into the following main steps.

1. Selection of the general model type and a suitable identification method corresponding to the desired accuracy and using a-priori knowledge about the system.
2. Generation of suitable test signals and sampling of the system response. This collection must be done with respect to the chosen identification method and the desired accuracy.
3. Definition of the model structure (e.g. a polynomial of third degree) using perhaps a pre-analysis of the sampled system data or knowledge of theoretic system analysis.
4. Determination of the parameters of the model (e.g. the coefficients of the polynomial) applying the estimation technique to the model.
5. Validation of the determined model. Two considerations have to be made: First the applicability of the identification method has to be checked and second the degree of correspondence between the system and model input-output behaviour has to be compared with the desired one. In case of falsification one has to reenter the process in step one with another model type and/or estimation method, in step two using other test data or in step three using another model structure.

When high correspondence between system and model is required, one has to select a model type having the same properties as the system in step one. Especially for nonlinear systems the chosen model type must be capable to represent a wide class of functions. Feedforward neural networks with at least one hidden layer have been proven to be capable to estimate any Borel-measurable function to any desired degree of accuracy [3]. This generalization property makes neural networks well suited candidates for the modelling of nonlinear systems.

The model parameters are the weights and biases of the network which can be adjusted using the famous backpropagation algorithm [7] in step four. Here we see a first problem when introducing neural networks in the identification process. There are no hints or at least rules of thumb to determine the best network structure in step three for a given nonlinear function. The number of hidden layers and units for each layer have to be chosen arbitrary hoping not to underfit the model structure. Even overfitting does not guarantee to reach an optimal result.

Other occurring problems are due to the nature of iterative gradient-descent methods, to which backpropagation belongs. Depending on the chosen stepwidth η and the starting-point on the error surface the backpropagation method might stick in a local minimum or oscillate between different values. The next section briefly gives the theoretic background necessary for the further reading.

Section III presents novel approaches to overcome the mentioned problems mainly by the introduction of global numeric optimization techniques like genetic algorithms [2] and simulated annealing [4] into the identification process. Subjects of the global optimization are the network topology, the stepwidth η, the steepness of the activation functions of each neuron and the network weights. Another approach tries to get better results by the increase of the number of free parameters to adjust by the backpropagation method. Here beneath the weights and the biases also the steepness of all activation functions is changed towards the steepest negative gradient.

In section IV an example system is introduced to get an illustrative representation of the identification processes. For comparison three validation criteria are mentioned showing the improvement or deterioration of the results.

The last section closes this paper summarizing the important topics and pointing out further worthwhile research areas.

II. BASIC CONCEPTS AND NOTATION

1. Multilayer Feedforward Neural Networks and Backpropagation

When the units of a neural network are grouped to succeeding layers and the output of each unit is just the input to each unit of the following layer, then the network is called a multilayer feedforward neural network. The first or input layer holds the incoming data whereas the last or output layer exits the results of the network calculations. All other layers are called hidden layers. Each hidden and output unit can mathematically be described by the following formula:

$$o_{k,j} = f_{act}\left(u_{k,j}\right) \tag{2}$$

where

$$u_{k,j} = \sum_{i=1}^{n_{k-1}} w_{k,i,j} o_{k-1,i}$$ input of unit j in layer k;

$k \in \{2,...,m\}$ index for all layers;

$i, j \in \{1,...,n_k\}$ index for a unit in a layer;

$o_{k,j}$ output of unit j in layer k;

f_{act} function with sigmoid activation function;

$w_{k,i,j}$ weight of the connection between unit i in layer k-1 and unit j in layer k.

The input units are proposed just to pass through the network inputs u_j with $j \in \{1,...,n_1\}$.

We can completely describe the structure of a multilayer feedforward neural network using the following notation:

$$o = N_{n_1,...,n_m}(u) \tag{3}$$

N is a functional symbol representing the overall mapping of input vector u to output vector o. The subscripts $n_1,...,n_m$ denote the structure of the network giving the number of units per layer respectively. The parameter estimation technique most common for feedforward neural networks is the backpropagation method. In the so called online case all patterns are learned one by one. For each pattern the procedure is first to apply the input part of the pattern to the input units (yielding vector u) and then propagate the data forward. The network output $o=N(u)$ and the desired output of the pattern o_p are used to calculate the instantaneous network error

$$e = \frac{1}{2}(o_p - o)^2 \tag{4}$$

Backpropagation tries to minimize this function through stepwise altering the weights and biases of all units in the direction of the steepest descent. With a given stepwidth η the rule for each pattern and each weight is

$$\Delta w_{k,i,j} = -\eta \frac{\partial e}{\partial w_{k,i,j}} \tag{5}$$

For weights of hidden layers this amount of change can not be computed directly, so the strategy of backpropagation is to propagate an error term δ backwards through the net using the equation

$$\Delta w_{k,i,j} = -\eta o_{k-1,i} \delta_{k,j} \tag{6}$$

where

$$\delta_{k,j} = \begin{cases} f'_{act}(u_{k,j})(o_p - o_{k,j}) & \text{for } k = m \\ f'_{act}(u_{k,j}) \sum_{i=1}^{n_{k+1}} \delta_{k+1,i} w_{k,j,i} & \text{for } 2 < k < m \end{cases} \tag{7}$$

f'_{act} is the derivative of f_{act} with respect to the unit input $u_{k,j}$ as defined in eq. (2).

The main problem of learning with feedforward neural networks is to find the best fitting structure for a given set of test pattern. It is proven, that a network with at least one hidden layer is capable to approximate any Borel-measurable function with any degree of accuracy, but the number of necessary hidden units in this layer is not given.

Further on the backpropagation method brings along a few problems due to its gradient descent approach. First the parameter η is pattern and structure dependent and must be chosen at hand. A small value might intolerably increase the learning speed whereas for great values the algorithm might miss or escape from small valleys in the error curvature holding perhaps the global optimum.

Despite the value of η it is always possible that backpropagation traps into a local optimum or leads to oscillation.

2. Neural Networks in System Identification

The integration of neural networks into the identification process is straight forward. If the maximal time regression of the system input and output is known (that is p and q of eq. 1 are given), the identification system can be represented as in figure 1.

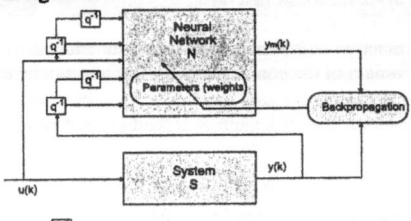

$\boxed{q^{-1}}$ one time step delay

Figure 1: Identification System

The actual and delayed input and output values at time step k are given as input to the neural network. After forward propagation the network and system output can be used by backpropagation to calculate the error and so to adjust the weights (and biases) of the network.

The main problem in this identification procedure is the knowledge of the maximal input and output time delays. In this work we assume that they are well defined since we are just interested in the problems resulting from the usage of feedforward neural networks and the backpropagation method.

3. Genetic Algorithms

Genetic algorithms [2] belong to the class of parametric search-methods which are able to find the global optimum in a highly deteriorated objective function due to their probabilistic character. Since the independent parameters are binary encoded, genetic algorithms are problem independent. The most important property regarding the success of genetic algorithms is the parallel treatment of individuals (each one a special rating of the variables), which are collected in a population. The population evolves in generations, hoping to contain a near optimum individual after a predefined amount of time. Every generation the population is worked on in some steps:

a) *Evaluation*: Each member is evaluated using the objective function to get the member's fitness.
b) *Selection*: A new population is mixed up from the old one, where individuals with high fitness have greater chance to survive than other ones.
c) *Crossover*: Parts of some randomly chosen members are exchanged, taking care that new individuals are incorporated to the population. Together with selection this is the mechanism guaranteeing a search towards regions with better fitness.
d) *Mutation*: A few randomly chosen bits of the whole population are flipped in order to avoid the loss of information.

4. Simulated Annealing (SA)

Another optimization method copied from nature like genetic algorithms is simulated annealing. The process of annealing brings a melted substance with high energy to a crystal at lower energy. The more gently the temperature is reduced, the more likely it is to find the global energy minimum, the perfect crystal. This process was transferred into a general optimization method to find the global optimum of combinatorial problems with large solution spaces [4].

Given an objective function $e: W \rightarrow R^+$ over the solution space W the procedure can be outlined as follows:

1. Start with time step counter $n=0$ and an initial temperature T_0 at high value.
2. Select an initial state w_0 at random and calculate the corresponding cost $e(w_0)$.
3. Generate a new state w_{n+1} with respect to the actual state w_n and temperature T_n using function $f_{gen}(w_n, T_n)$.
4. Calculate the objective function due to w_{n+1}. If the result is improved, retain the new state, otherwise choose a random number r uniformly from interval $[0,1]$ and accept the worsening if $r < f_{acc}(\Delta e, T_n)$, where f_{acc} is a given distribution and Δe is the change of cost.
5. Decrease temperature T_n using function f_T.
6. Increase n by 1 and restart at step 3 as long as a given criterion is not fulfilled. This may be the goodness of the cost reached so far or the value of n, which is proportional to the amount of time used for the computation.

From the theory of (inhomogenous) Markov chains there are some considerations on how to choose the functions f_{gen}, f_d and f_T in order to reach a global optimum with probability one. The main problem appearing hereby is the amount of time necessary for good convergence. In [10] is given a way to overcome this problem using the following functions: For the generation of a new solution function

$$w_{n+1} = f_{gen}(w_n, T_n, r) = w_n + T_n \tan(r) \tag{8}$$

is used. The number r here is chosen from a uniform distribution over the open interval $]-\pi/2, \pi/2[$.

The acceptance function f_{acc} is given by the Boltzmann distribution, and the function for temperature reduction

$$f_{acc} = e^{-\frac{\Delta e}{kT_n}} \tag{9} \qquad\qquad f_T(T_0, n) = \frac{T_0}{1+n} \tag{10}$$

This is the main improvement in comparison to the methods developed so far, since the reduction rate is no longer inversely logarithmic, but inversely linear.

1056

III. NOVEL APPROACHES

This chapter introduces three approaches to improve the identification of dynamic systems with neural network models and backpropagation. This is mainly done by comprising the identification process into a global search method (subsection 1) or vice versa (subsection 3). The attempt of subsection 2 is a little bit different trying to get better results by increasing the number of free parameters encountered in the backpropagation method.

1. The Use of Genetic Algorithms in the Identification Process

Some parameters hardly affecting the results of identification have to be defined in advance. From the model point of view these are the topology of the network (number of hidden layers and units per layer) and the steepnesses of the activation functions as well. From the point of view of the identification method the stepwidth η can be considered. Genetic algorithms as universal optimization methods will be used to get best fitting values for these parameters as initializations for the identification process [9].

The reason of choosing these parameters is to outline the optimization potential left in systems identification with feedforward neural networks and backpropagation. For the topology and the stepwidth there is no other chance than parametric search methods as a straight forward way to obtain near best values. Both parameters are problem dependent and cannot be derived from the available information. The optimization of the steepnesses can instead be included into the backpropagation algorithm as shown in the next subsection. Here it is considered to see if it is worth further investigation and to get a first insight into the distribution of good values. The hope to improve the results by altering the steepnesses is related [1], where is founded that the majority of units in the hidden and output layers of a trained network work predominantly in the linear part of their activation functions. Changing the linear region in advance (see fig. 2) gives backpropagation the chance to find weight combinations mapping the unit inputs for more test patterns into the linear region.

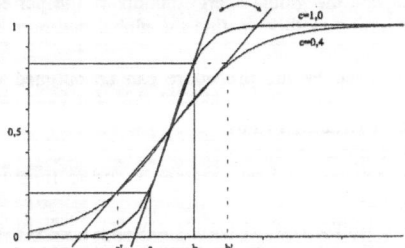

Figure 2: Slope optimization

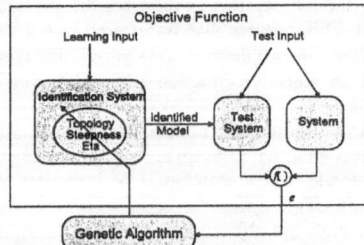

Figure 3: Diagram of the optimization of the identification process.

Beneath the last statement the consideration of steepnesses gives more degrees of freedom to the objective function, though allowing the error surface to hold deeper minima than with constant values.

Figure 3 gives an overview of the embedded identification process into genetic algorithms. The identification system can be considered to be the same as shown in figure 1. Before each calculation of the objective function the genetic algorithm set the parameters to optimize (topology, eta or steepnesses). Then the learn input is taken to identify the system with the backpropagation method. The computed model is tested in comparison to the system using the same test input data and the deviation between both outputs is the basis for an error criterion (e.g. m.s.e.) giving the output of the whole objective function.

2. Extension of Backpropagation with Respect to the Steepnesses of all Activation Functions

Similar to the introduction of biases it is possible to extend backpropagation to the variation of the steepnesses of the activation functions of all hidden and output units. The amount of change $\Delta c_{k,i}$ for the steepness of unit i in layer k and a fixed pattern can be defined by

$$\Delta c_{k,i} = -\eta \frac{\partial e}{\partial c_{k,i}} \tag{11}$$

Analogously to the derivation of the amount of change of the network weights (see eq. 6 and 7) we can formulate the update rules for the steepnesses in the following equations:

$$\Delta c_{k,i} = \eta \gamma_{k,i} \tag{12}$$

where,

$$\gamma_{k,i} = \begin{cases} \dfrac{\partial}{\partial c_{k,i}} f_{act}(u_{k,i}, c_{k,i})(o_p - o_{k,i}) & \text{for } k = m \\ \dfrac{\partial}{\partial c_{k,i}} f_{act}(u_{k,i}, c_{k,i}) \sum\limits_{j=1}^{n_{k+1}} \delta_{k+1,j} w_{k+1,i,j} & \text{for } 2 < k < m \end{cases} \tag{13}$$

$\delta_{k,j}$ is the error term defined in eq. (7), and can be taken from standard BP.

3. Incorporation of Simulated Annealing Principles into Backpropagation

In subsection 1 backpropagation was embedded into a parametric search procedure. Here it is just the other way: The universal optimization technique, namely simulated annealing, is included into backpropagation (see [10]) to combine the advantages of both principles. Simulated annealing has a drawback in the amount of time it might take, but is capable to find the global optimum on the error surface. Against that, backpropagation is quick in operation but suffers in getting to the best solution.

The general idea of the combined algorithm is to add a value calculated by eq. (8) to the amount of change each time a weight is updated in backpropagation (eq. 6). The new update rule is then defined by

$$w_{k,i,j} = w_{k,i,j} + \alpha \eta o_{k-1,i} \delta_{k,j} + (1-\alpha)(T_n \tan(r)) \tag{14}$$

where α controls the relative amounts of both parts (setting α to 1 gives the pure backpropagation algorithm). After updating each weight, steps 4 and 5 of the simulated annealing procedure described above are executed. That is, the network error is calculated in order to check increased deterioration if the weight change should be retained or withdrawn. Then the temperature is decreased and the time step counter increased before continuation with the next weight. To speed up this procedure the calculations can be made once after all weights of a whole layer have been updated.

IV. SIMULATION RESULTS

To show the capabilities of the described methods we use a dynamic system given by the following difference equation (see [6])

$$f[x_1, x_2, x_3, x_4, x_5] = \frac{x_1 x_2 x_3 x_5 (x_3 - 1) + x_4}{1 + x_3^2 + x_2^2} \tag{15}$$

The input is bounded to the interval [-1,1] in order to get a BIBO system, which is necessary for proper identification. The experiments with genetic algorithms are carried out using as learning data a white sequence (uniform distribution over the input interval) over 10^4 steps. All testing is done with a white sequence over 1000 steps. With the parameters produced by genetic algorithms an additionally identification process is started using 10^5 uniformly distributed input patterns. This is the same input as used for the other simulations if not mentioned otherwise. For comparison purposes showing the robustness of the identified model, we look at two validation criteria, the m.s.e. which is used as a cost function for the genetic algorithm and the Theilian inequality coefficient [11].

The underlying model structure for all experiments is the one proposed by Narendra, who used a neural network with topology $N_{5,20,10,1}$ and a stepwidth $\eta = 0.25$. The validation results for such an identified model (abbr. StandardBP) can be seen in the first row of table 1. All error graphs for the identification process using the new approaches are presented in comparison with the one of StandardBP.

$$e_{mse} = \frac{1}{n_p}\sum_p \left(y_p - y_{m,p}\right)^2$$

$$e_{the} = \frac{\sqrt{\dfrac{1}{n_p}\sum_p \left(y_p - y_{m,p}\right)^2}}{\sqrt{\dfrac{1}{n_p}\sum_p y_p^2} + \sqrt{\dfrac{1}{n_p}\sum_p y_{m,p}^2}}$$

where y_p is the system output, $y_{m,p}$ the model output for pattern p and n_p the number of input patterns.

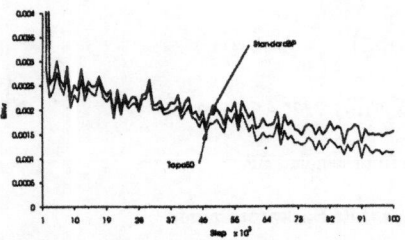

Figure 4: Learning error for topology optimization

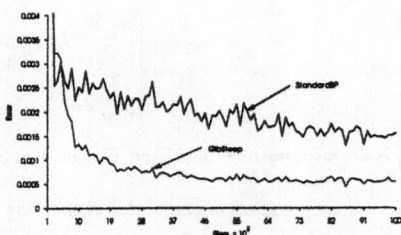

Figure 5: Learning error for global steep optimization

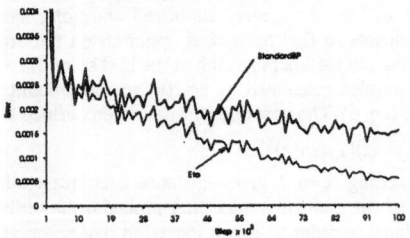

Figure 6: Learning error for η optimization

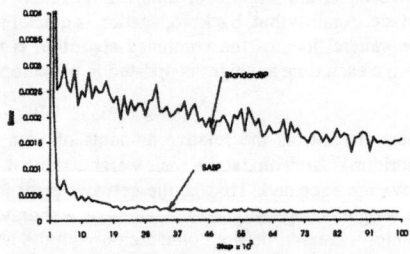

Figure 7: Learning error for embedded SA optimization

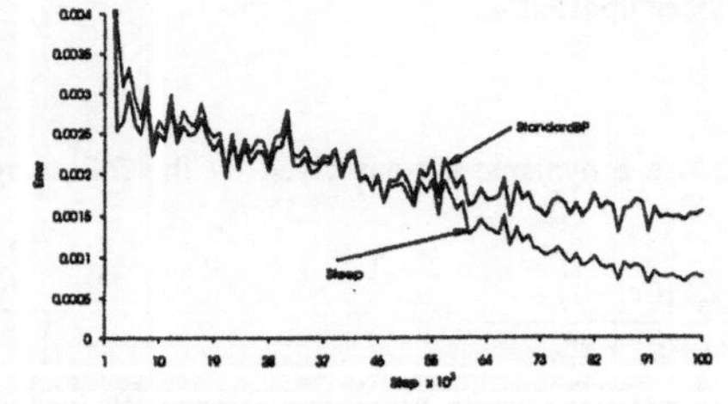

Experiment	e_{mse}	e_{the}
StandardBP	0,0065	0,0776
Topo50	0,00483	0,0668
GlbSteep	0,0022	0,0445
Eta	0,0024	0,0472
Steep	0,003917	0,0599
SABP	0,001345	0,0349

Figure 8: Learning error for steep optimization   **Table 1:** Validation results

In the following the results are shortly described [7]:

Topo50: Genetic search for the best unit numbers for two hidden layers, thereby given an upper bound of 50 units per layer. The topology found best in genetic algorithm is $N_{5,48,32,1}$, which leads to the assumption, that better results could be reach by increasing the limit of 50 units. Due to amount of time this is impractical. Since the weights of the improved network are initialized randomly it might be possible, that the decreased network error results from a better starting point on the error surface. To retain this possibility we make 1000 trials with different weight initializations and 10000 learning inputs for both, Topo50 and StandardBP. The average mean squared error for StandardBP is calculated to 0.008108 and for Topo50 to 0.007723, so demonstrating the robustness of the model found by genetic algorithms.

GlbSteep: Optimization of all steepnesses by genetic algorithms. The best obtained results found values for the steepnesses in the range of 0 to 100 for a logistic activation function. The validation criteria make clear, that the steepnesses contain a great optimization potential, much better than the one given by the network topology. The learning error exhibits fast convergence and a smooth trend.

Eta: Searching for good values of the stepwidth using genetic algorithms. The best value obtained is 0.692. Here the validation presents similar good results like GlbSteep, although the learning error converges slower and seems not to reach saturation during the first 100000 input patterns.

Steep: Altering the slopes of the activation functions in the same way as the network weights during backpropagation learning. For the initialization of weights given in the StandardBP experiment this method shows better convergence since circa the 50000th input step, which results in better validation criteria. For other initializations we found divergences beginning with the same step, but with a worse course of the new method. The averaged mean squared error for 1000 experiments with 100000 inputs each proved to be for both, StandardBP and Steep, the same value of about 0.0048. The difference between the methods can be seen in the results of the best and worst experiments, which were 0.003189 and 0.006661 for StandardBP and 0.001537 and 0.008833 for Steep. So the Steep approach gives the chance of getting as good results as promised in GlbSteep, but not with great security.

SABP: Embedded simulated annealing principles into backpropagation. This approach brings up the most promising results, fully justifying the hopes that the method combines the advantages of both, the simulated annealing schedule and the backpropagation algorithm. The introduced algorithm is able to find a much deeper minimum than backpropagation in a faster time than simulated annealing alone would need.

V. CONCLUSIONS

With the presented methods, we can build reliable and robust models for any nonlinear system by using parametric models, like connectionist models, in combination with genetic optimization methods. We solve the characteristic problems due to the parameter design of neural networks and their learning algorithm BP by embedding the identification process into a general optimization techniques. Additionally we enhance the degree of freedom of the model by the introduction of another parameter, the steepnesses of all activation functions. This approximation improves the performance of the best known model [6].

Literature

[1] Burrows, T.L.; Niranjan, M.: The use of feed-forward and recurrent neural networks for system identification. *Technical Report TR-158*, University of Cambridge, U.K., 1993.
[2] Goldberg, D.: *Genetic Algorithms in Search, Optimization & Machine Learning*, Addison-Wesley, 1989.
[3] Hornik, K.: Multilayer feedforward networks are universal approximators. *Neural Networks*, Vol. 2, pp. 359-366, 1989.
[4] Kirkpatrick, S; Gelatt, C.; Vecchi, M.: Optimization by Simulated Annealing. *Science* 220, pp. 671-680, 1983.
[5] Ljung, L., *System Identification*. Prentice-Hall, 1987.
[6] Narendra, K.S.; Parthasarathy, K.: Identification and control of dynamic systems using neural networks. *IEEE Trans. on Neural Networks*, Vol 1, N° 1, pp. 4-27, 1990.
[7] Rumelhart, D.E., Hinton, G.E., Williams R.J. Learning internal representations by error propagation. In *Parallel Distributing Processing*, Vol. 1, pp. 318-362, MIT Press, 1986.
[8] Sinne, S.: M.Sc. Thesis (in Progress), Dept. Computer Science I, University of Dortmund, 1994.
[9] Vergara, V., Moraga, C., Computational Intelligence for the Identification of Dynamical Systems, in *Proceedings of the 3dr Int. Conf. on Fuzzy Logic, Neural Nets and Soft Computing*, Iizuka, Japan, 1994.
[10] Wassermann, P.D.: *Neural computing: theory and praxis*. Van Nostrand Reinhold, 1989.
[11] Zwicker, E.: *Simulation und Analyse dynamischer Systeme in den Wirtschaft- und Sozialwissenschaften*, Berlin, 1981.

Recurrent Neural Networks
for Identification of Friction

M. DOMINGUEZ * **, J.M. MICHELIN**, J.M. MARTINEZ*

Commissariat à l'Energie Atomique (CEA) CEN Saclay, DMT/SERMA, 91191 Gif-sur-Yvette, FRANCE
***SFIM Industries Etablissement d'Asnières (SFIM EA) 7 rue Alphonse Kappler, 92600 Asnières, FRANCE*

Abstract

For high accuracy in mechanical control, the compensation of friction must be taken into account. When working at low velocities, the slip-stick phenomenon appears, introducing additive perturbation torque. Mechanical and robotics engineers use several models to simulate these torques. Here we select two dynamic friction laws (Dahl model and Reset-Integrator model), and we use Recurrent Neural Networks to model these dynamic systems in order to complete the «a priori» knowledge with what we don't know how to modelize. We use the «canonical» architecture to construct our network, and use a gradient based algorithm to train it. Results show that a general architecture for describing this family of friction laws can be obtained. This type of architecture may be used in regulation schemes.

1. Introduction

To achieve accurate results in tracking devices, it is often compulsory to model friction laws more elaborated than the well-known Coulomb model. Friction is a non-linear phenomenon depending on system's dynamics and caution to change during the mechanical system life (causes can be : age, lubrification, temperature).

Non-linear models have been proposed and used with success in some applications [1] [2] [3]. Generally compensation architecture is based on a specific friction model. In SFIM application no friction model found in the litterature is able to to represent the physical phenomenon. A more general framework capable of representing a set of friction models, and not only a precise model , can be useful.

Neural networks and their capacities of approximation and «black box» approach have been used in various attemps to identify and regulate non-linear dynamical systems [7], [4]. However, until now no attemp have been done in the identification of dynamic friction laws.

A neural network architecture may give a suitable frame to represent the friction behaviour without being specific of a friction model.

Then the first stage of the work is to define the neural network architecture in order to perform friction identification.

About identification with neural networks two general schemes are used and described in [7] : serie-parallel, or parallel methods.

If we have a direct access to the system outputs the first scheme can be easily achieved with MLP (Multi-Layered Perceptron). So we are then, working with a finite difference model as like :

$$\hat{y}_{k+1} = f\left(y_k, y_{k-1}, ..., y_{k-n_y}, u_k, u_{k-1}..., u_{k-n_u} \right)$$

where y_k denotes the systems outputs, and u_k the systems inputs.

When no access to the real values of the system to identify is possible, that is generally the case for friction torque, we use an estimation $\hat{y}_k$ of the real output, which lead us to a finite difference equation model as like:

$$\hat{y}_{k+1} = f\left(\hat{y}_k, \hat{y}_{k-1}, ..., \hat{y}_{k-n_y}, u_k u_{k-1}..., u_{k-n_u} \right)$$

This last system can only be identified by a recurrent network, where estimated outputs are fed into the input layer after being delayed an appropriated number of times.

In the second part of our paper we describe summarily the two friction laws we use to generate our data sets . Then we use the approach developped recently, by [8] to implement our recurrent neural network, and use the two above quoted identification schemes. The fourth part shows results we achieve and the conclusion we draw from this approach.

2. Frictions models

Friction can be described by a model in which we will take into account two areas:

○ a linear one, when the velocity is sufficiently high, corresponding to the asymptotic behaviour. It introduces in our tracking system a nearly constant bias of torque.

○ a non-linear area, when the velocity is low. It's called the «sticking» zone. High torques variations can be observed, which perturbates the mechanical control loop by introducing the own friction dynamics.

Among the numerous litterature of friction models we select the two following models because of their well-know applications and because of the very different dynamical behaviour when representing the sticking phenomenom.

The first model was proposed by Dahl in [5]. It is described by a non-linear differential equation, where a mechanism related to the representation of the zero velocity friction force is included.We select the common form for gimbals.

$$\frac{d\Gamma}{dt} = V \cdot \gamma \cdot (\Gamma - f_0 \cdot \text{sgn} V)^2 \quad \text{or} \quad \frac{d\Gamma}{dx} = \gamma \cdot (\Gamma - f_0 \cdot \text{sgn} V)^2 \quad \text{with } V = \frac{dx}{dt}$$

Where V is the relative velocity, Γ the friction torque, f_0 the maximum limit friction torque. γ is a damping coefficient which models the stiffness of the friction; the greater the stiffness is, the sharper the friction will be.

As $\Gamma = f_0 \text{sgn} V$,then $\frac{d\Gamma}{dx} \approx 0$, and the friction model describe a quasi-plastic behaviour.

When $\left|\frac{\Gamma}{f_0}\right| \ll 1$, then $\frac{d\Gamma}{dx} \approx \gamma f_0^2$, and this relation describes a quasi-elastic behaviour.

The second model we describe is the Reset-Integrator [6] model given in Figure.1. It was presented as an attemp to model the physical behaviour of «slip-stick» phenomenon by introducing an internal state variable, to relate the friction to the relative position.

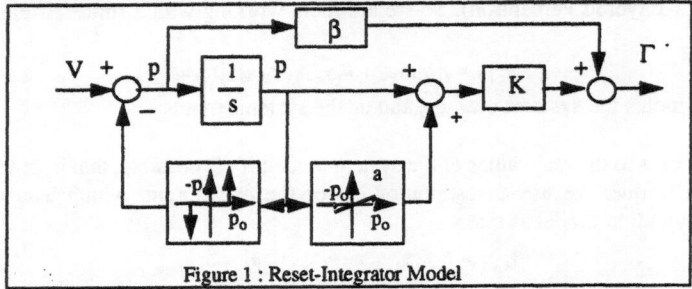

Figure 1 : Reset-Integrator Model

V represents the relative velocity, Γ the friction torque, β a viscous damping term, p_0 is a constant which defines the maximum amount of motion during sticking.

As long as $|p| \geq p_0$, then $\dot{p} = 0$, and the output of the integrator is constant; the friction torque value becomes $\Gamma = K \cdot p$, ($p = cst = p_0 \cdot sgn(p)$).

After a velocity change and while $|p| < p_0$, the output of the model is given by :

$$\Gamma = K \cdot p \cdot (1 + a) + \beta \cdot \dot{p}$$

These two models were implemented using the SIMNON software (SIMulator NON-linear); parameters of the simulation are given in the Appendice.

3. Identification using Neural Networks

3.1 Architecture

The basis of identification of non-linear process using neural networks were posed by [7] using two classical schemes in parametric estimation of systems.

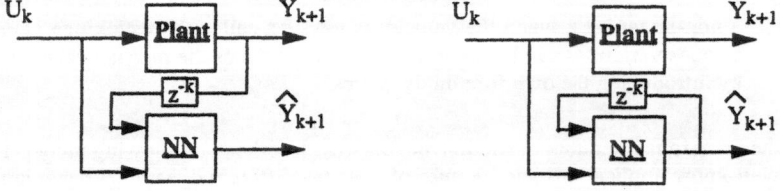

Figure 2 : Serie-Parallel Identification (a)   Parallel Identification (b)

Where U_k is a vector formed with the last n_u inputs of the plant. Y_{k+1} and $\widehat{Y}_{k+1}$ are vectors composed of the last n_y outputs and estimated outputs of the plant.

The first one, called serie-parallel method given in Figure 2.a, has the following advantages :

    O   If the system is stable in the sense : Bounded Inputs and Bounded Outputs (BIBO), then the inputs of the network are bounded.

    O   As there is no feedback in the model, we can use classical backpropagation algorithm with a MLP network. The complexity of the algorithm, and the computing time are lower than in parallel identification.

We can write then the finite difference equation of the model as :

$$\hat{y}_{k+1} = f\left(y_k, y_{k-1}, ..., y_{k-n_y}, u_k, u_{k-1} ..., u_{k-n_u} \right)$$

The main drawback in this architecture is its impossibility to represent the internal friction dynamics.

The second procedure, called parallel method given in Figure 2.b can not guarantee convergence of parameters , or even that the error of modelisation will tend to zero.
But, this scheme is useful when an internal friction dynamic has to be modelized. That is the case when working with these friction models and more genrally with the high accurate mechanical systems.
The main advantage of recurrent neural network is its capacity to have internal state variables, or when there is additive noise in the outputs [8]. So in this case we can write the finite difference equation of the model as :

$$\hat{y}_{k+1} = f\left(\hat{y}_k, \hat{y}_{k-1}, ..., \hat{y}_{k-n_y}, u_k, u_{k-1} ..., u_{k-n_u} \right)$$

It is shown in [8] that any recurrent discrete network of order N can be represented with a non recurrent network (feedforward) written in a suitable «canonical form» with N state variables. We used these results to design our network given in Figure 3.

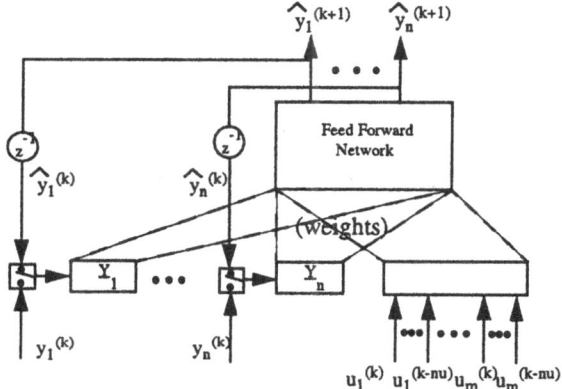

Figure 3 : General Architecture

Our network have as many outputs as states vectors $Y^{(k)}{}_i = \left(y_i^{(k)}, ..., y_i^{(k-n_y)} \right)$. Each one is of same length n_y. All state variables $y_i^{(k-nj)}$ are linked to the output of the network, with the corresponding numbers of delay units.The soft «switch» boxes in Figure 3 allow us to choose between serie-parallel or parallel approach. So, each state vector can be connected either to real output values $y_i^{(k)}$ for the serie-parallel approach, or to current outputs of the network $\hat{y}_i^{(k)}$ in parallel approach.

3.2 Algorithms

As there are no feedback in the serie-parallel approach, the minimisation over a (discrete) time interval Nt, makes calculus of gradients independent between them . The formula of weight θ_i

modification between 2 neurons at instant k is given by :

$$(\Delta\theta_i)^k_{total} = -\eta \cdot \frac{\partial J^k}{\partial\theta_i} = -\eta \cdot \sum_{p=1}^{Nt} (\Delta\theta_i)^k_p$$

where the Nt copies of network contributes independently in the total sum.

When using the parallel approach, and using a gradient based method to minimize our LMS criterion, calculus of gradients at a moment k needs the computing of all its preceeding values since the initialisation of the network [9], [10], [11]. The calculation of gradients of the criterion is given by :

$$\frac{\partial J^k}{\partial\theta_i} = \sum_{t=1}^{Nt} \left(Y^{(k+t)} - \hat{Y}^{(k+t)} \right) \cdot \frac{\partial \hat{Y}^{(k+t)}}{\partial \hat{Y}^{(k+t-1)}} \cdots \cdot \frac{\partial \hat{Y}^{(k+1)}}{\partial\theta_i}$$

In practice this calculus can not exceed a few number i.e Nt, due to time computation, and round-off errors. After several trials we select a Nt = 3 time window as a good compromise.

In the following figure we see how the modification of network at moment k, depends on the the contribution of the following Nt-1 copies retropropagated to k.

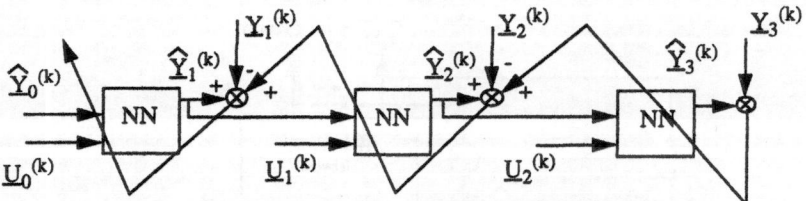

Figure 4 : Back Propagation Recurrent mode, Nt = 3

One essential point when propagating through the Nt copies of our network are the initial values of the $Y^{(k)}$ state vector . If we choose $\hat{Y}_0^{(k)} = 0, \forall (1 \le k \le Ncycles)$, then we observe an error three times superior than when initializing with $\hat{Y}_0^{(k)} = \hat{Y}_1^{(k-1)}, \forall (1 \le k \le Ncycles)$.
Then it is worth to keep the previous knowledge of the estimate .

4. Computer Simulation

We generate a filtered [0, 100 Hz], gaussian sequence with zero mean and covariance 1, as input velocity for our 2 models and create a training set of 2500 patterns and a test set of 500 patterns for the learning of our network.

As preprocessing the training data centered and reduced to optimize learning. It is shown in [12] that shifts in the input variables of the neurons introduce a preferred direction for weights changes, which slows down the learning. Data reduction enables us to equalize learning speed of input weights of the neurons. Indeed, the speed at which the output of a particular weight varies with gradient descent is proportional to the covariance of the input. In our case, these operations make the final RMS modelling error to be divided by two.

The input vectors we select, as a convenient choice to represent our two friction laws were : $(v_k, v_{k-1}, \Gamma_{k-1})$ for the Dahl model, $(v_k, v_{k-1}, \Gamma_{k-1}, \Gamma_{k-2})$ for the Reset-Integrator model. We select them after assuming that friction can be represented by a NARMAX (Non-linear Auto-Regressive Moving Average with eXogenous input) description as given in [13]. The selection is made by using a compromise between the numbers of inputs and the final validation RMS error.

Our network included of 10 neurons in hidden layer, and a single one in output layer.

We use then, the two schemes described above, and train the networks over a 2000 learning cycles, with a Nt = 3. With the following results for the validation errors :

RMS error	Serie-Parallel	Parallel
Dahl model	$1.14\ 10^{-3}$	$1.78\ 10^{-3}$
Reset integrator model	$7.18\ 10^{-3}$	$8.58\ 10^{-3}$

Here we present the simulation over the test set , for the two friction laws and their remaining approximation error in the parallel case as given in Figure 5.

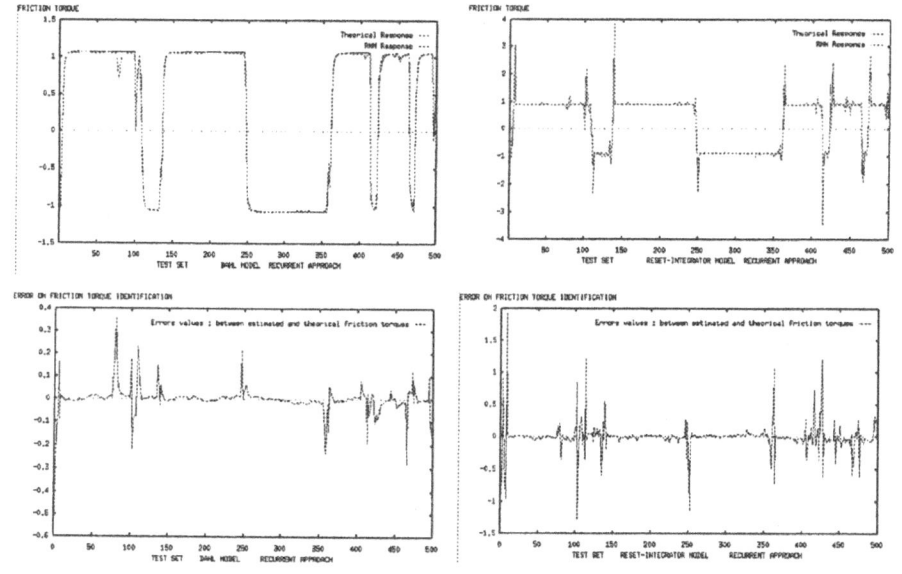

Figure 4

We can see that there is no sensible loss of performances between the parallel and serie-parallel approach, in both models. The differences in the two friction laws approximation, are mainly due to the more complex definition of the reset-integrator model, which have an extra torque on low velocities.

The ratio : RMS error / average |torque|, is here of 5% for parallel Dahl approximation, and 22% for parallel Reset-Integrator approximation. Classical approximations with a Coulomb model only give a 40-50% performances.

5. Conclusion

We have shown here that Recurrent Neural Network are able to modelize and identify a set of dynamics friction laws. This is possible due to their ability to represent internal state variables. Their use is particularly interesting when no physical model can be generated.

To achieve a small RMS error, preprocessing data and a well-chosed initialisation method were used.

Next step in our work will be to introduce the network in a closed mechanical system. In order to identify and to compensate the undesirable torque friction for an accurate tracking device.

APPENDICE

Parameters for the friction simulations :
Dahl : $\gamma = 10^6$, $f_0 = 0.035$ N
Reset-Integrator : K = 1 N/m, $p_0 = 10^{-4}$m, $\beta = 3.0$ N/(m/s), a = 0.25

REFERENCES

[1] J. Gilbart, G. Winston (1974) : «Adaptative Compensation for an Optical Tracking Teles cope» Automatica Vol 10, pp 125-131

[2] C.Walrath (1984) : «Adaptative Bearing Friction Compensation Based on Recent Knowledge of Dynamic Friction», Automatica Vol 20, N°6, pp717-727

[3] «Compensation de frottements secs sur un système de stabilisation 3 axes » SFIM Industries, Internal Report 1992.

[4] R.Murray-Smith, D.Neumerkel (1992) : « Neural Networks for Modelling and Control of a Non-Linear Dynamic System» Proc. of the 1992 IEEE Int. Symp. on Intelligent Control, pp404-409.

[5] P.R.Dahl (1968) : «A solid friction model» AFO 4695-67-C-0158, Aeropace Corporation, El Segundo, CA

[6] D.Haessig,B.Friedland (1991) : «On the Modeling and Simulation of Friction» Trans. of ASME Jrnl of Dyn. Syst., Meas. and Contr. Vol 113, pp-362.

[7] K.S.Narendra, K.Parthasarathy (1990) : «Identification and Control of Dynamical Systems Using Neural Networks» IEEE Trans. on Neural Networks, March 1990, Vol 1, Nb 1, pp 4-27.

[8] O.Nerrand, P.Roussel-Ragot, D.Urbani, L.Personnaz, G.Dreyfus (1994) : «Training Recurrent Neural Network : Why and How ? An illustration in Dynamical Process Modeling» IEEE Trans. on Neural Networks, March 1994, Vol 5 , Nb 2, pp 178-185.

[9] D.E. Rumelhart, G.E.Hinton, R.J.Williams (1986) : « Learning internal representations by error propagation» Parallel Distributed Processing: Explorations in the Microstructures of Cognition, Vol. 1, MIT Press.

[10] P.J. Werbos (1990) : «Backpropagation Through Time : What It Does and How to Do it» Proceedings of the IEEE, October 1990, vol 78, no 10, pp1550-1560.

[11] J.M. Martinez, C.Parey, M.Houkari (1992) : « Learning Optimal Control using Neural Networks» Proceedings of Neuro-Nimes, pp 431-441.

[12] Y. Le Cun (1994) : «Efficient Learning and Second Order Methods» Numerical Analysis Summer School, Neural Networks and Applications, June 13th-24th 1994 CEA, INRIA, EDF FRANCE

[13] Leontaritis, Billings, (1985) : «Input-output parametric models for non-linear systems », Int. J. Control, vol 41, n°2, 303-328

Solving an End-Effector Positioning Problem
by Hopfield Neural Network

S.Cavalieri, M.Martini

Universita' di Catania, Facolta' di Ingegneria
Istituto di Informatica e Telecomunicazioni
Viale A.Doria, 6 95125 Catania (ITALY)
email:cavalieri@iit.unict.it, fax:+39 95 338280, tel: +39 95 339449

Abstract.
The paper proposes application of a Hopfield network to optimization of the movement of a 3R planar robot. More specifically, the network is used to solve a typically complex problem from the computational point of view - determination of the positions of the robot along a certain trajectory - in such a way as to minimize the final end-effector positioning error. The paper illustrates the methodology followed to solve this problem and discusses the results that can be obtained by using the neural solution proposed.

1.Introduction.

Neural optimization networks such as Hopfield, Boltzman, etc. networks [1][2] offer evident advantages in solving problems which feature a high level of computational complexity. There are certain categories of problems which have no solution other than trying all the possibilities. As such an exhaustive search is generally impractical, heuristic methods are applied to find solutions that are acceptable, if not optimal. Neural optimization networks represent an excellent alternative to traditional heuristic methods as they allow an optimal or quasi-optimal solution to be reached in finite periods of time. This paper proposes application of a Hopfield network [3][4][5] to the problem of optimizing the movement of a 3R planar robot [6]. The mechanical characteristics of the actuators in the joints, the accuracy of the angle position sensors, and dimensional errors in the mechanical elements which make up the end-effector the robot is equipped with, all cause differences between the theoretical behaviour of the robot and its actual behaviour. In particular, the placing of the end-effector in a certain final position is greatly affected by these factors, causing errors which are unacceptable for some applications. The three degrees of freedom the robot is endowed with make it possible to choose the position of the arms while in movement so as to minimize the error in positioning the end-effector. Choosing the position of the arms is a typically onerous problem from a computational viewpoint, and involves a number of steps.

Once the trajectory the robot has to follow has been established, all the positions it could possibly occupy along this trajectory are identified. Then the combinations of all these possible positions are explored until one corresponding to a minimal positioning error is found.

In this paper the authors deal with the problem of determining the positions of a robot along a certain trajectory by using a Hopfield network. This network has the interesting property of being able to find a solution (optimal or nearly) to any problem, provided that the conditions surrounding the problem are correctly established. As far as the problem at hand is concerned, the surrounding conditions are the robot's trajectory and the constraints relating to unacceptable positions along this trajectory. On the basis of these conditions, the network has to choose the angle positions, for each point in the trajectory, that minimize the final end-effector position error.

The network used is the continuous Hopfield network model [3], while the methodology the authors adopt to solve the problem is the one proposed in [4], which consists of constructing an energy function in the form given by the Liapunov function, starting from the surrounding conditions of the problem itself. On the basis of a comparison between the energy function constructed and the Liapunov function, it is possible to obtain the parameters of the neural network, i.e. the weights of the connections and external bias currents.

Below, after a brief overview of the problem of positioning error minimization, the method will be applied to the problem of end-effector positioning. The results obtainable with the neural solution proposed will then be illustrated and discussed.

2. A Brief Overview of the Problem of Positioning Error Minimization.

Fig.1 is a schematic representation of the robot in the space x,y,z. It comprises three joints, each of a length L_i (i=1,2,3), and can rotate around the z axis of an angle θ. The end-effector is forced to move in the plane x,y along the generic trajectory shown in the figure. The radial distance, ρ, of the end-effector is calculated from the origin of the axes. Fig.2 is a projection of the robot onto a plane orthogonal to x,y. As said above, the distance ρ marks the position of the end-effector, while a set of three angles $\beta = (\beta_1, \beta_2, \beta_3)$, refer to the relative position of the robot's joints.

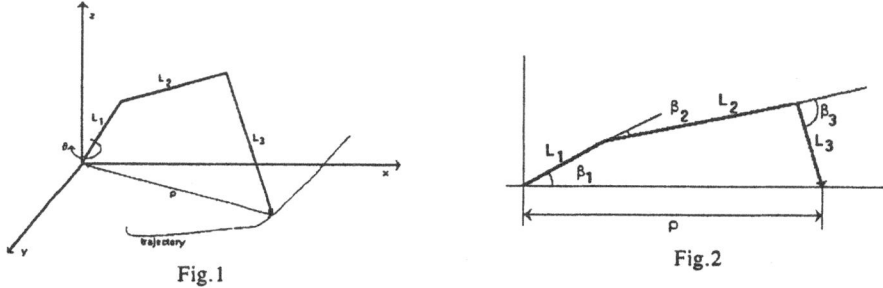

Fig.1   Fig.2

The position of the end-effector is described by the kinematic function:

$$\delta(\beta) = \frac{\rho}{\sum_i L_i}$$

where the radial distance, ρ, is given, with reference to Fig.2, by the following system of equations:

$$\begin{cases} \rho = L_1 \cdot \cos\beta_1 + L_2 \cdot \cos(\beta_1 + \beta_2) + L_3 \cdot \cos(\beta_1 + \beta_2 + \beta_3) \\ 0 = L_1 \cdot \sin\beta_1 + L_2 \cdot \sin(\beta_1 + \beta_2) + L_3 \cdot \sin(\beta_1 + \beta_2 + \beta_3) \end{cases}$$

If ϕ indicates the set of all the admissible values of the angles $\beta = (\beta_1, \beta_2, \beta_3)$, the robot's workspace is defined by the set:

$$\text{Workspace} = \{\delta | \delta = \delta(\beta), \beta \in \phi\}$$

Fig. 3 shows the possible state space of the robot, with a geometric configuration of the arms of $L_1 = 1.1$ m, $L_2 = 0.9$ m, $L_3 = 1.0$ m. The abscissa refers to the position δ of the end-effector, while the ordinate is related to β_2.

The presence of errors $\Delta\beta_i$ in angle positioning, which depends on the quality of the actuators used, means that positioning of the end-effector will be affected by an error that can be calculated by:

$$e_p = \sqrt{\Delta\rho^2 + \Delta z^2}$$

where $\Delta\rho$ is the positioning error in the radial direction, and Δz represents the positioning error in the direction orthogonal to the plane x,y. In [7] the authors plotted a maximum end-effector positioning error, e_p, in the state space, obtained by randomly varying the error $\Delta\beta_i$ within realistically foreseeable extremes. Fig.3 also shows a map of the errors in the state space again for the geometric configuration of the arms of $L_1=1.1$ m, $L_2=0.9$ m, $L_3=1.0$ m. As can be seen, four regions are identified in which the ranges of positioning error are comprised in the intervals [0.0000-0.00010], [0.00010-0.00020], [0.00020-0.00030], [0.00030-1.33417].

According to the neural solution here proposed, the error map is given to the Hopfield network, so that it can determine the sequence of sets of angles $\beta = (\beta_1, \beta_2, \beta_3)$ in such a way as to minimize the overall positioning error along the whole trajectory.

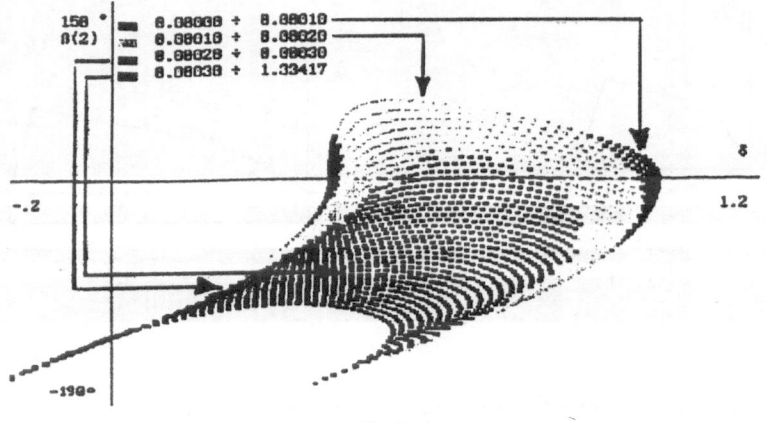

Fig.3

3. The Neural Approach to Error Minimization.

The trajectory the arm has to follow can be represented by the set of values δ, such that:

$$\text{Trajectory} = \{(\delta,\theta)|\delta = \delta(\beta), \beta \in \phi\}$$

The aim is to identify, for each value of δ along the trajectory, a set of three angles, $\beta = (\beta_1, \beta_2, \beta_3)$, such as to minimize the overall end-effector positioning error along the whole trajectory.

Let m indicate the number of values of δ which determine the trajectory to be followed, and k the number of all the possible sets of angles $\beta = (\beta_1, \beta_2, \beta_3)$ admissible for the m values of δ. This number, k, of sets can be logically decomposed into m groups. To each i-th (i=1..m) group g_i belong all the k_{gi} possible sets of angles $\beta = (\beta_1, \beta_2, \beta_3)$ which correspond to the i-th value of δ. Obviously we get $\sum_{i=1}^{m} k_{gi} = k$. On the basis of the hypothesis, the output neurons of the Hopfield network were logically subdivided into k groups of m neurons each. Henceforward we will identify each neuron with a double index, xi, (where the index x=1..k refers to the group and the index i=1..m refers to the neurons in each group), its output with OUT_{xi}, the weight for neurons x_i and y_j with $W_{xi,yj}$, and the external bias current for neuron x_i with I_{xi}. The variability of the index x=1..k can be subdivided into the subintervals $[1..k_{g1}],[1+k_{g1}..k_{g2}],[1+k_{g2}..k_{g3}], ...,[1+k_{gm-1}..k_{gm}]$. Each subinterval $[1+k_{gi-1}..k_{gi}]$ represents the group g_i referring to all the sets of angles $\beta = (\beta_1,\beta_2,\beta_3)$ corresponding to the value δ_i. According to this convention, the Liapunov energy function can be expressed by:

$$E = -\frac{1}{2} \bullet \sum_x \sum_i \sum_y \sum_j W_{xi,yj} \bullet OUT_{xi} \bullet OUT_{yj} - \sum_x \sum_i I_{xi} \bullet OUT_{xi} \qquad (1)$$

If the output of the generic neuron xi, OUT_{xi}, assumes a value of 1, it indicates that corresponding to δ_i in the trajectory being considered, the set of angles with the index x has to be considered in order to minimize the overall positioning error. Obviously, this set of angles with index x has to be in the interval $[1+k_{gi-1}..k_{gi}]$, i.e., it has to belong to the group g_i. If it did not, the neural network would have chosen an inadmissible set of angles corresponding to the particular value of δ_i. The matrix shown in Fig. 4 provides a clearer representation of the output state.

	δ_1	δ_2	δ_3	δ_4
$\beta_1 (\delta_1)$	1	0	0	0
$\beta_2 (\delta_1)$	0	0	0	0
$\beta_3 (\delta_1)$	0	0	0	0
$\beta_4 (\delta_2)$	0	0	0	0
$\beta_5 (\delta_2)$	0	1	0	0
$\beta_6 (\delta_3)$	0	0	0	0
$\beta_7 (\delta_3)$	0	0	1	0
$\beta_8 (\delta_4)$	0	0	0	0
$\beta_9 (\delta_4)$	0	0	0	0
$\beta_{10} (\delta_4)$	0	0	0	1

Fig.4

The example shown refers to quite a simple scenario: it was assumed that the path is identified by only four values of δ ($\delta_1,\delta_2,\delta_3,\delta_4$). Each of these values corresponds to a very limited number of β sets. More specifically, it was assumed that group g_1 (referring to position δ_1) comprised the sets β_1,β_2,β_3, group g_2 (referring to position δ_2) the sets β_4,β_5, group g_3 (referring to position δ_3) the sets β_6,β_7, and, finally, group g_4 (referring to position δ_4) the sets $\beta_8,\beta_9,\beta_{10}$.

Each column in the matrix refers to a δ_i (i=1..4, in the example), whereas each row refers to the set β_x belonging to the group g_i.

Fig.4 also shows an example of a correct neural network output. As can be seen, columns 1, 2, 3 and 4 of the matrix have a 1 corresponding to rows 1, 5, 7 and 10 respectively. This corresponds to identification of the set β_1 for δ_1, β_5 for δ_2, β_7 for δ_3 and β_{10} for δ_4.

For the neural network to be able to provide correct solutions, the conditions surrounding the problem have to be defined. Below we will describe these surrounding conditions and the relative energy function terms.

As said above, each group, g_i, of sets of angles in the interval $[1+k_{gi-1}..k_{gi}]$ refers to the value δ_i. In each of these groups of sets there can only be one that corresponds to the value δ_i. In other words, in each group g_i, the total number of 1s present in the column for δ_i has to be unitary. This condition is expressed by the energy function term:

$$\frac{A}{2} \bullet \sum_{x \in g_i} \sum_i \sum_{j \neq i} OUT_{xi} \bullet OUT_{xj}$$

No set belonging to the group g_i can correspond to a δ_j value where $j \neq i$. This means that each row in the output matrix can have at most one 1. The corresponding energy function term is:

$$\frac{B}{2} \bullet \sum_i \sum_x \sum_{y \neq x} OUT_{xi} \bullet OUT_{yi}$$

Each 1 in the output matrix identifies the β set which corresponds to the generic value δ_i. As the total number of δ_i values is m, the number of 1s in the output matrix has to be m. The corresponding term in the energy function is:

$$\frac{C}{2} \bullet (\ \sum_x \sum_i OUT_{xi} - m)^2$$

Another important surrounding condition concerns connectivity. Let us assume that β_x is the set of angles corresponding to δ_i and β_y the set corresponding to δ_j; we hypothesize that δ_i and δ_j are two consecutive values along the path. The sets β_x and β_y have to assume adjacent angle values. In order to establish the adjacency between all the possible sets of angles β, a matrix named Conn was built, comprising only 0s and 1s. It is a square matrix and both the rows and the columns correspond to the set β. A value of Conn[x,y]=0 (x,y=1..k) indicates that the sets β_x and β_y are adjacent and that δ_i and δ_j are two consecutive values along the path; a value of Conn[x,y]=Dxy indicates the opposite. Dxy assumes quite a high value so as to prevent choice of not adjacent sets of angles that correspond to consecutive δ_i distances. The surrounding condition referring to adjacency between the β sets can be expressed as follows:

$$\frac{D}{2} \bullet \sum_i \sum_{\substack{x \in g_i \\ y \in g_j}} \sum_{y \neq x} Conn[x,y] \bullet OUT_{xi} \bullet (OUT_{yi+1} + OUT_{yi-1})$$

As can be seen, the fact that the δ_i distances are consecutive is expressed by the condition $j=i+1$ or $j=i-1$. The condition of adjacency between the β sets is expressed by the matrix Conn.

The network may choose sets of angles which are adjacent but one or both of which may not belong to groups corresponding to consecutive values of δ_i and δ_j. It is therefore necessary to introduce a further term to prevent this possibility. This is achieved by using the term Dxy, through the following surrounding conditions:

$$\frac{D}{2} \bullet \sum_i \sum_x \sum_{\substack{y \neq x \\ x \in g_i, y \notin g_j}} D_{xy} \bullet OUT_{xi} \bullet (OUT_{yi+1} + OUT_{yi-1})$$

$$\frac{D}{2} \bullet \sum_i \sum_x \sum_{\substack{y \neq x \\ x \notin g_i, y \in g_j}} D_{xy} \bullet OUT_{xi} \bullet (OUT_{yi+1} + OUT_{yi-1})$$

$$\frac{D}{2} \bullet \sum_i \sum_x \sum_{\substack{y \neq x \\ x \notin g_i, y \notin g_j}} D_{xy} \bullet OUT_{xi} \bullet (OUT_{yi+1} + OUT_{yi-1})$$

which refer respectively to the various circumstances in which at least one of the sets chosen does not belong to the corresponding group, i.e.:

$$(x \in g_i \text{ and } y \notin g_j) \text{ or } (x \notin g_i \text{ and } y \in g_j) \text{ or } (x \notin g_i \text{ and } y \notin g_j)$$

The last surrounding condition refers to minimization of the error made in positioning the end-effector. As said, in a previous work the maximum error positioning was obtained for each value of δ and for different geometric configuration of the arms. In this work a matrix called Err contains these error for the geometric configuration considered: each single set β_x corresponds to a positioning error expressed by the matrix Err[x] (x = 1..k). The surrounding condition which minimizes the overall error is as follows:

$$\frac{E}{2} \bullet \sum_i \sum_x \sum_{y \neq x} (Err[x] + Err[y]) \bullet OUT_{xi} \bullet (OUT_{yi+1} + OUT_{yi-1})$$

From a comparison between the Liapunov function expressed by (1) and the energy function constructed by summing all the terms seen previously, the weights of the neural network were determined according to the optimization problem being considered. They are expressed by:

$$
w_{xi,yj} = \begin{cases}
\text{if } (i \neq j) \text{ and } x,y:((x,y \notin g_i) \text{ or } (x,y \notin g_j)) \text{ then } - = A \\
\text{if } x \neq y \text{ and } i = j \text{ then } - = B \\
\forall\ x,i,y,j\ -\ = C \\
\text{if } x \neq y \text{ and}((j = i \dot{+}1) \text{ or } (j = i - 1)) \text{ then} \\
\qquad \text{if}(x \in g_i \text{ and } y \in g_j) \text{ then } - = D \cdot Conn[x,y] \text{ else } - = D \cdot D_{xy} \\
\text{if } x \neq y \text{ and } ((j = i +1) \text{ or } (j = i - 1)) \text{ then } - = E \cdot (Err[x] + Err[y])
\end{cases}
$$

4. An Evaluation of the Neural Network Solution.

As we said in the introduction, the optimization problem being dealt with in this paper is of the NP Hard type, that is, it is one of a class of problems to which there is no solution other than trying all the possibilities. With reference to the specific problem at hand, an exhaustive search would involve all the possible sequences of β sets in order to find the sequence that minimizes the overall positioning error along the whole trajectory. Because such a search is generally impractical, heuristic methods have been applied to find solutions that are acceptable, if not optimal. The solution using a Hopfield recurrent network is typical of this: solutions are rapidly provided by the network, but there is no guarantee that they will be optimal. The aim of this section is to evaluate the validity of these solutions. To do this, we examined the state space and error map shown in Fig. 3 and considered different end-effector trajectories. For each of them the Hopfield network provided sequences of sets of angles. The tests were performed using an Anza Plus [8] neural simulator. Figs.5.a,b,c,d show the overall error for the sequences of sets identified for four different paths. As can be seen, the network makes choices that minimize the end-effector positioning error, which is always confined within a range of minimal errors of [0.00000-0.00010] and [0.00010-0.00020].

Final Remarks.

The paper has dealt with an NP Hard problem - that of minimizing the positioning error for the end-effector of a robot equipped with three degrees of freedom. The authors have given a mathematical definition of the conditions surrounding the problem, which are necessary to determine the energy function to be minimized. In order to evaluate the validity of the approach, some of the most significant experimental results obtained have been illustrated. They show that the Hopfield network provides solutions with extremely low positioning errors. This represents a further demonstration of the effectiveness of the network as a valid alternative to classical heuristic methods to solve real problems featuring non-deterministic polynomial complexity.

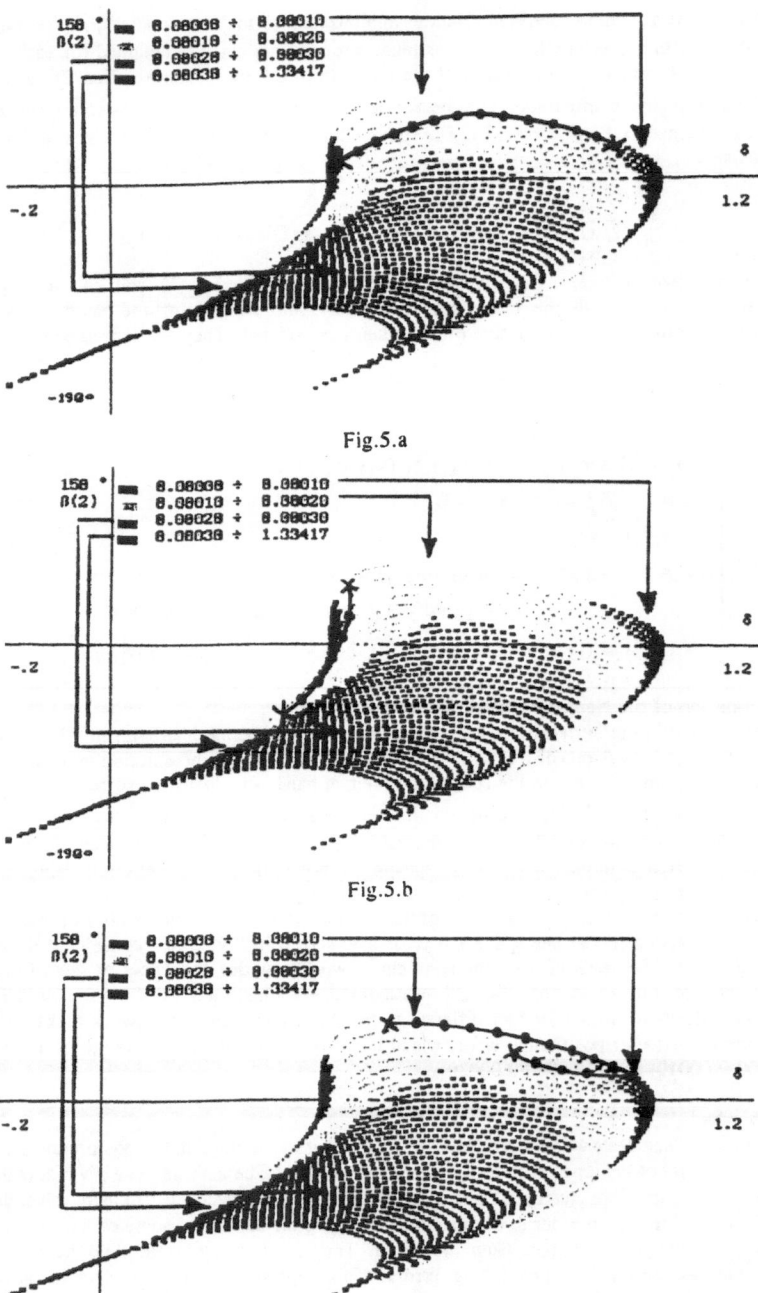

Fig.5.a

Fig.5.b

Fig.5.c

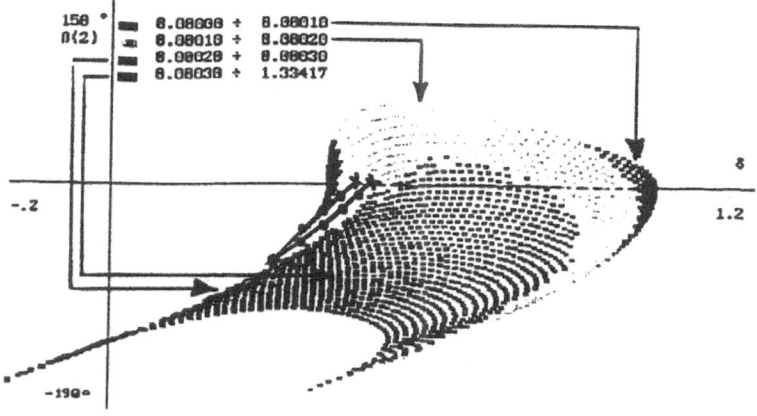

Fig.5.d

References.

[1] P.D.Wasserman, "Neural Computing - Theory and Practice". Van Nostrand Reinhold Editor, pp.106-109.

[2] R.P.Lippmann, "An Introduction to Computing with Neural Nets", IEEE ASSP Magazine, April 1987, pp.2-22.

[3] J.J.Hopfield, "Neurons with Graded Response Have Collective Computational Properties Like those of two-state Neurons", proceedings National Academy of Sciences 81:3088-3092, May 1984.

[4] J.J.Hopfield, D.W.Tank, "Neural Computation of Decision in Optimization Problem", Biological Cybernetics, vol.52, pp.141- 152, July 1985.

[5] J.J.Hopfield, "Neural Networks and Physical Systems with Emergent Collective Computational Abilities", proceedings National Academy of Sciences 79:2554-2558, April 1982.

[6] S.Cavalieri, M.Martini, F.Petrone, R.Sinatra, "A Neural Network Approach for Position Error Minimization Problem in Redundant Robot", Ninth World Congress on the Theory of Machines and Mechanisms, Politecnic of Milan, Italy, August 30-September 2, 1995.

[7] V.Marchis, F.Petrone, R.Sinatra, "Analisi dello Spazio di Lavoro di Robot Ridondanti", XI National Congress AIMETA, 28 September-2 October, 1992, Trento, Italy.

[8] Anza Plus User's Guide and Neurosoftware Documents Release 2.2 15 May, 1989.

Visuomotor Control Using an Artificial Neural Network

P.F. McGuire and G.M.T. D'Eleuterio

Institute for Aerospace Studies
University of Toronto
Downsview, Ontario, Canada M3H 5T6
Phone: (416) 667-7731, Fax: (416) 667-7799
e-mail: mcguire@utias.utoronto.ca, gde@utias.utoronto.ca

Abstract

An approach to visuomotor control using an artificial neural network (ANN) is presented. The architecture of the controller is founded on the Cerebellar Model Arithmetic Computer (CMAC) and consists of two principal elements: the Image CMAC (ICMAC), which provides a visual representation of a target object, and a Differential Image CMAC (DICMAC), which supplies rate information on the object. The approach differs from the conventional use of artificial vision in control in that the visual images are not merely reduced to position and orientation of the object but rather are employed integrally. Thus the controller would be able to generalize across a set of images and their corresponding objects. Learning is accomplished by means of a reinforcement scheme. Computer simulation results are presented for an inverted pendulum system.

1 Introduction

Many researchers have attacked the problem of visuomotor control using artificial neural networks (ANNs), (Miller (1987b), Martinetz *et al.* (1990), Kuperstein (1990), Hashimoto (1992)) but these have reduced the vision component to merely identifying targets or extracting parameters via task-specific preprocessing of the visual image. While this approach may work in isolated instances, each application requires redesign and/or recalibration before the ANN can begin to learn. When targets are used, the issue of general applicability is raised; occultation of targets, specific lighting conditions, and camera focal parameters are all factors to be considered.

One exception known to this author is the ALVIN project (Pomerleau (1993)). This project uses a back-propagation network that operates on low resolution images directly, letting each pixel represent a neuronal input. The goal is to develop a system that can effectively steer a vehicle through a variety of road conditions. Supervised training is provided by human drivers while the network associates the video input directly with steering commands. The system performs well given appropriate training and some bootstrapping techniques.

The ANN presented here, based on Albus's (1981) Cerebellar Modeled Arithmetic Controller (CMAC), achieves similar video-to-control mapping. Owing to its simple form, the Image-CMAC (ICMAC) processes images quickly, allowing high-resolution images to be processed in real time. While the ICMAC does not interpret images by recognizing objects, it provides a basis for development of ANNs that do. This aspect, however, is not discussed here.

The ICMAC structure maps greyscale pixel images to memory cell locations such that 'similar' images are mapped to overlapping sets of memory locations, and 'distinct' images are mapped to nonoverlapping sets. The term 'similar' in this context implies that for equivalent lighting conditions, contents of the image may only change by small amounts in terms of spacial location, orientation, and scale. Computer simulation utilizing a reinforcement learning scheme in combination with the ICMAC demonstrates the ability to control dynamic systems using direct visual cues instead of reduced position and orientation data.

2 The ICMAC Design

2.1 Coarse Coding

Let us begin our description of the ICMAC by relating it to a biological mechanism (see Figure 1). In nature it is often found that precision measurements are not made by precision sensors, but rather through the combination of many poor resolution sensors. One example of such a system may be found in the proprioceptive feedback detector network that allows the brain to control motor functions. No one single sensor cell is capable of encoding a sufficient estimate of muscle contraction (joint angle), but the unique stimulation of many of these cells can adequately encode this information for use by the brain. This recovery of precise data from multiple poor resolution sources is known as coarse coding.

Coarse coding was used as a basis in the development of Albus's (1981) Cerebellar Modeled Arithmetic Computer (CMAC), which has since been used in a variety of learning controller applications. The CMAC is basically a multi-dimensional coarse coding algorithm based on the cerebellar encoding properties of higher primates, although examples of coarse coding may be found throughout nature. For more information on CMAC and it's applications see Albus (1981), Miller *et al.* (1987a), Graham & D'Eleuterio (1991), and McGuire & D'Eleuterio (1992).

The ICMAC utilizes the coarse-coding principle in reducing image data. An image is broken down into regular grids of large 'paxels' or 'coarse cells' (see Figure 2). Each of the grids are shifted by one pixel in both the x and y directions. For the purposes of this paper we will assign each paxel a value that corresponds to the average greyscale value of the pixels that fall within its 'receptive field' (the pixels falling directly above the cell). Although the averaging process destroys detailed

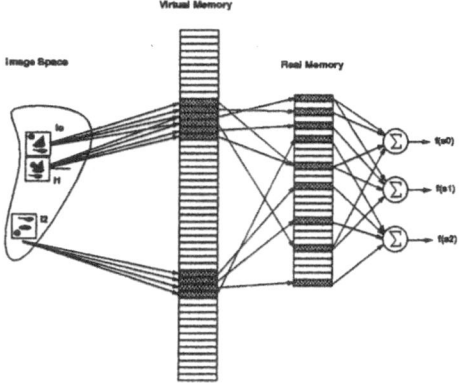

Figure 1: Mapping similar images to overlapping groups of memory cells.

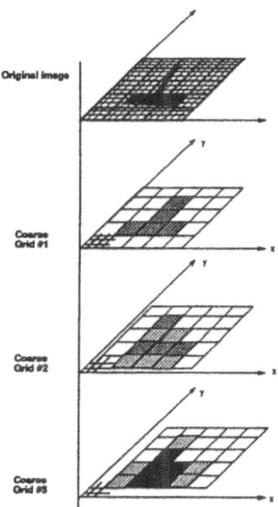

Figure 2: An illustrative example of coarse coding images with averaging paxels. Each image has a unique representation provided there are enough grids. Grayness of coarse cells represent level of activation, with darker cells representing a higher level of activation.

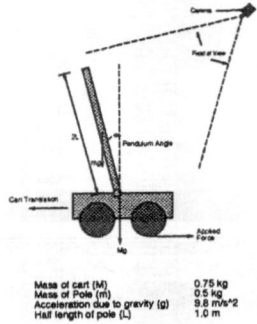

Figure 3: Inverted pendulum dynamic system with external camera monitoring the progress

information at each grid layer, every unique image can be differentiated through the unique coarse cell activations provided there are a sufficient number of grids.

At this point, we cannot easily extract or classify images in a sensible fashion since the information is distributed over the activation of the coarse cells in each grid. We must distill this coarse-coded data into a tractable form for our particular application. For visuomotor control, where the images consist of a visual representation of a dynamic system in various states of activity, we would like to group together images that show the system in similar states of activity.

Consider an inverted pendulum system with the camera viewing the system in action from the outside (see Figure 3). The state of the system (given by the pendulum angle and rate) is similar when the pole has moved by only a small amount (also when the rate has changed little, as will be discussed later). What we require here is a method of grouping such that small changes in the image group together. One way of doing this is to define a set of higher-level computational cells, analogous to the method by which the brain interprets the proprioceptive sensors, that are activated when specific images appear. If properly organized, we can build a structure that activates one set of cells for each image with the constraint that 'similar' images map to intersecting sets of cells.

The desired mapping can be achieved yet again by coarse coding, this time dividing up the greyscale dimension. This not only helps achieve our goal, but will create a more robust mapping. Each of the cells in each of the coarse grids will be split up according to greyscale. In essence we will be creating a new subgrid of coarse cells, one for each range of greyscale discriminated. In each subgrid, cells will be shifted by both x and y position and greyscale (see Figure 4), thus each image will map to a unique set of cells provided that enough greyscales are discriminated.

Let us say that the coarse grids divide the image evenly into Q divisions along both the x and y axis (the regular division is not a requirement but is convenient for computational purposes). Let us also say that there are k grids, then if we shift each grid by one pixel, we will get a unique set of coarse cells for each of the $N \times N$ pixels in the original image when $kQ = N$.

Now suppose that each grey scale axis in each subgrid complex is divided into Q_g and that these cells are shifted in a regular fashion from grid to grid. If each pixel can take on any of N_g distinct grey values then each paxel must take on any of $N_g(N/Q)^2$ distinct values in order to retain all information. We will obtain a unique representation for every possible image if a sufficient number of subgrids allowing the greyscales to be fully represented, that is if:

$$kQ_g = N_g(N/Q)^2 \tag{1}$$

But we also have the requirement $kQ = N$, which implies:

$$kQ_g = k^2 N_g \tag{2}$$

Thus

$$Q_g = kN_g \tag{3}$$

This means that we will have a space defined by:

$$kQ^2Q_g = k(N/k)^2 kN_g = N^2N_g \tag{4}$$

which is exactly the size of the original space. Thus if we wish to retain all information we will save nothing in terms of memory requirements. Fortunately, due to the way the information is now structured, we can selectively ignore certain aspects in creating the mapping we seek.

If we look at two images that are similar in scale, translation, and rotation, we see that the differences occur in local regions and that the rest of the image remains unaltered (see Figure 5). Our coarse coding method will select one set of cells for each image. Since the change occurs locally, most of the cells will be the same, but in the regions where the change does occur the local greyscale averages will be different, hence some of the subgrid cells that are activated will be different. By reducing the value of of Q_g, ie. reducing the number of grey scale distinctions, we can control how many cells will be different between the two images as each subgrid coarse cell will activated by a wider range of grey scale values. This will also reduce the sensitivity of the coding scheme to random pixel noise that occurs in every practical application. In essence, Q_g defines the closeness of two images under this encoding scheme.

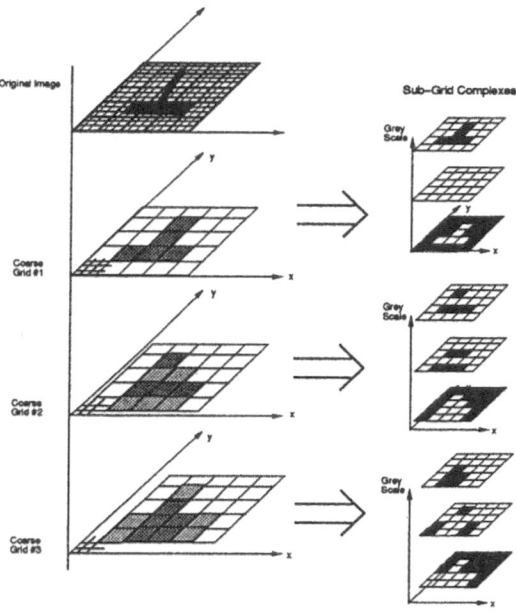

Figure 4: Coarse grids are divided into a number of subgrids that coarse code greyscale cell activations.

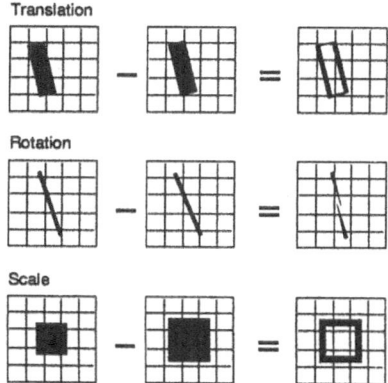

Figure 5: Differences in similar images with respect to rotation, translation, and scale changes. The differences occur in local regions of the image. The divisions shown are coarse cells before averaging.

Each image is represented by the activation of kQ^2 cells. If $kQ = N$ it is conceivable that we will be talking about a very large number of cells for even moderate sized images. We can reduce this number by mapping each of the subgrids to a higher level cell, or even by mapping each grid to a single cell according to the activations contained within these structures. If a more complex generalization is desired then subgrid cross comparison may be used to extract higher order characteristics, but that issue is beyond the scope of this paper.

Once this cell reduction has been performed, we have distilled the image down to a set of only k cells. Note that each step of combining cells incorporates global changes into the mapping process, thus a change anywhere in the image can affect the selection of a high level cell. Fortunately, the mechanism for detecting changes locally is still in place, and the desired generalization is still intact. The effect of combining cells becomes apparent when the background changes, since a larger percentage of cells will be affected even when the intended difference between two images is small. This then limits the ICMAC to situations with constant or predictable backgrounds. While this is a significant limitation, the ICMAC is still applicable to a wide range of tasks. On-going research into more intelligent strategies for combining cells will hopefully resolve this problem.

3 Generalization and Parameter Effects

There are a number of design parameters that must be chosen when employing the ICMAC. Suitable parameters depend on the application, fortunately the algorithm is robust with respect to changes in these parameters. This robustness implies that the designer need not spend extraordinary amounts of time searching for the 'workable' set of parameters, since performance will not vary significantly over large ranges.

The most important parameters are the variables k, Q, and Q_g, which define the number of grids, the number of coarse cells per dimension in the grids, and the number of subgrids respectively. These variables affect the generalizing property of the ICMAC with respect to rotation, translation, and relative size of the objects depicted in the input image.

Throughout this paper we will insure that each pixel in the input image is represented by a unique set of coarse cells, so that:

$$kQ = N \tag{5}$$

Whereas the value of Q_g will be chosen so as to neglect greyscale information as discussed earlier.

Also important in the design is the size of the hashing memory. For applications involving control of simple dynamic systems, the hashing memory can be chosen much larger than is required, thus insuring 'collision' effects (different images mapping to the same set of memory locations in a random fashion) to be negligible. In more complex tasks, where a significant portion of the image space is used, further steps must be taken to insure that hashing collisions do not hinder performance. We will not be considering these tasks here as it is not clear that the ICMAC would be applicable.

Finally, one must consider the image size (N) and grey scale resolution (N_g). While often these will be set by the hardware of the task, software transformations may always be used to set N and N_g. Typically, images will be reduced by averaging pixels and combining greyscales, but caution must be used here for important information may be lost through absentmindedly chopping data which appears negligible (see section 3.1). Since this software reduction costs computational time, it may be easier to let the ICMAC process the raw image with parameters set to ignore data. This processing guarantees that the data reduction is consistent with the ICMAC input requirements.

3.1 Generating Images

As mentioned above, one must be careful when performing operations on greyscale images. For the purposes of this paper we will be generating images based on internally computed state variables characterizing a dynamic system. Since we are working with finite resolution images, it is imperative that we generate the images in the same fashion that an actual CCD camera would capture them. For instance, Figure 7 shows an example where a boundary falls part way between two pixel edges. In this case, the half covered pixels will register half the grayness that they would otherwise register had the object completely covered the pixels. When generating images it is important not to limit the movement of objects to only whole pixel jumps, but to allow objects to traverse their domain smoothly. Thus objects will often have a 'soft' or 'fuzzy' border. This characteristic of digital imaging has been used to determine that locations of targets with sub-pixel accuracy, and is quite important to the mapping process of the ICMAC. One can easily imagine how a whole edge of half intensity greyscale pixels can affect the average of a coarse cell.

3.2 The Task

The dynamic system that we will be considering will be that of the inverted pendulum of Figure 3. In order to simplify the task we will ignore translation of the cart and concentrate on controlling angle of the pole. The camera is affixed to the cart, looking up the length of the pole at a target object. For convenience we use a translating square to represent the target object, although any shape(s) could be used.

3.3 DICMAC: Extracting Differences from Images

It is a well known result from control theory that both the pendulum angle φ and the rate $\dot{\varphi}$ must be measured in order to stabilize the inverted pendulum plant with a constant gains feedback controller. Thus it is imperative that we take into consideration velocity information represented in the video images.

This can be accomplished with a slight modification to the ICMAC. We simply coarse code the difference between two images by performing a pixel-wise subtraction. As a result of the subtraction process, the range of coarse cell values changes, but this is corrected by a simple scaling operation so that the original ICMAC mapping process can be reused. This modified ICMAC will be henceforth referred to as a DICMAC (Difference Imaging Cerebellar Modeled Arithmetic Computer).

Actual Object superimposed on a pixel grid.   **Pixel Representation of object.**

Figure 6: Typical digital imaging process. Pixels that contain a sharp boundary end up registering only a portion of the actual the greyscale over the whole pixel. This effect can be countered by averaging surrounding pixels to determine the exact location of the edge.

The result of subtracting two successive images creates a new image that highlights the changes in the system over short periods of time. The DICMAC operating on these new images groups similar changes together when the rest of the state variables are the same. When the ICMAC is used in conjunction with the DICMAC, an unique set of memory cell pointers are generated for every unique combination of φ and $\dot{\varphi}$, that is, for every state.

3.4   Reinforcement Learning

Consider a task with real valued states at discrete time intervals (x_t), and a controller generating continuous valued actions (a_t) performed by the controller. While the plant variables are real values, the generalization and the discretization of the CMAC allow us to reduce the problem to that of a deterministic finite-state Markovian process with unknown transition probabilities. Note that we are making an assumption about the behavior of the plant when we describe the process as deterministic, that is, at state x_t, performing action a_t will always bring the plant to the same next state x_{t+1}. This assumption ignores the effects of noise and external perturbations, which are always present to some degree in a real system.

We will be considering a task in which the goal is to bring the plant to a desired state (x_d) in minimum time and keep it there for the remainder of an allotted run interval. In addition, the reinforcement learning scheme to be described operates on a delayed-consequences reward system. Rewards are given only after the allotted run interval has elapsed, based on the performance of the controller during the interval. The reinforcement evaluation signal z_i, (where ()$_i$ indicates the interval number), will be taken as a single scalar number where the greater the value the stronger the reinforcement intended. The only requirement on the reinforcement signal is that it be ordered, that is, it rates 'better' performances with larger signals. The more information the reinforcement signal contains, the faster the reinforcement technique can discover an appropriate solution. In a sense, the reinforcement signal is used to direct the exploration, thus reducing the amount of undirected 'random' searching that needs to be done.

In contrast to supervised learning where the task is to determine input-output relationships based on labeled data sets, reinforcement learning is based on a punishment/reward system. Recent development of reinforcement learning schemes using ANNs have led to a number of learning controllers for dynamic systems. Most notable in this category are the developments of Barto, Sutton, and Anderson's (1983) Adaptive Heuristic Critic, Watkin's (1989,1992) Q-learning, and the formal Temporal difference TD(λ) family of reinforcement learners detailed by Sutton (1988).

The reinforcement learner utilized here is most closely related to Q-learning and TD(0) but uses a somewhat unique search technique. In essence, an external reinforcement signal provided by the environment is used to rank the performance of the controller after a task is attempted. Based on this signal, the reinforcement algorithm modifies the parameters of the ANN (contents of memory cells in the case of the ICMAC) and the task is attempted again. Repeated trials result in overall improvement of the controller.

The algorithm proceeds as follows: There are 2 ICMAC modules used, one (ICMACz) for predicting the end of interval reinforcement signal at each time step (z_i), and the other (ICMACμ) for determining the action to be taken at each time step (a_t). The input to ICMACz consists of an image depicting the plant in its current state and a bar graph at the edge of the image that indicates the value of the action to be performed by the controller. The input to ICMACμ is simply an image of the plant in its current state. Note that an ICMAC module comprises one ICMAC and one DICMAC operating on the same input, the output of the module is the average of the two outputs.

At each time step M actions $a_t(1), a_t(2), ..., a_t(M)$ are selected by using a normal random number generator with fixed standard variation σ_t and a mean μ_t computed by ICMACμ. These actions are then displayed in the input of ICMACz and used to compute M predictions of the final reward $\hat{z}_t(1), \hat{z}_t(2), ..., \hat{z}_t(M)$. The action that produces the highest predicted final reward is then chosen as the action for that time step a_t. This action is then applied to the plant which after a time interval has evolved to x_{t+1}. The procedure is then repeated, while the action a_t, and maximal prediction $\hat{z}_t = max\{\hat{z}_t(1), \hat{z}_t(2), ..., \hat{z}_t(M)\}$ are recorded for training later.

By looking at M different options at each step we are balancing exploitation with exploration. Small values of M mean that we are choosing actions with less regard for what the predictor ICMACz is telling us thus performing random actions, whereas if M is large, we are taking the advice of ICMACz serious and looking for the optimal actions at each time step. If the predictor is accurate then a large value of M is desirable, since pure exploitation may be evoked, on the other hand, if the predictor is inaccurate, a small value of M is necessary to effect exploration. Since we are assuming that the mean generator

ICMACμ learns optimal actions as the predictor improves, a large value of M will never be required as the exploration will be centered on areas known to bring about a favorable evaluation. This condition on ICMACμ, which is enforced during learning, may limit the accuracy of the predictor in regions of poor evaluations, hence the classic trade off between exploration and exploitation.

In addition to M, σ affects the exploration. Typically, we take the value of σ to be constant while M is large, but when ICMACz begins to predict accurately, σ can be reduced towards zero assuming that ICMACμ has been trained appropriately.

At the end of each run interval, both ICMACs are trained. Training ICMACz is a straightforward matter, but ICMACμ must be trained selectively so as to ensure that the selected actions continually improve. ICMACz is trained at each state accessed during the last run interval via the 'delta rule' to recall the final reward signal z_i:

$$\Delta w_{active_{i,t}} = \frac{\beta}{k}(z_i - \hat{z}_t) \qquad \forall\, t \in interval\ i \tag{6}$$

Where the learning parameter β is chosen from the range $\beta \in (0,1]$, and $\Delta w_{active_{i,t}}$ refers to the change in the values stored at all memory locations accessed at time step t during run interval i.

ICMACμ is likewise trained using the delta rule, but is only trained when the final reward z_i is greater than any reward yet achieved (z_{max}), and then only if the final reward is greater than or equal to the predicted reward $\hat{z}_t$. In this case, the delta rule becomes:

$$\Delta w_{active_{i,t}} = \frac{\beta}{k}(a_t - \mu_t) \qquad \forall\, t \in interval\ i \tag{7}$$

If $z_i \leq \hat{z}_t$, but is still larger than z_{max}, then the mean previously computed by CMACμ is re-learned.

The variable r_{max} can occasionally be corrupted by a lucky guess before full exploration has been completed, thus it is periodically reset to zero to insure complete learning. It is this variable that insures that ICMACμ is learning adequately during the exploration.

Initially, ICMACz is set to predict the median reinforcement value for every state-action pair (typically zero). The first run attempted will be a random attempt, and will likely fail (state variables leaving their allowable range). If it does fail, each of the state-action pairs accessed during that attempt will receive a negative final value prediction. During the next run, a random walk will again be performed, only this time the predictor values will steer the system away from any of the state-action pairs accessed during the first run since they will have a lower final value prediction than any other pair. As attempts are continued, the predictor will continue to identify poor state-action pairs, and drive the system towards those that show some success. All the while, CMACμ is monitoring the progress, learning only the best actions to perform at any given state. Eventually, ICMACμ will build up an action generating map that always leads to a success, and the predictor will be no longer necessary.

3.5 Computer Simulations

We now present results of a computer simulation of the the reinforcement algorithm using Image CMACs operating on images of the inverted pendulum system. The task interval was chosen to be three seconds long, and one thousand trails were performed. The images used were of poor quality (60×60 pixels, ie. $N=60$) and the size of the time steps were chosen to be 0.03 seconds.

The ICMAC parameters k, Q, Q_g were chosen to be 15, 4, and 2 respectively, with the learning parameter β set at 0.07 for ICMACmu training and 0.1 for ICMACz training. The size of the hash memory was taken to be much larger than required, thus the effects of hashing collisions were unimportant.

The reinforcement learner parameter M was set equal to 10 while the value of the random number generator was fixed at 0.25 throughout the simulation. The variable r_{max} was reset every 100 trials.

In our example presented here, the reinforcement signal is computed by scaling a sum squared error of position over the entire trajectory to the range $(0,1)$ with 0 being the worst possible performance, and 1 being the best. A negative valued reinforcement was computed if the system state variables left the predefined ranges of $\varphi \in (-28.6\,deg, 28.6\,deg)$ and $\dot{\varphi} \in (-171.9\,deg/s, 171.9\,deg/s)$. The reinforcement signal was set to -1 if the state variables left their defined ranges after the first time step. This punishment signal was scaled linearly with the number of time steps taken to failure. A failure at the last time step in the run interval would receive a zero reinforcement signal.

Figure 11 shows the performance of the controller at various stages of learning. The examples were chosen to reflect the characteristic stages in learning. In the first graph the controller allows the system state variables to leave the prespecified ranges, while in the second graph the controller has over-compensated for the problem. In the third graph, the controller demonstrates that the basic principle behind the task has been discovered, and in the final graph the performance has been fine-tuned. This result was found to be repeatable for many variations of the problem including random starting positions, reinforcement signal computation, and interval length. The final run in this simulation exhibited less than perfect performance. This result is believed to be caused by poor resolution in the generated images, and could easily be improved. The present goal, however, has been to demonstrate the applicability of the ICMAC to visuomotor control, and establish a proof-of-concept simulation. Future research involving a real-time experiment is planned, with update rates of up to 100 Hz anticipated.

4 Conclusions

This paper has served to introduce the Image CMAC and show how it may be used in the control of a dynamic system. The simulation results are not intended to reflect state-of-the-art performance, but rather demonstrate that such a neural network can indeed be used in control applications.

The approach taken to the task of visuomotor control using reinforcement learning is fairly novel and represents the general philosophy behind out research. Our approach is rooted in the belief that biological systems do not adhere to the strict functionalist view that have constituted traditional control formulation. Rather, we believe that the development of parallel

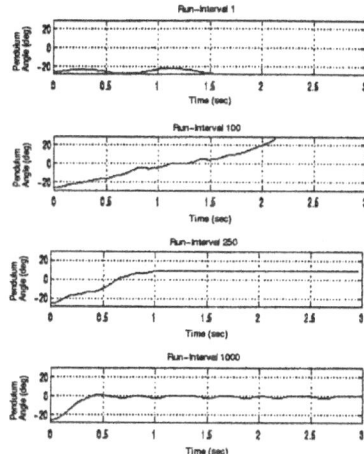

Figure 7: Performance of ICMAC-Reinforcement controller at various stages of learning the dynamics of an inverted pendulum system. The goal is to reduce the pendulum angle magnitude to zero in minimum time.

distributive computing approaches that tie all of the aspects of the problem into one whole will, if properly understood, can lead to much more versatile systems.

5 References

Albus J.S. (1981), *Brain, behavior and robotics*, Peterborough, N.H.: BYTE Books.

Barto A.G., Sutton R.S., Anderson C.W. (1983), Neuronlike elements that can solve difficult learning control problems. *IEEE Transactions on Systems, Man, and Cybernetics*, 13:835-846.

Graham D.P.W., D'Eleuterio G.M.T. (1991), MOVE-A Neural-Network Paradigm for Robotic Control, *Canadian Aeronautics and Space Journal*, V37, N1, pp 17-26.

Hashimoto H., Kubota T., Kudou M., Harashima F. (1992), Self-Organizing Visual Servo System Based on Neural Networks, *IEEE Control Systems*, April, pp 31-36.

Kuperstein M. (1990), INFANT Neural Controller for Adaptive Sensory-Motor Coordination, *Neural Networks*, V4, pp 131-145.

Martinetz T., Ritter H.J., Schulten K.J. (March 1990), Three-Dimensional Neural Net for Learning Visuomotor Coordination of a Robot Arm, *IEEE Transactions on Neural Networks*, V1, N1, pp 131-136.

McGuire P.F., D'Eleuterio G.M.T. (1992), Active Control of Interference in CMAC/MOVE Neural Networks for Robotic Applications, *Seventh CASI Conference on Astronautics*, Ottawa, ON, 4-6.

Miller W.T., Glanz F.H., Kraft L.G. (1987a), Application of a General Learning Algorithm to the Control of Robotic Manipulators, *The International Journal of Robotics Research*, V6, N2, pp 84-98.

Miller W.T. (1987b), Sensor based Control of Robotic Manipulators Using a general Learning Algorithm., *IEEE Journal of Robotics and Automation*, V3, N2, pp 157-165.

Pomerleau (1993), *Neural network perception or Mobile Robot Guidance*, Carnegie Mellon University: Kluwer Academic Publishers.

Sutton R.S. (1988), Learning to predict by methods of temporal difference, *Machine Learning*, V3, pp 9-44.

Watkins C.J.C.H. (1989), *Learning from Delayed Rewards*, PhD thesis, Kings College.

Watkins C.J.C.H. and Dayan P. (1992), Q-learning, *Machine Learning Journal*, 8(3/4), May 1992 Special Issue on Reinforcement Learning.

Learning the Visuomotor Coordination of a Mobile Robot by Using the Invertible Kohonen Map

C. Versino,* L.M. Gambardella

IDSIA, Corso Elvezia 36, 6900 Lugano,Switzerland

cristina@idsia.ch, luca@idsia.ch, http://www.idsia.ch

Abstract

In this paper we present an experiment of the *visuomotor coordination* of a simple mobile robot. Our approach to sensorimotor modelling belongs to the category of *learning by doing*. First, < *perception, action* > pairs are collected by observing the robot behavior during operation. Then, a learning by examples method is used to estimate the parameters of the model. In our experiment the learning machine is an *Extended Kohonen Map*. We have observed that this neural network model has the property of being naturally *invertible*. Given an input pattern, the network output value is retrieved by competition on the neuron fan-in weight vectors. This is the standard use of the input-output mapping that we call *forward* mode. Viceversa, given an output value, a corresponding input pattern can be obtained by competition on the neuron fan-out weight vectors. We call this use of the network *backward* mode. The invertibility property makes the Extended Kohonen Map worth considering for sensorimotor modeling. Our experiment shows that by training the network on the robot direct kinematics (the forward mode), one obtains at the same time a solution to the inverse kinematics problem (the backward mode). The experiment has been performed both in a computer simulation and by using a real robot.

1 Introduction

Robotics research devotes considerable attention to the study of sensorimotor coordination [BDM94, BGG93, M94, RMS91, ZGLC94]. This term refers to the association of signals coming from various sensory modalities to motor commands in view of a given task. A widely studied instance of sensorimotor association is the *visuomotor coordination* of a robot arm [RMS91, BGG93]. The task is to automatically position the end effector of a robot arm at an arbitrary target location (τ_1, τ_2, τ_3) in a 3D environment (Figure 1, left). The information about the location of the target is provided to the robot system by two cameras which observe the workspace. The robot arm moves when a valid set of angles (in Figure 1, $(\vartheta_1, \vartheta_2, \vartheta_3)$) is selected and transmitted to the joint motors. The end effector positioning task requires visuomotor coordination because one needs to relate the visual information about the target location, as it is perceived by the camera system, to the movements of the arm.

Studies on sensorimotor coordination deal with the *bidirectional association* existing between sensory perception and motor action. On one side, there is a transformation from the space of motor actions to the space of resulting sensory situations: this part is commonly referred to as the *forward* association. On the other side, one needs to know how to generate motor commands to reach a target sensory perception: this part is referred to as the *inverse* association. In the robot arm example, the forward relationship corresponds to the ability of predicting the position of the end effector when a set of joint angles is transmitted to the arm motors. On the other side, the inverse association maps positions in the workspace into joint angles configurations that allow the robot to reach these positions.

From a methodological point of view, the study of robot sensorimotor coordination (both forward and inverse) may be addressed in two different ways. The classical approach consists of assuming a *a priori* model of the interaction between sensors and actuators. By *a priori model*, we mean, for instance, a model which is built on the basis of physical laws. In this way, the forward association, which predicts the robot arm position for a joint configuration, can be expressed as a direct kinematics equation, while the end effector positioning task is viewed as the solution to an inverse kinematics equation. The alternative approach consists of building a *a posteriori* description of the interaction between sensors and actuators. By the term *a posteriori* description, we mean a model built on data which implicitly

*Supported by the No. 21-36365.92 project of the Fonds National de la Recherche Scientifique, Berne, Suisse.

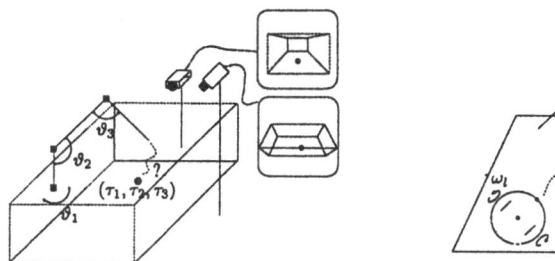

Figure 1: Experimental setup for the visuomotor coordination of a robot arm (left) and of a mobile robot (right).

describes the association between sensory perceptions and motor commands. First, $< perception, action >$ pairs are collected by observing the robot behavior during operation. Then, a *learning by examples* method is used to estimate the parameters of the model. Advocates of the *a posteriori* methodology stress the realism of their approach, in that no fixed model is assumed to be valid before experimenting with the real robot; on the contrary, the *a priori* model describes ideal interactions which are met in simulated environments only[1] [BDM94].

In this paper we study the *visuomotor coordination for a simple mobile robot by using the learning by examples scheme*. The task is to automatically guide a mobile robot to an arbitrary target location (τ_1, τ_2) in a 2D environment (Figure 1, right). As in the arm example, the target location with respect to the robot's current position is observed by a camera located above the workspace. The robot movement is controlled by choosing the angular velocities (ω_l, ω_r) for its left and right driving wheels. Hence, the task is to find a sequence of wheel velocities that would eventually bring the robot from any initial position in the workspace to any target location; in order to do that, the visual information provided by the camera is used. Although the main focus is put on a typical inverse task (the inverse kinematics of a mobile robot), we show that both forward and inverse associations can be addressed at the same time by an Artificial Neural Network (ANN) model, which has the property of being *invertible*.

The paper is organized as follows. In the second section, we review related work on visuomotor coordination addressed by the learning by examples approach, in particular those based on ANN. In the third section, we present the ANN model we have chosen for our experiment, namely an Extended Kohonen Map (EKM). We motivate the choice of this model by observing an interesting property of its input-output behavior, namely the invertibility, which makes it worth considering for the study of forward and inverse associations. In the fourth section, we present the experiment on the visuomotor coordination of a mobile robot addressed by the previously introduced ANN. We describe the $< perception, action >$ data collection, the ANN training and the performance phase, when the trained network guides the mobile robot to target locations in the workspace. Solution paths obtained in a computer simulation will be presented, as well as a description of an experiment performed on a real robot. Finally, we will draw some conclusion.

2 Related work on visuomotor coordination

A widely studied case of visuomotor coordination is the automatic positioning of the end effector of a robot arm in the experimental setup depicted in Figure 1. Following the learning by examples approach, Ritter, Martinetz and Schulten have realized an ANN, typically an EKM, which learns to gradually solve the arm inverse kinematics problem [RMS91]. Their methodology is referred to as *direct inverse modeling* and can be summarized as follows. During the network training phase, target positions are chosen at random in the workspace. Each target point is supplied as input data to the ANN, which decides a joint configuration. This is transmitted to the joint motors to move the end effector to a new position, which is recognized by the camera system. The difference between the target position and the one actually reached by the robot allows the computation of an "improved" joint angle configuration which can be taught to the network as target output value. Strictly speaking, the learning rule is *supervised*, because the network weights are modified to reproduce the improved joint configuration. However, one should notice that there is no *a priori* decided configuration for any given target position in the workspace: the improved joint angle configuration is derived by letting the robot operate and by observing the results of its action. Therefore, the authors claim that the robot represents an autonomous system which learns in a closed-loop mode: it receives all the information needed for adaptation from the cameras. This style of learning is correctly called *learning by doing*.

Arm trajectory generation has also been studied by Bullock, Grossberg and Guenther [BGG93]. They have used a neural architecture called DIRECT, which learns a transformation between spatial and velocity coordinates through

---

[1]While we essentially agree with this statement, we wish to note that *a priori* knowledge is also present in the *a posteriori* approach. The decision of which variables are to be observed, the choice of a particular class of models as a suitable skeleton for sensorimotor description are all examples of *a priori* knowledge currently used to construct *a posteriori* models.

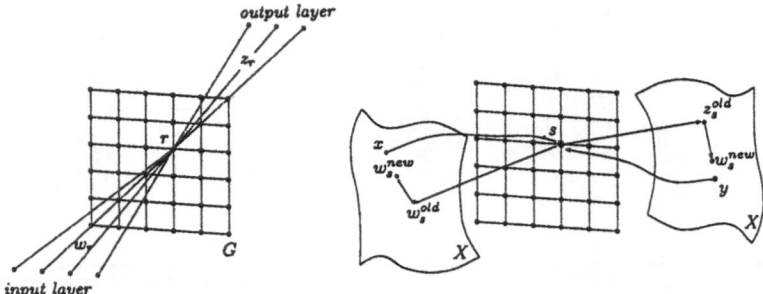

Figure 2: Extended Kohonen Map architecture (left) and learning step (right).

a sequence of spontaneously generated random movements.

Recently, following the same approach, Zalama, Gaudiano and López Coronado have proposed a neural network model (NNETMORC) for the control of a mobile robot in a 2D environment [ZGLC94]. Their model combines VAM learning and associative learning within an architecture similar to the DIRECT model.

In this paper we study the visuomotor coordination problem for a mobile robot as it is addressed in [ZGLC94], i.e. in the experimental setup of Figure 1. We have found that a very straightforward solution can be expressed in terms of an EKM. Although our work takes inspiration from the experiment on end effector positioning realized by Ritter, Martinetz and Schulten (in that we use the same neural model), *we don't use the direct inverse modeling* approach. Instead, our departure point is the observation of an interesting property of the EKM that greatly eases our task.

3 An interesting property of the Extended Kohonen Map

We assume that the reader is familiar with the Kohonen Map (KM) neural model [K84]. To describe our experiment we briefly introduce the required extension to the basic KM algorithm (EKM) proposed by Ritter and Schulten in [RS87], which enables us to train on output values the KM network by supervised learning. From the architecture point of view, the KM network is augmented by adding to each neuron r on the competitive grid G a fan-out weight vector z_r to store the neuron output value (Figure 2, left). The computation in the EKM network proceeds as follows. When an input pattern x is presented to the input layer, the neurons on G compete to respond to it. The competition involves the neurons fan-in weight vectors w_r, and consists of the computation of the *distance*[2] between x and each w_r. The neuron s, whose fan-in vector w_s is the closest to x, is declared the winner of the competition, and its fan-out vector z_s is taken as the network output answer to x.

During the training phase, both the input pattern x and the desired output value y proposed by a teacher are learned by the winning neuron and by its neighbors on the grid. The learning step consists of moving the fan-in weight vectors of the selected neurons closer to x, and their fan-out weight vectors closer to y (Figure 2, right). This learning style has been described as a *competitive-cooperative* training rule [RMS91]. It is *competitive* because the neurons compete through their fan-in weight vectors to respond to the presented input pattern. As a consequence, only that part of the network[3] which is relevant to deal with the current input data undergoes the learning process. Moreover, neighboring locations on the grid will correspond to fan-in weight vectors that are close to each other in input data space: this property is referred to as the *data topology-conserving* character of the network. The rule is also *cooperative* in that the output value learned by the winning neuron is partially associated to the fan-out weight vectors of its neighbors. The authors claim that if the input-output function to be learned is a continuous relationship, spreading the effect of learning an output value to the neighborhood of the winner represents a form of generalization which accelerates the course of learning [RMS91].

We are now in a position to make an observation about an interesting property of the input-output mapping realized by the EKM. We have noticed that this mapping is naturally *invertible*. To explain this point, suppose you have trained an EKM on a set of examples $< x, y >$, where x is a point in the input data space X and y is a point in the output data space Y. We see two possibile ways of using the map. The first use, which we call *forward*, consists of presenting an input pattern x to the network input layer to retrieve the corresponding output value, that will be an approximation to y. This is accomplished in the usual way, namely by letting the fan-in weight vectors w_r compete on x, and by taking the fan-out weight vector z_s of the winner s as the network output value. The second use, which we call *backward*, consists of presenting an output pattern y to the network output layer to retrieve a corresponding input value, that will be an approximation to x. This is achieved by letting the fan-out weight vectors z_r compete on

---

[2] The distance measure may be, for instance, the Euclidean distance.
[3] As defined by the neighborhood function.

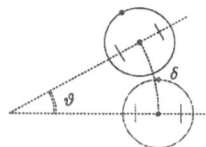 

Figure 3: Displacement values for the robot (left side), the Khepera robot (right).

y, and by taking the fan-in weight vector w_s of the winner s as the network input value. The possibility of retrieving output values, through forward mode, and input values, through backward mode, is due to the fact that the EKM is completely symmetrical with respect to its input-output behavior, as the neuron activation is based on a distance measure [4]. Moreover, input and output patterns are learned by the network in exactly the same way (Figure 2).

Why is this property so interesting? Remember that our objective is to solve an inverse problem, namely the inverse kinematics of a mobile robot. Inverse tasks are often characterized by the existence of a well-defined forward task. In our case, the forward task corresponds to the robot direct kinematics. On the contrary, inverse tasks may be ill-posed, or may have multiple solutions. The observation above provides us with a rudimentary but effective approach to tackle forward and inverse problems at the same time. It suggests that, first, a model for the forward task may be constructed by training an EKM on examples of the forward association; these examples should be easily available. Second, by using the trained network in backward mode, we are provided with a solution to the inverse task. Thus, to apply this idea to our instance of inverse problem, it will be sufficient to train the network on the robot direct kinematics. The backward use of the network will be the solution to the inverse kinematics problem. In the next section we describe this procedure in detail.

4 Visuomotor coordination of a mobile robot by the invertible EKM

4.1 Example collection

To model the robot direct kinematics according to the *a posteriori* sensorimotor modeling approach, one has to generate a set of examples < *action, perception* >. In our case the relevant variables to record are the robot wheel velocities, for the action side, and the corresponding robot displacement, for the perception side. To be more specific, an example is a pair < $(\omega_l, \omega_r), (\vartheta, \delta)$ >. (ω_l, ω_r) are the angular velocities of the left and right wheel, while (ϑ, δ) are, respectively, the robot heading change and the distance travelled by its axle mid-point during a fixed Δt time (Figure 3, left).

The example generation phase amounts to repeatedly selecting a velocity pair at random[5], applying it to the wheels for a fixed Δt time, and measuring the corresponding robot displacement by using the camera system. These examples, which implicitly describe the real robot kinematics, are the starting point for modeling the forward association between the space of motor commands and the space of resulting sensory situations. The advantage of the learning by doing approach to sensorimotor modeling lies in the fact that if the real robot em consistently deviates from the ideal behavior as predicted by a kinematics equation (if the robot is "malfunctioning"), this deviation will be included in the model[6].

In our experiment, examples of direct kinematics have been generated by using Khepera, a miniature robot [MFI93] (Figure 3, right). Khepera is equipped with two wheels, each controlled by an independent motor. The motor speed is expressed in integer units, one unit corresponding to 8 millimeter of advancement per second. With Khepera we generated a training set consisting of 1000 < *velocity, displacement* > pairs, with velocity chosen in the range [0, 5], and applied to the wheels for approximately one second.

4.2 Network training

Having collected a set of examples, the EKM training can take place. The network architecture is a 6×6 competitive grid, with two input units and two output units (Figure 4, left). The input units are used to encode the velocity pair, while the output units represent the robot displacement. The competitive grid dimension (6×6) was chosen to reflect the fact that each wheel may assume six speed values only. The network was trained as explained in the third section.

The result of the training phase is graphically depicted in Figure 4, right side. Each square is a neuron on the competitive grid. For each neuron, the learned fan-in and fan-out weight vectors have been represented. The fan-in weight vector is a velocity pair indicated by the position of the neuron on the (ω_l, ω_r) reference system. The fan-out

---

[4]This would not be possible, for instance, in a feed-forward network.
[5]With uniform probability in the space of valid speed range.
[6]Infact it is hard to talk about a deviation as there is no *a priori* expectation on how the robot should behave.

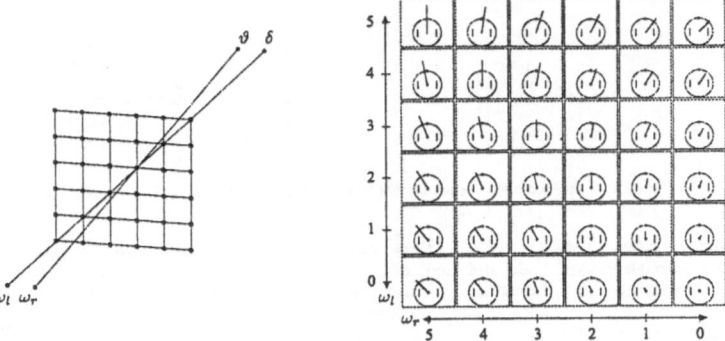

Figure 4: The network architecture of the experiment (left), representation of the network weights after training (right).

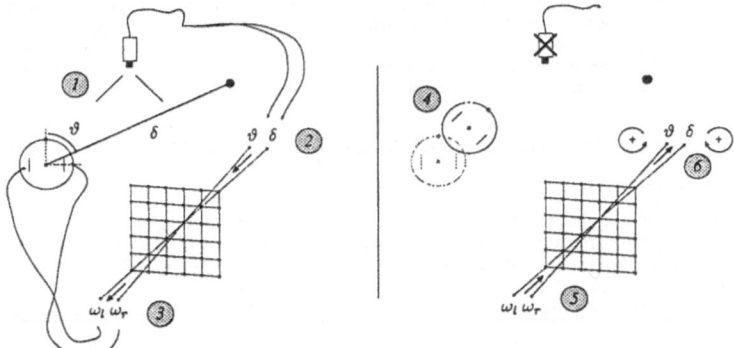

Figure 5: The way the trained network is used to guide the mobile robot to a target location.

weight vector is the robot displacement and has been represented as a vector: the vector orientation indicates the robot heading change, while the vector length is the distance travelled by the its axle mid-point. By looking at the overall aspect of the grid, one may appreciate the *data topology-conserving* character of the network. Velocity pairs are ordered along the grid horizontal and vertical axes, so similar velocity pairs are mapped onto neighboring locations on the grid. A continuous variation is also present in the displacement values. So similar velocity pairs will lead to a similar displacement. We argue that the EKM model is especially well-suited to learn the robot kinematics, thanks to the cooperative aspect the learning rule, which naturally tends to associate similar output values to similar inputs. This turns out to be a desirable generalization property, as the function to be learned is a continuous relationship between velocity and displacement.

4.3 Network performance

The EKM has been trained on the forward mode, namely on a transformation from the space of motor commands to the space of visual perceptions. We will now explain how the trained network is used in backward mode to compute the inverse function, which transforms a visual perception into a motor command. The task is to guide the robot to a target location placed at arbitrary angle ϑ and distance δ in the workspace (Figure 5). The angle and distance information is provided by the camera system, which observes both the robot and the target (step 1). To be more specific, ϑ is now defined as the angle between the robot heading direction and the vector connecting the robot axle mid-point and the target, while δ is the Euclidean distance between the robot axle mid-point and the target location.

The observed ϑ and δ values are supplied to the EKM in backward mode to retrieve a velocity pair (steps 2 − 3). For this particular application, the competition on the pattern (ϑ, δ) has been designed to consider its components *in sequence*. First, ϑ is processed: the competition is restricted to the weight vector component of the neuron which

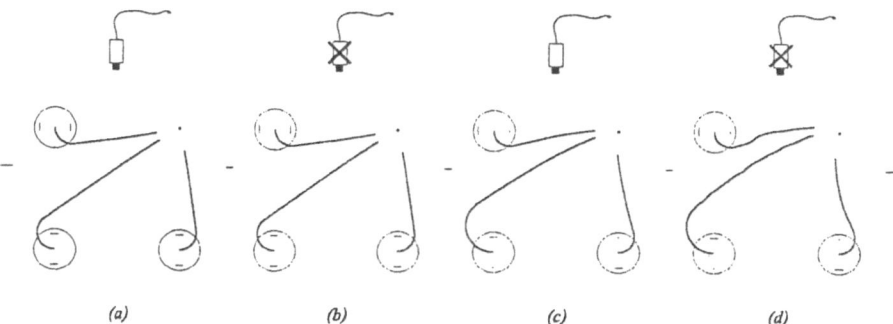

Figure 6: Trajectories performed by the robot with the camera information (a), without camera feedback (b), with left wheel radius reduced and visual information (c), with left wheel radius reduced and sporadic visual feedback (d).

stores the angle information. As the result of this preliminary step, a subset of grid neurons are selected: these are those neurons which match ϑ equally well. Second, δ is processed, but only those neurons selected at the previous step are allowed to compete; among these, the competition is restricted to the weight vector component of the neuron which stores the distance information. The overall competition process leads to the selection of a velocity pair for the robot wheels, namely the fan-in weight vector of the winning neuron. The rationale for splitting the competition in two steps is twofold. First, consider that, while the angle varies in a limited range of values, the distance value can be arbitrarily large. Therefore, if both data were processed by the network in one step, the distance component would have had more importance in the determination of the winning neuron. This can lead to undesirable effects, such as always selecting the neuron corresponding to the highest speed for both wheels[7]. Second, it seems natural to give priority to the angle information, so that the robot's first goal is to turn to face the target. The distance information becomes really relevant when the robot is close to the target so as to gradually reduce the wheel speed[8].

After having applied the speed pair to the wheels for a fixed time Δt, the robot arrives to a new position (step 4). Its new position with respect to the target is again derived by the camera system, and the whole procedure is repeated until the robot reaches the target. However, it is interesting to note that, as an alternative to employing the camera system to update the robot positional information, the network direct kinematics model can be used instead. This means that, after having applied a velocity pair to the robot wheels (step 4), the same pair is processed by the network in forward mode to retrieve a displacement pair (steps 5–6) and to update the new position of the robot with respect to the target. So "in principle" the robot is able to reach the target position blindly, i.e. without the visual feedback provided by the camera.

4.4   Results on network performance

The network performance has been tested both in a computer simulation and on the Khepera robot.

Figure 6 refers to the computer simulation. It depicts the trajectories performed by the robot to reach a target starting from several initial positions and under different experimental conditions. In Figure 6a the robot reaches the target by using the visual feedback from the camera system to update its positional information during motion. In other words, the EKM is just employed in backward mode to retrieve velocity pairs when the positional information is supplied by the camera system (steps 1 to 4 of Figure 5). Figure 6b shows the paths followed by the robot when the network kinematic model is used to update the robot position. In other words, the network is supplied with the target position at the beginning of the path only, then steps 2 to 6 are repeated until the target is reached. So the network is able to reach the target position blindly by using both forward and backward modes. No significant deviation from the previous trajectories is observed. Finally, Figures 6c–d show the trajectories of the robot after having reduced its left wheel radius to half of the original size. In this way, the robot kinematics dramatically changes. Nevertheless, when supplied with the camera visual feedback continuously (Figure 6c) or at least sporadically (Figure 6d), the robot is still able to reach the target, although the trajectories are now different. This last experiment suggests that the system exhibits a kind of "noise resistance". However, it should be made clear that the compensation ability to perturbations (in this case, the reduction of a wheel radius), is due to the fact that the robot is provided with the information of how it is moving with respect to the target. This information gives the robot a signal that something is going wrong. In other words, at each step of its trajectory, the robot can compare the actual position to the goal to the one calculated by the network kinematic model. Eventually, if the difference between the two is too large, the

---

[7]This is the neuron that machtes the longest distance, which corresponds to the speed pair $(5,5)$.

[8]For example, by selecting the speed sequence $[(5,5),(4,4),(3,3),(2,2),(1,1),(0,0)]$, which corresponds to one diagonal of the competitive grid.

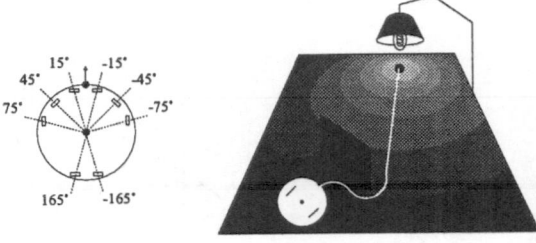

Figure 7: The Khepera infrared proximity sensors (left), the robot moving towards the light (right).

robot may decide that the model is no longer accurate enough and start a new training phase. We want to stress once again that this decision is induced by the camera system. If the robot is not given the visual information, at least sporadically, there is no way to compensate for noise. A real robot which moves blindly will get lost, sooner or later, as noise is always present and unpredictable.

We conclude by briefly describing the experiment with the Khepera robot. The experiment was organized by taking advantage of the robot's sensing capabilities. Khepera is provided with eight infrared proximity sensors (Figure 7, left). Each sensor contains a light receiver which allows the measurement of normal ambient light along a particular direction. We identified the target position with a light source (Figure 7, right). In this way, the angle between the robot forward direction and the target can be defined, for example, as the weighted mean of the angles corresponding to the two most activated sensors. This gives a rough approximation of the ϑ value. As far as the distance value δ is concerned, we kept it fixed to a chosen value[9]. With ϑ and δ defined in this way, Khepera is able to reach a fixed source of light or to track a moving light. The behavior of the robot clearly depends on the value to which δ is fixed. It the distance value is high[10], the network selects high velocities and the robot moves fast. If a small distance value is selected, smaller velocities are preferred. As δ is always set to a value other than 0, the robot never stops; rather, after having reached the area of maximal light, it keeps moving in that area. Finally, a nice effect we observed, is the robot's ability of avoiding obstacles while moving towards the light (Figure 7). As the obstacles project shadows, the robot, which always moves in the direction of the strongest light, takes the shadow as an indication of which areas are to be avoided to prevent collisions. In a certain sense, this experiment can be regarded as the physical implementation of a robot moving in a potential field generated by an attraction force towards the goal (the light) and by repulsion forces exerted by the obstacles (the shadows).

5 Conclusions

We have presented an experiment on the *visuomotor coordination* of a mobile robot. Our approach to sensorimotor modeling belongs to the category of *learning by doing*. First, $<$ *perception, action* $>$ pairs are collected by observing the robot during operation. These data are the starting point to describe the real robot behavior. Second, a learning by examples method is used to estimate the parameters of the model. The advantage of the *a posteriori* approach to sensorimotor modeling lies in the fact that, in the case where the robot consistently deviates from the ideal behavior as predicted by its kinematics equation, this deviation will been included in the model.

In our experiment the learning machine is an *Extended Kohonen Map*. We have observed that this neural network model has the property of being *invertible*. Given an input pattern, the network retrieves the output value by competition on the fan-in weight vectors of the neurons. We have called this use of the input-output mapping the network *forward* mode. Viceversa, given an output pattern, a corresponding input data is retrieved by competition on the neuron fan-out weight vectors. We have called this use of the map *backward* mode.

We argue that the invertibility property makes the Extended Kohonen Map a suitable architecture for sensorimotor modelling.. The experiment of the visuomotor coordination of a mobile robot shows that by training such a network on the robot direct kinematics (the forward mode), one obtains at the same time a solution to the inverse kinematics problem (the backward mode). The trained network can be used to guide the robot to any target location in a 2D workspace. To update its position with respect to the target during motion, the robot may take advantage either of the visual feedback provided by a camera system, or of the network direct kinematics model. However, experimentation in the presence of noise has stressed the importance of providing the robot with at least some sporadic visual information about the target position. In this way, the system exhibits a kind of noise resistance. In our opinion, this issue becomes really relevant in path finding problems, where the robot is expected to keep track

---

[9]A sensible definition of distance is hard to give.
[10]Here "high" means high with respect to the distance range learned by the network.

of its position with respect to the goal autonomously. A real robot which moves blindly will get lost sooner or later. Rather, the definition and recognition of landmarks on the path could greatly help the robot task. This will be one of the subjects of our future research.

Acknowledgements

We would like to thank *Neuristique* (France) for having provided the SN neural network simulator to support the experimental part of the work. Thanks to Vicente Ruiz de Angulo, Thomas Barbas and Domenico Perrotta for their comments on early drafts of this paper.

References

[BDM94] Béssiere, P., Dedieu, E., Mazer, E. *Representing Robot/Environment Interactions Using Probabilities: the "Beam in the Bin" Experiment.* Proc. "From Perception to Action", Lausanne, Switzerland.

[BGG93] Bullock, D., Grossberg, S., Guenther, F. *A Self-Organizing Neural Network Model for Redundant Sensory-Motor Control, Motor Equivalence and Tool Use.* Journal of Cognitive Neuroscience, 5, 408-435.

[K84] Kohonen, T. *Self-Organization and Associative Memory.* Springer Series in Information Sciences, 8, Heidelberg.

[M94] Massone, L. L. E. *Sensorimotor Learning.* To appear in "The Handbook of Brain Theory and Neural Networks", M. A. Arbib Editor, MIT Press.

[MFI93] Mondada, F., Franzi, E., Ienne, P. *Mobile Robot Miniaturization: a Tool for Investigation for Control Algorithms.* Proc. of the Third International Symposium on Experimental Robotics, Kyoto, Japan.

[RMS91] Ritter, H., Martinetz, T, Schulten, K. *Neural Computation and Self-Organizing Maps. An Introduction.* Addison-Wesley.

[RS87] Ritter, H., Schulten, K. *Extending Kohonen's Self-Organizing Mapping Algorithm to Learn Ballistic Movements.* Neural Computers, R. Eckmiller and E. von der Marlsburg (eds.), Springer, Heidelberg.

[ZGLC94] Zalama, E., Gaudiano, P., López Coronado, J. L. *A Real-Time, Unsupervised Neural Network Model for the Control of a Mobile Robot in a Nonstationary Environment.* Boston University Technical Report CAS/CNS-94-002.

Neural Networks
for Automatic Fuzzy Control System Design

Jesús Villadangos[1], J. R. González de Mendívil[2], C. F. Alastruey[1], J. R. Garitagoitia[2],

[1] University of the Basque Country
Dpt of Electricity and Electronics
P.O. BOX 644
48080 Bilbao, SPAIN
e.mail: fitxi@we.lc.ehu.es

[2] Public University of Navarra
Dpt of Automatica, Electronics and System Engineering
Campus Arrosadia s/n
31006 Pamplona, SPAIN
e.mail: jrmen@we.lc.ehu.es

Abstract. In this paper, a method for the automatic design of Fuzzy Control Systems is introduced. The method is based on the identification of the inverse model of the process to be controlled by using a Neural Network. The Neural Network which models the inverse process, is used again to obtain a set of tuples representing the fuzzy variables of the fuzzy controller. In order to obtain the fuzzy linguistic variables involved in the fuzzy controller, a Neural Network is used with the DCL algorithm. Finally, the fuzzy controller is implemented by a decision table. The method has been applied to the automatic development of a fuzzy controller for a highly non linear process.

1. Introduction

A fuzzy control system consists of a set of fuzzy variables and a fuzzy inference system. Usually the hardest part of designing fuzzy control systems is selecting which fuzzy sets best represent the controlled and controlling variables, namely, the tunning of the controller. The fuzzy controler design is totally made by experts who know the process to be controlled. A trial-error method is followed to implement the best controller for the process.

Neural Networks (NNs) theory has been applied to identification and control design. Many works have been reported concerning non linear and linear plants [5,7,8]. A NN which models a process could be used as a process expert, so the NNs learning capability can model an expert knowledge. The goal of this paper is to present a possible symbiosis between Fuzzy Systems and NNs to obtain a fuzzy controller based on NNs.

A generic method based on NNs is introduced in order to automatically obtain a fuzzy control system for any process, either linear or non linear. This method substitutes the role of the experts by the NNs. The method follows four steps.

First, the behaviour of the process is obtained with a Pseudo Random Binary Signal (PRBS), as process input. This signal is usually utilized in adaptive control to process identification [2]. The backpropagation learning algorithm is used to invert process identification. The trained NN which models the inverse process is called *NN-process*.

Second, a *trajectory* for the process is defined as a sequence of process outputs that could be followed by the process to reach the reference. The trained NN-process together with a predefined trajectory are used to obtain values of the universe of the controller fuzzy variables; the obtained data are to be called *virtual process data* because they are calculated from artificial data.

Third, an NN with the unsupervised Differential Competitive Learning algorithm (DCL) [3] is utilized to classify the virtual process data into linguistic terms for each fuzzy variable [1,3,9,10].

Fourth, the identified fuzzy controller variables are used to implement the decision table of the fuzzy control system. The decision table implements the control actions for the controller. If the information about the process was not enough, then a backpropagation learning algorithm is used to complete the decision table[1,11].

The rest of the paper is organized as follows: Section 2 describes the method for automatic fuzzy control system design. Section 3 presents the process identification and the generation of data for the fuzzy variables identification. In section 4, the fuzzy inference system design will be shown. In section 5, an example of the use of the method is provided. Finally, conclusions end the paper.

2. Fuzzy control system design method

The fuzzy controller follows the general scheme presented in figure 1. The process output is fuzzified [4,6] into the input controller variables E and EC. E represents the error between the reference and the process output. EC represents the error variation between consecutive sample times. The defuzzification process for the variable U, represents the process control signal. These variables (E, EC, U) are fuzzy variables, and based on them, the fuzzy inference system must be defined. The values "ge", "gec" and "gu" are gains for the controller variables, in this case their value is permanently one because the design process automatically sets the fuzzy controller variables.

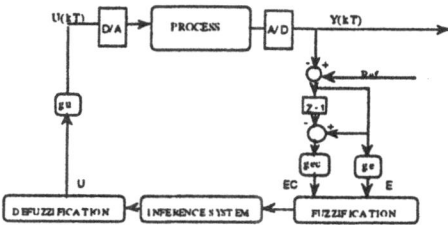

Figure 1: Fuzzy control system.

The experts should define the fuzzy variables and the inference system based on their knowledge about the process to be controlled. Here the experts role is substituted by the NNs, which will drive completely the fuzzy controller design. By considering an unknown process, all later development could be considered general for any process. The proposed method follows the next scheme:

Step 1   Modeling the inverse dynamic of the process by a backpropagation learning algorithm. The NN-process simulates the experts knowledge about the process.

Step 2   Obtaining the virtual process data. A trajectory, as defined above, is applied to the NN-process to yield the control signals for such trajectory. The data about the trajectory and the control signal which constitute the fuzzy variables universes are called *virtual process data*

Step 3   Classification of the virtual process data by the unsupervised Differential Competitive Learning (DCL) algorithm to obtain the linguistic terms for each fuzzy controller variables E, EC and U.

Step 4   The virtual process data and the identified fuzzy variables are used to define the inference system. For each pair of virtual process data (E, EC) a value of the control signal U from the virtual data is assigned.

This method will be used in section 5 to identify a fuzzy control system for a highly non linear process.

3. Process identification

The process is modelled with an NN, the NN-process. A previously defined trajectory for the process and the NN-process will generate the virtual process data. Such data will be later used to identify the memberships functions (linguistic terms) of the fuzzy control variables (E, EC, U).

3.1 Identification of a SISO process by using NNs

The inputs to the NN are delayed-process outputs, and the desired output for the NN is the input to the process. The identification of a SISO process could follow the scheme in figure 2.

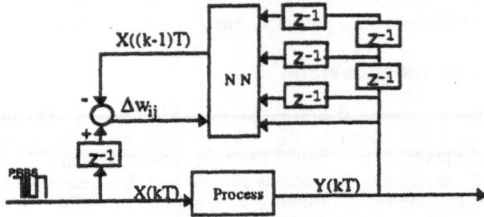

Figure 2: NN process modelling.

The error between the desired input process and the obtained output from the NN is used to modify the weights of the neurones (backpropagation algorithm). This is represented in figure 2 by the signal Δw_{ij}.

3.2 Obtaining the virtual process data

A trained NN simulates the experts knowledge, the NN-process. We define possible trajectories for the process output to reach the reference, as shown in figure 3.

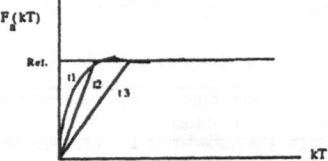

Figure 3: Possible outputs trajectories for the process.

The trajectory will be presented as input to the NN-process, and so will be stored the virtual process data for E, EC and U. The error E (E= $Y_{reference}$ - Y(kT)), the difference between the reference and the process output signal. EC, is defined as EC=E(kT) - E((k-1)T), the difference between errors at consecutive sample times and U is the obtained output from the NN-process.

The virtual process data are used, in the next section, to generate the fuzzy variables, and eventually the fuzzy inference system.

4 Fuzzy inference system identification

Firstly the linguistic terms of the fuzzy variables E, EC and U will be generated, and then the fuzzy inference system.

4.1 Fuzzy variables identification

The sets of values obtained for the variables E, EC and U based on the NN-process and the desired trajectory to reach the reference will be classified in separate ways, although with the same method. This method is the unsupervised Differential Competitive Learning (DCL) algorithm, that groups similar data into zones corresponding to those values.

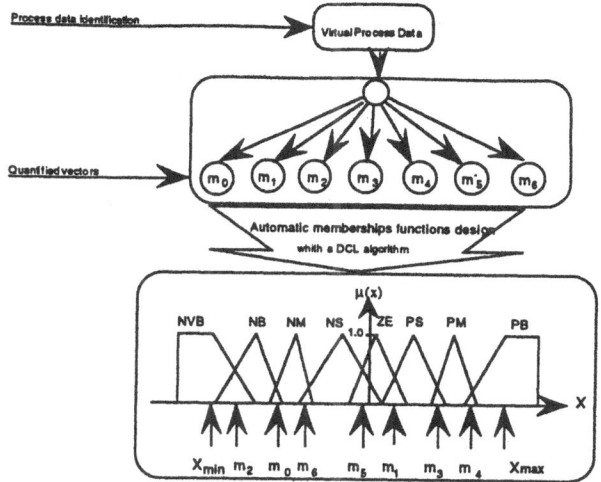

Figure 4: Linguistic terms identification for any fuzzy variable.

The figure 4 shows the virtual data classification result for some fuzzy variable using the DCL learning algorithm. m_i points to different concentration zones of values. These pointers will define the cross points of the membership functions that compose each fuzzy variable. The pointers allow to define triangle or trapezoidal membership functions, functions that flexibilize the tunning of the fuzzy control system. Notation:ZE for zero, P for positive, N for negative, B for big, S for short, M for medium and V for very, so NVB denotes the linguistic term "Negative Very Big", all are used to name the linguistic terms or membership functions of the fuzzy variables.

The identified fuzzy variables allow to generate the fuzzy inference system as shown in the sequel.

4.2 Fuzzy inference system generation

The fuzzy inference system is a set of if-then rules. For two variables it can be represented by a table. The virtual process data will be used to generate the output for the controller, as shown in figure 5.

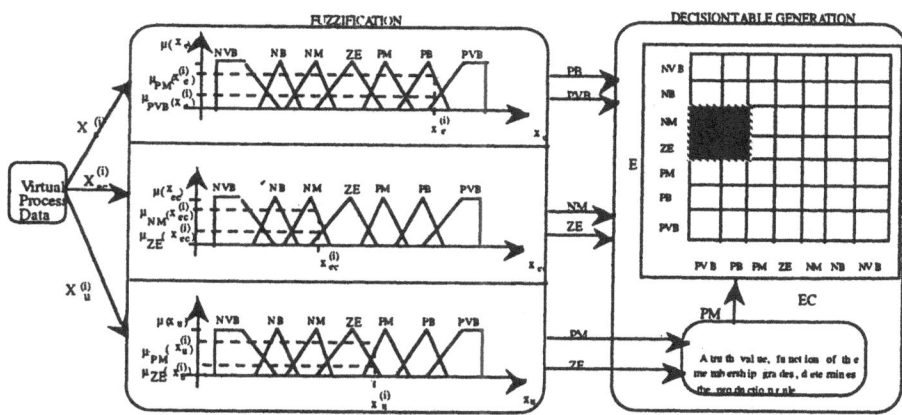

Figure 5: Inference system identification.

All the virtual process data should be presented for the fuzzy variables. The fuzzy sets of the variables were obtained previously. All the possible generated fuzzy rules are used to fill the blank spaces in the table. For every place in the table a conflict could exist, to say, different rules could be found for the same input values. For example, all the possible rules that can be extracted from figure 5 are:

IF	E_1 = PB	and	E_2 = ZE	THEN	U = PM
IF	E_1 = PB	and	E_2 = NM	THEN	U = PM
IF	E_1 = PVB	and	E_2 = ZE	THEN	U = PM
IF	E_1 = PVB	and	E_2 = NM	THEN	U = PM
IF	E_1 = PB	and	E_2 = ZE	THEN	U = ZE
IF	E_1 = PB	and	E_2 = NM	THEN	U = ZE
IF	E_1 = PVB	and	E_2 = ZE	THEN	U = ZE
IF	E_1 = PVB	and	E_2 = NM	THEN	U = ZE

For each rule a truth value [11] is assigned to determine which rule is the correct one. The truth value depends on the membership value for the fuzzy set. In this example this factor is calculated as follows:

$$D_{truth} = \mu_{PB}(x_e) \times \mu_{NM}(x_{ec}) \times \mu_{PM}(x_u) = 0,70 \times 0,60 \times 0,80 = 0,336$$

In this case the chossed rule for E1=PB and E2=NM is U=PM.

If the virtual process data could not contain enough information for generating the complete fuzzy inference system, then the table won't be full. The backpropagation algorithm will be used to fullfill it. The filled spaces on the table (U), and the corresponding fuzzy variables E and EC, are used to train an NN. Quantified values of E and EC are used as inputs for the NN and the corresponding U is considered the output to learn.

When the training process is finished, the quantified fuzzy sets for E and EC, which didn't defined a value for U, are presented to the NN trained before. The NN output for such combination E and EC is considered the corresponding U value to be placed in the table.

The next section shows an application example for a non linear process.

5. Problem application

The process to be controlled is a polish machine. The machine approximates the emery to the processing piece with a radial velocity v_r, the controlling signal. The output signal of the process is the normal force (F_n) over the piece, due to the contact between the piece and the emery. The reference following is the goal for the fuzzy control system, and in this method the reference should be known previously to the fuzzy control system identification. The reference is a force, as well as the controlled output signal. The controlling signal is the radial approximation velocity of the emery to the working piece.

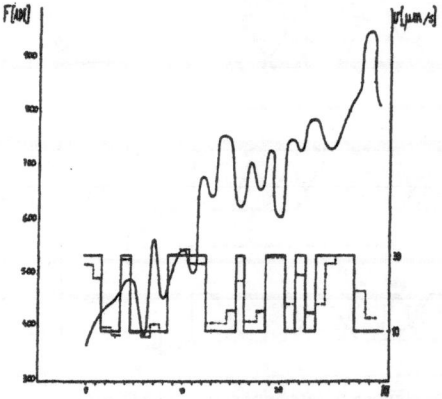

Figure 6: Behaviour of the process and inverse process identification (dotted line).

The behaviour of the process is obtained by means of a Pseudo Random Binary Signal (PRBS). The process input (v_r) is the PRBS and the process output is a force (F_n) as illustrated in figure 6. The PRBS varies between: -10 μm/s < v_r < 30 μm/s, and the normal force: F_n<2000[ADU] where ADU are the Analog Digital Units. The sample time is 9 ms.

The results of the automatic fuzzy control system generation for the process illustrated in figure 6 are now presented. The control parameteres are: the reference is F_n=500 [ADU]; the gains "ge" an "gec" are equal one and the gain "guc" takes different values.

Firstly, the process is modelled by an NN. The activation function for the neurones is tanh(x). Figure 6 represents the identification results for the process. The dotted line denotes the modelled process.

Learning characteristics:

number of layers:	3
number of input neurones:	4
number of hidden neurones:	10
number of output neurones:	2
learning error:	0.3
learning cycles:	25(X)

Secondly, the virtual process data are obtained based on an ideal trajectory similar to t2 represented in figure 3. Then, these data and the DCL algorithm help to find the fuzzy sets for the linguistic terms E, EC and U. These ones are represented in figure 7.

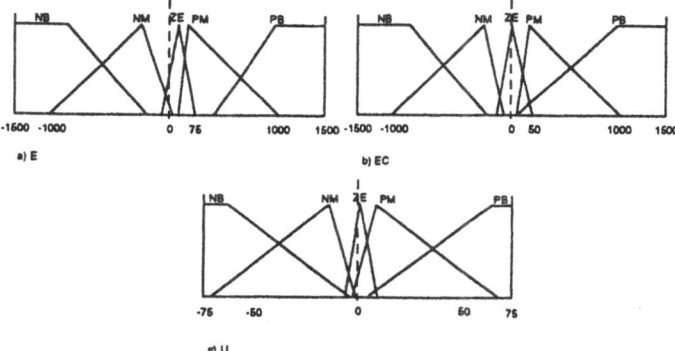

Figure 7: Generated fuzzy variables for the example fuzzy control system. a) E, b) EC, c) U.

The fuzzy inference system is following represented in a table in figure 8.

		EC				
		NB	NM	ZE	PM	PB
E	NB	Œ	NM	NM	Œ	Œ
	NM	PM	NM	NM	Œ	Œ
	ZE	PM	PM	PM	Œ	Œ
	PM	PM	PM	Œ	PM	PM
	PB	PM	PM	Œ	PM	PM

Figure 8: Generated fuzzy inference system for the example process.

The following graphics present the control results for different parameters. Figure 9 illustrates the result for different gain values for the controlling signal U.

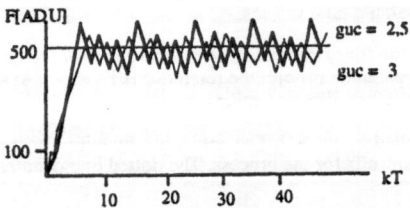

Figure 9: Control for different "guc" values.

The oscillation in the output signal is due to the use of a PRBS. This signal is useful in adaptive control but for this method a Pseudo Random Multivalued Signal. could be more suitable.

Conclusion

An approach to the automatic fuzzy control system generation and adaptable to any unknown process are presented. It is useful for every not completely know process, and flexibilize the tunning of the fuzzy control system, due to the possible definition of triangular or trapezoidal memberships functions.

This method could be extended to MIMO systems, which will utilise the complete powerful of the method. Another way to improve the method could be include it in an adaptive fuzzy control system.

References

[1]   Faßmer, J. "Adaptive Generierung der Produktionsregeln eines Fuzzy-Regler mit Hilfe eines Neuronalen Netzwerkes". Diplomarbeit IFW Hannover, December 1992.

[2]   Isermann, R. "Prozeidentification". Band 1, 2. Springer Verlag 1988.

[3]   Kosko. B. "Neural networks and fuzzy systems". Prentice-Hall International 1992.

[4]   Lee, C.C. "Fuzzy logic in control systems: Fuzzy logic controler part I-II". IEEE Transactions on Systems, Man and Cybernetics, Mar/April 1990, vol. 20, n. 2, pp. 1320-1336.

[5]   Narendra, K.S.; Parthasarathy, K. "Identification and control of dynamical system using neural networks". IEEE Transactions on Neural Networks, March 1990, vol. 1, n. 1, pp. 4-27.

[6]   Procyk, T.J.; Mamdami,E.H. "Alinguistic self-organizing process controller". Automatica 1979,vol. 15, pp. 15-30.

[7]   Quin, S.-Z.; Su, H.T.; Mc Avoy, T.J. "Comparation of four neural net learning methods for dynamic system identifjcation". IEEE Transactions on Neural Networks, Jan 92, vol. 3, n. 1, pp. 122-130.

[8]   Reynold Chu, S.; Shoureshi, R.; Tenorio, M. "Neural networks for system identification". IEEE Control System Magazine, April 1990, pp. 31-34.

[9]   Takagi, H. "Fusion technology of fuzzy theory and neural networks suvey and future directions". Proceedings of the International Conference on Fuzzy Logic and Neural Networks., July 1990, pp. 13-26.

[10]   Takagi, H.; Hayashi, I. "Fuzzy reasoning". International Journal of Approximate Reasoning, Mai 1991, vol. 5, n. 5, pp. 191-212.

[11]   Wang, L-X.; Mendel, J.M. "Generating fuzzy rules by learning from examples". Proceedings of the 1991 IEEE International Symposium on Intelligent Control. Arlinton, California 1991.

Supervised Classification
with Variable Kernel Estimators

Pierre COMON   and   Yves CHENEVAL [†]

THOMSON-SINTRA, BP157, F-06903 Sophia Antipolis Cedex
comon@asm.thomson.fr
[†] *LAMI, EPFL, CH-1015 Lausanne*
cheneval@di.epfl.ch

Abstract

Assuming uniform losses, Bayesian classification leads to the lowest possible misclassification rate, by definition. In order to carry out supervised classification in the Bayes sense, kernel density estimators with variable width are utilized. Nevertheless, in their standard form, they would require each learnt pattern to be stored, which is often beyond the hardware specifications. For this reason the variable kernel algorithm proposed is based on clusters, determined so as to minimize the final misclassification rate. This rate is evaluated by a cross-validation type algorithm in order to avoid overfitting. Experimental results show that it is possible to determine the number of clusters and their parameters, and to take into account hardware constraints, unavoidable in neural implementations.

1   Probabilistic framework

The classification problem consists of building a mapping from a set of patterns (observations) to a set of classes. It is assumed throughout this paper that patterns are real valued and of dimension d. In other words, any pattern $x \in \mathbf{R}^d$ can be associated with a class ω_j by this mapping. In the context of supervised classification, a set of examples $\{(x(n), \omega_{j(n)}), 1 \leq n \leq N\}$ is given, so that the mapping is known at a finite number of points. This set of input-output pairs is the *learning set*.

In a probabilistic framework, it is assumed that all patterns belonging to the same class follow the same underlying distribution. In this paper, it is assumed that this distribution admits a density, denoted $p(x|\omega_j)$.

Assuming uniform losses, the Bayesian approach allows to build the mapping that minimizes the total number of misclassifications, provided the conditional densities $p(x|\omega_j)$ and the priors $P_j = P(\omega_j)$ are known. More precisely, any observation $x(k)$ is assigned the class $\omega_{j(k)}$, where

$$j(k) = Arg \operatorname*{Max}_i \{ P_i \, p(x(k)|\omega_i) \}. \tag{1}$$

Now, it is clear that knowing a mapping on a finite set will never provide the complete definition of the mapping on $\mathbf{R}^d$ without further information. That's why supervised classification is usually carried out by assuming –sometimes implicitly– a parametric model, either on the classifying rules (as in neural networks), or on the conditional densities (as in Bayesian approaches).

Part of this work has been funded by the ESPRIT-BRA project 6891, supported by the Commision of the European Communities.

One of the most interesting estimator for densities is known as the "Kernel" estimator. It has not only nice consistency properties, but also can provide continuous estimates regardless of the number of patterns available in the learning set [9], which is of great practical interest, as opposed to histograms for instance. Denote A_i the subset of the learning set that contains patterns belonging to class ω_i, and N_i its cardinality. Then the kernel estimate of $p(x|\omega_j)$ takes the form:

$$\hat{p}(u|\omega_i) = \frac{1}{N_i} \sum_{x(n) \in A_i} \frac{1}{h(n;i)^d} K\left(\frac{u - x(n)}{h(n;i)}\right), \tag{2}$$

where $h(n)$ is positive and $K(\cdot)$ is the kernel function. The choice of the kernel gives the estimator the basic finite sample properties; for instance, if $K(\cdot)$ is positive and C^∞, then so is $\hat{p}$.

If $h(n;i)$ depends only on N_i and not on n, then the estimator is said to have a fixed width, or to be a fixed kernel estimator, in short. This estimator was originally proposed by Parzen [10], and Cacoullos [2] extended it to the multichannel case. The suggestion of a variable width has been proposed independently by Wagner [11] and Breiman [1]. Thus the variable kernel estimator should not be called a Parzen estimator, for the sake of clarity.

For every class index i, sufficient conditions that yield consistency of $\hat{p}$, at every point where p is continuous, are [11] [9]:

$$K(u) > 0, \quad and \quad \int K(u)\,du = 1, \tag{3}$$

$$\|u\|^d K(u) \to 0 \quad as \quad \|u\| \to \infty, \tag{4}$$

$$h(n) \to 0 \quad as \quad N \to \infty, \tag{5}$$

$$N\,h(n)^d \to \infty \quad as \quad N \to \infty. \tag{6}$$

Actually, this last assumption ensures a stronger convergence (*e.g.* in quadratic mean). As a by-product, more accurate properties on the $h(n)$ series are obtained, mainly in the fixed kernel case [3] [9].

The exact shape of the kernel function has little effect on the convergence rate even if it has an influence on the asymptotic variance [9] [3]. In the multivariate case ($d > 1$) there is no reason to take a non isotropic kernel in estimator (2), since each kernel is contributing on behalf of a single pattern. Therefore, the function $K(\cdot)$ will denote in the remaining a radial function, that is, a function depending only on the norm of its argument. However, non isotropic kernels will be subsequently used when they represent clusters of patterns.

Variable width kernel estimators are recognized to perform much better than the fixed width ones in the finite sample case, although they are both consistent [1]. One of the key difficulty in using the variable width estimator is eventually to determine the widths that lead to the best performance. The choice of the width factor has been seen to be very critical, even for the fixed kernel estimator. One first possibility is to base the width value on the distance to the nearest neighbor, as proposed by Breiman. Another one is to minimize a local quadratic error, for every class ω_i:

$$e(u;i) = E\{[\hat{p}(u|\omega_i) - p(u|\omega_i)]^2\} \tag{7}$$

This can be the starting point of a refinement procedure [5] [3].

Here, the approach is different: (i) the number of parameters in the model is reduced by the means of clustering, and (ii) the parameters are chosen so as to minimize the final misclassification rate instead of a density quadratic error.

2 A density estimator based on clusters

Since each vector from the learning set is involved in (2), it is clear that the model becomes very heavy to utilize whenever some N_i's are large. Neural networks avoid overfitting because they perform in some way data compression. Here, overfitting does not appear to be a crucial problem, but data compression is relevant to reduce both memory storage and computational time during the test phase.

Assume that the data in every class ω_i has been clustered into Q_i groups, $\Gamma_{q;i}$, each containing $N[q;i]$ patterns. With our notation, we thus have: $\sum_{q=1}^{Q_i} N[q;i] = N_i$. Now from the Bayes rule, the density in class ω_i can be decomposed into:

$$p(u|\omega_i) = \frac{1}{P_i} \sum_{q=1}^{Q_i} P(\Gamma_{q;i}) \, p(u|\Gamma_{q;i}). \tag{8}$$

This relation suggests the following reconstruction formula, if $P(\Gamma_{q;i})$ is estimated by the ratio $N[q;i]/N$, and P_i by $\hat{P}_i = N_i/N$:

$$\tilde{p}_I(u|\omega_i) = \frac{1}{N_i} \sum_{q=1}^{Q_i} \frac{N[q;i]}{\sigma(q;i)^d} K\left(\frac{u - C(q;i)}{\sigma(q;i)}\right), \tag{9}$$

where a single width factor $\sigma(q;i)$ has been used within each cluster. This relation can equivalently be obtained by replacing $x(n)$ by its centroid $C(q;i)$ in (2).

Another more accurate reconstruction procedure involves a positive definite matrix $L[q;i]$ that accounts for the cluster shape:

$$\tilde{p}_A(u|\omega_i) = \frac{1}{N_i} \sum_{q=1}^{Q_i} \frac{N[q;i]}{h(q;i)^d} K\left(\frac{L[q;i]^{-1}(u - C(q;i))}{h(q;i)}\right). \tag{10}$$

It remains to estimate centroids $C(q;i)$, shape factors $L[q;i]$, and width factors $\sigma(q;i)$ or $h(q;i)$. We describe below one reasonable solution. Assume the clusters are sufficiently well separated so that the kernel tails of neighboring clusters vanish. Then, with the description above, the density estimate within a cluster reduces to a single mode. Yet, in reconstructions (9) and (10), only moments of order 1 and 2 are used, so that the density within a cluster may be approximated by a Gaussian density. In other words, $K(\cdot)$ can be assumed to be a radial Gaussian kernel for this calculation:

$$K(u) = (2\pi)^{-d/2} exp\{-||u||^2/2\}.$$

With this approximation, maximum likelihood estimates can be easily computed. If $\tilde{p}_I(u|\omega_i)$ is maximized with respect to $C(q;i)$ and $\sigma(q;i)$, we obtain:

$$C[q;i] = \frac{1}{N[q;i]} \sum_{x(n)\in\Gamma_{q;i}} x(n), \tag{11}$$

$$\sigma(q;i)^2 = \frac{1}{d} \frac{1}{N[q;i]} \sum_{x(n)\in\Gamma_{q;i}} ||x(n) - C[q;i]||^2. \tag{12}$$

Next, if $\tilde{p}_A(u|\omega_i)$ is maximized with respect to $C(q;i)$, $h(q;i)$ and $L[q;i]$, we obtain:

$$L[q;i]\,L[q;i]^T = A[q;i], \quad \text{with} \tag{13}$$

$$A[q;i] = \frac{1}{N[q;i]} \sum_{x(n)\in\Gamma_{q;i}} (x(n) - C[q;i])(x(n) - C[q;i])^T, \tag{14}$$

$$h(q;i)^d = \det L[q;i]. \tag{15}$$

Thus, $L[q;i]$ is any positive square root of $A[q;i]$, for instance its lower triangular Cholesky factor. Of course, this estimate is biased. To remove the bias, one can replace $N[q;i]$ by $N[q;i] - 1$, as usual.

However, it is clear that these solutions are not the best possible, neither with respect to criterion (7), nor with respect to the misclassification rate. Since the final goal is actually classification, a criterion measuring deviations from the true densities, like (7), is not the most appropriate, especially if memory resources are strongly limited.

Thus, we propose in the next sections to find the best parameter set, $\{Q_i, C(q;i), \sigma(q;i), h(q;i), L[q;i]\}$, in the Bayes sense. The best solution obtained may –or may not– yield good density estimates, it does not matter, but it will lead to the best classification rate. Since this search is quite complicated, we shall reduce it to a local search, starting from initial guesses given by equations (11) to (15). Initially, there will be a single cluster per class (*i.e.* $Q_i = 1, \forall i$), and the numbers Q_i will increase until the space complexity bound is reached, $\sum_i Q_i = Q$.

3 Direct minimization of the Bayes risk

3.1 Clustering.

Two clustering procedures are used in this paper. The first one was proposed by Diday, and is sometimes referred to as the Dynamic Clusters algorithm [6]. Each iteration consists of two stages: assessment and assignment. *Assignment:* Each pattern x is assigned the cluster $\Gamma_{q;i}$ whose centroid is the closest. *Assessment:* Each cluster $\Gamma_{q;i}$ is characterized by its centroid, $C(q;i)$, the arithmetic mean of the patterns it contains, and its size $N[q;i]$.

Initially, centroids are arbitrarily fixed, on a grid for instance. We do not recommend so much to fix them in a random manner, for reproducibility. Iterations are run until clusters do not change. This algorithm terminates after a finite number of iterations.

The second procedure considered here is a modification of the previous one, and can be described as follows. Each cluster is characterized not only by $C(q;i)$ and $N[q;i]$, but also by its shape matrix, $L[q;i]$. *Assignment:* Each pattern x is assigned the cluster $\Gamma_{q;i}$ whose centroid is the closest, in the sense of the Mahalanobis distance, $\|L[q;i]^{-1}(x - C(q;i))\|$. *Assessment:* Each cluster is updated by computing its new centroid $C(q;i)$, and its new covariance matrix, $A(q;i) = L[q;i]\,L[q;i]^T$.

When the density estimator $\tilde{p}_I$ is used, the first procedure is utilized, whereas the second one is required for estimator $\tilde{p}_A$.

3.2 Total misclassification rate.

The set of examples is used for two purposes at the same time: first to build the classifier (learning), and second to evaluate its performances (testing). In the present case, these two aspects cannot be separated because clusters parameters required in learning are determined from the classification rate.

It is well known that if no particular care is paid, the performances will be strongly optimistic (resubstitution, overfitting). The only way we can perform this optimization without overfitting is by cross validation [7]. Among those rating methods, one is more attractive because every pattern is used once in performance evaluation and $N - 1$ times in learning. This method is the Totally Averaged Leave-One-Out (TALOO) procedure [7] [8] [4]. Beside the fact that the performance evaluation is unbiased, there are two key advantages in using TALOO that are not shared by other cross validation techniques such as the Holdout, or the Averaged Holdout (AH): the patterns are completely symmetrically utilized, and there is no loss of information.

Actually, TALOO may be seen as a particular AH where partitions are all of the form $\{1\}\{N - 1\}$, and are all generated. In practice, AH cannot be totally averaged except in that particular case, so that patterns cannot in general play all the same role.

Assume n ranges in $\{1, .., N\}$. Once $\tilde{p}(u|\omega_i)$ has been computed for every pattern u of the database, our algorithm first computes a LOO estimate of the density, leaving pattern $x(n)$ out:

$$\tilde{p}_I(u|\omega_i; n) \stackrel{def}{=} \tilde{p}_I(u|\omega_i) - \frac{1}{N_i} \frac{\bar{\delta}(i, i(n))}{\sigma(q(n); i)^d} K\left(\frac{u - C(q(n); i)}{\sigma(q(n); i)}\right), \qquad (16)$$

where $i(n)$ and $q(n)$ denote the class and the cluster of $x(n)$, respectively, and where $\bar{\delta}(a, b)$ is 1 if $a = b$ and zero otherwise. Here we used the trick proposed by Fukunaga to compute the LOO estimate from the Resubstitution one [8].

Next, the classification error is evaluated for that sample $x(n)$ as follows. Pattern $x(n)$ is assigned the class $\omega_{\hat{i}(n)}$, with:

$$\hat{i}(n) \stackrel{def}{=} Arg \max_i \{\hat{P}_i \tilde{p}_I(x(n)|\omega_i; n)\}. \qquad (17)$$

The elementary error is thus:

$$\epsilon(n) = \bar{\delta}(i(n), \hat{i}(n)). \qquad (18)$$

Last, all elementary errors are averaged to form the class- and the total- misclassification rates:

$$\epsilon(\omega_i) = \frac{1}{N_i} \sum_{x(n) \in \omega_i} \epsilon(n), \qquad (19)$$

$$\epsilon = \sum_i \hat{P}_i \epsilon(\omega_i) = \frac{1}{N} \sum_{n=1}^{N} \epsilon(n). \qquad (20)$$

A very similar procedure is defined if the second clustering algorithm is used. It is indeed only needed to substitute $\tilde{p}_A$ for $\tilde{p}_I$ everywhere, and to replace equation (16) by:

$$\tilde{p}_A(u|\omega_i; n) \stackrel{def}{=} \tilde{p}_A(u|\omega_i) - \frac{1}{N_i} \frac{\bar{\delta}(i, i(n))}{h(q(n); i)^d} K\left(\frac{L[q; i]^{-1}[u - C(q(n); i)]}{h(q(n); i)}\right). \qquad (21)$$

3.3 Optimization of parameters.

Clearly, ϵ is an implicit function of $\{Q_i, C(q; i), \sigma(q; i), 1 \leq q \leq Q_i\}$, or $\{Q_i, C(q; i), h(q; i), L[q; i], 1 \leq q \leq Q_i\}$. In other words, it can be minimized for any given set of examples. This is precisely how the memory and computational resources of a neural network can be used in the best way for the purposes of classification, as shown in the next section.

In order to cope with this optimization, one starts with $Q_i = 1$, and $L[q; i] = I$, and run one of the two clustering algorithms. This gives a first set of initial guesses. From that point, a polytope algorithm is run [13] and provides the best clusters parameters in the neighborhood (local minimum). Next, it is run again for the d cases where a single Q_i is increased by one, and the best is selected. And so forth.

This algorithm defines a sequence of tuples $(Q_1, Q_2, .., Q_c)$ of the same size as the number of classes, for which the best misclassification rate ϵ should decrease monotically. Of course, if the total number of clusters, Q, reaches N, ϵ reaches zero. But it is not relevant to go that far. When looking at the curve giving ϵ as a function of Q, it can be noticed a breaking point: this is the value of Q that should be retained.

If the data set is very large, or if the dimension d is large, then the algorithm above may be time consuming. A suboptimal algorithm can possibly be run instead, and consists of the following. Find $(Q_1, .., Q_c)$ by computing suboptimal solutions (11) to (15), at each iteration. Then fix the set of Q_i obtained, and find the best clusters parameters in a second stage. In this simplified algorithm, there is a single polytope optimization.

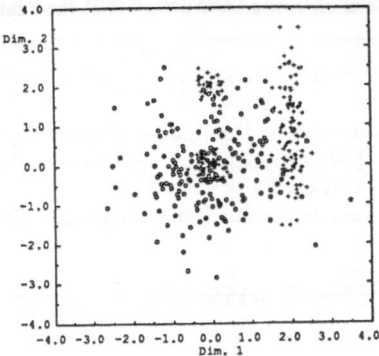

Figure 1: The *clouds* database

4 Simulation results

4.1 Description of the database

- Bidimensional distributions, two classes.
  400 patterns, 200 in each class (50% in each class).

- Class 0 : sum of three different Gaussian distributions with $P_{w_0} = 0.5$ and:

$$p(x|w_0) = \frac{1}{2}\left(\frac{p_1(x)}{2} + \frac{p_2(x)}{2} + p_3(x)\right),\qquad(22)$$

with

$$p_j(x) = \frac{1}{2\pi}\frac{1}{\sigma_{jx}\sigma_{jy}}e^{-\left(\frac{(x-m_{jx})^2}{2\sigma_{jx}^2} + \frac{(y-m_{jy})^2}{2\sigma_{jy}^2}\right)}\qquad(23)$$

where m_{jx} and m_{jy} are the x and y means of the j's distribution while σ_{jx}, σ_{jy} are their x and y standard deviations:

j	σ_{jx}	σ_{jy}	m_{jx}	m_{jy}
1	0.2	0.2	0.0	0.0
2	0.2	0.2	0.0	2.0
3	0.2	1.0	2.0	1.0

- Class 1 : a single normal distributions with $P_{w_1} = 0.5$ and:

$$p(x|w_1) = \frac{1}{2\pi}e^{-\frac{x^2+y^2}{2}}\qquad(24)$$

- The graphical representation of this dataset can be seen in figure 1.

4.2 Results

All the simulations have been carried out with the suboptimal algorithm as described in section 3.3, using both clustering algorithms. By using table 1, we can find the optimum trajectory in the (Q_1, Q_2) space. For the $\tilde{p}_I$ estimator, we follow the path $(1, 1) \rightarrow (1, 2) \rightarrow (1, 3) \rightarrow (1, 4) \rightarrow (1,5)$. For the $\tilde{p}_A$ estimator, we follow the route $(1, 1) \rightarrow (1, 2) \rightarrow (1, 3) \rightarrow (2, 3) \rightarrow (2, 4)$. Due

class 0 → / class 1 ↓	1	2	3	4	5
1	30.00	32.50	32.25	31.75	32.25
2	30.00	33.25	32.50	29.25	36.25
3	18.00	16.75	18.25	17.25	19.25
4	16.00	16.00	15.00	14.75	18.25
5	15.25	16.25	15.00	15.50	17.75

class 0 → / class 1 ↓	1	2
1	30.00	29.50
2	16.25	17.00
3	13.25	13.25
. 4	13.25	12.25
5	13.75	13.00

Table 1: Total misclassification rate according to $C(q; i)$ using $\tilde{p}_I$ (left) and $\tilde{p}_A$ (right)

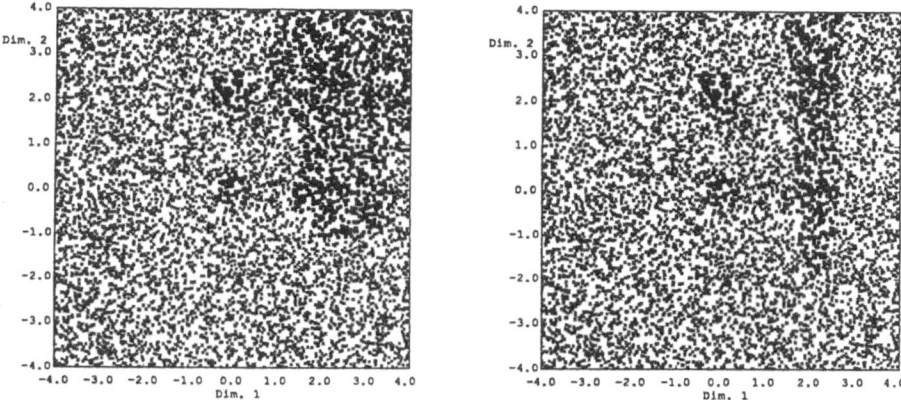

Figure 2: Decision surfaces for $(Q_1, Q_2) = (1, 4)$ using $\tilde{p}_I$ (left) and $\tilde{p}_A$ (right)

to the unstability of the clustering algorithm using the $\tilde{p}_A$ estimator, it was impossible to have more than 2 clusters for class 0.

We can immediately remark that the $\tilde{p}_A$ estimator gives much better result than $\tilde{p}_I$. The figure 2 shows that the clusters for the class 2 are narrower and fit better the data of figure 1 the same clusters using the $\tilde{p}_I$ estimator.

4.3 Choice of (Q_1, Q_2)

If we plot the total misclassification rate for the chosen path, we can clearly see a break located at $(1, 3)$ for $\tilde{p}_I$ and $(1, 2)$ for $\tilde{p}_A$. We have chosen the next point after the break with a small derivative value, so that we are sure that ϵ has finished to decrease. This gives us the point $(1, 4)$ for $\tilde{p}_I$ and $(1, 3)$ for $\tilde{p}_A$.

4.4 Polytope algorithm

We have run the polytope algorithm for the two chosen points. We have optimized using the centroid $C(q; i)$ and the sigma $\sigma(q; i)$ of the clusters as optimization parameters for $\tilde{p}_I$. We have reached 13.75% using point $(1, 4)$, which is less than any result of table 1.

We have optimized using only the centroid $C(q; i)$ of the clusters as optimization parameters for $\tilde{p}_A$ (i.e. $h(q; i)$ and $L[q; i]$ were each time recalculated with the new $C(q; i)$). However, we have not found any difference in misclassification error using the point $(1, 3)$ with these parameters (still 13.25%) This is probably due to the fact that $L[q; i]$ was not optimized.

All the simulations and figures have been produced by the Packlib environment [12] developed in the frame of the ESPRIT-BRA project 6891.

5 Conclusion and future work

This work aimed at using optimally finite hardware constraints. As a consequence, efficient computing of probability density functions for Bayesian classification requires sub-optimal methods, avoiding to compute as many kernels functions as there are vectors in the learning set. We have shown that polytope optimization used with different clustering algorithms can reach this objective under certain stability conditions. Although costly in terms of computing time, this optimization must be done only during learning. Results for the $\tilde{p}_A$ density estimator are encouraging but stability problems prevent from using it with a large number of clusters.

Future work consists of optimizing all parameters instead of only centroids, and finding a stable clustering algorithm with shape matrices. One idea is to use $\tilde{p}_I$ during the clustering, then $\tilde{p}_A$ during polytope optimization.

References

[1] L. BREIMAN, W. MEISEL, E. PURCELL, "Variable kernel estimates of multivariate densities", *Technometrics*, vol. 19, no. 2, pp. 135–144, May 1977.

[2] T. CACOULLOS, "Estimation of a multivariate density", *Annals of Inst. Stat. Math.*, vol. 18, pp. 178–189, 1966.

[3] P. COMON, "Supervised classification, a probabilistic approach", in *ESANN-European Symposium on Artificial Neural Networks*, Verleysen, Ed., Brussels, Apr 19-21 1995, D facto Publ.

[4] P. COMON, C. JUTTEN et al., "ELENA, Axis A: Theory", Esprit Basic Research Project 6891, CEC, June 1993.

[5] P. COMON, J. L. VOZ, M. VERLEYSEN, "Estimation of performance bounds in supervised classification", in *ESANN-European Symposium on Artificial Neural Networks*, M. Verleysen, Ed., April 20-22 1994, pp. 37–42, D facto Publ.

[6] E. DIDAY, "The dynamic clusters method in non hierachical clustering", *Int. Jour. Computer Inf. Sciences*, vol. 2, pp. 61–88, 1973.

[7] R.O. DUDA, P.E. HART, *Pattern Classification and Scene Analysis*, Wiley, 1973.

[8] K. FUKUNAGA, R.R. HAYES, "Estimation of classifier performance", *IEEE Trans. PAMI*, vol. 11, no. 10, pp. 1087–1101, Oct. 1989.

[9] D.J. HAND, *Kernel Discriminant Analysis*, RSP press, 1982.

[10] E. PARZEN, "On the estimation of a probability density function and the mode", *Ann. Math. Stat.*, vol. 33, pp. 1065–1076, 1962.

[11] T. J. WAGNER, "Nonparametric estimates of probability densities", *IEEE Trans. on Inf. Theory*, vol. 21, no. 4, pp. 438–440, July 1975.

[12] Y. CHENEVAL, "Packlib, an interactive environment to develop modular software for data processing", *to appear* in *IWANN-International Workshop on Artificial Neural Networks*, Mira, Ed., Madrid, Jun 7-9 1995, Springer-Verlag.

[13] J.A. NELDER, R. MEAD, "A simplex method for function minimization", in *Comput. J.*, vol. 5, pp 308-313, 1965

Daily Electrical Power Curves: Classification and Forecasting Using a Kohonen Map

Marie Cottrell [1], Bernard Girard [1], Yvonne Girard [1], Corinne Muller [2], Patrick Rousset [1]

(1) SAMOS Université Paris 1 90, rue de Tolbiac 75634 PARIS Cedex 13 FRANCE	(2) Electricité de France Direction des Etudes et Recherches 1, avenue du Général de Gaulle 92141 CLAMART FRANCE

Abstract : This paper addresses an extensively studied problem : how to forecast the daily half-hour electrical power curve. Many methods have been developed, classical linear methods (like ARIMA methods) as well as neural ones. In this paper, we present a very simple method : the past daily curves are normalized and one considers the corresponding profile (with mean 0 and variance 1). These profiles are classified using a Kohonen map. Then, for some future point, a strategy is defined in order to compute its typical profile, the mean and the variance are forecast and the expected power curve is computed. This method uses little computation time and is easy to develop. The first results are satisfactory and promising.

1 The Problem

The goal is to predict the complete half-hour electrical power curve of the following day in order to help Electricité de France satisfy consumption at any time and regulate production in function of the demand. For each day, the curve is composed of 48 half-hour power loads.

The idea which first comes to mind is to extend the tried and tested one-step forecast, by means of Multilayer Perceptrons as in [1], [5], [6] for example. But, we know ([3], [8]) that it is not possible due to the non-linearity of the recurrence equation defining the model. In fact, when the predicted values are successively taken as inputs for the long term forecast, the limit values of the output depend strongly upon the weights and of the initial values. The reason for this phenomenon is the existence of several fixed points, some of them which attract, while others repulse.

It would also be possible to consider a Multilayer Perceptron with 48 output units in order to associate the complete following daily curve to the input variables (exogenous variables like meteorological variables, nature of the day, past values of the power, etc). But this method does not work well. For example, in [4], it is shown that a previous classification is necessary and the authors propose to use as many perceptrons as classes that have been identified. In [7], another mixed method is developed : firstly, a classification of the curves is achieved, then, 48 Multilayer Perceptrons specialized within each half-hour are defined inside each class, so that 48 by 4 different Multilayer Perceptrons are necessary. Even if the use of the Statistical Stepwise Method (SSM) [2],[3], allows to reduce as much as possible the number of parameters, this method is somewhat complex and not very easy to put into practice.

In this paper, we propose another method to forecast the next curve, which is essentially based on a classification. It is a semi-parametric method, parametric for the prediction of the mean and variance of the next day, non parametric for the normalized profile.

2 The data

We have power curves for 2188 days dating from January 2, 1986,. Among them, 2084 are ordinary days (weekdays or week-end days) and 104 are public holidays or extra days.

The first remark is that the power curves are quite different from one day to another. See Figures 1, 2.

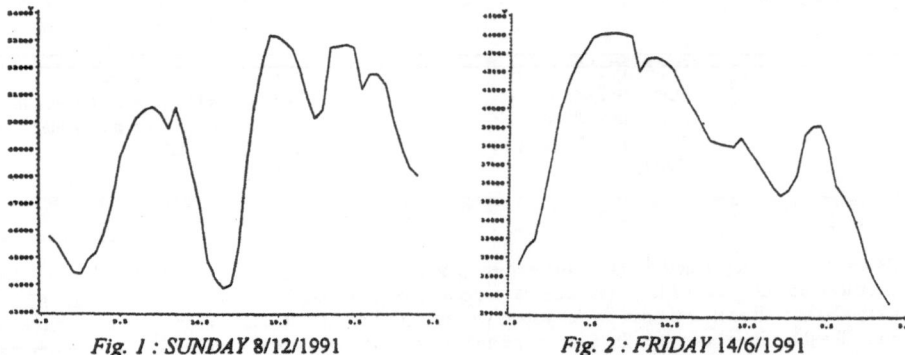

Fig. 1 : SUNDAY 8/12/1991   *Fig. 2 : FRIDAY* 14/6/1991

The second remark is that the shape of the curve leads us to fix the origin of the day at 4H30 a.m.. The values which correspond to this half-hour seem to be those least dispersed and are the minima or approximately minima. So, from now, the days begin at 4H30 (as in Figures 1 and 2). For a day j, the power curve is denoted by $C(j) = (C(j,h)$, $h = 1, ..., 48)$ and is composed of the observed values of the power at 4h30, 5H, 5H30, ..., etc.

We split up the curve $C(j)$ into three characteristics : the mean $m(j)$ of the day j , its variance $\sigma^2(j)$, and the profile P(j) defined by

$$P(j) = \left(P(j,h), h = 1,...,48\right) = \left(\frac{C(j,h) - m(j)}{\sigma(j)}\right)$$

The methodology that we propose here consists of three steps :

1) *predict the mean and the variance* of a day by means of any method (ARIMA, Multilayer Perceptron, for example),
2) *make up a classification of the profiles* just defined,
3) for any given day needed to be forecast, build *its typical profile* and compute the power curve by *redressing it* (multiply by the standard deviation and add the mean).

In this paper, we do not address the problem 1), that is the prediction of the mean and of the variance. See for example [] for this point. In the rest, we consider these two parameters as if they were known.

3 Classification of the profiles

The Kohonen algorithm is principally used with some specific features. First of all, to take into account the annual periodicity (January is between December and February which belong to two distinct years), we consider a cylindrical 2-dimensional network where the left and right sides are attached, but not the top and bottom, as is showed in Figure 3.

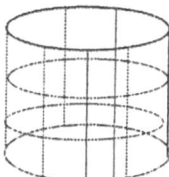

Fig. 3 : Topology of the network

Since the variability remains almost stable according to the half-hour, the distance is the Euclidean distance with equal weighting for each coordinate in $\mathbf{R}^{48}$. Finally, as all input vectors are profiles with a norm equal to 1, the weights are constrained to remain on the unity sphere. Let us denote N the number of units.

Let us write down the Kohonen algorithm which is used to train the network :

- initialize the weights at random on the unity sphere

- at each step t, if the current values of the weights are denoted by $w_i(t)$,
      - present a profile P
      - look for the winner unit i_0, that is the unit whose weight vector is the closest to the profile P
      - modify the weights of i_0 and of its neighbors on the network by

$$w_i(t+1) = w_i(t) + \varepsilon(t)(P - w_i(t)), \text{ for } i = i_0 \quad \text{or} \quad i \text{ neighbor of } i_0$$

      where $\varepsilon(t)$ is a decreasing to 0 adaptation parameter
      - normalize again the vectors w_i

After training the Kohonen map, each unit is represented by its weight vector and the classification of the 2188 profiles consists in associating to each profile the class defined by the number of the winner unit. The Kohonen algorithm allows us to classify the profiles into N classes, with the usual property : two similar profiles belong to the same class or to neighbor classes. See in Figure 4, the result of the classification for a 10 by 10 network. Each class is represented by the weight of its unit.

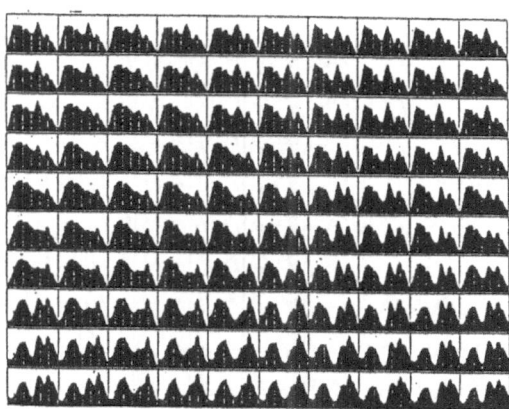

Fig. 4 : the 10 by 10 classification

We can observe that neighbor classes have very similar typical profiles. To facilitate the interpretation of this classification in terms of days and seasons, we can divide into 13 classes, in a hierarchical classification, these 100 classes making them easier to identify. See Figure 5 the new big classes with their name.

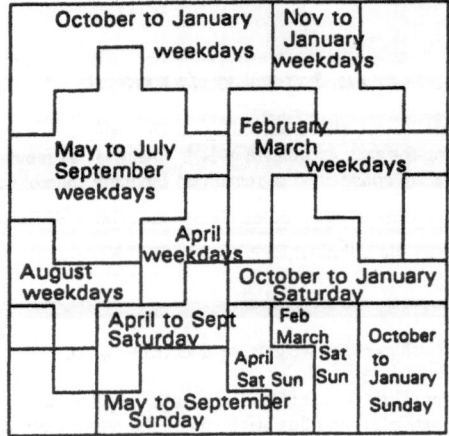

Fig. 5 : Super-classes and their identification

Do not forget that the right side is attached to the left one. We can see that the top of the network is occupied by the week-days while the bottom corresponds to the Saturdays and the Sundays. We also observe that there is a continuity from the winter days to the summer days, via the spring and autumn days.

Finally, let us represent the contents of the classes. The Figure 6 shows all 2188 profiles drawn into its corresponding class. We can see that the dispersion is small and that there are few badly represented profiles.

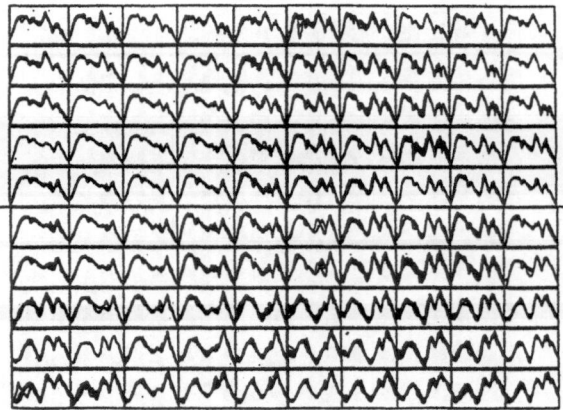

Fig. 6 : Contents of the classes

4 Forecast

Remember that we suppose that the mean and the variance of the following day are known. Let us have to forecast the profile of a day *j*. We build a table which gives for each day, defined by its month and its name, all the classes where such a day has been classified, together with the number of its instances. See for example in Figure 7, that the Tuesdays of October were 6 times in the class 1, 10 times in the class 2, and so on.

Month	Day	Class	Number
October	Tuesday	1	6
October	Tuesday	2	10
October	Tuesday	11	4
October	Tuesday	12	6
October	Tuesday	21	1

We observe that the concerned classes are adjacent in the network, so that even if several profiles could be chosen for this day, all these profiles are very similar.

Fig. 7 : Distribution of the Tuesdays of October

So the predicted profile of a Tuesday of October (for example) will be computed as the weighted mean of all the typical profiles of the concerned classes.

More generally, if *j* is a day, if *(aⱼᵢ , i=1,...,N)* are the number of instances of this day in all the classes (*aⱼᵢ* is 0 if day *j* never was in the class *i*), if *Wᵢ* is the weight of the unit *i*, the estimated profile of the day *j* is given by :

$$\hat{P}(j) = \frac{\sum_{i=1}^{N} a_{ji} W_i}{\sum_{i=1}^{N} a_{ji}}$$

Then the profile is redressed (multiplied by the standard deviation and increased by the mean). The predicted curve for a day *j* is given by

$$\hat{C}(j) = \sigma(j)\hat{P}(j) + m(j)$$

See in Figures 8 and 9 two forecast curves drawn in comparison with the real ones.

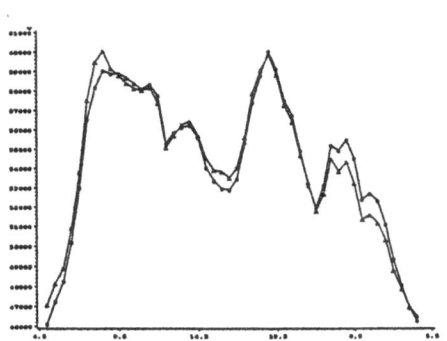

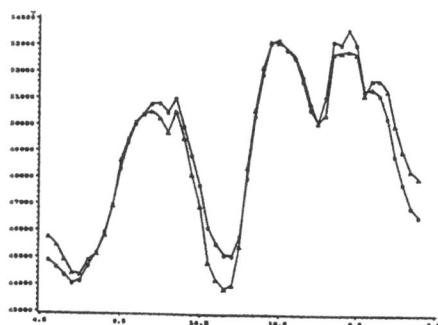

Fig. 8 :
*Observed (A) and forecast (O) curve
for the Tuesday 15/1/9191*

Fig. 9:
*Observed (A) and forecast (O) curve
for the Sunday 8/12/1991*

5 Conclusions

We conclude this paper by studying the averaged relative errors. The relative error for a curve C is defined by $E = \frac{1}{48} \sum_{k=1}^{48} \frac{|C_k - \hat{C}_k|}{C_k}$, where C (resp. $\hat{C}$) denote the observed curve (resp. the predicted one).

The next tables (Figures 10, 11, 12) represent the global averaged relative error, the same by month, the same by year.

	Number	Mean
Extra days and public holidays	104	0.0189
Others	2084	0.0151
All the days	2188	0.0152

Fig. 10 : Averaged relative error

Year	Number	Mean
1986	346	0.02
1987	346	0.017
1988	348	0.014
1989	350	0.013
1990	346	0.013
1991	348	0.013

Fig. 11 : Averaged relative error by year, without the extra days and the public holidays

Month	Number	Mean
January	176	0.016
February	169	0.016
March	162	0.018
April	197	0.017
May	148	0.015
June	178	0.013
July	176	0.011
September	157	0.015
October	208	0.014
November	163	0.016
December	174	0.017

Fig. 12 : Averaged relative error by month without the extra days and the public holidays

One can observe that the results are better since 1988, what seems to indicate a change of structure (change of the rate timetable for example). One can also verify that the modes (the most frequent values) of the relative errors distributions are equal to 0.12 for the ordinary days as well as for the extra days and public holidays, even if the averaged value is greater for the extra days and public holidays.

In addition to these results, we apply the proposed method to the year 1992, which was not used for the classification, and we get an averaged relative error of about 0.015 as for the global result. This fact is really promising and encouraging.

The performances cannot be really compared for the moment to the performances of the linear model which is generally used, because we do not estimate the mean value and the variance. However, we see that the precision is about 1.5%, while the usual models give about 2.4%. The main advantage of the methodology that we propose, are its simplicity, its small computation cost and its adaptability.

We are now working to incorporate the prediction of the mean and the variance together with our methodology and the first results are very satisfactory.

References

[1] R.Capone, S.Kimbrough, Using a neural network to predict electricity generation, *Proc. WCNN 94*, San Diego, pp. I-324-329, 1994.

[2] M.Cottrell, B.Girard, Y.Girard, M.Mangeas, Time serires and neural networks : a statistical method for weight elimination, *Proc. of ESANN 93*, M.Verleysen Ed., Editions Quorum, (ISBN 2-9600049-0-6), 1993.

[3] M.Cottrell, B.Girard, Y.Girard, M.Mangeas, C.Muller, Neural modeling for time series : a statistical stepwise method for weight elimination, *IEEE Transactions on Neural Networks, in press* , Prepublication SAMOS No. 20., 1993.

[4] A.Garcia Tejedor, M.Cosculluela, C.Bermejo, R.Montes, A neural system for short-term load forecasting based on day-type classification, *to appear in Proc. ISAP 94*, 1994

[5] K.L.Ho, Y.Y.Hsu, C.C.Yang, Short term forecasting using a multilayer neural network with an adaptative learning algorithm, *Transactions on Power Systems*, Vol. 7, No. 1, pp. 141-149, 1992

[6] K.Y.Lee, Y.T.Cha, J.H.Park, Short-term load forecasting using an artificial neural network, *Transactions on Power Systems*, Vol. 7, No. 1, pp. 124-131, 1992.

[7] C.Muller, M.Cottrell, B.Girard, Y.Girard, M.Mangeas, A neural network tool for forecasting french electricity consumption, *Proc. WCNN 94*, San Diego, pp. I-360-365, 1994.

[8] Q.Yao, H.Tong, Quantifying the influence of initial values on nonlinear prediction, *Technical Report*, No. UKC/IMS/S92/5c, University of Kent, U.K., 1992.

CMAC Real-Time Adaptive Control Implementation on a D.S.P. Based Card

G.MERCIER, K.MADANI CMTF/LIIA Centre de Sénart Université Paris 12

Adresse : G.MERCIER IUT Sénart, département MI
Avenue Pierre POINT
77127 LIEUSAINT France
Tel : 33-1-64-13-44 80   Fax : 33-1-64 13 45 01

Abstract :

Numerous paper on CMAC and adaptive control, have been published. However few of them relate effective hardware real-time implementation. G.Kraft [Kraft 90] exposes an adaptive scheme, based on a first order processes (i.e. integrator). This application brings interesting simulation results, but first order process is not useful with real processes.

First we describe briefly a modified CMAC paradigm, with an on-line computation of connections to reduce memory size and allowing real-time process control. In a second step, we present the DSP architecture car, we have developed, and give comparisons results in learning and transfer phase between different processors.

Then in a third phase, we expose the modified Kraft structure, with a second order unknown process and we detail the result of a manipulation and give experimental results.

I The CMAC paradigm.

The CMAC algorithm belongs to a class of neural networks that consists of an adjustable combination of fixed basis functions and a linear update rule. The action of this rule is described by equations (1) .

$$h_o = \phi\left(\|x - \gamma_o\|\right)$$
$$O_h = \sum_{j=1}^{M} w_{jk} * h_o \quad (1)$$

$\emptyset = \emptyset(r)$ is a basis function, γ_j is the j th basis function centre and h_o is its heights at the input x, ω_{jk} is the contribution of the basis function to the h_{th} component of the output $\hat{O}_h$ and M is the total number of basis function and hence is proportional to the memory requirements.

The normalized output is obtained by relations (2):

$$\emptyset_r \begin{cases} 1, r < G/2 & (2) \\ \\ 0, r > G/2 \end{cases}$$

G is the generalization factor of the network and $r = \| x' - \gamma_j \|$.

where

$$\hat{O}_h = \frac{\sum_{j=1}^{M} W_{ij} * h_o}{\sum_{j=1}^{M} h_o} \quad (3)$$

This allows the network output to be written in a simplified way (4) for the selected weights.

$$\widehat{O_h} = \frac{\sum_{j=1}^{G} W^*_{ok}}{G} \qquad (4)$$

The network is trained in a supervised manner. It is forced to emulate an **(x,y)** by applying the update Widrow rule to such pair. The training set of data consists of all such pairs may be presented to the network several times.

$$\omega^n_{ok} = \omega^{n-1}_{ok} + \eta \varepsilon^n$$
$$\varepsilon^n = \widehat{O}_k - O^n_k \qquad (5)$$

where **n** is iteration index, η is the learning rate (usually comprised between 0.2 and 0.8). The learning rate value affects only the learning time and has no effected on the accuracy of the solution.

Physical Realisation :

The main problem is to find an iterative relation between binary input vectors and activated memory cells performing the size output. We have developed an on line relation between a N dimensional input vector{ X,Y,W;... }.

An appropriate choice of basis function centers gives rise to G regular N dimensional lattices (N=input dimension), in which N-d cube is of G side, G is quantisation intervals (or generalization factor).The lattices are so arranged that any input falls within exactly G*N d-cubes, one from each lattice observation. For a two dimensional input vector, relation between the G activated memory cells and inputs vectors is given by equations (6) and (7).

$$K_{ij} = [(X_i/G) * (2^r + G-1) + (Y_j/G)] -1 \quad (6)$$

$$O = \sum U_{ij} = K_{ij} * [1- (sup\ K_{ij} - inf\ K_{ij})] \quad (7)$$

where Uij are the activated cells addresses for a binary coded pair input vector {X,Y }for i and j varying from 0 to G. K_{ij} and $U_{ij} \in N^+$ and **O** the output computed value.

In classical CMAC architecture, a connection-matrix is computed at the initialisation operation phase, this matrix occupies several k-bytes memory proportional to the squared bit of the resolution of the input vectors. The use of this "on line "relations spares the maximum memory and avoid "hashing" techniques. In practice for real time applications, because of the limitation of the memory size, the solution for a 16 bit CPU is limited to a 3 dimensional 8 bit resolution input vector.

II Physical neural support : The DSP card.

The Figure 1 gives the block diagram, of hardware implementation

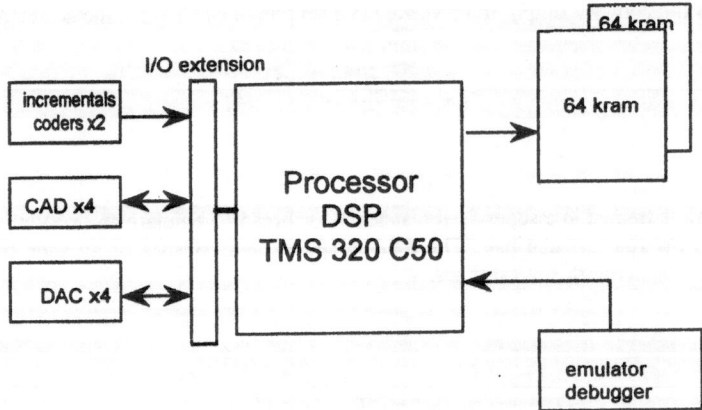

Figure 1: DSP Hardware implementation

Analogue I/O are connected to ADC and DAC converters throw the I/O expansion connector of the DSP. For the position controller application, a 1 ms sampling time (input and output) was the first objective for the neural network, and is widely under the performances of the processor. Incremental coders inputs are also available for position and velocity information. Memory map is shared between Data and Program. External memory is 64 Kbytes (50ns access time). The DSP has 9 k bytes internal memory (35 ns access time) useful for recurrent instructions and temporary data storage. The maximum external memory space is 224k bytes.

A J-TAG controller allows emulation with a PC for the software development and debugging.

III Application : Second order adaptive control .

The adaptive schematic proposed by G. Kraft is given by *Figure 2*.

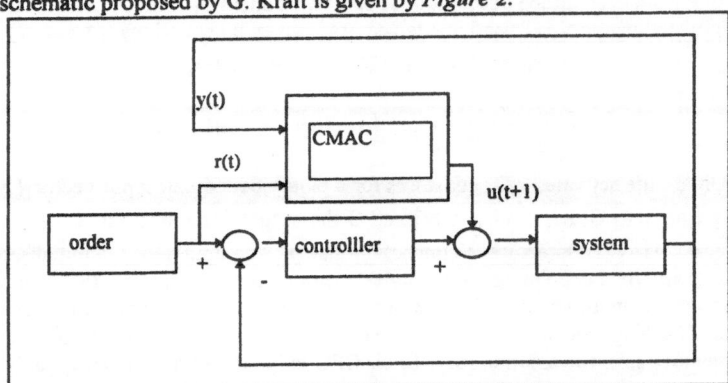

Figure 2 : Adaptive schematic

If u(t) is the input of a first order system and y(t)its output, the discrete equation of this system can be written under the form:

$$y(t+1)= a*y(t) + b*u(t). (8)$$

where a and b are unknown parameters and t the discretisation time.

Then we can define a predictive element which determines the command input u(t) of the system, function of the actual output y(t) and the desired future value y(t+1).

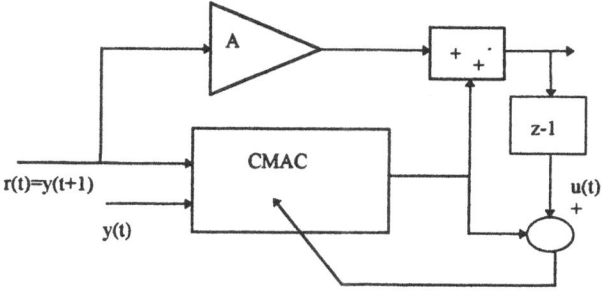

Figure 3 : CMAC learning.

If the parameters a and where known, it would be sufficient to store predifined values into the CMAC memory and to use it as a look-up table. In the present case, a and b parameters are unknown, the area surface $\phi\{y(t),y(t+1),u(t)\}$ must be learned. The modification of weights in the memory is obtained by the Widrow learning rule.

$$m(t+1)=m(t) + \beta*[u(t)-m(t)]. \qquad (9)$$

where m(t) represents the contents of the cells memory for the contribution of the output calculation such as

$$m = \sum_{i=1}^{g}\sum_{j=1}^{g} m_{ij}(t)$$
, (g = generalisation factor)   (10)

The CMAC output u(t+1) is compared to its previous value u(t). The error comparison modifies the weights using the Widrow rule (8).The principle of the CMAC learning is given by the *Figure 3*.

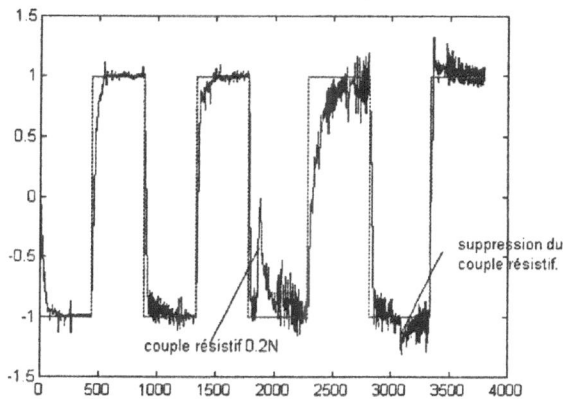

Figure 4 : simulation results for a second order DC motor

We applied this model to a second order DC motor, results of the simulation are given by the *Figure 4*.

The network converges in the first iteration, but oscillations appear in the following periods. If we applied a resistive torque, we notice that the compensation occurs immediately during the same cycle, and if this torque is suppressed, the reverse compensation is immediate.

To prevent these oscillations due to the second order term of the DC motor transfer function, we have introduced a first order function G(p) between the order r(t) and the CMAC input. This function is smoothing the rectangular input order and expand the activated input sensors of the CMAC. The real schematic is presented by the *Figure 5*

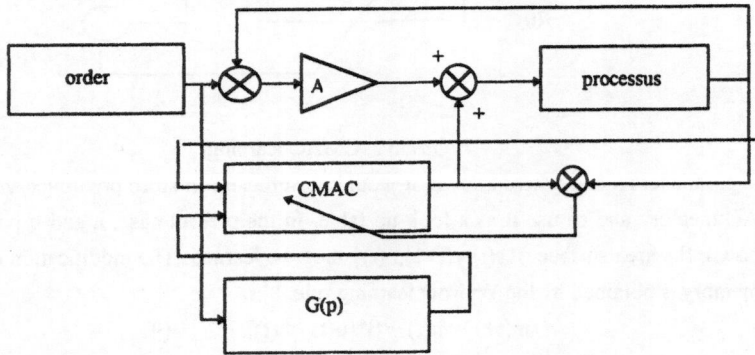

Figure 5:_Modified structure with G(p) transfer reference function

Practical mounting .

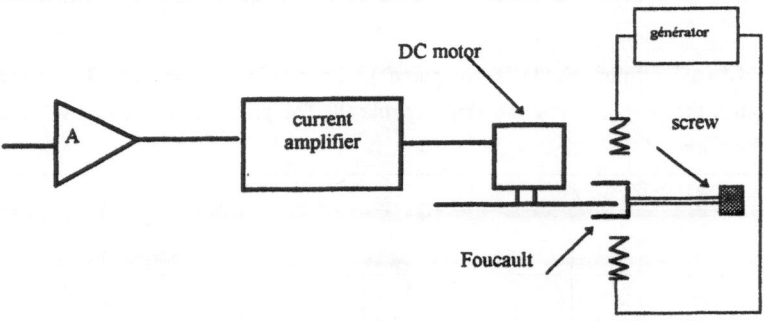

Figure 6: Experimental platform including an electric motor with a Foucault effect braking system.

Experimental results.

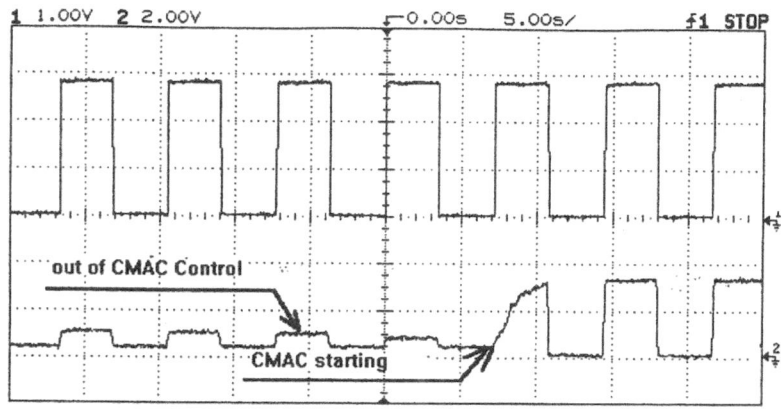

Figure 7 : Network learning

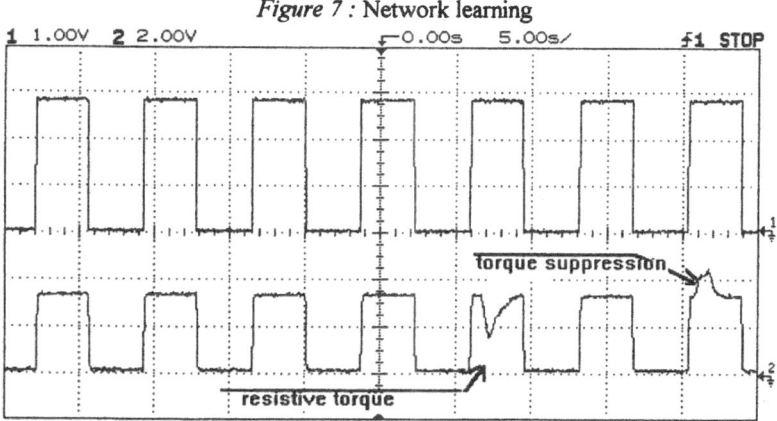

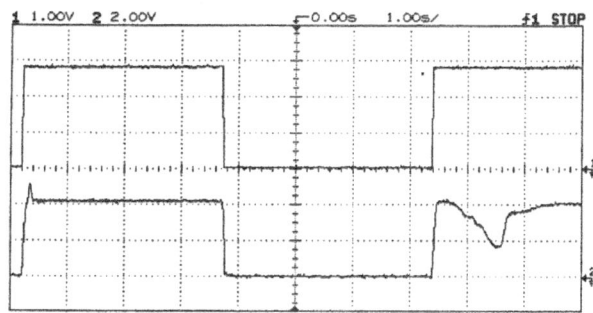

Figure 8 a-b : application and suppression of resistive torque.

IV Conclusion .

We have demonstrate that the CMAC network is interesting because its convergence speed in learning and in transfer time (0.24 ms with the DSP card).Most of publications concerning CMAC applications are only simulated, because this network requires a lot of memory, hashing code is usually used to compress virtual memory because of limitation of the physical memory. We have retained an on-line calculation of the connections, this method spares a lot of physical memory;and the hashing coding is no longer necessary.

The material support has been developed with a based DSP card, that we have entirely defined and used for a real application. This application consists in the adaptive control of a real second order DC motor, with unknown parameters by introducing a modification to the classical first order application of G.Kraft. We have presented the measurements of this application.

V : Bibliography .

[HOR 89]   Hornik K *Multilayer feedforward networks ar e universal approximators,* Neural Networks, 2, pp 359-366 -1989.

[JOR 89]   Jordan MI. *Generic constraints on underspecified target trajectories.* Proceedings of International Joint Conference On neural Networks. Washington Vol I, pp 217-225.

[KRA 90]   Kraft G & Campagna D.*A Comparison between CMAC Neural Network Control and two traditionnal Adaptive Control Systems.* American Control Conference Pittsburg 1989 pp 21-23.

[MER 94-1]   G.Mercier,K. Madani , C.Barret *Various Neural Network Implementations on a low cost and versatile DSP card., EUFIT 94 pp1482-1488.*

[MER 94-2]   G.Mercier, K Madani *A new on-line CMAC Algorithm for Real tTme Applications, R.I Informatics congress St petersbourg, May 94.*

[NAR89]   Narandra KS & Partharsaraty K *Stable Adaptative Systems, Englewood Cliffs ,NJ Prentice Hall 1989.*

[NAR 90]   Narandra KS & Partharsaraty K. *Identification and control of dynamic systems using neural networks.* IEEE Transactions on Neural Networks, Vol 1 pp 4-27.

[PSA 87]   Psaltis,. D, Sideris A., &Yamamura A., *Neural controllers,*Proceedings of IEEE first International Conference on Neural Networks,San Diego,Vol 4, pp 551-558.

[SAE89]   Saerens M.&Soquet. M, *A neural controller,* Proceedings of IEEE first International Conference on Neural Networks,London pp 211-215.

[SAE91]   Saerens M.&Soquet. M, *Neural Controller based on Backpropagation Algorithm.*IEEE peoceeding F-138, pp 55-62.

Neural Networks in Digital Data Transmission[1]

Inma Ortuño Ortín
Joan Serra Sagristà

UCCD, Dpt.Informàtica, Fac.Ciències
Universitat Autònoma de Barcelona
08193 Bellaterra (Barcelona)
SPAIN

Phone: +34 3 581 30 12
Fax: +34 3 581 24 78
Email: joans@melq.uab.es

January 1995

Abstract

This paper applies to both the applications of artificial neural networks and to communications systems. To ensure a reliable digital data transmission over noisy insecure channels, information must undergo a coding process. Neural nets can be used as both coders and, much more interesting, decoders, outperforming by far performance achieved with algebraic methods when data is sent over an *AWGN* channel. In particular, we show that a three-layer feed-forward neural network is sufficient to decode any *Reed-Solomon* (*RS*) code. The simulation results for the $RS[8, 4, 3]$ code over $GF(2^2)$ show that soft decision neural network decoding after transmission over an *AWGN* channel could give $1.75dB$ coding gain relative to hard decision.

---

[1]Submitted to International Workshop on Artificial Neural Networks (IWANN'95)

1 Introduction

This paper presents results of an investigation of soft decision neural network decoding as applied to *Reed-Solomon* codes.

Any communications system consists of a data source sending out a message, a channel through which the message is sent, and a receiver who gets the message. Quite often the channel introduces noise and perturbation. In order to minimize its harming effect on the information sent, messages undergo a coding process before entering the channel: some amount of redundancy is added in a certain way. At the output of the channel the receiver is faced with the problem of retrieving the original message as reliable as possible. This decoding process can then be done in two different ways: in hard decision, when only the quantified symbol most probably sent is chosen; or in soft decision, when the perturbed data together with probabilities, mean and standard deviation of the channel, are all forwarded to the next step of the decoding algorithm. Soft decision decoding methods are only suitable for those codes allowing the use of a minimum euclidean distance, for instance, convolutional codes, but not for linear codes as *Reed-Solomon* codes.

In recent papers, [TaSe 94, ShSw 94] new methods for soft decision decoding of *Reed-Solomon* codes are presented, both trying to reduce the complexity of the decoder. In this work, instead of the standard methods, we develop a new soft decision decoding method by means of neural networks that further reduces the complexity of the decoder, but at the price of not being too computationally efficient when the code length is large.

Despite the well known result stating that *RS* codes are bad in the sense of Gilbert-Varshamov bound, this family of codes is among the most widely used in digital data transmission, from spacecraft and militar communications to Compact Disk (CD) industry.

Reed-Solomon codes, a subset of multilevel cyclic block codes (in particular non-binary *BCH* codes) with powerful error correcting capabilities (since they are maximum distance separable they have largest possible minimum distance and thus maximum error correcting capability), are very efficient in computational aspect when algebraic decoding is applied (see [MaSl 77] for a general introduction to the theory of error correcting codes). They may, however, give weaker performance compared to convolutional codes, at least at moderate bit error rates (around 10^{-5} to 10^{-6}) on the *Additive White Gaussian Noise (AWGN)* channel. This disadvantage mainly results from the lack of a general applicable method for soft decision decoding of linear codes.

We describe a neural net model that can be used to decode words belonging to a *Reed-Solomon* code and to correct noisy patterns that it receives from an *AWGN* transmission channel. A simulation model incorporating an *AWGN* channel has been used to establish the performance of our decoding method.

2 Neural networks as decoders

Since the decoding process of any code, even in the linear case, can be classified as a *NP-Complete* problem (see for instance [BeMcTi 78, GaJo 78]); and since, on another side, as soon as 1985 John Hopfield [HoTa 85] proved that neural networks could be useful to solve some *NP-Complete* problems, the step to take was quite clear: try to relate the two fields.

From the work by Bruck and Blaum [BrBl 89] to that by Rampone et al. [RaEsTa 94], numerous authors have concentrated on this subject. Most of them do not try to develop new decoding algorithms, but simulate a maximum likelihood decoding with the help of neural networks, benefiting from their inherent advantages, i.e., parallel computing capability, graceful degradation under hardware failure, generalisation, robustness in the presence of noise, distributed information storage,...

Different kinds of neural network models have been proposed to implement the decoding process in a communications system. Depending on the topology, the unit processor and the learning rule, some methods give better performance than others. For instance, Hopfield networks proposed to decode linear codes and convolutional codes suffer from their low storing capability. Recurrent neural networks would be a good choice it it were not for the engagement between synchronous or asynchronous updating rule, difficult to argue in the context of error correcting codes. And last, hard-limiter units or threshold logical units are best suited for *Binary Symmetric Channels (BSC)*, but not for *AWGN* channels.

3 Neural network model

The model presented here is designed to decode a particular extended $RS[N = 4, K = 2, \delta = 3]$ code over $GF(q = 4)$; the same construction can be further enlarged to design a neural network decoder for a general $RS[N, K, \delta]$ code over $GF(q)$. For the sake of simplicity, the given code is first mapped onto a binary $RS[n = 8, k = 4, \delta = 3]$ code over $GF(2^2)$, with the special mapping $0 \longrightarrow -1$ and $1 \longrightarrow 1$. That is, codewords have "binary" length 8. The minimum distance between two codewords is $\delta = N - K + 1 = 3$, and the error correcting capacity is given by $t = \lfloor \frac{\delta-1}{2} \rfloor = 1$.

The neural network is a fully interconnected three-layer feedforward net with $n = 8$ units in the input layer, $q^K = 4^2 = 16$ units in the hidden layer, and $k = 4$ units in the output layer. For a general $RS[N, K, \delta]$ code over $GF(q)$, again a fully interconnected three-layer feedforward net is appropiate; the method requires first a mapping onto a "binary" $RS[n = Nm, k = Km, \delta' \geq \delta]$ code over $GF(2^m)$, and then build the neural network with n neurons in the input layer, the binary codeword length; q^K neurons in the hidden layer, as many neurons as codewords; and k neurons in the output layer, the binary length of the information vector sent. This means the neuron number required to carry away the decoding process grows exponentially with the number of codewords, which is not a practical solution for codes with large parameters.

The basic model of neuron is that of figure 1.

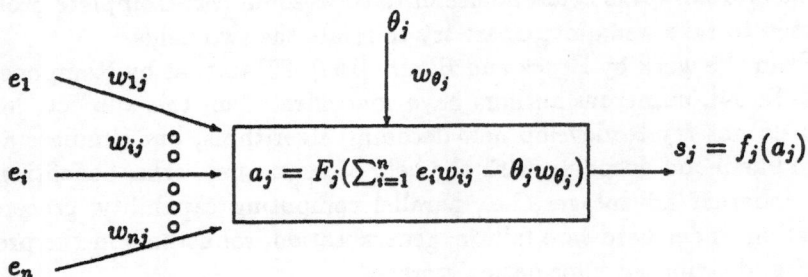

Figure 1: Diagram for neuron j

Neurons in the input layer only forward towards the network the received perturbed vector at the end of the transmission channel, they have an identity transfer function (concatenation of the activation function F_j and the output function f_j).

Neurons in the hidden layer can be of two different types: perceptron-like units, that is

$$s_j(t) = a_j(t) = \begin{cases} 1 & \text{if } \sum_{i=1}^n (e_i w_{ij}) \geq \theta \\ 0 & \text{if } \sum_{i=1}^n (e_i w_{ij}) < \theta \end{cases}$$

or, with a sigmoidal activation function

$$s_j(t) = \frac{1}{1 + e^{-\sum_{i=1}^n (e_i w_{ij}) - \theta}}$$

Neurons in the output layer are thus of also two different types, with an identity transfer function or with a hard-limiter transfer function respectively. In both cases, the neural network output is a binary vector of length k.

3.1   Training Phase

The training phase depends on the activation function used by neurons in the hidden layer.

If they use a perceptron-like function, then weights can be properly initialized and no training is required at all. Each one of the hidden neurons is associated with a codeword. When presented an input vector ∇, we have to choose that codeword that best approximates to ∇, that is, we want only a hidden neuron to be active. This goal can be met by setting the initial (and final) weights between the input and the hidden layer to the same "bit" values of the expected input vector bound to a right codeword, so that the neuron closest to the input vector will have the maximum activation value, and by setting all hidden neuron biases to $\theta = n - \delta + 1$, this will be the winner neuron. Weights from the hidden to the output layer are set accordingly to the codeword the hidden neuron is associated with.

If neurons from the hidden layer use a sigmoidal activation function, then learning is required and can be done with the *backpropagation* algorithm. Weights and biases must be randomly initialized.

4 Simulation results

We have performed simulations for the *Reed-Solomon RS*[8, 4, 3] code over $GF(2^2)$ both for the *BSC* and the *AWGN* channel. In the first case (not exposed here) coding gain achieved by the neural network decoder relative to plain transmission is the same to that achieved by algebraic methods, but this avoids the multiplication of the received vector by the parity check-matrix.

The whole communications system has been simulated. Data source, coding process and channel transmission has been implemented with Maple V.2, meanwhile for the neural network decoder we have used the SNNS v3.2 package (Stuttgart Neural Network Simulator [2]).

When transmission takes place over an *AWGN* channel, simulations show a coding gain of $5.6dB$ and $1.75dB$ relative to plain transmission and hard decision respectively for a bit error rate of 10^{-5}. Work by Shin and Sweeney [ShSw 94] showed a coding gain of $2dB$ relative to hard decision when applied to a larger code, a *RS*[15, 13, 1] code, but their method demands more computational resources. Furthermore, our neural network has never wrongly decoded, but always either decoded properly, or detected the error to be larger than its correcting capacity. Figure 2 show these results.

5 Conclusion

This work is devoted to an application of neural networks for error correcting decoding of *Reed-Solomon* codes. We have designed a fully interconnected three-layer feedforward neural network that performs the decoding process of a *RS*[8, 4, 3] code over $GF(2^2)$; hints have been given to extend this design to more general *RS*[N, K, δ] codes over $GF(q)$.

The idea of connecting neural nets and error correcting codes was first proposed by Bruck and Blaum, where, as in our case, one can not expect

---

[2]Free Software under the aegis of GNU avalaible via anonymous ftp at ftp.informatik.uni-stuttgart.de, /pub/SNNS.

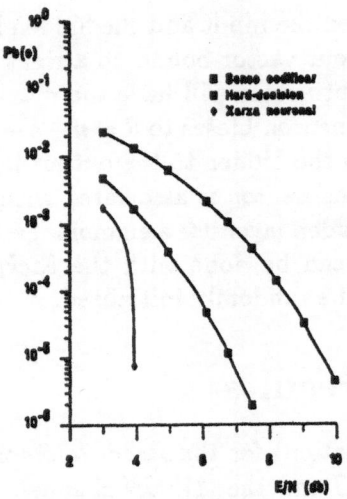

Figure 2: Neural network decoding performance of $RS[8,4,3]$ code

the network to converge to the optimal solution unless the neuron number increases exponentially (in a sense, this large neuron number meets the bound given by Bulsari [Bul 93]). In contrast, after learning the code, our neural network reduces the error decoding probability to zero for a noise version of codewords with a number of errors in the range of the error correcting capacity of the code, with the further advantage of a fast and simple training.

Two different activation functions have been used, one requiring a training phase, and another that needs a proper weight initialization but no learning, both achieving similar performance. This may suggest that theoretical studies of the case given may dramatically diminish learning time. Simulation results show that soft decision decoding of the $RS[8,4,3]$ code over $GF(2^2)$ would give $1.75dB$ coding gain over hard decision decoding.

References

[BeMcTi 78] Berlekamp,E.R., McEliece,R. & Tilborg,H. "On the inherent intractability of certain coding problems", *IEEE Trans. on Inf. Theory*, Vol. 24: 384-386, 1978.

[BrBl 89] Bruck,J. & Blaum,M. "Neural Networks, Error-Correcting Codes, and Polynomials over the Binary n-Cube", *IEEE Trans. on Inf. Theory*, Vol. 35, No. 5: 976-987, 1989.

[Bul 93] Bulsari,A. "Some analytical solutions to the practical form of the general approximation problem for feedforward neural networks", *Technical Report Abo Akademi*, March 1993.

[GaJo 78] Garey,M. & Johnson,D. *Computers and Intractability: A Guide to the Theory of NP-Completeness*, Freeman 1978.

[HoTa 85] Hopfield,J.J. & Tank,D.W. "Neural Computing of Decision in Optimization Problems", *Biological Cybernetics*, Vol. 52: 141-152, 1985.

[MaSl 77] MacWilliams,F.J. & Sloane,N.J.A. *The Theory of Error-Correcting Codes*, North-Holland Mathematical Library, 1977.

[RaEsTa 94] Rampone,S., Esposito,A. & Tagliaferri,R. "A Neural Network for Error Correcting Decoding of Binary Linear Codes", *Neural Networks*, Vol. 7, No. 1: 195-202, 1994.

[ShSw 94] Shin,S.K. & Sweeney,P. "Soft decision decoding of Reed-Solomon codes using trellis methods", *IEE Proc.-Commun.*, Vol. 141, No. 5: 303-308, October 1994.

[TaSe 94] Taipale,D.J. & Seo,M.J. "An Efficient Soft-Decision Reed-Solomon Decoding Algorithm", *IEEE Trans. on Inf. Theory*, Vol. 40, No. 4: 1130-1139, July 1994.

A New Neural Network Approach to the Floorplanning of Hierarchical VLSI Designs

M. Saheb Zamani and G. R. Hellestrand

Computer and Systems Technology Laboratory, School of Computer Sc. and Eng.
The University of New South Wales, Sydney 2052 Australia

Abstract: In this paper, we introduce a new neural network approach to the floorplanning of VLSI (Very Large Scale Integrated circuit) designs. The network used is a Kohonen self-organising map. An abstract specification of the design is converted to a set of appropriate input vectors fed to the network at random. At the end of the process, the map shows a 2-dimensional plane of the design in which the modules with higher connectivity are placed adjacent to each other, hence minimising total connection length in the design. The approach can be extended to consider external connections and is able to floorplan a rectilinear boundary. These features makes the approach capable of floorplanning hierarchically specified designs. Since only the global ordering phase is needed for the floorplanner, the process is fast.

1   INTRODUCTION

Floorplanning is an early process in the layout design of VLSI circuits. The floorplanning problem in known to be NP-complete [10]. Given an abstract specification of a circuit, a floorplanner is to find the placement and shape of all modules in the design and the port positions on the modules, in a 2-dimensional plane so that an objective function (usually a combination of connection length and wasted area) is minimised. The specification normally consists of a list of connectivities[1] between the modules in the form of a matrix, called a connectivity matrix, an estimation of the size of modules, and sometimes restrictions on the module port positions and the shapes of the modules.

Floorplanning is particularly useful for hierarchically specified circuits (Figure 1) where, in a top-down process, the shape and the port positions of the submodules are not fixed and must be determined by the floorplanning process. In a top-down strategy, at the highest level of the specification hierarchy, the rectangular area of the chip is partitioned into a set of regions corresponding to the submodules. These submodules are, in turn, partitioned further and the process is iterated recursively according to the topology of the hierarchy tree (Figure 1). The shape of the submodules is usually taken to be rectangular but there is no reason for this assumption other than simplifying algorithms.

Even rectangular floorplanning is an NP-hard problem [10]. One of the most common approaches to floorplanning is rectangular dualisation in which a graph representing the module connectivities is converted to another graph from which the rectangular dissection of a rectangle can be derived [2, 5, 6]. Since the connectivity graph should be planarised, triangulated and simplified to have no complex triangles[2], the time complexity of such algorithms is non-polynomial due to the fact that the planarisation and complex triangles elimination processes are NP-complete [11]. In all the above approaches, rectangular shapes are assumed for the submodules. This may restrict the solution space and may result in large wasted spaces and as well long connections in the design.

In this paper, we present an algorithm based on the Kohonen self-organising map [4]. The self-organising principle has been applied to placement problem [3, 8, 1] which is a subproblem of floorplanning in which the shape, size and port positions of all modules in the design are fixed. However, to our knowledge, no approach to floorplanning based on neural networks has been reported.

In our approach, a set of input vectors corresponding to each submodule is calculated and presented to

---

[1] Connectivity is often considered as the number of connections between modules. It can also be defined as a number indicating the importance of connections to be minimised.

[2] A complex triangle is a 3-edge cycle which is not an inner face.

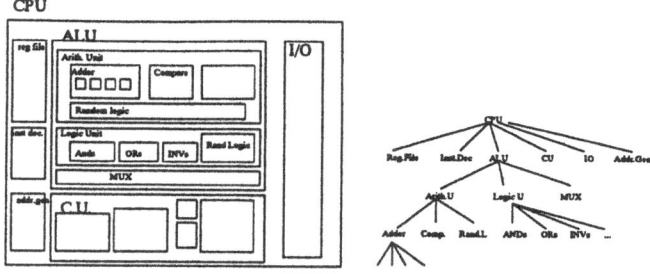

Figure 1: Design Hierarchy

the map at random. At the end of the process, the map will show a dissection of a rectilinear shaped module into a set of rectilinear submodules. In the following section, the idea is explained in detail.

2 BASIC CONCEPT

A rectangular grid of $w \times h$ nodes is to be dissected into n regions (corresponding to n modules). The connectivity information is given in the form of a matrix whose entry, C_{ij} $i, j \in [0, n-1]$, is the connectivity between the module pair i and j. To minimise the total connection length in the design, the heavily connected modules should be placed adjacent to each other. Since in general, satisfying all the adjacency requirements may not be possible in a 2-dimensional plane, a self-organising map is used to map this higher dimensional proximity information onto a 2-dimensional plane.

The Kohonen self-organising process starts with a set of distance vectors representing module connectivities and assigns the vectors with highest connectivities (least distances) to the neurons which are geographically close in the map after a number of iterations. To apply the self-organising principle to the floorplanning problem, we calculate a vector for each module so that the distance between each pair of them reflects the final desired proximity in the 2-dimensional map. The desired proximity can be derived from the connectivity information; the heavier the connectivity C_{ij}, the closer (in distance) the modules i and j should be placed. Around these main vectors, a set of random vectors is generated near to the main vector of each module. These vectors are then fed to the network and recycled at random during successive iterations. At the end of the process, the distance vectors belonging to a module are assigned to the neurons which end up being close to each other (hence producing the contiguous rectilinear shape of the module on the grid) and the distance vectors belonging to the modules heavily connected to this module are assigned to the neurons geographically close to its grid nodes (neurons).

3 DISTANCE VECTORS

The objective is to generate a set of vectors whose similarities (distances) correspond to the connectivities between n modules. The following equation holds for all vector pairs $\vec{V}_i = (v_{0i}, v_{1i}, ..., v_{n-1,i})$ and $\vec{V}_j = (v_{0j}, v_{1j}, ..., v_{n-1,j})$:

$$\sum_{k=0}^{n-1}(v_{ik} - v_{jk})^2 > \sum_{k=0}^{n-1}(v_{ik} - v_{lk})^2 \qquad iff \quad C_{ij} < C_{il}$$

or if the vectors are normalised vectors:

$$\forall i, j = 0, ..., n-1 \qquad \vec{V}_i . \vec{V}_j = K.C_{ij} < 1 \qquad if \quad i \neq j \tag{1}$$

$$\vec{V}_i . \vec{V}_i = 1 \qquad if \quad i = j \tag{2}$$

where K is a constant factor to keep the v_{ij}s within the domain of real numbers[3].

This results in a system of $\frac{n(n+1)}{2}$ equations with n^2 variables, so we can select the $\frac{n(n-1)}{2}$ variables arbitrarily to generate a solution. To simplify the solution process, we take

$$v_{ij} = 0 \qquad \forall i > j.$$

The vectors resulting from the above equations are assigned to each module as main vectors. For the floorplanning problem, in the case where the number of output neurons is greater than n and extra distance vectors

---

[3] K must be so as to make v_{ik} small enough to satisfy equation 2.

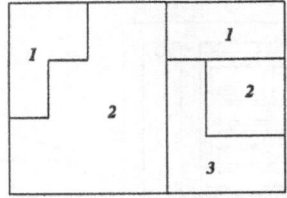

Figure 2: Disintegration of modules where the random distance vectors are not close enough to the main vectors.

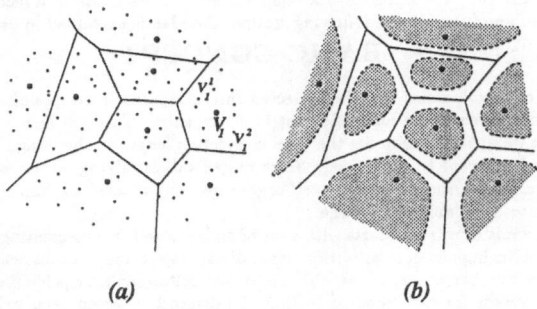

(a)   *(b)*

Figure 3: An illustration of n main vectors and their random vectors. (*a*) Totally random vectors, (*b*)random vectors in shaded area.

around the main vectors specifying a module are needed, a set of random vectors whose distances to the main vectors are within a certain range (half of the distance to their second closest main vector), is generated. If these randomly generated vectors are too far from the main vectors, the neurons assigned to a particular module may become separated and cause the disintegration of that region (Figure 2), which is usually undesirable in VLSI layout design. On the other hand, if the randomly generated vectors around a main vector are too close to the main vector, then a number of random distance vectors may be assigned to a single neuron and a local minimum may easily result.

Where the disintegration problem is not important, a network may be trained for any n-module problem by a set of totally random vectors in n-dimensional space. Then the main distance vectors are presented to each neuron to assign the modules to the grid nodes. In this case, a network can be trained for any n-module design only once, and thenceforward, according to the connectivity information of a particular problem, the nodes are assigned to modules properly.

In order to illustrate the above discussion, we show n vectors as n points in 2 dimensions (Figure 3), eventhough they are n-dimensional. If we generate totally random distance vectors regardless of the positions of the n main vectors (Figure 3.a), since the network may not be able to consider the proximity of any two vectors, say $\vec{V_1^1}$ and $\vec{V_1^2}$, it may position them at widely separated neurons on the 2-dimensional map, and this may result in dis-contiguous regions for some modules. To counteract this problem, random vectors associated with each main vector are generated inside the shaded area shown in Figure 3.b, where the solid lines form a Voronoi diagram[4] constructed from the n-points and the dashed borders contain the regions for which the random inputs are generated. For example, for the point i, the dashed border contains all the points which have a distance to the point i which is less than half of its distance to the second closest main point. Instead of finding the position of the Voronoi regions and hence the borders, a "generate and test" strategy is used in

---

[4]Voronoi diagram of n points is a partitioning of a space into n regions in such a way that a region $i \in [0, n-1]$ contains all the points which are closest to the point i.

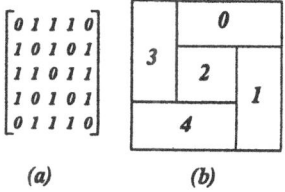

Figure 4: 5-module wheel. (a): Connectivity matrix, (b):an optimum floorplan.

$$\begin{array}{cccccccccc}
0 & 0 & 0 & 0 & 0 & 3 & 3 & 3 & 3 & 3 \\
0 & 0 & 0 & 0 & 2 & 2 & 3 & 3 & 3 & 3 \\
0 & 0 & 0 & 0 & 2 & 2 & 3 & 3 & 3 & 3 \\
0 & 0 & 2 & 2 & 2 & 2 & 2 & 3 & 3 & 3 \\
0 & 0 & 2 & 2 & 2 & 2 & 2 & 3 & 3 & 3 \\
1 & 1 & 2 & 2 & 2 & 2 & 2 & 4 & 4 & 4 \\
1 & 1 & 2 & 2 & 2 & 2 & 2 & 4 & 4 & 4 \\
1 & 1 & 1 & 1 & 2 & 4 & 4 & 4 & 4 & 4 \\
1 & 1 & 1 & 1 & 4 & 4 & 4 & 4 & 4 & 4 \\
1 & 1 & 1 & 1 & 4 & 4 & 4 & 4 & 4 & 4 \\
\end{array}$$

Figure 5: The result of floorplanner for the wheel example.

which a vector is generated randomly and checked against the following relation:

$$|\vec{V_p} - \vec{V_{fc}}| < \frac{|\vec{V_p} - \vec{V_{sc}}|}{2} \tag{3}$$

where $\vec{V_p}$ is the generated random vector, $\vec{V_{fc}}$ is the closest main point to $\vec{V_p}$ and $\vec{V_{sc}}$ is the second closest main point to it. If the relation holds, the vector is accepted; it is rejected otherwise. The same number of random vectors are generated as there are output neurons.

4 EXPERIMENTAL RESULTS

The floorplanner[5] was first applied to a simple example known as the 5-module wheel whose connectivity matrix is given in Figure 4.a. One of the optimal floorplans, if the modules are taken as rectangles, is shown in Figure 4.b. The map used was a grid of 10×10 neurons which were connected to 5 input neurons. For each module, $20 = \frac{10 \times 10}{5}$ distance vectors were generated around the main vectors and were fed to the network at random, for 1000 iterations. Only the global ordering phase [4] was used to train the network since for floorplanning there is no need for a fine-adjustment phase. The initial neighbourhood was set to 7 to cover most of the neurons, and was then decreased linearly with time to 0 after 1000 iterations. The gain factor was initialised to 0.4 and decreased linearly with time to 0.04 after 1000 iterations, as recommended by Kohonen [4]. The result is shown in Figure 5 which is optimum with respect to meeting all adjacency requirements. (The number at a neuron corresponds to the main vector to which that neuron responds best).

The second example was a real VLSI design adopted from [6](Figure 6.a). The design consists of 7 modules with the connectivity matrix shown in Figure 6.b. The network parameters were the same as described above. The result is shown in Figure 7. Almost all module proximity requirements are met except those with a small number of connections. (The dualisation floorplan in [6] fails to meet the important proximity requirement between the modules 0 and 5 with 17 connections but satisfies all other requirements).

The last example is of a rectangular floorplan not realisable in a 2-dimensional plane (Figure 8). The floorplanner automatically ignored some of the adjacency requirements (between modules 1 and 2, and to some extent between modules 0 and 1) to realise the floorplan in 2 dimensions (Figure 9.a). The parameters were the same as above. In another attempt when the gain factor was decreased to 0.01 after 1000 iterations, a better result was obtained (Figure 9.b). In this experiment, only one adjacency requirement was ignored, and this is the least number of connectivity relations which can be ignored to make the floorplan realisable.

The time required by the algorithm on a SUN4 workstation for all the above examples was less than 7

---

[5]The algorithm used is exactly as given in [7]

(a)

$$\begin{bmatrix} 0 & 3 & 2 & 2 & 11 & 17 & 3 \\ 3 & 0 & 0 & 0 & 0 & 0 & 16 \\ 2 & 0 & 0 & 16 & 0 & 10 & 0 \\ 2 & 0 & 16 & 0 & 0 & 10 & 0 \\ 11 & 0 & 0 & 0 & 0 & 0 & 32 \\ 17 & 0 & 10 & 10 & 0 & 0 & 16 \\ 3 & 16 & 0 & 0 & 32 & 16 & 0 \end{bmatrix}$$

(b)

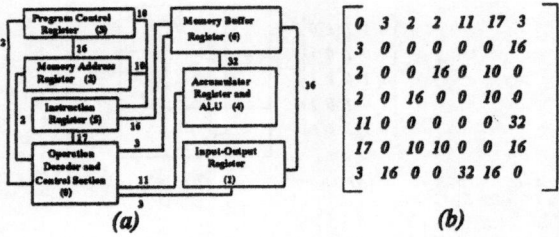

Figure 6: (*a*): Schematic diagram of a real design. (*b*): Connectivity matrix.

0	0	0	5	5	5	2	2	2	2
0	0	0	5	5	5	2	2	2	2
0	0	0	5	5	5	3	3	3	2
0	0	0	5	5	3	3	3	3	3
4	4	4	4	3	3	3	3	3	3
4	4	4	4	3	3	3	3	3	3
4	4	4	4	6	6	3	3	3	3
4	4	4	6	6	6	6	1	1	1
4	4	6	6	6	6	6	1	1	1
4	4	6	6	6	6	6	1	1	1

Figure 7: The result of the floorplanner for the design in Figure 6.

$$\begin{bmatrix} 0 & 1 & 1 & 0 & 0 \\ 1 & 0 & 1 & 1 & 1 \\ 1 & 1 & 0 & 1 & 1 \\ 0 & 1 & 1 & 0 & 0 \\ 0 & 1 & 1 & 0 & 0 \end{bmatrix}$$

(a)

(b)

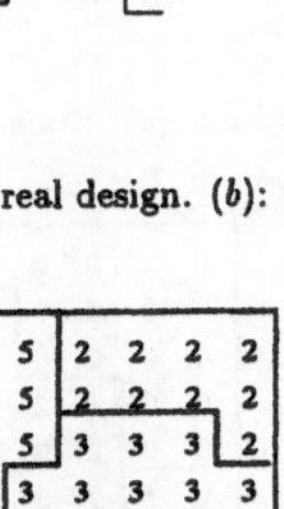

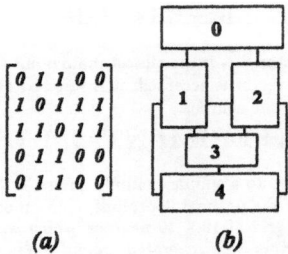

Figure 8: (*a*): Schematic diagram of the third example. (*b*): Connectivity matrix.

4	4	4	4	4	4	2	2	2	2
4	4	4	4	4	4	2	2	2	2
4	4	4	4	3	3	3	2	2	2
4	4	4	3	3	3	3	2	2	2
4	4	1	3	3	3	3	3	2	2
1	1	1	3	3	3	3	3	2	2
1	1	1	3	3	3	3	3	0	0
1	1	1	3	3	3	3	0	0	0
1	1	1	3	3	3	0	0	0	0
1	1	1	3	3	0	0	0	0	0

4	4	4	4	4	4	2	2	2	2
4	4	4	4	4	2	2	2	2	2
4	4	4	4	2	2	2	2	2	2
4	4	4	4	2	2	2	2	2	2
3	3	3	1	1	1	2	2	2	2
3	3	3	1	1	1	1	2	2	2
3	3	3	1	1	1	1	0	0	0
3	3	3	1	1	1	1	0	0	0
3	3	3	1	1	1	1	0	0	0
3	3	3	3	1	1	1	0	0	0

Figure 9: The result of floorplanner for the design in Figure 8: (*a*) with the previous parameters, and (*b*) with modified parameters.

	No of modules	grid	Time (sec.)
example 1	5	10 × 10	5.0
example 2	7	10 × 10	7.0
example 3	5	10 × 10	5.0
example 3	5	5 × 5	1.8

Table 1: The time required by the algorithm.

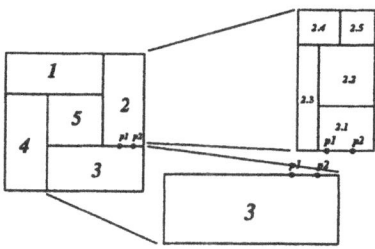

Figure 10: Setting I/O ports position at one level of hierarchy to guide the floorplan operating at a lower hierarchy level.

seconds which shows the approach is practical. The summary of the results in terms of efficiency is given in Table 1.

5   CONCLUDING REMARKS

In this work, we have only considered the internal connections between modules. This is reasonable for flat designs (i.e. consisting of only one level of hierarchy) and where there are no I/O pad constraints. For hierarchically specified circuits, the floorplan of the upper hierarchy levels can guide the floorplanning process of the lower levels by applying constraints to the submodules' ports positions [9]. For example, the floorplan obtained in Figure 10 can guide the process of finding the floorplan of module 2 to place the submodules connected to the ports $p1$ and $p2$ (i.e. submodule 2.1) close to the bottom part of the module to reduce connection length. These constraints can be considered in our floorplanner by initialising the weight vectors of the neurons at the desired port positions to values close to the main vectors of the submodules connected to the ports, depending on the number of connections between them. In other words, the ports are considered as single point submodules whose positions are fixed. In this way, the floorplanner can handle complex cases where a port is connected to many modules with different connectivities and can place the modules with more connections closer to that port.

Another strength of this approach is its capability to work with rectilinear shapes[6]. The complexity of rectilinear shapes results in current floorplanners employing rectangular dissection approaches. Especially in hierarchical floorplanning, even if a rectilinear floorplan is obtained, the floorplanning of modules within a rectilinear boundary at lower hierarchy level, can result in high complexity. In our approach, this can be done simply by considering a rectilinear map with the shape obtained from the upper level floorplanning process.

The relative size of submodules can be controlled by considering different neighbourhood sizes for different submodules according to their relative sizes. During the process of decreasing the neighbourhoods, the relative sizes of neighbourhoods should be preserved.

References

[1] R-I Chang and P-Y Hsiao. Arbitrarily sized cell placement by self-organizing neural networks. *Proceedings of IEEE International Symposium on Circuits and Systems*, pages 2043–2046, 1993.

[2] W. R. Heller, G. Sorkin, and K. Maling. The planar package planner for system designers. *Design Automation Conference*, pages 253–260, 1982.

[3] A. Hemani and A. Postula. Cell placement by self-organisation. *Neural Networks*, 3:377–383, 1990.

[4] T. Kohonen. The self-organizing map. *Proceedings of the IEEE*, 78(9):1464–1480, September 1990.

[5] K. Kozminski and E. Kinnen. An algorithm for finding a rectangular dual of a planar graph for use in area planning for VLSI integrated circuits. *Design Automation Conference*, pages 655–656, 1984.

---

[6]The grid size should not be smaller than the size of leafcells (e.g. gates in gate array design style).

[6] S. M. Leinwand and Y-T Lai. An algorithm for building rectangular floorplans. *Design Automation Conference*, pages 663–664, 1984.

[7] R. Lippmann. An introduction to computing with neural nets. *IEEE ASSP Magazine*, pages 4–22, 1987.

[8] R. Sadananda and A. Shrestha. Topological maps for VLSI placement. *International Joint Conference on Neural Networks*, pages 1955–1958, 1993.

[9] M. Saheb Zamani and G. R. Hellestrand. An integrated approach to the floorplanning, placement and routing of hierarchical designs. *Australasian Computer Science Conference*, 1995.

[10] S. Sahni. The complexity of design automation problems. *Design Automation Conference*, pages 402–411, 1980.

[11] S. Tsukiyama, K. Koike, and Shirakawa. An algorithm to eliminate all complex triangles in a maximal planar graph for use in VLSI floor-plan. *Proceedings of IEEE International Symposium on Circuits and Systems*, pages 321–324, 1986.

A Neural Network Approach to Quality Control Charts

Thomas STÜTZLE

Department of Statistics and O.R.
University Complutense, Madrid

Abstract

In this paper Quality Control Charts without memory are compared to neural networks trained with the Backpropagation algorithm. Neural networks are used to decide whether a process is under statistical control or out of control. As only the last sample is used to decide upon the state of the production process, a comparison to Shewhart-control charts leads automatically to a comparison between statistical tests and neural networks. By using a combined control chart to control the process mean and the process variability, the kind of classifications of the kind of change is considered explicitly. Finally neural networks are used to classify the kind of change occurred, considering only the last sample.

1 Introduction to quality control charts

Control Charts are used as an aid to check whether a production process is under or out of statistial control. A process is said to be out of statistical control when some parameters like the process mean or variability have changed. In order to control the paramters of a production process at constant intervals of time, a sample of the process is taken and the realization of a quality characteristic is measured. Let $x_t = (x_{t1}, x_{t2}, .., x_{tn})^T$ be the vector of these measurements. This vector can be interpreted as a realization of the random vector $X_t = (X_{t1}, X_{t2}, .., X_{tn})^T$. The running of a control chart can be considered as a periodically repeated statistical test, where the null hypothesis is H_0: *The process is under statistical control* and the alternative hypothesis is H_1: *The process is out of statistical control.* As with every test errors of type 1 and type 2 can be commited. The possible errors are **Error of type 1:** *Some action is taken although the process is under control* and **Error of type 2:** *No action is taken although the process is out of control.* The errors of type 1 are also called false alarms. False alarms are to be avoided because an interruption of the production process and an unnecessary search for possible errors cause avoidable costs.

Often Shewhart control charts are used to control the mean and the variability of a production process. In this paper a $\bar{x}$-chart, that uses the sample mean as sample statistic, is chosen to control the process mean and a s-chart, that makes use of the sample standard deviation, is taken to control the process variability. These control charts do not consider past samples and only the most recent sample is used to test whether some parameter has changed.

To control the mean and the variability of a production process at the same time, often a $\bar{x}$-chart and a s-chart are used in parallel. This possibility was also considered here. The interesting point for a combined control chart is, that a sole increase of the process variability results in an increase of the probability that the mean chart gives an out of control signal, independently of the sample size n chosen. Otherwise a change of the process mean has no effect on the s-chart. This may lead to false classifications of the kind of process change. These false classifications are to be avoided because a correctly classified change may give an indication for possible faults and facilitate the search for them.

One possibility to avoid these false classifications would be to use past samples as inputs for some classification procedures and to use them to distinguish the different kinds of change.

2 Applications of neural networks in quality control

Recently there have been some attempts to use neural computing for special tasks in the Statistical Process Control (SPC). One approach is to compare control charts to neural networks. Pugh (1991) tried to imitate a Shewhart Control Chart for the process mean with neural networks. In Smith (1994) a neural network was trained to replace a combined $\bar{x}$-R-chart.

Another kind of application of neural networks was considered in the article of Guo, Dooley (1992). When a shift of the parameters by two control charts for the process mean and the process variability was detected, they used neural networks to determine whether the out-of-control signal was actually owed to a shift of the process mean or of the process variability. They compared the results obtained with neural networks to a quadratic discriminant function and some obvious heuristics that consider the control chart that first signals a change in the parameters. Here it is tried to improve the evaluation of a Quality Control System to give a better indication for possible errors.

Another possibility is the interpretation of control charts to detect unnatural patterns on the charts. This approach was described in the articles of Hwarng, Hubele (1993) and Pham, Oztemel (1994). Some unnatural patterns on control charts are described for example in Montgomory (1985). These unnatural patterns also indicate an out-of-control situation. In Hwarng, Hubele the Backpropagation-algorithm was used, whilst Pham, Oztemel make use of some Learning Vector Quantizer algorithm (LVQ).

In the last two possible approaches, neural networks were used as a possible classification tool for patterns on control charts. This may be the more usefull application of a nonparametric method like neural networks. Nevertheless in this paper neural networks are compared to Shewhart Control Charts especially considering the different errors of type 1 and type 2 as in the articles of Pugh and Smith different weights for this two types of errors were not considered. Here the results of the neural networks are compared to control charts for the process mean and the process varibility. Next neural networks are trained to imitate a combined $\bar{x}$-s control chart.

Basically, in this paper neural networks are therefore used to imitate a special statistical test. Usually neural networks can be used as a special classification procedure, but a statistical test might be interpreted also as some kind of classification.[1] In a statistical test the null hypothesis and the alternative hypothesis may be interpreted as two different classes. In terms of control charts these two classes can be interpreted as *the process is under statistical control* and *the process is out of statistical control*.

It is well known, that the output of neural networks when used for classification approximates under certain conditions bayesian a posteriori probabilities, see e.g. Richard, Lippman (1992). This has also some implications for statistical tests, what will be investigated in the first experiment. Actually statistical tests may be formulated by a bayesian approach where the posterior probabilities are used.

3 Comparison of Neural Networks to Shewhart Control Charts

3.1 Basic assumptions and practical considerations

3.1.1 Assumptions for the comparison

For the comparisons a sample size of $n = 6$ was chosen. For the generation of the samples by Monte Carlo simulation it was supposed that $X_{t1}, X_{t2}, .., X_{tn}$ are independent, identically $N(\mu_t, \sigma_t^2)$ distributed. If the process is under statistical control the parameters are chosen $\mu_0 = 0$ and $\sigma_0 = 1$. For the different control charts a significance level of $\alpha = 0, 01$ is chosen. This results in an overall level of error of type 1 of $\alpha = 0.0199$ for the combined $\bar{x}$-s control chart.

3.1.2 Partitioning of the samples into sets

Three disjunctive sets of 1500 samples each were generated. The first set was used as the training set. As the different weights of errors of type 1 and type 2 had to be considered, a second set of samples was used to determine the interpretation limits for the network outputs in a proper way, as explained in section 3.1.4. The last set was used as a test set, to obtain the final results for the comparison.

3.1.3 Preprocessing of the samples

It was also one objective of this investigation to examine the effect of the different possibilities for the preprocessing of the samples. In the presentation of the results differences caused by the preprocessing will be discussed only if they lead to obviously different results.

One possibility is, not to preprocess the sample vector $x = (x_1, x_2, .., x_n)^T$ and take it as input for the neural net.

Additionally sample statistics like the sample mean and the sample standard deviation can be calculated and used as additional inputs for the neural net. This should not be necessary, since the neural networks should be able to imitate a control chart without additional sample statistics.

Another possibility is to order every sample in increasing order, so that $x_{(1)} < x_{(2)} < .. < x_{(n)}$. Such a preprocessing has the advantage that the jth input neuron always gets the j-largest sample value as input. With this kind of preprocessing e.g. the sample range, that has a direct relation to the process variability, can easily be calculated as a linear function.

A further possibility is to consider only the upper half of the $\bar{x}$-chart and to 'reflect' the sample realizations falling beneath the middle line of the $\bar{x}$-chart with respect to this middle line. This can easily be done, since the $\bar{x}$-chart is symmetrical with respect to the middle line. So only one control limit, i.e. two classes, has to be considered.

3.1.4 Classification considering errors of type 1

One possibility for clasification with neural networks is to choose one output neuron for each class. This corresponds to a vectorial codification of the objective outputs. In this case the jth class may be represented by a basis vector whose jth component is 1.

---

[1] Or every classification might be interpreted as some special kind of statistical test.

To interpret the outputs, one way is to assign the input to the class that is represented by the output neuron with the highest output value or to assign a class only if a specified limit is exceeded. For the comparison between neural networks and quality control charts it was considered essentially, that also with the neural networks a rate of error of type 1 comparable to control charts can be observed. Therefore here another possibility was used to determine proper limits for the interpretation of the outputs.

An independent set of samples was used to determine these interpretation limits. The number of errors of type 1 and type 2 made by the corresponding control chart were compared to the number of errors of type 1 and type 2 made by the neural network for different values of the output limits. Then the interpretation limits for the outputs were chosen so, that approximately the same number of errors of type 1 as with the control charts were made. If there were several possible values, also the errors of type 2 were considered to choose adequate limits. The values for the output limits were examined over a fixed range of values with a stepsize of 0.01. The final results for the comparison between neural networks and control charts are then based on the evaluation of the test set with the fixed limits.

Another way for the representation of the outputs is to take one output neuron and to assign each class a certain value as the objective output. This can be considered as a scalar codification of the outputs. Here interpretation limits have to be assigned to divide the output into different intervals. No noticeable difference between the two types of representation of the objective outputs were found in the experiments.

3.1.5 Training of neural networks

For the comparisons to the different control charts feedforward neural nets, that were trained with the Backpropagation algorithm, were used. As activation functions for the hidden and the output neurons a sigmoid function was used. To determine the time when to stop the training, a validation set was used, as it is proposed in Hecht-Nielson (1990).

3.2 Discussion of the experimental results

3.2.1 Approximation of posterior probabilities and statistical tests

The approximation of the posterior probabilities of neural networks and the resulting relation to statistical test is examined here, considering a simple test with $H_0: \mu_0 = 0$ and $H_1: \mu_1 = 1$. Here σ_0 is assumed to be 1.[2]

For this test, the posterior probabilities for the two parameters can be calculated explicitly and can be compared to the outputs of a neural network. It can be checked also, whether the outputs sum 1 as they should, if posterior probabilities are approximated. This test can be formulated as a Neyman-Pearson test or as a bayesian test. The both test formulations can be shown to be equivalent. For the Neyman-Pearson test a level $\alpha = 0.01$ is chosen. For the bayesian test prior probabilities of the parameters have to be given or estimated. Here it is assumed that $P(\mu_0) = P(\mu_1) = 0.5$, so for each set 750 samples with μ_0 and 750 samples with μ_1 were generated. With these prior probabilities and the assumptions made here, the posterior probability for the parameters can be calculated and for the parameter μ_1 a value for its posterior probability can be determined, from that one should choose parameter μ_1 to get a bayesian test equivalent to a Neyman-Pearson test with a fixed level α. In this example, this probability is approximately $P(\mu_1|x) = 0.937$. The interpretation limit of the output of the neural nets trained for this test should be therefore close to this value.

For the neural networks two output neurons were used to represent the two different parameters or classes. Nets with and without a hidden layer were trained and then the errors of type 1 and type 2 for the test set were compared to the performance of the corresponding statistical test. The statistical test corresponds to the application of an one-sided control chart with the upper control limit of $\tau_{0.99} \cdot 1/\sqrt{6} = 0.9497$.

In table 1 the interpretation limits for the node that represents parameter μ_1, the maximum and the minimum of the summed outputs, that should be near 1, and the mean absolute deviation of the output of the neuron that represents parameter μ_0 from the correct posterior probability were calculated. Here e.g. 6–3–2 refers to a neural net with 6 input, 3 hidden and 2 output neurons.

Architecture	errors of type 1/type 2	interpretation limits	max. of summed outputs	min. of summed outputs	mean absolute deviation from posterior prob.
QRK	5 / 354	/	/	/	/
6-2	6 / 354	0.92	1.00002	0.99998	0.020986
6-3-2	6 / 343	0.91	1.01231	0.98492	0.030631
6-5-2	6 / 385	0.93	1.0264	0.97846	0.026766
6-7-2	7 / 341	0.90	1.00354	0.99674	0.027840

Table 1 Results for the approximations of posterior probabilities

---

[2]This test can be seen also as a typical classification problem, see e.g. Duda, Hart (1973).

With respect to the approximation to the posterior probabilities, the net with architecture 6-2 gives the best performance. This can be best seen taking account of the mean absolute deviation of the posterior probabilities from the network outputs. Also the maximum and the minimum of the sum of the outputs are very close to 1. Compared to the net 6-2, the nets that use hidden units give quite large values for this maximum and minimum. The limits for the interpretation of the output are for all networks close to the theoretical value of 0.937. This should be this way, as with a good approximation of the posterior probabilities, the interpretation limits should also be close to this theoretical value. Also the number of errors of type 1 and type 2 should be close to the number of errors for the statistical test, as with the posterior probabilities a bayesian test, equivalent to a Neyman-Pearson test, can be formulated and this bayesian test depends on the posterior probabilities. The experimental results show, that neural networks can be trained to imitate some special statistical tests also with respect to a certain rate of error of type 1 that has to be considered. So e.g. the net 6-2 gives nearly the same results as the corresponding statistial test. Only one error of type 1 more is made. The nets with additional hidden units give similar results, e.g. net 6-3-2 gives even less errors of type 2 as net 6-2.

3.2.2 Comparison to a $\bar{x}$-chart

In this section a more general comparison to a $\bar{x}$-chart is carried out. The samples were generated so, that approximately for one third of the samples the mean was $\mu_0 = 0$, for approximately one third a negative shift, uniformly distributed in $[-3\sigma_0; -0.5\sigma_0]$, and for the last third a positive shift, uniformly distributed in $[0.5\sigma_0; 3\sigma_0]$, was supposed.[3] For the comparisons only the number of errors of type 1 and type 2 were considered. Another possibility would have been to estimate the average run length for some selected shifts. One objective of the experiments was also to investigate the effect of different kinds of preprocessing for the samples. The preprocessing and the determination of correct outputs can be chosen in such a form, that theoretically no hidden layer is necessary to imitate a $\bar{x}$-chart. This is the case when the samples are 'reflected', as described in section 3.1.3 or when an increase and a decrease of the mean are considered explicitly as classes. As there were only slight differences between the different kinds of output representation and preprocessing, these will not be discussed here.

In the test set, where 501 samples did not undergo a change of the mean, 5 errors of type 1 and 203 errors of type 2 were made with the $\bar{x}$-chart. The results of the neural networks have to be compared with these numbers of errors.

In table 2 some results of neural networks for different types of preprocessing of the data are given. Here 'reflected' refers to the use of only one half of the control chart, ordered refers to the ordering of the samples and with codification the presentation of the objective output is meant. The different number of output nodes is due to the representation of the objective outputs.

Architecture	errors of type 1	errors of type 2	remarks on preprocessing an output representation
6-2	5	206	reflected, ordered data, vectorial codification
6-2	5	200	not reflected, ordered data, vectorial codification
6-1	5	196	not reflected, ordered data, scalar codification
6-3	5	207	not reflected, not ordered, vectorial codification

Table 2 Results for the neural networks for the comparison to a shewhart control chart for the process mean

Generally it can be said, that neural networks are able to imitate a $\bar{x}$-chart and may obtain quite the same results with respect to errors of type 1 and type 2. This can be seen comparing the results for the neural networks given in table 2 to the results of the test set for a $\bar{x}$-chart. Accidentally all the neural nets give 5 errors of type 1 with little differences in the number of errors of type 2. The kind of preprocessing of the samples had only very little influence on the results. This seems reasonable, since the kinds of preprocessing considered here do not influence on the decisions made by the $\bar{x}$-chart either.

Another outcome is, that no hidden layer is necessary to obtain results, that compare to a $\bar{x}$-chart. But this is only the case if no hidden units are necessary from theoretical considerations.[4] It is also not necessary to use additional inputs like the sample mean.

Finally it should be noticed that the results presented here could be expected to be this way since the problem considered here is basically a linear problem. This is so because the sample mean, that is a sufficient statistic for the parameter μ of a normal distribution, is a linear function, that should be easily approximated by neural networks.

3.2.3 Comparison to a s chart

For the comparison between a standard deviation chart and neural networks only an increase of the process variability was considered. For the generation of the samples it was supposed, that approximately half of the samples were subject to a shift of the standard deviation, that was taken as uniformly distributed in $[1.2\sigma_0; 3\sigma_0]$. For the test set, where

---

[3]Here a special interpretation of the prior class probabilities was supposed. This interpretation corresponds to three classes, that have equal prior probabilities. Another possibility would have been to choose for one half of the samples a shift that can be positive or negative and for the other half no shift at all.

[4]Hidden units are only necessary when the samples are not 'reflected' and only the classes shift/no-shift are distinguished. For a justification of this result, the problem may be compared to the well known XOR-Problem.

742 samples did not undergo a change of the standard deviation, 7 errors of type 1 and 325 errors of type 2 were made. In the following table some results obtained with neural networks are summarized.

Architecture	errors of type 1	errors of type 2	remarks
6-2	8	297	reflected, ordered data
6-9-2	9	278	reflected, ordered data
6-15-2	9	398	reflected, unordered data
6-13-5-2	11	397	reflected, unordered data

Table 3 Results for the neural networks for the comparison to a shewhart control chart for the process variability

The most salient result is, that neural networks can give comparable results to a s-chart, if the ordered samples are taken as inputs. Then even a network without hidden layer suffices. If the unordered samples are taken as inputs, a considerably higher number of errors of type 2 is made. This means, that here the kind of preprocessing of the samples has a noticeable influence on the results.

It should be observed here, that the sample standard deviation actually is a nonlinear function of the sample. If no ordering of the sample is made, hidden units are necessary to obtain reasonable results. The worse performance of neural nets, that take unordered samples as inputs, shows, that neural networks were not able to approximate good enough the decision rule of a s-chart. One reason may be, that not enough training examples were used as inputs for the neural net. Another reason may be, that the architecture of the neural net is not complex enough to approximate the desired function. Here no further experiments were made to investigate the reasons of this worse performance. If the sample standard deviation or the sample range are used as additional inputs to the unordered samples, similar results to a s-chart are obtained. This seems to be natural as the sample standard deviation is also the statistic used by the s-chart and the neural network only has to 'learn' to interpret this statistic correctly.

With the ordered samples, similar results to a s-chart could be obtained. It makes no sense to talk about an overall error rate, if the two different kinds of errors are not considered. We will limit us here to state, that neural nets and the s-chart give approximately the same result, if the samples are ordered. It might be interesting to investigate the slightly different results in more detail.

The reason for the good performance of neural nets with ordered inputs may be explained. The greatest and the smallest sample value suffice to calculate the sample range, that can also be used to control the process variability. Here the input for the neural network is the vector whose components are the order statistics of the sample, so that the sample range could be calculated as a linear function of these inputs. If no hidden unit is used, basically a monotone transformation of the weighted sum of the order statistics is calculated in the output neurons. If one examines the weights for these networks, it can be seen that the smallest and the greatest value have also the greatest weights associated with them.

3.2.4 Comparison to a combined $\bar{x}$-s chart

If one uses a combined $\bar{x}$-s-chart, there are three classes that are to be distinguished. These classes may be interpreted as *The process is under statistical control*, *A change of the process mean occurred* and *A change of the process variability occurred*. A further class like *Both parameters changed* might be considered.

As a consequence of the results obtained in the last two sections, the ordered samples were used as inputs. Here the samples were generated so, that approximately one third of the samples had no shift, approximately the second third had a shift of the parameter μ, uniformly distributed in $[\mu_0 - 3\sigma_0; \mu_0 - 0.5\sigma_0] \cup [\mu_0 + 0.5\sigma_0; \mu_0 + 3\sigma_0]$, and the last third had a shift of the standard deviation, uniformly distributed in $[1.2\sigma_0; 3\sigma_0]$. In table 4 the evaluation of the test set with a combined control chart is given and explained below.[5]

position of sample points	samples without shift	samples with shift of μ	samples with shift of σ
on both charts inside c.l.	495	99	175
outside c.l. on $\bar{x}$-chart	5	387	28
outside c.l. on s-chart	6	1	232
outside c.l. on both charts	0	2	70
sum	506	489	505

Table 4 Results of the test data set for a combined $\bar{x}$-s-chart. Errors of type 1: 11 Errors of type 2: 274 False classifications: 31

With a $\bar{x}$-s-chart a shift of a parameter is indicated, if at least one chart gives an out of control signal. E.g. a shift of μ may be indicated on the $\bar{x}$-chart, on the s-chart or on both charts at the same time. No error is only made if a shift of σ is indicated on the s-chart, a shift of μ is indicated on the $\bar{x}$-chart and if no shift occurred, both charts do not indicate an out of control situation.

The control charts made $11 (= 5 + 6)$ errors of type 1 and $274 (= 99 + 175)$ errors of type 2. Another kind of error occurs when the kind of shift is classified in a wrong way. It is striking here, that a lot of times (70) a shift of σ is indicated on both charts. This is mainly due to the possible values for the shifts of the standard deviation up to

---

[5]In the table c.l. means control limit.

$3\sigma_0$. To determine the number of false classifications it should be considered that the $\bar{x}$-chart reacts to an increase of the process variability with a higher probability of false alarms whereas the s-chart is not sensible to changes of μ. If both charts give an out of control situation, this should be therefore interpreted actually as a shift of σ. So the total number of false classifications amounts to $31(= 28 + 1 + 2)$.

Next an exemplar result of a neural net, that was trained to distinguish between the three classes is given. This result is representative of others that were obtained with other architectures of neural networks that were trained for this problem.

network 6–5–3	samples without shift	samples with shift of μ	samples with shift of σ
no change shown	491	120	194
shift of μ shown	8	367	14
shift of σ shown	7	2	297
shift of μ and σ shown	0	0	0
sum	506	489	505

Table 5 Results of the test data for a neural net, architecture 6–5–3. Errors of type 1: 15 Errors of type 2: 314

False classifications: 16

It can be seen, that for the neural network the number of errors of type 1 and of type 2 is higher than for the combined control chart. Only the number of false classifications $(16 = 14+2)$ is smaller.[6] In general neural nets could not reach the lower numbers of errors of type 1 and type 2 of the $\bar{x}$-s-chart. They only made less false classifications. Nevertheless the false classifications are mainly of that kind, that a shift of σ is interpreted as a shift of μ. It should also be noticed, that in this example it never occurred, that a shift is indicated on both output neurons that indicate shifts of μ or σ.

It should be stated here, that the main problem is, that neural nets do not reach the number of errors of type 1 or type 2 of the control chart. Although they offer an advantage with respect to a lower number of false classifications, this should be no reason to use neural nets to replace a combined control chart, as the essential thing are the lower numbers of errors of type 1 or type 2. The false classifications could also be handeled by a further interpretation of the control charts considering past samples. Nevertheless neural nets could offer an advantage in a combination with control charts as it is discussed in the next section. One reason for neural networks not to reach this low numbers of errors of type 1 or type 2 may be, that they also minimize explicitly false classifications because of the objective function they use, whereas with the application of a combined $\bar{x}$-s-chart the classification of the kind of change is not considered.

3.2.5 Classification of the kind of change

It is interesting, that the number of false classifications with neural networks is smaller than with the combined control charts. This leads to the idea, that neural networks may only be used as a classification procedure to classify the kind of change, once that such a change is indicated by a combined control chart. To investigate whether with such a combined Shewhart-chart/neural-net scheme the number of false classifications can be reduced, a neural network was trained to distinguish between shifts of the mean and shifts of the standard deviation. A test set of 2000 samples was then evaluated by a combined $\bar{x}$-s-chart. If the combined $\bar{x}$-s-chart indicated an out of control situation, the kind of shift was classified by a neural network. The number of false classifications given by the neural network was then compared to the number of false classifications made by the combined $\bar{x}$-s-chart.

The samples were generated so, that only shifts of μ, uniformly distributed in $[\mu_0 + 0.5\sigma_0; \mu_0 + 2\sigma_0]$, or shifts of σ, uniformly distributed in $[1.2\sigma_0; 2\sigma_0]$, were considered.

The evaluation of the test data with a combined $\bar{x}$-s-chart is given in table 6.

Results for the test data for the combined $\bar{x}$-s-chart	samples with shift of μ	samples with shift of σ
on both charts inside c.l.	706	688
only outside c.l. on the $\bar{x}$ chart	256	64
only outside c.l. on the s-chart	9	240
Outside c.l. on both charts	7	30
sum	978	1022

Table 6 Results of the test data for a combined control chart. False classifications: 80

With the combined control chart $80(= 64 + 9 + 7)$ false classifications would be made. In table 7 now the results of the test of the kind of classification with the neural network is given. Here only the 606 samples where an out of control situation was detected are considered.

---

[6]This holds in general for all other neural nets, that were trained for this problem.

Classification of shift with neural nets	neural net shows shift of σ	neural net shows shift of μ
on both charts outside c.l., shift of σ	30	0
on both charts outside c.l., shift of μ	2	5
outside c.l. on the $\bar{x}$ chart, shift of σ	38	26
outside c.l. on the $\bar{x}$ chart, shift of μ	4	252
outside c.l. on the s chart, shift of σ	238	2
outside c.l. on the s chart, shift of μ	5	4

Table 7 Classification of the kind of change with neural nets. False classifications: 39

With neural nets $39(= 2 + 26 + 4 + 2 + 5)$ false classifications are made. This means, that the number of false classifications has approximately been halfed. This seems to indicate, that by using neural nets for only this special classification task, an improvement could be achieved. Anyway these result should be carefully interpreted because in this experiments only the last sample was considered.

4 Summary and final comments

In this paper it was essentially investigated whether neural networks can imitate a certain model, like it is assumed for the control charts. This problem can be interpreted as a special kind of statistical test.

Generally it can be pointed out, that neural networks can be trained to imitate the control charts considered here. This holds for the $\bar{x}$-chart as well as for the s-chart. Neural nets can be trained to consider the different kinds of error of type 1 and type 2. If both kind of changes are considered, the results of neural networks with respect to errors of type 1 and type 2 are worse than those of a combined $\bar{x}$-s-chart. Only for the number of false classifications an improvement could be found. So it seems to be more reasonable to use neural networks for special tasks such as the classification of the kind of shift. A correct classification of the kind of change of the process parameters is important as it may facilitate the search for possible errors.

The main drawback of the problems treated here is, that past samples are not considered. Past samples are used e.g. by the CUSUM-charts that are able to detect small shifts of the parameters faster than Shewhart charts. There are also additional rules for the interpretation of the Shewhart charts that give them an artificial memory. Here it would be one possibility to investigate whether neural networks are also able to imitate or even to give an improvement over these charts. Considering the patterns of the past samples, neural networks can also be used to detect special patterns on control charts like mixtures or a gradual increase/decrease of the mean or the process variability. These special patterns indicate also changes of the parameters. Another possibility is, that different patterns may indicate certain errors occurred during the running of a machine. As the use of computers increases also in the Statistical Process Control, adequate algorithms or classification procedures may be used to obtain a better evaluation of the production process. This evaluation has to be made in a real time manner. One possibility here is the use of neural networks for this task, but still further investigation is needed.

References

Duda, R. and Hart, P. *Pattern Classification and Scene Analysis,* John Wiley & Sons, New York, 1973.

Guo, Y. and Dooley, K.J. *Identification of Change Structure in Statistical Process Control,* International Journal of Production Research, Vol.30, 1992, 1655–1669.

Hecht-Nielson, R. *Neurocomputing,* Addison-Wesley Publishing Company, 1990.

Hwarng, H.B. and Hubele, N.F. *Back Propagation Pattern Recognizers for X Control Charts: Methodology and Performance,* Computers & Industrial Engineering, Vol. 24, 1993, 219–235.

Montgomery, D.C. *Introduction to Statistical Quality Control,* John Wiley & Sons, 1985.

Pham, D.T. and Oztemel, E. *Control Chart Pattern Recognition Using Learning Vector Quantization Networks* International Journal of Production Research, Vol.32, 1994, 721–729.

Pugh, G.A. *A Comparison of neural Networks to SPC Charts,* Computers & Industrial Engineering, Vol.21, 1991, 253–255

Richard, M.D. and Lippman, M.P. *Neural Network Classifiers Estimate Bayesian a posteriori Probabilities,* Neural Computation, Vol.3, 1991, 461–483.

Smith, A.E. *X-bar and R control Chart Interpretation Using Neural Computing,* International Journal of Production Research, Vol.32, 1994, 309–320.

Bankruptcy Prediction
with Artificial Neural Networks

Eugenio Fernández[*,**] and Ignacio Olmeda[*]

[*]Dpto. de Fundamentos de Economía e Historia Económica
[**]Dpto. de Matemáticas
Universidad de Alcalá
Alcalá de Henares 28802 Madrid SPAIN

Abstract: *In this paper we compare the forecasting accuracy of feedforward neural networks against various competing models (C4.5, MARS, Discriminant Analysis and Logit) on the problem of predicting bankruptcy. The neural network model is found to provide generally better results, though the computational effort is several orders of magnitude higher. We also consider mixtures of the methods and show that many of these are always more accurate than any single method. We suggest that an optimal system for risk rating should include two or more of the models considered.*

1. Introduction.

Financial agents are increasingly interested on the use of Artificial Neural Networks (ANN´s), as well as some other appealing techniques such as Genetic Algorithms or Machine Learning, for modelling and forecasting purposes. The reason for this is quite obvious, if these "high-tech" tools were truly more powerful, the competitive advantage from using them would be decisive, at least until these technologies were used by any agent so that diferential benefits were fully arbitraged. The number of successful applications of ANN's reported has been so high that a "folk-theorem" asserts their universality and superiority against any other procedure. Considering that the process of developing a ANN-based Decision Support System (DSS) is relatively much more costly than a traditional statistical one, it is crucial, from the economic point of view, to determine the soundness of this belief.

Comparisons on the forecasting accuracy of ANN´s against various models in classification problems are relatively common in the literature. Most of these comparisons consider only a competing model (such as Discriminant Analysis) so that the appropriateness of ANN´s in a general forecasting context is not resolved. Another salient feature of the mentioned studies is that they only employ simple models as the aternative, and not the combination of two or more of them. In this paper we consider both questions by employing recently developed methods as alternatives and well as mixtures of them. We will show that though ANN´s models can be near optimal (under a forecasting criterion) when compared against the traditional or sophisticated models considered, a combination of the methods provides in general better results.

2. Methods compared.

The models chosen for comparison include a standard feedforward neural network with a single hidden layer trained with backpropagation (NN), two classical statistical techniques: Discriminant

Analysis (DA) [Fisher, 1936] and Logit (Logit), and two recent extensions of the CART algorithm of Breiman et al. (1984): Multivariate Adaptive Regression Splines (MARS) and C4.5 (C4.5). The first three approaches are already well known so that we will only briefly describe the last two ones.

MARS [Friedman, 1991] is a nonparametric technique wich exploits the ideas of stepwise regression and recursive partitioning. The model begins with a simple structure (say, a linear regression in the explicative variables) and sucessively adds new terms of higher order (basis functions), for example, cross products between variables. Whenever the new function fails to improve the fit to the data, the model splits the region into different subregions and repeats the procedure. The regions are subdivided in the following manner: for each dependent variable x_i fixed at a determined level (called *knot*), MARS considers two subregions, one including the data points with a value higher than the knot and its complementary. After performing a regression in each of these subregions, the fit is computed and then, a different spliting variable or a different level are tried. This procedure is repeated and the best parameters are chosen.

The C4.5 algorithm combines some improvements of the well-known ID3 [Quinlan, 1983]. C4.5 generates a decision tree by evaluating the information gain of further partitioning the tree at a certain stage. The algorithm begins with a minimal tree and evaluates the attribute which produces the most informative partition of the training cases (which is equivalent to minimizing the entropy of the partition). Each of the leaves generated is treated again as a new tree and the procedure iterates until there are no missclassifications in the training data. The resulting tree is "pruned" to produce a minimal tree by reducing its complexity while conserving its generalizing properties.

Though conceptually very similar these methods differ both in their structure (the MARS algorithm uses truncated cubic polynomials as basis functions while C4.5 uses step functions) as well as in their performance criterion (the MARS algorithm minimizes a cross-validated error while C4.5 maximizes an information criterion), consequently they can provide different conclusions.

3. Database used and results.

From 1977 to 1985 the Spanish banking system suffered the worst crisis of its whole history, affecting 52% of the 110 banks that were operative at the beginning of this period. Such concentration in time offers the oportunity to compare alternative methods for bankruptcy prediction, since the economic conditions can be considered stable enough to assess the significance of the financial ratios used. Following previous studies (see Pina, 1989 and references therein), we employ a database consisting on 66 banks (29 failed and 37 non-failed) and 9 financial and economic ratios (working capital/total assets, sales/total assets, etc.). This database was randomly splited in two sets, Set 1 consisted on 34 banks (15 failed and 19 non-failed) and Set 2 on 32 banks (14 failed and 18 non-failed).

We tried a variety of especifications for each of the models (number of basis functions for MARS, number of leaves for C4.5, number of hidden nodes for the NN, etc.), always using all the attributes. For reasons of brevity we give only the results for the best model found (full results are available upon request). First we estimated the models on Set 1 and use them for predicting on Set 2, then we reversed the procedure, estimating on Set 2. The criterium employed to evaluate accuracy is the percentage of correct classifications, note that under this criterion the learning algorithm of any of the methods could be suboptimal [Olmeda, 1993].

In Table 1 we have computed the number of correct predictions, as well as the relative percentage of successes. We also indicate when a model is at least as good as any other on a particular set by a small asterisk (*). As we can see, the best model is NN, performing slightly better than logit. These models are followed by MARS, C4.5 and, finally DA. The bad behavior of DA is not surprising, since it assumes a gaussian distribution of the variables and equal variances for the attributes. None of these hypotheses hold: a standard Kolmogorov-Smirnov test rejected the normality assumption for 6 of the 9 attributes and the variances were also significantly different. It is also noticeable the excellent performance of the logit model which, considering its simplicity, computational efficiency and the existence of a sampling theory to test the consistence of its parameters, can be chosen as the "best" model in this application.

Table 1
Performance of alternative models

Method	Training Set 1	Testing Set 2	Overall	Training Set 2	Testing Set 1	Overall	Total Overall
D.A.	30 (88.23%)	26 (81.25%)	56 (84.85%)	25 (78.12%)	26 (76.47%)	51 (77.27%)	81.06%
Logit	32* (94.12%)	28 (87.50%)	60* (90.91%)	32* (100%)	29 (85.29%)	61 (92.42%)	91.66%
MARS	31 (91.18%)	24 (75.00%)	55 (83.33%)	31 (96.87%)	28 (82.35%)	59 (89.39%)	86.36%
C4.5	29 (85.29%)	28 (87.50%)	57 (86.36%)	31 (96.87%)	27 (79.41%)	58 (90.62%)	85.60%
N.N. (8 hidd)	31 (91.18%)	29* (90.62%)	60* (90.91%)	31 (98.87%)	31* (91.17%)	62* (93.93%)	92.42%

Computationally, the NN model is clearly the worst, the computing time required is several orders of magnitude higher than for any other method: for example, the learning phase of the NN required more than 12 minutes on a 486 66Mhz., while the computing time of MARS on the same platform is around 3 seconds and less than one second for C4.5 on a Sun IV.

It is possible, though, that the random selection of training and testing sets could have biased the results in favor of a particular method. For this reason we employed a resampling procedure to robustfy our inferences. A very convenient way to estimate the expected performance of any particular method consists on dividing the training data (N) into v equal subsamples of size $n_v = N/v$, estimating the model using $(v-1)n_v$ examples and predicting the remaining n_v ones (*v-fold cross validation*) [Stone, 1974]. We repeat this procedure for each of the v subsamples, computing the mean prediction error. The Set 1 was divided into 6 approximately equal subsamples, and each of the 5 methods was estimated using these subsamples (which gives 30 different models), then we computed the mean prediction error for each of the methods. The results are shown in Table 2.

Table 2
Cross-validated performance

	DA	Logit	MARS	NN	C4.5
Training	90.00%	95.29%	94.11%	97.05%*	84.70%
Test	61.76%	76.47%	79.41%	82.35%*	79.41%
Total	85.29%	92.15%	91.66%	94.60%*	83.82%

As one can see, the NN is again the best model both in terms of in-sample fitting and ou-of-sample prediction. This robustness is remarkable, since the reduction of the size of the training set (to 28 patterns) could have induced to overfitting the data. The second best model is logit, closely followed by MARS. It is also interesting to note the inversion of ranking position of DA and C4.5 (fourth and fifth, respectively), this seems to indicate that the sample size reduction has been particularly perverse for the last method which it is known to work better for a large number of examples.

In any case, from a DSS perspective the optimal system may not be an individual method but the combination of several of them. In fact, this is the usual way to proceed to evaluate projects in many financial contexts: the opinions of a comitee of human experts (each of them representing a particular and relevant aspect of the problem) are aggregated to give and optimal decision.

Based on this, we consider several possible mixtures of 2, 3, 4, and 5 of the above methods. In every combination we include the two best models (NN and logit), since none of them dominates the other in the sense of Pareto; for example (see Table 1) the logit exhibits better behavior in the training sets while the NN does in the testing sets. We employ the vote of the majority as the forecast (for example predicting bankruptcy if two out three models do). In the case of combining an even number of methods a tie is possible. As this was the case for the model combining NN and logit we used a probabilistic rule by randomly chosing any of the predictions with equal probability.

Table 3
Accuracy of mixed models

Methods combined	Training Set 1	Testing Set 2	Overall	Training Set 2	Testing Set 1	Overall	Total Overall
NN + logit	32* (94.12%)	29* (90.62%)	61* (92.42%)	32* (100%)	31* (91.17%)	63* (96.97%)	93.93%*
NN+logit+C4.5	31 (91.18%)	31* (96.87%)	62* (93.94%)	31 (96.87%)	31* (91.17%)	62* (93.94%)	94.94%*
NN+logit+DA	32* (94.12%)	29* (90.62%)	61* (92.42%)	32* (100%)	29 (85.29%)	61 (92.42%)	92.42%*
NN+logit+MARS	31 (91.18%)	28 (87.50%)	59 (89.39%)	31 (96.87%)	30 (88.23%)	61 (92.42%)	90.91%
NN+logit+C4.5+MARS	31 (91.18%)	31* (96.87%)	62* (93.94%)	31 (96.87%)	30 (88.23%)	61 (92.42%)	93.18%*
NN+logit+C4.5+DA	32* (94.12%)	31* (96.87%)	63* (95.45%)	32* (100%)	32* (94.11%)	64* (96.97%)	96.21%*
NN+logit+MARS+DA	32* (94.12%)	30* (93.75%)	62* (93.94%)	32* (100%)	31* (91.17%)	63* (95.45%)	94.69%*
All methods	31 (91.18%)	29 (90.62%)	60 (90.91%)	31 (96.87%)	30 (88.23%)	61 (92.42%)	91.66%

As one can see from Table 3, many mixtures produce more accurate forecasts than any single method. In fact 3 mixtures (NN+logit, NN+logit+C4.5+DA and NN+logit+MARS+DA) dominate any other single alternative in the sense of Pareto, so that they should be preferred in any situation. Also note that there is a certain trade-off between the number of methods used and the accuracy of predictions: including too many marginally inefficient models conduce to suboptimal systems (last row of Table 3).

4.   Conclusions and future research.

In this paper we analyzed the performance of ANN's models on a problem of bankruptcy prediction. Our results are by no means conclusive but they suggest that ANN's models could de superior to both clasical and recently developed statistical and Machine Learning classifiers.

The main finding is that when one combines two or more of the methods in a simple manner, the predictions are generaly more accurate than the ones obtained by applying any single method. In this sense, ANN's are not the best alternative available. Our results can be extended in a number of ways, for example, taking into account the asymetric costs of missclassification (Type I and II errors), using a weighted cross-validated measure of the predictions, or by considering a the classification problem as a sequential decision problem in which one desires to minimize the value of marginal information.

References.

Breiman, L.; Friedman, J.; Olsen, J. and Stone, C. (1984): *Classification and Regression Trees*. Wadsworth International, CA.

Fisher, R.A. (1936): The use of multiple measurements in taxonomic problems. *Ann. Eugenics*, **7**, pp. 179-188.

Friedman, J.H. (1991): Multivariate Adaptive Regression Splines. *The Annals of Statistics*, **19**, pp. 1-141.

Olmeda, I. (1993): Aprendizaje y generalización, in I. Olmeda and S. Barba-Romero (eds.): *Redes Neuronales Artificiales: Fundamentos y Aplicaciones*. Servicio de Publicaciones de la Universidad de Alcalá.

Pina, V. (1989): La información contable en la predicción de la crisis bancaria. *Revista Española de Contabilidad y Finanzas*, **18**, pp. 309-338.

Quinlan, J.R. (1983): Learning Efficient Classification Procedures and their Application to Chess End Games, in R.S. Michalski, J.G Carbonell and T.M. Mitchell (eds.): *Machine Learning: An Artificial Intelligence Approach*. Tioga Publishing Co., Palo Alto, CA.

Stone, M. (1974): Cross-validatory Choice and Assessment of Statistical Predictions. *Journal of the Royal Statistical Society B*, **2**, pp. 111-147 (with discussion).

Authors Index

Acevedo-Sotoca, M.I. 882
Adamo, J.M. 642
Adams, R. 308
Aguiló, J. 121
Aizenberg, I.N. 389
Aizenberg, N.N. 389
Alastruey, C.F. 1092
Albizuri, F.X. 144
Aleksander, I. 566
Alexandre, F. 24
Almagro León, A. 948
Almaraz, J. 486
Álvarez, J.R. 15, 137, 1036
Andrade, M.A. 108
Andreu, E. 85
Anguita, D. 642
Anguita, M. 736
Arbib, M.A. 412
Arcas Túnez, F.J. 906
Arroyo, F. 987
Arroyo, V. 1036
Asensi Muñoz, D. 948
Augereau, B. 231
Avellana, N. 800
Ayoubi, M. 1045

Bandera, C. 979
Baradona da Fonseca, I. 260, 268
Barahona da Fonseca, J. 260, 268
Barro, S. 137
Baumela, L. 993
Beiu, V. 822
Benítez, J.M. 744
Benítez-Díaz, D. 527
Bernard, J. 231
Bernier, J.L. 838
Blasco-Alberto, J. 712
Blayo, F. 771, 781
Bluff, K. 53
Bojarczak, P. 1009
Bolea, S. 85
Bray, A. 189
Butchart, K. 308
Buti, M. 114

Caballero, A. 634
Cabello, D. 860

Cabestany, J. 752, 761
Cabruja, E. 114
Calvet, J.M. 627
Calvet, S. 114
Cano-Izquierdo, J.M. 934
Capillas, R. 800
Carrasco, R.C. 605
Carrillo-Menéndez, S. 166
Casacuberta, F. 433
Castaño, M.A. 433
Castillo, F. 752, 761
Català Mallofré, A. 478
Cavalieri, S. 1068
Cembrano, G. 1016
Cervera, E. 345
Cheneval, Y. 673, 1099
Chentouf, R. 519
Cheung, W.P. 868
Chiaberge, M. 666
Cho, J.-W. 926
Christodoulou, C. 223
Clarkson, T. 223, 441
Cloete, I. 374, 382, 611
Colodro, F. 789
Comon, P. 1099
Conde, I.M. 979
Corbacho, F.J. 412
Corredor Lacha, D. 942
Cottrell, M. 1107
Crestani, F. 597

D'Anjou, A. 144
D'Eleuterio, G.M.T. 1076
Daffertshofer, A. 76
Davey, N. 308
de la Hera, A. 144
De Wilde, P. 202, 584
Del Moral Hernandez, E. 216
del Pobil, A.P. 345
Delgado, A.E. 15, 1036
Di Bene, G. 666
Di Pascoli, S. 666
Díaz-Otero, F. 634
Dimitriadis, Y.A. 934
Domínguez, J. 696
Dominguez, M. 1060
Dorronsoro, J.R. 291

Dracopoulos, D.C. 315
Duro, R.J. 31

Echanobe, J. 913
Engelbrecht, A.P. 374, 382
Ennaji, A. 330

Farhat, N.H. 216
Feng, J. 353
Fernández, E. 1142
Fernández, F.J. 728, 736
Fernández, J.A. 800
Fernández, M.A. 137
Fernández, R.A. 1036
Ferreiro Garcia, R. 448
Fiesler, E. 535, 807
Flanagan, J.A. 322
Flexer, A. 454
Forcada, M.L. 605
Franquelo, L.G. 789

Gambardella, L.M. 1084
García, F. 505
García, I. 144
García, J.A. 752
García-Chamizo, J.M. 942
García-Quesada, J. 527
Garitagoitia, J.R. 913, 1092
Garrido, P. 114
Gaudiano, P. 471
Gedeon, T.D. 543, 551
Geldenhuys, J. 374
Giménez, V. 252
Girard, B. 1107
Girard, Y. 1107
Göppert, J. 419
Gómez, A. 31
Gómez-Vilda, P. 252, 963
González, A.M. 291
González, M. 979
González De Mendívil, J.R. 913, 1092
Gonzalo, A. 987
Goser, K. 338, 513
Grabec, I. 158
Graña, M. 144
Grossberg, S. 1
Guan, Y. 441
Guedes, K. 1001
Guérin-Dugué, A. 771, 781, 956

Guyot, F. 24

Haken, H. 76
Hamilton, A. 815
Han, J. 195
Harris, D. 551
Hasler, M. 322
Heit, B. 231
Hellestrand, G.R. 1128
Hérault, J. 845
Higuchi, T. 396
Hilera, J.R. 987
Ho, A.M.C.-L. 202
Hoekstra, J. 45
Hofacker, G.L. 180
Hofman, P.M. 166
Holgado, R. 800
Hombrados, M.A. 963
Hruby, P. 276

Ibarra-Picó, F. 942, 948

Jahnke, A. 720
Jaramillo-Morán, M.A. 882
Jašić, T. 239
Jerez, J. 979
Jones, A.J. 315
Joya, G. 283
Jutten, C. 361, 519

Kanstein, A. 513
Klar, H. 720
Köppen, M. 875
Kosak, A. 338
Krivosheev, G.A. 389
Kruizinga, P. 90
Kussinger, M. 37
Kuzmina, M.G. 246

Lambert, R. 666
Lang, E.W. 37
Lazzerini, B. 666
Lecourtier, Y. 330
Lee, C.K. 868
Lee, S.-Y. 926
Legat, J.-D. 404, 696, 704
Leisenberg, M. 462
Leibovic, K.N. 209
Letelier, J.-C. 130

Li, K.C. 868
Lis, J. 498
López, M.T. 137
López, V. 166, 291
López-Aligué, F.J. 882
López Coronado, J.L. 471, 934
Lorenzo, A. 634
Lourens, T. 61

Macías-Macías, M. 882
Madani, K. 1114
Madrenas, J. 696, 761
Maggiore, A. 666
Manjarrés, A. 15, 137
Manykin, E.A. 246
Maouli, M. 45
Maravall, D. 993
Maria, N. 771, 781
Marín, F.J. 505
Martín-del-Brio, B. 712
Martin-Smith, P. 728
Martinez, J.M. 1060
Martini, M. 1068
Maturana, H. 130
Mayes, D.J. 815
McGuire, P.F. 1076
Menéndez de la Prida, L. 7
Mercier, G. 1114
Merelo, J.J. 838
Michelin, J.M. 1060
Millán, J. del R. 1016
Mira, J. 15, 137, 658, 1036
Miranda, E. 830
Miró, J. 69
Miura, K.-i. 589
Monte, E. 627
Moraga, C. 195, 1052
Morán, F. 108
Morcego Seix, B. 478
Moreno, J.M. 752, 761, 771
Moreno-Díaz jr., R. 209
Mosquera, A. 860
Mpodozis, J. 130
Muller, C. 1107

Nagano, T. 589
Natowicz, R. 890
Navarro, X. 114
Nejad, A.F. 543

Nieto, V. 963
Nishii, J. 151
Noda, I. 971

Olmeda, I. 1142
Ortega, F. 486, 979
Ortega, J. 744, 838
Ortuño Ortín, I. 1121
Osowski, S. 1009
Otero, R.P. 658

Pacheco, M.A. 1001
Palagi, P.M. 956
Park, C.H. 926
Paton, J.F.R. 100
Pelayo, F.J. 728, 736
Perez Castelo, F.J. 448
Pérez-Castellanos, M. 252
Pérez-Maroto, J. 934
Petkov, N. 90
Piera Carreté, N. 478
Poh, H.L. 239
Prem, E. 619
Prieto, A. 728, 736, 838, 898
Puntonet, C.G. 898
Puzenat, D. 559

Requena, I. 744
Reyneri, L.M. 666, 830
Rizo-Aldeguer, R. 942
Rodellar, V. 963
Rodríguez, F.B. 166
Rodríguez-Alvarez, M. 898
Ros, E. 728
Ros, S. 15
Rosenstiel, W. 419
Roth, U. 720
Rousset, P. 1107
Ruiz García, A. 906
Ruiz-del-Solar, J. 875
Rybak, I.A. 100

Sainz, L. 1036
San Anselmo, S. 761
Sánchez-Andrés, J.V. 85, 174
Sandoval, C. 427
Sandoval, F. 283, 486, 505
Santa Cruz, C. 291
Santos, J. 31, 658

Saxena, I. 807
Schäfer, M. 1045
Schels, A. 37
Schreter, Z. 492
Schwaber, J.S. 100
Seeger, E. 37
Serra Sagristà, J. 1121
Shams, S. 919
Signes Pont, M.T. 174
Sigüenza, J.A. 166, 634, 650
Simões da Fonseca, J. 260, 268
Simon, T. 231
Sinne, S. 1052
Sinsel, S. 1045
Soria, B. 85
Stetter, M. 37
Stocker, E. 330
Stone, J.V. 189
Strey, A. 800
Stützle, T. 1135
Surina, I.I. 246

Taylor, J.G. 441, 822
Thimm, G. 535
Thissen, P. 404, 696, 704
Tirozzi, B. 353
Torralba, A. 689, 789
Torrano, E. 252
Torras, C. 1016
Trejo, L.A. 427

Valderrama, E. 114, 800
Varona, P. 650
Vellasco, M. 1001
Vergara, V. 1052
Verleysen, M. 404, 696, 704
Vermesan, O. 794
Versino, C. 1084
Viana, L. 298
Vico, F.J. 486, 979
Vidal, E. 433
Viktor, H.L. 611
Vilariño, D.L. 860
Vilarrubla, S. 627
Villa, R. 114, 121
Villadangos, J. 1092
Voz, J.-L. 404

Wallace, J.G. 53
Waszczuk, R. 1009
Weitze, M.-D. 180
Weitzenfeld, A. 683
Wells, G. 1016
Wong, P.M. 551

Zalama, E. 471
Zamani, M.S. 1128
Zebulum, R.S. 1001
Zhao, Q. 396
Zurada, J.M. 374, 382

Springer-Verlag
and the Environment

We at Springer-Verlag firmly believe that an international science publisher has a special obligation to the environment, and our corporate policies consistently reflect this conviction.

We also expect our business partners – paper mills, printers, packaging manufacturers, etc. – to commit themselves to using environmentally friendly materials and production processes.

The paper in this book is made from low- or no-chlorine pulp and is acid free, in conformance with international standards for paper permanency.

Lecture Notes in Computer Science

For information about Vols. 1–854
please contact your bookseller or Springer-Verlag

Vol. 855: J. van Leeuwen (Ed.), Algorithms – ESA '94. Proceedings, 1994. X, 510 pages.1994.

Vol. 856: D. Karagiannis (Ed.), Database and Expert Systems Applications. Proceedings, 1994. XVII, 807 pages. 1994.

Vol. 857: G. Tel, P. Vitányi (Eds.), Distributed Algorithms. Proceedings, 1994. X, 370 pages. 1994.

Vol. 858: E. Bertino, S. Urban (Eds.), Object-Oriented Methodologies and Systems. Proceedings, 1994. X, 386 pages. 1994.

Vol. 859: T. F. Melham, J. Camilleri (Eds.), Higher Order Logic Theorem Proving and Its Applications. Proceedings, 1994. IX, 470 pages. 1994.

Vol. 860: W. L. Zagler, G. Busby, R. R. Wagner (Eds.), Computers for Handicapped Persons. Proceedings, 1994. XX, 625 pages. 1994.

Vol: 861: B. Nebel, L. Dreschler-Fischer (Eds.), KI-94: Advances in Artificial Intelligence. Proceedings, 1994. IX, 401 pages. 1994. (Subseries LNAI).

Vol. 862: R. C. Carrasco, J. Oncina (Eds.), Grammatical Inference and Applications. Proceedings, 1994. VIII, 290 pages. 1994. (Subseries LNAI).

Vol. 863: H. Langmaack, W.-P. de Roever, J. Vytopil (Eds.), Formal Techniques in Real-Time and Fault-Tolerant Systems. Proceedings, 1994. XIV, 787 pages. 1994.

Vol. 864: B. Le Charlier (Ed.), Static Analysis. Proceedings, 1994. XII, 465 pages. 1994.

Vol. 865: T. C. Fogarty (Ed.), Evolutionary Computing. Proceedings, 1994. XII, 332 pages. 1994.

Vol. 866: Y. Davidor, H.-P. Schwefel, R. Männer (Eds.), Parallel Problem Solving from Nature - PPSN III. Proceedings, 1994. XV, 642 pages. 1994.

Vol 867: L. Steels, G. Schreiber, W. Van de Velde (Eds.), A Future for Knowledge Acquisition. Proceedings, 1994. XII, 414 pages. 1994. (Subseries LNAI).

Vol. 868: R. Steinmetz (Ed.), Multimedia: Advanced Teleservices and High-Speed Communication Architectures. Proceedings, 1994. IX, 451 pages. 1994.

Vol. 869: Z. W. Raś, Zemankova (Eds.), Methodologies for Intelligent Systems. Proceedings, 1994. X, 613 pages. 1994. (Subseries LNAI).

Vol. 870: J. S. Greenfield, Distributed Programming Paradigms with Cryptography Applications. XI, 182 pages. 1994.

Vol. 871: J. P. Lee, G. G. Grinstein (Eds.), Database Issues for Data Visualization. Proceedings, 1993. XIV, 229 pages. 1994.

Vol. 872: S Arikawa, K. P. Jantke (Eds.), Algorithmic Learning Theory. Proceedings, 1994. XIV, 575 pages. 1994.

Vol. 873: M. Naftalin, T. Denvir, M. Bertran (Eds.), FME '94: Industrial Benefit of Formal Methods. Proceedings, 1994. XI, 723 pages. 1994.

Vol. 874: A. Borning (Ed.), Principles and Practice of Constraint Programming. Proceedings, 1994. IX, 361 pages. 1994.

Vol. 875: D. Gollmann (Ed.), Computer Security – ESORICS 94. Proceedings, 1994. XI, 469 pages. 1994.

Vol. 876: B. Blumenthal, J. Gornostaev, C. Unger (Eds.), Human-Computer Interaction. Proceedings, 1994. IX, 239 pages. 1994.

Vol. 877: L. M. Adleman, M.-D. Huang (Eds.), Algorithmic Number Theory. Proceedings, 1994. IX, 323 pages. 1994.

Vol. 878: T. Ishida; Parallel, Distributed and Multiagent Production Systems. XVII, 166 pages. 1994. (Subseries LNAI).

Vol. 879: J. Dongarra, J. Waśniewski (Eds.), Parallel Scientific Computing. Proceedings, 1994. XI, 566 pages. 1994.

Vol. 880: P. S. Thiagarajan (Ed.), Foundations of Software Technology and Theoretical Computer Science. Proceedings, 1994. XI, 451 pages. 1994.

Vol. 881: P. Loucopoulos (Ed.), Entity-Relationship Approach – ER'94. Proceedings, 1994. XIII, 579 pages. 1994.

Vol. 882: D. Hutchison, A. Danthine, H. Leopold, G. Coulson (Eds.), Multimedia Transport and Teleservices. Proceedings, 1994. XI, 380 pages. 1994.

Vol. 883: L. Fribourg, F. Turini (Eds.), Logic Program Synthesis and Transformation – Meta-Programming in Logic. Proceedings, 1994. IX, 451 pages. 1994.

Vol. 884: J. Nievergelt, T. Roos, H.-J. Schek, P. Widmayer (Eds.), IGIS '94: Geographic Information Systems. Proceedings, 1994. VIII, 292 pages. 19944.

Vol. 885: R. C. Veltkamp, Closed Objects Boundaries from Scattered Points. VIII, 144 pages. 1994.

Vol. 886: M. M. Veloso, Planning and Learning by Analogical Reasoning. XIII, 181 pages. 1994. (Subseries LNAI).

Vol. 887: M. Toussaint (Ed.), Ada in Europe. Proceedings, 1994. XII, 521 pages. 1994.

Vol. 888: S. A. Andersson (Ed.), Analysis of Dynamical and Cognitive Systems. Proceedings, 1993. VII, 260 pages. 1995.

Vol. 889: H. P. Lubich, Towards a CSCW Framework for Scientific Cooperation in Europe. X, 268 pages. 1995.

Vol. 890: M. J. Wooldridge, N. R. Jennings (Eds.), Intelligent Agents. Proceedings, 1994. VIII, 407 pages. 1995. (Subseries LNAI).

Vol. 891: C. Lewerentz, T. Lindner (Eds.), Formal Development of Reactive Systems. XI, 394 pages. 1995.

Vol. 892: K. Pingali, U. Banerjee, D. Gelernter, A. Nicolau, D. Padua (Eds.), Languages and Compilers for Parallel Computing. Proceedings, 1994. XI, 496 pages. 1995.

Vol. 893: G. Gottlob, M. Y. Vardi (Eds.), Database Theory – ICDT '95. Proceedings, 1995. XI, 454 pages. 1995.

Vol. 894: R. Tamassia, I. G. Tollis (Eds.), Graph Drawing. Proceedings, 1994. X, 471 pages. 1995.

Vol. 895: R. L. Ibrahim (Ed.), Software Engineering Education. Proceedings, 1995. XII, 449 pages. 1995.

Vol. 896: R. N. Taylor, J. Coutaz (Eds.), Software Engineering and Human-Computer Interaction. Proceedings, 1994. X, 281 pages. 1995.

Vol. 897: M. Fisher, R. Owens (Eds.), Executable Modal and Temporal Logics. Proceedings, 1993. VII, 180 pages. 1995. (Subseries LNAI).

Vol. 898: P. Steffens (Ed.), Machine Translation and the Lexicon. Proceedings, 1993. X, 251 pages. 1995. (Subseries LNAI).

Vol. 899: W. Banzhaf, F. H. Eeckman (Eds.), Evolution and Biocomputation. VII, 277 pages. 1995.

Vol. 900: E. W. Mayr, C. Puech (Eds.), STACS 95. Proceedings, 1995. XIII, 654 pages. 1995.

Vol. 901: R. Kumar, T. Kropf (Eds.), Theorem Provers in Circuit Design. Proceedings, 1994. VIII, 303 pages. 1995.

Vol. 902: M. Dezani-Ciancaglini, G. Plotkin (Eds.), Typed Lambda Calculi and Applications. Proceedings, 1995. VIII, 443 pages. 1995.

Vol. 903: E. W. Mayr, G. Schmidt, G. Tinhofer (Eds.), Graph-Theoretic Concepts in Computer Science. Proceedings, 1994. IX, 414 pages. 1995.

Vol. 904: P. Vitányi (Ed.), Computational Learning Theory. EuroCOLT'95. Proceedings, 1995. XVII, 415 pages. 1995. (Subseries LNAI).

Vol. 905: N. Ayache (Ed.), Computer Vision, Virtual Reality and Robotics in Medicine. Proceedings, 1995. XIV, 567 pages. 1995.

Vol. 906: E. Astesiano, G. Reggio, A. Tarlecki (Eds.), Recent Trends in Data Type Specification. Proceedings, 1995. VIII, 523 pages. 1995.

Vol. 907: T. Ito, A. Yonezawa (Eds.), Theory and Practice of Parallel Programming. Proceedings, 1995. VIII, 485 pages. 1995.

Vol. 908: J. R. Rao Extensions of the UNITY Methodology: Compositionality, Fairness and Probability in Parallelism. XI, 178 pages. 1995.

Vol. 909: H. Comon, J.-P. Jouannaud (Eds.), Term Rewriting. Proceedings, 1993. VIII, 221 pages. 1995.

Vol. 910: A. Podelski (Ed.), Constraint Programming: Basics and Trends. Proceedings, 1995. XI, 315 pages. 1995.

Vol. 911: R. Baeza-Yates, E. Goles, P. V. Poblete (Eds.), LATIN '95: Theoretical Informatics. Proceedings, 1995. IX, 525 pages. 1995.

Vol. 912: N. Lavrac, S. Wrobel (Eds.), Machine Learning: ECML – 95. Proceedings, 1995. XI, 370 pages. 1995. (Subseries LNAI).

Vol. 913: W. Schäfer (Ed.), Software Process Technology. Proceedings, 1995. IX, 261 pages. 1995.

Vol. 914: J. Hsiang (Ed.), Rewriting Techniques and Applications. Proceedings, 1995. XII, 473 pages. 1995.

Vol. 915: P. D. Mosses, M. Nielsen, M. I. Schwartzbach (Eds.), TAPSOFT '95: Theory and Practice of Software Development. Proceedings, 1995. XV, 810 pages. 1995.

Vol. 916: N. R. Adam, B. K. Bhargava, Y. Yesha (Eds.), Digital Libraries. Proceedings, 1994. XIII, 321 pages. 1995.

Vol. 917: J. Pieprzyk, R. Safavi-Naini (Eds.), Advances in Cryptology - ASIACRYPT '94. Proceedings, 1994. XII, 431 pages. 1995.

Vol. 918: P. Baumgartner, R. Hähnle, J. Posegga (Eds.), Theorem Proving with Analytic Tableaux and Related Methods. Proceedings, 1995. X, 352 pages. 1995. (Subseries LNAI).

Vol. 919: B. Hertzberger, G. Serazzi (Eds.), High-Performance Computing and Networking. Proceedings, 1995. XXIV, 957 pages. 1995.

Vol. 920: E. Balas, J. Clausen (Eds.), Integer Programming and Combinatorial Optimization. Proceedings, 1995. IX, 436 pages. 1995.

Vol. 921: L. C. Guillou, J.-J. Quisquater (Eds.), Advances in Cryptology – EUROCRYPT '95. Proceedings, 1995. XIV, 417 pages. 1995.

Vol. 923: M. Meyer (Ed.), Constraint Processing. IV, 289 pages. 1995.

Vol. 924: P. Ciancarini, O. Nierstrasz, A. Yonezawa (Eds.), Object-Based Models and Languages for Concurrent Systems. Proceedings, 1994. VII, 193 pages. 1995.

Vol. 925: J. Jeuring, E. Meijer (Eds.), Advanced Functional Programming. Proceedings, 1995. VII, 331 pages. 1995.

Vol. 926: P. Nesi (Ed.), Objective Software Quality. Proceedings, 1995. VIII, 249 pages. 1995.

Vol. 927: J. Dix, L. Moniz Pereira, T. C. Przymusinski (Eds.), Non-Monotonic Extensions of Logic Programming. Proceedings, 1994. IX, 229 pages. 1995. (Subseries LNAI).

Vol. 928: V.W. Marek, A. Nerode, M. Truszczynski (Eds.), Logic Programming and Nonmonotonic Reasoning. Proceedings, 1995. VIII, 417 pages. 1995. (Subseries LNAI).

Vol. 929: F. Morán, A. Moreno, J.J. Merelo, P. Chacón (Eds.), Advances in Artificial Life. Proceedings, 1995. XIII, 960 pages. 1995 (Subseries LNAI).

Vol. 930: J. Mira, F. Sandoval (Eds.), From Natural to Artificial Neural Computation. Proceedings, 1995. XVIII, 1150 pages. 1995.

Vol. 931: P.J. Braspenning, F. Thuijsman, A.J.M.M. Weijters (Eds.), Artificial Neural Networks. IX, 295 pages. 1995.

Vol. 932: J. Iivari, K. Lyytinen, M. Rossi (Eds.), Advanced Information Systems Engineering. Proceedings, 1995. XI, 388 pages. 1995.

Vol. 933: L. Pacholski, J. Tiuryn (Eds.), Computer Science Logic. Proceedings, 1994. IX, 543 pages. 1995.